ELECTRONICS, COMMUNICATIONS AND NETWORKS IV

PROCEEDINGS OF THE 4TH INTERNATIONAL CONFERENCE ON ELECTRONICS, COMMUNICATIONS AND NETWORKS, 12 – 15 DECEMBER 2014, BEIJING, CHINA

Electronics, Communications and Networks IV

Editor

Amir Hussain

Division of Computing Science and Mathematics, University of Stirling, Stirling, UK

Mirjana Ivanovic

Faculty of Sciences, Department of Mathematics and Informatics, University of Novi Sad, Novi Sad, Serbia

VOLUME 1

CRC Press

Taylor & Francis Group

Boca Raton London New York Leiden

CRC Press is an imprint of the Taylor & Francis Group, an **informa** business

A BALKEMA BOOK

CRC Press/Balkema is an imprint of the Taylor & Francis Group, an informa business

Typeset by diacriTech, Chennai, India
Printed and bound in Great Britain by CPI Group (UK) Ltd, Croydon, CR0 4YY

Published by: CRC Press/Balkema
P.O. Box 11320, 2301 EH Leiden, The Netherlands
e-mail: Pub.NL@taylorandfrancis.com
www.crcpress.com – www.taylorandfrancis.com

ISBN: 978-1-138-02830-2 (set of two volumes Hardback)
ISBN: 978-1-138-02868-5 (Volume 1)
ISBN: 978-1-138-02869-2 (Volume 2)
ISBN: 978-1-315-68210-5 (eBook PDF)

Electronics, Communications and Networks IV – Hussain & Ivanovic (eds)
© 2015 Taylor & Francis Group, London, ISBN: 978-1-138-02830-2

Table of contents

VOLUME 1

Communications and networks

Computer technology

VOLUME 2

Electronic application

Electronics, Communications and Networks IV – Hussain & Ivanovic (eds)
© 2015 Taylor & Francis Group, London, ISBN: 978-1-138-02830-2

Preface

On behalf of the conference organizing committee, I am pleased to welcome you to join the 4th International Conference on Electronic, Communications and Networks (CECNet2014), which is held from December 12 to 15, 2014, in the Sunworld Hotel, Beijing, China. It inherits the fruitfulness of the past three conferences and lays a foundation for the coming CECNet2015 at the same time next year in Shanghai.

CECNet2014 is hosted by Hubei University of Science and Technology, China, with the main objective of providing a comprehensive global forum for experts and participants from acadamia to exchange ideas and presenting results of ongoing research in the most state-of-the-art areas of Consumer Electronics Technology, Communication Engineering and Technology, Wireless Communications Enginneering and Technology, and Computer Engineering and Technology.

In this event, 13 famous scholars and Engineers are invited to deliver the keynote speeches on their latest research, including Prof. Vijaykrishnan Narayanan (a Fellow of the Institute of Electrical and Electronics Engineers), Prof. Han-Chieh Chao (the Director of the Computer Center for Ministry of Education Taiwan from September 2008 to July 2010), Prof. Borko Furht (the founder of the Journal of Multimedia Tools and Applications), Prof. Kevin Deng (who served as Acting Director of Hong Kong APAS R&D Center in 2010), and Prof. Minho Jo (the Professor of Department of Computer and Information Science, Korea University). Special thanks to all the plenary speakers.

Special acknowledgements should also go to the tpc members, session organizers, the reviewers and all the authors, with whom the CECNet 2014 is more successful.

Finally, special thanks to CRC Press / Balkema (Taylor & Francis Group) for the support of publication services.

Prof. Zhongming Li
Committee Chair of CECNET 2014

Electronics, Communications and Networks IV – Hussain & Ivanovic (eds)
© 2015 Taylor & Francis Group, London, ISBN: 978-1-138-02830-2

Organizing committee

General Co-Chairs

Prof. Zhongming Li, *Hubei University of Science and Technology, China*
Hung-Yuan Chung, *Ph.D., Distinguished Professor, FIET (U.K.), SMIEEE (U.S.A.), AE of IEEE TNNLS (U.S.A), Department of Electrical Engineering,National Central University, Taiwan*
Dr. Srikanta Patnaik, *Professor of SOA University & I.I.M.T., Bhubaneswar, India*

Technical Program Committee Co-Chairs

Prof. Rugang Zhong, Hubei *University of Science and Technology, China*
Mr. Joseph V. Lillie, *2009-2010 IEEE Foundation Director and Treasurer, USA*
Prof. Lei Shu, *Guangdong University of Petrochemical Technology, china*
Prof. Chengwu Liu, *Hubei University of Science and Technology, China*
Prof. Jianfeng Mao, *Hubei University of Science and Technology, China*
Dr. Stefan Mozar, *Vice President of the IEEE Consumer Consumer Electronics Society, USA*
Prof. Dai Li, *Hubei University of Science and Technology, China*
Prof. Zhengze Cheng, *Hubei University* of Science and Technology, *China*

Publication Co-Chairs

Dr. Nurul I Sarkar, *School* of *Computer & Mathematical Sciences (D-75), Auckland University of University, New Zealand*
Chairman of IEEE Joint NZ North, *Central, and South ComSoc Chapter*
Simone Fiori, *PhD, Dipartimento di Ingegneria dell'Informazione, Università Politecnica delle Marche, Via Brecce Bianche, 60131 Ancona - Italy*

Technical Program Committee

Dr. Sharon Peng, *Vice President of IEEE Comsumer Electronics(During 2013~2015), USA*
Prof. Feng Shu, *School of Electronic and Optical Engineering, Nanjing University of Science and Technology, Nanjing, China*
Prof. Aliakbar Montazer Haghighi, *Department of Mathematics,Prairie View A&M University, Prairie View, Texas, USA*
Prof. Jun-Juh Yan, *Department of Computer and Communication, Shu-Te University,Taiwan*
Prof. Chaman Lal Sabharwal, *Missouri University of Science & Technology, USA*
Prof. ByungKwan Lee, *Kwan-Dong University, Korea*
Prof. Benfdila Arezki, *Faculty of Electrical and Computer Sciences, University M. Mammeri, Algeria*
Prof. Sangho choe, *The Catholic Univ. of Korea, Bucheon-si, Gyeonggi-do, Korea.*
Prof. Chongfu Zhang, *University of Electronic Science and Technology of China, China*
Prof. Ana C.L.Cabeceira, *Department of Electricity and Consumer Electronics, University of Valladolid, Spain*
Dr. Tim French, *Department of Computer Science and Technology, Faculty of Creative Arts and Technologies,University of Bedfordshire, UK*
Prof. Hai-Han Lu, *Department of Electro-Optical Engineering, National Taipei University of Technology, China (Taiwan)*
Dr. Ritaban Dutta, *CSIRO Digital Productivity and Services Flagship, Hobart 7001, Tasmania, Australia*

Communications and networks

Electronics, Communications and Networks IV – Hussain & Ivanovic (eds)
© 2015 Taylor & Francis Group, London, ISBN: 978-1-138-02830-2

Novel scheduling techniques for multiuser diversity MU-MIMO TDD communication system

Joyatri Bora* & Md. A. Hussain
North Eastern Regional Institute of Science and Technology, Nirjuli, Itanagar, Arunachal Pradesh

ABSTRACT: Here, we propose a new scheduling scheme for multiuser multiple input-multiple output (MU-MIMO) system, considering Multiuser Diversity (Mu-Div) in highly scattered Rayleigh channel. The scheduling scheme is based on non-uniform quantization slots and thresholding of Gaussian distributed spectral efficiency using techniques similar to Single User MIMO Spatial Multiplexing (SU-MIMO-SM) system in TDD where all the BS antenna resources are allocated to a single user. The scheme may be implemented in two ways. In one way, the BS selects users blindly from the set of users in the high slot with blind allocation of a BS antenna to each user. On the other way, the users selected from the slot may be based on their feedback on antenna index, the index indicating the BS antenna to which it has maximum channel gain with any of its receiving antennas. The proposed scheduling scheme seems to be simpler so far as feedback information processing is concerned.

1 INTRODUCTION

MIMO has proved to be an efficient technique of providing large spectral efficiency by exploiting spatial multiplexing of parallel data stream through different channels of same frequency bandwidth in a rich scattering environment (Paulraj et al. 2004, Telatar 1999, Goldsmith et al. 2003). However, in a multiuser system increased spectral efficiency is achieved due to multiuser diversity property of the channel, and data transmission is scheduled to the user with the best channel (Knopp & Humblet 1995), adopting a TDMA protocol and various scheduling techniques. In (Chen & Wang 2006), the authors report a fair scheduling scheme where BS assigns its all transmit antenna to a single user with the best channel condition at one time slot in response to limited CSI feedback that increases the system capacity and coverage area of a multiuser MIMO system.

Recently, there has been a lot of research work on multiuser MIMO where multiple users are served simultaneously along spatially multiplexed channels (Gesbert et al. 2007). The capacity of MU-MIMO BC can be increased by transmitting to a set of users simultaneously having better channel than to a single user with best channel (Caire & Shamai 2003). Although Costa's dirty paper coding (DPC) (Costa 1983) achieves the maximum theoretical sum-rate capacity of MU-MIMO BC, its large computational complexity and feedback requirement of full CSIT leads to adapt low complexity suboptimal precoding technique such as block diagonalization (BD) (Spencer et al. 2004) proposed for downlink transmissions of MU-MIMO system. In a multiuser network where users suffer from different level of interference and also because the numbers of users are more than the antennas at the BS, scheduling is necessary in the downlink of MU-MIMO system to serve the best users maximizing the system performance. Scheduling requires CSI feedback from users. Some scheduling schemes require full CSI feedback (Yoo & Goldsmith 2006, Zhang et al. 2007). A distributed scheduling algorithm with partial CSI feedback for MU-MIMO is proposed in (Li et al. 2009) where only a vector is feedback as CSI, and adaptively selecting a set of user. To reduce the feedback load transmission scheme based on quantized CSI has been developed (Huang et al. 2009).

We reported three novel scheduling techniques based on quantized mutual information (MI) and SNR dependent feedback for multiuser MIMO-SM system (Bora & Hussain 2014b). Adopting these techniques, we propose a new scheduling scheme where multiple users equal to $M_T=M_R=M$ are simultaneously fed with data transmissions from M BS antennas, one antenna for one user. To select M users from a total of K active users and allot BS antenna resource, the active users are selected from those that are in a slot above a threshold value determined by the spectral efficiency of M parallel channels of a highly scattering wireless medium between users and BS, having a multi user diversity advantage. The technique is based on TDMA protocol and do not require separate

*Corresponding author: *bjoyatri@yahoo.com*

channels for data transmission to users and collecting channel state information (CSI) from users to BS.

2 SYSTEM MODEL AND METHODOLOGY

We consider a multiuser diversity (Mu-Div) MIMO down link data transmission system with highly scattered Rayleigh fading channel and TDD. The BS has M_T transmitting antennas and K users each equipped with M_R receiving antennas where $M_T \leq M_R$. Each transmitting antenna of the BS sends the independent data stream to M_T users simultaneously, the base station (BS) selecting its user for each transmitting antenna based on blind selection amongst the users who are in a given slot above the threshold, or utilizing antenna identification bits integrated with the feedback bits sent by the users above the threshold for a SNR slot.

The users process the received SNRs in two ways. In one way, as in SU-MIMO-SM, each user derives its instantaneous mutual information for $M_T=M_R=M$ and sends feedback bits if it is in a slot above a threshold as instructed by the BS. M_T numbers of users above the threshold are blindly selected by the BS for down link transmission, allocating one transmitting antenna to one user blindly, and each user with $(1/M_T)$ times the data rate of the spatial multiplexing channel. In the other way, each user determines if it is in a slot above the threshold as in the first way, and also finds the transmitting antenna to which it has maximum channel gain with any one of its M_R receiving antennas. The feedback bits indicating the slot above the threshold and the feedback bits indicating the maximum channel gain antenna are integrated such that a minimal number of bits are required in the comprehensive feedback which is further elaborated later in this section. The BS selects M_T numbers of users above the threshold, such that each transmitting BS antenna may be allocated to a user which recorded a maximum channel gain from it, as is revealed by the comprehensive feedback bits and each user with $(1/M_T)$ times the data rate of the spatial multiplexing channel.

For the second way cited above, we assume that the M_T numbers of users selected from those above the threshold are such that each one has recorded maximum channel gains from different transmitting antenna. Otherwise a blind allocation of the antennas is the choice. The sum data rate or the sum capacity of this down link transmission system is equal to the data rate of the spatial multiplexing channel and hence differs from the ideal sum capacity where allocations of transmit antennas to the users are based on maximum SNR reporting through separate channels as in the multiuser antenna selection (MUAS).

Assuming the channel between the BS and each user to be rich scattering and i.i.d. Rayleigh flat fading, the instantaneous mutual information (MI) in b/s/Hz of the kth user in the Multiuser diversity SU-MIMO-SM system is expressed as

$$I_k = \log_2\left(\det\left(I_M + \frac{\rho}{M} H_k^* H_k\right)\right) \tag{1}$$

where ρ is SNR per receive antenna, H_k is channel gain matrix between k^{th} user and BS which is random and $(\bullet)^*$denotes the complex conjugate operator.

Assuming Gaussian approximation of MI in Equation1, the asymptotic Gaussian distributed spectral efficiency of the SU-MIMO-SM channel between the base station and the kth user is expressed as (Hochwald et al 2004),

$$S_K = \mu + \sqrt{2\sigma^2 \ln K} \tag{2}$$

Where $\mu=E$ $(\log_2 \det (I_M + (\rho/\mathrm{M})\ H_k^* H_k))$ and $\sigma^2 = (\log_2 e)^2\ [\log M + 1.58]$ are the ergodic spectral efficiency and the variance of the single user open loop MIMO channel. The second term on the right hand side of Equation 2 is the additional spectral efficiency due to multiuser diversity advantage in a rich scattering environment.

Three novel scheduling techniques are reported in (Bora & Hussain 2014b) for SU-MIMO-SM, utilizing multiuser diversity, for high data rate transmissions in high scattering channel. We explain below our proposal for MU-MIMO scheduling using TDD is taking the case of system parameters of scheduling technique1 in (Bora & Hussain 2014b), which are: Mean of instantaneous spectral efficiency *(MI)*, *M_IH* $=\mu$ of users over 10000 simulations of Rayleigh fading high scattering channel, Standard deviation *STD_H*$=\sigma_s$ of instantaneous spectral efficiency of the simulated channels, between BS and each user, and the number of active users K=70. MIMO configurations M_Tx M_R considered here are 2x2, 3x3, 4x4, 5x5, 6x6, 7x7, and 8x8, to receive *SNR (SINR)* per antenna = 20 dB. With L_b=1, 2 bit feedback indicating the slot above a threshold, all the users above threshold for the highest slot in scheduling technique1, over 100 simulations are considered. The scheduling scheme is as follows.

For implementation of the system practically, an estimate of S_k is used as $S_k'=\mu + 2\sigma_s$ and non-uniform slotting of $S_k'$is taken as shown below in Figure 1, for L_b= 2 bit feedback. Although $2\sigma_s$ is quantized into equal slot lengths, the last slot of quantized $2\sigma_s$ is added with mean μ to make the last slot of quantized S_k'. If the number of slots above μ is taken such that always a few number of users are available in the highest slot, then size of L_b appears to be 1-3 bits.

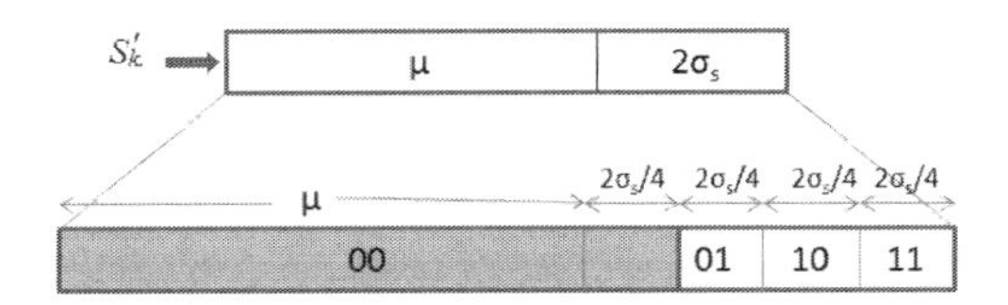

Figure 1. Unequal quantization of S_k, the estimate of S_k, for scheduling technique 1.

A user sends feedback bits '11' if its MI ≥ (μ+3/4.(2σ_s)), '10' if its MI is in the range (μ+1/2.(2σ_s))≤ MI < (μ+3/4.(2σ_s)), '01' if its MI is in the range (μ+1/4.(2σ_s))≤ MI <(μ+1/2.(2σ_s)), and '00' for $\mu \leq$ MI < (μ+1/4.(2σ_s)). We consider here that the BS instructs the users to send feedback '11' for the highest slot only and K_{TH} is the number of users respond, then the downlink sum data rate to be provided to M_T selected users blindly is expressed as in Equation 4. This Equation 4 we get from the downlink data rate expressed in Equation 3 (Rajanan & Jindal 2012) for the MIMO-SM multiuser diversity system with K users, utilizing TDMA protocol and TDD system.

$$R(K) = B_{data}(K).S_K = \left(1 - \frac{KL_b}{T_b \log_2(1+\rho)}\right).S_K \tag{3}$$

where $R(K)$ is the downlink data rate to the scheduled user, B_{data} is downlink data bandwidth, T_b is the total block length of the channel, L_b is the number of bits used for feedback, and ρ is received SNR per antenna. We estimate the S_k as $S_k'$and considering ρ=20 dB, T_b= 100 symbols, the downlink data rate will be

$$R(K) = \left(1 - \frac{KL_b}{T_b \log_2(1+\rho)}\right).S_K'$$

$$= \left(1 - \frac{K_{TH}L_b}{T_b \log_2(1+100)}\right)(\mu + n.2\sigma_s) \tag{4}$$

where n=1/2, 3/4, 7/8 and 15/16 for L_b=1, 2, 3, 4.

It may be noted that, to collect feedback for individual slots at a time, one bit feedback is sufficient. We consider here two bit feedback, assuming that the technique may also be implemented by taking feedback for all slots simultaneously.

As is clear now, the M_T antennas are allocated blindly to M_T users, selected blindly from those found in the high slot. In the other case where antenna preferences are taken from users, the comprehensive feedback system is designed as below.

Let M_T=2 to 8, and L_b is the number of feedback bits for the slot, the number of comprehensive feedback bits C_b= L_b + A_b, where A_b is additional bits needed so that the binary sequence using the C_b bits indicates the antenna index of a user sending feedback. As an example, if M_T=2, 4, 8, then A_b=0, 1, 2 for L_b=1, A_b=0, 0, 1 if L_b=2, and A_b=0, 0, 0 for L_b=3. The antenna index for the 4th antenna, when M=8, L_b=2, C_b=3 is 100 as an example. It may be noted that for finding sum data rates using TDD and C_b, C_b replaces L_b in Equation 4.

3 RESULTS AND DISCUSSIONS

As cited above, we consider 2x2, 3x3, 4x4, 5x5, 6x6, 7x7 and 8x8 MIMO configuration with received SINR per antenna ρ=20 dB for blind antenna allocations and blind user selection above threshold (TDD and blind selection) as well as in the case where the comprehensive feedback system is used for antenna allocations (TDD and C_b), and K=70 active users with L_b=1, 2 bit in the system. For explaining the results we take the case of 4x4 MIMO and L_b=2 bits. Figure 2, show the numbers of times the different users found above threshold for the highest slot over 100 simulations, with L_b=2, M_T =M_R=4, and SINR per antenna ρ=20 dB. We observe from Figure 2, that there are a number of users like 15, 66, 30, and 58 which occur in most of the 100 simulations and may be taken as the users blindly selected for antenna allocations, randomly assigning one user to one antenna with capacity equal to 6.23 b/s/Hz. The downlink sum data rate of the MU-MIMO is 24.9 b/s/Hz. It may be noted that we assume simultaneous data transmission to four users by four respective antennas and hence received SINR per antenna is ρ=20 dB.

A variation of this may be each antenna transmitting for a duration of ((1 /M_T) times the total time bandwidth available for data transmission, and since one antenna transmits at a time, SNR per receive antenna is ρ=20 dB.

For comprehensive feedback case, every user also finds the BS antenna to which it detects highest channel gain with any one of its receive antenna, and we call it antenna index. In the comprehensive feedback system, as explained above the users above the threshold also send this index. We simulated the scheme for 100 runs, and record the contenders for each BS transmission antenna in every run, the contender being the user with highest channel gain for a given antenna in a run.

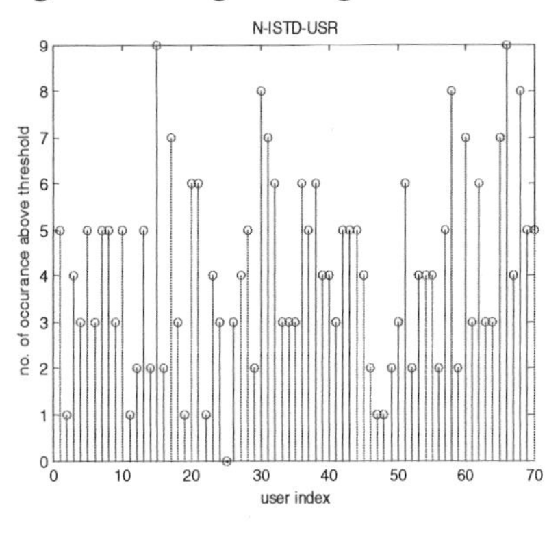

Figure 2. Number of occurrence of users above threshold.

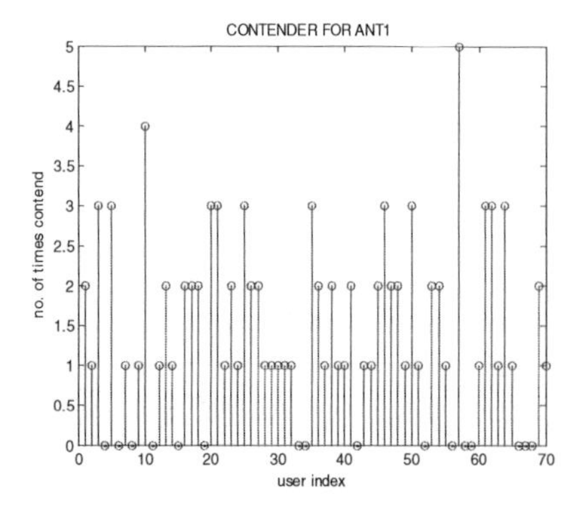

Figure 3. Contender for BS's antenna1 over 100 simulations using MUAS algorithm.

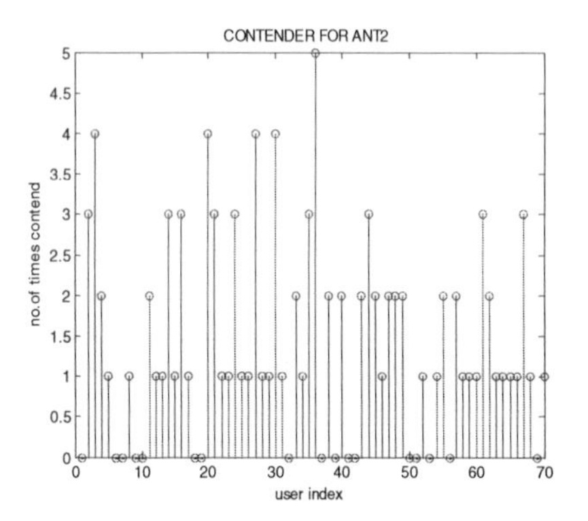

Figure 4. Contender for BS's antenna2 over 100 simulations using MUAS algorithm.

Figures 3–6, show the number of times a user contends for a given antenna over 100 runs. From the above figures, we see that some users never contend for any antenna in 100 simulations. User 57, 36, 63, and 30 contends maximum times for antenna1, 2, 3, and 4 respectively.

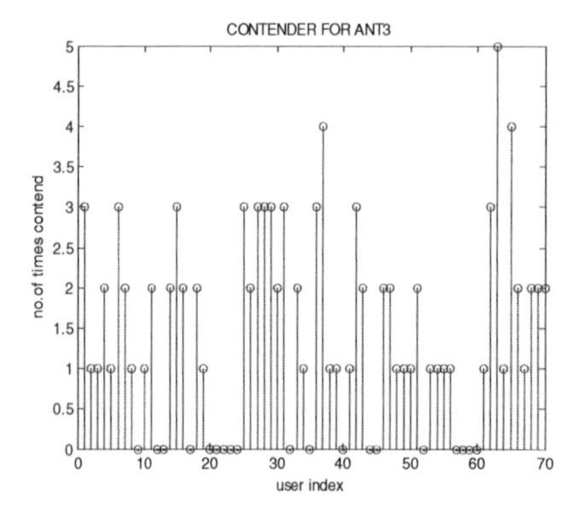

Figure 5. Contender for BS's antenna3 over 100 simulations using MUAS algorithm.

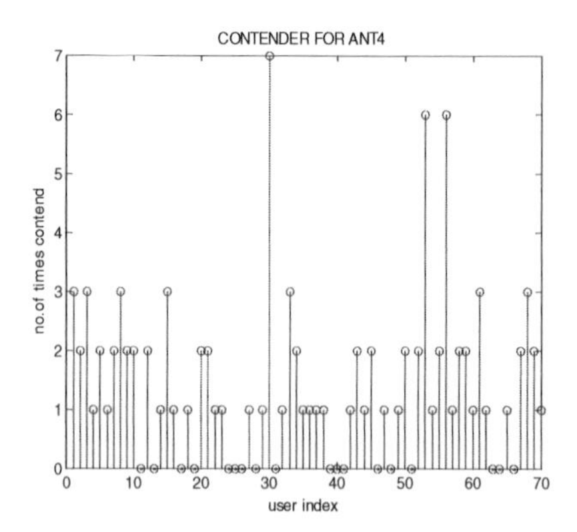

Figure 6. Contender for BS's antenna4 over 100 simulations using MUAS algorithm.

Hence, using Figure 2, of the users above threshold we may say that the comprehensive feedback system is most likely to select user 57, 36, 63, and 30 for antenna 1, 2, 3, and 4 respectively. BS will now allot equally divided capacity equal to 6.23 b/s/Hz to each selected user. We checked the maximum data rates with each one of BS antennae and any one of receiving antenna of users, occurring most of the times over 100 simulations. The values are 7.3, 7.2, 7.1 and 7.1 b/s/Hz respectively, and here the sum capacity of the users would be 28.7 b/s/Hz. We assume this value as the sum data rate where separate channels are used for feedback. Since our scheme provides downlink sum data rate of 24.9 b/s/Hz, a difference of around 3.8 b/s/Hz is observed.

For M=6(8) with L_b =2, the comprehensive feedback needs one extra bit as A_b, the downlink sum data rates are 34.4(44.5) b/s/Hz (TDD and blind selection) and 33.8 (43.7) b/s/Hz (TDD and C_b).

Table 1 shows simulation results for different sum data rates for TDD and blind selection, TDD and C_b,

Table 1. Sum data rate of the scheduling scheme with 1 and 2 bit feedback.

L_b	A_b	M	Sum data rate using TDD & blind selection	Sum data rate using TDD & C_b	Sum data rate using separate feedback channel
Bits	Bits		b/s/Hz	b/s/Hz	b/s/Hz
	0	2	12.73	12.73	17.7
	1	4	23.7	23.2	28.7
1	2	6	33.15	30.9	40.0
	2	8	43.3	40.5	51.2
	0	2	13.7	13.7	17.7
	0	4	24.9	24.9	28.7
2	1	6	34.4	33.8	40.0
	1	8	44.5	43.7	51.2

and using separate channels for collecting feedback, L_b=1, 2, and different A_b and M values.

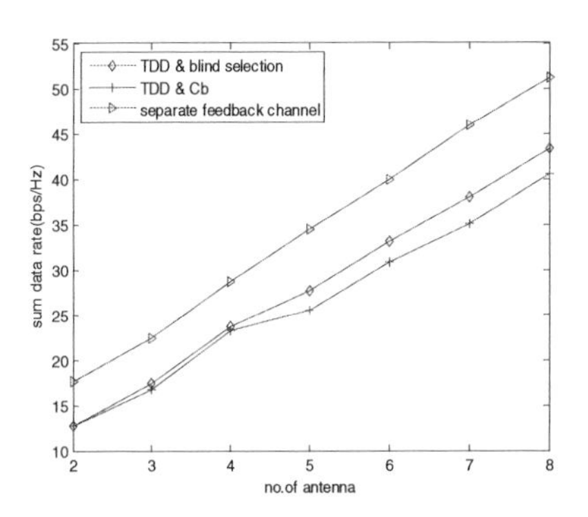

Figure 7. Downlink sum data rate with no. of antennas.

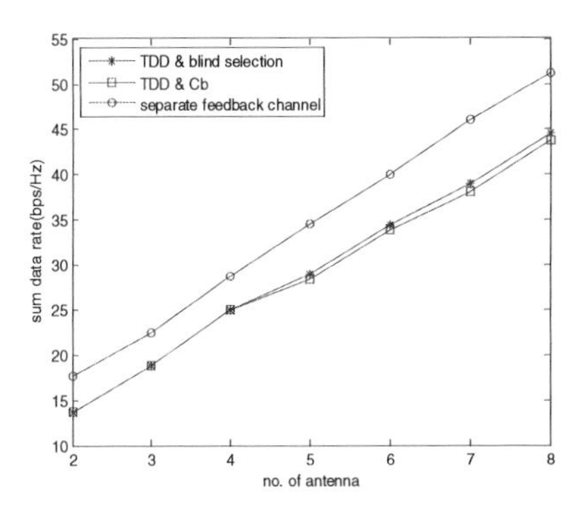

Figure 8. Downlink sum data rate with no. of antennas.

We also show these results in the Figure 7 and Figure 8, with L_b=1 and 2, respectively.

We observe from Figure 7 and Figure 8, that difference between sum data rates using TDD and blind selection, and TDD and C_b, are not remarkable for the M values shown, although for L_b =1 in Figure 7 this difference is noticeable for M=5 to 8.

4 CONCLUSION

We propose here scheduling schemes for scheduling simultaneously to $M_T = M_R = M$ numbers of downlink transmissions to users, in a rich scattering environment and utilizing the property of multi user diversity in the channel, for MU-MIMO and using TDD system. To explain our scheme, we have considered the scheduling technique 1 of our previously proposed techniques for SU-MIMO-SM system (Bora & Hussain 2014b). To choose M users from K active users, we consider two ways. In one we choose the users blindly from those reporting to be in the slot and above a threshold, and allocate a BS antenna blindly to one selected user. In the other, we use comprehensive feedback and choose the users in the slot above a threshold and try to allocate BS antennas to selected users as per antenna indices using C_b bits. The downlink data rates to each user are *(1/M).R(K)* b/s/Hz with appropriate L_b(C_b replaces L_b in case of TDD and C_b) values in equation 3, with the threshold value corresponding to the highest slot of $S_k' = \mu + 2\sigma_s$. We compare the respective down link sum data rates with the scheduling technique where separate channel is used for sending feedback. Although the sum data rates using TDD and blind selection, and TDD and C_b differ considerably from the sum data rates where separate feedback channel is used, the difference between them is not remarkable except for L_b=1 with M = 5 to 8. We are exploring utility of our scheme for the other two scheduling techniques of(Bora & Hussain, 2014b).

REFERENCES

Bora J. & Hussain A. Md. 2014a. Scheduling techniques for high data rate downlink in multiuser diversity systems using MIMO-spatial multiplexing transmission. *Int. Conference on Control, Instrumentation, Energy and Communication (CIEC14), University of Calcutta, India, January 31-February 2, 2014.*

Bora J. & Hussain A. Md. 2014b. Novel scheduling techniques with minimal feedback load for multiuser diversity MIMO-SM systems. In *Proc. of IEEE Int. Conference on Information Science and Applications (ICISA), South Korea, Seoul, 6–9 May, 2014.*

Caire G. & Shamai S. 2003. On the achievable throughput of a multi antenna Gaussian broad cast channel. *IEEE Trans. Inform. Theory* 49: 1691–1706.

Chen C.J. & Wang L.C. 2006. Enhancing coverage and capacity for multiuser MIMO systems by utilizing scheduling. *IEEE Trans.Wireless Comm* 5(5): 1148–1157.

Costa M. 1983. Writing on dirty paper. *IEEE Trans. Inform. Theory*: 439–441.

Gesbert D., Kountouris M., Heath R., et al. 2007. From single user to multiuser communications: shifting the MIMO paradigm. *IEEE Signal Process. Magazine* 24(5): 36–46.

Goldsmith A., Jafar S.A., Jindal N., et al. 2003. Capacity limits of MIMO channels. *IEEE J. Select. Areas Commun* 21: 684–702.

Hochwald B.M., Marzetta T.L. & Tarokh V. 2004. Multiple-antenna channel hardening and its implication for rate feedback and scheduling. *IEEE Trans. Inform. Theory* 50(9): 1893–1909.

Huang K., Andrews J.G. & Heath R.W.Jr. 2009. Performance of orthogonal beamforming for SDMA with limited feedback. *IEEE Transactions on Vehicular Technology*58(1): 152–164.

Knopp R. & Humblet P. 1995. Information capacity and power control in single cell multiuser communications. In *IEEE Proc. Int. Conf. on Commu.*: 331–335.

Li Z., Yang J. & Yao J. 2009. Distributed scheduling algorithm for multiuser MIMO downlink with adaptive feedback. *Journal of Communications* 4(3): 164–169.

Paulraj J., Gore D.A., Nabar R.U., et al. 2004. An overview of MIMO communications-A key to gigabit wireless. *In Proc. of IEEE* 92(2): 198–218.

Rajanan A. & Jindal N. 2012. Multiuser diversity in downlink channels : when does the feedback cost outweigh the spectral efficiency gain? In *IEEE Trans on Wirelees Comm.* 11(1): 408–418.

Spencer Q. H., Swindlehurst A. L. & Haardt M. 2004. Zero-forcing methods for downlink spatial multiplexing in multiuser MIMO channels. *IEEE Trans. Signal Process* 52: 461–471.

Telatar I. 1999. Capacity of multi-antenna Gaussian channels. In *European Trans. Telecomm* 10(6): 585–595.

Yoo T. & Goldsmith A. 2006. On the optimality of multi-antenna broadcast scheduling using zero-forcing beamforming. *IEEE Journal Selected Areas Commun* 24: 528–541.

Zhang X., Lee J. & Liu H. 2007. Low complexity multi-user MIMO scheduling with channel decomposition. In *proc. of IEEE Wireless Comm. and Networking Conference*: 2452–2456.

Electronics, Communications and Networks IV – Hussain & Ivanovic (eds)
© 2015 Taylor & Francis Group, London, ISBN: 978-1-138-02830-2

A cloud computing solution for medical institutions

Xiao-jun Chen
Information Department, Affiliated Hospital of Jiangsu University, Zhenjiang, Jiansu, China
Computer Science and Telecommunication Engineering, Jiangsu University, Zhenjiang, Jiangsu, China

Jia Ke
Computer Science and Telecommunication Engineering, Jiangsu University, Zhenjiang, Jiangsu, China
Information Management Department, Management of Jiangsu University, Zhenjiang, Jiangsu, China

ABSTRACT: The Information System in Hospital (HIS) is a necessary component in a medical institution, providing key messages to sustain medical care for virtually all the branches in the installation. Right now HIS does not have sufficient technique to sustain the several functions of the wireless access and Virtual Private Network (VPN). The main problem is how to use the Virtual machine effectively. The solution is designed to solve this problem. This cloud computing system uses VMware as a virtualized technology and works on virtual private network. A cloud-based technology enables software easily accessible, with abundant and convenient extended capabilities for the expansion of storage capacity, and making the information available in the clinical field. By using this cloud computing system, medical staffs can quickly and easily access resources, significantly saving the time of searching and managing. It can enhance the competitiveness of hospitals and improve patients' satisfaction.

1 INTRODUCTION

The Information System in Hospital (HIS) is a means for the software system which is used in clinical medicine management and the financial management of the hospital. It manages the operation of all departments and equipments (Cao & Wang 2013). HIS consists of Electronic Medical Record (EMR), Laboratory Information System (LIS), Anesthesia Information Management System (AIMS), Nosocomial Infection Surveillance (NIS), Radiology Information System (RIS), and Physical Examination Information System (PEIS), etc. The HIS component which follows the health level 7 standards (HL7) is shown as Fig. 1.

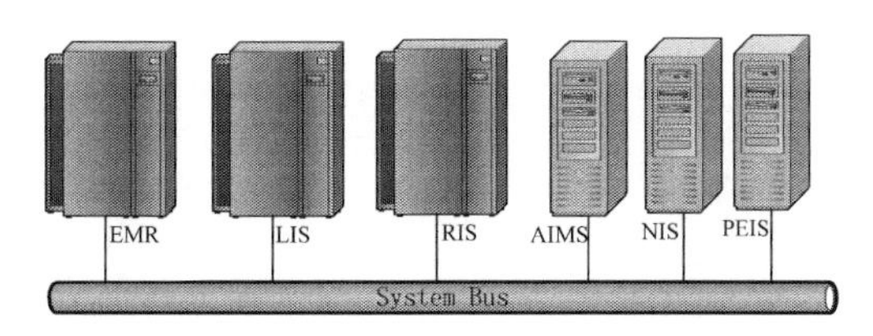

Figure 1. Traditional HIS architecture.

In our country, HIS construction is involving all aspects of matters such as hospital digital collection, reporting the data, saving and transmitting the medical information (Lee et al. 2013). By now, the Hospital Service Bus (HBS), small scale hospital information sharing platform (Chan et al. 2014) and other sophisticated solutions (Miroslav et al. 2013) have been widely used. But, we should also realize that HIS being in the initial stage, the lack of technical innovation will lead HIS to an inefficient, cumbersome and resource wasting system (Tsumoto et al. 2013). It can also result in the problems in the hospital information system's data storage requirement of security, efficiency and realtime (Fattouh et al. 2014).

A cloud is defined as a place over network infrastructure where information technology (IT) and computing resources such as computer hardware, operating systems, networks, storage, databases, and even entire software applications are available instantly, on-demand as described in (Rajkumar et al. 2011). Cloud computing uses cloud resources (hardware and software), that are delivered as a service crossing the whole local net or internet. While cloud-based system is not involving a lot of new technologies, it certainly is a new way of managing IT. In many cases, it will not only change the workflow within the IT organization, but also result in a complete reorganization of the IT department. Cost savings and scalability can be highly achieved by cloud computing.

Cloud computing is often compared with Service Oriented Architectures (SOA), Grid, Utility and Cluster computing as described in (Avram 2013). Cloud-based system and SOA could run separately or simultaneously in the cloud-based operating system. The storage server can supply the value-added support for SOA (Dai 2012).

The cloud-based system has become a new model for the host to provide services on the local net ware and the internet. It is favored to the business, because users do not need to configure any hardware environment, and allows enterprises to establish a new application with minimal resources. It can add storage easily for users according to their requirement. The application of cloud-based technology can effectively integrate EMR, LIS, RIS, AIMS, NIS, and PEIS. We transfer the comprehensive doctors shift system (CDSS) which needs to collect comprehensive data from EMR, LIS and NIS to the cloud-based software platform. After it migrates to the cloud-based software platform, the CDSS reduces the retrieval time, improving operation efficiency. In this way, the operation of query and manipulating data can be more convenient and efficient.

2 MOTIVATION

2.1 *Shortage of traditional HIS*

With the development of in-depth hospital information construction and clinical information systems on the market specialized products, more and more hospitals buy a lot of business subsystems. The data structures of each subsystem, the development environment, the dictionary have different standards. For instance, most of patient registration information cannot be effectively shared and used. Duplication of information input and information asymmetry can result in the blocked information flow. There are some typical shortages as follows.

2.1.1 *The low access efficiency*

Traditional HIS is a large and complex system, and involves various departments of the hospital. With the continuous generation of new business, installing an operating system in the traditional one server, running an application, and configuring for each application deployment model under separate storage devices cannot meet the hospital's needs. In addition, medical staffs only use the traditional PC accessing to HIS via the intranet, so this limited the workplace and effective working hours of medical staffs.

2.1.2 *Strong coupling connection*

Hospitals use the traditional point to point connections among the various specialized subsystems. With the gradual increase in the deployment of software, and coupled increased subsystems connected, a huge number of interfaces integrate in-depth, resulting in maintenance difficulties, and the enhancement of the cooperation of the subsystem developers and service attitude dependent. Hospital restrictions currently take advantage of emerging applications and innovative new technologies.

Meanwhile, a large number of medical devices, such as electronic blood pressure monitors, ECG, temperature gauge, PETCT, ultrasound instruments and endoscopies, produce large amounts of unstructured data storage format of the different data structures in the hospital system. Although there are interfaces between different systems, it is very difficult to realize the integrated subsystems. Therefore, it directly causes the information unable to be shared effectively across different hospitals, even between different depart-mends in the same hospital, thus resulting in duplications and waste of resources.

2.1.3 *A lack of the whole process management about doctor's orders*

Currently, most hospitals lack the whole process management about doctor's orders. The whole process management of orders is the core to make the whole process of medical information digitized. To implement various business data collection of the hospital, it needs to process safety warning medical knowledge to help medical service and medical management assessment work. It makes the hospital achieve clinical center and provide patients with excellent medical services. The hospital offers a comprehensive and accurate source of clinical information to improve health care quality and service levels.

2.1.4 *A lack of the definition about the standard in hospital resources*

It has not always been fully applied to financial accounting, costing, accounting, and logistics management as the main line of the management systems in hospital. Some of the hospital managers do not realize the centralized logistics and financial management controlling. There is no business to achieve financial integration, comprehensive budget or expense control. Hospital resources (human, material, equipment, etc.) are not using a definition uniform and standard data format.

2.2 *Why the cloud-based system is applied in the medical system?*

Many community hospitals and large hospitals have adopted a basic electronic medical record system, but most of them have saved in different medical records databases. Under normal circumstances, the patient may go to many different hospitals for different diseases. For example, there are dental specialist hospitals, maternity hospitals and pediatric hospitals serving for different diseases. And different doctors in different hospitals want to see the whole medical records with the standard of HL7.This problem must be solved by cloud-based system

Currently, the hospital information system software design company has its own database of electronic medical records (EMR). It needs to design exchanging interface to share information between different information systems. To consume a lot of shared system resources, the cost of this approach is high. Such sharing is already the biggest obstacle to the exchange of information. Cloud-based platform can reduce the electronic medical record system interface development costs and make connecting and sharing in whole platform.

HL7 standards-based cloud computing platform centrally stores patient medical data, outpatient data, inpatient data and community hospital data in virtual storage. How to comprehensively and rationally manage the patient information and how to share information are concerned by the medical staff. The cloud-based system can bring change to the traditional medical information system.

To make the health information be accessed at any time and any place will not only help us to improve medical services, but also help us to reduce costs significantly. The development of medical data to the cloud is a first fundamental step to success. And it is a deep understanding and effective implementation of the security and privacy of cloud-based system.

3 SYSTEM VIRTUALIZATION MODEL

3.1 *Design of architecture*

The application of integration service bus characteristics with the HSB and SOA can be a concept to build a business integration platform. EMR, LIS, RIS, AIMS, NIS, and PEIS systems are distributed in business platforms of different institutions. So the purpose of the virtual integration platform is to realize the exchange of information, establish the combination process control, make each independent business system of hospital internal form an organic whole, and serve the patients better. At the same time, since the current medical system between each module coupling degree is high, the system will be difficult to fully interact with and to maintain the development in the future, we should make a new system to join the business, establish HL7 as the standard of the data exchange platform, connect the independent systems, and realize the loose coupling between various systems. Make the business application easily to be expanded and be reused, and more integrated.

3.2 *Virtual resource management architecture*

The cloud-based technology may theoretically be seen as a different set of services; therefore, it is divided into three layers of cloud-based architecture, which are infrastructure layer, network layer, application layer. Only the application layer is applied in hospital. There are simple applications and softwares applied to the application layer, and they provide some interface for using. Therefore, we call this type of service at the application layer software as a service (SaaS). We provide the application software of the hospital for medical personnel using.

CDSS is a practical system which provides doctors required patient information; it needs to collect comprehensive data from EMR, LIS and NIS. We transfer the CDSS to the cloud-based software platform as shown in Fig.2 (a).

3.3 *System hardware, construction and connection*

In our hospital, the physical cloud is constructed by the IBM virtual machine and EMC storage system. The IBM server allows the system to achieve maximum consolidation ratio, also supports MAX5 memory expansion unit and FlashPack solid state drive, in order to increase data throughput. When an IBM server works in conjunction with VMware vSphere, it can simultaneously provide excellent and innovative hardware, software technology and optimization capabilities, that are ideal for this platform. The hospital will be able to create dynamic, highly flexible, cost-effective virtualization infrastructure. The system is shown schematically in Fig.2 (b)

3.4 *Cloud-based tools*

The VMware ESXi installation system runs on each server so that the server works as the ESXi host. ESXi virtual machine manages the underlying program vSphere virtualization platform, and provides hardware resources on the physical host for virtual machines. vSphere is a single virtualized server management tool, needing to manage the entire cloud computing center. The virtualization server, which has deployed a single vCenter server to a centralized management is a console to manage the

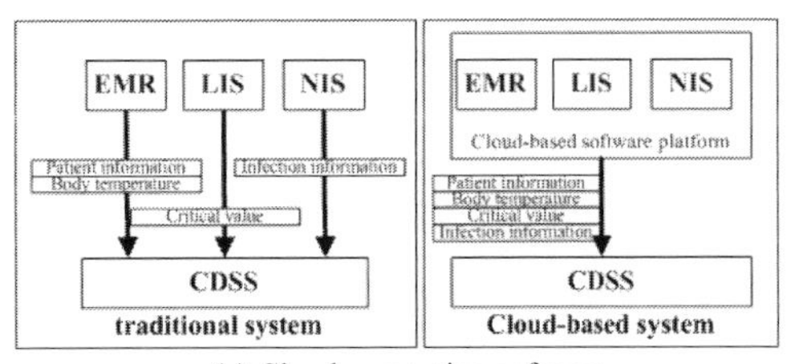

(a) Cloud computing software

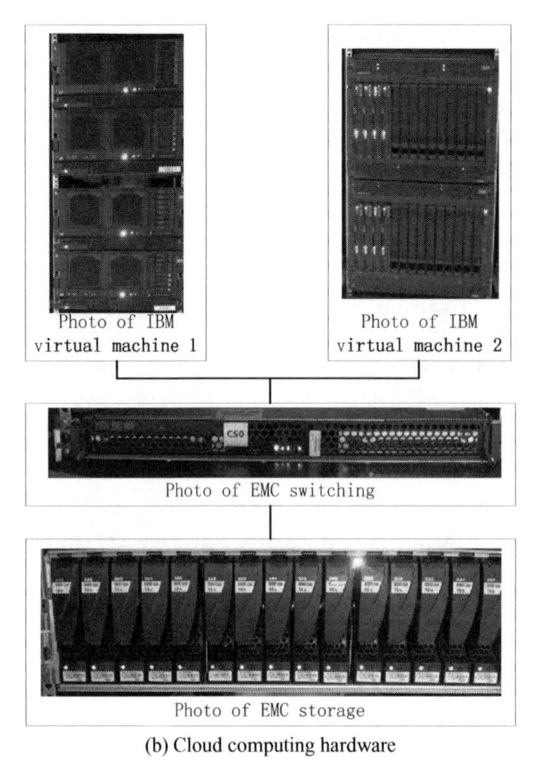

(b) Cloud computing hardware

Figure 2. Cloud computing.

entire vSphere virtualized environments. vCenter multiple ESXi hosts in a cluster management, can achieve migration, cloning, and other functions of the virtual machine between servers, further realization vMotion, HA and DRS and other advanced features. After the vCenter server deployment completed, you can use the vCenter management console to manage the entire vSphere virtualization environment. The VMware vSphere software tools are shown as Fig. 3

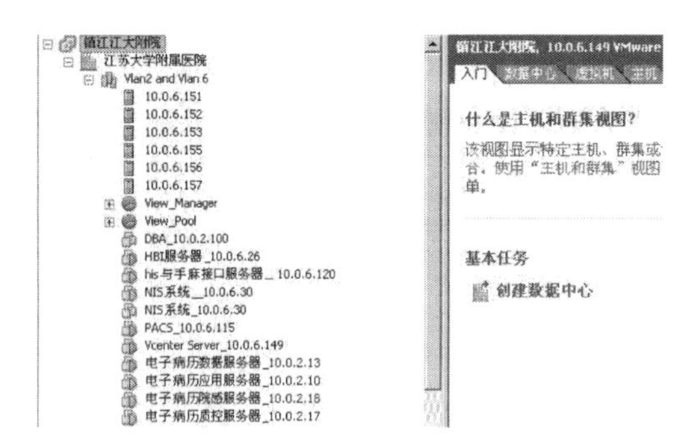

Figure 3. VMware vSphere software tools.

Applied in clinical practice, desktop workstation virtualization is realized. Every health care is only one common terminal device, but everyone has a separate desktop. Any operation would not affect others. Health care workers can use Virtual Private Network software tools authorized by the internal and external network access network, while supporting smart handheld devices, the system administrator can configure different security for different desktops. The Virtual Private Network software tool is shown as Fig. 4

Figure 4. Virtual private network software tools.

4 RESULTS

After server virtualization transformation, Affiliated Hospital of Jiangsu University changed the application system complexity, variety of operating platform, and complexities caused by isolated backstage IT support infrastructure. Numerous subsystems EMR, LIS, RIS, AIMS, NIS, PEIS and others run on a unified virtualization architecture. A large number of servers from the original decentralized management, now uses vCenter unified management console. You can easily manage the entire configuration server by vCenter management console. The administrator console can control all the vCenters. All kinds of data computing centers, storage, network resources can be managed by each virtual machine of the vCenter. And each virtual machine can adjust its own configuration.

4.1 *Improve the system efficiency*

After the virtual transformation, all independent types of business systems integrate as background data center server clusters, and several actual physical servers share their computing resources, in accordance with the requirements of the different types of

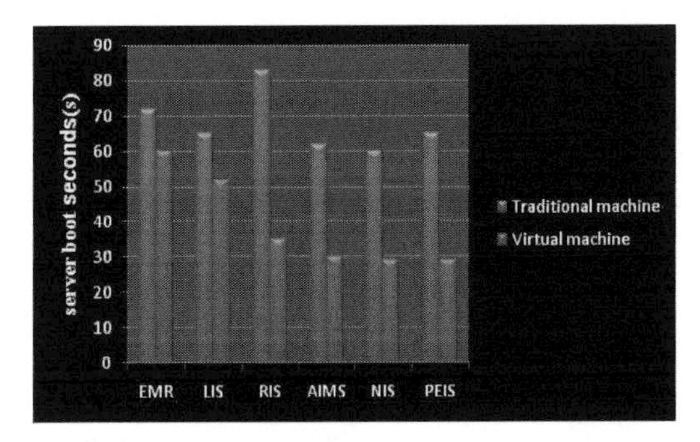

Figure 5. Server boot seconds.

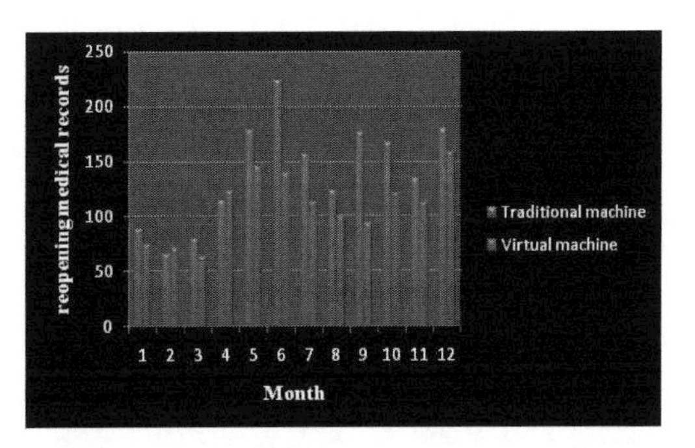

Figure 6. Reopening medical records.

applications assigned to the service center. After the entire virtualization server management tends to be simple and convenient. Unified management interface reduces the workload of IT management staff. Using virtualization technology to build a unified cloud computing platform to solve the security control, management tools, backup, disaster recovery and a series of deficiencies that existed before, reliability of the overall business operation of the hospital has been greatly improved. With a restart problem, especially for hardware systems, virtualization device can greatly improve the efficiency of operating system restarts, and reduce the start time, as shown in Figure 5, except for EMR and LIS, because the use of Oracle database, is only slightly faster at boot time, other systems are significantly faster at boot time.

4.2 *Improve the operational efficiency of medical personnel*

By implementing desktop virtualization, the doctor, nurse workstations, and fee collectors are classified into general access type, general operational, supervisory staff and other personnel to collect different types of applications. It will extract the same type of user application requirements and CPC has been installed in a virtual machine template. When new users join in, it simply generates user virtual machine from a template. As a result, user virtual machine and VPN technology improve the efficiency of the medical personnel to operate the hospital information system. The Internet network in smart handheld devices can successfully access the hospital information system, shown in Figure 6. The cloud computing application virtual machines connected with network could compare year medical care re-opening record. Health care working can be easier due to the operation of information systems. The situation that the medical record is asked to reopen because of not being completed in time has reduced a lot. Except for February and April, the reopening medical records in other months have reduced.

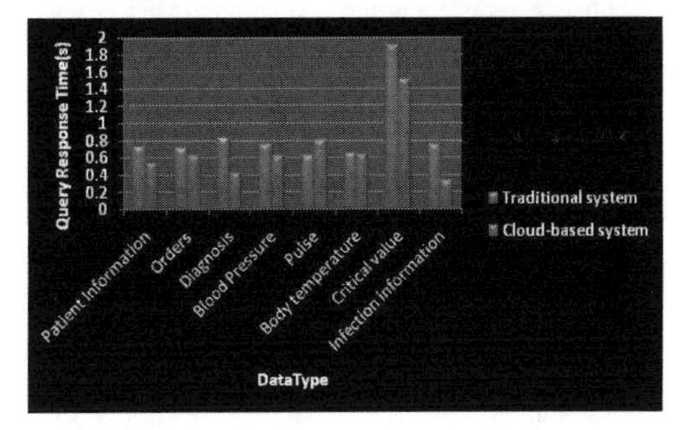

Figure 7. CDSS data query response time.

4.3 *Comparative analysis*

Comparison of CDSS in traditional systems and in cloud-based system is shown in Figure 7, we found that the cloud-based system's query response time is much less than the traditional system. Because the cloud-based system is based on VMware ESXi, it can run faster.

5 CONCLUSIONS

This paper introduces a series of problems existing in hospital information systems. It summarizes the advantages of cloud-based system and analyzes the feasibility of using the hospital information system which is based on cloud-based system. We achieve a flexible cloud-based hardware and scalable software platform in our hospital. Adoption of the cloud computing information platform, not only significantly reduces construction costs, but also provides a wide range of clinical applications to achieve information sharing. Meanwhile, medical personnel can quickly and easily access resources. In addition, we show that the cloud computing platform can improve the operating speed of the system and reduce the time of physicians query by experiments. Based on the new platform and relying on powerful servers, it can achieve a handheld device and access the external network. The cloud infrastructure does not only realize the convenience of updating and the application process, but also achieve a unified basis for the construction of the hospital system.

ACKNOWLEDGMENTS

This research has partially been supported by the National Natural Science Foundation of China under Grant No. 60673190 and 61203244, College Natural Science Research of Jiangsu Province under Grant No. 14KJB520008, Senior Technical Personnel of Scientific Research Fund of Jiangsu University under Grant No. 13JDG126.

REFERENCES

Cao, X. J. & Wang, X. F.. 2013. Application of business intelligence in hospital information management. *Applied Mechanics and Material* 373: 1098–1101.

Lee, K., Wan, T. T. & Kwon, H. 2013. The relationship between healthcare information system and cost in hospital. *Personal and Ubiquitous Computing* 17(7): 1395–1400.

Chan, Y. S., Liang, H. J. & Lin, Y.H. 2014. Using wireless measuring devices and Tablet PC to improve the efficiency of vital signs data collection in hospital. *IEEE International Symposium on Bioelectronics and Bioinformatics (ISBB), Chung Li: 1–4 April 2014.* IEEE.

Miroslav, B., Lenka, L., Vaclav, C., et al. 2013. Visualization in information retrieval from hospital information system. Advances in Intelligent Systems and Computing, *Proc. The 7th International Conference, SOCO'12, Ostrava, Czech Republic, 5–7 September 2013.*

Tsumoto, S., Hirano, S. & Iwata, H. 2013. Mining nursing care plan from data extracted from hospital information system. *Proc. The 2013 IEEE/ACM International Conference on Advances in Social Networks Analysis and Mining, Niagara Falls, Canada, 25–28 Auguest 2013.* IEEE.

Fattouh, I. L., Suzan, S., Shahd, H., et al. 2014. Enhanced hospital information system by cloud computing: SHEFA'A. *In Design, User Experience, and Usability. User Experience Design for Everyday Life Applications and Services*: 56–62. Springer International Publishing.

Rajkumar, B. & Rajiv, R. 2011. Federated resource management in grid and cloud computing systems. *J Future Gener Comput Syst* 26(5): 1189–1191.

Avram, M. G. 2013. Advantages and challenges of adopting cloud computing from an enterprise perspective. *Proceedings of The 7th International Conference Interdisciplinarity in Engineering, INTER-ENG 2013*: 529–534.

W. Dai, Rubin. 2012. Service-oriented knowledge management platform. *In: Proceedings of the 13th IEEE international conference on information reuse and integration* 15(4): 255–259.

Electronics, Communications and Networks IV – Hussain & Ivanovic (eds)
© 2015 Taylor & Francis Group, London, ISBN: 978-1-138-02830-2

A multicast transmission scheme based on block acknowledgment with network coding in wireless LANs

Chun-Xiang Chen*, Yuto Izumi & Jianfei Ai
Department of Management and Information Systems, Prefectural University of Hiroshima, Hiroshima, Japan

ABSTRACT: In this paper, we apply the network coding (NC) technology to multicast transmission in wireless LANs and propose a retransmission scheme to recover packets from transmission error. We focus on the feedback channel and the period of the block acknowledgment (ACK) phase in which receivers inform the sender of the reception status of data packets by sending back an ACK packet. In order to improve the system performances in terms of the reception ratio of the data packet and the average number of packets decoded at each receiver, the receiver embeds the packets which have been received successfully into the acknowledgment packet with network coding. The performances are evaluated by computer simulation, and the results show that the proposed scheme achieves better performances than the existing scheme.

1 INTRODUCTION

With the development of wireless LAN (WLAN), multicast communication via WLAN is widely used not only for streaming-content delivery (such as voice, video etc.), but for data delivery service with high reliability. However, in multicast communication based on the IEEE Std 802.11 (2007) DCF (Distributed Coordination Function), error control mechanism is not implemented for the recovery from data loss.

To recover the packets from transmission error, the ARQ (automatic repeat request) schemes are widely employed in which each receiver sends a request to the transmitter for retransmission of lost packets. Compared with the unicast communication, multicast transmission have multiple receivers. Thus, the retransmission of the lost packet will increases the transmission delay of packets and will affect the quality of multicast communication.

Network coding (NC) is a new approach for enhancing the quality of wireless communication (Ahlswede et al. 2000, Katti et al. 2008, Elloumi et al. 2014). NC has been proposed originally for maximizing the network information flow in broadcast/multicast transmission (Ahlswede et al. 2000, Matsuda et al. 2011). It has been studied and applied in various areas such as throughput enhancement, robustness enhancement, security etc (Matsuda et al. 2011). The core concept of NC is to encode several packets from a single link or multiple links into one packet (called NC packet) at the transmitter or at an intermediate node. At each receiver, the packets which have not been received are decoded from the NC packet based on the packets which have been received successfully. Wu & Zheng (2011) presented an NC based multicast retransmission scheme in mobile communication networks and analyzed the optimal number of packets encoded into a retransmission packet. Nakagawa et al. (2012) proposed a selective retransmission method of lost packets with small packet loss for multimedia data, and evaluated the reception ratio and transmission delay by computer simulation.

In this paper, we propose a network coding based retransmission scheme for multimedia data in multicast communication over WLAN. The core idea of the proposed scheme is similar to that of Nakagawa's scheme (Nakagawa et al. 2012). However, we focus on the block acknowledgment period, and introduce an NC based acknowledgment scheme for enhancing the system performance in terms of the reception ratio and the average number of packets decoded at each receiver. We clarify the performances by computer simulation.

The rest of this paper is organized as follows. Section 2 describes the system model and the operations of the proposed scheme. Section 3 shows the evaluation of the system performances in terms of reception rate of packets and the average number of packets decoded at each receiver. Conclusion remarks are given in Section 4.

*Corresponding Author: *chen@pu-hiroshima.ac.jp*

2 NETWORK CODING BASED ACKNOWLEDGMENT

In this section, we describe the network model proposed in this paper and give its operations by comparing with the existing scheme (Nakagawa et al. 2012).

Figure 1 illustrates an example of NC based retransmission of a multicast communication system which consists of a base station S (multicast sender) and four multicast receivers (A, B, C and D). In Figure 1, S has consecutively sent packets P1, P2, P3, P4 and P5 to the multicast group. At receiver A, packets P1, P2, P4 and P5 have been received correctly, but packet P3 has not. Each receiver accesses its WLAN according to CSMA/CA of 802.11 DCF and sends back an acknowledgment packet (ACK) to the sender (base station) to acknowledge the sender that which packet has been received correctly or not.

In the traditional ARQ (Automatic Repeat re-Quest) scheme, the sender must retransmit packets P1, P2, P3 and P5 in sequence at the cost of 4 slots, where slot is defined as the time period of the transmission time of a fixed-size packet. On the contrary, with the retransmission scheme based on a simple network coding (Wu and Zheng 2011, Katti et al. 2008) , each receiver stores the packets received correctly for a short time for decoding the lost packet. At the situation shown in Figure 1, for example, base station S simply mixes packet P1 with packet P2 in exclusive OR (XOR), generates a mixed packet PX, and broadcasts PX to the multicast group, then generates the next NC-ed packet X by XOR-ing packet P3 with packet P5 and broadcasts it to the multicast group. Each receiver attempts to decode the packet which has not been received correctly yet. At receivers A and D, they ignore packet PX because they have successfully received packets P1 and P2. At receivers B (or C), the receiver gains packet P2 (or P1) by XOR-ing PX with P1 (or P2) which has been received correctly. In the same way as described above, receivers A, C and D can get packets P3, P3 and P5 by decoding packet X. Obviously, base station S retransmits the four packets in 2 slots, and each receiver obtains

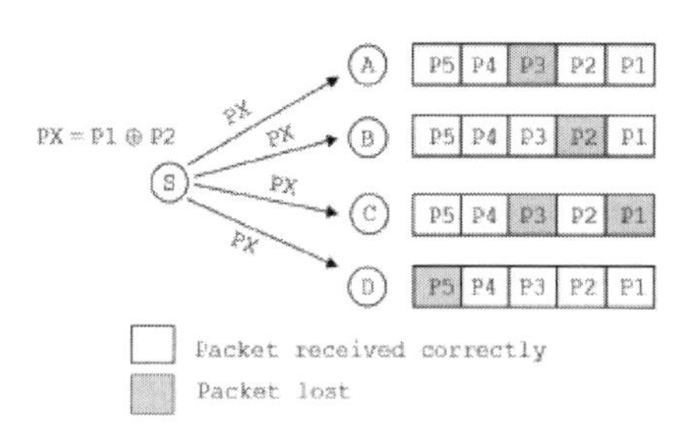

Figure 1. An example of NC based retransmission.

the packets which are lost in the first transmission. The NC based retransmission scheme greatly reduces the retransmission time.

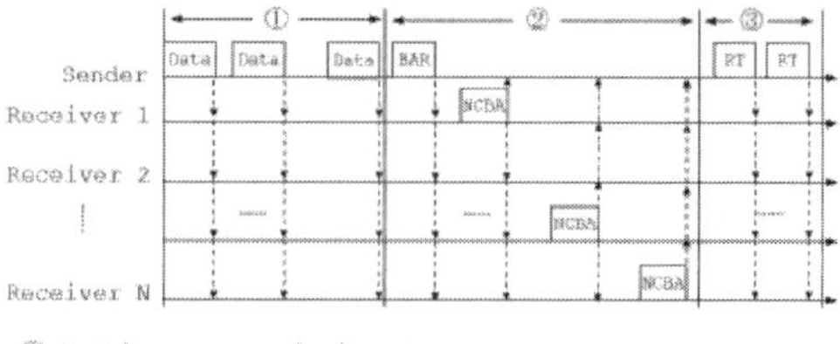

Figure 2. One cycle of NC based retransmission.

It is not always so favorable as described above. NC based retransmission largely depends on the situation of lost packets. At a receiver, it can not decode the packet if the NC-ed packet includes two or more packets which have not been received correctly at that receiver. In Nakagawa et al. (2012), a selective retransmission scheme based on NC is presented. The system performance in terms of reception ratio of packets and packet delay is evaluated by computer simulation, and the effectiveness of the scheme is shown by comparing with other schemes.

The idea of the scheme proposed in this paper is similar to that of Nakagawa et al. (2012). However, we apply the network coding to the block ACK packet in the acknowledgment period. This revision characterizes several features in our scheme. Figure 2 shows one cycle of the operation of the proposed scheme. It consists of a transmission phase of data packets, a block acknowledgment phase and a retransmission phase. The operation in each phase is described in the following subsections.

2.1 *Transmission phase of data packets*

When the base station seizes the transmission right according to IEEE 802.11 DCF (i.e. the first packet are transmitted without collision), it successively transmits its data packets stored in its buffer for transmission. The number of packets which can be transmitted successively is assumed to be N_{burst}. Once a packet is transmitted, it is stored for a short time for retransmission until the next transmission cycle. The sender inserts a SIFS (Short Inter Frame Space) between two packets. When the number of packets successively transmitted reaches N_{burst}, or all the packets waiting for transmission in the buffer are transmitted, the sender sends out a BAR (Block ACK Request) informing the receivers that the transmission of data packets is finished, then switches to block acknowledgment phase.

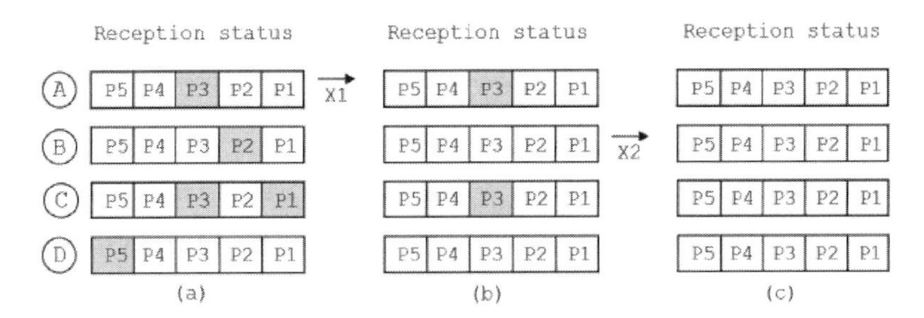

Figure 3. Operation of NC based retransmission.

At each receiver of the multicast group, the receiver collates the received packets and stores the packet without error in its receiving buffer and discards the packet received in error.

2.2 *Block ACK phase with network coding*

When the receiver receives the BAR packet, it understands that the transmission of data packets in burst has finished, and switches to block acknowledgment phase. In this phase, each receiver examines the received packets and creates the corresponding block acknowledgment (BA) with network coding (network-coded BA: NCBA). Unlike the existing scheme (Nakagawa, Tanabe, Tanigawa, & Tode 2012), the receiver encodes all the data packets into the data field of this BA by simply XOR-ing all packets which have been received correctly and sends this NCBA to the multicast group. This scheme adroitly makes use of the wireless channel by broadcasting the NCBA not only to the sender but to all receivers. Therefore, some of the receivers are able to gain the data packet by decoding this NCBA in block acknowledgment phase.

Figure 3 illustrates an example of operation in the block acknowledgment phase. Figure 3(a) shows the reception status at each receiver at the end of multicast transmission phase. The sender has transmitted continuously 5 data packets to its 4 receivers, and the packets lost at each receiver is painted. In the block acknowledgment phase, when a receiver (i.e. A) gets the right of transmission, it sends its NCBA X1 to the multicast group. In the case here, receiver A sends out its NCBA which includes the reception status and the encoded data of packets (P1, P2, P4 and P5). Sender S and other receivers B, C and D receive this NCBA, and attempt to extract the packets which have been lost in the multicast transmission phase by XOR-ing this NCBA with the packets received correctly. Therefore, receivers B, C and D are able to recover packets P2, P1 and P5, respectively. After the transmission of the NCBA of receiver A, the reception status at each receiver is shown in Figure 3(b). We assume that receiver B gains the transmission right to send its NCBA packets X2 successfully at this time. At this moment, receiver B has received all the 5 data packets correctly, its NCBA includes the data packet made from the 5 packets (P1, P2, P3, P4 and P5). When receivers A and C receive the NCBA from receiver B, receiver A and B are able to recover packet P3 by decoding the NCBA with the packets correctly received so far. The reception status at each receiver is shown in Figure 3(c) after the transmission of B's NCBA. Obviously, it is possible to enhance reception rate of data packets at each receiver by introducing the network coding into the acknowledgment packet.

2.3 *Retransmission phase*

After the sender receives all NCBAs from all the receivers, it knows completely the reception status at each receiver. In retransmission phase, the sender consists of the retransmission data packet according to the following procedures.

- If at least two packets are lost (received incorrectly) at different receivers, the sender determines the lost packets which are lost at the most receivers, and chooses the one with the smallest sequencing number as the first packet (say packet L_1), then chooses another lost packet (say packet L_2) which satisfies that $P_m = L_1 \oplus L_2$ can be decoded at as many receivers as possible, sends P_m to the multicast group.
- If there are no packets satisfying $L_m = L_1 \oplus L_2$, the sender retransmits these lost packets individually.
- The sender repeats the procedures described above until all of the packets needed retransmission are retransmitted.

In multicast communication for multimedia, introducing the retransmission for error control will increase the delay of data packet. Therefore, an upper bound limit of the interval of the block acknowledgment phase (defined as D_{BAP}) and/or the number of the retransmission packets in retransmission phase may be introduced.

3 SIMULATION RESULTS

To clarify the effectiveness of the proposed scheme, we investigate the system performances in terms of the reception ratio of the packets and the average number of packets decoded at each receiver by computer simulation. Table 1 gives the simulation environment. The topology of the multicast communication system consists of a sender (base station of WLAN) and M multicast receivers which are put in the circle around the sender within 100-meter radius. Size of the data packet is 512 bytes and the arrival interval of data packets is 5 msec. The transmission rate of the WLAN is 2Mbps.

Table 1. Simulation environment.

CPU	Intel Core(TM) i7-2600, 3.4GHz
Memory	8.00GB
OS	Windows 7 Pro. 64bit
Simulator	QualNet 6.1

Compared with the existing scheme (Nakagawa et al. 2012), our scheme introduces the network coding of the data packets into the acknowledgment packet. Therefore, it is unavoidable that the block acknowledgment phase will become longer. In our simulation, we assume that $N_{burst} = 8$ ($N_{burst} = 16$ in Nakagawa et al. (2012)) and $D_{BAP} = 70$ msec. If the interval of the block acknowledgment phase exceeds the time limit D_{BAP}, the block acknowledgment phase will be terminated, and the operation turns to retransmission phase.

In the retransmission phase, the number of retransmission packets with network coding is limited to be 2 because some of the lost packets may be retransmitted in block acknowledgment phase.

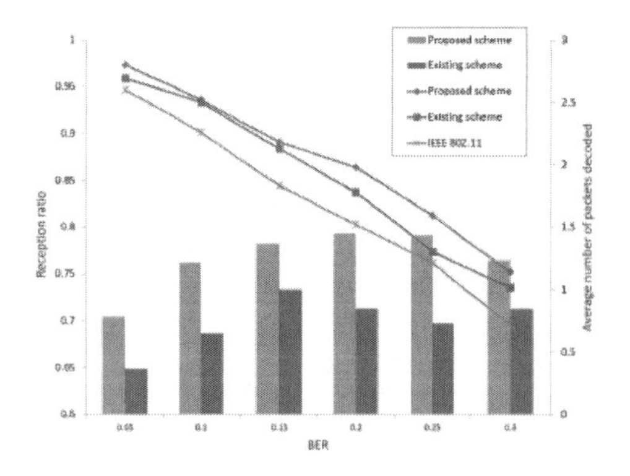

Figure 4. Reception ratio of data packet and average number of packets decoded at each receiver ($M = 3$).

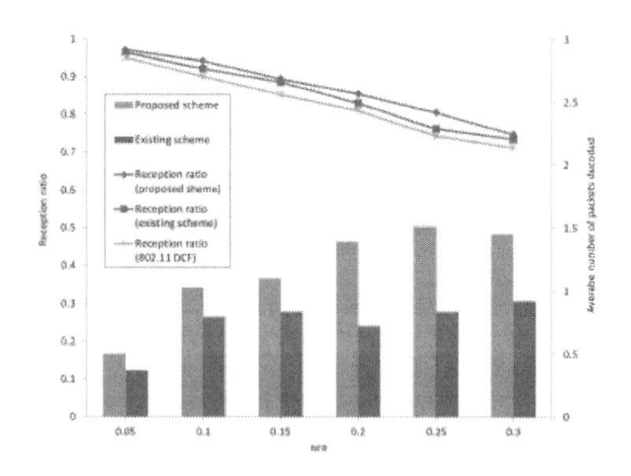

Figure 5. Reception ratio of data packet and average number of packets decoded at each receiver ($M = 5$).

Figure 4 shows the reception ratio of data packets (line graph) and the average number of packets decoded at each receiver (bar graph) as the PER (Packet Error Rate) changes. Figures 5 and 6 show the cases at $M = 5$ and $M = 10$, respectively. By comparing the proposed scheme with the existing scheme and 802.11 DCF, it is clear that (1) the reception ratio of data packet decreases as the PER increases; (2) the reception ratio of the proposed scheme is the largest among the three schemes; (3) the average number of packets decoded at each receiver is superior to that of the existing scheme.

Figure 7 illustrates the reception ratio and the average number of packets decoded at each receiver of our scheme, where $M = 2; 5;$ 10 and 15. The reception ratio decreases as the packet error rate increases. However, the amount of decrement of the reception ratio becomes smaller as the number of the receivers (M) increases. The result shows that the suitable number of the receivers is 10.

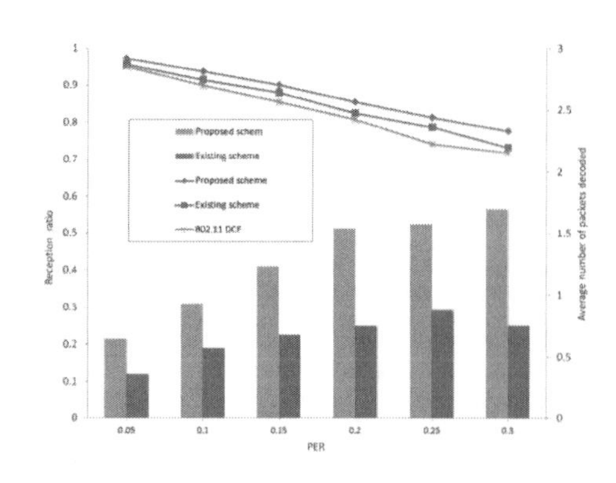

Figure 6. Reception ratio of data packet and average number of packets decoded at each receiver ($M = 10$).

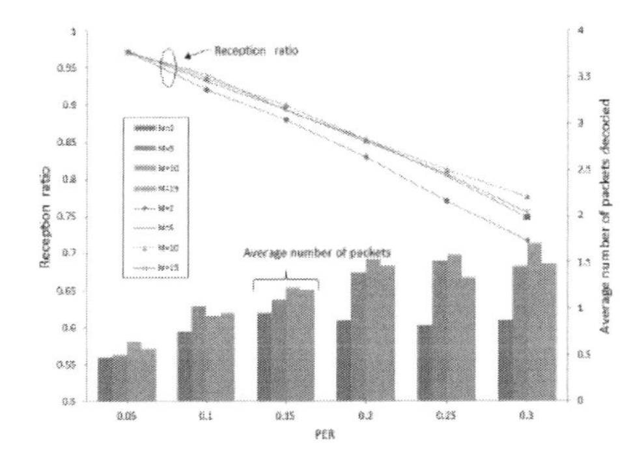

Figure 7. Reception ratio of data packet and average number of packets decoded at each receiver.

4 CONCLUDING REMARKS

In this paper, we focus on the block acknowledgment phase in the multicast transmission system and have presented a novel multicast transmission scheme with network coding in the block ACK phase. Taking the feature of radio wave into account, our scheme makes use of the wireless channel by embedding the data packet into the acknowledgment packet and broadcasting it to the multicast group. To restrict the increment of packet delay, the number of packets with NC in the retransmission phase is limited to two.

The numerical results of simulation show that the performances in terms of the reception ratio and the average number of packets decoded at each receiver are superior to those of the existing scheme. As the future works, the performances of the proposed scheme should be evaluated in practical environment with confliction and/or unicast communication.

REFERENCES

Ahlswede, R., Ning, C., Li, S.-Y. & Yeung, R. 2000. Network information flow. *IEEE Trans. Inf. Theory, 46*(4), 1204–1216.

Elloumi, S., Gourdin, E. & Lefebvre,T. 2014. Does network coding improve the throughput of a survivable multicast network? *10th International Conference on the Design of Reliable Communication Networks (DRCN)*, 1–8.

IEEE Std 802.11. 2007. *Wireless LAN Medium Access Control (MAC) and Physical Layer (PHY) Specifications*. IEEE.

Katti, S., Rahul, H., Hu, W., Katabi, D., Medard, M. & Crowcroft, J. 2008. Xors in the air: practical wireless network coding. *IEEE/ACM Transaction on Networking, 16*(3), 497–510.

Matsuda, T., Noguchi, T. & Takine, T. 2011. Survey of network coding and its applications. *IEICE Trans. Commun. E94-B*(3), 698–717.

Nakagawa, M., Tanabe, S., Tanigawa, Y. & Tode, H. 2012. Multicast transmission method based on selective retransmission with network coding in wireless lans. *IEICE Technical Report, NS2011–236*, 319–322.

QualNet. Qualnet simulator version 6.1. *Scalable Network Technologies*, http://www.scalable-network.com/content/qualnet.

Wu, H.&Zheng, J. 2011. Coret: A network coding based multicast retransmission scheme for mobile communication networks. *Proceedings of IEEE ICC2011*, 2011–015–05.

Electronics, Communications and Networks IV – Hussain & Ivanovic (eds)
© 2015 Taylor & Francis Group, London, ISBN: 978-1-138-02830-2

An enhanced TCP for satellite network with intermittent connectivity and random losses

Qiongbing Chen, Yong Bai* & Liang Zong
College of Information Science & Technology, Hainan University, Haikou, Hainan, China

ABSTRACT: When mobile users transmit data over heterogeneous satellite network through satellite gateway, there are random packets losses due to bit errors on satellite links, and intermittent connectivity may also occur when the mobile users disconnect from the satellite gateway. TCP performance degrades drastically if invoking the congestion control mechanism when encountering intermittent connectivity and random losses. This paper proposes a TCP scheme, called TCP M-Veno, which incorporates the mechanism of M-TCP at the gateway to deal with the intermittent connectivity, and refines TCP Veno at TCP sender to mitigate the effects of random losses and long propagation delay. The performance improvement of our proposed TCP M-Veno is demonstrated by simulations by comparing our proposed TCP scheme to other three TCP variants, NewReno, Veno, and M-TCP.

1 INTRODUCTION

With the development of the satellite network tech nology, satellite network has gradually become an indispensable part of the Internet. The satellite network can be integrated with other wireless networks to support new services through satellite gateway. For instance, a mobile ad hoc network (MANET) can be established between ocean fishery ships, and the satellite network can be integrated with the maritime MANET and terrestrial network to form a heterogeneous network, which can support ship-to-ship and ship-to-shore communications (Du et al. 2010, Du & Bai 2013, Zhou et al. 2013). To support environmental and ecological research, the architecture of a sensor network that uses satellite communication to transfer data from remote sensors was presented in (Ye et al. 2008).

Transmission Control Protocol (TCP) provides reliable delivery of data stream at the transport layer (Allman & Stevens 1999, Floyd & Henderson 1999). The original design of TCP protocol mainly considered the terrestrial wireline networks. When the TCP protocol is applied over satellite network, the characteristics of satellite network pose challenges on TCP. Firstly, high latency occurs in the data delivery mainly due to long propagation delay over GEO satellite link. When the connections traverse Geostationary (GEO) links, the Round Trip Time (RTT) for sending one packet and receiving its ACK can be above 500ms. Secondly, high bit error rate (BER) can lead to random packet losses. Without the information about the reason of losses, TCP treats each packet loss as an indication of network congestion, and the TCP sender will reduce its sliding window size to avoid congestion collapse, which can substantially reduce the overall throughput. Thirdly, the mobile nodes in the network can experience intermittent connectivity with the network due to temporary link outages, signal path blockages, interference, and moving away from the satellite gateway. Such disconnections result in packet losses, TCP still considers them as the signs of congestions and shrinks its congestion window to limit the amount of data being sent into the network. In addition, TCP retransmission time out (RTO) mechanism employs the exponential backoff algorithm where the retransmission interval is doubled for each retransmission. Repeated losses lead to smaller sliding window size and larger RTO, which means fewer amounts of data being sent into network and longer intervals between each attempt to probe network connectivity.

To deal with the above negative characteristics of satellite network, a number of solutions have been proposed for improving TCP performance in satellite networks (Allman et al. 1999, Allman et al. 2000, Taleb et al. 2011, Botta & Pescape 2014, Pirovano & Garcia 2013). The proposed solutions can be roughly divided into the following three categories: end-to-end improvement solutions (Brakmo & Peterson 1995, Fu & Liew 2003, Omueti & Trajkovi 2007, Roseti et al. 2010, Luglio et al. 2012), link-layer solutions (Kuhn

*Corresponding author: *bai@hainu.edu.cn*

et al. 2014, Dunaytsev et al. 2011), and split connection solutions (Balakrishnan et al. 1995, Balakrishnan et al. 1997). Specifically, to address the issue of long propagation delay, TCP Veno (Fu & Liew 2003) adopts the same methodology of Vegas (Brakmo & Peterson 1995) to estimate the backlogged packets in the network and select the minimal RTT as a reference to derive the optimal throughput the network can accommodate. To address the issue of random loss, Veno further improves Vegas with the abilities of distinguishing random loss and congestion loss and performing different sending reactions. TCP ADaLR (Omueti & Trajkovi 2007) introduces adaptive window increase and loss recovery mechanisms to address TCP performance degradation in satellite networks. On the other hand, M-TCP (Balakrishnan et al. 1997) was proposed mainly to improve TCP performance in the case of intermittent connectivity. In M-TCP, the sender will not suffer the reduction of sliding window by keeping the size of congestion window unchanged when the mobile node disconnects from the network. TCP Noordwijk was proposed to efficiently transmit short amount of data (i.e., Web traffic) by adoption of a burst based paradigm instead of window based paradigm of TCP standard (Roseti et al. 2010, Luglio et al. 2012). It has been evaluated over DVB-RCS satellite links for high speed trains with frequent interruptions in millisecond level, but it has not been evaluated for the connections with both random losses and long disconnection period (e.g., in second level) or for the non-Web traffics (e.g., large file download traffic using FTP).

In the existing studies for improving TCP performance over satellite network, there is still lack of solutions that can deal with the intermittent connectivity and high BER effectively at the same time. In this paper, we propose an enhanced TCP scheme to fulfill this dual task. Our proposal combines the benefits of M-TCP (which can deal with frequent disconnections) and Veno (which can deal with random packet losses). In the sender side, TCP Veno is enhanced such that the slow start phase is divided into four subphases, which enables TCP to better utilize the satellite link; the gateway is improved with the M-TCP mechanism to deal with intermittent connectivity of mobile users. The performance improvement of our proposed TCP M-Veno is demonstrated by simulations by comparing it with other three TCP variants, NewReno, Veno, and M-TCP.

This paper is organized as follows. Section 2 provides an overview of TCP NewReno, Veno, ADaLR, and M-TCP. Section 3 describes our proposed TCP M- Veno scheme. The performance evaluation of TCP M- Veno, NewReno, Veno, and M-TCP is presented in Section 4. Finally, the conclusions are drawn in Section 5.

2 REVIEW OF RELATED TCP VARIANTS: TCP NEWRENO, VENO, ADALR, AND M-TCP

2.1 *TCP NewReno*

TCP uses a mechanism called slow start (SS) to increase the congestion window (*cwnd*) after a connection is initialized and after a timeout. It starts with a window of one (or two) times the maximum segment size (MSS). For every packet acknowledged, *cwnd* increases by 1 MSS so that the *cwnd* effectively doubles for every RTT. When *cwnd* exceeds a threshold, *ssthresh*, the algorithm enters a new state, called congestion avoidance (CA). In the CA phase, *cwnd* is additively increased by one MSS every RTT as long as non-duplicate ACKs are received. When a packet is lost, the likelihood of duplicate ACKs being received is very high. In TCP Reno (Allman & Stevens 1999), if three duplicate ACKs are received, Reno will halve the *cwnd*, set *ssthresh* to new *cwnd*, perform a fast retransmit, and enter a phase called fast recovery. In the state of fast recovery, TCP retransmits the missing packet that was signaled by three duplicate ACKs, and waits for an ACK of the entire transmit window before returning to CA. If there is no ACK, TCP Reno experiences a timeout and enters the SS state. If an ACK times out, slow start is used. TCP NewReno (Floyd & Henderson 1999) is the most widely adopted TCP version. TCP NewReno further modifies TCP Reno within the fast recovery algorithm to solve the timeout problem when multiple packets are lost form the same window. When multiple packets are lost, a partial ACK acknowledges some but not all packets that are outstanding at the start of a fast recovery. TCP NewReno will result in TCP performance degradation since it treats random packet loss as congestion loss.

2.2 *TCP Veno*

TCP Veno (Fu & Liew 2003) was proposed with mechanism to distinguish between congestion loss and random packet loss due to bit errors, and provide different measures to deal with them. Veno makes use of a mechanism to estimate the state of the connection and deduce what kind of packet loss, congestion loss or error loss, is most likely to have occurred, rather than to pursue preventing packet loss. If the number of backlogged packets is below a threshold, the loss is considered to be random. Otherwise, the loss is said to be congestion induced.

First, it estimates *BaseRTT* as the round trip time of a segment when the connection is not congested, in practice, the minimum of all measured round trip times. Next, the sender measures the so-called Expected and Actual rates:

$Expected = cwnd/BaseRTT \; Actual = cwnd/RTT$

and obtain the difference of them:

$DIFF = Expected - Actual.$

When $RTT > BaseRTT$, there is a bottleneck link where the packets of the connection accumulate. The backlog at the queue, denoted by N, can be calculated as:

$N = Actual \times (RTT - BaseRTT) = DIFF \times BaseRTT.$

Using N as an indication of whether the connection is in congestive state. The connection is in the congestive state if $N \geq \beta$. When a packet loss is detected in this state, it assumes a congestion loss; the connection is not in the congestive state if $N < \beta$. When a packet loss is detected in this state, it assumes an error loss. β is a threshold value for deciding whether the connection is in the congestive state, and its value is suggested to be determined by experimentation. In Vegas, *BaseRTT* is continually updated throughout the TCP connection. In Veno, *BaseRTT* is reset whenever packet loss is detected either due to timeout or duplicate ACKs.

2.3 *TCP ADaLR*

TCP ADaLR (Omueti & Trajkovi 2007) introduces the following modifications at the sender: a scaling component ρ, adaptive window (*cwnd* and *rwnd*) increase, and loss recovery mechanisms. A scaling factor ρ is calculated by the measured RTT as $\rho = (sampleRTTs/1s) \times 60$. The *sampleRTT* is normalized by 1s. The value of 60 is the minimum recommended value for the maximum RTO normalized by 1s.

In ADaLR, the SS phase is divided into four subphases based on the current *cwnd*, *ssthresh*, *rwnd* (receiver advertised window), and *flightsize* (total unacknowledged byte in the network). A breakpoint $\rho = 15$ is selected to define the transition for the adaptive *cwnd* mechanism. The breakpoint value ($\rho = 15$) corresponds to a *sampleRTT* of 250 ms. The *cwnd* is incremented exponentially for $\rho < 15$ as the default TCP SS phase. For $\rho \geq 15$, *cwnd* is increased as $(\sqrt{\rho}/4) \times MSS$. If losses occur or if the conditions of the four sub-phases do not hold, *cwnd* is increased as in conventional TCP SS phase.

The loss recovery mechanism of ADaLR modifies the fast recovery phase to enable the back-to-back transmission of two segments. The default maximum ACK delay of 200 ms is added to the current time used in the computation of the subsequent RTO value.

2.4 *M-TCP*

M-TCP (Balakrishnan et al. 1997) is a modified version of TCP, which takes the split connection approach and divide the connection into wireline and wireless domains. Fig. 1 illustrates M-TCP in a simplified network model. TCP connections are split into two at the Super Host (SH), which acts as a gateway. The TCP sender at the Fixed Host (FH) uses standard TCP to send data to the SH. The TCP client at the SH (called SH-TCP), receives data packets transmitted by the sender, and sends these packets to the M-TCP client at the Mobile Host (MH). The ACKs received by M-TCP at the SH are forwarded to SH-TCP for delivery to the sender.

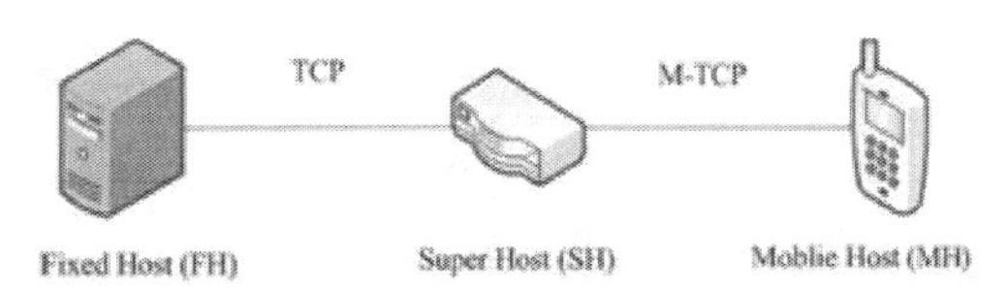

Figure 1. Illustration of M-TCP.

The M-TCP approach is to choke the TCP sender when the MH is disconnected, and allow the sender to transmit at full speed when the MH reconnects by manipulating the TCP sender's window. In case of disconnections, the sender is forced into *persist* mode by receiving an indication via special ACK from M-TCP. This ACK contains a TCP window size update that sets the sender's window size to zero. In *persist* mode, the sender preserves the size of its congestion window, and freezes its timer such that it will not suffer retransmit timeout and will not exponentially back off its retransmission timer. When the connection is regained, M-TCP at the MH sends a specially marked ACK to M-TCP at the SH which contains the sequence number of the highest byte received thus far. It unfreezes M-TCP timers to allow normal operation to resume.

To implement M-TCP, the SH needs to handle packets from server and client as follows:

2.4.1 *Handling packets from server*

Upon receiving a segment from the FH, the SH forwards that segment to the MH. It does not ACK this data segment until receiving an ACK from the MH. This ensures that the end-to-end TCP semantics are ensured. When the MH temporarily disconnects, the sender at the FH is forced into *persist* mode. The mechanism is performed as follows: if *rwnd* contains w bytes. Assume that the client has ACKed bytes up to $w' \leq w$. The SH sends an ACK for bytes up to $w' - 1$ in the normal way. When the MH temporarily disconnects from the network because it stops receiving ACKs for bytes transmitted after w'. M-TCP sends an indication of this fact to the SH which then sends an ACK for the w' th byte to the sender. This ACK contains a TCP window update (*rwnd* =0).

Because the sender's window size, *wnd*, is the minimum of *cwnd* and *rwnd*, i.e., *wnd = min(rwnd, cwnd)*, the sender's window size is set to zero when receiving an ACK with *rwnd=0*. In addition, this ACK triggers the sender to enter into the *persist* mode. In the *persist* mode, the sender preserves the size of its congestion window, and freezes its timer. Thus, it will not suffer retransmit timeout and will not exponentially back off its retransmission timer.

When the MH regains its connection, it sends a reconnection ACK to the SH. When the SH receives that packet, it forwards it ACK to the sender. Upon receiving reconnection ACK, the SH will retransmit all the packets that have not been acknowledged by the client. When receiving the reconnection ACK, the sender resumes its transmission and exits the persist state. Since the sender never times out, it never performs TCP congestion control. Thus, the sender can resume transmission at the rate before the packet loss occurred.

2.4.2 *Handling packets from client*

M-TCP maintains a packet queue in order to retransmit packets from the SH, as opposed to retransmit from the sender when packets are lost between the SH and the MH. A packet is removed when its ACK is received from the client. If the packet contains a duplicate ACK number, M-TCP checks the packet queue. If the queue is empty, M-TCP forwards the packet to the sender. If this is a new ACK packet which indicates that the MH could have reconnected and replied to the packet sent by the SH. If the packet contains a new ACK number, the newly acknowledged packet in the queue is purged, and the last ACK number reduced by one is sent to the sender.

3 PROPOSED TCP M-VENO

To deal with the challenges of both intermittent connectivity and random losses over satellite network, we propose TCP M-Veno by taking advantages of M-TCP and TCP Veno. TCP M-Veno combines the benefits of M-TCP and Veno. It incorporates the mechanism of M-TCP to let the sender not invoking congestion control during disconnections of the mobile nodes. To further mitigate the effect of long delay and high BER, we enhance the TCP Veno by refining the slow start phase of Veno, and call it Veno+. In our proposed TCP M-Veno, TCP connections are split into two at the gateway as shown in Figure 2. The sender implements TCP Veno+, and the gateway implements the M-TCP mechanisms. Next, we describe the improvements on TCP sender for supporting TCP M-Veno.

In the sender side of TCP M-Veno, TCP Veno is enhanced and called Veno+ to enable TCP to better utilize the satellite link. Specifically, our TCP M-Veno refines the SS phase of TCP ADaLR. In the CA phase, TCP M-Veno employs the window update algorithm for loss recovery as that in TCP Veno. Next, we describe the refined adaptive cwnd increase algorithm in the SS phase and the window update algorithm for loss recovery in the CA phase.

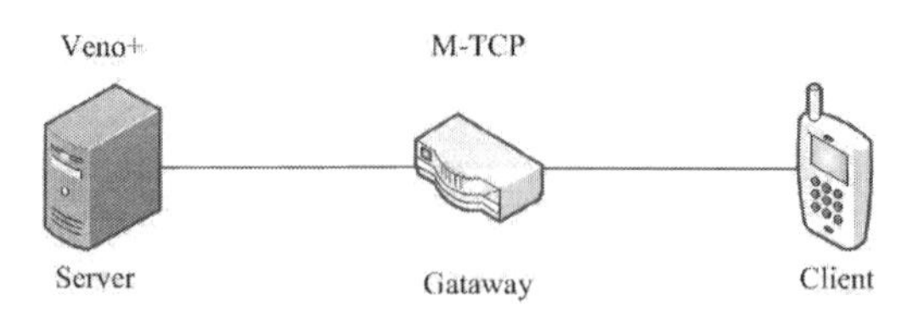

Figure 2. Illustration of TCP M-Veno.

3.1 *Refined adaptive cwnd increase algorithm*

As in TCP ADaLR, the SS phase is still divided into four sub-phases based on the current *cwnd*, *ssthresh*, *rwnd*, and *flightsize*. In our proposal, the M-Veno has a different increase rate in four sub-phases for *cwnd* in order to be adaptive to different conditions. In the sub-phase 1, the values of *cwnd* and *flightsize* are very small, which means that there are not much sent and unacknowledged data in the network. Therefore, thesent rate is increased as $(\sqrt{\rho}/4) \times MSS$. In the subphase 2 and 3, the values of *cwnd* and *flightsize* are in the intermediate range. In TCP M-Veno, we propose to refine the increase of *cwnd* in the second and third sub-phases. In the second and third sub-phases of TCP M-Veno, *cwnd* is increased more aggressively by $(\sqrt{\rho}) \times MSS$ instead of at $(\sqrt{\rho}/4) \times MSS$ TCP ADaLR. In the last sub-phase, *cwnd* and *flightsize* have relatively larger values. It is more likely that large backlogs exist in the network. In order to keep the network stability, it is appropriate to increase *cwnd* less aggressively by $(\sqrt{\rho}/4) \times MSS$.

3.2 *Window update algorithm for loss recovery*

In the CA phase, the window update algorithm of TCP Veno for loss recovery is used in M-Veno instead of using the loss recovery in ADaLR. In Veno, the window update algorithm is as follows,

if ($N < \beta$) // available bandwidth underutilized
set *cwnd = cwnd + 1/cwnd*
when each new ACK is received;
else if ($N \geq \beta$) // available bandwidth fully utilized
set *cwnd = cwnd + 1/cwnd* when every other new ACK is received;

When the sender receives three duplicate ACKs, the window update algorithm in Veno is as follows,

if ($N < \beta$) // random loss due to bit errors is most likely to have occurred
set *ssthresh = cwnd ∗ 4/5*;

else // congestion loss is most likely to have occurred

set *ssthresh* = *cwnd*/2;

4 PERFORMANCE EVALUATION

The simulation for performance evaluation is conducted using OPNET software. Our simulated network scenario is an integrated heterogeneous network for maritime communications. The simulated network model is shown in the Figure 3. The networking model consists of terrestrial network, satellite network, and MANET of ships. MANET is established between ocean ships which can connect via satellite for remote communications with terrestrial network. The users (clients) on ships connect to access point first, and then connect to shipborne satellite gateway to download data from the terrestrial network from the server. The link rate of the server connecting to the terrestrial gateway is set to 10Mb/s. The downlink data rate of satellite network is 2048kb/s, and uplink data rate of satellite network is 256kb/s. The one-way link propagation delay of satellite is 250ms. The BER of the satellite link is from 10–9 to 10–5. The client has 20 seconds disconnections for every 5 minutes. The application used in this connection is FTP. The server sends a 50Mbytes file size to the client. The terrestrial gateway implements the M-TCP mechanism. The Veno+ is implemented on the TCP clients. We evaluate and compare the performance of TCP M-Veno with M-TCP, TCP Veno, and TCP NewReno in such a networking scenario.

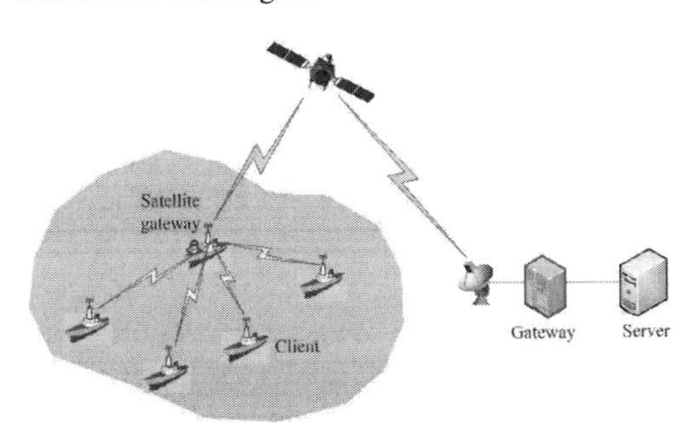

Figure 3. Simulated networking scenario.

4.1 *Simulation results*

4.1.1 *Update of cwnd*

Figure 4 shows the update of cwnd with intermittent interruptions at a low BER (10–9). It can be seen that the TCP Veno performs very closely to the TCP NewReno. TCP M-Veno outperforms other three TCP variants. The *cwnd* of TCP Veno with disconnections decreases drastically during the disconnection period and is unable to reach the value in TCP M-Veno. During the disconnections of the mobile terminal, the TCP M-Veno is forced into the persist mode to keep its *cwnd*. Hence, it avoids unnecessary decrease of *cwnd* and helps improving the transmission performance.

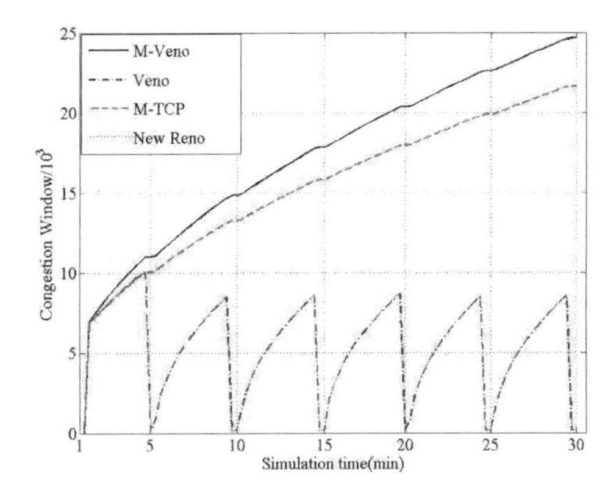

Figure 4. Update of *cwnd* of TCP variants at BER=10–9.

4.1.2 *Throughput*

Figure 5 shows the satellite link throughput of four TCP versus BERs. With the increase of BER, the satellite network throughput of four TCP variants decreases. When the BER changes from 10–7 to 10–5, the throughput decreases rapidly. At higher BER, (10-5) the throughput of TCP M-Veno is higher than that of other three TCP variants. Under low BER regime (10–9 to 10–8), TCP M-Veno and M-TCP has higher throughput than that of TCP NewReno. This is because the TCP M-Veno and M-TCP tends to have larger congestion window size that enables more data to be transmitted. When BER continues to increase, the M-TCP's throughput is close to that TCP NewReno. TCP M-Veno's throughput is higher than that of other TCP variants at all BERs.

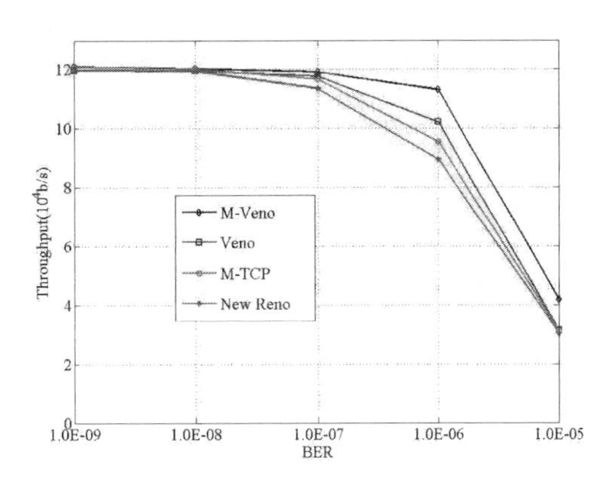

Figure 5. Throughput of TCP variants.

5 CONCLUSION

In the paper, we presented the TCP M-Veno algorithm for improving TCP performance over satellite networks in the presence of frequent disconnections and random losses. Our proposal combines the benefits of two previous TCP variants, M-TCP (which can handle intermittent connectivity) and TCP Veno (which can handle random packet losses). TCP M-Veno incorporates the M-TCP mechanism to deal with the disconnections of mobile users. The TCP Veno is further improved with adaptive window increase mechanism in the slow start phase to mitigate the effect of long propagation delay. To implement TCP M-Veno, the required improvements on TCP sender and gateway are described in this paper. Our proposed TCP scheme and related three TCP variants (NewReno, Veno, and M-TCP) are evaluated by simulations over an integrated heterogeneous networking scenario where satellite network is integrated with a mobile ad hoc network. The simulation results showed that our proposed TCP M-Veno outperforms other three TCP variants for such a networking scenario with intermittent connectivity and random losses.

ACKNOWLEDGEMENT

This paper was supported by the National Natural Science Foundation of China (Grant No. 61062006 and Grant No. 61261024) and the Special Social Service Project Fund of Hainan University, China (Grant No. HDSF201301).

REFERENCES

Allman, M. & Stevens, W. 1999. TCP congestion control. *RFC:* 2581.

Allman, M., Dawkins, S., Glover, D., et al. 2000. Ongoing TCP research related to satellites. *RFC:* 2760.

Allman, M., Glover, D. & Sanchez, L. 1999. Enhancing TCP over satellite links using standard mechanisms. *RFC:* 2488.

Balakrishnan, H., Srinivasan, S. & Katz, R. H. 1995. Improving reliable transport and handoff performance in cellular wireless networks. *Wireless Networks* 1(4): 469–481.

Balakrishnan, H., Srinivasan, S. & Katz, R. H. 1997. M-TCP: TCP for mobile cellular networks. *ACMSIGCOMM Computer Communications Review* 27 (5): 19–42.

Botta, A. & Pescapè, A. 2014. On the performance of new generation satellite broadband internet services. *IEEE Commun. Mag* 52 (6): 201–209.

Brakmo, L. S. & Peterson, L. L. 1995. TCP Vegas: end to end congestion avoidance on a global internet. *IEEE J. Sel. Areas Commun* 13: 1465–1480.

Du, W. & Bai, Y. 2013. Supporting maritime VoIP service over integrated heterogeneous wireless/wireline network. *In Proc. PIMRC, London, UK, 3461–3465 July 2013.* Piscataway: IEEE Communications Society.

Du, W., Ma, Z., Bai, Y., et al. 2010. Integrated wireless networking architecture for maritime communications. *In Proc. SNPD, London, UK, 134–138.* Piscataway: IEEE.

Dunaytsev, R., Dmitri, M., Yevgeni, K., et al. 2011. Modeling TCP SACK performance over wireless channels with completely reliable ARQ/FEC. *Int. J. Commun. Syst.* 24 (12): 1533–1564.

Floyd, S. & Henderson, T. 1999. The NewReno modification to TCPs fast recovery algorithm. *RFC 2582.*

Fu, C. P. & Liew S. C. 2003. TCP Veno: TCP enhancement for transmission over wireless access networks. *IEEE J. Sel. Areas Commun* 21 (2): 216–228.

Kuhn, N., Lochin, E., Lacan, J., et al. 2014. On the impact of link layer retransmission schemes on TCP over 4G satellite links. *International Journal of Satellite Communications & Networking* 1: 1–24.

Luglio, M., Roseti, C. & Zampognaro, F. 2012. Transport layer enhancements on a satellite-based mobile broadband system for high speed trains. *International Journal of Satellite Communications & Networking* 30 (6): 235–249.

Omueti, M. & Trajkovi, L. 2007. TCP with adaptive delay and loss response for heterogeneous networks. *In Proc. Wireless Internet Conf.(WICON), Austin, TX, USA 5.2007.* Brussels, Belgium.

Pathmasuntharam, J. S., Kong, P. Y., Zhou, M. T., et al. 2008. TRITON: High speed maritime mesh networks. *IN 2008 IEEE 19th International Symposium on Personal, Indoor and Mobile Radio Communications, Cannes, France 1-5 September 2008.* IEEE Communications Society.

Pirovano, A. & Garcia, F. 2013. A new survey on improving TCP performances over geostationary satellite link. *Network & Communication Technologies* 2(1): 1–18.

Roseti, C., Luglio, M. & Zampognaro, F. 2010. Analysis and performance evaluation of a burst based TCP for satellite DVB RCS links. *IEEE/ACM Trans. Networking* 18(3): 911–921.

Taleb, T., Hadjadj-Aoul, Y. & Ahmed, T. 2011. Challenges, opportunities, and solutions for converged satellite and terrestrial networks. *IEEE Wireless Commun* 18(1): 46–52.

Ye, W., Silva, F. DeSchon, A., et al. 2008. Architecture of a satellite-based sensor network for environmental observation. In *Proc. Earth Science Technology Conference (ESTC), Maryland, USA, June 2008.*

Electronics, Communications and Networks IV – Hussain & Ivanovic (eds)
© 2015 Taylor & Francis Group, London, ISBN: 978-1-138-02830-2

Study on the propagation characteristics of UWB signal waveform distortion

Feng Chen*, Yuanjian Liu, Xi Yan & Xingyu Qi
Nanjing University of Posts & Telecommunications, Nanjing, China

ABSTRACT: In recent years, the propagation distortion problem of Ultra-Wideband (UWB) signal waveform has become one of the most popular research topics to the short-range high-speed wireless communications with the advantages of good confidentiality, anti-multipath, anti-interference and anti-electromagnetic interference. First of all, the law of ultra-wideband signal waveform distortion is discussed in theory. Then the UWB signal propagation in the typical indoor environment is simulated based on time-domain ray-tracing method. The factors of UWB signal waveform propagation distortion are analyzed.

KEYWORDS: Ultra-wideband signal; time-domain ray-tracing method; waveform distortion.

1 INTRODUCTION

In recent years, the short-range and high- speed ultra-wideband technology is rising increasingly, and becoming the hot topic in wireless communication area. Ultra-wideband technology is not sensitive to channel fading advantages, and it has low power spectral density, weak interference and centimeter-level position accuracy. Besides, its system is not complex. As UWB pulse signal covers a frequency range of several GHz, UWB has complex frequency behaviors and the frequency selective fading that it experienced is far more serious than the general narrowband signal. It will lead to a serious distortion of the received waveform in the time domain. Therefore, the study on UWB signal propagation characteristics and the channel model is the main part of UWB's key technology.

The problems of UWB signal propagation are usually solved by high-frequency approximation method, ray-tracing method. Field strength has been predicted by time domain ray-tracing method. Effective ray channel model has been established, and time domain reflection coefficients and diffraction coefficients have been given (Barnes &Tesche 1991, Yao et al. 2003).The reflection of transient plane waves in finite conducting half-space has been analyzed(Landron et al. 1996). The reflection coefficient of the typical indoor wall has been measured, and its theoretical formulas and empirical formulas have been compared using the Laplace transform (Manteuffel &Kunisch 2004). In addition, field strength prediction is made by FDTD, but the obvious defect is the large number of calculation (Rappaport 1996). A completely random approach based on experimental measurements could analyze a variety of channel characteristics using a variety of statistical tools, but the inadequacy of the random method needs a large number of test data (Yao 1997). So far, the study on indoor propagation characteristics of UWB pulse waveform about how to change by using time domain ray- tracing is rare.

In this paper, the indoor environment propagation characteristics of ultra-wideband signal and its waveform distortion are researched by time domain ray-tracing technology, and an effective ultra-wideband channel model is established. The factors of UWB signal waveform propagation distortion are analyzed(Wu 2008, Yao et al. 2013) .

2 TIME DOMAIN REFLECTION COEFFICIENT

The time domain reflection field amplitude is determined by the reflection coefficient. The contribution of reflection field to the impulseresponse in the improved time domain UTD is e_r^+ .

where

$$e_r^+(t) = E_0^r \left| A_r(s^r) \right| \overset{=+}{r} (\tau_r) \tag{1}$$

$$\tau_r = t - \frac{s^i}{c} - \frac{s^r}{c} \tag{2}$$

$$\overset{=+}{r} (\tau_r) = r_{hp}^+(t) e_\perp^i e_\perp^r + r_{vp}^+(t) e_{//}^i e_{//}^r \tag{3}$$

$r(\tau_r)$ is the time-domain reflection coefficient.

*Corresponding author: *276196701@qq.com*

$r_{hp}^{+}(t)$ and $r_{vp}^{+}(t)$ are respectively horizontal polarization coefficient and vertical polarization coefficient.

For the signal reflections caused by the loss media surface, P. R. Barnes (Barnes &Tesche 1991) has described the reflection of time-domain coefficient in detail.

When it is the horizontal polarization

$$r^{+}(t)=-[K\delta(\tau_r)+\frac{4\kappa}{1-\kappa^2}\frac{e^{-a\tau_r}}{\tau_r}\sum_{n=1}^{\infty}(-1)^{n+1}nK^{n}I_n(a\tau_r)] \quad (4)$$

When it is the vertical polarization

$$r^{+}(t)=[K\delta(\tau_r)+\frac{4\kappa}{1-\kappa^2}\frac{e^{-a\tau_r}}{\tau_r}\sum_{n=1}^{\infty}(-1)^{n+1}nK^{n}I_n(a\tau_r)],\cos^2\varphi/\varepsilon_r<<1 \quad (5)$$

The above formula can be abbreviated as

$$r^{+}(t)=\pm[K\delta(t)+\frac{4\kappa}{1-\kappa^2}e^{-at}F] \quad (6)$$

Where

$$F=\sum_{n=1}^{\infty}(-1)^{n+1}\frac{nK^{n}}{t}I_n(at) \quad (7)$$

By using sum of series to represent it, the reflection coefficient is simplified to:

$$\begin{aligned}r^{+}(t)=\pm\{K\delta(t)+\frac{4\kappa}{1-\kappa^2}e^{-at}[e^{-Kat/2}(\frac{a^3}{16K}t^2+\frac{1-K^2}{4K^2}a^2t+\frac{K^2-1}{2K}a)\\+\frac{a^3}{16K}t^2-\frac{a^2}{4K^2}t+\frac{a}{2K}]\}\end{aligned} \quad (8)$$

Ordering the formula in the second term $W(t)$, that is:

$$\begin{aligned}W(t)&=\frac{4\kappa}{1-\kappa^2}e^{-at}[e^{-Kat/2}(\frac{a^3}{16K}t^2+\frac{1-K^2}{4K^2}a^2t+\frac{K^2-1}{2K}a)\\&+\frac{a^3}{16K}t^2-\frac{a^2}{4K^2}t+\frac{a}{2K}]\\&=\frac{4\sin\varphi\sqrt{\varepsilon_r-\cos^2\varphi}}{\varepsilon_r-1}e^{-at}[e^{-Kat/2}(\frac{a^3}{16K}t^2+\frac{1-K^2}{4K^2}a^2t+\frac{K^2-1}{2K}a)\\&+\frac{a^3}{16K}t^2-\frac{a^2}{4K^2}t+\frac{a}{2K}]\end{aligned} \quad (9)$$

where

$$K=(\varepsilon_r\sin\varphi-\sqrt{\varepsilon_r-\cos^2\varphi})/(\varepsilon_r\sin\varphi+\sqrt{\varepsilon_r-\cos^2\varphi}) \quad (10)$$

$$a=60\pi\sigma\ c/\varepsilon_r \quad (11)$$

Taking positive sign, that is

$$r^{+}(t)=K\delta(t)+W(t)=\frac{\varepsilon_r\sin\varphi-\sqrt{\varepsilon_r-\cos^2\varphi}}{\varepsilon_r\sin\varphi+\sqrt{\varepsilon_r-\cos^2\varphi}}\delta(t)+W(t) \quad (12)$$

As can be seen in the above formula, when the incident angle of the signal is fixed, the reflection coefficient is mainly associated with ε_r and $\frac{\sigma}{\varepsilon_r}$.

3 SIMULATED RESULTS

In order to verify the validity of the analysis, simulation according to the literature (Yao et al. 2013) is made, and the results in the literature (Yao et al. 2013) are compared to verify its correctness.

The simulation environment is an empty room and its length, width and height are respectively 4.5 m, 3m and 2.5m, as shown in Figure 1. The back wall is a glass wall, and other walls and the floor are cement materials. Besides, the ceiling is gypsum material. In addition, all electromagnetic parameters of the materials are shown in Table 1 below. The height of transmitting antenna is 1.6m, and the height of receiving antenna is 1m. In this paper, the transmitting and receiving antennas are assumed to be the ideal antennas. By establishing a Cartesian coordinate system, the coordinates of the transmitting and receiving antennas are shown in Figure 2. The frequency of transmitting signal is 7000MHz, and the transmission power is 30dBm.

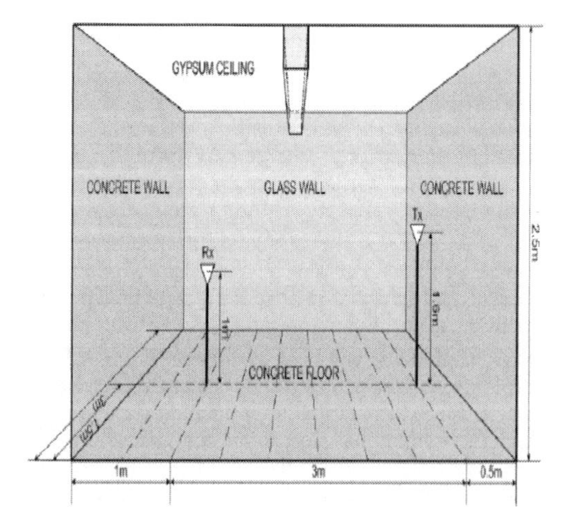

Figure 1. Simulation room layout.

The problem of ultra-wideband signal waveform transmission distortion is built on the specific propagation channel model. Therefore, study on the waveform distortion, in fact, is the study on the model of the propagation channel. The simulation and analysis of UWB

Table 1. Electromagnetic parameters of walls.

Wall materials	Electromagnetic parameters	
	ε_r	σ
Dry concrete	5	0.7
Glass	6	0.001
Gypsum board	2.8	0.15

signal waveform are made. UWB signal waveforms in transmitter and receiver are compared. The waveform distortion is discussed, according to the size of the incident, the angle of incident ray, and the electromagnetic properties of the reflecting surface material.

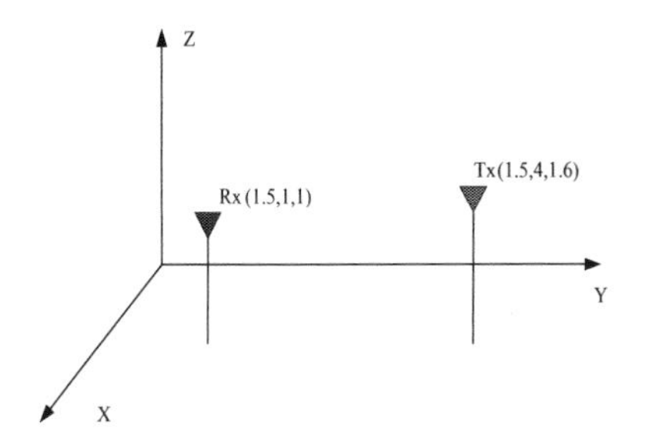

Figure 2. The coordinates of antennas.

In order to analyze the results of simulation conveniently, a baseline is introduced to compare the results. To study the angle of incidence and electromagnetic characteristics parameters, only one of the parameter values is changed each time in the baseline condition while other parameter values remain unchanged. Under the basic conditions, the incident angle is $\theta = 45°$and the relative permittivity of the reflecting material is $\varepsilon = 10$ .Moreover, the conductivity is $\sigma = 8.854187818 \times 10^{-12}$. The waveforms of transmitter and receiver at this time are shown in Figure 3.

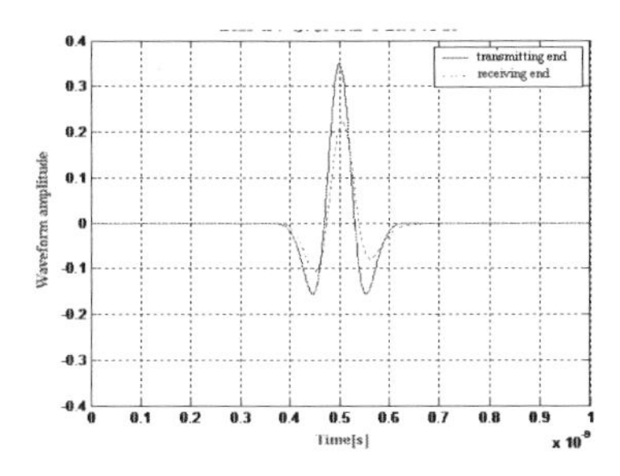

Figure 3. The waveforms at $\theta = 45°$.

When the incident angle is $\theta = 5°$, the waveforms of transmitter and receiver are shown in Figure 4.

As shown in Figure 3 and Figure 4, when the incident angle changes, the waveform suffers from distortion. And when the incident angle is smaller, the degree of distortion is larger. Changing the size of incident angle and the degree of distortion is not the same, which indicates that the changes in size of the incident angle affect the distortion. Therefore, the incident angle is a factor which can influence the waveform distortion.

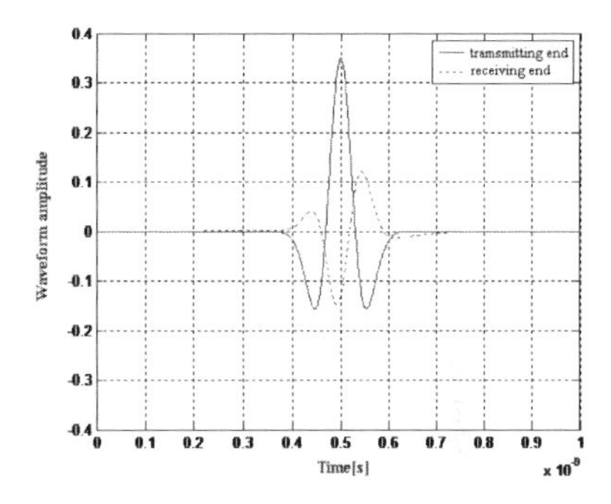

Figure 4. The waveforms at $\theta = 5°$.

When the material relative permittivity is 3.9, the waveforms of transmitter and receiver are shown in Figure 5.

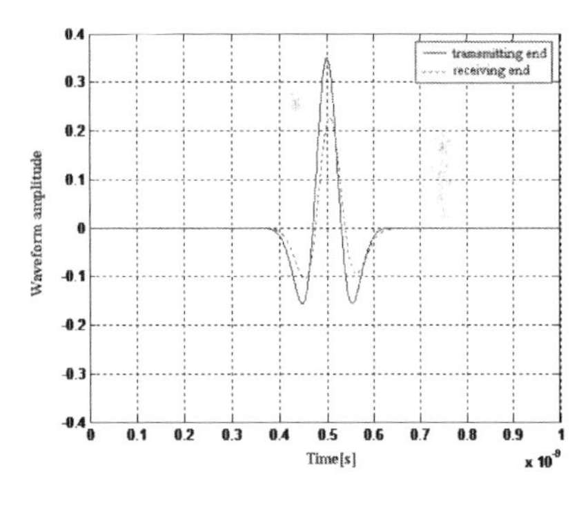

Figure 5. The waveforms at $\varepsilon = 3.9$.

As shown in Figure 5, when the material relative permittivity is 10, the waveform suffers from a very large degree of distortion. When the material relative permittivity is 3.9, the degree of waveform distortion is very small. It shows that if the material relative permittivity changes, the degree of waveform distortion will also change. So the relative permittivity is also one of the factors affecting waveform distortion.

When the material conductivity is $\sigma = 8.8\times10^{-14}$, the waveforms of transmitter and receiver are shown in Figure 6.

As shown in Figure 6, when the material conductivity is $8.854187818\times10^{-12}$, the waveforms suffer from a very large degree of distortion. When the material conductivity is 8.8×10^{-14}, the degree of waveform distortion is very small. It shows that if the material conductivitychanges, the degree of waveform distortion will also change.So the conductivity is also one of the factors affecting waveform distortion.

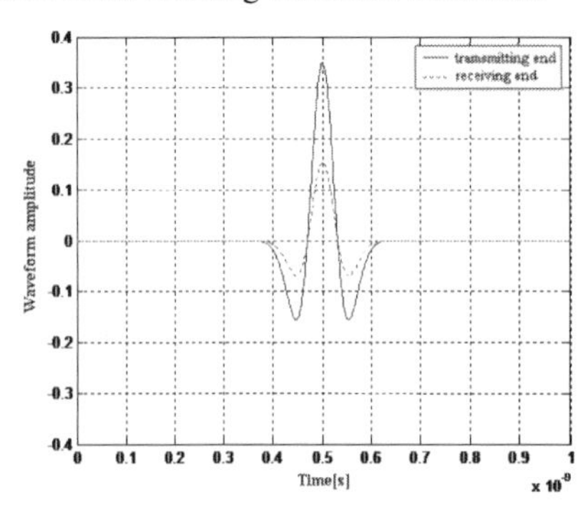

Figure 6. The waveforms at $\sigma = 8.8\times10^{-14}$.

4 CONCLUSION

From the above simulation and analysis, we can see that the factors, which affect the distortion from the Gaussian second order derivative of the pulse waveform, include the incident angle of the incident ray, the relative permittivity of material and the conductivity of the reflecting surface.

The analysis above is based on simple indoor channel model, even only the incident angle, relative permittivity and conductivity etc. A small number of model parameters are considered. For the complex indoor and outdoor models which are much closer to the actual application, a further study needs to be made, and the other parameters in the model ought to be taken into consideration, such as the type of reflection coefficient (vertically polarized reflection coefficient, horizontally polarized reflection coefficient) and the polarization type in the electric field etc.

ACKNOWLEDGEMENTS

This work has been supported by the National Natural Science Fund under Grant No. 61372045 and by the Ministry of International Journal of Antennas and Propagation and Education of Higher Specialized Research Fund for the Doctoral Program under Grant No. 20123223120003.

REFERENCES

Barnes, P. &Tesche, F.M. 1991.On the direct calculation of a transient plane wave reflected from a finitely conducting half space.*IEEE Transactions onElectromagnetic Compatibility*33 (2): 90–96.

Landron, Feuerstein, M. &Rappaport, T.S. 1996.A comparison of theoretical and empirical reflection coefficients for typical exterior wall surfaces in a mobile radio environment.*IEEE Transactions on Antennas and Propagation*44(3): 341–351.

Manteuffel, D. &Kunisch, J.2004. Efficient Characterization of UWB antennas using the FDTD method.*IEEE Antennas and Propagation Society International Symposium*:1752–1755.

Rappaport, T.S. 1996.*Wireless communications principles and practice*.New Jersey: Prentice-Hall.

Wu dong. 2008. Study on Indoor Channel and Waveform Nonlinear-Distortion of UWB Signal.*Master dissertation*, Nanjing University of Posts and Telecommunications.

Yao, R. 1997. A time-domain ray model for predicting indoor wideband radio channel propagation characteristics. *The conference of Universal Personal Communication*: 848–852.

Yao, R., et al. 2013. UWB multipath channel model based on time-domain UTD technique.*IEEE Global Telecommunications Conference*: 1205–1210.

Yao, R., Zhu, W. &Chen, Z. 2003.An efficient time-domain ray model for UWB indoor multipath propagation channel.*IEEE Vehicular Technology Conference*:1293–1297. 2003.

Electronics, Communications and Networks IV – Hussain & Ivanovic (eds)
© 2015 Taylor & Francis Group, London, ISBN: 978-1-138-02830-2

Research on ultra-wideband dual-polarized quadruple-ridged horn antenna

Lijia Chen, Hao Li & Nannan Wang
School of Electronics and Information Engineering, Harbin Institute of Technology, Heilongjiang, China

ABSTRACT: The paper presents a quadruple-ridged horn antenna for EMC tests and measurements. Parameters which have a big influence on the characteristics of the horn antenna are discussed and compared. The ridges of the antenna are redesigned to decrease reflection loss over a wide frequency band, and a dielectric lens is loaded. The antenna is dual-polarized and its VSWR is less than 2 in the frequency band 1-20GHz with almost 20dB gain, and at the same time, the cross polarization is less than -40dB. The simulation results and analyses of the quadruple-ridged horn antenna are presented.

KEYWORDS: horn antennas, broadband, ridges, antenna feeds, lens.

1 INTRODUCTION

With the development of electronic technology, the millimeter-wave technology begins to be used in the standard test, electronic warfare and radar systems, and the antennas with good performance covering the millimeter-wave range are needed by more and more people, and the polarization of the antenna is asked to be variable in a lot of situations (Chung et al. 2003). Due to its wide frequency bandwidth, high gain and good directivity, horn antenna has been developed and widely used in these fields, usually as a separate antenna or feed source. To allow horn antenna work in a wide band, the ordinary horn must be improved, and usually loaded with ridges. Quadruple-ridged horn antenna can not only meet the requirements of a wide band but also the requirements of the variable polarization (Chung 2010).

In this paper, a novel quadruple-ridged horn antenna is designed for the 1-20GHz frequency band. Accordingly, a new design is proposed to reduce the reflect wave in the back cavity of the horn by cutting the end of the ridge. The simulated results show that the improved dual-polarization quadruple-ridged horn antenna exhibits a low VSWR as well as good radiation pattern over 1-20GHz frequency. It meets the need of time consuming measurement that the source antenna is always required to accommodate different frequency bands and to change a different polarized antenna.

2 HORN ANTENNA PARAMETERS

The general form of quadruple-ridged circular horn antenna is shown in Figure 1. The design of ridged horn antenna is mainly decomposed into three parts: the design of the horn, feeding section and the shape of the ridge.

Normally, the open angles and caliber sizes of the horn are limited by the specifications. The feeding section contains a part of the ridges and an N-type input connector with a coaxial line. Actually, it is a circular waveguide, in which TE_{10} mode wave can transform. The ridges are tapered so that they will not touch each other, as shown in Figure 2. The end of the ridges is placed in a cavity. This cavity can suppress the TE_{20} mode in the waveguide (Rodriguez & Vicente 2009). The coaxial line passes through one ridge, and its shield is connected to this ridge, while the core of the coaxial line is connected to the other ridge.

The bandwidth of ridge horn is mainly decided by these factors: the distance between ridges, shape and size of the ridges. The shape of the ridges must be able to keep the resistance well-matched between the feed point and the open space (Turk & Keskin 2012). It is demonstrated that ridges with the shape of exponential function work well for this task (Chung et al. 2004). Since the size of the horn is decided, the curve of the ridge can be calculated easily. Meanwhile, by chamfering the end of the ridge at the caliber of the horn, the characteristic of the low frequency band can be reduced. Since the distance between the ridges and

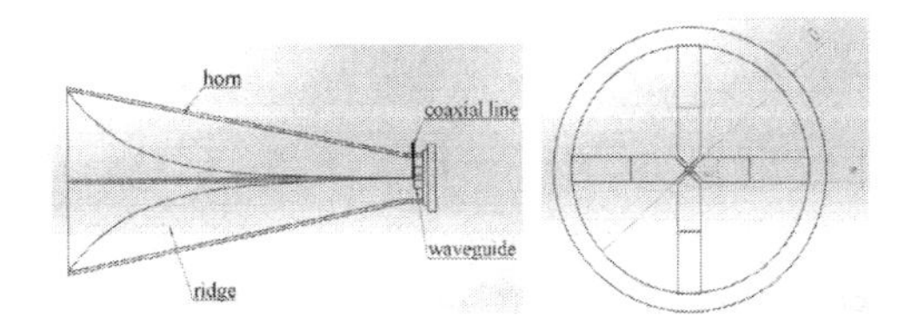

Figure 1. Ridged horn.

Figure 2. Tapered ridge.

the width of the ridges are equivalent to a microstrip line, the impendence at the feed point can be calculated. When the ratio of feed spacing and ridge width is 1:4, the impendence is rounded to 50 Ohms. The simulation shows that when the impendence is a constant, the bandwidth of the horn will be larger if the feed spacing is smaller as shown in Figure 3.

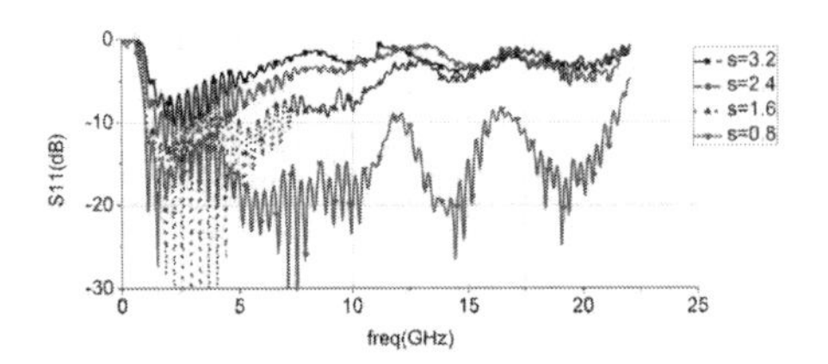

Figure 3. Relationship between feed spacing and reflection loss.

Generally, the end of the ridge in the feed section is a rectangle. To improve the bandwidth of the horn antenna, the shape of the ridge end is redesigned to a ladder-shape, a round arc type, and a polyline type, as shown in Figure 4. Simulation proves the polyline type can reduce the reflection loss in the band of 15GHz-18GHz, as shown in Figure 5.

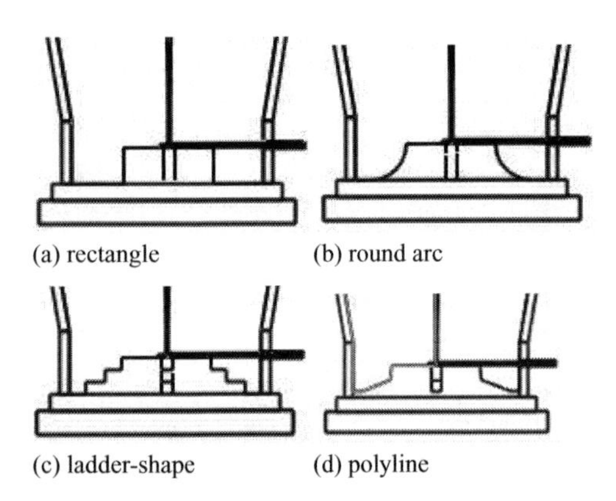

Figure 4. Shapes of ridges in the waveguide.

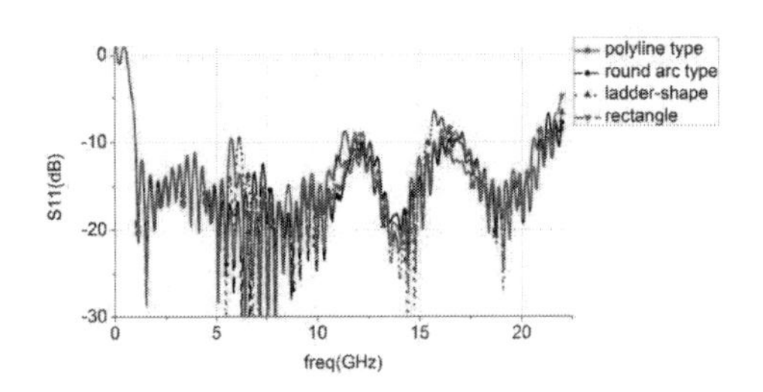

Figure 5. Reflection loss of different shapes of ridge.

3 ANTENNA DESIGN AND SIMULATION

Based on the design ideas above, a new quadruple-ridged dual-polarized horn antenna with dimension of 176*176*356 mm^3 is built and simulated. By changing the shapes of the end of the ridges, the antenna works well in the band from 1GHz to 20GHz with VSWR<2. However, it's observed that there are some distortions in the shape of the radiation pattern at higher frequency bands. To deal with such distortions, a dielectric lens is loaded on the caliber of the horn. The cone-shaped lens can adjust the phase of waves at the caliber to the same. Then the radiation pattern shape can be improved, and at the same time, the gain of the antenna increases. Figure 6 shows the directivity of the horn before and after loading the lens.

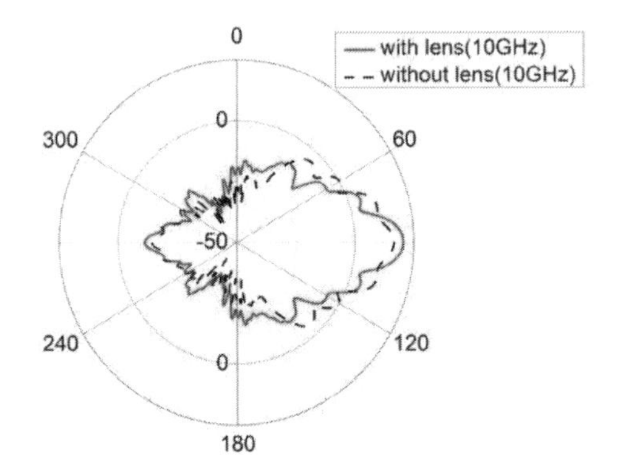

Figure 6. Radiation pattern of horn.

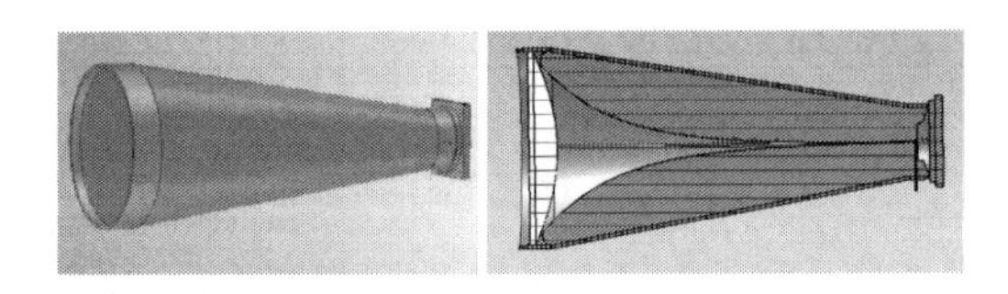

Figure 7. Designed antenna and its cross-view section.

The final designed antenna is shown in Figure 7. In the frequency band 1-20GHz, VSWR is less than 2 as shown in Figure 8. With the lens, the beamwidth is narrower, and the bandwidth is the same as without the lens. The beam width of the horn decreases with the increasing frequency as visible on radiation patterns in Figure 9. The gain of the antenna is shown in Figure 10. It is above 9dB from 2GHz to 20GHz, except for being 3.7dB in 1GHz.

For dual-polarized antenna, small cross-polarized property is significant. Generally, it is desired that cross-polarization is less than -20dB. Figure 11 shows that the cross-polarization of this horn antenna is below -40dB from 1GHz to 20GHz.

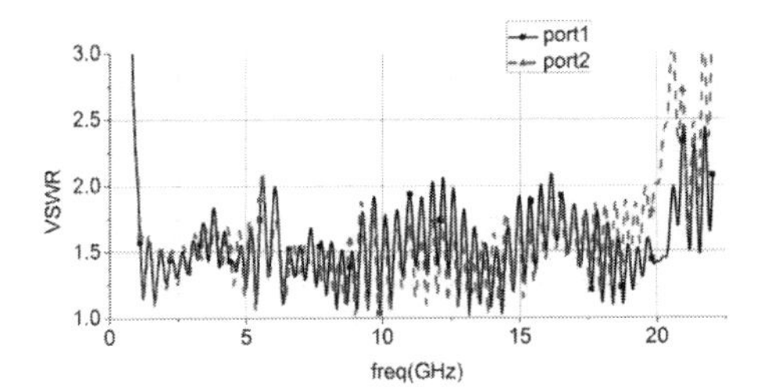

Figure 8. VSWR of two ports of the horn.

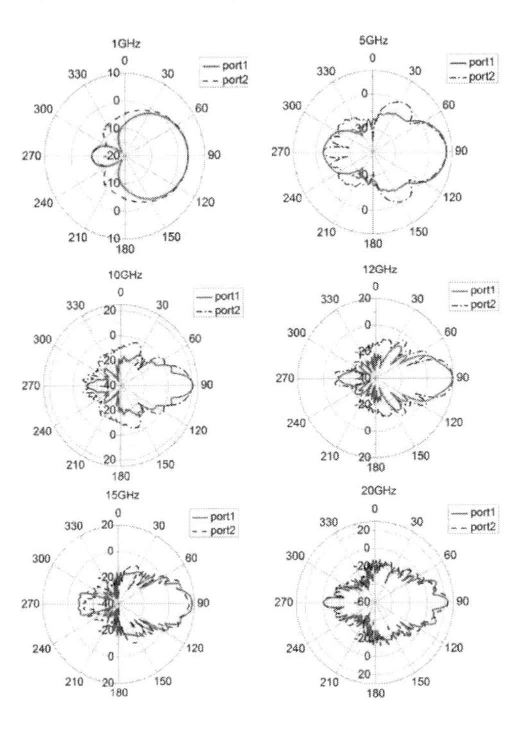

Figure 9. Radiation patterns of quadruple-ridged horn.

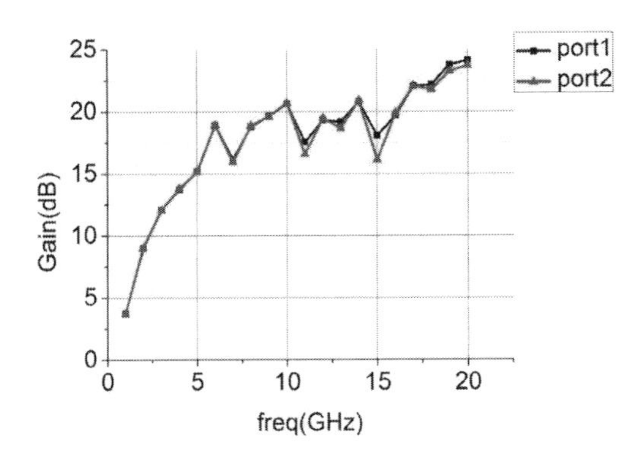

Figure 10. Gain of quadruple-ridged horn.

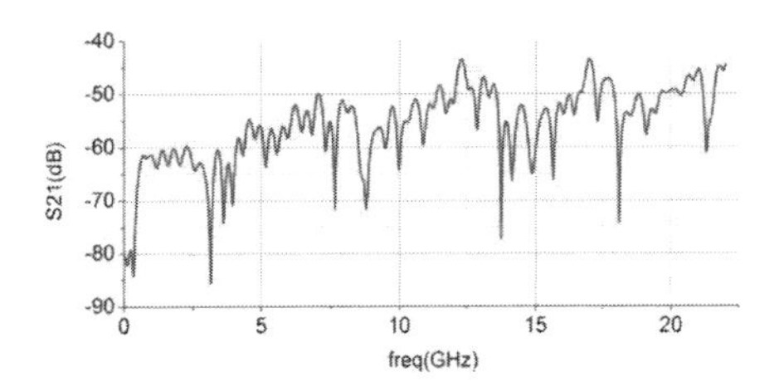

Figure 11. Cross polarization of quadruple-ridged horn.

4 CONCLUSION

In this work, different factors of quadruple-ridged horn antenna are discussed, and the results are described. An ultra-wideband dual-polarized quadruple-ridged horn antenna is designed and simulated. It performs well in the 1-20GHz frequency band with VSWR<2, and at the same time, it has advantages of good directivity, high gain and low cross-polarization. Thus, the antenna is applicable for the UWB test and measurement purposes.

ACKNOWLEDGMENT

The authors wish to acknowledge the assistance and support from the National Natural Science Foundation of China under Grant Nos. 61171063 and the International S&T Cooperation Program of China under Grant Nos.2011DFR10890.

REFERENCES

Chung, K.H. Pyun, S.H. Chung, S.Y. & Choi, J.H. 2003. Design of a wideband TEM horn antenna. *IEEE Trans. AP-S*. 1: 229–232.

Chung, J.Y. 2010. Ultra-wideband dielectric-loaded horn antenna with dual-linear polarization capability. *Prog. Electromagn.* Res. 102: 397–411.

Rodriguez & Vicente. 2009. Recent improvements to dual ridge waveguide horn antennas: The 200MHz to 2000MHz and 18GHz to 40GHz models. *Electromagnetic Compatibility*, 2009. *EMC 2009. IEEE International Symposium on*:24–27. IEEE.

Turk, A.S. & Keskin, A.K. 2012. Partially Dielectric-Loaded Ridged Horn Antenna Design for Ultrawideband Gain and Radiation Performance Enhancement. *Antennas and Wireless Propagation Letters. IEEE*. 11: 921–924.

Chung, K.H. Pyun, S.H. & Choi, J.H. 2004. The design of a wideband TEM horn antenna with a microstrip-type Balun. *Antennas and Propagation Society International Symposium IEEE*. 2:1899–1902.

Electronics, Communications and Networks IV – Hussain & Ivanovic (eds)
© 2015 Taylor & Francis Group, London, ISBN: 978-1-138-02830-2

An uplink scheduling mechanism based on user satisfaction in LTE networks

Kuo-Chih Chu*, Tzu-Chi Huang & Shu-Hua Cheng
Lunghwa University of Science and Technology, Taoyuan, Taiwan

Wen-Chung Tsai
Chaoyang University of Technology, Taichung, Taiwan

ABSTRACT: In recent years, LTE networks are quickly deployed everywhere in the world to resolve the congestion problem in 3G networks. Since various new network services are emerging, however, we can expect that the bandwidth of LTE networks will not satisfy requirements of users. In this paper, we propose an uplink scheduling mechanism based on user satisfaction in order to meet user satisfaction even though the bandwidth is not enough. In experiments, we simulate our mechanism and other mechanisms to verify that our mechanism indeed can meet user satisfaction in comparison with other mechanisms.

KEYWORDS: LTE; Scheduling Mechanism; User Satisfaction.

1 INTRODUCTION

Mobile devices and communication technologies emerge quickly in recent years. More and more users use mobile devices to access the Internet through mobile communication networks. In order to have more bandwidth to meet user satisfaction, many countries in the world gradually replace their 3rd-Generation (3G) network infrastructures with Long Term Evolution (LTE) (Ekstrom et al. 2006) network infrastructures. Up to September 2014, there are 112 countries and 331 LTE network service providers in the world (Global mobile Suppliers Association 2014). Although LTE networks have the bandwidth, multiple times more than 3G networks, we can expect that LTE networks still will fail to satisfy the requirements of users in the future.

Insufficient bandwidth will lower user satisfaction. Because of this, users probably change to other network service providers to make the original network service provider lose incomings. In order to resolve the problem of insufficient bandwidth in LTE networks, besides the upgrade to LTE-A networks, it is an alternative to design a good scheduling mechanism capable of efficiently utilizing the limited bandwidth. Because uplink scheduling mechanisms currently are not many, we propose a new uplink scheduling mechanism based on user satisfaction in this paper for resolving the problem of insufficient bandwidth in LTE networks.

This paper is composed of five sections as follows. Section 2 introduces LTE networks. Section 3 presents our proposal. Section 4 simulates the performance of our proposal in user satisfaction. Section 5 concludes this paper.

2 LTE NETWORK OVERVIEW

LTE networks have the Evolved Universal Terrestrial Radio Access (E-UTRAN) networks and the Evolved Packet Core (EPC) networks. The E-UTRAN network has User Ends (UEs) and an Evolved Node B (eNodeB). LTE networks have a scheduling mechanism in an eNodeB to allocate bandwidth in units of Resource Blocks (RBs) to the UEs. When a UE has data for transmission, it sends a Scheduling Request (SR) to the eNodeB. When the eNodeB receives the SR, it allocates bandwidth to the UE for sending back a Buffer State Reporting (BSR) that can tell the eNodeB the requested bandwidth. After that, the eNodeB executes the scheduling mechanism to allocate bandwidth to a UE according to the bandwidth requested by the UE. Finally, the eNodeB uses a downlink to tell the UE the scheduling result.

LTE networks use the Single-Carrier FDMA (SC-FDMA) for the uplink in order to decrease the Peak to Average Power Ratio (PAPR) and the power consumption. When the SC-FDMA is activated, an eNodeB has to allocate adjacent RBs to a UE if more

*Corresponding author: *kcchu@mail.lhu.edu.tw*

than 1 RB will be given to the UE. Due to the working principle of the SC-FDMA, it is difficult to design an uplink scheduling mechanism.

Currently, there are certain works related to uplink scheduling mechanisms in LTE networks (Yaacoub & Dawy 2012). Abu-Ali et al. (2014) divide basic LTE scheduling mechanisms into three categories, i.e. Best-Effort Schedulers, Schedulers Optimizing QoS, and Schedulers Optimizing QoS and Power. Best-Effort Schedulers mainly seek the maximum throughput or fairness. For example, the Recursive Maximum Expansion (RME) mechanism Ruiz de Temino et al. (2008) use a greedy algorithm to achieve the maximum throughput in the system by allocating no bandwidth to users who have bad signal quality, but it will lower satisfaction of the users. The Mean Enhanced Greedy (MEG) mechanism Nwamadi et al. (2009) allocate the same quantity of RBs to different UEs in order to achieve the fairness by avoiding the difference of bandwidth gotten by different UEs. Schedulers Optimizing QoS allocate RBs to UEs according to different QoS requirements requested by UEs. For example, Wen et al. (2008) propose a mechanism to make a scheduler, both support QoS and avoid the starvation problem due to insufficient available bandwidth. Schedulers Optimizing QoS and Power are designed for reducing the consumption of power in UEs. For example, Li et al. (2009) delay transmission of a little data in UEs, so they can send more data in a single transmission operation and reduce the power consumption. However, the related works seldom discuss issues about the guarantee of User Satisfaction. The User Satisfaction Balance (USB) mechanism Wu et al. (2012) give user satisfaction more priority and tries to give all users high satisfaction no matter whether the network is congested or not. Because of lacking an admission control mechanism, however, the USB mechanism may lower user satisfaction quickly when many users are in the network. Against the problem resulting from the lack of an admission control mechanism, we introduce a User Satisfaction Oriented (USO) mechanism to our proposal in this paper, which both have an admission control mechanism not available on the USB mechanism and uses a scheduling mechanism different to the USB mechanism. We hope that our proposed mechanism can maintain high user satisfaction no matter in what network condition users are working.

3 USER SATISFACTION ORIENTED (USO) MECHANISM

The USO mechanism is composed of an admission control mechanism and a scheduling mechanism. According to the traffic load on networks, the admission control mechanism controls the number of UEs that can enter the scheduling mechanism, in order to avoid overloading the system with too many UEs to congest networks. The scheduling mechanism allocates RBs to UEs according to the requirements of UEs. We introduce the two mechanisms, respectively, as follows.

3.1 *Admission control mechanism*

The admission control mechanism in Figure 1 works by the following steps:

1 Determining the total number of UEs scheduled this time: First, we define the total number of UEs accepted by the admission control mechanism as S_t, while S_{t-1} is the total number of UEs accepted last time. If the scheduling result gets 100% user satisfaction last time, the admission control mechanism accepts and increases the number of UEs by 1 this time; otherwise, it decreases the number of UEs by 1 this time.
2 Determining whether a UE can enter the scheduling mechanism: When a new UE wants to enter the scheduling mechanism, we determine whether the total number of UEs in the system, i.e. U_t, is less than S_t or not. If it is, the new UE is allowed to enter the scheduling mechanism; otherwise, it is rejected. If there is no other new UEs or the allowed number of UEs is reaching the limitation (i.e. U_t equal to S_t), the next step is proceed.

3.2 *Scheduling mechanism*

The scheduling mechanism in Figure 2 works by the following steps:

1 Building Transmission Matrix: Because the signal quality between each UE and its eNodeB is different, different UEs will not get the same bandwidth in the corresponding RB. According to the response of each UE, the base station can build a transmission matrix that has a relationship mapping all UEs to all RBs as shown in Figure 3. In Figure 3, we assume that the schedule this time needs to allocate 6 RBs to 3 UEs. RB_1 gives a bandwidth of 20bps, 40bps and 30bps to UE_1, UE_2 and UE_3. Similarly, we can look up Figure 3 to know bandwidth allocated to each UE by other RBs.
2 Calculating User Satisfaction: In this step, we define the User Satisfaction Index (USI) in Equation (1) as a metric to evaluate user satisfaction. In Equation (1), the USI refers to the amount of bandwidth allocated to the ith UE this time as $alloc_i$, the amount of bandwidth requested by the ith UE this time as req_i and the user satisfaction of the ith UE as USI_i.

$$USI_i = (alloc_i / req_i) \quad (1)$$

Next, we assume that UE_1, UE_2 and UE_3 want to transmit data of 45bps, 50bps and 60bps. According to Figure 3 and (1), we can get the USI of each UE corresponding to each RB. According to all USIs in Figure 4, we find that the USI of UE_1 is 1 if it uses RB_6, which implies that UE_1 can be satisfied with RB_6. According to Figure 4, we also can calculate the average USI of each UE, i.e. USI=0.78 in UE_1, USI=0.65 in UE_2 and USI=0.54 in UE_3.

3 Enumerating RB Combinations to Satisfy UEs: In this step, we define two parameters, i.e. USI_{th} and USI_{Ti}. We define the USI threshold as USI_{th} that will be initiated to 1 for each schedule and define the summation of all USIs for all RBs chosen to the ith UE as USI_{Ti}. For example, we choose to allocate RB_1 and RB_2 to UE_1, so its USI_{T1} is 1.11 (i.e. 0.44+0.67=1.11). From Figure 4, accordingly, we can enumerate for each UE all RB combinations that can meet $USI_{Ti} \geqq USI_{th}$.

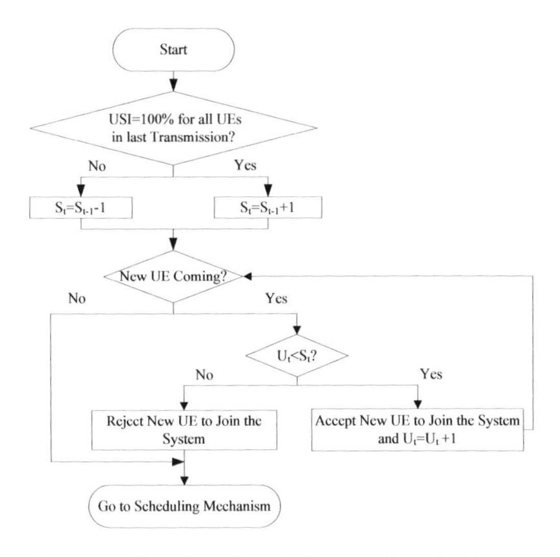

Figure 1. Flow chart of admission control mechanism.

We take UE_1 for example. The scheduling mechanism begins searching USIs from RB_1 to the end and concludes that more RBs should be allocated to UE_1 because the USI of UE_1 corresponding to RB_1 is 0.44 that cannot make USI_{T1} larger than USI_{th}. Next, we have to allocate adjacent RBs to UE_1 due to the working principle of the SC-FDMA. Accordingly, we add RB_2 to RB_1 and calculate whether the combination of RB_1 and RB_2 can make USI_{T1} larger than USI_{th} or not. Because USI_{T1} now is 1.11 and larger than USI_{th}, allocating the combination of RB_1 and RB_2 to UE_1 can satisfy UE_1. We cease to add RB_3, because our mechanism is designed to locate the

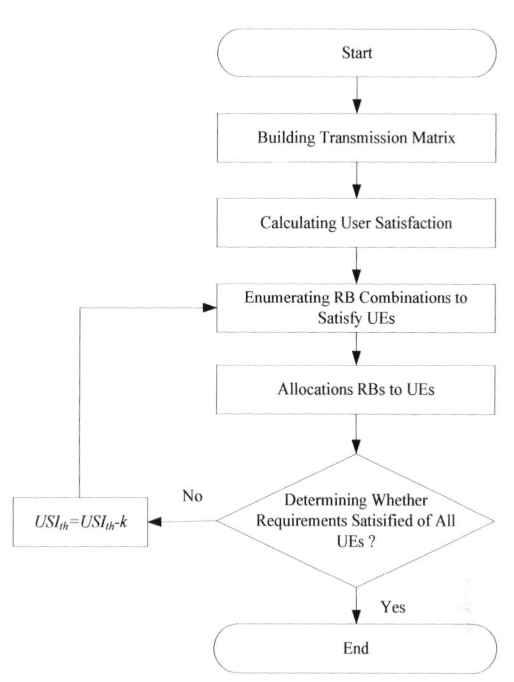

Figure 2. Flow chart of scheduling mechanism.

	RB_1	RB_2	RB_3	RB_4	RB_5	RB_6
UE_1	20	30	35	40	40	45
UE_2	40	35	30	25	30	35
UE_3	30	30	30	35	35	35

Figure 3. Transmission matrix.

	RB_1	RB_2	RB_3	RB_4	RB_5	RB_6
UE_1	0.44	0.67	0.78	0.89	0.89	1
UE_2	0.8	0.7	0.6	0.5	0.6	0.7
UE_3	0.5	0.5	0.5	0.58	0.58	0.58

Figure 4. USIs among UEs and RBs.

	RB_1	RB_2	RB_3	RB_4	RB_5	RB_6
UE_1	0.44	0.67	0.78	0.89	0.89	1
UE_2	0.8	0.7	0.6	0.5	0.6	0.7
UE_3	0.5	0.5	0.5	0.58	0.58	0.58

Figure 5. Scheduling result.

RB combination satisfying UE_1 as soon as possible and because the combination of RB_1 and RB_2 already can satisfy UE_1. Accordingly, we repeat the step and begin searching USIs from RB_2 to the end. Finally, we can enumerate 6 cases of combinations that can satisfy UE_1, i.e. (RB_1+RB_2), (RB_2+RB_3), (RB_3+RB_4), (RB_4+RB_5), (RB_5+RB_6) and RB_6. We can enumerate 5 cases of combinations that can satisfy UE_2, i.e. (RB_1+RB_2), (RB_2+RB_3), (RB_3+RB_4), (RB_4+RB_5) and (RB_5+RB_6). We can enumerate 5 cases of combinations that can satisfy UE_3, i.e. (RB_1+RB_2), (RB_2+RB_3), (RB_3+RB_4), (RB_4+RB_5) and (RB_5+RB_6).

4 Allocating RBs to UEs: The scheduling mechanism will begin locating RBs to a UE that has the average minimum USI. In Step 2, we find that UE_3 has the average minimum USI, so we allocate RBs to UE_3 first. From the 5 cases of combinations enumerated for UE_3 in Step 3, we locate the combination whose USI_{T3} is closest to USI_{th}. We find the combination of (RB_1+RB_2) because its USI is 1 and equal to USI_{th} and allocate it to UE_3. Next, we target UE_2 because it has the second average minimum USI. From the 5 cases of combinations enumerated for UE_2 in Step 3, we find that the combination of (RB_3+RB_4) has USI=1.1 which is closest to USI_{th} and then allocate it to UE_2. After that, only UE_1 has not been allocated and we have RB_5 and RB_6 that are not allocated to a UE. We find that the RB_6 has USI=1 which is equal to USI_{th}, so we give RB_6 to UE_1.

   After our scheduling mechanism satisfy all UEs with RBs, we may improve the throughput of the system by allocating adjacent RBs to UEs if there are RBs not used. We may continue the previous case and allocate RB_5 to a UE because it is not used. After comparing the USIs of UE_2 that uses the combination of (RB_3+RB_4) and the USI of UE_1 that uses RB_6, we find that UE_1 can get more bandwidth than UE_2 if UE_1 uses RB_5. Accordingly, we allocate RB_5 to UE_1 instead of UE_2 and get the scheduling result in Figure 5.

5 Determining Whether Requirements of All UEs are Satisfied or Not: After we allocate all RBs to UEs, we determine whether all UEs can meet $USI_{Ti} \geq USI_{th}$ or not. If it is positive, the scheduling mechanism ends its entire process and notifies all UEs of the result. If it is negative, which implies that the total bandwidth in networks is not enough, the scheduling mechanism proportionally decreases requested bandwidth of all UEs in order to execute the scheduling operation. By proportionally decreasing the requested bandwidth of all UEs, the scheduling mechanism can use the insufficient bandwidth to satisfy all UEs without lowering much user satisfaction. Currently, the scheduling mechanism decreases USI_{th} by k (0.05 in default) each time and repeats Step 3 to execute the scheduling operation.

4 SIMULATION RESULTS

In this section, we develop a program to do simulations. In simulations, we assume that each channel occupies 5 MHz and each sub-frame has 25 RBs. According to the location and the signal quality, each UE can choose a certain modulation mechanism to transmit data to an eNodeB. First, a UE transmits data at the size of 500 bits and then gradually increases the size until 5000 bits. We simulate the USO, RME and USB mechanisms, and compare their Average User Satisfaction and Throughput when transmitting different sizes of data.

First, we observe the average user satisfaction of the three mechanisms. When calculating the average user satisfaction, we consider it 100% if it is larger than 100%. According to the average user satisfaction in Figure 6, we observe that our USO mechanism maintains the 100% average user satisfaction when transmitting data ranging from 500 bits to 2500 bits. After data is transmitted more than 2500 bits, our USO mechanism makes its average user satisfaction higher than other mechanisms due to the work of its admission control mechanism. Although the USB mechanism is designed for user satisfaction, it lacks an admission control mechanism and lowers the

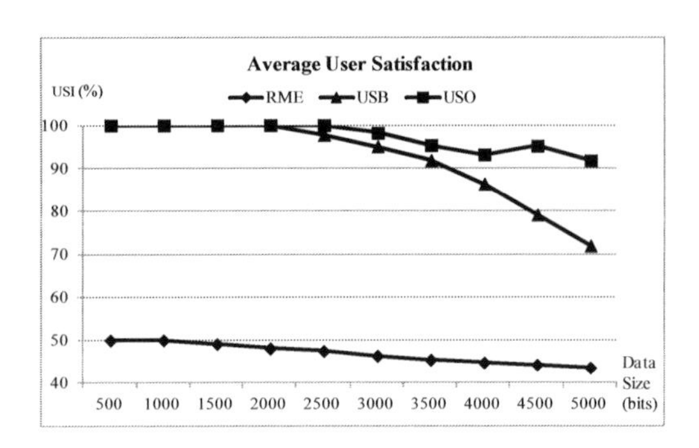

Figure 6. Average user satisfaction comparisons.

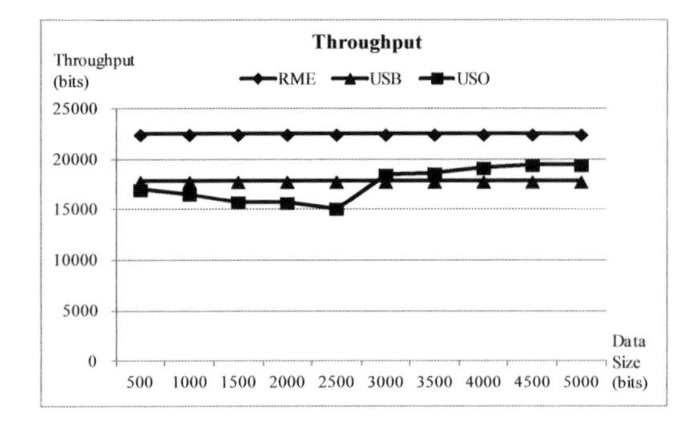

Figure 7. Throughput comparisons.

average user satisfaction when a UE transmits more than 2500 bits to congest the networks. Because the RME mechanism allocates most bandwidth to UEs that have better signal quality, starving UEs that lowers worse signal quality the average user satisfaction.

Figure 7 shows the throughput differences among the three mechanisms. The throughput expresses the summation of bandwidth allocated to all UEs in the scheduling operation this time. If the throughput is high, the network gets more bandwidth for data transmission. Because the RME mechanism is an algorithm designed for approaching the maximum throughput, we observe that its throughput is the highest among the simulation. Although our USO mechanism Although our USO mechanism and the USB mechanism both show different curves in throughput, the curves are not distant from that of the RME mechanism, which implies that the three mechanisms all have a similar capability in throughput.

According to above two simulation observations, we conclude that our USO mechanism can give all users good user satisfaction all the time without decreasing throughput of the entire system.

5 CONCLUSIONS

Since the demand for bandwidth increases quickly day by day, an LTE network gradually shows its difficulty in meeting the bandwidth requirement of users and inevitability lowers user satisfaction. Against the problem, we propose a USO mechanism in this paper. We design the USO mechanism to use an admission control mechanism that can control the number of UEs entering the scheduling mechanism. After that, the scheduling mechanism considers the requested bandwidth of all UEs and tries to make all UEs have user satisfaction equal or close to 100%. In experiments, we simulate an LTE network that uses the USO, USB and RME mechanisms, respectively, and observe the performances of the three mechanisms. We observe that our USO mechanism can outperform other mechanisms in user satisfaction. When LTE networks meet the challenge of network congestion in the future, we believe that our USO mechanism can provide much help by efficiently decreasing the negative impact of network congestion on user satisfaction.

ACKNOWLEDGEMENT

This work was supported by the National Science Council under contract: NSC102-2221-E-262-010.

REFERENCES

Abu-Ali, N., Taha, A.-E.M., Salah, M., et al. 2014. Uplink Scheduling in LTE and LTE-Advanced: Tutorial, Survey and Evaluation Framework. *IEEE Communications Surveys & Tutorials* 16(3): 1239–1265.

Ekstrom, H., Furuskar, A., Karlsson, J., et al. 2006. Technical Solutions for the 3G Long-Term Evolution. *IEEE Communications Magazine* 44(3): 38–45.

Global mobile Suppliers Association. 2014. 1800MHz band fuels LTE growth: GSA confirms 331 commercially launched LTE networks, including 150 LTE1800 deployments. *http://www.gsacom.com/news/gsa_414.php.*

Li, Z., Yin, C. & Yue, G. 2009. Delay-Bounded Power-Efficient Packet Scheduling for Uplink Systems of LTE. *International Conference on Wireless Communications, Networking and Mobile Computing ,China,24-29 Sept. 2009.* Beijing: IEEE.

Nwamadi, O., Xu, Z. & Nandi, A.K. 2009. Enhanced Greedy Algorithm Based Dynamic Subcarrier Allocation for Single Carrier FDMA Systems. *Wireless Communications and Networking Conference, Hungary, 5-8 April 2009.* Budapest: IEEE.

Ruiz de Temino, L., Berardinelli, G., Frattasi, S., et al. 2008. Channel-aware Scheduling Algorithms for SC-FDMA in LTE Uplink. *International Symposium on Personal, Indoor and Mobile Radio Communications, France, 15–18 Sept. 2008.* Cannes: IEEE.

Wen, P., You, M. & Wu, S. 2008. PFGBR Scheduling for Streaming Services over Wideband Cellular Network. *Vehicular Technology Conference, Singapore, 11–14 May 2008.* Singapore: IEEE.

Wu, J.-S., Yang, S.-F., Hsu, C.-M., et al. 2012. User Satisfaction-Based Scheduling Algorithm for Uplink Transmission in Long Term Evolution System. *International Conference on Parallel and Distributed Systems, Singapore, 17–19 Dec. 2012.* Singapore: IEEE.

Yaacoub, E. & Dawy, Z. 2012. A Survey on Uplink Resource Allocation in OFDMA Wireless Networks. *IEEE Communications Surveys & Tutorials* 14(2): 322–337.

Electronics, Communications and Networks IV – Hussain & Ivanovic (eds)
© 2015 Taylor & Francis Group, London, ISBN: 978-1-138-02830-2

Building of multilingual software simulation platform based on education network private cloud

Qing Cui*, Gang Shi & Chongguo Wang
School of Information Science and Engineering, XinJiang University, Urumqi, China

ABSTRACT: Focusing on analyzing the shortcomings and limitations of Xinjiang University Common Computer Test (CCT), this paper puts forward construction scheme of software simulation platform based on the private cloud. It can effectively improve students' mastery of basic knowledge and basic skills and the level of computer education. The research content mainly includes:building tool to function grab, architecture for the target system between the state of directed graph, and system environment variables set; cloud storage based on the distributed users for Hadoop.HTML5 drawing technology and desktop theme meta users in browser are utilized to reproduce the target system; and multilingual translation information is combined to achieve the goal of software multilingual interfacestruct private cloud exam platform based on Xinjiang cernet.

KEYWORDS: PaaS, Target software, HTML5, Xinjiang CCT.

1 INTRODUCTION

Xinjiang, located in northwest of China, enjoys vast territory and abundant resources, rich products, where many ethnic groups like Uygur, Han, Kazak and Kirgiz live and work together. For thousands of years, people of all ethnic groups have formed a social environment with multi languages coexisting and developing together. Improving backward situations of economic development is of great significance to China's Western Development Program, especially to raising the education level of the minority groups and people living in the backcountry, and local economic development. Therefore, it is the priority among priorities to develop education and cultivate talents in exploiting Xinjiang.

It has been the focus of the training system construction for universities in Xinjiang to improve the students' basic computer skills and practical ability. The examination as a teaching, training and evaluation system is valued by universities as a link education and evaluation National Computer Rank Examination (NCRE) earlier used network test system, while this system generally uses LAN. The exam provides software of the related registration, and examination arrangements will be provided by the specialized websites before the exam, and then the item bank will be installed on each examination site server. CCT has conducted thirty-six examinations since 2000, and the participants of CCT now cover all the universities from earlier, more than twenty, which amounts up to nearly one hundred thousand each year. Many universities use the CCT to evaluate the student situation, mastering the basic computer skills, and to evaluate the effect of teaching with it.

2 SITUATION OF CCT

However, CCT based on the IBT (Internet-Based Test) has been facing the following difficulties.

First, the type of questions is increasingly single. In order to facilitate machine scoring, CCT uses objective test, and subjective items uses the way of matching results, though the grading policy has been continuously improved, it is very difficult to be fair and accurate on the whole.

Second, the scale of question bank is restricted. It is difficult to expand question types of operation of the computer. Because of restriction of examination mode, they emphasize more basic knowledge, and involve less operating computer skills. In the process of realization, the computer operating exam can only realize the function of the file and item operations in Windows XP system. And it can't involve the test in the configuration management and the system maintenance. Replacing the test operating system, such as an update to Windows 7 or Windows Sever 2008, is associated with the whole test procedure and need store form computer rooms of all universities in Xinjiang. But these requirements are a major problem that the students will face after graduation.

*Corresponding author: *luanyq2003@163.com*

Third, there are a large number of concurrent burst characteristics in the process of computer operating exam and teaching, such as in the annual examination, tens of thousands of people begin answering at the same time. There is a great restrict in fund and technology to solve the question of the wide range of concurrent requests having the character of a longer period of time so as to maintain the features of interaction of the state by universities. At present, universities have inserted to CERNET, which can ensure the net speed and bandwidth though the net speed of the common internet is lower. It will extremely increase the level of Xinjiang computer education to construct the private-cloud flat based on Xinjiang CERNET, and realize the simulation platform of multi-language software open to universities.

3 TARGETS OF SYSTEM CONSTRUCTION

A software simulation platform based on multilingual target system, can provide a platform for simulation test at ordinary times for students. For CCT in Xinjiang, it can continually increase the level of basic knowledge and skills. This has greater application value in the western minority regions without experimental conditions.

For the requirement of the simulation platform based on the multilingual software, the main goal of system constructions is to study its key technology, and to construct the Tele -Education Private Cloud based on the XERNET (Xinjiang Education and Research Network), and as node taking Xinjiang universities.

First, to develop simulation modeling tools, researching target software, the basic operating object of Windows 7 should be grasped and analyzed according to the requirements of NCT syllabus, and finally a target software model can be established. It is difficult to avoid hard coding for target system when establishing test surroundings of the invented target system, because the workload of hard coding is too large to realize the system software such as DOS, and it will be renovated once every few years, let alone the simulation objectives. So it is necessary to develop grasping tools based on the Windows API's programming technology to grasp every state under an operating simulation object (it is the window appearing every state rather than recording screen photo). Under grasping from different routes realizing operational, it will realize objects, sub-objects, and there's attributes, to create a vector diagram through analyzing objects, sub-objects and there's attributes from different routes by navigation, to store the vector diagram in cloud-terminator by XML, and to become common data meeting user's requirements.

Second, cloud computing technology is to store all the data of the simulation learning system and experimentation compiling running used for information education on Virtual Machines.And user terminators realize all of operation functions reappearing target software on IE for large concurrency users by using the function of browser graphics rendering of HTML5, and record changes of user operation and surroundings information. Cloud computing technology can be also used to provide a software simulation environment, and to accomplish tasks of teaching and examination by reappearing, and to simulate learning and experiment in the education environment with high-performance and multi-user (Lin et al. 2011).

Finally, to research and develop theengine and service of the multi language software simulation platform based on the cloud, the distributed architecture used to store large server cloud computing is set up to realize WEB application making use of HTML5 technology for client presentation layer, to realize the multi-language interface switching of the target software, and to explore how to realize teaching of the remote virtual at the least cost, and then to promote IT development in the minority areas (Wushour·&Riyiman2005).

4 KEY TECHNOLOGY OF SYSTEM BUILDING

4.1 *Virtual simulation technology*

It requires different experiments and test environments for various courses, for example, course A requires the environment ofWindows+DirectX+MPI, and course B requires the environment of The RedhatLinux+MySQL+Apache. The environment and process of the experiment can be simulated using computer and simulation software.The virtual simulation technology (Virtual Reality, abbreviated VR) is a kind of method of processing graph and reproducing image relying on computer.At present, generally hard code is used for target system using VR at home and abroad.But it is difficult as the workload of realizing.Dos using hard code is very large, and it needs updating once very few years, let alone the objects in need of simulation(Li2011).

4.2 *Technology of distributed cloud storage*

In the data center, the cloud services provided by the cloud computing include: computing resource, storage resource, and software resources, etc. Cloud of computing resources is equivalent to a virtualized computing resource pool. To accommodate a wide variety of work patterns, these patterns can be rapidly deployed to physical facilities. Cloud storage integrates a lot of storage devices and coordinating works, and provides data store service together. Cloud of Software service based on the application of virtualization technology, is a virtualization of cloud application, and is a brand-new mode of software utilization and leads into personal user domain. Users do not need to install, to reinstall, thus there is no waste produced (Sunet. al. 2011).

4.3 *HTML5-Reconstruction technique of the client WEB*

HTML5 standard is designed to help developers write all kinds of Web applications more easily, and to adapt to the latest trend of the current SaaS, cloud computing and RIA technology. The new HTML5 standard will include more powerful label and application programming interface (API) for interaction, multimedia and localization. Many modifications of the core elements are made to let users control data displaying on the screen more effectively. In the past if users wanted to develop RIA applications, they might be forced to use Java plug-in technology. The HTML5 limits to support new rich-experience elements named CANVAS, let users realize RIA experience without plug-in implementation. Appearance of HTML5 canvas elements fills the shortage; developers can use JavaScript scripting language to carry on graphic operations based on a series of commands on canvas. It allows users to draw the 2D image programmatically. The new HTML updated will free the browser, and make the browser no longer rely on proprietary plug-in such as Flash, QuickTime and Silverlight and others, and realize Web morestandardly (Liu&Yang2011).

4.4 *Multilingual interface switching technology*

Based on state property, the target system can realize the automatic extraction of software language resource files, and the multi-lingual interface switching combined with the translation of multilingual information. Multi-language resources base is a target software resource extracted based on the target software model of XML. Multi-language resources base contains two groups of service. The first one automatically extracts the list of resources from the target software resources. Developers carry on translation of character information efficiently. The second one submits the translation results in the multi-language resources base. When switching language library for user simulation plug-in, the corresponding language library resources are provided (Wushour·&Riyiman2005).

At present, universities in Xinjiang haven't achieved the multilingual virtualization teaching at cloud technology, based on the simulation platform of multilingual target system, to realize the switching scheme of the multilingual interface in application using the dynamic loading method and the object-oriented programming ideas, to provide students training simulation test maximally, and to provide a platform for Xinjiang's CCT examination, thus effectively improving the level of students'basic knowledge and skills (Miuet al.2004.).

5 SYSTEM CONSTRUCTION

PasS (Platform as a Service) platform is a branch of the cloud computing technology service. The technology does not directly provide software, but provide the functional components for developing software. The assembly is deployed to the cloud, and provided to the programmer with Web Service or App Engine. Developers construct own application system taking platform as a basic framework. Research on the key technology of multi-language software on the simulation platform based on the cloud platform, has functions to set up a simulator, re-appear target system, simulate operating processes, and trace the changes of the character data of the target system in the field of software simulation. etc. And these functions are encapsulated into different types of software or services. The platform can be divided into: the capture modeling tool for target software, the simulation engine, the services of the target software library, the storage services for the user, the services of multi-language resources, etc.

The hierarchical architecture of the whole platform and its application are shown in Figure 1:

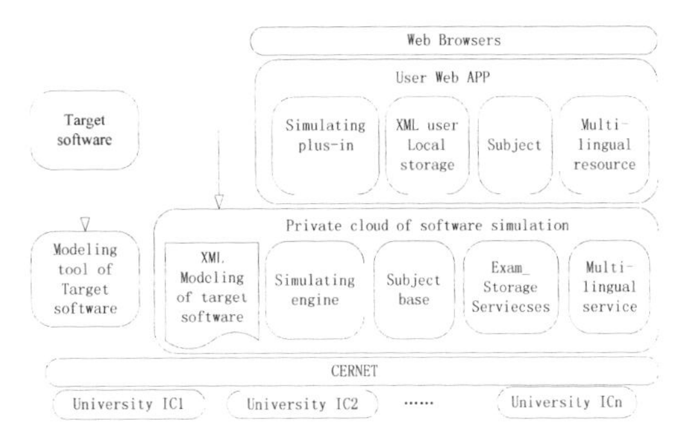

Figure 1. Hierarchical architecture.

Modeling tools of the target software: simulation of the target software, if only hard coding method is used, will lead to a set software having the same function, then the workload of the whole development process will be very large, and will develop continuously because of the different target software. This scheme is not desirable. This paper uses simulation modeling tools to realize the process of all the target software, which, through the operation to the user of the target software in a specific environment, dynamically grasps all the windows of target software, decomposes the classes, attributes and navigation, constructs the target operating state directed graph, and then stores in the cloud at XML, composes the model of the XML target software.

The simulation engine includes two parts: one is the user simulation plug-in based on HTML5, this plug-in is downloaded by users, and applied to own-software based on browser, which is used for the rapid development of the respective application process; the other one is the simulation service of cloud, this service will treat the user request of the simulation plug-in on the cloud, and analyze the behavior of the target software from the model of the XML object software, and then feed back to the user simulation plug-in, where plug-in realizes the simulation to the target software (Liu &Yang 2011).

The software interface base is an assembling to return to the desktop subject elements based on the Web Service. In fact, the subject is drawn, requiring pictures, font and style, user simulation plug-in in the simulation engine, when reappearing the behavior of the target software in the user's browser, if using the picture mode, the number of transmission data between networks will be increased greatly, when storing the dialog box rather than the picture of the dialog box, but the property is also needed, it can realize the re-appearing process to the dialog box using the drawing tools of the HTML5 in client. The reappearing attributes required, is an XML document like tree structure, various storage to digraph construct a public data needed from users, and data is downloaded from the cloud to the client memory, so the drawing with JS in the browser should be very fast (Liu&Yang2011).

This study takes the fonts, pictures, rendering, interface style of the target software subject as a whole using subject mode of software interface; draw interface effect consistent with the target software in the Canvas of the user's browser with the user simulation plus-in. The modeling environment of software can greatly reduce the flow of the user access network.

Multi-language resources base is the target software attributes, resource extracting based on the XML target software.

The storage service of the end user simulation instance, is to store and send the process of the software simulation operation from the end user to the cloud distributing computing and storing. Based on this service, programmer can join the function of storing and continuing of the simulation operation process in own applications software (Lin et al. 2011).

6 CONCLUSIONS

The feature of the project is to research the improvement of the education and teaching to backward equipment using cloud computing technology, through the combination of the cloud computing technology and HTML5 technology, so as to realize the virtual tele-education environment, and realize analyzing and treating process, and solve the actual problems and difficulties, lay the foundation for the development of information technology of Xinjiang' university.

This subject can also be widely used for the software test and the analyses of software user behavior habit. When software is developed, it will create the software state and navigation graph during set up the process of the software simulation.Using the directional graph and the digraph can further research the different operational strategies of automatic generation of traversal of this graph, thus providing support for this software automatic test. When software is in the sales process, the simulation products can also be generated, and can provide experience to the potential users. And this simulator can also be used to record every user operation function, realize the analysis to the operating habits of users, and help the development and upgrade of this software.

ACKNOWLEDGEMENT

The research work was supported by the National Natural Science Foundation of China (Grant No. 61262023) and the National High - Tech Research and Development Program of China (Project No. 2013AA13702).

REFERENCES

Li,Gangjian. 2011. Research on the framework of the cloud computing platform based on the virtual technique. *Journal of Jilin Architectural Engineering Institute*:89–92.

Lin,Kun, Li,AiJu&Dong, Longjiang. 2011. Research and realize on the cloud memory. *Microcomputer Information* 07 :221–223.

Liu,Huaxing&Yang,Geng. 2011. Research on the next generationWEB developing standard HTML5. *Computer Technology and Development*08 :171–174.

Miu,Cheng, Yuan,Baoshe, Wushour·SI_lamu, et al. 2004. Design and implementation of processing platform for uighur, Kazak, Kirgiz, Chinese and English.*Computer Engineering*30(10) :92–94.

Sun,Fuquan, Zhang,Dawei, ChengXu, et al. 2011. Establishment of enterprise private cloud storage platform based on Hadoop.*Journal of Liaoning Technical University*30(6):913–916.

Wushour·SI_lamu&Riyiman-Tursun. 2005. Key techniques for uighurscreen translation and its realization. *Journal of Xinjiang University*22(3):75–78.

Electronics, Communications and Networks IV – Hussain & Ivanovic (eds)
© 2015 Taylor & Francis Group, London, ISBN: 978-1-138-02830-2

Data transmission of deep sea GPS wave buoy based on the BeiDou satellite system

Chaoqun Dang*, Zhanhui Qi, Mingbing Li & Suoping Zhang
National Ocean Technology Center, Tianjin, China

ABSTRACT: The GPS wave buoy designed by the national ocean technology center requires only a GPS receiver without any extra sensors. It has the advantages of simple hardware, low power consumption, small size, lightweight, easy deployment, highly precise measurements, and is suitable for large-scale and meticulous wave measurements. In order to solve the problem of deep sea GPS wave buoy data transmission, this device employs the BeiDou satellite system to communicate. This paper introduced three satellite systems, Argo, Iridium, and BeiDou, and analyzed the advantages and shortcomings of each when used for marine environment observation. Then, the system structure, circuit design, data measurement, and transmission theory of each satellite system were described in detail. Finally, with laboratory and field tests, it was proved that the BeiDou system could be used for deep-sea data transmission. Specifically, when the GPS wave buoy experienced typhoon conditions, it was able to accomplish normal communication transmissions, which also verified the reliability of the BeiDou satellite system.

1 INTRODUCTION

The focus of China's marine policy is on understanding the deep-sea, and the data transmission method is increasingly important for marine environmental protection. However, acquiring deep-sea marine environmental factors in a convenient, accurate, and rapid way has proved to be a difficult task.

Currently, wireless communication is commonly used in the acquisition of marine environmental factors. Digital radio and radio data communication methods widely used near receivers are convenient, real-time, and economical. However, due to the limitations of distance and communication conditions, the methods used near the coast are not appropriate for deep-sea data communications.

Satellite system networks provide a potential solution for deep-sea data communication. Because of the outstanding performance, satellite communication has emerged in the deep-sea data communication field. The Argo, Iridium, and BeiDou systems have gradually become preferred choices and are widely used for marine environmental observation.

2 SATELLITE SYSTEMS

2.1 *Argo satellite system*

The Argo satellite system is a satellite communication system for data collection and positioning established by France and the United States. It uses polar-orbiting satellites to transmit various environmental monitoring data and to locate the positionsof instrument carriers. Due to the high latitudes (Arctic and Antarctic circles) geosynchronous satellites cannot be used to communicate; the Argo system provides a good method of communication for hydrometeorological monitoring instruments in these areas. (Li et al. 2011)

The Argo system is widely used in the field of oceanography as a method for locating buoys. The latest Argo-3platform messaging transceiver(PMT) was designed specifically for the constellation of Argo, and overcomes the shortcomings of the Argo-2 platform transmitter terminal(PTT) with characteristics of bidirectional communication, superior data capacity, and high speeds (Guigue 2010).

The PMT of the Argo system is not a real-time system and must wait for the satellites above it. Therefore, it is only suitable for transmissions that don't require real-time data. Typical Argo system applications include Argo surface drifters and Argo profile measurement buoys.

2.2 *Iridium satellite system*

The Iridium satellite system is a global mobile satellite communication system consisting of 66 LEO satellites. The most notable features are the interplanetary links and polar orbit. The interplanetary links guarantee continuous satellite communication through a gateway. And the polar orbit provides unobstructed

*Corresponding author: *dangchaoqun06@163.com*

communication service between the north and south poles. Iridium can achieve high signal intensity and reliable communication anywhere on earth without the effects of weather, altitude, ionosphere, distance, or any other factor. Therefore, the Iridium satellite system is especially suitable for areas in which other communication methods cannot reach.

The cost of Iridium SBD communication is relatively high compared to other terrestrial mobile communication systems. However, Iridium SBD data transfer is still quite economical in certain areas, industry, and circumstances.(Ledesma 2012, Zhang et al. 2006, Ju et al. 2010)

The next-generation Iridium communication module is the Iridium 9602, which boasts a small size and is light-weight. It is widely used in the fields of marine buoys and hydrological monitoring as a SBD transceiver in an embedded application.

2.3 *BeiDou satellite system*

According to the plan of the BeiDou satellite system construction, the coverage of the system expanded to the Asia-Pacific region in December of 2012. With the advantages of being the latecomer, it offers a short message communication service that GPS and other satellite navigation systems do not have except in navigation and timing services. The BeiDou satellite system is better than international maritime satellites communication in the aspects of the addressing method, flow rate of channel, capacity of users, performance of real-time communication, and price. In addition, the BeiDou satellite system is broader than the coverage of GPRS/CDMA, more stable than VHF, and more mature than the Chinese Area Positioning System (CAPS). The current coverage of the BeiDou system basically includes the entire sea area that is under the jurisdiction of China.

The service of short message communication is a considerable feature of the BeiDou system, in that it not only can know its own position, but can also pass its position to others. The BeiDou user machine which relies on the capability of bidirectional communication can achieve three primary functions: (1) Upload monitoring data: a GPS wave buoy uses the GPS sensor to collect wave data, sends the data to the receiving terminal, and then the buoy monitoring center uses the BeiDou commander machine to receive and store the buoy data. (2) Point to point communication: In cases of emergency situations or when there is a need to send commands to a single buoy, the buoy monitoring center can send commands to the BeiDou user machine that is fixed on the intended buoy via the BeiDou commander machine. The buoy then executes the appropriate commands according to the instructions. (3) Broadcast communication of common information: For common information, the buoy monitoring center can send the same message to all subordinate BeiDou user machines. (Li et al.2012)

Currently, the second generation of the BeiDou system has completed coverage of the Asia-Pacific region, and offers high security and real-time communication. The capability of short message bidirectional communication is suitable for large-scale data control and transmission. In addition, the communication module has characteristics of small size, low price, low communication cost, and large system capacity.

3 GPS WAVE BUOY

Because the data transmission of the Argo and Iridium systems are first sent through foreign data processing centers and then distributed to domestic users after processing, the security of data cannot be guaranteed. In addition, all of the key technologies are mastered by others. Therefore, GPS wave buoys adopt the communication system of the BeiDou satellite system considering the aspects of data security, tariffs, and other practical maritime applications.

3.1 *System structure*

GPS wave buoys consist of three main parts, including the GPS sensor, the hardware circuit board, and the BeiDou communication module. The structure of the system is shown in Figure 1 where the hardware circuit board consists of the power and control system.

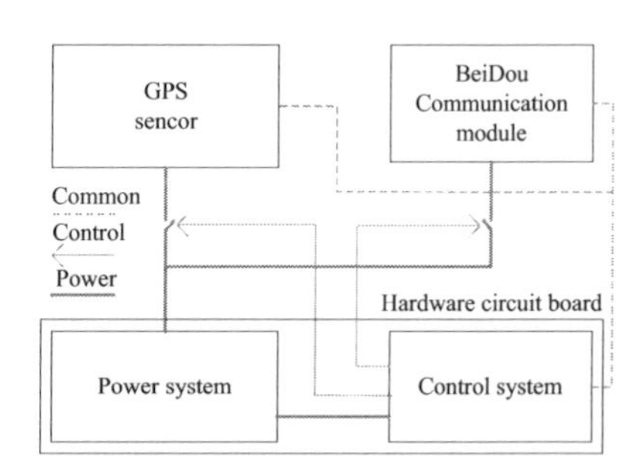

Figure 1. Structure of GPS wave buoy system.

After deploying the GPS wave buoy, the hardware circuit uses the power of the GPS sensor in accordance with a predetermined sampling interval. Then, the ARM microcontroller begins to read the raw data collected from the GPS sensor and stores it in the expansion SRAM. Next, the system shuts down the power of the GPS sensor until achieving the predetermined number of points. Then, the ARM microcontroller reads the raw data from the SRAM and calculates the wave parameters. The system uses the power of the BeiDou communication module after completing the calculation to send the results to the receiving terminal in a short message.

3.2 *Design of hardware circuit*

In order to meet the requirements of GPS wave buoy signal processing, data communication, and power control of each module, the hardware circuit board mainly consists of the ARM microcontroller, system power, external clock, SRAM, power control of external sensors and communication modules, interfaces of external sensors and communication, system reset, and the level conversion module. The overall design of the hardware circuit is shown in Figure 2.

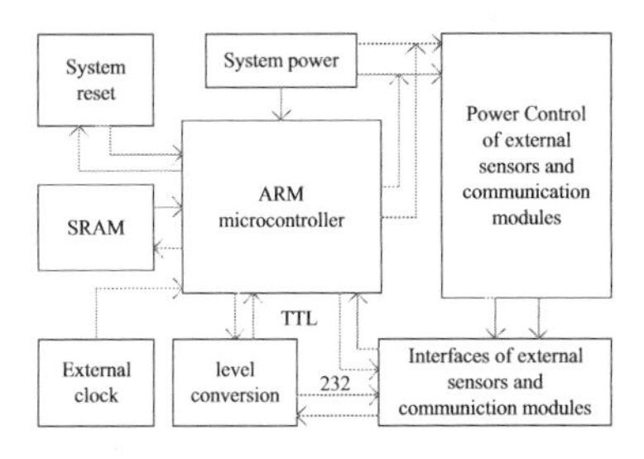

Figure 2. Overall design of hardware circuit board.

The ARM microcontroller is the core of the entire system and plays a dual role of calculator and controller. It controls normal operation of each module. System power provides power for the microcontroller, external sensors, and the communication modules. The external clock provides RTC and the system clock for the microcontroller. SRAM is used to store the raw data of the GPS sensor. The microcontroller controls the power of the external sensors and communication modules via the power control module. System reset is used to generate a reset signal when it is mandatory. The system achieves normal communication between the ARM microcontroller and external devices via the level conversion, external sensors, and communication interfaces.

The system passed the environmental experiment, and can work normally at -5°C - 45°C.

3.3 *Measuring principle and data transmission process*

GPS wave buoy stake advantage of the Doppler shift principle of the GPS satellite signal to measure the motion of the buoy, obtains the wave spectrum from the motion information, calculates the characteristic value of the wave data from the wave spectrum, and then stores and sends the data.

In order to reduce power consumption of the system, the power control module was added to optimize the power consumption of each module and make each module work in a timesharing and segmented way. The process of measurement and data transmission is shown in Figure 3.

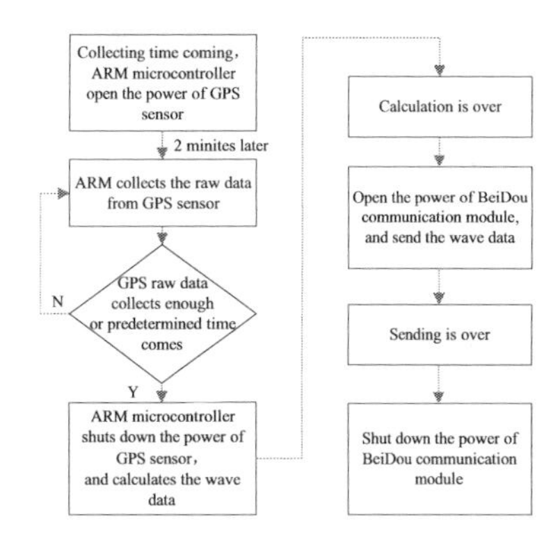

Figure 3. Processes of measurement and data transmission.

The ARM microcontroller uses the switch control module to provide power to the GPS sensor after the power of the system is stable. Two minutes later, it begins to collect GPS raw data until it's collected enough data or reaches the pre-set time. Then, the ARM microcontroller shuts down its power and begins calculating the wave data.

The ARM microcontroller reads the GPS raw data and calculates the wave data. The ARM microcontroller turns on the power of the BeiDou communication module after the calculation is completed and begins to send wave data two minutes later. After a successful interactive transmission, the ARM microcontroller shuts down the power of the BeiDou communication module.

The GPS wave buoy uses the RTC clock to control hourly measurements when the system is powered. In order to avoid program crashes, the system adopts two strategies: a software watchdog within the code and an increase in the system power-on reset circuit in the design of the hardware.

4 TEST

4.1 *Laboratory test*

A 15-day laboratory experiment was carried out in September 2013. During the test, the system experienced inclement weather conditions like clouds, thunderstorms, and rains, which had significant impacts on the satellite data transmissions. During the duration of the test the wave buoy transmitted 360 sets of hourly wave data, and the receiving terminal received 360 sets of data; therefore, the efficiency of the data transmission was 100%.

4.2 *Field test*

During the voyage of the Northwest Pacific in late October 2013, a GPS wave buoy was launched from a R/V ship. During the test, two typhoons, KROSA and Petrels, passed through the testing area. The KROSA was used as an example to analyze the communication system of the GPS wave buoy in typhoon conditions. Figure 4 shows the path of typhoon KROSA (blue line) and the GPS wave buoy drifting track (red line).

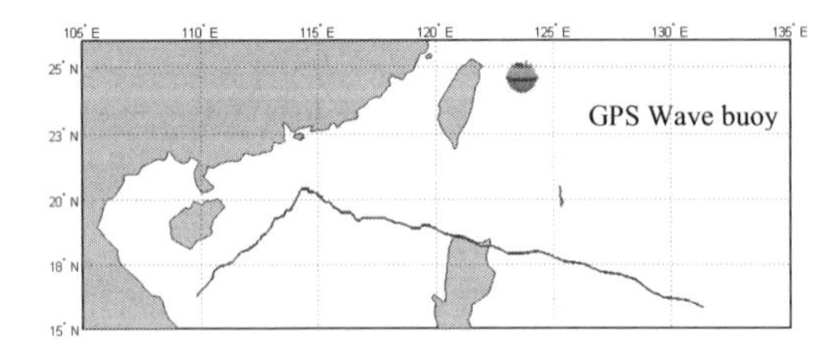

Figure 4. Path of typhoon KROSA (blue) and the GPS wave buoy drifting track (red).

The wave data was analyzed from Oct. 29th to Nov. 4th during which time Typhoon KROSA passed. This period of time included 141 sets of data. The wave heights are shown in Figure 5.

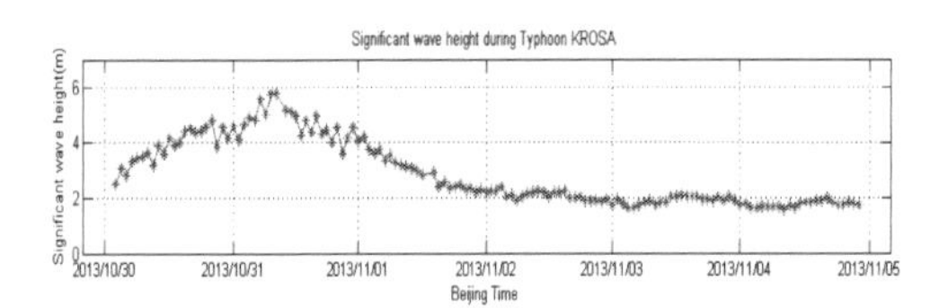

Figure 5. GPS wave buoy data transit through BeiDou satellite system during Typhoon KROSA.

The receiving terminal of the GPS wave buoy system received 140 sets of wave data, having lost just one set of data at 5:00 on Nov. 1st. As such, the efficiency of the data transmission was 99.3%.

5 CONCLUSIONS

The GPS wave buoy introduced in this essay not only works normally under bad sea conditions, but also has an outstanding reliability in data transmission. Field and laboratory tests proved the feasibility of the BeiDou satellite system for use in deep-sea environments. Due to limitation of the coverage of BeiDou, at this stage, the system only works in a limited sea area. After the BeiDou satellite system is completed, it will have a broad application range. In order to improve the system performance, topics discussed in this paper were as follows:

1. Ensure the validity of data transmissions by setting an appropriate buoy sampling interval, and take advantage of the BeiDou bidirectional communication capability, which has a high validity of data transmission.
2. Reduce power consumption using the buoy's internal control board module for power control and management, which can control the work mode of each module to optimize power distribution and reduce the system power consumption overall.
3. Take advantage of the bidirectional communication capability. The work mode of the wave buoy can be set remotely to meet the requirements of different observation points in the sea.
4. Use the concealed observation capability to remain discrete in the deep sea. The wave buoy has a small size, and the structure contains a crash cushion design, which ensures the buoy's concealment and safety in the deep sea.

ACKNOWLEDGEMENTS

A sincere thanks to the support of the National Natural Science Foundation (Number:41406114), Special Fund for Marine Renewable Energy Projects of China (Number: GHME2012ZC02), the Natural Science Foundation for Young Scholars of Tianjin, China(Number: 13JCQNJC03800) and the Natural Science Foundation for Young Scholars of Tianjin, China (Number: 12JCQNJC02400).

REFERENCES

Guigue, M. 2010. PMT RFM-YTR-3000 USER MANUAL.

JJ de Vries. 2007. Designing a GPS-based mini wave buoy. *International ocean system*: 21–23.

JuYuqiang, Liu Jingbiao & Yu Haibin. 2010. A remote monitoring and data transmission system of buoy based on Iridium satellites. *Embedded technology* 26(12): 36–39.

Ledesma Miguel Martínez. 2012. *Development of an oceanographic drifter with IRIDIUM bidirectional communication capability*:1–51 . University of Balearic Islands: Master of science in electronic engineering.

Li Wenbin, Zhang Shaoyong, & Shang Hongmei, et al. 2011. Design of surface drifting buoys based on Argos satellites system. *Ocean Technology* 30 (1): 1–4.

Li Wenqing, Fu Xiao, Wang Wenyan, et al. 2012.Application of BeiDouand generation satellites navigation system in marine data buoy supervision and management. *Shandong Science* 25(6): 21–26.

Peng Wei, Xu Junchen, Du Yujie, et al. 2009. Design of marine monitoring data transmitting system based on BeiDou satellitess system. *Ocean Technology* 28(3): 13–15.

Zhang Shuwei, Wang Xiufen & Qi Yong. 2006. Application on ocean data buoy of Iridium data communication. *Shandong Science* 19(5): 16–19.

Electronics, Communications and Networks IV – Hussain & Ivanovic (eds)
© 2015 Taylor & Francis Group, London, ISBN: 978-1-138-02830-2

A layered interconnected graph model based algorithm for traffic grooming in optical transport network

Xijie Dong & Liangrui Tang
A State Key Laboratory of Alternate Electrical Power System with Renewable Energy Sources (North China Electric Power University), Changping District, Beijing, China

Jiangyu Yan
B School of control and computer engineering (North China Electric Power University), Changping District, Beijing, China

ABSTRACT: In order to solve the problem of low wavelength resource utilization caused by the huge gap between the capacity need by the service and the single wavelength capacity in OTN, a Layered Interconnected Graph Model (LIGM) based algorithm for Traffic grooming is presented. The LIGM is based on the optical network physical topology, and the costs of every link represent the optical network resources. The initial solution set is calculated with the grooming path optimization selection strategy and the optimal grooming route is chosen from the set with the O/E/O number of each path. By adapting the load balancing mechanism, the algorithm can perfectly solve the problem of traffic grooming in optical networks. The simulation shows that, compared with other traffic grooming algorithms, the algorithm LIGA effectively reduces the blocking probability and makes full use of the optical network resource at the same time.

KEYWORDS: Optical Transport Network; Layered Interconnected Graph Model; Traffic grooming

1 INTRODUCTION

With the rapid growth of the numbers of data service types and bandwidth requirements in communication network, a flexible and efficient optical transmission network built with the technology of OTN becomes the inevitable development results (Zhang Haiyi 2013). However, by adapting DWDM technology, a single wavelength capacity has grown to 100 Gb/s, and there is a huge gap between the capacity requirement of the service and the single wavelength capacity. In order to avoid the waste of bandwidth resources, the research on traffic grooming has become one of the most important topics in the construction of OTN(Yang Xiaolong 2009).

The traditional way of traffic grooming is to establish the static virtual topology by adapting wavelength routing and then these low rate services are routed on the virtual topology. However, service connections change dynamically in the actual network. The traditional way for low-speed connection routing would cause low resource utilization. Layered graph model is proposed to solve the problem of route and wavelength assignment (RWA) in ref.(Zhang Yu et al. 2004). But when applied to the traffic grooming, it can't effectively solve the problem of service multiplex. The strategy of adapting the virtual topology, layered graph and integration graph in turn is proposed in ref.(Wen Haibo et al. 2004). It perfectly solves the problem of service multiplex and RWA, but it ignores the network load balance problem, so that it cannot make full use of the network resource. The strategy of shortest path algorithm is adapted as the grooming path in ref.(Li Jia et al. 2008). This strategy makes the key link in the congestion situation, and the network blocking probability is relatively high.

According to the above analysis and comprehensive consideration of the dynamic information of optical network resources, a Layered Interconnected Graph Model (LIGM) based algorithm for traffic grooming in optical transport network is proposed in this paper. The cost of each link is dynamically updated according to the available network resources in the LIGM, so as to realize the goal of load balancing, low blocking probability and large network throughput. At the same time, under the quantitative restrictions of optical transceivers, the strategy of grooming path optimization selection would choose the path with less O/E/O conversion to groom the service in order to fulfill the transmission reliability requirement.

2 LAYERED INTERCONNECTION GRAPH MODEL

Establishing a light path in the optical transmission network, all the links of the path should be assigned the same wavelength. This is known as the Wavelength-Continuity Constraint. Previous researches prove that wavelength conversion technology improves the performance of all-optical network to a large extent. However, it is expensive to configure the wavelength converter for the node of OTN. With the consideration that OTN nodes have the ability of O/E/O conversion, all-optical wavelength conversion function could be realized in the way of O/E/O conversion to transfer the service from wavelength λ_i to λ_j, in order to meet the requirements of network construction cost and Wavelength-Continuity Constraint. Based on the above consideration, the Layered Interconnection Graph Model which is extended from the original layered graph is put forward to be suitable for the OTN in this paper.

The steps to build the LIGM from the original physical topology are as follows. According to the number of wavelength $|W|$ in a physical topology link, make $|W|$ copies of the physical topology to form the layered graph with $|W|$ layers. Each layer is just one wavelength plane. Each physical node has corresponding virtual nodes in each layer of the layered graph, for example, node V_k^i is the virtual node of node $V_k(K= 1,2, \dots |V|)$ in the wavelength plane λ_i. And then the corresponding virtual nodes are connected by the layered interconnect sides. The physical link $E_{mn} \in (m, n = 1,2 \dots |V|)$ is corresponding to wavelength link E_{mn} in each wavelength plane λ_i.

When a service request R(s, d, bw) arrives, the layered interconnect sides can be divided into two classes. They are the layered interconnect sides of source, and destination node (s-d node) and the one of intermediate node (i node). If there is a direct link formed by the free wavelength link in a wavelength plane, build this direct link and then delete those free wavelength links to complete the network resources update, thus, concludes the bandwidth update of the grooming path link and the optical transceivers of the grooming path nodes. Finally, the LIGM consists of wavelength links, direct links, layered interconnect sides (LIS) and virtual nodes. The process of building the LIGM after grooming the service request $R(1, 3, b)$ is showed in Figure 1.

In the LIGM, the cost of each link is updated dynamically according to the resources usage of the link. Only in this way, the goal of load balancing could be achieved. The definition of the cost of each link is as follows.

$$w_e = \begin{cases} \frac{|W|}{c}, & e \text{ is the wavelength link.} \\ 1, & e \text{ is the lightpath link.} \\ |V|-1, & e \text{ is the LIS of } i \text{ node.} \\ 0, & e \text{ is the LIS of } s-d \text{ node.} \end{cases} \quad (1)$$

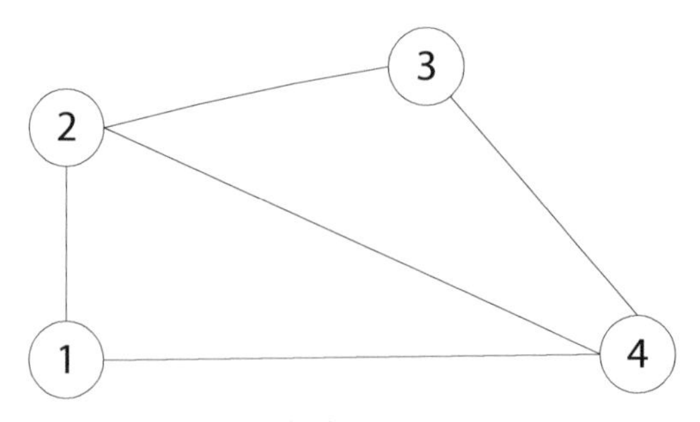

(a) Physical topology

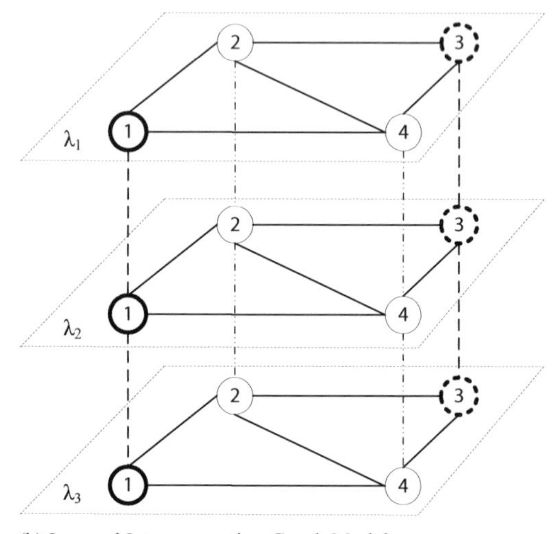

(b) Layered Interconnection Graph Model

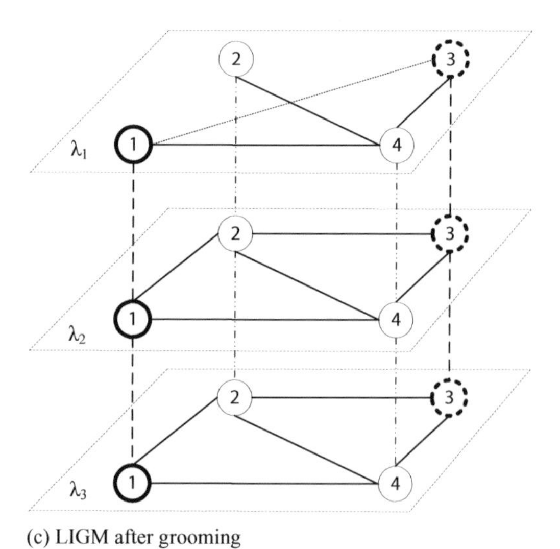

(c) LIGM after grooming

Figure 1. Process of building the LIGM.

For the wavelength link e, c represents the numbers of available wavelength in the corresponding physical link in physical graph. The initial value of w_e is 1. When there are no service requests, the number of available wavelength is $c = |W|$. For the LIS, the cost is defined with the comprehensive consideration of the numbers of optical transceiver and O/E/O conversion. $|V|$ represents the number of nodes in the physical topology, and $|V|-1$ is just the theoretical maximum path value in any wavelength plane.

In the LIGM, the cost of each link is defined with the comprehensive consideration of load balancing mechanism, the numbers of optical transceiver, and O/E/O conversion. Based on the LIGM, the problem of traffic grooming in OTN is simplified to the extent of finding out the available optimal light path in the LIGM to groom the service.

3 ALGORITHM DESCRIPTIONS

By adapting the LIGM, the complexity of dynamic traffic grooming in OTN can be significantly reduced. It can be seen from the LIGM that the network scale grows $|W|$ times, but the problem of RWA and services multiplex can be solved at the same time. How to get an optimal path with less O/E/O conversion average times, low blocking probability and high network resource utilization becomes the main purpose of the algorithm presented in this paper.

3.1 *Traffic grooming strategy*

Comprehensively considering the traffic grooming targets such as balancing network load, reducing the average numbers of O/E/O conversion and reducing the network blocking probability, the algorithm based on the LIGM and presented in this paper is achieved as follows.

Step 1: Establish the service connection path with the direct link or the layered interconnection sides which meet the bandwidth requirement of service in the LIGM. The direct link has occupied the node optical transceiver resources. So it can save the optical transceiver resources by adapting the existing direct link to build single-hop or multi-hop connection.

Step 2: When step 1 failed, establish a direct link with free wavelength link in a wavelength plain in the LIGM. An optical transceiver should be occupied in the source and destination nodes of the new direct link by this strategy.

Step 3: When step 1 and 2 both failed to build grooming path, establish the grooming path with the combination of free wavelength links, direct links and layered interconnection sides to build multi-hop grooming path. It could make full use of the OTN resources by this strategy.

Step 4: If all the steps above failed to establish the grooming path, the request of the service is refused.

3.2 *Optimization selection strategy of traffic grooming path*

Traditional traffic grooming algorithms adapt the shortest path algorithm such as Dijkstra algorithm to select the grooming path. However, it will accelerate the congestion degree of key links in the network in this way. Therefore, in order to balance the network load, not only the cost of each link should be updated according to the network resources, but also the grooming path optimization selection strategy should be introduced. When the service connection request arrives, the cost of each link is updated dynamically according to the network resources. In this process, b represents the bandwidth requirement of the service connection. Then find out the route from the source node s to the destination node d. h represents the hops of the route and wid represents the cost of link from node i to node d. represents the link cost of h hops from source node to destination node. The function is given as

$$f_{(s,d)}(h) = min\left\{f_{(s,d)}(h-1), \min_{i\in K}\{f_{(s,i)}(h-h_{id}) + w_{id}\}\right\}, s \neq d \tag{2}$$

where h = 1,2,...|V|−1, and K represents the set which is consist of the nodes which are just one hop from the destination node *d*. Let $f(s,d)(h) = 0$ when nodes *i* and *s* are the mapping nodes of the same physical node, which means the minimum link cost of a node to itself is always 0. Let $f(s,d)(0) = \infty$ when the nodes *i* and *s* are the mapping nodes of different physical nodes, which means that the minimum path cost of any two nodes is infinite while the link hop between them is zero. The minimum path cost between source-destination node pairs can be calculated by (2) with the value of h increasing gradually. When $f(s,d)(h)$ steadily comes to a minimum value eventually, the traffic grooming path is got whenever it achieves the minimum for the first time, with the initial solution set of optimal path between node *s* and *d*. Let *S* representing the set of all optimal paths obtained.

If the initial solution set *S* contains only one path, the path is the output seen as the optimization solution. Otherwise, the optimized solution can be got by putting the initial solution set into optimization objective function.

When service is transported in an optical network, the service delay consists of transmission delay and conversion delay. Ref.(Fu Gang & Wu Yibo 2008) presents that compared with transmission delay, conversion delay is the mainly part of the service delay. Transmission delay can be neglected. Therefore, the O/E/O conversion number of the service is chosen as the optimization goal of the initial solution set. The optimization objective function is expressed as

$$f_{opt} = min\{f_{(s,d)}(h)*hop\} \quad (3)$$

where *hop* represents the O/E/O conversion number of the path in the set *S*. By calculating the value of f_{opt}, we can obtain the optimal solution, then groom the service request $R(s,d,bw)$ on the corresponding appropriate route, and occupy the network resources. With the optimization selection strategy, the service request can be groomed on the path with less O/E/O conversions to maintain the reliability of the service transmission.

4 SIMULATIONS AND ANALYSIS

The simulation is conducted on 14 nodes with 21 bi-directional physical links of the NSFNET which is showed in Figure 2 in this paper. The simulation process is based on the events simulation environment. The number of optical transceiver on each node is |T|. It is supposed that there are 14 wavelengths in each fiber and the single wavelength capacity is 10. The service requests $R(s,d,bw)$ are all two-way services, and the source and destination nodes are randomly chosen from the node set. The bandwidth requirement of each service is uniformly distributed in 1-6 integer units. The arrival of service requests follows the Poisson distribution with mean β, and the duration of service requests follows the exponential distribution with mean 1/μ. So the service load strength is defined as $\rho = \beta/\mu$ (Erl). There is no queuing mechanism for a service request. It will be rejected as soon as the service connections fail to establish.

The EMNZ - BCP algorithm presented in ref. (Fan Tao 2012) and traditional SLRA presented in ref. (K. Zhu & B. Mukherjee 2002) are chosen as the comparison algorithms to compare with the LIGM-based algorithm (LIGA) presented in this paper. The blocking probability (BP) and service average O/E/O conversion numbers respectively under the condition of different numbers of optical transceivers and load strengths are the primary indicators to measure the traffic grooming algorithms above.

Figure 3 shows the BP in the condition of different numbers of optical transceivers in each node. It can be seen from the figure, when the number of transceivers is relatively less, the BP of LIGA is less than those of SLRA and EMNZ-BCP. This is because the LIGA firstly selects existed single or multi-hop direct links as the grooming path. It can save the resource of optical transceivers so as to decrease the blocking probability. It can also be seen that the BP decreases with increasing of the number of optical transceivers by using the three algorithms. However, when it increases to a certain extent, the trend of BP decrease becomes flat. The number of wavelengths in the fiber now turns into the main constraint.

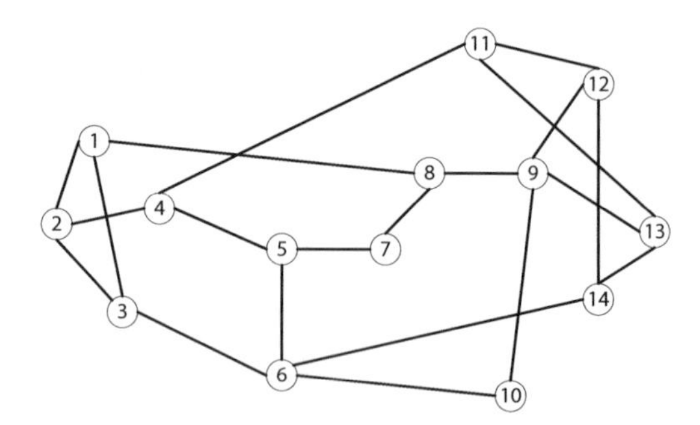

Figure 2. The NSFNET network topology.

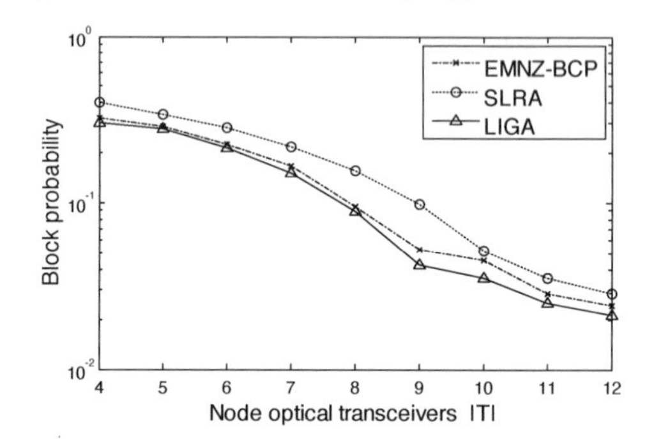

Figure 3. The BP performance with different number of optical transceivers.

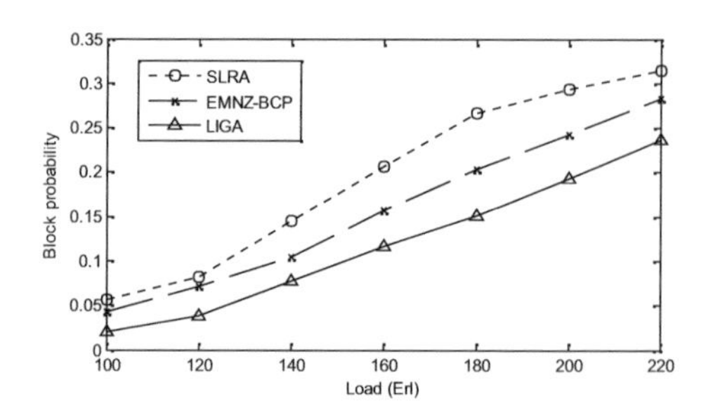

Figure 4. The BP performance with different network load.

Figure 4 shows the BP in different network load strengths of the three algorithms. It can be seen from the figure that the BP performances of EMNZ-BCP and LIGA are lower than that of the traditional SLRA. EMNZ-BCP and LIGA respectively consider the network resources usage and optical transceivers of each node in grooming path selection. But when it comes to the SLRA, the congestion situation of network key links is accelerated with the reason of adapting the Dijkstra as the grooming path selection algorithm. On the contrary, EMNZ-BCP and LIGA respectively adapt the KSP and Optimization selection strategy to avoid the BP increasing sharply. Compared with EMNZ-BCP, LIGA takes the number of O/E/O conversion as the optimization objective function. The performance of BP improved obviously in this way in this by saving the wavelength resources for the following access service requests.

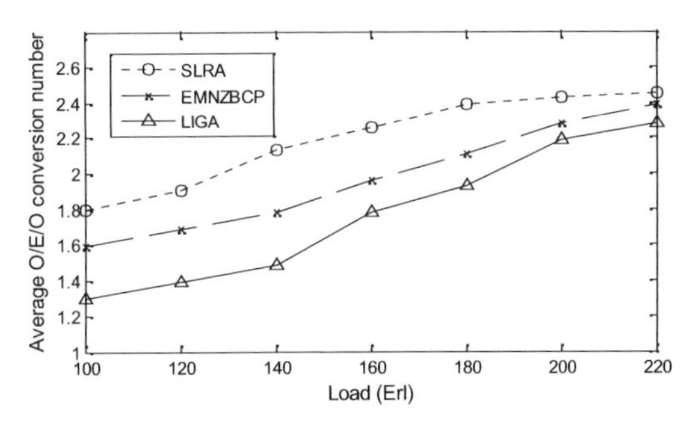

Figure 5. Average O/E/O conversion number of service.

Figure 5 shows the average O/E/O conversion number of service in each algorithm with different load strengths. It is seen that the average O/E/O conversion number of service of these three algorithms tends to increase with growth of the service strength. After the service strength increases to a certain degree, the average O/E/O conversion number of service begins to be flat, that is due to the large percentage of existed direct link with high service strength that makes every service request be groomed via direct links of single or multi-hops. As a result, the difference of O/E/O conversion number decreases gradually. It also can be seen that the average O/E/O conversion number of LIGA is less than those of the other two algorithms. With the result of low optical transceiver usage of each service, LIGA achieves the goal of full use of the network resources and indirectly decreases the BP. The LIGA algorithm makes better use of wavelength resources based on the LIGM and outperforms the EMNZ-BCP a lot in optimization selection strategy.

Based on the above analysis, LIGA is obviously better than EMNZ-BCP and SLRA in the performance of BP and service average numbers of O/E/O conversion. By grooming on the LIGM, the LIGA saves the optical transceiver resources, and improves the performance of BP. On the other hand, the network wavelength resources utility improves by adapting the load balancing mechanism which updates the cost of each link according to the real-time situation of resources. For real-world NSFNET optical network, the LIGA is proven to have better performance than the EMNZ-BCP and SLRA.

5 CONCLUSIONS

The rapid development of multimedia services causes higher requirements on the security and reliability of the optical communication network. The OTN transmission technology, with the characteristics of flexibility, efficiency, reliability, and large wavelength capacity, becomes one of the important technologies in the area of future optical network development. For the characteristic of large single wavelength capacity, the problem of traffic grooming in OTN is deeply studied in this paper. With comprehensive consideration of the dynamic information of optical network resources, a Layered Interconnected Graph Model (LIGM) based algorithm for traffic grooming in optical transport network is presented in this paper. The initial solution set of the optimal path is obtained by the optimization selection strategy, and the optimized solution is determined by using optimization objective function, which takes O/E/O conversion number into consideration. The comparison studies by computer simulations indicate that the algorithms presented in this paper have better performance on reducing the block probability and making full use of the network resources.

ACKNOWLEDGMENT

This work is financially supported by "the Fundamental Research Funds for the Central Universities" (2014QN14).

REFERENCES

Fan Tao. 2012. Research on Algorithm for the dynamic traffic grooming of neighbor zone expansion in IP over WDM networks. Xi'an: Xidian university.

Fu Gang & Wu Yibo. 2006. Analysis of End to End Delay of NGN Service. *Designing techniques of posts and telecommunications* (2): 22–26.

K. Zhu & B. Mukherjee. 2002. On-line approaches for provisioning connections of different bandwidth granularities in WDM mesh networks. *Optical Fiber Communication Conference and Exhibit*: 549–551.

Li Jia, Wang Jianshe, Jiao Fangyuan. 2008. A new integrated grooming algorithm based on integrated graph. *Information technology* (8): 139–142.

Wen Haibo, Li Lemin, Yu Hongfang, et.al. 2004. A different reliability requirements supported traffic grooming algorithm for WDM mesh networks. *Journal of china institute of communications* 25(3): 1–10.

Yang Xiaolong, Liu Xiao & Zheng Huan. 2009. A Novel Advance Reservation-Based RWA with Less Resource Fragmentations Based on Layered Graph for Lambda-Grids. *Journal of University of Electronic Science and Technology of China* 38(4): 529–532.

Zhang Haiyi, Zhao Wenyu & Wu Bingbing. 2013. Considerations on transport networks technologies evolution. *China Communication* 10(4): 7–18.

Zhang Yu, Li Zhengbin, Xu Anshi, et.al. 2004. A routing and wavelength assignment algorithm in multi-granularity switching optical networks. *Acta electronica sinica* 32(12): 93–97.

Electronics, Communications and Networks IV – Hussain & Ivanovic (eds)
© 2015 Taylor & Francis Group, London, ISBN: 978-1-138-02830-2

"Intelligence aids to navigation cloud" based on cloud computing technology

Zhixiu Du* & Guojun Peng
Research Center of Ship Navigation Service, Fujian, China
Marine Navigation & Pilot Technology Research Centre of Jimei University, Xiamen, Fujian, China

ABSTRACT: To solve the problem in aids to navigation informatization construction, this paper studies the applications of cloud computing technology in internal management organization level, remote monitoring and external service informatization of aids to navigation. The solution to overall architecture of "intelligent aids to navigation cloud" by using cloud computing technology is proposed, and the logical architecture and network architecture of "intelligent aids to navigation cloud" are described. The research can provide a reference for aids to navigation informatization by using cloud computing technology.

KEYWORDS: Cloud Computing Technology; Intelligence; Aids to navigation Cloud.

1 INTRODUCTION

In compliance with the general requirements of the administrative system reform and the national strategy on ocean development, the Ministry of Transport (MOT) established three navigation guarantee centers in the North China Sea, the East China Sea and the South China Sea respectively in 2012. By 2013, a total of 12,228 aids to navigation have been deployed. China MSA undertakes the management and maintenance of 7,331 aids to navigation. Among them, there are 396 shore-based AIS stations, 35 VTS centers, 146 radar stations and 22 marine RBN/DGPS stations(MSA 2014). How to realize the efficient management of aids to navigation? Informatization is the key. After years of informatization construction of aids to navigation, several business subsystems have been completed, and a lot of data information has been accumulated. Those are valuable information to aids to navigation informatization management, and play active roles in the management of aids to navigation department. But with the development of shipping market and aids to navigation informatization, the aids to navigation information system becomes more and more complex, and the development mode of traditional independent system becomes the bottleneck of data exchange and data sharing. Cloud computing is a new method of sharing data architecture, and a virtualized resource pool of super computer, which becomes a new network sharing model(Zheng 2012) . Introducing cloud computing improves the ability of data processing, resource sharing, collaboration, network security and so on. Internal management, remote monitoring and other external services of aids to navigation informatization are comprehensively improved.

2 THE APPLICATION ANALYSIS OF CLOUD COMPUTING TECHNOLOGY IN AIDS TO NAVIGATION INFORMATIZATION MANAGEMENT

Cloud technology is a generic term of network technology, information technology, integration technology, management platform technology, application technology, etc. Based on the business model of cloud computing, the cloud technology can form a resource pool, which is on-demand, and flexible(Baidu 2014). Cloud computing is the core of the cloud technology. Cloud computing which is a computing model can access all shared resources pool, such as computing facilities, storage devices, application degree, and so on by using the Internet anytime and anywhere(Mell, P 2011). Cloud computing uses the distributed computing technology that will enlarge more computing resources, and uses the redundant resources to process fault tolerant. It has obvious advantages in terms of network resource sharing because of the features of supercomputing power and low cost, high safety, user-center, and so on. Cloud computing technology can provide more environmentally friendly, green,

*Corresponding author: *duzhixiu_0219@126.com*

energy-saving, intelligent, quick and powerful informatization support, and more favorable conditions for the construction of integrated information management system.

Aids to navigation informatization uses cloud computing technology to build a unified data center as a cloud computing center. It centrally managed based IT resources through virtualization, the Internet, e-Navigation and other related technologies and concepts. And it provides basic IT resources for aids to navigation business application in the form of sharing and on-demand. Informatization of aids to navigation builds a unified data standard to ensure data's integrity, consistency, availability, and forms a unified data sharing service. It has distributed and paralleled computing capabilities and a rich variety of applications of delivery mechanism. Cloud computing technology promotes the management way of aids to navigation informatization, which is transformed from technology-driven to the business-driven. Business applications and data realize centralized management from decentralized deployment. Finally, aids to navigation informatization realizes the high fusion of the management business and information(Liu 2013).

3 THE OVERALL ARCHITECTURE OF "WISE AIDS TO NAVIGATION CLOUD"

Wise aids to navigation cloud architecture transmits perceptual information to basic information communication network through a various ways of connection, and establishes an automatic resource scheduling allocation mechanism without human's intervention to provide resource services for all kinds of applications. The resource services mainly include data resources and computing resources(Zhang 2012). Data resource refers to the data resources of all kinds of information systems, such as telemetry and telecontrol, base station electric and kinetic data, basic database of aids to navigation, etc. The computing resource refers to all kinds of hardware resources, such as the server's CPU, memory, disk space, etc. It mainly uses virtualization to integrate resources, and establish a resource pool by combining virtualization and management to achieve elastic supply. The overall architecture of cloud computing platform of aids to navigation mainly includes the infrastructure-as-cloud-service (IacS), platform-as-cloud -service (PacS), software-as-cloud-service (SacS) technology architecture, management and service system architecture and function architecture.

3.1 *The technical architecture analysis*

Intelligent aids to navigation cloud platform technologies include

IacS layer: virtualization technology, data center construction technology, resource integration and optimization technology, network communication technology.

PacS layer: huge numbers of data storage and processing technology, resource management and scheduling technology, security mechanism design technology, etc.

SacS layer: web services technology, internet technology, application development technology, etc.

In addition, in order to safeguard the application services of the IacS, PacS and SacS in the intelligent aids to navigation cloud platform, the secure, stable and efficient operation, a secure management and operation guarantee mechanism based on the whole system need to be established. The security management and service guarantee mechanism mainly include software service quality guarantee mechanism, data security and privacy protection technology.

3.2 *The management service architecture system analysis*

IacS management service architecture mainly includes cloud platform data center hardware resource management, cloud computing resource allocation management, cloud platform network resource management and allocation, cloud computing platform storage resource management and allocation, daily data center platform maintenance and management operations of aids to navigation.

PacS management service architecture mainly includes the unified development language, database and component, basic platform framework service, corresponding platform application service mechanism.

SacS management service architecture mainly includes comprehensive business information management platform, telemetering and remote control system, video inspection, e-government systems, navigation information distribution system, lifecycle management of aids to navigation, and other kinds of SacS. According to actual needs, aids to navigation departments can develop SacS software by using PacS or entrusting platform builder, conforming to the requirements of the practical applications.

3.3 *The functional architecture analysis*

The cloud platform of intelligent aids to navigation mainly includes three functions, namely, the internal management operation function, external services and extended functions. The internal management function mainly includes completing the operation, storage, management of basic business data and related affair data of aids to navigation, management of OA data within the aids to navigation department and so on. The external service function mainly includes completing the building and publishing of website information of the aids to navigation department, the e-government platform of department, related information distribution of aids to navigation, etc. The extended function refers to the newly added function according to the actual needs of each aids to navigation department. The " Intelligent aids to navigation Cloud " platform overall architecture is shown in Figure 1.

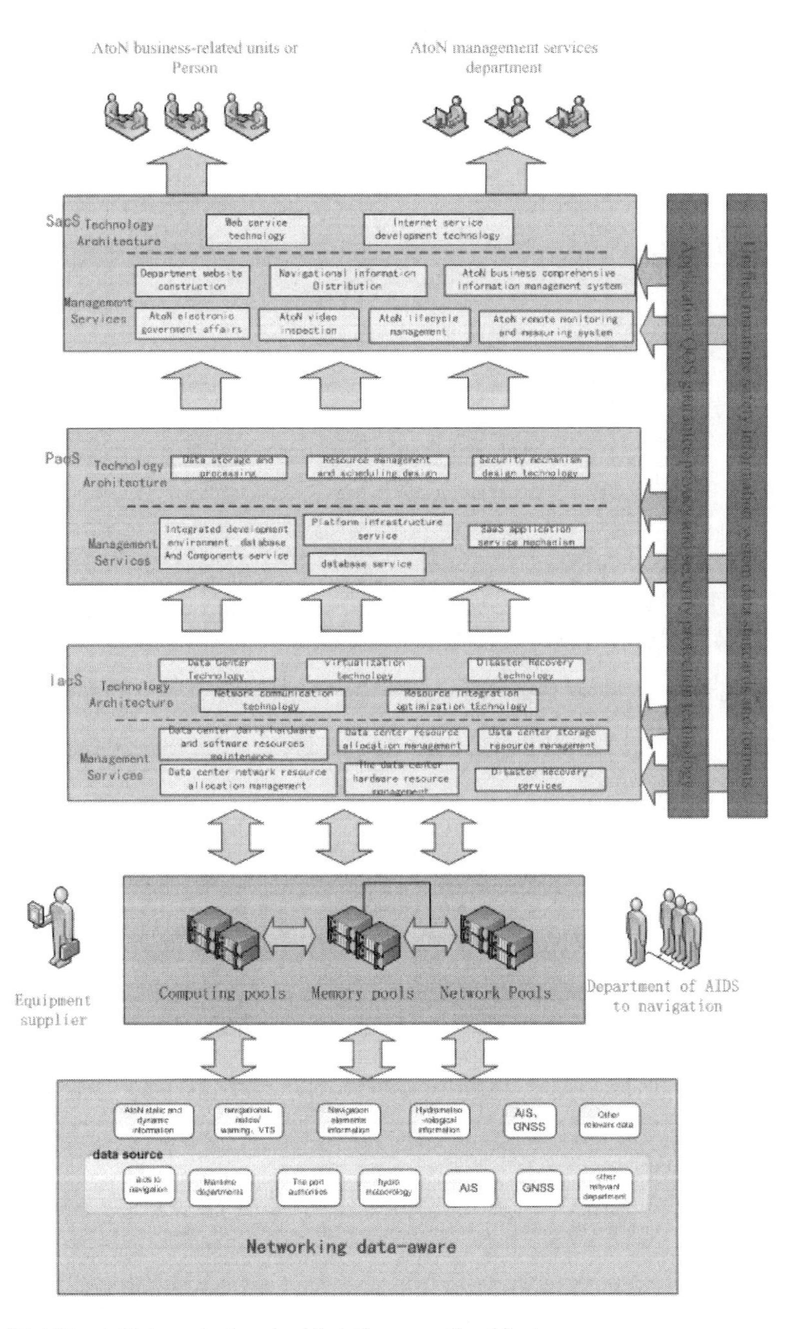

Figure 1. The " Intelligent aids to navigation cloud " platform overall architecture.

4 THE LOGICAL ARCHITECTURE OF "INTELLIGENT AIDS TO NAVIGATION CLOUD "

The platform constructs aids to navigation cloud platform data center relying on the China Maritime Safety Administration, and part of the data centers rely on three maritime navigation guarantee centers and each aids to navigation department. Under the fit e-Navigation technical conditions, the platform unifies network, data and communication infrastructure system, realizing data sharing and high-speed transmission between the data center and sub-center, sub-center and sub-center. Specific logical data center architecture model is shown in Figure 2. Logical intelligent aids to navigation cloud architecture can be divided into the production resource pool, which includes operation management, professional application layer, integration architecture layer, unified resource management scheduling layer, infrastructure layer and the physical environment layer, the development and test resource pool and the disaster recovery resource pool. The basically logical structure of development and test resource pool and disaster recovery resource pool is the same as the production resource pool.

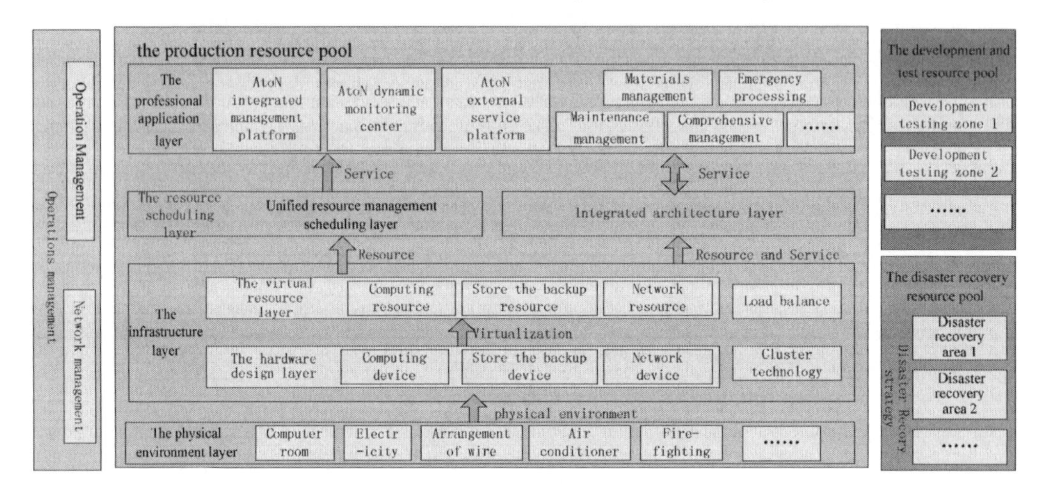

Figure 2. The "intelligent aids to navigation cloud" platform logical architecture.

5 THE "INTELLIGENT AIDS TO NAVIGATION CLOUD" NETWORK ARCHITECTURE

Network is the key of aids to navigation informatization. Cloud computing service depends on the network. Building the "intelligent aids to navigation cloud" network architecture is a focus in this paper. According to the function of network, aids to navigation information communication network architecture is divided into two parts, namely, data communications network (DCN) and data-aware network (DSN).

1 Data Communications Network (DCN). DCN is a network system for each level maritime administration departments to carry out the information exchange and network service, and also is an important channel to provide information services to the public. Aids to navigation information system network architecture is divided into the bureau backbone network, navigation guarantee center DCN backbone network and the every access network of aids to navigation. The detailed information as shown in figure 3. The aids to navigation management information DCN builds a dedicated information network for all levels of maritime apartments to improve the existing backbone network, expand regional and wireless network, increase the network transmission capacity, meet the transmission needs of data, image and video conference system. It could provide a safe, convenient, and manageable multiply access method for application systems and terminals through the unified management network boundaries interface, which could guarantee the high-speed, smooth, real-time and accurate information transmission, and meet all kinds of service requirements of end-to-end business.
2 Data Sensor Networks (DSN). DSN is a network system based on data collection technology and data access technology, which provides the basic network service to perceive overall work condition, current, environment and other factors of aids to navigation. DSN is divided into the self-built network and other-built network. The self-built

network mainly includes the dedicated network of aids to navigation video, the aids to navigation RFID network, AIS network, VHF communications network and radar network. The other-built network includes mobile communication network of the operator and maritime satellite network. Aids to navigation management information DSN network is to build a full range of network-sensor platforms, and form a 360 ° real-time monitoring of live video. The telemetry and terminal remote control sense the power source, lights, position and other work status in real time to realize the overall perception of the work condition, current, the environment and other aids to navigation management elements, to realize the safe, real-time, accurate and effective information collection of aids to navigation.

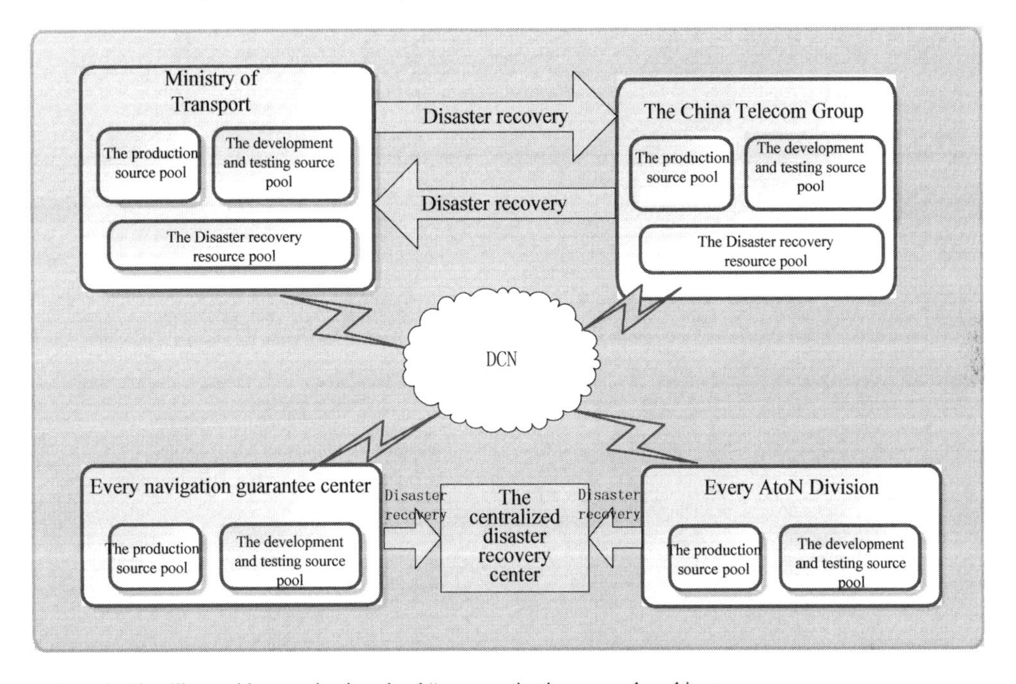

Figure 3. The "intelligent aids to navigation cloud " communication network architecture.

6 CONCLUSION

In recent years, the cloud computing technology has been rapidly developing in all industries as a new technology trend. The application of cloud computing technology in aids to navigation information management has transformed the way of working unprecedentedly. But cloud computing technology application still has many problems to be solved. For instance, data security can't be ensured and each information system of aids to navigation lacks uniform technology standard, interface standards and other issues. But these cannot prevent the application and promotion of cloud computing technology in the aids to navigation informatization process, and all of these will be our directions of future research.

REFERENCES

Baidu Encyclopedia. 2014. Cloud Computing. http://baike.baidu.com/view/2108643.htm.

Liu Shijiang. 2013. Use Cloud Computing Technology to Build "Intelligence Maritime Cloud". *Systems and Programs* (01): 53–56.

Maritime Safety Administration of the People's Republic of China(MSA). 2014. Annual Report on Navigation Guarantee Service in 2013. http://www.aton.gov.cn/showInnerN.aspx?Type=000601.

Mell, P. & Grance, T. 2011. *The NIST Definition of Cloud Computing*. National Institute of Standards and Technology.

Microsoft TechNet. 2014. About virtual machine templates. http://technet.microsoft.com/library/bb740838.aspx.

Zheng Xuhui. 2012. Cloud computing applications in the future Intelligent Transportation System. *Light Industry Technology* (03): 88–89.

Zhang Yunyong et al. 2012. Cloud Data Center and Energy Efficiency Communication. *Information and Communication Technology* (01): 49–53.

Zhang Yuchi. 2012. Cloud Computing Technology and Application Analysis. *Computer CD-ROM Software and Applications* (21): 126–128.

Electronics, Communications and Networks IV – Hussain & Ivanovic (eds)

Multi-objective nodes placement problem in large regions wireless networks

Mahmoud Gamal*
School of Information Science and Engineering, Hunan University, Changsha, Hunan, China
Department of Mathematics, Suez Canal University, Ismailia, Egypt

Ehab Morsy
Department of Mathematics, Suez Canal University, Ismailia, Egypt

Ahmad Salah
School of Information Science and Engineering, Hunan University, Changsha, Hunan, China
College of Computers & Informatics, Zagazig University, Sharkia, Egypt

ABSTRACT: In this paper, we concern with the problem of node placement in wireless communication networks. Given a set of nodes and a set of communication devices, the node placement problem requires to choose positions from a set of designated candidate sites to place these nodes such that a set of conflicting objectives are met. In particular, we focus on minimizing construction cost and network interference as well as maximizing network coverage and total bandwidth. This problem is formulated as a multi-objective optimization problem using the well-known algorithm Multi-Objective Evolutionary Algorithm Based on Decomposition (MOEA/D) algorithm. Furthermore, we apply an adapting version of this algorithm to solve our problem. Our experimental results show that MOEA/D has a good performance in a reasonable running time. Moreover, comparative results show that our algorithm is effective in all the desired objectives.

KEYWORDS: Wireless Networks, Node Placement, Genetic Algorithm, Optimization Problems, Multi-objective Evolutionary Algorithm Based on Decomposition.

1 INTRODUCTION

Planning for heterogeneous wireless networks comprises a lot of important issues that should be taken into consideration. Typically, many researchers attempt to propose approaches to increase the performance and efficiency of the network by finding the best planning of cellular networks (St-Hilaire et al. 2006, St-Hilaire et al. 2008), by finding the best placing for base stations and nodes in the networks using multi-objectives (Chuan-Kang et al. 2009, Jin-Kyu et al. 2001, Lee and Kang, 2000) and by selecting locations and types of access points in wireless local area networks (WLANs) (Unbehaun & Kamenetsky 2003, Bahri & Chamberland 2005, Gast 2005) In this paper, we study the placement of nodes in Wireless networks. The node placement problem requires choosing the appropriate positions from sets of nodes and communication devices. The main target is to place these nodes and devices such that the construction cost and the overlap are minimized and the coverage and the total bandwidth are maximized. Each node can connect to different communication networks through several communication devices. At each candidate site, at most one node is placed. We also assume that receivers are uniformly distributed through the given region. Each node has possible relationships with other nodes of the network under prescribed constraints. This model is closely related to that presented recently in (Abdelkhalek et al. 2011). Additionally, we apply an adapting version of the (MOEA/D) algorithm proposed by (Jan, 2010) to solve the placement problem.

Various algorithms have been introduced for the node placement problem to extend the network lifetime or to maintain better connectivity. Authors in (Pandey et al. 2009) presented a two-tier hierarchical heterogeneous wireless sensor network with the objective of placing the minimum number of nodes to handle the underlying traffic. The model in (Yu et al. 2007) addressed the problems of deciding the number and positions of nodes which should be deployed in the wireless sensor network. Nevertheless, this paper provides a solution to optimize placement of heterogeneous nodes in an arbitrary network for maximizing network performance.

*Corresponding author: *mgamal_sci@hnu.edu.cn*

In (Abdelkhalek et al. 2011)[1], same problem was covered using multi-objective genetic algorithm for heterogeneous network infrastructure. The problemis to find the optimal placement of nodes from a set of candidate sites and their optimal connections to an existing hierarchical network structure.

The rapid growth of the number of users in the wireless network simplifies the problem of base station placement. (Chuan-Kang et al. 2009) concerned with datacom networks, specially, Wi-Fi and Wi-MAX networks, in which a given set of transmitters should be placed such that the coverage and the capacity satisfaction are maximized, and the cost and the overlap are minimized. The heterogeneity in this problem derived from the reality of including two different node type small and large power radius nodes. (Jin-Kyu et al. 2001) They proposed a new representation that describes a base station placement with real numbers, as well as a new genetic operators are used to it.

This paper is organized as follows. In the next section, we describe the underlying problem. In Section 3, we develop a genetic algorithm for the problem and the proposed algorithm is presented. Our experimental results are discussed in Section 4. Finally, we conclude the paper in Section 5.

2 PROBLEM FORMULATION

Given a set $E=\{e_1,e_2,......, e_E\}$ of candidate sites with positions $e_u=(\alpha_u, \beta_u)$. We define a set of nodes $N=\{n_1,n_2,......,n_N\}$ to be placed on the candidate sites along with a set of communication devices denoted as $D=\{d_1,d_2,......,d_D\}$. Each device d_v has power p_v, capacity s_v, cost c_v, type t_v and power range w_v. Besides a set of receivers $R=\{r_1,r_2,......,r_R\}$ are placed in positions (α_u, β_u). Each receiver r_g has a signal threshold θ_g^d associated with a communication device d_v. The signal threshold, together with the signal strengths of the devices, determines the area in which the nodes can cover it. σ_g^d is the data rate demand of the receiver r_g from communication device d_v. The main goal is to place nodes n_i in the service area such that all receivers r_g associated with the node n_i can be covered with a small interference occurred.

2.1 *Objectives*

One major factor that affects installing a wireless network is the high expenses incurred by the network operator. For this reason, a major concern of this paper is to optimize the number of nodes with concerning coverage, cost, capacity, and overlaps.

2.1.1 *Maximizing the communication coverage*

The first objective is to maximize the number of receivers covered by the network. A node n_i covers receiver r_g only if the signal strength $S^v_{g,i,u}$ is greater than the signal threshold θ_g^d of the receiver. The communication coverage can be calculated by:

$$\text{Max } Z_1(X) = \sum_{v=1}^{D}\sum_{i=1}^{N}\sum_{u=1}^{E} a_{uv} x^v_{iu}, \tag{1}$$

where x^v_{iu} is a decision variable defined as

$$x^v_{iu} = \begin{cases} 1 & \text{if } r_g \text{ is assigned to } n_i \text{ connected with } d_v \\ 0 & \text{otherwise} \end{cases} \tag{2}$$

and

$$a_{uv} = \begin{cases} |\{r_g\}| & \forall g \text{ where } S^v_{g,i,u} > \theta^v_g \\ 0 & \text{otherwise} \end{cases} \tag{3}$$

At this point, $S^v_{g,i,u}$ is the signal strength between receiver r_g and node n_i that are associated with a communication device d_v and located on candidate site e_u is calculated as(Pahlavan and Krishnamurthy)

$$S^v_{g,i,u} = \frac{p_v G_g G_i \lambda^2}{(4\pi)^2 d^2_{ug}}, \tag{4}$$

where p_v is the power for each communication device d_v, G_i and G_g are the antenna gains of node and receiver r_g, respectively, λ is the carrier wavelength, and d_{ug} is the Euclidean distance from receiver r_g to candidate site e_u

2.1.2 *Minimizing the total cost*

This objective aims to minimize the network installation cost (i.e., the cost C_{iv} of setting up node n_i and communication device d_v that associated with it). In our problem, we also solve the problem of minimizing the number of nodes placed which achieve full coverage with minimum overlap occurred that leads to minimize placement cost. We compute this objective as follows.

$$Min\ Z_2(X) = \sum_{v=1}^{D}\sum_{i=1}^{N}(C_{iv})\sum_{u=1}^{E} x^v_{iu} \tag{5}$$

Where C_{iv} represents the cost of installation for communication device d_v in node n_i.

2.1.3 *Minimizing the maximum bandwidth*

A network design model has to provide sufficient data rate (bandwidth) for users. The main issue of this objective is to maximize the minimum capacity bandwidth of the network. In other words, we require minimizing the absolute difference between the communication device dv bandwidth and the sum of the data rate demands of its covered receivers. This can be described formally as in equation 6.

$$Min\ Z_3(X)=\sum_{v=1}^{D}|\ s_v-\sum_{i=1}^{N}\sum_{u=1}^{E}\sum_{g=1}^{R}\sigma_g^d\alpha_{ig}^v\ x_{iu}^v\ | \quad (6)$$

Where s_v is the data rate provide by communication device d_v, $\sigma_g{}^d$ is the data rate demand of receiver r_g from a communication device , and

$$\alpha_{iu}^{v}=\begin{cases}1 & if\ r_g\ \ is\ assigned\ to\ n_i\ connected\ with\ d_v\\ 0 & otherwise\end{cases} \quad (7)$$

2.1.4 *Minimizing the overlap*

Decrease the network interferencedue to overlappingamong communication devices is our last objective (Bosio et al. 2007). This can be determined by es timating the number of receivers covered by more than one communication device as in equation8.

$$Min\ Z_4(X)=\sum_{g=1}^{R}\sum_{i=1}^{N}\sum_{u=1}^{E}\psi_{uv}\ \alpha_{ig}^v x_{iu}^v\ \forall v \quad (8)$$

Where

$$\psi_{uv}=\begin{cases}|\{d_v\}| & \forall g\ \ where\ S_{g,i,u}^v>\theta_g^v\\ 0 & otherwise\end{cases}$$

2.2 *Constraints*

In our model, we define the following constraints that must be met in any feasible solution of the problem. The sum of the total data rate demand for all receivers connected to node n_i with communication device d_v should be at most the capacity s_v of d_v:

$$\sum_{u=1}^{E}\sum_{g=1}^{R}\sigma_g^d\alpha_{ig}^v\ x_{iu}^v\leq\ s_v\qquad\forall\ i\in N,\forall\ v\in D \quad (10)$$

For each node n_i at least one communication device dv is assigned to the node:

$$\sum_{v=1}^{D}x_{iu}^v P_{iv}\geq 1\qquad\forall\ i\in N \quad (11)$$

$$\text{where } P_{iv}=\begin{cases}1 & if\ \ d_v\ \ in\ node\ n_i\\ 0 & Otherwise\end{cases}$$

Each node n_i is assigned to one candidate site:

$$\sum_{u=1}^{E}x_{iu}^v\ =1\qquad\forall\ i\in N,\ \forall v\in D \quad (12)$$

At most one node is assigned to each candidate site:

$$\sum_{i=1}^{N}x_{iu}^v\leq 1\qquad\forall\ u\in E\ ,\exists v\in D \quad (13)$$

Each receiver r_g can be connected to at most one node n_i with communication device d_v:

$$\sum_{i=1}^{N}\alpha_{ig}^v\ \geq 1\qquad\forall\ v\in D \quad (14)$$

3 APPLYING CMOEA/D-DE-ATPALGORITHM IN NODES PLACEMENT PROBLEM

Decomposition approaches is a good choice for solving many-objective optimization problems. In (Zhang & Li 2007), the authors proposed multi-objective evolutionary algorithm based on decomposition (MOEA/D). The MOEA/D tackles the problem of approximation of the Pareto front by decomposing a multi-objective optimization problem into a number of scalar optimization subproblems and optimizing them concurrently. We apply CMOEA/D-DE-ATP to optimize the sufficient number of nodes and their positions in our problem under the given constraints. This algorithm includes the standard procedures of genetic algorithms: constructing an initial population, representing different individuals of the population, selecting the fittest individuals, applying a crossover operation that mates individuals, and applying the mutation operator on the resulting children. We first construct an initial population.

Population initialization is the first and the primary task for all evolutionary algorithms. If a priori information about the solution is absent, we normally start generating the assignment of substrings in a chromosome randomly. Each candidate solution is encoded into a chromosome. Namely, each node is represented as a substring of length n composed of the position X that represents the candidate site, CD that represents communication device, and Y the nodes connected to the underlying nodes.

The selection operator selects chromosomes of the highest finesses. The method of normal selection causes those individual that encode successful structures to create copies more regularly. In our algorithm we use mating selection. During the crossover operation, the two parent's output from the selection procedure are recombined to a new offspring chromosome. The new chromosome is fitter than its parents if it gets the best characteristics from them. There are several approaches for the crossover operator. MOEA/D approach uses the single point crossover. MOEA/D-DE uses DE operator (see Step 2.1 in (Jan 2010)). With a pre-described probability rate, the crossover changes node's positions associated with candidate sites and communication devices and mate them to generate a new solution.

After that, mutation operator is applied to individual solutions to get a new solution by complementing some genes randomly. Mutation is generally performed as a bit flip to maintain diversity in the population and inhibit premature convergence. In this algorithm, we perform a mutation with rate P_m=1/(Parameters dimension), i.e., each bit has a probability of P_m to be flipped.

The mind of heuristic operator increases the node's individual utilization in crossover and mutation operators, then it creates solution Z which is (Jan 2010) used for updating the populations by using a DE operator, after that check if an element of y is outside the boundary, reset its value by selecting values inside the boundary which achieves all our constraints then Evaluate y by using equations in step 2-4 in (Jan 2010). Afterwards update the populations for each solution Z in the set of the best value for the objective considering all the neighbors of the sub-problem, Algorithm computes the new function of values y as in (Jan 2010) equation (4) or (5), if y perform better than $xj$with consider the j_{th} sub-problem, it replaces x_j with y. Finally, termination criterion is checked to decide whether the search should stop or continue, in order that our algorithm will stop after a definite number of generation increase generation $gen=gen+1$.

4 COMPUTATIONAL RESULTS

The performance of the algorithm has been evaluated through a wide range of different test problems which generated randomly as provided in (Raisanen and Whitaker, 2005, Abdelkhalek et al. 2011). Each test problem has two issues to be resolved, 1) determining placement for nodes in heterogeneous network infrastructure. 2) Simultaneous optimization concerning our four objectives with their constraints. In our experiential application, we use some parameters that we need as shown in Table 1 These parameters are defined in (Abdelkhalek et al. 2011). There are three communication devices d_v with a power radius of 13, 15, 17 km and capacity150,160, 170 Mb/s Respectively, random data rate demands σ_g^d and only one type exists for all devices, but with different cost C_v as 3500, 4500,5500 Respectively. The gains G_i, G_g of node n_i and receiver r_g are set to 1 and the bandwidth between two nodes bv is $500MHz$ with wavelength λ is $0.025\ km$. The threshold θ_g^d for all receivers r_g is 1 and the cost C_{iv} for all node n_i different from 50 to 2300.

Table 1. Problem parameters.

Parameters	Values
Region size **(Km²)**	(40×40)
Distribution distance between test points **(Km²)**	2
A number of candidate sites	17
Number of nodes	46
Number of communication devices	3

4.1 *Computational results*

Test problems are categorized in two approaches: the size of the area and the density of receivers r_g where they are placed, as indicated in Table 2. A total of 9 test problem cases are conducted by combining the density of the receivers r_g with the size of the region (see Table 2). All problem instances are available and are implemented 30 independent times. After the maximum number of iterations is reached (i.e., 300 iterations), the algorithm stops running. However, if there isn't enhancement of the objective functions after 100 iterations is performed with a population size of 30, it also stops. Additionally, starting populations randomly have been used in our problem class.

We implemented our algorithm in MATLAB and run the experiments on Intel® Core™ i5-2450QM CPU, 2.50GHz, with 4GB of RAM running Windows 7 platform. Our results are illustratedin Table 2 which contains lists of the average CPU time,the number of feasible solutions in the final population P_f and the best values for our four objective functions according our proposed algorithm.

In Table 3,when the number of test points and area increase, the CPU Time dramatically increases to find the optimal solution. Obviously, the CPU time is positively correlated with the problem size.

Table 2. Number of receivers.

Region size	Distribution distance of test points		
Km²	Km²- Number of test points		
	2	3	4
40 x 40	400	178	100
45 x 45	506	225	127
50 x 50	625	278	156

Table 3. Objectives results in different test problem.

Problem	CPU Time (s)	P_f	Z_1	Z_2	Z_3	Z_3
C_1	4.113	3	2685	2641	327101	762
C_2	2.168	1	11717	258050	382962	366
C_3	1.508	10	6131	252050	404202	207
C_4	5.251	3	34913	261050	312571	929
C_5	2.950	4	15597	258050	372715	451
C_6	1.833	9	7743	250050	396230	242
C_7	7.094	6	34329	263050	305727	898
C_8	3.217	10	19107	261050	354341	548
C_9	2.061	3	9769	251050	393133	295

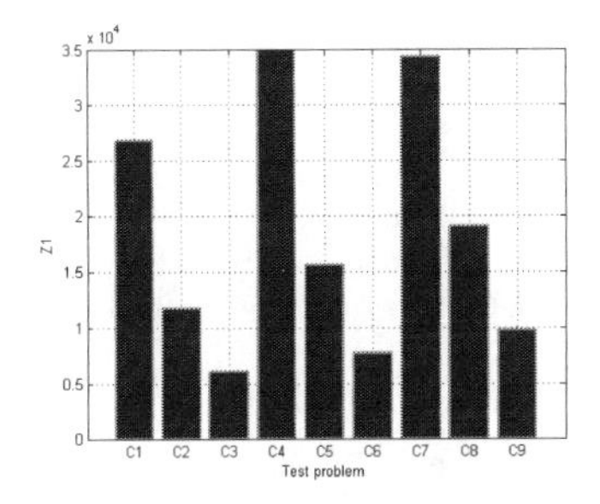

Figure 1. A relationship of 9 test problems with Z1.

As shown in Figure 1,with the increasing of test points and region size, first objective will increase to can cover almost of test points in the networks.

In Figure 2, the cost is proportional to the size of the coverage area. We can notice that by means of increasing test points to cover, the more expensive is the cost of our placement due to communication devices' cost.

In Figure 3 all test problem instances with *Z3*, almost all test points are covered and their demands are satisfied. The Bandwidth is Inversely proportional with the size of the coverage area. Noticeably, the more of the test points we have to cover, the less bandwidth we have for our placement due to bandwidth limitation.

The overlap associated with the additional test points and area with *Z4* is illustrated in Figure 4. It is clear that with the increase of test points and region size, the fourth objective (overlap) also will increase.

4.2 *Comparison results*

In this section, we compare our solution with the same problem instance introduced in (Abdelkhalek et al. 2011) with same objective functions. Moreover, we add overlap objective function and propose 5 vital constraints for our problem. The results are in Table 4 shows that our algorithm is extremely better than MOGA algorithm applied in (Abdelkhalek

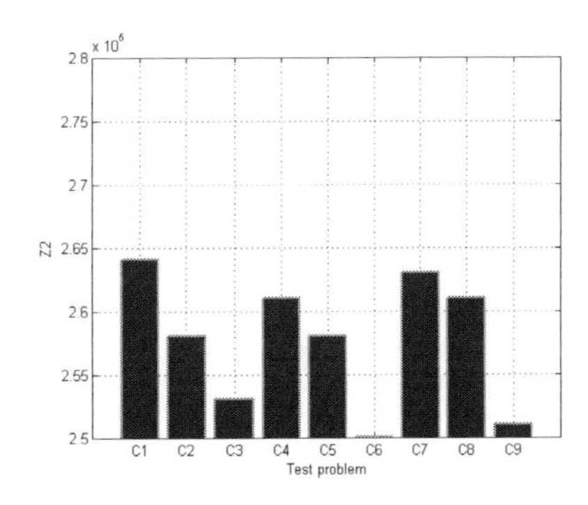

Figure 2. A relationship of 9 test problems with *Z2*.

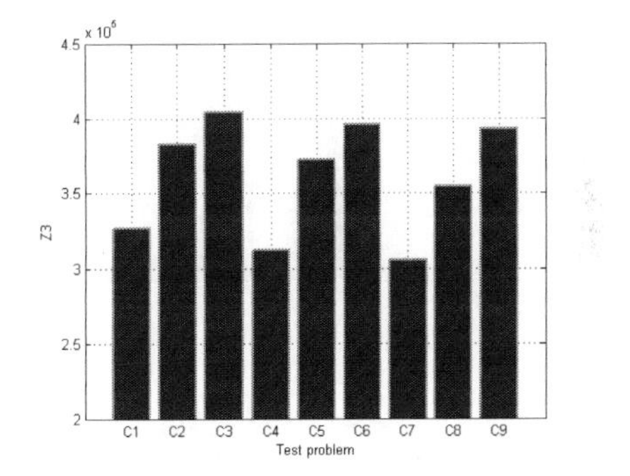

Figure 3. A relationship of 9 test problems with Z3

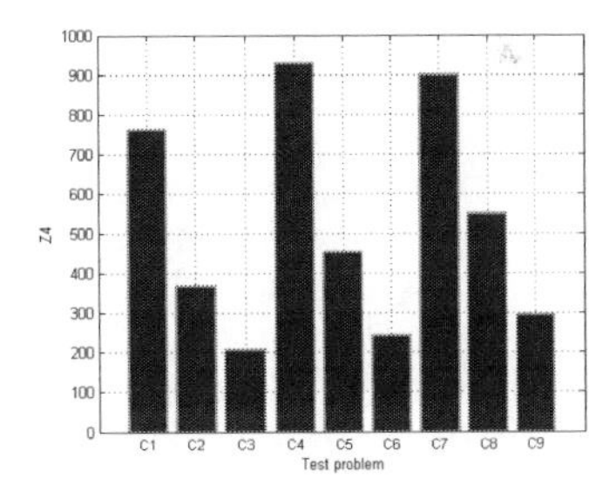

Figure 4. A relationship of 9 test problems with Z4.

et al. 2011) for all the 3 objectives. Additionally, we present another objective value (overlap). Since the CPU time is positively associated with number of objectives and the problem size, then compared to (Abdelkhalek et al. 2011) the CPU time is slightly large which is only 0.983039 second.

Figure 5 illustrates the comparison and results which is listed in Table 4.The red color represents CMOEA/D-DE-ATP algorithm which we applying in this paper based on the fourobjectives.However, the blue color is for MOGA algorithm applied in (Abdelkhalek et al. 2011) with only three objctives.

Table 4. Comparision results.

RSULTS	CPU Time (s)	P_f	Z_1	Z_2	wZ_3	Z_4
CMOEA/ D-DE-ATP	4.1130	3	2685	2641	327101.08	762
MOGA	3.175	4	400	505815	1.05E8	---

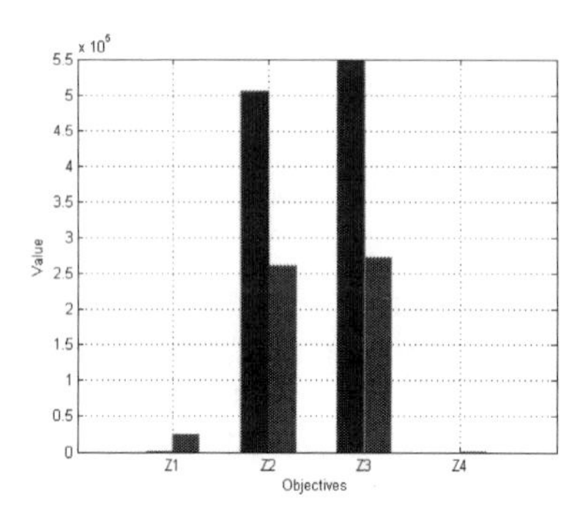

Figure 5. Comparison diagram.

5 CONCLUSION

In this paper, we formulate a model for node placement in Wireless Network.We propose a number of constraints and penalty components focusing on the specific features of the wireless environment. The model is efficiently solved using an adapting version of MOEA/D algorithm to analyze our problem. Experimental results show that our algorithm has a good performance in a reasonable running time. Furthermore, This model achievebetter performance than MOGA ALGORITHM. Finally, the results presented in this paper can be easily applied to any existing network as to expand it by placing some new nodes. Assuming that we know demand distribution through the time horizon for already existing network and we want to deploy new nodes and combine them through the existing infrastructure. This existing infrastructure is a set of homogenous or heterogeneous nodes that is associated with settled and hierarchical networks and connected to each communication device type. We placed these additional nodes only if there is a necessary need for it.

ACKNOWLEDGMENT

This research is supported partially by Alexander von Humboldt Foundation

REFERENCES

Abdelkhalek, O., Krichen, S., Guitouni, A. & Mitrovic-Minic, S. 2011. A genetic algorithm for a multi-objective nodes placement problem in heterogeneous network infrastructure for surveillance applications. Wireless and Mobile Networking Conference (WMNC), 2011 4th Joint IFIP, 2011. IEEE, 1–9.

Bahri, A. & Chamberland, S. 2005. On the wireless local area network design problem with performance guarantees. Computer Networks 48(6): 856–866.

Bosio, S., Capone, A. & Cesana, M. 2007. Radio planning of wireless local area networks. IEEE/ACM Transactions on Networking, 15(6): 1414–1427.

Chuan-Kang, T., Chung-Nan, L., Hui-Chun, C. & Jain-Shing, W. 2009. Wireless Heterogeneous Transmitter Placement Using Multiobjective Variable-Length Genetic Algorithm. IEEE Transactions on Systems, Man, and Cybernetics, Part B: Cybernetics, 39(4): 945–958.

GAST, M. 2005. 802.11 wireless networks: the definitive guide, O'Reilly Media, Inc.

Jan, M. A. 2010. MOEA/D for constrained multiobjective optimization: Some preliminary experimental results. Computational Intelligence (UKCI).

Jin-Kyu, H., Byoung-Seong, P., Yong Seok, C. & Han-Kyu, P. 2001. Genetic approach with a new representation for base station placement in mobile communications. Vehicular Technology Conference, VTC 2001 Fall. IEEE VTS 54th 4:2703–2707.

Lee, C. Y. & Kang, H. G. 2000. Cell planning with capacity expansion in mobile communications: a tabu search approach. IEEE Transactions on Vehicular Technology, 49(5): 1678–1691.

Pahlavan, K. & Krishnamurthy, P. 2002. Principles of Wireless Networks: a Unified Approach. New Jersey: Prentice Hall PTR.

Pandey, S., Dong, S., Agrawal, P. & Sivalingam, K. M. 2009. On performance of node placement approaches for hierarchical heterogeneous sensor networks. Mobile Networks and Applications 14(4):, 401–414.

Raisanen, L. & Whitaker, R. 2005. Comparison and Evaluation of Multiple Objective Genetic Algorithms for the Antenna Placement Problem. Mobile Networks and Applications, 10(1–2): 79–88.

St-Hilaire, M., Chamberland, S. & Pierre, S. 2006. Uplink UMTS network design—an integrated approach. Computer Networks 50(15): 2747–2761.

St-Hilaire, M., Chamberland, S. & Pierre, S. 2008. A tabu search algorithm for the global planning problem of third generation mobile networks. Computers & Electrical Engineering 34(6): 470–487.

Unbehaun, M. & Kamenetsky, M. 2003. On the deployment of picocellular wireless infrastructure. Wireless Communications, IEEE 10(6): 70–80.

Yu, L., Wang, N., Zhang, W. & Zheng, C. 2007. Deploying a heterogeneous wireless sensor network. International Conference on Wireless Communications, Networking and Mobile Computing, 2007. (WiCom 2007): 2588–2591. IEEE.

Zhang, Q. & Li, H. 2007. MOEA/D: A multiobjective evolutionary algorithm based on decomposition. IEEE Transactions on Evolutionary Computation 11(6): 712–731.

Electronics, Communications and Networks IV – Hussain & Ivanovic (eds)
© 2015 Taylor & Francis Group, London, ISBN: 978-1-138-02830-2

Performance modelling of turbo-coded non-ideal single-carrier and multi-carrier waveforms over wide-band Vogler-Hoffmeyer HF channels

Fatih Genç, Mustafa Anil Reşat, Asuman Savaşçhabeş & Özgür Ertuğ*
Department of Electrical and Electronics Engineering, Gazi University, Ankara, Turkey

ABSTRACT: The purpose of this paper is to analyze and compare the turbo coded Orthogonal Frequency Division Multiplexing (OFDM) and turbo coded Single Carrier Frequency Domain Equalization (SCFDE) systems under the effects of Carrier Frequency Offset (CFO), Symbol Timing Offset (STO) and phase noise in wide-band Vogler- Hoffmeyer HF channel model. Hence, in coded SC-FDE and coded OFDM systems; a very efficient, low complex frequency domain channel estimation and equalization are implemented in this paper. Also, Cyclic Prefix (CP) based synchronization synchronizes the clock and carrier frequency offset. The simulations that analyzes average BER versus bit signal to noise ratio parametrized by the CFO, STO and phase noise show that nonideal turbo-coded OFDM has better performance with greater diversity than non-ideal turbo-coded SC-FDE system in wide-band HF channel.

1 INTRODUCTION

Spectral and power efficiency of terminal in the limited bandwidth and transmit power have been developing continuously for the new generation of wireless communication systems. To meet the new user demands, new air interfaces are needed to be enhanced. In OFDM systems, one Inverse Fast-Fourier Transform (IFFT) block is used at the transmitter and also one FFT block is used at the receiver sides of the link. In the IFFT block, OFDM transmitter multiplexes the information into many low-rate streams which are transmitted parallelly instead of sending the information as a single stream (Weinstein & Ebert 1971). The modulated signals in an OFDM system have high peak values in time domain since many sub-carriers are added via an IFFT operation. Therefore, OFDM systems are known to have high Peak-to-Average Power Ratio (PAPR). Due to the limited battery life in mobile terminals, the PAPR problem is a main disadvantage of the OFDM system for the up-link (Myung 2006). On the other hand, Single-Carrier Frequency-Domain Equalization (SC-FDE) is a desirable alternative to OFDM systems. In the case of SC-FDE technique, no IFFT and FFT blocks existed on the transmit side while FFT and IFFT operators are performed at the receiver side of the link. SC-FDE experiences lower PAPR levels than OFDM because no IFFT is performed at the transmitter to precode the signal. In order to mitigate the PAPR problem, Single-Carrier (SC) transmission uses single-carrier modulation instead of many sub-carriers (Weinstein & Ebert 1971, Pancaldi 2008). Low-complexity channel equalization and estimation in the frequency-domain are used to mitigate the inter-symbol interference (ISI) (Hassa 2009). For this purpose, the frequency domain MMSE equalizer is generally used to minimize the attenuations of the fading channel. For wide-band channels, conventional time domain equalizers are impractical because of the very long channel impulse response in the time domain. Frequency domain equalization is more practical for such channels because the FFT size does not grow linearly with the length of the channel response and the complexity of the FDE is much lower than the time domain equalizer. At the same time, frequency domain MMSE estimation is preferred with the comb-type pilot tone arrangement to predict the multi-path channel coefficients (Shen & Martinez 2008, Van de Beek 1995). In order to estimate the channel characteristics, the comb-type pilot symbols are placed as periodically as possible in coherence time. An additional way to eliminate ISI almost completely, is to use a guard interval which is called cyclic-prefix (CP). The CP is the replica of the last L symbols of the block as shown in Figure 4. The guard time L must be larger than the expected channel delay spread. At the receiver, the received CP is discarded before processing the block. By doing so CP also prevents inter-block interference. CP is also used in CP-based channel synchronization to compensate the inter-carrier interference (ICI) caused by the Doppler effect. CP-based synchronization enables CFO estimation

*Corresponding author: *ertug@gazi.edu.tr*

without need of additional redundant pilots. In fact, the key point is that CP already contains sufficient information to perform synchronization. Without CFO, the sub-channels do not interfere with one another. The impact of frequency offset is loss of orthogonality between the tones. Hence, the received signal is not a white process because of its probabilistic structure and it contains information about the timing offset and carrier frequency offset (Sandell 1995). Estimations of timing offset θ and frequency offset $\hat{\varepsilon}$ are achieved by the relation of the CPs of consecutive frames. In this paper, the performances of the turbo-coded SC-FDE and OFDM systems in wide-band Vogler-Hoffmeyer HF channels are compared. In practice, a wide-band radio channel has time-variant, frequency-selective and noisy properties. Most commonly used HF channel model is recommended by CCIR and ITU-R that is called as a Watterson HF channel (Wattersen 1970, ITU 1986). The main restriction of the Watterson model is that the model is designed and tested for narrow-band channels but not for ones having more than 12kHz bandwidth. In the design of high data speed wide-band HF communication systems, exact modeling and simulation of HF channel are needed. Therefore, Vogler-Hoffmeyer HF channel model is used in this paper (Hoffmeyer & Vogler 1990, Guao & Wang 2009).

The remainder of this paper is organized as follows. Section II gives an overview of the wide-band Vogler-Hoffmeyer HF channel model. Section III overviews OFDM and SC-FDE structures. In Section IV., channel equalization/estimation and synchronization methods are defined in detail. Numerical results and discussions are given in Section V, and finally, conclusions are drawn out from the results in Section VI.

2 WIDE-BAND HF CHANNEL MODEL

HF channel characteristics are directly shaped by the ionosphere behavior because HF channels utilize the ionospheric reflections in order to provide long-distance communications. The wide-band HF channel can be modeled as an FIR filter where the taps are time-variant and have complex values. This model can be described by the following equation:

$$y_t = \sum_{i=0}^{L-1} h_i x_t + n_i \tag{1}$$

where y_t is the complex output of the channel, L is the length of the channel, h_i is one of the L taps of the time-varying transversal filter, x_t is the complex input to the channel and n_i is Additive White Gaussian Noise (AWGN). This type of a complex-valued FIR filter can be formed easily by convolving the input signal with the channel impulse response. Thus, the coefficients of the filter can be defined as the samples of the HF channel impulse response which is given as:

$$h(t,\tau) = \sqrt{P(\tau)}D(t,\tau)\psi(t,\tau) \tag{2}$$

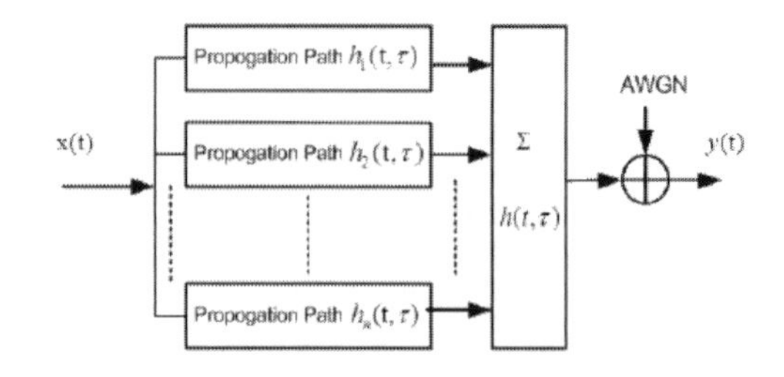

Figure 1. HF channel model.

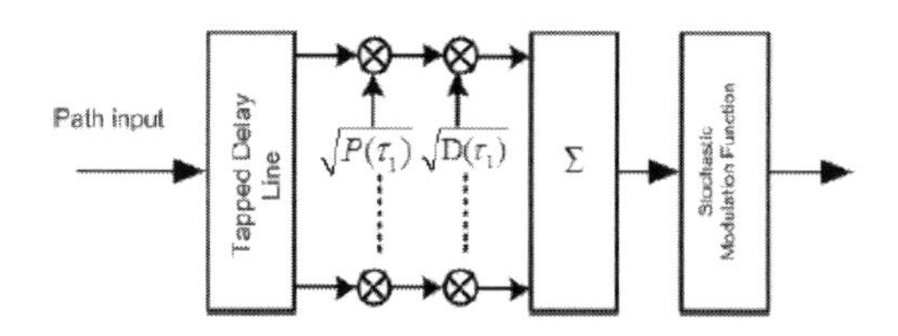

Figure 2. Single propagation path.

where $P(\tau)$ is the delay power profile, and its square root $\sqrt{P(\tau)}$ describes the shape in delay dimension; $D(t,\tau)$ is the deterministic phase function showing each path's Doppler shift, and $\psi(t,\tau)$ is the stochastic modulation function which describes the fading value of the impulse response.

The Doppler Effect can also be given by the formula:

$$D(t,\tau) = e^{j2\pi f_D t} \tag{3}$$

where f_D is the Doppler shift value. The stochastic modulation function $\psi(t,\tau)$ can be stated as random variables with an auto-correlation function that possesses a Gaussian shape. Whilst Figure 1 shows the structure of the wide-band HF channel model with propagation paths, Figure 2 shows the model of a single propagation path (Guao & Wang 2009). It is important to specify the main difference between the narrow-band Watterson model and the wide-band channel model here. At the Watterson model time delay spread is neglected and the time delay of each path has a single value. On the other hand, the wide-band model has a delay power profile symbolized with $P(\tau)$ and relates the Doppler effect with the time delay of each path.

3 SYSTEM MODELS

3.1 *OFDM system*

The system structure is illustrated in Figure 3. First, each binary source data are encoded in non-punctured, $R = 1/3$ code-rate turbo encoder and Log-Map algorithm is chosen for best decoding performance. At this time, N sub-carriers X_k for k=0,1,....N-1 are modulated by a signal alphabet $A = \{\pm 1, \pm 3, \pm j, \pm 3j\}$ used for transmitting the information for 16-QAM. After base-band modulation, pilot tone symbol insertion is applied for the channel estimation where the pilot pattern is shown in Figure 4. Pilot arrangement in OFDM system issue is discussed in more detail under the Section IV. Here, the output of the IFFT after the serial to parallel conversion can be expressed as:

$$x(n) = \frac{1}{\sqrt{N}} \sum_{k=0}^{N-1} X(k) e^{j2\pi kn/N} \tag{4}$$

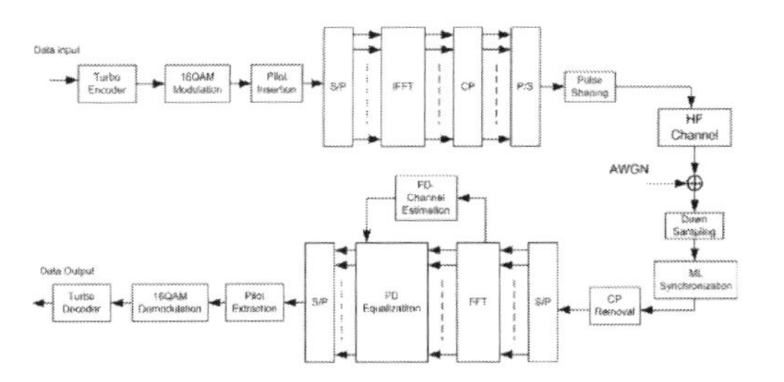

Figure 3. OFDM structure.

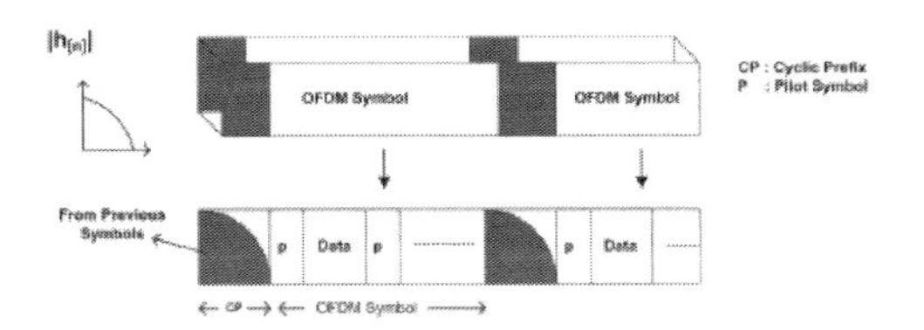

Figure 4. Frame structure.

where the constant $\frac{1}{\sqrt{N}}$ normalizes the power N is the sub-carrier number and $X(k)$ is the modulated input symbols. Cyclic prefix (CP) of length N_c is added at the beginning of the frame which must be greater than the maximum channel delay spread then the composite symbols are transmitted through the HF channel. In order to eliminate inter-carrier interference (ICI), this guard time includes the cyclic extended part of the OFDM symbol. Next, Root-Raised-Cosine (RRC) pulse shaping filtering is used to reconstruct on the data symbols. After FFT is applied at the receiver, the received signal is given by:

$$Y[k] = H[k]X[k] + Z[k] \tag{5}$$

where $X[k]$ denotes the kth sub-carrier frequency components transmitted symbol, $Y(k)$ is received symbol, $H[k]$ is channel frequency response and $Z[k]$ is noise in the frequency domain, respectively. At the receiver, after passing to discrete-time domain through A/D converter and pulse shaping filter, CP-based ML synchronization is applied to compensate the Carrier Frequency Offset (CFO) which is mentioned in section IV then guard time is removed:

$$y[n] = \{ y_g(n) \quad M < n < N \} \tag{6}$$

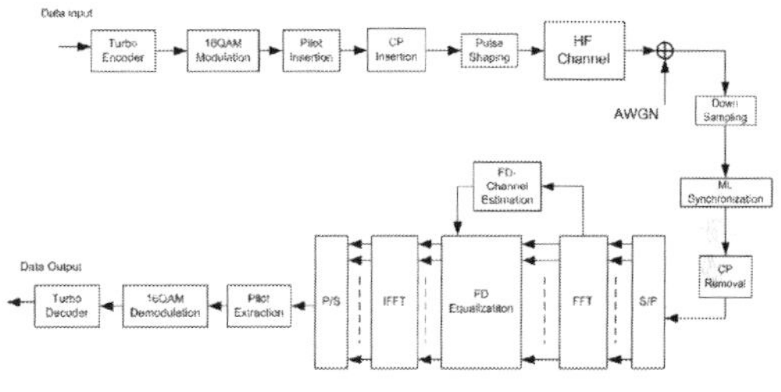

Figure 5. Frame structure.

where N is the sub-carrier, M is CP length, y_g received signal that have guard interval insertion. Then y_n is received to FFT block for the following operation:

$$\begin{aligned} Y[k] &= FFT\{y(n)\},\ k,n = 0,1,...N-1 \\ &= \frac{1}{N} \sum_{n=0}^{N-1} y(n) e^{-j2\pi kn/N} \end{aligned} \tag{7}$$

Next *Least Square* estimated $\hat{H}_{LS}^{p}[k] = \frac{Y^p[k]}{X^p[k]}$ is obtained by extracting the pilot signals $Y^p[k]$. The interpolated $\hat{H}[k]$ for all data sub-carriers is obtained in MMSE channel estimation. Then, in the Frequency domain equalization (FDE) block the transmitted data is equalized by MMSE equalizer as below:

$$\hat{X}_n = IFFT\{Y_k C_k\} = y_n \otimes c_n \tag{8}$$

where C_k represents the equalizer correction term, which is computed according to the FDE as follows:

- MMSE Equalizer:

$$C_k = \frac{\hat{H}_{FFT}^{*}}{\left|\hat{H}_{FFT}\right|^2 + (E_b/N_o)^{-1}} \tag{9}$$

where $(.)^*$ denotes conjugate. MMSE equalizer in R9 makes an optimum trade-off between noise enhancement and channel correction term, while using the signal-to-noise ratio (SNR) value R4. Finally, the binary information data are obtained back in 16-QAM modulation and turbo decoding respectively.

3.1.1 *SC-FDE system*

OFDM and SC-FDE are similar in many ways. However, there are explicit differences that make the two systems perform differently. As shown in Figure 5, the main difference between OFDM and SC systems is the placement of the IFFT block. In SC systems, it is placed on the receiver side to transform the frequency domain equalized signals, thus compensating for channel distortion, bringing back to the time domain R3. All the other blocks are formed with the same manner like OFDM system at the both sides of the transmission. In the OFDM system, symbols are exposed to an additional transformation by using the IFFT, $x(n) = IFFT\{X[k]\}$, but in the SC-FDE system no transformation is used. The frame of SC-FDE is transmitted during the time instant after the Turbo encoder, 16-QAM modulation, pilot insertion and CP insertion are applied, respectively, and the receiver maps received data into the frequency domain in order to equalize. When the channel delay spread is large it is more efficient computationally to equalize in the frequency domain.

4 CHANNEL ESTIMATION & SYNCHRONIZATION

4.1 *Channel estimation*

Comb-type pilot arrangement which is used for frequency domain interpolation to estimate channel frequency response R9,R10. In Comb-type pilot arrangement, every OFDM and SC symbol has pilot tones where are periodically located at the each sub-carriers. Notice that $S_f = \frac{1}{\sigma_{\max}}$ the periods of pilot tones in frequency domain must be placed at the coherence bandwidth. The coherence bandwidth is determined by an inverse of the *maximum delay spread* $\sigma_{\max}$. Let consider the $\hat{H}_{LS} = X^{-1}Y\Delta = \tilde{H}$, using the weight matrix W channel estimate $\hat{H}\Delta = W\tilde{H}$ is defined and MSE of the channel estimate is calculated as below:

$$J(\hat{H}) = E\{\| e \|^2\} = E\{\| H - \hat{H} \|^2\} \tag{10}$$

The MMSE estimation method finds a better estimate that minimizes the MSE in R10. For the derivation of MMSE channel estimation, the cross-correlation R_{eH}, error vector e with channel estimate H is forced to zero.

$$\begin{aligned} R_{e\tilde{H}} &= E\{e\tilde{H}^H\} \\ &= E\{(H - \hat{H})\tilde{H}^H\} \\ &= E\{(H - W\tilde{H})\tilde{H}^H\} \\ &= R_{H\tilde{H}} - WR_{\tilde{H}\tilde{H}} = 0 \end{aligned} \tag{11}$$

where $(.)^H$ *Hermitian* operator. Solving equation R11 for W yields:

$$W = R_{H\tilde{H}} R_{\tilde{H}\tilde{H}}^{-1} \tag{12}$$

Using R12 the MMSE channel estimate follows as:

$$\hat{H} = W\tilde{H} = R_{H\tilde{H}} R_{\tilde{H}\tilde{H}} \tilde{H} \tag{13}$$

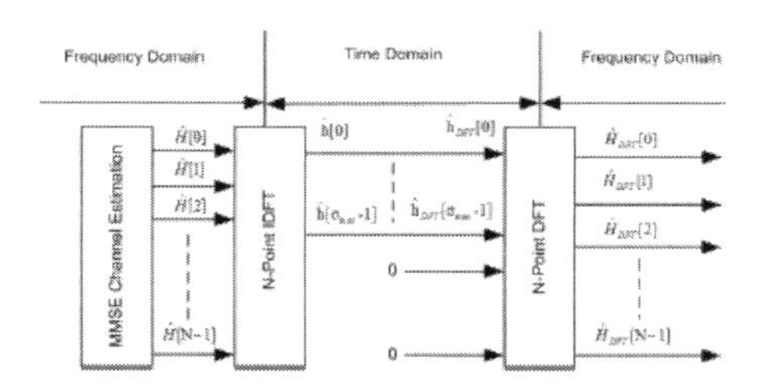

Figure 6. FFT-based channel estimation.

Figure 6 shows the block diagram of FFT-based channel estimation, given the MMSE channel estimation. An important point is that $\sigma_{\max}$ must be known formerly to remove the effect of noise outside the channel delay. Taking the IFFT of the MMSE channel estimate $\hat{H}$ to get into the time domain, that the coefficients contain the noise are ignored with zero padding and then transform the remaining $\sigma_{\max}$ elements back to the frequency domain to achieve $\hat{H}_{FFT}$. Finally $\hat{H}_{FFT}$ is used in R9 at the Frequency Domain MMSE Channel Equalizer block.

4.2 *Synchronization*

In general, there are two types of distortion related to the carrier signal. One is the Phase Noise due to the Voltage Control Oscillator (VCO) and the other is Carrier Frequency Offset (CFO) caused by Doppler Frequency shift f_d. Let define the normalized CFO, $\varepsilon = \frac{f_d}{\Delta_f}$, as a ratio of the CFO to sub-carrier spacing Δ_f. Where f_d is the Doppler Frequency. Here Δ_f is the ratio of the bandwidth to subcarrier number $\left(BW / N_{\Delta f}\right) = \frac{24000}{256} = 93.75Hz$. CP-based channel synchronization estimates the time and carrier-frequency offset. This algorithm exploits the cyclic prefix preceding the OFDM and SC symbols, thus reducing the need for pilots. The received data in the time domain $e^{j2\pi\varepsilon k/N}$, where ε denotes the difference in the transmitter and receiver oscillators as a fraction of the inter-carrier spacing, that is calculated in (14). Notice that all sub-carriers are affected by the same shift ε is shown as:

$$r(k) = s(k-\theta)e^{j2\pi\varepsilon k/N} + n(k) \tag{14}$$

where $r(k)$ is the received data, $s(k-\theta)$ is the unknown arrival time transmitted signal and $n(k)$ is the AWGN. Hence, $r(k)$ contains information about the time offset θ and carrier offset ε. From the observation shown in Figure 7 that the estimation of frequency offset and the estimation of timing offset are calculated as below:

$$\begin{aligned}
\gamma(m) &\triangleq \sum_{k=m}^{m+L-1} r(k)r^{*}(k+N) \\
\Phi(m) &\triangleq \frac{1}{2}\sum_{k=m}^{m+L-1} |r(k)|^{2} + |r(k+N)|^{2} \\
\hat{\theta}_{ML} &= \arg\max_{\theta}\left\{|\gamma(\theta)| - \frac{SNR}{SNR+1}\Phi(\theta)\right\}, \\
\hat{\varepsilon}_{ML} &= -\frac{1}{2}\angle\gamma\left(\hat{\theta}_{ML}\right)
\end{aligned} \tag{15}$$

where L is the CP length, m is the index of samples, $\gamma(m)$ is the correlation coefficient and $\Phi(m)$ is an energy term R11,R12. Figure 7 shows the estimation of timing offsets and frequency offsets. The index values of the peaks of the timing estimate give the estimates of carrier frequency offset $\hat{\varepsilon}$ values:

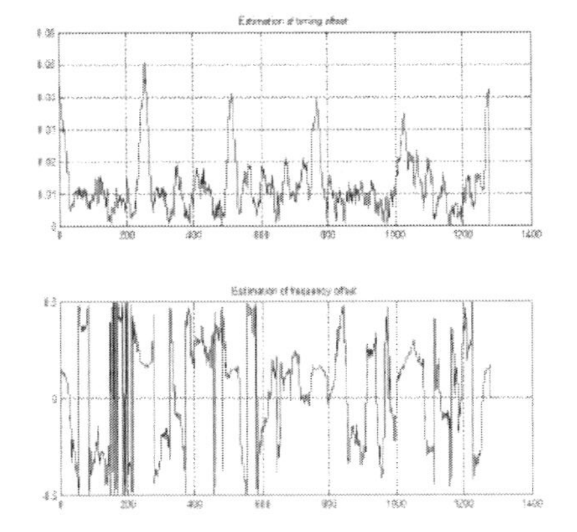

Figure 7. STO and CFO estimates.

$$\hat{\varepsilon} = \hat{\gamma}(maxindexvalues(\hat{\Phi}(m))) \tag{16}$$

Finally, these estimates are used in channel synchronization block to compensate the carrier frequency offset as:

$$\hat{s}(k) = r(k)\cdot e^{-j2\pi\hat{\varepsilon}k/N} \tag{17}$$

where $\hat{s}(k)$ is the synchronized signal.

5 SIMULATION RESULTS

In this section, the BER performance of the proposed systems for CFO, STO and phase noise are shown. The simulation parameters are compliant to the wide-band HF channel model: 24 kHz bandwidth, 16 QAM constellation, 256 sub-carriers, 210 occupied sub-carriers, 16 cyclic prefix length and the pilot tone number is equal to 30. The code rate of the turbo code is 1 / 3, the interleaver is 512 block interleaver and 10-iteration log-MAP decoding is used. In all of the simulations, normalized frequency offset of each system is a constant value between 0.1 and 0.5. For the channel model, multipath Rayleigh fading channel is used which can be modeled as a tapped-delay line with $L_{ch} = 3$ delay taps are [3 7 10] ms. Furthermore the channel gains of the taps are [0 -3 -8] dB.

In Figure 8, the effect of the CFO is analyzed. Hence, normalized CFO, ε. Channel delay spread and phase noise are neglected. As can be seen from the Figure 8, as the frequency offset of the channel increases, BER performances decrease as well. This is because of the way that CFO increments the ICI without the CP-based channel synchronization. Both OFDM and SC-FDE systems experience the impacts of severe frequency-selective fading channels even so there are certain contrasts between the performance of their decoders. For lower code rates, such as $R = 1/3$ Turbo code; OFDM out-performs SC-FDE. For SC-FDE, the noise amendment loss increases with the average input SNR. When the channel is ineffective and the SNR is high, the equalizer tries to invert the nulls and as a result, the noise in these ineffective locations is amplified. Conversely, OFDM combines the useful energy across all sub-carriers through turbo-coding and interleaving.

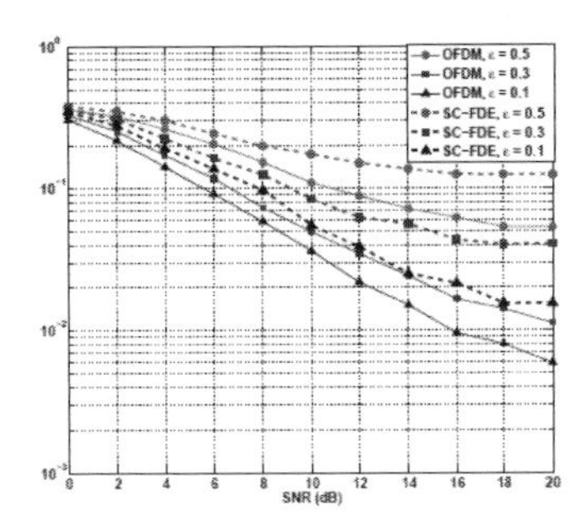

Figure 8. BER performance versus SNR parametrized by the CFO.

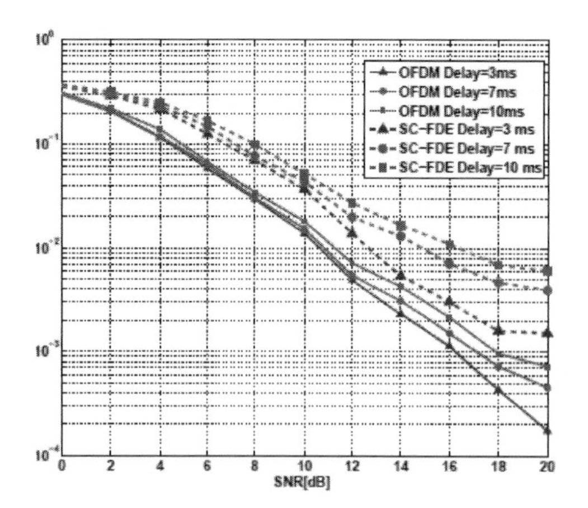

Figure 9. BER performance versus SNR parametrized by STO.

In Figure 9, the effect of the channel delay spread, that is modeled with zero padding in each propagation path is analyzed. It is assumed that no CFO and phase noise exist. The $R = 1/3$ rate, 4 state (7,5) convolutional turbo encoder has d_{free}. Therefore, a coded OFDM system with this turbo code can achieve a diverse order of 5 without implementing any additional transmit/receive antennas, or using any other diversity techniques. Hence, especially when the channel order is larger, lower rate codes are required to achieve full diversity in OFDM systems. When this is the case, OFDM gives better performance than the SC-FDE system because of the reduced effect of ISI. In Figure 10, simulation, the effect of random fluctuations in the phase of a waveform due to the VCO at the -140dBc/Hz, -100dBc/Hz, and -70dBc/Hz values is analyzed.

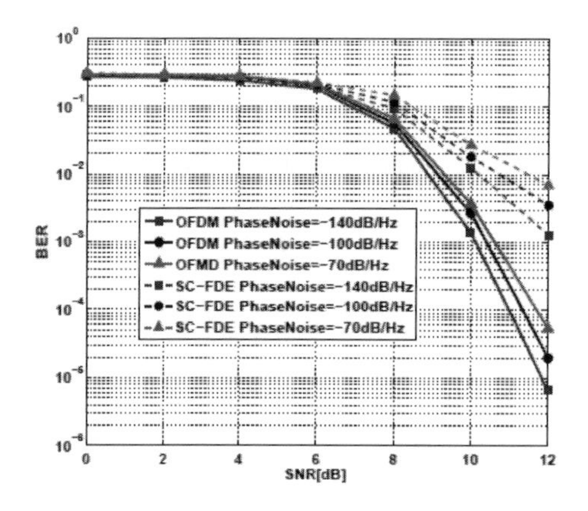

Figure 10. BER performance versus SNR parametrized by phase noise.

6 CONCLUSION

In this paper, the performances of the turbo-coded SC and OFDM systems using FDE, MMSE channel estimation, CP-based synchronization over the Wideband Vogler-Hoffmeyer HF channel are simulated. The performance of the proposed systems was compared under only CFO, STO and phase noise effects. The simulation results confirm that turbo-coded OFDM performs significantly better than turbo-coded SC-FDE in HF channel model with the large diversity.

REFERENCES

Coleri, S., e. a. (2002). Channel estimation techniques based on pilot arrangement in ofdm systems. *IEEE Transactions on Broadcasting* 48, 223–229.

Guao, Y. & K. K. Wang (2009). A real-time software simulator of wideband hf propagation channel. *International Conference on Communication Software and Networks*, 304–308.

Hassa, E. S., e. a. (2009). Enhanced performance of ofdm and single-carrier systems using frequency domain equalization and phase modulation. *National Radio Science Conference*, 1–10.

Hoffmeyer, J. A. & L. E. Vogler (1990). A new approach to hf channel modeling and simulation. *Military Communications Conference 3*, 1199–1208.

ITU (1986). Hf ionospheric channel simulators. Report 549-2, Recommendations and Reports of the CCIR,.

Myung, H. G., e. a. (2006). Peak-to-average power ratio of single carrier fdma signals with pulse shaping. *The 17th Annual IEEE International Symposium on Personal, Indoor and Mobile Radio Communications (PIMRC)*, 1–5.

Pancaldi, F., e. a. (2008). Single-carrier frequency domain equalization. *IEEE Signal Processing Magazine 25*, 37–55.

Sandell, M., e. a. (1995). Timing and frequency synchronization in ofdm systems using the cyclic prefix. *International Symposium on Synchronization*, 16–19.

Shen, Y. & E. Martinez (2008). Channel estimation in ofdm systems. Report rev. 1/2006, Free scale Semiconductor.

Van de Beek, J.J., e. a. (1995). On channel estimation in ofdm systems. *Vehicular Technology Conference, 1995 IEEE 45th 2*, 815–819.

Wattersen, C. C., e. a. (1970). Experimental confirmation of an hf channel model. *IEEE Trans. On Comm. Tech. 18*, 792–803.

Weinstein, S. & P. M. Ebert (1971). Data transmission by frequency-division multiplexing using the discrete fourier transform. *IEEE transactions on communication technology 19*, 628–634.

Electronics, Communications and Networks IV – Hussain & Ivanovic (eds)
© 2015 Taylor & Francis Group, London, ISBN: 978-1-138-02830-2

Reconfigurable transmission with wideband spectrum sensing using GNU radio and USRP

Kai Gu, Jun Tan & Yong Bai*
College of Information Science & Technology, Hainan University, Haikou, Hainan, China

ABSTRACT: To evaluate the feasibility of cognitive radio in practical systems, this paper establishes an experimental platform that supports reconfigurable transmission with wideband spectrum sensing using GNU Radio and Universal Software Radio Peripheral (USRP). The established experimental platform can perform wideband spectrum sensing by using multiple steps of narrow band energy detection. Based on the spectrum sensing results, the platform can then determine the available vacant transmission bandwidth and reconfigure the OFDM transmission parameters for data delivery. The experimental results of the implemented platform are presented in this paper.

1 INTRODUCTION

With the dramatically increasing demand for wireless communication services, the spectrum resource becomes much scarcer in supply. Nevertheless, a spectrum hole can appear which is a band of frequencies assigned to a Primary User (PU), but is not utilized by the user at a particular time and specific geographic location (Haykin 2005). The concept of cognitive radio, introduced by Dr. Joseph Mitola (Mitola 1999), enables the capabilities of spectrum sensing, spectrum management, and spectrum access (Hossain et al. 2009). Spectrum sensing determines the vacant frequency bands of PUs to support opportunistic spectrum access of Secondary Users (SUs), and may automatically switch frequency to avoid interference to PUs when there is a conflict. Spectrum management analyzes the idle spectrum detected, and then takes certain strategies to allocate transmission parameters, such as bandwidth. Spectrum access dynamically changes the spectrum to realize the communications between users on the basis of spectrum sensing. The functions of reconfiguration intend to use the real-time knowledge of its environment to allocate transmission parameters such as transmission frequency, bandwidth, modulation, and coding methods dynamically without causing harmful interference to PUs (Leaves et al. 2004, Maldonado et al. 2005).

GNU Radio is an open source software development toolkit that provides signal processing blocks to implement software-defined radios and signal processing systems. USRP (Universal Software Radio Peripheral) serves as the digital baseband and the intermediate frequency part of a radio communication system, making the ordinary computer work as software radio equipment. USRPs are commonly used with the GNU Radio software suite to create software-defined radio systems. A motherboard of USRP provides the following subsystems: clock generation and synchronization, FPGA, ADCs, DACs, host processor interface, and power regulation. These are the basic components required for baseband processing of signals. A modular front end of USRP, called a daughterboard, is used for analog operations such as up/down-conversion, filtering, and other signal conditioning. The application of GNU Radio and USRP in the establishment of an experimental platform was introduced in (Tucker & Tagliarini 2009). A spectrum sensing with energy detection based on GNU radio and USRP was given in (Sarijari et al. 2009). An experimental study of OFDM implementation utilizing GNU radio and USRP was presented in (Marwanto et al. 2009).

To evaluate the feasibility of cognitive radio in practical systems, the experimental platform needs to be established. This paper introduces our work on reconfigurable transmission with wideband spectrum sensing using GNU Radio and USRP. The established experimental platform can perform wideband spectrum sensing by using multiple steps of narrow band energy detection. Based on the spectrum sensing results, the platform can then determine the available vacant transmission bandwidth and reconfigure the OFDM transmission parameters for data delivery.

The rest of the paper is organized as follows. Section 2 describes considered system scenario of cognitive transmissions. Then, wideband spectrum

*Corresponding author: *bai@hainu.edu.cn*

sensing procedure is proposed in Section 3. The reconfigurable OFDM transmission is described in Section 4. Section 5 reports the developed testbed system and the experiment results. Section 6 makes the conclusions.

2 SYSTEM SCENARIO OF COGNITIVE TRANSMISSIONS

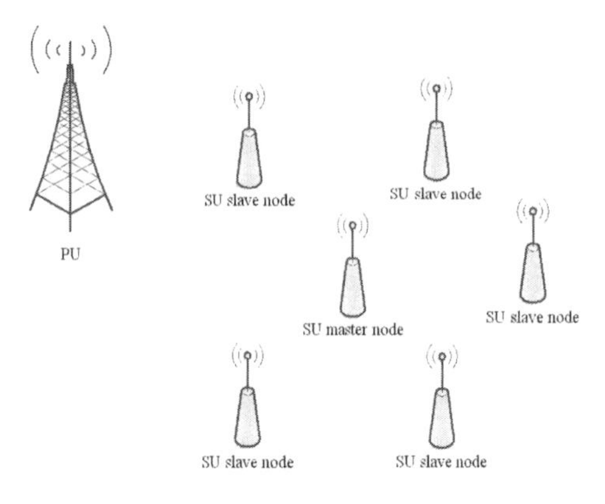

Figure 1. System scenario of cognitive transmissions.

In this paper, a centralized CR system for flexible spectrum usage is considered. As shown in Figure 1, the considered system scenario consists of Primary Users (PU), Secondary User (SU) master node, and SU slave nodes. The PU is the authorized user for spectrum utilization, and should be guaranteed no interference from the SU network. Figure 2 shows the time cycle of cognitive transmissions in the SU system. The cognitive SU master node is enabled with the abilities of spectrum sensing and spectrum decision. After real-time perception and decision of available spectrum, the information on selected suitable channels is sent to SU slave nodes in the signaling phase. The SU network can detect the idle channels of PU for opportunistic spectrum access. Due to the adoption of centralized networking in the SU system, a common control channel is required in the SU system. The SU slave nodes receive the reconfiguration signaling frame in the monitoring stage via the common control channel, and then send back acknowledge frames for confirmation. Then, the SU network nodes reconfigure the transmission parameters and start to transmit data in the SU network. This process recycles in the system. The spectrum sensing can also be cooperated by the SU master node and slave nodes. In the cooperative spectrum sensing, the slave nodes sense the spectrum and send periodic reports to the master node informing it about what they sense; the master node, with the information gathered, evaluates whether a change is necessary in the channel used, or on the contrary, whether it should stay transmitting and receiving in the same one.

3 WIDEBAND SPECTRUM SENSING PROCEDURE

The frequency domain energy detection strategy is adopted in USRP (Cabric et al. 2006). The sampling signals are transformed into frequency domain via FFT. The modulus square of transforming results is calculated, and then compared with a threshold value for decision. Since the A/D sampling rate of the USRP for the experiment is 64M Hz, the maximum detection bandwidth of USRP is adc rate-decim = 4MHz with decimation ratio (decim) of 16. Thus, the USRP can not detect wide frequency band. We propose to perform multiple steps of a single-band spectrum sensing in the SU master node as shown in Figure 3. The multiple frequency bands are sensed in steps, and single-band detection is performed in each step until all frequency bands of PU are sensed. The principle of energy detector is to find the energy of the received signal and to compare that with a predefined threshold for deciding whether the primary signal exists or not. The PU is decided to be presented if the energy is above the threshold, or else it will be absent.

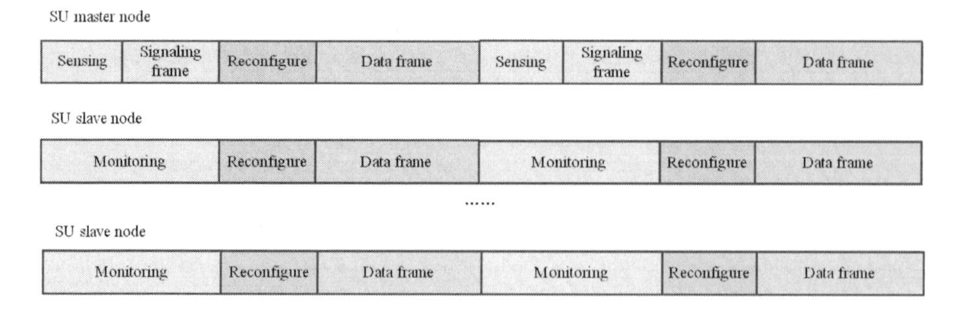

Figure 2. Time cycle of cognitive transmissions in the SU system.

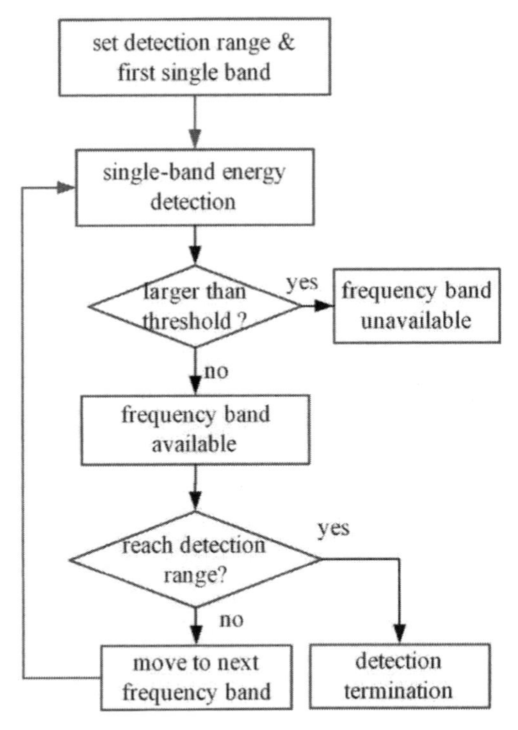

Figure 3. Multi-band spectrum sensing procedure in the SU master node.

4 RECONFIGURABLE OFDM TRANSMISSION

With the signaling from the SU master node, the SU nodes can reconfigure their transmission channels. In our experimental platform, an OFDM PHY layer is employed with the SU nodes. The functional block diagram of the transmitter and receiver for the OFDM transmission is shown in Figure 4. The binary data intended for transmission is supplied to the PHY layer from the MAC layer. This input is sent to a channel coding processor. The coded data are mapped to data constellations according to the modulation schemes such as QPSK, 16-QAM, and 64-QAM. The subcarrier allocator assigns the data constellations to the corresponding subchannels. The pilots are inserted at fixed positions in the frequency domain within each OFDM data symbol. Preambles and pilots can support the synchronization, channel estimation and tracking process. The resultant stream of constellations is subsequently input to an IFFT after a serial-to-parallel (S/P) conversion. In order to prevent inter-symbol interference (ISI) eventually caused by the channel delay spread, the OFDM symbol is extended by a cyclic prefix (CP) that contains the same waveform as the corresponding ending part of the symbol. The CP duration T_{CP} could be one of the following derived values: $T_{FFT}/32$, $T_{FFT}/16$, $T_{FFT}/8$, and $T_{FFT}/4$, where T_{FFT} represents the time duration of the IFFT output signal. Finally, the OFDM signal is transferred to the RF transmission modules via a digital-to-analog converter. Based on the sensed vacant channels and incumbent channels, the master node can make decisions on channel usage, and determine the ON/OFF status of subcarriers. The subcarriers used by incumbent users are off, while the unoccupied subcarriers for usage are on. The data is only assigned to the subcarriers whose status information is ON.

The OFDM receiver roughly implements the same operations as performed by the transmitter in reverse order. The RF signal is yielded to the baseband signal for processing. Then, the signal is converted into parallel streams by using S/P converter; the CP is removed; and the FFT is applied to transform the time domain data into the frequency domain data. Then, the data in the active subcarriers are multiplexed by using a P/S converter. Finally, the time-domain signals are demodulated into a reconstructed version of the original input at the transmitter. In addition to the data processing, synchronization and channel estimation must be performed at the receiver. With a channel estimation, adaptive selection of channel coding and modulation can be enabled in the OFDM transceivers.

5 EXPERIMENT RESULTS

In our experiment, an USRP acts as a cognitive master node for spectrum sensing, and transmit with another USRP which acts as a cognitive slave node. The maximum bandwidth for single band detection is determined by adc_rate (analog-digital conversion rate whose default is 64MHz/s) and decim (FPGA decimation rate whose default value is 16). The number of FFT bins (FFT size) affects the detection resolution, which is set to 128, and the detection resolution is 781.25Hz. We use the USRP of PU to transmit two signals at 407MHz and 413MHz, and the bandwidth of the transmitted signal is 1.875MHz. Figure 5 shows the signal spectrum at the PU node.

The energy detection range is from 400MHz to 420MHz, and Figure 6 shows the multi-band spectrum detected by the cognitive master node. Note that in Figure 6, the green line indicates the result of energy detection without PU, and the red bar chart shows the result of energy detection when there is PU. In Figure 6, we can observe two high energy bands between 405MHz and 415MHz, which correspond to 407MHz and 413MHz as shown in Figure 5. Because of small step frequency, the burst noise and noise uncertainty have a great impact on the results of each detection. For instance, one detection value is close to 73 occurs at 403MHz due to burst noise. With a larger step frequency, the impact of burst noise on detection will be reduced. Because of noise uncertainty, a suitable detection threshold needs to be determined to reduce the interference. The signaling frame transmits on a fixed common control channel, which is implemented with GMSK modulation at frequency 650MHz. The transmission bandwidth is notified in the reconfiguration signaling frame.

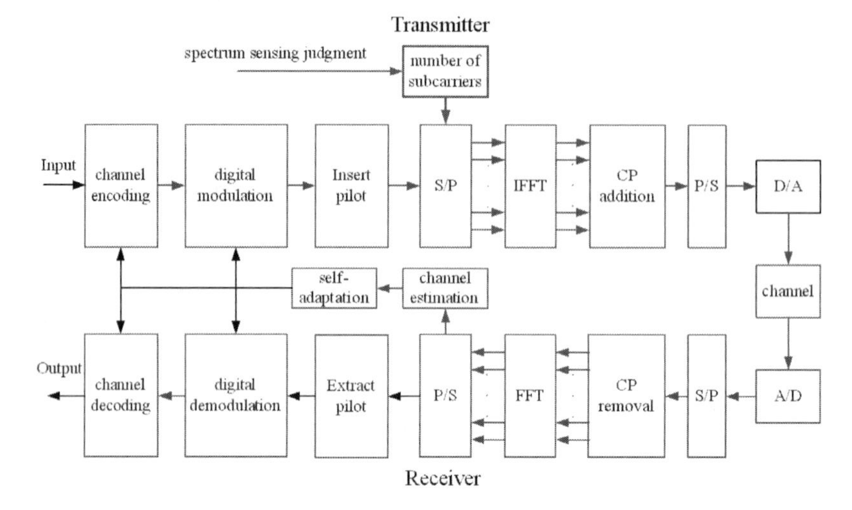

Figure 4. Reconfigurable OFDM transmission.

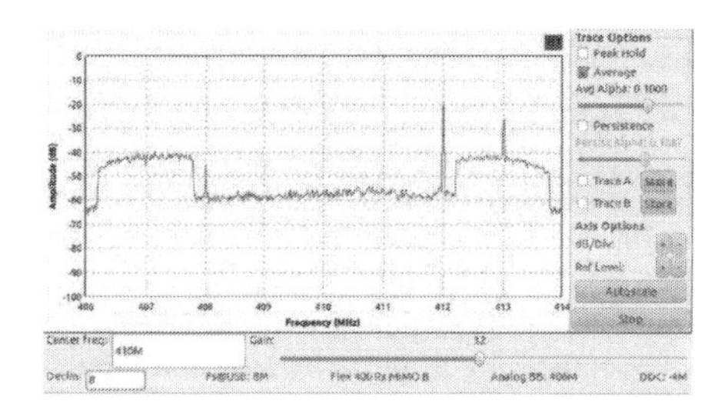

Figure 5. PU signal spectrum.

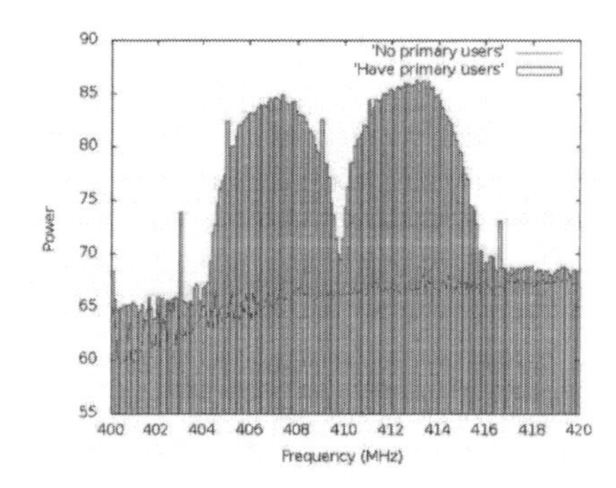

Figure 6. Energy detection result at SU master node.

The data frame transmissions use OFDM technology. In our experiment, USRP contains digital to analog conversion, and the OFDM transmission bandwidth is defined by $dac\_rate / i / fft\_length * occupied\_tones$. In the formula, dac_rate is the D/A conversion rate which is 128M samples per second; i is the FPGA interpolation rate with default value 32; fft_length is the number of FFT bins with default value 512; $occupied\_tones$ is the number of occupied FFT bins with default value 200; dac_rate / i is the maximum bandwidth of sending a signal. Using this formula, the actual bandwidth of OFDM can be calculated. Based on the sensed vacant bandwidth, the parameters such as the FFT subcarrier number can be adjusted, and then the bandwidth for OFDM transmission can be configured accordingly.

Figure 7 shows the reconfigurable OFDM transmission with 2MHz bandwidth. The transmission carrier frequency of data frames is configured to 1011MHz, and the transmission bandwidth is 2MHz. The FPGA interpolation rate is still set to the default value; the number of FFT bins and the number of occupied FFT bins are configured to 1024 and 512, respectively.

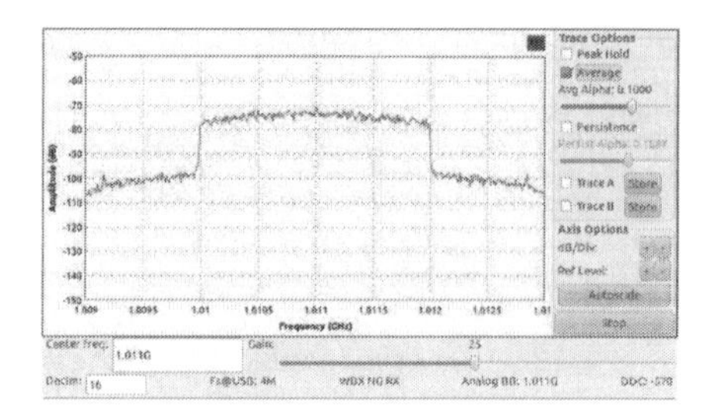

Figure 7. Reconfigurable OFDM transmission with 2MHz bandwidth.

In the next experiment, the OFDM transmission parameters at the transmitter and receiver can be reconfigured according to the sensed PU vacant channels. Figure 8 shows the reconfigurable OFDM transmission with 4MHz bandwidth. The transmission carrier frequency of data frames is reconfigured to 760MHz with 4MHz transmission bandwidth. In this case, the number of FFT bins and the occupied FFT bins remain unchanged, while the FPGA interpolation rate is changed from 32 to 16.

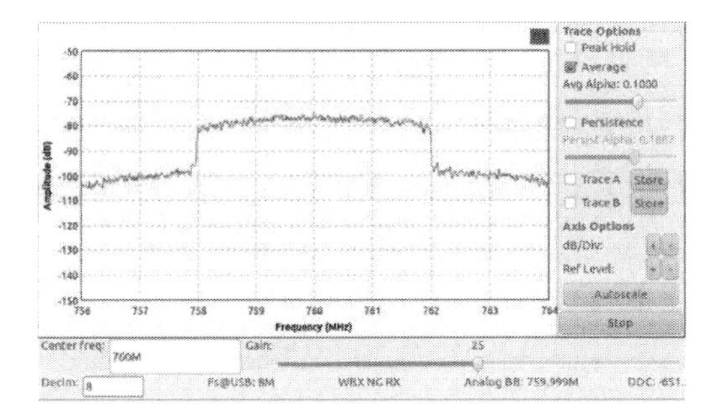

Figure 8. Reconfigurable OFDM transmission with 4MHz bandwidth.

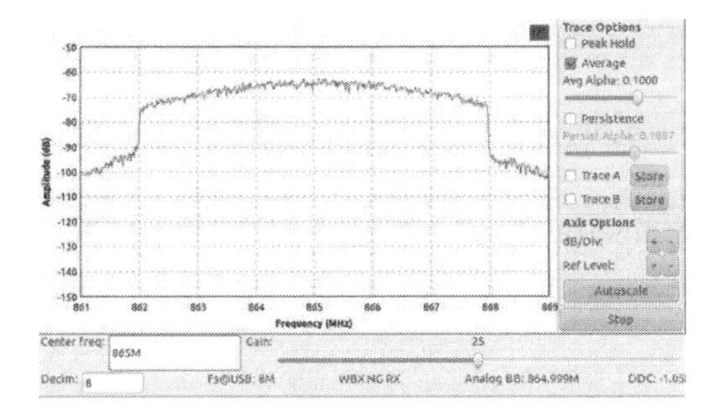

Figure 9. Reconfigurable OFDM transmission with 6MHz bandwidth.

Figure 9 shows the reconfigurable OFDM transmission with 6MHz bandwidth. The transmission carrier frequency of data frames is reconfigured to 865MHz with 6MHz transmission bandwidth. In this case, the FPGA interpolation rate and the number of FFT bins are 16 and 32 respectively, and the number of occupied FFT bins is reconfigured to 760. Theoretical calculated bandwidth is 5.94MHz.

Table 1 shows the experimented transmission performance with different modulation and transmission frequencies. In the table, pktno represents the total number of sent packets; n rcvd represents the number of packets received; and n crct is the number of packets received correctly. PER (Packet Error Rate) is calculated with the missing frames and CRC error frame contained. It is observed that BPSK has higher reliability compared with QPSK at the same transmission conditions.

Table 1. Experimented transmission performance.

Modulation	Frequency (MHz)	pktno	n rcvd	n crct	PER
BPSK	760	9977	9881	9871	1.06%
QPSK	760	9960	9939	9836	1.24%
BPSK	865	9977	9957	9796	1.81%
QPSK	865	9959	9928	9726	2.34%

6 CONCLUSIONS

The reconfigurable system with cognitive radio can take full advantage of dynamic characteristics of radio channel to improve spectrum efficiency by adjusting the parameters of transceivers. In our established platform, the available vacant transmission bandwidth can be sensed by using a multi-step wideband frequency-domain energy detector, and reconfigurable end-to-end OFDM transmission is implemented. Further study can be continued on this established experiment platform, for example, how to deal with the effect of noise uncertainty in energy detection, and how to reconfigure more transmission parameters in the practical systems.

ACKNOWLEDGEMENT

This paper was supported by the National Natural Science Foundation of China (Grant No. 61062006 and Grant No. 61261024) and the Special Social Service Project Fund of Hainan University, China (Grant No. HDSF201301).

REFERENCES

Cabric, D., Tkachenko, A. & Brodersen, R.W. 2006. Spectrum sensing measurements of pilot, energy, and collaborative detection. *Proc. Military Communications Conference, 23–25 Oct. 2006, Washington, D.C.,USA*: 1–7.

Haykin, S. 2005. Cognitive radio: brain-empowered wireless communications. *IEEE J. Sel. Areas Commun.* 23 (2): 201–220.

Hossain, E., Niyato, D. & Han, Z. 2009. *Dynamic spectrum access and management in cognitive radio networks.* United Kingdom: Cambridge University Press.

Leaves, P., Moessner, K., Tafazolli, R., Grandblaise, D., et al. 2004. Dynamic spectrum allocation in composite reconfigurable wireless networks. *IEEE Commun. Mag., May 2004,* 42 (5): 72–81.

Maldonado, D., Le, B., Hugine, A., Rondeau, T. & Bostian, C. 2005. Cognitive radio applications to dynamic spectrum allocation: a discussion and an illustrative example. *Proc. First IEEE International Symposium on New Frontiers in Dynamic Spectrum Access Networks, Baltimore, Maryland, 8–11 Nov. 2005, USA:* 597–600.

Marwanto, A., Sarijari, M.A., Fisal, N., Yusof, S.K.S. & Rashid, R.A. 2009. Experimental study of ofdm implementation utilizing gnu radio and usrp-sdr. *Proc. IEEE 9th Malaysia International Conference on Communications (MICC), 2009, Kuala Lumpur, Malaysia:* 132–135.

Mitola, J. 1999. Cognitive radio for flexible mobile multimedia communications. *Proc. IEEE International Workshop on Mobile Multimedia Communications (MoMuC'99), 15 Nov 1999, Sandiego, California, USA:* 3–10.

Sarijari, M. A., Marwanto, A., Fisal, N., Yusof, S.K.S., Rashid, R.A. & Satria, M.H. 2009. Energy detection sensing based on gnu radio and usrp: An analysis study. *Proc. IEEE 9th Malaysia International Conference on Communications (MICC), Kuala Lumpur, Malaysia 15–17 Dec. 2009:* 338–342.

Tucker, D. C. & Tagliarini, G.A. 2009. Prototyping with gnu radio and the usrp-where to begin. *Proc. IEEE SOUTHEASTCON'09, Atlanta, GA, USA 5–8 March 2009:* 50–54.

Electronics, Communications and Networks IV – Hussain & Ivanovic (eds)
© 2015 Taylor & Francis Group, London, ISBN: 978-1-138-02830-2

IPv6 network virtualization architecture for smooth IPv6 transition

Dujuan Gu
Computer Network Information Center, Chinese Academy of Sciences
University of Chinese Academy of Sciences, Beijing, China

Xiaohan Liu, Ze Luo & Baoping Yan
Computer Network Information Center, Chinese Academy of Sciences

ABSTRACT: IPv6 transition has become inevitable and fairly urgent. Facing such slow IPv6 adoption, we attempt to propose IPv6 network virtualization architecture (VNET6) to facilitate IPv6 transition in a seamless way for applications and networks. While user's experience remains transparent and seamless, the extended network virtualization converts IPv6 into an incrementally deployable protocol. VNET6 dynamically fulfills different heterogeneous application's requirements of end-to-end communication. The evaluation of our prototype implementation demonstrates that: (1) Available applications through IPv6 connectivity are more than those through IPv4 in VNET6. (2) This architecture can provide sufficient incentives to trigger IPv6 adoption until IPv6 achieves full adoption, as IPv6 virtual networks are deployed gradually and seamlessly. Hence, VNET6 can greatly accelerate the pace of transition to IPv6 along with the continuing development of Internet applications.

1 INTRODUCTION

The current Internet has exposed limitations of IPv4 (Huston). IPv4 address exhaustion has occurred in Asia and Europe, and is coming soon to North America. However, IPv6 has not achieved widespread adoption even after over a decade of existence. According to Google's statistics about IPv6 adoption, only 2.30% of users (Google, 2014) would use Google's services over IPv6 on December 20, 2013. The lack of IPv6 adoption causes resistance to IPv6 transition, although most operating systems, devices, and networks are IPv6-ready.

Internet Service Providers (ISPs) and Internet Content Providers (ICPs) have gone their own way to study IPv6 transition. From ISPs' perspective, previous studies of various IPv6 transition mechanisms unfortunately have not gone into that widespread deployment, because they cannot solve various applications interoperation problems (Hiromi et al. 2012, Skoberne & Ciglaric 2011). Accordingly, the end-to-end communication of IPv4/IPv6 heterogeneous applications cannot be ensured. From ICPs' perspective, a number of researchers have upgraded specific applications to IPv6(Hirorai & Yoshifuji 2006). It is not realistic to expect all applications to be upgraded to IPv6 any time soon (Hamarsheh et al. 2011). Meanwhile, numerous applications have found it difficult to adapt them to underlying heterogeneous networks of IPv4 and IPv6 coexisting. Therefore, ISPs and ICPs should be bound together to upgrade everything to IPv6. However, it has become clear that upgrading everything from IPv4 to IPv6 is quite a messy affair.

In fact, few researchers have been engaged in studying an overall architecture for IPv6 transition. Future network architectures (Roberts 2009, Paul et al. 2011, McKeown 2009, McKeown et al. 2008, Jacobson et al. 2009) are attracting interest to deal with different Internet ossification problems; however they are complete grounds-up re-designs. Network virtualization (Sahoo et al. 2010, Chowdhury & Boutaba 2009, Anderson et al. 2005) has been proposed as a diversifying attribute of future paradigm, yet how to enable the seamless integration between IPv6 and current Internet remains unexplored. Particularly, there is no research on adaptive provision of virtual paths to cope with the end-to-end communication of various applications in the complex environment of IPv4 and IPv6 coexisting.

In order to propose a comprehensive solution to IPv6 transition, we borrow from the innovative insights of network virtualization and propose a simple architecture of IPv6 network virtualization (VNET6), which is more closely aligned to the incumbent IPv4 Internet. The major challenge in this architecture is how dynamically fulfill heterogeneous application's different requirements of end-to-end communication during IPv6 transition. Our main contributions in this paper are as follows:

- VNET6 facilitates IPv6 transition by a new decoupling mechanism, which effectively shields the application layer IP operations from underlying network in a seamless way. Thus, IPv6 network deployment remains transparent to all applications.
- VNET6 extends the research of network virtualization to flexibly cope with heterogeneous communication in the environment of IPv4 and IPv6 coexisting. With incremental scalability, networks can be gradually upgraded from IPv4 to IPv6 through the seamless integration of IPv6 network resources and IPv6 transition mechanisms.

Our prototype implementation results show that user's experience remains transparent and seamless during IPv6 network deployment. In VNET6, more accessible applications through IPv6 connectivity must provide an incentive for users to adopt IPv6. With incremental scalability, IPv6 virtual networks coexist with IPv4 networks over the shared infrastructures. VNET6 indeed accelerates growth in terms of IPv6 network deployment and user's IPv6 adoption. Hence, VNET6 also greatly accelerates the pace of transition to IPv6 along with the continuing development of Internet applications.

The rest of the paper is organized as follows. Section 2 outlines the architecture of VNET6. The main topic of section 3 is the design of IPv6 network virtualization. Section 4 implements and evaluates VNET6 in comparison with mainstream IPv6 transition mechanisms. In section 5, the conclusion and future work are presented.

2 ARCHITECTURE AND NOTATIONS

2.1 *An overview of VNET6*

For the coexisting IPv4 and IPv6 environment between applications and networks, VNET6 virtualizes the mixed IPv4 and IPv6 physical networks and various IPv6 transition mechanisms into IPv6 virtual networks. IPv6 virtual networks provide abstract IPv6 transition services for the end-to-end communication of IPv4 and IPv6 applications. This novel architecture can be logically viewed in three layers depicted in Figure.1.

- Physical infrastructure layer: a weighted indirection graph for a physical network topology. This layer is the current complex physical infrastructure of IPv4 and IPv6 coexisting; there are native IPv4 networks, native IPv6 networks or dual stack networks, and various IPv6 transition mechanisms. A physical node may be a physical entity with network resources and IPv6 transition mechanisms.
- Virtual network layer: a weighted indirection graph for a virtual network topology. VNET6 extends network virtualization technology, so as to provide network abstractions and meet application-specific requirements. On one hand, network abstractions enable IPv6 virtual networks to be built with incremental scalability and without any modifications of existing networks. A virtual node on-demand integrates various network resources and IPv6 transition mechanisms; a virtual link between two virtual nodes is a logical Softwire (Li et al. 2007) by tunneling the data packet. This virtual consolidation mechanism has been implemented as described previously (Gu et al. 2013a). On the other hand, IPv6 transition mechanisms are dynamically adapted to a service profile of an application. The service profile is a combination of application-specific heterogeneous interoperation requirements between IPv4 and IPv6. Specifically, the optimal paths ensure application's end-to-end communication in the mixed IPv4 and IPv6 environment. Hence this IPv6 virtual network is a scale-out architecture (Vahdat et al., 2010) that supports incremental scalability and flexible services.
- Application layer: This layer includes IPv6 applications and IPv4 applications. A novel mechanism Decoupling API effectively shields the application layer IP operations from underlying networks. Thus, incumbent IPv4 applications are able to benefit from IPv6, without any modifications.

UCM is a virtualization orchestrator (Gu et al., 2013a), which unifies the control and management of applications and networks. To satisfy the evolving demands of network, UCM maintains virtual network topologies and supports dynamic virtual consolidation mechanism for networks. Therefore UCM simplifies both network and application management of the environment of IPv4 and IPv6 coexisting.

2.2 *Decoupling application ip operations from networks*

To effectively decouple applications from networks, VNET6 proposes Decoupling API to identify double IP addresses for application IP operations and network connectivity. Figure.2 shows the difference

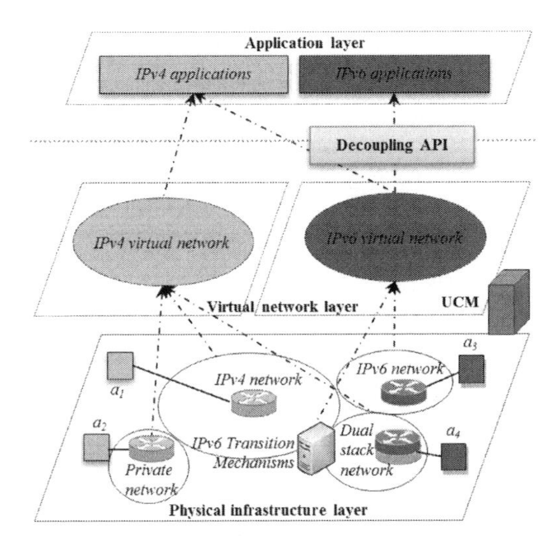

Figure 1. VNET6 architecture overview.

between TCP/IP and VNET6. In TCP/IP, the underlying networks and various upper applications depend on a unique IP address, which would result in a very high degree of coupling. In VNET6, applications and networks are provided with *IP_A* and *IP_N* respectively; double IP addresses may belong to different IPs. Here VNET6 like HIP (Nikander et al., 2010), LNA (Balakrishnan et al., 2004), and Serval (Nordstrom et al., 2012) offers two addresses, so that networks and applications could independently change IP addresses. By contrast, the main difference is that VNET6 need not address mapping management, because algorithmic translation can be used in the mapping information between *IP_N* and *IP_A*.

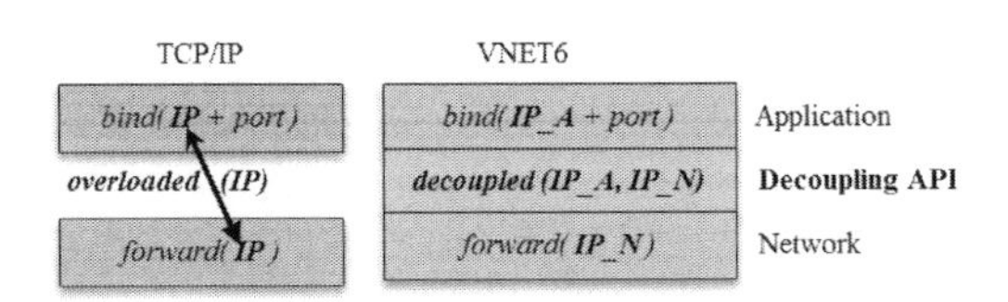

Figure 2. TCP/IP stack versus VNET6.

This algorithmic translation self-manages the mapping information of double IP addresses. A destination address in *IP_A* is obtained via DNS Name Resolver, while other addresses are algorithmically translated as below.

- A mapping IPv4 address *L_4add* is assigned to local IPv4 applications over IPv6 network.
- A mapping IPv6 prefix *L_Prefix* is added to *L_4add* to construct IPv4-embedded IPv6 address (Bao et al., 2010) for IPv6 network connectivity of IPv4 applications.
- A mapping IPv6 prefix *R_Prefix* and the IPv4 addresses of remote IPv4 applications are used to construct IPv4-embedded IPv6 address to communicate with local IPv6 applications over IPv6 network.
- A mapping IPv4 address *R_4add* acts as a dummy address for remote IPv6 applications with global IPv6 addresses (Hinden and Deering, 2003).

Practically, the mapping information is required for successful inbound and outbound communication for heterogeneous applications. There are *L_Prefix* and *L_4add* for applications of a local end, which are assigned by an access network. Meanwhile, applications of a remote end are assigned with *R_Prefix* or *R_4add*, which should agree with the resource pool of IPv4-IPv6 translation, used to translate network address. Hence, address routing needs to do nothing for these application's addresses and prefixes, which are consistent with network's ones. The mapping information is saved in Decoupling API for algorithmic translation.

This mechanism retains standard network sockets for applications, and invokes new sockets to adapt to network connections. Practically it retains standard network sockets for applications and networks without introducing new characteristics. By communicating directly on Decoupling API, IPv4 applications without any modifications can go on running over IPv6 networks as before. For example, an IPv4 application in IPv6 network initiates a connection with a remote IPv4 end on the Internet. According to the above mapping information, *L_4add* in *IP_A* is translated to an IPv4-embedded IPv6 address *L_Prefix+L_4add* in *IP_N*; the IPv4 destination address in *IP_A* is combined with *R_Prefix* in *IP_N*. Particularly, IPv4 addresses in application payload don't need to be translated; IPv6 addresses of *IP_N* are translated to IPv4 addresses reducing IPv6 prefixes by the IPv4-IPv6 translator of *R_Prefix*. Hence this mechanism effectively shields the application layer IP operations from underlying networks and remote ends.

During IPv6 transition, the interoperation IP capability with remote ends is essentially determined by applications, rather than network IPs. Hence VNET6 fends off the tight coupling between applications and networks.

3 DYNAMIC VIRTUALIZATION PROCESS

Figure 3 outlines our IPv6 dynamic virtualization process to set up virtual networks according to application-specific requirements in above architecture.

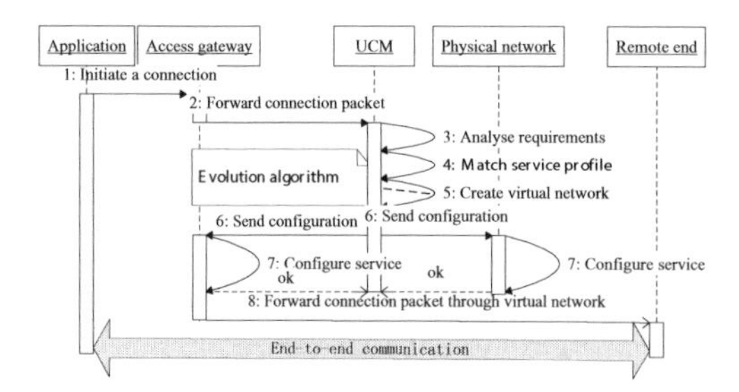

Figure 3. Dynamic virtualization process.

This process effectively ensures optimal end-to-end paths in IPv4/IPv6 heterogonous environment, when one application on a local end initiates a connection to another application on a remote end.

1 Decoupling API has helped the local application to obtain addresses of *IP_A* and *IP_N* through algorithmic translation. Then the application initiates a connection to the remote application.
2 Upon receiving the application's connection packet, the access gateway of this local end only needs to forward this initial connection packet to UCM. The default route of this access gateway guides this packet with the next hop to UCM.
3 According to the address values in IP header of this packet, UCM easily identifies the address attributes of *IP_N*.
4 UCM matches the address attributes with a service profile, which complies with IPv6 transition services specified by end-to-end communication requirements for this application's connection.
5 UCM identifies the candidate physical resources of existing network resources and IPv6 transition mechanisms for this service profile. Especially an evolution algorithm is applied to ensure an optimal end-to-end path with the necessary IPv6 transition mechanisms. UCM parses the specified service profile, and then evaluates the suitability of any matching network resources and transition mechanisms. The suitability is the distance from a matching one to the request host. The closest one is assigned among these matching ones. It is a simple evaluation strategy to keep IPv6 transition mechanism simple enough to be hosted in performance constrained entities. A virtual path is bound to the specific physical resources. The binding information is recorded by UCM. Hence, the UCM assigns the set of physical nodes and physical links to a virtual network. Accordingly, all binding information is maintained by UCM. UCM binds the most suitable network resources and transition mechanisms with the optimal path.
6 Finally, UCM sends configuration commands via standard network management interface to those specific physical nodes, including the access gateway. These configuration commands are related to the necessary IPv6 transition mechanisms for this service profile.
7 After receiving the configuration commands, the specific physical nodes process the related configuration commands and report the processing results to UCM. As long as all configuration commands of IPv6 transition services are configured, the virtual network will be created completely. Especially, the first node in this virtual network is bound with the access gateway. The access gateway needs to configure at least a new route for the destination of connection and the mapping information that will be learnt and saved in Decoupling API.
8 According to the added new route, the access gateway guides this connection request to the specific remote end through the built virtual path.

The following packets will be directly transported through the path, on which IPv6 transition mechanisms have been configured to ensure application's end-to-end communication.

4 IMPLEMENTATION AND EVALUATION

4.1 *Implementation*

As a proof of concept, we provide an overview of our implementation of VNET6. This implementation extends the implementation of preview studies in (Gu et al., 2013b) and (Gu et al., 2013a). This prototype of VNET6 is composed of commodity entities and two novel components: Decoupling API and UCM.

- Decoupling API should be installed on IPv6-only hosts that possibly run not only IPv6 applications, but also IPv4 applications. This component runs natively in the Linux kernel as a module.
- UCM is a control and management software tool, which is implemented to enhance the management capabilities of existing network management software tools for topology. The next step of implementation will extend Flowvisor (Sherwood et al., 2009) to support UCM function as a virtualization orchestration, which unifies the control and management of applications and networks.

The behavior is identified for applications running in VNET6. As long as IPv4 applications run over IPv6 networks, these applications should advertise their extra DNS records with IPv4-embedded IPv6 addresses for inbound communication.

4.2 *Evaluations and discussions*

While performance is not a first-order consideration of VNET6, the actual application value should be the accessibility of applications on account of the emphasis on IPv6 adoption. Meanwhile, the ultimate goal of new approaches is to replace all of IPv4 with IPv6.

As compared with dual stack and IPv4-IPv6 translation mechanism, VNET6 provides higher IPv6 application accessibility than these two mainstream transition mechanisms, because IPv6 networks in VNET6 call for unimpeded access to not only IPv6-only applications, but also all incumbent IPv4 applications and heterogeneous applications. The higher IPv6 application accessibility is, the more users adopting IPv6 will be. As a result of application's intercommunication over time, IPv6 can achieve full adoption through VNET6. By contrast, IPv6 could not achieve full adoption through dual stack mechanism(Cho et al., 2004). Meanwhile, IPv6 may achieve full adoption through IPv4-IPv6 translation mechanism, but with very slow speed and severe initial conditions. Therefore, VNET6 has certain advantages over mainstream IPv6 transition mechanisms.

5 CONCLUSION AND FUTURE WORK

The above verification and evaluation demonstrate VNET6's feasibility. As might be expected, IPv6 in VNET6 provides immediate benefit: high IPv6 application accessibility, apart from larger address space. Thus VNET6 generates the necessary motivation to adopt IPv6 in end-users. Eventually, IPv6 can achieve full adoption, as incremental IPv6 network deployment is speeded up. VNET6, as a promising approach, not just can sustain Internet incumbent applications, but also can accelerate the pace of transition to IPv6.

This paper only proposed a preliminary IPv6 network virtualization architecture. In future research, we will focus on the improvement of VNET6, especially how to explore the overall system performance facing large-scale applications, how to refine the design of IPv6 network virtualization architecture and algorithm, and how to revise our implementation in the wide range of networks. The further investigation of network virtualization, future network and IPv6 transition technologies is also our concern.

ACKNOWLEDGMENT

This work is jointly supported by the Natural Science Foundation of China under Grant NO. 61361126011, the Special Project of Informatization of Chinese Academy of Sciences in "the Twelfth Five-Year Plan" under Grant No. XXH12504-1-06 and Around Five Top Priorities of "One-Three-Five" Strategic Planning, CNIC(Grant No. CNIC_PY-1408 and No. CNIC_PY-1409). Thank our colleagues who contributed towards these projects for their collaborations.

REFERENCES

Anderson, T., Peterson, L., Shenker, S. & Turner, J. 2005. Overcoming the Internet impasse through virtualization. *Computer* 38:34–41.

Balakrishnan, H., Lakshminarayanan, K., Ratnasamy, S., Shenker, S., Stoica, I. & Walfish, M. 2004. A layered naming architecture for the Internet. *In ACM SIGCOMM Computer Communication Review* 34(4): 343–352.

Bao, C., Huitema, C., Bagnulo, M., Boucadair, M. & Li, X. 2010. IPv6 addressing of IPv4/IPv6 translators. *Work in Progress*. RFC 6052,October.

Cho, K., Luckie, M. & Huffaker, B. 2004. Identifying IPv6 network problems in the dual-stack world. *Proceedings of the ACM SIGCOMM workshop on Network troubleshooting: research, theory and operations practice meet malfunctioning reality*: 283–288.

Chowdhury, N. M. K. & Boutaba, R. 2009. Network virtualization: state of the art and research challenges. *Communications Magazine, IEEE*, 47(7): 20–26.

Google. 2014, May. Global IPv6 statistics-measuring the current state of IPv6 for ordinary users. http://www.google.com/ipv6/statistics.html#.

Gu, D., Liu, X., Qin, G., Yan, S., Luo, Z. & Yan, B. 2013. VNET6: IPv6 virtual network for the collaboration between applications and networks. *Journal of Network and Computer Applications* 36: 1579–1588.

Gu, D., Liu, X., Wu, C., Qin, G., Luo, Z. & Yan, B. 2013. A Transparent Solution for Legacy Applications between Heterogeneous Networks. *Global Telecommunications Conference (Globecom 2013), 2013b. IEEE*: 2176–2181. IEEE.

Hamarsheh, A., Goossens, M. & Alasem, R. 2011. Decoupling Application IPv4/IPv6 Operation from the Underlying IPv4/IPv6 Communication (DAC). *American Journal of Scientific Research*:101–121.

Hinden, R. M. & Deering, S. E. 2003. Internet protocol version 6 (IPv6) addressing architecture. RFC 3513, April.

Hiromi, R., Nakamura, O., Hazeyama, H. & Ishihara, T. 2012. Experiences from IPv6-Only Networks with Transition Technologies in the WIDE Camp Autumn 2012.

Hirorai, R. & Yoshifuji, H. 2006. Problems on IPv4-IPv6 network transition. Applications and the Internet Workshops, 2006. *SAINT Workshops 2006. International Symposium on, 2006. IEEE*, 5 pp.-42.

Huston, G. 2014. IPv4 address report. http://www.potaroo.net/tools/ipv4/index.html.

Jacobson, V., Smetters, D. K., Thornton, J. D., Plass, M. F., Briggs, N. H. & Braynard, R. L. 2009. Networking named content. *In Proceedings of the*

5th international conference on Emerging networking experiments and technologies, December 2009:1–12. New York :ACM.

Li, X., Dawkins, S., Ward, D. & Durand, A. 2007. Softwire problem statement. *IETF RFC4925. Internet Engineering Task Force, Fremont, CA.*

Mckeown, N. 2009. Software-defined networking. *INFOCOM keynote talk, Apr.*

Mckeown, N., Anderson, T., Balakrishnan, H., Parulkar, G., Peterson, L., Rexford, J., Shenker, S. & Turner, J. 2008. OpenFlow: enabling innovation in campus networks. *ACM SIGCOMM Computer Communication Review,* 38: 69–74.

Nikander, P., Gurtov, A. & Henderson, T. R. 2010. Host identity protocol (HIP): Connectivity, mobility, multi-homing, security, and privacy over IPv4 and IPv6 networks. *Communications Surveys & Tutorials, IEEE* 12: 186–204.

Nordstrom, E., Shue, D., Gopalan, P., Kiefer, R., Arye, M., Ko, S., Rexford, J. & Freedman, M. J. 2012. Serval: An end-host stack for service-centric networking. *Proc. 9th USENIX NSDI*: 85–98.

Paul, S., Pan, J. & Jain, R. 2011. Architectures for the future networks and the next generation Internet: A survey. *Computer Communications* 34: 2–42.

Roberts, J. 2009. The clean-slate approach to future Internet design: a survey of research initiatives. *Annals of telecommunications,* 64: 271–276.

Sahoo, J., Mohapatra, S. & Lath, R. Virtualization: A survey on concepts, taxonomy and associated security issues. 2010. IEEE: 222–226.

Sherwood, R., Gibb, G., Yap, K.-K., Appenzeller, G., Casado, M., Mckeown, N. & Parulkar, G. 2009. Flowvisor: A network virtualization layer. *OpenFlow Switch Consortium, Tech. Rep.*

Skoberne, N. & Ciglaric, M. 2011. Practical Evaluation of Stateful NAT64/DNS64 Translation. *Advances in Electrical and Computer Engineering* 11: 49–54.

Vahdat, A., Al-Fares, M., Farrington, N., Mysore, R. N., Porter, G. & Radhakrishnan, S. 2010. Scale-out networking in the data center. *Micro, IEEE* 30: 29–41.

Electronics, Communications and Networks IV – Hussain & Ivanovic (eds)
© 2015 Taylor & Francis Group, London, ISBN: 978-1-138-02830-2

The evolution of business supporting network in the environment of cloud computing

Wei Guo*
CMCC Shanxi Ltd., Taiyuan, China

Zhihong Yuan
Shanxi University, Taiyuan, China

ABSTRACT: In this paper the current situation of business supporting network of operators is introduced. Combined with the characteristics of cloud computing, the direction of the evolution of business supporting network has been studied via virtual network equipment, enterprise-class private cloud and software defined network. It has been proved by the evolution of business supporting network that the enterprise-class cloud network is capable of increasing the efficiency in the usage of IT infrastructure and decreasing the cost of equipment.

1 CURRENT SITUATION OF BUSINESS SUPPORTING NETWORK

As one of the core business systems, the major point of the system in the security of business supporting network lies upon its usability, confidentiality and integrity.

In the typical business supporting network for operators, the conventional network framework is used as its main body, and the classification and development are applied to the infrastructure (host, memory, network etc.) of supporting networks (See Figure 1), in accordance with different security levels of various business types. From the top of the hierarchy of the level of security, network security domains (SD) would be the core domain, internal domain and external domain. It can be seen that the group server of the core domain is of the highest reliability while the group server of the external domain is of the lowest. And the internal parts of all security domains are of the same credibility (Zhang et al. 2010).

With the carrying out of numerous applications, the basic network is in need of continuous amendments and expansion in order to adapt to the fast-paced business types and changing operation modes, which have brought huge pressure to the running and maintenance of the system. Conventional network planning design, which is generally thought of as highly reliable, has a mesh network that contains complicated redundancy. Yet the physical topology of structural mesh network has significant advantages due to its high-reliability maintenance, fault tolerance and performance improvements. Thus it becomes a general rule of design. The pre-designed planning and configuration status will be altered by various faults and errors including accessing of multiple loops in layer 2, interconnection of routers in Full Mesh, variations of link status in network, and running errors of points. Thus, the diagnosis of operation and maintenance will be rather complicated. Besides, expansion and movement of applications will lead to network changes in varying degrees, and these changes are likely to have negative influences over the normal operation of business systems.

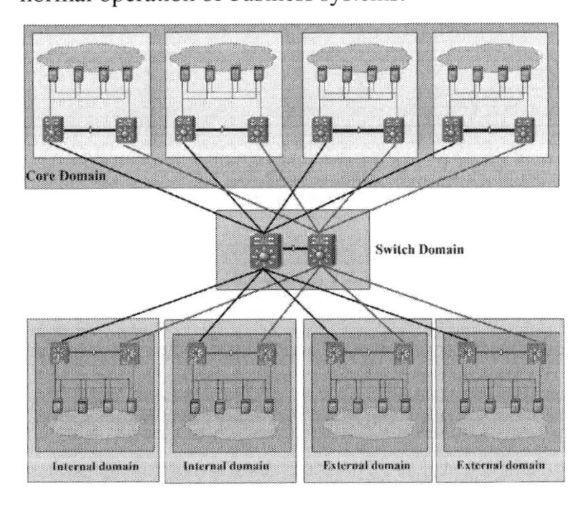

Figure 1. Typical classification of the security domain of business supporting network.

*Corresponding author: *weiwei3547@163.com*

Nowadays, the virtual resource pool deployed in the business supporting network makes the platform much easier to be delivered, and its advantage of higher flexibility is gradually seen. However, with the expansion of cloud computing platform, the structure and operation of conventional basic network are insufficient for the flexible cloud computing. Firstly, the classification of network security domain becomes a limit on the virtual resource pool, which leads to the decrease of the efficiency of cloud computing. Secondly, for the conventional firewall, it is not possible to transfer the security policy with virtual machine, thus frequent human intervention will keep the usability efficiency of virtualized resources at a lower level and the cost of maintenance high. Therefore, the structural evolution of business supporting networks in the environment of cloud computing has become a major issue in current network operation.

2 EVOLUTION OF STRUCTURE OF BUSINESS SUPPORTING NETWORK

Because of various factors including the multiple layer structures, security domain, security level, strategy deployment, router control, VLAN classification, layer 2 loop, redundancy design and so on, the structure of traditional business supporting network is rather complicated, which makes it fairly difficult to run and maintain the basic network of data center (DC). Combined with the characteristics of network structure in the environment of cloud computing and special cases of business supporting network, a thought of evolution of business supporting network structure has been developed, with significant application in practice.

2.1 *Virtualization of network equipment*

Virtualization of network equipment leads to the horizontal integration of network structure. Hence, the main network equipment will form a logic unit. The network structure can be simplified by management simplification, configuration simplification and virtual port channel between devices and these will also increase the reliability of redundancy (Cisco 2010). And if the network structure is vertically separated, there will be independent data paths in the logic network which is divided in a virtual way. Therefore the existence of the logic network will not be perceived by either the end users or the upper level applications.

In Figure 2, it can be seen that the conventional business supporting network structure is of high complexity but of low stability. Specifically, there are multiple loops level 2 accesses, complicated high-reliability design "VRRP+MSTP", and complex router design due to intertwined links. What's more, a router oscillation or convergence will take place when there are errors at points or links.

It can be seen in Figure 3 that through network element virtualization, several network points will be virtualized into one single point, and intertwined links will be bound to form a single logic link, thus the complicated MSTP/VRRP is no longer needed and the using efficiency of link bandwidth is improved in the process. In addition, planning of routes and VLAN in the data center will be dramatically simplified. Finally, malfunction of individual physical device or link will have no influence on the upper-level routers, which enables the stability of the structure to be sharply improved.

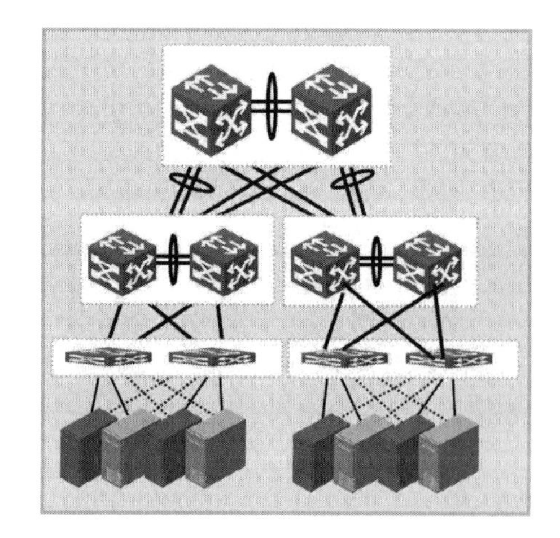

Figure 2. Conventional business supporting network structure.

Figure 3. Business supporting network structure with a network element virtualization.

2.2 *Enterprise-class private cloud network structure*

A proper platform structure and protocol will be defined in accordance with the real situation in Shanxi province, combined with the establishment of data center via combing and summarizing the Business Support System (BSS), Operation Support System (OSS) and Management Support System (MSS). The mutual backup between systems and mutual disaster recovery between two DCs will be realized among multiple DCs. A DC-class core switch between working DC will be established, see Figure 4. And the enterprise-class security domain including core domain, internal access domain, external access domain, internet access domain and terminal access domain, will be covered in accordance with different system security levels in BSS OSS and MSS (the number of security domain is adjustable according to real situations). Meanwhile, an enterprise-class resource pool of the security domain is to be established for proper resource sharing and allocation inside the working stations.

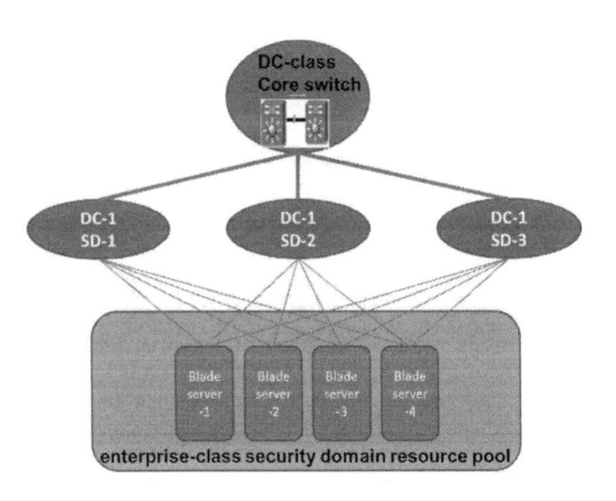

Figure 4. Deployment of enterprise-class security domain resource pool.

The mutual network between each working station will be realized via the establishment of packet optical transport network (POTN) among DC-class core switches (Cheng et al. 2012). The ring network of working station constituted by POTN will not only integrate the transmission of each business network among different DCs, expand the bandwidth in a flexible way and increase the transmission efficiency, but also simplify the network structure (namely the integration of transmitting equipment and data network equipment) , which can make a carrier-class switching protection less than 50ms. See Figure 5.

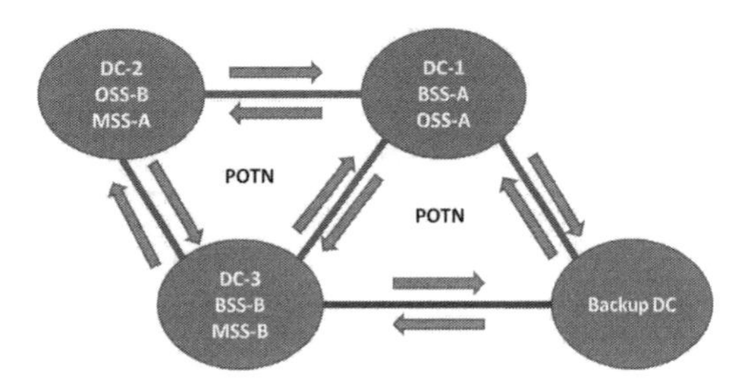

Figure 5. Deployment of data center mutual connection.

2.3 *Programmable supporting network*

As a new generation of communication network, software defined network (SDN), mostly realized by automatic control, is featured by its openness and simplicity (Open networking 2011). In SDN, the controlling plane and data plane of the network will be separated completely, that is to say, the behavior of underlying points will be controlled by a controller through the unified network operation system (NOS), where NOS communicates with underlying equipment via "South interface", while it communicates with various applications via "North interface", namely the application programming interface (API).

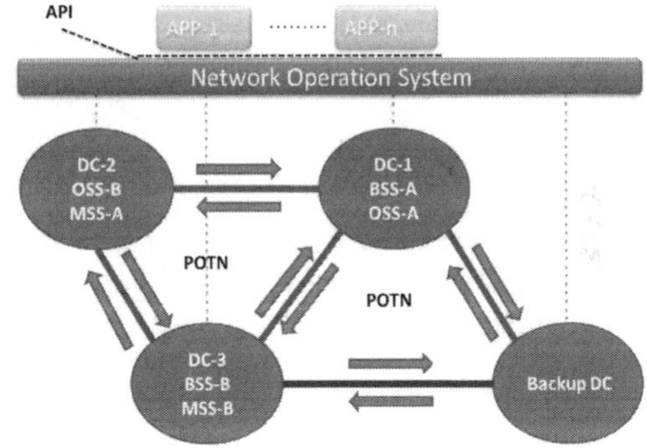

Figure 6. SDN network structure of business supporting network.

It can be seen in Figure 6 that a business supporting network of the SDN structure will have both programmable data plane and controlling plane. Between or inside each working DC, the adding and deleting of virtual equipment will be managed when necessary. By writing an application that can be named as "firewall", the resources between virtual equipment will be thoroughly separated and the data package will be processed. And by writing an application which can be named as "router & switch", various types of virtual equipment will be routed and switched. Furthermore, through the SDN controlling platform deployed in the virtualized resource pool, the administrator will be able to schedule network resources easily and freely, also change the network resources and security strategies flexibly when necessary.

3 PRACTICE OF STRUCTURAL EVOLUTION

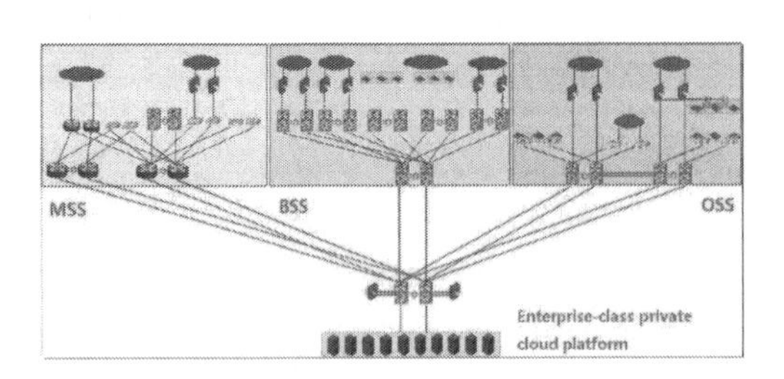

Figure 7. Enterprise-class private cloud platform.

See Figure 7. Through the establishment of enterprise-class private cloud network structure, the BSS, OSS and MSS can acquire relevant services via network in an expandable way when necessary. Meanwhile, resources will be instantly released after using, which significantly increases the utilization rate of IT infrastructure resources and reduces the cost at the same time. Private cloud can be accessed either in the way of business designation or private cloud security domain. The core switch of BSS, OSS and MSS will be accessed with the switch of private cloud platform resource pool respectively, and then resources will be allocated to each supporting network by resource pool in way of virtualization in accordance with needs, and the gateway of resource pool will be set on the virtual firewall of private cloud. Network security domain isolation will be realized by the private cloud platform switch with the application of layer 2 VLAN and different VLAN will be allocated to different supporting networks, in order to meet the demand of current network security domain regulations.

In addition, with the new structure of date sharing access of each global business module at the storage level, essentially, the applications and business flow can be separated from specific host platforms, ensuring that the processing ability of each business module is independent from the physical limit of individual host, so that any single business module of the whole business system can run seamlessly on any point under cluster circumstances. This innovative structure enhances the operating efficiency of the business system, which ensures that the development of business is not affected by the bottleneck of processing ability of specific host platform.

4 CONCLUSION

Taking the IT structure as a whole, network is regarded as the infrastructure carrying the platform communications, such as server and storage etc. Its development is subject to the deployment and development of the platform and business it carries. Therefore, in the formation and creation of industry chain of cloud computing, the development of network virtualization has always been falling behind that of system virtualization and server virtualization. Being the most important core network of operator, the study on its network structure evolution under the circumstances of cloud computing plays a positive role in minimizing equipment cost and reducing the complexity of network operating for the operator. And this study is of great assistance in improving network efficiency and flexibility, which will speed up the promotion of new businesses as a consequence.

REFERENCES

Cisco. 2010. Virtual Port Channel Quick Configuration Guide. http://www.cisco.com/.

Cheng, M., Zhang, J. & Shen, C. 2012. Evolution of OTN Technology in Metropolitan Area Network. Telecommunications Technology8:40–44.

Open networking. 2011. Soft defined network. https://www.opennetworking.org/.

Zhang, B., Feng, M. & Jing, X. 2010. Study on Security Protection System of Enterprise Network based on Security domain, Computer Security:36–38.

Electronics, Communications and Networks IV – Hussain & Ivanovic (eds)
© 2015 Taylor & Francis Group, London, ISBN: 978-1-138-02830-2

An ant-based MP2P network robustness-enhanced method

Fangfang Guo, Shuang Yang*, Guangsheng Feng & Xiaogang Wang
College of Computer Science and Technology, University of Harbin Engineering, Harbin, Heilongjiang, China

ABSTRACT: Ant colony optimization achieves a global optimal effect on the resource search when operating distributed parallel search in MP2P (Mobile Peer to Peer) network. However, network connectivity and resource availability can hardly be guaranteed because of several problems, such as ant scale, network load and node-mobility. To avoid heavy network load and long search delay, the paper puts forward an ant dynamic intelligent survival management mechanism. To ensure the sustainability of resource search of the strong dynamic in the network during resource search, the paper proposed a period T detection mechanism. Aiming at the target resource failure in the process of resource transfer, the paper also puts forward a different target resource QoS (Quality of Service) backup mechanism so that it can improve the recall rate of the target resource.

1 INTRODUCTION

With the development of MP2P network, ensuring its applications and security are more and more important. At present, resource search of unstructured MP2P network based on ant colony optimization has achieved large achievements in terms of efficiency and success rate of resource search (Bocek et al. 2010, Zhang 2006, Veijalainen 2011). However, resource search based on ant colony optimization also brings certain challenges to the MP2P network in security.

Some articles (Hadi et al. 2011, Li 2004) put forward that the classic ant colony optimization is leaded in the resource search of unstructured P2P (Peer to Peer) network. By updating pheromone and calculating re-broadcasting probability, this algorithm presents an intelligence resource search based on breadth-first search. Compared with the traditional breadth-first search algorithm, this algorithm has obvious effects on both controlling the resource search scale and alleviating the network load. However, this algorithm becomes a flooding-based search when it is applied to the MP2P network. As it leads to the surge of network pressure, search ability is significantly decreased. Therefore, performance and availability of the network are severely affected. In (Zhou& Sun2010, Tsai et al. 2010), taking account of message numbers and time delay in existing P2P network resource search algorithm, the algorithm utilizes positive feedback mechanism of pheromone to guide search message packets to target resource area. The algorithm could enhance the success rate of resource search, but the pheromone updating methods are too simple. Therefore, it is easy to fall into local optimization path in the process of distributed search that will increase the network load.

Ant colony optimization insures the diversity of search solution through feedback mechanism and multi-path search. Distributed search enhances the robustness, but it also has some defects: ant scale affects network load and search time; searching may easily trap into local optimization, which leads to premature stagnations. Thus, by analyzing the networking storm, this paper proposes a dynamic and intelligent survive management mechanism in order to achieve lighter network load and shorter search delay. In order to solve the problem of strong dynamic in MP2P network, a period T detection mechanism is proposed to ensure the sustainability of resource search. Last but not least, the paper also puts forward a different target resource QoS backup mechanism for solving the failure problem of target resource to enhance the recall rate of the target resource.

During searching resource in MP2P network, effectiveness and usability of resource are severely affected by a high level of dynamic in topology which results from the node mobility of MP2P network. Both search and response of data packets, increase network load and consume network bandwidth, meanwhile, it also affects searching efficiency and success rate. Thus, it is significant to promote the availability of the MP2P network by searching and recalling resource to enhance the robustness of research method in MP2P network.

1.1 *Networking storm*

Networking storm occurs in the process of resource search. When a request resource node is searching

*Corresponding author: *yangshuang900512@163.com*

and transmitting the ants, the number of search ants will exponentially grow by reason of a series of factors such as dynamic changes, mobile nodes, transmitting delay, re-generating search ants and so on. The situation severely increases the network load and nodes as well as consumes bandwidth sources. What is worse, there are terrible effects on the normal operation of the network, which may lead to the paralysis of the whole net. Therefore, how to prevent the emergence of broadcast storms is very important to the application and maintainability of MP2P network.

1.2 *MP2P network search loop*

MP2P network is a kind of covering layer net. In fact, it is not only finishing the search of request resources, but also transferring resources as the ultimate goal. In the search process, the repeated transmission of search ants occurs in the mobility and high dynamic of network topology. Either the changes of node location or calculation errors will cause the loops in the network. The situation not only reduces the efficiency of resources search, but also consumes the network bandwidth. As a result, it leads to a sharp decrease of network performance. Therefore, it is essential for resources search and network availability to prevent loops during the search process.

1.3 *Failure of intermediate node*

As the relay node of resource search ants, the intermediate node is important for resource search of network and transfer. However, both volatility of a network and efficiency of resource search will be seriously impacted by the node login, logout and abnormal failures in MP2P network. The path of searching cannot be reused during the process of recalling, it affects the real-time of resource delivery.

1.4 *Failure of target resource node*

In MP2P network, the target resource node is a provider for the request node. Due to the ant colony optimization is a distributed parallel search algorithm, the failure of resource search only happens at all of the search ants get back failure messages. If a target resource node fails in the transmission process, invalid redundancy information would occupy network bandwidth and it will increase network load, which also cannot meet the need of users' requesting resources. Therefore, it is very important to deal with the target resource node for users' requirement and MP2P network load.

2 ENHANCED THE ROBUSTNESS OF MP2P NETWORK BASED ON ANT COLONY OPTIMIZATION

2.1 *Treatment of MP2P network storm*

The paper puts forward the intelligent life management mechanism of ant search in MP2P network resource search. It fundamentally reduces the network load, saves the network bandwidth and improves the efficiency of search, even avoids the redundant ants.

Based on probability calculation, search ants carry requested messages for heuristic resource search and ants transmission in the course of resource search. If the network is not requested resource, the stranded ants will be useless. Thus, the paper proposes dynamic survival time ($TTL_{dynamic}$) to intelligently manage the search for ants.

The conception of the frequently, transmission message of the node is put forward, aiming at the dynamic, intelligent management of $TTL_{dynamic}$ that is the count of node transmitted packet (search packet and respond package) in unit time. According to analyze of network statistics, the frequent degree of node transmits packet directly affects the network load and change of performance. Thus, two threshold values of node transmit packet Fd_1 and Fd_2 are come up with the subject (simulation results shows that $Fd_1<Fd_2$) to intelligently manage the ant $TTL_{dynamic}$.

Calculation formula of $TTL_{dynamic}$ as follows :

$$TTL_{dynamic} = \begin{cases} TLL_{dynamic} & Fd<Fd_1 \\ aTTL_{dynamic} & Fd_1<Fd<Fd_2, 0<a<1 \\ 0 & Fd>Fd_2 \end{cases} \quad (1)$$

Figure 1. shows the concrete algorithm as follows :

```
//initialization of ant search TTLdynamic
Initialization(TTLdynamic)=15;
Calculate(Fd);
SelectMIN(FD);
if(FD<FD1)
TTLdynamic=TTLdynamic-1;
else  if(FD1<FD<FD2)
      FD= TTLdynamic-t;
else  if(FD>FD2)
      TTLdynamic=0;
Generate(message,N);
Destruct(SearchANT);
Return;
```

Figure 1. Concrete algorithm.

The intelligent survival management of search ants has many advantages, such as reducing useless data redundancy, ensures the effectiveness of network connectivity, improves utilization rate of bandwidth and whole performance of the network.

2.2 *MP2P network avoids the search loop mechanism*

Due to the mobility of nodes and the high dynamic changes of whole network topology, transmitting search ants in the MP2P network may account many repeated relay nodes (caused by changes of node location and calculating probability). Thus, search ant may generates loops, even it becomes circle phenomenon in the network(Huang et al. 2010).

In order to resolve node mobility and loops, the article adds Forward NodeID. When search ant is transmitted by net node, it records ID of transmitting node that it goes through, and these data form transmitting a node sequence. Every time search ant chooses nodes, it retrieves the transmitting node sequence by itself. If the next hop transmitting node has existed in the sequence, search ant will calculate transmit rate again. Then it chooses the transmitting node, and performs a heuristic resource search. Figure 2 shows the specific node as follows:

```
//generate search ant
Generate(SearchANT);
Initialization(ForwardNodeID)=NULL;
   if(NodeID ∉ Forward NodeID)
   Forward(SearchANT);
   Add Forward NodeID(NodeID);
 else
   Calculate(NodeIDnext);
 Forward(SearchANT);
 Add Forward NodeID(NodeIDnext);
 Return;
```

Figure 2. The specific node.

2.3 *MP2P network node mobility management mechanism*

The highly dynamic of the MP2P network mainly expresses in terms of node mobility. Normal login, logout and abnormal interrupt may occur at any time during a resource search and transmission. This article uses segmented thought to deal with login, logout, move and breakdown of intermediate nodes, which occurs during a search and recall resource.

First of all, the intermediate node announces its behavior to surrounding neighbors, when it normally login or logout MP2P network. During resource search, if the leaving node chooses by the search ant, according to the calculating probability, then the ant will have to recalculate the probability. If the reused path node quits network in the process of resource transmitting, search ant will find its neighbor node according to Route Information of response ant, then skip the breakpoint and go on transmitting.

For abnormal failure of node: if the chosen next-hop node is searching resources and happens to break down, then executes the routing algorithm again and selects a new next-hop node. If the resource transmitting is performed, then firstly sets up detecting time t and detecting number n. If it receives no response to detecting achieve number n, it considers the node as failures and then moving on to detect neighbors of this node until the link is established successfully.

2.4 *Failure to process mechanism of the target resource node*

As the diversity development of network application, users have different levels of QoS requirements to different services. The target resource QoS is severely affected by the frequently occurring of transmission delay, node movement and node jitter. The traditional ant-based MP2P network resource search algorithm can be simply divided into minimum resource search and maximize resource search when searching for the target resource nodes.

Minimum resource search will be finished when request ants find the first resource, thus to save search time. However, the obtained resource may not necessarily satisfy request ants, and the target node may fail at any time.

Maximum resource search requires search ants to traverse active nodes as many as possible, providing a lot of target resource for request nodes which ensure diversity of search. Thought the global optimal solution will be found, the search process needs longer time delay.

Request resource nodes also expect to get many target resources. It is not only increases the selections of resource, but also ensures backup nodes to cope with the failure of target resource nodes. It ensures the resource transmission can be continued. Therefore, the subject tries best to find more target resources, which is applied for different QoS at certain constraint condition. It guarantees the robustness of resource search. QoS constraint condition as follows:

$$Q=\alpha*n+\beta*t \quad 0<\alpha<1, 0<\beta<1, \alpha+\beta=0 \quad (2)$$

n is visited number of resource Q, indicating its heat degree of resources; t is time from recent visit to the present time, showing a freshness degree of resource Q. The subject orders resource Q, and recalls required data according to request nodes' indeed request (QoS requires high-demanding or real-time) and data distribution.

3 EXPERIMENT SIMULATION AND ANALYSIS

3.1 *Experimental parameter settings*

Normal application of MP2P network is guaranteed the robustness of MP2P network. The paper will evaluate performance on ant-based resource search (referred to as A) and robustness-enhanced mechanism (referred to as B) from search delay and utilization of network resources.

3.1.1 *Search delay*

Search delay is a time interval, which from the request node sending out search ant packets to receive response successfully. The average search delay is the summation of successful research delay time divided by the summation of delay time. In the same network environment, the lower search delay means higher success rate of resource search and higher effective utilization of network.

3.1.2 *Network utilization*

Network utilization is a key parameters which measures bandwidth occupation and network congestion. If utilization is too high, it indicates the network is overloaded. On the contrary, if utilization is low, it shows the network is spare. It stands for the degree of the busy network. Try best to improve the effective utilization of network on the premise of network congestion not occur.

3.2 *Experimental environment settings*

This paper uses OverSim as simulation platform. First of all, 100 mobile node is set to make a random waypoint model motion within the range of 1000m×1000m, the movement speed random changes between 0-20m/s. The covering diameter of wireless node is 250 meters. Single node stored 100 file, and every file has 5 to 10 keywords. Each query carries 3 keywords at most, and the network bandwidth is 2Mb/s. During the simulation, 30s suspend time is needed. Specific simulation time is set depending on different needs.

In the beginning of the experiment, we initialize the information about the network and parameters. The paper comparatively analyze different resource search method by accumulating experiment data and visual results interface diagram. Figure 3 shows MP2P network random movement topology, as follows:

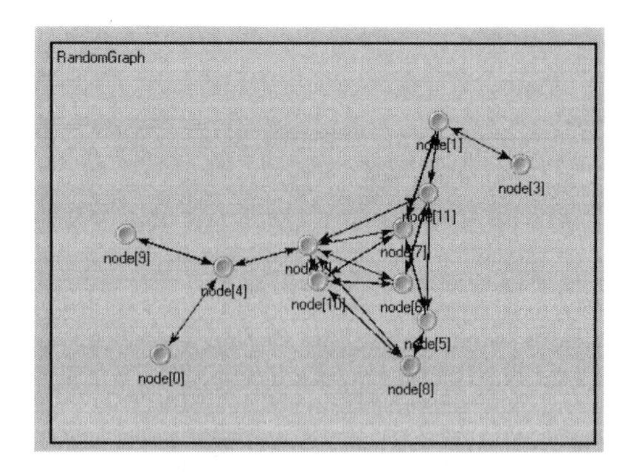

Figure 3. OverSim running diagram.

3.3 *Simulation results analysis*

3.3.1 *Simulation results about search delay*

MP2P network resource search delay directly responses the effect of search ants dynamic, intelligent survive management mechanism to reduce network load. The paper from node movement speed, and node scale to simulation, the results show in Figure 4 and Figure 5.

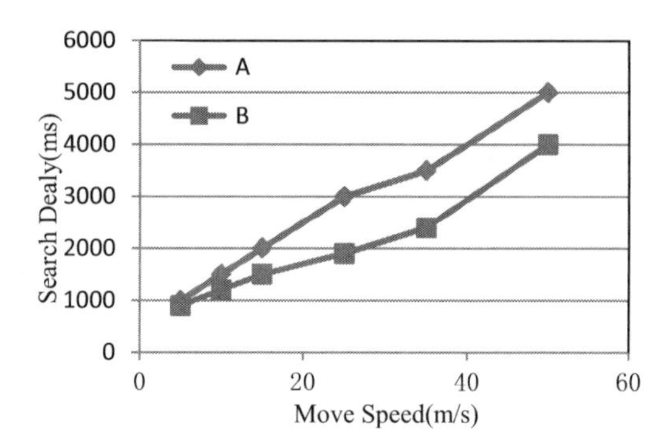

Figure 4. Search delay changes along with node movement speed.

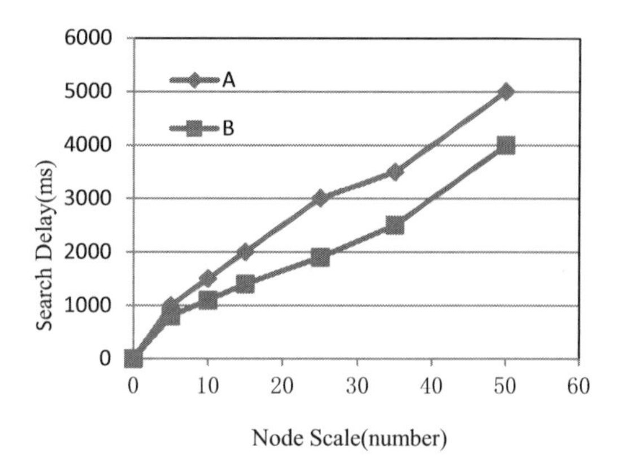

Figure 5. Search delay changes along with node scale.

3.3.2 *Simulation results about Network utilization*

Network utilization is a very important reference index in network health examination and network maintenance. By testing MP2P network utilization, we can realize the operation status of MP2P network. From the simulation results, we can see that after using robustness-enhanced mechanism, MP2P network utilization improves without increasing network load, as it shows in Figure 6:

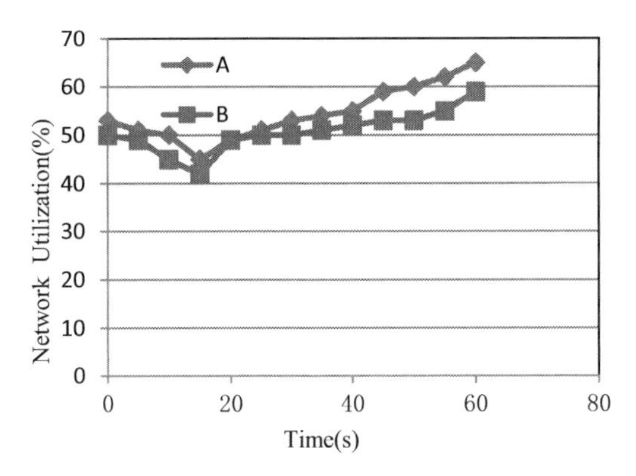

Figure 6. Network utilization.

4 CONCLUSION

Due to the strong network topology dynamic and limited node capacity, which restricts application diversification of MP2P network(Huang et al. 2010). The paper proposes an ant dynamic, intelligent survival time management mechanism, a period T detecting mechanism and a QoS target resource backup mechanism to secure the persistence and robustness of the MP2P network resource search and transmission in the aspects of networking storm, node mobility and node failure. The simulation experiment shows that the robustness-enhanced method has stronger applicability and fault tolerance than others, and it also satisfies the demand of the highly unstructured dynamic of MP2P network. The future work will be explored high match of resource search, physical link and logical link. It is very important to decrease the network load and improve the robustness of MP2P network .

ACKNOWLEDGMENT

The paper is partially supported by the National Science Foundation of China (61370212), the Research Fund for the Doctoral Program of Higher Education of China (20122304130002), the National Natural Science Foundation of China (61402127), the Fundamental Research Fund for the Central Universities (HEUCFZ1213) and the Fundamental Research Fund for the Central Universities (HEUCF100601).

REFERENCES

Bocek, T. Hunt, E. & Stiller, B. 2009. Mobile P2P Fast Similarity Search. *CCNC IEEE Consumer Communications and Networking Conference. Las Vegas, USA. 10–13 January 2009*. Piscataway: IEEE.

Hadi, G., Salavati, A.H. & Pakravan M.R. 2011. An ant-based rate allocation algorithm for media streaming in peer to peer networks: Extension to multiple sessions and dynamic networks. *Journal of Network and Computer Applications* 34(1):327–340.

Huang, Y.R., Zhong, C. & Li, Z. 2010. Location aided proactive search mechanism for mobile Peer-to-Peer network. *Computer Engineering* 34(19):127–129.

Li, S.Y. 2004. *Ant colony algorithms with application.* China: Harbin Institute of Technology.

Tsai, K.Y., Hsu, C.L. & Wu, T.C. 2010. Mutual anonymity protocol with integrity protection for mobile peer-to-peer networks. *International Journal of Security and Networks* 5(1): 45–52.

Veijalainen J. 2011. Autonomy, heterogeneity and trust in mobile P2P environments. *International Journal of Security and Its Applications* 1(1):57–72.

Zhang, X.M. 2006. *Research of data replication in hierarchical P2P Systems.*Hunan: Graduate School of National University of Defense Technology.

Zhou, L. & Sun, L. 2010. A P2P resource search model based on Ant Colony Optimization. *Computer and Automation Engineering (ICCAE), Singapore, 26–28 February, 2010*. Piscataway: IEEE.

Electronics, Communications and Networks IV – Hussain & Ivanovic (eds)
© 2015 Taylor & Francis Group, London, ISBN: 978-1-138-02830-2

Shape classification based on multiscale bispectral invariants

Hengguang Guo* & Jun Qv
Department of Airborne Vehicle Engineering, Naval Aeronautical Engineering Institute, Yantai, Shandong, China

ABSTRACT: Shape is a basic and an important visual feature used for image classification. There are many shape descriptors, and multiscale description is a promising method for shape classification. When using multiscale description method, features obtained at different scales, can be combined to increase the classification power, therefore increase the correct classification rate. In this paper, a novel multiscale bispectral shape invariants are presented. First, the image is filtered by a low-pass Gaussian filter or a high-pass Gaussian filter to obtain multiscale images. Then, Radon transform of different directions is used to map the images to a set of one dimensional projections. At last, bispectral analysis is used to obtain multiscale bispectral shape invariants. The result shows that the multiscale bispectral invariants using HPGF perform better than multiscale bispectral invariants using LPGF, elliptic Fourier descriptors, Zernike moments, and wavelet-based multiscale contour Fourier descriptors.

1 INTRODUCTION

Object classification and recognition is an important goal in image analysis. Object can be classified by several ways, such as color, texture and shape. Shape, one basic feature, is an important visual feature in image classification and recognition. In shape classification, many shape features have been proposed in the past, and there are two main shape description methods: contour-based methods and region-based methods. In contour-based methods, the shape boundary points are used to obtain shape features. The most famous contour-based shape descriptors are Fourier descriptors (Zahn et al. 1972, Persoon et al. 1977), wavelet descriptors (Chang et al. 1996), wavelet-Fourier descriptors (Kunttu et al. 2006) and curvature scale space (Mokhtarian et al. 1992). In region-based methods, the shape features are obtained by using all the shape interior points. Popular region-based methods include moments (Prokopet al. 1992) and generic Fourier descriptors (GFD) (Zhang et al. 2002).

High order spectra (HOS) are spectral representations of high order moments of a random process. In (Chandran et al. 1993), a set of translation, scale change and amplification invariant features defined from high order spectra was proposed, and they were used for one dimensional pattern recognition. It was extended for application to two dimensional images and recognition of object in (Chandran et al. 1997). The two-dimensional version has been used for classification of viruses in electron microscope images (Ong et al. 2005), classification of sea mines in sonar images (Chandran et al. 2002), detection of eye abnormalities from digital fundus images (Tan et al. 2008), and breast cancer detection from thermal images (Tavakol et al. 2013).

2 MULTISCALE HIGH ORDER SPECTRA INVARIANTS

2.1 *LPGF and HPGF*

In this paper, the multiscale shape features are obtained by using the low-pass Gaussian filter (LPGF) and the high-pass Gaussian filter (HPGF) separately. When using LPGF, the object is smoothed and as scale (standard deviation of the filter) decreases, the contour and exterior regions are lost. On the other hand, when using HPGF the object contour can be detected and as standard deviation of the filter decreases the exterior regions are represented. Therefore using the LPGF of different scales, instead of object contour, the inner and central part are emphasized. On the other hand, when using the HPGF of different scales, the object contour and exterior parts are focused more than the central part.

2.2 *The Radon transform*

Let $f(x,y)\in \mathbf{R}^2$ be a two-dimensional function, $L(\theta,\rho)$ be a straight line in $\mathbf{R}^2$ represented by:

$$L=\{(x,y)\in \mathbf{R}^2 : x\cos\theta + y\sin\theta = \rho\} \qquad (1)$$

where θ is the angle L makes with the y axis and ρ is the distance from the origin to L.

*Corresponding author: *guohengguang@126.com*

For an image $f(x,y)$, the Radon transform for a particular angle θ is the line integral (projection) of the image intensity along the specified direction. The radon function computes the projection of image matrix along parallel paths oriented perpendicular to the rotation angle θ. Radon transform for an image $f(x,y)$ at an angle θ is defined as:

$$R(\rho,\theta)=\int_{-\infty}^{+\infty}\int_{-\infty}^{+\infty} f(x,y)\delta(\rho - x\cos\theta - y\sin\theta)\,dx\,dy \quad (2)$$

where (x,y) are the spatial coordinates and ρ is the perpendicular distance from origin to the line oriented at angle θ.

Some useful properties on translation, rotation, and scaling of the Radon transform are described as follows:

Translation: when the f is translated by a vector $\vec{u}=(x_0,y_0)$, a shift of its transform in the variable ρ by a distance $d=x_0\cos\theta+y_0\sin\theta$ is produced, which equals to the projection of $\vec{u}$ on the line $x\cos\theta+y\sin\theta=\rho$.

Rotation: when the image is rotated by an angle θ_0, the transform is shifted by θ_0 in the variable θ.

Scaling: when the f is scaled by a factor α, the ρ coordinate of the transform is scaled by a factor α and the amplitude of the transform is scaled by a factor $1/\alpha$.

2.3 *Bispectrum and bispectral invariants*

The bispectrum of a random signal $x[k]$ can be defined in the frequency domain as

$$B(f_1,f_2)=E\left[X(f_1)X(f_2)X^*(f_1+f_2)\right] \quad (3)$$

where $X(f)$ is the Fourier transform of $x[k]$, f is normalized frequency (devided by one half of the sampling frequency) that is between 0 and 1, $*$denotes the complex conjugate and $E[\;]$ denotes the statistical expectation operation over an ensemble of possible realization of the signal. For a deterministic signal there is no need for expectation.

$$B(f_1,f_2)=X(f_1)X(f_2)X^*(f_1+f_2) \quad (4)$$

Bispectrum is often calculated by a direct or indirect estimation method. The direct method starts with the one-dimensional discrete Fourier transform of $x[k]$, and then applying the triple product by Equation (4). In order to extract the bispectra invariants, the FFT technology is used when applying the direct method.

The bispectral invariant feature is the phase of the integrated bispectrum along a straight line of given slope,a,in bi-frequency space . In (Chandran et al. 1993), the bispectral invariant feature $\phi(a)$ is defined by

$$I(a)=\int_{f_1=0^+}^{\frac{1}{1+a}} B(f_1,af_1)\,df_1 = I_{\text{Re}}(a)+jI_{\text{Im}}(a) \quad (5)$$

$$\phi(a)=\arctan\left(\frac{I_{\text{Im}}(a)}{I_{\text{Re}}(a)}\right) \quad (6)$$

2.4 *Multiscale bispectral invariants extraction*

Multiscale spectral invariants are extracted from the image and used for classification according to the following procedure

1 The size of an object in an image is normalized by bilinear interpolation. The object size, which is the sum of intensities over the image, is normalized to be 2500. The image size is 151×151.
2 The normalized images are filtered by LPGF and HPGF separately, to generate scale space. When using LPGF for multiscale description, the selected scales are: $\sigma_1=25$, $\sigma_2=20$, $\sigma_3=15$, $\sigma_4=10$, $\sigma_5=5$. On the other hand, when using LPGF, the selected scales are: $\sigma_1=15$, $\sigma_2=10$, $\sigma_3=5$, $\sigma_4=3$, $\sigma_5=1$.
3 The filtered images are mapped to a set of one dimensional projections using Radon transform. Radon projections were computed at 0, 45, 90, and 135. Each projection is zeros padded to length 256 before taking the FFT.
4 Bispectral invariant features are computed from each projection at four equal slopes in the bi-frequency plane given by the tangent values of 1/4, 1/2, 3/4 and 1. The bispectrum need not to be computed over the entire bi-frequency space, only along the line of integration. Bilinear interpolation is used to map bispectral values from the grid of frequencies to points along the line.

3 SHAPE CLASSIFICATION

The nearest mean algorithm (Glowacz et al. 2012, 2014) is used for shape classification, and the similarity between objects is measured by Euclidean distance (*Ed*). Euclidean distance (*Ed*) at each scale is calculated by the following equation:

$$Ed^s(\mathbf{T},\mathbf{D})=\sqrt{\sum_{x=1}^{C}\sum_{y=1}^{E}\left(\mathbf{OD}_{\mathbf{T}}^{\mathbf{s}}(x,y)-\mathbf{OD}_{\mathbf{D}}^{\mathbf{s}}(x,y)\right)^2} \quad (7)$$

where $Ed^s(\mathbf{T},\mathbf{D})$ is the Euclidean distance between the object features, $\mathbf{OD}_{\mathbf{T}}^{\mathbf{s}}$, of the test image $\mathbf{T}$ and object features $\mathbf{OD}_{\mathbf{D}}^{\mathbf{s}}$ of an image from database $\mathbf{D}$, at scale index s. Then average distance (*Ad*) is computed as:

$$Ad=\frac{1}{Y}\sum_{s=1}^{Y} Ed^s \quad (8)$$

where Ad is the average distance and Y is the number of scales. When using the average distance for classification, the result performs better than single scale distance, and correct classification can be increased.

The validation of shape classification is made using leave-one-out method. When using this method, each image is left out in turn and used as test image. The correct classification rate (CCR%) is computed as follows:

$$CCR(\%) = \frac{c_0}{t_0} \times 100 \tag{9}$$

where c_0 is the total number objects that are correctly classified and t_0 is the total number objects that are classified.

4 EXPERIMENTS AND RESULTS

For shape classification experiments, MPEG-7 CE-Shape-1 Part B database (Latecki et al. 2000) is used, which consists of shapes acquired from real world objects. There are pre-segmented and binary form1400 images in the database,. The objects are classified into 70 classes, and with 20 images in each class.

In experiments, first, results of representation using HPGF and LPGF of single scales (filtering at different scales) and average distance (with method given in section 3) are investigated and compared. Original result, which represents the classification result without any filtering operation, is also compared with single scale and average distance results. Second LPGF-based and HPGF-based multiscale bispectral features (average performance) are compared with other shape description methods.

4.1 *Original, single scale and average distance results*

When using LPGF for multiscale representation, five differ-ent scales are selected. The selected scales are: σ1=25, σ2=20, σ3=15, σ4=10, σ5=5. On the other hand, when using HPGF for multiscale representation, the selected scales are: σ1=15, σ2=10, σ3=5, σ4=3, σ5=1.

Table 1 shows the CCR% of the original, selected single scales using LPGF and average distance of selected scales. When the database is applied without any LPGF, the highest CCR% can be observed, which is 92.2%. The CCR% decreases with the scales decreases, and that means the object become smoother. The result of average distance is 90.1%, and the orginal result and some single scale results are higher than this. Therefore it is not effective when using LPGF, when the database is without noise.

Table 1. CCR% of the original, single scale and average distance using LPGF and HPGF.

	LPGF	HPGF
Original	92.2%	92.7%
σ_1	91.9%	92.3%
σ_2	91.6%	93.1%
σ_3	90.2%	94.5%
σ_4	89.7%	96.0%
σ_5	88.4%	94.1%
Average distance	90.1%	96.7%

4.2 *Comparison with other techniques*

In this section, the performance of LPGF-based multiscale bispectral invariants and HPGF-based multiscale bispectral invariants is compared with each other, and with other methods, such as elliptic Fourier descriptors (EFD) (Kuhl et al. 1982), Zernike moments (ZM) (Kim et al. 2000) and wavelet transform-based multiscale contour Fourier descriptors (Kunttu et al. 2003).

When using elliptic Fourier descriptors (EFD) for shape description, the boundary is represented by complex coordinates, and the EFD is obtained by using the Fourier expansion. Zernike moments (ZM), which makes full use of boundary information, can obtain higher classification result. By applying the Fourier transform to the coefficients of the multiscale complex wavelet transform, the wavelet transform-based multiscale contour Fourier descriptors is obtained.

Table 2 shows the correct classification rate (CCR%) of the HPGF-based multiscale bispectral invariants, LPGF-based multiscale bispectral invariants, EFD, Zernike moments, and wavelet-based multiscale contour Fourier descriptors. It is observed that multiscale bispectral invariants using HPGF perform better than multiscale bispectral invariants using

Table 2. CCR% of HPGF-based and LPGF-based multiscale bispectral invariants, wavelet-based multiscale Fourier descriptors, EFD, and ZM.

Descriptors	CCR%
Multiscale bispectral invariants using HPGF	96.7%
Multiscale bispectral invariants using LPGF	90.1%
Elliptic Fourier descriptors (EFD)	82.0%
Zernike Moments (ZM)	90.2%
Wavelet-based multiscale Fourier descriptors	94.1%

LPGF, elliptic Fourier descriptors, Zernike moments, and wavelet transform-based multiscale contour Fourier descriptors. The CCR% of HPGF-based multiscale bispectral invariants is 96.7%, whereas the CCR% of LPGF-based multiscale bipsectral invariants is 90.1%, wavelet-based multiscale descriptors is 94.1%, EFD is 82% and ZM is 90.2%.

5 CONCLUSION

Multiscale technology is a very efficient method for shape classification. When using multiscale description method, features obtained at different scales, can be combined to increase the classification power, therefore increase the correct classification rate.. In this paper, a novel multiscale bispectral invariants is presented, applying a low-pass Gaussian filter (LPGF) and a high-pass Gaussian filter (HPGF), separately.

First, the image object size is normalized, and the normalized image is filtered by a low-pass Gaussian filter or a high-pass Gaussian filter to obtain multiscale images. Then the filtered images are mapped to a set of one dimensional projections using Radon transform of different direction. At last, bispectral invariant features are computed from each projection at different slopes in the bi-frequency plane. For shape classification, the Euclidean distance at each scale is computed, and the the average distance of each object is calculated. HPGF-based multiscale bispectral invariants, which focus the contour and exterior parts of an object more than the central part, perform better than LPGF-based multiscale bispectral invariants, and three other shape description methods compared in this paper. Therefore, Classifying objects with this new HPGF-based multiscale bispectral invariants can increases immunity to noise and classification power.

REFERENCES

Chandran, V. & Elgar, S. 1993. Pattern recognition using invariants defined from higher order spectra–one dimensional inputs. *IEEE Transactions on Signal Processing* 41: 205–212.

Chandran, V. & Elgar, S. 2002. Detection of mines in acoustic images using higher order spectral features. *IEEE Journal of Oceanic Engineering* 27: 610–618.

Chandran, V., Carswell, B., Boashash, B. & Elgar, S. 1997. Pattern recognition using invariants defined from higher order spectra-2-D image inputs. *IEEE Transactions on Image Processing* 6: 703–712.

Chang, G.C.H. & Kuo, C.C.J. 1996. Wavelet descriptor of planer curves: theory and application. *IEEE Transactions on Image Processing* 5: 56–70.

Glowacz, A. & Korohoda, P. 2014. Recognition of monochrome thermal images of synchronous motor with the application of binarization and nearest mean classifier. *Archives of Metallurgy and Materials* 59(1): 31–34.

Glowacz, A. Glowacz, A. Glowacz, Z. 2012. Diagnostics of direct current generator based on analysis of monochrome infrared images with the application of cross-sectional image and nearest neighbor classifier with Euclidean distance. *Przeglad Elektrotechniczny* 88(6): 154–157.

Kim, W.Y. & Kim, Y.S. 2000. A region-based shape descriptor using Zerike Moments. *Signal Processing: Image Communication* 16: 95–102.

Kuhl, F.P. & Giardina, C.R. 1982. Elliptic Fourier features of a closed contour. *Computer Graphics and Image processing* 18: 236–258.

Kunttu, I. Lepisto, L. Rauhamaa, J. 2006. Multiscale Fourier descriptors for defect image retrieval. *Pattern Recognition Letters* 27(2): 123–132.

Kunttu, I., Lepisto, L., Rauhamma, J. & Multiscale 2003. Fourier descriptor for shape classification. *IEEE International Conference on Image Analysis and Processing, 17–19 September 2003*: 536–541. IEEE.

Latecki, L.J., Lakamper, R. & Eckhardt, U. 2000. Shape descriptors for non-grid shapes with a single closed contour. *IEEE International Conference on Computer Vision and Pattern Recognition, Hilton Head Island, SC, 13–15 June 2000*: 424–429. IEEE.

Mokhtarian, F. & Mackworth, A.K. 1992. A theory of multiscale, curvature-based shape representation for planar curves. *IEEE Transactions on Pattern Analysis and Machine Intelligence* 14(8): 789–805. Ong, H. & Chandran, V. 2005. Identification of gastroenteric viruses by electron microscopy using higher order spectral features. *Journal of Clinical Virology* 34: 195–206.

Persoon, E. & Fu, K. 1977. Shape discrimination using Fourier descriptors. *IEEE Transactions on System, Man, and Cybernetics* 7: 170–179.

Prokop, R.J. & Reeves, A.P. 1992. A survey of moment-based techniques for unoccluded object representation and recognition. *CVGIP: Graphical Models and Image Processing* 54(5): 438–460.

Tan, T.G., Acharya, U.R. & Ng, E.Y.K. 2008. Computer-based identification of anterior segment eye abnormalities using higher order spectra. *Journal of Mechanics in Medicine and Biology* 8: 121–136.

Tavakol, M.E., Chandran, V., Ng, E.Y.K. & Kafieh, R. 2013. Breast cancer detection from thermal images using bispectral invariant features. *International Journal of Thermal Sciences* 69: 21–36.

Zahn, C.T. & Roskies, R.Z. 1972. Fourier descriptors for plane close curves. *IEEE Transactions on Computers C* 21: 269–281.

Zhang, D.S. & Lu, G. 2002. Shape-based image retrieval using generic Fourier descriptor. *Signal Processing: Image Communication* 17:825–848.

Electronics, Communications and Networks IV – Hussain & Ivanovic (eds)
© 2015 Taylor & Francis Group, London, ISBN: 978-1-138-02830-2

Exploring the limited feedback schemes for 3D MIMO

Zheng Hu* & Rongke Liu
School of Electronic and Information Engineering, Beihang University, Beijing, China

Shaoli Kang & Xin Su
State Key Laboratory of Wireless Mobile Communications, China Academy of Telecommunications Technology (CATT), Beijing, China

Hongtian Li
Beijng Institute of Mechanical and Electrical Engineering, Beijing, China

ABSTRACT: The three dimensional (3D) MIMO is a key technology to improve system performance. In this paper, based on the 3D channel model developed by 3GPP, three limited feedback schemes are proposed. In vertical domain, scheme I makes a sectorization based on the users' Zenith of Departure angle (ZoD) to construct the precoding vector. Each user's ZoD is selected to construct the precoding vector for the vertical domain in scheme II. Scheme III calculates the optimal angle for the precoding vector through the uplink pilot training. System-level simulation is conducted to evaluate the performance of the 3D MIMO under 3D-UMa and 3D-UMi scenarios. Simulation results reveal that the three schemes have a better performance than the traditional 2D MIMO scheme. Also scheme III performs the best, followed by scheme II and scheme I.

1 INTRODUCTION

The rapidly increasing demand for wireless data services, such as smartphones, mobile TV and tablets, relative to the finite frequency bandwidths in cellular network forms a new challenge for mobile communication. Multiple input multiple output (MIMO) is a key technology in Long Term Evolution (LTE) and LTE-Advanced.

Recently, massive MIMO as a key technology, has become a promising and sophisticated technology for the 5th generation (5G) of mobile communication (Marzetta 2010). Restricted to the actual physical space, the number of antennas aided the BS can not go to infinity. The uniform planar array (UPA) draws a lot of attentions. With the use of active antenna systems (AAS), such two dimensional (2D) antenna array can provide more spatial degrees of freedom (DoFs) in both elevation and azimuth domains relative to the uniform linear array (ULA), and the BS can form beams in 3D space adaptively (Lu et al. 2011). In the traditional 2D MIMO system, the users can be just served simultaneously in different horizontal directions. However users located in the same azimuth angel can not be served at the same time because all beams in vertical domain have the same downtilt (Xie et al. 2013). 3D MIMO technology which can distinguish users in vertical domain attracts more attentions. Vertical sectorization can provide the higher gain of the sector patterns in vertical domain. User-specific elevation beamforming can steer transmitted energy in elevation and spray less interference to adjacent cells (Song et al. 2013). In Li et al. 2013, a 3D beamforming technique is proposed which can improve the throughput of the cell edge users and the cell center users compared to the 2D beamforming algorithm.

Actually, signals from the array naturally experience 3D channel. However, most channel models are just restricted to 2D propagation in the horizontal plane and the impact of elevation angle is not considered. The assumption of 2D propagation collapses when in some propagation environments the elevation angle distribution is significant (Shafi et al. 2006). In order to be compatible with the actual radio wave propagation, a large effort in channel model has to be made accounting for the impact of the channel component in the elevation direction (Kammoun et al. 2014). In order to evaluate 3D MIMO, 3D channel model must be taken into account. 3GPP has developed a 3D channel model. In the stage of 3GPP Release 12, 3D urban micro cell (3D-UMi) and

*Corresponding author: *huzheng2008168@sina.com*

3D urban macro cell (3D-UMa) scenarios have been defined. The 3D channel model is more compatible for the practical wireless communications (Zheng et al. 2014).

The channel state information (CSI) can help the BS to eliminate the multi-user interference and enhance the MIMO system performance. The CSI can be obtained through channel reciprocity in the time division duplexing (TDD) operation. In frequency division duplex (FDD) operation, the user can select the appropriate precoding matrix from a finite set of precoders, named codebook, shared by the BS and the users (Inoue & Heath 2009). And then based on the limited feedback strategy, the user provides the codebook index to the BS.

In actual deployment, there are many cellular networks supporting FDD. To retain a backward compatibility, in this paper, based on the characteristics of 3D MIMO, three limited feedback schemes are proposed and evaluated. In the horizontal domain, the existing codebook of LTE-Advanced is still utilized. We concentrate on the construction of precoding vector in vertical domain under the 3D-UMa and 3D-UMi scenarios. The 3D channel model developed by 3GPP is introduced for the system performance evaluation.

This paper is organized as follows. In section II, the system model is presented. The limited feedback schemes are described in section III. In Section IV downlink transmission of MU-MIMO is described. The system-level simulation results and analysis are given in section V. At last, section VI concludes this paper.

Notation: We use upper bold letters to denote matrix and lower bold letters to denote vectors. CN (***m***, ***R***) denotes the circular symmetric complex Gaussian distribution with mean ***m*** and covariance matrix ***R***. The superscripts T, H stand for the transpose and conjugate transpose. $\otimes$ denotes the Kronecker product.

2 SYSTEM MODEL

2.1 *3D antenna model*

The 3D antenna model proposed by 3GPP is introduced here (3GPP 2014, TR 36.873). The horizontal and vertical radiation patterns are listed below:

$$A_{E,H}(\varphi) = -\min[12(\frac{\varphi}{\varphi_{3dB}})^2, A_m], \quad \varphi_{3dB} = 65^\circ, A_m = 30dB \tag{1}$$

$$A_{E,V}(\theta) = -\min[12(\frac{\theta - 90}{\theta_{3dB}})^2, SLA_v], \quad \theta_{3dB} = 65^\circ, SLA_v = 30dB \tag{2}$$

The 3D radiation pattern is

$$A_E(\theta,\varphi) = -\min\{-[A_{E,V}(\theta) + A_{E,H}(\varphi)], A_m\} \tag{3}$$

where φ is the angle between the boresight of the horizontal antenna and the direction of interest and θ is the angle between the vertical direction and the direction of interest. A_m and SLA_V denote the front-to-back attenuation and sidelobe attenuation respectively. The horizontal and vertical half-power beamwidth (HPBW) are φ_{3dB} and θ_{3dB}, respectively.

2.2 *3D channel model*

Relative to the traditional 2D channel model, the 3D channel model considers the radio propagation in the vertical dimension. Here the 3D channel model developed by the 3GPP is recommended (3GPP 2014, TR 36.873).

The process of channel generation includes the scenario selection, determination of user parameters and generation of channel coefficient. The channel coefficients include large scale parameter and small scale parameter. The large scale parameter such as pathloss and shadow fading can be referred to 3GPP 2014, TR 36.873. The small scale channel coefficient is generated by summing the contribution of some rays. Assume that one ray is composed of M sub-paths. The coordinate system for 3D channel model can be seen from Figure 1. $ray_{m,n}$ means the m-th sub-path in the n-th ray. The global coordinate system defines the zenith angle $\theta_{ZoD/ZoA}$ and the azimuth angle $\varphi_{AoD/AoA}$. $\theta_{ZoD/ZoA}$ =90° points to the horizontal direction and $\theta_{ZoD/ZoA}$ =0° points to the zenith direction. $\tilde{\boldsymbol{n}}$ is the given direction of $ray_{m,n}$. $\hat{\boldsymbol{e}}$ and $\hat{\boldsymbol{u}}$ are the spherical basis vectors.

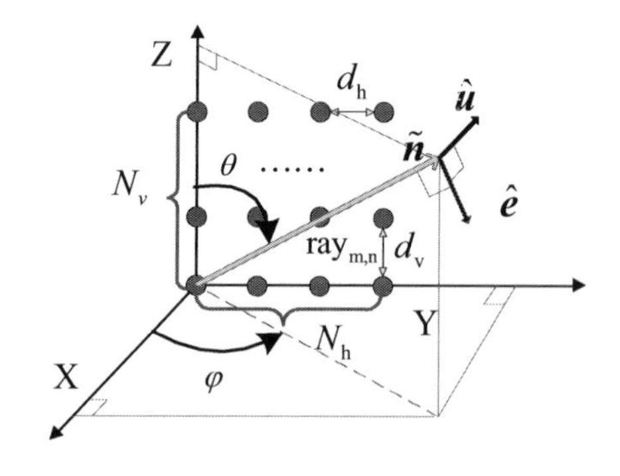

Figure 1. The coordinate system for 3D MIMO and antenna configuration in the BS.

The channel coefficient from transmitter element s to receiver element u for the n-th ray is modeled as: for non-line-of-sight (NLOS) path,

$$H_{u,s,n}(t)=\sqrt{P_n/M}\sum_{m=1}^{M}\begin{bmatrix} F_{rx,u,\theta}\left(\theta_{n,m,ZOA},\varphi_{n,m,AOA}\right) \\ F_{rx,u,\varphi}\left(\theta_{n,m,ZOA},\varphi_{n,m,AOA}\right)\end{bmatrix}^T$$

$$\begin{bmatrix} \exp\left(j\Phi_{n,m}^{\theta\theta}\right) & \sqrt{\kappa_{n,m}^{-1}}\exp\left(j\Phi_{n,m}^{\theta\varphi}\right) \\ \sqrt{\kappa_{n,m}^{-1}}\exp\left(j\Phi_{n,m}^{\varphi\theta}\right) & \exp\left(j\Phi_{n,m}^{\varphi\varphi}\right)\end{bmatrix} \quad (4)$$

$$\begin{bmatrix} F_{tx,s,\theta}\left(\theta_{n,m,ZOD},\varphi_{n,m,AOD}\right) \\ F_{tx,s,\varphi}\left(\theta_{n,m,ZOD},\varphi_{n,m,AOD}\right)\end{bmatrix}\exp\left(j2\pi\lambda_0^{-1}\left(\hat{\boldsymbol{h}}_{rx,n,m}^T.\boldsymbol{d}_{rx,u}\right)\right)$$

$$\exp\left(j2\pi\lambda_0^{-1}\left(\hat{\boldsymbol{h}}_{tx,n,m}^T.\boldsymbol{d}_{tx,s}\right)\right)\exp\left(j2\pi v_{n,m}t\right)$$

for LOS path,

$$H_{u,s,n}(t)=\sqrt{\frac{1}{K_R+1}}H_{u,s,n}^{'}(t)$$

$$+\delta(n-1)\sqrt{\frac{K_R}{K_R+1}}\begin{bmatrix} F_{rx,u,\theta}\left(\theta_{LOS,ZOA},\varphi_{LOS,AOA}\right) \\ F_{rx,u,\varphi}\left(\theta_{LOS,ZOA},\varphi_{LOS,AOA}\right)\end{bmatrix}^T$$

$$\begin{bmatrix} \exp\left(j\Phi_{LOS}\right) & 0 \\ 0 & -\exp\left(j\Phi_{LOS}\right)\end{bmatrix} \quad (5)$$

$$\begin{bmatrix} F_{tx,s,\theta}\left(\theta_{LOS,ZOD},\varphi_{LOS,AOD}\right) \\ F_{tx,s,\varphi}\left(\theta_{LOS,ZOD},\varphi_{LOS,AOD}\right)\end{bmatrix}.$$

$$\exp\left(j2\pi\lambda_0^{-1}\left(\hat{\boldsymbol{h}}_{rx,LOS}^T.\boldsymbol{d}_{rx,u}\right)\right).\exp\left(j2\pi\lambda_0^{-1}\left(\hat{\boldsymbol{h}}_{tx,LOS}^T\cdot\boldsymbol{d}_{tx,s}\right)\right)$$

$$.\exp\left(j2\pi v_{LOS}t\right)$$

where $F_{rx,u,\theta}$ and $F_{rx,u,\varphi}$ are the antenna radiation patterns for element u in the direction of the spherical basis vectors, $\hat{\boldsymbol{e}}$ and $\hat{\boldsymbol{u}}$ respectively. $F_{tx,s,\theta}$ and $F_{tx,s,\varphi}$ are the antenna radiation patterns for element s in the direction of the spherical basis vectors, $\hat{\boldsymbol{e}}$ and $\hat{\boldsymbol{u}}$ respectively.

$$\hat{\boldsymbol{h}}_{rx,n,m}=\begin{bmatrix}\sin\theta_{n,m,ZOA}\cos\varphi_{n,m,AOA} \\ \sin\theta_{n,m,ZOA}\sin\varphi_{n,m,AOA} \\ \cos\theta_{n,m,ZOA}\end{bmatrix}^T$$

is the unit vector about the azimuth of arrival angle (AoA) $\varphi_{n,m,AoA}$ and the zenith of arrival angle(ZoA) $\theta_{n,m,ZOA}$. And also $\hat{h}_{tx,n,m}$ is the counterpart at the transmit side. $\boldsymbol{d}_{tx,s}$ and $\boldsymbol{d}_{rx,u}$ are the location vectors of the transmit and receive elements, respectively. $\kappa_{n,m}$ denotes the cross-polarization power ratio in linear scale. $\{\Phi_{n,m}^{\theta\theta}, \Phi_{n,m}^{\theta\varphi}, \Phi_{n,m}^{\theta\varphi}, \Phi_{n,m}^{\varphi\varphi}\}$ are the random initial phases for sub-path m of ray n. λ_0 is the wave length of the carrier frequency. K_R is the Ricean K-factor. $v_{n,m}$ is the doppler frequency component. Due to space limitations, more detailed description about the generation of 3D channel model can be referred to 3GPP 2014, TR 36.873.

3 LIMITED FEEDBACK SCHEME

Here assume that the number of transmission spatial layers to each desired user is 1 and the user can achieve the perfect channel estimation.

The BS is equipped with 2D antenna array. The number of antennas in horizontal direction is N_h and in vertical direction is N_v.So the total number of antennas is $N=N_v \cdot N_h$. Figure 1 depicts the configuration of 2D antenna array in the BS. d_v and d_h are the antenna spaces in the vertical direction and horizontal direction respectively. Each user has N_r receiving antennas.

To describe the limited feedback scheme, take the limited feedback of user k in cell j for example. When the transmitted antenna elements are row-wise indexed, the channel matrix between user k and the respective BS in cell j is $\boldsymbol{H}_k=[\boldsymbol{H}_{k1}\,\boldsymbol{H}_{k2}\ldots\boldsymbol{H}_{kNv}]\in C^{Nr}\times^{N}$. $\boldsymbol{H}_{ki}\in C^{Nr}\times^{Nh}$ $(i=1,2,\ldots N_v)$denotes the channel matrix from the i-th row transmitting antenna elements in the BS to user k.

The procedure of feedback scheme is as follows:

Step 1: Calculate the average correlation matrix in the horizontal domain by the user k.

$$\boldsymbol{R}=\frac{1}{\boldsymbol{N}_v}\sum_{i=1}^{N_v}(\boldsymbol{H}_{ki})^{\mathrm{H}}\boldsymbol{H}_{ki} \quad (6)$$

Implement the eigenvalue decomposition of $\boldsymbol{R}$ and get the eigenvector corresponding to the largest eigenvalue. Assume that the feedback overhead is B-bit. The set of codebook is $\boldsymbol{C}=\{\boldsymbol{c}_1, \boldsymbol{c}_{1\ldots}\boldsymbol{c}_M\}$, $M=2^B$. So the optimal codeword can be selected from the codebook set through the principle as follows:

$$\boldsymbol{w}_{\mathrm{h}}=c_n, n=\arg\max_{i=1\cdots M}\left|\boldsymbol{v}_{\mathrm{k}}^{\mathrm{H}}\boldsymbol{c}_{\mathrm{i}}\right| \quad (7)$$
$$\left[\boldsymbol{v}_k,\lambda_{\mathrm{k}}\right]=eig(\boldsymbol{R})$$

where $eig()$ denotes the eigenvalue decomposition. λ_k and $\boldsymbol{v}_k$ are the largest eigenvalue and the corresponding eigenvector, respectively. $\boldsymbol{w}_{\mathrm{h}}\in C^{N\mathrm{h}}\times^{1}$is the optimal quantized precoding vector for the horizontal domain.

Step 2: In the BS, the precoding vector for vertical domain is constructed as follow:

$$\boldsymbol{w}_v=\begin{bmatrix} w_1 & w_2 & \cdots & w_{N_v}\end{bmatrix}^{\mathrm{T}}\in C^{N_v\times 1} \quad (8)$$

where

$$w_m=\frac{1}{\sqrt{N_{\mathrm{V}}}}\exp(j\frac{2\pi}{\lambda}(m-1)\mathrm{d}_v\cos(\theta_{ketilt})),$$

$m=1,2\ldots N_v$. λ is the carrier wavelength. θ_{ketilt} is the angle for precoding vector in the vertical space. Here we adopt three schemes to select θ_{ketilt} for user k.

Scheme I:

Based on the distribution of users' ZoD in 3D-UMi and 3D-UMa scenarios, here we will split the cell into four sectors in vertical domain. Figure 2 depicts the definition of user's ZoD. In the figure, θ_{kZoD} is the ZoD. It is the vertical angle measured between the zenith direction and the line connecting the user k to the BS in cell j. Figure 3 shows the split sectors in vertical domain. Based on the reference signal received power (RSRP), the user determines which sector it is attributed to. The vertical precoding vector of the users attributed to the same vertical sector will be the same, namely θ_{ketilt} is assigned to the same value. There are four angle values for the four vertical sectors respectively. For 3D-UMi scenario, θ_{ketilt} is assigned to 102°, 95°, 90°, and 85° corresponding to sector 1, sector 2, sector 3 and sector 4 respectively. For 3D-UMa scenario, θ_{ketilt} is assigned to 102°, 99°, 95°, and 93°corresponding to sector 1, sector 2, sector 3 and sector 4 respectively.

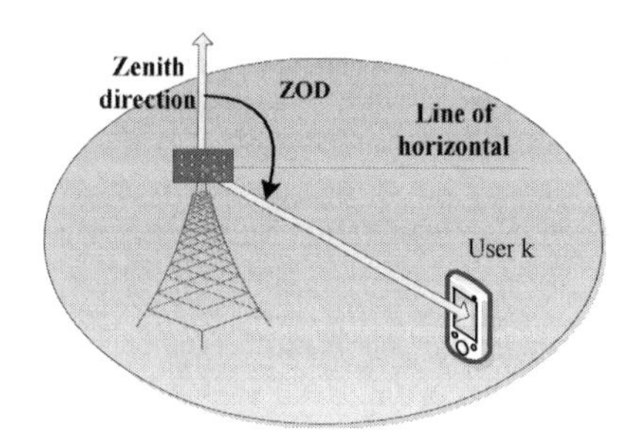

Figure 2. The definition of ZoD.

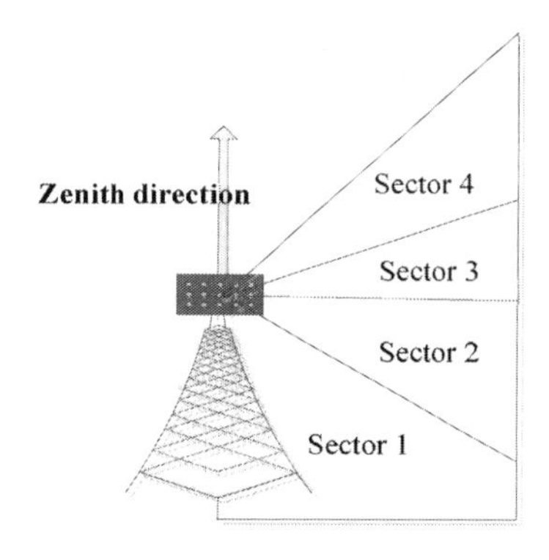

Figure 3. The split sectors of one cell in vertical domain.

Scheme II:

We assume that the ZoD of each user can be obtained through the uplink pilot training. The BS selects θ_{kZoD} for θ_{ketilt} to construct the vertical precoding vector, namely

$$\theta_{ketilt} = \theta_{kZoD} \tag{9}$$

Scheme III:

In this scheme, we will exploit uplink pilot training to calculate the optimal angle for θ_{ketilt}. Assume that the antenna elements in the BS are column-wise indexed and the uplink channel matrix between the BS and user k is $\boldsymbol{H}_k^{*}=[\boldsymbol{H}_{k1}^{*}\boldsymbol{H}_{k2}^{*}\ldots\boldsymbol{H}_{kN\mathrm{h}}^{*}]^{\mathrm{T}}\in C^{N}\times^{Nr}$ when utilizing the uplink pilot. $\boldsymbol{H}_{ki}^{*}\in C^{Nr}\times^{Nv}$ is the channel matrix between the i-th (i=1, 2...N_h) column antenna elements in the BS and the receiving antennas of user k.

We can calculate the average correlation matrix in the vertical domain.

$$\boldsymbol{R}_\mathrm{v} = \frac{1}{\boldsymbol{N}_\mathrm{h}}\sum_{i=1}^{N_\mathrm{h}}(\boldsymbol{H}_{ki}^{*})^{\mathrm{H}}\boldsymbol{H}_{ki}^{*} \tag{10}$$

Then the optimization problem can be expressed as:

$$\begin{aligned}&\boldsymbol{max}\quad \boldsymbol{w}_\mathrm{v}^{H}\boldsymbol{R}_\mathrm{v}\boldsymbol{w}_\mathrm{v}\\ &subject\ to: \theta_0 \le \theta_{ketilt} \le 180^{\circ}\end{aligned} \tag{11}$$

where θ_0 equals to 90° for 3D-UMa scenario and θ_0 is assigned to 0°for 3D-UMi scenario. In 3D-UMa scenario, the BS is above the surrounding buildings. So the ZoD of each user in 3D-UMa scenario is more than 90°.

Through (11), we can get the optimal θ_{ketilt} to construct the vertical percoding vector. But in FDD operation, the uplink channel and downlink channel generally occupy different frequency resources. So the optimization problem (11) can just roughly get a reasonable θ_{ketilt} for the vertical percoding vector in the downlink.

Step 3: The BS can get the precoding vector $\boldsymbol{w}_\mathrm{h}$ for the horizontal domain through the feedback of user k. And then $\boldsymbol{w}_\mathrm{v}$ can be calculated by obtaining θ_{ketilt} in the BS. So the 3D precoding vector of user k can be calculated by:

$$\boldsymbol{w}_k = \boldsymbol{w}_\mathrm{v} \otimes \boldsymbol{w}_\mathrm{h} \tag{12}$$

4 DOWNLINK TRANSMISSION OF MU-MIMO SYSTEM

The received signal of the users can be expressed as:

$$\boldsymbol{x}_j = \sqrt{\rho_f}\boldsymbol{H}_{jj}\boldsymbol{W}_j\boldsymbol{s}_j + \sum_{l=1,l\neq j}^{L}\sqrt{\rho_f}\boldsymbol{H}_{jl}\boldsymbol{W}_l\boldsymbol{s}_l + \boldsymbol{v}_j \tag{13}$$

where $\boldsymbol{W}_j = 1/(K)^{1/2} \times [\boldsymbol{w}_1\ \boldsymbol{w}_2\ \ldots\boldsymbol{w}_K]\in C^{N}\times^{K}$ is the matrix composed of K users' precoding vectors calculated by (12). K is the number of co-scheduled users. $\boldsymbol{H}_{jj}=[\boldsymbol{H}_1^{\mathrm{T}}\boldsymbol{H}_2^{\mathrm{T}}\ldots\boldsymbol{H}_K^{\mathrm{T}}]^{\mathrm{T}}$ is the downlink channel between the users in cell j and the BS transmitter in cell j. $\boldsymbol{s}_j=[s_1\,s_2\ldots s_K]^{\mathrm{T}}\in C^{K}$ is the transmitted signal for the K users in cell j. The second term in the right side denotes the interference from adjacent cells. $\boldsymbol{v}_j \sim CN(0,\sigma^2\boldsymbol{I}_K)$ is the additive noise.

5 SIMULATION RESULTS

In this section, system-level simulation is performed to evaluate the 3D MIMO system. In the simulation, the cellular network layout adopts wrap-round technique. The network is comprised of 19 hexagonal sites. Each site is split into 3 cells.10 active users are distributed uniformly in each cell, namely 570 users are distributed in 57 cells. Each user is equipped with cross-polarized antennas and N_r=2. The BS adopts the 2D planar ±45°cross-polarized antenna array.

In the simulation, there are 4 columns of cross-polarized antennas and each column has 10 pairs cross-polarized antennas. Each column contains 2 columns of co-polarized antennas. So N_v=10 and N_h=2×4=8. And the number of antennas is N=80. For simplicity, detailed simulation parameters are listed in table 1. In (7), the 8Tx codebook of LTE-A is utilized for horizontal domain (3GPP 2014, TS 36.213). To make a comparison, here the baseline is the 2D MIMO scheme.

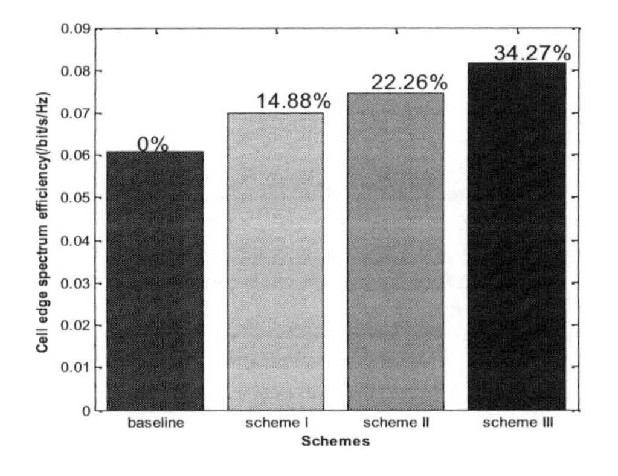

Figure 4. Cell edge spectrum efficiency of 3D MIMO under 3D-UMa scenario.

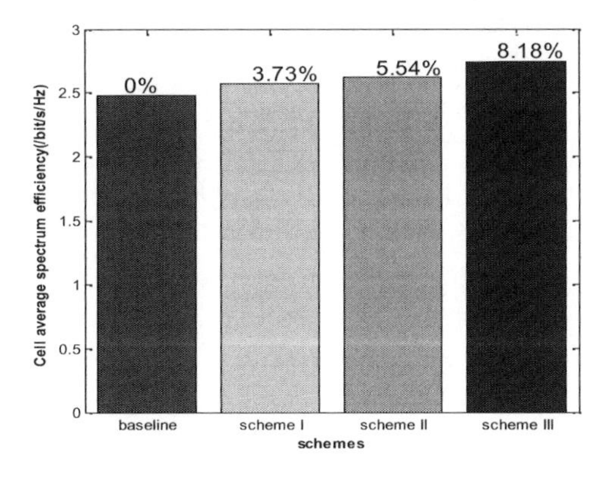

Figure 5. Cell average spectrum efficiency of 3D MIMO under 3D-UMa scenario.

Figure 4 and Figure 5 show the cell edge spectrum efficiency and the cell average spectrum efficiency of the three proposed schemes in MU-MIMO system under the 3D-UMa scenario respectively. In this paper, the maximal number of co-scheduled users is 2. We can see the proposed schemes perform better than the baseline scheme. The cell edge spectrum efficiency and the average spectrum efficiency of the 3 schemes have a performance gain compared to the baseline scheme. Among the three schemes, scheme III has the best performance and scheme I performs the worst. There are some scatters in the radio propagation, so the energy will not be completely concentrated on the direction of the each user's ZoD. Scheme III selects the optimal angle for the vertical precoding vector and scheme II just selects the ZoD of each user for the vertical precoding. So scheme III outperforms scheme II. In scheme I, the users select the same angle for the vertical precoding vector when they are in the same vertical sector. Of course, the interference among the users in scheme I is larger than the interference in scheme II. So scheme I has a lower gain than scheme II.

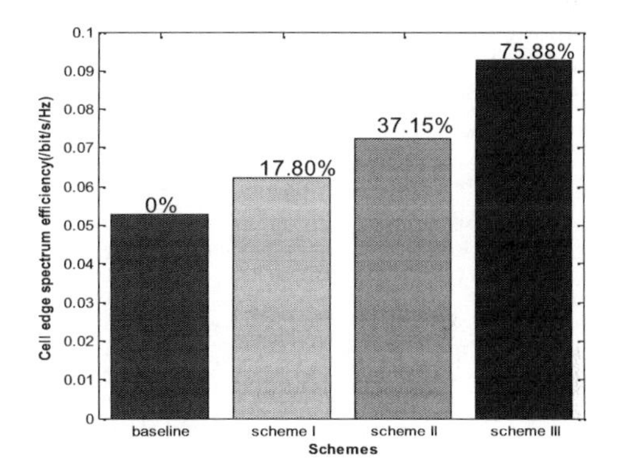

Figure 6. Cell edge spectrum efficiency of 3D MIMO under 3D-UMi scenario.

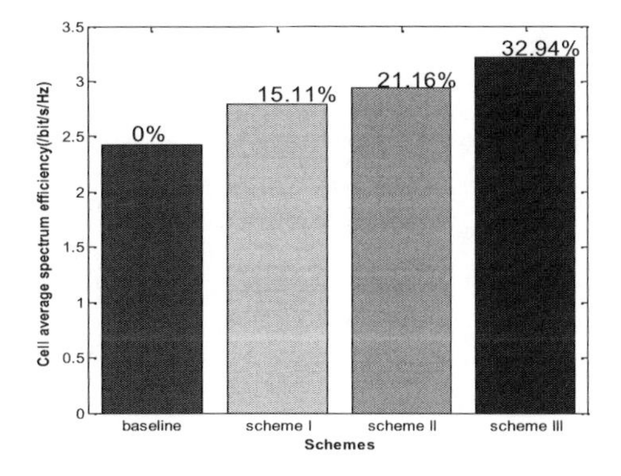

Figure 7. Cell average spectrum efficiency of 3D MIMO under 3D-UMi scenario.

Figure 6 and Figure 7 present the simulation results of the three schemes under the 3D-UMi scenario. The similar observations can be made for 3D-UMi scenario. The performance gain of scheme III is the highest, followed by scheme II and scheme I. The distribution range of users' ZoD in 3D-UMi scenario is wider than that in the 3D-UMa scenario. The inter-beam interference of the co-scheduled users in vertical domain will be smaller in 3D-UMi scenario than that in 3D-UMa scenario. So we can find that the performance gain of the three schemes relative to the performance of the baseline scheme in 3D-UMa scenario is lower than that in 3D-UMi scenario.

Table 1. 3D-MIMO simulation parameters.

Parameters	Settings
Scenarios	3D-UMa/3D-UMi
Bandwidth	10MHz
Antenna element interval	0.5λ both in horizontal and vertical direction
Carrier frequency	2GHz
User speed	3Km/h
Schedule	proportional fair
Channel estimation	ideal
Traffic mode	full butter
HARQ	the maximal number of retransmission is 4
Wrapping method	geographical distance based

6 CONCLUSION AND FUTURE WORK

In this paper, the 3D channel model calibrated by 3GPP is introduced. The BS is equipped with 2D antenna array. According to the characteristics of 3D MIMO, the existing codebook of LTE is used in the horizontal dimension, while in the vertical dimension we propose three schemes to construct the precoding vector. In the 3D-UMa and 3D-UMi scenarios, system-level simulation is implemented. The three schemes achieve a better performance than the baseline scheme. The performance gain of scheme III is the highest, followed by scheme II and scheme I

ACKNOWLEDGEMENT

This research has been supported by 863 Project of China ((No.2014AA01A705)) and the State Key Laboratory of Wireless Mobile Communication of China Academy of Telecommunication Technology (No.2007DQ305156).

REFERENCES

3GPP. 2014. TR 36.873 v12.0.0(2014-09), Study on 3D channel model for LTE (Release 12). *http://www.3gpp.org/ftp/Specs/2014-12/Rel-12/36_series/.*

3GPP. 2014. TS 36.213 v12.4.0 (2014-12) Physical layer procedures (Release12). *http://www.3gpp.org/ftp/Specs/2014-12/Rel-12/36_series/.*

Inoue, T. & Heath, R.W. 2009. Kerdock codes for limited feedback precoded MIMO systems. *IEEE Transactions on* Signal *Processing* 57(9):3711–3716.

Kammoun, A., Khanfir, H, Altman, Z, et al. 2014. Preliminary Results on 3D Channel Modeling: From theory to standardization. *IEEE Journal on Selected Areas in Communications* 32(6):1219–1229.

Li Yan, Ji Xiaodong , Liang Dong , et al. 2013. Dynamic beamforming for three-dimentional MIMO technique in LTE-Advanced networks. *International Journal of Antennas and Propagation* 2013.

Lu Xiaojia, Tolli, A., Piirainen, O., et al. 2011. Comparison of antenna arrays in a 3-D multiuser multicell network. 2011 IEEE International Conference on *Communications (ICC), Kyoto, 5-9 June, 2011.* Piscataway: IEEE.

Marzetta, T.L. 2010. Noncooperative cellular wireless with unlimited number of BS antennas. *IEEE Transactions on* Wireless Communications 9(11): 3590–3600.

Shafi, M., Zhang Min, Smith, P.J., et al. 2006. The impact of elevation angle on MIMO capacity. 2006 IEEE International Conference on *Communications (ICC), Istanbul, June 2006.* Piscataway: IEEE.

Song Yang, Yun Xiang, Nagata, S, et al. 2013. Investigation on elevation beamforming for future LTE-Advanced. *2013 IEEE International Conference on* Communications Workshops (ICC), *Budapest, 9-13 June 2013.* Piscataway: IEEE.

Xie Yi, Jin Shi, Wang Jue, et al. 2013. A limited feedback scheme for 3D multiuser MIMO based on Kronecker product codebook. *2013 IEEE 24th International Symposium on* Personal Indoor and Mobile Radio Communications, *London, 8-11 Sept. 2013.* Piscataway: IEEE.

Zheng Kan, Ou Suling, Xuefeng Yin. 2014. Massive MIMO channel models: a survey. *International Journal of Antennas and Propagation* 2014.

Electronics, Communications and Networks IV – Hussain & Ivanovic (eds)
© 2015 Taylor & Francis Group, London, ISBN: 978-1-138-02830-2

Depth map processing method based on edge information for DIBR

XueRui Hu*, Ying Yu, Yan Shi & Bo Wang
Department of Automation, Communication University of China, Beijing, China

ABSTRACT: Depth Image Based Rendering (DIBR) is one of the key technologies in Free-viewpoint Television (FTV) system. DIBR is used to render virtual view in FTV system, and the quality of depth map is very important in the processing of view synthesis. To improve the quality of depth map and reduce the holes in virtual view, a depth map processing method is proposed in this paper. In the proposed method, we take advantage of the edge information and use a new smoothness term based on Graph Cuts to smooth the boundary areas of depth map. The experimental results show that the proposed method improves the quality of depth maps and synthesized views.

1 INTRODUCTION

Free-viewpoint television (FTV) is an innovative 3DTV that enables the users to view a 3D scene in different viewpoints freely (Tanimoto 2012). In the video encoder, a real world 3D scene is typically captured by an array of cameras that produce large amounts of data. To greatly reduce the amount of data during encoding and transmission, some views can be synthesized by depth image based rendering (DIBR) (Zhu 2011, Fehn 2004). DIBR is an effective approach for view synthesis, which takes a reference image and its associated depth map as an input to generate a virtual view by 3D warping.

However, there is an essential problem in DIBR that holes and artifacts appear in the synthesized view, particularly in the boundary areas of objects because of occlusion and disocclusion (Tran & Harada 2013, Mori et al. 2008). Although the occlusion problem can be handled by estimating the depth map from three camera views and blending the virtual views, there still remain some holes and artifacts after rendering. These remaining holes and artifacts are caused by the discontinuity and inaccuracy of depth values, especially at the boundary areas or disocclusion areas. Our goal is to smooth the boundaries of depth map and generate synthesized view with few holes and artifacts.

In this paper, the edge information of objects is considered as a very important factor influencing the quality of the depth map. A new edge dependent depth map processing method is proposed. We use a contour detection algorithm to detect edge information. Then, we propose a smoothness term based on Graph Cuts to handle the boundary problem of depth map.

* Corresponding author: *ruixue_hu@cuc.edu.cn*

2 RELATED WORKS

2.1 *The depth estimation reference software*

The depth map is a 2D image that indicates the distance between a world point and the camera. Once a depth map is obtained, we can synthesize a virtual view by 3D warping. The depth value can be calculated from the disparity value by using (1), where v is the intensity of depth map, d is the disparity value of a pixel, d_{max} and d_{min} represent the maximum and minimum disparity respectively.

$$v = 255 \cdot \frac{d - d_{min}}{d_{max} - d_{min}} \tag{1}$$

The depth estimation reference software (DERS) is developed to help the research of depth estimation (Tanimoto et al. 2008). DERS estimates the disparity value by stereo matching (Lee et al. 2012). We use a global matching algorithm in the proposed method, because it is more accurate than the local method, and the computation speed is acceptable (Scharstein & Szelisk 2002).

DERS supports 3 main models of depth estimation, automatic mode, segmentation mode, and semi-automatic mode. Take the automatic mode for example, the simplified diagram is shown in Figure 1.

Three camera views are inputted, and the smallest matching cost is calculated by matching the right and the center view, and the left and the center view. In the Update Matching Cost step, the static pixels detected as static areas are updated to perform good temporal performance. Finally, the Graph Cuts is used to optimize the depth map.

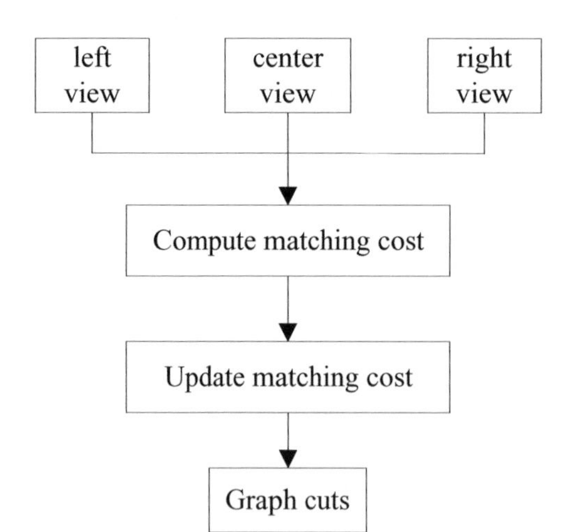

Figure 1. Diagram of the automatic mode.

2.2 *Graph Cuts*

Graph Cuts is a useful and popular energy minimization algorithm. It can be used to find the best disparity value for all pixels. Kolmogorov & Zabih (2004) described how Graph Cuts can be used in stereo and multi-camera scene reconstruction, and the standard form of the energy function is (2). (3) is the data term, while (4) is the smoothness term.

$$E(f)=\sum_{p\in P} D_p(f_p)+\sum_{p,q\in N} V_{p,q}(f_p,f_q) \tag{2}$$

$$E_{data}(f)=\sum_{p\in P} D_p(f_p) \tag{3}$$

$$E_{smooth(f)}=\sum_{p,q\in N} V_{p,q}(f_p,f_q) \tag{4}$$

P represents all pixels of the current image. N is a neighborhood system on pixels. Label f_p and label f_q represent the disparity of pixel p and pixel q, respectively. $D_p(f_p)$ represents the matching cost when the disparity value of pixel p is f_p. $V_{p,q}(f_p,f_q)$ is used to impose spatial smoothness, and it includes the disparity difference between neighboring pixels p and q.

Wildeboer & Fukushima (2010) proposed a semi-automatic method based on Graph Cuts. In their method, the edge map is provided by manually tracing an edge, and the smoothness term is defined as follows:

$$V_{(p,q)}=\beta\lambda\left|f_p-f_q\right| \tag{5}$$

where β is a scaling factor, and λ is an empirically chosen smoothing factor. If p is on the edge, then β=0.1, or else it is 1.0. This approach requires people to draw the edge map manually, which may cause wrong edge information and produce big holes in boundary areas. And while changing the smoothness term, the different changes of disparity around the edge areas are not taken into consideration, which could cause wrong depth values.

In the next section, we will describe our method and propose a new smoothness term based on Graph Cuts.

3 PROPOSED METHOD

As mentioned in previous sections, most of the holes and artifacts appear at the boundaries of objects where the occlusion and disocclusion happen, and the depth value changes greatly. The method in this paper is proposed to smooth the boundary areas by taking advantage of edge information. The detected edge map is used in Graph Cuts to change the smoothness term. The process of the proposed method can be simply displayed in Figure 2. We use left view, center view and right view as input to estimate the depth value of center view. During the step of Graph Cuts, we will optimize the depth map with the proposed smoothness term according to the edge information.

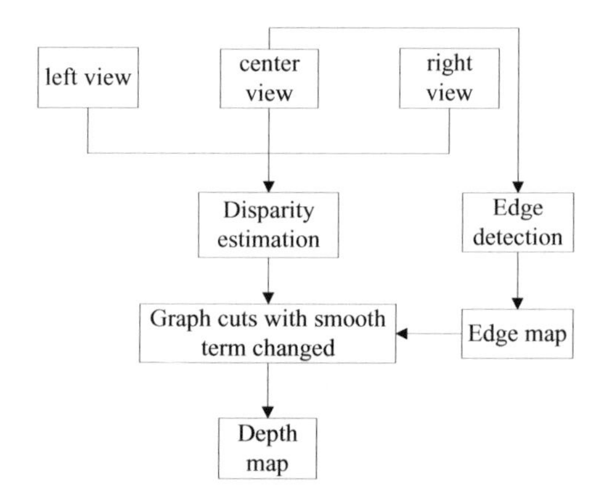

Figure 2. Process of the proposed method.

First, we need to obtain an edge map of the center view. The edge image indicates the location of the boundary areas where the depth value may be inaccurate and discontinuous. Although Canny can be used to detect the edge information effectively, the detected edge map contains too many texture edges which would consume more computing cost. So the contour detection algorithm is preferred to generate the edge map. The contour detection algorithm can obtain the boundaries of objects with little texture edges. Figure 3 is the color image and Figure 4 is the edge map.

Figure 3. Color image.

Figure 4. Edge map of contour detection.

If an edge pixel is defined, and then the smoothness term can be defined as (6), where q is a weight. The value of θ varies with the changes of disparity difference between an edge pixel and its neighboring pixels.

$$E_{smooth(f)} = \sum_{p,q \in N} \theta \cdot V_{p,q}(f_p, f_q) \tag{6}$$

Actually, not all detected edges are useful in the proposed method. Our goal is to smooth the areas where the depth value is discontinuous. The circular area in Figure 4 indicates the texture edges where the depth value hardly changes, so there is no need to smooth these texture edges. The square area on Figure 4 indicates the boundary areas where the background is occluded by foreground, and the depth value may be discontinuous.

As shown in Figure 5, p is a pixel on the edge; b and d are pixels in the foreground; a and c are pixels on the background. Let label f be the disparity value of pixels. The disparity values of pixels are compared to decide whether pixel p is on the boundary edges or not. Pixel p and d are both on the foreground, and their disparity could be similar or the same, so we compare f_c and f_d instead of f_p and f_d. We also compare f_a and f_b instead of f_p and f_b to smooth the depth value in the vertical direction. On the other hand, the value of q should be greater than zero. If θ is equal to zero, there will be no smoothness term at the edge pixel. Therefore, taking the horizontal direction, for example, we propose that the weight value can be defined as follows:

$$\theta = 1 - \frac{|f_c - f_d|}{f_{\max} - f_{\min}} \tag{7}$$

where $f_{\max}$ is the maximum disparity value of all pixels, and $f_{\min}$ is the minimum disparity value; $f_{\max}$ and $f_{\min}$ can be calculated by stereo matching. If p is a pixel on the texture edge, then $|f_c - f_d|$ will be equal to 0, and q will be equal to 1. In this case, the pixel p will be smoothed by the regular method. If p is a pixel on the boundary of the object, θ will be a small value. By changing the smoothness term according to the edge information, the depth value at boundary areas can be more accurate, and the quality of synthesized view can be better. The experimental results will be discussed in section 4.

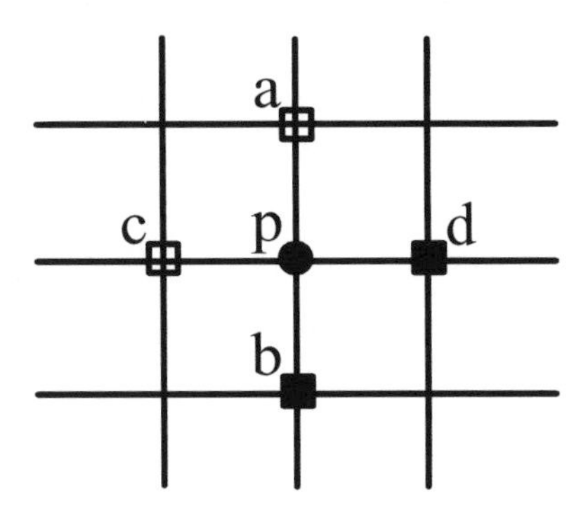

Figure 5. Smoothing between pixels in edge areas.

4 EXPERIMENTAL RESULTS

We use video test sequences "Champagne_tower" and "Akko&Koya" in our experiment to verify the proposed method. Both of these two sequences are captured by fixed cameras with 50mm interval, and they are provided by Nagoya University.

"Champagne_tower" has a resolution of 1280×960, and "Akko & Koya" has a resolution of 640×480. For each sequence, we obtain depth maps by using automatic method, semi-automatic method (Wildeboer & Fukushima 2010) and the proposed method. We use the View Synthesis Reference Software (VSRS_3.5) to synthesize the virtual view. The results of view synthesis are evaluated by objective evaluation measures, such as PSNR and SSIM.

Figures 6 and 7 are the results of depth estimation, and the results show that the proposed method generates better depth maps than automatic method does. Although the semi-automatic method achieves better boundary lines in depth map, it causes some wrong depth values at boundary areas which generate the holes in Figure 8c.

Figures 8 and 9 show the results of view synthesis, and the circular areas indicate the holes and artifacts appear in the boundary areas. According to the depth maps in Figure 6 and 7, we can see that these holes and artifacts are caused by inaccurate depth values. From Figure 8d and Figure 9d we can see that the proposed method reduces the holes and artifacts and achieves better quality of virtual view.

PSNR and SSIM results are shown in Table1 and Table2. The proposed method has a better performance on both PSNR and SSIM. Our method obtains 30.739dB of PSNR for "Champagne tower", which is about 2.2dB higher than automatic method and 0.8dB higher than semi-automatic method. For "Akko & Koya", the PSNR of the proposed method achieves about 0.45dB higher than semi-automatic method. The proposed method also obtains higher SSIM value than automatic method and semi-automatic method do.

Table 1. Experimental results of Champagne tower.

Evaluation measures	Automatic method	Semi-automatic method	Proposed method
PSNR(dB)	28.537	29.885	30.739
SSIM	0.953	0.96	0.963

Table 2. Experimental results of Akko & Koya.

Evaluation measures	Automatic method	Semi-automatic method	Proposed method
PSNR(dB)	36.811	36.8604	37.3296
SSIM	0.9581	0.9582	0.9689

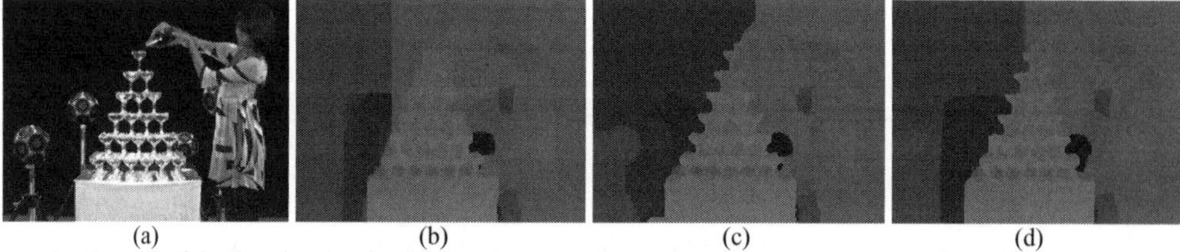

Figure 6. Results of depth estimation for "Champagne_tower". (a) camera view; (b) automatic method; (c) semi-automatic method; (d) proposed method.

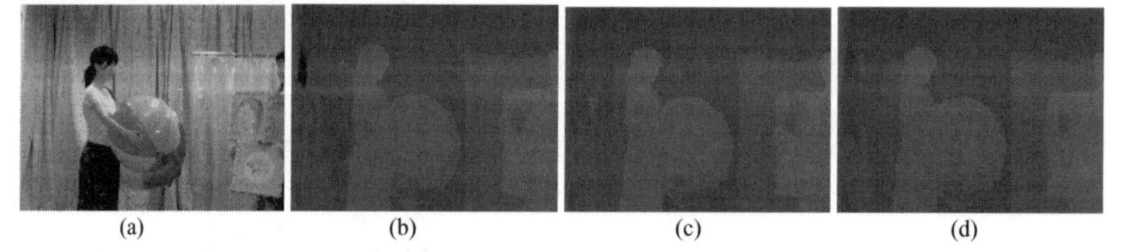

Figure 7. Results of depth estimation for "Akko & Koya". (a) camera view; (b) automatic method; (c) semi-automatic method; (d) proposed method.

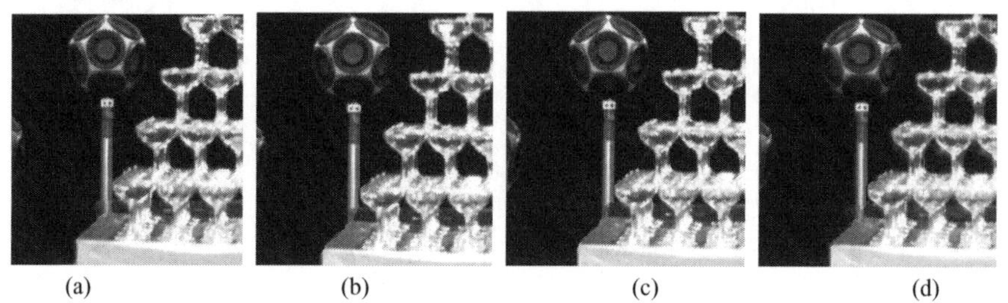

Figure 8. Results of view synthesis for "Champagne_tower". (a) camera view; (b) automatic method; (c) semi-automatic method; (d) proposed method.

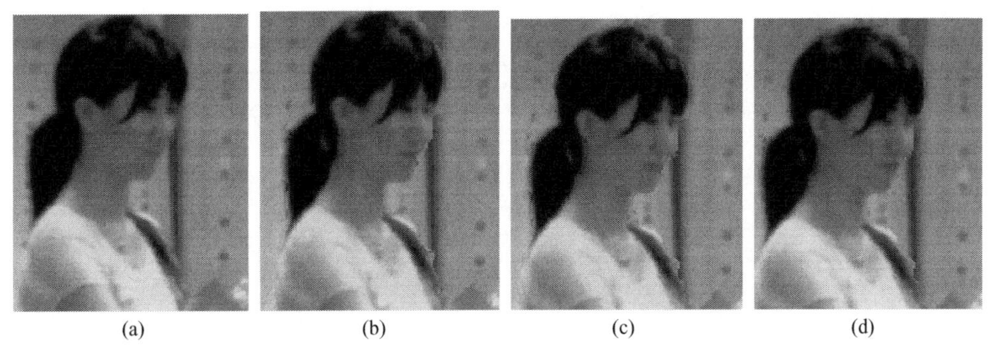

Figure 9. Results of view synthesis for "Akko & Koya". (a) camera view; (b) automatic method; (c) semi-automatic method; (d) proposed method.

5 CONCLUSIONS

In this paper, we propose a depth map processing method based on edge information for DIBR. An edge map is generated by using a contour detection algorithm, and we take advantage of edge information to smooth the boundary areas of depth map. The smooth method in this paper is based on Graph Cuts, and the smoothed depth map achieves better quality in the experiments. The experimental results of view synthesis show that the proposed method gets higher PSNR and SSIM value than related works, and reduces the holes and artifacts.

REFERENCES

Conference Report. 2010. Report on Experimental Framework for 3D Video Coding. *Doc W11631*. Guangzhou, China: ISO/IEC JTC1/SC29/WG11.

Fehn, C. 2004. Depth-image-based rendering (DIBR), compression and transmission for a new approach on 3D-TV. In *Proceedings of SPIE Stereoscopic Displays and Virtual Reality Systems* XI: 93–104.

Kolmogorov, V. & Zabih, R. 2004. What Energy Functions Can Be Minimized via Graph Cuts?. *IEEE transactions on pattern analysis and machine intelligence* 26(2): 147–159.

Lee, P.J., Yu, W.J. & Chuang, C.F. 2012. Watershed-based stereo matching algorithm for depth map generation. *2012 IEEE 1st Global Conference on Consumer Electronics (GCCE), Tokyo, 2–5 October 2012*. IEEE.

Mori, Y., Fukushima, N., Fuji, T. & Tanimoto, M. 2008. View generation with 3D warping using depth information for FTV. *Proc. IEEE 3DTV-CON*: 229–232.

Scharstein, D. & Szeliski, R. 2002. A taxonomy and evaluation of dense two-frame stereo correspondence algorithms. *IJCV* 47(1–3): 7–42.

Tanimoto, M. 2012. Free viewpoint Television. *Signal Processing: Image Communication* 27: 555–570.

Tanimoto, M., Fujii, T. & Suzuki, K. 2008. Reference software for depth estimation and view synthesis. *MPEG Report M15377*. Archamps, France: ISO/IEC JTC1/SC29/WG11.

Tran, A.T. & Harada, K. 2013. Depth Based View Synthesis Using Graph Cuts for 3DTV. *Journal of Signal and Information Processing* 4: 327–335.

Wildeboer, M.O. & Fukushima, N. 2010. A semi-automatic multi-view depth estimation method. *Visual Communication and Image Processing 2010, Proc. SPIE7744, Huangshan, 11 July 2010*. International Society for Optics and Photonics.

Zhu, L. 2011. A New Virtual View Rendering Method based on Depth Map for 3DTV. *Procedia Engineering* 15: 1115–1119.

Electronics, Communications and Networks IV – Hussain & Ivanovic (eds)
© 2015 Taylor & Francis Group, London, ISBN: 978-1-138-02830-2

Design and implementation of mobile money system using Near Field Communication (NFC)

Emir Husni* & Adrian Ariono
School of Electrical Engineering & Informatics, Institut Teknologi Bandung, Bandung, Jawa Barat, Indonesia

ABSTRACT: Nowadays smartphone is widely used in every part of the world. Trend that is happening now is mobile phone penetration tends to decrease while the smartphone tend to increase. More and more newly launched smartphone offers support for Near Field Communication (NFC). NFC technology enables simple and safe two-way interaction between electronic devices. This research has developed a smartphone-based electronic payment system using NFC. Besides being used to make a payment, the system can be expanded to other applications such as time and attendance machine for a community. To compensate the lack of physical secure element, registration, synchronization, top-up, and attendance record process is done online, while the transaction is done offline. Various cryptographic algorithms are used, such as password-based key derivation function and key wrap algorithm to maintain the security of the system. Devices used in this research are Android smartphones and NFC reader module CN-370S-2.

1 INTRODUCTION

Mobile phone penetration in Indonesia, according to Roy Morgan Research, has reached 84 % in March 2013. Although this number is still dominated by conventional mobile phones, the penetration of smartphones is doubled from the previous year, from 12% to 24% (Chadha 2013). Today' trends are mobile phone growth slowly decreasing while smartphone growth rapidly increasing. In addition, the computational resources in a smartphone are increasing every year so that more and more applications can be integrated into it. Because of various applications, smartphones are most commonly used electronic device in everyday life, according to a survey conducted by Forrester Research (Tsirulnik 2014).

Near Field Communication (NFC), according to the definition from the NFC-Forum, is a technology that enables simple and safe two-way interactions between electronic devices with a single touch. Global Industry Analysts predict that by 2018 the number of NFC-enabled smartphone will reach 1.6 billion units.

Electronic money is a payment instrument that is growing rapidly in Indonesia. The number of electronic money issued in Indonesia, according to Bank of Indonesia, showed significant growth every year. Nowadays, the majority of electronic money issuer is a bank, and the media is a smartcard.

From the data mentioned above, there are opportunities that can be utilized with an electronic payment system using NFC-enabled smartphone. Smartcard-based electronic money does not offer other features except for the transaction. Electronic money using NFC-enabled smartphone can be used for various needs with help from smartphone capabilities. This research will offer an alternative for developing an electronic payment system based on NFC-enabled smartphones. The electronic payment system is designed to be used in a small to medium community, e.g. office or school, because it can also be used for community member's attendance log.

2 SECURITY DESIGN

2.1 *Security specification*

Before designing security specification, hardware limitations must be taken into consideration. One of the most limiting factors when designing security specifications is a USB driver for NFC reader used in Linux-based computer can only send 100bytes of data in one sequence. Other limiting factors are the unavailability of secure module and offline transaction. After limitations are known, we can define the security specifications. Below are specifications that must be taken into consideration when designing security feature, as shown in Figure 1:

- Total length of the transaction data packet must not exceed 100 bytes.
- Encryption key for transaction must be securely stored in the device.
- Transaction data log stored in encrypted form in the device.

* Corresponding author: *ehusni@lskk.ee.itb.ac.id*

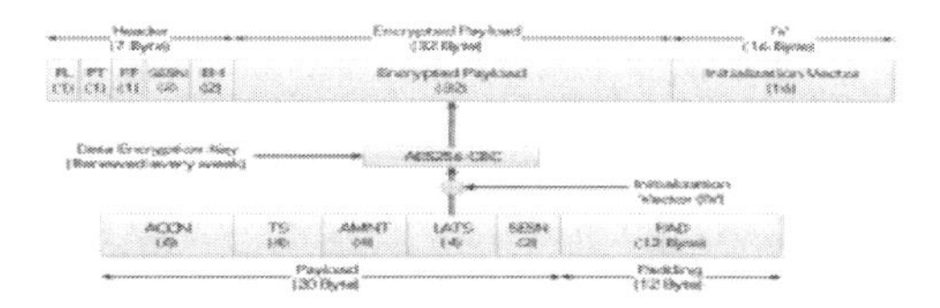

Figure 1. Transaction data packet.

2.2 *Transaction security design*

To comply with the specifications described above, transaction data packet length must be shorter than 100 bytes. In order to get a short data packet with good confidentiality, the chosen encryption scheme is symmetric encryption. Symmetric encryption is used because of the low cyphertext overhead. A symmetric encryption algorithm that will be used is AES because AES is considered the most secure and practical symmetric encryption algorithm today (Paar & Pelzl 2010).

2.3 *Key management*

Key management must be taken into consideration because key for encrypting transaction data packet is stored in the device. In most of the payment system that already established, payment assets (e.g. keys, payment data) are stored in secure elements. Due to unavailability of secure elements that can be used, payment assets in this system will be stored on the user's device.

Transaction key is used by all devices in the system, therefore the storing mechanism of said key must carefully designed. The best practices for software-based symmetric key management are password-based key derivation function, and AES key wrap (Burnett 2001). Using the AES key wrap, AES key that used for encrypting transaction data is stored in wrapped state. The key used for wrapping AES key is derived using password-based key derivation function. By using this mechanism, the key stored in every device will be in different state depending on the user's password.

2.4 *Application data security*

Transaction data logs and e-Money data are stored as application data. Transaction data logs are stored using database while e-Money data stored as SharedPreferences (Android) or configuration files (Linux). There are various attributes stored in e-Money data, e.g. account number, hashed application password, and not all of those attributes need to be a secret. Table 1 shows attributes in e-Money data and storing method for each attribute.

Table 1. E-money data attributes and storing method for each attribute.

Attribute	Storing method
Account Number (ACCN)	Plaintext
Hardware Identification (HWID)	Plaintext
Password	Hashed with salt using SHA256
Last Sync Timestamp	Plaintext
Last Transaction Timestamp	Plaintext
Balance Unverified	Encrypted with key derived using password-based key derivation function
Balance Verified	Encrypted with key derived using password-based key derivation function
Transaction Key	Wrapped using an AES Key Wrap algorithm with key derived using password-based key derivation function

Transaction data logs are the data that represent transaction done by the payer and merchanst in the past, as shown in Figure 2. E-Money issuer redeems the money to the merchant and deducts payer balance on the server based on transaction data logs sent by both payer and merchant to the server. Transaction data logs need to be encrypted because if the merchant can create this data manually, server needs to pay nonexistent transaction to the server.

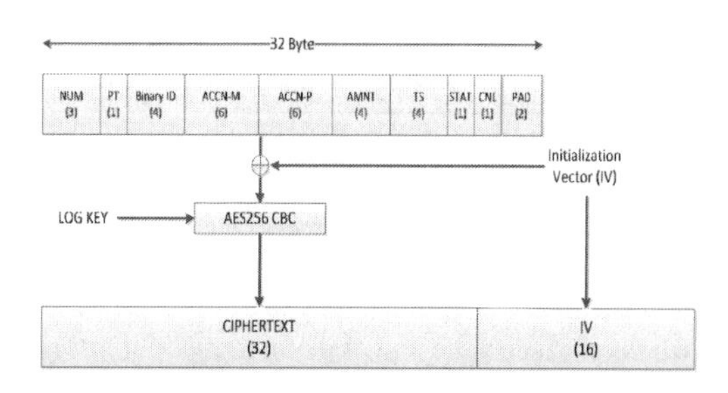

Figure 2. Transaction data log format stored in the database.

3 SYSTEM DESIGN

3.1 *System requirement*

3.1.1 *Functional requirement*

There are four major processes in this system. Those processes are: registration, transaction, synchronization, and top-up. Each of these processes has their own requirement.

a. *Registration*
Registration is done both by the system administrator and user. Registration assigns an account number to the user's device. User needs to report to the system administrator if they want to change their device and use the same account number.

b. *Transaction*
Transaction processing is done in offline state. A transaction process' sequence that involving users need to be as simple as possible.

To avoid automated attack, payers need to input three digit random numbers from merchant device (EDC) before transaction starts.

Transaction between the two smartphones can be done if one acts as the payer and the other as a merchant. Prior to the transaction, smartphone that act as a merchant need to send merchant requests. A transaction data log is stored in an encrypted manner.

c. *Synchronization*
Synchronization data security must be taken into consideration. Time needed to do synchronization must be as fast as possible.

d. *Top-up*
Balance's top-up must be done on the server by authorized personnel. To receive topped-up the balance, user needs to synchronize their device.

3.1.2 *Technical requirement*

In order to implement this system, devices needed are:

- NFC-enabled Android smartphone
- NFC reader with PC connectivity
- Linux-based PC
- Web Server.

3.2 *Top-level system design*

As stated before, there are four main processes in this system. Transaction processing is done in offline state while the other process is done in the online state.

Besides four main processes, there are three main actors that have an active role in the system. The three actors are: payer, merchant, and server (administrator).

3.3 *Registration process*

There are two actors that involved in the registration process, payer / merchant and server administrator. First, server administrator needs to create new random account number. Newly created account number needs to be inputted into the user's device, and then the user's device will send its hardware identification (IMEI for smartphone or serial number for NFC reader) to the server. The registration process sequence diagram is shown in Figure 3.

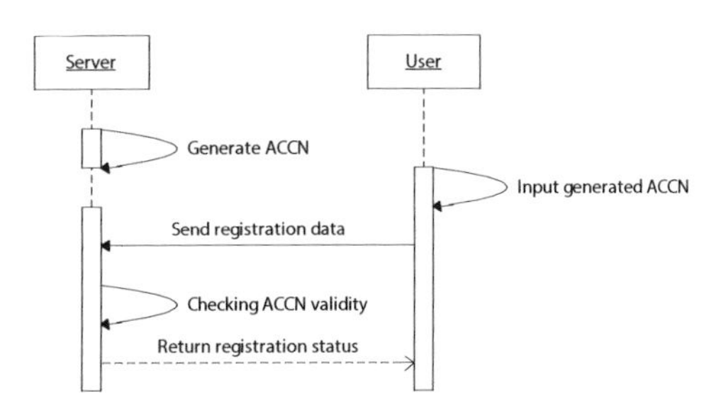

Figure 3. Registration process sequence diagram.

3.4 *Transaction process*

The transaction process is divided into two categories: transaction between smartphone and smartphone, transaction between smartphone and EDC. The transaction process between smartphone and smartphone utilizes NFC P2P Simple NDEF Exchange Protocol (SNEP), while the transaction process between smartphone and EDC utilizes card emulation using Logical Link Communication Protocol (LLCP) implementation.

In Android API, Android Beam is implemented on top of SNEP. Android Beam features "tap to beam" to initiate communication between two devices. After users tap their device, the communication link is established between two devices prior to sending data. Android Beam only supports one-way NDEF transfer. If the receiving device already received NFC data, communication link will be disconnected, regardless the received data validity. Because of this nature, transaction processing between smartphone and smartphone needs additional sequence. Before the transaction is done, the merchant needs to send merchant request. Merchant request is used to ensure both payer and merchant holds the same transaction key.

In transactions between smartphone and EDC, merchant request can be omitted because programming in EDC is done from LLCP layer. Programming in LLCP layer gives more control of the data received or sent by an NFC reader in EDC. The data received from smartphone can be checked first. If there's an error in the data checking process, NFC reader will send "command not supported" APDU to smartphone. If there's no error in the data checking process, NFC reader will send "command normally completed" APDU to smartphone. In this transaction process, NFC reader emulates NFC tag. Tag emulation is used because P2P cannot be implemented perfectly by the CN-370S-2 NFC reader.

Both transaction process between smartphone and smartphone, and transaction process between smartphone and NFC reader have similar sequence, transaction data transfer. In the transaction process between smartphone and smartphone, the transaction data transfer is done after merchant request completed. In the transaction process between smartphone and NFC reader, transaction data transfer is done after payer inputted amount and three digit random numbers. Transaction data transfer contains encrypted transaction data packet as shown in Figure 4.

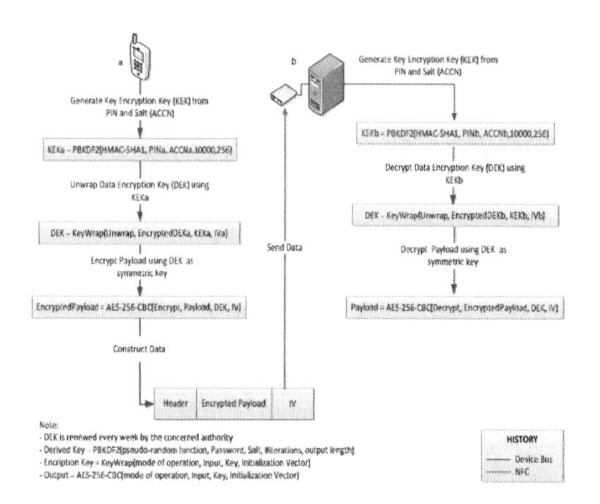

Figure 4. Transaction data transfer.

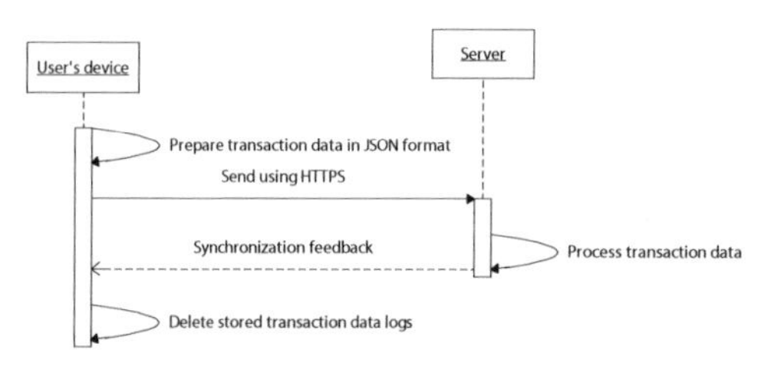

Figure 5. Synchronization process sequence diagram.

3.5 *Data synchronization process*

The synchronization process is done in the online state. Transaction data logs are sent in plain text form. HTTPS protocol is used to ensure that transaction data logs are sent to a destination address, and to ensure that data is not sniffed in the middle (Gourley & Totty 2002), as shown in Figure 5.

Transaction data logs are sent in plain text form so that the server will need less time to process transaction data sent by the user.

3.6 *Top-up, and key change process*

Top-up and key change process is done in a similar way. First, the system administrator needs to make changes in the server. The changes can be key replacement or topping up the balance for the payer account in the server. Users need to synchronize their device to get these changes. After synchronization, users will get a new key or new balance value. The top-up and key change process sequence diagram is shown in Figure 6.

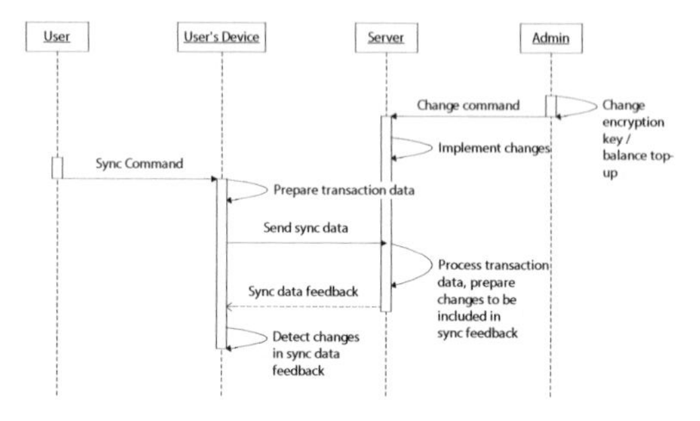

Figure 6. Top-up and key change process sequence diagram.

3.7 *Attendance process*

By following the existing pattern, in the attendance process, the seller is replaced by attendance machine. Attendance machine is used to replace the role of the merchant in the attendance process. Security feature used here is in the form of three-digit random numbers to prevent automated attendance filling.

The format of attendance data packet sent to the attendance machine is similar to transaction data packet, as shown in Figure 7.

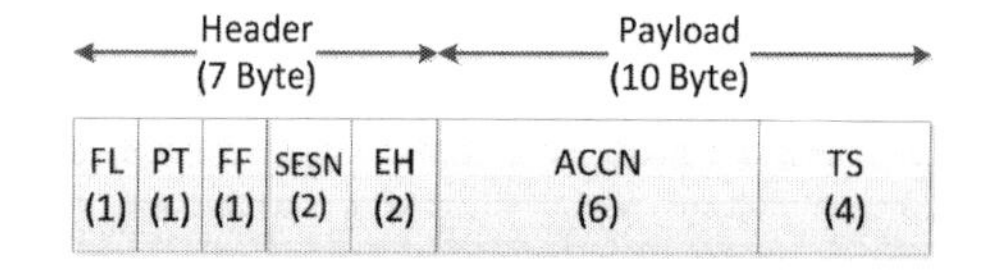

Figure 7. Attendance data packet format.

4 SYSTEM TESTING

4.1 *Key deriving time*

Password-based key derivation function is an algorithm that's very CPU-intensive. It has iteration as its function parameter. The more iteration used, the longer time needs to finish deriving key. When the algorithm is first introduced, 1000 iterations are a recommended minimum number of iterations. Google Nexus S, as the first NFC-enabled Android smartphone, needs almost a second to derive three keys with a number of iterations 1000, 900, and 800 while the newer device like Samsung Galaxy Note II only needs a quarter of a second to derive three keys with the same number of iterations, as shown in Table 2.

Table 2. Average time for deriving key.

Device	Average time (ms)
Sony Xperia M	469
Samsung Galaxy Note II	262.8
Google Nexus S	974.03
Asus X200CA	157.5

4.2 *Transaction time*

Transaction time measured is the transaction time between smartphone and smartphone, and transaction time between smartphone and EDC. The transaction time between smartphone and EDC takes more time than the transaction time between smartphone and smartphone, as shown in Table 3.

In transactions between smartphone and EDC, as stated before, NFC tag emulation is used. Using this approach, there are more steps required prior to sending data than using P2P approach.

Table 3. Average transaction time.

Transaction	Average time (ms)
Smartphone – Smartphone	141.50
Smartphone – EDC	270.87

4.3 *Synchronization time*

The amount of time required to synchronize transaction data depends on the number of synchronized transaction data. From the results obtained, the change of time is almost consistent with the number of synchronized transaction data change. When the number of synchronized transaction data gets bigger, the time required to synchronize is getting more unpredictable. This likely occurs because other processes that require an internet connection running in the device's operating system get higher priority due to long waiting time.

Table 4. Average synchronization time.

# transaction data	Average time (ms)
0	1354,80
20	2203,47
60	4351,43
100	7329,17

5 CONCLUSION

Development of integrated mobile e-money system can be done by designing the registration, synchronization, top-up, and attendance process in online state while transaction process is done in offline state and the data are in encrypted form.

Transaction between two Android smartphone only requires 141.50 msecs to complete, while transaction between Android smartphone and NFC reader requires 270.87 msecs to complete.

Transaction data packet consists of three parts: header, payload, and initialization vector. The header is used to provide information on the overall data packet. The payload contains encrypted transaction data. The initialization vector is used to decrypt the payload.

A key for encrypting transaction data is stored using AES Key Wrap algorithm. The key to wrapping key is derived using password-based derivation function.

Time required to derive key using password-based key derivation function depends on the device's computational resources. Google Nexus S, as the first NFC-enabled Android smartphone, needs almost a second to derive three keys with a number of iterations 1000, 900, and 800.

A synchronization process uses Hyptertext Transfer Protocol Secure (HTTPS) to ensure the transaction data received by the server in safely manner. Time required to synchronize the transaction data is directly proportional to the number of synchronized transaction data.

REFERENCES

Burnett, S. 2001. *RSA Security's Official Guide to Cryptography.* Berkeley: McGraw-Hill.

Chadha, R. 2013. Indonesia Online: A Digital Economy Emerges, Fueled by Cheap Mobile Handsets. *http://www.slideshare.net/slideshow/ embed_code/21155895*

Gourley, D. & Totty, B. 2002. *HTTP: The Definitive Guide.* Sebastopol: O'Reilly.

Paar, C. & Pelzl, J. 2010. *Understanding Cryptography: A Textbook for Students and Practitioners.* Berlin Heidelberg: Springer-Verlag.

Tsirulnik, G. 2014. Mobile Phone Ranked Most Used Electronic Device: Forrester. *http://www.mobilemarketer.com/cms/news/research/7473 .html*

Electronics, Communications and Networks IV – Hussain & Ivanovic (eds)
© 2015 Taylor & Francis Group, London, ISBN: 978-1-138-02830-2

A simple optimized joint routing and scheduling algorithm for multi-hop wireless network

Md. Anwar Hussain* & Arifa Ahmed
North Eastern Regional Institute of Science & Technology, Nirjuli, Itanagar, Arunachal Pradesh

ABSTRACT: A simple optimized routing and scheduling technique is reported here for routing data packets through wireless multi-hop networks. The technique optimizes end-to-end throughput or delay and is based on Minimum Angle Routing with Distributed (MA-DS) or Centralized (MA-CS) Scheduling between a Base Station (BS) and a user (U) node through a set of Intermediate (IM) nodes. One extra IM node is included for each U-IM, IM-IM or IM-BS node pair of MA-DS /MA-CS technique of route construction if it provides improved data rate between the pairs. Considerable increase of end-to-end throughput and decrease of delay are observed. Three network topologies with 39, 77, and 47 nodes are considered, including the users and the BS. The route computations require simple algorithm, hence the newly proposed algorithm is attractive for networks with large number of nodes as computational power involved is much less and particularly suitable for battery operated mobile devices.

1 INTRODUCTION

Nowadays, broadband wireless access networks are designed with cutting edge technologies for high data rate communication and connecting wired internet for global access. The wireless links now carry multimedia data of high data rate with enhanced and sophisticated signal processing and receiver design. For a wireless network consisting of large number of nodes with links and nodes scheduling requirements, efficient routing of data packets from a user node to a destination node is required, and the design of such routing and scheduling is guided by two main objectives: the efficiency of the route construction with joint scheduling in mind, and the optimal end to end throughput or delay performance.

Wireless mesh networks such as IEEE WiMax/802.16 standard (Fu et al. 2005) consist of fixed nodes that act as routers connecting mobile or fixed user nodes to a wired network through a base station (BS) which may act as a Gateway. The networks are designed to carry big data volume through its wireless links. The concerns in designing such broadband links are the signal interference at nodes (Ramamurthi et al. 2008) and the quality of service (Hong et al. 2007) transmitted to a user.

Sophisticated signal processing with suitable signal design and complex signal transmission and reception techniques are the challenges for research in this field for allocation of resources to users (Nahle et al. 2008). Route computation algorithms (Nahle et al. 2007) that guarantee optimal data rates may require scheduling technique design with formation of transmission groups (El-Najjar et al. 2010) for efficient utilization of the network resource. Joint centralized routing and scheduling (Tang et al. 2009, Du et al. 2007) and concurrent transmission (Tao et al. 2005) are designed for improving the performance of wireless mesh networks. A comprehensive work on the different routing functions for wireless multi-hop networks with their unifying and distinguishing features is seen in (Hamid et al. 2011).

Chakravarty & Hussain (2012) and Hussain et al.(2012) reported simple joint routing and scheduling techniques based on minimum angle routing between a user node U and BS through a set of intermediate nodes (IM) with centralized and distributed schemes. The techniques, called MA-CS and MA-DS, require an IM be found, which is within the transmission range of the starting node S (U, IM or BS) and at minimum angle with the reference line drawn from S to BS (U). A set of links are selected by the route computation from U to BS (BS to U). The route computations are simple and consume less computational power under the assumption that the locations (x-y coordinates) of all nodes including BS and U are known. The minimum angle node may be termed as terminating node T of a link constituting the route.

In the present work, one IM node between S (U, IM, or BS) and T (U, IM, or BS) is included, S and T are obtained from MA-CS or MA-DS technique. The extra IM is included between S and T provided it increases throughput in the path between S and T. Else the route remains as the direct link connection

*Corresponding author: *bubuli_99@yahoo.com*

between S and T as in MA-DS (MA-CS). The IM node that is included between S and T must be from the common neighbor set of S and T and is the one that provides the highest end to end throughput in the path from S to T. It is called MAIM-DS (MAIM-CS) route construction algorithm.

We consider two scheduling schemes for link or node scheduling to download or upload data packets. In the first scheme, when only one user node U (or the BS) is ready with data packets for sending them to BS (or a user U), the route from the U (BS) to the BS (U) is constructed by MAIM-DS (MAIM-CS) and links are activated sequentially. In the second scheme, when more than one user U (or the BS) is ready with data packets to send them to BS (or to more than one user U), respective routes are constructed by the algorithm; routes are activated sequentially, senders take turns to transmit. In the latter case, a scheduling frame is constructed with time slots for every route. Each slot of time is the total uploading (downloading) time for each data through respective routes.

2 ROUTING TECHNIQUE

2.1 *Minimum angle based routing using one extra im node*

In a wireless mesh network, a user U connects to BS through IM nodes for uploading data to wired network, whereas for downloading, BS connects to U. As reported in the research of Chakravarty & Hussain (2011) and Hussain et al. (2012), a simple route construction is possible from the U to BS (BS to U) by selecting minimum angle IMs between U and BS (BS and U), provided the U (BS) knows the x, y coordinates of all the nodes in the network. We modify the route construction algorithm of Chakravarty & Hussain (2011), Hussain et al. (2012) as explained below.

To route data packets, we need an end-to-end route consisting of different links. A link is a wireless connection between two nodes, a starting node S and a terminating node T. The MA-DS routing protocol draws a reference line from S (U, IM) to B and selects an IM from the neighbor set of nodes of S (IM nodes which are at one transmission range of S), which is at minimum angle, i.e. the S-T (T is the selected IM) line makes minimum angle with the reference line. Next, the selected IM acts as the S node, draws a reference line from it to BS, selects a minimum angle T node from its neighbor and the next link of the route is obtained. The IM (IMs) selected in the previous link (links) is (are) excluded. This process continues until the last IM is the BS. In the case of MA-CS, the route constructions start from the BS and end through different IMs, in the same way as in MA-DS, until the last IM is the user U. The two routes, from U to BS and from BS to U, may be different.

We modify both MA-DS and MA-CS routing protocols to optimize end-to-end throughput between U and BS. As shown in Figure 1 (a), the MA-DS would select the link U1–N4 (first user to 4th node) as the first link of the route between U1 to BS. This link has data rate as per the link length and transmission range and other parameters are shown in Table 1. Now in order to optimize end-to-end throughput, data rate possibility in every pair of S-T is optimized by searching for an alternate path which takes only one IM between S and T and the IM needs to be a common neighbor of both S and T. We call this routing protocol as MAIM-DS. As shown in Figure 1(a), MAIM-DS selects N3 as the extra IM, the earlier link U1-N4 now converts to a path of two links: U1-N3 and N3-N4, and the data rate of the path U-N3-N4 is more than the link data rate of U1-N4 link. The Figure 1 (b) shows the complete route from the U1 to BS: U1-N4-N7-BS as per MA-DS, and U1-N3-N4-N5-N7-N9-BS as per MAIM-DS routing protocol.

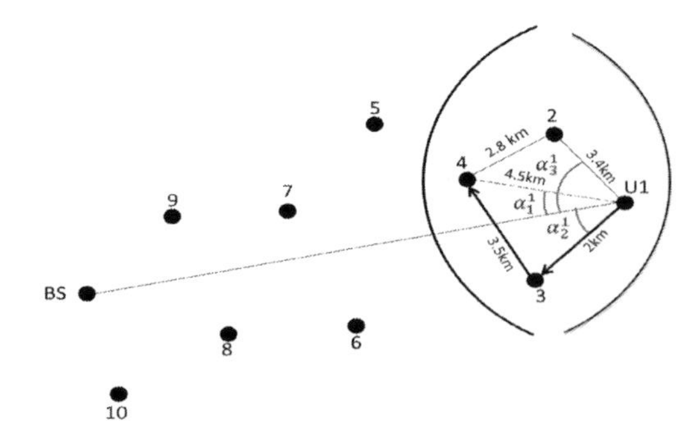

Figure 1(a). MIMA- DS using one extra IM node.

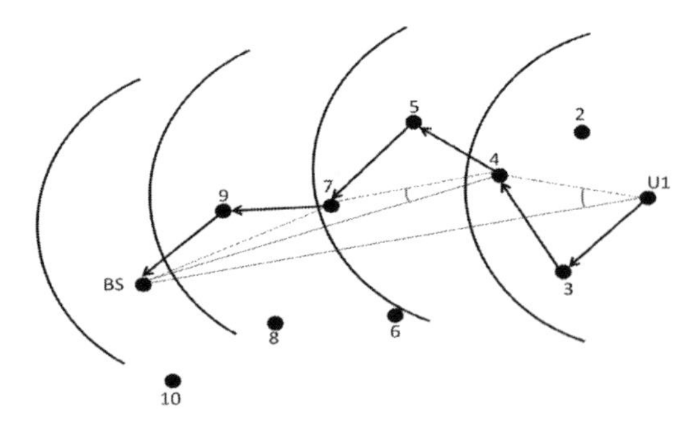

Figure 1(b). Complete route using MAIM- DS.

For MA-CS and MAIM-CS routing protocols, the complete route selection is explained in Figure 2(a). As said above, the route computation here starts from the BS end and the first link of the route is BS-N8 as per MA-CS and the first path is BS-N10-N8 as per MAIM-CS, the path provides more data rate than the link. The principle of route construction is same as in MA-CS, for every link between S and T one extra IM being found and IM being the common neighbour of S and T. The other links or paths of the complete route are derived in the same way until the last T is a user U. Figure 2(b) shows the complete route between the BS and U1, BS-N8-N6-N3-U1 as per MA-CS, and BS-N8-N6-N4-N3-U1 as per MAIM-CS routing protocol.

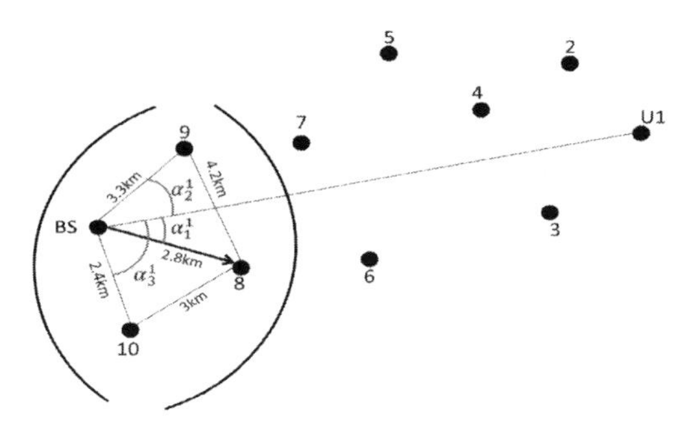

Figure 2(a). MAIM-CS using one extra IM node.

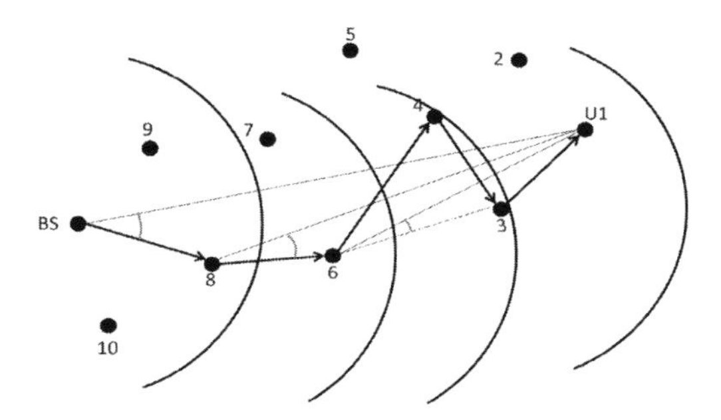

Figure 2(b). Complete route using MAIM-CS.

3 METHODOLOGY

The performance of multi-hop wireless networks that we consider here is evaluated in terms of the end-to-end throughput *ETH* from a user (BS) to the BS (user). We consider a fixed data frame of size 11 Mb, 5 Km. Transmission range of each node and the burst profiles over various transmission distances (link lengths) are shown in Table 1.

Table 1. Bit rates and transmission ranges.

Modulation	Coding Rate (*R*)	Link Range (Km)	Bit Rate (*BR*) (Mbps)
QPSK	1/2	$3.5 < l_i \leq 5$	2
16-QAM	1/2	$2 < l_i \leq 3.5$	5.5
64-QAM	3/4	$l_i \leq 2$	11

The table shows link data rates for different transmission ranges and the modulation and coding rate to be used by a transmitting node. As an example, over a link length of 5 Km. a data rate of 2 Mbps is possible with QPSK modulation and data coding with rate ½.

As seen clearly from Table 1, nodes connected with smaller wireless link distances have better data rates. This data rate can be further enhanced if an alternative route, involving other neighbor nodes (neighbor of both S and T) between the *S* and *T* node, is possible, which requires smaller time durations TN_i than T_i. Else the link remains the link between *S* and *T* directly as per MA-DS (MA-CS). Only one extra *IM* between S and T is included. We calculate the time durations T_i of different slots through different wireless links l_i required to deliver the fixed length data frame *F* Mb from the user (BS) to the BS (user) by using both MA-DS and MA-CS algorithms. The corresponding value is TN_i in MAIM-DS and MAIM-CS, here the link implies the path from S to T through the extra IM (if one is needed) included. The end-to-end throughput of the networks is derived by using both MA-DS (MA-CS) and MAIM-DS (MAIM-CS) route construction algorithms for comparison. To transmit a data file of size *F* Mb through the ith link of a network, the link having bit rate BR_i Mbps for code rate R_i, the MA-DS (MA-CS) algorithm requires time slot T_i (second) as given by the following equation.

$$Ti = (F/Ri)/BRi \tag{1}$$

The end-to-end throughput *ETH* over a route is obtained as

$$ETH_{MA} = F/sum\ (T_i) \tag{2}$$

where $sum(T_i)$ is the total time required through all the links to reach the destination. For the case of MAIM-DS (MAIM-CS) routing, ETH_{MAIM} is calculated as follows. As explained above, one *IM* between the *S* and *T* is considered, the selection of *IM* is explained above, and the time duration T_{SN} between *S* and *IM* and T_{NE} between IM and T are expressed as follows, we have then

$$TN_i = T_{SN} + T_{NE} \tag{3}$$

$$T_{SN} = (F/R_{SN})/BR_{SN} \tag{4}$$

$$T_{NE} = (F/R_{NE})/BR_{NE} \tag{5}$$

where R_{SN} and R_{NE} are the respective coding rates, and BR_{SN} and BR_{NE} are the respective link bit rates. The end-to-end throughput by using MAIM-DS (MAIM-CS) is calculated as

$ETH_{MAIM} = F/sum\ (TN_i)$ (6)

where *sum (TN_i)* is the total time required through all the paths between U (BS) and BS (U) to reach the destination.

In the second scheme of scheduling where every U-BS (BS-U) takes turns in uploading or downloading, the throughput performance of the whole network is given as

$TH_{MAIM\text{-}NET} = F/sum\ (sum\ (TN_i))$ (7)

where the second sum takes care of all routes constructed by MAIM-DS (MAIM-CS) algorithm. In MA-DS (MA-CS) algorithm, the corresponding expression is

$TH_{MA\text{-}NET} = F/sum\ (sum\ (T_i))$ (8)

It is assumed that the nodes have burst profiles with modulation and coding rates to obtain different data rates over various transmission distances (link lengths), as shown in Table 1. We consider three network scenarios as mentioned in Table 2.

Table 2. Various network parameters.

Network Scenario	No of Users	No of IMs	No of BS	Transmission Range of nodes (Km.)	Area of Network
1	6	32	1	5	20x20
2	5	71	1	5	40x40
3	9	37	1	5	25x25

4 PERFORMANCE EVALUATION

4.1 *Route construction, scheduling and performance analysis: Case 1*

Figure 3(a) shows the routes for data transmissions. In the first network from different users to the BS by using MA-DS and MAIM-DS algorithm, dotted lines indicate routes using MA-DS, and solid lines indicate routes using MAIM-DS. Similarly, Fig 3(b) shows the routes for data transmissions from BS to different users by using MA-CS and MAIM-CS algorithm, dotted lines indicate routes using MA-CS, and solid lines indicate routes using MAIM-CS.

To compare the routing performance of the two algorithms, different elements are tabulated as Table 3(a) and it can be found that with an increase of total IMs in the route the MAIM-DS, the sufficient increase in end-to-end throughput *ETH* from different users to BS is provided, compared with MA-DS. Table 3(b) shows the corresponding values for MAIM-CS and MA-CS, and the same trend as in Table 3(a) is also shown.

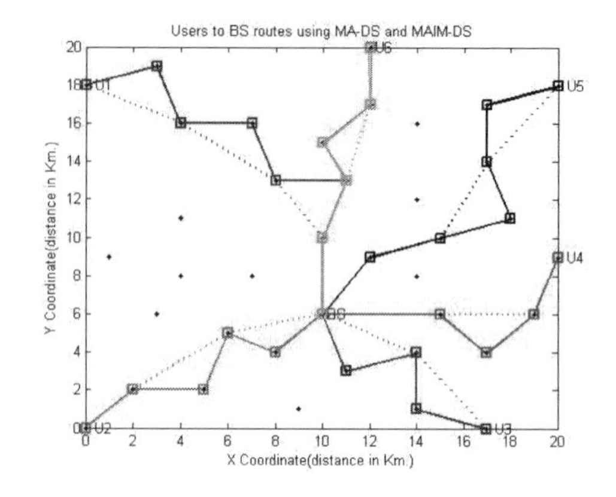

Figure 3(a). Routes constructed by using MA-DS and MAIM-DS for the first network.

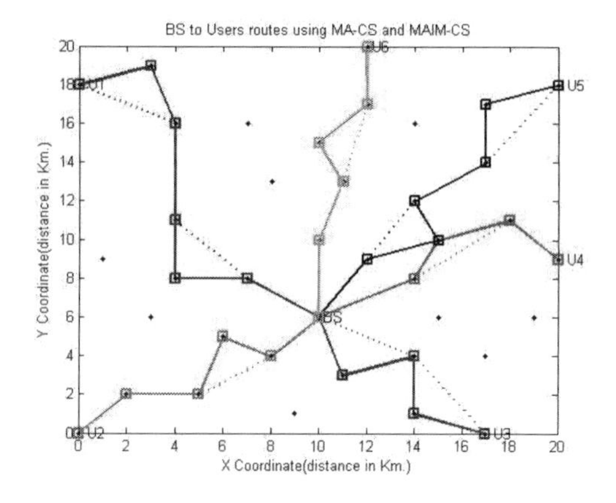

Figure 3(b). Routes constructed by using MA-CS and MAIM-CS for the first network.

Table 3(a). End-to-end throughput for first network.

	MA-DS/ MAIM-DS			
Usr	ETH_{MA}/ ETH_{MAIM} (Mbps)	$Sum(T_i)$/ $Sum(TN_i)$ (s)	No. of IMs / No. of IMs	% increment of *ETH* by MAIM-DS
U1	0.25/ 0.314	44/ 35	3/6	25.72
U2	0.423/ 0.55	26/ 20	2/4	30
U3	0.5/0.69	22/ 16	1/3	37.50
U4	0.423/0.478	26/ 23	2/3	13.05
U5	0.297/0.35	37/ 31	3/5	19.34
U6	0.367/0.407	30/ 27	3/4	11.10

Table 3(b). End-to-end throughput for first network.

User	MA-CS/ MAIM-CS ETH_{MA}/ ETH_{MAIM} (Mbps)	$Sum(T_i)$/ $Sum(TN_i)$ (s)	No. of IMs/No. of IMs	%increment of ETH by MAIM-CS
U1	0.25/0.29	44/38	3/5	15.8
U2	0.478/0.55	23/20	3/4	15
U3	0.5/0.69	22/16	1/3	37.50
U4	0.423/0.478	26/ 23	2/3	13.05
U5	0.25/0.29	44/38	3/5	15.8
U6	0.367/0.407	30/27	3/4	11.1

In Figure 4(a), the route constructions for the second network using MA-DS (dotted lines), MAIM-DS (solid lines) algorithm, and in Figure 4(b), the constructions by using MA-CS (dotted lines), MAIM-CS (solid lines) algorithm are shown. Performance comparisons are shown in Table 4(a) and Table 4(b), respectively. Sufficient improvement in the end-to-end throughput performance by using MAIM-DS and MAIM-CS is observed. Also it is noted that the improvement is obtained with a sufficient increase in the number of IM nodes from the U (BS) to BS (U) node routes.

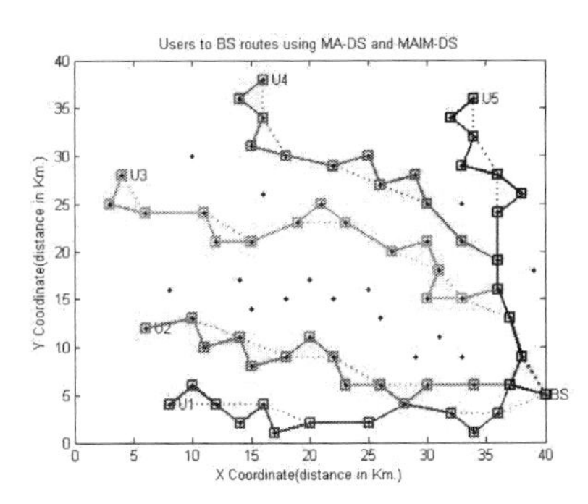

Figure 4(a). Routes constructed by using MA-DS and MAIM-DS for the second network.

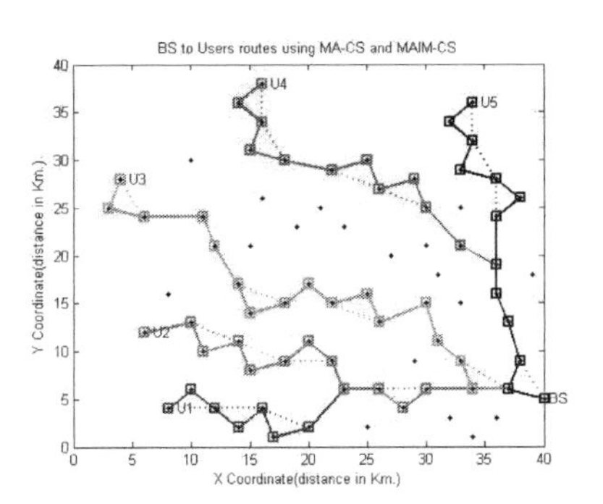

Figure 4(b). Routes constructed by using MA-CS and MAIM-CS for the second network.

Table 4(a). End-to-end throughput for second network.

User	MA-DS/MAIM-DS ETH_{MA}/ ETH_{MAIM} (Mbps)	$Sum(T_i)$/ $Sum(TN_i)$ (s)	No. of IMs / No. of IMs	% increment of ETH by MAIM-DS
U1	0.125/0.151	88/73	7/12	20.56
U2	0.129/0.157	85/70	8/13	21.41
U3	0.091/0.11	121/100	10/17	21.01
U4	0.103/0.12	107/92	10/15	16.34
U5	0.149/0.177	74/62	7/11	19.38

Table 4(b). End-to-end throughput for second network .

User	MA-CS/ MAIM-CS ETH_{MA}/ ETH_{MAIM} (Mbps)	$Sum(T_i)$/ $Sum(TN_i)$ (s)	No. of IMs/ No. of IMs	% increment of ETH by MAIM-CS
U1	0.14/0.167	78/66	8/12	18.23
U2	0.134/0.157	82/70	9/13	17.15
U3	0.102/0.115	108/96	12/16	12.46
U4	0.103/0.12	107/92	10/15	16.34
U5	0.149/0.177	74/62	7/11	19.38

The route constructions for the third network, as is done for the first and second network, are shown in Figure 5(a) by using MA-DS and MAIM-DS, and in Fig 5(b) by using MA-CS and MAIM-CS algorithms. The respective performance comparison tables are Table 5(a) and Table 5(b). The tables show a clear advantage of MAIM-DS (MAIM-CS) over MA-DS (MA-CS). The improvement of end-to-end throughput is remarkable if we construct routes by using MAIM-DS and MAIM-CS algorithms. Here, as in the case of the first and second network, increase in the total number of IM nodes between U (BS) and BS (U) is also notable.

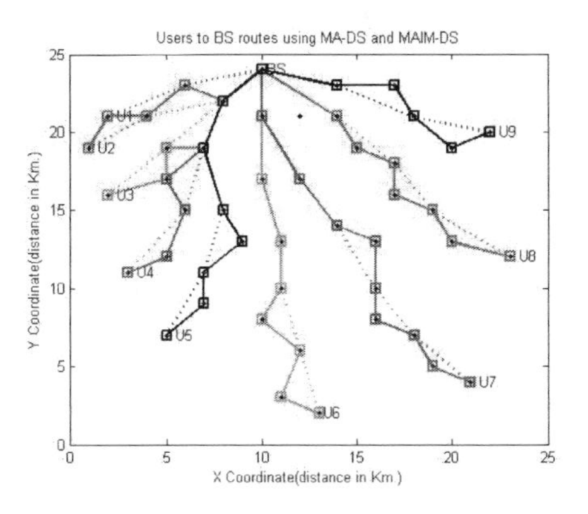

Figure 5(a). Routes constructed by using MA-DS and MAIM-DS for the third network.

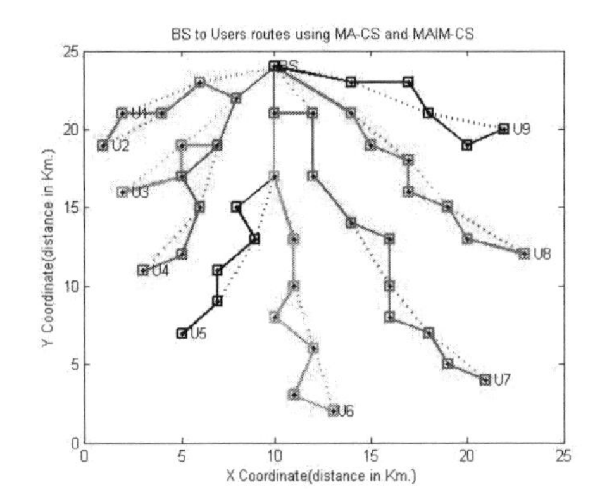

Figure 5(b). Routes constructed by using MA-CS and MAIM-CS for the third network.

Table 5(a). End-to-end throughput for third network.

	MA-DS/ MAIM-DS			
User	ETH_{MA}/ ETH_{MAIM} (Mbps)	$Sum(T_i)$/ $Sum(TN_i)$ (s)	No. of IMs /No. of IMs	% increment of *ETH* by MAIM-DS
U1	0.5/0.825	22/13.3	1/3	65.04
U2	0.423/0.635	26/17.3	2/4	50.01
U3	0.423/0.75	26/14.66	2/4	77.33
U4	0.367/0.458	30/24	3/5	24.98
U5	0.268/0.34	41/32.33	4/6	26.80
U6	0.212/0.239	52/46	5/7	13.05
U7	0.186/0.232	59/47.3	5/8	24.68
U8	0.25/0.340	44/32.3	3/6	36.08
U9	0.333/0.407	33/27	2/4	22.23

Table 5(b). End-to-end throughput for third network.

	MA-CS/MAIM-CS			
User	ETH_{MA} / ETH_{MAIM} (Mbps)	$Sum(T_i)$/ $Sum(TN_i)$ (s)	No. of IMs / No. of IMs	% increment of *ETH* by MAIM-CS
U1	0.5/0.825	22/13.33	1/3	65.04
U2	0.423/0.635	26/17.33	2/4	50.01
U3	0.423/0.75	26/14.66	2/4	77.33
U4	0.367/0.458	30/24	3/5	24.98
U5	0.268/0.34	41/32.33	4/6	26.80
U6	0.212/0.24	52/46	5/7	13.05
U7	0.167/0.226	66/48.66	5/9	35.63
U8	0.25/0.34	44/32.33	3/6	36.08
U9	0.333/0.407	33/27	2/4	22.23

4.2 *Route construction, scheduling and performance analysis: Case 2*

As explained in the introduction, in this case, data packets are scheduled through various links from the nodes via IMs for every source-destination pair, routes constructed by MAIM-DS (MAIM-CS) algorithm and every source-destination pair takes turn in their uploading or downloading. Scheduling frames are constructed and the throughput performance of the whole network is given by

$$TH_{MAIM\text{-}NET} = F/sum\ (sum\ (TN_i)) \tag{8}$$

where the first sum takes care of all routes. Here although throughput obtained by each pair is less, all pairs get fair treatment by the network. In MA-DS (MA-CS) algorithm, the corresponding expression is

$$TH_{MA\text{-}NET} = F/sum\ (sum\ (T_i)) \tag{9}$$

Table 6. Frame schedule length & throughputs.

Routing Algorithm	Network	$sum(sum(TN_i))$ (seconds)	$TH_{MAIM\text{-}NET}$ (Kbps)
MA-DS	1	185	59.46
	2	475	23.16
	3	333	33.03
MAIM-DS	1	152	72.37
	2	397	27.71
	3	254.31	43.25
MA-CS	1	189	58.20
	2	449	24.50
	3	340	32.35
MAIM-CS	1	162	67.90
	2	386	28.50
	3	255.64	43.03

Table 6 shows the results in different networks for MA-DS, MAIM-DS, MA-CS and MAIM-CS algorithms. From Table 6, it is observed that the achieved throughput performance of the whole network by using MAIM-DS and MAIM-CS algorithms is better than that by using MA-DS and MA-CS, respectively, and the trend is same in all the networks. Thus we may say that MAIM-DS and MAIM-CS algorithms perform better than MA-DS and MA-CS algorithms, respectively.

In the above discussed schemes and algorithms, the transmission times are only considered and propagation times, processing times for coding, decoding, modulation and demodulation of data signals are ignored.

5 CONCLUSION

In this paper, simple optimized routing and scheduling algorithm is proposed for improving the performance of a wireless multi-hop mesh network. Two algorithms are discussed here, namely MA-DS (MA-CS) and MAIM-DS (MAIM-CS). In the second algorithm, one IM between every starting and minimum angle node, which is a common neighbor to both the nodes, is included such that it enhances the throughput between the starting and minimum angle node. If there is no enhancement, the direct link between the node pairs remains. Routing paths by using these two algorithms are constructed. End-to-end throughput for each pair of User-BS route of a network and throughput of a complete network are calculated for both the algorithms and they are compared. Here the first mentioned algorithm is proposed earlier and the second mentioned algorithm is proposed newly here which is an improvement of the earlier proposed algorithm. The performances of the two routing algorithms with two different schemes of scheduling on the throughput of three network topologies are shown by simulations. From simulation results, it is seen that the proposed new algorithm works better than the earlier one in giving much higher throughput for the two different cases of scheduling. Here both distributed and centralized computations for constructing routes by using the two algorithms are considered. Both the route construction algorithms are simple and require only position (x-y co-ordinate) information and the burst profiles of the nodes. The distributed or centralized route computations refer to taking the starting node as the user node or the BS, respectively. The only information needed as the input to both the algorithms is the node's positions and burst profile.

REFERENCES

Chakraborty, I. & Hussain, M. A. 2012. A simple joint routing and scheduling algorithm for a multihop wireless network. *ICCSII12*, AUS, Dubai.

Du, P., Jia, W., Huang, L. & Lu, W. 2007. Centralized scheduling and channel assignment in multi-channel single-transreceiver WiMax mesh network. *Wireless Communications and Networking Conference, Proc. of WCNC '07, Kowloon, 11–15 March 2007* : 1734–1739. IEEE.

El-Najjar, J., Assi, C. & Jaumard, B. 2010. Maximizing the network stability in mobile WiMAX mesh networks. *Mobile New Applications published in Springer Science & Business Media*15(2): 253–256.

Fu, L., Cao, Z. & Fan, P. 2005. Spatial reuse in IEEE 802.16 based wireless mesh networks. *In Proc. IEEE International Symposium on Communications and Information Technology, 12–14 October 2005*: 1358–1361. IEEE.

Hamid, S. A., Hassonein, H. & Tkahara, G. 2011. Routing for wireless multi hop networks-unifying and distinguishing features. *Technical Report2011*: 583, Telecommunication Research Lab, Queen's University.

Hong, C.-Y., Pang, A.-C. & C.Wu, J.-L. 2007. QoS routing and scheduling in TDMA based wireless mesh backhaul networks. *In proc. of IEEE conference on Wireless Communications and Networking (WCNC '07), Kowloon, 11–15 March 2007*: 3232–3237. IEEE.

Hussain, M. A, Faiz, M. F. I, & Chakravarty, I. 2013. Simple Routing Algorithm for Multi-hop Wireless Network. *Journal of Emerging Trends in Computing and Information Sciences* 4: 58–65.

Nahle, S. & Malouch, N. 2008. Joint routing and scheduling for maximizing fair throughput in WiMAX mesh network. *In Proc. of IEEE International Symposium on Personal, Indoor and Mobile Radio Communication, Cannes, 15–18 September 2008: 1–5*. IEEE.

Nahle, S., Iannone, L., Donnet, B. & Malouch, N. 2007 Dec. On the construction WiMAX mesh tree. *IEEE Communication Letters* 11(12): 967–969.

Ramamurthi, V., Reaz, A. & Mukherjee, B. 2008. Optimal capacity allocation in wireless mesh networks. *In proc. of IEEE Global Telecommunications Conference, , New Orleans, LO, 30 Nov.– 4 Dec. 2008*: 1–5. IEEE.

Tang, Y., Yao, Y. & Yu, J. 2009. A novel joint centralized scheduling and channel assignment scheme for IEEE 802.16 mesh networks. *In proc. of IEEE international conference on Computer Science & Education (ICCSE '09), Nanning, 25–28 July 2009*: 289–293. IEEE.

Tao, J., Liu, F., Zeng, Z. & Lin, Z. 2005. Throughput enhancement in WiMAX mesh networks using concurrent transmission. *In Proc. of IEEE International Conference on Wireless Communications, Networking and Mobile Computing, 23–26 September 2005*: 871–874. IEEE.

Electronics, Communications and Networks IV – Hussain & Ivanovic (eds)
 ISBN: 978-1-138-02830-2

Shadowing in M2M channels at 2100 and 700 MHz in dense scattering environments

Y. Ibdah & Y. Ding
Department of Electrical Engineering and Computer Science, Wichita State University, USA

Y. H. Ding*
School of Electronics Information Engineering, Tianjin Key Laboratory of Film Electronic and Communication Devices, Tianjin University of Technology, Tianjin, China

ABSTRACT: This paper presents the shadowing in Mobile-to-Mobile (M2M) channel from the measurements at 2:1GHz and 700MHz bands in suburban and forest areas with dense scattering environments. Four test scenarios are considered for the transmitter (Tx) and receiver (Rx) placed inside traveling vehicles or on a test cart pushed at a walking speed. Channel models are studied for path loss, shadowing, small-scale fading based on measurements. The variance of shadowing varies from 1.8 to 4.9.

1 INTRODUCTION

In mobile ad hoc wireless networks and intelligence transportation systems, the nodes or terminals are no longer stationary (Buccio et al. 2005). The channels in the M2M hops are different from those in cellular communication systems, since the Tx and Rx are both in motion, in addition, the antennas at the Tx and Rx are closer to ground levels. Small-scale M2M channel models are studied for isotropic and nonisotropic scattering in Rayleigh, cascaded-Rayleigh, and Rician fading Patel et al. (2005). Intensive measurements for vehicle-to-vehicle (V2V) channels are conducted in various highway, suburban, rural roads, and some application-specific scenarios such as traffic congestion at 5GHz and 700MHz, e.g. (Matolak, Sen, Xiong, & Yaskoff 2005, Sen & Matolak 2008). In most V2V channel measurements, transmitter and/or receiver antennas are usually placed on the roof of test vehicles. However, in many application of M2M communications, the mobile units (and their antennas) are often inside a vehicle traveling on roads; or inside a pocket or bag carried by a pedestrian walking at side roads, parking lots, or woody/forest areas. Available channel measurements and modelings for these types of M2M channels are found limited, although the vehicle penetration loss for antennas inside a vehicle is studied by measurements in 100 ~ 2400MHz(Hill & Kneisel 1991, Tanghe, Joseph, Verloock, & Martens 2008).

In this paper, we conduct measurements for the M2M channels in woody/forest and suburban areas with dense scattering environments at 2.1 GHz and 795 MHz frequency bands, and explore the channel models for large scale and small-scale fading. Suburban areas usually include streets with one or two lanes in each traffic direction, and buildings or houses set back from the curb; while urban areas have wider streets with buildings or houses closer to the curb (Mecklenbrauker, Molisch, Karedal, Tufvesson, Paier, Bernado, Zemen, Klemp, & Czink 2011). Depending on the tests routes and areas, the Tx and Rx travel towards or away from each other in an angle (e.g. perpendicularly), or in convoy and opposite directions as in many V2V measurements (e.g. (Matolak, Sen, Xiong, & Yaskoff 2005, Sen & Matolak 2008, Paier, Karedal, Czink, Hofstetter, Dumard, Zemen, Tufvesson, Molisch, & Mecklenbrauker 2007, Maurer, Fugen, & Wiesbeck 2002, Paier, Karedal, Czink, Dumard, Zemen, Tufvesson, Molisch, & Mecklenbruker 2009, Karedal, Czink, Paier, Tufvesson, & Molisch 2011)). Two antenna placements are considered: a). inside-vehicle-to-inside-vehicle (IV2IV) where the antennas of both Tx and Rx are placed inside the test vehicles; b). inside-vehicle-to-walk (IV2W) where the antenna of Tx is inside the vehicle and antenna of Rx is placed on a cart pushed at a walking speed.

*Corresponding author: *lucydyh@163.com*

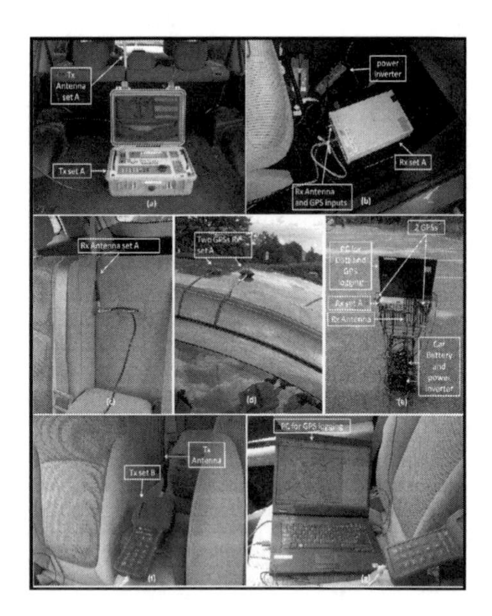

Figure 1. (a) and (b). Tx and Rx of Set A inside test vehicle; (c). Set A Rx antenna inside test vehicle; (d). GPSs for Set A Rx on roof of vehicle; (e). Set A Rx on test cart; (f). Set B Tx inside test vehicle; (g). Laptop and GPS for Set B Rx.

2 MEASUREMENTS SETUP

The channel measurements were conducted in northern New Jersey in summer 2012 when the trees were in their full growth. Test areas were chosen in woody/forest and suburban areas with a mix of dense trees, story buildings, street signs and corners, where heavy diffracting and dense scattering to radio signals were likely to occur (Rappaport 2001). The areas were scanned while the Tx was turned off before conducting the tests. A low received power at −151dB was recorded at the testing frequencies, 2110:2MHz and 795MHz, to ensure that the selected radio frequencies were not used by other carriers. The receiver was configured to, respectively, 50kHz bandwidth with −165dB sensitivity/collection threshold for frequency 2110:2MHz (2:1GHz frequency band); and 100kHz bandwidth with −140dB sensitivity for frequency 795MHz. The measurements were collected using test vehicles and a mobile cart. Laptop computers were used to log the data. A Global Position System (GPS) was connected to the laptop to record the location, elevation, and speed. One GPS is used for the Tx and two GPSs were used at the Rx for a higher accuracy.

2.1 *Test scenarios*

Four scenarios in frequencies 2110:2MHz and 795MHz are characterized by different tests areas. The test routs are shown in Fig. 1.

2.1.1 *Scenario 1 - IV2IV Forest at 2110:2MHz*

The test routes for this scenario are inside and at boundary of a woody/forest area. The Tx and Rx travelled in the same directions on traffic roads (convoy) Rx → Tx → and ← Rx ← Tx, they also traveled away or towards each other at an angle, such as 90° out ← ↑ and 90° in ← ↑, marked by the dotted and dash dotted lines. About 2600 data samples were collected at 2110:2MHz. The Tx and Rx antennas were vertically placed inside the test vehicles at heights of, respectively, 1.5m and 1.25m above the ground level. The Tx and Rx traveled at speed 3 ~ 6 miles/hr with a separation distance 60 ~ 310m between them. The visibility from the Tx to Rx was guaranteed in most of the measurements collected, except in those due to the trees between the Tx and Rx when the test vehicles traveled in 90° in/out at the roads shown by the dotted and dash dotted lines.

2.1.2 *Scenario 2 - IV2IV Suburban at 2110:2MHz*

In this scenario, the test equipment and antenna positions were configured the same as in Scenario 1. The test areas were in suburb streets. Although suburban areas are determined by the geography of the land, the visibility from the Tx to Rx are not always guaranteed. We collected two sets of measurements: visibility existed and not existed, in order to obtain more accurate channel models.

Scenario 2a -Without LOS Three test routes were configured, by dotted, dashed, and dash-dotted lines, respectively. The Tx traveled at a T-shaped street at street speed along with the traffic and the Rx travelled in a smaller street in a parking lot at speed 5 ~ 20 miles/hr. The Tx and Rx traveled the same or opposite direction with a separation 15 ~ 460m, and there was no visibility between them. Around 6000 samples were collected for Scenario 2a.

Scenario 2b - With LOS The Tx and Rx travelled in a two-way suburb street with two-lanes in each direction. Groups of two- or three-story buildings were found 20 ~ 40m away from the curbs. A 10m wide divider or median strip of trees, bushes, and grass was in the middle of the street. The LOS between the Tx and Rx existed when they were moving in the same direction, convoy ← Tx ← Rx, and partially not existed in opposite directions, Tx → ← Rx, due to the trees and grass in between.

2.1.3 *Scenario 3 - IV2W Suburban at 2110:2MHz*

The Tx was inside the test vehicle with antenna 1.5m above the ground level. The Rx was placed on a cart and the antenna was 1m above ground. In the first area, the Tx traveled at 20 ~ 35 miles/hr in the streets marked by dash-dotted and dotted lines, and the Rx was pushed in the cart at pedestrian speed 2 ~ 3 miles/hr in the side walks marked by dash-dotted and dotted lines. The separation between Tx and Rx was 20 ~ 300m. Around 3200 samples were collected in this test area. The LOS between the Tx and Rx existed for most of samples collected, excepts for those when the LOS was blocked or partially blocked by the traffic, the trees, and the person pushing the cart.

2.1.4 *Scenario 4 - IV2W Suburban at 795MHz*

The Tx was placed inside the vehicle and the Rx on the cart. The heights of both Tx and Rx were 1m above ground level. The Tx traveled in a two-way suburb street marked by dash-dotted lines and a side street marked by dotted lines. The two-way street did not have divider strip. Groups of 2- or 3-story buildings were found 20m away from the curbs in the side street. The Rx was pushed in the parking lot marked by dash-dotted lines, and at a side walk marked by dotted lines. The separation between Tx and Rx was 4 ~ 150m, and the visibility between them was obstructed by trees, bushes. Around 1430 samples were collected for this scenario at 795MHz.

3 SHADOWING CHANNEL MODELING

In general, the received signal power at the receiver is determined by the transmitted power, path loss, large scale (shadowing), and small-scale fading (Rappaport 2001). The received signal power in (dB) can be generally expressed as $P_r = P_t - PL + S + \Omega$, where P_t is the transmitted power, PL is the path loss, S is the power in shadowing, and Ω is the power in smallscale fading (Rappaport 2001). In this section, we analyse the channel models for shadowing based on the measurements obtained in scenarios 1-4. We remove the effect path loss and obtain models for shadowing from the measurements. Table 1 provides a summary of the proposed shadowing models for all scenarios.

Shadowing characterizes the local average variation of the envelope around the path loss. Typically, it is estimated by a moving average window over distance (Rappaport 2001). Depending on the test conditions and surrounding environments, the window size is usually 5 ~ 40 times of the wavelength in the radio frequency (Mecklenbrauker, Molisch, Karedal, Tufvesson, Paier, Bernado, Zemen, Klemp, & Czink 2011). Fig. 2 illustrates four moving average windows with different sizes using the measurements in Scenario 1. The window sizes are 1m, 2m, 3m, and 4m, respectively. These window sizes correspond to 7, 14:1, 21:1, and 28:1 times of the wavelength in frequency 2:1GHz; and 2:7, 5:3, 8, and 10:6 times wavelength in 795MHz. It can be observed that 4m window fails to follow the local average of shadowing in the received signal; the 1m and 2m windows seem too small, as a result, picking up part of fast variations (small-scale fading) components; the *3m* window appears to be a reasonable choice to extract shadowing for the measurements.

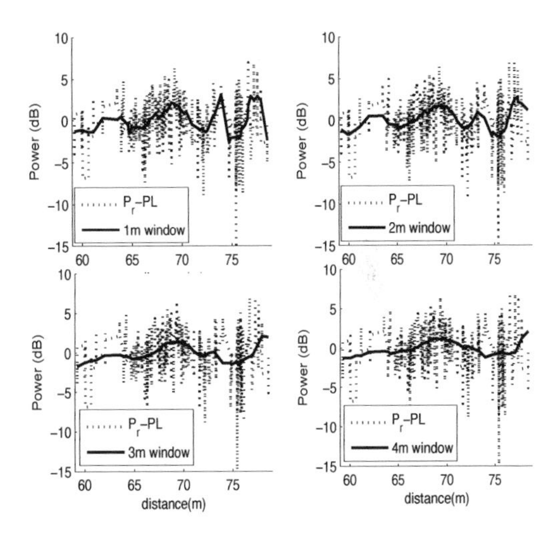

Figure 2. Moving average windows in shadowing estimation, Scenario 1.

Let S represent the local average power (in dB) after removal of the path loss component. The lognormal probability density function (pdf) for shadowing is given by (5.59) (Molisch 2010): $f_s(s) = \frac{10}{\sqrt{2\pi}ks\ \ln(10)} \exp\left(\frac{-(10\log s-\mu)^2}{2k^2}\right)$, where ln is the natural logarithm of base e, mean $\mu = \mathbb{E}[S]$, and variance $k^2 = \mathbb{V}[S]$. Fig. 3 show the distributions of shadowing in Scenario 1. The probability density function of log-normal are also plotted with values of mean μ and variance k estimated from the measurements. As listed in Table 1, the means of shadowing are close to zero for all scenarios, except for Scenario 2b in which there is no visibility between the Tx and Rx. For frequency 2:1GHz, the variance of shadowing in the forest environment in Scenario 1 is lower (2:37dB) than those in Scenario 2 (4:85 and 4:87dB) and in Scenario 3 (3:92dB). For frequency 795MHz, the variance of shadowing is 1:80dB, which is lower than those for 2:1GHz.

Table 1. Scenarios summary.

Scenario	Shadowing
Scenario 1	2.37
Scenario 2a	4.8536
Scenario 2b	0.1042
Scenario 3	3.9178
Scenario 4	1.7980

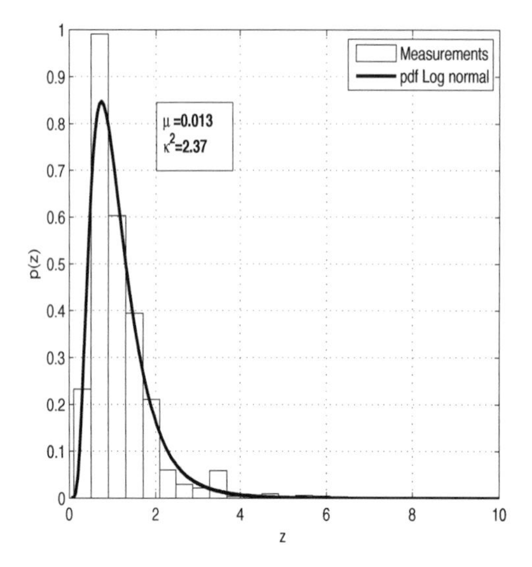

Figure 3. Distribution of the shadowing in Scenario 1.

4 CONCLUSION

We presented IV2IV and IV2W M2M channel measurements in suburban areas with dense scattering environments in 2.1GHz and 795MHz. Empirical models for shadowing are studied based on the measurements. The measurements suggest that the variance varies from 1.8 to 4.9.

ACKNOWLEDGEMENT

This work was supported by the National Natural Science Foundation of China (No. 51101113), Tianjin Natural Science Foundation (No.14JCYBJC16200).

REFERENCES

Buccio, P. Masala, E. Kawaguchi, N., Takeda, K. & Martin, J. C. D. 2005. Performance evaluation of H. 264 video streaming over inter-vehicular 802.11 ad hoc networks. In *Proc. IEEE Symp. on Personal, Indoor and Mobile Radio Commun.*, Volume 3, Berlin, Germany, pp. 1936–1940.

Hill, C. & Kneisel, T. 1991. Portable radio antenna performance in the 150, 450, 800, and 900 MHz bands outside and in-vehicle. *IEEE Trans. Veh. Technol. 40*(4), 750–756.

Karedal, J., Czink, N., Paier, N., Tufvesson, F. & Molisch A. F. 2011. Path loss modeling for vehicle-to-vehicle communications. *IEEE Trans. Veh. Technol. 60*(1), pp. 323–328.

Matolak, D., Sen, I., Xiong, W. &Yaskoff, N. 5 GHz wireless channel characterization for vehicle to vehicle communications. In *Proc. IEEE Military Commun. Conf.*, Atlantic City, NJ, 2005. pp. 3016–3022.

Maurer, J., Fugen, T. & Wiesbeck, W. Narrow-band measurement and analysis of the inter-vehicle transmission channel at 5.2 GHz. In *Proc. IEEE Veh. Technol. Conf.*, Volume 3, Birmingham, AL, 2000 pp. 1274–1278.

Mecklenbrauker, C. F., Molisch, A. F., Karedal, J., Tufvesson, F., Paier, A., Bernado, L., Zemen, T., Klemp, O. & N. Czink. Vehicular channel characterization and its implications for wireless system design and performance. *Proc. IEEE 99*(7), 2111, pp.1189–1212.

Molisch, A. F. 2010. *Wireless Communications* (2nd ed.). John Wiley & Sons Ltd.

Paier, A., Karedal, J., Czink, N., Dumard, C., Zemen, T., Tufvesson, F., Molisch, A. F.& Mecklenbruker, C. F. 2009. Characterization of vehicle-to-vehicle radio channels from measurements at 5.2 GHz. *Wireless Pers. Commun. 50*(1), pp. 19–32.

Paier, A., Karedal, J., Czink, Hofstetter, H., Dumard, C., Zemen, T., Tufvesson, F., Molisch, A. F.& Mecklenbruker, C. F. First results from car-to-car and car-to-infrastructure radio channel measurements at 5.2GHz. In *Proc. IEEE Int. Symp. on Personal Indoor and Mobile Radio Commun.*, Volume 1, Athens, Greece, pp. 1–5.

Patel, C., G. St¨uber, & Pratt, T. 2005. Simulation of Rayleigh-faded mobile-to-mobile communication channels. *IEEE Trans. Commun.* 53, 1876–1884.

Rappaport, T. S. 2001. *Wireless Communications: Principles and Practice* (2nd ed.). Prentice Hall.

Sen, I. & Matolak, D. 2008. Vehicle-vehicle channel models for the 5-GHz band. *IEEE Trans. Intell.* Transp. Syst. 9(2), pp. 235–245.

Tanghe, E., Joseph, W., Verloock, L. & Martens, L. 2008. Evaluation of vehicle penetration loss at wireless communication frequencies. *IEEE Trans. Veh. Technol.* 57(4), pp. 2036–2041.

Electronics, Communications and Networks IV – Hussain & Ivanovic (eds)
© 2015 Taylor & Francis Group, London, ISBN: 978-1-138-02830-2

An improved multi-path TCP scheme over handover between Wi-Fi and cellular networks

Sunghyun Im, Seung Ki Park, Byoungkwan Kim & Ju Wook Jang*
Dept. of Electronic Engineering, Sogang University, Seoul, Korea

ABSTRACT: We consider Multipath TCP (MPTCP) which allows simultaneous use of Wi-Fi and cellular interfaces for a single TCP connection, presenting a standard TCP socket API to the application. In previous work, Full-MPTCP mode and Back Mode have been proposed. In Full mode, both Wi-Fi and cellular are used while, in backup mode, only Wi-Fi is used when available and cellular is used only when Wi-Fi is not available. Full mode enables seamless TCP connection with the expense of cellular resource consumption and faster battery exhaustion while Backup mode may involve disconnection during handover with saving in cellular resource and battery. We propose Hybrid scheme which enables seamless connection while minimizing cellular resource waste and battery consumption. Simulation results show that our scheme exhibits seamless handover (on leaving Wi-Fi coverage) with similar throughput as Full mode while consuming cellular resource and battery as little as Backup mode.

1 INTRODUCTION

Currently, wireless networks like Wi-Fi, 3G, and LTE coexist and user can make simultaneous connections to different radio access networks. To deal with the seamless handover problem, schemes like MPTCP (Passch et al. 2012, Raiciu et al. 2012) are suggested. MPTCP is one of TCP extensions which can make numerous connections with heterogeneous networks at the same time and was standardized in IETF (Ford et al. 2013, Ford et al 2011, Raiciu et al. 2011).

Handover algorithms using MPTCP are proposed in (Passch et al. 2012). One is Full-MPTCP Mode (Full Mode). This scheme refers to the regular MPTCP operations where all flows are used. For example, MS shown in Figure 1 transmits packets through 3G network and Wi-Fi network at the same time. MPTCP is mostly intended for users who want to obtain the best data transfer rates. Also, this scheme supplies seamless handover service to the user.

Another method is Backup Mode. This technique also maintains connections to all paths, but only uses a subset of all paths. Also, this algorithm has a longer battery lifetime than Full Mode's and saves 3G network resource. So, more users can be accommodated for giving cellular resource.

However, proposed handover schemes have some problems. First, Full Mode is not the best option for users willing to save battery consumption. Moreover, MS uses both networks, although MS can only use its WI-Fi connection, so the 3G network's radio resource is wasted.

Meanwhile, in case of Backup Mode, MS transmits data with an initial congestion window when MS implements handover and sends data to 3G networks. Thereby, the early part of the 3G network's throughput is low and user who uses a real time service can experience delays (Can't support seamless handover service).

The remainder is organized as follows. Our scheme is explained in section 2 and we evaluate our algorithm with MATLAB in section 3. And Section 4 concludes this paper..

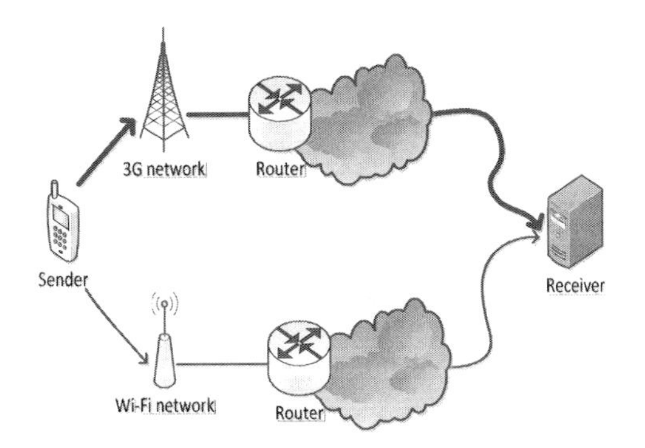

Figure 1. MPTCP scenario in wireless network.

*Corresponding author: *jjang@sogang.ac.kr*

2 PROPOSED SCHEME

In this section, we propose an MP-TCP handover algorithm called Hybrid Mode. This scheme aims to decrease the usage of 3G in Wi-Fi network and to provide seamless handover.

When MS is within the coverage of a Wi-Fi AP, MS only uses Wi-Fi network resource (Backup Mode). If MS is expected to leave Wi-Fi with high probability, it changes its mode to Full Mode. In this case, if a measured Received Signal Strength (RSS) from the 3G networks satisfies equation (1), MN sends packets to 3G networks during N RTTs.

$$10\log M_{new} - \alpha \geq 10\log M_{old} \tag{1}$$

M_{new} is the RSS from the 3G networks and M_{old} is the RSS from the current associated Wi-Fi network. We set decision value (α) which is smaller than the handover threshold $(offset - H)$. So, MS temporarily operates like Full Mode and sends packets during N RTTs before being unavailable the Wi-Fi connection, as shown in Figure 2.

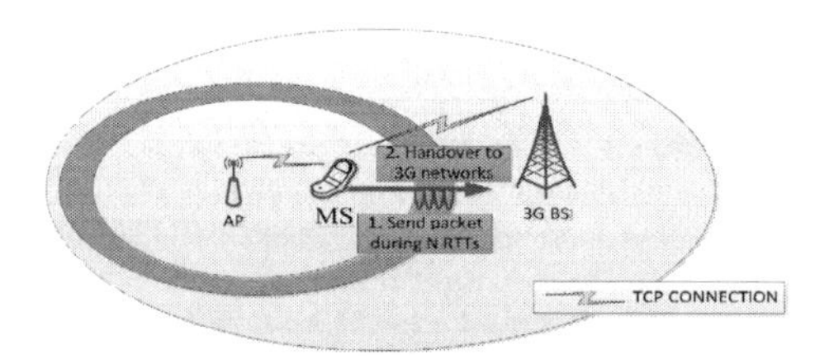

Figure 2. Network environment.

If we determine the value of N and there is no congestion, the average throughput of 3G network (T) is estimated by using equation (2).

$$T = \begin{cases} \dfrac{\sum_{k=1}^{N} 2^{k-1}}{N \times RTT\_3G} \times packetsize (N \leq 4) \\ \dfrac{15 + \sum_{k=5}^{N} (k-4)}{N \times RTT\_3G} \times packetsize (otherwise) \end{cases} \tag{2}$$

Otherwise, we set coverage of Wi-Fi is d, so equation (1) is satisfied when the distance from Wi-Fi is $d - (N \times RTT \times v)$. So, we set that α is the difference between the RSS from 3G and the RSS from Wi-Fi in $d - (N \times RTT \times v)$.

After sending data during N RTTs, MS saves information about the last value of the congestion window. If MS gets away from Wi-Fi again and measured RSS is satisfied following equation (3), MS implements handover.

$$10\log M_{new} - (offset - Hysteresis) \geq 10\log M_{old} \tag{3}$$

So MS stops transmitting data to Wi-Fi network and sends data through the 3G network. MS uses the congestion window value which was saved by MS before. Flow chart shows all process in Figure 3.

After handover, MS sends packets with initial congestion window value when it uses Backup Mode algorithm. However, Hybrid Mode uses a half of the congestion window value which has been saved before handover. So, the throughput of Hybrid Mode is higher than Backup Mode and Hybrid Mode can provide seamless handover service like Full Mode. Reason of using a half of the congestion window value is to decrease transmission failure probability. When MS sends data through the network, other users also use that network. So, network congestion changes frequently. Hence, Hybrid Mode uses a half value to decrease probability of collision.

Also, Hybrid Mode can save MS's battery consumption and 3G radio sources because MS in Wi-Fi only sends packets through Wi-Fi like Backup Mode.

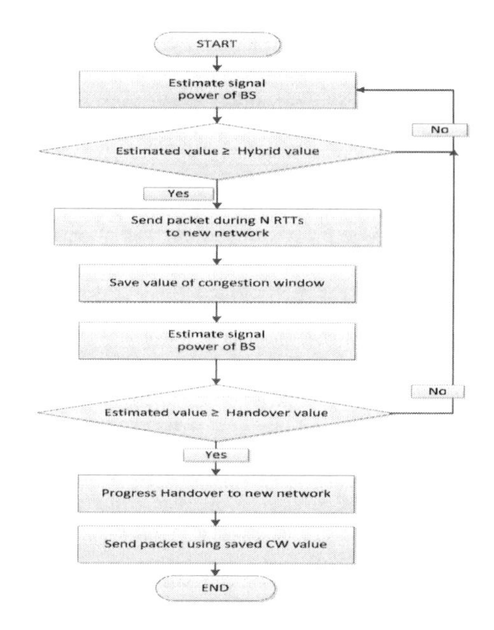

Figure 3. Flowchart about Hybrid Mode.

3 EVALUATION

Figure 4 shows simulation results of our scheme. Wi-Fi average throughput is 10Mbps and that of 3G is 1Mbps. In this paper, we assume a free space environment and set both value of antenna receive gain and transmit gain to 1. Also, we set RTT of 3G network is 0.06 sec and N is 9.

In case of Hybrid Mode, initially, only the 3G inter-face is active, and all traffic is received through this interface. If MS is expected to leave Wi-Fi with high probability, Hybrid Mode transmits data to the 3G networks. So, the early part of the 3G network's throughput is similar to Full Mode after handover, as shown in Figure 5. This means that the Hybrid Mode algorithm is able to provide a seamless handover and is more efficient than Full Mode, Backup Mode.

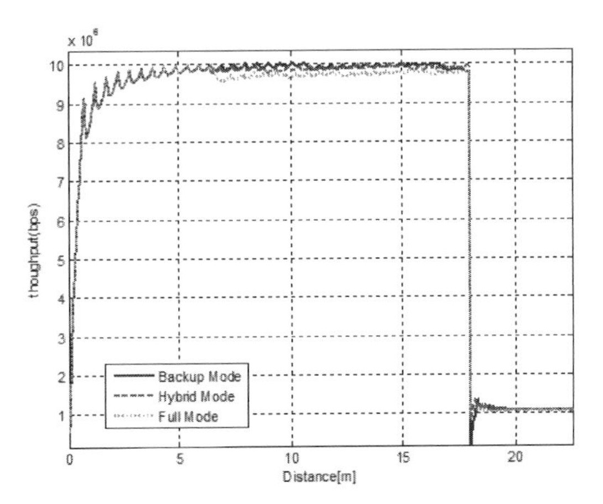

Figure 4. Throughput of Backup Mode, Hybrid Mode and Full Mode.

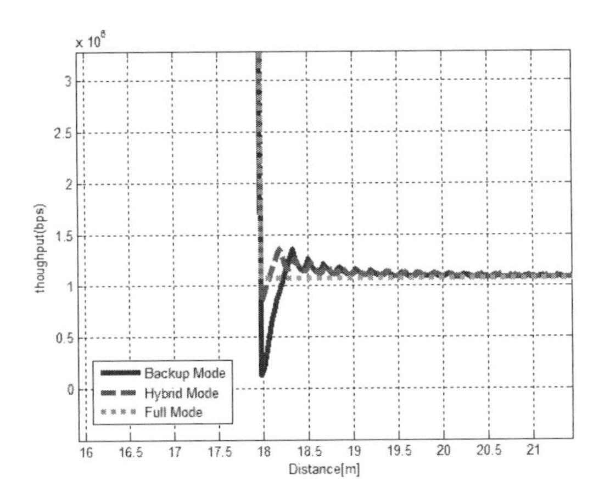

Figure 5. Throughput after handover.

4 CONCLUSION

Seamless handover are important aspects because of guarantee higher quality of service for end-users. On the other hand, mobile devices are equipped with multiple wireless interfaces and can be connected simultaneously to different radio access networks. Also, users can transmit to different networks at the same time by using MPTCP scheme. In this paper, we propose a new handover scheme with MPTCP which can service seamless handover and keep battery life-time. So, this scheme can be solved about seamless data offload in communication networks (e.g., smart-phone network, vehicular network).

Since, packets belonging to the same connection are distributed across multiple interfaces, packet reordering issues might arise (Leung et al. 2007). However, this issue might be resolved by developing an efficient packet synchronization mechanism, for example, at an anchor access node like MPTCP proxy.

ACKNOWLEDGEMENT

This research was supported by the MSIP(Ministry of Science, ICT and Future Planning), Korea, under the ITRC (Information Technology Research Center) support program (NIPA-2014-H0301-14-1006) supervised by the NIPA (National IT Industry Promotion Agency).

REFERENCE

Ford, A., et al. 2011. Architectural guidelines for multipath TCP development. *RFC 6181, IETF.*

Ford, A., et al. 2013. TCP Extensions for Multipath Operation with Multiple Addresses. *RFC6824, IETF.*

Leung, K.-C., et al. 2007. An overview of packet reordering in transmission control protocol (TCP): problems, solutions and challenges. *IEEE Transactions on Parallel and Distributed Systems* 18 (4): 522–535.

Paasch, C., et al. 2012. Exploring Mobile/Wi-Fi Handover with Multipath TCP. *ACM SIGCOMM Workshop on Cellular Networks (CellNet).*

Raiciu, C., et al. 2011. Coupled Congestion Control for Multipath Transport Protocols. *RFC6356, IETF.*

Raiciu, C., et al. 2012. How Hard Can It Be? Designing and Implementing a Deployable Multipath TCP. *In Networked Systems Design and Implementation (NSDI '12).*

Electronics, Communications and Networks IV – Hussain & Ivanovic (eds)
© 2015 Taylor & Francis Group, London, ISBN: 978-1-138-02830-2

A three-dimensional dynamic allocation scheme for WDM-OFDM-PON

Jun Jiang, Min Zhang*, Yang Yang, Zhuo Liu & Xue Chen
State Key Laboratory of Information Photonics and Optical Communications, Beijing, China

ABSTRACT: A Three-Dimensional Dynamic Allocation algorithm (TDDA) with scheduling of wavelength, frequency and time domain is proposed, It makes full use of the advantages of WDM-OFDM-PON. The algorithm meets the characteristics of each service priority and makes a flexible bandwidth allocation. Simulations were conducted to study the performance of TDDA algorithm.

1 INTRODUCTION

WDM-PON will be one of the key technology of broadband access network in the future.

However, the major disadvantages of WDM-PON are its high cost of passive devices and low bandwidth utilization, the main reason is that comparing with other networks, access network has a wealth of business types and granularity, while the granularity of a single wavelength is too large. However, OFDM has a flexible granularity, high spectrum efficiency, and it can support more users. However, On the one hand, the total uplink bandwidth of OFDM is limited, on the other hand, it is extremely dependent on high speed DSP when the subcarrier number is large, this leads to the expensive cost. Combing WDM, OFDM and TDM technologies, which is a promising solution to the future broadband PONs.

A subband access scheme for multiband high-speed OFDM-PON has been proposed (Cheng et al. 2013), where the granularity of the bandwidth allocation is as small as the bandwidth of just one subcarrier and the ONUs only receives the desired subcarriers according to the bandwidth allocation strategy from the whole downstream OFDM signal without band limitation. This provides the possibility of implementing the algorithm proposed in this paper.

A three-dimensional dynamic allocation algorithm (TDDA) with scheduling of wavelength, frequency and time domain is proposed, It makes full use of the advantages of WDM-OFDM-PON. The algorithm meets the characteristics of each service priority and makes a flexible bandwidth allocation.

2 SCHEME OF THREE-DIMENSIONAL DYNAMIC ALLOCATION-TDDA

The proposed TDDA performs the joint scheduling of wavelength, frequency and time. The operation of it is as follows.

Step1: The time of first packet with low priority arrives OLT as the primary factor to determine the order of bandwidth allocation for each ONU, and the secondary factors is the request bandwidth of each ONU.

Step2: The period of each poll is divided into EF subinterval and AF/BE subinterval. First, allocate bandwidth for high-priority traffic. if $R_{1,k}<B_1^{max}$, $W_{1,k}=R_{1,k}$, else $W_{1,k}=B_1^{max}$, wherein B_1^{max} is the threshold of the allocated bandwidth for high-priority traffic of a ONU, R and W matrix is respectively the requested bandwidth and allocated bandwidth for different SLAs and ONUs, $R_{1,k}$, $R_{2,k}$, $R_{3,k}$ is respectively the requested bandwidth of high-priority, medium-priority and low-priority traffic of ONU_k, $W_{1,k}$, $W_{2,k}$, $W_{3,k}$ is respectively the allocated bandwidth for high-priority, medium-priority and low-priority traffic of ONU_k, $W_{2,3,k}$ is the sum of the allocated bandwidth for medium-priority and low-priority traffic of ONU_k, $R_{2,3,k}$ is the sum of the requested bandwidth of medium-priority and low-priority traffic of ONU_k, $R_{2,3}$ is the sum of $R_{2,3,k}$ of all ONU, E_k^{AF} is the average cache of medium-priority traffic of ONU_k during the idle period. Then allocate the combined bandwidth for medium-priority and low-priority traffic of each ONU. Calculating the total allocated bandwidth for high-priority traffic as W_{EF} and the

*Corresponding author: *mzhang@bupt.edu.cn*

remaining bandwidth can be allocated as W_{rest}. If $R_{2,3}$ is less than W_{rest}, allocate $W_{2,3,k}$ the sum of $R_{2,3,k}$ and the value calculated in proportion to remaining bandwidth of W_{rest} minus $R_{2,3}$, Else, allocation for each ONU according to the principle of giving priority to the ONU with the minimum of requested bandwidth (see Figure 1). And then allocate bandwidth for medium-priority and low-priority traffic of each ONU respectively,

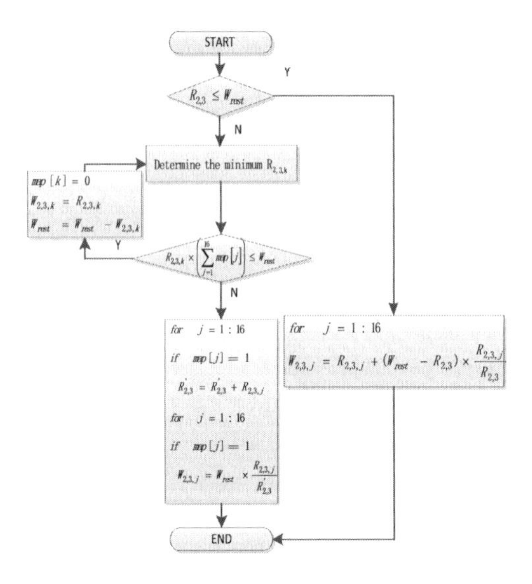

Figure 1. Bandwidth allocation for AF/BE.

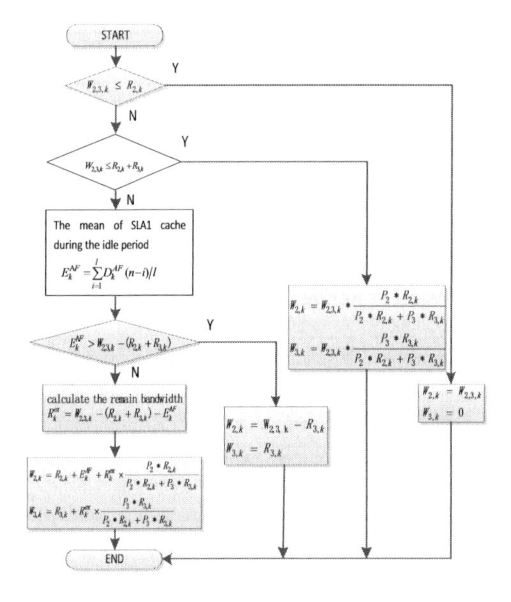

Figure 2. Bandwidth allocation for AF and BE.

Select the appropriate allocation methods in accordance with different relationships of $W_{2,3,k}$, $R_{2,3,k}$, $R_{2,k}$ and E_k^{AF} (see Figure 2).

Step3: The wavelength is allocated for each ONU according to the descending of ONU traffic load. Calculate the synthetic cost S_{ijk} use Equation (1). wherein C_{ijk} is the wavelength tuning cost and Ld_k is the maximum load difference between wavelengths when tuning wavelength i to j at ONU_k. We allocate the wavelength with the least synthetic cost to the current ONU.

$$S_{ijk} = w_{tk} * C_{ijk} + w_{lk} * Ld_k \tag{1}$$

Step4: Slots Allocation for the SLA0 of all ONU first, and then do it for SLA1 and SLA2(see Figure 3). Slot allocation for SLA0 is according to the principle of from bottom to top and from left to right, and that for SLA1 and SLA2 is to allot the subcarrier with the least allowable bandwidth to the current packet, then update the load on this subcarrier and continue to the next allocation.

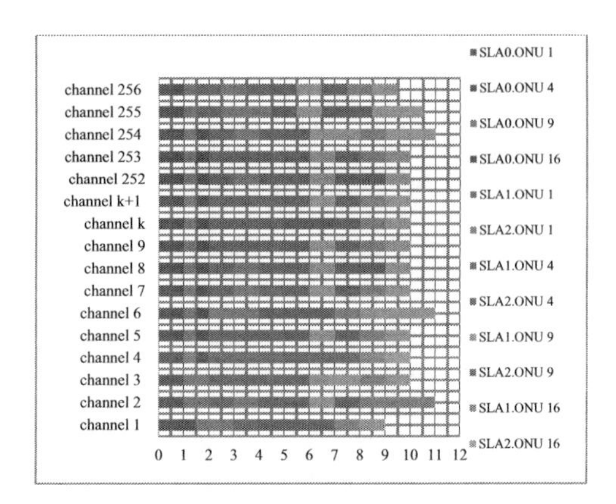

Figure 3. Slots allocation for each ONU.

3 SIMULATION BASED OPNET AND RESULT

A WDM-OFDMA-PON network topology is built by the OPNET (see Figure 4), which is composed by an OLT, an ODN, 16 ONUs (see Figure 5-7) and a link containing four wavelengths. Wherein each transmitter/receiver corresponds to a wavelength, and contains 64 channels, each channel represents an OFDM subcarrier. In the OLT, the total downstream signal is still divided into several bands and generated, respectively, to release the device requirement, but in the ONU,

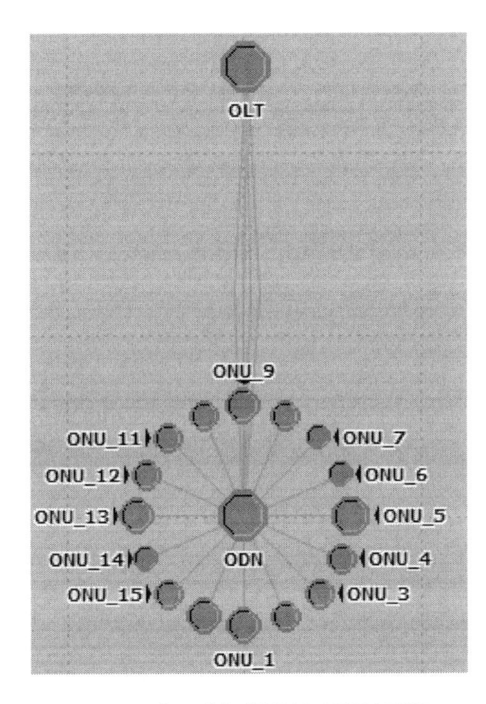

Figure 4. Network model of WDM-OFDM-PON.

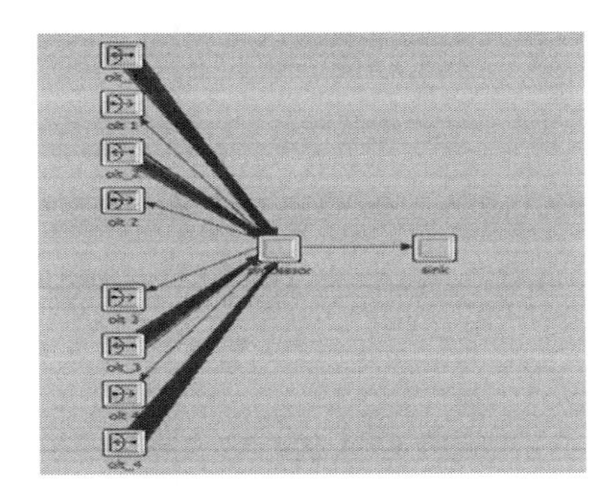

Figure 5. Node model of OLT.

there will be no OFDM-band concept, and the ONUs are able to access any subcarriers from the whole downstream OFDM signal, both inside an OFDM band and across adjacent OFDM bands. What is more, focusing only on the desired subcarriers allows the ONUs to use low-speed devices to realize higher system capacity, which is very helpful to reduce the system cost in the access network. The high-priority business is generated by a traffic source of a Poisson distribution, and the medium-priority and low-priority business of self-similar distribution are generated respectively by 10 overlapped-tail traffic source.

Since the evaluation parameters of previous algorithms for WDM-OFDM-PON system is not the

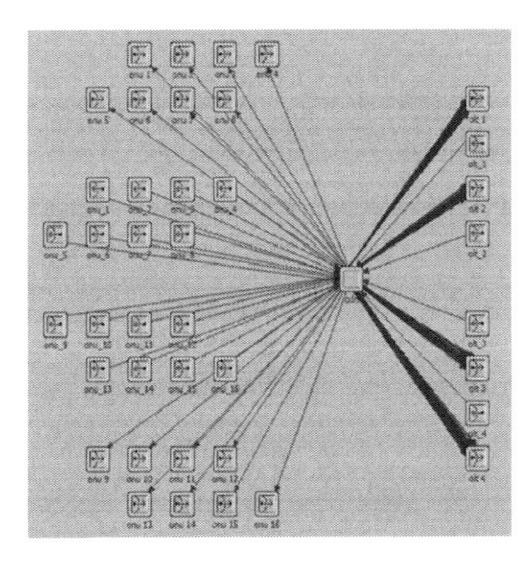

Figure 6. Node model of ODN.

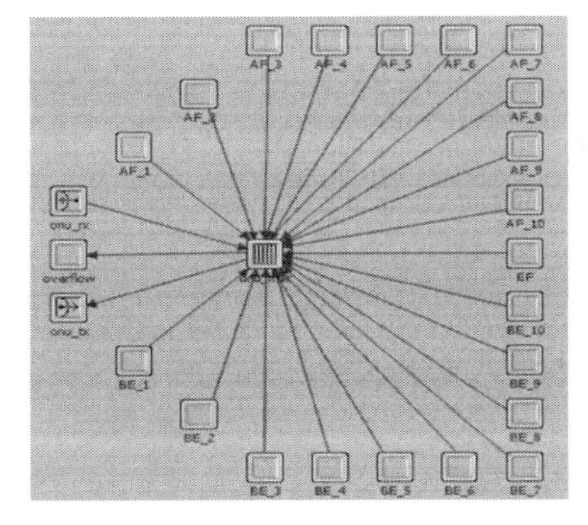

Figure 7. Node model of ONU.

network throughput, packet delay, wavelength tuning cost and load balance between wavelengths, the performance of new algorithms for WDM-OFDM-PON system is compared with that of the algorithms for OFDMA-PON system and TWDM-PON system with the same uplink rate and the same transmission distance respectively.

To evaluate the proposed algorithm in terms of the network throughput and end-to-end packet delay, A WDM-OFDM-PON system with 100 Gbps, 40 km extended-reach is developed to evaluate the new algorithm, comparison with the status based sequential dynamic subcarrier allocation (SDSCA) algorithm (Lim et al. 2013). The obtained figure of 80Gbps, exhibits an improvement in channel utilization rate by 3.9%, compared to the status based SDSCA that stalls at around 77 Gbps. The packet delay of the SLA0 is only 600 us when the traffic load of ONU is lower than the threshold ONU loads of 0.7, comparison with the status based SDSCA algorithm, it reduces 400us, and 1.6ms packet delay which is far less than that of the status based SDSCA for SLA0 even at maximum ONU load. It can be observed from the Figure 6, there is a sharp increase in packet delay of the SLA1 and SLA2 with the status based SDSCA, and that of TDDA is lower and flatter when exceeds the threshold.

To evaluate the algorithm in terms of the wavelength tuning cost and load difference between wavelengths, A WDM-OFDM-PON system with 40 Gbps, 10 km reach is developed to evaluate the new algorithm compared with the algorithm A and algorithm B (Luo et al. 2012). As shown in Figure 7 and Figure 8, the new algorithm generates less than 160 units tuning cost, and the maximum load difference between wavelengths fluctuates between 30 and 42 units. This means that the new algorithm makes a trade-off between wavelength tuning cost and system load balance compared with that of algorithm A and algorithm B.

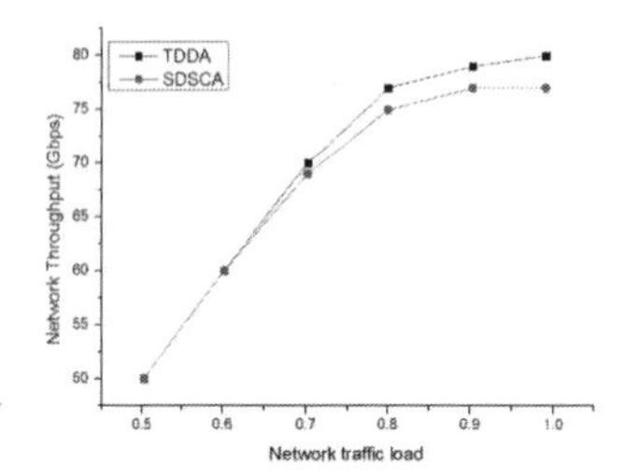

Figure 5. Network throughtout.

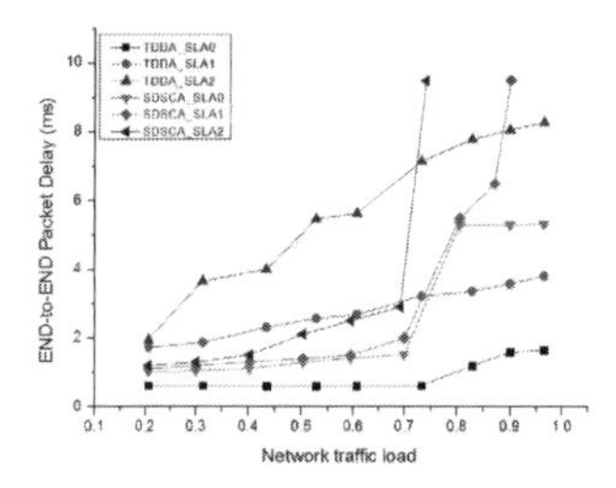

Figure 6. Packet delay for TDDA and SDSCA.

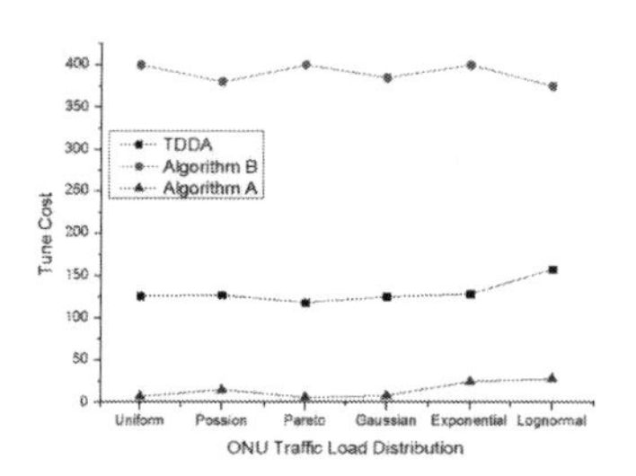

Figure 7. Tuning cost comparison.

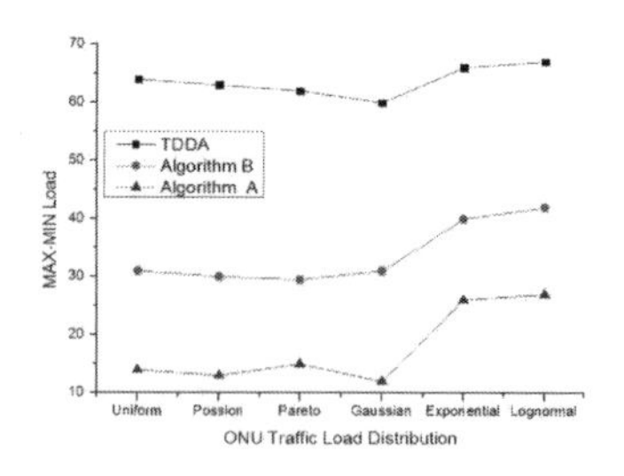

Figure 8. Load balance comparison.

4 CONCLUSION

The proposed TDDA algorithm achieves a significant improvement in throughputs, packet delay, tuning cost and load balance. The increased performance is mainly due to three factors. First of all, the multi-threaded polling algorithm based on SLA0 makes both the current and the next polling operations can be performed simultaneously in the current polling period, so the transmission delay caused by report frame and gate frame is avoided, secondly, the maximum bandwidth priority algorithm and the monitoring of the cache of SLA1 during the idle period makes the packet delay keep flat with increasing traffic loads. And finally, the optimal real-time load balance makes the congestion rate on all wavelengths lower, and benefits the network management and control with a stable status.

REFERENCES

Cheng Lei, Minghua Chen & Hongwei Chen. 2013. 1-Tb/s WDM-OFDM-PON system with subband access scheme and flexible subcarrier-level bandwidth allocation. *IEEE Photonics Journal:* 5(1).

D. Qian, N. Cvijetic, J. Hu, et al. 2010. 108 Gb/s OFDMA-PON with polarization multiplexing and direct detection. *J. Lightw. Technol* 28(4): 484–493.

Lim Wansu, Pandelis Kourtessis, et al. 2013. Dynamic subcarrier allocation for 100 Gbps, 40 km OFDMA-PONs with SLA and CoS. *IEEE Journal of Lightwave Technology:* 31(7).

Luo Yuanqiu, Effenberger Frank & Sui Meng. 2012. Wavelength management in time and wavelength division multiplexed passive optical networks (TWDM-PONs). *Globecom 2012-Optical Networks and Systems Symposium.*

M. Yuang, P. Tien, D. Hsu, et al. 2012. A high-performance OFDMA PON system architecture and medium access control. *J. Lightw. Technol* 30(11):1685–1693.

N. Cvijetic. 2012. OFDM for next-generation optical access networks. *J. Lightw. Technol* 30(4): 384–398.

Electronics, Communications and Networks IV – Hussain & Ivanovic (eds)
© 2015 Taylor & Francis Group, London, ISBN: 978-1-138-02830-2

DCS1800 and LTE B39 co-existence problem in SGLTE mobile terminal

Haiying Jiang *& Yougang Gao
College of Electronic Engineering, Beijing University of Posts and Telecommunications, Beijing, China

Dan Zhang
Beijing Art and Media College, Beijing, China

ABSTRACT: In this paper, the co-existence problem of TDD-LTE band 39 and DCS 1800 in SGLTE (Simultaneous GSM and TDD-LTE) mobile terminal is studied for the first time. The noise figure that the LTE band 39 receiver is interfered by DCS1800 transmitter is calculated, and it is found to be substandard. Therefore, a new design is invented by modifying DCS1800 transmitter circuit and adding special SW configuration. Then, the noise figure of the LTE B39 receiver with the new design transmitter can meet the requirement with good margin. The engineering samples with the new design are built, and the practical test is performed. It can be found that the calculating result is well in accord with the test result. At last, through the same method, a new LTE B39 transmitter is designed to suppress LTE B39 transmission noise with DCS 1800 receiver and is verified by test on the engineering built samples.

1 INTRODUCTION

SGLTE (Simultaneous GSM and TDD-LTE) is the simultaneous work of GSM and TD-SCDMA/TDD-LTE, while GSM is used for voice service, and TD-SCDMA/TDD-LTE is used for data service in one mobile terminal. According to CMCC's network configuration, GSM900/DCS1800 is for voice and TD-SCDMA or TDD-LTE is for data service during the simultaneous work. In order to meet the requirement, two radio systems are required to be supported in one mobile terminal: one radio system supports GSM900/DCS1800 for voice call, which is designated as the second radio system, and one radio system supports TD-SCDMA/TDD-LTE for data service, which is designated as the first radio system. Compared with ordinary terminals that do not support simultaneous work, this kind of terminals provide the convenience that enables us to have a call without disconnection of data transmission or receiving disconnection; however, they also introduce co-existence problems, which are new and bigger challenges in SGLTE mobile terminal design. For example, the transmission (TX) power or TX noise from the transmitter of one radio system can interfere other system's Receiving (RX), block the receiver, or degrade receiver's sensitivity. In the ordinary design, there are no such problems. Without study on co-existence problems, the simultaneous work cannot be realized, or it will be realized with big performance degradation. As a result, co-existence problems of the two systems should be carefully studied as early as possible.

*Corresponding author: *jianghaiying198@aliyun.com*

2 CO-EXISTENCE PROBLEMS IN SGLTE MOBILE TERMINAL

According to the system working configuration, series of co-existence problems are listed in Table 1 (3GPP TS 51.010-1 version 5.6.0 Release 5, 3GPP TS 36.521-1 V10.5.02013) (Ludwig & Bretchko 2002). In addition to these series, there is also a problem of receiver blocking that the first Radio TX signal may block the second Radio receiver, and the second Radio TX signal may block first Radio receiver (Ludwig & Bretchko 2002). Moreover, there is a problem of GSM900 third harmonic to degrade B41 receiver (3GPP 2003, 3GPP 2013, Ludwig & Bretchko 2002). In one paper, the series of the co-existence problems cannot be covered, and in this paper we will only focus on the co-existence problem between DCS1800 and LTE band 39.

Table 1. Series of co-existence problem

	Transmitter TX noise as Interference	Receiver as victim
1	GSM900	TD-SCDMA band34
2	GSM900	TD-SCDMA Band F/LTE band39
3	GSM900	TDD LTE band38
5	GSM900	TDD LTE band40
6	GSM900	TDD LTE band41
7	GSM1800	TD-SCDMA band34

8	GSM1800	TD-SCDMA Band F/LTE band39
9	GSM1800	TDD LTE band38
11	GSM1800	TDD LTE band40
12	GSM1800	TDD LTE band41
13	TD-SCDMA band34	GSM900
14	TD-SCDMA Band F/ LTE band39	GSM900
15	TDD LTE band38	GSM900
17	TDD LTE band40	GSM900
18	TDD LTE band41	GSM900
19	TD-SCDMA band34	GSM1800
22	TD-SCDMA Band F/ LTE band39	GSM1800
21	TDD LTE band38	GSM1800
23	TDD LTE band40	GSM1800
24	TDD LTE band41	GSM1800

3 STUDY ON DCS1800 AS INTERFERENCE TO LTE B39

3.1 *Frequency range of DCS1800 and LTE B39*

The channels of CMCC DCS1800 are from Ch.512 to Ch.635 (CMCC 2013), but they are not the same with those of globe DCS1800 which are from channel Ch. 512 to Ch. 885. Frequency range of globe DCS1800 and CMCC DCS1800 is shown in Figure 1. Also, in Figure 1, the narrow frequency space between DCS1800 highest TX/RX channel and LTE B39 lowest RX/TX channel is shown.

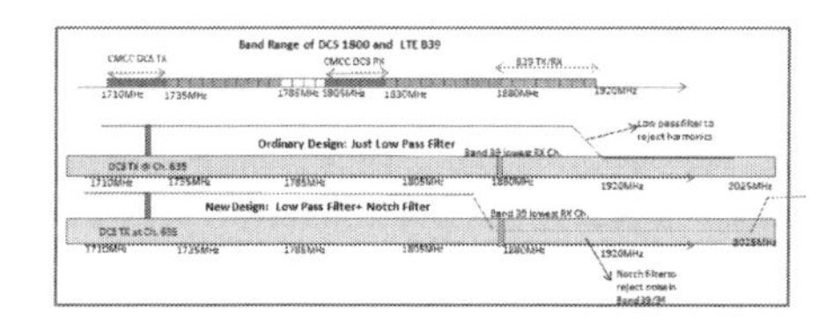

Figure 1. Frequency range of DCS1800 and LTE B39

3.2 *Noise figure of LTE Band 39 due to interference from DCS1800 transmitter*

The diagram of DCS1800 transmitter of the ordinary design is shown in Figure 2 And on DCS1800 path, there is only a Low Pass Filter (LPF). The frequency range of LPF can be found in Figure 1.

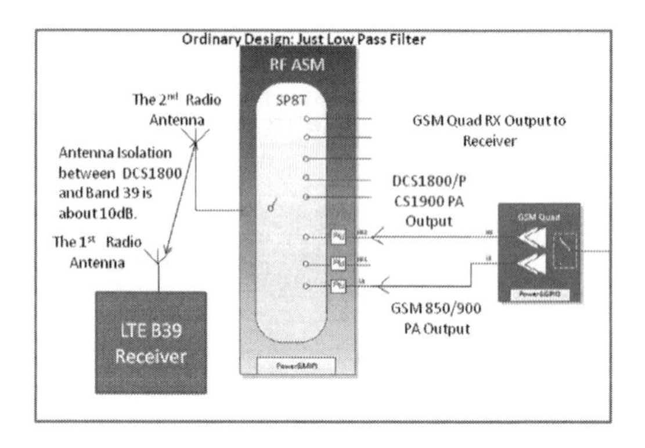

Figure 2. DCS1800 transmitter block diagram for ordinary design

The TX noise of DCS1800 transmitter falling to the LTE B39 RX band is generated by two parts: the transceiver (specified as N_1 dBm/Hz) and the Power Amplifier (PA). For the noise generated by PA, it is specified as follows: for white noise (specified as N0 dBm/Hz) input and gain (specified as G1 dB), its output noise is N_2 dBm/Hz. According to Figure 2, the loss on the path can be specified as follows: L_3 dB is specified as LPF loss; L_4 dB is the loss of print circuit board (PCB) tracing between antenna switch and antenna feeding point; and L_5 dB is specified as antenna isolation. The noise figure that the LTE B39 receiver is not interfered by DCS1800 transmitter is F dB.

The total noise after PA output is specified as N_{PA}, and calculated as (1). The total noise after LPF output is specified N_{LPF}, and calculated as (2). The total noise before antenna feeding point is specified as N_{AFP}, and calculated as (3). The total noise adding to B39 receiver is specified as N_{Total}, and calculated as (4). The receiver noise figure due to DCS1800 TX noise is specified as $F_{with\_Tx\_noise}$, and calculated as (5). The noise figure degradation due to DCS1800 TX noise is specified as $F_{Degradation}$, and calculated as (6).

$$N_{PA} = 10\times\log(10^{0.1\times N_0} + 10^{0.1\times N_1}) + G_1 + 10\times\log[10^{0.1\times N_0}\times(10^{0.1\times(N_2 - N_0 - G_1)} - 1)/(10^{0.1\times N_0} + 10^{0.1\times N_1}) + 1](\text{dBm/Hz}) \quad (1)$$

$$N_{LPF} = N_{PA} - L_3 + 10\times\log[10^{0.1\times N_0}\times(10^{0.1\times L_3} - 1)/10^{0.1\times N_{PA}} + 1](\text{dBm/Hz}) \quad (2)$$

$$N_{AFP} = N_{LPF} - L_4 + 10\times\log[10^{0.1\times N_0}\times(10^{0.1\times L_4} - 1)/10^{0.1\times N_{LPF}} + 1](\text{dBm/Hz}) \quad (3)$$

$$N_{Total} = N_{AFP} - L_5 + 10\times\log[10^{0.1\times N_0}\times(10^{0.1\times L_5} - 1)/10^{0.1\times N_{AFP}} + 1](\text{dBm/Hz}) \quad (4)$$

$$F_{with\_Tx\_nosie} = 10 \times \log[10^{0.1 \times N_0} \times (10^{0.1 \times F} - 1) + 10^{0.1 \times N_{Total}}] - N_0(\mathrm{dB}) \quad (5)$$

$$F_{Degradation} = 10 \times \log[10^{0.1 \times N_0} \times (10^{0.1 \times F} - 1) + 10^{0.1 \times N_{Total}}] - N_0 - F(\mathrm{dB}) \quad (6)$$

The parameters of the components listed above are shown in Table 2, and the antenna isolation L_5 between the first radio antenna and the second radio antenna is about 10dB. According to equation (1) to (6), we can get:

$$F_{with\_Tx\_nosie} = 36.6(\mathrm{dB}) \quad (7)$$

$$F_{Degradation} = 29.6(\mathrm{dB}) \quad (8)$$

The ordinary noise figure degradation is much bigger than it is required, so it cannot meet the requirement from CMCC.

Table 2. Component S' parameter Value for ordinary design.

N_0 and N_1 (dBm/Hz)	White noise (dBm/Hz)	PA Spurious (dBm/Hz)	PA Gain (dB)	Low pass filter loss (dB)	RF trace loss (dB)	Receiver Noise Figure (dB)
N_{0+} N_1 -154	N_0 -174	N_2 -133	G_1 29	L_3 2	L_4 1	F 7

3.3 *The invention of new design*

In Figure 2, in order to reduce the noise figure degradation, we can devote our efforts in three aspects: the first one is to reduce the interference level from the transmitter; the second one is to increase the coupling loss between the transmitter and the victim; and the last one is to improve the victim's receiving sensitivity to suppress higher level interference.

However, with specified PCB size, antenna position of the two radio systems, the antenna isolation cannot be bigger, so the coupling loss is almost fixed. Also, for certain hardware platform used in terminals, the flexibility to select the transceiver and PA with lower TX noise is restricted, and the noise generation is determined and cannot be reduced. Also, for a certain hardware platform, the receiver sensitivity cannot be improved to be higher than its specification. The only way we can choose is to increase band rejection of the filter to suppress the TX noise that falls into TDD-LTE B39 RX band.

In Figure 2, after PA LPF is integrated in antenna switch, its loss cannot be big enough because the antenna switch is selected, and the only method we can select is to add additional notch filter after DCS1800 PA output or before LTE B39 receiver input. However, it can be found that if the noise filter is inserted after PA output or before receiver input, because of the insertion loss of the filter, it will also introduce extra loss to useful DCS1800 and PCS1900 TX signal or useful LTE B39 RX signal. According to system link budget calculation, this loss introduced to PCS1900 TX or to LTE B39 RX signal cannot be accepted. At last, without loss of PCS1900 TX signal or TDD-LTE B39 RX signal, a new design with design change on DCS1800 transmitter and special SW configuration is introduced.

The block diagram of DCS1800 transmitter is changed as shown in Figure 3. There are two paths for DCS1800 TX signal after PA output to go to antenna switch, one with notch filter and one without notch filter. When phone works in CMCC network, by controlling SW configuration, DCS1800 TX is switched to the path with notch filter; if it works at PCS1900 or globe DCS 1800, it is switched to the path without notch filter.

Compared with Table 2, the only difference in Table 3 is the loss of L_3, which is the total loss of the LPF and notch filter, and its value is 30dB, and the antenna isolation L_5 is also about 10dB. Also, from equation (1) to (6) we can get

$$F_{with\_Tx\_nosie} = 10.9(\mathrm{dB}) \quad (9)$$

$$F_{Degradation} = 3.9(\mathrm{dB}) \quad (10)$$

According to system link budget calculating result, it can satisfy CMCC requirement with a good margin with sensitivity degradation of 3.9dB.

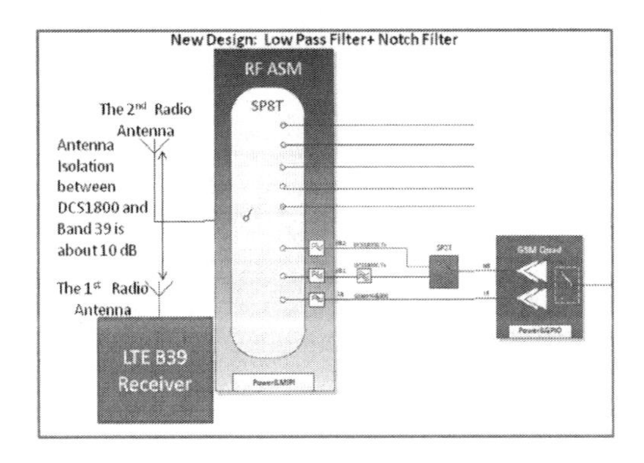

Figure 3. DCS1800 transmitter block diagram for new design

Table 3. Component S' parameter value for new design.

N_0 and N_1 (dBm/Hz)	White noise (dBm/Hz)	PA Spurious (dBm/Hz)	PA Gain (dB)	Low pass filter loss (dB)	RF trace loss (dB)	Band 39 Receiver Noise Figure (dB)
N_{0+} N_1	N_0	N_2	G_1	L_3	L_4	F
-154	-174	-133	29	30	1	7

3.4 *Verification*

The test is performed on real built samples in the full wave anechoic room, according to CTIA (Certification program test requirements) (CTIA 2012), and the test setup is as shown in Figure 4. Two rounds of LTE Band39 free space TIS (Total Isotropic Sensitivity) are performed. The first round test is taken as the reference TIS, and the second round test is with DCS1800 transmitter's max TX. The test result is shown in Table 4.

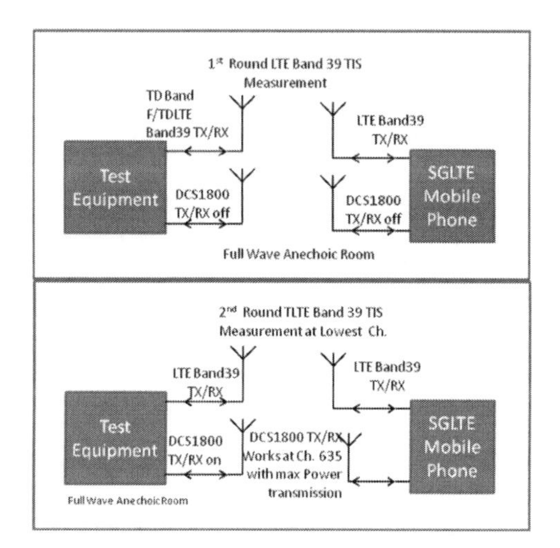

Figure 4. Test setup for design verification.

4 STUDY ON TDD-LTE B39 AS INTERFERENCE TO DCS1800

Through the same calculating method from (1) to (6), the DCS1800 receiver noise figure and noise figure degradation due to the ordinary TDD-LTE B39 transmitter design, whose block diagram is shown in Figure 5, are presented as (11) and (12).

$$F_{with\_Tx\_nosie} = 44.6(\text{dB}) \tag{11}$$

$$F_{Degradation} = 37.6(\text{dB}) \tag{12}$$

According to system link budget calculation, the degradation fails to meet CMCC requirement. Through the same method, the notch filter should be added after LTE B39 PA output to suppress the TX noise that falls into DCS1800 RX band. Block diagram of new design LTE B39 transmitter is shown in Figure 5. From equation from (1) to (6), we can get

$$F_{with\_Tx\_nosie} = 11.2(\text{dB}) \tag{13}$$

$$F_{Degradation} = 4.2(\text{dB}) \tag{14}$$

The test result of the engineering samples can be found in Table 4

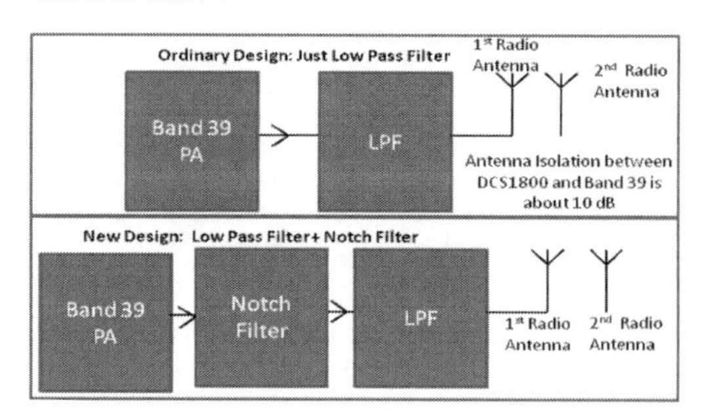

Figure 5. LTE B39 transmitter block diagrams for ordinary design and new design.

Table 4. Verification result.

Interference	Victim		
DCS 1800	TDD LTE TIS (free space) Band 39 Ch. 38350 , Band width 20MHz		
	Test Items	Test data	limit
	Reference	-93.3dbm	
	DCS1800 with max level transmission on Ch.635	-90.0 dbm	-85
	Desense	3.3 dB	
LTE Band 39	DCS 1800 TIS (BHR:Beside Head Right Side (Head Phantom Only)) at Ch. 635		
	Test Items	result	limit
	Reference	-105.0dbm	
	LTE Max transmission at Ch. 38350, Band width 20 MHz	-100.4dbm	-94.5
	Desense	4.6 dB	

REFERENCES

3GPP 2003. TS 51.010-1 V5.6.0 Digital cellular telecommunications system (Phase 2+); Mobile Station (MS) conformance specification; Part 1: Conformance specification .

3GPP 2013. TS 36.521-1 V10.5.0 Evolved Universal Terrestrial Radio Access (E-UTRA); User Equipment (UE) conformance specification; Radio transmission and reception; Part 1: Conformance testing.

CMCC 2013. v120130507 LTE-TDSCDMA-GSM Multimode terminal technology specification-dual standby part.

CTIA 2012. OTA Test Plan version 3.2 Test Plan for Wireless Device Over the-Air Performance: Method of Measurement of radiated RF power and Receiver Performance.

Ludwig, R. & Bretchko, P. 2002. *RF Circuit Design: Theory and Applications*, Simplified Chinese translation edition. Copyright Beijing Pearson Education North Asia Limited and Publishing House of Electronics Industry.

Electronics, Communications and Networks IV – Hussain & Ivanovic (eds)
© 2015 Taylor & Francis Group, London, ISBN: 978-1-138-02830-2

Research in congestion-resistant frequency allocation scheme for HF IP network

Yuan Jing*
School of Information and Navigation, Air Force Engineering University, Xi'an, Shaanxi, China
Unit 95876, Zhangye, Gansu, China

Guoce Huang, Qilu Sun & Peng Cao
School of Information and Navigation, Air Force Engineering University, Xi'an, Shaanxi, China

ABSTRACT: HF IP network is a novel modern communication network, which is the important direction of HF communication in future. In an HF IP network, the rationality of the frequency allocation scheme would have great influence to the performance of the network, especially the congestion resistance performance. Therefore, based on complex network theory, a kind of frequency allocation algorithm was designed based on edge betweenness. Firstly the algorithm calculates the edge betweenness according to the network topology. Secondly, the best frequency allocation results could achieve the best conditions in the network transmission capacity, which can be obtained through the edge betweenness. Research results show that the scheme with different routing policy, and network scale could significantly improve the network capacity and ascend the ability to resist congestion.

1 INTRODUCTION

The frequency management of HF (High Frequency) has a great influence to the network running status, as well as to the network congestion characteristics. Frequency allocation scheme directly affects the efficiency of the physical layer data transmission. And then, the transmission delay of the user data affects the transmission performance of the entire network and congestion characteristics. Therefore, reasonable distribution of frequency based on available frequency settings, during HF IP (Internet Protocol) network operation process, can ensure to avoid different frequency interference and also efficient frequency resources using.

At present, the focus of the frequency management research for HF, mainly concentrated in high quality ALE (Automatic Link Establishment), frequency detection (Yehuda 2014, Furman et al 2013), frequency planning and distribution (Arthur 2010, Eliardsson & Karlsson 2012). And the frequency allocation study has been less studied. While with the development of the HF IP network, it has been found that the reasonable planning of frequency between different nodes would have great influence on the performance of network transmission and congestion.

2 HF IP NETWORK

The HF IP network is a data transmission network through HF channel, which is put forward by U.S. (United States) and NATO (North Atlantic Treaty Organization) aimed to realize the IP connectivity between different physical networks (Serinken 2012).

2.1 *Summary of HF IP network*

The HF IP network contains lots of nodes and covers a large area and is suitable for using a fixed frequency allocation scheme.

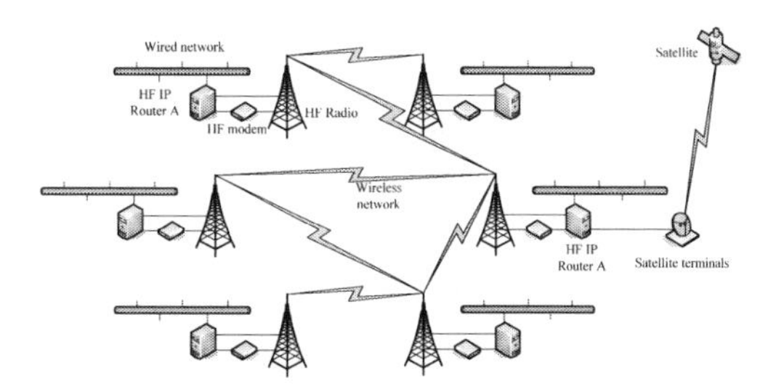

Figure 1. HF IP network diagram.

Figure 1 is a HF IP network diagram. Each HF node provides channel access to local users and wireless link to the LAN (Local Area Network) users in different nodes.

In order to guarantee the reliability of the wireless link, stop and wait ARQ (Automatic Repeat-reQuest) protocol was used between different nodes. Therefore, in a data frame transmission period, for a wireless link

*Corresponding Author: *gmt_jingyuan@163.com*

there is only one sending node and receiving node. Nodes which connect by a wireless link queue up the data according to the arriving order. Any node in the network is not connected directly. And the user data transmission is helped by routing nodes.

2.2 *Characteristics of frequency allocation*

As the transmission efficiency of the HF link influence the network transmission performance, each link should choose the frequency with high transmission efficiency. But the frequencies with high quantity are limited, how to make reasonable allocation is a worthy problem to research. At the same time, there are some basic principles in HF frequency allocation for HF IP network:

1. Usable frequencies must deduct the disable and protection frequencies which are not allowed by the government and army.
2. The frequencies for different radios in a single node should ensure a certain interval, to avoid the co-site interference.
3. The spectrum characteristics of HF would be affected by the ionosphere changes in day and night. It needs to distinguish the frequency between day and night.

3 FREQUENCY ALLOCATION IMPACT THE NETWORK CONGESTION PERFORMANCE

The traditional frequency allocation scheme is mainly based on statistics and experience. It is an artificial random distribution method on the basis of general allocation principles. And this method can be equivalent to a frequency of random distribution (Li & Liu 2005). And the following researches of network congestion performance are combined with a different frequency allocation scheme and different routing policy.

N_{node} is the node number of the network. The adjacency matrix of the network is $\mathbf{A}_{net}$, and $\mathbf{A}_{net}$ is $N_{node} \times N_{node}$ adjacency matrix.

$$\begin{cases} \mathbf{A}_{net}(i,j) = \mathbf{A}_{net}(j,i) = 0 \\ \mathbf{A}_{net}(i,j) = \mathbf{A}_{net}(j,i) = a_f \quad f = 1,2,3, \end{cases} \tag{1}$$

If $\mathbf{A}_{net}(j,i) = 0$ is true, it means there is no wireless link between node i and node j. If $\mathbf{A}_{net}(j,i) = a_f$ means that node i and node j was connected with an HF link. The a_f is the normalized transmission rate of an HF link, and the footnote f is the code name of the frequency.

HF data transmission requires minimizing link routing number (namely the link hop). The best routing policy is planned according to the principle of minimum routing hops. Assuming that matrix $\mathbf{A}'_{net}$ is the normalized matrix of $\mathbf{A}_{net}$.

$$\mathbf{A}'_{net}(i,j) = \begin{cases} 0 & \mathbf{A}_{net}(i,j) = 0 \\ 1 & \mathbf{A}_{net}(i,j) > 0 \end{cases} \tag{2}$$

Under the condition of A'_{net} the shortest path between two nodes (maybe more than one path) can be calculated. Here the method of literature(Arenas et al 2001) will be used, which can get the shortest path solution set easily.

The betweenness of edge $i \leftrightarrow j (i \neq j)$ can be expressed as:

$$b\_link_{ij} = \sum_{e1,e2 \in V, e1 \neq e2} \frac{n\_path_{e1e2,i \leftrightarrow j}}{n\_path_{e1e2}} \tag{3}$$

n_path_{ele2} is the number of the shortest path between node $e1$ and node $e2$, and $n_{ele2,i \leftrightarrow j}$ is the number of the shortest path which through edges $i \leftrightarrow j$.

Under the topology condition, the edge betweenness of b_link_{ij} can be get by the shortest path set. And the network edge betweenness can be expressed as the matrix **B_link**. Through previous analysis, we can get the edge betweenness relations as followed:

$$\mathbf{B\_link}(i,j) = \mathbf{B\_link}(j,i) \tag{4}$$

Therefore, the edge betweenness matrix **B_link** is determined by the topology matrix $\mathbf{A}'_{net}$, and has nothing to do with the frequency allocation scheme.

3.1 *Shortest path routing*

Matrix $\mathbf{A}'_{net}$ and **B_link** could be calculated by the network topology structure. Under the condition of the shortest path routing, the edge betweenness size determines the number of data, which would be transmitted by the edge(Hao et al 2013).

$\mathbf{F}_{hf}\{a_f\}$ is the available frequency set for the network. Edge frequency allocation according to certain rules, then we could get the adjacency matrix $\mathbf{A}_{net}$.

N_{data_net} is the amount of data produced by the network in a single unit time, which was a data frame transmission period. And the source and destination node of the data are selected at random. Meanwhile the normalized data size is 3. So the amount of data for edge during a unit of time can be represented as:

$$\begin{aligned} &N_{data\_node}(i,j) = N_{data\_node}(j,i) \\ &= F_{data\_node}\left(N_{data\_net}, \mathbf{B\_link}(i,j)\right) \end{aligned} \tag{5}$$

F_{data_node} describe the relationship between **B_link** and N_{data_net} for the edge in a unit time. During the network congestion status, with N_{data_net} and formula

3, it can calculate the amount of data arriving at the edge per unit time as follows:

$$F_{data\_node}\left(N_{data\_net}, \boldsymbol{B\_link}(i,j)\right) = \frac{2N_{data\_net}\boldsymbol{B\_link}(i,j)}{N_{node}\left(N_{node}-1\right)} \tag{6}$$

When the N_{data_net} increases to the critical value, the network congestion would show up. And the critical value is that there was at least one edge which had too much data to deal with and beyond its ability.

$$\begin{aligned} & N_{data\_node}(i,j) \geq \mathbf{A}_{net}(i,j) \\ & \Rightarrow F_{data\_node}\left(P_{data}, \mathbf{B\_link}(i,j)\right) \geq \mathbf{A}_{net}(i,j) \\ & \Rightarrow \frac{2N_{data\_net}\mathbf{B\_link}(i,j)}{N_{node}\left(N_{node}-1\right)} \geq \mathbf{A}_{net}(i,j) \end{aligned} \tag{7}$$

Through the formula 6, $N_{data\_net}(i,j)$ and $\mathbf{B\_link}(i,j)$ present approximate proportional relationship. When any edge's betweenness and frequency transmission efficiency meet the relationship of formula 7, the network congestion phenomenon would be occurred.

a. Frequency Randomly Distribution

The matrix $\mathbf{B\_link}$ always exists a maximum betweenness, and the frequency random distribution does not guarantee the maximum a_f assigned to the corresponding $\mathbf{A}_{net}(i,j)$, even when N_{data_net} is small and may exist an edge meet the formula 7.

b. Frequency Allocation Scheme Based on Edge Betweenness

With the positive relationship between $N_{data\_net}(i,j)$ and $\mathbf{B\_link}(i,j)$, the edge which have a greater amount of data would have got higher data processing ability, as the frequency had been assigned according to the edge betweenness. Meanwhile, the influence of N_{data_net} to the network congestion could be reduced through this novel frequency allocation scheme.

Therefore, the frequency allocation principle prompts the biggest N_{data_net} with all edges on the basis of formula 8.

$$\frac{2N_{data\_net}\mathbf{B\_link}(i,j)}{N_{node}\left(N_{node}-1\right)} \leq \mathbf{A}_{net}(i,j) \tag{8}$$

3.2 *The routing policy based on the load change rate of the edge*

The shortest path routing is one of the simplest routing policies, which is not suitable for the network congestion. Here the frequency allocation scheme based on edge betweenness is researched under a new routing policy, which was based on the load change rate of the edge. This routing policy has a characteristic with both deterministic (the length of shortest path l_node_{ij} between node i and node j) and randomness (the load of edge q_edge_{ei} and its rate of change Δq_edge_{ei}). The node e is the current node for the data, and its destination node is j. And the node i is any one of the nodes which connected to the node e. Then, the different frequency allocation scheme would be studied under the different network conditions.

Formula 9 is the expression of the new routing policy. If $\mathbf{A}_{net}(i,j) > 0$, δ_{ij} is 0, otherwise δ_{ij} is 1. While α and λ are the weights of different elements. Through formula 9, we will know the smallest d_route_{ei} which decides the next hop routing nodes for the data.

$$d\_route_{ei} = \begin{cases} \left(1-\delta_{ij}\right)\left(\alpha l\_node_{ij} + (1-\alpha) q\_edge_{ei} + \lambda \Delta q\_edge_{ei}\right) & \Delta q\_edge_{ei} > 0 \\ \left(1-\delta_{ij}\right)\left(\alpha l\_node_{ij} + (1-\alpha) q\_edge_{ei}\right) & \Delta q\_edge_{ei} \leq 0 \end{cases} \tag{9}$$

The following analysis of network transmission state change processing is under the routing policy based on the load change rate of the edge. State of the network is divided into three parts. Firstly, if the N_{data_net} is small, the most of the data would be transmitted following the shortest path. And the edge load is small; the network current state is defined as the free state. Secondly, if the N_{data_net} is increased, the edge which effective betweenness became larger appears a balance of data arrival and processing ability, and the network state is defined as the busy state. Finally, if the N_{data_net} is increased continuously, some of the edges have a situation that the arriving data was so much and the edges could not deal with, and the network state is defined as the congestion state.

a. free state

When the N_{data_net} is small, the edges would have a small load. At this point, the data routing process has given priority to the shortest path routing (q_edge_{ei} and Δq_edge_{ei} are both very small). Comparing with two frequency allocation schemes, we found that the frequency random distribution increases the probability of edge satisfy formula 7. That means under the condition of N_{data_net}, the frequency random distribution is more likely to make the network in the busy state.

There is a certain N_{data_net} makes network in a busy state under the condition of frequency random distribution, while the network is in the free state under the condition of the frequency allocation based on edge betweenness. In the free state, a data in a network would have fewer hops and shorter transmission delay. Whereas in a busy state, a data would meet a long transmission delay and choose the suboptimal path for transmission in order to reduce the transmission pressure of busy link.

b. busy state

When the network is in the busy state, the routing policy based on the load change rate of the edge would have an advantage that the data would choose the best route through

$$(1-\alpha)q\_edge_{ei}+\lambda\Delta q\_edge_{ei}.$$

Compared with the two kinds of frequency allocation scheme, once the edge with large $\mathbf{B\_link}(i,j)$ is not assigned to the frequency with high transmission efficiency, it may force much data to choose suboptimal path. On the one hand, this would increase the routing hops of the data. On the other hand, it would increase the data survival time in the network. Under the same condition of N_{data_net}, there would be more instantaneous existing data in the network. Meanwhile, along with the increasing of N_{data_net}, frequency random distribution would cause the network getting into congestion state earlier.

c. congestion state

When the network is in congestion state, the edge would have so much arriving data greater than its processing ability. The two kinds of routing policy are not able to meet the optimal operation of the resources in the network. With the increasing of N_{data_net}, the amount of the existing data in the network will increase rapidly.

4 FREQUENCY ALLOCATION SCHEME BASED ON EDGE BETWEENNESS

It has already been analyzed if all edges can meet the formula 8 and make the largest N_{data_net}, the network can maximize the transmission capacity.

With further analysis, it has been found that each frequency spectrum characteristic of the changes can assume as a markov process. Therefore, the network always has a certain number of good frequencies. After t_{freq} time, we should operate the frequency allocation scheme again according to the formula 8 to ensure a high betweenness edge assigned a better frequency.

5 SIMULATION ANALYSIS THE NETWORK CONGESTION PERFORMANCE

Network congestion resistance is an important index of network transmission performance. Use the parameter H to describe the phase transition properties of the network congestion (Yang et al. 2013).

$$H\left(N_{data\_net}\right)=\lim_{t\to\infty}\frac{1}{N_{data\_net}}\frac{\Delta W_{net}(t)}{\Delta t} \tag{10}$$

$\Delta W_{net}(t)$ is the total data variation for the network within a certain time range Δt.

If $H(N_{data\_net})$ is equal to zero or less than zero, the total amount of data in a network would keep in constant, and the network is in the free state or busy state and without congestion. When the $H(N_{data\_net})$ is larger than 0, the total amount of data in the network shows an increase. That means the network gets into the congestion state. Use formula 10 to identify the critical state of $H(N_{data\_net})$, then we can get the maximum N_c, it is the largest N_{data_net} that the network can handle. The N_c referers to the network congestion phase transition point.

5.1 *Shortest path routing*

The network congestion characteristics would be analyzed with different frequency allocation scheme under the condition of the shortest path routing.

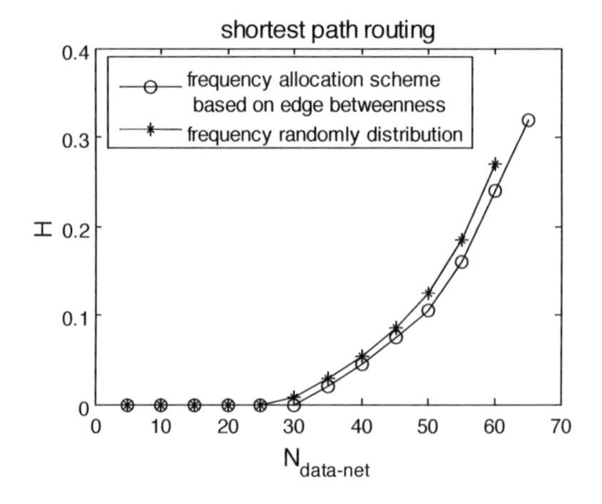

Figure 2. The congestion phase transition curves with two kinds of frequency allocation scheme.

The network includes 100 nodes, and the network model is an ER random graph. The connectivity probability p_{er} of different nodes is 0.1. The normalized transmission rate of edge is 1, 2 and 4 respectively. The frequency allocation uses two kinds of schemes. The simulation result is the average of the simulation of 100 times.

Figure 2 shows the congestion phase change characteristics of the network under the different frequency allocation scheme. It can be seen that when use the frequency allocation scheme based on edge betweenness, the network congestion phase change effect is better than the frequency randomly distribution. The N_c changed from 30 to 25. The results are consistent with theoretical analysis.

5.2 *Routing policy based on the load change rate of the edge*

The routing node is calculated through formula 9. Network simulation environment remains the same.

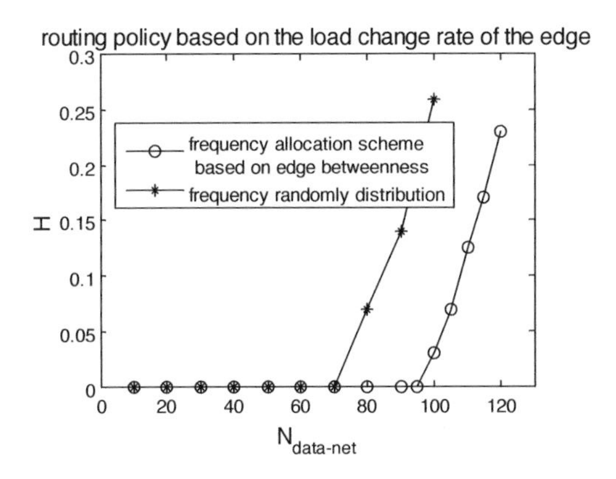

Figure 3. The congestion phase transition curves with two kinds of frequency allocation scheme.

From Figures 2 and 3, it can be found that the routing policy based on the load change rate of the edge could make the phase transition point obviously moves to the right compared with shortest path routing, namely the network capacity increase significantly. When using the routing policy based on the load change rate of the edge, frequency allocation scheme based on edge betweenness will lead to better network transmission performance and the ability to resist congestion. The N_c changed from 70 to 95. The results are consistent with theoretical analysis.

5.3 *Under different network scale*

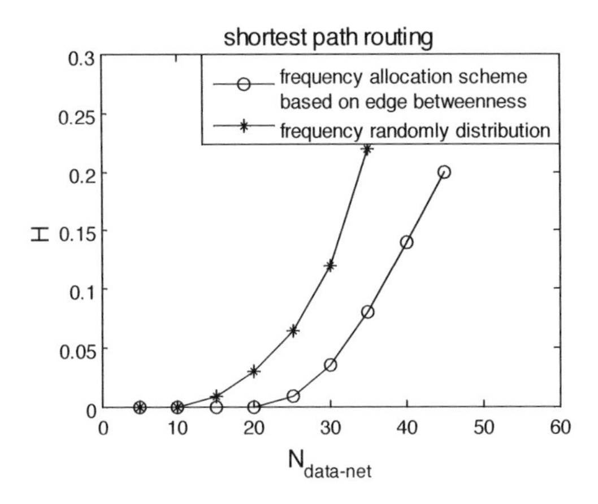

Figure 4. The congestion phase transition curves with two kinds of frequency allocation scheme (shortest path routing).

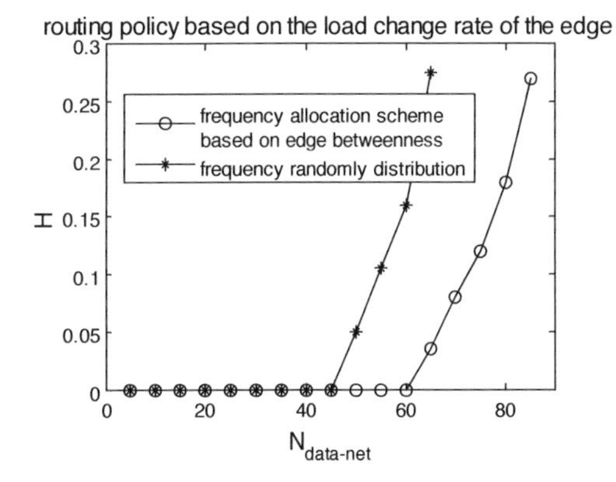

Figure 5. The congestion phase transition curves with two kinds of frequency allocation scheme (routing policy based on the load change rate of the edge).

Through the previous research, it has been found that the integrated use of a frequency allocation scheme based on edge and routing policy based on the load change rate of the edge can effectively improve the network transmission capacity and enhance the ability to resist congestion of the network. What is more, we will analysis the congestion resistance performance with different network scale under two kinds of frequency allocation scheme.

The network includes 50 nodes, and the network model is still an ER random graph. The connectivity probability p_{er} of different nodes is 0.1. For edges, the normalized transmission rate of edges is 1, 2 and 4 respectively. The final results are the average 100 times of simulation results.

Figures 4 and 5 show the congestion phase transition curves with two kinds of frequency allocation scheme, when the network has 50 nodes. The simulation results consistent with the network scale of 100 nodes. Compared with frequency random distribution, this new frequency allocation scheme can significantly enhance the ability of the network, which is to resist congestion and improve the transmission performance of the network with different network scale and different routing policy.

6 SUMMARY

Traditional HF IP network frequency allocation scheme can not meet the needs of the network congestion resistance optimal. Through the theoretical analysis of the reason for network congestion problem, we can find the relationship between the frequency allocation scheme and the edge betweenness, then designed a frequency allocation scheme based on edge betweenness. After the theoretical analysis and simulation comparison under different network scale and different routing policy, the new frequency allocation scheme can improve the network congestion resistance, compared with the traditional frequency allocation scheme.

REFERENCES

Arenas A, Guilera A D & Guimeµa R. 2001. Communication in networks with hierarchical branching. *Phys Rev Lett.* 86: 3196–3199.

Eliardsson P & Karlsson J. 2012. Dynamic frequency allocation in impulsive noise environments. *Electronics Letters* 48: 657–658.

Furman W. N., Koski E. & Nieto J. W. 2013. Design Concepts for a Wideband HF ALE capability. *HFIA, San Diego California, USA.*

Li Wenjie & Liu Bin. 2005. Preemptive Shor-t Packe-t First Scheduling in Input Queueing Switches. *Acta ElectronicaSinica* 33(4): 577–583.

Nigel Arthur. 2010. Analysing the performance of dynamic frequency selection as a HF frequency management technique. *HFIA. San Diego, USA.*

Nur Serinken. 2012. Experimental Wide Band HFIP System SkyNet. *HFIA. York, UK.*

Yang Hanxin, Wu Zhixi & Wang Binghong. 2013. Suppressing traffic-driven epidemic spreading by edge-removal strategies. *Physical Review* 87: 1–5.

Yehuda Eder. 2014. Adaptive Multiple ALE Networking. *HFIA, San Diego California, USA.*

Yu Hao, Zhou Yu-Cheng, Jing Yuan-Wei, et al. 2013. Traffic dynamics of the complex networks with the heterogeneous bandwidth allocation. *Acta Phys. Sin.* 62(8): 1–7.

Electronics, Communications and Networks IV – Hussain & Ivanovic (eds)
© 2015 Taylor & Francis Group, London, ISBN: 978-1-138-02830-2

Uncertainty research of remote sensing image classification using the Modified Rough Entropy Model

Xia Jing, ZeFei Liu & Min Yang
College of Geomatics, Xi'an University of Science and Technology, shannxi, Xi'an, China

Yan Bao
The Key Laboratory of Urban Security and Disaster Engineering, Ministry of Education, Beijing, China

ABSTRACT: Remote sensing classification is an effective technical means to obtain ground cover information, and its quality directly influences the application effectiveness of remote sensing data. Therefore, it is very important to research the precision of remote sensing image classification. However, sampling method and sample size have a relatively large impact on accuracy evaluation using the confusion matrix. In the paper, classification uncertainty of remote sensing image was evaluated based on rough set theory. Firstly, the remote sensing image was classified using SVM (Support Vector Machine) method on the basis of image processing. And then the probability vector after classification is used in this paper as the basis in defining upper and lower approximation sets for remote sensing image classification uncertainty evaluation model. Finally, MRED (Modified Rough Entropy Model) is applied to evaluating the uncertainty of classification results. The results showed the MRED measured the classification uncertainty at the scale of land cover classes are more reasonably and accurately.

1 INTRODUCTION

Remote sensing image classification is an effective technology to obtain ground cover information, special category information to interpret the resulting classification is widely used in various fields has become an important part of Earth data (Sun 2003). We evaluate the quality of the information by accuracy or error normally or also called uncertainty in its broadest definition. The process of classification won't be completed unless the accuracy was evaluated (Li et al. 2012). Remote sensing classification accuracy test is one of the most important parts in the remote sensing image classification. With the accuracy test of remote sensing image classification, on the one hand, we can effectively evaluate the performance of the classifier, and thus transform the classifier; on the other hand, we can also make a final assessment on the quality of the final results of remote sensing classification (Tong 2008). Accuracy evaluation of remote sensing images has experienced a process which is from non-location to location and from qualitative to quantitative (Bo &Wang 2005). The error matrix evaluation is the most stringent method to evaluate classification accuracy statistically and most widely used, what's more, there is some ones who propose that taking it as the assessment criteria of remote sensing image classification accuracy (Smits 1999). But in fact, it cannot express the quality of classification information completely. The basic assumption of the traditional evaluation method is that every pixel can one hundred percent be classified into a certain category, in which the assumption has certain limitation since it ignores the uncertainty during the classification process. It is gradually realizing that, in the process of classification in remote sensing image, the characteristic of uncertainty is ubiquitous, only the uncertainty itself is determined, which is the main factor to affect classification quality (Bo &Wang 2005). In this context, the uncertainty of remote sensing classification has earned widespread attention, and the study begins with booming trend.

In this paper, we take the rough set theory as the theoretical framework to measure the uncertainty of the remote sensing image classification, make full use of modified rough entropy evaluation model to express the quality of remote sensing image classification at the scale of land cover classes; and take advantage of visualization method for remote sensing image classification uncertainty, which show the value , distribution and spatial structure of remote sensing image classification uncertainty by gray-scale, to describe the uncertainty of the classification of remote sensing image completely.

*Corresponding author: *jingxia1001@163.com*

2 REMOTE SENSING IMAGE PREPROCESSING

2.1 *Atmospheric correction*

Using FLAASH (Fast Line-of-sight Atmospheric Analysis of Spectral Hypercubes) model based on the atmospheric transmission model of atmospheric to correct remote sensing image. FLAASH model can effectively eliminate the effects caused by light and atmosphere and other factors on the reflectance, and obtain the physical model parameters more precisely ground radiance and reflectance and surface real temperature (Hao 2008). In the paper, we choose IKONOS imaging for further research, whose center located at longitude 44°35 '33.51 "N, latitude 86°5' 8.71" E, and imaging time for imaging date at 5:21 on July 28, 2008 (GMT), select the mid latitude summer as the atmospheric model.

2.2 *Geometric correction*

In this paper, we use the high-precision GPS field survey ground control points as control points, there are 27 control points evenly laid in the IKONOS image, and the majority of which are selected near the intersection of main road and pathway; to improve the accuracy of field positioning, the 18 control points which using polynomial correction method for geometric precision correction of IKONOS image, and the rest 6 control points as the registration accuracy testing to ensure that the geometric correction error control in 0.5 pixel.

2.3 *Image fusion*

This paper selects Shihezi region IKONOS multispectral image (Figure 1) and panchromatic image of the same area using Gram-Schmidt spectral sharpening method fusion, as shown in Figure 2 fused images.

Figure 1. Regional remote sensing image (red band, near-infrared, green band).

Figure 2. The fused image (a part of the image).

In order to reflect the change before and after the fusion of spatial resolution, the fusion of the same region of the image before and after local amplification, the comparison results are shown in Figure 2. From the comparison of the two images can be seen in the fused image not only retains the high spectral resolution of the original multispectral image, but also to improve the spatial resolution to 1m, which can greatly reduce the error produced in the process of choosing training samples, and improve the accuracy of classification

3 REMOTE SENSING IMAGE CLASSIFICATION

In this paper, the selection of Support Vector Machine (SVM) algorithm for remote sensing image classification study area. SVM is a novel learning method to establish in statistical learning theory, it makes the confidence interval reaches a minimum, at the same time based on the structural risk minimization has maintained the empirical risk fixed, eclectic experience risk and confidence interval, thereby obtaining the minimum risk decision function (Suresh 2012).

According to the field survey data and study area, the surface study area covered is divided into cotton, grapes, water, building land and bare land five types and using a Support Vector Machine algorithm to research regional coverage classification. First of all, respectively, choose a certain number of training samples, and then, by calculating the J-M distance (Jeffreys-Matusita Distance) to determine the ground between categories of separability. The J-M distance J_{ij} can be used to evaluate separability between class features. When $0.0 < J_{ij} < 1.0$, there have no the spectral separability between samples; when $1.0 < J_{ij} < 1.9$, there has a spectral separability between samples, but a greater degree of overlap; when $1.9 < J_{ij} < 2.0$, the spectral separability between samples is better

(Ma et al. 2010). The selected J-M distance between the training samples as shown in the following table:

Table 1. Each category of training samples between J-M distances.

	Cotton	Water	Grapes	Building land	Bare land
Cotton	-	-	-	-	-
Water	2.0000	-	-	-	-
Grapes	1.9871	1.9999	-	-	-
Building land	2.0000	1.9985	1.9970	-	-
Bare land	2.0000	2.0000	1.9999	1.8579	-

As can be seen from Table 1, in addition to building land and bare weak separability, the separability among other samples are better, therefore, images can be classified using the training samples, the classification results as shown in Figure 3.

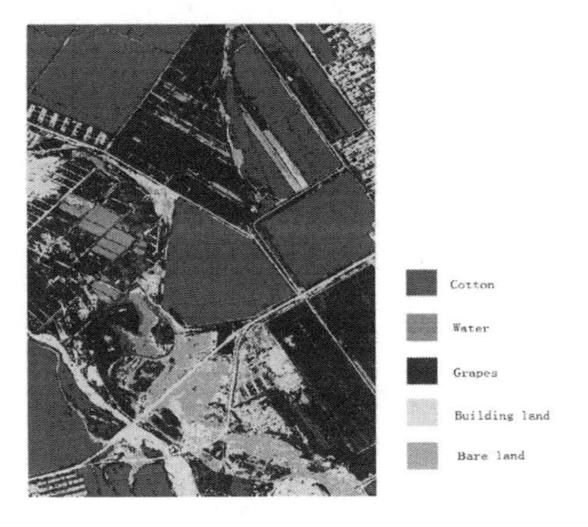

Figure 3. Remote sensing image classification results.

4 EVALUATING THE CLASSIFICATION UNCERTAINTY OF REMOTE SENSING IMAGE BASE ON MODIFIED ROUGH ENTROPY MODEL

4.1 *Modified Rough Entropy Model*

Rough Entropy is a kind of Entropy based on rough set theory, which could evaluate the classification uncertainty at the scale of land cover classes are more reasonably and accurately (Liang et al. 2009). In a remote sensing image we suppose that S=(U,A) is a information system built up by images, where U is standing for all pixels in the image, and A is constituted with gray value, texture, geometry characteristics of different bands, $U/A = \{R_1, R_2, \cdots, R_m\}$is a way to classify remote sensing. suppose that $S = (U,A)$ is a information system , and $X \subseteq U$,U and A should be equal,

$U/R = \{X_1, X_2, \cdots, X_n\}$ means that R is one of the classifications of A, so that we can define that the rough entropy R about rough set A:

$$E_R(X) = -\rho_R(X)[\sum_{i=1}^{n} \frac{|X_i|}{|U|} ln \frac{1}{|X_i|}] \tag{1}$$

where, $|X_i|$and $|U|$ means to the number of pixels in X_i and U separately (Bastin et al. 2002).

Researching the definition (1) of rough entropy further, due to $0 \le \rho_R(X) \le 1$ we can get that:

$$-\rho_R(X)[\sum_{i=1}^{n} \frac{|X_i|}{|U|} ln \frac{1}{|X_i|}] \le -(\sum_{i=1}^{n} \frac{|X_i|}{|U|} ln \frac{1}{|X_i|}) \tag{2}$$

We can realize from the formula (2) that it is not suitable for reality, which the rough entropy is much smaller when we taken two kinds of uncertainty into consideration than one. So Modified Rough Entropy Model had been put forward to solve this problem by He Yaqun ect. (He et al. 2006):

$$E_R(X) = \frac{|BN_R(X)|}{|U|} ln|U| - [\sum_{i=1}^{n} \frac{|X_i|}{|U|} ln \frac{1}{|X_i|}] \tag{3}$$

Where $BN_R(X) = \overline{R}(X) - \underline{R}(X)$ is standing for boundaries, and $\frac{|BN_R(X)|}{|U|} ln|U|$ is used to measure the uncertainty caused by boundaries, if the boundaries enlarged, the value will be larger. From the formula(3), it is easy to see that the modified rough entropy had not put the two uncertainties into account, but also had fixed the rough entropy raised by the formula (1).

4.2 *Evaluate the classification uncertainty at the scale of land cover classes*

We can figure out the classification uncertainty of land cover classes on the image thought the probability vector after classification according to rough set theory. First of all, we must figure out the number of pixels in an upper approximation set or lower approximation set about different classifications. And then evaluate the rough entropy of every classification (Table 2) according to formula (3) and show it out by grayscale form, the final result had been shown in Figure 4. We can know that the rough entropy value will enlarge with the classification uncertainty rising, the color will be light on rough entropy gray image correspondingly.

Table 2. Rough entropy of classifications.

Classes of land cover classes	Cotton	Building land	Bare land	Grapes	Water	$\sum$
Upper approxim ation set	2487442	3938588	287063	1772434	733878	
Lower approxim ation set	2308900	3205111	166225	934484	294297	
Boundary area	178542	733477	120838	837950	439581	
classification	2401402	3655875	189520	1369557	512114	8128468
$-\sum_{i=1}^{n}\frac{\lvert X_i\rvert}{\lvert U\rvert} ln \frac{1}{\lvert X_i\rvert}$	4.3403	6.7967	0.2833	2.3807	0.8283	14.6294
Measurement of boundary area	0.349	1.4357	0.2365	1.6402	0.8604	
Rough entropy	14.9789	16.0651	14.8659	16.2696	15.4899	

Figure 4. Grayscale of rough entropy.

From Table 2, we can see that the rough entropy of building area is highest which means the classification uncertainty was most; the following is the classification uncertainty of grape and bare area; the rough entropy of cotton and water are lower and the uncertainty about the classes is less.

The difference of rough entropy in every classes is mainly due to the difference of their spatial distribution and spectral characteristics, which make the phenomenon happen there will be a varying degrees of mixed pixels in different classes as well as " the same objects with different spectral values "or" the different objects with same spectral values ", the combined effect of these two aspects have different effects on the uncertainty caused by boundary region and indiscernible relation, thus the final rough entropy is different.

Building mostly act as some small pattern spots scattered, and there is too much green vegetation interspersed in, so the proportion of pure pixel in the whole category is very small, the situation of mixed pixels become worse, and the uncertainty caused by boundary region getting larger; meanwhile building land and bare are easily mixed, so the phenomenon of " the different objects with same spectral values" and the uncertainty caused by indiscernible relation getting worse. In a word, the uncertainty of this class is the highest.

Although grapes are massive distributed on the ground overall and has obviously texture features different from the other types, but there must be some mixed pixels for there is much bare soil among grape trellis which have long distance, the uncertainty caused by boundary region has higher value, thus the classification uncertainty of grapes has higher value too.

Although the distribution of bare land is fragmented, but there is no other land type interludes in and the mixed pixel is not serious, the classification uncertainty of boundary region is small; however, the spectral characteristics of bare land and building land is closer, the phenomenon of " the different objects with same spectral values" is more serious, the uncertainty caused by indiscernible relation become large. Therefore, the classification uncertainty of bare land is mainly caused by the indiscernible relation.

Cotton contains more blocks and every block almost act as flakiness, the pixel in the block are very pure and the mixed pixels only exist in the boundary area of each block, the uncertainty of boundary area is small; meanwhile the texture features of cotton and grapes are quite different, the uncertainty of indiscernible relation is small. Thus, the total uncertainty of the class will be small.

The distribution of water is more concentrated, and each block showed a planar. So the intersections region of water and other classes take a small proportion of the whole region, and which cause less mixed pixels; at the same time, the spectral characteristics are different from others obviously, and it's not easy to be mixed. So the uncertainty of water caused by boundary area and indiscernible relation are small, the final uncertainty must be smallest.

However, we can see that from the grayscale of rough entropy the rough entropy of the piece of water on the upper left is high, now we extract the grayscale of rough entropy, classification and original images of this place for further research individually (Figure 5). From the picture we can see that there is some place was classified into grapes and some was classified into building in this classification. Corresponding to the original image, there may be some aquatic plants live in the region classified into grapes, which make the spectral characteristics of the region similar to the spectral characteristics of grapes, and the region was classified into the building may due to the less water level and underwater sediment exposed, which make the spectral characteristics of the region similar to the spectral characteristics of building. The two reasons above increase the uncertainty caused by indiscernible relation, which make the classification uncertainty of the water increase and make the rough entropy greater finally.

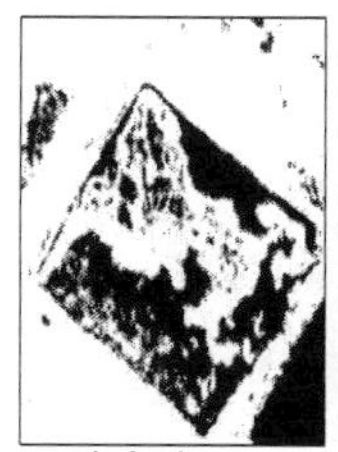
grayscale of rough entropy

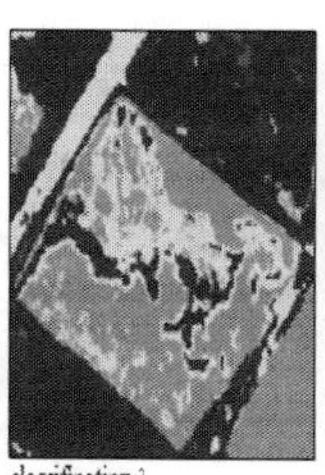
classification

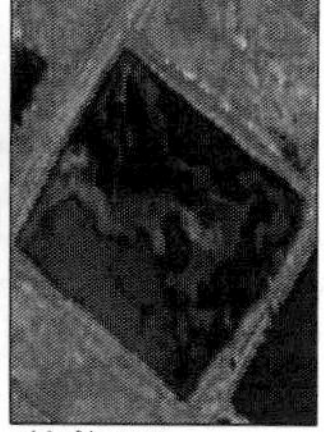
original images

Figure 5. The grayscale of rough entropy, classification and original images of the special water

5 CONCLUSIONS AND DISCUSSION

In this paper, basing on remote sensing classification, we adopt MRED to evaluate the classification uncertainty at the scale of land cover classes, the results showed that the MRED measured the classification uncertainty at the scale of land cover classes are more reasonably and accurately.

There are many methods to evaluate the classification uncertainty of remote sensing image, and each method has its own advantages and disadvantages, how to combine several methods to evaluate classification uncertainty remains be further studied.

In this paper, the measured index of uncertainty is mainly aimed at the pixels attribute information in remote sensing image uncertainty evaluation, which does not take the uncertainty of position information into consideration. It needs further study to analyze and discuss how to make full use of the two uncertainty evaluation index conjunctively.

ACKNOWLEDGEMENTS

We thank the anonymous reviewers for their comments. The Project supported by Chinese National Natural Science Fund (No. 51208016) and Beijing Natural Science Fund (No. 8122008), and Beijing Education Commission Fund (No. KM201310005023)

REFERENCES

Bastin, L., Fisher, P.F. & Wood, J. 2002. Visualizing Uncertainty in Multispectral Remotely Sensed Imagery. *IEEE: Computers & Geosciences* 28(3): 337–350.

Bo, Y.C. & Wang, J.F. 2005. Assessment on Uncertainty in Remotely Sensed Data Classification: Progresses, Problems And Prospects. *Advances in Earth Science* 2(11): 1218–1225.

Hao, J.T., Yang, W. N., Li, Y.X. & Hao, J.Y. 2008. Atmospheric Correction of Multi-spectral Imagery ASTER. *Remote Sensing Information* (1): 78–79.

He, Y.Q., Hu, S.S. & Zhu, J. 2006. Modified Rough Entropy Method for Measuring Uncertainty in Rough Sets. *Journal of Naval University of Engineering*18(4): 26–29.

Li, W.J., Qiu, J. & Long, Y. 2012. Application of Rough Set Theory in Map Generalization. *Acta Geodaetica et Cartographica Sinica* 41(1): 298–301.

Liang, J.Y., Wang, J.H. & Qian, Y.H. 2009. A New Measure of Uncertainty Based on Knowledge Granulation for Rough Sets. *Information Sciences* 179: 458–470.

Ma, N., Hu,Y.F., Zhuang, D.F. & Wang, X.S. 2010. Determination on the Optimum Band Combination of HJ -1A Hyperspectral Data in the Case Region of Dongguan Based on Optimum Index Factor and J-M Distance. *Remote Sensing Technology and Application* 25(3): 358–365.

Smits, P.C., Dellepiane, S.G. & Schowengerdt, R.A. 1999. Quality Assessment of Image Classification Algorithm for Land Cover Mapping: A Review and Proposal for a Cost-based Approach. *International Journal of Remote Sensing* 20(8): 1461–1468.

Sun, S. 2003. Earth Data-An Important Resources for Geoscience Innovation. *Advances in Earth Science* 18(3): 334–337.

Suresh, G.V., Reddy, E.V. & Reddy, E.S. 2012. Uncertain Data Classification Using Rough Set Theory. *Advances in Intelligent and Soft Computing* 132: 869–877.

Tong, Y.E. 2008. *Uncertain Measure in Remote Sensing Image Classification Based on Information Theory and Rough Set*. Fuxin: Liaoning Technical University.

Electronics, Communications and Networks IV – Hussain & Ivanovic (eds)
© 2015 Taylor & Francis Group, London, ISBN: 978-1-138-02830-2

Formal modeling and verification for SDN firewall application using pACSR

Miyoung Kang & Jin-Young Choi
Secure Software Research Center, Korea University, Seoul, Korea

Hee Hwan Kwak
SOLiD, Seongnam, Korea

Inhye Kang
University of Seoul, Seoul, Korea

Myung-Ki Shin & Jong-Hwa Yi
ETRI, Deajeon, Korea

ABSTRACT: The main purpose of this paper is to describe the formal modeling using the process algebra language called pACSR and then suggest a method to verify the firewall application running on SDN using pACSR. In order to detect the violation of firewall rules in case of SDN network topology changes, we propose a verification framework that can check the deadlock through parallel composition of the specification (SPEC) and its implementation(IMPL). If any mismatches or inconsistencies between SPEC and IMPL occur, they could be detected within the formal framework. This framework provides in advance verification for consistency in SDN before critical error might occur by SDN controlling.

1 INTRODUCTION

Recently, various platforms that control networks based on SDN (OpenFlow 2013) have been commercialized by service providers or communicative enterprises such as Google, Cisco, and HP. Since SDN is a software programming that can set, control and manage the network path, administrator's fault or incorrect programming can cause the whole network errors. Therefore, a number of related studies (Canini et al. 2012, Foster et al. 2011, Forster et al. 2014urrently underway to guarantee safety and consistency for SDN network topology.

Our goal is to suggest one of the formal verification methods to assure safety and consistency of the SDN framework. SDN is a management technology that separates the network device control component from the data transfer component using openAPI, such as OpenFlow. Through the SDN controller that deals with the control component, forwarding and packet processing rules are decided and forwarding rules are transferred to the lower SDN switch. For example, various applications such as Security Policy, QoS Policy, and Firewall Policy in [Figure 1] can transfer the rules to the controller. Even if each independent application program does not cause any error, collisions can occur due to rule conflicts among SDN switches when various the application programs are executed in one controller. Since incorrectly written SDN applications can cause errors on the whole network, verification is necessary.

Therefore, we are developing the veriSDN (Shin et al. 2013a), a tool that verifies the rule conflict in such applications. VeriSDN provides a basic formal theory and framework for pACSR(Shin et al. 2013a). It introduces the use of formal modeling and verification techniques in the early stage to describe design sketches of OpenFlow. The model specifed pACSR uses a model checker as a verification tool to ensure that the predicted behavior of the designed OpenFlow satisfies certain properties. The formal model formalizes its correctness in logical formulas as safety properties, then the model is verified against the safety properties by the model-checker VeriFM.

This paper suggests a verification method for veriSDN that can detect the firewall rule violation due to the SDN network topology changes. The verification framework checks deadlocks through the parallel composition of specification (SPEC) and its implementation(IMPL). If any mismatches or inconsistencies between SPEC and IMPL happen, they can be detected within the framework. In this way, the

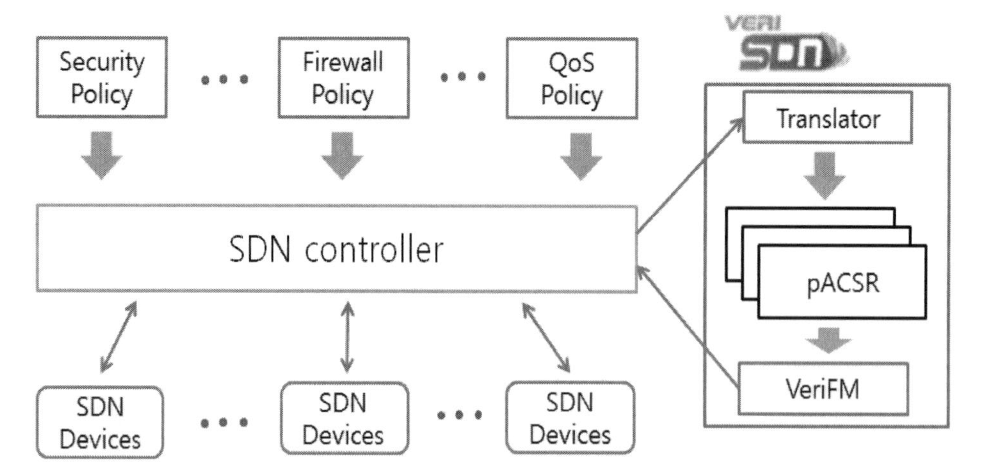

Figure 1. VeriSDN conceptual diagram.

framework provides in advance verification for consistency in SDN before critical errors might occur by SDN controlling.

2 PACSR

A formal description is a specification expressed in a language whose semantics and syntax are formally defined. The need for a formal semantic definition means that the specification language must be based on logic, mathematics, etc., and on natural languages. Formal methods imply the act of proving the correctness of designs or implementations with respect to requirements and properties with which they must satisfy, using formal techniques (Clarke et al. 1996, Kwak et al. 1998a).

ACSR is a formal description language for behavior modeling using concepts of processes, resources, events, and priorities, which is originally developed for the formal verification of real-time embedded systems and Cyber Physical Systems (CPS) (Kwak et al. 1998b). In both CPS and SDN, software is the key because it's the software that determines system and network complexity. So, CPS and SDN have the same issues on how to model and abstract their real-time, concurrent behaviors and feedback controls. ACSR(Kwak et al. 1998b), like other process algebras, consists of a set of operators and syntactic rules for constructing process; a semantic mapping which assigns meaning or interpretation to processes; a notion of equivalence or partial order between process; and a set of algebraic laws that allows syntactic manipulation of processes. ACSR uses two distinct action types to model computation: time and resource-consuming actions, and instantaneous events.

In our research, pACSR extends the process algebra ACSR by allowing network packets to communicate with network ports to describe SDN's behaviors. In pACSR, values passed through ports are network packets (packet passing) and packets can also be passed as parameters for process.

$P := NIL \mid A{:}P \mid e.P \mid be \rightarrow P \mid P_1 + P_2 \mid P_1 \parallel P_2 \mid P\backslash F \mid$
$P \backslash\backslash I$
$e := (T,ve) \mid (l?,ve) \mid (l!,ve) \mid (l?x\ ,ve) \mid (l!ve_1,ve)$
$A := \emptyset \mid \{S\}$
$S := (r,ve) \mid (r,ve), S$

The syntax of process term, *NIL,* represents a deadlock process that does nothing. Process terms $e.P$ and $A{:}P$ are event and actin prefixes, respectively. $P_1 + P_2$ defines a nondeterministic choice between P_1 and P_2 which is subject to priority arbitration. $P_1 \parallel P_2$ represents the parallel composition of P_1 and P_2 in which the events from P_1 and P_2 synchronize or interleave while the timed actions from P_1 and P_2, if they do not share any resource, reflect the synchronous passage of time.

The following pACSR description shows a simple example of semantics with packet passing and parameterized process with packet, which could be seamlessly added in the global flow table's behaviors managed by OpenFlow controller.

P(x) := if matchSrcIP(x,sourceIP)ch!x.NIL
S := ch?y.NIL
Sys := (P(x) || S)/{ch}

P(x) represents a process that has packet x as a parameter. Once process P gets a packet x, it checks with a predefined predicate matchSrcIP(x,source_IP). If the condition in a match field of packet x is satisfied, process P sends packet x through channel which is represented as ch?x. After the egress of packet x, process P becomes NIL. Otherwise, it becomes NIL also in this example. In case of process S, it gets packet through port and names it as y and becomes NIL. Process Sys represents parallel composition of P(x) and S, that is process P(x) and S are running in parallel. pACSR has other features such as predefined predicates (PP) and predefined functions (PF). The term matchSrcIP () in example above is the PP. There could be many predicates predefined as need as long as they don't have side effects.

In this paper, formal verification is accomplished as follows. Let IMPL represent a pACSR process for implementation, and let SPEC represent a pACSR process for the property that IMPL should satisfy. SPEC is described in the following manner: if IMPL violates the property, SPEC prevents the progress of IMPL; otherwise SPEC does not interfere with the progress of IMPL. Then the parallel composition of SPEC and IMPL runs in forever if the property is satisfied. The violation of the property raises a deadlock. Therefore, we can check the correctness as follows:

((SPEC||IMPL)\Sync)\\ALLR » IDLE

where Sync is the set of synchronization channels between SPEC and IMPL and ALLR is the set of all resources in IMPL. In other words, we show that ((SPEC||IMPL)\Sync) produces no deadlock and thus does not reach to NIL.

3 VERIFICATION OF FIERWALL ON SDN

In this section, we will highlight the use case of our verification method using firewall application running on SDN. We show that the violation of firewall rules by the change of network topology should be detected, and propose the framework that can detect the firewall rule violation due to the network topology change.

3.1 *Firewall Application*

We suppose that a firewall application is running in SDN with 4 switches and 3 firewall rules in a policy server that the network should satisfy. Table 1 shows the 3 firewall rules: R1, R2, and, R3. The first rule R1 states that if source IP of a packet is 1 (in this example, we simplify the IP address format as an integer) the packet should be dropped. The second rule R2 states that if the destination IP of a packet is 2 it should be allowed in the network. Rule 3 states that other packets should not be allowed in the network.

Let's assume that firewall rule R2 and R3 are applied to switch SW1. Similarly R1, R2, and R3 are applied to SW2, R2 and R3 to SW3, R2 and R3 to SW4, respectively. With this firewall rules assignment, it is clear that a packet with source IP equals 1 should be dropped at SW2 by the rule R1. Switches in SDN can be connected in many different topology and for this example, for simplicity, we start with the chain topology of 4 switches in the network as shown in Figure 2.

Table 1. Firewall rules.

	rule	action
R1	srcIP==1	drop
R2	dstIP==2	allow
R3	*	drop

Figure 2. Switches in chain topology.

Once firewall rules shown in Table 1 are applied to switches with the knowledge of network topology shown in

Figure 2, OpenFlow table for each switch will be generated by SDN controller as shown in Table 2.

Table 2. OpenFow table of switches in Figure 2.

DP-ID	Match Fields – Flow entry	Instructions
SW1	dstIP == 2	out_port : 2
	*	drop
SW2	srcIP == 1	drop
	dstIP == 2	out_port : 2
	*	drop
SW3	dstIP == 2	out_port : 2
	*	drop
SW4	dstIP == 2	out_port : 2
	*	drop

Now we have firewall rules, R1, R2, and R3, and OpenFlow table (Table 2) for switches, SW1, SW2, SW3, and SW4 connected in a chain topology (Figure 2). A packet with source IP equals 1 and destination IP equals 2 (i.e., srcIP==1, dstIP==2) will be dropped at SW2. In consequence it cannot be delivered to the SW4 or outside of network.

3.2 *Specification and implementation*

There are two elements in our proposed verification framework. The first element is a specification (SPEC) that defines the correct behavior of some parts of the system. The second element is an actual implementation (IMPL). Given the SPEC and IMPL, we can check if there is any inconsistency between SPEC and IMPL, that is, if IMPL follows SPEC. If there are any mismatches or inconsistencies between SPEC and IMPL, they can be detected within the formal framework proposed in this paper using pACSR language. As illustrated in this paper, the firewall rules in Table 1 can be considered as SPEC and OpenFlow table in Table 2 can be considered as IMPL.

3.2.1 *SPEC*

Given the firewall rules and OpenFlow table, pACSR descriptions for both models (i.e., firewall rules and OpenFlow table) can be generated. For instance, firewall rules in Table 1 can be translated into the pACSR description as shown in Figure 3.

```
RSW = {}*: in?x. RSW1(x)
RSW1(x) = if srcIP(x)==1 → RSW                   // R1
          else if dstIP(x)==2 → {}*: out? RSW// R2
          else RSW                               // R3
```

Figure 3. pACSR description for firewall rules.

RSW process represents a set of firewall rules. It waits for an event "in" to synchronize. Once event is synchronized, it becomes RSW1 process. RSW1 process represents firewall rule R1, R2, and R3. The first condition for 'if statement' represents R1 and the second one for R2 and so on. RSW1 process becomes RSW process if source IP of a packet "x" (i.e., srcIP(x)) equals 1. That is, it expects the network to drop a packet of which srcIP(x) is 1. The second if condition (i.e., for R2) waits for event "out" to synchronize if destination IP of a packet "x" (i.e., dstIP(x)) equals 2. Once it is synchronized, it becomes RSW process. The event "out" will be used to notify the fact that packet egresses the network. That is, when event "out" synchronized, SPEC (i.e., firewall rules) knows that a packet in the IMPL (i.e., OpenFlow table) egresses the network. The notation "{}*" represents zero or more time consumption.

For the purpose of verification, RSW process, which is a SPEC, will be composed parallel with IMPL, where SPEC and IMPL will communicate along with event "in" and "out". The pACSR description (i.e., Path) for IMPL will be generated from OpenFlow table and will be explained later.

3.2.2 *IMPL*

Flow entries in OpenFlow table shown in Table 2 is IMPL. They can be represented as pACSR process as shown below.

```
Path(x) = {}*: in!x. {SW1_P1}: SW1(x)
```

"Path" process gets a packet "x" that ingress the network and waits to be synchronized with the event "in". This event "in" is also defined in SPEC, which is represented as RSW process. When SPEC (i.e., "RSW") and IMPL (i.e., "Path") are synchronized via the event "in", we know that a packet enters the network. That is, by the synchronization of the event "in", IMPL notifies to SPEC (i.e., firewall rules) that a packet ingresses. After the synchronization, "Path" process consumes a resource "{SW1_P1} for one time unit and pass the packet "x" to SW1 process.

```
SW1(x) = if dstIP(x)==2 → {}*: {SW1}: SW2(x)
         else {}*: {SW1}: IDLE
```

When SW1 process gets a packet "x", it becomes SW2 process if the destination IP of the packet "x" equals 2. Otherwise, it becomes an IDLE process. IDLE process simply consumes one time unit and becomes IDLE itself again. That is, when a packet "x" moves to IDLE, it can be considered as being dropped. IDLE process is denoted as "IDLE = {}: IDLE".

```
SW2(x) = if srcIP(x)==1 → {}*: {SW2}: IDLE
         else if dstIP(x)==2 → {}*: {SW2}: SW3(x)
         else {}*: {SW2}: IDLE
```

```
SW3(x) = if dstIP(x)==2 → {}*: {SW3}: SW4(x)
         else {}*: {SW3}: IDLE
```

Switch SW2 and SW3 can be modelled similarly as SW1 and shown below.

```
SW4(x) = if dstIP(x)==2 → {}*: {SW4}: NWOUT
         else {}*: {SW4}: IDLE
NWOUT = out! IDLE
```

Once a packet "x" arrives at switch SW4, it will notify to SPEC via the event "out" that a packet egresses the network if the destination IP equals 2. Otherwise, packet will be dropped.

3.2.3 *Verification*

Once SPEC (i.e., RSW) and IMP (i.e., Path) are defined in pACSR description, they can be composed parallel to represent the SDN system as shown below.

```
SDN = RSW(PACKET) || Path(PACKET)
```

"PACKET" represents an instance of a specific packet. If any mismatches or inconsistencies between SPEC and IMPL occur, there will be synchronization failure between SPEC (i.e., RSW) and its IMPL (i.e., Path).

In pACSR description, the synchronization failure between RSW (i.e., SPEC) and Path (i.e., IMPL) happens when the SDN process reaches a deadlock. Given the example with a chain topology shown in

Figure 2, we can see that SDN process never reaches a deadlock. That implies that a packet enters the network and is routed by OpenFlow tables in each switch will follow the firewall rules. For instance, when the packet "x" with srcIP(x)==1 && dstIP(x)==2 enters the network, Path process sends the packet via event "in" to RSW process and will be synchronized with RSW. After synchronization RSW becomes RSW1 process and Path becomes SW1 process. This packet continues to move to SW2 since dstIP(x)==2 and will be dropped at SW2 since srcIP(x)==1 the condition satisfies. That is, SW2 becomes IDLE after consumes one time unit of resource "{SW2}". Another process, RSW1, becomes RSW by the rule R1 and idles forever. Since both RSW and Path processes become idle, we can conclude that there is no deadlock.

We note that the other case such as source IP does not equal to 1 and destination IP equals 2, also do not cause any deadlock. A packet will not be dropped at SW2 and moved to SW3 since its source IP does not equal to 1 and destination IP is 2. This packet will eventually reach SW4 and then egress to the out of the network. Once a packet goes out of network, NWOUT process synchronize the event "out" with RSW1 process and both process will become idle. As we have shown, these two different packets does not cause any deadlock.

4 VERIFICATION OF MODIFIED NETWORK

In this section, we consider the case when the topology is changed from chain to diamond shape as shown in Figure 4. Even though the same firewall rules are applied to each switch, the behavior of packets in the diamond topology will be different from chain topology due to the topology difference.

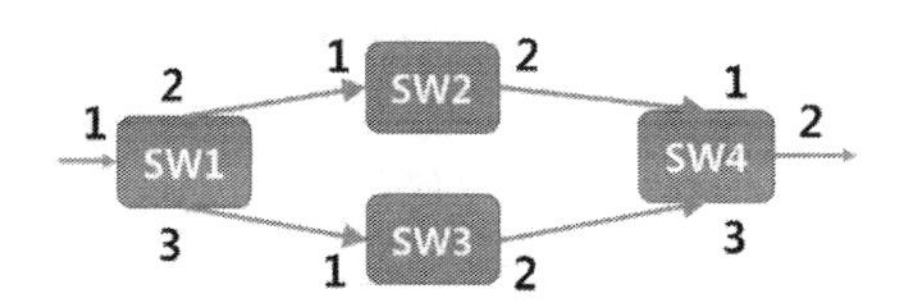

Figure 4. Switches in diamond topology.

For instance, any packet with source IP equals to 1 could not egress to the outside of network since it will be dropped at SW2. However, with diamond topology, it can be delivered to SW4 and can be egressed to the outside of network.

We show that the violation of firewall rules by the change of network topology should be detected, and propose the framework that can detect the firewall rule violation due to the network topology change.

In the following sections, we first explain how a packet with source IP equals 1 and destination IP equals 2 can reach to SW4 and egress the diamond topology network. This packet has been dropped at SW2 in the chain topology network. We then highlight how our formal framework detects this problem.

There will be the same firewall rules and their assignment for each switch. Only the OpenFlow table will be updated by the SDN controller due to the change of topology. Newly updated OpenFlow table is shown in Table3.

4.1 *SPEC of modified topology*

There will be no changes in the firewall rules. However, the pACSR description for the firewall rules should be updated to reflect the topological changes in order to have correct synchronization of events. New RSW process is shown below.

Table 3. OpenFlow table of switches in Figure4.

DP-ID	Match Fields – Flow entry	Instructions
SW1	dstIP == 2	out_port : 2 out_port : 3
	*	drop
SW2	srcIP == 1	drop
	dstIP == 2	out_port : 2
	*	drop
SW3	dstIP == 2	out_port : 2
	*	drop
SW4	dstIP == 2	out_port : 2
	*	drop

RSW = {}*: in?x. RSW1(x)
RSW1(x) =
 if srcIP(x)==1 → RSW // R1
 else if dstIP(x)==2 → {}*: out? {}*: out? RSW// R2
 else RSW // R3

For firewall rule R2, RSW1 process waits for the event "out" twice since a packet at switch SW1 has been copied into two packets at switch SW2 and SW3.

4.2 *IMPL of modified topology*

There will be no changes in the firewall rule assignment for each switch. However, to reflect the topological changes, OpenFlow table will be updated by the SDN controller. Newly updated OpenFlow table is shown in Table3.

From an updated OpenFlow table, we can generate pACSR descriptions shown below.

Path(x) = {}*: in!(x). {SW1_P1}: SW1(x)

SW1(x) = if dstIP(x)==2 → {}*: {SW1}: SW1_1(x)
 else {}*: {SW1}: IDLE
SW1_1(x) = SW2(x) || SW3(x)

We note in a diamond topology in Fiqure4., a packet in SW1 is copied into SW2 and SW3. This is denoted as SW1_1(x) process, which spawns two processes SW2 and SW3. That is, a packet in SW1 is copied into two packets, one in SW2 and another one in SW3. Other pACSR processes are identical to the ones shown in a chain topology.

4.3 *Verification of modified topology*

Same as the chain topology case, RSW and Path process will be parallel composed to represent the SDN system as shown below.

SDN = RSW(PACKET) || Path(PACKET)

In a diamond topology, when a packet with source IP equals 1 and destination IP equals 2 enters the network, there is a path that this packet can reach to SW4 such as SW1 $\rightarrow$ SW3 $\rightarrow$ SW4. This is the violation to firewall and should be detected. This packet was dropped in the chain topology.

When to consider the SDN process with PACKET, a packet with source IP equals 1 and destination IP equals 2. The firewall rule process RSW becomes RSW1 and checks if source IP equals 1. Since source IP is 1, it becomes RSW again and idles forever until new packet comes in. However, Path process becomes SW1 and two processes SW2 and SW3 are spawned. SW3 becomes SW4 and tries to synchronize "out" with RWS. Unfortunately, since there is no other process ready to synchronize "out" event, it comes NIL (i.e., deadlock). This implies that there is a firewall violation with diamond topology.

As we mentioned, our verification framework consists of SPEC and IMPL. We can check if IMPL matches SPEC by checking the deadlock in the parallel composition of SPEC and IMPL. SPEC states the correct behavior of firewall. IMPL models the actual instances of OpenFlow table. OpenFlow table of each switch are described in terms of pACSR processes. It turns out to be quite natural to model OpenFlow table with pACSR language. Each OpenFlow entry can be a pACSR process and each matching conditions can be a condition, and action field of OpenFlow entry can be modeled as pACSR with argument.

5 CONCLUSION

This paper proposed the verification methods for SDN network topology that performs in veriSDN, the formal verification tool. VeriSDN verifies the rule conflict among various SDN applications such as firewall, QoS, and routing. The methods in veriSDN detect the firewall rule violation caused by the SDN network topology changes. The verification framework checks deadlocks through parallel composition of SPEC and its IMPL. If any mismatches or inconsistencies between SPEC and its IMPL occur, they can be detected within the formal framework. That is, the framework provides in advance verification for consistency in SDN before critical errors might occur by SDN controlling.

In the future, we plan to investigate on how scalable our formal method will be with a number of flow table entries, topology, and properties. We are currently in the process of generate whole pACSR description. This approach is redundant in a sense that most of pACSR description remains intact. Hence, we will investigate incremental verification methods to reduce this overhead and only the altered part of flow table should be translated into pACSR.

ACKNOWLEDGMENT

This research was funded by the MSIP(Ministry of Science, ICT & Future Planning), Korea in the ICT R&D Program 2014, supported by the MSIP(Ministry of Science, ICT and Future Planning), Korea, under the ITRC(Information Technology Research Center) support program (NIPA-2014-H0301-14-1023) supervised by the NIPA(National IT Industry Promotion Agency) , and was supported by Basic Science Research Program through the National Research Foundation of Korea(NRF) funded by the Ministry of Education, Science and Technology(2012R1A1A2009354).

REFERENCES

Canini, M. , Venzano, D. , Peresini, P. Kostic, D. & Rexford, J. 2012. A NICE way to test OpenFlow applications. *NSDI*: 127–140.

Clarke, E. & Wing, J. 1996. Formal methods: State of the art and future directions. *ACM Computing Surveys* 28(6): 626–643.

Foster, N., Harrison, R., Meola, M. L. & Freedman, M. J. 2014. NetKAT: Semantic Foundations for Networks. *In Proceedings of the 41st annual ACM SIGPLAN-SIGACT symposium on Principles of programming lanuages* 49(1): 113–126.

Foster, N., Harrison, R., Meola, M. L., Freedman, M. J., Rexford, J. & Walker, D. 2011. Frenetic: A high-level language for OpenFlow networks. *In Proceedings of the Workshop on Programmable Routers for Extensible Services of Tomorrow*: 6.

Kang, M., Kang, E., Kwak, H., *Hwang, D.,* Kim, B., Nam, K. Choi, J. & Shin, M. 2013. Formal Modeling and Verification of SDN-OpenFlow. *2013 IEEE Sixth International Conference on Software Testing, Verification and Validation (ICST), Luxembourg. 18–22 March 2013.* IEEE.

Kwak, H., Choi, J., Lee, I. & Philippou, A. 1998a. Symbolic weak bisimulation for value-passing calculi. *Technical Report, MS-CIS-98–22, Department of Computer and Information Science*, University of Pennsylvania.

Kwak, H., Choi, J., Lee, I. & Philippou, A. 1998b. Symbolic Schedulability Analysis of Real-time Systems, *In IEEE Real-Time Systems Symposium. The 19th IEEE. Madrid 2–4 December 1998*. IEEE.

OpenFlow Specification 1.3.1 2013. https://www.opennetworking.org/sdn-resources/onf-specifications/openflow

Shin, M., Kwak, H., Choi, J. & Kang, M. 2013a. Process Algebra based Symbolic Verification for Software-Defined Networking(SDN). *Telecommunication Review, SK Telecom* 23(5): 583–593.

Electronics, Communications and Networks IV – Hussain & Ivanovic (eds)
© 2015 Taylor & Francis Group, London, ISBN: 978-1-138-02830-2

Fast recovery of TCP congestion window on a backup 3G path for MPTCP via occasional probing

Byoungkwan Kim, Sunghyun Im, Seung Ki Park & Ju Wook Jang*
Department of Electronic Engineering, Sogang University, Seoul, Korea

ABSTRACT: Mobile operators forecast continuous increase of mobile data traffic generated by their customers. Offloading to Wi-Fi is being considered as a solution to reduce the burden of the cellular networks. Multipath TCP (MPTCP) supports the simultaneous use of multiple interfaces (Ford 2011). Therefore, mobile operators who use MPTCP for mobile data transmission can provide more bandwidth and achieve smooth handover from Wi-Fi to cellular network for user.In this paper we analyze existing MPTCP optional modes (Full mode, Backup mode)'s problem and propose complemented MPTCP mode. Finally, we experimentally prove benefits of our schemes.

1 INTRODUCTION

In recent years mobile data traffic is increased sharply, and mobile operators have a hard time to provide mobile data service for their customers. According to Cisco's report (Cisco 2013), this trend is going to deepening year after year. Offloading to Wi-Fi is being a main option to solve the problem, and MPTCP is considered as key technology to offloading mobile data traffic from Wi-Fi to cellular network. MPTCP allows simultaneous use of multiple interfaces, and supports smooth handover from Wi-Fi to cellular network.

For moving traffic away from congested links, two modes of operation for MPTCP were suggested. One is Full-MPTCP mode and the other is Backup mode (Paasch et al. 2012). Full-MPTCP mode refers to the regular MPTCP operations where all sub-flows are used. Such mode allows seamless handover to the mobile node when it moves to a cellular network from Wi-Fi network. But in the Full-MPTCP mode, a mobile node uses too much cellular data. In the contrary, the mobile node uses only Wi-Fi network's data when stayed Wi-Fi coverage in Backup mode. But, Backup mode can't support seamless handover as many as a Full-MPTCP mode because the CW(congestion window) on the cellular sub-flow is not large like Full-MPTCP mode. Our aim is to develop these MPTCP modes and to propose new MPTCP operation algorithms. We propose a new MPTCP mode (Section 2) and evaluate our scheme through ns-2 simulation results for proving the advancement from previous works (Section 4). Finally, Section 4 concludes this paper.

2 PROPOSED SCHEME

In this paper, we propose a new several algorithms which are developed from Full MPTCP mode and Backup mode. The mode of operation for MPTCP which includes those algorithms is the H-mode.

H-mode enables the mobile node to spend a few cellular data when the TCP sub-flow on the Wi-Fi interface is established. Also, when Wi-Fi sub-flow is disconnected, mobile node's cellular sub-flow CW is set to a reasonable value and, not initial value.

CW pre-checking and adaptation: In Backup mode (Paasch et al. 2012), mobile node's good-put is very low and slowly recovered immediately after vertical handover (by Wi-Fi loss) because the CW on the cellular network sub-flow still has initial value. Therefore, we propose a new MPTCP mode, which contains an effective solution to this issue. The first proposed algorithm is performed following process; cellular sub-flow's CW is pre-checked, recorded, updated when mobile node had maintained Wi-Fi connection, and that will be adapted to the cellular sub-flow after Wi-Fi loss.

*Corresponding author: *jjang@sogang.ac.kr*

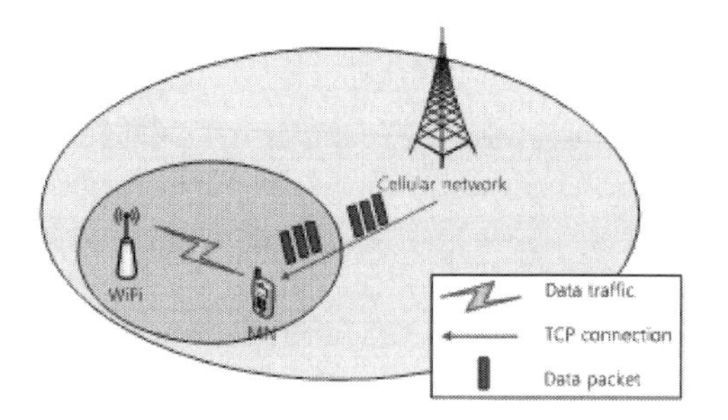

Figure 1. Cellular sub-flow's CW pre-check, save, and update.

As shown in Figure 1, mobile node pre-checks connected cellular sub-flow congestion by transmitting some data packets to cellular networks. This transmission is maintained before enough data is gathered for estimating congestion and calculating CW magnitude.

Then calculated CW magnitude is recorded at cellular network's server and this measurement process is repeated periodically time t for upgrading data. At this time R_H, cellular data consumption by the mobile node is calculated as

$$R_H = R_F \times \left(\frac{\sum t_{t0}}{\sum T} \right) \tag{1}$$

where R_F is cellular data consumption by MS in full-mode, t_{t0} is data transmission time for estimating congestion in H-mode, t_{min} is a measurement cycle in H-mode, T is sum of t_{t0} and t_{min}.

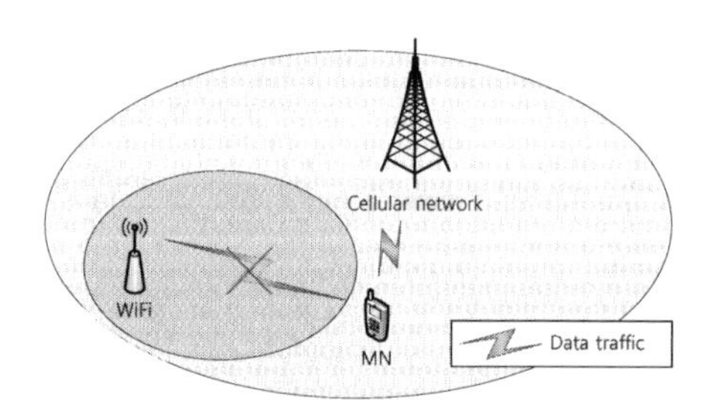

Figure 2. Data transmission by cellular network after Wi-Fi loss.

As shown in Figure 2, mobile node emerged from Wi-Fi coverage (Wi-Fi loss), and is only connected to the cellular network. Then, cellular network adapt CW_A to the mobile node, and mobile node's good-put is maintained more that value. $CW_{A,}$ which is adapted at cellular sub-flow is calculated as

$$CW_A = \frac{1}{2} \times CW_R \tag{2}$$

where CW_R is recorded CW magnitude when the mobile node is in Wi-Fi coverage. The reason of not adoption CW_R to cellular network is to avoid a timeout. Cellular network congestion is changed frequently, and CW_A is not accuracy value at the time of the change. And SSth(Slow start threshold) value is also changed to CW_A. Before MS emerged from Wi-Fi coverage, the cellular sub-flow's SSth value is the initial value. It doesn't reflect cellular network congestion. So, if SSth is not updated when MS emerged from Wi-Fi coverage, time out is occurred immediately.

The flowchart of the *CW pre-checking and adaptation* algorithm is illustrated in Figure 3, respectively.

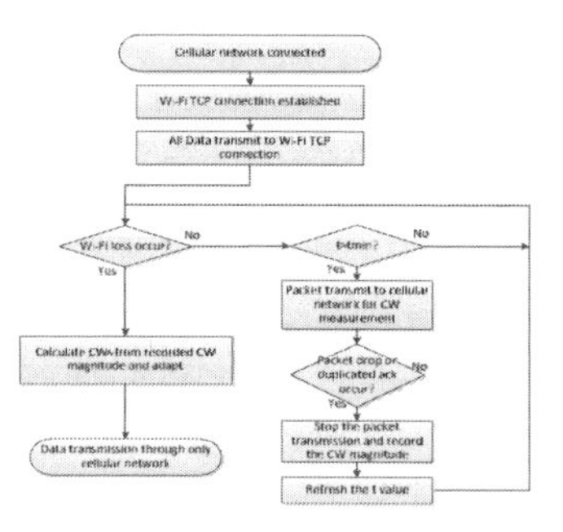

Figure 3. CW pre-checking and adaptation.

3 EVALUATION

In this section we provide an evaluation of the performance by ns-2 (Fall et al. 2007) when using the two MPTCP modes, Backup mode and H-Mode. The simulation is designed to demonstrate the merits of H-mode over the Backup mode. As shown in Figure 4, cellular sub-flow maintains similar CW values. H-mode is operated for measuring to cellular sub-flow CW value and applying value for reconnection.

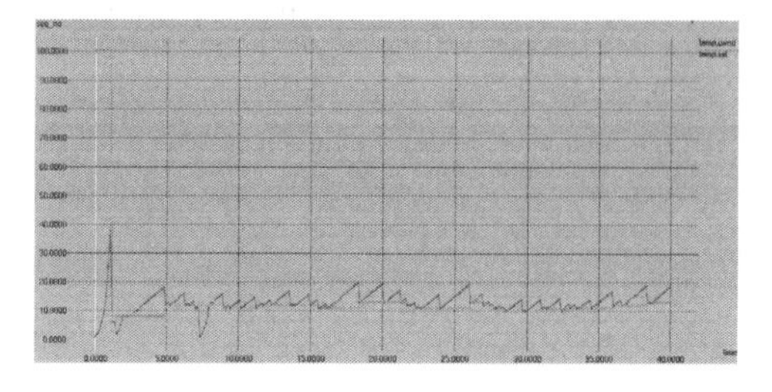

Figure 4. The congestion window of cellular sub-flow.

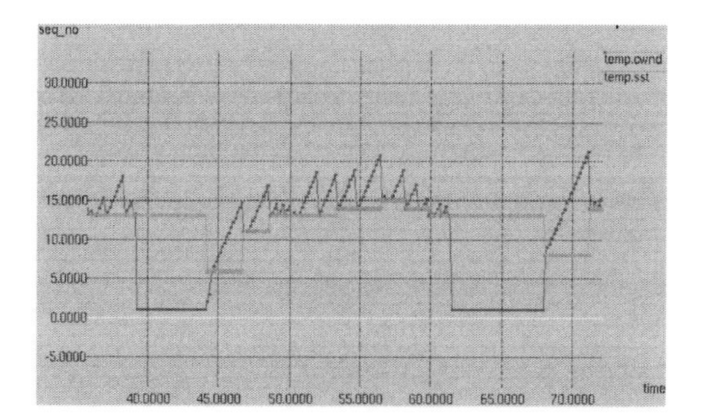

Figure 5. The congestion window adaptation scenario.

Figure 5 depicts the congestion window adaptation scenario and Figure 6 depicts the details of Figure 5's graph.

As shown in Figure 6a, after tcp connection is disconnected, the cellular sub-flow's congestion window is increased by slowstart (Floyd 1999), and it takes 0.3 seconds. In contrast, as shown in Figure 6b, When H-mode is operated and recorded congestion window is adapted to cellular sub-flow, sub-flow's congestion window recovery delay is almost 0 seconds.

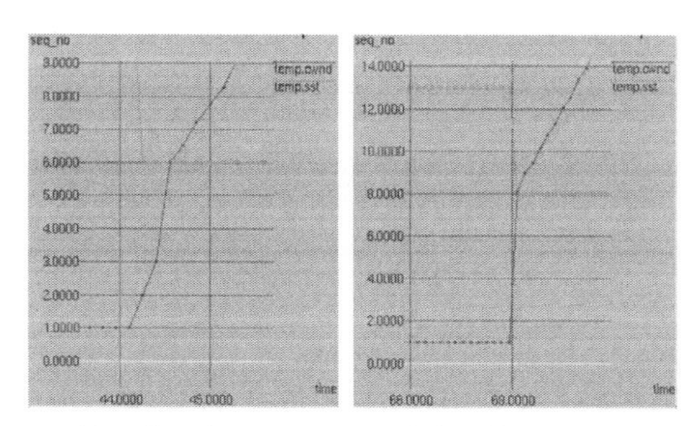

a. Not adaptaion b. Adaptaion

Figure 6. CW recording and CW_A adaptation.

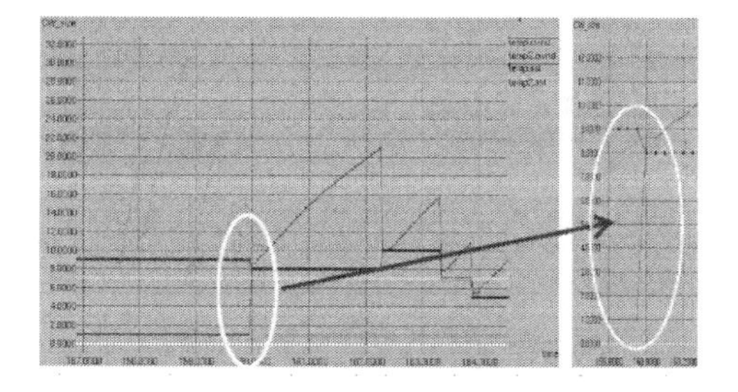

Figure 7. The congestion window adaptation of H-mode.

Figure 7 depicts the congestion window adaptation of H-mode when handover from Wi-Fi to cellular network is happened. As shown in Fig. 7, when Wi-Fi loss is occurred, H-mode supports high CW magnitude for cellular sub-flow immediately.

4 CONCLUSION

Offloading to Wi-Fi is an important solution to reduce the burden of the cellular networks. Multipath TCP (MPTCP) is primary scheme which can support the simultaneous use of multiple interfaces. Thus, it means that user can transmit to mobile data by different networks at the same time using MPTCP scheme. In this paper, we propose a new algorithm, H-mode, which is developed from previous MPTCP's modes (Full MPTCP mode, Backup mode). H-mode enables the mobile node to spend a few cellular data when the mobile node exists on Wi-Fi coverage. When mobile node move from Wi-Fi coverage to outside and Wi-Fi sub-flow is disconnected, mobile node's cellular sub-flow CW is set to a reasonable value. That means we can provide smooth handover to mobile users. Finally, we can prove the usefulness of H-mode by ns-2 simulations.

ACKNOWLEDGEMENT

This research was supported by the MSIP(Ministry of Science, ICT and Future Planning), Korea, under the ITRC(Information Technology Research Center) support program (NIPA-2014-H0301-14-1048) supervised by the NIPA(National IT Industry Promotion Agency)

REFERENCES

Cisco, T. 2013. Cisco Visual Networking Index: Global Mobile Data Traffic Forecast Update *Cisco Public Information*: 2012–2017.

Fall, Kevin, et al. 2007. The network simulator (ns-2). *URL: http://www. isi. edu/nsnam/ns.*

Floyd, Sally et al. 1999. The NewReno modification to TCP's fast recovery algorithm. RFC 2582, IETF.

Ford, Alan, et al. 2011. TCP extensions for multipath operation with multiple addresses. *IETF MPTCP proposal. URL: http://tools. ietf. org/id/draft-ford-mptcp-multiaddressed-03. txt.*

Paasch, Christoph, et al. 2012. Exploring mobile/WiFi handover with multipath TCP. In *Proceedings of the 2012 ACM SIGCOMM workshop on Cellular networks: operations, challenges, and future design*: 31–36. ACM.

Electronics, Communications and Networks IV – Hussain & Ivanovic (eds)
© 2015 Taylor & Francis Group, London, ISBN: 978-1-138-02830-2

Design of a compact frequency-reconfigurable notched UWB antenna based on meander lines

Yuanyuan Kong, Yingsong Li* & Wenhua Yu
College of Information and Communications Engineering, Harbin Engineering University, Harbin, Heilongjiang, China

ABSTRACT: This paper proposes a printed reconfigurable band-notched Ultra-Wideband (UWB) antenna, which has a bandwidth ranging from 3.1 GHz to 10.6 GHz. The antenna consists of a fan-shaped radiating patch, a trapezoidal ground plane and two Meander Line (ML) structures. By engraving two snake-shaped meander lines on the patch and ground plane, respectively, two notch bands are generated in order to filter undesired signal from WiMAX frequency bands (3.3 GHz~3.6 GHz) and WLAN frequency bands (5.15 GHz~5.825 GHz). Additionally, two ideal radio frequency switches are utilized on the snake-shaped MLs to realize the required notch reconfiguration. Simulated results obtained from the HFSS showed that the designed antenna can provide flexible band-notched and attractive reconfigurable characteristics, making the proposed antenna suitable for multi-mode UWB communication applications.

1 INTRODUCTION

With the increasing developments of wireless communication technologies, ultra-wideband (UWB) technology has been widely investigated for its high data rate, large communication capacity, low detection rate and interception rate (Win et al. 1998), etc. In particular, the UWB antenna has been greatly studied in industrial and academic fields and lots of UWB antennas have been developed to meet the requirements for indoor applications. A half-sized unbalance dipole antenna with fan-shaped and trapezoidal elements is designed for ultra-wideband communication application (Hiraguri et al. 2013), which can provide a good bandwidth and omnidirectional radiation characteristics over the desired operating band ranging from 3.1GHz to 10.6GHz.

However, the bandwidth of the UWB communication band designated by FCC overlapped with several existing narrowband communication systems such as worldwide interoperability for microwave access (WiMAX: 3.3GHz~3.6GHz) and wireless local area network (WLAN: 5.15GHz~5.825GHz), which will interfere with the UWB system, vice visa. The UWB system faces a challenging problem to deal with these potential interferences. One of effective techniques is to design a band-notched UWB antenna to filter unwanted narrowband signals, which may give interference to the UWB systems. After that, a great number of UWB antennas with band-notched characteristics have been reported to suppress these potential interference signals. Band-notched characteristics can be generated by many methods, such as cutting various slots on the radiation patch and the ground plane or adding different resonators along the feed signal lines for the microstrip fed UWB antennas. By etching a quarter-wavelength slot on the coplanar waveguide antenna (Nguyen et al. 2012), surface current distribution of the radiation patch is changed and a notch band is produced, which might be used to suppress undesired signal. Rostamzadeh et al. (2012) has designed a V-shaped slot and two rod-shaped parasitic structures to give a frequency band-stop function, which gives frequency filtering characteristics. In addition, such band-notch function and filtering bandwidth can be controlled by adjusting the dimensions of the slot and the rod-shaped parasitic elements. Another effective method is to settle different shapes of resonators near the microstrip feed signal strip line, which is to filter unwanted signal interferences. By exploiting this technique and reducing the potential interference from used narrowband communications, interdigital capacitance loading loop resonator (IDCLLR) structure is utilized to reject designated bands within the passband of the UWB antenna (Li et al. 2012). Recently, a compact multiple band-notched UWB antenna with four pairs of meander lines (MLs) that work as desired resonators to produce quadruple band notch characteristics has been reported. In addition, embedding U-shaped parasitic strips along the feed line can also be used to design narrowband notching characteristic (Weng et al. 2012, Jiang & Che 2012). Although these

*Corresponding author: *liyingsong@ieee.org*

designed band-notched UWB antennas can provide frequency filtering functions to filter unwanted narrowband signal interferences, the band-notched UWB antennas cannot provide entire bandwidth for conventional UWB applications. In order to solve the problem, reconfigurable UWB antennas have been invented to provide a notched function when there is an interference located within the UWB band, while they can act as UWB antennas without any potential interference. On the basis of the concepts, a band-notched UWB antenna with reconfigurable characteristics has been studied by mounting electronic switches across or along these resonators, which serve to active or deactivate their corresponding band notches (Al-Husseini et al. 2010). Similarly, Li et al. (2012), Li et al. (2013) designed reconfigurable UWB antennas by the integration of stepped impedance resonators (SIRs) and ideals switches. A reconfigurable UWB antenna has been discussed by the use of PIN-diode switches, which can provide an UWB state, three narrowband state, or a dual-band state (Boudaghi et al. 2012).

In this paper, a compact frequency-reconfigurable notched UWB antenna is proposed by using ideal switch and meander line techniques. The proposed dual-band characteristics are realized by means of meander lines (MLs), while the reconfigurable function is realized by using two ideal radio frequency switches that are integrated on the designed two MLs. Simulated results obtained from the HFSS demonstrated that the proposed antenna can give desired notch characteristic when the antenna is used as a band-notched UWB antenna, while it can cover the entire bandwidth of the UWB when all the switches are ON.

2 STRUCTURE OF THE PROPOSED RECONFIGURABLE BAND-NOTCHED UWB ANTENNA

The configuration of the proposed antenna is shown in Figure 1, which is based on the fan-shaped UWB antenna (Hiraguri et al. 2013). The proposed notch bands are obtained by using the meander lines and the ideal switches that are proposed by Weng et al. (2012) and Li et al. (2012), Li et al. (2013). It is found that the proposed antenna consists of fan-shaped radiating patch that is connected with the 50-Ohm microstrip feed signal line and is printed on the top side of a substrate with a relative permittivity of 2.65, the thickness of 0.4mm, and the dielectric loss tangent is 0.001, a partial ground plane that is printed on the opposite side of the substrate, two snake-shaped MLs that are embedded in the radiating patch and the partial ground plane with trapezoidal shape, respectively, and two ideal switches, namely switch-1 (SW1) and switch-2 (SW2), which are integrated into the two MLs. The 50-Ohm microstrip feed structure is comprised of the microstrip feed signal line and the trapezoidal ground plane that is used to enhance the bandwidth of the proposed antenna. The ML etched on the radiating patch is to produce a notch at the WiMAX band, while the ML embedded in the trapezoidal ground plane aims to give the WLAN band notch. The center frequencies of these two notch bands can be controlled by adjusting the dimensions of the MLs, while the reconfigurable characteristics can be designed by controlling the ON/OFF states of the SW1 and SW2. To implement these switches, the presence of a metal bridge represents the ON state, while its absence represents the OFF state (Li et al. 2012, Li et al. 2013).

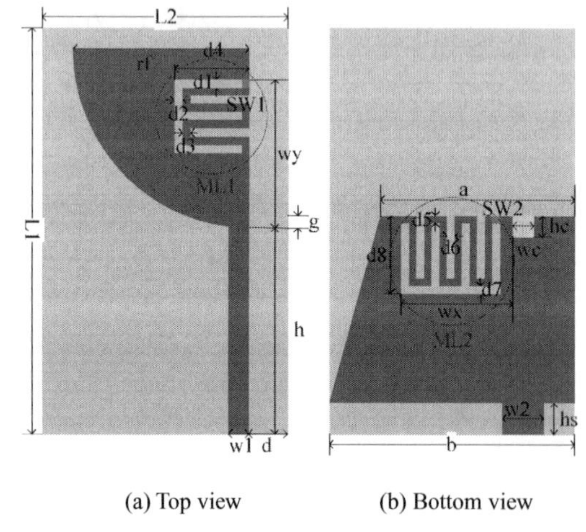

(a) Top view (b) Bottom view

Figure 1. Configuration of the proposed reconfigurable notched UWB antenna.

It can be seen that the proposed antenna has a length of L1=41.6mm and a width of L2=20mm. The microstrip feed line width is w1=1.05mm and the gap between the trapezoidal ground plane and the fan-shaped radiating patch is g=0.4mm. the proposed antenna is optimized by using the HFSS and the optimized parameters are listed as follows: w2=6mm, wc=3mm, hc=1.3mm, hs=1mm, d=4mm, wx=10mm, wy=11mm, d4=5.5mm, a=16mm, b=20mm, rf=15mm, h=23mm, d1=d2=d3=0.59mm, d5=d6=d7=0.2mm and d8=2.6 mm.

3 ANTENNA PERFORMANCE BANDWIDTH

In this section, the effects of the MLs, the parameters of the MLs and the effects of the integrated switches are extensively investigated.

3.1 *Effects of the MLs on the impedance bandwidth*

Figure 2 shows the effects of the MLs on the impedance bandwidth of the proposed antenna. It is observed that the proposed antenna without MLs is a UWB antenna that covers the entire bandwidth designated by FCC. The proposed antenna with only ML1 gives a notch band at the 3.3 GHz~3.6 GHz, which is to filter the interferences from WiMAX system. When the two MLs are integrated into the UWB antenna at the same time, the proposed antenna is a dual-band-notched UWB antenna with two notch bands located at the WiMAX band and the WLAN band, respectively. In this case, the proposed antenna can filter the potential interferences from both the WLAN and WiMAX bands. In addition, we can see that the WiMAX band notch characteristics are generated by the ML1, while the WLAN notch band is given by the ML2. In this design, the resonance length of the MLs is about $\lambda_g/4$, where λ_g is the guided wavelength approximately given by

$$\lambda_g \approx \lambda_0 \Big/ \sqrt{(\varepsilon_r + 1)/2} \quad (1)$$

with λ_0 being the free space wavelength and ε_r being the relative permittivity.

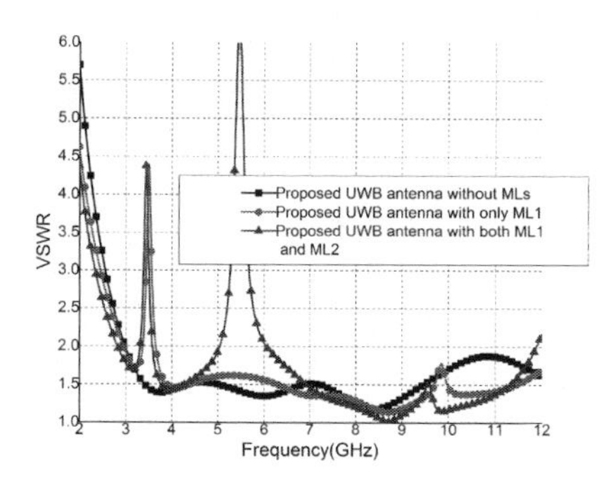

Figure 2. VSWR characteristics of proposed UWB antenna with MLs.

3.2 *Effects of the parameters on the impedance bandwidth*

From the discussion in section 3.1, we can see that the two notch bands are produced by ML1 and ML2, respectively. Key parameters d4 and d8 are selected to investigate the effects of the dimensions of the MLs and both the switches are set as OFF. In this case, the MLs act as open-circuit stubs that can prevent the unwanted signals from the narrowband systems. The effects of the d4 are shown in Figure 3. We can see that the center frequency of the lower notch band moves from 3.5 GHz to 3 GHz with an increment of d4, which is caused by the increased resonance length of the ML1. When the d4 increases from 5.2mm to 5.8mm, the current path along the snake-shaped ML is also increased, which result in lengthened effective resonance length. Additionally, the resonance length is approximately equal to a quarter-wavelength, which is according to the center frequency. From the Figure 3, it is found that the proposed notch can provide immunity to the unwanted 3.5 GHz WiMAX signal.

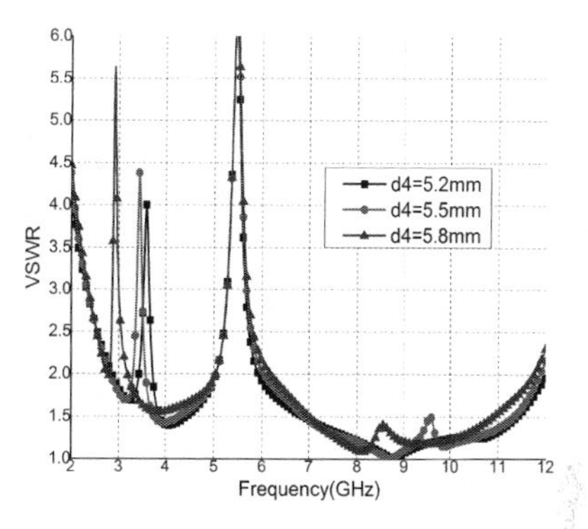

Figure 3. VSWR characteristics of proposed reconfigurable UWB antenna with variant d4.

Figure 4 depicts the effects of d8 on the impedance bandwidth. With the increment of d8, the center frequency of the higher notch band shifts to the low frequency because of the increased resonance length of ML2. It is found that the trend of center frequency of higher notch band is inversely proportional to d8, while the center frequency of the lower notch band remains constant. As d8 increases from 2.4mm to 2.8mm, the center frequency of the higher notch band moves from 5.8 GHz to 5.2 GHz. Thus, the antenna can filter the potential interference signals from the WLAN communication systems. In addition, there is a mismatch around the 9.5 GHz, which may be caused by the high-order spurious of the 5.5 GHz WLAN band.

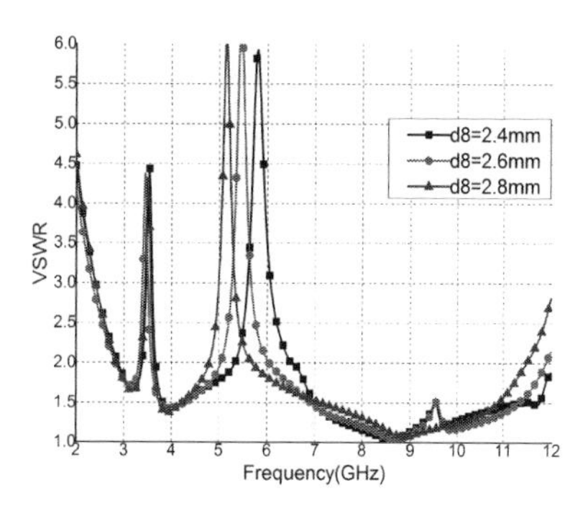

Figure 4. VSWR characteristics of proposed reconfigurable UWB antenna with variant d8.

3.3 *Effects of the switches on the impedance bandwidth*

From the discussions above, we found that the proposed antenna with both the switches OFF is a dual-band notch UWB antenna and the two notch bands are located at the WiMAX and the WLAN bands. However, when a UWB antenna is needed, it needs to design a UWB antenna with no notches. Luckily, reconfigurable antenna can change the modes to make the antenna suitable for different systems. On the basis of the previous knowledge of the reconfigurable antennas and the radio switch techniques, a reconfigurable notch band UWB antenna is proposed and investigated based on the proposed dual-band UWB antenna discussed in sections 3.1 and 3.2. To make the proposed antenna reconfigurable, two switches are integrated on the MLs, which are shown in Figure 1. In the simulation, the presence of a metal bridge represents the ON state, while its absence represents the OFF state. The performance of the proposed reconfigurable notch band UWB antenna is shown in Figure 5.

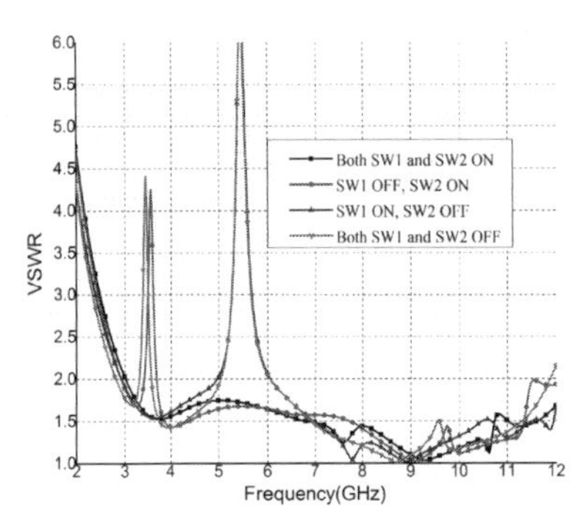

Figure 5. Effects of SW1 and SW2 on the proposed UWB antenna.

It can be seen that the proposed reconfigurable antenna with both the switches ON is a UWB antenna that covers the entire bandwidth designated by the FCC in 2002. When SW1 is OFF and SW2 is ON, the proposed reconfigurable antenna has a notch band at the WiMAX band, which helps to prevent the potential interferences from WiMAX systems. When SW1 is ON and SW2 is OFF, the proposed reconfigurable antenna can provide a notch function that is located in the WLAN band. In this case, the proposed antenna can filter the undesired narrowband signal from the WLAN systems. In addition, the antenna can also be used as a dual band antenna. When both SW1 and SW2 are OFF, the antenna has two notch bands at the WiMAX and WLAN bands, respectively. In this case, the antenna can also be used as a triple band antenna, which can work in 2.9 GHz~3.2 GHz, 3.7 GHz~5 GHz and 6 GHz~12 GHz.

3.4 *Current distribution*

To better understand the principle of the proposed antenna, the current distributions of the proposed antenna are investigated at 3.5 GHz and 5.5 GHz and are shown in Figure 6. It is found that the current distribution of the proposed dual-band notched UWB antenna at 3.5 GHz shown in Figure 6(a) mainly focus on the ML1, microstrip feed signal line, while the current distribution on the ML2 is small. From the Figure 6(b), we can see that the current concentrate on the ML2, microstrip feed signal line. Thus, we can conclude that the lower notch band is produced by the resonance of ML1, while the higher notch band is obtained by the use of the ML2.

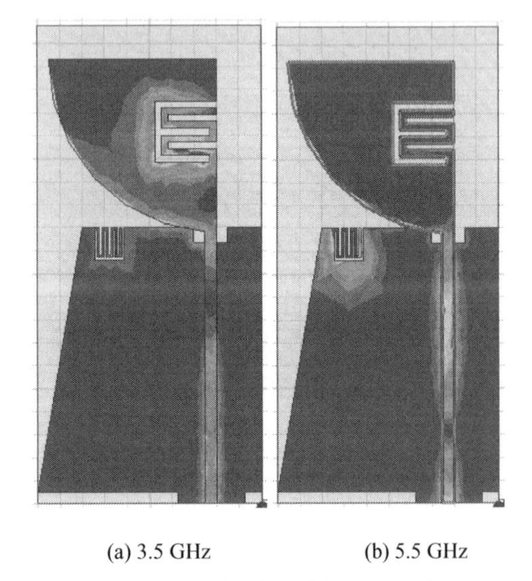

(a) 3.5 GHz (b) 5.5 GHz

Figure 6. Current distribution of the proposed antenna at 3.5 GHz and 5.5 GHz.

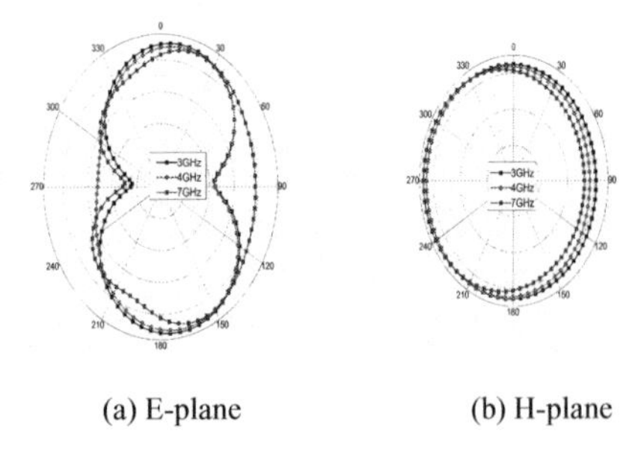

(a) E-plane (b) H-plane

Figure 7. Radiation pattern of antenna at three different frequencies.

3.5 *Radiation pattern*

The radiation patterns at the 3 GHz, 4 GHz and 7 GHz are obtained from the HFSS and are shown in Figure 7. Generally, omnidirectional radiation characteristic is obtained at H-plane, while quasi-omni-directional radiation patterns are produced at the E-plane. Thus, the proposed reconfigurable UWB antenna is suitable for UWB and band-notched UWB communication applications.

4 CONCLUSION

In this paper, a UWB antenna with a reconfigurable band-notched characteristic has been proposed and investigated in detail. The proposed notch functions are realized by the use of the MLs, and the wide notch tunable characteristic is achieved by adjusting the dimensions of the MLs. In addition, the reconfigurable band-notched characteristics are designed by means of the ideal switches that are integrated on the proposed MLs. The simulated results demonstrated that the proposed antenna can be used as a UWB antenna, a band-notched UWB antenna, a dual band-notched UWB antenna, a dual-band antenna or a triple band antenna. From the results discussed above, we can conclude that the proposed UWB antenna reconfigurable band-notched characteristic is suitable for multi-mode wireless communication applications. With the project ongoing, our team will fabricate the proposed antenna by the use of PIN diodes or switch controllable network to make the proposed antenna suitable for practical engineering applications.

ACKNOWLEDGMENT

This work was partially supported by Fundamental Research Funds for the Central Universities (HEUCFD1433).

REFERENCES

Al-Husseini, M., J. Costantine, Christodoulou, C.G., et al. 2010. A reconfigurable frequency-notched UWB antenna with split-ring resonators. *Proceedings of Asia-Pacific Microwave Conference*: 618–621.

Boudaghi, H., Azarmanesh, M. & Mehranpour, M. 2012. A frequency-reconfigurable monopole antenna using switchable slotted ground structure. *IEEE Antennas and Wireless Propagation Letters* 11: 655–658.

Hiraguri, K., Koshiji, K. & Koshiji, F. 2013. A wideband antenna with fan-shaped and trapezoidal elements on printed circuit board for ultra wideband radio. *IEEE 2nd Global Conference on Consumer Electronics*: 267–268.

Jiang, W. & Che, W.Q. 2012. A novel UWB antenna with dual notched bands for WiMax and WLAN applications. *IEEE Antenna and Wireless Propagation Letters* 11: 293–296.

Li, T., Zhai, H. Q., Li, G.H., et al. 2012. Compact UWB band-notched antenna design using interdigital capacitance loading loop resonator. *IEEE Antenna and Wireless Propagation Letters* 11:724–727.

Li, Y., Li, W. & Ye, Q. 2012. Compact reconfigurable UWB antenna integrated with SIRs and switches for multimode wireless communications. *IEICE Electronics Express* 9: 629–635.

Li, Y.S., Li, W. & Ye, Q. 2013. A reconfigurable wide slot antenna integrated with SIRs for UWB/Multiband communication applications. *Microwave and Optical Technology Letters* 55: 52–55.

Nguyen, D.T., Lee, D.H. & Park, H.C. 2012. Very compact printed triple band-notched UWB antenna with quarter-wavelength slots. *IEEE Antenna and Wireless Propagation Letters* 11: 411–414.

Rostamzadeh, M., Mohamadi, S., Nourinia, J., et al. 2012. Square monopole antenna for UWB applications with novel rod-shaped parasitic structures and novel V-shaped slots in the ground plane. *IEEE Antenna and Wireless Propagation Letters* 11: 446–449.

Weng, Y.F., Cheung, S.W. & Yuk, T.I. 2012. Design of multiple band-notch using meander lines for compact ultra-wide band antennas. *IET Microwaves, Antennas &Propagation* 6: 908–914.

Win, W.Z. & Scholtz, R.A. 1998. Impulse radio: how it works. *IEEE Communications Letters* 2: 36–38.

Electronics, Communications and Networks IV – Hussain & Ivanovic (eds)
© 2015 Taylor & Francis Group, London, ISBN: 978-1-138-02830-2

Joint antennas selection and power optimization for energy efficient MIMO systems

Peng Kou, Xiaohui Li* & Yongqiang Hei
State Key Laboratory of Integrated Service Network, Xidian University, Xi'an, Shaanxi, China

ABSTRACT: Antenna selection is an effective way to improve the Energy Efficiency (EE) of massive MIMO systems. In this paper, we try to figure out the optimal number of antennas for massive MIMO systems, and the optimal transmit power is considered. With respect to the form of energy efficiency, we propose a high rate joint optimization algorithm of the number of selected antennas with transmit power. With the basis of ergodic capacity, the proposed algorithm will reduce the complexity by a large scale since it avoids the frequent search for the optimal number of selected antennas and transmit power. The proposed algorithm is suitable for both the Zero-Forcing (ZF) and Maximum-Ratio Transmission (MRT) precoding, and in this paper, we give the details of the proposed algorithm with the ZF precoding. The simulation results show that the proposed algorithm could achieve a nearly maximum energy efficiency under both uncorrelated and correlated channels.

1 INTRODUCTION

Massive MIMO system is known for its high spectrum efficiency and energy efficiency. However, using so many antennas is not optimal for energy efficiency because of the additive power consumption by additive radio frequency chains, thus antenna selection is an important issue in massive MIMO systems. Besides, transmit power can also be optimized to improve energy efficiency.

There have been a lot of antenna selection methods considering the norm of channel vector, antenna correlation coefficient, capacity and so on. Norm-based antenna selection method was proposed in (Pei et al. 2012), which intended to select the L antennas with the biggest norms. Another method for antenna selection by optimizing the precoding matrix was investigated in (Mehanna et al. 2013), where the number of selected antennas was predefined and the optimal precoding matrix should be the one with the minimum norm while achieving a reasonable signal-to-interference plus noise ratio (SINR) with the predefined number of selected antennas. Antenna correlation coefficient based selection method was proposed in (Zhou et al. 2014b), which aimed to select the antenna set with the smallest antenna correlation coefficients. In (Zhou et al. 2014a), the authors selected the antenna set considering the capacity with constrained total power consumption, and proved that capacity increased first and then decreased with the growing number of selected antennas under this method, similar conclusion was also given in (Li et al. 2013). Most works on antenna selection focused on which kind of antenna set was likely to be optimal considering capacity or energy efficiency, and little attention was given to the optimal number of selected antennas. Some work was done in (Lee et al. 2013), with the limitation of narrowing down the search space of number of selected antennas, in (Pei et al. 2012), the authors tried to give optimal number of selected antennas but the effect of power costed by radio frequency chains was ignored. An energy efficient power allocation algorithm was given in (Zhao et al. 2013), which became complicated in time-varying channel. A scheme of code diversity is introduced in (Wu & Calderbank 2009), which enables pretty good performance with low complexity decoding and only a small number of feedback bits, which is very valuable in large-scale MIMO systems.

Based on the conclusion that if the number of selected antennas is large enough, the random antenna selection method can achieve an energy efficiency close to that achieved by the optimal antenna selection method (Lee et al. 2013). In this paper, we give a joint optimization algorithm of the number of selected antennas with transmit power on energy efficient, and the proposed algorithm is expected to be nearly optimal for more effective antenna selection methods. Power consumption in radio frequency

*Corresponding author: *xhli@mail.xidian.edu.cn*

chains is considered in this paper, and the converge rate is higher than binary search algorithm in (Li et al. 2013) by utilizing the characteristic of the expression of egodic energy efficiency. Simulation results verify the performance of the proposed algorithm under both uncorrelated and correlated channels.

The rest of the paper is organized as follows. The system model is introduced in Section 2. In Section 3, both the ergodic expression of energy efficiency of very large multi-user MIMO (MU-MIMO) systems with ZF and random antenna selection method are derived, and the joint optimization algorithm of the number of selected antennas with transmit power is established in Section 4. Simulation results are shown in Section 5. Section 6 summarizes this paper. And related proofs are included in the Appendices.

2 SYSTEM MODEL AND ERGODIC ENERGY EFFICIENCY

2.1 *System model*

We consider the downlink of a single cell of very large MU-MIMO system, which consists of one BS equipped with N antennas and K single-antenna users that are scheduled in the same time-frequency resource, where $N\to\infty$, $K\to\infty$, $N>>K$, and the BS select L antennas from all the N antennas for data transmission. The system model is shown in Figure 1.

The received signal can be represented as

$$\boldsymbol{y}=\sqrt{\frac{\rho}{\gamma}}\boldsymbol{HWx}+\boldsymbol{n} \tag{1}$$

where $\boldsymbol{y}$ is the K-component vector of received signals, $\boldsymbol{x}$ is the K-component vector of transmitted signals. $\boldsymbol{H}=(\boldsymbol{h}_1^T,\boldsymbol{h}_2^T,\ldots,\boldsymbol{h}_K^T)^T$ is the $K\times L$ propagation matrix of complex-valued channel coefficients. The elements of $\boldsymbol{H}$ are independent identically distributed complex Gaussian random variables with zero mean and unit variance. $\boldsymbol{W}=\boldsymbol{H}^H(\boldsymbol{HH}^H)^{-1}$ is the precoding matrix, $\gamma=\mathrm{tr}(\boldsymbol{W}^H\boldsymbol{W})/K$ is the normalization factor, which normalizes the total power by averaging fluctuations in power based on $\boldsymbol{W}$. $\boldsymbol{n}$ is the K-component additive noise vector of received noise, the components of which are independent identically distributed complex Gaussian random variables with zero-mean and unit-variance. Without loss of generality and to minimize notations, we take the noise variance to be one.

2.2 *Ergodic energy efficiency*

With the ZF precoding, the signal received at the kth user is given by

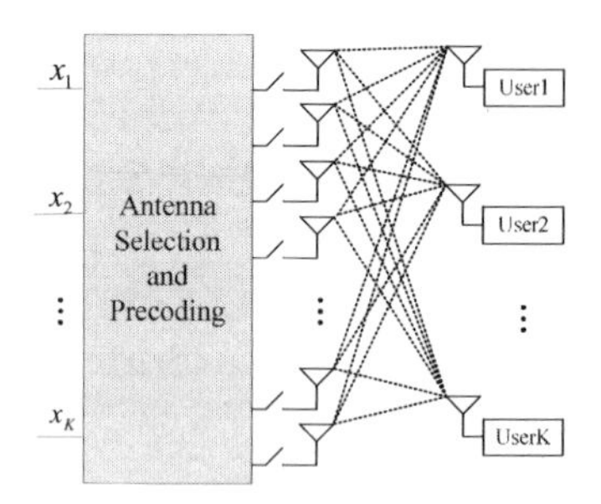

Figure 1. System model of very large MU-MIMO.

$$y_k=\sqrt{\frac{\rho}{\gamma}}x_k+n_k \tag{2}$$

from which we can derive the instantaneous received SINR per user as shown below

$$SINR_k=\frac{\rho}{K\gamma}=\frac{\rho}{\mathrm{tr}(\mathrm{W}^H\mathrm{W})}=\frac{\rho}{\mathrm{tr}\left((\mathrm{H}^H\mathrm{H})^{-1}\right)} \tag{3}$$

hence, the data rate of user k is

$$r_k=\log_2\left(1+\frac{\rho}{\mathrm{tr}\left(\left(\boldsymbol{H}^H\boldsymbol{H}\right)^{-1}\right)}\right) \tag{4}$$

When both the number of users K and the number of selected antennas L grow large, we can obtain (Rusek et al. 2013)

$$1\Big/\mathrm{tr}\left(\left(\boldsymbol{H}^H\boldsymbol{H}\right)^{-1}\right)\approx(L-K)/K \tag{5}$$

thus, the ergodic data rate of user k can be expressed as

$$r_k=\log_2\left(1+\rho\left(\frac{L}{K}-1\right)\right) \tag{6}$$

Energy efficiency is defined as the ratio of data rate to the total power consumption, based on the ergodic analysis of data rate with the ZF precoding, therefore, the ergodic energy efficiency can be written as

$$\eta=\frac{K\log_2\left(1+\rho\left(\frac{L}{K}-1\right)\right)}{P_{total}} \tag{7}$$

where P_{total} is the total power consumption comprising transmit power and circuit power consumption. We use the power consumption model in (Cui et al. 2004)(Cui et al. 2005), in which the total power consumption is given as $P_{total}=\rho+P_1+P_2L$, divided into three parts: the transmit power ρ, the part independent of the number of antennas P_1 and the part proportional to the number of antennas P_2L. Besides $P_1=2P_{syn}+P_{LNA}+P_{mix}+P_{IFA}+P_{filr}+P_{ADC}$, where P_{syn}, P_{LNA}, P_{mix}, P_{IFA}, P_{filr}, and P_{ADC} refer to the power consumption in frequency synthesizer, low noise amplifier, mixer, intermediate frequency amplifier, receiving filter, and A/D converters respectively, and $P_2=P_{DAC}+P_{mix}+P_{filt}$, where P_{DAC}, P_{mix}, and P_{filt} refer to the power consumption by D/A converters, mixer and receiver filter respectively.

3 OPTIMIZATION METHOD FOR NUMBER OF SELECTED ANTENNAS AND TRANSMIT POWER WITH ZF PRECODING

The optimization of ergodic energy efficiency given by Eq.(7) is a fractional program, in which the objective function is a ratio of two real-valued functions. The fractional program can be transformed into an equivalent convex program in some conditions, which is supposed to be an effective method for this kind of problem. In this section, the conditions and process of this method are introduced first, afterwards, the optimization methods for the number of selected antennas and transmit power with the ZF precoding are also investigated.

3.1 *Fractional programming method*

An energy efficiency optimization program has the general form as

$$\max_{x \in S} \eta = \frac{m_1(\boldsymbol{x})}{m_2(\boldsymbol{x})} \tag{8}$$

where $S \subseteq \mathbf{R}^n$, m_1, m_2: $S \rightarrow \mathbf{R}$ and $m_2(x)>0$. Problem (8) is called a concave-convex fractional program if m_1 is concave, m_2 is convex, and S is a convex set, this kind of fractional program can be transformed to an equivalent convex program, which may be solved more efficiently (Long Zhao et al. 2013)(Isheden et al. 2012).

Considering the following equivalent form of the fractional program (8)

$$\begin{aligned} &\max_{x \in S, \eta \in \mathbf{R}} \quad \eta \\ &s.t. \quad \frac{m_1(\boldsymbol{x})}{m_2(\boldsymbol{x})} - \eta \geq 0 \end{aligned} \tag{9}$$

Upon rearranging the constraint, can we obtain

$$\begin{aligned} &\max_{x \in S, \eta \in \mathbf{R}} \quad \eta \\ &s.t. \quad m_1(\boldsymbol{x}) - \eta m_2(\boldsymbol{x}) \geq 0 \end{aligned} \tag{10}$$

As m_1 is concave and m_2 is convex, $m_1(\boldsymbol{x})$-$\eta m_2(\boldsymbol{x})$ is concave with a fixed value of η. Thus the constraint is rearranged as

$$\max_{\mathrm{x}} m_1(\boldsymbol{x}) - \eta m_2(\boldsymbol{x}) > 0 \tag{11}$$

Considering the function

$$F(\eta) = \max_{\mathrm{x}} m_1(\boldsymbol{x}) - \eta m_2(\boldsymbol{x}) \tag{12}$$

This function is convex, continuous and strictly decreasing in η (Isheden et al. 2012). Let q^* be the optimal solution of the objective function in problem (8), therefore the following statements are equivalent:

$$\begin{aligned} &F(\eta) > 0 \Leftrightarrow \eta < q^* \\ &F(\eta) = 0 \Leftrightarrow \eta = q^* \\ &F(\eta) < 0 \Leftrightarrow \eta > q^* \end{aligned} \tag{13}$$

Thus, problem (8) can be transformed as solving the nonlinear function as below

$$F(\eta) = \max_{\mathbf{x}} m_1(\boldsymbol{x}) - \eta m_2(\boldsymbol{x}) = 0 \tag{14}$$

We can solve Eq.(14) using Newton's method. Noticing that $F(\eta_n)=m_1(x_n^*)-\eta_n m_2(x_n^*)$, $F'(\eta_n)=-m_2(x_n^*)$, thus the update in each iteration can be derived as below

$$\eta_{n+1} = \eta_n - \frac{F(\eta_n)}{F'(\eta_n)} = \frac{m_1(\boldsymbol{x}_n^*)}{m_2(\boldsymbol{x}_n^*)} \tag{15}$$

where $\boldsymbol{x}_n^*$ is optimal for $F(\eta_n)$.

And $\boldsymbol{x}_n^*$ converges to the optimal $\boldsymbol{x}^*$ with a super-linear convergence rate. The initial condition of Newton's method $F(\eta_0) \geq 0$ should be satisfied when we set the initial point η_0.

3.2 *Optimization method for number of selected antennas*

Referring to Eq.(7), the problem of the number of selected antennas optimization is described as

$$L^* = \arg\max_{L} \frac{K \log_2(1+\rho(L/K-1))}{\rho + P_1 + LP_2} \tag{16}$$

problem (16) is a fractional problem. As the function $g_1(L)=K\log_2(1+\rho(L/K\text{-}1))$ is concave and function $g_2(L)= \rho+ P_1+P_2L$ is convex, it can be solved using the method in section 3.1, and the problem is equivalent to

$$F(\eta) = \max_{L} g_1(L) - \eta g_2(L) = 0 \tag{17}$$

We can solve Eq.(17) using Newton's method with the update in each iteration as

$$\eta_{n+1} = \eta_n - \frac{F(\eta_n)}{F'(\eta_n)} = \frac{g_1(L_n^*)}{g_2(L_n^*)} \tag{18}$$

where the initial point η_0 should be a value that satisfies $F(\eta_0) \geq 0$, and L_n^* is optimal for $F(\eta_n)$ with the following form

$$L_n^* = \left(\frac{1}{\eta_n P_2 \ln 2} - \frac{1}{\rho} + 1\right) K \tag{19}$$

When η converges to the optimal point, we get the optimal L^*.

3.3 *Optimization method for transmit power*

Referring to Eq.(7), the problem of transmit power optimization is described as

$$\rho^* = \arg\max_{\rho} \frac{K \log_2\left(1 + \rho(L/K - 1)\right)}{\rho + P_1 + LP_2} \tag{20}$$

problem (20) is a fractional problem. As the function $f_1(\rho) = K\log_2(1+\rho(L/K-1))$ is concave and function $f_2(\rho) = \rho + P_1 + P_2 L$ is convex, therefore it can be solved using the method in section 3.1, the problem is equivalent to

$$F(\eta) = \max_{\rho} f_1(\rho) - \eta f_2(\rho) = 0 \tag{21}$$

We can solve Eq.(21) using Newton's method with the update in each iteration as

$$\eta_{n+1} = \eta_n - \frac{F(\eta_n)}{F'(\eta_n)} = \frac{f_1(\rho_n^*)}{f_2(\rho_n^*)} \tag{22}$$

where the initial point η_0 should be a value that satisfies $F(\eta_0) \geq 0$, and ρ_n^* is optimal for $F(\eta_n)$ with the following form

$$\rho_n^* = \left(\frac{L-K}{\eta_n \ln 2} - 1\right) \Big/ (L/K - 1) \tag{23}$$

When η converges to the optimal point, we get the optimal ρ^*.

4 JOINT OPTIMIZATION ALGORITHM

4.1 *Joint optimization algorithm*

Optimization methods for the number of selected antennas and transmit power are given in Section 3. In order to obtain the maximum EE considering both the number of selected antennas and transmit power, we propose a joint optimization algorithm. Firstly, we find the optimal number of selected antennas with a fixed transmit power. Next find the optimal transmit power with the obtained number of selected antennas. Then, we find the optimal number of selected antennas with the obtained transmit power, ..., until $\Delta\eta = \eta_n - \eta_{n-1} < \varepsilon$, where ε can be adjusted to make a trade-off between accuracy and complexity. The termination condition of fractional programming method in this algorithm is $|F(\eta)| \leq \delta$, where δ is a proper positive number. The detail of the joint optimization algorithm is given in Table 1.

Table 1. Joint optimization algorithm.

Initialization: $\eta 0 = \eta_{ini}(\eta_{ini} > 0, \eta_{ini} \to 0)$, ρ=0dB, compute Ln^* according to equation (19), n=1;
Iteration Process: **Step 1:** Update η_n according to equation (18); **Step 2:** Compute L_n^* according to equation (19), $n=n+1$. If ($|F(\eta_n)| > \delta$), go to **Step 1**. Otherwise compute ρ_n^* according to equation (23); **Step 3:** Update ηn according to equation (22); **Step 4:**Compute ρ_n^* according to equation (23). If ($|F(\eta_n)| > \delta$), $n=n+1$, go to **Step 3**. Otherwise go to **Step5**. **Step 5:** If ($\Delta\eta > \varepsilon$), $n=n+1$, go to **Step 1**. Otherwise $L^* = L_n^*$ and $\rho^* = \rho_n^*$.

4.2 *Converge analysis*

The initial point of η should satisfy $F(\eta) \geq 0$ when the fractional programming method is used. In the joint optimization method, we set the initial η_0 as $\eta_{ini}(\eta_{ini} > 0, \eta_{ini} \to 0)$. Because η_n converges to the optimal η by one side, and the endpoint of the fractional programming method is smaller than the optimal η, $F(\eta) \geq 0$ is always satisfied during the iterative process of the joint optimization algorithm, and the algorithm will converge to the optimal η with a superlinear rate.

The joint optimization algorithm is based on ergodic energy efficiency, thus the nearly maximum energy efficiency can always be achieved with the optimal number of selected antennas and transmit power obtained by this algorithm in time-varying channel. The proposed algorithm can avoid the frequent search for the optimal number of antennas and transmit power when the channel changes. Noticing that Newton's method converges at a high rate, the proposed algorithm can reduce the complexity by a large margin.

5 SIMULATION RESULTS

5.1 *Ergodic energy efficiency*

The comparison between the simulated and the analytical ergodic energy efficiency with random antenna selection method are shown in Figure 2, the channel model is the Rayleigh channel model, N=200, K=10, and the simulation results are based on 1000 runs. We can see that with different L and ρ, they all match perfectly, and it is not wise to use all the antennas for data transmission. The simulated points match the analytical results so well even when the number of selected antennas is small, for example from 20 to 40. The reason of this phenomenon is that the points shown in Figure 2 are the ergodic energy efficiency while Eq.(7) describes the almost precise energy efficiency with L growing large and the ergodic energy efficiency with L being not so large.

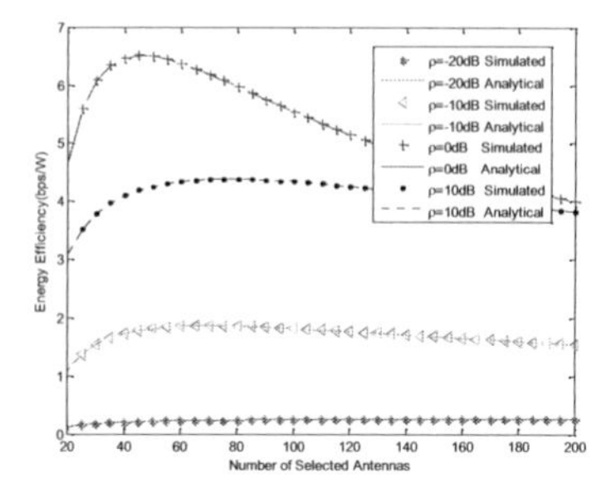

Figure 2. Comparison between analytical and simulated results.

5.2 *Converge rate of the proposed algorithm*

In Figure 3, we investigate the convergence of the proposed algorithm, the number of users K=8, 10, 12 and 14 respectively, the initial η is set as 0.1, and the initial transmit power is set as 0dB. We can see that the proposed algorithm converges to the maximum energy efficiency after about 7 comparisons, and this number barely differs among different K. The optimal number of selected antennas and transmit power for different numbers of users are shown in Table 2. We can find in Table 2 that, the optimal number of selected antennas increases with the number of users, so does the transmit power.

An antenna selection algorithm called binary search algorithm is proposed in (Hui Li et al. 2013). The complexity of the binary search algorithm is $O(\log_2 N)$, which is comparable with that of the proposed algorithm when transmit power optimization is not considered. However, if optimal transmit power is searched with a high accuracy based on binary search as it implies during the analysis of transmit power's effect on energy efficiency, especially when joint optimization with number of selected antennas is considered, the complexity will surely be higher than the proposed algorithm in this paper.

Table 2. The optimal number of selected antennas and transmit power with ZF.

Number of users (K)	Number of Selected antennas (L)	Transmit Power (ρ)
8	41	2.0506dB
10	48	2.5791dB
12	54	3.0182dB
14	59	3.3944dB

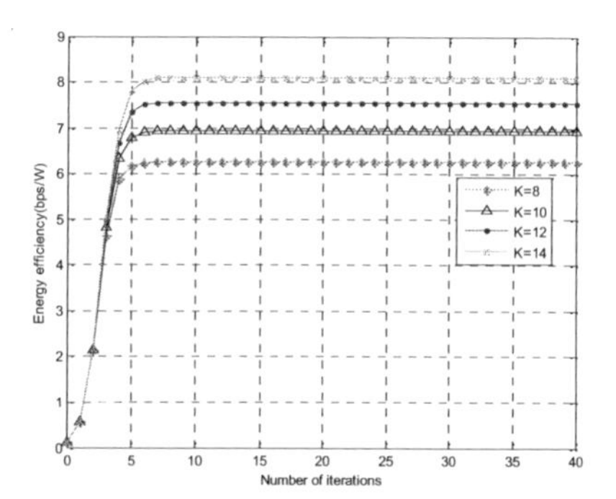

Figure 3. Converge rate of the proposed algorithm.

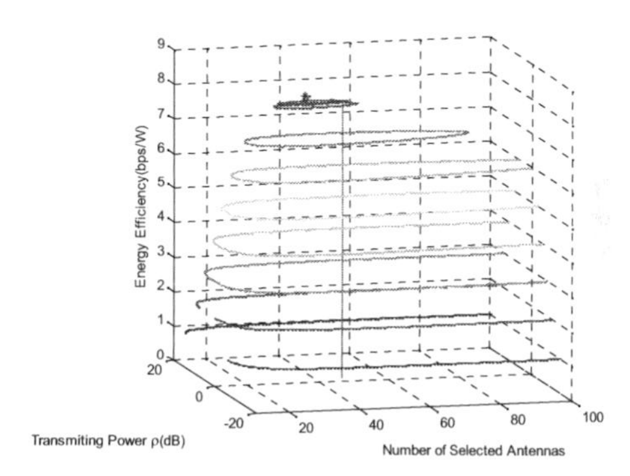

Figure 4. Performance of the proposed algorithm with ZF under uncorrelated channel.

5.3 *Performance of joint optimization algorithm*

In very large MU-MIMO systems, complexity of optimal antenna selection algorithm becomes a big problem due to the massive antennas. Known for low complexity and fairly well performance norm-based selection is widely used in very large MU-MIMO systems, and is also used as the antenna selection method in simulations.

The performance of the proposed algorithm under both uncorrelated channel (UCC) and correlated channel (CC) is shown in Figure 4 and Figure 5. The UCC model we use is the Rayleigh channel model, and the CC model is the popularly used exponential correlation channel model, more details can be seen in (Loyka 2001), the correlation coefficient in which is set as 0.7 in simulations, just as it was set in (Hui Li et al. 2013). N=200, K=10, simulation results are based on 1000 runs. The energy efficiency reached by exhaustive search algorithm and the joint optimization algorithm is shown in Table 3, and the proposed algorithm can almost achieve the maximum energy efficiency under both uncorrelated channel and correlated channel.

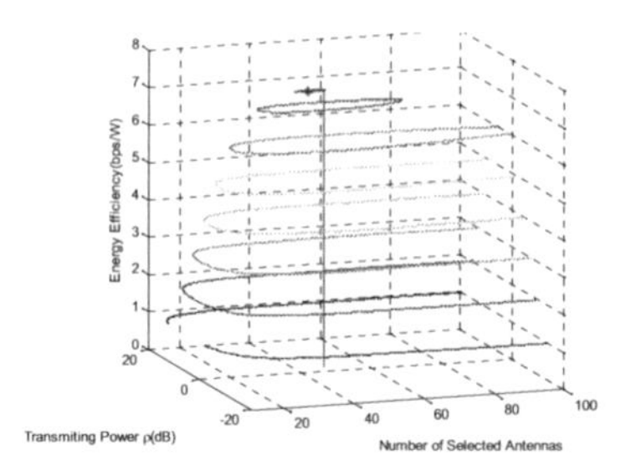

Figure 5. Performance of the proposed algorithm with ZF under correlated channel.

Table 3. Energy efficiency(bps/W) with exhaustive search algorithm and joint optimization algorithm.

Precoding/ Channel model	Exhaustive search algorithm	Joint optimization algorithm
ZF/CC	7.4591	7.4380
ZF/UCC	8.2524	8.0767
MRT/CC	5.0731	4.8037
MRT/UCC	6.2445	6.1258

We can further see that, if the solution is a smaller number of selected antennas, higher energy efficiency can be achieved, which means that the number of selected antennas obtained by the proposed algorithm is a little larger than the optimal one, because of the antenna selection effect, and that the capacity gain of an extra antenna is decreasing as the number of selected antennas increases, which differs from that in random antenna selection. Further work should be done to compensate this residual.

6 CONCLUSIONS

In this paper, based on the idea that if the number of selected antennas is large enough, the random antenna selection method can reach an energy efficiency approximate to that achieved by the optimal antenna selection method, we give the joint number of selected antennas and transmit power optimization algorithm based on the ergodic energy efficiency of large scale MU-MIMO with random antenna selection method. Simulation results show that the proposed algorithm can achieve the nearly maximum energy efficiency under both uncorrelated and correlated channels with more effective antenna selection method.

ACKNOWLEDGEMENT

This paper is supported by the national natural science foundation (61201134, 61201135), 111 Project of China under Grant (B08038), the fundamental research funds for the central universities (K50510010016) and the state major projects of the next generation broadband wireless mobile communication networks (2012ZX03001027).

REFERENCES

Cui, Shuguang A. J. G. & Bahai, A. 2005. Energy-constrained modulation optimization. *IEEE Transactions on Wireless Communications* 4(5): 2349–2360.

Cui, Shuguang, A. J. G. & A. Bahai. 2004. Energy-efficiency of MIMO and cooperative MIMO techniques in sensor networks. *IEEE Journal on Selected Areas in Communications* 22(6): 1089–1098.

Isheden, C., Chong, Z., E. J., et al. 2012. Framework for link-level energy efficiency optimization with informed transmitter. *IEEE Transactions on Wireless Communications.* 11(8): 2946–2957.

Lee, Byung Moo, Choi Jin Hyeock, J. B., et al. 2013. An energy efficient antenna selection for large scale green MIMO systems. *IEEE International Symposium on Circuits and Systems, Beijing, 19–23 May 2013.* Piscataway: IEEE.

Li, Hui, Song, Lingyang, D. Z., et al. 2013. Energy efficiency of large scale MIMO systems with transmit antenna selection. *IEEE International Conference on Communicatons, Budapest, 9–13 June 2013.* Piscataway: IEEE.

Loyka, S. L. 2001. Channel capacity of MIMO architecture using the exponential correlation matrix. *IEEE Communications Letters* 5(9): 369–371.

Mehanna Omar, N. D. S. & Giannakis, G. B.. 2013. Joint multicast beamforming and antenna selection. *IEEE Transactions on Signal Processing* 61(10): 2660–2674.

Pei, Yiyang, T.-H. P. & Liang, Y.-C.. 2012. How many RF chains are optimal for large-scale MIMO systems when circuit power is considered?. *IEEE Global Communications Conference, Anaheim, 3–7 December 2012.* Piscataway: IEEE.

Rusek, F., Persson, D., B. K. L., et al. 2013. Scaling up MIMO: Opportunities and challenges with very large arrays. *IEEE Signal Processing Magazine* 30(1): 40–60.

Wu, Yiyue & Calderbank. R.. 2009. Code Diversity in Multiple Antenna Wireless Communication. *IEEE Journal of Selected Topics in Signal Processing* 3(6): 928–938.

Zhao, Long, Zhao, Hui, F. H., et al. 2013. Energy Efficient Power Allocation Algorithm for Downlink Massive MIMO with MRT Precoding. *IEEE Vehicular Technology Conference, Las Vegas, 2–5 September 2013.* Piscataway: IEEE.

Zhou, Xingyu, B. B. & Chen, W. 2014a. Iterative Antenna Selection for Multi-Stream MIMO under a Holistic Power Model. *IEEE Wireless Communications Letters* 3(1): 82–85.

Zhou, Zhiqiang N. G. & Lin, X. 2014b. Reduced-Complexity Antenna Selection Schemes in Spatial Modulation. *IEEE Communications Letters* 18(1): 14–17.

Electronics, Communications and Networks IV – Hussain & Ivanovic (eds)
© 2015 Taylor & Francis Group, London, ISBN: 978-1-138-02830-2

Optimal training sequence for OFDM based two-way relay networks in the presence of phase noise

Xin Li
China University of Geosciences, Wuhan, Hubei, China

ABSTRACT: In this paper, we investigate the problem of optimal training sequence design for channel estimation in two-way amplify-and-forward relay networks. Specially, we consider a practical scenario that phase noise appears at the transmitter and receiver antennas. Orthogonal frequency-division multiplexing (OFDM) technique is employed to combat the frequency selective fading. Assuming the full phase noise information can be obtained, we propose a design of training sequences to minimize the mean-square error (MSE) of channel estimation. Based on this design, we also propose robust training sequences for the case that no phase noise information is obtained. Simulation results are presented to show that we designed sequences outperform random sequences.

KEYWORDS: two-way relay networks, OFDM, channel estimation, optimal training design, phase noise.

1 INTRODUCTION

Research on two-way relay network (TWRN) has attracted much attention for its capability of enhancing the overall communication rate compared to one-way relay network (Rankov & Wittneben2007). It was firstly investigated by Shannon (1961), where inner and outer bounds on the capacity region were derived. Generally, both amplify-and-forward (AF) and decode-and-forward (DF) protocols developed for one-way relay networks (OWRN) can be extended to TWRN The achievable rate region for amplifying-and-forward (AF) and decode-and-forward (DF) based TWRN has been developed in (Rankov & Wittneben 2006). Optimal relay strategies were reported in (Cui et al. 2009) to decrease the symbol error rate. For broadband TWRN, tone permutation and power allocation algorithm has been proposed in (Ho et al. 2008).To cope with the channel frequency-selectivity, a frequency domain equalization scheme based on OFDM and single carrier radio access was reported in (Gacanin & Adachi 2010).

Most of the previous work assumed perfect channel state information at both the relay node and/or the two source nodes, which makes the channel estimation be an important issue for TWRN. Channel estimation for TWRN employing AF protocol has been extensively investigated in. The optimal training design was proposed in for frequency flat fading channels. Specially, fast fading channels were considered in (Wang et al. 2011) and the optimal training was designed to estimate the basis expanding model coefficients. For OFDM based TWRN, (Yang, w. et al. 2010) proposed the optimal training sequence to minimize the mean square error (MSE) of channel estimation. Block and pilot tone based channel estimation algorithms were designed in (Gao, F. et al. 2009) to recover the individual frequency selective fading channel between a source node and relay node. As carrier offset (CFO) is inevitable, nulling-based least square (LS) algorithm was developed to jointly estimate CFO and channel coefficients (Wang et al. 2011). In addition, (Wang et al. 2010) proposed a superimposed training sequences scheme to estimate CFO and channel coefficients. Recently, the effect of channel estimation error was also investigated (Gacanin et al. 2011). However, the effect of phase noise on channel estimation for TWRN has not been well considered in previous works.

This paper is organized as follows. Section 2 introduces the two-way relay OFDM system and channel model. Optimal training sequence design with and without phase noise information are presented in Section 3. Section 4 discusses the numerical results. Finally, Section 5 concludes the paper

Notation: throughout the paper, all the vectors and matrices are represented by boldface **a** and **A**, respectively. **F** denotes the FFT matrix with $F_{ij}=(1/N)exp(-j2\Pi(i-1)(j-1)/N)$.The operators $(\cdot)^T$ and $(\cdot)^H$ denote the transpose and complex conjugate transpose operations. $(\cdot)_N$ is the modular arithmetic of N. The function diag(x) returns a diagonal matrix with the vector x on its diagonal. $E(x)$ represents the expectation of a random variable x.

2 CHANNEL AND SYSTEM MODEL

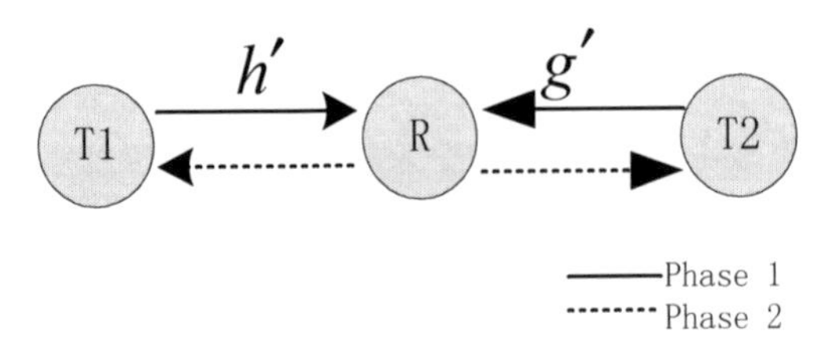

Figure 1. A two-way relay network.

Consider a TWRN with two source nodes $T1$, $T2$ and a relay node R, see Figure 1. Each node is mounted with a half-duplex antenna. One round of information exchange consists of two phases. In the first phase, both source nodes transmit information to relay node simultaneously. In AF mode, the relay scales the received signal and retransmits to source nodes. The channels from $T1$ and $T2$ to R are assumed to be static frequency-selective fading, namely, they do not change during one frame of transmission. Let $\boldsymbol{h}'=[h_0',h_1'...h_{L-1}']^T$ and $\boldsymbol{g}'=[g_0', g_1'...g_{L-1}']$ denote the discrete-time baseband equivalent impulse response vectors of frequency selective fading channels from $T1$ and $T2$ to R, respectively. Though the same number of taps from $T1$ and $T2$ to R is assumed, the results in the following can be extended to more general scenarios easily. As time-division duplex (TDD) is a generally adopted assumption, the channel from R to $T1$ is still $\boldsymbol{h}'$ and the channel from R to $T2$ is still $\boldsymbol{g}'$. The elements in $\boldsymbol{h}'$ and $\boldsymbol{g}'$ are modeled as zero-mean circularly symmetric complex Gaussian random variables and are independent from each other. Let $\sigma^2_{h',l}, l=0,1,\cdots L-1$ denote the variance of elements in $\boldsymbol{h}'$ and $\sigma^2_{g',l}, l=0,1,\cdots L-1$ denote the variance of elements in $\mathbf{g}'$. We define $\sigma^2_{h'}=\sum_{l=0}^{L-1}\sigma^2_{h',l}$ and $\sigma^2_{g'}=\sum_{l=0}^{L-1}\sigma^2_{g',l}$ for later expression simplicity.

In the first phase, the input data bits are firstly mapped to complex symbols drawn from signal constellation. Let $\dot{\mathbf{d}}_i=[\dot{d}_{i,0},\dot{d}_{i,1},\cdots \dot{d}_{i,N-1}]^T$ be the frequency domain signal to be transmitted from $Ti, i=1,2$. The power constraint of the transmitted signal is $E\{\dot{\mathbf{d}}_i^H\dot{\mathbf{d}}_i\}=P, i=1,2$, where P is the average transmitting power of $T1$ and $T2$. After an inverse fast Fourier transform (IFFT), $\dot{\mathbf{d}}_i$ is transformed into a time domain signal $\mathbf{d}_i=\mathbf{F}^H\dot{\mathbf{d}}_i$. To combat inter-symbol interference (ISI), a cyclic prefix of length L_{CP} is appended to, where L_{CP} is larger the length of the channel impulse response L. Assuming perfect time and frequency synchronization, the received signal at the relay, after cyclic prefix removal, is derived as

$$\mathbf{r}=\mathbf{E}_{R,1}\mathbf{H}'\mathbf{E}_{T1,1}\mathbf{d}_1+\mathbf{E}_{R,1}\mathbf{G}'\mathbf{E}_{T2,1}\mathbf{d}_2+\mathbf{n}_r, \tag{1}$$

where $\mathbf{E}_{s,1}=diag(e^{j\phi_{s,1}(0)},e^{j\phi_{s,1}(1)},\cdots e^{j\phi_{s,1}(N-1)}), s\in\{T1,T2,R\}$ is a $N\times N$ diagonal matrix representing the phase noise in the first phase. Phase noise is mainly caused by phase fluctuation of the local oscillator. Whenever a local oscillator is used, phase noise is assumed to be present. Assuming free-running oscillator is used, the phase noise is modeled as Winner process (Demir et al. 2000) and $\phi_{s,1}(n)=\phi_{s,1}(n-1)+\varepsilon_{s,1}$. $\varepsilon_{s,1}$ is a Gaussian random variables with zeros mean and variance $2\pi\beta_{s,1}T_s$, where T_s is the sampling interval and $\beta_{s,1}$ is the two-sided 3dB phase noise linewidth, i.e. frequency spacing between 3dB points of its Lorentzian power spectral density function. H' and G' are $N\times N$ column-wise circulant matrices with the first column $[\mathbf{h}'^T,0,\cdots 0]^T$ and $[\mathbf{g}',0,\cdots 0]^T$, respectively. $\mathbf{n}_r$ is a white Gaussian noise vector with zero mean and covariance $\mathbf{R}_{n_rn_r}=\sigma_n^2\mathbf{I}$.

In the second phase, the relay amplifies its received signal and retransmits to both source nodes after adding a cyclic prefix. For symmetry, we only consider the signal received at $T1$. After cyclic prefix removal, the received signal at $T1$ can be represented as

$$\begin{aligned}\mathbf{y}_1&=\alpha\mathbf{E}_{T1,2}\mathbf{H}'\mathbf{E}_{R,2}\mathbf{r}+\mathbf{n}_1\\&=\alpha\mathbf{E}_{T1,2}\mathbf{H}'\mathbf{E}_{R,2}\mathbf{E}_{R,1}\mathbf{H}'\mathbf{E}_{T1,1}\mathbf{d}_1+\alpha\mathbf{E}_{T1,2}\mathbf{H}'\mathbf{E}_{R,2}\mathbf{E}_{R,1}\mathbf{G}'\mathbf{E}_{T2,1}\mathbf{d}_2+\underbrace{\alpha\mathbf{E}_{T1,2}\mathbf{H}'\mathbf{E}_{R,2}\mathbf{n}_r+\mathbf{n}_1}_{\mathbf{n}}\end{aligned} \tag{2}$$

Where α denotes the power amplifying factor, and is expressed as

$$\alpha=\sqrt{\frac{P_r}{\sigma^2_{h'}P+\sigma^2_{g'}P+\sigma^2_n}}.$$

$\mathbf{n}$ is the effective noise with zero mean and covariance $\mathbf{R}_{\mathbf{nn}}=\sigma_n^2(\alpha^2\mathbf{H}'\mathbf{H}'^H+\mathbf{I})$. $\mathbf{E}_{s,2}, s\in\{T1,T2,R\}$ denotes the phase noise diagonal matrix in the second phase.

According to the results in (Liu et al.2006), for $\beta_iT_s<<1$ and a slow-fading channel, (2) can be approximated as

$$\begin{aligned}\mathbf{y}_1&\approx\alpha\mathbf{E}_{e1}\mathbf{H}'\mathbf{H}'\mathbf{d}_1+\alpha\mathbf{E}_{e2}\mathbf{H}'\mathbf{G}'\mathbf{d}_2+\mathbf{n}\\&=\alpha\mathbf{E}_{e1}\mathbf{H}\mathbf{d}_1+\alpha\mathbf{E}_{e2}\mathbf{G}\mathbf{d}_2+\mathbf{n}\end{aligned}, \tag{3}$$

Where $\mathbf{E}_{e1}$ and $\mathbf{E}_{e2}$ are effective phase noise matrices with effective 3dB bandwidths $\beta_{e1}=\beta_{T1,2}+\beta_{R,1}+\beta_{R,2}+\beta_{T1,1}$ $\beta_{e2}=\beta_{T1,2}+\beta_{R,1}+\beta_{R,2}+\beta_{T2,1}$, respectively. Let a length of $K=2L-1$ the vector $\mathbf{h}$ be the convolution of $\mathbf{h}'$ and $\mathbf{h}'$, a length of K vector $\mathbf{g}$ be the convolution of $\mathbf{h}'$ and $\mathbf{g}'$. $\mathbf{H}$ and $\mathbf{G}$ are $N\times N$ column-wise circulant matrices with the first column $[\mathbf{h}^H,0,\cdots 0]^T$ and $[\mathbf{g}^T,0,\cdots 0]^T$, respectively.

The approximation in (3) simplifies receiver design significantly. Therefore, our discussion in the following will be based on a signal model in (3).

For coherent detection, the source node should acquire the channel state information $\mathbf{H}, \mathbf{G}$ and effective phase noise $\mathbf{E}_{e1}, \mathbf{E}_{e2}$. In the next section, we adopt LS algorithm for channel estimation and investigate the problem of optimal training design with and without phase noise information.

3 OPTIMAL TRAINING SEQUENCE DESIGN

We assume block training based channel estimation in this paper, which happens at the beginning of transmission over slow fading channels. To relax the signal processing burden, the low complexity LS algorithm is employed for channel estimation.

3.1 *Optimal training design with perfect phase noise information*

In this subsection, we assume perfect phase noise information can be acquired. Due to commutativity, (3) can be reformed as

$$\mathbf{y}_1 = \alpha \mathbf{E}_{e1}\mathbf{D}_1\mathbf{h} + \alpha \mathbf{E}_{e2}\mathbf{D}_2\mathbf{g} + \mathbf{n} = \alpha \underbrace{\left(\mathbf{E}_{e1}\mathbf{D}_1 \quad \mathbf{E}_{e2}\mathbf{D}_2\right)}_{\mathbf{P}} \underbrace{\begin{pmatrix}\mathbf{h}\\ \mathbf{g}\end{pmatrix}}_{\boldsymbol{\theta}} + \mathbf{n}, \tag{4}$$

where $\mathbf{D}_1$ and $\mathbf{D}_2$ are $N \times K$ circulant matrices with first column d_1 and d_2, respectively. Then, the least-square estimation of the composite channel $\boldsymbol{\theta}$ is given by

$$\hat{\boldsymbol{\theta}} = \frac{1}{\alpha}(\mathbf{P}^H\mathbf{P})^{-1}\mathbf{P}^H\mathbf{y}_1 = \boldsymbol{\theta} + \frac{1}{\alpha}(\mathbf{P}^H\mathbf{P})^{-1}\mathbf{P}^H\mathbf{n}, \tag{5}$$

And the MSE of estimation error is defined as

$$MSE_0 = \frac{1}{2K} tr\{E[(\hat{\boldsymbol{\theta}} - \boldsymbol{\theta})(\hat{\boldsymbol{\theta}} - \boldsymbol{\theta})^H]\}, \tag{6}$$

Substituting (5) into (6), we can obtain that

$$MSE_0 = \frac{\sigma_n^2(\alpha^2\sigma_{h'}^2 + 1)}{2K\alpha^2} tr\{(\mathbf{P}^H\mathbf{P})^{-1}\}, \tag{7}$$

Where

$$\mathbf{P}^H\mathbf{P} = \begin{pmatrix} \mathbf{D}_1^H\mathbf{D}_1 & \mathbf{D}_1^H\mathbf{E}_1\mathbf{E}_2\mathbf{D}_2 \\ \mathbf{D}_2^H\mathbf{E}_2\mathbf{E}_1\mathbf{D}_1 & \mathbf{D}_2^H\mathbf{D}_2 \end{pmatrix}, \tag{8}$$

From the majorization theory in (Golub & Van 1996) it can be shown that the MSE is lower bounded as

$$MSE_0 \geq \frac{\sigma_n^2(\alpha^2\sigma_{h'}^2 + 1)}{2K\alpha^2} \sum_{m=1}^{2(2L-1)} \frac{1}{[\mathbf{P}^H\mathbf{P}]_{m,m}}, \tag{9}$$

Where the equality holds if and only if $\mathbf{P}^H\mathbf{P}$ is a diagonal matrix. Using Cauchy-Schwartz inequality, the MSE is further lower bounded as

$$MSE_0 \geq \frac{1}{\alpha^2} \sqrt[2(2L-1)]{\prod_{m=1}^{2(2L-1)} \frac{1}{[\mathbf{P}^H\mathbf{P}]_{m,m}}}, \tag{10}$$

Where the equality holds if and only is the elements in $\mathbf{P}^H\mathbf{P}$ are identical. From (8, 9, and 10), we derive the conditions to achieve minimum MSE:

A1. $\mathbf{D}_1^H\mathbf{E}_1\mathbf{E}_2\mathbf{D}_2 = \mathbf{0}$ and $\mathbf{D}_2^H\mathbf{E}_2\mathbf{E}_1\mathbf{D}_1 = \mathbf{0}$

A2. $\mathbf{D}_1^H\mathbf{D}_1$ and $\mathbf{D}_2^H\mathbf{D}_2$ are diagonal matrices with equal diagonal elements.

In the following, we provide a way to design $\mathbf{D}_1$ and $\mathbf{D}_2$. We assume that $\mathbf{D} = [\mathbf{D}_1\ \mathbf{D}_2]$ is a $N \times 2K$ circulant matrix and its first column is denoted by $\mathbf{d} = [d_0, \cdots d_{N-1}]^T$. With this assumption, it is obvious that condition A1 will be satisfied, when $\mathbf{D}^H\bar{\Lambda}\mathbf{D}$ is a diagonal matrix for any $N \times N$ diagonal matrix $\bar{\Lambda}$. The $(i, j)th$ element of $\mathbf{D}^H\bar{\Lambda}\mathbf{D}$ can be expressed as

$$(\mathbf{D}^H\bar{\Lambda}\mathbf{D})_{i,j} = \sum_{k=0}^{N-1} d_k\bar{\Lambda}_k d^*_{(k+|i-j|)_N}, \tag{11}$$

where $\bar{\Lambda}_k$ is the kth diagonal element of $\bar{\Lambda}$. Because $(\mathbf{D}^H\bar{\Lambda}\mathbf{D})$ is a $2K \times 2K$ matrix, the maximum value of $|i-j|$ in (11) is $2K-1$. With (11), we find that the off-diagonal element in $(\mathbf{D}^H\bar{\Lambda}\mathbf{D})$ will be zero if the space between any nonzero elements in $\mathbf{d}$ is equal or larger than $2K-1$. From previous discussion, the first column of $\mathbf{D}$ has the following form

$$d_n = \sum_{i=0}^{c-1} A_i\delta(n - l_i) \quad 0 \leq n \leq N-1, \tag{12}$$

where $l_{i+1} - l_i \geq 2K-1, N + l_0 - l_{c-1} \geq 2K-1$. The number of nonzero elements should be less or equal than N/(2K).The power of $A_i, i = 0, \cdots c$ is normalized to P, namely $\sum_{i=0}^{c}|A_i|^2 = P$. When $\bar{\Lambda}$ is an identity matrix, the diagonal element of $\mathbf{D}^H\bar{\Lambda}\mathbf{D}$ is

$$(\mathbf{D}^H\bar{\Lambda}\mathbf{D})_{i,i} = \sum_{k=0}^{N-1}|d_k|^2 = \sum_{j=0}^{c}|A_j|^2 = P, \tag{13}$$

Interestingly, (13) indicate that condition A2 is also satisfied. With the discussion above, the optimal training sequences $\mathbf{d}_1$ and $\mathbf{d}_2$ for channel estimation can be designed in the following steps:

Step1. Construct a sequence in the form of (12)

Step2. Construct a$N \times 2K$circulant matrix $\mathbf{D}$ with the first column $\mathbf{d} = [d_0, \cdots d_{N-1}]^T$

Step3.The optimal training sequences $\mathbf{d}_1$ and $\mathbf{d}_2$ are the first and $(K+1)th$columns of $\mathbf{D}$, respectively.

3.2 *Channel estimation without phase noise information*

In this subsection, we consider the case that channel estimation is performed before phase noise estimation. Because phase noise matrices $\mathbf{E}_{e1}$ and $\mathbf{E}_{e2}$are unknown, the LS estimation of channel $\boldsymbol{\theta}$ is given by

$$\hat{\boldsymbol{\theta}} = \frac{1}{\alpha}\begin{pmatrix} P_1\mathbf{D}_1^H\mathbf{D}_1 & \sqrt{P_1P_2}\mathbf{D}_1^H\mathbf{D}_2 \\ \sqrt{P_1P_2}\mathbf{D}_2^H\mathbf{D}_1 & P_2\mathbf{D}_2^H\mathbf{D}_2 \end{pmatrix}^{-1}\begin{pmatrix} \sqrt{P_1}\mathbf{D}_1^H \\ \sqrt{P_2}\mathbf{D}_2^H \end{pmatrix}\mathbf{y}_1 = \boldsymbol{\theta} + \Delta\boldsymbol{\theta}, \tag{14}$$

The estimation error $\Delta\boldsymbol{\theta}$ can be expressed as

$$\Delta\boldsymbol{\theta} = \begin{pmatrix} \mathbf{D}_1^H\mathbf{D}_1 & \mathbf{D}_1^H\mathbf{D}_2 \\ \mathbf{D}_2^H\mathbf{D}_1 & \mathbf{D}_2^H\mathbf{D}_2 \end{pmatrix}^{-1}\begin{pmatrix} \mathbf{D}_1^H \\ \mathbf{D}_2^H \end{pmatrix}\begin{pmatrix} (\mathbf{E}_1-\mathbf{I})\mathbf{D}_1 & (\mathbf{E}_2-\mathbf{I})\mathbf{D}_2 \end{pmatrix}\begin{pmatrix} \mathbf{h} \\ \mathbf{g} \end{pmatrix} + \begin{pmatrix} \mathbf{D}_1^H\mathbf{D}_1 & \mathbf{D}_1^H\mathbf{D}_2 \\ \mathbf{D}_2^H\mathbf{D}_1 & \mathbf{D}_2^H\mathbf{D}_2 \end{pmatrix}\begin{pmatrix} \mathbf{D}_1^H \\ \mathbf{D}_2^H \end{pmatrix}\mathbf{n}, \tag{15}$$

And the MSE is obtained as

$$MSE_1 = MSE_0 + \frac{1}{2K}tr\{E\underbrace{\begin{pmatrix} \underbrace{\mathbf{D}_1^H(\mathbf{E}_1-\mathbf{I})\mathbf{D}_1\mathbf{R}_h\mathbf{D}_1^H(\mathbf{E}_1-\mathbf{I})^H\mathbf{D}_1}_{\mathbf{M}_1} & \mathbf{D}_1^H(\mathbf{E}_1-\mathbf{I})\mathbf{D}_1\mathbf{R}_{hg}\mathbf{D}_2^H(\mathbf{E}_2-\mathbf{I})^H\mathbf{D}_2 \\ \mathbf{D}_2^H(\mathbf{E}_2-\mathbf{I})\mathbf{D}_2\mathbf{R}_{gh}\mathbf{D}_1^H(\mathbf{E}_1-\mathbf{I})^H\mathbf{D}_1 & \underbrace{\mathbf{D}_2^H(\mathbf{E}_2-\mathbf{I})\mathbf{D}_2\mathbf{R}_g\mathbf{D}_2^H(\mathbf{E}_1-\mathbf{I})^H\mathbf{D}_2}_{\mathbf{M}_2} \end{pmatrix}}_{\mathbf{M}}\}, \tag{16}$$

where $\mathbf{R}_h = E\{\mathbf{hh}^H\}$, $\mathbf{R}_g = E\{\mathbf{gg}^H\}$, $\mathbf{R}_{hg} = E\{\mathbf{hg}^H\}$ and $\mathbf{R}_{gh} = E\{\mathbf{gh}^H\}$. As channel vectors $\mathbf{h}$ and $\mathbf{g}$ are zero mean and independent, we have $\mathbf{D}_2^H(\mathbf{E}_2-\mathbf{I})\mathbf{D}_2\mathbf{R}_{gh}\mathbf{D}_1^H(\mathbf{E}_1-\mathbf{I})^H\mathbf{D}_1 = \mathbf{0}$ and $\mathbf{D}_1^H(\mathbf{E}_1-\mathbf{I})\mathbf{D}_1\mathbf{R}_{hg}\mathbf{D}_2^H(\mathbf{E}_2-\mathbf{I})^H\mathbf{D}_2 = \mathbf{0}$. In addition, $\mathbf{M}_1$ and $\mathbf{M}_2$ are diagonal matrices because $\mathbf{R}_h$ and $\mathbf{R}_g$ are diagonal matrices.

4 SIMULATION RESULTS

In this section, we provide simulation results to verify the effectiveness of our design. The size of training vectors is$N = 64$. The OFDM signal bandwidth is 20MHz corresponding to $T_s = 50\,ns$.The channel response length is $L = 4$. We further assume that the power delay profile of each channel is uniform, i.e., each tap of $\mathbf{h}'$or $\mathbf{g}'$is modeled as a zero-mean Gaussian random variable with variance$1/L$. We assume all the 3dB phase noise bandwidths are the same, i.e., $\beta = \beta_{T1,2} = \beta_{R,1} = \beta_{R,2} = \beta_{T,1} = \beta_{T2,1}$. In addition, the phase noise level percentage (PLP) is defined as $N\beta T_s$ (Rabiei et al. 2011). The SNR in all the simulations is defined as P/σ_n^2. The relay transmitting power is set to$P_r = P$. For comparison, we also simulated the MSE performance of random training sequence, which is generated by QPSK modulated random bits.

Figure 2 shows the MSE performance of channel estimation with perfect phase noise information. It is shown that our designed training sequences outperform the random training sequences from low to high SNR regime. That type1 and type2 sequences having the same performance verifies our optimal training sequences design in subsection 3.1. Namely, if the generating sequence has the form of (12), designed sequences from step1 to step3 will have the optimal performance. It is also observed from Figure 2 that the phase noise will not affect the channel estimation error, which verifies that the MSE in (9) will not be affected by phase noise if training sequences are properly designed. Figure 3 shows the MSE performance of channel estimation without phase noise information. This paper investigates block training based channel estimation for OFDM base two-way relay networks in the presence of phase noise. Most of the existing joint channel and phase noise estimation algorithms can be categorized into two cases. One is channel estimation after phase noise compensation, and the other is channel estimation before phase noise compensation.

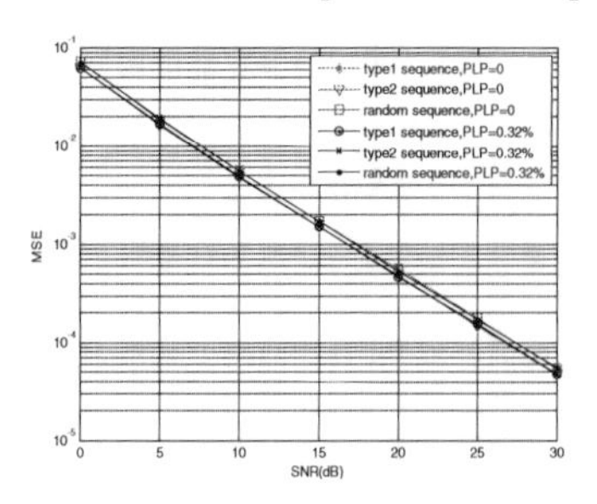

Figure 2. MSE performance of channel estimation with perfect phase noise information.

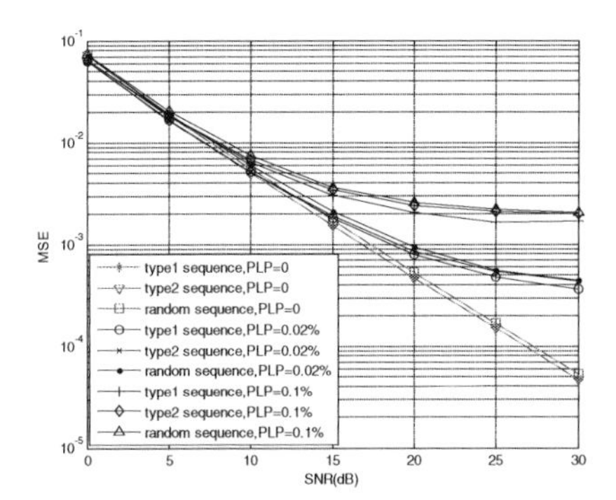

Figure 3. MSE performance of channel estimaiton without phase noise information.

REFERENCES

Cui, T., Ho, T. & Kliewer, J. 2009. Memoryless relay strategies for two-way relay channels. *IEEE Trans. Commun.* 57(10): 3132–3143.

Demir, A., Mehrotra, A. & Roychowdhury, J. 2000. Phase noise in oscillators: a unifying theory and numerical methods for characterization. *IEEE Trans. Circuits Systems I* 47(5): 655–674.

Gacanin, H. Adachi, F. 2010. Broadband analog network coding. *IEEE Trans. Commun.* 9(5): 1577–1583.

Gacanin, H., Sjödin, T. & Adachi, F. 2011. On channel estimaiton for analog network coding in a frequency-selective fading channel. *EURASIP J. Wireless Commun. Netw.*: 1–12.

Gao, F., Zhang, R. & Liang, Y. C. 2009. Channel estimation for OFDM modulated two-way relay networks. *IEEE Trans. Signal Process* 57(11): 4443–4455..

Golub, G. H. & Van Loan, C. F. 1996. *Matrix computations*,. Baltimore: JHU Press.

Ho, C. K., Zhang, R. & Liang, Y. C. 2008. Two-way relaying over OFDM: optimized tone permuation and power allocation. In Proc. *IEEE ICC, Beijing, China, May 2008*: 3908–3912. IEEE.

Jiang, B., Gao, F., Gao, X. & Nallanathan, A. 2010. Channel estimation and training design for two-way relay networks with power allocation. *IEEE Trans. Commun.* 9(6): 2022–2032.

Liu, P., Bar-Ness, Y. & Zhu, J. 2006. Effects of phase noise at both transmitter and receiver on the performance of OFDM systems. In Proc. *IEEE Inf. Sciences Syst., Princeton, USA, November 2006*: 312–316. IEEE. .

Rabiei, P., Namgoong, W. & Al-Dhahir, N. 2008. Frequency domain joint channel and phase noise estimaiton in OFDM WLAN systems. *In Proc. IEEE Signals System Computers, Pacific Grove, CA, October 2008*: 928–932. IEEE..

Rabiei, P., Namgoong, W. & Al-Dhahir, N. 2011. On the performance of OFDM-based amplify-and-forward relay networks in the presence of phase noise. *IEEE Trans. Commun.* 59(5): 1458–1466.

Rankov, B. & Wittneben, A. 2006. Achievable rate region for the two-way relay channel. *In Proc. IEEE ISIT, Seattle, USA, July 2006:* 1668–1672. IEEE.

Rankov, B. & Wittneben, A. 2007. Spectral efficient protocols for half-duplex fading relay channels. *IEEE J. Sel. Aeras Commun.* 25(2): 379–389,

Shannon, C. E. 1961. Two-way communication channels. *In Proc. 4th Berkeley Symp. Math. Stat. Prob.*: 611–644.

Wang, G., Gao, F., Chen, W. & Tellambura, C. 2011. Channel estimation and training design for two-way relay networks in time-selective fading environments. *IEEE Trans.Wireless Commun.* 10(8): 2681–2691.

Wang, G., Gao, F., Wu Y. C. & Tellambura, C. 2011. Joint CFO and channel estimation for OFDM-based two-way relay networks. *IEEE Trans. Commun.* 10(2): 456–465.

Wang, G., Gao, F., Zhang, X. & Tellambura, C. 2010. Superimposed pilot based joint CFO and channel estimation for CP-OFDM modulated two-way relay networks. *EURASIP J. Wireless Commun. Netw.*: 1–9.

Yang, W., Cai, Y., Hu, J. & Yang, W. 2010. Channel estimation for two-way relay OFDM networks. *EURASIP J. Wireless Commun. Netw.*: 1–6.

Zou, Q., Tarighat, A. & Sayed, A. H. 2007. Compensation of phase noise in OFDM wireless systems. *IEEE Trans. Signal Process* 55(11): 5407–5424.

Electronics, Communications and Networks IV – Hussain & Ivanovic (eds)
© 2015 Taylor & Francis Group, London, ISBN: 978-1-138-02830-2

Outage probability of dual-hop multiple antenna fixed-gain AF relaying with CCI in Nakagami-*m* fading channels

Cong Li, Yuming Zhang, Yunpeng Cheng* & Yuzhen Huang
College of Communications Engineering, PLA University of Science and Technology, Nanjing, Jiangsu, China

ABSTRACT: In this paper, we present an analytical investigation on the outage performance of dual-hop multiple antenna fixed-gain amplify-and-forward relaying systems in the presence of Co-Channel Interference (CCI). Assuming that all channels follow Nakagami-*m* distribution, an exact closed-form analytical expression for the outage probability of the systems is derived. Furthermore, we look into the asymptotical outage performance at the high signal-to-noise ratio regime, and characterize the diversity order and the coding gain achieved by the system. Our results reveal that the diversity order of the system is solely determined by the channel fading severity parameters and number of antenna, while CCI degrades the outage performance by affecting the coding gain of the system. Moreover, numerical simulations are provided to corroborate our theoretical results.

1 INTRODUCTION

Due to the ability of significantly improving the coverage, reliability, and throughput of wireless communication systems, there has been a growing interest in investigating dual-hop relaying transmissions (Hasna & Alouini 2004a, b). One of the simplest relaying protocols is the amplify-and-forward (AF) which generally falls into two categories, i.e., fixed-gain relaying and channel state information (CSI)-assisted relaying (Laneman et al. 2004). Parallel to the development of relaying transmissions, multiple antenna technique has been identified as another efficient approach to considerably improve the system performance and allow for higher throughput of wireless links. Hence, adopting multiple antennas in relaying systems can further bring substantial performance gains (see (Yooh et al. 2011) and references therein).

Unfortunately, these remarkable performance gains are only promised in an ideal noise-limited environment. However, in practical wireless communication systems such as cellular networks, a key feature, namely, co-channel interference (CCI), must be considered in the performance analysis due to aggressive frequency reuse. In (Ding et al. 2012), the outage probability of dual-hop fixed-gain AF multiple-input multiple-output (MIMO) relaying systems with interference at both the relay and destination nodes was derived. In (Phan et al. 2012), the authors investigated the effect of feedback delay on the performance of dual-hop beam forming AF relaying systems with an interference-limited relay node. In (Zhong et al. 2013), the outage performance of dual-hop multiple antenna AF relaying systems with both fixed-gain and variable-gain relaying schemes under a single Rayleigh interferer at the relay was investigated. In (Huang et al. 2013), the authors analyzed the ergodic capacity of dual-hop multiple antenna AF relaying systems employing transmit beam forming and maximum ratio combining with feedback delay and CCI.

It is worth pointing out that most of prior works assume the interference-limited scenario, the joint impact of CCI and noise on the outage performance of dual-hop multiple antenna AF relaying systems has not been well-understood. In addition, all the prior works consider Rayleigh fading channels, hence, the effect of channel fading severity level on the relaying system employing multiple antennas with CCI remains unknown. Only in the recent works (Al-Qahtani et al. 2011, Suraweera et al. 2012), the authors studied the performance of dual-hop AF relaying systems employing variable-gain and fixed-gain relaying schemes with CCI in Nakagami-*m* fading channels, respectively. However, these works are limited to the single antenna system. To fill these important gaps, we extend the analysis of (Suraweera et al. 2012) to multiple antennas scenario. Specifically, we derive exact and asymptotic expressions for the outage probability of dual-hop multiple antenna fixed-gain AF relaying systems with interference and noise at the relay, which provide an efficient means to investigate the impact of key system parameters such as CCI, number of antennas and channel fading severity on the outage performance of the system.

*Corresponding author: *chengyp2000@sina.vip.com*

2 SYSTEM AND CHANNEL MODELS

Let us consider a dual-hop multiple antenna AF relaying network, which consists of a source S equipped with N_t antennas, a destination D equipped with N_r antennas, and a single antenna relay R assisting the communication of source node. We assume that the relay node is subjected to a single dominate interference[1] and additive white Gaussian noise (AWGN), while the destination node is corrupted by AWGN only. The entire communication consists of two slots. During the first slot, the source S sends the signal symbol x_s to the relay R with the transmit power Ps. Therefore, the received signal at the relay R can be expressed as

$$y_r = \sqrt{P_s}\mathbf{h}_1^\dagger \mathbf{w}_1 x_s + \sqrt{P_I} h_I x_I + n_1 \tag{1}$$

where $\mathbf{h}_1 = \left[h_{11}, h_{12}, \ldots, h_{1N_t}\right]^T$ is the channel vector for the source-relay link with $(\cdot)^T$ being the transpose operator, the amplitude of $\{h_{1i}\}_{i=1}^{N_t}$ follows a Nakagami-*m* distribution with fading severity parameter m_{sr} and average power Ω_{sr}, $(\cdot)^\dagger$ is the conjugate transpose operator, $\mathbf{w}_1$ is the $N_t \times 1$ transmit beamforming vector defined as $\mathbf{w}_1 = \frac{\mathbf{h}_1^\dagger}{\|\mathbf{h}_1\|_F}$, P_I is the average power of interference signal, h_I is the channel for interference-relay link, and its amplitude also follows a Nakagami-*m* distribution with fading severity parameter m_I and average power Ω_I, x_I is the interference symbol satisfying zero-mean and unit energy, n_1 is the AWGN noise at the relay node with $E\left[|n_1|^2\right] = \sigma^2$, where $E[\cdot]$ is the expectation operator.

At the second phase, the relay R retransmits its received signal to the destination D after applying an amplification factor. Consider that it is difficult to obtain the instantaneous interference link information at the relay node, hence, the fixed-gain relaying scheme, where the relay simply forwards a scaled version of the received signal, becomes more appealing for such case. Correspondingly, the amplification factor G is given by (Zhong et al. 2013)

$$G = \sqrt{\frac{1}{N_t P_s \Omega_{sr} + P_I \Omega_I + \sigma^2}} \tag{2}$$

Hence, the received signal at the destination D after MRC is given by

$$y_D = \sqrt{P_r} G \mathbf{w}_2^\dagger \mathbf{h}_2 y_r + \mathbf{w}_2^\dagger \mathbf{n}_2 \tag{3}$$

where P_r is the transmit power at the relay, $\mathbf{h}_2 = \left[h_{21}, h_{22}, \ldots, h_{2N_r}\right]^T$ is the channel vector for the relay-destination link, the amplitude of $\{h_{2i}\}_{i=1}^{N_r}$ follows a Nakagami-*m* distribution with fading severity parameter m_{rd} and average power Ω_{rd}, $\mathbf{w}_2$ is the $N_r \times 1$ transmit beamforming vector defined as $\mathbf{w}_2 = \frac{\mathbf{h}_2}{\|\mathbf{h}_2\|_F}$, $\mathbf{n}_2$ is the $N_r \times 1$ AWGN vector with $E\left[\mathbf{n}_2 \mathbf{n}_2^\dagger\right] = \sigma^2 \mathbf{I}$.

To this end, by substituting (1) and (2) into (3), the end to-end signal-to-interference-plus-noise ratio (SINR) can be derived as

$$\gamma = \frac{\gamma_1 \gamma_2}{(1+\gamma_3)\gamma_2 + C} \tag{4}$$

where $\gamma_1 = \frac{P_s\|\mathbf{h}_1\|_F^2}{\sigma^2}$, $\gamma_2 = \frac{P_r\|\mathbf{h}_2\|_F^2}{\sigma^2}$, $\gamma_3 = \frac{P_I\|\mathbf{h}_I\|_F^2}{\sigma^2}$ and $C = \frac{1}{G^2\sigma^2}$.

3 END-TO-END PERFORMANCE ANALYSIS

In this section, we analyze the outage performance of the system. We first derive the exact closed-form expression for the outage probability, and then investigate the asymptotical outage behavior at high signal-to-noise ratio (SNR) regime to obtain two key design parameters, i.e., diversity order and coding gain.

3.1 *Outage probability*

The outage probability is defined as the probability that the instantaneous end-to-end SINR γ falls below a predefined threshold γ_{th}, or mathematically

$$P_{\text{out}}(\gamma_{\text{th}}) = \Pr\{\gamma < \gamma_{\text{th}}\} = F_\gamma(\gamma_{\text{th}}) \tag{5}$$

It is noted from (5) that to evaluate the outage probability of the system, the statistical behavior of the end-to-end SINR γ is required, which can be expressed as

$$F_\gamma(x) = \int_0^\infty \int_0^\infty F_{\gamma_1}(\theta) f_{\gamma_2}(y) dy f_{\gamma_3}(z) dz \tag{6}$$

where $\theta = x(1+z) + \frac{Cx}{y}$. To obtained the cumulative distribution function (CDF) $F_\gamma(\cdot)$, we need the CDF of γ_1 and the probability density functions (PDFs) of γ_2 and γ_3.

[1] We assume a single interferer at the relay node for the mathematical tractability. However, the analysis can be easily extended to multiple interferers. In addition, in a practical cellular network, the relay at the edge of cell may be subjected to a single dominant interferer due to frequency reuse. Hence, the assumption has been broadly adopted in the published works(Al-Qahtani et al. 2011, Suraweera et al. 2012, Poon & Beaulieu 2004).

As all the channels undergo Nakagami-m fading, the CDF of γ_1 is given by (Yang et al. 2012)

$$F_{\gamma_1}(x) = 1 - e^{-\frac{m_{sr}x}{\bar{\gamma}_1}} \sum_{k=0}^{m_{sr}N_t-1} \frac{1}{k!}\left(\frac{m_{sr}x}{\bar{\gamma}_1}\right)^k \tag{7}$$

where $\bar{\gamma}_1 = \frac{P_s\Omega_{sr}}{\sigma^2}$ is the average SNR of the first hop. In addition, the PDF of γ_2 is given by

$$f_{\gamma_2}(y) = \left(\frac{m_{rd}}{\bar{\gamma}_2}\right)^{m_{rd}N_r} \frac{y^{m_{rd}N_r-1}}{(m_{rd}N_r-1)!} e^{-\frac{m_{rd}y}{\bar{\gamma}_2}} \tag{8}$$

where $\bar{\gamma}_2 = \frac{P_r\Omega_{rd}}{\sigma^2}$ is the average SNR of the second hop. In addition, the PDF of γ_3 is given by

$$f_{\gamma_3}(z) = \left(\frac{m_I}{\bar{\gamma}_3}\right)^{m_I} \frac{z^{m_I-1}}{(m_I-1)!} e^{-\frac{m_I z}{\bar{\gamma}_3}} \tag{9}$$

where $\bar{\gamma}_3 = \frac{P_I\Omega_I}{\sigma^2}$ is the interference-to-noise ratio (INR) of interference signal.

Now substituting the CDF/PDFs of (7), (8) and (9) into (6) and applying the binomial expansion, the CDF of the end-to-end SINR γ can be further expressed as (10).

$$\begin{aligned} F_\gamma(x) &= 1 - e^{-\frac{m_{sr}}{\bar{\gamma}_1}x} \sum_{k=0}^{m_{sr}N_t-1} \frac{1}{k!}\left(\frac{m_{sr}x}{\bar{\gamma}_1}\right)^k \\ &\times \sum_{n=0}^{k} \binom{k}{n} \left(\frac{m_{rd}}{\bar{\gamma}_2}\right)^{m_{rd}N_r} \left(\frac{m_I}{\bar{\gamma}_3}\right)^{m_I} \frac{C^{k-n}}{\Gamma(m_{rd}N_r)\Gamma(m_I)} \\ &\times \int_0^\infty y^{m_{rd}N_r+n-k-1} e^{-\frac{m_{sr}Cx}{\bar{\gamma}_1 y}-\frac{m_{rd}}{\bar{\gamma}_2}y} dy \int_0^\infty z^{m_I-1}(1+z)^n e^{-z\left(\frac{m_{sr}x}{\bar{\gamma}_1}+\frac{m_I}{\bar{\gamma}_3}\right)} dz. \end{aligned} \tag{10}$$

To this end, by utilizing (Gradshteyn & Ryzhik, 2007, eq. (3.471.9)) and (Gradshteyn & Ryzhik, 2007, eq. (9.211.4)), the CDF of γ can be derived as (11) after some mathematical manipulations, where $K_n(\cdot)$ and $\Psi(\cdot)$ denote the nth order modified Bessel function of the second kind and the confluent hypergeometric function (Gradshteyn & Ryzhik, 2007), respectively.

$$\begin{aligned} F_\gamma(x) &= 1 - e^{-\frac{m_{sr}}{\bar{\gamma}_1}x} \left(\frac{m_I}{\bar{\gamma}_3}\right)^{m_I} \sum_{k=0}^{m_{sr}N_t-1} \frac{1}{k!}\left(\frac{m_{sr}m_{rd}xC}{\bar{\gamma}_1\bar{\gamma}_2}\right)^{\frac{m_{rd}N_r+k}{2}} \\ &\times \sum_{n=0}^{k} \binom{k}{n} \left(\frac{m_{sr}\bar{\gamma}_2 x}{m_{rd}\bar{\gamma}_1 C}\right)^{\frac{n}{2}} \frac{2}{\Gamma(m_{rd}N_r)} K_{m_{rd}N_r+n-k}\left(2\sqrt{\frac{m_{sr}m_{rd}Cx}{\bar{\gamma}_1\bar{\gamma}_2}}\right) \\ &\times \Psi\left(m_I, m_I+n+1; \frac{m_{sr}x}{\bar{\gamma}_1}+\frac{m_I}{\bar{\gamma}_3}\right) \end{aligned} \tag{11}$$

Hence, the outage probability can be directly obtained by substituting $x = \gamma_{th}$ into (11). Note that the expression (11) only involves standard functions which allows for fast evaluation in popular mathematical software such as Matlab or Mathematica, thereby providing an efficient means to assess the impact of various key system parameters such as number of antennas, CCI and channel fading severity on the outage performance of the system.

3.2 *Asymptoticala analysis*

The closed-form expression in (11) provides an efficient means to evaluate the exact outage probability of the system for any given SINR rather than the network diversity behavior. Motivated by this, we hereafter pursue an asymptotic analysis of the outage probability to investigate the impact of key parameters on the outage performance and we have the following key results. Recall the definition of the diversity order in a noise-limited environment, only the asymptotically large SNR regime is of interest, hence we assume that the powers of the interferers are fixed in the diversity analysis.

Theorem 1. Let $\bar{\gamma}_2 = \mu\bar{\gamma}_1$ and $\bar{\gamma}_1 \to \infty$, the outage probability can be approximated as

$$P_{\text{out}}(\gamma_{\text{th}}) \approx \begin{cases} \theta_1\left(\frac{m_{sr}\gamma_{\text{th}}}{\bar{\gamma}_1}\right)^{m_{sr}N_t}, & m_{sr}N_t < m_{rd}N_r \\ (\theta_2+\theta_3)\left(\frac{\gamma_{\text{th}}}{\bar{\gamma}_1}\right)^{L}, & m_{sr}N_t = m_{rd}N_r = L \\ \theta_4\left(\frac{m_{rd}\gamma_{\text{th}}}{\mu\bar{\gamma}_1}\right)^{m_{rd}N_r}, & m_{sr}N_t > m_{rd}N_r \end{cases} \tag{12}$$

where θ_1, θ_2, θ_3 and θ_4 are given in (16), (18), (19) and (24).

Proof: See Appendix A.

As can be observed from Theorem 1, the diversity order G_d of the considered system is only determined by the fading severity parameters and number of antennas, i.e., $\min\{m_{sr}N_t, m_{rd}N_r\}$. It is also noteworthy that CCI only affects the coding gain G_c, which is given by

$$G_c = \begin{cases} \frac{1}{m_{sr}\gamma_{\text{th}}}\theta_1^{-\frac{1}{m_{sr}N_t}}, & m_{sr}N_t < m_{rd}N_r \\ \frac{1}{\gamma_{\text{th}}}(\theta_2+\theta_3)^{-\frac{1}{L}}, & m_{sr}N_t = m_{rd}N_r = L \\ \frac{\mu}{m_{rd}\gamma_{\text{th}}}\theta_4^{-\frac{1}{m_{rd}N_r}}, & m_{sr}N_t > m_{rd}N_r \end{cases} \tag{13}$$

4 SIMULATION RESULTS

In this section, we present numerical results analytical expressions and investigate the impact of channel fading severity level, number of antennas and CCI on the outage probability of the system. Unless otherwise specify, we set $\gamma_1 = \gamma_2$, $m_I = 3$ and $\gamma_{th} = 0$dB in all simulations.

Figure 1 plots the outage probability of the system under different channel fading severity level with $N_t = N_r = 2$ and $\gamma_3 = 15$dB. As can clearly be seen from this figure, the analytical and simulation curves are in an excellent agreement and the asymptotic curves match very well with the exact curves in the high SNR regime, which corroborates the accuracy of our derivation. Moreover, the outage performance significantly improves when m_{sr} and m_{rd} increase from1 to 3, which gives an intuitive indication that the outage performance heavily depends on channel fading severity level. In addition, the curves show that the full diversity order of the system under three scenarios can be achieved, which validates our analysis.

Figure 2 investigates the joint impact of number of antennas and interference power on the outage probability of the system with $m_{sr} = m_{rd} = 1$. As can be expected, increasing the interference power degrades the outage performance of the system while increasing the number of antennas can significantly improve the outage performance of the system. It is also seen that increasing the interference power results in higher outage probability, the curves remain parallel to each other which means CCI does not affect the diversity order. This validates our proposed analysis.

5 CONCLUSION

In this paper, we have studied the outage performance of dual-hop multiple antenna fixed-gain AF system with interference in Nakagami-*m* fading channels. Specifically, we derived the exact closed-form expression for the outage probability, which provides a fast and efficient means to evaluate the impact of number of antennas, channel fading severity level and interference on the outage performance of the system. In addition, simple and informative high SNR approximations for the outage probability were also provided, which accurately identify the diversity order and the array gain as two key design components of the system. The findings of this paper suggest that the diversity order is equal to $\min\{m_{sr}N_t, m_{rd}N_r\}$, and that CCI degrades the outage performance by reducing the coding gain of the system.

6 APPENDIX A PROOF OF THEOREM 1

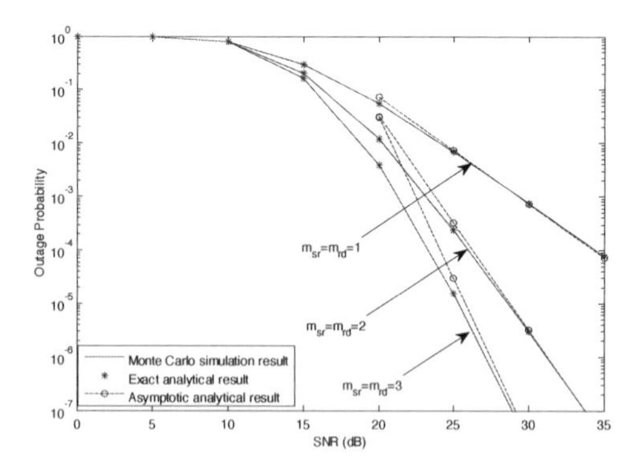

Figure 1. Impact of fading severity parameters on the outage probability of multiple antenna fixed-gain AF relaying system with CCI.

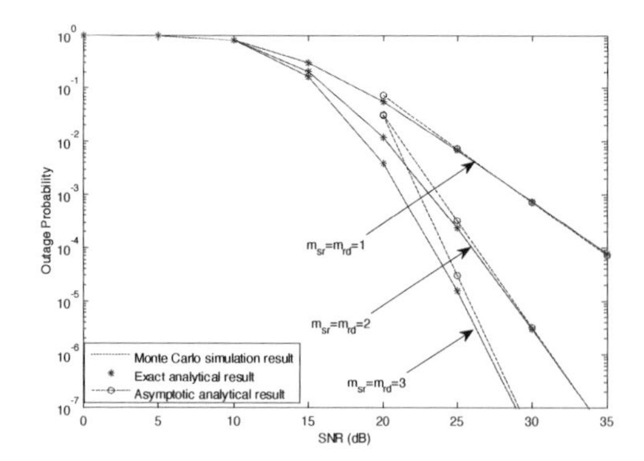

Figure 2. Impact of interference on the outage probability of multiple antenna fixed-gain AF relaying system with CCI.

6.1 *Case 1:* $m_{sr}N_t < m_{rd}N_r$

When $\overline{\gamma}_1 \to \infty$, the CDF of γ_1 can be approximated as

$$F_{\gamma_1}(x) = \frac{\gamma\left(m_{sr}N_t, \frac{m_{sr}x}{\overline{\gamma}_1}\right)}{\Gamma(m_{sr}N_t)} \approx \left(\frac{m_{sr}x}{\overline{\gamma}_1}\right)^{m_{sr}N_t} \frac{1}{(m_{sr}N_t)!} \tag{14}$$

where we have used the asymptotic property of the incomplete gamma function $\gamma(\cdot,\cdot)$ near zero (Gradshteyn & Ryzhik, 2007, eq.(8.354.1)). Then, substituting (14) into (6) and utilizing (Gradshteyn & Ryzhik, 2007, eq. (3.351.3)) and (Gradshteyn & Ryzhik, 2007, eq. (9.211.4)), the CDF of the end-to-end SINR γ can be approximated as

$$F_\gamma(\gamma_{th}) \approx \theta_1 \left(\frac{m_{sr}\gamma_{th}}{\overline{\gamma}_1}\right)^{m_{sr}N_t} \tag{15}$$

where θ_1 is shown in (16).

$$\theta_1 = \sum_{k=0}^{m_{sr}N_t-1} \binom{m_{sr}N_t}{k} \left(\frac{m_{rd}N_t}{\mu}\right)^k \frac{\Gamma(m_{rd}N_r - k)}{\Gamma(m_{rd}N_r)\Gamma(m_{sr}N_t+1)} \times \left(\frac{m_I}{\bar{\gamma}_3}\right)^{m_I} \Psi\left(m_I, m_I + m_{sr}N_t - k + 1; \frac{m_I}{\bar{\gamma}_3}\right) \tag{16}$$

6.2 *Case 2:* $m_{sr}N_t = m_{rd}N_r = L$

Due to the fact the resulting integration for γ_2 does not converge by following the same way of case 1, we directly expand the modified Bessel function $K_n(\cdot)$ to derive the approximate outage probability. Hence, by utilizing a Maclaurin series expansion of the exponential function and (Gradshteyn & Ryzhik, 2007, eq. (8.446)), we have

$$F_\gamma(\gamma_{\text{th}}) \approx (\theta_2 + \theta_3)\left(\frac{\gamma_{\text{th}}}{\bar{\gamma}_1}\right)^L \tag{17}$$

where θ_2 and θ_3 are shown in (18) and (19), respectively.

$$\theta_2 = \sum_{n=0}^{L-1} \frac{1}{\Gamma(n+1)} \sum_{k=0}^{n} \binom{n}{k} \left(\frac{m_{rd}m_I N_t}{\mu\bar{\gamma}_3}\right)^k \times \sum_{p=0}^{L-k-1} \frac{(-1)^{L-n+1}\Gamma(L-k-p)\Gamma(L-k-p+m_I)}{\Gamma(L)\Gamma(p+1)\Gamma(L-n-p+1)\Gamma(m_I)} \left(\frac{\bar{\gamma}_3}{m_I}\right)^{L-p} \left(\frac{m_{rd}L}{\mu}\right)^p \tag{18}$$

$$\theta_3 = \sum_{i=0}^{L-1} \left(\frac{m_{sr}m_{rd}N_t}{\mu}\right)^i \frac{(-1)^{N-i}\left\{\ln\frac{Lm_{sr}m_{rd}\gamma_{\text{th}}}{\mu\bar{\gamma}_1} - \psi(1) - \psi(L-i+1)\right\}}{\Gamma(i+1)\Gamma(L-i+1)} \tag{19}$$

6.3 *Case 3:* $m_{sr}N_t > m_{rd}N_r$

For this case, we adopt the following alternative approach to derive the approximate outage probability. We note that the end-to-end SINR γ can be rewritten as

$$\gamma = \frac{\rho_1\rho_2\|\mathbf{h}_1\|_F^2\|\mathbf{h}_2\|_F^2}{\left(1+\rho_3|h_I|^2\right)\rho_2\|\mathbf{h}_2\|_F^2 + C} \tag{20}$$

where $\rho_1 = \frac{P_s}{\sigma^2}$, $\rho_2 = \frac{P_r}{\sigma^2}$ and $\rho_3 = \frac{P_I}{\sigma^2}$. Hence, the CDF of γ can be alternatively computed as

$$F_\gamma(\gamma_{\text{th}}) = \Pr\left\{\frac{\|\mathbf{h}_1\|_F^2\|\mathbf{h}_2\|_F^2}{\left(1+\rho_3|h_I|^2\right)\|\mathbf{h}_2\|_F^2 + C/\rho_2} < \frac{\gamma_{\text{th}}}{\rho_1}\right\} \tag{21}$$

At the high SNR regime, the CDF of γ can be approximated by

$$F_\gamma(\gamma_{\text{th}}) \approx \Pr\left\{\min\left(\frac{\|\mathbf{h}_1\|_F^2\|\mathbf{h}_2\|_F^2}{N_t/\mu}, \frac{\|\mathbf{h}_1\|_F^2}{1+\rho_3|h_I|^2}\right) < \frac{\gamma_{\text{th}}}{\rho_1}\right\} \tag{22}$$

To this end, by involving (Zhong et al. 2013, Lemma 1) and utilizing the asymptotic property of the incomplete gamma function $\gamma(\cdot,\cdot)$, we have

$$F_\gamma(\gamma_{\text{th}}) \approx \theta_4\left(\frac{m_{rd}\gamma_{\text{th}}}{\mu\bar{\gamma}_1}\right)^{m_{rd}N_r} \tag{23}$$

$$\theta_4 = (m_{sr}N_t)^{m_{rd}N_r} \frac{\Gamma(m_{sr}N_t - m_{rd}N_r)}{\Gamma(m_{sr}N_t)\Gamma(m_{rd}N_r+1)} \tag{24}$$

REFERENCES

Al-Qahtani, F. S., Yang, J., Radaydeh, R. M., et al. 2011. Performance analysis of dual-hop AF systems with interference in Nakagami-*m* fading channels. *IEEE Signal Processing Letters* 18(8): 454–457.

Ding, H., He, C. & Jiang, L. 2012. Performance analysis of fixed gain MIMO relay systems in the presence of co-channel interference. *IEEE Communications Letters*. 16(7): 1133–1136.

Gradshteyn, I. S. & Ryzhik, I. M. 2007. *Table of Integrals, Series, and Products. 7th edition.* California: Academic Press.

Hasna, M. O. & Alouini, M.-S. 2004a. A performance study of dual-hop transmissions with fixed gain relays. *IEEE Transactions on Wireless Communications* 3(6): 1963–1964.

Hasna, M. O. & Alouini, M.-S. 2004b. End-to-end performance of transmission systems with relays over Rayleigh-fading channels. *IEEE Transactions on Wireless Communications* 2(6): 1126–1131.

Huang, Y., Li, C., Zhong, C., et al. 2013. On the capacity of dual-hop multiple antenna AF relaying systems with feedback delay and CCI. *IEEE Communications Letters* 17(6): 1200–1203.

Laneman, J. N., Tse, D. N. C. & Wornell, G.W. 2004. Cooperative diversity in wireless networks: efficient protocols and outage behavior. *IEEE Transactions on Information Theory* 50(12): 3062–3080.

Phan, H., Duong, T. Q., Elkashlan, M., et al. 2012. Beamforming amplify-and-forward relay networks with feedback delay and interference. *IEEE Signal Processing Letters* 19(1): 16–19.

Poon, T. & Beaulieu, N. 2004. Error performance analysis of a jointly optimal single co-channel interferer BPSK receiver. *IEEE Transactions Communications* 52(7): 1051–1054.

Suraweera, H. A., Michalopoulos, D. S. & Yuen, C. 2012. Performance analysis of fixed gain relay systems with a single interferer in Nakagami-*m* fading channels. *IEEE Transactions on Vehicular Technology* 61(3): 1457–1463.

Yeoh, P. L., Elkashlan, M. & Collings, I. B. 2011. Selection relaying with transmit beamforming: A comparison of fixed and variable gain relaying. *IEEE Transactions on Communications* 59(6): 1720–1730.
Yang, N., Elkashlan, M., Collings, I. B. 2012. Multiuser MIMO relay networks in Nakagami-m fading channels. *IEEE Transactions on Communications.* 60(11): 3298–3310.
Zhong, C., Suraweera, H. A., Huang, A., et al. 2013. Outage probability of dual-hop multiple antenna AF relaying systems with interference. *IEEE Transactions on Communications* 61(1): 108–119.

Electronics, Communications and Networks IV – Hussain & Ivanovic (eds)
© 2015 Taylor & Francis Group, London, ISBN: 978-1-138-02830-2

A cross-layer protocol in molecular communication nanonetworks

Zuopeng Li* & Xiangguo Chen
School of Information and Electrical Engineering, Hebei University of Engineering, Handan, Hebei, China

ABSTRACT: A molecular communication nanonetwork is a new distributed computer network which operates at nanoscale. The simple single nanomachine and the new communication paradigm, which are completely different from the traditional communication technologies such as electromagnetic wave, all pose challenges for the development of communication protocols in nanonetworks. In this paper, a new architecture of molecular communication nanonetworks is first provided, and then on this basis, a Concentration-Aware (CA) routing protocol in nanonetworks is proposed, which utilizes the concentration gradient of diffused molecules to establish the path of data relay in multi-hop nanonetworks. Finally, to achieve the reliable data relay, a CA based cross-layer protocol is further proposed, which concerns improvement of the data link quality, and then an optimal path of data relay in multi-hop nanonetworks is established. The experimental results show that the concentration aware routing protocol is viable for molecular communication nanonetworks and has an excellent performance of the network load.

1 INTRODUCTION

The rapid development of nanotechnology has opened many new research fields in the academic world, for example, the nanonetwork is bringing human intelligence into the microscopic world. A nanomachine is a device that consists of components on the nanoscale and the whole dimension is at the nanoscale or microscale, which is able to perform functions of computing, storing data, communicating, sensing and actuation (Li et al. 2013). Since the single nanomachine is able to only perform very simple tasks, they should be interconnected through data exchange to comprise a distributed nanonetwork which can perform more complex tasks in the larger space.

Molecular communication (MC) is a new short-range communication paradigm that uses biochemical molecules as information carriers in a fluid medium (Ian et al. 2008), and the carrier molecules are called information molecules. Because of the appropriate transceiver size and very low power consumption, the bio-inspired MC is considered as one of the most promising communication technologies of nanonetworks (Kamal & Attahiru 2013).

With high bio-compatibility as well as high-resolution sensing and actuation at the molecular level, molecular communication nanonetworks can be widely applied into not only biomedical area (e.g., body area nanonetworks), but also environmental, military and industrial areas (e.g., nanosensor networks) and so on. So it is attracting wide attention of the academic. In the future, molecular communication nanonetworks will become important parts of Internet of Things (IoTs), and even Cyber-physical System (CPS).

At the early research stage of nanonetworks, the current works mainly focus on the component and supporting technologies of nanonetworks, especially the physical layer of molecular communication protocol, e.g., development of various MC technologies involving short-range (nm-μm) MC, medium-range (μm-mm) MC and long-range (mm-m) MC(Lluis & Ian 2009), exploration of channel characteristics and noise modeling of MC (Dogu 2011), design of encoding technologies of MC(Tadashi 2010). The achievement of a reliable nanonetwork needs high-performance communication protocols. Despite the very simple capability of the single nano-machine and the unique characteristics of MC all pose challenges for developing communication protocols in nanonetworks, they also provide wider research opportunities at the same time.

In this paper, we concerned with using the characteristics of molecular channel and the bio-inspired methods to solve problems of developing communication protocols in nanonetworks. Firstly, a new architecture of nanonetwork is proposed, and then on this basis the concentration-aware routing protocol is proposed for short-range MC in nanonetworks and this protocol is further extended based on integrating MAC layer with network layer to enhance the availability of the concentration aware routing protocol. To our knowledge, this is the first specialized cross-layer protocol of short-range MC in nanonetworks. Finally, the performance of the protocol by simulation is verified.

*Corresponding author: *lizuopeng@hebeu.edu.cn*

2 A NEW ARCHITECTURE OF MOLECULAR COMMUNICATION NANONETWORKS

This architecture of molecular communication nanonetworks is composed of two types of nodes, namely, nanomachine and nanosink. Nanomachines serve as the basic entities to perform tasks at the nanoscale, which can only use short-range MC to send information due to their simplicity. As a special type of nanomachines, nanosinks have larger sizes and more powerful capabilities such as processing and communicating, which are used to extend the scale and enhance the reliability of nanonetworks. As shown in Figure 1, a nanosink and several nanomachines consist of a sub-nanonetwork in which nanomachines will exchange information with nanosink based on short-range MC and multi-hop methods. Several sub-nanonetworks consist of a nanonetwork through medium-range MC between nanosinks.

Because of the capability of fast transmitting information molecules in a short distance as well as no need of extra communication infrastructure, the free diffusion based short-range MC has become the focus of researchers and achieved many research results. Essentially, the free diffusion based MC is a broadcast communication by which information molecules emitted from nanomachines diffuse in any direction. Similar to the amplitude modulating (AM) and frequency modulating (FM) technologies in traditional communication such as electromagnetic wave, the concentration level or rate of concentration change of information molecules can be used as the carrier of information in MC.

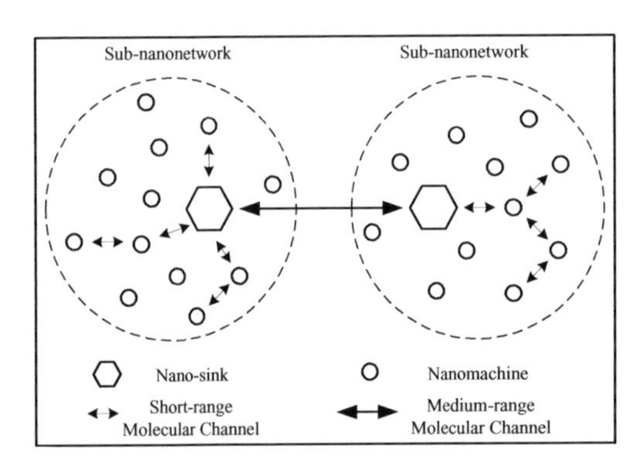

Figure 1. The architecture of molecular communication nanonetwork.

3 A CROSS-LAYER PROTOCOL IN MOLECULAR COMMUNICATION NANONETWORKS

In this paper, we focus on the information transmission from source (nanomachine) to destination (nanosink) in one sub-nanonetwork through free diffusion based MC.

3.1 *Theoretical foundation*

The concentration-aware (CA) routing protocol will take advantage of the concentration information of diffused molecules to enable the data relay in a multi-hop nanonetwork. Furthermore, a bio-inspired method, namely, multiple communication interface architecture, is also used to support the achievement of an efficient CA routing protocol.

1 The diffusion is a process that molecules move from a region of high concentration to a region of low concentration, ultimately achieving the state of uniform concentrations. Molecular diffusion can be modeled by Fick's law (Jean 2006):

$$J = -D \times \frac{dC(x,t)}{dx} \tag{1}$$

where x is the position, J is the diffusion flux along x, D is the diffusion coefficient of the medium, C is the concentration and t is time. Equation (1) demonstrates that the diffusion can create a concentration gradient around the sender nanomachine (or sender for short) and the location around sender has the highest value of concentration. CA routing protocol will choose the path of data relay based on the concentration gradient of a specific type of molecules.

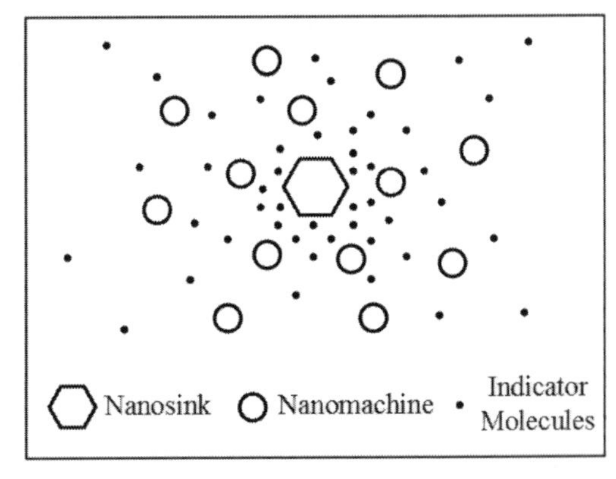

Figure 2. Diffusion of indicator molecules.

In nanonetworks, the sender diffuses information molecules into propagation medium. The receiver nanomachines (or receivers for short) are equipped with the specific type of receptors and only when the concentration of appropriate type of information molecules reaches a sufficient level, then information molecules start to be received. As shown in Figure 2, nanosink has a larger size. So it can release a relatively larger number of specific types of molecules whose concentration gradients indicate which nanomachines have located more closely to the nanosink, and then the next-hop node to relay data is determined.

2 The multiple communication interface architecture commonly exists in bio-cells, which have hundreds of receivers and are able to communicate by using multiple types of information molecules simultaneously. The bio-inspired nanomachines can take advantage of this multiple channel access technique to enhance interconnection between nanomachines and then support implementation of various communication protocols (Stephen 2010).

3.2 *Simple CA routing protocol*

The basic steps of the CA routing protocol are as follows:

a. Nanosink releases indicator molecules *I*, and the diffusion of *I* creates a concentration gradient;
b. Nanomachine *i* which sensed *I* records the maximum concentration of I as *CIi*;
c. Source nanomachine s encodes information and then broadcasts a data packet which contains *CIs* to one-hop neighbors, where *CIs* denotes the maximum concentration of *I* sensed by source *s*;
d. Nanomachine *i* which received the data packet from the last hop nanomachine *j* compares *CIi* with *CIj*, and if $CIj \geq CIi$, then discards this packet, otherwise sends a feedback packet which contains *CIi* to *j*;
e. Nanomachine *j* receives feedback packets, and if the number of feedback packet is more than 1, then chooses the *CIc* = max *CIi*, and sends a confirmation packet which contains *CIc*;
f. If nanomachine k which received the confirmation packet can confirm CIc = CIk, then replaces CIj with CIk and sends the data packet forward, until it reaches nanosink.

3.3 *Concentration-Aware Contention based Routing protocol (CACR)*

The CA routing protocol just describes a strategy of establishing the path of data relay based on the local concentration information on indicators, which does not concern the impact of channel state, e.g., signal collision caused by using the same type of information molecules. For keeping availability and reliability of routing protocol, a concentration-aware contention based routing protocol (CACR) is proposed to expand the CA routing protocol in this section, which takes advantage of the multiple communication interface architecture of nanomachines and first uses CA protocol to determine nodes available for data rely, and then further chooses the optimal next-hop node using molecular channel characteristics-based contention mechanism.

The basic steps of CACR cross-layer protocol are as follows:

Assumption: nanomachines can use *N* types of information molecules to communicate with each other, *Mp* {*p*=*1*,......, *N*} denotes a type of information molecules.

a. Nanosink releases indicator molecules *I*, and the diffusion of *I* creates a gradient of concentration;
b. Nanomachine *i* which sensed *I* records the maximum concentration of *I* as *CIi*;
c. Source nanomachine s encodes information and then broadcasts a requirement packet which contains *CIs* to one-hop neighbors, where *CIs* denotes the maximum concentration of I sensed by source s;
d. Nanomachine *i* which received the requirement packet from the last hop nanomachine *j* compares *CIi* with *CIj*, and if $CIj \geq CIi$, then discards this packet, otherwise chooses the information molecule *Mq* {*q*=*1*,......, *N*} which has a larger storage volume than other types of information molecules stored in nanomachine *i* and then uses *Mq* to send a feedback packet which contains *CIi* to *j*;
e. Nanomachine *j* receives feedback packets, and if the number of feedback packet is more than 1, then *j* chooses next hop based on the following rules:
    - If without collusion detected, then chooses *Mr* {*r*=*1*,......, *N*}, which contains a maximum concentration value and uses *Mr* to send a data packet;
    - If collusion of *Mt* {*t*=*1*,......, *N*} is detected, then namomachine *j* sends requirement packet again using *Mt*. Nanomachine *k* which confirmed that *Mt* is same as the its information molecules carrying feedback packet receives requirement packet again, and then random chooses *Mv* {*v*=*1*,......,*N*} and sends feedback packet using *Mv*, which contains *CIk* to *j*. Nanomachine *j* receives feedback again, chooses *Mw* {*w*=*1*,......, *N*} which contains a maximum concentration value and uses *Mw* to send data packet;
f. If nanomachine l which received the data packet confirms that Mr or Mw is same as its information molecules carrying feedback packet, then sends the data packet forward, until it reaches nanosink.

4 EXPERIMENT ANALYSIS

4.1 *Experimental setup*

When a nanosink wants to initiate the information transmission in its sub-nanonetwork, it instantaneously releases a pulse of indicator molecules, and the propagation of this pulse can be analytically modeled by solving Fick's law of diffusion. If the nanosink releases Q molecules at the instant t = 0, the molecular concentration at any point in space is given by (Bossert & Wilson 1963):

$$c(x,t)=\frac{Q}{(4\pi Dt)^{3/2}}e^{-x^2/4Dt} \tag{2}$$

where D is the diffusion coefficient, t is time and x is the distance from the nanosink location. In this paper, the diffusion coefficient is set to D = 1 nm2/ns, similar to the diffusion coefficient of ionic calcium in cytoplasm. In this section, the CA protocol will be evaluated based on verifying both feasibility and performance of the protocol through computer simulation.

1 Feasibility of protocol

The feasibility of the protocol will be verified by the simulation of CA and CACR protocol. We set that the CA and CACR routing protocol are carried out by a molecular communication nanonetwork which is composed of 1 nanosink and 50 nanomachines in area of 500*500 μm2, and the simulation process involves the operational results of both CA and CACR routing protocols.

2 Performance of protocol

Performance of the protocol will be verified by an analytic evaluation of the network load for CA and CACR routing protocol, which show the average number of packets, delivered for a data transmission from source to nanosink at different network scale.

The simulation environment is set as follows:

Nanosink's operations are not taken into account in protocol evaluation due to its relatively more powerful capability such as computing and energy carried;

After an initialization phase of the protocols, 50 times data transmissions are performed based on randomly selecting a source node, and every functional packet (e.g., conformation packet and feedback packet) is considered as a half of the data packet in length.

4.2 *Experimental results*

1 Feasibility of protocol

As shown in Figure 3(a), CA protocol can construct the path of data rely form source to nanosink, while Figure 3 (b) illustrates that CACR protocol can also construct the path of data rely form source to nanosink, where pentagon, circle, diamond and dot denote nanosink, nanomachine, data source and nodes chosen to relay data packet, respectively. Since two intermediated nodes have the same concentration of indicators, there are two paths constructed and the redundant path can ensure the reliability of data rely. Since signal collision can increase the packet loss ratio and several intermediated nodes may have the same concentration of indicators, CA routing protocol establishes more than one path to relay data packet. While based on channel contention to choose next-hop node, CACR routing protocol can establish only one optimal path to relay packet.

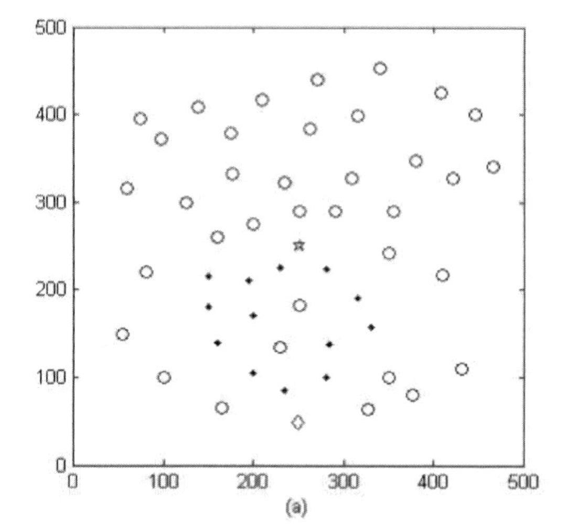

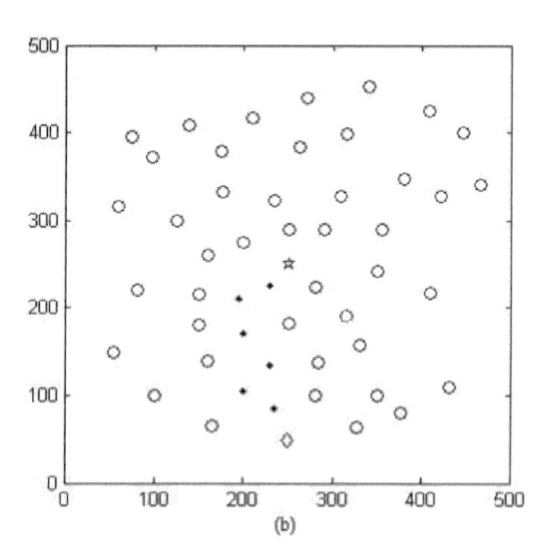

Figure 3. Operational result of CA (a) and CACR (b) protocol.

2 Performance of protocol

CA and CACR algorithm can choose a small number of nodes to relay data packet, although the establishment of a relay path needs a relatively more packet (e.g., feedback, confirmation packet).

Through executing 50 times of CA/CACR protocols, respectively, Figure 4 shows that CACR protocol is more appropriate for nanonetworks, and the advantage of reducing network load becomes more obvious with the increasing network scale.

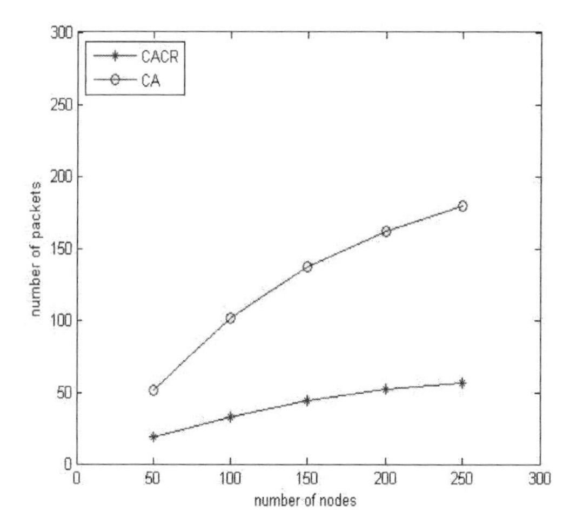

Figure 4. Average network load of CA and CACR protocol.

5 CONCLUSION

At the early research stage of molecular communication nanonetworks, the effective methods for developing communication protocols need to be found, which should be able to meet the characteristics of both nanonetworks and molecular communication. In this paper, a new architecture of molecular communication nanonetworks is proposed first, and then based on characteristics of molecular channel and bio-inspired mechanisms, the concentration aware (CA) routing protocol, which utilizes the concentration gradient of diffused molecules to establish the path of data relay in multi-hop nanonetworks is proposed. Then, for achieving the reliable data relay, a CA based cross-layer protocol is further proposed, which concerns improvement of the data link quality, and then an optimal path of data relay in multi-hop nanonetworks is established. The simulation results show the CA routing protocol has an excellent performance and is appropriate for the molecular communication nanonetworks. In the future, based on the development of related technologies such as energy model of molecular communication, the performance of the CA routing protocol will be further verified and improved.

REFERENCES

Bossert, W. & Wilson, E. 1963. The analysis of olfactory communication among animals. *J. theoretical biology* 5(3): 443–469.

Dogu, A. 2011. Capacity Analysis of a Diffusion-based Short-range Molecular Nano-communication Channel. *Computer Networks* 55(6): 1426–1434.

Ian, F., Fernando, B. & Cristina, B. 2008. Nanonetworks: A new Communication Paradigm. *Computer Networks* 52(12): 2260–2279.

Jean, P. 2006. One and a Half Century of Diffusion: Fick, Einstein, before and beyond. *Diffusion Fundamentals* 4(1): 1–19.

Kamal, D. & Attahiru, A. 2013. Molecular communication via microtubules and physical contact in nanonetworks: A survey. *Elsevier Journal on Nano Communication Network* 4(2): 73–85.

Li, Z., Zhang, J., Cai, S., Wang, Y. & Ni, J. 2013. Review on Molecular Communication. *Journal on Communications* 34(5):152–167.

Lluis, G. & Ian, F. 2009. Molecular Communication Options for Long Range Nanonetworks. *Computer Networks* 53(16): 2753–2766.

Stephen, B. 2010. Nanoscale Communication Networks. Boston :Artech House.

Tadashi, N. 2010. In-sequence molecule delivery over an Aqueous Medium. *Nano Communication Networks* 1(3): 181–188.

Electronics, Communications and Networks IV – Hussain & Ivanovic (eds)
© 2015 Taylor & Francis Group, London, ISBN: 978-1-138-02830-2

Probabilistic state machine mining method of wireless unknown protocol

Fen Li, Weiyan Zhang* & Kuilin Tao
Computer Application Institute, China Academy of Engineering Physics, Mianyang

ABSTRACT: In order to identify and analyze wireless unknown communication protocol, based on protocol bit stream data after protocol discovery and frame segmentation, corresponding solution was proposed based on four different input data. In order to resolve the third situation which is nearest to the research group background of protocol reverse engineering, the Protocol Informatics Project was applied to the identification and analysis of wireless unknown communication protocol, and reconstructing algorithm of the protocol probabilistic state machine was designed and implemented. The experimental results show that unlayered finite state machine can be reconstructed in the condition without prior knowledge.

1 INTRODUCTION

Protocol reverse engineering refers to extracting protocol grammar, syntax and semantics, through monitoring and analyzing the network input-output, system behavior and instruction execution flow of the protocol, without depending on protocol description. Protocol reverse engineering includes two interrelated aspects: one is the identification of protocol message formats, including correctly identifying the message field layout and its functional meaning; the other is the identification of the protocol behavioral model, from a practical sense, to extract the finite state machine model consistent with the protocol source program (Tian et al. 2011). Due to the inherent complexity of the protocol, constructing its state machine model, the advanced logic model, is important for accurate and in-depth program analysis.

Depending on different analysis objects, the existing protocol reverse engineering techniques fall roughly into two categories: network trace-based and execution trace-based. A network trace-based analysis mainly focuses on sniffed network traffic, and its feasibility is: 1) the data stream of single packet sample is an example of the message format, and multiple message samples corresponding to the same message format have similarities; 2) session is the complete interactive process of protocol entities, and the state transition sequence of a protocol entity in a session is a subset of the protocol state machine. The sequential relationship of packets within the same session contains part of protocol state transition information. Among the typical work on the protocol reverse engineering of packet sequence (Beddoe 2004, Wu et al 2010, Cui et al 2007) the most representative is the PI project launched by Beddoep in 2004. It introduces sequence alignment algorithm of Bioinformatics, and attempts to analyze the configuration information of the target protocol. Literature (Wu et al 2010) proposed complete solution of automatic network protocol fuzzy testing, but it does not extract state machine information and is not available to provide the fuzzy testing of the entire packet exchange process. Cui et al (Cui et al 2007) proposed a protocol reverse scheme Discover focusing on recursive classification. It achieves initial domain division based on segmentation and sequence alignment based on a domain as an element, which is more targeted than PI project with bytes as its element.

Execution trace-based analysis mainly focuses on instruction execution sequence in the process of data analysis, and its feasibility is: 1)the process of protocol entity accepting packets is the process of packets parsing, while the expressions and attributes of protocol format can be obtained by analyzing domain boundary and domain usage mode to procedures; 2) instruction sequence of complete session can be divided into arrangement of single packet instruction sequence, and the sequential relationship between subsequences contains state transition information(Pan et al 2011). In the exiting execution trace-based work(Jcaballero et al 2007, Lin et al 2008, Wondracek et al 2008), Jcaballero et al. (2007) proposed a protocol reverse method based on dynamic taint analysis and designed a prototype system Polyglot, using off-line analysis to identify separator, locator and keyword in packet format. Literature Lin et al. (2008) proposed domain structure realization scheme AutoFormat based on taint data analysis, using heuristic strategy to identify predefined field properties and dependencies,

*Corresponding author: *zwy18@caep.cn*

but it cannot handle complex structure and unknown semantics. Wondracek et al. (2008) proposed an improved scheme based on Polyglot, which carries out semantic information integration on packets with the same format to extract packet structure with better versatility.

In this paper, a state machine reconstruction method of wireless unknown communication protocol is proposed based on one of the inputs of protocol reverse analysis as its background, which can build a probabilistic state machine of wireless unknown protocol.

2 PROBLEM DESCRIPTION AND SOLUTION

As can be seen from the research status of protocol reverse engineering, most of the research focuses on the protocol message structure, but there is little work on directly reconstructing the state machine model without depending on any design information. The existing research on protocol state machine reconstructing is mainly based on instruction execution sequence, and work based on packet sequence analysis is less, in which most focus on application protocol state machine model. In this paper, protocol reverse method based on network trace is employed to reconstruct a state machine model of wireless unknown communication protocol.

The identification and analysis of wireless unknown communication protocol need to implement signal acquisition, protocol discovery and protocol identification of airborne wireless information. The first step is a signal acquisition process, the captured wireless signal is transformed to bit stream data after parameter identification, correct demodulation and decode. The second process is protocol discovery, to accurately separate data belonging to a communication process from bit stream data with noise. The third procedure is protocol identification, to analyze separated bit stream, identify real communication control fields and message formats, and reconstruct a protocol state machine model, which includes data preprocessing, frame segmentation, frame format parsing, protocol layering, protocol state analysis and a series of possible steps.

This paper focuses on wireless unknown communication protocol bit stream data after protocol discovery and data frame segmentation, to address its state machine reconstruction. The analysis results of each front-end module act as the input of back-end state analysis module. Due to protocol complexity, the prior knowledge and different analysis solutions, the front-end analysis results also vary, thus proposing appropriate solutions respectively based on four different inputs.

2.1 *Ideal condition*

Assuming that the front acquisition module can obtain a data frame with a preamble with evident fixed pattern, and data frame can be accurately segmented based on a frequent pattern search algorithm to find the preamble; the boundary location and semantic of each domain can be obtained on the basis of abundant prior knowledge, to clearly divide each layer of the protocol. In this ideal condition, the data part of the corresponding layer of each frame needed can be extracted to analyze, and abstract the final finite state machine model according to the semantics of every domain.

2.2 *Less ideal condition*

Assuming that the front acquisition module can accurately segment data frame; the boundary location of every domain can be obtained based on certain prior knowledge; part of the layer can be divided according to the structure and semantic of partial domain, and the data part of the corresponding layer of each frame needed to analyze can be extracted, then the finite state machine model of identifying layer can be obtained according to following scheme.

The core idea of this program is that the convention of protocol state is determined only by certain fixed domain in a frame, and different values of this domain result in each finite state of the protocol, thus the focus of the scheme is to find the state field. Taking into account that the protocol state is limited, the variation of state field is the smallest compared with the rest of the frame, thus the search of state field can be changed to find the filed with the smallest variation in frame. Therefore the variance is introduced to find the field with the smallest variance.

Assuming that each data frame comprises N fields, and the boundary division of fields is shown in Figure 1, to separately find the variance of each field σ_i^2, where $i = 1...N$, taking the field with the smallest variance as the state field.

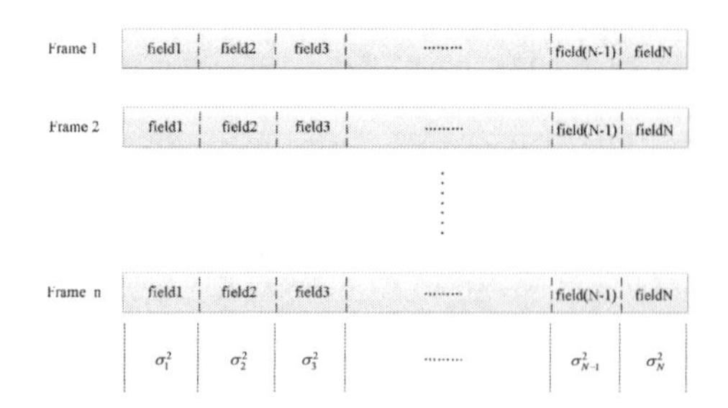

Figure 1. Analyzing protocol state machine model in less ideal condition.

Taking into account that certain field has fixed value, for example, some protocol version number and address field, whose variance is 0. Due to the noise during transmission, some bits in fixed field may also change, resulting in the variance value not being 0. Therefore the threshold is set to be a small value, the field corresponding to the minimum variance σ_i^2 greater than ε is set to be state field. After it is determined, state field with different values is set to be different protocol state, which results in the protocol state machine model.

2.3 *Not ideal condition*

Assuming that captured data frame has no preamble with evident fixed pattern, and the protocol frame boundary can only be roughly identified. If layer can be divided based on the structure and semantics of part of the field, and the data part of corresponding layer of each frame needed to analyze can be extracted, the finite state machine model of some layer can be obtained according to the following scheme; If the layer cannot be divided, the whole protocol state transition diagram can be reconstructed according to the following scheme.

1 Use clustering algorithm to cluster two classes with a minimum distance to a new class, and classify each frame to different classes by calculating the longest common substring of two frames;
2 Assign each class an ID number, thus each frame is marked by a cluster ID, and each data stream is converted to a cluster ID sequence;
3 Extract state with the same cluster ID in each stream as the same state ID;
4 Use common substring searching algorithm to find other states with a length greater than 2 in the remaining sub - stream; and remaining frame is set as a separate state;
5 Create state diagram for each stream; merge all of these figures as a finite state model; use "pull-out" and "pull-in" to extract common prefix state; emerge the two states for precursor state with only one successor state, finally obtain the simplified state diagram.

3 ALGORITHM DESIGN

In the above four different conditions, the research background of our project team of wireless unknown protocol reverse analysis is close to the third state. The front-end module can accurately segment data frame and divide each byte, thus in this paper the corresponding algorithm is proposed according to this situation.

The protocol state machine is a formal means for describing protocol, which can check a variety of errors in protocol specification. The state information provided by network protocol for intrusion detection can be utilized, and generated state diagram is compared with a state diagram defined in an RFC, in order to find the untested features. The state in the diagram is composed of the values of all variables, in most cases, for the purpose of analysis, large number of states can be divided into a small quantity of groups.

For known protocols, protocol semantic can be analyzed, and a few main states can be extracted to obtain the protocol finite state machine; against unknown protocol, only some bit stream data can be obtained rather than semantic information. Taking into account that the frame related to the same state will have similar behavior and thus similar structure, therefore frames with the same structure are clustered as one of the states. The cluster numbers from frame clustering can be used as the state numbers of finite state machine, with which as the core idea, the state machine reconstructing algorithm is obtained as shown in Figure 2.

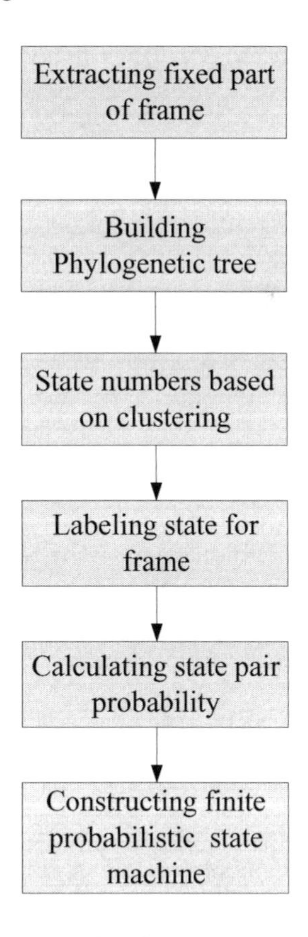

Figure 2. Algorithm flow diagram.

3.1 *Extracting protocol state*

Since the protocol state is only related to frame header, the only frame fixed part can be extracted as the analysis object according to experience. For frame clustering, PI algorithm is utilized to build Phylogenetic tree, based on which corresponding threshold and clustering numbers, as well as the state number of finite state machine are obtained.

The obtained central point of each class is regarded as state, represented by type π_i, which results in k states, $\pi_1,...,\pi_i,...,\pi_k \in S$, where S is the aggregation of all protocol states. The distance between each frame (where the i-th frame is presented by f_i) and state π_i is calculated. If meeting $\arg\min d(f_i, \pi_i)$, the i-th frame f_i is labeled by state π_i, where $i \in [1,k]$. $\Delta(C_i)$ denotes the diameter of the i-thC_i, and $d_{\max}$ can be defined as $d_{\max} = \max_{1 \le i \le k}\{2\Delta(C_i)\}$. C_h denotes the nearest cluster with frame f', if $d(\pi_h, f') > d_{\max}$, the frame f' can be assigned to an unknown state π_0.

3.2 *Constructing probabilistic protocol state machine*

The above steps label all frames in the protocol stream as a state, therefore protocol stream can be denoted as $F = (\pi_1,...\pi_h)$. The probability of every state type pair(π_i, π_{i+1}) is calculated to get the sequence of different state types and the probability from state type π_i to π_{i+1}. On the real transmission environment, the received data frame sequence may be wrong, therefore a threshold is taken to filter the state type pair out of order. In this paper, the threshold is set as ε, smaller than the probability of state type pair in probabilistic protocol state machine.

Based on the obtained state type pair, oriented diagram can be used to draw the link between every type pair, and obtains the final finite probabilistic protocol state machine based on obtained state and probabilistic transition condition between states.

4 THE EXPERIMENTAL DATA AND RESULT ANALYSIS

In order to verify the validity of the probabilistic state machine reconstructing algorithm of wireless unknown protocol, a network packet capturing tool Wireshark is utilized to capture internet traffic, with which as the algorithm input to verify the validity of this reconstructing algorithm. The experimental environment configuration is shown in Tables 1 and Table 2.

Wireshark is utilized to capture data in a complete communication process as the experimental data to verify the probabilistic state machine reconstructing algorithm of wireless unknown protocol, and ultimately get the finite probabilistic state machine with no layer division shown in Figure 3.

Table 1. The hardware configuration list.

Device name	Quantity	Configuration introduction
Lenovo K27	1	CPU: i5-2450M-2.5G, Memory: 4G, HardDisk: 500G

Table 2. The software configuration list.

Software name	Version	Software category	Testing purpose
Windows XP professional SP3	Service Pack 3	System software	Windows OS system platform
Microsoft Visual Studio 2010 Ultimate	Version 10.0.30319.1	Integrated development environment	The integrated environment testing program development and debugging

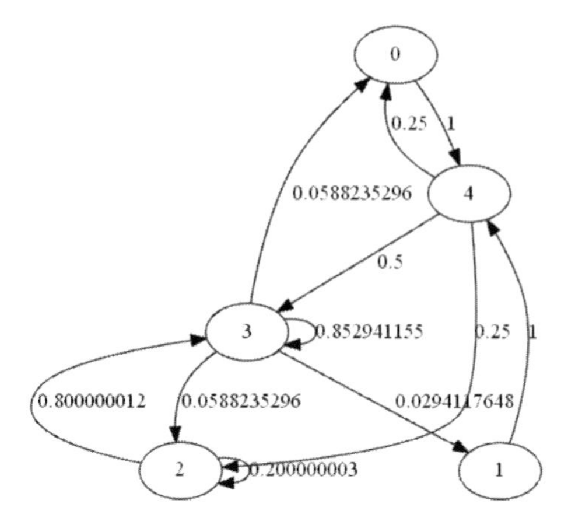

Figure 3. Constructed probabilistic protocol state machine.

5 CONCLUSION

As the network application types gradually increase, automated protocol reverse technology will get more attention, and the corresponding achievement will be widely used in intrusion detection, fuzzy testing, protocol reuse, and other areas. As can be seen from the research status, protocol reverse engineering is not a new technology, but the integration and expansion of multiple sequence alignment, grammar inference, dynamic taint analysis and data flow analysis. In this paper, based on one of the inputs of protocol reverse analysis as its background, a state machine reconstruction method of wireless unknown protocol is proposed based on PI algorithm, which can reconstruct the probabilistic state machine of unknown protocol.

As a result of analysis of unknown protocol, temporarily protocol semantic can't be obtained, thus the protocol state machine in this scheme does not contain semantic, and the core idea of it is the assumption that the frame related to the same state will have similar behavior and similar structure. Therefore, this scheme is effective to unknown protocol analysis meeting the assumption. The follow-up work will identify the semantic of unknown protocol state field, and further improve the state machine constructing algorithm of unknown protocol.

ACKNOWLEDGMENT

We are grateful to these anonymous reviewers for their helpful suggestions and comments. This work has been supported by the advanced research of China Academy of Engineering Physics.

REFERENCES

Beddoe M. Protocol informatics project [EB/OL]. (2004-10-05) http://www.4tphi.net/~awalters/PI/PI.html.

Cui Wei-dong, Kannan J & WangH J. 2007. Discoverer: automatic protocol reverse engineering from network traces. *Proc of the 16th USENIX Security Symposium*: 199–212.

Jcaballero, Yin Heng, Liang Zhen-kai, et al. 2007. Polyglot: automatic extraction of protocol format using dynamic binary analysis. *Proc of the 14th ACM Conference on Computer and Communications Security*: 317–329.

Lin Zhi-qiang, Jiang Xu-xing, Xu Dong-yan, et al.2008. Automatic protocol format reverse engineering through context-aware monitored execution. *Proc of the 15th Symposium on Network and Distributed System Security*.

Pan Fan, Wu Lifa, Du Youxiang, et al. 2011. Overview of protocol reverse engineering. *Application Research of Computers* 28(8): 2801–2806.

Tian Yuan, Li Jianbin & Zhang Zhen. 2011. Effective approach to re-engineering of protocol's state machine model. *Computer Engineering and Applications* 47(19): 63–67.

Wondracek G, Ccomparetti M P, Kruegel C, et al. 2008. Automatic network protocol analysis. *Proc of the 16th Symposium on Network and Distributed System Security*.

Wu Zhiyong, Wang Hongchuan & Sun Lechang. 2010. Survey of Fuzzing. *Application Research of Computers* 27(3):829–832.

Electronics, Communications and Networks IV – Hussain & Ivanovic (eds)
© 2015 Taylor & Francis Group, London, ISBN: 978-1-138-02830-2

A ultra-wideband antenna with tunable and frequency reconfigurable filtering characteristics

Yingsong Li*, Yuanyuan Kong & Wen Zhang
College of Information and Communications Engineering, Harbin Engineering University, Harbin, Heilongjiang, China

ABSTRACT: In this paper, an Ultra-Wideband (UWB) antenna with tunable and frequency reconfigurable filtering function is proposed and verified on the basis of the HFSS. The proposed UWB antenna possesses two notch bands that are realized by the use of a pair of Arc-L-shaped Slot (ALS) and an L-Shaped Stub (LSS). The tunable filtering function is achieved by adjusting the dimensions of the ALS and LSS, while the frequency reconfigurable filtering characteristic is obtained by controlling the ON/OFF states of the ideal radio frequency switches integrated on the proposed antenna. The experiments obtained from the HFSS showed that the proposed UWB antenna can be used as a conventional UWB antenna or a filtering UWB antenna, making it flexible and practical for UWB communication applications.

1 INTRODUCTION

To cater the rapid growing demands for high data rate communication systems, ultra-wideband technique is becoming a proper candidate to meet the high data rate requirements. Since the frequency 3.1–10.6 GHz has been allocated for commercial UWB communications by FCC in 2002, the UWB communication system has attracted much attention and been widely studied because of its many inherent advantages such as high data rate, less power consumption, good range measurement accuracy and range resolution, low system complexity and cost (Li et al. 2012a, Chu et al. 2008, Li et al. 2010, Li et al. 2013). UWB antenna is one of the key components to realize a practical UWB system (Chu et al. 2008, Li et al. 2010, Li et al. 2013). Therefore, many UWB antennas have been reported and analyzed in the recent years. However, some of these UWB antennas are embarrassed in their large size or their complex structures. To develop UWB antennas with simple structure and broad bandwidth, a great number of printed UWB antennas have been proposed and studied to meet the requirements for UWB communications (Chandel et al. 2014, Ojaroudi et al. 2012, Li et al. 2012b, Li et al. 2012c, Wu et al. 2013, Koohestani et al. 2010). Among these proposed printed UWB antennas, microstrip fed and coplanar waveguide (CPW) fed UWB antennas have been widely developed and investigated in recent years (Li et al. 2012b, Li et al. 2012c, Wu et al. 2013, Koohestani et al. 2010). However, the microstrip fed UWB antennas are difficult to integrate with the radio frequency front. Thus, the CPW-fed UWB antennas have attracted much more attention and have been widely studied. CPW-fed wide slot UWB antennas, having wide bandwidth and simple configuration, have been proposed and investigated (Koohestani et al. 2010).

Although these CPW-fed UWB antennas can provide wide bandwidth, which meets the requirements of the released FCC standard, it might interfere with the existing narrowband systems such as wireless local area networks (WLANs) and X-band signals. Thus, designing a UWB antenna with a filtering characteristic is necessary to filter these potential interferences from the WLANs and X-band narrowband systems (Li et al. 2012b, Li et al 2012c, Wu et al. 2013). In a sequence, plenty of UWB antennas with single or multiple notch bands have been proposed and investigated. However, the previously reported band-notched UWB antenna can only serve as a UWB antenna or a band-notched UWB antenna. When a UWB antenna is needed to cover the entire band of the designated UWB bandwidth, these band-notched UWB antennas are difficult to meet this requirement (Li et al. 2012b). Thus, design a UWB antenna with a reconfigurable function and wide tunable notch characteristics is desired to solve the above requirements.

In this paper, an UWB antenna with simple structure is proposed to provide a wide tunable band-notch characteristic and a reconfigurable function, which aims to deal with the above mentioned drawbacks. The designed band-notched functions are realized by the use of an ALS and an LSS, while the reconfigurable characteristics are carried out by the integration

*Corresponding author: *liyingsong@ieee.org*

of the ideal radio frequency switches. By controlling the ON/OFF states of these ideal switches integrated on the UWB antennas, the proposed antenna can act as a UWB antenna or a band-notched UWB antenna even or a multi-band antenna. The experimental results demonstrated that the proposed antenna has wide impedance bandwidth, good omnidirectional radiation patterns, and flexible reconfigurable characteristics, which is suitable for indoor UWB communications.

2 ANTENNA CONFIGURATION

Here, we will discuss the configuration of the proposed wide tunable and reconfigurable band-notched UWB antenna. Figure 1(a) describes a dual band-notched UWB antenna, which is comprised of a circular ring radiating patch, a circular shaped slot etched on the CPW ground plane, a pair of ALSs which is cut on the CPW ground to produce the X-band notch band, an LSS inserted inside of the circular ring radiating patch and a CPW-fed 50-Ohm structure together with the CPW ground plane. The proposed dual band-notched UWB antenna is printed on a substrate with a relative permittivity of 2.65, loss tangent of 0.002, and substrate thickness of 1.6mm. The 50 Ohm CPW feed structure is comprised of a CPW-fed transmission signal strip line with a width of WF=3.6mm, and the gap g between the CPW-fed transmission signal strip line and the CPW ground plane is 0.2mm. The center frequencies of the proposed notch bands can be tuned by adjusting the dimensions of the proposed ALS and LSS. In addition, three radio frequency switches, namely switch-1 (SW1), switch-2 (SW2) and switch-3 (SW3), are integrated into the proposed antenna to generate the reconfigurable characteristics and the proposed reconfigurable band-notched UWB antenna is shown in Figure 1(b). It is found that the three switches control the states of the ALS and the LSS. In this paper, the three switches are implemented by the use of the ideal switches concepts used in Li et al. (2012b). The reconfigurable function is realized via controlling the states of these three switches. To obtain the symmetrical radiation characteristics, the SW2 and SW3 are turned ON/OFF at the same time.

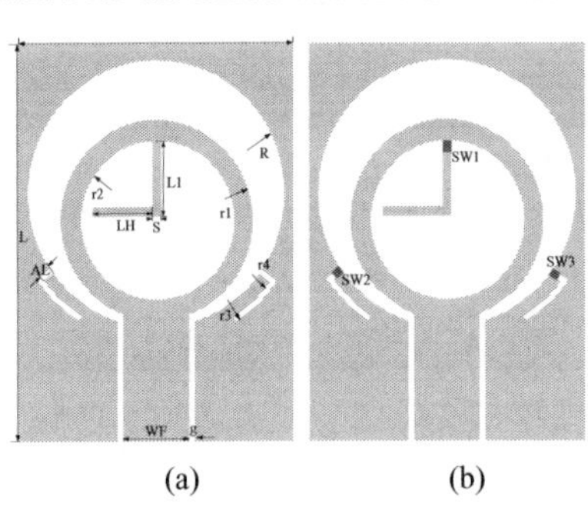

Figure 1. Configuration of the proposed antenna.

3 EFFECTS ON THE PERFORMANCE OF THE ANTENNA

In this section, we will discuss the performance of the proposed dual band-notched antenna by means of the HFSS. Figure 2 shows the effects of the ALS and LSS on the impedance bandwidth. It is found that the proposed antenna without the ALS and the LSS is a UWB antenna, which can cover the entire UWB bandwidth designated by FCC in 2002. On the other hand, the proposed antenna with only the LSS is a band-notched UWB antenna with a notch located at the WLAN band, while the proposed antenna with both the ALS and the LSS is a dual band-notched UWB antenna that can filter the unwanted signal interferences from the WLAN- and X- bands.

From the Figure 2, we can see that the two notch bands are generated by using the ALS and the LSS. Here, the parameters LH and the AL are selected to investigate the performance of the proposed antenna. Figure 3 describes the effects of LH on the impedance bandwidth of the proposed antenna. We can see that the lower notch band moves to the low frequency because of the increased resonance length. With an increment of the LH, the resonance length of the LSS is increased and hence the resonance frequency of the lower notch band shifts to the low frequency. Additionally, the center frequency of the higher notch band remains constant. However, the notch depth of the higher notch band is reduced.

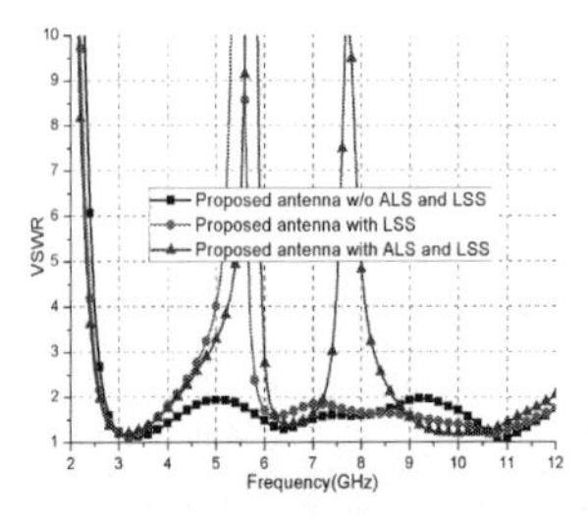

Figure 2. Effects of the ALS and LSS.

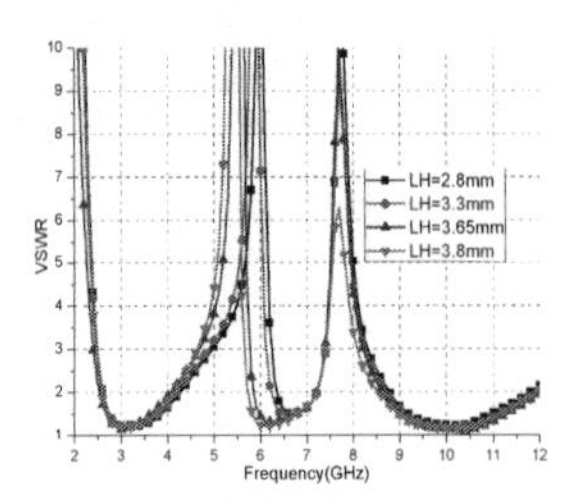

Figure 3. Effects of the impedance bandwidth with varying LH.

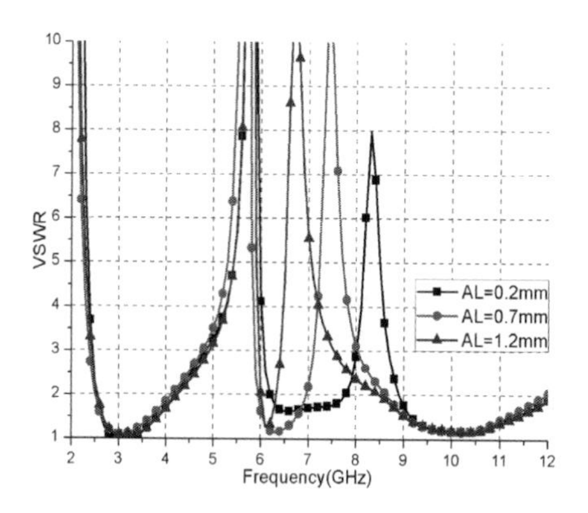

Figure 4. Effects of the impedance bandwidth with varying AL.

Figure 4 demonstrates the effects on the impedance bandwidth with varying AL. It is observed that the center frequency of the higher notch band shifts towards the low frequency with an increase of the AL, which is caused by the increased resonance length of the ALS. When the AL increases from 0.2mm to 1.2mm, the resonance length of the ALS is increased. Thus, the center frequency of the higher notch band moves to the low frequency. From the discussions above, we can say that the notch bands are generated by the use of the ALS and LSS, while the wide tunable notch characteristics are realized by adjusting the dimensions of the ALS and LSS.

4 RESULTS AND DISCUSSIONS

In this section, the frequency reconfigurable filtering characteristics and the principle of the proposed antenna are discussed to further understand this antenna. In this paper, the optimized parameters are L=32, W=24, R=11.6, r1=6.7, r2=5, r3=12.3, r4=12.5, L1=5.1, LH=3.45, S=0.8, AL=0.7 (unit: mm). Figure 5 gives the frequency filtering characteristics of the proposed antenna with different switch modes. Here, to implement these radio frequency switches, the presence of a metal bridge represents the ON state, while its absence represents the OFF state (Li et al. 2012b). It is found that the proposed antenna with all the switches OFF is a band-notched UWB antenna. In this case, the notch band located at the X-band, which aims to filter the potential interferences from X-band satellite communications. In addition, this antenna can also be used a dual band antenna. When SW1 is ON and SW2 and SW3 are OFF, the proposed antenna is a dual notch band and the two notch bands are settled at WLAN band and the X-band, respectively. Thus, the proposed antenna may filter the unwanted signals from the WLAN- and X- band. Furthermore, it can be also worked as a triple antenna. When the radio switch SW1 is OFF and the switches SW2 and SW3 are ON, the proposed antenna can be used as a UWB antenna that covers the entire band designated by the FCC. The proposed antenna is also a band-notched UWB antenna that has a notch band at the WLAN band when all the switches are ON. Additionally, the antenna is also a dual band antenna. Therefore, by controlling the ON/OFF states of the proposed three switches, the proposed antenna can be used as a dual band-notched antenna, a band-notched antenna, a UWB antenna, dual band antenna or a triple band antenna.

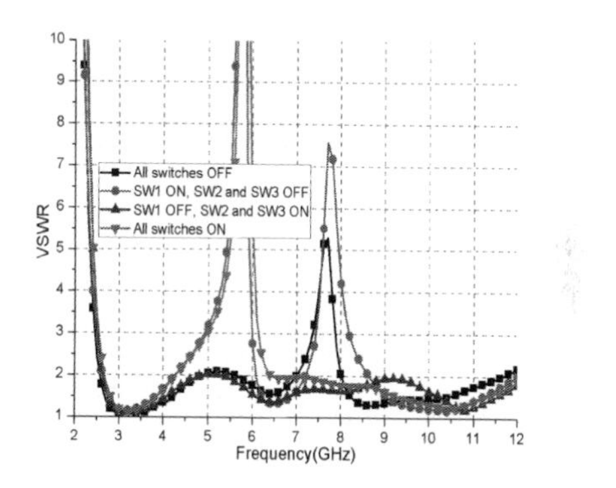

Figure 5. Frequency reconfigurable filtering characteristics.

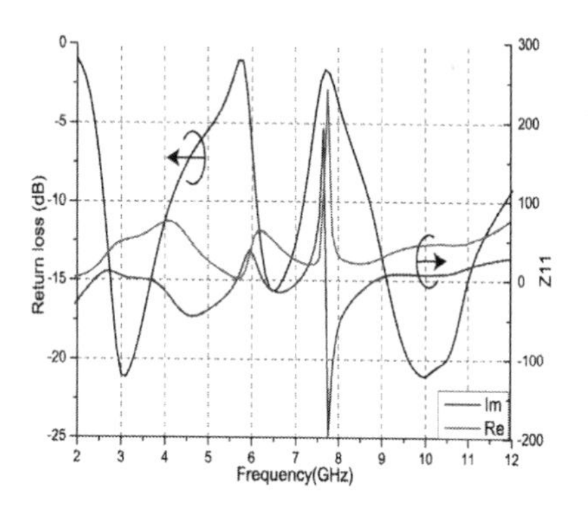

Figure 6. Impedance characteristics of the proposed antenna.

To further understand the principle of the proposed antenna, the impedance bandwidth and the current distributions of the proposed antenna are investigated by using the HFSS. Figure 6 shows the impedance bandwidth of the proposed antenna. We can see that there are two notch bands located at WLAN- and X- bands to reduce the potential narrowband interferences. Moreover, the real of the proposed antenna drops to zero at the lower notch band, while the real increase to 200-Ohm sharply at the higher notch band. Thus, the proposed antenna generates two mismatches at these two bands. This is because that the ALS and the LSS can act as resonance filters that are to filter the potential narrowband interferences.

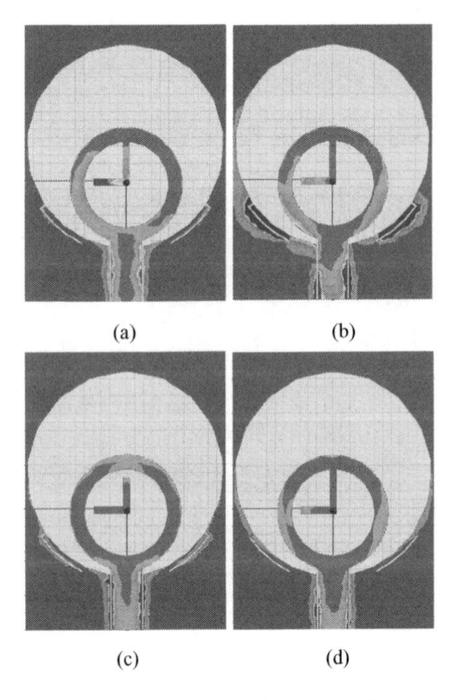

Figure 7. Current distribution of the proposed antenna.

On the basis of the discussions mentioned above, we can see that the ALS and the LSS play an important role in designing the two wide tunable notch bands, while the switches mainly affects the reconfigurable function of the proposed antenna. Here, the current distributions of the proposed antenna with SW1 ON and SW2, SW3 OFF, all the switches ON/OFF are obtained by HFSS, which are shown in Figure 7. Figures 7(a) and (b) show the current distribution with the mode of SW1 ON and SW2, SW3 OFF at 5.5 GHz and 7.9GHz, respectively. It is observed that the current mainly flows on the LSS at the 5.5 GHz band. In this case, the antenna has a strong current at the LSS, while the current on the CPW feed structure and the circular radiating patch are small. As for the 7.9 GHz, the current distribution of the proposed antenna focused on the ALS while the current distribution at the LSS is small. Thus, the two notch bands are manly affected by the dimensions of the LSS and ALS which is the same as that of the discussion in section 2. The current distribution at 5.5 GHz of the proposed antenna with all the switches OFF is shown in Figure 7(c). It can be seen that the current distribution mainly flows along the CPW feed structure and the edge of the circular ring patch, while the current on the ALS and the LSS are small. In this case, the strong current on the LSS is disappeared owing to the open switch-1, which cut off the current path of the LSS. Figure 7(d) is the current distribution of the proposed antenna at 7.9 GHz when all the switches are ON. We can see that the current on the ALSs are disappeared because of the closed switches, SW2 and SW3, on the ALS. From the analysis, we found that the band-notched principle of the proposed antenna is obtained by means of the ALS and LSS, while the tunable band-notch characteristic and the reconfigurable function are achieved by adjusting the dimensions of the ALS and the LSS and controlling the states of the integrated switches.

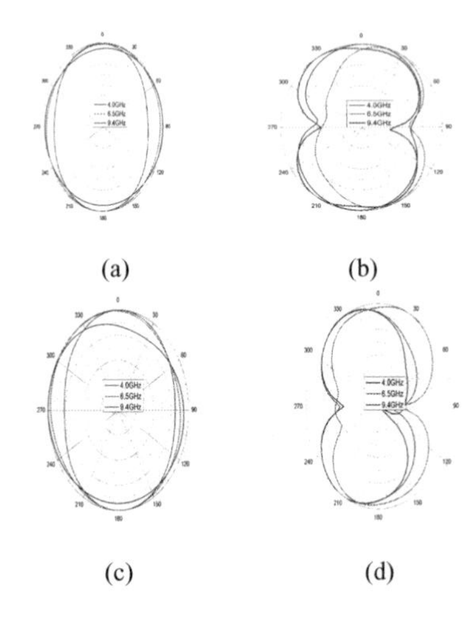

Figure 8. Radiation patterns of the proposed antenna. (a) (b) SW1 OFF and SW2, SW3 ON; (c) (d) SW1 ON and SW2, SW3 OFF.

The radiation patterns of the proposed antenna at the 4.0 GHz, 6.5 GHz and 9.4 GHz are shown in Figure 8. It is found that it has a nearly omnidirectional radiation patterns in the H-plane and a figure-of-eight radiation pattern in the E-plane, indicating that the proposed antenna is suitable for UWB communication applications.

5 CONCLUSION

In this paper, an ultra-wideband antenna with wide tunable and frequency reconfigurable filtering characteristics is proposed and analyzed by the use of the HFSS. The simulated results demonstrated that the filtering function can be realized by using the ALS and LSS, while the reconfigurable filtering function is achieved by controlling the ON/OFF states of the three ideal radio frequency switches. From the performance of the proposed antenna, we found that the antenna possesses wide bandwidth covering the entire UWB band and the antenna has a wide tunable filtering characteristic and reconfigurable filtering function to filter the undesired signals from the WLAN- and X- bands. In addition, the proposed antenna can also provide a good omnidirectional radiation patterns, making it suitable for UWB communications. In the future, we'll fabricate and design a switch network or use PIN diodes to verify the wide tunable and frequency reconfigurable filtering antenna.

ACKNOWLEDGMENT

This work was partially supported by Fundamental Research Funds for the Central Universities (HEUCFD1433).

REFERENCES

Chandel, R., Gautam, A.K. & Kanujia, B.K. 2014. Microstrip-line FED beak-shaped monopole-like slot UWB antenna with enhanced band width. *Microwave and Optical Technology Letters* 56(11): 2624–2628.

Chu Q.X. & Yang Y.Y. 2008. A compact ultra wideband antenna with 3.4/5.5 GHz dual band-notched characteristics, *IEEE Transactions on Antennas and Propagation* 56(12): 3637–3644.

Koohestani M. & Golpour M. 2010. Compact rectangular slot antenna with a novel coplanar waveguide fed diamond patch for ultra wideband applications. *Microwave and Optical Technology Letters* 52: 331–334.

Li, Y., Li, W. & Ye, Q. 2013. Miniaturization of asymmetric coplanar strip-fed staircase ultra wideband antenna with reconfigurable notch band. *Microwave and Optical Technology Letters* 55(7): 1467–1470.

Li, Y., Li, W. & Yu W. 2012b. A switchable UWB slot antenna using SIS-HSIR and SIS-SIR for multimode wireless communications applications. *Applied Computational Electromagnetics Society Journal* 27(4): 340–351.

Li, Y., Yang, X,. Liu, C. & Jiang, T. 2012c. Miniaturization cantor set fractal ultrawideband antenna with a notch band characteristic. *Microwave Opt. Technol. Lett.* 54: 1227–1230.

Li, Y.S., Li, W.X. & Ye Q.B. 2012a. Compact reconfigurable UWB antenna integrated with stepped impedance stub loaded resonators and switches. *Progress In Electromagnetics Research* 27: 239–252.

Li, Y.S., Yang, X.D., Liu, C.Y., et al. 2010. Compact CPW-fed ultra-wideband antenna with dual band-notched characteristics. *Electron. Lett.* 46: 967–968.

Ojaroudi, N., Ojaroudi, M. & Ghadimi, N. 2012. UWB omnidirectional square monopole antenna for use in circular cylindrical microwave imaging systems. *IEEE Antennas and Wireless Propagation Letters* 11: 1350–1353.

Wu A. & Guan B. 2013. A Compact CPW-fed UWB antenna with dual band-notched characteristics. *International Journal of Antennas and Propagation* 2013.

Electronics, Communications and Networks IV – Hussain & Ivanovic (eds)
© 2015 Taylor & Francis Group, London, ISBN: 978-1-138-02830-2

The influence of hidden stations on throughput of unsaturated WLANs

Fu Li*
School of Information Science and Engineering, Shandong University, Jinan, China

ABSTRACT: In the system with imbalanced traffic, hidden stations are in different states when the data is transmitted. Based on the unsaturated two-dimensional Markov chain model, this paper proposes an algorithm for calculating collision probability and unsaturated throughput of the IEEE802.11 wireless local networks (WLANs) with hidden stations. The validity and efficiency of the presented algorithm are demonstrated according to the simulation results of a specific network environment by NS2.

1 INTRODUCTION

IEEE802.11 (IEEE STD 802.11 2012) DCF (Distributed Coordination Function) protocol is the fundamental media access mechanism of the IEEE802.11 WLAN, which is based on the CSMA/CA protocol. According to the CSMA/CA protocol, the station perceives the channel before sending the data. If the state of the current channel is free, the station will send data to channel. Otherwise, the station will enter the back-off process. The station will not access channel until the size of the back-off counter is diminished to 0. In WLANs, because the transmitting power of stations is limited, the signal can only cover a limited distance (Pal & Nasipur 2011). The stations outside the coverage of the transmitting station's signal are unable to perceive the existence of the transmitting station. We can say that the transmitting station and stations outside the coverage of the signal are hidden from each other. For a system with hidden stations, hidden stations can add the potential collision threat in channel (Lee & Ikjun 2009), and then the collision probability will be increased. The hidden station problem cannot be avoided. An accurate analysis of the influence of hidden stations provides a basis for the system optimization and the improvement of the DCF protocol.

Nowadays, most existing methods for analyzing the throughput of WLANs with hidden stations only consider the saturated throughput (Kim & Lim 2009, Huang et al. 2010). These methods cannot satisfy the actual network environment because the traffic of a system is not balanced, and the queue buffer cannot have data waiting for the transfer at any time (Zheng et al 2012). A few papers put forward methods, analyzing the unsaturated performance of a system with hidden stations, but the application range of these methods is limited. The paper (Ekici & Yongacoglu 2008) assumes that the length of the queue buffer is 1 when it calculates the probability that a station has no data waiting in the queue. Because the assumption brings about severe data loss, the usefulness of the paper is limited.

Based on the existing unsaturated two-dimensional Markov chain model, the paper proposes a method for calculating the unsaturated throughput of a system with hidden stations. The method considers the different status of hidden stations to analyze the collision probability caused by hidden stations in unsaturated case. The paper calculates the unsaturated throughput and the saturated throughput of the system, and uses NS2 simulation tool to check the validity of this method.

2 MODEL ANALYSIS

Assuming the transmission channel is ideal, the system contains hidden stations, the number of total stations is Nt, which includes f hidden stations and c contending stations, the traffic is unsaturated, which means the station may have no data waiting in the queue to be sent after it completes the previous data transmission, and the capacity of the queue is large enough not to overflow on the unsaturated condition. The maximum retransmission times and the maximum back-off stage are set to be m, the size of the minimum contention window is $w0$, the back-off stage i is between 0 and m, and a denotes the parameter in the exponential distribution followed by the packet arrival interval of each station.

Based on the IEEE802.11 DCF protocol, in normal case, the station will enter the back-off process before sending the data. Let two-dimensional random process $[u(t),v(t)]$ be the back-off state of the stations at t moment, in which $u(t)$ denotes back-off stage at time

*Email: *fuliwuli@163.com*

t, and $v(t)$ denotes the size of the back-off time counter at time t. Let $b_{i,r} = \lim_{t\to\infty} P\{u(t)=i, v(t)=r\}$ be the steady state probability for the station in i back-off stage and the size of the back-off time counter be r.

The back-off state transition diagram of stations on the unsaturated conditions is shown in Figure 1. The E represents the waiting state; the ellipses in Figure 1 represent the station's back-off state. The first number in every ellipse is the station's back-off stage, the second number denotes the value of the back-off time counter in the current back-off stage, and wi is the size of contention window in the i back-off stage. The p is the collision probability, not affected by the back-off stages, p_a is the probability of packet arriving at an average slot time in the waiting state, and p_f represents the queue idle probability.

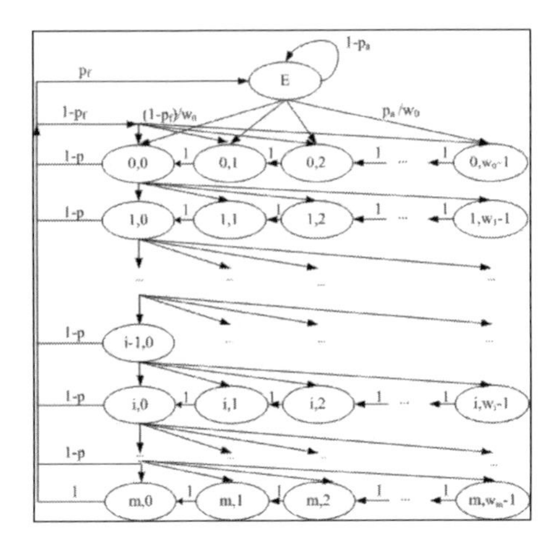

Figure 1. The back-off state transition diagram of a station on the unsaturated conditions.

The steady state probability of each state can be calculated by Figure 1 as (ZHANG et al 2011):

$$b_{0,0} = \frac{2(1-2p)(1-p)}{(1-p)(1-(2p)^{m+1})w_0 + (1-2p)(1-p^{m+1})} \tag{1}$$

$$b_{i,r} = \frac{w_i - r}{w_i} p^i b_{0,0} \tag{2}$$

$$b_E = p_f b_{0,0} / p_a \tag{3}$$

and it is constrained by:

$$\sum_{i=0}^{m} \sum_{r=0}^{w_i-1} b_{i,r} + b_E = 1. \tag{4}$$

The probability that a station sends data at any time slot can be calculated by (1), (2), (3) and (4) (ZHANG et al 2011):

$$\tau = \sum_{i=0}^{m} b_{i,0} = \frac{2(1-2p)(1-p^{m+1})p_a}{2(1-2p)(1-p)p_f + p_a[(1-p)(1-(2p)^{m+1})w_0 + (1-2p)(1-p^{m+1})]} \tag{5}$$

Let T_{suc} be the average time for successful transmitting; T_{col} is the average time for conflict; δ is the transmission delay; σ represents the physical time slot; t_{ep} is the transmission time of the packet; t_H is the time needed for transmitting the physical layer header and the MAC layer header; t_{SIFS}, t_{DIFS}, t_{RTS}, t_{CTS} and t_{ACK} are the duration time for various kinds of frame.

The paper (Zhang et al 2011) gets the probability p_a by dividing the time into time slots when the station is in the waiting state, and figures out p_f by calculating t_{js}, which denotes the average service time of grouping. Let t_I represent the average length of time slots when the station is in the waiting state, t_{BO} represents the average length of time slots when the station is in the back-off state. Because hidden stations have an impact on t_I and t_{js}, the calculation methods of p_a and p_f will be affected. Calculation expressions of t_I and t_{BO} are:

$$t_{BO} = (1-\tau)^{c-1}\sigma + [1-(1-\tau)^{c-1}][(1-p)T_{suc} + pT_{col}] \tag{6}$$

$$t_I = t_{BO} \tag{7}$$

Expressions of p_a and p_f are:

$$p_a = 1 - e^{-at_I} \tag{8}$$

$$p_f = 1 - at_{js} \tag{9}$$

where,

$$t_{js} = \{\frac{(1-p)[1-(2p)^{m+1}]}{1-2p} w_0 - \frac{p(1-p^m)}{2(1-p)} + 2^m w_0 p^{m+1} - \frac{w_0+1}{2}\} t_{BO} + [\frac{p(1-p^m)}{1-p} + p^{m+1}]T_{col} + (1-p^{m+1})T_{suc}$$

is the average packet service time.

In the unsaturated system, the station will enter the ideal state with the probability p_f when no packet is waiting in the queue. When the data arrival rate tends to be saturated, p_f tends to be zero.

The steady state probability bE representing the ideal state can be calculated by (1), (3), (8) and (9):

$$b_E = \frac{2p_f(1-2p)(1-p)}{2p_f(1-2p)(1-p)+p_a[(1-p)(1-(2p)^{m+1})w_0+(1-2p)(1-p^{m+1})]}. \quad (10)$$

For a system with hidden stations, since the contending stations within the signal range of the transmitting station can cause conflicts, the potential interference caused by hidden stations also needs to be considered. Then, the expression of the collision probability is derived respectively under the below two mechanisms.

2.1 *The expression of collision probability under the basic mechanism*

Under basic mechanism, the station transmits data directly when it senses that the channel is idle. Hidden stations may cause conflicts during the packet transmission time (Kumar &.Krishnan 2010). Let h be the number of physical time slots needed for transmitting a packet, and $h=tep/\sigma$; l be the minimum back-off stage satisfying $2_l w0\text{-}1 \geq h$. In the unsaturated system, hidden stations may be in the back-off state or the waiting state when the transmitting station starts to send the data. First, we calculate the probability that a hidden station does not affect the transmitting station when the hidden station is in the back-off state. If a hidden station's back-off time counter decreases to 0 before the end of the transmission, a collision will be caused. In such case, only if the back-off time counter's value is greater than h when the transmitting station starts transmitting data, won't the hidden station terminate its back-off process before the end of the transmission, or disturb the current transmission. So the probability that the hidden station will not cause a collision can be calculated by:

$$\begin{aligned} p_v = \sum_{i=l}^{m}\sum_{r=h+1}^{w_i-1} b_{i,r} = \frac{b_{0,0}}{2}\{p^m(w_0 2^m + \frac{h(h+1)}{w_0 2^m} \\ -(1+2h)) + \frac{w_0((2p)^l-(2p)^m)}{1-2p} \\ + \frac{h(h+1)}{w_0(1-\frac{p}{2})}((\frac{p}{2})^l-(\frac{p}{2})^m) - \frac{1+2h}{1-p}(p^l-p^m)\} \end{aligned} \quad (11)$$

Next we consider the condition that a hidden station is in the waiting state when the transmitting station starts to transmit. In this case, only if the hidden station did not have packet arrivals before the last time slot of the transmission, won't the hidden station interfere with the transmission. So the probability that the hidden station will not cause collision can be given by:

$$\begin{aligned} p_k &= b_E[1-\sum_{i=1}^{h-1}(1-p_a)^{i-1}p_a] \\ &= \frac{2p_f(1-2p)(1-p)}{2p_f(1-2p)(1-p)+p_a[(1-p)(1-(2p)^{m+1})w_0+(1-2p)(1-p^{m+1})]}(1-p_a)^{h-1} \end{aligned} \quad (12)$$

Considering the above two cases, the probability p_h that a hidden station will not cause a collision in the unsaturated system can be expressed as:

$$p_h = p_v + p_k \quad (13)$$

In conclusion, for the unsaturated system with hidden stations, the collision probability p is:

$$p = 1-(1-\tau)^{c-1}p_h{}^f. \quad (14)$$

Then, τ and p can be calculated by expression (5) and (14).

2.2 *The expression of collision probability under the RTS/CTS mechanism*

Under the RTS/CTS mechanism, a station transmits the RTS frame when the channel is detected to be ideal, then it sends data only when it receives the CTS frame from the purpose station (Kumar &.Krishnan 2010). By the CTS frame, hidden stations can know that a station outside the range they can perceive will take up the channel. In the mechanism, because hidden stations cause conflicts only when the RTS frame is transmitted, the transmitting station will not be affected when it starts transmitting data. Let k_R be the number of physical time slots needed for transmitting an RTS frame, and $k_R=t_{RTS}/\sigma$, v be the minimum back-off stage satisfying $2^v w_0\text{-}1 \geq k_R$. Let v and k_R respectively replace l and h in the expression (11) and (12), the expression of the probability that the hidden station will not cause collision under the RTS/CTS mechanism can be obtained. Then, the collision probability can be calculated.

2.3 *The throughput*

The throughput of the system can be calculated by (Pal & Nasipur 2011),

$$S = \frac{P_{tra}P_s t_{ep}}{(1-P_{tra})\sigma + P_{tra}P_s T_{suc} + P_{tra}(1-P_s)T_{col}} \quad (15)$$

where $P_{tra} = 1-(1-\tau)^{N_t}$ is the probability that there are stations transmitting packets at any time slot; $P_s = N_t\tau(1-\tau)^{c-1}(p_h)^f/(1-(1-\tau)^{N_t})$ is the probability that a packet is being sent successfully.

The T_{suc} and T_{col} are different when the station adopts different access mechanisms. Under the basic mechanism, the expressions of T_{suc} and T_{col} are:

$$T_{suc} = t_H + t_{ep} + \delta + t_{SIFS} + t_{ACK} + \delta + t_{DIFS} \quad (16)$$

$$T_{col} = t_H + t_{ep} + \delta + t_{SIFS} + t_{ACK} + t_{DIFS} \quad (17)$$

Under the RTS/CTS mechanism, the expressions of T_{suc} and T_{col} are:

$$\begin{aligned} T_{suc} &= t_{RTS} + \delta + t_{SIFS} + t_{CTS} + \delta + t_{SIFS} \\ &+ t_H + t_{ep} + \delta + t_{SIFS} + t_{ACK} + \delta + t_{DIFS} \end{aligned} \quad (18)$$

$$T_{col} = t_{RTS} + \delta + t_{SIFS} + t_{CTS} + \sigma \quad (19)$$

3 NUMERICAL AND SIMULATION RESULTS

We use NS2 (Network 2011) to verify the accuracy of the algorithm. Topology structure is set to the symmetrical loop, the receiving station is in the center, and transmitting stations are evenly located on the loop. The carrier sense range (Cai & Miao 2012) and the effective transmission range are set as a fixed value (250m). We can get different numbers of hidden stations through changing the radius of a loop. The simulation parameters adopt the parameters in protocol IEEE802.11b (IEEE80211b 1999), shown as Table1. We set the total amount of stations as 32. The amount of hidden stations is set as 0, 1 and 3.

Table 1. Parameters in the system.

Parameters	number
The total number of stations	32
W_0	32
The maximum condition window	1024
The maximum back-off stage	5
The maximum retransmission times	5
The physical layer header	192 bits
The Mac layer header	272 bits
The size of Packet	8000 bits
ACK frame	304 bits
RTS frame	352 bits
CTS frame	304 bits
Transmission delay	1 us
Length of the physical time slot	20 us
t_{SIFS}	10 us
t_{DIFS}	50 us
Data rate	11 Mbit/s
Basis rate	1 Mbit/s

For the basic mechanism and the RTS/CTS mechanism, the comparison of analysis results and simulation results are shown respectively in Figure 2 and Figure 3. Figures show that analysis results match simulation results well.

The results indicate that hidden stations can cause a throughput decline, and in a system, the more the number of hidden stations is, the less the throughput will be. They also indicate that the hidden stations decrease the saturation point of a system, based on which the throughput is approximately proportional to the packet arrival rate.

Compared with the basic mechanism, the reduction of the throughput caused by the same amount of hidden stations under the RTS/CTS mechanism is smaller. This is mainly because the RTS/CTS mechanism lowers the adverse effect brought by hidden stations, but this mechanism cannot eliminate collisions caused by hidden stations.

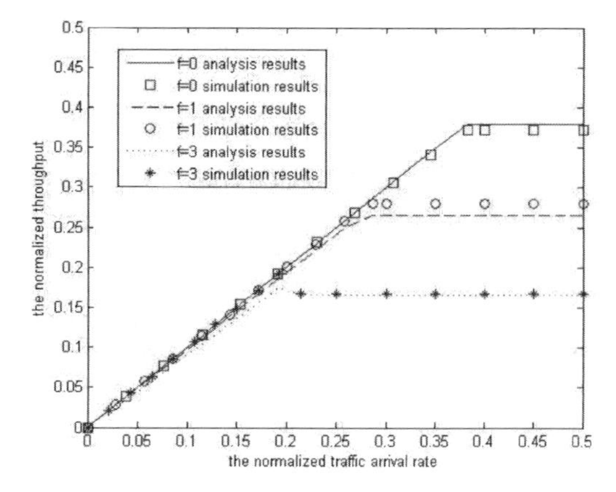

Figure 2. Basic access methods, the relationship of normalized throughput and normalized traffic arrival rate.

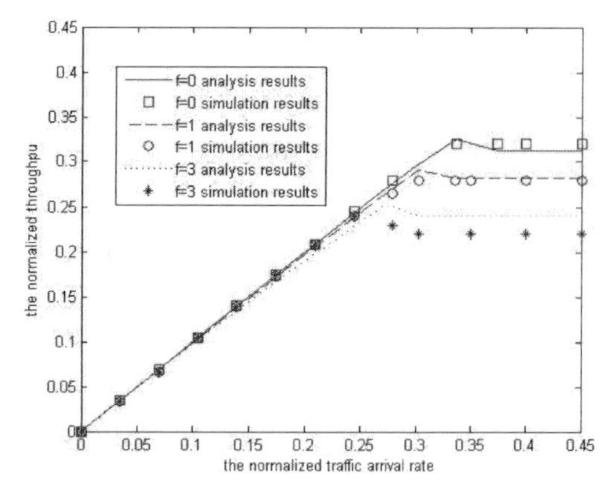

Figure 3. RTS/CTS access methods, the relationship of normalized throughput and normalized traffic arrival rate.

4 SUMMARY

In this paper, an algorithm analysis is proposed for calculating collision probability and unsaturated throughput of the IEEE802.11 WLANs with hidden stations, and it is verified with NS2 simulation. For designers and researchers of networks, the algorithm can give them a basis for the optimization of the system.

REFERENCES

Amitangshu Pal & Asis Nasipuri. 2011. Performance Analysis of IEEE 802.11 Distributed Coordination Function in Presence of Hidden Stations under Non-saturated Conditions with Infinite Buffer in Radio-over-Fiber Wireless LANs. *Local and Metropolitan Area Networks*: 1–6.

Cai Baoguo & Miao Xuening. 2012. Saturation throughput analysis of IEEE 802.11 DCF under capture effect. *2012 2nd International Conference on Consumer Electronics, Communications and Networks*: 3489–3492.

Ekici Ozgur & Yongacoglu Abbas. 2008. IEEE 80211a Throughput Performance with Hidden Nodes. *IEEE Communication Letters* 12(6):465–467.

Huang Xiaodi, Zhang Xiaomin & Zhang Haitao. 2010. Effect of Hidden Nodes on Performance of IEEE802.11 WLAN. *Collected papers of the 17th annual Information theory academic meeting of Chinese Institute of Electronics*: 387–391.

IEEE STD 802.11 2012. Wireless LAN Medium Access Control (MAC) and Physical Layer (PHY) Specifications *IEEE STD*: 818–972.

IEEE80211b. 1999. Wireless LAN Medium Access Control (MAC) and Physical Layer (PHY) Specifications: High-Speed Physical Layer Extension in the 2.4GHz Band:1–96.

Kim Taejoon & Lim Jong-Tae.2009. Throughput Analysis Considering Coupling effect in IEEE 802.11 Networks with Hidden Stations. *IEEE Communications Letters* 13(3): 175–177.

Kumar Ponnusamy & Krishnan A. 2010 Throughput Analysis of the IEEE 802.11 Distributed Coordination Function considering Capture Effects. *Third International Conference on Emerging Trends in Engineering and Technology*: 836–841.

Lee, Jinkyu & Ikjun Yeom. 2009. Avoiding Collision with Hidden Nodes in IEEE 802.11 Wireless Networks. *IEEE Communications Letter* 13(10): 743–745.

Network Simulator (NS2). (2011–11–04). http://www.isi.edu/nsnam/ns/ns-build.html.

Zhang Haitao, Zhang Xiaomin & Chen Gaoming. 2011. Performance Analysis of IEEE802.11 DCF in Non-saturated Conditions. *The International Conference on Business Management and Electronic Information* 4(1):495–498.

Zheng qin, Xiano JinSheng & Xie Honggang. 2012. A novel model for non-saturated performance analysis of IEEE802. 11 DCF. *2012 2nd International Conference on Consumer Electronics, Communications and Networks. IEEE*: 1552–1556.

Electronics, Communications and Networks IV – Hussain & Ivanovic (eds)

Analysis of online user's behavior: A case of leading e-business websites in China

Huang Li*, Qiujian Lv & Jun Liu
School of Information and Communication Engineering, Beijing University of Posts and Telecommunications, Beijing, China

ABSTRACT: As e-business gains mainstream popularity, this field has drawn increasing attention. Analysis of user's behaviors can bring e-business Service Providers (SPs) a better understanding of user's behavior pattern, which is of great importance for the growth of e-business. In this paper, we address three basic questions to depict the behaviors of e-business website users. To solve these questions, a Hadoop-based analysis system is designed to delve the real-world big data of two leading e-business websites (website T and J) collected from a northern city of China. This system enables to provide the quantitative pictures of the user's access and purchase behaviors, including purchase rate and temporal distribution. Furthermore, our study reveals the internal and external factors that determine the user's behaviors and provides three possible improvements from the perspective of e-business SPs.

KEYWORDS: user's behavior; Hadoop; big data; e-business.

1 INTRODUCTION

E-business has been rocketing these years in China. According to the data monitoring report of Chinese e-business market, Chinese e-business market transaction amount has reached 10.2 trillion RMB by the end of 2013, which is an increase of 29.9% over the year 2012. Moreover, the number of Chinese online shopping users has reached 312 million by the end of 2013.

With the rapid growth of e-business market and the huge amount of users, the study of user's behaviors has become an interesting topic and greatly draws our attention. This research is motivated by our desire to find out the patterns of users' browsing e-business websites. In particular, we are interested in three questions: How do users access e-business websites? How many individuals eventually purchase items? What is the temporal distribution of the user's access and purchase behavior in one day? To answer the above questions, we collect data from a northern city of China on 2014.03.11, and analyze the behavior pattern with the real-world data of two leading e-business websites (website T and J). Due to the humongous size, it is inefficient to handle the collected data in traditional ways. Thus, a Hadoop-based system is developed to solve this big data problem.

The main purpose of this paper is to investigate the factors that affect the user's behaviors and provide possible improvements from the perspective of the e-business service providers (SPs). Toward this end, we quantitatively analyze the access pattern, purchase pattern, and the relationships between access and purchase behavior with real-world web traffic data.

The rest of the paper is organized as follows. Section 2 outlines the dataset used for our analysis. In section 3, the analysis method is discussed. We present the results of the analysis in section 4. Related work is reviewed in Section 5. Section 6 concludes the paper and presents the future work.

2 DATA SET

In this section, we depict the approach for data set collection, followed by the description of the data format.

2.1 *Data collection*

Our study is based on the data collected by our self-developed Traffic Monitoring System (TMS), which contains hardware and software. The hardware-based probe processes at the speed of 10G/s and has the capability of traffic capturing and classification. The deployment of TMS in the network system is presented in Figure 1.

*Corresponding author: *lihdesty@126.com*

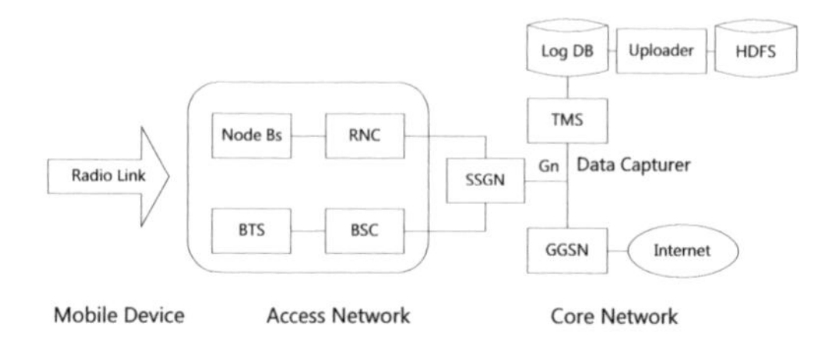

Figure 1. Deployment of TMS in the network system.

The TMS is deployed at the *Gn* interface between SGSN (Serving GPRS Support Node) and GGSN (GPRS Gateway Support Node) to capture all downlink and uplink IP packets.

2.2 *Data format*

The data set produced by TMS for the following analysis is comprised of a sequence of records, which are defined as a one-dimensional vector of six fields and these fields are respectively:

[Client IP, Request Time (s), Request Time (ms), URL, Host, Referer]

Client IP is considered as the identity of an individual. Request Time composed of Request Time (s) and Request Time (ms) represents the time when access behavior occurs. By mining the knowledge contained in URL, Host and Referer, we characterize different user's behaviors.

Moreover, we have cleaned up invalid flow logs, which have formatting errors on IP address and time.

3 ANALYSIS METHOD

In this section, we introduce the message flows when viewing e-business websites firstly. Next, the characteristics of different kinds of behaviors are presented. Then, we introduce the Hadoop-based analysis system in detail.

3.1 *Message flow analysis*

As shown in Figure 2, we split the interaction between user and e-business website into three steps: (1) access, (2) browse and (3) purchase. During the first step, the user accesses an e-business website through different source websites, such as search engine, SNS website, etc. When the user has accessed e-business website, the step (2) starts. This step includes searching, viewing the items and deciding which item to buy. The third step occurs after (2). The step (3) composes of adding the item to cart, checking out and confirming the order. In this paper, we mainly focus on the discussion of the step (1) and (3). From Figure 2, information conveyed in the request URLs enables us to determine the characteristics of user's behaviors, which will be depicted in 3.2.

3.2 *Characteristics of user's behavior*

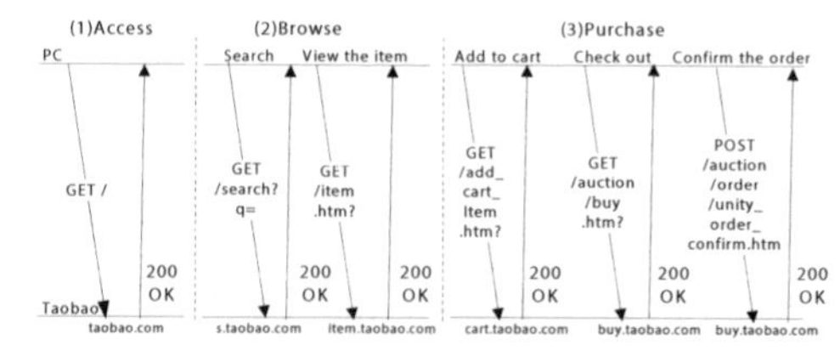

Figure 2. Message flow of user's behaviors in website T.

As shown in Figure.2, the URL in step (1) is "GET/", which means no available information exists in the request URL for the analysis of access behavior. However, the Host and Referer respectively contain knowledge about the source website (from which an individual accesses e-business websites) and target website (in this paper, the target website is website T or J), as the examples below describe.

Example1

*Host: my.**taobao.com***

*Referer: http://www.**baidu.com**/s?wd=taobao&rsv_spt=1&issp=1&rsv_bp=0&ie=utf-8&tn=baiduhome_pg&rsv_sug3=4&rsv_sug=0&rsv_sug1=3&rsv_sug4=296&inputT=10962*

Example2

*Host: shi.**taobao.com***

*Referer: http://**weibo.com**/mygroups?gid = 3682907912753324&wvr = 5&leftnav=1*

In example1, the Host contains "taobao.com" and "baidu.com" appears in Referer, which shows that the user accesses taobao.com from baidu.com (search engine). Example 2 indicates that the user accesses taobao.com from weibo.com (SNS website), as the Host displays "taobao.com" and "weibo.com" is included in the Referer.

Due to the fact that the user may cancel the order in every step of the described three procedures, so the completion of the confirming the order procedure in the purchase behavior is defined as the realization of purchase behavior. Information contained in the URL and Host of the confirming the order procedure represents an users' purchase behavior, as the example3 describes.

Example3

*URL:POST /auction/order/**unity_order_confirm**.htm*

*Host: buy.**taobao.com***

As can be seen from example3, the appearance of "unity_order_confirm" in the URL is the realization of purchase behavior. Moreover, the Host is "buy.taobao.com", meaning the purchase behavior occurs in taobao.com.

It should be noted that the characteristics are not the same for different e-business websites. Thus, in the following, the analysis of the characteristics for website T and J are respectively depicted. Based on these characteristics, we discover the domains of source websites and the IPs of the users who have purchased items.

3.3 *Hadoop-based analysis system*

To process the collected big data, an analysis system based on a distributed platform called Hadoop (Abe et al. 2014)is designed. This platform enables the storage and processing of the huge size of the packet data efficiently.

3.3.1 *Introduction of Hadoop*

Hadoop is a software framework designed for data-intensive distributed applications. It is well known for its two main components: Hadoop Distributed File System (HDFS) and MapReduce.

The HDFS is a distributed file system designed to run on common hardware. HDFS is highly fault-tolerant and designed to be deployed on low-cost hardware. It provides high throughput to access the application data and is suitable for the applications that have large data sets (Abe et al. 2014). The architecture of HDFS is presented in Figure 3.

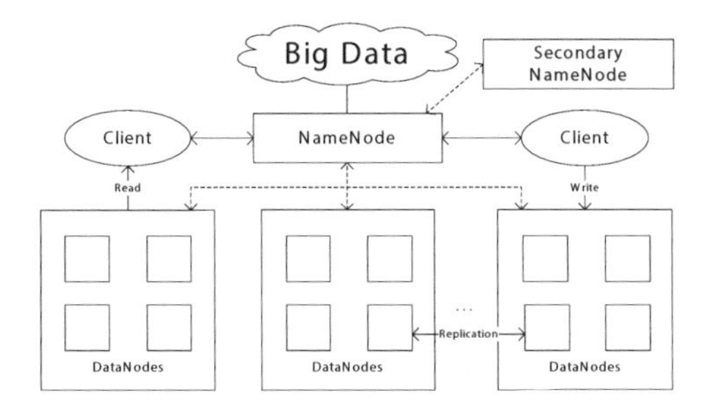

Figure 3. Architecture of HDFS.

We access the files through NameNode, which is a master server managing the file system namespace. Moreover, there are a number of DataNodes, which manage storage attached to the nodes that they run on.

MapReduce is a programming model and an associated implementation for processing and generating large data sets. Users specify a map function that processes a key/value pair to generate a set of intermediate key/value pairs, and a reduce function that merges all intermediate values associated with the same intermediate key (Dean & Ghemawat 2008).

3.3.2 *Source domain matching by RegEx expression*

The characteristics of access behavior in the Referer discussed earlier determine how customers access e-business websites. RegEx expressions are used in MapReduce programs to extract the domains of the source websites contained in the Referers. The self-designed regular expression suits three kinds of domains:

domain format 1: http://.{com,cn,net}*
domain format 2:http://.*.{com,cn,net}*
domain format 3: http://.*.*.{com,cn,net}*
("" represents any word that contains 1~20 letters)*

However, based on the regular expression, more than one domain can be extracted from the Referer if the Referer contains several domains, as the Referer below presents. So, the disturbance may be induced.

http://strip.taobaocdn.com/*tfscom/T1z1c9FX4hXXbMsGbX.html?name = itemd-sp&url =* ***http://www.163.com/****&iswt = 1&pid = tt_16143929_2253494_8923336&refpos=,n,i&adx_type = 0&pvid = 0aed0a27000053196 fa85965 0040407c_0&ps_id = d4aca751c88842be34713f3 8fad1dad1& tanxdspv =* ***http://rdstat.tanx.com****......*

Specifically, this Referer is found in the packet of access behavior from 163.com to taobao.com. Thus We add a "source domain mark" in the RegEx expression to eliminate the disturbing domains. For the above Referer, the string "url=" or "u=" is treated as the marks to locate the source domain.

Moreover, another kind of disturbed domains is the generated data when users jump from e-business website to itself. Empirically, these scenarios happen frequently. The domain of this website can be the biggest part of the matched source domain. To avoid this disturbance, we filter the source domains of website T and J.

3.3.3 *Classification of the source domains*

The classification of the matched source domains we defined is presented as follows:

Video app: apps that provide access to videos of any form.

Video website: websites that provide access to videos of any form.

Information portal website: websites that allow users to access to a wide range of contents like news, sports.

Vertical portal website: specialized websites that let users to access a specific subject area.

SNS: websites that offer Social Networking Services.

Navigate website: websites that gather multiple links on the webpage and give users the convenience of finding a website quickly.

Music app: apps that provide access to music of any form.

Entertainment website: websites that offer literal entertaining contents.

Business website: websites that contain shopping information (other than the target e-business website itself).

Search engine: websites that provide searching services.

Other: websites that are not mentioned above.

3.3.4 *Time conversion*

The time when users entering e-business websites is considered as the Request time. In the initial data, the time is presented in the structure, which specified in seconds and millisecond separately. We convert the time into UTC(Universal Time Coordinated) format.

3.3.5 *Calculation of purchase rate*

In former discussion, we depict the characteristics of purchase behavior. Using these characteristics, the users identified by the user IP who have purchased can be extracted, and the time of their access and purchase can be found out. Specially, the customer may access an e-business website in the morning but purchase at night. This kind of connection is weak and unpersuasive. Thus we exclude the IPs who purchase more than one hour after this IP has accessed the e-business website. The formula of purchase rate in every hour is presented below.

$$purchase\ rate = \frac{\sum IP_{purchased\ within\ one\ hour}}{\sum IP_{accessed\ in\ an\ hour}} \quad (1)$$

4 ANALYSIS RESULT

In this section, we present our analysis results firstly. Secondly, some available improvements that can be implemented by e-business SPs are introduced.

4.1 *User's access & purchase behaviors*

4.1.1 *Matched source domains*

By running a self-developed MapReduce program on HDFS, 1511 source domains has been extracted for website T; website J corresponds to 147 source domains. Sorting the domains by the number of accessing IPs in descending order, we select .orting g cending order bysed onchase.s and rementthe first 48 domains for website T, which account for 90% of the accessed IPs, to represent the entirety. Similarly, for website J, the first 28 domains account for 92% of the accessed IPs are selected. The selected domains are classified by our definition mentioned in 3.3.3, and the accessed IPs of those domains are used to calculate the ratio of each type of deep analysis of user behavior.

4.1.2 *Classification of the source domains*

Based on the classification discussed earlier, we create a figure presenting the ratio of classified accessed IPs for website T and J.

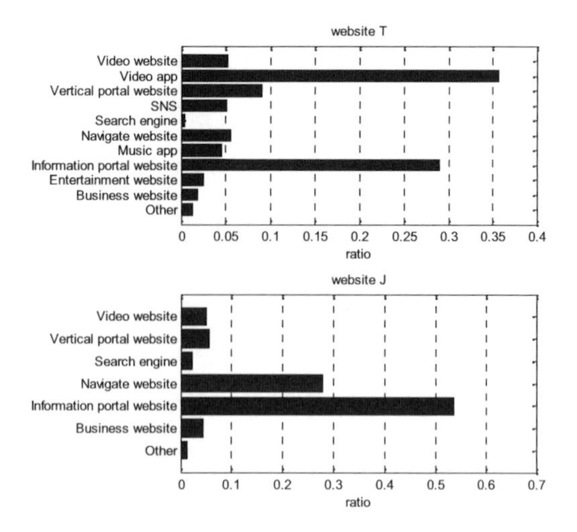

Figure 4. Ratio of the classification for website T and J.

The Figure 4 illustrates the ratio of two significant source domains for website T and J. For website T, Video app (35.72%) and Information portal website (28.88%) are shown to be the first and second biggest parts; with regard to website J, Information portal website (52.62%) and Navigate the website (27.45%) are the first two biggest parts. Other than the classifications mentioned above, all ratios of classifications for website T and J are lower than 10%.

4.1.3 *All-day purchase rate*

For website T, 24977 IPs are found accessed and 1786 IPs purchased. 4881 IPs have accessed the website J, and 201 IPs finally purchased. Figure 5 illustrates the purchased ratio of the two websites.

As can be seen in Figure 5, the purchase rate for both of these two websites are lower than 10%. It indicates that having access to e-business websites from outside links, few people do actual purchase.

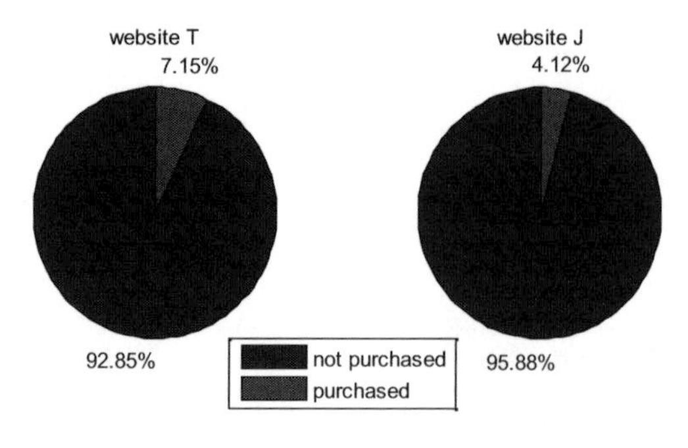

Figure 5. All day purchased rate for website T and J.

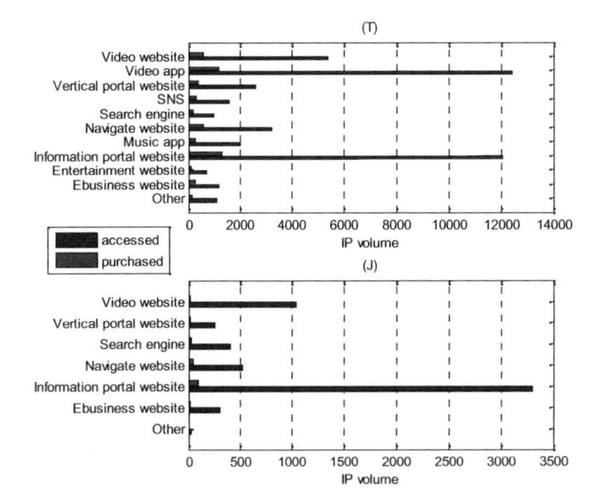

Figure 6. Accessed & purchased IP volume of different classifications.

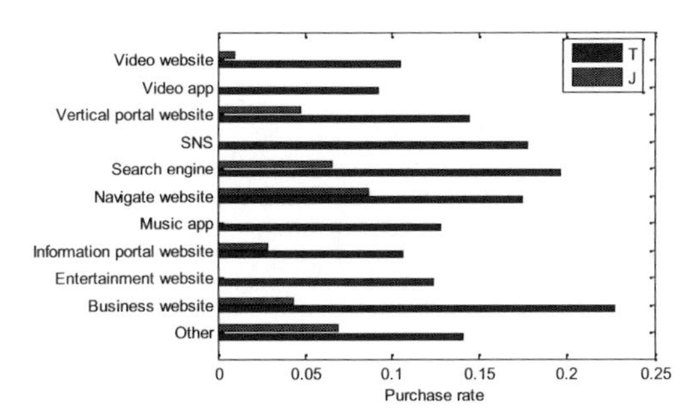

Figure 7. Purchase rate of different classifications.

4.1.4 *Classified purchase rate*

Figure 6 shows the volume of accessed IPs and purchased IPs with regard to different classifications of website T and J. Figure 7 shows the purchase rate of different classifications for the two e-business websites.

Obviously, for website T, business website and video app, respectively hold the maximum and least purchase rate. For website J, navigate a website contributes the maximum purchase rate, while video website makes up the least.

Based on the combination analysis of the Figure 6 and Figure 7, we investigate the user's access and purchase behavior in detail: Video websites and apps present the two lowest purchase rates for both website T and J (10.50%, 9.22% and 0.96%, 0%). They share the characteristic of extremely high volume of accessed IPs and low volume of purchased IPs. One reason for this phenomenon is the massive quantity of advertisements deployed on video websites and apps. For example, lots of advertisements jump out from the popup banners when opening up the app or show up when pausing the video. Another reason is the attitude of the user. When browsing the video websites or watching videos with the apps, getting information from the video other than buying items is user's main purpose. Thus, although the advertisements draw a lot of the user's attention, few people do real purchase.

Other than video website and app, although the information portal website contributes the maximum amount of accessed IPs for website J and second largest number of accessed IPs for website T, this kind of website has the lowest purchase rate (10.61% and 2.82%). It indicates that the ability to draw attention is indeed effective, similar to video websites and apps. Moreover, due to the massive amount of other information, the user's purposes are not always purchasing items.

Music app, entertainment website and vertical portal website present similar purchase rates for website T (12.81%, 12.43%, 14.45%). The amount of access IPs from the Music app and entertainment website to website J are both almost zero. The possible reason is that there were few links of website J on music apps and entertainment websites. Due to the fact that less information is conveyed from the vertical portal websites than from the information portal websites, there are smaller amount of accessed IPs and purchased IPs for vertical portal websites.

Business website has the most purchase rate (22.72%) for website T, but for website J, it is the third-last purchase rate (4.35%). The possible reason for this phenomenon is website T has lots of friendly business websites that provide links to website T. Moreover, these outside websites are the source where users start to purchase a certain item, thus a high purchase rate is achieved.

SNS, navigate website, search engine provide high purchase rate for both website T and J. For SNS websites, the information (such as discount information, group purchase information, etc.) provided by friends and other link provider is more effective than other ways and create a desire for the goods, which leads to the promotion of purchase rate. The high purchase rate of navigating website and search engine is due to the fact that users have the firm purpose to purchase certain items.

In summary, the user's attitude towards online browsing and the layout of the webpage links are two elements that effectively affect the user's purchase behavior. Respectively, the user's attitude is generated from the users themselves, like purchase desire, while webpage link layout is provided not by users, such as the position of the advertisement. We will discuss these factors in 4.3.

4.2 *Temporal distribution*

4.2.1 *Temporal distribution of normalized access & purchase volume*

Firstly, we discuss the access volume of a day over time, followed by the discussion of the purchase volume. The IP normalized volume is calculated hour-by-hour, and respectively for website T and J. The formula of IP normalized volume is presented below:

$$IP\ normalized\ volume = \frac{IP\ volume\ within\ one\ hour}{Total\ IP\ volume} \quad (2)$$

The Figure 8 presents the accessed IP normalized volume of website T and J. As is shown in the figure, the temporal distribution of both websites present a similar trend that is stable at a low level before 6:00; starts growing at about 6:00 and becomes stable at a high level at about 9:00; declining at about 21:00. In the period of 9:00 to 21:00, the average volume of the accessed IPs shows difference for these two websites and is about 25000 for website T and 3000 for website J. However, the normalized volume in this period is almost the same.

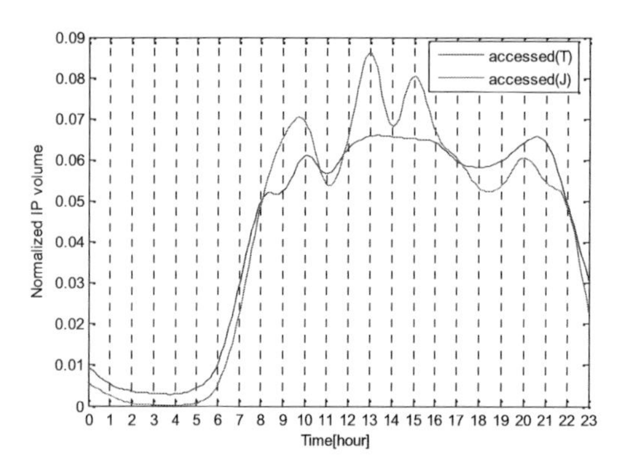

Figure 8. Accessed IP volume over time for website T and J.

The Figure 9 presents the purchased IP normalized volume of website T and J. The trends of both web site's purchased IP volume are similar to the accessed IP volume over time. Moreover, it is worth to notice that there are two peak volumes for both website T and J. The peak purchased IP volume of both websites occur at around 21:00, which illustrates most customers prefer to purchase items in this time period. For website T, the second biggest volume appears at around 13:00; website J shows its second biggest volume at around 10:00.

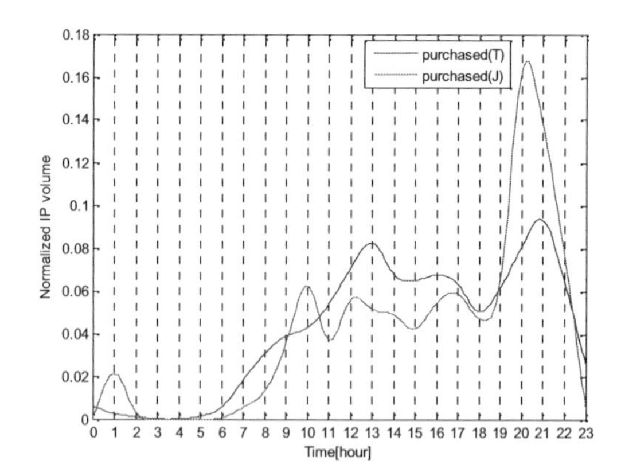

Figure 9. Purchased IP volume over time for website T and J.

4.2.2 *Purchase rate over time*

The purchase rate of website T and J over time is shown in the Figure 10. It can be observed that both the peak volume and the second biggest volume of the purchase rate for each website appears slightly earlier than the peak volume and the second biggest volume of purchased IP. For example, the peak volume of purchase rate for website T appears at around 20:00, but the peak volume of purchased IP volume appears at around 21:00.This is of great significance for SPs to predict the "convergence zone" of purchase behavior. When the purchase rate is rising up, massive incoming customers intending to purchase items can be predicted by SPs.

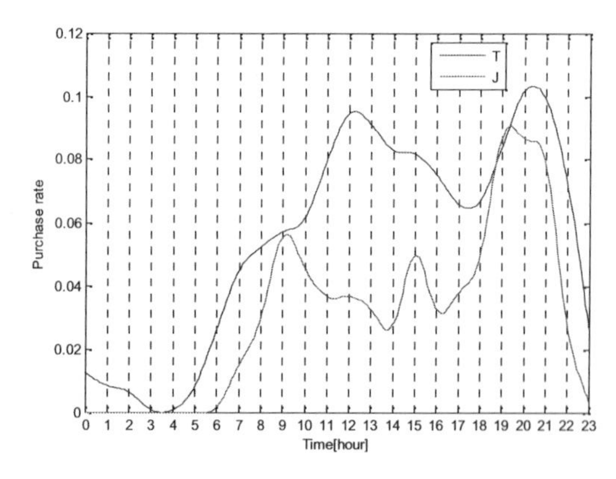

Figure 10. Purchase rate over time for website T and J.

4.3 *Analysis of e-business service improvement*

From these above results, internal and external factors that significantly determine the user's access and purchase behavior are proposed. Based on these factors, possible improvements for e-business SPs are analyzed.

4.3.1 *Internal factor*

The internal factor is the user's browsing attitude. For example, if the user has a firm purpose to buy a certain item, the purchase behavior can be displayed. However, if the user browses the websites aimlessly, the purchase behavior may not occur frequently.

Alternatively, as the links or advertisements may draw the user's attention when websites browsing, access behavior occurs frequently, whereas the purchase behavior seldom happens.

4.3.2 *External factor*

The external factor includes the webpage link layout and the time period.

The link layout plays an important role in the user's accessing process. There are many factors about link layout, such as the amount of the advertisements or the location of the link on the webpage. For example, the video app "baofeng" has a popup banner that contains advertisements for website T at a special zone, which contributes the highest volume of accessed IPs.

The time distribution provides us knowledge about what period of time draws the biggest amount of users who are willing to purchase. The peak volume of purchase rate appears at around 20:00 to 21:00. This fact is easy to be perceived. Individuals usually work in the morning and the afternoon, which leads to fewer purchase behavior. From this perception, we can speculate that most users of e-business websites are workers.

4.3.3 *Possible improvements*

In my opinion, there are three possible improvements for e-business SPs. The first one is providing personalized services for users. Such as analyzing the user's web traffic to find out the user's preferences helps altering the user's browsing attitude. Secondly, improve the quality of advertisement. To draw more user's attention, the advertisements need to be designed more attractive and appear at a better place in the web pages. Thirdly, having a good knowledge of access and purchase behavior enables to implement attractive services at the proper time. For example, starting a discount in the night when there are more potential purchasers.

5 RELATED WORK

E-business has been growing rapidly and changing business patterns over the past several years. Many researches have been done in this particular field. (Gupta & Bhushan 2012) presented an architectural framework for personalized e-business service recommendation system using data mining techniques. The impact of e-service quality, customer perceived value, and customer satisfaction on customer loyalty have been examined in an online shopping environment in (Chang & Wang 2011). Moreover, the analysis of consumer behavior is a key aspect for the growth of e-business. (Kumar & Tomkins 2010) undertook a large-scale study of online user behavior based on search and toolbar logs. In (Hostlera et al. 2011), a theoretical model has been developed to illustrate the impact of recommendation agents on online consumer behavior.

Besides, the purchasing behavior is what we concern most in this paper and several articles are related to it. (Wang & Sarwar 2011) presented a methodology in designing a post-purchase recommender system and evaluated the performance. (Hernández et al. 2010) analyzed the perceptions which induce customers to purchase over the Internet, and found out the perceptions that induce individuals to purchase online for the first time may not be the same as those that produce repurchasing behavior. (Guo & Barnes 2011) developed and tested a conceptual model of purchase behavior in virtual worlds using a combination of existing and new constructs. (Jøsang et al. 2007) gave an overview of existing and proposed systems that can be used to derive measures of trust and reputation for Internet transactions. (Banerjee & Saha 2012) mainly concerned what triggers impulse buying and how does sensory marketing aid impulse buying behavior.

6 CONCLUSION AND FUTURE WORK

In this paper, we presented an analyze system based on Hadoop platform, where HDFS and MapReduce enables the storage and parallel computing of big data sets. This system leverages the collected real-world data about two popular e-business websites to discover the relationship between user's access and purchase behavior in e-business websites. Through investigation, the analysis results answered the three major questions: How customers access e-business websites; how many of the accessed customers eventually purchase items; what is the temporal distribution of the user's access and purchase behavior in one day. We found out that individuals entering e-business websites from external links rarely end up purchasing items. In addition, different kind of websites show different purchase rate. Moreover, we provided the purchase rate trend over time, and found out the highest purchase rate appears at around 20:00 to 21:00. On the basis of these results, factors that determine the user's behavior are analyzed and we presented three possible improvements for the e-websites SPs, including providing more personalized services, attractive advertisement, and special promotion policy in proper time period.

In the future, we plan to analyze bigger data collected from different cities on our system and provide more useful insights. Moreover, we will extend our system by stepping into the realm of mobile purchasing, which has become an interesting topic nowadays.

REFERENCES

Abe S et al. 2014. Apache Hadoop. http://hadoop.apache.org/.

Banerjee, S., & Saha, S. 2012. Impulse buying behaviour in retail storesn r behaviour in rrviou Asia Pacific Journal of Marketing & Management Review 1(2): 1–21.

Chang H.H., & Wang H.W. 2011. The moderating effect of customer perceived value on online shopping behaviour. Online Information Review 35(3): 333 - 359.

Dean J. & Ghemawat S. 2008. MapReduce: simplified data processing on large clusters. Communications of the ACM 51(1): 107–113.

Guo Y. & Barnes S. 2011. Purchase behavior in virtual worlds: An empirical investigation in Second Life. Information & Management 48(7): 303–312.

Gupta P. & Bhushan B. 2012. Data mining Techniques in E-Business. International Journal of Advances in Engineering Sciences 2(1): 13–15.

Hernández B., Jiménez J. & Martín M.J. 2010. Customer behavior in electronic commerce: The moderating effect of e-purchasing experience. Journal of Business Research 63(9–10): 964ppears

Hostlera R.E., Yoonb V.Y., Guo Z., Guimaraesd T. & Forgionne G. 2011. Assessing the impact of recommender agents on on-line consumer unplanned purchase behavior. Information & Management 48(8): 336–343.

Jøsang A., Ismail R. & Boyd C. 2007. A survey of trust and reputation systems for online service provision. Decision support systems 43(2): 618–644.

Kumar R. & Tomkins A. 2010. A characterization of online browsing behavior. Proceedings of the 19th international conference on World wide web, Raleigh, USA, 26–30 April 2010: 561–570. New York: ACM.

Wang J. & Sarwar B. 2011. Utilizing Related Products for Post-Purchase Recommendation in E-commerce. Proceedings of the fifth ACM conference on Recommender systems, Chicago, USA, 23–27 October 2011: 329–332. New York: ACM.

Electronics, Communications and Networks IV – Hussain & Ivanovic (eds)
© 2015 Taylor & Francis Group, London, ISBN: 978-1-138-02830-2

A fast traffic classification method based on SDN network

Wei Li*, Guojun Li & Xiufen Yu
Key Lab of Beijing Network Technology, School of Computer Science and Engineering, Beihang University, Beijing, China

ABSTRACT: With the development of mobile Internet, the functionality and portability of mobile device have changed greatly. Many users gradually adapt to using a variety of applications on mobile devices. Mobile applications are combined with other technologies, such as cloud computing and CDN, which make the traffic classification become more and more complex. Application traffic classification is the basis of network quality of service management, security and intrusion detection. A current classification method based on machine learning technology can identify app offline. Subjected to traditional network architecture, online classification is less efficient, and cannot meet the new business demand. This paper proposed a novel method that integrates SDN architecture and nearest application based cluster classifier, then designed a prototype system on a multi-core processor platform. A number of experiments are carried out on a real-world traffic dataset to demonstrate the effectiveness and robustness of the proposed approach.

1 INTRODUCTION

Network traffic classification plays an important role in the modern network security and management (LIU 2008) especially in the application classification field. Classification results are widely used in network planning, service quality analysis, intrusion detection, user fees and other network management. On the other hand, it can be applied to the user's behavior analysis. For service providers, they can better understand user behaviors in order to provide more personalized service to enhance user satisfaction using flow classification techniques.

Traditional application classification mainly relies on the conventional TCP/UDP port number in the early days of the Internet. With the evolution of technology and wide use of cloud computing and CDN, the amount of applications deployed on the Internet is quickly increasing and many applications adopted encryption techniques. For instance, some uses the end-to-end encryption channel such as HTTPS and SRTP. Traditional deep packet inspection technology can not deal with such situation (Zhang et al. 2013). Recently, the statistical characteristics of the flow level are mainly used in the traffic classification (Karagiannis et al. 2005, Bernaille et al. 2006, Moore & Zuev 2005). After that, the focus shifted to the methods which handling the flow features based on the machine learning technology. However, in a production environment, traffic identification's performance based on the technology is still unsatisfied.

On the other hand, the study mainly focused on the coarse-grained division of traffic, such as according to the protocols or application type. While fine-grained division research is relatively less, especially to identify the application name. Some papers (Zhang et al. 2013, Karagiannis et al. 2005) indicated that application can be identified relied on a single feature vector, but recent analysis of mobile application showed that one app usually has several behaviors. In this paper, we put forward that application can be characterized as a matrix consisted of several feature vectors.

SDN/OpenFlow (McKeown et al. 2008) is a new kind of network architecture, traditionally the packet forwarding is controlled by the switches and routers while now in this architecture, it's controlled by the SDN/OpenFlow and controller, so as to realize the separation of data forwarding and routing control. The controller can control the flow table on the OpenFlow switches through the API. With this character, we integrated the NACC as a module into the controller and realized the fast online classification avoiding classifying the identified app repeatedly.

The major contributions of this work are summarized as follows:

- We proposed an identification method called nearest application based cluster classifier (NACC), and integrate it into the controller module.
- It can put the learned classification results to the switches, so as to achieve fast identification purpose. Our experiment shows identification accuracy can reach 90%.

*Corresponding author: *liw@buaa.edu.cn*

2 RELATED WORK

2.1 *Related work of traffic classification*

In this section, we described the current research situation of traffic classification.

The purpose of the traffic classification is to classify the applications from the flows which generating from the apps. Current studies about the traffic identification mainly focused on the classification methods with the characteristics of flows based on machine learning. In the early stage, Roughan et al. (2004) first applied the k-Nearest Neighbor, linear discriminant analysis and quadratic discriminant analysis, classify the network traffic on coarse-grained level. Zuev & Moore (2005) proposed to use naïve Bayes techniques to classify network traffic based on the flow statistical features. For this method, all attributes involved in the classification are of conditional independence and obey Gaussian distribution. Later on, Auld et al. (2007) proposed a neural network model based on Bayesian methods for further improving the accuracy and stability of the classification model. Alshammari (2012) reviewed the past traffic classification methods.

Unsupervised machine learning methods do not need the priori category information of samples and have the potential to deal with the unknown applications. It applies the unsupervised clustering algorithms to categorize a set of unlabeled training samples and uses the produced clusters to construct a traffic classifier. (Bernaille et al. 2006) were the first to put forward the real-time traffic classification model using spectral clustering, k-means and Gaussian mixture model. Then (Yang 2008) proposed an improved k-means clustering algorithm and its performance is better compared with simply using k-means clustering over the traffic flows. This method doesn't need the pre-labeled training data, and the amount of clusters and traffic classes is not equal, so it can't reach the performance of the supervised classification method.

However, unsupervised clustering algorithms have problems of mapping from clusters of real applications without knowing any information about applications. (Erman et al. 2007) proposed to use a small set of supervised training set in an unsupervised approach to address the problem of mapping from clusters to real applications. But it suffers from low accuracy about predicting unknown flows. Liu (2008) proposed a new feature selection method by firstly introducing combination of entropy function. For the problem of supervised method couldn't identify the new traffic type, they carried out a semi-supervised method to classify traffic flows. Rates of detection of the experimental results show that the algorithm is better than the k-means method. In the case of small samples of tags, as unlabeled sample increases, the detection rate is on the rise.

Zhang (2013) proposed a novel nonparametric approach for traffic classification, which can improve the classification performance, effectiveness by incorporating correlated information into the classification process. The results show the traffic classification performance can be improved significantly even under the extreme difficult circumstance of very few training samples. Then Zhang (Zhang et al. 2013) proposed a new method to tackle the problem of unknown applications in the crucial situation of a small supervised training set. The proposed method possesses the superior capability of detecting unknown flows generated by unknown applications and utilizing the correlation information among real-world network traffic to boost the classification performance.

2.2 *Related work of SDN*

Qazi et al. (2013) presented a framework, Atlas, which incorporates application-awareness into a Software-Defined Networking (SDN). Atlas enables fine-grained, accurate and scalable application classification in SDN. It employs a machine learning (ML) based traffic classification technique, a crowd-sourcing approach to obtain ground truth data and leverages SDN's data reporting mechanism and centralized control. They prototype Atlas on HP Labs wireless networks and observe 94% accuracy on average, for top40 Android applications.

Traffic classification is a core problem underlying efficient implementation of network services. Kogan et al. (2013) drew from experience in classifier design for commercial systems to address this problem in SDN and OpenFlow and proposed an efficient design of packet classifiers. Their proposed abstractions and design patterns can significantly reduce requirements on network elements and enable deployment of functionality that would be infeasible in a traditional way.

3 SYSTEM MODEL

In this section, we present a framework, SSTC, as shown in Figure 1, which is short for SDN based semi-supervised traffic classifier and incorporates the traffic classification into the SDN. When designing our classification method, we are interested in such problem: application today has become more and more complex and one application can't be represented by one feature. In order to solve the classifying flow under SDN architecture, this paper introduces a NACC classifier as a controller module deployed in the controller. When a packet enters into the SDN, first it will be stored in the OpenFlow enabled switches, and then the switches will match that packet with own flow table to decide next action, if it found an entry in the table, the switch will directly do as the action told, otherwise, this packet will be sent to the controller. Classification model will deal with the unknown packet and get the class result, then deliver the flow entry according to the flow class and corresponding strategies.

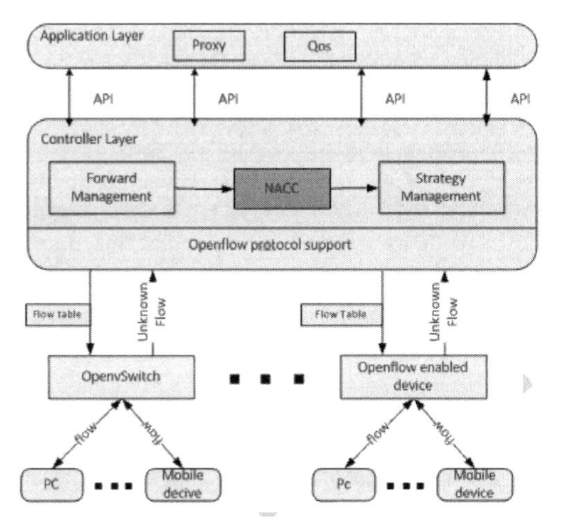

Figure 1. The SDN system model.

As shown in Figure 2, in system design, the "Input Control" module will put the flow feature vectors into the feature pool, concretely, this module will use "Receive" function to receive the unknown openflow packets from switches and get the raw packets and then "Extract flow feature vector" module extract the flow 5-tuple information, if such flow exists in the flow feature pool, put it into that queue, otherwise, create a new flow queue and put it into the feature pool.

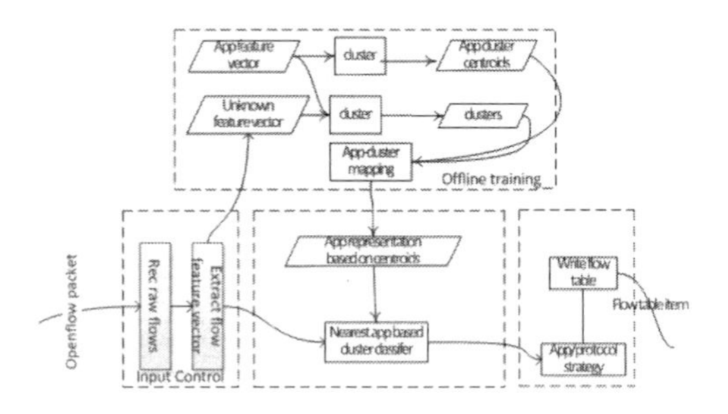

Figure 2. The structure of NACC classifier.

The NACC classifier is integrated into the SDN controller as a module and handle with unknown flow feature vector.

4 METHODOLOGY

4.1 *Nearest application based cluster classifier*

In this section, we firstly present the details of the proposed traffic classification method with unknown flow. When designing our classification method, we are interested in such problem: application today has become more and more complex and one application can't be represented by one feature. To solve this problem, our idea is to construct a matrix with several cluster centroids. That is the representation of the app.

Now we introduce notations and terminology to describe the problem. Firstly, we start with a small number of labeled flows to construct our supervised data set which is used to map app to clusters. Let $x^{(i)} = \left\{x_1^{(i)},\cdots,x_j^{(i)},\cdots,x_n^{(i)}\right\}$ is a set of flows. $x^{(i)}$ is a flow instance characterized by a vector of attribute values, such as $x^{(i)} = \left\{x_1^{(i)},\cdots,x_j^{(i)},\cdots,x_n^{(i)}\right\}$. Where n is the number of attributes, and m is the number of instances, $x_j^{(i)}$ is the value of j-th attribute i-th flow. Also, let $Y = \left\{y^{(1)},\cdots,y^{(q)}\right\}$ be the set of traffic classes, where q is the number of classes of interest. For example, $y^{(q)}$ could be classes such as "weixin", "qq", "weibo", or be " video", "pic" type in coarse-grained type. Our goal is to learn a mapping from m-dimensional variable X to Y. Along with the Erman's idea, collect the unlabeled flows in the target network for some time, remembered as $X_u = \left\{x_u^{(1)},x_u^{(2)},\cdots\right\}$, also, let $X_T = \left\{x_T^{(1)},x_T^{(2)},\cdots\right\}$ be labeled data set and then merge the data together, we get the training set X for traffic clustering, where $X = X_T \cup X_u$. Also, at the merging time, we can extend the labeled flow set by searching the correlated flows between X_u and X_L for correlated flows have same 2-tuples:{destination_ip, transport_protocal}.This method has been applied in (Zhang et al. 2013).For example, when we use weibo, it will generate traffic flows. This app having many behaviors can't be characterized by one feature. The source ip will change with user's geographical location, but the server ip will not change, that means one app's 2-tuples is the same in different time apart from the port number. If there are two flows x_a and x_b, if we know the x_a is generated by weibo, and x_a and x_b have same 2-tuples, then we can determine that x_b is also generated by weibo.

In this paper, we chose the k-means algorithm to construct the clusters due to its good performance. The k-means clustering first to partition the pure app traffic flows into k clusters. Here's how to choose k is a problem for different apps having different behaviors, so the k is not fixed. We need to determine the k value automatically. According to the experiments, the cluster number is inversely proportional to within-cluster sum of squares, when k is equal or greater than real cluster number, the sum rising trend will be slow, but once the k value is less than the real number, this index will rise sharply. So first carry out k-means with k=1,2,4,8,……, we will eventually find two value v and 2v, the index changes slow in such range, use binary search algorithm will find out best k for special pure app traffic flows data in O(2logv) time. After determining the best k value, the k-means algorithm will partition the flows into k clusters, $C = \{c_1, c_2, \cdots, c_k\}$. The cluster centroids set is the representation of app. Notice that, here the app's k value is different, so the size of the app's representation set is different. For ease of calculation, we chose $K = \max\{k\}$ while K=15. Suppose some app's cluster number is less than K, it will be filled with zero vector. For fixed k value, so as to minimize the within-cluster sum of squares:

$$\arg\min \sum_{i=1}^{k} \sum_{x_j \in C_i} \| x_j - m_i \| \tag{1}$$

Where m_i denotes the centroids of C_i. So the representation of app is:

$$y = \{c_1, c_2, \cdots, c_K\} \tag{2}$$

Input : Megered dataset X ;
Output : Semi—supervised classifier f(x)
Create k cluster $C = \{c_1, c_2, \cdots, c_k\}$ get their centroids $M = \{m_1, m_2, \cdots, m_k\}$ by k-means; get the app representation $y = \{c_1, c_2, \cdots, c_K\}$ by performing k-means on the dataset of app $y^{(j)}$.
For i=1 to k // is the number of labeled flows in cluster i
If $n_i == 0$
$c^{(i)}$ is labeled as unknown
Else
For j=1 to q
// $n_i^{(j)}$ is the number of flows labeled with $y^{(j)}$ in cluster i
Compute $P(Y=y^{(j)} \mid c^{(i)}) = n_{ij}/n_i$
$y = \arg\min_j P(Y=y^{(j)} \mid c^{(i)})$
label $c^{(i)}$ with y ;
end
end
foreach app traffic class $y^{(i)}$ representation model $M^{(i)}$
$M^{(i)} = \{m_j : c_j \in y^{(i)}\}$
Construct the NACC
$f(x) = \arg\min_j \left(\sum_{m \in y_j} \|x - m\| \right)$

Figure 3. NACC algorithm.

If k is small than K, $c_{k+1}, \ldots, c_K$ will be zero vector. Then cluster the training set and map the clusters generated by k-means to the different application-based traffic classes.

$$y = \arg\min_j \left(\sum_{m \in C_j} \| x - m \| \right) \tag{3}$$

To support application-oriented classification, we apply a probabilistic assignment mechanism to the mapping procedure. $P(Y = y^{(j)} \mid c^i)$, where $y^{(j)}$ means an app, $P(Y = y^{(j)} \mid c^i)$ denotes the probability of correctly mapping cluster c^i to the j-th traffic class $y^{(j)}$. We use the flows in X_E to estimate the probabilities. $P(Y = y^{(j)} \mid c^i) = n_{ij} / n_i$. Where n_{ij} is the number of flows that were assigned to $y^{(j)}$, n_i is the total number of labeled flows assigned to c^i.The clusters that don't have any labeled flows assigned to unknown .

So the decision function for mapping cluster is:

$$y = \arg\max_j P(Y = y^{(j)} \mid c^i) \tag{4}$$

That means all flows in c^i will be labeled as j, classified into j-th traffic class.

After the app-cluster mapping procedure, the key point to generate the application representation method, we have discussed before, for each application, we use the flow feature which is also called the cluster centroid to represent the app, it can be described by a set of cluster centroids,

$$M^{(i)} = \{m_j : C_j \in y^{(i)}\} \tag{5}$$

So, the final classification rule is,

$$y = \arg\min_j \left(\sum_{m \in Y_j} \| x - m \| \right) \tag{6}$$

Then for the flows have been classified, the controller will change the rule to a flow entry and send it to the switches. Online classification means if a flow has a flow entry in the switches, it will be delivered immediately. Otherwise, the flows will send to controller to be classified. The algorithm is shown in the Figure 3.

5 EXPERIMENTAL EVALUATION

This section describes the data set used in this work and experiments.

To facilitate our work, we collected the traces from the Network Centre of Beihang University. Depending on the specific subnets traced, the collected traces are categorized as Campus and Residential Wireless Lan.

Although our classification approach uses only flow statistics, application-layer information is helpful for training the classifier and required for validating the results. Thus, we decided to collect full packet traces. But limited by the disk space and writing speed and the campus traffic speed peak can reach 980MB/s, we have to reduce our acquisition time. According to the experiment demand, we avoid the peak period to collect traffic. The device writing speed is 300MB/s and disk space is 500GB. We collect the trace for 15 minutes at On May 26, 2014 and gathered 202GB traffic. Apart from that, we also collect 101GB wireless traffic flows of Residential Wireless Lan.

The Campus traffic mainly consists of flows generated by teachers and students, and school doesn't forbidden p2p application such as utorrent, so p2p traffic makes up a large proportion of the dataset. This is the unsupervised training data set. We also collect a few of popular android application's traffic flows, in details, we build a pure traffic capture system that generates one application traffic at a time in a short period. In this way, we got the pure application traffic respectively, and merge them together. This is how our supervised training dataset come from.

From Table 1, there are almost 0.4 billion IP packets. UDP accounts for about 80.1% and TCP is 8.4%, the leftover's protocols are ICMP, GRE and so on. Campus trace is 202GB, 81.5% is UDP, 14.5% is TCP, the other is 4%;

Table 1. Campus traffic trace constitution.

Protocol/app	Percentage	Protocol/app	Percentage
HTTP	2.63%	ARP	0.74%
TCPDATA	41.48%	IPV6	0.29%
uTorrent	0.04%	ICMP	0.29%
QICQ	0.1%	DNS	0.35%
UDPDATA	51.25%	Teredo	0.01%
SSL	2.3%	NetBIOS	0.01%
Other	0.02%	Bootstrap	0.13%

While Table 2 demonstrates the Residential Wireless Lan trace, it's about 101GB, where 40.0% IP packets use UDP and 4.9% use TCP, 54.1% use GRE protocol. In our ground truth, about 52.3% IP packets use UDP and 46.2% use TCP, note that SSL and HTTPS use TCP as their carrier, 1.5% use other protocols.

In our experiment, we chose 34 features of flow characterization to describe a traffic flow, the attributes are shown in Table 3. After extracted from the data set, we got about 10.79 million flow feature vectors where there are 22000 supervised features, 106650 features of Residential WLAN trace and 10.66 million features of Campus trace.

Table 2. WLAN traffic trace constitution.

Protocol/app	Percentage	Protocol/app	Percentage
HTTP	5.52%	XMPP	0.01%
TCPDATA	37.89%	IPV6	0.09%
uTorrent	0.34%	BitTorrent	0.01%
GRE	0.92%	DNS	0.44%
UDPDATA	26.84%	Teredo	0.08%
OTHER	18%	QICQ	0.01%
eDonkey	0.03%	NetBIOS	0.03%

Table 3. Statistical feature.

Feature type	Description	Num.
Flow feature	Source_ip, source_port, dest_ip, dest_port, protocol num	5
Payload feature	Payload length in byte	10
Packet interval	Packets interval times in undirected	9
Packet feature	Packet length in byte	10

From the Figure 4, we can conclude that the TCP has taken place a large percentage in the mobile apps protocols distribution.

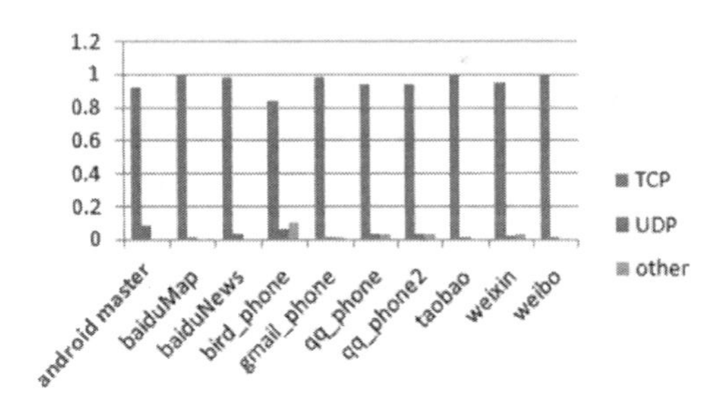

Figure 4. Mobile apps protocols distribution in supervised data set.

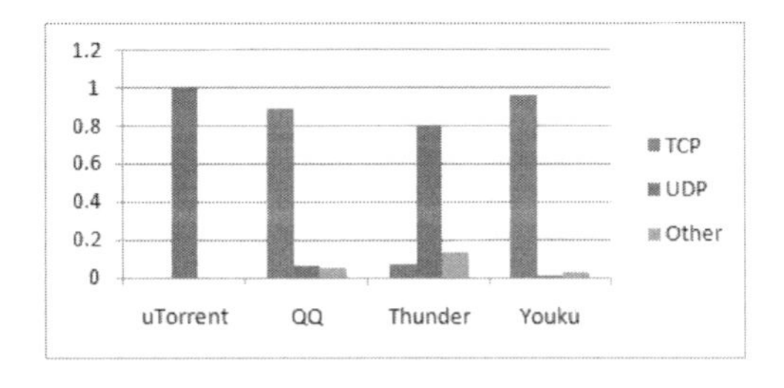

Figure 5. PC apps protocols distribution in supervised data set.

As shown in the Figure 5, not like the mobile apps, pc apps protocol usage depends on the specific application.

We select the top 15 popular apps, including the mobile apps and pc apps, in order to test the classifier's accuracy; we took 40% supervised feature vectors as our testing set, and use F-score as our metrics. F-score is calculated as below:

$$F-score = \frac{2 \times precision \times recall}{precision + recall} \tag{7}$$

Where the precision is the ratio of correctly classified flows over all predicted flows in a class and recall is the ratio of correctly classified flows over all ground truth flows in a class. F-score is used to evaluate the per-class performance.

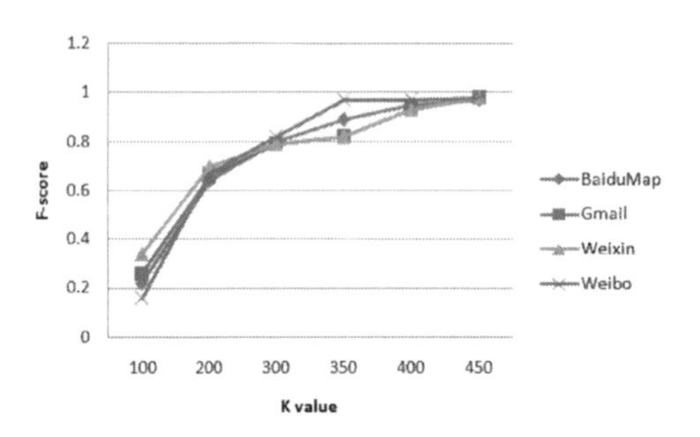

Figure 6. Accuracy of apps distribution with different k values.

Figure 6 describes the our experiments with choosing the best k value and the accuracy of apps identification with different k, the figure reports the best accuracy can be reached when k equals to 400.

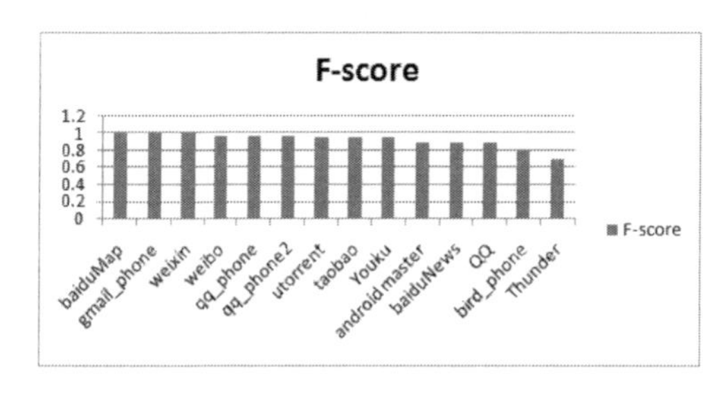

Figure 7. PC apps protocols distribution in supervised data set.

Figure7 reports the overall accuracy of 15 apps. The identification accuracy of weibo and weixin can reach nearly 100%, and QQ mobile's identification accuracy is higher than QQ on PC. Compared to Qazi's job (Qazi 2013), we got higher identification accuracy. Almost all apps identification accuracy can reach 90%, and the mean value is 92%. As can be see, our proposed approach identification accuracy can reach 90% using the training data set.

However, Thunder and Angry bird reach a low identification accuracy about 80% due to the third party ads traffic. This is because those applications traffic only accounts for a fraction of the whole. In addition, the low identification rate of thunder is caused by less usage in the campus. Compared with utorrent, which has high identification accuracy, the most important reason is that the packets are passed through in the local network and the network is stable.

There are mainly two reasons to explain this:

- Some mobile apps nearly have no ads while such apps on windows platform always have ads.
- The traffic purity of mobile traffic is higher than that on PC.

6 CONCLUSION

In this paper, we proposed a novel method called NACC for fine-grained application classification, which implemented the prototype in SDN controller to achieve online network traffic classification. Modern applications become more and more complex, however, traditional methods employ one feature vector to represent the app. Traditional methods based on one feature vector can lead to low accuracy when deal with complicated application classification. To address this problem, we put forward to utilize the application's cluster center to characterize the application. Thus, classifying traffic flows is to find out the minimum distance sum between flows feature vector and application matrix. Firstly, cluster the training set and map the cluster of the app class. Secondly, we use the flow features or cluster centroids to represent each application. In this way, identifying the flows equals to assign the flows to cluster with the highest probability. Finally, for the flows which have been classified, the controller will generate corresponding flow entry and deliver it to the switches. Online classification means that flow will be delivered immediately if it has a flow entry in the switches. Otherwise, the flow will be transmitted to controller to be classified.

In future work, we will expand our data set for further experiment. Considering enormous dataset, we can take advantage of multi-core processor to speed up the offline training. Meanwhile, we will improve the k-means on multi-core server in parallel for better performance.

ACKNOWLEDGMENT

Thanks to Beijing Municipal Education Commission for sponsoring our research project.

REFERENCES

Alshammari, R. 2012. *Automatically generating robust signatures using a machine learning approach to unveil encrypted voip traffic without using port numbers, ip addresses and payload inspection*. Halifax, Nova Scotia: Dalhousie University.

Auld, T., Moore, A. W., & Gull, S. F. 2007. Bayesian neural networks for internet traffic classification. *IEEE Transactions on Neural Networks* 18(1): 223–239.

Bernaille, L., Teixeira, R., Akodkenou, I., Soule, A. & Salamatian, K. 2006. Traffic classification on the fly. *ACM SIGCOMM Computer Communication Review* 36(2): 23–26.

Erman, J., Mahanti, A., Arlitt, M., Cohen, I. & Williamson, C. 2007. Offline/realtime traffic classification using semi-supervised learning. *Performance Evaluation* 64(9): 1194–1213.

Karagiannis, T., Papagiannaki, K. & Faloutsos, M. 2005. BLINC: multilevel traffic classification in the dark. *In ACM SIGCOMM Computer Communication Review* 35(4): 229–240.

Kogan, K., Nikolenko, S., Culhane, W., Eugster, P., & Ruan, E. 2013. Towards efficient implementation of packet classifiers in SDN/OpenFlow. *HotSDN "13 In Proceedings of the second ACM SIGCOMM workshop on Hot topics in software defined networking, New York, 16 August 2013*: 153–154. ACM.

Liu Ying-Qiu, Li Wei & Li Yun-Chun. 2008. Study of network traffic classification and application identification. *Application Research of Computers* 25(5): 1492–1495.

McKeown, N., Anderson, T., Balakrishnan, H., Parulkar, G., Peterson, L., Rexford, J. & Turner, J. 2008. OpenFlow: enabling innovation in campus networks. *ACM SIGCOMM Computer Communication Review* 38(2): 69–74.

Moore, A. W. & Zuev, D. 2005. Internet traffic classification using bayesian analysis techniques. In ACM SIGMETRICS Performance Evaluation Review 33(1): 50–60.

Qazi, Z. A., Lee, J., Jin, T., Bellala, G., Arndt, M. & Noubir, G.2013. Application-awareness in SDN. *In Proceedings of the ACM SIGCOMM 2013 Conference on SIGCOMM, New York,August 2013*: 487–488. Hong Kong. ACM SIGCOMM Computer Communication Review

Roughan, M., Sen, S., Spatscheck, O. & Duffield, N. 2004,. Class-of-service mapping for QoS: a statistical signature-based approach to IP traffic classification. In *Proceedings of the 4th ACM SIGCOMM Conference on Internet Measurement, New York, 25 October 2004*: 135–148. ACM.

Yang, C. & Huang, B. 2008. Traffic classification using an improved clustering algorithm. *International Conference on Communications, Circuits and Systems, 2008. ICCCAS 2008*. Fujian, 25–27 May 2008: 515–518. IEEE.

Zhang, J., Chen, C., Xiang, Y., Zhou, W. & Vasilakos, A. V. 2013. An effective network traffic classification method with unknown flow detection. *IEEE Transactions on Network and Service Management* 10(2): 133–147.

Zhang, J., Xiang, Y., Wang, Y., Zhou, W., Xiang, Y. & Guan, Y. 2013. Network traffic classification using correlation information. *IEEE Transactions on Parallel and Distributed Systems* 24(1): 104–117.

Electronics, Communications and Networks IV – Hussain & Ivanovic (eds)
© 2015 Taylor & Francis Group, London, ISBN: 978-1-138-02830-2

Application of DPX image in situational awareness of frequency hopping signal

Po Li*, LiNi Zhou & Huang Zhang
College of Continuing Education, National University of Defense Technology, Changsha, Hunan, China

ABSTRACT: Digital phosphor technology (DPX) has been a recently crucial innovation of the real-time spectrum analysis technology. It is widely used in electronic reconnaissance for time-varying signals, such as frequency hopping signal. As a start, the basic working theory of DPX is introduced. Then based on real-time spectrum analysis, the typical hardware system generating the DPX spectrum is given. In the following part, the spectrum situation of frequency hopping signal is illustrated respectively through the waterfall diagram and DPX image. Last but not the least, the simulation results indicate that DPX image combined with the method of time-frequency analysis can be effectively applied in situational awareness of frequency hopping signal.

1 INTRODUCTION

Frequency hopping signal with short dwell time at each frequency point and the random appearance time is not easy to be revealed and captured. In the increasingly complex electromagnetic environment, it is not only crucial and urgent, but also more difficult for electronic reconnaissance to completely capture frequency hopping signal by the higher interception probability.

At present, electronic reconnaissance often focuses on signal detection, identification and pays more attention to the detailed information of the signal spectrum. However, electromagnetic spectrum situational awareness is inadequate as a whole. In addition to signal characteristics, the changes of the electromagnetic spectrum over a period of time from the electronic reconnaissance information output are also among our concerns (Chen et al. 2011). Thus the electromagnetic spectrum awareness technology related to time adapts more to the need of modern electronic reconnaissance and the application fields is much wider.

The development of digital receiver provides an excellent platform for the improvement and application of electromagnetic spectrum situational awareness. With the rapid development of digital technology, the real-time spectrum analysis technology has also witnessed a number of significant and increasing widespread progresses (Su et al. 2007, Zhang & Ding 2007). When this technology is used in the digital receiver, the signal parameter may be measured flexibly in multiple domains according to various needs.

The commonly used electromagnetic spectrum situational awareness technology is a waterfall diagram based on the time-frequency analysis technology.

Time-frequency analysis is the most frequently employed electronic reconnaissance technology combining the spectrum analysis with the time parameter. After more than sixty-year development, the time-frequency analysis technology has gained great achievements in many aspects and has been widely applied in many fields such as signal processing (Ma 2008). "Waterfall diagram" converts two-dimensional spectrum generated by FFT transform to the three-dimensional display changing with time, and can effectively express the changes of signal in frequency domain during a continuous period of time. It contains more abundant information than simple analysis in time domain or in frequency domain, which is often applied in the study of non-stationary random signal. The waterfall diagram in the application of electromagnetic spectrum situational awareness is based on Short Time Fourier Transform (STFT) (Cabal-Yepez et al. 2012). The spectrum charts are shown at a time axis with a certain direction (such as from bottom to up or from left to right and so on) frame by frame in waterfall diagram. This technology adds time cumulative effect of multiframe spectrum and obtains the spectrum change trend over a certain time through the accumulation of time which cannot be found in a single spectrum. It is a kind of simple understanding and observation of the electromagnetic spectrum reconnaissance technology.

*Corresponding author: *lipo_lipo@nudt.edu.cn*

The DPX has been an innovative technology employed in real-time spectrum analysis recently distinguished from the waterfall diagram. The name 'Digital Phosphor' is derived from the phosphor coating on the inside of cathode ray tubes (CRTs) used as displays in televisions, computer monitors and older test equipment. DPX technology developed by the Tektronix Company applies the persistence and ratio into the real-time spectrum analysis. The DPX spectrum image can demonstrate multiple signals at different times in the same frequency range. Moreover, it can also employ intensity level, color scheme, and track statistics and other digital enhancements to highlight the variety of different information of the signal. The DPX technology can generate real-time spectrum situation image over a period of time, and greatly improve the signal acquisition and observation capabilities of the equipment, which makes it well suited for application in signal detection. The methods of digital image processing, such as the trend analysis, the statistics track detection and the image segmentation methods, can also be used to process the DPX image and realize the automatic signal detection and separation (Li et al. 2008, Guo et al. 2013c).

2 THE PRINCIPLE OF DPX AND SYSTEM DESIGN

2.1 *The principle of DPX*

The DPX image is a three dimensional spectrum that gathers all the signal spectrums during a certain time to integrate one spectrum for display as well as show the appearance probability of various signals. 3D information contained in the DPX image includes frequency, spectrum amplitude and the hit count (hit probability) in this period that can fully show a variety of characteristics of each signal. The concrete process of generating DPX includes three steps: signal digitization, graphic and display. First of all, by continuous real-time FFT operation the signal is transformed from the time domain into the frequency domain. Then the spectrums are written in the storage space and are accumulated as well. In the end, visual, dynamic picture is generated.

Figure 1 illustrates the process that generates the DPX images. The digitized data stream is divided into data records, and then FFT is performed on each record, continually producing spectral waveforms. The spectral waveforms are plotted onto a grid of counting cells called the "bitmap database", shown in Figure1a. As a dense grid, the bitmap database is created by dividing a spectrum graph. The columns represent the trace of amplitude values and rows as points on the frequency axis. A waveform contains one amplitude value for each frequency. The actual 3-D database on a DPX system for real-time spectrum analyzer may consist of many columns and rows. But a 9 x 8 matrix is used to illustrate the concept. Therefore the spectral waveforms will contain 9 x 8 points respectively.

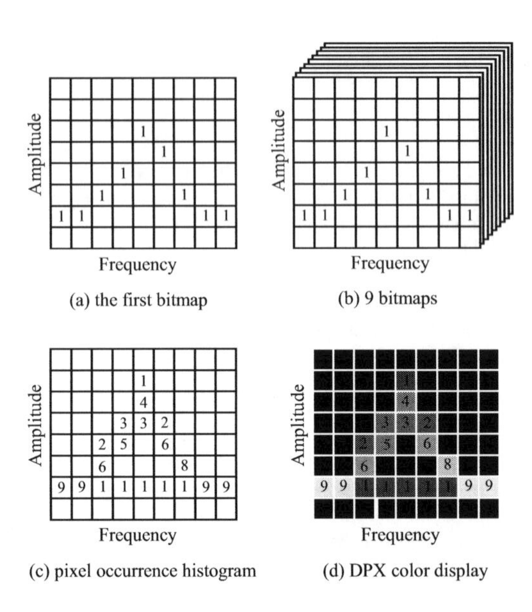

Figure 1. Example for the process of generating the DPX image.

Spectrum calculation and digitization are continuous. As waveforms are plotted to the grid, the cells boost their values each time they receive a waveform point. Each cell in this grid includes the number of times hit by the incoming spectrum. Figure1b demonstrates what the database cells might contain after a single spectrum has been mapped into it. Blank cells represent the value zero, referring to that no points from a spectrum have fallen into them yet. Values that the simplified database might contain are shown in Figure1c. After additional eight spectral transforms have been performed. The results have been stored in the cells. One of the nine spectrums happens to be sampled at a time when the signal is absent, which can be seen by the string of "1" values at the noise floor.

When the number of occurrence value is mapped to a color scale, the accumulated data turn into intuitively visual information. Many color mapping algorithms can be the potential choices and generally warmer colors (red, orange, yellow) indicate more occurrences (Guo et al. 2013b). In order to display them in the text gray mapping is used, for example, as illustrated in Figure1d.

The color displayed in the DPX images is proportional to the occurrence counts. Thus we can visually distinguish rare transients from normal signals and background noise.

It can be seen from DPX image process that DPX is related to the parameters of frequency domain, spectrum amplitude and time domain. The parameter which changes in each domain has a direct effect on DPX. DPX image demonstrates a large number of spectrums accumulated in a certain period of time and the cumulative effect is represented by bitmap colors. This method is highly effective to grasp the rule of signal changes.

2.2 *System design based on real-time spectrum analysis*

The DPX technology has been constantly making progress on the base of the real-time spectrum analysis technology. Therefore DPX can be realized through the real-time spectrum analysis system. On one hand, Figure2a. illustrates a typical block diagram of DPX system. Excluding the antenna, the system consists of the radio frequency processing module (RFPM) and the data processing module (DPM). These two hardware modules receive the signal, calculate the frequency waveform and accumulate the spectrum for DPX. The host which may be a computer or an industrial control computer, receives and displays the DPX 3-D spectrum, as well as sending out the control commands though the high-speed interface.

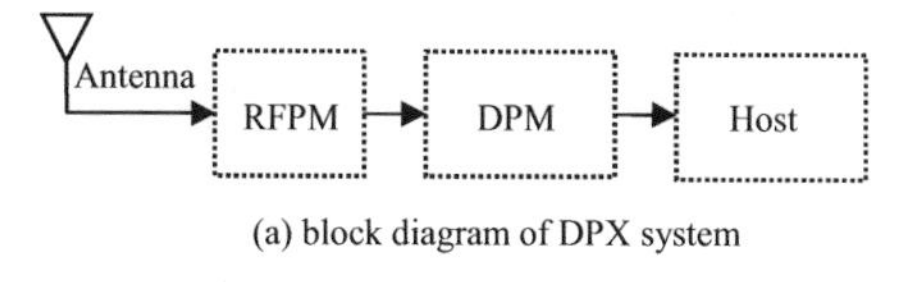

(a) block diagram of DPX system

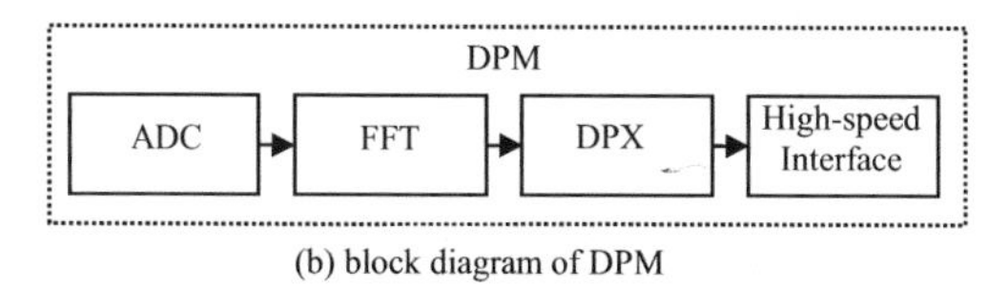

(b) block diagram of DPM

Figure 2. A topical DPX system.

On the other hand, Figure2b demonstrates the block diagram of data processing module which mainly generates DPX. After the process in RFPM the input signal enters DPM. Through ADC (Analog-to-Digital Converter) the signal is transformed into digital sampling value which is helpful to the follow-up digital signal processing. Then these data are divided into frames to do real-time flow FFT and continuous signal spectrums are obtained frame by frame. These spectral data are stored in a two-dimensional matrix bitmap in which the rows stand for frequency points and the columns are for spectrum amplitude. The stored value represents the hit count. DPM sends the data in the cache to the host through a high-speed interface at regular intervals in correspondence with the refresh rate of the display. Subsequently, the accumulated bitmap can be color matched and displayed in the host according to the preset color scheme. At the same time, DPM can also perform other tasks such as reconnaissance or measurement in a parallel way.

DPX image owns some advantages as follows: no loss in the long time of data accumulation , and visualize the data as well as make them more graphic. These merits have been applied in some electronic reconnaissance systems (Li et al. 2011, Guo et al. 2013a).

3 THE APPLICATION IN SITUATIONAL AWARENESS OF FREQUENCY HOPPING SIGNAL

Electronic reconnaissance technology concentrates on analyzing the rules of changes varying with time in a comparatively short period. Waterfall diagram and DPX technology are both based on FFT algorithm, but with different focuses. Waterfall diagram emphasizes on the change of the time while DPX technology focuses on the accumulative effect of spectrum towards the time and analyzes the trend of spectrum through statistics. DPX technology presents the rules of the changes of spectrum with time from another angle, which is a beneficial supplement to electronic spectrum situational awareness technology.

3.1 *The situational awareness of frequency hopping signal based on waterfall diagram*

Waterfall diagram has been very widely applied in electromagnetic spectrum situational awareness as it can count and demonstrate the changes of the spectrum during a certain time. Thus the capability of description especially the effective detection of frequency hopping signal is extremely strong. Waterfall diagram presents the spectrum during a certain time integrally. From the diagram it is obvious to see the existence and disappearance of frequency. Therefore, the movement of frequency hopping signal can be observed visually. Figure3a is the waterfall diagram resulting from a frequency hopping signal experiment, from which the distribution of frequency hopping signal at each point is apparent.

3.2 *The situational awareness of frequency hopping signal based on DPX image features*

DPX can calculate the spectrum in a certain time, and analyze the accumulated data as time goes by, which is remarkably adaptive in the detection of frequency hopping signals.

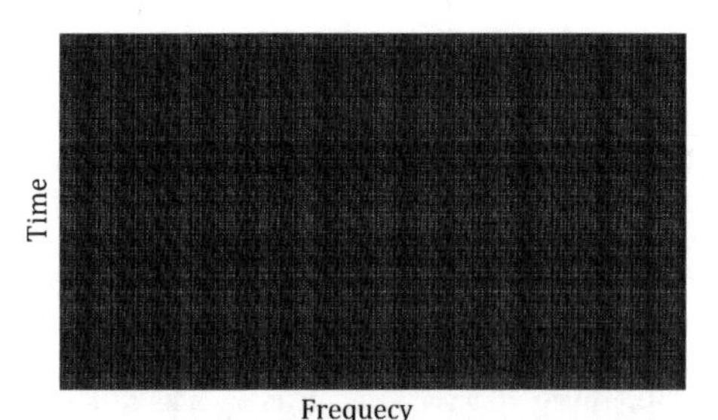

(a) A waterfall diagram of a frequency hopping signal

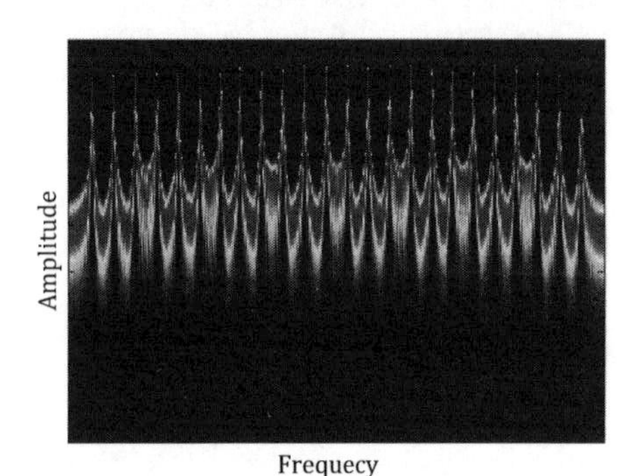

(b) A DPX image of a frequency hopping signal

Figure 3. Frequency hopping signal in waterfall and DPX.

DPX image of frequency hopping signal at certain moment is shown in Figure3b. Through the comparison of these two diagrams it can be seen that DPX concentrates the spectrum changes of frequency hopping signal during a certain time on one diagram for demonstration. From this diagram, it can be seen that the signal shares 24 uniformly distributed hopping points. Then combined with the analysis of axis, the frequency value of each hopping point can be gained. Compared with waterfall diagram, DPX presents the characteristics of hopping signals from the angle of frequency points.

3.3 *The situational awareness of frequency hopping signal integrated multi-methods*

DPX and waterfall diagram, each have their merits. If the electronic spectrum is stable, the results of the analysis through these technologies are the same as that of traditional Fourier transformation analysis. However, only when the spectrum is changing, the merits and advantages of these technologies can be reflected. Electromagnetic spectrum situational awareness faces a complex electronic magnetic environment, where electronic magnetic spectrum varie all the time. The integrated application of these technologies offers an effective method of electromagnetic spectrum situational awareness.

Take a hopping signal as an example. When the sampling rate is certain, the length of signal continual points can represent time. Therefore, in this example the continual points take the place of time concept. The single frequency of the signal continues 64sampling points, and the rule of hopping is [1; 2; 3; 4; 5; 6; 7; 8]. In a hopping period there are 8 hopping points, which hop 32 times. The purpose of the design is to highlight the non-uniformity of hopping points of hopping signals. This phenomenon is apparent in DPX. Thus, according to the energy of various hopping points from DPX, we can estimate which points exist more frequently.

In electromagnetic spectrum situational awareness, the result of detection relates to detection time, detection methods and parameters. In this example, the detection time is fixed to the end of a hopping period, FFT calculation points on DPX diagram, waterfall diagram and spectrum diagram are 128, and the data length of spectrum analysis is 128 points. The calculation result is shown in Figure4.

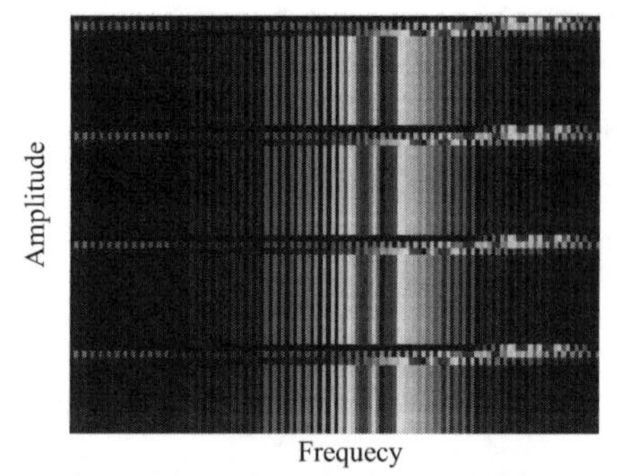

(a) waterfall diagram of the example in the text

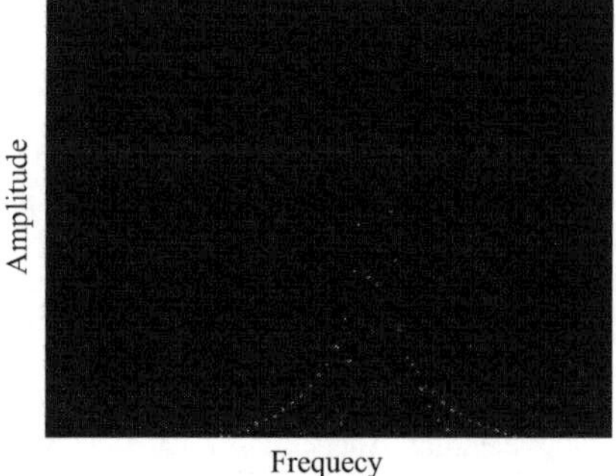

(b) DPX image of the example in the text

Figure 4. The result of the example in the text.

After analyzing DPX diagram, it can be that DPX diagram demonstrates 8 frequency points, among which the previous two frequency points are brighter. This means a high rate of existence, which conforms to the characteristics we have set at the very beginning. The waterfall diagram illustrates the frequency changes during a certain time. The hopping rule can also be seen from the diagram. However, since the calculation time contains two hopping points(which can also be seen from the spectrum diagram), only the existing rule of each two points can be estimated. At this moment, the spectrum analysis method focusing on delicate analysis of the signals, displays its advantage to estimate that the two hopping points are changing from low at first to high afterwards. Besides the lasting time for each hopping point is the same. Real electronic detection result can be obtained by integrating these technologies.

Some researchers have studied the method of analyzing hopping frequency through spectrum diagram (Guo 2008). In practice, upon the influence of the change of hopping signals, the positive analysis result can only be obtained in an ideal situation. It is suggested that integration of various kinds of analysis methods should be taken into consideration in practice. After all, the result closest to correctness can only be achieved after a sufficiently long period.

4 CONCLUSION

As a new technology DPX provides an innovative method to electromagnetic spectrum situational awareness. As has shown in this paper, DPX combined with frequency analysis method such as waterfall diagram can be used to conduct effectively situational awareness of hopping signals.

We can probe in DPX image to find more information. Further studies will focus on the method and application of digital image processing technology in automatic detection and recognition of signal. Through automatic processing of DPX diagram integrated with image processing, feature extraction and other methods, automatic recognition of hopping signals can be realized.

REFERENCES

Cabal-Yepez, E., Garcia-Ramirez, A. G., Romero-Troncoso, R.J., et al. 2012. Reconfigurable Monitoring System for Time-Frequency Analysis on Industrial Equipment Through STFT and DWT. *IEEE Transactions on Industrial Informatics* 9(2): 760–771.

Chen, Z.P., Wu, J.H., Su, S.Y., et al. 2011. The Technology and Application of Wideband Real-time Spectrum Analysis. *Signal Processing* 28(2): 152–155.

Guo, J.T. 2008. Ambiguity function of random frequency-hopping signals and its time-frequency analysis. *Computer Engineering and Applications* 44(8): 121–123.

Guo, S.J., Su, J., Sun, G., et al. 2013. Design of DPX System for Real-time Spectrum Analysis and Signal Detection. *IEICE Electronics Express* 10(19): 1–8.

Guo, S.J., Sun, G., Li, J., et al. 2013. A New Pseudo-coloring Coding Method for DPX. *2013 Fourth International Conference on Digital Manufacturing and Automation* : 54–57.

Guo, S.J., Sun, G., Li, P., et al. 2013. Application of DPX in Wideband Signal Detection. *2013 Fourth International Conference on Digital Manufacturing and Automation*: 283–286.

Li, P., Liu, W.Q., Su, S.Y., et al. 2008. Implementation of Digital Phosphor Technique in Electronic Reconnaissance. *Journal of Electronic Measurement and Instrument* 22(s2): 342–346.

Li, P., Yang, J., Zhang, Y., et al. 2011. Application of the Digital Phosphor Technology in Radar Reconnaissance. *Computer Engineering & Science* 33(4): 168–172.

Ma, Q.M., Wang, X.Y. & Du, S.P. 2008. Research of the Method for the Weak Signal Detection Based on the Amplitude Fluctuation Property of the Frequency Spectrum. *Journal of Electronics & Information Technology* 30(11): 2642–2645.

Su, S.Y., Liu P. & Chen Z.P. 2007. Design and Implementation of Wideband Real-time Spectrum Analyzer. *Journal of Electronic Measurement and Instrument* 21(5): 113–117.

Zhang, P. & Ding, Z. 2007. Application of Real-time Spectrum Analyzer in Countermeasure. *Modern Radar* 29(3): 84–89.

Electronics, Communications and Networks IV – Hussain & Ivanovic (eds)
 ISBN: 978-1-138-02830-2

An improved mechanism for semantic web service discovery

Hui Li & Yunfeng Hu
School of Economics and Management, Xidian University, Xi'an, Shaanxi, China

Jianfeng Ma
School of Computer Science and Technology, Xidian University, Xi'an, Shaanxi, China

ABSTRACT: Recently with the rapid development of web service technology, the number of web services increases dramatically. How to efficiently locate functionality-desired web services among lots of web services and how to select the best one among large functionality-similar web services become the hot topics of research. In order to discover the most optimal web services for users, a hybrid mechanism is proposed in this paper. Firstly, a list of candidate services is generated in this approach through computing similarity between the users' requests and a web service's profile. Then the integrated QoS values of candidate Web services are calculated. The candidate Web services are ranked for users according to QoS values. Finally, we demonstrate the feasibility and effectiveness of our discovery mechanism by validating it on a set of web services.

1 INTRODUCTION

In the past few years, web services have become widely distributed source used for providing services on the network. With the rapid growth of the quantity of Web services, the discovery process is more and more difficult. In general, the discovery process is the process of finding a service or several services which can best satisfy a user's demand.

UDDI is one of the widely used services, matchmaking technology, which discovers proper services at the syntactic level based on keywords. However, this keywords-based service discovery method is obviously insufficient, because it requires the same understanding of words in an application domain between the service providers and requesters. This means the requester must input the "right" keywords to search for relevant services and locate which candidate services are more likely to be used in the context.

With the increasing quantity of web services, it becomes harder to find the most suitable service. Thus, semantic service matchmaking is developed to solve this problem, because the description language of semantic web service could be understood by machines (Fenza et al. 2008). The combination of semantic web (Hendler 2005) and web service realizes intelligent and automatic service discovery. Semantic web technology could add semantic information to web service's description (McIlraith & Son 2001). So, semantic web service based on the semantic concept can be searched. It is efficient for service discovery to solve the problem of keywords matching. Now, this has become a hot topic of research. And one of the most widely used languages for semantic web service description is OWL-S (W3C 2003).

In this paper, a hybrid matching algorithm is presented to discover more appropriate web services for requesters. The hybrid algorithm computes the similarity degrees between the Web services and the requests based on semantic concept and syntax tree. Meanwhile, QoS values are used to sort the selected Web services.

The remainder of this paper is organized as follows: Section II summarizes the related essential works. Section III presents the proposed approach for semantic Web service discovery and describes a prototype of the mechanism as well as its operations. Section IV analyzes and evaluates the matching results. Section V summarizes this work.

2 RELATED WORK

This section presents the technologies to describe and discover semantic Web service, which are widely used.

2.1 *Semantic web service description language*

Semantic Web service description languages, for example Ontology Web Language for Service (OWL-S) (Martin & Burstein 2004) and Web Service Modeling Ontology (WSMO) (Roman & Keller 2005), are proposed as abstractions of syntactic Web service description languages e.g. WSDL. OWL-S describes the inputs, outputs, categories and consequences of

Web services by using concepts defined in the ontology. And it also provides the low-level framework for specialization into a WSDL frame to be compatible with existing Web services.

The above is used to build Web service description ontologies. Other semantic Web service description languages are used to add semantic annotations to WSDL, such as WSDL-S (Akkiraju & Farrel 2005) and SAWSDL (Farrell & Lausen 2007). SAWSDL is derived from the WSDL. It is a simple extension of WSDL through using three additional attributes, i.e. model Reference, lifting Schema Mapping and lowering Schema Mapping. It annotates existing Web services excluded semantic information with semantics in an intuitive way.

2.2 *Semantic web service discovery*

Currently, to improve the precision rate is mainly applied to semantic Web service discovery mechanism, especially to calculating the similarity between service request and service advertisement in semantic Web service. Up to now, a number of SWS (Semantic Web Service) discovery methods have been published, which can be divided into four categories: OWL-S based, WSMO based, SAWSDL and WSDL-S based, other ontology based. The approach presented in (Farrag & Saleh 2013) is similar to this paper, which uses the distance between the users' requests and the tested services to discover the proper Web services, but the QoS information is not considered, and it needs plenty of calculation time.

In (Lu & Wang 2012), a Web service discovery method based on the domain ontologies is proposed, which calculates the semantic similarity by ontology concept and distance of service request and service advertisement. However, this Web service discovery algorithm only calculates the semantic similarity of the function descriptions. It ignores the semantic similarity of the descriptions of the input/output parameters.

2.3 *QoS-aware discovery*

Since many functionally similar Web services are presented in the Web, it is an urgent requirement to distinguish them by using a series of non-functional criteria such as Quality of Service (QoS).

Many researchers provide various methods for evaluating QoS of the Web services and the results are used to select services. Yu 2014proposed a collaborative filtering scheme for evaluating the QoS of large-scale Web services. However, this scheme is difficult to implement when the users' features are hard to collect. The work described in (Lin & Lai 2014)[12] presents a trustworthy two-phase Web service discovery mechanism based on QoS and collaborative filtering. It could recommend the needed services to users, but the ontology is not considered.

The similarity measure is proposed in this study to address the drawbacks of ontology-based approaches, namely the ignorance of input/output semantic information. It combines syntax similarity with semantic similarity to promote the accuracy of results. Our work also differs from other efforts in service matching in its goal and methodology. Our goal is to satisfy the needs of requesters' individuality. It is realized by calculating the QoS values of the candidate services and ranking them.

3 SEMANTIC WEB SERVICE DISCOVERY MECHANISM

Our mechanism finds similar services to meet the consumers' requests based on semantic and syntax similarity, and integrated QoS values. The whole process of discovering service is shown in Figure 1. Each step uses the results of the previous steps, and aims to rank the selected Web services according to the integrated QoS values.

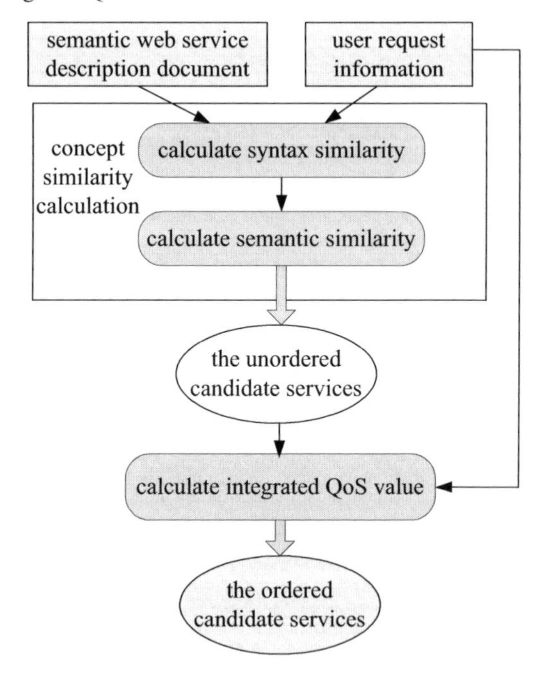

Figure 1. The service discovery process.

3.1 *Syntax similarity computation*

Calculating the semantic similarity is a complex and time-consuming work. To save time, the Web services are filtered based on keywords before calculating the semantic similarity. In Web service description documents, we extract the service name and text description information. The formula of calculating the syntax similarity is as follows:

$$Gras = \frac{2 * T_3}{T_1 + T_2} \tag{1}$$

where T_1 and T_2 represent the number of keywords in service description and request information. T_3 represents the number of the same keywords. If $Gras > \gamma$, where γ is a threshold, the service continues as the following steps. Service providers provide the service introduction information in a text. This step can filter out irrelevant services by using this text.

3.2 *Semantic similarity computation*

In this step, the user's request is firstly converted into a set of concepts according to the domain ontology. This set is denoted as $CR = \{c^k, c^{in}, c^{out}\}$, where c^k, c^{in} and c^{out} are subsets. c^k is keywords' subset, $c^k = \{c_1^k, c_2^k, ..., c_l^k\}$. c^{in} is the user defined input parameters' subset, $c^{in} = \{c_1^{in}, c_2^{in}, ..., c_m^{in}\}$. c^{out} is the user defined output parameters' subset, $c^{out} = \{c_1^{out}, c_2^{out}, ..., c_m^{out}\}$. Meanwhile, the WSDL files of Web services are transformed into OWL-S format. The concepts in descriptions and input/output parameters of the Web service are extracted and saved in a set. The set is denoted as $CS = \{c^d, c^{in\_s}, c^{out\_s}\}$, where c^d, c^{in_s} and c^{out_s} are concept subsets of the Web service's description, input and output parameters respectively, $c^d = \{c_1^d, c_2^d, ..., c_u^d\}$, $c^{in\_s} = \{c_1^{in\_s}, c_2^{in\_s}, ..., c_v^{in\_s}\}$, $c^{out\_s} = \{c_1^{out\_s}, c_2^{out\_s}, ..., c_w^{out\_s}\}$. Then the sum of concepts similarities is considered as the similarity between user request and Web service.

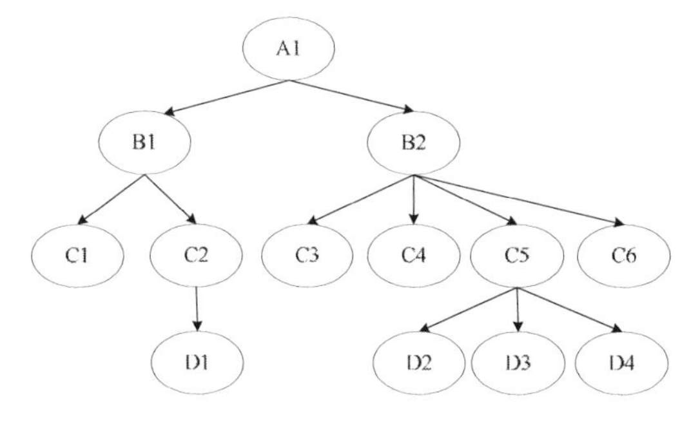

Figure 2. The service discovery process.

The structure of concepts in ontology can be represented as a tree, as shown in Figure 2. Algorithm design mainly considers the following design principles: First, the semantic similarity is quantifiable. The calculation result must be numeral which could be compared. Secondly, the semantic similarity is adjustable. The algorithm needs to have adjustable parameters, which can adjust the needs of different ontology. Finally, semantic similarity calculation results should be as accurate as possible. It has to meet people's understanding. By the summary of the previous studies, the influence factors of concept semantic similarity based on ontology include the depth of a concept node, the short path between two concepts, the number of the edges of the shortest path and the density of a concept (Zeng & Benatallah 2004).

With a view to the depth of a concept and the short distance between any two concepts, the depth of the hierarchy indicates the level of detail of the concept classification. When the distance between two concepts is certain, the deeper they located, the more similar they are. The formula for ontology similarity of concepts is shown below:

$$Lev(c_i, c_j) = \begin{cases} \left(\frac{Sur(c_i)}{Lev} + \frac{Sur(c_j)}{Lev}\right) * t & c_i = c_j \\ \left(\frac{Sur(c_i)}{Lev} + \frac{Sur(c_j)}{Lev}\right) * \frac{t}{|Sur(c_i) - Sur(c_j)|} & c_i \neq c_j \end{cases} \tag{2}$$

where the $Sur(c_i)$ and $Sur(c_j)$ are the depths of c_i and c_j. Lev is the depth of the concept tree. t is an adjustment coefficient. When the two concepts are at the bottom level and when the level difference is 0, they should be relatively most similar. In this case, Lev should be 1. Therefore, t is recommended to be 1/2.

As for the short distance between any two concepts, the semantic distance of two concepts becomes larger with the increasing number of the shortest paths between them. The distance between them is computed as follows:

$$Dis(c_i, c_j) = \prod_{k=1}^{n} x \tag{3}$$

x is a predefined weight in the ontology. n is the number of the edges of the path from c_i to c_j. The weights of all the edges are tired multiply. Since the weight value is less than 1, with the increase in path length, $Dis(c_i, c_j)$ decreases. This algorithm takes into account both the length of the path and edge weight.

In existing studies on the concept density, they utilize the number of sub-concepts which are next to the certain concept and the number of sub-concepts which are next to the minimum parent concept of two certain concepts to measure the density of concepts. But this kind of methods is not an accurate measurement. A density measurement of a concept Ci has nothing to do with the number of sub-concepts of chi. And if the minimum common parent concept is added into the measurement, in some cases, this function may be not monotone. As shown in Figure 2, the density at (C1, C2) is smaller than (C5, C6). But if the formula (4) is used to calculate concept similarity, the value at (C1, C2) is larger than (C5, C6), which is unreasonable.

$$S(Ci,Cj)=\frac{2\log^{n_{f(Ci,Cj)}}}{\log^{n_{Ci}}+\log^{n_{Cj}}} \tag{4}$$

where $n_{f_{(Ci,Cj)}}$ is the number of sub-concepts of the minimal father concept of Ci and Cj. n_{ci} is the number of sub-concepts of concept Ci, which is the same as Cj. So in this paper, the density similarity measure formula of concepts is as follows:

$$\rho(c_i,c_j)=\frac{Bro(c_i)+Bro(c_j)}{TotBro} \tag{5}$$

where the $Bro(c_i)$ represents the number of brother concepts which have the same father with c_i. *TroBro* is the number of concepts in the last layer of ontology tree. To conclude, considering the various influencing factors of similarity, we define the semantic similarity formula between two concepts as follows:

$$Sim(c_i,c_j)=\alpha Lev(c_i,c_j)+\beta Dis(c_i,c_j)+\delta\rho(c_i,c_j) \tag{6}$$

where α, β and δ are weight coefficients within range [0,1], and $\alpha+\beta+\delta=1$.

Finally, the semantic similarity between CR and CS can be computed as follows:

$$Sim(\text{CR, CS})=\frac{\sum_{i=1,j=1}^{i=l,j=u} Sim(c_i,c_j)+\sum_{k=1,t=1}^{k=m,t=v} Sim(c_k,c_t)+\sum_{r=1,s=1}^{r=n,s=w} Sim(c_r,c_s)}{N} \tag{7}$$

$c_i \in c^k$, $c_j \in c^d$, $c_k \in c^{in}$, $c_t \in c^{in\_s}$, $c_r \in c^{out}$, $c_s \in c^{out\_s}$, N is the number of non zero similarity values between CR and CS. If Sim(CR, CS)> π, where π is a threshold and π [$\in$0, 5, 1) , this web service can be the candidate service.

3.3 *QoS computation and matching*

In this step, QoS attributes and values are processed. Sometimes, maximum requirements or minimum requirements on some QoS attributes have been restricted by service consumer. For instance, a consumer may want a service with a price not surpassing 10$. So, candidate services whose prices are over this threshold will be eliminated. Therefore, this paper uses a QoS-based services filtering algorithm. We can avoid selecting the Web services that don't meet the consumer's expectation with this algorithm. Algorithm 1 is taken as follows: input a set of candidate services obtained in the last step and a set of QoS based constraints, then output the eligible Web services.

Algorithm 1: QoS filtering for candidate Web services

Input: candidate Web services list, QoSParameter, QoSRequest

Output: eligible Web services list

```
Begin
for each service S in Candidate list do
if (QoSParameter.name=QoSRequest.name)and
(QoSParameter.monotony="increase")and
(QoSParameter.value<QoSRequest.threshold)then
Eliminate (service S)
end if
if (QoSParameter.name=QoSRequest.name)and
(QoSParameter.monotony="decrease")and
(QoSParameter.value>QoSRequest.threshold)then
Eliminate (service S)
end if
end for
End
```

It is necessary to normalize every QoS value to have a value in the range of [0,1]. To normalize the QoS values, each QoS value is considered as monotonically decreasing or increasing. Monotonically increasing QoS values are normalized by equation (8) and monotonically decreasing QoS values are normalized by equation (9). Besides, q_{max} and q_{min} are the maximum and minimum values of the QoS attributes in all candidate services.

$$Q_i=\begin{cases}\frac{q-q_{\min}}{q_{\max}-q_{\min}} & \text{if}(q_{\max}\neq q_{\min})\\ 1 & \text{if}(q_{\max}=q_{\min})\end{cases} \tag{8}$$

$$Q_i=\begin{cases}\frac{q_{\max}-q}{q_{\max}-q_{\min}} & \text{if}(q_{\max}\neq q_{\min})\\ 1 & \text{if}(q_{\max}=q_{\min})\end{cases} \tag{9}$$

In order to calculate all the QoS scores for the services, every normalized QoS attribute Q_i is multiplied by the corresponding weight provided by a service consumer, as shown by formula (10). The integrated QoS value can be used to sort the output set of algorithm 1.

$$QoS(S)=\frac{\sum_1^n Q_i * w_i}{\sum_1^n w_i} \tag{10}$$

4 EXPERIMENTS

To evaluate the performance of our approach, the experiments are conducted in this section and compared with other approaches (e.g., the one proposed in (Lu & Wang 2012) and (Plebani & Pernici 2009)). SWSDDO represents the method in (Lu & Wang 2012). URBE represents the approach in (Plebani & Pernici 2009). Firstly, only the functional performance of the Web services is considered. In this evaluation process, we use the data provided by wsdream (Zhang & Zheng 2010) for validating our proposal. This collection consists of 3,378 WSDL files from 69 countries and 15,811 operations covering five application domains (i.e., business, education, science, weather, media).

Recall and precision are adopted as metrics to evaluate the performance of different discovery approaches. Recall and Precision can be calculated by:

$$Recall = \frac{|A \cap B|}{|A|} \tag{11}$$

$$Precision = \frac{|A \cap B|}{|B|} \tag{12}$$

where A is the relevant set of Web services for a query, and B is a set of Web services search results.

We change the WSDL files into OWL-S format, and create ontology in Pretege 3.4. Meanwhile the concepts of user requests are extended by using WordNet. 5 test queries are done on each domain of the dataset. In the syntax similarity computation step, according to the number of Web services in the corpus, the threshold γ is set to be 0.625 to filter plenty of irrelevant services. By increasing the capacity of the Web service library, the value of γ can be linearly increased. α and β in formula (6) are set to be 0.4 and 0.3 respectively. Figure 3 shows the experimental results of our approach, the URBE approach and the SWSDDO approach. In Fig 3(a), the recall values of our approach are higher than those of URBE. In Figure 3(b), the precision values of our approach are considerably higher than those of the other two approaches, indicating that the recommended services by our approach are very proper to user's request.

Then we use the QoS dataset of wsdream (Zhang & Zheng 2011) to evaluate our approach. The dataset consists of real-world QoS evaluation results from 339 users on 5,825 Web services. The quality attributes include response-time and throughput. Ten requests with QoS threshold are conducted, and the averages of recall and precision values are 0.893 and 0.932 respectively. The results verify our approach is useful for QoS based Web service discovery.

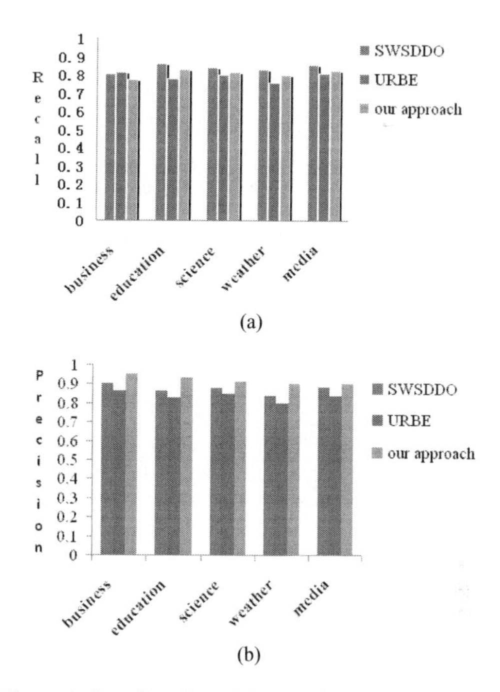

Figure 3. Recall and precision performance.

5 CONCLUSION AND FUTURE WORK

In this paper, an improved Web service discovery mechanism is presented to select the desired Web services. Syntax similarity and semantic similarity are both used to produce a list of the candidate services for users. The non-functional requests are realized through QoS filtering. And the final results are sorted by QoS values. Experiments on real-world dataset are conducted to study the performance of our discovery mechanism prototype. The results show that our approach outperforms related works.

In future work, we will apply semantic reasoning on Web service discovery to realize intelligent search and increase the precision degree.

ACKNOWLEDGMENT

This work is supported by the National Natural Science Foundation of China (No.71203173), the Fundamental Research Funds for the Central Universities, China (No. K5051306004).

REFERENCES

Akkiraju, R. & Farrell, J. 2005. Web service semantics – wsdl-s. http://www.w3.org/Submission/WSDL-S/.

Farrag, T. A. & Saleh, A. I. 2013. Semantic Web Services Matchmaking: Semantic Distance-based Approach. *Computer and Electrical Engineering* 39(2): 497–511.

Farrell, J. & Lausen, H. 2007. Semantic annotations for wsdl and xml schema. http://www.w3.org/TR/sawsdl/.

Fenza, G., Loia, V. & Senatore, S. 2008. A Hybrid Approach to Semantic Web Services Matchmaking. *Int. J. Approx. Reasoning* 48(3): 808–828.

Hendler, J.A. 2005. Knowledge Is Power: A View from the Semantic Web. *AI Magazine* 26(4): 76–84.

Lin, S. Y. & Lai, C.H. 2014. A Trustworthy QoS-based Collaborative Filtering Approach for Web Service Discovery. *Journal of Systems and Software* 93: 217–228.

Lu, G. & Wang, T. 2012. Semantic Web Services Discovery Based on Domain Ontology. *In Puerto Vallarta, World Automation Congress, Proc. Mexico, 24–28 June 2012*. IEEE.

Martin, D. & Burstein, M. 2004. OWL-S: semantic markup for Web services. http://www.w3.org/Submission/OWL-S.

McIlraith, S.A. & Son, T.C., 2001. Semantic Web Services. *IEEE Intelligent Systems* 16(2): 46–53.

Plebani, P. & Pernici, B. 2009. URBE: Web Service Retrieval Based on Similarity Evaluation. *IEEE Transactions on Knowledge and Data Engineering* 21(11): 1629–1642.

Roman, D. & Keller, U. 2005. Web service modeling ontology. *Applied Ontology* 1(1): 77–106.

The OWL Services Coalition. 2003. OWL-S: Semantic Markup for web Services. http://www. daml-s.org/owl-s/1.0/owl-s.pdf

Yu, Q. 2014. QoS-aware Services Selection Via Collaborative QoS Evaluation. *World Wide Web* 17(1): 33–57.

Zeng, L. & Benatallah, B. 2004. QOS Aware Middleware for Web Service Position. *IEEE Transactions on Software Engineering* 30(5): 311–327.

Zhang, Y. & Zheng, Z. 2010. WSExpress: A QoS-aware Search Engine for Web Services. *In Miami Florida, International Conference on Web Services, Proc. USA, 5–10 July 2010*. IEEE.

Zhang,Y. & Zheng Z. 2011. Exploring Latent Features for Memory-Based QoS Prediction in Cloud Computing. *In Madrid, IEEE Symposium on Reliable Distributed Systems, Proc. Spain, 4–7 October 2011*. IEEE

Electronics, Communications and Networks IV – Hussain & Ivanovic (eds)
© 2015 Taylor & Francis Group, London, ISBN: 978-1-138-02830-2

Privacy preserving range search based on comparable encrypted index in wireless sensor networks

Shiyang Li & Xiaoming Wang
Department of Computer Science, Jinan University, Guangzhou, Guangdong, China

ABSTRACT: With the flourish of wireless sensor networks, how to execute privacy preserving range search in the encrypted sensory data becomes a significant issue of wireless sensor network security. In this paper, a privacy preserving range search scheme is proposed based on a comparable encrypted index of sensory. In our scheme, an encryption mechanism of sensory data and query trapdoor to preserve the privacy of query is put forward. Moreover, a specific encrypted index of sensory data is proposed which can direly compare with the query trapdoor without the auxiliary data structure like prefix set and search tree. Thus, compared with the existing schemes based on auxiliary data structure, our scheme can remarkably improve the search efficiency and reduce communication traffic of data collection and query. The security of our scheme is illustrated by security analysis. Performance evaluation shows that our tool is practical and deployable.

1 INTRODUCTION

With the surging advancement of Micro-Electro Mechanism, Embedded Technology and Wireless Communication Technology, the wireless sensor network technology, proposed at the end of 20 century, has been widely applied to a variety of fields, such as industrial automation, real time monitoring and environmental information collection and analysis. In the wireless sensor network, the traditional trust model assumes that the sensor and the system owner of the network are the collector and consumer of sensor readings. However, faced with the requirement of large scale data monitoring, the traditional trust model is insufficient, because the amount of computing power, storage and energy of sensor is less. Therefore, the two-tiered sensor network based on the cloud storage service is proposed. In the two-tiered sensor network, a multitude of storage nodes with high computing power and storage in the cloud environment serve as an intermediate tier between sensor and system owner. The storage nodes store the data collected by sensors and provide data query service for system owner. Nevertheless, since the storage nodes, which store massive data collected from sensors may be attacked by adversaries, there are risks of confidential data leakage and privacy loss.

Data encryption is the fundamental methodology to prevent confidential data leakage. The data collected from sensors are encrypted before uploaded to storage nodes. The encryption mechanism can guarantee that the confidential data are secure since attackers can't find one way to access even capture the encrypted data. However, a fundamental problem is how to retrieve the encrypted data upon a query from the system owner. Recently, there are some schemes (Sheng & Li 2008, Shi et al. 2009, Zhang et al. 2009, Shi et al. 2007, Chen & Liu 2012, Tsou et al. 2012) proposed in order to solve the problem of privacy preserving range query in the wireless sensor network. The schemes are based on bucket partitioning technology. The scheme (Sheng & Li 2008) first is proposed to utilize the encrypted data bucket to realize the encrypted range query, and schemes improve the accuracy and effectiveness of the query method based on the bucket partitioning. However, the borderline information of the bucket may be exposed to the adversaries to attack the storage nodes. Indeed, the precision and efficiency of these schemes are low, because the result of the query may contain some useless data. The schemes put forward the idea that storage nodes adopt an auxiliary data structure (Bit-map, search tree, prefix) to preserve the data sequence and use the features of auxiliary data structure to realize the function of query. In the schemes (Tsou et al. 2012, Shi et al. 2007), the storage nodes need to store the Bit-map or search tree which contains global data so that the space efficiency and computational efficiency is quite low. Meanwhile, the query method based on Bit-map or search tree only can be adopted in the integer data query. The scheme (Chen & Liu 2012) based on a prefix set can enforce the search efficiency and improve accuracy of query. In the scheme (Chen & Liu 2012), taking the prefix set of index and trapdoor to compute the intersection can solve the query matching. However, the principle of the scheme (Chen & Liu 2012) is computing the auxiliary set intersection as the schemes (Tsou et al. 2012, Shi et al. 2007), there is still room for improving the computational complexity and space efficiency.

As above, the schemes(Tsou et al. 2012, Shi et al. 2007, Chen & Liu 2012) use some auxiliary data structure (prefix set, Bit-map, search tree) and compute the auxiliary set intersection to realize the range search over the encrypted sensory data. Thus, the schemes (Tsou et al. 2012, Carbunar et al. 2010, Chen & Liu 2012) need to allocate extra computing resources and storage space for the auxiliary data structure. Faced with the efficiency issue of existing encrypted range query in wireless sensor network, a preserving range query scheme with higher practicability and performance efficiency is proposed. Comparable encrypted indexes of the sensor data to achieve encrypted range query are also put forward. In our scheme, the comparable encrypted index can directly compare with the encrypted trapdoor by utilizing the property of logarithmic function and the homomorphic encryption. Our scheme doesn't need to use the auxiliary data structure. Consequently, the computational efficiency and storage efficiency can be improved obviously. Moreover, a key generation policy to realize the forward security and backward security is put forward. Accordingly, if the key of sensor node is exposed in the time slot t, the adversaries only can query the data in the time slot t, but cannot execute a value query operation to search the data collected in the other time slot. The forward security and backward security properties of an enacrypted query search play a significant role in the actual query application in wireless sensor network.

2 RELATED WORK

2.1 *Searchable encryption*

In 2000, Song et al. (Song et al. 2000) proposed the concept of searchable encryption and gave an encrypted keyword search scheme, which stimulates scientific researchers to investigate encrypted search. In order to improve the query efficiency, Goh (Goh 2003) used Bloom filters to screen the valid query trapdoor. In 2006, R. Curtmola, et al (Curtmola R,et al. 2006) proposed a searchable symmetric encryption. Their schemes adopted sequence list and chain table to structure a special query index. They ameliorated the encrypted keyword index construction to improve the efficiency and security of encrypted search. To adapt to the public-key cryptosystems, the public key encryption with the keyword search scheme was proposed by Boneh et al. (Boneh et al, 2004). Data owner encrypts data files using the receiver's public key and only the receiver can generate a valid query trapdoor by his private key. In their schemes, the property of bilinear maps and the secure hash function can prevent the content of encrypted data from leaking during the query phase. This construction is later improved by several researchers (Yang et al. 2011, Bao et al. 2008, Chang & Mitzenmacher 2005, Fuhr & Paillier 2007, Gu & Zhu 2010, Baek et al. 2008, Ballard et al. 2005, Bellovin & Cheswick 2007, Wang et al. 2010, Zheng et al. 2013). These schemes modify the encryption method of keyword index and the query parameters to improve function and security of query. The schemes (Yang et al. 2011, Bao et al. 2008) realize the encrypted search schemes for the multi-user query and the management of the query permission in a cloud environment. With the development of searchable encryption, many new schemes are proposed in succession, such as the schemes with encrypted range searches (Sheng & Li 2008, Shi et al. 2009, Zhang et al. 2009, Tsou et al. 2012, Carbunar et al. 2010, Chen & Liu 2012), encrypted conjunctive keyword searches (Boneh & Waters 2007, Golle et al. 2004) and encrypted fuzzy keyword search (Li et al. 2010).

2.2 *Privacy protecting query search in wireless sensor network*

Owing to the actual application requirement, most of the existing searchable encryption schemes are based on public key encryption system and focus on the issue of encrypted keyword search. Nevertheless, restricted by the low computing power and energy problem of wireless sensors, the query scheme in wireless sensor network would rather adopt the symmetric cryptographic than public key encryption to encrypt the index and data. Indeed, because the wireless sensor is routinely used in data monitoring and analysis, the range query instead of the keyword search, is the main research direction of a privacy preserving query in wireless sensor network. Recently, there are some schemes (Sheng & Li 2008, Shi et al. 2009, Zhang et al. 2009, Tsou et al. 2012, Carbunar et al. 2010, Chen & Liu 2012)proposed in order to solve the problem of privacy preserving range query in the wireless sensor network. The schemes (Sheng & Li 2008, Shi et al. 2009, Shi et al. 2009, Zhang et al. 2009) are based on bucket partitioning technology. The scheme (Sheng & Li 2008) first is proposed to utilize the encrypted data bucket to realize the encrypted range query, and schemes (Shi et al. 2009, Zhang et al. 2009) improve the accuracy and effectiveness of the query method based on the bucket partitioning. To improve the query precision, the schemes (Tsou et al. 2012, Shi et al. 2007) put forward that storage nodes adopt an auxiliary data structure (Bit-map, search tree) to preserve the data sequence and utilize the features of data structure to realize the range query. The scheme (Chen & Liu 2012) based on a prefix set can improve the accuracy of query and reduce time and space. Taking the prefix set of index and trapdoor and computing the intersection can solve the accurate query matching. In addition, verifying completeness is also a research focus of the wireless sensor network query, the schemes (Shi et al. 2009, Zhang et al. 2009, Tsou et al. 2012, Chen & Liu 2012, Pang et al. 2005) adopt data structure combined with digital signature to ensure the completeness of query results.

3 MODELS AND PROBLEM STATEMENT

3.1 *System model*

As is illustrated in Figure 1, a two-tiered sensor network consists of three levels: the sensing level, the storage level, and the user level. The sensing level is comprised of the sensor nodes, which are commonly distributed in a field to collect sensory data. Because of the limited storage and computing power of the sensors, each sensor node sends the sensory data collecting its nearby storage node to achieve data storage and communion. The storage level is the intermediate tier between the sensor and the system owner. The storage nodes which make up the storage level are the powerful wireless devices with high storage and computing capacity. The task of the storage level is storing the data collected by sensors and realizing the range query from system owner. The user level is the system owner who owns this wireless sensor system, the demand of the system owner is acquiring the sensor data which he is concerned with.

Because the storage nodes sprinkle in each query district and storage massive sensor data, they may be attacked by adversaries and exist the risks of confidential data leakage. Therefore, the sensory data are encrypted before uploaded to storage nodes. The range query takes over the encrypted sensory data. The encrypted range search scheme contains four steps. Firstly, the system owner draws up the key generation policy and the encrypted query mechanism of range search scheme. Secondly, the sensor in sensing level collects sensory data, encrypts and sends to the storage level to achieve encrypted data storage. Thirdly, the system owner in user level generates the encrypted query trapdoor as a query requirement and sends it to the storage level. Finally, the storage node matches the encrypted data and the encrypted trapdoor to actualize query and return the query result to the system owner.

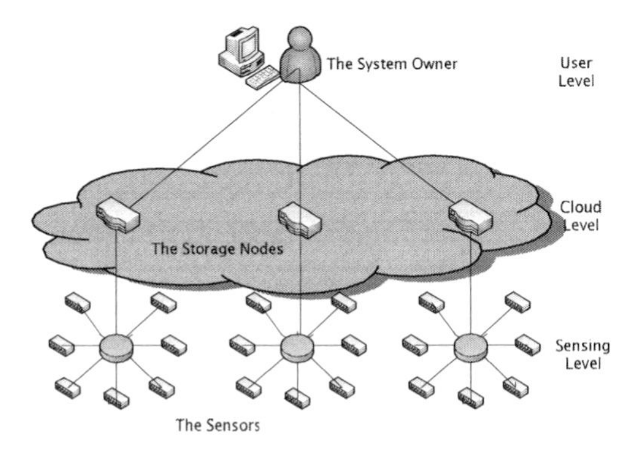

Figure 1. The system model of our searchable encryption scheme.

3.2 *Threat model*

In the wireless sensor two-tiered network model, it's assumed that the system owner and the sensor nodes are trusted, while the storage nodes which store massive data collected by sensors are incredible. As is known to all, the sensor nodes and the storage nodes can be attacked by adversaries. When the adversaries sacrifice the sensor node, adversaries can acquire the data collected by this sensor, forge sensory data and send them to the storage level. But the data from one sensor only constitute a small fraction of the collected data of the whole sensor network, while the storage node stores the data collected from its district. Sacrificing a storage node can cause more serious damages to the wireless sensor network than sacrifice a sensor. If a storage node is sacrificed by adversaries, the large number of data stored on the storage node will be exposed. Moreover, the trapdoor also will be acquired and the falsified query result will be forged by adversaries. Consequently, the main security issue of wireless sensor network is protecting the data stored in storage nodes.

3.3 *Formulate definition of our encrypted search scheme*

Definition of privacy preserving range query scheme in wireless network is described below. The scheme consists of the following four polynomial time randomized algorithms:

1 $KeyGen(s) \rightarrow (K_i, qk_i, qk_s, qk_u, qk_{i,u\rightarrow s})$:
Take a security parameter, generate the query key qk_i, qk_s, qk_u for the system owner, the storage nodes, the sensors and generate an auxiliary query key $qk_{i,u\rightarrow s}$ used for query. Indeed, generate an encryption key K_i used to encrypt index and generate trapdoor.

2 $Buildindex(K_I, qk_I, \{I, T, (d_1, ..., d_n)\}, params)$
$\rightarrow \{I, T, [E(qk_I, Ind(d_1)), ..., E(qk_I, Ind(d_n))]\}$:
The sensor i collects sensory data $\{I, T, (d_1, ..., d_n)\}$ in time slot T. Sensor i uses an encryption key K_i to generate searchable encrypted indexes $\{I, T, [Ind(d_1), ..., Ind(d_n)]\}$ for sensory data $\{I, T, (d_1, ..., d_n)\}$, utilize query key qk_i to encrypt indexes again. The encrypt indexes are.

$\{I, T, [E(qk_I, Ind(d_1)), ..., E(qk_I, Ind(d_n))]\}$.

3 $Trapdoor(I, T, qk_U, K_I, [a,b])$
$\rightarrow \{I, T, E(qk_U, TD(a)), E(qk_U, TD(b))\}$:

The system owner U desires to query the sensory data collected by the sensor i in the time slot t between the range [a,b]. The system owner U uses the encryption key K_i to generate the searchable encrypted query trapdoor $(TD(a), TD(b))$ and utilizes query key qk_u to encrypt the trapdoor

$\{I, T, E(qk_U, TD(a)), E(qk_U, TD(b))\}$.

4 $Test(I, T, qk_S, qk_{I,U\rightarrow S}, E(qk_I, ind(d_j)), E(qk_U, T(a)),$
$E(, qk_U T(b)) \rightarrow T / F$:

The storage node utilizes its query key qk_u and auxiliary query key $qk_{i,u\rightarrow s}$ to compare the encrypted index $E(qk_i, ind(d_j))$ and the encrypted query trapdoor $E(qk_u, TD(a))$, $E(qk_u, TD(b))$. If the sensory data are in the range [a,b], returns are true. Otherwise, returns are false.

4 OUR PRIVACY PRESERVING RANGE SEARCH SCHEME

Our privacy preserving range search scheme consists of the following four polynomial time randomized algorithms:

1 $KeyGen(1^\lambda) \rightarrow (\{K_I\}, qk, params)$
The system owner U runs this algorithm to generate the encryption key, query key and the public parameter.

Take the security parameter λ as input. Let Z_p be the Galois Field of prime order p. Define two cryptographic hash functions: $f_K : \{0,1\}^* \rightarrow Z_P$, $h_K : \{0,1\}^* \rightarrow Z_P$, where K is the key of hash function.

Choose query keys $qk_U, qk_S \in Z_p$ uniformly at random for the system owner U and storage node S, where $qk_U < qk_S$.

Generate query key for each sensor. For instance the sensor I, choose query key $qk_I \in Z_P$ uniformly at random, where $qk_I < qk_S - qk_U$. Compute the auxiliary query key $qk_{S\rightarrow I,U} = qk_S - qk_i - qk_U$. Choose the encryption key $K_i \in Z_p$ uniformly at random.

The algorithm outputs the encryption key $\{K_I\}$ for the sensor I in the system, the query key $qk = (qk_U, qk_S, qk_I, qk_{S\rightarrow I,U})$ for the system owner U, the storage node S and the each sensor I. In the system, the public parameter is $params = (p, h_K(\cdot), f_K(\cdot))$.

The system owner U possesses his query key qk_U and saves the encryption key $\{K_I\}$ for the sensor I belong to him. Distribute the query key qk_i and the encryption key K_i to each sensor I in the system. For the storage node S, the system owner allocates the query key qk_S to it and builds the auxiliary query key table for all the sensors managed by it. The auxiliary query key table for the storage node is shown by Table1.

Table 1. The auxiliary query key table.

Sensor	Auxiliary query key
1	$qk_{S\rightarrow 1,U}$
2	$qk_{S\rightarrow 2,U}$
...	...
I	$qk_{S\rightarrow I,U}$
...	...

2 $Buildindex(K_I, qk_I, \{I, T, (d_1, ..., d_n)\}, params)$
$\rightarrow \{I, T, [E(qk_I, Ind(d_1)), ..., E(qk_I, Ind(d_n))]\}$:

The sensor I runs the Buildindex() algorithm to generate the encrypted indexes of the sensory data $\{I, T, (d_1, ..., d_n)\}$ collected in the time slot T.

The sensor I computes the encryption parameters of the time slot T: $\alpha_{I,T} = h_{K_I}(T), EK_{I,T} = f_{K_I}(T)$.

For each sensor data d_j collected in the time slot T, generate searchable encrypted index:

$$Ind(d_j) = \log_{\alpha_{I,T}}(d_j + EK_{I,T})$$

Utilize the simple homomorphic encryption $E(K, M) = M + K \bmod p$ and the query key qk_i of sensor I to encrypt the index again:

$$E(qk_I, Ind(d_j)) = \log_{\alpha_{I,T}}(d_j + EK_{I,T}) + qk_I \bmod p$$

The algorithm outputs the encrypted indexes $\{I, T, [E(qk_I, Ind(d_1)), ..., E(qk_I, Ind(d_n))]\}$ of the sensory data $\{I, T, (d_1, ..., d_n)\}$ collected by the sensor I in the time slot T.

The sensor I adopts the secure symmetric encryption $Enc(K_I, d)$ to encrypt the sensory data $\{I, T, (d_1, ..., d_n)\}$, and sends the encrypted indexes and the encrypted sensory data to the storage node S.

3 $Trapdoor(I, T, qk_U, K_I, [a,b])$
$\rightarrow \{I, T, E(qk_U, TD(a)), E(qk_U, TD(b))\}$
The system owner U executes this algorithm to submit query request to search the sensory data collected from sensor I in time slot T between the range $[a,b]$.

The system owner U computes the encryption parameters of the sensor I and time slot T:

$\alpha_{I,T} = h_{K_I}(T)$, $EK_{I,T} = f_{K_I}(T)$.

Generate the searchable encrypted range trapdoor:

$$TD(a) = \log_{\alpha_{I,T}}(\frac{1}{a + EK_{I,T}})$$

$$TD(b) = \log_{\alpha_{I,T}}(\frac{1}{b + EK_{I,T}})$$

Utilize the simple homomorphic encryption $E(K, M) = M + K \bmod p$ and the query key qk_U of the system owner U to encrypt the trapdoor again:

$$E(qk_U, TD(a)) = \log_{\alpha_{I,T}}(\frac{1}{a + EK_{I,T}}) + qk_U \bmod p$$

$$E(qk_U, TD(b)) = \log_{\alpha_{I,T}}(\frac{1}{b + EK_{I,T}}) + qk_U \bmod p$$

The algorithm outputs the encrypted range trapdoor $\{I, T, E(qk_U, TD(a)), E(qk_U, TD(b))\}$.

The system owner U sends the encrypted range trapdoor to the storage node S as the query request.

4 $Test(I, T, qk_S, qk_{I,U \to S}, E(qk_I, ind(d_j)), E(qk_U, T(a)), E(,qk_U T(b)) \to T/F$:

The storage node S receives the query request of the system owner U, which runs this algorithm to execute encrypted search operation. According to the time slot T and the sensor I, The storage node S can locate the encrypted index $\{I, T, \{E(qk_I, Ind(d_j))\}\}$, for each $E(qk_I, Ind(d_j))$ execute the algorithm $Test(I, T, qk_S, qk_{I,U \to S}, E(qk_I, ind(d_j)), E(qk_U, T(a)), E(,qk_U T(b)) \to T/F$.

Because the simple homomorphic encryption $E(K, M) = M + K \bmod p$ adopted in our scheme has the homomorphic property $E(K_1, M_1) + E(K_2, M_2) = E(K_1 + K_2, M_1 + M_2)$, compute $E(qk_I, ind(d_j)) + E(qk_U, T(a))$:

$$= \log_{\alpha_{I,T}}(d_j + EK_{I,T}) + qk_I + \log_{\alpha_{I,T}}(\frac{1}{a + EK_{I,T}}) + qk_U \bmod p$$

$$= \log_{\alpha_{I,T}}(\frac{d_j + EK_{I,T}}{a + EK_{I,T}}) + qk_U + qk_I \bmod p$$

$$= E(qk_U + qk_I, \log_{\alpha_{I,T}}(\frac{d_j + EK_{I,T}}{a + EK_{I,T}}))$$

As the generation rule of query keys is $qk_{S \to I,U} = qk_S - qk_i - qk_U$, the storage node S uses its query key qk_S the auxiliary query key $qk_{S \to I,U}$ which can be found in the auxiliary query key table saved in the storage node to decrypt the data $E(qk_U + qk_I, \log_{\alpha_{I,T}}(\frac{d_j + EK_{I,T}}{a + EK_{I,T}}))$:

$$D(qk_S - qk_{S \to I,U}, E(qk_U + qk_I, \log_{\alpha_{I,T}}(\frac{d_j + EK_{I,T}}{a + EK_{I,T}}))$$

$$= \log_{\alpha_{I,T}}(\frac{d_j + EK_{I,T}}{a + EK_{I,T}})$$

Although, the storage node S can obtain the value of $\log_{\alpha_{I,T}}(\frac{d_j + EK_{I,T}}{a + EK_{I,T}})$, the storage node S can't compute the range boundary a and the sensory data d_j. The simple homomorphic encryption E() is applied to realize the privacy preservation and authenticate the storage node S which solely can compute the decryption key by its query key qk_S the auxiliary query key $qk_{S \to I,U}$.

According to the property of logarithmic function, if $\log_{\alpha_{I,T}}(\frac{d_j + EK_{I,T}}{a + EK_{I,T}}) \ge 0$, we can judge that $d_j \ge a$; otherwise $d_j < a$. In like manner, the cloud storage S can decrypt $E(qk_I, ind(d_j)) + E(qk_U, T(a))$ and obtain $\log_{\alpha_{I,T}}(\frac{d_j + EK_{I,T}}{b + EK_{I,T}})$. The cloud storage S can judge that $d_j > b$ or $d_j \le b$. Therefore, by the comparable encrypted index and trapdoor, the storage node S can compare the sensory data d_j with the range boundary a,b without knowing the value of them.

If the value of sensory data d_j is between the range [a, b], the algorithm outputs truth and the storage node S returns the corresponding encrypted data $Enc(K_I, d)$ to the system owner U; otherwise outputs false.

5 SECURITY ANALYSIS AND PERFORMANCE ANALYSIS

5.1 *Security analysis*

1 Privacy preservation of search content. Our range search scheme can enable the storage to perform privacy preserving range searches on an encrypted database. The comparable index and sensory data are encrypted by the sensors. The query trapdoor is encrypted by the system owner. The storage node executes query operation over the encrypted data. The encryption mechanism can guarantee that the sensory data is secure since storage node and attackers can't find one way to access the encrypted data.
2 Specify the query executor. According to the specific query key generating rule in our scheme, only the storage node possessing the query key and the auxiliary query key of the subordinate sensor can

execute the rightful encrypted query. Even acquiring rightful query trapdoor from system owner, the attacker and the other storage node not belonging to this sensor district don't have query permission and cannot execute the rightful encrypted query.

3 Control query permission. In our scheme, only the system owner can utilize his query key and encryption key of the corresponding sensor to generate rightful query trapdoor. The attacker and the storage node cannot generate rightful query trapdoor. Therefore, they cannot match the encrypted index in the storage node.

4 Comparable encrypted index. Different from the most existing encrypted range search scheme based on computing set intersection, the scheme of the encrypted range search by the specific comparable index is realized. The search method in our scheme remarkably improves the search efficiency and reduces communication traffic of data collection and query. Therefore, our scheme realizes the encrypted range search with higher practicability and performance efficiency.

5.2 *Performance analysis*

In this section, the performance of our scheme in terms of both asymptotic complexity and actual execution time is evaluated. Our scheme is compared with the scheme (Chen & Liu 2012) based on the prefix set and the scheme (Shi et al. 2007) based on the search tree. The performance comparison is analyzed in two aspects: the computation cost and communication cost.

Table 1 shows that the computation cost and communication cost of BuildIndex, Trapdoor and Test algorithm in our scheme and the schemes (Carbunar et al. 2010, Chen & Liu 2012). Asymptotic complexity is measured in terms of three kinds of operations: L denotes the operation of logarithm, E denotes the public key encryption, T denotes the matching operation, |D| denotes the size of sensory data. In our system, the number of sensory data saved in storage node is N, the average prefix set size of each data interval is K, the data space of sensory data is M.

From Table 2, it can be observed that the computational complexity and communication traffic of scheme (Shi et al. 2007) and scheme (Chen & Liu 2012) are much more expensive than our scheme. The comparable index of our scheme can directly compare with the query trapdoor, so our scheme does not need the extra auxiliary data structure to achieve query. Therefore, the computation complexity and communication traffic in our scheme are low. Our encrypted range search scheme providing with higher practicability and performance efficiency is more suitable for the wireless sensor network with low computing capacity.

our encrypted range search scheme in Java based on Java Pairing Based Cryptography library (jPBC) is also implemented. A computer that has an Intel Pentium Dual E2140 1.5GHz and 1G RAM and runs under Windows XP is used. Table 3 and Figure 2, 3, 4 show the execution time of our scheme.

6 CONCLUSION

The problem of privacy preserving range search in the wireless sensor networks is focused. Directly against the query efficiency issues of existing searchable encryption schemes, an encrypted range search scheme is proposed based on a specific comparable encrypted index. In our scheme, the encryption mechanism of sensory data and query trapdoor can preserve the privacy of query. The specific encrypted index of sensory data which can direly compare with the query trapdoor without the auxiliary data structure like prefix set and search tree is put forward. Thus, compared with the existing schemes based on auxiliary data structure, our query scheme remarkably can improve the search efficiency and reduce communication traffic of data collection and query.

Table 2. The asymptotic complexity of our scheme.

Algorithm	BuildIndex		Trapdoor		Test
	Computational complexity	Communication traffic	Computational complexity	Communication traffic	Computational complexity
Our scheme	N*L	N*|D|	2*L	2*|D|	N*T
(Chen & Liu 2012)	K*(N-1)*E	K*N*|D|	2*E	2*|D|	K*N*(N-1)*T
(Shi et al. 2007)	Log(M)*N*E	Log(M)*N*|D|	Log(M)*E	Log(M)*|D|	Log(M)*Log(M)*N*T

Table 3. The actual execution time of our scheme.

Number of index items		10	20	50	100	200	500	1000
Our scheme	Buillndex	0.906	1.875	4.468	9.968	17.906	44.266	89.062
	Trapdoor	0.096	0.094	0.093	0.094	0.094	0.097	0.110
	Test	0.016	0.022	0.046	0.079	0.156	0.408	0.797
(Chen & Liu 2012)	Buillndex	5.248	9.835	26.808	54.769	102.064	234.609	489.841
	Trapdoor	0.134	0.136	0.134	0.134	0.135	0.136	0.134
	Test	0.089	0.127	0.243	0.418	0.842	2.273	4.463
(Shi et al. 2007)	Buillndex	8.643	17.257	36.978	87.718	160.103	392.822	782.789
	Trapdoor	0.696	0.683	0.702	0.704	0.699	0.703	0.705
	Test	0.143	0.173	0.277	0.446	0.998	2.611	4.861

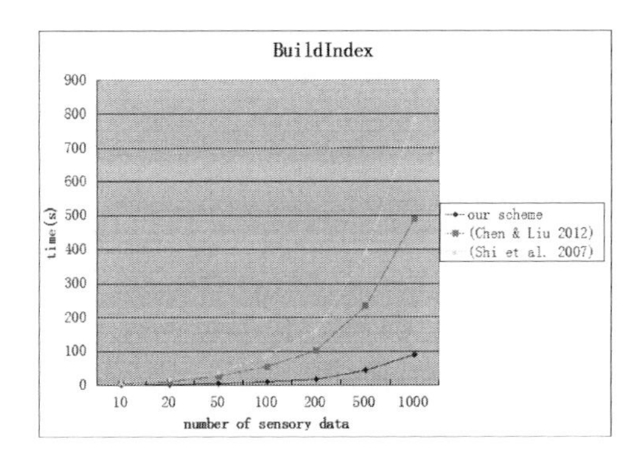

Figure 2. The actual execution time of BuildIndex() algorithm.

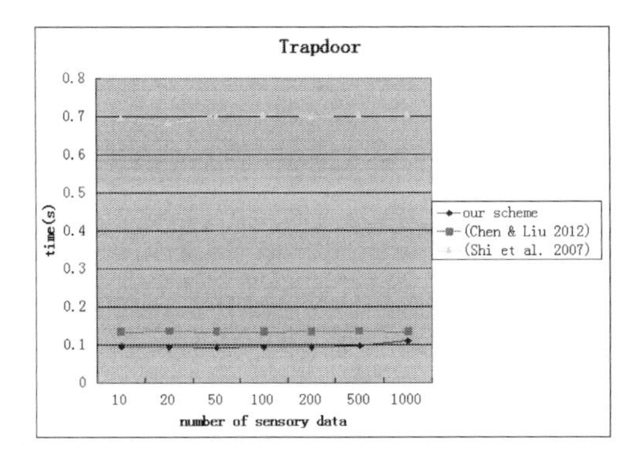

Figure 3. The actual execution time of Trapdoor() algorithm.

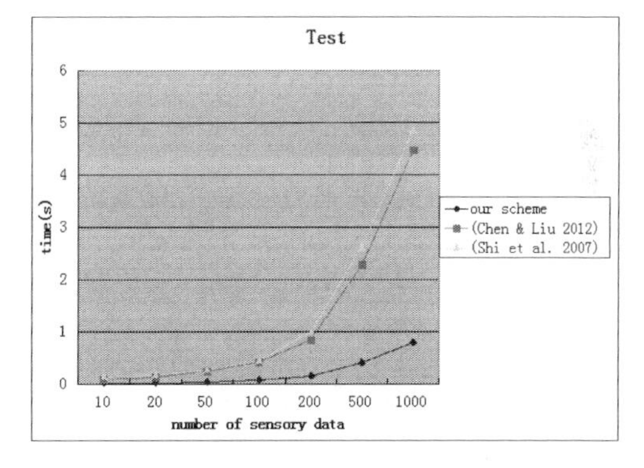

Figure 4. The actual execution time of Trapdoor() algorithm.

ACKNOWLEDGMENT

This work is supported by the National Natural Science Foundation of China (Grant No. 61272415, 61070164, 61272413). It is also supported by Science and Technology Planning Project of Guangdong Province, China (Grant No. 2013B010401015, 2012B091000136), and Natural Science Foundation of Guangdong Province, China (Grant No.S2012010008767, 815106 32010000022), and Science and Technology Planning Project on Guangdong City, China (Grant No. 12C542071906).

REFERENCES

(Boneh, D., Di Crescenzo, G., Ostrovsky, R., et al. 2004. Public key encryption with keyword search. *Advances in Cryptology-Eurocrypt*: 506–522.

Baek, J., Safavi-Naini, R. & Susilo, W. 2008. Public key encryption with keyword search revisited. *Computational Science and Its Applications–ICCSA*: 1249–1259.

Ballard, L., Green, M., de Medeiros, B., et al. 2005. Correlation-Resistant Storage via Keyword-Searchable Encryption. *IACR Cryptology ePrint Archive*: 417.

Bao, F., Deng, R. H., Ding, X., et al. 2008. Private query on encrypted data in multi-user settings. *Information Security Practice and Experience*: 71–85.

Bellovin, S. M. & Cheswick, W. R. 2007. Privacy-enhanced searches using encrypted bloom filters.

Boneh, D. & Waters, B. 2007. Conjunctive, subset, and range queries on encrypted data. *Theory of cryptography*: 535–554.

Chang, Y. C. & Mitzenmacher, M. 2005. Privacy preserving keyword searches on remote encrypted data. *Applied Cryptography and Network Security*: 442–455.

Chen, F. & Liu, A. X. 2012. Privacy-and integrity-preserving range queries in sensor networks. *IEEE/ACM Transactions on Networking* 20(6): 1774–1787.

Curtmola, R., Garay, J., Kamara, S., et al. 2006. Searchable symmetric encryption: improved definitions and efficient construction. *Proceedings of the 13th ACM conference on Computer and communications security*: 79–88.

Fuhr, T. & Paillier, P. Decryptable searchable encryption.2007. *Provable Security*: 228–236.

Goh, E. J. 2003. Secure Indexes. *IACR Cryptology ePrint Archive*: 216.

Golle, P., Staddon, J. & Waters. B. 2004. Secure conjunctive keyword search over encrypted data. *Applied Cryptography and Network Security*: 31–45.

Gu, C. & Zhu, Y. 2010. New Efficient Searchable Encryption Schemes from Bilinear Pairings. *Network Security* 10(1): 25–31.

Li, J., Wang, Q., Wang, C., et al. 2010. Fuzzy keyword search over encrypted data in cloud computing. *INFOCOM, 2010 Proceedings IEEE*: 1–5.

Pang, H. H., Jain, A., Ramamritham, K., et al. 2005. Verifying completeness of relational query results in data publishing. *Proceedings of the 2005 ACM SIGMOD international conference on Management of data*: 407–418.

Sheng, B. & Li, Q. 2008. Verifiable privacy-preserving range query in two-tiered sensor networks. *IEEE INFOCOM*: 457–465.

Shi, E., Bethencourt, J., Chan, T. H. H., et al. 2007. Multi-dimensional range query over encrypted data. *Security and Privacy, 2007. SP'07. IEEE Symposium on*: 350–364.

Shi, J. Zhang, R. & Zhang, Y. 2009. Secure range queries in tiered sensor networks. *IEEE INFOCOM*: 945–953.

Song, D. X., Wagner, D. & Perrig, A. 2000. Practical techniques for searches on encrypted data. *Security and Privacy, 2000. S&P 2000. Proceedings. 2000 IEEE Symposium on. IEEE*: 44–55.

Tsou, Y. T., Lu, C. S., Kuo, S. Y. 2012. Privacy-and integrity-preserving range query in wireless sensor networks. *Global Communications Conference (GLOBECOM)*: 328–334.

Wang, C., Cao, N., Li, J., et al. 2010. Secure ranked keyword search over encrypted cloud data. *Distributed Computing Systems (ICDCS), 2010 IEEE 30th International Conference on*: 253–262.

Yang, Y., Lu, H. & Weng, J. 2011. Multi-user private keyword search for cloud computing. *Cloud Computing Technology and Science (CloudCom), 2011 IEEE Third International Conference on*: 264–271.

Zhang, R. Shi, J. & Zhang, Y. 2009 Secure multidimensional range queries in sensor networks. *10th ACM MobiHoc*: 197–206.

Zheng, Q., Xu, S. & Ateniese, G. 2013. VABKS: Verifiable Attribute-based Keyword Search over Outsourced Encrypted Data. *IACR Cryptology ePrint Archive*: 462.

Electronics, Communications and Networks IV – Hussain & Ivanovic (eds)
© 2015 Taylor & Francis Group, London, ISBN: 978-1-138-02830-2

A design of UWB planar antenna with band-notched characteristic

J.J. Liao, W.B. Zeng*, X.D. Wu & S.L. Li
School of Electric and Information Engineering, Guangxi University of Science and Technology, Liuzhou Guangxi, P. R. China

ABSTRACT: In this thesis, a planar antenna for Ultra-Wide Band (UWB) application is designed. A spade-shaped patch and an E-shaped slot being used, an excellent band-rejection characteristic is achieved. Measurements show that the antenna gains a frequency bandwidth of 2.03–10.8 GHz with a good impedance matching (return loss ≤ –10 dB), which meets the requirement of UWB application well. Besides, a notched band of 4.98 – 6.32 GHz is designed to avoid the electromagnetic interference caused by the WLAN and HIPERLAN/2 systems. Moreover, the antenna shows a good performance of radiation patterns.

1 INTRODUCTION

In recent years, ultra-wideband technology has attracted significant attention in the field of networking, wireless communication, radar detection, and other applications. Small size, high capacity, low cost, and ease of fabrication are the advantages that have been widely applied in ultra-wideband antenna design (Chuang et al. 2012). The FCC (Federal Communications Commission) announced that the spectrum from 3.1 to 10.6 GHz is for UWB application in 2002; however, there still remains wireless local area network (WLAN) as well as HIPERLAN/2 which are used at 5.15 – 5.825 GHz. They could make severe frequency disturbance to the UWB communication application (Zhang et al. 2012). In order to avoid and solve the effect caused by the frequency band, antennas with band-notched function at these existing bands for UWB applications are required.

Recently, many of ultra-wideband antennas with one or more rejected-bands have been considered. The effective and conventional way to realize the characteristic of band-notched is inserting one or more slots or stubs on the patch, such as embedding L-shaped slot (Liu et al. 2013), T-shaped slot (Zeng et al. 2012), U-shaped slot (Mishra & Mukherjee 2012), tapered-shape slot (Azim et al. 2011) and etc. Besides, adding parasitic structures is also a feasible design. The sizes of these parasitic structures are nearly $\lambda / 2$ or $\lambda / 4$ resonant lengths, corresponding to their notched frequencies.

Firstly, a CPW-fed planar antenna without notch-band for UWB applications is presented in this thesis. The antenna is studied in simulations and verified through experiments. Then, a band-notched antenna design is presented. The band-notched characteristic in the 5 GHz frequency bandwidth is obtained by embedding an E-shaped slot to the original antenna design. Moreover, omnidirectional patterns in the E-plane as well as relatively steady gains have also been obtained from the antenna.

2 CONFIGURATION OF THE DESIGNED ANTENNA

The configuration and a prototype photograph of the designed antenna are shown in Figure 1. The substrate of the proposed antenna is single-side FR4 PCB, and the antenna profile is etched on the metal surface of the substrate. The dimension of the antenna is 30×28×1 mm^3, and the dielectric constant εr and the dielectric loss tangent $\tan\delta$ of the substrate are 4.4 and 0.02, respectively. The design consists of a spade-shaped patch and a 50-Ω step to 100-Ω CPW transmission line. In addition, both the CPW feeding structure and the spade-shaped patch are symmetrical to the y-axis.

Ansoft HFSS software was employed to design and optimize the antenna. Simulation shows that the dimensions of the patch and the distance between the patch and the ground determine the main resonant frequency of the proposed antenna, so it is easy to realize the UWB operating bandwidth simply by optimizing the main dimensions of the antenna. The dimensions (in millimeter) of the optimized antenna are: $W_1 = 12.5$, $W_2 = 4.2$, $W_G = 0.5$, $W_S = 2$, $L_1 = 4$, $L_2 = 6$, $L_3 = 8$, $L_G = 2$, $L_s = 2.5$, $h = 1$.

*Corresponding author: *ZWB@gxut.edu. cn*

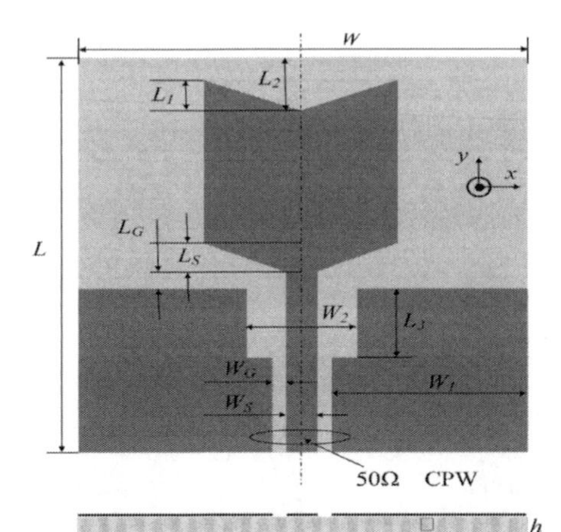

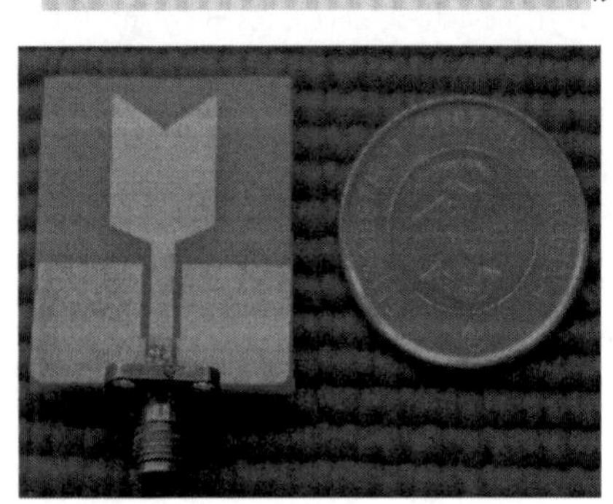

Figure 1. The geometry of the designed antenna.

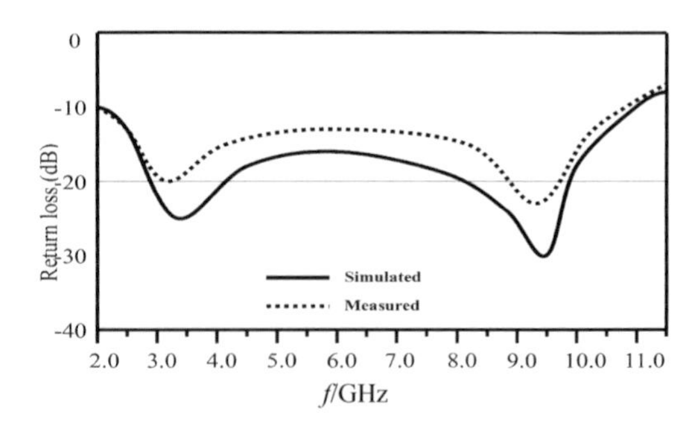

Figure 2. Simulation and measurement return loss of the antenna.

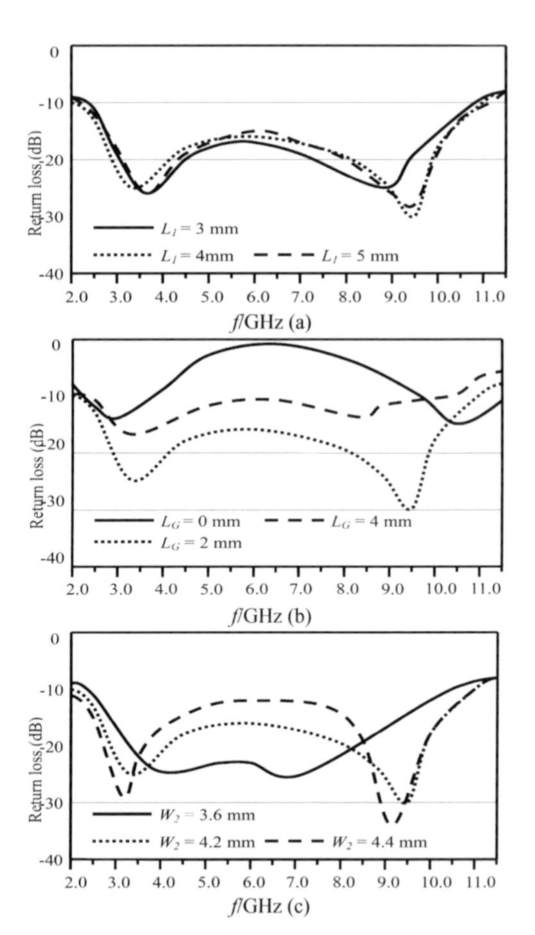

Figure 3. Simulation of the return loss for different (a) L1, (b) LG, and (c) W2.

3 RESULTS AND PARAMETREIC ANALYSE

HP8510C network analyzer is applied to measure the impedance matching of the antenna. The result is shown in Figure 2. Obviously, within the frequency band of 2.03– 10.80 GHz, the return loss is not more than -10 dB, and this frequency band covers the frequency range of 3.1 - 10.6 GHz, which is for UWB applications. Besides, one can observe two nulls in the operating frequency band in Figure 2, which means that the designed antenna can work in the mode of double resonances.

In order to explore the effects of geometrical parameters on the impedance matching, a parametric study has been conducted. Through simulation work the key dimensions are L_1, L_G, and W_2, which determine the main impedance performance of the antenna in the operation band. The other parameters do not affect the impedance of the antenna so much; however, they can be adjusted and optimized to improve the other performance of the antenna.

The return loss curves of the antenna with different L_1 are indicated in Figure 3(a). The bandwidth in which the return loss is less than -10 dB has changed when L_1 increases. In consideration of the bandwidth of the UWB application, the height of $L_l = 4$ mm is the best choice because the impedance bandwidth of the antenna achieves its maximum value of about 6.77 GHz (2.03 – 10.80 GHz) in this case.

Figure 3(b) shows that the distances between the patch and the ground plane have a significant effect on the return loss. When increases from 0 mm to 2 mm with other parameters being fixed, the operation bandwidth of the designed UWB antenna increases. On the contrary, the bandwidth of the proposed UWB antenna decreases as increases from 2 mm to 4 mm.

Furthermore, from Figure 3(c), one can see that the operating bandwidth of the proposed UWB antenna increases by varying from 3.6 mm to 4.4 mm.

4 BAND-NOTCHED DESIGN AND RESULT

The band-notched characteristic of the antenna is obtained by etching an E-shaped slot on the spade-shaped patch; the simulation works show that it is easy to control the band-notched performance by adjusting the size of the E-shaped slot. Figure 4 shows the configuration and a photograph of the band-notched design antenna. The optimized parameters (in millimeter) are as follows : = 6, = 5, = 1, = 8, = 0.5.

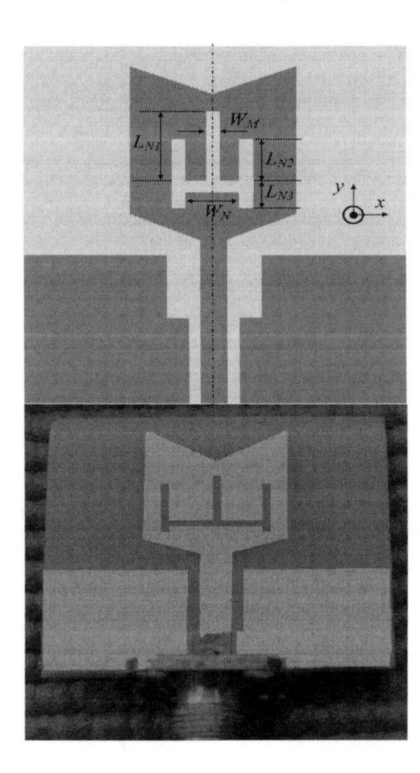

Figure 4. The configuration and a photograph of the band-notched antenna.

The curves of return loss of the band-notched design antenna are shown in Figure 5. The simulation impedance bandwidth (return loss ≤ -10 dB) is about 9.07 GHz (1.98 – 11.05 GHz), and the notched frequency band (return loss ≥ -10 dB) is from 4.82 to 5.88 GHz. Moreover, by measuring, an operating frequency bandwidth of 2.03 – 10.80 GHz and a notched frequency bandwidth of 4.98 – 6.32 GHz are also obtained. Both of the simulated and measured results agree well. The above results indicate that the proposed band-notched design antenna can not only operate in UWB frequency band well but also in filter 5 – 6 GHz interference signal caused by WLAN as well as HIPERLAN/2 systems.

The simulation radiation patterns at different frequencies are depicted from Figure 6 (a), (b) and (c). The band-notched design antenna achieves an omni-directional radiation characteristic in the H-plane. Besides, the antenna becomes more directive when the frequency increases.

Figure 7 exhibits the simulated surface electric current distribution in three typical frequencies of 3.5, 5.5 and 9 GHz. As shown in (a) and (c), the density of the surface current on the feeding point as well as the area between the ground plane and the patch is stronger than that around the E-shaped slot relatively. This phenomenon indicates that the introducing of E-shaped slot can not influence the characteristic of the original antenna within the UWB operation band significantly. At the frequency of 5.5 GHz, as shown in Figure 7 (b), the surface current intensity increases obviously around the E-shaped slot, and this will cause the sudden change of the input impedance, which can result in characteristic of the band-notched.

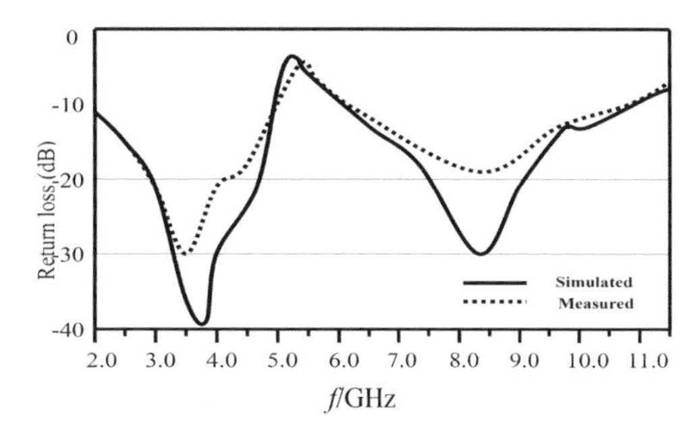

Figure 5. Simulation and measurement return loss of the band-notched antenna.

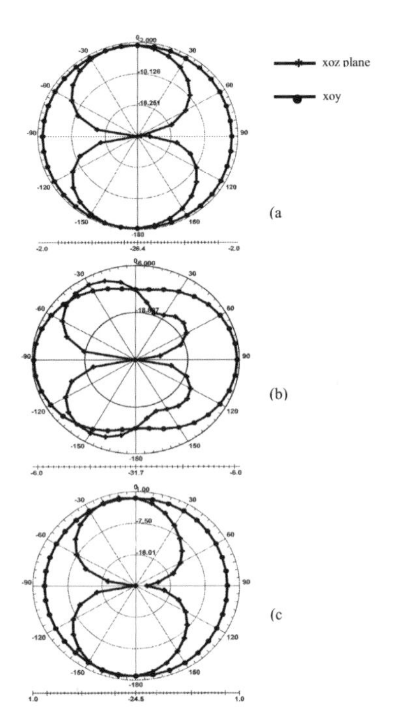

Figure 6. Simulation radiation patterns at (a) 3.5 GHz, (b) 5.5GHz and (c) 9 GHz.

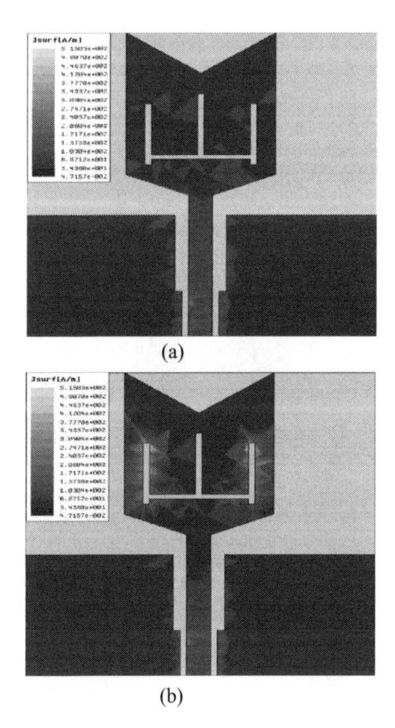

(a)

(b)

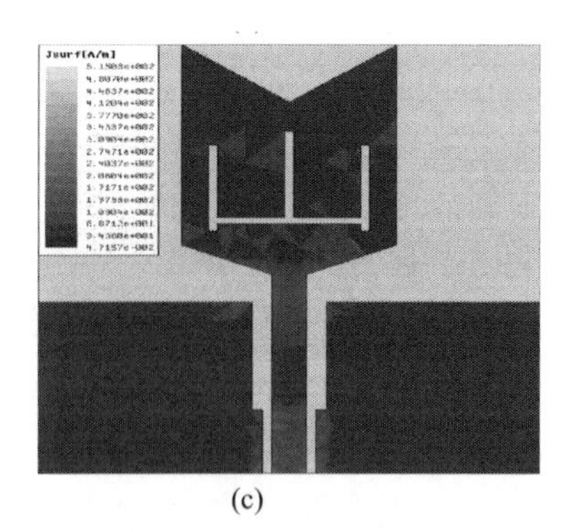

(c)

Figure 7. Simulated surface electric current distributions at (b) 5.5 GHz, and (c) 9 GHz.

5 CONCLUSION

A compact UWB planar antenna is reported in this thesis. The antenna was designed and optimized by using the software of HFSS and then tested experimentally. The size of the antenna prototype is 30×28×1 mm^3. Simulation and measurement results indicate that the antenna can achieve a good impedance matching and a stable radiation characteristic over the UWB frequency band. Parametric studies have been conducted to explore the effects of key geometrical parameters on the impedance characteristics of the designed antenna. Moreover, a band-notched design is also presented, and by etching an E-shaped narrow slot to the original design, the notched-band of 5– 6 GHz is achieved.

ACKNOWLEDGEMENT

This study is partially supported by the Research Projects for higher education of Guangxi, China (No. 2010JGZ028).

REFERENCES

Azim, R., Islam, M.T. & Misran, N. 2011. Compact tapered-shape slot antenna for UWB applications. 10: 1190–1193.

Chuang, C.T., Lin, T.J. & Chung, S.J. 2012. A band-notched UWB monopole antenna with high notch-band-edge selectivity. 60(10): 4492–4499.

Liu, X.L., Yin, Y.Z., Liu, P.A. et al. 2013. A CPW-fed dual band-notched UWB antenna with a pair of bended dual-L-shape parasitic branches. 136: 623–634.

Mishra, S.K. & Mukherjee, J. 2012. Compact printed dual band-notched U-shape UWB antenna. 27: 169–181.

Zeng, W.B., Zhao, J. & Wu, Q.Q. 2012. Compact planar ultra-wideband wide-slot antenna with an assembled band-notched structure. 54(7): 1654–1659.

Zhang, S.M., Zhang, F.S., Li, W.Z. et al. 2012. A compact UWB monopole antenna with WiMAX and WLAN band rejections. 31: 159–168.

Electronics, Communications and Networks IV – Hussain & Ivanovic (eds)
© 2015 Taylor & Francis Group, London, ISBN: 978-1-138-02830-2

TOC: Lightweight event tracing using online compression for Wireless Sensor Networks

Wen Liu*
School of Software Technology, Zhejiang University, Hangzhou, China

Chenhong Cao, Yi Gao & Jiajun Bu
College of Computer Science, Zhejiang University, Hangzhou, China

ABSTRACT: Several trace based diagnostic techniques have been proposed for abnormal detection and fault diagnosis in Wireless Sensor Networks (WSNs). Such techniques usually incur large runtime overhead. However, massive trace collection at runtime is a non-trivial task in resource-constraint WSN nodes. To reduce the trace size, several compression algorithms have been applied to WSN applications, yet the implementation of such methods is still challenged by constraining resource of sensor nodes. To address these problems, we propose TOC, a new event Tracing technique using Online Compression, which combines periodical pattern mining and efficient token assignment. TOC is able to effectively reduce the trace size with acceptable runtime overhead. We implement our method based on TinyOS 2.1.2 and evaluate its effectiveness by case studies in sensor network applications. Results show that TOC reduces the size of event trace to 39.9% and 47.8% on average, compared with the straightforward tracing method and state-of-the-art method.

1 INTRODUCTION

Wireless Sensor Networks (WSNs) are being increasingly deployed in a variety of fields to help monitoring and understanding of the micro-behavior of the physical world. Some of the latest WSNs projects include GreenOrbs (Liu et al. 2011), CitySee (Mao et al. 2012), LOFAR-agro (Langendoen et al. 2006), and etc.

Unfortunately, WSNs are very susceptible to failures after being deployed in the severe environment, such as an urban field (Du et al. 2014), volcanoes (Werner-Allen et al. 2006) and mountains (Hasler et al. 2008). For deployed projects, an effective way to troubleshoot the root cause of failures is to trace important system events. Many techniques have been recently proposed for system diagnosis based on log analysis to help developers to localize the potential bugs of the system (Khan et al. 2008, Luo et al. 2006, Krunic et al. 2007). However, due to the highly limited storage and computational resource of sensor nodes, it is a non-trivial task to obtain and store system logs in a lightweight manner.

Some logging and tracing techniques have been proposed recently to reduce the run-time computational and storage overhead of logging (Shea et al. 2010, Tsiftes et al. 2008, Goeman et al. 2001, Welch. 1984). However, they do not take more efficient logging mechanism into consideration, and also the high run-time computational overhead, which would probably introduce the Heisenbugs into the system.

We notice that, the behaviors of WSNs are highly repetitive. In most cases, a small number of system events occur frequently, while the rest of the events occur rarely. Based on these observations, we propose TOC (Tracing using Online Compression), a new event tracing method using online compression, which combines periodical pattern mining and efficient token assignment. TOC employs Fast Fourier Transform (FFT) method to capture the systematic periodical pattern and further allocates the number of bits of each token based on the times each event appears during system periods. Due to the consideration of event frequency in system behavior periods, TOC could efficiently reduce the trace size.

We implement our method based on TinyOS 2.1.2 (Levis et al. 2005) and evaluate its effectiveness by case studies in sensor network applications. Results show that our method incurs an acceptable overhead in terms of space and computation while reducing the size of event trace to 39.9% and 47.8% on average compared with the straightforward tracing method and the state-of-art method.

*Corresponding author: *liuwen@emnets.org*

The contributions of this work are summarized as follows.

- *We propose a novel event tracing method, which could dynamically optimize event trace size based on runtime system behaviors.*
- *We propose a lightweight method to mine the systematic periodical pattern of the event sequence and provide an effective accounting unit based on event frequency which is used by event token reassign.*
- *We implement our method and demonstrate its effectiveness in reducing trace size in real world sensor networks.*

The rest of this paper is structured as follows. Section 2 describes the related work. Section 3 introduces the design principles. Section 4 shows the evaluation results, and finally, Section 5 concludes this paper and gives directions of future work.

2 RELATED WORK

There are numerous works focusing on the reduction of system runtime trace. We classify existing works into two main categories: static methods and dynamic methods.

LIS (Shea et al. 2010) is a typical static method. It provides a language to gather runtime information efficiently by using local namespaces and bit-aligned logging. To achieve an effective reduction of log size, LIS defines three kinds of tokens (global, local and point). The global token is the mark of the start of a function and the point token is the mark of the end of a function. These two tokens constitute a function namespace scope. In such a scope, local type token could be parsed into a unique token in the whole system space. Due to the reuse of local type token, the bits of each local token, as well as global token, could be saved, and a significant reduction in trace size would be achieved compared to assign all system events a global token.

On the other hand, dynamic methods handle the trace at run-time. One method in this category is to directly employ a compression algorithm in runtime trace once trace buffer is full. Such methods include two widely used UNIX utility programs gzip (LZ77 (Tsiftes et al. 2008)) and compress (LZW (Welch 1984). These two methods also been used for trace compression. However, these techniques would incur considerable computational and storage overhead during compression, therefore, are inapplicable for WSNs due to extreme resource constraints.

To address such challenges, several compression algorithms such as FCM (Goeman et al. 2001) and SLZW (Sadler & Martonosi 2006) are proposed in the purpose of reducing the amount of calculation requirements at runtime. However, these two methods need massive training data to learn a model to predict value or look up the encoded index during the processing of the trace. When the model cannot be obtained or the generated model fail to reflect the comprehensive pattern of trace, the performance of such algorithms would drop rapidly in some extreme cases, the compressed trace may consume more space than the original one.

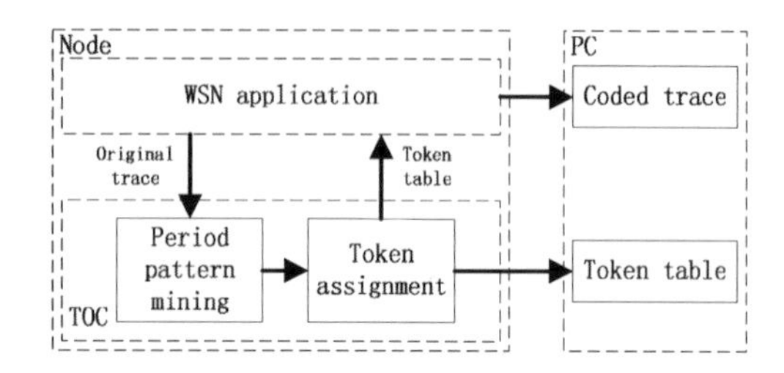

Figure 1. TOC overview.

3 DESIGN

In this section, we present the design of TOC. Section 3.1 presents an overview of TOC. Section 3.2 describes how TOC obtains event log at runtime. Section 3.3 describes how TOC discovers the periodical pattern of system behaviors by mining event log. Section 3.4 demonstrates how TOC reassigns token for each event based on their frequency in the log.

3.1 *Architecture overview of TOC*

Figure 1 shows an overview of TOC. First, TOC obtains the runtime event trace of the system and stores them into RAM of WSN nodes. After collecting enough information, TOC employs FFT to capture the periodical pattern of the event sequence and further calculate the frequency of each event. Finally, based on the different frequencies, TOC assigns each event a new token with variable-length to perform efficiently tracing at runtime, so the system trace could be decoded from the coded trace and the token table.

3.2 *System event sequence collection*

The first step in TOC workflow is to get the system event sequence.

Typically, WSN application is composed of two phases: the initialization phase and the functional phase. In most cases, the abnormal system behaviors occur during the functional phase. Therefore, the monitoring of system events behavior of the functional phase in resource-constraint WSN node is of high value compared to the initialization phase. After the boot of WSN node, TOC first holds on a time interval to make sure the initialization phase is finished. When the node entering the functional phase, TOC will monitor the system event trace, and buffer the event ID sequence into node RAM. Once the collected event ID sequence is long enough, TOC will stop event collection and perform periodical pattern mining.

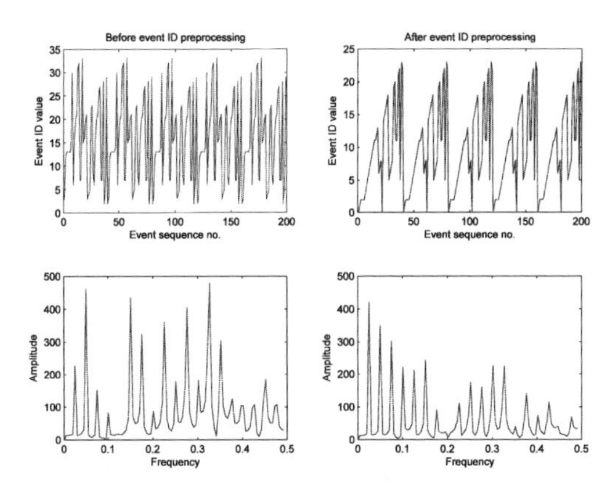

Figure 2. Event sequence and result of FFT before and after event ID preprocessing.

3.3 *Periodical pattern mining*

After obtaining the event ID's sequence during the functional phase, we use a periodical pattern mining technique to find out the period of the sequence. Further, we calculate the frequency of each event.

We employ Fourier Transform method to process the event ID sequence. Note that, Fourier Transform is designed to analyze the frequency feature of the signal in time domain, and the amplitude of signal wave in time domain is meaningful in the process of Fourier analysis. However, the number of event ID indicates nothing except distinguish itself among others in the event ID sequence, which would make Fourier Transform unlikely to capture the real periodical feature in event sequence. To make Fourier Transform meaningful in this scenario, we adopt preprocessing on the ID of each event. The key idea of event ID preprocessing is to reassign each event a new ID value based on the order of the first appearance in the sequence of each event. An example is shown in Figure 2. The figure on the left and right is the event ID sequence and the result of Fourier Transform on this sequence before and after preprocessing, respectively. It is easy to find that, the original event ID sequence is relatively disorder compare to the preprocessed one. In this example, the dominated frequency of this event sequence is 0.025, which means the period of this sequence is 40.

After preprocessing the event ID in collecting sequence, we could calculate the sequence period. We employ an FFT method to find the dominated frequency component, so the periodical pattern of event traces is the reciprocal of this frequency. To get the most likely bias-free result, we take the longest period among the components with a high ranking of amplitude as the system periodically.

3.4 *Event trace encoding*

When we got the period of event sequence, we could perform Huffman coding based on the different frequencies of each event to achieve space saving purpose. To make full use of the buffered data to obtain an accurate result, we calculate occurrence frequency of each event in max integral multiple of periods in the buffered event sequence. Then we perform Huffman coding method to calculate the variable-length code of each event. We use one extra flag byte to indicate whether the encoding step is performed. When the encoding phase is executed, the flag is set. When an event happens in WSNs application, the system first check the flag, then trace the event's corresponding code based on the system state (before or after Huffman encoding).

4 EVALUATION

In this section, we present an evaluation of TOC in terms of overhead and compression ratio.

4.1 *Benchmarks*

In order to evaluate the overhead of TOC on real sensor nodes, we investigate four typical benchmarks on TelosB node with 8MHz MSP430f1611 processor (Msp430f1611 Datasheet), 128KB program size, 10KB memory size, 1MB external flash size, and 250Kbps CC2420 radio.

All four benchmarks are in the TinyOS 2.1.2 distribution: RadioCountToLeds, TestDip, Test Dissemination, and TestNetwork. We instrumented main, LEDs, radio or network layer, and timer components in these benchmarks to get runtime system trace.

4.2 *Overhead*

4.2.1 *RAM overhead*

TOC requires storing a string of event ID on RAM to perform periodical pattern mining. In addition, TOC needs extra RAM to perform Huffman coding and store Huffman code of each event.

We first studied the event trace of the four benchmarks offline. We found that the periods of event trace at the normal level of instrumentation varied from 40 to 100.

To capture the periodical pattern of event trace accurately while consuming RAM consumption as less as possible, we choose to buffer 256 events into RAM for further pattern mining which will occupy 2K bytes. Consider the effectiveness of FFT, we build a sine/cosine table for later looking up, which would take additional 256 bytes. Due to Huffman coding is a light weight entropy coding algorithm, the RAM consumption at Huffman coding step is only a few hundred bytes.

Table 1. Program flash size (K bytes).

Benchmarks	Without TOC	With TOC
RadioCountToLeds	11.35	16.02
TestDip	18.79	23.47
TestDissemination	13.19	17.87
TestNetwork	27.10	31.77

Note that when the periodical pattern mining is done, TOC could release the RAM, which occupied by FFT, so the maximum RAM consumption during the workflow of TOC is only the RAM required by periodical pattern mining which is approximately 2.3K. Overall, the RAM overhead is acceptable since TelosB has a total of 10K RAM.

4.2.2 *Program flash*

Table 1 shows the program flash size of the four benchmarks before and after adding TOC module. We could easily find that TOC introduces program flash size increase 4.7K bytes on average. Consider the program flash of TelosB is 48K bytes, the overhead of TOC is acceptable.

4.2.3 *CPU consumption*

We evaluate the CPU consumption on TelosB node. As shown in Figure 3, the time consumption difference among each benchmark is very limited. Consider the mechanism of TOC, the dominated CPU consumption incurs at periodical pattern mining step and the event trace encoding step to perform FFT and Huffman coding algorithms correspondingly.

4.3 *Compression ratio*

In this section, we evaluate the performance of TOC on compression ratio.

We implement three kinds of approaches to compare with TOC.

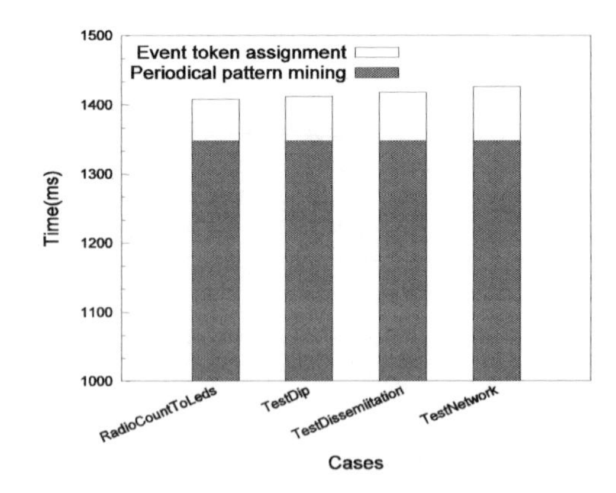

Figure 3. Time consumption of TOC.

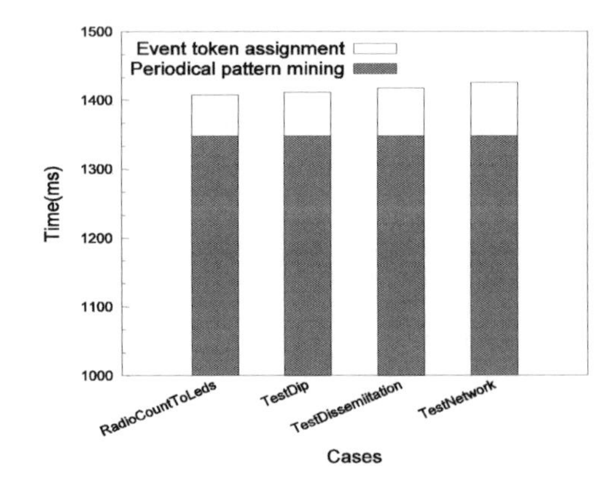

Figure 4. Compression ratio.

- *LIS: utilize local namespace to reduce trace size.*
- *TOC: adopt our method on trace which is obtained by Naive approach.*
- *TOC on local token of LIS: perform TOC method based on local token of LIS.*

Figure 4 shows that LIS could reduce the trace size 20% on average. This number is relatively small compared to the evaluation on (Shea et al. 2009), it is because in our cases, the sum of all traces points and the number of local type trace point is less. In such cases, the trace size can hardly get benefit from LIS mechanism. Compared to LIS, TOC could reduce the trace size aggressively. TOC could achieve an aggressive compression ratio 35% to 55%. When we implement TOC on local token of LIS, TOC further reduces the trace size to 40% on average.

5 CONCLUSION

We propose a novel online trace compression technique, TOC, which exploited the fact that WSNs computations are highly repetitive and do not evolve much over time. TOC combines periodical pattern mining and efficient token assignment, which could effectively reduce trace size while incurring acceptable runtime overhead. Compared to existing methods, TOC could adapt to various WSN applications online based on their system behaviors. To mining the pattern of WSNs application, TOC only needs a small set of training data and only incurs limited calculation and space consumption while processing the training data to get the periodical pattern of event sequence and assigning each event a new token. What's more, TOC does not incur additional calculation overhead at run time once the token assignment step complete.

We further implement TOC based on TinyOS 2.1.2 and evaluated its effectiveness by case studies in sensor network applications. Results show that TOC could reduce the size of event trace to 39.9% and 47.8% on average compared to straightforward tracing method and the state-of-art methods with optimizing the bandwidth of event trace.

However, there are still several aspects could be further considered to improve the performance and scalability of TOC: TOC can utilize sequence pattern to further optimize online trace compression; TOC could improve its scalability in compression on more generic traces, such traces include not only event ID, but also variable value and timestamp etc.

ACKNOWLEDGMENTS

This work is supported by National Key Technology R&D Program (2012BAI34B01).

REFERENCES

Du, W., Xing, Z., Li, M., et al. 2014. Optimal Sensor Placement and Measurement of Wind for Water Quality Studies in Urban Reservoirs. In *Proc. of ACM/ IEEE IPSN.*

Goeman, B., Vandierendonck, H. & Bosschere, K. 2001. Differential FCM: increasing value prediction accuracy by improving table usage efficiency. In *Proc. of HPCA.*

Hasler, A., Talzi, I., Beutel, J., et al. 2008. Wireless sensor networks in Permafrost research - concept, requirements, implementation and challenges. In *Proc. of NICOP.*

Khan, M., Le, H., K., Ahmadi, H., et al. 2008. Dustminer: Troubleshooting interactive complexity bugs in sensor Networks. In *Proc. of ACM SenSys.*

Krunic, V., Trumpler, E. & Han, R. 2007. NodeMD: Diagnosing node-level faults in remote wireless sensor. In *Proc. of ACM MobiSys.*

Langendoen, K., Baggio, A. & Visser, O. 2006. Murphy loves potatoes: Experiences from a pilot sensor network deployment in precision agriculture. In *Proc. of WPDRTS.*

Levis, P., Madden, S., Polastre, J., et al. 2005. TinyOS: An operating system for wireless sensor networks. *Ambient Intelligence.* Berlin Heidelberg: Springer.

Liu, Y., He, Y., Li, M., et al. 2011. Does wireless sensor network scale? A Measurement Study on GreenOrbs. *In Proc. of IEEE INFOCOM.*

Luo, L., He, T., Zhou, G., et al. 2006. Achieving repeatability of asynchronous events in wireless sensor networks with EnviroLog. In *Proc. of IEEE INFOCOM.*

Mao, X., Miao, X., He, Y., et al. 2012. CitySee: Urban CO2 Monitoring with Sensors. In *Proc. of IEEE INFOCOM.*

Msp430f1611 Datasheet. *http://focus.ti.com/docs/prod/folders/ print/msp430f1611.html.*

Sadler, C. M. & Martonosi, M. 2006. Data compression algo-rithms for energy-constrained devices in delay tolerant net-works. In *Proc. of SenSys.*

Shea, R., Srivastava, M. & Cho, Y. 2009. Optimizing bandwidth of call trace for wireless embedded systems. In *IEEE Embedded Systems Letters.*

Shea, R., Srivastava, M. & Cho, Y. 2010. Scoped identifiers for efficient bit aligned logging. In *Proc. of DATE.*

Tsiftes, N., Dunkels, A. & Voigt, T. 2008. Efficient Sensor Network Reprogramming through Compression of Executable Modules. In *Proc. of IEEE SECON.*

Welch, T. A. 1984. Technique for high-performance data compression. In *IEEE Computer* 17 (6).

Werner-Allen, G., Lorincz, K., Johnson, J., et al. 2006. Fidelity and yield in a volcano monitoring sensor network. In *Proc. of OSDI.*

Electronics, Communications and Networks IV – Hussain & Ivanovic (eds)
© 2015 Taylor & Francis Group, London, ISBN: 978-1-138-02830-2

Research on location-based service in distributed interactive multimedia system

Shan Liu* & Jianping Chai
School of Information Engineering, Communication University of China, Beijing, China

ABSTRACT: This paper presents the importance and feasibility of the location-based services with personalized, distributed and interactive multimedia system for digital campus. Our goal is to achieve the multimedia content distribution and sharing information in the wireless roaming network by designing system function modules and terminal server application modules. The results demonstrate that the designed system can provide a channel for external communication for school management level, and a platform for internal resource sharing. The designed system can also provide a channel for independent study and information exchange platform for school staff and visitors.

1 INTRODUCTION

With the development of new media technologies and widely spread use of new mobile media, such as smart phones and tablet computers, the people's daily life have greatly changed. Recently, the concepts of the digital city, digital home, digital campus, etc. have become more and more popular (Wang 2011). In addition, the information and communication technology together with the location-based services enable us to acquire the personalized information and resources conveniently (Reed & Rappaport 1998).

Nowadays, the mobile positioning technology has been widely used in our daily life. The mobile positioning technology makes use of mobile communications network technology, by measuring and computing the wireless signal that received via mobile terminal, to acquire the location information of the mobile device or a particular person. In this way, we can provide various services related to the location information, which is also called "Location-Based Service" (LBS).

The distribution of the multimedia content over the Internet mainly uses the content distribution network (CDN) technology, which builds the overlay network between the applications and Internet (Song & Tao 2005). The CDN technology makes up many defects of current IP network, which provides a flexible and efficient content distribution service for multimedia applications (Yin et al. 2012). This technology has been widely used in the applications of multimedia network and has brought convenience to the large-scale video and audio content distribution over the Internet.

Related works can refer to the services of digital city and the digital home, which also aims to improve the people daily life by using information and communication technology. In addition, the users tend to get various kinds of services in a certain range and access to a variety of resources using mobile devices recently. Especially for the campus, how to make the teachers and students teaching and learning effectively has become more and more important.

2 SYSTEM ARCHITECTURE

2.1 *System architecture overview*

The system platform uses multimedia content as the sources of information, and uses the computer network as information dissemination channels, by which the sources of information are distributed to the user terminal. The information sources and system equipment, together with the management of terminal users are maintained and consolidated by the system maintenance personnel in the background management module. The user's terminal requests the multimedia content via the applications, which the server has already prepared.

The whole system is separated with three different layers: central server, sub-network and user group. The central server provides service to terminal user and administrator, also dispatches the database and controls the network. The sub-network is a wireless local area network mainly consists of routers, which accesses the backbone network and working as a data bridge. The user group mainly consists of applications in user's terminal. The system architecture is shown in Figure 1.

*Corresponding author: *liushan@cuc.edu.cn*

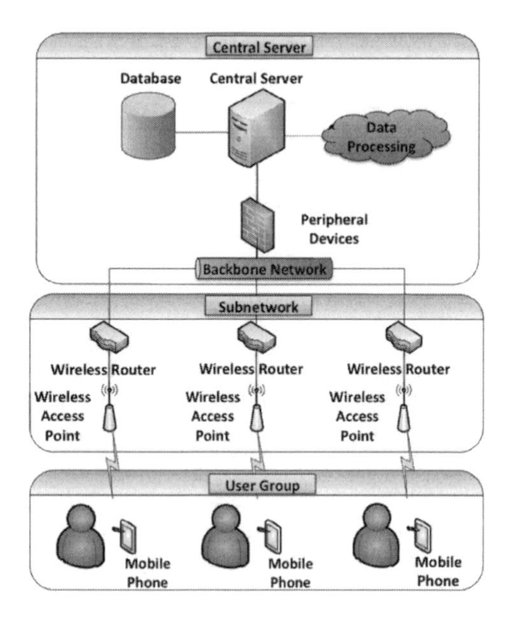

Figure 1. System architecture.

2.2 *System mode design*

With the development of network and computer technology, the current architecture of the system can be generally categorized into three modes: single mode (Zhang et al. 2007), mode of client/server (C/S), mode of browser/server (B/S).

In this paper, the B/S mode is chosen to design and develop our system. The B/S mode stores the data on a network server where the user uses the browser to access the data through the Web server via the Internet. The Web server plays the role of intermediaries in the whole process, which the Web server receiving the request from a browser and then forward the request to the application server. The application server processes the request and returns the result to the Web server. Finally the result will be presented in the user's browser through the Web server via the Internet.

2.3 *System architecture design*

The system is divided into the central server module and terminal application modules. The central server module consists of the background management module and a terminal access module, which the background management module including the functional modules of the system maintenance personnel. The detail structure of the terminal application module and central server module are shown in Figure 2 and Figure 3, respectively.

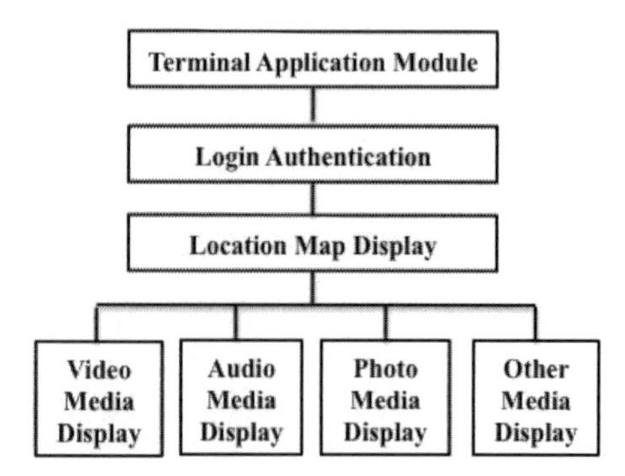

Figure 2. Structure of terminal application module.

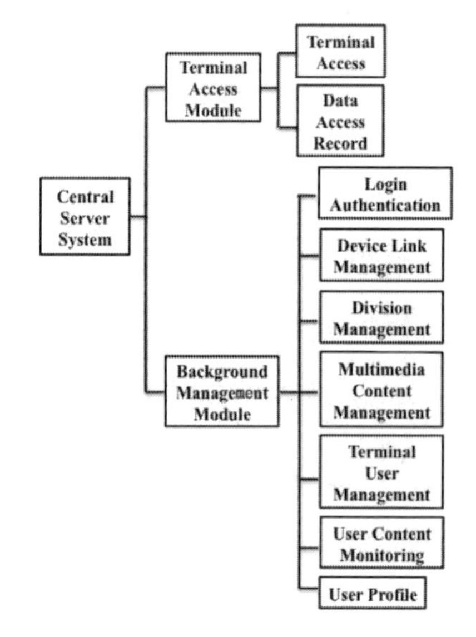

Figure 3. Structure of central server module.

3 WIRELESS LOCAL AREA NETWORK DESIGN

In our system, the handheld device of the terminal user will access the wireless network and connect to the system server. The divided regions will sometimes exceed the coverage of the single wireless device, so we need to expand the coverage of the wireless network by wireless extension. The common method is to increase the wireless access point in a certain area to expand wireless coverage, but there still will be a big disadvantage. When a user using a handheld device moves among different access points, he/she needs to find the network first and reconnect to the network again, or even reset the device each time. To solve this problem, the technologies of wireless network roaming can be used.

When there are multiple access points in a particular network and their micro units have a range of overlap with each other, the wireless user can move between the wireless coverage areas where the wireless network card can automatically find the access point with the maximum signal strength and choose to send and receive data via this access point. By this way, the wireless user will cut off the previous access point and maintain the network connection with the later access point without being interrupted. It appears to be that the user is among an uninterrupted network environment, which is called wireless roaming.

4 EVALUATION

4.1 *Background management system test*

In our system, we used black box test method to see if our goal of system functional design can be achieved. The specific information on test performance and results for each case are shown in Table 1. From the test we can see that the background management system can work correctly in each case and the various functional modules are able to realize the expected effect, which also can accept a large number of client accesses.

4.2 *Network environment test*

The test is focused on testing of the wireless roaming function in the campus wireless network. The specific information of test performance and results for each case are shown in Table 2. And we can see that the realization of wireless roaming function of the designed wireless network environment can be achieved. And the designed network provides the terminal users with a high-speed multimedia access rate.

4.3 *Terminal application test*

The test uses black box test method. We will test each function of the terminal applications in different cases. The results of test performance and results are listed in Table 3. From the test result, we can see that the terminal applications can provide the reliable services, and can run smoothly on the mobile phone.

Table 1. Description of test performance and results.

Case No.	Test Name	Test Description	Test Result
01	System access	Access the server from the client.	Success. Access succeed.
02	System login	Login with different users and test the user information authentication.	Success. User login succeed and the authentication operation can be used to warning the unauthorized user .
03	Hotspot management	Monitor, change and filter information of hotspots.	Success. The action of monitoring, filtering, addition and deletion succeed.
04	Device management	Monitor, change and filter information of devices.	Success. The action of monitoring, filtering, addition and deletion succeed.
05	Multimedia management	Monitor, change and filter multimedia resources.	Success. The action of monitoring, filtering, addition and deletion succeed.
06	Binding of hotspot and multimedia	Bind and unbind of hotspot and multimedia.	Success. The action of binding and unbinding succeed.
07	User management	Monitor, change and filter information of users.	Success. The action of monitoring, filtering, addition and deletion succeed.
08	User profile management	Monitor and record the user's action and provide password reset..	Success. The user's action is recorded and the password can be reset.
09	Stability test	Test the system stability.	Success. The system keeps running in stable states and can respond immediately.
10	Stress test	Increase the connections of server and test the system response.	Success. The response of the server is stable when increase the connections.

Table 2. Description of test performance and results.

Case No.	Test Description	Test Result
01	First, run the server program and open the WiFi of the mobile. Then run the terminal application to map interface, using the mobile close to one wireless router A. Finally, slowly move to another wireless router B.	Success. When the mobile moves from wireless router A close to wireless router B, the location of the terminal application re-located to the wireless router B.
02	Using the terminal application installed on the mobile to access a video and keep playing. Then move from wireless router A to wireless router B.	Success. When moving close to wireless router B, the network connection switches from wireless router A to wireless router B and the video play smoothly while moving.

Table 3. Description of test performance and results.

Case No.	Test Item	Test Description	Test Result
01	System Access	Login the application to access the server.	Success. Access succeed.
02	Map Resources	Map display, hotspots display and position display.	Success. The map and hotspots can be displayed and the position is correct.
03	Video Media	Request to video media display and download video resources.	Success. All the video resources display correctly and can be downloaded.
04	Audio Media	Request to audio media display and download audio resources.	Success. All the audio resources display correctly and can be downloaded.
05	Photo Media	Request to photo media display and download photo resources.	Success. All the photo resources display correctly and can be downloaded.
06	Other Media	Request to other media display and download resources.	Success. All the other resources display correctly and can be downloaded.
07	Access Record	Record the connections information, user login and logout information and the access information of multimedia resources.	Success. All the actions including the connections information, user's action and access information of multimedia can be recorded.

5 CONCLUSION

In this paper, we have introduced an approach to design a location-based service with personalized, distributed and interactive multimedia system for digital campus. The study of the key technologies and the module design provide us an overview of the system structure and the main work of this paper. We have analyzed the system characteristic and demand, proposes a feasible scheme. The study of the key technologies of the system implementation helped us to choose a suitable development technology to develop the system. The analysis of the system based on the demand of the design, system structure, system function and database structure has been pointed out. The Java Web technology and Android application development technology are used to realize each part performance. Based on the system performance and results, we can conclude that system can provide a communication channel for the school management, a platform for sharing resources on digital campus. Future work will expand our work to larger network environment and more users.

REFERENCES

Reed J. & Rappaport, T. 1998. An Overview of the Challenges and Progress in Meeting the E-911 Requirement for Location Service. *IEEE Communication Magazine*: 30–37.

Song, J. & Tao, H. 2005. The Development and Application of Content Distribution Network Technology. *Television Technology* 6: 75–77.

Wang Y. 2011. Survey of Digital Campus Construction in China. *Modern Distance Education on Research* 4(112): 39–50.

Yin, H., Zhan, T. & Lin, C. 2012. Multimedia Network: From Content Distribution Network to Future Internet. *Chinese Journal of computers* 6: 1120–1130.

Zhang, X., Peng, G. & Wei, W. 2007. Research of Marine Traffic Information Sharing Technology Based on WEB. *Institute of Navigation China 2007 Annual Symposium*, 2007: 44–48.

Electronics, Communications and Networks IV – Hussain & Ivanovic (eds)
© 2015 Taylor & Francis Group, London, ISBN: 978-1-138-02830-2

Adaptive multi-user resource allocation with partial information

Lihan Liu* & Hong Wu
School of Economics and Management, Beijing University of Posts and Telecommunications, Beijing, People's Republic of China

ABSTRACT: In this paper, the effect of imperfect Channel State Information (CSI) on the system performance of radio resource management for a downlink multi-user CDMA system was investigated. First, the Symbol Error Rate (SER) was proven to be constellation-point dependent due to its imperfect CSI, which correspondingly yielded an error floor on the SER. An iterative bisection algorithm, which would minimize the total transmit power under a certain quality requirement, was proposed under consideration that the throughput function is always nonconvex and nondifferentiable in real communication systems.

1 INTRODUCTION

In recent years, due to its ability to achieve high spectral efficiency, the CDMA system has captured growing interest in the wireless communication field. However, efficient radio resource management among multiple users is still a challenge in the design of CDMA systems. Adaptive resource management, in terms of assigning powers, channels, and rates in CDMA systems, has been extensively investigated over the past several years (Meshkati et al. 2007, Jafar & Goldsmith 2003, Lee et al. 2005, Musku et al. 2010). Various optimization algorithms, with objective functions such as throughput/rate maximization and power minimization under certain constraints, have been discussed in detail based on the availability of perfect channel state information (CSI). Unfortunately, in real systems, perfect CSIs are not always available or practical.

Many researchers have investigated the effect of imperfect CSIs on the downlink and uplink transmission of CDMA systems (Wong & Evans 2009) (Mokari et al. 2010, Awad et al. 2010, Wang et al. 2011). Most of these studies assumed that the noise variance was constant and not affected by the imperfect CSI. The first effort to analyze the effect of the signal-and-power dependent noise caused by imperfect CSIs is shown in works (Wang & Liu 2014, Wang et al. 2014). However, only the OFDMA system was considered to achieve the optimal radio resource management.

In this paper, the result of (Wang & Liu 2014, Wang et al. 2014) was extended to include downlink multi-user CDMA transmission. In addition, under consideration that the throughput function in practical communication systems is nonconvex and nondifferentiable, an iterative bisection algorithm was proposed in order to solve the optimization problem of adaptive resource allocation for downlink CDMA systems with imperfect CSIs.

In Section 2, the system model and problem formulation are presented. The effects of imperfect CSIs are discussed in Section 3. In Section 4, the iterative bisection algorithm for the optimization problem is derived. The numerical results and conclusions are given in Sections 5 and 6, respectively.

2 SYSTEM DESCRIPTION AND PROBLEM FORMULATION

A downlink multi-user CDMA system with K users communicating with the same base station was considered for the purposes of this paper. The available frequency band was assumed to be divided into equi-width independent subchannels, and the fading of each subchannel was assumed to be constant during a frame period. This study primarily focused on square M-QAM modulation. Hence, each symbol that included the in-phase and quadrature components, which were assigned PN sequences to spread the desired signal. The received signal, after de-spreading and demodulation, is given by (Abu-Rgheff 2007)

$$y_{k,n} = \sqrt{P_{k,n}}h_{k,n}x_{k,n} + \omega_{k,n} + I_{k,n} \tag{1}$$

In this equation, $x_{k,n}$ is the complex transmitted signal, $P_{k,n}$ is the transmit power, $h_{k,n}$ is the channel fading, and $\omega_{k,n}$ and $I_{k,n}$ are the Guassian noise and interference of user k on n-th subchannel, respectively. The effect of the spreading factor was included in the interference term.

*Corresponding author: *lihanliu.nyu@gmail.com*

The decision signal is given by

$$z_{k,n} = y_{k,n}\frac{h_{k,n}^{*}}{\left|h_{k,n}\right|^{2}} = \sqrt{P_{k,n}}x_{k,n} + \omega_{k,n} + I_{k,n}\ , \tag{2}$$

where $\omega_{k,n}$ and $I_{k,n}$ are the noise and interference, respectively.

In the paper, the utility function is defined as the throughput in the research of Wang & Liu (2014) and Wang et al.(2014). That is,

$$u_{k,n} = T_{k,n}\left(P_{k,n}\right) = \max_{M}\left\{T_{k,n}^{M}\left(P_{k,n}\right)\right\}, \tag{3}$$

where $T_{k,n}\left(P_{k,n}\right)$ is the throughput of user k on n-th subchannel, and $T_{k,n}^{M}\left(P_{k,n}\right)$ is the throughput with constellation size M, which is given by (Meshkati et al. 2007).

$$\begin{aligned} T_{k,n}^{M}\left(P_{k,n}\right) &= L\log_2 M \\ &\times\left[1-\frac{2\left(\sqrt{M}-1\right)}{\sqrt{M}}Q\left(\sqrt{\frac{3P_{k,n}\left|h_{k,n}\right|^{2}}{(M-1)\left(\sigma_{\omega,k,n}^{2}+\sigma_{I,k,n}^{2}\right)}}\right)\right]^{2L} \end{aligned} \tag{4}$$

In this equation, L is the frame size of symb.ols, and $\sigma_{\omega,k,n}^{2}$ and $\sigma_{I,k,n}^{2}$are the variance of the noise and interference, respectively.

The objective of multi-user resource management is to maximize the utility function by optimally assigning each user's transmit power and constellation size under power constraints, which are given by

$$\begin{aligned} &\max_{P_{k,n}} \sum_{k=1}^{K}\sum_{n=1}^{N} T_{k,n}\left(P_{k,n}\right) \\ &s.t.\ \sum_{k=1}^{K}\sum_{n=1}^{N} P_{k,n} \le P_{tot} \qquad , \\ &\qquad 0 \le P_{k,n} \le P_{tot},\ \ \forall k,\ n \end{aligned} \tag{5}$$

where P_{tot} denotes the power constraint.

3 THE EFFECT OF THE IMPERFECT CSI

In this section, the effect of an imperfect CSI on the system performance of a multi-user CDMA system was investigated.

In general, an imperfect CSI can be expressed as

$$h_{k,n} = \hat{h}_{k,n} + e_{k,n}, \tag{6}$$

where $h_{k,n}$ and $\hat{h}_{k,n}$ are the real and estimated channel fading, respectively, and $e_{k,n}$ is the Gaussian distributed estimation error with a mean of zero and a variance of (Baissas & Sayeed 2002)

$$\sigma_{e,k,n}^{2} = \frac{p(0)}{1+SNR_{k,n}^{pilot}\,t_{co}/t_{s}} \tag{7}$$

In this equation, $SNR_{k,n}^{pilot}$ denotes the pilot SNR, $p(0)$ is the fading process power, and t_{co} and t_s are the coherence time and symbol duration, respectively.

With an imperfect CSI, only $\hat{h}_k$ is known in the receiver. Thus, the decision signal after the matched filtering operation is equal to

$$\begin{aligned} \hat{z}_{k,n} &= y_{k,n}\frac{\hat{h}_{k,n}^{*}}{\left|\hat{h}_{k,n}\right|^{2}} \\ &= \sqrt{P_{k,n}}x_{k,n} + \frac{\sqrt{P_{k,n}}x_{k,n}e_{k,n}}{\hat{h}_{k,n}} + \frac{\omega_{k,n}}{\hat{h}_{k,n}} + \frac{I_{k,n}}{\hat{h}_{k,n}} \end{aligned} \tag{8}$$

The imperfect CSI introduced a new power-and-signal dependent noise $\sqrt{P_{k,n}}x_{k,n}e_{k,n}\big/\hat{h}_{k,n}$ resulting in a constellation-point dependent SER. In previous studies (Wong & Evans 2009, Mokari et al. 2010, Awad et al. 2010, Wang et al. 2011), this property was ignored so that the uncertainty would usually associate with the channel gain, and the performance, such as throughput and data rate, could be directly derived. However, the power-and-signal dependent noise caused an error floor on the SER shown below.

If $\Pr\left(x_{k,n}\middle|\hat{h}_{k,n}\right)$ is defined as the conditional probability of $x_{k,n}$ received correctly, then $\Pr\left(x_{k,n}\middle|\hat{h}_{k,n}\right)$ is constellation-point dependent:

$$\Pr\left(x_{k,n}\middle|\hat{h}_{k,n}\right)= \begin{cases} \left[1-2Q\left(\eta_{k,n}\right)\right]^{2}, & \text{internal constellation points} \\ \left[1-Q\left(\eta_{k,n}\right)\right]^{2}, & \text{corner points} \\ \left[1-2Q\left(\eta_{k,n}\right)\right]\left[1-Q\left(\eta_{k,n}\right)\right], & \text{constellation border points} \end{cases} \tag{9}$$

where

$$\eta_{k,n} = \frac{d_{\min}\sqrt{P_{k,n}}\left|\hat{h}_{k,n}\right|}{\sqrt{2\sigma_{\omega,k,n}^{2}+2\sigma_{I,k,n}^{2}+2P_{k,n}\sigma_{e,k,n}^{2}\left|x_{k,n}\right|^{2}}} \tag{10}$$

In this equation, $d_{\min}$ is the minimum distance among the constellation points.

Thus, the average SER under an imperfect CSI is given by

$$\widehat{SER}_{k,n} = 1-\frac{1}{M}\sum_{\{x_{k,n}\}}\Pr\left(x_{k,n}\middle|\hat{h}_{k,n}\right), \tag{11}$$

where $\{x_{k,n}\}$ denotes the set of all available constellation points. When $P_k \to \infty$, $\eta_{k,n} \to d_{\min}\left|\hat{h}_{k,n}\right|\Big/\sqrt{2\sigma_{e,k,n}^{2}\left|x_{k,n}\right|^{2}}$, which is constant, an error floor is introduced to the SER. This property was completely neglected in previous studies (Wong & Evans 2009, Mokari et al. 2010, Awad et al. 2010, Wang et al. 2011).

Under an imperfect CSI, the throughput is derived as

$$\hat{T}_{k,n}^{M}\left(P_{k,n}\right)=L\log_2 M\left(1-\widehat{SER}_{k,n}\right)^{L}$$
$$=L\log_2 M\left[\left(\frac{1}{M}\sum_{\{x_{k,n}\}}\Pr\left(x_{k,n}\middle|\hat{h}_{k,n}\right)\right)^{L}\right]. \qquad (12)$$

Thus, with an imperfect CSI, the optimization problem for a multi-user CDMA system could be rewritten as

$$\max_{P_{k,n}} \sum_{k=1}^{K}\sum_{n=1}^{N}\hat{T}_{k,n}\left(P_{k,n}\right)$$
$$s.t. \quad \sum_{k=1}^{K}\sum_{n=1}^{N}P_{k,n}\le P_{tot}, \qquad , \qquad (13)$$
$$0\le P_{k,n}\le P_{tot}, \quad \forall k,n$$

where $\hat{T}_{k,n}\left(P_{k,n}\right)=\max\limits_{M}\left\{\hat{T}_{k,n}^{M}\left(P_{k,n},\hat{h}_{k,n}\right)\right\}$.

4 ITERATIVE BISECTION ALGORITHM

Since $\hat{T}_{k,n}\left(P_{k,n}\right)$, which is same as $T_{k,n}\left(P_{k,n}\right)$, is non-convex and nondifferentiable, in order to solve the optimization problem in (13), the dual optimization problem is

$$g^{*}=\max_{\mu\ge0}\hat{g}(\mu)$$
$$s.t. \sum_{k=1}^{K}\sum_{n=1}^{N}\hat{P}_{k,n}(\mu)\le P_{tot}, \qquad (14)$$

where

$$\hat{g}(\mu)=-\mu P_{tot}+\sum_{k=1}^{K}\sum_{n=1}^{N}\min_{0\le P_{k,n}\le P_{tot}}\left(-\hat{T}_{k,n}\left(P_{k,n}\right)+\mu P_{k,n}\right) \qquad (15)$$
$$\hat{P}_{k,n}(\mu)=\arg\min_{0\le P_{k,n}\le P_{tot}}\left(-\hat{T}_{k,n}\left(P_{k,n}\right)+\mu P_{k,n}\right)$$

Similar to *Propositions 1* and *2* in (Wang et al. 2011), $\hat{g}(\mu)$ and $\hat{P}_{k,n}(\mu)$ were both determined to be the non-increasing functions of variable μ. Based on the properties of $\hat{g}(\mu)$ and $\hat{P}_{k,n}(\mu)$, the bisection algorithm proposed in (Wang et al. 2011) could also be used to solve the optimization problem in (13), which is summarized in Figure 1.

The above bisection algorithm only works for a constant SNR for each user. However, when the new transmit power is allocated to a certain user, the interferences to other users changes, thereby changing the SNR of each user. In this case, the iterative process should be employed to adaptively mange the radio resources in order to make the algorithm converge, as shown in Figure 2.

5 SIMULATION RESULTS

Numerical simulations are presented to illustrate the effect of an imperfect CSI on the system performance of multi-user resource management in a downlink CDMA system. 4-QAM, 16-QAM, and 64-QAM were selected for the modulations, and the number of users and subchannels were designated as $K=5$ and $N=2$, respectively.

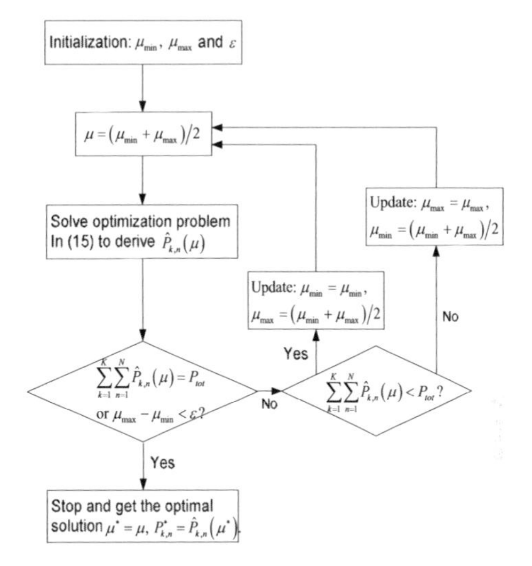

Figure 1. Bisection algorithm.

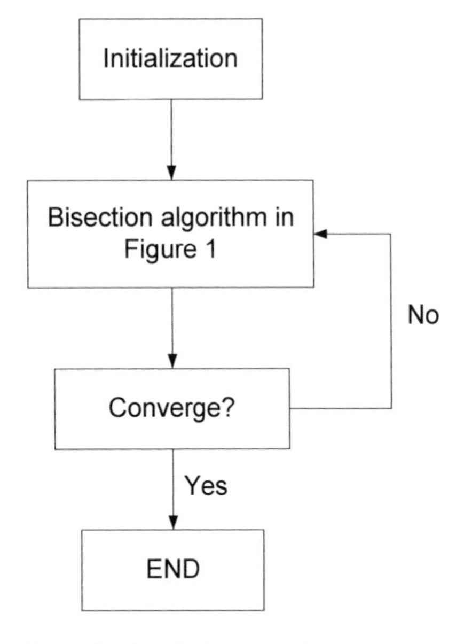

Figure 2. Iterative bisection algorithm.

Figures 3-5 compare the performance of the proposed optimal algorithm in multi-user resource management considering the power-and-signal dependent noise and the optimal algorithm in multi-user resource management with imperfect an CSI (Wang et al. 2011). Two cases employ the same iterative bisection algorithm. In Figure 3, the total power constraint is equal to $P_{tot} = 40$ and the channel error ratios (CER), the ratios between the variance of channel estimation error and noise, are 0dB, -10dB, and -30dB. In Figures 4 and 5, the CER is -10dB, and the total transmit powers are 40, 15, and 5.

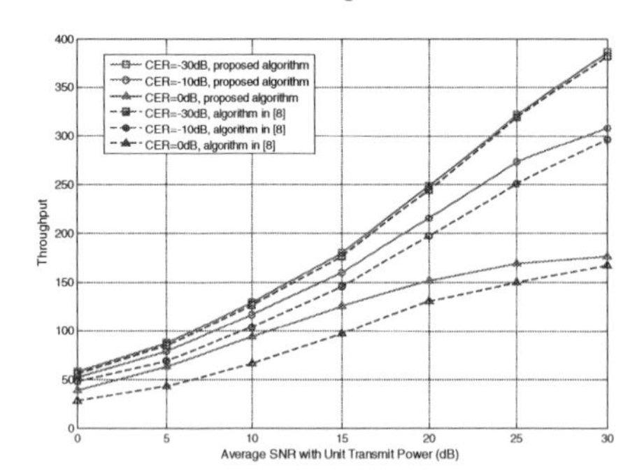

Figure 3. Performance comparison of various CER.

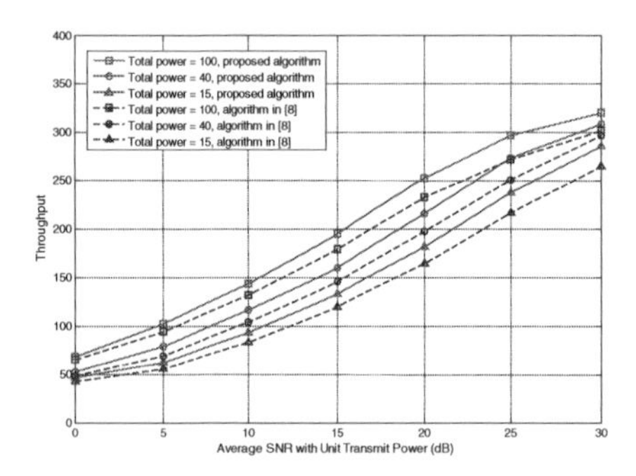

Figure 4. Performance comparison of various total transmit powers.

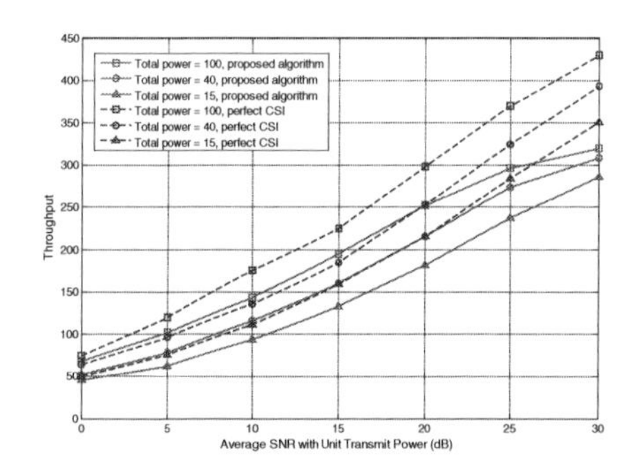

Figure 5. Performance comparison of various total transmit powers.

From Figures 3-5, the proposed and optimal algorithms in (Wang et al. 2011) were both observed to show the performance degradation under an imperfect CSI. In addition, the performance of the proposed algorithm was better than that of the optimal algorithm in (Wang et al. 2011), especially when either the CER or the total power constraint was large. This is reasonable because, as shown in (8), CER and transmit power are both able to enhance the effect of the power-and-signal dependent noise. Figures 3 and 4 reveal that the effect of an imperfect CSI has to be considered in research concerning multi-user resource management.

6 CONCLUSIONS

This paper addressed the effect of imperfect CSI on the optimal resource management of a downlink multi-user CDMA system. The imperfect CSI introduced an extra signal-and-power dependent noise term, which was neglected in existing studies and yielded significant performance degradation.

REFERENCES

Abu-Rgheff M. 2007. *Introduction to CDMA wireless communications*, Academic Press.

Awad M. K., Mahinthan V., Mehrjoo M., Shen X. and Mark J. W. 2010. A dual-decomposition-based resource allocation for OFDMA networks with imperfect CSI. *IEEE Trans. Vehic. Techn.* 59(5): 2394–2403.

Baissas M.-A. R. and Sayeed A. M. 2002. Pilot-based estimation of time varying multipath channels for coherent CDMA receivers. *IEEE Trans. Signal Process.* 50(8): 2037–2049.

Jafar S. A. & Goldsmith A. 2003. Adaptive multirate CDMA for uplink throughput maximization. *IEEE Trans. on Wireless Commun.* 2(2): 218–228.

Lee J.-W., Mazumdar R. R. & Shroff N. B. 2005. Downlink power allocation for multi-class wireless systems. *IEEE/ACM Trans. Netw.* 13(4): 854–867.

Meshkati F., Goldsmith A. J., Poor H. V. & Schwartz S. C. 2007. A gametheoretic approach to energy-efficient modulation in CDMA networks with delay QoS constraints" *IEEE J. Sel. Areas Commun.* 25(6): 1069–1078.

Mokari N., Javan M. R. & Navaie K. 2010. Cross-layer resource allocation in OFDMA systems for heterogeneous traffic with imperfect CSI. *IEEE Trans. Vehic. Techn.* 59(2): 1011–1017.

Musku M. R., Chronopoulos A. T., Popescu D. C. & Stefanescu A. 2010. A game-theoretic approach to joint rate and power control for uplink CDMA communications. *IEEE Trans. Commun.* 58(3): 923–932.

Wang Z. and Liu L. 2014. Adaptive resource management for a downlink OFDMA system with imperfect CSI, *Electronics Letters* 50(7): 554–556.

Wang Z., Liu L., Wang X. and Zhang J. 2014. Resource allocation in OFDMA networks with imperfect channel state information. *IEEE Communications Letters* 18(9): 1611–1614.

Wang Z., Peng Q. and Milstein L. B. 2011. Multiuser resource allocation for downlink multi-cluster multicarrier DS CDMA system. *IEEE Trans. Wireless Commun.* 10(8): 2534–2542.

Wong I. C. & Evans B. L. 2009. Optimal resource allocation in the OFDMA downlink with imperfect channel knowledge. *IEEE Trans. Commun.* 57(1): 232–241.

Electronics, Communications and Networks IV – Hussain & Ivanovic (eds)
© 2015 Taylor & Francis Group, London, ISBN: 978-1-138-02830-2

Design and implementation of the postgraduate information management system based on ASP.NET

Zhihai Liu*
College of Transportation, Shandong University of Science and Technology, Qingdao, Shandong, China

Kaidi Yang, Shoubo Lu, Su Yang & Ronghua Zhang
College of mechanical and Electronic Engineering, Shandong University of Science and Technology, Qingdao, Shandong, China

ABSTRACT: In this paper, the problems of slow response and imperfect function in operations of postgraduate information system were analyzed. The function structure and E-R diagram were drawn. And ASP.NET was selected as the main development tool of the designed system after that several developing tools were compared. At last, the postgraduate information management system was developed combined with MySQL database. The result shows that each function module in the system is more harmonious and unified, which can further promote the process of information construction of colleges and universities.

1 INTRODUCTION

With the continuous expansion of postgraduate enrollment, how to improve the efficiency and quality of postgraduate management has become a priority. Compared with developed countries, our country's information level lags behind obviously, and traditional information management system for postgraduates has many defects such as low efficiency, incomplete functions, the untimely updated database, etc. For example, Chang Bao-ying took JSP as the development platform of the Postgraduate Information Management System. Because of the need to take up large amounts of memory and disk space when JSP webpage runs, the burden on the computer is increased and the system has a lower work - efficiency (Chang 2008). The system designed by Lu Cheng-jun could solve the former problems such as response speed and information interaction, but it only consisted of register information management module and basic information query function for students(Lu 2009). Therefore, the system was inadequate to perform the tasks that generated by universities. As previously mentioned, these systems have yet to be perfected. In the viewpoint of the foreign countries, this kind of management system also faces many challenges, including backward technology, inflexible operations, and narrow contents of design. Each item is to be solved at present.

In order to solve the defects that exposed in JSP development technology, the ASP.NET technique was introduced into the Postgraduate Information Management System. A simple ASP.NET page looks just like an ordinary HTML page, and the combinations of HTML language, scripting language and ActiveX controls could produce dynamic, interactive and efficient web-based applications. At the same time, ASP.NET applies ADO (Active Data Objects) method of database access, and makes the process of operations such as adding, modifying and deleting the relevant data more convenient (Li & Liang 2009). Thus, it has all the advantages of simpler system deployment, simpler construction process, and less investment of system development.

By comparison with other management systems, the management systems are more imperfect from the functional perspective. So some modules such as the management of theses, patents, scientific researches, and training plans were added to the system. Through continuous improvement, this system provides a new comfortable working environment for postgraduates training and management, and lays a solid foundation for the realization of the digital campus.

2 DEMAND ANALYSIS

The system can display the teachers' name, teachers' number and login account information when teachers enter the system. At the same time, teachers can view the campus bulletin and students' announcement information as well as carry out the maintenance and management of personal information, students' achievements and the papers. Among them, the students' achievements module can achieve uploading, modifying, viewing

*Corresponding author: *zhihliu@126.com*

performance and other functions (Cao 2011). The module structure of this system is shown in Figure 1.

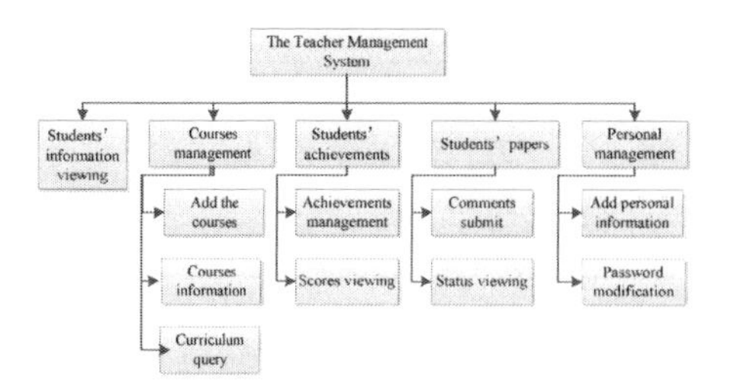

Figure 1. The module's structure diagram of this system.

3 DESIGN AND IMPLEMENTATION

3.1 *Database design*

Database design was mainly composed of logical design and physical design. In the logical database design, each entity is converted into a corresponding table while the property is converted into a corresponding column (Liu 2007). Finally, the relationships between entities are converted into relationships of primary key and foreign key between the tables. Logical database design was the source of the physical database design and provided very important information for the design of the database. In the physical database design phase, it was necessary to determine that the logical design in the target DBMS (data base management system) was realized physically. The diagrams of data flow and main entity relationships are respectively shown in Figure 2 and Figure 3.

According to the database design process, a postgraduates' information database was established firstly, and then some tables were created to store all important information. Once the system was run, the user could obtain the corresponding data in each table, thus to realize the data interaction. The structure of these nine tables is shown in Figure 4.

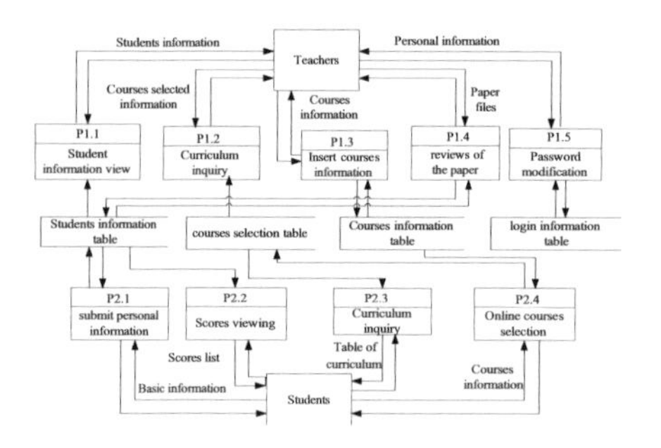

Figure 2. The diagrams of data flow.

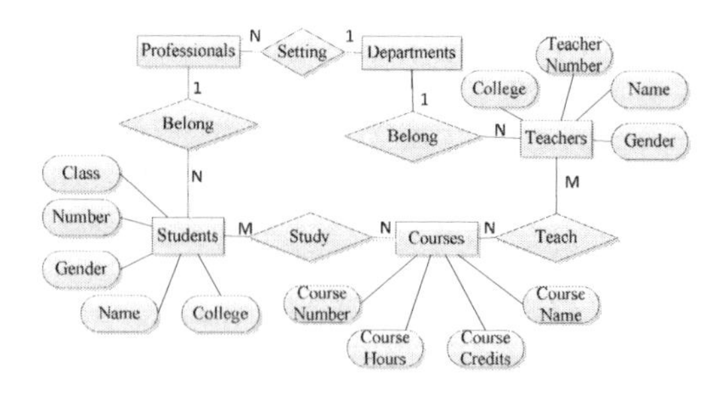

Figure 3. E-R diagram of main entity relationship.

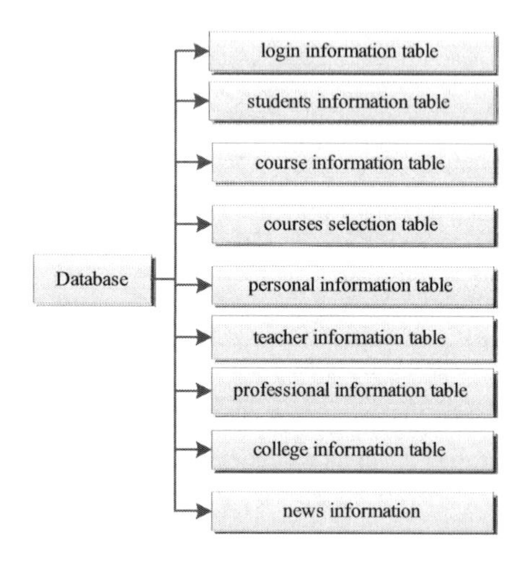

Figure 4. The structure of nine tables in the database.

3.2 *Design of all the interface functions*

3.2.1 *Login interface*

The system login module is the main entrance of the system; it adopts unified-interface login method which is used to implement unified authentication development standards. In more detail, it sets the permissions fields respectively as 1, 2, 3 but logins through the same interface. The users can be authorized to login according to the permissions fields when getting into the system, which ensures that the different users have the corresponding permission. The login information of the administrator, teachers and students is stored in the login table.

3.2.2 *Students' information viewing*

From this page, teachers can view students' basic information, such as name, gender, student number, political affiliation, register information, tutor's name, etc. Therefore, it can be more convenient for teachers to understand the students and promote the interaction between teachers and students. All the information is stored in the personal information table.

3.2.3 *Course selection management*

There are three contents, including courses' adding, courses' information viewing and curriculum query in this function module. In the interface by adding courses, the teacher can insert the course type, course name, and class time cycle, class venue and the other important information into the course information table. The course information can be displayed in the course information interface, and can achieve adding, modifying and deleting operations. In curriculum query interface, teachers can conveniently inquire their corresponding semester curriculum. The course information is stored in the course information table while the course selection information is stored in the course selection table.

3.2.4 *Students' achievements*

This function part includes performance management and score viewing. In the performance management interface, teachers can entry students' related-course score in the course selection table, but it should be noted that the score must be modified through the administrator once it is submitted. The teachers don't have permissions of modificatory.

3.2.5 *Students' papers*

There are two contents of this function module, one is comments submission and the other is state viewing. In comments submission interface, teachers can conveniently review students' theses and submit their opinions while in the state viewing interface, teachers can view submitted comments state. Similar to the student performance module, a revision of opinions must be submitted through the administrator, and the teachers still don't have permissions of modificatory.

3.2.6 *Personal information management*

In order to achieve a more humanized management, a teachers' personal information modify module has become an indispensable part of this system. This function module includes adding personal information and password modification. Teachers can input their names, numbers, departments, photographs and login information into the teachers' information table, and can modify these above information and the login passwords. All the teachers' personal information is stored in the teacher information table.

3.2.7 *The tree structure of all function interfaces*

According to the systematic analysis and the functional requirements of the system, the overall structure of the system was designed and shown in Figure 5.

3.3 *The implementation of main interface*

Framework plays the role of splitting the webpage into several different regions and realizing the display of multiple HTML pages in a browser window, so it is frequently used in webpage design (Cai & Wang 2008).In this way, the computation number is reduced and the burden of server is lightened when a browser refreshes. The biggest characteristic is the consistent website style, so in the process of system design, the same part of all the site pages was made into a separate page which was regarded as a child frame structure of the whole website content.

The main interface of the system guides the users to read all the contents by the directory index, so the users can switch webpages more conveniently. The interface is shown in Figure 6.

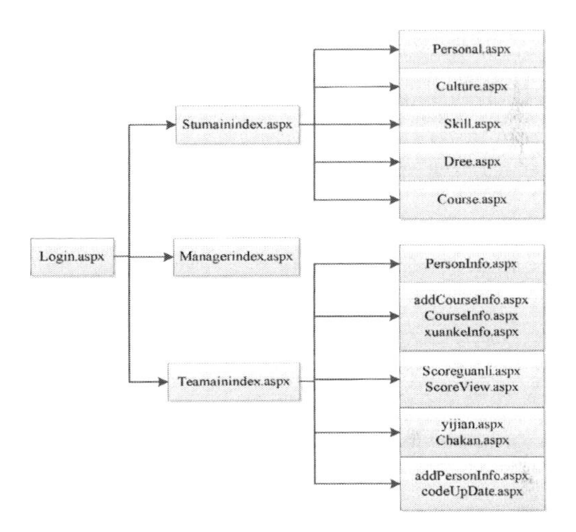

Figure 5. Overall structure of the system.

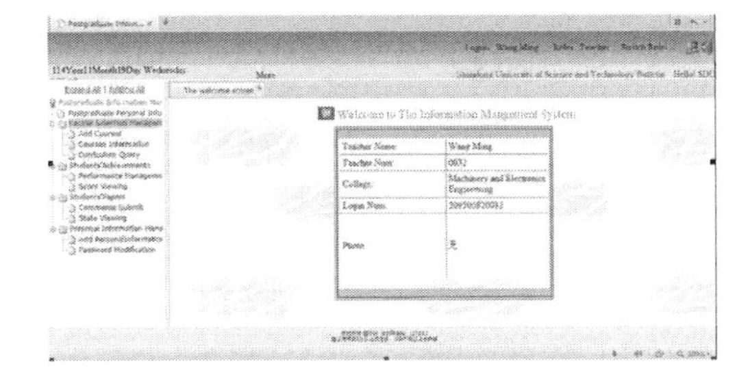

Figure 6. The main management interface.

3.4 *The implementation of adding courses interface*

The data-adding function of postgraduates' information management is one of the indispensable parts in data management system. It requires teachers and administrators to input related information in the database timely and to facilitate students' course selection process.

In the system, the data such as teachers' curriculum information should be added to the database completely. In order to solve these issues such as connection, query, addition, deletion, etc., ASP.NET provides a rich database operation functions such as INSERT, SELECT, UPDATE and DELETE SQL commands for database access.

It's also usually necessary to detect the validity of input data in the system, that is to say, the system should give a friendly reminder when the user input undesirable data types or data contents. In order to solve these problems, this design made full use of the ASP.NET test controls, such as Required Field Validator control in mandatory fields, Regular Expression Validator control in testing expression and the Range Validator control in checking the value scope, etc. The use of these elements greatly improved the design efficiency and ensured the security and correctness of data (Yu & Du 2006). The interface of adding courses is shown in Figure 7.

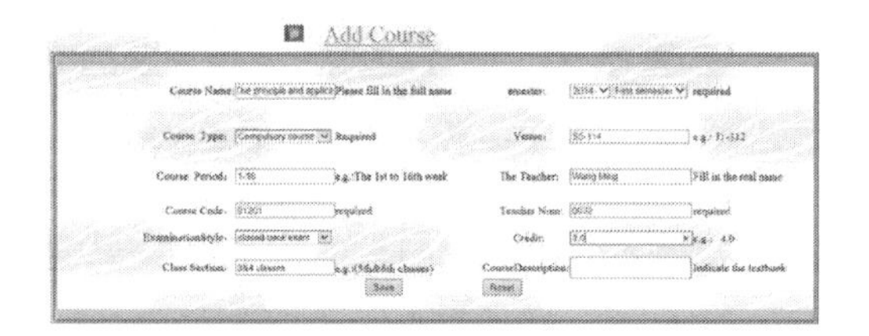

Figure 7. Diagram by adding course interface.

4 CONCLUSION

The paper was presented based on the demand analysis and functions design. The whole development process mainly includes the data flow analysis, the structure and function diagram design, database design, functional interface design, and coding a design. Finally, it passed the test of security, stability and functional operation successfully.

The implemented system is of prime importance not only in improving the efficiency and management level, but also in saving money and streamlining. It also has certain actual application value in the information management system of colleges and universities, and further promotes the process of informatization construction.

ACKNOWLEDGEMENT

The paper is supported by the Postgraduate Education Innovation Project of Shandong University of Science and Technology (No.KDYC13008).

REFERENCES

Cai, C.A. & Wang, Q. 2008. The design and implementation of students' information management system based on B/S model. *Computer engineering and design* 12: 98–100.

Cao, R. 2011. Research and Development of postgraduate information management system. *Information Science and Technology* No.S1.

Chang, B.Y. 2008. Research and Development of postgraduate information management system. *DA ZHONG KE JI* 103(3): 60–62.

Li, K.M. & Liang, X.M. 2009. The analysis and design of student information management system based on ASP technology. *Computer knowledge and technology* 12: 98–100.

Liu,N.L. 2007. *Proficient in ASP.NET 2.0 & SQL Server 200 projects development*. Beijing: People's posts and telecommunications press.

Lu,C.J. 2009. Design of college student information management system with ASP.NET 2.0 Technology. *Journal of ChongQing University of Arts and Sciences(National science Edition)*28(4): 38–42.

Yu,G.F. & Du,W.L. 2006. Object-Oriented Programming Method in Design of Web Pages. *Computer Technology and Development* 16:58–60.

Electronics, Communications and Networks IV – Hussain & Ivanovic (eds)
© 2015 Taylor & Francis Group, London, ISBN: 978-1-138-02830-2

D2D Cooperative Communication Technology in TD-LTE-Advanced Systems

Jingan Liu, Wei Wu* & Xuejun Sha
Harbin Institute of Technology, Harbin, Heilongjiang Province, China

ABSTRACT: TD-LTE-Advanced system coverage is limited, and the communication quality of cell edge users can not be guaranteed. In this paper, in order to solve this problem, based on D2D communication and cooperative communication technology, D2D cooperative communication was proposed. Cell edge users could receive statistically independent multiple copies from the base station and another UE serving as a relay. By combining the copies, diversity gain could be obtained, better cell coverage could be achieved, and system performance could be improved. The simulation results indicated that D2D cooperative communication could increase the SINR of cell edge users and improve the throughput of TD-LTE-Advanced systems.

1 INTRODUCTION

In recent years, Device-to-Device (D2D) communication has been considered as a technology that could improve the performance of LTE-Advanced systems. Numerous conventional technologies have achieved device communications, such as WLAN and Bluetooth technology, but D2D communication is significantly different. It uses the licensed frequency band of cellular systems to avoid the uncertainty of non-licensed frequency use. Introducing D2D communication into LTE systems was first proposed by Doppler (Doppler et al. 2009). Based on his study, Koskela and Seppala proposed the cluster multicast concept (Koskela et al. 2010, Seppala et al. 2011). Lei designed operator controlled D2D communications in LTE-Advanced systems (Lei et al. 2012), and Yue proposed secrecy-based access control for D2D Communication (Yue et al. 2013).

Fixed relay cooperation has been proven to be able to improve system spectral efficiency and system overall performance (Laneman et al. 2004) and (Janani et al. 2004). However, fixed relay cooperation is not flexible enough and is high in cost. As a possible solution to this problem, based on the user cooperation concept (Sendonaris et al. 2003), cellular users could be chosen as relays to achieve cooperative communication, and users at the edge of the cell could obtain diversity gain by receiving statistically independent multiple copies.

D2D cooperative communication, proposed in this paper, is based on D2D communication and cooperative communication. Its purpose is to improve the communication quality of edge users and improve the overall system performance.

The rest of this paper is organized as follows. Section 2 provides the workflow for D2D cooperative communication. The power control algorithm and the improved resource allocation algorithm, which would suppress interference and select relay users, respectively, are presented in Section 3. The simulation results are shown in Section 4. The conclusions are presented in Section 5.

2 PROCESS OF D2D COOPERATIVE COMMUNICATION

When a user's communication quality can not be guaranteed, it can be defined as an edge user; then, the base station selects a relay node so that it can achieve D2D cooperation, thereby improving its communication quality, as shown in Figure 1.

In Figure 1, a relay UE (rUE) reuses the resources of a cellular user (cUE) resources. A terminal UE (tUE) is an edge user. When eNodeB sends data to a certain tUE, the tUE received two copies via the direct link and relay link. By combining two copies from eNodeB and the rUE, the tUE can obtain diversity gain.

Based on the work of Doppler and Lei, the communication process was designed for D2D cooperative communication in TD-LTE-Advanced systems.

D2D cooperative communication is launched and ended by SINR judgment rule with two thresholds, which can be described by equation (1) and equation (2):

*Corresponding author: *kevinking@hit.edu.cn*

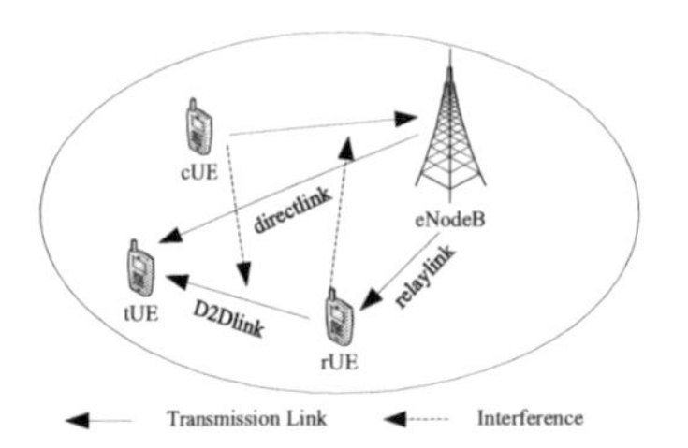

Figure 1. Downlinks in D2D cooperative communication.

$$SINR_{UE} \le SINR_{mean} * cooperationfactor_{low} \quad (1)$$

$$SINR_{UE} > SINR_{mean} * cooperationfactor_{high} \quad (2)$$

where $SINR_{mean}$ is the average SINR of all users, $cooperationfactor_{low}$ is the lower cooperation threshold, and $cooperationfactor_{high}$ is the higher cooperation threshold.

When a user's SINR satisfies equation (1), the system begins D2D cooperation. When the communication quality is improved and satisfies equation (2), the system completes the cooperation.

TheD2D cooperative communication process in TD-LTE-Advanced systems is shown in Figure 2.

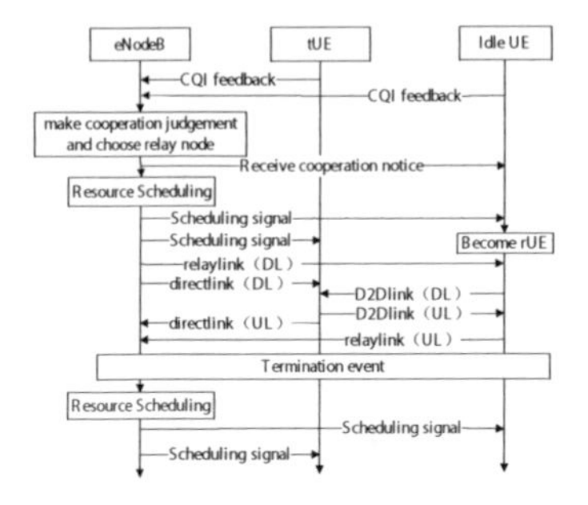

Figure 2. Process of D2D cooperative communication in TD-LTE-Advanced systems.

As shown in the process details, D2D cooperative communication is based on the original process, and only little overheads are required.

3 RESOURCE ALLOCATION OF D2D COOPERATIVE COMMUNICATION

3.1 *Power control algorithm of D2D cooperative communication*

In this paper, the rUE reused the cUE's resources in order to communicate with the tUE. Although reusing resources can introduce interference, it has higher resource utilization. When the downlink relay reuses uplink slot resources, the cUE interferes with D2D cooperative communication, and the D2D cooperation users interfere with the base station, as shown in Figure 1.

To suppress interference, the rUE's transmission power should be minimized based on the achievement of D2D cooperative communication. The minimum power is mainly determined by three aspects: First, according to the 3GPPstandards, the UE's transmission power cannot be higher than 23dBm. In addition, the received signal must satisfy the edge user's minimum received power resolution. Furthermore, the communication quality of relaylink and D2Dlink should be equal.

According to the 3GPP standards, the path loss model in the D2D cooperative relay scenario could be described by equation (3) and equation (4):

$$PL_{ER}(\text{dB}) = 22.0\lg(d_{ER}) + 28.0 + 20\lg(f) \quad (3)$$

$$PL_{RT}(dB) = 16.9\lg(d_{RT}) + 32.8 + 20\lg(f) \quad (4)$$

Where PL_{ER} is the path loss between eNodeB and the rUE, PL_{RT} is the path loss between the rUE and the tUE, d_{ER} is the distance between eNodeB and the rUE, and d_{RT} is the distance between the rUE and the tUE.

The relaylink and D2Dlink were assumed to have the same additive Gaussian white noise, P_N. To match them, the received signal at the rUE and tUE should have the SNR.

$$SNR_{rUE} = SNR_{tUE} \quad (5)$$

$$P_{RrUE} / P_{N(ER)} = P_{RtUE} / P_{N(RT)} \quad (6)$$

$$(P_{eNodeB} - PL_{ER}) / P_N = (P_{rUE} - PL_{RT}) / P_N \quad (7)$$

$$P_{rUE} = P_{eNodeB} - 22.0\lg(d_{ER}) + 16.9\lg(d_{RT}) + 4.8 \quad (8)$$

where P_{RrUE} is the rUE's received power from eNodeB, P_{RtUE} is the tUE's received power from the rUE, and P_{eNodeB} and P_{rUE} are the transmission powers of eNodeB and the rUE, respectively.

P_{rUE} can be calculated as P_{rUE1} according to equation (8). The tUE's received signal power from the rUE is:

$$P_{RtUE} = P_{rUE} - PL_{RT} - P_S \quad (9)$$

where P_S is the shadow fading. According to the 3GPP standards, in TD-LTE-Advanced systems, the UE's minimum received power is -94dBm (P_{RtUE}); thus, P_{rUE} can be calculated as P_{rUE2} according to equation (9).

Finally, P_{rUE} must be less than 23dBm, the rUE's transmission power:

$$P_{rUE} = \min(P_{rUE1}, P_{rUE2}, 23) \tag{10}$$

3.2 *Relay selection and resource allocation algorithm*

The optimal resource allocation algorithm uses the rUE and the reused resources of each edge user as two variables of the user SINR. Then, the system throughput can be calculated according to the SINRs. By maximizing the system throughput, each edge user's rUE and reused resources can be confirmed. However, this algorithm is too complex to function properly. In this paper, an improved algorithm was proposed in order to make D2D cooperation more practical.

Several sets were defined as follows: $\mathbf{T}_i$ (i=1,2,3) was the set of edge users and $\mathbf{C}_i$ (i=1,2,3) was the set of cellular users, where i denoted three sectors; **U** was the set of selected rUEs; **S** was the set of reused resources of the cUEs; and $\mathbf{R}_1$, $\mathbf{R}_2$, and $\mathbf{R}_3$ were the temp sets of the rUEs.

Step.1: Idle cellular users who are in the same sector with the current edge user were selected to work as the rUEs, The first α percent was picked up as $\mathbf{R}_1$ according to the communication quality with the base station.

Step.2: The first β percent was picked up as $\mathbf{R}_2$ according to the communication quality with the current edge user from $\mathbf{R}_1$.

Step.3: The rUEs that will not interfere with the base station were selected as $\mathbf{R}_3$ from $\mathbf{R}_2$.

Step.4: The edge user's SINRs with different rUEs in $\mathbf{R}_3$ were calculated when reusing resources of different cellular users in the other two sectors; Then the final rUE and s (the reused resources) with the best SINRs were selected.

Step.5: Whether the rUE belongs to **U** was obtained to ensure that the rUE has not been used. If true, the process ignored this rUE and returned to Step.1; otherwise, the process continued to Step.6.

Step.6: Whether s belong to **S** was obtained to ensure that the cUE's resources have not been reused. If true, the process ignored the resources and returned to Step.1; otherwise, the process continued to Step.7.

Step.7: The reused resources were allocated to the rUE to reuse. The rUE was placed into **U** and s was placed into **S**.

Step.8: Whether the allocation was performed for edge users in this sector was obtained. If true, the process moved to Step.9; otherwise, the allocation for the next edge user was started.

Step.9: Whether the allocation was performed for all three sectors was obtained. If true, the process was end up; otherwise, the process returned to Step.1 and began the next sector's allocation.

The specific relay selection and resource allocation process is shown in Figure 3. This algorithm's complexity was greatly reduced by group selection.

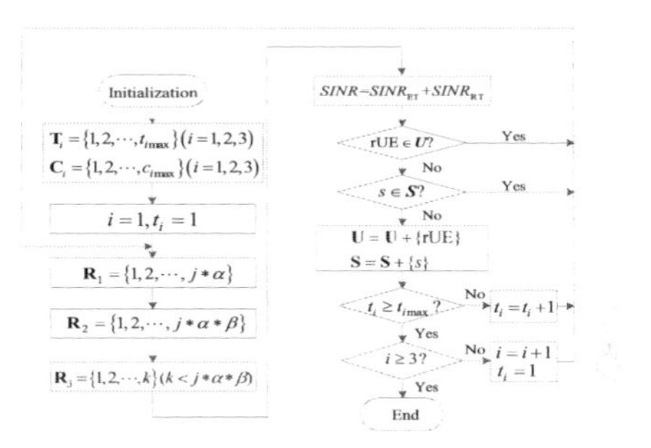

Figure 3. Relay selection and resource allocation algorithm for D2D cooperative communication in a TD-LTE-Advanced system.

4 SIMULATIONS

In this section, a system-level simulation for D2D cooperative communication in a seven-cell TD-LTE-Advanced system model was conducted. There were 48 users per cell. The rUEs adopted amplify and forward (AF). The tUEs adopted maximum ratio combining (MRC).The carrier frequency was 2.14GHz, the bandwidth was 20MHz, *cooperation-factor*$_{low}$ and *cooperationfactor*$_{high}$ were set to 0.6 and 0.7, respectively, and α and β were both set to 50. The simulation results are shown in Figures 4 through 6.

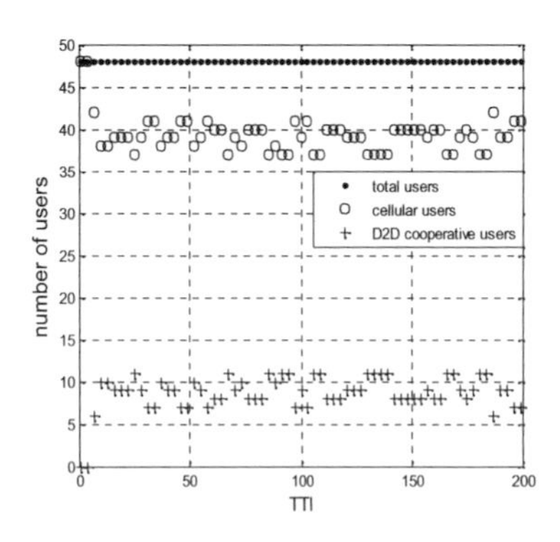

Figure 4. Number comparison of different users.

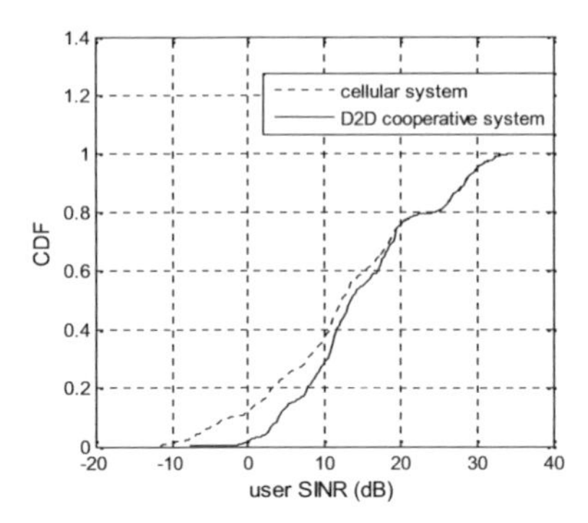

Figure 5. CDF of user SINR.

As indicated by Figure 4, in each TTI, about 8 to 9 edge users adopted D2D cooperative communication due to the poor communication quality.

As indicated in Figure 5, with D2D cooperative communication, the SINR distribution of users and communication qualities of edge users were improved.

As shown by Figure 6, the system throughput increased by approximately 20% after introducing D2D cooperative communication due to the improvement of the edge users' communication quality. Using the power control algorithm and proper resource allocation algorithm, the interference was suppressed. As a result, the overall system performance was improved.

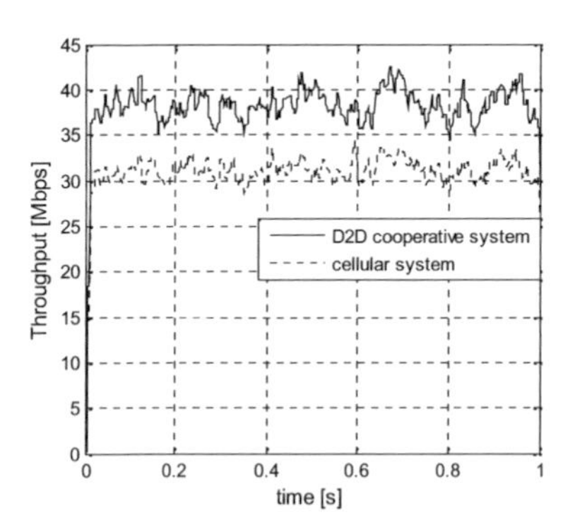

Figure 6. Throughput comparison of the original cellular system and D2D cooperative system.

5 CONCLUSIONS

In this paper, D2D cooperative communication was designed based on D2D communication and cooperative communication. The designed process of the D2D cooperative communication, power control algorithm, relay selection, and resource allocation algorithm were given. According to the simulation result, the communication qualities of edge users were improved, and the overall performance of the system increased by approximately 20%.Thus, D2D cooperative communication technology could be used to improve the performance of TD-LTE-Advanced systems.

ACKNOWLEDGEMENT

This research was supported by the Next Generation Wireless Mobile Communication Network of China under Grant No. 2012ZX03001031-003 and Heilongjiang Postdoctoral Financial Assistance under Grant No.LBH-Z11153.

REFERENCES

Doppler, K., Rinne, M., Janis, P., et al 2009. Device-to-Device Communications; Functional Prospects for LTE-Advanced Networks. *IEEE International Conference on Communications: 1–6, 14–18 June 2009*. Piscataway: IEEE.

Doppler, K., Rinne, M., Wijting, et al. 2009. Device-to-device communication as an underlay to LTE-advanced networks. *IEEE Communication. Magzine* 47(12): 42–49.

Janani, M.A., Hedayat, T.E. Hunter, et al. 2004. Coded cooperation in wireless communications: space-time transmission and iterative decoding. *IEEE Transaction on Signal Process* 52(2): 362–371.

Koskela, T., Hakola, S., Chen, T. et al. 2010. Clustering Concept Using Device-To-Device Communication in Cellular System. *Wireless Communications and Networking Conference 18–21 April 2010,* 1–6. Piscataway: IEEE

Laneman, J.N., Tse, D.N.C. & Wornell, G.W. 2004. Cooperative Diversity in Wireless Networks: Efficient Protocols and Outage Behavior. *IEEE Transaction on Information Theory* 50(12): 3062–3080.

Lei, L., Zhangdui, Z., Chuang, L., et al. 2012. Operator Controlled Device-to-Device Communications in LTE-Advanced Networks. *IEEE Wireless Communication* 19(3): 96–104.

Sendonaris, A., Erkip, E. & Aazhang, B. 2003. User Cooperation Diversity, Part I & Part II. *IEEE Transaction on Communication* 51(11): 1927–1948.

Seppala, J., Koskela, T., Tao, C., et al. 2011. Network controlled Device-to-Device (D2D) and cluster multicast concept for LTE and LTE-A networks. *Wireless Communications and Networking Conference), 28–31 March 2011*: 986–991. Piscataway: IEEE

Yue, J., Ma, C., Yu, H., et al. 2013. Secrecy-Based Access Control for Deviceto-Device Communication Underlaying Cellular Networks. *IEEE Communication. Letter* 17(11): 2068–2071.

Electronics, Communications and Networks IV – Hussain & Ivanovic (eds)
© 2015 Taylor & Francis Group, London, ISBN: 978-1-138-02830-2

A random early detection based active queue management algorithm in power optical communication network

Zhao Liu & Liangrui Tang
State Key Laboratory of Alternate Electrical Power System with Renewable Energy Sources (North China Electric Power University), Changping District, Beijing, China

Jiangyu Yan
School of Control and Computer Engineering (North China Electric Power University), Changping District, Beijing, China

ABSTRACT: According to the different QoS requirements of services in power communication network, a random early detection based active queue management algorithm combined with the differentiated service model is proposed in this paper. The congested queue can take advantage of other idle queues to ensure the QoS requirement of the high priority queue. The simulation shows that the algorithm presented in this paper has better performance in reducing the overall packet loss rate and enhancing the utilization ratio of buffer.

KEYWORDS: Power Optical Communication Network; Differentiated Services; IP QoS; Active Queue Management

1 INTRODUCTION

With the rapid development of smart grid, the power communication services present the features such as IP, broadband and diversity (Zhao & Zhang 2011). As a typical representative technology of the next generation of optical network, Optical Transport Network (OTN) possesses of the features including high capacity, safety and efficiency. Thus OTN is the future trend of the power transmission network (Feng et al. 2012).

In recent years, with the development of IP network, IP services have become the main form of services. As a technical means of end-to-end congestion control, Active Queue Management (AQM) has been always an emphasis concerned by researchers at home and abroad. Key indicators of AQM include packet loss rate, link utilization, throughput, delay, stability, and equity, etc. By using packet loss events and idle link events to manage congestion, a congestion control with the relatively short buffer is proposed to reduce the end-to-end delay and improve the throughput (Feng et al. 2002). Improved algorithms are proposed to identify and punish the non-adaptive flows, provide better protection for adaptive flows and improve the fairness of the network (Huang et al. 2010, Tian & Wu 2012, Gong & Wu 2010). The SBlue algorithm is proposed which can effectively maintain the stability of queue length, reduce the overflow or idle phenomenon (Wu & Jiang 2005).

The algorithms above manage queues in accordance with the condition that data flows occupy the router and link, they only consider from the perspective of adaptive and non-adaptive flows. However, the Qualities of Services (QoS) of different communication services in the power grid are quite different. Power production dispatching services usually need higher security and stability, because they are core services to ensure that the power grid can run safely and stably. Nevertheless, the QoS requirement of management information services is much lower. Therefore, it is necessary to divide the flows into multiple priorities.

According to the characteristics of power communication services, an improved active queue management algorithm Length Change Random Early Detection (LC-RED) combined with the differentiated service model is proposed in this paper. By predicting and changing the buffer length of different queues, the congested queues can take advantage of other idle queues. Then we can ensure the QoS requirement of the high priority queue, reduce the overall packet loss rate and enhance the utilization ratio of buffer.

2 PROPOSED ALGORITHM

IP QoS network model is divided into Integrated Services (IntServ) model and Differentiated Services

(DiffServ) model, DiffServ model is widely used for its simplicity, flexibility, and expansibility. According to the different QoS requirements of power communication services, data flows are divided into three types: Expedited Forwarding (EF) service, Assured Forwarding (AF) service and Best Effort (BE) service. EF service, which has the highest priority, can obtain high-quality service even in the case of network congestion. AF service has lower priority than EF service. When the network is congested, AF service can obtain a certain amount of QoS assurance. As for BE service, it is the only best effort service provided by the network, which is also the main mode of services provided by the network nowadays.

For the convenience of description, the total buffer length of the router is defined as L_0. In accordance with the long-term average service conditions of EF, AF and BE service, the total buffer is divided into three parts as the initial buffers of the three services. The buffer length of EF, AF and BE service is defined as L_1, L_2, and L_3. In this paper, L_0=500 packets, L_1=100 packets, L_2= L_3=200 packets. Two thresholds are set in each queue: high threshold *maxi* (i=1,2,3) and low threshold *mini*(i=1,2,3), α_i and β_i represent low threshold coefficient and high threshold coefficient, $0 \le \alpha_i < \beta_i \le 1$.

When a new packet arrives at the router, the type of packet and affiliated queue should be determined according to the Differentiated Services Code Point (DSCP) of arriving packet (Zhang 2012). Then we can use the exponentially weighted moving average (EWMA) to calculate the average length of the affiliated queue at the current time. avg_i represents the average queue length, which can be described just as followed:

$$avg_i = \begin{cases} (1-w_q)avg_i + w_q q_i & q_i > 0 \\ (1-w_q)^m avg_i & q_i = 0 \end{cases}, \text{ i} = 1, 2, 3 \quad (1)$$

w_q is the weight coefficient, $0 < w_q < 1$. In general, w_q is determined by the magnitude and duration of burst service that the router allows, typically the value of w_q is 0.002[8]. q_i is current queue length when the new packet arrives, and m is the duration of the empty queue.

Starting from 0 seconds, points are marked every t seconds. These points are referred to as the buffer update points, t=1 in this paper. Compare L_i (i=1,2,3) and L_{Ei} (i=1,2,3), L_{Ei} is the predicted length of the buffer, which represents the length that the buffer needs to reach after conversion. The initial $L_{Ei}=L_i$, L_{Ei} updates when the time arrives at update points. If $L_{Ei}=L_i$, it indicates that the existing buffer can satisfy the requirement of forecast, the queue has the ability to accommodate the packets arrived in this period. So the newly arrived packets can enter the buffer queue without packet loss. If $L_{Ei}>L_i$, it indicates that the predicted length of buffer still does not reach after the conversion last time, so the queue cannot accommodate all the packets arrived in this period. According to Random Early Detection (RED) algorithm, avg_i is compared with a high threshold max_i and low threshold min_i:

1 if $avg_i<min_i$, do not reject arriving packet,
2 if $avg_i>max_i$, reject arriving packet,
3 if $min_i \le avg_i \le max_i$, reject arriving packet with probability P_{ai}(i=1,2,3), which is calculated as follows:

$$P_{bi} = max_p(avg_i - min_i)/(max_i - min_i) \quad (2)$$
$$P_{ai} = P_{bi}/(1 - count_i \times P_{bi}) \quad (3)$$

max_p is the largest drop probability, $count_i$ indicates the number of packets newly arrived that are not dropped.

When the time arrives at update points, change the lengths of buffers. Process is as follows:

1 Determine the buffer's status of each queue. If $avg_i<min_i$, this indicates that the spare area of the buffer in the queue is too long, which means the queue is idle, the length of the buffer can be shortened. If $avg_i>max_i$, this indicates that the spare area of the buffer in the queue is too short, which means the queue is tense, the length of the buffer should be increased. If $min_i \le avg_i \le max_i$, this indicates that the spare area of the buffer in the queue is appropriate, so the length conversion is not necessary for this queue.

The value of α_i and β_i should be different. In order to ensure the highest QoS requirements, the high priority queue should be more stringent to shorten the length of the buffer; meanwhile the requirements of increasing the buffer can be relatively loose. On the contrary, the low priority queue should be relatively loose to shorten the length of buffer, meanwhile the requirements of increasing the buffer can be more stringent, thus $\alpha_1<\alpha_2<\alpha_3$, $\beta_1<\beta_2<\beta_3$. In this paper, (α_1, β_1)=(0.5,0.8), (α_2, β_2)=(0.6,0.9), (α_3, β_3)=(0.8,0.95).

2 Calculate the predicted length of the buffer L_{Ei} and determine the length change of buffer. L_{Ei} is calculated as follows:

$$L_{Ei} = avg_i / \gamma_i \text{ (i} = 1, 2, 3) \quad (4)$$

γ_i represents the predicted coefficient, $\alpha_i<\gamma_i<\beta_i$. The purpose of the equation (4) is to make the average queue length appropriate, which is located between the high and low thresholds after the buffer conversion. Meanwhile, if the buffer is too short, the queue would be sensitive to the new packets, which may result in additional loss. Therefore, a lower limit of buffer length l_i (i=1,2,3) is provided in each queue, the buffer length cannot be less than the lower limit, $l_1 = l_2 = l_3 = 60$ packets in this paper.

Each queue has two variates: buffer idle length $Brrrow_i$ (i=1,2,3) and buffer tense length $Loan_i$ (i=1,2,3). For the tense queue, $Brrrow_i$=0, $Loan_i$ is calculated as follows: if $L_{Ei} \leq l_i$, $Loan_i = L_i - l_i$; if $L_{Ei} > l_i$, $Loan_i = L_i - L_{Ei}$. For the idle queue, $Loan_i$=0, $Brrrow_i = L_{Ei} - L_i$. For the appropriate queue, $Brrrow_i = Loan_i = 0$.

3 Change the length of the buffer. Starting from EF queue, the states of three queues from high to low priority can be determined successively. If the queue is in the tense state, find an idle queue from low to high priority, then increase the buffer length of tense queue and shorten the same length in the idle queue at the same time until the buffer length of tense queue reaches its predicted buffer length or the idle buffer is exhausted. If three queues are all idle from high to low priority, recover the queue that is shorter than its initial length to its initial length or the idle buffer is exhausted.

In terms of queue scheduling, in order to simplify the problem, packets are set as the same size. Thus the Weighted Round-Robin scheduling algorithm (WRR) is adopted.

3 SIMULATION AND ANALYSIS

During the actual power grid operation, the amount of different priority services is also different: The amount of high priority services, such as Power production dispatching services, is smaller. On the contrary, the low priority services such as management information services, has a larger amount. Therefore, the ratio of packet number of three priority services is set to 1: 2: 7. The arrival rate of three priority services is assumed to obey Poisson distribution for the parameter λ_i(i=1,2,3), λ_1=200 packets/s, λ_1: λ_2: λ_3=1:2:7. MATLAB is adopted for the calculation of the simulation, and the simulation time is 50s. At the 0s, 20s and 40s, the burst of packets occurs in EF, AF and BE queue, the burst size is 50 packets/s, 100 packets/s and 200 packets/s. The duration is 6s, 6s and 10s, as is shown in Figure 1. The leaving total rate of service is 2000 packets/s, the weight ratio of WRR scheduling algorithm is also 1:2:7.

Figure 2 shows the buffer length changes of three priority queues with LC-RED algorithm. It can be seen from the figure that the packet number of EF queue has begun to mutate since 0s. The predicted buffer length of EF exceeds its original length at about 4s, thus EF buffer begins to increase, while the buffers of AF and BE are shortened. AF is shortened more than BE for its longer idle buffer. Mutation disappears at 6s, after a period of cooling, EF, AF and BE resume their initial buffer length at about 16s.

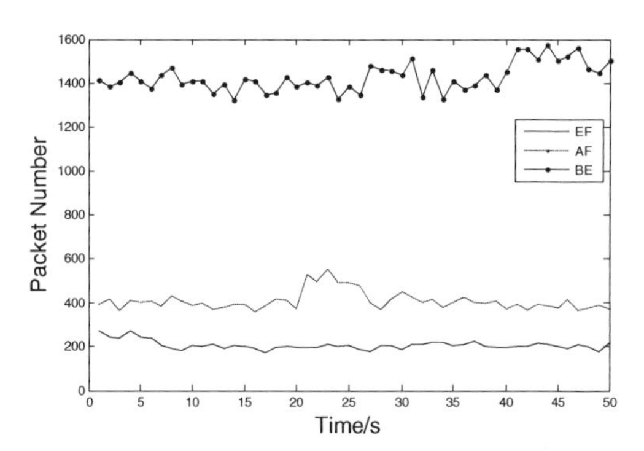

Figure 1. Number of packets.

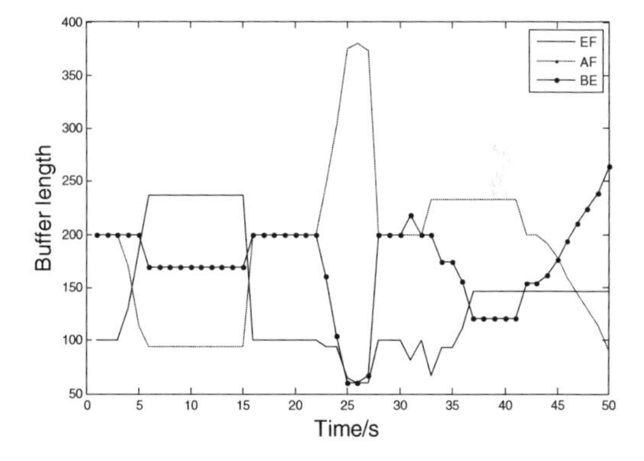

Figure 2. Change of the buffer length with LC-RED.

The packet number of AF begins to mutate since 20s. The predicted buffer length of AF exceeds its original length at about 23s, thus AF buffer begins to increase, while the buffers of EF and BE are shortened. AF buffer continues to increase until the buffers of EF and BE reach their lower limit. Similarly, the packet number of BE begins to mutate since 40s, EF buffer is appropriate at this time, so only the lengths of AF and BE buffers are converted.

Figure 3 shows the loss rate of each priority queue with LC-RED and RED algorithms. Figure 3 (a) shows that the packet loss rate of an EF queue with LC-RED is far less than that with the traditional RED when the packet number mutates. It is because the EF queue with LC-RED predicts and increases its buffer in time, enhances the ability to accommodate packets. Figure 3 (b) shows that the loss rates of AF queue with two algorithms have a high situation at about 24s, which is because of the large amount of mutations and insufficient predicted length. Then the loss rate of RED is still increasing until mutations disappear while LC-RED can quickly reduce the loss rate. Figure 3 (c) shows that the loss rate of BE queue with LC-RED has a high situation at about 26s. It is

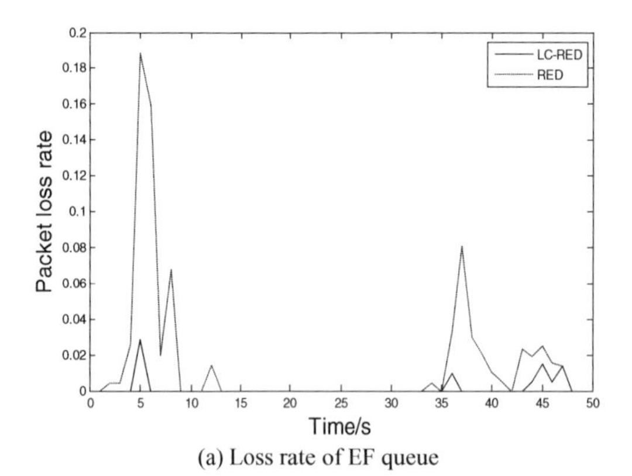

(a) Loss rate of EF queue

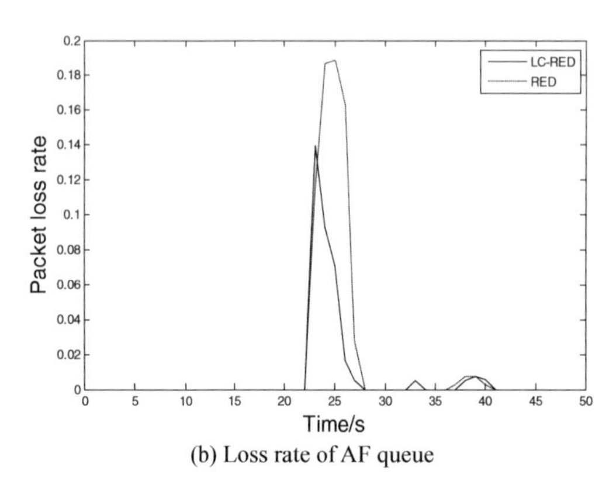

(b) Loss rate of AF queue

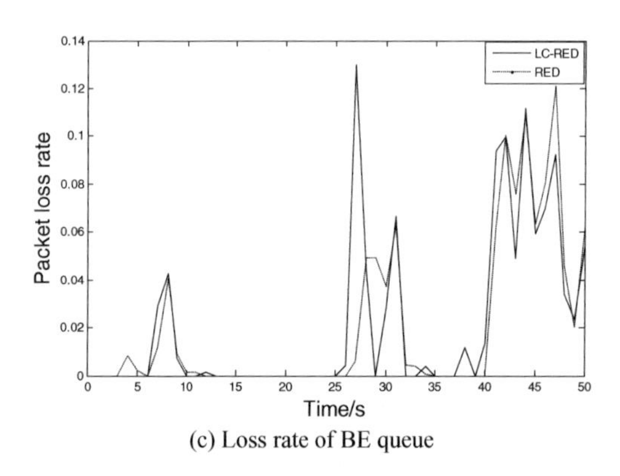

(c) Loss rate of BE queue

Figure 3. Loss rate of each priority queue with LC-RED and RED.

because BE buffer is short after conversion, so the average queue length of BE can reach the high threshold easily, which result in additional loss.

Figure 4 shows the overall loss rate with LC-RED and RED algorithms. Combined with figure 3 and figure 4, it indicates that when the congestion occurs in the high priority queue but it is not serious,

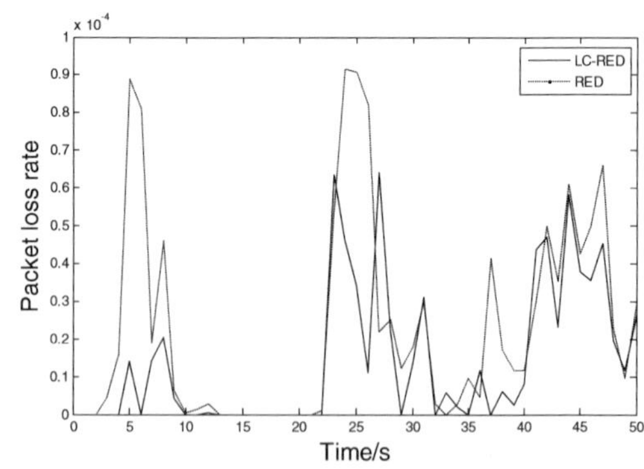

Figure 4. Overall loss rate with LC-RED and RED.

LC-RED algorithm can effectively reduce the packet loss rate. When the congestion is serious, LC-RED algorithm can reduce the packet loss rate of high priority queue. However, this comes at the expense of reducing the service quality of low priority queue.

Figure 5 shows the buffer utilization rate with LC-RED and RED algorithms. As can be seen from the figure, the buffer utilization rate of the two algorithms is basically the same at the earliest stage of packet mutation. It is because the initial buffer can meet its growth without transforming the buffer length. Then the buffer utilization rate with LC-RED is higher than that with the RED. It is because that RED can only maintain the queue length by dropping packets when initial buffer is insufficient, so the utilization rate is low. Meanwhile, LC-RED can reduce packet loss by changing the length of the buffer, thus improving the utilization rate of overall buffer.

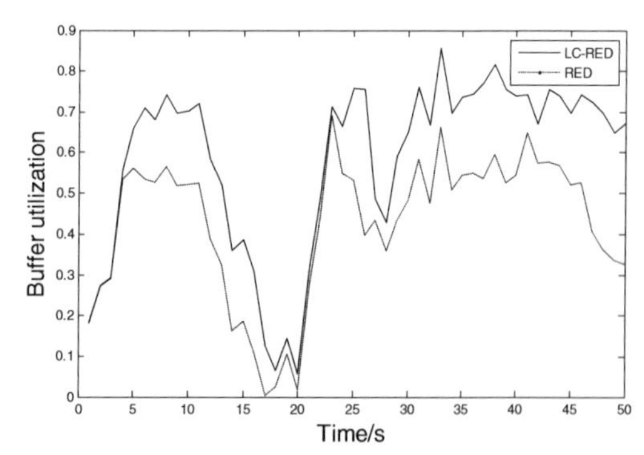

Figure 5. Buffer utilization rate with LC-RED and RED.

Based on the above analysis, LC-RED is obviously better than RED in the performance of packet loss rate and buffer utilization. By predicting and changing the buffer length, LC-RED can effectively alleviate the loss caused by burst packets. The loss rate of the high priority queue and the overall loss rate can be greatly reduced. Meanwhile the buffer utilization rate also increases significantly.

4 CONCLUSION

The rapid development of multimedia services causes higher requirements on the security and reliability of the optical communication network. The OTN transmission technology with the characteristics of flexibility, efficiency, reliability, and large wavelength capacity becomes one of the important technologies in the area of future optical network development. According to the different QoS requirements of multiple services in OTN network, the problem of AQM is deeply studied in this paper. An improved active queue management algorithm LC-RED combined with the differentiated service model is presented in this paper. The congested queues can take advantage of other idle queues to ensure the QoS requirement of the high priority queue. The comparison studies by computer simulations indicate that the algorithm presented in this paper have better performances on reducing the overall packet loss rate and enhancing the utilization ratio of buffer.

ACKNOWLEDGMENT

This work is financially supported by "the Fundamental Research Funds for the Central Universities" (2014QN14).

REFERENCES

Feng Bao, Fan Qiang, Li Yang, et al. 2012. Research on Framework of the Next Generation Power Communication Transmission Network. *2012 International Conference on Computer Science and Electronics Engineering, Hangzhou, 23–25 March 2012*: 414–417. IEEE Computer Society Press.

Feng Wuchang, Dilip D. Kandlur & Debanjan Saha. 2002. The BLUE Active Queue Management Algorithm. *IEEE/ACM Transactions on Networking* 10(4): 513–527.

Gong Jing & Wu Chunming. 2010. S-CHOKe: An AQM Algorithm for Enhances the Fairness of the CHOKe. *Acta Electronica Sinica* 38(5): 1100–1104.

Huang Lei, Wu Chunming, Jiang Ming, et al. 2010. REDu: A New Active Queue Management Algorithm for Detection and Punishment of Unresponsive Flows. *Acta Electronica Sinica* 38(8): 1759–1762.

Tian Quan & Wu Bin. 2012. ML-XCHOKe: Malicious Level based-improved XCHOKe AQM. *The 9th Annual Conference of China Institute of Communications, Beijing, 17 August 2012*: 44–48. Beijing: Beijing University of Post and Telecommunications Press.

Wu Chunming & Jiang Ming. 2005. SBlue: stabilized Blue. *Journal on Communications* 26(5): 68–74.

Zang Lili. 2012. *The research of priority checking RED algorithm based on IPV6 network*. Jilin: Jilin University.

Zhao Ziyan& Zhang Dawei. 2011. Analysis on the Requirement of SGCC on Telecommunication Services in the "12th Five-Year Plan" Period. *Telecommunications for Electric Power System* 32(233): 56–60.

Electronics, Communications and Networks IV – Hussain & Ivanovic (eds)
© 2015 Taylor & Francis Group, London, ISBN: 978-1-138-02830-2

Atmosphere impact and ground station selection of satellite to ground laser communication

Yan Lou, Yiwu Zhao, Chunyi Chen, Shoufeng Tong & Huilin Jiang
National and Local Joint Engineering Research Center of Space Optoelectronics Technology, Changchun University of Science and Technology, Changchun, Jilin, China

Zhipeng Ren
National Asset Institute, Changchun University of Science and Technology, Changchun, Jilin, China

ABSTRACT: A satellite to ground laser communication scheme of multiple ground stations which take the climatic geographical distribution and circumstance characteristics into consideration is presented in this paper. The downlink characteristics of a GEO satellite to five ground stations are analyzed according to the STK software. The results show that Tibet's Ali has the best longitude and latitude with 52° horizontal angle of satellite to ground laser communications. The power attenuation shows a decreasing trend as well as the wavelength increases due to scattering. Optical power attenuation intensifies as visibility reduces. The atmospheric attenuation caused by the average power is small along with the rise of the horizon dip. The wavelength λ increases as the fluctuation index decreases. As the scintillation index decreases, the receiver-aperture diameter increases. However, the scintillation index decreases more rapidly with the high altitude. These results will be helpful to the experiment of satellite to ground laser communication.

1 INTRODUCTION

From 2003 to 2009, ESA's involvement in the development of optical communication terminals for high data rate links between satellites already resulted in the successful demonstration of the SILEX program using optical inter-satellite links and the ARTEMIS satellite in the GEO orbit, and experimental program is like the OPEL. The Optical Ground Station (OGS) facility on the Canary Islands and communication links through the Earth's atmosphere have demonstrated successfully that coherent laser communication links are operational in-orbit (Sodnik 2009) since February 2008. With the transmitting data at a rate of 5.625Gbps, they verify the capability of laser communication exemplarily in LEO-LEO and Ground-LEO constellations (Smutny et al. 2009, Heine et al. 2010). Japanese Engineering Test Satellite and four ground stations carried out communications in 2009. The Satellite will communicate with one of the four ground stations, DLR, ESA, JPL and NICT, depending on the weather situation. (KOISHI et al. 2011, Toyo et al. 2011). In 2011, Harbin Institute of Technology (China) first actualized the direct detection of high-speed satellite to ground laser communication. Changchun University of Science and Technology in China has also reached the international advanced level in space and satellite to ground laser communication (Wenhe et al.2010)

2 THE OPTICAL GROUND STATION LOCATION

2.1 *Optical ground station address location*

Earth's atmosphere affects the optical signal from deep space in two ways. First, when the optical signals pass through the atmosphere, it is partially attenuated. Second, to limit the effects of atmospheric turbulences, the ground station should be located higher in altitude. Therefore, implications of atmospheric turbulence effects on the deep space downlink analysis need to be included. A disadvantage of deep space downlink is the fluctuation of atmospheric attenuation caused by a number of phenomena in the atmosphere, such as scattering, absorption, and turbulence. Turbulence, in particular, will experience the effects of Scintillation, Beam wander, Angle-of-arrival fluctuations and Beam expansion. These phenomena may result in a loss of power and intensity fluctuations at the receiver, and in the worst case, may limit the availability of the deep space downlink.

The availability calculation in the present work is based on the power budget analysis of deep space optics and the statistical analysis of atmospheric attenuation. The availability is defined as the time in which the atmospheric attenuation is lower than the power margin of deep space optics in this paper. The available probability is given by

$$P_0 = 1 - \prod_{i=1}^{N} p_i,$$

here p_i is the probability that the ith ground station is not available.

The available probability attains 60% with more than 5km ground visibility through the optimization design of the selected ground station addresses.

2.2 *GEO satellites orbit planning*

Figure 1 of GEO satellite diversity or distribution is provided with the east longitude 10.5°, 77°and 176.8°. The link of satellites with east longitude 77° can achieve full coverage of the Chinese territory and is best suited for GEO satellite to ground communication with small channel's zenith according to the simulation of STK.

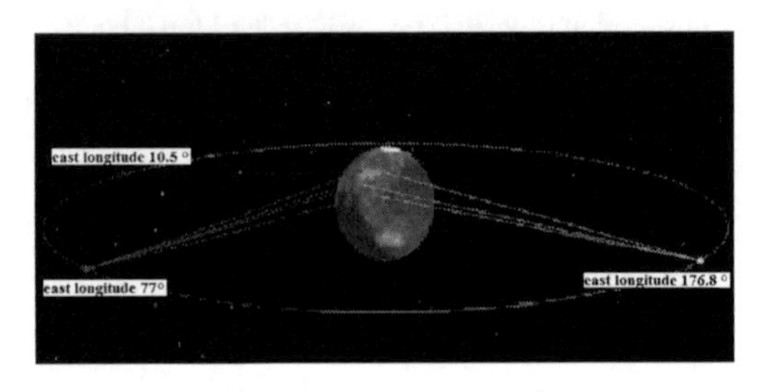

Figure 1. Map of GEO-ground communication by STK.

2.3 *The ground stations selection*

Clouds, fog, haze, mist, and other precipitation in the atmosphere cause strong attenuation of the laser signal and communication outages. Therefore, ground receiving telescopes need to be located on sites where cloud coverage is low and statistically predictable. Fortunately, weather diversity and the global cloud distribution patterns can provide, with a reasonably high likelihood, a cloud line of sight for the optical link to at least one location on the ground. This chapter describes the effects of Earth's atmosphere upon laser communication beams. In China, there are several astronomical observatories, whose geographical information and satellite stations are as below:

Changchun: north latitude43°48'52.54", east longitude:125°19'38.68", altitude:211m.

Kunming: north latitude25°02'03.80", east longitude:102°43'16.62", altitude:1899m.

Hainan: north latitude20°04'48.18", east longitude:110°17'21.48", altitude:20m.

Urumqi: north latitude43°49'23.92", east longitude:87°36'44.64", altitude:846m.

Ali: north latitude32°27'15.63", east longitude:81°11'50.41", altitude:5022 m.

Ali station is a new national observatory, which has a unique altitude advantage of above 5km with very low average cloud that makes it suitable for optical observation and tests. The ground station is suitable for the establishment of a link with longitude 77° of GEO satellites.

2.4 *The satellite to ground link characteristics simulation*

Table 1. Simulation of GEO satellites and five ground stations Link with longitude 77 deg by STK.

Link	Chang chun	Urumuqi	Kunming	Hainan	Ali
Visibility Times/%	100	100	100	100	100
Duration/s	86400	86400	86400	86400	86400
Ground station Azimuth / degree	21°	39°	49°	46°	52°
Communication Distance /Km	39471	37869	37121	37356	36935

2.5 *The simulation of satellite to ground link characteristics*

Table 1 shows simulation results of GEO satellites and five ground stations Link with longitude 77 degrees by STK. The following conclusions can be drawn from the table:

Among the five ground stations, the lower the latitude is, the higher the Ground station azimuth of the visual axis is. The Ground station azimuth must be improved in order to reduce atmospheric effect; The Ground station of Changchun with azimuth 21° is unfavorable with longitude and latitude; The Ground station of Urumqi with azimuth 39° is ideal with longitude and high-latitude; The Ground station of Hainan with azimuth 46° is ideal with latitude and high longitude; The Ground station of Kunming with azimuth 49°is favorable with latitude and longitude; The Ground station of Ali with azimuth 52° is the most Optimal with latitude and longitude, which is best suited for the development of satellite to ground laser communication.

3 ANALYSIS OF EFFECT ON ATMOSPHERIC CHANNEL OF SATELLITE TO GROUNDS LINKS

The impact on the link of satellite to grounds slant atmospheric channel in this paper includes average

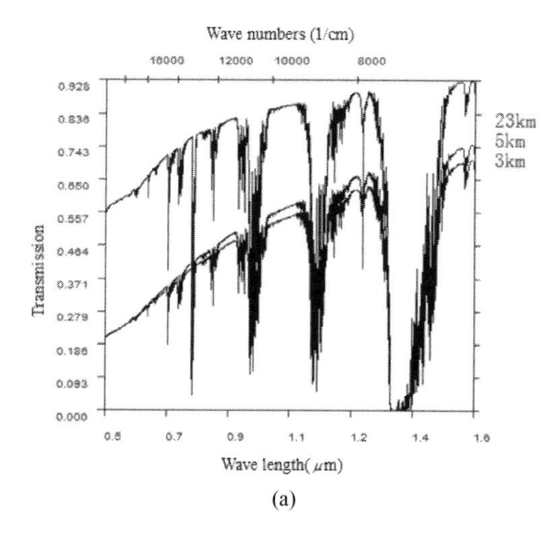

(a)

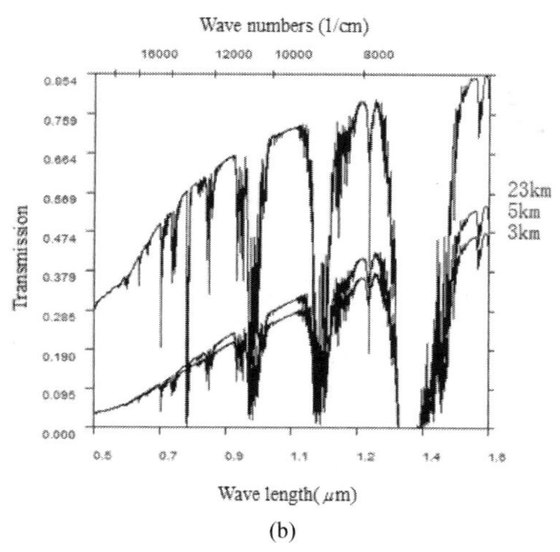

(b)

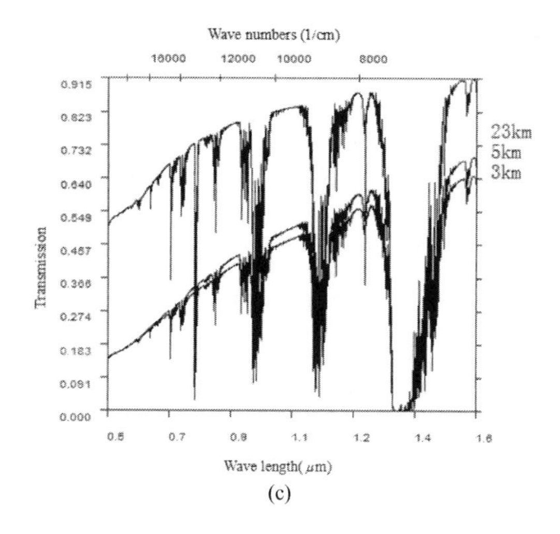

(c)

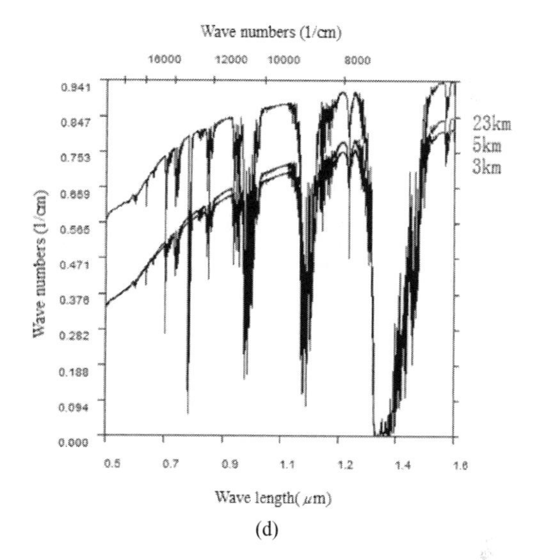

(d)

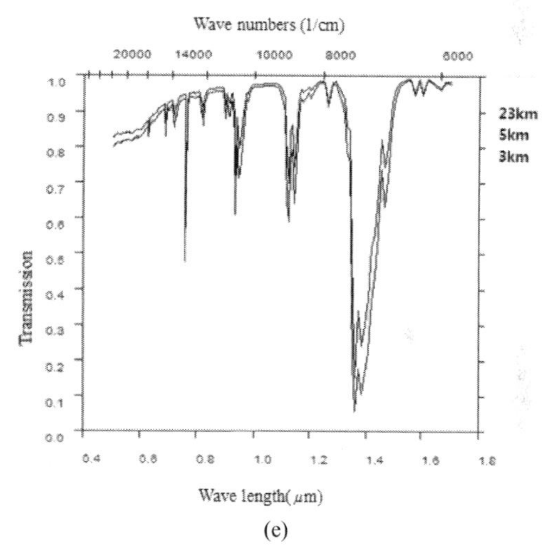

(e)

Figure 2. Transmittance spectrum of the atmosphere for a space-to-ground link.

power attenuation, transient power flicker and jitter caused by atmospheric turbulence.

3.1 *Atmospheric absorption and scattering effects*

Earth's atmosphere, via absorption and scattering, attenuates the propagating electromagnetic wave. The atmospheric transmittance related to atmospheric attenuation is described as :

$$T_{atm}(\lambda) = \exp\left[-\sec\varphi \int_0^H k_e(\lambda, h)\,\mathrm{d}h \right] \quad (1)$$

where λ is the laser wavelength; ke=ks+ka is the extinction or the attenuation coefficient of atmosphere; H is the vertical height of the atmospheric channel. The parameter ϕ is the zenith angle. Analyzing the transmittance spectrum, one can clearly notice that the spectrum itself is modulated between transmission bands with high transmittance and the bands where the transmittance is close to zero (forbidden bands). Atmospheric transmittance in over 5 km above sea level after MODTRAN simulation and attenuation caused by longitude 77° GEO satellite to five ground stations for slant atmospheric scattering is shown in Figure 2. Providing the condition that the zenith angle is 0°, the altitude is low, and the visual range is 5Km in mid-latitude, the transmittance spectrum between longitude GEO 77° GEO satellite and five ground stations is analyzed.

Several observations can be derived from these figures. For simplicity, all of these MODTRAN plots are restricted to the case of rural aerosol model and located at mid-latitudes in China continent. Tab. 2 shows the data of the atmospheric transmittance simulation result which is summarized in the different links of satellite to ground stations at the 800nm band, 1060nm band and 1550nm band.

Table 2. Different link of transmission simulation results.

				Transmission		
Ground station	Altitude	Horizon angle	Visibility	800 nm	1060 nm	1550 nm
Chang chun	211m	21deg	5km	0.15	0.25	0.43
Kunming	1899m	49deg	5km	0.63	0.70	0.73
Hainan	20m	45deg	5km	0.34	0.48	0.60
Uru muqi	846m	39deg	5km	0.40	0.60	0.65
Ali	5022m	57deg	5km	0.91	0.93	0.92

4 RESULTS

Under the same weather conditions, with the increasing wavelength, the slant path atmosphere transmission increases, and the corresponding power attenuation due to scattering shows a decreasing trend. For Changchun station, when the visibility is 3Km, the transmittance of 800nm band is 0.15 and of the 1550nm band is up to 0.43. It means that the longer the wavelength of laser light is, the smaller the loss of atmospheric scattering is. If the beacon light uses the 800nm band wavelength and signal laser uses the 1550nm band, the optical power attenuation will be greater than the communication beacon light.

Along with the reduced visibility, the optical power average attenuation is greatly intensified by atmospheric scatter. The power loss caused by the slant path channel is only 0.7dB for the Changchun ground station when the visibility is 23km and the wavelength is 1550nm; the power loss caused by the slant path channel reaches 4dB when visibility is 3km. Therefore, the higher the visibility is, the lower power attenuation caused by atmospheric is. So, the single ground station with better available probability and link loss margin must be optimized first.

Along with rises of horizontal angle, the average power of atmospheric attenuation is getting smaller. The horizontal angle of GEO to Urumuqi ground station and Changchun ground station respectively, are 39degree and 21degree with 3km visibility, and transmittances are 0.65 and 0.43. Therefore, the bigger the horizontal angle is, the smaller the loss caused by atmospheric is.

Along with the increasing altitude ground stations, the slant atmospheric scattering loss decreases. The horizontal angle of GEO to Kunming ground station and Hainan ground station respectively, are 49deg and 45deg with altitude 1899m and 20m. The transmittances are 0.73 and 0.60 of 1550nm under 3km visibility condition for two ground stations. For Ali Station, the power loss is less than 1dB of GEO to ground station with 5022m altitude, which is under the best condition and with better visibility. Laser beam propagation is perturbed by random refractive index fluctuations of the Earth's atmosphere. Therefore, compared with a plane wave, when the laser beam propagates from deep space, the receiver will undergoes wavefront phase distortion. Experimental results of laser beam propagation through horizontal atmospheric paths are included. And even though these are not true representatives of deep space optical link configurations, they are indicative of the relevance of the theory. Therefore, we use the plane wave model to calculate the scintillation index near the center of the beam and the differences of the results of model calculations using the Gaussian beam can be ignored. Flashing simplified by a plane wave Rytov variance index is defined as [8]

$$\sigma_I^2 = 2.606k^2 \sec\varphi \int_{h_0}^{H} C_n^2(h) \int_0^{\infty} \kappa^{-8/3} \times$$

$$\left\{1-\cos[\frac{(h-h_0)\kappa^2 \sec\varphi}{k}]\right\} \mathrm{d}\kappa \mathrm{d}h =$$

$$2.25\mu_1 k^{7/6}(H-h_0)^{5/6} \sec^{11/6}\varphi \text{ ,} \tag{2}$$

$$\mu_1 = \int_{h_0}^{H} C_n^2(h)(\frac{h-h_0}{H-h_0})^{5/6} \mathrm{d}h \tag{3}$$

Eq. (3) shows that, σ_I^2 represents the fluctuations which are characterized by a scintillation index variance. Parameter H is the height of the satellite to Parameter ground station;

Parameter h is the altitude; h_0 is ground elevation;

Parameter k=2π/l is the wave number;

Parameter l is the scale of turbulent eddies;

Parameter φ represents the zenith angle.

Experimental data of Cn^2 are not readily available. However, a number of parametric models have been formulated to describe the $Cn^2(h)$ profile, and among those, one of the most used models is the Hufnagel-Valley, given by:

$$C_n^2(h) = 0.00594(v_v/27)^2(10^{-5}h)^{10}\exp(-h/1000) + 2.7\times10^{-16}\exp(-h/1500) + A_{hv}\exp(-h/100), \tag{4}$$

Parameter h is the altitude.

Parameter n and A_{hv} are to be set by the user defined as below:

$$v_v = [1/15000\int_{5000}^{20000} v_h^2(h)\mathrm{d}h]^{1/2} \tag{5}$$

Parameter $v_h(h)$ is the height of the horizontal orthogonal component of the wind speed, here v_v=21m/s, $A_{hv}=1.7\times10^{-14}\mathrm{m}^{-2/3}$.

Space-to-ground aperture averaging factor expression allows taking the Cn^2 profile into account and is given by:

$$A = \frac{1}{1+A_0^{-1}(\frac{D^2}{\lambda h_0 \sec\varphi})^{7/6}}, \tag{6}$$

where $A_0 \approx 1.1$; D is the aperture diameter; altitude h_0 is given by

$$h_0 = [\frac{\int_{\text{path}} \mathrm{d}hC_n^2(h)h^2}{\int_{\text{path}} \mathrm{d}hC_n^2(h)h^{5/6}}]^{6/7}. \tag{7}$$

Fluctuation variance is usually associated with the atmospheric refractive index structure constant, the link distance, receiving optical aperture and other factors. The results based on the east longitude of 77.0 degree GEO satellite to the two ground stations in Kunming and Hainan is calculated as below:

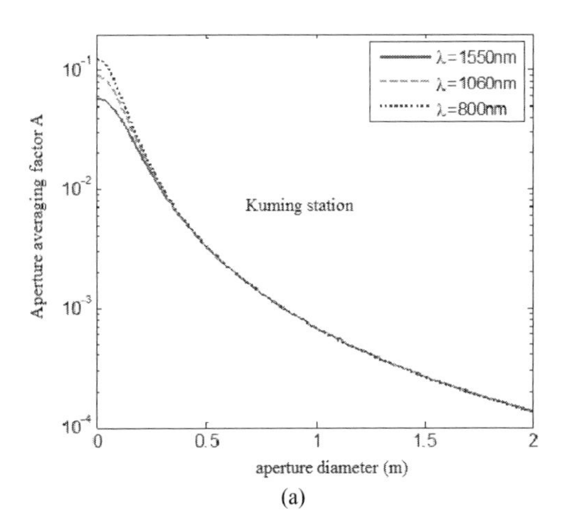

(a)

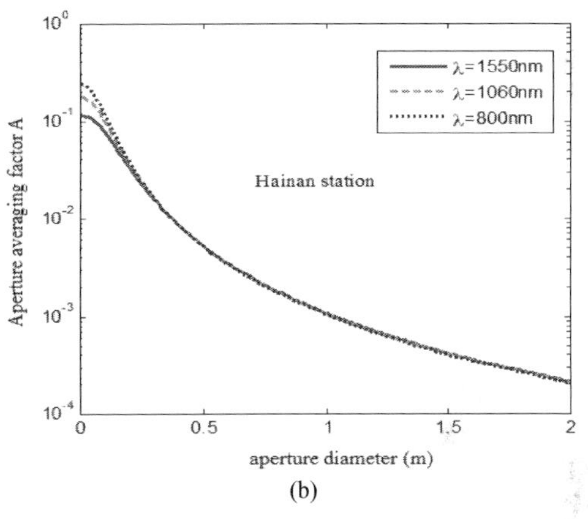

(b)

Figure 3. Fluctuation index of satellite communicates to two ground stations with different receiving apertures. Aperture averaging factor depends on the aperture diameter with wave-length such as 1550nm, 1060nm and 800nm of two stations.

We can see the results from Figure 3:

a. The longer the wave-length λ is, the smaller the scintillation index is.
b. With the increase of the receive aperture diameter, scintillation index decreases rapidly. When the diameter is 0.5m, the scintillation variance can be less than 0.01. If the receiving aperture increases bigger (more than 1m), aperture averaging effect is weakened.

With the increase of altitude, scintillation variance is improved. Under the sub-caliber conditions (abscissa close to 0), Kunming station's altitude is

1899m, and the corresponding scintillation variance is 0.07; while Hainan station's altitude is 20m, and corresponding scintillation variance is 0.11.

5 CONCLUSIONS

The satellite communication to five ground stations is analyzed in this paper by STK. The results show that five ground stations in the lower latitudes as well as the higher visual axis of the communication horizontal angle; The horizontal angle must be higher in order to reduce atmospheric effects. Tibet Ali has the best longitude and latitude and maximum contribution to satellite laser communication with horizontal angle 52°.

Considering above situations, when the power attenuation showed a decreasing trend, the wavelength increases due to scattering. Optical power attenuation intensifies as the visibility is reduced. The atmospheric attenuation caused by the average power is small along with the rise of horizon dip. The wavelength λ increases as the fluctuation index decreases. As the scintillation index decreases, the receiver-aperture diameter increases. However, the scintillation index decreases more rapidly with the high altitude. These results will be helpful for the experiment of satellite to ground laser communication.

ACKNOWLEDGMENTS

This work was financially supported by the National Natural Science Foundation (61007046), Innovation Program of Jilin Municipal Education Commission (2014052011 5JH).

REFERENCES

Deng, W., Tan, L., Ma, J., Yu, S., et al. 2010. Measurements of angle-of-arrival fluctuations over an 11.8 km urban path. *Laser Part. Beams* 28(01): 91–99.

Hamid, H. 2005. *Deep Space Optical Communications*. California: Jet Propulsion Laboratory California Institute of Technology Press.

Heine, F., Kmpfner, H., Czichy, R., et al. 2010. Optical Inter-Satellite Communication Operational. *2010 Military Communications Conference, San Jose, CA, Oct. 31 -Nov. 3, 2010*. IEEE.

Jing, M., Yi, J., Li, T. & Si, Y. 2008. Influence of Beam Wander on Bit-error Rate in a Ground-to-Satellite Laser Uplink Communication System. *Optics Letters* 33(22): 2611–2613.

Koishi, Y., Suzuki, Y., Takahashi, T., et al. 2011. Research and Development of 40 Gbps Optical Free Space Communication from Satellite/Airplane. Space Optical Systems and Applications (ICSOS), 2011 International Conference on, Santa Monica, CA, 11–13 May 2011: 88–92. IEEE.

Smutny, B., Kaempfner, H., Muehlnikel, G., et al. 2009. 5.6Gbps Optical Inter Satellite Communication Link. *Proceedings of SPIE, San Jose, CA, 24 February 2009*. International Society for Optics and Photonics.

Sodnik, Z., Perdigues, J., et al. 2009. Adaptive Optics and ESA's Optical Ground Station. *Proc. SPIE Bellingham: SPIE7464, Free-space communications IX, San Diego, CA, 02 August 2009,* International Society for Optics and Photonics.

Toyo Shi MAM, Sasaki, T., Takenak AH, et al. 2011. Research and Development of Free Space Laser Communications and Quantum Key Distribution Technologies at NICT. 2011 International Conference on Space Optical Systems and *Applications (ICSOS), Santa Monica, CA , 11–13 May 2011*. IEEE.

Electronics, Communications and Networks IV – Hussain & Ivanovic (eds)
© 2015 Taylor & Francis Group, London, ISBN: 978-1-138-02830-2

A content dissemination model for mobile internet to minimize load on cellular network

Xiaofeng Lu
Beijing University of Posts and Telecommunications, Beijing,. China

Pietro Lio
Computer Laboratory, University of Cambridge, Cambridge, UK

Pan Hui
Hong Kong University of Science and Technology, Clear Water Bay, Hong Kong

ABSTRACT: Cellular networks do not have enough capacity to accommodate the exponential growth of mobile data requirement. Data can be delivered between mobile terminals through peer-to-peer WiFi communications (e.g. WiFi direct), but contacts between mobile terminals are frequently disrupted because of the user's mobility. This paper proposes a Subscribe-and-Send architecture, under which a user subscribes contents on the Content Service Provider (CSP) but does not download the subscribed contents. Some users who have downloaded the contents deliver the contents to the subscribers through WiFi opportunistic peer-to-peer communications. Numerical simulations provide an evaluation of the traffic offloading performance of Subscribe-and-Send. Simulation results show that Subscribe-and-Send architecture can offload 72%~90% cellular network data traffic load.

1 INTRODUCTION

With the proliferation of smart mobile terminals (SMTs) such as smart phones, laptops, pads and so forth, people use these devices to access Internet more and more frequently. According to Global Mobile Data Traffic report, global mobile data traffic grew 81 percent in 2013. Currently, a large percentage of mobile Internet data traffic is generated from the SMTs and mobile broadband-based PCs. Cisco forecasts that the compound annual growth rate (CAGR) of mobile data traffic will grow by 61 percent from 2013 to 2018. Current cellular networks do not have enough capacity to accommodate such an exponential growth of data. Cellular networks operators are able to increase bandwidth to meet the increasing bandwidth requirement. However, increasing the infrastructure capacity is costly, and the infrastructure capacity cannot increase unlimitedly.

People found that many applications such as email or file transfer can afford to delay data transfers without significantly hurting user experience. So researchers began to use delay tolerant networking (DTN) and opportunistic routing technologies to transfer bulk data instead of cellular network (Laoutaris et al. 2009, Fall 2003). However, in delay tolerant networks, a continuous end-to-end path between a source-destination pair is not guaranteed. As SMT users keep moving, the wireless links between SMTs are highly prone to disruption. Hence, opportunistic routing between SMTs begins to play a more and more important role in saving the bandwidth of cellular network infrastructure.

In this paper, the way to offload cellular network traffic through WiFi opportunistic peer-to-peer communications is studied. This paper proposes a Subscribe-and-Send architecture, which composes of an application and an opportunistic routing protocol. The application deals with the subscribing and delivery. With Subscribe-and-Send, a mobile user downloads some contents through the cellular network infrastructure and routes the contents to subscribers.

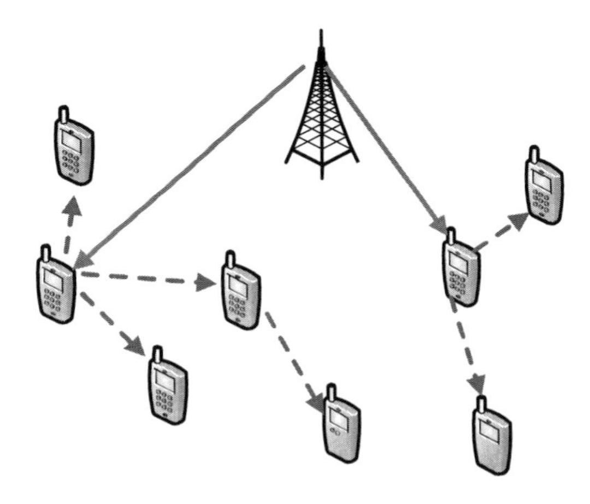

Figure 1. Opportunistic WiFi communication.

The mobile data may include multimedia newspapers, videos, weather forecasts, and movie trailers etc., generated by Content Service Providers (CSPs). In fact, lots of SMT users are interested in a common content, so a SMT user can download the content through the cellular network infrastructure, and then send it to other users opportunistically through WiFi peer-to-peer communication as Figure 1 shows.

2 RELATED WORK

There are already several cellular traffic offloading solutions and applications proposed in both academic and industrial fields.

Balasubramanian designed a system, called Wiffler, to augment mobile 3G capacity(Balasubramanian et al. 2010). By using a simple method to predict future WiFi throughput, it waits until 3G savings being expected within the application's delay tolerance. For applications that are extremely sensitive to the delay or loss (e.g., VoIP), Wiffler quickly switches to 3G if WiFi is unable to successfully transmit the packet within a small time window. Balasubramanian studied how to switch interface between WiFi and 3G smoothly. In our study, both 3G and WiFi interfaces are in work mode. Only the content subscribed will be transferred through WiFi peer-to-peer.

Han proposed to intentionally delay the delivery of information and offload it to the free opportunistic communications with the goal of reducing cellular data traffic (Han et al. 2010). They studied the target-set selection problem in the cellular traffic offloading for information delivery in MoSoNets. They proposed three algorithms called Greedy, Heuristic, and Random to solve the target-set selection problem. Although Greedy performs the best in simulation, the Greedy algorithm is not practical in real world. In comparison, every user can download data through cellular network infrastructures proposed in this paper.

Haddadi proposed a scalable, location-aware, personalized and private advertising system for mobile platforms, called MobiAd (Haddadi et al. 2010). Its key benefit of using mobile phones is to take advantage of the vast amount of information on the phones and the users' interests in order to provide personalized ads. MobiAd would perform a range of data mining tasks in order to maintain an interesting profile on the user's phone, use the infrastructure network to download, display relevant ads, and report the clicks via a DTN protocol.

Dimatteo proposed an integrated architecture exploiting the opportunistic networking paradigm to migrate data traffic from cellular networks to metropolitan WiFi access points (APs) (Dimatteo et al. 2011). To quantify the benefits of deploying such architecture, they considered the case of bulk file transfer and video streaming over 3G networks and simulated data delivery using real mobility data set of 500 taxis in an urban area. Their results gave the numbers of APs needed for different requirements of quality of service for data delivery in large metropolitan area. We think the WiFi infrastructures could be in favor of the traffic offloading of our architecture as well.

Whitbeck proposed Push-and-track, a content dissemination framework that harnesses ad hoc communication opportunities to minimize the load on the wireless infrastructure while guaranteeing tight delivery delays (Whitbeck et al. 2012). Push-and-track achieves this through a control loop that collects user-sent acknowledgements to determine if new copies need to be re-injected into the network through the 3G interface. In our study, the number of copies of a file is not determined by the architecture, but by the network condition.

Hongyi Wu proposed an integrated cellular and ad hoc relaying system, iCAR (Hu et al. 2001). The basic idea of iCAR is to place a number of ad hoc relaying stations (ARSs) at strategic locations, which can be used to relay signals between mobile hosts (MHs) and base transceiver stations (BTSs). Our study is to reduce the traffic load on cellular network and avoid diverting traffic from one cell to another cell.

3 SUBSCRIBE-AND-SEND ARCHITECTURE

In this section, we introduce the Subscribe-and-Send architecture and how it offloads the mobile Internet data traffic from cellular networks through WiFi opportunistic peer-to-peer communications.

3.1 *Premise: transmission model*

We are concerned with the issue of distributing content to a variable set of SMTs. The SMTs is equipped with wireless broadband connectivity (e.g., 3G) and is also able to communicate in WiFi. Both the wireless broadband connectivity and WiFi interface are in working mode. A software is installed on the SMT to subscribe contents on CSP and to send files to encountering nodes. The software can get a file through both the wireless broadband connectivity and the WiFi connectivity.

We suppose the SMTs are WiFi Direct devices. WiFi Direct devices have a new capability that allows the creation of peer-to-peer connections between WiFi client devices without requiring the presence of a traditional WiFi infrastructure network (i.e., AP or router) (Alliance W F 2010).

3.2 *Overview of Subscribe-and-Send architecture*

There are two stages in Subscribe-and-Send: subscribe and send.

1 In the subscribe stage, a user accesses the CSP and subscribes some interested contents on it. The subscription composes of the name of subscribed content, the user's ID and the deadline of the subscription. Before the deadline of the subscription, the software does not download the subscribed content through the cellular network. It prefers to receive the content sending by others through WiFi connections.
2 Some users would like to download the interested contents through the cellular network. These users can be called source nodes. A source node could be a member of a CSP, or the one who would like to pay for downloading contents through 3G. When a source node accesses the CSP, if the content that the node has is subscribed by others, it starts the sending process and delivers the content to the subscriber through opportunistic WiFi connections. When two nodes meet, the source node will deliver the subscribed content to the other node. The opportunistic routing protocol working in the routing layer determines how a node delivers its files. Different opportunistic forwarding protocols can be employed in the routing layer (Lindgren et al. 2003, Spyropoulos et al. 2005, Burgess et al. 2006, Lu 2010).

For example, node_1 contacts the CSP and downloads movie_1 through 3G, and node_7 subscribes movie_1 on the CSP as Figure 2 shows. Node_1 accesses the CSP and knows that node_7 subscribes movie_1. Then node_1 begins to send movie_1 to node_7 through opportunistic WiFi connections. As the WiFi connection cannot be always created, a relay has to carry the content and waits for the future WiFi connections. In Figure 2, the solid line means the connection created by 3G interface and the dot line means peer-to-peer WiFi communication. After node_3 gets movie_1 from node_1, as it has no other WiFi connection, node_3 has to carry movie_1 all the time. When node_3 meets node_7 at some place, it sends movie_1 to node_7.

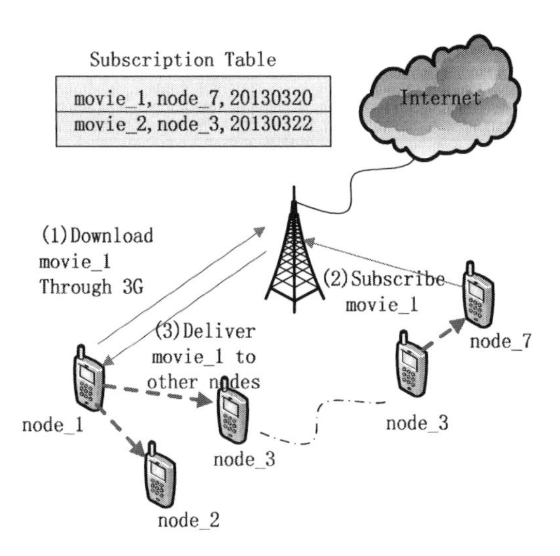

Figure 2. Subscribe-and-Send architecture. A user subscribes a movie, and then the source node reads the subscription table and sends the movie to the subscriber one by one.

3.3 *Subscription management*

1 When a subscriber receives the content subscribed on the CSP, it sends a response message to the CSP and removes the subscription of this content from the subscription table.
2 If a subscriber does not receive the subscribed content after the deadline of the subscription, the CSP prompts the user of the failure of the subscription. Then the user can download the content through 3G, or extend the deadline of the subscription. If the user does not extend the deadline, the CSP removes the subscription of the content from the subscription table on CSP. Meanwhile, the subscription is removed from the subscription table on the user's device also.
3 As the deadline of a subscription is created by the subscriber, the deadlines of different subscriptions are different. And, it is not necessary that the clocks on all devices are synchronous because the CSP maintains the total subscription table.
4 Each node accesses the CSP to check the subscription table every 10 minutes or 30 minutes through 3G, so the 3G traffic load of an SMT is very low.

3.4 *Cellular network traffic offloading analysis*

During the data dissemination from the source to the destination, some relays might be interested in the content as well. When a node receives a content that it also subscribes, it sends a response message to the CSP and removes itself from the subscription table. As the node does not download the content through 3G, the load on the cellular network is released. In Figure 3, node_2, node_3, node_4, node_5 and node_6 subscribe movie_1, and only node_1 downloads movie_1 through 3G. Node_1 accesses the subscription table on the CSP and knows that some other nodes subscribe movie_1. It sends movie_1 to node_2 and node_3 when the WiFi connections exist. Node_2 and node_3 continue to send the content to others when WiFi connections exist. Finally, all nodes get movie_1 before the deadline of each subscription. Since only one node downloads movie_1 through 3G, 85.7% of the cellular network traffic is offloaded.

4 SIMULATION

4.1 *Simulation setup*

We employ the Opportunistic Network Environment (ONE) simulator to evaluate the forwarding performance and traffic offloading performance. In the simulator, nodes only travel on the roads of Helsinki on a city may. In this paper, we focus on the area surrounding Helsinki's city center, covering 15.3 km^2. The simulation considers six groups: two groups of people walking, three groups of cars whose speeds are between 7 to 10 m/s and another group of cars whose speeds are between 2.7 to 13.9 m/s. In each round of simulation, the numbers of nodes of each group are different from 20 to 60. We assume the transmission range of all nodes to be the same at 100 meters. The transmission speed is 10MB per second. The simulation time is 5000 seconds (83 minutes). The size of the content is between 5000KB and 10MB. We also assume that all nodes participate in the data dissemination.

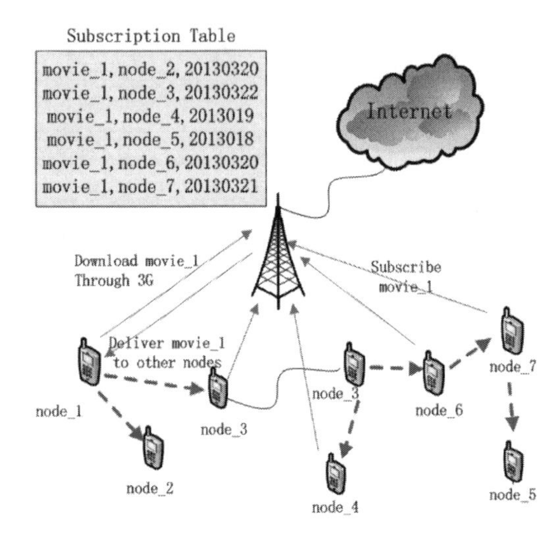

Figure 3. When a relay receives a content that is subscribed by itself, the relay removes itself from the subscription table.

4.2 *Result: traffic offloading ratio*

We assume that all nodes participate in the data dissemination, but only parts of all nodes are interested in each data. Subscriber ratio means the number of subscribers over all nodes. Figure 4 shows the traffic offloading ratios under different subscriber ratios. It indicates that the traffic offloading ratio increases with the increasing of subscriber ratio. In the simulation, when the subscriber ratio is equal to 1, the traffic offloading ratio is 99.4%. Even when the subscriber ratio is only 2%, Subscribe-and-Send can offload 72% of the cellular network traffic.

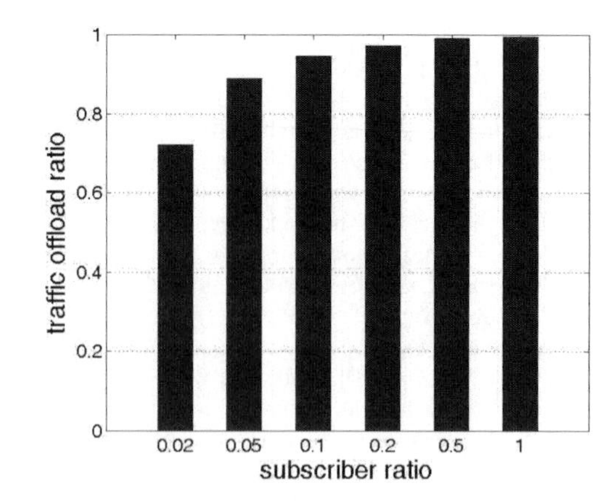

Figure 4. Traffic offloading ratios under different subscriber ratios.

4.3 *Compare with 3G*

Figure 5 shows about 53% of the contact time is between 10 to 20 seconds, and the probability of the contact time being shorter than 10 seconds is 12.8%. So, about 87.2% of contact time is longer than 10 seconds. Here we assume the bandwidth of WiFi is 10MB/s, so a node can finish sending 100MB data to a receiver in a contact time. Even if we assume the maximum size of a bulk file is 100MB, the probability of transferring the file to a receiver within all contacts is higher than 87.2%.

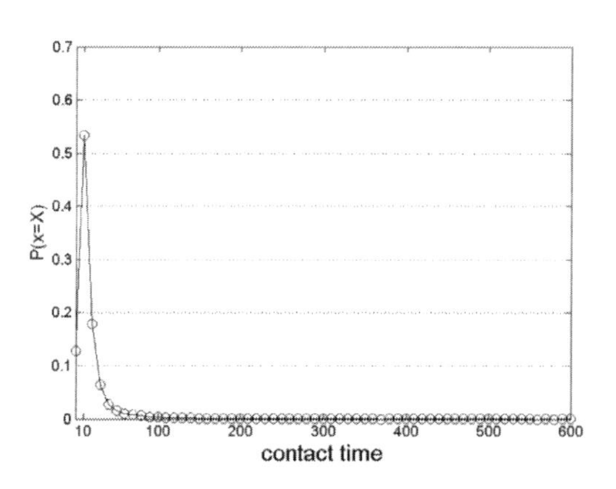

Figure 5. Contact time distribution.

In this paper, two message latency deadlines are tested: a tight 1 min delay and a more relaxed 10 min delay. Figure 6 shows the cumulative probability of the latency of MaxProp. It shows about 95% of delays are shorter than 600 seconds (10 min) and about 26.8%

of delays are shorter than 1 minute. Applications that can tolerate some delays in the delivery process (e.g., file transfers), such as 10 minutes, can take advantage of the WiFi ad hoc communication. Indeed, some data do not have to be downloaded at the instant they are used, and they can be smoothly pre-fetched into mobile devices. The delay is between the time when a source node sends a message and the time when the destination receives the message.

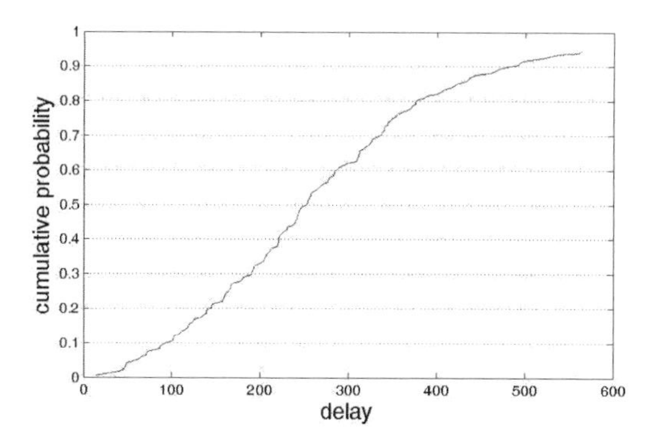

Figure 6. Cumulative probability of the latency of MaxProp.

We assume nodes in Subscribe-and-Send check the subscription table on the CSP every 10 minutes. As about 95% of delivery delays are shorter than 10 minutes, the probability that a subscriber can receive a 100MB file in 20 minute is. If the subscriber downloads the 100M file from the infrastructure, it will cost the subscriber 83.3 minutes to finish the downloading at a practical 2G bandwidth 20KB/s and cost the subscriber 8.3 minutes at an average practical 3G bandwidth 200KB/s and 30 RMB. As can be seen, when people download a bulk file through 3G, they have to wait for minutes and pay much money for it. For most users, 20 minutes or even 1 hour delivery delay is tolerable if the file can be received freely.

5 DISCUSSION

To consider the range of users, different kinds of users could profit from Subscribe-and-Send. In some cities, users pay for downloading data through the cellular network. Firstly, the more files a user downloads or uploads through the cellular network, the more money the user pay to the operator. In this scenario, operators profit from users' data traffic, so users would like to use Subscribe-and-Send to get data and to save the money. Secondly, some Internet CSPs, such as Youtube Company, have to pay a lot for the Internet bandwidth they rent from the cellular network operators every year. They are eager to apply the offloading technology to release the bandwidth load and save the bandwidth fee.

6 CONCLUSION

Currently, cellular networks do not have enough capacity to accommodate the exponential growth of mobile data requirement. A solution is to apply delay tolerant networking technology to transfer bulk data through free wireless communication, e.g. WiFi. We propose Subscribe-and-Send architecture and an opportunistic routing protocol for it. In Subscribe-and-Send, a user subscribes contents through the cellular network infrastructure but does not download the subscribed content. Some users who have downloaded the contents deliver the contents to other subscribers through WiFi with opportunistic protocol. The simulations demonstrate that Subscribe-and-Send architecture can offload 72%~99% of cellular network traffic. The Subscribe-and-Send architecture can be used by SMT users and CSPs to save the cellular networks traffic.

ACKNOWLEDGMENTS

This work was supported by National Natural Science Foundation of China (Grant No.61100208, 61472046, 61372109).

REFERENCES

Alliance Wi-Fi. 2010. Wi-Fi CERTIFIED Wi-Fi Direct. White Paper: http://www. wi-fi. org/news_articles. php

Balasubramanian Aruna, Ratul Mahajan & Arun Venkataramani. 2010. Augmenting Mobile 3G Using WiFi. *In Proc. of MobiSys'10, San Francisco, 15–18 June 2010*. ACM.

Burgess John, Brian Gallagher, David Jensen, and Brian Neil Levine. 2006. MaxProp: Routing for vehicle-based disruption-tolerant networks. *In Proc. of IEEE INFOCOM, Spain, 23–29 April 2006*. IEEE.

Cisco. 2013. Cisco Visual Networking Index: Global Mobile Data Traffic Forecast Update, 2013-2018. http://www.cisco.com/c/en/us/solutions/collateral/service-provider/visual-networking-index-vni/white_paper_c11-520862.html.

Dimatteo Savio, Pan Hui, Bo Han & Victor O.K. Li. 2011. Cellular Traffic Offloading through WiFi Networks. *In Proc. of Mobile Adhoc and Sensor Systems (MASS), Spain, 17–21 October2011*. IEEE.

Fall, K. 2003. Delay tolerant networking architecture for challenged internets. *In Proc. of SIGCOMM '03: 2003 conference on Applications, technologies, architectures, and protocols for computer communications, Germany, 25–29 August 2003*. ACM.

Haddadi Hamed, Pan Hui & Ian Brown. 2010. MobiAd: Private and Scalable Mobile Advertising. *In Proc. of MobiArch'10, USA Chicago. 20–24 September 2010*. ACM.

Han Bo, Pan Hui, V. S. Anil Kumar, Madhav V. Marathe, Guanhong Pei & Aravind Srinivasan. 2010. Cellular Traffic Offloading through Opportunistic Communications: A Case Study. *In Proc. of CHANTS'10, USA. 20–24 September, 2010*. ACM.

Laoutaris, N. Smaragdakis, G. Sundaram, R. & Rodriguez, P. 2009. Delay-Tolerant Bulk Data Transfer on the Internet. *In Proc. of ACM SIGMETRICS 2009, USA. 15–19 June 2009*. ACM.

Lee Kyunghan, Injong Rhee, Joohyun Lee, Yung Yi & Song Chong. 2010. Mobile Data Offloading: How Much Can WiFi Deliver? *In Proc. of SIGCOMM'10. India, 30 August 2010*. ACM.

Lindgren A., Doria, A. & Schelen, O. 2003. Probabilistic routing in intermittently connected networks. *SIGMOBILE Mobile Computing and Communications Review, Vol 7, October, 2003*. ACM.

Lu Xiaofeng & Hui, P. 2010. An energy-efficient n-epidemic routing protocol for delay tolerant networks. *In Proc. of Networking, Architecture and Storage (NAS), Macau, 15–17 July 2010*. IEEE.

Spyropoulos T., Psounis, K. & Raghavendra, C. 2005. Spray and wait: An efficient routing scheme for intermittently connected mobile networks, *In Proc. of SIGCOMM 2005 workshop on Delay Tolerant Networking, USA. 22–26 August 2005*. ACM.

Whitbeck John, Yoann Lopez,Jérémie Leguay,Vania Conan & Marcelo Dias de Amorim. 2012. Push-and-track: Saving infrastructure bandwidth through opportunistic forwarding. *Pervasive and Mobile Computing* 8: 2012.

Wu, H., Qiao, C., De, S., et al. 2001. Integrated cellular and ad hoc relaying systems: iCAR. *IEEE Journal on Selected Areas in Communications* 19(10): 2105–2115.

Electronics, Communications and Networks IV – Hussain & Ivanovic (eds)
© 2015 Taylor & Francis Group, London, ISBN: 978-1-138-02830-2

Testability evaluation method based on multi-source information fusion technique

Jin Luo*, ZiHua Kong & Chen Meng
Department of Missile Engineering, Ordnance Engineering College, Shijiazhuang, China

ABSTRACT: To promote the confidence of the testability evaluation conclusion, this paper studied on the multi-source testability information fusion in the framework of fuzzy theory. Information fusion technique takes full advantage of the multi-source testability information all through the product life cycle and gets a more precise evaluation conclusion. The description method of fuzzy testability information is promoted, the multi-source information fusion strategy based on fuzzy information possibility distribution, and testability evaluation based on the prior possibility distribution fusion result are discussed.

1 INTRODUCTION

Testability verification test and evaluation are the foundation for the equipment purchase management and scientific decision making. How to make a high confidence evaluation conclusion is a pending problem in the testability theory and engineering. Information fusion technique provides an effective way to handle the scarcity of testability verification test data. Testability evaluation based on multi-source information fusion technique can take full advantage of the multi-source testability-related information which be acquired in different phase of the product life cycle and get a more precise evaluation of the equipment testability (Fang 2006).

Considering lots of prior testability information is fuzzy, this paper describes the multi-source fuzzy testability information fusion based on fuzzy set theory. Focuses on how to comprehensively utilize the multi-source information fusion result and improve the confidence level of the testability evaluation result, this paper gives out a new testability evaluation method based on Bayesian theory and illustrates its application in the end. The testability evaluation model based Bayesian inference conduces to a higher confidence of the testability evaluation conclusion with the prior distribution and complete fusion of growth test data in the development stage and field test data.

2 DESCRIPTION OF THE FUZZY TESTABILITY INFORMATION BASED ON FUZZY SET THEORY

Prior information that utilized in multi-source testability information fusion includes testability test data of equipment replaceable units, expert experience, and the growth test data in development stage (Liu et al. 2007). Considering there are lots of imprecise prior information is described in fuzzy language, such as expert experience, the inaccurate statistic result, etc. we can utilize the fuzzy set to describe the fuzzy information in the testability process (Feng 2006).

Let V be an object space, $\forall x \in V$, $A \subseteq V$. A characteristic function $\mu_A(x)$ is defined to study whether x belongs to A or not. Thus u, together with $\mu_A(x)$, constitutes a coupled pair $\{x, \mu_A(x)\}$. Fuzzy subset A in V may be defined as $A=\{x, \mu_A(x) | x \in V\}$, where $\mu_A(x)$ is called as a fuzzy membership function of u to A, and $\mu_A(x) \in [0,1]$. We can choose an appropriate fuzzy set model to describe the imprecise testability information. Trapezoid fuzzy set and triangle fuzzy set model are two kinds of fuzzy subset be utilized frequently.

Regard the unknown testability index R as a variable, We can use fuzzy number A to denote the expert evaluation result for testability index R whose membership function is $\mu_A(x)$, and construct a possibility distribution function $\pi_R(x)$ to describe the expert judgment quantificationally, $\pi_R(x) = \mu_A(x)$, $\forall x \in V$. If experts consider the testability index R is in the zone $[S_l, S_x]$, then the possibility distribution can be denoted as follows.

$$\begin{cases} \pi_R(x) = 1 & x \in [S_l, S_x] \\ \pi_R(x) = 0 & else \end{cases}$$

If experts give out the least possibility value x_l and the most possible value x_s for index R, and the most possible value of R is x_0, then the possibility

*Corresponding author: *zoe_luojin@163.com*

distribution of R based on triangle fuzzy set model can be denoted as follows.

$$\pi_R(x) = trian(x; x_l, x_0, x_s) = \begin{cases} \dfrac{x - x_l}{x_0 - x_l} & x_l \le x \le x_0 \\ \dfrac{x_s - x}{x_s - x_0} & x_0 \le x \le x_s \\ 0 & else \end{cases}$$

3 MULTI-SOURCE TESTABILITY INFORMATION FUSION STRATEGY

The above research indicates that the multi-source possibility distribution function is a fuzzy set, therefore, different possibility distribution can be operated according to the fuzzy set algorithm and reduce to a new possibility distribution. Two kinds of multi-source information fusion strategy are discussed in this paper. One is the serial fusion strategy that fuses information one by one, and the other is the parallel fusion strategy that fuses all information one time.

3.1 *Serial fusion strategy*

Serial fusion strategy is fit to the information fusion in equipment research and manufacture phase that can access the new information continuously. Supposed we accessed the decision information from two experts whose possibility distribution is denoted by π_1 and π_2. We can fuse the information from two experts by fusing their possibility distributions into a new one that is expressed by π_c. Then accesses another expert's information π_3 and fuses it with π_c. Because π_c is the fused possibility distribution, we consider the confidence of π_c is higher than π_3, then we can promote a priority fusion strategy according to the coherence between π_c and π_3 as follows.

$$\forall x \in V, \pi(x) = \min(\pi_c(x), \max(\pi_3(x), 1 - h(\pi_c, \pi_3)))$$

Where $h(\pi_c, \pi_3)$ is the coherence between π_c and π_3, and

$$h(\pi_c, \pi_3) = \sup_{u \in V}(\pi_c(x), \pi_3(x))$$

Expression $h(\pi_c, \pi_3)=0$ means that π_c and π_3 are contrary, and only π_c will be kept. Expression $h(\pi_c, \pi_3)=1$ means that π_c and π_3 are accordant. Compute the new possibility distribution by fusion operation, $\pi = \min(\pi_c, \pi_3)$. We can take the information fusion strategy, according to the above steps.

3.2 *Parallel fusion strategy*

A common parallel fusion operation for multi-source information is weight fusion as follows.

$$\pi(x) = \sum_{i=1}^{n} w_i \pi_i(x) \tag{1}$$

Where $w_i \in [0,1]$, $\sum_{i=1}^{n} w_i = 1$.

How to decide the weight of the information source is the key to weight fusion method. The fusion weight is difficult to decide because it is under the influence of lots of factors in the testability evaluation process, such as the weightiness, the reliability of the information source, etc. this paper uses a self-adaptive weight fusion method to solve the weight problem.

Testability evaluation always requires the intersection of the information possibility distribution. If two possibility distributions aren't intersect, we can consider they are conflict and reduce to two conflict values for testability index R. that is to say, these two possibility distributions are unfit for fusion operation. On the contrary, if two possibility distributions are intersect, the abetment between two information sources will be improved according to their intersection. The abetment $S(\pi_i, \pi_j)$ or $S(\pi_i, \pi_j)$ between two possibility distributions $\pi_i(x)$ and $\pi_i(x)$ are defined as follows.

$$S(\pi_i, \pi_j) = S(\pi_j, \pi_i) = \frac{\int_x (\min\{\pi_i(x), \pi_j(x)\})dx}{\int_x (\max\{\pi_i(x), \pi_j(x)\})dx}$$

Regard to the fusion condition that have more than two information sources, the first operation step should be done is computing the abetment by two information sources and constructing the abetment matrix AM. The abetment matrix $AM=S_{n\times n}$ where $S_{ij} = S(\pi_i, \pi_i)$, and $S_{ij}=1$ when $i=j$.
Defined

$$A(E_i) = \frac{1}{n-1} \sum_{j=1, j\neq i}^{n} S_{ij}$$

Then we get the fusion weight as follows.

$$w_i = \frac{A(E_i)}{\sum_{j=1}^{n} A(E_j)}$$

4 TESTABILITY EVALUATION WITH FUZZY PRIOR INFORMATION

Based on the above research, we get a final possible distribution result from the multi-source information fusion. How to comprehensively utilize the multi-source information fusion result to improve the confidence of the testability evaluation conclusion is another problem in the testability evaluation process. Bayesian method provides a good ideal for testability evaluation with fuzzy prior information(Zhang 2007). Regarded the final possibility distribution reduced from multi-source information fusion as the prior possibility distribution that is denoted by $\pi(x)$, we can update this prior possibility distribution to the posterior possibility distribution by test data, and evaluate the equipment testability based the posterior possibility distribution. Because the posterior possibility distribution combined fuzzy prior information with test data information, the confidence of the evaluation conclusion is higher than the conclusion which is educed from test data only (Zhou 2008).

Let $v=(v_1, v_2,\ldots, v_n)$ be the sample of Y, and $f(v, u)$ expresses the probability density function, where u is the distribution parameter. Then we can get the likelihood function as follows.

$$L(v \mid u) = \prod_{i=1}^{n} f(v_i, u)$$

Likelihood function reflects the appearance probability of the fact value of the sample when the distribution parameter is u. A condition possibility distribution can be educed from the likelihood function as follows.

$$\pi(v \mid u) = \frac{L(v \mid u)}{\max(L(v \mid u))}$$

Computes the posterior possibility distribution according to the reasoning rule of Bayesian method which is described by

$$T(\pi(v), \pi(u|v)) = \pi(u|v)$$

Let choose an appropriate norm which is described as $T(a,b)=ab$, then we can get the posterior possibility distribution as follows.

$$\pi(u|v) = \pi(u|v)/\pi(v)$$

Based on the posterior possibility distribution, we can compute the evaluation value, and access the confidence region of the testability index such as FDR.

5 EXAMPLES

There are three experts gave out the testability evaluation conclusions of index FDR. First expert considered FDR is between 0.7 and 0.95, and the most possible value is 0.86. The second expert considered FDR is between 0.72 and 0.96, and the most possible value is 0.88. The last expert considered FDR is between 0.8 and 0.98, and the most possible value is 0.9. We can use the triangle fuzzy set to describe the fuzzy testability evaluation conclusions of three experts as follows.

$$A_1 = (0.70, 0.86, 0.95)$$

$$A_2 = (0.72, 0.88, 0.96)$$

$$A_3 = (0.80, 0.90, 0.98)$$

Then we can reduce three possibility distributions from the above testability information as $\pi_1(x)=A_1$, $\pi_2(x)=A_2$ and $\pi_3(x)=A_3$.

Let apply the parallel fusion strategy as follows. First, computing the abetment by twos and constructing the abetment matrix AM.

$$AM = \begin{bmatrix} 1 & 0.76 & 0.38 \\ 0.76 & 1 & 0.51 \\ 0.38 & 0.51 & 1 \end{bmatrix}$$

Then we have

$A(E_1)=0.76+0.38=1.14$

$A(E_2)=0.76+0.51=1.27$

$A(E_3)=0.38+0.51=1.14$

So, we get the fusion weights of different information sources as follows.

$w_1=0.345$, $w_2=0.385$, $w_3=0.270$

According to the express (1), the possibility distribution fusion result is described as $\pi(x)=(0.73, 0.88, 0.96)$. Let regard the most possible value of $\pi(x)$ as the most possible value of FDR whose upper and lower limits are 0.8 and 1 respectively, describe the prior information as a triangle fuzzy set model, and suppose there are 51 faults has been detected and one is the false alarm, then

$$L(1 \mid r) = C_{51}^{1} r^{50} (1-r)^{1}$$

We can compute the max likelihood estimation r is 0.98 and the max value of the likelihood function is 0.37, then we can get

$$\pi(1 \mid r) = \frac{L(1 \mid r)}{\max(L(1 \mid r))} = \frac{L(1 \mid r)}{0.37}$$

Compute the posterior possibility distribution which is described as

$$\pi(r\mid 1)=\begin{cases}\dfrac{r^{50}(1.85r-r^2-0.85)}{0.00042} & 0.8\le r\le 0.88\\ \dfrac{r^{50}(r^2-2r+1)}{0.00021} & 0.88\le r\le 1\\ 0 & else\end{cases}$$

Educed by the prior expert information and test data, we access the most possible value of FDR that is 0.875.

6 CONCLUSIONS

When the field usage data for equipment is limited, the evaluation sample is small, and it is difficult to enhance the confidence of the testability evaluation conclusion. A multi-source information fusion strategy is brought forward in this paper to take full advantage of the multi-source testability information all through the product life cycle. With the prior distribution and complete fusion of growth test data in the development stage and field test stage, the testability evaluation model based Bayesian inference is studied. The example illustrates that a testability evaluation method based on Bayesian inference can improve the confidence level of the testability evaluation result.

REFERENCES

Fang, G H. 2006. *Research on the multi-source information fusion techniques in the process of reliability assessment.* Hefei: Hefei University of Technology.

Feng, J. 2006. Fusion of Information of mulitple sources based on ML-Π theory in bayesian analysis. *Mathematics in Practice and Theory* 36(6): 142–145.

Liu, Q, et al. 2004. The fusion method for prior distribution based on expert's information. *Chinese Space Science and Technology* (3): 68–71.

Zhang, J H. 2007. *Bayes method in the test analysis.* Changsha: National University of Defense Technology Press.

Zhou, T. 2008. A new Bayesian mehtod for estimation of reliability with fuzzy prior information. *Journal of Projectiles, Rockets, Missiles and Guidance* 28(1): 216–218.

Electronics, Communications and Networks IV – Hussain & Ivanovic (eds)
 ISBN: 978-1-138-02830-2

Window adaptive cost aggregation method for stereo correspondence

Jing Luo, Zuren Feng & Na Lu

State Key Laboratory for Manufacturing Systems Engineering, Systems Engineering Institute, Xi'an Jiaotong University, Xi'an, Shaanxi, China

ABSTRACT: Stereo matching is an active research area in computer vision field, and a lot of excellent researchers have focused on this challenging topic. For stereo matching, Census is proven to be a simple matching algorithm with high accuracy. But how to choose the aggregation window remains a critical problem. This paper presents a new approach to choose the size of aggregation window based on texture and the disparity boundary near the pixel to be matched, which does not increase much extra computational complexity. Experiment results show that the proposed method is effective to improve the matching result.

1 INTRODUCTION

In the field of computer vision, stereo matching has attracted the attention of many researchers (Klaus et al. 2006, Mei et al. 2011, Veksler 2003). The goal of stereo matching is to determine the disparity between two images that shoot the same scene but from different viewpoints.

At present, there are still a lot of stereo matching problems need to be solved. In general, stereo matching algorithms can be classified into global matching algorithms and local matching methods (Klaus et al. 2006). Global algorithms usually take the stereo matching as an optimization problem which can be solved by minimizing certain energy functions. Global methods usually consume more computation, but could produce maps with accurate disparity. Local algorithms, on the other hand, employ the intensity or gradient within a local region to compute the disparity of each pixel. Compared with the global algorithms, the local ones have the simpler mechanism and are more time efficient. However, it is difficult to identify the correct matching points, especially in the flat areas (Lee 2013). The research in this paper mainly focuses on the local stereo matching method.

In addition, the local methods could be further categorized into two classes, i.e., sparse stereo methods and dense stereo methods. The sparse stereo method is usually feature-based, which could be employed in tasks like image stitching through matching robust key points or features (Wang et al. 2008, Zheng et al. 2008, Zheng et al. 2008). On the other hand, our study mainly focuses on dense disparity computation, which aims at obtaining the depth information of the whole image.

According to the taxonomy by Scharstein & Szeliski (2002), local dense two-frame stereo correspondence algorithms usually perform a subset of the following four steps:

1 matching cost computation,
2 cost aggregation,
3 disparity computation,
4 disparity refinement.

Every step of the algorithm has effects on the matching performance, among which, the matching cost computation and the cost aggregation are the two most important procedures.

Matching cost computation methods are widely applied, including normalized cross correlation (NCC), sum of absolute difference (SAD), sum of squared difference (SSD) and Census and so on, among which, the Census is turned out to be a simple matching cost method with high accuracy for stereo matching (May et al. 2011). Census computes the matching cost by evaluating the relative ordering distinction of the neighbor pixels, rather than the intensity difference. Therefore, the Census is relatively more robust to radiometric distortion (Lee 2013), which is the reason for choosing Census as the matching cost computation method in our algorithm.

Cost aggregation is another critical procedure. Zhang et al. (2009) adaptively constructed an upright cross local support window employed as the aggregation window with four varying arm lengths which are determined by the color similarity and connectivity constraints. Yoon & Kweon (2006) presented a new window-based method for correspondence search using varying support-weights. They adjusted the support-weights of the pixels in a given support window based on intensity similarity and geometric proximity to reduce the image ambiguity. However, one common deficiency of these methods is that they are all time-consuming.

Our work in this paper mainly focuses on window adaptive aggregation method. The information from the edge feature is employed to guide the selection of the aggregation window size. Specifically, the regions with rich texture or with segment boundary prefer smaller aggregation windows and vice versa. The experiments on Middlebury benchmark have verified that our approach can effectively improve the stereo matching performance.

2 ALGORITHM

2.1 *Aggregation window problem*

Through quantities of experiments, we have observed that some images prefer smaller aggregation windows (e.g. Figure 1), while some images prefer larger aggregation windows (e.g. Figure 2), and others prefer middle size aggregation windows. The experimental result is given in Table 1, which summarizes the disparity accuracy employing Census cost and fixed size aggregation windows that could well support this observation. So how to choose suitable aggregation windows is an important issue which needs to be studied.

Figure 1. Image rock. An example image which prefers a small aggregation window.

Specifically, the regions which lack texture require relative big aggregation windows; while for the other regions with rich texture, increasing the size of aggregation windows cannot help to improve the matching result but would add computation burden. In addition, the pixels near the boundary of different disparity prefer small windows, because large window size could easily induce edge blurring effects in the disparity results. As can be seen in Figure 1,

Figure 2. Image bowling. An example image which prefers a large aggregation window.

the Rock image is rich of texture everywhere and has abundant disparity boundaries, which explains why it could obtain better matching performance employing small aggregation window. Figure 2, on the contrary, is lacking of texture in most of the regions and has fewer disparity boundaries, which thus prefer larger aggregation window.

As a conclusion, it is necessary to develop a new approach to select the size of aggregation window adaptively. In this paper, the information from the edge feature is employed as the evidence to select the size of the aggregation windows.

Figure 3 gives some image patches which could obtain better disparity accuracy with small aggregation window applied; the image patches in Figure 4 are turned out to be better matched with large aggregation window. It could be clearly seen that the images are turned out to be better matched with large aggregation window. It could be clearly seen that the images in Figure 3 include high textures and many of them have a depth discontinuous area inside. On the other hand, in Figure 4, the images are obviously smoother than those in Figure 3.

Figure 3. Image patches preferring small aggregation windows.

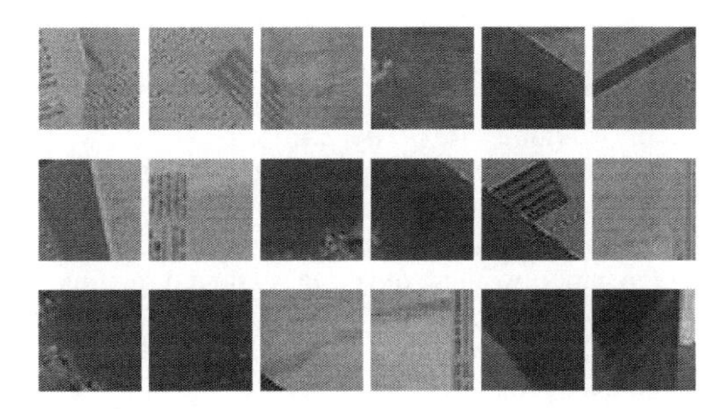

Figure 4. Image patches preferring large aggregation windows.

2.2 *Matching cost computation*

As mentioned above, the Census is employed to compute the matching cost. In details, we process each pixel in the image using Census transform within a 5×5 sub-window around the pixel (Zabih & Woodfill 1994), which results in a binary string such as 10111 00001 00110 10111 0100. In this string, '1' means the intensity of the center pixel is higher than that of the corresponding neighbor pixel; while '0' indicates the intensity of the center pixel is lower than the neighbor pixel's. The Census transforms of two pixels respectively from the stereo image pairs are computed, and the similarities of which are then measured using Hamming distance, i.e. the number of different bits.

2.3 *Edge detection using Sobel operator*

One common property of the rich texture region and the areas close to the boundaries is that abundant edges could be detected. Therefore, it is natural to use the edge feature to select the size of aggregation windows.

In our study, Sobel operator is employed to detect the edge. Technically, Sobel operator is a discrete differentiation operator which can compute an approximation of the gradient of the image. The operator includes two 3×3 kernels, which execute convolution with the original images to get approximations of the derivatives. The computations are as follows:

$$G_x = \begin{bmatrix} -1 & 0 & +1 \\ -2 & 0 & +2 \\ -1 & 0 & +1 \end{bmatrix} * I, \tag{1}$$

$$G_y = \begin{bmatrix} -1 & -2 & -1 \\ 0 & 0 & 0 \\ +1 & +2 & +1 \end{bmatrix} * I, \tag{2}$$

where * denotes the 2D convolution operation. I is the source image; G_x and G_y are two images that contain the vertical and horizontal derivative approximations at each point. Then we combine G_x and G_y to get G as the indicator to select the aggregation windows by

$$G = \sqrt{(\omega \times G_x)^2 + G_y^2} \tag{3}$$

where ω is a weight to stress the influence of the horizontal gradient. In our experiments, ω is set as 1.5. The reason why G_x is emphasized by a weight larger than 1 is that, compared with the vertical derivative, the horizontal derivative is more helpful to decide the disparity due to the epipolar geometry in stereo vision.

2.4 *Aggregation window selection*

The window adaptive aggregation method based stereo matching mainly includes four steps. At first, the matching cost of each pixel at different disparities in the left image is computed. Secondly, the size of the aggregation windows is selected from 9×9, 13×13 or 17×17. Thirdly, the sum of the cost of the selected aggregation windows is obtained. And finally, the disparity of each pixel is simply computed by the WTA (Winner-Takes-All) method.

The regions with rich texture would have a lot of pixels with high value of G, and these regions prefer smaller aggregation windows; the regions where the disparity boundary exists would also have high value of G, and prefer smaller aggregation windows. On the other hand, the regions that are lacking of texture and without disparity boundary will have small value of G. So we can decide which aggregation window to choose based on the value of G.

Specifically, two thresholds (p_1, p_2) are predetermined empirically with p_1 set as 1000, and p_2 as 1500 in our experiments. The thresholds are applied to the sum of G in a 9×9 sub-window, which is computed as

$$S_1(i,j) = \sum_{x,y \in [-4,4]} G(i+x, j+y), x, y \in Z, \tag{4}$$

where i and j are the coordinate of the pixel in question. If S_1 is larger than the threshold p_1, which means that the region around this pixel has enough texture or has the disparity boundary, we select the size of the aggregation window as 9×9.

Furthermore, if S_1 is smaller than the threshold p_1, then we compute the sum of G in a 13×13 sub-window as

$$S_2(i,j) = \sum_{x,y \in [-6,6]} G(i+x, j+y), x, y \in Z. \tag{5}$$

If S_2 is larger than the threshold p_2, which means that the region around this pixel has enough texture or has the disparity boundary, we chose the size of aggregation window as 13 × 13.

And if S_2 is smaller than p_2, which means the region around this pixel does not have enough texture and will not cross the boundary of disparity, we can use the largest aggregation window of 17 × 17.

2.5 *Determine the disparity*

After we have selected the size of the aggregation windows, the cost in each aggregation window at different disparity is computed by

$$S_{\cos t}(i,j,d) = \sum_{pixel(x,y)\in W} \cos t(x,y,d), x, y \in Z, \quad (6)$$

where W is the aggregation window for pixel at (x, y), and $\cos t\,(x, y, d)$ is the matching cost computed by the Census difference. Then the WTA algorithm is employed to determine the disparity by detecting the minimum of S_{cost} at different disparity values.

3 EXPERIMENTAL RESULT

In our experiments, the criterion of accuracy is employed to evaluate the experimental result. The pixels that cannot be processed by the algorithms (such as the boundary of the image) and the pixels without ground truth have been excluded from the result. Pixels with disparity error greater than one pixel are supposed to be errors. The experimental result of four data sets from the Middlebury benchmark is given in Figure 5.

In addition, to verify that our algorithm can improve the matching performance, the result of our algorithm is compared with three fixed aggregation window calculations. 9 × 9 window means the fixed aggregation window of 9 × 9 is applied to compute the disparity at each pixel, as the same to 13 × 13 and 17 × 17 windows. The results are given in Table 1, where the "Adaptive" is the proposed method.

It could be seen from Table 1 that our algorithm is effective to improve the stereo matching performance. For the images Bowling and Flowerpot, when larger aggregation window is employed, the matching accuracy is higher. While in case of the images Cloth and Rock, smaller aggregation windows may obtain better results. Compared with the algorithms using fixed aggregation windows, the accuracy of the proposed algorithms has been always competing to be the highest. Particularly, in the datasets of Bowling and Flowerpot, the proposed method has achieved the best performance.

Table 1. Compare of the experimental result.

Window size	9 × 9	13 × 13	17 × 17	Adaptive
Bowling	88.62%	90.92%	91.45%	91.47%
Cloth	98.43%	97.57%	96.37%	97.74%
Flowerpot	88.44%	89.16%	89.23%	89.24%
Rock	99.13%	98.82%	98.38%	98.90%

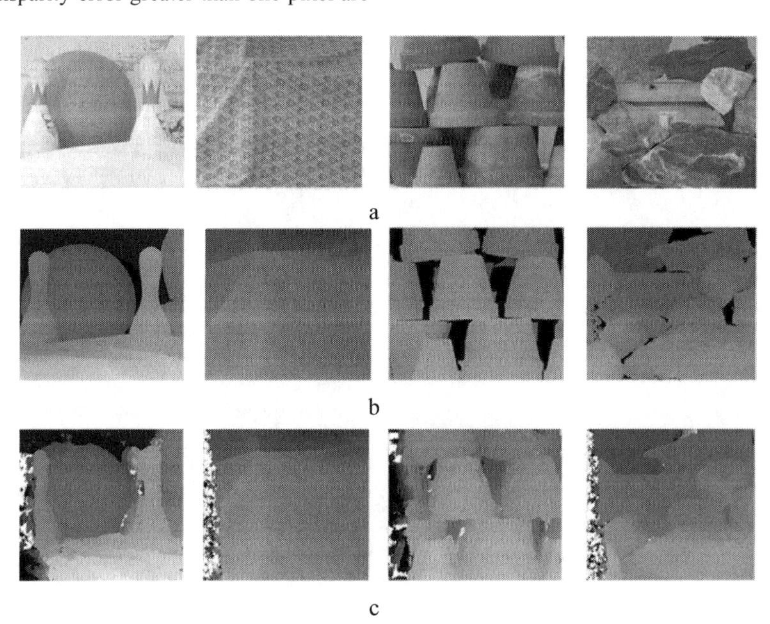

Figure 5. Experimental result using our algorithm. (a) Reference images. (b) Ground truth. (c) Our results.

4 CONCLUSION

This paper develops a new stereo correspondence method which mainly makes use of edge feature information to select the size of the aggregation windows by double thresholds. The rationale behind is that the regions with rich texture or with boundary prefer smaller aggregation windows and vice versa. Experiments on the Middlebury benchmark have verified the effectivity of the proposed algorithm.

ACKNOWLEDGMENT

This work is supported by the National Natural Science Foundation of China grant 61105034, China Postdoctoral Science Foundation grant 20110491662, 2012T50805.

REFERENCES

Klaus, A., Sormann, M. & Karner, K. 2006. Segment-based stereo matching using belief propagation and a self-adapting dissimilarity measure. *The 18th International Conference on Pattern Recognition, August 2006.* IEEE.

Lee, Z., Juang, J. & Nguyen, T. 2013. Local disparity estimation with three-moded cross census and advanced support weight. *IEEE Transactions on Multimedia* 15(3):1855–1864.

Mei, X., Sun, X., Zhou, M., Wang, H. & Zhang, X. 2011. On building an accurate stereo matching system on graphics hardware. *2011 IEEE International Conference on Computer Vision Workshops. November 2001:* 467–474 IEEE.

Scharstein, D. & Szeliski, R. 2002. A taxonomy and evaluation of dense two-frame stereo correspondence algorithms. *International journal of computer vision* 47(1–3): 7–42.

Veksler, O. 2003. Fast variable window for stereo correspondence using integral images. *2003 IEEE Computer Society Conference on Computer Vision and Pattern Recognition, June 2003.* 1: I–556. IEEE.

Wang, B.Q., Guo, L. & Zheng, M. 2008. Sub-pixel image registration algorithm in image panoramic mosaic. *Journal of Computer Engineering And Applications* 44(17):191–200.

Yoon, K. J., & Kweon, I. S. 2006. Adaptive support-weight approach for correspondence search. *IEEE Transactions on Pattern Analysis and Machine Intelligence* 28(4):650–656.

Zabih, R., & Woodfill, J. 1994. Non-parametric local transforms for computing visual correspondence. *ECCV, 1994*: 151–158. Springer Berlin Heidelberg.

Zhang, K., Lu, J. & Lafruit, G. 2009. Cross-based local stereo matching using orthogonal integral images. *IEEE Transactions on Circuits and Systems for Video Technology* 19(7): 1073–1079.

Zheng, M., Chen, X. & Guo, L. 2008. Stitching Video from Webcams. *Advances in Visual Computing* : 420–429. Springer Berlin Heidelberg.

Zheng, M., Guo, A., Zhong, W. & Guo, L. 2008. Image Stitching of Scenes with Large Misregistration. *International Conference on Computer Science and Information Technology, August 2008* : 803–807. IEEE.

Electronics, Communications and Networks IV – Hussain & Ivanovic (eds)
 ISBN: 978-1-138-02830-2

Demonstration of image sensor communication

Lan Lv, Rongzhao Wu, Jiang Liu, Peng Liu & Song Liu
North China Electric Power University, Beijing, China

ABSTRACT: In this paper, a Visible Light Communication (VLC) system utilizing image sensor is devised. The structure of the proposed system is introduced. Capture, tracking and extraction of Light Emitting Diode (LED) signal are achieved by applying the locating algorithm to the visible light communication system. Edge algorithms, region growing algorithm and frame difference method are used to find out the location, and these algorithms are compared. Moreover, on the basis of this system, a simple, character-oriented ASCII communication protocol is achieved, and relevant results are given. The results show that a practicable communication speed can be achieved by using the proposed system.

1 INTRODUCTION

With the rapid development of solid state lighting technology, LED is widely used in our daily life. Compared with the traditional lighting source, white LED has lower energy consumption, longer lifetime, smaller size, better performance, more environmental friendliness, faster switching speed and greater reliability (Eason et al. 1955). Based on these advantages, the physicist Herald Haas and his team at the University of Edinburgh in England invented a new patented technology, which used the flash of the light source to transmit digital information. This proposal is called VLC (Maxwell 1881). The rapid development of VLC promotes another kind of communication technology based on visible light called image sensor communication (ISC), which uses LED as transmitter and image sensor as receiver.

In order to achieve high speed optical signal, high speed cameras or newly developed image sensors have been used in the receiving system. Region of Interest (ROI) function of the camera can effectively improve the frame rate. By using the ROI function, the target region which contains the light source can be selected and only the selected image will be processed. Consequently the computational complexity can be reduced; the frame rate can be greatly improved. If the whole image needs to be processed, other noise sources will probably be included, but if only the effective region is analyzed, communication errors can be reduced. In order to minimize the interference of other sources in the field of view, misjudgment ratio can be reduced by diminishing the area of the processed image. After the target source is located, a small window including LED light signal will be opened.

In this paper, edge algorithms, region growing algorithm and frame difference method are used to find out the location. And the simulation results are compared. After the receiving terminal receives the signals from the camera, the information of LED will be separated, detected, and stored. Thus the communication from the transmitting terminal to the receiving terminal is achieved. In this paper, only static image processing is considered without taking the situation into account when the recipient is moving. The correspondence of the mobility and real-time processing will be addressed in future research.

The rest of the article is organized as follows: In section 2, the theory and system of ISC are introduced; Section 3 describes the image processing algorithms, including edge algorithms, region growing algorithm and frame difference method; Section 4, conclusions and future work are presented.

2 THE STRUCTURE OF ISC SYSTEM

Image sensor communication is a good candidate for short message transmission. Since the frame rate of an image sensor is below a few kilohertz, the ISC data rate for one transmitter is limited by the frame rate. Therefore, it is very important to improve the communication rate. For VLC, some researches tried to improve the communication rate by using parallel LED transmitters or some equalization technology (Tanaka & Nakagawa 2000, Tanaka 2001, Afgani et al. 2006, Sugiyama & Nakagawa 2006, Sugiyama & Masao 2007, Elgala 2007, Linnartz & Schenk 2008, Lin & Hirohashi 2009, Vucic 2010, Vucic 2011, Khalid & Ciaramella 2012, O'Brien 2008b, O'Brien 2008a, Le Minh 2009). In this paper, image processing algorithms are introduced to improve communication speed.

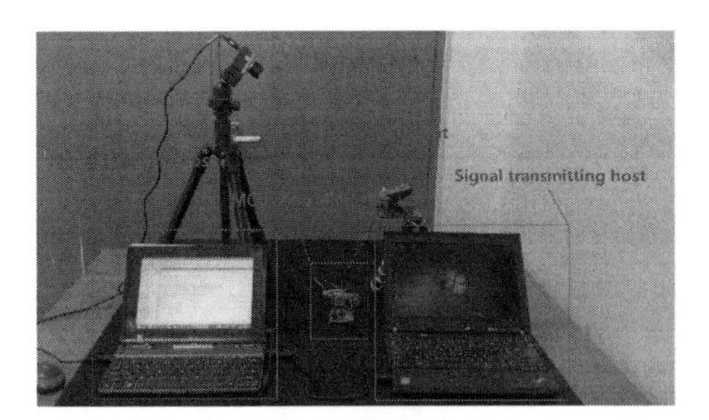

Figure 1. Structure of ISC system.

The structure of proposed ISC model is shown in the Figure 1. Different from the conventional VLC system, the receiving apparatus of the system is the camera instead of the detection device such as photodiode. The system is composed by two parts: the transmitting part and the receiving part. As shown in Fig. 1, the transmitting part consists of a Micro Control Unit (MCU), a LED light and a signal transmitting host. The receiving part consists of an image sensor and a signal receiving host. Signal transmitting and signal receiving host are ordinary personal computers. The image sensor is a camera device. On-Off Keying (OOK) is used as modulation scheme. In the captured images, only useful information is desired and unnecessary information should be cut. There may be other light sources, which will confuse the judgment of target source, then how to distinguish the useful light source from other sources is very important. Here we use edge algorithms, region growing algorithm and frame difference method to distinguish the light source.

As the ISC system is an asynchronous communication system, the head need to be added into the information, and the information needs to be packaged. After the frame encoder, the signal is modulated for the optical signal transmission. OOK and Pulse-Position Modulation (PPM) are two widely-used modulation schemes for optical communication. Since the received time of image sensor is hard to be synchronized in the ISC system, the OOK modulation is chosen in our paper. In ISC system, the LED driver is controlled by a computer and transmits information through the flash of LED light. The specified character code is sent by computer serial debugging assistant window. At the receiver side, the camera is used to acquire the images. At first, the full frame of the image sensor is read out and the readout data is transmitted to a processor. Then the light source is detected by the processor using image processing algorithms. For the ISC system, since the contrast of the desired object and the background is always sharp, edge algorithms, region growing algorithm and frame difference method can be used to distinguish the object region from the background. After detection, the target region can be achieved. After the judgment of the signal position, the tracking and communication period begin. After locating, a small window will be opened which contains the LED. After acquisition of location, demodulation is followed and then asynchronous communication can be realized.

From the whole system, it can be noticed that the processing of the picture is very important for the location in the system. Without this step, the following steps cannot continue. In the next part, image processing algorithms will be introduced.

3 THE IMAGE PROCESSING ALGORITHM

The images must be captured and processed by the ISC system. If the speed of camera is not high enough then it is difficult to be applied to real life. A large amount of information is contained in a picture. If the pictures are processed directly, a lot of useless information will be included, thus the cost will be increased and the processing speed will be very slow. If the pictures are pretreated, the cost can greatly be saved and processing speed can be improved. What we need to do is to find out the location of the light source. By locating the LED, a lot of time and storage space will be saved.

Image matching is used by some researchers to find out the light source, but the simulation to verify the feasibility of this method is not presented. The edge algorithms, region growing algorithm and frame difference method are introduced to locate the light source in this paper. The different location methods are compared, and from the simulation results we can conclude that, the frame difference method is the most appropriate for our system.

3.1 *Edge algorithm*

In order to locate the LED, edge algorithms are used. The simulation results of edge algorithms are shown as follows. The original image is shown as Figure 2(a). Robert's operator adopts partial difference operator to find the edges. The simulation result of Roberts is shown as Figure 2(b). The key of Prewitt operator is the convolution of each pixel in an image with the two convolution kernels by moving the two templates of a small region in the image. And the final output of the edge magnitude can be used for image edge detection. The simulation result of Prewitt is shown as Figure 2(c). There exist two Sobel operators, one is the horizontally detected edges, and the other is the vertically detected edges. Its mask templates are 3x3 extremely similar to Prewitt operator. The simulation result of Sobel is shown as Figure 2(d). The operator of Laplace of Gaussian (LoG) is a second-order differential edge detection method, which detects the edges by finding out the zero-crossing point value from the second derivative of gray level of the image. The operator is sensitive to noise. The simulation result of LoG is shown as Figure 2(e). The essence of Canny is to make a smooth operation using a standard Gaussian function, and then to locate the maxima of the first-order differential with the direction. The simulation result of Canny is shown as Figure 2(f).

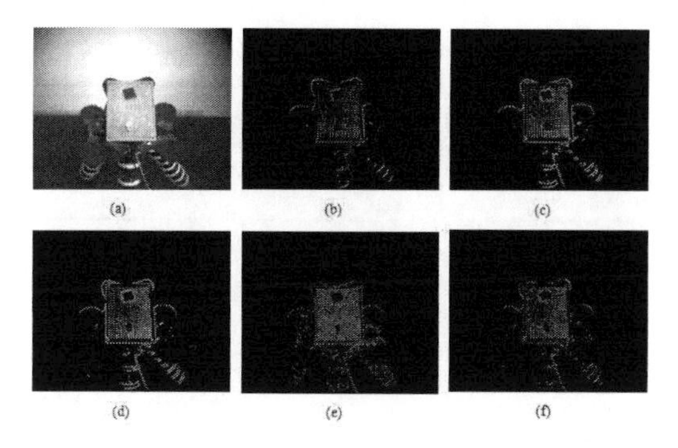

Figure 2. The edge algorithm simulation results.

The edge extraction result of Robert's operator is not accurate; for smooth edges of the image, the extraction of Prewitt operator is complete; Aside from the complete edges, Sobel operator has particularly desirable effects on concentrated parts of the picture. As can be seen from the simulation results, the effects of extraction of Sobel operator and Prewitt operator are almost the same; the result of LoG operator for edge detection is fuzzy; the effect of Canny operator is also not so good. From the simulation results above, it can be clearly seen that, there are so many edges in the images, so it is difficult to find the edge of light source unless we use shape matching. And shape matching is complex and it will take much more time.

3.2 *Region growing algorithm*

Region growing method, mainly takes relations between pixels and their spatial domain pixels into consideration, whose basic idea is to get similar pixels together to constitute regions. Specifically, one or more seed pixels are defined as the starting point of growth for the area of each needed to be split, and then areas grow according to some similarity criterion, gradually forming a region of space with certain uniformity. Pixels or regions around the field of seed pixels with similar properties of the seed pixels are merged into seed pixel areas, until there are no more pixels or other small areas that can be merged into. In this paper, the seed is selected manually. When region growing algorithm is used, one point should be clicked as the starting point, one point of the light source is selected as the seed and then the Enter key is pressed. Take the pixel as the growing center, then check its adjacent pixels, compare the neighborhood pixel with it, if the gradation difference is smaller than a threshold value determined in advance, then merge them. After repeated experiments, we find out that, when the threshold is 0.02, the area of growing is obvious and large; when the threshold is 0.03, the area of growing is close to the original shape. In our paper, finding out the light source is the purpose, so we choose 0.03 as the final threshold. The simulation result is shown in Figure 3(b), and Figure 3(a) is the original image.

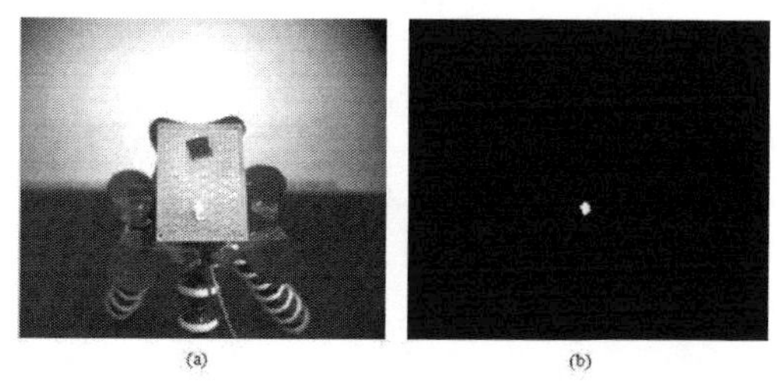

Figure 3. The region growing algorithm simulation result.

From the simulation result, we can see that, the location of LED can be found out accurately. But the time of running the algorithm is relatively long.

3.3 *Frame difference method*

Frame difference method, as the name suggests, is the subtraction between multi-frame images which is almost the same, and changes of the images are achieved. Thus, the frame difference method is usually used to detect moving targets and as the basis for the next steps of image processing. The process of frame difference method is as follows: First, two frames of pictures are achieved by the camera; subtraction is introduced to process these two pictures. Then, binarization is used, and after binarization, denoising is followed; finally, the connected components are achieved. The main advantage of the frame difference method to detect the dynamic between multi-frame images is that background changes a little so the changes are easily separated from the relatively static background. Difference between two consecutive frames is utilized to locate the LED light source by the frame difference method. The process of location which uses frame difference method is shown as Figure 4. First, every variable should be initialized, then two frames of pictures will be captured by the camera, after that, the images will be processed by the frame difference method. If the location of LED cannot be achieved, the programming will return to the previous two steps; if the location is achieved then another image will be gotten with the camera, after tracking and extraction; if the signal can be extracted, then the location of LED will be displayed; if not, the programming will return to the previous three steps.

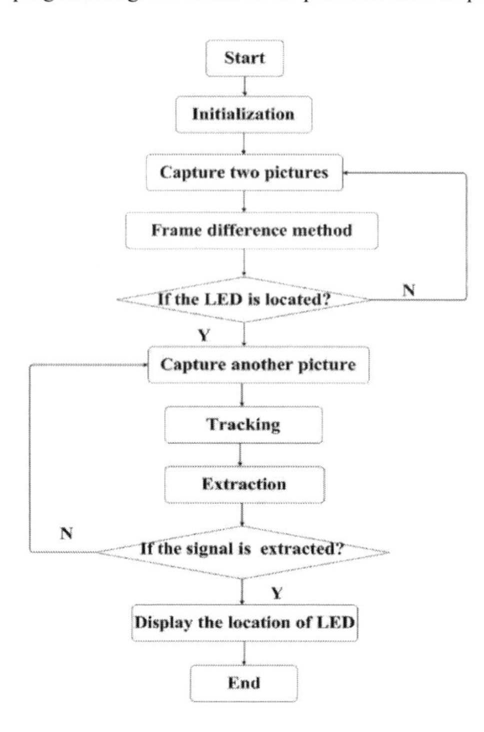

Figure 4. The process of the location of LED.

4 EXPERIMENT RESULTS

The frame difference method simulation result is shown as Figure 5.

As can be seen from the simulation results, the edge algorithms can just find out the edges, if we want to locate the light, other steps should be followed, such as shape matching mentioned in (Liu & Shimamoto 2011), but it is very complex to do so and it will take much more time. The region growing algorithm and frame difference method can find out the location of LED, but region growing algorithm is an iterative method, the overhead of space and time is large. Noise and uneven grey level may result in void and excessive segmentation, and the effect of shadow processing is often not very good. Selecting the seed manually needs more time. And it can be seen from Table 1 that, the runtime of frame difference method is the shortest. We can save time if frame difference method is used.

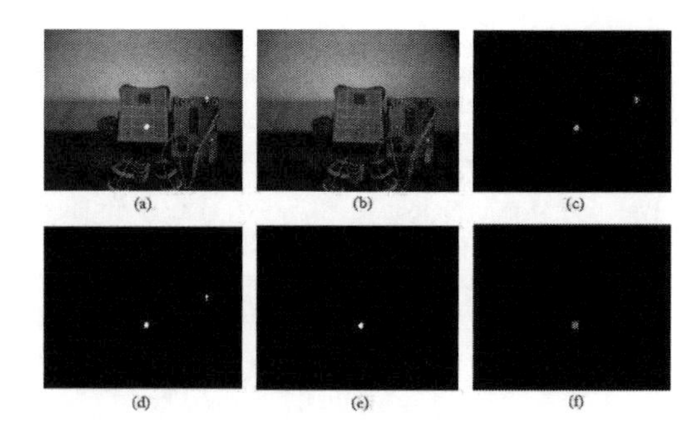

Figure 5. The frame difference method simulation results.

Table 1. Comparison of runtime of different image processing algorithms.

Algorithm	Runtime(s)
Roberts	6.847788
Prewitt	7.476628
Sobel	7.125441
LoG	1.318209
Canny	20.526323
Region growing algorithm	5.468019
Frame difference method	0.535738

However, the result of frame difference method is complete and clear. So, in this paper, frame difference method is chosen as our final image processing algorithm. Figure 5 (a) is the first frame of the image captured by the camera; Figure 5(b) is the second frame; Figure 5(c) is the processing result of frame difference method; Figure 5(d) is the result of binarization; Figure 5(e) is the result of denoising; Figure 5(f) is the final result of algorithm, which is the location of LED. The location of LED is marked by a green rectangle, the red star is the central part of this green rectangular. From the result, it can be seen that even a noise source is added next to the light source, the noise can be filtered by the algorithm, leaving the LED only. After the image denoising processing, the LED is completely separated from the background. The simulation results show that the frame difference method can accurately locate the light source, it is very appropriate for the ISC system.

After LED is located, the next step is the judgment of the state of LED, whether the light is on or off. After the data is received, parity check will be followed. After the image processing algorithm is used in the ISC system, the speed and bit error rate are improved greatly. In this experimental system, the distance between LED and camera lens is 35 cm. In theory, as long as the distance can satisfy the conditions; the area projected on LED camera sensor is larger than the unit pixel, the signal reception hosts are able to capture, track, and extract the LED signals. In order to send and receive a number of ASCII codes, a simple ASCII character-oriented synchronous serial communication protocol is proposed. Within the constraints of communication protocol, ASCII information can be transmitted from the transmitter to the receiver reliably and effectively. So far, the communication of one character and five characters is achieved, and the result is pretty good. The communication result is shown in Figure 6.

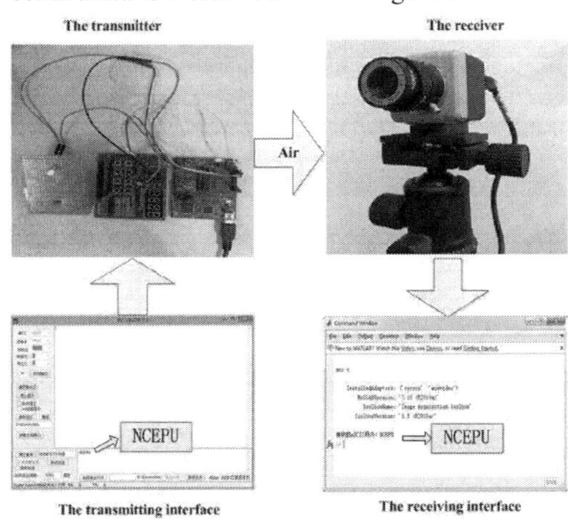

Figure 6. The transmitter, receiver and their operating interfaces.

5 CONCLUSION AND FUTURE WORK

In this paper, image processing algorithms to locate the LED are used. From the simulation results, it can be seen that the algorithms especially the frame difference method can indeed find out the location of the light source, make full preparations for the later work. In this way, the speed can be improved greatly and bit error rate can be reduced effectively. A simple communication system has been set up, and the communication has been achieved successfully though the speed is rather low. In our future work, the improvement of speed will be our focus.

ACKNOWLEDGMENT

The project is sponsored by the Fundamental Research Funds for the Central Universities, the Scientific Research Foundation for the Returned Overseas Chinese Scholars, State Education Ministry and State Key Laboratory of Alternate Electrical Power System with Renewable Energy Sources (Grant No. LAPS14015).

REFERENCES

Afgani, M. Z., Haas, H., Elgala, H. & Knipp, D. 2006. Visible light communication using OFDM. *The 2nd IEEE International Conference on Testbeds and Research Infrastructures for the Development of Networks and Communities (TRIDENTCOM), Barcelona, Spain, 2006*. IEEE.

Eason, G., Noble, B. & Sneddon, I. N. 1955. On certain integrals of Lipschitz-Hankel type involving products of Bessel functions. *Philosophical Transactions of the Royal Society of London. Series A, Mathematical and Physical Sciences* 247(935): 529–551.

Elgala, H., Mesleh, R., Haas, H. & Pricope, B. 2007. OFDM visible light wireless communication based on white LEDs. *Proceedings of the 65th IEEE Vehicular Technology Conference, Dublin, Ireland, 22-25 April 2007*: 2185–2189. IEEE.

Haruyam, H., Haruyam, S. & Nakagawa, M. 2006. Experimental investigation of modulation method for visible-light communications. *IEICE transactions on communication* 89(12), 3393–3400.

Jiang, L., Noonpakdee, W., Takano, H. & Shimamoto, S. 2011. Foundational analysis of spatial optical wireless communication utilizing image sensor. *Proceedings of 2011 IEEE International Conference on Imaging Systems and Techniques (IST), Batu Ferringhi, Penang, Malaysia, 17–18 May 2011*: 205–209. IEEE.

Khalid, A. M., Cossu, C., Corsini, R., Choudhury. P. & Ciaramella. E. 2012. 1 Gbit/s transmission over a phosphorescent white led by using rate adaptive discrete multitone modulation. *Photonics Journal* 4(5): 1465–1473.

Lin, K. I. X. & Hirohashi, K. 2009. High-speed full-duplex multiaccess system for led-based wireless communications using visible light. *Proceedings of the International Symposium on Optical Engineering and Photonic Technology (OEPT) July 2009*: 1–6. IEEE.

Linnartz, J.-P.M.G., Feri, L., Yang Hongming, Colak, S.B. & Schenk, T.C.W. 2008. Communications and sensing of illumination contributions in a power led lighting system. *Proceedings of 2008 IEEE International Conference on Communications, Beijing, China, 19–23 May 2008*: 5396–5400. IEEE.

Maxwell, J. C. 1881. *A treatise on electricity and magnetism*. Oxford: Clarendon Press.

Minh, H. L., O'Brien, O., Faulkner, G., et al. 2009. 100 Mbit/s NRZ visible light communications using a post equalized white led. *IEEE Photon. Tech-nol. Lett.* 21(15): 1063–1065.

O'Brien, D., Faulkner, G., Zeng, L., Minh, H. L., Lee, K., Jung, D. & Oh, Y. 2008. 80 Mbit/s visible light communications using pre-equalized white led. *The 34th European Conference on Optical Communication, Brussels, Belgium, 21–25 September 2008*: 1–2. Piscataway: IEEE.

O'Brien, D., Minh, H. L., et al. 2008. Indoor visible light communications: challenges and prospects. *Proceedings of the International Society for Optics and Photonics (SPIE), San Diego, CA, USA, 10 August 2008*. IEEE.

Sugiyama, H., Haruyam, S. & Nakagawa, M. 2007. Brightness control methods for illumination and visible-light communication systems. *Proceedings of the 3rd International Conference on Wireless and Mobile Communications (WMC), Guadeloupe, French Caribbean, 4–9 March 2007*: 78. IEEE.

Tanaka, Y., Haruyam, S. & Nakagawa, M. 2000. Wireless optical transmissions with white colored led for wireless home links. *Proceedings of the 11th IEEE International Symposium on Personal, Indoor and Mobile Radio Communication (PIMRC), London, UK, 18–21 September 2000*: 1325–1329. IEEE.

Tanaka, Y., Komine, T., Haruyama, S. & Nakagawa, M. 2001. Indoor visible communication utilizing plural white leds as lighting. *Proceedings of the 12th IEEE International Symposium on Personal, Indoor and Mobile Radio Communication (PIMRC), San Diego, California, USA, 30 September-3 October, 2001*: 81–85. IEEE.

Vucic, J., Kottke, C., Habel, K. & Langer, K.-D. 2011. 803 Mbit/s visible light WDM link based on DMT modulation of a single RGB led luminary. *The Optical Fiber Communication Conference on Optical Society of America (OSA), Los Angeles, CA, USA, 6–10 March 2011*: 1–3. IEEE.

Vucic, J., Kottke, C., Nerreter, S., Langer, K.-D. & Walewski, J.W. 2010. 513 Mbit/s visible light communications link based on DMT-modulation of a white led. *Journal of Lightwave Technology* 28(24): 3512–3518.

Electronics, Communications and Networks IV – Hussain & Ivanovic (eds)
 ISBN: 978-1-138-02830-2

Study of intelligent agriculture system based on IOT technology

Xuefen Ma
Jingchu University of Technology, Jingmen, Hubei, China

ABSTRACT: A new agricultural system based on IOT technology has been designed using wireless sensors, ZigBee, and wireless networking. The system consists of four organizational structures, each with their own function. The system serves to collect, analyze, and store date, perform remote monitoring tasks, and find errors. The wireless sensor network system of intelligent irrigation sub-system is discussed in the paper.

KEYWORDS: intelligent agriculture; IOT; zigbee; wireless sensor technology.

1 INTRODUCTION

After the computer and the internet, the IOT, or Internet of Things comes to our life. It has Great prospects for development, and it becomes the focus of the global information industry. IOT is useful to governments, academia, and enterprises due to its broad application prospects in industry, agriculture, health and education (ITU 2005).

2 THE CONCEPT OF AN INTERNET OF THINGS

Massachusetts Institute of Technology (MIT) established the automatic identification (Auto-ID) in 1999 (Cai 2011), proposing that "everything is interconnected through the network". IOT connects all objects as a means of communication, relying on an embedded system of RFID (Yao 2010). Single objects can become some LAN intelligent systems by key technologies. Then the LAN systems of specific function will connect to the Internet with wireless network technology, the system of IOT. The IOT report published by the International Telecommunication Union (ITU-T) in 2005 (Cai 2011) proposed that everything in the world can be connected through technology, including sensors, RFID, ZigBee, Wireless location and robotics et.

On February 17, 2009, the United States signed the "American Recovery and Reinvestment Act 2009". Once it took effect, $30 billion was invested in the field of smart grid and health information technology. In July 2009, the IT Strategy Headquarters of Japan issued an "I-Ja-pan Strategy 2015" (Lu & You 2012), whose goal is "to realize a digitized and smart society." Japan's strategy focuses on the development of IOT services including transportation, environment, education, and telecommuting. On August 7, 2009, Premier Wen Jiabao visited the Wuxi Micro Sensor Network Engineering Technology Research and Development Center (Cai 2011), noting that the importance of IOT, and we should study the key technology as soon as possible. This is best achieved by developing wireless network technology and sensing technology. He also proposed that the system would " develop IOT, improve the productivity of science and technology" (Cai 2011). In 2009, IOT technology was one of the top five emerging strategic industries, allowing us to solve many of China's agricultural problems.

3 INTELLIGENT ARGICULTURAL SYSTEM BASED ON THE TECHNOLOGY OF IOT

The reconstruction of our province, also referred to as China's agricultural valley, is taking modern agricultural science and technology in order to modernize. To achieve this goal, the application of IoT is essential. We can gather information such as rainfall regime, water in all directions, allowing us to provide the technical means for reasonable irrigation in real time and monitor the weather. Vehicle positioning in real time would also be possible. An operation condition in real time in sowing and harvesting stage makes farm equipment operating efficiently. IOT is becoming a lot more widespread as a way advance sensing, communications and data processing (Lin & Long 2013). It is a new trend in the world's agricultural development to apply IoT technology.

3.1 *System function design*

The intelligent agriculture system contains data collection terminals, properties for data analysis and storage, the ability to remotely monitor and control, and the property of error alarm.

3.1.1 *Data collection terminal*

Conditions such as crop production environment temperature, humidity, light intensity and CO_2 concentration, leaf humidity, and dew point temperature can be gathered in real time by wireless sensors .We can also gather video signal in the process of production in real time via live cameras. The data collection points can be monitored a ZigBee network. This agricultural condition data acquisition system can provide a scientific basis for comprehensive agricultural automatic monitoring of ecological information, environmental automatic control and intelligent management.

3.1.2 *Data transfer and analysis*

The data are transmitted to the gateway via a wireless network it is acquired. The gateway is responsible for gaining access to the transmission network and sending the date to the control or management center. The service management platform of the system stores and analyzes the data by the relevant software, obtaining a variety of visual results. Knowing and solving problems in time through real-time tracking crop growth, serving as the basis for agricultural operations and management.

3.1.3 *False alarms*

The system allows the user to create a custom range of the data according to the actual situation. If the collected data is out of range, it will be marked in the system as an alarm. These markers can be used for reference to make some operation.

3.1.4 *Remote monitoring and control*

The user can view a variety of real-time data transmitted by the sensor by any device with Internet access at any time. Specifically, he or she can remotely control a variety of productions.

The system usually practices unified certification, centralized management and control, including user management, device management, rights management and authentication management in order to achieve high efficiency of IOT.

3.2 *System module*

In support of IOT technology, users can adjust the parameters of various operations to improve work efficiency. Modules of the intelligent agriculture system are shown in Figure 1.

An agricultural system based on intelligent networking is composed by the four-tier structure, The intelligent agricultural system based on IOT consists of four-tiered architecture, each with their own function. They serve to collect, analyze, and store date, perform remote monitoring tasks, and find errors. The first one is the perception layer, monitoring various planting information through the sensor. The second is the network layer, sending data to the control center and forming the Internet of things. The third is the application layer, the intelligent agricultural service platform for data collection, storage and processing. The fourth one is the application of terminal access layers, which manages the production process in real time.

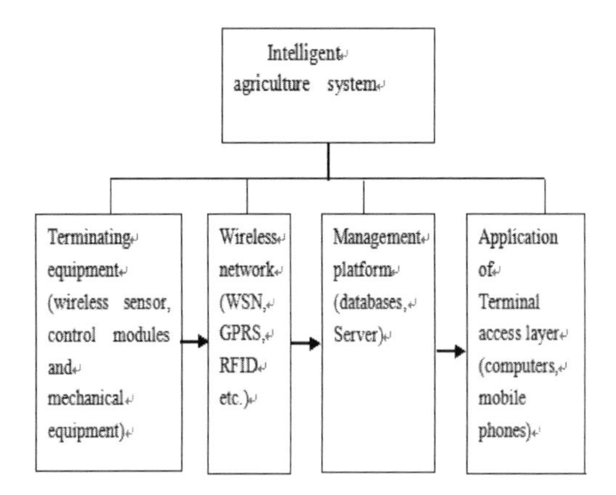

Figure 1. Figure module of the intelligent agricultural system.

4 DESIGN OF THE INTELLIGENT AGRICULTURAL SYSTEMS

The structure of Intelligent agricultural systems is very huge and complicated.

4.1 *Structure diagram of subsystem*

Intelligent agricultural systems can be divided into various subsystems according to different function: lighting control system, integrated heating system, intelligent irrigation (fertilization) systems, and agricultural deployment systems. The structure and principles of each subsystem are similar. The structure diagram of an intelligent irrigation system is shown in Figure 2. It uses sensors to sense temperature, humidity and nutrient information of the soil and it controls the intensity and extent of irrigation as needed, combining with weather information. If fertilization is needed, the exact formula water and fertilizer will be sprayed, building a water-saving irrigation and fertilization system of high efficiency and low consumption. Fertilization systems like this not only improves the yield and quality of crops, but also reduces the environment pollution. It is conducive to the sustainable use of land resources(Gao 2010).

4.2 *Network architecture of subsystem*

The system is a hybrid network structure, mainly consisting of monitoring and controlling devices, ZigBee networks, wireless gateways, and management centers. The Wireless sensor network coverage area is composed of the master node and individual irrigation units. Every irrigation unit is provided with some sensor nodes and a sink node. Sensor nodes are deployed largely in the soil at a representative position, which can detect soil humidity, ambient temperature, and nutrient data in real time. The sink node sends the collected data information to the wireless gateway. The sink node is often equivalent to a special routing equipment. They form a data acquisition system via multiple ZigBee monitoring network. Because the spatial variability characteristics of weather conditions are clear, the system can be equipped with a weather monitoring instrument. The meteorological data collected can be as a supplement to their information.

Each monitoring network has the collects data, and receives the control command transmitted by a higher nodes. There are two control valves; one is an open valve of irrigation equipment and the other is a control valve of the fertilizer coming into the pipe. The monitoring network uses a Y-shaped structure, a parallel distributed processing system, to ensure the independence of each node. The tasks of wireless gateway are to establish and maintain the entire communications network, to record the data collected by monitoring network and weather stations, and communicate with the monitoring center (Cao 2010). The management center handles the received data and stores them in the database. Then it can discover and solve problem timely and accurately , by use of some corresponding software and expert decision-making system. The system should have the capacity to self-organize and adapt, since the network topology may change if a node fails.

The system organizes a network automatically. If the point of failure arrives, the network can automatically find and readdress it using the wireless gateway. Therefore, the maintainability and scalability of the system is strong, satisfying the practical production.

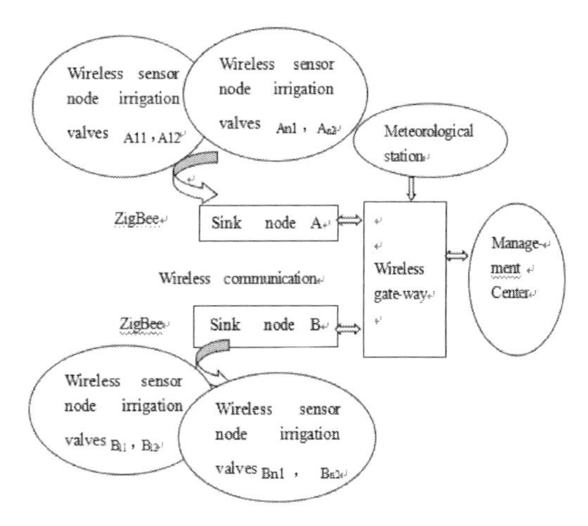

Figure 2. Structure diagram of the intelligent irrigation system.

4.3 *The design of system software*

The wireless sensor network software program for intelligent irrigation system includes three part (Den & Chen 2013). The sensor nodes programs serve to collect data.The sink nodes programs serve to store and transmit the data.The master node programs are used to analyze and communicate.

The structure is shown as Figure 3.

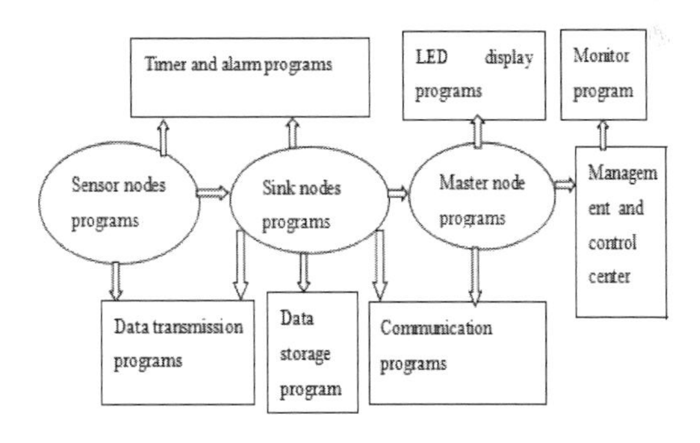

Figure 3. Program structure of the system.

There are data transmission programs, communication programs, external data storage programs and peripheral clock with LED display programs in every part. All data and information are transmitted to the management and control center, so there must be a monitor program for the system. The communication protocol can use TDMA.

5 SUMMARY

Through the use of a variety of data acquisition sensors organized into a terminal sensor network, the user can view a variety of agricultural information and control of agricultural equipment in real time by using mobile terminals. It can reduce administrative staff, environmental resource depletion and waste, while improving production quality and efficiency to a certain extent. This is part of the intelligent agriculture based on IOT networking technology (Den & Chen 2013).

These principles can also be fully applied in transportation, sales, and distribution management of agricultural products. Although the functions and purposes are different in various fields, they are closely related in technical level, allowing for the modernization of agricultural production. If they combine each other, then can realize intelligence of the overall agricultural field, the modernization of agricultural production ultimately.

REFERENCES

Cai, Rimei. 2011. Overview of the internet of things. *Electronic Product Reliability and Environmental Testing* 29(1): 59–63.

Cao, Yuanjun. 2010. The research of wireless sensor network irrigation system based on changes of crop canopy temperature. *The research of Agricultural Mechanization* 32(9): 126–129.

DenYun. & Chen Xiaohui. 2013. Design of the Wireless Sensor Network for Smart Irrigation System. *Process Automation Instrumentation* 34(2): 80–83.

Gao, Jun. 2010. Water saving irrigation control system based on Wireless Sensor Network. *Modern electronic technology* 312(1): 204–206.

ITU. 2005. Internet Reports. *The Internet of Things*. Tunis. www.itu.int/internet of things/.

Lin, Yuanguai & Long, Shunyu. 2013. Intelligent agricultural application system based on IOT technology. *Intelligent processing and application* 2013(3): 71–73.

Lu, Tao & You, Anjun. 2012. Europe-U.S.-Japan-South-Korea IOT development strategy and Its Enlightenment to China. *Science &Technology Progress and Policy* 29(4): 47–51.

Yao, Wanhua. 2010. The concept and basic connotation of Internet of things. *Information China* 141(5): 22–23.

Electronics, Communications and Networks IV – Hussain & Ivanovic (eds)
 ISBN: 978-1-138-02830-2

Research and implement bidirectional OFDM-PON based on recycling residual Raman pump

Mingzhi Mao, Caixia Kuang*, Qianwu Zhang, Rujian Lin, Yingxiong Song, Min Wang & Jun Yu
Shanghai University, Shanghai, China

ABSTRACT: A novel OFDM-PON architecture based on a distributed Raman amplifier is implemented which amplifies the downstream optical signal and reuses the residual Raman pump as an upstream optical carrier. The error-free 16QAM downstream and 16QAM-OFDM upstream transmission over a 60km SMF are realized.

OCIS codes: (060.4510) optical communication; (120.4570) Optical design of instruments; (190.5650) Raman Effect.

1 INTRODUCTION

Orthogonal Frequency Division Multiplexing (OFDM) Passive Optical Network (PON) has been considered as an alternative scheme to realize next generation access network for increasing data bandwidth, security, and spectrum efficiency (Deng 2014). However, the performance of an OFDM - PON is severely limited by the low sensitivity of optical receiver and the high link loss caused by an optical splitter (Alavi et al. 2014). The distributed Raman amplifier has been used as optical amplifier to solve the link budget problem in optical trunk systems for many years. The Raman amplifier can improve system power, data rate and deployment distance (Gunning 2012). At the same time, the reflective semiconductor optical amplifier (RSOA) has been adopted as an upstream optical modulator to achieve cost-effective optical network units (ONUs) in OFDM-PON (Wei et al. 2010).

Therefore, it is greatly beneficial if use is made of the distributed Raman amplifier incorporating RSOA as a reflective modulator in ONU to construct an intensity modulation-direct detection (IMDD) single-mode fiber (SMF) uplink system for OFDM-PONs. However, a number of crucial issues still remain unsolved. The residual Raman pump is often ignored and discarded directly. Huge link loss is caused by an optical splitter in the traditional upstream transmission system. At the same time, the optical signal strength injecting to RSOA is always weak, and the low optical power makes the RSOA modulation bandwidth is narrow (Wei et al. 2010).

In this paper, we propose a high performance, fully bidirectional OFDM-PON scheme that utilizes the distributed Raman amplification and the residual Raman pump power recycling to inject RSOA for upstream modulation by a 16QAM-OFDM signal. The proposed system realized error-free downstream and upstream 16QAM-OFDM transmission over a 60 km single-mode fiber (SMF) link.

2 TRANSMISSION SYSTEM

The bidirectional OFDM transmission system was constructed and depicted in Figure 1. At the transmitter, there is an ECL at 1550nm with 10-MHz linewidth. The continuous-wavelength (CW) lightwave from ECL is modulated in an intensity modulated Mach-Zehnder modulator (IM-MZM) driven by an electrical 16QAM signal. The distributed Raman amplifier boosts the optical signal and the pumping power (24dBm-32dBm) can be controlled by a variable optical attenuator. At the receiver, optical signal can be demodulated directly after a 23dB attenuator instead of a 1×128 splitter and a 60km SMF. Then the residual pumping light at 1450 nm wavelength is separated from the downstream signal via a tunable optical filter and recycled as a seeding beam to RSOA for upstream data transmission.

*Corresponding author: *rosy99@shu.edu.cn*

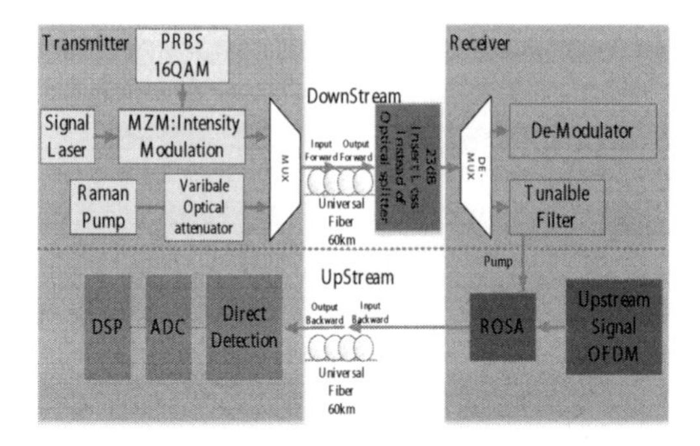

Figure 1. Schematic diagram of the proposed bidirectional OFDM link. ADC: analog-to-digital converter. MZM: Mach-Zehnder modulator. DSP: digital signal process.

3 SIMULATION DEMONSTRATION

3.1 *60km downstream transmission*

The downstream transmission system uses a Raman pump laser diode operating at 1450nm with 10-MHz line-width. The pump power can be adjusted from 350mW~1.5W by a variable optical attenuator to find an appropriate pump power for Raman amplification. The power can meet the downstream optical signal gain requirement and ensure the residual pump power to play a role as an upstream optical carrier. Taking account together, higher pump power, the relative intensity noise, the fiber linear and nonlinear effects, it is necessary to find the pump power to make system BER below 10e-3. The signal laser is operating at 1550nm with 10-MHz linewidth and 0dBm output power.

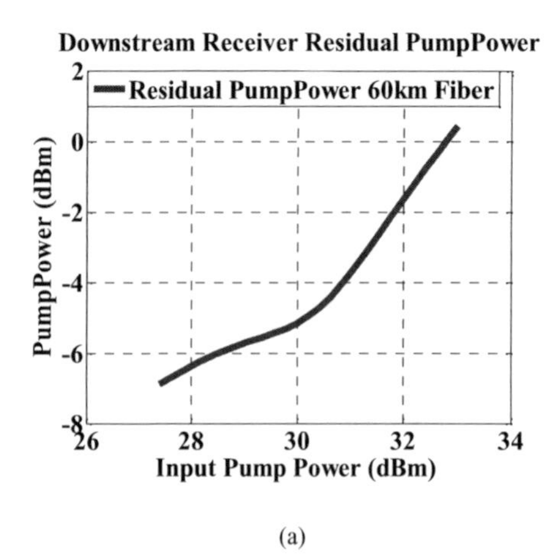

(a)

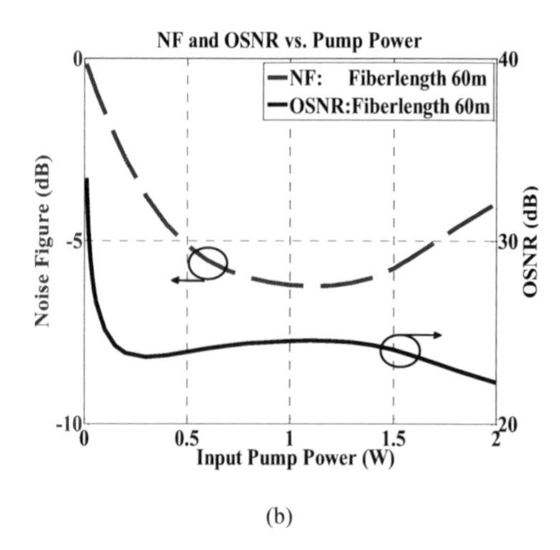

(b)

Figure 2. (a) Downstream residual Raman pump power; (b) NF and OSNR vs. input pumppower calculated downstream NF and OSNR. NF: noise figure. OSNR: optical signal noise rate.

A bidirectional transmission experiment is performed simultaneously with downstream and upstream signals. For downstream signal, the nonlinear noise and transmission loss are the main limiting factors. The main link loss is caused by 1×128 splitter and amounts to 23dB. Raman amplification is implemented to compensate the loss. Some researchers studied the effect of system noise in SMF transmission and analyzed the influence of nonlinear noise in detail (Ma et al. 2013, Faralli & Pasquale 2003). It is clear that the residual pump power is increased by adding the input pump power. After Raman amplification over a 60km SMF the residual pump power was measured to be ~0.99 mW (Figure 2a). Figure 2(b) shows the equivalent NF and OSNR performances versus input pump power, respectively. By increasing the input pump power, NF and OSNR reduced first and then, for high pump power, the NF and OSNR with Rayleigh influence will degrade due to the superposition of Rayleigh backscattering noise. The threshold pump power of 60 km SMF for Raman effect was measured about 25dBm in accordance with the conclusion in the related literature (Faralli & Pasquale 2003). The Rayleigh noise is the most important limiting factors. It is found that the nonlinear transmission impairments could be neglected if the pump power is below 1W.

In order to confirm a power quantitative evaluation of distributed Raman amplifier, the downstream BER versus pump power is drawn in Figure 3. The downstream signal bit rate is set from 8Gbit/s up to 32Gbit/s with 16QAM data format. Figure 3 shows that with the increase of pump power, the overall trend of BER declines. However, each curve has a pump power threshold, above which the BER curve flattens in spite of the pump power increase. Thereby, the input Raman pump power has to be appropriately taken to satisfy the downstream signal bit rate. The higher transmission bit rate is more sensitive to fiber linear and nonlinear effect and need higher pump power to amplify the downstream signal. However, if the pump power is too high, the residual pump beam will suffer from severe effects of Rayleigh noise. In order to ensure the residual pump power can satisfy the upstream requirement, the paper gives the following simulation is given.

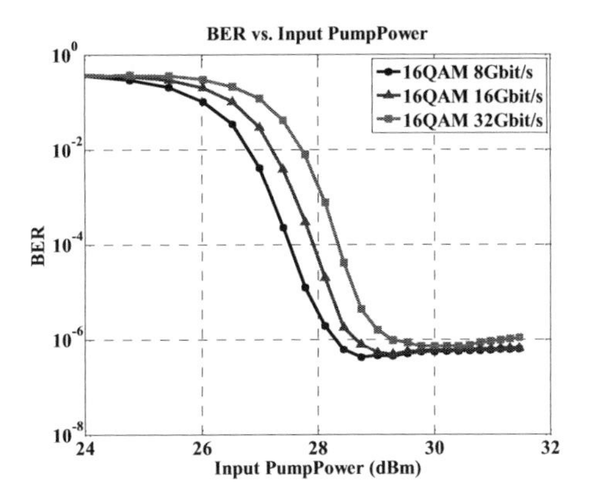

Figure 3. Downstream different line rate signal BER.

3.2 *60km upstream transmission*

Through the analysis in chapter 3.1, the pump power is recognized to exceed 27dBm at least to ensure downstream BER below 10e-3 and the residual pump power should exceed 1mW as shown in Figure 2(a). Therefore, recycling of the downstream residual Raman pump as the upstream optical carrier is feasible for the upstream transmission. The RSOA and SMF parameters in upstream system are taken from reference (Wei et al. 2010). The upstream OFDM signal has 64 carriers in the Hermitian conjugate arrangement and Cyclic Prefix (CP) is 12.5% (Wei et al. 2010). The RSOA bias current is 120 mA in the 60 km SMF upstream system. The bit rate of upstream signal is 10 Gbit/s.

Because high pump power causes high noise, the evaluation of upstream 16QAM-OFDM signal BER versus downstream input pump power is carried out resulting in Figure 4(a). Through measurement, the optimal transmit pump power is found to be 30 dBm as shown in Figure 4. Upstream constellation with respect to 30dBm/31.5dBm downstream input pump power is shown in Figure 4 (b) and (c).

On the other hand, the modulation bandwidth can be improved by adjusting the appropriate input pump power, as shown in Figure 4(a). It can be seen that the upstream modulation bandwidth of 3GHz can be achieved with the input pump power range from 23dBm to 30dBm. This is because the signal line rate depends on input optical power (Wei et al. 2010). Accordingly, the appropriate input pump power is 26dBm to 30dBm.

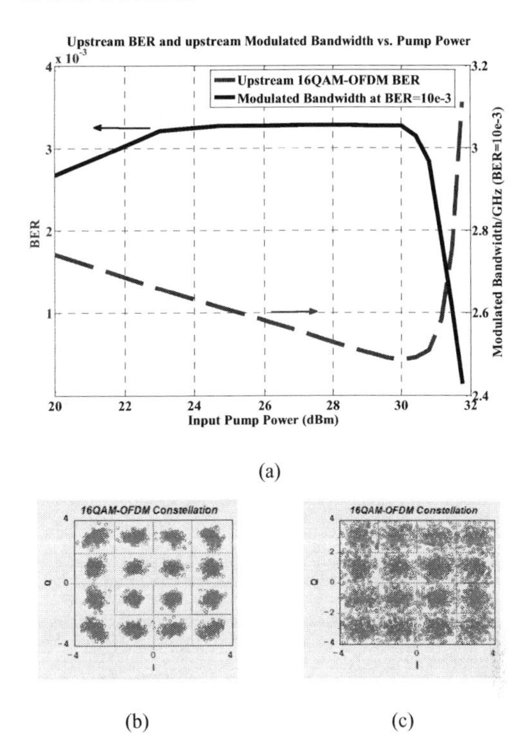

Figure 4. (a) Upstream BER vs. downstream input Raman pump power; (b) Upstream constellation at the power 30dBm/31.5dBm.

4 CONCLUSIONS

A novel long reaches OFDM-PONs architecture based on a distributed Raman amplifier is implemented in which the downstream optical signal is amplified by Raman amplification to compensate the link loss and the upstream optical carrier is generated by recycling the residual Raman pump beam which injects into RSOA to carry the OFDM signal. The scheme is fully bidirectional with reduced cost of the upstream system since distributed Raman amplification and residual Raman pump power recycling are easily implemented. The proposed scheme is proving to be a potential OFDM-PON solution for future access network.

ACKNOWLEDGMENTS

This work is supported by Program of Natural Science Foundation of China (No.61132004, No.61275073) and Shanghai Science and Technology Development Funds (No. 14511100100, No.13JC1402600).

REFERENCES

Alavi, S.E. Skudai, Malaysia, Amiri, I.S. Idrus, S.M. & Supa'at, A.S.M. 2014. All-Optical OFDM Generation for IEEE 802.11a Based on Soliton Carriers Using Microring Resonators. *IEEE Photonics Journal* 6(1).

Deng, M.L. Ling, Y. Chen, X.F. R.P. Giddings, Y.H. Hong, Yi, X.W. Qiu, K. & Tang, J.M. 2014. Self-seeding-based 10Gb/s over 25km optical OFDM transmissions utilizing face-to-face dual RSOAs at gain saturation. *Opt. Express* 22(10): 11954–11966.

Faralli, S. & Pasquale, Di F. 2003. Impact of Double Rayleigh Scattering Noise in Distributed Higher Order Raman Pumping Schemes. *Photon. Technol. Lett. IEEE, 21 May 2003*: 804–806. Piscataway: IEEE.

Gunning, F.C.G. Frascella, P. Antony, C. Fabbri, S.J. Rafique, D. Sygletos, Stylianos. Gunning, P. Reidy, D. McAuliffe, W. Cassidy, D. Ellis. & Andrew, D. 2012. All-optical OFDM and distributed Raman amplification: Challenges to enable high capacities and extend reach.) *Transparent Optical Networks (ICTON), 2012 14th International Conference on, Coventry, 2–5 July 2012*:1–5. IEEE.

Ma, L. Tsujikawa, K. Hanzawa, N. & Yamamoto, F. 2013. 3.5 W optical power delivery over 5 km single-mode fiber using C-band amplified spontaneous emission light source. *Optical Fiber Communication Conference and Exposition and the National Fiber Optic Engineers Conference (OFC/NFOEC), Anaheim, CA 17–21 March 2013*: 1–3 IEEE.

Wei, J. L. Hamié, A Gidding, R. Hugues-Salas,P. E. Zheng, X. Mansoor, S. & Tang, J. M. 2010. Adaptively modulated optical OFDM modems utilizing RSOAs as intensity modulators in IMDD SMF transmission systems. *Opt. Express* 18: 8556–8573.

Wei, J.L. Sanchez, C. Giddings, R. P. Hugues-Salas, E. & Tang, J.M. 2010. Significant improvements in optical power budgets of real-time optical OFDM PON systems. *Opt. Express* 18: 20732–20745.

Electronics, Communications and Networks IV – Hussain & Ivanovic (eds)
© 2015 Taylor & Francis Group, London, ISBN: 978-1-138-02830-2

An improved Rwgh algorithm for node localization in wireless sensor network

Qingmin Meng*
School of Informatics, Linyi University, Linyi, Shandong, China

ABSTRACT: Determination of the position accurately is an important issue in Wireless Sensor Networks (WSN). When the signal transmission path between two nodes is obstructed by some obstacles, Non-Line-of-Sight (NLOS) propagation error comes into existence, and becomes the major reason to affect the localization precision. Based on studies on NLOS error restraining techniques, an improved residual weighted (Rwgh) algorithm is presented. Relative to the traditional Rwgh algorithm, the improved one reduces the number of distance measurements gradually in generating combinations of measurements, and selects the combinations with the smaller residual values as the basis of next calculation step. Thus, using the improved method can reduce the calculation amount and improve the localization precision. Simulations show that the improved algorithm can inhibit the impact of NLOS to some extent relative to the traditional Rwgh algorithm.

1 INTRODUCTION

Node localization technology is one of the main supporting technologies, and the premise of many applications in WSN. Data obtained from the sensor, only combined with the position information, will have practical significance. Node localization is realized by extracting the characteristic parameters of the signal, and then by a certain algorithm.

Depending on different algorithm ways used, localization methods can be divided into two types: range-based localization and range-free localization (Tian et al. 2003). In range-based localization algorithm, node position is determined by triangulation method or least square (LS) method through calculating the measurements of the point-to-point distance or angle information. This technique can achieve precise localization, but it needs higher hardware requirements, and its energy consumption is large. For this type, the main algorithms are Received Signal Strength Indicator (RSSI) (Lewis 2002, Xiang 2004, Sun 2005), AOA (Angle of Arrival) (Nissanka 2001, Dragos 2001), TOA (Time of Arrival) (Andy 1997, Lewis et al. 2001), TDOA (Time Difference of Arrival) (Lewis et al. 2001, Patwari 2003), etc. For range-free localization algorithm, it does not need the distance information or angle information, and only depends on network connectivity to realize location. For this type, the main algorithms are Centroid Localization Algorithm (Bulusu 2002, Nissanka 2000), Convex Programming (Lance 2001), APS (Ad-hoc Positioning System) (Dragos et al. 2001), MDS- MAP(Yi 2003), APIT(Tian et al. 2003), Amorphous (Radhika et al. 2003), etc. In this paper, the study focuses on a range-based localization.

Line of sight (LOS) transmission is a necessary condition for getting accurate signals eigenvalue. But in fact, signals do not always propagate along the path of LOS. When signals encounter obstacles such as mountains or buildings, they will be spread by scattering, reflection, etc., and their transmission delay will increase, so non-line-of-sight (NLOS) errors are resulted. Localization accuracy is greatly reduced because of the presence of NLOS errors. Thus the performance of the application system is affected too. Therefore, research on WSN localization algorithm under the NLOS environment has great practical significance.

According to the effects of NLOS, there exist some algorithms to identify and eliminate or mitigate the effects of NLOS error, including Wylie's NLOS propagation reconstruction algorithm (Wylie et al. 1996), Kalman filter algorithm (Li & Liu 2005), residual weighted algorithm (Rwgh)(Chen 1999), iterative minimum residual estimation algorithm (IMR)(Li 2006), position residual testing (PRT) algorithm(Chan et al. 2006), etc.

All these algorithms mentioned above have their advantages and disadvantages. The method proposed by Wylie requires adequate sample values or historical information, and needs multiple communications to get more distance measurements to obtain the statistical properties of the error distribution, therefore, most of nodes energy will be consumed, so it

*Corresponding author: *qingmin_meng@163.com*

does not comply with the principle of energy conservation. Kalman filtering method is an iterative method which is more suitable for fast-moving target, and also requires repeating communication. During the localization process, calculation of some of the parameters depends on the application environment. However, to obtain and determine these parameters is very difficult, because the application environment of part of the sensor network cannot be predicted in advance, which will directly affect the performance of the algorithm. The Rwgh algorithm can reduce the impact of NLOS propagation without the statistical characteristics of distance measurement error or without distinguishing NLOS propagation, but the computational complexity increases rapidly with the increasing number of anchor nodes. The IMR algorithm has the smallest amount of computation, but its localization accuracy is affected, because its estimation results are obtained from the minimum residual estimation or its suboptimal solution. PRT algorithm is not suitable for sensor networks because of greater algorithm complexity, though it can exclude NLOS error to a certain extent (Meng 2011).

This paper proposes one improved algorithm with simplified computational complexity based on the discussion about Rwgh algorithm, which chooses the combination with minimum residuals from many residual combinations at each step for the next step, by decreasing the number of distance measurements to combine gradually from maximum to minimum. Finally, the final node coordinates are got by weighting all the coordinates with the minimum residuals.

2 RWGH ALGORITHM

In order to reduce the adverse effects of NLOS error, and improve localization accuracy, localization results are weighted by using localization residual in this algorithm.

Let $X=[x,y]^T$ and $X_i=[x_i,y_i]^T$ be an unknown node coordinate and an anchor node coordinate, respectively, and then the minimum mean square estimation (MMSE) is expressed by

$$\hat{X}=\arg\min_X \sum_{i\in S}\left(r_i-\|X-X_i\|\right)^2 \tag{1}$$

where, $r_i=\sqrt{(x-x_i)^2+(y-y_i)^2}$, and S is the combination of different anchor nodes, $r_i-\|X-X_i\|$ is the residual to a specific X value, and $R_{es}(X;S)$ is used to express the square sum of residual to a specific X value:

$$R_{es}(X;S)=\sum_{i\in S}\left(r_i-\|X-X_i\|\right)^2 \tag{2}$$

The minimum X value, which let $R_{es}(X;S)$ be the minimum value, is obtained by using the LS estimation method to unknown nodes, and is denoted by

$$R_{es}(\hat{X};S)=\min_X R_{es}(X;S) \tag{3}$$

Anchor nodes are divided into different groups, with some groups having NLOS signals and some not. The weighted sum may reduce the effect of NLOS by increasing the weight of position estimation in groups without NLOS signals or with less NLOS signals. The normalized squares sum of residual is defined as below:

$$\bar{R}_{es}(X;S)=\frac{R_{es}(X;S)}{size \quad of \quad S} \tag{4}$$

The computational steps of the Rwgh localization algorithm are as follows:

Step 1: Assume that M (M>3) measurements are got from different anchor nodes in the premise of the statistical properties of the NLOS error unknown. There will be $N=\sum_{i=3}^{M}C_M^i$ different combinations of distant measurements through grouping to all M measurements, and all the groups are expressed as S_k, where $k=1,2,\cdots,N$.
Step 2: Calculate coordinates estimated values $\hat{X}_k$ and normalized square sum of residual $\bar{R}_{es}(\hat{X}_k;S_k)$ for each combination of the unknown node position, where $k=1,2,\cdots,N$.
Step 3: Do residual weighting for all $\hat{X}_k$ to get the final coordinates estimation, and use the reciprocal of it as the weighted value, that is:

$$\hat{X}=\frac{\sum_{k=1}^{N}\hat{X}_k\left(\bar{R}_{es}(\hat{X}_k;S_k)\right)^{-1}}{\sum_{k=1}^{N}\left(\bar{R}_{es}(\hat{X}_k;S_k)\right)^{-1}} \tag{5}$$

It can be seen that Rwgh algorithm does not need the optimal choice after obtaining all the measurements, but lists all the combinations of measurements, and calculates coordinates estimation to all of them, then does the residual weighted average. Because measurement combinations of smaller residuals have a higher localization accuracy, while the combinations of larger residuals have lower localization accuracy, so the impact of NLOS for localization accuracy can be inhibited to some extent using the reciprocal of the residuals as weight to obtain the estimated coordinates of unknown nodes (Chen 1999). Because of listing exhaustive combination of the distance measurements, a lot of time and energy are spent in calculating on the one hand, and plenty of combinations with larger residuals may occur in the other. Although the residual weighted is used, there will still be some impact on the final result, thus affecting the accuracy of localization estimation.

3 IMPROVED ALGORITHM ABOUT RWGH

The ideal of the improved algorithm is to decrease the number of distance measurements being included in combinations gradually at each step when the combinations of measurements are generated, then to get the combinations with smaller residuals, and exclude the combination with larger residuals. Using those combinations as the basis for the calculation in the next step, the final coordinates of codes are obtained by residuals weighted to all the coordinates of the minimum residual values finally. Thus, the aim is achieved, that is, reducing the calculation work and improving the localization precision.

Suppose that all the measurements come from all M anchor nodes. The steps of the improved algorithm are as follows:

1 Calculate coordinates estimation values $\hat{X}_k$ and normalized square sum of residual $\bar{R}_{es}(\hat{X}_k;S_k)$ using LS method to all the measurements if i=M. Here, the combination number, $C_M^M = 1$, so k=1.
2 If i=M-1, then the combination number is $C_M^{M-1} = M$, calculate coordinates estimation values $\hat{X}_k$ and normalized square sum of residual $\bar{R}_{es}(\hat{X}_k;S_k)$ using LS method ($k = 1,2,\cdots,M$).
3 If i>3, then let i=i-1, and repeat step 2); otherwise, go to step 4).
4 Do residuals weighted to the coordinates estimated values, which are obtained according to the minimum residuals at each step, as the final coordinates of the unknown nodes using the following formula:

$$\hat{X} = \frac{\sum_{k=1}^{p} \hat{X}_k \left(\bar{R}_{es}(\hat{X}_k;S_k)\right)^{-1}}{\sum_{k=1}^{p}\left(\bar{R}_{es}(\hat{X}_k;S_k)\right)^{-1}} \tag{6}$$

The improved algorithm flow chart is shown in Figure 1.

4 ANALYSIS OF ALGORITHM COMPLEXITY

For the position estimation of the unknown nodes using the LS method, if the number of distance measurements is greater than or equal to 3 in Rwgh algorithm, the combined number of all distance measurements, $N = \sum_{i=3}^{M} C_M^i$.

Therefore, the calculation amount becomes very large with the increasing number of the anchor nodes M. The minimum residual can be obtained by iteration of IMR algorithm. If $M \geq 4$, then the possible maximum iteration times are M-3, and times of LS estimation are:

$$M + 1 \leq C_i \leq 1 + \frac{1}{2}(M - n + 1)(M + n) \tag{7}$$

where, $n = \max\left(M - M_i + 1, 4\right)$, M_i is the predefined iteration times.

It can be seen from the above algorithm description, that the computed amount of the improved algorithm and the iterative IMR algorithm is very similar. The difference is that the improved algorithm uses the residual weighted method to obtain the final results through weighting all optimized combinations, and the purpose of the IMR algorithm is to find the position estimated values with the smallest residuals, or to get the corresponding position estimation values with the combination of measurements needed.

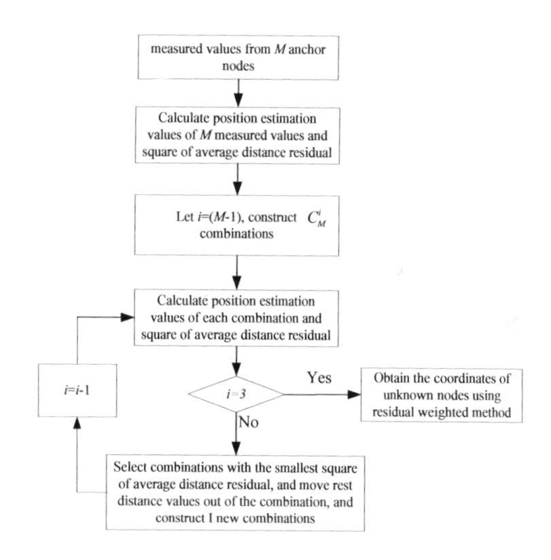

Figure 1. The improved algorithm flow.

This algorithm begins with the combination including all the M measured values. When one time preferred choice is finished, the combinations with M-1 measurements will be generated, until 4 combinations including 3 measurements are generated. So all the calculation times of LS estimation are:

$$C = \frac{1}{2}M(M+1) - 5 \tag{8}$$

The comparative results of LS estimation times for Rwgh algorithm, IMR algorithm and the improved algorithm are shown in Table 1.

Table 1. Comparison results of LS estimation times for Rwgh algorithm, IMR algorithm and the improved algorithm.

Anchor node amount	4	5	6	7
Rwgh Algorithm	5	16	42	99
IMR Algorithm	5	6–10	7–16	8–23
Improved Algorithm	5	10	16	23

From Table 1, it can be seen that the improved algorithm can save LS calculation amount compared with Rwgh algorithm. Though the calculation amount of IMR algorithm is the lowest, and its calculation times may be set according to the iteration times, but there is a little difference between it and the improved algorithm, or even the same when calculation times of LS are selected to be the maximum value of IMR algorithm.

5 ALGORITHM SIMULATION AND RESULTS ANALYSIS

Simulation scene is set as follows: using MATLAB as simulation tool; Nodes are randomly distributed in a 2000m×2000m area; Measurement errors obey Gaussian distribution with zero mean and standard deviation from 0.05m to 30m; take the mean value of 100 times independent estimation as the mean square error(MSE).

$$MSE = \sqrt{\frac{\sum_{i=1}^{100}\left[(x-\hat{x})^2+(y-\hat{y})^2\right]}{100}} \tag{9}$$

where, $(\hat{x},\hat{y})$ is the final coordinate of one unknown code.

Suppose that the measurements can be got for the target nodes from all the anchor nodes, for convenience, let m/n express that n NLOS errors exist in m measured values. The MSE curves of LS, Rwgh, IMR and the improved algorithm, changing with noise standard deviation, are shown from Figures 2~5 in the case of 4/1, 7/1, 7/2 and 7/3, respectively.

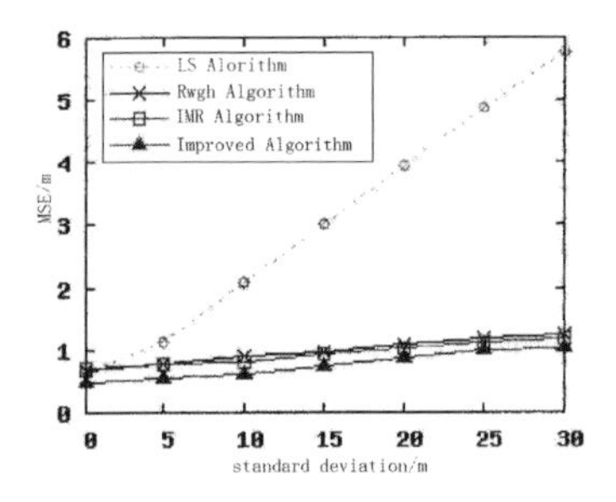

Figure 2. Performance comparison of 4 algorithms in case of 4/1.

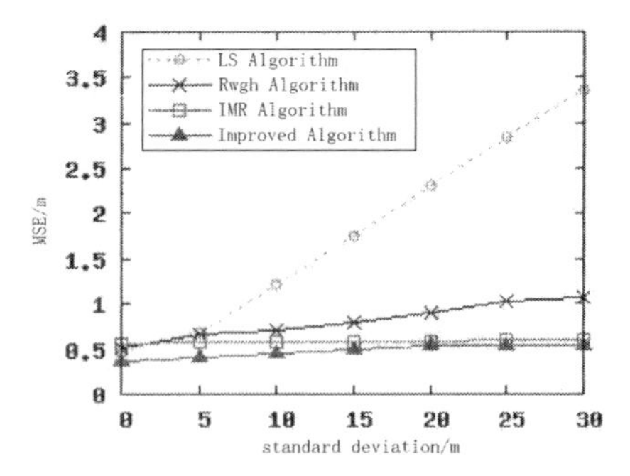

Figure 3. Performance comparison of 4 algorithms in case of 7/1.

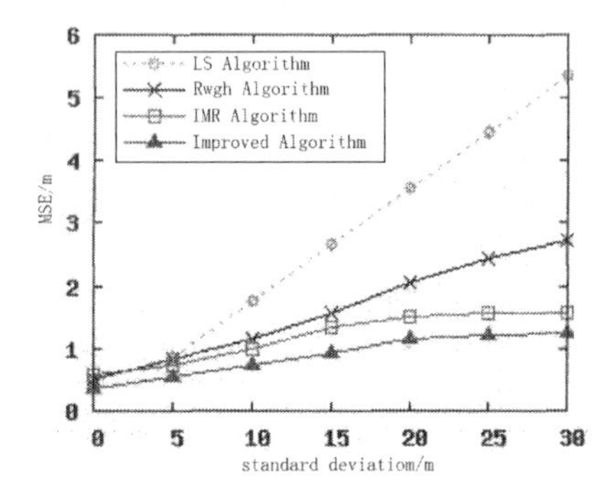

Figure 4. Performance comparison of 4 algorithms in case of 7/2.

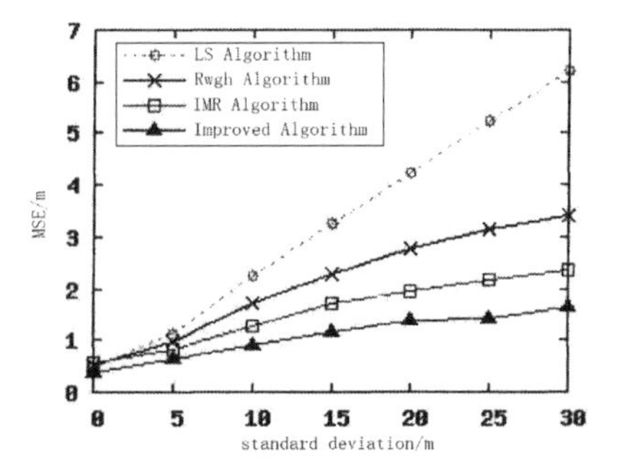

Figure 5. Performance comparison of 4 algorithms in case of 7/3.

It can be seen that the localization error of LS algorithm increases rapidly with the increasing NLOS error in the case of 4/1, but the localization error of other three algorithms increases relatively small. This indicates that the three algorithms have a certain degree of inhibition on the NLOS error, though there is not obvious difference in localization performance among them, localization accuracy of the improved algorithm is slightly higher than that of Rwgh and IMR. Though the localization accuracy of all the 4 algorithms can be improved with the increase of LOS anchor nodes by comparing Figures 2~5, and there are the same trends with the increasing NLOS anchor nodes (see Figures 3~5), the localization performance with the increase of measurement noise of the improved algorithm is obviously higher than other three algorithms.

6 CONCLUSIONS

Through the theoretical analysis and simulation above, it can be seen that the MSE of the LS algorithm increases linearly with the increase of the NLOS error, and the other three algorithms have certain robustness to NLOS error. But the MSE of the proposed algorithm is less than that of the Rwgh algorithm and IMR algorithm obviously, and the computational complexity is greatly reduced relative to the traditional Rwgh algorithm.

ACKNOWLEDGMENT

The project is supported by the Special Funds of the National Natural Science Foundation of China (Grant No. 41340028).

REFERENCES

Andy Ward, Alan Jones & Andy Hopper. 1997. A New Location Technique for the Active Office. *IEEE Personal Communications* 4(5):42–47.

Bulusu N. 2002. Self-Configuring localization systems. *Ph.D. Thesis, Los Angeles, University of California.*

Chan Yiu-tong, Tsui Wing-yuc, So Hing-Cheung, et al. 2006. Time-of-Arrival Based Localization under NLOS Conditions. *IEEE Transactions on Vehicular Technology* 55(1): 17–24.

Chen Pi-Chun. 1999. A Non-line-of-sight Error Mitigation Algorithm in Location Estimation. *Proceedings of Wireless Communication and Networking Conference*: 31–320. Piscataway: IEEE.

Dragos Niculescu & Badri Nath. 2001. Ad hoc positioning system(APS). *Global Telecommunications Conference. San Antonio,TX, USA:IEEE*: 2926–2931.

Ji Xiang & Zha Hongyuan. 2004. Sensor positioning in wireless ad-hoc sensor networks using multidimensional scaling. *23th Annual Joint Conference of the IEEE Computer and Communications Societies, Hong Kong, China*: 2652–2661.

Lance Doherty, Kristofer S. J. Pister & Laurent El Ghaoui. 2001. Convex Position Estimation in Wireless Sensor Networks. *Proceedings of 20th Joint Conference of the IEEE 32 Computer and Communications Societies Anchorage,AK,USA:IEEE Computer and Communications Societies* 3: 1655–1663.

Lewis Girod & Deborah Estrin. 2001. Robust Range Estimation using Acoustic and Multimodal Sensing. In:*Proceedings of IEEE/RSJ International Conference on Intelligent Robots and Systems (IROS' 01),Maui,Hawaii,USA: IEEE Computer Society* 3: 1312–1320.

Lewis Girod, Vladimir Byehovskiy, Jeremy Elson, et al. 2002. Locating tiny sensors in time and space: *A case study. Proceedings of the 2002 IEEE International Conference on Computer Design:VLSI in Computers and Processors, Freiburg: IEEE Computer Society* : 214–219.

Li Jing & Liu Ju. 2005. NLOS error mitigation in TOA using Kalman filter. *Journal of Communications* 26(1): 32–137, 143.

Li X. 2006. An iterative NLOS mitigation algorithm for location estimation in sensor networks. *IST Mobile and Wireless Commun. Summit, Myconos, Greece, June 2006.*

Meng Qing-min. 2011. Improvement on non-line-sight nodes localization for wireless sensor networks. *Computer engineering and applications* 47(21):223–226.

Nissanka B. Priyantha, Anit Chakraborty & Hair Balakrishnan. 2000. The Cricket Location-Support System. *Proc. 6th Int'l Conf. Mobile Computing and Networking*:32–43. New York: ACM Press.

Patwari, N., Nero III A. O., Perkins, M., et al. 2003. Relative location estimation in Wireless Sensor Networks. *IEEE Transactions on Signal Processing* 51(8): 2137–2148.

Priyantha N B, Miu A K L, Balakrishnan H, et al. 2001. The cricket compass for context-aware mobile applications. *Proc. of the 7th Annual Int'l Conf. on Mobile Computing and Networking,Rome, Italy* 1–14.

Radhika Nagpal, Howard Shrobc & Jonathan Bachrach. 2003. Organizing a Global Coordinate System from Local Information on an Ad Hoc Sensor Network. *Proceedings of the 2nd International Workshop on Information Processing in Sensor Networks,Palo Alto, CA,USA* 333–348.

Sun Li-min, Li Jian-zhong & Chen Yu. 2005. *Wireless sensor networks*. Beijing: Tsinghua university press.

Tian He, Chengdu Huang, Brian M. Blum, et al. 2003. Range-Free Localization Schemes for Large Scale Sensor Networks. MobiCom: *Proceedings of the 9th Annual International Conference on Mobile Computing and Networking, San Diego, CA, USA* 81–95.

Wylie M.P. & Holtzman J. 1996. The non-line-of-sight problems in mobile location estimation. *Proc IEEE Int Conf Universal Personal Communications*: 827–831. Cambridge:Piscataway Publishers.

Yi Shang, Wheeler Ruml & Ying Zhang. 2003. Localization from mere connectivity. In: *Proc.of the 4th ACM International Symp.on Mobile Ad-Hoc Networking & Computing. Annapolis, Maryland*:201–212.

Electronics, Communications and Networks IV – Hussain & Ivanovic (eds)
© 2015 Taylor & Francis Group, London, ISBN: 978-1-138-02830-2

Design of cattle health monitoring system using wireless bio-sensor networks

Myeong-Chul Park
Department of Biomedical Electronics, Songho College, Hoengseong, Gangwon, Republic of Korea

Hyon-Chel Jung
Department of Biomedical Engineering, KunKuk University, Chungju, Republic of Korea

Tae-Koon Kim
Medical Supply Co., Ltd., Wonju, Gangwon, Republic of Korea

Ok-Kyoon Ha*
Engineering Research Institute, Gyeongsang National University, Jinju, Republic of Korea

ABSTRACT: The spread of highly contagious livestock diseases can cause economic damages by limiting cattle farm productivity, decreasing tourism, and decreasing the exports of trading companies. However, there are also many social costs associated with recovering from these economic damages. It is still difficult to prevent livestock diseases using current monitoring systems that track livestock activity and the environmental conditions of livestock barns. In this paper, we design a cattle health monitoring system to help prevent livestock diseases, such as foot-and-mouth disease, using bio-sensors. We also design a sensing module to collect different bio-signals of each animal, such as the heartbeat, the breath rate, and the momentum, and use a Wireless Sensor Network module (WSN) to transmit this biometric data. We implement both sensing and WSN modules as an integrated device on a single board.

1 INTRODUCTION

The highly contagious livestock diseases lead to not only economic damages, such as lower productivity of cattle farms, fall in tourism income, and fall in exports of trading companies, but also much social costs to escape from the damages. It is important to isolate livestock diseases, such as foot-and-mouse disease, through the prevention of infectious diseases, because these epidemic diseases may incur damages in a country. Especially, the forecasting systems are urgently needed to effectively control these epidemic diseases for Korean farms which have heavy density of livestock.

Previous forecasting systems (Ju & Kim 2006, Hwang et al. 2010, Lee et al. 2011, Kim et al. 2012) monitor the environment of livestock barns and the activities of cattle using a pedometer or thermal imaging, making it difficult to produce clear predictive information describing the occurrences of livestock diseases. In this paper, we design a cattle health monitoring system that uses biosensors to collect vital information from each individual cow in order to help forecast livestock diseases. We developed a sensing module to measure cattle bio-signals, such as the heartbeat, the breath rate, and the momentum, and a wireless sensor network module to transmit the biometric data to the forecasting system.

2 RELATED WORK

The jog trot system (Fujitsu 2011) uses a pedometer installed on the foot of each individual cow to analyze movement patterns and activity. The system provides cattle management and milking services through the monitoring of the activity of each entity and breeding status. This active sensing system identifies the health condition of each entity and detects special activities in advance to allow swift measures such as artificial insemination on time. The jog trot system transmits the collected activity data to a graphical information for user PC or smart phone using WSN.

The intelligent management system for livestock farms (Nare Trends Inc. 2011) can be connected to prior livestock farms automation system, and uses

*Corresponding author: *jassmin@gnu.ac.kr*

smart phones to remotely control different aspects of the operating facilities, such as temperature, humidity, self-feeders, and ventilation units. This system is practical for monitoring the emergency state of cattle sheds because it employs a closed circuit television (CCTV) system for collecting video information of the livestock farm and its facilities.

The livestock disease forecasting system (Kim et al. 2012) is an integrated management system that collects data on the activity and body temperatures of each livestock using acceleration sensors and thermal imaging cameras. The system compares the collected information with control data according to livestock disease collected by challenge test. This forecasting system can identify livestock diseases before it spreads in order to minimize associated damages. However, it is difficult to apply to individual animals such as cattle, because it employs general sensors to collect the biometric data of livestock.

3 DESIGN OF THE PROPOSED SYSTEM

In this paper, we design a cattle health monitoring system that collects vital information on each entity of cattle farms, such as heart rate, respiration rate, and quantity motion, to help forecast livestock diseases using wireless bio-sensors. For the purposed system, we also design a sensor module to measure the bio-signals of cattle, such as the heartbeat, the breath rate, and the momentum, and a wireless sensor network module to transmit biometric data to the forecasting system. Figure 1 shows the architecture of the cattle health monitoring system.

This cattle health monitoring system uses an Electrocardiogram (ECG), a Thermistor, and a Gyro-sensor to measure the bio-signals of each individual cattle. The raw data is filtered using a Band Pass Filter (BPF), a Low Pass Filter (LPF), and a High Pass Filter (HPF) to create digital biometric data. The biometric data is then transmitted to an integrated management system that stores the vital cattle information using a Zigbee WSN module after the signals are amplified using a processing amplifier. Finally, the cattle health monitoring system graphically presents the vital information to help understand the symptoms of each individual through a clinical decision support system (CDSS).

3.1 *Bio-sensor module*

The bio-sensor module consists of three parts: an ECG component which detects the pulse frequency of cattle, a Respiration component that uses a force sensing resistor (FSR) to detect the breath rate, and an Accelerometer component for measuring the momentum of each individual. The ECG component employs a processing amplifier as the BPF, LPF, and HPF signal filter algorithms are applied to each part.

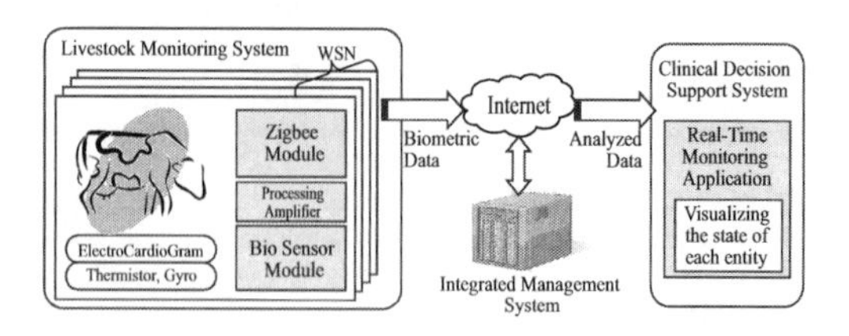

Figure 1. The overall architecture of the proposed cattle health monitoring system.

3.1.1 *Detecting heartbeat*

The ECG component detects the pulse frequency of the cattle in order to monitor heartbeat activity. We employed the Einthoven triangle method to measure a potential difference in the ECG. This ensures that the separation distances of the electrodes are suitable for cattle trunks. We estimated heartbeat activity using measured ECG signals through the three electrodes. Figure 2 depicts the heartbeat detection process for the ECG component. To convert the ECG analog signals to digital signals, the process applies BPF to the raw data of the ECG and differentiates the signals to analyze any variation in quantity. It then generates the digital signals by applying an LPF. This process applies the active threshold level according to the peak of the signal to effectively detect heartbeat.

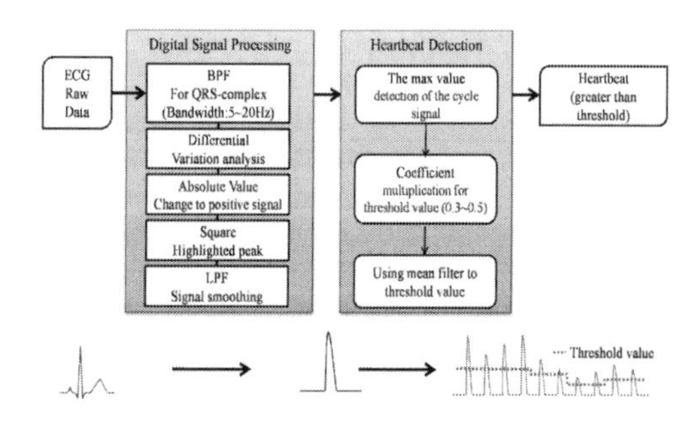

Figure 2. The heartbeat detection process for the ECG part.

3.1.2 *Measuring breath rate*

In general, healthy cattle breathe from the abdomen and the chest at the same time. Cattle with respiratory or digestive diseases do not simultaneously use both the abdomen and chest to breath and experience increased breath rates. The Respiration component records the breath patterns and calculates breath rate using the FSR, which measures the change of resistance value according to the pressure change on a sensor. We installed a FSR between a band and the cattle trunk to measure respiration. Figure 3 depicts the measurement process for the breath rate of cattle based on the FSR. The LPF is applied on the raw data from the FSR (Cut-off frequency: 1Hz) because the signals based on cattle breath are low in frequency. Finally, the Respiration component counts a breath to detect the breath of cattle if the signal of the FSR is less than a threshold level immediately after the signal is larger than the threshold.

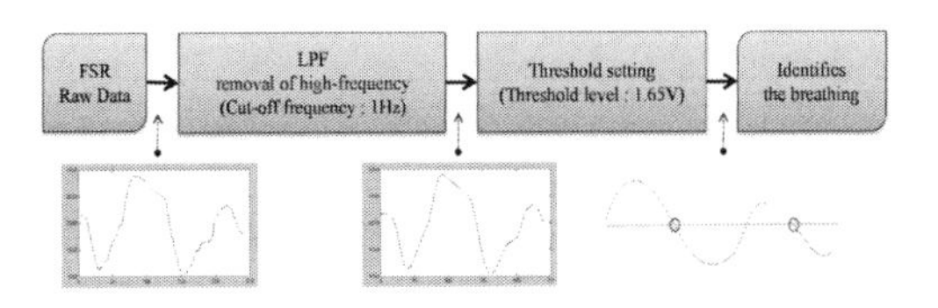

Figure 3. The process for measuring the breath rate.

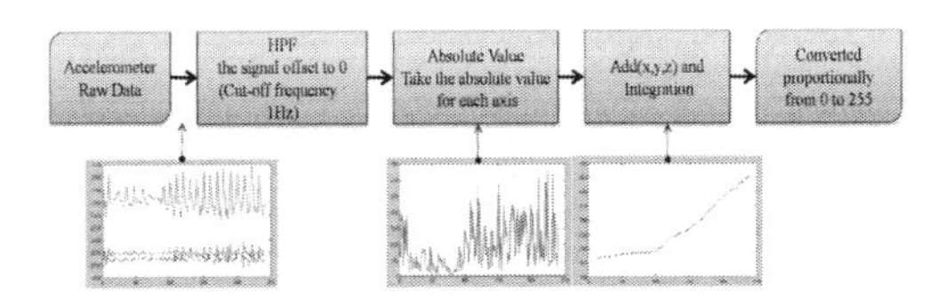

Figure 4. The process for measuring the momentum.

3.1.3 *Measuring momentum*

An accelerometer was used to measure the momentum of the cattle. The Accelerometer component measures the displacement of the accelerometer according to the activity of cattle, and calculates the numerical value of the momentum (0-255) by integrating the activity value during a unit of time. Figure 4 shows the measurement process for the momentum of cattle using the accelerometer. During measurement, HPF (Cut-off frequency: 1Hz) is used to detect available signals from raw data, and the values of each sensor axis are changed to absolute values. Finally, we detected scalable values from 0 (for inactive) to 255 (for continuous active) after the integration of the sum of the absolute values.

3.2 *Zigbee WSN module*

We configured a WSN that uses a Zigbee sensor module based on IEEE 802.15.4 to transfer biometric data from the biosensor module to the integrated management system. For the Zigbee WSN module, we employed a machine control unit (MCU) that provides low-power consumption to control data flow and rapid signal processing, and designed a network protocol for the data packet. The data packet consists of a start character, an entity identifier, heartbeat data, breath data, momentum data, an error checker, and a terminal character. The error checker includes various error states, such as the heartbeat signal error, the breath signal error, the momentum signal error, and the power states of the integrated device. The structure of the data packet and the description of each bit for error checker are shown in Figure 5.

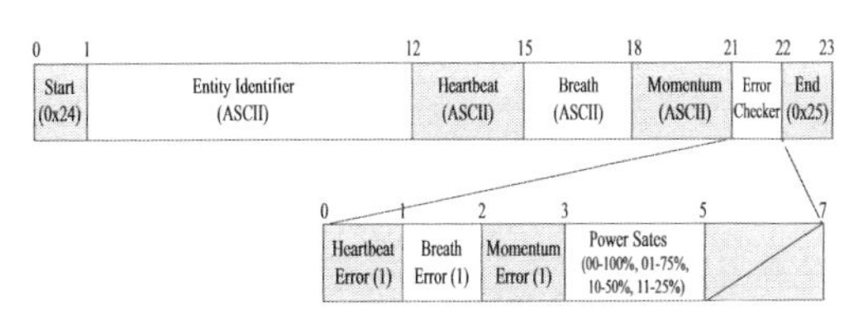

Figure 5. The protocol of a data packet for WSN.

4 IMPLEMENTATION

We implemented both the bio-sensor module and the Zigbee WSN module as an integrated device on a single board for an easy installation on the trunk of each cattle. Figure 6 illustrates implemented hardware. In the figure, the ECG component used for heartbeat detection (① in Figure 6), the Respiration component used for the breath measurement (② in Figure 6), and the Accelerometer component used for the momentum measurement (③ in Figure 6). Our integrated device includes a MCU for signal processing and a Zigbee WSN module (④ and ⑤ in Figure 6, respectively). The integrated device was protected by an aluminum box, and was fixed on the trunk of cattle using a band which includes a FSR to measure the breath rate.

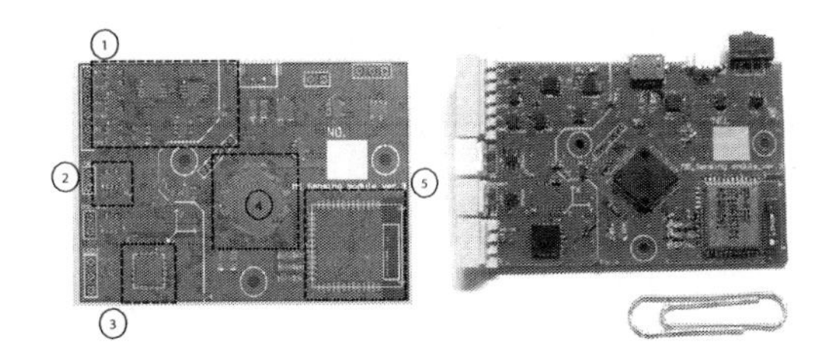

Figure 6. Implementation of the integrated device.

In our cattle health monitoring system, the biometric data of each entity are collected using a biosensor module and are stored in an integrated management system. The vital information is provided to use on a PC or a smart phone through our clinical decision support system (CDSS) when a permitted user wishes to monitor the states of each entity. Figure 7 shows the user interface of the application in the CDSS for real-time monitoring. The application provides the options for selecting entities and reports vital information on the selected entities for every hour. We can easily manage each device because the state of each sensor and batteries is provided graphically.

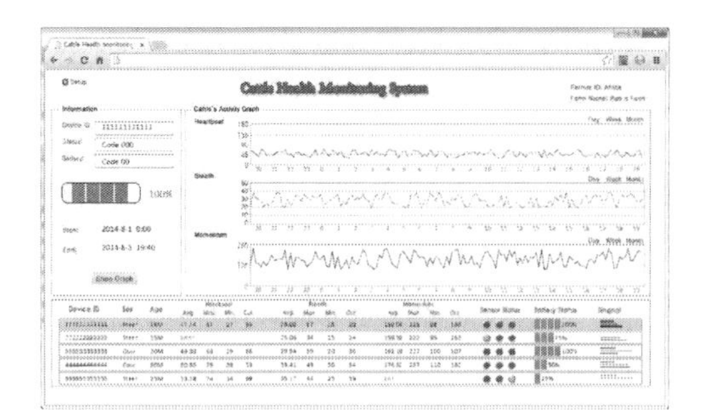

Figure 7. User application of the CDSS for real-time monitoring.

5 CONCLUSION

It is important to isolate the livestock diseases, such as foot-and-mouse disease, because these epidemic diseases can incur significant economic damages. We designed a cattle health monitoring system that collects vital information of each cattle, such as the heartbeat, the breath rate, and the momentum, to help forecast livestock diseases using biosensors based on WSN.

We implemented an integrated device including a biosensor module and a Zigbee WSN module on a single board for easy installation on the trunk of cattle. In the system, the biometric data collected by the integrated device is transmitted to an integrated management system that stores vital information. Finally, the cattle health monitoring system graphically provides vital information to help understand the symptoms of each entity, such as livestock disease, through a CDSS.

ACKNOWLEDGMENTS

This work was supported by the Livestock Disease Control for Smart Green Bio Security System Development Program (R0000575) funded by the Ministry of Trade, Industry and Energy, and also was supported by Basic Science Research Program through the National Research Foundation of Korea (NRF) funded by the Ministry of Education (NRF-2013R1A1A2011389), Republic of Korea.

REFERENCES

Fujitsu 2011. A jog trot system. http://www.fujitsu.com/kr/sustainability/agriculrure/.

Hwang, J. H., Lee, M. H., Ju, H. D., Kang, H. J. & Yoe, H. 2010. Implementation of swinery integrated managementsystem in ubiquitous agricultural environments. *The Journal of Korea Information and Communications Society 35*(2B): 252–262.

Hwang, J. H., Shin, C. S. & Yoe, H. 2010. Study on an agricultural environment monitoring server system using wireless sensor networks. *Sensors (Basel)* 10(12): 11189–11211.

Ju, M. & Kim, S. 2006. Logistic services using RFID and mobile sensor network. *International Journal of Multimedia and Ubiquitous Engineering* 1(2): 25–29.

Kim, H., Yang, C. & Yoe, H. 2012. Design and implementation of livestock disease forecasting system. *The Journal of Korea Information and Communicatons Society* 37(12): 1263–1270.

Lee, J., Hwang, J. H. & Yoe, H. 2011. Design of integrated control system for preventing the spread of livestock diseases. *Lecture Notes in Computer Science* 7105: 169–173.

Nare Trends Inc. 2011. Xspark: An intelligent management system. http://www.xspark.co.kr/system/cattleshed.php.

Electronics, Communications and Networks IV – Hussain & Ivanovic (eds)
© 2015 Taylor & Francis Group, London, ISBN: 978-1-138-02830-2

Dispersion relations of saw propagating under periodical gratingon langasite

Xiaolan Qian, Fangqian Xu*, Yixiang Chen, Xuelan Zou & Zhenfei Zhao
Zhejiang University of Media and Communications, Hangzhou, Zhejiang, China

ABSTRACT: Properties of surface acoustic waves propagated under Pt grating on langasite cut (at 0°, 138.5°,and 26.6°)substrate were studied. The dispersion relations of Pt strips of different thicknesses were given. It was discovered that there are two stop-bands on the dispersion curves, which is very unusual in piezo-electric crystal substrates and theoretically supports the design of relevant SAW filters.

1 INTRODUCTION

The features of the piezoelectric langasite (LGS), such as high electromechanical coupling coefficients, the existence of temperature compensated cuts, and the absence of a structural phase transition below its melting point of about 1400°C, make it a desirable piezoelectric substrate for BAW and SAW devices and sensors (Weihnacht et al. 2012, Naumenko 2011, Naumenko 2012, Kenny 2006)

To predict the characteristics of a SAW device, exact theory work is required to adequately describe its propagation in the analyzed piezoelectric substrate.

Hashimoto et al. (1993,1994) proposed the concept of discrete Green function and effective permittivity for grating on the basis of Blφtekjær's theory (Blφtekjær et al. 1973), which is of great use for characterizing wave propagation in infinite periodic metal grating. According to finite element method-boundary element method (FEM-BEM), the dispersion relation was found by searching the pole of the admittance of waves at different frequencies, from which the COM parameters were extracted.

Because Hashimoto's program was designed to calculate only the dispersion relations for strips of gold and aluminium, this paper proposed a new method to investigate characteristics of SAW behavior in platinum grating on langasite cut with Euler angles (0°, 138.5°, and 26.6°), by which the dispersion relation on Pt electrodes with different thicknesses on langasite substrate was obtained.

2 THEORY ANALYSIS

Figure 1 shows an infinite periodic metal grating with the period p, electrode width w, and thickness h. Surface acoustic waves supposedly propagate in the x_1 direction, so Newton's equation is

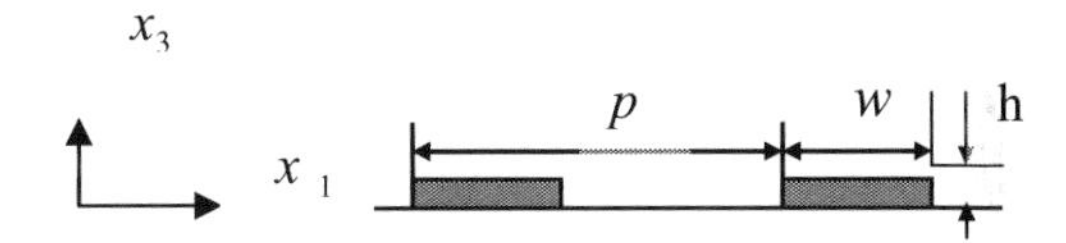

Figure 1. Substrate and strip.

written as:

$$-\rho\omega^2 u_n = \frac{\partial T_{n1}}{\partial x_1} + \frac{\partial T_{n3}}{\partial x_3} \quad (n = 1,2,3) \tag{1}$$

According to Blφtekjær's original theory, the electric charge $q(x_1)$ on the electrodes and electric field $e(x_1)$ within the electrode gaps are as follows:

$$q(x_1) = \sum_{m=-\infty}^{+\infty} \frac{A_m \exp(-j\beta_{m-1/2}x_1)}{\sqrt{\cos(\beta_g x_1) - \cos\Delta}} \tag{2}$$

$$e(x_1) = \sum_{m=-\infty}^{+\infty} \frac{B_m \operatorname{sgn}(x_1)\exp(-j\beta_{m-1/2}x_1)}{\sqrt{-\cos(\beta_g x_1) + \cos\Delta}} \tag{3}$$

where, $\beta_m = \beta_g(m+s)$, $\beta_g = 2\pi/p$ and $\Delta = \pi w/p$. $Q(\beta_n)$ and $E(\beta_n)$ were used to denote the Fourier transform for $q(x_1)$ and $e(x_1)$, respectively, thus:

$$Q(\beta_n) = p^{-1}\int_{-p/2}^{+p/2} q(x_1)\exp(+j\beta_n)dx_1 \tag{4}$$

$$E(\beta_n) = p^{-1}\int_{-p/2}^{+p/2} e(x_1)\exp(+j\beta_n)dx_1 \tag{5}$$

*Corresponding Author: *xufangqian2005@163.com*

Substituting Eq.(2) and (3) into (4) and (5) yields the following:

$$Q(\beta_n) = 2^{-0.5} \sum_{m=-\infty}^{+\infty} A_m P_{n-m}(\cos\Delta) \tag{6}$$

$$E(\beta_n) = 2^{-0.5} \sum_{m=-\infty}^{+\infty} S_{n-m} B_m P_{n-m}(\cos\Delta) \tag{7}$$

where $P_m(\theta)$ is the m-th order Legendre function. According to the relation between $q(x_1)$ and $e(x_1)$, the following is obtained:

$$\sum_{m=-\infty}^{+\infty} \left(\frac{jS_n \varepsilon(\infty)}{\varepsilon(\beta_n/\omega)} A_m + S_{n-m} B_m \right) P_{n-m}(\cos\Delta) = 0 \tag{8}$$

For all m, the following should come into existence:

$$B_m = -j\varepsilon(\infty)^{-1} A_m \tag{9}$$

Therefore:

$$\sum_{m=M_1}^{M_2} A_m \left(S_{n-m} - S_n \frac{\varepsilon(\infty)}{\varepsilon(\beta_n/\omega)} \right) P_{n-m}(\cos\Delta) = 0 \tag{10}$$

By solving the linear equations with respect to $(M_2 - M_1)$ the unknowns, the coefficient A_m can be obtained by:

$$Q = \int_{-w/2}^{w/2} q(x_1)\,dx_1 = 2^{-0.5} p \sum_{m=M_1}^{M_2} A_m P_{m+S-1}(\cos\Delta) \tag{11}$$

$$\Phi = -\int_{-w/2}^{w/2} e(x_1)\,dx_1 = -\int_{-p+w/2}^{-w/2} \frac{e(x_1)dx_1}{1-\exp(2\pi js)}$$
$$= \frac{2^{-1.5} p}{\varepsilon(\infty)\sin(s\pi)} \sum_{m=M_1}^{M_2} (-1)^m A_m P_{m+S-1}(-\cos\Delta) \tag{12}$$

The admittance $Y(s,\omega)$ is determined by calculating the ratio between Q and Φ; that is:

$$Y(s,\omega) = \frac{j\omega Q}{\Phi} = 2j\omega\sin(s\pi)\varepsilon_g(s,\omega) \tag{13}$$

In the above equation, $\varepsilon_g(s,\omega)$ denotes the effective permittivity.

$$\varepsilon_g(s,\omega) = \varepsilon(\infty) \frac{\sum_{m=M_1}^{M_2} A_m P_{m+S-1}(\cos\Delta)}{\sum_{m=M_1}^{M_2} (-1)^m A_m P_{m+S-1}(-\cos\Delta)} \tag{14}$$

For short-circuited (SC) gratings, the electric potential $\Phi = 0$ and the admittance $Y(s,\omega)$ take extreme value. By solving $Y(s,\omega)^{-1} = 0$ with different frequencies to extract the velocity of the surface acoustic wave, the dispersion relation can be obtained for waves in short-circuited metal gratings.

3 DISPERSION AND COM PARAMETER

According to the above-mentioned theoretical analysis of the dispersion relation for SAW propagating in periodic short-circuited metal gratings, parts of the Hashimoto program were modified to calculate the dispersion relation for Pt grating on langasite cut at different angles (0°, 138.5°, and 26.6°) in a different ration between strip thickness and grating period $h/p = 0.005 - 0.07$ (see Figure 2).

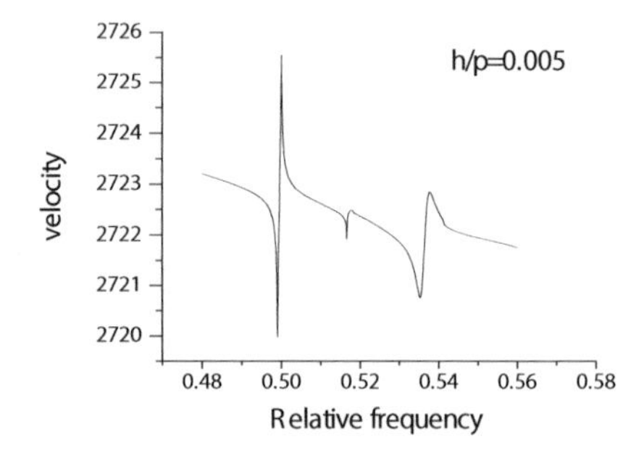

a. $h/p = 0.005$

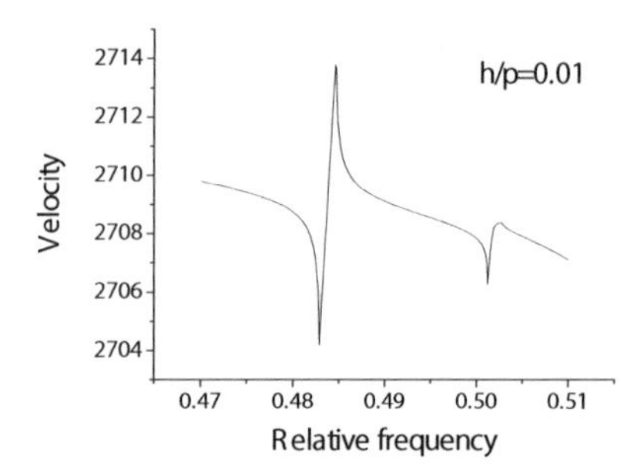

b. $h/p = 0.01$

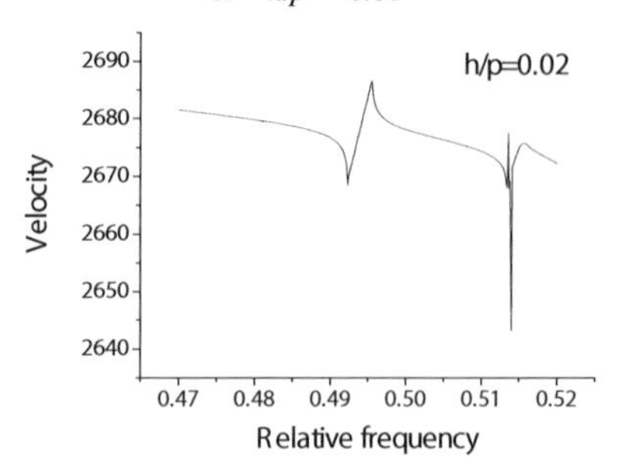

c. $h/p = 0.02$

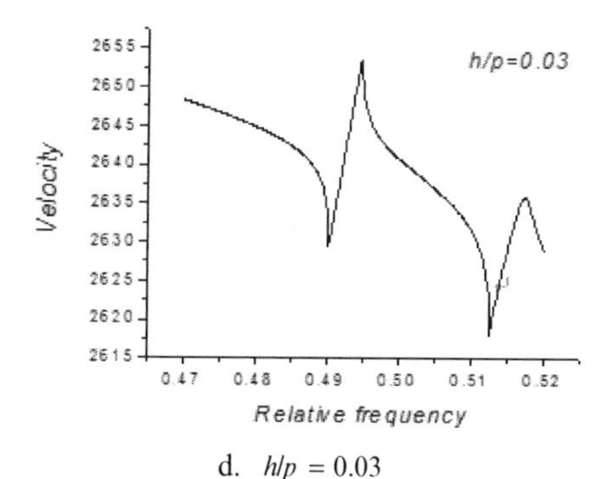

d. h/p = 0.03

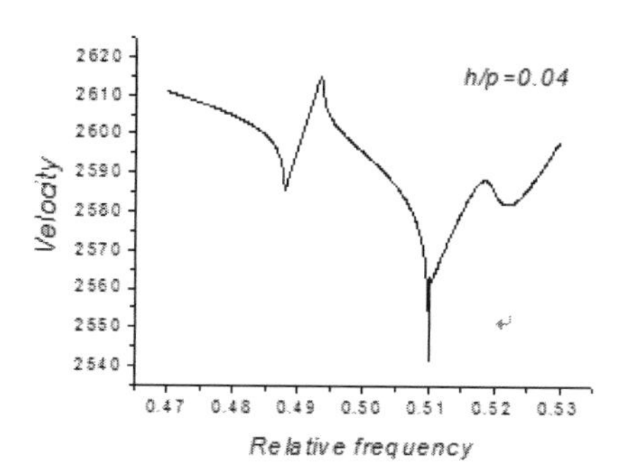

e. h/p = 0.04

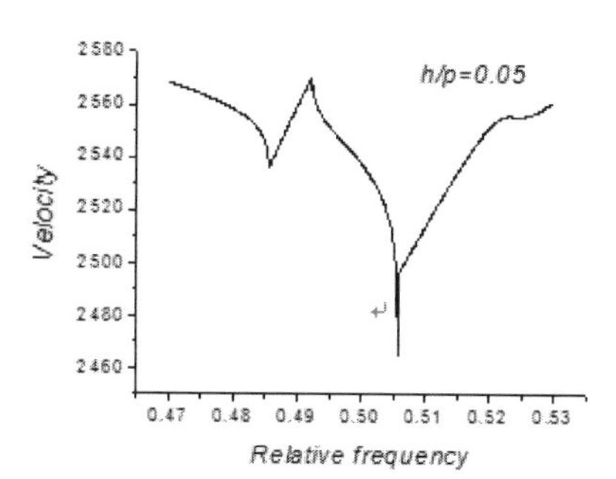

f. h/p = 0.05

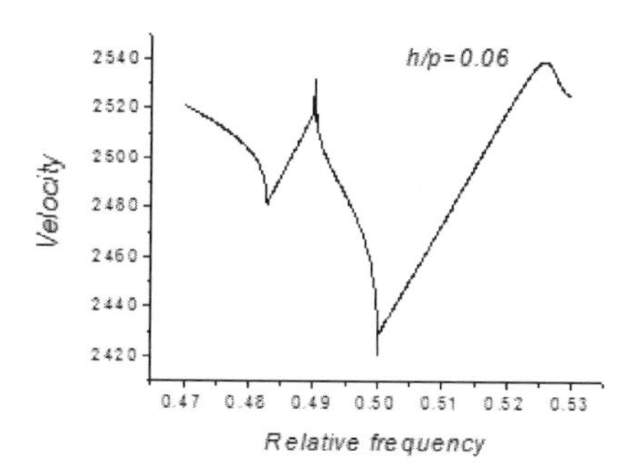

g. h/p = 0.06

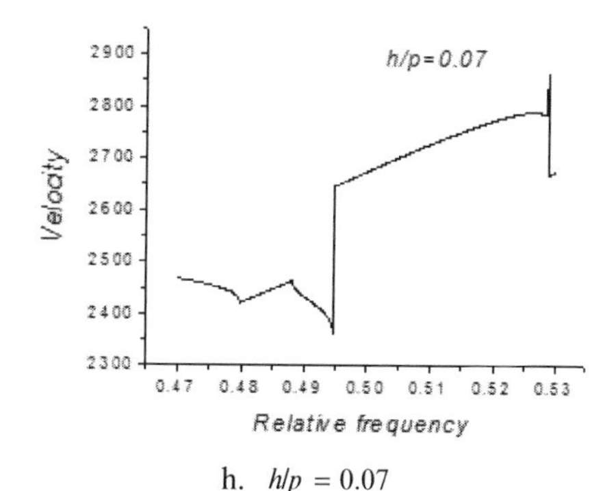

h. h/p = 0.07

Figure 2. Dispersion relation on different strip thicknesses.

From the dispersion relations, some conclusions could be drawn:

1 There were two stop-bands on the dispersion curves for Pt grating on the langasite cuts (at 0°, 138.5°, and 26.6°) in Figure 2. This is very unusual on piezo electric crystal substrates.
2 With an increased Pt thickness, the second and third stop-bands merged. They produced a very wide detuned stop-band.

4 CONCLUSION

Because the Hashimoto program was designed to calculate the dispersion relation for two kinds of metal grating, gold, and aluminium, it was improved upon to calculate SAW propagating in Pt gratings on langasite cuts (at 0°, 138.5°, and 26.6°). From the calculated results, it was found that there are two stop-bands on the dispersion curves. Moreover, the second and third stop-bands merged when the Pt thickness was further increased.

ACKNOWLEDGMENTS

This work was carried out with the program developed by Hashitomo, etc. at Chiba University and supported by the National Nature Science Foundation of China (No. 11374254). The authors express their sincere gratitude.

REFERENCES

Analysis for Measured LGS SH-SAW Devices. *IEEE trans. on UFFC* 53(2): 402–411.

Blϕtekjær, K., Ingebrigesen, K.A. & Halvor Skeie. 1973. Acous-tic Surface Waves in Piezoelectric Meterials with Periodic Metal Strip on the Surface. *IEEE trans. Electron. Device* 20(12): 1133–1138.

Hashimoto, K. & Yamaguchi, M. 1993. Analysis of excitation and propagation of acoustic waves under periodic me-tallic-grating structure for SAW device modeling. in *Proc. IEEE Ultrason. Symp*:143–148.

Hashimoto, K. & Yamaguchi, M. 1994. Precise simulation of surface transverse wave devices by discrete Green function theory. In *Proc. IEEE Ultrason. Symp*:2 53–258.
Kenny, T. D. 2006. FEM/BEM Impedance and Power.
Naumenko, N. F. 2011. Analysis of Interaction Between Two SAW Modes in Pt Grating on Langasite Cut (0°, 138.5°, 26.6°).*IEEE Trans on UFFC* 58(11): 2370–2377.
Naumenko, N. F. 2012. Effect of Anisotropy on Characteristics and Behavior of Shear Horizontal SAWs in Resonators Using Langasite. *IEEE Trans on UFFC* 59(11): 2515–2521.
Weihnacht, M., Sotnikov, A., Schmidt, H., et al.2012. Langasite: High Temperature Properties and SAW Simulations. In *Proc. IEEE Ultrason. Symp*: 243–248.

Electronics, Communications and Networks IV – Hussain & Ivanovic (eds)
© 2015 Taylor & Francis Group, London, ISBN: 978-1-138-02830-2

A technical scheme of secure communication based on controlled projective synchronization method

Hui Qian* & Hongjie Yu
Department of Engineering Mechanics, Shanghai Jiao Tong University, Shanghai, China

ABSTRACT: This paper proposes a technical scheme of chaotic masking for secure communication based on the controlled projective synchronization method. A nonlinear drive vector function is constructed with a scaling factor for chaotic synchronization by a proper separation of the linear and nonlinear terms in a chaotic system. The transmitted message is injected into one element of the state vector. Thus, this scheme increases the complexity of masking messages and the difficulty of decoding. A Lorenz attractor is used to simulate the application of sinusoidal information. A simple and optimal chaotic masking scheme is given through comparative analysis. The results of the simulation show that our system is highly secure.

1 INTRODUCTION

Since drive-response synchronization was first created by Pecora & Caroll (1990) in the early 1990s research on chaotic synchronization method and chaotic communication scheme has been on the rise around the world over the past twenty years. Researchers have proposed a variety of chaotic synchronization methods, such as identical synchronization (Pecora & Carroll 1990, Carroll & Pecora 1993), phase synchronization (Rosenblum et al. 1996), generalized synchronization (Kocarev & Parlitz. 1996) and projective synchronization (Xu & Chee 2002, Fink et al. 2000, Yu et al. 2006 Practical applications of chaotic synchronization, specifically securing communication systems have been reported in Ref. (Cai et al. 2010). Research on chaotic synchronization is the basis of secure communication. Many aspects should be considered when performing chaotic communication, such as the transmission channel, circuit matching, the encryption method, and anti-jamming measures.

Various kinds of chaos-based secure communication systems have been proposed (Wang & Gao 2010, Lvarez & Montoya 2004, Wang & Zhang 2006, Li & Zhang 2011, Liao & Hu 2012, Zhu 2010, Chen & Yang 2011, Mou et al. 2013). These can roughly be classified into the following categories: chaotic making, chaotic switching, chaotic modulation, digital communication, and the inverse system approach. A simple method is to add the information signal directly to the chaotic carrier, using the mixed signal to drive the receiver. In such circumstances, the magnitude of the useful information signals must be small enough to successfully execute chaotic masking. Although there is a linear relationship between coupling systems, projective synchronization cannot be classified as generalized synchronization because of the non-negative maximum conditional Lyapunov exponent. In 2006, Yu, Peng and Liu proposed a new method of projective synchronization for unidentical chaotic systems based on stability criterion of linear systems (Yu et al. 2006). Two unidentical chaotic systems can synchronize up to any desired scaling factor by a suitable separation of the systems. This method is suitable not only for three-dimensional coupled partially linear systems, but also for higher dimensional, or even hyper-chaotic systems. In this paper, we present a technical scheme of secure communication based on the Controlled Projective Synchronization Method (Yu et al. 2006). The linear and nonlinear terms of the driving system are separated and configured properly with all eigenvalues of the Jacobian matrix **A** having negative real parts. Then, the transmission sinusoidal information is injected into the nonlinear part with the scaling factor α as carrier signal. The responding system receives the masking information, decoding the sinusoidal information by projective synchronization at the receiving end. Different from traditional secure communication schemes, the transmitted chaotic signal at the sending end contains multiple signal vectors. Therefore, what we receive in the practical circuit equipment is a vector output containing useful information turned into any component of the chaos carrier wave. The new added scaling factor α is an arbitrary nonzero constant, adding complexity to the decoding information. As a result, the masking

*Corresponding author: *huiqian91@hotmail.com*

message is difficult to hack, thus achieving the goal of secure communication. Numerical simulations are carried out to confirm the feasibility and effectiveness of the proposed secure communication scheme in this paper.

2 CONTROLLED PROJECTIVE SYNCHRONIZATION BASED ON STABILITY CRITERION

Consider a chaotic system described by

$$\dot{\mathbf{x}} = \mathbf{f}(\mathbf{x}(t)) \tag{1}$$

where $\mathbf{x}(t) \in \mathbf{R}^n$ is an n-dimensional state vector of the chaotic system, and $\mathbf{f}: \mathbf{R}^n \to \mathbf{R}^n$ is defined as a vector field in n-dimensional space. The function $\mathbf{f}(\mathbf{x}(t))$ is decomposed as

$$\mathbf{f}(\mathbf{x}(t)) = \hat{\mathbf{A}}\mathbf{x}(t) + \hat{\mathbf{h}}(\mathbf{x}(t)) \tag{2}$$

where the function $\hat{\mathbf{A}}\mathbf{x}(t)$ is a linear term of $\mathbf{f}(\mathbf{x}(t))$. $\hat{\mathbf{h}}(\mathbf{x}(t))$ is the nonlinear term of $\mathbf{f}(\mathbf{x}(t))$. Then $\hat{\mathbf{A}}\mathbf{x}(t)$ is suitably disposed as

$$\hat{\mathbf{A}}\mathbf{x}(t) = \mathbf{A}\mathbf{x}(t) + \bar{\mathbf{A}}\mathbf{x}(t).$$

Here, it is required that $\mathbf{A}$ is a full rank matrix and all of its eigenvalues have negative real parts, and $\bar{\mathbf{A}}\mathbf{x}(t)$ is the remainder of the linear function $\hat{\mathbf{A}}\mathbf{x}(t)$.

Let the function

$$\mathbf{g}(\mathbf{x}(t)) = \mathbf{A}\mathbf{x}(t), \mathbf{h}(\mathbf{x}(t)) = \bar{\mathbf{A}}\mathbf{x}(t) + \hat{\mathbf{h}}(\mathbf{x}(t)).$$

Then, the system (1) as a driving system can be rewritten as

$$\begin{aligned} \dot{\mathbf{x}}(t) &= \mathbf{g}(\mathbf{x}(t)) + \mathbf{h}(\mathbf{x}(t)) \\ &= \mathbf{A}\mathbf{x}(t) + \mathbf{h}(\mathbf{x}(t)) \end{aligned} \tag{3}$$

For the given chaotic system (3), we construct a new system, which will act as a responding system:

$$\begin{aligned} \dot{\mathbf{w}}(t) &= \mathbf{g}(\mathbf{w}(t)) + \mathbf{h}(\mathbf{x}(t))/\alpha \\ &= \mathbf{A}\mathbf{w}(t) + \mathbf{h}(\mathbf{x}(t))/\alpha \end{aligned} \tag{4}$$

where $\mathbf{w}(t) \in \mathbf{R}^n$ is an n-dimensional state vector of the new created system. The scaling factor α is an arbitrary nonzero constant. We define the synchronization error as $\mathbf{e}(t) = \mathbf{x}(t)/\alpha - \mathbf{w}(t)$, the evolution of which is determined by the following equation

$$\begin{aligned} \dot{\mathbf{e}}(t) &= \dot{\mathbf{x}}(t)/\alpha - \dot{\mathbf{w}}(t) \\ &= \mathbf{A}(\mathbf{x}(t)/\alpha - \mathbf{w}(t)) \\ &= \mathbf{A}\mathbf{e}(t) \end{aligned} \tag{5}$$

Obviously, the zero point of $\mathbf{e}(t)$ is the equilibrium of the linear error system (5). Since all eigenvalues of the matrix $\mathbf{A}$ have negative real parts, according to the stability criterion of linear system, the zero point of scaling synchronization error $\mathbf{e}(t)$ is asymptotically stable and $\mathbf{e}(t)$ tends to zero when $t \to \infty$. Then, the state vectors $\boldsymbol{x}(t)$ and $\mathbf{w}(t)$ of driving system (3) and responding system (4) synchronize up to a constant factor α, such that

$$\lim_{t\to\infty} \|\mathbf{x}/\alpha - \mathbf{w}\| \to 0, \text{or} \frac{\mathrm{x}_i}{\mathrm{w}_i} = \alpha, i = 1, 2, \ldots, n \tag{6}$$

where x_i and w_i are variable components of the state vectors $\boldsymbol{x}(t)$ and $\boldsymbol{w}(t)$ respectively.

The configuration of $\mathbf{A}$ has a wide range, so due to its simplicity and agility, the method becomes simple and effective.

3 CHAOTIC MASKING TECHNIQUE

We then turn the nonlinear chaotic signal $\mathbf{h}(\mathbf{x}(t))$ of driving system (3) into $\mathbf{h}(\mathbf{x}(t))/\alpha$ with the processing of signal amplifier, injecting the transmission sinusoidal information into $\mathrm{h}_j(\mathrm{x}(t))/\alpha$ directly to obtain the hybrid driving signal $\boldsymbol{s}(t)$ mixed with chaotic signals and useful information

$$\begin{aligned} \mathbf{s}(t) = \mathbf{h}(\mathbf{x}(t))/\alpha + \mathbf{i}(t) &= \begin{bmatrix} \mathrm{s}_1 \\ \vdots \\ \mathrm{s}_j \\ \vdots \\ \mathrm{s}_n \end{bmatrix} \\ &= \begin{bmatrix} \mathrm{h}_1(\mathrm{x}(t))/\alpha \\ \vdots \\ \mathrm{h}_j(\mathrm{x}(t))/\alpha \\ \vdots \\ \mathrm{h}_n(\mathrm{x}(t))/\alpha \end{bmatrix} + \begin{bmatrix} 0 \\ \vdots \\ \mathrm{i}_m(t) \\ \vdots \\ 0 \end{bmatrix} \\ &= \begin{bmatrix} \mathrm{h}_1(\mathrm{x}(t))/\alpha \\ \vdots \\ \mathrm{h}_j(\mathrm{x}(t))/\alpha + \mathrm{i}_m(t) \\ \vdots \\ \mathrm{h}_n(\mathrm{x}(t))/\alpha \end{bmatrix} \end{aligned} \tag{7}$$

where $\mathbf{s}(t) \in \mathbf{R}^n$ is an n-dimensional vector of chaotic signal. Any attacker who wants to decode the information will run into difficulties when trying to discover where the useful information is hidden in a chaotic carrier. The information could be encrypted in any component $s_j(t)$. The chaotic carrier may scale up or down by a certain factor to make the cracking with a conventional synchronization method much more difficult. The projective synchronization is achieved between the driving system and the responding system when the chaotic signal $s(t)$ is transmitted to the receiver through the channel. After the responding system outputs the synchronous chaotic signal $\mathbf{h}'(\mathbf{w}(t)) = \mathbf{h}(\alpha\mathbf{w}(t))/\alpha$, the transmitted useful information $\mathrm{i}_m(t)$ can be regained at the receiver through $s(t)$ minus $\mathbf{h}'(\mathbf{w}(t))$. Figure 1 shows the encryption and decryption process of sinusoidal information.

The differential equation driven by the chaotic signal is

$$\dot{\mathbf{w}}(t) = \mathbf{A}\mathbf{x}(t) + \mathbf{s}(t) \tag{8}$$

The decoded information is

$$\begin{aligned} \hat{\mathrm{i}}_m(t) &= \mathbf{s}(t) - \mathbf{h}'(\mathbf{w}(t)) \\ &= \mathbf{s}(t) - \mathbf{h}(\alpha\mathbf{w}(t))/\alpha \end{aligned} \tag{9}$$

According to the analysis above, the hybrid signal $s(t)$ contains the nonlinear term scaled up or down by a factor α, an arbitrary nonzero constant of the driving system. Compared to the former scheme with the linear transmission signal, the new scheme increases the complexity of the decryption process, and therefore makes decoding more difficult.

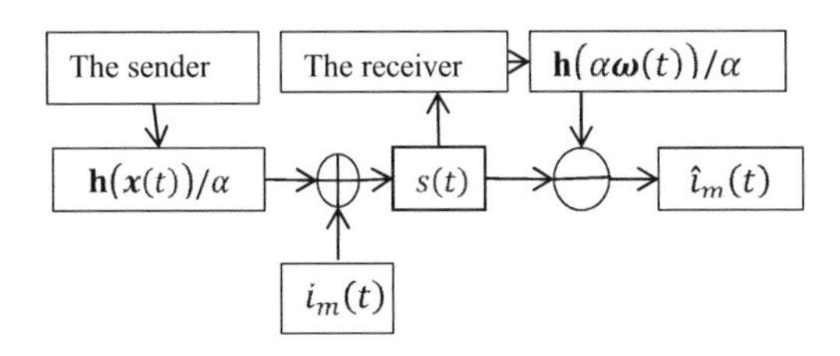

Figure 1. The encryption and decryption process through chaotic masking.

4 APPLICATION TO THE NUMERICAL EXAMPLE

The nonlinear differential equations of the Lorenz attractor are

$$\begin{aligned} \dot{x}_1 &= \sigma(x_2 - x_1) \\ \dot{x}_2 &= -x_1 x_3 + \gamma x_1 - x_2 \\ \dot{x}_3 &= x_1 x_2 - b x_3 \end{aligned} \tag{10}$$

The system (10) has chaotic behavior at the parameter values $\sigma = 16, \gamma = 45.92, b = 4$, decomposing Eq.(10) according to (3) into $\mathbf{g}(\mathbf{x}(t))$ and $\boldsymbol{h}(\boldsymbol{x}(t))$ as

$$\begin{aligned} \mathbf{g}(\mathbf{x}(t)) &= \mathbf{A}\begin{bmatrix} x_1 \\ x_2 \\ x_3 \end{bmatrix} = \begin{bmatrix} -\sigma & & \\ & -1 & \\ & & -b \end{bmatrix}\begin{bmatrix} x_1 \\ x_2 \\ x_3 \end{bmatrix}, \\ \mathbf{h}(\mathbf{x}(t)) &= \begin{bmatrix} \sigma x_2 \\ -x_1 x_3 + \gamma x_1 \\ x_1 x_2 \end{bmatrix} \end{aligned} \tag{11}$$

the diagonal matrix $\mathbf{A}$ has negative real eigenvalues $(-16, -1, -4)$. We take sinusoidal information as $\mathrm{i}_m(t) = a\sin(wt)$, injecting the information signal $\mathrm{i}_m(t)$ into the first component $h_1(x(t))/\alpha$ of nonlinear transmission signal $\mathbf{h}(\mathbf{x}(t))/\alpha$ in order to get the hybrid driving signal $s(t)$

$$\mathbf{s}(t) = \begin{bmatrix} \sigma x_2/\alpha \\ (-x_1 x_3 + \gamma x_1)/\alpha \\ x_1 x_2/\alpha \end{bmatrix} + \begin{bmatrix} \mathrm{i}_m(t) \\ 0 \\ 0 \end{bmatrix} \tag{12}$$

Then, we construct the receiving end:

$$\begin{aligned} \dot{w}_1 &= -\sigma w_1 + s_1(t) \\ \dot{w}_2 &= -w_2 + s_2(t) \\ \dot{w}_3 &= -b w_3 + s_3(t) \end{aligned} \tag{13}$$

The expression of the decoding information is:

$$\hat{\mathrm{i}}_m^{(1)}(t) = s_1(t) - h_1(\alpha w(t))/\alpha = \mathrm{i}_m(t) + \sigma e_2 \tag{14}$$

If the information signal is injected into the second component $h_2(x(t))/\alpha$ or the third component $h_3(x(t))/\alpha$, the decoding information at the receiving end is respectively as follows

$$\begin{aligned} \hat{\mathrm{i}}_m^{(2)}(t) &= s_2(t) - h_2(\alpha w(t))/\alpha \\ &= \mathrm{i}_m(t) - x_1 x_3/\alpha + \alpha w_1 w_3 + \gamma e_1, \\ \hat{\mathrm{i}}_m^{(3)}(t) &= s_3(t) - h_3(\alpha w(t))/\alpha \\ &= \mathrm{i}_m(t) + x_1 x_2/\alpha - \alpha w_1 w_2 \end{aligned} \tag{15}$$

The differential equation of synchronization error according to (5) and (12) is

$$\begin{bmatrix} \dot{e}_1 \\ \dot{e}_2 \\ \dot{e}_3 \end{bmatrix} = \begin{bmatrix} -\sigma & & \\ & -1 & \\ & & -b \end{bmatrix} \begin{bmatrix} e_1 \\ e_2 \\ e_3 \end{bmatrix} = \mathbf{A} \begin{bmatrix} e_1 + i_m(t) \\ e_2 \\ e_3 \end{bmatrix} \tag{16}$$

The synchronization error between the transmitter (drive) system and the receiver (response) system has asymptotic stability. This is based on the stability criterion of linear systems with all eigenvalues of a Jacobian matrix **A** having negative real parts.

Take parameters and initial conditions as:

$a = 15, \omega = 0.5, \alpha = 0.5$,

$(x_1(0), x_2(0), x_3(0)) = (6.0, 3.5, 1.5)$

and $(w_1(0), w_2(0), w_3(0)) = (2.0, 1.0, 0.5)$.

Figure 2 shows the time history of synchronization error $e_1(t), e_2(t), e_3(t)$ of the state variables $\mathbf{w}$ and $\mathbf{x}/\alpha$. It can be seen that $e_2(t)$ and $e_3(t)$ quickly converge to zero, while $e_1(t)$ is a trigonometric function curve in the time domain. The reason for the above situation is that the sinusoidal message $i_m(t)$ has a large amplitude which leads to the destruction of the synchronization result when the message is hidden in the first component $s_1(t)$ of the chaotic carrier $\mathbf{s}(t)$. However, the original information signal can still be recovered accurately according to the expression (14). This does not contain w_1 and x_1/α. The first component $s_1(t)$ of the transmitted hybrid driving signal $\mathbf{s}(t)$ is depicted in Figure 3. It shows that $s_1(t)$ is a chaotic signal and the sinusoidal message is well hidden, achieving a successful encryption. In Figure 4, it can be found that the recovered information $\hat{i}_m(t)$ stabilizes to the original information $i_m(t)$ after a short process. The time history of synchronization error between the original sinusoidal signal and the recovered information signal when the sinusoidal signal is injected into three components of $\mathbf{s}(t)$ respectively are shown in Figure 5. When the sinusoidal information is injected into the second component $s_2(t)$ of $\mathbf{s}(t)$, the recovered information $\hat{i}_m(t)$ has the shortest convergence time of the three situations. As a result, under the condition of not affecting decoding accuracy, it is optimal to place the sinusoidal information into the second component $s_2(t)$.

According to (16), the three synchronization error equations are independent of each other which means a group of state variables w_j and x_j/α $(j = 1,2,3)$ cannot synchronize due to the transmission sinusoidal signal, while the remaining two groups' projective synchronizations are not affected. From (12), each of three components $s_j(t)$ of $\mathbf{s}(t)$ does not contain a corresponding group of state variables w_j and x_j/α that are not synchronous, so the convergence and accuracy of the recovered information are not affected, no matter which transmission signal that we inject the sinusoidal message into. We then take the amplitude a, frequency ω and scale factor α as control parameters. With $a \in [0.01, 100], \omega \in [0.01, 100], \alpha \in [-100, 100]$, we then create a numerical simulation analysis of the encryption and decryption processes. It turns out the figure of signal synchronization error is in accordance with Figure 5. It is concluded that any of the sinusoidal messages, despite the frequency or the amplitude, can be chosen as the transmission information. The scale factor α can be chosen arbitrarily, and add complexity to the expression of decoding information, improving security.

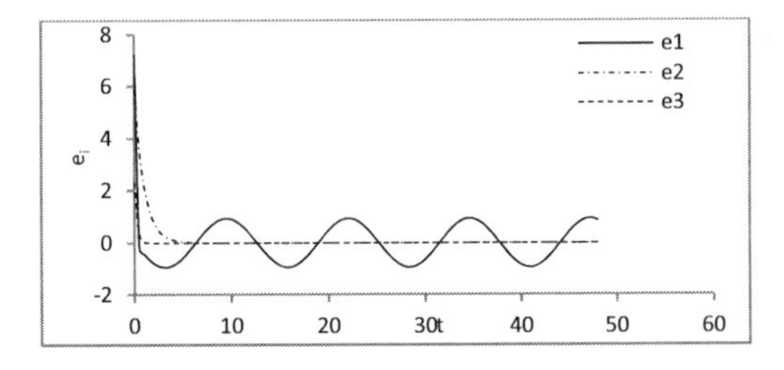

Figure 2. The projective synchronization error.

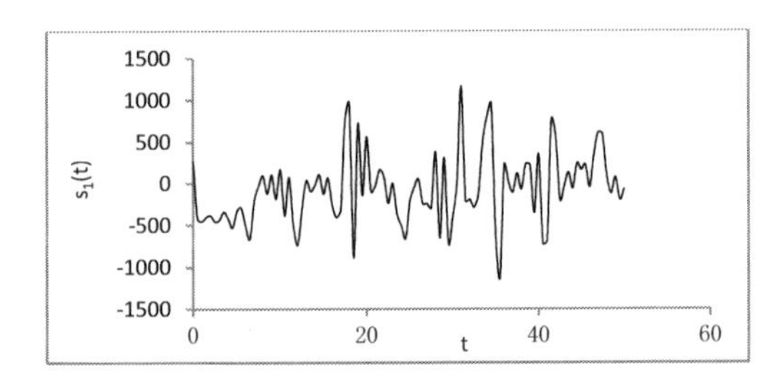

Figure 3. Chaotic signal $s_1(t)$.

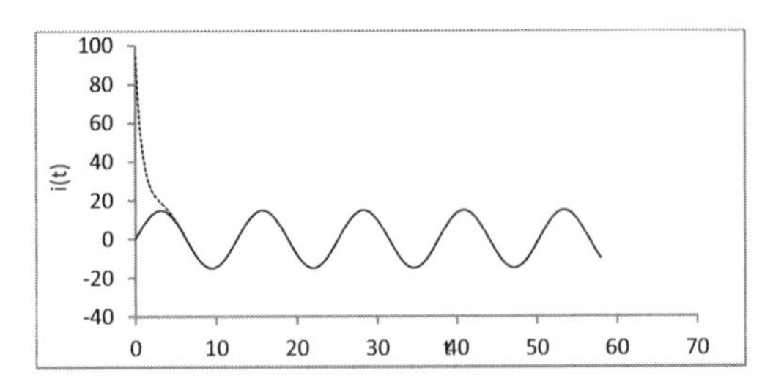

Figure 4. $\hat{i}_m(t)$ and $i_m(t)$.

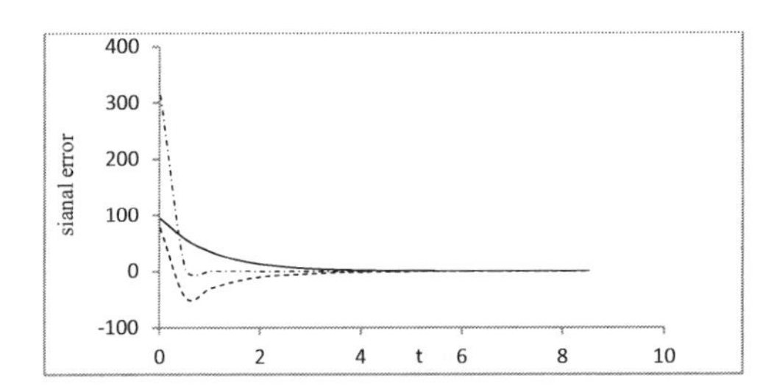

Figure 5. The signal error $\hat{i}_m(t) - i_m(t)$.

5 CONCLUSIONS

A new scheme of chaotic masking for secure communication is proposed in this paper. Out method of projective synchronization is based on the stability criterion of linear systems. The linear term and nonlinear terms of the driving system are separated and properly configured. The transmission sinusoidal information is injected into the nonlinear part with the scaling factor α as the carrier signal. The responding system receives the masked information, and the transmission sinusoidal information will be decoded through projective synchronization at the receiving end. Simulation results show that in the Lorenz system, the optimal configuration of the Jacobian matrix **A** is diagonal and the amplitude and frequency of the sinusoidal signal can be chosen from a wide range. Besides, the new added scale factor α makes the decryption process more complex, thus improving communication security. Numerical simulations with the Lorenz system confirm the feasibility and effectiveness of the method.

ACKNOWLEDGMENT

The research is supported by the National Natural Science Foundation of China (Grant No.10802030).

REFERENCES

Cai, J., Lin,M. & Yuan, Z. 2010. Secure communication using practical synchronization between two different chaotic systems with uncertainties. *Mathematical and Computational Applications*. 15: 166–175.

Carroll, T.L. & Pecora, L.M. 1993. System for producing synchronized signals. *Physica D* 67:126–140.

Chen Qi-shuan & Yang Li-na. 2011. Chaos synchronization of secure communication based on adaptive inverse control. *IEEE*: 525–527.

Fink,K.S., Johnson,C., Carroll,T., et al. 2000. Three coupled oscillators as a universal probe of synchronization stability in coupled oscillator arrays. *Phys.Rev.E* 61: 5080–5090.

Kocarev, L. & Parlitz ,U. 1996. Generalized synchronization and equivalence of unidirectionally coupled dynamical systems. *Phys. Rev.Lett.*76: 1816–1819.

Li Zhang-guo & Zhang Zheng. 2011. Secure communication scheme with hyper-chaotic Lorenz-Stenflo system. *IEEE*: 2272–2275.

Liao Ni-huan & Hu Zhi-hong. 2012. A hybrid secure communication method based on synchronization of hyperchaos systems. *IEEE*: 289–292.

Lvarez G & Montoya F. 2004. Cryptanalyzing a diecrete-time chaos synchronization secure communication system. *Chaos, Solutions and Fractal.* 21: 689–694.

Mou Jun, Li Peng, Zu Long-qi, et al. 2013. Control of chaos observer synchronization and the application in secure communication. *IEEE*: 568–572.

Pecora, L.M. & Carroll, T.L. 1990. Synchronization in Chaotic System. *Phys. Rev. Lett.* 64: 821–824.

Rosenblum, M.G., Pikovsky, A.S. & Kurths, J. 1996. Phase Synchronization of Chaotic Oscillators. *Phys Rev. Lett.*76: 1804–1807.

Wang Xing-yuan & Gao Yong-feng. 2010. A switch-modulated method for chaos digital secure communication based on user-defined protocol. *Commom Nonlinear SCI Number Simulat.* 15: 99–104.

Wang XM & Zhang JS. 2006. Chaotic secure communication based on nonlinear autoregressive filter with changeable parameters. *Phys.Lett A* 357: 323–329.

Xu, D. & Chee, C.Y. 2002. Secure digital communication using controlled projective synchronisation of chaos. *Phys.Rev.E* 66: 046218-1-5.

Yu, H.J, Peng, J.H. & Liu, Y.Z. 2006. Projective synchronization of unidentical chaotic systems based on stability criterion. *International Journal of Bifurcation and Chaos*. 16: 1049–1056.

Zhu Zi-qi. 2010. A highly robust chaotic synchronization scheme and its application in secure communication. *IEEE*: 4, 24–27.

Electronics, Communications and Networks IV – Hussain & Ivanovic (eds)
© 2015 Taylor & Francis Group, London, ISBN: 978-1-138-02830-2

The constellation condensing for signal space alignment in MIMO Y channel

Jiaju She, Xinling Wu & Liang Geng
Beijing GuoDianTong Network Technology Co. Ltd, Beijing, China

ABSTRACT: This paper proposes a procedure consisting of three standard steps to find the Constellation Condensing (CC) design for both symmetrical and asymmetrical transmission. CC simplifies the demodulation at the receiver in the classic MIMO Y channel with Signal Space Alignment for Network Coding (SSA-NC), which aligns two signals from different users into a signal spatial dimension at the relay. The modulation of the sum signal, which consists of a signal in M order modulation from one user and the other signal in N (M < N) order modulation from another user, can be condensed to an N order through easy and effective steps. The performance gain brought by the proposed design was validated through Monte-Carol simulations. The analysis shows that the CC not only ensures low-complexity, but also reduces the bit error rate.

1 INTRODUCTION

Relaying technique is the technique provides a wireless communication system with uniform high coverage and potential energy savings. It has been thoroughly investigated in the academia and widely employed in the industry. The relaying model is originally introduced by van der Meulen (1971) and substantially developed by Cover &El Gamal (1979). Two-way relaying (TWR) (Rankov & Wittneben 2007, Weng & Murch 2007)communication is a spectral-efficient relaying technique that completes the information exchange between one or more pairs of users through two phases: multiple accessing channel (MAC) phase and broadcasting channel (BC) phase (Chen & Yener 2009). However, both the multi-user interference (MUI) and the self-interference (SI) impair the desired signals at the receiving end. Owing to that wireless communication, systems are broadcast systems, signals from various users share the wireless medium concurrently and frequently create interferences to one another, leading to the MUI problem. A variety of signaling schemes have been designed to handle the MUI problem and improve the network transmission rate. Conventional zero-forcing algorithms, such as the block diagonalization (BD) scheme (Spencer et al. 2004) and the eigen-value based beam-forming (EBB) scheme (Fezali & proakis 1990), are usually employed to suppress MUI. SI refers the previously transmitted information about a user that returns to itself and can be removed by self-interference cancelation (SIC). The network coding (NC) approach is usually applied to suppress SI.

At present, NC in the TWR model is a focus. It provides a new perspective on the utilization of interferences. Physical-layer network coding (PNC) (Zhang et al. 2008) and analog-network coding (ANC) (Katti et al. 2007,2008) have been employed to boost the TWR channel (Rankov & Wittneben 2006, Hausl & Hagenauer 2006). Both the schemes allow two independent signals to be transmitted in the same dimension, and let the relay jointly process the sum of two signals interfering with each other. In the MAC phase, the relay receives and utilizes those interfering signals, and then broadcasts them to the end users of the side information to cancel the interference signals in the BC phase. The authors of (Zhao et al. 2012) studied PNC with spatial modulation and evaluated the average symbol error probability. A performance analysis of adaptive PNC is presented in (Muralidharan& Rajan2012).

Lee, Lim and Chun (Lee et al. 2010) proposed the signal space alignment for network coding (SSA-NC) in the MIMO Y channel with three users and an intermediate relay. As there is no direct link between users, each user is equipped with M antennas and communicates with one another via the relay equipped with N antennas. The degrees of freedom (DoF) were investigated to characterize the Gaussian channel capacity. The SSA-NC aligns two signals from different users into a signal spatial dimension. The capacity of the model is $3M\log(SNR)+o(\log(SNR))$, if $N \geq \lceil 3M/2 \rceil$. Subsequently, various extended channel models with SSA-NC have emerged. In (Lee et al. 2011) and (Ganesan et al. 2011), the DoF of different channel models with SSA-NC is investigated. In (She et al. 2012), we made a generalization of the applications of SSA-NC in general MIMO Y channel with K ($K \geq 2$) users. Assuming that each user is equipped with Mk antennas and user*i*and user *j* are to exchange nij signals with the help of the relay, the DOF of $\sum_{k=1}^{K} m_k$ could be achieved, when

$$N > \tfrac{1}{2}\sum_{k=1}^{K} m_k \left(m_k = \sum_{i \neq k}^{K} n_{ki}\right)$$

and $\left(M_i + M_j\right) \geq \left(N + n_{ij}\right)$. The authors of (Liu& Yang 2011) studied the signal alignment (SA) for the multi-carrier code division multiple (MC-CDMA) user TWR system. By exploiting the unique features of both TWR systems and MC-CDMA channels, they proposed a spectral-efficient SA signaling for MC-CDMA TWR systems, where each pair of users employs a maximal ratio transmitter of their counterpart to align their signal sat the relay. Asymptotic analysis and simulation results showed that the SA signaling supported more users with higher spectral efficiency comparing with NSA signaling. Wang et al. (2011) proposed a CC design at the relay for SSA-NC to ensure the low-complexity detection at the receiver in the symmetrical transmission. Exact expressions of symbol error rate (SER) were developed. The literature indicated that the CC designs are capable of enhancing the system performance. However, most of the designs are dedicated to the symmetrical transmission case.

This paper analyzes the constellation condensing design (Wang et al. 2011), and proposes three standard steps to find the CC design is proposed for both symmetrical transmission and asymmetric transmission. In a TWR system that uses SSA-NC, the relay receives a sum signal composed by a signal in M order modulation from one user and another signal in N order modulation from another user. To find the general CC designed, the first step is to arrange the observing constellation of the sum signal at the relay in a table; then, the observing constellation is grouped; the last step is to remap them to N order modulation constellation. Following the three steps, the modulation of the sum signal can be condensed into N order. The CC operation on the sum signal at the relay reduces the demodulation complexity and improves BER performance at the users. The remainder of this paper is arranged as follows. In section 2, the conventional 3-user MIMO Y channel with SSA-NC is reviewed. In section 3, the design is introduced and described in details. In section 4, the simulation results presented to show the performance gain. Finally, section 5 concludes this paper.

2 SYSTEM MODEL

In the conventional3-user MIMO Y channel every user exchanges a message, i.e., $n_{ij}=n_{ji}=1$, with each of the others via an intermediate relay. SSA-NCis employed to improve the spectral efficiency of the wireless resource. For example, user i($i \in \{1; 2; 3\}$) has $M_i = 2$ antennas, and the relay has $N = 3$ antennas. The transmission is split into two slots: the MAC phase and the BC phase as shown in Figure 1.

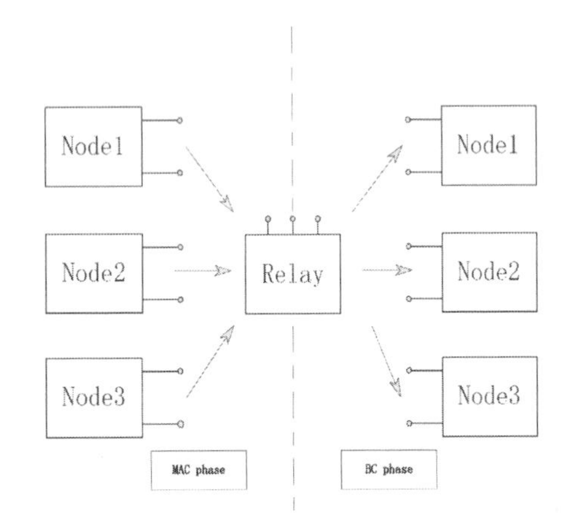

Figure 1. Channel model.

In the MAC phase, user i transmits the pre-coded signals to the relay. The received signals at the relay are given as

$$Y^{[R]} = \sum_{j \neq i, i=1}^{3} H^{[R,i]} V^{[j,i]} S^{[j,i]} + \eta^{[R]} \tag{1}$$

Where $H^{[R,i]}$ denotes the $N \times M_i$ channel matrix from user i to the relay, $S^{[j,i]}$ denotes the $n_{ji} \times 1$ transmitted signal vector from user i to user j, and $\eta^{[R]}$ denotes the additive white Gaussian noise (AWGN)matrix. The key concept of SSA-NC is to choose specific pre-coding vectors that allow the two paired signal streams from two users to be aligned to jointly perform detection and encoding of network coding at the relay. The total signal spatial dimensions are halved. Thus, the utilization efficiency of the wireless resource is improved.

$$span(H^{[R,i]} V^{[j,i]}) = span(H^{[R,j]} V^{[i,j]}) \tag{2}$$

Assuming that the intersection subspaces are denoted by Q_{ij}^{is}, $\left\|Q_{ij}^{is}\right\|^2 = 1$, an alignment algorithm for the design of the pre-coding matrices is given as

$$\begin{bmatrix} I_N & -H^{[r,j]} & 0 \\ I_N & 0 & -H^{[r,i]} \end{bmatrix} \cdot \begin{bmatrix} Q_{ij}^{is} \\ V_n^{[i,j]} \\ V_n^{[j,i]} \end{bmatrix} = 0 \tag{3}$$

When equation (2)stands, 6 independent signals are aligned into 3 spatial dimensions. The relay receives3 sum signals in MAC phase. With W_{ij} denoting the sum signal of $S^{[j,i]}$ and $S^{[i,j]}$, W_{12}, W_{13} and W_{23} are the three signals received by the relay.

In the BC phase, the relay condenses the constellation of the signals obtained in the last phase, and broadcasts them to the desired users by applying the zero-forcing(ZF) beam-forming scheme. The received signal sat the user are:

$$y^{[j,i]} = H^{[j,R]} F^{[j,i]} X^{[R]} + \eta^{[j,i]} \tag{4}$$

where $H^{[j,R]}$ denotes the $M_j \times N$ channel matrix from the relay to user j, $F^{[j,i]}$ denotes the MU-MIMO zero-forcing matrix, and $\eta^{[j,i]}$ denotes the $n_{ij} \times 1$ AWGN vector. The $X^{[R]}$is

$$X^{[R]} = \begin{bmatrix} W'_{12} \\ W'_{13} \\ W'_{23} \end{bmatrix}, \tag{5}$$

where W'_{12} denotes the sum signal after CC. The ZF algorithm used here is BD (Spencer et al. 2004), and $F^{[j,i]}$is obtained by:

$$F^{[j,i]} = nullspace\{H^{[k,R]}\} \; k \neq i, j \tag{6}$$

Then, user j receives W'_{ij}, by eliminating the self-interference, user j obtains the desired information transmitted from user i. In a similar way, every user obtains information from other users.

3 CONSTELLATION CONDENSING DESIGN

In (Wang et al. 2011), the authors propose an "XOR" CC design for symmetrical transmission. Take QPSK for example, if {00; 01; 11; 10} corresponds to $\left\{\frac{1+j}{\sqrt{2}}, \frac{1-j}{\sqrt{2}}, \frac{-1-j}{\sqrt{2}}, \frac{-1+j}{\sqrt{2}}\right\}$ the constellation points observed at the relay are $\sqrt{2}(1+j)$, $\sqrt{2}$, $\sqrt{2}(1-j)$, $\sqrt{2}j$, 0, $-\sqrt{2}j$, $-\sqrt{2}(-1+j)$, $-\sqrt{2}$, $-\sqrt{2}(-1-j)$ corresponding to {00 ⊕00}; {00 ⊕01; 01 ⊕00}; {01 ⊕01}; {00 ⊕10; 10 ⊕00}; {11 ⊕00; 00 ⊕11; 01 ⊕10; 10 ⊕01}; {11 ⊕01; 01 ⊕11}; {10 ⊕10}; {10 ⊕11; 11 ⊕10} and {11 ⊕11} respectively. Therefore, the 16 pairs are divided into 9 groups that are mapped to 9 points. Based on the "XOR" result, the nine groups are re-grouped to four, i.e.{00 ⊕00; 11 ⊕11; 10 ⊕10; 01 ⊕01}; {01 ⊕00; 00 ⊕01; 10 ⊕11; 11 ⊕10}; {11 ⊕00; 00 ⊕11; 01 ⊕10; 10 ⊕01} and {10 ⊕00; 00 ⊕10; 01⊕11; 11⊕01)}. After re-grouping, if the group and bit information of one signal is selected, the bit information of another signal is determined uniquely. Thus, the four groups can be mapped to points in QPSK, and the constellation is condensed at the relay. The receiver demodulates and decodes the received signal with self-information. This type of CC design is advantageous in that it simplifies the observing constellation and prevents the prohibitively high computational complexity of the demodulation at users. In addition, it brings BER performance gain. The number of the points in a constellation is reduced, and the Euclidean distances between them increase while the transmitting power at the relay stays the same. However, the authors did not consider the asymmetrical transmission case, where the paired users use different constellations. In this paper, we propose a general method to find the CC design for both symmetrical and asymmetrical transmission.

Table 1. 8PSK+2PSK.

8PSK\2PSK	0 (0.7071+0.7071i)	1 (–0.7071–0.7071i)
000 (–0.3827–0.9239i)	0.3244–0.2168i	–1.0898–1.6310i
001 (–0.9239–0.3827i)	–0.2168+0.3244i	–1.6310–1.0898i
011 (–0.9239+0.3827i)	–0.2168+1.0898i	–1.6310–0.3244i
010 (–0.3827+0.9239i)	0.3244+1.6310i	–1.0898+0.2168i
100 (0.3827–0.9239i)	1.0898–0.2168i	–0.3244–1.6310i
101 (0.9239–0.3827i)	1.6310+0.3244i	0.2168–1.0898i
111 (0.9239+0.3827i)	1.6310+1.0898i	0.2168–0.3244i
110 (0.3827+0.9239i)	1.0898+1.6310i	–0.3244+0.2168i

The "XOR" design illustrated above reveals that a pair in a group mapped to a point possesses a certainty for the receiver. When the receiver demodulates the symbol, it determines the bit information transmitted from the paired user with self-information. Taking the case of BPSK+8PSK as an example, the bit information of BPSK {0; 1} correspond to {0.7071+0.7071i;–0.7071–0.7071i}, and the bit information of 8PSK {000; 001; 011; 010; 100; 101; 111; 110}correspond to {–0.3827 –0.9239i;–0.9239–0.3827i;–0.9239+0.3827i;–0.3827+0.9239i;

0.3827−0.9239i; 0.9239−0.3827i; 0.9239+0:3827i; 0.3827+0.9239i}. The constellation of the sum signal is listed in Table 1, which shows no overlapped pairs. Grouping any two pairs from different columns and rows to be mapped to a point in 8BPSK can ensure the certainty. As different points in8PSK have the same theoretical BER, different grouping schemes have the same BER performance. For convenience, an approach like "XOR" that simplifies the decoding procession at the receiving user should be applied. For example, "XOR" the bit of BPSK with the first bit of 8PSK, and the groups are{000&0; 100&1}; {001&0; 101&1}; {011&0; 111&1},{010&0; 110&1}; {100&0; 000&1}; {101&0; 001&1},{111&0; 011&1}, and {110&0; 010&1}.

To generalize the grouping process into discrete steps, the case of QPSK+QPSK is chosen. The bit information of QPSK, {00; 01; 11; 10} corresponds to{1; 1j;−1;−1j}. The actual transmission hardly uses such constellation. However, the sum signal includes various possible conditions that may appear in the CC design. The grouping of this case is detailed below as three steps.

Step I. List the observing constellation of the sum signal and scale the pairs into different groups, as shown in Table 2. Obviously, there are cases where four pairs or two pairs overlap together. We marked the observing groups in the table, and 7 groups appear. For instance, G{A} denotes the group A.

Table 2: QPSK+QPSK.

QPSK\QPSK	00	01	11	10
00	2 G {F}	1+1i G{B}	0 G{A}	1−1i G{C}
01	1+1i G{B}	2i G{G}	−1+1i G{D}	0 G{A}
11	0 G{A}	−1+1i G{D}	−2 G{H}	−1-1i G{E}
10	1−1i G{C}	0 G{A}	−1−1i G{E}	−2i G{I}

Step II. Regroup the observing constellation. Each symbol denotes 2 bits of information for the receiver. QPSK was selected to be the condensed constellation. Since QPSK has only four points, the number of groups has to be reduced, i.e. some observing groups must be merged into one. Each four pairs selected to be grouped together should be from different rows and columns. As the observing groups consisting of more pairs have more restrictions, the re-grouping process should start with them. G{A} already have four pairs from different rows and columns, so we move to G{B}. The Two pairs in G{B} occupy two rows and columns. Thus the other two must be selected form the intersection of the other two rows and columns, leaving us two options, G{E} or G{H}&G{I}. They can be selected randomly, for example G{H}&G{I}. Then, G{F} and G{C} are left in the first column. Since G{C} occupies two rows and two columns and has more restrictions, only G{D} is suitable. At last, the remaining G{F} , G{G} and G{E} are grouped together naturally. In total3 grouping schemes exist for this case. As different points in QPSK have the same theoretical BER, different grouping schemes lead to different decoding process but result in the same BER performance at the receiver.

Step III. Map the re-grouped pairs to the QPSK constellation. This step is simple, and no more detail needs to be addressed here.

Generally given two signals, S_a and S_b in M and N order modulation respectively, the constellation points in them are denoted as $\{a_1,...,a_{2^M}\}$ and $\{b_1,...,b_{2^N}\}$, respectively. There are $2^M$and 2^N different points in constellations, $a_1 \neq a_2 \neq ... \neq a_{2^M}$ and $b_1 \neq b_2 \neq ... \neq b$. S_c denotes the sum signal, and $ci = a_m + b_n; (i \in \{1,2,...,2^M \times 2^N\})$. Although, $c_p = a_{m_p} + b_{n_p} = c_q = a_{m_q} + a_{n_q}$, $a_{m_p} = a_{m_q}$ and $b_{n_p} = b_{n_q}$ cannot be derived. The contradiction easily proves that 2^M pairs overlapped in the observing constellation at the relay. Assuming that more than 2^M pairs overlap, there must be more than 2^M distinguishing points in the M order modulation to move more than 2^M distinguishing points in the N order modulation to the same position. However, the M order modulation has only 2^M different points, hence that assumption is impossible. Naturally, we can conclude that at most 2^M pairs overlap in the observing constellation at the relay. Considering the extreme condition where every 2^M pairs overlap together, and points in the observing constellation of S_c can be scaled into 2^N groups, an N order modulation is required. Therefore, the sum signal, which is composed by an M order modulation signal from one user and an $N(M \leq N)$ order modulation signal from another user, can be condensed to an N order modulation.

The derivation of the BER is similar to that described in (Wang et al. 2011), thus it is not detailed here.

4 SIMULATION RESULTS

In this section, the Monte-Carlo simulation results are provided to assess the BER performance of the proposed CC design. We assume that the transmitting power in the MAC phase and in the BC phase is $P_{S^{[j,i]}} = P_{S^{[j,i]}} = P_{W_{ij}} = P$, and the AWGN at the relay and the users have the same variance, i.e.

$\sigma_r^2 = \sigma_{W_{ij}}^2 = \sigma_n^2$. The simulation results are displayed in terms of the ratio of the transmitting signal power to the noise variance ($SNR = P/\sigma_n^2$). In the simulation, all wireless channels are assumed as i.i.d. Rayleigh fading. The sum signals, each composed by an *M* order modulation signal and an *N* order modulation signal are condensed and mapped to the *N* order modulation.

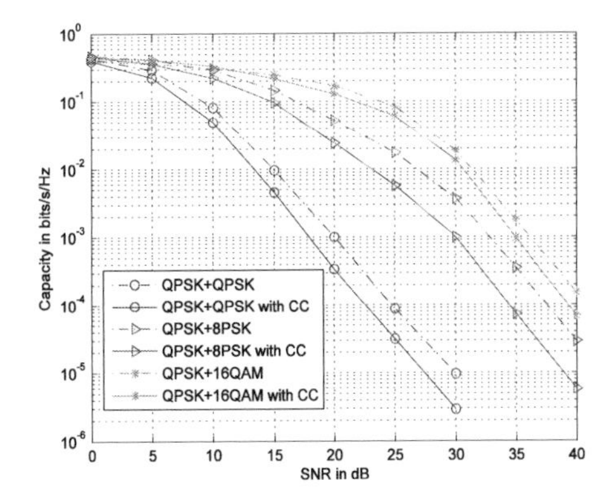

Figure 2. BER performance.

Figure 2 illustrates the simulation results for three sets of QPSK related assumptions, i.e. QPSK + QPSK, QPSK + 8PSK and QPSK +16QAM. Take QPSK + 8PSK as example, it indicates that one user employs the QPSK modulation, and the other one employs the 8PSK modulation. Each BER curve demonstrates the BER in one signal spatial dimension, and BER is calculated as

$$BER=\frac{E_N+E_M}{T_N+T_M}, \tag{7}$$

Where E_N and E_M denote the number of error bits in the N order modulation signal and the *M* order modulation signal, respectively, while TN andTM denote the total number of bits in them. As seen from the curves, the system BER performance is improved when CC is employed at the relay.The QPSK + 16QAM has almost 1dB gain, and QPSK + QPSK has almost 2dB gain, while QPSK + 8PSK has more than 2dB gain. The simulation result verifies that the CC design can indeed improve BER performance. The corresponding CC rules of the three simulation assumptions are detailed in the Table 3, 4 and 5. Due to the fact that there is no overlapping in the QPSK+8PSK and QPSK+16QAM,the CC designs are in a relatively casual manner.

The 16QAM modulation mapper can be referred to in (3GPP TS 36.211 2011). Different points in 16QAM have different BERs, thus different CC designs maybe lead to different BER performance. A more suitable grouping approach can be formulated to reduce the BER by following certain rules.

5 CONCLUSION

In this paper, a procedure consisting of three standard steps to find the CC design for any sum signal at the relay in the classic MIMOY channel with SSA-NC is proposed. Through the proposed steps, the observing constellation of synthetic signals can be condensed at the relay. The CC not only simplifies the demodulation, but also reduces the BER at the receiver. The simulation results are provided to verify the performance gain brought by the approach.

Table 3. QPSK+QPSK CC rule.

QPSK\QPSK	00	01	11	10
00($e^{\pi/4}$)	G{D}	G{B}	G{A}	G{C}
01 ($e^{7\pi/4}$)	G{B}	G{D}	G{C}	G{A}
11 ($e^{5\pi/4}$)	G{A}	G{C}	G{D}	G{B}
10 ($e^{3\pi/4}$)	G{C}	G{A}	G{B}	G{D}

Table 4. QPSK+8PSK CC rule.

8PSK\QPSK	00	01	11	10
000 ($e^{11\pi/8}$)	G{A}	G{B}	G{C}	G{D}
001($e^{9\pi/8}$)	G{B}	G{C}	G{D}	G{E}
011($e^{7\pi/8}$)	G{C}	G{D}	G{E}	G{F}
010($e^{5\pi/8}$)	G{D}	G{E}	G{F}	G{G}
100($e^{13\pi/8}$)	G{E}	G{F}	G{G}	G{H}
101($e^{15\pi/8}$)	G{F}	G{G}	G{H}	G{A}
111($e^{\pi/8}$)	G{G}	G{H}	G{A}	G{B}
110($e^{3\pi/8}$)	G{H}	G{A}	G{B}	G{C}

ACKNOWLEDGMENT

This article is supported by the science and technology project of State Grid Corp., "key technology research and application of smart grid supporting smart city".

Table 5. QPSK+16QAM CC rule.

16QAM\QPSK	00	01	11	10
0000	G{A}	G{B}	G{C}	G{D}
0001	G{B}	G{C}	G{D}	G{E}
0011	G{C}	G{D}	G{E}	G{F}
0010	G{D}	G{E}	G{F}	G{G}
0100	G{E}	G{F}	G{G}	G{H}
0101	G{F}	G{G}	G{H}	G{I}
0111	G{G}	G{H}	G{I}	G{J}
0110	G{H}	G{I}	G{J}	G{K}
1100	G{I}	G{J}	G{K}	G{L}
1101	G{J}	G{K}	G{L}	G{M}
1111	G{K}	G{L}	G{M}	G{N}
1110	G{L}	G{M}	G{N}	G{O}
1000	G{M}	G{N}	G{O}	G{P}
1001	G{N}	G{O}	G{P}	G{A}
1011	G{O}	G{P}	G{A}	G{B}
1010	G{P}	G{A}	G{B}	G{C}

REFERENCES

3GPP TS 36.211, V10.2.0, June 2011, Chapter 7.

Chen,M. &Yener, A. 2009.Multiuser Two-way Relaying: Detection and Interference Management Strategies. *IEEE Trans.Wireless Commun.* 8(8): 4296–4305.

Cover, T. &Gamal, A.E. 1979.Capacity Theorems for the Relay Channel.*IEEE Trans. Inform. Theory* 25: 572 –584.

Fezali,W. &Proakis, J. G. 1990.Adaptive SVD Algorithm for Covariance Matrix Eigenstructure Computation. *Proc. Int. Conf. Acoustics, Speech, Signal Processing, Albuquerque, NM, 3–6 AprIL 1990*:2615–2618.

Ganesan, R. S., Weber, T. & Klein A. 2011. Interference alignment in multi-User two way relay Networks. *Vehicular Technology Conference (VTC Spring), 2011 IEEE 73rd. Yokohama. May 2011*: 1–5. IEEE.

Hausl, C.& Hagenauer, J. 2006. Iterative network and channel decoding for the two-way relay channel. *IEEE International Conference on Communications. Istanbul. June 2006* (vol. 4):1568–1573. IEEE.

Katti, S., Gollakota, S.&Katabi, D. 2007. Embracing wireless interference: Analog network coding.*ACM SIGCOMM Computer Communication Review* 37(4): 397–408.

Katti, S., Rahul, H., Hu, W., Katabi, D., Medard, M. & Crowcroft, J. 2008. XORs in the air: Practical wireless network coding. *IEEE/ACM Transactions on Networking* 16: 497–510.

Lee K., Lee N. & Lee I. 2011. Feasibility Conditions of Signal Space Alignment for network coding on K-user MIMO Y channels. *Communications (ICC), IEEE International Conference on. Kyoto. June 2011*: 1–5. IEEE.

Lee N., Lim J. B. & Chun J. 2010. Degrees of Freedom of the MIMO Y Cahnnel: Signal Space Alignment for Network Coding. *IEEE Transactions on Information Theory* 56(7): 3332–3342.

Liu T. & Yang C. 2011. Signal Alignment for multicarrier Code Division Multiple User Two-way Relay Systems. *IEEE Transactions on Wireless Communications* 10(11): 3700–3710.

Muralidharan V. T. & Rajan B. S. 2012. Performance analysis of adaptive physical layer network coding for wireless two-way Relaying. *Personal Indoor and Mobile Radio Communications (PIMRC), 2012 IEEE 23rd International Symposium on. Sydney, NSW. Sept. 2012*: 596–602. IEEE.

Rankov, B. & Wittneben, A. 2006. Achievable rate regions of for the two-way relay channel.*IEEE International Symposium on Information Theory, Seattle, WA, 9–14 July 2006*: 1668–1672. IEEE.

Rankov,B. &Wittneben, A. N. 2007. Spectral efficient protocols for halfduplex fading relay channels.*IEEE J. Sel. Areas Commun.*25(2):379–389.

She J., Chen S., Hu B., Wang Y.&Su X. 2012. Practical conditions of signal space alignment for generalized MIMO Y channel. *Science China Information Sciences* 55(10): 1–13.

Spencer, Q. H.,Swindlehurst, A. L.&Haardt, M. 2004.Zero-Forcing Mthods for Downlink Spatial Multiplexing in Multiuser MIMO Channels.*IEEE Trans. Signal, Process.*52(2): 461–471.

Van der Meulen, E.C. 1971.Three-terminal Communication Channel.*Adv. Appl. Prob.*3: 120–154.

Wang N., Ding Z., Dai X. & Vasilakos A. V. 2011. On Generalized MIMO Y Channels: Precoding Design, Mapping, and Diversity Gain. *IEEE Transactions onVehicular Technology* 60(7): 3525–3532.

Weng,L. &Murch, R. D. 2007.Multi-user MIMO relay system with selfinterference cancellation.*Proc. IEEE WCNCKowloon, 11–15 March 2007*: 958– 962.

Zhang, S., Liew, S. C.&Lam, P. P.2006. Physical layer network coding. Available: arXiv: 0704.2475v1.

Zhang, S., Liew, S. C.&Lu, L. 2008.Physicallayer network coding schemes over finite and infinite fields. *Global Telecommunications Conference, IEEE GLOBECOM,New Orleans, LO, 30 Nov. – 4 Dec. 2008*: 1–6. IEEE.

Zhao Z., Peng M. & Wang W. 2012. Spatial modulation in two-way network coded channels: Performance and mapping optimization. *Personal Indoor and Mobile Radio Communications (PIMRC), 2012 IEEE 23rd International Symposium on. Sydney, NSW. Sept. 2012*: 72–76. IEEE.

Electronics, Communications and Networks IV – Hussain & Ivanovic (eds)
 ISBN: 978-1-138-02830-2

Performance of TCP variants over integrated satellite network and Multi-Hop MANET

Lina Shen, Yong Bai* & Liang Zong
College of Information Science & Technology, Hainan University, Haikou, Hainan, China

ABSTRACT: Geostationary (GEO) satellite networks can be integrated with multi-hop Mobile Ad hoc Networks (MANETs) to support new services such as local and remote marine communications. High latency and transmission errors are two main challenges when transporting data efficiently over such an integrated heterogeneous network. Different TCP performances need to be investigated in this new networking architecture. In this paper, TCP performances of four TCP variants (NewReno, SACK, Veno, and Hybla) were evaluated by simulations, and it was discovered that TCP Hybla exhibited the best performance in terms of throughput and download response time in such a networking environment.

1 INTRODUCTION

Geostationary (GEO) satellite communication has characteristics of wide coverage and easy user access without adversely affecting the natural environment. Satellite communications are an important means of communication in areas where it is difficult to construct terrestrial network facilities. Specifically, satellite communication systems have a unique superiority in maritime communications. A ship's communication relies primarily on the Inmarsat (international maritime satellite) system for communications out at sea. A MANET (mobile ad hoc network) is a continuously self-configuring, infrastructure-less network of mobile devices. It is applicable for conditions that were previously impossible or inconvenient for laying network infrastructure where there is a need for rapid network establishment. With a limited transmission power of network nodes, MANETa usually employ multi-hop transmissions for communication between its network nodes. An integrated heterogeneous network composed of a satellite network and multi-hop MANET can support both local wireless transmissions between mobile nodes and remote data delivery to terrestrial networks by using the satellite network as the transit network. The TRITON project in Singapore developed a low-cost and high-speed maritime ship-to-ship/shore IEEE 802.16-based mesh network to complement or replace satellite communications in narrow water channels or traffic lanes close to shorelines (Pathmasuntharam et al. 2008). To facilitate the communications of mobile users on fishing vessels, an integrated wireless networking system consisting of MANETs, cellular mobile networks, and satellite networks was established (Du et al. 2010; Bai & Du, 2013). In such an integrated heterogeneous network, the maritime MANET was set up for ship-to-ship communications in the fishing ship fleet, and the ship-to-shore communications were further supported by integrating the maritime MANET with a ship-borne satellite gateway in one ship of the fleet. In support of environmental and ecological research the architecture of a sensor network that uses satellite communication to transfer data from remote sensors was presented by Ye (2008).

The original design of the TCP transport protocol design primarily considered wired networks, where the bit error rate (BER) is low and packet losses are generally caused by congestion. When there is a packet loss, TCP adjusts its congestion window to reduce the data transmission speed and the network load to alleviate congestion on an RTT (round-trip time) basis. In a satellite link, the RTT significantly increases due to long-distance transmission and long propagation times. In a GEO network, the RTT can be 600 ms if both the forward and return links are via a satellite network. A satellite link has a high BER (between 10^{-9} and 10^{-5}), thus the packet error rate increases and seriously affects the performance of the transport layer. In the absence of knowledge about why a packet was lost, a loss event in TCP is always interpreted as an indication of congestion. When this happens, a TCP sender will reduce its sliding window size drastically to avoid congestion collapse, which can substantially reduce overall throughput.

There have been studies conducted to improve TCP performance over MANETs and satellite networks. TCP DOOR (detection of out-of-order and response)

*Corresponding author: *bai@hainu.edu.cn*

determines whether network routing changes occur and takes corresponding actions to detect both the sender and receiver sides of the TCP out-of-order packets (Wang & Zhang 2002). TCP-F (feedback) relies on RFNs (route failure notifications) and RRNs (route reestablishment notifications) to distinguish the routing failures and network congestion to improve TCP performance in MANETs (Chandran et al. 2001). TCP-ELFNs (explicit link failure notifications) make a distinction between a link disconnection and route disconnection with feedback from the lower layer (Holland & Vaidya 1999). A number of alternative congestion control algorithms have been proposed to improve TCP performance over GEO satellite networks. Veno monitors the network congestion level by RTT variations and uses this information to decide whether packet losses are likely to be due to congestion or random bit errors (Fu & Liew 2003). Other algorithms try to solve the problems of wireless channels. Specifically, Hybla mainly deals with long RTTs (Caini & Firrincieli 2004), and Westwood mainly deals with wireless link losses (Casetti et al. 2002).

The TCP transmission performance needs to be further investigated in the new networking architecture where GEO satellite networks are integrated with multi-hop MANETs. This paper includes comparisons of the performance of four TCP variants: TCP NewReno, SACK, Veno, and Hybla. The throughput and download response time of the four TCP variants versus the number of multiple hops at different BERs were evaluated. It was discovered that TCP Hybla exhibited the best performance in terms of throughput and download response time in such a networking environment.

The rest of this paper is organized as follows. Section 2 provides an overview of TCP Reno, NewReno, SACK, and Veno. Section 3 describes TCP Hylab over the satellite network. The performance of the four TCP variants by simulations is evaluated and discussed in Section 4. Finally, the conclusions are drawn in Section 5.

2 OVERVIEWS OF TCP VARIANTS

2.1 *TCP Tahoe and Reno*

TCP uses a mechanism called slow start (SS) to increase the congestion window (*cwnd*) after a connection is initialized and after a timeout. It starts with a window of two times the maximum segment size (MSS). For every packet acknowledged, the *cwnd* increases by one MSS so that the *cwnd* effectively doubles for every RTT. When the *cwnd* exceeds a threshold (*ssthresh*), the algorithm enters a new state called congestion avoidance (CA). In the CA phase, *cwnd* is additively increased by one MSS every RTT as long as non-duplicate ACKs are received. When a packet is lost, the likelihood of duplicate ACKs being received is very high. The behavior of Tahoe and Reno differ in how they detect and react to packet loss. In Tahoe, triple duplicate ACKS are treated the same as a timeout. Tahoe will perform fast retransmit, set *ssthresh* to half the current *cwnd*, reduce *cwnd* to 1 MSS, and then reset to the SS state. In Reno, if three duplicate ACKs are received, Reno will halve the *cwnd* (Tahoe sets it to 1 MSS), set the *ssthresh* equal to the new *cwnd*, perform a fast retransmit, and enter a phase called fast recovery (Allman & Stevens 1999). In the fast recovery state, TCP Reno retransmits the missing packet that was signaled by three duplicate ACKs, and waits for an ACK of the entire transmit window before returning to CA. If there is no ACK, TCP Reno experiences a timeout and enters the SS state. If an ACK times out, slow start is used as it is with Tahoe. Both algorithms reduce *cwnd* to 1 MSS in a timeout event. Hence, both TCP Tahoe and Reno treat the packet loss as a manifestation of network congestion, and reduce the transmission rate to maintain the stability of the network. This is reliable for wired networks because BERs are very low in wired links. However, wireless networks can cause random packet loss, especially in a long-delay satellite link where BERs can be as high as 10^{-4}. TCP Tahoe and Reno will result in TCP performance degradation because random packet loss is treated as congestion.

2.2 *TCP New Reno and TCP SACK*

TCP NewReno is the most widely adopted TCP variant (Floyd & Gurtov 1999). TCP NewReno further modifies TCP Reno within the fast recovery algorithm to solve timeout problems when multiple packets are lost from the same window. When multiple packets are lost, a partial ACK recognizes some but not all packets that are outstanding at the start of a fast recovery. Because it takes the sender out of fast recovery, the sender has to wait until timeout occurs. In the presence of multiple losses, TCP Reno and NewReno retransmit a maximum of one lost packet per RTT. To overcome this problem, the selective acknowledgment (SACK) option for TCP was proposed (Mathis et al. 1996). With this option, the sender can recover more than one packet per RTT. The receiving TCP sends back SACK packets to the sender informing the sender that non-continuous blocks of data have been received. The sender can then retransmit only the missing data segments. This ability is extremely useful when dealing with large *cwnds* (where multiple losses are frequent) and long RTTs in satellite networks.

2.3 *TCP Veno*

TCP Veno was proposed containing a mechanism to distinguish between congestion loss and random packet loss due to bit errors, and provides different measures to deal with them (Fu & Liew 2003). Veno makes use of a mechanism to estimate the state of the connection that deduces what kind of packet loss, congestion loss, or error loss is most likely to have occurred rather than preventing packet loss. If packet loss is detected while the connection is in the congestive state, Veno assumes the loss is due to congestion; otherwise, it assumes the loss is due to an error.

First, Veno estimates *BaseRTT* as the RTT of a segment when the connection is not congested: the minimum of all measured round trip times. Next, the sender measures the so-called Expected and Actual rates:

$$Expected = cwnd/BaseRTT \tag{1}$$

$$Actual = cwnd/RTT \tag{2}$$

and obtains the difference between them:

$$Diff = Expected - Actual \tag{3}$$

When $RTT > BaseRTT$, there is a bottleneck link where the packets of the connection accumulate. The backlog in the queue, denoted by N, can be calculated as:

$$N = Actual\times(RTT\text{-}BaseRTT) = DIFF\times BaseRTT \tag{4}$$

N is used as an indication of whether the connection is in a congestive state. When packet loss is detected, if $N < \beta$ (available bandwidth underutilized), it assumes an error loss; if $N \geq \beta$ (available bandwidth fully utilized), it assumes a congestion loss. The value of β is set at 3. In Veno, *BaseRTT* is reset whenever a packet loss is detected either due to timeout or duplicate ACKs.

3 TCP HYBLA IN SATELLITE NETWORK

In 2004, Carlo Caini, et al. proposed a new TCP algorithm, TCP Hybla, with the main purpose of eliminating penalization of TCP connections that incorporate a high-latency satellite radio link with long RTTs (Caini & Firrincieli 2004).

TCP Hybla adopts a congestion window growth factor ρ, which is proportional to the RTT. The Hybla aims at attaining the same instantaneous segment transmission rate $B(t)$ for long RTT connections (e.g., satellite, wireless, multi-hop networks, etc.) as a relatively fast reference TCP connection (e.g., a wired one).

$$B(t) = W(t)/RTT \tag{5}$$

In order to achieve this goal, the transmission rate of $B(t)$ is independent of the RTT; and the congestion window, $W(t)$, needs a faster growth rate. Its congestion window adjusts according to a normalized round trip time, ρ, which represents the ratio between the actual RTT and the round trip time of the reference connection, denoted by RTT_0.

$$P = RTT/RTT_0 \tag{6}$$

Such normalization helps to remove the performance dependence on RTTs. It has been proved that the TCP Hybla congestion control mechanism using the following algorithm alternative to the standard congestion control algorithm can obtain the same instantaneous segment transmission rate $B(t)$ of ideal connection:

$$W_{i+1} = \begin{cases} W_i + 2^{\rho} - 1, & \text{SS} \\ W_i + \rho^2 / W_i, & \text{CA} \end{cases} \tag{7}$$

where W_i denotes the *cwnd* with the reception of the i-th ACK.

TCP Hybla makes the congestion window $W(t)$ independent of the RTT, so it can effectively avoid the impact of long RTTs on throughput.

4 SIMULATION RESULTS

Apart from the modification of the congestion control rules, Hybla includes several other enhancements, including the mandatory adoption of the SACK policy, the use of timestamps, the adoption of Hoes channel bandwidth estimates, and the implementation of packet spacing techniques. The SACK policy is mandatory because a larger average *cwnd* is expected for long RTT connections. Large *cwnd*s also frequently cause severe inefficiencies of the "exponential back-off" RTO policy. These can be avoided by resorting to timestamps. It is possible to have fresh RTT estimates after timeouts, thus limiting any unnecessary use of the RTO exponential back off algorithm. Bandwidth estimates are used in order to appropriately set the initial *ssthresh*. Packet spacing helps counteract the bursts associated with large *cwnd*s.

4.1 *Topological and simulation setup*

The simulated network scenario was an integrated heterogeneous network for maritime communications. The network topology diagram is shown in Figure 1. The networking model consisted of satellite networks, multi-hop MANET of ships, and a terrestrial network. The MANET was composed of ships far away from shore that could only be connected via satellite for remote terrestrial communications. Users (clients) on the ships first connected to an access point, and then connected to the terrestrial network for data delivery from the server via the ship-borne satellite gateway.

The above networking scenario was modeled by OPNET. The one-way propagation delay of the GEO satellite link was 250 ms with a satellite link bandwidth of 1.54 Mbps. The BERs of the satellite links were 10^{-9} to 10^{-5}. The terrestrial server sent a 12.5 Mb file to the remote multi-hop MANET. Four TCP variants (NewReno, SACK, Veno & Hybla) were compared. The maximum transmission unit (MSS) was set to 536 bytes. The RTT_0 of Hybla was set at 75 ms. The throughput and download response times were evaluated with respect to different satellite link BERs and the number of hops.

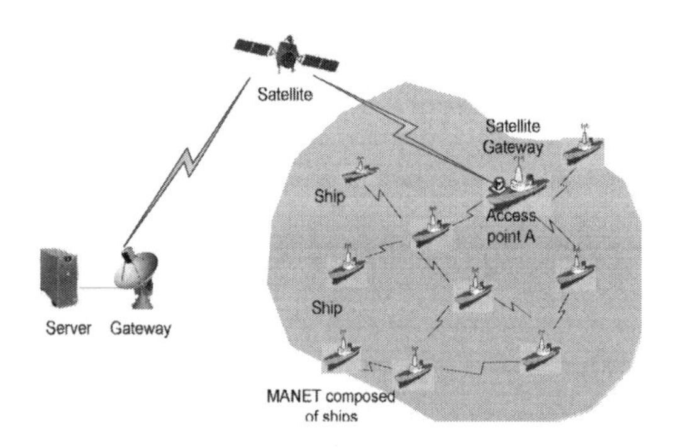

Figure 1. An integrated heterogeneous network consisting of the satellite network, MANET, and terrestrial network.

4.2 *Simulation results*

Figure 2 shows the throughputs of four TCP variants (NewReno, SACK, Veno, & Hybla) versus BERs from 10^{-9} to10^{-5} with different hops. It was observed that the throughput decreased with the increase of the BER. The throughputs of Hybla and Veno were considerably higher than that of SACK and NewReno. With the increase of the number of hops, the TCP throughput was reduced with all TCP variants. The throughputs of Hybla and Veno were significantly reduced; the throughput of Hybla dropped to about 70% of the original performance and the throughput of Veno dropped to about one-third to half of the original performance. The throughputs of SACK and NewReno dropped to about 80% of the original performances. Among the four TCP variants, the throughput of TCP Hybla was the highest with respect to the different number of hops and BERs.

Figure 3 illustrates the throughputs of the four TCP variants for three hops and various BERs (10^{-9} to 10^{-5}). When the BER was 10^{-9} to 10^{-7}, the throughput of Hybla was 4% to 11% higher than that of Veno. At a higher BER range (10^{-6} to 10^{-5}), the performance of Hybla was 30% higher than that of Veno. It is interesting to note that, at a higher BER (10^{-5}), the throughput of Hybla was two times that of Veno and four times that of TCP SACK and NewReno. When the BER increased, the throughput of Hybla increased more drastically than the other TCP variants, and TCP Hybla outperformed all the other algorithms.

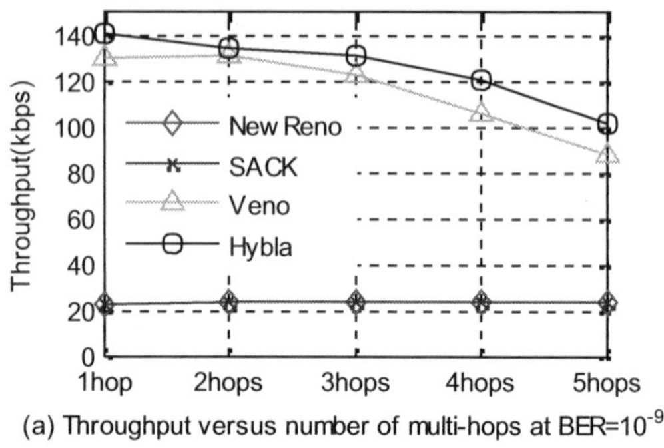

(a) Throughput versus number of multi-hops at BER=10^{-9}

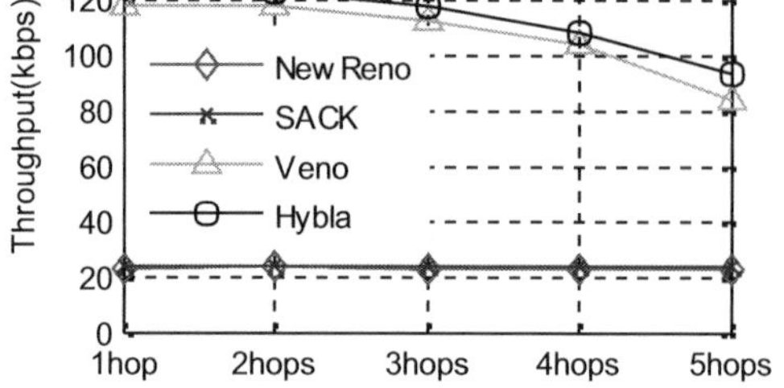

(b) Throughput versus number of multi-hops at BER=10^{-7}

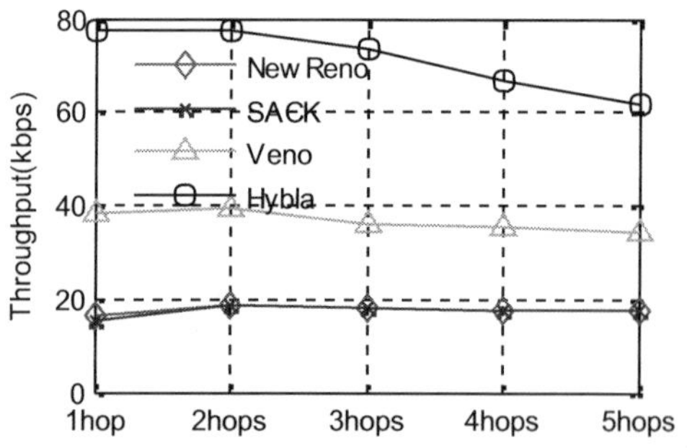

(c) Throughput versus number of multi-hops at BER=10^{-5}

Figure 2. Throughputs versus different number of hops and BERs.

Figure 4 shows the download response time of the four TCP variants with various BERs (10^{-9} to 10^{-5}) and the number of hops. When the BER increased gradually, the download response time increased gradually.

With an increase in the number of hops, the download response time increased slightly. The performance difference of the four TCP variants was not significant when the BER changed from 10^{-9} to 10^{-7}, but the download response time increased rapidly when the BER changed from 10^{-6} to 10^{-5}. This showed that the download response times of Hybla and Veno exhibited only 15%-28% that of SACK and NewReno. When BER was 10-5, the download response time of Hybla was half of Veno, and one-quarter that of SACK and NewReno. TCP Hybla achieved the best performance in terms of download response times.

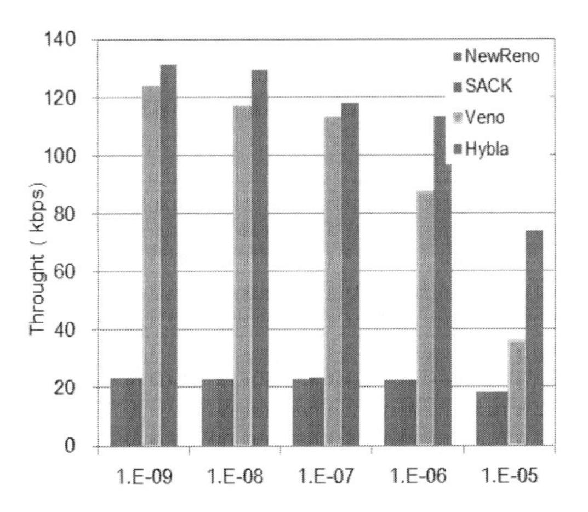

Figure 3. Throughputs of TCP variants at different BERs.

5 CONCLUSIONS

Though there are many options of TCP variants for wireless networks, their effectiveness needs to be further investigated in regards to GEO satellite networks integrated with multi-hop MANETs. In this paper four TCP variants, NewReno, SACK, Veno, and Hybla, were evaluated by simulation for such a networking scenario. The simulation results revealed that the TCP performance of the Hybla algorithm was better than the other three TCP variants in terms of throughput and download response times. When the BER was changed from 10^{-9} to 10^{-7}, the performance of Hybla was slightly higher than that of Veno, and drastically higher than that of NewReno and SACK. With high BERs (10^{-6} to 10^{-5}), the performance of Hybla was considerably higher than that of Veno, NewReno, and SACK. Hence, Hybla was relatively effective in the integrated heterogeneous network, and future research on improving TCP performance over such an integrated network should be based on this TCP variant. Future work will concentrate on a novel TCP scheme that is able to deal with challenges (long propagation delay and random losses over satellite links, multi-hop delays, lost channels, routing failures, and network partitions in MANETs) in such an integrated heterogeneous networking scenario.

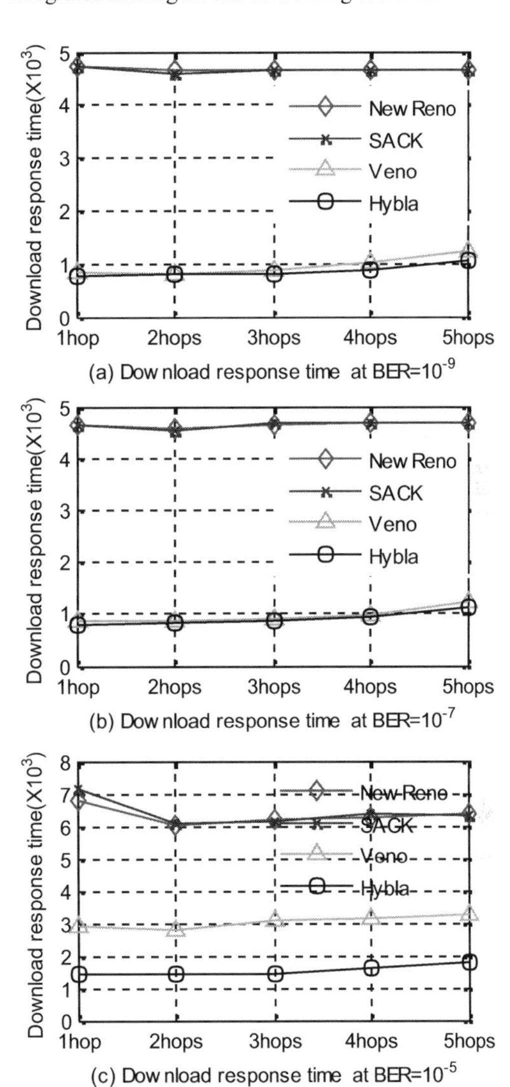

Figure 4. Download response time of four TCP variants for various BERs (10^{-9} to 10^{-5}) and different number of hops.

ACKNOWLEDGMENT

This paper was supported by the National Natural Science Foundation of China (Grant No. 61062006 and Grant No. 61261024) and the Special Social Service Project Fund of Hainan University, China (Grant No. HDSF201301).

REFERENCES

Allman, M. & Stevens, W. 1999. TCP congestion control. *IETF RFC*: 2581.

Bai, Y. & Du, W. 2013. VoIP services for ocean fishery vessels over integrated wireless and wireline networks. *In 2013 IEEE 24th International Symposium on Personal Indoor and Mobile Radio Communications (PIMRC)*: 3461–3465.

Caini, C. & Firrincieli, R. 2004. TCP Hybla: a TCP Enhancement for Heterogeneous Networks, *International Journal of Satellite Communications and Networking*22(5): 547–566.

Casetti, C., Gerla, M. Mascolo, S., Sanadidi, M. Y. & WangR. 2002. TCP Westwood: end-to-end congestion control for wired/wireless networks. *Wireless Networks* 8(5): 467–479.

Chandran, K., Raghunathan, S. Venkatesan, S. & Prakash, R. 2001. A feedback based scheme for improving TCP performance in ad-hoc wireless networks. *Proceedings of the 18th International Conference on Distributed Computing Systems, Amsterdam, USA*: 472–479.

Du, W., Ma, Z. Bai, Y. Shen, C. Chen, B. & Zhou, Y. 2010. Integrated Wireless Networking Architecture for Maritime Communications. In *Proc. 11th IEEE ACIS International Conference on Software Engineering Artificial Intelligence Networking and Parallel/ Distributed Computing (SNPD), London, 9–11 June 2010*: 134–138.IEEE.

Floyd, S., Henderson, T. & Gurtov, A. 1999. The NewReno modification to TCP's fast recovery algorithm. *IETF RFC*: 2582.

Fu,C. P. & Liew, S. C. 2003. TCP Veno: TCP enhancement for transmission over wireless access networks. *IEEE J. Sel. Areas Commun.* 21(2):216–228.

Holland, G. & Vaidya, N. H. 1999. Analysis of TCP performance over mobile ad hoc networks. *Proceedings of the 5th Annual ACM/IEEE International Conference on Mobile Computing and Networking, Seattle*: 219–230.

Mathis, M., Mahdavi, J., Floyd, S. & Romanow, A. 1996. Romanow TCP selective acknowledgment options. *IETF RFC*: 2018.

Pathmasuntharam, J. S., Kong, P. Y., Zhou, M. T., Ge, Y., Wang, H., Ang, C. W. & Harada, H. 2008. TRITON: High speed maritime mesh networks. *In Proc. 19th IEEE International Symposium on Personal, Indoor and Mobile Radio Communications, Cannes, 15–18 September 2008*: 1–5.IEEE.

Wang, F. & Zhang, Y. 2002. Improving TCP performance over mobile ad-hoc networks with out-of-order detection and response. *Proceedings of the 3rd ACM International Symposium on Mobile Ad Hoc Networking & Computing. Lausanne, New York, June 2002*: 217–225. ACM.

Ye, W., Silva, F., DeSchon, A. & Bhatt, S. 2008. Architecture of a Satellite-Based Sensor Network for Environmental Observation. *In Proc. of the Earth Science Technology Conference (ESTC), Maryland, June 2008*: 1560–1565.

Electronics, Communications and Networks IV – Hussain & Ivanovic (eds)
© 2015 Taylor & Francis Group, London, ISBN: 978-1-138-02830-2

Dual-band circularly polarized L-shaped dielectric resonator antenna

Wenhui Shen, Jie Liu, Jian Wu & Kang Yang
School of Communication and Information Engineering, Shanghai University, Shanghai, China

ABSTRACT: A dual-band Circularly Polarized (CP) Dielectric Resonator Antenna (DRA) is presented. The structure of the antenna is that L-shaped Dielectric Resonator (DR) is fed by a Y-shaped microstrip line. Each arm of the DR generates a resonant frequency, so we can adjust each of the resonant frequency by adjusting the sizes of the arm. The simulated impedance bandwidths (S11 ≤ -10dB) are 11.9–13.7GHz and 14.9–16.4GHz, which suit for satellite broadcasting, aeronautical radio navigation service, and radio positioning. Circularly Polarized design is achieved by adjusting the two orthogonal feeding stubs of Y-shaped microstrip line. Each of the simulated axial ratio (AR ≤ 3dB) bandwidth is 11.8–12.3GHz and 14.3–14.8GHz. The DRA can be used as uplink and downlink bands linking the mobile station antenna on the ground and the satellite communication system.

1 INTRODUCTION

In recent years, the demand for better antenna performance, such as wide bandwidth, low loss and miniaturization is increasing with the development of modern wireless communication.

DRA has been widely studied because of a number of advantages, such as small size, light weight, low cost, ease of excitement and high radiation efficiency (Luk & Leung 2003, Petosa 2007). Early studies of DRA mainly concentrated on linearly polarized operation. But in order to establish a good satellite communication links, most satellite communications and navigation systems have paid more attention to the CP DRA. A number of CP DRAs are proposed with shapes like a cylinder, rectangle (Pan& Leung 2012, Fakhte et al. 2014, Motevasselian 2013), annulus (Li & Leung 2013), trapezoid (Pan & Leung 2010), etc. Several dual-band CP DRAs have been designed using different DR shapes and feeding structures. In (Fang & Leung 2012), a doubly-fed dual-band circularly polarized DRA is introduced. It uses a dual-band 90° coupler to provide two quadrature signals. It significantly increases the size and complexity of the system. In (San et al. 2012), a dual-band slot-coupled CP DRA is investigated. Although this design has a quite narrow AR with 1.4% only in the upper band, its feeding network is simple.

In this paper, an L-shaped DRA with a simple Y-shaped microstrip line is proposed. The L-shaped DRA is composed of two rectangular dielectric resonators of the same widths but different lengths. The dual-band is achieved by adjusting the lengths and widths of these two DRs. The CP radiation is obtained by adjusting the lengths and widths of two feeding stubs of Y-shaped microstrip line to excite two orthogonal modes of equal amplitudes and 90° phase difference (Li et al. 2009). Good radiation performance suitable for satellite communication system, radio positioning and aeronautical radio navigation service are obtained, and details of the antenna design are discussed.

2 ANTENNA CONFIGURATION

The configuration of the proposed antenna is shown in Figure 1. If the size of a known rectangle DR is the length of a, widths of b, height of h and relative dielectric constant of ε_r, the resonance frequency can be calculated as follows (Mongia 1992):

$$K_x = \frac{\pi}{a} \tag{1}$$

$$K_z = \frac{\pi}{2h} \tag{2}$$

$$K_y \tan\left(K_y b/2\right) = \sqrt{(\varepsilon_r - 1)K_0^2 - K_y^2} \tag{3}$$

$$K_0 = \frac{2\pi}{\lambda_0} \tag{4}$$

$$f_0 = \frac{C}{2\pi\sqrt{\varepsilon_r}}\sqrt{K_x^2 + K_y^2 + K_z^2} \tag{5}$$

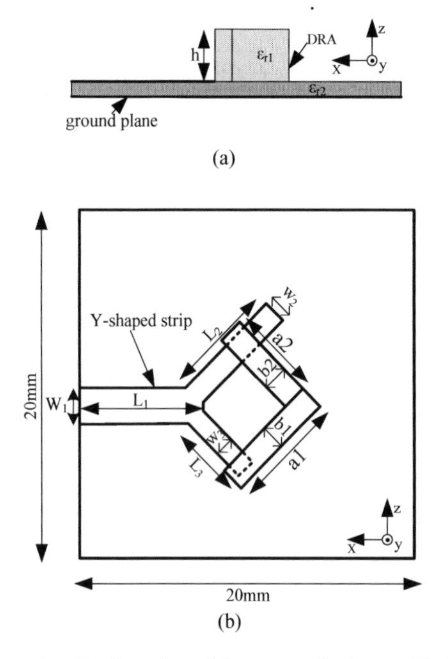

Figure 1. Configuration of the proposed antenna: (a) side view and (b) top view.

According to these formulas, given $\varepsilon_{r1} = 16$, h = 4.9mm, a1 = 10mm and b1 = 1.5mm, the resonance frequency of the DRA is fr = 16GHz. If a2 = 8mm, b2 = 3mm, then fr = 13GHz. Adjusting the lengths and widths of the two arms, the L-shaped DRA can operate in different frequency. We simulated the antenna with Ansoft HFSS Microwave Studio software. Good performance was obtained when a1 = 10.3mm, b1 = 2mm; a2 = 7mm, b2 = 2mm, h = 4.9mm.

The DR was mounted on an FR4 substrate with relative dielectric permittivity of $\varepsilon_{r2} = 4.4$ and size of 20mm × 20mm × 1.6mm. The DRA is fed by a Y-shaped microstrip line printed on the FR4 substrate. Two orthogonal modes of equal amplitudes and 90° phase difference are mainly controlled by adjusting the lengths and widths of two feeding stubs. After optimization and analysis, an example is selected to show the performance of the proposed antenna. In this example, the size of Y-shaped microstrip is $L_1 = 14$mm, $W_1 = 4.7$mm, $L_2 = 10$mm, $W_2 = 2.3$mm, $L_3 = 6$mm, $W_3 = 2$mm.

3 RESULTS AND DISCUSSIONS

Figure 2 displays the simulated return loss of the antenna. The antenna resonates at 12.9GHz (that the simulated return loss under -10dB ranges 11.9 ~ 13.7GHz, 14% bandwidth), which covers downlink band of satellite broadcasting in ITU3 area. In the upper band, it resonates at 15.8GHz (that the simulated return loss under -10dB ranges 14.9 ~ 16.4GHz, 9.6% bandwidth), which covers uplink band of satellite broadcasting, aeronautical radio navigation service band, and radio positioning band.

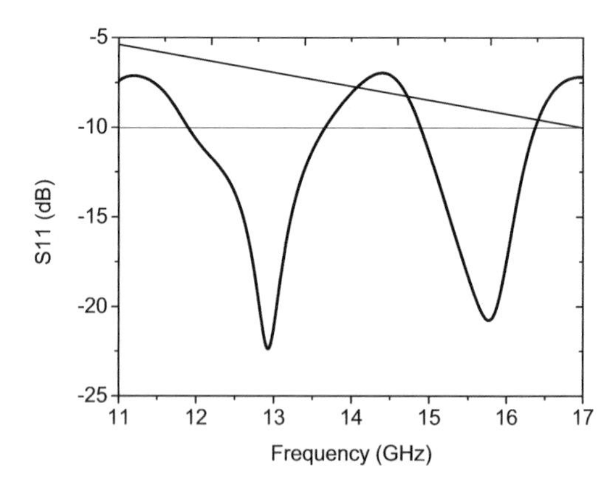

Figure 2. Simulated return loss of the proposed antenna.

In order to obtain a dual-band design with good performance, various design parameters were adjusted carefully. It has been mentioned that the lengths of two arms of L-shaped DRA are mainly responsible for the upper band, while the widths of two arms have an impact on the lower band as well as the upper band. The effects of various a1, a2, b1 and b2 on the antenna performance were studied here, with only one parameter variable at a time.

Figure 3 shows the simulated return loss for different a1 and a2. As expected, the upper band is more sensitive to a1 and a2 than the lower band. So the lengths of a1 and a2 are useful to optimize the performance of the upper band without affecting the lower one.

The effects of b1 and b2 on the return losses are investigated in Figure 4. In contrast to a1 and a2, both bands change significantly when b1 and b2 change. With the increase of b1 and b2, both bands move to

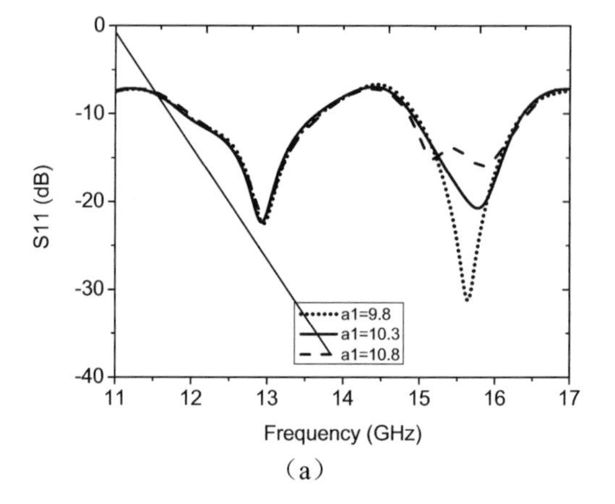

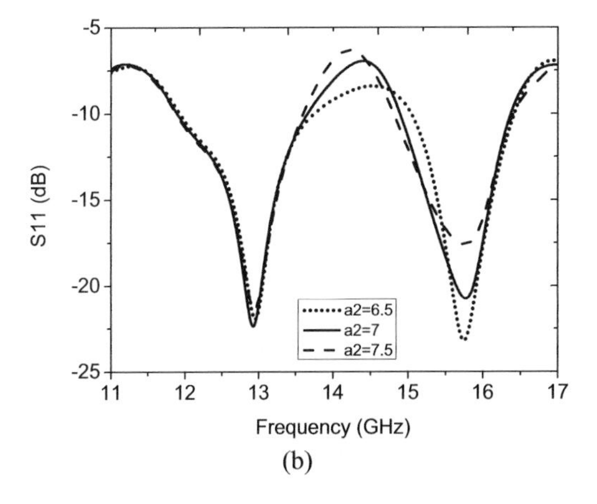

Figure 3. Simulated return loss of the proposed antenna against frequency for various a1 (a) and a2 (b).

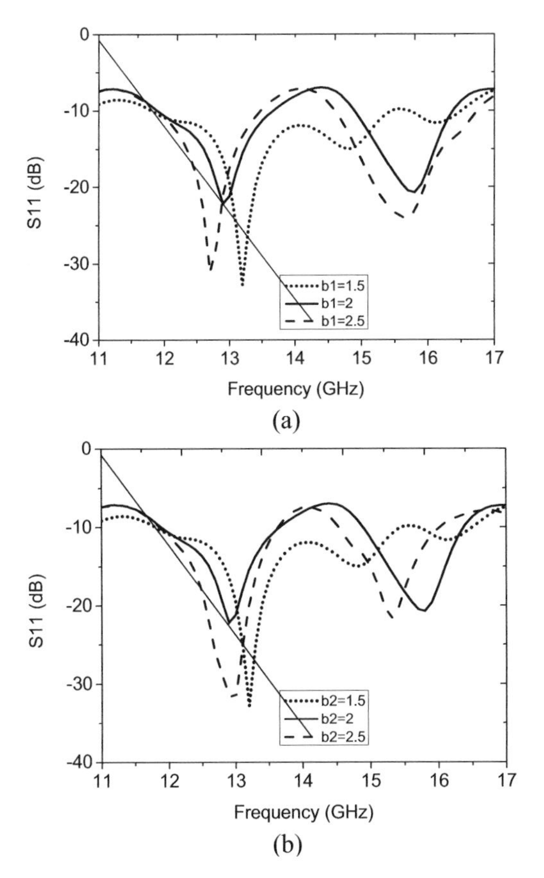

Figure 4. Simulated return loss of the proposed antenna against frequency for various b1 (a) and b2 (b).

a lower frequency. If the value of b1 or b2 was too small, the dual-band feature would disappear. So b1 and b2 are the critical parameters which decide the dual-band operation.

Figure 5 shows the simulated axial ratio of this antenna. From the simulated results, it is clear that, the antenna has two CP radiation frequencies: the lower CP frequency is 11.8–12.3GHz, referred to 3dB axial ratio. The upper CP frequency is 14.3–14.8GHz. The minimum AR is 1.18 dB at 12.1GHz and 0.08 dB at 14.6GHz respectively.

The simulated radiation patterns are shown in Figure 6. The co-polarization is found to be right-hand CP (RHCP) in the lower band. In this band, the RHCP fields of XZ and YZ plane are at least 37dB higher than the left-hand CP (LHCP) fields in the

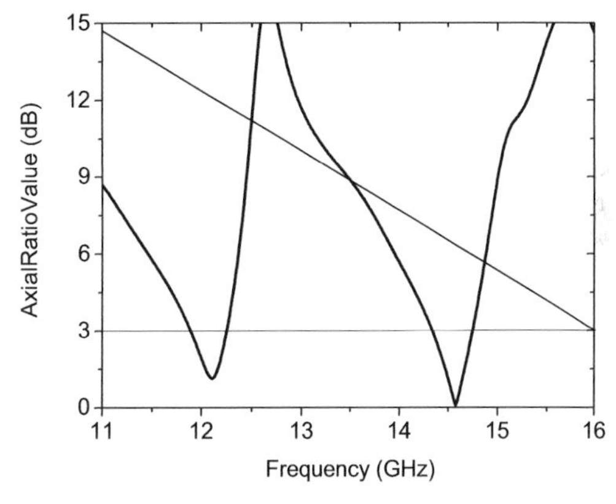

Figure 5. The simulated axial ratio of the L-shaped DRA.

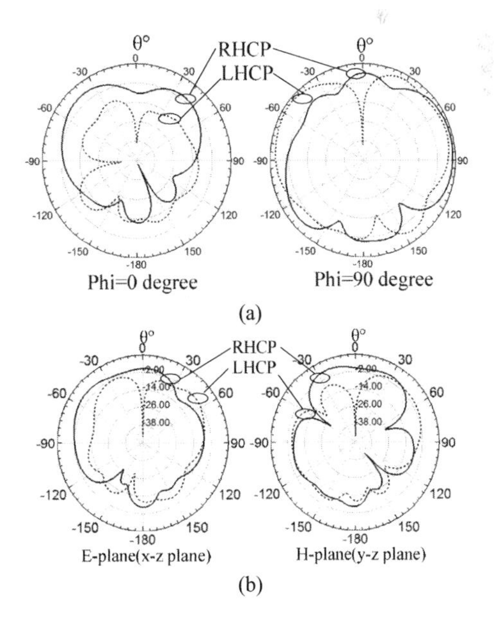

Figure 6. Simulated radiation patterns of the proposed antenna: (a) 12.1GHz and (b) 14.6GHz.

boresight direction ($\theta = 0°$). In the upper band, the co-polarization is also RHCP. The RHCP fields of XZ and YZ plane are at least 46dB higher than the LHCP fields in the boresight direction ($\theta = 0°$).

4 CONCLUSION

A dual-band circularly polarized L-shaped DRA is investigated in this paper. Dual-band is achieved by adjusting the two arms of the L-shaped DR. By using a Y-shaped microstrip line, two sets of orthogonal modes can be excited simultaneously to generate dual-band CP waves. The effects of several key parameters are studied in details. The proposed dual-band circularly polarized L-shaped dielectric resonator antenna is suited for the satellite broadcasting, aeronautical radio navigation service and radio positioning.

ACKNOWLEDGMENT

The paper is supported by the National Natural Science Foundation of China: 61171031.

REFERENCES

Fakhte S. et al. 2014. A Novel Low-Cost Circularly Polarized Rotated Stacked Dielectric Resonator Antenna. *Antennas and Wireless Propagation Letters, IEEE* 13: 722–725.

Fang X. S. & Leung K. W. 2012. Linear-circular-polarization designs of dual-wide-band cylindrical dielectric resonator antennas. *Antennas and Propagation, IEEE Transactions on* 60(6): 2662–2671.

Li L. X. et al. 2009. Circularly polarized ceramics dielectric resonator antenna excited by Y-shaped microstrip. *Microwave and Optical Technology Letters* 51(10): 2416–2418.

Li W. W & Leung K. W. 2013. Omnidirectional circularly polarized dielectric resonator antenna with top-loaded alford loop for pattern diversity design. *Antennas and Propagation, IEEE Transactions on* 61(8): 4246–4256.

Luk K. M. & Leung K. W. 2003. *Dielectric resonator antennas*. London: Research Studies Press.

Mongia R. K. 1992. Theoretical and experimental resonant frequencies of rectangular dielectric resonators. *IEE Proceedings H (Microwaves, Antennas and Propagation). IET Digital Library* 139(1): 98–104.

Motevasselian A. et al. 2013. A circularly polarized cylindrical dielectric resonator antenna using a helical exciter. *Antennas and Propagation, IEEE Transactions on* 61(3): 1439–1443.

Pan Y. & Leung K. W. 2010. Wideband circularly polarized trapezoidal dielectric resonator antenna. *Antennas and Wireless Propagation Letters, IEEE* 9: 588–591.

Pan Y. M. & Leung K. W. 2012. Wide band omnidirectional circularly polarized dielectric resonator antenna with parasitic strips. *Antennas and Propagation, IEEE Transactions on* 60(6): 2992–2997.

Petosa A. 2007. *Dielectric resonator antenna handbook*. Artech House.

San Ngan. H. et al. 2012. Design of dual-band circularly polarized dielectric resonator antenna using a higher-order mode. *In Antennas and Propagation in Wireless Communications, 2012 IEEE-APS Topical Conference on, IEEE*: 424–427.

Electronics, Communications and Networks IV – Hussain & Ivanovic (eds)
 ISBN: 978-1-138-02830-2

Compact ultra-wideband F-shaped dielectric resonator antenna integrated with an narrow band slot antenna

Wenhui Shen, Jian Wu, Jie Liu & Kang Yang
School of Communication and Information Engineering, Shanghai University, Shanghai, China

ABSTRACT: A compact Ultra-Wideband (UWB) antenna integrated with a Narrow Band (NB) slot antenna is presented. The UWB antenna consists of an F-shaped dielectric resonator fed by a microstrip line. In addition, the proposed structure integrates a NB slot antenna which is excited by a strip with the same polarization. The simulated results demonstrate that the UWB and NB antenna provide a 3.05 ~ 10.7GHz and 5.72 ~ 5.91GHz impedance bandwidth, respectively. The UWB antenna can be used for sensing signals while the NB slot antenna can be used for the high performance LAN or WLAN. Moreover, the two antennas have a very good isolation. The size of the narrow band antenna is smaller than the conventional slot antenna by covering one arm of the DR on it, causing the reduction of the volume of the entire antenna.

1 INTRODUCTIONS

In recent years, the UWB communication systems have attracted more and more attentions due to high transmission rates, large space capacity, low cost and power consumption, and any other advantages that can ease the current tension radio spectrum resources (Powell 2004, Schantz 2005). The flexibility of the design offered by dielectric resonator antennas (DRAs) makes them suitable for wireless communication applications and one of the potential candidate antennas for UWB systems (Liang & Denidni 2008, Denidni & Weng 2009, Denidni et al. 2010, Azari et al. 2013, Ozzaim et al. 2013). For this purpose, several designs of UWB DRAs have been proposed. For instance, in the work of Ryu & Kishk (2011), a simple rectangular dielectric resonator is excited by a bevel-shaped patch connected to a coplanar waveguide (CPW), which has an impedance bandwidth covering the frequency range from 3.1 ~ 10.6GHz. Other examples with a strip feeding for the different shape of DRAs to achieve the ultra-wideband (Gao et al. 2012, Kumari & Behera 2013, Dhar et al. 2013, Messaoudene et al. 2013).

In this paper, a compact UWB F-shaped DRA integrated with a narrow band antenna which is fed by a microstrip line through a slot is presented. The hybrid structure is composed of two antennas: an F-shaped dielectric resonator antenna that is used to sense the signals in 3.05 ~ 10.7GHz, and a narrow band antenna whose working frequency is 5.8GHz that can be used for the high performance LAN or WLAN. That the polarizations of the two antennas are the same ensures that the sensing can measure the interference of the operation. Simulations results show the good isolation between the two antennas. With these features, the antenna is a proper candidate for radar, medical imaging and cognitive radio systems.

2 ANTENNA CONFIGURATION

The geometry of the antenna proposed is shown in Figure 1. Figure 1(a), 1(b) and 1(c) show the front view, top view and bottom view of the antenna.

The substrate of the antenna is Rogers RT588LZ with relative permittivity $\varepsilon_{r1} = 1.96$, and its volume L × W × h = 30 × 20 × 0.762 mm^3. And the F-shaped DRA is Arlon AD1000(tm) with relative permittivity $\varepsilon_{r2} = 10.2$ and thickness h1 = 5.08 mm.

As shown in Figure 1(c), the L-shaped ground plane is printed below the substrate. On the surface of the substrate, a slot which is excited from port 2 by a microstrip transmission line is under one arm of F-shaped DR. Port 2 is mounted on the opposite side of the DR to ensure the two ports have the same polarization. The parametric analysis has been carried out to determine the optimal parameters for the proposed antenna. The optimal design parameters of the proposed antenna are as follows: L = 30 mm, W= 20 mm, L1 = 16.5 mm, L2 = 15 mm, L3 = 12 mm, L4 = 11 mm, L5 = 4 mm, L6 = 9.6 mm, L7 = 5.5 mm, W1 = 20 mm, W2 = 0.35 mm, W3 = 1.5 mm, W4 = 1.5 mm, W5 = 1 mm, W6 = 2mm, W7= 6.25 mm, b1 = 10.6 mm, b2=8 mm and a1 = 3 mm , a2 = 3.65 mm, a3=10.95mm.

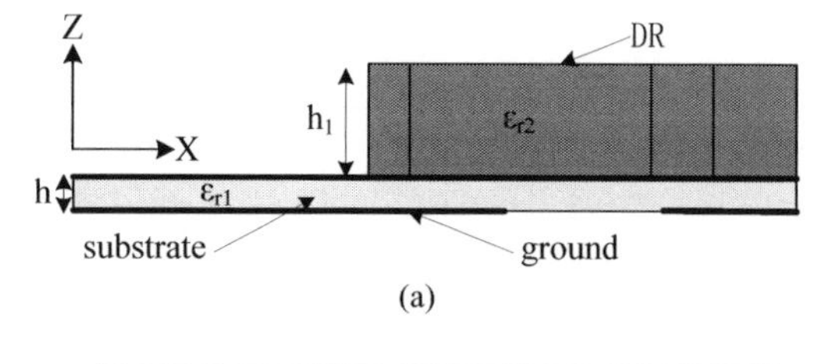

(a)

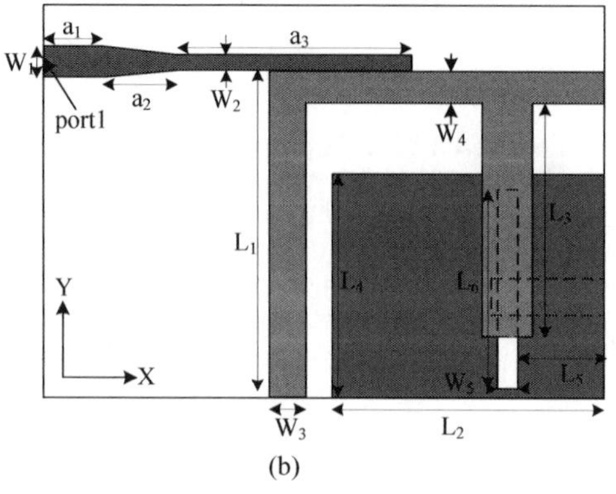

(b)

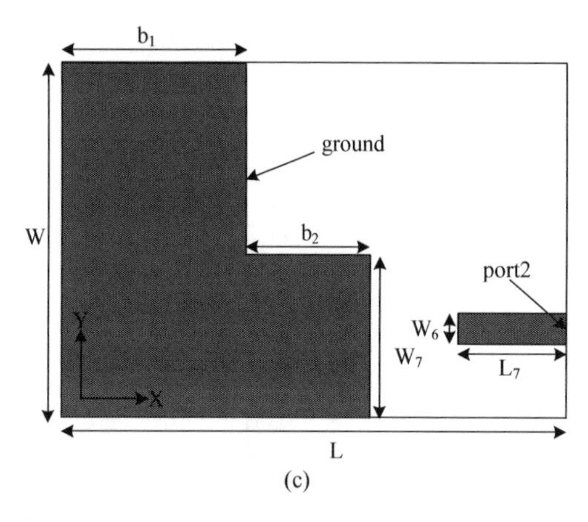

(c)

Figure 1. Configuration of the proposed antenna: (a) front view (b) top view and (c) bottom view.

3 RESULTS AND DISCUSSIONS

To investigate the characteristics of the antenna, the software Ansoft HFSS is used. Figure 2(a) shows the S11 of the antenna (the solid line is for the DRA antenna and the dash line is for the slot antenna). It can be seen that the simulated impedance bandwidth for the UWB DRA is 3.05 ~ 10.7GHz for S11 ≤ -10 dB which largely covers the 3.1 ~ 10.6GHz ultra-wide band. And the narrow band slot antenna operates at 5.8GHz and provides an impedance bandwidth of 190MHz (5.72 ~ 5.91GHz for S22 ≤ -10dB) suitable for the high performance LAN or WLAN. Figure 2(b) is the curve of transmission coefficient S12 of the antenna. As it can be noticed, S12 $\leq$ -20dB across UWB except 5 ~ 6GHz (S12 < -15dB), which ensures good isolation between the two antennas.

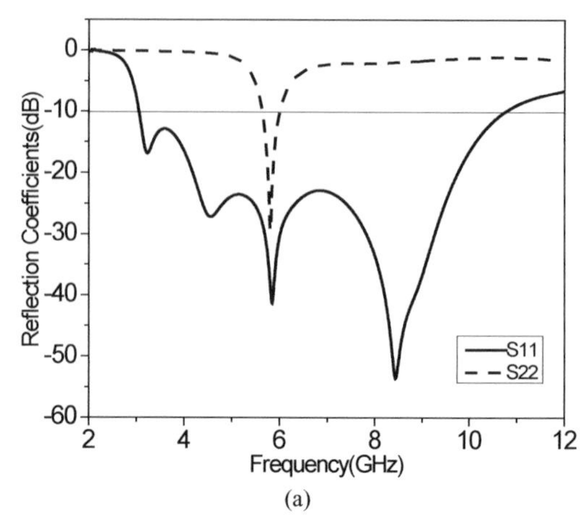

(a)

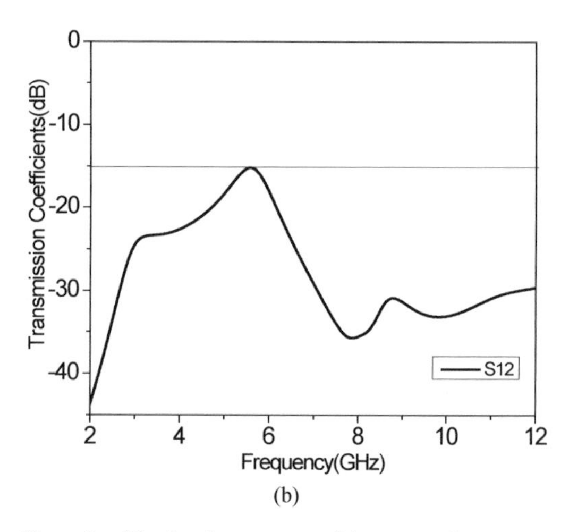

(b)

Figure 2. Simulated parameters of the proposed antenna.

W4 and L6 are important in adjusting the resonant frequency of the DRA and the slot antenna, respectively. Figure 3 shows the impedance bandwidth against the frequency, Figure 1(a) is obtained by adjusting W4 and Figure 1(b) is by adjusting L6.

As it can be seen from Figure 3(a), the frequency of the DRA antenna moves to the left as W4 increases, while the frequency of the slot antenna almost keeps the same. So we can adjust W4 to achieve suitable UWB bandwidth.

Figure 3(b) shows the simulated reflection coefficients for different L6. Also, it can be discovered that the longer L6 is, the lower the resonate frequency is. So the length L6 is useful to optimize the resonant frequency of the slot antenna.

From Figure 3, the two parameters W4 and L6 are independent of each other, so that we can adjust one parameter without causing unexpected results of the other one. This is important because we can adjust the antenna easily to achieve the performance.

Other parameters of cause can also affect the antenna, but the effects are not regular and they are correlative to each other.

Figure 4 shows the simulated E(yz)-plane and H (xz)-plane radiation patterns of the proposed antenna at 4, 6.5 and 9 GHz. It can be observed that the antenna radiates at broadside direction. The radiation pattern is almost symmetrical in the operating frequency range in the E-plane. However, the unsymmetrical radiation patterns in the H-plane are caused by the asymmetry of the structure and the excitation. The simulated radiation patterns for narrow band antenna at 5.8GHz are shown in Figure 5.

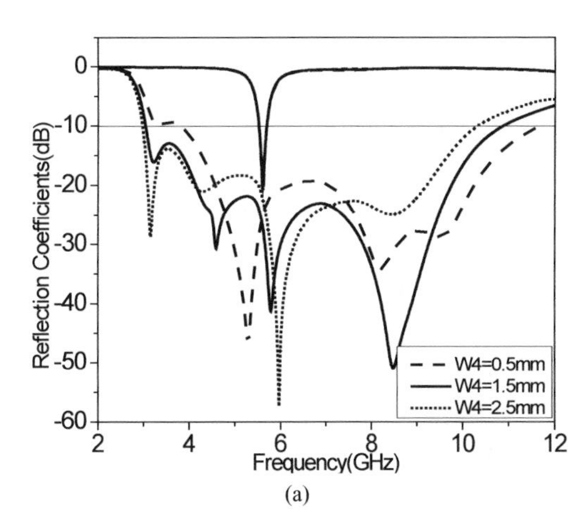

(a)

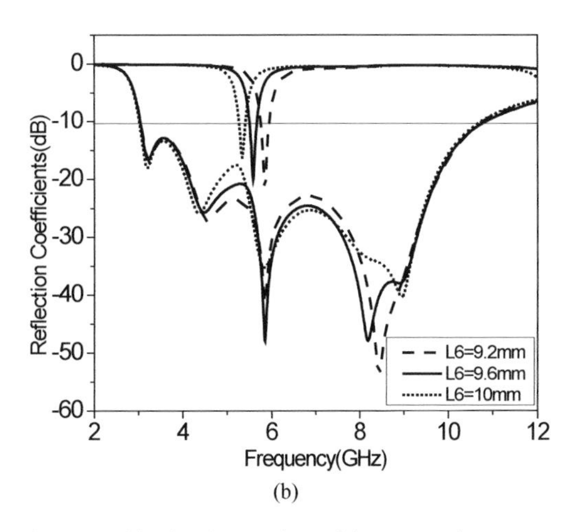

(b)

Figure 3. Simulated return loss of the proposed antenna against frequency for various W4 (a) and L6 (b).

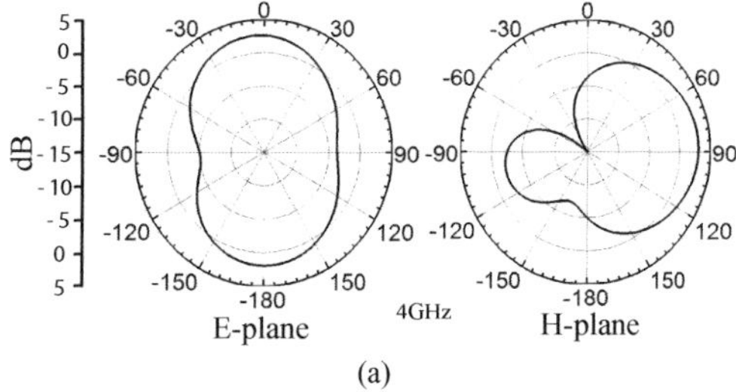

(a)

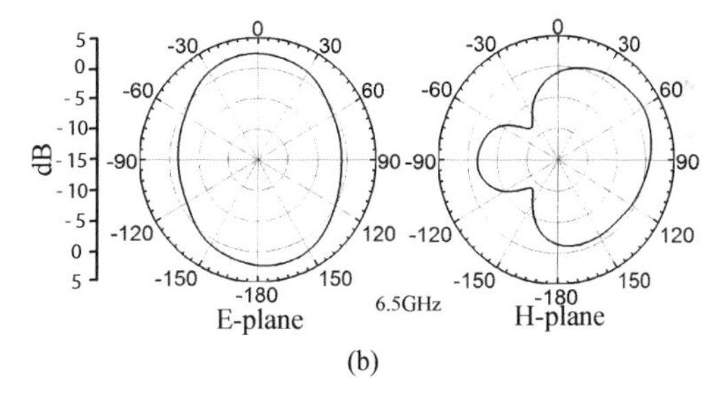

(b)

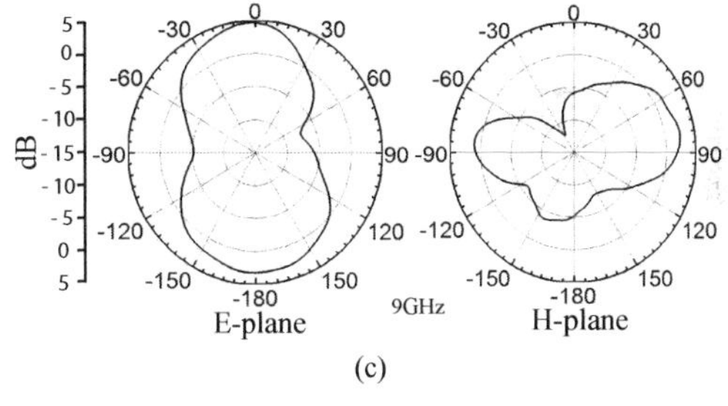

(c)

Figure 4. Simulated radiation patterns for UWB DRA: (a) 4GHz (b) 6.5GHz and (c) 9GHz.

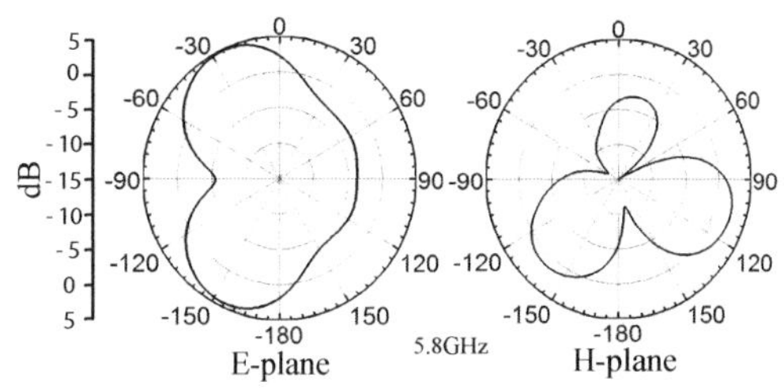

Figure 5. Simulated radiation patterns for narrow band antenna at 5.8GHz.

The simulated gains of the ultra-wideband antenna are presented in Figure 6. It is found that the UWB antenna gains ranges from 2 to 7.8dB and is stable around 3 dB which is suitable for UWB systems.

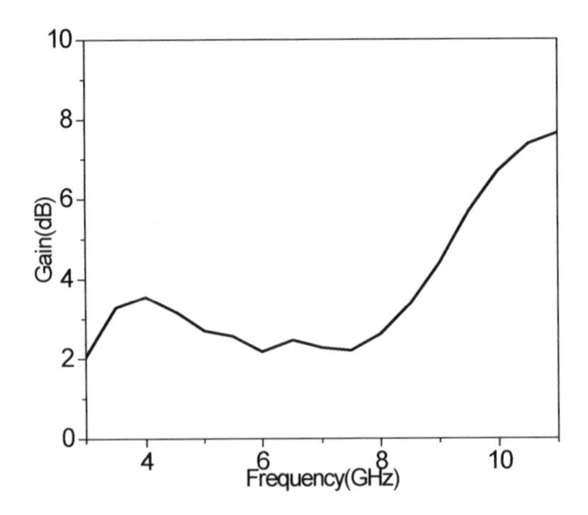

Figure 6. Simulated gain of UWB antenna.

4 CONCLUSION

A compact UWB DRA integrated with a narrow band slot antenna is presented. The proposed UWB antenna has been applied with an F-shaped DR, which is excited by a microstrip line. And the NB antenna is applied with a slot, which is fed by a short microstrip line in the opposite side of the port 1. The simulated results exhibit that the proposed UWB DRA provides a wide impedance bandwidth of 111% (3.05 ~ 10.7GHz) with the gain being stable around the 3dB and the narrow band antenna operating around 5.8GHz with 5.1dB in resonant frequency, which can be used in wireless transmission system. Moreover, the proposed design provides good isolation between the two antennas. With these features, the antenna can be used for radar, medical imaging and cognitive radio systems.

ACKNOWLEDGMENTS

The paper is supported by the National Natural Science Foundation of China: 61171031.

REFERENCES

Azari. A. et al. 2013. *A New Super Wideband Fractal Monopole-Dielectric Resonator Antenna.*

Denidni. T. A. & Weng. Z. 2009. Rectangular dielectric resonator antenna for ultrawideband applications. *Electronics letters* 45(24), 1210–1212.

Denidni. T. A. et al. 2010. Z-shaped dielectric resonator antenna for ultrawideband applications. *Antennas and Propagation, IEEE Transactions on* 58(12), 4059–4062.

Dhar. S. et al. 2013. A Wideband Minkowski Fractal Dielectric Resonator Antenna. *Antennas and Propagation, IEEE Transactions on* 61(6), 2895–2903.

Gao. Y. et al. 2012. Compact asymmetrical T-shaped dielectric resonator antenna for broadband applications. *Antennas and Propagation, IEEE Transactions on* 60(3), 1611–1615.

Kumari. R. & Behera. S. K. 2013. Wideband log-periodic dielectric resonator array with overlaid microstrip feed line. *Microwaves, Antennas & Propagation, IET* 7(7).

Liang. X. L. & Denidni. T. A. 2008. H-shaped dielectric resonator antenna for wideband applications. *Antennas and Wireless Propagation Letters, IEEE* 7, 163–166.

Messaoudene. I. et al. 2013. Experimental investigations of ultra-wideband antenna integrated with dielectric resonator antenna for cognitive radio applications. *Progress In Electromagnetics Research C* 45, 33–42.

Ozzaim. C. et al. 2013. Stacked conical ring dielectric resonator antenna excited by a monopole for improved ultrawide bandwidth. *Antennas and Propagation, IEEE Transactions on* 61(3), 1435–1438.

Powell. J. 2004. Antenna design for ultra wideband radio. *Doctoral dissertation, Massachusetts Institute of Technology.*

Ryu. K. S. & Kishk. A. A. 2011. UWB dielectric resonator antenna having consistent omnidirectional pattern and low cross-polarization characteristics. *Antennas and Propagation, IEEE Transactions on* 59(4), 1403–1408.

Schantz. H. 2005. *The art and science of UWB antennas.* Artech House.

Electronics, Communications and Networks IV – Hussain & Ivanovic (eds)
© 2015 Taylor & Francis Group, London, ISBN: 978-1-138-02830-2

Secure design of VMI-IDS

Jiangyong Shi*, Chengye Li, Yuexiang Yang & Kun Jiang
College of Computer, National University of Defense Technology, Changsha, Hunan, China

ABSTRACT: This paper analyzed the typical Virtual Machine Introspection based Intrusion Detecting System (VMI-IDS) in the past ten years, including Livewire and so on. Through comparison and analyzing, we concluded that a trend of non-intrusive, anomaly-based and distributed IDS will be the future direction of VMI-IDS development. Furthermore, two central problems in designing of VMI-IDS, namely the semantic gap problem and security problem are discussed. Potential solutions including trust computing, integrity checking and rootkit detection is discussed.

1 INTRODUCTION

With the development of cloud computing and virtualization technology, more and more enterprises migrate their IT infrastructures into the cloud, individuals are also benefiting from the application of cloud in areas like cloud storage, cloud apps, etc. The security of virtual machines plays a significant role in this trend. To protect the virtual machines, traditional security methods like IDS are adjusting their way to adapt to a cloud environment.

Traditionally, there are two main kinds of IDS, namely Host Based IDS (HIDS) and Network Based IDS (NIDS). HIDS detects intrusion for the machine by collecting information such as file system used, network events, system calls, etc. HIDS has an excellent view of what is happening in that host's software, but is highly susceptible to attack. Besides, the efficiency is greatly dependent on choosing system characteristics of the host. While NIDS avoid the problem of being accessible by malware, it has a poor view of what's happening inside the host. Moreover, NIDS cannot detect any intrusion if the network traffic is encrypted, so it is impossible to detect intrusion in highly precise.

Considering the architecture of virtualization in Figure 1, we can see that Virtual Machine Monitor (VMM, also called hypervisor) has a higher privilege than Guest Oss and can introspect almost everything happened in VMs, including virtual memory, registers and storage. It's possible to implement security measures in VMM to monitor the VMs. Such measures may include IDS, Antivirus scanning, Firewall, etc. By doing so, there are at least two benefits. Firstly, it can protect all the VMs on the same VMM in spite of the types of Guest OSs. Secondly, by implementing security measures in VMM, a lot of attacks to security software can be avoided. Other benefits may include an on-demand Antivirus scanning without occupying the resources of VMs, a uniform disposition of firewall rules to VMs of the same service or demand, and so on. More properties of VMM including Isolation, Inspection and Interposition can be used to design a better IDS than host-based IDS as is sum in (Garfinkel & Rosenblum 2003). Examples and the detailed development of such IDSs will be presented in part two. The problems of these IDSs and corresponding solutions will be offered in part three.

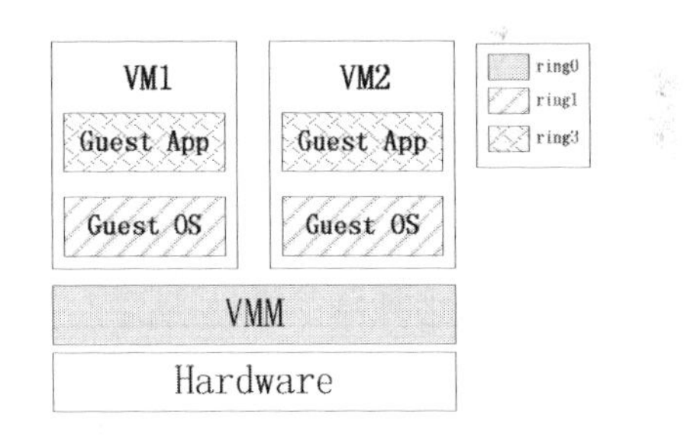

Figure 1. Architecture of virtualization.

2 RELATED WORKS

The first model of VMI-IDS is proposed by Garfinkel in 2003. Over the past ten years, VMI-IDS which takes the advantage of the virtualization technology to protect VMs are researched widely, mainly consists of Livewire (Garfinkel & Rosenblum 2003), HyperSpector (Kourai & Chiba 2005), IntroVirt (Joshi et al. 2005), Collabra (Bharadwaja et al. 2011), HA-CIDS (Hisham et al. 2013), VMwatcher

*Corresponding author: *shijiangyong@nudt.edu.cn*

(Kittel 2010)0, vProbes (Dehnert 2012), etc. Based on whether an agent-like information gatherer is installed in the VM, we divided them into two types, namely intrusive VMI-IDS and non-intrusive VMI-IDS. The main reason to do this is the big differences in implementation and security between them. What's more, VMM-Based IDS can be divided into another two types based on the type of detection rules, namely misuse-based VMI-IDS and anomaly-based VMI-IDS. Besides that, there are VMI-IDSs such as HA-CIDS trying to distinguish between different Guest Oss to improve the detection efficiency and Collabra trying to apply distributed thought into VMI-IDS to improve the detection accuracy.

Table 1. Development of VMI-IDSs.

VMI-IDSs	Year	Special Properties
Livewire	2003	Misuse based
HyperSpector	2005	Intrusive, Distributed
IntroVirt	2005	Anomaly based
VMwatcher	2007	Non-intrusive
Collabra	2011	Hypercall inspection
vProbes	2012	Non-intrusive,Distributed
HA-CIDS	2013	Hierarchical, autonomous

Table 1 is a collection of current VMI-IDS, from which we can see that VMI-IDSs have developed from misuse-based to anomaly-based, from intrusive to non-intrusive, from single node to distributed, and from static to dynamic and adaptive. These developments have greatly increased the quality of VMI-IDSs, including detection of novel attacks, transparency to attackers, more accurate detection, more efficiency, etc. Firstly, an anomaly-based VMI-IDS is more able to detect novel attacks since any deviation from normal behaviors can be treated as intrusions. Secondly, a non-intrusive VMI-IDS is more robust in resisting attacks with its transparency to VMs. Thirdly, a distributed VMI-IDS is more suitable to apply in cloud environment which contains a large number of compute nodes. Lastly, dynamic and adaptive VMI-IDS can best adapt to the dynamic feature of cloud and variety of Guest OSs so as to minimize the performance cost of introducing VMI-IDS.

3 PROBLEMS & SOLUTIONS

Even though VMI-IDSs have evolved a lot in the past ten years, none of the above VMI-IDSs have come into a product. This is because there are several challenges in designing such an integrated IDS into the VMM while keep the good qualities we mentioned in the last part. In this part, we will discuss these problems and make a clear way to the future research.

3.1 *Semantic gap*

Semantic gap is the first challenge in designing a non-intrusive VMI-IDS. Semantic gap is the difference between what we see from the VMM view and what we see from the VM view. Typically, a VMM monitors the events in VMs through interpreting hypercalls and introspecting resources like memory and disks. To understand what happened in VMs through interpreting hypercalls and introspecting resources, there must be a translation process. A lot of researches have been done to solve this problem.

Livewire uses OS Interface Library to translate the Hardware State of VMs into OS-level view. The OS Interface Library is built by modifying the Linux crash dump examination tool crash. The crash tool can provide access to monitored host's memory. Since Livewire is designed for VMware Workstation which is a hosted VMM, this means the VM's memory is accessible by crashing. However, the crash is designed to debug a kernel or analyze a dump file. It requires details about the kernel such as symbol table, data types, etc. What's more, it's very slow in analyzing a whole kernel for specific information. So its efficiency and effectiveness should be reconsidered.

Different from Livewire, VMwatcher utilizes a technique named guest view casting to minimize the semantic gap. Guest view casting uses guest device drivers and related file system drivers reconstruct VM information such as files and directories from the virtual disk. The memory state reconstruction is achieved by translating the virtual memory address into the corresponding physical memory address. Firstly, from System.map file exporting Init_task_union which contains a list of task_structs. Each task_structs represents a process in VM. Then we can sequentially get the page directory entry from the task structure, the page table entry from the mm structure and finally the physical address from the pgd structure.

Even though VMwatcher can detect intrusions and scan malware files in VMs with commercial anti-virus software, there are still several problems to solve. The view casting process is based on knowledge of the related drivers which commercial systems such as Windows don't offer. What's more, what VMwatcher can watch through introspection and guest view casting is limited to files and processes while advanced malware can hide itself from these detections by technologies such as DKOM. Introspecting to registers should be further studied.

A similar concept of view casting is called kernel map in (Lin et al. 2012). By resolving the kernel map of VM in VMM layer, the services of the VM can be identified and the required detection rules can be adopted. Take Windows as an example, kernel map first obtains the VM OS information through TIB (Thread Information Block), PEB (Process Environment Block) and OSMajorVersion & OSMinorVersion. By choosing the corresponding detection rules of specific Windows version, it can be more efficient compared with detecting all kinds of OSs. Then the kernel map extracts the service information through key structures, including ETHREAD, EPROCESS and PID. It's actually the same way as VMwatcher does in obtaining physical memory address. The problem is that how to find TIB or ETHREAD from a raw binary bit of VMM.

Instead of being entangled with the problem of semantic gap and trying to find malware behaviors or files through checking files and processes, there are also solutions trying to interpret the hypercalls to identify intrusion behaviors. One example of these is Collabra. Hypercall is a concept in Xen, which is used by the VMs to access sensitive resources and request privileged operations. Collabra gives each hypercall an anomaly score according to related elements. By comparing the anomaly scores with a threshold, Collabra divides hypercalls into two categories as normal and anomalous (Bharadwaja et al. 2011). This method is relatively unreliable since both legitimate and malicious behaviors could cause a suspicious hypercall with high anomaly scores. The author states this can be solved by reducing the influence of an element in the scoring process. However, what a related element is and how it affects the anomaly scores haven't been addressed. To improve it and make it practicable, methods in machine learning such as classifying and clustering can be used to analyze the hypercall, so as to find anomalous behavior. This method has the benefit of privacy preserving without introspection of details of virtual machine memory or the file system.

3.2 *Security of the VMI-IDS*

General attacks to IDS are summed up in (Corona et al. 2013). As the document states, there are six types of attacks, including Evasion, Overstimulation, Poisoning, DOS, Response Hijacking and Reverse Engineering.

These attacks belong to three phases: Measurement, Classification and Response, which separately represents the function of the event generators, event analyzers and response units. These attacks are partly avoided by introducing VMI-IDS. However, some of them are still vulnerable to these attacks.

When the VMs are compromised by an attacker, it is possible to directly attack the VMI-IDS. The attacker can use a passive attack (Serjantov & Sewell 2005) against the IDS to subvert the VMM. In this way an attacker can directly attack the IDS by manipulating the results of requests send by the IDS, thus resulting in a compromised IDS. An attacker is also able to attack the IDS through host system which provides the virtual machine environment. This is especially true to hosted VMI-IDS. For example, an attacker is able to tamper with the VMM interface of Livewire when he has access to the host machine.

What's more, all of the IDS in the above ignored one fundamental question, which is the security of the VMM where IDS stays. They may be able to protect the virtual machines well with malwares only implemented in the VM and knowing nothing about the existence of IDS. But if this is not true, such as when the host or the VMM is attacked by rootkits, the detection results of IDS will be unreliable. To solve this problem, we should first make sure that the VMM itself is safe, so that the IDS in it can keep from being attacked. Secondly, we should be able to figure out the rootkits when the hypervisor is already affected.

To address the above problems and protect the VMI-IDS, following access can be considered:

3.3 *Trusted computing*

SHype (Sailer et al. 2005) is an enforcement access control mechanism used to control information flow between virtual machines sharing a single hardware platform. This is especially important to VMI-IDS, because one compromised virtual machine may affect other VMs through resource sharing, thus bypassing the IDS.

The main structure of sHype contains hook functions, the access control module (ACM), and security policy manager. Hook functions are used to interpret the VM's access requests to share resources. After the requests in interpreted, they will be delivered to ACM. Then the ACM checks if the process has access attributes of the resource by looking inside the Policy. Policy Manager is used to configure the secure policy of ACM, such as BLP, Chinese Wall (Brewer & Nash 1989).

Terra (Garfinkel et al. 2003) is a virtual machine-based platform for trusted computing, the core part of Terra is a trusted virtual machine monitor (TVMM) which isolates and protects independent virtual machines (VMs). Closed box VMs, shown in gray, are protected from eavesdropping or modification by anyone but the remote party who has supplied the box. The TVMM can identify the contents of the closed box to remote parties, allowing them to trust it.

TVMM not only provides the traditional benefits of VMM, namely isolation, extensibility, efficiency, compatibility and security, but also has additional capabilities such as Root Secure, Attestation and Trusted Path. Root Secure means even the system manager cannot destroy the privacy and isolation of Closed-box VMs. Attestation enables local applications running in Closed-box VMs identify itself to a remote party so that the party can trust these applications. Trusted Path is a secure channel between the users and applications which can prevent malicious code from snooping.

By using the architecture, a VMI-IDS like Livewire can keep a secure communication between the monitored VMs and the IDSVM from being sniffed or modified by a rootkit.

3.3.1 *Integrity checking*

A fundamental assumption of the current virtualization technology is the presence of a trustworthy hypervisor. Unfortunately, the large code base of commodity hypervisors and recent successful hypervisor attacks (e.g. VM escape) seriously question the validity of this assumption.

HyperSafe (Wang & Jiang 2010) is a lightweight approach that endows VMMs with a unique self-protection capability to provide lifetime control flow integrity. HyperSafe mainly uses two key techniques. The first one is non-bypassable memory lockdown which reliably protects the hypervisor's code and static data from being compromised even in the presence of exploitable memory corruption bugs (e.g., buffer overflows), therefore successfully providing VMM code integrity. The second one is restricted pointer indexing which introduces one layer of indirection to convert the control data into pointer indexes. These pointer indexes are restricted such that the corresponding call/return targets strictly follow the control flow graph, hence expanding protection to control-flow integrity. A prototype is built and used to protect two open-source VMMs: BitVisor and Xen. The experimental results with synthetic hypervisor exploits and benchmarking programs show HyperSafe can reliably enable the VMM self-protection and provide the integrity guarantee with a small performance overhead.

The methods HyperSafe used are not specified to VMI-IDS, but can be treated as a defensive measure to protect the VMM and the IDS in it from being compromised.

3.4 *Rootkit detection*

However, what if rootkits have already infected the hypervisor? This demands the hypervisor with abilities to detect existing rootkits. Rootkits like BLUEPILL (Rutkowska 2006), VMBR (King & Chen 2006) and DKSM (Bahram et al. 2010) that utilize hardware-assisted virtualization (HAV) technology is hard to detect using traditional ways like data integrity comparison. This is because the effectiveness of these ways is dependent on a higher privilege of detection codes than rootkit. While rootkits like BLUEPILL can hook the functions which query system files, processes, registry table or services, thus cheating the detection. Even the Direct Kernel Object Manipulator (DKOM) can be cheated since BLUEPILL has a higher privilege than the virtual machine kernel. But this doesn't mean rootkits like BLUEPILL cannot be detected, since its interpretation of instructions introduced a delay and a change of the memory. Based on this, some methods are listed in Table 2.

Table 2. Detection methods of HAVR.

Detection Methods	Merits	Demerits
Time differences on privilege instructions' execution (Fritsch 2008)	Many timers, easy to implement	The timers are easy to be intercepted and modified, unreliable
Profile differences on memory resources (Ferrie 2007)	Don't rely on timer	Cannot detect rootkits using AMD-V
CPU anomaly and error detection (Myers & Youndt 2007)	Based on the intrinsic anomaly or errors of CPU	Specified on the type of CPU, not general
Comparison of instructions counter (Barbosa 2007)	Don't rely on time, hard to cheat	Only capable on dual-core or multi-core CPU

In fact, the methods listed above can detect only the existence of the hypervisor, rather than the malicious of the hypervisor. Since there have been a lot of legal hypervisors being used, it is not sufficient in detecting hypervisor-based rootkits or HAVR. To do that, we can add an IDS module with HAVR detection function to pre-installed legal hypervisor to prevent the install of rootkits. Or we can use hardware-based detection method (Bulygin & Samyde 2008) to detect the existing rootkits which is more reliable since no code will be more privileged than hardware.

4 CONCLUSION & FUTRUE WORK

This paper analyzed the typical VMI-IDS in the past ten years, including Livewire, HyperSpector, IntroVirt, VMwatcher, Collabra, vProbes, HA-CIDS, etc. Results show that a trend of non-intrusive, anomaly-based and distributed IDS will be the future direction of VMI-IDS development. Besides, we discussed two central problems of VMI-IDS, namely the semantic gap problem and security problem. While many ways to minimize the semantic gap have been proposed, many problems remained. These problems include the requirement for symbol files and other OS-related information, the limited ability of VMI-IDS, the entry point of Guest OS related structures in VMM memory. Although method of identifying IDS with hypercalls is proposed in Collabra, the precision is very low.

The security problem of VMI-IDS is often ignored by VMI-IDS designers. As we discussed in part 3, there are many attacking methods to VMI-IDS such as cheat attack, passive attack, and VMM-based rootkit. To address the security problem, we discussed three technologies, namely trusted computing, integrity checking and rootkit detection. Many instances are given in the discussion. Our future works mainly focus on implementing these technologies in designing secure VMI-IDS which can not only detect intrusions to virtual machines but also protect it from being compromised. Nevertheless, we hope our job can help VMI-IDS designers better understand the requirements and trends of VMI-IDS designing by inspecting the problem of current implementations.

ACKNOWLEDGEMENT

This work is supported by the National Natural Science Foundation of China (Grant No.61170286 and NO.61202486). We would like to thank the anonymous reviewers for their insightful comments and suggestions.

REFERENCES

Bahram, S., Jiang, X .& Wang, Z. DKSM: Subverting virtual machine introspection for fun and profit. *Reliable Distributed Systems, 2010 29th IEEE Symposium on, New Delhi , Oct. 31 -Nov. 3* 2010: 82–91. IEEE.

Barbosa E. 2007. Detecting Bluepill. *Presentation on SyScan Conference, Singapore, 2007.*

Bharadwaja, S., Sun, W. & Niamat, M. 2011. Collabra: a xen hypervisor based collaborative intrusion detection system. *Information Technology: New Generations (ITNG), 2011 Eighth International Conference on, Las Vegas, 11-13 April 2011*: 695–700. IEEE.

Brewer, D. F. C. & Nash, M. J. 1989. The chinese wall security policy. *1989 IEEE Symposium on Security and Privacy. IEEE*: 206–214.

Bulygin, Y. & Samyde, D. 2008. *Chipset based approach to detect virtualization malware.* USA: BlackHat Briefings.

Corona, I., Giacinto, G. & Roli, F. 2013. Adversarial attacks against intrusion detection systems: Taxonomy, solutions and open issues. *Information Sciences*: 205–225.

Dehnert Alex. 2012. Intrusion detection using VProbes. *VMware Technical Journal.*

Ferrie, P. 2007. Attacks on more virtual machine emulators. *Symantec Technology Exchange.*

Fritsch, H. 2008. *Analysis and detection of virtualization-based rootkits*. Technische Universitat Munchen.

Garfinkel, T. & Rosenblum, M. 2003. A virtual machine introspection based architecture for intrusion detection. *NDSS* 3: 191–206.

Garfinkel, T., Pfaff B & Chow, J. 2003. Terra: A virtual machine-based platform for trusted computing. *ACM SIGOPS Operating Systems Review* 37(5): 193–206.

Hisham, A., Kholidy Abdelkarim Erradi & Sherif Abdelwahed. 2013. HA-CIDS: a hierarchical and autonomous IDS for cloud systems. *Fifth International Conference on Computational Intelligence, Communication Systems and Networks, Madrid, 5-7 June 2013*: 179–184. IEEE.

Joshi, A., King, S. T. & Dunlap, G. W. 2005. Detecting past and present intrusions through vulnerability-specific predicates. *ACM SIGOPS Operating Systems Review* 39(5): 91–104.

King S T. & Chen, P M. 2006. SubVirt: Implementing malware with virtual machines. *Security and Privacy, 2006 IEEE Symposium on. IEEE, Berkeley/Oakland, CA, 21-24 May 2006*: 14.-327. IEEE.

Kittel Thomas. 2010. Design and implementation of a virtual machine introspection based intrusion detection system. *Technical University Munich* 3: 191–206.

Kourai, K. & Chiba, S. 2005. HyperSpector: virtual distributed monitoring environments for secure intrusion detection. *Proceedings of the 1st ACM/USENIX International Conference on Virtual Execution Environments, June 2005*: 197–207. ACM.

Lin Chih-Hung, Tien Chin-Wei & Pao Hsing-Kuo. 2013. Efficient and effective NIDS for cloud virtualization environment. *IEEE 4th International Conference on Cloud Computing Technology and Science, Taipei, 5–7 June 2013* . IEEE.

Myers, M. & Youndt, S. 2007. An Introduction to Hardware-Assisted Virtual Machine (HVM) Rootkits. *White Paper of Crucial Security.*

Rutkowska, J. 2006. *Subverting VistaTM kernel for fun and profit*. Black Hat Briefings.

Serjantov A, Sewell P. 2003. Passive attack analysis for connection-based anonymity system. *Computer Security–ESORICS 2003*: 116-131. Sailer R, Valdez E & Jaeger T. 2005. sHype: Secure hypervisor approach to trusted virtualized systems. *Techn. Rep.* RC23511.

Wang, Z. & Jiang, X. 2010. Hypersafe: A lightweight approach to provide lifetime hypervisor control-flow integrity. *2010 IEEE Symposium on Security and Privacy (SP), Oakland, CA, USA, 16-19 May 2010*: 380–395. IEEE.

Electronics, Communications and Networks IV – Hussain & Ivanovic (eds)
© 2015 Taylor & Francis Group, London, ISBN: 978-1-138-02830-2

A novel TD-LTE private network working at discrete narrow band for power industry

Zhan Shi
Power Grid Control Center of Guangdong, China South Power Grid, Guangzhou, China

Xiaobin Wei* & Jianming Zhang
School of Electronic Engineering, Beijing University of Posts & Telecommunications, Beijing, China.

ABSTRACT: In order to build "Strong Smart Grid", it is important to establish a private communication network first. With the advantages of spectrum utilization and asymmetric transmission, the 4th wireless communication technology, TD-LTE, has attracted a lot attentions from the private network builders. However, the scarce spectrum resource and the high loan cost have limited the growth of the TD-LTE private network's construction. In this paper, we will take the power wireless private network for example, make use of the "Own Frequency Resource" to explore the construction of a TD-LTE private network, and also will carry on a simulation to analyze the performance of the Distribution and Utilization Information Acquisition System, in order to provide references for the construction of the TD-LTE private network.

1 INTRODUCTION

During the 12th Five-Year-Plan period, China State Grid puts forward a goal to build "Strong Smart Grid" in an all-round way. Strong smart grid is an important platform for supporting the new energy strategy and also a pushing hand to promote the integration of a variety of advanced technology and innovation. However, to realize the plan, a power communication private network must be built at first for three reasons: 1st, to guarantee the stability, when a sudden emergency happens at the public places, it may issue in the communication cut, and also may lead the information platform to paralysis, which will cause a secondary hazard with much more damage; 2nd, to guarantee the safe differentiated control design to operation authority, different permission should be set in the power industry manufacturing platform according to the jobs, and cross-tier communication is forbidden; 3rd, to realize cluster scheduling, which is essential for the multimedia communication (CATR 2009).

Besides, according to the "Regulations of the electric secondary system's safety protection" and "Overall scheme of the electric secondary system's safety protection", the electric secondary system's safety protection must follow the principles of security division, dedicated network, horizon segregation and the originality of vertical authentication. Therefore, it is necessary to build a power private network.

And so far, a phase objective has been achieved: the 110kV and above voltage level power transmission network has been 100% optical fiber, and the 35kV and 10kV mid-voltage distribution network has been part-fiber(Kang 2010). However, at some places of the city, to build a fiber network needs too much resource, which would not be a cost-effective way. In this case, to build a wireless private network to extend the wired fiber network would be a good option.

In 1995, China Radio Regulatory Commission allocated 40 frequency points during 223.025~235.000MHz to the power industry. The wireless digital broadcasting station used to be used for the power wireless communication. However, with the rapid development of the mobile multimedia services, now it is hard for the wireless digital broadcasting station technology to meet the needs of the broadband services.

In this paper, we propose a novel construction project for the power wireless private network. It applies the technologies of TD-LTE to realize the information acquisition and system remote control. Now a number of mainstream operators, such as Japan's Softbank, Vodafone, Deutsche Telekom, France Telecom, SK Telecommunications, India Yota, American Clearwire, as well as those new operators who possess TDD band, have accepted TD-LTE technology, which would provide a complete production chain for the construction of the TD-LTE private network.

The rest of the paper is organized as follows: In Section 2, we present the usage of carrier aggregation.

*Corresponding author: *manjusaka33@163.com*

Section 3 analyzes the capacity of 230MHz TDLTE private network. In Section 4 we introduce the business requirements of Distribution and Utilization Communication System. Section 5 analyzes the performance through simulation. Finally, in Section 6, we conclude our paper.

2 CARRIER AGGREGATION

According to the regulations of China Radio Regulatory Commission, there are 40 frequency points during 223.025 235.000MHz, 1MHz in total, allocated to the power industry in China. The spectral distribution is shown in Figure 1.

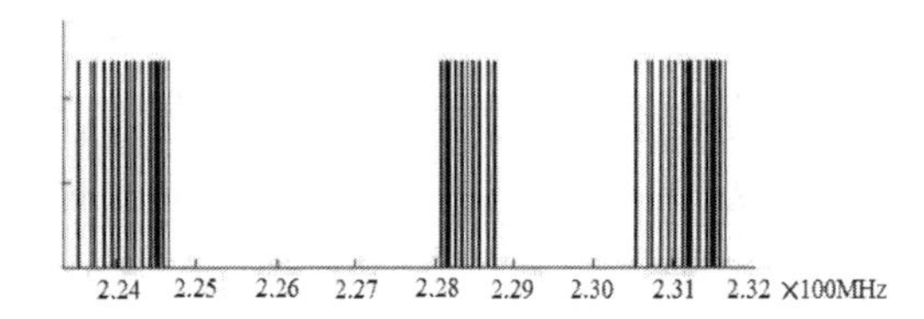

Figure 1. The spectral distribution.

The wireless digital broadcasting station technology has been used in the power industry. However, as wireless digital broadcasting station can only use one frequency point for data transmission, it can hardly meet the need of the power wireless broadband communication. By using carrier aggregation, TD-LTE can effectively make use of those discrete frequency bands as one continuous broadband, and get a higher data transmission speed. All of these operations could be realized by a base band processing module.

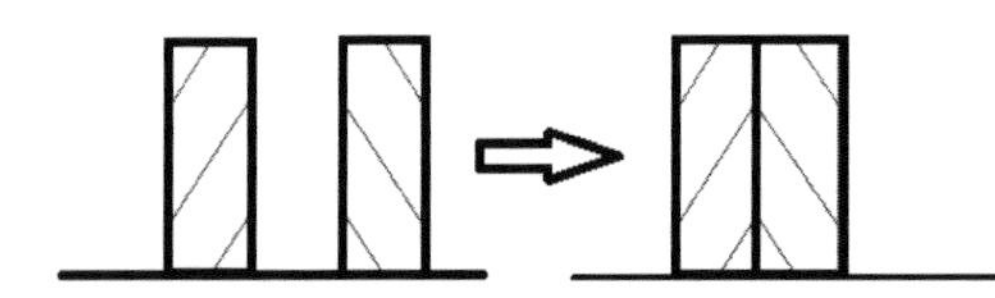

Figure 2. Discrete carrier aggregation.

What is shown in Figure 2 is Discrete Carrier Aggregation, which can effectively make use of those discrete narrow frequency bands, without any modifications to the system physical channel and the coding modulation scheme. However, in order to reduce the interference caused by the leakage at the adjacent frequency points, there is a need to reverse 10% of the frequency for the protection band, so after carrier aggregation, we can get 0.9MHz band for the private network(Cao et al. 2013).

3 THE CAPACITY OF 230MHZ TD-LTE PRIVATE NETWORK

The capacity of the TD-LTE system is decided by many factors: system bandwidth, scheduling algorithm, transmitter power, length of the cyclic prefix, resource allocation, sub-carrier spacing, time slot schedule, link overhead, MIMO, interference cancellation and so on. But in this paper, we only focus on the effects of the system bandwidth, resource allocation and link overhead(Xiao et al. 2012).

3.1 *System bandwidth*

The TD-LTE system has no special filter in the time domain, as well as frequency domain, but needs to set a transitional band to eliminate the impact of "waveform broadening" and "waveform oscillation" in the time domain. So not only a guard band should be set for carrier aggregation, but also a part of the band needs to be reversed for the protecting bandwidth in the TD-LTE system. Even with more guard bands, the leakage of energy out of the system bandwidth will be smaller. But considering a balance between the frequency spectrum efficiency and the system protection, we assign 10 percent of the frequency resource for the guard band. As a result, we would get 20 kHz per discrete frequency point at 230 MHz for the power wireless communication. In TD-LTE system, the subcarrier spacing Δf is 15kHz. So the total number of subcarriers in 230MHz TD-LTE system would be:

$$N_{sc} = \frac{2000}{\Delta f} \times 40 = 53 \tag{1}$$

The number of symbols per time slot would be:

$$N_{symb} = \frac{0.5}{T_{symbol}} \times 2 = 14 \tag{2}$$

The symbol period is: $T_{symbol} = 66.67$ μs, and the number of the Resource Blocks per time slot would be:

$$N_{RB} = \left\lfloor \frac{N_{sc}}{12} \right\rfloor = 4 \tag{3}$$

3.2 *The configuration of uplink-and-downlink time slot and special sub-frame*

The TD-LTE system supports two switching period: 5ms and 10ms, in total 7 kinds of the uplink-and-downlink time slot configuration. Meanwhile, in order to save the cost of the network, special time slot such as DwPTS and UpPTS can be used to transfer data and the system control information.

3.3 *Resource grid*

In the TD-LTE system, whether uplink or downlink, the minimum resource unit is Resource Element (RE), the physical channels correspond to a range of REs to carry the high-level information. The signals, transferred in a slot, can be described as a resource grid. Taking the uplink for instance, there are N_{RB}^{UL} N_{sc}^{RB} subcarriers and N_{symb}^{UL} SC-FDMA symbols. N_{RB}^{UL} is the number of uplink RBs, decided by the configuration of the transmission bandwidth in the cell, and matching the formula: $N_{RB}^{min,UL} \le N_{RB}^{UL} \le N_{RB}^{max,UL}$, the number of SC-FDMA symbols per time slot, is decided by the high-level CP configuration.

3.4 *Cyclic prefix*

In order to avoid serious Inter-Symbol Interference (ISI) and Inter-Carrier Interference (ICI), the length of CP should be much larger than the maximum allowable delay spread of the channel. In TD-LTE system, when it is normal CP, the number of symbols in uplink each slot N_{symb}^{UL} would be 7, when it is Extended CP, N_{symb}^{UL} would be 6.

3.5 *Downlink overhead*

In TD-LTE system, the downlink overhead includes Reference Signal (RS), Synchronization Signal, Physical Broadcast Channel (PBCH), Physical Control Format Indicator Channel (PCFICH), Physical HARQ Indicator Channel (PHICH) and Physical Downlink Control Channel (PDCCH). This article assumes that RS, PCFICH, PHICH and PDCCH completely occupies the first one or two OFDM symbols of the DwPTS, so PDSCH data will not be transmitted at the first one and two OFDM symbols of DwPTS.

3.6 *Uplink overhead*

In TD-LTE system, uplink overhead includes DMRS (Demodulation Reference Signal), PUCCH (Physical Uplink Control Channel) and SRS (Sounding Reference Signal).

3.7 *Transmission rate of 230MHz TD-LTE power special network*

In this paper, the system is configured as follows: 1MHz system bandwidth, 2×2 antenna, normal CP, the uplink and downlink configuration-NUM0, the uplink and downlink sub-frame ratio of 3: 1, the special sub-frame configuration of 10: 2: 2, 2 OFDM symbols in DwPTS for the transmission of control information(Deng & Zhang 2011).

3.7.1 *Downlink transmission rate*

Downlink transmission speed is related to the number of REs for PDSCH in downlink sub-frame and the number of RE for PDCCH in DwPTS. Each frame has two sub-frames for downlink, in total 1344 REs. Except the first two OFDM symbols in DwPTS for PDCCH, there are 768 REs left. And, 744 REs is assigned for the link spending (192 REs for RS, 288 REs for P-SCH, 288 REs for S-SCH, 264 REs for PBCH), so there are 1080 REs left for PDSCH. With different channel environments, the downlink transmission rate of 230MHz TD-LTE private networks would be:

$$Rate_{DL} = N_{RE} \times ModType \times \frac{CodeRate}{1024} \tag{4}$$

In the formula, NRE is the number of REs for PDSCH per frame, ModType is the modulation scheme, CodeRate/1024 is the validity of the modulation coding.

3.7.2 *Uplink transmission rate*

The Uplink transmission rate is related to the number of REs for PUSCH in the uplink sub-frame. Each frame has six sub-frames for uplink. At the edge of the frequency band, 2 RBs are assigned to PUCCH per time slot, so 2 RBs are left for PUSCH. Besides, 1 OFDM symbol would be assigned for transmitting the uplink RS per time slot, so in total 1728 REs can be used for PUSCH. With different channel environments, TD-LTE uplink transmission rate of 230MHz TD-LTE private networks would be:

$$Rate_{UL} = N_{RE} \times ModType \times \frac{CodeRate}{1024} \tag{5}$$

In this formula, N_{RE} is the number of REs for PUSCH per frame, ModType is the modulation scheme, CodeRate/1024 is the validity of the modulation coding.

What is shown in Figure 3 is the uplink maximum transmission rate of 230MHz TD-LTE private network system through an AWGN channel with different modulation and coding schemes, and we can see that: with the QPSK modulation mode, the uplink maximum transmission rate can reach 198.4 kbps; with the 16QAM modulation mode, the uplink maximum transmission rate can reach 406.05kbps; with the 64QAM modulation mode, the uplink maximum transmission rate reaches 959.9kbps.

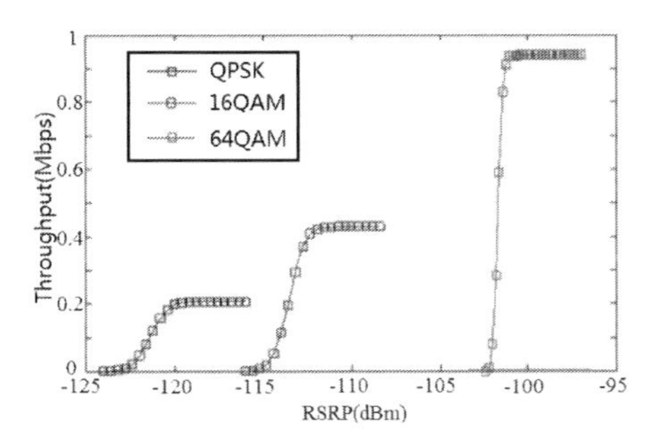

Figure 3. Transmission rate of uplink with different MCS.

4 THE BUSINESS OF POWER WIRELESS NETWORK

The power wireless network is at the end of the power communication network, and contacts with the users directly. It has many characteristics, such as: a large number of terminal nodes, widespread distribution, unbalance density of the nodes and low communication traffic per node. Distribution and utilization communication, as the primary application, is mainly responsible for the transmission of the power distribution production business data and marketing business data. Power distribution production business includes tele-control, tele-command and telemetering of the distribution equipment. And marketing business mainly includes the electric energy data acquisition and load management(Gong et al. 2013).

4.1 *Distribution business*

The main devices of the distribution network are PTB(Pole Top Breaker), RMU(Ring Main Unit), SS(Switching Station), PT(Pole Transformer) and BT(Box Transformer). The data volumes of these devices are shown in the Table 1.

Table 1. Distribution service.

	Business Data(Byte)			
Terminal	Tele-command	Tele-meter	Tele-control	Degree of Ammeter
PTB	2	22	10	0
SS	10	80	85	112
RMU	3	32	45	0
BT	2	28	60	0
PT	1	26	0	0

4.2 *Utilization business*

According to the property of the power users and the requirement of marketing business, the power users can be divided into six types: class A is a large-scale transformer user, class B is Small and medium-scale transformer user, class C is a three-phase commercial user, Class D is single-phase commercial user, class E is residential users, class F is public assessment measurement points.

According to Power User Electric Energy Data Acquire System Functional Specification, the data acquisition of the power user power-expenditure information mainly includes the functions of automatic information acquisition, interrogation, information report, tele-control and parameter setting. The data volumes of the power users are shown in Table 2:

Table 2. Utilization service.

Terminal	Periodic Service (bps)	Events Upload	DL Control	Parameter Setting
A	145.92	240bit	87bit	400bit
B	86.19	220bit	87bit	400bit
C	8.53	227bit	87bit	400bit
D	8.53	193bit	87bit	400bit
E	14.93	218bit	87bit	400bit
F	21.1	210bit	87bit	400bit

4.3 *Business requirements*

In this paper, we assume that each distribution area has K output lines, and each distribution line has 15 pole top breakers, 2 switching stations, 8 ring main units, 30 box type substations, 50 pole transformers. And there are N electrical terminals are distributed in the distribution area, including 10 terminals of Class A, 90 terminals of Class B, 9 terminals of Class C, 81 terminals of Class D, 730 terminals of Class E, 60 terminals of Class F per 1000 utilization terminals. In order to simplify the calculation, K and N satisfy the following relationship:

$$K = N\%1000 \tag{6}$$

In the acquisition system, the uplink terminal number *NEvent*, who will transfer event information to the station per minute, submits to Poisson Distribution:

$$P\left(N_{Event} = k\right) = \frac{e^{-\lambda}\lambda^{k}}{k!}, \lambda = 0.001 \times N \tag{7}$$

Figure 4 shows the uplink traffic distribution of the distribution and utilization information acquisition system with different numbers of the distribution and utilization terminals, it can be calculated as follows:

$$Tr_{UL} = Per_{Dis} + Per_{Uti} + E \times Event \tag{8}$$

In the formula, Per_{Dis} is the uplink periodic service of the distribution terminals, including tele-command, tele-metering and degrees of ammeters. Per_{Uti} is the uplink periodic service of the utilization terminals. E is the number of the utilization terminals who upload event information.

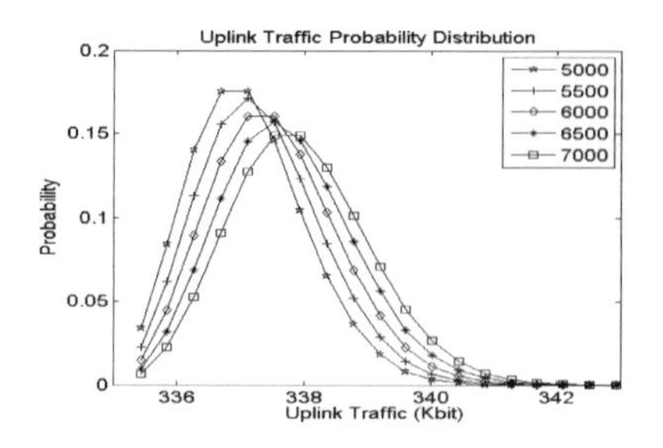

Figure 4. Uplink traffic probability distribution.

5 THE PERFORMANCE ANALYSIS OF 230MHZ TD-LTE SYSTEM

5.1 *Maximum urgency first*

In this simulation, we use Maximum Urgency First (MUF) algorithm to schedule the users. MUF takes the static priority scheduling algorithm and dynamic priority scheduling algorithm into account, and makes use of the dynamic priority and static priority to ensure the effective transmission of the burst traffic(Wang 2010).The priority of each task is called as the urgency, which consists of a static priority, a dynamic sub-priority and a static sub-priority, which submits to Poisson distribution.Event is the data size of the event information of one utilization terminal.

5.2 *Simulation environment*

The simulation is based on TD-LTE system dynamic simulation. The sampling interval is a TTI (1ms), and the simulation time is 600s. Each simulation has a different number of the terminals. Then we will get the success rate of Distribution and Utilization information acquisition with different conditions. The system parameters are shown in Table 3.

Table 3. System configuration.

Parameter	Setting
Scene	Dense urban
RB	180kHz
Channel	AWGN
MCS level	29 levels
Scheduling	MUF
CQI Delay	5ms
UE Transmit power	17 dBm
PathLoss	ITUR P1546 model
OFDM symbol number	14
Sub wave interval	15kHz
eNB Transmit power	46 dBm
Height of eNB antenna	30m
Height of UE antenna	1.5m
DL Timeslot	0,5
UL Timeslot	2,3,4,7,8,9

5.3 *Analysis of the simulation result*

In the Distribution and Utilization Information Acquisition System, the uplink capacity is limited. Therefore, the simulation focuses on the uplink channel.

When there are 6000 utilization terminals and 6 distribution outlines, we can see that 230MHz TD-LTE power wireless private network can effectively meet the needs of the distribution and utilization information acquisition. And it is in accordance with Power User Electric Energy Data Acquire System Functional Specification, that the success rate of the periodic information acquisition should be more than 99.5.

The simulation result is shown in Figure 5:

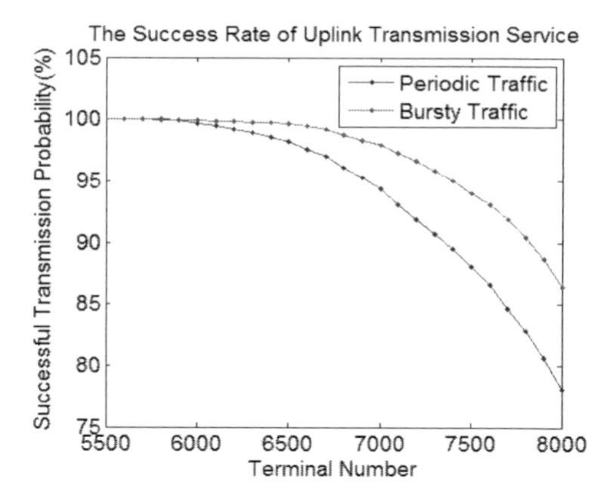

Figure 5. The success rate of uplink transmission service.

6 CONCLUSIONS

This paper proposed a novel network, 230 MHz TD-LTE private network, to build the wireless private network, which can provide a guarantee for the Strong Smart Grid. By using carrier aggregation, we can make effective use of those discrete frequencies in 230MHz frequency band. And through the analysis of the system simulation, we get the performance of the private network, which can provide references for the following construction of the private network.

REFERENCES

Cao Jinping, Liu Jianming & Li Xiangzhen. 2013. Carrier aggregation technology on 230 mhz dedicated spectrum of power systems. *Automation of Electric Power Systems* 37(12): 63–68.

Catr, M. 2009. Research on private network technology and development stratagem. *Technical report, SCDMA Wireless Broadband Industry Alliance*.

Deng Xu, Z. J. 2011. Analysis of td-lte capacity. *Mobile Communications* 35: 49–52.

Gong Gangjun, Xiong Chen, X. G. 2013. Research of communication business model of power distribution and utilization. *Power System Protection and Control* (22): 19–24.

Wang Xibo, T. X. 2010. A modified maximum urgency first algorithm and its implement. *Control and Automation* 26: 40–48.

Xiao Qinghua, Mao Zhuohua, Li Wenjie, et al. 2012. Comprehensive analysis of td-lte capacity cpability. *Designing Techniques of Posts and Telecommunications* 4: 36–40.

Zhiyi, K. 2010. Telecom service and communication equipment. *Technical report*. Donghai Securities.

Electronics, Communications and Networks IV – Hussain & Ivanovic (eds)
© 2015 Taylor & Francis Group, London, ISBN: 978-1-138-02830-2

Dependence on track pitch for focusing error signals in the land-groove-type optical disk

M. Shinoda*
Kyoto Works, Mitsubishi Electric Corporation, Kyoto, Japan

H. Nakatani
M-TEC Company Limited, Osaka, Japan

ABSTRACT: The behavior of the focusing error signals on the land and the groove in the land-groove-type optical disk is calculated theoretically, which is known for having peculiar offsets at the focus point. The dependence of the behavior of the focusing error signal on the track pitch is investigated by using the format of Digital Versatile Disc-Random Access Memory (DVD-RAM) as the land-groove-type optical disk model. The calculation results show that the focusing error signals strongly depend on the relationship among the elements of the optical intensity profile of the focus spot, in particular, spot size, defocus condition, and track pitch. The calculation method and results described in this paper will be reflected in the next generation of land-groove-type optical disks.

1 INTRODUCTION

An enormous amount of data is gradually distributed through the Internet because of the improvements in its traffic capacity and speed. However, optical disk is still an important means of distributing big data such as movie files and application software. Therefore, technological development for realizing high recording density has been progressing steadily. An optical disk, with a track pitch narrower than the previously used one, is one of the approaches for realizing high recording density. The technique of recording on both the land and the groove substantially meets this approach, since the recording density can be doubled without changing the track pitch and the optical parameters of the optical pickups. Many studies on land-groove-type optical disks have been reported (Miyagawa et al. 1993, 1996, Morita et al. 1997). In these studies, the track pitch, which is the most important parameter for realizing high recording density, is considered from the perspective of the qualities of reproduced signals, such as the amount of cross-talk or cross-erase. Of this type of optical disk, only the digital versatile disc-random access memory (DVD-RAM) (Yoshizawa et al. 1998) has been commercialized by virtue of the above issues related on the track pitch being overcome.

On the other hand, Kondo (2007) analyzed the focusing error signals on the land and the groove in DVD-RAM, and pointed out that peculiar offsets occurred at the focus point in the focusing error signals.

The authors of this paper are also interested in the above characteristics and have calculated the slope sensitivity of the focusing error signals in DVD-RAM (Shinoda et al. 2014). The calculation results show that the gain difference in the slope sensitivity of the focusing error signals on the land and the groove strongly depends on the optical intensity distribution in the optical pickup. Therefore, all the factors related to the gain difference should be restricted to ensure that its value is lower than that allowed value from the viewpoint of stable focusing servo control.

In the calculation mentioned above, general purpose optical disk modeling software "DIFFRACT" (Mansuripur 2002) which is based on the quasi-vector diffraction theory, is used. The applicability of this software is verified in many papers and also in the authors' papers, which address the theoretical calculations related to the tracking error signals in writable optical disks (Shinoda 2011, Shinoda et al. 2010, 2011).

In this paper, we calculate the focusing error signals under the condition of a track pitch narrower than that of the previously used one. This calculation is intended to realize a higher recording density than that of DVD-RAM, and investigate the dependence of the behavior of the focusing error signals on the track pitch.

2 CALCULATION CONDITIONS

Figure 1 shows an optical model of this calculation. The laser diode radiates diversely, and its far-field pattern is elliptic. This characteristic is specified as rim intensity which is the normalized optical intensity at the aperture of the objective lens. In this calculation, the values of

*Corresponding Author: *Shinoda.Masahisa@aj.MitsubishiElectric.co.jp*

the rim intensity for the tangential and radial directions are set to be 0.70 and 0.31, respectively.

The physical format of DVD-RAM is applied except for the track pitch which is the parameter addressed in this study. The configuration of the groove is binary, and its duty-cycle ratio is 0.5. The depth of the groove is set to be $l/6$. This means that the phase difference between the land and the groove in a single path is $\pi/3$. A cylindrical lens is used to generate the focusing error signal FE by the astigmatic method (Cohen et al. 1984). However, in this calculation, the Zernike aberration coefficient is introduced to generate the astigmatic aberration and therefore the linear zone of the focusing error signal becomes 7 mm. In order to simplify the calculation, the optics from the laser diode to the objective lens is considered to have no aberration.

Table 1 summarizes the disk parameters and optical parameters of the optical pickup used in this calculation.

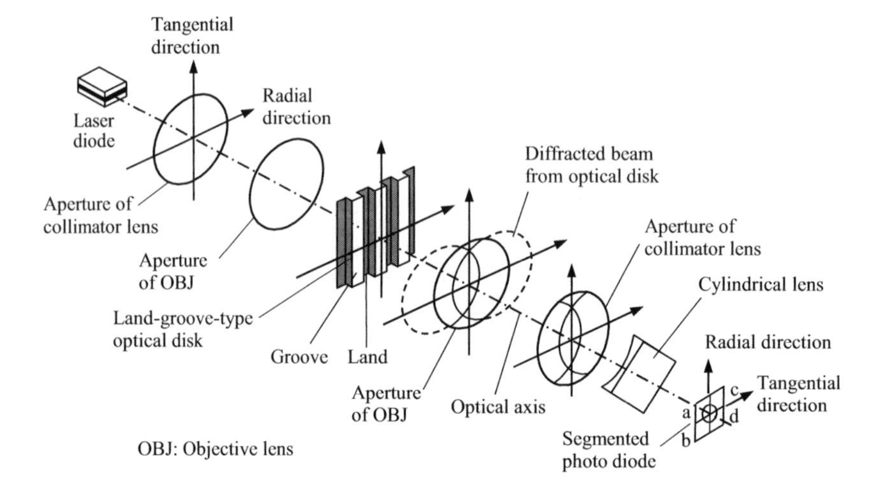

Figure 1. Optical model of this calculation.

Table 1. Disk and optical parameters in calculation.

Item		Symbol	Value	Unit
Optics	Wavelength	λ	0.65	µm
	Polarization		Circular	
Objective lens (OBJ)	Focal length		1.9	mm
	Numerical aperture	NA	0.65	
Rim intensity at the aperture of OBJ	Tangential		0.70	
	Radial		0.31	
Disk	Track pitch	T_p	Parameter	
	Groove shape		Binary	
	Duty-cycle ratio		0.5	
	Phase difference (single path)		$\pi/3$	rad

Figure 2 depicts the principle of the astigmatic method, which is introduced in this study, for detecting focusing errors. The focusing error signal FE is obtained from the outputs of a four-segment photo diode by using the operation:

$$FE = (A + C) - (B + D), \qquad (1)$$

where A-D are output signals from segments a-d, respectively.

When the laser beam is focused correctly on DVD-RAM, the optical intensity distribution on the photo diode becomes circular, and therefore, all signals, A-D, are equal. As a result, the focusing error signal FE is zero. When the laser beam is focused at the side near to DVD-RAM, the optical intensity distribution becomes elliptic by the astigmatic aberration of the cylindrical lens. Therefore, all the signals change, such that A, $C <$ B, D. This results in $FE < 0$. In contrast, when the laser beam is focused at the side far from DVD-RAM, all the signals change such that A, $C > B$, D. This results in $FE > 0$. According to the above characteristics, the operation of a focusing-servo-control circuit is such that the focusing error signal FE remains zero.

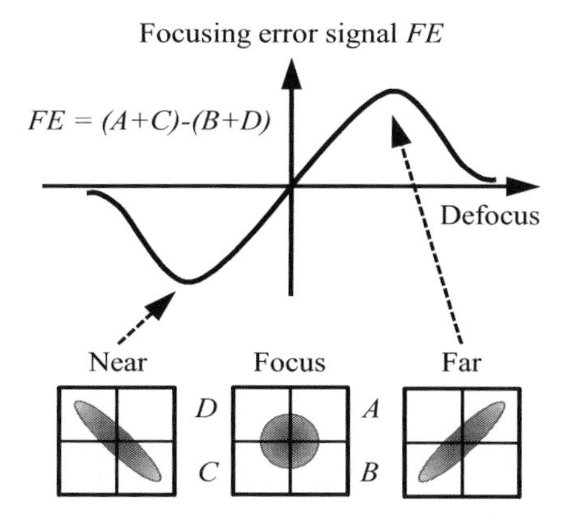

Figure 2. Principle of focusing error detection by the astigmatic method.

3 CALCULATION RESULTS

Figures 3(a)-(f) depict the calculated focusing error signals, in which the track pitches T_p are 1.23, 1.1, 1.0, 0.8, 0.7, and 0.5 mm, respectively. The focusing error signals strongly depend on the track pitch. The case of DVD-RAM in which the track pitch is 1.23 mm is shown in Figure 3(a). The focusing error signal from the land does not coincide with that from the groove. The result of this characteristic is that each focusing error signal has an offset around the focus point, as described by Kondo (2007).

When the track pitch is 1.1 mm, the focusing error signal from the land almost coincides with that from the groove but not perfectly; whereas, when the track pitch is 1.0 mm, the focusing error signals from the land and the groove coincide perfectly at the focus point. If the track pitch is smaller than 1.0 mm, the focusing error signals again do not coincide near the focus point. However, when the track pitch is 0.5 mm, the focusing error signals coincide perfectly at any defocus points including the focus point.

One more interesting behavior of the focusing error signals is that the relationship between the two signals in terms of the signal intensity near the focus point is reversed with a track pitch of 1.0 mm.

4 DISCUSSIONS

The objective of this study is to investigate the behaviors of the focusing error signals from the viewpoint of the relationship between the size of the focus spot and the track pitch. The optical intensity profile of the focus spot varies with the defocus. Figure 4 shows the spot size and the peak intensity of the focus spot versus the defocus value. Here, the spot size for the radial direction is defined as the diameter for which the optical intensity is e^{-2} at the focus spot peak. The peak intensity value is normalized as that at the focus point. If defocus occurs, the peak intensity decreases and the spot size increases. This means that the optical intensity distribution spreads toward a marginal region.

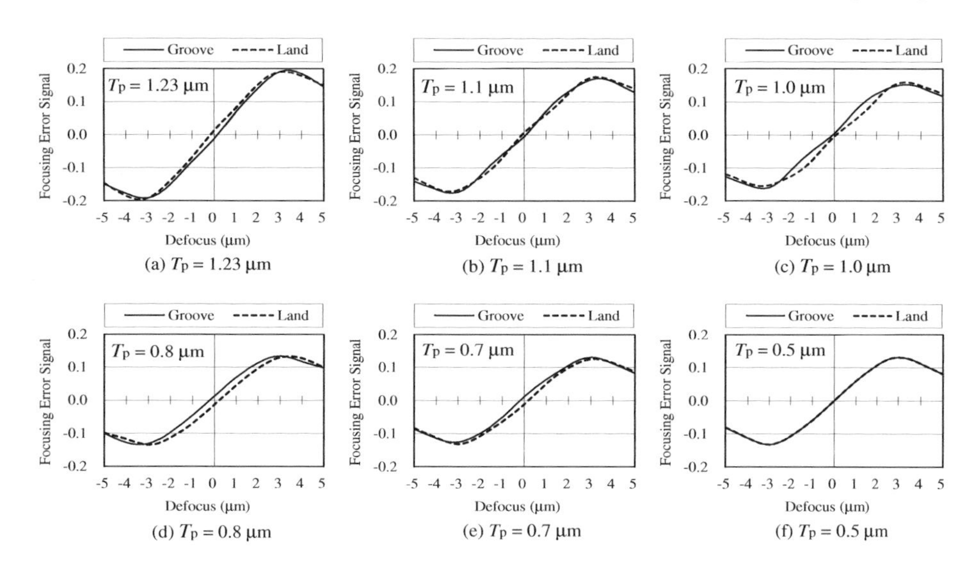

Figure 3. Calculated focusing error signals. The track pitches in Figures 3(a)-(f) are 1.23, 1.1, 1.0, 0.8, 0.7, and 0.5 μm, respectively.

Now, a resolution limit is introduced, RL, which is defined as

$$RL = \lambda / NA / 2. \qquad (2)$$

Here, λ is 0.65 mm and NA is 0.65, from Table 1, and therefore, the resolution limit RL in this calculation model is 0.5 mm. When the track pitch is 1.0 mm, as shown in Figure 3(c), its length is exactly double of the resolution limit. It is assumed that it is this relationship that causes the focusing error signals from the land and the groove to coincide perfectly at the focus point.

Figures 5(a) and (b) depict the relationship between the optical intensity profile of the focus spot and the land-groove disk size. The track pitch is 1 mm. The values of the defocus of the focus spot in these figures are 0 and 1.0 mm, respectively. When there is no defocus, as shown in Figure 5(a), the center portion of the optical intensity profile is irradiated on the land and both sides of its profile are irradiated on the groove. When there is a defocus of 1.0 mm, as shown in Figure 5(b), the center portion of the optical intensity profile decreases, which is irradiated on the land; while both sides of its profile increase, which are irradiated on the groove.

A similar study is applied in the case that the track pitch is 0.7 mm, shown in Figures 6(a) and (b). The defocus conditions are the same as those for the case shown in Figure 5. In this case, whereas the focus spot is located on the land, both sides of the optical intensity profile larger than those in Figures 5(a) and (b) are irradiated on the groove. In addition, the marginal portion is irradiated on the lands next to the center land.

When the track pitch is 0.5 mm, as shown in Figure 3(f), the focusing error signals from the land and the groove coincide perfectly at any defocus point. Since the value of the track pitch is the same as the resolution limit RL, the focus spot cannot discriminate between the land and the groove. Therefore, for practical use, this condition is not good.

In our calculations, the track pitch of 1.1 mm shown in Figure 3(b) rather than that of 1.23 mm in DVD-RAM standard has the advantage from the viewpoint of stable focusing servo control, because the focusing error signal from the land almost coincides with that from the groove. Contrary to this, the track pitch narrower than that of DVD-RAM makes the reproduced signal poor. For examples, the cross-talk from the adjacent land or groove will increase.

The possibility of erasing the signals will increase when the other signals are recorded on the adjacent tracks (cross-erase). The signal to noise ratio (SNR) of the reproduced signals will decrease. Therefore, the best value of the track pitch should be investigated from the viewpoint of both the reproduced signal qualities, such as the amount of cross-talk, cross-erase, and SNR as well as the stability of the focusing servo control.

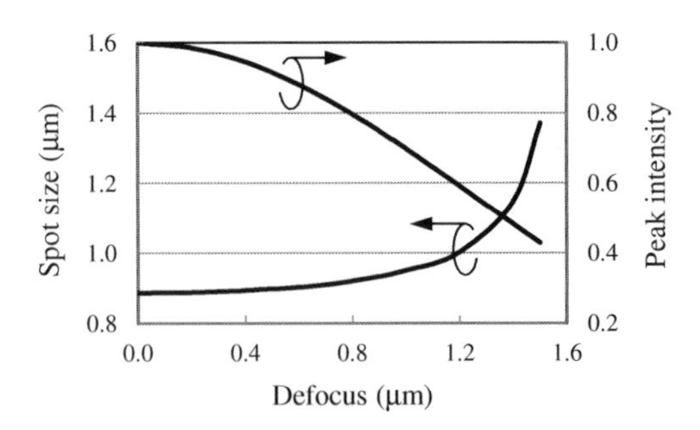

Figure 4. Spot size and peak intensity of the focus spot versus defocus.

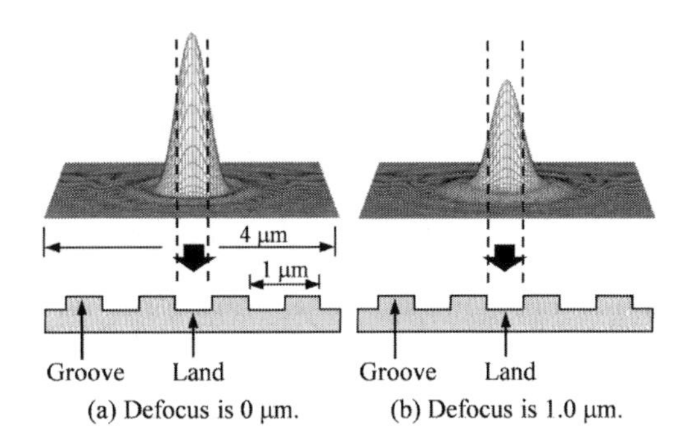

Figure 5. Relationship between the optical intensity profile of the focus spot and the land-groove disk size. The track pitch is 1.0 mm. The defocus in Figures 5(a) and (b) is 0 and 1.0 mm, respectively.

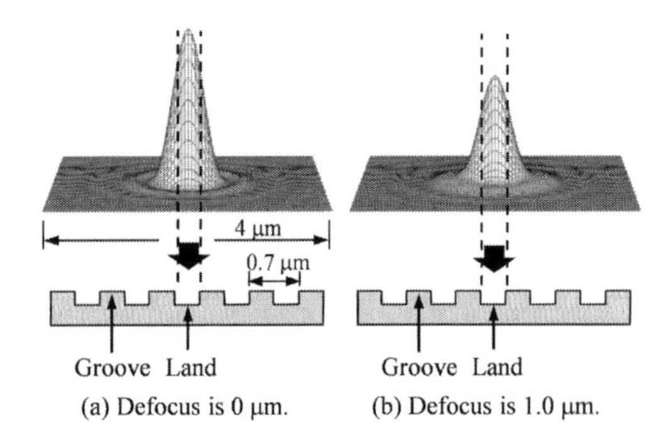

Figure 6. Relationship between the optical intensity profile of the focus spot and the land-groove disk size. The track pitch is 0.7 mm. The defocus in Figures 6(a) and (b) is 0 and 1.0 mm, respectively.

5 CONCLUSIONS

We calculated the focusing error signals under the condition of the narrower track pitch for realizing the higher recording density than that in DVD-RAM.

As a result, the behavior of the focusing error signals strongly depends on the relationship among the optical intensity profile of the focus spot, especially the spot size, the defocus condition, and the track pitch.

It becomes clear that the narrower track pitch of 1.1 mm rather than that of 1.23 mm in DVD-RAM standard has the advantage from the viewpoint of stable focusing servomechanism.

The calculation method and results described in this paper will be applicable to the next generation of optical disk with different track pitches and different land or groove widths as well as DVD-RAM.

REFERENCES

Cohen, D. K., Gee, W. H., Ludeke, M. & Lewkowicz, J. 1984. Automatic focus control: the astigmatic lens approach. *Appl. Opt.* 23: 565–570.

Kondo, K. 2007. Analysis of origin of offset seen in astigma method when we read land-groove disk. *Jpn. J. Appl. Phys.* 46: 229–231.

Mansuripur, M. (ed.). 2002. *Classical Optics and its Applications*. Cambridge: Cambridge University Press.

Miyagawa, N., Gotoh, Y., Ohno, E., Nishiuchi, K. & Akahira, N. 1993. Land and groove recording for high track density on phase-change optical disks. *Jpn. J. Appl. Phys.* 32: 5324–5328.

Miyagawa, N., Ohno, E., Nishiuchi, K. & Akahira, N. 1996. Phase change optical disk using land and groove method applicable to proposed super density rewritable disc specifications. *Jpn. J. Appl. Phys.* 35: 502–503.

Morita, S., Nishiyama, M. & Ueda, T. 1997. Super-high-density optical disk using deep groove method. *Jpn. J. Appl. Phys.* 36: 444–449.

Shinoda, M. & Nakai, K. 2011. Dependence of tracking error characteristics of in-line-type differential push-pull methods on arrangement error of segmented gratings. *Opt. Rev.* 18: 367–373.

Shinoda, M. 2011. In-line-type differential push-pull methods for tracking error detection. *IEEE Trans. Magn.* 47: 532–538.

Shinoda, M., Nakai, K. & Ohmaki, M. 2014. Analysis of behavior of focusing error signals in astigmatic method in the scheme of land-groove recording. *IEEE Trans. Magn.* 50: 3501904.

Shinoda, M., Nakai, K. & Takeshita, N. 2010. Dependence of tracking error characteristics on objective lens shift for in-line-type differential push-pull methods using segmented gratings. *Opt. Rev.* 17: 360–366.

Yoshizawa, T., Satoh, H. & Osawa, H. 1998. Digital versatile rewritable disc (DVD-RAM) complementary allocated pit address (CAPA) signal and tracking error signal detection method; Improving objective lens radial shift immunity. *JpnJ. Appl. Phys.* 37: 2255–2256.

Electronics, Communications and Networks IV – Hussain & Ivanovic (eds)
© 2015 Taylor & Francis Group, London, ISBN: 978-1-138-02830-2

Performance evaluation of SNMP, NETCONF and CWMP management protocols in wireless network

M. Słabicki & K. Grochla
Institute of Theoretical and Applied Informatics of the Polish Academy of Sciences, Gliwice, Poland

ABSTRACT: This paper compares the performance of three popular network management protocols: SNMP, Netconf and CWMP (TR069) applied to automate the configuration change and monitoring of wireless network equipment. The OMNeT++ discrete event simulator was used, with models of these three protocols. Two use cases were used for the evaluation: execution of a simple transaction (e.g. configuration change) and statistic collection of devices connected through a wireless network with different packet loss ratio. The results show that SNMP has the smallest time required to perform typical management operations, because of the smallest transmission overhead. Among the three protocols analyzed the CWMP requires the largest amount of time on average and has a highest standard deviation, what shows that the time required varies significantly. The reconfiguration of the devices for SNMP protocol required more time than the monitoring of the device statistics, but in NETCONF and CWMP it is heavily influenced by the link speed.

1 INTRODUCTION

The number of devices used to build a network constantly grows. The increasing size and complexity of the network makes the management of the network more complex and demand more time and effort. A few network management protocols have been proposed to automate remote management of network devices. They are used by the network management systems to communicate with the devices and perform operations related to reconfiguration, monitoring, fault detection, accounting and security. It allows to decrease the operational expenses of the network operator, by decreasing the human work related to the everyday management tasks of the network administrator. The complexity of the management of a network also grows as more devices are behind wireless channels. Currently the administrators need to control not only infrastructure devices, connected through a wired link to the network operation center, but also routers and bridges connected through IEEE 802.16 or IEEE 802.11 links.

The aim of our research was to evaluate, how popular management protocols behave in typical operations executed over a wireless network. Several methods to achieve this aim can be used. In (Carlsson et al. 2002) authors measured performance of the SNMP protocol for real devices, and the impact of a protocol implementation on different hardware. Authors in (Grochla & Naruszewicz 2012) proposed framework to emulate network devices. We decided to simulate network and all devices. In our research, we analyzed the only influence of the network features (delay, packet loss etc.). We chose three widely used management protocols: SNMP (Stallings 1998), Netconf (Enns et al. 2011) and CWMP (also known as TR.069) (Blackford & Digdon 2012). The expected outcome of the experiments was statistics of the time necessary to execute request realized by the management protocol. In our research, we used tree topology of the network, and two management procedures: to get data from network devices, and set value in network devices. To evaluate these protocols we provided models of protocols and determined the test environment which contained several different scenarios.

To the best of our knowledge, so far there is no comparison of the three selected protocols in wireless network scenario. Chen et al. (2013) evaluated the SNMP protocol used to manage Wireless Sensor Network. In the paper (Hedstrom et al. 2011) authors performed a quantitative analysis of the performance of the SNMP and Netconf protocols. They developed a simple test environment in which they investigated basic requests of protocols. Their results show that NETCONF is a reasonable alternative for recent complex networks. However, they did not take into the consideration delays related to transmission messages through the network, which could be important. On the other hand, paper (Sehgal et al. 2012) considers the usage of resources in SNMP and Netconf protocols implemented on embedded devices, but there is no evaluation of the influence of the network conditions on the transmission efficiency. There are quite a few papers analyzing a single protocol performance over wireless channels, such as (Kantorovitch & Mahonen 2002) or (Johnson 2009) for SNMP, but they do not provide the comparison to other network management protocols.

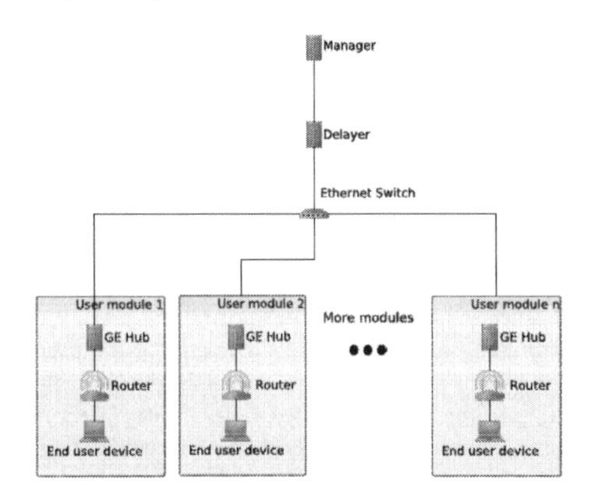

Figure 1. Network schema.

The rest of the paper is organized as follows: the simulation model of the network, proposed scenario and the model of protocols are described in Section 2. Results are shown in Section 3. The work is summarized in Section 4.

2 MODEL DESCRIPTION

In our research, we decided to use OMNeT++ (Varga et al. 2001). It is a well known discrete event simulator, designated to simulate networking protocols. To achieve a reasonable model of network behavior we used INET Framework, which adds to OMNeT++ support for simulating network based on TCP/IP.

2.1 *Network description*

Simulated network had a tree topology. The schema of the simulation scenario is shown in Figure 1. The central element of the network was a network manager. It was connected through an Ethernet switch to user modules. Between the manager and Ethernet switch we inserted a module "Delayer". Its aim was to simulate a delay of the connection between the end-user and backbone (ADSL or Ethernet). In the Ethernet connection, we assumed that data rate is symmetric, and is equal 1 *Gbps*. In ADSL case the throughput to devices was equal 8 *Mbps*, and to manager was equal 0.5 *Mbps*. The delay in "Delayer" module was calculated as a function of throughput of the channel and packet length. Moreover, we added delay of the core network. The delay of the core network was constant, and was equal 10 ms.

Because in our considerations was the performance of the network management protocols cooperation with equipment located in user private networks, we assumed that the user equipment is installed in private homes (for example behind a router with WLAN). The features of the WLAN were added by a module called "GE Hub". Wireless channel model in WLAN was based on two separated models. To model packets loss in wireless channel we implemented a Gilbert-Elliot model (Gilbert 1960), (Elliott 1963). In our simulations each of the end user devices had their own Gilbert-Elliot module. Separate Gilbert-Elliot modules generated for each device independent periods of error-free communication. The implemented model used data from (Aráuz & Krishnamurthy 2003). We used parameters related to two WLAN throughput: 5.5 *Mbps* any 11 *Mbps*; and frames with a length of 1000 *bytes*. These parameters were marked respectively as a GEI and GEII. Our implementation gave similar Packet Error Rate (Aráuz & Krishnamurthy 2003), respectively, 1.35 % and 7.98 %. We compared these GEI and GEII results with the simpler model based on percentage packet loss. They are labeled PERI and PER II. The packet loss level was the same as in the GEI and GEII. To model the propagation delay in the WLAN we developed our new model. The main idea of this model was that the biggest impact on the delay has the backoff time involved with collisions in the MAC layer. In the paper (Issariyakul et al. 2005) authors investigated the distribution of the delay time. We used their data on delay probability in our model. To sum up we can expect the delay of the simple packet sent from a manager to user device should be the sum of the elementary delays as follows:

- constant delay of the backbone rate – 10 *ms*,
- delay of the connection between the end-user and backbone (depends on link datarate set in Delayer and the packet length),
- delay related to backoff time in WLAN MAC layer,
- delay related to datarate in WLAN channel (we assumed that efficient datarate of this channel is 3 *Mbps* or 6 *Mbps* respectively if theoretical throughput is 5.5 *Mbps* or 11 *Mbps*).

2.2 *SNMP protocol – model*

SNMP (Stallings 1998) is a management protocol which uses UDP as a transport layer. Typical communication looks as follows:

- *Manager* sends a UDP packet with GET/SET request,
- *Device* sends a UDP packet with GET/SET response.

In our consideration, we do not define the exact version of SNMP protocol. There is no significant difference between protocol versions in simple requests evaluated in this research. Packet sizes used in the simulations were determined by capture via Wireshark example packets of SNMP protocol and averaged sizes from the series of the same packets. Simulation parameters are shown in Table 1.

Table 1. Parameters of the SNMP and Netconf simulations.

Protocol	SNMP		Netconf	
Scenario	GET	SET	GET	SET
Request packet size	82 *B*	140 *B*	407 *B*	509 *B*
Response packet size	100 *B*	155 *B*	506 *B*	337 *B*
Timeout to resend	2 *s*	2 *s*	2 *s*	2 *s*
Time to next procedure	10 *s*	10 *s*	10 *s*	10 *s*
Request processing time	0.1 *s*	0.2 *s*	0.1 *s*	0.2 *s*

2.3 *Netconf protocol – model*

Netconf enns2011netconf is a management protocol proposed by IETF. It is based on remote procedure call (RPC) paradigm. The network manager creates an RPC request, code it in the XML and sends to a network device using a secure, connection-oriented session. The network device sends response in a similar way. The connection between devices is set and finished by the network manager. In our simulation we used raw TCP sockets, but it is possible to use more sophisticated protocols e.g. SSH. The message trace in our implementation looked as follows:

- *Manager* begins a TCP connection,
- *Manager* sends a GET/SET request,
- *Device* sends a GET/SET response,
- *Manager* closes a TCP connection.

Packet sizes used in the simulations were determined by capture via Wireshark example packets of Netconf protocol and averaged sizes from series of the same packets. Parameters of the simulation are shown in Table 1.

2.4 *TR-069 protocol – model*

CWMP, which is commonly called TR-069 (Blackford & Digdon 2012), is a protocol described by the Broadband Forum. The protocol gives availability to secure auto-configuration, management of the firmware and software, monitoring and diagnostics of the network devices. The main difference between TR-069 and previous protocols is that the connection is begun by the user device. In many cases the user devices are hidden, e.g. behind NAT or firewall, and it is impossible to begin connecting to the user device from the network manager. TR-069 messages are sent as an HTTP requests and responses. HTTP server are running on the network manager. In our simulations we used raw TCP sockets and size of all requests were increased by the size of the HTTP overhead. A typical sequence in our TR-069 implementation looked as follows:

- *Device* begins a TCP connection,
- *Device* sends Inform request (HTTP Post),
- *Manager* sends Inform reply (HTTP Response),
- *Device* sends empty message (HTTP Post),
- *Manager* sends GET/SET request (HTTP Response),
- *Device* sends GET/SET response (HTTP Post),
- *Manager* sends empty message (HTTP Response),
- *Device* closes a TCP connection.

Table 2. Parameters of the TR-069 simulation.

Scenario	GET	SET
Inform request size	2508 *B*	2508 *B*
Inform response size	831 *B*	831 *B*
Empty request size	196 *B*	196 *B*
Request size	500 *B*	700 *B*
Response size	1023 *B*	400 *B*
Empty request size	196 *B*	196 *B*
Inform proc. time	0.1 *s*	0.1 *s*
Empty packet proc. time	0.1 *s*	0.1 *s*
Message proc. time	0.1 *s*	0.2 *s*
Time to next procedure	10 *s*	10 *s*
Waiting time after failure	3 *s*	3 *s*

Packet sizes used in simulations were determined by capture via Wireshark example packets of TR069 protocol and averaged sizes from series of the same packets. Parameters used in the simulation are shown in Table 2.

2.5 *Monitored data*

In our simulation we assumed that network manager intends to execute request for two simple commands: GET (to receive the value of a variable), and SET (to set value of a variable). In SNMP and Netconf scenario the procedure is started by manager, in TR069 communication is started by the user's device. Both requests were simulated in separate scenarios. We wanted to check the distribution of the time that is necessary to perform the whole request. To have comparable results we defined the way to gather data. In SNMP the time of the request was measured from sending the request to receiving the response. If the response was lost, the time expands on the request resend. In Netconf and TR-069 the time of the request was measured from the start of the TCP initializing to receive a Response packet. The new procedure was not started before the end of the last one or before the procedure was interrupted by a failure event (in TCP) or a timeout (in UDP). All cases were repeated 50 times and results were averaged.

2.6 *Evaluated scenarios*

We decided to evaluate parameters as follows:

- backbone datarate (ADSL or Ethernet),
- different WLAN channel (5 *Mbps* or 11 *Mbps*),
- two management procedures (GET or SET).

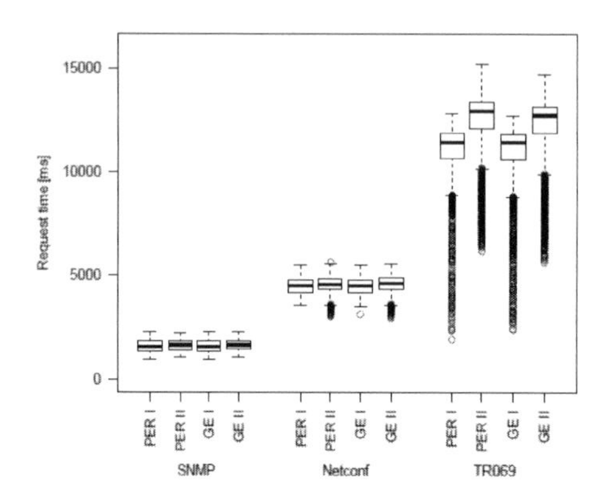

Figure 2. GET request time for 200 nodes with ADSL connection.

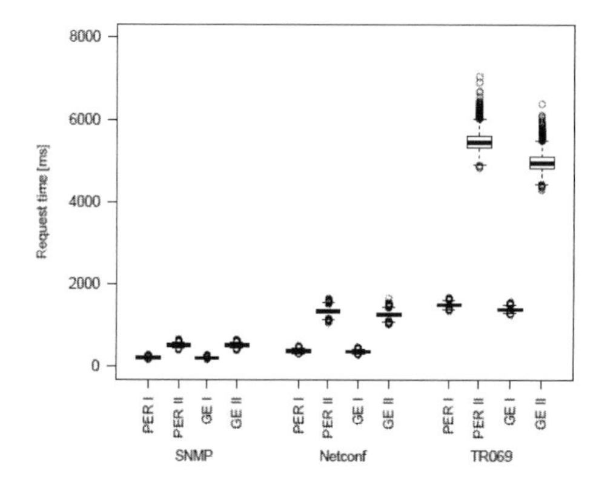

Figure 3. GET request time for 200 nodes with ethernet connection.

3 RESULTS

We present results from the network with manager and 200 network elements connected via ADSL or Ethernet. GE model used in simulation to pretend connection via WLAN with 5.5 *Mbps* (GEI and PERI) or 11 *Mbps* (GEII and PERII). The network manager sends 1000 GET (or SET) request to each network element. After finishing all requests, we repeated the simulation (50 repetitions).

3.1 *Request time distribution*

Figures 2 and 3 show the distribution of GET request time in different protocols. Respectively Figure 4 and 5 shows the distribution in SET request. Time achieved in network with ADSL is generally longer than in network with Ethernet. In all cases, SNMP achieved the smallest results and TR069 the highest one. The loss rate (in both loss models) affects achieved results – higher level gave longer mean time. Our results show that TR069 is sensitive to delays and packet losses.

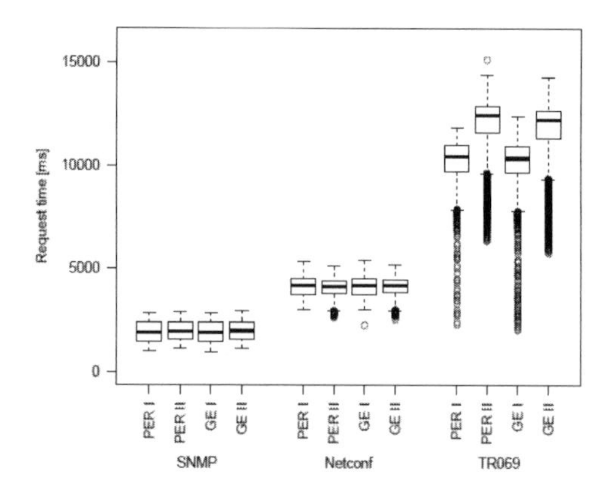

Figure 4. SET request time for 200 nodes with ADSL connection.

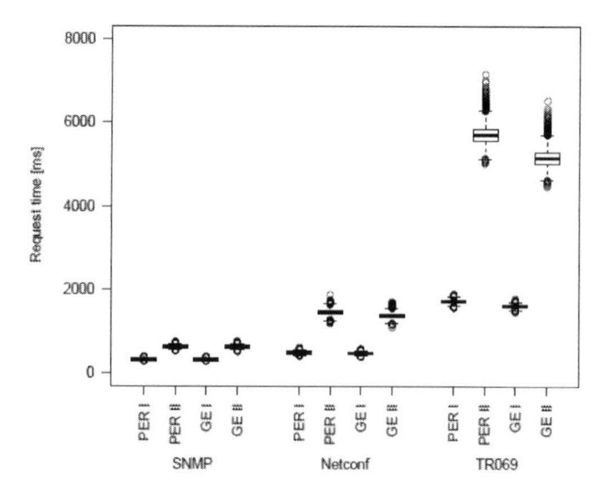

Figure 5. SET request time for 200 nodes with ethernet connection.

3.2 *Histograms*

To show the time of request distribution in all protocols, we summarized the histograms in all nodes for each simulation. Figure 6 shows a histogram of the GET requests in a network with an ADSL link. It can be noticed, that in SNMP protocol requests are finished in turns – they are shown as distinct peaks. It is due to the simplicity of the SNMP protocol. It contains only request and reply packets. If the packet is lost the next try is started after timeout. That is why the peaks are clear. Next protocols are more complex. They use TCP (so there is retransmission in the lower layer) and they use more packets to establish connection and send a request. Because of that fact, the histogram does not have clear peaks. There is the significant difference between Netconf and TR069 – Netconf needs less packets to finish the whole request. That is why the Netconf had wider peaks, and TR069 do not have peaks, can be also observed, but they are less visible.

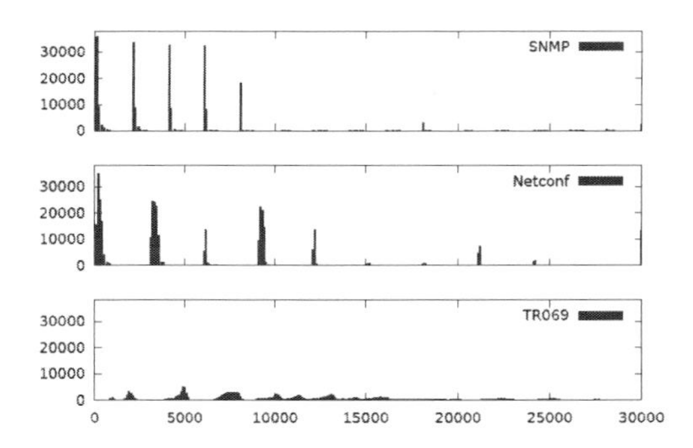

Figure 6. Histogram of requests time (ms) in all nodes (sum of node histograms) [ADSL, 200 nodes in network].

Table 3. Aggregated request execution time [in milliseconds].

		GET		SET	
		mean	std.dev	mean	std.dev
	GE I	1608	1371	1948	1432
SNMP	GE II	1676	1524	2023	1620
ADSL	PER I	1608	1372	1948	1432
	PER II	1673	1522	2020	1618
	GE I	212	329	316	328
SNMP	GE II	521	921	625	920
Eth.	PER I	214	334	318	334
	PER II	521	920	626	920
	GE I	4513	3719	4174	3748
Netconf	GE II	4633	4463	4195	4344
ADSL	PER I	4508	3707	4168	3734
	PER II	4590	4444	4116	4257
	GE I	371	686	471	688
Netconf	GE II	1274	2234	1373	2232
Eth.	PER I	388	726	487	727
	PER II	1352	2313	1453	2315
	GE I	11109	8815	10194	7308
TR069	GE II	12381	15540	11894	14257
ADSL	PER I	11157	8766	10285	7330
	PER II	12627	15319	12125	14167
	GE I	1403	1200	1601	1201
TR069	GE II	4968	6321	5169	6425
Eth.	PER I	1508	1311	1711	1317
	PER II	5471	6627	5714	6662

3.3 *Numerical results*

In Table 3, we show aggregated numerical results from simulations. The output of one simulation is a mean time needed to finish a request on all network devices. The table contains a mean values and standard deviations of these outputs (averaged from 50 simulation repetitions). In SNMP protocol the GET procedure is on average faster than SET procedure. In Netconf and TR069 it depends on major factor – if more important is the link speed (in ADSL scenario, GET gives longer values) or execution time (in Ethernet scenario, GET gives smaller results).

4 SUMMARY

Research shows that evaluated management protocols have different protocol overhead. This overhead influences the expected time necessary to finish the request. Although the Netconf and TR069 protocols are more sophisticated (they use TCP as a transport layer), the SNMP protocol achieved the smallest average time needed for transactions in all assumed scenarios. It is related to the small protocol overhead. Among the three protocols analyzed the TR069 requires largest amount of time on average and has highest standard deviation, what shows that the time required variates significantly. SNMP protocol requires more time for reconfiguration than for the monitoring of the device statistics. In NETCONF and TR069 it depends on the link speed.

ACKNOWLEDGMENTS

This work was funded by the Polish National Center for Research and Development, grant no LIDER/10/194/L-3/11.

REFERENCES

Aráuz, J. & Krishnamurthy, P. 2003. Markov modeling of 802.11 channels. Proceedings of the IEEE Vehicular Technology Conference 2: 771–775.

Blackford, J. & Digdon, M. 2013. TR-069 CPE WAN Mangement Protocol Issue: 1 Amendment 5 version 1.4.

Carlsson, P. et al. 2002. Obtaining reliable bit rate measurements in SNMP-managed networks. Proceedings of the 15th ITC Specialists Seminar on Traffic Engineering and Traffic Management: 114–123.

Chen, W. et al. 2013. Wireless Sensor Network Management Based on SNMP Protocol. Applied Mechanics and Materials 303: 292–296.

Elliott, E. 1963. Estimates of Error Rates for Codes on Burst-Noise Channels. Bell system technical journal 42(5): 1977–1997.

Enns, R. et al. 2011. Network Configuration Protocol (NETCONF), IETF RFC 6241.

Gilbert, E. N. 1960. Capacity of a Burst-Noise Channel. Bell system technical journal 39(5): 1253–1265.

Grochla, K. & Naruszewicz, L. 2012. Testing and Scalability Analysis of Network Management Systems Using Device Emulation. In A. Kwiecien et al. (eds.), Computer Networks 19th International Conference, CN 2012, Szczyrk, Poland, June 19–23, 2012. Proceedings: 91–100. Berlin: Springer.

Hedstrom, B. et al. 2011. Protocol Efficiencies of NETCONF versus SNMP for Configuration Management Functions. Masters thesis, University of Colorado.

Issariyakul, T. et al. 2005. Exact Distribution of access delay in IEEE 802.11 DCF MAC. Proceedings of the Global Telecommunications Conference 5: 2534–2538.

Johnson, R.B. 2009. Evaluating the use of SNMP as a wireless network monitoring tool for IEEE 802.11 wireless networks. Masters thesis, Clemson University.

Kantorovitch, J. & Mahonen, P. 2002. Case studies and experiments of SNMP in wireless networks. Proceedings of the IEEE Workshop on IP Operations and Management: 179–183.

Sehgal, A. et al. 2012. Management of resource constrained devices in the Internet of Things. IEEE Communications Magazine 50(12): 144–149.

Stallings, W. (ed.) 1998. SNMP, SNMPv2, SNMPv3, and RMON 1 and 2. Boston: Addison-Wesley Longman Publishing Co., Inc.

Varga, A. 2001. The OMNeT++ discrete event simulation system. Proceedings of the European Simulation Multiconference 9.

Electronics, Communications and Networks IV – Hussain & Ivanovic (eds)

FPGA based physical layer protocol and rapid synchronization in FSO communication

Wanxin Su
Changchun Institute of Optics, Fine Mechanics and Physics, Chinese Academy of Sciences, Changchun, China

ABSTRACT: A small factor FPGA based Free-Space-Optical communication platform is demonstrated. To reduce the bandwidth cost in the handshaking procedure, a rapid synchronization method is applied. Since the system is targeted at terrestrial communication use, the link channel is greatly influenced by the transmission media, i.e. the atmosphere. Firstly, the atmospheric channel causes an attenuation of the carrier power due to the absorption and the scattering effect of the atmosphere molecules and aerosols. Secondly, the turbulence of the atmosphere causes the laser power to scintillate, which may cause logic discrimination error leading to a much higher bit error rate and frequent resynchronization. To overcome this problem, this paper uses the locally generated clock as the sampling clock and the multi-phase sampling technique to improve the synchronization speed. As proved by experiments, the presented technique realizes more efficient synchronization than PLL-based clock recovery and saves the bandwidth cost when channel failure happens.

KEYWORDS: FSO; Physical Layer; Synchronization; FPGA

1 INTRODUCTION

Terrestrial Free Space Optical Systems (Scott et al. 2003) uses a narrow laser beam that travels through the atmosphere as the link carrier. The atmosphere link greatly influences the carrier transmission by attenuating the transmitting power and scintillating the received power, which make the channel more challenging to be a transmission media.

To construct a usable FSO system, the atmospheric channel must be analyzed, according to which the hardware is designed and the application specified protocol is developed. The hardware and the protocol demonstrated here are specifically designed for terrestrial FSO use, whereas they could also be used in any other communication media. To reduce the bandwidth waste in the frequent resynchronization, special technique is implemented to speed up the synchronization process.

The experiments prove that the demonstrated hardware and the protocol can be used in an FSO link and can improve the bandwidth efficiency due to the rapid synchronization technique.

2 THE UNSTABLE LINK CHANNEL

The main influence of the atmosphere on the laser carrier can be summarized by the atmospheric absorption, the atmospheric scattering and the atmospheric turbulence (Muhammad et al. 2005).

The atmospheric absorption includes the gas absorption and the water absorption caused by the atmospheric molecules and water molecules in the atmosphere. The commonly used FSO frequency band is the near-IR band, in which the gas absorption is much lower than the water absorption. So the gas absorption is usually not taken into account. By choosing the carrier laser band in the "atmospheric windows", the water absorption is so weak that the power loss by water absorption can be ignored. There are mainly two atmospheric windows in the near-IR frequency band, 780–850 nm and 1520–1600 nm. In the demonstrated platform, the 1550nm laser is adopted because it's safer to human eyes under the same output power (Isaac et al. 2001).

The atmospheric scattering can be described by Rayleigh scattering, Mie scattering and geometric scattering. In the atmospheric windows commonly adopted in FSO communication systems, the atmospheric molecules are the main cause of Rayleigh scattering. If the system chooses the 1550ns band as the carrier, the influence caused by Rayleigh can be low enough not to be considered.

In near-IR terrestrial FSO system, the main cause of Mie scattering is aerosols in the atmosphere with diameters of 0.03 um – 2000 um. These aerosols include the solid or liquid particles, among which the fog and haze are the most important sources of the attenuation (Penndorf. 1957). Equ. (1) gives an empirical expression of the Mie scattering coefficient, in which V_M equals the visibility of the environment, and q is given by Equ.(2) as an empirical constant and it's a function of the visibility.

When the size parameter of the atmospheric particles becomes larger than 50, the molecules are large enough to consider their scattering as geometric scattering and to use geometric optics to analyze the effect. Such particles include rain, snow, heavy fog, hail, etc.; refer to (Shaban et al. 2010, Tatarko et al. 2012).

Another influence of the atmosphere on the link is variation of the refractive index caused by the atmospheric turbulence. The result of the variation is the scintillation of the received optic power, the drifting and the spreading of the beam and the speckle effect.

$$\sigma_a = \frac{3.912}{V_M}\left[\frac{\lambda}{550nm}\right]^{-q} \tag{1}$$

$$q = \begin{cases} 1.6, & (V_M > 50km) \\ 1.3, & (6km < V_M < 50km) \\ 0.16V_M + 0.34, & (1km < V_M < 6km) \\ V_M - 0.5, & (0.5km < V_M < 1km) \\ 0, & (V_M < 0.5km) \end{cases} \tag{2}$$

3 HARDWARE STRUCTURE

The demonstrated system is a point-to-point system with two terminals, each of which has a transmitter and a receiver. The system uses an FPGA as the central unit and has full-duplex capability, as shown in Fig.1.

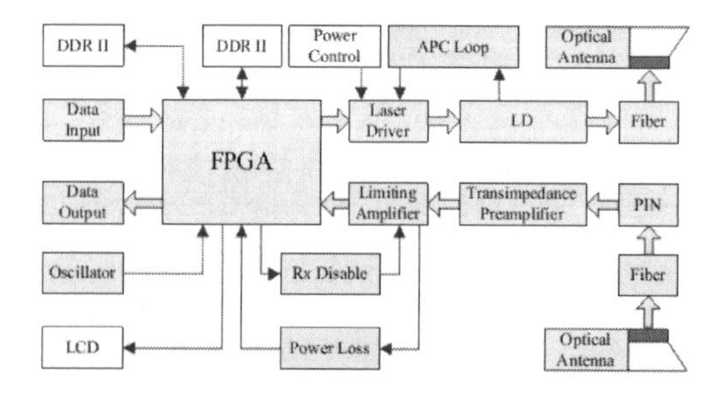

Figure 1. FSO system hardware structure.

The transceiving protocol resides in a single FPGA chip, including the transmitting protocol and the receiving protocol. The transmitting protocol receives data from an external source through a dedicated parallel interface. Then it buffers the data in an FIFO, encodes the data, and serializes the data to output them to the laser driving subsystem. LVDS technology is used between the laser driving subsystem and the FPGA chip to improve signal integrity and the communication data rate.

For the receiving subsystem, the collected optical signal feeds the coupling fiber before it is sent to the detector sensor. Then the amplifier and signal shaper amplify the weak current signal and convert it to a voltage signal, which is also sent to the receiving protocol in the FPGA chip through an LVDS interface. The protocol samples the input stream and deserializes it, after which it determines if the received signal is data or a command. If the received content is data, the protocol buffers the data in an FIFO so that the external data sink can read the data through a parallel interface. If the content is a command, the protocol executes it.

3.1 *Auto power control loop*

Fig.2 demonstrates the laser driving subsystem. To prevent the output power drift due to temperature change and aging of the laser, there must be an auto power control mechanism. In the demonstrated system, a power monitor PIN is used to realize the power stabilization. The output optic power is defined as two parts, the high power and the low power. When the output is logic 0, the optic power is the low power; otherwise the optic power is the combination of the low power and the high power if the output is logic 1. The PIN collects part of the output power and feeds the low pass filter. Then the signal is compared with the low power set value and modifies the output of A2 to keep the average optic power constant over time. The low power set value and the high power set value can be determined by two requirements, the required output power and extinction ratio. To use this method to keep the average output power constant, there should be a balance of the possibilities of the high optic power and the low optic power, or the baseline excursion will happen.

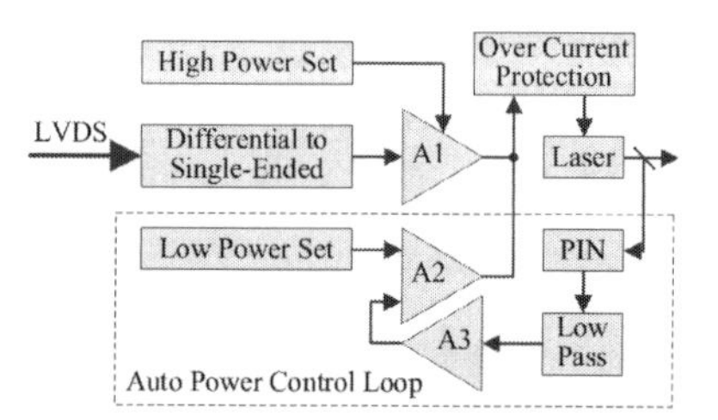

Figure 2. Laser driver subsystem

3.2 *Signal amplifying and shaping*

Fig.3 demonstrates the signal amplifying and shaping procedure in the receiving subsystem. It consists of a transimpedance amplifier and a limiting amplifier. The transimpedance amplifier is a max3665 chip used to transform the weak signal of the detector to a voltage signal and cancel the DC component of the signal. The limiting amplifier is a max3748 chip used to amplify the weak voltage and shape the output to fit the LVDS bus.

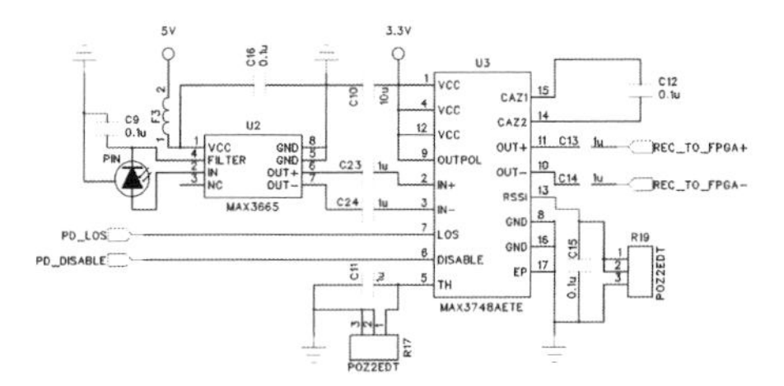

Figure 3. Signal amplifier and shaping.

The MAX3665 chip has got a DC cancellation feature(MAX3665 datasheet). It implements a low-frequency feedback to remove the DC component of the received signal, thus to center the signal within the dynamic range of the chip. Because of this DC cancellation feature, the transceiving protocol should also balance the possibilities of the high optic power and the low optic power to prevent baseline drift, as was mentioned in previous auto power control section.

3.3 *Transceiving protocol*

The physical layer protocol is shown in Fig.4. It is divided into two main parts, the transmitting protocol and the receiving protocol. For the transmitting protocol, it interfaces with an external data source, buffers the data in a FIFO, encodes the incoming data, serializes the data, and interfaces with the external laser driver through an LVDS pair. For the receiving protocol, it interfaces with the receiving hardware through an LVDS pair, filters and synchronizes the received stream, deserializes and decodes the data, and performs error monitoring of the receiving procedure. The receiving protocol also contains an exception handling feature of determining the link status and scheduling the transceiving activity. The received data is buffered in the FIFO for external data sink to read them out. Since the transceiving protocol could reside in a single FPGA chip, the platform has got improved system integrity and reduced system cost.

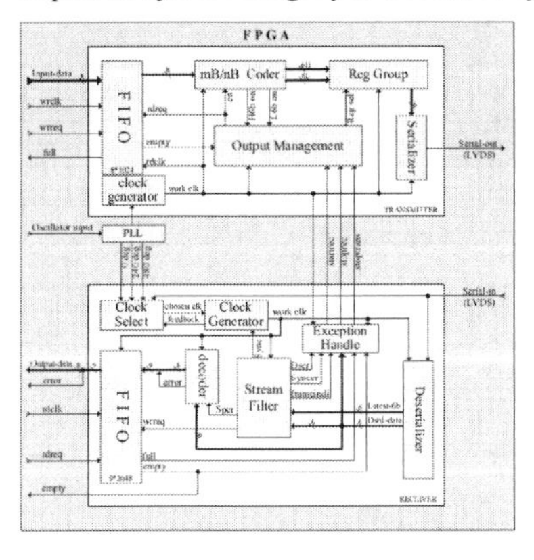

Figure 4. Physical layer protocol structure.

4 RAPID SYNCHRONIZATION

Most high-speed serial communication systems use the PLL-based clock-and-data recovery circuitry to generate synchronized sample clock. It is a highly reliable method in almost all kinds of serial link with stable channel. However, it takes too long before the PLL's got a stable clock output, usually hundreds or thousands of received bits. For an FSO link, especially the terrestrial link, the channel availability varies much with the environment it's in, taking weather for example. When the availability of the link gets very poor, the channel frequently fails, and the PLL-based method takes too much of the bandwidth to resynchronize, thus reduces the effective data bandwidth.

In the demonstrated physical layer protocol, a multi-phase clock sampling mechanism is used. The sampling clock is generated locally rather than from the incoming data stream, so two problems must be solved. One is the clock alignment with the incoming data, and the other is the frequency drift between two terminals. Commercial oscillators often have a 10^{-4} ~ 10^{-6} frequency drift, which means the system can never have a BER performance better than 10^{-6} if the sample clock is generated locally and will lose synchronization quite frequently, increasing the bandwidth waste in resynchronization, as demonstrated in Fig.5.

To eliminate the frequency difference (Barua et al. 2011) between the two terminals, multi-phase sampling is used, as shown in Fig.6 with the shaded clock phase as the working phase. Fig.7 and Fig.8 give the timing chart of the sampling clock generation. Signals named Cx are three input clocks with a 240°phase difference one another. Signals named smpl_clk[n] are three monitor sampling clocks generated from Cx with a 120° phase difference one another. Ser_in is the input data stream. Inclk is the chosen input clock used to generate a working sampling clock. Smplclk is the final generated working sampling clock.

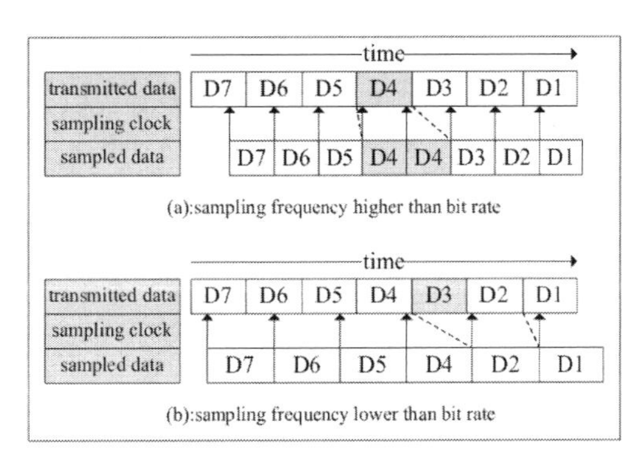

Figure 5. Frequency mismatch between terminals.

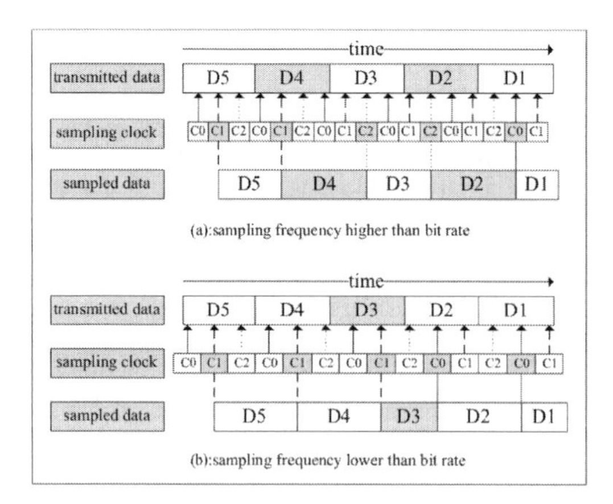

Figure 6. Multi-phase sampling.

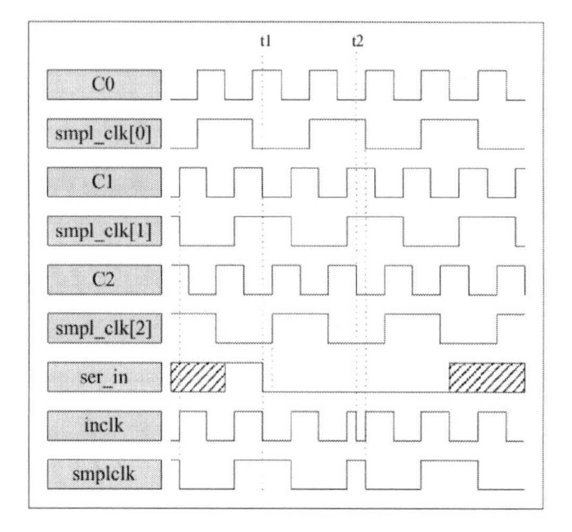

Figure 7. Sampling clock generation ($f_R<f_T$).

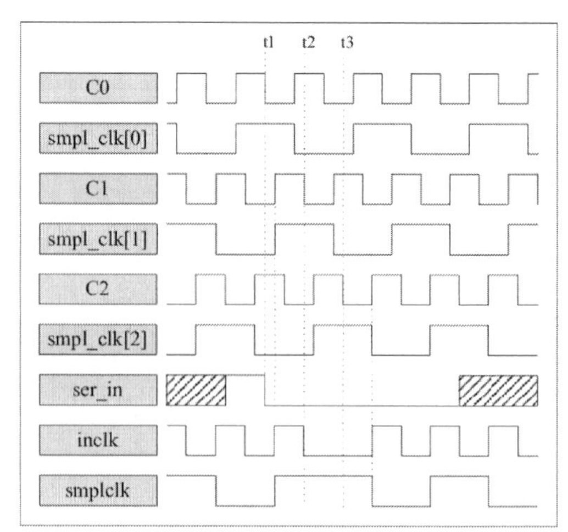

Figure 8. Sampling clock generation ($f_R>f_T$).

Using this method, the frequency difference between the two will not affect the data transmission in case that the frequency difference is within a certain scale given by Equ. 3, in which f_R is the sampling clock frequency of the receiver; f_T is the bit rate of the transmitter; max_{1s} is the maximum number of possible long 1s, and max_{0s} is the maximum number of possible long 0s.

$$f_T \cdot [1 - \frac{1}{3(\max_{1s} + \max_{0s})}] < f_R < f_T \cdot [1 + \frac{1}{3(\max_{1s} + \max_{0s})}] \quad (3)$$

Fig.9 shows the working wave when the transmitter clock is 50MHz and the receiver clock is 51.67MHz. Fig.10 shows the working wave when the transmitter clock is 50MHz and the receiver clock is 48.44MHz. Fig.9 and Fig.10 are read from the FPGA chip in real time when the system is working, and show that the presented method is a

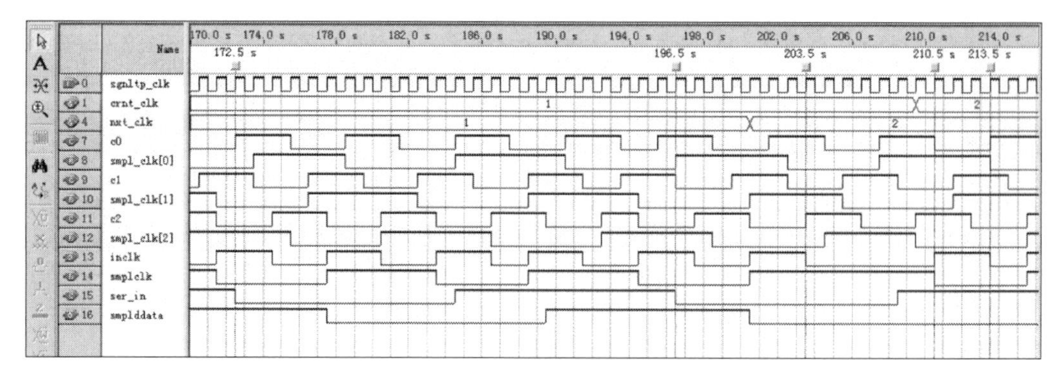

Figure 9. Sampling clock generation working wave (f_T=50MHz, f_R=51.67MHz).

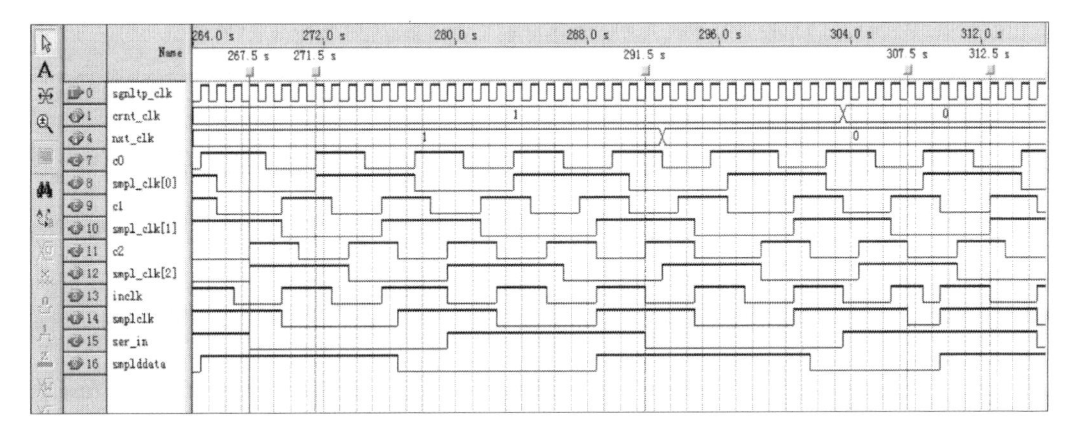

Figure 10. Sampling clock generation working wave (f_T=50MHz, f_R=48.44MHz).

choice for eliminating frequency difference between two FSO terminals. Since the sampling clock is locally generated, the method saves effective data bandwidth over PLL-based clock recovery circuitry, with no need to wait until the PLL is stable, thus realizes the rapid synchronization and resynchronization after channel failure.

5 CONCLUSIONS

In this paper, the low-cost hardware structure of an FPGA based FSO system is introduced. The physical layer protocol and the rapid synchronization method to save bandwidth cost in resynchronization by using locally generated sampling clock are demonstrated. The overall hardware and the physical layer protocol are tested under a bit rate of 300Mbps, during which the protocol synchronizes rapidly after channel failure. It's proved that locally generated sampling clock together with multi-phase sampling is a feasible data sampling technique for terrestrial FSO system. It solves the data and clock alignment problem by using locally generated sampling clock, eliminates the frequency difference between two terminals, and contributes to rapid synchronization, thus improves the bandwidth efficiency by reducing the handshaking bandwidth cost.

REFERENCES

Barua B & Barua D. 2011. Evaluate the performance of FSO communication link with different modulation technique under turbulent condition. *Proc. ICCIT*:22–24.

Isaac I. Kim, Bruce McArthur & Eric Korevaar. 2001. Comparison of laser beam propagation at 785 nm and 1550 nm in fog and haze for optical wireless communications. *Proc. SPIE* 4214: 26–37.

MAX3665 datasheet, *www.maxim-ic.com*.

Muhammad S. Sheikh, Kohldorfer P. & Leitgeb E. 2005. Channel Modeling for Terrestrial Free Space Optical Links. *ICTON*. (Tu.B3.5):407–410.

Penndorf R. 1957. Tables of the refractive index for standard air and the Rayleigh scattering coefficient for the spectral region between 0.2 and 20.0um, and their application to atmospheric optics. *JOSA*. 47:119–129.

Scott Bloom, Eric Korevaar, John Schuster, et al. 2003. Understanding the performance of free-space optics. *JOURNAL OF OPTICAL NETWORKING*. 2(6):178–200.

Shaban H, El Aziz S.D.A & Aly M.H. 2010. Probability of error performance of free space optical systems in severe atomspheric turbulence channels. *Proc. Microwave Symposium (MMS). Mediterranean*:25–27.

Tatarko M, Ovsenik L & Turan J. 2012. Availability and reliability of FSO links estimation from measured fog parameters. *Proceedings of the 35th International Convention*:21–25.

Electronics, Communications and Networks IV – Hussain & Ivanovic (eds)
© 2015 Taylor & Francis Group, London, ISBN: 978-1-138-02830-2

Deployment of IPsec based on IPv6 campus network

Yantao Tao
Jingchu University of Technology, Jingmen, Hubei, China

ABSTRACT: As a new generation of Internet protocol, IPv6 has solved, to some extent, the problem that the number of IPv4 addresses would be consumed soon. The IPv6 protocol enhances the support for network security and uses, in a compulsory manner, the IPSec protocol to implement the security functions including authentication and encryption. This paper introduces the deployment of IPSec using OpenSWAN according to the topological structure of IPv6 campus network.

KEYWORDS: Campus Network, IPv6, IPSec.

1 INTRODUCTION

1.1 *Background*

As a fundamental new-generation protocol, IPv6 makes great improvements to IPv4 regarding IPv4. Its encryption and authentication are implemented through IPSec on the network layer(Bell 1986), thus solving the defects of IPv4 in terms of encryption and authentication. However, most of the IPv6-related work are still in research and experiment phases(Diffie & Hellman 1976). Thus, it is especially important that deploying the Internet Protocol Security (IPSec) protocol in an actual campus network environment is of great significance.

1.2 *Current situation*

IPSec is composed of a series of protocols that can provide complete and secure solutions for IP network(Metcalfe 1976). The combination of these protocols offers various protective measures for the application entity and constitutes the system architecture of IPSec(Lin 2010). Although IPSec improves the security of IPv6, its deployment process is limited to theoretical scientific research.

2 LINUX OPERATING SYSTEM

Linux gives excellent support to the protocol stack NETKEY module of its kernel. For Linux operating system, OpenSWAN is the most representative implementation approach of IPSec(Kastas 1998).

3 DEPLOYMENT OF IPSEC IN IPV6 CAMPUS NETWORK

3.1 *Experimental environment and network topological structure*

The topology structure of the network used for experiment is shown in Figure 1. The environment consists of four hosts, one pair installed with CentOS release 5.3 operating system and the other pair Windows XP. The two hosts installed with CentOS release 5.3 are referred to as "Left" and "Right" (or left gateway server and right gate server) respectively and the two hosts with Windows XP are referred to as Hosts A and B. The two network servers Left and Right are also installed with OpenSWAN. The installation of OpenSWAN is detailed in section 3.2.

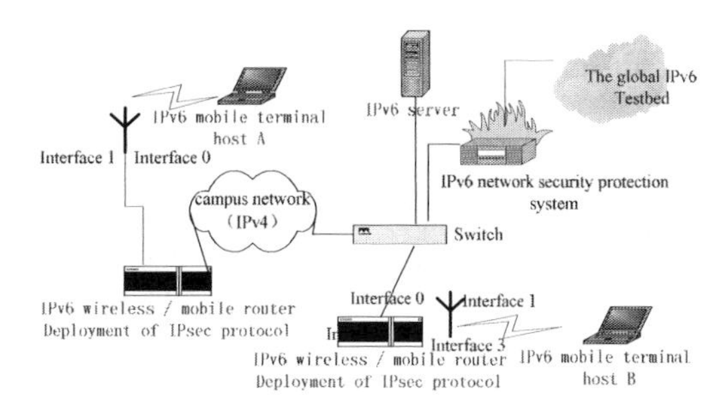

Figure 1. Topological structure of IPSec protocol experimental environment.

3.2 *Installation of openSWAN*

For the experiment, the OpenSWAN software was installed with source package downloaded prior to installation. The installation process is illustrated below with the Left gateway server as an example.

3.2.1 *Association package*

Instructions: yum install make gcc gmp-devel bison flex //install association package online

3.2.2 *Download the source code online*

wget http://www.openswan.org/download/openswan-2.6.24.tar.gz

3.2.3 *Decompress the file downloaded*

Instructions: tar zxvf openswan-2.6.24.tar.gz

3.2.4 *Modify directory*

Instructions: cd openswan-2.6.24

3.2.5 *Install OpenSWAN*

Instructions: make programs install

3.2.6 *Set shared key*

Instructions: vi /etc/ipsec.secrets

3.2.7 *Restart IPSec*

In order to verify whether IPSec is installed, the following instructions may be used:ipsec verify

The execution output is shown in Figure 2.

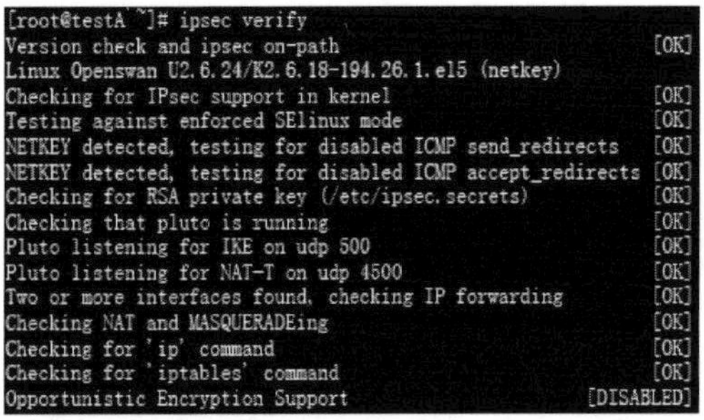

```
[root@testA ~]# ipsec verify
Version check and ipsec on-path                              [OK]
Linux Openswan U2.6.24/K2.6.18-194.26.1.el5 (netkey)
Checking for IPsec support in kernel                         [OK]
Testing against enforced SElinux mode                        [OK]
NETKEY detected, testing for disabled ICMP send_redirects    [OK]
NETKEY detected, testing for disabled ICMP accept_redirects  [OK]
Checking for RSA private key (/etc/ipsec.secrets)            [OK]
Checking that pluto is running                               [OK]
Pluto listening for IKE on udp 500                           [OK]
Pluto listening for NAT-T on udp 4500                        [OK]
Two or more interfaces found, checking IP forwarding         [OK]
Checking NAT and MASQUERADEing                               [OK]
Checking for 'ip' command                                    [OK]
Checking for 'iptables' command                              [OK]
Opportunistic Encryption Support                       [DISABLED]
```

Figure 2. Normal Starting of IPSec.

The information shown in Figure 2, or similar information, indicates successful IPSec installation. With the above steps, the installation of OpenSWAN-2.6.23 software on the (Left) gateway server is completed. The installation on the Right Gateway server follows the same procedure.

3.3 *Configuration of IP addresses*

The gateway servers and IP addresses of hosts were configured according to the topology structure shown in Figure 1. In addition, the default routes for the two gateway servers were configured and two sets of gateway servers opened to implement the IPv6 data forwarding function. The configuration is detailed as follows:

3.3.1 *Configure the IPv6 addresses of the Left gateway server*

ifconfig eth0 add 2001:250:4005:1002::10/64
ifconfig eth1 add 2001:250:4005:1001::1/64

3.3.2 *Default route of the Left gateway server and instructions for IPV6 data forwarding:*

route–A inet6 add default gw 2001:250:4005:1002::11
echo 1 > /proc/sys/net/ipv6/conf/all/forwarding

3.3.3 *Configure the Right gateway sever*

ifconfig eth0 add 2001:250:4005:1002::11/64
ifconfig eth1 add 2001:250:4005:1003::1/64

3.3.4 *Default route for Right gateway server and Instructions for IPv6 data forwarding:*

route–A inet6 add default gw 2001:250:4005:1002::10
echo 1 > /proc/sys/net/ipv6/conf/all/forwarding

3.3.5 *Configure Ipv6 addresses of Host A:*

IPv6 adu 4/2001:250:4005:1001::8

3.3.6 *Configure the default route of Host A:*

ipv6 rtu ::/0 5/2001:250:4005:1001::1

3.4 *Configuration of openSWAN*

The configuration of OpenSWAN includes two steps.

3.4.1 *Generate a new RSA public key on the Left gateway server through the following instructions:*

ipsec newhostkey --output /etc/ipsec.secrets
ipsec showhostkey --left>>/etc/ipsec.conf

3.4.2 *Copy the IPSec.conf on the Left gateway server to the directory of /etc/ for the Right gateway server and create the secrete key of the Right gateway server through the following instructions:*

ipsec newhostkey --output /etc/ipsec.secrets
ipsec showhostkey --Right >> /etc/ipsec.conf

Complete the above /etc/ipsec.conf file of the Right gateway server and copy to the directory of /etc/ for the Left gateway service. The exchange of secrete keys between the two gateway servers is completed according to the above procedures.

3.5 *Starting up of openSWAN software*

3.5.1 *Start up OpenSWAN on both Left and Right gateway servers:*

Service ipsec restart

3.5.2 *Start up IPSec on the Right gateway server; the instructions and execution result are shown in Figure 3.*

```
"ipv6-to-ipv6" #1:STATE_MAIN_I1:initiate
"ipv6-to-ipv6" #1:received Vendor ID payload [Openswan]
"ipv6-to-ipv6" #1:received Vendor ID payload [Dead Peer Detection]
"ipv6-to-ipv6" #1:received Vendor ID payload [RFC3497] method set to=109
"ipv6-to-ipv6" #1:STATE_MAIN_I2:sent MI2, expecting MR2
"ipv6-to-ipv6" #1:NAT-Traversal:Result using RFC 3497(NAT-Traversal):
no NAT detected
"ipv6-to-ipv6" #1:STATE_MAIN_I3: sent MI3, expecting MR3
"ipv6-to-ipv6" #1:received Vendor ID payload [CAN-IKEv2]
"ipv6-to-ipv6" #1:STATE_MAIN_I4:ISAKMP SA established {auth=OAKLEY_RSA_
SIG cipher=aes_128 prf=Oakley_sha group=modp2048}
"ipv6-to-ipv6" #1:STATE_QUICK_I1:initiate
"ipv6-to-ipv6" #1:STATE_QUICK_I2: sent QI2, IPsec SA established tunnel
 mode {ESP=>0x8e6fd86 <0x0508b899 xfrm=AES_128-HMAC-SHA1 NATOA=none NATD
=none DPD=none}
```

Figure 3. Information upon starting up of IPSec on the Right gateway server.

The information in Figure 3 demonstrates the IPSec exchange process. The "ISAMP SA established" (successful establishment of ISAKMP) displayed in Line 9 indicates that the first-phase SA negotiation has been completed between the Right and Left gateway servers. Line 11 shows "IPSec SA established" (successful establishment of IPSec SA) in tunnel mode, indicating that IPSec SA has been successfully established on both Right and Left gateway servers. The information demonstrates that the IPSec connection has been successfully established in tunnel mode between Right and Left gateway servers.

3.6 *IPSec connection test*

Data capturing software Ethereal was installed on the Left gateway server. The IPSec connection was tested through the following steps.

Ping the Left gateway server on Host A (2001:250:4005:1001::1/64)

Execute the instructions: ping6 2001:250:4005:1001::1 and the information shown in Figure 4 can be obtained:

Figure 4 indicates that the communication between Host A and the Left gateway server has been established.

```
Pinging 2001:250:4005:1001::1
from 2001:250:4005:1001::8 with 32 bytes of data:
Reply from 2001:250:4005:1001::1: bytes=32 time=1ms
Reply from 2001:250:4005:1001::1: bytes=32 time=1ms
Reply from 2001:250:4005:1001::1: bytes=32 time=1ms
Reply from 2001:250:4005:1001::1: bytes=32 time<1ms
Ping statistics for 2001:250:4005:1001::1:
    Packets: Sent = 4, Received = 4, Lost = 0 (0% loss),
Approximate round trip times in milli-seconds:
    Minimum = 0ms, Maximum = 1ms, Average = 0ms
```

Figure 4. Communication data information.

Ping the Right gateway server, Right gateway server, Host B and Host A on Hosts A, A, A and B, respectively, using the same method. Information similar to that shown in Figure 4 can be obtained. Such information indicates that Hosts A and B are able to communicate through the two gateway servers.

3.7 *Data capture and the analysis of data packets in IPSec experiment*

The data packet exchange in the two phases is further discussed in this section. Figure 5 shows the first message sent from the Right gateway server to the Left gateway server in primary mode.

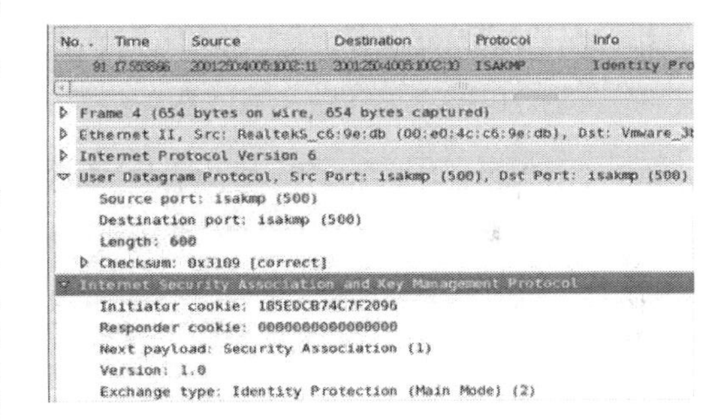

Figure 5. First message in primary mode.

As seen in Figure 5, the first message contains the ISAKMP protocol, the header (HDR) of which contains the Cookies of the initiator and the responder. Since it is the first message, the Cookie of the responder is 0. The Next payload field is the SA payload, and the proposal payload is encapsulated in this field. In addition, several exchange payloads is encapsulated in the proposal payload. The exchange payloads contain information such as exchange IDs, encryption algorithm, authentication method, etc.

In the primary transmission mode from the Left to the Right gateway server, the initiator's cookie in the HDR of the second message's ISAKMP protocol becomes the cookie of the initiator in the first message. Meanwhile, the responder generates a cookie through a certain way as the cookie of the responder in the second message. In addition, based on the message from the initiator, the responder chooses one of the exchange payloads encapsulated in the proposal payload. The SA property of the chosen payload is one of the many exchange payload of the first message. The responder uses this as the SA payload to send to the initiator, i.e. the Right gateway server. Therefore, the negotiation regarding the ISAKMP SA property is completed. Through the later six messages of the first phase, ISAKMP SA is established.

Upon the completion of the above operations, the IPSec strategy of IPV6 could be implemented on any Linux node of a campus network.

3.8 *Network performance of IPv6 campus network before and after applying IPSec strategy*

3.8.1 *Influence of IPSec strategy on the network performance*

In order to assess the changes of network performance and security before and after applying IPSec strategy in the IPv6 campus network, the Ttcpw tool was used. Through the tool, the data to be transmitted was generated by Host A and then sent to Host B, where the change of the data is monitored. Figure 6 shows the comparison of network performance when a 12 MB TCP file is transmitted before and after applying IPSec strategy. In the diagrams, *KB/sec* denotes the transmission rate, *calls/sec* denotes the number of calls processed per second, *real msec* denotes the total time cost to receive the data, and *A* and *B* denote the IPv6 campus network environment before and after applying IPSec strategy.

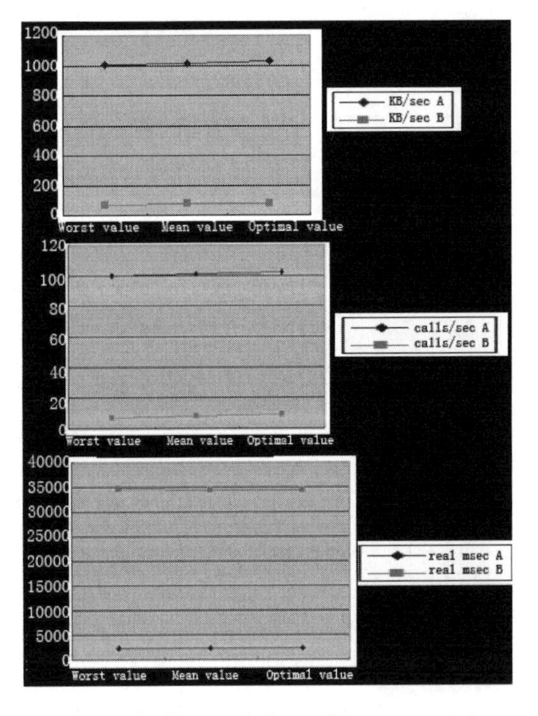

Figure 6. TCP file transmission performance comparison.

As seen in Figure 6, before IPSec strategy was applied, the transmission rate almost reached the ideal value, while the average transmission rate decreased a little after applying IPSec (with AH transmission mode). The values of other parameters also decreased to a certain extent, indicating that the implementation of network transmission security strategy had an adverse influence on the actual network performance.

Similarly, a 4 MB HTTP file was transmitted and the data change at the receiver was monitored. The transmission mode was ESP. Figure 7 shows the observed data variation.

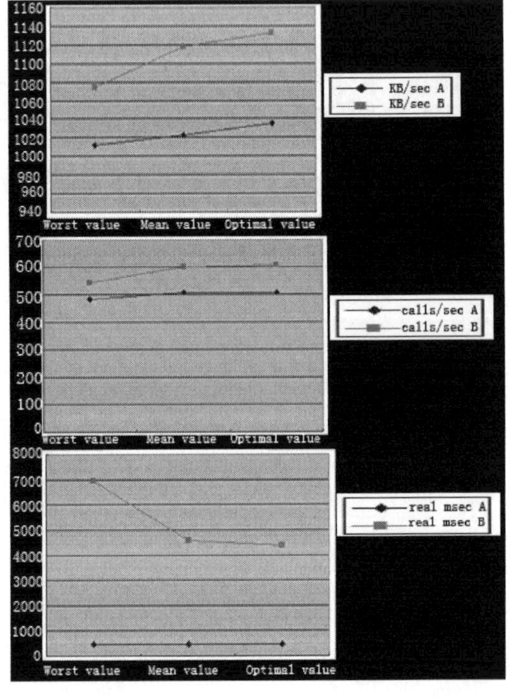

Figure 7. HTTP file transmission performance comparison.

As seen in Figure 7, the network performance before and after applying IPSec barely varied for transmitting HTTP file and TCP file. However, the choice of AH authentication or ESP authentication has significant influence on network performance.

3.8.2 *Influence of IPSec strategy on network security*

This section analyzes the network security performance change before and after applying IPSec strategy. The performance was evaluated based on network throughput and data packet forwarding, as shown in Figs. 8 and 9, where A and B denote the IPv6 campus network environment before and after IPSec strategy is applied.

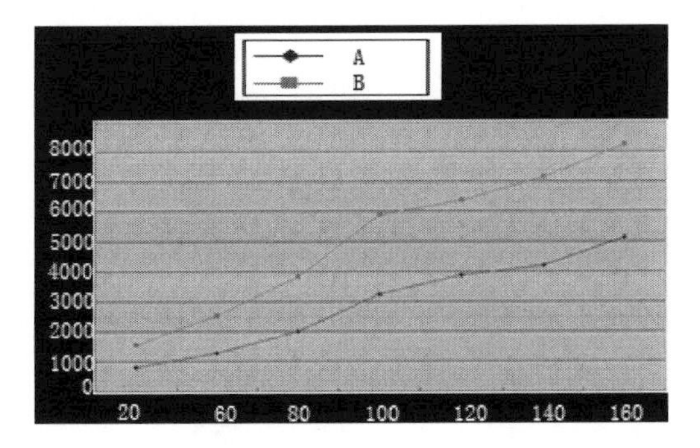

Figure 8. Network throughput.

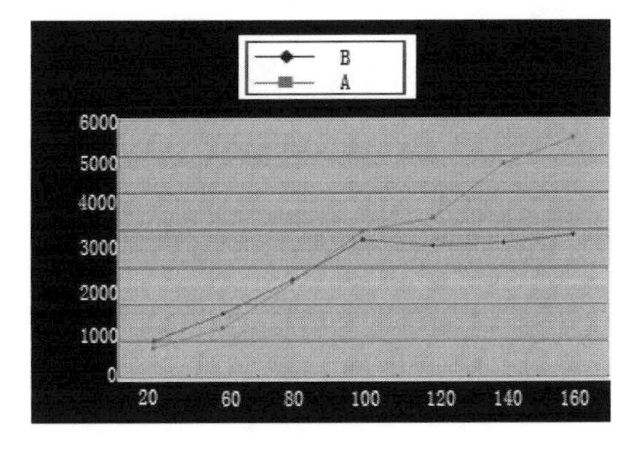

Figure 9. Forwarded packets.

As seen in Figure 8, after IPSec strategy was applied, the network throughput increased significantly with time. By continuously sending broadcast messages to Host B using Tcpw tool, the number of received packets was monitored. As seen in Figure 9, the number gradually decreased with time after IPSec is applied, comparing with that before IPSec was applied. We can conclude that the application of IPSec strategy to a certain extent made the access to Host B less smooth, whereas the network security was improved relatively.

ACKNOWLEDGMENT

Thank University-level Scientific Research Project "ASP Online Examination System" (ZR201024) of Jingchu University of Technology in 2010 to provide financial support.

REFERENCES

Bell, P.R. 1986. Review of point-to-point network routing algorithms. *IEEE Communication Magazine* 24(1):34–38.

Diffie, W. & Hellman, M. 1976. New Directions in Cryptography. *IEEE Trans.* IT-22(6): 644–654.

Kastas, T. J. 1998. Real-Time Voice Over Packet-Switched Networks. *IEEE Network* 12(1):18–27.

Lin Chien-Chuan. 2010. Network Intrusion Detection Using Particle Filter and Colored Petri Nets. *2010 International Symposium on Computer, Communication, Control and Automation Proceedings* 2(C):198–202.

Metcalfe, R. M. 1976. Ethernet:Distributed Packet Switching for Local Computer Networks. *Commun. ACM* 19(7): 395–404.

Electronics, Communications and Networks IV – Hussain & Ivanovic (eds)
© 2015 Taylor & Francis Group, London, ISBN: 978-1-138-02830-2

On the number of BPC permutations admissible to k-extra-stage Omega networks

G. Veselovsky
Assumption University, Bangkok, Thailand

ABSTRACT: A $N \times N$ k-Omega network is obtained by adding k extra stages to an original Omega with $logN$ stages. A permutation is a very common communication pattern in parallel computations. Among various classes of regular permutations, the class of BPC (bit-permute-complement) permutations is one of the most important. Admissibility of a permutation means that it can be realized in a given network without blocking. In this paper the algorithm for BPC permutations admissibility, introduced by us before, is used for determining the number of BPC permutations admissible to k-Omega networks of various sizes and with the various numbers of extra stages. The number of additional stages needed for non-blocking realization of some frequently used BPC permutations is also determined. The results are true for multistage and recirculating versions of aforesaid communication topology and for the electronic and optical implementations as well.

1 INTRODUCTION

The present-day approach to satisfying growing demands for computer performance consists of the realization of massively parallel computations which include thousands or even hundreds of thousands small and relatively inexpensive processing nodes. The performance of such systems today is limited mainly by their interconnections rather than their logic or memory. Shuffle-exchange multistage networks (SENs) represent one of the most common interconnecting topologies. In such networks, binary addresses of input terminals to the next stage are produced by circularly shifting one position out of the previous stage, which is called "perfect shuffle" connection (Stone 1971). In the case of a SEN, the uniformity of an interconnection pattern allows it to be implemented as a multistage interconnection network (MIN) or as a recirculating network as well.

A SEN of size $N \times N$, with $logN$ stages and N/2 2×2 switches in each Omega network. (Lawrie 1975). An 8×8 Omega network is shown in Figure 1. An Omega network provides a single route for any input-output pair. It refers to generalized-cube topology, which is used for many networks, including indirect binary n-cube, baseline, STARAN flip network etc. (Siegel 1990). To make it fault-tolerant and increase the permutation capability of the original Omega network, k extra stages can be added in front of it with $1 \leq k \leq n-1$ (Shen 1995a). In this case the number of disjoint routes for any input-output pair is 2^kpaths. Such configuration is called k-Omega network. In what follows an Omega network without extra stages is referred to as a 0-Omega.

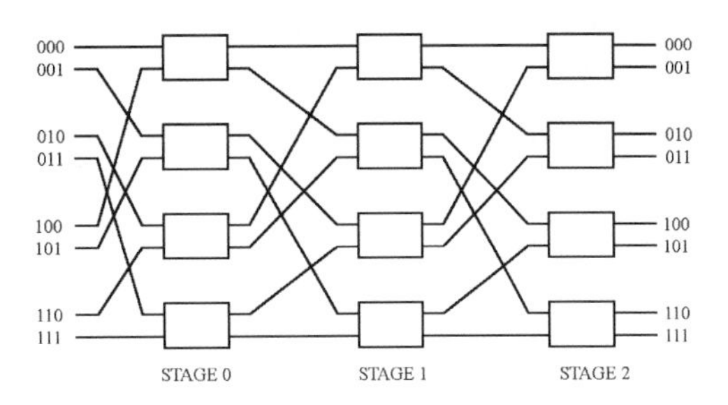

Figure 1. An 8×8 Omega network.

The uniformity of interconnection pattern between stages makes it possible to implement a recirculating version of SEN with one stage only (Figure 2). Such network can pass data from inputs to outputs for some number of steps through itself. (Wilkinson 1996). Additional k steps (beyond $logN$) are equivalent to additional stages for 0-Omega.

Optical multistage interconnection networks (OMINs) of the shuffle-exchange type are possible (Al-Shabi & Othman 2008), (Yang et al. 2000), including variable-stage ones (Das et al. 2003). Recirculating optical networks are also being studied intensively (Wong & Cheng 2000), (Kang & Zhang 2001).

*Corresponding author: *gveselovsky@au.edu*

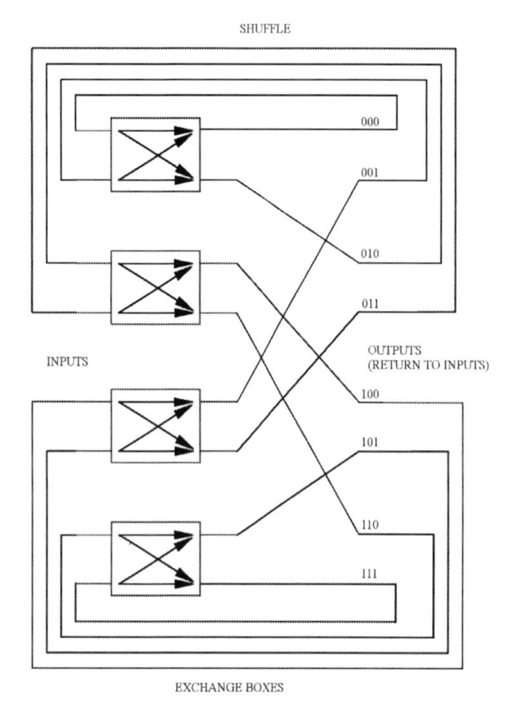

Figure 2. An 8 × 8 recirculating SEN.

Various communication patterns are demanded among the processing nodes of a parallel computing system: such as one-to-one, broadcasting, multicast patterns, and permutations. The latter means parallel transfer of data from all inputs to all outputs, with a distinct output for each of the inputs. A permutation is admissible to a given network, if it can be realized without blocking one pass through the network. When permutations are realized, two possible states of a switch are needed. One is *straight*, with the upper input connected to the upper output and the lower input connected to the lower output. The other is *cross*, with the upper input connected to the lower output and the lower input connected to the upper output. The ratio of the number of admissible to a network permutations to the total number of possible permutations defines the network's combinatorial power. It can be proved that any permutation can be realized on a multistage SEN with 2n – 1 stages (Grammatikakis et al. 2000). Whereas only $N^{N/2}$ of $N!$ are admissible to a 0-Omega network (Shen et al. 1995b). Thus, the availability of redundant routes in the network not only provides fault tolerance, but also improves its permuting capability. However, the quantitative evaluation of the improvement is needed. If a permutation does not possess any regularity, it is called an arbitrary permutation; but if there are some general rules for producing a destination address from a source address, a permutation is called a regular one. There are different known classes of regular permutations, among which the BPC (bit-permute-complement) class is one of the most important. BPC permutations are very common in scientific applications, including digital signal processing. Moreover, the nature of this class of permutations allows the use of a simple elegant approach for checking their admissibility to SENs in the wide range of the number of stages. Finally, it gives an insight into more general trends, which are not available in works devoted to exploring admissibility of arbitrary permutations.

In (Gazit & Malek 1989) the impact of adding one extra stage to multistage interconnection networks on their permuting capability was explored. Thereafter an $O(N \lg N)$ admissibility algorithm for 1-Omega network was proposed (Shen 1995b). In both aforesaid works, arbitrary permutations were considered. In (Bashirov & Karanfiller 2005) an approach based on modeling with Petri nets was proposed. It allows determining the necessary minimum number of stages in a variable-stage OMIN for an arbitrary permutation also. Our paper concentrates on the admissibility of BPC permutations to the aforesaid communication topology. In (Das et al. 2003) an $O(n^2)$ algorithm is proposed that allows determining the minimum number of stages required to pass a given BPC permutation through a SEN. In this paper, we use an approach that allows doing the same in $O(n)$ time only (Veselovsky 2011a). Statistics about the number of BPC permutations admissible to SENs of variable sizes and with various numbers of stages presented here have been tabulated with its use. Although our exploration concentrates on BPC permutations, the obtained information allows observing the tendency of improving permutation capability along with increasing the number of stages in general. For the convenience of the reader in what follows we give a concise description of the algorithm. The number of extra stages necessary for admissibility of some frequently used BPC permutations is ial

2 PRELIMINARIES AND DEFINITIONS

A permutation is called a bit-permute (BP) permutation if the destination address for all source-destination pairs can be obtained by permuting its components in accordance with some general rules, so for any n there are $n!$ BP permutations. A permutation is called a bit-permute-complement (BPC) permutation if the destination address can be produced from the source address by permuting bits in the source address or complementing some or all of its bit positions. It is evident that any *parental* BP permutations can generate 2^n *descendant* BPC permutations including the parental one. Thus for a given n the total number of BPC permutations equals to $2^n n!$. Among BP/BPC permutations there are frequently used ones. Such permutations are usually referred to by names (Grammatikakis et al. 2000). Some of frequently used BP and BPC permutations in parallel computing are listed below, with each equation showing the mapping of a source to the destination. In all foregoing cases a source address is supposed to be as follows:

$$s_0 s_1 \ldots s_{n-2} s_{n-1}$$

1 *Perfect shuffle*

$$\pi_{PSH} = s_1 s_2 \ldots s_{n-1} s_0$$

2. *Unshuffle*

$$\pi_{USH} = s_{n-1} s_0 \ldots s_{n-3} s_{n-2}$$

3. *Vector reversal*

$$\pi_{VR} = \overline{s_0}\,\overline{s_1} \ldots \overline{s_{n-2}}\,\overline{s_{n-1}}$$

4. *Butterfly*

$$\pi_{BF} = s_{n-1} s_1 \ldots s_{n-2} s_0$$

5. *Exchange*

$$\pi_{EXCH} = s_0 s_1 \ldots s_{i-1} \overline{s_i} s_{i+1} \ldots s_{n-1} s_{n-2}$$

6. *Bit reversal*

$$\pi_{BR} = s_{n-1} s_{n-2} \ldots s_1 s_0$$

7. *Matrix transpose*

$$\pi_{MT} = s_l s_{l+1} \ldots s_{2l-1} s_0 s_1 \ldots s_{l-1} \text{ if } n = 2l$$
$$\pi_{MT} = s_l s_{l+1} \ldots s_{2l} s_0 s_1 \ldots s_{l-1} \quad \text{if } n = 2l + 1$$

8. *Bit shuffle*

$$\pi_{BSH} = s_0 s_2 \ldots s_{n-2} s_1 s_3 \ldots s_{n-1} \text{ if } n = 2l$$
$$\pi_{BSH} = s_0 s_2 \ldots s_{n-1} s_1 s_3 \ldots s_{n-2} \text{ if } n = 2l+1$$

9. *Shuffle row major*

$$\pi_{SHRM} = s_0 s_l s_1 s_{l+1} \ldots s_l s_{2l-1} \text{ if } n = 2l$$
$$\pi_{SHRM} = s_0 s_{l+1} s_1 s_{l+2} \ldots s_{l-1} s_{2l-1} s_l \qquad \text{if} = 2l + 1$$

The *transition matrix* for a BP permutation can be presented in a symbolic form by $S_0 S_1 \ldots S_{n-2} S_{n-1} S_{\pi(0)} S_{\pi(1)} \ldots S_{\pi(n-2)} S_{\pi(n-1)}$ (Shen 1995a).

The only BP permutation admissible to the 0-Omega network, which also means n-stage $N \times N$ SEN, is *identity* permutation when all destination addresses are the same as corresponding source addresses. In (Das et al. 2003) it is proved that, if a *parental* BP permutation is admissible to an m -stage SEN, $n < m \leq 2n-1$, all its 2^n*descendant* BPC permutations, including the *parental* one itself, are also admissible. So if a given BPC permutation contains complements, it is enough to find the necessary number of extra stages for its parental BP permutation.

In earlier works, increasing permutation capabilities of multistage interconnection networks (MINs) was considered in the assumption of adding extra stages to the basic n-stage network, i.e. for fixed-stage networks. In this paper, as has been mentioned already, we also consider a variable-stage and recirculating SENs. The possible architecture of a variable-stage hybrid OMIN can be found in (Das et al. 2003). Variable-stage OMINs and recirculating networks, optics and electronics as well, can vary the number of stages m in the range $1 \leq m \leq 2n-1$, but permuting properties of SENs should be examined separately for two cases: namely, for $1 \leq m \leq n$, and for $n < m \leq 2n-1$. It is so, because in the first case a SEN does not hold full accessibility and in the second case, adding extra stages, multiple routes for any input/output pair is available. The first case was explored by us in many details in (Veselovsky & Agrawal 2014b), so in this paper we concentrate on the second case, in other words, on exploring k-Omega permuting ability. However, despite that not too many BPC permutations are admissible in the first range, among them there are at least two very important stages. We mean here, mentioned above *perfect shuffle* (one stage is needed) and *matrix transpose* (*N/2* stages are needed). The *perfect shuffle* plays a fundamental role in parallel processing. It is efficient for executing fast Fourier transform (FFT) on a parallel processor and it provides evaluating polynomials in minimum time. It is also very efficient for various kinds of sorting. The *matrix transpose* is needed in many applications of linear algebra.

3 A MINIMAL ALGORITHM FOR BPC PERMUTATIONS ADMISSIBILITY TO K-OMEGA NETWORKS

The case with extra stages was considered for fixed-stage electronic and variable-stage optical MINs in (Shen 1995a) and (Das et al. 2003) respectively. The new solution of a very similar problem was presented by us in (Veselovsky 2011a). Its novelty consists in the analysis of a symbolic transition matrix rather than a binary one for a given BPC permutation. In what follows the reader will find a concise description of our algorithm. In the case under consideration all outputs of the network are reachable from any inputs. Moreover, redundant disjoint routes with the number of such routes which equal to 2^k are available.

For a BPC permutation π, the symbolic *transition matrix* now looks as follows:

$$S_0S_1...S_{n-2}S_{n-1} * ... * S_{\pi(0)}S_{\pi(1)}...S_{\pi(n-2)}S_{\pi(n-1)}.$$

In the foregoing expression, "*" means extra bits. The number of those bits equals to the number of extra stages. Hereafter components of a *transition matrix* given in the above form will be called symbolic components in distinction from binary components in a binary transition matrix which consists of N rows. For checking permutation admissibility we here apply a modified window method, based on comparing only indices of symbolic components within a window, instead of inspecting binary combinations among a large number of rows within a binary transition matrix by shifting a window along it, as was done traditionally. Recall that when checking admissibility, each window of the size n defines the terminal numbers in one stage of the network. So in order not to block each window, the forming of different terminal numbers should be permitted in a given stage for all input-output pairs. However, it is possible only if there are no symbolic components with the same subscripts, or like components inside the window. Hence, the aforesaid modification of the window method provides fast and straightforward way to check the admissibility of any BPC permutation to a k-Omega network with the number of extra stages k in the range $0 < k \leq n-1$. Begin the check with assigning $k = 1$, and then increment it until aforesaid condition is satisfied.

An illustrative example of applying our approach for determining the necessary number of extra stages k for 8 × 8 k-Omega, when *bit reversal* permutation is realized, is given below. This permutation falls to the worst-case category from the viewpoint of routing it on 0-Omega and on topology similar to the networks. (The *bit reversal* is used for rearranging the results of the FFT procedure as its last step). From Figure 3 it can be seen that there are symbolic components with the same indices in the windows W_1 and W_2, which means blocking between the first and the second stages and between the second and the third stages as well in the 0-Omega. Similarly there are like components in the window W_2 in Figure 4 (1-Omega), so there will also be blocking. However in Figure 5 all windows are independent and so in this case the permutation in question is admissible to 2-Omega. As to assigning extra bits X_0 and X_1 in the transition matrix for example, this issue is beyond the scope of our paper. We can recommend an approach which is proposed in (Veselovsky & Ackarachalanonth 2012b).

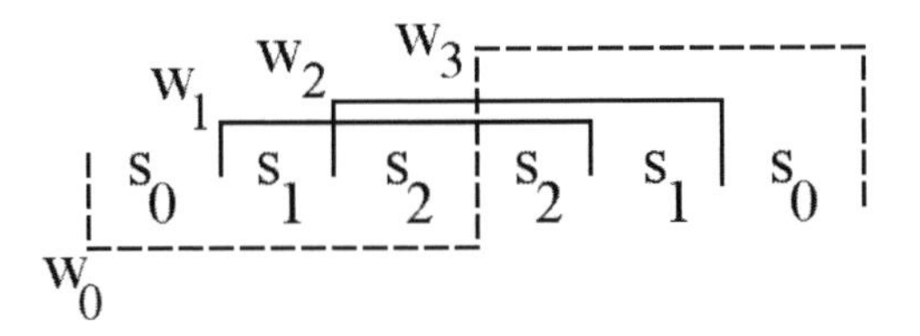

Figure 3. Windows for *bit reversal* on 8 × 8 0-Omega network.

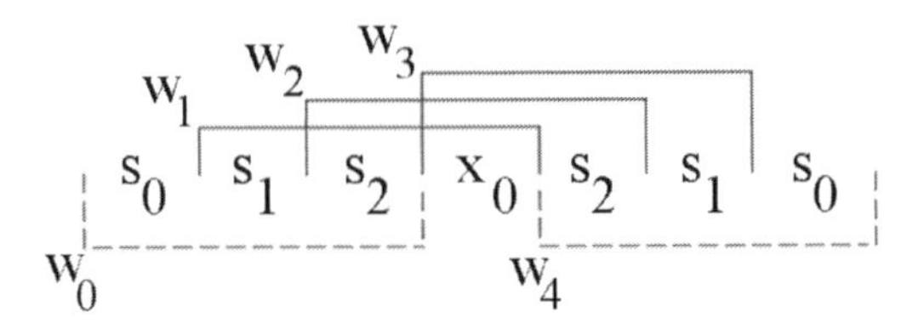

Figure 4. Windows for bit reversal on 8 × 8 1-Omega network.

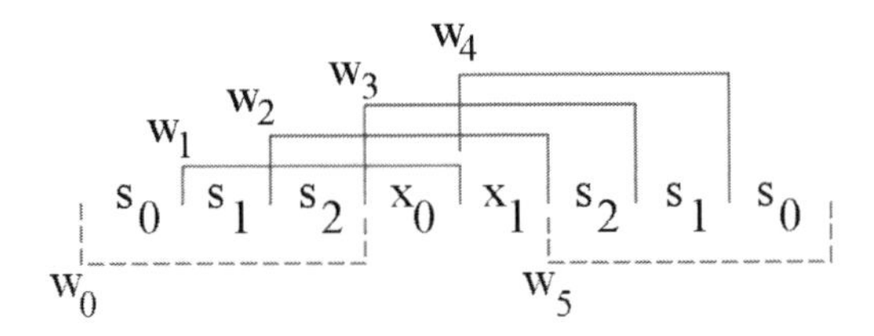

Figure 5. Windows for bit reversal on 8 × 8 2-Omega network.

Algorithm Determining the minimal number of extra stages ($0 < k \leq n-1$)

Step 1: Initialize N, n.

Step 2: Initialize a transition sequence.

Step 3: Compute the distance *L[I]* between components of the transition sequence with the same indices in the destination tag and in the source tag.

FOR I=0 TO I=n-1 DO

L[I]:=$x_{d,I} - x_{S,I}$

Step 4: Procedure *min L[0], L[1], ... L[n-1]: integers*

Min L[0]

FOR I=1 TO n-1 DO

Imin ≥ L[I] THEN min:= L[I]

Step 5: Compute the number of extra stages k

k:=n-min L[I]

It is easy to see that the above algorithm essentially includes two FOR-DO loops and hence requires around *2n* steps, so its time complexity is $O(n)$. This is the lower boundary.

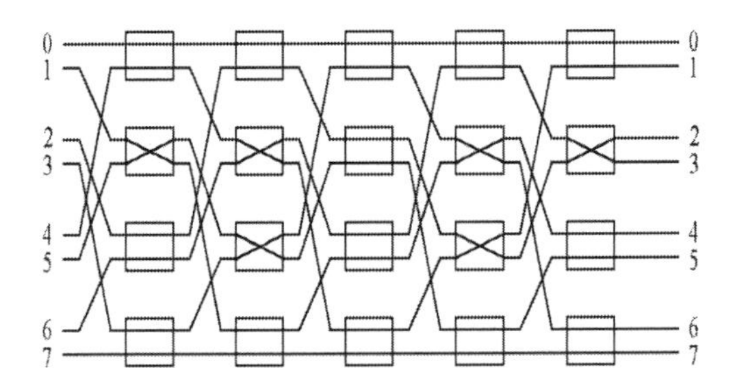

Figure 6. Routing *bit reversal* permutation on 8 × 8 5-stage SEN

By using foregoing approach, it is not difficult to find that in general *bit reversal* for admissibility needs *n – 1* extra stages, and the same is suitable for *butterfly*. On the contrary, *bit shuffle* and *shuffle row major* need only one extra stage to become admissible. The reader is encouraged to try the rest frequently used BPC permutations. Routing *bit reversal* permutation on 8 × 8 2–Omega is shown in Figure 6.

4 EVALUATION THE NUMBER OF ADMISSIBLE BPC PERMUTATIONS

The foregoing algorithm is implemented in MATLAB. The results obtained by the program for various values N with the various values of the number of extra stages k are presented in the tables given below, so now the influence of extra stages added to the basic n-stage structure on permutation capability of k-Omega can be inspected. It will be recalled that the total number of BPC permutations equals to $2^n n!$ where $n = \log_2 N$, and $k = m - n$.

Table 1. BPC Permutations admissibility for N = 16.

Number of extra stages	Status	Number of permutation
k = 0	Admissible	16 (4.167%)
	Blocked	368
k = 1	Admissible	128 (33.333%)
	Blocked	256
k = 2	Admissible	288 (75%)
	Blocked	96
k = 3	Admissible	384 (100%)
	Blocked	0

($2^n n! = 384$)

As can be seen from the tables1-3, for larger values of N with a smaller number of extra stages, $k = 2$ e. g. and the gain in permutation capability of the network, expressed as a percentage, is insignificant. However, two extra stages provide four routes in a network for each source-destination pair, which allows us in many cases to meet the requirements for fault-tolerance.

Table 2. Permutations admissibility for $N = 256$.

Number of extra stages	Status	Number of permutation
k = 0	Admissible	256 (0.00248%)
	Blocked	10321664
k = 1	Admissible	32768 (0.317%)
	Blocked	10289152
k = 2	Admissible	373248 (3.616%)
	Blocked	9948672
k = 3	Admissible	1572864 (15.238%)
	Blocked	8749056
k = 4	Admissible	3840000 (37.202%)
	Blocked	6481920
k = 5	Admissible	6635520 (64.286%)
	Blocked	3686400
k = 6	Admissible	9031680 (87.5%)
	Blocked	1290240
k = 7	Admissible	10321920 (100%)
	Blocked	0

($2^n n! = 10321920$)

Table 3. Permutations admissibility for N = 1024.

Number of extra stages	Status	Number of permutation
k = 0	Admissible	1024 (0.000028%)
	Blocked	3715890176
k = 1	Admissible	524288 (0.014%)
	Blocked	3715366912
k = 2	Admissible	13436928 (0.36%)
	Blocked	3702454272
k = 3	Admissible	100663296 (2.71%)
	Blocked	3615227904
k = 4	Admissible	384000000 (10.33%)
	Blocked	3331891200
k = 5	Admissible	955514880 (25.71%)
	Blocked	2760376320
k = 6	Admissible	1770209280 (47.64%)
	Blocked	1945681920
k = 7	Admissible	2642411520 (71.11%)
	Blocked	1073479680
k = 8	Admissible	3344302080 (90%)
	Blocked	371589120
k = 9	Admissible	3715891200 (100%)
	Blocked	0

($2^n n! = 3715891200$)

From the aforesaid, we can conclude that problems concerning improving the permuting capability of the network and providing the fault tolerance are essentially different. Even though our reasoning stems from exploring the admissibility of BPC permutations rather than arbitrary ones, beyond all doubts, the tendency is general.

5 CONCLUSION

In this paper our fast algorithm for determining the number of extra stages needed for admissibility of a BPC permutation developed before is used for evaluating the number of BPC permutations admissible to k-Omega networks of various sizes and with various numbers of extra stages. Similar statistics are known for MINs with only one extra stage and for arbitrary permutations. Although our investigation concerns admissibility of a special class of permutations, it allows to judge, as a first approximation, the impact of increasing the number of extra stages in a wide range on permuting capability of MINs. In our opinion the paper contributes to the knowledge about fundamental properties of multistage interconnection networks and their recirculating counterparts. Our predecessors (Gazit & Malek 1989) explored the influence of one extra stage on permutation capability of very small size networks (up to $N = 32$) and got optimistic results. However, our investigations show that with increasing the size of a network, the gain due to one extra stage becomes negligible and realizes significant values with the number of extra stages close to their upper boundary ($n - 1$).

REFERENCES

Abed, M. & Othman. M. 2008. Fast method to find conflicts in optical multistage interconnection networks. *Intl. J. of the Computer,The Internet Management* 16(1): 18–25.

Adams, G. B. & Siegel, H. J. 1982. The extra-stage cube: a fault-tolerant interconnection networks for supersystems. IEEE Trans. Computers C-31(5): 443–454.

Al-Shabi, M. A. & Othman, M. 2008. A new algorithm for routing and scheduling in optical Omega networks. 2008. *Intl. J. of the Computer, The Internet and Management* 16(1): 26–31. 16(1): 26–31.

Bashirov, R. & Karanfiller, T. 2005. On path dependent loss and switch crosstalk reduction in optical networks. *Information Sciences* 180: 1040–1050.

Das, N. & Bhattacharya, B. & Bezrukov S. 2003. Permutation routing in optical MIN with minimum number of stages. *Systems Architecture* 48: 311–323.

Gazit, I. & Malek, M. 1989. On the number of permutations performable by extra-stage multistage interconnection networks. *IEEE Trans. Computers* 38(2): 297–302.

Grammatikakis, M. D.& Frank, Hsu D. & Kraetzl M. 2000. *Parrallel System Interconnections and Communication.* CRC Press.

Kang, W.-S. & Zhang, E. P. 2001. Optical Switching Device Having Recirculating Structure and Optical Switching Method Therefore. US Patent 6211979.

Lawrie, D. H. 1975. Access and alignment of data in an array processor. *IEEE Trans. Computers* 24(12): 1145–1155.

Leighton, F. Th. 1992. *Introduction to parallel algorithms and architectures: arrays, trees , hypercubes.* San Mateo, CA: Morgan Kaufmann Publishers.

Mohammad, A. A. 2005. New algorithm to avoid crosstalk in optical multistage interconnection networks. In *Proc. MICC-ICON* (1): 501–504.

Raghavendra, C. S. & Varma A. 1986. Fault-tolerant multiprocessors with redundant path networks. *IEEE Trans. Computers* 35(4): 307–316.

Shen, X. & Xu, M. & Wang, X. 1995. An optimal algorithm for permutation admissibility to multistage interconnection networks. *IEEE Trans. Computers* 44(4): 604–608.

Shen, X. 1995. An optimal algorithm for permutation admissibility to extra stage cube-type networks. *IEEE Trans. Computers* 44(9): 1144–1149.

Shen, X. 1995. Optimal realization of any BPC permutation on k-extra-stage Omega networks. *IEEE Trans. Computers* 44(5): 714–719.

Siegel, H. J. 1990. *Interconnection Networks for Large-Scale Parallel Processing. Theory and Case Studies.* 2nd ed. McGraw-Hill International Editions.

Stone, H. S. 1971. Parallel processing with the perfect shuffle. *IEEE Trans. Computers* 20(2): 153–161.

Veselovsky, G. & Ackarachalanonth, S. 2012. Constructing transition matrices for routing BPC permutations on shuffle-exchange recirculating networks. In *Proc. DICTAP2012*: 378–383.

Veselovsky, G. & Agrawal A. 2014. On crosstal-free BPC permutations routing in an opticl variable-stage shuffle-exchange network. *Proc. IASTED Intl. Conf. Parrallel and distributed computing and networks PDCN:* 232–238.

Veselovsky, G. 2011. On BPC permutations admissibility to variable-stage hybrid optical shuffle-exchange networks. In *Proc. PDCN 2011*, IASTED 44(5): 714–719.

Yang, Y. & Wang J. & Pan Y. 2000. Permutation capability of optical multistage interconnection networks. *J. Parallel Distrib. Computing* 60(1): 72–91.

Electronics, Communications and Networks IV – Hussain & Ivanovic (eds)
© 2015 Taylor & Francis Group, London, ISBN: 978-1-138-02830-2

A practical robust efficient data-retrieving architecture for wireless sensor networks

Dingcheng Wang*, Bo Tang*, Beijing Chen, Xi Liu, Yujia Ni & Zhili Cao
College of Computer & Software, Nanjing University of Information Science & Technology, Nanjing, Jiangsu, China

ABSTRACT: At present, Wireless Sensor Networks (WSNs) are now applied in many fields like environment monitoring, healthcare and military intelligence, but most of them are simple-constructed, costly and low-efficiency. Therefore, how to build economic, reliable practical network systems becomes extremely significant. So far, scholars proposed various mechanisms and algorithms to improve network's efficiency. However, they as a whole suffer several shortages, e.g., hard to implement, exorbitant economic expenditure. In this paper, we investigate many real-world application scenarios of WSNs and develop a practical, robust efficient data-retrieving mechanism named "Practicability-oriented Multi-subsink Energy & Link-quality Aware (PMELA)" architecture. It includes a multi-subsink network structure and a novel Energy & Expected-transmissions aware Data Collection Protocol (EEDCP). Besides, it reserves the features of ad hoc like easy-to-deploy, adaptive, self-organized and multi-hop. The real-world simulation results show that our proposed architecture performs much better than Collection Tree Protocol (CTP) in network lifetime (about double of CTP's), energy efficiency and latency (about 1/4 of CTP's).

KEYWORDS: adaptive and self-configuring, practical data-retrieving network architecture, wireless sensor networks.

1 INTRODUCTION

A typical wireless sensor network (WSN) is composed of hundreds or thousands of low-cost wireless sensor nodes which are densely deployed in the area, we concern and one or multi sinks used to retrieve data from sensor nodes (Fowler 2009). Nowadays, WSNs have indeed helped people to solve many research issues and coped with a lot of realistic problems. And its fields can be categorized with several headings include military applications, environmental monitoring, health care, home networking and so on (Arampatzis et al. 2005). For most of WSNs, their tasks are sensing data in the phenomenon area and storing the data in the base station (BS) or sink. Due to the limits of wireless devices and the low-efficiency of current WSNs, researches of practical, robust efficient data-retrieving architecture become hot (Al-Karaki & Kamal 2004).

At present, there is none uniform data-retrieving architecture for all application scenarios because every of them somehow has unique requirements, e.g., applications in military information collection usually ask for data-retrieving speed and data security while applications in wild environment monitoring commonly wish the networks could work effectively as long as possible. Our proposed architecture suits to all scenarios with one accessible path in sensing area. In terms of the network structure, the WSN architectures can be categorized by flat structures and hierarchical structures. Familiar flat protocols include WRP (Murphy & Aceves 1996), TBRPF (Ogier et al. 2004), TORA (Park & Corson 1997), Gossiping (Hedetniemi & Liestman 1998), Flooding (Lim & Kim 2001), RR (Braginsky & Estrin 2002), etc. and famous hierarchical protocols include LEACH (Heinzelman et al. 2000), LEACH-C (Heinzelman et al. 2002), GAF (Roychowdhury & Patra 2010), HEED (Younis & Fahmy 2004), etc. Our proposed architecture is half-flat half-hierarchical. It divides the sensing area into several clusters (or zones) by subsinks and in each cluster it is flat.

Nearly all proposed WSN architectures or protocols are to cope with two main issues:, resource allocation (energy, bandwidth, etc.) and routing schemes (Gatzianas & Georgiadis 2008). An ideal protocol should have three features, namely energy efficient, robust and low latency. In theory, hierarchical architectures perform much better than flat ones, and complex advanced protocols perform better than simple ones generally. However, the majority of commonly used protocols in real world is those simple ones, e.g., direct transmission (DT) and CTP (Gnawali

*Corresponding author: *dcwang2005@126.com, tangbo_nuist@163.com*

et al. 2009), which belong to flat protocols. In our investigation, we found there are a lot of difficulties to construct those hierarchical or advanced architectures. And some of these difficulties are too hard to overcome, at least for the present. We consider that those architectures who suffer a lack of universality or practicability are "NOT practical". Not only hierarchical protocols, e.g., LEACH, SISR (Baek et al. 2010) and LEACH-C, but also some complex flat protocols, e.g., TSA-MSSN (Song & Hatzinakos 2007) and QAZP (Cheng & Chang 2012), belong to "not practical" architectures.

In part II we try to find out the reasons why those excellent data-gathering architectures, e.g., LEACH, GAF, QAZP and TSA-MSSN, are hard to implement or popularize in real-world. And then, in part III.A and III.B, we propose our architecture "Practicability-oriented Multi-subsink Energy & Link-quality Aware (PMELA)" which is easy to construct and suits to present technique conditions. It reserves the features of Ad-hoc like adaptive, self-organized and multi-hop. The basic idea of PMELA is to use one accessible trajectory, e.g., road, bridge and alley, to deploy and manage some wireless nodes (NOT sensor nodes) to function as subsinks to gather data from other sensor nodes. After that, users can get result data from those subsinks easily. In part III.C, we illustrate PMELA's working procedure in detail, and then discuss related parameters by mathematic analysis. In part IV, we imitate the real-world environment and make lots of experiments to validate PMELA's dominance by the commonly-used collection tree protocol (CTP) (Gnawali et al. 2009).

2 RELATED WORK

At present, there are mainly two structures of data gathering architectures for WSNs, namely flat structures and hierarchical structures. Since the famous hierarchical routing architecture named low-energy adaptive clustering hierarchy (LEACH) (Heinzelman et al. 2000)was proposed in 2000, many excellent clustering algorithms were proposed to improve the efficiency of WSNs. Adding those energy-efficient flat ones, there are so many advanced architectures (protocols) now (Pantazis et al. 2012). However, we found in our investigation that most of them are not used in real-world applications. To explain the reasons why those excellent methods are not practical or popular, we list several protocols (architectures) as instances: LEACH, LEACH-C, HEED, TTDD (Luo et al. 2005), PEGASIS (Lindsey & Raghavendra 2002), GAF, QAZP (Cheng & Chang 2012), SISR (Baek et al. 2010) and TSA-MSSN (Song & Hatzinakos 2007).

We divide the possible factors into three categories: "NOT Hardware-supported", "NOT Robust" and "OVER Technique-relied". The first one includes two potential problems caused by hardware: (a) the ideal sensor nodes are too expensive to be not suitable for large-scale densely deployed WSNs; (b) the standard of current wireless-node technique does not meet the requirements of those advanced methods. E.g., GAF, TSA-MSSN and TTDD require sensor nodes having the ability of precise positioning. It can be imagined that if the sensor nodes are assembled with GPS, it would be a highly expensive to build such a WSN. Similarly, if it uses range-based positioning algorithm, it also asks for other expensive hardware like distance or angle measurement system. In LEACH, LEACH-C, HEED and PEGASIS, the sensor nodes are assumed to communicate with BS directly and change radio range freely. In real-world scenarios, the long distance transmission is commonly accomplished by multi hops. If the sensor nodes assemble high capacity of radio and high capacity of the battery, it means high investment of sensor nodes. Therefore, in terms of this factor, most of highly centralized protocols (LEACH-C, TSA-MSSN, etc.) are not practical because they usually ask for precise locating systems or long distance transmission modules.

The second difficulty "NOT Robust" refers to that some advanced protocols are too ideal in consideration and they are not so robust in the real world. For instance, LEACH utilized (1) to choose cluster heads (CHs) in a certain round.

$$P(i)=\begin{cases}\dfrac{k}{N-k*(r \bmod \frac{N}{k})} & : \textit{node i has not been a CH}.\\ 0 & : \textit{otherwise}.\end{cases} \quad (1)$$

In (1), k refers to the number of clusters, N refers to the number of total nodes, r refers to the number of rounds and $P(i)$ refers to the probability of being the CH for node i. We simulated LEACH in MATLAB platform (N=100, k=5, in an area of 100*100) and the results revealed that k is not 5 in a large number of rounds, and in many rounds k is zero. In the real world, this situation would cause serious error. QAZP assumes a mobile anchor can move freely in the sensing area, which is impossible for most ground-based applications. As for TTDD, sensor nodes are assumed to be always stationary, which is also hard to guarantee in outdoor. In LEACH-C, nodes need to send the information about location and energy level to the BS directly in each round.

This procedure would squander much energy, especially when we cannot guarantee that the size of result data is much larger than that of information.

The last factor "OVER Technique-relied" means that the good performance of those advanced methods mainly rely on certain techniques like data fusion technique, precise radio range control technique, distance estimating technique and reliable transmission control protocol. In terms of data fusion technique, it seems to be not generally accepted in real-world applications. Because in most of real-world scenarios, all the data sensed by sensor nodes is valuable, and loss compression (e.g., high-fusion-rate methods) is not acceptable. On the other hand, package loss is not considered in some protocols. This makes them low fault-tolerant and fragile in real-world. Like for those centralized protocols (LEACH-C, TSA-MSSN, etc.), the BS needs to communicate with numerous nodes using reliable links at the same time. However, at present, there is no uniform, reliable transmission control protocol for WSNs like TCP/IP for traditional networks guarantee the communication quality. Table 1 shows the detailed drawbacks (√) and dominance (—) of CTP, PMELA and architectures mentioned above in terms of practicability.

Table 1. Reasons of unserviceable for each architectures.

Architecture	NOT Hardware Supported	NOT Robust	OVER Technique-relied
LEACH	√	√	√
LEACH-C	√	√	√
TTDD	√	√	√
PEGASIS	√	—	√
GAF	√	—	√
HEED	√	—	√
QAZP	√	√	—
SISR	—	√	√
TSA-MSSN	—	√	—
CTP	—	—	—
PMELA	—	—	—

3 OUR PROPOSED PMELA

3.1 *Application scenarios & features*

In many realistic scenarios of WSNs, the sensing fields are not totally inaccessible, such as the forests, factory area, animal farm, greenhouse area, zoo, etc. and usually there are some trajectories for people to access. Fig. 1 shows the three scenarios can apply our proposed Practicability-oriented Multi-subsink Energy & Link-quality Aware (PMELA) architecture.

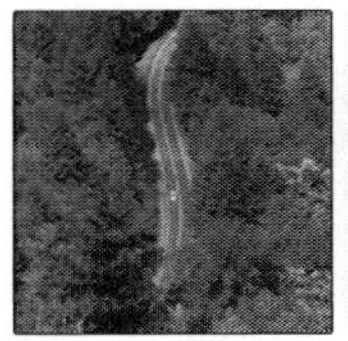

Forests Greenhouse Factory area

Figure 1. Three scenarios suit to use PMELA.

A PMELA network is composed of numerous random-deployed sensor nodes, several subsinks (powerful wireless nodes) and one data-retrieving terminal. It has three main features below compared with other advanced WSNs architectures.

- Easy to construct: This is same to traditional ways. Our proposed architecture reserves the advantages of original Ad hoc networks, namely randomly deployed, easy to manage, adaptive and self-configuring.
- New Structure: A PMELA networks is constructed along an accessible trajectory like a road in the sensor field. We deploy numerous cheap sensors in the sensing area randomly, and then deploy several wireless nodes along the trajectory to work as subsinks. We assume those subsinks have infinite power, because their batteries can be recharged or replaced easily.
- New Routing Algorithm: In PMELA, a new practical routing strategy named energy & expected-transmissions-aware data collection protocol (EEDCP) is proposed. To utilize it in real-world, we suggest modifying the relevant codes based on the CTP which is already implemented in TinyOS.

3.2 *Basic idea*

Our motivation is to use common-used cheap wireless sensors to build practical robust efficient networks. We argue to build a network along an accessible trajectory like a road, alley or creek. Firstly, users deploy their sensor nodes in the area near the trajectory randomly (by plane or other ways). Then users deploy several enhanced wireless nodes along the trajectory to work as recipients. After the network is built, users can retrieve data from those wireless recipients from mobile vehicles assembled wireless receiving device. In PMELA, we utilize EEDCP which is adaptive and self-configuring routing protocol. It aims at making tradeoff between residual energy of nodes and the data transmitting success rate to improve efficiency and robustness of the network.

3.3 *Details & problem formulation*

3.3.1 *Working principles of "EEDCP"*

In PMELA, we propose a new routing protocol named energy & expected-transmissions-aware data collection protocol (EEDCP). Its working procedure can be divided into rounds. Each round begins with a set-up phase when the path tables are formed, followed by a steady-state phase when sensor nodes send their result data to subsinks.

To guarantee the efficiency of PMELA, the steady-state is set much longer than set-up phase. There are three operations to form the path table (Figure 2) in set-up phase. Here to elaborate the complex procedure, we use a simplified instance. As is shown in the Figure 2-a, the first step is to set down the father-subsink for sensor nodes. In this sub-procedure, every subsink intensifies its radio amplifier and broadcasts a message contained their own ID. Every sensor node receives one message at least. If receiving several messages, sensor chooses the one with highest signals as its father-subsink. By this operation, the network is divided into several clusters or zones.

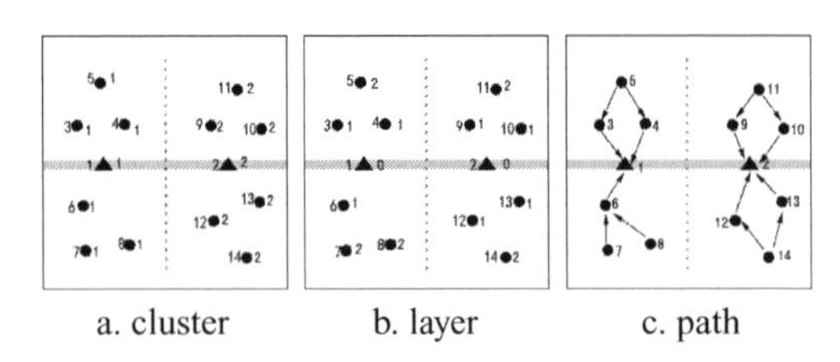

a. cluster b. layer c. path

Figure 2. Three steps to establish path table.

The second and third operations are intra cluster. Firstly, subsinks decrease the radio amplifier down to the level same as sensor nodes, and then broadcast a message to settle down the layer 1(as shown in Figure 2-b). After that, the nodes in layer 1 repeat this procedure. This procedure iterates until all nodes are settled, when each node can calculate their own transmission path table which records the available paths to forward data packets (Figure 2-c).

In steady phase, the data packets flow from the higher layer to lower layer generally. For node "*i*" who has n_p next hops $(1,2,3,\ldots,n_p)$, it chooses the next hop "*k*" subjected to (2). *Next(i)* refers to the next hop node of node "*i*" and *layer(i)* refers to the layer of node "*i*".

$$\begin{aligned} & Next(i) = node(k) \\ & s.t. \\ & \quad layer(k) = layer(i) + 1; \\ & \quad W(i \sim k) = Max\{W(i \sim 1),\ldots,W(i \sim m),\ldots,W(i \sim n_p)\}; \\ & \quad EETX_{i/j} \geq 0. \end{aligned} \tag{2}$$

In (2), $EETX_{i/j}$ refers to the extra expected transmissions of node "*j*" to node "*i*", it is defined by Link Estimation Extension Protocol. It can be calculated by

$$EETX_{i/j} = \begin{cases} -1; & data\_success_{j\text{-}to\text{-}i} = 0; \\ (\dfrac{data\_total_{j\text{-}to\text{-}i}}{data\_success_{j\text{-}to\text{-}i}} - 1) \times 10; & otherwise. \end{cases} \tag{3}$$

And *W(i~j)* is the weight of the transmission from node "*j*" to node "*i*", which represents the ability of node "*j*" receiving packets from node "*i*". To guarantee the efficiency of PMELA, the calculating formula should be designed scientifically. First, it is considerate that the path to high residual power is preferred. Besides, in terms of transmission success rate, the node with low expected transmissions (ETX) is favored to guarantee the network's robustness. These two aspects can be balanced by

$$\phi(j) = \frac{\overline{AEL_j}^{\,x}}{\overline{AETX_j}^{\,y}}, \tag{4}$$

where

$$\overline{AEL_j} = \frac{E_{rest\_j} + \sum_{s=1}^{n_p} AEL_s}{(n_p + 1) \times (E_{max} - Round * E_{min\_cost})}, \tag{5}$$

$$\overline{AETX_j} = \frac{\sum_{s=1}^{n_p} (AETX_s + EETX_{j/s})}{n_p \times AETX_{max}}. \tag{6}$$

In (5), E_{max} stands for the initial power of each node, and E_{min_cost} stands for the minimal energy cost of each node in each round. The *AEL*, *AETX* of subsinks are 0. Due to the fact that we do not know the extent of contribution of these two parameters, we add two other variables (x,y) to control the impact of them, as is shown in (4).

In EEDCP, the steady phase is much longer than set-up phase and forwarding packets to only one node would result in hotspots in the long run, so it is wise to assign the packets to different nodes in the path table fairly. Assigning meanly is a choice, but different path has different ability (ϕ) of forwarding packets. We argue to assign the packets according to different paths' transmission ability quotient. E.g., at a certain moment, the numbers of packets have been delivered by node "*i*" via different paths are $(P_1,P_2,\ldots,P_m,\ldots,P_{np})$. We assume the ideal number of packets forwarded via the m-th path is $\chi(m)$ which can be calculated by

$$\chi(m) = \varphi(m) \times \sum_{s=1}^{n_p} P_s, \tag{7}$$

where $\varphi(m)$ is the ratio of packets delivered from node "*i*" to node "*m*" in theory and it is decided by

$$\varphi(m) = \frac{\phi(m)}{\sum_{s=1}^{n_p} \phi(s)}. \tag{8}$$

Therefore, we can get the weight of the path (next hop "m") for node "i" by

$$W(i \sim m) = \chi(m) - P_m = \frac{\dfrac{\overline{AEL_m}^x}{\overline{AETX_m}^y}}{\sum_{s=1}^{n_p} \dfrac{\overline{AEL_s}^x}{\overline{AETX_s}^y}} \times \sum_{s=1}^{n_p} P_s - P_m . \quad (9)$$

So far, node "i" knows whom it sends the current packet to. In a realistic situation, there are multiple packets waiting for being forwarded at certain times. We suggest calculating the total number of packets for each path together, and then sending corresponding numbers of packets to next-hop node. By doing this, frequent switching between computation module and radio module can be avoided.

3.3.2 *Optimum number of subsinks*

In PMELA, the number of subsinks is decided by balancing the economic cost of hardware and the contribution they make.

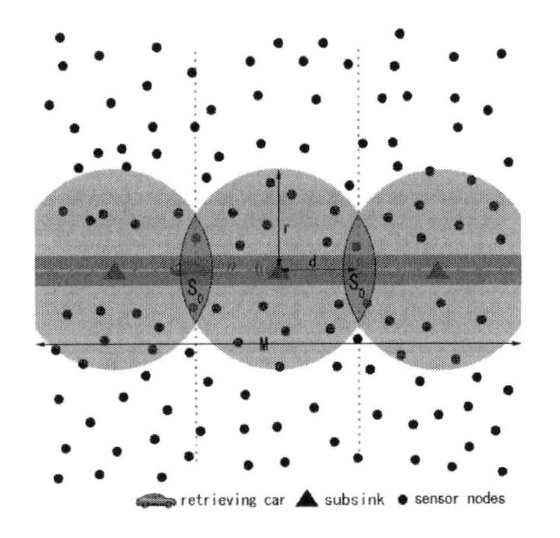

Figure 3. Network structure of PMELA. Note that the height of the map should be larger than "r".

In terms of efficiency, one way to relieve the impact of hotspot issue is to mean energy dissipation among those nodes that connect with subsinks directly and we denote such nodes as "key nodes". For the total number of packets is fixed and all packets need to be forwarded by key nodes, the number of key nodes should be as many as possible. For easy analysis, we assume the density of nodes is uniform. Therefore, one way to increase the number of key nodes is to maximize the radio cover area by using a large number of subsinks. Unfortunately, the cost of subsinks is much more than that of sensor nodes both in price and management. So we wish the number of subsinks is as little as possible in terms of cost. In addition to that, mass of subsinks cause much waste. As shown in Figure 3, the S_0 areas are repeated twice by each two adjacent subsinks and one area is useless. Hence, we make a tradeoff between number of subsinks and their waste, and give a limitation to the number of subsinks in order to avoid the extra cost of numerous subsinks at the same time.

For the reason of page limitation, we omit the derivation procedure. We can find the optimal k by

$$k_{opt} = \left[\frac{M}{\mathrm{NSD} \times r} \right], \quad (10)$$

where the "[]" means the half-adjusting operation. NSD refers to the decision-making coefficient of number of subsinks. According to our derivation, NSD=1.8296.

3.3.3 *Principle of setting value for "x" and "y"*

These two parameters are used to deal with the disorders caused by dynamic real-world environment and improportional numbers of packets via different paths in the long run. In addition to disorders, they also can be used when the user demand is changed. In real-world running, if the energy levels of key nodes are out of synchronization, we can increase x to reduce the number of packets forwarded by low-energy nodes. And if we want to improve the data collection success rate, we increase y.

In general situations, to extend the lifetime of the network as long as possible, we increase the value of "x" regularly when the collection success rate is beyond the expected value which is an application-specific index. We denote the expected data collection success rate is $Threshold_{data\_col}$. Then we can regulate the value of "x" and "y" by

$$x = \begin{cases} x + 0.1; & Rate_{suc} > Threshold_{data\_col}; \\ x - 0.1; & otherwise. \end{cases}, \quad y = \begin{cases} y - 0.1; & Rate_{suc} > Threshold_{data\_col}; \\ y + 0.1; & otherwise. \end{cases} \quad (11)$$

where $Rate_{suc}$ stands for the data collection success rate of last round. And the initial value of "x" and "y" is 1.

4 NUMERICAL RESULTS

To evaluate our architecture, we simulated CTP and PMELA using MATLAB tools, and then compared their capabilities in four aspects, namely network lifetime (by rounds), energy utilization ratio, data collection success rate and communication delay (by hops), under the same condition: 200 sensor nodes were randomly deployed in the zone from [*0, 0*] to [*100, 100*] with an accessible path along the line (*y=50*). To be fair and precise, we run the simulating codes ten times and use the medial one as the final result.

As shown in Table 2, the network lifetime of the PMELA network is 2400 rounds, about 1.8 times of that of the CTP network (1320 rounds). Figure 4 shows the detailed number of valid nodes in each round in GAF and PMELA networks. This graph indicates PMELA is more effective than GAF in dealing with the hotspot issue.

Figure 5 shows us the energy utilization ratio of CTP and PMELA in each round. It is clear that the energy utilization ratio of the CTP network in each round before the 1000th round was almost double of that of the PMELA network. However, their final energy utilization ratios were similar, which indicates CTP networks consume the energy more quickly and it is less energy efficient than PMELA networks.

Figure 6 shows us the collecting success rate figures in each round using CTP and PMELA. Before the 1320th round, the data collecting success rate of PMELA was almost 4 times of that of CTP. The straight line during the fore 1700 rounds implies the impact of "*x*" and "*y*" in (4). We should notice that if we decrease the required success rate for PMELA to CTP's level, we can further improve the network lifetime dramatically.

Table 2. Network lifetime.

Protocol Architecture	CTP	PMELA
Lifetime (rounds)	1320	2400

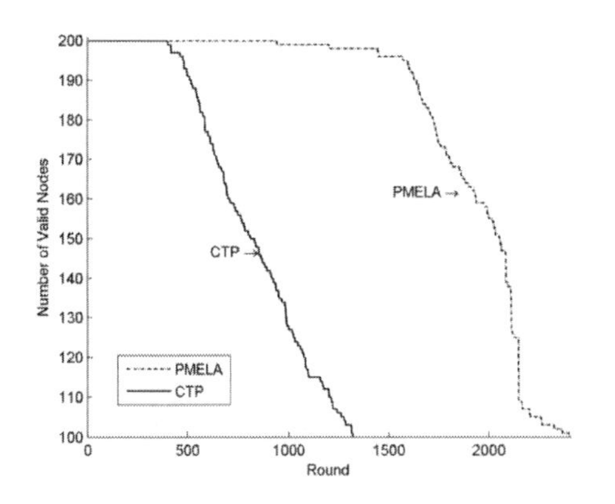

Figure 4. Number of valid nodes in each round.

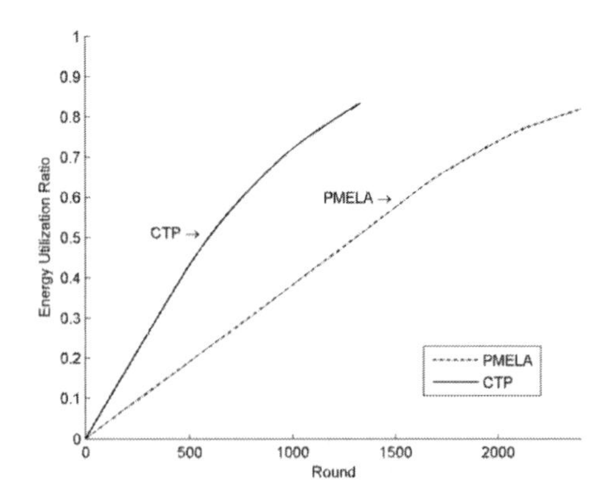

Figure 5. Energy utilization ratio.

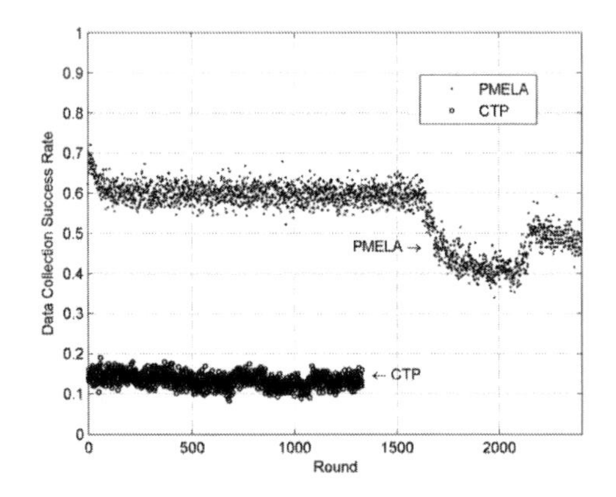

Figure 6. Data collection success rate in each round.

Figure 7 records the average data-delivering hops of nodes in each round using different protocols. We can see the latency of PMELA was steady during its lifetime. It was less than 1/4 of that of CTP during the fore 500 rounds. On the other hand, the decreasing latency in the CTP network after the 500th round was due to the early death of sensor nodes.

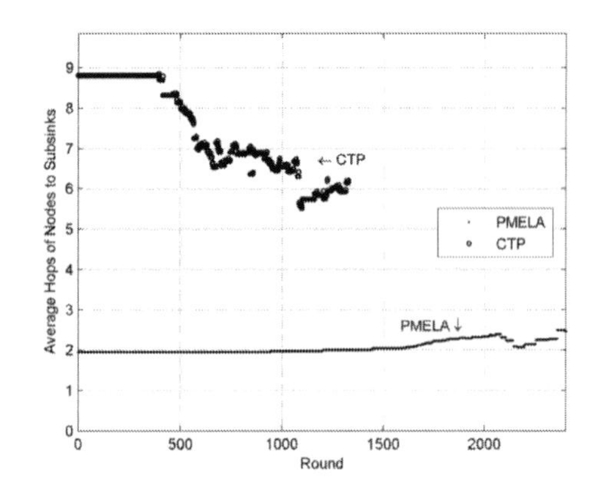

Figure 7. Average hops for nodes to transmit data to subsinks.

5 CONCLUSION

When designing practical protocol architectures for wireless sensor networks, it is necessary to consider the present conditions, requirements of the application, the need for ease of deployment, robustness and efficiency of the protocol and cost of implementation. These conditions impulse us to design PMELA, a data collection architecture for WSN, where a novel routing protocol (EEDCP) is proposed and used to make the architecture practical, robust and efficient. It reserves the advantages of original Ad hoc networks, e.g., easy-to-construct, low investment and robust. Besides, the final real-world simulations prove that our PMELA performs much better than CTP in efficiency, latency and network lifetime.

ACKNOWLEDGMENT

This work was supported in full by the Natural Science Foundation of JiangSu Province NO. BK2012858, and supported in part by the National Natural Science Foundation of China under grant numbers 61103141.

REFERENCES

Al-Karaki, J.N. & Kamal, A.E. 2004. Routing techniques in wireless sensor networks: a survey. *Wireless Communications. IEEE* 11(6): 6–28.

Arampatzis, Th. Lygeros, J. & Manesis, S. 2005. A Survey of Applications of Wireless Sensors and Wireless Sensor Networks. *Proceedings of the 13th Mediterranean Conference on Control and Automation, Limassol, Cyprus, 27-29 June 2005*. IEEE.

Baek Jinsuk, An Sun Kyong, & Fisher, P. 2010. Dynamic cluster header selection and conditional re-clustering for wireless sensor networks. *IEEE Transactions on Consumer Electronics* 56(4): 2249–2257.

Braginsky, D. & Estrin, D. 2002. Rumor Routing Algorithm for Sensor Networks. *Proceedings of the 1st ACM international workshop on Wireless sensor networks and applications, Atlanta, USA, 2002*. New York: ACM.

Cheng, S.T. & Chang, T.Y. 2012. An adaptive learning scheme for load balancing with zone partition in multi-sink wireless sensor network. *Expert Systems with Applications* 39(10): 9427–9434.

Fowler, K. 2009. Sensor Survey Results. *Instrumentation & Measurement Magazine, IEEE* 12(2): 40–44.

Gatzianas, M. & Georgiadis, L. 2008. A Distributed Algorithm for Maximum Lifetime Routing in Sensor Networks with Mobile Sink. *IEEE Transactions on Wireless Communications* 7(3): 984–994.

Gnawali, O., Fonseca, R., & Jamieson, K. 2009. Collection tree protocol. *Proceedings of the 7th ACM Conference on Embedded Networked Sensor Systems, Berkeley, California, 4-6 Nov 2009*. New York: ACM.

Hedetniemi, S.M & Liestman, A. 1998. A Survey of Gossiping and Broadcasting in Communication Networks. *Networks* 18(4): 319–349.

Heinzelman, W.R., Chandrakasan, A., & Balakrishnan, H. 2000. Energy-efficient communication protocol for wireless microsensor networks. *Proceedings of the 33rd Hawaii International Conference on System Sciences, 4-7 Jan 2000*. IEEE.

Heinzelman, W.R., Chandrakasan, A., & Balakrishnan, H. 2002. An application-specific protocol architecture for wireless microsensor networks. *IEEE Transactions on Wireless Communications* 1(4): 660–670.

Lim, H. & Kim, C. 2001. Flooding in Wireless Ad Hoc Networks. *Computer Communications* 24(3): 353-363.

Lindsey, S. & Raghavendra, C. 2002. PEGASIS: Power-Efficient GAthering in Sensor Information Systems. *In Proc. IEEE Aerospace Conference, Montana, USA, 2002*. IEEE.

Luo, H., Ye, F., Cheng, J., et al. 2005. TTDD: Two-Tier Data Dissemination in Large-Scale Wireless Sensor Networks. *Wireless Networks, Springer Netherlands* 11(1): 161–175.

Murphy, S. & Aceves, L. 1996. An Efficient Routing Protocol for Wireless Networks. *Mobile Networks and Applications* 1(2): 183–197.

Ogier, R., Templin, F. & Lewis, M. 2004. Topology Dissemination Based on Reverse-Path Forwarding. RFC Editor 2004. *http://www.hjp.at/doc/rfc/rfc3684.html*.

Pantazis, N., Nikolidakis, S., & Vergados, D. 2012. Energy-Efficient Routing Protocols in Wireless Sensor Networks: A Survey. *Communications Surveys & Tutorials, IEEE* 15(2): 551–591.

Park, D. & Corson, S. 1997. A Highly Adaptive Distributed Routing Algorithm for Mobile Wireless Networks. *Proceedings of the 16th Conference on Computer and Communications Societies, Kobe, Japan, 7–12 Apr 1997*. IEEE.

Roychowdhury, S. & Patra, C. 2010. Geographic Adaptive Fidelity and Geographic Energy Aware Routing in Ad Hoc Routing. *Special Issue of International Journal of Computer & Communication Technology* 1(2, 3, 4): 309–313.

Song, L. & Hatzinakos, D. 2007. Architecture of Wireless Sensor Networks With Mobile Sinks: Sparsely Deployed Sensors. *IEEE Transactions on Vehicular Technology* 56(4): 1826–1836.

Younis, O., & Fahmy, S. 2004. HEED: A Hybrid, Energy-Efficient, Distributed Clustering Approach for Ad Hoc Sensor Networks. *IEEE Transactions on Mobile Computing* 3(4): 366–379.

Electronics, Communications and Networks IV – Hussain & Ivanovic (eds)
© 2015 Taylor & Francis Group, London, ISBN: 978-1-138-02830-2

A hybrid one dimensional optimization

Zhengyuan Wang
High-Tech Institute of Xi'an, Xi'an, P.R. China

Li Gao
Peking University, Peking, P.R. China

Huizhen Wang
High-Tech Institute of Xi'an, Xi'an, P.R. China

ABSTRACT: In this paper, we propose a hybrid method for one-dimensional unimodal function optimization. An approximating function to the given function is fitted by five points selected on its curve. In order to improve the precision approach, using heuristic rules to select the structure of approximate function automatically. The minimum point of the approximation function is an approximation to the minimum point of the given unimodal function. Repeated iteration strategy is used to improve the approximating solution. Proposed termination condition effectively improves the search efficiency. Hybrid optimization method has been successfully solved many function optimization problems, such as convex function, non-convex function, non-smooth function and no expression functions. The hybrid method can be shown to have almost quadratic convergence. Numerous experiments show that the algorithm is efficient.

1 INTRODUCTION

We consider the following minimization problem:

$$\min_{a \le x \le b} f(x) \tag{1}$$

where $f(x)$:$[a,b] \rightarrow R$ is continuous and unimodal. The function f (x) may be not differentiable, not to mention its convexity. This problem generally uses the line search method for solving.

There are many line search methods. Each method has its own advantages and applications. Newton's method is a classical one dimensional optimization method. For optimization problem of a convex function, it's proved that Newton's method has quadric convergence (Wang 2010, Abraham et al. 2009, Huang et al. 2009, Crina & Ajith 2007, Zhao & Gao 2005). Newton's method is invalid to the minimization problem of non-smooth function, even to the function without explicit expression. For the objective function $f(x)$, Direct line search methods such as gold section method, interpolation method, six point method and double secant method, has less restricted conditions than Newton's method (Tan & Qin 2000, Yuan & Sun 2005, Yin 1989, Zhou 1984). These direct methods may be used to obtain the extreme value of a continuous function and the function without expression also. The direct method is time-consuming, usually (Crina & Ajith 2009, Hu et al. 2007, Ioannis 2006, Shi et al. 2005). The gold section method is used to get the minimum of continuous function and has a linear convergence rate. Double secant method is the most effective direct method, but it's difficult to determine the argument. The interpolation method assumes that function $f(x)$ is similar to a polynomial. The minimum point of interpolation function is used to approximate the minimum point of $f(x)$. Interpolation method has superlinear convergence if the approximation error to $f(x)$ is very small. The interpolation method applies only to a convex function optimization problems.

In many fields, efficient method for solving general unimodal function optimization problems is important. More domain knowledge for obtaining the minimum, higher efficiency of the solving method.

In order to improve efficiency of the search process, a hybrid method is proposed which is combined Thiete type continued fraction with five-point line search. Firstly, we obtain five key points on the curve of function $f(x)$ and then calculate the approximation of the extreme point. Finally an approximate solution is obtained by repeated iteration. Experimented results show that five-point line search method is faster than the gold section method and the interpolation method generally. Five-point method proposed here can obtain the extreme value of convex functions. But Five-point method is invalid for solving optimization

*Corresponding Author: *zywang1999@163.com*

problems of non-convex unimodal function. Thiete type continued fraction is used to approximate the given function $f(x)$ in this case and five-point method can use to determine whether the solution obtained meets the accuracy requirements or not. In this way we can significantly improve the search efficiency. Experiments show that the hybrid method proposed here was successful in solving unimodal function optimization problems.

2 FIVE-POINT LINE SEARCH METHOD TO CONVEX FUNCTION OPTIMIZATION

Suppose that the minimization problem is described by (1), $f(x)$ is a lower convex function and maybe an explicit function, implicit function, even to a function without expression. The value of $f(x)$ can be worked out with x given.

2.1 *Five-point line search method to the minimization problems of lower convex function*

Suppose that function $f(x)$ is a lower convex function. Five points on the curve of $y=f(x)$ are given in XOY platform (Figure 1(a)). Let $P_i(x_i, y_i)$ satisfies

$$x_1 < x_2 < \ldots < x_5, y_i = f(x_i) \tag{2}$$

$$y_1 \geq y_2 \geq y_3, y_5 \geq y_4 \geq y_3 \tag{3}$$

where $x_1=a$ and $x_5=b$ initially.

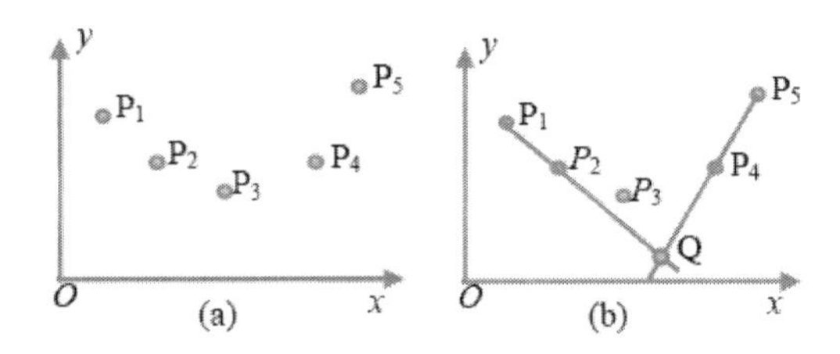

Figure 1. Five key points and new search point.

Suppose that Q is the intersection of line P_1P_2 and P_4P_5 (Figure 1(b)). X-coordinate of point Q is

$$x_a = (y_4 - y_2 + r_2 x_2 - r_4 x_4)/(r_2 - r_4) \tag{4}$$

$$r_2 = (y_1 - y_2)/(x_1 - x_2) < 0, r_4 = (y_5 - y_4)/(x_5 - x_4) > 0 \tag{5}$$

x_a is updated by a number close to x_3 if it equals to x_3. For example, x_a is updated by Eq.(6).

$$x_a = \begin{cases} 0.1x_2 + 0.9x_3, 2x_3 \geq x_4 + x_2 \\ 0.9x_3 + 0.1x_4, 2x_3 < x_4 + x_2 \end{cases} \tag{6}$$

X-coordinate x_a is an approximating solution while $y_a=f(x_a)$ is an approximation of the minimum of the function $f(x)$. In order to obtain a solution which meets the tolerance, repeated iteration strategy is used. If $f(x_a)<f(x_3)$, the new five-point is as follows:

$$(P_1', P_2', P_3', P_4', P_5') = \begin{cases} (P_1, P_2, (x_a, f(x_a)), P_3, P_4), x_a < x_3 \\ (P_2, P_3, (x_a, f(x_a)), P_4, P_5), x_a > x_3 \end{cases} \tag{7}$$

Otherwise

$$(P_1', P_2', P_3', P_4', P_5') = \begin{cases} (P_1, P_2, P_3, (x_a, f(x_a)), P_4), x_a > x_3 \\ (P_2, (x_a, f(x_a)), P_3, P_4, P_5), x_a < x_3 \end{cases} \tag{8}$$

This method to obtain the extreme value of convex function is named five-point line search method. Generally, a lower convex function $f(x)$ satisfies the inequality condition (9) and (10).

$$r_2(x - x_2) + f(x_2) \leq f(x) \text{ where } x \geq x_2 \tag{9}$$

$$r_4(x - x_4) + f(x_4) \leq f(x) \text{ where } x \leq x_4 \tag{10}$$

2.2 *Convergence*

By analysis in section 2.1, it is found that a series of the approximation of minimum of $f(x)$ is a monotonically decreasing series. So we obtain a theory for minimization problem of a lower convex function by five-point line search method.

Theory 1(Convergence theory of lower convex function) Let function $y=f(x)$ is continuous and lower convex in $[a, b]$. The Point $P_i^{(k)}\left(x_i^{(k)}, y_i^{(k)}\right)$ is on the curve of the function $y=f(x)$ and satisfies that

$$\begin{aligned} & a \leq x_1^{(k)} < \ldots < x_5^{(k)} \leq b, y_i^{(k)} = f\left(x_i^{(k)}\right) \\ & y_1^{(k)} > y_2^{(k)} > y_3^{(k)}, y_5^{(k)} > y_4^{(k)} > y_3^{(k)} \end{aligned} \tag{11}$$

The new points are produced by Eq.(12).

$$\left(P_1^{(k+1)}, \ldots, P_5^{(k+1)}\right) = \begin{cases} \left(P_1^{(k)}, P_2^{(k)}, Q^{(k)}, P_3^{(k)}, P_4^{(k)}\right), x_a^{(k)} < x_3^{(k)}, y_a^{(k)} \leq y_3^{(k)} \\ \left(P_2^{(k)}, Q^{(k)}, P_3^{(k)}, P_4^{(k)}, P_5^{(k)}\right), x_a^{(k)} < x_3^{(k)}, y_a^{(k)} > y_3^{(k)} \\ \left(P_2^{(k)}, P_3^{(k)}, Q^{(k)}, P_4^{(k)}, P_5^{(k)}\right), x_a^{(k)} > x_3^{(k)}, y_a^{(k)} \leq y_3^{(k)} \\ \left(P_1^{(k)}, P_2^{(k)}, P_3^{(k)}, Q^{(k)}, P_4^{(k)}\right), x_a^{(k)} > x_3^{(k)}, y_a^{(k)} > y_3^{(k)} \end{cases} \tag{12}$$

where $Q^{(k)}\left(x_a^{(k)}, y_a^{(k)}\right)$ satisfies Eq.(13).

$$x_a^{(k)} = \begin{cases} 0.9x_3^{(k)} + 0.1l_k, \left|t^{(k)} - x_3^{(k)}\right| \le \varepsilon, 2x_3^{(k)} \ge l_k + u_k \\ 0.9x_3^{(k)} + 0.1u_k, \left|t^{(k)} - x_3^{(k)}\right| \le \varepsilon, 2x_3^{(k)} < l_k + u_k \\ 0.1x_3^{(k)} + 0.9l_k, t^{(k)} \le l_k \\ 0.1x_3^{(k)} + 0.9u_k, t^{(k)} \ge u_k \\ t^{(k)}, \text{otherwise} \end{cases} \tag{13a}$$

$$\begin{aligned} & y_a^{(k)} = f\left(x_a^{(k)}\right), \varepsilon = 0.01\min\left\{x_3^{(k)} - x_2^{(k)}, x_4^{(k)} - x_3^{(k)}\right\} \\ & t^{(k)} = \left(r_4^{(k)}x_4^{(k)} - r_2^{(k)}x_2^{(k)} + y_2^{(k)} - y_4^{(k)}\right)/\left(r_4^{(k)} - r_2^{(k)}\right) \\ & r_2^{(k)} = \frac{y_1^{(k)} - y_2^{(k)}}{x_1^{(k)} - x_2^{(k)}}, r_4^{(k)} = \frac{y_5^{(k)} - y_4^{(k)}}{x_5^{(k)} - x_4^{(k)}} \\ & l_k = \left(y_3^{(k)} - y_2^{(k)}\right)/r_2^{(k)} + x_2^{(k)}, u_k = \left(y_3^{(k)} - y_4^{(k)}\right)/r_4^{(k)} + x_4^{(k)} \end{aligned} \tag{13b}$$

Then

$$\lim_{k\to\infty} y_3^{(k)} = \lim_{k\to\infty} f\left(x_3^{(k)}\right) = \min_{x\in[a,b]} f(x) \tag{14}$$

Proof: Function $f(x)$ will obtain the minimum at $[a, b]$ because it is continuous in $[a, b]$. $P_i^{(k)}$ is on the curve of $f(x)$ and satisfies

$$\begin{aligned} & a \le x_1^{(k)} < \ldots < x_5^{(k)} \le b, y_i^{(k)} = f\left(x_i^{(k)}\right) \\ & y_1^{(k)} > y_2^{(k)} > y_3^{(k)}, y_5^{(k)} > y_4^{(k)} > y_3^{(k)} \end{aligned}$$

From Eq.(13) we know that $\left\{x_2^{(k)}\right\}$ is an increasing series with upper bound b and $\left\{x_4^{(k)}\right\}$ is a decreasing series with a lower bound a. Let

$$\lim_{k\to\infty} x_4^{(k)} = U, \lim_{k\to\infty} x_2^{(k)} = L$$

Then

$$\begin{aligned} & x_2^{(k)} \le L \le x_3^{(k)} \le U \le x_4^{(k)} \\ & y_2^{(k)} \ge f(L) \ge y_3^{(k)}, y_3^{(k)} \le f(U) \le y_4^{(k)} \end{aligned}$$

In the other hand

$$l_k < x_3^{(k)} < u_k, l_k < x_a^{(k)} < u_k$$

$$x_4^{(k)} - \max\left\{x_3^{(k)}, x_a^{(k)}\right\} > \frac{y_4^{(k)} - y_3^{(k)}}{r_4^{(k)}} > \frac{f(U) - y_3^{(k)}}{r_4^{(1)}}$$

$$\min\left\{x_a^{(k)}, x_3^{(k)}\right\} - x_2^{(k)} > \frac{y_3^{(k)} - y_2^{(k)}}{r_2^{(k)}} > \frac{y_3^{(k)} - f(L)}{r_2^{(1)}}$$

$$\begin{aligned} & \left(x_4^{(k)} - x_2^{(k)}\right) - \left(x_4^{(k+1)} - x_2^{(k+1)}\right) \\ & = \begin{cases} x_4^{(k)} - x_3^{(k)}, x_a^{(k)} < x_3^{(k)}, y_a^{(k)} \le y_3^{(k)} \\ x_a^{(k)} - x_2^{(k)}, x_a^{(k)} < x_3^{(k)}, y_a^{(k)} > y_3^{(k)} \\ x_3^{(k)} - x_2^{(k)}, x_a^{(k)} > x_3^{(k)}, y_a^{(k)} \le y_3^{(k)} \\ x_4^{(k)} - x_a^{(k)}, x_a^{(k)} > x_3^{(k)}, y_a^{(k)} > y_3^{(k)} \end{cases} \end{aligned}$$

$$\begin{aligned} & \left(x_4^{(k)} - x_2^{(k)}\right) - \left(x_4^{(k+1)} - x_2^{(k+1)}\right) \\ & \ge \min\left\{\left[f(U) - y_3^{(k)}\right]/r_4^{(1)}, \left[y_3^{(k)} - f(L)\right]/r_2^{(1)}\right\} \end{aligned}$$

$$\lim_{k\to\infty}\left(x_4^{(k)} - x_2^{(k)}\right) - \left(x_4^{(k+1)} - x_2^{(k+1)}\right) = 0$$

$$\lim_{k\to\infty} \min\left\{\left(f(U) - y_3^{(k)}\right)/r_4^{(1)}, \left(y_3^{(k)} - f(L)\right)/r_2^{(1)}\right\} = 0$$

By equation (13) we can obtain

$$\lim_{k\to\infty} y_3^{(k)} = f(L) = f(U)$$

Similarly, this line search method is also suitable to the maximization problem of upper convex function.

Theory 2 Function $y=f(x)$ is continuous and upper convex in $[a, b]$. Point $P_i^{(k)}$ satisfies inequality (15).

$$\begin{aligned} & y_1^{(k)} < y_2^{(k)} < y_3^{(k)}, y_5^{(k)} < y_4^{(k)} < y_3^{(k)} \\ & y_i^{(k)} = f\left(x_i^{(k)}\right), a \le x_1^{(k)} < \ldots < x_5^{(k)} \le b \end{aligned} \tag{15}$$

The new points are produced by Eq.(16).

$$\begin{aligned} & \left(P_1^{(k+1)}, \ldots, P_5^{(k+1)}\right) \\ & = \begin{cases} \left(P_1^{(k)}, P_2^{(k)}, Q^{(k)}, P_3^{(k)}, P_4^{(k)}\right), x_a^{(k)} < x_3^{(k)}, y_a^{(k)} \ge y_3^{(k)} \\ \left(P_2^{(k)}, Q^{(k)}, P_3^{(k)}, P_4^{(k)}, P_5^{(k)}\right), x_a^{(k)} < x_3^{(k)}, y_a^{(k)} < y_3^{(k)} \\ \left(P_2^{(k)}, P_3^{(k)}, Q^{(k)}, P_4^{(k)}, P_5^{(k)}\right), x_a^{(k)} > x_3^{(k)}, y_a^{(k)} \ge y_3^{(k)} \\ \left(P_1^{(k)}, P_2^{(k)}, P_3^{(k)}, Q^{(k)}, P_4^{(k)}\right), x_a^{(k)} > x_3^{(k)}, y_a^{(k)} < y_3^{(k)} \end{cases} \end{aligned} \tag{16}$$

where $Q^{(k)}$ is determined by Eq.(13). Then

$$\lim_{k\to\infty} y_3^{(k)} = \lim_{k\to\infty} f\left(x_3^{(k)}\right) = \max_{x\in[a,b]} f(x) \tag{17}$$

Only one point is produced once in the process for solving the optimization problem. So five-point method, gold section method and interpolation method have the same calculation each step. But five-point line search method is the most efficient methods.

2.3 *Termination condition*

It is unnecessary to obtain the exact solution to the optimization problem usually. It is important to determine whether a solution is satisfactory. Generally the optimization tolerance ε is given and we can give a simple rule to determine satisfactory solution.

Theory 3 If inequality (18) or (19) holds,

$$r_2(x_4-x_2)+f(x_2)\geq f(x_4)-\varepsilon \tag{18}$$

$$r_4(x_2-x_4)+f(x_4)\geq f(x_2)-\varepsilon \tag{19}$$

Then $f(x_a)$ is an estimate to minimum of lower convex function $f(x)$ with the optimization tolerance ε given.

$$\left|f(x_a)-\min_{a\leq x\leq b} f(x)\right|\leq\varepsilon \tag{20}$$

where

$$x_a=\frac{y_4-y_2+r_2x_2-r_4x_4}{r_2-r_4}, r_2=\frac{y_1-y_2}{x_1-x_2}, r_4=\frac{y_5-y_4}{x_5-x_4} \tag{21}$$

$$y_1>y_2>y_3, y_5>y_4>y_3, y_i=f(x_i), x_1< \quad <x_5$$

Proof: Function $f(x)$ is lower convex, then

$$f(x)-r_2(x-x_2)-f(x_2)\geq 0, x\geq x_2$$
$$f(x)-r_4(x-x_4)-f(x_4)\geq 0, x\leq x_4$$

By inequality (18), we can get that

$$f(x)-r_2(x-x_2)-f(x_2)$$
$$\leq f(x_4)-r_2(x_4-x_2)-f(x_2)\leq\varepsilon, x_2<x<x_4$$

By inequality (19), we can get that

$$f(x)-r_4(x-x_4)-f(x_4)$$
$$\leq f(x_2)-r_4(x_2-x_4)-f(x_4)\leq\varepsilon, x\in(x_2,x_4)$$

Let

$$g(x)=\begin{cases} r_4(x-x_4)+f(x_4), x_a\leq x<x_4 \\ r_2(x-x_2)+f(x_2), x_2\leq x<x_a \end{cases}$$

then

$$g(x)\leq f(x), x\in[x_2,x_4]$$
$$g(x_a)=\min_{x_2\leq x\leq x_4} g(x)$$

Suppose that $y^*=f(x^*)$ is the minimum of $f(x)$ where x belongs to the interval $[x_2, x_4]$, then

$$g(x_a)\leq g(x^*)\leq f(x^*)\leq f(x_a), x_2<x^*<x_4$$

If inequality (18) or (19) holds with tolerance ε given, then

$$f(x_a)-g(x_a)<\varepsilon, 0\leq f(x_a)-f(x^*)<\varepsilon$$

Similarly, we can get condition to determine the satisfactory solution for maximization problems of upper convex function.

Theory 4 If inequality (22) or (23) holds, then $f(x_a)$ is an approximation to the maximum of upper convex function $f(x)$ with the optimization tolerance ε given.

$$r_2(x_4-x_2)+f(x_2)\leq f(x_4)+\varepsilon \tag{22}$$

$$r_4(x_2-x_4)+f(x_4)\leq f(x_2)+\varepsilon \tag{23}$$

where

$$x_a=\frac{y_4-y_2+r_2x_2-r_4x_4}{r_2-r_4}, r_2=\frac{y_1-y_2}{x_1-x_2}, r_4=\frac{y_5-y_4}{x_5-x_4} \tag{24}$$

$$y_1<y_2<y_3, y_5<y_4<y_3, y_i=f(x_i), x_1< \quad <x_5$$

To the minimization problem of lower convex function, a new five-point is produced according to Eq.(12) and (13) if inequality(18) and (19) don't hold. Repeat the iteration until inequality (18) or (19) holds and obtain an approximation which meets the optimization tolerance at last. The algorithm is described as follows:

Algorithm 1 Five point line search method to obtain the minimum of a lower convex function

Step1 Set k=1, Input $P_i^{(k)}$which satisfies Eq.(11). Initialize the tolerance ε.

Step2 Eq.(13) yields $x_a^{(k)}$ and then get$f\left(x_a^{(k)}\right)$.

Step3 If inequality (18) or (19) holds, then $f\left(x_a^{(k)}\right)$ is the answer which meets the optimization tolerance given, output the result and exit.

Step 4 Eq.(12) and (13) yield $P_i^{(k+1)}$. Set k=k+1 and go to Step 2.

3 HYBRID OPTIMIZATION METHOD

Five-point line search method is used to get the minimum of a convex function. It is also valid to determine whether the approximation is satisfactory.

We select Thiete type continued fractions to approximate function $f(x)$. Continued fraction $u(x)$ is formulated by Eq. (25) and satisfied with (26).

$$u(x)=y_1^{(k)}+\left(x-x_1^{(k)}\right)\frac{a_0x+a_1}{x^2+b_1x+b_2} \tag{25}$$

$$u\left(x_i^{(k)}\right)=f\left(x_i^{(k)}\right), i=1,2,\ldots,5 \quad (26)$$

Parameters a_0, a_1, b_1 and b_2 are determined by Eq.(26)(Tan J.Q. 2007). Let $u'(x)=0$ and then we get the minimum point of $u(x)$ by Eq.(27).

$$x_a^{(k)}=\begin{cases}0.9x_3^{(k)}+0.1x_2^{(k)}, \left|t^{(k)}-x_3^{(k)}\right|\le\varepsilon, 2x_3^{(k)}\ge x_2^{(k)}+x_4^{(k)}\\ 0.9x_3^{(k)}+0.1x_4^{(k)}, \left|t^{(k)}-x_3^{(k)}\right|\le\varepsilon, 2x_3^{(k)}< x_2^{(k)}+x_4^{(k)}\\ t^{(k)}, \text{otherwise}\end{cases} \quad (27)$$

Where

$$\varepsilon=0.01\times\min\left\{x_3^{(k)}-x_2^{(k)}, x_4^{(k)}-x_3^{(k)}\right\} \quad (28)$$

$$t^{(k)}=\begin{cases}\dfrac{-a_0b_2-a_1x_1^{(k)}+v^{(k)}}{a_0b_1-a_1+a_0x_1^{(k)}}, a_0b_1-a_1+a_0x_1^{(k)}>0\\ \dfrac{-a_0b_2-a_1x_1^{(k)}-v^{(k)}}{a_0b_1-a_1+a_0x_1^{(k)}}, a_0b_1-a_1+a_0x_1^{(k)}<0\end{cases} \quad (29)$$

$$v^{(k)}=\sqrt{\left(a_0b_2+a_1x_1^{(k)}\right)^2-\left(a_0b_1+a_0x_1^{(k)}-a_1\right)\left(a_1b_2+\left(a_1b_1-a_0b_2\right)x_1^{(k)}\right)} \quad (30)$$

Sometimes the approximation to function $f(x)$ don't exist, i.e. parameters a_0, a_1, b_1, b_2 satisfied with Eq.(25)-(26) don't exist. If it happens, continued fractions $u(x)$ is replaced by an interpolation function.

It is proved that $x_3^{(k)}$ almost quadratic converges to the minimum point of function $f(x)$ if the condition (11) holds(Wang1993). Thiete type continued fraction may approximate arbitrary continuous functions. Integrated Thiete type continued fraction approximation methods and Algorithms 1, we get hybrid optimization methods.

Algorithm 2 Hybrid optimization method to minimization problems of unimodal function

Step1 $k\leftarrow 1$. Input $P_i^{(k)}$which satisfies Eq.(11) and the optimization tolerance ε.

Step2 Get $x_a^{(k)}$ by Eq.(27) and get $f\left(x_a^{(k)}\right)$.

Step3 If inequality (18) or (19) holds, $f\left(x_a^{(k)}\right)$ is the minimum, then exit. Otherwise, update $P_i^{(k)}$ by Eq. (12), (27) -(30). Go to step 2.

4 EXPERIMENTS AND RESULT

Five methods are used to solve problems in Table 1. Table 2 shows the number of iterations.

Table 1. Functions tested.

Number	Function	Interval $[a,b]$
1	$y=x^2+3x+2$	[-5,-1.7]
2	$y=x^2+3x+2$	[-5,-1]
3	$y=x^3+2x^2+x+5$	[-0.5,6]
4	$y=x^3+2x^2+x+5$	[0, 6]
5	$y=\sin x$	[3.16,5]
6	$y=-\sqrt{\sqrt{18x^2+20.25}-x^2-4}$	[-3, 0]
7	$y=\begin{cases}e^{-x}, x\le 1.3065586410393501\\ \dfrac{1}{5-x}, x>1.3065586410393501\end{cases}$	[0.5,4.5]
8	$y=100(1-x^2)^2+(1-x)^2$	[0, 10]
9	$y=(1-x^2)^2+(1-x)^2$	[0, 10]
10	$y=(1-x^2)^2+100(1-x)^2$	[0, 10]
11	$y=100(1-x^3)^2+(1-x)^2$	[0, 10]
12	$y=2.545664x^2-11.928x-14.20315$	[0, 10]
13	$y=100(1+x^2)^2+(1-x)^2+9176$	[0, 10]
14	$y=(x-10)^2+10(x-1)^4+6$	[0, 10]
15	$y=(e^x-2)^4+x^2+1$	[0, 10]
16	$y=(x^2-10)^2+(x-6)^2$	[0, 10]
17	$y=x^2(1-x)^4[1-(1-x)^4]^2$	[0, 10]
18	$y=(x-2)^2+1-0.16/(x^2+12)+5(x-3)^2$	[0, 10]
19	$y=\cosh x$	[0, 10]

Table 2. Number of iterations by different methods.

Number	Gold section	Quadratic interpolation	Cubic interpolation	Five-point line search	Hybrid method
1	39	66	23	7	6
2	40	271	6	25	7
3	41	14	6	27	12
4	41	/	/	8	7
5	38	>2000	10	20	11
6	40	46	10	22	12
7	40	44	29	19	23
8	42	23	15	/	15
9	42	21	15	32	14
10	42	10	11	31	11
11	42	56	22	/	16
12	41	31	6	27	7
13	41	>2000	/	24	13
14	42	48	16	34	12
15	42	33	42	47	14
16	41	48	12	/	13
17	42	>2000	41	11	8
18	41	9	8	28	8
19	43	/	/	11	8

*Tolerance is 0.0000001; '/' denotes 'unsolvable'.

Experimented results in Table 2 show that the hybrid method is valid. For functions with symmetric curve, interpolation method almost has the same iterations number with hybrid method. For functions with asymmetric curve, the hybrid optimization method is the best.

5 CONCLUSIONS

One dimension search method is very important to improve the efficiency of multidimensional search method. Hybrid one dimensional optimization method is proposed for solving the optimization problem of convex function, non-convex function, non-smooth function and even to the function without expression. Experimented results show that hybrid optimization method is valid.

REFERENCES

Abraham D., Rafael M., Fred G., et al. 2009. Hybrid scatter tabu search for unconstrained global optimization. *Annual of operation research*.

Crina G. & Ajith A. 2007. Modified line search method for global optimization. *Proceedings of 1th Asia International Conference on Modeling & Simulation*.

Crina G. & Ajith A. 2009. A Novel Global optimization technique for high dimensional functions. *International journal of intelligent systems* 24: 421–440.

Hu J.Q. et al. 2007. A model reference adaptive search method for global optimization. *Operations research* 55(3): 549–568.

Huang Z. H., Hu S.l. & Han J.Y. 2009. Convergence of a smoothing algorithm for symmetric cone complementarity problems with a nonmonotone line search. *Science in China series A: mathematics* 52(4): 617–630.

Ioannis G. Tsoulos & Isaac E. Lagaris. 2006. Genetically controlled random search: a global optimization method for continuous multidimensional functions. *Computer Physics Communications* 174: 152–159.

Shi Z. J. et al. 2005. New inexact line search method for unconstrained optimization. *Journal of optimization theory and applications* 127(2): 425–446.

Tan J.Q. 2007. *Continued fraction theory and its application*. Peking: Science Press.

Tan H.W. & Qin X.Z. 2000. *Practical optimization*. Dalian: Dalian University of Technology Press.

Wang X.D. 1993. On the convergence of the continued fraction algorithm. *Journal of Fuzhou university* 21(4): 26–31.

Wang Z.Y. 2010. Combinatorial optimization based on state transition. Xi'an: Xi'an Jiao Tong University Press.

Yin Y.M. 1989. A double tangent method and double secant method used in line search. Journal of East China institute of technology (4): 21–27.

Yuan Y.X. & Sun W.Y. 2005. Optimization theory and method. Beijing: Science press.

Zhao H. & Gao Z.Y. 2005. Equilibrium algorithms with nonmonotone line search technique for solving the traffic assignment problems. *Journal of science and complexity* 18(4): 543–555.

Zhou L. 1984. New improvements in direct search techniques. *Journal of Tianjin University* (4): 62–68.

Electronics, Communications and Networks IV – Hussain & Ivanovic (eds)
© 2015 Taylor & Francis Group, London, ISBN: 978-1-138-02830-2

Research of routing algorithm based on levels for wireless sensor networks

Meng-Jiao Wang & Yong-Zhen Li*
Department of Computer Science & Technology Yanbian University Yanji, China

ABSTRACT: In order to save energy consumption and prolong lifetime of the network, the dissertation presents a new routing algorithm named LEACH-LMT (LEACH-Level Multi-hop Transmission) based on LEACH (Low–Energy Adaptive Clustering Hierarchy) algorithm. This paper uses MATLAB to simulate the operation of networks, and to compare LEACH-LMT with LEACH and LEACH-EE algorithm through the number of surviving nodes and energy consumption. The result shows that LEACH-LMT algorithm performed better than LEACH and LEACH-EE algorithm.

1 INTRODUCTION

WSN (Wireless Sensor Network) (Lin 2012) technology has been researched with hundreds or thousands of sensor nodes to analyze the sensing abilities, computing abilities and power abilities. At present, Wireless Sensor Network has been used in an amount of areas such as environment, military, energy, medical treatment and so on.

The Wireless Sensor Network has some matters such as low computing speed, transmission range, limited energy power and stable position. To achieve better performance, it is essential to improve the lifetime of the entire network, and to save the energy that is focused on a few sensor nodes.

2 RELATED WORKS

Hierarchical routing, originally processed in wireless networks, is a well-known structure with extra advantages in scalability and efficient transmitting. Hierarchical routing is also used to achieve energy efficient clustering algorithm in WSNs (Qing & Zhang 2010, Manjeshwar & Agrawal 2001).In a hierarchical routing protocol, higher energy nodes can be used to merge data and transmit the message while lower energy nodes can be used to perform the sensing. Hierarchical routing protocol achieves to decrease energy consumption of the whole network and prolong the lifetime of the network by data aggregation and mergence. Hierarchical routing protocol usually has two stages. One stage is to choose cluster heads, and the other stage is for transmitting data.

The typical algorithms of the hierarchical routing protocol include LEACH (Heinzelman & Chandrakasan 2002, Qing & Zhang 2010), TEEN (Threshold sensitive Energy Efficient sensor Network protocol) (Manjeshwar & Agrawal 2001), and APTEEN (Adaptive Periodic Threshold-sensitive Energy Efficient sensor Network) (Manjeshwar & Agrawal 2002). LEACH is a clustering protocol in which the cluster head collects data from member nodes, merges data, and directly transmits the merged data to the base station. The fusion of data received from member nodes can decrease energy consumption of normal nodes. The performance of LEACH proves that it ensures a fixed probability to be a cluster head node, but the amount of clusters is not stable. So it has some problems to be solved.

TEEN processes like LEACH. Besides that, sensor nodes do not have data to be periodically sent. TEEN operates time-critical data. Time-critical data are required in applications such as earthquake forecast, and the critical value is broadcasted in cluster forming stage. When the sensed value of data does not reach the critical value, it cannot be sent.

APTEEN processes a hybrid network that combines the virtues of proactive and reactive sensor networks, and minimizing their limits. In APTEEN, sensor nodes not only periodically send data, but also react to the sudden change of sense data. Moreover, when a sensor node does not send data during certain time for proactive operation, it senses data and sends the data to the cluster head, therefore improves the shortage of TEEN.

LEACH-EE (Li & Zhang 2007) algorithm is an effective clustering routing algorithm based on LEACH algorithm. The main idea of the LEACH-EE is that building a multi-hop chain (MHC) that goes through all cluster-head nodes in every round. The data collected will be transmitted to the base

*Corresponding author: *lyz2008@ybu.edu.cn*

station according to the chain. This multi-hop method reduces the energy consumption of nodes. At present, there are many improved algorithms for cluster-head nodes multi-hop based on LEACH. A LEACH-EE algorithm is shown in the following steps.

1 LEACH-EE algorithm uses the same method to select cluster-head nodes as LEACH algorithm does.
2 Setting up a cluster and building the multi-hop chain. The cluster-head nodes broadcast their messages of being cluster-head, including loading position and TDMA scheduling information
3 The data are transmitted from the cluster head that is farthest from the sink node.

The advantage of LEACH-EE is that it can prevent the node with little residual energy from dying early through multi-hop path between cluster head nodes and base station. The shortcoming is that it may go around in circles to transmit information if the nodes are distributed intensively in the monitor area.

3 PROPOSED PROTOCOL

In this paper, it proposes the multi-hops routing algorithm according to level method. Based on the assumption, the selection of the cluster head nodes in LEACH algorithm is random (Anastasi & Conti 2009, Sharma M & Sharma K 2012), resulting in an unbalanced distribution of cluster heads. The LEACH-EE algorithm produces a circuitous data transmission. So this paper introduces an algorithm named LEACH-LMT based on LEACH algorithm, which can improve the lifetime of the network compared with LEACH and LEACH-EE algorithm. The improved routing algorithm is divided into cluster setup stage and data transmission stage.

At cluster setup stage, sensor nodes are chosen from all the nodes to be the cluster head nodes with a specific probability. The interesting data are broadcasted from the cluster head, according to multi-level, and all the sensor nodes produce the routing table information of the property information on their own. The data transmission step is that the data gathered from the nodes communicates with the cluster head and then the data is fused by the cluster head nodes. Finally, the merged data communicates with the sink node. Then the current round is finished.

3.1 *Initialization of the node properties*

The proposed protocol sets level value of all network sensor nodes. On this stage, every sensor node sets up the routing table information according to the property information of nodes from the sink node to the last node.

1 Firstly, all nodes are set to be level 0. At the beginning of the sink node, all the sensor nodes will be set to the corresponding levels. According to the range of the transmission radius of the general nodes, the base station broadcast message (node identifier, level information, residual energy and so on) to every node in the range of radius R. R is the transmission radius of normal nodes. So the nodes within a radius R will receive the broadcasted message. Nodes that accept the message from the base station set its level value as one and add themselves to the level_ID list, while the rest nodes turn to sleep mode and add them to non_level ID list. The nodes will be activated as they receive the next broadcast message. If the message is from the node that has a level (i.e. not included in the non_level_ID_list) (e.g. L) they make their level as L + 1.
2 The nodes with the level of 1 broadcast the useful message to the second level in the range of R. The nodes that receive a key message will firstly see if they have been set level. If not, they will be set to level 2 and write a new interest message, and then broadcast the interest message to the next level. In this paper, we divide the monitoring area into four levels. After the fourth level is complete, the stage of setting level is finished. The supposed level is finished, and the routing table is ended, according to the received message.
3 After step (1) and (2) the interest message is broadcasted gradually, and every sensor node initializes the routing table information and divides the level value.

3.2 *Network clustering*

Based on this paper, we assume that there are more cluster heads in the level that are remote from the sink node and fewer cluster heads in the area that are close to the sink node. Thus, the cluster heads are evenly distributed. The selection of the cluster head in the proposed algorithm is shown as the following procedures:

1 Setting up the levels of all the nodes in the monitoring area and setting the radius of the level.
2 Nodes in different levels are set to a different value of p (the percentage of the cluster heads within the total nodes) as varying levels. The value of p is set $p*(1-2x)$, $p*(1-x)$, p and $p*(1+3x)$, with $0<x<1$. x is a random value between 0 and 1; it is just a variable that controls the probability of cluster heads being among all nodes. In this paper, we divide the monitor area into four levels to simulate. The four values of p match with four levels. The higher the level is, the larger the value of p will be. So the threshold would be larger in high level, and the amount of the cluster heads in the high level would be more. Thus, energy consumption of transmitting messages to the sink node can be reduced.

3 Each sensor node chooses a random number between 0 and 1. If the number is less than or equal to the threshold T (n), the sensor node is chosen as a cluster head node. If there is no node whose random number is less than or equal to the threshold T(n), the node with most residual energy will be chosen as the cluster head.

The proposed algorithm considers residual energy and connectivity when selecting the cluster head nodes compared to LEACH. The threshold T (n) of LEACH-LMT algorithm is produced through the following procedure as (1).

$$T(n) = \begin{cases} \frac{p}{1-p(r \bmod (1/p))}(\alpha * \frac{E(i)_{res}}{E_{ave}} + \beta * \frac{S(i).n}{S(i).\ln}), n \in G \\ 0, else \end{cases} \quad (1)$$

P is the percentage of the amount of cluster heads in the total sensor nodes. p is usually set to 5%; r is the current rounds; G is the nodes that are not chosen as cluster nodes at present; E(i)res is the residual energy of sensor node i; Eave is the network average energy at present; S(i).n is the alive neighbor node number of sensor node i; S(i).ln is the alive sensor nodes in which level node i in; α is the weight factor of energy; β is the control weight factor of connectivity; and α+β=1.

Suppose that the amount of the sensor node in the whole network is n, p*n is the average amount of the cluster heads that will be chosen. The larger the value of E(i)res/Eave is, the larger the value of the threshold T(n) will be. And then the probability of the node to be a cluster head will be greater. Because of the strategy of the level, if the ratio of S(i).n/S(i).ln is larger, the neighbor nodes of node i is more and the probability of the node to be clustered head will be greater. Considering connectivity can avoid high energy alone node being selected as cluster head. When a sensor node is chosen as a cluster head in the level by the previous method, the node broadcasts a message that declares from a cluster head to the sensor nodes in the range of the communication radius. All sensor nodes that accept the message see if they are clustered heads, if not, these nodes will be the cluster members that are nearest to the cluster head node that broadcast the message.

The distribution of the cluster head nodes equals the energy decrease of sensor nodes in different levels in the proposed algorithm. The method balances the energy decrease of the nodes in high levels, and it performs better than the cluster head selection of the LEACH algorithm does.

3.3 *Data transmission*

On this stage, the proposed algorithm applies multi-hop transmission of the communication of the cluster heads and the sink node. Thanks to the idea of the level, we should decide whether the cluster head can be relayed node in terms of the weight of the cluster head in the high levels. Setting up a spanning tree between cluster heads and the base station can decrease the distance of data transmission, save the energy consumption of the cluster head nodes at high levels, and improve the lifetime of the network. The proposed algorithm communicates the data to the cluster head based on the routing table information stored in the node. The setup of the multi-hop spanning tree using LEACH-LMT algorithm is shown as follows.

1 Firstly, header information of the data packet is updated as the node's property information. And the next node broadcasting the data is selected from the routing table. A data transmission begins from the cluster head existing at the highest level.
2 Calculating the weight of the cluster head nodes under the previous level through the following formula (2).
3 The method to select a relay node is choosing the cluster node with the biggest weight (the result of step2). Moreover, if there are several cluster nodes with the same weight, select the node with the shortest distance with the sink node.
4 If there is no cluster head node in the current level, turn to step 2.
5 The cluster head node that receives the data packet continues the procedure 2 to 4 and communicates the data with a node at level 1. Data packet of cluster head nodes at level 1 is transmitted to the base station directly, and the data transmission is finished.

The weighted model based on levels in LEACH-LMT is:

$$w_{(\text{CH\_i, CH\_j})} = \gamma * (1 - \frac{S(j).t}{S(i).t}) + (1-\gamma) * k_i \quad (2)$$

Formula (2) illustrates the weight value of a cluster head CH_i based on to select cluster head CH_j as a relay node. γ is the value between 0 and 1. S(j).t/S(i).t is the level ratio of cluster head CH_j and CH_i. The ki in formula (2) is produced through the following equation.

$$k_i = \frac{(E_{to\sin k})_i / (E_{resided})_i}{(E_{toCHj})_i / (E_{resided})_i + (E_{to\sin k})_j / (E_{resided})_j} \quad (3)$$

k_i is the energy control portion for a multi-hop path. It avoids data to warp transmission. $(E_{to\ sink})_i$ is the energy consumption of the cluster head CH_i sending data to the base station directly. $(E_{resided})i$ is the residual energy of cluster head CH_i. $(E_{toCHj})_i$ is the energy consumption of cluster head CH_i sending data to CH_j. $(E_{tosink})_j$ is the energy consumption of CH_j sending messages to sink node. $(E_{resided})_j$ is the residual energy of cluster head CH_j.

From equation (2) we can see if the residual energy of a cluster head node is more, the value of level is smaller, the cluster head node is closer to the sink node, and it consumes less energy. So the cluster head node is more suitable to be a relay node.

4 SIMULATION AND ANALYSIS

4.1 *Initialization*

In the paper, we simulated the proposed algorithm using MATLAB and compared the proposed algorithm with LEACH and LEACH-EE algorithm to analyze the performance of the proposed algorithm. The simulated sensor field size is 200m * 200m, and the amount of the sensor nodes is 200. We suppose that each node has the same initial energy, and the energy is consumed when the data is transmitted. The weight coefficient of the spanning tree is set 0.5.

4.2 *Analysis of the result*

Figure 1 illustrates the comparison of the energy consumptions using three different algorithms. In the beginning (0–682 rounds) the energy consumptions of LEACH and LEACH-EE are almost the same. That is because the energy of the nodes is enough at first, and we cannot see the advantages of LEACH-EE algorithm. With the increase of the round, LEACH-EE algorithm performs better than LEACH. LEACH-LMT always performs best among the three different algorithms. It saves energy consumptions in the selection of the cluster head nodes and multi-hop method.

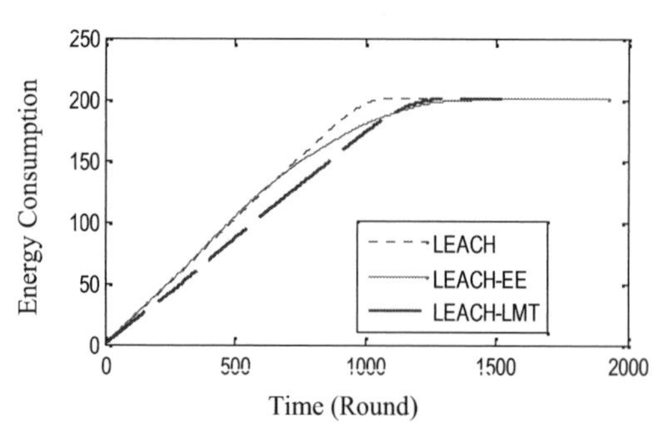

Figure 1. Comparison of three algorithms in energy consumption.

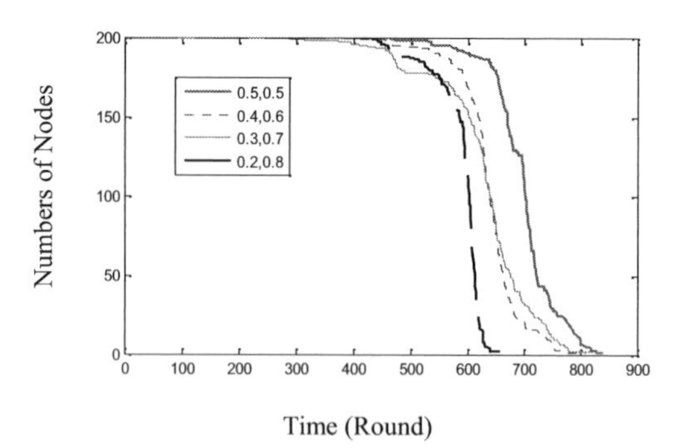

Figure 2. Comparison of LEACH-LMT in threshold factors.

Figure 2 illustrates the comparison of the lifetimes with different threshold factors. We simulate different threshold factors to find out the best values of the threshold factors. The lifetimes of the first nodes dying with different threshold factors are very different. We can see that the lifetime of nodes performs best when α is set 0.5 and β is also set 0.5.

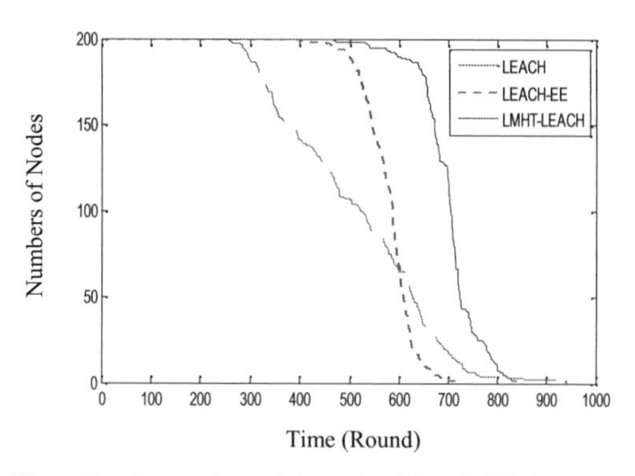

Figure 3. Comparison of three algorithms in lifetime.

Table 1. The results of a comparative standard.

Algorithm	FND	HND
LEACH	263 rounds	520 rounds
LEACH-EE	406 rounds	577 rounds
LEACH-LMT	458 rounds	708 rounds

Figure 3 illustrates the comparison of FND (the time of First Node Dies) and HND (the time that Half of the Node Dies) (Rajesh & Sunil 2011) using three different algorithms. We can see that the performance of LEACH-LMT is better than LEACH and LEACH-EE according to FND.

Table 1 illustrates the specific values of FND and HND using three different algorithms. The proposed algorithm improves the FND much more than LEACH algorithm. Compared with LEACH algorithm, the FND of LEACH-LMT algorithm improves about 74%, and the HND improves about 36%. Compared with LEACH-EE algorithm, the FND of LEACH-LMT algorithm improves about 10%, and the HND improves about 23%. That is because LEACH-LMT algorithm considers the residual energy and connectivity when choosing cluster head nodes. In this way, energy consumption of transmission from cluster heads to the base station can be reduced. And multi-hop between cluster head nodes can make transmission paths optimized.

5 CONCLUSIONS

Based on this paper, we simulated the energy consumption and the lifetime of three different algorithms. We use a multi-level method to achieve multi-hop in the cluster communication. The proposed algorithm not only considers residual energy, but also connectivity to select cluster head nodes. The proposed algorithm makes sure that the cluster head nodes using levels to transmit data to the base station saves more energy than transmitting directly to sink node. In general, the performance of LEACH-LMT algorithm is better than LEACH and LEACH-EE according to FND and HND. It can make sure load balancing and equal distribution in energy consumption of every node.

In this paper, we analyzed the proposed method and the performance compared with the existing algorithms. Then we concluded that the proposed method improved the lifetime span and saved the energy consumption.

REFERENCES

Anastasi, G. & Conti, M. 2009. Energy conservation in wireless sensor networks: A survey. *Ad Hoc Networks*: 537–568.

Heinzelman, W. B. & Chandrakasan, A. P. 2002. An application-specific protocol architecture for wireless microsensor networks. *IEEE Transactions on Wireless Communications* 1(4): 660–670.

Li Yan & Zhang Nuanhuang. 2007. LEACH-EE-Efficient clustering routing algorithm based on LEACH protocol. *Computer Application* 27(5): 1103–1105.

Lin Yangwu. 2012. Research on the Secure Authentication of Web of Things. *Wireless Internet Technology* 20(10): 6–7.

Manjeshwar, A. & Agrawal, D. P. 2001. *TEEN*: A routing protocol for enhanced efficiency in wireless sensor networks. *Parallel and Distributed Processing Symposium, Proc.15th International*: 23–27.

Manjeshwar, A. & Agrawal, D. P. 2002. *APTEEN*: A Hybrid Protocol for Efficient Routing and Comprehensive Information Retrieval in Wireless Sensor Networks. *Parallel and Distributed Processing Symposium, Proceedings International*: 195–202.

Qing Bian & Zhang Yan. 2010. Research on Clustering Routing Algorithms in Wireless Sensor Networks. *Intelligent Computation Technology and Automation*: 1110–1113.

Rajesh, P. & Sunil, P. 2011. Energy and Throughput Analysis of Hierarchical Routing Protocol for Wireless Sensor Network. *International Journal of Computer Applications* 20(5): 32–36.

Sharma, M. & Sharma, K. 2012. An Energy Efficient Extended LEACH. *Communication Systems and Network Technologies*: 377–382.

Electronics, Communications and Networks IV – Hussain & Ivanovic (eds)
© 2015 Taylor & Francis Group, London, ISBN: 978-1-138-02830-2

A subcarrier suppression based opportunistic network coding approach for uplink cooperative OFDM transmission system

Fei Wang*, Dongmei Zhang, Kui Xu & Wei Xie
College of Communications Engineering, PLA University of Science and Technology, Nanjing, Jiangsu, China

ABSTRACT: This paper proposes a strategy of subcarrier suppression and opportunistic network coding for uplink cooperative OFDM transmission system. Under the actual channel propagation conditions, the performance quickly degrades for the high decoding error probability. With the aim of counteracting this drawback, specifically, two subcarriers suppression methods are introduced and jointly designed with opportunistic network coding scheme. Simulation results show that the two methods have a better performance than the scheme with no suppression on the term of Bit Error Rate (BER). Meanwhile, we obtained an interesting result from the simulation as depicted in the paper.

1 INTRODUCTION

In recent years, the mobile telephone service, data service and multi-media service of the wireless communication are booming rapidly. The future wireless network must afford the high quality service. The aim of the cooperative communication is to help the nodes having the demand of communication reach a high and reliable communication by the use of other nodes' resources.

Based on the propagation character of the wireless channel (Tse et al. 2004), the cooperative communication allows the single-antenna terminal device sharing the antennas of other users' to form a virtual antenna array. Therefore, the same information can reach to the receiver through different independent wireless channels. In recent years, cooperative communication has been accepted by several standards such as IEEE 802.11s, IEEE 802.16j and LTE-Advanced as a powerful technique to provide spatial diversity in wireless systems. For the cooperative communication, the relay can choose either Amplify-and-Forward (AF) (W et al. 2001) or Decode-and-Forward (DF) (Lam et al. 2006) to forward the information.

For the AF protocol, the relay retransmits an amplified version of the received signal including noise and channel degradations, without reading its content. Usually, AF protocol leads to a lower system performance since the noise level is amplified jointly with the useful signal. In particular, the AF protocol is effective only under high Signal-to-Noise Ratio (SNR) conditions. Usually, we talk more about the DF, a strategy used by the relay. For the DF protocol, the relay decodes the received information and forwards a newly encoded signal to the destination nodes. This protocol eliminates the interference of the noise, and is in favor of the final judgment at the destination receiver, so the reliability performance is improved.

One of the shortages of cooperative communication is the requirement of extra resources which reduces the system spectral efficiency. An efficient way to deal with this problem is to apply network coding in wireless cooperative system (Xu et al. 2013). It is shown in (Kishore et al. 2006) that applying network coding in wireless networks with cooperative communication could also achieve diversity gain.

However, network coding may not always be helpful (Kompella et al. 2012). Actually, there is situation that network coding does harm to the performance of cooperative communication. As a result, it is of great importance to use opportunistic network coding (Letaief et al. 2007). When no opportunity for network coding exists at an intermediate node, it is beneficial to send an uncoded packet (Hsu et al. 2012).

In this paper, we apply NC to the cooperative uplink communication system, where two users transmit the information to the base station with the aid of the relay. As an interesting solution for performance improvement of the system, (Popovski et al. 2009) proposed an ideal pre-equalization scheme of the wireless channel for DF protocol. The primary drawback of the pre-equalization scheme is that the transmitting side performs amplification for pre-equalization the wireless channel whenever the signals are affected by deep fading attenuations, however, it is hard to achieve in reality. And it is not a good choice in some condition. A blind signal separation based physical layer network coding (PLNC) approach at the relay node is proposed in (Bing 2009). By using of the complex matrix inversion and diagonalization, it achieves good performance in multipath fading channels. In (Liang et al.

*Corresponding Author: *wfzd01@sina.com*

2008), power control is applied to lower the influence of the multipath fading channel on the system performance. The main drawback of the scheme proposed in (Liang et al. 2008) is the significant increase in signaling overhead. This paper aims to improve the performance of the uplink transmission system through the use of two subcarrier suppression methods, named discrete subcarriers suppression (DSS) and combined subcarriers suppression (CSS). Simulation results show that both the proposed two methods outperform the uplink transmission scheme with no suppression (Bhargava et al. 2010) in the term of BER.

The rest of the paper is organized as follows: Section 2 describes the model of the system. The suppression of the subcarriers in the systems is introduced in section 3. Then Section 4 analyzes BER of the proposed scheme and also provides the numerical simulation comparison. At last, conclusions are offered in Section 5.

2 SYSTEM MODEL

We focus on an uplink transmission scenario in Figure 1 Two source nodes U_1 and U_2, one relay node R and one destination node D. Two users are transmitting information to a base station D with the help of the relay R. The information transmitted by two users are denoted as s_1 (by U_1) and s_2 (by U_2), as a result of the nature of wireless transmission, the relay and the base station can receive the information from the users simultaneously. Each node is equipped with an omni-directional antenna. The channel is half duplex so that transmission and reception of a node must occur in different time slots. In this paper the communication channel is assumed to be affected by multipath fading with Rayleigh distribution and unchanged during a system period.

The transmission of the messages completed in three time slots. In the first time slot, the user U_1 sends s_1 to R and D. In the second time slot, the user U_2 sends s_2 to R and D, and in the last time slot, the relay *R* send the processed signal to the base station D. The relay node R chooses the DF strategy to process the information.

Each node exploits the Orthogonal Frequency Division Multiplexing (OFDM) modulation with *K* subcarriers, each one modulated through a BPSK scheme. Then we consider the transmission of the signal. The channel links U_1, U_2 to R and R to D are modeled as an FIR filter Signals with *L* taps.

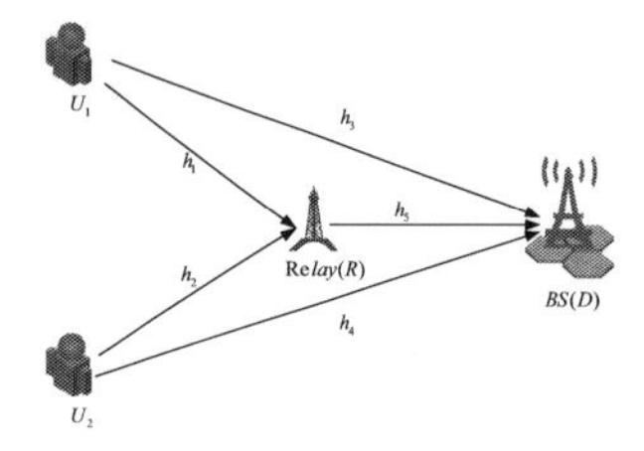

Figure 1. Structure of the system model.

Let h_i be a vector whose entries $h_i(k)$, with k=0,1,…,L–1, denotes the tap coefficients of the propagation channel between the nodes, as:

$$h_i=\left[h_i(0),h_i(1),\cdots,h_i(L-1)\right]^T,(i=1,2,3,4,5) \quad (1)$$

Where the symbol *T* denotes the transposition, we consider h_i constant in the period of three time slots, equal to the duration of three OFDM symbols. The coefficient of each channel is a Gaussian random variable, and $h_i[l]=(P_t[l]/dmn^a)a_{mn}[k],l=0,1,\ldots,L\text{-}1$, in which, $Pt[k]$represents the transmission power of the *kth* subcarrier. α is the path loss exponent in 2~5. $amn[k]$is the Rayleigh-distributed fading factor. *dmn* is the distance between node *m* and *n*, with $m,n\in(U_1,U_2,R,D)$. Hence, the frequency response for the *kth* subcarrier of the two users and the Relay are given by:

$$H_i[k]=\sum_{l=0}^{L-1}h_i(l)e^{-2j\pi\frac{k}{K}l} \quad (2)$$

So the signal arriving at the receivers (Relay or Base station) can be expressed as:$n_0[k]\sim CN(0,\sigma_n^2)$

$$y[k]=|H_i[k]|s_t[k]+n_0[k],t=1,2. \quad (3)$$

Where $n_0[k]$ ~ $CN(0,n_2)$ represent the additive noise at the terminal with zero mean and variance n_2.

2.1 *In the first time slot*

Only the user U_1 send signals with normalized transmitting power P_{U1}, as a result of the broadcast nature in a wireless channel, R and D can receive the signal at the same time. The information bits “0” or “1”is mapped into BPSK symbols “–1” or “1”.The signal received by R is:

$$y_{r1}[k]=H_1[k]s_1[k]+n_0[k] \quad (4)$$

And then the relay decodes the signal yr_1, then, the decoded signal is modulated to $ydec\_r_1$. The message received by D is:

$$y_{d1}[k]=H_3[k]s_1[k]+n_0[k] \quad (5)$$

2.2 *In the second time slot*

Only the user U_2 sends signals, also the transmitting power P_U is normalized. R and D receive the signals with white additive Gaussian noise at the receiver:

$$y_{r2}[k]=H_2[k]s_2[k]+n_0[k] \quad (6)$$

$$y_{d2}[k]=H_4[k]s_2[k]+n_0[k] \quad (7)$$

The relay decodes the signal yr_2, then, the decoded signal is modulated to $ydec\_r_2$.

2.3 *In the third time slot*

R transmits the mixed signal $ydec\_r_1$ and $ydec\_r_2$ to D, with normalized power P_R. The specific mixture of the signal is discussed in the next section.

3 SUBCARRIERS SUPPRESSION METHODS IN THE SYSTEM

It is straightforward to note that the performance of the system is due to the most attenuated subcarriers which make the mean power level at the R side very low. We analyzed if omitting one or more subcarriers having the low SNRs may sufficiently improve BER performance of the uplink transmission system.

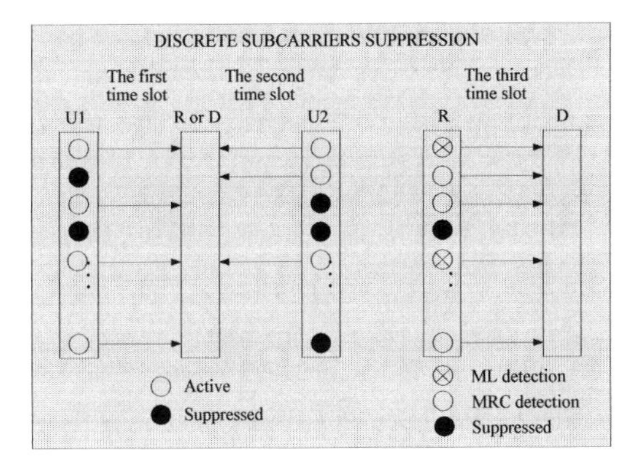

Figure 2. Proposed discrete subcarriers suppression scheme.

Firstly, we use λ denotes the suppression threshold value. If the following constraint is valid, the subcarriers are turned off:

$$|\hat{H}_i[k]|^2 < \lambda \Rightarrow P_t[k] = 0 \quad for\ k = 0, \cdots, K-1. \tag{8}$$

This can improve the use of power and allows fully allocate the power on the active subcarriers (Fantacci et al. 2006). When λ increases, the number of suppressed subcarriers grows also. Both the relay and the base station need to know the full knowledge of which subcarrier is suppressed and they can avoid detecting on it. Hence, additional signaling is needed for the subcarriers suppression scheme.

In the following two subchapters, two subcarriers suppression methods are introduced. Also, the communication reliability (BER), and signaling overhead are taken into account.

3.1 *Discrete subcarriers suppression*

The first method is Discrete Subcarriers Suppression (DSS). We choose a common threshold value λ for each of the two users and the relay node in relation to a target BER. When a subcarrier has a channel gain below the threshold value, it'll be suppressed. The information of the subcarriers is expressed by a vector whose dimension is identical as the number of the subcarriers. The vector's element is "1" or "0", and "1" indicates the active subcarriers, on the contrary, "0" denotes the suppressed ones. We use *Vi* to express the vectors used by the five channels. The specific method of the selectivity of the subcarriers is introduced as the formula (9). During the first and the second time slot, U_1 and U_2 transmit its message only on the active subcarriers, at the relay node, the typical BPSK detection has been used. Then, the relay sends the sum of the two modulated signals to the base station. For the subcarrier used by U_1 and U_2, R allocates the available power to the two modulated signals in proportion. Figure 2 describes the DSS.

$$\begin{aligned}
&|\hat{H}_1[k]|^2 > \lambda \Rightarrow V_1[k] = 1, |\hat{H}_1[k]|^2 < \lambda \Rightarrow V_1[k] = 0;\\
&|\hat{H}_2[k]|^2 > \lambda \Rightarrow V_2[k] = 1, |\hat{H}_2[k]|^2 < \lambda \Rightarrow V_2[k] = 0;\\
&|\hat{H}_3[k]|^2 > \lambda \Rightarrow V_3[k] = 1, |\hat{H}_3[k]|^2 < \lambda \Rightarrow V_3[k] = 0;\\
&|\hat{H}_4[k]|^2 > \lambda \Rightarrow V_4[k] = 1, |\hat{H}_4[k]|^2 < \lambda \Rightarrow V_4[k] = 0;\\
&|\hat{H}_5[k]|^2 > \lambda \Rightarrow V_5[k] = 1, |\hat{H}_5[k]|^2 < \lambda \Rightarrow V_5[k] = 0.
\end{aligned} \tag{9}$$

The vector for the user U_1 to select the subcarrier is Vs_1,

$$V_{S1}[k] = V_1[k] \cdot V_3[k] \cdot V_5[k] \tag{10}$$

Similarly, the vector for U_2 is Vs_2,

$$V_{S2}[k] = V_2[k] \cdot V_4[k] \cdot V_5[k] \tag{11}$$

The process of the information for the base station has three cases:

3.1.1 *The subcarrier used by both U1 and U2*

Three signals are received by the base station, so the Maximum Likelihood (ML) detection has to be used to retrieve the information. In the third time slot, the signal received by the base station is denoted as:

$$y_{DSS\_d\_U1}[k] = y_{d1}[k] H_3[k] + y_{d\_U1}[k] H_5[k] \tag{12}$$

$$
\begin{aligned}
P[k](-1,-1) &\propto \exp\left(-\frac{|y_{d1}[k]+H_3[k]|^2+|y_{d2}[k]+H_4[k]|^2+|y_{d3}[k]+(\sqrt{r}+\sqrt{1-r})H_5[k]|^2}{\sigma_n^2}\right)\\
P[k](-1,+1) &\propto \exp\left(-\frac{|y_{d1}[k]+H_3[k]|^2+|y_{d2}[k]-H_4[k]|^2+|y_{d3}[k]+(-\sqrt{r}+\sqrt{1-r})H_5[k]|^2}{\sigma_n^2}\right)\\
P[k](+1,-1) &\propto \exp\left(-\frac{|y_{d1}[k]-H_3[k]|^2+|y_{d2}[k]+H_4[k]|^2+|y_{d3}[k]+(\sqrt{r}-\sqrt{1-r})H_5[k]|^2}{\sigma_n^2}\right)\\
P[k](+1,+1) &\propto \exp\left(-\frac{|y_{d1}[k]-H_3[k]|^2+|y_{d2}[k]-H_4[k]|^2+|y_{d3}[k]-(\sqrt{r}+\sqrt{1-r})H_5[k]|^2}{\sigma_n^2}\right)
\end{aligned}
\tag{13}
$$

Then the base station computes the probability in (13) (at the top of the page), where $yDSS\_r_1$ stands for the signal send by U_1, and $yDSS\_r_2$ send by U_2. We think the two BPSK symbols corresponding to the largest one of the four probabilities are the information send by the two users.

As for the case, an interesting question must be emphasized. The BPSK symbols are -1 or 1, at the relay node, if the two symbols are equivalent probability. The sum of the two signals will be zero, so the ML detections will not accuracy. For this, we add a coefficient r to the two modulated signals, in other words, the relay node allocate its power to the two modulated signals non-equivalent probability. In addition, the relay and the base station must know the knowledge of which subcarriers are active, so additional signaling information is needed.

3.1.2 *3.1.2 The subcarrier used by either U_1 or U_2*

For the case, we can use the Maximum Ratio Combining (MRC) detection to get the information. In the third time slot, the information send by the relay belongs to U_1 can be expressed as:

$$y_{d\_U1}[k] = y_{DSS\_r3}[k]H_5[k] + n_0[k] \tag{14}$$

Where $y_{DSS\_r}3$ is the signal sent by U_1, y_{d_U1} is the signal received by the base station. We can use (5) and (14) to get the signal we want.

$$y_{DSS\_d\_U1}[k] = y_{d1}[k]H_3[k] + y_{d\_U1}[k]H_5[k] \tag{15}$$

Where $y_{DSS\_d\_}U1$ is the received signal of U_1 after MRC.

3.1.3 *The subcarrier used by neither U1 nor U2*

None of the user nodes has transmitted, so the detection is not needed.

3.2 *Combined subcarriers suppression*

Another proposed subcarriers suppression method applied in the cooperative uplink system named Combined Subcarriers Suppression (CSS). The two users and the relay choose a common threshold value, and the principle of the selectivity of the subcarrier is similar to DSS. However, the transmitting of the information takes place only on the subcarriers selected by the three nodes. Figure 3 describes the proposed CSS scheme.

As a result, the vector for the two users to select the subcarrier is *Vall*,

$$V_{all}[k] = V_1[k]\cdot V_2[k]\cdot V_3[k]\cdot V_4[k]\cdot V_5[k] \tag{16}$$

So, the process of the information for the base station has two cases:

3.2.1 *The subcarrier used by both U1 and U2*

The ML detection has put into use to retrieve the information.

3.2.2 *The subcarrier used by neither U1 nor U2*

No information have transmitted by the two users, therefore, the detection is avoid.

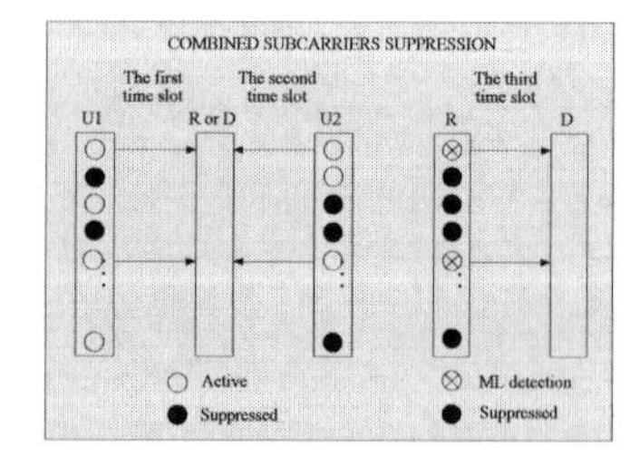

Figure 3. Proposed combined subcarriers suppression scheme.

Just as the DSS, at the relay node, CSS also need a coefficient added to the two modulated signals, it can avoid the appearance of the two symbols "-1"and "1" in the same probability. But the transmission rate of

CSS scheme is lower than the DSS method, because the transmission takes place only on the subcarriers which are active for the three terminals.

For the CSS scheme, the complexity of the data detection is simpler compared with DSS. Meanwhile, this method has a considerable improvement in terms of data reliability, but a reduction of the transmission rate. And also, the additional signaling information is essential.

4 NUMERICAL RESULTS

The subsequent presented analytical and simulation results assume perfectly synchronized for the uplink transmission system. We assumed equal power allocation between the S and R station, and the path loss exponent α=4. The distance between the two users and the relay is 1, which is the same as the distance between the relay and the base station. In the meantime, the variable *is* for different situations is simulated. The performance of the proposed two subcarriers suppression methods has been evaluated in terms of BER.

Figure 4 shows the comparison of BER versus for different SNR for the DSS and CSS schemes. From Figure 4, the employ of DSS in the System do a significant improvement in term of BER. In particular, BER improves a little compare to the situation of No suppression for the low SNR and λ. We can see that the proposed two subcarriers suppression schemes obtain a significant BER performance improvement when compared with tradition no suppression scheme (Bhargava et al. 2010). Meanwhile, the BER performance of the CSS scheme is better than that of the DSS scheme.

As the former case, the lower SNR give rise to the worsen condition of the propagation channel, and the fewer number of suppressed subcarriers will not bring about obvious improvement of the system performance. For the later, due to the high SNR and λ, the gain of the channel will be better. Therefore, the link between the relay and the base station do a little affect to the preciseness of the ML detection.

Figure 5-7 show the BER performance as a function of the variable r at different SNR and λ. From Figure 5, the curve rise spiky when the coeffic-ient r has a value of 0.5. The reason is that the two

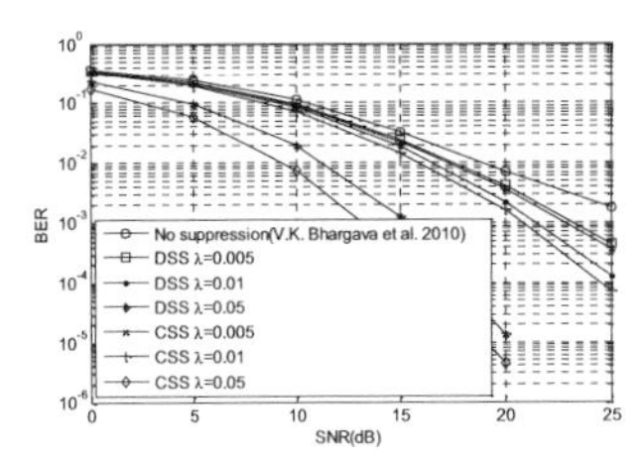

Figure 4. The BER of the DSS and CSS schemes for different threshold values, which as a function of the mean SNR.

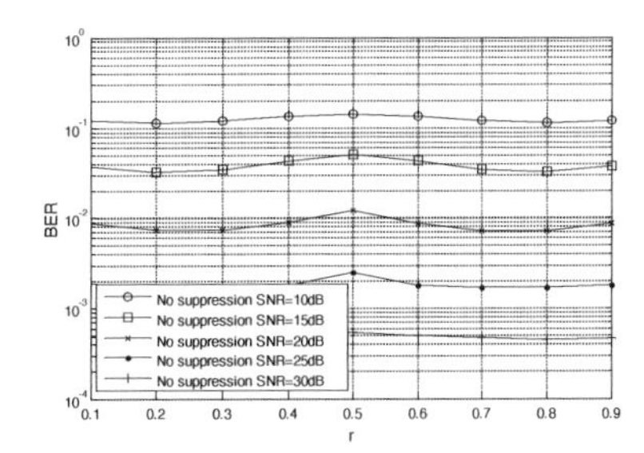

Figure 5. BER as a function of the variable r for different SNR values in the situation of No suppression.

BPSK symbols "1"and "-1" are with equivalent probability, i.e.,1/2 for each symbol. The sum of the two modulated signals will have three values, 2, 0 and -2. The probability of 0 is1/2, so the ML detection will have a fixed error1/4. Also we can find that the curve is nearly symmetrical. Meanwhile, when r is set to 0.2 or 0.8, the system gets an optimized BER performance.

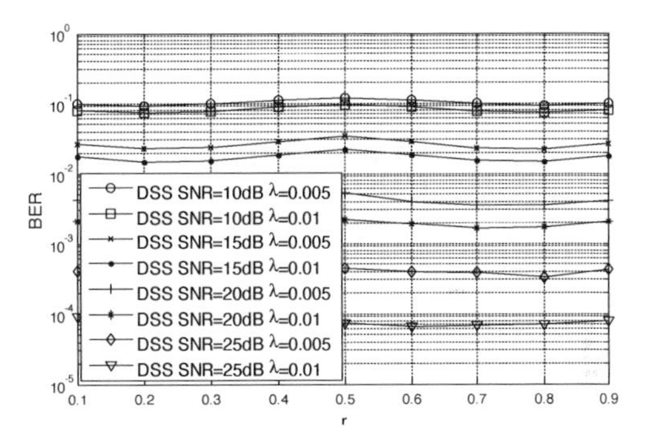

Figure 6. BER as a function of the variable r for different SNR and λ values in DSS.

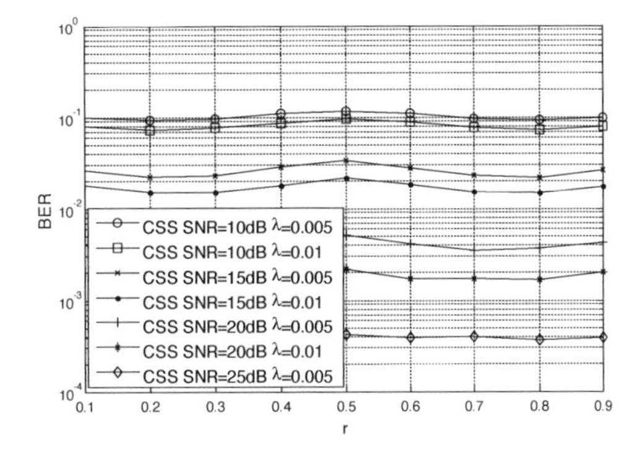

Figure 7. BER as a function of the variable r for different SNR and λ values in CSS.

5 CONCLUSIONS

In this paper, we consider the uplink cooperative transmission system, where two users transmit its information to the base station in multipath fading channels with the help of the relay. And we take into account the decoding BER of the systems, for the purpose, two methods are applied to evaluate the effect to the performance of the system. There are DSS and CSS. The result of the simulation reveals that both DSS and CSS can improve the performance of the system in terms of BER compared to the case of no suppression. In the meantime, CSS does better than DSS in the aspect in the cost of the reduction of the transmission rate, since the larger number of subcarriers has been suppressed. Both the two methods need additive signaling for the sake of letting the relay and the base station nodes know which subcarriers have been suppressed.

Further on, we discuss the coefficient r. The selection of the scheme depends on the system requirements. In other words, if we ask for the higher reliability of the data, then the CSS scheme will be the best choice for the efficient use of the subcarriers suppression. The DSS scheme reaches a moderate result in terms of BER compared to the CSS and the case of no suppression.

All the presented BER performance have approved that Uplink Cooperative OFDM Transmission System with subcarrier suppression, represents very interesting solution for the wireless communication systems.

ACKNOWLEDGEMENT

This work was supported by the National Natural Science Foundation of China (No.61371123), Jiangsu Province Natural Science Foundation for Young Scholar under Grant (BK2012055), National Natural Science Foundation of China for Young Scholar (No.61301165), and Jiangsu Province National Science Foundation under Grant (BK2011002), China Postdoctoral Science Foundation (2014M552612).

REFERENCES

Bhargava. V. K, Kundan Kandhway & Mohammad Mamunur Rashid. 2010. Cooperative communication in wireless uplink transmissions using random network coding. *Vehicular Technology Conference Fall (VTC 2010-Fall), IEEE 72nd, Ottawa, ON, 6–9 Sept. 2010*: 1–5. IEEE.

Bing, D. & Jun, Z. 2009. Physical-layer network coding over wireless fading channel. *Proc. International Conference on Information, Communications and Signal Processing. Macau, 8–10 December 2009*: 1–5. IEEE.

Fantacci. R. S., Morosi, D., Marabissi, E. D. Re & DelSanto, N. 2006. A rate adaptive bit-loading algorithm for in-building power-line communications based on DMT-modulated systems. *IEEE Trans. Power Del.*21(4): 1892–1897,

Hsu, Y.P. & Sprintson, A. 2012. Opportunistic network coding: Competitive analysis. *Network Coding (NetCod), International Symposium, Cambridge, Massachusetts, USA, 2012:* 191–196.

Kishore. S., Chen, Y. & Li, J. 2006. Wireless diversity through network coding. *In wireless communications and networking conference, WCNC 2006, Las Vegas, United States, 2006*: 1681–1686. IEEE.

Kompella. S., Sharma, S., Yi Shi, Jia Liu, Hou Y.T. & Midkiff, S. F. 2012. Network coding in cooperative communications: friend or foe. *Mobile Computing, IEEE Transactions on* 11: 1073–1085.

Lam, P. P, Zhang, S., Liew, S. C. 2006. Hot topic: physical-layer network coding. *In Proc. Int. Conf. Mobile Computing and Networking, Los Angeles, California, USA, 2006:* 358–365. ACM.

Letaief, K. B., Chen, W. & Cao, Z. 2007. Opportunistic network coding for wireless networks. *Communications. ICC'07. IEEE International Conference, Moscow, Russia, 2007*: 4634–4639. IEEE.

Liang. Y. C, E. C. Y. Peh & Guan, Y. L. 2008. Power control for physical-layer network coding in fading environments. *Proc. IEEE Int. Symp. Personal, Indoor Mobile Radio Commun.,Cannes, 15–18 September 2008*: 1–5. IEEE.

Popovski, P. T., Koike-Akino & Tarokh, V. 2009. Adaptive modulation and network coding with optimized precoding in two-way relaying. *Proc. IEEE Global Telecommun. Conf., Hilton, Hawaiian, Village, Hnolulu, Hawaii, USA, 30 Nov. 4 - Dec. 2009:* 1–6.

Tse. D., Laneman, J. &, Wornell, G. 2004. Cooperative diversity in wireless networks: efficient protocols and outage behavior. *IEEE Trans. Inf. Theory* 50(12): 3062–3080.

Wang Xingwei, Wang Shiqiang, Song Qingyang & Jamalipour, A. 2011. Power and rate adaptation for analog network coding. *IEEE Trans. Veh. Technol.* 60(5): 2302–2313.

Xu, K., Xu,Y., Xia, X.C. & Chen, Y. 2013. Symbol error rate of two-way decode-and-forward relaying with co-channel interference. *Personal Indoor and Mobile Radio Communications (PIMRC), 2013 IEEE 24th International Symposium on, London, 8–11 Sep 2013*. IEEE.

Electronics, Communications and Networks IV – Hussain & Ivanovic (eds)
© 2015 Taylor & Francis Group, London, ISBN: 978-1-138-02830-2

BER and distance adaptive spectrum resource allocation in FWM-based wavelength convertible spectrum-sliced elastic optical path network

Fazong Wang, Min Zhang*, Danshi Wang, Jiahui Wu, Zhiguo Zhang & Shanguo Huang
State Key Laboratory of Information Photonics and Optical Communications, Beijing University of Posts and Telecommunications, Beijing, China

ABSTRACT: A network layer and physical layer were combined to propose a wavelength conversion scheme and a spectrum resource allocation algorithm for a wavelength-convertible spectrum-sliced elastic optical path network while simultaneously considering the BER and distance. The feasibility of FWM-based wavelength converter was demonstrated, and the algorithm simulation results indicated that the algorithm was able to decrease the blocking rate and improve spectrum utilization.

KEYWORDS: Routing and spectrum allocation, Wavelength conversion, BER-distance adaptive, FWM.

1 INTRODUCTION

Optical networks are undergoing significant changes, fueled by the exponential growth of high-speed Internet traffic. Recently, the spectrum-sliced elastic optical path network (SLICE) has been proposed to mitigate this problem by adaptively allocating spectral resources according to client traffic demands (Jinno et al. 2009). By taking advantage of bandwidth-variable modulation technologies such as optical orthogonal frequency-division multiplexing (O-OFDM), it can allocate spectrums with speeds ranging from Gb/s to Tb/s (Jinno et al. 2009).

Traditional routing, wavelength assignment, and spectrum allocation (RWSA) algorithms must satisfy the spectral consistency, spectral continuity and spectral conflict constraints (Charbonneau et al. 2009). Thus, despite the availability of sufficient resources, connections could still be blocked. Some algorithms utilize the wavelength converter to increase the network efficiency and decrease the blocking probabilities. Therefore, the wavelength conflicts along the route of a channel can be resolved by changing the wavelength of a transit connection from one incoming wavelength to another wavelength. In (Patel et al. 2012, Subramaniam et al. 1996), using a spare wavelength converter, a relatively small number of converters can achieve the same benefits as that of a full wavelength conversion capability with a full conversion range.

In current optical networks, multi-level modulation formats are widely used to increase spectrum efficiency, such as QPSK (quadrature phase shift keying), and n-QAM (quadrature amplitude modulation). A novel distance-adaptive (DA) modulation was proposed to allocate different spectrum slices to an end-to-end path according to different path distances, in which the modulation level was chosen with respect to the transmission characteristic for an example optical signal noise rate (OSNR). Thus the network resources could be used effectively. By selecting a set of subcarriers and modulation levels, according to path distance, this approach was employed in a DA routing algorithm (Jinno et al. 2010), which adopted a complex modulation format for short distances and a robust, simple modulation format for long distances.

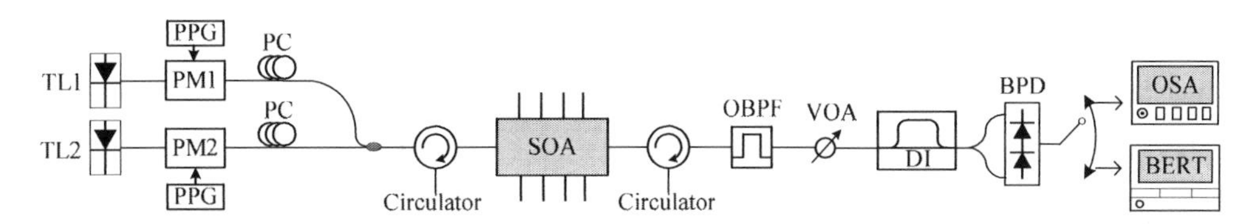

Figure 1. Experimental setup: tunable laser(TL); phase modulator(PM); pulse pattern generator(PPG); polarization controller(PC); semiconductor optical amplifier(SOA); optical bandpass filter(OBPF); variable optical attenuator(VOA); delay interferometer(DI); balanced photonics detector(BPD); optical spectrum analyzer(OSA); bit-error rate tester(BERT).

*Corresponding Author: *mzhang@bupt.edu.cn*

As the growth of bandwidth, some users want to get a better quality of service, but in DA algorithm, it may choose the complex modulation to decrease the number of subcarriers while the bit-error rate (BER) is increased so that it cannot meet the users demand. The BER factor uses more frequency slots than the DA algorithm. Although the bit-error rate and DA are considered in (Wang et al. 2012), the spectrum efficiency improvement and the blocking rate decline are from the last step. In this type of situation the quality of service could not meet the user's requirements.

In this paper, a dynamic routing and spectrum allocation algorithm was proposed while considering multiple factors, such as DA modulation, BER requirement, and wavelength conversion in SLICE. Using this approach, the users can obtain a better service quality and higher spectrum efficiency. The blocking rate and the total throughput were evaluated by comparing the proposed BER-DA-WC to BER-DA in SLICE. The goal of this study was to achieve a minimum request blocking rate with reasonable service. A new spectrum allocation algorithm called BER-DA-WC was proposed considering three factors: the wavelength conversion, BER, and distance. The simulation results indicated that it improved resource utilization.

2 THE LOWER SOLUTION BASED ON INTERPOLATION

2.1 *BER-DA-WC routing and frequency allocation algorithm*

For the same data rate 16-QAM used half the bandwidth of QPSK since there were twice the number of bits per symbol. Thus, spectrum bandwidth could be conserved by reducing the symbol rate and increasing the number of bits per symbol to transmit the same data.

In order to eliminate the spectrum consistency restriction and improve the spectrum efficiency, a new wavelength conversion algorithm was used. The procedure of the proposed BER-DA-WC algorithm is described as follows:

Step 1. When the client request arrives, select the route that offers the highest numbers of available frequency slots from the candidate paths of the source-destination node pair.

Step 2. If the required BER of the service is low, select a low-level modulation such as QPSK, without considering distance; otherwise, consider the distance. If the hop number of the chosen route exceeds the threshold, select QPSK; otherwise, select a high-level modulation, such as 16QAM, instead.

Step 3. Select frequency slots using the WC algorithm described as follows.

Step 4. Remove the connections with expired holding times and release the resource.

Suppose that a physical topology G (V, E) exists, where V is a set of nodes, E is a set of fibers, and C GHz is the total network spectrum. Each slot is referred to as a frequency slot, with a spectral width of σ GHz, A wavelength converter of $\pm M$ GHz. $M = C$ indicates that the incoming frequency can be converted to any outgoing frequency.

The WC algorithm is described as follows, where Wij is the link between node i and node j, and wij is the first slot that satisfies the request.

Step 1: Collect the available continuous slots for every link of the path and put the lowest index of the continuous slots in the set Wij.

Step 2: At the first link of the route, use the First-Fit algorithm to choose the frequency slot in Wij that meets the requirements, and record the lowest index of the allocated frequency slots as wij.

Step 3: In the middle link of the route, check the slots with previously chosen links; if those slots are available on this link, choose them; otherwise implement Step 4.

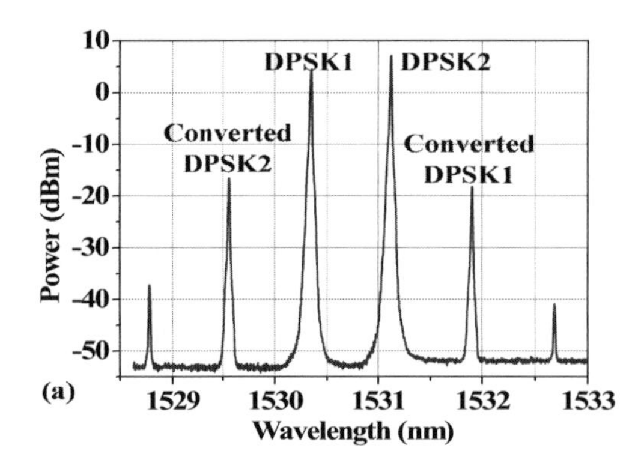

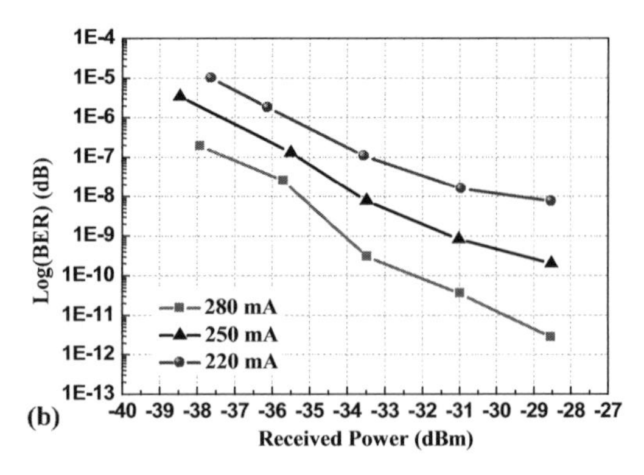

Figure 2. (a) The optical spectrum of wavelength conversion; (b) the BER curve as a function of received power under different biascurrents.

Step 4: In the range of the wavelength conversion, use the First-Fit algorithm to select the appropriate spectrum in W_{ij}, and record the lowest index of the allocated frequency slots as *wij*. If there are insufficient slots available, drop the lowest slots in W_{ij} for every link, and continue to Step_2.

Step 5: If there are no available slots and the set of a link is null, then block the request.

2.2 *Experimental setup*

In order to demonstrate the feasibility of our scheme, a physical experiment was set up, as shown in Fig. 1. Two coupled tunable lasers (TLs) were sent into two phase modulators (PM1 & PM2), which were driven by the 10 Gbps pseudo-random binary sequence (PRBS) from two pulse pattern generators (PPG). Two DPSK signals were generated at 1530.3 nm and 1531.1nm, both with 5dBm. The polarization states of the two signals were adjusted with two polarization controllers (PCs). The two circulators before and after the SOA were used as isolators to decrease the influence of the reflection. In SOA, with a bias current of 280 mA, four-wave mixing could be achieved to realize the wavelength conversion. After FWM, the newly generated idlers were separately filtered by an optical bandpass filter (OBPF). The selected signal was demodulated by the fiber-based delayed interferometer (DI) and detected by a balanced photonics detector (BPD). Finally, the eye-diagrams were obtained from an oscilloscope (OSC) and BERs are measured with a BER tester (BERT).

The optical spectrum measured at the output of SOA is shown in Fig. 2(a). As shown in this figure two new idlers at 1529.5 nm (Converted DPSK2) and 1531.9 nm (Converted DPSK1) had the large optical signal to noise rates (OSNRs) of more than 30 dB and high conversion efficiencies (CEs) of more than -23 dB, which indicated a good converted signal performance.

Then, other conditions were left unchanged, and the bias current of SOA was from 220 mA and 250 mA to 280 mA. The BER of the converted DPSK1 as a function of received power under different bias currents is shown in Fig. 2(b). The BER of the converted signal was improved as the bias current and received power increased. The error-free performance demonstrated the feasibility of the FWM-based wavelength conversion scheme.

3 NUMERICAL SIMULATIONS AND DISCUSSIONS

We performed numerical simulations using C++ in order to compare the number of blocked services of 10,000 and the total throughput of BER-DA with those of BER-DA-WC. The whole number of Frequency Slots (FSs) on each link was set to 128, and the required slots for a connection request was uniformly distributed from the smallest granularity 1 FS to 8 FSs. The proposed algorithm was simulated on the 14-node NFSNET topology. Requests were randomly distributed over the network. Requests inter-arrival time was exponentially distributed, and the holding time exhibited a negative exponential distribution. The source-destination node pair was generated randomly from all the nodes, and the frequency slot width was set as 12.5Hz. "Unchanged" was used to denote a traffic load of 400Erlang, a hop number of 5, and a high BER service by 50%.

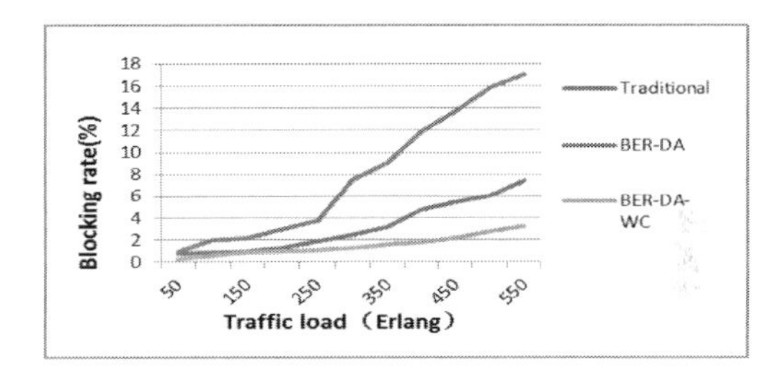

Figure 3. Blocking rate at different traffic loads while BER and hop threshold remaind unchanged.

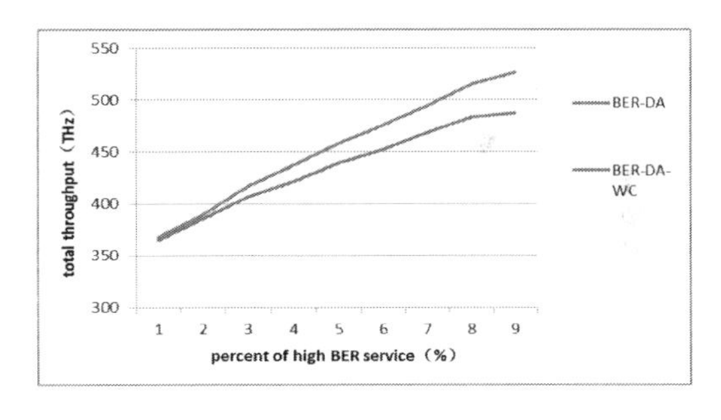

Figure 4. Total throughput at different BER distributions while the traffic load and hop threshold remained unchanged.

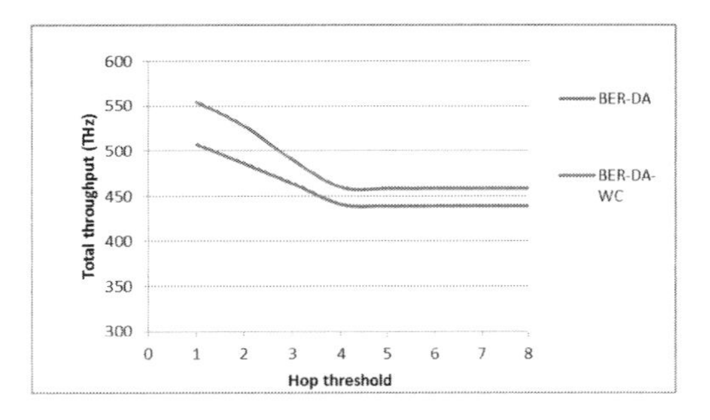

Figure 5. Total throughput at different hop thresholds while the traffic load and BER distributions remained unchanged.

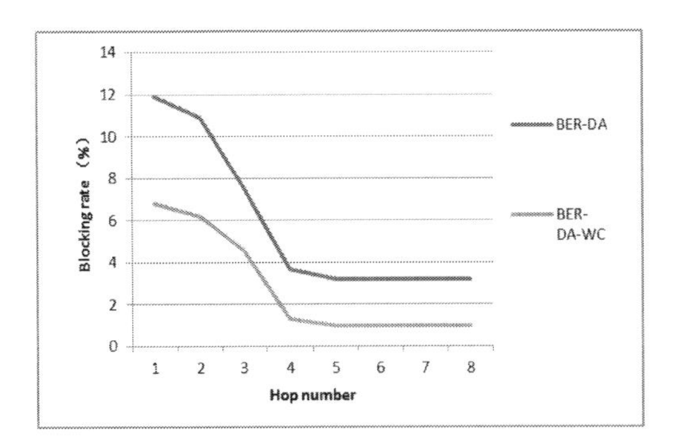

Figure 6. Blocking rate at different hop thresholds while the others conditions remain unchanged.

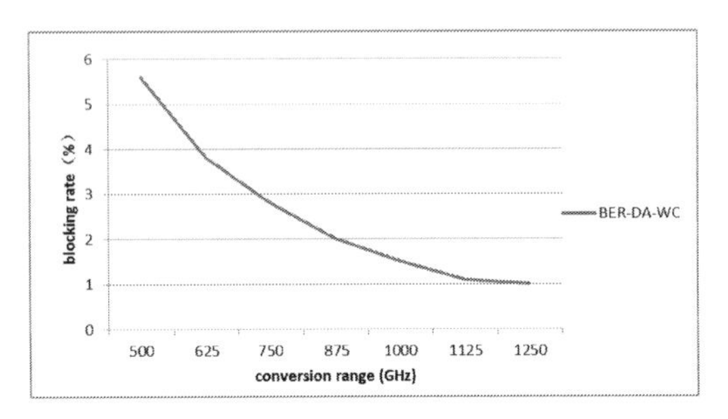

Figure 7. Blocking rate at different conversion ranges.

Requests were divided into two categories according to their characteristics, i.e., low BER requests, and high BER requests. The BER value was randomly generated to simplify the simulation.

In Figure 3 the BER-DA-WC algorithm is compared to the BER-DA algorithm and the traditional algorithm that does not consider BER and distance. As is shown in Figure 3 as the traffic load increased, the blocking rate of all algorithms increased, however, the BER-DA outperformed the traditional algorithm due to the DA. By choosing the high-level modulation format for the services with hop numbers less than the threshold, DA algorithm occupied fewer spectrums than the traditional algorithm. BER-DA-WC outperformed BER-DA primarily because the wavelength conversion obtained a higher frequency utilization.

Figure 4 shows that, under the same conditions, when the proportion of high BER services increased, the total throughput was increased. This was because high-level modulation formats (such as QPSK) were selected for the high BER service, so that more requests were able to be served. While BER-DA-WC improved the spectrum utilization efficiency through wavelength conversion, an incoming wavelength can be converted to any outgoing wavelength with a wavelength converter. The same situation is shown in Figure 5; with BER-DA-WC more requests can be served, so the total throughput was much greater than the BER-DA.

Figure 6 displays that the blocking rates decreased initially, then stabilized as the hop threshold increased since, in the NFSNET topology, nearly all paths include less than five hops. When the network load was 400 Erlangs, as the hop threshold increased, more services chose high-level modulation, and BER-DA-WC obviously outperformed BER-DA.

Each wavelength converter had a limited conversion range, and all the results mentioned above were in full frequency conversion range. As shown in Figure 7, as the conversion range increased the blocking rates decreased initially and stabilized. Thus, by employing optical wavelength converters with limited ranges, benefits similar to those with full ranges were achieved.

4 CONCLUSION

The feasibility of wavelength conversion from the network layer and physical layer was demonstrated, The physical layer was based on the FWM effect in order to achieve the dual-channel wavelength conversion. In addition to a dynamic routing and spectrum allocation algorithm in WC-SLICE networks while simultaneously considering the BER and distance. The feasibility of a FWM-based wavelength converter was demonstrated, and the simulation results indicated that the proposed algorithm could largely reduce the blocking rate and improve the resource utilization efficiency while providing better quality of service.

ACKNOWLEDGMENTS

This study is supported by 863 Program No. 2012AA011302, 973 Programs No. 2012CB315604, a Doctoral Scientific Fund Project of the Ministry of Education of China (No. 20120005110010), BUPT Excellent Ph.D. Students Foundation.

REFERENCES

Charbonneau, N. 2010. *Dynamic non-continuous advance reservation over wavelength-routed networks.* University of Massachusetts Dartmouth.

Jinno, M, Takara, H & Kozicki, B. 2009. Concept and enabling technologies of spectrum-sliced elastic optical path network (SLICE). *In Asia Communications and Photonics Conference and Exhibition (p. FO2). Optical*

Society of America. Shanghai, China, 2–6 November 2009. OSA, IEEE Photonics Society.
Jinno, M., Takara, H., Kozicki, B., et al. 2009. Spectrum-efficient and scalable elastic optical path network: architecture, benefits, and enabling technologies. *Communications Magazine, IEEE* 47(11): 66–73.
Jinno, Masahiko, et al. 2010. Distance-adaptive spectrum resource allocation in spectrum-sliced elastic optical path network. *Communications Magazine, IEEE* 48 (8): 138–145.
Patel, A. N, Ji, P. N, et al. 2012. Routing, wavelength assignment, and spectrum allocation in wavelength-convertible flexible optical WDM (WC-FWDM) networks. In *National Fiber Optic Engineers Conference* . OSA.
Subramaniam S S, Azizoglu M & Somani A K. 1996. All-optical networks with sparse wavelength conversion. *Networking, IEEE/ACM Transactions on* 4(4): 544–557.
Wang, L, Zhang, M, et al. 2012. Dynamic frequency allocation in SLICE considering both BER and distance. *TELKOMNIKA Indonesian Journal of Electrical Engineering* 10(7): 1537–1540.

Electronics, Communications and Networks IV – Hussain & Ivanovic (eds)
© 2015 Taylor & Francis Group, London, ISBN: 978-1-138-02830-2

The research on M-learning based on network streaming media in higher vocational education

Ge Wang
Kaifeng Culture and Arts College, Kaifeng, Henan Province, China

ABSTRACT: With the development of 4G and wireless network, M-learning has revolutionary changes. We can study audio and video at any time and in any place with streaming media. By questionnaire, this paper makes the conclusion of necessity and situation of higher vocational students in M-learning from network streaming media, discusses the course composition, course features, course requirement analysis and design strategy in higher vocational education by micro-video M-learning.

1 INTRODUCTION

At present, there exists a little study and application for M-learning in higher vocational education. Advanced network technology and hardware equipment is mainly used in non-educational aspects, such as business, entertainment, life, etc. On the one hand, M-learning in higher vocational education is lack of support from relevant departments. On the other hand, intelligent mobile phone media are popular among higher vocational students. It's an important problem about how to improve the learners' interests. The theories and practices are poor. This paper lists related knowledge, analyzes the present situation of higher vocational education in M-learning, lists the advantages of applying M-learning from streaming media in 4G or wireless network environment. Through the results of the questionnaire, discuss the course composition, course features, course requirement analysis and design strategy in higher vocational education by micro-video M-learning.

1.1 *4G and wireless network*

The application of 4G mobile communication network and wireless multimedia communication in the field of streaming media, mainly includes online news, streaming media messages, video telephone, video conference, mobile phone, video on demand, interactive games, car navigation and so on, which make people communicate more conveniently, more richly and colorfully. (Khan et al. 2009, Krenik et al. 2008) Most important, it provides a broader platform and more resources for M-learning in streaming media. (Liu et al. 2013)

1.2 *Streaming media*

The process of streaming media is that the coding information, such as video, audio and data, is transformed from server to client through real-time transport in continuous flow. The client can immediately play video or audio after receiving information data in the cache. (Gan et al. 2013)

1.3 *M-learning*

M-learning is based on the present mature wireless network, Internet and multimedia technology. Teachers and students use mobile devices (such as the portable computer, IPAD, telephone, etc.) to realize interactive learning. (Prapaipis et al. 2009)

2 THE ANALYSIS ON STUDY SITUATION OF HIGHER VOCATIONAL COLLEGE STUDENTS

Nowadays, in basic education, the students of weak foundation are put into vocational education. . These students need more effective teaching methods, more attractive teaching environment, more training about characteristic learning habits. Higher vocational education is formed for the cultivation of students' work habits, social responsibilities. In this process, if you give the students a more effective learning environment, more self-expression space, it will be more conducive to the cultivation of the students' learning habits, including the habit of independent thinking, innovative thinking, the correct expression of self.

With the development of computers' popularity, fewer students can use nothing. Maybe students can't use computers only on some knowledge points. So more and more students use search engines to search knowledge points. If teachers express the knowledge of a specific operation in micro video class, it will be more conducive for students to learn anytime or anywhere. This kind of study is about intuition, short time consumption and strong characteristics. It is more conducive to making fragmentary learning and achieving complete knowledge in a short time. (Wang et al. 2013)

The characteristics of mobile learning: firstly, we can learn anywhere and anytime. Make classroom, extra-curricular environment, family connect together; secondly, encourage the students to learn from all aspects, so that the depth and width of knowledge are all increased; thirdly, develop cooperation and communication abilities. Mobile technology provides a possibility to narrow the distance between school and the real world. The teacher may integrate class activities and resources outside class into the classroom through the network, and the higher vocational students can apply the skills in reality by using mobile devices, wireless connection and application, improving the students' interest in learning skills (Giousmpasoglou et al. 2013, Yuan et al. 2012).

3 ANALYSIS OF THE SURVEY RESULT

According to "The 34th China Internet development statistics report"(Gao et al. 2014) from Chinese Internet Network Information Center, in the first half of 2014, China's Internet users who use the mobile Internet continue to maintain growth, rising from 81% to 83.4%; the number of mobile phone users has reached 527000000 in China in 2014, an increase of 26990000 compared with the end of 2013. The proportion of Internet access via the desktop and notebook computers has slightly decreased. This year our country's Internet users using the mobile Internet exceed traditional PC online for the first time as the ratio (80.9%).

A survey of "university students' needs for network streaming, media and M-learning" is aimed at the computer majored students from the author's vocational college. It received a total of 85 valid questionnaires, in which 58 copies are submitted by webpage, 27 copies are submitted by mobile phone (Figure 1). Investigation shows that only about 2.35% of the students have no mobile network or wireless network equipment (Figure 2), nearly 80% of the students use Internet over 3 hours every day (Figure 3). And the number of students whose main access time are used to download data or study only accounts for 15.29%, and the students who understand wireless or 4G network only accounts for 12.94%, nearly 50% of the students hope to study through the video case or teaching software, 35.29% of the students want to study through a short video on certain knowledge (Figure 4).

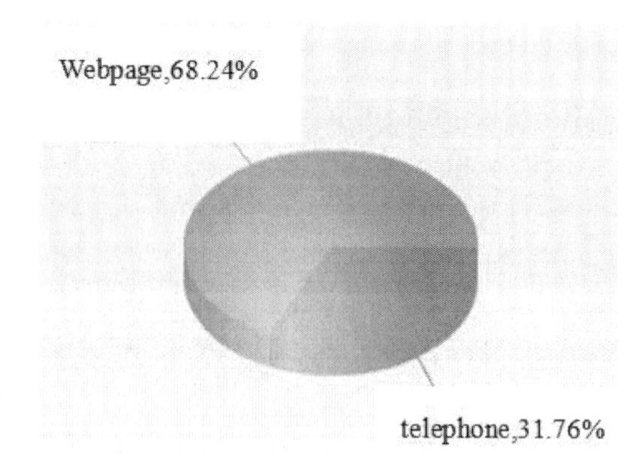

Figure 1. Analysis of survey source.

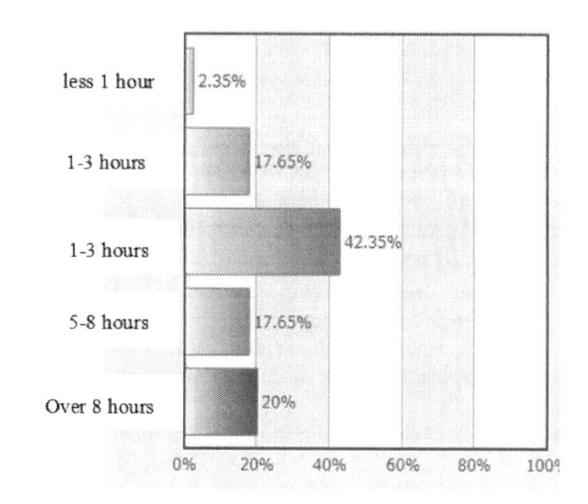

Figure 2. Daily use of Internet.

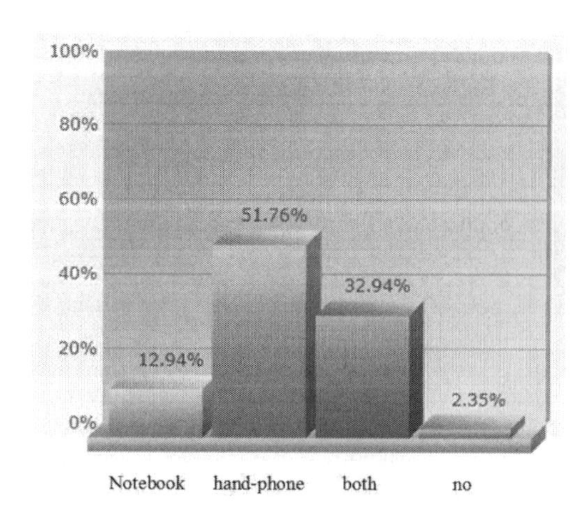

Figure 3. Mobile or wireless equipment

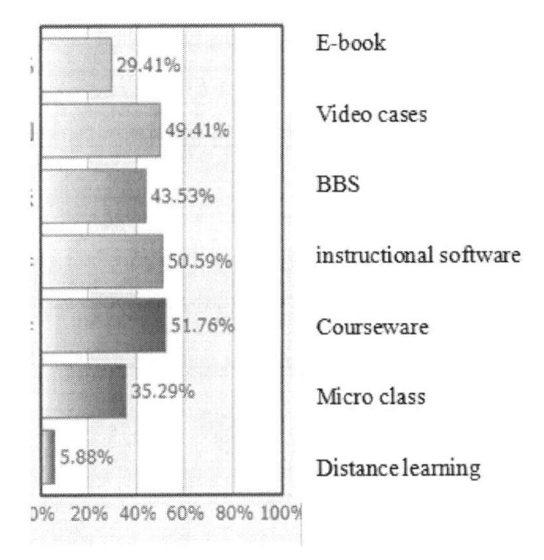

Figure 4. The most interest forms of learning.

Therefore, as a higher vocational educator, it is important to arrange students to study through M-learning with streaming media outside the classroom for more than 3 hours. Mobile Internet technology is becoming more and more mature, such as micro-message, micro-blog, micro class, micro video, etc. "Micro" elements are changing people's cognition and learning habits. Learners can interact with teachers, experts, learners anywhere and anytime. Learners can make full use of the advantages of 4G or WiFi network, watching instructional videos or fragments through the mobile terminal or computer terminal. Learners can browse the questions and answers on the screen freely. Teachers can give the teaching information through the Internet, and learners can receive the information anytime or anywhere. Making full use of the "micro" element can active vocational college students' after-school life, cultivate professional accomplishment, enhance the professional skills, create more higher vocational students of high-level comprehensive qualities. (Bai et al. 2009)

4 THE CURRICULUM MODEL AND THE DESIGN STRATEGY OF M-LEARNING USING STREAMING MEDIA IN HIGHER VOCATIONAL COLLEGE

4.1 *The composition of network streaming media M-learning*

According to the characteristics of M-learning, the fragmentation, the learning content is fragmented and miniaturized. It is no longer complete, linear, curing. Traditional big size module of learning content cannot meet the learners' needs of M-learning better. Therefore, the resources based on M-learning are fragmented, miniaturized and flexible. It is easy to obtain, disseminate and share. (Zhang et al. 2011)

Network streaming media video course benefits from the rapid development of network technology and video technology. The research suggests that network streaming media course is the sum of the teaching content and teaching activities for the implementation of a certain subject through micro video display. It consists of two parts: the teaching content supported by teaching target, teaching strategy and the teaching activities. One of the main carriers of teaching content is the teaching micro video. Teaching supporting environment refers to teaching software tools, teaching resources and the implementation of the teaching activities on the platform.

Therefore, the content of network streaming media micro video course is composed of a plurality of micro video teaching according to the teaching aims and teaching strategies, composed of a sequence of some structure. The content structure of the curriculum is related to curriculum content capacity. If a micro video course is of small capacity (such as lectures on short-term training, etc.), it can be directly formed by a plurality of micro video teaching. If the course contents are more systematic and complete, the network streaming media micro video courses may be constituted by a number of the thematic units, or formed by a plurality of thematic units and fragmented micro videos.

4.2 *The features of network streaming media micro course*

According to its properties and connotation and definition, this kind of courses has its own characteristics in the curriculum structure level, curriculum design level, knowledge content level and resource acquisition level.

1 Curriculum structure level—loose affinity.
   The knowledge module in the course is made up of the adjacent loose knowledge module, and has structural features of the course of study. At the same time, some modules need to organize knowledge in a meaningful way and constitute a learning unit to achieve specific learning objectives.
2 Curriculum design level—modular theme.
   The course design is based on the knowledge module, learning unit, learning theme. For the realization of learners' desires of choosing learning content according to the needs and interests, and the implementation of reuse and regeneration of

video courses. Video knowledge module's theme unit is independent knowledge points, which help to break the curing and strict structure of traditional video course and achieve the course construction and sharing.

3 Knowledge content level—miniature fragmentation.

Attention decentralized in fragments period, sustained attention time is shorter. Miniaturized resources are easier for learners to use and share fragmented time learning. Through the knowledge module reuse, construct new course.

4 Resource access level—association.

Network streaming media M-learning resource is the fragmentation of loose coupling. For learners of fragmented study, they need to associate these fragments of the coupling body to become meaningful things. Not only do they consider the relative independence of micro-content, but also some association hidden behind the loose content, and a latent continuous structure is formed gradually in learning an experience.

5 Video attribute level —art.

The main characteristics of network streaming media M-learning course is visual resources, and video appearance and knowledge content are very important for the availability of the curriculum, which directly influences the learner' motivation and efficiency.

4.3 *Demand analysis of network streaming media micro video*

Based on the change of the content of fragmentation learning and the curriculum connotation and characteristics, this study suggests that the network streaming media micro course can meet the learning needs of fragmentation. Main content teaching micro video can make fragmented learning effective and feasible. Mainly based on the following aspects:

1 Network streaming media micro video course has a wide application market and is favored by the majority of learners

Video resources as the network learning resource have the following advantages:

1. Meet the various learning styles of learners;
2. Improve the transfer efficiency of situation knowledge, make learners accept more easily;
3. Compared with the abstract text resources, its content is more intuitive;
4. Make present knowledge and forthcoming knowledge connected, associated better.

Therefore, network streaming media micro video courses play a greater role. For example, a well-known university public class and all kinds of network video resources have become one of the most important channels for learning; video resources will be a main teaching form in the future open course. Teachers and students pay more attention to it. In the field of education, it has produced influence in reality.

2 The length and content meet the demand of fragmentation learning

Network streaming media video belongs to miniature resources, its learning content is short, learning relative to the traditional video course is shortened obviously. It can be used as suitable learning resources of micro learning activities. Therefore, it can be used as a suitable resource in fragmentation situation, meeting the needs of learners' resources.

3 Network streaming media micro video course structure is conducive to the study of debris to polymerization

The teaching content of micro video course is based on the teaching objectives, teaching strategies. But it is not disorderly, scattered teaching micro video collection. Therefore, micro video course targets to help the learners to learn to make the scattered teaching micro video aggregated and associated to realize knowledge content, promote the self-perfection of thinking and learning of structured knowledge.

4.4 *Design strategy for the knowledge contents of network streaming media M-learning*

1 According to the law of short attention, design effective teaching strategy to attract learners.

The attention length is different from people in different times with different media. Adult attention duration is 20 minutes when they are listening to the speech report. The average attention length of watching video is 2.7 minutes; the average length of human attention is 12 seconds in 2000, is 8 seconds in 2012. The attention length is very short. Therefore, it is necessary to design the attention contents according to the needs of the learner's short attention rules.

In view of the short attention learners, we need to pay attention to the following contents:

(1) Instant useful content; (2) valuable content; (3) be quickly retrieved content; (4) short content; (5) content with index; (6) provide an overview of the contents of short generalized introduction for learners; (7) the loose structure content. Independent module content is flexible and easy to transplant.

Learners can quickly browse little module of the course; (8) display video content length, in order to help learners grasp the rhythm of learning; (9) the immediate dialog. The instant, quick, short dialog can help learners understand the learning content better.

2 Adopt WPW teaching process

Research on the psychological experiments has found that the learner's keeping attention degree will show a certain fluctuation in the unit of learning time, and the degree of keeping attention will directly affect the memory of the short term and long term. In the WPW teaching process, the learners' best memory effect appears the earliest, followed by the stimulation of the last occurrence. So, the importance of beginning and summarizing part of teaching maintain a higher degree of attention.

3 Increase the authenticity of knowledge content, establish a connection between knowledge and life, promote knowledge transfer of learners.

From selecting knowledge content level, learners should consider and pay attention to the authenticity, connectivity, and significance of knowledge content. So learners can cultivate the ability to solve problems by using the knowledge content, and obtain the learning satisfaction and a sense of achievement. Through sampling analysis of network streaming media micro video, finding case, problems and the story is a commonly used strategy in micro video teaching. It can put the learner into real situations to increase the authenticity of knowledge content.

5 CONCLUSIONS

In summary, the network streaming media M-learning is an open teaching system, teaching information interactivity, personalized study characteristics, rich teaching resources. Through the questionnaire survey, in higher vocational education it has become more and more necessary to enhance the streaming media, M-learning, wireless network, 4G network. By using the network environment, develop M-learning of higher vocational students effectively, and make it become a part of school education, make the professional learning more effective, private, acceptable, and with stronger skills.

ACKNOWLEDGMENTS

This paper is supported by the Education Department of Henan Province, which is the key scientific and technological project, project number: 13B520132

REFERENCES

Bai Jian & Li Zhijun. 2009. Exploring into application of mobile education based on mobile streaming media technology in higher vocational education. *Technique on line* 2009(18): 102–103.

Gan Lu, Xie Donglan & Huang Lan. 2010. Research on the construction of remote video education system based on Streaming Media Technology. *Journal of Yulin Normal University* 31(2): 143–146.

Gao Kang & Luo Yufan. 2014. China's mobile phone Internet users have a higher proportion of computer Internet users. *http://news.xinhuanet.com/fortune/2014-07/21/c_1111720117.htm*. 2014.07.21

Giousmpasoglou, C. & Marinakou, E. 2013. The Future Is Here: M-Learning in Higher Education. *Computer Technology and Application* 4(2013): 317–322.

Khan, A.H. 2009. Future Computer and Communication, 4G as a Next Generation Wireless Network. *International Conference on ICFCC 2009*: 334–338.

Krenik, B. 2008. 4G wireless technology: When will it happen? What does it offer?. *Solid-State Circuits Conference A-SSCC* '08: 141–144.

Liu Dongli, He Kai & Yang Yingxiang. 2013. Mobile HD streaming media application based on 4G wireless network. *Journal of electronic technology and software engineering* 20: 59.

Prapaipis, G., Xu, G.Z. & Supanee, S. 2009. M-Learning: a New Stage of d-Learning in Higher Education of Thailand. *Journal of Yunnan Normal University* 2009(29): 37–40.

Wang Mi. 2013. On Designing the Content of Micro-Video Courses in a Fragmented Learning Age. *ECNU Doctoral Dissertation*: 1–236.

Yuan Jinying & Ma Xiufeng. 2012. Interaction design of network courses based on inquiry learning. *Software Guide* 2012.2: 71–73.

Zhang Tianyun & Cui Lingling. 2011. Research on micro learning process and structure design theory basis. *Research in Teaching* 34(1): 60–61.

Electronics, Communications and Networks IV – Hussain & Ivanovic (eds)
© 2015 Taylor & Francis Group, London, ISBN: 978-1-138-02830-2

M-SVM solving the phase rotation of CAP access network

Yida Wang & Min Zhang
State Key Lab. of Information Photonics and Optical Communications, Beijing University of Posts and Telecommunications, Beijing, China

ABSTRACT: Carrier-less Amplitude and Phase (CAP) modulation is a kind of modulation format that is similar to quadrature amplitude modulation. It has great potential to be used in positive optical access network. In this paper, a machine learning algorithm named M-Support Vector Machine (M-ary SVM) is used to solve different phase rotation problem in the real transmission caused by digital filters in the CAP modulation system. M-ary SVM is an M-dimension classification algorithm with logN classifiers that is superior to other classification algorithms with N complexity. The impact of phase rotation on system performance and algorithm is demonstrated. The BER performance of different transmitting length over Standard Single Mode Fiber (SSMF) is measured. It demonstrates M-ary SVM algorithm in short positive optical access network, further expand to multi-level and multi-band CAP modulation system.

1 INTRODUCTION

Recent year, large transmission capacity and low power consuming become more and more important in access network. Using high order modulation format such as M-quadrature amplitude modulation(M-QAM) or orthogonal frequency division multiplexing(OFDM) to enlarge the capacity, and employ positive device in transmission system like positive optical network to reduce the power overhead scored tremendous achievements. However, the complexity and cost raise have limited the scale of these advanced modulations in short range communication. Nowadays, a modulation mode named carrier-less amplitude and phase arouses great interest because it allows high data rate to be transmitted in limited bandwidth with no complex modulate and demodulate components. It generated by combining two outputs of the filters. The filter impulse responses are a pair of Hilbert transforms, which are utilized to shape the in-phase and quadrature signals respectively. CAP is quite similar to QAM because the two filters take the place of the carriers in QAM modulation. There is no need to use radio frequency (RF) source or IQ modulator in CAP system so that it reduces the equipment cost greatly. Then the speed of DAC and ADC becomes the main factor constraining the transmitting capacity of the high speed CAP system (Ingham et al. 2011).

CAP-based system (Chen et al. 1992) is very sensitive to symbol timing offsets and jitters. It has been showed that 1ps timing jitters for 100 Gbps CAP system incurred around 4dB power penalty (Wei et al. 2012). Moreover, the flat channel frequency response is necessary to cause non-flat channel will decrease the system performance significantly. In real transmitting system, the different DAC sample timing points of ONU will produce different phase rotation of the received signal, which leads to higher difficulty to demodulation and increases the bits error rate. Especially in multi-band or multi-level communication system such as WDM for the impurity of transmitting length to each ONU. In this letter, M-SVM algorithm is used to demodulate directly with no need to solve the phase problem and reach an adequate BER performance. The solution is not only verified in MATLAB and VPI simulations, but also in 10 Gbps CAP-based system from back to back to 40km standard single mode fiber. Further simulated in multi-level of difference SNR in the standard plural channel.

2 OPERATION PRINCIPLE

Support vector machine is a general and effective machine learning algorithm to solve problems as small samples, dimension trouble and nonlinear disasters. It has extensive use for the classify of low-dimensional linear inseparable problem through a kernel function that transforms the issue into a high dimensional linear classifiable (Kreßel 1999). The SVM could be converted into the following

$$\text{argmax}_{\alpha}: \sum_{i=1}^{L}\alpha_i - \frac{1}{2}\sum_{i=1,j=1}^{L} y_i \cdot y_j \cdot \alpha_i \cdot \alpha_j \cdot \langle \mathbf{v}_i, \mathbf{v}_j \rangle \tag{1}$$

$$\textbf{s.t. } 0 \le \alpha_i \le C,\ i = 1, 2, \ldots\ldots L \tag{2}$$

$$\sum_{i=1}^{L} \alpha_i y_i = 0 \tag{3}$$

equivalent (1–3) with KKT conditions of quadratic programming problem.

In N category case, SVM usually classifies one class to the other N-1 classes, respectively, so called 'one against all' which requires the N classifier (Sebald & Bucklew 2001). M-ary SVM makes it *logN* by labeling each signal case in binary format with each bit modeled by conventional SVM. Figure 1 illustrates the core unit of the M-ary SVM detector in the 16-CAP system (Li et al. 2013).

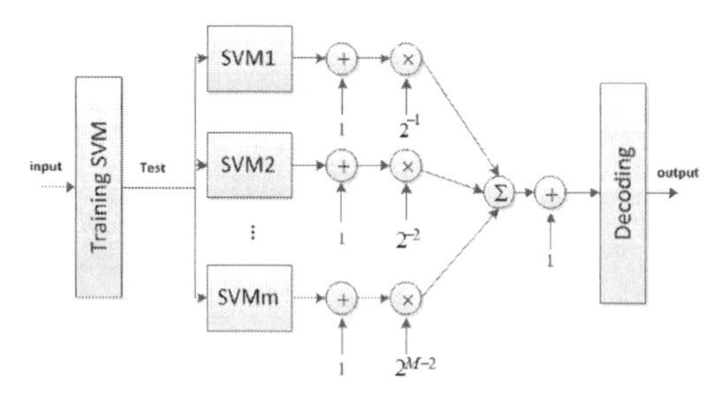

Figure 1. Processing structure of M-ary support vector machine detector used for the 16-CAP optical access network.

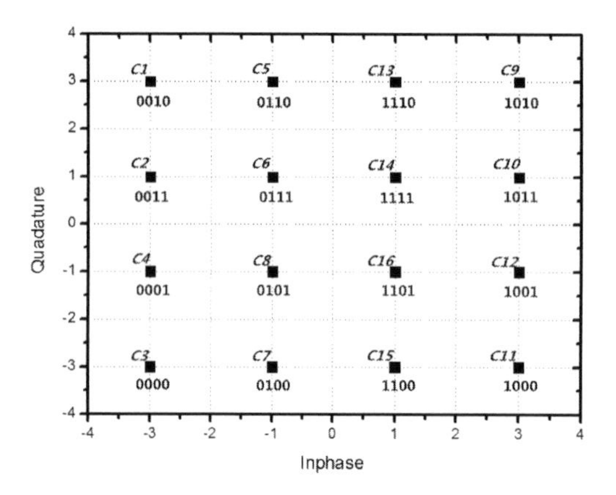

Figure 2. Gray coding and class assignment strategy for every constellation point in 16-CAP systems.

The whole structure includes two main steps training and testing. Given a classification strategy to soft margin binary classifiers and construct the feature vector by in-phase and quadrature CAP symbol. Training stage is devoted to establishing the binary SVM's separating hyperplane. Every symbol is mapped into one of the 16 classes. According to Figure 2, to make the demodulation much more concise that integrates the gray-code decoding, PAM signal demodulating, serial to parallel converting, reaching bits stream finally. Because these constellation mappings can be changed based on the different real transmitting system, so it provides flexibility to the expansion of the system. Although receiving signals located not in an ideal position caused by the imperfect sample timing, clock jitters and noise influence, training data can be easily identified by the receiver. Function (4) gave the *ith* SVM strategy.

$$D_i^+ = \left\{ n \in S \middle| mod\left((n-1) \cdot 2^{-(i-1)}, 2 \right) = 0 \right\}$$
$$D_i^- = \left\{ n \in S \mid mod\left((n-1) \cdot 2^{-(i-1)}, 2 \right) \neq 0 \right\} \tag{4}$$

Where n is training data label, S = {1,2......,16} is set of the classes. D_i^+ and D_i^- are judgment results of *ith* SVM. Finally, every training data is assigned to *logN* numbers of {−1,+1} represent *logN* SVM decision answers. By function (5), *ith* SVM classify

$$\hat{c}^{(i)} = sign\{\sum_{k \in V} \alpha_k^{(i)} \cdot y_k^{(j)} \cdot \langle v, v_k^{(i)} \rangle + b_i\} \tag{5}$$

$$\hat{c} = 1 + \sum_{i=1}^{4} (\hat{c}^{(i)} + 1) \cdot 2^{i-2} \tag{6}$$

the unknown symbols after each support vector machine established. Then the function (6) helps to calculate the specific category before decoding, if all SVMs classify correctly, the final detection results decoded with the mapping of Figure 2.

3 SIMULATION AND EXPERIMENTS

Figure 3 shows the fundamental system structure. OLT side, 10 Gb/s pseudo-random bits sequence is mapped into 4 level. After PAM modulation, the symbols are shaped separately by a pair of FIR filters with 32 taps whose impulse responses are represented by

$$h_1(t) = g(t)\cos\omega t$$
$$h_2(t) = g(t)\sin\omega t$$

Where $g(t)$ is square root raised cosine function, f_c is the carrier frequency. The computational complexity of an ideal system to generate and demodulate CAP signals is directly related to the number of taps of the shaping and matched filters. Setting the up-sample factor 4, f_c is 3Ghz and roll-off factor be 0.2. The in-phase and quadrature output is combined into AWG, whose sample rate is set to 10Gsa/s. The electric signal is transformed to optical signal by MZM, which is driven by laser with continuous wave fixed at 1550nm, then transmitted through SSMF about 20km on launch power 0 dBm. In ONU, after PD the received signal is sampled by an oscillator at a sampling rate of 40 Gsa/s and is processed offline, then the M-ary SVM works.

Usually, high-frequency components have large power attenuation. There are two reasons for this problem one is the bandwidth limitation from AWG and the other is frequency fading induced by chromatic

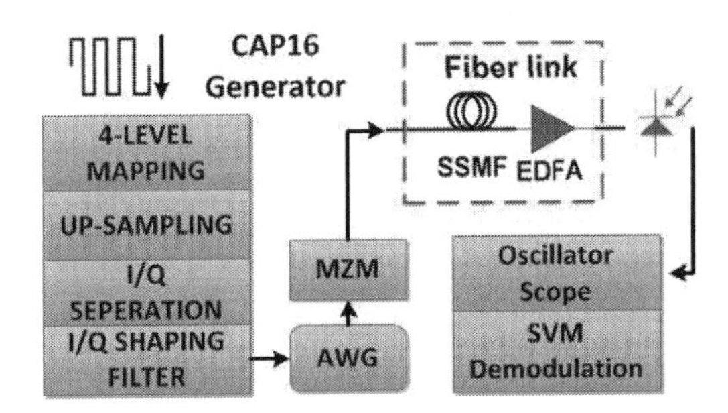

Figure 3. Experiment fundamental structure of carrier-less amplitude & phase access network with M-SVM demodulation.

dispersion. It will degrade the system performance heavily. As a result, conventional experiment adds pre-emphasis to solve this problem. In the experiment, a blind equalization of DD-LMS schedule is employed to compensate the distortion from timing offsets and channel frequency response. Figure 4 is the received signal constellation diagrams of experiment at −15dbm received optical power. Firstly 500 symbols are sent to training support vector machines which sets the penalty factor C = 3 and the kernel parameter s = 2. Secondly transmitting 2e4 bits optical signal under the same transmit length and launch optical power, finally receiver got different phase rotation as we expected.

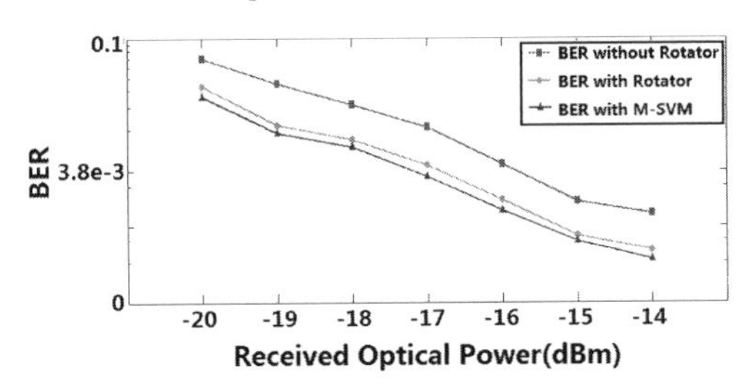

Figure 4. Experiment BER performance of 10Gbps 16 CAP system and constellation over 20km SSMF.

Despite the sample timing, approximate BER under 3.8e-3 got. The BER performance of different received optical power with M-SVM is reflected in Figure 4 which indicates the feasibility of M-SVM in solving the phase rotation of CAP-based system. It found that phase rotation aggravate the judgment effect, traditional demodulation with rotator lower the BER about 3%, meanwhile M-ary SVM demodulation could reach a little bit better cause the Un-Sensibility to the timing jitters of clock. The BER performance of different transmitting length is investigated in Figure 5, setting 0 dBm launch power and 10Gbps transmitting data rates to ensure having objective comparison. It nearly identical in short length distance cause the fewer jitters influence and small phase rotation are not impacting greatly. As the longest of transmitting length, there is a trend that demodulation with M-ary SVM has a lower BER than the conventional with rotator. Higher transmitting length's performance is mainly caused by FIR filter taps. FIR filters with higher taps could get much better transmitting performance. The experiment is concentrate on short distance transmit which is mainly applied to access network, especially positive optical network whose equipment devices are required lower power consuming.

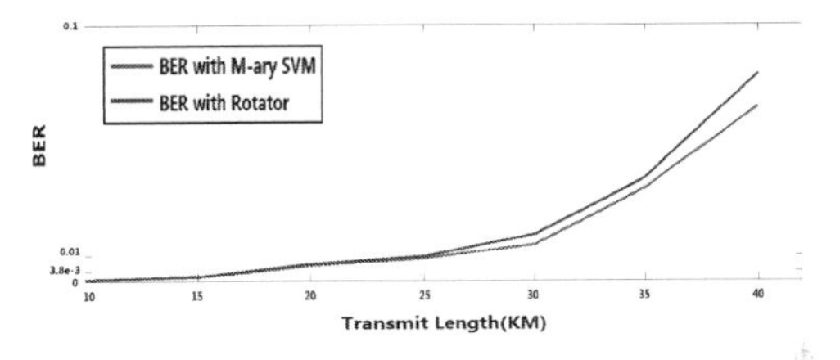

Figure 5. Different transmit distance on same launch power, M-ary SVM is better than conventional demodulation with rotator.

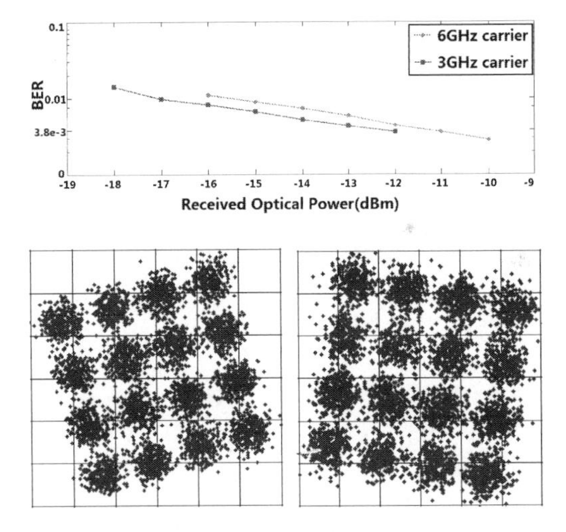

Figure 6. Simulation BER performance of multi-level and multi-band CAP system and received constellation.

The decoding solution expands to multi-level CAP signals. It is quite easy to move the baseband signal to high frequency using only different carrier multiply with function $g(t)$ listed before. The carriers of two pairs of FIR filters are set on 3GHz and 6GHz. Under the same parameters, Figure 6 shows the BER curve and received constellation diagrams. As single-level experiment, 500 symbols sequence is sent to train, support vector machines and define the transmit distance random, but under 50km.Through emulation, although different transmitting length and sample timing lead to unknown phase rotation, the M-ary SVM decode the signal accurately. For a practical implementation, a trade-off will be made between complexity and performance.

4 CONCLUSION

In this letter, the feasibility of a CAP-based system with Mary SVM to demodulate without phase recovery is demonstrated. M-ary SVM is a method to perform multiple category classification. The BER performance of 0dBm launch optical on different receives optical power is investigated. Compared with the conventional demodulation, former one can reach the same performance or better. Different transmit distance system performance is investigated that M-ary SVM fulfill the access network demand. Then multi-level CAP positive optical access network is simulated, indicating that multi-level CAP-based system with M-ary SVM demodulation can reach the demand in an ideal situation. The experiment and simulation show that the low complexity SVM algorithm allows the realization of high-speed short distance CAP based positive optical access network.

REFERENCES

Chen, W. Y., Im, G. H. & Werner, J. J. 1992. Design of digital carrier-less AM/PM transceivers. *AT&T and Bellcore contribution to ANSI T1E1.4*: 92–149.

Ingham, J. D., Penty, R. V. & White, I. H. 40 Gb/s carrier-less amplitude and phase modulation for low-cost optical data communication links. *Proc. OFC, Los Angeles, CA, USA, Mar.* 2011:1–3. Optic Society of America.

Kreßel, U. H. G. 1999. Pairwise classification and support vector machines in *Advances in Kernel Methods: Support Vector Learning*: 255–268. Cambridge, MA: MIT Press.

Li Minliang, Yu Song, Yang Jie, Chen Zhixiao, Han Yi & Gu Wanyi. 2013. Nonparameter nonlinear phase noise mitigation by using M-ary support vector machine for coherent optical system. *Photonics Journal* 5(6).

Li Tao, Wang Yiguang, Gao Yuliang, Alan Pak Tao Lau, Chi Nan & Lu Chao. 2013. 40 Gb/s CAP32 system With DD-LMS equalizer for short reach optical transmissions. *IEEE Photonics Technology Letters* 25(23): 2346–2349.

Sebald, D. J. & Bucklew, J. A. 2001. Support vector machines and the multiple hypothesis test problem. *IEEE Transactions on Signal Processing* 49(11): 2865–2872.

Wei, J. L., Ingham,J. D., Penty,R. V., White, I. H., & Cunningham, D. G. 2012. Update on performance studies of 100 gigabit Ethernet enabled by advanced modulation formats. *Proc. IEEE Next Gen 100G Opt. Ethernet Study Group, Sep. 2012* IEEE.

Electronics, Communications and Networks IV – Hussain & Ivanovic (eds)
© 2015 Taylor & Francis Group, London, ISBN: 978-1-138-02830-2

A buffer management policy combining token bucket with WFQ for DTN

H. Q. Wang, J.M. Zhu & G.S. Feng
College of Computer Science and Technology, University of Harbin Engineering, Harbin, Heilongjiang, China

ABSTRACT: Messages in Delay Tolerant Networks (DTN) are forwarded in the store-carry-forward manner. Aiming at the situation of each node's limited buffer and the requirement of different messages with different priorities in DTN, how to efficiently make use of each node's buffer and meanwhile provide differentiated services to the messages with different priorities, hereby not only improving the communication ability of the network but also keeping fairness of message forwarding, is a significant problem in DTN multiple-copy routing protocols (such as Epidemic routing). In this paper, we propose a new buffer management policy based on token bucket and Weighted Fair Queuing (WFQ), named BMP_T&WFQ, in which each node's buffer is divided into several queues according to the mechanism of combining token bucket with WFQ. The simulation results show that our policy realizes the fairness of queue scheduling and at the same time provides balanced sizes of each node's queues with the higher utilization of node's buffer.

1 INTRODUCTION

DTN is a new kind of network, which is proposed in order to overcome serious obstacles to IPN (interplanetary network) communication (Fall 2003). The characteristics of the communications in DTN mainly include intermittent connections, high latency, low transmissibility, inexistence of end-to-end path (Fu et al. 2013). On account of these characteristics, traditional routing protocols can't directly be applied. DTN deploys a store-carry-forward style routing (Xiao & Huang 2009). Meanwhile, for the sake of improving delivery ratio, multiple-copy routing protocols are generally adopted.

However, the limited buffer space of each node in DTN has imposed restrictions on network performance to a great extent. Thus how to management buffer of nodes has been one of the most important research topics (Rashid et al. 2013, Zhao & Chen 2013). Many buffer management policies for DTN are proposed from the perspective of the buffer replacement. When the buffer of one node is full, the proposed buffer management policies replace messages according to messages' certain characteristics. Both the differentiated services to the messages with different priorities and the fairness of queue scheduling were not taken into consideration by the above policies However, it is a necessary requirement that messages are classified and differentiated services are provided according to message importance under certain environments such as battlefield and disaster scene (Xie & Han 2014, Sun & Zhang 2012).

According to the above observations, we propose BMP_T & WFQ algorithm, a new buffer management policy based on token bucket and WFQ (Kurose 2005), in which each node's buffer is divided into several queues according to message priorities. Accordingly each queue is deployed a token bucket with a weight which varies as the number of tokens in the token bucket. When a node attempts to forward messages, the transmission right of the node is assigned to the queue which has the maximum weight. In order to realize differentiated services to the messages with different priorities and fairness of queue scheduling, higher priority queues have a higher probability to gain more tokens. At the same time, in order to improve utilization of node's buffer, BMP_T&WFQ also provides a dynamic balancer among each node's buffer queues, which performs balancing according to the historical state of each queue and the frequency of creating messages of different priorities.

2 RELATED WORKS

In consideration of the importance of the buffer management for DTN, many works have been done. Lindgren A (Lindgren & Phanse 2006) proposed FIFO, MOFO, and SHIL. They dropped messages according to the received time, the forwarded times and the lifetime of messages. The buffer management algorithm EBMP estimated the number of message replications by three message features (Shin & Kim 2011), and then replaced the message marked with the largest estimated number of replications. However, its complicated calculation processes brought some obvious side effects, such as large network overhead and high latency. Rashid S proposed MDC-SR algorithm which intended to control the number of the messages in each node. In order to reduce the number of unnecessary messages relayed and discarded, the algorithm defined a threshold to control the number of dropping messages (Rashid et al. 2013). However, for a static threshold, the network performance is unstable when the environment changed.

The algorithm DPMQ generated three dynamic queues for each node according to the estimated delivery ratio of messages (Rashid et al. 2013). And messages were buffered to different queues with different cases. But the algorithm only could be used in PROPHET routing and the applicable scenes were limited. Buffer replacement was performed on the basis of matching probabilities between the destination node and the node of carrying the message (Ye et al. 2010). The policy replaced the message with the minimum matching probability firstly. The paper proposed an adaptive buffer management mechanism based on message redundancy estimating (Wu et al. 2012). The mechanism made use of the node's activity degree and the number of message copies to predict the probability of successful delivering the message and replaced the message with the minimum probability. Yao Liu et al proposed a buffer management scheme based on estimated status of messages (Liu et al. 2011). The total number of copies in the network and the dissemination speed of a message were estimated on the basis of encounter histories.

Although many buffer management policies for DTN have been proposed, but most of them mainly utilize the characteristics of nodes or messages to perform buffer management. For example, DPMQ generated three dynamic queues, but it only considers the dynamic priorities of messages. Differentiated services to the messages with different priorities and the fairness of queue scheduling had been not taken into account by the above works.

3 BMP_T&WFQ BUFFER MANAGEMENT POLICY

3.1 *The main ideas of BMP_T & WFQ*

Token bucket mechanist can be used to limit the sending rate of network and WFQ can provide the fairness of message forwarding. Given that, token bucket and WFQ are applied to BMP_T&WFQ with a little modification. BMP_T&WFQ performs buffer management aiming at Epidemic routing (Mundur et al. 2008). The policy divides each node's buffer into *n* queues according to the priorities of messages. Each queue is assigned a token bucket with a weight which varies as the number of tokens in the token bucket. When a node attempts to forward a message, the transmission right of the node is assigned to the queue which has the maximum weight. BMP_T&WFQ buffer management policy is shown in Figure 1.

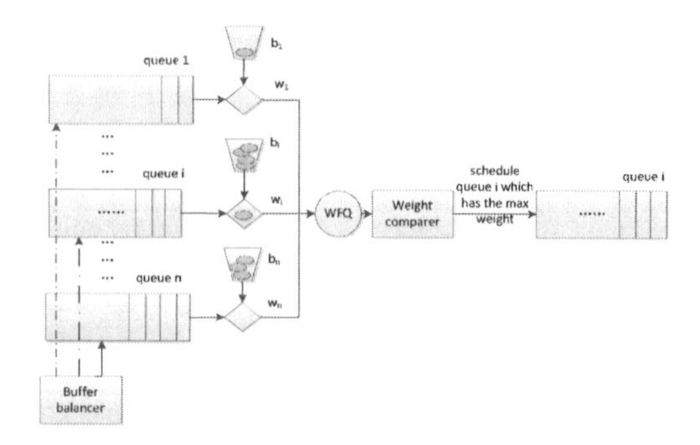

Figure 1. BMP_T&WFQ buffer management policy.

(1) Token allocation of each queue

$$token_i = \begin{cases} token_i - 1 & w_i > 0 \cap w_i > w_j, j \in \forall(1, i-1) \cup (i+1, n)) \\ b_i & w_j = 0, j \in \forall(1, n)) \end{cases} \tag{1}$$

(2) WFQ weight of each queue

$$w_i = 1 - (b_i - token_i) / (b_i - b_{i-1}) \tag{2}$$

(3) Transmission right allocation of queues in each node

The condition of queue j getting transmission right: wj>0 and wj is the maximum.

(4) Adaptive sizes of each node's queues

$$bS_i = \begin{cases} bS_i - \lambda_i * bSH_i & bSH_i > \theta_{threi} \\ bS_i + \sum_j \lambda_j * bSH_j & bSH_j > \theta_{threj}, j \in \forall(i+1, n) \cup \exists(1, i-1) \end{cases} \tag{3}$$

Note that λi is:

$$\lambda_i = 1 - nrofmessage_i / \sum_{j=1}^{n} nrofmessage_j \tag{4}$$

where bS_i is the size of queue i. bSH_i is the historical size of queue i and it is the average value of recording ten times its sizes. $qtre_i$ is the threshold of queue i carried out dynamic equilibrium, $nrofmessage_i$ is the number of messages in queue i.

3.2 *Algorithm*

The buffer management algorithm is shown in Figure 2 and Figure 3.

```
process weightingToken()
    begin
       for each weight of queue i
          do
            weight_i =1-(b_i-token_i)/(b_i-b_i-1);
          end
if(weight_0==0&&..weight_i==0..&& weight_n==0)
          for each weight of queue i
                do
                   token_i ← b_i;
                end
              // queue n get the transmission right
       token_n← token_n -1;
     end if
     else
           //get the max weight
            Weightcomparar();
// queue i who has the max weight gets the
transmission right
           token_i ← token_n -1;
     end else
   end
```

Figure 2. Weight computing algorithm.

```
process Bufferbalancer()
    begin
       for each queue i in this router
          do
λ_i=1-nrofmessage_i/sum(nrofmessage);
          end
       if(bSHidtory_i>θ_trei)
            bS_i= bS_i-λ_i* bSHistory_i;
       end if
       else
          from n start, find the first queue i whose
bSHistory_i<θ_trei;
          use formula(3) and (4)to balance each
queue buffer;
       end else
    end
```

Figure 3. Buffer balancer algorithm.

4 EMPIRICAL STUDY AND SIMULATIONS

In this empirical study, for the purpose of brevity, we assume each node's buffer be divided into three queues, named queue1,queue2 and queue3 ,which buffer the messages with low priority, middle priority and high priority respectively. Each queue is deployed a token bucket with a parameter of b_i. Note that $i=1...3$ and b_1=20, b_2=30 b_3=40. The token bucket of each queue is filled after initialization, that is to say token$_1$=20, token$_2$=30, token$_3$=50. For briefness to calculate, we assume b_0=15, b_4=0. Note that b_0 and b_4 are not calculated.

4.1 *Demonstration of differentiated services and fairness*

The procedures of changing of token and weight of each queue are presumed according to formula (1) and (2) as shown in Table 1.

Table 1. The procedures of changing of token and weight of each queue.

	queue1		queue2		queue3		Tran. right
	token	w1	token	w2	token	w3	
0	20	1	30	1	50	1	queue3
1	20	1	30	1	49	19/20	queue2
2	20	1	29	9/10	49	19/20	queue1
…							
…							
30	16	1/5	21	1/10	33	3/20	queue1
31	15	0	21	1/10	33	3/20	queue3
32	15	0	21	1/10	32	2/20	queue3
33	15	0	21	1/10	31	1/20	queue2
34	15	0	20	0	31	1/20	queue3
35	15	0	20	0	30	0	queue3
0	20	1	30	1	50	1	queue3

From Table 1, we can observe that queues3, queue2, queue1 get twenty times transmission right, ten times transmission right, five times transmission right, respectively. In consequence, the probability of scheduling queue3 is 0.57, queue2 is 0.29 and queue1 is 0.14. Hence, BMP_T&WFQ can realize differentiated services to the messages with different priorities. And transmission right allocation is on a round robin, which can avoid the queue with the same priority holding transmission right all the time. Therefore differentiated services and fairness can be achieved by BMP_T&WFQ.

4.2 *Simulations and results*

4.2.1 *Simulation environment*

In this paper, we adopt ONE (Opportunistic Networking Environment) (Keranen 2008) to evaluate our proposed buffer management policy BMP_T&WFQ. In this scenario, the simulation area is 4500m*3400m, simulation time is 12 hours and the total number of nodes is 42. The transmission speed of nodes is 250kBps and the transmission range is 10 meters. The buffer size of nodes is 5M. We place three groups of nodes that respectively simulate the move of pedestrians, cars and trains. Group1, Group2 both hold 20 nodes and Group3 hold 2 nodes. The movement speeds in Group1 and Group2 are 0.5m/s~1.5m/s and 2.7m/s~13.9m/s with shortest path map based movement model. The movement speeds in Group3 are 7~10m/s with map route model. And the message size is 400k.

4.2.2 *Simulation experiment*

In our simulation, the following metrics are used.

A relay ratio is defined as the ratio of the number of relayed messages to the number of created messages.

A delivery ratio is defined as the ratio of the number of delivered messages to the number of created messages.

1. The influence of simulation parameters on relay ratios

We configure the buffer size of each queue is one third of buffer size of each node. In Figure 4, queue1, queue2 and queue3 represent low, middle, and high priority queue respectively. Figure 4 and Figure 5 establish the influence of buffer size and message size on relay ratios without buffer balancer. As shown in Figure 4, the averages of relay ratios of three queues are 0.53, 0.29, 0.18 respectively, coinciding with the summary in section 4.1. We can detect that the relay ratios have no obvious change when the buffer size increasing It can be seen from Figure 5, the relay ratio of queue3 is the highest among three queues and the relay ratios of three queues have reduced as the message size increasing. This is because the more message size increased, the less quantity of stored messages in each node, which results in the lower probability of relaying messages correspondingly.

2. The influence of buffer balancer on delivery ratios

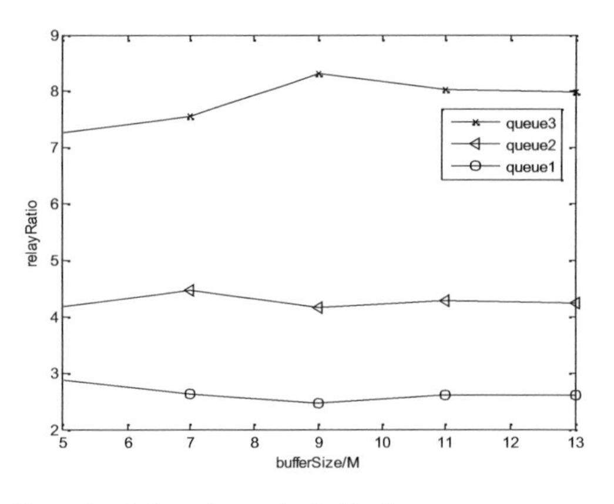

Figure 4. Relay ratio varying buffer size.

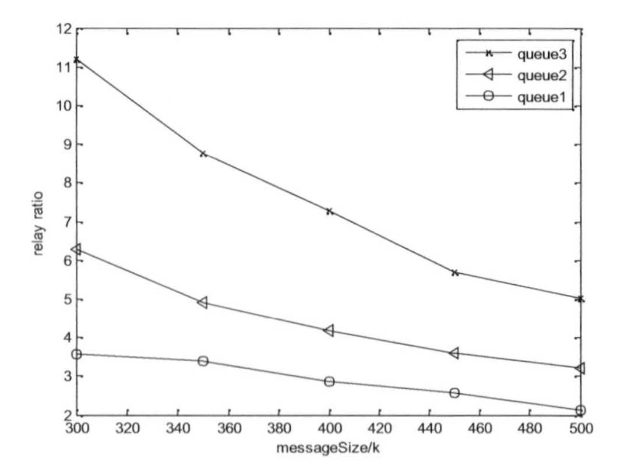

Figure 5. Relay ratio varying message size.

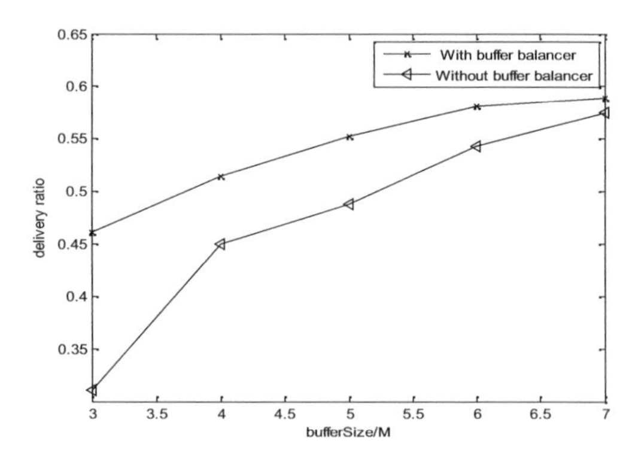

Figure 6. The total delivery ratio varying buffer size.

In order to verify the influence of buffer balancer on delivery ratios, we assume an extreme case where there are only two kinds of messages in the network, the messages with middle priorities and the messages with high priorities. In this case, we compare the delivery ratios of the networks with and without buffer balancer. The conditions of comparing are shown in Table 2.

Note that *bS* is the buffer size of each node in Table 2.

Figure 6 and Figure 8 show that the total delivery ratios with/without buffer balancer. Figure 7 and Figure 9 show that the delivery ratio of each queue

Table 2. Sizes of each queue with/without buffer balancer.

	With buffer balancer		Without buffer balancer
	Sizes of queue	θtre_i	Sizes of queue
queue1	1/3**bS*	1/4**bS*	0
queue2	1/3**bS*	1/2**bS*	1/2**bS*
queue3	1/3**bS*	3/4**bS*	1/2**bS*

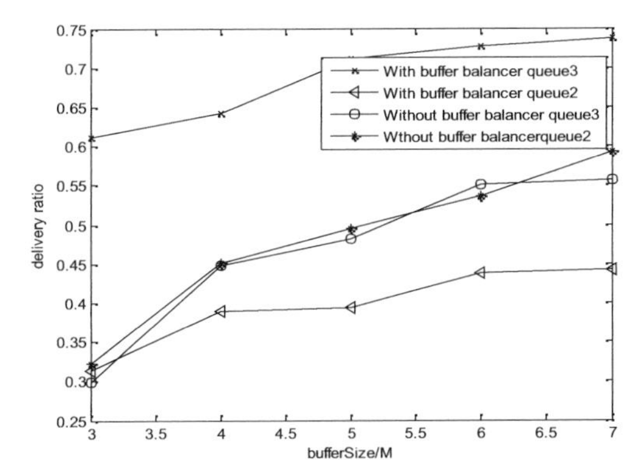

Figure 7. The delivery ratio of each queue varying buffer size.

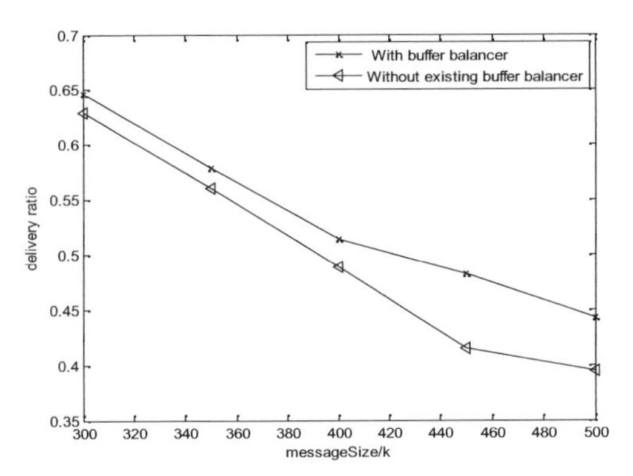

Figure 8. The total delivery ratio varying message size.

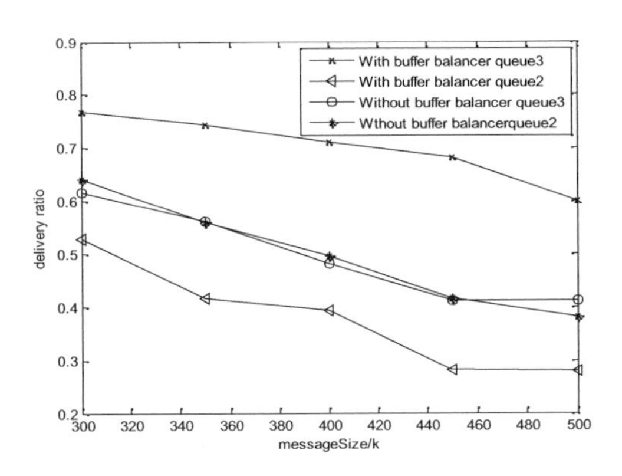

Figure 9. The delivery ratio of each queue varying message size.

with/without buffer balancer. From Figure 6, we can observe that the delivery ratios of two cases increase as buffer sizes increasing. And the delivery ratio with buffer balancer is higher. From Figure 7 and Figure 9, we can find out that the delivery ratio of queue 3 is the highest when buffer balancer is used. From Figure 8, it can be seen that the delivery ratios of two cases reduce as message size increasing. And the delivery ratio with buffer balancer is higher than that without buffer balancer.

5 CONCLUSION

This paper proposes a buffer management policy named as BMP_T & WFQ based on token bucket and WFQ. In the empirical study in section 4.1, we have proven that the queues with higher priorities hold more transmission rights and the transmission right allocation is on a round robin. The simulation results show that the queues with higher priorities can obtain higher relay ratios and the delivery ratio of the network with buffer balancer is higher than that without buffer balancer. Hence, the empirical study and simulation results both show that BMP_T & WFQ realizes the differentiated services to the messages with different priorities and fairness of queue scheduling and at the same time balances the sizes of each node's buffer queues with the higher utilization of node's buffer.

ACKNOWLEDGMENT

The paper is partially supported by the National Science Foundation of China (61370212), the Research Fund for the Doctoral Program of Higher Education of China (20122304130002), the National Natural Science Foundation of China (61402127), the Fundamental Research Fund for the Central Universities (HEUCFZ1213) and the Fundamental Research Fund for the Central Universities (HEUCF100601).

REFERENCES

Eyuphan Bulut & Boeslaw K. Szymanski. 2013. Secure Multi-copy Routing in Compromised Delay Tolerant Networks. *Wireless Personal Communications* 73(1): 149–168.

Fall, Kevin. 2003. A delay-tolerant network architecture for challenged Internets. *Proc of the ACM SIGCOMM, New York, August 2003*. ACM.

Fu Kai, Xia Jing-Bo & Yin Bo. 2013. Network situation-aware probabilistic routing algorithm in DTN. *Journal of Chinese Computer Systems* 34(1): 145–149.

Keranen, A. 2008. *Opportunistic Network Environment Simulator.* Helsinki: Helsinki University of Technology.

Kurose, J. F. 2005. *Computer networking: a top-down approach featuring the Internet*. India: Pearson Education.

Lindgren, A. & Phans,e K. S. 2006. Evaluation of queueing policies and forwarding strategies for routing in intermittently connected networks. *Communication System Software and Middleware, Comsware, January 2006*. IEEE: 1–10.

Liu, Y., Wang, J. & Zhang, S. 2011. A buffer management scheme based on message transmission status in delay tolerant networks. *Global Telecommunications Conference (GLOBECOM 2011), December 2011:* 1–5. IEEE.

Mundur, P., Seligman, M. & Lee, G. 2008. Epidemic routing with immunity in delay tolerant networks. *Military Communications Conference, MILCOM, San Diego, CA, 16-19 November 2008:*1-7. Baltimore: IEEE.

Rashid, S., Ayub, Q. & Zahid, M. S. M. 2013. Message drop control buffer management policy for DTN routing protocols. *Wireless personal communications* 72(1): 653–669.

Shin, K. & Kim, S. 2011. Enhanced buffer management policy that utilises message properties for delay-tolerant networks. *IET communications* 5(6): 753–759.

Sulma Rashid, Abdul Hanan Abdullah & Qaisar Ayub. 2013. Dynamic Prediction based Multi Queue (DPMQ) drop policy for probabilistic routing protocols of delay tolerant network. *Network and Computer Applications* 36(5): 1395–1408.

Sulma Rashid, Qaisar Ayub & M. Soperi Mohd Zahid. 2013. Message drop control buffer management policy for DTN routing protocols. *Wireless Personal Communications* 72(1): 653–669.

Sun Jianzhi & Zhang Yingxin & Chen Dan. 2012. Routing scheme of opportunistic network with Epidemic based on priority strategy. *Computer Engineering and Applications* 48(20): 108–111.

Wu Da-peng, Zhou Jian-er & Wang Ru-yan. 2012. Message-redundancy Estimating Adaptive Buffer Management Mechanism for Opportunistic Network. *Journal of Electronics & Information Technology* 1: 18.

Xiao Ming-jun & Huang Liu-sheng. 2009. Delay Tolerant Network Routing Algorithm. *Journal of Computer Research and Development* 46(7): 1065–1073.

Xie Lingjie & Han Xuedong. 2014. DTN routing protocol research based on multiple queues. *Computer Engineering And Design* 35(2): 376–380.

Ye, H., Chen, Z. G. & Zhao, M. 2010. ON-CRP: Cache replacement policy for opportunistic networks. *Journal of China Institute of Communications* 31(5): 98–107.

Zhao Guang-song & Chen Ming. 2013. Congestion control mechanism based on accepting threshold in delay tolerant networks. *Journal of Software* 24(1) : 153–163.

Electronics, Communications and Networks IV – Hussain & Ivanovic (eds)
 ISBN: 978-1-138-02830-2

Credibility assessment on agricultural information website

Xiaoqiao Wang, Kexi Wang* & Zhenjun Zhao
Management Department, Hunan University of Science and Technology, XiangTan 411201, China

ABSTRACT: Website credibility measurement is the basis and important content of network information evaluation so as to provide reliable information for users. This paper presents a web credibility evaluation index system with all indexes being integrated in fuzzy comprehensive evaluation. With some rape information websites being the experimental objects, the collection and analysis of experimental data are made. Methods to evaluate the website's credibility of comprehensive type and professional type of agriculture are obtained through calculating credibility of 43 agricultural comprehensive and professional websites containing rape information. Web site credibility evaluation model proposed in this paper can provide reference for others with the indexes and percentages being determined respectively in different actual situations.

1 INTRODUCTION

Agricultural information website is an important window and main source for users in agricultural industry, especially in rural areas to get information. It also contributes a lot to "agriculture" service and agricultural informatization. Due to a relative low level of rural users' knowledge and their inability to identify the authenticity of information, their demand for reliable information is urgent. Users' doubt in agricultural information service restricts the development of agricultural informatization, thus credibility evaluation is very important. Web site credibility measurement is the basis and important content of network information evaluation.

So far, credibility research has had more than 80 years of history. There is a continuing and systematic attention in academia in China, which attaches importance to the credibility research. However, there is still a big gap from the Western. Western media credibility research is divided into three major categories: source credibility, information / content credibility and channel credibility. The study on source credibility is the start and focus of credibility research. The current domestic research on credibility is mainly divided into two directions: one focuses on the credibility and reputation of information publisher such as websites and authors; the other is directed at the truth and accuracy research on text, image, audio and video objects. According to different objects, text fields include blog, email, web page and Wikipedia, etc. (Zhang & Lin 2011).

Information credibility research in China is an emerging and challenging research topic. There are not many scholars and papers in this aspect, and very little research has addressed the field of agricultural information. This paper applies the definition of literature (Iding et al. 2008), that is, the degree of faith or trust of users towards websites. On the basis of relevant research, this paper puts forward a kind of index system of website credibility evaluation, and searches for a comprehensive evaluation method of website credibility by collecting and analyzing experimental data of rape information website through fuzzy comprehensive evaluation method.

2 MATERIALS AND METHODS

2.1 *Evaluation standard of information credibility*

Metzger M summarized the influence factors of internet information credibility, which can be divided into four categories: "website characteristics", "information characteristics", "author characteristics" and "user characteristics".

The authors of this paper start the research from the angle of "website characteristics" and give the evaluation indexes and corresponding methods, then fuse the important indexes by fuzzy comprehensive evaluation method, finally obtain the comprehensive credibility rank of 43 agricultural information websites and summarize the assessment results.

2.2 *Evaluation index and method of website credibility*

In the study on trusted attributes, generally credibility should contain attributes such as "reliability",

*Corresponding author: *truewxq_2004@126.com*

"usability" and "safety" etc. (Kawai et al. 2008). According to the researchers' analysis, evaluation index system of website credibility in this paper is constructed by three main characteristics, "Reliability", "Usability" and "Safety".

1 Reliability. It refers to the level of satisfaction or trust of users towards the website. A second index is used to describe it such as "Importance", "Authority", "Popularity", "Webpage Quality", "Domain Suffix", "Connectivity", "Usage" and "Organization".
2 Usability. It refers to the websites' validity, efficiency and subjective satisfaction. A second index is utilized to describe it such as "Comprehensiveness", "Professional Relevancy", "Novelty", "Navigation", "Page design", "Friendly Level", "Searching" and "Advertising".
3 Safety. It refers to the system's ability to avoid being potentially dangerous or unstable state. A second index is selected to describe it such as "Credibility Sign" and "Confidentiality".

The detailed index and evaluation content of website credibility are shown in Table 1.

Table 1. Credibility evaluation index system.

The first index	The second index	Evaluation content
Reliability	Importance	PR Value
	Authority	Reverse Link Number
	Popularity	Alexa Ranking
	Webpage quality	Bounce Rate
	Domain suffix	Expert evaluation, standard: level of domain suffix
	Connectivity	Disconnected Chain Rate
	Usage	Daily IP Visits, Daily PV Views
	Organization	Expert evaluation, standard: level of website's organization
Usability	Professional Relevancy	Expert evaluation, standard: Professional Relevancy
	Comprehensiveness	Page Index Number
	Novelty	Expert evaluation, standard: update frequency
	Navigation	Expert evaluation, standard: navigation usability
	Page Design	Expert evaluation, standard: Page Friendly Level
	Friendly Level	Speed (Download Time, Response Time)
	Searching	Expert evaluation, standard: searching function in station
	Advertising	Expert evaluation, standard: Advertising
Security	Credibility Sign	Expert evaluation, standard: whether from the third-certification; signs of recommendation and credibility
	Confidentiality	Expert evaluation, standard:privacy protection strategy

2.3 *Assessment of website credibility based on the fuzzy comprehensive evaluation method*

Fuzzy comprehensive evaluation method converts qualitative evaluation into quantitative evaluation on the basis of Membership Theory in fuzzy mathematics.

2.3.1 *Establishment of factor sets*

1 Establish first class evaluation factors. The factor set U of evaluation objects is divided into m factor subsets, U={U1,U2,…,Um}; for the website credibility, the first class evaluation factors are set as Reliability, Usability and Security, U={U1,U2,U3}.
2 Establish secondary evaluation factors. According to the first class evaluation factors, the secondary evaluation factors are set as U1={U11,U12,…,U1n},…, Um={Um1,Um2,…,Umn}.

"Reliability" contains eight indexes, U1={Importance(U11), Authority(U12), Popularity(U13), Webpage Quality(U14), Domain Suffix(U15), Connectivity(U16), Usage(U17), Organization(U18)};

"Usability" contains eight indexes, U2={Comprehensiveness(U21), Professional Relevancy (U22), Novelty(U23), Navigation(U24), Page Design(U25), Friendly Level(U26), Searching(U27), Advertising(U28)};

"Security" contains two indexes,

U3={CredibilitySign(U31),Confidentiality(U32)}

2.3.2 *Establishment of evaluation rules*

Determine the evaluation value and relationship of evaluation factors:

1 Establish the evaluation set V. V={V1,V2,…, Vp} is the collection which results from evaluation.

The website credibility can be divided into four levels, V={ no credibility low credibility middle credibility high credibility}.

2 Establish a weight set W. W is the weight vector of the evaluation matrix with first class factors, reliability, usability and safety, $W=\{W_1, W_2, W_3\}$, Wi is the weight of the secondary factor W_{ij}, $W_i=\{W_{i1},W_{i2},\ldots,W_{ij}\},0\leq W_{ij}\leq 1$;

$$\sum_{j=1}^{ki} W_{ij} = 1,\ i=1, 2, 3; j=1, 2,\ldots,ki.$$

2.3.3 *Fuzzy comprehensive evaluation*

Combined with the secondary evaluation factors, each kind of factors Umn, the membership level of the four levels (rij1,rij2,rij3,rij4) and the evaluation results of ki factors, they can be expressed as ki×4 orders matrix.

$$R_i = \begin{bmatrix} R|u_1 \\ R|u_2 \\ \vdots \\ R|u_k \end{bmatrix} = \begin{bmatrix} r_{i1} & r_{i2} & \cdots & r_{i4} \\ ri & r_{i2} & \cdots & r_{i4} \\ \vdots & \vdots & & \vdots \\ r_{iki1} & r_{iki2} & \cdots & r_{iki4} \end{bmatrix}_{ki\times 4}$$

i=1,2,3 j=1,2,...,ki Ri is the first class evaluation factor of Ui, rijm is the membership level of Uij. According to the weight set W, one can get the first class fuzzy evaluation matrix of Ui:

$$B_i = W_i \circ R$$

$$= (w_1, w_2, \cdots, w_{ij}) \begin{pmatrix} r_{i11} & r_{i12} & \cdots & r_{i14m} \\ r_{i21} & r_{i22} & \cdots & r_{i24} \\ \vdots & \vdots & & \vdots \\ r_{iki1} & r_{iki2} & \cdots & r_{iki4} \end{pmatrix}$$

$$= (b_1, b_2, \cdots, b_{iki})$$

The fuzzy judgment matrixes of first class factors Bi(i=1,2,3) constitute the comprehensive evaluation matrix R. The secondary fuzzy comprehensive evaluation set can be obtained in a similar way. According to the principle of maximum membership level or other methods, the evaluation result can be achieved.

3 CASE ANALYSIS

3.1 *Evaluation objects*

Rape information websites are divided into two categories: the comprehensive agricultural websites containing rape information and professional rape information websites. The evaluation objects of this experiment are 43 websites containing rape information (Table 2) and there are 38 websites belonging to the first category (the agricultural websites of a province or direct-controlled municipality in China) and 5 websites belonging to the second category. Two groups of test indexes are designed in this paper, named as experiment I and experiment II.

3.2 *Selection of evaluation index*

In experiment I, four indicators such as "PR Value of website", "Reverse Link Number", "Alexa ranking" and "Page Index Number" are selected as important indexes to measure the website credibility by fuzzy comprehensive evaluation., and five indicators such as "PR Value of website", "Reverse Link Number", "Alexa ranking", "page Index Number" and "Professional Relevancy" are selected in experiment II.

It is found that the stability of indexes influences the evaluation. In the process of data collecting, the objective indexes such as "Bounce Rate", "Disconnected Chain Rate", "Usage", "Download Time" and "Response Time" fluctuate across time largely, and will influence the stability of the result, but the objective indexes such as "PR Value", "Reverse Link Number", "Alexa Ranking" and "Page Index Number" bring small fluctuations, so it will make the result more accurate and reliable.

Therefore, the influence factors of credibility mentioned in literatures (Du et al. 2010, Iding et al. 2008, Kawai et al. 2008, Lee et al. 2008, Li 2011, Ma & Yuan 2011, Staddon & Chow 2008, Wanas et al. 2008, Zhang & Lin 2011) and the stability of indexes during the collection process are taken into consideration. Experiment shows the credibility ranking of each website, and these indexes only reflect the comprehensive strength of websites without considering "Professional Relevancy" when evaluating the website credibility. So the "Professional Relevancy" index is added in experiment II and the page's amount gotten from searching the keyword 'rape' is approximately regarded as it. If the webpage related to the rape is few, even though the comprehensive strength index is perfect, then the website can't meet the professional requirements and its credibility declines accordingly. Therefore the weight of "Professional Relevancy" index should be relatively large.

3.3 *Data source*

1 Alexa website. In this paper, "Comprehensive Ranking" of a website and "Page Index Number" in Baidu are provided by Alexa.
2 Chinaz website. The "PR value" and "Reverse Link Number" in this paper are provided by Chinaz.

Table 2. Objects of website credibility evaluation.

Category	Website address	Organization
Comprehensive Agricultural information website	www.farmers.org.cn	Chinese Academy of Agricultural Information Institute
	www.moa.gov.cn	Chinese Ministry of Agriculture
	www.caas.net.cn	Chinese Academy of Agricultural Information Institute
	www.aweb.com.cn	Beijing agricultural fair Digital Technology Co., Ltd.
	www.new9e.com	Beijing nine hundred million net new agriculture information technology Co., Ltd.
	www.agri.gov.cn	Ministry of Agriculture Information center
	www.heagri.gov.cn	Hebei province agriculture information center
	www.haagri.gov.cn	market and Economic Information division of agriculture Department, Henan Province
	www.ynagri.gov.cn	Agricultural information center of Yunnan Province
	www.lnjn.gov.cn	Information center of Rural Economy Committee ,Liaoning Province
	www.hljagri.gov.cn	agriculture committee of Heilongjiang province
	www.hnagri.gov.cn	Agriculture Department of Hunan Province
	www.hnagri.com	science and Technology Information Research Institute of Hunan agricultural Academy of sciences
	www.ahny.gov.cn	Rural business management office of Anhui province agriculture committee
	www.sdny.gov.cn	Agriculture Department of Shandong Province
	www.xj-agri.gov.cn	Information Center of Xinjiang Agriculture Department
	www.jsagri.gov.cn	Agricultural Information Center of Jiangsu Province
	www.zjagri.gov.cn	Agriculture Department of Zhejiang Province
	www.jxagri.gov.cn	Agriculture Department of Jiangxi Province
	www.hbagri.gov.cn	Hubei Rural Information Propaganda Center
	www.gxny.gov.cn	Agricultural Information Center of Guangxi Zhuang Autonomous Region
	www.gsny.gov.cn	agriculture Department of Gansu Province
	www.sxnyt.gov.cn	Agriculture and animal husbandry Information Center of Shanxi Province
	www.nmagri.gov.cn	agriculture and Animal Husbandry department Information Center of Inner Mongolia Autonomous Region
	www.sxny.gov.cn	Agricultural Information Station of Shaanxi Province
	www.jlagri.gov.cn	agriculture committee of Jilin province
	www.fjagri.gov.cn	Agriculture Department of Fujian Province
	www.qagri.gov.cn	Agriculture Department of Guizhou Province
	www.gzaas.com.cn	Guizhou Academy of Agricultural Sciences
	www.qhagri.gov.cn	agriculture and animal husbandry Department of Qinghai Province
	www.scagri.gov.cn	Agriculture Department Information Center of Sichuan Province
	www.nxny.gov.cn	The Ningxia Hui Autonomous Region Department of agriculture and animal husbandry
	www.tjagri.gov.cn	Tianjin agriculture department
	www.cqagri.gov.cn	Chongqing Agricultural information center
	www.cdagri.gov.cn	Chengdu Agriculture Commission
	www.hznky.com	Hangzhou Research Institute of Agricultural Sciences
Professional rape Information website	www.d288.com	Chen Bingzhu (individual)
	www.bamudi.com	Liu Bing (individual)
	www.lvfangzhe.com	individual
	www.ymt360.com	Yi Cun Tong Da (Beijing) Network Technology Co.,Ltd.
	www.rapeseed.cn	Oil Crops Research Institute of Hunan Agricultural University
	www.zgycjjw.com	industry economic research Department of rape modern industrial technology system
	www.cnrapeseed.com	Jingzhou Fubang feed grain and Oil information Co., Ltd.

3 Manual work. Rape industry experts who are engaged in the research of agricultural information technology are invited to mark the "Professional Relevancy" index and weight of each factor. The highest and lowest scores of data are removed and then the average of the remaining data is taken as the index score.

4 EVALUATION RESULT AND ANALYSIS OF THE RAPE WEBSITE CREDIBILITY BASED ON FUZZY COMPREHENSIVE EVALUATION

Through statistical analysis of expert scoring tables, the index weight matrix in experiment I and II are obtained respectively as follows: W={0.4, 0.4, 0.1, 0.1} W={0.24, 0.24, 0.06, 0.06,0.4}. The credibility results of rape information website provided by Matlab are shown in Table 3 and all the index values are measured in December 19, 2012.

5 CONCLUSION

1 Experiment II shows that when considering website comprehensive strength, the better ones are "www.aweb.com.cn", "www.agri.gov.cn", and "www.ymt360.com", and their utilization is comparatively high. In general, the credibility of national websites is higher than local ones. The forefront of provincial websites ranking is as follows: Yunnan, Liaoning, Shandong, which shows that the credibility is related to the level of local agriculture informatization. The website "www.ymt360.com" got the highest score on the professional website, so it can be a reference for professional rape websites. The other professional rape websites are not good and have a large gap with the provincial websites.
2 Experiment shows that when considering website comprehensive strength and professional needs, better ones are listed as follows: "www.aweb.com.cn", "www.agri.gov.cn", "www.moa.gov.cn" and professional rape websites. The forefront of provincial websites is as follows: Yunnan, Liaoning, Shandong, Heilongjiang, Jiangsu, Hunan, Chongqing. The ranking for Hunan, Qinghai and Guizhou changes a lot (ranking up) in the two experiments. It shows that Hunan, Qinghai and Guizhou pay more attention to rape relatively.
3 The evaluation indexes in experiment are suitable for the credibility evaluation of comprehensive website. The comprehensive strength index is in favor of the comprehensive website evaluation, because the comprehensive website covers a large range of information and gets page-links easily

Table 3. Website credibility.

Website addresse	Exp I Composite score	Exp II Composite score
www.aweb.com.cn	0.08547	0.06000
www.agri.gov.cn	0.07692	0.06000
www.ymt360.com	0.06838	0.06000
www.moa.gov.cn	0.05128	0.05319
www.caas.net.cn	0.04282	0.04282
www.ynagri.gov.cn	0.03855	0.04787
www.bamudi.com	0.03855	0.03855
www.lnjn.gov.cn	0.03846	0.03846
www.sdny.gov.cn	0.03846	0.03846
www.hljagri.gov.cn	0.03778	0.03778
www.jsagri.gov.cn	0.03778	0.03778
www.cqagri.gov.cn	0.03628	0.03628
www.farmers.org.cn	0.03419	0.03419
www.zjagri.gov.cn	0.03419	0.03419
www.gxny.gov.cn	0.03419	0.03419
www.sxny.gov.cn	0.03401	0.03401
www.haagri.gov.cn	0.03275	0.03275
www.gsny.gov.cn	0.03175	0.03175
www.fjagri.gov.cn	0.03175	0.03175
www.hbagri.gov.cn	0.03023	0.03023
www.ahny.gov.cn	0.02913	0.02913
www.heagri.gov.cn	0.02771	0.02771
www.jxagri.gov.cn	0.02771	0.02771
www.hnagri.gov.cn	0.02589	0.03723
www.hnagri.com	0.02589	0.02589
www.xj-agri.gov.cn	0.02589	0.02589
www.sxnyt.gov.cn	0.02589	0.02589
www.nmagri.gov.cn	0.02589	0.02589
www.jlagri.gov.cn	0.02589	0.02660
www.qagri.gov.cn	0.02589	0.02660
www.qhagri.gov.cn	0.02589	0.03191
www.scagri.gov.cn	0.02589	0.03191
www.nxny.gov.cn	0.02589	0.02589
www.tjagri.gov.cn	0.02589	0.02589
www.new9e.com	0.02265	0.02265
www.gzaas.com.cn	0.02265	0.02265
www.cdagri.gov.cn	0.02265	0.02265
www.d288.com	0.02265	0.02265
www.hznky.com	0.01942	0.01942
www.cnrapeseed.com	0.01942	0.05319
www.rapeseed.cn	0.00647	0.05319
www.lvfangzhe.com	0.00504	0.05319
www.zgycjjw.com	0.00227	0.05319

compared with other type web sites. According to a certain question, the information on the single professional website is even more valuable than the comprehensive one. In addition, the comprehensive strength index is with a bias to the website existing for a long time, because these indexes need time to accumulate and the rank of new-built websites may be relatively lower, so it is unfair to them. In fact, the information on newly built website may be newer, and the users may be more interested in new information.

4 The evaluation indexes in Experiment II are suitable for the credibility evaluation of professional website. After adding the "Professional Relevancy" to it, the result changes greatly because of its large weight. In experiment II, two factors are considered: comprehensive strength of a website and professional requirements. Compared with the result of experiment I, the ranking of professional website has a big change, namely, the final position in experiment I become the forefront in experiment II. When considering the importance of user demand during the evaluation process, even if the comprehensive strength index of website is perfect, but it is seldom related to rape information, the website does not meet the professional requirements and its credibility is lower accordingly. Taken together, the user demand is added to the influence factor of credibility, and it will lay a foundation to evaluate the credibility of rape information.
5 For some ties in the result caused by the limitation of the fuzzy comprehensive evaluation method, it is suggested to take the tiles together to re-evaluate by evaluation method again, and the result will be more accurate. The fuzzy comprehensive evaluation method is more effective within ten objects.
6 During the process of data gathering, it is found that the existing professional rape website is few and some are unable to open. It shows that rape service platform is poor and the comprehensive strength is relatively low. It is imperative to construct a professional and credible rape service platform.

Taken together, the website credibility evaluation model proposed in this paper provides reference and basis for the construction of the credible rape information platform. All this work will lay a foundation for the subsequent evaluation of the rape information and it will promote the healthy development of the rape website.

ACKNOWLEDGEMENT

This research is supported by the project of Humanites & Social Sciences Planning Foundation of the Ministry Education of China (11YJA630124), project of Hunan Provincial Social Science Foundation (13JD20) and project of Hunan Provincial Natural Science Foundation (2015JJ4025).

REFERENCES

Du Jing, Li Daoliang & Li Hongwen. 2010. Research on evaluation system of agricultural information websites in China. *Jiangxi Journal of Agricultural Sciences*(3): 191–193.

Iding, M., Auernheimer, B. & Crosby, M. E. 2008. Towards a meta-cognitive approach to credibility determination. *Proc of the 2nd ACM Workshop on Information Credibility on the Web*: 75–80.

Kawai, Y., Fujita, Y. & Kumamoto, T. 2008. Using a dentiment map for visualizing credibility of news sites on the web. *Proceedings of WICOW 2008, California: ACM*: 53–58.

Lee, R., Kitayama, D. & Sumiya, K. 2008. Web-based evidence excavation to explore the authenticity of local events. *Proceedings of WICOW 2008, California: ACM*: 63–66.

Li Luyang. 2011. Network Text Oriented Research on Information Credibility. Harbin: Master thesis of Harbin Industrial University.

Ma Weiyu & Yuan Fang. 2011. Improved evaluation method of webpage information credibility based on PageRank. *Journal of Zhengzhou University: Science Edition* 43(2): 43–47.

Staddon, J. & Chow, R. 2008. Detecting reviewer bias through web-based association mining. *Proceedings of WICOW 2008, California: ACM*: 5–9.

Wanas, N., El-Saban, M., Ashour, H., et al. 2008. Automatic scoring of online discussion posts. *Proceedings of WICOW 2008, California: ACM*: 19–25.

Zhang Tianyu & Lin Hongfei. 2011. The multiple representation of military information credibility research. *Computer Engineering and Science* 33(9): 109–116.

Electronics, Communications and Networks IV – Hussain & Ivanovic (eds)
© 2015 Taylor & Francis Group, London, ISBN: 978-1-138-02830-2

An ultra-wide band CPW-fed antenna with WiMax/WLAN dual frequency band-notch function

X.D. Wu, W.B. Zeng*, J.J. Liao & S.L. Li
School of Electric and Information Engineering, Guangxi University of Science and Technology, Liuzhou, Guangxi, P.R. China

ABSTRACT: A planar Ultra-Wideband (UWB) antenna is proposed in this paper. Two U-shaped narrow slots are etched on the radiating patch to achieve dual notched-bands. By using the software of Ansoft HFSS, the proposed antenna has been analyzed and optimized, and then the antenna prototype is tested experimentally. Measurement results confirm that the proposed antenna obtains a good impedance matching along 2.2 – 11.08 GHz frequency band(return loss ≤ -10 dB) and is with two rejection bands over the frequency range of 3.2 – 3.7 GHz as well as 4.8 – 6.0 GHz, respectively. Besides, a relatively stable radiation performance and an appropriate antenna gain within the operating band are also achieved. Moreover, it is easy to control the notch-band characteristic of the antenna by varying the size of the U-shaped narrow slot.

1 INTRODUCTION

Since the FCC (Federal Communications Commission) announced that the spectrum 3.1 – 10.6 GHz is for ultra-wideband application, the increasing UWB application demands have stimulated many researches into the design of UWB antenna. The main factors for UWB antenna design include impedance matching, stability of the radiation patterns, and EMI (Electromagnetic Interference) problems, especially the compact size design. The UWB system can be interferenced by some wireless communication systems, such as WLAN (Wireless Local Area Networks which operates in the frequency range of 5.150 – 5.825 GHz) (Ren et al. 2007), WiMax (Worldwide Interoperability for Microwave Access, 3.40 – 3.69 GHz) (Jang et al. 2007). To overcome this problem, antenna with band-notched characteristic is suggested. Recently ,a lot of methods to design band-notched antenna have been proposed, these methods include attaching one or more slots to the antenna patch (Zhou et al. 2008), using U-slot defected ground structure , employing a split ring slot (Liao et al. 2010) and so on.

In this thesis, a compact planar UWB antenna as well as its dual notch-bands design is proposed. Within the frequency range of 2.2 –11.08 GHz, the measurement results have indicated that the antenna achieved a good impedance matching (with return loss not more than -10 dB) .In addition, two frequency bandwidths of 3.2 – 3.7 GHz and 4.8 – 6.0 GHz with return loss lager than -10 dB have been obtained. Simulation works confirm that two U-shaped narrow slots embedding in the radiation patch result in the dual notch-bands characteristic of the proposed antenna, and the performance of the band-notched can be controlled simply by adjusting the location and the size of the narrow slots.

2 CONFIGURATION OF THE ANTENNA

The configuration and a photograph of the antenna prototype are depicted in fig 1. The antenna is etched on the surface of a single-side FR4 PCB substrate with an overall size of 36×36×1 mm^3. The dielectric constant of FR4 PCB ε_r is 4.4, and the dielectric loss tangent tan δ is 0.02. A 50Ω CPW is introduced here to excite the radiation patch. Both the CPW feeding structure and the patch are symmetrical to the y-axis. Besides, a couple of rectangle wide slots are etched on the ground plane.

Two U-shaped narrow slots etched on the radiation patch are the key to achieve the two notch-bands for the proposed antenna. The antenna is designed and optimized by the software of Ansoft HFSS. Simulation work results indicate that the two center notched frequencies can be controlled simply by changing the total lengths of L_1 and L_2, as well as the widths of W_a and W_b, where L_1 is the sum of $2L_a$ and L_b, and L_2 represents the sum of $2L_c$, $2L_d$, $2L_e$, L_f. The optimized size of the antenna is: $L_a = 4$, $L_b = 8$, $L_c = 5$, $L_d = 4$, $L_e = 1$, $L_f = 10$, $L_g = 2$, $L_h = 4$, $L_i = 16.5$, $L_j = 3.5$, $L_k = 6$, $L = 36$, $W_a = 0.25$, $W_b = 0.25$, $W_c = 0.5$, $W_d = 2$, $W = 36$, $h = 1$ (in millimeter).

*Corresponding author: *ZWB@gxut.edu. cn*

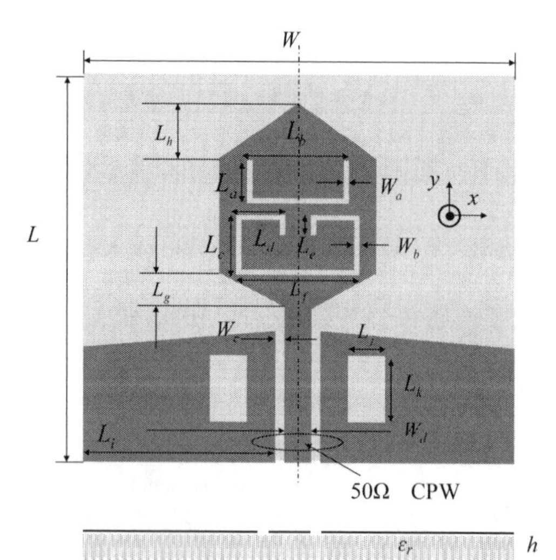

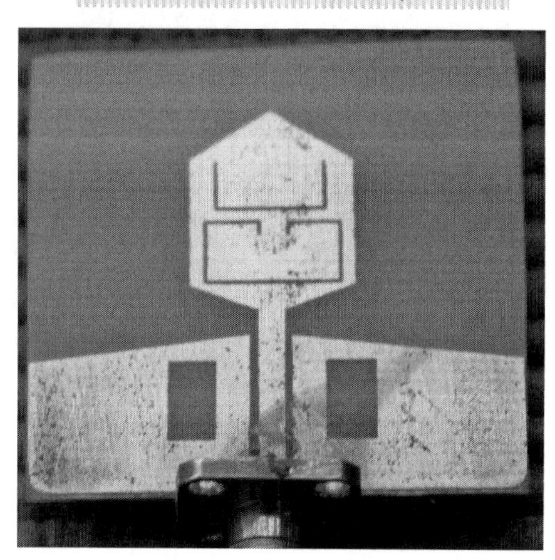

Figure 1. The geometry structure and a photograph of the antenna prototype.

3 RESULTS AND DISCUSSIONS

HP8510C network analyzer is applied to measure the impedance matching of the antenna. The results are illustrated in Figure 2. One can see clearly that the simulation bandwidth (return loss $\leq$ -10 dB) of the proposed antenna is 2.0 – 11.5 GHz with two notched bands (return loss $\geq$ -10 dB) of 3.2 – 3.7 GHz as well as 4.9 – 6.1 GHz. Moreover, an operation frequency bandwidth of 2.2 – 11.1 GHz (return loss $\leq$ -10 dB) with two notched frequency bandwidths (return loss $\geq$ -10 dB) of 3.2 – 3.7 GHz as well as 4.8 – 6.0 GHz are achieved by measurement. Both the simulation and measurement results agree well. The above results indicate that the proposed antenna can not only operate in UWB frequency band well but also reject two frequency bands (5.15 –5.825 GHz and 3.4 – 3.69 GHz), which are WLAN and WiMax operating frequency band, respectively.

The simulation radiation patterns at frequencies of 3.20, 3.50, 5.50 and 9.00 GHz are depicted in Figure 3. One can see clearly from (a) and (d) that the antenna is Omnidirectional radiation in the H-plane as well as a suitable gain at the frequencies of 3.2 and 9 GHz (which are within the UWB operating band). But, as shown in (b) and (c), the radiation patterns become worse and the gain decreases at the frequencies of 3.5 and 5.5 GHz, which are within the notched bands.

In order to study the effects of geometrical parameters on the band-notched characteristics as well as the impedance matching, parametric studies have been conducted. The geometrical parameters considering here are W_a, L_1, W_b and L_2. As shown in Figure 4, the bandwidth of the higher notched band increases with an increasing W_a, while the lower one does not vary significantly. Figure 5 shows that the central frequency of notched band goes up with a decreasing L_1. Furthermore, as it is shown in Figure 6, the bandwidths of both the upper and lower notched bands increase as W_b increases, and the total notched bands shift toward the higher frequencies with a decreasing L_2 as shown in Figure 7.

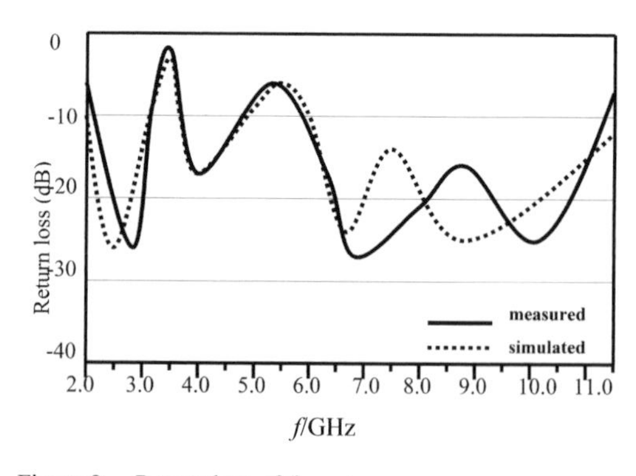

Figure 2. Return loss of the antenna.

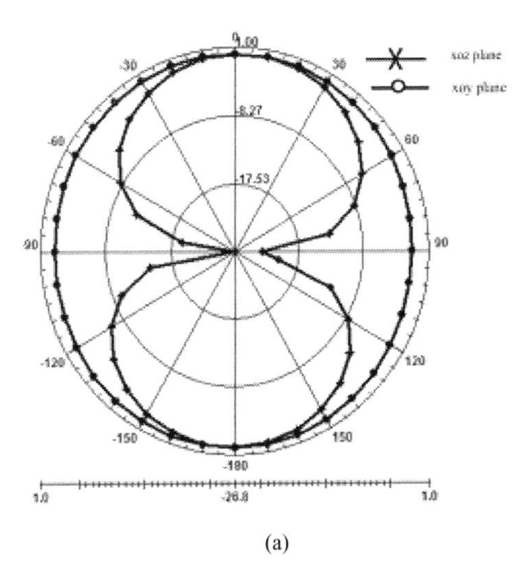

(a)

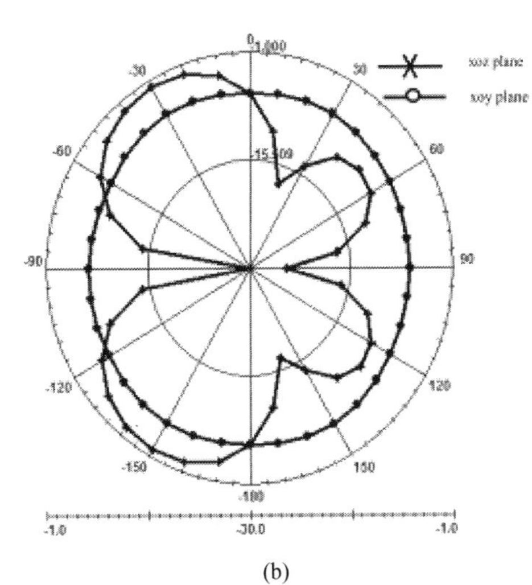

(b)

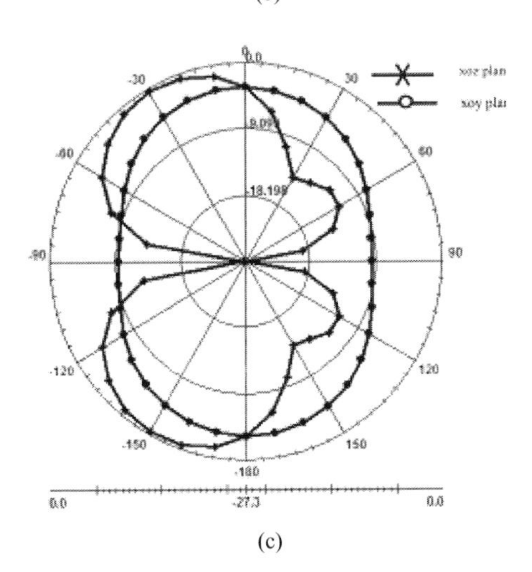

(c)

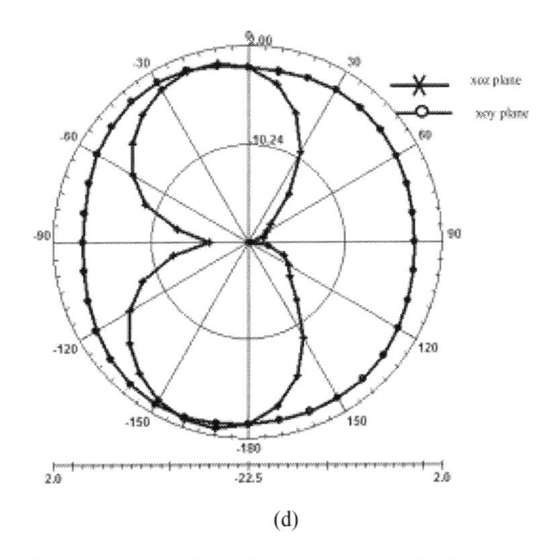

(d)

Figure 3. Simulation radiation patterns at the frequencies of (a) 3.2 GHz, (b) 3.5 GHz, (c) 5.5 GHZ and (d) 9 GHz.

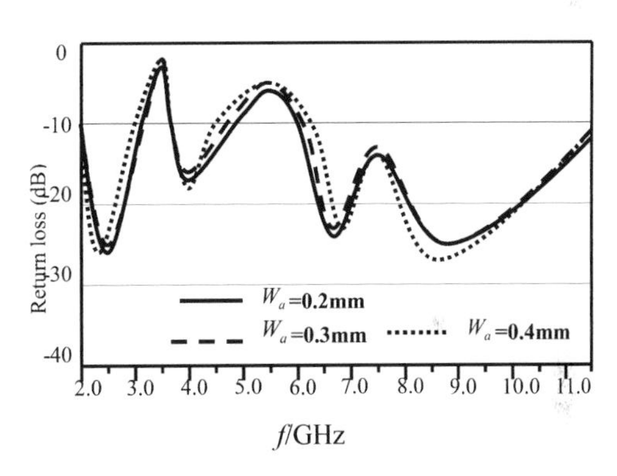

Figure 4. Simulation return loss with different W_a.

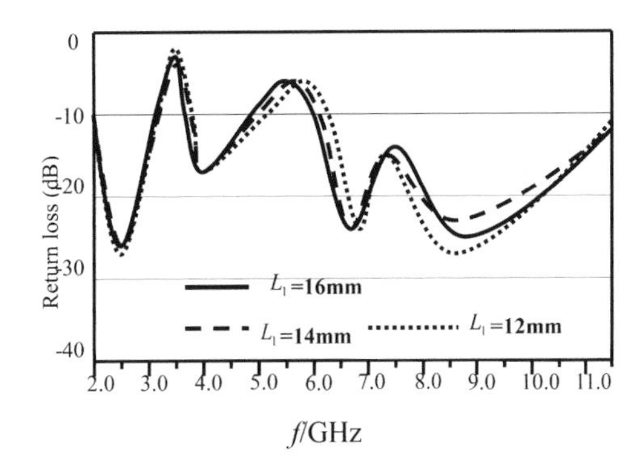

Figure 5. Simulation return loss with different L_1.

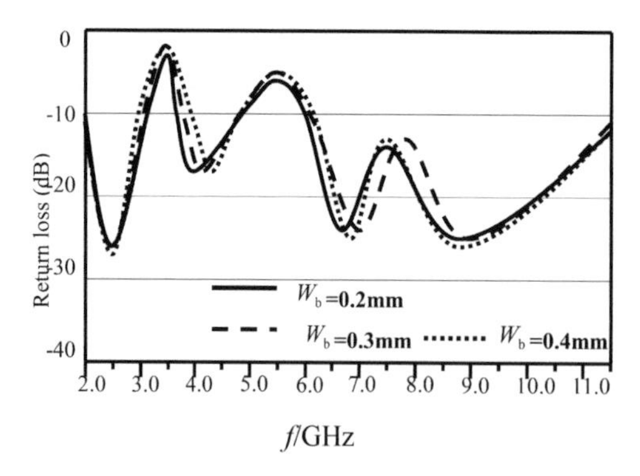

Figure 6. Simulation return loss with different W_b.

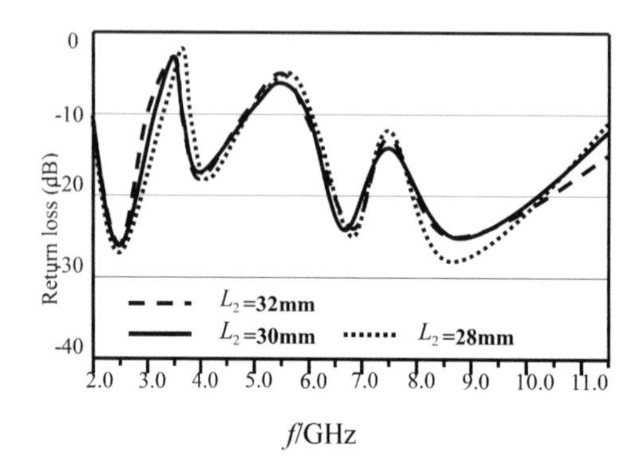

Figure 7. Simulation return loss with different L_2.

4 CONCLUSION

A planar antenna with two notch-bands for UWB applications is proposed in this paper. The antenna prototype achieves good matching along 2.2 – 11.08 GHz frequency band and rejects two bands of 3.2 – 3.7 GHz as well as 4.8 – 6.0 GHz. The dual notched-band characteristic is obtained by etching two U-shaped narrow slots on the radiating patch. Besides, a relatively stable radiation patterns in y-z plane as well as an appropriate antenna gain within the operating band are also obtained. The antenna is compact and easy to be produced, and it is a good solution for UWB wireless communication applications.

ACKNOWLEDGEMENT

This research work is supported by the research projects for higher education of Guangxi, China (NO.2010JGZ028).

REFERENCES

Jang, Y., Park, H., Jung, S. & Choi, J. 2007. A Compact band-selective filter and antenna for UWB application. *PIERS Online* 3(7): 1053–1057.

Liao, X.J., Yang, H.C., Han, N., et al. 2010. UWB antenna with dual narrow band notches for lower and upper WLAN bands. *Electronics Letters* 46(24): 1593–1594.

Ren, W., Deng, J. Y. & Chen, K. S. 2007. Compact PCB monopole antenna for UWB applications. *Journal of Electromagnetic Waves and Applications* 21(10): 1411–1420.

Zhou, H.J., Sun, B.H., Liu, Q.Zh., et al. 2008. Implementation and investigation of U-shaped aperture UWB antenna with dual band-notched characteristics. *Electronics Letters* 44(24): 1387–1388.

Electronics, Communications and Networks IV – Hussain & Ivanovic (eds)

 ISBN: 978-1-138-02830-2

Periodic continuous attractors of a population decoding model with external input

Weigen Wu*
School of Electronic Information Engineering, Pan Zhi Hua University, Panzhihua, Sichuan, China

Xing Yin
School of Mathematics and Computer Science, Pan Zhi Hua University, Panzhihua, Sichuan, China

ABSTRACT: The population decoding model is a good way to describe the encoding of external information in neural activity patterns of the brain. For further understanding the periodic information processing, this paper proposes decaying periodic continuous attractors of a population decoding model with external inputs. The model can obtain the decaying periodic continuous attractors if the synaptic connections are the decaying periodic functions, and some certain conditions are provided. An example is finally carried out to illustrate the developing theory.

1 INTRODUCTION

Recently, the population coding has attracted extensive attention because it can be used to understand how the external information is encoded in the neural system by populations. In the dynamic of neural system, population coding acts as the core of using the mathematical tool to model brain functions (Amari & Nakahara 2005) In mathematical model, continuous stimuli, such as eye position (Seung 1998), head direction (Zhang 1996), moving direction (Seung & Lee 2000), and the spatial location of objects can be encoded as continuous attractors in the brain. In population coding, continuous attractors of a population decoding model are generally to be of a Gaussian shape as the synaptic connections among neurons and the responses of the neurons are assumed to have a Gaussian distribution (Wu et al 2005, 2012, 2014). Moreover, continuous attractors with periodic activity have been studied in the population decoding model (Yu & Li 2013) based on a phenomenon that periodic membrane oscillations are typical for various brain regions (Cheu et al 2012).

However, initial vectors are regarded as external inputs and periodic continuous attractors don't have a center shape, such as the Gaussian shape in (Yu & Li 2013). Based on previous works (Yan 2014), this paper proposes decaying periodic continuous attractors of a population decoding model with external inputs.

The remainder of this paper is organized as follows. Some preliminaries are introduced in Section 2. Section 3 proposes some certain conditions for continuous attractors of RNNs with infinite neurons. An experiment is presented in Section 3. Finally, we draw conclusions in Section 4.

2 PRELIMINARIES

For convenience, $x\ (\xi,\ t)$ is denoted by the internal state of neuron ξ at time t. In order to study the dynamics of the population decoding model, a simple model is demonstrated by the following dynamic equation (3):

$$\dot{x}(\xi,t) = -x(\xi,t) + \frac{\int_{-\infty}^{\infty} \omega(\xi,\eta)x^2(\eta,t)d\eta}{1+\rho\int_{-\infty}^{\infty} x^2(\eta,t)d\eta} + h(\xi) \qquad (1)$$

for $t \geq 0$, where $\xi \in R$; ρ is a positive constant; $\dot{x}(\xi,t)$ denotes the first derivative of $x(\xi,\ t)$ with respect to t; $\omega(\xi,\ \eta) > 0$ denotes the synaptic connection between neuron ξ and neuron η. A neuron ξ can be fired by increasing the input from the neighboring neuron η through synaptic connection $\omega\ (\xi,\ \eta)$. Its gain is controlled by a synaptic connection of strength ρ. $h\ (\xi)$ denoting an external input to neuron ξ.

$|\cdot|$ is denoted by the absolute value of a real number and some definitions are listed.

Definition 1: A set $\{x^*(\xi) \mid \xi \in R\}$ is called an equilibrium point of (1), if it satisfies:

$$-x^*(\xi) + \frac{\int_{-\infty}^{\infty} \omega(\xi,\eta)x^{*2}(\eta)d\eta}{1+\rho\int_{-\infty}^{\infty} x^{*2}(\eta)d\eta} + h(\xi) = 0$$

for all $\xi \in R$.

*Corresponding author: *weigenwu@126.com*

Definition 2: An equilibrium $\{x^*(\xi) \mid \xi \in R\}$ of (1) is said to be stable, if given any $\varepsilon > 0$, there exists a $\delta > 0$, such that

$$| x(\xi,0) - x^*(\xi) | < \delta, \xi \in R$$

implying that

$$| x(\xi,t) - x^*(\xi) | < \varepsilon, \xi \in R$$

for all $t \geq 0$. And a real equilibrium is called unstable if it is not stable.

Definition 3: A set of equilibria C is said to be a stable continuous attractor of (1) if it is a connected set and each point $x^* \in C$ is stable. A set of equilibria C is said to be an unstable continuous attractor of (1) if it is a connected set and each point $x^* \in C$ is unstable.

3 CONTINUOUS ATTRACTORS

In this section, we will study the decaying periodic continuous attractors of a population decoding model (1). The decaying periodic connection between neuron ξ and neuron η is denoted by:

$$\omega(\xi,\eta) = \omega_{\max} \cdot \exp\left(2 \frac{\frac{3\sin(\xi-\mu)}{\xi-\mu} - 3}{\delta^2} \right) \quad (2)$$

where ω is a positive constant. Obviously, the connection has a center shape and Figure 1 shows the synaptic connection.

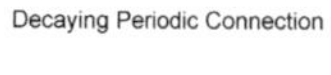

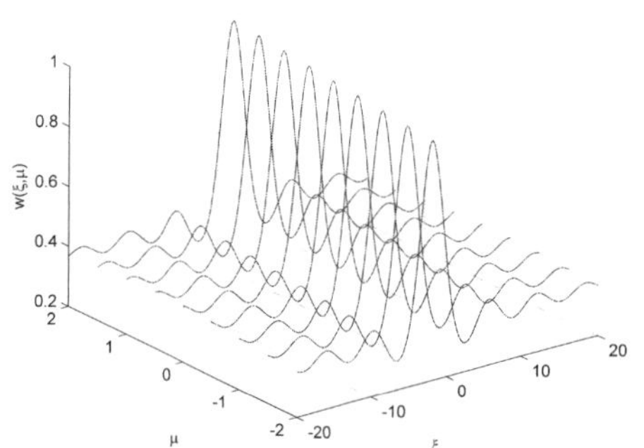

Figure 1. A decaying periodic connection between $\omega_{\max} = 1$ and $\rho = \sqrt{6}$.

The external stimulus is denoted by:

$$h(\xi) = h \cdot \exp\left(\frac{\frac{3\sin(\xi-\mu)}{\xi-\mu} - 3}{\delta^2} \right) \quad (3)$$

where h is a nonnegative constant and $\mu \in R$ is a free parameter corresponding to the memory stored in the network.

Lemma 1: Define $\gamma = 1 + \delta\sqrt{2\pi} x_{\max}^2(t) > 0$, $a = \sqrt{2\pi}\delta$, $b = -(\delta^2 \omega_{\max} + \sqrt{2\pi}\delta h)$, $c = 1$, and $d = -h$, each trajectory of RNN (1) can be represented as

$$x(\xi,t) = x_{\max}(t) \cdot \exp\left(\frac{\frac{3\sin(\xi-\mu)}{\xi-\mu} - 3}{\delta^2} \right), t \geq 0 \quad (4)$$

where $x_{\max}(t)$ is a differentiable function, which satisfies

$$\dot{x}_{\max}(t) \approx -\gamma(a x_{\max}^3(t) + b x_{\max}^2(t) + c x_{\max}(t) + d) \quad (5)$$

for all $t \geq 0$.

Proof of Lemma 1: Substituting (2-4) in the right-hand side of (1), it follows that:

$$-x(\xi,t) + \frac{\int_{-\infty}^{\infty} \omega(\xi,\eta) x^2(\eta,t)\, d\eta}{1 + \rho \int_{-\infty}^{\infty} x^2(\eta,t)\, d\eta} + h(\xi)$$

$$= -x_{\max}(t) \cdot \exp\left(\frac{\frac{3\sin(\xi-\mu)}{\xi-\mu} - 3}{\delta^2} \right) + \omega_{\max} x_{\max}^2(t) \cdot$$

$$\frac{\int_{-\infty}^{\infty} \exp\left(2 \frac{\frac{3\sin(\xi-\mu)}{\xi-\mu} - 3}{\delta^2} \right) \exp\left(2 \frac{\frac{3\sin(\eta-\mu)}{\eta-\mu} - 3}{\delta^2} \right) d\eta}{1 + \rho x_{\max}^2(t) \int_{-\infty}^{\infty} \exp\left(\frac{\frac{3\sin(\eta-\mu)}{\eta-\mu} - 3}{\delta^2} \right) d\eta} +$$

$$h \cdot \exp\left(\frac{\frac{3\sin(\xi-\mu)}{\xi-\mu} - 3}{\delta^2} \right)$$

By Taylor expansion

$$\exp\left(\frac{\frac{3\sin(\xi-\mu)}{\xi-\mu} - 3}{\delta^2} \right) \approx \exp\left(-\frac{(\xi-\mu)^2}{2\delta^2} \right)$$

we can have

$$\int_{-\infty}^{\infty} \exp\left(2 \frac{\frac{3\sin(\xi-\mu)}{\xi-\mu} - 3}{\delta^2} \right) \exp\left(2 \frac{\frac{3\sin(\eta-\mu)}{\eta-\mu} - 3}{\delta^2} \right) d\eta$$

$$\approx \int_{-\infty}^{\infty} \exp\left(-\frac{(\xi-\eta)^2 - (\eta-\mu)^2}{\delta^2} \right) d\eta$$

$$= \delta^2 \pi \exp\left(-\frac{(\xi-\mu)^2}{2\delta^2} \right)$$

and

$$\int_{-\infty}^{\infty} \exp\left(\frac{\frac{3\sin(\eta-\mu)}{\eta-\mu}-3}{\delta^2}\right) d\eta$$

$$\approx \int_{-\infty}^{\infty} \exp\left(-\frac{(\xi-\mu)^2}{2\delta^2}\right) d\eta = \delta\sqrt{2\pi}$$

then

$$-x(\xi,t) + \frac{\int_{-\infty}^{\infty} \omega(\xi,\eta) x^2(\eta,t)\, d\eta}{1+\rho\int_{-\infty}^{\infty} x^2(\eta,t)\, d\eta} + h(\xi)$$

$$\approx \left(-x_{max}(t) + \frac{\omega_{max} x_{max}^2(t)}{1+\delta\sqrt{2\pi} x_{max}^2(t)} + h\right) \cdot \exp\left(\frac{\frac{3\sin(\xi-\mu)}{\xi-\mu}-3}{\delta^2}\right)$$

thus, $x_{max}(t)$ must satisfy that

$$\dot{x}_{max}(t) \approx -x_{max}(t) + \frac{\delta^2 \pi \omega_{max} x_{max}^2(t)}{1+\delta\sqrt{2\pi} x_{max}^2(t)} + h \quad (6)$$

By the factorization, (6) can be written as:

$$\dot{x}_{max}(t) \approx -\gamma(a x_{max}^3(t) + b x_{max}^2(t) + c x_{max}(t) + d)$$

The proof is complete.

Based on the root formulas of cubic equation (Shengjin's formulas) (Fan 1989), the following theorem provides sufficient conditions for network (1) to possess stable and unstable continuous attractors.

Theorem 1: *Define* $A=b^2-3ac=(\delta^2\omega_{max}+\sqrt{2\pi}\delta h)^2 - 3\sqrt{2\pi}\delta$, $B = bc-9ad = 8\sqrt{2\pi}\delta h-\delta^2\omega_{max}$, $C = c^2-3bd = 1-3\delta^2\omega_{max}h-3\sqrt{2\pi}\delta h^2$, $\Delta = B^2-4AC$, $Y_1 = -A(\delta^2\omega_{max}+\sqrt{2\pi}\delta h)+\frac{3\sqrt{2\pi}\delta}{2}(B+\sqrt{B^2-4AC})$, $Y_2 = -A(\delta^2\omega_{max}+\sqrt{2\pi}\delta h)+\frac{3\sqrt{2\pi}\delta}{2}(B-\sqrt{B^2-4AC})$, $\gamma = B/A$, $\kappa = \arccos\left(\frac{-2A(\delta^2\omega_{max}+\sqrt{2\pi}\delta h)+3\sqrt{2\pi}\delta B}{2\sqrt{A^3}}\right)$ and $\Lambda_j =$

$$\left\{\lambda_j \cdot \exp\left(\frac{\frac{3\sin(\xi-\mu)}{\xi-\mu}-3}{\delta^2}\right) \mid \xi,\mu\in R\right\}, j=1,2,3$$

- **Case 1:** Denoted by $\lambda 1 = 3h$, if $A = B = 0$, then the set Λ_1 is a stable continuous attractor of (1);
- **Case 2:** Denoted by $\lambda 2 = \frac{(\delta^2\omega_{max}+\sqrt{2\pi}\delta h)-(\sqrt[3]{Y_1}+\sqrt[3]{Y_2})}{3\sqrt{2\pi}\delta}$, if $B^2-4AC > 0$, then the set $\Lambda 2$ is a stable continuous attractor of (1);
- **Case 3:** *Denoted by* $\varphi = \left\{\frac{(\delta^2\omega_{max}+\sqrt{2\pi}\delta h)}{\sqrt{2\pi}\delta}+\gamma, -0.5\gamma\right\}$, $\lambda 3 = max(\varphi)$ *and* $\lambda 4 = min(\varphi)$, if $B^2-4AC = 0$, then the set $\Lambda 3$ is a stable continuous attractor of (1) and the set $\Lambda 4$ is an unstable continuous attractor of (1);
- **Case 4:** Denoted by $\Phi = \{\frac{(\delta^2\omega_{max}+\sqrt{2\pi}\delta h)-2\sqrt{A}\cos(\frac{\kappa}{3})}{3\sqrt{2\pi}\delta}$, $\frac{(\delta^2\omega_{max}+\sqrt{2\pi}\delta h)-\sqrt{A}(\cos(\frac{\kappa}{3})+\sqrt{3}\cos(\frac{\kappa}{3}))}{3\sqrt{2\pi}\delta}$, $\frac{(\delta^2\omega_{max}+\sqrt{2\pi}\delta h)-\sqrt{A}(\cos(\frac{\kappa}{3})-\sqrt{3}\cos(\frac{\kappa}{3}))}{3\sqrt{2\pi}\delta}\}$, $\lambda^5 = max(\Phi)$, $\lambda^6 = \Phi - \{\lambda^5, \lambda^7\}$ and $\lambda^7 = min(\Phi)$, *if* $B^2-4AC < 0$, then the sets Λ^5 and Λ^7 are stable continuous attractors of (1) and the set Λ^6 is an unstable continuous attractor of (1).

Proof of Theorem 1 Similar to (Yu et al 2013, 2014), the proof is composed of two parts for proving that Λ^j is stable or unstable continuous attractor. In first part, we will prove that Λ^j is a continuous attractor. Clearly, Λ^j is a connected set. We only prove that each point of Λ^j is an equilibrium point.

Given any $\mu\in\boldsymbol{R}$, $x^*(\xi,\mu)$ can be represented as:

$$x^*(\xi,\mu) = \lambda_j \cdot \exp\left(\frac{\frac{3\sin(\xi-\mu)}{\xi-\mu}-3}{\delta^2}\right), j=1,\dots,7$$

for $\{x^*(\xi,\mu)|\xi\in\boldsymbol{R}\}\in\Lambda j$ ($j = 1,\dots, 7$). Substituting $x^*(\xi,\mu)$ into the right-hand side of (1), based on Shengjin's formulas (Fan 1989), we have

$$-\gamma(a\lambda_j^3 + b\lambda_j^2 + c\lambda_j + d) = 0$$

then

$$-x^*(\xi) + \frac{\int_{-\infty}^{\infty} \omega(\xi,\eta) x^{*2}(\eta)\, d\eta}{1+\rho\int_{-\infty}^{\infty} x^{*2}(\eta)\, d\eta} + h(\xi) = 0$$

By Definition 1, $\{x^*(\xi,\mu)|\xi, \mu\in\boldsymbol{R}\}$ is an equilibrium point. Clearly, Λ^j is a continuous attractor.

In second part, we prove whether each equilibrium of Λ^j is stable or unstable. Based on **Lemma 1**, the RNN (1) is rewritten as (5), so we take (5) into consideration.

The proof of **Case 1** is as following:

By the linearization, (5) at $\lambda 1$ is given by

$$d(x_{max}(t)-\lambda_1) = H_1 \cdot (x_{max}(t)-\lambda_1)^3 dt \quad (7)$$

where $H1 = -\gamma$ and $\gamma > 0$ is defined in **Lemma 1**.

Then it follows from (7) that

$$x_{max}(t)-\lambda_1 = (x_{max}(0)-\lambda_1)\cdot\sqrt{\tfrac{1}{2\gamma t+1}} \quad (8)$$

Because $\gamma > 0$, we have $0 < \sqrt{\frac{1}{2\gamma t+1}} < 1$ for all $t \geq 0$.

Given any $\varepsilon > 0$, there exists a $\sigma = \varepsilon$ for ξ, $\mu\in R$, and supposed $|x_{max}(0)-\lambda_1| < \varepsilon$, then we have that

$$|x(\xi,0)-x^*(\xi,\mu)|$$

$$= \left|x(\xi,0)-\lambda_1\cdot\exp\left(\frac{\frac{3\sin(\xi-\mu)}{\xi-\mu}-3}{\delta^2}\right)\right|$$

$$= \left| x_{max}(0) - \lambda_1 \right| \cdot \exp\left(\frac{\frac{3\sin(\xi-\mu)}{\xi-\mu} - 3}{\delta^2} \right)$$

$$\le \varepsilon$$

$$= \sigma$$

Thus, it holds that

$$\left| x(\xi,t) - x^*(\xi,\mu) \right|$$

$$= \left| x_{max}(t) - \lambda_1 \right| \cdot \exp\left(\frac{\frac{3\sin(\xi-\mu)}{\xi-\mu} - 3}{\delta^2} \right)$$

$$\le \left| x_{max}(t) - \lambda_1 \right|$$

$$= \left| x_{max}(0) - \lambda_1 \right| \cdot \sqrt{\tfrac{1}{2\gamma t+1}}$$

$$\le \varepsilon$$

$$= \sigma$$

By **Definitions 3**, $\Lambda 1$ is a stable continuous attractor. Similar to **Case 1**, the results of **Case 2**, **Case 3** and **Case 4** are easily obtained. The proof is complete.

4 EXAMPLE

Let us consider the RNN (1)

$$\dot{x}(\xi,t) = -x(\xi,t) +$$

$$\frac{16\pi}{27} \cdot \frac{\int_{-\infty}^{\infty} \exp\left(\frac{2\pi}{3} \cdot \left(\frac{\sin(\xi-\mu)}{\xi-\mu} - 1\right)\right) x^2(\eta,t)\, d\eta}{1 + 0.1\int_{-\infty}^{\infty} x^2(\eta,t)\, d\eta} +$$

$$\exp\left(\frac{2\pi}{3} \cdot \left(\frac{\sin(\xi-\mu)}{\xi-\mu} - 1\right)\right)$$

where $\delta = \frac{3}{\sqrt{2\pi}}$, $\omega(\xi,\eta) = \frac{16\pi}{27} \cdot \exp\left(\frac{2\pi}{3} \cdot \left(\frac{\sin(\xi-\mu)}{\xi-\mu} - 1\right)\right)$, $\rho = 0.1$, $h = \frac{1}{9}$.

Combined with ***Lemma 1***, $A = B = 0$, the (1) is rewritten as:

$$\dot{x}_{max}(t) \approx -x_{max}(t) + \frac{\frac{16\pi}{27} \cdot x_{max}^2(t)}{1 + 3x_{max}^2(t)} + \frac{1}{9} \tag{9}$$

for all $t \ge 0$. By the factorization, (9) can be written as:

$$\dot{x}(t) = -(1 + 3x_{max}^2(t))\left(x(t) - \frac{1}{3} \right)^3 \tag{10}$$

for all $t \ge 0$. This 1-D equation has four equilibrium points: $x_{max}^* = \frac{1}{3}$ (stable). Figure 2 shows the stability of the equilibrium point.

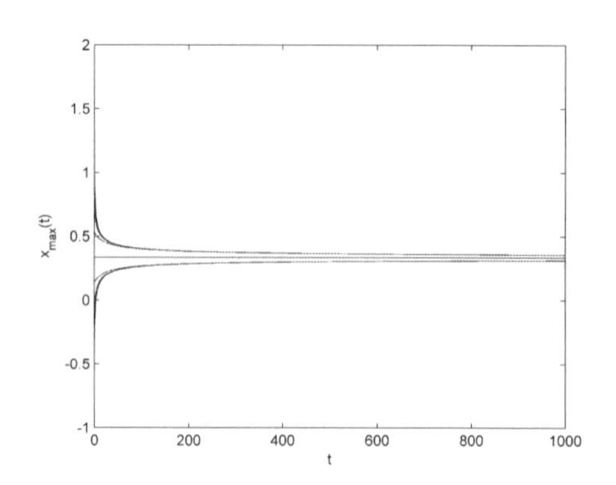

Figure 2. A stable equilibrium point of (9). $x_{max}^* = \frac{1}{3}$ is stable.

By **Definition 3**, the network possesses two stable continuous attractors $\Lambda 1$, where

$$\Lambda_1 = \left\{ \tfrac{1}{3} \cdot \exp\left(\tfrac{2\pi}{3} \cdot \left(\tfrac{\sin(\xi-\mu)}{\xi-\mu} - 1\right)\right) \mid \xi, \mu \in R \right\}$$

Then Figure 3 shows the continuous attractor Λ_1 (stable).

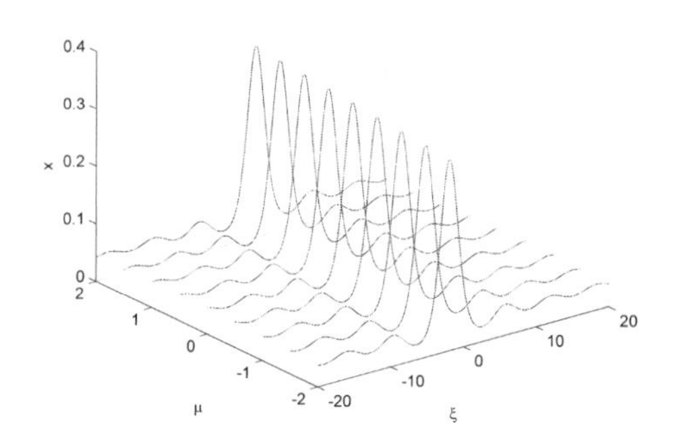

Figure 3. Continuous attractor of example. The red surface is the stable continuous attractor Λ_1. It is a decaying continuous attractor.

5 CONCLUSIONS

In this paper, we propose the decaying periodic continuous attractors of a population decoding model with external inputs. First, the population decoding model is written as a 1-D ordinary differential equation. Second, under certain conditions, this model can obtain some stable and unstable equilibrium points by the linearization. Finally, one example is carried out to illustrate the developed results.

REFERENCES

Amari, S. & Nakahara, H. 2005. Difficulty of singularity in population coding. *Neural Comput.* 17(4): 839–858.

Cheu, E. Y., Yu. J., Tan, C. H., & Tang, H. 2012. Synaptic conditions for auto-associative memory storage and pattern completion in Jensen et al.s model of hippocampal area CA3. *J. Comput. Neurosci.* 33(3): 435–447.

Fan, S.J. 1989. A new extracting formula and a new distinguishing means on the one variable cubic equation. *Natural Science Journal of Hainan Teacheres College* 2(2): 91–98.

Fung, C. A., Wong, K. M. & Wu, S. 2008. Dynamics of neural networks with continuous attractors. *Europhys. Lett.* 84, 18002.

Li, J., Yang, J., Yuan, X. & Hu, Z. 2014. Continuous attractors of higherorder recurrent neural networks with infinite neurons. *Neurocomputing* 131: 388–396.

Machens, C. & Brody, C. 2008. Design of continuous attractor networks with monotonic tuning using a symmetry principle. *Neural Computation* 20: 452–485.

Seung, H. S. & Lee, D. D. 2000. The manifold ways of perception, *Science* 290(5500): 2268–2269.

Seung, H. S. 1998. Continouous attractors and oculomotor control. *Neural Netw.* 11(7–8): 1253–1258.

Wu, S. & Amari, S. 2005. Computing with continuous attractors: Stability and online aspects. *Neural Comput. l.* 17(10): 2215–2239.

Wu, S., Hamaguchi, K. & Amari, S. 2008. Dynamics and computation of continuous attractors. *Neural Computation* 20: 994–1025.

Yan, L. 2015. Dynamics Analysis of A Class of Recurrent Neural Networks. *ICIC Express Letters* 9:(7).

Yu, J., Tang, H. & Li. H, 2013. Dynamics Analysis of a Population Decoding Model, *IEEE Trans. Neural Networks and Learning systems* 24(3): 498–503.

Zhang, K. C. 1996. Representation of spatial orientation by the intrinsic dynamics of the head-direction cell ensemble: A theory. *J. Neurosci.* 16(6): 2112–2126.

Electronics, Communications and Networks IV – Hussain & Ivanovic (eds)
© 2015 Taylor & Francis Group, London, ISBN: 978-1-138-02830-2

Reliability of narrow-band TD-LTE network transmission

Zanhong Wu
Power Grid Control Center of Guangdong, Guangzhou, China

Chao Cheng* & Jianming Zhang
Beijing University of Posts and Telecommunications, Beijing, China

ABSTRACT: Power communication service has widely used various communication technologies. With the development of the power grid to the smart one and the further development of enterprises informatization, the higher reliable transmission requirements on power communication is put forward. In the TD-LTE network, it can provide different quality of service (QoS) according to the user's level, business type and access location, to protect the reliable transmission of business. This article describes in detail the main module of PCC in the EPC network, specifically introduces each module function of PCC and QoS mechanisms in the EPS load, fully considers the business characteristics of the special power network, to improve the PCC architecture, ensuring the implementation of each QoS business. eNode B is adopted to control the functions of the RRC layer, and achieve the reliable transmission of each business.

1 INTRODUCTION

With the development of the construction of the smart grid, power business demand for reliability and security continues to improve, and power wireless private network construction attracts more and more attention. With the rapid development of wireless broadband communication technology, as a complement to the power cable of optical fiber communication, wireless communication's ability to support power with electricity side business has been on a large increase, more and more power electrical distribution service considers using a wireless communications carrier.

TD-LTE is the next-generation wireless broadband technology of China's independent intellectual property rights, using a variety of advanced wireless communication technologies such as Orthogonal Frequency Division Multiplexing (OFDM), Multi - Input Multi-Output (MIMO), a multi-channel smart antenna technology and so on, which effectively enhances the data throughput rate, coverage and the number of users online (Li 2014). Based on power private 230 MHz spectrum resources, spectrum aggregation technology is used to achieve the purpose of discrete narrowband broadband transmission, expanding the scope of application of wireless communication in the power industry, and carrier realizes the power information collection, distribution automation, video transmission and many other smart grid services and so on, at the same time, guarantees the reliable ability of power wireless broadband communication system service smart grid.

2 SMART GRID BUSINESS NEEDS

TD-LTE 230 power wireless communication system is faced with communications network applications for smart grid, primarily distribution electricity side of electricity information collection, distribution automation business.

Electricity terminal collecting device according to the type of the user is roughly divided into two kinds, one is single-phase resident and single-phase Industrial and Commercial users collector, the other is a three-phase Industrial and Commercial users special collector, the two different collecting devices own the same business type, but the corresponding business data volume is different. There are five types of business: real-time data acquisition, collecting, downlink control commands, parameter setting, event upload (Nian 2012).

Our distribution automation business originated in the late 1970s began electricity load control business, and distribution automation communications possess multi-point, dispersion, the characteristics of fewer amounts of communication data of each node. To reduce investment, optimize channel structure, usually a distribution substation is used to transmit the communication data of a region of distribution

*Corresponding author: *guangguang@bupt.edu.cn*

terminal, and distribution substation is set in power substation and switching station. However, laying fiber to each node will face the problem of the long construction period and investment cost, and singly laying fiber will not bring more economic benefit to the remote area (Wang 2010, Ou et al. 2012). The TD-LTE230 communication system only installs the private wireless communication terminal in distribution automation equipment, which can achieve the basic functions of distribution master station data collecting and monitoring and the expansion functions of analytical applications. In the scene of the distribution network, all kinds of terminals include Feeder Terminal Unit (FTU), Distribution Terminal Unit (DTU), Transformer Terminal Unit (TTU), Remote Terminal Unit (RTU). Using optical fiber and auxiliary wireless way realizes a complete coverage of the distribution automation network. Distribution automation involves the distribution equipment types, including column switch, switching stations, ring main unit, box-type substation and pole top transformer. According to business type, the very distribution equipment business can be divided into four categories: remote control, remote communication, remote measuring and power measurement (Quan 2012, Li 2012).

3 TD-LTE QOS MECHANISM

3.1 *PCC architecture*

In 3GPP TS 23.203 protocol, PCC architecture is divided into roaming and non-roaming scene, while different scenes can be further divided, but in the power LTE private network, the information collection device is fixed and will not move between different regions. Therefore, this paper mainly considers non-roaming scene, with its architecture shown in Figure 1:

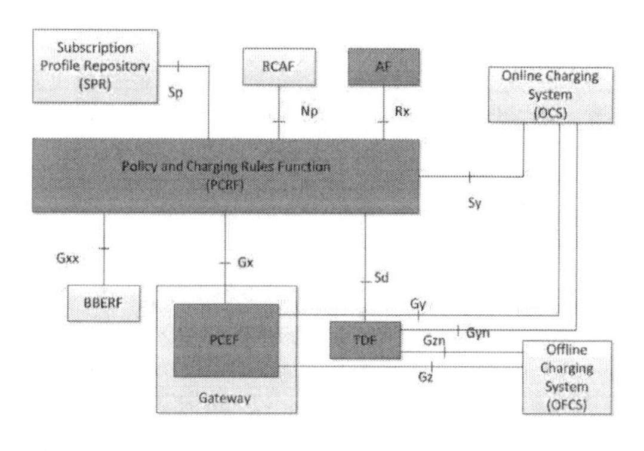

Figure 1. Non-roaming PCC architecture.

Architecture main function modules will be described in detail.

The PCRF module full name is Policy and Charging Rules Function, namely policy and charging rule function unit, and the functional entity contains the policy control decision and flow-based charging control functions. PCRF accepts input from Policy and Charging Enforcement Function (PCEF), SPR and AF, and provides PCEF for network control functions of service data flow detection, gating, QoS-based and flow-based accounting (except credit control). Combining the PCRF custom information makes PCC policy.

The PCEF module full name is Policy and Charging Enforcement Function, namely Policy and Charging Enforcement Unit. PCEF is a logical functional unit which implements QoS policies, and it accepts the QoS policy from PCRF (Kong 2011, Yuan 2012). The strategy is embodied in the radio access network air interface resource allocation command, so that the QoS policy of PCRF decision is implemented in the wireless side, in addition, the QoS policy will be mapped on IP packet header DSCP by PCEF, thus the uplink data stream QoS on the external IP network can be controlled. In the EPS system, GGSN completes the corresponding PCEF logic functions (Yuan 2012).

AF is the upper application part of the PS domain, depending on the application-related information to the PCRF by sending dynamic DIAMETER based on the Rx interface QoS request, the upper layer, through the API service provider interface, plays a role in AF, according to specific business requirements, QoS request issued to PCRF.

For power LTE private network, because the type of businesses almost changes, and there is no need to consider the accounting problems and every information collector should be treated equally, there is no distinction between user level security and other requirements. Therefore, the appropriate adjustments can be made to the PCC architecture according to the characteristics of the power LTE network. Detailed architecture is shown in Figure 2.

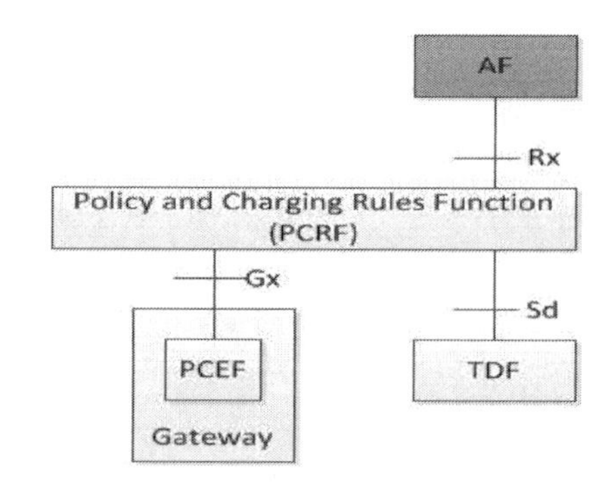

Figure 2. Electricity private network dedicated PCC architecture.

In dedicated PCC architecture, only AF module, PCRF module, PCEF module and TDF module are kept. According to 3GPP TS 23.401 protocol, modified dedicated PCC architecture is implemented as shown in Figure 3 in the power LTE private network, where PCEF is achieved by SGW / PGW NE, and TDF module is integrated into the PDW module, the AF server module is set up and configured by power companies.

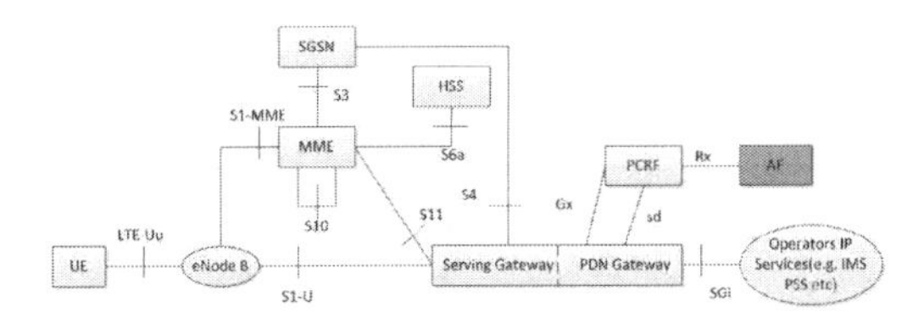

Figure 3. Electricity dedicated network element Figure.

3.2 *EPS radio bearer management*

RB management is mainly completed in the RRC signaling connection, where RB includes SRB0, SRB1, SRB2 and DRB on Uu mouth. Then the related content of RRC connection is introduced.

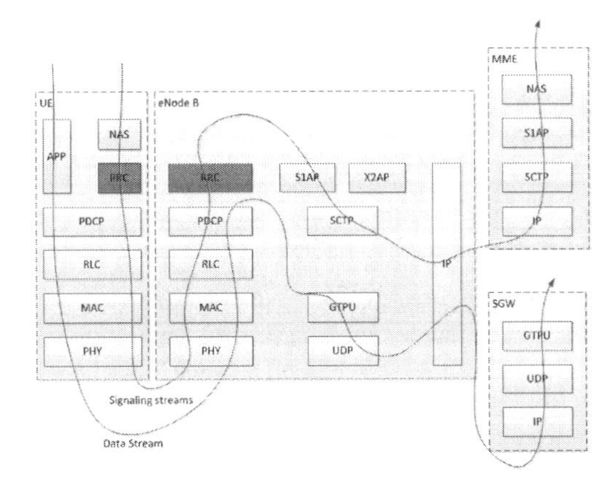

Figure 4. RRC protocol architecture.

Figure 4 shows that RRC is protocol entity which manages radio bearer, through the interaction of RRC signaling to realize the function of establishment, modification and release of RB. Popular speak RRC connection is established SRB1 between UE and eNode B (Zhang 2009). The RRC signaling control UE transmits the corresponding data stream.

Whether the RRC connection setup message, RRC connection reconfiguration message or RRC connection reconstruction message, eNode B carries out the same configuration for the radio resource configuration in an Information Element (IE). When adding or modifying the list of DRB in private radio resource configuration, RRC will assign a radio bearer's identity, and also gives it a logical identity at the same time. The various business types on the radio bearer transmissions are known. Meanwhile, RRC also through signaling LogicalChannelConfig priority fields determines logical channel priority and prioritized BitRate field determines the value of PBR, so that different wireless carriers provide a logical channel priority level.

4 POWER TD-LTE PRIVATE NETWORK RELIABILITY SECURITY SCHEMES

4.1 *DSCP and QCI mappings*

DSCP owns 64 priorities, and network transport layer nodes (routers, SGW PGW) can be controlled QoS by DiffServ system. In the base station side, for the uplink and downlink data stream, just the wireless RB mapping S1 interface GTP tunnels are made to pass through, which does not resolve the IP packet, its QoS classification by QCI class (Yuan 2012).

According to the different QoS, EPS Bear can be divided into two categories: GBR (Guranteed Bit Rate) and Non-GBR. The so-called GBR is that network allocates a "permanent" constant bit rate for the bearer requirements, even in the case of nervous network resource, and the corresponding bit rate can be maintained. Maximum Bit Rate (MBR) parameter defines that GBR Bear can achieve the speed limit under the condition of adequate resources. MBR value is likely to be greater than or equal to GBR value. In contrast, Non-GBR, the traffic (or bearing) needs to withstand reduction rate requirements under network congestion, each of the Non-GBR traffic can be preset as the minimum guaranteed bit stream on the base station side, in order to ensure that the Non - GBR business won't be starved to death because access GBR business takes too much resources. Where the MBR is set in the AF, the minimum GBR can use the method of presetting at the base station.

In the access network, the air interface bearer QoS is controlled by the eNode B, and every bearer has a corresponding QoS parameter QCI (QoS Class Identifier) and ARP (Allocation And Retention Priority).

In the LTE system, lower QCI value is corresponding to better reliability. QCI values corresponding to the QoS guarantee the reliability is respectively shown in Table 1.

Table 1. QCI index.

QCI Index	Resource Type	Priority	Packet Delay Budget	Packet Error Loss Rate
1	GBR	2	100ms	10^{-2}
2		4	150ms	10^{-3}
3		3	50ms	10^{-3}
4		5	300ms	10^{-6}
5	non-GBR	1	100ms	10^{-6}
6		6	300ms	10^{-6}
7		7	100ms	10^{-6}
8		8	300ms	10^{-6}
9		9	300ms	10^{-6}

Where QCI index 4 corresponds to the business of remote control, remote communication, remote measuring and downlink control commands, index 5 corresponds to the parameter settings, index 6 corresponds to collecting, index 7 corresponds to the real-time data acquisition and degrees of ammeter, index 8 corresponds to the event upload. Business should be the same in DSCP and QCI rank order. This ensures a high priority real-time requirements business access to a network, and business can enjoy better resource allocation, and the priority forward right at the core network.

4.2 *Access success rate*

4.2.1 *Business satisfaction rate calculation*

1 GBR business satisfaction rate calculation method

Among the indicators of QoS, the delay is generally guaranteed by the MAC scheduler, L2 packet loss rate is guaranteed by ARQ and HARQ mechanisms, so for GBR business, GBR is the measure indicator of the best satisfaction, when data are transmitted, the business rate is higher than the guaranteed rate of GBR, whether business QoS is satisfied, or is not satisfied.

$$S_{GBR} = 100\% - \frac{1}{Sum_{GBR}} \times \sum \frac{D_i}{GBR_i} \qquad (1)$$

Where S_{GBR} is the GBR business satisfaction rate (expressed as a percentage), and D_i is no-empty GBR business total number of i-th RLC buffer queue. Formula concepts are to calculate the non-satisfaction rate of each GBR business of data transmission (the ratio of the GBR rate difference and GBR rate), then find the average of non-satisfaction rate, and use 100% satisfaction rate to subtract above value. The GBR business satisfied rate will then be obtained. For simplicity, the above calculation S_{GBR} formulas for all GBR business are the same, and are used for summation averaged, and without using a weighted averaging, the formulas can be optimized if necessary.

D_i is calculated as follows:

$$D_i = \begin{cases} 0 & , R_i \geq GBR_i \\ GBR_i - R_i, & R_i < GBR_i \end{cases} \qquad (2)$$

Where R_i is the i-th GBR traffic throughput of the statistical time period, the formula means that when the GBR traffic statistics throughput is higher than the GBR rate in the period, this business is the satisfaction, the GBR difference rate is zero; otherwise GBR difference rate should be calculated. Note that the above formula is only for RLC buffer queue and is not an empty business which is effective, the RLC buffer queue is empty business without a corresponding D_i value, nor being involved in the calculation.

2 Non-GBR business satisfaction rate calculation method.

Introduced in non-GBR business minimum rate MinBR≠0, the rate is defined when all non-GBR traffic data is transmitted, and the least MinBR rate service will be provided for each non-GBR business. When the non-GBR business does not drop, it means that business is satisfied.

Non-GBR business satisfaction rate is calculated as follows:

$$S_{non-GBR} = \begin{cases} 100\% & , \sum R_i \geq R \\ \dfrac{\sum R_i}{MinBR \times Sum_{non-GBR}} & , \sum R_i < R \end{cases} \qquad (3)$$

Where $S_{non-GBR}$ is non-GBR business satisfaction rate, R_i is the no-empty non-GBR traffic throughput of i-th RLC buffer queue in a time period, $Sum_{non-GBR}$ is the total number of the no-empty non-GBR business of the RLC buffer queue. Formula means: when the average rate of data transmission of all non-GBR businesses is over MinBR, satisfaction rate is 100%; otherwise, their satisfaction rate is that the average rate of the MinBR radio.

In practical applications, there are two methods of user satisfaction and the time curve can be obtained. In the actual operating curve, it is recommended, at the base station architecture, to use the base of user satisfaction snooping statistics GBR and non-GBR service business satisfaction rate curve.

4.2.2 *Access judgment rule*

It is known that both GBR business judgment rule and non-GBR business judgment rule are the same. Therefore, this introduces access GBR judgment rule.

Downlink access GBR business judgment rule is as follows:

Definition downlink GBR business access two thresholds: $T_{DL\text{-}HO\text{-}GBR}$ (which allows switching service access downlink GBR threshold) and $T_{DL\text{-}New\text{-}GBR}$ (GBR allows new businesses downstream access threshold). And $T_{DL\text{-}HO\text{-}GBR} < T_{DL\text{-}New\text{-}GBR}$, is made to ensure priority access GBR switching operations. $T_{DL\text{-}HO\text{-}GBR}$ and $T_{DL\text{-}New\text{-}GBR}$ can be preset at the base station.

When $S_{GBR} \geq T_{DL\text{-}New\text{-}GBR}$:

If the calculated level of other QoS (such as business or NON-GBR business GBR) satisfaction level is higher than its corresponding QoS switching threshold, the case of a downlink GBR new business and switching operations are allowed access.

If any one of the calculated QoS satisfaction levels of service (such as GBR traffic or NON-GBR service) has its own QoS level lower than the switching threshold, in this case, the new service is not allowed access to the downlink GBR switching operations allow access.

When $T_{DL\text{-}New\text{-}GBR} > S_{GBR} \geq T_{DL\text{-}HO\text{-}GBR}$:

The new business at this time that does not allow access to the downstream GBR switched services can access.

When $T_{DL\text{-}HO\text{-}GBR} > S_{GBR}$:

At this time of the downlink GBR new business and switching operations are not allowed access.

GBR Uplink business access: principle above, just replace the threshold: $T_{UL\text{-}HO\text{-}GBR}$ and $T_{UL\text{-}New\text{-}GBR}$, and $T_{UL\text{-}HO\text{-}GBR} < T_{UL\text{-}New\text{-}GBR}$.

4.2.3 *Data collection success rate*

When a business needs to meet a specified time collecting success rate m% actually, it should meet that the business service satisfaction rate is greater than the percentage of time quasi-entry limit which is greater than or equal to m%. This is done through setting reasonable limits and business MBR, QCI, GBR equivalent. MinBR and GBR should try to follow the basic needs of the business to set the bandwidth, making the fullest use of the resource as follows:

Set this business template information, MBR value (GBR type of business), DSCP value in the AF module,

Set the MinBR value of different services in the base station.

The business satisfaction rate curve is set to meet the acquisition success rate which is greater than threshold.

4.3 *Radio resource allocation*

Through a dedicated PCC architecture, eNode B can know business type of each data packet, while RRC assigns logical channel priorities for the radio bearer business. Figure 4 shows that the data flows through the PDCP layer, RLC layer, MAC layer and the PHY layer to transmit, our main concern is the MAC layer multiplexing. MAC layer is responsible for the plurality of logical channels (logic channel) multiplexed onto the same transport channel (transport channel). Every TTI UE can send a MAC PDU, but RLC SDU from a plurality of logical channels can put in the same MAC PDU, which is the origin of the MAC multiplexed. MAC layer multiplexing aims to achieve, through the logical channel priorities for each radio bearer to provide the service, in order to determine the generated MAC PDU which contains the data of the logical channel, and various logical channels included in the total.

MAC PDU only has one, but there are many multiplexed logical channels, which requires the assignment of a priority level for each logical channel. The highest priority logical channel data is preferentially contained in the MAC PDU, then the second highest priority logical channel data, and so on, until the allocated MAC PDU is full, or no more data can be sent. However, this allocation may make a high priority of the logical channel which has always occupied a wireless resources that are allocated to the UE by eNode B, resulting in a low-priority logical channel being "starved to death." To avoid the above situation, MAC layer uses a similar token bucket (token bucket), which achieves MAC multiplexed algorithm, and the basic idea of the algorithm is based on whether there is a token in the token bucket and a token to determine whether the number of transmitting data of a logical channel controls the amount of the logical channel data in the MAC PDU. Therefore, LTE introduces PBR (Prioritized Bit Rate) concept, namely the allocation of resources to the logical channel prior, the data rate of each logical channel is configured, thus providing a minimum data rate for each logical channel, avoiding the low priority logical channel being "starved to death."

5 CONCLUSIONS

At present, the domestic power grid has built reliable electric power communication network, formed mainly optical fiber communication. The coexistence of means of microwave communication, carrier, satellite, etc, self-healing network at different levels is the main feature of power private communications network architecture. Construction of power TD-LTE wireless broadband private network, will greatly promote the development of smart grid. The paper improves the PCC architecture using QoS mapping to achieve end to end mapping, which ensures the reliability of the power transmission business.

REFERENCES

Kong, Xiangyi. & Zhao, Jihong. 2011. Study of the mapping of qos parameters in lte-a systems. *ZTE Technology Journal* 17(4): 43–47.

Li Jin You, Yan Lei & Qi Huan. 2014. Research and practice of power wireless communication network based on lte 230system. *Electrical Engineering*: 132–134.

Li, Wenwei., Chen, Baoren., Wu, Qian. et al. 2012. Applied research of td-lte power wireless broadband private network. *Telecommunications for Electric Power System* 33(11): 82–87.

Ou Qing Hai, Xie Jie Hong. et al. 2012. Application of td-lte technology in power distribution and utilization system. *Modern Electronics Technique* 35(23): 27–31.

Quan Nan, Lei Yu Qing, Huang Biyao. et al. 2012. Research on architecture of terminal communication access network in smart grid. *Telecommunications for Electric Power System* 33(1): 57–59.

Wang Yong, Li Shao Cong & Chen Baoren. 2010. Analysis on the application and development of power communication services. *Telecommunications for Electric Power System* 31(11): 44–52.

Xu Guonian. 2012. Construction and application of 230 mHz power wireless broadband communication system. *Telecommunications for Electric Power System* 33(7): 58–62.

Yuan Penghui, Long Biao & Chen Jie. 2012. Analysis of the policy of pcc deployment for service control in lte. *Telecommunication Science* 28(11): 29–31.

Zhang Xincheng & Zhao Xiaojin. 2009. *LTE air interface and performance*. Beijing: Posts & Telecom Press.

Electronics, Communications and Networks IV – Hussain & Ivanovic (eds)
© 2015 Taylor & Francis Group, London, ISBN: 978-1-138-02830-2

An improved resource allocation algorithm of D2D communication for TD-LTE-advanced systems

Wei Wu*, Jingan Liu, Xiuzhi Guan & Jianzhong Li
Harbin Institute of Technology, Harbin, Heilongjiang, China

ABSTRACT: D2D communication was introduced to the cellular system to improve the coverage and performance of the edge user. In this paper, in order to reduce the interference caused by D2D communication, an improved resource allocation algorithm of D2D communication for TD-LTE-Advanced systems was proposed. In the proposed algorithm, the system throughput was improved by introducing greedy algorithm, and the computational complexity was decreased by reducing traversal times. The simulation result indicated that, using the proposed resource allocation algorithm, the system throughput increased approximately 20% with a ratio of D2D pairs to cellular users of 1:5.

1 INTRODUCTION

With the popularization of smart phones and tablet PCs, the demands of mobile data transmission have increased greatly. Context-aware applications and some bandwidth-hungry applications, such as high-quality multimedia, require high data rate communication among neighbors (Zhang et al. 2014, Tehrani et al. 2014). The Long-Term Evolution Advanced (LTE-A) system could provide a high rate of communication in licensed cellular bandwidth (Kaufman & Aazhang 2008). The users in LTE-A are required to transmit data via the eNodeB. However, the users located at the edge of a cell experience difficulty in communicating with eNodeB due to poor channel conditions. In order to resolve this problem, Device-to-Device (D2D) communication was introduced in the cellular system (Doppler et al. 2009). D2D communication technology could extend cell coverage and improve QoS and system throughput.

Resource allocation is one of the key technologies of D2D communication (Wang et al. 2011). D2D communication works in licensed cellular bandwidth. In other words, D2D users and cellular users share the same time-frequency resource. A suitable resource allocation method could ensure that cellular communication is used as a primary mode and D2D communication is used as a secondary mode. The goal of resource allocation is to improve the communication quality of D2D users with the precondition that cellular users communicate correctly. In this way, the system capacity could be greatly enhanced. In recent years, some resource allocation algorithms have been proposed for D2D communication. Most of these have aimed to maximize the system throughput by traversing all reusable resources in the system to search for suitable resources for D2D users (Zulhasnine et al. 2010, Xu et al. 2012). In these algorithms, the system throughput was calculated according to the SINRs of cellular users and D2D users, and optimal resource allocation was achieved by solving the throughput optimization problem. However, these algorithms are computationally complexity and exhibit poor real-time performance.

In this paper, an improved resource allocation algorithm based on the throughput greedy algorithm was proposed. This algorithm improved the throughput by introducing greedy algorithm and decreased the computational complexity by reducing traversal times.

The rest of this paper is organized as follows. Section 2 describes Fractional Frequency reuse in D2D communication. In Section 3, the detailed procedure of the proposed D2D communication resource allocation algorithm is presented. In Section 4, the simulation parameters are given and the results are analyzed. Section 5 presents the conclusion.

2 FRACTIONAL FREQUENCY REUSE

In the traditional cellular system, the entire frequency band is allocated to the entire cell. The eNodeB emits and receives signal with equal power. However, in D2D communication using the TD-LTE-Advanced system, the traditional resource allocation algorithm could result in co-channel interference (CCI) and a decrease in throughput, especially when D2D users

*Corresponding author: *kevinking@hit.edu.cn*

reuse the frequency. Therefore, Fractional Frequency Reuse (FFR) was employed in the proposed resource allocation algorithm (Hyang Sin et al. 2011).

In the FFR scheme, the entire frequency band is divided into three parts, and each part is allocated to one sector of the cell. By doing so, the spectrum efficiency is greatly increased, and the FFR improvement increases the gain by avoiding the interference in D2D communication. The spectrum allocation of FFR is shown in Figure 1.

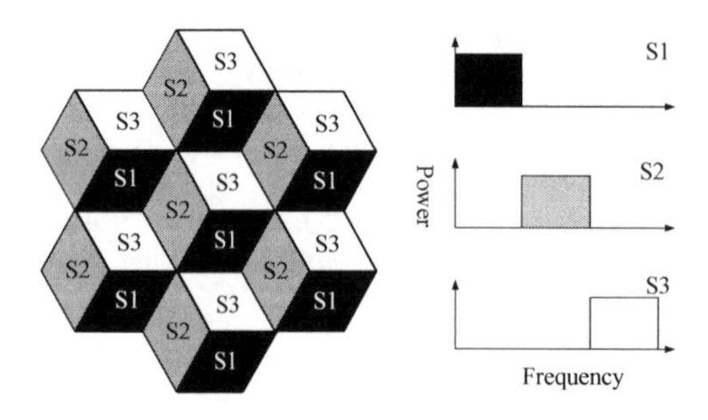

Figure 1. Fractional frequency reuse.

In Figure 1, the sector with different colors uses the different part of the frequency band. D2D users located in sector S1 or S2 reuse the frequency resource of S3, while the cellular users located in sector S1 or S2 use the frequency resource of S1 or S2 respectively. In this way, the co-channel interference is reduced significantly, and the spectrum efficiency and the throughput of the cell are improved.

3 IMPROVED RESOURCE ALLOCATION ALGORITHM OF D2D COMMUNICATION

If the D2D users are far from eNodeB, the reuse of the uplink resources of the cell would be rational. In this way, the interference within the cellular system would be reduced, and the SINRs of the cellular and D2D users would both increase significantly. As a consequence, the system throughput would be greatly improved. In contrast, reusing the downlink resource of the cell would be a better choice for D2D users near eNodeB. Since the traditional algorithm has a heavy computational burden, it would be difficult to complete the task in a scheduling period (1ms in the TD-LTE-Advanced system). In this paper, an improved resource allocation algorithm based on the greedy algorithm was proposed to improve the system throughput and decease the computational complexity by reducing traversal times.

In the proposed resource allocation algorithm, $\mathbf{C}_i=\{1,2,\ldots c_{imax}\}(i=1,2,3)$ was the set of cellular users and $\mathbf{D}_i=\{1,2,\ldots d_{imax}\}(i=1,2,3)$ was the set of D2D users, where i denoted the three sectors. In addition, $\mathbf{G}_{cidj}$ was the channel gain matrix between the cellular users and D2D users, where ci denoted the cellular users in the sector i, and d_j denoted the D2D users in sector j. The reuse procedure of downlink resources by D2D users was as follows.

Step 1: The CQIs of cellular users in each sector of one cell were sorted in descending order. The matrix of the sorted CQI could be acquired. D2D users reuse the cellular users' resource with the maximum CQI. In this way, the interference of cellular users was decreased.

Step 2: The D2D users in the cell into pairs were divided, and the resource could be reused by a certain D2D user pair is determined. The D2D user pair reused the resource belonging to the other sectors.

Step 3: The eNodeB collected the information of all users in a cell, estimated and stored the channel gain matrix, **G**, between the cellular users and D2D users.

Step 4: The resource of sector 1 for D2D users to reuse was selected first.

Step 5: The decision criterion of whether $\mathbf{D}i\neq\emptyset$ or $ci\neq ci_{max}$ was true or not was obtained to judge whether all the D2D pairs were served and whether there were resources available for D2D users to reuse. If true, the procedure moved to Step 6; otherwise, the procedure continued to Step 11.

Step 6: The first element of the matrix, $\mathbf{C}i$ was chosen and this cellular user's frequency resource was allocated to corresponding D2D users for reuse.

Step 7: The channel gain matrix, **G** was searched to determine which D2D pair had the minimum channel gain with the cellular user selected in Step 6. Then, this D2D pair was set as an alternative pair to reuse the spectrum resource.

Step 8: The cellular user's downlink SINR, γ_c^{DL}, and the D2D users' received SINR, γ_d^{DL}, were calculated according to equation (1) and equation (2), respectively.

$$\gamma_c^{DL} = \frac{P_B G_{Bc}}{N_0 + I + P_d G_{cd}} \tag{1}$$

$$\gamma_d^{DL} = \frac{P_d G_{dd}}{N_0 + I + P_B G_{dB}} \tag{2}$$

where PB and Pd are the transmission power of the eNodeB and a D2D user, respectively; GBc and Gcd are the channel gains between the cellular user and the eNodeB and the cellular user and the D2D user, respectively; Gdd and GdB are the channel gains between two D2D users and between the D2D user and the eNodeB, respectively; N_0 is the thermal noise; and I accounts for the inter cell interference.

Step 9: Whether $\gamma c^{DL} > \gamma c,tgt^{DL}$ and $\gamma d^{DL} > \gamma d,tgt^{DL}$, were found out, where $\gamma c,tgt^{DL}$ and $\gamma d,tgt^{DL}$ are the minimum SINR of cellular users and D2D users to work properly. If true, the procedure moved to step10; otherwise, the process returned to Step 5 and restarted the allocation.

Step 10: The selected spectrum resource was allocated to the cellular user and the D2D users for reuse. **D** and c_i was set as $\mathbf{D}=\mathbf{D}-\{d_i\}$, $c_i=c_i+1$. Then, the procedure returned to Step 5 and began the next round of allocation.

Step 11: Whether the allocation was complete for all three sectors was determined. If true, the procedure was ended; otherwise, the procedure returned to Step 6 and began the next sector's allocation until the allocation was complete for all three sectors.

The reuse procedure of the downlink resource of cellular users by D2D users is shown in Figure 2.

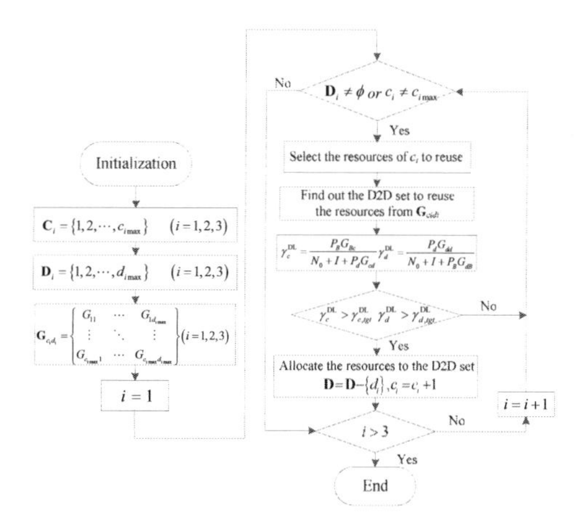

Figure 2. The reuse procedure of D2D communication downlink resources.

The reuse procedure of the uplink resources of cellular users by D2D users is similar to the reuse of the downlink resources. The difference is the interfered targets. When reusing downlink resources, the cellular users are interfered by D2D pairs, and the D2D users are interfered by eNodeB. When reusing uplink resources, eNodeB will be interfered by D2D pairs, and D2D users will be interfered by the cellular users. As a result, when reusing uplink resources, Step 9 is different, and the eNodeB SINR γB^{UL} and D2D user SINR γd^{UL} can be calculated using equation (3) and equation (4).

$$\gamma_B^{UL} = \frac{P_c G_{Bc}}{N_0 + I + P_d G_{dB}} \tag{3}$$

$$\gamma_d^{UL} = \frac{P_d G_{dd}}{N_0 + I + P_c G_{cd}} \tag{4}$$

where Pc is the transmission power of the cellular user.

4 SIMULATION OF THE PROPOSED ALGORITHM

A system-level simulation for the resource allocation algorithm of D2D communication in TD-LTE-Advanced systems was conducted. According to the 3GPP (3rd Generation Partnership Project), the simulation parameters were configured as shown in Table 1.

Table 1. Simulation parameters.

Simulation parameters	Configuration
Cell structure	Hexagonal 7 quarters, 3sectors per cell
Carrier Frequency	f_c=2.14GHz
System bandwidth	B=20MHz
Resource block size	f_{RB}=180kHz
Transmission power of eNodeB	36dBm
Transmission power of cellular users	20dBm
Transmission power of D2D users	17dBm
Number of cellular user	60
D2D distance	≤20m

The throughput performance of the proposed resource allocation algorithm for D2D communication in TD-LTE-Advanced systems is shown in Figures 3 through 5.

The relationship between the system throughput and the number of cellular users and D2D numbers is shown in Figure 3, and the average throughput per user is shown in Figure 4.

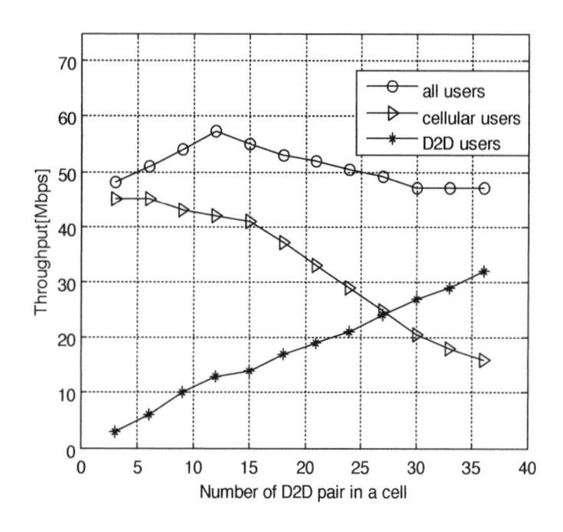

Figure 3. Throughput versus the number of D2D pairs.

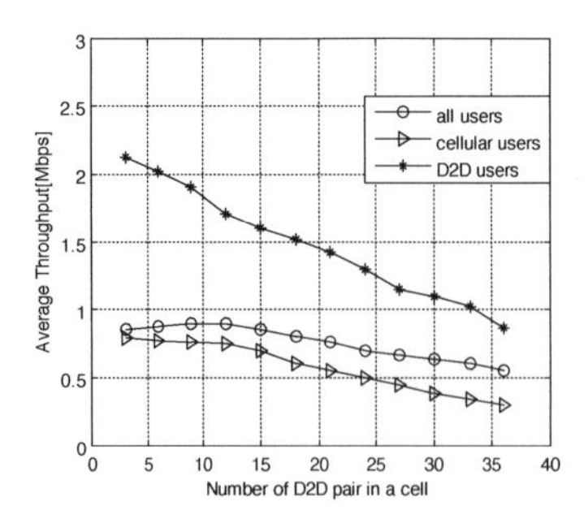

Figure 4. Average throughput per user versus the number of D2D pairs.

As shown in Figures 3 and 4, the cellular users throughput decreased as the number of D2D pairs increased. This was because the reused resources increased was the number of D2D pairs increased. In this way, an increasing number of cellular users could experience interference. As a result, the throughput of cellular users decreased. In addition, the throughput and average throughput were at their maximum values when there were 12 D2D pairs in the simulation scenario. In other words, the throughput performance was best when the ratio of the number of D2D pairs to the number of cellular users was 1:5. The system throughput comparison of the TD-LTE-Advanced system with and without D2D communication is shown in Figure 5.

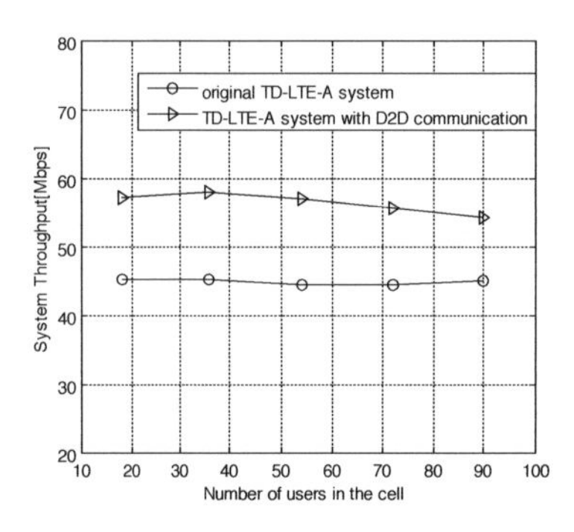

Figure 5. Throughput comparison of the TD-LTE-Advanced system with and without D2D communication.

As shown in Figure 5, the system throughput of the TD-LTE-Advanced system with D2D communication was approximately 20% higher than the system without the D2D communication system.

5 CONCLUSIONS

In this paper, an improved resource allocation algorithm of D2D communication for TD-LTE-Advanced systems was proposed in order to reduce the interference caused by D2D communication and the computational complexity of resource allocation. The system throughput could be improved by introducing the greedy algorithm and the computational complexity could be decreased by reducing traversal times. As indicated by the simulation result, the throughput was best when the ratio of the number of D2D pairs to the number of cellular users was 1:5, and the throughput could increase by approximately 20%.

ACKNOWLEDGEMENT

This research was supported by the Next Generation Wireless Mobile Communication Network of China under Grant No. 2012ZX03003011-004 and Heilongjiang Postdoctoral Financial Assistance under grant No. LBH-Z11153.

REFERENCES

Doppler, K., Rinne, M., Wijting, C., et al. 2009. Device-to-device communication as an underlay to LTE-advanced networks. *IEEE Communications Magazine* 47(12): 42–49.

Hyang Sin, C., Jaheon, G., Bum-Gon, C, et al. 2011. Radio resource allocation scheme for device-to-device communication in cellular networks using fractional frequency reuse. *The 17th Asia-Pacific Conference on Communications: 58–62, 2–5 Oct 2011.* Piscataway: IEEE.

Kaufman, B. & Aazhang, B. 2008. Cellular networks with an overlaid device to device network. *The 42nd Asilomar Conference on Signals, Systems and Computers: 1537–1541, 26–29 Oct 2008.* Piscataway: IEEE.

Tehrani, M. N., Uysal, M., Yanikomeroglu, H. 2014. Device-to-device communication in 5G cellular networks: challenges, solutions, and future directions. *IEEE Communications Magazine* 52(5): 86–92.

Wang Bin, Chen Li, Chen Xiaohang, et al. 2011. Resource Allocation Optimization for Device-to-Device Communication Underlaying Cellular Networks. *IEEE the 73rd Vehicular Technology Conference*: 1–6 *15–18 May 2011.* Piscataway: IEEE.

Xu Chen, Song Lingyang, Han Zhu, et al. 2012. Interference-aware resource allocation for device-to-device communications as an underlay using sequential second price auction. *IEEE International Conference on Communications: 445–449, 10–15 June 2012.* Piscataway: IEEE.

Zhang, Y., Pan, E., Song, L., et al. 2014. Social Network Aware Device-to-Device Communication in Wireless Networks. *IEEE Transactions on Wireless Communications* 14(1): 177–190.

Zulhasnine, M., Changcheng, H. & Srinivasan, A. 2010. Efficient resource allocation for device-to-device communication underlaying LTE network. *IEEE the 6th International Conference on Wireless and Mobile Computing, Networking and Communications: 368–375, 11–13 Oct 2010.* Piscataway: IEEE.

Electronics, Communications and Networks IV – Hussain & Ivanovic (eds)
© 2015 Taylor & Francis Group, London, ISBN: 978-1-138-02830-2

Betweenness centrality and its application in the power grid

Runze Wu & Xiao Yang
School of Electrical & Electronic Engineering, North China Electric Power University, Changping District, Beijing, China

ABSTRACT: According to the definition and characteristic of betweenness centrality, its application in the power grid was investigated in this paper. Taking loads into consideration, a revised parameter was proposed. To find the important edge of the network, a change of the betweenness centrality was conducted. Adding up the changes in the neighbor node betweenness centrality before and after the edge was deleted, the importance of the edge could be known. Results show that a strong correlation existed among the betweenness centrality changes of the neighbor edges and the cascading failure amount the deleted edge caused, which made it feasible to identify the important lines of the power grid by betweenness centrality changes at the same time.

KEYWORDS: complex network; betweenness centrality; cascading failure.

1 INTRODUCTION

Complex networks, including the internet, transportation systems, and power grids,function based on their inherent interdependencies. Complex structures of the networks determine its complex characteristics (Shi 2008). In order to discover the important lines in the power grid, betweenness was used to measure the importance of the lines (Wu&Wen 2013, Deng 2014, Bompard&Wu 2010). The results showed that lines with high betweenness played an important role in the network. Limitations still existed in simulation when using an un-weighted graph to analyze the power grid to solve the problem, and inputting weights into consideration was a proper way to investigate the characteristics of the power grid (Ding&Han 2008).

After doing research on network models, such as IEEE models, the necessity and feasibility of the models were questioned and ways to improve power grid models were proposed. So as to make a proper way to simulate the power grid, electrical betweenness was used (Shi&Luo 2008). Researchers have used electrical betweenness to identify important lines in the power grid. Especially in recent years, major blackout events have frequently occurred, bringing huge economic losses and marked impacts on social stability as well. As mentioned above, identifying important lines is paramount in the research of power grids (Xu&Wang 2010).

The paper firstly gives the definition and characteristics of betweenness centrality, and then presents an improved way to count the betweenness centrality. When an accident occurs, the load of the broken edge would change its way to another neighbor edges, which would bring great impact on other edges at the same time. Analyzing the power grid showed a strong correlation between changes in the betweenness centrality of neighboring edges and the amount of cascading failure the deleted edge caused. According to the delta betweenness centrality of the edge, the important lines of the power grid were identified.

2 DEFINITION

A complex network can be simplified into a simple agent model G = (V, E) based on the complex network theory. V_i denotes node i in graph G and E_{st} denotes the edge between node s and node t. The value of E_{st} denotes the distance between node s and node t. The shortest path D_{st} represents the shortest path between node s and node t. The value that all the shortest paths between node$s$ and node t pass through the edge E_{st} in the network divided by the shortest path between node s and node t as $\mu_{st}(E_{st})$ was defined, specifically:

$$\mu_{st}(E_{st}) = \frac{\lambda_{st}(E_{st})}{\lambda_{st}} \tag{1}$$

where λ_{st} is the number of the shortest paths between node s and node t and $\lambda_{st}(E_{st})$ is the number of the shortest path between node s and node t pass through the edge E_{st}. Betweenness centrality Q is defined as the summation of all the $\mu_{st}(E_{st})$ in the network,specifically:

$$Q(E_{st}) = \sum_{s,t} \mu_{st}(E_{st}) \tag{2}$$

Betweenness centrality can be used to measure the importance of a node or an edge in the network. An edge with high betweenness centrality plays an important role in the network. At the same time, betweenness centrality shows the importance of an edge to some extent.

Putting weights to the edges and nodes into consideration, $W(E_{st})$ represents the weight of edge E_{st}. The value of $W(E_{st})$ shows the maximum load that edge E_{st} can bear. For a weighted graph, betweenness centrality of edge $Q(E_{st})$ is defined as:

$$Q(E_{st}) = \sum_{s,t} \mu_{st}(E_{st}) \tag{3}$$

At the same time,the shortest path in the weighted graph is redefined as:

$$D(E_{st}) = \sum \frac{P(E_{st})}{W(E_{st})}[A] \tag{4}$$

where $P_{s\,t}$ denotes the load on edge E_{st} , W_{st} denotes the maximum load that E_{st} can bear, and matrix $[A]$ is the revise parameter matrix. For different networks, matrix $[A]$ has different parameters. In the case where the network is a power grid:

$$[A] = \frac{\max|\theta| - \theta_{st}}{P(E_{st})} \tag{5}$$

where θ is the inverse Jacobian matrix result of *P-Q* decomposition when all the *PV* nodes are seen as *PQ* nodes. The change of betweenness centrality shows the importance of the edge in the network, using the revised parameter of the shortest path to conduct the betweenness centrality identifies the important edges of the network more accurately.Changes of betweenness centrality shows the effect of the edge when cascading failures occur. The change of betweenness centrality of neighbor edges is defined as $\Phi(EST)$ when the edge is deleted by the network, specifically:

$$\Phi(E_{st}) = \sum_{i=s,j\neq t} \Delta\Psi(E_{ij}) + \sum_{i\neq s,j=t} \Delta\Psi(E_{ij}) \tag{6}$$

where $\sum_{i=s,j\neq t}\Delta\Psi(E_{ij})$ is the betweenness centrality changes in neighbor edges of node *s* when the edge E_{st} is deleted and $\sum_{i\neq s,j=t}\Delta\Psi(E_{ij})$ is the betweenness centrality changes in neighbor edges of node *t* when the edge E_{st} is deleted. According to the definition of $\Phi(E_{st})$, it is learned that the value of $\Phi(E_{st})$ is affected by the neighbor edges of node s and node *t*. When an edge in the network is deleted, loads on the edge change its way to other neighbor nodes and edges, the shortest paths of the loads in the network are changed. Because of the changes in the shortest paths, the betweenness centrality of these neighbor edges is changed as well. The betweenness centrality changes in neighbor edges affect the amount of cascading failure to some extent. If the average load on the edges is very high, a failure on edge E_{st} would quite likely cause a cascading failure. It is a proper way to use the parameter $\Phi(E_{st})$ to identify important edges in the network.

3 SIMULATION

In order to analyze the case where an edge of the network was accidentally deleted,a power grid model was used in this paper. When a line of the power grid was deleted, a blackout occurred. According to the amount of the blackout, a parameter of the amount of the blackout could be used to identify the importance of the edges. To simulate this, a simplified model of a power grid with 86 nodes and 113 lines was created, as shown in Figure 1.

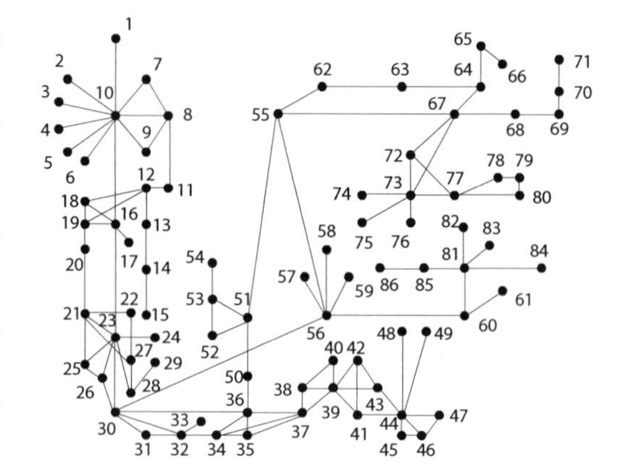

Figure 1. Simplified model of power grid.

To make the parameters of the edges more clear, the betweenness of the edges were reordered by ascending sequence, as shown in Figure 2.

The abscissa shows the number of the edges that were reordered by ascending sequence. The ordinate shows the betweenness centrality of the edges.

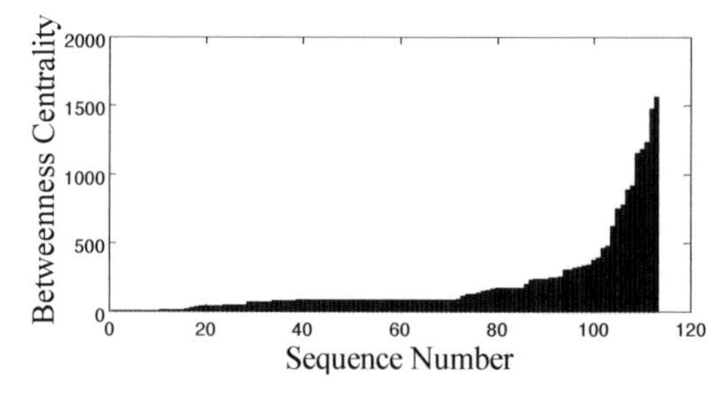

Figure 2. The distribution of the betweenness centrality of the edges.

The distribution of the betweenness centrality shows that a small part of the edges assumed most of the betweenness centrality in the network. Edges with high numbers of betweenness centrality played an important role in the power transmissions; when those edges were deleted, a great change happened to the power grid. To identify the influence of the accident, the amount of the changes in the betweenness centrality is shown in Figure 3.

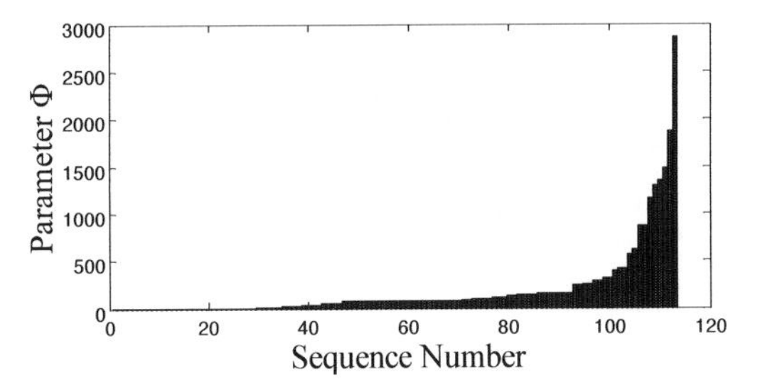

Figure 3. The distribution of changes in betweenness centrality.

Table 1. Changes in betweenness centrality of the edges.

Φ	Sequence	Staring node, Ending node
2869	1	23, 30
1876	2	30, 56
1480	3	55, 56
1353.33	4	10, 16
1300.03	5	56, 60
1162	6	67, 73
874	7	60, 81
865.67	8	16, 23
616	9	51, 55
570	10	37, 38

According to the date shown in Table 1, the importance of the edges could be found in the network. The results showed that edges with high changes of betweenness centrality played important roles in the network. Node 23 and Node 30 were located in the center of the network, and most of the transmissions had to change their routes when the edge (23-30) was accidentally deleted; at the same time, the loads changed their way from the edge (23-30) to edge (23-26) and edge (26-30). The two edges had to undertake the loads of edge (23-30). If edge (23-26) or edge (26-30) lacked the ability to undertake those loads, the power grid would be divided into two unconnected parts, which would bring considerable energy losses. As mentioned above,the edge (23-30) had a great importance in the network, as shown in Figure 2. When a power grid runs in a similar situation and an edge is accidentally deleted, the loads of the network are re-routed immediately. After a series of activities, some nodes in the network go wrong, increasing the numbers of the failure nodes, and the influence of the deleted edge is noticed. The result is shown in Figure 4.

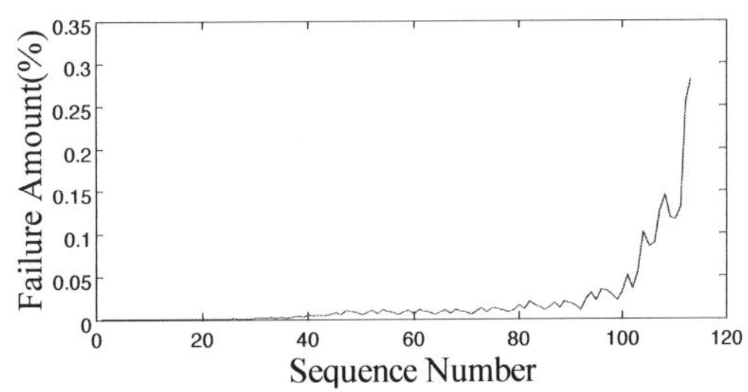

Figure 4. Distribution of the amount of cascading failure.

In Figure 4, the reordered number of edges by ascending sequence was used in the x-axis to show the failure amount of the deleted edges. Because of the sort of the edges, the bigger the edge had, the bigger the changes of betweenness centrality would take place when it was deleted. However, as shown in Figure 4, although the trend of the changes was alike,differences still existed. Figure 5 illustrates the failure amount and the betweenness centrality changes.

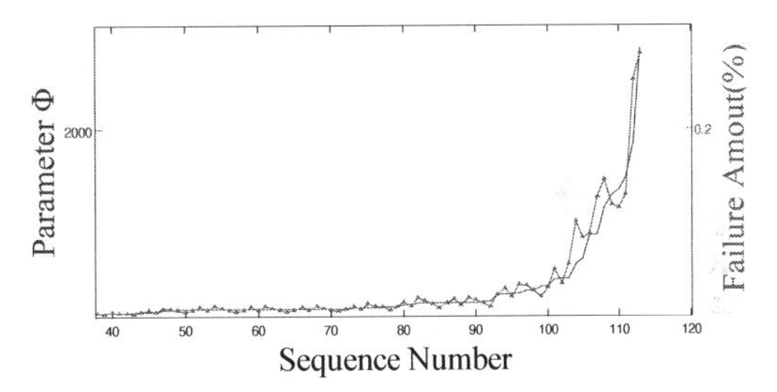

Figure 5. Collation of the failure amount and the betweenness centrality changes.

The higher the betweenness centrality of the deleted edge, the larger the amount of cascading failure the deleted edge caused. As shown in Figure 5, the amount of the cascading failure had a great correlation with the parameter Φ. The parameter Φ was affected by the loads on the edge, the position of the edge of the network, and the number of the shortest paths passing through the edge. If the markov process was used to simulate the cascading failure in the power grid, the initial conditions of the markov process would be affected by the value of parameter Φ.As a consequence, the cascading failure amount showed great correlation to parameter Φ, making it possible to know the cascading failure amount caused by the edge by comparing the Φ of the edge and to identify the important edges of the network.

4 CONCLUSION

Based on the complexity theory, loads were taken into consideration and changes to neighbor edges betweenness centrality parameter Φ was proposed in this paper. The parameter Φ was used to identify the important edges in the network. A power grid model was used to analyze the parameter Φ and its effect on the cascading failure. Results showed that the amount of cascading failure had a great correlation to parameter Φ, making it possible to identify the important edges of the network by the value of parameter Φ. Although the amount of the cascading failure could be judged by the value of parameter Φ, the influence of the cascading failure was not only assessed by the amount of the cascading failure, but also the power loss of the cascading failure, the change in frequency of the power system, and the node voltage and so on. Many shortcomings still existed when judging the importance of the edge by the amount of the cascading failure. That will be the next focus in important line identification research.

ACKNOWLEDGMENTS

This work was financially supported by the Fundamental Research Funds for the Central Universities (13MS01).

REFERENCES

Bompard, E. & Wu, D. 2010. The Concept of Between ness in the Analysis of Power Grid Vulnerability. *Complexity in Engineering, 2010. COMPENG'10 Rome, 22–24 February 2010*. IEEE.

DengChunlan. 2014.*Vulnerability research for cascading failures blackout in power system based on complex network theory*. Chengdu: Southwest Jiaotong University.

Ding Ming&HanPingping. 2008. Vulnerability Assessment to Small-world Power Grid Based on Weighted Topological Model. *Proceedings-Chinese Society of Electrical Engineering* 28(10): 20.

Liu Yaonian&ZhuXi. 2011. Identification of vulnerable lines in power grid based on the weighted reactance between ness index. *Power System Protection and Control*39(23).

ShiJin&LuoGuangyu. 2008. Complex Network Characteristic Analysis and Model Improving of the Power System. *Proceedings of the CSEE*(25): 018.

ShiJin. 2008. *Network model and structure performance analysis of Powersystem based on complex network theory*.Huazhong University of Science & Technology.

WuKewen & WenFushuan. 2013. A Markov Chain Based Model for Forecasting Power System Cascading Failures. *Automation of Electric Power Systems* 37(5): 29–37.

XuLin & WangXiuli. 2010. Electric Between ness and Its Application inVulnerableLine Identification in Power System. *Proceedings of the CSEE*(1): 007.

Electronics, Communications and Networks IV – Hussain & Ivanovic (eds)
© 2015 Taylor & Francis Group, London, ISBN: 978-1-138-02830-2

A resource allocation approach based on a simulated annealing algorithm in electric power communication networks

Fei Xia*, Zongze Xia & Xiaobo Huang
State Grid Liaoyang Electric Power Supply Company, Liaoyang, Liaoning, China

Xiao Gao
State Grid Liaoning Electric Power Company Limited, Shenyang, Liaoning, China

ABSTRACT: This paper studied the problem of resource allocation in electric power communication networks. A novel resource allocation approach was proposed based on a simulated annealing algorithm. In detail, after considering the different network users, the resource allocation problems in a link were first investigated, and then an optimization resource allocation model was built. The optimization problem described a utility function via the social distance and link bandwidth, which was the objective function of the proposed optimization model. The proposed optimization problem was solved with the simulated annealing algorithm, and then a bandwidth allocation solution with an optimal utility function was obtained. Finally, numerical results were provided to validate the performance of the proposed model. The simulation results state that the proposed approach is promising.

1 INTRODUCTION

With the rapid development of smart grids, the electric power communication networks are regarded as a crucial technical support system. They provide services for the operation of our smart grids such as data transmission, video, and voice etc. These services are useful for the allocation of power grids. Moreover, office automation can be realized with electric power communication networks. Generally, electric communication networks need to provide several services simultaneously in a link. Electric power communication networks use several communication technologies consisting of wired and wireless networks. However, most existing research does not consider wireless network technologies for electric power communication network environments. Hence, it is necessary to bring wireless network technologies into electric power communication networks, and research the problem of resource allocation to guarantee the quality of service (QoS) in a wireless link (Chen et al. 2014, Feng et al. 2014, Gortzen & Schmeink 2014, Kim et al. 2014, Ruder et al. 2014).

To date, the issue of resource allocation has been extensively studied. The authors (Liang, et al. 2014) studied resource allocation problems of heterogeneous relay networks based on the game theory and proposed a Stackelberg model-based hierarchical game to address resource allocation in heterogeneous relay networks. Xu, et al. (2014) studied non-stationary resource allocation problems and proposed a new solution that focuses on the problem without assuming detailed packet-level knowledge which is unavailable at resource allocation time. Taking into account the energy-efficiency of the networks, the authors (Xu et al. 2014) addressed the resource allocation problems in heterogeneous orthogonal frequency division multiple access downlink networks. Transmitter energy consumption and receiver energy consumption were taken into account at the same time and a mixed combinatorial and nonconvex optimization model was proposed. For this nonconvex optimization problem, a heuristic method was developed to obtain the optimal solution. The authors (Minarolli & Freisleben 2014) addressed the problem of the distributed resource allocation in cloud computing networks. This problem was dealt with using an optimization model with a utility function, and the proposed optimization problem was solved with an artificial neural network algorithm. Most current technologies model the resource allocation problem with heuristic algorithms, statistical methods, and stochastic knapsack problem. Different from the previous methods, Pandit et al. (2014) used a modified multi-dimensional bin packet to handle resource allocation issues in cloud computing networks. This model was solved with a

*Corresponding author: *empire_1009@163.com*

simulated annealing algorithm. However, due to the development of current networks, it is still difficult to allocate the bandwidth with an efficient utility.

Motivated by the above issue, in this paper focused on the problem of resource allocation. The utility of each link was first considered, and a utility function was given. The novel utility function with respect to the bandwidth was described by a social distance function and a traditional utility function (Crucitti et al. 2006, Daly & Haahr 2007, Tsiropoulos et al. 2011, Shi et al. 2008). In order to attain the best bandwidth for each service request, an optimization problem was constructed where the novel utility function was regarded as the objective function, which was the main contribution of this paper. However, it was difficult to calculate the optimal bandwidth via the proposed optimization model. As a result, the simulated annealing algorithm was utilized in this paper to deal with this optimization model. Then the bandwidth of each service request could be obtained with an optimal utility.

The rest of this paper is organized as follows: In Section 2, the resource allocation problem and the novel method are introduced. Section 3 provides numerical results to assess the properties of the proposed approach. Finally, this work is concluded in Section 4.

2 PROBLEM STATEMENT AND METHODOLOGY

2.1 *The resource allocation problem*

Figure 1 illustrates the construction of the resource allocation problem for a link. There are N service requests in the source node. For each service request, there are N_i users. They will be sent for the source node to the destination node. Thereby, a sufficient bandwidth should be provided in order to guarantee the QoS of each user. The same bandwidth is used for each user belonging to the same service request.

2.2 *Our methodology*

Assuming there are N service requests, and each service request contains N_i users, the amount of users is:

$$C = \sum_{i=1}^{N} N_i. \tag{1}$$

In this method, a utility function is used to measure the holistic property of the electric power communication network. This utility function consists of a social distance function and a traditional utility function. For each service request, its utility function can be denoted by the following equation (Tsiropoulos et al. 2011):

$$u(b_i, \eta_i) = u(b_i) S(\eta_i), \tag{2}$$

where $u(b_i)$ is the traditional utility of the i-th service request, and $S(\eta_i)$ is the social distance function. Bringing in the social distance function can exploit the information of the social network in order to obtain efficient resource allocation. In detail, the traditional utility function can be expressed by (Tsiropoulos, et al., 2011):

$$u(b_i) = \frac{\ln(b_i - B_{min})}{\ln(B_{max}) - \ln(B_{min})} \frac{\text{sgn}(b_i - B_{min}) + 1}{2}, \tag{3}$$

where B_{max} and B_{min} are the upper bound and lower bound of bandwidth, respectively. Besides, sgn(*) is a sign function.

Considering the social distance function, η_i describes the social distance of all users in the service request i. It can be calculated by:

$$\eta_i = \frac{1}{N_i} \sum_{n=1}^{N_i} \gamma_n, \tag{4}$$

where γ_n is the social distance of the users in a social network (i.e., a service request in this paper). The social distance function should be monotonically increasing, thus in this paper the social distance function is denoted as follows:

$$S(\eta_i) = 1 - e^{\alpha(1-\eta_i)}, \tag{5}$$

where α is a constant (Tsiropoulos et al. 2011). According to Equation (2), the novel utility function can be denoted by:

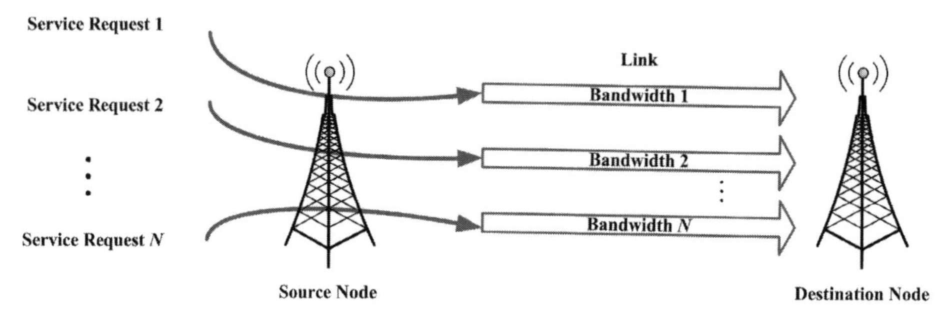

Figure 1. Resource allocation construction.

$$u(b_i,\eta_i)=u(b_i)S(\eta_i)$$
$$=\frac{\ln(b_i-B_{min})}{\ln(B_{max})-\ln(B_{min})} \quad (6)$$
$$=\frac{\text{sgn}(b_i-B_{min})+1}{2}(1-e^{\alpha(1-\eta_i)})$$

So the amount of network utility is:

$$U_\Sigma=\sum_{i=1}^{N}N_i u(b_i,\eta_i). \quad (7)$$

In this method, the utility of the electric power communication network is as large as possible. Consequently, it will be used as the objective function in the optimization resource allocation model. Meanwhile, the bandwidth b_i has an upper bound and a lower bound. In this paper, they are denoted by B_{max} and B_{min}, respectively. Assuming that the bandwidth of the link is B, then:

$$\sum_{i\in\{N\},n\in\{N_i\}} b_{i,n}\le B \quad (8)$$

where $b_{i,n}$ denotes the bandwidth of the n-th user in the i-th service request. Finally, considering Equations (7) and (8), an optimization resource allocation model can be built as follows:

$$\max\ U_\Sigma+\lambda\sum_{i\in\{N\},n\in\{N_i\}} b_{i,n}$$
$$s.t. \quad (9)$$
$$B_{min}\le b_{i,n}\le B_{max}$$

where λ is a small constant. However, it is still difficult to solve this optimization problem directly. Thereby, the simulated annealing algorithm is involved in the proposed method.

The simulated annealing algorithm, which is a general and effective approximation algorithm for solving a large-scale combinatorial optimization problem, is a novel random search method. The simulated annealing algorithm can obtain an optimal solution efficiently and simply. Compared to other random search algorithms, the simulated annealing algorithm is much more efficient and exact. The detail steps of the proposed method are as follows:

Step 1: Initialize the parameters, i.e. the social distance η_n;
Step 2: Let the initial temperature be t_0 and decreasing temperature step be Δt. Denote the initial bandwidth of each user by $\{b_{i,n}^0\}$. Set the number of iterations $k=1$, and let the maximum number of iterations be K. Set the variable that is not updated $p=0$, and the maximum number of times that is not continuously updated as P;
Step 3: From the solution set $\{b_{i,n}^{k-1}\}$, attain a new solution set $\{b_{i,n}^k\}$ according to the simulated annealing algorithm and the constraints in Equation (9). Let the current solution set $\{b_{i,n}^c\}=\{b_{i,n}^{k-1}\}$;
Step 4: According to Equation (7), calculate the difference of the utility function with respect to the new solution set $\{b_{i,n}^k\}$, and denote this difference by ΔU_Σ;
Step 5: If $\Delta U_\Sigma>0$, let the current solution set $\{b_{i,n}^c\}=\{b_{i,n}^k\}$ and $p=0$, or obtain the solution set $\{b_{i,n}^k\}$ and set $p=p+1$;
Step 6: If $p>P$, go to Step 9, otherwise, continue to Step 7;
Step 7: Set $k=k+1$, and $t=t-\Delta t$;
Step 8: If $k<K$, go back to Step 3. Otherwise, continue to Step 9;
Step 9: Set the final solution set $\{b_{i,n}^f\}=\{b_{i,n}^c\}$, and exit.

Then, an optimal solution set of the resource allocation model in Equation (9) can be obtained.

3 NUMERICAL RESULTS

In this section, the properties of the proposed method will be assessed. The parameters of the simulation are listed in Table 1. It is assumed that there are two kinds of service requests. Service Request 2 consists of 1000 users. Meanwhile, it is supposed that the bandwidth of a link is 10 Gbps. The constant λ in this method is 0.01 for achieving an optimal solution. The parameter α in social distance function is 5. The social distance is set at η_2=0.5, and $\eta_1=0.25$, 0.5, and 0.75, respectively.

Table 1. Parameters in the simulation.

Parameter	Value
B	10 Gbps
B_{min}	64 Kbps
B_{max}	10 Mbps
N	2
N_2	1000
λ	0.01
α	5

In this case, the bandwidth of Service Request 1 is plotted in Figure 2. The x-axis denotes the number of the users of Service Request 1. The y-axis denotes the bandwidth of each user in Service Request 1. From Figure 2, it can be seen that the bandwidth increases as the number of users in Service Request 1 increases. The different bandwidths under the various social distance parameters were also plotted. Obviously, the social distance parameter did not have an effect on the bandwidth allocation of this method. In other words, this algorithm can obtain an optimal bandwidth and is not sensitive to the social distance function. Besides, the bandwidth has large fluctuations versus the number of users. That is because this method only solves an optimal utility, and does not consider the stability of the bandwidth.

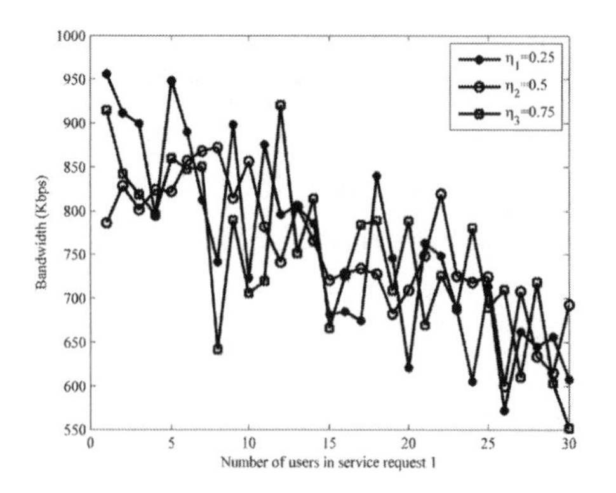

Figure 2. The bandwidth of serivce request 1.

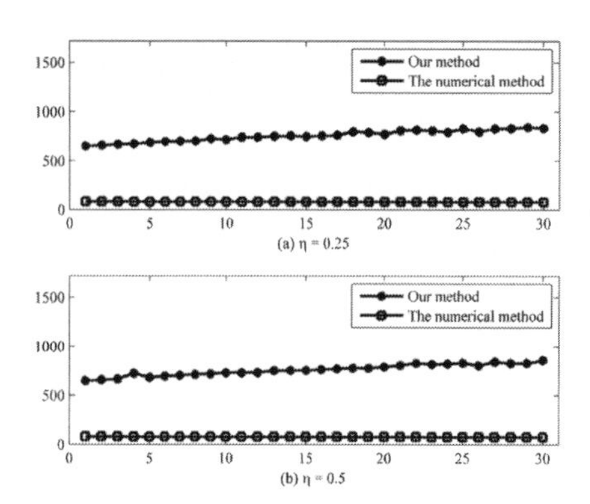

Figure 3. The utility of two methods.

In the simulation, the proposed algorithm was compared to a numerical resource allocation method (Tsiropoulos et al. 2011). First, the utilities of two methods were plotted, as shown in Figure 3. In Figures 3a and 3b, the utilities of two methods are described when the social distance parameter η_1 is 0.25 and 0.5, respectively. The x-axis denotes the number of users in Service Request 1. Obviously, the proposed method was outstanding in utility.

In Figure 4, the average utility of two methods is given when the number of users in Service Request 1 increased from 1 to 30. The averages were potted when the social distance parameter η_1 was 0.25, 0.5, and 0.75, respectively. The average values of the proposed method were 756.23, 760.21, and 755.20, respectively. In addition, the average values of the numerical method (Tsiropoulos et al. 2011) were 79.59, 79.50, and 79.20.

Finally, the improvement ratios of algorithm performances were used to validate the holistic property of the proposed method. The improvement ratio was denoted as follows:

$$\Phi = \left|\sum u_a - \sum u_b\right| / \left|\sum u_a\right|, \tag{10}$$

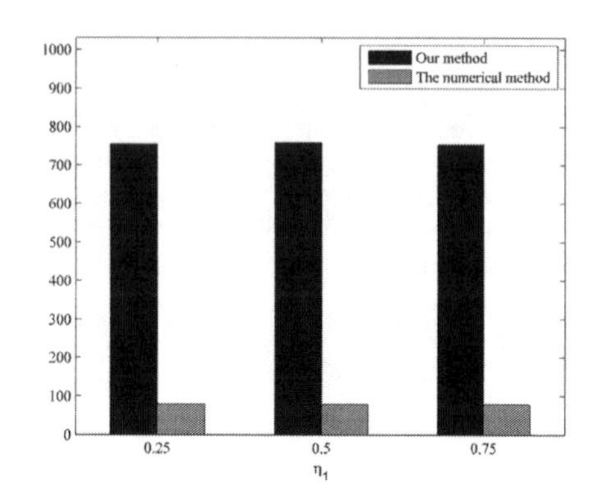

Figure 4. The average utility.

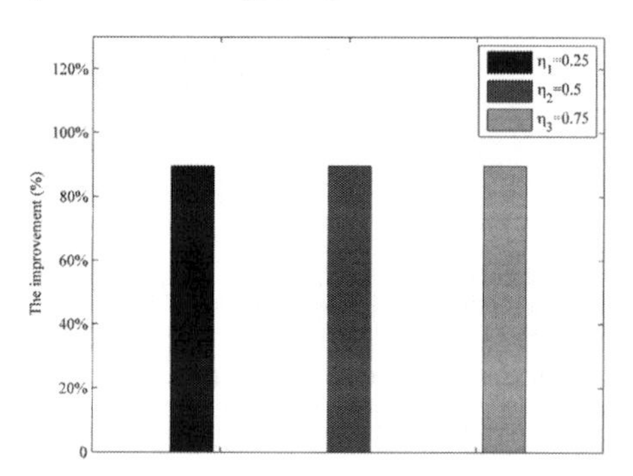

Figure 5. The improvement ratio of performance.

where u_a and u_b are the utilities from algorithms a and b, respectively. It states the improvement ratio of algorithm a with respect to algorithm b. As shown in Figure 5, the improvement ratios were 89.48%, 89.54%, and 89.52%. This demonstrates that the proposed method was outstanding compared to the numerical method.

4 CONCLUSIONS

This paper studied the problem of resource allocation for links in electric power communication networks. A new resource allocation approach was proposed based on a simulated annealing algorithm. After considering different network users, first the resource allocation problems in a link were investigated, and then an optimization resource allocation model was built. The optimization problem described a utility function via the social distance and the link bandwidth, which was the objective function of the proposed optimization model. The proposed optimization problem via the simulated annealing algorithm was dealt with, and then a bandwidth allocation solution with the optimal utility function was achieved. The simulation results stated that the proposed method was successful.

REFERENCES

Chen, F., Vleeschouwer, C. & Cavallaro, A. 2014. Resource allocation for personalized video summarization. *IEEE Transactions on Multimedia* 16(2): 455–469.

Crucitti, P., Latora, V. & Porta, S. 2006. Centrality in networks of urban streets. *Chaos* 16(1): 1–9.

Daly, E. & Haahr, M. 2007. Social network analysis for routing in disconnected delay-tolerant MANETs. In *Proc. MOBIHOC*. September 2007: *32–40*. ACM.

Feng, T., Bi, J. & Wang, K. 2014. Joint allocation and scheduling of network resource for multiple control applications in SDN. In *Proc. 2014 IEEE Network Operations and Management Symposium*: 1–7.

Gortzen, S. & Schmeink, A. 2014. Non-asymptotic bounds on the performance of dual methods for resource allocation problems. *IEEE Transactions on Wireless Communication* 13(6): 3430–3441.

Kim, S., Lee, B.G. & Park, D. 2014. Energy-per-bit minimized radio resource allocation in heterogeneous networks. *IEEE Transactions on Wireless Communications* 13(4): 1862–1873.

Liang, L., Feng, G. & Jia, Y. 2014. Game-theoretic hierarchical resource allocation for heterogeneous relay networks. *IEEE Transactions on Vehicular Technology* PP(99): 1.

Minarolli, D. & Freisleben, B. 2014. Distributed resource allocation to virtual machines via artificial neural networks. In *Proc. 22nd Euromicro International Conference on Parallel, Distributed, and Network-Based Processing, Torino, 12-14 February 2014*: 1–10. IEEE.

Pandit, D., Chattopadhyay, S., Chattopadhyay, M. & Chaki, N. 2014. Resource allocation in cloud using simulated annealing. In *Proc. 2014 Applications and Innovations in Mobile Computing, Kolkata, 27 Feb. -1 March 2014*: 1–7. IEEE.

Ruder, M., Meyer, R., Obernosterer, F., Kalveram, H., Schober, R. & Gerstacker, W. 2014. Receiver concepts and resource allocation for OSC downlink transmission. *IEEE Transactions on Wireless Communications* 13(3): 1568–1580.

Shi, L., Liu, C., & Liu, B. 2008. Network utility maximization for triple-play services. *Computer Communications* 31(10): 2257–2269.

Tsiropoulos, G., Stratogiannis, D., Mantas, N. & Louta, M. 2011. The impact of social distance on utility based resource allocation in next generation networks. In *Proc. 2011 3rd International Congress on Ultra Modern Telecommunications and Control Systems and Workshops* : 1–6.

Xu, J., Andrepoulos, Y., Xiao, Y. & Schaar, M. 2014. Non-stationary resource allocation policies for delay-constrained video streaming: application to video over Internet-of-things-enabled networks. *IEEE Journal on Selected Areas in Communication* 32(4): 782–794.

Xu, Q., Li, X., Ji, H. & Du, X. 2014. Energy-efficient resource allocation for heterogeneous services in OFDMA downlink networks: systematic perspective. *IEEE Transactions on Vehicular Technology* 63(5): 2071–2081.

Electronics, Communications and Networks IV – Hussain & Ivanovic (eds)
© 2015 Taylor & Francis Group, London, ISBN: 978-1-138-02830-2

Simulation and analysis of AODV routing protocol in MANETs with different network connectivity

Huihui Xiang*, Jincheng Huang & Jun Gao
Department of Electronic and Information Engineering, Yancheng Institute of Technology, Jiangsu, China

ABSTRACT: Mobile Ad hoc Networks (MANETs) are composed of a cluster of wireless mobile appliances (nodes) and can be quickly deployed without the aid any centralized administration and predetermined infrastructure. Routing is an essential factor due to dynamic change of the network topology of MANETs. Various protocols for routing have been proposed to efficiently transfer data among different mobile nodes by researchers for MANETs in the recent past. This paper mainly deals with AODV that is a reactive routing protocol and presents performance evaluation and comparison of AODV with different network connectivity. Random Waypoint (RWP) mobility model has been used to design networks and the simulations have been performed by using OPNET Modeler 17.5. The simulations are carried out by varying the network coverage, number of nodes and transmission range of each node. Performance of AODV is evaluated based on the number of hops per route, route discovery time, packet delivery ratio, average end-to-end delay.

KEYWORDS: MANETs; AODV; *Connectivity; Random Waypoint Mobility Model; OPNET.*

1 INTRODUCTION

Mobile Ad Hoc Networks (MANETs) are autonomously self-organized and self-configured mobile wireless and infrastucture-less communication networks, where mobile nodes can freely enter, leave or move around the network (Wang et al. 2014, Saeed et al. 2012). MANETs are independent of any infrastructure and centralized administration, which makes MANETs suitable for rapid deployment in response to application requirements (Shivashankar et al. 2014).

Nodes in MANETs are not all within the direct transmission range of each other and require other nodes to forward data (Wang et al. 2014, Zebra et al. 2012). Therefore, nodes act as hosts and routers and should have complete information about the network at any time to find the optimum route to forward data to the specific nodes till the destination. Due to the mobility, entrance and leave of nodes, which leads to dynamic change in network topology and configurations of MANETs, there are many challenges in routing and management.

Routing is one of the active research topics for MANETs. MANETs require specific routing protocols to solve the issues such as lack of central administration and management, mobility of nodes, constraints of resources (battery life, processing power, bandwidth), etc. (Maurya et al. 2014). Routing data efficiently is essential and complicated (Venkatesan et al. 2014, Gupta & Mathur 2014). Furthermore, routing algorithms for MANETs should be adaptive and robust to change in network topology and configurations in a decentralized and self-organizing way.

Routing protocols in MANETs can be classified into different categories based on different criteria. The mostly used criterion to categorize MANET routing protocols is based on the way how nodes acquire, update and maintain routing information. Based on this criterion, routing protocols in MANETs are classified into three types, i.e. proactive (periodic, table-driven) routing, reactive (on-demand) routing and hybrid routing (Seon et al. 2014, Gupta et al. 2013). In proactive routing protocols, every node should periodically exchange routing information with other nodes to maintain the whole network topology information through. This type of routing protocol includes DSDV, WRP, OLSR, FSR, GSR, HSR, TBRF, STAR and CGSR, etc. (Maurya et al. 2014, Roye & Chai-Keong 1999). In reactive routing protocols, nodes search or maintain a routing path when route is needed or required (on-demand). AODV (Perkins et al. 2003), DSR, TORA, ABR, CBRP, LAR, ARA, DYMO, ANODR, SSA and PLBR, etc. (Shivashankar et al. 2014) are reative routing protocols. Hybrid routing protocols combine qualities of proactive and reactive routing protocols while overcome their shortcomings. ZRP, ZHLS, DST, DDR and CEDAR, etc. are some hybrid routing protocols (Maurya et al. 2014).

*Corresponding author: *xianghh@ycit.cn*

AODV (Ad-hoc On-demand Distance Vector) combines the concept of route discovery and maintenance of DSR with the concept of the sequence number and sending of periodic hello messages from DSDV. Therefore, AODV can maintain multiple paths and minimize the number of active paths from the source to destination.

Nodes that are within the transmission range of the one another are called neighbors. The number of neighbors of one node is defined as node degree. In MANETs, due to limits on power and interference, each node has a certain transmission range and can only communicate with the neighbors. When a node needs to send data to non-neighboring nodes, the data will be transferred according to routing protocols through a sequence of multiple hops, which are called intermediate nodes. One intermediate node is the neighbor of the previous and back node in the route. Therefore, network connectivity is an essential requirement for successful routing and communication among nodes. Network connectivity depends on the coverage area, number of nodes and transmission range of each node as shown in (Bettstetter 2002, Santi & Blough 2003, Santi & Blough 2002).

Most of the work has been done by researchers to evaluate the performance of different routing protocols by changing network load and size (Goyal 2012), mobility and density of nodes (Thriveni et al. 2013), and QoS (quality of service) metrics (Bagwari et al. 2012). In this paper, we evaluate the performance of AODV as a function of network connectivity by changing the area size, number of nodes and transmission range of the nodes.

The remainder of this paper is organized as follows. Section II describes design and implementation details of the simulation environment. The computer simulations using OPNET Modeler 17.5 and comparisons are presented in Section III. Section V concludes this paper with a discussion of future research work followed by the references.

2 SIMULATION ENVIRONMENT

We performed simulations on OPNET Modeler 17.5 for the performance comparison and evaluation of AODV with different network connectivity.

The simulations consider the space of 1000m *1000m or 2000m*2000m in which 50, 100 or 200 nodes are placed randomly to form a network. Each packet begins the transmission from a random source node to a random destination.

The simulation runs for 1 hour with mobility patterns. For simulation, random waypoint (RWP) mobility model (Bettstetter 2001) is used. The simulation parameters are summarized in Table 1. Figure 1 shows the network scenario with randomly placed nodes moving around according to RWP mobility model in OPNET Modeler 17.5.

Table 1. Simulation parameters.

Parameter	Value
Total Number of Nodes	50,100 and 200
Size of Area	1000m*1000m, 2000m*2000m
Transmission Range	100-600 (m)
Routing Protocol	AODV
Standard Ad hoc Speed	10m/s
Pause Time	100s
Simulation Time	1 hour
Node Movement Model	RWP
Data Rate	11 Mbps
Node Placement Strategy	Random
Packet Size	1024 bits(average)

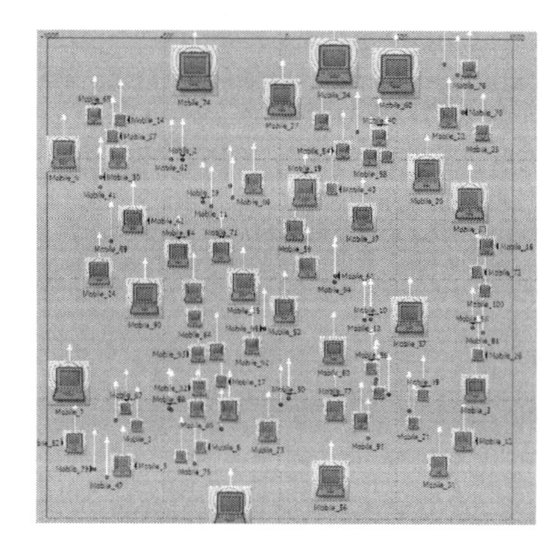

Figure 1. The snapshot of network scenario.

3 SIMULATION RESULTS AND DISCUSSION

3.1 *Performance metrics*

The following metrics are used: minimum node degree, the average number of hops per route, average route discovery time, average End-to-end delay, delivery ratio.

- Minimum Node Degree: This is the minimum number of neighbors of all nodes when the network is deployed.
- Average End-to-end Delay: This is the average time from the creation of a data packet at the source to the destruction at the destination of all nodes.
- Average Number of Hops per Route: This is the average number of hops in each route to every destination listed in the route table.
- Average Route Discovery Time: This is the average time from sending a route request to the reception a route reply of all nodes.

- Delivery Ratio: This is the ratio of the total of data packets successfully transferred to the destinations and the total number of data packets generated.

3.2 *Minimum node degree*

The minimum node degree of each scenario is shown in Table 2. The results show that the network performs higher connectivity with more number of nodes or larger transmission range, which coincides with the results in (Bettstetter 2002). Moreover, when the transmission range of each node is 100 meters, there are isolated nodes in the deployed network, which means that the deployed network is not connected practically. Nodes should have more power for larger transmission range.

Table 2. Minimum node degree (Area: 1000m*1000m).

Transmission	Number of Nodes	
Range(m)	50	100
100	0	0
200	1	4
300	5	11
400	8	18
500	10	22
600	17	36

3.3 *50,100,200 nodes deployed in the area 1000m*1000m, transmission range 100m*

The delivery ratio of 50, 100, 200-node scenario are 20.41%, 53.61% and 55.15% respectively. The results of average end-to-end delay and routing performance are shown in Figure 2. All though there are isolated nodes, the left nodes can still communicate with each other through routing. With more nodes in the area, it has lower delay and shorter route discovery time, while with more hops per route. The overall performance is better when network connectivity grows higher.

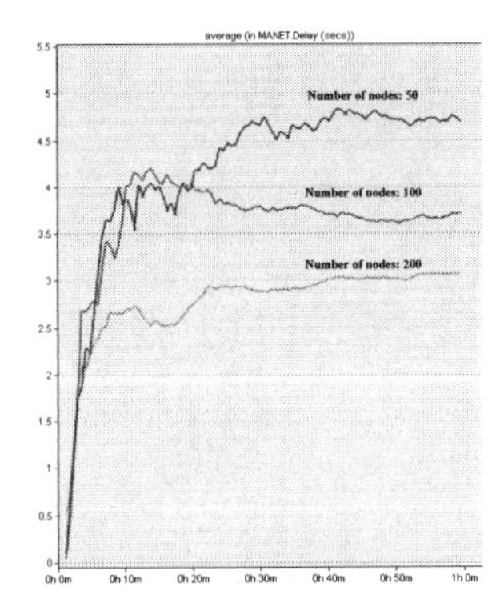

(a) Average End-to-end delay (seconds)

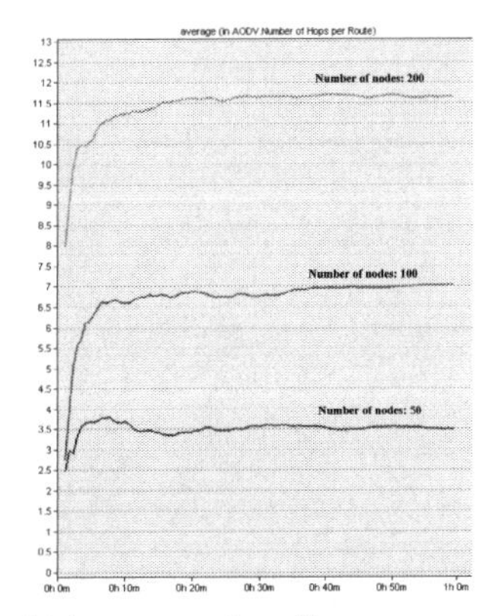

(b) Average number of hops per route

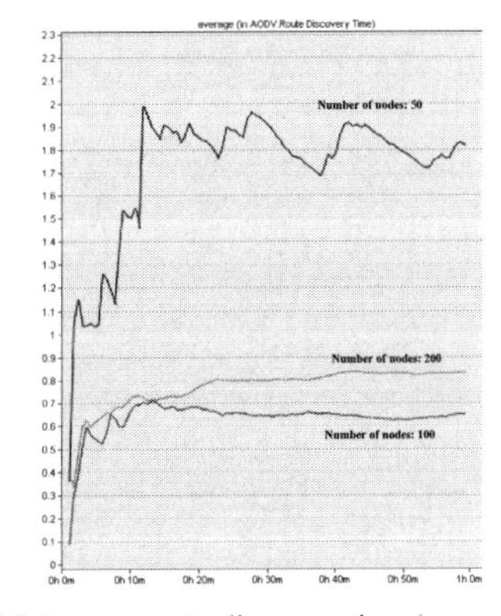

(c) Average route discovery time (seconds)

Figure 2. Comparisons between performance with varying number of nodes.

3.4 *50, 100 nodes deployed in the area 1000m * 1000m, transmission range from 100m-600m*

The delivery ratio is shown in Table 3 and the average end-to-end delay and routing performance are shown in Figure 3 and Figure 4. When there are 50 nodes, the overall network performance becomes better with higher connectivity. However, when there are 100 nodes, the performance becomes worse with higher connectivity due to interference caused by higher power of each node.

Table 3. The delivery ratio with different connectivity.

Transmission	Number of Nodes	
Range (m)	50	100
100	20.41%	53.61%
200	89.8%	85.57%
300	97.96%	47.42%
400	97.96%	28.87%
500	97.96%	16.49%
600	100%	14.43%

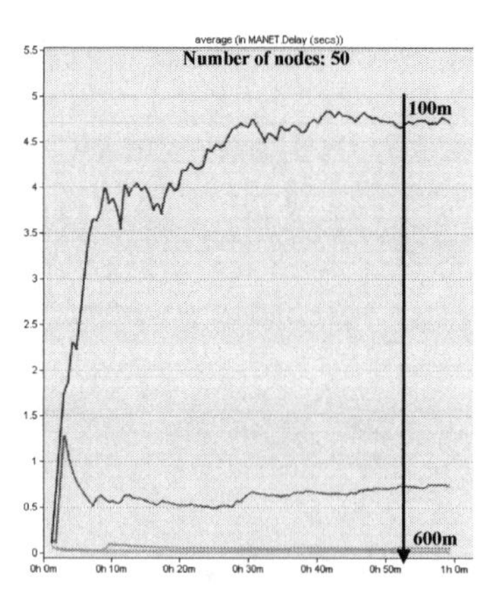

(a) Average End-to-end delay (seconds)

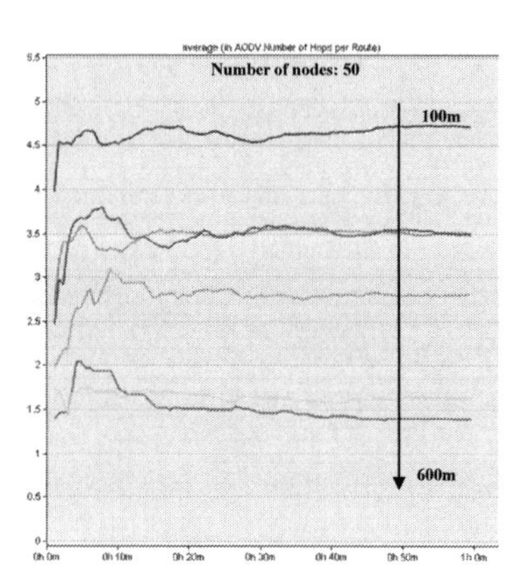

(b) Average number of hops per route

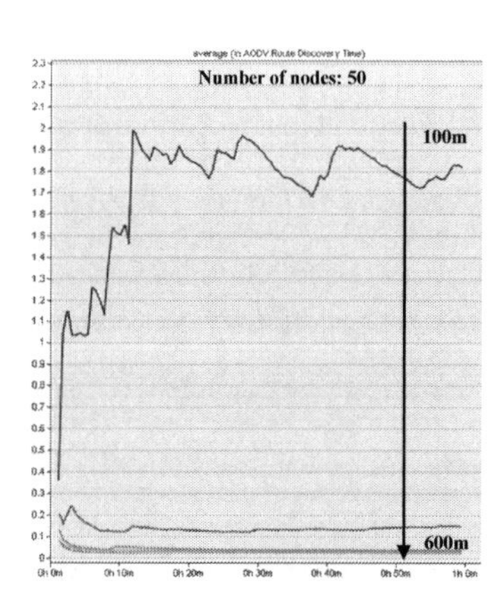

(c) Average route discovery time (seconds)

Figure 3. Network performance of 50-node scenarios.

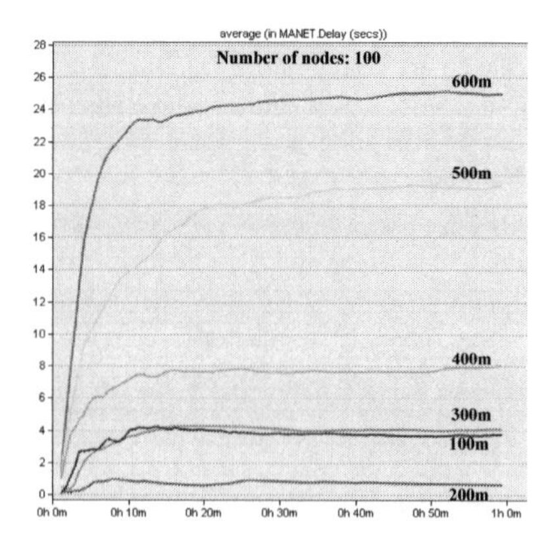

(a) Average End-to-end delay (seconds)

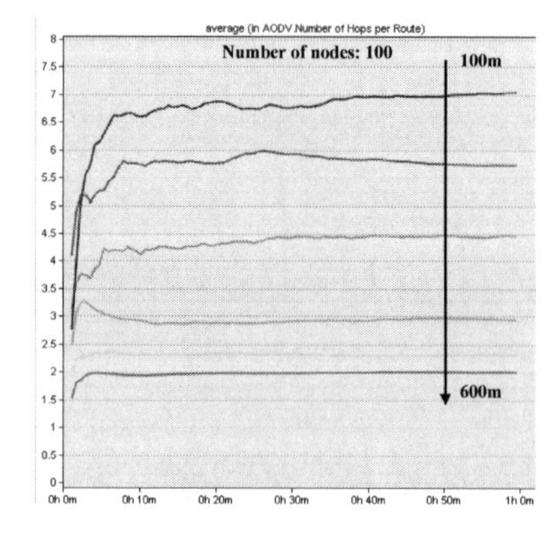

(b) Average number of hops per route

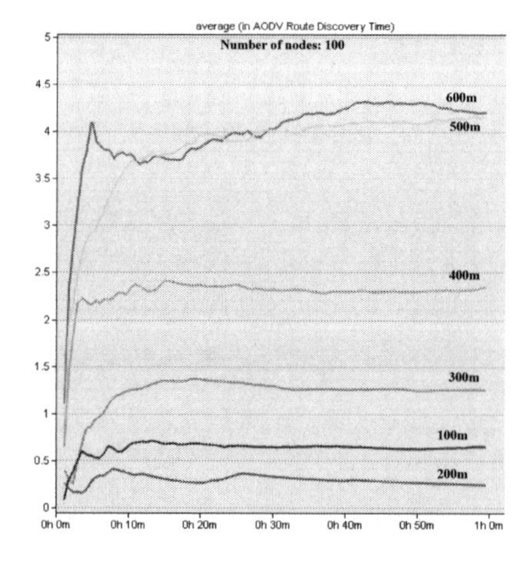

(c) Average route discovery time (seconds)

Figure 4. Network Performance of 100-node scenarios.

3.5 *100 nodes deployed in the area 1000m*1000m, 2000m*2000m, transmission range 600m*

The minimum node degree is 2, and the ratio is delivery 67.01% when 100 nodes with transmission range 600m are deployed in the area 2000m *2000m. The average end-to-end delay and routing performance are shown in Figure 5. When the area grows larger with the same number of nodes with the same transmission range, the network connectivity becomes lower which leads to more hops per route. However, the interference decreases dramatically within larger area, and the other three metrics are better which means that the overall network performance is better compared with that in a smaller area.

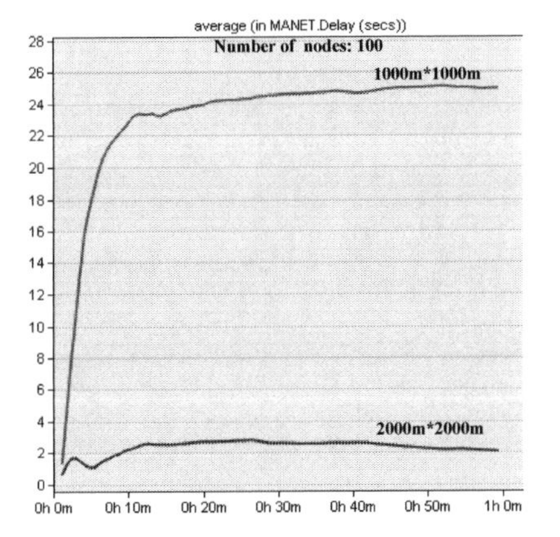

(a) Average End-to-end delay (seconds)

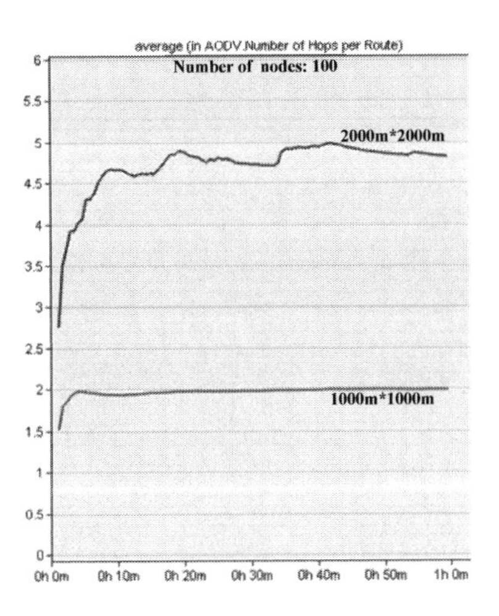

(b) Average number of hops per route

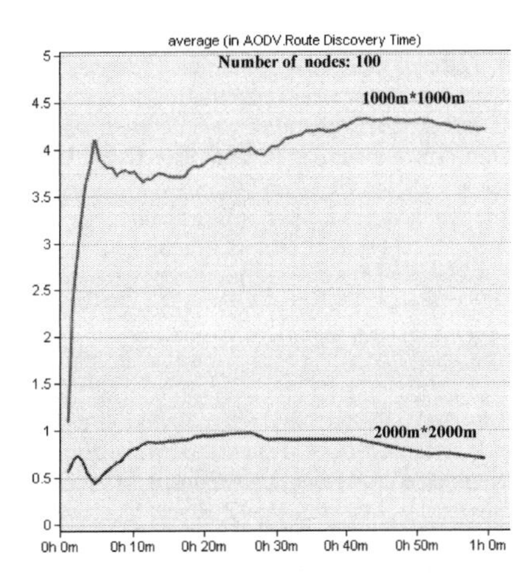

(c) Average route discovery time (seconds)

Figure 5. Network performance of 100-node scenarios deployed in different areas.

4 CONCLUSION AND FUTURE WORK

In this paper, we have studied and analyzed the performance of AODV with different network connectivity by varying network parameters through simulations. The simulations and analyses are useful for choosing the appropriate and optimum network design parameters while deploying MANETs for practical application. The simulation results show that more nodes or higher power of each node will not surely bring better performance in a specific area. This is because more nodes or higher power may lead to severer interference between nodes and congestion between packets, which decreases the network performance to a great extent. In the future, the research work will be extended to evaluate the effect of network connectivity on the performance of proactive and hybrid routing protocols of MANETs.

ACKNOWLEDGMENT

This work is supported by the Department of Electronic and Information Engineering, Yancheng Institute of Technology, China. We thank Yunping Li and Qian Liu for reading the draft of this paper and their valuable comments and suggestions.

REFERENCES

Bagwari, A. et al. 2012. Performance of AODV routing protocol with increasing the MANET nodes and its effects on QoS of mobile ad hoc networks. *International*

Conference on Communication Systems and Network Technologies, Rajkot, 11–13 May 2012. Piscataway: IEEE.

Bettstetter, C. 2001. Mobility modeling in wireless network: categorization, smooth movement, and border effects. *ACM SIGMOBILE Mobile Computing and Communications Review* 5(3): 55–67.

Bettstetter, C. 2002. On the minimum node degree and connectivity of a wireless multihop network. *The 3rd ACM International Symposium on Mobile Ad Hoc Networking &Computing, Boston, 9–11 June 2002*. Lausanne: ACM.

Goyal, P. 2012. Simulation study of comparative performance of AODV, OLSR, FSR & LAR, routing protocols in MANET in large scale scenarios. *World Congress on Information and Communication Technologies, Trivandrum, Oct. 30 - Nov. 2 2012*. Piscataway: IEEE.

Gupta, S. & Mathur, A. 2014. Enhanced flooding scheme for AODV routing protocol in mobile ad hoc networks. *International Conference on Electronic Systems, Signal Processing and Computing Technologies, Nagpur, 9–11 Jan. 2014*. Piscataway: IEEE.

Gupta, S. K. et al. 2013. Simulation and analysis of reactive protocol around default values of route maintenance parameters via NS-3. *International Conference on Information Systems and Computer Networks, Mathura, 9–10 March 2013*. Piscataway: IEEE.

Maurya, A. K. et al. 2014. Random waypoint mobility model based performance estimation of On-Demand routing protocols in MANET for CBR applications. *International Conference on Computing for Sustainable Global Development, New Delhi, 5–7 March 2014*. Piscataway: IEEE.

Perkins, C. et al. 2003. Ad hoc On Demand Distance Vector Routing. IETF RFC3561.

Royer, E. M. & Chai-Keong, T. 1999. A review of current routing protocols for ad hoc mobile wireless networks. *IEEE Personal Communications* 6(2): 46–55.

Saeed, N. H. et al. 2012. MANET routing protocols taxonomy. *International Conference on Future Communication Networks, Baghdad, 2–5 April 2012*. Piscataway: IEEE.

Santi, P. & Blough, D. M. 2002. An evaluation of connectivity in mobile wireless ad hoc networks. In *Proceedings of International Conference on Dependable Systems and Networks, 89–98, 2002*. Piscataway: IEEE.

Santi, P. & Blough, D. M. 2003. The critical transmitting range for connectivity in sparse wireless ad hoc networks. *IEEE Transactions on Mobile Computing* 2(1): 25–39.

Seon Yeong, H. et al. 2014. An application-driven path discovery mechanism for MANET routing protocols. *International Conference on Communications, Sydney, NSW, 10–14 June 2014*. Piscataway: IEEE.

Shivashankar, G. et al. 2014. Implementing a new power aware routing algorithm based on existing dynamic source routing protocol for mobile ad hoc networks. *IET Networks* 3(2): 137–142.

Shivashankar, H. N. et al. 2014. Designing energy routing protocol with power consumption optimization in MANET. *IEEE Transactions on Emerging Topics in Computing* 2(2): 192–197.

Thriveni, H. B. et al. 2013. Performance evaluation of routing protocols in mobile ad-hoc networks with varying node density and node mobility. *International Conference on Communication Systems and Network Technologies, Gwalior, 6–8 April 2013*. Piscataway: IEEE.

Venkatesan, T. P. et al. 2014. Overview of proactive routing protocols in MANET. *Fourth International Conference on Communication Systems and Network Technologies, Bhopal, 7–9 April 2014*. Piscataway: IEEE.

Wang, Zehua. et al. 2014. PSR: A lightweight proactive source routing protocol for mobile ad hoc networks. IEEE Transactions on Vehicular Technology 63(2): 859–868.

Zebra, W. et al. 2012. CORMAN: A novel cooperative opportunistic routing scheme in mobile ad hoc networks. IEEE Journal on Selected Areas in Communications 30(2): 289–296.

Electronics, Communications and Networks IV – Hussain & Ivanovic (eds)
© 2015 Taylor & Francis Group, London, ISBN: 978-1-138-02830-2

Network coding based MMF routing

Guan Xu, Bin Dai*, Jun Yang & Benxiong Huang
Department of Electronics and Information Engineering, Huazhong University of Science and Technology, Wuhan, China

Peng Qing
China Academy of Electronics and Information Technology, Beijing, China

ABSTRACT: To solve the fair maximum flow problems in multi-unicast networks, general solutions can fairly allocate available bandwidth to network flows according to the Max Min Fairness (MMF). However, existing solutions only consider available links on paths of existing flows directly, ignoring other links that may provide potential of higher fairness. In this paper, we propose a network coding based MMF routing algorithm to address the fair maximum flow problems by considering potential idle links. With potential links, we extend the fairness solutions by encoding flows according to the network configuration of inter-session network coding before bandwidth allocation. As a result, the intersection flows will be encoded into a set of new flows without the intersection link, shifting the offered load of links to idle links. The simulations with the OpenFlow protocol, verify that our algorithm can get higher fairness than MMF algorithm with little sacrifice in throughput.

1 INTRODUCTION

In communication networks, most network flows are type of best-effort traffic, transmitted individually with the flow bandwidth elastic to the variation of admitted bandwidth, which bring the maximum flow problem in traffic engineering. With the max-min fairness (MMF) (Nace & Pi´oro 2008) as a commonly theoretical support, the fair maximum flow problem can be solved by selecting flow paths and allocating flow bandwidth with the objective function of maximum fairness. To solve this issue, a number of studies on MMF routing algorithms for splittable flows or unsplittable flows have presented in literature.

For unsplittable flows with static routing, the fair flow optimization problem can be solved (Pi´oro & Medhi 2004) easily by choosing the minimal bandwidth per link along each path, as the flow amount and the average bandwidth of each link can be measured directly. However, the static routing brings low link utilization as the flows can be routed in alternative paths. For unsplittable flows with dynamic routing, the flow optimization problem can be optimized by designing the path selection functions on the length and cost of all candidate paths. With these routings, flows can be fairly installed on available links with the multi-paths diversity of each flow. However, the only links on feasible paths would be considered, which ignore other links that may be used to construct structured routings for existing flows. Therefore, we focus on the possible increment of fairness from fully utilizing the ignored links.

Compared with unsplittable flows, the splittable ones allow multi-path routing per flow, which makes the fair maximum flow problem computationally tractable (Hartman et al. 2012, Nace et al. 2006). Formulating the fair maximum flow problem, solvable algorithms for each node of general networks was proposed based on linear programming. In this work, we solve this problem with network-wide visibility of link resources in a centralized manner.

By the traditional methods, flows share links linearly on their flow weights. Alternatively, flows can be compressed in the network elements to share the links with higher efficiency. Among existing flow compress techniques, inter-session network coding routing (Khreishah et al. 2010) is promising for multiple unicast sessions with flow encoding on the shared links.

In the structure of inter-session network coding (NC) routings, encoding and decoding are determined by finding feasible paths of duplicate packets to decode the encoded packets. The chosen paths of duplicate packets from the source of a session to thesink of the others are extra paths compared to non-NC routings, also called the remedy paths. For the fairness maximum flow problem, we formulate the problem with the constraints of remedy paths,

*Corresponding author: *daibin@hust.edu.cn*

which consume additional links (Kiss et al. 2011) in the intersession network coding routings, compared to single path routing.

In routing algorithm design and implementation, there are some challenges. Firstly, the network coding routing algorithms are more practical for centralized flow control, as the key operations of tracking every possible path of sessions and inter-sessions would be implemented with less cost in control packets and algorithm complexity to synchronize the global information of topology. As most MMF algorithms are distributed, the stability of flows bandwidth will depend on the complexity of the convergence. Secondly, its incompatible to implement encoding and decoding operations for packets on general switches as the hardware architecture is not programmable.

Fortunately, the Software Defined Network (SDN) is a typical network for our algorithms, as it brings new programmatic abstraction for network management in the logical centralized control plane and the programmatic flow tables in the flow forwarding plane, leading to innovative approaches for traffic engineering (Egilmez et al. 2013, Curtis et al. 2011). In this paper, the main contributions are as follows,

We propose a network coding based centralized MMF routing algorithm (MMF-NC) in networks of multiple unicast flows, addressing the fair maximum flow problems by network coding routings with further utilization of available links. The results show that in fair maximum flow problems, the inter-session network coding routings can increase the fairness of existing MMF routings with affordable sacrifice of throughput.

We design the MMF-NC algorithm as a new application of the OFC s routing management, creating a bandwidth reservation (BR) API and extending the OpenFlow API in platform of OMNeT++. The fairness, throughput, and link utilization are measured and compared to the WaterFilling algorithm (WF) (Marbach 2003).

This paper is organized as follows. The related work is introduced in Section 2. In Section 3, we introduce the network setting and the formula of fairness routing. The MMF-NC algorithm is explained in Section 4. The simulation is shown in Section 5 and discussions in 6. Finally, we make the conclusion in Section 7.

2 RELATED WORK

To increase the fairness of bandwidth allocation, the flow splittability can provide multiple paths to guarantee fair bandwidth allocation. With multiple paths of minimal cost, the algorithms in optimizing splittable flows (Hartman et al. 2012) can balance the fairness and throughput for a set of flow demands in a given network. Danna proposed IEWF (Danna et al. 2012), a variation of WF algorithm for fairness optimization in multipath multi-commodity problems with a natural value vector converge. However, in practice, the optimal splittable flow solutions are shown to be mostly unsplittable (Danna et al. 2012), only a small proportion of flows has to be splittable. Our MMF-NC algorithm search network coding structures for saturated links, depending on the topology as well as flows of this link, which would only be used for saturated links with network coding opportunities.

The WF algorithm (Marbach 2003) solves the MMF problems, with the most saturated link detected and allocated in iterators until all flows are served or all links are used up. The main performance is evaluated by the lowest bandwidth allocation, the link utilization, and the throughput. We follow their design method, and design our extension by adding network coding aware checking and network coding structure searching in each iterator, as encoding flows decrease the flow number of the most saturated link.

With data flow reservation providing individual or groups of data flows QoS guarantees from the Northbound API, the OpenFlow network can change the routing and bandwidth of each flow (Egilmez et al. 2013) in OpenFlow Controller (OFC). We also consider the flow reservation request from northbound API as the input of our algorithm. With OFSs statistically detecting elephant flows, DevoFlow (Curtis et al. 2011) only has to response elephant flow requests with routings by a decreasing best-fit bin packing algorithm, with the input of flow visibility and global topology. Adding bandwidth guarantee to the OFHosts flow demands as an additional input, our centralized algorithm finds out an optimal guaranteed routing with bandwidth allocation for OFHost.

The detection of the most saturated link and its largest flow has been implemented as a NOX application (Mann et al. 2012), based on which the VM scheduling decisions made. Similar to this design, we also make use of the detection of links and flows, and the information of application layer to calculate an optimal allocation result.

3 SETTINGS AND FORMULATION

We study the fair maximum flow problem in the undirected network G=(E,V), with each link e of capacity c_e. We assume that all flows are unsplittable, so each flow f_d will be routed by a single path routing from source s_d to destination t_d with a flow bandwidth x_d. The flow bandwidth of existing flows, denoted by $\mathbf{x}=\{x_d\}$, can be arranged in non-decreasing order, which is also called lexicographic order(Nace & Pi´oro 2008). The lexicographically maximum objective is shown in the following link path formula,

$$\text{lexmax } \mathbf{x}(t)$$

$$\sum_{i,j} \delta_{eij} x_{ij} + \sum_{d} \delta_{ed} x_d < c_e$$

$$e = 1, 2, ..., E \quad (1)$$

where δ_{ed} stands for whether link e is on the path of flow d, δ_{eij} means the same for link e to the network coding routing of different flows f_i *and* f_j, as the original flows will be encoded into a new flow with a single demand and a network coding routing. This objective function can be optimized with a unique solution as the objective function $\mathbf{x}(t)$ is concave.

4 NETWORK CODING BASED MAX MIN FAIR ROUTING ALGORITHM

For the fair maximum flow problems, the WF algorithm was used to find a routing path and allocate bandwidths for existing flows fairly, to get a lexicographical maximization (Marbach 2003). For unsplittable flows with fixed paths, the WP algorithm can be implemented with the pseudocode shown in algorithm 1, as an example of a simple centralized algorithm.

Algorithm 1: WaterFilling

Require: link capacities C_e, flow paths $P = P_d$, x_{ed}, **return** flow bandwidth x_d.
1: Initialization: $\mathbf{x}^0 = 0$, $k = 0$
2: **while** $P \neq 0$ **do**
3: $k = k + 1; x_d^0 = P(x_{sat})$
4: $c_e = c_e - x_d^0, \forall e \in \{e \mid \delta_{ei} = 1\}, P = P - P_d$
5: **if** $x_d \neq 0, \forall f \in F$ **then** break;
6: **end if**
7: end **while**

This algorithm consists of iterations of finding the maximum of the minimal bandwidth of existing links, as a linear programming problem with formula,

$$\max x_{sat} \quad (2)$$

$$\sum_{e \in P_d} x_d \leq C_e \quad (3)$$

$$x_d \geq x_{sat} \quad (4)$$

where (3) shows that the flows bandwidth of a link would not exceed the link capacity, and (4) guarantees the order of fairness that all unallocated flows bandwidth x_d will be no less than x_{sat}, the one to be allocated.

In the fair bandwidth allocation, each flow cannot get more bandwidth without any sacrifice of the lower flows bandwidth. In the case of unsplittable flow and fix path, the problem $P(x_{sat})$ can be calculated directly by calculating the average bandwidth of each link.

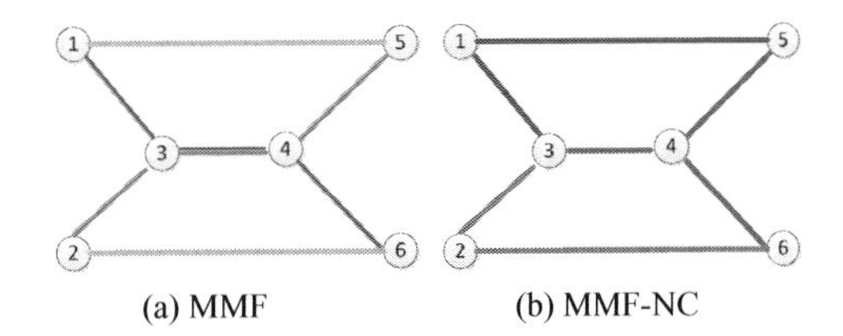

(a) MMF (b) MMF-NC

Figure 1. In a butterfly structured topology of 6 nodes, one flow is generated from node 1 to node 6 in purple, and the other is from node 2 to node 5 in blue. The encoded flow is in green. (a) With MMF algorithm, the link (3,4) is the bottleneck; (b) with MMF-NC algorithm, the bottleneck doesn't exist.

With the inter-session network coding routings (Wang & Shroff 2007), the fairness may be improved. For example, in a directed acyclic network of butterfly structure with link capacity 1 in the Figure 1, two flows with unlimited demands are routed with general MMF and MMF-NC algorithm separately. The fairness for each flow is 0.5 for MMF algorithm, shown in Figure 1(a), with the saturated link (3, 4) shared by the two flows and other links "idle"; while the fairness of each flow will be 1, in the latter algorithm, as the two flows are suitable for inter-session network coding routing, shown in Figure 1(b), replacing two original flows with an encoded flow on the link (3, 4), which eliminate the saturated link. With flows on shared links encoded, the pairwise intersession network coding routings for two flows can increase the fairness of both flows with addition cost of extra remedy paths. As a result, the average of flow bandwidth the most saturated link e_i will change from c_i/a_i *to* $c_i/(a_i-1)$, as encoding flows decreases the flow amount in the shared links.

For MMF-NC routing, the fairness problem $P(x_{sat})$ in step 3 will be modified as,

$$\max x_{sat} \quad (5)$$

$$\sum_{e \in P_d} x_d + \sum_{e \in P_i, e \in P_j, P_i, P_j exists} x_{ij} \leq C_e \quad (6)$$

$$x_d \geq x_{sat} \quad (7)$$

The link-path constraints show the network coding routings P_{ij} as the original flows are uniformly assigned by the bandwidth of encoded flow x_{ij}. Based on this formula, the WF algorithm can be modified with network coding (shown in algorithm 2), as network coding routings on saturated links can compress the load of existing flows.

For unallocated flows, the most saturated link is selected as the lowest value of average bandwidths for existing available links in step 3. After the saturated link e_i is selected, all pairs of flows on e_i are checked for network coding routings.

Recall that in a fair solution, each flows bandwidth cannot be increased unless sacrificing the flows with lower bandwidth. For flows routed by network coding routing, the remedy paths may bring fairness sacrifice to other flows. So to minimize the fairness sacrifice, the remedy paths would be chosen with a guarantee that other unallocated flows can be routed at the same bandwidth. With this guarantee, all unallocated flows are pre-allocated the same bandwidth of $c_i/(a_i-1)$ from the residual graph in step 4, so that the network coding routing will not decrease the bandwidth fairness of the other flows, especially the flows with lower bandwidth.

By inter-session network coding routings, two different flows are encoded to be a new flow with the bandwidth calculated individually. So in step 6 and 13, a pair of different flows is randomly chosen from the unallocated flows on link e.

In step 8 inter-session network coding routing is searched in a centralized way, with structures similar to the butterfly or grail structures (Wang & Shroff 2007), proved to be feasible structures for 2 flow inter-session network coding routing. Once the network coding routing is feasible, the encoded flow will replace the original two flows with the last saturated link relaxed. As a result the algorithm will go to step 3, checking for another saturated link. If no network coding routing exists, the unallocated flows on saturated links will be allocated with x_{sat} and labeled as allocated flow. For the algorithmic complexity and scalability, only encoded flows will not to be encoded again for another network coding routing in MMF-NC algorithm to reduce the complexity.

After the algorithm, the routing and bandwidth for each flow will be optimized. As an example of successful instance, the MMF and MMF-NC algorithm are compared for the example of the Figure 1. The admitted flow bandwidth for each flow can be increased about 100%.

5 SIMULATION

In this section, we evaluate the fairness of bandwidth allocations for multiple unicast sessions in a packet level simulation in OMNeT++. With topology and traffic given in the following subsection, different algorithms are compared.

5.1 *Implementation*

To implement MMF-NC algorithm in OpenFlow networks, a set of services for MMF-NC routing will be created on the OpenFlow Controller, while the OpenFlow switches are NC enabled.

Flows are divided into guaranteed flow with reserved bandwidth and normal flows without band width allocation. The guaranteed flows are generated by the OpenFlow supported hosts (OFhosts) with a bandwidth guarantee through the Northbound Routings (NR) requests to OFC. The unlimited flows are directly routed by the shortest path algorithm via the PACKET IN[1] messages.

Algorithm 2: MMF-NC

Require: link capacities c_e, flow paths $P=\{P_d\}$, x_{ed}
return flow bandwidth x_d
1: Initialization:x_0=0, k=0;
2: **while** $P \neq 0$ **do**
3: k=k+1; x_d^0=P(x_{sat}) with link e
4: Pre-allocate all unallocated flows a bandwidth $c_e/(a_e-1)$
5: Set **F** equal to all unallocated flows on link e
6: Select different flows $f_i, f_j \in$ **F**
7: **while** f_i and f_i exist **do**
8: try finding the network coding routing
9: **if** network coding routing is feasible **then**
10: Set x_{eij}, δ_{eij} ;
11: BREAK
12: **else**
13: Choose another pair of different flows for $f_i, f_j \in$ **F**;
14: CONTINUE
15: **end if**
16: **end while**
17: **if** there is a network coding routing **then** CONTINUE
18: **end if**
19: $c_e = c_e - x_d^0, \forall e \in \{e \mid \delta_{ei} = 1\}, P = P - P_d$
20: **if** $x_d \neq 0, \forall f \in F$ **then** break;
21: **end if**
22: **end while**

According to the concept of northbound API (Ferguson et al. 2013), we implement communication between OFhosts and OFC. OFhosts reserve a guaranteed flow by sending OFC a NR request message with request with the routing destination and the guaranteed bandwidth, and then start the flow on receiving the corresponding admission from OFC. The OFC will calculate an available routing for receiving NR request, install routing into routing tables of switches along the routing, and then reply with the allocated bandwidth. The OFC also can start the routing algorithms at the link bandwidth variation, which can be detected by the OFS. The OpenFlow protocol message OFPMP PORT DESCRIPTION can be used to inform OFC current and max speed of existing OFS ports.

---

[1] http://www.opennetworking.org

A routing path can be installed or updated by configuring each OFS from the end to the head along the path, and can be removed only by removing flow entry in the head OFS of the path. A network coding routing is a directed graph, which can be installed from the two end OFSs, to the encode/decode OFS, and then to the head OFSs of the graph. The removal of a network coding routing can be achieved in the opposite order.

The basis modules we add are Topology Management (TM), OFS Management, MMF Routing and MMF-NC. TM discovers and maintains network connectivity through controller-to-switch messages and asynchronous messages; OFS Management detects congested links by collecting link utilization counted per port on the OFS periodically; MMF Routing calculates fairly allocated routing paths for existing flows. If MMF Routing finds no feasible solution, MMF-NC will be used.

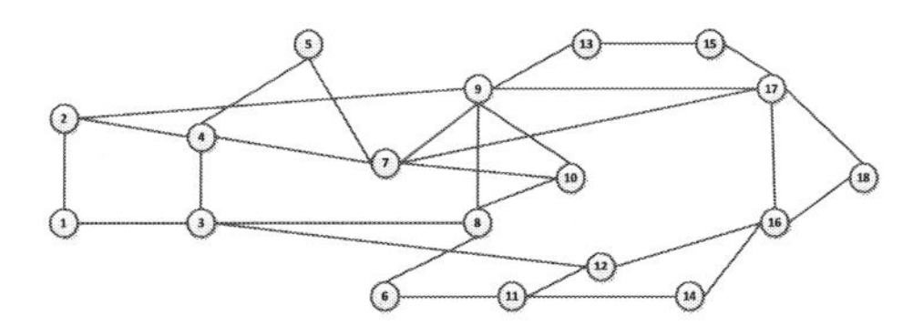

Figure 2. This is a scenario similar to the topology of the MCI core network.

To implement the MMF-NC function on OFC, we implement modules of Residual Topology and Flow management for existing flows and spare link capacities. To support NC actions on OFS, flow tables are programmed, NC buffer is created for each NC flow.

5.2 *Simulation setup*

The simulation is built on an OpenFlow protocol network platform with OpenFlow switches and OpenFlow controller implemented in an OpenFlow Extension for the OMNeT++ INET framework (Klein & Jarschel 2013). As a baseline algorithm, the WF algorithm is also implemented on the same platform, with the same simulation setup for MMF-NC routing as follows,

Topologies: To evaluate our routing algorithms, we build a butterfly structured topology, the original structure for inter-session network coding routings, and a general bi-directed topology similar to the MCI internet backbone topology (Figure 2), which general for traffic load analysis (Ma & Steenkiste 1997). End hosts are randomly attached to a switch, a node in the topology, in the initialization stage.

Traffic: Each traffic flow is assumed to be random in location and demand, with an unsplittable routing path. The bandwidth of each flow is controlled by OFC. The unlimited bandwidth flows are generated at the same probability. The flow demands are always set greater than the capacity of the link.

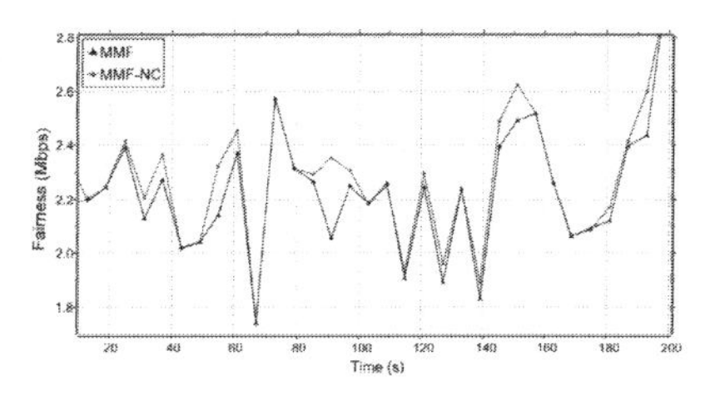

(a) Fairness

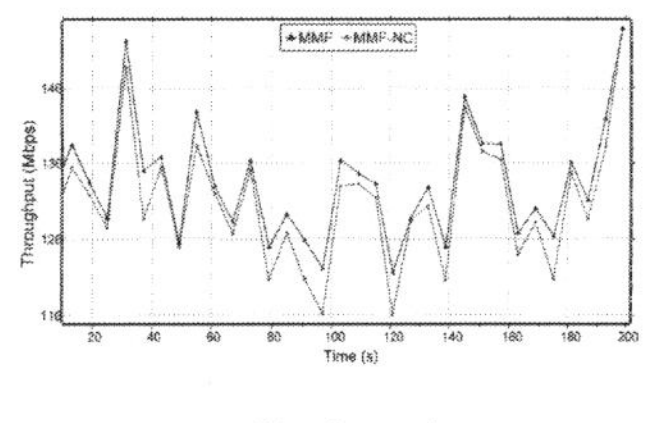

(b) Throughput

Figure 3. The performance of MMF and MMF-NC algorithms in general networks. (a) shows the fairness of the network, while (b) shows the throughput of the same simulation.

Link channel: Assume that all saturated links drop packets randomly. Hosts are sending flows randomly with a flow rate measured periodically and with the lowest sending rate for the network fairness.

5.3 *Result*

For general networks, we consider the MCI structured topology of 29 switches, on which 31 sending hosts and 31 receiving hosts are randomly connected to a switch. Each sending host transmits data to a randomly selected receiving host, with a bandwidth demands from 8Mbps to 9.6Mbps, which is closed to the link capacity. When this transmission is closed, the sending host will sleep from 4ms to 40ms, and then select a receiving host to start a new flow so that there are totally at most 31 flows in the network.

To compare the performances of different routing algorithms, the fairness and throughput are measured periodically per 6s in each run. For fairness, we just compare the minimal flow rate, which is a sufficient condition for the lexicographical comparison of two sets of flow rates. The results in Figure 3 show that the fairness in MMF-NC is no less than the one in MMF all the time, with the fairness increased by 2.91% and the throughput decreased by 1.26%. The fairness improvement of MMF-NC is not significant especially at time around t=70, as bandwidth allocation for network coding routings may bring unfairness to higher rate for uniformly distributed traffic when the extra paths increase the load of some links. The link utilization increments in MMF-NC are no more than 3% higher than the one in MMF for most links, which means that the MMF-NC is seldom used in the MCI topology with random flows.

For traffic of real-time networks, we consider a butterfly structured network similar to the one in Figure 1, which consists of 1 controller, 8 OpenFlow switches, from OFS to OFS7, 10 sending hosts and 10 receiving hosts (Figure 4), with the same traffic pattern for the last experiment which consist of at most 10 flows coexisting in the network.

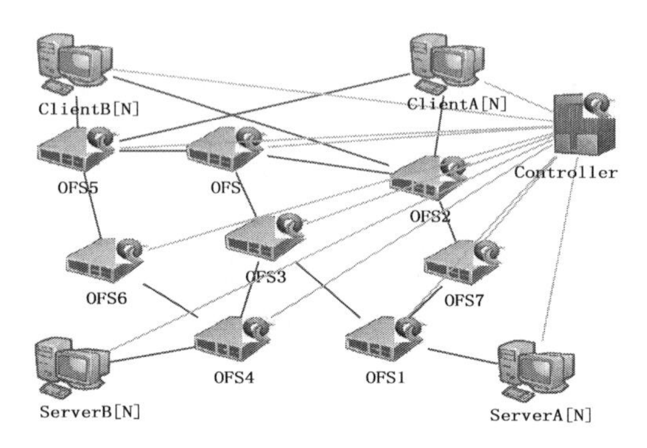

Figure 4. The simulation topology of butterfly structure with 2N flows.

In Figure 5, the fairness of MMF-NC is increasing 24.4% with a throughput sacrifice of 24.2% in average, which shows that network coding is a feasible way of achieving fairness with the additional cost of extra traffic load on idle links. The throughput of MMF can steadily approach to 2.8Mbps, as each flow requires unlimited bandwidth in our assumption. The network throughput is not so stable in MMF-NC, which may be caused by the possibility that the remedy paths would conflict with other flows paths. Compared to examples in Figure 1, the fairness gain is not significant, because sometimes the link for remedy paths may be subscribed by other flows.

The link utilization (in table 1) shows that 4 links of the original paths are increasing about 30%, which provide network coding gains for admitted flow bandwidth; other links are increasing no more than 5%, which means a small extra cost for remedy paths. It is considered that the improvement is mainly due to the load reduction on saturated links (3,0). From the first two experiments, MMF-NC algorithm is shown to bring significant fairness for resource allocation in simple butterfly topology.

The last two experiments show that there will be a high probability of network coding gain for MMFNC algorithm in networks with some links saturated, which is consistent with the fact that MMF-NC can improve the fairness.

Table 1. Link utilization of each link for the second experiment.

Link	MMF-NC	MMF	Growth Ratio
1,3	0.65	0.49	0.32
1,7	0.98	0.93	0.053
7,2	0.98	0.93	0.053
0,2	0.67	0.51	0.31
3,0	0.98	0.98	0
4,3	0.68	0.53	0.29
4,6	0.98	0.97	0.02
6,5	0.98	0.97	0.02
0,5	0.64	0.47	0.35

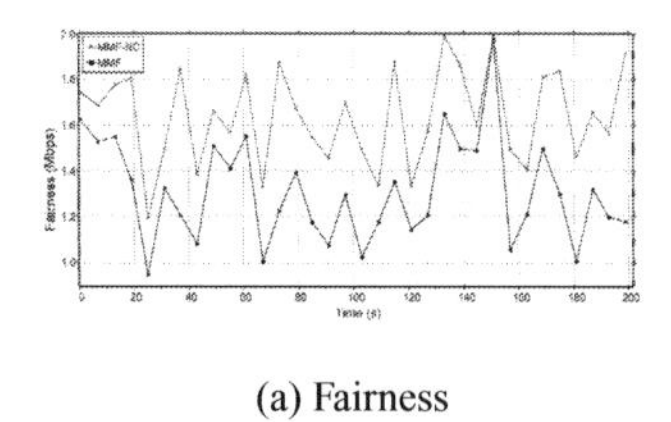

(a) Fairness

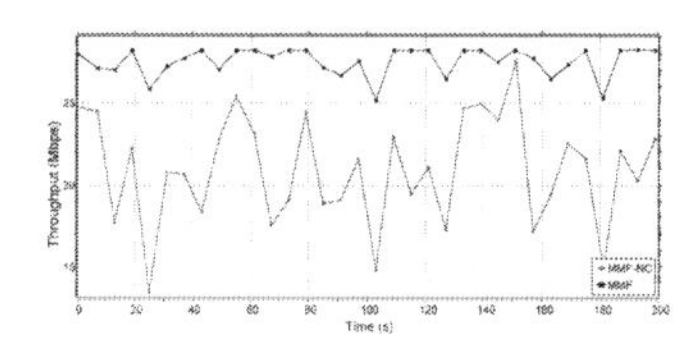

(b) Throughput

Figure 5. The performance of MMF and MMF-NC algorithms in butterfly structured networks. (a) shows the fairness of the network, while (b) shows the throughput of the same simulation.

6 DISCUSSION

The simulation results show that our algorithm provides a better, fairer than a general MMF algorithm, several improvements can be employed for future studies. Making use of inter-session paths instead of paths for each session, network coding routings intuitively bring no sacrifice to the bandwidth of original flows; however, it may bring sacrifice to other flows bandwidth. Network coding routings consume links more than original routings, so in MMF-NC algorithm, the cost of network coding routings should be considered in future. For example, network coding routings would be used as urgent schemes for heterogeneous traffics, where there are enough links for network coding routings besides the bandwidth of links for original flows.

The network coding routing is used as a relaxation on bandwidth allocation, which might change the order of flow bandwidth vector and bring fluctuation for calculation. In our work network coding routing is chosen after pre-allocating the bandwidth to all unallocated flows, so that the selected network coding routing would not influence the flows with lower bandwidth. As a result, the network coding routing would not exist since that the per-allocating bandwidth is too high to find a feasible remedy path. In future work, the pre-allocated bandwidth can be varied from the value $c_i/(a_i\text{-}1)$ to c_i/a_i, with a certain schedule method to get the network coding routing with optimal bandwidth allocation.

When searching for network coding routings with the butterfly and grail structure (Wang & Shroff 2007), remedy paths are chosen by the shortest path algorithm, with the constraints that new network coding routings would not decrease the flows of lower flow bandwidth allocations. In future work, the remedy paths can be chosen by selecting the path of links with low flow density, which would be a maximum flow problem for both network coding and normal flows.

For weighted MMF problems, flows with different weight would share links proportional with the weight. If a network coding routing is feasible to eliminate the bandwidth allocation for flows of different weight, then the shared link would allocate bandwidth to the original flows as an encoded flow, which is different from normal MMF solutions. And the encoded flows will have a new weight between the weights of original flows, and can be used for optimization with other flows. That also applies to the splittable flows. So for future work, the weighted flow fairness problem will be solved by network coding based MMF algorithms.

7 CONCLUSION

In this paper, we propose a network coding based MMF algorithm to fairly allocate bandwidth for multiple unicast networks, with implementation and verification in OpenFlow protocol. With a better fairness compared to the traditional MMF, the MMF-NC is shown to be an efficient improvement of traditional MMF algorithms. In the future, we will consider the optimization in the extra cost of network coding routings, and the details in the implementation of network coding routing in the discussion.

ACKNOWLEDGEMENTS

This work was supported by the National Key Technology Research and Development Program of the Ministry of Science and Technology of China under Grant no. 2012BAH93F01, the Innovation Research Fund of Huazhong University of Science and Technology, No. 2014TS095 and the National Science Foundation of China under Grant no. 60803005.

REFERENCES

Curtis, A. R., Mogul, J. C., Tourrilhes, J., et al.2011. Devoflow: scaling flow management for high-performance networks. *SIGCOMM Comput. Commun. Rev.*41(4): 254–265.

Danna, E., Hassidim, A. Kaplan, H., et al. 2012. Upward max min fairness. *INFOCOM, 2012 Proceedings IEEE, Orlando, 25–30 March 2012.* IEEE.

Egilmez, H., Civanlar, S. & Tekalp, A. 2013. An optimization framework for QoS-enabled adaptive video streaming over openflow networks. *IEEE Transactions on Multimedia* 15(3): 710–715.

Ferguson, A. D., Guha, A. Liang, C. Fonseca, R. & Krishnamurthi, S. 2013. Participatory networking:An API for application control of sdns. *SIGCOMM Comput. Commun. Rev.* 43(4): 327–338.

Hartman, T., Hassidim, A., Kaplan, H., Raz,D. & Segalov, M. 2012. How to split a flow? *INFOCOM, 2012 Proceedings IEEE, Orlando, 25–30 March 2012.*IEEE.

Khreishah, A., C.-C. Wang, & N. B. Shroff. 2010. Rate control with pairwise intersession network coding. *IEEE/ACM Transactions on Networking* 18(3): 816–829.

Kiss, Z. I., Polgar, Z. A. Giurgiu, M. & Dobrota, V. 2011. Resource efficient network coding based congestion control for streaming applications. *2011 34th International Conference on Telecommunications and Signal Processing (TSP), Budapest, 18–20 August 2011.* IEEE.

Klein, D. & Jarschel, M. 2013. An openflow extension for the omnet++ inet framework. *Proceedings of the 6th International ICST Conference on Simulation Tools and Techniques, Cannes, 5–7 March 2013.* Brussels: ICST.

Li, L., Widmer, J. & Yin, H.(Eds.). 2012. *NETWORKING 2012*, Volume 7289 of *Lecture Notes in Computer Science*: 190-204. Berlin: Springer.

Ma, Q. & Steenkiste, P. 1997. On path selection for traffic with bandwidth guarantees. *Network Protocols, 1997. Proceedings., 1997 International Conference on. Atlanta, 28 October 1997.* IEEE.

Mann, V., Gupta, A., Dutta, P., Vishnoi, A., Bhattacharya, P., Poddar, R. & Iyer, A. 2012. Remedy:Network-aware steady state vm management for data centers. *11th International IFIP TC 6 Networking Conference, Prague, Czech Republic, May 21–25, 2012.* Berlin: Springer.

Marbach, P. 2003. Priority service and max-min fairness. *IEEE/ACM Transactions on Networking* 11(5): 733–746.

Nace, D. & Pi´oro, M. 2008. Max-min fairness and its applications to routing and load-balancing in communication networks: A tutorial. *IEEE Communications Surveys and Tutorials* 10(4): 5–17.
Nace, D., Doan, N.-L., Gourdin, E. & Liau, B. 2006. Computing optimal max-min fair resource allocation for elastic flows. *IEEE/ACM Transactions on Networking* 14(6): 1272–1281.
Pi´oro, M. & D.Medhi. 2004. *Routing, flow, and capacity design in communication and computer networks*. San Francisco:Elsevier.
Wang, C.-C. & N. B. Shroff. 2007. Beyond the butterfly - a graph-theoretic characterization of the feasibility of network coding with two simple unicast sessions. *IEEE International Symposium on Information Theory, Nice 24–29 June 2007*. IEEE.

Electronics, Communications and Networks IV – Hussain & Ivanovic (eds)
© 2015 Taylor & Francis Group, London, ISBN: 978-1-138-02830-2

Designing low rate LDPC codes with iterative threshold close to the channel capacity

Ming-Yang Xu*, Xiang-Yu Wang, Shulong Han & Song Yu
State Key Laboratory of Information Photonics and Optical Communications, Beijing University of Posts and Telecommunications, Beijing, China

ABSTRACT: Low rate LDPC codes are required to ensure the reliability of channel transmission in the low signal-to-noise ratio regime. In this paper, we design three low rate LDPC codes by employing density evolution and progressive-edge-growth algorithm. The code efficiencies of the LDPC codes are larger than 90%, and can be widely used in the low signal-to-noise ratio regime.

1 INTRODUCTION

Low rate codes ensure the reliability of channel transmission in the low signal-to-noise ratio regime (Andriyanova & Tillich 2012), such as wireless sensor network. In the communication system with low signal-to-noise ratio and precious channel resource, we hope to improve the performance of the codes, meaning that we can reduce the redundant codes to ensure the reliability of channel transmission. We choose low density parity check codes (Gallager 1962) as the channel coding as a result of the LDPC codes with high performance (Luby et al. 1998, Brink 1999) and low decoding complexity (Zhang & Paihi 2001).

Although regular and irregular LDPC codes have gained very good results, they behave very poorly in the low signal-to-noise regime. In the low signal-to-noise regime, we need to design a good low rate LDPC code with an iterative threshold close to the channel capacity. According to the analysis of LDPC codes, we have to satisfy two conditions (Andriyanova & Tillich 2012) to design low rate LDPC codes with high performance. One condition is to make LDPC code with high performance use only low degrees and the other is to make minimum distance large, which is necessary to obtain a low error floor. The two conditions are in conflict with each other (Tillich & Zemor 2006). If we satisfy the requirement of the first condition, we need to design a large proportion of variable nodes of degrees 1 and 2. But it doesn't meet the second condition for a regular or irregular LDPC code with a large number of variable nodes of low degrees. We introduce multi-edge type LDPC codes to circumvent this contradiction.

The paper is organized as follows. In section 2, we review the multi-edge type LDPC codes and the cascaded constructions for degree one variable nodes. In section 3, we design three degree distributions of low rate LDPC codes and analyze their performances. We conclude the main content of this paper in section 4.

2 MULTI-EDGE TYPE LDPC CODES

2.1 *Parameterization of multi-edge type LDPC codes*

In order to understand multi-edge type LDPC codes, we review the parameterization of multi-edge type LDPC codes and introduce the meaning of the various parameters.

Supposing that we have mastered the regular and irregular LDPC codes completely, we review multi-edge LDPC codes further. Regular and irregular LDPC codes are only defined by degree distributions, so all variable nodes or constraint nodes are statistically interchangeable (Richardson & Urbanke 2004). In other words, there is only one class of edges in the Tanner graph representation of regular and irregular LDPC codes. Now we consider the LDPC codes with multiple classes of edges. In general regular and irregular LDPC codes are defined from the edge perspective. In the multi-edge type LDPC codes, the edge perspective can't be used because there is no single edge type connected to every node type. Therefore, in the multi-edge type LDPC codes, it is more convenient for us to define it from a node perspective (Richardson & Urbanke 2004).

In order to simplify the analysis, we assume the low rate LDPC codes are applied to channel codes in the communication system with BPSK modulation and Gaussian channel. To represent multi-edge type LDPC codes from a node perspective, we introduce the following multinomial representation. Exponents

*Corresponding Author: *xumingyang316@126.com*

are defined as degrees. Let $x=(x_1, x_2, \ldots, x_n)$ denote various classes of edges and let $d=(d_1, d_2, \ldots, dn)$ be a degree distribution (Richardson & Urbanke 2004). In order to simplify the description, the following expression is introduced (Richardson & Urbanke 2004)

$$\vec{x}^{\vec{d}} = \prod x_i^{d_i} \tag{1}$$

This expression means a type node connected by various types of edges. A Tanner graph is defined by two polynomials. One polynomial is associated to variable nodes and the other is associated to constraint nodes. We denote these polynomials by (Richardson & Urbanke 2004)

$$v(\vec{x}) = \sum v_{\vec{d}} \vec{x}^{\vec{d}}, u(\vec{x}) = \sum u_{\vec{d}} \vec{x}^{\vec{d}} \tag{2}$$

Now we explain these coefficients. Let N be the length of the codeword. The quantity vN is the number of variable nodes in the Tanner graph. Similarly, the quantity uN is the number of constraint nodes in the Tanner graph.

According to the definition of multi-edge type LDPC codes, we achieve code rate by (Richardson & Urbanke 2004)

$$k = v(\vec{1}) - u(\vec{1}) \tag{3}$$

In order to understand the expression, let us multiply the rate by N then the result is the number of variable nodes minus the number of constraint nodes. This is the number of independent variable nodes. According to the definition of code rate, dividing this by N, we can achieve the code rate.

Let me introduce some additional notation. We introduce (Richardson & Urbanke 2004)

$$v_{x_i}(\vec{x}) = \frac{d}{dx_i} v(\vec{x}), u_{x_i}(\vec{x}) = \frac{d}{dx_i} u(\vec{x}) \tag{4}$$

As each type of edges is connected between variable nodes and constraint nodes, so the number of edges of each type is equal from variable node perspective and constraint node perspective. We introduce the following formula to explain the relationship (Richardson & Urbanke 2004).

$$v_{x_i}(\vec{1}) = u_{x_i}(\vec{1}), i = 1, 2, \cdots n \tag{5}$$

where $\vec{1}$ denote the vector of all 1's and the length of vector is n.

2.2 *Density evolution of multi-edge type LDPC codes*

Density evolution describes the behavior of the decoder with the infinite block length. When decoding with MP algorithm, messages are passing between variable nodes and constraint nodes. Density evolution of regular and irregular LDPC codes has been studied in some papers (Wei & Akansu 2001). We just discuss density evolution of multi-edge type LDPC codes in this paper. The main difference between the two is that message densities emerging from the different node types are averaged. In regular and irregular LDPC codes, they are averaged across the different degrees. In multi-edge type LDPC codes, they are averaged across node types independently for each edge type. In the regular and irregular LDPC codes, messages are represented by a single density function, since all edges are statistically equivalent. In the multi-edge type LDPC codes, messages are represented by n density functions, one for each edge type (Richardson & Urbanke 2004).

Density evolution is represented by the following recursion (Richardson & Urbanke 2004)

$$P^{l+1} = \lambda(R, \rho(P^l)) \tag{6}$$

Here, the multiplication operation inside $\rho(P^l)$ is the convolution of densities in the constraint node domain and inside $\lambda(R,Q)$ it is the convolution of densities in the variable node domain (Richardson & Urbanke 2004).

After introducing density evolution theoretically, we describe how to implement practically. Gaussian approximation of density evolution is an efficient method (Chung et al. 2001). In order to simplify operation, we replace probability density function by the means of probability density function. Thus we can turn multidimensional operation into one-dimensional operation.

2.3 *Cascaded constructions for degree one variable nodes*

Through the understanding of multi-edge type LDPC codes, now we focus on the cascaded constructions for degree one variable nodes (Richardson & Urbanke 2004), which is a special construction of multi-edge type LDPC code.

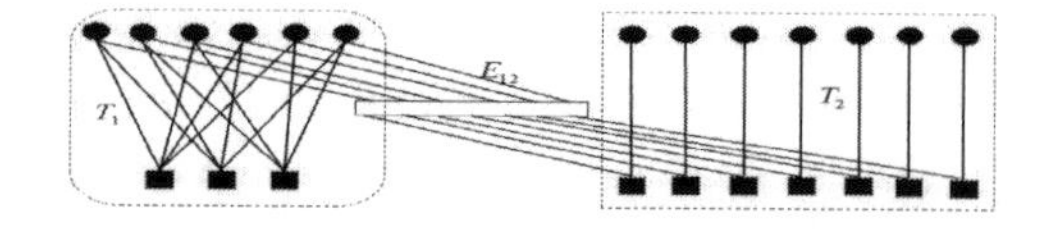

Figure 1. Cascaded constructions for degree one variable nodes Tanner graph. The structure of Tanner graph T_1 is designed to make minimum distance big enough.

This structure contains three types of edges, we can refer to Figure 1. The Tanner graph T_1 is a standard irregular LDPC codes structure and it is favorable for the minimum distance to increase with growth of codeword. Tanner graph T_2 only contains all of variable nodes of degree 1 and each variable node corresponds to each constraint node. The two structures satisfy the conditions to design low rate LDPC codes. Last but not least, there is only $E1_2$ edges connected by between Tanner graph T_1 and Tanner graph T_2. Therefore, this structure can get both advantages of Tanner graph T_1 and Tanner graph T_2, so we can use it to design low rate LDPC codes.

In general, the number of variable nodes in right of Tanner graph should become larger as the code rate becomes lower.

Obviously, we can use more complex constructions to design low rate LDPC codes, but this sample structure has met our need.

3 DEGREE DISTRIBUTION OF LOW RATE LDPC CODES

In order to obtain low rate LDPC codes, we need three steps. Firstly, according to the code rate, we use density evolution to design degree distributions of low rate LDPC codes with threshold close to channel capacity (Richardson et al 2001). Secondly, we generate LDPC code with PEG algorithm according to the degree distribution. Lastly, we obtain the performance of the LDPC code through system simulation. If the performance is actually close to channel capacity, we will eventually obtain the low rate LDPC code. But if not, we will need to design degree distribution again.

Density evolution is one of the methods to analyze the performance of degree distribution with the result of threshold. In order to simplify the computational complexity, we use Gaussian approximation of density evolution (Chung et al. 2001). Simultaneously, we find the threshold with Gaussian approximation is nearly equal to channel capacity. Gaussian approximation is a very efficient method.

In order to describe the performance of low rate LDPC code, we review some parameters as follow.

Threshold: Channel parameter is the standard deviation σ of Gaussian noise in the communication system with BPSK modulation and Gaussian channel. We can compute a threshold σ^* through density evolution when degree distribution is determined (Richardson & Urbanke 2001). Note that I use Gaussian approximation of density evolution in order to simplify the calculation.

SNR: We can achieve the minimum of signal-to-noise ratio when decoding process can be performed successfully with system simulation.

Code Efficiency: This parameter is used to describe the performance of LDPC code close to channel capacity quantitatively. Dividing code rate by channel capacity is the code efficiency.

After hard work, we have designed three LDPC codes with code rate of 0.04, 0.06 and 0.08 respectively. Taking an example of code rate 0.04, I analyze the process.

We design the degree distribution of code rate 0.04 with Gaussian approximation of density evolution. According to the degree distribution on Table 1, we generate check matrix of LDPC code with progressive-edge-growth algorithm (Hu et al. 2001). We gain the maximum value of Gaussian noise by simulating the coding process with BPSK modulation and Gaussian channel. Further we compute the channel capacity (Kabashima & Saad 2004) and code efficiency. We display these results on Table 4.

The degree distribution of rate code 0.04 is shown on Table 1 and is also described with polynomials as follow.

Table 1. Degree distribution of code rate 0.04.

v	d			u	d		
0.03	2	37	0	0.01	7	0	0
0.03	3	37	0	0.01	8	0	0
0.94	0	0	1	0.60	0	2	1
				0.34	0	3	1

Table 2. Degree distribution of code rate 0.06.

v	d			u	d		
0.05	2	30	0	0.01	9	0	0
0.03	3	30	0	0.01	10	0	0
0.92	0	0	1	0.36	0	2	1
				0.56	0	3	1

Table 3. Degree distribution of code rate 0.08.

v	d			u	d		
0.07	2	25	0	0.01	11	0	0
0.03	3	25	0	0.01	12	0	0
0.90	0	0	1	0.20	0	2	1
				0.70	0	3	1

$$v(\vec{x}) = 0.03x_1^2x_2^{37} + 0.03x_1^3x_2^{37} + 0.94x_3 \quad (7)$$

$$u(\vec{x}) = 0.01x_1^7 + 0.01x_1^8 + 0.6x_2^2x_3 + 0.34x_2^3x_3 \quad (8)$$

As the analysis processes of code rate 0.06 and 0.08 are similar to that of code rate 0.04, so we only present the results of degree distributions and performances as follow.

We optimize degree distribution with density evolution. According to my experience, the degree of edges E_{12} in Tanner graph T_2 is the combination of two and three. In Tanner graph T_1, the ratio of the number of constraint nodes and that of variable nodes range from 0.2 to 0.6. We adjust the degree distribution of each type edge in turn to achieve the biggest threshold. It takes a long time to optimize degree distribution. In this paper, the degree distributions presented may not be the best and they are only good enough to be used in practical.

We use PEG algorithm to generate LDPC codes according to the optimized degree distribution. In order to obtain good LDPC codes, we need to make the girth of LDPC codes as big as possible. The PEG algorithm is one of effective way to generate LDPC codes. You can use another way to generate LDPC codes only if the girth is big enough.

As the results are shown on Table 4, we know that the code efficiencies are very high, meeting our expectation. In order to show the performance of low rate LDPC codes, it is convenient to draw a figure with MATLAB. So we can display the performance of error correction intuitively in Figure 2.

Table 4. Performance of low rate LDPC codes.

Code rate	Threshold	SNR	Channel capacity	Code efficiency
0.04	4.05	0.062	0.0435	92.0%
0.06	3.31	0.094	0.0645	93.0%
0.08	2.84	0.129	0.0875	91.4%

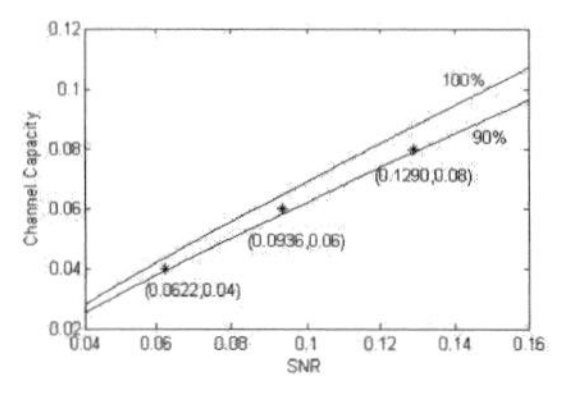

Figure 2. The relationship between channel capacity and SNR. The code efficiencies of three examples are more than 90%. Note that the results are related to the simulation accuracy of Gaussian noise. So we should make an effort to simulate Gaussian noise in performance analysis.

In this section, we describe the normal steps to design low rate LDPC codes. We conclude that the way to design low rate LDPC codes is very good.

4 CONCLUSIONS

In this paper, with analysis of the conditions to design low rate LDPC codes, we introduce multi-edge type LDPC codes and focus on the cascaded constructions for degree one variable nodes, which is used to design low rate LDPC codes. Then, we show three examples of low rate LDPC codes with density evolution and PEG algorithm. We analyze the performances of LDPC codes and conclude that it is an efficient method to design low rate LDPC codes in the end.

As multi-edge type LDPC codes combine the advantages of various structures, so low rate LDPC codes with the cascaded constructions for degree one variable nodes can also enjoy many advantages, such as low decoding complexity and fast convergence. In the future, we believe that low rate LDPC codes will be widely used in low signal-to-noise ratio regime.

ACKNOWLEDGMENTS

This work was supported in part by the National Basic Research Program of China (973 Pro-gram) under Grants No. 2012CB315605 and No. 2014CB340102, in part by the National Natural Science Foundation under Grant No. 61271193, in part by the Fund of State Key Laboratory of Information Photonics and Optical Communications.

REFERENCES

Andriyanova, I. & Tillich, J. P. 2012. Designing a Good Low-Rate Sparse-Graph Code. *IEEE Transaction on Communication* 60(11): 3181–3190.

Chung, S. Y., Richardson, T. J. & Urbanke, R. L. 2001. Analysis of Sum-product Decoding of Low-density Parity-check Codes using a Gaussian Approximation. *IEEE Transactions on Information Theory* 47(2): 657–670. Brink, S. T. 1999. Convergence of iterative decoding. *Electron. Lett.* 35: 806–808.

Gallager, R. G. 1962. Low-density parity-check codes. *IRE Transactions on Information Theory* 8(1): 21–28.

Hu, X. Y., Eleftheriou, E., & Arnold, D. M . 2001. Progressive Edge-growth Tanner Graphs. *Global Telecommunications Conference, 2001. GLOBECOM'01. San Antonio, TX, 25–29 November 2001*: 995–1001. IEEE.

Kabashima, Y. & Saad, D. 2004. Statistical mechanics of low-density parity check codes. *Journal of Physics A: Mathematical and General* 37(6): R1-R43.

Luby, M. G., Mitzenmacher, M., Shokrollahi, M. A. & Spielman, D. A. 1998. Analysis of low density codes and improved designs using irregular graphs. In Proc. of *the 30th Annual ACM Symposium on Theory of Computing*: 249–258.

Richardson, T. J. & Urbanke, R. L. 2001. The Capacity of Low-density Parity-check Codes Under Message-passing Decoding. *IEEE Transaction on Information Theory* 47(2) 599–618.

Richardson, T. J. & Urbanke, R. L. 2004. Multi-edge type LDPC codes. Available: http://wiiau4.free.fr/pdf/Multi-Edge%20Type%20LDPC%20Codes.pdf.

Richardson, T. J., Shokrollahi, M. A. & Urbanke, R. L. 2001. Design of Capacity-approaching Irregular Low-density Parity-check Codes. *IEEE Transaction on Information Theory*,47(2): 619–637.

Tillich, J. P. & Zemor, G. 2006. On the minimum distance of structured LDPC codes with two variable nodes of degree-2 per parity-check equation. *In Proceedings of ISIT 2006, Seattle, USA, 9–14 July 2006*. IEEE.

Wei, X. & Akansu, A. N. 2001. Density Evolution for Low-density Parity-check Codes under Max-Log-MAP Decoding. *Electronics Letters* 37(18): 1125–1126.

Zhang T. & Paihi K. K. 2001. High-performance, Low-complexity Decoding of Generalized Low-density Parity-check Codes. *Global Telecommunications Conference, 2001. GLOBECOM'01. San Antonio, TX, 25–29 November 2001*: 181–185. IEEE.

Electronics, Communications and Networks IV – Hussain & Ivanovic (eds)
© 2015 Taylor & Francis Group, London, ISBN: 978-1-138-02830-2

A Mobile Relay prioritized and fairness guaranteed resource allocation for HSR network

Hua Yang, Jian Xiong* & Lin Gui
Department of Electronic Engineering, Shanghai Jiao Tong University, China

Bo Rong
Communications Research Centre Canada, Canada

ABSTRACT: Broadband access system for High Speed Railway (HSR) network attracts huge attention, due to the wide deployment of HSR at 350 kilometers per hour. Mobile Relay (MR) is introduced in the Long Term Evolution Advanced (LTE-A) of 3GPP Release 12 technical report to overcome some of the problems caused by high speed mobile scenario. However, resource allocation still remains a crucial and a challenging task in HSR network. This paper proposes a resource allocation algorithm for HSR relay User Equipments (UEs) and macro UEs based on Modified Largest Weighted Delay First (MLWDF). To address the time-varying channel and transmission delay, we further apply one step linear prediction to estimate the Channel Quality Indicator (CQI) of MR. Simulation results show that our proposed resource allocation algorithm with perfect CQI feedback improves the throughput of relay UEs in HSR as well as the whole system by assuring better fairness among UEs. Moreover, compared with a conventional MLWDF algorithm, our proposed algorithm with the predicted CQI method can significantly boost the data rate and fairness factor of relay UEs though not much gain in system capacity.

1 INTRODUCTION

High speed railway (HSR) networks are being deployed at an increasing rate worldwide. More and more people are choosing to travel by high speed train, because of its convenience and low carbon footprint. Recently, a survey found that about 72\% business travelers have a huge demand for on-board wireless access (Zhang et al. 2010). Therefore, providing cellular services to onboard passengers is important. However, due to the fast moving and well shield carriage, HSR network faces severe Doppler frequency shift, high penetration loss, fast and frequent handover, together with increasing power consumption of user equipments (UEs). To overcome these problems, mobile relay (MR) is introduced in the Long Term Evolution Advanced (LTE-Advanced) of 3rd Generation Partnership Project (3GPP) Release 12 technical reports. MR is a base access point mounted on the high speed train, which is connected to Donor evolved node B (DeNB) in wireless way and can provide wireless connectivity service to users inside the train.

With MR assisted, high speed mobility still remains a challenge environment for HSR networks. The typically rapid and significant variations in the instantaneous channel conditions are highly undesirable for multimedia communications given the stringent demand on data bandwidth and continuity. Therefore, a proper resource allocation mechanism is needed to accomplish both efficient service provision and better resource utilization.

Up to now, a few studies have addressed resource scheduling in HSR scenario. In (Gao et al. 2013), a Quality of Service (QoS) guaranteed scheduling algorithm for the heterogeneous converged network is proposed. Taking time-varying channel into consideration, (Karimi et al. 2012) proposes a resource allocation mechanism based on the periodical signal quality changes. (Zhu 2012) divides each data stream into two types with different bit error rate requirements, and proposes a chunk-based resource allocation approach. In (Zhao et al. 2013), a multi-dimensional resource allocation scheme for downlink transmission is introduced. (Aida & Kambori 2008) expands the use of heterogenous wireless links to provide continuous and intermittent fast connections. In these literatures, the designed resource scheduling schemes are generally based on user's channel status information (CSI) and QoS requirements of each user. Moreover, most literatures assume that the CSI can be perfectly estimated at the receiver and reliably fed back to DeNB with no delay. And the goal of these scheduling schemes is to find a balance between maximizing the system's throughput and providing fairness among users.

*Corresponding author: *xjarrow@sjtu.edu.cn*

As far as the authors know, few works have considered the resource scheduling between users in high speed train and users moving at pedestrian speeds in the same cellular. Moreover, such scenario occurs quite frequently, especially when a train moves across or stays in a station. Therefore, there is a strong motivation to propose a kind of resource scheduling strategy to maximize system throughput, guarantee QoS requirements together with fairness among users.

In this work, a novel allocation scheme is proposed for users in high speed train and users in cellular to achieve throughput gain and guarantee users' fairness. The contributions of this work are two-folds. First, based on Modified Largest Weighted Delay First (MLWDF) algorithm, we propose a resource reservation scheme for users in high speed train. Second, concerning the time-varying wireless channel conditions of HSR as well as the transmission delay, we employ a prediction method for Channel Quality Indicator (CQI) of the HSR network.

The remainder of this paper is organized as follows. Section 2 gives an elaborate description of the system model, and the problem is also formulated in Section 2. The proposed resource allocation mechanism for users in high speed train and users in cellular follows in section 3. The simulation and performance analysis are provided in Section 4. Finally, we conclude the paper in section 5.

2 SYSTEM MODEL AND PROBLEM FORMULATION

2.1 *System model*

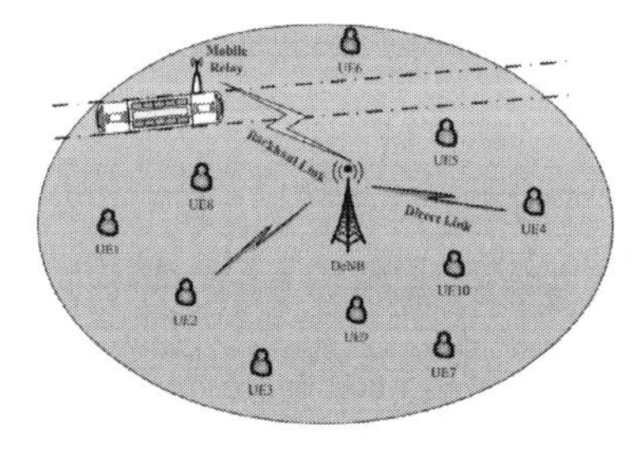

Figure 1. Illustrations of high speed mobile system architecture.

The system model is illustrated in Figure.1. According to the introduction in Section 1, users in high speed train, communicate with DeNB through MR, while users in cellular communicate with DeNB directly. We define UEs in cellular as macro UEs and UEs in high speed train as relay UEs. Suppose the scenario comprises K_1 macro UEs and one MR, within MR there are K_2 relay UEs. We define them as

$$U = \left\{u_{macro}^{1}, u_{macro}^{2}, \cdots, u_{macro}^{K_1}, u_{relay}\right\}$$

$$u_{relay} = \left\{u_{relay}^{1}, u_{relay}^{2}, \cdots, u_{relay}^{K_2}\right\}$$

In our work, we focus on the downlink data transmission in a single LTE-A cell. We only consider resource allocation between macro UEs and MR. The detail scheduling in MR among relay UEs is not considered in this paper.

In 3GPP LTE standard, a Resource Block (RB) is the smallest radio resource allocation unit. The size of a RB is 12 adjacent subcarriers of 15KHz in the frequency domain, and the time of a RB is the time slot consisting of 6 or 7 symbols. Every Transmission Time Interval (TTI) contains two time slots, and TTI equals 1ms in the time domain. Besides, each RB can be independently modulated and assigned to one user. Meanwhile, the Modulation and Coding Scheme decides the number of bits to be transmitted in each RB. And the selection is dynamic, depending on the channel's changes over time. In this study, we consider LTE-A system with 100MHz of bandwidth and 500 RBs per time slot. It is assumed that all RBs are used for data transmission for simplification.

2.2 *Problem formulation*

Resource allocation in wireless communication considers the CQI between UEs and DeNB. CQI is a wireless communication channel quality measurement standard, which is calculated and sent back to DeNB by UEs according to the downlink reference signals. Higher CQI indicates better channel quality, higher modulation technique, and higher data transmission rate.

$$TR = BW \cdot (1 - P_{BW}) \cdot efficiency \tag{1}$$

where TR represents transmission rate, BW is bandwidth, *efficiency* means information bits per resource element (Gui et al. 2010). P_{BW} is short for protection bandwidth, and each LTE-A data symbol saves 10% of the bandwidth as guard band.

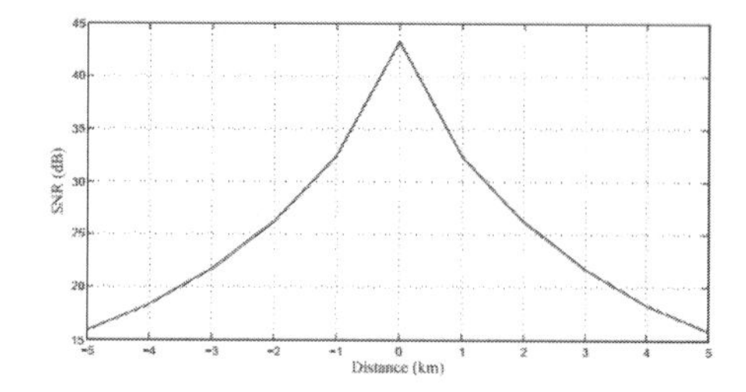

Figure 2. Time-varying SNR measured in HSR network.

In normal mobility patterns, changes of the channel conditions are not fast enough to change the channel quality from the last CQI report to the transmission time. Therefore, scheduling is based on the last reported CQI of users. However, this assumption is not suitable for HSR network. Due to the time-varying channel conditions, there is almost no published paper studying CQI in the high speed mobile scenario.

Figure.2 shows the time-varying signal to noise ratio (SNR) in the HSR network (GUI et al. 2010), which is measured in Shanghai-Wuxi experimental section. Moreover, Mehlführer et al. (2009) has statistics to show that the SNR is roughly proportional to CQI. In the resource scheduling mechanism, every UE sends a $\overline{CQI}=\{CQI_u(t,n)\}^{1\times N}$ feedback vector containing supported $CQI \in \{1,\cdots,CQI_{\max}=15\}$ values for each RB to the DeNB after receiving a scheduled RB, where N means the total number of RBs. In this paper, we assume frequency selective CQI, where CQI is reported for each RB.

Therefore, resource allocation in one LTE-A cell, containing both relay UEs and macro UEs, can be formulated as follows

$$\begin{aligned}
\max \quad & \sum_{u=1}^{U}\sum_{n=1}^{N}\sum_{t=1}^{T} TR_{u,n}(t)\cdot x_{unt} \\
\text{s.t.} \quad & \sum_{u=1}^{U}\sum_{n=1}^{N} x_{unt} \le N, \forall t \in \{1,\cdots,T\} \\
& x_{unt} \in \{0,1\} \\
& \gamma \le F_{index} \le 1
\end{aligned} \tag{2}$$

Where U equals the total amount of UEs (including MR), T is the number of scheduling periods. The F_{index}, ranging from 0 to 1, is based on Jain's fairness index as defined in (3). A system with a bigger F_{index} is considered to be fairer. And $\gamma \in [0,1]$ is a parameter that represents how fairness the system intends to achieve.

$$F_{index} = \frac{(\sum x_u)^2}{U\cdot\sum x_u^2} \tag{3}$$

and

$$x_u = \frac{a_u}{d_u} \tag{4}$$

where a_u denotes the actually acquired bandwidth of UE u, while d_u corresponds to the required bandwidth of UE u.

3 RESOURCE SCHEDULING ALGORITHMS

The presented scheduling formulation for relay UEs in HSR and macro UEs in cellular is a 0-1 integer programming problem. Since scheduling decisions must be carried out in each scheduling period, the presented NP-hard formulation can hardly be implemented in real time. Therefore, we need to find out some other solutions.

In a mobile communication network, the location and network environment of each UE won't be the same, resulting in the received signal strength is different. The best CQI scheduling algorithm always selects the UE with the best CQI according to the feedback index from UEs. Round Robin (RR) algorithm stems from the concept of fairness and considers nothing of CQI. Proportional Fair (PF) scheduling algorithm can be thought as the hybrid of Best CQI and Round Robin, which means all UEs may assure a certain amount of resource in time and no starvation would possibly occur. However, PF algorithm does not differentiate service types, but UE channel quality.

3.1 *Modified Largest Weighted Delay First algorithm*

The well-known scheduling algorithm, referred to as Modified Largest Weighted Delay First (MLWDF) is introduced by Andrews in (Andrews et al. 2004). MLWDF algorithm considers QoS priority and attempts to balance the weighted delays of terminals, while trying to make use of channel resources efficiently. MLWDF algorithm calculates priority according to the ratio of the instantaneous transmission rate and the average transmission rate, the queuing delay of the Head of Line, packet loss rate constraints and delay constraints, which is expressed as

$$i = \arg\max_u[-\frac{\lg\delta_u}{T_u}\cdot\frac{TR_{u,n}(t)}{\overline{TR}_{u,n}(t)}\cdot W_u(t)] \tag{5}$$

where, i is the selected user in RB n at time t, δ_u is the upper limited loss rate of terminal u, T_u is the upper limited delay of terminal u, $W_u(t)$ is the waiting time in the head of terminal u at time instant t. $TR_{u,n}(t)$ is the transmission rate of terminal u for the resource block n at time t, and $\overline{TR}_{u,n}(t)$ is the average transmission rate of terminal u over a certain period in the past. The average transmission rate of each UE for each RB n is updated according to the following expression

$$\overline{TR}_{u,n}(t+1) = \begin{cases} (1-\frac{1}{t_c})\overline{TR}_{u,n}(t)+\frac{1}{t_c}TR_{u,n}(t) & u=i \\ (1-\frac{1}{t_c})\overline{TR}_{u,n}(t) & u\ne i \end{cases} \tag{6}$$

where t_c is the window size of the average data rate and can be adjusted to maintain fairness. Normally, t_c should be limited in a reasonable range so that terminals cannot notice the quality variation of the channel.

3.2 *Proposed resource scheduling algorithm*

As Figure.2 shows in section 2, the SNR of HSR changes periodically. When a high speed train comes into a cellular, its channel condition is rather poor. Then, as it moves on, the distance between the train and DeNB becomes shorter and the channel condition gets better. When the distance reaches the lowest point, the channel condition achieves the best value. After that, as the train moves away, the channel condition becomes worse and worse. Take the above mentioned circumstance into consideration, both PF and MLWDF algorithms are not suitable for resource allocation between macro UEs and relay UEs within MR in HSR network.

Based on MLWDF algorithm, we propose a resource reservation strategy with perfect CQI feedback. When the CQI of MR is poor, the proposed strategy will keep relay UEs' real-time traffic scheduled, but significantly lower theirs' non-real-time data rate. On the contrary, if the CQI of MR is good enough, the proposed strategy will raise the priority of relay UEs within MR. In the proposed resource allocation algorithm with perfect CQI feedback, we preset a CQI value of MR, named *threshold*. We also define $D_u(t)$ as the bandwith that UE u requires at time t, $A_u(t)$ as the bandwith that UE u acquires at time t. At the beginning of each scheduling period, we test whether $CQI_{relay}(t,n)$ is greater than *threshold*, where $CQI_{relay}(t,n)$ is the perfect feedback CQI of MR in RB n at time t. If so, our proposed resource reservation scheme with perfect CQI feedback would raise the priority of MR remarkably, such as reserving all RBs for relay UEs within MR. Otherwise, we just employ MLWDF algorithms to accomplish the resource allocation.

However, the time-varying channel and transmission delay would make the outdated CQI unreliable. To address the fast changing CQI in the high speed mobile scenario, we employ the simple one step linear prediction to predict the values of MR's CQI on the transmission time

$$CQI'_{relay}(t,n) = CQI_{relay}(t-1,n) + [CQI_{relay}(t-1,n) - CQI_{relay}(t-2,n)], t \geq 3 \quad (7)$$

where $CQI'_{relay}(t,n)$ is the predicted CQI of MR in RB n at time t, $CQI_{relay}(t-1,n)$ and $CQI_{relay}(t-2,n)$ are the last reported CQI of MR in RB n at time $t-1$ and $t-2$ respectively.

Based on the one step linear prediction, we propose a resource reservation algorithm with predicted CQI method, as shown in Figure 3. At each scheduling period, we use (7) to predict the current CQI of MR. Then, we regard the predicted $CQI'_{relay}(t,n)$ as $CQI_{relay}(t,n)$ in the following allocation algorithm. The rest of this scheme is the same as the resource reservation scheme with perfect CQI feedback.

Algorithm 1: Resource Reservation Based on MLWDF algorithm with predicted CQI method (On Demand Traffic)

Input: N, U, T, $Threshold$, $D_u(t)$, $A_u(t)$.

1 Initialization: Set each UE's $\hat{TR}_{u,n}$ in each RB at the first scheduling period as $TR_{u,n}$, initialize CQI'_{relay} for each RB at time $t = 1, 2$;
2 **for** $t \leftarrow 1$ *to* T **do**
3   **for** $n \leftarrow 1$ *to* N **do**
4   **if** $t \geqslant 3$ **then**
5   Calculate $CQI'_{relay}(t,n)$ according to (7);
6   **if** $CQI'_{relay}(t,n) \geqslant$ *Threhold* **then**
7   Select RB n for u_{relay};
8   **else**
9   **for** $u \leftarrow 1$ *to* U **do**
10   **if** $D_u(t) \geq A_u(t)$ **then**
11   Calculate $[-\frac{\lg \delta_u}{T_u} \cdot \frac{TR_{u,n}(t)}{\hat{TR}_{u,n}(t)} \cdot W_u(t)]$;
12   **else**
13   Drop u from U;
14   Select RB n for user i according to (5);
15   Update $A_i(t)$ by adding $R_i(t)$;
16   **for** $u \leftarrow 1$ *to* U **do**
17   Update $\hat{R}_{u,n}(t)$ according to (6);
18   **for** $u \leftarrow 1$ *to* U **do**
19   Update $W_u(t)$ by adding 1;

Figure 3. Resource reservation algorithm based on MLWDF with CQI prediction.

4 SIMULATION AND RESULTS

4.1 *Simulation conditions*

In our simulation, we assume that when UEs send requested bandwidths to a DeNB, the DeNB would know the number of UEs in the cell and the service type of each UE. These services are then inserted into different service queues waiting for a resource.

Table 1. QoS parameters for typical traffics.

Traffic	Latency (ms)	Loss Rate	macro	relay
Real time Gaming	50	10^{-3}	10%	10%
Conversational Voice	100	10^{-2}	25%	25%
Interactive Gaming	100	10^{-3}	10%	10%
Live Streaming Video			7.5%	5%
Conversation Video	150	10^{-3}	2.5%	5%
FTP			10%	10%
E-mail			5%	5%
p2p File Sharing	300	10^{-6}	5%	0%
Web Browsing			10%	15%
Buffered Streaming Video			15%	15%

3GPP classifies LTE services into the Guaranteed Bit Rate (GRB) and Non-GRB resource types with different QoS Class Identifier. In this simulation, we classify data transmission services into 10 types (Alasti et al. 2010) as shown in Table 1. Table 1 also indicates the percentage of each traffic type that macro UEs and relay UEs account for, respectively.

Table 2. Simulation parameters.

Parameter	Value
System bandwidth	100MHz with 500RBs
Radius of a DeNB	5km
Transmission direction	Downlink
Mobility of high speed railway	350km/h
Number of MR	1
Number of macro UEs	400
Number of relay UEs in MR	400
Scheduling algorithm	Proposed in section 3
Traffic model	As shown in Table 1
Scheduling period	10ms
Time constant t_c in (6)	100ms
Threshold of MR's CQI	10

Table 2 describes the simulation parameters adopted throughout this paper. In the simulation, we consider LTE-A system with 100MHz bandwidth and 500 RBs. It is assumed that all RBs are used for data transmission for simplification. Moreover, we focus on downlink transmission. In our system model, there are 400 active macro UEs and one active MR in a single cellular. Within the MR, there are 400 active relay UEs. We also assume that the CQIs of macro UEs are uniformly distributed within 3~15, while the CQI of MR are set as section 2 describes. The moving speed of HSR is 350km/h. And the scheduling period is 10ms.

4.2 *Simulation results and comments*

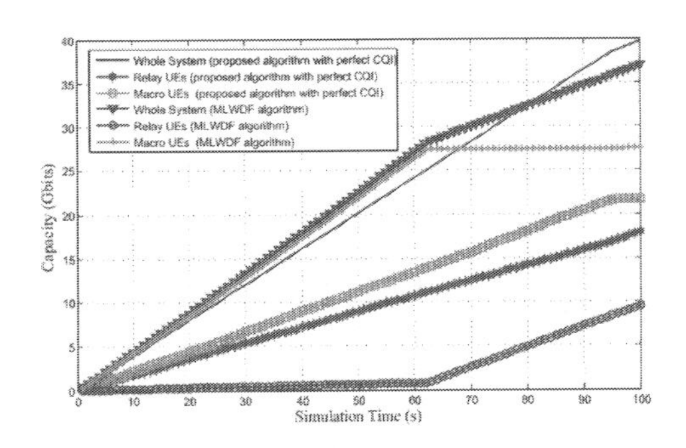

Figure 4. Capacity comparison of proposed resource allocation algorithm with perfect CQI feedback and MLWDF algorithm.

Figure.4 shows the capacity performance of different algorithms against simulation time. Our proposed algorithm with perfect CQI feedback significantly increases the capacity of relay UEs, without sacrificing the capacity of the whole system. Since the channel conditions of relay UEs are worse than that of macro UEs, macro UEs are more likely to be scheduled in the conventional MLWDF algorithm. In our proposed algorithm with perfect CQI feedback, we raise the priority of relay UEs once the feedback CQI of MR is beyond the predetermined threshold. In Figure.4, we also find that there is an inflection point in MLWDF, when the simulation time equals 64s. This is because the waiting time in the head of relay UEs is long enough to obtain the resource. Compared with MLWDF, our proposed algorithm also avoids allocating resource to relay UEs just because of their increasing waiting time, especially when their channel conditions are poor.

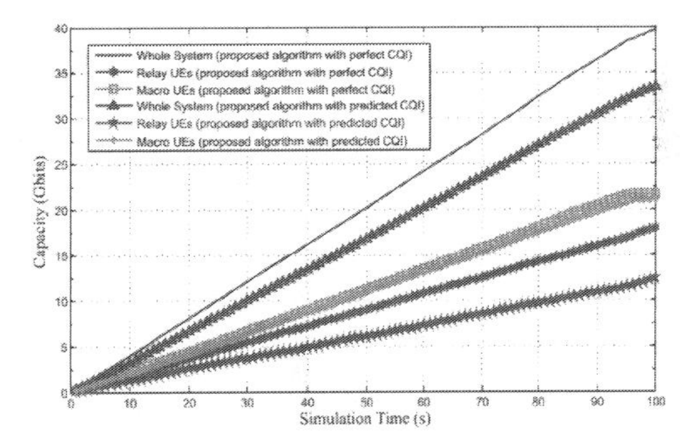

Figure 5. Capacity comparison of proposed resource allocation algorithm with perfect CQI feedback and predicted CQI.

The effects of using one step linear prediction scheme on system capacity are also compared with simulation. Figure.5 shows the capacity of our proposed algorithm with perfect CQI feedback as well as predicted CQI. In Figure.5, we can see that there is a gap between perfect CQI feedback algorithm and predicted CQI method, concerning the capacity of relay UEs as well as a whole system. This is because of the inaccuracy of the prediction scheme we applied. To deal with the time-varying channel and transmission delay, we introduce the simple one step linear prediction. However, one step linear prediction scheme is not accurate enough. As a result, there exists a capacity loss.

Traditionally, the fairness index F_{index} is calculated with (3), where U equals the number of macro UEs together with the number MR. However, since one MR may include lots of relay UEs, the fairness calculation mentioned above is not suitable for our system. Consequently, we redefine U as the number of macro UEs together with the number of relay UEs within MR. In addition, we suppose that every relay UE shares the allocated resource impartially within MR.

Figure.6 shows the fairness performance of different algorithms, where the fairness index of MLWDF, our proposed algorithm with perfect CQI and predicted CQI are 0.8199, 0.9643, 0.8896, respectively. The proposed resource reservation algorithm with perfect CQI feedback outperforms MLWDF. This is because our proposed algorithm would allocate more resources to relay UEs when the channel condition of MR is beyond the predetermined threshold. In an HSR network, relay UEs have worse channel states than macro UEs. MLWDF algorithm doesn't take this particular case into consideration. Hence, MLWDF has a worse fairness index. Our proposed resource reservation algorithm with the predicted CQI method has a little worse performance than that of perfect CQI feedback, but better than that of MLWDF. This is because one step linear prediction is simple, but not accurate enough. As a result, there may exist, an error gap between the predicted CQI and the perfect CQI feedback. So is the fairness index.

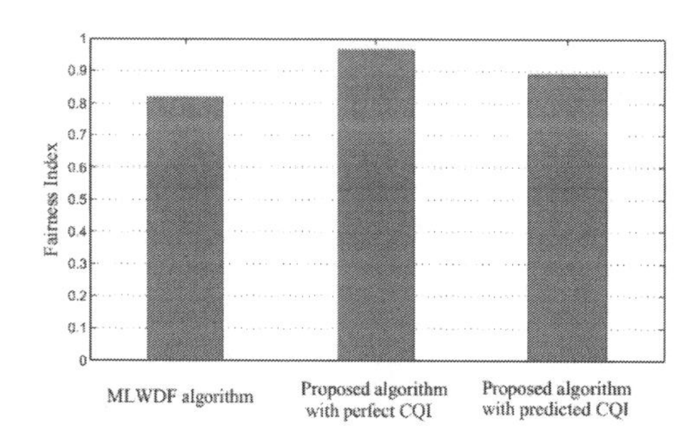

Figure 6. Fairness Index of MLWDF, proposed algorithm with perfect CQI, Proposed algorithm with predicted CQI.

5 CONCLUSION

In this paper, we formulate the optimal resource allocation problem between relay UEs and macro UEs in high speed railway network with mobile relay assisted. Since the problem is a nonlinear constrained optimization problem, we first propose a resource reservation algorithm with perfect CQI feedback. Concerning the time-varying channel and transmission delay, we propose a resource reservation algorithm with the predicted CQI method. The simulation results show that our proposed resource allocation algorithm with perfect CQI feedback improves the capacity for relay UEs in HSR together with the whole system, and achieves a better fairness among UEs. Moreover, compared with MLWDF algorithm, our proposed algorithm with predicted CQI method cannot gain much in system capacity, but can significantly improve the capacity for relay UEs in HSR and also the fairness.

ACKNOWLEDGMENT

This work was supported in part by the Major State Basic Research Development Program of China (973 Program) (No. 2009CB320403), National Natural Science Foundation of China (61201222), the 111 Project (B07022) and the Shanghai Key Laboratory of Digital Media Processing and Transmissions. It was also partially supported by the funds of MIIT of China (Grant No. 2011ZX03001-007-03).

Jian Xiong is the corresponding author.

REFERENCES

Aida, H. & Kambori, S. 2008. Effective use of heterogeneous wireless links in high speed railways by predictive scheduling. In *Applications and the Internet: 459–462; International Symposium, Saint, July 2008*. IEEE.

Alasti, M., Neekzad, B. & Hui, J., et al. 2010. Quality of service in WiMAX and LTE networks: Topics in Wireless Communications. *Communications Magazine, IEEE 48*(5): 104–111.

Andrews, M., Kumaran, K., Ramanan, K., et al. 2004. Scheduling in a queuing system with asynchronously varying service rates. *Probability in the Engineering and Informational Sciences* 18(02): 191–217.

Gao, H., Ouyang, Y., Hu, H., et al. 2013. A QoS-guaranteed resource scheduling algorithm in high-speed mobile convergence network. In *Wireless Communications and Networking Conference Workshops IEEE*: 45–50.

Gui, L., Ma, W., Liu, B., et al. 2010. Single frequency network system coverage and trial testing of high speed railway television system. *Broadcasting IEEE Transactions on 56*(2): 160–170.

Karimi, O. B., Liu, J., & Wang, C. 2012. Seamless wireless connectivity for multimedia services in high speed trains. *Selected Areas in Communications, IEEE Journal on 30*(4): 729–739.

Mehlführer, C., Wrulich, M., Ikuno, J. C., et al. 2009. Simulating the long term evolution physical layer. In *Proc. of the 17th European Signal Processing Conference, Glasgow, Scotland* 27: 124.

Zhang, J. Y., Tan, Z. H., Zhong, Z. D., et al. 2010. A multi-mode multi-band and multi-system-based access architecture for high-speed railways. In *Vehicular Technology Conference Fall IEEE 72nd* : 1–5.

Zhao, Y., Li, X., Li, Y., et al. 2013. Resource allocation for high-speed railway downlink MIMO-OFDM system using quantum-behaved particle swarm optimization. In *Communications, 2013 IEEE International Conference on* : 2343–2347.

Zhu, H. 2012. Radio resource allocation for OFDMA systems in high speed environments. *Selected Areas in Communications, IEEE Journal on 30*(4): 748–759.

Electronics, Communications and Networks IV – Hussain & Ivanovic (eds)
© 2015 Taylor & Francis Group, London, ISBN: 978-1-138-02830-2

An optimization model based on radial basis function neural networks

Biyuan Yao, Jianhua Yin*, Zhen Guo & Wei Wu
College of Information Science and Technology, Hainan University, China

ABSTRACT: The total energy consumption increases dramatically huge with the propelling of Chinas development. This paper focused energy structure data of China trying to analyze the past situation and future trends. By comparing weka application and Matlab optimization, Radial Basis Function Neural Networks receive excellent prediction results, it verifies the effectiveness of high performance both of convergence speed and optimization data. Finally, we proposed a specified direction of the proper energy structure of China.

1 INTRODUCTION

It signifies profound theoretical and practical significance to conduct research on China's energy structure. Energy is the preliminary crucial material basis for human survival and socioeconomic development, a fit energy supply and consumption system saves tremendous resource, thus contributes to a sustainable society, especially China economic viewpoint emphasizes on quality rather than speed.

Guiyang Wang proposed optimal control and energy saving operation algorithm based on RBFNN, received high performance in horizontal ground-coupled heat pump systems. RBFNN pattern recognition is another useful tool to automatically classify audio clips and other multimedia tech., the studies provided constructive contributions on how to optimize data model and prediction (Chen et al. 2006). In this paper, in consideration of all beneficial suggestions are depend on precise data analysis and prediction, we focused on the energy structure and Gross Domestic Production data from 1980 to 2012, introduced two methods to build a model reflecting the potential relation between energy and social consumption, based on this pattern to make a prediction of future energy trends and provide reasonable directions. It is organized as follows: in Section 2 we introduce theoretical knowledge (energy and economy, weka application and mathematic theory, radial basis function neural networks (RBFNN)); we focus on data analysis in Section 3; conclusions and some advices will be shown in Section 4; Section 5 is our future study direction.

2 PRELIMINARY RESULTS

2.1 *Relationship between energy structure and economy*

Total energy structure refers to the proportion relationship between all kinds of primary and secondary energy composition from total energy production or consumption. Energy structure is an important study area of energy system engineering, it influences finally a utilization way of the national economics department directly, and reflect people's living standard (Glvez et al. 2013). The study of production structure and consumption structure of energy can grasp important information and lay foundation for the energy supply and demand balance system. Figure out energy production resources, variety, quantity as well as consumption product provides a scientific theory for mining investment, allocation and utilization (Paul et al. 2005). Meanwhile, to make a prediction of future status according to these data significantly, especially energy structure, adjustment is one of the important tasks facing China (Zhang et al. 2008).

2.2 *Weka applications and regression*

WEKA is an open platform for data mining. As a collection of big data mining task, it integrates machine learning algorithms, data preprocessing, classification, regression, clustering, association rules, and new interactive interface visualization, etc.

As is mentioned above, it's well acknowledged exist a strong relationship between energy and economy, for

*Corresponding author: *yinjh@hainu.edu.cn*

energy consumption directly reflects the social production capacity and people's spending power, which values are an important measure of economic area, thus an effective regression model can be achieved, we try to establish an energy structure model and energy economy relational model.

2.3 *Radial basis function neural networks*

The basic form of a neuron is shown in Figure 1. The neural network consists of three layers, where each layer plays a different role. The input layer is composed of some of source input points (sensing unit), the second layer (hidden layer) converts nonlinear data from the input space to the hidden space (Ong et al. 1996). The output layer is linear and provides the response of activation mode of the input layer (Gang 2011).

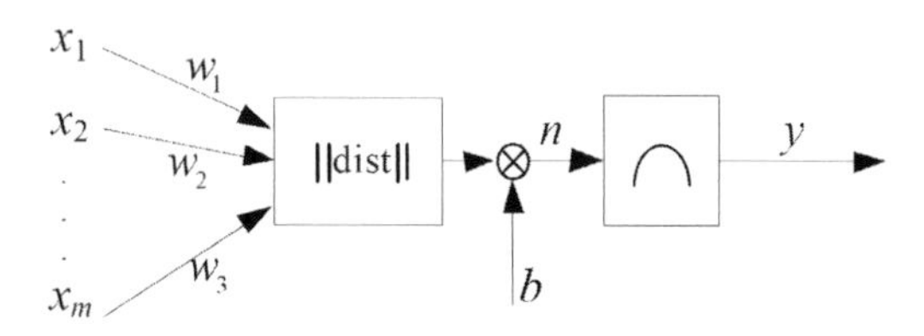

Figure 1. Structure of a single neuron.

A general form can be expressed as $h_N(x) = W^T S(x) = \sum_{i=1}^{m} w_i s_i(x)$, where its input vector is a bounded compact set, $h_N(x)$ is the output of neuron N, $W = [w_1\ w_2 \cdots w_m]^T \in R^m$ is the weight vector, m is the number of neural network central nodes, $S(x) = [s_1(x), s_2(x), \cdots, s_m(x)]^T \in R^m$, $s_i(x)$ is radial basis function, the most commonly used radial basis function is Gaussian function in the following expression:

$$s_i(x) = e^{-\|x-c_i\|/\eta_i^2}, i = 1, 2, \cdots, m \tag{1}$$

Where $c_i \in \Omega_x, i = 1, 2, \cdots, m$ is the center nodes set, η_i is the normalized parameter of the hidden node i which determines the scope of the base functions to the central node. When the center nodes are enough and distributed reasonably, NN is seen to approach a continuous function $h(x)$ on a bounded compact set Ω_x, and the approximation error can be arbitrarily small.

Let there are m neurons in hidden layer, when weights w_i is determined, from Figure 2, the output is:

$$y = \sum_{j=1}^{m} w_j h_j(x) + b_j \tag{2}$$

3 DATA PREPARATION

China implemented its reform and open-up policy in 1978, we begin our data collection from 1980 to achieve more representative and accuracy in Table 1. where:

x_1: oil consumption (million tonnes);

x_2: gas consumption(million tonnes oil equivalent,mtoe);

x_3: coal consumption (mtoe);

x_4: nuclear energy consumption (mtoe);

x_5: hydro consumption (mtoe);

x_6: solar consumption (mtoe);

x_7: wind consumption (mtoe);

x_8: Geo biomass (mtoe).

x_9: Primary energy(oil/natural gas/coal/nuclear / hydro electric/renewables(mtoe).

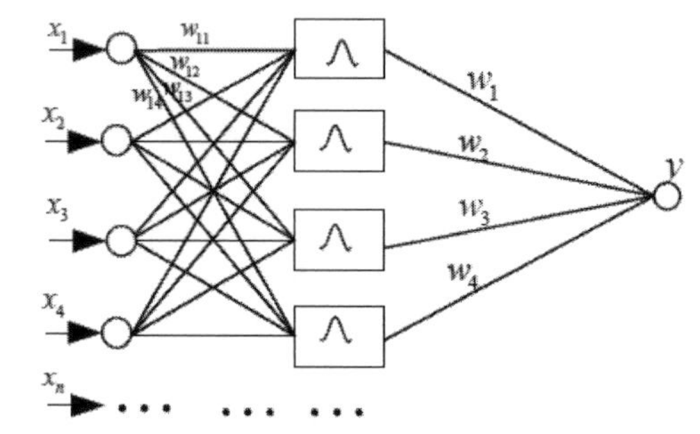

Figure 2. Structure of neural networks.

x_1, x_2, x_3 constitute the main part of the energy consumption and carries crucial impact on economic data, some missing data (partial from $x_4, x_{6,7,8}$) should be fixed with zero. These new types of energy resources occupy tiny proportion and cannot be exactly confirmed separated accurately, a summation operation was conducted to finish this part. Secondly, to find out those variables with tight correlations to economic contribution is important. Regression analysis studies influence of variable(s) (dependent) on another (more) explanatory variable(s) and theoretical calculation methods. It starts from data samples to determine the mathematical relationship and identify the impact of a specific variable from the many variables which affect the significant variables, which are not significant.

4 DATA MODELING

In this section, we try to use two different methods for modeling Gross Domestic Product (GDP) prediction. Based on forecasting data of each energy type, we adopt Neural Networks to predict GDP of 2013.

Table 1. Energy structure of China (1980-2012).

year	x_1	x_2	x_3	x_4	x_5	$\sum x_{6,7,8}$	x_9
1980	85.4	12.8	305.1	-	13.2	-	416.4
1981	81.1	11.5	302.9	-	14.8	-	410.4
1982	80.1	10.7	320.6	-	16.8	-	428.4
1983	81.8	11	343.6	-	19.6	-	455.9
1984	84.6	11.2	374.8	-	19.6	-	490.3
1985	89.8	11.6	408	-	20.9	-	530.4
1986	95.7	12.4	430.1	-	21.4	-	559.5
1987	101.5	12.5	464	-	22.6	-	600.7
1988	108.8	12.9	496.8	-	24.7	-	643.2
1989	114	13.5	517.1	-	26.8	-	671.4
1990	112.9	13.7	509.3	-	28.7	-	664.6
1991	121.9	14.3	527.2	-	28.2	-	691.6
1992	132.4	14.3	545.2	-	29.6	-	721.4
1993	145.8	15.1	574.3	0.4	34.4	-	770
1994	148.1	15.6	612	3.3	37.9	0.2	817.2
1995	160.2	16	663.5	2.9	43.1	0.8	886.5
1996	175.7	16.6	677.4	3.2	42.5	0.3	915.8
1997	193.9	17.6	672.6	3.3	44.4	0.6	932.4
1998	197.1	18.2	652	3.2	47.1	0.7	918.2
1999	209.3	19.3	672.8	3.4	46.1	0.7	951.7
2000	224.2	22.1	679.2	3.8	50.3	0.7	980.3
2001	228.4	24.7	692.8	4	62.8	0.8	1013.3
2002	247.5	26.3	728.4	5.7	65.2	0.8	1073.8
2003	271.7	30.5	868.2	9.8	64.2	0.8	1245.3
2004	318.9	35.7	1019.9	11.4	80	0.6	1466.8
2005	327.8	42.1	1128.3	12	89.8	1	1601.2
2006	351.2	50.5	1250.4	12.4	98.6	1.4	1764.7
2007	369.3	63.5	1320.3	14.1	109.8	1.8	1878.7
2008	376	73.2	1369.2	15.5	132.4	0.9	1969.9
2009	388.2	80.6	1470.7	15.9	139.3	6.9	2101.5
2010	437.7	96.2	1609.7	16.7	163.4	14.2	2338
2011	459.4	117.5	1760.8	19.5	158.2	25.4	2540.8
2012	483.7	129.5	1873.3	22	194.8	31.9	2735.2

4.1 *Regression on each energy*

Suppose variables $x_i (i = 1, 2, \cdots, 8)$ all contributed to GDP measured by energy consumption, from table 1, use curve fitting toolbox we can build function of increasing trends of each energy. The oil consumption curve fitting is shown in Figure 3, and new types of energy sum are shown in Figure 4.

From the data point of Figure 3, in consideration of bend area, it is apparent to define the degree of the function is 2. The residual graph is shown bellow. Linear model Poly2:

$$x_1 = p_1 \cdot t^2 + p_2 \cdot t + p_3 \tag{3}$$

where t is normalized ($t = year - 1980$), and evaluation index coefficients (95% confidence bounds):

$$p_1 = 0.4305(0.394, 0.4669)$$

$$p_2 = -2.265(-3.542, -0.9885)$$

$$p_3 = 87.32(77.9, 96.73)$$

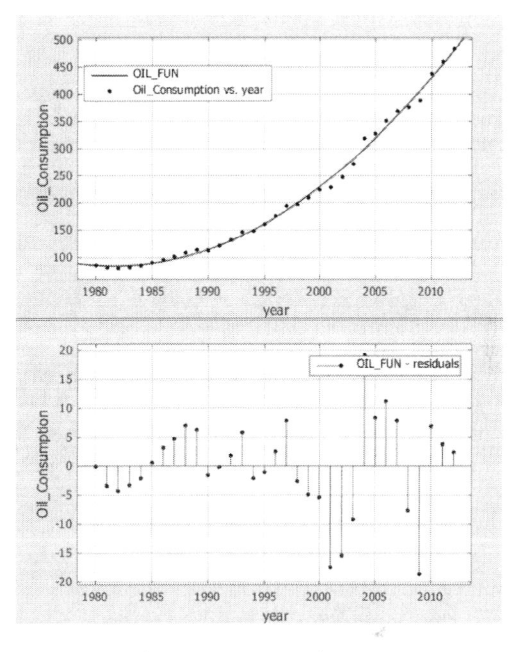

Figure 3. Coal consumption curve fitting

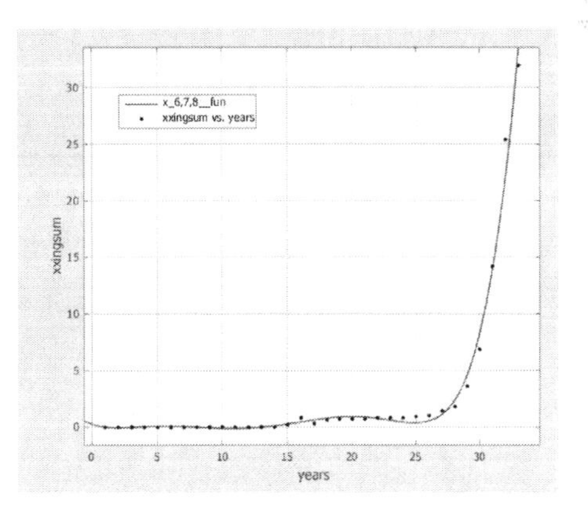

Figure 4. Fluctuate x^* curve fitting.

Similarly to other energy forecasting data, functions approached as follows:

$$x_2 = 0.0118 \cdot t^3 - 0.3922 \cdot t^2 + 3.862 \cdot t + 3.297 \tag{4}$$

$$x_3 = 0.10670 \cdot t^3 - 3.600 \cdot t^2 + 54.45 \cdot t + 206.6 \quad (5)$$

$$x_4 = 0.0334 \cdot t^2 - 0.5013 \cdot t + 1.376 \quad (6)$$

$$x_5 = 0.0113 \cdot t^3 - 0.3360 \cdot t^2 + 4.346 \cdot t + 6.379 \quad (7)$$

x_6 can be fixed by gaussian function:

$$x_6 = a_1 \cdot e^{-(\frac{t-b_1}{c_1})^2} \quad (8)$$

where $a_1 = 3.112, b_1 = 36.7, c_1 = 3.471$).

Goodness of fit evaluation index can be comparatively small, but it cannot reflect t because center $b_1 = 36.7$ while actual $t \in (-\infty, +\infty)$. Beside x_4, fluctuate variables $x_{6,7,8}$ partly represent clean energy, it is not obvious to fit function, from table 1 we can infer that these energy forms strongly lie on policy makers and carried out in late 1990s, to build such function is not reasonable, with a view to w_4 represents nuclear investment, this clear energy need to be evaluated for Japan nuclear plant accident, and in consideration of its investment scale is huge and lasts for a long period, according to latest authority survey, we just add $x_{6,7,8}$ as one variable x^* instead of each single one x^*, function can be approached.

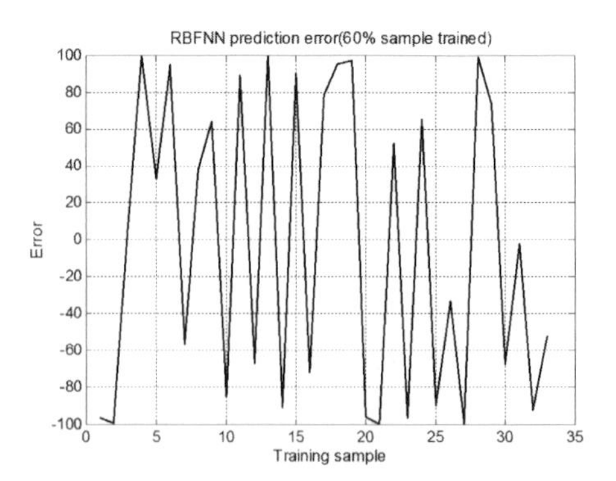

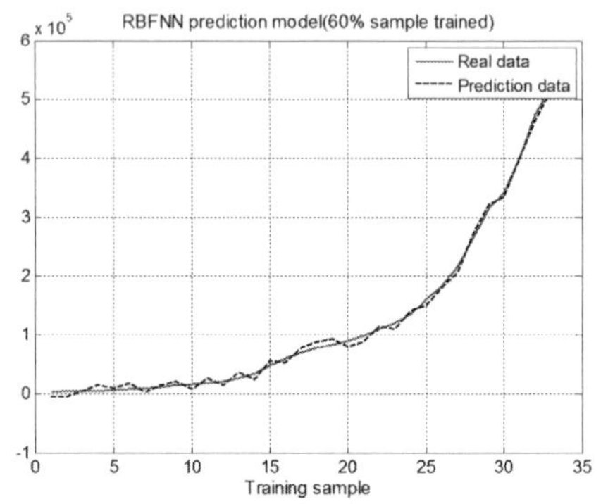

Figure 5. RBFNN prediction results 1.

4.2 *RBF prediction*

Let a matrix $P_{(7\times33)}$ as input data, and matrix $T_{(1\times33)}$ as output data, then we can train the two data set in radial basis function neural networks, from regression on each energy, let $t = 34$, prediction data achieved as testing matrix M_{pre}, and use this matrix as testing matrix M_{test}:

$$M_{pre} = M_{test} = \begin{pmatrix} t \\ x_1 \\ x_2 \\ x_3 \\ x_4 \\ x_5 \\ x^* \end{pmatrix} = \begin{pmatrix} 34 \\ 507.9680 \\ 144.6160 \\ 2.0900 \times 10^3 \\ 22.9306 \\ 209.4692 \\ 50.2176 \end{pmatrix} \quad (9)$$

Prediction model results based on RBFNN are shown in Figure 5 and Figure 6, the spread speed is 0.5 and target goal is 1.

As is shown above, 60% percentage data set was used to build network and adapted to simulate all data sets, the error curve is not so satisfied.

$$M_{pre1} = M_{test} = \begin{pmatrix} t \\ x_1 \\ x_2 \\ x_3 \\ x_4 \\ x_5 \\ x^* \end{pmatrix} = \begin{pmatrix} 34 \\ 574.7145 \\ 139.2217 \\ 2.1500 \times 10^3 \\ 24.7841 \\ 22.7784 \\ 50.2241 \end{pmatrix} \quad (10)$$

The prediction results simulated the real data well with acceptable error, this situation also verifies that suitable data sets are helpful to a network training.

The results predicted is shown above, and comparisons between the curve fitting forecast and RBFNN method shares some approximate value.

5 CONCLUSIONS AND RELATED WORKS

In Section IV, subsections A and B, we analyzed the energy prediction and the economy prediction, respectively, based on the current trends. The results shown verify the correctness of the prediction model based on the Radial Basis Function Neural Network. The curve fitting results represent the predicted inputs of the model rather well (Li et al. 2014).

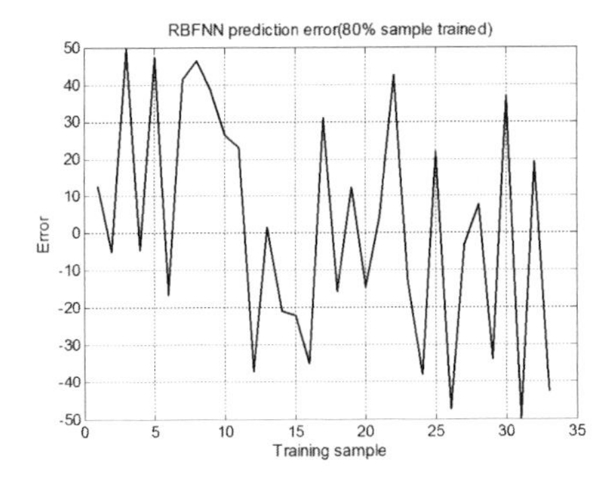

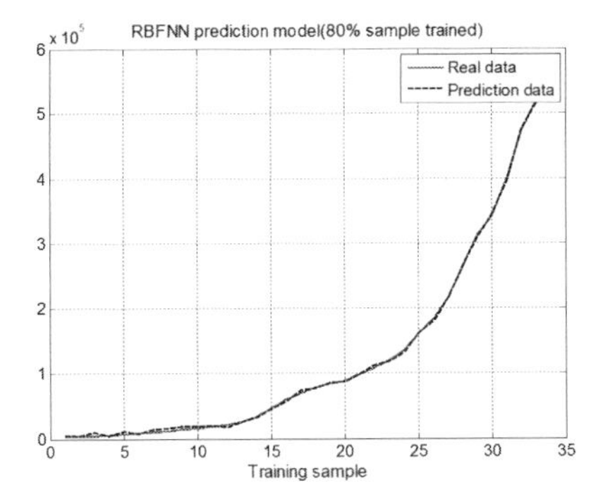

Figure 6. RBFNN prediction results 2.

Table 2. Energy investment of China (1980–2012).

year	w_1	w_2	w_3	w_4	w_5	$w_{6,7,8}$
2003	946.0	322.0	436.4	-	-	380.3
2004	1112.3	637.9	690.4	-	-	506.4
2005	1463.6	801.3	1162.9	-	-	677.7
2006	1822.2	939.3	1459.0	-	-	760.5
2007	2225.5	1415.4	1804.6	-	-	825.3
2008	2675.1	1827.5	2399.2	-	-	944.3
2009	2791.5	1839.8	3056.9	-	-	1178.9
2010	2928.0	2035.1	3784.7	-	-	1287.9
2011	3022.0	2268.5	4907.3	-	-	1284.7

Actually, all energy exploiting closed to funds invested in its area, which this factor is crucial while has not been taken into account. Even for any organization, its financial capacity is limited, this phenomenon confronted us with a problem of optimization (Yang 2013).

In Table 2, for some missing data of w_5, its investment data can not reflect the actual situation for the same reason with w_4, use curve fitting to fix it relevant to the year is accessible.

Let the relationship between total primary energy (E_{ng}) and GDP (G_{dp}) is linear, from table 1, linear model Poly1:

$$G_{dp} = 246.1 \cdot E_{ng} - 1.756 \times 10^5 \tag{11}$$

Functions of the contribution rate value of investment I_{nv} and each type proportion of total energy retained:

$$x_i = \alpha_i \cdot I_{nv}, \sum \alpha_i = 1 \tag{12}$$

$$x_i = \beta_i \cdot E_{ng}, \sum \beta_i = 1 (i = 1, 2, \cdots, 8) \tag{13}$$

where α_i, β_i represents proportionality coefficient from total investment and energy of x_i.

As total investment has been fixed to a constant number C, from the equations above obtain:

$$G_{dp} \propto I_{nv} \tag{14}$$

this problem mathematically conveyed:

$$G_{dp} = f_{max}(x_1, x_2, x_3, x_4, x_5, x_6; t) \tag{15}$$

subject to

$$I_{nv} = C, x_i = \alpha_i \cdot I_{nv}, \sum \alpha_i = 1, x_i = \beta_i \cdot E_{ng}, \sum \beta_i = 1.$$

From this equation set, achieve maximum G_{dp} via optimization. According to BP prophecy, China will remain the world's largest energy consumer and become the world's largest energy consuming entity, it's worth to focus on how to push development direction, this also can be defined as a valuable investment index (Dhanalakshmi et al. 2009).

This paper builds relationship model between energy and economic with polynomial regression, RBFNN, the extremum value of multivariate functions, there will need more demonstration of these equations, also may be other factors has not been considered in, which is another variable definition problem, but as high dimension the variables matrix, the results would be more actuate.

REFERENCES

C. J. Ong, et al. 1996. An optimization approach for biarc curve-fitting of B-spline curves. *Computer Aided Design* , 28(12): 951–959.

Dhanalakshmi P., et al. 2009. Classification of audio signals using SVM and RBFNN. *Expert Systems with Applications* , 36(3): 6069–6075.

Glvez A. & Iglesias A. 2013. A new iterative mutually coupled hybrid GACPSO approach for curve fitting in manufacturing. *Applied Soft Computing* , 13(3): 1491–1504.

H. Chen & Kim A. S. 2006. Prediction of permeate flux decline in crossflow membrane filtration of colloidal suspension: a radial basis function neural network approach. *Desalination*, 192(1): 415–428.

Paul Crompton, et al. 2005. Energy consumption in China: past trends and future directions. *Energy Economics* , 27(1): 195–208.

Wen Gang. 1996. Knowledge-based Artificial Neural Network Applied to Subdividing control of Stepping Motors. *Modern Electronics Technique* ,34(7): 190–199.

X. L. Yang & Q. He. 2013. Influence of Modeling Methods for Housing Price Forecasting. *Advanced Materials Research* , 798: 885–888.

Z. Li, et al. 2014. An ontology-based Web mining method for unemployment rate prediction. *Decision Support Systems*, 66: 114–122.

Z. Zhang, et al. 2008. Asian energy and environmental policy: Promoting growth while preserving the environment. *Energy Policy* , 36(10): 3905–3924.

Electronics, Communications and Networks IV – Hussain & Ivanovic (eds)
© 2015 Taylor & Francis Group, London, ISBN: 978-1-138-02830-2

Novel excitation of dielectric resonator antenna for wideband applications

Lin Yi, He Yan, Gao Ge & Hu Yang
College of Electronics Science and Engineering,
National University of Defense Technology, Changsha, Hunan Province, China

ABSTRACT: A novel excitation method using a circular patch was employed to excite a Dielectric Resonator Antenna (DRA). The proposed method offered a flexible approach to design a wideband antenna for monopole-like radiation. A circular patch-fed annular column DRA was designed to demonstrate the applicability of this method. The radiation performance of the proposed antenna was presented.

1 INTRODUCTION

Dielectric resonator antennas (DRAs) (Long 1983) have been studied extensively due to their advantages such as compact size, low loss, high radiation efficiency, and high flexibility in excitation. Many practical and novel excitation methods for DRAs have been proposed including the SIW feed (Abdel-Wahab, 2011), coplanar-waveguide feed (Gao 2011), aperture coupled feed (Nikkhah 2013), coaxial probe feed (Pan 2012), and direct-microstripline feed (Huitema, 2011). However, most of them focused on broadside radiation patterns. It is well known that a DRA excited in its $TM_{01\delta}$ mode has monopole-like radiation (Verplanken 1976), and this mode has been investigated widely for various practical antenna applications (Mongia 1993, Guo 2005, Ozzaim 2013). However, the method to excite this mode is restricted to the probe feed.

In this paper, a novel method to excite DRAs is proposed, which makes their application for monopole-type radiation more flexible, and provides a new approach to design a wideband DRA. This method employs a circular patch to excite a dielectric resonator antenna. In order to demonstrate the availability and flexibility of the proposed method, it was used to excite an annular column DRA operating in its $TM_{01\delta}$ mode. The simulated return loss and radiation patterns are reported in this paper.

2 THEORY AND ANTENNA GEOMETRY

In order to excite the $TM_{01\delta}$ mode of the DRA, the circular patch was designed to operate on the TM_{02} mode in which the field is invariant in the φ-direction (Garg 2001). Since this mode is not the fundamental mode, the size of the patch was relatively large (about 1 λ_g).

The resonator frequency of the DRA excited at the $TM_{01\delta}$ was given as (Verplanken 1976):

$$f = \frac{c}{2\pi\sqrt{\varepsilon_r}} \cdot \sqrt{\left(\frac{\pi}{2h}\right)^2 + \left(\frac{x_o}{a}\right)^2} \tag{1}$$

$$\frac{J_1(x_0)}{Y_1(x_0)} = \frac{J_1(\frac{b}{a}x_0)}{Y_1(\frac{b}{a}x_0)} \tag{2}$$

where a, b, and h are the outer radius, inner radius, and height of the DRA, respectively, c is the speed of light in free space, and x_0 is the root of Formula (2) and its value depends on the ratio of b/a.

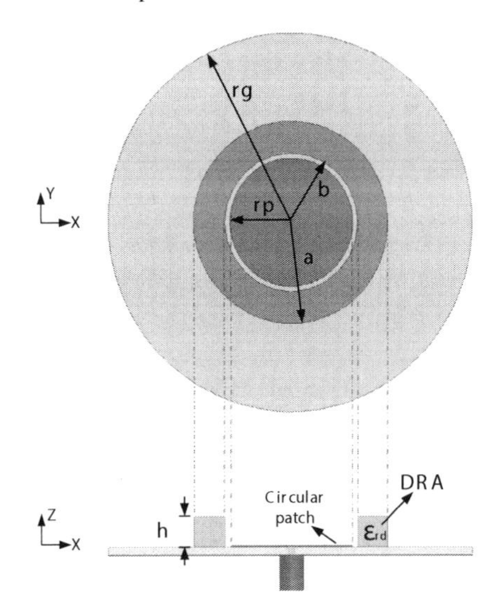

Figure 1. Geometry of the circular patch-fed annular column DRA.

The geometry of the circular patch-fed annular column DRA is shown in Figure 1. The circular patch had a radius of rp = 25.6 mm and was printed on a substrate of relative permittivity of 2.2 and thickness of 2 mm. The patch was center-fed by a coaxial probe. The annular column DRA had an inner radius b = 26 mm, outer radius a = 38 mm, height h = 6 mm, and dielectric constant $\varepsilon_{rd} = 16$. The ground plane radius rg was 75 mm.

3 SIMULATED RESULTS AND DISCUSSION

The radiation performance of the DRA was simulated using Ansoft HFSS software. Figure 2 shows the return loss of the circular patch-fed annular column DRA. A wide impedance bandwidth from 4.2 GHz to 5.2 GHz (about 21.7%) was obtained. The theoretical resonance frequency of $TM_{01\delta}$ was 4.6 GHz using Formula (1). However, it was obvious that there were two resonator frequencies (4.3 GHz and 4.9 GHz) in the working bandwidth. By changing the value of the patch radius, as shown in Figure 3, the frequency changed. If the radius of the patch increased, the lower resonator frequency decreased. Simultaneously, the higher resonator frequency was stable in this process. Therefore, it could be assumed that the lower resonator frequency was generated by the circular patch and the higher one was linked to the annular column DRA. The discrepancy between the theoretical resonator frequency of the $TM_{01\delta}$ mode of DRAs and the simulated one was caused by the excitation method and imprecise model of the DRA.

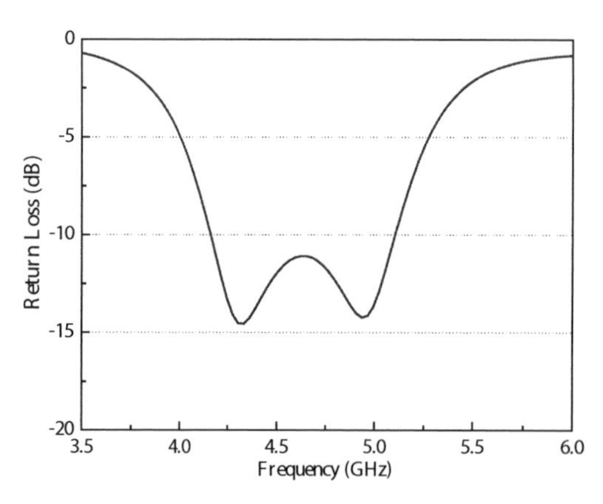

Figure 2. Return loss of the circular patch fed annular column DRA.

Figure 4 shows the radiation patterns for the 4.3 GHz and 4.9 GHz frequencies. According to the E-plane radiation patterns, a deep null at zenith was obtained and the maximum radiation beam appeared at around $\theta = \pm 40^0$. The H-plane radiation patterns showed that the radiation pattern was omnidirectional. Therefore, a monopole-like pattern was realized in the pass band. The simulated antenna gain at 4.9 GHz was 5.6 dB. Meanwhile, the cross-polarization of this antenna was extremely low, about 40 dB below the co-polarization level both in the E-plane and H-plane.

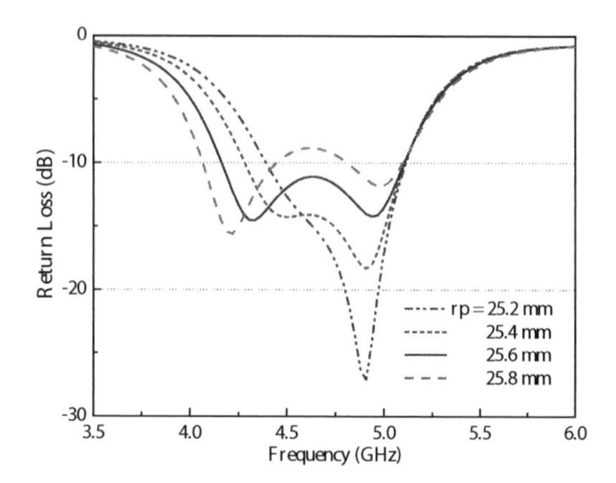

Figure 3. Return loss for the circular patch-fed annular column DRA for different values of patch radius.

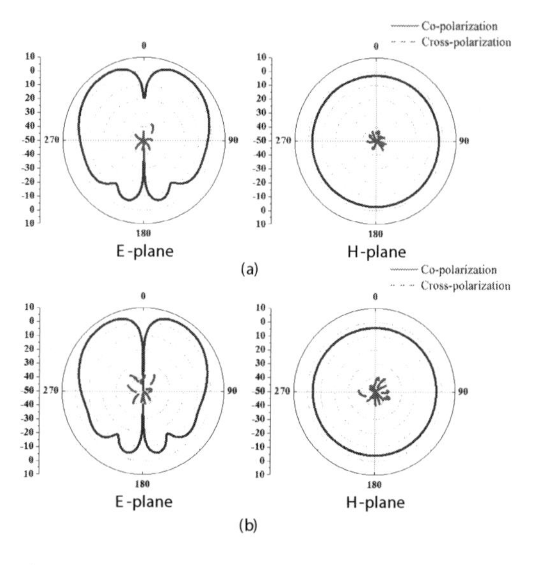

Figure 4. Simulated radiation patterns of the circular patch-fed annular column DRA: (a) $f = 4.3GHz$; (b) $f = 4.9GHz$.

4 CONCLUSIONS

By using a circular patch to excite the $TM_{01\delta}$ mode of a dielectric resonator antenna, a novel excitation method was introduced in this paper. A circular patch-fed annular column DRA was designed to demonstrate the applicability of this excitation method. The results confirmed that the proposed excitation method also provided a flexible approach to achieve a wideband antenna for monopole-like radiation.

REFERENCES

Abdel-Wahab, W. M. 2011. Millimeter-wave high radiation efficiency planar waveguide series-fed dielectric resonator antenna (dra) array: Analysis, design, and measurements. *J. IEEE Transactions on Antennas and Propagation* 59(8): 2834–2843.

Garg, R. 2001. *Microstrip antenna design handbook*. Artech House.

Gao, Y 2011. Compact CPW-fed dielectric resonator antenna with dual polarization. *J. Antennas and Wireless Propagation Letters, IEEE.* 10: 544–547.

Guo, Y. X. 2005. Wide-band stacked double annular-ring dielectric resonator antenna at the end-fire mode operation. *IEEE Transactions on Antennas and Propagation* 53(10): 3394–3397.

Huitema, L. 2011. Compact and multiband dielectric resonator antenna with pattern diversity for multistandard mobile handheld devices. *J. IEEE transactions on antennas and propagation* 59(11): 4201–4208.

Long, S. A. 1983. The resonant cylindrical dielectric cavity antenna. *J. IEEE Transactions on Antennas and Propagation* 31: 406–412.

Mongia, R. K. 1993. Electric-monopole antenna using a dielectric ring resonator. *J. Electronics Letters* 29(17): 1530–1531.

Nikkhah, M. R. 2013. High-Gain Aperture Coupled Rectangular Dielectric Resonator Antenna Array Using Parasitic Elements. *J. IEEE transactions on antennas and propagation* 61(7): 3905–3908.

Ozzaim, C. 2013. Stacked conical ring dielectric resonator antenna excited by a monopole for improved ultrawide bandwidth. *IEEE Transactions on Antennas and Propagation* 61(3): 1435–1438.

Pan, Y. M. 2012. Wideband omnidirectional circularly polarized dielectric resonator antenna with parasitic strips. *J. IEEE transactions on antennas and propagation* 60(6): 2992–2997.

Verplanken, M. 1976. The Electric-Dipole Resonances of Ring Resonators of Very High Permittivity. *J. IEEE transactions on Microwave Theory and Techniques* 24(2): 108–112.

Electronics, Communications and Networks IV – Hussain & Ivanovic (eds)
© 2015 Taylor & Francis Group, London, ISBN: 978-1-138-02830-2

Dual-routing-engine based satellite optical network with periodically changed topology

Yanan Yue, Min Zhang*, Yan Long, Dahai Han & Shanguo Huang
State Key Laboratory of Information Photonics and Optical Communications, Beijing University of Posts & Telecom, Beijing, China

ABSTRACT: With the explosive growth of data business, the traditional satellite microwave's bandwidth is more crowded, so there is an inevitable trend to develop satellite optical communication. In this paper, we put forth a new satellite model based on periodically dynamic changed and the dual-routing-engine model. The two different route modules: Group Engine (GE) and Unit Engine (UE) are the key for the satellite network, they separate the topology to two-layer and multi-domain, determine the information path across the Optical Inter-Satellite Link (OISL) in the same orbit and Optical Inter-Orbit Link (OIOL). Then we simulate the structure aimed at computing and analyzing the resource blocking rate and the time blocking rate of different service models. The simulation proves that optical satellite has the higher time blocking rate, so it is necessary to re-route and reallocation in transmission. The result shows optical satellite network under dual-routing-engine model performs better than monolayer satellite network.

1 INTRODUCTION

With the ever increasing of data traffic in the satellite network, traditional satellite communication has obviously reached a bottleneck in speed owing to its adoption of the microwave frequency band (such as Ku, Ka frequency band). Therefore, it is increasingly important to build an optical satellite network.

Currently, there are some researches about networking and routing in the monolayer of low-earth-orbit (LEO) (Ekici et al. 2001); there are also some studies about two layers between middle-earth-orbit (MEO) and LEO, or three-layered network models of geosynchronous-earth-orbit (GEO)/MEO/LEO (Wang et al. 2006). Comparing with microwave link, optical link greatly improves the transmission bandwidth, but it raises new demand on the laser-links switching and route designing. The acquisition, tracking and pointing (ATP) in OISL/OIOL has great difficulties, and the links switching will also bring an amount of overhead. Here, the OISL means optical communication links of adjacent satellite in the same orbit, they are stable links can be established permanently if needed. The OIOL presents optical communication links in different orbit or different layer, they aren't stable links due to motion of satellites. For the past few years, the wavelength division multiplexing (WDM) of the optical satellite network has been put forward and researching on satellite networks (Yang et al. 2010, Karafolas et al. 2000).

According to the topological structure and service flow characteristics in the satellite network, we proposed a two-layered and multi-domain satellite network, which can change the topology periodically and dynamically, the dual-engine of GE and UE are responsible for the services transmission of the whole network. Then we use the simulation tools: Satellite Tool Kit (STK) and C++ to simulate the network performance under different service models: the topology constitutes of LEO/MEO, the optical link has been set to a coarse wavelength division multiplexing (CWDM) mode. At last, the results of the analysis based on simulations have been given.

2 THE DUAL-ROUTING-ENGINE STRUCTURE

2.1 *Two-layered & multi-domain*

The LEO layer has a larger number of satellites, a short orbital period, a fast changing network topology and a small coverage area, so it needs frequent communication path switching. However, the satellites in the MEO layer have a relatively stable, steady relationship and a large coverage. On the basis of those features, we proposed a two-layered, multi-domain optical satellite network infrastructure based on the dual-routing-engine. The topology relationship is constructed and maintained by changing periodically and dynamically.

As depicted in Figure 1, the solid lines represent OISL, and the dotted lines represent OIOL. Closed curve composed of permanent links is in the same orbit satellite, and there is no inter-orbit laser link in the LEO layer.

*Corresponding author: *mzhang@bupt.edu.cn*

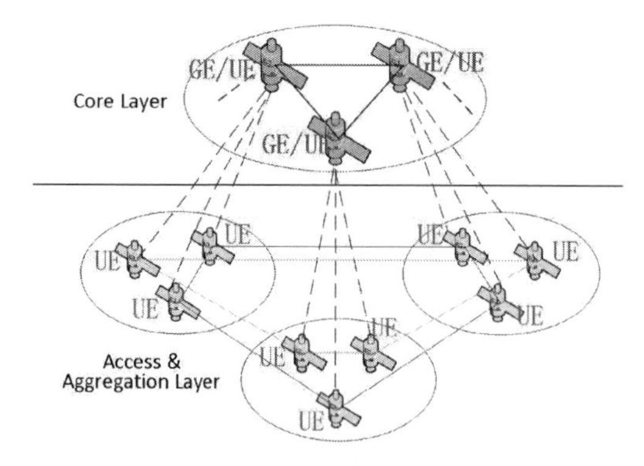

Figure 1. Two-layered & multi-domain optical satellite network infrastructure.

The Core Layer is composed of MEO satellites, and the Access & Convergence Layer consists of LEO satellites which has a small mass and relatively simple function and configuration. The nodes in the Core Layer are not only the management nodes in each domain but also the transmission nodes. They have the global topological information and can route across different domain, which reflects the relatively concentrated arbitration and coordination. The nodes in the Access & Convergence Layer take charge of the access and flow convergence for services and applications. They have independent routing capabilities of the intra-orbit services, but have to apply for complicated routing through the core layer. This two-layered structure possesses a superior extension ability and management ability for the whole network, as well as reducing the routing overhead.

The satellites in the same layer and orbit constitute a routing domain. It can construct a stable OISL between them, which is the best link to conduct data transmission. MEO satellites and the LEO satellites which are in the vision of the former form a logic domain. The MEO manage the subordinate LEO, and maintain the routing tables of this domain. In addition, it undertakes the forwarder for the services of the different orbit, avoiding frequent laser link switching.

2.2 *Dynamically periodic replacement topology*

Although the satellite keeps moving with high speed constantly, but in fact it is a regular periodic motion, so we can classify the satellite network structure as the periodically dynamic, changing structure, and construct a topology which is real-time changing but stable and simple. Then we are able to analyze the network performance.

Based on the Walker constellation topology (Mortari et al. 2011), we can design the network to a quasi periodic dynamic topology. Under the quasi periodic dynamic topology, different levels of satellites conform to the hooping structure as the following Formula (1), in which m and n are positive integer and T_m refers to the period of the satellite in the MEO, T_L refers to the period of the satellite in the LEO, and N_L refers to the number of satellites in the monorail of the LEO.

$$T_m = nT_L \pm mT_L / N_L \tag{1}$$

When the T_m period over, the satellites in the LEO go back to the origin place, but the location of them has changed. At this time it can exchange the topology rule table and adapt it to the next cycle in order to complete the circulation state.

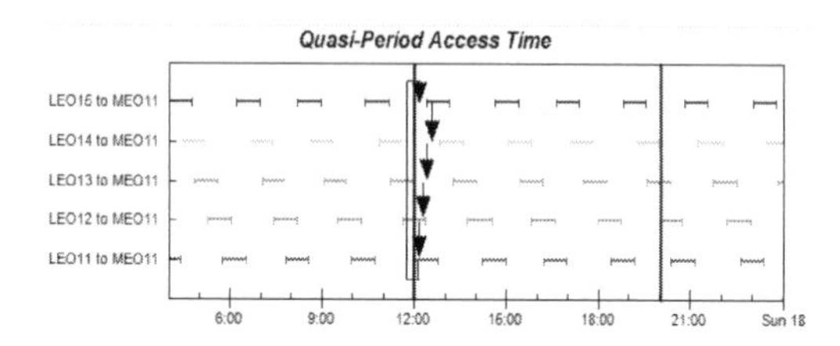

Figure 2. Quasi-period access time.

For example in Figure 2: T_m=480min, T_L =100min and n=5, m=1. The vertical coordinate means the links between one MEO with different LEO. The horizontal ordinate means the running time. Every time after T_m, we can observe that the link relation between LEO and MEO is not fixed, but LEO11 becomes LEO15, and LEO12 becomes LEO11 and so on.

2.3 *Dual-routing-engine module*

We designed an operating mode based on dual-routing-engine: GE and UE.

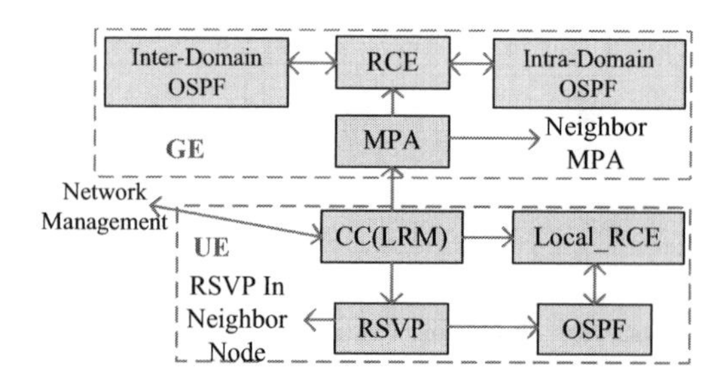

Figure 3. GE and UE module.

As depicted in Figure 3, GE is configured in the core nodes, and adopts the extended Path Computation Element (PCE) protocol to realize (Gonzalez et al. 2013). It has a global field vision and is responsible for the path calculation for services' transmission of inter-layer and inter-domain. While UE is configured with all nodes in the core layer and access convergence layer. It mainly takes charge of the domain path computation, reflects distributed intelligence, supports that each domain's ability of self-organization, and limits the network information flooding in one domain, which greatly reduces the convergence time of intra-orbit routing.

When service arrives, it will be sent to the UE unit for processing firstly. Then the routing engine selector (RES) judges whether the service can be calculated in UE. If it is judged to be an intra-orbit service, it will not be reported to the upper layer. Instead, it will be sent to the local routine computation engine (RCE). And the local RCE will conduct routing computing and resource reserving in the local, whose specific function realization relying on OSPF and Resource Reservation Protocol (RSVP). OSPF provides local RCE to prestored local routing table and real-time routing information by continuous flooding, and link-resource management (LRM) provides resource information. Then local RCE will get the shortest effective and available path. RSVP conducts detailed backward reservation according to the routing information, and attempts to establish the optical link through a link traversal. If the link establishment succeeds, RSVP will report to connect-controller (CC) module and the upper layer, informing each node that the resource are used; if it fails, the service will be blocked or rerouted. When the service is judged to be the inter-orbit or inter-domain service by the RES unit, it will be sent to the GE module, in which Multi-Protocol Analyzer (MPA) analysis the received data packets, including establishment request, resource information and topology information etc., providing the basis of the calculation for the RCE module. Then RCE realizes the calculation of a path between any two stars relying on the information from inter-domain OSPF and intra-domain OSPF as well as MPA, then conduct resource reservation of the services. All processes shown in Figure 4.

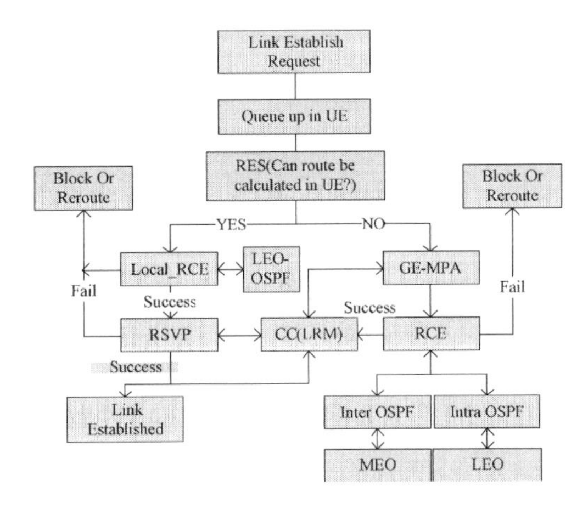

Figure 4. Dual-routing working principle.

3 SIMULATION RESULTS AND DISCUSSION

3.1 *Simulation data*

According to the above designing idea, we designed a typical simplified satellite optical model based on a two-layered and multi-domain, the parameters are listed in the Table 1, then we simulated by STK and C++.

Table 1. Basic satellite parameters.

	LEO	MEO
Constellation Type	walker delta	walker delta
Number of Planes	2	1
Number of Sats per Planes	6	3
Phase factor	1	0
Orbit Altitude	1680km	13929km
Eccentricity	0	0
Orbit Inclination	45°	0°
Intra Orbit Range	8059km	35173km

Then we need to get the distance of MEO vision, this is the main parameters to decide the domain's division. We can image the vision to a cone, then use geometric equations to calculate the below parameters:

$$\begin{cases}\theta_{1/2} \geq 23.4^\circ \\ d \geq 18639Km\end{cases}, \text{ so we set } \begin{cases}\theta_{1/2} = 25^\circ \\ d = 19000Km\end{cases}$$

where $\theta_{1/2}$ = half-apex angle; d = length of cone bus.

Finally, we simulated in STK, we can obtain a single MEO logical domain at a time and the constellation figure in this domain is displayed in Figure 5.

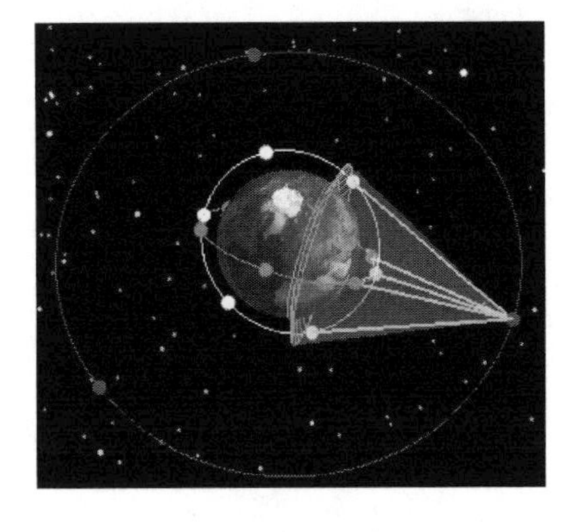

Figure 5. Simulate network model.

According to this kind of method to build an inter-orbit link, we can calculate the link construction circumstance, which is depicted in Figure 6. The vertical coordinate means the number of OIOL in one logic domain. As it is shown, there are at least 12 links between MEO and LEO, which assure that LEO can access to the management scope of the MEO at any time.

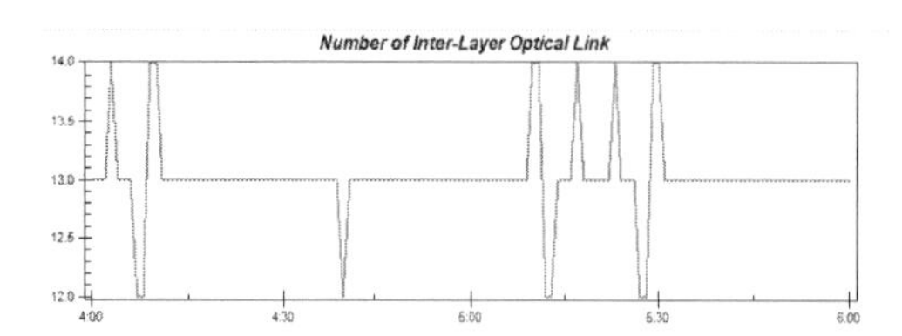

Figure 6. Number of inter-layer optical link.

3.2 *Resource block & time block rate*

Here, we assume that the service distribution model is Poisson. There are 4, 6 or 8 wavelengths in the network, and each service lasts 1-30 minutes. We use first-fit (FF) algorithm to assign the wavelength for services; then use backward resource reservation in RSVP of the simulation.

Figure 7 shows the resource blocking rate for the dual-routing-engine network configured of different wavelength number. Once the optical link resources be saturated, the blocking rate would increase rapidly. When the wavelength number is 8, the network is in a state of resources excessively and won't occur a situation of resources, lack mostly; But when the number of wavelengths to reduce to 4, a low service arrival rate could remain a low blocking rate, when the arrival rate achieves 29%, the OISL and OIOL stay in a saturated state, most multi-hop service be blocked., causing resources rate level presents a pattern of linear.

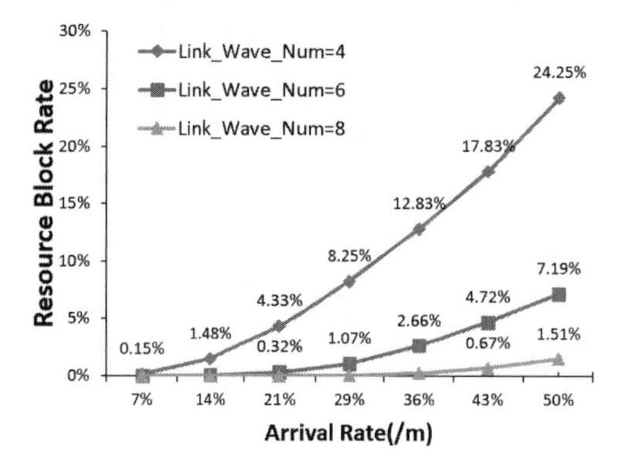

Figure 7. Resource blocking rate and arrival rate.

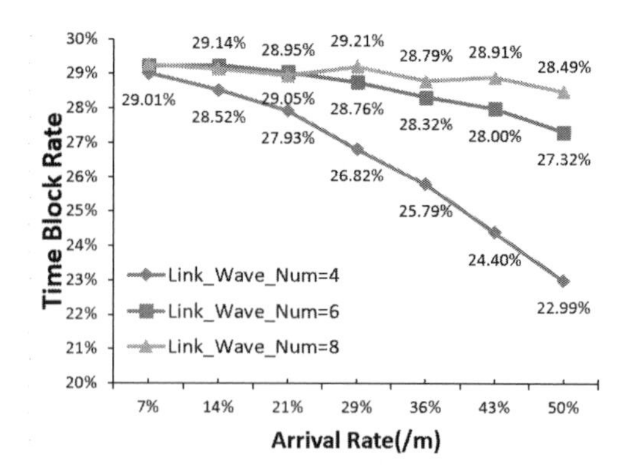

Figure 8. Time blocking rate and arrival rate.

The Figure 8 shows the time blocking rate. Drawing from the graph, we could see that in the majority of the time, the time blocking rate of service is stably at a 25%-30% comparatively high level. The reason is that the time blocking rate mainly depends on the duration of the stable link. For an established two-layered satellite topology, the duration of the stable link and its distribution are fixed without a topological change. The service only uses OISL calculated by UE will not be blocked at any time. In the model, the generating probability of LEO and MEO service only transfer by OISL is as Formula (2):

$$P(\text{LEO}) + P(\text{MEO}) = \frac{N_{Lo} \times N_L \times (N_L - 1) + N_{Mo} \times N_M \times (N_M - 1)}{N \times (N-1)} = 0.3 \quad (2)$$

where the N_{Lo} and N_{Mo} refer to the orbit number of LEO and MEO layer, N_L, N_M refer to the satellite number of each orbit. Thus the blocking rate of inter-orbit service is:

$$P = \frac{P(\text{TimeBlock})}{1 - P(\text{LEO}) - P(\text{MEO})} \approx 40\% \quad (3)$$

That 40% service will be in the state of re-routine or waiting due to the block of link establishment time. Certainly we could lower the time blocking rate to a large extent by allowing a link switching in the service transmission process. However, it needs the superior and swift switching and ATP technology of the space-borne laser. Meanwhile, in the same circumstance the time blocking rate and the resource blocking rate have an inverse relation, which is a contradiction relationship.

3.3 *Compare with the monolayer satellite network*

We have computed and analyzed the package level for the network between dual-routing-engine and monolayer network. Here, the monolayer satellite network means a simple network with single-layer. All the nodes have the same function. When a service generated, the route computation depends on the flooding of the route-request message.

Under dual-routing-engine, each domain has a core routing node, so the level of sending packets P_D is composed of three parts: the obtaining the routing information p_r, the resource reservation p_s and the global resource flooding p_f:

$$P_D = p_r + p_s + p_f = \sum_{Source}^{GE} R + 2\sum_{Source}^{Dist} R + (N_{Meo} + \frac{N_{Leo}}{O_{Leo}} - 1) \quad (4)$$

where R is path information, N is the number of the satellites and O is the number of the orbits.

In the structure of the monolayer and hop-by-hop routing pattern, the level of sending packets is also made up of three parts, flooding for obtaining the routing information P'_f, for obtaining routing information P'_r, and for resource reservation P'_s. We assume that the network is a 2-D Torus topology consisted by Walker constellation, LEO on each orbit can consist of at least 4 links of two inter-plane links and two intra-plane links, so we can get the following formula:

$$P'_D = p'_f + p'_r + p'_s = \sum_{Source}^{Dist} H_p + \sum_{Dist}^{Source} R + 2\sum_{Source}^{Dist} R \quad (5)$$

Where R is path information, H_p is the flooding packets at each path node. Then we can infer that:

$$\sum_{Source}^{Dist} H_p = 4 \times (1 + \frac{3H(H-1)}{2}) \quad (6)$$

where H is the sending hops of flooding packets when the addressing succeeds.

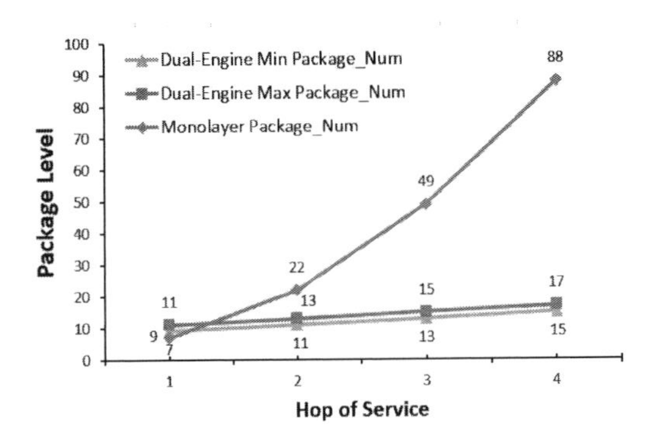

Figure 9. Package level of different hops' service.

According to the above formula and the network model in the simulation, the signal overhead of establishing links (the level of sending packets) under different hops can be depicted as the below Figure 9. In the dual-routing-engine model, the overhead of signal packet is always at a low level, and it appears slowly linear change with the hop increasing. But in the monolayer and hop-by-hop model, the signal packets of establishing service show an exponential growth trend. Once the service is over two hops, the overhead of signal is apparently more than a dual engine with two-layered structure.

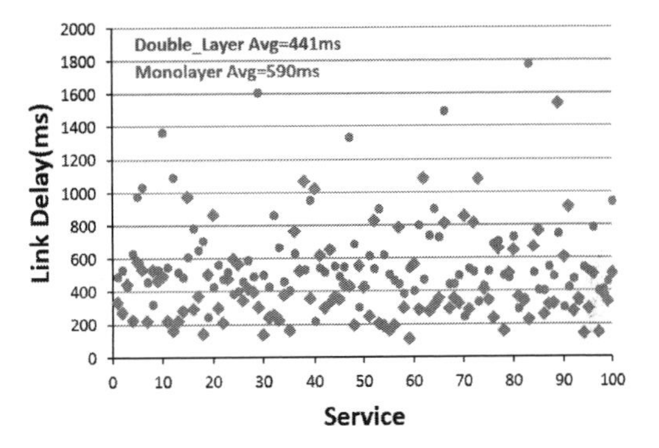

Figure 10. Link establishment delay.

Follow on, we make comparison of link delay between dual-routing-engine with two-layer network and a monolayer satellite network under the same scale. The graph (Figure 10) is the link establishment delay by generating 100 services randomly. We could obviously see that the link establishment delay is focused on the range of 200-400ms. Certainly it stays in a comparatively high level because of the exclusion of the satellite process delay and the laser alignment delay. The reason is that the service needs a backward-resource-reservation after the route acquisition of UE and GE. Only after the success of backward reservation, the route establishment can be regarded as succeeding. After the optical link established, the data transmission delay will be approximately 150ms.In addition, if we take the ascending and descending link time of the access layer into consideration, the 400ms transmission delay standard can be satisfied. If we could solve the time blocking rate problem of service, then this structure of the optical satellite network absolutely can bear high real-time services, such as satellite video live broadcast, etc.

4 CONLUSION

In this paper, we proposed a dual-routing-engine based optical satellite networks. It aims at solving the bandwidth requirements. Through the simulation, we found the time block is a main issue that we should improve. The comparison results show the dual-routing-engine have a better performance than monolayer network relatively.

REFERENCES

Ekici, E. et al. 2001. A distributed routing algorithm for datagram traffic in LEO satellite networks. *Networking,* IEEE/ACM Transactions on 9(2): 137–147.

Gonzalez de Dios, O et al. 2013. Experimental demonstration of a PCE for wavelength-routed optical burst-switched (WR-OBS) networks. Optical network design and modeling, 17th International Conference on. *IEEE*: 269–274.

Karafolas, N. et al. 2000. Optical WDM networking in broadband satellite constellations. *Lasers and Electro-Optics Society 2000 Annual Meeting. LEOS 2000, 13th Annual Meeting, IEEE* 1: 100–101.

Mortari, D. et al. 2011. Design of Flower Constellations for Telecommunication Services. *Proceedings of the IEEE* 99(11): 2008–2019.

Wang Zhenyong et al. 2006. Analysis on connectivity of inter-orbit-links in a MEO/LEO double-layer satellite network. *Chinese Journal of Aeronautics* 19(4): 340–345.

Yang Qinglong et al. 2010. Analysis of crosstalk in optical satellite networks with wavelength division multiplexing architectures. *Journal of Lightwave Technology* 28(6): 931–938.

Electronics, Communications and Networks IV – Hussain & Ivanovic (eds)
© 2015 Taylor & Francis Group, London, ISBN: 978-1-138-02830-2

Dual-band B-shaped antenna for wireless communication technology

Wen-Jie Zeng, Dong-Mei Cai* & Jia Peng
College of Physics and Optoelectronics, Taiyuan University of Technology, Taiyuan, Shanxi, China

Jian-Xia Liu
College of information engineering, Taiyuan University of Technology, Taiyuan, Shanxi, China

ABSTRACT: A dual-band B-shaped antenna was proposed in this paper. This antenna covered the frequency bands of 4G frequency division duplex long-term evolution (FDD-LTE) and worldwide interoperability for microwave access application (WiMAX). The proposed antenna was fed by 50 Ω micro-strip line on a 1.6-mm-thick FR4 material substrate. The antenna was designed and analyzed using HFSS15.0. The simulation results showed that the impedance bandwidth (return loss < -10 dB) was 180 MHz (1.73-1.91 GHz) for FDD-LTE and 480 MHz (3.46-3.94 GHz) for WiMAX 3.5 GHz. There was a maximum simulated gain of 3.76 dB and 4.24 dB at the lower and higher resonance frequencies, respectively. This paper illustrated that the antenna showed stable omnidirectional radiation patterns.

1 INTRODUCTION

In recent years, there has been a rapid development of wireless communication technology. The dual-band micro-strip antenna has become one of the most important circuit devices in modern wireless communication systems. Because of its advantageous properties such as small size, light weight, ease of integration, and low cost (Ullah et al. 2012, de Oliveira et al. 2012), the micro-strip antenna has attracted much attention. With the developments of network technology, the demand for fast mobile network speeds has increased among worldwide users. The 4G (FDD-LTE 1.755-1.785 GHz and 1.85-1.88 GHz) networks, with the advantages of maximum download speeds of 100 Mbps and upload speeds of 50 Mbps, are subject to greater favor. Meanwhile, the WiMAX (3.5 GHz band of 3.4-3.69 GHz) technology has been used in the commercial, medical, and industrial fields. A network may not require a wide bandwidth to prevent the impact of other frequencies, so it is imperative that the dual-band antenna does not produce any parasitic bands. Researchers have studied a variety of methods to create a dual band. Some multi-frequency slot antennas have been designed, achieving multi-frequency characteristics through different combinations of slot antennas with micro-strip antennas (Lv et al. 2006, Tao et al. 2007), but in these cases, the high frequency omnidirectional pattern was poor. An S-shaped antenna with a rectangular ring was developed by Huang and Zhang (2014). It loaded a slot in the ground plane to achieve tri-band. A small size 20×20 mm^2 dual-band inverted A-shaped antenna for WiMAX and C-band was achieved by loading a slot on the patch (Ahsan et al. 2014). Abdul Salam et al. 2014), proposed a dual-band micro-strip antenna with a simple structure covering the WLAN 2.4-2.5 GHz and 5.7-5.9 GHz. (Zhang et al. 2008) proposed a monopole antenna that loaded a micro-strip stub at the rectangular ring terminal to obtain a high wideband frequency. A wide dual-band antenna, using a combination of circular and rectangular shapes, was designed (Chu & YeE 2010), but it interfered with bands of certain frequencies. All the above methods achieved a dual band and an expansion of bandwidth; however, they have the potential to produce interference.

Based on the above reviews, this paper proposed a B-shaped dual-band micro-strip antenna coving FDD-LTE and WiMAX with resonant frequencies of 1.82 GHz and 3.67 GHz. The minimum return loss at the lower and higher resonance points were -32.4825dB and -37.4815dB, respectively. The antenna was dual-band and did not produce any other frequencies. Stable omnidirectional radiation patterns and gains were thus achieved.

2 ANTENNA DESIGN

The geometric structure and configuration of the proposed B-shaped patch antenna are shown in Figure 1. The device was designed on a 1.6-mm-thick fiberglass

*Corresponding author: *zwjwhxz@126.com or dm_cai@163.com*

polymer resin (FR4) substrate with a relative dielectric constant ε_r=4.4 and a loss tangent tanδ=0.02 fed by 50 Ω micro-strip line. The HFSS15.0 program was used for the design and simulation of the proposed B-shaped antenna. The B-shaped patch had three parts: a rectangle with a width of *ws* and two semi-circles with radius of r_1 and r_2 and widths of d_1 and d. The same B-shaped slot, the same size as the substrate, was loaded on the ground. The radiation and ground patches were located on both sides of the substrate.

The antenna had a B-shaped patch to excite the two resonant modes. The entire upper circular ring and the bottom half of the lower ring produced the higher resonant frequency, while the upper half of the lower circular ring and bottom half of the upper ring excited the lower resonant frequency. Figure 2 shows the current distributions on the surface of the patch at higher and lower resonant frequencies. Stronger impedance matching was created by loading a B shape slot on the ground plane. The antenna's final parameters are shown in Table 1.

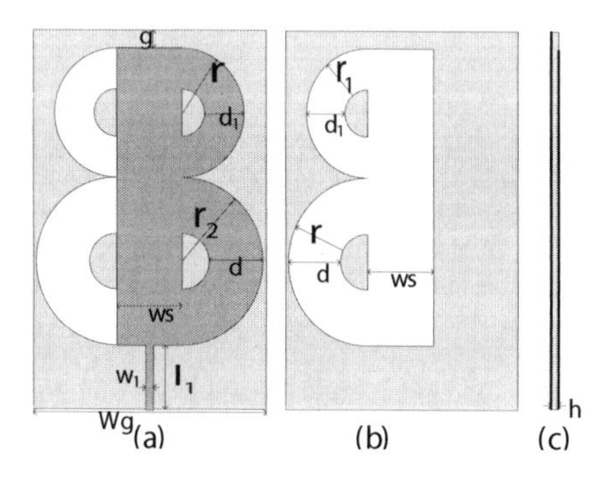

Figure 1. Geometry of the antenna: (a) Top view; (b) Ground plane; (c) Side view.

Table 1. The parameters of the proposed antenna. (mm).

Wg	r_1	r_2	*d*	d_1
30	8	10.5	7	5
ws	l_1	w_1	*g*	*h*
8.5	8	0.9	2	1.6

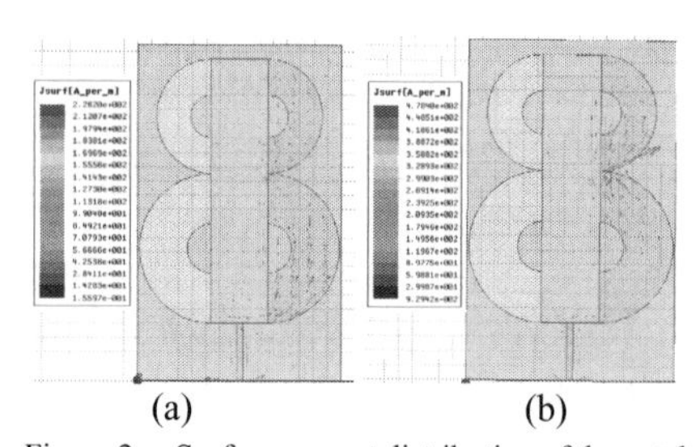

Figure 2. Surface current distribution of the patch: (a) 3.67 GHz; (b) 1.82 GHz.

3 SIMULATION RESULTS AND DISCUSSION

The return loss of the B-shaped antenna was simulated with the HFSS15.0 program; the results are shown in Figure 3. There were two bandwidths for the developed antenna if there was no B-shaped slot on the ground, but the impedance matching was poor. After loading a B-shaped slot on the ground plane, the antenna still had two bandwidths and the impedance matching improved significantly. Moreover, the antenna did not produce any other parasitic bands. In Figure 3, it can be seen that there were two bands for the final B-shaped antenna. The relative impedance bandwidths were 10% and 13%, respectively.

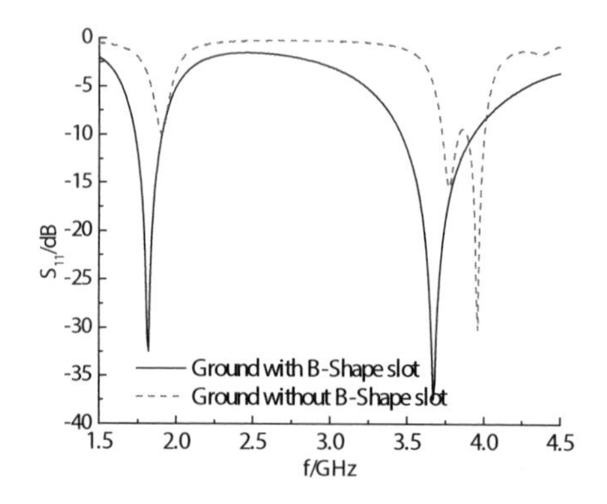

Figure 3. Simulated return loss of the antenna.

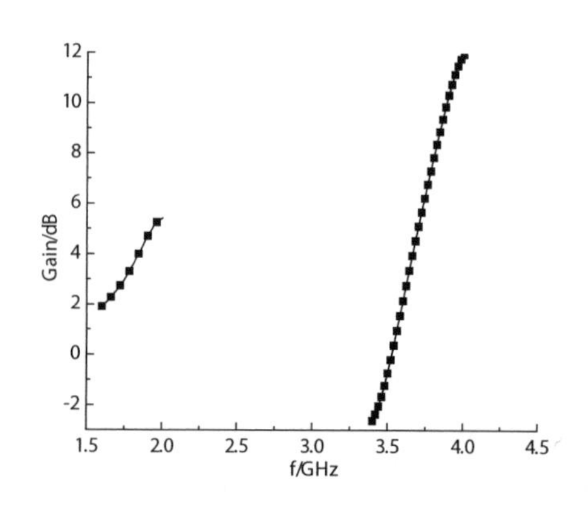

Figure 4. Gain of the antenna.

The simulated radiation patterns, including the E-plane (x-z-plane) and H-plane (y-z-plane), for the B-shaped antenna at the operating frequencies of 1.82 GHz and 3.67 GHz are shown in Figure 5. The radiation patterns at all given frequencies were stably bi-directional and predominantly vertically polarized. The antenna had stable omnidirectional radiation patterns for the two bandwidths.

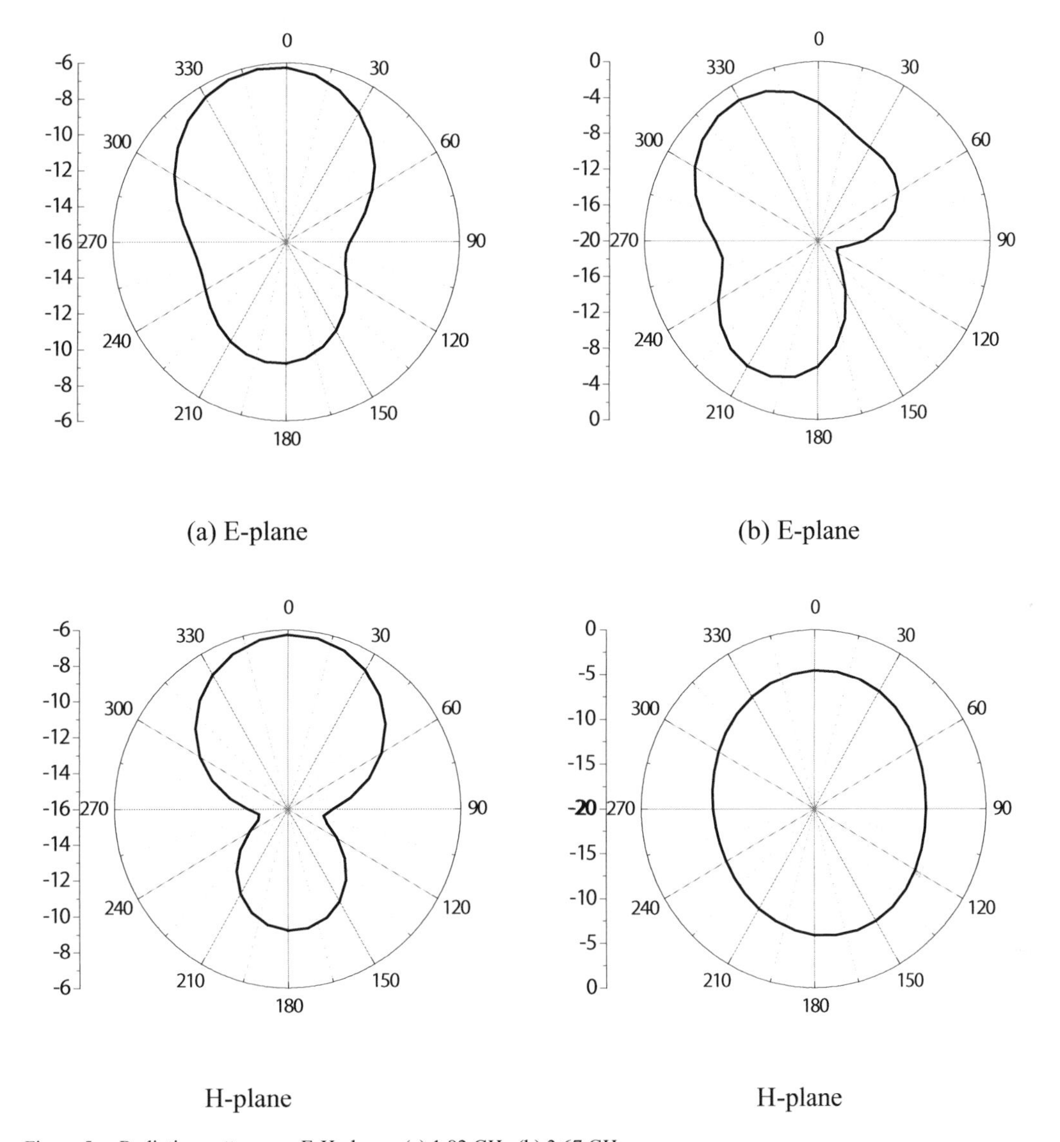

Figure 5. Radiation patterns on E-H planes: (a) 1.82 GHz (b) 3.67 GHz.

The gains for the two frequency bands are shown in Figure 4. The average gains for 1.82 GHz and 3.67 GHz were 3.76 dB and 4.24 dB, respectively.

4 CONCLUSION

A B-shaped structure dual-band antenna for 4G FDD-LTE and WiMAX 3.5 GHz was proposed. The simulated results showed that the proposed antenna had two bandwidths and did not produce other frequency bands. Strong impedance matching was obtained by loading a B-shaped slot on the ground plane. The relative impedance bandwidths were 10% (1.73-1.91 GHz) and 13% (3.46-3.94 GHz) at the lower and higher resonant points, respectively. The antenna had better omnidirectional radiation patterns. The average gains for 1.82 GHz and 3.67 GHz were 3.76 dB and 4.24 dB, respectively. Due to the properties stated above, the proposed device could be widely used in 4G networks and WiMAX 3.5 GHz dual-band systems.

ACKNOWLEDGMENTS

This work has been supported by Natural Science Foundation of Shanxi Province of China, No.2013011006—4, and supported by the State Key lab of Optics Technologies for Micro-fabrication, Institute of Optics and Electronics and Science Foundation of Chinese Academy of Science No.KFS4.

REFERENCES

Ahsan Md Rezwanul, Mohammad T. Islam & Habib Ullah M. 2014. A Compact Multiband inverted A-shaped patch antenna for WiMAX and C-band. *Microwave and Optical Technology Letters* 56(7): 1540–1543.

Chu Qing-xin & Ye Liang-hua. 2010. Compact dual band antenna for WLAN/WiMAX applications. *Chinese Journal of Radio Science* 25(5): 893–898.

de Oliveira, E.E.C., D'Assunc, A.G., Oliviera, J.B.L. & Cabral, A.M. 2012. Small size quasi-fractal microstrip antenna M1 EBG-GP. *Int. J. Appl. Electromagn Mech.* 39(4): 645–649.

Huang Hui-Fen & Zhang Shao-Fang. 2014. Compact Multiband Monopole antenna for WLAN/WiMAX Applications. *Microwave and Optical Technology Letters* 56(8): 1809–1812.

Lv Wenjun, Cheng Yong & Cheng Chonghu. 2006. Design of a novel multi-frequency microstrip slot antenna. *Chinese Journal of Radio Science* 21(2): 265–269.

Salam, A., Khan, A. A. & Hussain, M. S. 2014. Dual Band Microstrip antenna for wearable applications. *Microwave and Optical Technology Letters* 56(4): 916–918.

Sanz-Izquierdo, B., Batchelor, J.C. & Sobhy, M.I. 2010. Button antenna on textiles for wireless local area network on body applications. *IET Microwaves Antennas Propag.* 4(11): 1980–1987.

Tao, J., Cheng, C. H. & Zhu, H. B. 2007. Compact dual band slot antenna for WLAN applications. *Microwave Opt. Technol. Lett.* 49(5): 1203–1204.

Tiang, J. J., Islam, M. T., Misran, N. & Singh, J. M. 2014. Wideband L-probe circular patch antenna for dual-frequency RFID application. *Microwave and Optical Technology Letters* 56(6): 1421–1424.

Ullah, M.H., Islam, M.T., Mandeep, J.S., Misran, N. & Nikabdullah, N. 2012. A compact wideband antenna on dielectric material substrate for K band. *Electron Elect Eng.* 23(7): 75–78.

Zhang, L., Jiao, Y. C. & Zhao, G. 2008. Broadband dual band CPW- fed closed rectangular ring monopole antenna with a vertical strip for WLAN operation. *Microwave Opt. Technol. Lett.* 50(7): 1929–1931.

Electronics, Communications and Networks IV – Hussain & Ivanovic (eds)
© 2015 Taylor & Francis Group, London, ISBN: 978-1-138-02830-2

Evaluation method for node importance based on service in electric power communication network

Ying Zeng & Xingnan Li
Power Dispatch and Control Center of Guangdong Grid, Guangzhou, Guangdong, China

Boren Deng*, Runze Wu & Yingjie Zou
State Key Laboratory of Alternate Electrical Power System with Renewable Energy Sources (North China Electric Power University), Beijing, China

Yun Luo
Sichuan Enrising Information Technology Co. Ltd, Chengdu, Sichuan, China

ABSTRACT: A new evaluation method for node importance is proposed with the combination of the node contraction method in complex networks and the functional analysis of the importance evaluation for power service in practical operation. The definite application of this algorithm features the particularity of the power industry. Firstly, we classify the power services, then relate the importance degree of power service to the edge weight which is used to calculate the node weight and its importance. Finally, the simulation example, we have testified corresponds ultimately to the practical situation.

KEY WORDS: Importance of power service, complex network, node contraction, node importance

1 INTRODUCTION

As a private network of electric power system, the electric power communication network operates all services of power production and management. With the development of the smart grid, variety and amount of power services are growing rapidly, the structure of the power communication network is becoming more and more complex. Studies showing the importance of each node have an evidential distinction in a complex network. The network will be heavily damaged, once 5% of core nodes of the scale-free network are attacked (Wang et al. 2005). The evaluation of the node importance is significant for the research of network vulnerability and has a certain practical significance in the reliability and the risk management of the electric power communication network.

The evaluation methods of node importance in the complex network include Betweenness Method, Node Contraction Method and so on (Kermarrec et al. 2011). Betweenness Method is a classical method, but has the disadvantage of higher computation complexity (Manteach et al. 2010). (Tan et al. 2006) proposes the conception of aggregation of network. Node Contraction Method has low computation complexity, yet it can't tell the difference between the nodes which share the same location in the looped network. Based on the method above, (Chen et al. 2009) put forward an improved method which takes the extent of the interaction among nodes into consideration. (Zeng et al. 2013) use bandwidth as network weight, and the weight of the node as correction coefficient to evaluate the importance of the node in the electric power communication network. However, the method neglects the effect of power service in the electric power communication network; the result can't entirely reflect the actual operation situation of the network.

On the premise of the previous studies about node importance, this paper proposes a node importance evaluation method based on the importance of power service. Firstly, the power service is analyzed from safe operation requirements aspect and QoS (Quality of Service) requirements aspect to evaluate the importance of power service. Then, we analyze the link weight from the perspective of power service and the node weight obtained. The importance of a node can be calculated by using Node Contraction Method finally. The proposed method analyzes the importance of a node in the electric power communication network from the perspective of power service. Combining the power service with the topology

*Corresponding author: *dengboren@ncepu.edu.cn*

of the network, the proposed method emphasizes the impact of the node on the actual operation status of the network and provides a significant reference for the risk management of the electric power communication network.

2 ANALYSIS OF SERVICES

2.1 *Importance of power service*

Electric power services are divided into two categories including Production Control Services and Management Information Services. Production Control Services have two parts: Security Zone I and Security Zone II. Management Information Services can also be classified into two parts of Security Zone III and Security Zone IV based on the real-time requirement of power services. The power services in Production Control Services have high real-time requirement and are directly related to the survival of the power grid. Services in the Security Zone I such as the transmission line protection and wide-areas measurement system are the core services of the electricity production. Services in the Security Zone II such as the Dispatching Automation System ensure the normal operation of electric power system. Management Information Services are the ancillary services in electric power production. Services in the Security Zone III demand high real-time performance, such as the Substation Video Monitoring Service. Services like the Telephone System for Administrative Department insensitive to the delay requirements belong to Security Zone IV.

The importance of power service is a mathematical weight, which is determined by the harm degree on the basis of service interruption. The greater the importance value of a service is, the greater harm it will bring about the power grid when it is interrupted. So the importance of power service is a vital indicator for evaluating the operational risk of the electric power communication network.

2.2 *The evaluation of power service importance*

According to the characteristics of electric power production, we analyze the service importance based on the demand of operational security and QoS.

The demand of operational security reflects the safety requirements of power services and includes three indicators: Security Zone, Operation mode, Transmission channel. The indicators indicate the service importance by describing the actual operation safety state of running services from different angles.

The demand of QoS reflects the requirements of the communication quality of power services. Broadly speaking, the services related to the operation of the power system have a deeper effect towards the security and stability of the power grid, and a higher demand of communication quality. Therefore, they attach greater service importance.

According to the analysis of power service importance and the relevant regulation of electric power company, we get the importance of typical power services shown in Table 1.

Table 1. Importance of power services.

Label	Services	Importance $/r_i$
S_1	500kV Line relay protection	0.9503
S_2	220kV Line relay protection	0.9399
S_3	Stability and Safety System	0.9044
S_4	Telephone System for Dispatching Department	0.7474
S_5	Dispatching automation System	0.8422
S_6	Information System of Management and protection	0.6043
S_7	Wide-area measuring system	0.7442
S_8	Lightning location monitoring system	0.3524
S_9	The substation video monitoring service	0.3595
S_{10}	Video conference system	0.3424
S_{11}	Telephone System for Administrative Department	0.2436

3 THE DEGREE OF THE NODE IMPORTANCE BASED ON THE SERVICE

3.1 *The link weight analysis*

Power communication network runs plenty of power services, which are direct factors affecting the security and stability of the power grid. The traditional weighted network uses link length or link capacity as the edge weight.

However, for the power communication networks in which optical transmission technology is widely used, the main factors affecting information transmission are not distance or capacity, but services. As the transmission carriers of services and links are important factors affecting network operation. The more important the running services are, the greater the implications of the link on the operational security of the power grid are. So we consider the service as an evaluation factor of the link weight.

$G=(V,E)$ is a weighted complex network, $V=[v_i]$ is the node set of the network, and n is the total number of nodes, $E=[e_i]$ is the link set of the network, m is the total number of links. The weight of the link e_k is:

$$w_k = \sum_{i=1}^{x_k} n_{ki} \times v_i \tag{1}$$

where x_k represents the category number of the running service on the link e_k, n_{ki} represents the number of the i–th category running service on the link e_k, v_i represents the importance of the i–th category service.

The link weight w_k describes the relative importance of the link e_k. The greater the weight of the link is, the more important the carrying service is, and the greater the influence is on the network.

3.2 *The node weight based on service*

In a complex network, the node degree describes the tightness between the node and the neighboring nodes. According to the definition of the node degree and the analysis from a service perspective we can get the link weight and the node weight.

The node weight s_i of node v_i is:

$$s_i = \sum_{w_k \in N_i} w_k \tag{2}$$

where N_i is the set of links connected with the node v_i.

Node weight s_i takes both the number of connections with neighboring nodes and the running services into account, which is a comprehensive reflection of the node local information.

3.3 *The evaluation of the node importance*

(Wang et al. 2005) introduces the concept of aggregation of network, which uses the differences of variation in the aggregation of the network between after using the node contraction method and before that to evaluate node importance. However, the problem is that it cannot distinguish the importance between nodes of the same location conditions in the ring network.

In this paper, the node weight is gotten by the analysis of services and the results obtained can reflect the actual network operation better.

Analyzed from the service perspective, the aggregation of the weighted network:

$$\partial(WG) = \frac{1}{s * l} \tag{3}$$

where s represents the average node weight that based on service of all the nodes in the network; l represents the average distance of the un-weighted network transferred by the weighted network.

After the node contraction, the higher the network aggregation is, the more important the node is. Therefore, the node importance is:

$$I(v_i) = 1 - \frac{\partial(WG)}{\partial(WG * v_i)} \tag{4}$$

where ∂(WG) is the aggregation of the original network, $\partial(WG^* V_i)$ represents the aggregation of the network after the contraction of the node v_i.

The node importance $I(v_i)$ not only describes the node status in the network topology, but also reflects the node service information. So it measures the importance of a node in the network more comprehensively, which has practical applications in power communication network.

4 SIMULATION AND ANALYSIS

The network model proposed by (Fan et al. 2014) is used for simulation and Matlab is adopted to the calculation of the simulation.

4.1 *The analysis of network*

There are 14 nodes and 16 links in the simulation network. Node v_1 is the Provincial Electric Power Dispatching Center, Node v_{13} is the Regional Electric Power Dispatching Center, Node v_{14} is a 220KV substation and the remaining nodes are 500KV substations. The link set e={e_1, e_2, e_3, e_4, e_5, e_6, e_7, e_8, e_9, e_{10}, e_{11}, e_{12}, e_{13}, e_{14}, e_{15}, e_{16}}, the power Services set working in the network S={ S_1, S_3, S_4, S_5, S_6, S_7, S_8, S_9, S_{10}, S_{11}}, the classification of power services is shown in Table 1, the operation arrangement of the power services is shown in this model .We define that service matrix S(e) represents the number of the power services running in each link, and the element $S(e)_{ij}$ represents the number of the service S_j in the link e_i.

4.2 *Calculation of the node importance*

According to the importance of the value of the power services in Table 1, we obtain the vector of the power services as follows:

v=[0.9503,0.9044,0.7474,0.8422,0.6043,0.7442,0.3524,0.3595,0.3424,0.2436]^T

Then the weight vector of links calculated by equation (1) is:

$$\mathbf{w} = \mathbf{S}(\mathbf{e}) \cdot \mathbf{v} \tag{5}$$

Therefore the weight vector w={0.2872, 0.3042, 0.2574, 0.0776, 0.0419, 0.0655, 0.0074, 0.1712, 0.218, 0.0542, 0.0074, 0.0074, 0.1478, 0.218, 0.2574, 0.0702}.

Then the network node weight vector s can be calculated by equation (2). Therefore the node weight vector s={0.8488, 0.4722, 0.085, 0.3966, 0.2328, 0.0616, 0.1552, 0.3732, 0.4754, 0.5616, 0.4754, 0.0419, 0.1357,0.0702}. So that the average weight of nodes in the network is s=0.3133.

Besides, the calculated average shortest path length of the network is l=2.7747. And the aggregation of the network is:

$$\partial(WG)=\frac{1}{s*l}=1.1505 \tag{6}$$

Similarly, the aggregation of the network after the node v_1 contracted is $\partial[WG*v_1]$=3.4363 .

According to (4), the importance of the node v_1, $I(v_i)$ is:

$$I(v_1)=1-\frac{\partial(WG)}{\partial(WG*v_1)}=1-\frac{1.1505}{3.4363}=0.6652 \tag{7}$$

In the same way the importance of the each node in the network can be calculated, and compared with the results calculated by Betweenness Method and Node Contraction Method. The calculated results are shown in Table 2 and Figure 1.

Node	Betweenness Method	Node Contraction Method	Proposed Method			
	Importance	Rank	Importance	Rank	Importance	Rank
v1	44	1	0.4887	6	0.6652	1
v2	39	2	0.5150	4	0.5469	6
v3	0	10	0.5380	3	0.4978	11
v4	3	9	0.6282	1	0.5491	5
v5	13	6	0.5701	2	0.5099	8
v6	0	10	0.4578	9	0.4663	14
v7	0	10	0.4464	12	0.5016	10
v8	27	5	0.4564	10	0.5401	7
v9	30	4	0.4752	7	0.5784	3
v10	34	3	0.4611	8	0.5995	2
v11	7	8	0.4422	13	0.5659	4
v12	0	10	0.3890	14	0.4805	13
v13	12	7	0.4945	5	0.5092	9
v14	0	10	0.4513	11	0.4942	12

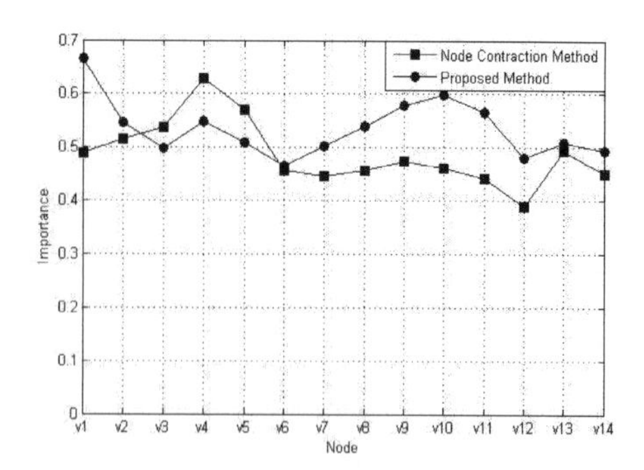

Figure 1. Comparison of the results of the Node Contraction Method and the Proposed Method.

4.3 *The result of simulation*

The node v_1 is the Provincial Electric Power Dispatching Center, which is the core of the entire network, and it plays a critical role in the daily operation of the power grid. Therefore the importance value of the node v_1 is supposed to be the maximum. According to the results of Table 2 and Figure 1, the importance of node v_1 of Betweenness Method and the proposed method are satisfactory and in agreement with the practice. Furthermore, with the proposed method, the importance value of node v_1 has a larger degree of difference which reflects its status and significance.

Compared with Node Contraction Method, the proposed method also considers the operation of power service. It can take advantage of power service factors to distinguish the importance between the network nodes with similar location. For example, the results of node v_6 and node v_7 are zero with Betweenness Method and are 0.4578 and 0.4464 with the Node Contraction Method. The two nodes share a similar position in the topology and both the Betweenness Method and the Node Contraction Method cannot tell the difference of importance between the two nodes. However, according to the calculated node weight vector s on the basis of the analysis of power service, the node weights of the two nodes are 0.0616 and 0.1552. The node v_7 has a greater impact on the operational security of the power grid. The important values of node v_6 and node v_7 in the proposed method are 0.4663 and 0.5016 respectively. Obviously, the proposed method reflects the actual operation of a network and is more rational.

In a word, compared with Betweenness Method and Node Contraction Method, the proposed method combines the factors of node power service with network topology to evaluate node importance comprehensively. The evaluation results not only reflect the influence degree of nodes on the connectivity of network topology, but also reflect the importance of nodes of protecting the operation of power service. The results can reflect the operational status of the nodes in the network and have a certain practical significance in the electric power communication network.

5 CONCLUSION

The importance of nodes in the network is analyzed from the perspective of service. The exact values of the node importance are obtained while substituting the value of distance for the one of the services. It is the joint consideration, both nodes' contribution degree to the service and the impact degree of network connectivity, which makes the proposed method the comprehensive evaluation. Compared with the results of past researches in the simulation, this algorithm shows the superior rationality and lower computational complexity. Hence, we conclude the paper as a major reference for practical application and risk assessment in the electric power communication network.

REFERENCES

Chen Jing & Sun Linfu. 2009. Evaluation Method for Node Importance in Complex Networks. *Journal of Southwest Jiaotong University* 44(3): 426-429.

Fan Bing & Tang Liangrui. 2014. Vulnerability analysis of power communication network. *Proceeding of the CSEE* 34(7): 1191 -1197.

Jungwon, H. & Aehintya, H. 2011. A Novel Risk Assessment for Complex Structural Systems. *IEEE Transactions on Reliability* 60(l): 210-218.

Kermarrec, A.M., Merrer E.L., Sercola, B., et al. 2011. Second order centrality: Distributed assessment of nodes criticity in complex networks. *Computer Communication* 34(5): 619-628.

Manteach, A., Yen, L., Callut, J., et al. 2010. The sum-over-paths covariance kernel: A novel covariance measure between nodes of a directed graph. *IEEE Transaction on Pattern Analysis and Machine Intelligence* 32(6): 1112-1 126.

Q/CSG110020-2011. 2012. Application of communication network interface technical specifications of CSG.

SERC. 2006. Secondary System Security Protection Scheme of Power System.

Tan Yuejin, Wu Jun & Dong Hongzhong. 2006. Evaluation method for node importance based on node contraction in complex networks. *Systems Engineering–Theory & Practice* 11(11): 79- 83.

Wang Xun, Ling Yun & Fei Yulian. 2005. Personalization recommendation system based on web log& cache data mining . *Journal of the China Society for Scientific and Technical Information* 24(3): 324-328.

Zeng Ying, Wang Ying, Dong Xijie, et al. 2013. Node importance evaluation strategy on electric power communication backbone network. *Journal of North China Electric Power University* 40(5): 65-69.

Electronics, Communications and Networks IV – Hussain & Ivanovic (eds)
© 2015 Taylor & Francis Group, London, ISBN: 978-1-138-02830-2

A novel P2P information retrieval framework using locality-sensitive hashing and B+ tree

Zengrong Zhan
College of Information Engineering, Guangzhou Panyu Polytechnic, Guangzhou, Guangdong, China

ABSTRACT: This paper proposes a novel information retrieval framework in P2P overlay network storage system. The framework uses Vector Semantic Module to represent the feature that extracted from each data element, and employs LSH similar technique to build indexing of the vector. By mapping the indexing to P2P overlay Chord, similarity searching in the framework is efficient. B+ tree is introduced to the framework to enhance the querying accuracy and improve the efficiency of final local filtering process. The results show that the proposed framework is accurate and efficient in retrieving relevant answers of queries. The system was also tested for different network sizes to show its scalability.

KEYWORDS: information retrieval; peer-to-peer; B+ tree; locality-sensitive hashing.

1 INTRODUCTION

With the rapid growth of online information, massive data are generated by users because of the convenience of the internet in both professional services and entertainment. A recent study from International Data Corporation (IDC) found 2.8 ZB of data was created and replicated in 2012 and the total amount of data doubled every two years, which means there are about 5TB of data for everyone on earth in 2020 (Gantz & Reinsel 2012). Moreover, users normally post queries across a large amount of data stored on their PC or data network. Therefore, an efficient, distributed, and scalable data management architecture with its information retrieval framework has become an urgent need for users.

Peer-to-Peer (P2P) networks, in which any two nodes can directly share information without intermediate servers, are widely used for facilitating the sharing of large amounts of data that are distributed in large-scale networks. The scalability, reliability and small infrastructure costs of P2P networks make them a perfect architecture to support file sharing applications in the rapidly growing, large-scale, and naturally distributed information ecology (Cholez et al. 2013). However, a successful system for these demands not only a huge amount of storage capacity, but also the ability of low-latency and scalable queries, since the data generated from users are in rich representation of different type and the numbers of features extracted from the data is often very high. Hence, it is critical to have a comprehensive and efficient means of information retrieval for P2P networks. This topic has recently aroused researcher's interest and leads to the development of novel technologies and solutions. For example, Distributed Hash Table (DHT) like Chord (Stoica et al. 2001), CAN (Ratnasamy et al. 2001), and Pastry (Rowstron & Druschel 2001) used in Structured P2P networks can provide scalable, load balanced, and robust distributed applications, and allow efficient key lookups in a logarithmic number of routing hop, but are limited to exact or range queries. Thus, Similarity search in high dimensional data has been a popular research topic in the last years, and many index structures for approximate similarity search have been proposed such as Gionis Aristides et al. (1999), Yu et al. (2001), Datar et al. (2004) and Bhattacharya et al. (2005). Among the newest methods, Locality Sensitive Hashing (LSH) (Indyk & Motwani 1998, Andoni & Indyk 2006) has proven to be efficient on indexing high-dimensional data. Euclidean LSH (Datar 2004) is a successful one which guarantees the query accuracy by invoking too many hash tables. Other examples like Panigrahy (2006), Lv et al. (2007) and Joly & Buisson (2008) are also proposed and show the improvement by reducing the memory consumption, and all of them are based on the same structure as Euclidean LSH.

These existing approaches mentioned above either focus on centralized setting or fail at providing both low-latency and high quality search results. Although those solutions based on LSH are efficient to organize and query large-scale data with high-dimensional features, they need to do an exact similarity measure on the final filtering process. When the data store on the peer is large-scale, which is very common in practice, the cost for the exact similarity measure will reduce performance significantly. Besides, in many cases, the query is mapped to the peer which is close to the one that stores the expect data, and it causes the query return inaccuracy results.

In this paper, we present an information retrieval framework that includes feature exaction, indexing creation, P2P overlay network and local indexing creation. We use the SVM as the feature representation of the data, and we also consider using the similarity search method based on LSH over high dimensional data in structured P2P overlay networks where N peers P_1, P_2, …, P_N are connected by a DHT that is organized in a cyclic ID space such as Chord. Mostly, we introduce the B+ tree to accelerate the querying process in local similarity measure and enhance the accuracy for the query.

2 RELATED WORK

Due to the limitation for search provided by the structured P2P network, queries within the overlays are not sufficient and even impossible in complex request. Therefore, many systems architectures are proposed to build support for different type of queries like keyword based with wild-cards query, range query, multi-dimensional query, similarity query, and so on.

2.1 *Keyword-based query*

For keyword-based query, data is associated with a set of keywords which are used as the feature that can be used to retrieve the data object. The simplest way to support keyword search is to index the pointer to the data in the P2P network like INS/Twine (Furness & Kolberg 2011), which extracts a unique sub-sequences of attributes and values from the data description and then hashed and store the sub-sequences to DHT. This method is suitable for any existing DHT without modifying the algorithms of routing, querying and replicating. The Open Distributed Search Engine Architecture proposed in Suel et al. (2003) also maintains a distributed index by hashing the extracted keywords and storing within the network, but it uses an inverted index rather than the keywords.

Instead of using keywords, some researchers, like eSearch (Tang & Sandhya 2004), use keyword sets as the value to be hashed and published within the DHT. Keyword-Set Search System (KSS) (Gnawali 2002) hashes the set of keywords and stores the inverted indexes to the appropriate peers. These hybrid indexing structures may need more storage to make the multi-keywords queries work.

Keyword based query with wild-cards is studied KISS-W in Joung & Yang (2006) and Joung & Yang (2007), which is a system extract key-tokens from keywords and supports the wild-card search over a hyper-cube based network.

2.2 *Tree-based search structures*

The Tree-based approaches perform very well when the feature dimensions are low, but degrade to linear search for high enough dimensions. For example, K-D trees (Bentley 1990) is used in Yi et al. (2013) to generate multi-dimensional keys for efficient search and Hua (2009) used R-trees (Guttman 1984) to facilitate fast membership queries.

The Skip Graph system (Aspnes & Shah 2007) uses a tree structure rather than consistent hashing to the overlay, and it adopts a graph structure into the system since the tree structure has a single point of failure. González-Beltrán et al. (2008) proposed a range query method by making use of the skip graphs. However, replacing the consistent hashing by a tree-based structure adds further complexity both in terms of theory and load balancing.

To support range queries, BATON system (Jagadish et al. 2005) describes a P2P network based on a balanced tree structure, and Distributed Segment Trees (Zheng et al. 2006) provides a way to support both range queries and cover queries.

Other approaches like SkipIndex (Zhang et al. 2004) and VBI-tree (Jagadish et al. 2006) both rely on tree-based approaches, but when the number of data dimensions is high, it degrades to linear search.

2.3 *Semantic vector indexing*

A representational approach for semantic search is known as pSearch (Tang et al. 2003) which uses the famous information retrieval techniques Vector Space Model (VSM) (Salton et al. 1975) and Latent Semantic Indexing (LSI) (Berry et al. 1999) to generate a semantic space. The space is built on the top of CAN known as eCAN (Xu & Zhang 2002). The process of this method is to compute a semantic vector from the document or service description by using VSM and LSI firstly and then directly mapped to a multidimensional CAN which basically has the same dimension of the space. Afterwards, it stores the data to peer in the DHT, which is denoted by S, resulting in similar data being close together. During the search process, a query will be computed to query vector Q, and flooded the query to those peers within a given radius. Although this approach efficiently solves the problem of semantic search, it is less practical in real applications since the dimension of the underlying P2P network depends on the dimension of the data, and hence the approach cannot be generalized to all DHTs. An improved method in Sahin et al. (2004)

follows pSearch by employing VSM and LSI, but map the resulting high dimensional space to a one dimensional Chord index, and therefore it is independent of the corpus size and the number of dimensions.

2.4 *Similarity search*

The similarity search algorithms aim to preserve query time guarantees, even for high-dimensional inputs. An efficient method for similarity is to find a way to reduce the dimensionality so that the query time will be drop dramatically.

One of the well-known methods is SFC indexing, which uses the Space Filling Curve to reduce d-dimensional coordinate space with their position defined by a set of keywords to one-dimensional P2P indexes. Squid (Schmidt & Parashar 2003 and Schmidt & Parashar 2004) and CRAP (Ganesan 2004) are the representatives which apply SFC indexing to Chord overlay and Skip graphs respectively. Besides, PHT (Chawathe 2005) also applies SFC indexing over generic DHTs, and Andrzejak & Xu (2002) give a very similar approach that makes use of a Hilbert SFC on top of the CAN overlay. The fatal weakness of SFC indexing is that the neighborhood in high dimensional space is not well preserved in the one dimensional SFC space, hence led to inefficient query.

An alternative approach is using Locality-sensitive Hashing mentioned in section 1, where several localities preserving hash functions are used to hash the high dimensional data, which is normally represented by a set of keywords, to one dimension such that with high probability the data that closed in high dimension are hashed to the same bucket. The method is very efficient in reducing the query time. Several follow-ups of this method exist and try to solve the problems associated with it (Andoni & Indyk 2006, Bawa et al. 2005, Lv et al. 2007 and Rina 2006). Moreover, Athitsos et al. (2008) presents a novel method that doesn't depend on the distance measured by hashing the distance. Approximate range search in high dimensions has also been the focused on some works such as Chazelle et al. (2008), but will not be considered in this work.

3 SYSTEM ARCHITECTURE

3.1 *Overview*

A view of our proposed system is shown in Figure 1, where the solid line indicates the indexing, creating flow and the dotted line shows a flow of querying process.

In our system, the data elements like documents are converted to vectors by using the VSM technology, and section 3.2 will describe this feature extraction process. The feature vectors generated from VSM are hashed and indexed, and details of the process will be discussed in section 3.3. Mapping the indexing to an existing overlay network like Chord are shown in section 3.4, and section 3.5 proposed an optimization method to improve the search process by constructing and storing a local B+ tree. The formal query process is given at the end of this section.

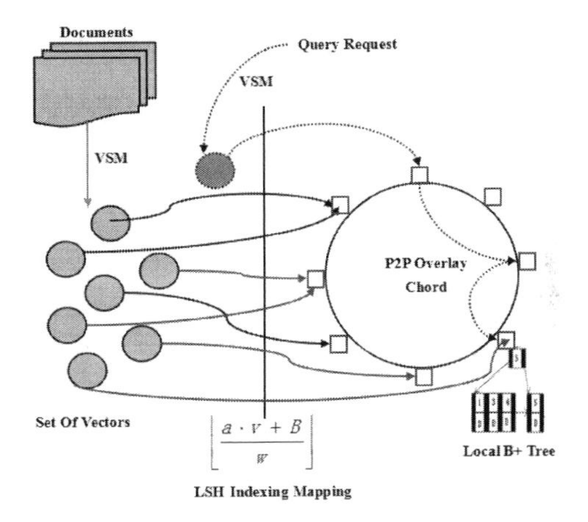

Figure 1. A general view of content indexing and querying.

3.2 *Feature representation*

A key component of a P2P system is to define the feature space, to which all the data elements are extracted. To support multiple keywords or similarity search in the system, each data element is associated with a sequence of keywords according the VSM in which a vector is used to reflect a particular concept with the given data element. For example, the sequences of keywords are converted into a vector in our case, and the value assigned to that component of the vector reflects the importance of the term or keyword in the data. Figure 2 shows an example of a data element that is described by three keywords "computer", "network" and "peer-to-peer" with weights 0.5, 2.5, 5.0 respectively.

By using the VSM, all the documents and queries can be treated as term vectors, and each element of the vector represents its importance or weights which are calculated by using term frequency multiply by inverse document frequency (Berry et al. 1999). Using inverse document frequency in weight calculating can avoid giving too high weight to some common terms since they are generally used in most documents.

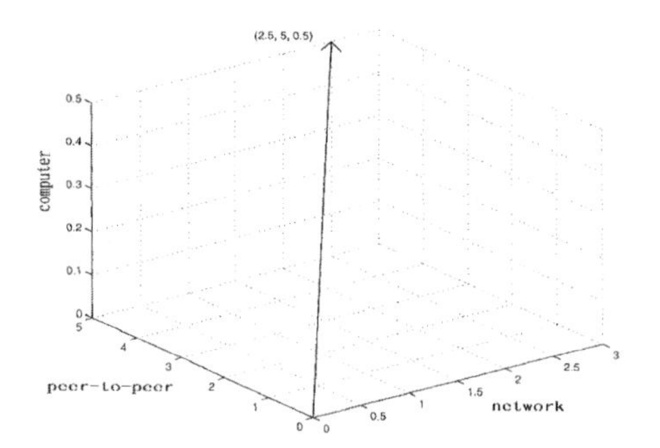

Figure 2. Example of a 3-Dimensional keyword space for storing computational resources, using the attributes: computer, network and peer-to-peer.

During query process, documents are ranked according to the similarity between the document vector and the query vector so that those with the highest similarity are returned. Jaccard Distance, Cosine Distance and *Lp* Distances are common similarity measures. In this work, we use L_2 Distance for the similarity measures. The L_2 distance is defined as follows:

$$d_2(a,b) = \|a-b\|_2 = \sqrt{\sum_{i=1}^{d}(a_i - b_i)^2} \quad (1)$$

where the non-negativity, identity, symmetry and triangle inequality are hold.

3.3 *Locality-sensitive hashing indexing (LSH)*

As the data is represented by a vector in Rd and the distance between vectors are defined, the left problem is how to perform indexing or similarity searching for querying objects. The key idea of locality-sensitive hash (LSH) is to hash the vector using many hash functions so as to ensure the probability of close objects are hashed together while apart objects are hashed distantly.

Formally, Let $M=(X,d)$ be any metric space, and $v \in X$. The ball of radius centered at v is defined as $B(v,r)=\{q \in X\} | d(v,q) \leq r\}$. Therefore, for a domain S of the points set with distance measure d, a LSH family is defined as follows:

$H=\{h:S \rightarrow U\}$ is called(r_1,r_2,p_1,p_2)-sensitive for d if for any $v,q \in S$

- if $v \in B(q, r_1)$ then $\Pr_H[h(q)=h(v)] \geq p_1$
- if $v \notin B(q, r_2)$ then $\Pr_H[h(q)=h(v)] \leq p_2$

and $p_1 > p_2$ and $r_1 > r_2$. In our work, we use LSH family based on p-stable distributions, which is defined as that for a distribution D over R if there exists $p_1 > 0$ such that for any n real numbers $r_1, r_2, \ldots, r_n$ and i.i.d. variables $X_1, X_2, \ldots, X_n$ with distribution D, the random variable$\sum_i r_i X_i$ has the same distribution as the variable $(\sum_i |r_i|^p)^{1/p} X$ where X is a random variable with distribution D. When $p=1$, it is the Cauchy distribution with density function $c(x)=1/\pi = 1/(1+x^2)$. And it is a Gaussian distribution with density function

$$g(x) = \frac{1}{\sqrt{2\pi}} e^{-x^2/2},$$

when $p=2$. This Gaussian distribution is used in our work since we used l_2 norm as our distant function. Formally, for each function $h_{a,b}(v): R_d \rightarrow N$ in H maps a d dimension vector v to an integers, and $a \in R^d$ is a random vector with entries chosen independently from a p-stable distribution and b is a real number chosen uniformly from the range $[0,w]$. Hence the random hash function can be given as

$$h_{a,b}(v) = \left\lfloor \frac{a \cdot v + b}{w} \right\rfloor \quad (2)$$

After defined the hashing function, LSH employs a function family $G=\{g:S \rightarrow U^M\}$, created by concatenating M hash functions from H, such that each $g \in G$ has the following form $g(v)=(h_1(v), h_2(v), \ldots, h_M(v))$ where $h_i \in H$ and $i \in [1,M]$. The idea behinds this is that by concatenating multiple function the probability of that the distant vector hashed together is decreased exponentially, but it will cause that the probability of that the close vector hashed apart is increased. Thus, it employs many $g_j \in G$, $j \in [1,L]$ to build L hash tables, in the hope that right answer will be found in at least one of them. In practical implementations, we normally give $L=3$ in order to limit the worst case of distances to be computed per query.

By using the p-stable LSH, vectors generated from data element are mapped into L dimension vector with each element is an integer, hence we can use every element of each vector of data to build the indexing or map data to exist DHT overlay like Chord.

3.4 *Mapping indexing to chord*

Chord uses a consistent hash function, e.g. SHA-1, to assign each node and each data element to an m-bit identifier which can be seen as an integer. Since each node has a unique identifier ranging from 0 to 2^m, these integers are ordered to form a ring structure module 2^m. Each node in Chord will also maintain routing state information, named finger tables, about several neighbors. Figure 3 from Stoica et al. (2001) shows an example of the overlay network where we can find each finger table contains the list of neighbor's information about their identifiers, intervals, and successor. To keep the finger tables up-to-date when nodes join and leave, it makes use of periodic maintenance.

In our system, node identifiers can be generated randomly or created by hashing the manual given like

IP address or some keywords that describe the node category. The management of data distributing, node joins and leaves, and query process is described below.

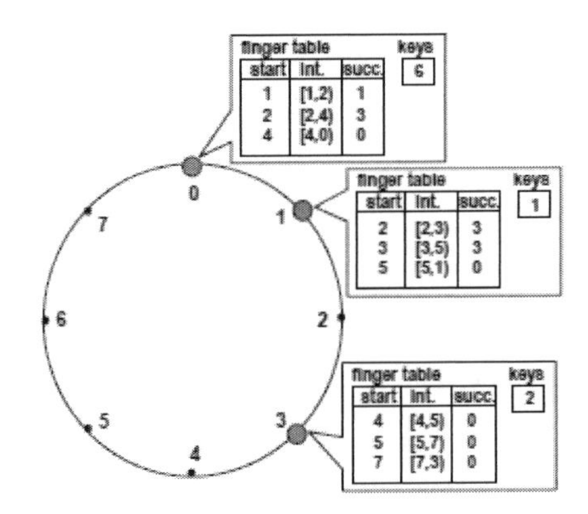

Figure 3. An example of finger tables and key locations in Chord.

Data Distribution: Each data element is converted into vector based on the VSM method mentioned in section 3.2, and the vector, then hashes to an integer vector with L dimension by using the LSH function based on 2-stable distribution that described in section 3.3. Every element of the vector will be treated as a data key, and the data will be distributed to the node whose identifier is equal to or follows the data key in the identifier space. The node is called the successor of the key. Consider an example by using the Chord ring in Figure 3 that is when the data element is hashed to a vector in which 6 is one element, and the node with key 1 received the request. When the node received the request it hashes the data element to 6 and looks up its finger table. When it finds out the key 6 is located in the item with 5 of the start, it will forwards the request to the successor which is the node with key 0 who will compare the key 6 in its finger table and find out it located in the item 4 with interval [4,0), and the successor is itself. Hence, the data is then distributed to the node.

Node Joins: The joining node has to generate an identifier according the predefined methodology, like using IP address as a label to be hashed, and then sends a join message with this identifier to a known node which will route this message across the overlay network to the successor of the new node based on the its identifier. The cost for joining is O (log_2^2N) messages. After joining the new node is inserted into the overlay network, it will take a part of the successor node's load.

Node Leaves: When a node leaves, the finger tables of the nodes that have entries pointing to the departing node update its successor list with O (log_2^2N) messages cost, ensuring lookups can always be succeed eventually.

Query Process: When a query request is reached, the query will be hashed to the integer vector of L dimensions by using VSM and LSH. Each element of the vector will be taken as the key to search the node whose identifier is equal to or follows the data key. Any node received the query request will normally compare the query with all the data elements in the local storage and return the result to the node that initiated the query. Since this local search process is slow and with a low accuracy rate, we will introduce a new method to optimize the local search algorithm in the following.

3.5 *Local search optimization using B+ tree*

Since a query received from the user is normally with a few keywords and sometimes ambiguous, the VSM with LSH function often map the query to the node near to the one that actually contains the data that user wanted. Moreover, it is impractical to do exact similarity measure for all data on the final filtering process. Therefore, we propose an optimization by using the B+ tree to reduce the search time and improve the query accuracy.

A B+ tree is an n-ary tree with a variable but often large number of children per node. The structure of a B+ tree consists of a root, many internal nodes and leaves. The root may be either a leaf or a node with two or more children. In a B+ tree each internal node only contains keys and the values are stored in the leaves. It is a dynamic, multilevel index, with maximum and minimum bounds on the number of keys in each index segment. Figure 4 shows an example of a B+ tree, where the key is integer range from 1 to 25, and the value are labelled with "D" stored on the leaf only.

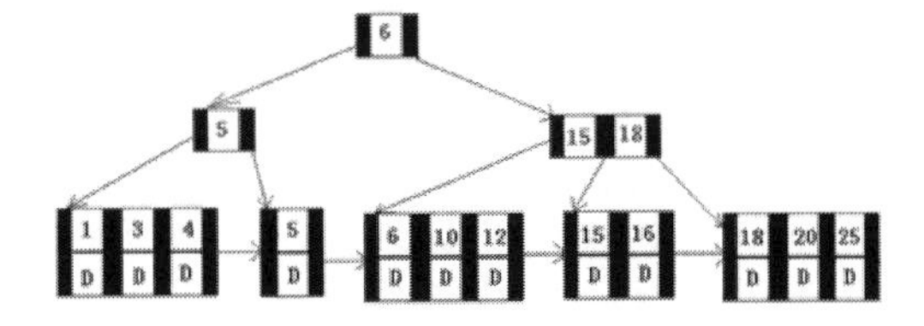

Figure 4. An example of B+ tree.

To construct a B+ tree for our system, we have to define the key and the data that are going to be maintained in the tree. Firstly, we select a pivot, which is a vector with the same dimension numbers as those generated from VSM, by a random or heuristic method. And then the distance between this pivot and the vector generated from VSM of each data element is calculated according to Equation (1). This distance is taken as the key to index for B+ tree. The data to which the key mapped are pointers (e.g. IP address with path) to the file or resource. There is a good chance that two vectors have the same distance to the pivot, therefore, a set of reference are stored in the data item.

Besides, each node in Chord not only maintains a B+ tree for local storage, but also maintains the storage information from successor and predecessor. The reason is that LSH has the highest probability to map the query vector to neighbors of the node with excepted data, hence gathering information of neighbors will accelerate the searching process and improve the accuracy.

During data distribution phase, the data with its vector generated from VSM are sending to the desired node and the node stores the data and compute the distance between the vector to the pivot vector. Hence, a pair of <key, reference> is provided after this progress.

The insertion process of B+ tree is performed as follows:

1. Perform a search to determine what bucket the new record should go into.
2. If the entries of this bucket are less than b-1 (b is 4 in our case), add the record to the bucket.
3. Otherwise, split the bucket. Allocate new leaf and move half the bucket's elements to the new bucket, and then insert the new leaf's smallest key and reference into the parent. If the parent is full, split it and add the middle key to the parent node. Repeat until a parent is found that need not split.
4. Send the <key, reference> pair to the neighbors. When the neighbors received this request, they repeat the step 1 to 3 to insert this record in their local tree.

When a data element is deleted, the <key, reference> pair is also required in the deletion process which is performed as follows:

1. Perform a search to find leaf L where the key belongs.
2. Remove the entry, and check if L is at least half-full. If it is then done. Otherwise, try to re-distribute by borrowing from a sibling. When re-distribution fails, it has to merge L and sibling and delete entries from parent of L.
3. Send the <key, reference> pair to the neighbors. When the neighbors received this request, they repeat the step 1-2 to delete this record in their local tree.

3.6 *Querying process*

In this part, we give a formal description of a query process. A query radius $r\in(0,1]$ and a set of hash function groups $G=\{g_1,g_2,...,g_L\}$ with each group $g_j=(h_1,h_2,...,h_M)$ is predefined.

1. A query will be converted to a vector $Q=(q_1,q_2,...,q_n)$ according to VSM.
2. Apply the hash function group G to Q, *and we got* $U=G(Q)=(u_1,u_2,...,u_n)$ *where* $u_i\in R$.
3. For each element u_i in U, we perform search operation by mapping u_i to the Chord ring, and we find the node to search locally.
4. Compare the distance d between query vector Q *and* local pivot vector in the B+ tree of the node that found in 3).
5. Query the B+ Tree by taking d as the key k. We first start from the root and look for the leaf which may contain the value k. At each node, we compare k with the node key to figure out where their distance is smaller than r. If it is, we follow the corresponding node by searching on the key values of the node. It is possible that two or more nodes will be followed at the same time. When a leaf is found, the data that contain a set of resource reference is returned. All the returns will be sent to the node that initiated this query request.
6. For all $u_k\,(k \geq i)$ in U, remove it from U if the predecessor's key of the node is smaller than u_k, and the upper bound of the first item in the finger table of this node is bigger than u_k.
7. Repeat step 3 to 6 until U is empty.

4 EXPERIMENTAL RESULTS

We implement a Java application for the P2P node, and use the Peersim (Montresor & Jelasity 2009) as the underlying P2P overlay Chord. Besides, the code in (Aaron 2013) is adopted for the implementation of B+ tree. A testing data set is from the Reuters 21578 collections. Basically, a network of 500 nodes is constructed, where each node is assigned an identifier of 128 bits.

4.1 *Query accuracy evaluation*

The search process was applied on 500 random queries with similarity radius 0.5, and evaluated according to the number of hash groups and the hash function in each group.

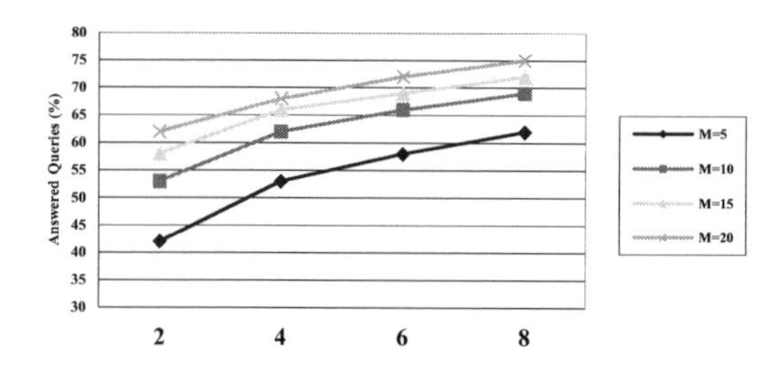

Figure 5. Query answer ratio vs the number of hash groups and hash functions.

Figure 5 shows the ratios of queries increased with the number of the hash group and hash function in each group. The result is 62% when M=10 and L=4, which is acceptable according to other methods. For example, Gupta et al. (2003) model is 38% by mapping scheme of 5 hash groups each of 20 functions and its average similarity range from 0.9 to 1.0, and Yee & Frieder (2005) returned answers about 1200 queries out of 10000, while Cohen et al. (2007) got about 22% of queries having relevant answers in case of 1% of nodes having data.

4.2 *Query performance evaluation*

Generally, the more the hash group uses the more answer returns, and by using the B+ tree for local search, the hops is less than the number of the groups. Figure 6, where the parameters are r=0.5 and M=10, shows that the answered queries increased with the hash groups, and the hops are normally half of the number of the hash groups. Zhou et al. (2008) got 55% average recall ratio by using an average of 8 hops, while our system got 66% of the answered queries with 3 hops needed or 6 Hash groups.

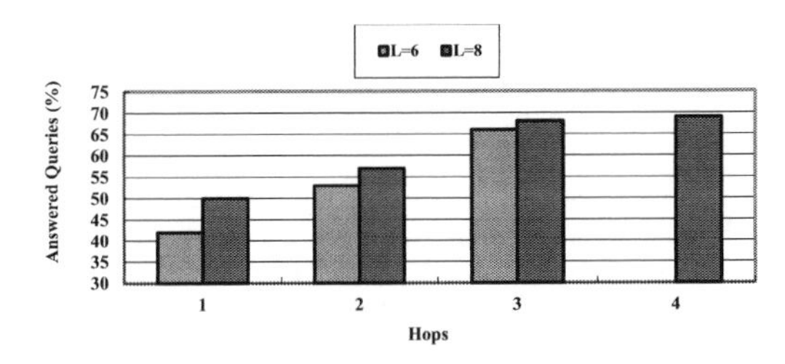

Figure 6. Query answer ratio vs hops and the number of hash functions.

4.3 *Scalability evaluation*

We test the system response by increasing the network size from 1000 to 10000. The experiments were performed using 4 and 8 hash groups of hash functions. Figure 7 shows that as the number of nodes increased the answered queries decreased, and we can enhance answer ratio by adding more hash group.

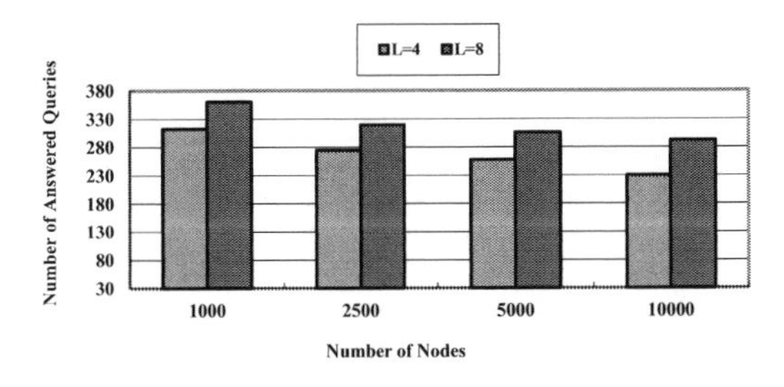

Figure 7. The Numbers of answered queries vs the number of nodes in overlay networks.

5 CONCLUSION

The information retrieval technology is one of the hottest topics in the research of Peer-to-Peer network, and the similarity search is the costly phases of query processing in such network. The feature generated from large data set by using VSM is high-dimensional, and when facing with high-dimensional data indexing, using LSH to map the indexing to overlay network like Chord is proved to be efficient. However, LSH indexing for large-scale and high-dimensional data set makes the exact similarity measure in the final filtering process. In this paper, we proposed an efficient and scalable P2P information retrieval framework by using VSM and LSH technology, and B+ tree is employed to improve the query accuracy and enhance filtering efficiency. Our proposed model could be improved by using cluster and predicting technology to optimize the parameters in the architecture.

ACKNOWLEDGEMENT

This work is supported by Education Science Planning Project of Guangzhou, China (2013A179).

REFERENCES

Aaron, W. 2013. B plus tree in C#, java and Python. *http://sourceforge.net/projects/bplusdotnet/.*

Andoni, A. & Indyk, P. 2006. Near-Optimal Hashing Algorithms for Approximate Nearest Neighbor in High Dimensions. *47th Annual IEEE Symposium on Foundations of Computer Science, FOCS '06. Berkeley, CA, USA, 21–24 October 2006.* IEEE.

Andrzejak, A. & Xu, Z. 2002. Scalable, Efficient Range Queries for Grid Information Services. *Second International Conference on Peer-to-Peer Computing (P2P 2002). Linkoping, Sweden, 7 September 2002.* IEEE.

Aspnes, J. & Shah, G. 2007. Skip Graphs. *ACM Trans. Algorithms* 3(4): 37.

Athitsos, V., Potamias, M., Papapetrou, P., et al. 2008. Nearest Neighbor Retrieval Using Distance-Based Hashing. *Proceedings of the IEEE 24th International Conference on Data Engineering. Cancun, Mexico, 7–12 April 2008.* IEEE Computer Society.

Bawa, M., Condie, T. & Ganesan, P. 2005. Lsh Forest: Self-Tuning Indexes for Similarity Search. *Proceedings of the 14th International Conference on World Wide Web. Chiba, Japan, May 2005.* ACM.

Bentley, J. L. 1990. K–D Trees for Semidynamic Point Sets. *Proceedings of the Sixth Annual Symposium on Computational Geometry. Berkley, California, USA, 07–09 June 1990.* ACM.

Berry, M. W., Drmac, Z. & Jessup, E. R. 1999. Matrices, Vector Spaces, and Information Retrieval. *SIAM Rev.* 41(2): 335–362.

Bhattacharya, I., Kashyap, S. R. & Parthasarathy, S. 2005. Similarity Searching in Peer-to-Peer Databases. *Proceedings of 25th IEEE International Conference on Distributed Computing Systems, ICDCS 2005. Columbus, Ohio, USA, 6–10 June 2005.* IEEE Computer Society.

Chawathe, Y., Ramabhadran, S., Ratnasamy, S., et al. 2005. A Case Study in Building Layered Dht Applications. *SIGCOMM Comput. Commun. Rev.* 35(4): 97–108.

Chazelle, B., Liu, D. & Magen, A. 2008. Approximate Range Searching in Higher Dimension. *Comput. Geom. Theory Appl.* 39(1): 24–29.

Chi, Z. & Arvind, K. 2004. Skipindex: Towards a Scalable Peer-to-Peer Index Service for High Dimensional Data. *Department of Computer Science, Princeton University. Tech. Rep.*: 703–704.

Cholez, T., Chrisment, I., Festor, O., et al. 2013. Detection and Mitigation of Localized Attacks in a Widely Deployed P2p Network. *Peer-to-Peer Networking and Applications* 6(2): 155–174.

Cohen, E., Fiat, A. & Kaplan, H. 2007. Associative Search in Peer to Peer Networks: Harnessing Latent Semantics. *Comput. Netw.* 51(8): 1861–1881.

Datar, M., Immorlica, N., Indyk, P., et al. 2004. Locality-Sensitive Hashing Scheme Based on P-Stable Distributions. *Proceedings of the Twentieth Annual Symposium on Computational Geometry. Brooklyn, New York, USA, 08–11 June 2004.* ACM.

Furness, J. & Kolberg, M. 2011. Considering Complex Search Techniques in Dhts under Churn. *Consumer Communications and Networking Conference (CCNC). Las Vegas, USA, 9–12 January 2011.* IEEE.

Ganesan, P., Yang, B. & Garcia-Molina, H. 2004. One Torus to Rule Them All: Multi-Dimensional Queries in P2p Systems. *Proceedings of the 7th International Workshop on the Web and Databases: colocated with ACM SIGMOD/PODS 2004. Paris, France, 17–18 June 2004.* ACM.

Gantz, J. & Reinsel, D. 2012. The Digital Universe in 2020: Big Data, Bigger Digital Shad-Ows, and Biggest Growth in the Far East. *IDC iView: IDC Analyze the Future. Tech. Rep.*

Gionis, A., Indyk, P. & Motwani, R. 1999. Similarity Search in High Dimensions Via Hashing. *Proceedings of the 25th International Conference on Very Large Data Bases. Edinburgh, Scotland, UK, 7–10 September 1999.* Morgan Kaufmann Publishers Inc.

Gnawali, O. D. 2002. *A Keyword-Set Search System for Peer-to-Peer Networks.* Massachusetts In-stitute of Technology. Tech. Rep.

González-Beltrán, A., Milligan, P. & Sage, P. 2008. Range Queries over Skip Tree Graphs. *Comput. Commun.* 31(2): 358–374.

Gupta, S. A., Gupta, A., Agrawal, D., et al. 2003. Approximate Range Selection Queries in Peer-to-Peer. *Conference on Innovative Data Systems Research Pacific Grove, California, USA, 5–8 January 2003.* VLDB Endowment.

Guttman, A. 1984. R-Trees: A Dynamic Index Structure for Spatial Searching. *SIGMOD Rec.* 14(2): 47–57.

Hua, Y., Xiao, B. & Wang, J. 2009. Br-Tree: A Scalable Prototype for Supporting Multiple Queries of Multidimensional Data. *Computers, IEEE Transactions on* 58(12): 1585–1598.

Indyk, P. & Motwani, R. 1998. Approximate Nearest Neighbors:Towards Removing the Curse of Dimensionality. *Proceedings of the Thirtieth Annual ACM Symposium on Theory of Computing. Dallas, Texas, USA, 24–26 May 1998.* ACM.

Jagadish, H. V., Beng Chin, O., Quang Hieu, V., et al. 2006. Vbi-Tree: A Peer-to-Peer Framework for Supporting Multi-Dimensional Indexing Schemes. *Proceedings of the 22nd International Conference on Data Engineering, ICDE '06. Atlanta, Georgia, USA, 3–7 April 2006.* IEEE.

Jagadish, H. V., Ooi, B. C. & Vu, Q. H. 2005. Baton: A Balanced Tree Structure for Peer-to-Peer Networks. *Proceedings of the 31st International Conference on Very Large Data Bases. Trondheim, Norway, 30 August 30 - 2 September 2005.* VLDB Endowment.

Joly, A. & Buisson, O. 2008. A Posteriori Multi-Probe Locality Sensitive Hashing. *Proceedings of the 16th ACM International Conference on Multimedia. Vancouver, British Columbia, Canada, 27–31 October 2008.* ACM.

Joung, Y.-J. & Yang, L.-W. 2007. Wildcard Search in Structured Peer-to-Peer Networks. *Knowledge and Data Engineering, IEEE Transactions on* 19(11): 1524–1540.

Joung, Y.-j. 2006. Kiss: A Simple Prefix Search Scheme in P2p Networks. *International Workshop on the Web and Databases (WebDB). Chicago, Illinois, USA, 30 June 2006.* ACM.

Lv, Q., Josephson, W., Wang, Z., et al. 2007. Multi-Probe Lsh: Efficient Indexing for High-Dimensional Similarity Search. *Proceedings of the 33rd International Conference on Very Large Data Bases. University of Vienna, Austria, 23–27 September 2007.* VLDB Endowment.

Montresor, A. & Jelasity, M. 2009. Peersim: A Scalable P2p Simulator. *IEEE Ninth International Conference on Peer-to-Peer Computing, P2P '09. Seattle, Washington, USA, 9–11 September, 2009.* IEEE.

Panigrahy, R. 2006. Entropy Based Nearest Neighbor Search in High Dimensions. *Proceedings of the Seventeenth Annual ACM-SIAM Symposium on Discrete Algorithm. Miami, Florida, USA, 22–24 January 2006.* Society for Industrial and Applied Mathematics.

Ratnasamy, S., Francis, P., Handley, M., et al. 2001. A Scalable Content-Addressable Network. *SIGCOMM Comput. Commun. Rev.* 31(4): 161–172.

Rowstron, A. & Druschel, P. 2001. Pastry: Scalable, Decentralized Object Location, and Routing for Large-Scale Peer-to-Peer Systems. *IN: MIDDLEWARE* 329–350.

Sahin, O. D., Emekci, F., Agrawal, D., et al. 2005. Content-Based Similarity Search over Peer-to-Peer Systems. *Proceedings of the Second International Conference on Databases, Information Systems, and Peer-to-Peer Computing. Toronto, Canada, 29–30 August 2004.* Springer-Verlag.

Salton, G., Wong, A. & Yang, C. S. 1975. A Vector Space Model for Automatic Indexing. *Commun. ACM* 18(11): 613–620.

Schmidt, C. & Parashar, M. 2003. Flexible Information Discovery in Decentralized Distributed Systems. *Proceedings 12th IEEE International Symposium on High Performance Distributed Computing. Seattle, WA, USA, 22–24 June 2003*. IEEE.

Schmidt, C. & Parashar, M. 2004. Enabling Flexible Queries with Guarantees in P2p Systems. *Internet Computing, IEEE* 8(3): 19–26.

Stoica, I., Morris, R., Karger, D., et al. 2001. Chord: A Scalable Peer-to-Peer Lookup Service for Internet Applications. *SIGCOMM Comput. Commun. Rev.* 31(4): 149–160.

Suel, T., Mathur, C., Wu, J.-w., et al. 2003. Odissea: A Peer-to-Peer Architecture for Scalable Web Search and Information Retrieval. *International Workshop on the Web and Databases (WebDB). San Diego, California, USA, 12–13 June 2003*. ACM.

Tang , C., Dwarkadas, S. & Dwarkadas, H. 2004. Hybrid Global-Local Indexing for Efficient Peer-to-Peer Information Retrieval. *Proceedings of the First Symposium on Networked Systems Design and Implementation (NSDI 2004). San Francisco, California, USA, 29–31 March 2004*. USENIX.

Tang, C., Xu, Z. & Dwarkadas, S. 2003. Peer-to-Peer Information Retrieval Using Self-Organizing Semantic Overlay Networks. *Proceedings of the 2003 conference on Applications, technologies, architectures, and protocols for computer communications. Karlsruhe, Germany, 25–29 August 2003*. ACM.

Xu, Z. & Zhang, Z. 2002. Building Low-Maintenance Expressways for P2p Systems. *HPL-2002-41, Hewlett-Packard Labs. Tech. Rep.*

Yee, W. G. & Frieder, O. 2005. On Search in Peer-to-Peer File Sharing Systems. *Proceedings of the 2005 ACM symposium on Applied computing. Santa Fe, New Mexico, USA, 13–17 March 2005*. ACM.

Yi, M. Y., Maw, A. H. & New, K. M. 2013. Usage of Kd-Tree in Dht-Based Indexing Scheme. *International Journal of Future Computer and Communication* 456–460.

Yu, C., Ooi, B. C., Tan, K.-L., et al. 2001. Indexing the Distance: An Efficient Method to Knn Processing. *Proceedings of the 27th International Conference on Very Large Data Bases. Roma, Italy, 11–14 September 2001*. Morgan Kaufmann Publishers Inc.

Zheng, C., Shen, G., Li, S., et al. 2006. Distributed Segment Tree: Support of Range Query and Cover Query over Dht. *In Electronic Publications of the 5th International Workshop on Peer-to-Peer Systems, IPTPS'06. Santa Barbara, CA, USA, 27–28 February 2006.*

Zhou, A., Zhang, R., Qian, W., et al. 2008. Adaptive Indexing for Content-Based Search in P2p Systems. *Data Knowl. Eng.* 67(3): 381–398.

Electronics, Communications and Networks IV – Hussain & Ivanovic (eds)
© 2015 Taylor & Francis Group, London, ISBN: 978-1-138-02830-2

A cross-layer AODV routing algorithm based on two-hop neighborhood information

Jinlong Zhang, Qinglin Hou, Lin Huang & Hong Chen*
Guangzhou GCI Science and Technology Company Limited, GuangZhou, China

ABSTRACT: To improve the performance of the AODV routing algorithm in wireless mobile ad hoc networks, a cross-layer AODV routing algorithm based on two-hop neighborhood information (CLAODV) was proposed in this paper. The CLAODV algorithm reduced the time required for building a routing table using a two-hop neighborhood information mechanism and was able to sense the status of links through a cross-layer design among the PHY, MAC, and network layers which decreased the routing overhead for link faults. Theoretical analysis verified the effectiveness of the CLAODV and simulation results showed that the CLAODV outperformed the traditional AODV routing algorithm in terms of end-to-end delay and success ratio.

1 INTRODUCTION

Ad hoc networking (Wang 2012, Li & Zhou 2012) is an increasingly important aspect of wireless communications and has been regarded as one of the key features of post-3G systems. In the 1970s and 1980s research on ad hoc networking was mainly for military purposes, when the primary use was for communication networks on battlefields where no infrastructures were available. Each node in an ad hoc network can work as a router to relay connections or data packets to their ultimate destination. There are two key issues of ad hoc networking, including MAC (medium access control) which shares common channel resources among wireless nodes, and routing which finds a path between the source and destination nodes across a number of possible relay nodes. Since the 1990s ad hoc networking has become more and more significant in commercial and residential areas. The reasons for that are as follows: (1) With the increase of small-size information processing devices, such as laptops, PDAs, and smart phones, the need to exchange digital information, such as video, music, and documents within a short range is ever-increasing. (2) Emerging wireless technologies, such as Bluetooth and IEEE 802.11, make ad hoc networking an important method of extending multimedia services from the internet to wireless environments.

Routing is one of the key components in the architecture of opportunistic networks. Routing in wireless mobile ad hoc networks have many challenges with multi-hop, dynamic topology, time-varying channels, and resource constraints. Some routing protocols which have been proposed in the last decade can be read in (Zhu et al. 2012, Liu & Li 2011) and the details are referenced therein. AODV routing protocol adopts the mechanisms of hop-to-hop routing and numbering by sequence in DSDV routing protocol and employs the mechanisms of route discovery and maintenance in DSR routing protocol. Because of its simple utility, the AODV routing protocol has garnered much attention in the last decade.

The reminder of this article is organized as follows: Section 2 contains the network model and problem description. Section 3 contains a cross-layer AODV routing algorithm based on two-hop neighborhood information described in detail and the related performance analysis. Section 4 analyzes the technology of the cross-layer design of this study. The simulation results are given in Section 5. Finally, Section 6 concludes the main results.

2 NETWORK MODEL AND PROBLEM DESCRIPTION

2.1 *Network model*

An ad hoc network based on a smart phone WiFi platform was established in this paper. Based on its characteristics and network system, a mobile community communication system was proposed which utilizes the wireless capability of a smart phone to establish an ad hoc network system to achieve communication between terminals. A mobile community is a communication model in which smart phones exchange information with other smart phones via WiFi in a

*Corresponding author: *chenhong@gcidesign.com*

certain area; it is free, convenient, and is used in various capacities such as entertainment, travel, emergency communication, and so on.

2.2 *Problem description*

Based on the authors' analysis, the following problems were found to exist in AODV routing algorithms when applied in the above-mentioned environments:

1. RREP packets are sent when destination or neighboring nodes receive RREQ packets.
2. The nodes are unable to sense the status of links without extra control packets.
3. When a link is broken, routing request messages are restarted, rather than repairing the broken link.

Based on the weakness of the AODV algorithm and routing overhead in mobile communities, the traditional AODV routing algorithm is ill-suited for the proposed environments where nodes can be mobile and the links are easily broken. To address these problems, a cross-layer AODV routing algorithm based on two-hop neighborhood information was proposed: the CLAODV.

3 AN IMPROVED AND EFFICIENT AODV ROUTING ALGORITHM BASED ON TWO-HOP NEIGHBORHOOD INFORMATION

Based on the above problems, an improved and efficient AODV routing algorithm, the CLAODV, was proposed which adopts the mechanism of two-hop neighborhood information to reduce routing time , improves the status sensing speed of links through cross-layers among the network layer, MAC layer, and physical layer, and then reduces control overhead for link breaks.

The CLAODV algorithm includes the following six steps:

Step 1: In the CLAODV algorithm, the nodes periodically broadcast the local (1-hop) HELLO packets which include the node's neighborhood information from their network layers. If a node receives a hello packet and extracts routing information, it renews its routing table.

Step 2: Nodes can sense the information from other nodes through carrier sensing of the PHY layer; if a node sends a frame which includes data frames, control frames, and expert broadcast frames (where the address of destination is -1), it will receive an ACK frame from the neighboring node.

Step 3: If a node sends a copy of a RREQ or RREP packet from the network layer it will send cross-layer information with "1" or "0" to the MAC layer. If it is sending a RREQ packet, the cross-layer information is "1."

Step 4: If the MAC layer of a node receives cross-layer information it obtains the value of that information. If it receives an ACK from the opposite, it will notify the network layer by cross-layer information sharing.

Step 5: After the network layer receives the cross-layer information, it will delete the copy of the RREQ or RREP packet or it will resend the RREQ or RREP packets with the referencing routing table.

Step 6: If the network layer of a node receives a RREQ packet from a neighboring node it will seek routing to reach the destination. If it searches the information, it will generate a packet in the network layer and send an RREQ packet.

4 CLAODV CROSS-LAYER DESIGN

An improved algorithm (CLAODV) which adopts a cross-layer design to utilize algorithm performance was created with the following details:

1. Nodes can deter information from other node sending through carrier sensing; at the same time, a node send each frame (except network layer send broadcast packets), the MAC layer of neighbor response an ACK frame.
2. If the network layer of a node sends a copy packet of RREQ or RREP, the network layer sends cross-layer information ("0" or "1") to the MAC layer. (If a node sends a RREQ packet, it will send cross-layer information as "1".)
3. If the MAC layer of the node receives the cross-layer information, it will report the network layer by sharing the cross-layer information when it sends a frame within the RREQ packet and receives an ACK frame from a neighbor.
4. If the network layer of the node receives the cross-layer information from the MAC layer, it will delete the copy of the RREQ or RREP packet; if not, it will refresh its routing table and resend the RREQ or RREP packets according to the routing table.

5 SIMULATION RESULTS AND PERFORMANCE ANALYSIS

5.1 *Performance analysis*

The cross-layer mechanism makes the CLAODV algorithm outperform the traditional AODV routing algorithm in terms of end-to-end delay, control overhead, success ratio, and so on.

5.1.1 *Algorithm convergence*

Because both of the algorithms utilize DV (distance vector), the CLAODV and AODV have the same convergence performance. If there is a path to the destination and packets are sent and received correctly, the CLAODV can find the path in a shorter time. Furthermore, there are no loops or counting to infinity in its routing process.

5.1.2 *Computing complexity*

Utilizing some new fields, the CLAODV increases the memory storage nodes. However, there is no transignification; the storage complexity is equivalent to that of the AODV which it is O(N) where N is the total number of nodes.

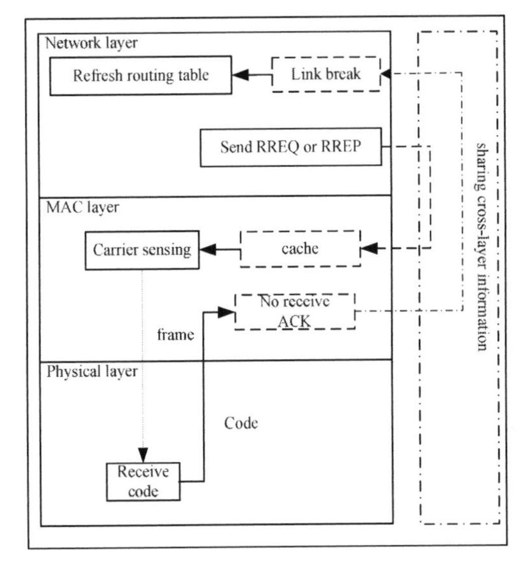

Figure 1. Cross-layer design.

5.1.3 *ETE (end-to-end) average delay*

$$\overline{T} = \sum_{i=0}^{N} (r_{ti} - s_{ti}) \Big/ N \tag{1}$$

where r_{ti} is the time when the i packet reaches the destination and s_{ti} is the time when the i packet is generated. Because the CLAODV algorithm utilizes two-hop neighbor information for reducing the routing time, it reduces the time needed for the packets to reach their destinations under the same load.

5.1.4 *Control overhead*

$$C = P_C / (P_C + P_D) \tag{2}$$

where P_C represents the bit number in all the control packets and P_D represents the bit number in the data packets that reach their destination. The CLAODV algorithm adopts a cross-layer design to reduce the number of RREQ and RREP packets to decrease the control overhead.

5.1.5 *Success ratio*

$$S = D_D / D_S \tag{3}$$

where D_D and D_S show the number of received and sent packets, respectively. The CLAODV algorithm reduces the time it takes packets to reach their destinations, so more data packets can reach their destinations in the same amount of time.

5.2 *Simulation environments and parameters*

Many simulations were executed to compare the performances of the CLAODV with the traditional AODV algorithm in terms of end-to-end delay, success ratio, and control overhead.

The simulation environments

(Fan et al. 2009) consisted of 40 mobile nodes evenly and randomly moving in a rectangle area of 400 m * 400 m. The nodes moved using a random waypoint model with a speed 0 m/s, 5 m/s, 10 m/s, 15 m/s, and 20 m/s; the transmission range of the nodes was 300 m. The traffic model was 1 packet/s and the destinations were random. The packet length was 1024 bits. The MAC protocol was 802.11b, and the channel rate was 1 Mbps. The simulation period was 600 s.

5.3 *Simulation results*

5.3.1 *Average end-to-end delay*

As illustrated in Figure 2, whether using the CLAODV or AODV, the ETE average delay increased as the velocity of the nodes increased. When the velocity of the nodes increased, the probability of the link breaking increased, too. Then, the number of retransmitting packets would also increase. Therefore, the ETE average delay increased. However, using the CLAODV resulted in less control packets and a higher success ratio of routing, which made transporting more fluent.

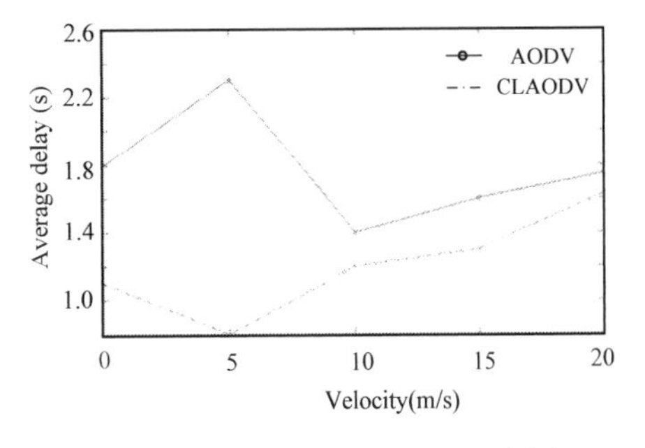

Figure 2. Comparison of average end-to-end delays.

5.3.2 *Control overhead*

As shown in Figure 3, there was an apparent difference in the protocol overhead between the CLAODV and AODV algorithms. The core idea of CLAODV was to decrease the control overhead in order to reduce the number of RREQ and RRER packets.

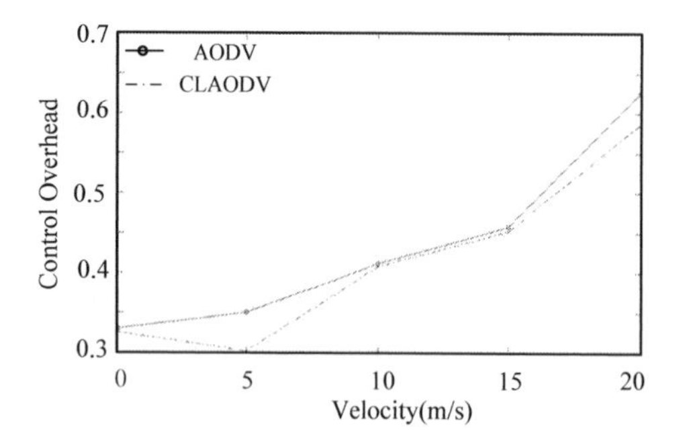

Figure 3. Comparison of control overhead.

5.3.3 *Success ratio*

As shown in Figure 4, the CLAODV showed a higher success ratio than the AODV. This was due to the fact that the CLAODV is able to make data packets reach their destinations more fluently.

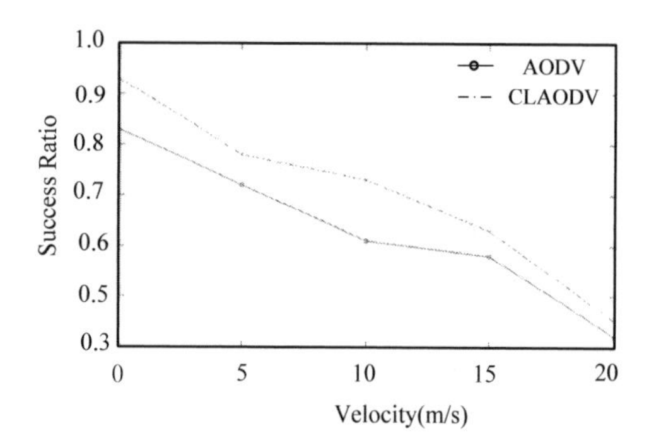

Figure 4. Success ratios of CLAODV and AODV algorithms.

6 CONCLUSION

A system of ad hoc networking based on smart phone Wi-Fi was proposed. Based on this system and the characteristics of smart phones, a cross-layer AODV routing algorithm was developed which adopted two-hop neighborhood information and the link status of cross-layer sensing to reduce route building times and to control overhead. In the future, further studies will be carried out to consider energy conservation for the AODV routing algorithm.

REFERENCES

Fan Ya-Qin, Wan Lin-Zhu & Sun Hui-Yin. 2009. Computer Simulation of AODV Routing Protocol in Ad Hoc Network Based on OPNET. *Journal of Jilin University (Information Science Edition)* 5: 534–538.

Li Jin & Zhou Ji-Peng. 2012. Ant colony optimization based energy aware routing algorithm for Ad Hoc networks. *Computer Engineering and Design* 33(4): 1315–1318.

Liu Hu & Li LuLu. 2011. Research and Improvement on AODV Routing Arithmetic in Mobile Ad Hoc Networks. *Journal of Wuhan Polytechnic* 31(11): 10–13.

Wang Yan-En. 2012. Smart phone sales. *Network and Information* 26(3): 5–5.

Zhou Qian. 2012. Disjoint multipath source routing in Ad hoc network. *Information Technology* 2: 94–96.

Zhu Xiao-Jian & Shen Jun. 2012. Minimum energy consumption multicast routing in ad hoc networks based on particle swarm optimization. *Journal on Communications* 33(3): 52–58.

Electronics, Communications and Networks IV – Hussain & Ivanovic (eds)
© 2015 Taylor & Francis Group, London, ISBN: 978-1-138-02830-2

Groundwater-level intelligent telemeter embedded platform based on GSM network

Xu Zhang & Qiong Liu
Key Laboratory for Highway Construction Technology and Equipment, Ministry of Education, Chang'an University, Xi'an Shaanxi, China

Aidi Huo*
School of Environmental Science & Engineering, Chang'an University, Xi'an Shaanxi, China

ABSTRACT: Groundwater monitoring is an important fundamental work for the sustainable development of society and economics. Groundwater pumping causes most of the urban environmental geology problems, so groundwater intelligent telemetry system has become an essential part of groundwater monitoring. In this paper, according to the characteristics of the environment of urban groundwater monitoring points, groundwater remote monitoring and management system has been designed, which is based on the SCM (Single Chip Microcomputer) and GSM network. The usage of short message service of GSM networks could complete the wireless remote and real-time transmission of the groundwater level, temperature and other parameters, which realizes the computer management and high-accuracy measurement of the results. The practical use shows that the system has higher stability and applicability, offering a scientific basis for the scientific development and rational use of groundwater resources, effectively preventing the development of land subsidence and ground fissures and government decision-making.

KEYWORDS: groundwater level; telemetry; GSM; embedded system.

1 INTRODUCTION

The groundwater level is an important indicator to measure the groundwater resources, the changes of which have a close relationship with groundwater extraction and ground subsidence, playing an important role in controlling land subsidence(Groom et al. 2000). In order to reasonably use, to fully understand groundwater resources and to achieve the automation of monitoring and remote management of the groundwater level, it is essential to make an implementation of real-time remote synchronous dynamic-monitoring via computers, automatic measuring instruments, common communication tools and so on (Weiming et al. 2009, Chen et al. 2009, Fu & Hu 2004, Anumalla et al. 2005).

Traditional groundwater monitoring methods mainly rely on manual and semi-artificial means of monitoring. With heavy workload, low efficiency and error-prone complex data processing, the transmission of information in a non-timely manner and poor timeless, they couldn't keep up with the needs of the development and modernization of information technology management. Besides, due to labor intensity and the low accuracy of measurement, the work is more difficult, especially in monitoring some of the relatively remote or dispersed monitoring sites.

With the comprehensive development of China's information process, there is an increased demand for information technology, as well as an enhanced understanding of information from a simple digital height to the height of being digital, network and wireless (Cox 1992, Sayed et al. 2005). In spite of the fact that the existing cable system could complete the digitalization and networking, its complex routing and high maintenance cost limit the spread of it. Therefore, there is a surge in demand for wireless data transmission(Paksuniemi et al. 2006).

The mobile network provides a lot of data service, covering almost every corner of the earth; it makes the wireless data transmission of many intelligent devices and instruments possible. SMS is of high reliability, the charge of which is relatively low. With the rapid development of mobile communications, the cost will be lower. As a communication medium, GSM can also simplify the ground facilities of the monitoring system. So using GSM as a communications medium is an effective way to achieve the groundwater level dynamic monitoring(Itaba et al. 2010, Huang & Liu 2006).

*Corresponding author: *huoaidi@163.com*

Two problems need to be solved in order to achieve a large area of automatic monitoring of the groundwater level. The first one is how to accurately measure the liquid level in harsh environmental conditions. The second is how to find a data communications means which is not only suitable for unmanned remote areas, but also in the populous cities with complex buildings to reach the remote wireless data and real-time communication, with less investment and large-scale adoption to facilitate the massive popularity and data communication means of easy-to-computer network management. Based on the GSM network, the embedded intelligence and remote sensing systems of groundwater level can achieve automated monitoring and remarkably improve the Technology and method of dynamic monitoring, which lays a solid foundation for the scientific and rational utilization of water resources and protecting ecological environment. The application of the system provides the necessary decision support and diversified services for sustainable urban development, disaster prevention and mitigation.

2 SYSTEM STRUCTURE AND WORKING PROCESS

The system uses the GSM network platform. SCM control technology and wireless network technology are the core. After intelligent information acquisition and information processing, the data arrives at the user terminal and is conversed to the water level. The system mainly consists of intelligent information collection terminal, the information server and the user terminal. The overall structure of the block diagram is shown in figure 1; the four parts of the intelligence information collection terminal are the SCM, GSM module, detection and control, which are mainly responsible for collecting information and sending it via the GSM module to the comprehensive information server. The comprehensive information server is made up of modules of the management, data receiving and sending and the terminal processing, aiming at receiving, storing and displaying the data. User terminal could be clarified into two kinds: computer terminals and GSM mobile phone user terminal. Computer end-user having installed the terminal application has access to the detailed water level data in the server of comprehensive information through the network, while mobile phone users authorized could obtain it by a phone from time to time.

The system works as follows. First the field level data could be transmitted in a standard signal that will be loaded into the digital data by the A/D convertor. Secondly the collected data is sent out by the GSM modem in a short message. Finally, after comprehensive information server receiving the message of the collection terminal, the water level elevation and other relevant information would be stored in the database for query and analysis. When needing them, users just start the transceiver module, then the water level information and database information can be sent to the phone or computer. Its basic functions are as follows: first, an automatic and timed collection of water level data and storage; second, manual entry to edit and modify the underlying data and the related stuff; third, statistical analysis of the collected data and tabulation drawing; the fourth, with computer and special software, users can directly setup, debug, and monitor the observation sub-station through SMS; then, remote data transmission.

3 THE DESIGN OF INTELLIGENT INFORMATION COLLECTION TERMINAL

3.1 *Intelligent information collection terminal hardware*

Field intelligence information collection terminal hardware is mainly made up of level transmitter, temperature transmitter, cable connection and groundwater intelligent monitor. Level transmitter is anti-corrosion under industrial grade, while temperature transmitter uses platinum resistance. Due to the fact that the water contains impurities in observation wells, the front end of the transmitters adopts stainless steel housing. The cable connection links the transmitter and groundwater intelligent monitor. With the measurement of liquid level in the pressure form, the dedicated cable with central ventilation duct is used as the connection cable. Following parts consolidate groundwater intelligent detector. One is SCM, which is the core of it.

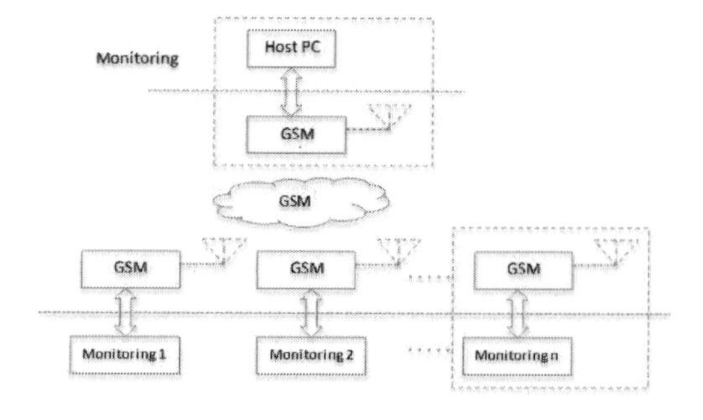

Figure 1. Overall structure of automatic monitoring system.

The other is peripheral including analog to digital conversion, data storage, calendar clock, data display, data communication, and backup power supply. The block diagram is shown in Figure 2.

The observed sub-station works as follows. The level and temperature transmitters, laid in the observation well, measure the liquid level and temperature signals. In order to reduce signal attenuation, 4-20 mA current signal transmission is used for sending signals to the groundwater intelligent monitor to make a conversion from analog to digital, which will be transmitted into the MCU to complete the data analysis and processing. The data storage part uses large-capacity memory chips to store the data processed. Calendar and clock part provides the data and clock signals. The data display part shows the site of the monitoring data. Data communication part achieves the data exchange between the site monitoring and computes via modem in the central monitoring station of the water sector at each level. In order to ensure that the groundwater intelligent monitoring works properly in the case of mains failure, the backup power is designed, which automatically switches between backup power and the mains.

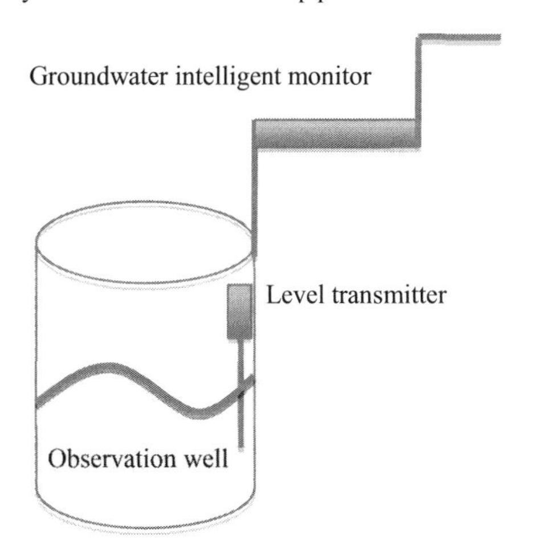

Figure 2. Field intelligent collection terminals.

3.2 *Intelligent information collection terminal software*

In order to improve the fidelity of the sampled signal of the water level, this system uses program to make digital filtering to external interference. When measuring the water level, the impact of waves may bring instantaneous and amplitude pulse interference to the sampled signal, which will disrupt the normal working of the system if it happens at the sampling time. Therefore, the data acquisition extension must filter the sampled data. Intelligent information collection terminal uses a method of the median filter. That is to say, continuously sampling the water level information in each five minutes, then ranging the sample values from small to large, finally taking the middle value as the real signal. In the design of software, to prevent interference, "redundant instruction" and "software trap" are arranged. Besides, it also considers the case of program disorder when the program should be allowed to reset to its normal state. The data acquisition, extension software includes the main program and data transfer subroutine. Intelligent data acquisition terminal data transfer subroutine process shown in Figure 3.

By using VC++ and the object-oriented programming techniques, intelligent information collection terminal software builds the main features of the application software in terms of component, improving system visibility, maintainability and scalability. It improves the system's reliability and operating efficiency by using database to save and process data.

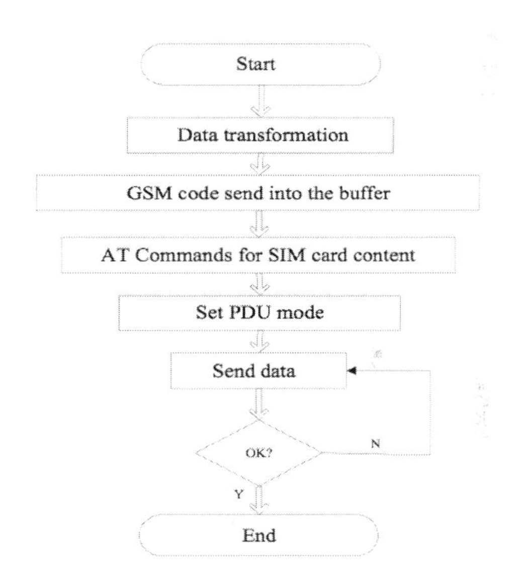

Figure 3. Data transfer subroutine.

4 THE DESIGN OF INTEGRATED INFORMATION SERVES

Groundwater remote monitoring network system is mainly made up of the sub-station of central stations and field observations, divided into two networking. It consists of three aspects: the field data collection and storage, remote data transmission, data analysis and database management. Figure 4 shows the block overall designing diagram of the groundwater resources telemetry management system.

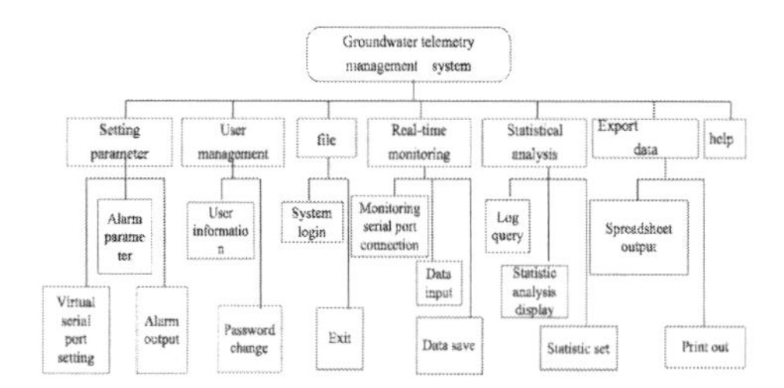

Figure 4. System design block diagram of the host computer.

Using visual C++ and net high-level language, groundwater resources telemetry management system adopts an overall design of the module from top to bottom. The system is an integrated information platform, including automatic remote acquisition, pre-processing, efficient data storage and management. With the wireless remote transmission of data, real-time access to monitoring wells' points of water resources and status information, data will be stored in the database of water resources in accordance with day, month and year. By drawing curves and histogram to analyze the data, the system displays the consolidated results to users and automatically generates reports in accordance with the requirements of users. The software management system of the host computer will be finished by three menus, which is made up of information management, real-time monitoring module, water resources decision support system, the output module and other auxiliary functions.

4.1 *Real-time monitoring module*

By using advanced wireless data transmission technology and serial communication technology, real-time monitoring system collects the real-time groundwater data and monitors the operation of the measuring point at any time. Real-time data acquisition includes real-time acquisition of the groundwater level and water temperature data. Water level and water temperature sensors in monitoring stations transmit the level and temperature into analog electrical signals and send them to forecasting instrument, which makes an A/D conversion and data processing on analog electrical signal and stores the water resources data. When water resources information is needed, telemetry terminal, according to the specified communication protocol, sends data to the connected modem, which issues the data in the form of short messages through GSM network. After receiving the short message, the modem of the monitoring center will send it to the linked host PC through the serial part. Similarly, when the monitoring center needs to collect real-time data; real-time data acquisition module of the water telemetry management system could send it in text messages to the monitoring station. Figure 5 shows the real-time data on the level of underground water.

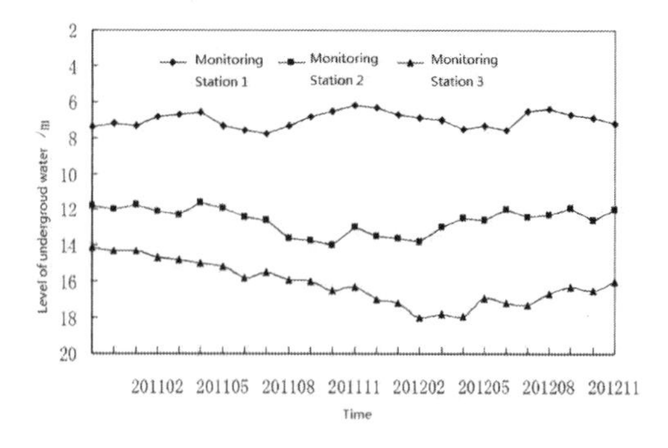

Figure 5. Real-time data of the level of underground water.

4.2 *Water resources decision support system*

The system uses the information fusion technology to make mathematical statistics and analysis of the data. Based on the information collected, it calculates each well's monthly and annual water level such as maximum, minimum, average and sum. In addition, according to the water environmental conditions during different periods, the status of regional development, the requirements of the regional master plan and specialized analysis models, it makes a comprehensive process and analysis of the data.

4.3 *Output module*

In order to facilitate data viewing and further process, the system allows users to import data into Excel or Word profile and automatically generates reports in the format that users require such as reporting daily, in three days or in five days. They can also be printed out.

5 THE FUNCTION OF THE SYSTEM EXPANSION AND INNOVATION

Although the system is currently only used for water level detection, due to the monitoring terminal providing multi-channel analog input interface, the system can simultaneously access to multiple signals. In other words, it can measure not only the water level, but also the physical quantities such as temperature, water quality and flow rate. As a result of the fact that the system enables remote data communication to break geographical constraints, the system can be not only used for remote monitoring of the groundwater level, but for remote measuring and reporting the levels of dams, river channels and reservoirs.

6 CONCLUSIONS

With the networking and technology of the hydrological automatic test system, hydrological information becomes more and more sophisticated. At the same time, hydrological observation projects and contents continue to increase. Besides, there are increasingly high requirements of the observation means and methods as well as the development and application of hydrological monitoring. Based on the GSM water level online monitoring, this paper designs a remote data acquisition system, which plays an important role in scientific research. It makes a combination of the remote data acquisition and GSM wireless data transmission technology, which gets rid of cable trouble, and makes it possible for the hydrological station to be unattended. The structure of the system is simple and easy to expand. With wide coverage of the network, the data is real-time. Moreover, communication costs are low.

Furthermore, without geographical restrictions, it can be widely used in geology, hydrology and other fields. It can also be applied in the remote mountain areas with inconvenience of transportation and electricity.

The successful development of this system is more than beneficial. It can change the status of low accuracy and poor reliability of the previous manual monitoring. Meanwhile, it achieves the purpose of the regional groundwater dynamic monitoring of automation and control, thus improving the level and efficiency of regional water resource management. In conclusion, it is of high value in the application and promotion.

ACKNOWLEDGEMENTS

The project was supported by the special fund for basic scientific research for central colleges, Chang'an University (No. 2014G1251030, No.2014G1251032).

REFERENCES

Anumalla S, Ramamurthy B, Gosselin DC, et al. 2005. Groundwater monitoring using smart sensors. *In IEEE* 6: 6.

Chen Q, Ding T, Li C, et al. 2009. Low-power wireless remote terminal design based on GPRS/GSM. *Journal of Tsinghua University*: 2.

Cox DC. 1992. Wireless network access for personal communications. *Communications Magazine, IEEE* 30(12): 96–115.

Fu M & Hu X. 2004. Groundwater inspecting system based on distributed bus. *Geo-information Science*: 1.

Groom BPK, Froend RH & Mattiske EM. 2000. Impact of groundwater abstraction on a Banksia woodland, Swan Coastal Plain, Western Australia. *Ecological Management & Restoration* 1(2):117–124.

Huang Z & Liu Z. 2006. The Application of Geophysical Technique in Investigating and Monitoring Groundwater Organic Contamination by DNAPLs. *Jiangsu Environmental Science and Technology*: 4.

Itaba S, Koizumi N, Matsumoto N, et al. 2010. Continuous observation of groundwater and crustal deformation for forecasting Tonankai and Nankai Earthquakes in Japan. *Pure and Applied Geophysics* 167(8):1105–1114.

Paksuniemi M, Sorvoja H, Alasaarela E, et al. 2006. Wireless sensor and data transmission needs and technologies for patient monitoring in the operating room and intensive care unit. *In IEEE*: 5182–5185.

Sayed AH, Tarighat A & Khajehnouri N. 2005. Network-based wireless location: challenges faced in developing techniques for accurate wireless location information. *Signal Processing Magazine, IEEE* 22(4): 24–40.

Weiming L, Defu L, Hui P, et al. 2009. The application of 3s technique in water pollution monitoring and forecasting. *In IEEE*: 822–826.

Electronics, Communications and Networks IV – Hussain & Ivanovic (eds)
© 2015 Taylor & Francis Group, London, ISBN: 978-1-138-02830-2

Multiple network coding of feedback for mobile P2P networks

Guoyin Zhang, Xu Fan*, Yongfeng Wang, Wei Gao & Yanxia Wu
College of Computer Science and Technology, Harbin Engineering University, Harbin, China

ABSTRACT: The data distribution scheme based on network coding can reduce the resource consumption of mobile P2P, but the data distribution efficiency is affected by the network topology. To improve the data distribution efficiency in mobile P2P networks and shorten the data distribution delay, this paper proposes multiple network coding, and designs multiple network coding of feedback to adjust data transmission scheme in mobile P2P. The scheme adjusts coding times dynamically, improves the amount of data received in the downstream nodes, and reduces the delay of data distribution in dynamic network environment. Experimental results are consistent with theoretical analysis. The experimental results show that the scheme can effectively improve the efficiency of data distribution.

1 INTRODUCTION

More recently, P2P technology has been employed to provide file sharing or live media streaming services over the mobile P2P due to the rapid improvements of wireless technologies and mobile devices capabilities. Because of the dynamic of its topology, existing routing protocols can not be directly used in mobile P2P.

Network coding, first proposed in information theory, has been introduced into P2P content distribution to improve system performance (Li & Niu 2011). It has been shown that network coding can improve the network efficiency in both unicast (Katti et al. 2008) and broadcast (Fragouli et al. 2006) wireless communication, by exploring the broadcast essence of the wireless network. Previous studies have optimized the network from the perspective of energy saving (Feng et al. 2013, Zhou et al. 2012) or throughput (Guo et al. 2011, Yan et al. 2012). With random network coding, each peer linearly encodes all the blocks with random coefficients, and then transmits the encoded blocks to its downstream peers.

A node can transmit any coded packet using network coding, since all of them can equally contribute to the eventual delivery of all data packets to the destination with high probability. This simplifies protocol design by avoiding block scheduling and enhancing block universal. However, only once network coding will increase the requirement and transmission of blocks in some network topologies. Multiple blocks are coded into one in every coding operation. Network coding reduces transmission times, at the same time reduces block amount of downstream peers. Compared with network coding, the paradigm of duplication and transmitting is another extreme that transmitting times equal to the number of blocks received by node. Obviously, two extremes are not good schemes. Unfortunately, encoder does not have precise global knowledge of how many coded blocks should be transmitted in mobile P2P. Motivated by our curiosity on choosing the "sweet spot" in coding times of various network topologies, the relation between coding times and in-degree in terms of graph theory is studied.

The number of linearly independent blocks in download node determines whether the node can decode a segment. To increase the chance of coding, coding node will wait before encoding operation until the block number in node buffer meets certain conditions. The waiting period provides opportunities to repeat coding for existing data, in order to increase block number in downstream nodes and reduce download time, and multiple coding has been proposed based on the basis of single coding.

The remainder of this paper is organized as follows. Sec. 2 presents the system model, including basic procedure of network coding and network architecture. Sec. 3 analyzes the relation of in-degree and coding times in terms of graph theory and presents feedback adjusting mechanism. In Sec. 4, simulations are carried out to corroborate the theoretical results. The final section concludes the paper.

2 OVERVIEW OF SYSTEM MODEL

2.1 *Random linear network coding*

A point-to-point communication network is represented by a directed graph $G=(V,E)$, where V is the

*Corresponding author: *fanxu@hrbeu.edu.cn*

set of nodes in the network and E is the set of edges in G which represent the point-to-point channels. The sets of input channels and output channels of a node v are denoted by *In(v)* and *Out(v)*, respectively.

This paper briefly describes procedures of encoding and decoding with random linear network coding (RLNC) (Li et al. 2003, Tracey et al. 2006). A file F is to be broadcasted to online peers. When segment-based network coding is applied, the file is divided into G segments, each of which is further broken into m blocks in source node. Assume that segment i has original blocks $\mathbf{B}^i = [B_1^i, B_2^i, \cdots, B_m^i]^T$, then a coded block b^i from segment i is a linear combination of $[B_1^i, B_2^i, \cdots, B_m^i]^T$ in the Galois field GF(q), i.e.

$$b^i = \sum_{j=1}^{k} c_k B_{k'}^i \quad 1 \le j \le m, c_k \in GF(q).$$

If a peer has $l (l \le m)$ coded or uncoded blocks of segment i $\mathbf{b} = [b_1^i, b_2^i, \cdots, b_l^i]^T$ when serving another peer p, it randomly chooses a set of coding coefficients from the GF(q) namely coding vector, denoted by $\mathbf{c} = [c_1, c_2, \cdots, c_l]^T$, and then encodes all its blocks from segment i, and produces one coded block

$$x^i = \sum_{j=1}^{l} c_j \cdot b_j^i = \mathbf{c} \cdot \mathbf{b}.$$

In a coded block x, the coding coefficients used to encode original blocks to x are embedded in the header of the coded block. As soon as peer p successfully receives a total of m coded blocks from segment i, $\mathbf{x} = [x_1^i, x_2^i, \cdots, x_l^i]^T$ which are linearly independent, it will be able to decode segment i with coding coefficients embedded in each of the m coded blocks. To decode segment i, the inverse of the coefficient matrix $\mathbf{A}_i$ should be computed first using Gaussian elimination. To obtain the original m blocks, it then needs to multiply

$\mathbf{A}_i^{-1}$ by $\mathbf{x}$, i.e.,

$$\mathbf{B}^i = \mathbf{A}^{-1} \times \mathbf{x}.$$

2.2 *Network architecture*

Mobile P2P content distribution system is organized as an application-layer overlay of peers, and each peer has a number of neighbors. The P2P platform supports centralized P2P network model, pure P2P network model and hierarchical P2P network model (Lua et al. 2005, Farooqi et al. 2009).

All of the peers in a centralized P2P network are managed by a control node. Each peer reports its existence and its neighbors to the control node. The control node can draw the topology of an entire network. It responses a request from a peer with related routing information, topology optimization and security functions according to the state of the desired P2P network. The centralized P2P network is suitable for a large well organized P2P network. On the other hand, the pure P2P network has no central server, and the content querying and content sharing are completed by broadcast between peers and their neighbors. The pure P2P network is well adapted to ad-hoc networks such as multi-hop wireless LANs and wireless sensor networks.

However, both of the two architectures are defective, especially for mobile P2P with high dynamic. The loose unstructured network framework design of pure P2P will decrease the accuracy of information transmission; while structured framework design leads to huge routing updating overhead due to frequent change of nodes. In this paper, centralized and pure P2P networks are integrated into a hierarchical model by several super nodes. Hierarchical P2P network model combines the advantages of the two models above. Nodes are divided into super node and common node by computing ability, power remaining, memory size and online time in hierarchical P2P.

Centered by super node as shown in Fig. 1, the network is divided into multiple subnetworks, each subnetwork is relatively independent. Super node is responsible for the management of subnetwork members and the query task between subnetworks. All nodes status is the same in terms of resource sharing. The super node stores the information of other nodes in the same subnetwork. When a request does not get response, super node will forward the request to other super nodes by discovery algorithm. Other super nodes will forward request to the common nodes which belong to the same subnetwork. The common node is responsible for data storage and transmission without distinguishing encoding node from non-coding node. In addition, common node processes network coding only once before transmission by default, and super node will adjust coding times of the common node based on network topology.

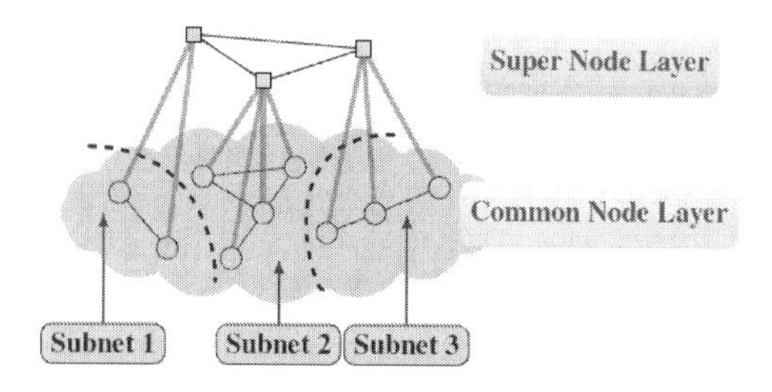

Figure 1. Network topology of hierarchical mobile P2P.

3 PROPOSED SCHEME

3.1 *Multiple network coding*

To take full advantage of network coding provided by the upstream nodes, this section presents multiple network coding. While the necessity, benefit and complexity of network coding are sensitive to the underlying graph structure of a network, existing research on network coding often focuses on reducing the time of coding operation and the number of coding nodes. However, peer encodes blocks it has received into a new block and then transmits the block to neighbors. Although coded blocks are useful to many peers, coding operation reduces the total number of blocks in downstream peers. The condition of recovering original blocks is to decode matrix of full rank, which means that peers must receive enough linear independent blocks. Thus once coding is not enough in some network topologies.

In Fig. 2, node h_4 encodes the blocks received from h_1, h_2 and h_3 into a new coded block and hen transmits it to r_1 and r_2. After once transmitting, r_1/r_2 receives two blocks from h_1 / h_3 and h_4. But r_1 and r_2 cannot decode the original blocks, therefore the twice encoding is needed. It can be found from Fig. 2 that twice encoding of node h_4 is the best solution.

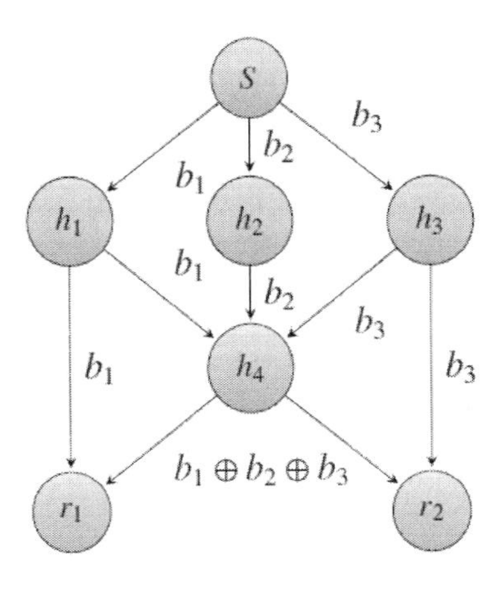

Figure 2. Extend network for network coding.

Definition 1 (Multiple network coding) *Encoder randomly generates k(k>1) coding vectors and encodes the received blocks into k coded blocks with these coding vectors respectively. The encoder forwards the k coded blocks finally.*

Receiver can decode the original blocks only when it has received $G \times m$ linearly independent blocks. The retransmission times of nodes in network depend on the network topology. Relay node does not need network coding when the max-flow (between source and every receiver) is no less than source transmitting rate and no link is reused. If some links are reused, relay node must network coding to achieve blocks forwarding in single transmission. In mobile P2P, sending rate of server is often greater than the max-flow, once transmission cannot forward all blocks, then multiple network coding can improve network performance by reducing forwarding times. Similar to network coding, multiple network coding can reduce the total times of transmission. Theoretically, coding operation will consume extra resource such as time and energy, and resource consumption of multiple network coding will be greater than direct transmission when the coding times reach some value. However, the coding consumption is far less than transmitting in reality, so this paper does not discuss the situation that the multiple network coding should be switched into duplication and transmitting. From the procedure of network coding, it can be found that multiple network coding can improve network efficiency if the in-degree of encoder is greater than that of downstream peers.

3.2 *Feedback adjusting scheme*

Encoders in mobile P2P network cannot have access to the topology details besides neighbors, so it is unable to determine the appropriate coding times. Too few or too many coding times will result in low network efficiency due to increased energy cost or number of forwarding. To select the appropriate coding times for encoders, the feedback mechanism is used to adjust the coding state in real time. Encoders and receivers will send information back to the super node, for which to make judgment and adjust coding schemes to ensure the whole system in the optimal state. The super node will periodically check the status of nodes in subnetwork. When the data distribution efficiency is low, common nodes with strong performance will be chosen into coding nodes.

Encoding tolerance rate of node v is denoted by $R(v) = \frac{|In(v)| - k_v}{|In(v)|}$, where |In(v)| is the in-degree of node v, and k_v is the coding times of this node. The global encoding tolerance rate is denoted by:

$$R = \frac{\sum_{v \in V} |In(v)| - \sum_{v \in V} k_v}{\sum_{v \in V} |In(v)|} \tag{1}$$

Success request rate of segment i is denoted by: $D_i = \frac{\sum_{v \in V} v_{Suc}^i}{\sum_{v \in V} v_{Req}^i}$, where v_{Suc}^i is node v success request times for segment i, v_{Req}^i is node v total request times for segment i. Success request rate for all segment is denoted by:

$$D = \frac{\sum_i \sum_{v \in V} v_{Suc}^i}{\sum_i \sum_{v \in V} v_{Req}^i} \tag{2}$$

Theorem 1 *D is inversely proportional to R.*

Proof The total number of successfully receiving is decided by transmit times and sum of receivers, i.e.

$$\sum_i \sum_{v \in V} v_{Suc}^i = N \cdot \sum_{v \in V} k_v \tag{3}$$

With (1) and (3), we have

$$R = 1 - \frac{\sum_i \sum_{v \in V} v_{Suc}^i}{N \cdot \sum_{v \in V} |In(v)|} \tag{4}$$

From (4), it can be found that R decreases when success request times $\sum_i \sum_{v \in V} v_{Suc}^i$ increase. Based on the definition of D, D is inversely proportional to R.

Negative feedback model is constructed from the relation between R and D. The adjusting operation of multiple network coding scheme of feedback is described as follows:

1 Set point is assumed as D_0. At the initial state, all the encoders process once network coding;
2 Super node periodically computes the D with feedback information based on Formula (2);
3 Super node activates adjusting factor T in the sub-network based on current D.

Adjusting function factor T is design as:

$$k_v = \begin{cases} \left\lfloor \dfrac{k_v}{2} \right\rfloor & D > D_0 \\ k_v + \left\lceil \dfrac{|In(v) - k_v|}{2} \right\rceil & D < D_0 \end{cases}$$

When $D < D_0$, it illustrates that the quantity of require data is greater than that of once transmission or coding times are not enough. Then multiple network coding operation of encoder will be activated by signal from super node. The sum of respond request will increase due to multiple network coding and D will also increase. When $D = D_0$, it implies that coding times are greater than ideal value of network, resource utilization is low in network. Then multiple network coding operation of encoder will be shutdown by signal from super node. Network coding times and D will decrease.

Theorem 2 Negative feedback adjustment mechanism can make D change within a small range, and when $D = D_0$ the transmission delay is the lowest.

Proof Based on cyclical check of super node, coding times will go down when $D > D_0$, and up when $D < D_0$. After a period adjusting, the blocks from encoder cover neighbors' request, the system comes into stable state. D changes in a small range, and when $D = D_0$ the total transmission times are the minimum in the network.

Next it is to be proved that the transmission delay is the lowest when $D = D_0$. In data distribution system with network coding, transmission delay includes request delay and coding delay. Request delay is the function of transmission rate $D_R = f(d)$ and $\frac{dD_r}{dD} = \frac{df(D)}{dD} < 0$, coding delay is the function of encoding tolerance rate $D_C = g(r)$ and $\frac{dD_C}{dD} = \frac{dg(R)}{dR} < 0$. There is a negative correlation between D_R and D_C from Theorem 1. Therefore, with the increase of network coding times, *D* will get improved, whereas successful request is reduced. And the increasing number of network coding can bring more transmission delay, so the improvement of successful request rate *D* will increase the transmission delay.

4 PERFORMANCE EVALUATION

This section investigates how much the node transmission efficiency can be improved by adopting the multiple network coding with feedback adjusting in mobile P2P, compared with the segmented network coding scheme and flooding scheme. First the simulation setting is described and then the simulation results are shown by event-driven simulator in C++ language.

4.1 *Simulation model*

Assume that there are *N* peers in a rectangular area of $100m \times 100m$. Let *N*= 30, 50, 80 and 100, where the node location is uniformly distributed in the area. Six super nodes are uniformly placed in the area. The maximum of neighbor is 8, the minimum is 4. The performance of dynamic and movement nodes is tested respectively. Node speed is $5m/s - 10m/s$. Since the network coding in this paper only considers the overlay, a network layer with AODV is built without considering the details of the physical and the MAC layer. It is assumed that there is no packet loss due to bit errors on wireless links and buffer overflows at receiver nodes.

4.2 *Performance results*

In the first set of experiments, the average in-degree $\overline{In(v)}$ is studied. From section 3.1, connections among nodes directly affect the network coding efficiency, and it is the essential condition to start multiple network coding. It shows in Fig. 3 that degree distribution is between 4 to 8 when the network comes into a stable state, and the average degree has decreased by different ranges after the node gets into a motion state. $\overline{In(v)}$ cannot reach the minimum of neighbor when N=30 and N=50. There is a broad range in $\overline{In(v)}$ of N=80 and N=100. All these results explain that node movement has an influence on average in-degree. In the second set of experiments, the optimal success request rate D' is searched for different $\overline{In(v)}$ in multiple network coding of feedback. For the best performance, the set point D_0 in the next set of experiments will be set to the corresponding value from approximate $\overline{In(v)}$ as shown in the Table 1. Super node will activate the adjusting factor T based on compared result between D and D_0.

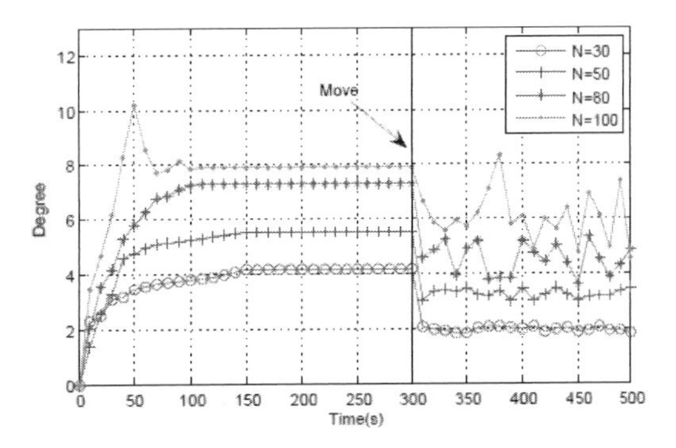

Figure 3. The change of $\overline{In(v)}$ with time and node movement.

Table 1. The optimal D' for different $\overline{In(v)}$.

$\overline{In(v)}$	2	2.5	3	3.5	4	4.5	5	6	7	8
D'/%	32	40	50	65	80	90	95	100	100	100

In the third set of experiments, the total network load is studied. Compared with network coding and multiple network coding, Figs. 4 and 5 show that network load of flood grows exponentially with the increasing number of nodes, but it is not sensitive to mobility. Network coding and multiple network coding are sensitive to mobility. When $\overline{In(v)} < 4$, multiple network coding has distinct advantages compared with network coding and can save 10-15% network load. When $4 < \overline{In(v)} < 5$, multiple network coding can save about 5% network load compared with network coding. Performance of multiple network coding is very close to network coding when $\overline{In(v)} \geq 5$.

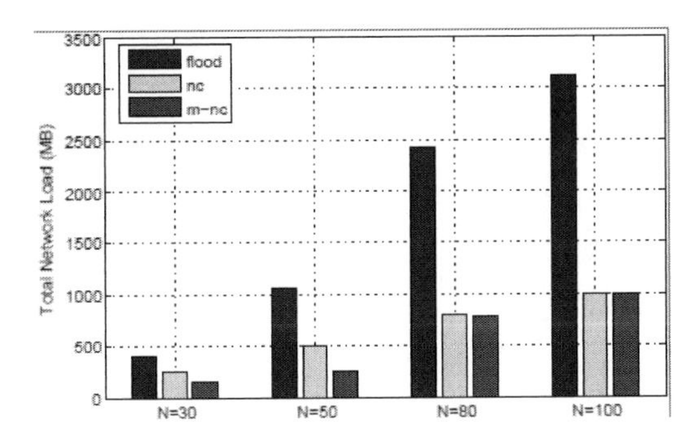

Figure 4. Network load for static network.

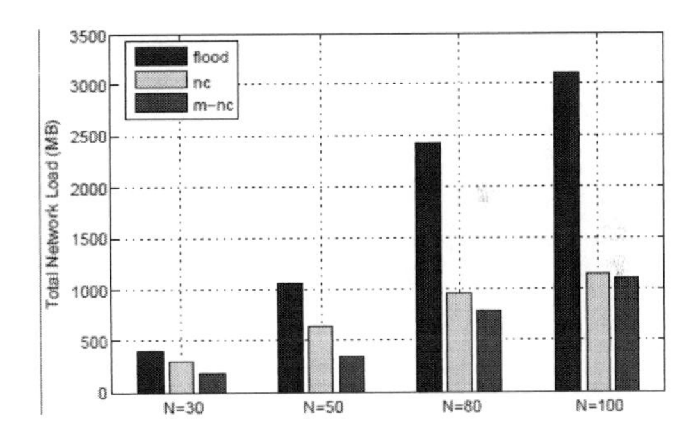

Figure 5. Network load for dynamic network.

5 CONCLUSIONS

This paper considers broadcasting with RNC in mobile P2P content distribution system. With several simulation scenarios, the multiple network coding of feedback for mobile P2P networks is evaluated. Our simulation results show that multiple network coding significantly reduces the network load in the network environment with low node in-degree or highly dynamic. However, multiple network coding is similar to network coding in a network environment with high node degree. Thus average in-degree is the key factor for multiple network coding performance.

ACKNOWLEDGMENT

This work was partially supported by the National Natural Science Foundation of China (No. 61073042 and No. 61170241) and Fundamental Research Funds for the Central Universities (No. HEUCF100606).

REFERENCES

Li Baochun & Niu Di. 2011. Random Network Coding in Peer-to-Peer Networks: From Theory to Practice. *In Proc. of the IEEE* 99(3): 513–523.

Katti, S., Rahul, H., Hu, W.J., et al. 2008. XORs in the Air: Practical Wireless Network Coding. *J. IEEE/ACM Transactions on Networking* 16(3): 497–510.

Fragouli, C., Widmer, J. & Le Boudec, J.Y. 2006. A Network Coding Approach to Energy Efficient Broadcasting: From Theory to Practice. In *Proc. 25th INFOCOM:* 1–11.

Guo Bin, Li Hongkun, Zhou Chi. et al. 2011. Analysis of General Network Coding Conditions and Design of a Free-Ride-Oriented Routing Metric. *J. IEEE Transactions on Vehicular Technology* 60(4): 1714–1727.

Feng Daquan, Jiang Chenzi, Lim Gubong, et al. 2013. A survey of energy-efficient wireless communications. *J. IEEE Communications Surveys & Tutorials* 15(1): 167–178.

Zhou, M., Cui, Q., Jantti, R. & Tao, X. 2012. Energy-Efficient Relay Selection and Power Allocation for Two-Way Relay Channel with Analog Network Coding. *J. IEEE Communications Letters* 16(6): 816–819.

Yan Qiben, Li Ming, Yang Zhenyu, et al. 2012. Throughput Analysis of Cooperative Mobile Content Distribution in Vehicular Network using Symbol Level Network Coding. *J. Selected Areas in Communications* 30(2): 484–492.

Lua, E. K., Crowcroft, J., Pias, M., et al. 2005. A survey and comparison of peer-to-peer overlay network schemes. *J. IEEE Communications Surveys & Tutorials* 7(2): 72–93.

Li, S. Y., Yeung, R. W. & Cai, N. 2003. Linear network coding. *J. IEEE Transactions on Information Theory* 49(2): 371–381.

Koetter, R. & Medard, M. 2003. An algebraic approach to network coding. *J. IEEE/ACM Transactions on Networking* 11(5): 782–795.

Tracey, H., Medard M., Koetter R., et al. 2006. A Random Linear Network Coding Approach to Multicast. *J. IEEE Transactions on Information Theory* 52(10): 4413–4430.

Farooqi, A. H., Kazmi, S. B. & Khan, F. A. 2009. Performance analysis of peer-to-peer overlay architectures for Mobile Ad hoc Networks. In *Proc. ICET*: 471–475.

Electronics, Communications and Networks IV – Hussain & Ivanovic (eds)
© 2015 Taylor & Francis Group, London, ISBN: 978-1-138-02830-2

Research on push-pull congestion control strategy in delay tolerant networks

Lihua Zhang* & Xiaoxu Cheng
Institute of Computer Science and Information Technology DaQing Normal University, Heilongjiang, China

Weimin Zhang
Oil production test branch of Daqing Oil Field Co, Heilongjiang, China

ABSTRACT: DTN network is a new network structure to obtain information under an extreme environment, with intermittent connectivity features; common control mechanism easily causes network congestion. Therefore, this paper researches Push-Pull congestion control algorithms aiming at congestion problems, and bases on the storage state's routing algorithm, as well as proposes Push-Pull congestion control algorithms based on the level of service. Moreover, the NS-2 simulation tool is used for simulation. Simulation results show that node congestion can be effectively controlled by the improved policy, which can improve information transfer rate.

KEYWORDS: DTN; Push-Pull; Congestion Control;Custody Transfer; NS-2.

1 INTRODUCTION

DTN is proposed stemming from applications called" restricted network", refers to a class of network (Durst et al. 1996) without stable end-to-end transmission paths due to node mobility, etc., reasons, and most of the time in the interrupt status.

Compared with other architectures, DTN architecture has two features, which make congestion control very difficult:

1 There may not have contact to arrive (Thus there are no chances for empty data accumulated in the nodes) in the next period of time.
2 The received message for custody transmission cannot be discarded (unless in extreme conditions cases or message expiration).

Given these constraints, possible measures to process congestion include reserving buffer space to different levels of service and so on. In extreme cases, the custody messages may be deleted, but such a move would be considered as a system error, so it should be avoided.

The current congestion control method is to use a shared priority queue to allocate custody storage. But this approach has two problems, namely priority inversion and the team head congestion.

2 SR ALGORITHM BASED ON THE STATE OF THE STORAGE

2.1 *Overview of SR algorithm*

SR(Jain et al. 2004) (Storage Routing) algorithm was proposed by Matthew Seligman, Kevin Fall and Padma Mundur in 2007, aiming at storage routing of DTN congestion control, it can use neighboring nodes with memory capacity to store information, which can effectively relieve congestion and maximize utilization of network resources.

SR algorithm (Daly & Haahr 2007) generally includes the following aspects:

2.1.1 *Information selection*

Choose according to the type of information (shown as Table 1).

*Corresponding author: *bearbear_318@126.com*

Table 1. Descripiton and application of push strategy.

Strategy name	Strategy description	Suitable occasion
Push Tail	The selection of the most late arrival information	When the data processing system using the continuous and orderly flow of information, such as voice service
Push Head	Select the earliest arrival information	When using a new data earlier arrival data is more important
Push Oldest Network Age	The choice has the most long survival time of information	When the system is more preferred to use the processing of all new data, for example Webpage browse
Push Latest Route Availability	The choice has the latest information available routing	Use when the link resource management system is very important and norms, the routing information to determine when, for example, satellite transmission services
Push Smallest	The choice of the length of the shortest information	The use of more important when large file transmission system or data, such as scientific data collection
Push Largest	The choice of the length of the longest information	Use when the system processing interactive application of small important information, such as a remote login service
Push Lowest Priority	Select the lowest priority information transfer	The high priority data transmission when the system is more important, such as the tactical network

2.1.2 *Node selection*

Select nodes closer to the congested nodes(Pan et al. 2008).

Under normal circumstances, the "hop count" is served as the measuring value, used to choose the destination nodes near the congested nodes. When SR invoking for a directed graph G=(V, E), k-hop neighbor node of the node v is defined as the node with k-hop distance from the node v(Leguay et al. 2005).

Definition 1:$N_v(k)$ is the collection of neighbor nodes with k-hop distance the node v in G, i.e.

$$Nv(k) = \{w | d(v,w) \leq k, w \in v\} \tag{1}$$

Where d(x,y) is the shortest path in all x→y paths, k refers to a range of for the "hop".

Definition 2: transfer cost from custody node v to transfer node c at the moment.

$$Mv,c(l,t) = Tv,c(l,t) + Sc(l,t)(1-\alpha) \tag{2}$$

Where, $T_{v,c}$ (l, t) is the transmission cost to move information with length l from node v to node c, S_c(l,t) is the cost of the node c to store information transferred from the node v with the length l.

Definition 3: transmission costs $T_{v,c}$ (l,t) is total time used to transfer information with the length l from the node v to the node c at the moment t; $T_{v,c}$(l,t) can be calculated by the link delay, bandwidth consumed and the on-off time spent by the transferred information.

$$Tv,c(l,t) = \log((Lv,c(t) + (l/Bv,c(t)))/(10\text{-}6))/10 \tag{3}$$

Where, $L_{v,c}$(t) represents the delay, and $B_{v,c}$(t) is the minimum bandwidth.

Definition 4: storage cost S_c(l, t) is the cost needed for storage of information with the length l transferred from the node c at the moment t. When each node receives and stores information from other nodes, it needs to determine the reserved available storage space(Zhang et al. 2007).

Storage cost normalization is calculated by the following formula:

$$Sc(l,t) = \begin{cases} 1-(Ax+Cx), l \leq Av \\ +\infty, l > Av \end{cases} \tag{4}$$

Where, Ax is the memory capacity of the node x at the moment t, C_x is the inherent memory capacity of the x.

Based on ERS rules and the formula, we calculate each of nodes c<N_v(k) in the k-hop range, calculating the transfer cost $M_{v,c}$(l,t).

2.1.3 *Selection of recovering*

Choose what information should be recycled by nodes.

3 CONGESTION CONTROL BASED ON PUSH-PULL STRATEGY

Currently, DTN custody model is based on the sender, which means all message transfer is initiated by providing Push operation. In order to avoid modifications to the overall routing algorithm, a Pull operation is used to extend the mechanism of bundle custody transfer, and form a new custody transfer mechanism. Under the new custody transport mechanism, the node holding the available storage resources sends request to other nodes to receive message custody, without affecting the whole network's routing system. Push-Pull custody transfer is implemented in the bundle layer, so this mechanism may fully take advantage of existing DTN custody transfer mechanism. Use Push-Pull custody transfer, it can transfer information to another storage node by Push, when congestion relief, news recycled Pull custody transfer operation when storage nodes occur congestion; after the congestion is released, Pull custody transfer operation is used to recycle the transferred messages,the original DTN routing is used to forward messages.

Study on the congestion control of the bundle layer mainly focuses on the congestion occurred at single nodes in the network, namely "node-level congestion control" in the traditional sense. Node-level congestion is judged by checking whether a single node may complete the custody transfer, and thus specific congestion control is completed by picking some bundles to transfer or discard, which conform to the conditions. Figure 1shows the time position of a node-level congestion occurred during the custody transmission process (Zhao et al. 2005).

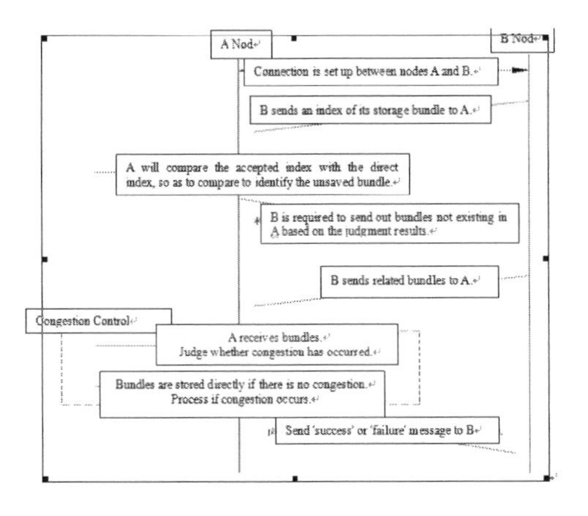

Figure 1. Congestion control in the custody transmission.

Push-Pull Congestion Control (PP) algorithm solves the congestion problem through the bundle transferring (Push Process) to neighboring nodes of the congested nodes, rather than discarded bundles at the nodes. When the bundle is required to forward, bundles are read back from the neighboring nodes through the process of reverse process Pull process Push read back from the bundle and forwarding.

4 CONGESTION CONTROL BASED ON THE SERVICE-LEVEL PUSH-PULL STRATEGY

4.1 *QoS service level*

QoS (Quality of Service) is a kind of network security mechanism, used to solve network latency and congestion. Service level refers to a group of end-to-end QoS capabilities; QoS usually supplies Best-Effort Service, IntServ (Integrated Service) and DiffServ (Differentiated Service) three service levels.

4.2 *Program design based on the Push-Pull congestion control algorithm in the service level*

To Push-Pull congestion control algorithm based on the service level (Tariq et al, 2006), first to the congestion detection. If there are no congestions, the bundles with high service level are directly stored in nodes by comparing with the threshold of service level, bundles with a low service level needs relative transfer operations, they are transferred to the smallest transferring cost in the node queue, and preset the storage space for high-level bundles received later; if congestion occurs, Push-Pull congestion control is performed. The basic process flow is shown in Figure 2.

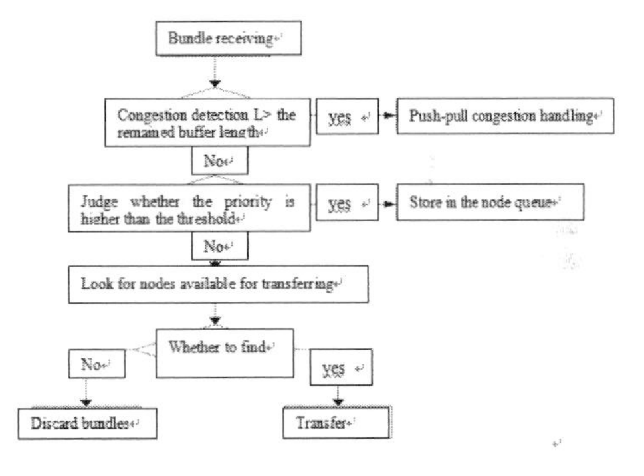

Figure 2. A flowchart of Push-Pull congestion control algorithm based on the service level.

4.3 *Network model*

Simulation of this paper uses a simple static routing in DTN routing(Shah et al. 2003). The simulation is based on the following assumptions:

The data source nodes have no flow of space mechanism; Each Bundle requires a custody transfer; Data can be discarded only when no other custody nodes are available; The storage capacity of the nodes is symmetric, that is any node may store data both in this node and transferred from other data;

Each link's bandwidth in the sending and receiving directions are symmetrical. The simulation topology is shown in Figure 1. There is a total of nine nodes divided into three domains, there are three nodes No. 0, No. 1and No. 2 in the Region 1; No. 3, No. 4 and No. 5 in the Region 2; No. 6, No. 7 and No. 8 in the Region 3. Link bandwidth is 1Mbps with 10ms delay.

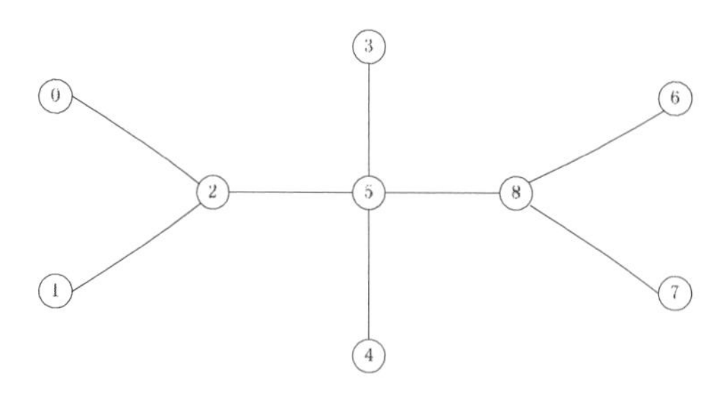

Figure 3. Topology.

4.4 *Simulation results*

The simulation uses the PushLowestPriority strategy. Simulations are made to the network transfer performance under two conditions with and without service level in the Push-Pull congestion control algorithm, from the influence of link availability rate to the information transfer rate.

Parameters are set as follows:

Traffic generator model: Poisson distribution model; the duration of data stream 1 is [1.5s, 9.5s], the priority is normal; duration of data stream 2 is [1.0s, 8.0s], the priority is expedited; load: 8000 bytes; link availability rate: 75%; the retransmission time is 0.5s; node cache size is 5000bytes; bundle size: 200 bytes. Bundle's valid survival time: 1s; time to terminal simulation: 50s; ML: 0.85 times of the node's maximum cache.

Simulation results are shown in figure 4 and figure 5.

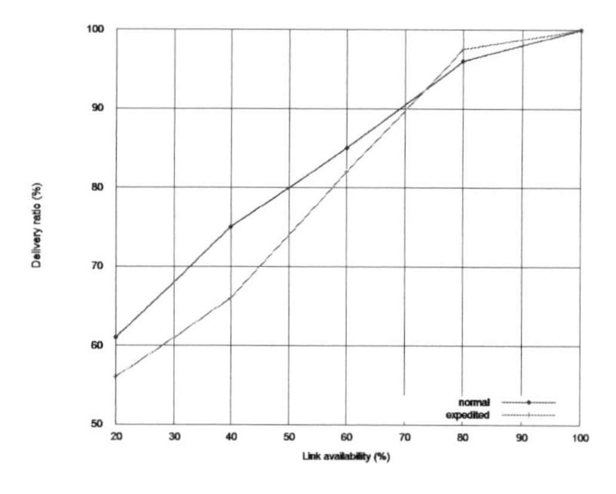

Figure 4. Comparison of the information transfer ratio with different priority to different link availability rate.

Figure 4 shows the information transmission rate with the expedited priority is higher than that with the normal priority, and greater amount of increase in the information transmission rate. When the link's availability rate reaches 100%, the likelihood of a node congestion is reduced along with the shortage of wait time, so that information transfer rates with different priorities all reach 100%.

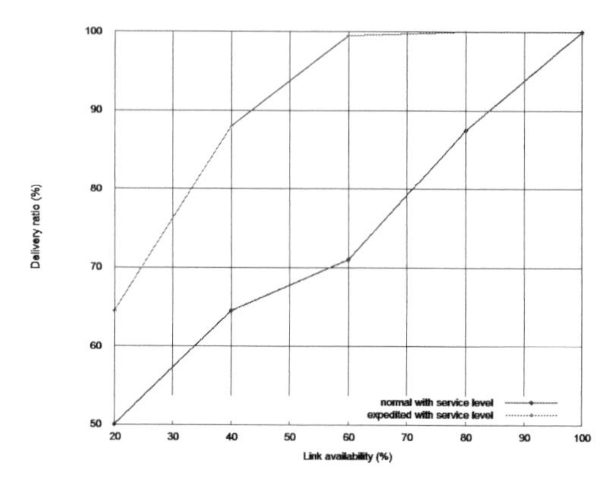

Figure 5 Change in the information transmission rate along with different link availability rate under the Push-Pull control algorithms with the service level.

Figure 5 shows when the link availability rate reaches 60%, information transfer rate with the expedited priority is close to 100%; the information transfer rate with normal priority relatively declines. When link available rates are up 100%, the transfer rates are getting closer and tend to be merged. This is mainly because the link is in the connected state, custody transfer can transmit information safely to the in the destination nodes without the impact of congestion.

5 ANALYSIS

According to the above simulation: the Push-Pull congestion control algorithm based on service level, can effectively reduce the transmission delay with high priority, and reduce the loss rate of network packets, as well as alleviate network congestion. However, bundle being discarded case still exists in the solution. Comparing with certain algorithms, the transmission delay has increased, so it needs to be improved.

There are obvious advantages in improving the end-to-end link information transmission rate and operational performance, especially for those with high priority. But on the contrast, it also reduces the performance of low priority.

6 CONCLUSION

This paper proposes Push-Pull control algorithm based on the service level aiming at deficiencies existing in SR algorithm. Through simulation analysis of the algorithm, it greatly reduces transfer delay of the high priority bundle, and delay of the low priority bundle is increased. The loss rate of this kind of bundles is increased under certain conditions, which is caused mainly due to the expiration of their life cycle.

Otherwise, the algorithm depends on availability of the network link, the connection nodes are needed to provide conditions for the transfer of bundles.

REFERENCES

Daly, E. & Haahr, M. 2007. Social network analysis for routing in disconnected delay-tolerant. *MANETs In: Proc of the ACM MOBIHOC.* New York: ACM.

Durst, R., Miller G. & Travis, E. 1996. TCP extensions for space communications. *In Proc. ACM MOBICOM 96*:15–26.

Jain, S., Fal,l K. & Patra, R. 2004 Routing in a delay tolerant network *Proc of the ACM SIGCOMM 2004.* New York : ACM.

Leguay, J., Friedman, T. & Conan, V. 2005. DTN routing in a mobility pattern space. *Proceeding of 2005 ACM SIGCOMM Workshop on Delay-Tolerant Networking.*

Pan Hui, Jon Crowerof & Eiko Yoneki. 2008. BUBBLE rap: Social-based forwarding in delay tolerant networks. *Proc of the ACM MOBIHOC*: 241–250. New York: ACM.

Shah, R., Roy, S. & Jain, S. 2003. Data MULEs: Modeling a three-tier architecture for sparse sensor networks *Proc of the 1st IEEE, 2003 IEEE Intl Workshop on Sensor Network Protocols and Applications*: 30–41.

Tariq, M.M.B., Ammar, M. & Zegura, E. 2006. Message ferry route design for sparse ad-hoc networks with mobile nodes. *Proc of the MobiHoc 200* Florence: ACM Press.

Zhang, X., Kurose, J. & Levine, B.N. 2007. Study of a bus-based disruption-tolerant network: Mobility modeling and impact on routing *Proc of the 13th Annual ACM Int'l Conf. on Mobile Computing and Networking*: 195–206 Montreal: ACM.

Zhao, W., Ammar, M. & Zegura, E. 2005. Controlling the mobility of multiple data transport ferries in a delay-tolerant network *Proceedings of IEEE INFOCOM 2005*, Miami, Florida.

Electronics, Communications and Networks IV – Hussain & Ivanovic (eds)
© 2015 Taylor & Francis Group, London, ISBN: 978-1-138-02830-2

Research on key management system in China Southern Power Grid

MingMing Zhang
Electric Power Research Institute.CSG, Guangzhou, China

Wenjun Gao*, Xin Xia & Fenglong Wang
Beijing Nari Smartchip Microelectronics Technology Company Limited, Beijing, China

ABSTRACT: This paper analyzes the metering safety status of China Southern Power Grid and proves that the establishment of the China Southern Power Grid safety, cost control protection system could guarantee the electricity information data transmission security. It is suggested that a three level key management system according to the power user measurement, automation safety system design and the construction be in line with the application based on the key safety management system. This will ensure the confidentiality of data public network transmission, the security of data storage and the safety and stability of the system operation of the system.

1 INTRODUCTION

The five provinces of China Southern Power Grid have all started the construction of measurement automation system. Wireless public network channels are used in the remote meter reading of special transformer, public transformer and low-voltage customers. While special lines, integrated data network, dialing and etc. are employed in the remote meter reading of measuring substation. In general, the safety protection system belongs to the unified management of the information management area III or control zone II of the power supply bureau. Currently, encryption is rarely applied in existing electric energy meters and measurement automation terminal technologies. Although encryption measures have been applied in some collection terminals, yet they are simple password authentication mechanisms based on the application layer protocol, which might lead to serious problems as follows.

1 It cannot provide effective protection of the important parameter data, the execution of control commands and etc.
2 There might exist an electricity tampering risk which would affect the accuracy of settlement of electricity.
3 There exists a high security risk in the remote controlling of electric power meter prepayment and large user peak load shifting management.
4 It is unable to adapt to the development requirements of intelligent electricity, intelligent home, new energy access and other smart grid technologies in the future.

Therefore, it is urgent to carry out the construction of energy metering automation cost control management system, which will implement remote electric data acquisition, remote control, remote setting of parameter and rate as well as other functions in the advanced meter secure management. Besides, it will not only provide convenient and efficient ways to achieve tiered electricity pricing, time-of-use electricity pricing and peak-valley electricity pricing, but also provide support for basic data in the field of dispatching and distribution as well as other related business field. As a result, high quality electrical services for controlling the electricity balance use and electrical safety can be eventually achieved.

2 SOUTHERN POWER GRID SAFETY COST CONTROL PROTECTION SYSTEM

The general framework of the Southern Power Grid security protection follows the international IEC62351 sanders, the requirement of smart grid network safety protection, the general requirements of "security protection for secondary system of power grid"(SERC order No. 5) by the State Electricity Regulatory Commission and "The guidance for grading of information system security levels in electric power industry". Combining with requirements of electricity measurement, automation system security protection applications (Jing et al. 1999), the electricity measurement security protection framework (Yun et al. 2010, Qiang et al. 2003) is proposed by analyzing the attacking mechanisms of each step, including

*Corresponding author: *gaowenjun@sgepri.sgcc.com.cn*

collection, processing, polymerization, communication, calculation. Security protection goals and technology development route have been clearly defined, which not only ensures the security of load controlling and data measurement, but also reduces the complexity and cost of security protection measures. Focusing the information protection on the key management system, a symmetrical algorithm encryption security module has been added to the cost controlling electric energy meter. At the same time, an asymmetrical one has been added to the station terminal, load control terminal and the concentrator. Eventually, the overall architecture of the complete cost control, security protection system ISO/IEC17799 & ISO/IEC27001 (2005) is as follows.

Southern power grid cost control security protection leveling system protects each system in its corresponding security domain. Furthermore, to implement security protection measures, it also divides each security domain into network boundary, network environment, host computer system and application environment. The Southern power grid will then provide technical and management support from four aspects, including standards, key systems, management systems and business processes, which ensures the security protection capability of the entire system.

3 KEY MANAGEMENT SYSTEM DESIGN

3.1 *Key management module design*

Southern Power Grid Key management system consists three levels (Gao 2012) and provides unified, secure key management of the application system. This system adopts a three-layer architecture, which implies the master key of the lower level is generated in the upper level and is encrypted by the cryptographic machine before being delivered to the next level. Considering the practical construction needs of provincial companies, the key management system will provide either two-level or three-level architecture according to hierarchical structure design.

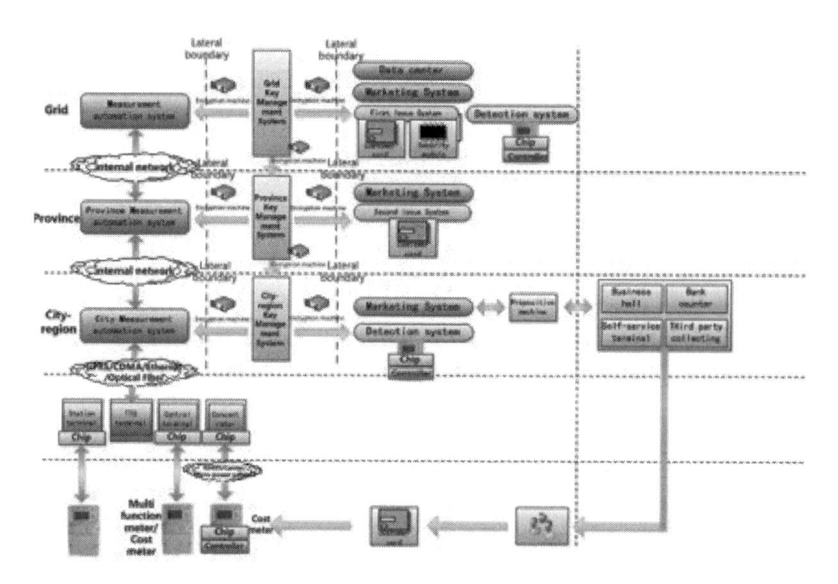

Figure 1. The entire framework of cost controlling security protection.

The provincial and municipal key is generated by the Grid key management system. These keys are encrypted and signed before being delivered to the lower key management system via IC cards or cryptographic machines. Boot cards and authorization cards are used to implement the security authentication and control mechanism during the key distribution. By confining the plaintext in the cryptographic machine, the keys are always secured during the distribution, which enhances the security of the electricity sales platform of the Southern Grid marketing system as well as the measurement, automation system. In the three-level architecture model, the first, second and third level are the grid, provincial and municipal key management system, respectively. While in the two-level model, the first and second level are the grid and provincial-municipal key management system The overall architecture of the system is shown in Figure 2.

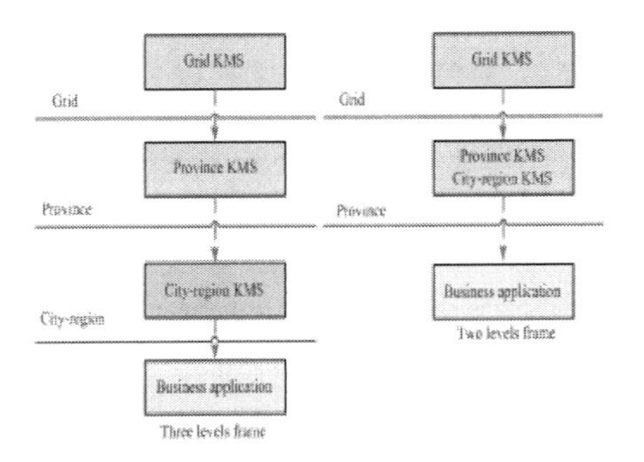

Figure 2. Overall architecture of the key management system.

Considering the system scalability and maintainability in the future, the provincial-municipal key management system of the two-level architecture model is merged from the provincial and municipal key management system of the three-level model.

3.1.1 *Three-level architecture model*

The grid key management system is the first level system of the Southern power grid electricity measurement information key management system, which includes the grid key management system software and hardware. It is responsible for key generation, distribution, transmission, storage, backup, restoration, management and maintenance. It generates the root key of the grid key management system as well as the, provincial root key management system based on the provincial code. With the generated keys, the grid key management system the initializes secure modules, user cards and cryptographic machines.

The provincial key management system is the second level system of the Southern power grid electricity measurement information key management system, which includes provincial key management system software and hardware. It is responsible for key acceptation, transmission, management and maintenance. It is built on cryptographic machines on the provincial key management system generated by the grid key management system. And it generates municipal key management system root keys based on the municipal code.

The municipal key management system is the third level system of the Southern power grid electricity measurement information key management system, which includes municipal key management system software and hardware. It is responsible for key acceptation, application, management and maintenance. It also generates business application keys and delivers them to the card issuing systems, detection systems, master station acquisition systems and electricity sales systems through the cryptographic machine.

3.1.2 *Two-level architecture model*

The grid key management system is the first level system of the Southern power grid electricity measurement information key management system, which includes grid key management system software and hardware. It generates the root key of the grid key management system as well as the, provincial root key management system based on the provincial code, same with the three-level model as above.

The provincial-municipal key management system is the second level system of the Southern power grid electricity measurement information key management system, which includes provincial and municipal key management systems software and hardware. It is responsible for key acceptation, transmission, management and maintenance. It builds provincial and municipal key management systems from the cryptographic machine of the provincial and municipal key management systems delivered by the grid key management system. Besides, it also generates business application keys and delivers them to the card issuing systems, detection systems, master station acquisition systems and electricity sales systems through the cryptographic machine.

3.2 *Key type design*

Keys are divided into master keys, root keys and application keys based on their importance and level architecture as shown in table 1.

Table 1. Key types of key management system.

Key type	Source	Algorithm	Purpose
Master key	Randomly generated by the cryptographic machine and backed up in the security card.	SM1/SM4 algorithm	To encrypt key transmission and back up, and import and export keys.
Root key	Generated by the random number of the cryptographic machine and the leading factor, backed up in the security card.	SM1/SM4 algorithm and SM7 algorithm	To generate application keys.
Application key	Directly distributed by root key.	SM1/SM4 algorithm and SM7 algorithm	To realize all kinds of supporting functions and act as a working key for business data encryption and decryption.

3.2.1 *Master key*

Master keys include transmission master keys of the grid, provincial and municipal levels, and are used for key distribution. The root key protection is realized by encrypting the cryptographic machine of root key import and export of the corresponding level. The transmission master key is randomly generated by the hardware noise source of the cryptographic machine, exported by cryptographic machine in the ciphertext and backed up. The transmission master key of the lower level is distributed by the corresponding regional dispersion code of master key of the upper level.

The grid master key is generated and backed up for the first time when the key management system is used, which can be recovered when necessary. If there is a key leakage, the grid master key can be either destroyed or updated by the grid key management system and cryptographic machine. The master keys of provincial and municipal levels are generated by the upper level key management system. They will be destroyed there is a key leakage and updated by the upper level key management system and cryptographic machine.

3.2.2 *Root key*

The root key includes business key and application key of all levels. The business key is mainly used to generate the business root key of the lower level and the application root key the same level, while the application root key will then generate the application keys. The grid business root key is generated by a random number and the grid leader factor, which is then distributed in provincial and municipal levels, so that the three level architecture of the business root key can be established as shown in Table 2.

Table 2. Root key types.

Key level	Business root key name	Source	Purpose
Grid	Grid level business root key	Distributed by a random number and the leader factor	Provincial level business root key generation
Provincial	Provincial level business root key	Distributed by the grid level business root key	Municipal level business root key generation
Municipal	Municipal level business root key	Distributed by the provincial level business root key	Same level business application root key generation
	Application root key	Distributed by the business root key	Application key generation

The grid level root key is generated by a random number and the grid level leader factor. It is backed up and recovered by the key management system, and gets destroyed and updated by the grid key management system and cryptographic machine when there is a key leakage. The root keys of provincial and municipal levels are distributed by the upper level key management system. The grid level root key generates the business root keys with the participation of the leading factor of the same level. The grid level root key will be destroyed by the cryptographic machine when it is leaked and updated by upper level key management system and the lead factor of the same level.

3.2.3 *Application key*

Application keys are working key for encryption and decryption of business system data. They are distributed by the dispersion code of the application root key, according to their functions and they are stored in the application cryptographic machine. In the process of business system application implementation, the corresponding application key distributes the application subkeys according to the meter number and the card serial number. As a result, the data encryption and decryption data using the application subkeys guarantees the "one meter one key, one card one key" configuration in the application process.

Application keys are distributed by the municipal key management system according to the intended business dispersion code, and then transmitted to each application system by the cryptographic machine, which can be destroyed and updated when it is leaked.

3.3 *Key management system design*

The Southern grid key management system consists consisting of the client, server and cryptographic machine. It adopts the B/S mode design of the distributed architecture, implements data communication between the server and cryptographic machine in an application layer protocol specified by the cryptographic machine. The technical architecture of the key management system is shown in Figure 3.

The Southern grid key management system is the core part of the safety protection system. It is centered on cryptography, and strictly follows the safety design standards. It is designed in accordance with the three-level key management system, and is based on the security module, smart card and cryptographic machine. This will guarantee the confidentiality and safety of data transmission and storage, and will provide security and stability to measurement, automation systems, electrical sales system and other business application systems.

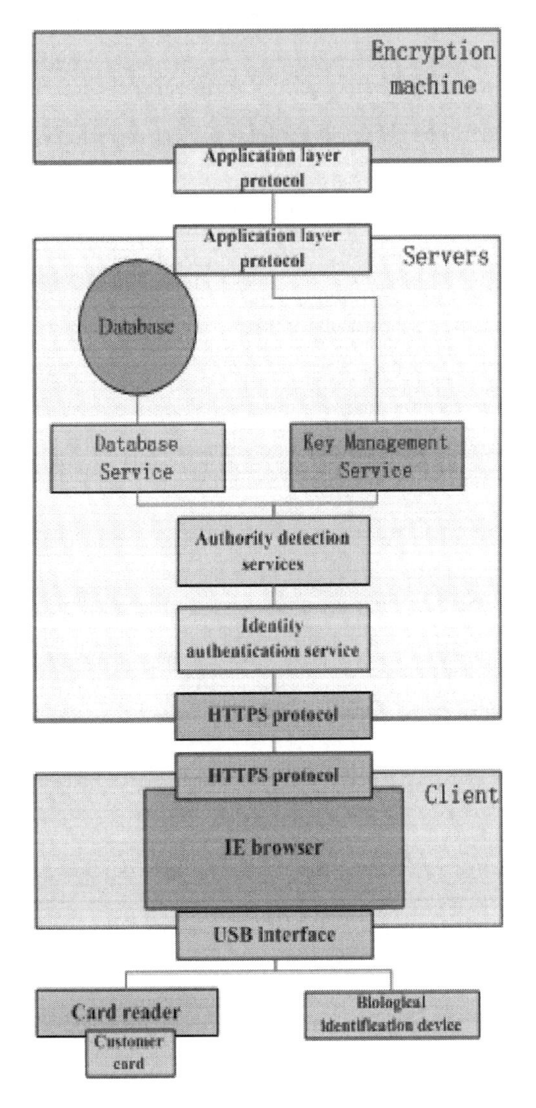

Figure 3. Technical architecture of the key management system.

4 CONSTRUCTION OF KEY MANAGEMENT SYSTEM

According to the province construction requirements of the China Southern Power Grid Key Management System, it adopts a three-level architecture pattern including grid, province and prefecture-level city. As described in Section 3.1, it deploys in the advanced security machine rooms of all patterns, respectively. There are no direct data interactions between the upper level and lower level key management or among the same level key management systems and the keys are passed by the cryptographic machine.

4.1 *Gird level key management system*

The Gird level key management system is mainly responsible for key generation, transmission, management and maintenance. It generates the root keys of the key management system of both the grid level and province level using the corresponding code. The Grid level key management system employs the business key to achieve the issue of the province level key management system, cryptographic machine, grid generation and test the cryptographic machine.

The Gird level key management uses the hardware noise source of the cryptographic machine to generate random numbers and the grid level root keys using specific algorithms. It adopts the leader seeds of the China Southern Power Grid to protect the backup of the grid root key. The system makes use of the input province code, generation code and test code to produce disperse province root key, generation and test keys, which are stored in the province, generation and test cryptographic machines, respectively. It also passes the province cryptographic machine to the province key management system.

4.2 *Province key management system*

The Province key management system is mainly in charge of key receiving, transmission, management and maintenance. By using the province system cryptographic machine, it can build province key management and generate disperse root keys of prefecture cities based on their codes. Meanwhile, the system also achieves the issue of the prefecture city cryptographic machine.

4.3 *Prefecture-level city key management system*

The functions of Prefecture-level city key management system include key receiving, transmission, management and maintenance. It builds Prefecture-level city key management system using the cryptographic machine. The system is based on the business needs of issuing corresponding keys into tool cards (including parameter preset card and local parameter card) and business application cryptographic machines (including metering automation, electricity sale, card generation and test cipher machines) by binding the applications of business root keys.

5 CONCLUSIONS

This paper conducts a study of metering equipment and corresponding system functions based on the security expense control construction needs and transformation technology requirements of the China Southern Power Grid. The research results can be applied in the processes of development, installation and maintenance of various terminal metering equipment and smart meters for companies. It can also be used in the development and construction of metering automation systems, sales systems, card issuing systems and key management systems. As a result, it lays a firm foundation for the comprehensive construction and application of security expense control management system in China Southern Power Grid.

REFERENCES

Jing, Jia, Yuan, Chen & Lina, Wang. 1999. *Security and confidentiality of information systems*. Beijing: Tsinghua university press.

Yun, Bai. 2010. Line of develepment and future trend of information system security architecture. *Journal of air force engineering university* 11(5): 75–80.

Qiang, Yan. 2003. *Research on the evaluation of information system security*. Beijing: Department of computer science and technology of Peking University.

ISO/ IEC 17799. 2005. Rules for the implementation of information technology, Security techniques, Information Security Management.

ISO/ IEC 27001. 2005. Information technology, Security techniques, information security management systems, Requirements.

Zhijiang, Gao. 2012. Design and implementation of key management system. *Software engineering of Beijing*. Beijing: University of Posts and Telecommunications.

Electronics, Communications and Networks IV – Hussain & Ivanovic (eds)
© 2015 Taylor & Francis Group, London, ISBN: 978-1-138-02830-2

Provable secure broadcast protocol for wireless sensor networks and RFID systems

Leyou Zhang*
School of Mathematics and Statistics, Xidian University, Xi'an, Shaanxi, China
School of Computer Science and Software Engineering, University of Wollongong, NSW, Australia

Zhuanning Wang
School of Mathematics and Statistics, Xidian University, Xi'an, Shaanxi, China

Yi Mu
School of Computer Science and Software Engineering, University of Wollongong, NSW, Australia

ABSTRACT: In Wireless Sensor Networks (WSNs), Broadcast Protocol (BP) can support their nodes to share their local information and transmit data by one-to-all or all-to-one (all). But the existing schemes are based on symmetric cryptosystem, which results in a single point of failure and lack of dynastic management. And it also results in serious redundancy and collision. How to overcome these limits is a main challenge problem. We introduce a new BP in this paper. This scheme is constructed based on an asymmetric cryptosystem which overcomes parts of the previous shortcomings. It has many advantages over the available, such as a good trade-off between the transmitted data size and the decryption key size, low amount of memory and storage space for the node. Finally, we also show it can be used in RFID systems.

1 INTRODUCTION

WSN has been used widely in the real life. It consists of many microelectronic devices where they are called sensor nodes. WANs need to broadcast some operations such as network queries, messages and command dissemination. In WANs, a node is a microelectronic device which has limited energy, processing power and memory. Hence a WSN has severe resource constraints. In addition, whenever the contents are broadcasted, all open devices can learn them. Protection for the content of the communication has widely recognized a crucial problem in the open environment. Furthermore, WSNs also need self-organizing (Sohrabi et al. 2000) protocols. These protocols support self healing, fault tolerance, and security features. However, as a limited hardware resource, a node in WSN cannot achieve all these features at the same time.

Broadcast is an essential feature in WSNs for critical operations such as network query, software updates, and network management. It will be helpful to transmit encrypted data which is recovered by a predefined node (or nodes). In addition, broadcast can support manager to control strictly selected nodes (Szalachowski et al. 2011) or configure them. Hence broadcast encryption (BE) protocol is an ideal proposition for WSNs (Shaheen et al. 2007). Two settings for BE are introduced, the symmetric key and public key settings. Most works focus on the symmetric BE (Szalachowski & Kotulski 2012, Durresi et al. 2007, Mann et al. 2007, Sheu et al. 2006). In a symmetric setting, a trusted center is needed. It takes all important roles such as extracting the secret key, transmitting the information to users. So a single point of failure (Park et al. 2008) appeared. But in an asymmetric setting (Public key system-PKS), any node can send the messages to the rest of the nodes, which overcomes the above shortcoming. Furthermore, in a PKS, the public parameters can be stored a public board (public device) which can be accessed by the users. It decreases the transmitting cost and storage cost. There are some others advantages over symmetric setting, such as public verification by Users et al. Recently, some works based on public key setting have been introduced such as (Li et al. 2010). Another disadvantage in symmetric BE is lack of dynamic user management (Mu et al. 2004). But in a PKS setting, a node encrypts a message using a unique key and only legal receiver can recover the message using the corresponding private key. Therefore, a PKS BE supports the dynamic management.

*Corresponding author: *xidianzhangly@126.com*

However, as a restricted device, a sensor has a limited amount of memory and storage. For constructing an efficient and secure system, it is needed to keep the size of the algorithm. Hence, a more natural idea is combining the advantages of two setting into one scheme. This motivates us to construct an efficient BE protocol based on symmetric and asymmetric settings.

Our Contribution. Combining the fast symmetric algorithm and the dynamic asymmetric algorithm, a new hybrid broadcast protocol is introduced. The broadcasting node encrypts the session key and this key will be used to encrypt a message with a symmetric algorithm such as AES. The new scheme achieves a good trade-off between the transmitted data size and the secret key size. It decreases the computation cost and storage cost for the code. In addition, it supports dynamic sensor networks. When a new node joins, it does not need to renew the system and only add partly computations in private keys. On the contrary, if some nodes are revoked, it also needs to make partly computations instead of reconstructing the system. Finally, we also show it can be used to an RFID system.

2 PRELIMINARIES

2.1 *Decision $q+1$-BDHE problem*

Given a tuple $(g, y_0, y_1, \cdots, y_q, T)$, where $y_i = g^{\alpha^i}$ and $y_0 = g^c$, decide T is random in G_1 or equal to $e(g,g)^{\alpha^{q+1}c}$.

Definition 1 We call a(t,ε)-decisional $q+1$-BDHE assumption holds if an adversary has only a negligible advantage ε in solving the game at most t-time.

2.2 *Identity-based broadcast encryption scheme*

Our protocol is based on the identity-based encryption (IBE) which is introduced firstly by Shamir (Shamir 1984). Broadcast Encryption (BE) (Zhang et al. 2012) allows a broadcaster to encode a message to some subset S of users and sends the ciphertexts to the users. Any legal receivers from *S*can use the received secret key to decrypt the ciphertexts. But illegal users could not recover the encrypted data. An IBBE with the maximal size$m$of the target set is given as follows.

Setup This algorithm takes in the security parameter, outputs public parameters and a master key.

Extract The algorithm takes in the master key, an identity and computes a corresponding private key.

Encrypt This algorithm takes in the public key and a set of included identities $S = \{ID_1, \ldots, ID_s\}$ $s \le m$, and outputs (Hdr, S, C_M).

Decrypt This algorithm takes in all corresponding parameters. If $ID_i \in S$, it recovers K at first, then decrypts the ciphertext C_M and finally outputs plaintexts.

2.3 *The BE security-model*

The theoretical analysis of the proposed protocols will use the security model of IBBE. Following (Szalachowski et al. 2012, Durresi et al. 2007, Mann et al. 2007, Sheu et al. 2006), we give the IBBE security model as follows:

Setup The challenger issues a call to *Setup* and obtains public keys *PK*. Then it sends *PK* to the adversary A.

Phase 1 In this phase, A adaptively takes some queries, where each query is given as follows.

Key query for ID_i. A takes a query for ID_i. Then the challenger issues a query to *Extract* on ID_i and gets corresponding secret keys. Finally, the challenger sends secret keys to A.

Challenge In this phase, A submits two messages M_0 and M_1 with the equal size. The challenger chooses a random bit $b \in \{0,1\}$ at first. Then it takes a call to *Encrypt* algorithm to get the ciphertext $C^*(M_h)$. The challenger sends the results to $\mathcal{A}$.

Phase 2 A continues to make others queries, and each one is described as follows.

Key generation query for ID_j as shown in phase 1, where $ID_j \notin S^*$.

Guess A outputs a bit $b' \in \{0,1\}$. If A succeeds, it means $b = b'$.

The advantage of A succeeding in the above game is defined as follows:

$$Adv_{IBBE}(t, m, A) = | 2P(b = b') - 1 |.$$

where t is the total time for key query.

If a scheme matches the above, we call it achieves $IND - full - CPA$ security.

3 THE PROPOSED BROADCASTING PROTOCOLS

3.1 *Constructions*

Initialization Let $S = (ID_1, \cdots, ID_n)$ denote the identity set of all notes in sensor networks, where $ID_i \in Z_n^*$. Select random generator $g \in G$ and some random elements $g_2, g_3, h, h_1, h_2 \in G$. Then pick randomly $\alpha \in Z_P^*$ and set $g_1 = g^\alpha$. Choose a function $f(x) = ax + b$ with the restriction $g_2 \ne g_3^{-a}$, $h_0 \ne g_3^{-b}$, $a, b \in Z_p^*$.The public keys are

$$PK = (g, g_1, g_2, g_3, h, h_1, h_2, f(x))$$

and the private keys for any ID_i S are

$$d_{ID_i} = (d_{i1}, d_{i2}, d_{i3}) = ((hg_2^r g_3^{f(r)})^\alpha (h_2 h_1^{ID_i})^{r'}, r, g^{r'})$$

where $r, r' \in Z_n^*$. It is worth noting that$\alpha$is kept secret by *PKG*.

The Initialization algorithm can be achieved before a WSN is formed.

Let $S_1(S_1 \subseteq S)$ denote a set nodes identity. We make BE protocol as follows.

Group key Generation Given the PK as described in Initialization algorithm, each nod of S_1 can complete the following computation. It selects randomly $t \in Z_n$ at first and computes $T_1 = g^t, T_2 = g_2^t, T_3 = h^t, T_4 = g_3^t$. Without loss of generality, we set $S_1 = (ID_1, \cdots, ID_s)$. Then for $i = 1, \cdots, s$, the member computes

$$x_i = F(\hat{e}((h_2 h_1^{ID_i})^t, g^{r'})),$$

Then it constructs

$$f(x) = \prod_{i=1}^{s}(x - x_i) = \sum_{i=0}^{s} a_i x^i$$

where a_i denotes the coefficient corresponding to x^i and $x \in Z_p$. Then he or she computes $(h_0, \cdots, h_s)$ as $h_0 = g^{a_0}, \cdots, h_s = g^{a_s}$. Hence, each member of S has an encryption key tuple $(h_0, \cdots, h_s, T_1, T_2, T_3, T_4)$. The keys $(h_0, \cdots, h_s, T_1, T_2, T_3, T_4)$ is fixed except that any node is deleted from S_1or new nodes join in.

Data Transmission If members in S_1 want to transmit a message each other, then the algorithm is constructed as follows:

Let K denote a session key. A broadcaster in S_1 selects a random $k \in Z_n^*$ and computes $C_0 = Kh_0^k$, $C_1 = h_1^k, \cdots, C_s = h_s^k$. Finally, the broadcaster outputs the header $Hdr = (C_0, C_1, \cdots, C_s, T_1, T_2)$ and broadcasts it to the rest of notes.

Data Recovering In order to retrieve the message encryption key K encapsulated in the header Hdr, the user with identity ID_i and the corresponding private key $d_{ID_i} = (d_{i1}, d_{i2}, d_{i3}) = ((hg_2^r g_3^{f(r)})^\alpha (h_2 h_1^{ID_i})^{r'}, r, g^{r'})$ computes (with identity $ID_i \in S_1$)

$$x_i = F(\hat{e}(d_{i1}, T_1) / (\hat{e}(g_1^{d_{i2}}, T_2)\hat{e}(g_1, T_3)\hat{e}(g_1^{f(d_{i2})}, T_4))),$$
$$K = C_0 \prod_{j=1}^{s} C_j^{x_i^j}.$$

Correctness. Assume the Hdr is well-formed for S_1. Then one can obtain

$$\begin{aligned} & F(\hat{e}(d_{i1}, T_1) / (\hat{e}(g_1^{d_{i2}}, T_2)\hat{e}(g_1^{f(d_{i2})}, T_4))) \\ &= F(\hat{e}((hg_2^r g_3^{f(r)})^\alpha (h_2 h_1^{ID_i})^{r'}, g^t) / \\ &\qquad (\hat{e}(g_1^r, T_2)\hat{e}(g_1, T_3)\hat{e}(g_1^{f(r)}, T_4))) \\ &= F(\hat{e}((h_2 h_1^{ID_i})^{r'}, g^t)) = x_i \end{aligned}$$

and

$$C_0 \prod_{j=1}^{s} C_j^{x_i^j} = Kh_0^k \prod_{j=1}^{s} g^{ka_j x_i^j} = Kg^{k\sum_{j=0}^{s} a_j x_i^j} = (g^{f(x_i)})^k K.$$

In fact $f(x_i) = 0$, and then we have $g^{f(x_i)} = 1$.

3.2 *Performance*

When a WSN is reformed, the proposed BE only deletes or adds the new member's identity to the beginning algorithm. It is more efficient than the existing works. Furthermore, it also achieves a good trade-off between the transmitted data size and the decryption key size. The detailed comparisons are given in Table 1.

Table 1. Comparisons between ours and the existing works.

Scheme	Pk	Storage	SM	Dyn.	Model
Shheen 2007	$O(\|S\|)$	$O(1)$	RO	No	Sy
Szalachowski 2012	$O(\|S\|)$	$O(1)$	RO	No	Sy
Li 2010	$O(1)$	$O(\|S\|)$	RO	No	Sy
Proposed scheme	$O(s)$	$O(s)$	SD	Yes	Asy

In Table 1, pk denotes private key. SM and Dyn denote security model and dynastic feature respectively. Sy and Asy are symmetric setting and asymmetric setting separately. OR and SD denote random oracle and standard model respectively.

In (Boneh & Boyen 2004, Canetti et al. 1998), authors have shown the issued schemes in the OR are insecure in the real life.

3.3 *Performance*

Now, we evaluate performance of the presented scheme. The new scheme is focused on the security and dynastic feature instead of single computation efficiency. Compared with the existing works, the new scheme is based on the asymmetric encryption system. So the computation cost will be somewhat higher than the available. In addition, a bilinear pair is used in the phase of group key generation. The computation cost of pairing operations is about 11110 multiplications. Let its security level be 128 bits. For tests we chose an assembler implementation (Szalachowski et al. 2010) of the AES (Daemen et al. 2002) block cipher. Let $n = 300, s = 100$. In (Szalachowski et al. 2011), the cost of all computation takes about 40 ms on each node for the block cipher. But the proposed scheme takes about 200 ms (the DBDH group was issued from the curve $E / F_{3^{163}}$). The experiments are run on a personal computer with core i5-3570 (3.4 GHz) and a maximum of 4.0 GB of the memory available.

4 SECURITY ANALYSIS

Next, we will show the proposed scheme is how to achieve the $IND-full-CPA$ security based on the hardness assumption. Let q be the maximum times of queries, τ be a multiplication performed time and ρ be an exponentiation performed time.

Theorem 4.1

Under the (t, ε) decision BDHE assumption, the introduced protocol is (t', ε, q)-$IND-full-CPA$ secure, where $t = t' + O(q\tau + q\rho)$.

Proof:

If there is an adversary A against the new protocol with advantage ε, then we will construct an algorithm B that has a same advantage to solve the hardness problem. It is described as follows.

Setup B constructs functions

$$f_i(x) \in Z_p^*[x], 1 \le i \le 3$$

with degree q, where

$$f_1(x) = \sum_{i=0}^{q} c_i x^i, f_2(x) = \sum_{i=0}^{q} b_i x^i,$$

$$f_3(x) = \sum_{i=0}^{q} c_i x^i, f(x) = -\frac{a_q}{c_q} x - \frac{a_q}{c_q}.$$

It sets $g_1 = g^{\alpha}, h = g^{f_1(\alpha)}, g_2 = g^{f_2(\alpha)}, g_3 = g^{f_3(\alpha)}$.

If $g_2 = g_3^{\frac{a_q}{c_q}}$ or $h = g_3^{\frac{a_q}{c_q}}$. Then it issues $f_i(x) \in Z_p^*[x], 1 \le i \le 3$ again. B continues to choose $\mu_0, \mu_1 \in Z_q^*$ and sets $h_1 = g^{\mu_0}, h_2 = g^{\mu_1}$. Finally, it transmits

$$PK = (g, g_1, g_2, g_3, f(x), h, h_1, h_2) \text{ to } A.$$

Queries The adversary A will issue private keys queries and B responds as follows.

Extract queries: Let A issue a query for an identity $ID_i \in S_1$, where S_1 is a set of notes. B sets

$$d_{i1} = (g^{\sum_{i=0}^{q-1}(a_i + rb_i + f(r)c_i)\alpha^{i+1}})(h_{l+1} \prod_{i=1}^{k} h_i^{v_i})^{r'},$$

$$d_{i2} = r, d_{i3} = g^{r'},$$

where $g_2^r g_3^{f(r)} \ne 1, r' \in Z_p^*$. It is simulated perfectly. In fact, $f(r) = -\frac{a_q}{c_q} r - \frac{a_q}{c_q}$ and $a_q + rb_q + f(r)c_q = 0$. And then

$$d_0 = (g^{\sum_{i=0}^{q-1}(a_i + rb_i + f(r)c_i)\alpha^{i+1}})(h_2 h_1^{ID_i})^{r'}$$

$$= (g^{\sum_{i=0}^{q}(a_i\alpha^i + rb_i\alpha^i + f(r)c_i\alpha^i)})^{\alpha}(h_2 h_1^{ID_i})^{r'}$$

$$= (h_0 g_2^r g_3^{f(r)})^{\alpha}(h_2 h_1^{ID_i})^{r'}.$$

It selects randomly $t \in Z_p$ at first and computes $T_1 = g^t, T_2 = g_2^t, T_3 = h^t, T_4 = g_3^t$. Without loss of generality, we set $S_1 = (ID_1, \cdots, ID_s)$. Then for $i = 1, \cdots, s$, the member computes $x_i = F(\hat{e}((h_2 h_1^{ID_i})^t, g^{r'}))$ and $f(x) = \prod_{i=1}^{s}(x - x_i) = \sum_{i=0}^{s} a_i x^i$, where a_i is the coefficient of x^i and $x \in Z_p$.

Finally it sets $(h_0, \cdots, h_s)$ as $h_0 = g^{a_0}, \cdots, h_s = g^{a_s}$. Hence, each node of S_1 has an encryption key tuple $(h_0, \cdots, h_s, T_1, T_2, T_3, T_4)$.

At the challenge phase, B will generate the challenge ciphertexts for $S^* = \{ID_1^*, \cdots, ID_s^*\}$ that the adversary A will attack. All the identities in S^* are not queried in the previous. B sets $T_1 = g^c$ and runs the Private Key Generating algorithm to compute $d_{ID_i^*} = (d_{i1}^*, d_{i2}^*, d_{i3}^*)$ as the previous method and sets $x_i = F(\hat{e}(d_{i1}^*, T_1)/T)$ and $g_2^{d_{i2}^*} g_3^{f(d_{i2}^*)} = 1$.

If B got a valid BDHE tuple, it means $T = e(g, g)^{\alpha^{n+1}c}$. In fact,

$$x_i = F(\hat{e}(d_{i1}^*, T_1)/(T\hat{e}(g^{\sum_{i=0}^{q-1} a_i\alpha^{i+1}}, T_1)))$$

$$= F(\hat{e}((hg_2^{d_{i2}^*} g_3^{f(d_{i2}^*)})^{\alpha}(h_2 h_1^{ID_i})^{r'}, g^c)/$$

$$(\hat{e}(g, g)^{\alpha^{n+1}c}\hat{e}(g^{\sum_{i=0}^{q-1} a_i\alpha^{i+1}}, T_1)))$$

$$= F(\hat{e}(g^{\alpha f_1(\alpha)}, g^c)\hat{e}((h_2 h_1^{ID_i})^{r'}, g^c)/$$

$$(\hat{e}(g, g)^{\alpha^{n+1}c}\hat{e}(g^{\sum_{i=0}^{q-1} a_i\alpha^{i+1}}, T_1)))$$

$$= F(\hat{e}((h_2 h_1^{ID_i})^{r'}, g^c)).$$

They yield valid encryption keys for S^*. Then B randomly picks $b \in \{0,1\}$, sets $K_b = K$. B makes a call to *Encrypt* algorithm and gets the Hdr^*. Hence Hdr^* is a valid simulation. If T is a random element of G_1, then B sets K_{1-b} to a random value. Finally, B returns Hdr^* to.

In the Phase 2, A will continues as in phase 1 with $ID_i \notin S^*$.

Finally, A outputs a bit $b' \in \{0,1\}$ and succeeds in tis the game when $b' = b$.

If A succeeds, it shows that B can decide $T = e(g, g)^{\alpha^{q+1}c}$ or a random element of G_1, which means B successfully solves the hardness assumption.

If T is a random element in G_1 then $\Pr[B(U, T) = 0] = \frac{1}{2}$. Otherwise $T = e(g, g)^{\alpha^{q+1}c}$, it means $|\Pr[b = b'] - \frac{1}{2}| \ge \varepsilon$. So, B has

$$|\Pr[B(U, e(g, g)^{\alpha^{q+1}c}) = 0] - \Pr[B(U, T) = 0]|$$

$$\geq |(\frac{1}{2} \pm \varepsilon) - \frac{1}{2}| \geq \varepsilon.$$

The time complexity of the algorithm B is dominated by the *multi.* and *expo.* performed in the queries phase. Hence the time complexity of B is $t = t' + O(q\tau + q\rho)$.

5 APPLICATIONS IN RFID

In a RFID tag encryption system, BE protocols encrypt the identity of an RFID tag (Jin et al. 2009). The BE stores corresponding information in the server, which makes the reader to read out the RFID tag by decrypting the encrypted ID based on the encrypted ID information.

The protocol in (Jin et al. 2009) cannot support dynamic setting. Our scheme overcomes this limit.

6 CONCLUSIONS

In this paper, a new broadcast protocol is proposed. It supports dynastic setting of wireless sensor networks, where some nodes join or are revoked, public parameters do not need to be renewed and only encrypted keys need to be computed. In addition, this protocol also supports RFID systems.

ACKNOWLEDGEMENT

Our work was supported partly by the NSF of China under Grant (61472307, 61100165, 61100231), NSF of Shaanxi (2014 JM8313, 2012JQ8044, 2010JQ8004), Foundation of Education, Department of Shanxi Province (2013JK1096, 2013JK0589), and the Fundamental Research Funds for the Central Universities of China and CSC Funds.

REFERENCES

Abdalla, M., Catalano, D., & Fiore, D. 2009. Verifiable Random Functions from Identity-Based Key Encapsulation. *Advances in cryptology-EUROCRYPT, Cologne, German, 26–30 April, 2009*: 554–571. Springer Berlin Heidelberg.

Boneh, D. & Boyen, X. 2004. Efficient selective-ID secure identity-based encryption without random oracles. *Proceedings of EUROCRYPT, Interlaken, Switzerland, 2–6 May 2004*: 223–238. Springer Berlin Heidelberg.

Canetti, R., Goldreich, O. & Halevi, S. 1998. The random oracle methodology, revisited (preliminary version). *Proceedings of ACM Symposium on Theory of Computing, TX, USA, 24–26 May 1998*: 209–218. New York: ACM.

Daemen, J. & Rijmen, V. 2002. *The Design of Rijndael.* Secaucus, NJ, USA: Springer Verlag.

Durresi, A. & Paruchuri, V. 2007. Broadcast protocol for energy-constrained networks. *IEEE Transactions on Broadcasting* 53(1): 112–119.

Jin, W., Sung, M. & Kim, D. 2009. Radio Frequency Identity Fication(RFID) Tag Encryption Method and System Using Broadcast encryption(BE) Scheme. Patent number: US7576651, USA.

Li, D., Yang, G. & Su, H. 2010. An ID-based Broadcast Encryption Scheme for Hierarchical Wireless Sensor Networks. *IEEE International Conference on Information Science and Engineering, Hong Kong, China, 11–13 December 2010*: 1–4, Piscataway: IEEE.

Mann, C.R., Baldwin, R.O., Kharoufeh, J.P. & Mullins, B.E. 2007. A trajectory-based selective broadcast query protocol for large-scale. *high-density wireless sensor networks, Telecommunication System* 35(1–2): 67–86.

Mu, Y., Susilo, W. & Lin, Y. 2004. Identity-Based Authenticated Broadcast Encryption and Distributed Authenticated Encryption. *Ninth Computing Science Conferenc(ASIAN), Chiang Mai University, Thailand, 8–10 Dec. 2004*: 169–181. Springer Berlin Heidelberg.

Park, J.H. , Kim, H.J. & Sung, M.H. 2008. Public Key Broadcast Encryption Schemes With Shorter Transmissions. *IEEE Trans on Broadcasting* 54(3): 401–411.

Shaheen, J., Ostry, D. & Sivaraman, V. 2007. Confidential and secure broadcast in wireless sensor networks. *The 18th Annual IEEE International Symposium on Personal, Indoor and Mobile Radio Communications (PIMRC), Athens, Greece, 3–7 September 2007*: 1–5. Piscataway: IEEE.

Shamir, A. 1984. Identity-based Cryptosystems and Signature Schemes. *Advances in Cryptology-CRYPTO. Santa Barbara, California, USA, 19–22 August 1984*: 47–53. Springer Berlin Heidelberg.

Sheu, J.P., Hsu, C.S. & Chang, Y.J. 2006. Efficient broadcasting protocols for regular wireless sensor networks. *Wireless Communications and Mobile Computing* 6(1): 35–48.

Sohrabi, K., Gao, J., Ailawadhi, V. & Pottie, G. J. 2000. Protocols for self-organization of a wireless sensor network. *IEEE Personal Communications* 7(5): 16–27.

Szalachowski, P. & Kotulski, Z. 2012. One-Time Broadcast Encryption Schemes in Distributed Sensor Networks. *International J. Distributed sensor networks* 2012: 1–9.

Szalachowski, P., Kotulski, Z. & Ksiezopolski, B. 2011. Secure position-based selecting scheme for wsn communication. *Computer Networks* 160: 386–397.

Szalachowski, P., Ksiezopolski, B. & Kotulski, Z. 2010. CMAC, CCM and GCM/GMAC: advanced modes of operation of symmetric block ciphers in wireless sensor networks. *Information Processing Letters* 110(7): 247–251.

Zhang, L.Y., Hu, Y. & Wu, Q. 2012. Adaptively Secure Identity-based Broadcast Encryption with constant size private keys and ciphertexts from the Subgroups. *Mathematical and computer Modelling* 55(1–2): 12–18.

Electronics, Communications and Networks IV – Hussain & Ivanovic (eds)
© 2015 Taylor & Francis Group, London, ISBN: 978-1-138-02830-2

Microwave frequency combs source based on Brillouin scattering

Peng Zhang*, Tianshu Wang, Wanzhuo Ma, Lizhong Zhang, Shoufeng Tong & Huiling Jiang
National and Local Joint Engineering Research Center of Space Optoelectronics Technology, Changchun University of Science and Technology, Changchun, Jilin, China

Mei Kong & Xin Liu
College of Science, Changchun University of Science and Technology, Changchun, Jilin, China

ABSTRACT: A simple photonic approach to generate microwave frequency combs is proposed and experimentally demonstrated. In this scheme, a multi-wavelength Brillouin Fiber Laser (BFL) with reflecting ring is used. The configuration includes a 7km long Single-Mode Fiber (SMF) as Brillouin gain, a circulator is used as reflecting ring for an increasing wavelength number. A multi-wavelength optical signal with Brillouin frequency shift is heterodyned at the high-speed photodetector (PD) to produce a microwave signal by adjusting Polarization Controller (PC). As a result microwave frequency combs at 10.69GHz, 21.43GHz, 32.1GHz, 42.79GHz can be obtained. The space of combs is about 10.7GHz. The linewidth of 40GHz microwave signal is about 5 MHz, and Side-Mode Suppression Ratio (SMSR) is more than 20 dB.

1 INTRODUCTION

Microwave signal sources generated by photonic approach are valuable in many potential applications which include wireless access networks, radars, remote sensing, navigation and satellite communication (Capmany & Novak 2007, Yao 2011). Microwave signal sources generated by photonic approach are gaining increasing interest for years, such as tunable microwave signal sources incorporating temperature controller (Wang et al. 2012) fiber strain controller (Liu et al. 2012), or tunable pump laser (R. Wang et al. 2013), ultra-narrow linewidth microwave signal sources based on dual-wavelength fiber laser under single-longitudinal-mode operation (SLM; Zhou et al. 2008, Feng et al. 2013, Sun et al. 2011, Feng et al. 2011, Liu et al. 2013) special shaped pulse generation and impulse ultra-wideband (UWB) signal generation (Muriel et al. 1999, Lancis et al. 2008, Wang & Yao 2008, Ye et al. 2010, 2011). Meanwhile, extra attention has been attracted to microwave frequency combs source. Compared to other microwave-generation approaches, stable and ultra narrow multi-wavelength lasers should be achieved to generate microwave frequency combs source. Multi-wavelength Brillouin laser with stable space may be a nice choose for microwave frequency combs.

In order to obtain a microwave frequency combs sources, a simple method to generate microwave frequency combs based on a BFL with reflecting ring is proposed and demonstrated in this paper. Based on the reflecting ring configuration, a multi-wavelength signal under SLM with space correspondent to Brillouin frequency shift through adjusting PC can be obtained. And the multi-wavelength output is heterodyned to produce a microwave frequency combs at 10.69GHz, 21.43GHz, 32.1GHz, and 42.79GHz by adjusting PC.

2 EXPERIMENTAL SETUP AND OPERATION PRINCIPLE

Figure 1 shows the schematic diagram of the experimental setup. The light from Brillouin pump (BP) and PC is injected into an optical circulator (OC1). The linewidth of the Brillouin pump (BP) is about 100 kHz. The OC is used to launch the pump into the fiber ring and extract the Stokes wave. The fiber ring is formed by a fiber coupler, 7km long SMF and a reflecting ring. The backward Brillouin laser beam and the incoming BP can be observed by optical spectrum analyzer (OSA) and injected into a PD through OC1. The beating signal is detected, and corresponds to the microwave signal. The signal is measured by an electrical spectrum analyzer (ESA).

*Corresponding Author: *zhangpeng@cust.edu.cn*

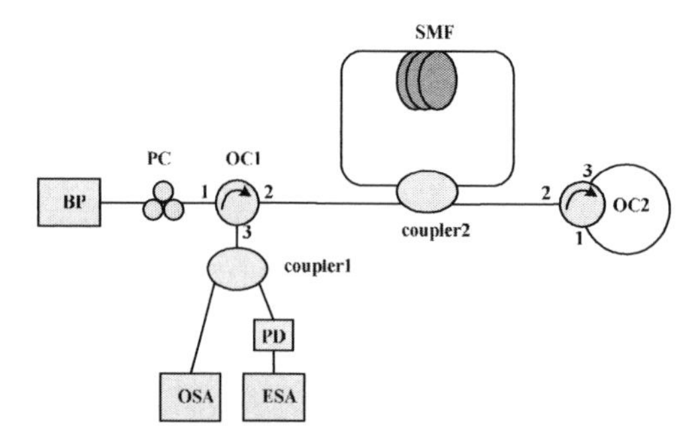

Figure 1. Experimental setup for generating microwave signal: **BP**, Brillouin pump; **PC**, polarization controller; **OC**, optical circulator; **SMF**, single-mode fiber; **OSA**, optical spectrum analyzer; **PD**, photodetector; **ESA**, electrical spectrum analyzer.

The Brillouin frequency shift *VB* can be expressed as,

$$v_B = \frac{2nV_a}{\lambda_P} \tag{1}$$

Where n is the refractive index, *Va* is the acoustic wave velocity in the fiber, and λp is the pump wavelength. When λp=1550nm, V_B is about 10GHz. As can be clearly seen, the beating microwave frequency combs with 10GHz space can be achieved.

3 RESULTS AND DISCUSSION

The pump light from PC, OC1 and coupler 2 is injected into SMF. As we know, the Stokes wave can be generated when the pump reaches the Brillouin threshold. The stoke wave also can generate the higher order stoke. And multi-wavelength Brillouin laser can be achieved by adjusting PC. The operation of multi-wavelength laser simultaneously with eight wavelength number was measured by an optical spectrum analyzer (OSA AQ6370C) with a resolution of 0.2 nm for optical spectrum measurement in Figure 2. The space of multi-wavelength signal is fixed and is about base on equation (1). Optical spectrum is measured by repeated scanning within 50 minutes. It can be found that multi-wavelength laser is relative stable as shown in Figure 3. And the optical signal to noise ratio (OSNR) of the lasers is more than 25dB. As we know, ~10GHz signal can be achieved mainly by beating pump and 1st Stokes. ~20GHz signal can be achieved mainly by beating pump and 2nd Stokes. ~30GHz signal can be achieved mainly by beating pump and 3rd Stokes. ~40GHz signal can be achieved mainly by beating pump and 4th Stokes. Multi-wavelength lasers with 10GHz space can be used for generating microwave frequency combs.

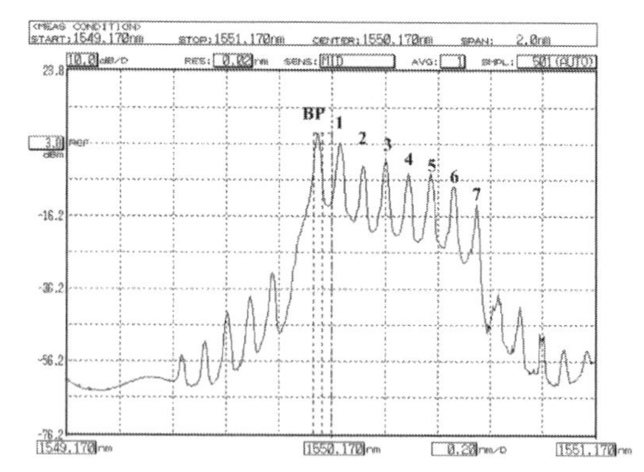

Figure 2. Multi-wavelength Brillouin fiber laser.

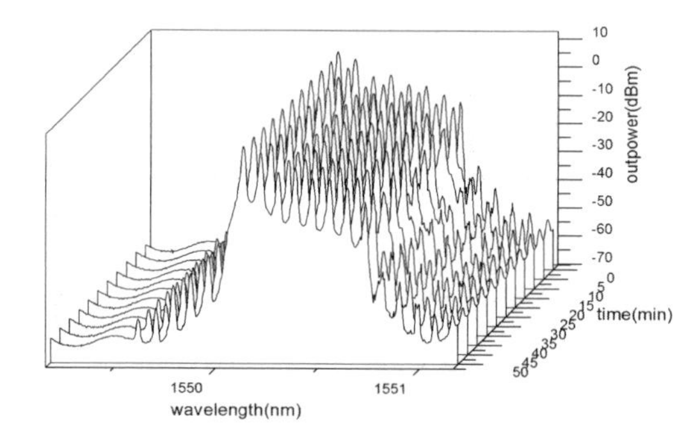

Figure 3. Stability of multi-wavelength Brillouin fiber laser.

And the output of multi-wavelength fiber laser is injected into a PD with a bandwidth of 50 GHz. Figure 4 illustrates that generated microwave beat signals are measured by ESA (Agilent PXA signal analyzer N9030A) with 44.36GHz span and 3.0 MHz resolution bandwidth. There are four microwave signals generated at 10.69GHz, 21.43GHz, 32.1GHz, and 42.79GHz. And the signal noise ratio (SNR) of microwave signals are about 20dB. The further verify the stability of microwave frequency combs. The output spectrum is measured over a ten-minute period within 40 min, shown in figure 5. It was found that the power of ~10GHz and ~20GHz microwave frequency signals is higher than other signals in figure 4. And the power of dual-wavelength lasers from the related scheme to beating frequency used for ~10GHz and ~20GHz microwave frequency signals are more than the lasers used for ~30GHz and ~40GHz microwave frequency signals.

The power fluctuations were recorded as in Figure 5. The power fluctuation is less ±2 dB from the average power. The stability of the microwave frequency combs power is relative worse than the results (Wang et al. 2012). The 3dB linewidth of microwave signals at 32.1GHz and 42.79GHz are estimated by Lorentz fitting at about 5 MHz in Figure 6 and Figure 7. The linewidth of ~30GHz and ~40GHz microwave frequency signals are relative bad.

The free spectral ranges (FSRs) of the cavity are written as

$$FSR=C/nL \tag{2}$$

Where C is the speed of light in vacuum, n is the refractive index of fiber, and L is the length of total cavity. L is about 7km, so the FSR of the cavity is about 0.03 MHz. And the range of Brillouin gain is more than 10MHz. So lasers were not under single-longitudinal-mode operation, leading to high phase noise and a broad bandwidth of the microwave signal. The special fiber and feedback fiber loop can be used to reduce the threshold of lasers under single-longitudinal-mode operation for ultra-narrow linewidth microwave signal at higher frequency (Liu et al. 2013).

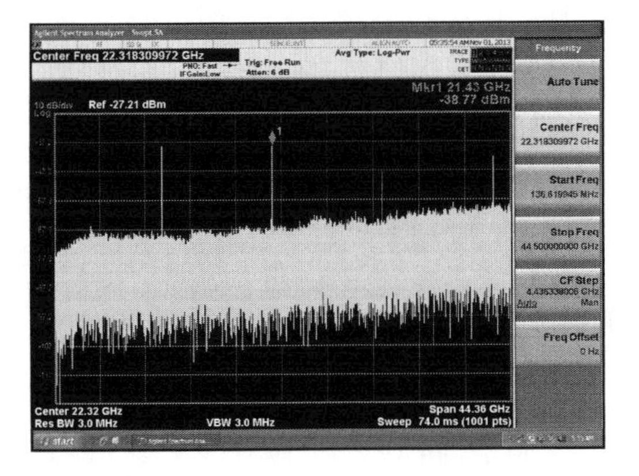

Figure 4. Microwave frequency combs generated from Brillouin scattering.

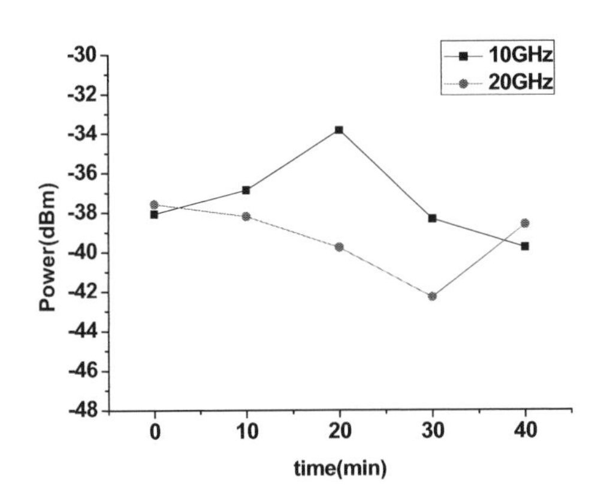

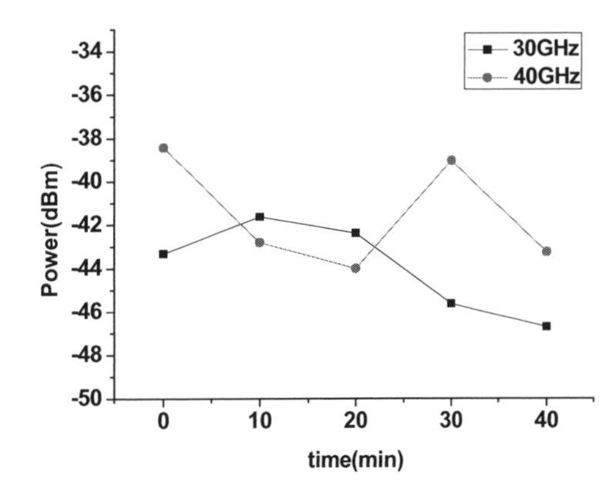

Figure 5. The power fluctuations of microwave frequency combs.

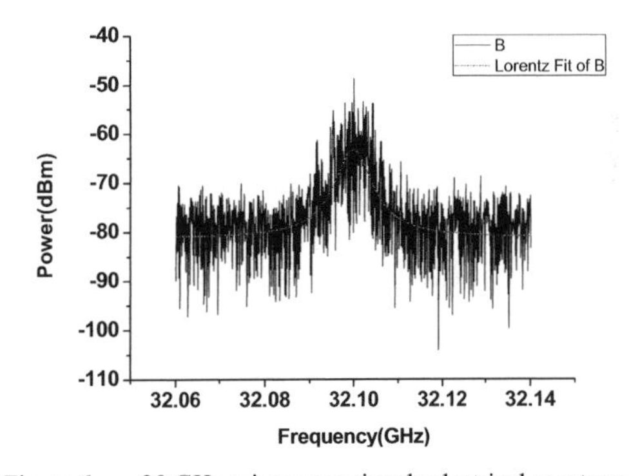

Figure 6. ~30 GHz microwave signals electrical spectrum.

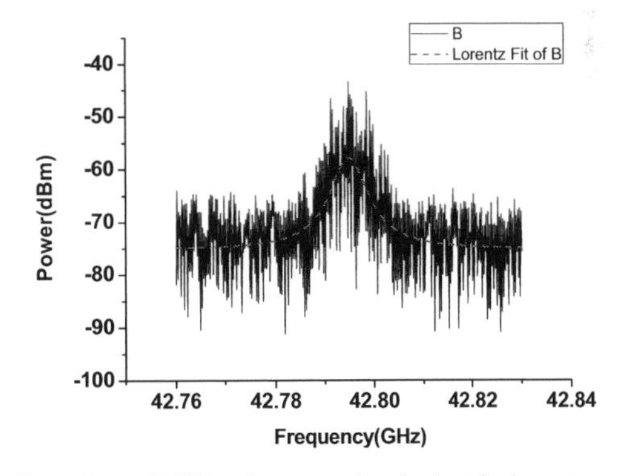

Figure 7. ~40 GHz microwave signals electrical spectrum.

4 CONCLUSION

The implementation of Brillouin lasers with a reflecting ring in the generation of microwave frequency combs by heterodyning multi-wavelength output is experimentally validated. ~10GHz, ~20GHz, ~30GHz and

~40GHz microwave signals about -40dBm have been obtained by heterodyning seven wavelength lasers. The power fluctuations of microwave frequency comb are less than ±2 dB from the average power. The line width of microwave signal at ~30GHz and ~40GHz from microwave frequency combs generated from the proposed method is about 5MHz. It may result from the lasers under multi-longitudinal-mode operation. Microwave frequency combs can be enhanced by the utilization of narrower linewidth pump, multi-wavelength lasers under single-longitudinal-mode operation and power stable mechanism. And the special fiber and feedback fiber loop can be used to reduce the threshold of lasers under single-longitudinal-mode operation for ultra-narrow linewidth microwave signal at higher frequency.

REFERENCES

Capmany, J. & Novak, D. 2007. Microwave photonics combines two worlds. *Nature Photonics* 1(6): 319–330.

Feng Suchun, Lu Shaohua, Peng Wanjing, Li Qi, Qi Chunhui, Feng Ting & Jian Shuisheng. 2013. Photonic generation of microwave signal using a dual-wavelength erbium-doped fiber ring laser with CMFBG filter and saturable absorber. *Optics & Laser Technology* 45: 32–36.

Feng Xinhuan, Cheng Linghao, Li Jie, Li Zhaohui & Guan Baiou. 2011. Tunable microwave generation based on a Brillouin fiber ring laser and reflected pump. *Optics & Laser Technology* 43(7): 1355–1357.

Lancis, J., Andrés, P. & Chen, L. R. 2008. Reconfigurable RF-Waveform Generation Based on Incoherent-Filter Design. *Journal of Lightwave Technology* 26(15): 2476–2483.

Liu Jinmei, Zhan Li, Xiao Pingping, Wang Gaomeng, Zhang Liang, Liu Xuesong, Peng Junsong & Shen Qishun. 2013. Generation of Step-Tunable Microwave Signal Using a Multiwavelength Brillouin Fiber Laser. *IEEE Photon. Technol. Lett.* 25(3): 220–223.

Liu Yi, Yu Jianglong, Wang Wenrui, Pan Honggang & Yang Enze. 2013. Narrow linewidth Single Longitudinal Mode Brllouin Fiber Laser Based on Feedback Fiber Loop. *ACTA Optica Sinca* 33(10): 1014003.

Liu, J., Zhan, L., Xiao, P., Shen, Q., Wang, G., Wu, Z., Liu, X. & Zhang, L. 2012. Optical generation of tunable microwave signal using cascaded Brillouin fiber lasers. *IEEE Photon. Technol. Lett.* 24(1): 22–24.

Muriel, M. A., Azaña, J. & Carballar, A. 1999. Real-time Fourier transformer based on fiber gratings. *Optics letters* 24(1): 1–3.

Sun Junqiang, Huang Yanxia, Li Hong & Jiang Chao. 2011. Photonic generation of microwave signals using dual-wavelength single-longitudinal-mode fiber lasers. *Optik* 122(9): 764–768.

Wang, C. & Yao, J. 2008. Photonic generation of chirped microwave pulses using superimposed chirped fiber Bragg gratings. *IEEE Photonics Technology Letters* 20(11): 882–884.

Wang, R., Chen, R. & Zhang, X. 2013. Two bands of widely tunable microwave signal photonic generation based on stimulated Brillouin scattering. *Opt. Commun* 287:192–195.

Wang, R., Zhang, X., Hu, J. & Wang, G. 2012. Photonic generation of tunable microwave signal using Brillouin fiber laser. *Appl. Opt.* 51(8): 1028–1032.

Yao, J. 2011. Photonic generation of microwave arbitrary waveforms. *Optics Communications* 284(15): 3723–3736.

Ye, J., Yan, L., Pan, W., et al. 2010. Two-dimensionally tunable microwave signal generation based on optical frequency-to-time conversion. *Optics letters* 35(15): 2606–2608.

Ye, J., Yan, L., Pan, W., et al. 2011. Photonic generation of triangular-shaped pulses based on frequency-to-time conversion. *Optics letters* 36(8): 1458–1460.

Zhou, J.L., Xia, L., Cheng, X.P., Dong, X.P., Shum, P. 2008. Photonic generation of tunable microwave signals by beating a dual-wavelength single longitudinal mode fiber ring laser. *Applied physics* B91(1): 99–103.

Electronics, Communications and Networks IV – Hussain & Ivanovic (eds)
© 2015 Taylor & Francis Group, London, ISBN: 978-1-138-02830-2

A TEM horn antenna with lens and tapered ridges for UWB radar application

Yichi Zhang, Jinghui Qiu & Nannan Wang
School of Electronics and Information Technology, Harbin Institute of Technology, Harbin, China

ABSTRACT: The paper presents a TEM double-ridged horn antenna with two laterally and longitudinally tapered ridges and a dielectric lens located at the antenna aperture. By tapering the ridges of a TEM horn antenna, a satisfied gain is got, the size of horn antenna is decreased, and the bandwidth is better for the practical operation. Meanwhile, the dielectric lens can increase the gains of antenna at high frequencies and narrow the main lobe, thus realize a better radiation pattern. Simulation results for the VSWR, radiation pattern, and the gain of the horn antenna are presented; the influence of both the lens and tapered ridges on the antenna properties is discussed as well. This paper introduces the results of the design, and realization and measurement of a TEM double-ridged horn antenna with a frequency band of 1-21GHz.

1 INTRODUCTION

The utilizations of ultra-wideband antennas for electromagnetic compatibility (EMC) measurement systems, the UWB radar systems, detection systems, and UWB communication systems in both military and civilian applications continue to increase. It is necessary for a UWB antenna to have a broadband frequency matching and a minimum signal distortion.

The TEM horn antenna is a travelling-wave, end-fire structure. Compared with Vivaldi antennas and other UWB antennas, the TEM double-ridged horn antenna has the advantages of wide bandwidths, directive performances, relatively easy constructions, high gains and no dispersions. However, its large size liemits its application. And at the high frequency of the working frequency band, the main lobe of the TEM horn antenna may be divided in several directions instead of focusing on the one direction in the antenna axis and thus the gain decreases.

For the purpose of improving the performance of the TEM horn antenna, many methods have been proposed. Shalger designed a new type of horn antenna with resistive sheet to reduce the distortion, but the resistive material may have a worse efficiency (Shalger 1996). Kanda introduced a TEM horn antenna with resistive sheet loaded to enlarge the bandwidth (Kanda 1982). An exponentially tapered TEM horn fed by microstrip balun is designed (Chung 2004) to increase the bandwidth of the TEM horn antenna; however it can not work in ultra band range. By adding metallic ridges to the waveguide, the parameters of TEM antennas can be improved (Taylor 1995), several kinds of tapered ridges and flared sections are adopted in TEM horn antenna designment. The next section describes the dielectric lens which improves the performance of TEM horn antenna. The inserted dielectric lens increases the antenna gain rapidly (Kiunho 2005); however it influences the impedance matching.

In this paper, a TEM double-ridged horn antenna with two tapered ridges and a dielectric lens is presented. Finally, the simulation and measurements are presented too.

2 TEM HORN ANTENNA DESIGN

2.1 *Design of the back cavity*

A ridged horn antenna consists of a waveguide part, a back cavity, double ridges, a cone part and a feeding element. The waveguide part is fed by a coaxial line. The outer conductor of the coaxial line is connected to the upper ridge while the inner conductor is connected to the lower ridge. Referring to a typical double-ridge horn antenna, the relevant dimensions are determined, and the main parameters influencing antenna performances includes: the dimensions of the back cavity, the shape of the ridge curve, the width of the ridge and the feeding distance.

The back cavity is actually a rectangular ridged waveguide connected to a reflection cavity, as shown in Figure 1. The rectangular ridged waveguide contributes to the transmission of the electromagnetic wave coming from the coaxial feeding line, while the reflection cavity is realized by cutting the end of the ridged waveguide, resulting in a stepped structure. According to the research results in section 2, the impedance and cutoff frequency of the waveguide are determined by the waveguide width a1 and height b1 along with the ridge dimensions including the a2 and b2.

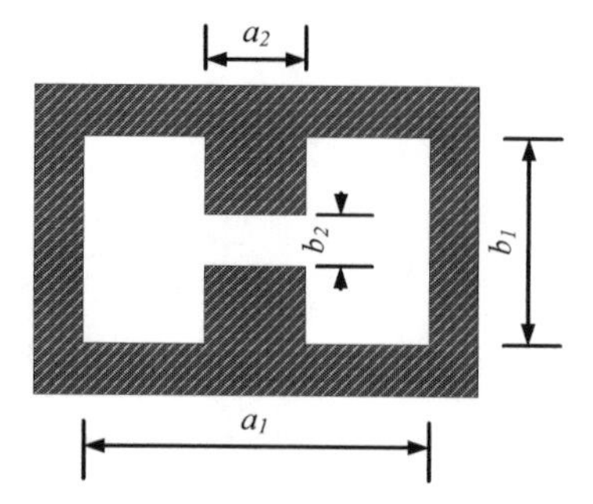

Figure 1. Reflection cavity in the ridged waveguide.

After the determination of the ridged waveguide dimensions, the effect of the reflection cavity at the end of the back cavity restrains the electromagnetic wave of TE_{20} mode, which could lead to a wider bandwidth of the dominant mode TE_{10}. The key parameters include the height of the step h and the length of last step l_s after cutting, which determine the cutoff frequency of the cavity. Simulation results prove that the antenna can achieve the best performances when h=7 and l_s=22.

2.2 *Design of the tapered double ridges*

According to the design principle of the ridge curve, an exponential curve $y = Ae^{pz}$ is usually adopted to realize a better impedance transition between the feed and the free space. An additional term of the first degree to the curve equation can make the impedance matching better in the low frequency range. Thus, the curve equation is changed into $y = Ae^{pz} + kz$. Where k is the term coefficient of the first degree and $A = b2/2$. In order to make the end of the ridge curve accord with the aperture, k and p should meet the following relationship:

$$p = \frac{1}{l}\lg\frac{2(b-kl)}{b_1} \tag{1}$$

where k is typically 0.02.

As shown in Figure 2, the thickness of the ridge is small near the feeding point and increases gradually when extending to the aperture, while the cross section is always the exponential-curve shape. Simulation is carried out referring to the improved ridged horn antenna and the ordinary double-ridge horn antenna, with the same ridge width of 4mm near the feeding point. The ridge thickness at the aperture of the

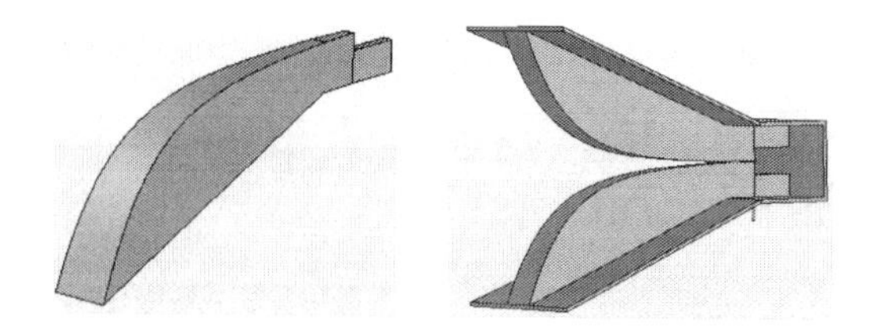

Figure 2. Tapered ridge and double-ridged horn antenna.

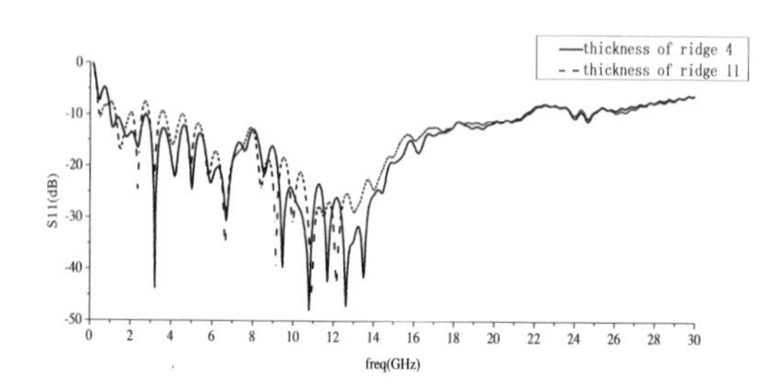

Figure 3. VSWR of tapered ridge and regular ridge.

improved antenna is optimized, generating the best reflection characteristic when it equals to 11mm. As shown in Figure.3, VSWR is less than 2.5 in the frequency range of 0.4-1GHz while the reflection coefficient is stable in the high frequency range. Therefore, the improved ridge structure can reduce the low-end frequency by 50%, leading to a double fractional bandwidth on the condition that the antenna dimension is unchanged, which means that the horn antenna can be miniaturized by 50% by the improved structure, referring to the same frequency range requirement.

2.3 *Design of antenna with a dielectric lens*

Radiation pattern distortion arises when the frequency is higher than 12GHz, which will lead to a lower radiation gain and worse radiation directivity. The reason is that at the aperture of the horn antenna, wave path difference results in phase difference (Walton 1964). When the frequency is higher, the wavelength is relatively small, so the wave path difference is enlarged, together with the phase difference, leading to radiation distortion. To solve this problem, phase should be adjusted consistently at the aperture. The traditional way is the addition of a lens (Cohn 1947).

Spherical wave can be transformed into a plane wave by a lens when the feed is located at its focus. Since horn antenna radiation can be approximated as a spherical wave, a general lens design principle can be applied here.

$$y = \sqrt{(n^2-1)x^2 + 2(n-1)fx} \tag{2}$$

Referring to this equation, n indicates the refractive index of the dielectric, and f is the distance between the intersection point of the two slopping walls and the aperture of the horn. After the determination of the corresponding parameters, the design can be carried out further, as shown in Figure 4.

The simulated VSWR of the TEM horn antenna with lens is shown in Figure 5, the calculated radiation patterns at the typical frequencies are shown in Fig.6, and the calculated gain of the antenna is in Fig.7. Gain is increased by 0.3dB by the addition of the lens at 1GHz with stable radiation beamwidth. When it reaches 12GHz, the radiation patterns of the antenna are improved significantly, and pattern distortion is eliminated. When the frequency is higher than 20GHz, the beamwidth is much smaller than that of the antenna without a lens. This proves that the addition of lens is an efficient method to enhance horn antenna radiation characteristics. Gain compassion is shown in Figure. 3-9, which shows the gain is typically increased by 5dB in the concerned frequency range.

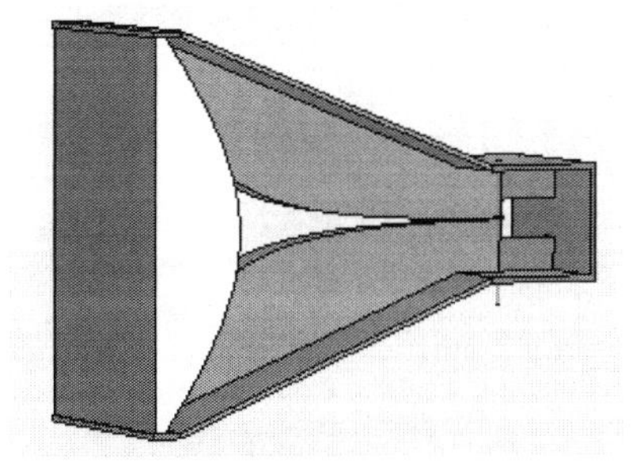

Figure 4. TEM horn with lens.

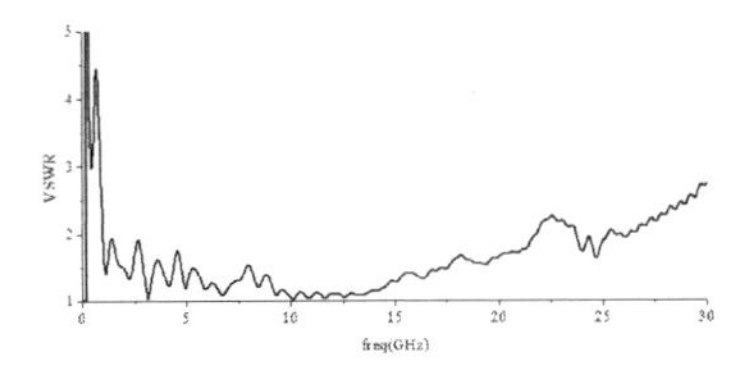

Figure 5. Simulated VSWR of antenna with lens.

3 ANTENNA REALIZATION

The TEM double-ridged horn antenna with two tapered ridges and a dielectric lens has been manufactured (Figure 8). The antennas are made from aluminum alloy. The antenna is fed through a coaxial cable with an SMA connector. The dimensions are the same as the designed model.

In Figure 9, the measured input reflection coefficients for the TEM horn antenna are presented. It can be seen that there is a pretty good agreement between simulation and measurement. However, because of the limitation of machining accuracy and process problems during welding connection, there is a large energy reflection in the high frequency.

The measured radiation pattern in E-plane and H-plane of manufactured TEM horn antenna with lens is shown in Figure.10 at 2, 10, 20GHz. The measured gains are shown in Figure.11. The measured gain

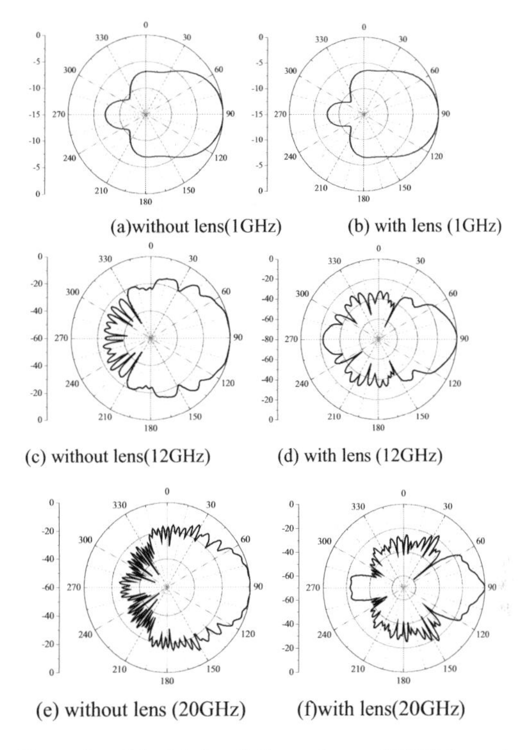

Figure 6. Calculated radiation patterns of TEM horn at selected frequencies.

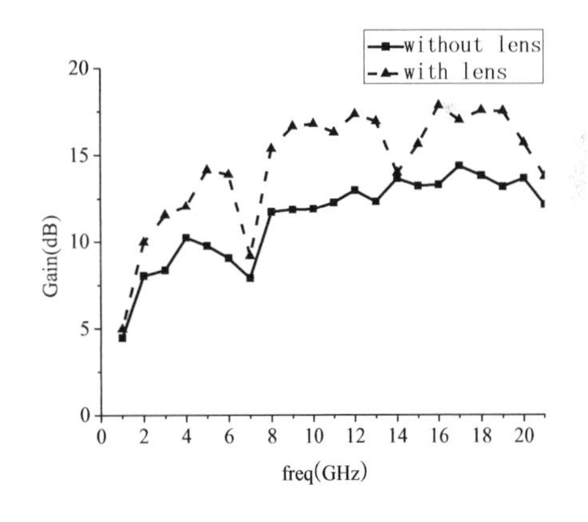

Figure 7. Simulated gain of antenna with/without lens.

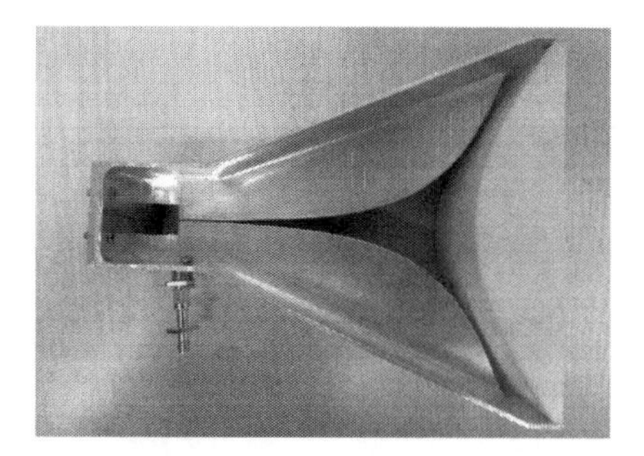

Figure 8. Manufactured TEM horn with lens.

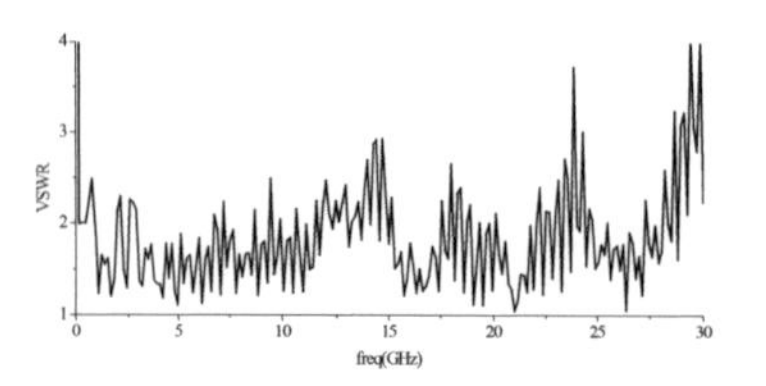

Figure 9. Measure impedance matching of TEM horn.

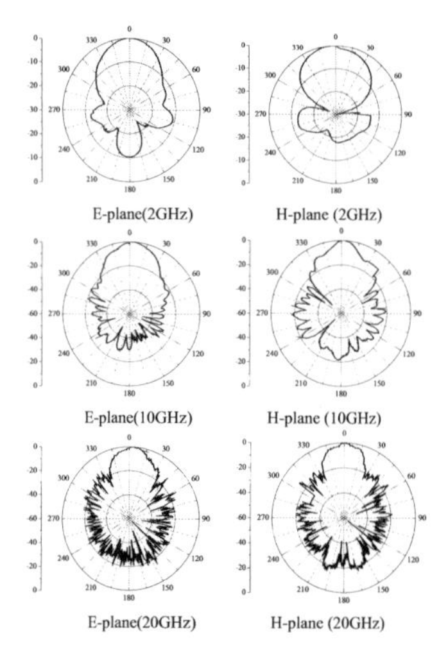

Figure 10. Measured radiation patterns of TEM horn.

curve shows that the lens increases antenna gain but also adds reflections to the TEM horn antenna.

4 CONCLUSION

Design and results of a TEM double-ridged horn antenna with two laterally and longitudinally tapered ridges and a dielectric lens are described. The antenna sample is manufactured and measured. The

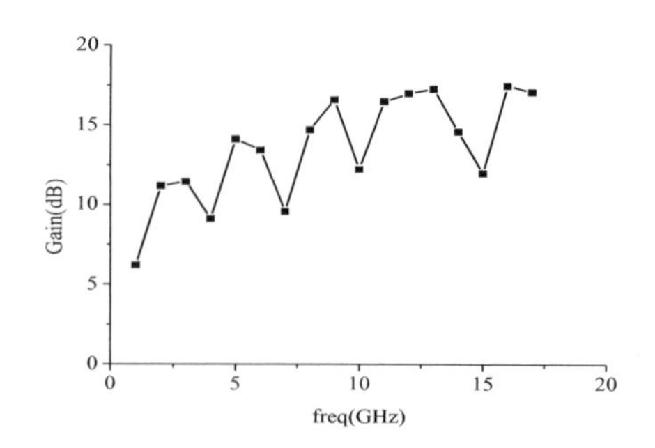

Figure 11. Measured gain of TEM horn antenna.

influences of lens and the tapered ridges are shown by the simulation. The design of tapered ridges leads to a significant size reduction of the device and satisfactory VSWR. The dielectric lens provides perfect far-field radiation characteristics and high gains over the practical bandwidth. Thus, this antenna works in 1-26GHz and is suitable for the UWB radar technology and also for measurement purpose.

ACKNOWLEDGMENT

The research is a part of international activities of ultra-wideband life detection radar technology import (2011DFR10890). The authors would like to express their sincere gratitude to CST Ltd., Germany, for providing the CST Training Center at our university with a free package of CST MWS software.

REFERENCES

Chung, K.H. 2004. The design of a wideband TEM horn antenna with a microstrip-type balun. *IEEE Antennas and Propagation Society International Symposium*: 1899–1902. Piscataway, New Jersey: IEEE.

Cohn, S.B. 1947. Properties of ridged waveguide. *Proc. IRE* 35: 778–783.

Kanda, M. 1982. The effects of resistive loading of "TEM" horns. *IEEE Transaction on Electromagnetic Compatibility* 24(2): 246–255.

Kiunho, C.S. 2005. Design of an ultrawide-band TEM horn antenna with a microstrip-type balun. *IEEE Transactions on Antennas and Propagation* 53(10): 3410–3413.

Shalger, K.L. 1996. Modified TEM horn antenna for wideband applications, *Microwave and Optical Technology letter* 12(2): 86–90.

Taylor, J.D. 1995. *Introduction to Ultra-Wideband Radar Systems*. CRC Press.

Walton, K.L. 1964. Broadband ridged horn design. *Microwave J* 4: 96–101.

Electronics, Communications and Networks IV – Hussain & Ivanovic (eds)
© 2015 Taylor & Francis Group, London, ISBN: 978-1-138-02830-2

On certain transformation semigroups

Jia Zhang
School of Mathematics and Computer Science, Northwest University for Nationalities, Lanzhou, Gansu, China

ABSTRACT: Let $T(M)$ denote the full transformation semigroup on a finite set M, where A is a nonempty subset of M,

$$F(M)=\{\ f\in\ T(M)\ \mid\ f(A)\subseteq A\ or\ \mid f(M)\mid=1\}. \tag{1}$$

Obviously, $F(M)$ is a subsemigroup of $T(M)$. And it is easy to see that if $A = M$, then $F(M) = T(M)$. In this paper, we mainly discuss the Green's relations of $F(M)$, describe the regular elements and completely regular elements of $F(M)$, and determine the regular and completely regular classes of $F(M)$.

1 INTRODUCTION

Let $T(M)$ denote the full transformation semigroup on a finite set $T(M)$, that is, the semigroup under composition of all mappings from M into itself. It is known that $T(M)$ is a regular semigroup. As customary, we denote composition of two transformations by juxtaposition and adopt a left mapping convention: for any f, $g\in T(M)$, gf denotes the mapping obtained by performing first f and then g. The kernel of f (the equivalence relation $\{(a, b)\in M\times M \mid f(a) = f(b)\}$) is denoted by $ker\ f$. Then for any $x\in M$, x' denotes the $ker\ f$-class containing x. Also, the mapping $x'\to f(x)$ is a bijection of $M/\ ker\ f$ onto $F(M)$. Hence, the set of equivalence class of $ker\ f$ and $F(M)$ have the same cardinality, that is $|M/\ ker\ f| = |F(M)|$.

Magill K.D (Magill 1966) studied an interesting subsemigroup of $T(M)$ is

$$K(M)=\{\ f\in\ T(M)\ \mid\ f(A)\subseteq A\ \}, \tag{2}$$

where $\varnothing\neq A\subseteq M$. He discussed its Green's relations. Elements of the semigroup $K(M)$ don't need to be regular. Regular elements of $K(M)$ are characterized by Nen-thein S., Youngkhong P. and Kemprasit Y. (Nenthein & Y. P. 2005). Moreover, $K(M)$ is shown to be an abundant semigroup by Higgins P. M. and Umar A. (Higgins & U. A. 1998).

In this paper, we consider a large semigroup $F(M)$, containing $K(M)$ and all constant mappings on M, that is

$$F(M)=\{f\in\ T(M)\mid\ f(A)\subseteq A\ or\mid f(M)\mid=1\}. \tag{3}$$

Obviously, $F(M)$ is a subsemigroup of $T(M)$, $K(M)$ is a subsemigroup of $F(M)$. And it is easy to see that if $A = M$, then $F(M) = T(M)$. In section 2, we characterize the Green's relations of $F(M)$. In particular, we consider a finite set M, so the relations D and J are same. In section 3, we describe the regular elements and completely regular elements of $F(M)$, and decide whether F(M) is a regular and completely regular semigroup. From this paper, we get a class of regular and completely regular semigroups. And, we know that the $K(M)$ is a special case in our discussion of section 2 and section 3.

2 GREEN'S RELATIONS

Let S be any semigroup and a, $b\in S$. We know that $(a, b)\in L$ if $S1a = S1b$, $(a, b)\in R$ if $aS1= bS1$, and $(a, b)\in J$ if $S1aS1= S1bS1$, where $S1$ is the semigroup S with an identity adjoined, if necessary. $H = L\cap R$ and $D = L\vee R$. These five equivalence relations on S are known as Green's relations (see Howie (Howie 1995)), which played a fundamental role in the development of semigroup theory. The relations L and R commute (Proposition 2.1.3, (Howie 1995)), and consequently $D = L\circ R = R\circ L$. In this section, we discuss the Green's relations of the semigroup $F(M)$, where $F(M)$ follows the section 1. If the Green's relations are very clear, we can know its characterization is regular and completely regular by the Green's relations. For convenience, we denote that the B is the complement set of A in M, that is $B = M \setminus A$.

Lemma 2.1 [Exercises 2.6.16, (Howie 1995)] Let $T(S)$ be the full transformation semigroup on any set S, and f, $g\in T(S)$. The

1 $(f, g) \in L$ if and only if $ker\, f = ker\, g$;
2 $f, g) \in R$ if and only if $f(S) = g(S)$;
3 $(f, g) \in D$ if and only if $|f(S)| = |g(S)|$;

And it is well-known that the $T(S)$ is regular.

Lemma 2.2 [Proposition 2.1.4, (Howie 1995)] Let S be a periodic semigroup. Then $D = J$.

2.1 *L relation*

We begin with the relation L.

Theorem 2.1.1 Let $f, g \in F(M)$. Then the following two conditions are equivalent

1 $(f, g) \in L$;
2 $ker\, f = ker\, g$ and if $|f(M)| = |g(M)| > 1$, then $f(X) \subseteq A$ if and only if $g(X) \subseteq A$ for any $X \subseteq M$.

Proof (1)⇒(2). Suppose that (1) holds. So $g = hf$ and $f = kg$ for some $h, k \in F(M)$. Clearly, $ker\, f = ker\, g$ and $|F(M)| = |g(M)|$. Now let $|f(M)| = |g(M)| > 1, X \subseteq M$. If $f(X) \subseteq A$, then $g(X) = hf(X) \subseteq h(A)$. Since $h \in F(M)$, if $h(A) \subseteq A$, then $g(X) \subseteq h(A) \subseteq A$. If $|h(M)| = 1$, then $|g(M)| = |hf(M)| = 1$, a contradiction. Thus we have $g(X) \subseteq A$. Similarly, $g(X) \subseteq A$ implies $f(X) \subseteq A$. Hence (2) holds.

(2)⇒(1). Suppose that (2) holds. So $|f(M)| = |g(M)|$. There are two cases to consider.

If $|f(M)| = |g(M)| = 1$, let $h = g$, then $g = hf$.

If $|f(M)| = |g(M)| > 1$. Define $h : M \to M$ by

$$h(x) = \begin{cases} g(f^{-1}(x)), & x \in f(M) \\ x, & otherwise. \end{cases} \quad (4)$$

Then h is well-defined since $ker\, f = ker\, g$ and, clearly, $g = hf$. It remains to verify that $h \in F(M)$. Let $f(M) \cap A = A1$. Obviously, $A1 \neq \varnothing$. Then $f(f^{-1}(A1)) = A1 \subseteq A$ implies $g(f^{-1}(A1)) \subseteq A$. Thus $h(A) = h(A1) \cup h(A \backslash A1) = g(f^{-1}(A1)) \cup (A \backslash A1) \subseteq A$. So we have $h \in F(M)$. Similarly, there exists some $k \in F(M)$ satisfying $f = kg$. Hence $(f, g) \in L$.

2.2 *R relation*

Next we consider the R relation.

Theorem 2.2.1 Let $f, g \in F(M)$. Then the following two conditions are equivalent

1 $(f, g) \in R$;
2 $F(M) = g(M)$ and $f(A) = g(A)$.

Proof (1)⇒(2). Suppose that (1) holds, so $g = fh$ and $f = gk$ for some $h, k \in F(M)$. Clearly, $f(M) = g(M)$. If $h(A) \subseteq A$, then $g(A) = fh(A) \subseteq f(A)$. If $|h(M)| = 1$, then $g(M) = fh(M) = f(x)$ for some $x \in M$ which implies $f(A) = g(A)$. So we have $g(A) \subseteq f(A)$. Similarly, we have $f(A) \subseteq g(A)$. Hence (2) holds.

(2)⇒(1). Suppose that (2) holds, so $|f(M)| = |g(M)| > 1$. We first construct $h \in F(M)$ with $g = fh$. Define $h : M \to M$ by

$$h(x) = \begin{cases} x1 & if\ g(x) \in g(A), \\ & where\ x1 \in f^{-1}g(x) \cap A \\ x2 & otherwise,\ where\ x2 \in f^{-1}(g(x)). \end{cases} \quad (5)$$

If $g(x) \in g(A) = f(A)$, then there exists some $a \in A$ such that $g(x) = f(a)$. This implies $f^{-1}(g(x)) \cap A \neq \varnothing$. Thus we could take some $x1 \in f^{-1}(g(x)) \cap A$. If $g(x) \notin g(A)$, then by $f(M) = g(M)$, we could take some $x2 \in f^{-1}(g(x))$. So h is well-defined and it is clear that $g = fh$. Furthermore by the definition of h, $h(A) \subseteq A$. Thus, $h \in F(M)$ as required. Similarly, there exists some $k \in F(M)$ satisfying $f = gk$. Hence $(f, g) \in R$.

As an immediate consequence of the previous two theorems we have

Theorem 2.2.2 Let $f, g \in F(M)$. Then the following two conditions are equivalent

1 $(f, g) \in H$;
2 $ker\, f = ker\, g, f(M) = g(M), f(A) = g(A)$ and
if $|f(M)| = |g(M)| > 1$, then $f(X) \subseteq A$ if and only if $g(X) \subseteq A$ for any $X \subseteq M$.

2.3 *D relation*

Now we consider the D relation.

Theorem 2.3.1 Let $f, g \in F(M)$. Then the following two conditions are equivalent

1 $(f, g) \in D$;
2 $|f(M)| = |g(M)|$, $|f(A)| = |g(A)|$ and if $|f(M)| = |g(M)| > 1$, then $|f(M) \cap A| = |g(M) \cap A|$.

Proof (1)⇒(2). Suppose that (1) holds. So $(f,h) \in L$ and $(h, g) \in R$ for some $h \in F(M)$. Clearly, $|f(M)| = |g(M)|$. And by Theorem 2.1.1 and 2.2.1, $|f(A)| = |h(A)| = |g(A)|$. Now let $|f(M)| = |g(M)| > 1$. Then $f(M) \cap A \neq \varnothing$, and $f(M) \cap A = f(f^{-1}(f(M) \cap A)) \subseteq A$. Thus $h(f^{-1}(f(M) \cap A)) \subseteq A$ by $(f,h) \in L$. That is $h(f^{-1}(f(M) \cap A)) \subseteq h(M) \cap A$. This implies $f^{-1}(f(M) \cap A) \subseteq h^{-1}(h(M) \cap A)$. Hence $|f(M) \cap A| \leq |h(M) \cap A|$ by $ker\, f = ker\, h$. So we have $|f(M) \cap A| \leq |g(M) \cap A|$ by $(h, g) \in R$. Similarly, we have $|g(M) \cap A| \leq |f(M) \cap A|$. Hence (2) holds.

(2)⇒(1). Suppose that (2) holds. If $|f(M)| = |g(M)| = 1$, then $(f, g) \in L$ by Theorem 2.1.1. Thus $(f, g) \in D$. If $|f(M)| = |g(M)| > 1$, then $f(A) \subseteq A$ and $g(A) \subseteq A$. We need to find some $h \in F(M)$ with $(f,h) \in L$ and $(h, g) \in R$. By condition (2), $|f(A)| = |g(A)|$, $|f(M) \cap A| = |g(M) \cap A|$ and $|f(M)| = |g(M)|$. So $|(f(M) \cap A) \setminus f(A)| = |(g(M) \cap A) \setminus g(A)|$ by $f(A), g(A) \subseteq A$, and $|f(M) \cap B| = |g(M) \cap B|$ by $A \cap B = \emptyset$. Thus we could take three bijections $h1$, $h2$ and $h3$, such that $h1 : f(A) \rightarrow g(A)$, $h2 : (f(M) \cap A) \setminus f(A) \rightarrow (g(M) \cap A) \setminus g(A)$ and $h3 : f(M) \cap B \rightarrow g(M) \cap B$. Now we define $h : M \rightarrow M$ by

$$h(x) = \begin{cases} h1(f(x)) & f(x) \in f(A) \\ h2(f(x)) & f(x) \in (f(M) \cap A) \setminus f(A) \\ h3(f(x)) & f(x) \in f(M) \cap B. \end{cases} \quad (6)$$

Clearly, h is well-defined and $h \in F(M)$. The following will show that $(f, h) \in L$ and $(h, g) \in R$. Obviously, $ker\, f = ker\, h$. Let $|f(M)| = |h(M)| > 1$, and $X \subseteq M$. If $f(X) \subseteq A$, then $h(X) \subseteq h1(f(X)) \cup h2(f(X)) \subseteq g(M) \cap A \subseteq A$. Conversely, if $h(X) \subseteq A$, then $X \subseteq h^{-1}(A) = f(M) \cap A$. Thus $f(X) \subseteq f(f(M) \cap A) \subseteq f(A) \subseteq A$. So we have $(f,h) \in L$. Furthermore by the definition of h, $h(M) = g(M)$ and $h(A) = h1(f(A)) = g(A)$. Thus $(h, g) \in R$. This completes the proof.

In this paper, we consider the $F(M)$ is a finite semigroup by the unity of M. So for the relation J, we have $D = J$ by Lemma 2.2.

In the end of this section, we characterize all the Green's relations of $F(M)$. Next we discuss regular and completely regular of $F(M)$.

3 REGULARITY AND COMPLETELY REGULARITY

An element a of a semigroup S is regular [completely regular] if $a = axa$ [$a = axa$ and $ax = xa$] for some $x \in S$. If all the elements of S are regular [completely regular], we say that S is a regular semigroup [completely regular semigroup]. Regular semigroup is a special class of semigroups which are particular amenable to be analyzed by using Green's relations. Completely regular semigroup is a special class of regular semigroups. The more other related information, the reader is referred to Petrich (Petrich & Reilly 1999). In this section, we first characterize the regular elements and completely regular elements of $F(M)$. Then we give a sufficient and necessary condition under which $F(M)$ is regular and completely regular. We first need some results that will be used later.

lemma 3.1 Let S be a semigroup, $a \in S$ and Ra denote the R-class containing a. Then a is a regular element if and only if Ra contains an idempotent.

Proof It is a straight consequence of Proposition 2.3.1 and 2.3.2 in (Howie 1995).

Lemma 3.2 Let S be a semigroup, $a \in S$ and Ha denote the H-class containing a. Then a is a completely regular element if and only if Ha contains an idempotent.

Proof It is a straight consequence of Lemma 1.7.9 and Proposition 2.1.2 in (Petrich & Reilly 1999).

Lemma 3.3 [Exercises 2.1.8, (Petrich & Reilly 1999)] Let $T(S)$ be the full transformation semigroup on any set S. Then $T(S)$ is completely regular if and only if $|S| \leq 2$.

3.1 *Regularity*

Proposition 3.1.1 Let $B \neq \emptyset$ and $f \in F(M)$. Then f is a regular element if and only if $f(B) \cap A \subseteq f(A)$.

Proof ($\Rightarrow$). To show the necessity we shall prove that if $f(B) \cap A \not\subseteq f(A)$, then f is not regular. By Lemma 3.1, we need only to show that the Rf contains no idempotents. Now if $|f(M)| > 1$, then let $f(A) = A1$ and $f(B) \cap A = A2$. Obviously, $A2 \setminus A1 \neq \emptyset$. If $g \in Rf$ is an idempotent, then $g(A) = A1$ by Theorem 2.2.1. And $g(a) = a$ for any $a \in A2 \setminus A1$. Thus $a \in g(A) = A1$. This is a contradiction. So there are no idempotents in Rf.

($\Leftarrow$). Suppose $f(B) \cap A \subseteq f(A)$. We shall show that Rf contains an idempotent, which implies f is a regular element. Now if $|f(M)| > 1$, then let $f(A) = A1$, and $f(B) \cap A = A2$. Obviously, $A2 \subseteq A1$. Fix some $a \in A1$, and define $g : M \rightarrow M$ by

$$g(x) = \begin{cases} x & x \in f(M) \\ a & otherwise. \end{cases} \quad (7)$$

Clearly, g is well-defined and $g \in F(M)$. Now we verify that g is an idempotent in Rf. If $x \in f(M)$, then $g(g(x)) = x = g(x)$; otherwise $g(g(x)) = g(a) = a = g(x)$. Thus g is an idempotent. Obviously, $f(M) = g(M)$. And $g(A) = g(A1) \cup g(A \setminus A1) = A1 \cup \{a\} = A1 = f(A)$. Thus, $g \in Rf$ as required and the proof is complete.

From the above proposition, we obtain

Theorem 3.1.2 $F(M)$ is regular if and only if $F(M)$ satisfies one of the following conditions

1 $B = \emptyset$;
2 $|A| = 1$ and $B \neq \emptyset$.

Proof ($\Leftarrow$). For convenience, we first proof the sufficiency. Suppose that (1) holds. So $F(M) = T(M)$. Thus $F(M)$ is regular. Now suppose that (2) holds. So for any $f \in F(M)$, $|f(M)| = 1$. Thus f is a regular element. So we have $F(M)$ is regular.

($\Rightarrow$). The proof of sufficiency has showed two cases of regular $F(M)$. Thus the proof of necessity needs to show that the regular $F(M)$ has only these two classes. So we claim that if (1) faults, then (2) holds.

Suppose that $F(M)$ is regular and $B \neq \emptyset$. If $|A| > 1$, then we could take some $f \in F(M)$ such that $f(B) \cap A \subseteq f(A)$ is not true. Thus by Proposition 3.1.1, f is not a regular element. This is a contradiction. So we have $|A| = 1$. The proof is complete.

3.2 *Completely Regularity*

Proposition 3.2.1 Let $B \neq \emptyset$ and $f \in F(M)$. Then f is a completely regular element if and only if f satisfies the following conditions

1 $f(B) \cap A \subseteq f(A)$;
2 $x' \cap f(M) \neq \emptyset$ for any $x \in M$.

Proof (⇒). Suppose that f is a completely regular element. By Proposition 3.1.1, (1) holds. And by Lemma 3.2, there exists an idempotent in Hf, denoted by g. So by Theorem 2.2.2, $f(M) = g(M)$. If (2) is not true, then there exists some $x0 \in M$ such that $x0' \cap f(M) = \emptyset$. Thus $x0' \cap g(M) = \emptyset$. But g is an idempotent, $g(g(x0)) = g(x0)$. Hence $g(x0) \in x0'$. This is a contradiction. So (2) holds.

(⇐). To show the sufficiency, we only need to prove that there exists an idempotent in Hf. We consider the following two cases to complete the proof.

If $|f(M)| = 1$, then $f^2 = f$. f is the idempotent as required.

If $|f(M)| > 1$, then $f(A) \subseteq A$. Define $g : M \to M$ by

$$g(x) = \begin{cases} a_{x'} & \text{if } x' \cap A \neq \emptyset, \text{ where } a_{x'} \in x' \cap f(A) \\ b_{x'} & \text{otherwise, where } b_{x'} \in x' \cap f(B). \end{cases} \quad (8)$$

We first show that g is well-defined. For any $x \in M$, if $x' \cap A \neq \emptyset$, then we claim that $x' \cap f(A) \neq \emptyset$. (If $x' \cap f(A) = \emptyset$, then by condition (1) and (2), there exists some $x0 \in B$ such that $x0 \in x' \cap f(B)$. And if $f^{-1}(x0) \cap A \neq \emptyset$, then there exists some $a \in f^{-1}(x0) \cap A$, such that $f(a) = x0$. This contradicts with $x' \cap f(A) = \emptyset$. Thus $f^{-1}(x0) \subseteq B$. By condition (2) again, there exists some $b1 \in B$ such that $b1 \in f^{-1}(x0) \cap f(M)$. And if $f^{-1}(b1) \cap A \neq \emptyset$, then there exists some $a1 \in f^{-1}(b1) \cap A$ such that $f(a1) = b1$. This contradicts with $f(A) \subseteq A$. Thus $f^{-1}(b1) \subseteq B$. Similarly, there also exists some $b2 \in B$ such that $b2 \in f^{-1}(b1) \cap f(M)$ and $f^{-1}(b2) \subseteq B$..... Since B is finite, there must exist some $bn \in B$ such that $f^{-1}(bn) \cap f(M) = \emptyset$. This is a contradiction. Thus $x' \cap A \neq \emptyset$.) Hence we could take a $x' \in x' \cap f(A)$, when $x' \cap f(A) \neq \emptyset$. If $x' \cap A = \emptyset$, then $x' \cap f(A) = \emptyset$. Thus we could take $b_{x'} \in B$ such that $b_{x'} \in x' \cap f(B)$. So g is well-defined. Next we will show that $(f, g) \in L$ and $(f, g) \in R$. By the definition of g, it is clear that $ker f = ker g$. Let $X \subseteq M$. If $f(X) \subseteq A$, then by condition (1), $x' \cap A \neq \emptyset$ for any $x \in X$. Thus $g(X) \subseteq f(A) \subseteq A$. Similarly, if $g(X) \subseteq A$, then we have $f(X) \subseteq A$. So by Theorem 2.1.1, $(f, g) \in L$. And it is obvious that $f(M) = g(M)$ and $f(A) = g(A)$. So by Theorem 2.2.1, $(f, g) \in R$. Finally, we show that g is an idempotent. For any $x \in M$, if $x' \cap A \neq \emptyset$, then $g(g(x)) = g(a_{x'}) = a_{x'} = g(x)$ by $a_{x'} \in x'$. Otherwise, $g(g(x)) = g(b_{x'}) = b_{x'} = g(x)$ by $b_{x'} \in x'$. Thus g is an idempotent as required. The proof is complete.

From the above proposition, we obtain

Theorem 3.2.2 $F(M)$ is completely regular if and only if $F(M)$ satisfies one of the following conditions

1 $|A| \leq 2$ and $B = \emptyset$;
2 $|A| = |B| = 1$.

Proof (⇐). For convenience, we first proof the sufficiency. Suppose that (1) holds. So $A = M$, $F(M) = T(M)$. Thus by Lemma 3.3, $F(M)$ is completely regular. Now suppose that (2) holds. So by the definition of $F(M)$, $F(M)$ contains only three idempotents. Thus $F(M)$ is completely regular.

(⇒). Since the proof of sufficiency has showed two cases of completely regular $F(M)$, the proof of necessity only needs to show that the completely regular $F(M)$ has only these two classes. Thus we proof that if (1) is not true, and then (2) must hold. Suppose that $F(M)$ is completely regular, and $B \neq \emptyset$. From Theorem 3.1.2, we have $|A| = 1$. Let $A = \{a\}$. If $|B| > 1$, then we could take $b1, b2 \in B$, $b1 = b2$. Now define $f : M \to M$ by

$$f(x) = \begin{cases} a & x = a \\ a & x = b_1 \\ b_1 & x = b_2 \\ x & \text{otherwise.} \end{cases} \quad (9)$$

Clearly, f is well-defined and $f \in F(M)$. But

$$(\exists b2 \in M)\ b_{2'} \cap f(M) = \emptyset. \quad (10)$$

Thus by Proposition 3.1.1, f is not a completely regular element. This is a contradiction. So we have $|B| = 1$. Thus (2) holds. The proof is complete.

4 CONCLUSIONS

Using the characterization of Green's relation, we describe the elements of regular and completely regular of a class of transformation semigroups. And we get a class of transformation semigroups which are regular and completely regular semigroups. Regular semigroups are a very important class of transformation semigroups. From this paper, we provide good examples for regular transformation semigroups. We know the abundant semigroups are also very important semigroups. From this paper, we can succeed in discussing the *-Green's relations and abundant elements of this transformation semigroups and deciding when these semigroups are abundant semigroups.

ACKNOWLEDGMENT

The author wishes to express his/her appreciation to the referee for some valuable comments and suggestions that help to improve the presentation of this paper.

REFERENCES

Fountain, J. 1982. Abundant Semigroup. *Proc. London. Math. Soc.* 44(1): 103–129.

Higgins, P. M. & Umar, A. 1998. *Semigroups of Weak V-stabilizer Mappings*. Technical Report, TR: 238. Technical Report Series.

Howie, J. M. 1995. *Fundamentals of Semigroup Theory.* New York: Oxford University Press.

Magill, K. D. 1966. Subsemigroup of S(x). *Math. Japan* 11: 109–115.

Nenthein, S., Youngkhong, P. & Kemprasit, Y. 2005. Regular Elements of Some Transformation Semigroups. *Pure Math. Appl.* 16(3): 307–314.

Petrich, M. & Reilly, N. R. 1999. *Completely Regular Semigroups*. New York: A Wiley-Interscience Publication.

Electronics, Communications and Networks IV – Hussain & Ivanovic (eds)
© 2015 Taylor & Francis Group, London, ISBN: 978-1-138-02830-2

Application of OTN technology in the provincial electric power backbone communication network

Liang Zhang
Information & Communication Company of State Grid Ningxia Electric Power Corp, Yin Chuan, China

ABSTRACT: With the rapid growth of optical communication technology and the power of IP data services, the traditional electric power backbone communication network is developing toward high capacity, multi granularity, and super-fast acceleration in the direction of the intelligent optical network evolution. Based on the analysis of the current provincial status and the access requirements of business power backbone communication network, and combined with application of OTN technology in electric power communication system, the provincial electric power backbone communication network of OTN network planning is analyzed in detail.

1 INTRODUCTION

With the rapid construction of smart grid, as an important support platform of strong and smart grid, the bearing capacity, the level of management, safety performance, the quality of the network of electric power backbone communication network are facing new challenges. At present, the provincial electric power backbone communication network is composed of many SDH networks. Although the safety performance of the richly protective mechanism is better, the network bandwidth can not meet the demand of integrated data access to the growing network traffic and video transmission business power of IP data services. While the dynamic scheduling capability and intelligent network management for large numbers business are also insufficient, therefore the scientific analysis of the existing problems and the future of the backbone communication network business developing trend, the provincial electric power backbone construction data, broadband, intelligent use of new technology of communication network is needed to meet the demand of electric power communication access to the growing business.

2 OTN TECHNOLOGY

OTN is based on wavelength division multiplexing technology in the transmission network of optical layer network, similar to the function of SDH, which is based on multi wavelength transmission as shown in Figure 1. Large numbers in the cross layer scheduling, light and electric layer realize wavelength operations, and achieve business then, mapping, multiplexing, cascade, protection / restoration, management and maintenance to form a large capacity with large granularity bandwidth service delivery for the characteristics of the transmission network. It solves the problem that the traditional WDM networks doesn't have wavelength scheduling service ability, the ability of network protection is weak and other issues to meet the requirements of high-definition video, the electric energy data acquisition and other large numbers business scheduling ability. The flexible networking, convenient extension, service scheduling ability, high reliability, are important directions of provincial electric power backbone communication network in the future(Huang et al. 2012).

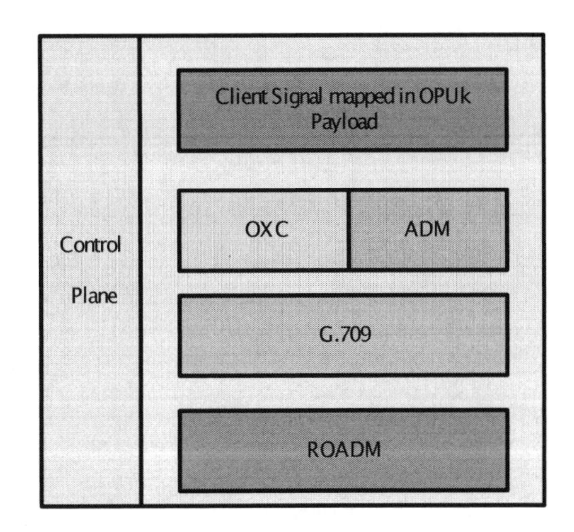

Figure 1. The function model of OTN.

Compared with the traditional backbone communication network technology, OTN technology has the following advantages:

1 OTN has the large numbers business scheduling function, very suitable for transmission of large capacity, high bandwidth data services.
2 OTN has the powerful ability of OAM. The overhead bytes that it defines can be rich, transparent to transmit different equipment's communication management information, enhance the operation management ability.
3 OTN can save fiber resources, bear power systems for all kinds of applications, recover disaster and replicate business information service.
4 OTN network with business ability and reliability of protection is good, which can withstand many failure tests, and the fault recovery time can meet the requirements of telecommunication level.

3 THE ANALYSIS OF PROVINCIAL ELECTRIC POWER BACKBONE COMMUNICATION NETWORK STATUS

At present, the provincial electric power backbone communication network is still based on SDH ring network. Although it has the simple network protection, short start-up time, mature technique, but in the early construction, speed-orientation and extensive form are mainly the priorities, and the overall planning is absent, which can't meet the requirements of multi service, elastic network construction of high bandwidth, high intelligence. Take the electric power backbone communication network as an example, its forms and needs are analyzed as shown in Figure 2.

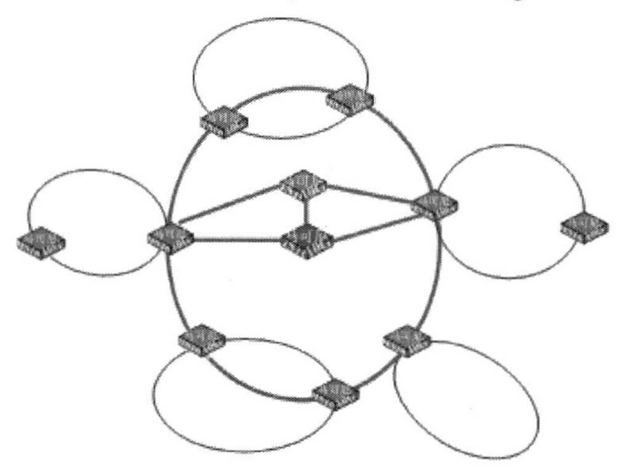

Figure 2. The topology of provincial power backbone communication network.

At present, the provincial electric power backbone communication network is 10G as the core network. All the local authorities through the core nodes ring networks converge to the backbone ring, communication service or to the traditional 2M aggregation business. Business flow is mainly composed of a transformer substation, power plant of the city, power supply bureau or company. The forms are as following:

1 With the continuing increase of high-definition video, the electric energy data acquisition and other large grain data services the bearing capacity of the existing equipment is limited, and interface mode also has some limitations.
2 With the establishment of various kinds of application systems, disaster recovery, centralized data center, information and communication have further integration trend, and strong backbone network demands greater carrying capacity. Reliability scheduling, security, business requirements are higher. QoS level which requires the power of the backbone communication network is also higher, and the level of protection is stricter.
3 As the types of business tend to be more diversified, more requirements for the various types of interface mode are put forward, and the exchange rate needs all kinds of businesses.
4 With the increase of the site, the scale of MSTP network is bigger, and the core resources become increasingly scarcer, which makes SDH network in network mode have certain limitations. The security of the network gradually decreases, and the cost expands.

So, the use of new communication technologies is required to solve the requirement of transmission, large capacity of fine scheduling, different QoS, and the OTN technology with its technical advantages. Currently the provincial trunk communication network construction is the most effective solution.

4 CONSTRUCTION SCHEME

4.1 *Construction principles*

1 Unified design standard, unified device model: Provincial Electric Power OTN (optical transmission network) backbone construction principle should strictly follow the requirements of unified technical standard, unified device model and so on. All the local authorities should coordinate the development, integrate network, complement each other, and make full use of the overall advantages of the network to ensure that the operation is not affected by the existing network.
2 Advanced technology and practice: in order to meet the needs of existing business access, it's considered that the future development needs three to five years of business, especially the actual requirements of the backbone communication network development of automation and information

of data service transmission technology and equipment. Advanced and matured technology is used to form a high capacity backbone communication network.

3 Network structure should take the evolution way of being active, steady and gradually promoted: the existing operating business should not be affected. The characteristics of the existing communication business are combined; equipment installation, commissioning, and service optical path guidance work are gradually completed.

4 Make full use of existing resources, effectively reduce the investment: the OTN of provincial electric power backbone communication network should be based on the business characteristics and existing optical fiber network, power, room and so on, which can save investment and adapt to requirements of different business.

4.2 *OTN overall planning*

Taking Ningxia provincial high capacity backbone communication network construction as an example, its main business is convergent around the city power supply bureau to gather a large number of IP business to the provincial company, so through the new OTN backbone communication network, it can serve for the IP service data channel, and effectively combine traditional TDM business. At the same time, on the basis of current optical fiber ring network, the mesh network development is gradually realized. Provincial networking company and each area bureau node combine MSTP, ASON and OTN technology. TDM business, low bandwidth channel are provided by the MSTP equipment, and the passage of large numbers (STM-64, GE, 10GE) is realized by the OTN equipment. According to the future business needs in the following 3-5 years, OTN Ningxia electric power backbone transmission network capacity for preliminary design is 40*10G, the 8-dimensional wave is WSS 40, and OUT 80 wave software is adjustable (Yu Xiaodong & Yu Fang 2010).

1 Cable routing organization schemes

The preliminary design should try to use cable lines which have been built or are newly built. Relying on the reliable backbone optical cable of 220kV and above, making substations with 220kV and above as the supporting point, a strong network frame is constructed, City Power Supply Bureau is the service access point to form the backbone transmission network access branch, and the built-in transmission capacity of OTN backbone transmission network is at least 40*10G, covering the city company.

2 OTN node selection

OTN network node selection should follow the following principles: (1) Cable is rich in resources. There are at least 2 OPGW cables, which are suitable for constructing the OTN ring network; (2) MSTP transmission of the first plane network structure of the backbone layer circuit site overlaps. On the condition of MSTP network fault, the business skips the emergency; (3) The center and information of the second distribution area should be contained, making it convenient for data communication backbone and main bearing transmission chain of SDH network routing switching to the high-capacity backbone optical transmission network.; (4) The communications room screen and power supply make it convenient to operate and maintain other geographical position. According to the Ningxia electric power site of the geography and cable resources, a node of 10 OTN is constructed, the main grid is mainly to achieve OTN. Two nodes of 7 OTN (including the city power supply bureau and an OLA site) are constructed in order to realize the service access of City Power Supply Bureau.

3 Topology of OTN network planning

OTN can support the type of optical layer connection based on single point to point, multipoint point to point, single point to multipoint, and can be- base on multiple kinds of topological network like linear type, generic type, star type and mesh type, etc. On account of the investment cost, OTN network construction can be constructed to be a ring form at the early stage and later be developed to be a mesh network. According to the geographical division, the whole OTN network can be divided into two management domains, in which A, B, C three cities constitute the southern domain, D, E, F three cities constitute the north field as shown in Figure 3. Inter domain node can adopt double equipment configuration to improve the safety and stability.

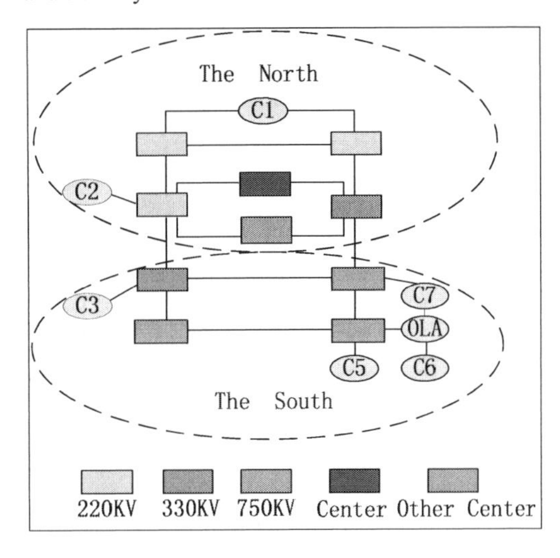

Figure 3. The topology of provincial OTN.

4 Network configuration

When the main channel with the different ODU 1+1 protection channels is configured, the bandwidth needs, business flow of the central station, second focal points and power supply bureau must be fully considered. To avoid the fault and save the panel, the different and shortest path should be selected. In the waveguide configuration, full consideration to the actual needs in each circuit should be given. Multiplex section should be configured with wave for redundancy protection and test. The service access of each site, and the statistics of the whole network average channel configuration and maximum bandwidth requirement should be considered. Taking Ningxia provincial backbone OTN construction as an example, the initial state can be configured with forty wave plates, and the whole network configuration of wavelength selection switch is based on multidimensional WSS, meeting the business requirements of current and future needs of expansion.

The OTN device configuration is composed of two parts, one part is the optical layer which configures and verifies transmission parameters and protection mode; the other part is the electric layer which rules and configures frame structure, principle and protection mode. In the rudimentary period of OTN construction, fiber Raman amplifiers or dispersion compensation fiber should be used to adapt to the requirements of MS-SNR such as the single direction OTN ,the double direction OTN, OLA and so on.

The whole OTN network mainly relates to the optical layer protection and electrical layer protection. Because the central station and city site cable route are susceptible to external damage, having high failure rate, two different routing sites are made to access zone by using optical multiplexing section protection mode. The protection of electric layer mainly uses SNCP 1+1 protection based on ODUK. For the section configured with both OMSP protection and ODUK 1+1 protection, in order to avoid the conflict of system protection, the collision avoidance mechanism can be used, namely the priority choice of optical multiplexing section protection, protection of ODUK 1+1 sub channel set to delay the trigger (Lan 2011).

At present, the provincial OTN network transmission business mainly includes 10G SDH services and 10GE information network service. Because the developing level of 4 communication transmission networks in each area is inconsistent, part of the region also converges to the central station taking the interface for 2.5G SDH and GE interface, therefore the access mode of 2.5G/10G SDH and GE/10GE interface is needed to consider. After being tested, the rate of 10GB/S (including STM-64, 10GE LAN/WAN and other business), the rate of 2.5Gb/s (including STM-1 and ODU1 particle based business) can realize service interworking between OTU2 interface and SNCP/N protection based on interconnection as shown in Table 1.

Table 1. Typical configuration of power business.

Setting	Bandwidth	Protection mode
Video VPN	8G	ODUK 1+1
SJW-A	10G	ODUK 1+1
SJW-B	2.5G	ODUK 1+1
Power Protection	2M	SNCP 1+1

Network management of OTN system adopts hierarchical management model which includes network layer, network element management layer and network management. According to the actual situation of provincial OTN, adjustable configuration of a network management system in the province can be reconciled, using geographic redundancy work mode. The OTN network contains two network management systems with two routers, which is double IP protection in the system (Sun & Yu 2010).

5 CONCLUSION

With further integration of surging and information communication network service power of the IP data, the electric power communication business develops towards diversity, and large numbers, therefore the scientific planning and construction of provincial electric power backbone communication network are needed by using OTN technology in order to make the current network move towards high bandwidth, high intelligence, high reliability of intelligent optical network, and meet the demands of construction of smart grid(Liang & Qin 2013).

REFERENCES

Huang Shan Guo, Zhang Jie, Luo Pei, et al. 2012. *The Planning and Optical Communication Nerwork* Beijing: Posts & Telecom Press.

Lan Qun. 2011. The Analysis of OTN Technology and Power Communication. *Communication Technology* 44(10): 80–82.

Liang Jing & Qin Miao. 2013. Application of Optical Transport Network(OTN)in Provincial Power Transmission Network. Power Construction 34 (3): 45–49.

Sun Hai Lian & Yu Fang. 2010. The Construction and Application of OTN Power Communication Network. *The Communication of Power System* 31(215): 24–27.

Yu Xiao-dong & Yu Fand. 2010. The Application of OTN Technology in Power Communication System. *The Communication of Power System* 31(217): 31–34.

Electronics, Communications and Networks IV – Hussain & Ivanovic (eds)
© 2015 Taylor & Francis Group, London, ISBN: 978-1-138-02830-2

Location-based AODVjr for wireless sensor networks in home automation

Bing Zhao
Metrology Department, China Electric Power Research Institute, Beijing, China

Xiaohui Li
College of Information Science and Engineering, Wuhan University of Science and Technology, Wuhan, China

Xiaobing Liang
Metrology Department, China Electric Power Research Institute, Beijing, China

Baojiang Cui
School of Computer Science, Beijing University of Posts and Telecommunications, Beijing, China

ABSTRACT: Applying wireless sensor networks to home automation is popular because of their characteristics of easy wiring, low cost, and self-organization. The popular standard of wireless sensor networks is ZigBee. The default routing algorithm of ZigBee, AODVjr, realizes route discovery by flooding route request packets (RREQs) to the whole network. However, flooding packets can result in more battery power consumption and shortening the lifetime of wireless sensor networks. Accordingly, this paper describes a location-based AODVjr that makes full use of the location information of the static nodes to limit route discovery to a cylindrical zone. We built up a simulation model using NS2 and analyzed the influence of zone size on packet delivery ratio in home automation applications using NS2 simulation. The simulation results showed that the proposed routing algorithm improved the packet delivery ration of AODVjr as well as residual energy ratio, routing overhead and packet average delay.

1 INTRODUCTION

Wireless sensor networks (WSN) are often used in home automation (HA) systems due to easy wiring, low cost, and self-organization (Biagioni 2012, Kazmi et al. 2014, Rajaraajeswari et al. 2014). The popular standard of WSN in HA (WSNHA) is the ZigBee specification. The IEEE 802.15.4/ZigBee/HA public application profile specifies how to realize the interoperability among various ZigBee HA devices. There are two routing algorithms in the network layer of ZigBee. One is AODVjr routing, a modified ad-hoc on-demand distance-vector (AODV) routing (Verma 2014). The other is Hierarchical Tree Routing (HTR). AODVjr is more frequently used than HTR is because AODVjr can get a more uniformed power consumption for a whole network. AODVjr starts route discovery from the source to the destination only when data packets must be sent. This kind of on-demand routing algorithm has a low routing overhead and memory requirement due to no routing table for all sensor nodes.

Although there are some excellent characteristics for ADOVjr, AODVjr faces certain challenges when it is applied to WSNHA. Firstly, the deployment of WSNHA depends on the home structure, so WSNHA has a nonuniform node distribution. Secondly, moving around of the home owners leads to instability of wireless link. On one hand, AODVjr adopts flooding route request packets (RREQs) to guarantee successful route discovery (Chakeres & Klein-Berndt 2002). On the other hand, frequent flooding RREQs results in a high possibility of broadcast storm and collision on MAC layer (Ortiz A. M. et al. 2014). Accordingly, the packet loss rate and power consumption are a little higher when using ADOVjr in WSNHA. In our previous study (Li et al. 2011), we did some research to improve the performance of AODVjr as well as to guarantee the reliability of data transmission in WSNHA. We called the modified AODVjr as a location-based AODVjr (AODVjr-LBR) algorithm which limits RREQ flooding to a round rectangle zone by using the location information of the sensor node. Restriction on RREQ flooding increases the performance of packet delivery ratio (PDR), meanwhile it decreases power consumption of WSNHA. However, the size of the round rectangle zone also affects the performance of AODVjr-LBR. This paper focused on how to select the size of the round rectangle zone and what kind of factors affect the performance of ADOVjr-LBR.

We firstly adopted the shadowing model to simulate the multipath effect of indoor wireless link, and predicted the indoor actual transmission distance according to the parameters of common wireless sensor chip. Then, we analyzed the influence of the zone size on the PDR and compared the performance of AODVjr-LBR and AODVjr in HA-specific application scenarios using NS2 (Fall & Varadhan 2009) simulation. The simulation results will help to guide engineering practice of HA system based on the ZigBee network.

The rest of the paper is organized as follows. Section II describes AODVjr-LBR algorithm. Section III details simulation model and discusses the simulation results. Section IV concludes the paper.

2 ROUTING ALGORITHM

WSNHA includes actuators, sensors, and controllers. We can view them as the nodes in WSN and most of the nodes are static. During network deployment, WSNHA nodes can easily get their location information. We make full use of location information to restrict flooding into a limited region and to reduce the number of RREQ.

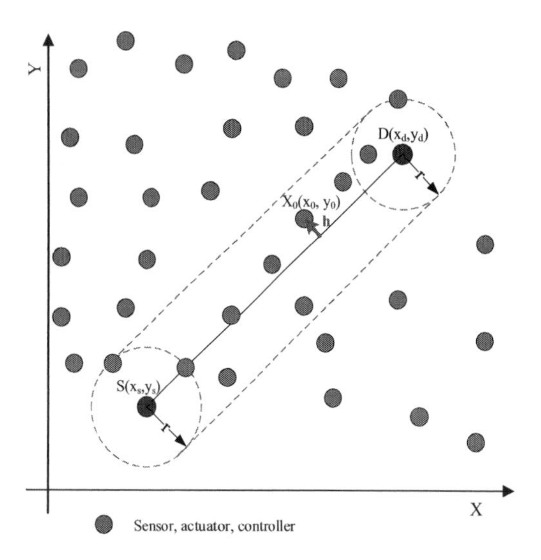

Figure 1. The round rectangle zone.

In Fig 1, consider Node S that needs to find a route to Node D. If there is no valid path for D in the routing table of S, S starts route discovery to find one. Before route discovery, S can establish a zone between S and D. The r is the radio signal distance. A circle with S as its center and r as its radius is the transmission range of the radio signal region; the transmission range of every node is assumed to be the same. Node S finds a route by broadcasting RREQs. The addresses of S and D are stored in the RREQ as source address and destination address, respectively. Each intermediate node $X0$ receives a RREQ, and then runs AODVjr-LBR, shown as Algorithm 1, to forward the RREQ. This algorithm uses the coordinates of S, D, and X_0 to restrict the flooding of RREQs to a limited zone.

Algorithm 1 AODVjr-LBR

Input *RREQ,X* 0
Output How to forward *RREQ*
if *RREQ* received before **then**
    discard *RREQ*
else if the destination of RREQ is X_0 **then**
    reply with Route Reply packet by the reverse link
else if the destination of RREQ is the X_0 's neighbor **then**
    directly forward *RREQ* to the destination
else if $0 \le h \le r$ **then**
    if X_0.*type* == *STATIC* **then**
        forward *RREQ*
    else
        discard *RREQ*
    end if
end if

The limited zone ban is a rectangle zone, a fan-shaped zone or a round rectangle zone. The performance of restricting RREQ in a rectangle zone is similar to flooding RREQ in the whole network because the coverage of WSNHA is not very big (Li et al. 2011). The performance of restricting RREQ in a fan-shaped zone is poor because the nonuniform distribution of WSNHA leads to scarce nodes in a fan-shaped zone which cannot guarantee successful rout discovery (Li et al. 2011). Consequently, AODVjr-LBR restricts the flooding of RREQs in a round rectangle zone shown as Fig 1. It is assumed that the coordinate of a node X_0 is (x_0, y_0) in WSNHA. The distance between X_0 and the line SD is h. The condition of judging whether X_0 locates in zone is $0 \le h \le r$. The calculation of h is the following. Suppose that the equation of a straight line $L(S, D)$ is $Ax + By + C = 0$. In the equation, A, B and C are constants. They can be calculated as follows.

$$A = (y_d - y_s)/(x_d - x_s), B = -1, C = y_s - Ax_s$$

So

$$h = \left| Ax_0 + By_0 + C \right| / \sqrt{A^2 + B^2}.$$

3 SIMULATION ANALYSIS

We developed the source code for AODVjr-LBR in the NS2 simulator. Our goals in conducting this simulation study were two-fold.

- Choose the proper round rectangle zone size. Flooding RREQs in a round rectangle zone that is too big results in the high collision possibility on the MAC layer which causes high packet loss rate. A round rectangle zone that is too small does not guarantee successful route discovery.

- Find the advantage of AODVjr-LBR by comparing the performance of AODVjr-LBR with AODVjr. We used NS2 to develop the source code for AODVjr and AODVjr-LBR, and compare the performance of these two routing algorithms.

We first analyzed the indoor wireless signal propagation distance using the shadowing model. Then we described the simulation scenarios referred the specific scenarios of WSNHA. Finally, we gave the simulation results.

3.1 *Indoor wireless signal propagation distance*

Considering multipath transmit effect of indoor wireless propagation model and the protocol stack of wireless sensor networks (Spadaciniet et al. 2014, Marcoet et al. 2013), it is necessary to use the shadowing model to simulate the multipath effect of indoor wireless link and to predict the indoor actual transmission distance according to the parameters of common wireless sensor chip. The shadowing model is a radio propagation model that estimates the path loss a signal encounters inside a home or a populated area over distance. The shadowing propagation model is described by (Rappaport 2002).

$$\left[\frac{P_r(d)}{P_r(\mathrm{d}_0)}\right]_{dB} = -10n\lg\left(\frac{d}{d_0}\right) + X_\sigma \tag{1}$$

where $P_r(d)$ denotes the received power when the distance is d. n is called the path loss exponent. $X_б$ is a Gaussian random variable with zero mean and standard deviation б. n and $X_б$ can be got by measurement. The shadowing propagation model indicates that the edge of communication range is based on a statistic model instead of an ideal circle. That is to say, the wireless nodes probabilistically receive packets at the edge of the communication range. The shadowing propagation model includes two parts. The first one is path loss model, and it adopts a close-in distance d_0 as a reference. The second part shows the variation of the received power at certain distance. It is a log-normal random variable, that is to say, it is of Gaussian distribution if measured in *dB*.

Supposed that

$$PL(\mathrm{d}) = 10\lg(\frac{P_t}{P_r(d)}) \tag{2}$$

where P_t is the transmitted power, then Equation (3) can be converted to

$$P_r(d) = P_t - PL(\mathrm{d}_0) - 10n\lg\left(\frac{d}{d_0}\right) - X_\sigma \tag{4}$$

Generally, d_0 is set to *1* m. Because d_0 is very close to the transmitter, the received power at d_0, $P_r(d_0)$, can be calculated by the free space propagation model. The free space propagation model only considers the ideal propagation condition. There is only one straight line between the transmitter and receiver in the free space propagation model. It is represented by

$$P_r(d) = \frac{P_t G_t G_r \lambda^2}{(4\pi)^2 d^2 L} \tag{5}$$

where G_t and G_r are the antenna gains of the transmitter and the receiver respectively. L ($L \geq 1$) is the system loss, and λ is the wavelength. Combined equation (2) with equation (3), $P_r(d)$ can be estimated by

$$P_r(d) = P_t - 10\lg(\frac{(4\pi)^2 L}{G_t G_r}(\frac{\mathrm{d}_0}{\lambda})^2) - 10n\lg\left(\frac{d}{d_0}\right) - X_\sigma \tag{6}$$

According to (4) and the hardware standard of the wireless sensor node, we can predict the received power changes with the distance in indoor wireless sensor networks, and estimate the actual valid distance of wireless signal. Assumed that CC2430 produced by TI is adopted to design the sensor node for indoor WSN, from the datasheet of CC2430, we can know that the transmitted power (P_t) is *0dBm*, and the sensitivity of the received power (S_r) is *-91dBm*. We used the parameters shown as Table 1 to simulate (4) in Matlab.

Table 1. Parameters used in indoor propagation model.

Parameter	Description	Value
P_t	The transmitted power	*0dB*
L	System loss	1
G_t	The antenna gains of the transmitter	1
G_r	The antenna gains of the receiver	1
λ	Wavelength	0.125m
n	The path loss exponent	4
б	The shadowing deviation	9.6

The concave curve in Fig 2 shows that the received power changes with the distance. The dot line around the concave curve shows the variation interval of the received power at certain distance. The variation interval is caused by the log-normal random variable. The horizontal thick line denotes the sensitivity of the received power (S_r). When the received power of a sensor node is greater than S_r, that sensor node can receive the stable wireless signal. As is shown in Fig 2, we can see clearly that almost all the dot line is over the horizontal thick line when the distance is less than 10m. Accordingly, when the distance is less than 10m, the sensor node can receive the stable wireless signal. In other words, the actual transmission range is 10m in actual WSNHA.

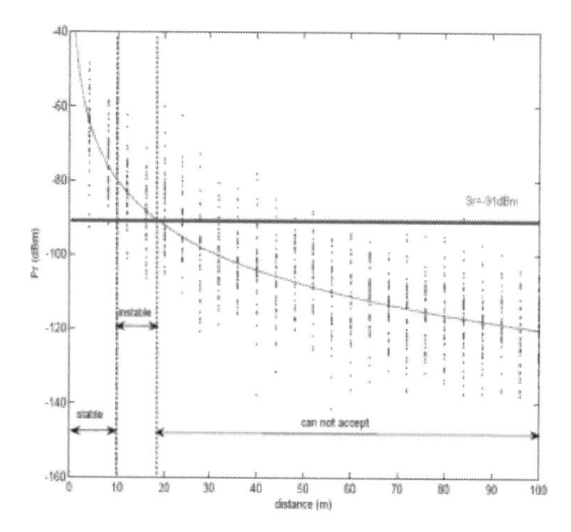

Figure 2. The received power changes with the distance.

3.2 *Design of the simulation scenarios*

Three simulation scenarios are designed for performance evaluation of AODVjr-LBR.

- Scenarios 1: The network is composed of one mobile node and twenty nodes, which are randomly distributed over an area of 16×6 m^2.
- Scenarios 2: The network is composed of one mobile node and thirty nodes, which are randomly distributed over an area of 16×9 m^2.
- Scenarios 3: The network is composed of one mobile node and forty nodes, which are randomly distributed over an area of 16×9 m^2.

IEEE 802.15.4 is set as the PHY and MAC layers in all the simulation scenarios. We assumed a signal propagation radius of 10 m on the basis of the analysis of indoor wireless signal propagation distance described in section 3.1. The movement speed of the mobile nodes was set as 0.5 m/s. The source node transmitted 1 data packet every 1 second. Each simulation ran 1000s. Because the number of the mobile nodes in HA is very low, the number of mobile nodes is one in all simulation scenarios. We use a fixed workload which consists of one sources-destination pairs. All sources and destinations are randomly selected from the deployed sensors within the sensor field. Table 2 shows the parameter specifications of our simulations.

Table 2. Parameters used in simulation.

Parameter	Value
Signal propagation radius	*10m*
Traffic type	CBR Traffic
Packet Size	70 Bytes
Data interval	1s
NO.of connections	1
NO.of mobile nodes	1
Velocity of the mobile node	0. 5m/s
Simulation time	1000
MAC protocol	IEEE 802.15.4

There are four metrics for analyzing the performance of AODVjr-LBR and AODVjr.

1. Packet delivery ratio (PDR). It is the ratio of the number of data packets received to the number originally sent. It is a critical performance measurement for a routing algorithm because it indicates whether this routing algorithm works correctly.
2. Residual energy ratio (RER). It is the ratio of the residual energy to the initial energy in the network. It describes the energy efficiency of the routing algorithm.
3. Routing overhead. It is the number of routing command packets. It represents overhead of the network layer. The main routing command packets of ADOVjr and ADOVjr-LBR are RREQ packets.
4. Average packet delay. It is the average one way delay for successfully transmitting a packet between the source and the destination. It measures the response time of the routing algorithm.

3.3 *Simulation results*

In this algorithm, carefully choosing the proper round rectangle zone can reduce power consumption caused by RREQ broadcast. Thus, the size of the round rectangle zone will affect the performance of the AODVjr-LBR algorithm. The value of R directly affected the performance of AODVjr-LBR because it determined the cylindrical zone size. Flooding RREQs in a zone that is too big increases the collision possibility of the frames transmitted by MAC layer and increases the packet loss rate. A zone that is too small does not guarantee successfully route discovery. Fig.3 shows PDR as a function of the round rectangle zone size for the three different scenarios. When R is greater than 10m, PDR changed little. It's worth noting that 10m is also the transmission range of the sensor node. Consequently, the effect of R greater than 10m is similar to flooding. AODVjr-LBR can maintain a relatively high PDR when R is in the range 2–4m.

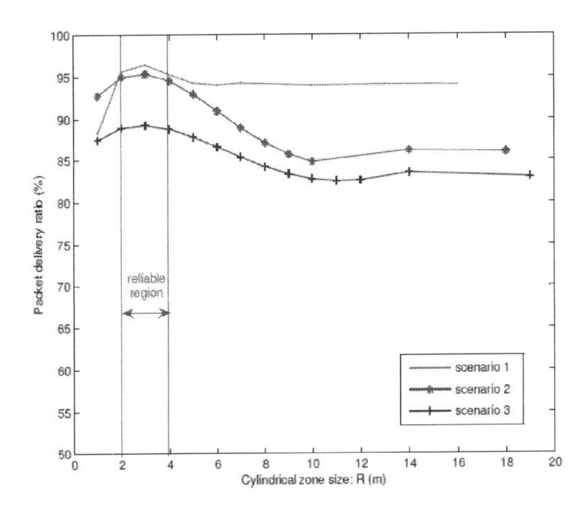

Figure 3. The packet delivery ratio as a function of cylindrical zone size for the three different scenarios.

Table 3 gives the simulation results for AODVjr-LBR when the radius of the round rectangle zone is 3 m. The PDR of AODVjr-LBR was better than that of AODVjr in all simulation scenarios because the round rectangle zone limited RREQ transmission. When achieving the high PDR, AODVjr-LBR can get the similar RER with AODVjr. The routing overhead and packet average delay of AODVr-LBR were a little lower than those of ADOVjr. Accordingly, AODVjr-LBR reduced the power consumption of RREQ transmission, lowered the packet transmission delay, and increased the data transmission reliability compared with AODVjr.

Table 3. Performance comparison: AODVjr-LBR vs. AODVjr.

	Performance measurement (unit)	AODVjr-LBR	AODVjr
scenario1	PDR(%)	97.00	89.98
	PER(%)	69.10	68.56
	Routing overhead(packets)	1010	1022
	Packet average delay(s)	0.015549	0.022737
scenario2	PDR(%)	96.93	85.84
	PER(%)	68.31	68.02
	Routing overhead(packets)	1013.00	1036
	Packet average delay(s)	0.019554	0.030255
scenario3	PDR(%)	91.00	79.94
	PER(%)	68.10	68.71
	Routing overhead(packets)	1012.00	1036
	Packet average delay(s)	0.023751	0.029452

4 CONCLUSIONS

We developed AODVjr-LBR, an improved version of AODVjr, for WSN in HA, and analyzed the influence of the round rectangle zone size on the PDR. The simulation results for four metrics showed that AODVjr-LBR performed better than AODVjr because of using the round rectangle zone to limit route discovery flooding.

The nodes are evenly distributed in our simulation model. If the node distribution is relatively dense or the nodes are evenly distributed, the round rectangle zone is a good choice to restrict RREQ flooding because it reflects stable performance in every performance measurements. If the node distribution is sparse or the nodes are not evenly distributed, the round rectangle zone may not work. Thus, flooding RREQ is the last resort in the proposed routing algorithm, because it can include more nodes to guarantee the routing in case of sparse node distribution.

ACKNOWLEDGMENT

This work has been supported in part by the National International Science and Technology Cooperation Project under Contract 2013DFG72850, and National Natural Science and Foundation of China under Grants 61105070.

REFERENCES

Biagioni, E. 2012. Topics in ad hoc and sensor networks. *IEEE Communications Magazine* 50(7): 120–121.

Chakeres, I. D., & Klein-Berndt, L. 2002. AODVjr, AODV simplified. *Mobile Computing and Communications Review* 6(3): 100–101.

Fall, K. & K. Varadhan. 2009. The ns Manual. A Collaboration between researchers at UC Berkeley, LBL, USC/ISI, and Xerox PARC.

Kazmi, A. H., M. J. O'grady, D. T. Delaney, et al. 2014. A review of wireless sensor-network-enabled building energy management systems. *ACM Transactions on Sensor Networks* 10(4): 325–349.

Li, X. H., S. H. Hong & K. L. Fang. 2011. Location-based self-adaptive routing algorithm for wireless sensor networks in home automation. *EURASIP Journal on Embedded Systems* 2011: 15 pages.

Marco, P. D., C. Fischione, F. Santucci, et al. 2013. Effects of Rayleigh-lognormal fading on IEEE 802.15.4 networks. *In 2013 IEEE International Conference on Communications (ICC)*: 1666–1671.

Ortiz, A. M., F. Royo, T. Olivares, et al. 2014. On reactive routing protocols in ZigBee wireless sensor networks. *Expert Systems* 31(2): 154–162.

Rajaraajeswari, S., R. Selvarani & P. Raj. 2014. Identification of multiple paths in smarter buildings networks. *In 2014 World Congress on Computing and Communication Technologies (WCCCT)*: 291–294.

Rappaport, T. S. 2002. *Wireless Communications: Principles and Practice (2nd Edition)*. Prentice Hall.

Spadacini, M., S. Savazzi & M. Nicoli. 2014. Wireless home automation networks for indoor surveillance: technologies and experiments. *EURASIP Journal on Wireless Communications and Networking* 6: 325–349.

Verma, V. K., S. Singh & N. P. Pathak. 2014. Analysis of scalability for AODV routing protocol in wireless sensor networks. *Optik - International Journal for Light and Electron Optics* 125(2): 748–750.

Electronics, Communications and Networks IV – Hussain & Ivanovic (eds)
© 2015 Taylor & Francis Group, London, ISBN: 978-1-138-02830-2

Cost functions and their application in designing blind equalizers based on BP neural networks

Juan Zhao
School of Electronics and Information Engineering, Jingchu University of Technology, Jing men, Hubei, China

ABSTRACT: In designing a blind equalizer based on Backward Propagation (BP) neural networks, the choice of the error function is fundamental to parameters that are updated online. Therefore, after some research on the properties of cost functions, this paper proposed a way to replace the error function of BP neural networks with cost functions, and the property of the equalizer based on BP neural networks with different p values of the constant modulus algorithm CMA (p, 2) were simulated. Decision-Direct algorithm (DD), hyperbolic tangent error, and the stop-and-go (SAG) algorithm and their combinations or revisions with the CMA were studied. Results showed that when $p = 2$, the CMA transferred a moderate value to update the parameters which leads to lower byte-error-rates (BERs) and good convergence trends. The combinations of CMA (2, 2) with the other three algorithms raised the behavior of the blind equalizer, while the HSCMA+CMA (2, 2) algorithm performed the best on the whole.

KEYWORDS: cost function; backward-propagation neural network; blind equalizer; constant modulus algorithm; decision direct algorithm; hyperbolic tangent error; stop-and-go algorithm.

1 INTRODUCTION

The blind equalization method is a kind of adaptive technology which does not need a training sequence as it reduces the inter-symbol interference according to the statistical characteristics of the received data. Thus, the utilization percentage of the channels is raised, the bandwidth resources are saved, the right decided ratio is improved, and the overall property of the signal transportation system is increased. These advantages all cause the equalizer to be an important part of a communication system. Since the adaptive recovering equalization method was proposed by Sato (Sato 1975), many algorithms have been developed such as the constant algorithm (Godard 1980) and multi-modulus algorithm (Joao et al. 2012), the decision directed algorithm (Papadias et al. 1997), the soft-decision directed algorithm (Chen et al. 2010), and the square-contour algorithm (Malik et al. 2010), among others. The fuzzy theory and the neural computing methods were also introduced to improve the usage and suitability of the equalizers.

The essence of blind equalization technology is the adaptive updating of the tap coefficients. It means that the memory-less nonlinear function would be a key work to design the equalizers. In appearance, it would be the choice of the cost functions. When designing an equalizer based on BP neural networks, the error function (Zhao 2011) is used it and can be treated as a constant modulus algorithm. Considering the good performances of other cost functions already developed, the author searched for a better error function for blind equalizers based on BP neural networks.

2 BASIC THEORIES

In designing blind equalizers, the cost functions can be chosen from various types of information rooted in the statistics (Rao 2011), the entropy, and the divergence (Mustafa 2012) of the received signals. They are derived from the low-order statistics, they can be easily calculated, designed in real world settings, and they can be combined with other theories to improve their properties. Most importantly, the equalizers based on the low-order statistics equalize quickly, meaning they are widely used and a large amount of research has been done.

2.1 *Properties of the cost functions*

The cost functions are used to tell the divergence of the signal sequences. This information can be traced back to the equalizers with backward propagation algorithms, and the equalizers then update the tap coefficients and the byte-error ratio (BER) is reduced. Therefore, cost functions must obey the following rules:

1 Non-negative. For each output signal $y(\text{n})$, the cost function values are $J[y(\text{n})] \geq 0$.
2 Symmetric. For each output signal $y(\text{n})$, there will always be $J[y(n)] = J[-y(n)]$.

3 Limited and unique. For each output signal y(n), J[y(n)] exists and is a sole value.
4 Continuous. For a given output signal y(n), there must be $\lim_{\Delta y \to 0} J[y(n)+\Delta y]=J[y(n)]$.
5 Concaved. For given output signals $y(n_1)$ and $y(n_2)$, there always exists a real $\lambda \in [0,1]$, which satisfies $J[\lambda y(n_1)+(1-\lambda)y(n_2)] \leq \lambda J[y(n_1)]+(1-\lambda)J[y(n_2)]$.
6 Consistent divergence. For given output signals $y(n_1)$ and $y(n_2)$, if its two-order divergences satisfy $D[y(n_1)] \geq D[y(n_2)]$, then its cost functions would satisfy $J[y(n_1)] \geq J[y(n_2)]$.

2.2 *Some proposed cost functions*

Cost functions based on the low-order statistics are commonly expressed in the following style:

$$J = E[e^2(n)] \tag{1}$$

The divergent function e(n) is:

$$e(n) = yd - y(n),\ yd = g[y(n)] \tag{2}$$

where $g(\cdot)$ is the memory-less nonlinear function of the output y(n).

Nowadays, there are several types of cost function, as seen below:

1 Decision directed (DD) cost functions

At the earliest, the memory-less nonlinear function is the decided function of the received sequence:

$$g[y(n)] = s_d = dec[y(n)] \tag{3}$$

The divergent function e(n) is:

$$e(n)_{DD} = s_d - y(n) \tag{4}$$

2 Constant modulus cost functions

It has been widely used and thoroughly studied since the constant modulus algorithm (CMA) was proposed by Godard. In CMA, the memory-less nonlinear function is as follows:

$$g[y(n)] = y(n)[1-|y(n)|^{2p-2}+R_p|y(n)|^{p-2}] \tag{5}$$

And the divergent function e(n) becomes

$$e(n)_{CMA} = [/y(n)/^p - R_p/]^{q/2} \tag{6}$$

where R_p is a low-order statistic and calculated by the following equation:

$$Rp=E[|s(n)|^{2p}]/E[|s(n)|^p] \tag{7}$$

In the above equations, p and q are integers and $q = 2$ is needed to act as a cost function.

Specifically, when $p = 1$ and $q = 2$, in unitary space, there exists:

$$|y(n)| = \sqrt{y(n)y^*(n)},\ \frac{\partial y^*(\,)}{\partial y(n)} = 0 \tag{8}$$

where $y^*(n)$ is the conjugation of the complex $y(n)$, so the backward propagated errors are divided into real and imaginary parts. It could then be called the dual-mode CMA or multi modulus algorithm (MMA) (Seok 2005).

2.2.1 *Some combined or improved cost functions*

The cost functions expressed in Equations (4) and (6) and others are often used to solve specific problems; while in engineering, the cost functions are usually combined to increase their capabilities. The most frequently met forms are listed below.

1 CMA-DD

The divergent function e(n) is a combination of Equations (4) and (6):

$$e(n)_{CMA\_DD} = \gamma(n)e(n)_{CMA} + \beta(n)\,e(n)_{DD} \tag{9}$$

where

$$\gamma(n) = f(|e(n)_{DD}|) \tag{10}$$

$$\beta(n) = [1 - {=} f(|e(n)_{DD}|)]|e(n)_{DD}|/|e(n)_{CMA}| \tag{11}$$

$$f(x) = \{1+\exp[-a(x-05)]\}^{-1},\ a>0 \tag{12}$$

2 CMA-HSCMA

In some studies (Guo et al. 2006), the cost functions are chosen according to the radius of the decided circle. Supposing r_{max} represents the radius of the positions of the outermost signals in the signal constellation sketch, then it satisfies:

$$r_{max} \geq \sqrt{10^{SNR/10}}\,\sigma_n = \sqrt{E[s^2(n)]/10^{SNR/10}} \tag{13}$$

In the above equation, SNR is the signal-to-noise ratio and σ_n is the standard error of the noise. Assuming D is the distance between two constellations and the decided radius, then it satisfies:

$$\frac{r_{max}}{\sqrt{10^{SNR/10}}} < d < \frac{D}{2} \tag{14}$$

Supposing the divergent function e(n) is

$$e(n)_{\text{HSCMA}} = \tanh e(n)_{\text{DCMA}} \quad (15)$$

where

$$e(n)_{\text{DCMA}} = \gamma e(n)_{\text{DCMA}} + (1-\gamma)\, e(n)_{\text{DCMA}} \quad (17)$$

and

$$\gamma = \begin{cases} 1 & |y(n)| < d \\ 0 & |y(n)| \geq d \end{cases} \quad (17)$$

3 SAG-CMA

In the SAG-CMA, the cost function is a combination of the stop-and-go algorithm (Xue et al. 2010) and the CMA. Then, the divergent function e(n) is as follows:

$$e(n)_{\text{SAG-CMA}} = \gamma e(n)_{\text{CMA}} \quad (18)$$

3 SIMULATIONS AND DISCUSSIONS

In the past few years, as the author was determined to find a better equalizer based on BP neural networks (Zhao 2013), a four-layer BP neural network was constructed and the variable step-size method was introduced as follows:

$$\eta(n+1) = \frac{\eta(n)}{1+\eta(n)\|e(n)\|^2} \quad (19)$$

At the beginning, the initialized step was the same and was set at $\eta = 0.0004$, the channel was $h = \{0.005, 0.009, 0.024, 0.854, 0.218, 0.049, 0.016\}$ and 4QAM signals were transported. For each simulation, the maximum iteration time is $n = 10{,}000$ and the results were averaged for N = 10 times. In simulations, the error function of the BP neural networks was changed by the cost functions mentioned above and their performance was verified.

3.3 *p value in CMA(p,2)*

The CMA is fundamental and thus attracts more attention. The CMA (1, 2), CMA (2, 2), and CMA (3, 2) are three types commonly met in research. For a given sequence of 4QAM signals, the cost functions have a relationship to the output y(n) (see Figure 1).

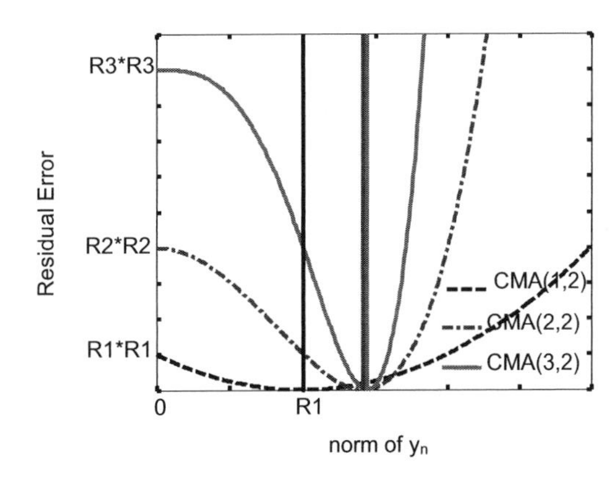

Figure 1. Relationship between y(n) and the CMA (p, 2) cost functions.

The larger values of p, the larger residual error of the signals far away from the center of the low-order statistic, and the larger the updated width of the tap coefficients, results in the equalizer converging more quickly but fluctuating violently. These can be easily demonstrated in simulations, as shown in Figure 2.

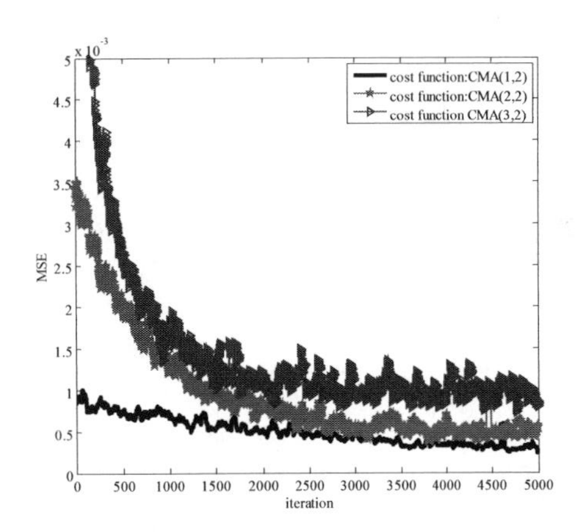

Figure 2. Windowed MSE with 50.

It can be seen that when $p = 1$, the residual error and the BER are both the smallest, but the distribution of the mean square error (MSE) before the equalizer was stabilized could not be clearly seen. When $p = 2$, the CMA performed better and acted as analyzed, had a better distribution of MSE, and its properties were better than that of the equalizer when $p = 3$. This is why the CMA (2, 2) is used in development and research, while the CMA (1, 2) and CMA (3, 2) remain in theory only.

3.4 *Simulations of the improved cost functions*

If the error function of the BP neural networks was replaced by Equations (9), (15), and (18), their performance as new equalizers was simulated (see Figure 3 and Figure 4).

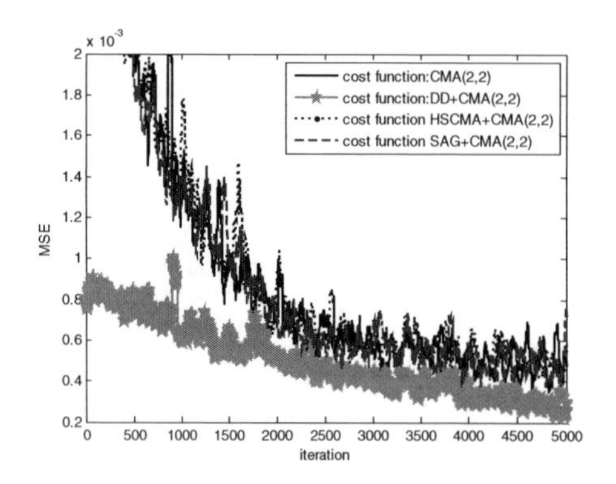

Figure 3. Windowed MSE of the equalizers with 50.

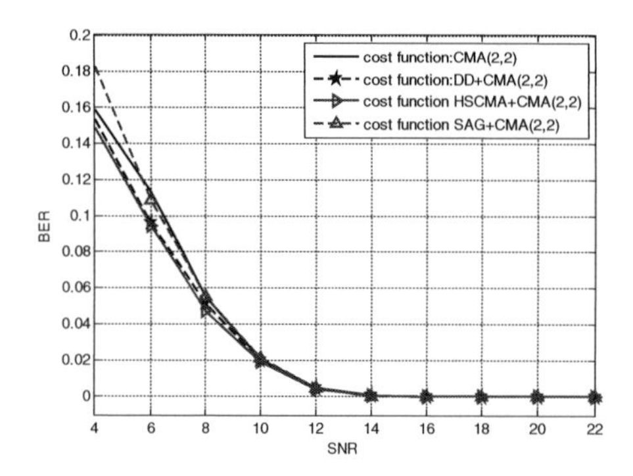

Figure 4. BER of the equalizers.

It could be seen that the new error functions improved by the combinations of the cost functions in the BP neural networks performed well. The introduction of the DD cost function decreased the residual error and the BER, but the distribution discipline was destroyed, while the introduction of the SAG and tangent function acted as expected. Comparatively, the HSCMA-CMA performed the best.

ACKNOWLEDGMENT

This work was supported in part by natural science foundation of Jingchu university of technology with grant No. ZR201022.

REFERENCES

Chen, S., Luk, B. L., Harris, C. J., et al. 2010. Fuzzy-logic tuned constant modulus algorithm and soft decision-directed scheme for blind equalisation. *Digital Signal Processing* (3): 846–859.

Godard, D.N. 1980. Self-recovering equalization and carrier tracking in two-dimensional data communication systems. *IEEE Transactions on Communications* (11): 1867–1875.

Guo Yecai, Zhou Qiaoxi & Duo Zhaofang. 2006. 3 decision circle based dual-mode constant modulus blind equalization algorithm. *8th international conference on singla processing, Beijing, 16-20 November 2006.* IEEE.

Gurcan, M. K., Weliwitegoda, D. & Chandra, G. 2012. Improved equalization and joint iterative detection. *Journal of the Franklin Institute* (1): 234–259.

Joao M. F., Maria, D., Miranda, M. & Silvax, T.M. 2012. A reginonal multimodulus algorithm for blind equalization of QAM signals: introduction and steady-state analysis. *Signal Processing* 11(11): 2643–2656.

Malik Muhammad Usman Gul, Shahzad Amin Sheikh. 2010. Design and implementation of a blind adaptive equalizer using Frequency Domain Square Contour Algorithm. *Digital Signal Processing* (6): 1697–1710.

Papadias, C. B. & Slock, D. T. M. 1997. Normalized sliding window constant modulus and decision-directed algorithms: a link between blind equalization and classical adaptive filtering. *IEEE Transactions on Signal Processing* (1): 231–235.

Rao Wei. 2011. Amplitude Transformation-Based Blind Equalization Part II: Suitable for High-Order QAM Signals. *Procedia Environmental Sciences* (0): 1282–1286.

Sato, Y. 1975. A Method of Self-Recovering Equalization for Multilevel Amplitude-Modulation Systems. *IEEE Transactions on Communications* 6(6): 679–682.

Seok Yoon, Choi SangWon, Lee Jumi, et al. 2005. *A Novel Blind Equalizer Based on Dual-Mode MCMA and DD Algorithm*//Yo-Sung Ho. Hyoung-Joong Kim. Advances in Multimedia Information Processing - PCM: 711–722. Berlin: Springer Berlin Heidelberg.

Xue Wei, Xiaoniu Yang & Zhaoyang Zhang. 2010 A Stop and Go Blind Equalization Algorithm for QAM Signals. *6th International Conference on Wireless Communications Networking and Mobile Computing (WiCOM), Chengdu, 23-25 September 2010.* IEEE.

Zhao Haiquan, Xiangping Zeng, Jiashu Zhang, et al. 2011. Pipelined functional link artificial recurrent neural network with the decision feedback structure for nonlinear channel equalization. *Information Sciences* 181(17): 3677–3692.

Zhao Juan. 2013 Application of the square contour algorithm in blind equalizers based on complex neural networks. *3rd International Conference on Consumer Electronics, Communications and Networks (CECNet), Xianning, 20-22. November 2013.* IEEE.

Electronics, Communications and Networks IV – Hussain & Ivanovic (eds)
© 2015 Taylor & Francis Group, London, ISBN: 978-1-138-02830-2

Suppression of Es layer clutter in HFSWR using a horizontally polarized antenna array

L. Zhao* & G. Yu
Civil Aviation Institute of Shenyang Aerospace University, Shenyang, China

Y.Z. Liu
Air Traffic Control Center of Northwestern Air Traffic Management Bureau of CAAC, Xian, China

H.K. Liu
Space Star Technology Co., Ltd, Beijing, China

ABSTRACT: Aiming at the problem of ionospheric Es layer clutter suppression in HF radar, a horizontally polarized antenna array was designed. Because of the difference of polarized patterns between the signal and ionospheric Es layer clutter, the horizontally polarized antenna arrays have a bigger gain on the ionospheric Es layer clutter and a smaller gain on the target signal. The simulation results of the horizontally polarized antenna array show that the clutter-to-signal ratios in the horizontally polarized antenna array output have improved obviously. So utilizing a horizontally polarized antenna array as an auxiliary antenna is a more effective way to so suppress the Es layer clutter in HF radar.

1 INTRODUCTION

There is an ionospheric Es layer clutter in high frequency surface wave radar because of the existence of the sporadic E layer in the ionosphere. Es layer is the ionized clouds that appear in the altitude of E layer, having a higher electron density than a normal E layer. The transmitted signal of the HFSWR is reflected by the Es layer, and received by the receiving antenna of the HFSWR. The Es layer clutter in the HFSER received signal that is stronger than the target signal. So the regular work of the HFSWR would be influenced.

The main antennas of the HFSWR are vertically polarized. So the target signal and Es layer clutter in the received signal of main antennas are vertically polarized. The Es layer clutter is a transmitted signal reflected by the ionosphere, so it is elliptically polarized (Stutzman 1998, Lee et al. 2000). Using the difference of polarized patterns between the signal and ionospheric Es layer clutter, a horizontally polarized antenna array is designed which can only receive Es layer clutter. So the Es layer clutter in the HFSWR received signal can be estimated by the Es layer clutter received by the horizontally polarized antenna array, and the estimated Es layer clutter can be removed from the HFSWR received signal (Wan et al. 2005, Wan et al. 2005, Gao et al. 2004).

In practice, the horizontally polarized antenna array also can receive little target signal. The suppression of the Es layer clutter in the HFSWR received signal is affected by the clutter-to-signal ratio in the horizontally polarized antenna array received signal. The Es layer clutter in the HFSWR received signal is suppressed more effectively when the clutter-to-signal ratio in the outputs of the horizontally polarized antenna array is higher (Leong et al. 1999).

2 STRUCTURE OF THE HORIZONTALLY POLARIZED ANTENNA ARRAY

The synthetic method given in literature (Liu et al. 2004) is used in the calculation of the weights for each antenna elements in the horizontally polarized antenna array. The Es layer clutter arrival direction which is around vertical direction is the main lobe direction for the horizontally polarized antenna array. So the direction of the target signal arrival which is around horizontally direction is a low side lobe for the horizontally polarized antenna array. It can be ensured that the clutter-to-signal ratio for the horizontally polarized antenna array received signal is high enough (Nuttall et al. 2001).

Two crosswise placed elements which are horizontally polarized constitute horizontally polarized

*Corresponding author: *zhaolong@sau.edu.cn*

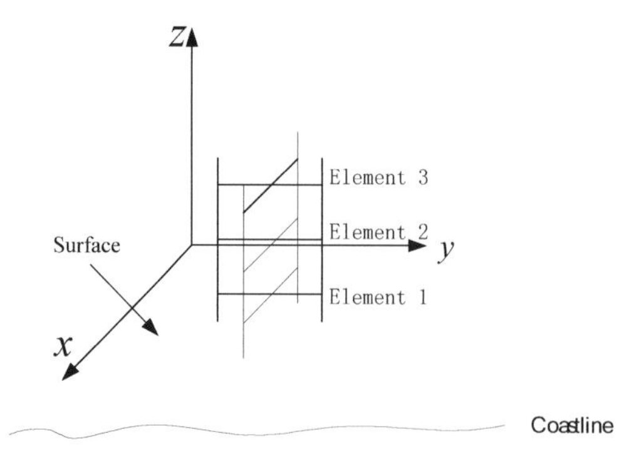

Figure 1. Horizontally polarized antenna array.

antenna. The two crosswise placed elements can receive all kinds of Es layer clutter.

Three horizontally polarized antennas constituted the horizontally polarized antenna array. The structure of the horizontally polarized antenna array is shown in Figure 1.

The antenna element which constitutes the horizontally polarized antenna array is dipole antenna. The interval between the horizontally polarized antennas is 5 meters. The length of the array elements which constitutes the horizontally polarized antenna is 10 meters. The work frequency of the horizontally polarized antenna array is from 4MHz to 8MHz.

3 PATTERNS OF THE HORIZONTALLY POLARIZED ANTENNA ARRAY

The weights for each antenna element in the horizontally polarized antenna array are calculated using the method given by the literature (Liu et al. 2004), when the Es layer clutter arrival direction which is around vertical direction is set to the main lobe direction for the horizontally polarized antenna array, and the direction of the target signal arrival, which is around horizontally direction is set to a low side lobe for the horizontally polarized antenna array.

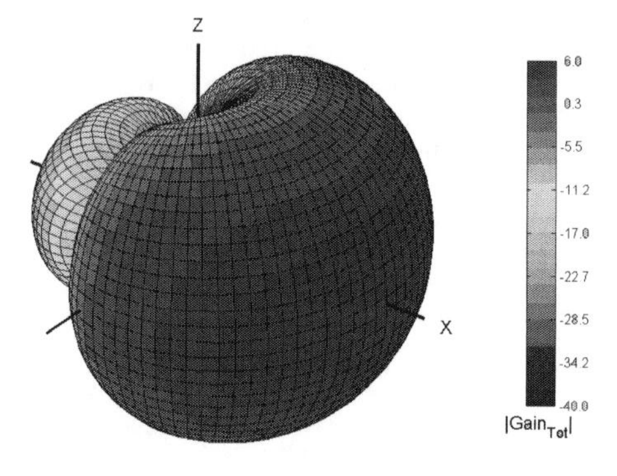

Figure 2. Pattern of the main antenna in HFSWR.

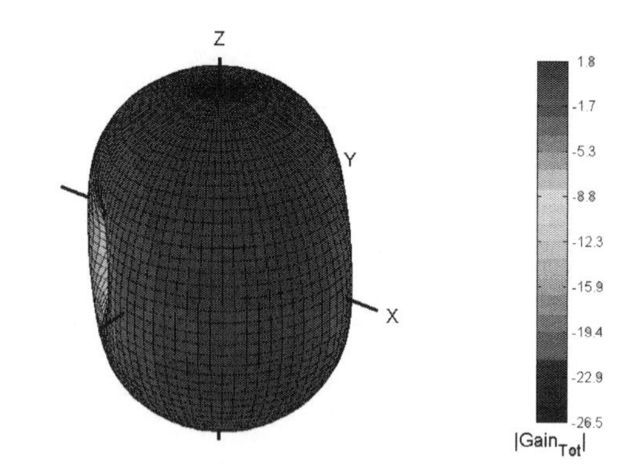

Figure 3. Pattern of the horizontally polarized antenna.

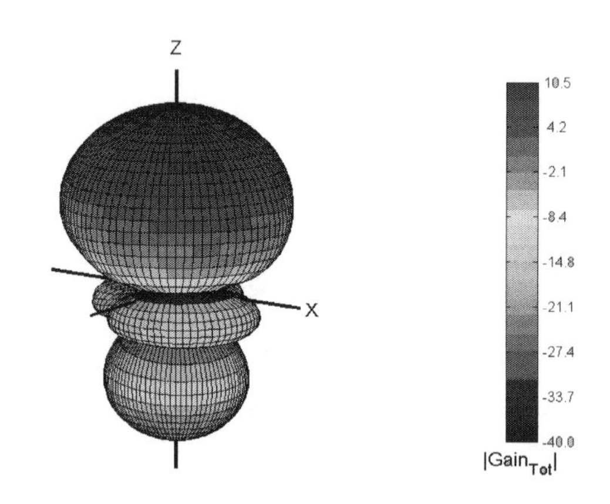

Figure 4. Pattern of the horizontally polarized antenna array.

The pattern of the main antenna in HFSWR is shown in Figure 2. The Pattern of the horizontally polarized antenna is in Figure 3. In this case, the work frequency of the horizontally polarized antenna array is 5MHz. In the figure, the X axis represents the signal arrival direction; Z axis represents the Es layer clutter arrival direction.

From the figure, it is can be seen that the vertically polarized main array has a big gain on the target signal. And the gain of target signals of the horizontally polarized antenna is smaller than the gain of the vertically polarized main array. At the mean time, the horizontally polarized antenna has a bigger gain on Es layer clutter than the vertically polarized main array.

From Figure 4, it also can be seen that the horizontally polarized antenna array has a bigger gain on Es layer clutter than a single horizontally polarized antenna on the X axis direction, and the horizontally polarized antenna array has a lower gain on the target signal than a single horizontally polarized antenna on the Z axis direction. So the clutter-to-signal ratio in the horizontally polarized antenna array output will be much bigger.

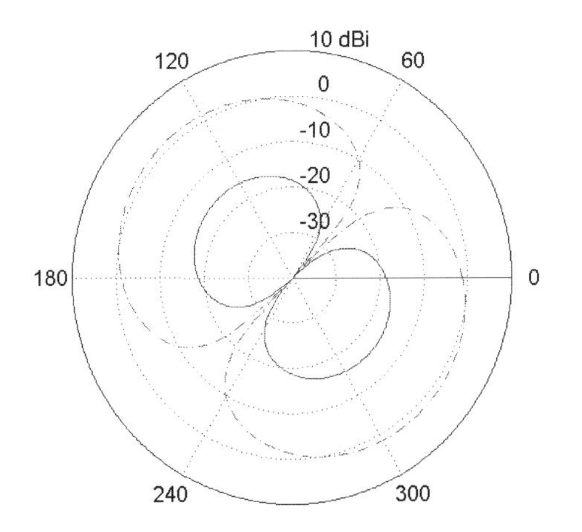

Figure 5. Horizontal plane pattern of the horizontally polarized array.

For further analysis, horizontal plane patterns of the horizontally polarized antenna array and a single horizontally polarized antenna are given in Figure 5. The solid line represents the horizontally polarized antenna array, and the dash line represents a single horizontally polarized antenna. The azimuth angle 0^0 represents the signal arrival direction; the azimuth angle 90^0 represents the Es layer clutter arrival direction.

The horizontally polarized antenna array has a 20dB bigger gain on target signal than a single horizontally polarized antenna.

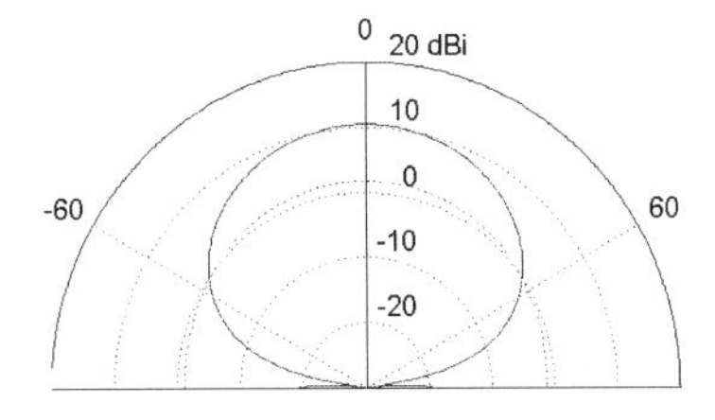

(a) Vertical plane pattern for angle 0^0.

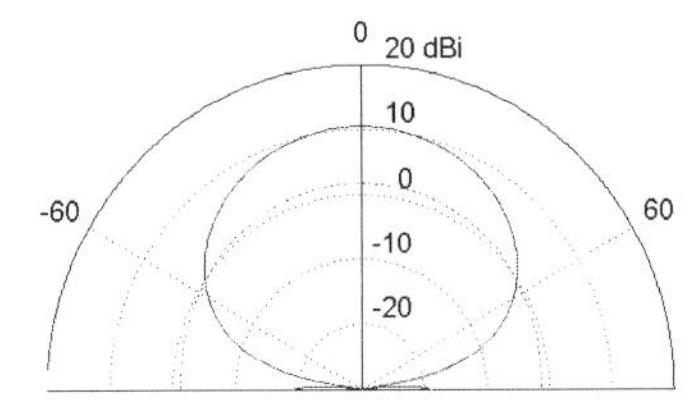

(b) Vertical plane pattern for angle 90^0.

Figure 6. Vertical plane patterns of the horizontally polarized antenna array.

The vertical plane patterns of the horizontally polarized antenna array and a single horizontally polarized antenna are given in Figure 6. The azimuth angle 90^0 represents the signal arrival direction; the azimuth angle 0^0 represents the Es layer clutter arrival direction. The solid line represents the horizontally polarized antenna array, and the dash line represents a single horizontally polarized antenna.

The horizontally polarized antenna array has a 10dB bigger gain on Es layer clutter than a single horizontally polarized antenna.

4 ANALYZE OF THE CLUTTER-TO-SIGNAL RATIO

Then the clutter-to-signal ratios in the received signal of the horizontally polarized antenna array are analyzed. For comparison purposes, the clutter-to-signal ratios in the received signal of a single horizontally polarized antenna are given at the same working condition.

In the analysis, A (dB) represents the magnitude of the target signal; B (dB) represents the magnitude of the Es layer clutter; $Gain_S$ (dB) represents the gain of the horizontally polarized antenna array on target signal; $Gain_C$ (dB) represents the gain of the horizontally polarized antenna array on Es layer clutter.

So the clutter-to-signal ratio in the received signal of the horizontally polarized antenna array is given by

$$\mathrm{CSR}_{\mathrm{ARRAY}} = \mathrm{B} + \mathrm{Gain}_{\mathrm{C}} - \mathrm{A} - \mathrm{Gain}_{\mathrm{S}} \tag{1}$$

Similarly, $G'ain_S$ (dB) represents the gain of the horizontally polarized antenna array on target signal; $G'ain_C$ (dB) represents the gain of the horizontally polarized antenna array on Es layer clutter.

So the clutter-to-signal ratio in the received signal of the horizontally polarized antenna is given by

$$\mathrm{CSR}_{\mathrm{ANTENNA}} = \mathrm{B} + \mathrm{G'ain}_{\mathrm{C}} - \mathrm{A} - \mathrm{G'ain}_{\mathrm{S}} \tag{2}$$

Then, the clutter-to-signal ratio in the received signal of the horizontally polarized antenna array is improved by

$$CSR_{IMPROVE} = Gain_C - Gain_S - \left(G'ain_C - G'ain_S\right) \tag{3}$$

The analysis in section 3 shows that the horizontally polarized antenna array has a 20dB bigger gain on the target signal than a single horizontally polarized antenna, and the horizontally polarized antenna array has a 10dB bigger gain on Es layer clutter than a single horizontally polarized antenna.

So the improvement of the clutter-to-signal ratio in the received signal of the horizontally polarized antenna array compared with a single horizontally polarized antenna is nearly 30 dB.

5 SIMULATION ANALYSIS

According to the results given in the last section, the Es layer clutter simulates the model given in the literature (Zhao. 2008). The same parameters are used in the simulation of Es layer clutter received by the main antenna and the horizontally polarized antenna array. Considering the different types of main antenna and auxiliary antenna used in the actual system, a fixed phase difference, a fixed amplitude attenuation and a small random phase fluctuation are added in the simulation of the Es layer clutter received by the auxiliary antenna compared with the Es layer clutter received by the main antenna. At the same time, considering the difference amplitude of the target signals received by the main antenna and auxiliary antenna, the target signal added in the reception of the main antenna is 30 dB bigger than the target signal added in the reception of the auxiliary antenna. The spectrums of the simulation Es layer clutter in the main antenna and auxiliary antenna are shown in Figure 7.

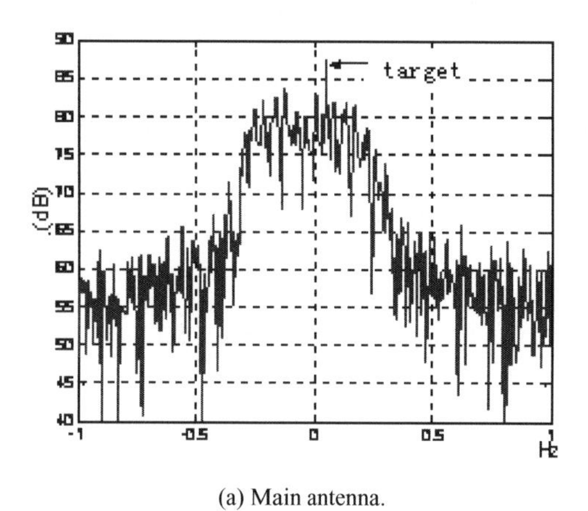

(a) Main antenna.

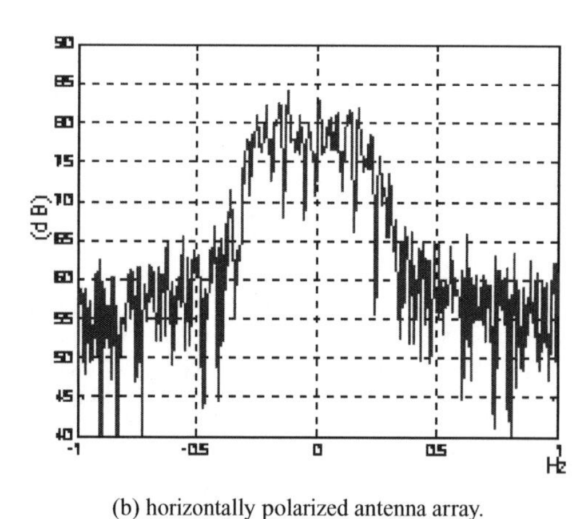

(b) horizontally polarized antenna array.

Figure 7. Spectrums of the main antenna and horizontally polarized antenna array.

The Es layer clutter received by the main antenna is suppressed using the adaptive cancellation algorithm given in the literature (Zhao et al. 2006). The cancellation result is shown in Figure 8.

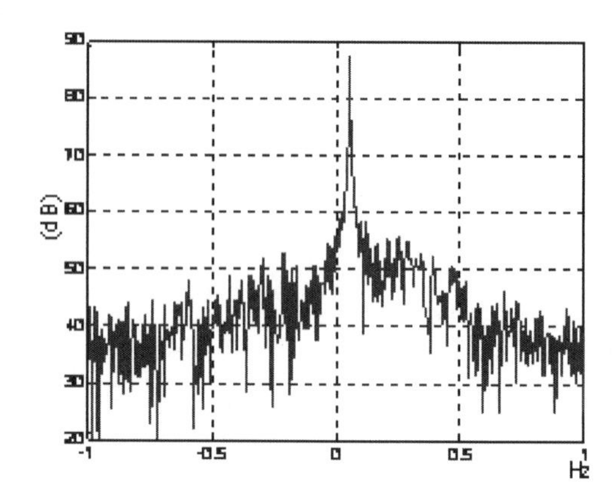

Figure 8. Spectrums of the main antenna after adaptive cancellation.

From Figure 8, it can be seen that after cancellation, the Es layer clutter amplitude has been significantly reduced, and the simulation target has been revealed obviously.

6 CONCLUSION

Aiming at the problem of ionospheric Es layer clutter suppression in HF radar, a horizontally polarized antenna array is designed. The synthetically method given in the literature (Liu et al. 2004) is used in the calculation of the weights for each antenna elements in the horizontally polarized antenna array. The simulation results of the horizontally polarized antenna array show that the clutter-to-signal ratios in the horizontally polarized antenna array output have improved obviously. And Es layer clutter amplitude after cancellation has been significantly reduced, and the simulation target has been added obviously.

ACKNOWLEDGMENT

This work is financially supported by "the National Natural Science Foundation of China (60939002, 61151002, U1433115)", "the Domestic civil aircraft project of Ministry of Industry and Information Technology (CXY2012SH16)", and "the Open Research Fund of The Academy of Satellite Application under grant NO. 2014_CXJJ-TX_12".

REFERENCES

Liu Yuan, Deng Weibo & Xu Rongqing 2004. A Study on Superdirectivity in HF band. *CJMW2004, Harbin, August, 2004*: 276–279.

Wan, X.R., Cheng, F. & Ke, H.Y. 2005. Sporadic-E Ionospheric Clutter Suppression in HF Surface-Wave Radar. *IEEE International Radar Conference, 9–12 May 2005*: 742–746. IEEE.

Wan, X.R., Ke, H.Y. & Wen, B.Y. 2005. Adaptive Cochannel Interference Suppression Based on Subarrays for HFSWR. *Signal Processing Letters, IEEE* 12(2): 162–165.

Gao, H.T., Zheng, X. & Li, J. 2004. Adaptive Anti-Interference Technique using Subarrays in HF Surface Wave Radar. *Radar, Sonar and Navigation, IEEE Proceedings* 151(2):100–104.

Nuttall,A.H. & Cray, B.A. 2001. Approximations to Directivity for Linear, Planar, and Volumetric Apertures and Arrays. *IEEE Journal of Oceanic and Engineering* 26(3): 383–398.

Stutzman, W.L.1998. Estimating directivity and gain of antennas. *IEEE Antennas and Propagation Magazine* 40(4): 7–11.

Lee, M.J., Song,L., Yoon, S. & Park, S.R. 2000. Evaluation of directivity for planar antenna arrays. *IEEE Antennas and Propagation Magazin*e 42 (3): 64–67.

Zhao Long. 2008. A Model for the Ionospheric Clutter in HFSWR Radar. International Conference on Information Management, Innovation Management and Industrial Engineering, 2008. ICIII '08, Taipei, 19–21 December 2008 :179–182. IEEE.

Zhao Long & Zhang Ning. 2006. Adaptive Cancellation of Es Layer Interference using Auxiliary Horizontal Antenna. *Journal of Systems Engineering and Electronics* (6): 313–315.

Leong, H. W.H. 1999. Adaptive Suppression of Skywave Interference in HF Surface Wave Radar Using Auxiliary Horizontal Dipole Antennas. *IEEE Pacific Rim Conference on Communications, Computers and Signal Processing, Victoria, BC, 22–24 August 1999*: 128–132. IEEE.

Electronics, Communications and Networks IV – Hussain & Ivanovic (eds)
© 2015 Taylor & Francis Group, London, ISBN: 978-1-138-02830-2

Speaker recognition based on support vector machine

Hai-jun Zhao, Hui Cao* Wan-jun Huang, Jia-ting Qiao & Jing Wei
Shaanxi Key Laboratory of Ultrasound, Shaanxi Normal University, Xi'an, China

ABSTRACT: Because a large number of redundant information exists in the Mel Frequency Cestrum Coefficients (MFCC) of speaker recognition programs, the method of Mean Impact Value (MIV) was introduced to select features of MFCC. Using this method, redundant features were removed and the MFCC had a greater impact on the recognition results taken as the input of SVM. Then, SVM was used to train input parameters and complete voice recognition. MFCC parameter variables of each dimension plus/minus were studied. After the verification of simulation experiments, MFCC parameters of each dimension plus/minus 90% were shown to improve the accuracy rate of classification to a greater extent, reduce recognition time, and improve operating efficiency

1 INTRODUCTION

Speaker recognition technology (also called voiceprint recognition technology) is a kind of biological authentication technology that can recognize a speaker automatically based on speech parameters in a speech waveform reflecting the speaker's physiological and behavioral characteristics (Wu & Yang 2009). Speaker recognition works by extracting the unique voice features of a speaker after obtaining samples of speech and storing them in a database; later the voice can be matched according to the features in the recorded samples and thus determining the speaker's identity. Speaker recognition has attracted attention throughout the world because of its economic convenience and accuracy.

The extraction of speech feature parameters is critical in speaker recognition. The main feature parameters presently used are linear predictive coding or LPC (Bouzid 2009), linear prediction cepstrum coefficient or LPCC (Zhang et al. 2007), and Mel frequency cepstrum coefficients or MFCC (Hosseinzadeh & Krishnan 2007). LPC and LPCC cannot describe the features of a speaker accurately because they do not use nonlinear frequency conversion or consider the characteristics of human ears. Therefore, the most commonly used parameter in speaker recognition is MFCC. All the speech samples studied for this method were clean speeches recorded in a quiet environment. This study used a recent machine learning method based on a statistical learning theory, SVM, to conduct the simulation experiments with accurate recognition performance of the MFCC parameters (Cao et al. 2013).Although a speech classification rate has been developed, there are large amounts of redundant information that can affect the classification rate in the MFCC feature parameters. MIV was used in this paper to select the MFCC parameters that have a greater effect on recognition results. It removed the MFCC parameters with negative impacts, reduced the computational complexity, and then developed a performance model.

2 SUPPORT VECTOR MACHINE

SVM has great adaptability and generalization capabilities and can resolve problems such as small samples, nonlinear and local minima, and other shortcomings (Wen et al. 2014). SVM is used to distinguish two kinds of samples with marks and has two components which include training and testing. For the linearly separable pattern, it considers the training samples$\{x_i, y_i\}_{i=1}^{N}$, where x_i is the sample of number I, and$y_i \in \{-1, 1\}$. Assuming that there is a hyper plane used to separate:

$$w \cdot x + b = 0 \tag{1}$$

where w is the normal vector of a hyperplane and b is the constant term. The main idea of SVM is to build a hyperplane as the decision surface and make the isolation edge between positive examples and negative examples become maximal. In other words, the optimal classification hyperplane is equivalent to obtaining the maximum interval. The special data point(x_i, y_i)

*Corresponding author: *caohui@snnu.edu.cn*

meeting the following conditions is called a support vector:

$$w \cdot x_i + b = -1, y_i = -1 \text{ or } w \cdot x_i + b = 1, y_i = 1 \quad (2)$$

Support vectors are the data points which are nearest to the decision surface. (Figure 1)

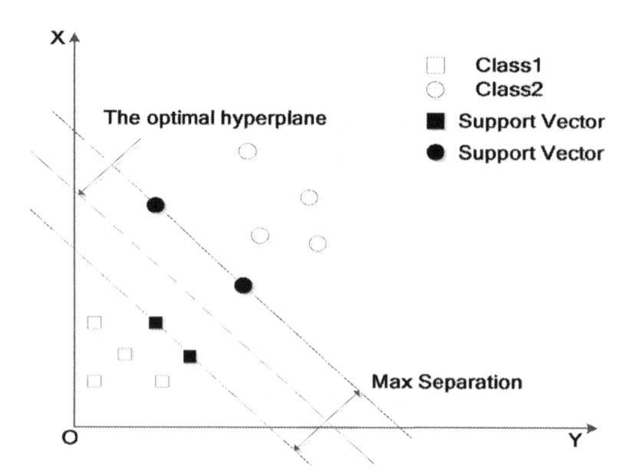

Figure 1. Optimal hyperplane.

Supposing there is two support vectors: $(x_1,1)$and $(x_2,-1)$, then the interval between the positive and negative examples would be:

$$dis = \frac{w}{\|w\|}(x_1 - x_2) \quad (3)$$

where $\frac{w}{\|w\|}$ is the unit normal vector. If $w \cdot x_1 + b = 1$ and $w \cdot x_2 + b = -1$, then $dis = \frac{2}{\|w\|}$.

If wanting to make $\frac{2}{\|w\|}$ the maximum, obtain the minimization of $\|w\|$ or $\frac{\|w\|^2}{2}$.

The basic method of SVM for nonlinear patterns that cannot be separated is to use kernel functions to map the sample points of the input features to the high-dimensional feature space. Then, data are divided by hyperplane and become linear and separable in the high dimensional feature space. So, the pattern is transformed to a linear separable one. Finally, it is corresponded to the nonlinear classification of low dimensional space (Liet al. 2013). Then, the slack variation $\xi \geq 0$ is introduced and should meet the following condition:

$$y_i(w \cdot x_i + b) + \zeta_i \geq 1 \quad (4)$$

This problem can be solved by obtaining the saddle point of the Lagrange function, which is:

$$J(w,b,a) = \frac{\|w\|^2}{2} + c\sum_{i=1}^{N}\zeta_i - \sum_{i=1}^{N} a_i\left[y_i(w \cdot x_i + b) - 1 + \zeta_i\right] \quad (5)$$

The auxiliary non-negative variable a_i is called the Lagrange multiplier and c is the penalty factor. By obtaining the partial derivative of w and b, setting them at zero and arranging the J, the dual problem of the initial problem is obtained. Then the optimal function with the use of the two planning methods is achieved:

$$f(x) = \text{sgn}(\sum_{i=1}^{N} a_i^* y_i K(x, x_i) + b) \quad (6)$$

3 MEAN IMPACT VALUE

As the index evaluating the importance of each independent variable's effect to a dependent variable, MIV's symbol represents the relevant direction and its absolute value represents the relevant importance of effect. The independent variables researched in this paper are dimensional MFCC feature parameters, and the dependent variable is the speech classification rate. The concrete process of calculation (Shi et al. 2010) is as shown in Figure 2:

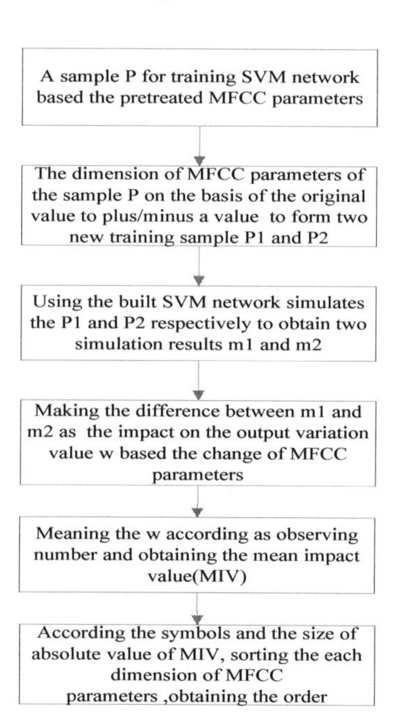

Figure 2. Flow diagram of MIV variable screening based on SVM.

4 SIMULATION EXPERIMENT

Simulation Environment: Windows XP operating system and MATLAB version 7.12.0. This experiment used the TIMIT voice database, which is a pure voice database in English libraries. In the TIMIT database, each speaker contained 10 segments of different voices, a different speaker said the same words, and the length of each speech was approximately 3-6s; voice sampling frequency was 8 KHZ, precision was 16 bit.

The TIMIT database selected six different male voices saying the same words. Each corresponding MFCC feature was extracted, and it selected the top 200 groups of data to transform parameters, including the first 150 samples of data as the experimental training samples. The latter 50 samples of data were selected as the experimental test samples. The recognition model was SVM; the dimension of MFCC features was 16; group of Mel filter was 24; the number of test: 2, 4, and 6 persons; Test Number: 2, 4, and 6 person; Testing times: each identification rans 10 times.

4.1 *The sorting of Mean Impact Value (MIV)*

The mean impact value (MIV) method was introduced to calculate each dimensional data of the MFCC feature parameters. The sorting of each dimensional MIV occurred when the variation was 10%. Each dimensional MFCC features was $c_i(i=1, 2\ldots16)$. (Table 1)

Table 1. The MIV sorting of six voices.

MFCC parameters	Sorting of s_1	Sorting of s_2	Sorting of s_3	Sorting of s_4	Sorting of s_5	Sorting of s_6
c_1	1	1	3	1	1	1
c_2	13	2	1	6	3	2
c_3	10	7	8	12	10	4
c_4	3	9	5	4	11	10
c_5	5	6	4	5	2	5
c_6	4	5	6	3	4	8
c_7	2	10	13	2	9	3
c_8	7	3	2	8	6	6
c_9	8	13	7	9	7	7
c_{10}	14	8	12	11	12	12
c_{11}	11	12	9	10	5	9
c_{12}	6	4	10	13	8	11
c_{13}	9	11	15	7	14	13
c_{14}	16	14	14	14	15	14
c_{15}	12	15	11	15	13	15
c_{16}	15	16	16	16	16	16

The highest five dimensions were selected to form new feature parameters: c_1, c_4, c_5, c_6, and c_7. Two speakers were designated as s_1 and s_2, four speakers were s_1, s_2, s_3 and s_4, and six speakers were s_1, s_2, s_3, s_4, s_5 and s_6.

Compared to the 30% variation and the 10% variation, the restructured 9th dimension and 12th dimension were changed in s_1, the restructured 2nd dimension and 13th dimension were changed in s_3, the restructured 4th dimension were changed in s_5, and the restructured 3rd dimension was changed in s_6. Compared to the 50% variation and the 30% variation, the restructured 3rd dimension and 4th dimension were changed in s_4, the restructured 6th dimension was changed in s_2, the restructured 8th dimension was changed in s_3, and the restructured 5th,6th,7th,8th, and 9th dimensions were changed in s_5. Compared to the 70% variation and the 50% variation the restructured 5th, 6th, and 11th dimensions were changed in s_1, the restructured 9th and 11th dimensions were changed in s_4, the restructured 2nd,3rd, and 4th dimensions were changed in s_4, the restructured 6th,10th,11th, and 12th dimensions were changed in s_5. Compared to the 90% variation and the 70% variation, the restructured 4th,5th,6th,7th,8th, and 9th dimensions were changed in s_1, the restructured 4th,7th, and 9th dimensions were changed in s_2, the restructured 1st,10th,11th, and 14th dimensions were changed in s_3, the restructured 5th and 12th dimensions were changed in s_4, and the restructured 11th and 13th dimensions were changed in s_5.

4.2 *Support Vector Machine (SVM) classification*

4.2.1 *Experiment 1*

According to the sorting of MIV values, redundant features were removed and the new feature parameters were restructured. The Figures (3-7) showed the relationships of speech recognition rates with recombined dimensions.

Each of the dimensional MFCC features of the training samples was made plus/minus 10%, 30%, 50%, 70%, and 90% on the basis of the original value, and their corresponding speech recognition rates are shown in Figures 3-7. As can be seen, when the number of speakers changed from 2 to 4 to 6, the recognition rate first increased and then decreased, and then leveled off. Every speech recognition rate changed according to the restructured dimension. Meanwhile, it was concluded that the speech recognition rates basically reached a peak when the restructured dimension was 10.

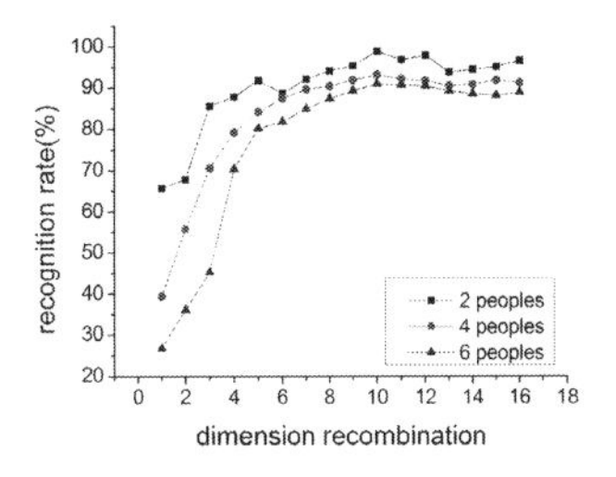

Figure 3. Plus/minus 10%.

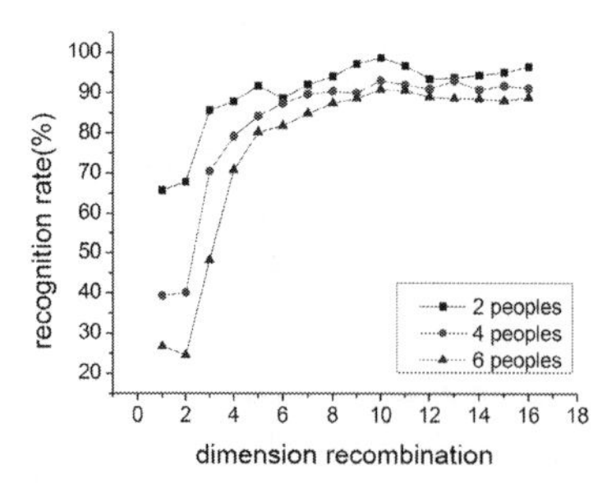

Figure 4. Plus/minus 30%.

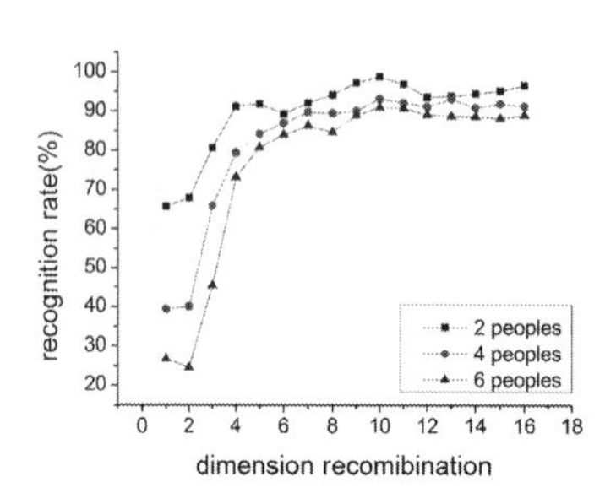

Figure 5. Plus/minus 50%.

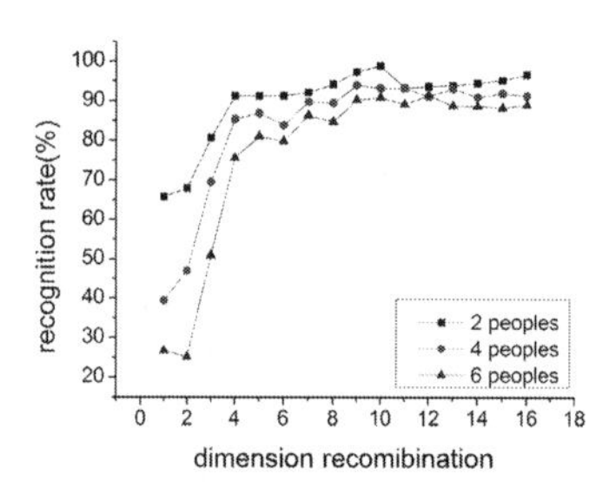

Figure 6. Plus/minus 70%.

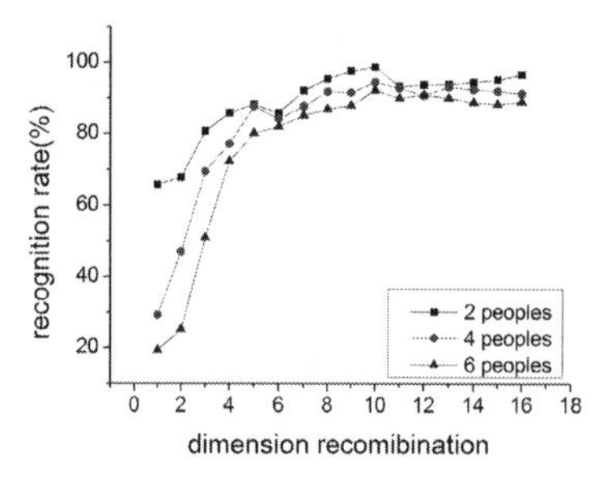

Figure 7. Plus/minus 90%.

4.2.2 *Experiment 2*

The restructured 10th-dimension feature was extracted separately to analyze the speech recognition rate and its running time.

By analyzing the data in Figure 8, when there were 2 speakers, they got the same recognition rate; when there were 4 speakers, the recognition rate of 90% variation was better than others; when there were 6 speakers, the recognition rate of 90% variation was also better than others.

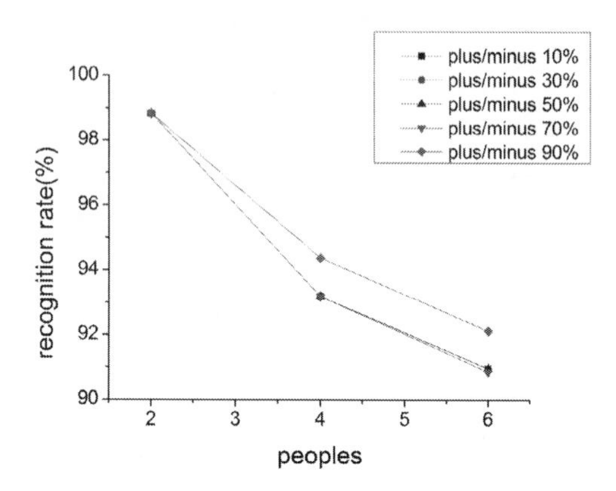

Figure 8. Comparison of recognition rate of the restructured 10 dimensions.

By analyzing the data in Figure 9, when there were 2 speakers, they took the same running time; when there were 4 speakers, the running time of 90% variation was less than others; when there were 6 speakers, the running time of 90% variation was also less than others.

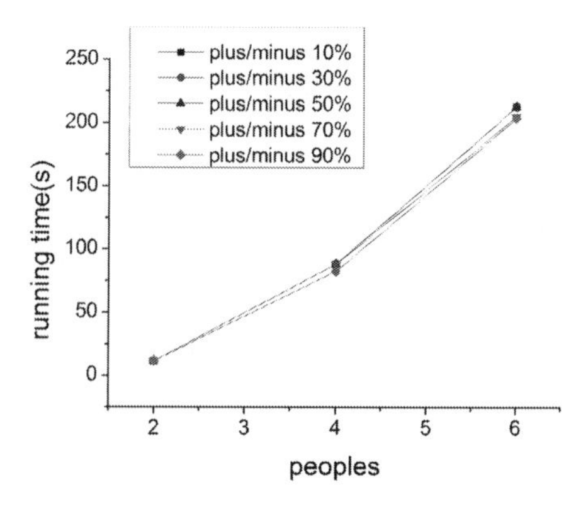

Figure 9. Comparison of running time of the restructured 10 dimensions.

In summary, when each of the dimensional MFCC features of training samples was made plus/minus 90% on the basis of the original value, the best speech recognition rate and the least running time were obtained.

4.2.3 *Experiment 3*

Next, the speech recognition rate and running time between the restructured 10 dimensions in the 90% variation and all 16 dimensions were analyzed.

By analyzing the data shown in Figure 10, when the number of speakers increased from 2 to 4 to 6, the speech recognition rate of the restructured 10 dimensions was better than all 16 dimensions.

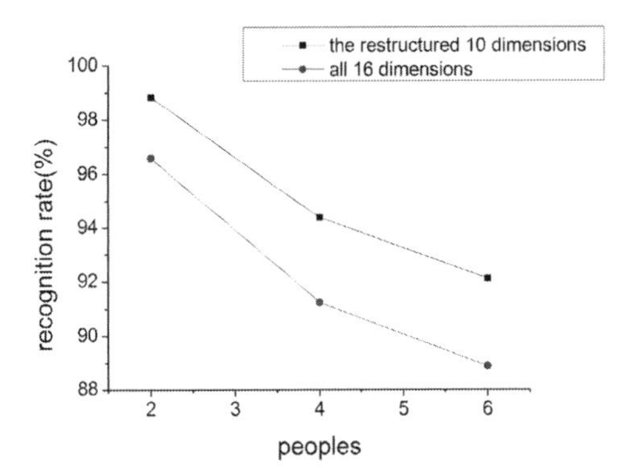

Figure 10. Comparison of recognition rate.

By analyzing the data in Figure 11, when the number of speakers increased from 2 to 4 to 6, the running times of the restructured 10 dimensions were all less than all 16 dimensions.

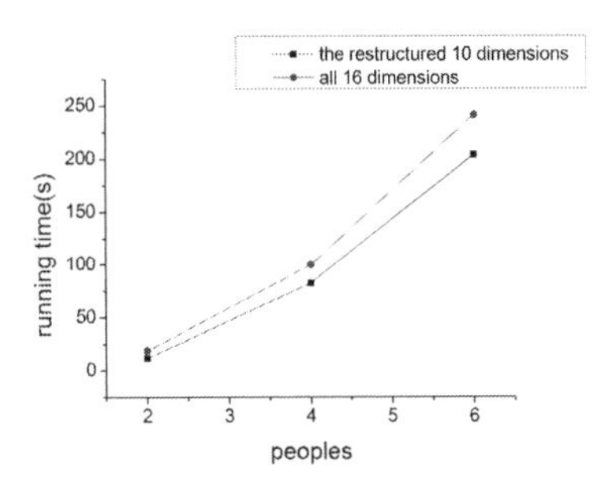

Figure 11. Comparison of running times.

Therefore, when the dimensional MFCC features were regrouped, it not only reduced the redundant features, but also shortened the running times and improved operation efficiency.

ACKNOWLEDGMENT

The completion of this paper was done under the National Natural Science Foundation of China(11074159 11374199), Thanks for people who supported and helped me.

REFERENCES

Bouzid, M. 2009. Robust Quantization of LPC Parameters for Speech Communication over Noisy Channel. *Proceedings of the 2nd International Conference on the Application of Digital Information and Web Technologies*: 713–718.

Hosseinzadeh, D. & Krishnan, S. 2007. Combining Vocal Source and MFCC Features for Enhanced Speaker Recognition Performance Using GMMs. *Proceedings of the IEEE 9th Workshop on Multimedia Signal Proceeding*: 365–368.

Hui, C. A. O., Chen, X. U., Xiao, Z. H. A. O., et al. 2013. The Mel-frequency Cepstral Coefficients in Speaker Recognition. *Journal of Northwest University* 43(2): 203–208.

Li, S., Liu, R., Zhang, L. & Liu, H. 2013. Speech emotion recognition algorithm based on modified SVM. *Journal of computer applications* 33(7): 1938–1941.

Shi, F., Wang, X. C., Yu, L. & Li, Y. 2010. *MATLAB Neural Network analysis of 30 Cases*: 183–188. Beijing: Beijing University of Aeronautics and Astronautics.

Wen Guoqiang et al. 2014. Fault Diagnosis Technology for Ball Screw based on Mean Impact Value and Support Vector Machine. *Machine tool & hydrauligs* 42(3): 173–176.

Wu Chaohui & Yang Yingchun. 2009. *Models and Methods for Speaker Recognition*. Beijing: Tsinghua University.

Zhang, X., Guo, Y. & Hou, X. 2007. A Speech Recognition Method of Isolated Words Based on Modified LPC Cepstrum. *Proceedings of the IEEE International Conference on Granular Computing*: 481–485.

Electronics, Communications and Networks IV – Hussain & Ivanovic (eds)
© 2015 Taylor & Francis Group, London, ISBN: 978-1-138-02830-2

Accurate link correlation measurement in wireless sensor networks

Zhiwei Zhao, Gaoyang Guan, Qin Zhang & Xiaofan Wu*
College of Computer Science, Zhejiang University, Hangzhou, Zhejiang, China

ABSTRACT: Recently, many works have been done to use link correlation for performance improvement. The effectiveness of these designs heavily rely on the accuracy of link correlation measurement. In this paper, we investigate state-of-the-art link correlation measurement and analyze the factors that impact the accuracy. We then propose a novel accurate and lightweight link correlation measurement (ALOM) based on the analysis of link correlation. We further implement ALOM as a stand-alone interface in TinyOS. Simulation and testbed results show that ALOM (1) achieves more accurate and lightweight link correlation measurement and (2) greatly improves the protocol performance of data collection via opportunistic routing.

1 INTRODUCTION

A wireless sensor network (WSN) contains a large number of small, inexpensive sensor nodes, which integrate sensing, computation, and wireless communication capabilities (Akyildiz et al. 2002). The key benefit of wireless broadcast is its ability to send a single packet to multiple receivers. Recently, some works (Guo et al. 2011, Wang et al. 2013, Srinivasan et al. 2010) have observed that link correlation can have a large impact on protocol performances.

Motivated by this observation, many works have been done to exploit link correlation for protocol optimization (Zhu et al. 2010, Guo et al. 2011, Basalamah et al. 2012, Zhao et al. 2014). Basically, the performance improvement of these works greatly relies on the accuracy of link correlation measurement. To measure link correlation, the widely used approach is beacon based measurement (We call this approach BCM). With this approach, each node periodically transmits beacon messages and records the received beacon receptions/losses in a bitmap (denoted as "1" and "0"). Each node then broadcasts its bitmaps to the corresponding beacon sender nodes. After receiving the bitmaps, a node can calculate the correlations between each pair of its outbound links.

With real-world experiments and analysis, however, we observe that beacon based approach have the following drawbacks: First, it lacks accuracy. Link correlation is often used to estimate performance metrics (e.g., expected number of transmissions) of data transmissions. More specifically, link correlation should be essentially the correlation of data packet receptions/losses. However, beacon messages are different with data packets in both data length and transmission rate, which, as we will discuss in Sec. II, are two important impacting factors of link correlation. Second, BCM relies only on the historical statistics and thus cannot provide timely (short-term) link correlation measurement. For traffic-dense applications, a timely link correlation measurement is strongly required (Dong et al. 2011). Third, the periodically exchanging beacons incur considerable measurement overhead.

To address these problems, in this paper, we propose an Accurate and Lightweight link cOrrelation Measurement (ALOM). Compared with BCM, ALOM has three salient features: First, instead of beacon messages, ALOM directly exploit data packet reception/loss traces instead of beacons, resulting the desired data packet level link correlation. Second, ALOM uses Received Signal Strength Indicator (RSSI) to calibrate the measurement results. Since RSSI indicates the immediate link behaviors, we can extract short-term link correlation with RSSI trace. By combining both RSSI-based result and the data reception trace based statistical result; we can get a timely yet accurate measurement of link correlation. Third, ALOM uses beacon messages only in the startup session. After that, ALOM depends only on data reception history and the immediately measured RSSI. RSSI measurements are fast and cheap (Fonseca et al. 2007), and the elimination of beacon messages greatly reduces the measurement overhead.

We implemented ALOM in TinyOS 2.1.2 with TelosB motes as a stand-alone interface. Then we incorporate ALOM into existing link correlation based opportunistic protocols to study its impact on end-to-end protocol performances (number of transmissions and delay). Simulation and testbed experimental

*Corresponding author: *wuxiaofan@zju.edu.cn*

results show that ALOM (1) provides more accurate link correlation measurement and (2) greatly reduces the measurement overhead. (3) ALOM-based protocols outperform their counterparts.

The contribution of this paper is summarized as follows:

We demonstrate the inaccuracy of beacon based measurement and give the reasoning (Sec. 2).

We combine link and physical layer information for accurate/lightweight link correlation measurement ALOM (Sec. 3).

We implement ALOM in TinyOS 2.1.2 and incorporate it into existing protocols. Experimental results show that ALOM provides more accurate link correlation measurement and ALOM-based protocols outperform existing BCM-based protocols (Sec. 4).

2 PRELIMINARIES AND RELATED WORKS

2.1 *Basics of link correlation*

Definition and metrics. We follow the common definition of link correlation as the correlation of packet reception/loss on different links. Among all these metrics, conditional packet reception/loss probabilities (CPRP/CPLP) are most widely used due to its meaningfulness and applicability for estimating upper layer performance such as expected number of transmissions (ETX). Therefore in this paper, we follow the metrics CPRP and CPLP.

Impacting factors. We first introduce the impacting factors of link correlation because it demonstrates the rationale of the proposed link correlation measurement. Previous works have concluded that link correlation is greatly affected by shadow fading and interference (Srinivasan et al. 2010, Wang et al. 2013). If two nodes have similar signal strengths and interference, they are more likely to lose packets at the same time. As a result, the link correlation between the two links would be strong.

With these two factors, however, we can still not decide link correlation. The reason is that packet transmission may not happen together with the fading/interference, i.e., although there is interference source (e.g.,Wi-Fi APs), the packet transmissions can possibly survive in the white space of WiFi communication. Intuitively, the fractions of interference captured by packet transmissions directly decide link correlation at different receivers. If the packets are transmitted when there is strong interference at both receivers, they are more likely to have common packet losses and link correlation is strong. Otherwise, link correlation is more likely to be weak.

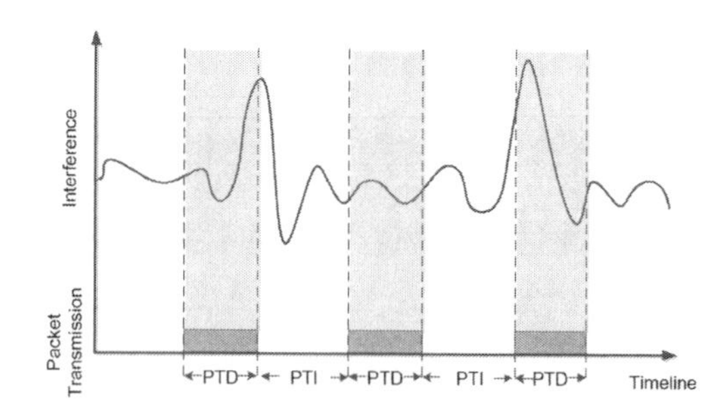

Figure 1. An example to illustrate the impacting factors of link correlation.

For given interference, clearly, data length and transmission rate are two factors that decide which fractions of interference have impact on packet transmissions and link correlation (It is worth noting that we mainly focus on the scenario where data packets are transmitted with fixed rate (Mo et al. 2009). We use a simple example with one sender and two nearly-placed receivers. The sender keeps broadcasting packets and we study the link correlation between the two links. As depicted in Figure 1, the x-axis denotes time and the y-axis denotes the environmental interference. The grey blocks denote broadcast packet transmissions. PTD means packet transmission duration, which is determined by packet length given the bit-rate. PTI means packet transmission rate. We can see that the three packets capture different fractions of interference, resulting certain link correlation values. Clearly, different PTIs and PTDs will lead to different interference impacts on the packet transmissions. This supports our conclusion that beacon-based measurement cannot obtain the desired data packet level link correlation.

2.2 *Measurement*

Existing works employing beacon messages for measuring link correlation works as follows: Each node periodically transmit beacon messages and record the beacon receptions/losses from other nodes with bitmaps, a "1" denotes a packet reception and a "0" denotes a packet loss. Then each node transmits the bitmaps to corresponding beacon senders. When a node obtains the bitmaps, it is able to calculate link correlation (CPRP/CPLP) between its outbound links. For example with one sender S and two receivers A and B, given two bitmaps indicating the packet receptions on S->A and S->B, "10011" and "11001", the link correlation can be calculated as the number of common packet receptions divided by packet receptions on S->B, i.e., $c_{A,B}=2/3$. This means S->A will receive a packet with probability of 2/3 given that S->B receives a packet.

The above beacon based link correlation measurement has two main drawbacks as follows: First and most important, it lacks accuracy. Beacon based measurement uses the beacon receptions to measure link correlation. As we analyzed in Sec.II.A, since both packet length and transmission rate of beacons are different with that of data packets, the fractions of interference captured by beacon based measurement are different with the interference that have real impact on data transmissions and the measurement result lacks accuracy. Second, The periodic beacons incur considerable overhead. In (Guo et al. 2011), beacons are transmitted every 10 seconds. However, in typical sensor networks, data packets are transmitted every 10 minutes (Mo et al. 2009). The beacon overhead is unacceptable.

Compared to the above common used approach, ALOM 1) uses data reception bitmaps for link correlation measurement, which is inherently accurate (the elimination of beacons also greatly reduces the measurement overhead) and 2) combines both physical layer parameter and network layer statistics for flexible link correlation measurement (considering both long-term and short-term link behaviors for correlation measurement).

3 MAIN DESIGN

In this section, we give the main design of our approach.

3.1 *ALOM measurement*

We use a simple example to describe the working progress of ALOM. Suppose node S transmits data packets to nodes A and B every t time with packet length pl. We would like to calculate the link correlation between S->A and S->B. Nodes A and B record the bitmaps as follows: When calling ALOM for data packet transmission, a sequence number (SN) is added to the data packet footer. The SNs are used to infer packet losses. For example, if node A receives packets 1, 2, and 5, it knows that 3 and 4 are lost and will use "11001" to denote the packet receptions/ losses from node S.

Before the real data packets are sent, ALOM first send a probing packet for obtaining the RSSI at its receivers. When node A and B receives the probing packet, they reply a respond packet that conveys the measured RSSI value and the historical bitmaps (Note that there is no ACKs under broadcast operation). After that, a node can calculates link correlation as:

$$c_{A,B} = \alpha \cdot ch_{A,B} + (1-\alpha) \cdot \quad (1)$$

where ch is the historical correlation value, ci is the instant correlation value and is a weighting factor.

When node S obtains the bitmaps, it can calculate the ch between S->A and S->B as the following Equation.

$$ch_{A,B} = \frac{\sum_{i=1}^{n}(b_A[i] \& b_B[i])}{\sum_{i=1}^{n} b_B[i]} \quad (2)$$

where $b_A[i]$ denotes the ith bit in A's bitmap b_A. Next, we use RSSI to infer the instant link quality of the two links, and further estimate the short term link correlation ci as follows:

$$ci_{A,B} = \frac{q_A^S * q_B^S}{q_B^S} = q_A^S \quad (3)$$

where q^S_A denotes the link quality of link S->A. We can see that the instant conditional packet reception ratio equals the link quality. It is reasonable that the instant packet reception is determined only by the signal to interference and noise ratio (SINR), while link correlation reflects the variations of the SINR in a period of time. After obtaining both $ch^S_{A,B}$ and $ci^S_{A,B}$, node S can calculate the expected link correlation as Eq. (1).

3.2 *Obtaining & processing necessary parameters*

Bitmaps and link quality are the two necessary parameters for link correlation calculation.

3.2.1 *Bitmaps*

As described above, the bitmaps are recorded at receivers and returned to the senders. There are two key issues: First, bitmaps at different receivers should be aligned with the same time offset. We use 32-bit length bitmaps to record the latest 32 received/lost data packets. However, the lost packet is indirectly identified (using the sequence number difference), i.e., when a packet loses a packet, it can identify this packet loss only when it receives a new packet in the future. As a result, at the time when a node loses a packet, its bitmaps are not updated. Further, bitmaps of different receivers may record different sets of packet transmissions; the calculated link correlation would be meaningless. For example, sender S transmits 10 packets, node A receives 1,2,3,5,7,9,10 and node B receives 1,2,3,6,7,8. If we set the bitmap length as 5 and collect the bitmaps after the 10th packet transmission, the bitmaps of A and B will be 01011 and 00111. Clearly, node B's bitmap does not record the latest 5 packets (6-10), instead, it records packets 4-8. The reason is that it is unaware of packets 9 and 10's transmissions. To solve this inconsistency, we add a initial offset at the beginning of each bitmap to indicate the start position of the bitmap. With the offset, the sender can align the bitmaps and calculate the link correlation. We use the same example as the above to illustrate how this works: when node A and B prepare to return the bitmaps, they add the start positions. Then S receives two vectors: 6—01011 and 4—00111. Now S uses the common parts of the bitmaps (010 and 111 for packets 6-8 respectively) to calculate link correlation ch according to Eq. (2).

The second issue is the length of bitmaps: when the bitmap is long, it can represent more long-term link behavior but requires more memory overhead. Considering the TelosB platform has only 10KB RAM and the packet payload length is no longer than 114 bytes, we set the length of bitmaps according to the neighbor size: When a node has many neighbors, it uses short bitmaps to save memory/transmission overhead. When a node has small number of neighbors, it uses large bitmaps for more accurate measurement. In our experiment, we use 32-bit length bitmaps to record the packet receptions.

3.2.2 *Link quality*

Compared to the bitmaps, obtaining link quality incurs less overhead. We use the relationship between SINR (signal to interference and noise ratio) and PRR (packet reception ratio) for instant link quality estimation. The receiver keeps sensing the environmental noise and records the signal strength when receiving a packet, so that a receiver can obtain its SINR. After that, it attaches the SINR to the feedback message (containing the bitmaps) and sends it back to the senders. When the sender node receives the SINR value, it extracts link quality as the following Equation:

$$q = g(SINR) \tag{4}$$

where $g()$ is the SINR-PRR conversion function in (Fonseca et al. 2007).

3.3 *Practical issues*

As discussed above, we use a weighting parameter to calculate the integrated correlation metric. When α is large, the history correlation is estimated with more weight; When α is small, the instant correlation is estimated with more weight. If the packet is received/lost at both receivers, we mark the ground truth of link correlation as 1; If the packet is received at only one receiver, we mark the ground truth as 0. We record the ground truth and adaptively tune for more accurate measurement. The intuition is that the correlation value close to the ground truth should be more weighted: If the ground truth of link correlation is 1, we increase the weight of $max(ch,ci)$. Otherwise, we increase the weight of $min(ch,ci)$. Finding a theoretical optimal will be studied in future works.

4 EVALUATION

We implemented a stand-alone interface of our approach in TinyOS on TelosB motes.We first test the accuracy of link correlation measurement in TOSSIM simulation with a 10x10 network (the "meyer-heavy" noise trace is used and the gain of each link is randomly set in range of [-95,-70] dbm). Next we incorporate our approach into existing network layer protocols (Basalamah et al. 2012) to study the performance improvements. Our 4x10 testbed is used for real motes evaluation.

4.1 *The accuracy of our approach*

We use the link correlation calculated with actual data traces as ground truth: If the two links receive/lose a packet at the same time, the instant correlation is 1. Otherwise, the link correlation is 0. We use our approach to measure link correlation before each packet transmission and study its accuracy compared to the above ground truth. For example, if the estimated link correlation is 0.8 and the packet is received by both links, the accuracy is 0.8. We calculate the average accuracy for all packet transmissions. Figure 2 depicts the measurement accuracy of our approach.

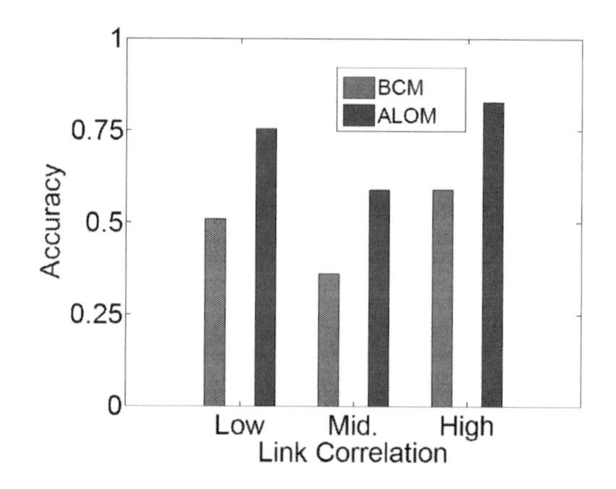

Figure 2. Measurement accuracy.

We can see that (1) for low and high link correlation, the approach accuracy is high; while for intermediate link correlation links, the approach accuracy is relatively low. The reason is that the conditional probability based link correlation metric has a PRR bias: when the two links are both with low/high PRR, the probability that they receive the packet at the same time will also be low/high. This is caused by the use of multiplication. As a result, when links are high or low, the estimated link correlation is more accurate. For intermediate links (PRR around 0.5), the estimated ci will be around 0.25. However, the actual link correlation with two 0.5 links can be any value from 0~1. The error is more likely to be large. (2) Our approach is more accurate than beacon-based link correlation measurement (BCM). The reason is that our approach extracts the link correlation directly from the data packet receptions. Due to the difference of data and beacon packets, the beacon based methods suffers from large measurement errors.

4.2 *Testbed results of protocol performance*

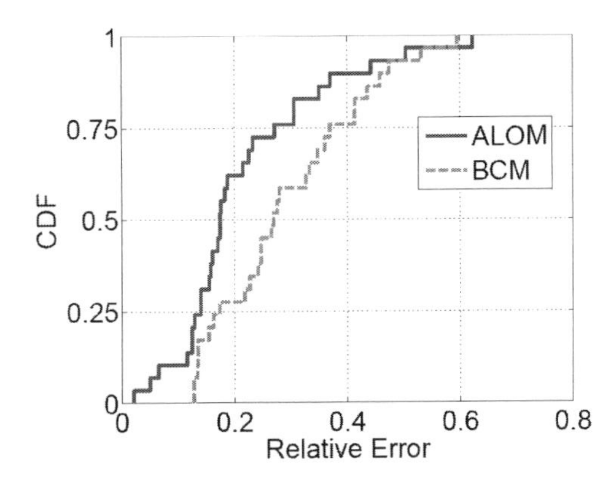

Figure 3. Measurement errors in opportunistic routing .

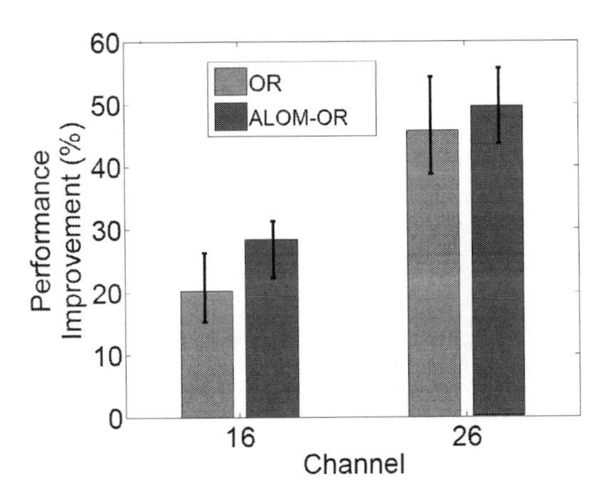

Figure 4. Performance improvement in opportunistic routing.

In this section, we study the protocol performance of opportunistic routing (Basalamah et al. 2012) with our approach. Figure 3 shows the measurement accuracy of our approach and the existing method. We can see that the measurement error rate of our approach is much less. The reason is the it uses data packets and RSSI for link correlation measurement; Figure 4 shows that ALOM-OR has more improvement than OR, compared with traditional routing. More specifically, the performance improvement is larger in channel 16 than in channel 26. The reason is that channel 16 is more severely interfered by WiFi traffic, and has stronger link correlation than channel 26. As a result, it has more neighborhoods with strong link correlation. A more accurate measurement can help OR to more accurate identify the strong link correlation links. Thus the performance improvement is larger in channel 16.

4.3 *Summary*

We can see that our approach-based protocols greatly outperform BCM-based protocols. The reason is that (1) BCM uses beacons receptions to measure the correlation of data receptions, which is rationally inappropriate to yield accurate link correlation based on data packets; (2) The use of periodic beacons incurs considerable overhead. According to the setting of (Guo et al. 2011), beacon overhead is even larger than the data transmission overhead; (3) BCM depends solely on historical receptions, which is not sufficient for timely correlation result. Differently, our approach combines both statistics and instantly measured RSSI for link correlation measurement, which can give a timely link correlation result.

5 CONCLUSION

In this paper, we investigated the state-of-the-art approaches of link correlation measurement, and proposed a novel link correlation measurement framework based on data flows (called our approach). Compared to existing beacon-based approaches, our approach 1) directly uses data packet receptions for link correlation calculation and 2) considers both historical data and the instant RSSI. Therefore, our approach is more accurate. Experiment results show that our approach achieves more accurate link correlation measurement and improves the performance of existing correlation-based works. Future direction lies in finding a theoretical optimal tradeoff between the historical correlation value and the instant value to meet various requirements from upper-layer protocols.

ACKNOWLEDGEMENT

This work is supported by National Key Technology R&D Program(2012BAI34B01).

REFERENCES

Akyildiz, I. F., Su, W., Sankarasubramaniam,Y., et al. 2002. Wireless sensor networks: A survey. *Computer Networks* 38: 393–422.

Basalamah, Kim, A., S., Guo, S., et al. 2012. Link Correlation Aware Opportunistic Routing. In *Proc. of INFOCOM, Orlando, FL, 25–30 March, 2012.* IEEE.

Dong, W., Liu, Y., Wang, C., et al. 2011. Link Quality Aware Code Dissemination in Wireless Sensor Networks. In *Proc. of ICNP, Vancouver, Canada, 17–20 Oct., 2011.* IEEE.

Fonseca, R., Gnawali, O., Jamieson, K., et al. 2007. Four-bit wireless link estimation. In *Proc. of HotNets, Atlanta, USA, 14–15 Nov., 2007.* ACM.

Guo, S., Kim, S., Zhu, T., et al. 2011. Correlated flooding in Low-duty-cycle Wireless Sensor Networks. In *Proc. of ICNP, Vancouver, Canada, 17–20 Oct., 2011.* IEEE.

Mo, L., He, Y., Liu, Y., et al. 2009. Canopy Closure Estimates with Greenorbs: Sustainable Sensing in The Forest. In *Proc. of SenSys, Berkeley, California, 4–6 November, 2009.* ACM.

Srinivasan, K., Jain, M., Choi, J., et al. 2010. The factor: Inferring protocol performance using Inter-link reception correlation. In *Proc. of MobiCom, Chicago, USA, 20–24Sep., 2010.* ACM.

Wang, S., Kim, S. M., Liu,Y., et al. 2013. Corlayer: a transparent link correlation layer for energy efficient broadcast. In *Proc. of MobiCom, Miami, USA, Sep. 30-Oct. 4, 2013.* ACM

Zhao, Z., Dong, W., Bu, J., et al. 2014. Exploiting link correlation for core-based dissemination in wireless sensor networks. In *Proc. of SECON, Singapore, June29 - July 3, 2014.* IEEE.

Zhu, T., Zhong, Z., He, T., et al. 2010. Exploring link correlation for efficient flooding in wireless sensor networks. In *Proc. of NSDI, San Jose, USA, 28–30 April, 2010.* USENIX.

Electronics, Communications and Networks IV – Hussain & Ivanovic (eds)
© 2015 Taylor & Francis Group, London, ISBN: 978-1-138-02830-2

The electromagnetic compatibility of a plasma antenna driven by HF power supplies

Jiansen Zhao, Xia Liu, Qinyou Hu*, Wei Liu & Hao Zhang
Merchant Marine College, Shanghai Maritime University, Shanghai, China

ABSTRACT: The Electromagnetic Compatibility (EMC) of a plasma antenna system generated by 13.56, 27.12, and 40.68 MHz High Frequency (HF) alternative current (ac) plasma power supplies was investigated using a field intensity meter, spectrum analyzer, and other sensitive devices. The results indicated that the strong EMI of the HF power supplies was 50 dB larger than the EMI of the direct current (dc) and low frequency ac power supplies. The interference included the induction field of the plasma, HF radiation, and harmonics. The induction field weakened as the distance between the antenna and field intensity meter increased. Moreover, the HF power supplies generated significant harmonic interference, which reached up to 10^{-3}times that of the power supply interference. Taking the spectrometer as the sensitive equipment, whose output spectrums shift seriously by the HF ac radiation. In addition, the EMI was suppressed by means of filtering and shielding.

1 INRODUCTION

It has been well known for decades that plasma antennas are antennas using plasma columns, usually produced by radio frequency (RF) power, instead of metal conductors or other dielectric materials as elements (Li 2007). They are comprised of insulating tubes filled with low pressure inert gas e.g. Neon, Argon (Rayner et al. 2004). In recent decades, plasma antennas have attracted more and more attention since they have numerous advantages over conventional metal antennas (Li et al.2010). For example, antenna parameters can be efficiently reconfigured by changing discharge conditions, which makes it possible to be used at wide wavebands and simplifies the matching network. Plasma can be rapidly created and destroyed by applying an electrical pulse to the discharge tube, hence it can be rapidly switched on and off. When it is on, it exhibits a good conductor providing a conducting medium for the applied RF signal. When it is off, the antenna is non-conducting and invisible to electromagnetic (EM) radiation, and the radar cross section (RCS) closes to 0.

The plasma antenna provides an integral and important part in both communication and plasma stealth technology.So far, many theoretical and experimental works have been conducted which studyimpedance, gain, phase, polar, radiation, dissipation, and many other properties. Kang and Alexeff (Alexeff et al. 2006) have successfully demonstrated both the operation of a plasma antenna in transmission and reception, as well as stealth features on a navy test range in San Diego. Borg has reported measurements of the efficiencies and radiation patterns of plasma column antenna elements used in communications. The radiation efficiency was reported to be in the range of 25 to 50% (Borg et al. 1999, 2000), and radiation patterns were found to be similar to those of antennas made from metal. The finite difference time-domain (FDTD) method (Qian et al. 2005, Li et al. 2009, 2010) was used to analyze the near and far fields of a plasma antenna and to determine the feasibility of constructing diverse antenna configurations at will. Kumar and Bora (Kumar et al. 2010, 2011) established a reconfigurable plasma antenna, and investigated the antenna properties of different plasma structures of a plasma column. By changing operating parameters such as the working pressure, drive frequency, input power, radius of the glass tube, length of the plasma column, and argon gas pressure, it was possible to transform a single plasma antenna into multiple small antenna elements. In addition, the effect on directivity with the number of plasma elements was also studied. Russo used a surfaguide device as plasma source for a plasma antenna. The surfaguide was optimized, realized, and used for the ignition of a plasma column which acted as a radiating structure. The coupling with the radiated signal network and plasma antenna efficiency were measured, demonstrating

*Corresponding Author: *qyhu@shmtu.edu.cn*

that a surfaguide can be effectively used to create and sustain the plasma conductive medium. The measurements highlighted that plasma antenna properties were strongly affected by the pump signal and that this signal required optimization in order to have the opitimal conductivity.

It is clear that the plasma antenna has gained increased attention throughout the world; however, the characteristics of EMC from a surface-wave-driven plasma antenna have not yet been extensively investigated. If EMC is not solved, the instrument of the same electromagnetic environment could be destroyed. In this paper, the experimental equipment was created in order to study the EMC of a plasma antenna driven by HF AC power supplies. In the experiment, we changed conditions such as discharge power, plasma source, gas pressure, and other parameters. The results may benefit the plasma application in modern engineering.

2 EXPERIMENTAL SETUP

The radiation equipment of different plasma sources is shown in Figure 1. .The equipment included two separate parts related to the characteristics of the power supplies: one for dc and low frequency (LF) ac(50 Hz and 5-20 kHz ac) power supplies (a) and another for 13.56, 27.12 and 40.68 MHz HF acones (b). We wanted to investigate the plasma driven induction field using various plasma power supplies. Hence, the signal transmitting and receiving devices were not used, while the field intensity meter, as shown in Figure 1 (b),was placed at a certain distance from plasma antenna and used to analyze the induction field of plasma.

The radiation field was analyzed by a spectrum analyzer (6) with a copper antenna (7). The copper antenna was fixed 20 m from the plasma antenna, which received the radiation signals generated by the different plasma power supplies. Finally, the EMIs from the plasma power supplies, especially that of the RF, were analyzed using a spectrometer. The fiber of the spectrometer was fixed near the discharge tube of the plasma antenna.

The discharge tubes in the dc and LF ac (50 Hz and 5-20 kHz ac) plasma antenna experimental systems were made of glass and filled with neon or argon; the gas cylinder, reducing valve strainer, and vacuum pump controlled the gas pressure. In the dc and low frequency ac systems, the discharge tubes were directly connected to the power supplies. In the HF ac plasma antenna system, the inductively coupling driving model was used with the inductive choke, which was winded around one side of the discharge tube.

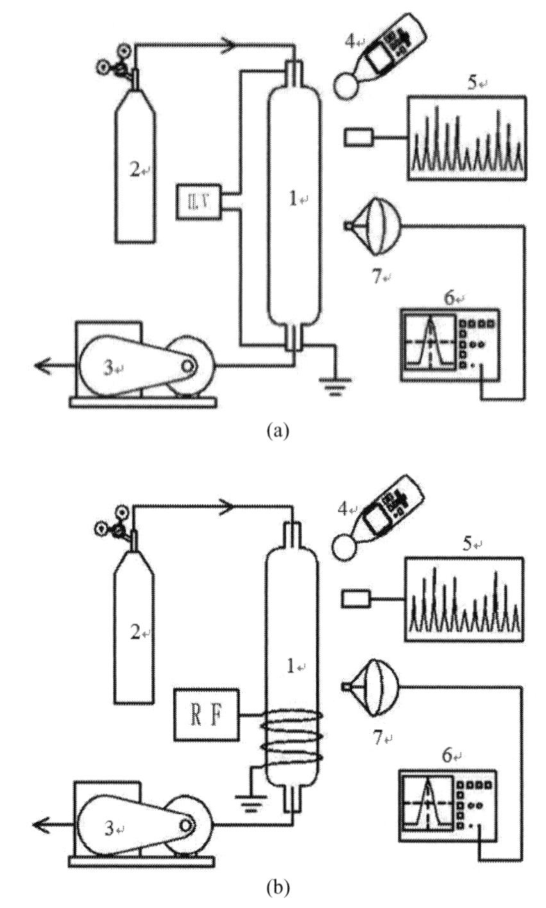

Figure 1. Experimental setup of the EMC of plasma antenna (a) Equipment for the dc and ac plasma antennae (b) Equipment for the RF plasma antenna.1.discharge tube 2.gas cylinder, 3.vacuum pump 4.field intensity meter 5.spectrometer, 6.spectrum analyzer 7.receiving antenna.

3 RESULTS

3.1 *Induction field of plasma*

EMIs must exist in plasma antennae systems due to the plasma sources. Plasma can also radiate electromagnetic waves with a given energy. The field intensity meter was used to measure the radiation of the plasma. First, the plasma power supply was switched off, and the distance between the field intensity meter and plasma antenna was fixed. The free-space field intensity, E_0 (V/m), was measured. Then, the power supply was switched on, and the total field intensity of the space,E_1 (V/m), was recorded. Hence, the induction field intensity of the plasma at the same location was E_1- E_0 (V/m). Next, the distance between the field intensity meter and plasma antenna was changed, and the field intensity was measured using the same method. The induction field of the dc plasma for every distance r is shown in Figure 2. The plasma was able to generate an induction field. When the plasma antenna was switched off, the field intensity, E_0, was approximately 0.2 V/m; while the plasma was on, the field intensity, E_1, was approximately 1.2 V/m. Thus, the induction field intensity from the plasma was 1 V/m. In addition, the induction field intensity decreased as the distance increased from 0.2 to 1 m. The plasma could have generated the induction field, but the intensity was limited, and could be weakened only by an increase the distance between the plasma and other devices.

Figure 2 also shows the radiation intensity of the plasma driven by a 5-20 kHz power supply. For a larger electron density(approximately$10^{17}m^{-3}$), the induction field intensity was larger than the plasma produced by the dc or 50 Hz ac sources (approximately 2×10^{16} m^{-3}). The reason the induction field intensity increased was verified. In the experiment, the discharge power was changed to analyze the changing of the induction field by the plasma.

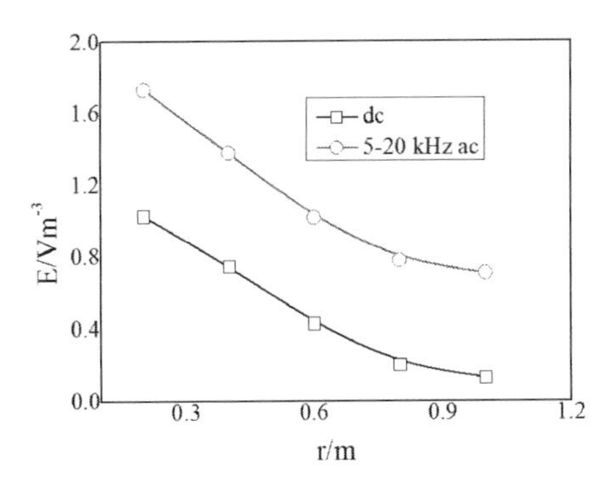

Figure 2. Electric plasma field driven by dc and low frequency ac discharge.

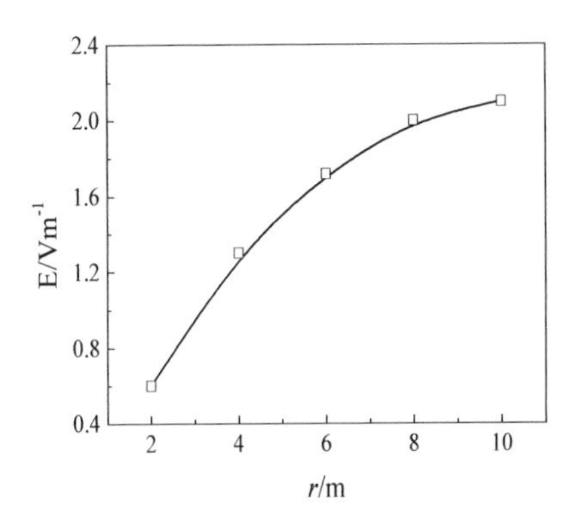

Figure 3. Electric plasma field at different discharge powers.

Figure 3 shows the induction field intensity of the plasma produced by a 5-20 kHz power supply at different discharge powers ranging from 2 to 10 W. The results indicated that the intensity increased as the discharge power increased. The electron temperature and density were diagnosed with a Langmuir probe. The electron temperature did not change significantly, but the density varied from $10^{16}m^{-3}$ at a2 W discharge power to approximately $10^{17}m^{-3}$ at a 10 W discharge power. Thus, the electron density was the most important factor that affected the induction field of the plasma.

3.2 *EMI of power supplies*

According to the experimental systems, the radiation of the plasma power supplies were investigated with the spectrum analyzer (Agilent N9310A, 9 kHz-3 GHz). In the experiment, the distance between the plasma and metal antennae was 10 m. The spectrums of the dc and LF ac power supplies are shown in Figure 4. The spectrums changed little, so the EMIs of the three power supplies were small and were unable to produce strong interference on the other devices under the same electromagnetic circumstance. Compared to the three power supplies above, the radiation values(13.56, 27.12, and 40.68 MHz) of the HF plasma power supplies were very large. As shown in Figure 5, unlike those of the three power supplies above, these signals appeared in the spectrum analyzer with several harmonic waves. The power of the signal was approximately2 dB with a white noise power of -55 dB, which was large enough to disturb or damage some devices nearby. Therefore, the HF plasma power supply could produce a large amount of radiation, which should not be neglected and could also affect other devices connected to plasma antennae.

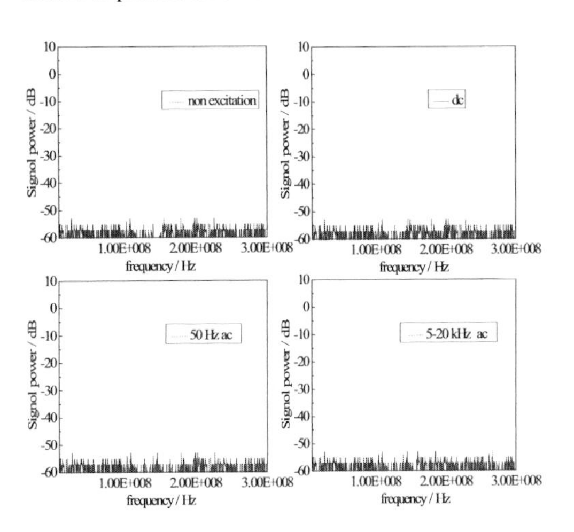

Figure 4. Spectrums of the dc and LF ac driven models.

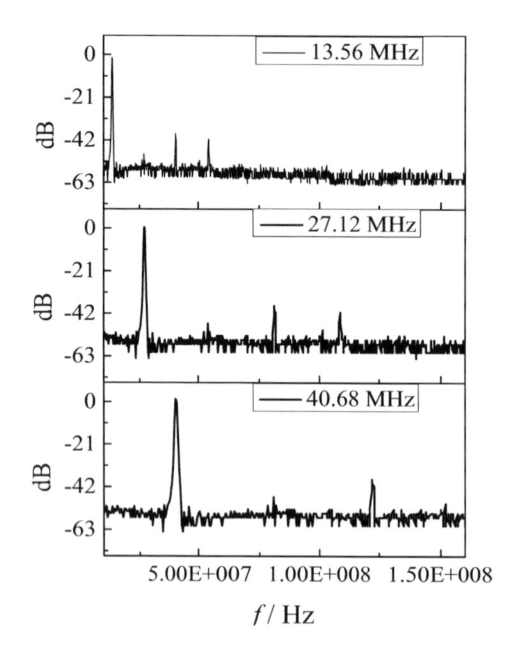

Figure 5. Spectrums of the HF plasma power supplies.

Figure 6 showed the plasma spectrums when the dc plasma antenna was on. For the low radiation from the dc power supply, there was little interference on the spectrometer. Compared to the dc power supply, when the 40.68 MHz HF plasma power supply was on, the spectrum line shifted from 585 to 887 nm in wavelength. Thus, the HF power supply produced very strong electromagnetic interference on the spectrometer. However, in the laboratory, the spectrometer and plasma antenna were connected to the same ground. The strong ground coupling also caused the EMI.

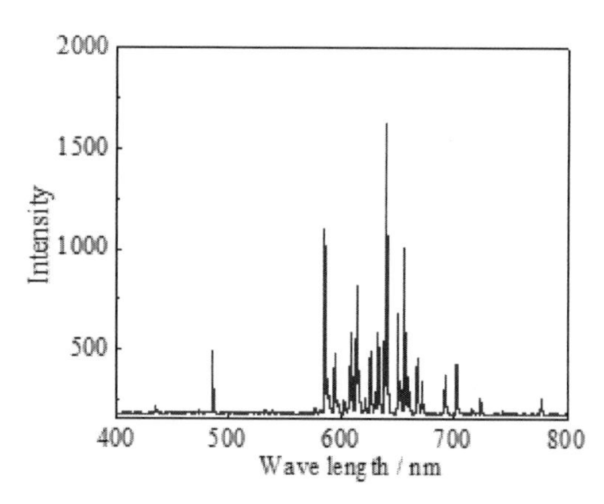

Figure 6. Normal line spectrum of Neon.

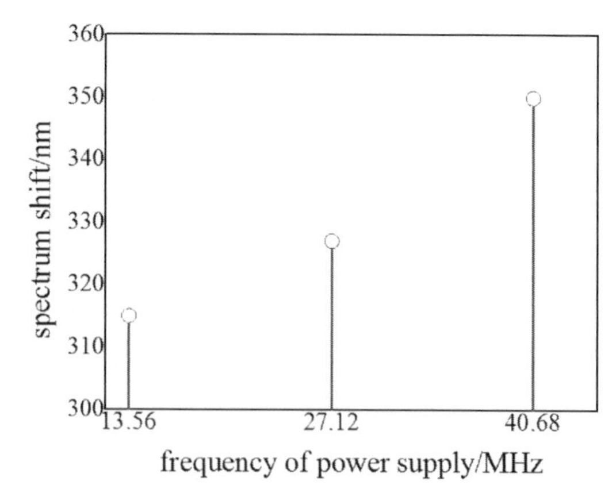

Figure 7. Line spectrum of HF plasma antenna at work.

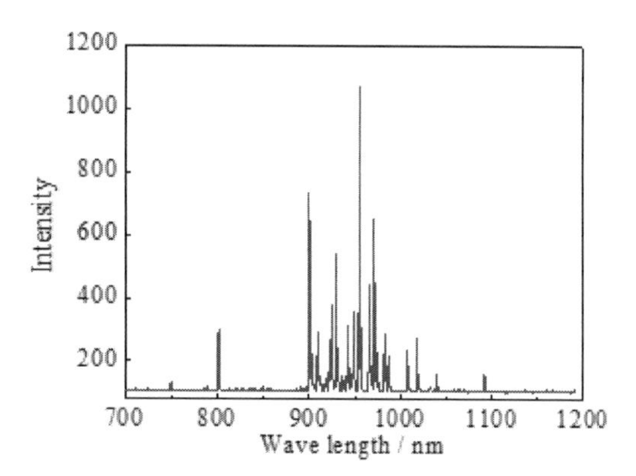

Figure 8. Spectrum shiftif the HF ac power supplies.

Figure 8 shows the spectrum shift of the different HF ac supplies, including 13.56 MHz, 27.12 MHz, and 40.68 MHz values. Strong EMI values from the power supplies were apparent. The average shift was approximately 315 nm with a13.56 MHz power supply frequency, and one was approximately 328 nm with a 27.12 MHz power supply frequency. Thus, the EMI clearly became stronger as the frequency of the power supplies increased.

3.3 *Optimization method of EMC*

The EMI was weakened through shielding and grounding. In the experiment, the distance between the spectrometer and plasma was increased, and they were not connected to the same ground. As shown in Figure 9, the spectrometer worked normally when the shielding technology was adopted.

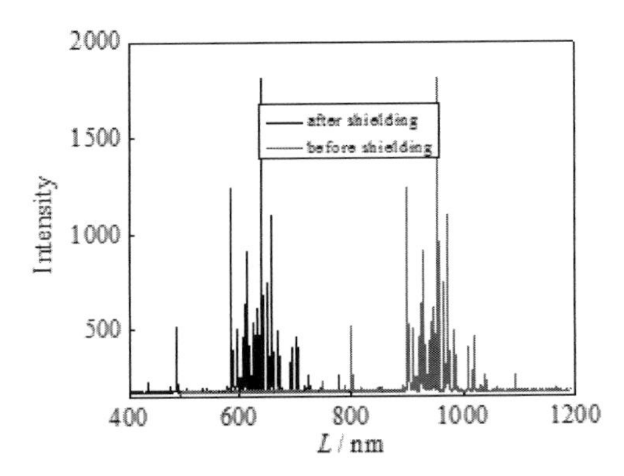

Figure 9. Working state of the spectrometer before and after shielding.

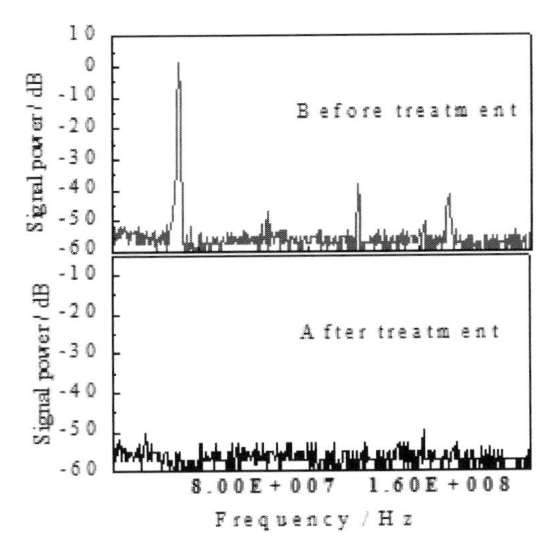

Figure 10. Spectrum received by the analyzer before and after treatment.

The harmonic waves were weakened with filtering. The band-pass filter was designed to prevent the harmonic waves from passing through the spectrum analyzer. The center frequency of the filter was 100 MHz and the bandwidth was 20 MHz (from 90 to 110 MHz), of which the insertion loss was less than 3 dB. Then, the filter was connected between the metal antenna and spectrum analyzer. The spectrums received by the analyzer connected to the band pass compared to the spectrum before treatment are shown in Figure 10. The 40.68 MHz HF signal and the harmonics were rejected effectively, the powers of which were nearly equal to the ground noise. The EMIs of the plasma antenna systems were able to be weakened through grounding, filtering, and shielding.

CONCLUSION

EMC prevents the use of plasma antennae in engineering, especially the type driven by HF ac power supplies. Plasma itself can generate an induction field, the intensity of which is too small to disturb surrounding devices. The interference signals, including the HF signal and its harmonic waves, could be produced by the plasma power supply. HF power supply interference could be eliminated through shielding and grounding, and harmonics could be weakened through filtering.

ACKNOWLEDGEMENTS

This project was supported by the development fund for Shanghai talents (NO.201436)

REFERENCES

Li Sheng.2007.Theoretical Analysis of Characteristics of Plasma Antennas Driven by Surface Wave. *Journal of University of South China (Science and Technology)*21(2): 37–40.

Rayner, J.P.,Whichello, A.P.&Cheetham, A.D. 2004. Physical Characteristics of Plasma Antennas. *IEEE Transactions on Plasma Science*32(1):269–281.

Li, X.S. Hu, B.J.&Li, H.Y.2010. Research and Application Progress of Plasma Antenna. *Modern Electronics Technious*(5): 67–69.

Alexeff, I., Anderson, T. & Parameswaran, S. 2006. Experimental and theoretical results with plasma antennas. *IEEE Transactions on Plasma Science*34(2): 166–172.

Borg, G.G. Harris, J.H. &Miljak, D.G. 1999.The application of plasma columns to radio frequency antennas. *Applied Physics Letters* 74(5): 3272–3274.

Borg, G.G.Harris, J.H. &Martin N.M.2000.Plasmas as antennas: Theory, experiment and applications. *Plasma Physics*7(5): 2198–2202.

Qian, Z.H. Chen, R.S. & Yang, H.W. 2005. FDTD analysis of a Plasma whip antenna. *Microwave and Optical Technology Letters*47(2): 147–150.

Qian, Z.H.Chen, R.S. &Leung, K.W.2005. FDTD analysis of microstrip patch antennacovered by plasma sheath. *Progress In Electromagnetic Research* 52(1):173–183.

Li X.S. Luo F. & Hu B.J.2009. FDTD Analysis of Radiation Performance of a Cylinder Plasma Antenna. *IEEE Antennas and Wireless Propagation Letters*8: 756–758.

Li, X.S.& Hu, B.J. 2010.FDTD Analysis of a Magneto-Plasma Antenna with Uniform of Ununiform Distribution.*IEEE Antennas and Wireless Propagation Letters*9: 175–178.

Kumar, R. & Bora, D.A.2010.Reconfigurable Plasma Antenna.*Journal of Applied Physics*107(5): 053303–053312.

Kumar, R. & Bora, D. 2011.Wireless Communication Capability of a Reconfigurable Plasma Antenna. *Journal of Applied Physics*109(6): 063303-1-9.

Electronics, Communications and Networks IV – Hussain & Ivanovic (eds)
© 2015 Taylor & Francis Group, London, ISBN: 978-1-138-02830-2

High-sensitivity gas sensor for chemicals leak in transportation

Dongjie Zhao, Jun Liu, Fang Yan, Yan Xu & Lei Wang
Beijing Wuzi University, Beijing, China

ABSTRACT: Concerned with chemicals leakage in transportation, there is an urgent need for a sensor which can quickly and accurately detect chemicals. In this study, an MEMS (Micro-electromechanical Systems) sensor is demonstrated, which is based on high-Field Asymmetric Waveform Ion Mobility Spectrometry (FAIMS) technology and can get spectra for chemical detection in one second. Corona discharge (CD) is used as the ionization source of FAIMS sensor. Three kinds of CWAs (Chemical Warfare Agents) stimulants are measured, and remarkable differences in spectra appear. Differential mobility of ions under high electric fields and low fields is calculated and the mobility increases with electric fields. DMMP is used as an example to investigate the capability of detecting CWAs by CD-FAIMS sensor, and the limit of detection is below $1 ng/L^3$. CD-FAIMS sensor is a promising method for tracing level of CWAs detection.

1 INTRODUCTION

Chemical warfare agents (CWAs) include nerve agents, blister agents, asphyxiants, and pulmonary irritants, which are harmful to human beings and the environment(Kolakowski et al. 2007). Enormous efforts have been made to detect CWAs via spectrophotometry, non-dispersive infrared spectroscopy, mass spectroscopy(MS), high performance liquid chromatography (HPLC), ion mobility spectroscopy (IMS) and high-field asymmetric waveform ion mobility spectrometry (FAIMS). Among various methods, FAIMS, as a new gas-phase analytical method, has attracted much attention due to small size, simple instrumentation, comparatively low costs and high sensitivity(Staubs & Matyjaszczyk 2008).

FAIMS sensor utilizes significantly higher electric fields and identifies the ion species based on the difference in their mobility coefficient in high and low strength of electric fields(Kolakowski & Mester 2007),which is based on the observation of Mason and McDaniel(Mason & McDaniel 1988), who reported that the mobility of an ion is affected by the strength of applied electric fields.. Above an electric field to gas density ratio (E/N) reaches 40 Td (E>10,700 V/cm at atmospheric pressure) the mobility coefficient K(E) has a non-linear dependence on the field. This dependence is believed to be specific for each ion species.

The initial step in FAIMS sensor response to CWAs is the formation of gas-phase ions from sample vapor. The traditional ionization source of the FAIMS is a 10 mCi radioactive ^{63}Ni foil with a half-life of about 85 years, which provides stable operation, and does not require an external power supply. However, radioactive ionization source is needed to make regular leak test and special safety regulations(Guharay et al. 2008). Several other alternative ionization methods have been reported, such as photoionization by ultraviolet light, laser multiphoton ionization, electrospray for liquid samples, corona discharge (CD), and so on. Corona discharge is one of the electrical discharges developed by the ionization gas around a sharp conductor, which occurs when the electric fields are strongly nonuniform and insufficient to cause completely electrical breakdown. CD in IMS is systemically designed and optimized by Tabrizchi(2000). According to the article, both better sensitivity and higher signal-to-noise ratio can be achieved. CD is a promising method to replace 63Ni, which is available in some producing instruments including IMS, MS (Eiceman & Karpas 2005).

In present work, an MEMS FAIMS sensor with corona discharge ionization source is fabricated, and the capability of detecting CWAs is investigated. Three kinds of CWAs stimulants are tested, including DMMP, 2_CEES and DEEA. Specifically, separation voltages effect on ion mobility is observed, and alpha function, which describes mobility's non-linear dependence on the field, is calculated.

2 EXPERIMENT

2.1 *FAIMS sensor*

The FAIMS sensor is fabricated by MEMS technology. A schematic sketch of the sensor is shown in Figure 1, which mainly consists of two planar ion filter electrodes(1), a deflector electrode(2) and a detector electrode(3). FAIMS drift tube is fabricated from two 7101 glass wafers and one DT263 glass wafer. Metal electrodes are formed on the top and bottom glass wafers defining the ion filter, deflector and detector plates. The metal electrodes are fashioned on the glass wafers by coating the glass wafers with AZ1500 photoresist first and patterning the resist by a photolithographic process. The patterned photoresist is then coated with metal (2000 Å gold on 300 Å titanium) by sputtering deposition. A lift-off process is used to remove the unwanted metal residing on the top of the photoresist, leaving behind the metal electrodes on the glass. Epoxy resin is used to assemble the sensor.

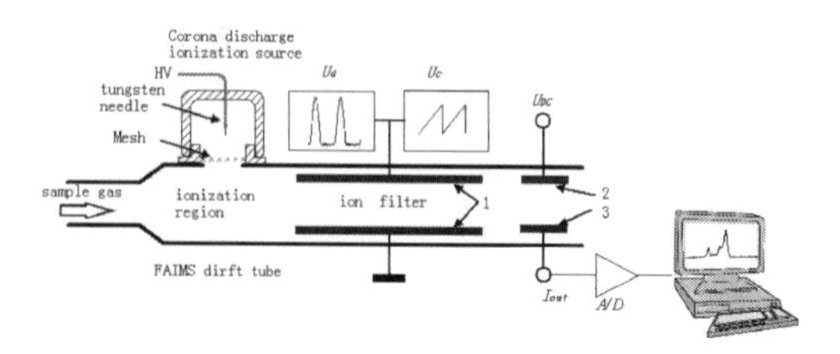

Figure 1. Schematic of FAIMS sensor.

The hole in the upper wafer provides a means for introducing the reaction ions. After ionization of the gas sample, the ions are carried by a carrier gas flowing through the ion filter region. Once through the ion filter, the ions are deflected onto a detector electrode. Depending on the polarity of the deflector electrode, either negative or positive ions are measured at the detector electrode. CD ionization source consists of a sharp tungsten needle, a target electrode of metal mesh and a DC high voltage (HV) power supply (30kV, 1mA), which is a kind of point-to-plain geometry.

The FAIMS sensor is operated by using specialized electronics containing a separation waveform (U_d) generator, a compensation voltage (U_c) generator, and a weak signal (I_{out}) amplifier. The separation waveform generator can output variable peak-to-peak amplitudes of the asymmetric waveform ranged from 0 to 1500V, and the operating frequency of the generator is 1MHz. The U_c is scanned between 20 to −20V in 1s period. The weak signal circuit is based on an ultra-low bias current (30fA) monolithic operational amplifier (Ti, OPA129).

2.2 *Gas supply and chemical vapor generation*

A schematic graph of the sample introduction system is shown in Figure 2. Clean nitrogen is provided by Praxair Inc. with gas scrubbing through 13X molecular sieve. Purified nitrogen split into three independently controlled parts with a flow of 0–2L/min used as the carrier gas flow(MFC1), 10–100mL/min used to control samples(MFC2), and 0–300mL/min used as waste flow(MFC3). Sample vapors are prepared by Teflon permeation tube which is located in a thermostat container at 35–70°C via a PID controller. This flow system allows a constant flow to be delivered to the drift tube while it permits changes in vapor concentration by adjusting the ratio of sample to diluents' flow. The diffusion of permeation tubes is weighed over several days to gravimetrically determine consumption of the sample (dm/dt), which can estimate concentration of the sample in the gas flow.

Three kinds of chemicals used in this work are not further purified. Diethylaminoethanol (DEEA, CAS 100-37-8) is purchased from Aldrich Inc. 2-Chloroethyl ethyl sulfide (2_CEES, CAS 693-07-2) is from Aladdin Inc. Dimethyl methylphosphonate (DMMP,CAS 756-79-6) is from Alfa Inc.

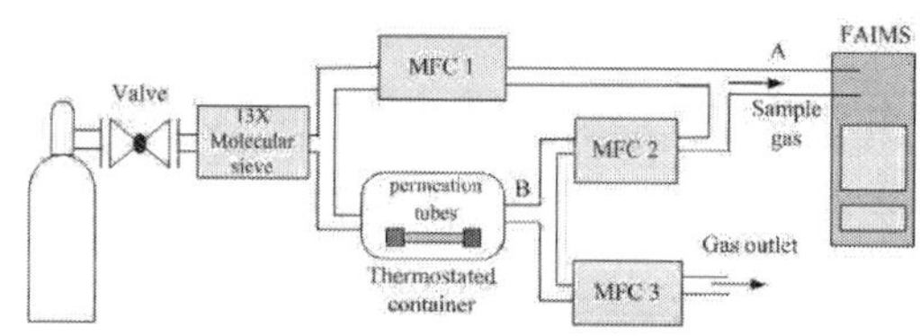

Figure 2. Schematic of sample introduction system.

3 RESULTS AND DISCUSSIONS

3.1 *Responses for CWAs stimulants*

CD-FAIMS spectra for background and three kinds of CWA stimulants at separation voltage (U_d) of 800V are shown in Figure 3. All spectra exhibite 2~3 peaks for compensation voltages range from −10 to +5 V. The absence of peaks between compensation voltages of −10~−20V suggests that little fragmentation of the product ions has been observed. Ion peaks of CWA stimulants in CD-FAIMS spectra suggest that the product ion intensity is concentration dependent, but compensation voltage (U_c) was independent of concentration.

The productions in positive polarity corona discharge contain N_2^+, N_4^+, $NH_4^+(H_2O)_n$ and $H^+(H_2O)_n$ (n is correlated with moisture). $H^+(H_2O)_n$ is commonly dominant ion and is called the reactant ions, which can be observed at U_d from 400V to 700V.

The top spectrum shown in Fig. 3 is a background spectrum without CWAs vapor. The low intensity for the ion peaks is at a compensation voltage of −1V ~ 3V(peak 1,2). Peak1 and peak 2 may be some impurities coming from corona discharge. The spectrum for DMMP shows peaks at compensation voltage of −6.9V, −0.5V and 2.5V. According to the reported result, peaks at -6.9V, 2.5V are protonated monomer

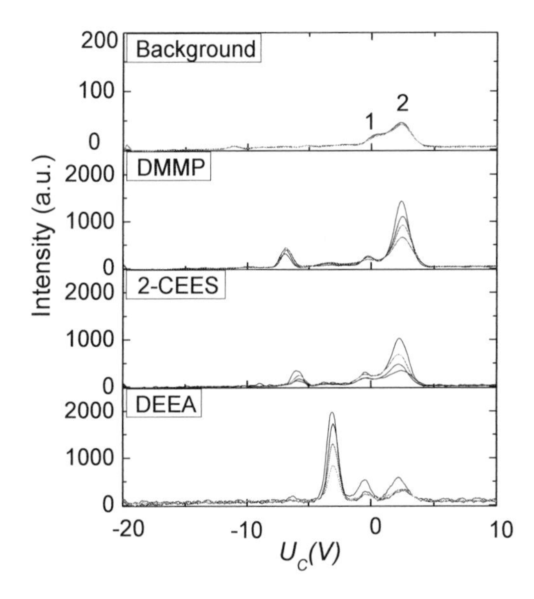

Figure 3. CD-FAIMS spectra of CWA stimulants under concentration of 80, 70, 60, 50.

and proton-bound dimmer of DMMP. Similar results are got in the spectrums of 2-CEES, DEEA. The right peaks and left peaks are thought to be protonated monomers and proton-bound dimmers, respectively. CWAs have higher proton affinity than most chemicals 6, which can minimize the impact of impurities.

3.2 *Separation voltage effect*

The separation voltages (U_d) have influence on ion mobility, and compensation voltages (U_c) for ion peaks are changed with ion mobility. Table 1 is given for U_d vs. U_c of CWA stimulants. Increased strength

Table 1. U_d vs U_c for CWAs.

U_d (V)	U_c (V)		
	DMMP	2-CEES	DEEA
450	−2.20	−1.23	−1.30
500	−2.55	−1.03	−1.71
550	−2.90	−1.35	−1.98
600	−3.20	−1.97	−1.88
650	−4.34	−2.59	−1.77
700	−5.50	−3.18	−1.97
750	−6.40	−3.74	−2.26
800	−6.90	−4.20	−2.54

of electric field in the asymmetric waveform can result in an increasing negative voltage that restores the transportation of ions through the drift tube. This polarity is consistent with ions that exhibit increased mobility with increased field strengths.

3.3 *Alpha functions dependence*

When electric field strength in the asymmetric waveform increases, ion mobility coefficient (K) will show the non-linear dependence on the field. K is characteristic structure of an ion and the ion-molecule reactions in gas, which is a form of perturbation propagation. K is influenced by frequency and energy obtained from the field by ions. The average energy of collision acquired from the electric field is determined by E/N. When E/N is small, it is considered to be negligible since any energy gained by the ion from the field is dissipated at high pressures by collisions with the neutral molecule. Under such conditions, K is a constant and independent of E/N. However, when E/N is large enough (E/N >40 *Td*), the mobility coefficient becomes dependent on electric field with increasing values of E/N as shown in eq1:

$$K(E/N) = K(0)[1 + \alpha(E/N)] \quad (1)$$

$$\alpha(E/N) = \sum_{n=1}^{\infty} \alpha_{2n}(E/N)^{2n} = \alpha_2(E/N)^2 + \alpha_4(E/N)^4 + \ldots \quad (2)$$

$K(0)$ is the mobility coefficient under low field conditions; $\alpha(E/N)$ is a function showing the nonlinear electric field's dependence of mobility for ions, and α_2, α_4, ... , α_{2n} are specific coefficients of even powers of the electric field; E/N is in the unit of *Td* (under normal conditions 1 *Td* corresponds to 268.67 *V/cm* since $E=N_0 10^{-17}$, where N_0 is Loschmidt's constant).

The alpha parameter is a function which contains the physical and chemical information about ion. The procedure of extraction of $\alpha(E/N)$ coefficient from experimental measurements (Table 1) of the electric field dependence on the mobility has been described. The alpha can be expressed as $\alpha(E/N) \approx \alpha_2(E/N)^2 + \alpha_4(E/N)^4$ with high accuracy as permitted measurements. The values of alpha coefficients are given in Table 2.

Table 2. Alpha coefficients for CWA simulants.

Compound	DMMP	2-CEES	DEEA
α_2 ($\times 10^{-6}$ Td^{-2})	21.55	4.24	8.13
α_4 ($\times 10^{-11}$ Td^{-4})	22.31	-7.39	10.07

The $\alpha(E/N)$ curves for CWAs stimulants are shown in Figure 4. In the whole range of E/N studied, $\alpha(E/N)$ grows in the order that DMMP > 2_CEES > DEEA. These plots are fundamental features of ions that are independent of the drift tube parameters. The difference in α function suggests that the effect of the electric field must be associated with molecular structure. One possible interpretation is that ions are heated so that the high field and the effect on the ion become striking.

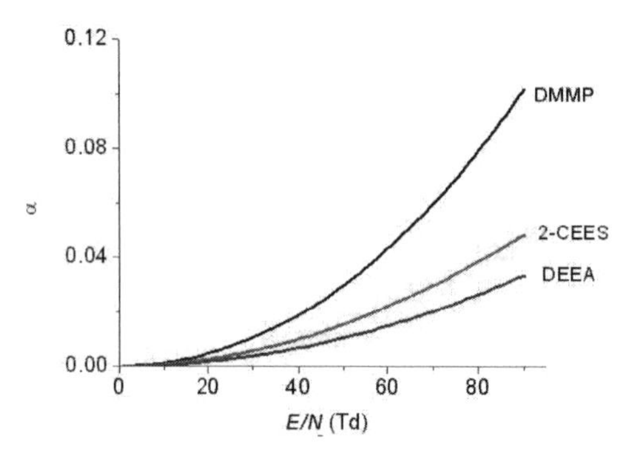

Figure 4. α(E/N) functions of the CWA stimulants.

In Figure 5, the calculated coefficients α for all the ions plot against experimental values of E_c/E_d (equivalent to U_c/U_d). The solid curve is the regression line. It is seen that the data points fit well the regression line and α shows a linear relation with the E_c/E_d (U_c/U_d). When U_d is same, the ration of different α values is equivalent to the ration of U_c values ($\alpha_1/\alpha_2= U_{c1}/U_{c2}$).

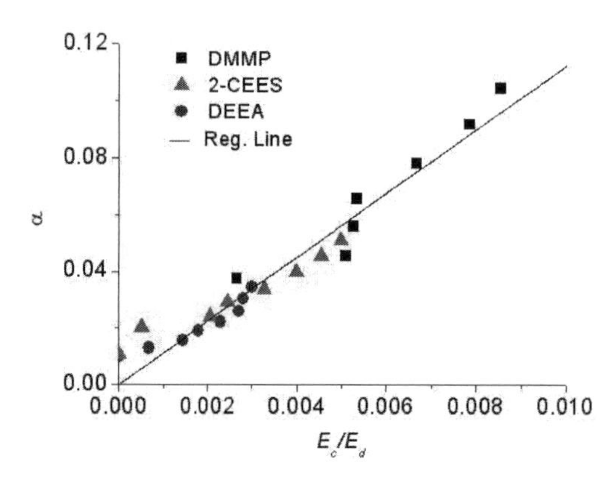

Figure 5. α versus Ec/Ed of the CWA stimulants.

3.4 *Limits of detection*

In order to investigate the sensitivity of CD-FAIMS for CWAs, a permeation tube of DMMP calibrated by VICI corp. (70°C, 25ng/min) is used to trace DMMP vapor ranged from 1~12.5ng/L^3 . In Figure 6, the response of CD-FAIMS has the linear relation with the concentration of DMMPand the limit of detection (LOD) is below 1ng/L^3, which is lower than the LOD accepted by ^{63}Ni-FAIMS.

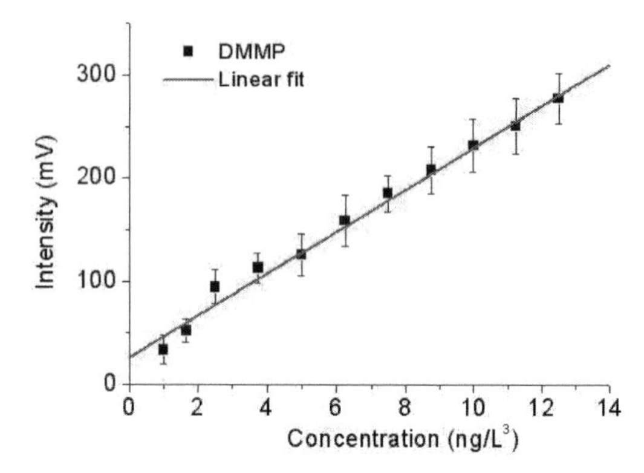

Figure 6. Intensity vs concentration for DMMP.

4 CONCLUSIONS

CD-FAIMS sensor has shown great potential for tracing CWAs detection. Three CWAs stimulants spectra exhibit remarkable differences. Ion mobility of CWAs stimulants increases with separation voltage ranged from 450V to 800V. The LOD of DMMP is below 1ng/L^3 by CD-FAIMS sensor.

ACKNOWLEDGMENT

The authors thank for the support from Beijing Logistics Engineering Research Center (No. BJLE2010).

The authors are also grateful for the support from Beijing Wuzi University (No.0351405702).The authors also appreciate the support from BEIJING WUZI UNIVERSITY (No. 2014XJQN15).

The authors thank for the help from the professor Jianping Li and his research team, too.

REFERENCES

Mason, E.A. & Mcdaniel, E.W. 1988. *Transport Properties of Ions in Gases*. New York: Wiley.

Eiceman, G. A. & Karpas, Z. 2005. *Ion Mobility Spectrometry*. Taylor & Francis.

Guharay, S. K., Dwivedi, P. & Hill, H. H. 2008. Ion mobility spectrometry: Ion source development and applications in physical and biological sciences. *IEEE Transactions on Plasma Science* 36: 1458–1470.

Kolakowski, B. M., Dagostino, P. A., Chenier, C. & Mester, Z. 2007. Analysis of chemical warfare agents in food products by atmospheric pressure ionization-high field asymmetric waveform ion mobility spectrometry-mass spectrometry. *Analytical Chemistry* 79: 8257–8265.

Kolakowski, B. M. & Mester, Z. 2007. Review of applications of high-field asymmetric waveform ion mobility spectrometry (FAIMS) and differential mobility spectrometry (DMS). *Analyst* 132: 842–864.

Staubs, A. E. & Matyjaszczyk, M. S. 2008. Novel trace chemical detection technologies for homeland security. *2008 IEEE Conference on Technologies for Homeland Security, Vols 1 and 2*: 199–204.

Tabrizchi, M., Khayamian, T. & Taj, N. 2000. Design and optimization of a corona discharge ionization source for ion mobility spectrometry. *Review of Scientific Instruments* 71: 2321–2328.

Electronics, Communications and Networks IV – Hussain & Ivanovic (eds)
 ISBN: 978-1-138-02830-2

Study on wavelet packet modulation-frequency hopping system in AWGN channel

Qin Zheng*
School of Communication Engineering, Hangzhou Dianzi University, Hangzhou, China

Xianghong Tang
School of Information Engineering, Hangzhou Dianzi University, Hangzhou, China

ABSTRACT: Authors of the paper propose a novel Multi-Carrier Modulation (MCM) technique which combines Wavelet Packet Modulation (WPM) technology with frequency hopping (FH) technology together, namely, Wavelet Packet Modulation-Frequency Hopping (WPM-FH) communication system. The theory analysis and deduction are given by means of the discussion about the relationship between the system BER performance and the number of users, carriers, hopping points in Additive White Gaussian Noise (AWGN) channel. The simulation results show that the system not only keep the advantage of wavelet packet modulation to improve the spectrum utilization, but also obtain the anti-jamming capability of frequency hopping technology, the communication system was better improved as a result.

KEYWORDS: wavelet packet modulation (WPM); frequency hopping (FH); multi-carrier modulation; Additive Gaussian noise channel (AWGN).

1 INTRODUCTION

Multi-carrier communication technology has received recent research attention in modern telecommunication areas. Wavelet packet modulation (WPM) is an alternative multi-carrier technology with frequency division multiplexing features as OFDM (Daoud 2012). On the other hand, WPM system can be flexible in the choice of wavelet functions and tree structures (Deng et al. 2007), and can improve spectrum utilization effectively. Since the wavelet packet functions' specialty of translational self-orthogonal and mutual orthogonality as well as the diversity of tree structures for wavelet packet modulation, the WPM based multi-carrier system can achieve strong ability of anti-carrier interference and inter-symbol interference (Akho et al. 2013).

The FH spread spectrum communication technology with the advantages of frequency hopping, concealment, and low spectrum density, etc, which ensures better anti-interference performance and application value. The schemes of direct sequence spread spectrum (DSSS) with multi-wavelet packet modulation techniques in (Shao et al. 2007), and the wavelet packet based M-ary multicarrier spread spectrum system (Ying et al. 2005) can improve the system performance effectively. Compared with the DSSS modulation techniques the FH technology is more suitable for military communication. Zhang & Zhang (2001) wavelet packet function modulation based frequency hopped WPDMA communication system has better performance of bit error rate (BER) in a variable multipath fading channel. However, the limited number of the basic functions as subcarriers in this method drawback the system performance. Moreover, the complexity of the system will be rising with the number of users and antennas increase. The theoretical BER derivation over different channels for the multi-user OFDM-FH system based phase modulation (Zeng & Peng 2010), has more general reference value. Unlike WPM multi-carrier system, the bandwidth of the OFDM multicarrier system based on Fast Fourier Transform (FFT) is not optional and has no Multi-rate diversity feature. What's more, the cyclic prefix in OFDM results in the decrease of band utilization.

Based on above described works, the WPM can effectively overcome these shortcomings. In this paper, we investigate the BER performance of the WPM-FH system with multi-user in AWGN channel. And present the derivation process of the theoretical BER equation. The simulation results prove that this scheme significantly enhance communication system performance.

*Corresponding author: *zheng_qin1@163.com*

The contents of this paper organized as follows: Section II firstly describes the multi–user WPM-FH system model, and then outlines the theoretical BER analysis and deduction of the system. Section III shows the simulation results with detail analysis in different elements. Finally, conclusions are drawn in section IV.

2 WPM-FH COMMUNICATION SYSTEM AND ANALYSIS

2.1 *WPM - FH communication system model*

Wavelet packet transform (WPT) based multi-carrier modulation system can provide the basis functions with orthogonality instead of subcarriers used in OFDM. In WPM system, the wavelet packet function with different node indices represent different sub-carriers, thus the modulation signal can be expressed (Yu et al. 2010)

$$s(t)=\sum_{(l,m)\in\Gamma}\sum_{i}x_l^m(i)w_l^m(t-iT_l), \tag{1}$$

where $x_l^m(i)$ is the digital signal based on the subcarrier function w_l^m, Γ is the set of wavelet packet tree's node label(l,m). Usually, the WPT realized by fast algorithm of wavelet packet, i.e. Mallat algorithm (Mallat 1991).

The transmitter of WPM-FH is shown in Figure 1. On the transmitting side, suppose the i-th information x_{ki} sent by user k. Firstly, the high speed bit stream x_{ki} for user k is serial-to-parallel(S/P) converted to M parallel sub-streams, then BPSK mapping for each sub-stream. The next step is to produce WPM signal by inverse discrete wavelet packet transform. At last, data symbols in M sub-channels superimposed to form a WPM symbol. Each user' data should be modulated onto the points of their corresponding frequency hopping sequence before being transmitted.

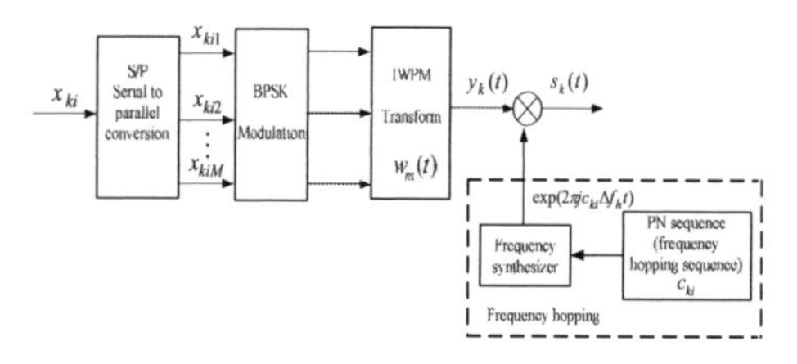

Figure 1. The transmitter of WPM-FH system.

Assuming the bit duration is T_b, after BPSK mapping the symbol period of sub-carrier is $T=T_b$, so the minimum band gap between two adjacent subcarriers is$\Delta f=1/T$, and the bandwidth for one WPM symbol is $W_{wpm}=M\Delta f$. To make the adjacent frequency points avoid interference and overlapping, the minimum interval of frequency hopping points is $\Delta f_h=M\Delta f$. When have N_h hopping points, the emission bandwidth of the WPM-FH system is $B=N_h\Delta f_h$. The transmission signal of the k-th user $s_k(t)$ can be expressed

$$s_k(t)=\sum_{i}\sum_{m=1}^{M}\sqrt{P}x_{kim}w_m(t-iT)$$
$$\cdot\exp(2\pi jc_{ki}M\Delta f(t-iT))\cdot p(t-iT), \tag{2}$$

where x_{kim} is the digital signal of the k-th user on m-th sub-channel, P is the average power of sub-carrier. T is the symbol duration of each sub-channel. $p(t)$ is the rectangular pulse function with unit amplitude, when $t\in[0,T]$, $p(t)$ has the value 1, otherwise 0. $\exp(2\pi jc_{ki}M\Delta ft)$ used to represent the complex envelope waveform of frequency hopping carriers, such said can bring convenience for later analysis. The frequency hopping sequence for the k-th user is $C_k=\{c_{ki}|c_{ki}\notin GF(N_h-1)\}$, $k=1,2,...,N_u$, N_u is the total number of users.

Suppose N_u users transmitting information simultaneously over the Gaussian channel, the signal on the receiver can be written as

$$r(t)=\sum_{i}\sum_{k=1}^{Nu}\sum_{m=1}^{M}[\sqrt{P}x_{kim}w_m(t-iT)$$
$$\cdot\exp(2\pi jc_{ki}M\Delta f(t-iT))\cdot p(t-iT)]+n(t), \tag{3}$$

where $n(t)$ is the AWGN interference signal, whose mean is 0 and the variance is $N_0/2$.

The receiver of the WPM-FH system is shown in Figure 2, mainly consists of three parts de-hopping part, demodulation part and sampling decision part. In order to facilitate the analysis, assume the system is completely synchronized, and user 1 is referenced as the primary user. The received signal should be de-hopping before the wavelet packet transform. Then generate the pre-judgment variable U_i, which can be expressed as follows

$$U_i=\sum_{m1=1}^{M}\int_{iT}^{(i+1)T}r(t)\cdot w_{m1}(t-iT)\cdot\exp(-2\pi jc_{1i}M\Delta f(t-iT))dt. \tag{4}$$

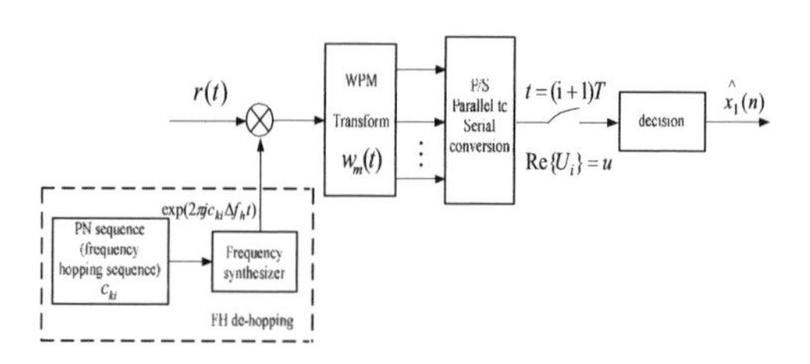

Figure 2. The receiver of WPM-FH system.

2.2 *BER performance analysis over AWGN channel*

For the convenience of description and analysis, let i=0, x_{ki} is the serial data generated by the repetition encoder (Lance & Kaleh 1997), namely $x_{kim} = x_k$, each sub-channel transmit the same data in WPM-FH system. The frequency hopping points are randomly selected from frequency set which contain N_h frequency points, i.e. c_{ki} is uniformly distributed over $[0, N_h - 1]$. Therefore, the (4) can be written

$$U_0 = \sum_{m1=1}^{M} \int_0^T \sqrt{P} \sum_{k=1}^{Nu} x_k \sum_{m=1}^{M} w_m(t) w_{m1}(t) \exp(2\pi j(c_{k0} - c_{10}) M\Delta f t) dt + \sum_{m1=1}^{M} \int_0^T n(t) w_{m1}(t) \exp(-2\pi j c_{10} M\Delta f t) dt$$
$$= \sqrt{P} MTx_1 + \sqrt{P} M \sum_{k=2}^{Nu} x_k \int_0^T \exp(2\pi j(c_{k0} - c_{10}) M\Delta f t) dt$$
$$+ \sum_{m1=1}^{M} \int_0^T n(t) w_{m1}(t) \exp(-2\pi j c_{10} M\Delta f t) dt \cdot \tag{5}$$

In equation (5), the first item is the data information that sent by user 1. The second item is multi-user interference caused by other $(N_u - 1)$ users, marked as I_u. As can be seen from the expression, the multi-user interference $I_u \neq 0$ when the frequency points collision occurs. Otherwise the $I_u = 0$. The third item is AWGN noise interference, marked as I_n, which meets the Gaussian distribution with mathematical expectation of 0 and variance of $MN_0 / 2$. So (5) can be simplified as

$$U_0 = \sqrt{P} MTx_1 + I_u + I_n \cdot \tag{6}$$

Take the real part u of the demodulation signal U_0, the u is as follows

$$u = \Re e\{U_0\}$$
$$= \sqrt{P} MTx_1 + \sqrt{P} M \sum_{k=2}^{Nu} x_k \int_0^T \cos[2\pi(c_{k0} - c_{10}) M\Delta f t] dt + \sum_{m1=1}^{M} \int_0^T n(t) w_{m1}(t) \cos(2\pi c_{10} M\Delta f t) dt$$
$$= \sqrt{P} MTx_1 + I_u^{'} + I_n^{'} \cdot \tag{7}$$

As can be seen from (7), in AWGN environment, both of multi-user interference and AWGN noise interference are the major factors that lead to the performance degradation of the WPM-FH system.

The average BER is a common method used to measure the communication system performance. In order to derive the WPM-FH system BER over AWGN channel, the probability density function of u is requested. According to (7), the interference value of the k-th $(k \neq 1,\ k > 1)$ user is

$$I_{uk} = \sqrt{P} M x_k \int_0^T \cos[2\pi(c_{k0} - c_{10}) M\Delta f t] dt \tag{8}$$

Consider if the k-th user sends the bit data "1", i.e. $x_k = 1$. If the frequency collision occurs between the k-th user and the primary user, i.e. $c_{k0} = c_{10}$, then the collision probability is $\alpha = 1/N_h$, and the interference value of the k-th user is $I_{uk} = \sqrt{P}MT$. If there is no frequency collision between the two users, i.e. $c_{k0} \neq c_{10}$, well, the probability with no collision is $\varepsilon = 1\text{-}\alpha$, and $I_{uk} = 0$. Therefore, the probability density function (PDF) is $f_{k1}(x)$ when the k-th interference user sends "1", $f_{k1}(x)$ can be written

$$f_{k1}(x) = \varepsilon\delta(x) + \alpha\delta(x - \sqrt{P}MT) \cdot \tag{9}$$

Similarly, if the k-th user sends"0", that means x_k = -1. In this case, the PDF $f_{k0}(x)$ is

$$f_{k0}(x) = \varepsilon\delta(x) + \alpha\delta(x + \sqrt{P}MT) \,. \tag{10}$$

Suppose transmitter send binary data "0" and "1" with equal probability, and different interference users are independent of each other (Li 2007). Thus the joint probability density of $(N_u - 1)$ interference users is

$$f(x) = [\frac{1}{2} f_{k1}(x) + \frac{1}{2} f_{k0}(x)]^{N_u - 1}$$
$$= [\varepsilon\delta(x) + \frac{1}{2}\alpha\delta(x - \sqrt{P}MT) + \frac{1}{2}\alpha\delta(x + \sqrt{P}MT)]^{N_u - 1} \,. \tag{11}$$

According to the central limit theorem, the multi-user interference item $I_u^{'}$ obeys Gaussian distribution when N_u is large enough. Obviously, the mathematical expectation of $I_u^{'}$ is $m_{I_{uk}} = 0$, and its variance is $\sigma_{I_{uk}}^2 = PM^2T^2\alpha(N_u - 1)$. Combining (7) and (11), it easy to get the mathematical expectation and variance of u, i.e. $m_u = \sqrt{P}MT$, $\sigma_u^2 = PM^2T^2\alpha(N_u - 1) + MN_0 / 2$.

Take the zero level as the decision threshold over AWGN channel, the theoretical BER of the multi-user WPM-FH system can be expressed

$$P_b = \int_{-\infty}^{0} \frac{1}{\sqrt{2\pi}\sigma_u} \exp\left(-\frac{(x - m_u)^2}{2\sigma_u^2}\right) dx \,. \tag{12}$$

Defining $v = (x - m_u) / \sigma_u$, so $v \sim N(0,1)$. Therefore (12) can be written as the form of $Q(\cdot)$ function

$$Q(x) = \int_x^{\infty} \frac{1}{\sqrt{2\pi}} \exp(-\frac{v^2}{2}) dv \,. \tag{13}$$

In equation (13), the integration limit $x = m_u / \sigma_u$. According to the above derivation process, the BER performance of the multi-user WPM-FH system is related to the number of users, frequency hopping points and the subcarriers.

3 SIMULATION ANALYSIS OF WPM-FH SYSTEM PERFORMANCE

Having derived the BER, the WPM-FH system performance is simulated over an AWGN channel in this section. Assume the WPM-FH system is completely synchronous, and the input source is binary data$\{0,1\}$. The effects of number of users, frequency hopping points, sub-channels and wavelet packet tree structures are tested. The simulation results are given below.

3.1 *Effect on the number of users N_u and theoretical BER*

Figure 3 shows the relationship between the WPM-FH system BER performance and the number of users ($N_u = 2,4,8$) when $M = 8$, $N_h = 32$. The theoretical BER curves corresponding to different number of users are also shown in the same figure. Simulation result shows that the system performance is decreasing with the increase of N_u. The influence of Gaussian white noise to the system is smaller while the SNR gradually increased. However, the multi-user interference becomes the major factor that results in the decrease of system performance. The probability of the occurrence of frequency collision is rising along with the increase of N_u, so the system performance slowdowns as the multi-user interference become more obvious. Compared with the simulation curves and theoretical curves, it can be seen that the BER theory derivation in this paper is reliable.

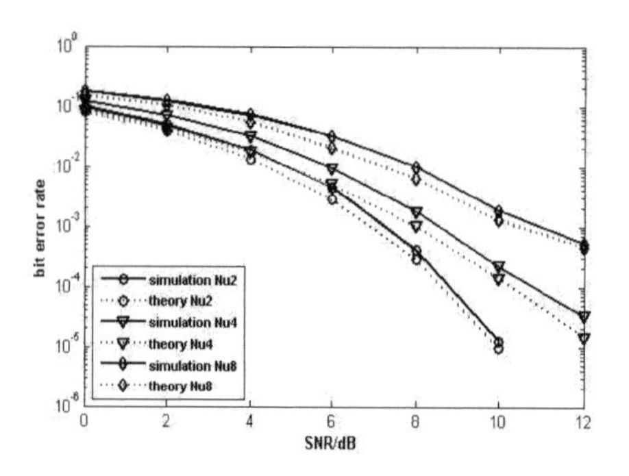

Figure 3. Number of users influence on WPM-FH system performance.

3.2 *Effect of number frequency hopping points N_h*

Figure 4 illustrates influence of the number of frequency hopping points to system performance. The number of users and subcarriers ($M = 8$) are certain. Take $N_h = 8,16,32$ separately, the simulation results show that the WPM-FH system gets better performance with the value of N_h increasing. It can be noted from the description of section II, the collision probability between frequency points is diminishing while N_h becoming larger, which means the multi-user interference becomes smaller. So the system performance improves naturally as a result.

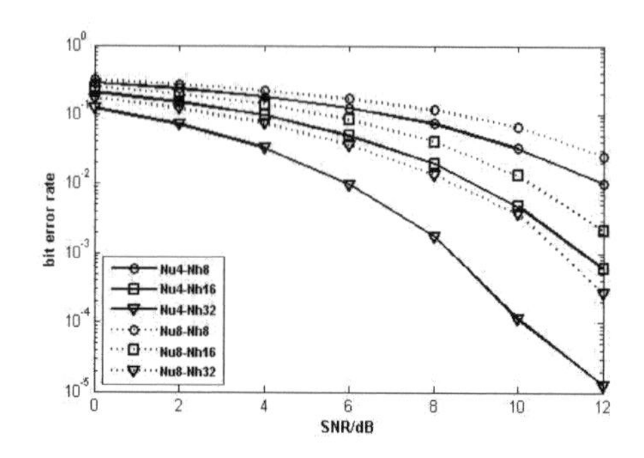

Figure 4. WPM-FH system performance under different hopping points.

3.3 *Effect of number of sub-channels M*

Fixed The number of users ($N_u = 4,8$) and frequency hopping points ($N_h = 32$). It is clear from Figure 5 that the WPM-FH system performance is advanced with the growing of M ($M = 8,16$). The value of the symbol period $T_s = MT$ gets lager as M increases. If the channel delay is far less than the symbol period, the code interference of the system will be smaller. In addition the waveform distortion can be reduced over channel transmission. Then the system can obtain better BER performance.

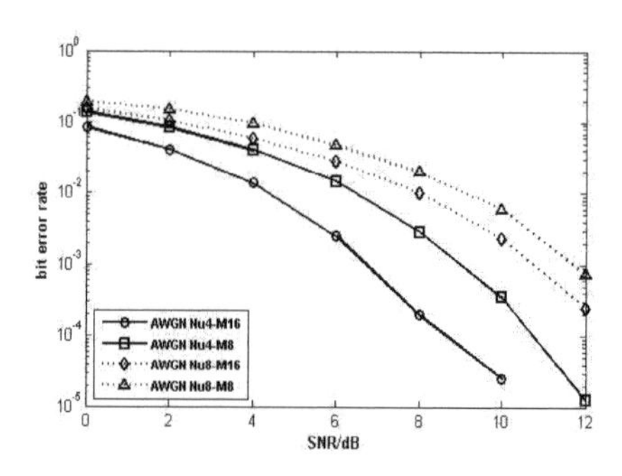

Figure 5. Influence of number of subcarriers to WPM-FH system.

3.4 *Effect of different tree structures*

Wavelet packet modulation has abundant tree structures, only 3-level wavelet packet modulation has 21 tree structures. In this paper, we take 3-level wavelet packet tree structure as an example, four kinds of modulation structures (showed in Fig. 6) performance in WPM-FH system is simulated, i.e. tree1, tree2 (Deng et al. 2007), tree3 (Li et al. 2014) and tree4. We can note that the simulation results in Figure 7, different tree structures with same level have very little influence on WPM-FH system performance. In this case, the simplified wavelet packet trees by pruning not only have simple structures and multi-rate modulation, but also keep a better system BER performance. The reliability of the WPM-FH system is increased since the diversity of wavelet packet tree structures. The characteristics and advantages of WPM modulation are also obvious that if the interference receivers want to demodulation a signal correctly, they must know both frequency hopping rules and the basis functions with corresponding tree structures.

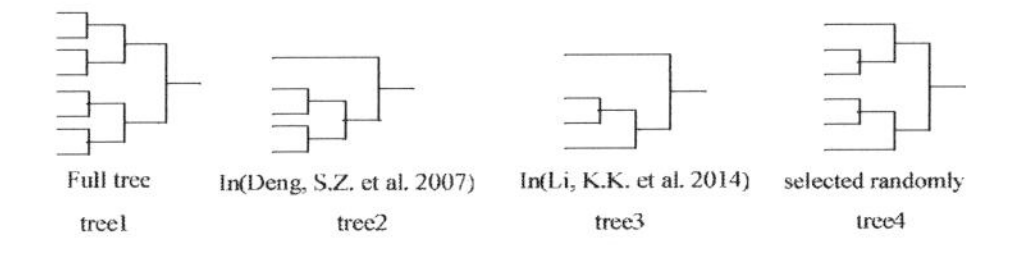

Figure 6. Four tree structures used in WPM-FH system.

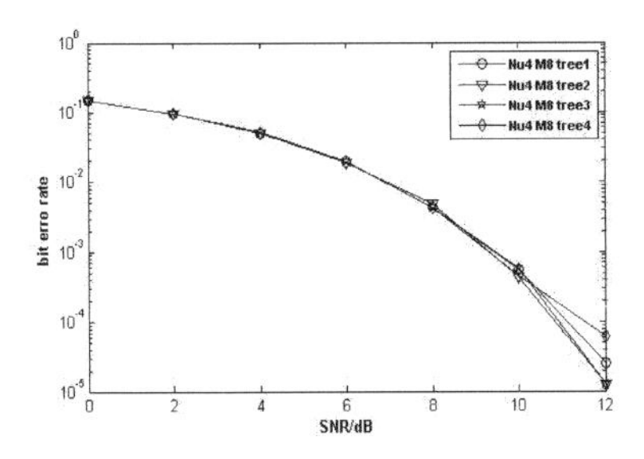

Figure 7. Performance of different tree structures in WPM-FH system.

4 CONCLUSION

This paper proposed the WPM-FH system based on the combination of frequency hopping spread spectrum and wavelet packet modulation. Performance analysis and simulation results about the number of sub-channels, users, frequency hopping points and tree structures are all given over AWGN channel. The simulation results verify the correctness of theoretical analysis. And the diversity of wavelet packet tree structures and basic functions rich the WPM-FH system's characteristics such as high security and flexibility as well as remarkable improvement of the anti-jamming performance. Although the WPM-FH system obtains the both advantages of WPM technique and FH technique, of course, there is also insufficient. The computation complexity and the requirements for hardware equipment are requested higher with the increase of tree structure levels.

ACKNOWLEDGMENT

The authors would like to thanks to the Project of postgraduate scientific research and innovation fund of Hangzhou Dianzi University (ZX130702308007).

REFERENCES

Akho, Z.M., Saleh, K. & Abdellatif, N. 2013. Performance of wavelet packets based multicarrier multi-code CDMA communication system in the presence of narrowband jamming. *International Conference on Communications, Signal Processing and their Applications*: 1–6.

Deng, S.Z., Ru, L., Du, X.M, et al. 2007. Optimal wavelet packet best-tree searching acquisition easy algorithm. *Journal of system simulation* 20(1): 4759–4761.

Kumbasar, V. & Kucur, O. 2009. Searching better wavelet acket tree for ISI and ICI reduction in WOFDM. *Wireless Telecommunications Symposium*: 1–4.

Lance, E. & Kaleh, G.K. 1997. A diversity scheme for a phase-coherent frequency-hopping spread-spectrum system. *IEEE Transactions on Communications* 45(9): 1123–1129.

Li, K.K., Tang, X.H. & Ma, D.D. 2014. Algorithm on optimization for wavelet packet modulation structures based on BER of sub-channel. *Journal of System Simulation* 26(2): 300–305.

Li, T.T., Ling, Q. & Ren, J. 2007. A spectrally efficient frequency hopping system. *Global Telecommunications Conference, IEEE*: 2997–3001.

Mallat, S. 1991. Zero-crossings of a wavelet transform. *IEEE Transactions on Information Theory* 37(4): 1019–1033.

Mohanty, M.N & Mishra, S. 2013. Design of MCM based wireless system using wavelet packet network & its PAPR analysis. *In processing of International Conference on Circuits, Power and Computing Technologies*: 821–824.

Shao, X.X., Tian, J.Q. & Tang, X.H. 2007. A Multi wavelet packets modulation based on DSSS. *Journal of Hangzhou Dianzi University* 1(1):38–41.

Ying, X.F., Yong, Z., Wang, C.Y, et al. 2005. Study on the wavelet packet based M-ary multicarrier spread

spectrum system. *Systems Engineering and Electronics* 11(27):1866–1870.
Yu, Z.W., Tang, X.H., Shen, C.P, et al. 2010. Research on structures of wavelet packet modulation based PAPR reduction. *Journal of Hangzhou Dianzi University* 8(2): 32–36.
Zeng, Q. & Peng, D.Y. 2010. Performance analysis of multiuser OFDM-FH communications system with PSK modulation scheme. *ACTA ELECTRONICA SINICA* 38(4): 943–948.
Zhang, G. & Zhang, X. 2001.Wavelet packet function modulation based frequency hopped WPDMA communication system. *Journal of Circuits and Systems* 6(1):13–16.

Electronics, Communications and Networks IV – Hussain & Ivanovic (eds)
© 2015 Taylor & Francis Group, London, ISBN: 978-1-138-02830-2

Analysis and design of the feistel structured S-box

Haoran Zheng*, Hongbo Wang & Kainan Zhang
Information Science and Technology Institute, Zhengzhou, Henan, China

ABSTRACT: S-box is the unique nonlinear component in many block ciphers. Its quality directly affects the security strength of the whole cryptographic algorithms. In this paper, we analyzed the nature of algebraic degree aspects of the 8 (×8 S-boxes, whose structure is a three-lap feistel such as CRYPTON and ZUC algorithm, etc. We proved that the algebraic degree of output component functions of such S-boxes cannot be higher than five times simultaneously. Therefore, such S-boxes are not ideal random nature. This paper presents the method of using five-lap feistel structure to construct 8 (×8 S-boxes. The differential uniformity, nonlinearity, algebraic number and item number distributions of these 8 (×8 S-boxes are in line with the characteristics of random distribution. Such S-boxes have the advantage of making such a structure constructed with a password and a random structure of the two construction methods. Accordingly, these S-boxes have higher security.

1 INTRODUCTION

S-box first appeared in Lucifer algorithm (Sorkin 1984)in the emergence of DES, the S-box is widely popular. S-box is the unique nonlinear component in many block ciphers. Its quality directly affects the security strength of the whole cryptographic algorithms. S-box is a good cryptographic algorithm that can resist differential attacks or linear attacks and other attacks to provide an effective guarantee. Precisely because of the extreme importance of the S-box, a series of design criteria (Liu & Feng 2000, Chen & Feng 2004, Wen et al. 2000, Filiol 2002, Fischer & Meier 2007, Chen & Feng 2007, Wu et al. 2009, Feng 2000) been presented on the basis of long-term research. Resist various attacks from the perspective of people on the S-box design made various demands, given the many design criteria, such as nonlinearity, differential uniformity, algebraic number and distribution of the number of items. Based on these design criteria, a number of constructs have been proposed a method of S-boxes, such as the randomly selected and tested with a password structural configuration, structure and other mathematical functions. The advantage of this construction method is a good indicator of cryptography and hardware to achieve lower costs. The disadvantage is that due to the restricted password structure and balance of efficiency, often in order to improve the efficiency of hardware implementation and appropriate indicators weakened cryptography. Randomly selected and tested method is usually the first use of a random way to generate random transformation, and then filtered through a set of indicators to test until you find to meet the requirements of S-boxes. The significant advantage is the obvious structural features, showing various property characteristics of the random transformation, not unexpected defects.

Using of password structure constructor usually makes use of some in-depth study of small-scale structures and passwords S-box construction of large-scale S - boxes. Such as CRYPTON (Chae & Hyo 1998) and ZUC (ETSI 2011) S-box is to use three-lap feistel structure consists of three 4 (4 small S-boxes constructed an 8 (8 S-boxes to achieve high performance on hardware. The advantage is that both the design of S-boxes with good indicators of cryptography, and a low implementation cost of hardware. Its disadvantage is that due to the restricted password structure and balance of efficiency, often in order to improve the efficiency of hardware implementation and appropriate indicators weakened cryptography. Randomly selected and tested method is usually the first use of a random way to generate random transformation, and then filtered through a set of indicators to test until you find to meet the requirements of S-boxes. A significant advantage is the obvious structural features, showing various property characteristics of the random transformation, not unexpected defects.

In this paper, we analyzed the nature of algebraic degree aspects of 8×8 S-boxes, whose structure is three-lap feistel such as CRYPTON and ZUC algorithm, etc. We proved that the algebraic degree of output component functions of such S-boxes cannot

*Corresponding Author: *haimozhang@163.com*

be higher than five times simultaneously. So these S-boxes on the performance of the random algebraic properties are not very good. This paper presents the method of using five-lap feistel structure to construct 8×8 S-box. The differential uniformity, nonlinearity, algebraic number and item number distributions of these 8×8 S-boxes are in line with the characteristics of random distribution. Such S-boxes have the advantage of making such a structure constructed with a password and a random structure of the two construction methods. Accordingly, these S-boxes have higher security.

2 BASIS KNOWLEDGE

Definition 2.1 Let $S(X)=(f_1(X),\cdots,f_m(X))$ $:F_2^n \to F_2^m$ be a multi-output Boolean functions, order. (Eli & Adi 1991)

$$\delta_s = \frac{1}{2^n} \max_{\substack{\alpha \in F_2^n/\{0\} \\ \beta \in F_2^m}} \#\{X \in F_2^n : S(X \oplus \alpha) \oplus S(X) = \beta\}$$

Here, $\#\{\bullet\}$ indicates the number of elements of the set $\{\bullet\}$. The δ_s is called for the differential uniformity of the $S(X)$.

Obviously, the $\delta_s \geq 2^{1-n}$ have established when $S(X): F_2^n \to F_2^m$.

Differential uniformity δ_s is an important indicator of cryptographic functions of the fight against differential cryptanalysis. From the angle against differential cryptanalysis, the smaller the difference δ_s, the better the uniformity, the smaller the difference uniformity δ_s, S-box input and output information, the more difficult the maximum differential leakage collected, resistance to differential cryptanalysis is stronger.

Definition 2.2 Let $S(X)=(f_1(X),\cdots,f_m(X))$ $:F_2^n \to F_2^m$ is a multi-output Boolean functions. L_n is recorded to all n-affine Boolean function.(Matsui 1994)Claimed that:

$$N_s = \min_{\substack{l \in L_n \\ 0 \neq u \in F_2^m}} d_H(u \cdot S(X), l(X))$$

is called to the nonlinearity of $S(X)$. Among them, $d_H(u \cdot S(X), l(X))$ means hamming distance between $u \cdot S(X)$ and $l(X)$.

Non-linearity N_s is an important indicator to measure the password function to fight linear cryptanalysis. Overcoming linear cryptanalysis stand -point, the better the non-linearity N_s, the greater the non-linearity N_s, the linear S-boxes, the more difficult the information collected leakage, resistance to linear cryptanalysis is stronger.

Definition 2.3 Let $f(X): F_2^n \to F_2$ is a n-affine Boolean function.(Jakobsen & Knudsen 1997) Then $f(X)$ can be uniquely means in the following form:

$$f(X) = a_0 \oplus \bigoplus_{1 \leq i_1 < i_2 < \cdots < i_k \leq n} a_{i_1 i_2 \cdots i_k} x_{i_1} x_{i_2} \cdots x_{i_k}$$

Among them, $X = (x_1, x_2, \quad, x_n)$ and $a_0, a_{i_1 i_2 \quad i_k} \in F_2$. And it is called to the algebra of regular type of $f(X)$.

Definition 2.4 Let $f(X): F_2^n \to F_2$ is a n-affine Boolean function. (Jakobsen & Knudsen 1997)Then the algebra of regular type of $f(X)$ in the following form:

$$f(X) = a_0 \oplus \bigoplus_{\substack{1 \leq i_1 < i_2 < \cdots < i_k \leq n \\ 1 \leq k \leq n}} a_{i_1 i_2 \cdots i_k} x_{i_1} x_{i_2} \cdots x_{i_k},$$ among

them, $X = (x_1, x_2, \cdots, x_n) \in F_2^n$, $a_0, a_{i_1 i_2 \cdots i_k} \in F_2$.

The $\max\{k \mid a_{i_1 i_2 \cdots i_k} \neq 0\}$ of $f(X)$ is called frequency. The number of items j times of $f(X)$ is the number of times a term that called the algebra of formal type j is $f(X)$'s. Several times that all $j (0 \leq j \leq n)$ of $f(X)$ is called for the item number.

From the perspective of anti-attack, if algebraic function of each component of the S-box output is too low, the corresponding block cipher vulnerable to higher order differential cryptanalysis attacks. If too few items each output component functions of S-box, it is possible to improve the success rate of the interpolation attack. Therefore, each output algebraic linear combination of S-boxes should be high enough number of items should be enough.

For a good 8×8 randomness bijective S-box, its algebraic output of each of its component functions best of 7, the expected value of the number of items to 127.5. Generally, the number of times each output component algebraic function of not less than 6. All components of the average number of output functions between 110 and 140, it is considered in line with the characteristics of a random distribution.

3 STRUCTURE AND PROPERTIES OF CRYPTON CATEGORY 8×8 S-BOXES

CRYPTON is one block cipher Advanced Encryption Standard (AES) algorithm candidate. Wherein the 8×8 S-box is to use three laps feistel structure which is constructed from three 4×4 S-boxes-S_0,S_1,S_2. Specifically, as shown in Figure 1:

S_0={15,9,6,8,9,9,4,12,6,2,6,10,1,3,5,15}
S_1={10,15,4,7,5,2,14,6,9,3,12,8,13,1,11,0}
S_2={0,4,8,4,2,15,8,13,1,1,15,7,2,11,14,15}

CRYPTON is one block cipher Advanced Encryption Standard (AES) algorithm candidate. Wherein the 8×8S-box is to use three laps feistel structure which is constructed from three 4×4 S-boxes S_0, S_1, S_2.

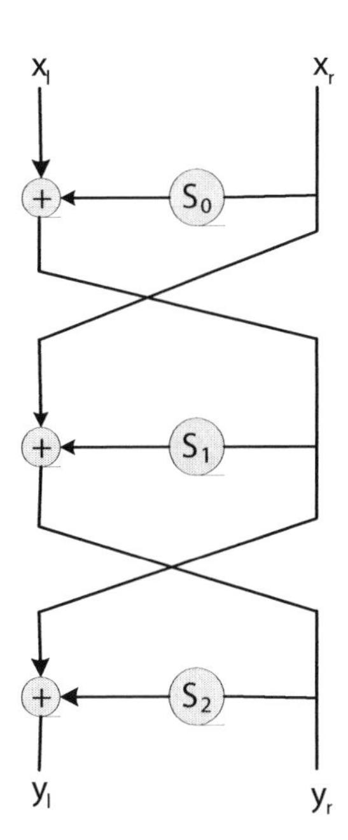

Figure 1. Three laps feistel structure.

The mathematical description of the S-boxes of CRYPTON is as follows:

$$(y_l, y_r) = S(x_l, x_r), x_l, x_r, y_l, y_r \in F_2^4$$

Wherein,

$$y_r = x_r \oplus S_1(x_l \oplus S_0(x_r)), \quad y_l = x_l \oplus S_0(x_r) \oplus S_2(y_r).$$

The characteristics of the S-box CRYPTON algorithm are analyzed in the literature (Chae & Hyo 1998).

In terms of security, the algorithm for CRYPTON S_0 and S_2, which are not required to-one mapping, but requested a degree of uniformity of their differential to be 2^{-3}, nonlinearity to be 4. For S_1, while requiring their differential uniformity of 2^{-2}, the nonlinearity of 4. In hardware implementation, when CRYPTON algorithm select S_0,S_1,S_2, in order to reduce the consumption of hardware, to use as little of the NAND gate circuit to achieve an S-box. CRYPTON algorithm will limit the number of output component algebraic functions S_0 and S_2 for 2, S_1 output component functions of algebraic number is limited to 3. The S-box of CRYPTON algorithm whose safety indicators better and good hardware implementation is efficiency, Such as the S-box differential uniformity of 2^{-5}, the non-linearity of 96, the number of all output components of algebraic functions are 5, the average number of 91, the hardware consumes only 107GE. Better security and lower hardware consumption indicators CRYPTON algorithm make its S-box to get a better application in the design of cryptographic algorithms, such as ZUC other algorithms in this class on the use of S-boxes. For convenience, we will call such kind of S-box for the CRYPTON category S-box. More generally, we define the following:

Definition 3.1 In Figure 1, the three-laps feistel structure utilizing the 8×8 S-boxes generated by the three 4×4 S-boxes, algebraic function if each output component 4×4 S-boxes of not more than 3, called the 8×8S-box for the CRYPTON category of S-boxes.

1 Such as CRYPTON, ZUC algorithm etc., they used the 8×8 S-boxes case for the CRYPTON category S-box;
2 As shown in Figure 1, the 8×8 S-boxes that generated by three-laps feistel structure consists of three one mapping 4×4 S-boxes, is also called for the CRYPTON category of S-boxes.

It can be seen from the safety indicators, differential non-linearity and uniformity of the S-boxes have reached the random 8×8 S-box requirements. However, the random nature of the distribution of algebraic number and the number of items are not very good performance. In fact, we can theoretically analyze the algebraic CRYPTON category S-boxes.

Theorem 3.1 Algebraic functions of the right four output components of CRYPTON category S-box is not more than five.

Proof: Let CRYPTON category S-box is:

$$(y_l, y_r) = S(x_l, x_r), x_l, x_r, y_l, y_r \in F_2^4$$

wherein, $y_r = x_r \oplus S_1(x_l \oplus S_0(x_r))$, $y_l = x_l \oplus S_0(x_r) \oplus S_2(y_r)$.

And let $x_l = (x_{l_1}, x_{l_2}, x_{l_3}, x_{l_4})$, $x_r = (x_{r_1}, x_{r_2}, x_{r_3}, x_{r_4})$, $x_{l_i}, x_{r_i} \in F_2$, $i = 1,2,3,4$.

The following study algebraic function y_r of output components.

Let $S_0(x_r) = (S_{0_1}(x_r), S_{0_2}(x_r), S_{0_3}(x_r), S_{0_4}(x_r))$, then $x_l \oplus S_0(x_r) = (x_{l_1}, x_{l_2}, x_{l_3}, x_{l_4}) \oplus (S_{0_1}(x_r), S_{0_2}(x_r), S_{0_3}(x_r), S_{0_4}(x_r)) = (x_{l_1} \oplus S_{0_1}(x_r),\ x_{l_2} \oplus S_{0_2}(x_r),\ x_{l_3} \oplus S_{0_3}(x_r),\ x_{l_4} \oplus S_{0_4}(x_r)) = S_1(x_{l_1} \oplus S_{0_1}(x_r),\ x_{l_2} \oplus S_{0_2}(x_r),\ x_{l_3} \oplus S_{0_3}(x_r),\ x_{l_4} \oplus S_{0_4}(x_r))$

Due to algebraic functions of the output component of S_1 is not more than three, algebraic function of the output component $S_1(x_l \oplus S_0(x_r))$ is not higher than the algebraic of

$(x_{l_i} \oplus S_{0_i}(x_r)) \cdot (x_{l_j} \oplus S_{0_j}(x_r)) \cdot (x_{l_k} \oplus S_{0_k}(x_r))$, $i, j, k \in \{1,2,3,4\}$ and $i \neq j \neq k$.

The algebraic of

$$(x_{l_i} \oplus S_{0_i}(x_r)) \cdot (x_{l_j} \oplus S_{0_j}(x_r)) \cdot (x_{l_k} \oplus S_{0_k}(x_r)),$$

$i, j, k \in \{1,2,3,4\}$ and $i \neq j \neq k$ $x_{l_i} \cdot x_{l_j} \cdot x_{l_k}$, $x_{l_i} \cdot x_{l_j}$

$\cdot S_{0_k}(x_r)$, $x_{l_i} \cdot S_{0_j}(x_r) \cdot S_{0_k}(x_r)$, $S_{0_i}(x_r)) \cdot S_{0_j}(x_r) \cdot S_{0_k}(x_r)$, $i, j, k \in \{1,2,3,4\}$ and $i \neq j \neq k$ This is four times higher in algebra.

Due to algebraic function S_0 component of not more than three, the algebraic of $S_{0_1}(x_r), S_{0_2}(x_r), S_{0_4}(x_r)$ is not more than three. So that, the algebraic of $x_{l_i} \cdot x_{l_j} \cdot S_{0_k}(x_r)$ is not more than five and the algebraic of $S_{0_i}(x_r) \cdot S_{0_j}(x_r) \cdot S_{0_k}(x_r)$ is not more than four.

So the algebraic of the output component functions of $S_1(x_l \oplus S_0(x_r))$ is not more than five.

So the algebraic of the output component functions of $y_r = x_r \oplus S_1(x_l \oplus S_0(x_r))$ is not more than five.

Scilicetly, algebraic function of the right four output components of CRYPTON category S-box is not more than five.

By Theorem 3.1, CRYPTON category S-box is just in line with the characteristics of random distribution, to construct a random feistel transformation properties better structured 8×8S-box, you need to increase the number of rounds feistel structure, but considering the hardware implementation efficiency, and the number of rounds in the feistel structure, the better. Thus, not only has the minimum number of rounds is given, and that it can meet the transformation properties of random feistel structure type 8×8S-box is very meaningful.

4 DESIGN METHOD ON 8×8 S-BOX BY FEISTEL STRUCTURE

This section gives a method for generating five-laps feistel structure 8×8S-box which is utilized by five randomly 4×4S-box.

The basic idea is to use each round 4×4S-box contains a feistel structure 5 as a candidate generation module S-box, the five by five 4×4S-boxes updated randomly, to achieve the S-box candidates the update, and then set targets using cryptography candidate S-box testing, screening out until one or more qualified S-boxes.

For using this method to generate qualified 8×8 S-boxes, we require uniformity of their difference is not more than 2^{-5}, the non-linearity of not less than 96, the number of output components algebraic function of not less than 6, the average number of output components function between 110 to 140, in order to ensure uniformity of its differential nonlinearity, algebraic number and item number distributions are in line with the characteristics of random distribution, making this kind of S-box structure of both the structure and the use of passwords randomly constructed two constructor advantage, which has higher security.

For candidate generation module S-box, we used the five-laps feistel structure shown in Figure 2.

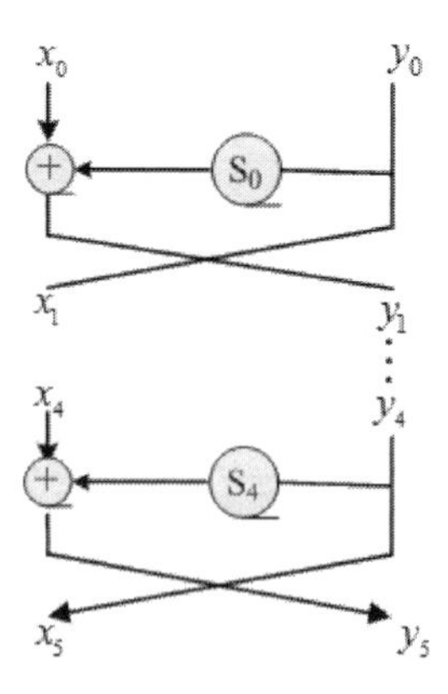

Figure 2. Candidate generation module structure of S-box.

Mathematical description of the candidate S-box generation module is as follows:

$(x_5, y_5) = S(x_0, y_0)$, $x_i = y_{i-1}$, $y_i = x_{i-1} \oplus S_{i-1}(y_{i-1})$, $x_i, y_i \in F_2^4$, $i \in \{1,2,3,4,5\}$.

Its construction algorithm is as follows:

Algorithm 4.1:

Step1 Initialization: m=0, randomly selected five 4×4S-boxes S_0, S_1, S_2, S_3, S_4;

Step2 Candidate generation S-boxes. 8×8 S-box which is five laps feistel structure is generated from the candidate 4×4S-boxes. Candidate 8×8 S-box as follows: $(x_5, y_5) = S(x_0, y_0)$. $x_i = y_{i-1}$, $y_i = x_{i-1} \oplus S_{i-1}(y_{i-1})$, $x_i, y_i \in F_2^4$, $i \in \{1,2,3,4,5\}$;

Step3 S-boxes in order to detect candidate differential uniformity, nonlinearity, algebraic degree, the average number of cryptography is set to meet targets. If the differential uniformity $> 2^{-5}$, then go to Step4; If the non-linearity <96, go to Step4; If certain algebraic S output component functions of <6, then go to Step4; If the average number of <110 or>140 then go to Step4; If the cryptography indicators meet the set requirements, go to Step5;

Step4 Update five 4×4S-boxes S_0, S_1, S_2, S_3, S_4, and return to Step2;

Step5 m=m+1, Output S_0, S_1, S_2, S_3, S_4 and Qualified S-box. If m<n, then return Step4; otherwise, the algorithm terminates.

According to the algorithm 4.1, we can construct these kind of 8×8 S-boxes whose differential uniformity is not more than 2^{-5}, the non-linearity of not less than 96, the number of output components algebraic function of not less than 6, the average number of Output component functions between 110 to 140.

$$
SS=\begin{pmatrix}
b4 & 0b & ff & 7a & c1 & 28 & 90 & d8 & 68 & 0f & 9e & be & 92 & f5 & 74 & 4c \\
21 & 55 & 86 & ab & b5 & 1b & e1 & 14 & 75 & 56 & 2f & 6f & f2 & f1 & 2b & d4 \\
30 & 6e & f8 & d3 & 1e & 87 & 36 & 52 & 31 & 15 & 6c & d2 & 32 & cb & 02 & 69 \\
45 & 59 & 17 & 63 & 41 & 3c & 05 & de & 60 & 9c & ea & 0e & 67 & a1 & 49 & b9 \\
4b & 4e & ed & 83 & 47 & 78 & 7c & 1c & a7 & 72 & 34 & b3 & 01 & 4f & e3 & 11 \\
c9 & 9a & 7b & d1 & c0 & a5 & b6 & 97 & d9 & 50 & dd & bf & da & 62 & 42 & 8c \\
94 & 79 & 51 & 23 & a8 & 82 & f6 & 88 & 9d & 33 & df & bb & 6d & 20 & 48 & 35 \\
fb & 7f & 37 & cd & 2a & fe & c3 & bd & af & ba & 27 & bc & 1a & c6 & a4 & 93 \\
29 & fa & 89 & 6a & dc & 57 & e5 & 44 & 9f & 61 & c8 & ae & e6 & 03 & f3 & 4d \\
09 & 10 & 13 & 70 & 6b & ce & 5c & ca & e9 & 99 & c5 & 2d & b0 & f4 & 4a & 22 \\
fd & a2 & a9 & ec & 0d & 16 & 3a & a0 & 3f & 38 & 54 & 3b & 00 & c7 & 65 & c4 \\
84 & 18 & cc & 80 & 85 & eb & 71 & 91 & 64 & 25 & d6 & 04 & 8e & 39 & ad & 76 \\
26 & 40 & 81 & 7e & 8f & 0c & 5b & 43 & f7 & e4 & cf & 66 & 1f & b2 & 9b & d7 \\
fc & ef & 12 & 07 & b8 & 7d & aa & 3d & b1 & 19 & db & 5f & d0 & 8b & 95 & a3 \\
a6 & 96 & b7 & e7 & 2e & 5d & 8a & f0 & 3e & 53 & d5 & 73 & 5a & 2c & 98 & e0 \\
08 & 46 & 77 & 0a & 1d & 8d & ac & e8 & 58 & 06 & c2 & 24 & f9 & ee & 5e & e2
\end{pmatrix}
$$

We give an example of a configuration, and examine their safety indicators and hardware consumption area.

Note one qualified S-box to SS, and specifically as follows:

$$S_0=\{4,12,14,5,11,2,0,3,9,7,1,6,15,10,13,8\}$$
$$S_1=\{12,13,9,10,8,11,7,2,1,15,0,6,14,5,4,3\}$$
$$S_2=\{15,6,11,4,5,0,7,1,10,8,14,12,2,9,13,3\}$$
$$S_3=\{14,12,8,2,7,10,13,4,1,9,0,11,6,5,3,15\}$$
$$S_4=\{4,11,8,7,9,13,0,5,15,14,3,10,1,2,12,6\}$$

To test SS-box for safety, get the differential uniformity of SS-box is 2^{-5}, the nonlinearity is 96, algebraic function of the number of output components of a six, seven, seven times the average number of 127.5, referral between 110 and 140. Therefore, the SS box has better security indicators, but also has the advantage of randomly constructed S-boxes.

In the 65nm (Weng 2006) process below, the SS box hardware implementation, the results obtained after comprehensive as shown in Table 1.

Table 1. Consolidated results.

Type	Circuit Delay	Clock frequency	Comprehensive Area
SS-box	1.1ns	909MHz	169Ge

As can be seen from the above results, the use of this design of 8×8 S-boxes constructed scheme implemented in hardware, the hardware consumes an area of 169GE, although the occupation of more hardware resources than CRYPTON category S-box. However the structure and random 8 (8 bijective S-boxes compared with more efficient hardware implementation to meet the requirements of resource-constrained environments.

5 CONCLUSIONS

S-box symmetric cipher algorithm design directly affects the safety performance of cryptographic algorithms and implementation of performance, design a good random transformation properties and have lower hardware implementation costs of both S-box cryptographic algorithms to ensure better resist password attacks, and able to meet the requirements of resource-constrained environment. In this paper, we analyzed the nature of algebraic degree aspects of 8×8 S-box, whose structure is three-lap feistel such as CRYPTON and ZUC algorithm, etc. We proved that the algebraic degree of output component functions of such S-boxes cannot be higher than five times simultaneously. Therefore, such S-boxes are not ideal random nature. This paper presents the method of using five-lap feistel structure to construct 8×8 S-box. The differential uniformity, nonlinearity, algebraic number and item number distributeions of these 8×8 S-boxes are in line with the characteristics of random distribution. Accordingly, the research results can be better applied to the block cipher algorithm coding aspects of design.

REFERENCES

Chae Hoon Lim & Hyo Sun Hwang. 1998. *A New 128-bit Block Cipher-Specification and Analysis*. Future Systems Inc.

Chen Hua & Feng Dengguo. 2004. An Evolutionary Algorithm to Improve the Nonlinearity of Self-Inverse S-Boxes. *International Congference on Information Security and Cryptology-ICISC 2004, Seoul, Korea, 2–3 December 2004*. Springer-Verlag.

Chen Hua & Feng Dengguo. 2007. An Evolutionary Algorithm to Improve the Nonlinearity of Self-Inverse S-Boxes. *International Conference on Information Security and Cryptology-ICISC. 2004. LNCS 3506, Seoul, Korea, 2–3 December 2004*. Springer-Verlag.

Eli Biham & Adi Shamir. 1991. Differential Cryptanalysis of DES-like Cryptosystems. *Journal of Cryptology* 4(1): 3–77.

ETSI/SAGE TS 35.221-2011. Specification of the 3GPP Confidentiality and Integrity Algorithms 128EEA3&128-EIA3. *Document 1:128-EEA3 and 128EIA3 Specification.*

ETSI/SAGE TS 35.222-2011. Specification of the 3GPP Confidentiality and Integrity Algorithms 128EEA3&128 EIA3. *Document 2: ZUC.*

Feng Dengguo. 2000. *Cryptanalysis.* Bejing: Tsinghua University Press.

Filiol Elle. 2002. A new statistical testing for symmetric ciphers and hash functions. *Information and Communications Security* 2513(1): 342–353.

Fischer, S. & Meier, W. 2007. Algebraic Immunity of S-Boxes and Augmented Functions. *Fast Software Encryption-FSE. 2007, LNCS 4593, Luxembourg, 26–28 March 2007.* Springer-Verlag.

Jakobsen, T., & Knudsen, L. R. 1997. The interpolation attack on block ciphers. *Proceedings of Fast Software Encryption: 4th International Workshop, Israel, 20–22 January 1997*: 28–40. Springer Berlin Heidelberg.

Liu Xiaochen & Feng Dengguo. 2000. Meet certain S-boxes constructed nature cryptography. *Journal of Software* 11(10): 1299–1302.

Matsui, M. 1994. Linear Cryptanalysis Method for DES Cipher. *Workshop on the Theory and Application of Cryptographic Techniques Lofthus, Norway, 23–27 May 1993.* Springer Berlin Heidelberg.

Sorkin, A. 1984. Lucifer: A cryptographic algorithm. *Cryptologia* 8(1): 22–41.

Wen Qiaoyan, Niu Xin Xin & Yang Yixian. 2000. Modern cryptography Boolean function. Beijing: Science Press.

Wen Shousong. 2006. 65nm technology and equipment. *Electronic equipment* 36(2).

Wu Wenling, Feng Dengguo & Zhang Wentao. 2009. *The design and analysis of block cipher.* Beijing: Tsinghua University Press.

Electronics, Communications and Networks IV – Hussain & Ivanovic (eds)
© 2015 Taylor & Francis Group, London, ISBN: 978-1-138-02830-2

SOA dynamic reconfiguration grid model of service scheduling and integration

Jiong Zheng
Information Science & Engineering College, XinJiang University, Urumqi, XinJiang, China

Zhixiang Zheng
Xinjiang Police College, Urumqi ,XinJiang, China

ABSTRACT: This paper proposes a Dynamic Integration and Schedule System for grid service (DISS) based on SOA, aimed at the problem of realizing dynamic selection and integration in the grid environment. We also introduce the idea of dynamic reconstruction, giving DISS a framework and implementation methods for dynamic reconfigurable grid applications. DISS achieves business integration quickly since the scheduling system has more flexibility through the use of existing resources.

1 INTRODUCTION

Service-Oriented Architecture is a component model for service-oriented. It distributes a deploy and combined service component, based on the demand for what separates computation programs from the framework (Li et al. 2012, Tsai et al. 2014). The famous grid architecture model-Open Grid Services Architecture (OGSA) is a server-centric model. Various service components of different worker nodes belong to different departments and people, which often results in the entering and evacuation of a component at any point in time (Karasavvas et al. 2005, Guo et al. 2004). It cannot ensure the continuity or the consistency of the applicable service. In order to satisfy dynamic changes in the grid environment and user requirements (Kakaševski et al. 2013), grid the system should possess dynamic reconfiguration capabilities and achieve integration and scheduling to support evolution from one application to another without interruption. This would provide users with a consistent service (Grimshaw et al. 2005, Zikos & Karatza 2008).

The idea of separating framework for computing is introduced in OGSA technology. It can achieve dynamic integration and scheduling functions, using loosely coupled SOA architecture components. If some components are down, this system will be invoked to replace them (Quan et al. 2007, Wu et al. 2011).

Based on the analysis of the grid system requirements of service-oriented architecture, we propose the dynamic service integration and scheduling system model (DISS) and its design framework. Finally, this paper outlines the grid environments of the development and the application of this grid system.

2 DYNAMIC SERVICE INTEGRATION AND SCHEDULING SYSTEM MODEL (DISS)

In the OGSA model, the kernel service layer provides some network service, such as the position of the grid system, discovery and binding, and grid virtual organization management. OGSA implements the share and cooperation of grid service, but it cannot provide a dynamic reconfiguration service. Based on the research of network functionality and OGSA framework, this paper's research and analysis serves as a combination of the SOA and the grid system, while introducing the dynamic reconfiguration service model onto the OGSA framework. This improves the grid system by giving it user-friendly properties and allowing it to meet requirements.

Definition 1

The connection between the components for the application is formed by service components and is named the grid system configuration. For the couple of (COMPONENT, CONNECTION), COMPONENT represents the set of component and CONNECTION represents the connection between different components.

Definition 2

The grid system configuration which adjusts in the running state is named dynamic reconfiguration.

Definition 3

The event that drives the grid system application dynamic reconfiguration in the service component under running conditions is named D-R event.

Based on the above analysis, this paper proposes a dynamic service integration and scheduling system model (DISS) based on SOA and the OGSA model. Figure 1 is the framework of the dynamic service integration and the scheduling system model (DISS). This model includes three layers: a grid system layer, an SOA organization layer, and a grid service resource layer.

1 Grid system layer
2 The grid system layer includes many applications. It builds on the higher layer of grid service and sends the requirements to the SOA organization before waiting for a response from there. SOA organization layer includes many modules, and is responsible for unification management and the scheduling of grid system service in order to achieve the share and cooperation of the grid system service. It includes the task management service, the DISS kernel service, and the service running environments: 1) task management service is responsible for analyzing the user task and getting the grid system requirements of task and monitor the status of the task. 2) The DISS kernel service includes a grid information service, a data management service, a virtual management service, a grid safe management service, and a D-R service. 3) The service running environments provide a mechanism of service representations, visits, and running.
3 Grid service resource layer

The grid service resource layer includes a physical resource, a logic resource, and framework. The physical resources are a cluster, a large database, a PC, storage, various special devices, and a network. The logic resource consists of a Web service, a special software system, an operation system, and so on. The framework is the infrastructure of DISS.

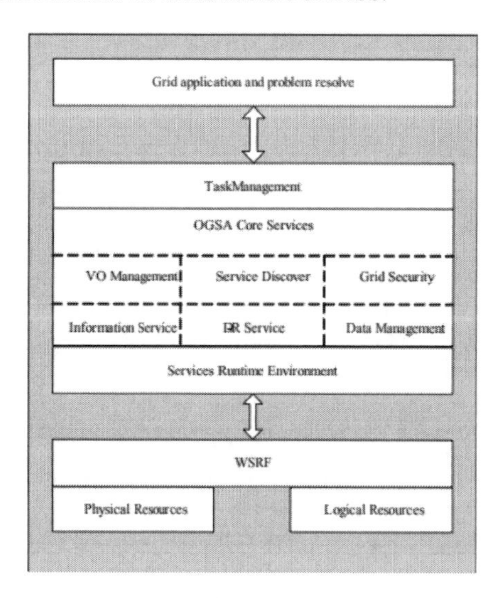

Figure 1. Framework of dynamic service integration and scheduling system model.

3 DISS MAIN MODULE DESIGN

3.1 *The service integration scheduling module implementation*

The service integration scheduling is provided by a D-R service. Figure 2 displays the function logic structure of the D-R service subsystem. It includes a grid system information service module, a service finding and binding module, a D-R service control module, service proxy, and web service resource framework.

The dynamic reconfiguration subsystem interaction process is as follows:

1 The service finding and binding module find a service component address in the grid information service.
2 The service binding is determined by the service proxy system and the grid system component.
3 During the application running period, specifically the service component launch reconfiguration event, control will be submitted to the D-R service module.
4 The D-R service control module generates a new requirement of service component finding and binding, achieving a reconfiguration of the D-R service control module.

In order to ensure that the grid system information service can reflect the new status of the service component, the grid system information service and grid system component keep a Time to Live (TTL).

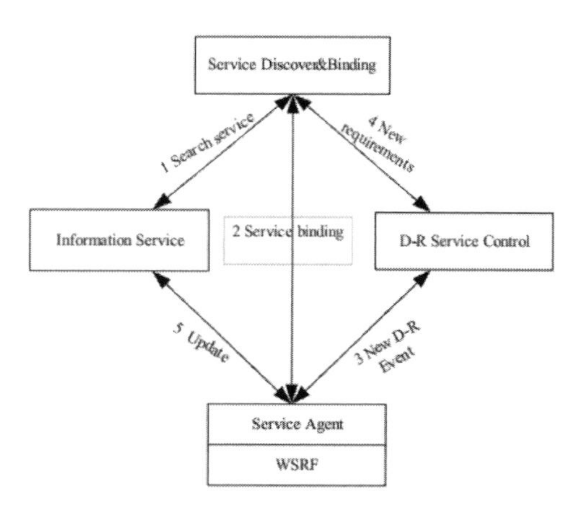

Figure 2. Function logic structure of D-R service subsystem.

3.2 *Description the rule of D-R*

This paper uses the Abstract State Machine (ASM) to describe the rule of D-R. ASM is a common method to create a system model, since it can be used in any system and depict a system from different abstract layers. This paper uses the ASM to show the grid system grammar rule, which includes the position of service component, the service component, and dynamic adding and dynamic replacement. ASM description shown as follows:

Pseudocode:

```
if componentX belong to Component and
Event(component)=replace
then
   replace(componentX) = true
   if replace(componentX)
   then
         if handle(component)=NULL
         then
                  leave(componentX)=true
                  component(componentX)=false
         else
                  pass(component)
                  requestUsable(handle
                  (componentX) = true
         endif
   endif
endif
```

4 DISS PROTOTYPE PLATFORM AND APPLICATION EXAMPLE

In order to verify the DISS model proposed by this paper, we implement the dynamic integration and schedule grid supporting (DISSEE) based on the above design framework. This environment supports a service-oriented application deployment, running, and monitor management. The kernel module includes a grid service system, a service proxy system, a D-R service system, an information service system, a safe, and a QOS maintain system. Figure 3 displays the logic framework.

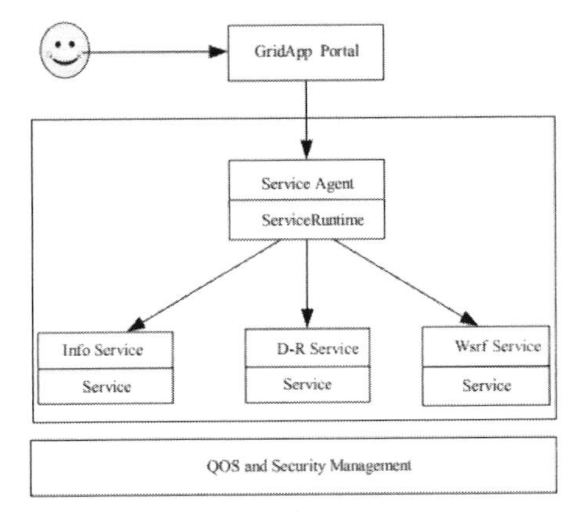

Figure 3. DISS system logic framework.

A user can login in a one-step service platform and submit their task by a GridApp page. When the service agent receives the requirements, it first distinguishes the service component of this requirement. It then finds the application example location of this requirement service by the query UDDI, and then sorting the application example and select the appropriate application. The service agent will send the query and book requirements. If the requirements do not receive a response of this service fault, the D-R control will use a SR table application backup information to send another service program requirement until all service programs finished. At last, the result will be saved in the DB-R database and give the user a result of a proxy service. Based on the dynamic integration and schedule grid supporting (DISSEE), it deploys a grid one-step service for a tour system (GOSS4T). Users can receive a group of travel services by sending a requirement in this system. For example, a travel service includes a weather query of sightseeing, booking tickets, and so on. Figure 4 is the logic framework of a grid one-step service for a tour system (GOSS4T).

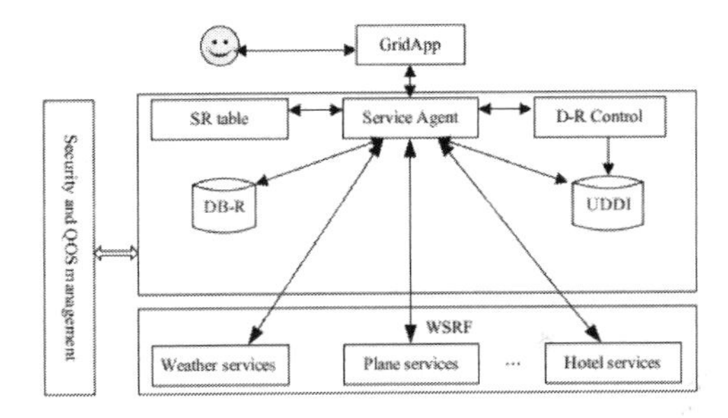

Figure 4. Grid one-step service for tour system (GOSS4T).

Compared with conventional architecture, SOA pays more attention to the role of service, providing extra values inside. It can also provide both customers and businesses with more suitable experiences, which do not appear in previous architecture. Hence, considering nowadays that the experience has drawn more attention, the development of SOA will be a main direction in the future.

5 CONCLUSION

Combining the grid system technology and SOA can improve the development, integration, and sharing of the grid system, At the same time, this meets the high requirements of using a grid system resource. This paper builds a D-R network model by using SOA technology and introduces the idea of a D-R service in OGSA framework. This model takes full advantage of the SOA framework to do the application component. We adjust it so it is now supported by grid infrastructure and fits into the dynamic change of the grid environment and user requirements. At last, this paper discusses the dynamic integration and schedule grid supporting (DISSEE), as well as the main module and the mechanism of implementation. This paper proposes a novel design idea and implementation method that builds a reliable and efficient grid service network.

ACKNOWLEDGMENTS

The research work was supported by the project of Education Sciences Planning of Xinjiang Uygur Autonomous Region of China (XJEDU2011S06).

REFERENCES

Grimshaw, A.S, Natrajan, et al. 2005. Legion: Lessons Learned Building a Grid Operating System. *Proceedings of the IEEE* (3) :589–603.

Guo-wei W, Zong-pu J & Wei Z. 2004. OGSA-based application study. *Journal of Jiaozuo Institute of Technology*.

Kakaševski, Mishev A, Krause A, et al. 2013. *ICT Innovations* : 151–160. Berlin Heidelberg: Springer.

Karasavvas K, Antonioletti M, Atkinson M, et al. 2005. Introduction to OGSA-DAI Services. *Lecture Notes in Computer Science*: 1–12.

Li X, Wang Z, Liu H. 2012. Optimizing initial chirp for efficient femtosecond wavelength conversion in silicon waveguide by split-step Fourier method. Applied Mathematics and Computation (24):11970–11975.

Quan L, Yang Y & Kai-jian L. 2007. Guarantee and control of quality of service on grid system: A survey. *Control and Decision*.

Tsai W, Zhong P, Bai X, et al. 2014. Dependence-Guided Service Composition for User-Centric SOA. *Systems Journal, IEEE,* 8(3):889–899.

Wu D, Liu C & Gao S. 2011. Coordinated control on a vehicle-to-grid system. *Electrical Machines and Systems (ICEMS), 2011 International Conference on. IEEE*:1–6.

Zikos S & Karatza H D. 2008. Resource Allocation Strategies in a 2-Level Hierarchical Grid System. *Simulation Symposium, ANSS 2008. 41st Annual. IEEE 2008*:157–164.

Electronics, Communications and Networks IV – Hussain & Ivanovic (eds)
© 2015 Taylor & Francis Group, London, ISBN: 978-1-138-02830-2

Research and design of 13.56 MHz RFID reader antenna

M.M. Zhou & W.P. Jing*
Jiangsu Key Laboratory of ASIC Design, Nantong University, Nantong, Jiangsu Province, China

ABSTRACT: Passive near field coupling antenna is widely used in Radio Frequency Identification (RFID). Antenna design is the key part of the RFID system design. The detailed design process of the antenna is derived in theory. A model can be established with the aid of High Frequency Structure Simulator (HFSS) and then the optimized antenna structure design can be achieved. It is too verbose and complex to design the matching network using theoretical formulas. The Smith chart method is simpler, easier and more intuitive to realize the matching process. The measured results show that $S_{11}$of the antenna is -14.299dB. The proposed antenna achieves a smaller reflection coefficient and a better efficiency.

1 INTRODUCTION

Radio frequency identification technology is a kind of automatic identification technology (Finkenzeller 2006). It can identify the objects and get the data via a non-contact and two-way data communication. Among those numerous identification technologies, RFID technology provides a precise and rapid identification for one or more targets. RFID technology can also work even in some relatively poor working environment, such as the situation of fast moving objects, etc. In addition, it has many other advantages, such as convenient application, no mechanical abrasion, long working life, high safety and so on. Considering the excellent performance and broad applicability of RFID technology, our country has applied this technology into the second generation identity card, transportation, the campus card and many other industry fields. Therefore, the study of RFID technology is really important.

An antenna has been put forward and designed in the paper (Huang & Luo 2007) for the existing radio frequency modules and cards. Circuit design of RF module and the antenna matching circuit are discussed. However, the antenna design lacks systematicness. The capacitance value is chosen through a lookup table for the antenna matching process. As a result, this design lacks flexibility. In consideration of RFID reader antenna, it puts forward a method by adding the open loop compensation at the bottom of the inductance coil to improve the intensity of magnetic field in this paper (Dai et al. 2013). However, after completion of the antenna matching, the main concerns of antenna design are reflection coefficient and energy transfer efficiency. The paper just explores the inductance value of the antenna rather than to explore the debugging of the antenna. Based on the working mode of passive inductance coupling RFID system, reference (Li 2007) analyzes the selection of the main parameters of the antenna and performance of the working environment and other issues affecting and a method for the near field of small loop antenna design is put forward. However the antenna design process is rather vague, there are no detailed numerical calculations and simulation debugging in that paper.

In this paper, the detailed antenna design and matching process in theory about the RFID reader antenna is discussed. After completion of the antenna design, a debugging for the antenna is achieved to ensure the compromise between the quality factor and the bandwidth. At last, the measurement results of the real, fabricated antenna show that the antenna has a smaller reflection coefficient and a better efficiency.

2 THE EQUIVALENT CIRCUIT AND THE CHOICE OF STRUCTURE ABOUT THE ANTENNA

The equivalent circuit diagram of the reader antenna coil is shown in Figure1. *L_ant* is the equivalent inductance of the antenna coil; *Rs_ant* is the resistance loss of the antenna coil and *C_ant* is the capacitance loss of the antenna coil. To make the antenna work at 13.56MHz, the antenna can be in series or parallel with a capacitance in the external. The capacitance and the antenna coil make up the L-C resonance circuit. It can ensure the resonance frequency of the working frequency on 13.56MHz by adjusting the

*Corresponding author: *13906294039@163.com*

value of the series or parallel capacitor. At this point, the reader will transfer energy to the RF cards and communicate with them.

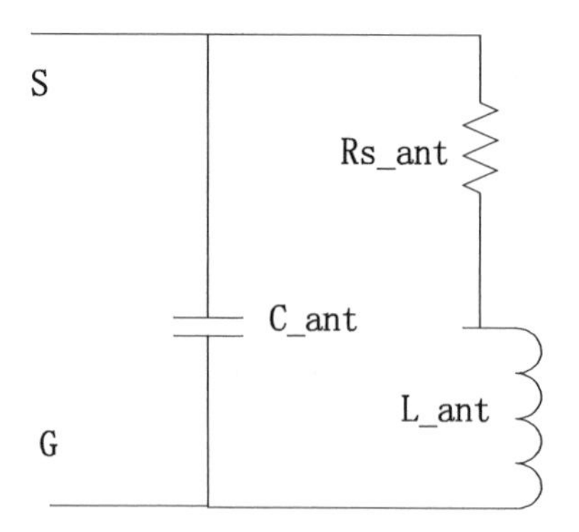

Figure 1. The equivalent circuit diagram of antenna.

Thomson formula is as follows:

$$f = \frac{1}{2\pi\sqrt{LC}} \tag{1}$$

where L is the inductance value and C is the capacitance value. The total capacitance value C is as follows:

$$C = \frac{1}{(2\pi f)^2 \times L} \tag{2}$$

where f is the frequency, taken as 13.56*MHz*, L is 5μH. Then C is approximately equal to 27.6pF. In engineering application, the value of capacitance loss is approximately 10pF and the matching about C will become very difficult. Therefore, the selections of the inductance values are range from 1μH to 3μH in general. According to the working environment, the antenna's shape is usually designed into a square, circular and rectangular. This paper introduces the design of the rectangular antenna, which deals with 13.56MHz RFID reader antenna as the PCB planar spiral inductors. Five key geometrical parameters of the rectangular coil are as follows: the peripheral length of the antenna coil L_x, the peripheral width L_y, the conductor width W, the coil spacing S and the coil number of turns N.

3 THE DESIGN OF THE ANTENNA

3.1 *The theoretical derivation of antenna design*

The Ampere's law is as follows:

$$B = \frac{\mu_0 i}{4\pi r}(\cos_{\alpha 2} - \cos_{\alpha 1}) \tag{3}$$

where B is the magnetic induction intensity, μ_0 is vacuum magnetic permeability, i is the current flowing through the conductor, r is the distance from B to the conductor and a_2 and a_1 are the included angle between B and the conductor. While current flows through the conductor, it can produce a magnetic field around the conductor. The magnetic induction intensity of the magnetic field is proportional to the current. But it is inversely proportional to the radius r.

Biot-Savart Law is expressed as (Xu & Cao 2006):

$$\underset{B}{\longrightarrow} = \int_{l_1}^{l_2} \frac{\mu_0 i}{4\pi} \frac{dl \times \vec{e_r}}{r^2} \tag{4}$$

where B, μ_0, i and r are the same as in equation 3, l_1 and l_2 are the integral path of the current in the conductor, dl is the tiny line element of the source current and e_r is the unit vector of current element towards magnetic field. It is concluded that the magnetic induction intensity is related to the vertical distance from the antenna coil's center. The result is as follows:

$$B = \mu_0 \frac{N_R^2}{2(R^2 + X^2)^{3/2}} i \tag{5}$$

where X is the vertical distance from the center of the antenna. R is the coil length. N is the number of turns. According to equation 5, B takes a derivation of R:

$$\frac{dB(R)}{dR} = \frac{\mu_0 iN}{2} \cdot \frac{2R\sqrt{(R^2+X^2)^3} - 3R\sqrt[3]{(R^2+X^2)}}{(R^2+X^2)^3} \tag{6}$$

The precondition is as follows:

$$\frac{dB(R)}{dB} = 0 \tag{7}$$

The conclusion is as follows:

$$R = \pm\sqrt{2}X \tag{8}$$

where R is equal to $\pm\sqrt{2}X$, the magnetic induction intensity has a maximum value. The plus-minus sign indicates that the magnetic induction intensity of the antenna is perpendicular to the two directions of propagation of the coil's axis.

3.2 *The determination of antenna's inductance value*

The approximate inductance value can be obtained by formula estimation, which can be expressed as:

$$L(nH) = 2T \times \{\ln(\frac{2 \times (Lx + Ly)}{W}) - K\} \times N^{1.8} \quad (9)$$

where L is the approximate inductance value of the antenna coil, L_x and L_y are the peripheral length and width, W is the conductor line's width and N is the coil number of turns. For the rectangular antenna coil, K is equal to1.47. According to the needs of the laboratory project engineering, the parameters are as follows: L_x=70mm, L_y=50mm, W=1mm, S=0.5mm. While N is equal to 4, the inductance value is more appropriate. A model which hasn't been matched (as shown in Figure 2.) is established in HFSS with these parameters.

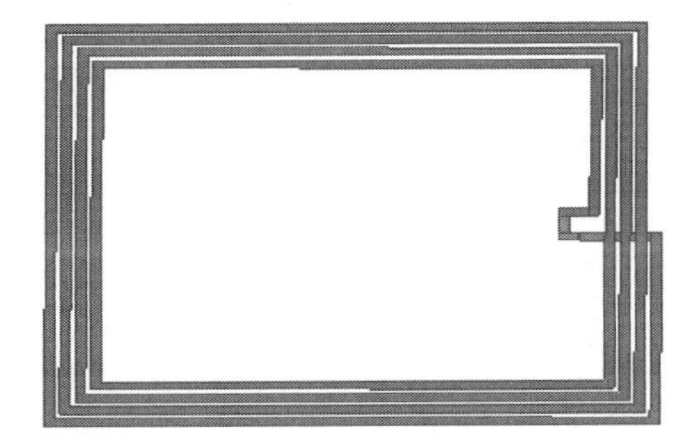

Figure 2. The antenna coil hasn't matched.

The parameter scanning analysis simulation function is used in HFSS, and the simulation parameters are L_x, W, S. L_y and L_x have the same influence on inductance, which determine the area of the coil. The simulation results are shown in Figure 3, Figure 4 and Figure 5.

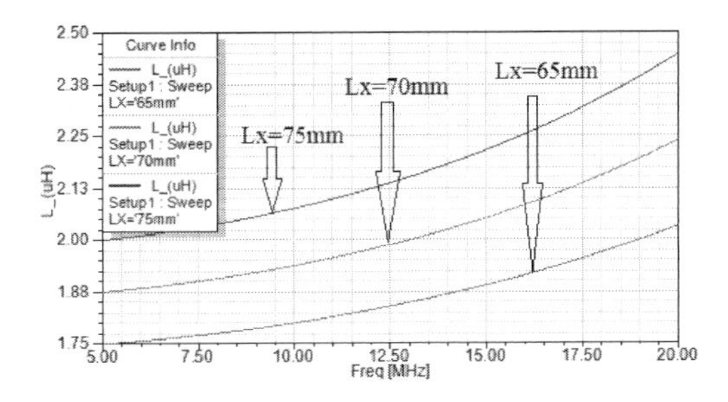

Figure 3. The effects of coil length on inductance.

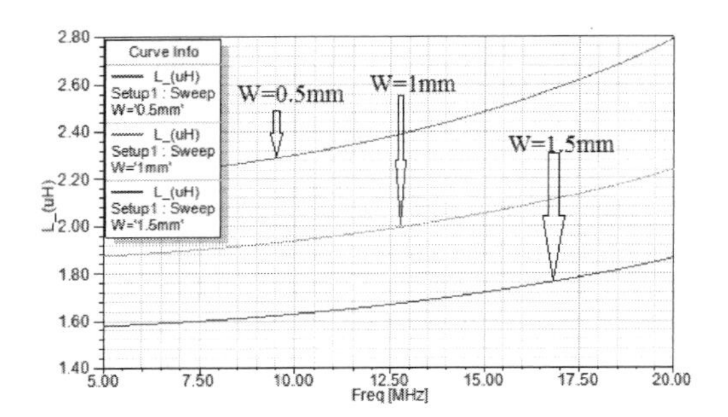

Figure 4. The width of the conductor on inductance.

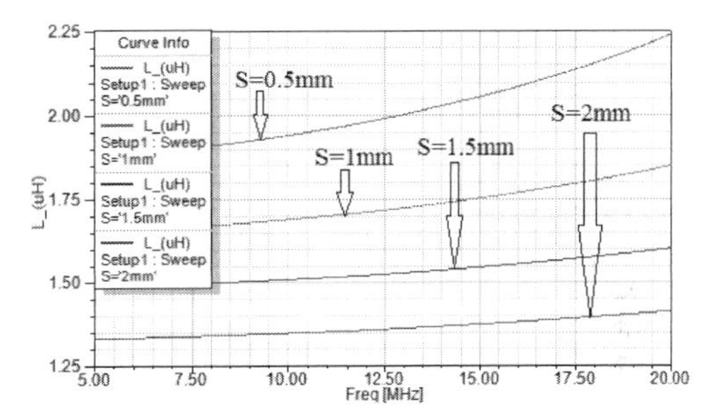

Figure 5. The coil spacing of the conductor on inductance.

The simulation results show that the impact of these parameters on the antenna inductance value is as follows: If the parameter L_x is increased, the inductance value will increase. If the parameter W is increased, the inductance value will decrease. If the parameter S is increased, the inductance value will decrease. According to equation 9, if the parameter N is increased, the inductance value will increase.

3.3 *The theoretical calculation and selection about antenna's quality factor*

In antenna theory, the quality factor (Q) represents the wastage performance of the inductance coil. In the resonance circuit, the higher the Q value becomes, the larger the output energy of the antenna will be. However, the higher quality factor will lead to a smaller transmission bandwidth. Then the useful subcarrier signal for data modulation won't be transferred out effectively by the antenna (Zhu et al.2008). When the quality factor is considerably smaller, the bandwidth will become wider. However, the antenna will consume more energy, which results in a shorter working distance of the antenna. Therefore, the appropriate quality factor becomes very important.

According to the ISO15693 air interface standard, the close coupling IC card system uses amplitude shift keying modulation mode. Based on the data transfer rate, the subcarrier frequency and the frequency offset, the required bandwidth can be calculated roughly.

According to the provisions in the ISO15693 standard, two subcarrier frequencies are 423.75 KHz and 484.29 KHz. The transmission rate is 26.69 Kbps. The frequency offset is 7KHz.

$$Bandwith = 2\times(423.75+26.69+7)\approx 0.914MHz$$

This is the minimum bandwidth we need to transfer the signal in engineering practice. Generally speaking, the antenna bandwidth must be greater than 1 MHZ. The quality factor is defined as follows:

$$Q=\frac{\omega\times L}{R}=\frac{2\pi f\times L}{R} \tag{10}$$

where f is the working frequency, taken as 13.56 MHz. L is the inductance of the antenna coil. R is the equivalent resistance of the antenna coil. According to the definition of bandwidth:

$$B=\frac{f}{Q} \tag{11}$$

The inequality ($B\times T\geq 1$) is obtained from engineering experience. T is the conduction time while the carrier system is modulating. In the ISO15693 standard, encoding is Miller coding. Pulse width is $3\mu s$. The result is as follows:

$$Q\leq f\times T\leq 13.56MHz\times 3\mu s\leq 40.68 \tag{12}$$

The relationship between antenna's quality factor and 3dB bandwidth is shown in Figure 6, which marks two subcarrier frequency points. It can be seen that when the reader's output impedance is 50Ω, the quality factor is not greater than 30. In engineering practice, the quality factor is chosen range from 10 to 30. In this paper, the quality factor is 20.

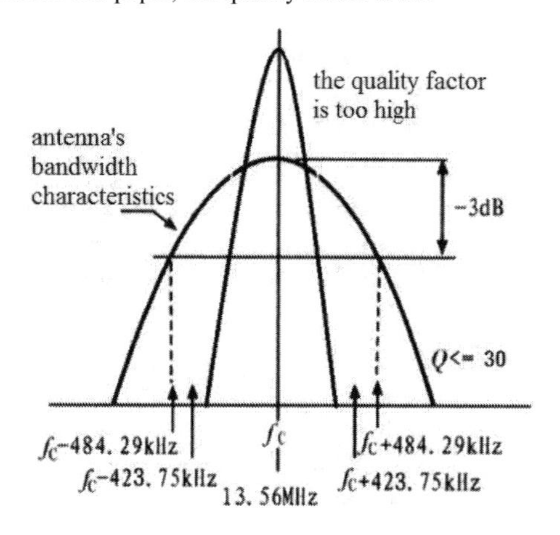

Figure 6. The relation between the antenna's quality factor and bandwidth.

After the completion of the preliminary design of the antenna, antenna's quality factor can be measured by the spectrum analyzer to calculate the antenna's bandwidth. If the antenna's bandwidth does not meet the using requirements, the series resistance in the matching circuit should be adjusted.

3.4 *The design of the antenna matching network*

In the radio frequency circuit design, the impedance matching problem between the circuits must be considered (Ludwig & Bogdanov 2013). The signal or energy should be transmitted from source to load reasonably. The reader antenna should also avoid signal reflection. Therefore, the antenna input impedance must be matched to 50Ω in the process of the antenna design (Rao et al. 2005). Different RF chips may have different impedance values. The impedance matching should be done by the matching network after the completion of the antenna design. Usually, there are two kinds of impedance matching methods. They are series-matching and parallel-matching. The series-matching design is introduced in this paper, which is based on the Smith chart method (Chen et al. 2012). The circuit diagram is shown in Figure 7. Smith chart is convenient to use. We need to set up some basic parameters in Options and DATAPOINT. It includes reference impedance, frequency and impedance. We should pay attention to these key parameters. Then electronic components can be in series or parallel into the circuit diagram. The impedance is divided into real part and imaginary part. In this paper, the reference admittance is 50Ω. The frequency is 13.56MHz. The impedance is the starting point. Firstly, we need to determine the starting point on the Smith chart. That is to say, we should determine the value of *Rs_ant* and *L_ant*. We can get the R_{s_ant} of the antenna coil in HFSS. The real part is 0.34Ω. The imaginary part is 171.6Ω. Then $L_{\_ant}$ is obtained by calculation. The calculation formula of Rs is as follows:

$$Rs=\frac{\omega R}{Q}-Rs\_ant \tag{13}$$

The impedance can be matched to 50Ω with the smith chart method. It is shown in Figure 8.

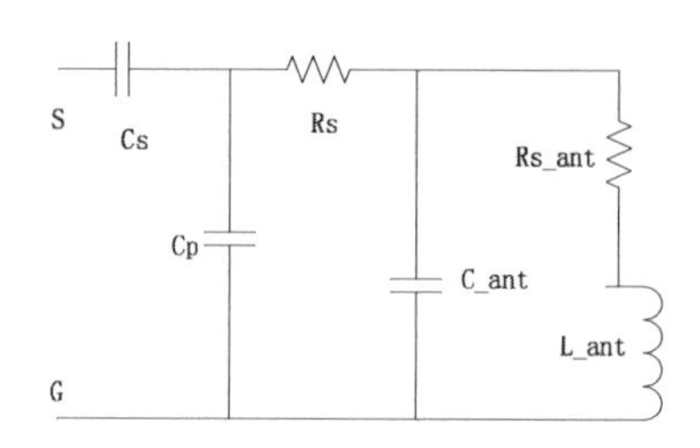

Figure 7. The antenna's matching circuit.

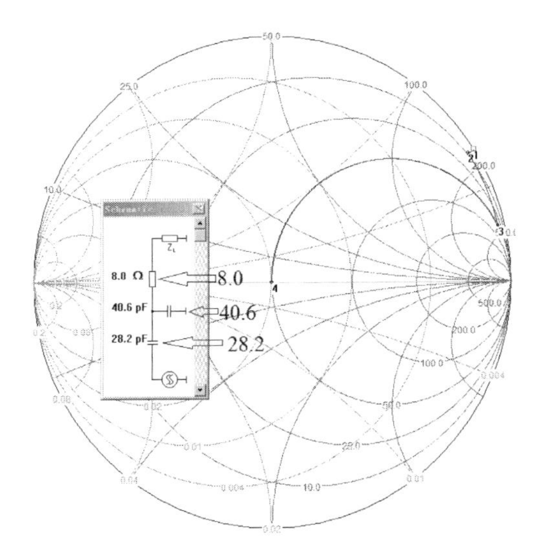

Figure 8. The smith chart after matching.

4 MODELING AND SIMULATION RESULTS ANALYSIS

The modeling and simulation of the proposed antenna can be tested by 3-D electromagnetism simulation software Ansoft HFSS (Li & Liu 2014). FR-4 which has high quality of electrical and mechanical performance is selected as substrate material with 1mm thickness in order to simulate practical PCB antenna effectively. The geometrical parameters of the coil are as follows: Lx=70mm, Ly=50mm, W=1mm, S=0.5mm, N=4, and the matched antenna model is as shown in Figure 9. The differences highlighted between Figure 2 and Figure 9 are shown in Figure 10. The gray part is the series resistance Rs. The blue part is series capacitor Cs. The yellow part is parallel capacitor Cp. Those electronic components are used for matching circuit.

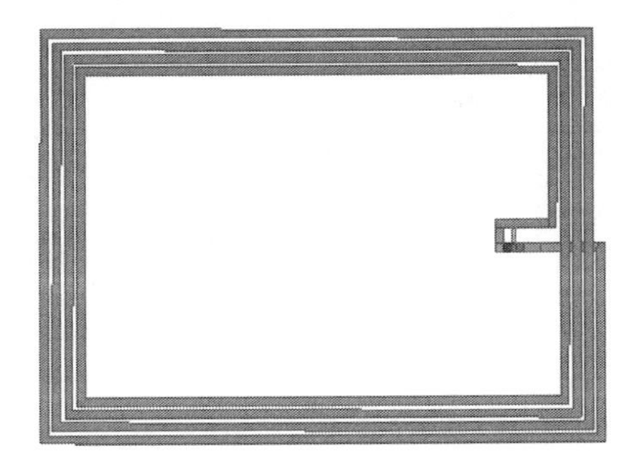

Figure 9. The antenna model after matching.

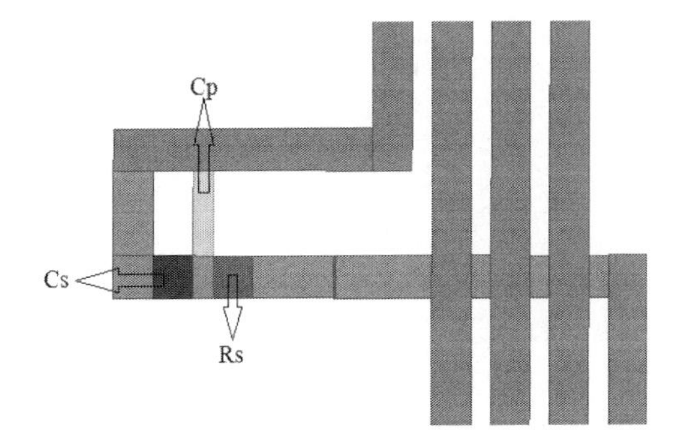

Figure10. The differences between Figure 2 and Figure 9.

According to the data of Figure 8 and the principle of the Smith Chart, the matching process can be simulated by adjusting series capacitor C_s and parallel resistance C_p, and the simulation result is shown in Figure 11. The device parameters are shown in Table 1.

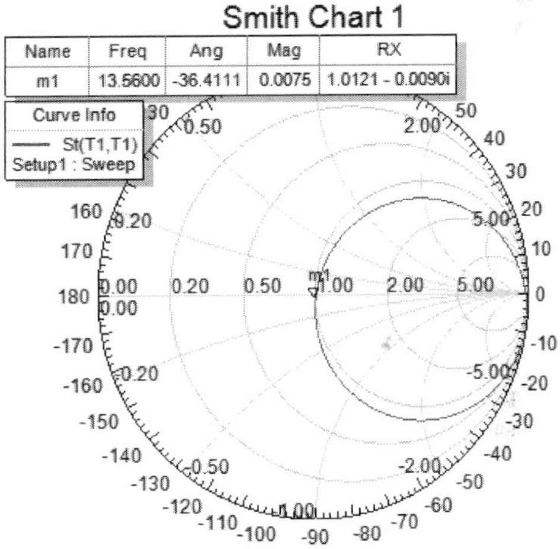

Figure 11. The final matching result.

Table 1. The matching circuit parameter value.

The Element's Name	Parameter value
Series resistance Rs	8.0Ω
Series capacitor Cs	26.9pF
Parallel capacitor Cp	9.4pF

Figure 12 illustrates the input impedance of the antenna, and it can be found that the real part is 50.6Ω and the imaginary part is close to zero at the frequency point of 13.56MHz. It meets the requirement of port impedance matching of RFID chip. S_{11} of the proposed antenna after matching is less than

-35dB and realizes well matching performance within the working frequency range as shown in Figure 13.

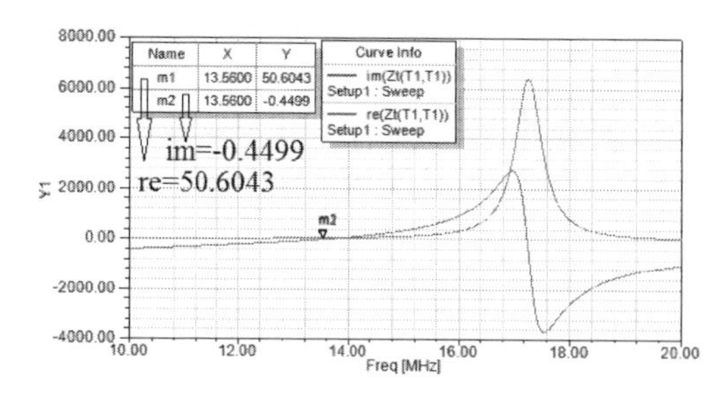

Figure 12. The input impedance value of the antenna.

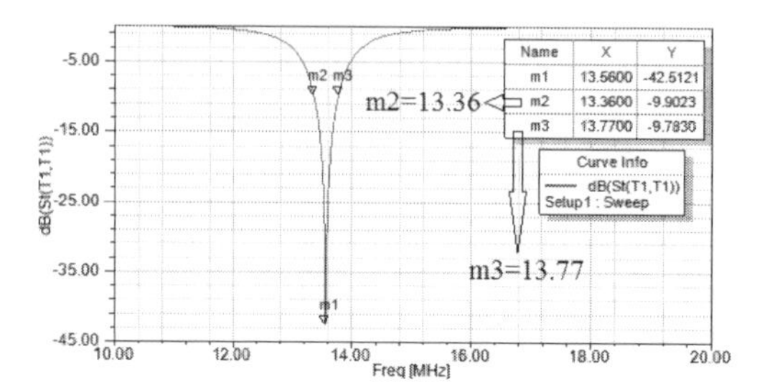

Figure 13. The characteristic curve S11.

It can be observed that the narrow bandwidth may not meet the actual requirement. $B = R/2\pi L$ can be derived from equation (10) and equation (11), which shows that the bandwidth can be enlarged by increasing the value of series resistance to 16Ω in the matching process, and the other matching steps are just as mentioned before. The parameters are shown in Table 2

Table 2. The matching circuit parameter value.

The Element's Name	Parameter value
Series resistance Rs	16Ω
Series capacitor Cs	38pF
Parallel capacitor Cp	29.3pF

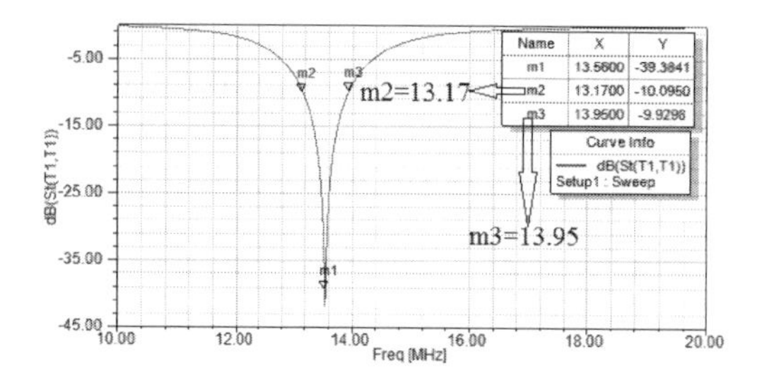

Figure 14. The characteristic curve S_{11} after increasing the value of series resistance.

The simulation results in Figure 14 show that the bandwidth is almost twice as large as the former one and achieves a high efficiency with the reflecting coefficient below -35dB. As the former design steps, it makes a good trade-off between bandwidth and the quality factor which can meet the practical engineering requirements properly.

5 TESE AND ANALYSIS OF THE REAL FABRICATED ANATENNA

A practical antenna which is shown in Figure 15 is fabricated after simulating, and the results of the real fabricated antenna are shown in Figure 16.

Figure 15. The real, fabricated antenna.

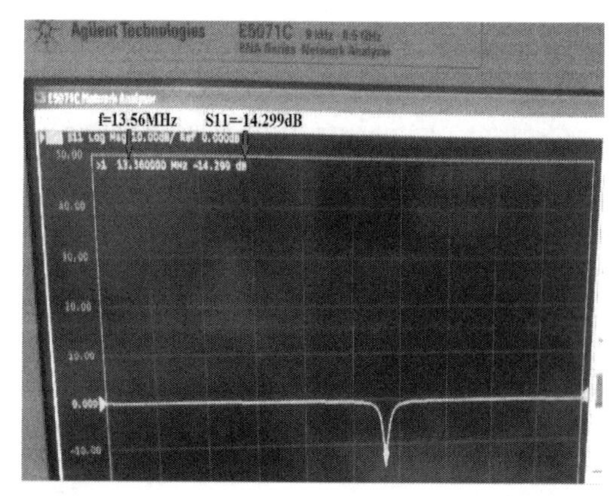

Figure 16. The reflecting coefficient measured by network analyzer.

According to Figure 16, we can find that the S_{11} is -14.299dB at 13.56MHz. Two copper wires in the external of the antenna's pin may cause parasitic effects, and the surrounding environment causes certain influence, which leads to S_{11} of the measurement results larger than the simulation value. Generally speaking, if the reflecting coefficient is less than -10dB in engineering practice, the real, fabricated antenna can work and read the tag antennas. Hence, the design has been successful not only theoretically but also practically.

6 CONCLUSIONS

This paper aims at the working principle of the 13.56 MHz inductance coupling RFID antenna system. It explores the detailed design and debugging methods about the RFID reader antenna. Based on the Smith chart method, the matching design of the antenna is simple and effective. The paper deals with 13.56 MHz RFID reader antenna as the PCB planar spiral inductors by the three-dimensional electromagnetic simulation software HFSS. The simulation and measurement results show that the antenna has a smaller reflection coefficient and a better efficiency. In the practical engineering, RFID reader antenna has many kinds of circuit structure, yet, they are the equivalent of the R-L-C resonant circuit in the final analysis. As a consequence, this paper has a good and general guiding value of the actual design of PCB planar reader antenna.

ACKNOWLEDGEMENT

The authors thank the industry, schools and research institutions, prospective joint research projects, the science and technology support program of Jiangsu Province-industrial part, and the key University Science Research Project of Jiangsu Province.

REFERENCES

Chen, Y.Y, Cai, J.Y, Dai, C.Y, et al. 2012. Rapid design of 13.56MHz antenna matching network based on smith chart. *Electronic Component and Device Applications* 14(12): 50–53.

Dai, C.Y, Cai, J.Y, Chen, Y.Y, et al. 2013. Design and HFSS simulation of 13.56 MHz RFID reader Antenna. *Radio Engineering* 43(1): 42–45.

Finkenzeller, K. 2006. *Radio frequency identification*. Beijing: Publishing House of Electronics Industry.

Huang, M. & Luo, Z.X. 2007. Design and realize for 13.56 MHz RFID reader Antenna. *Computer and Digital Engineering* 35(7): 151–153.

Li, B.S. 2007. The design for reader antenna on passive inductive RFID. *International RFID Technology Summit Forum.*

Li, M.Y & Liu, M. 2014. *Antenna design by HFSS.* Beijing: Publishing House of electronics industry.

Ludwig, R. & Bogdanov, G. 2013. *RF circuit design Theory and applications*. Beijing: Publishing House of Electronics Industry.

Rao, K.V.S, Nikitin, P.V & F.Lam, S. 2005. Impedance matching concepts in RFID transponder design. *4th IEEE Workshop on Automatic Identification Advanced Technologies*: 39–42.

Xu, L.Q. & Cao, W. 2006. *Field and wave electromagnetic.* Beijing: Science and Press.

Zhu, Y, Wang, G & Wang, H.J. 2008. Designof RFID Reader antenna at 13.56MHz frequency. *Journal of Microwaves* 24(5): 22–26.

Electronics, Communications and Networks IV – Hussain & Ivanovic (eds)
© 2015 Taylor & Francis Group, London, ISBN: 978-1-138-02830-2

Predicting student performances from access records on general websites

Qing Zhou
ChongQing University, ChongQing, China
Key Laboratory of Dependable Service Computing in Cyber Physical Society of Ministry of Education, College of Computer Science, Chong Qing University, ChongQing, China

Chao Mou*, Youjie Zheng & Yao Meng
ChongQing University, ChongQing, China

ABSTRACT: Predicting student performance using data mining tools is a difficult yet worthwhile undertaking for educational institutions. Previous studies on the student performance prediction restricted the data sources to one specific learning system. We show that it is possible to predict whether a student can pass Data Structure course from his or her access records on general websites. We first created a domain names catalog by assigning more than 12,000 domain names to 335 subcategories, 80 main categories and five top categories. This catalog was later used to calculate the frequencies that each student visited different categories of websites from about 20 million Internet access records. We then selected 14 categories of domain names as features to train a Naive Bayes Classifier. Experiments show that our method achieves a sensitivity (true positive rate) of nearly 90% and a specificity (true negative rate) of above 60%. Moreover, the proposed model has a high interpretability and conforms to our teaching experience.

1 INTRODUCTION

Data mining technology has now been successfully applied in education, which not only has emerged as a rising research area known as educational data mining (EDM) (Romero & Ventura 2013), but also provides us with a powerful tool to discover hidden knowledge and patterns in education.

Predicting student performance with EDM tools is a rewarding cause (Affendey et al. 2010). It can help administrators to make the right decisions, and teachers to adjust teaching strategies, and students to improve their final performance. For instance, Maria et al. used a logistic regression model to predict college enrollment from student interaction with a web-based intelligent tutoring system called ASSISTments (San Pedro et al. 2013). Natek & Zwilling (2014) worked on a small data set from a knowledge management system and found several key influencers to students' final grades. Perera et al. (2009) employed records of a collaborative learning system to improve the learning effect.

Students today were born in the digital age and have been in contact with digital technology since childhood, so they can be called "Digital Natives" (Arteaga Sánchez et al. 2014). The Internet has played a key role in these students' lives, involving their daily life as well as academic activity (Gurung & Rutledge 2014). Previous researches on student performance prediction focused on data set from a specific learning system. Is it possible to predict student performance from the pattern they access general websites? In order to verify this hypothesis, we trained a Naive Bayes Classifier with a data set collected from over 20 million Internet access records, and applied it to predict whether a student can pass a computer science course. Experiments indicate that our proposed method is promising concerning student performance prediction.

2 DATA SET

Our data set consists of Internet access records and course grade records. The Internet access records are obtained from the gateway server of the university, which accommodates more than 10,000 students. When a student requests a uniform resource locator (URL), a record (including student identifier, request time, and URL address, etc.) is written into a log file. Over 200, 000 log files and about one TB data are created every month. Each log file can be preserved

*Corresponding author: *mc8983@qq.com*

on the gateway for about six weeks before erased. With the permission of the university, we acquired about 500 GB of log files that were created during two weeks soon after the midterm of 2014 Spring Semester.

A course grade record (including student identifier and course grade) tells whether a student passed the *Data Structure*, one of the computer science courses delivered in the university. Altogether 300 students from 5 classes took the course. However, only 195 students were found in log files, and 23 (about 12%) of them failed the course. We use a binary vector $G_{195\times 1}$ to store student grades. If the ith student passed the examination, $G(i)$ is set to 1; otherwise, $G(i)$ is set to 0.

All student identifiers in Internet access records and course grade records were encrypted using the same cryptographic algorithm before these records are provided to us. Hence, we can associate both types of records from the same student together while keeping his or her identity secret.

3 PRE-PROCESSING

How to predict student academic performance from their Internet access records? As shown by the experience of many teachers, those who fail an examination are likely to visit some particular types of websites more often. Nevertheless, it is difficult for a computer program to determine the type of a website. Besides, we do not know which types of websites could be the most proper predictors. Solutions to these problems are decisive to the performance of the prediction model. The pre-processing is broken down into two stages: records extraction and frequency calculation.

3.1 *Records extraction*

From more than 10, 000 students' Internet access records, we drew a record of 195 students whose grades for Data Structure are known. Then we deleted all records in which URLs start from IP addresses other than domain names. A domain name can be either a registered domain name (e.g., yahoo.com) or a lower-level domain name (e.g., music.yahoo.com). Altogether, we got 21,003,921 extracted records, which involve 1,572 different registered domain names and 23,476 different lower-level domain names.

3.2 *Frequency calculation*

This stage aims to count the frequencies that different categories of domain names are visited by each student. Thus, we manually established a catalog of more than 12,000 domain names, and classified each domain name to one of 335 subcategories. Note that lower-level domain names with common registered domain (e.g., music.yahoo.com and shopping.yahoo.com) may belong to different subcategories.

In order to classify the domain names with a coarser granularity, we grouped 335 subcategories into 80 main categories, which were further merged into five top categories, namely, Entertainment & Leisure, Computer & Network, Life & Service, Culture & Education, Portals &Others. For example, the subcategory, main category and top category on music.yahoo.com are general music, music and Entertainment & Leisure, respectively. The outcomes of frequency calculation are three matrices S195×355, M195×80, and T195×5, where matrix element S(i, j) stores the number of times that the ith student have visited domain names belonging to the jth subcategory, while matrices M and T stores frequencies with regard to the main category and top category.

All elements of matrices S, M, and T were initialized to zeros before frequency counting. For each access record, we first extracted the lowest-level domain name from the URL and looked it up in the domain name catalog. If no match is found, we will search its higher-level domain names. The detection of a match would show us the subcategory, main category and top category to which current record belongs to. Then we added one to the values of the corresponding elements in matrices S, M, and T. Occasionally, a record may not be assigned to any category, for not all domain names can be embodied in our catalog. In this case, we simply discarded the record.

The above procedure is repeated until all records have been processed. Finally, the visiting frequencies of all sorts of websites for each student are stored in S, M, and T.

4 METHODOLOGY

The purpose of this study is to construct a student performance prediction model from students' online access records. After pre-processing, we knew the number of times each student accessed 440 categories (i.e., 355 subcategories, 80 main categories and 5 top categories) of websites. Hence, a student can be represented by 440 features, where each feature stands for the visiting frequencies of some category of websites. However, a model with 440 features is too complex for prediction and interpretation. Hence, we first extracted features that most relevant to student grade. Then we built a predictive model upon these features.

4.1 *Feature selection*

This stage intends to select a few categories as the features for our prediction model. The selected category must meet the following criteria:

1. The absolute value of the correlation coefficient between the course grade vector G and the visiting frequencies of such category of websites is larger than 0.12;
2. Such category of websites have been visited by at least 20 students;
3. Such category of websites have been visited by both groups of students, i.e., students passing and failing the course.

There are 14 categories meeting above requirements. Table 1 lists the names and granularities of these categories as well as their correlation coefficients with course grades.

Table 1. Detail about the selected categories.

Category Name	Category Granularity	Correlation Coefficient
Entertainment & Leisure	Top	–0.25
Culture & Education	Top	0.18
Movies & TV	Main	–0.26
News	Main	–0.18
Learning	Main	0.19
Movies	Sub	–0.24
Videos	Sub	–0.24
Online Games	Sub	–0.18
PC Games	Sub	–0.24
Game Company	Sub	–0.16
Journal	Sub	–0.22
Live Chat	Sub	–0.16
Documents	Sub	0.20
Programming	Sub	0.14

4.2 *Prediction model*

We use Naive Bayes Classifier (NBC) to model students' grades in Data Structure. NBC determines the most probable class given a set of features, assuming that all features are conditionally independent. Despite this oversimplified assumption, NBC is quite popular due to its high performance in classification effectiveness as well as computational efficiency. Another advantage of NBC is its interpretability, one of the factors valued by educational data mining. NBC is based on Bayes' theorem which can be stated in following form:

$$P(H \mid X) = \frac{P(X \mid H)P(H)}{P(X)} \tag{1}$$

where H denotes some class, e.g., passing (or failing) the course; X denotes the feature set, e.g., the frequencies that the student visits different categories of websites. As all features are considered conditionally independent by Naïve Bayes Classifier, we have

$$P(X \mid H) = \prod_i P(x_i \mid H) \tag{2}$$

where x_i denotes the ith feature. Normally, we assume each feature obeys Gauss distribution, i.e.,

$$P(x_i \mid H) = \frac{1}{\sqrt{2\pi}\sigma_i} e^{\frac{(x_i-\mu_i)^2}{2\sigma_i^2}} \tag{3}$$

where μ_i and σ_i^2 are estimated by the mean and variance of the ith feature.

Most NBC assumes that each feature follows the Gauss distribution. However, our experiment shows that the visiting frequencies conform better to the exponential distribution (See Figure 1 for the histogram for two typical categories of websites). Hence, we replaced Gauss distribution with exponential distribution in the revised method, and substituted equation (4) for equation (3).

$$P(x_i \mid H) = \frac{e^{-x_i/\mu_i}}{\mu_i} \tag{4}$$

where μ_i is estimated by the mean of the ith feature.

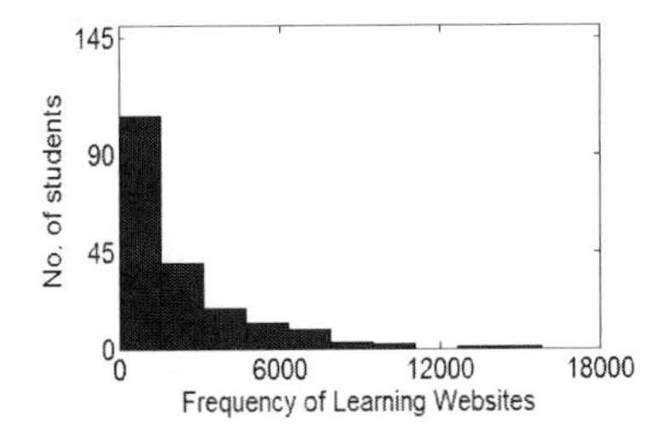

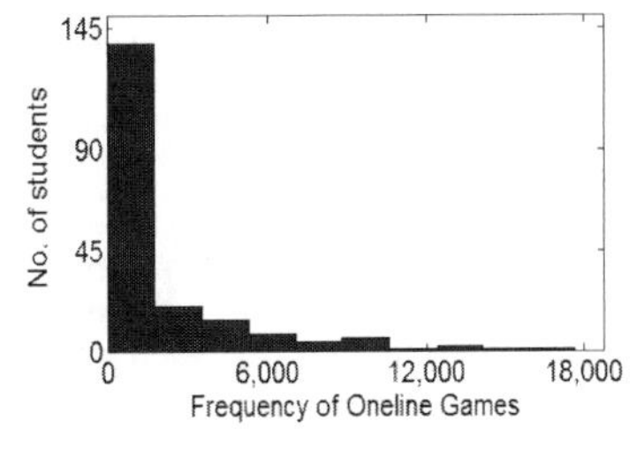

Figure 1. Caption of a typical figure. Photographs will be scanned by the printer. Always supply original photographs.

5 RESULTS AND DISCUSSION

We use leave one out cross-validation to test the effectiveness of our model. In each test, data from one student is used as the testing set, while data from the rest are used as the training set. This procedure is repeated until all the students are tested exactly once. Table 2 presents test results for our initial method (using Gauss distribution) and the revised method (using exponential distribution).

Sensitivity and specificity are two important measures of the performance of a classifier. In this experiment, sensitivity (a.k.a. true positive rate) is the ratio of true positive (the number of students who are correctly predicted as passing the course) to the number of students who actually pass the course. Our initial model achieves a high sensitivity of over 88% (152/172).

Table 2. Results of leave on out cross-validation for two methods.

Method	TP*	FP*	TN*	FN*	SE*	SP*
Initial method	152	20	11	12	88.37%	47.83%
Revised method	150	22	14	9	87.21%	60.87%

* TP=True Positive, FP=False Positive, TN=True Negative, FN=False Negative, SE=Sensitivity, SP=Specificity.

Nevertheless, we are more concerned with students who will probably fail the course, for they may need a higher attention or additional resources. Specificity is the ratio of true negative (the number of students who are correctly predicted as failing the course) to the number of students who actually fail the course. If a classifier simply predicts that all students pass the examination, the sensitivity will reach as high as 100%, while the specificity decreases to 0. The specificity in our method reaches 47.83%. That is to say, nearly half of the students who failed the course are recognized. It surpasses random guess significantly, considering that only about 12% of the students failed the examination.

The results of the revised method are presented in the second row of Table 2. The sensitivity values of the two methods are very close. However, the revised method improves a lot in specificity. Overall, the revised method achieves a better prediction performance than the initial one. This probably because the former assumes that the visiting frequencies follow exponential distribution, which approaches closer to the reality than Gauss distribution.

The experiments show that access to records on general websites can be used to predict students' grades in Data Structure. A more detailed investigation of the selected features indicates that those who browse more online games than learning websites are more likely to fail the course, which not only conforms to our teaching experience but also provides a rationale for our methods.

Although the sensitivity of our model reaches nearly 90%, its specificity is only slightly higher than 60%. Fortunately, our model can be further improved for the following reasons:

1. We only operated on online access records. Course-relevant information, such as students' demographic information, grades of course projects, and differences among five classes, also serve as good predictors of student performance;
2. We only used a straightforward feature, the visiting frequency of some category of websites. Some more complicated features, such as the duration of website access, may probably make our model more accurate;
3. We only analyzed two weeks' records, while the course lasted for sixteen weeks. More data are of help to improve the prediction performance.

6 CONCLUSION

In this paper, we predict whether students can pass the Data Structure course or not from the frequencies they accessed different types of websites. Even if none of the course-relevant information is available, the proposed model achieves a high sensitivity and a moderate specificity. Our future work is to upgrade the model by introducing a more complicated feature set and a larger amount of data.

ACKNOWLEDGEMENT

The work is supported by the National Nature Science Foundation of China under Grant 61472464, and the Fundamental Research Funds for the Central Universities under Grant CDJZR12.18.55.01.

REFERENCES

Affendey, L.S., Paris, I.H.M. & Mustapha, N. 2010. Ranking of influencing factors in predicting students' academic performance. *Information Technology* 9(4): 832–837.

Arteaga Sánchez, R., Cortijo, V. & Javed, U. 2014. Students' perceptions of Facebook for academic purposes. *Computers & Education* 70: 138–149.

Gurung, B. & Rutledge, D. 2014. Digital learners and the overlapping of their personal and educational digital engagement. *Computers & Education* 77: 91–100.

Natek, S. & Zwilling, M. 2014. Student data mining solution–knowledge management system related to higher education institutions. *Expert Systems with Applications* 41(14): 6400–6407.

Perera, D., Kay, J., Koprinska, I. et al. 2009. Clustering and sequential pattern mining of online collaborative learning data. *Knowledge and Data Engineering, IEEE Transactions on* 21(6): 759–772.

Romero, C. & Ventura, S. 2013. Data Mining in Education. *Wiley Interdisciplinary Reviews-Data Mining and Knowledge Discovery* 9(4): 832–837.

San Pedro, M.O.Z., Baker, R.S.J. & Bowers, A.J. et al. 2013. Predicting college enrollment from student interaction with an intelligent tutoring system in middle school. In D'Mello, S. K., Calvo, R. A., & Olney, A. (eds), *The 6th international conference on educational data mining; Proc., Memphis*, 6–9 July 2013.

Electronics, Communications and Networks IV – Hussain & Ivanovic (eds)
© 2015 Taylor & Francis Group, London, ISBN: 978-1-138-02830-2

SIP-based method to load friend lists in quantum communication network instant messaging system

Dexin Zhu, Shigang Wang, Jiawei Han, Jianan Wu & Rongkai Wei
Key Laboratory of Quantum Communication of Jilin Education Department, Changchun University, Changchun, Jilin Province, China
College of Computer Science and Technology, Changchun University, Changchun, Jilin Province, China

Nianfeng Li & Lijun Song
Key Laboratory of Quantum communication of Jilin Education Department, Changchun University

ABSTRACT: Quantum communication is an emerging interdisciplinary field of study that combines traditional communications and quantum mechanics. With its absolute security features that traditional communication methods lack, it has huge potential for application and great scientific significance. In this paper, a SIP-based solution for loading the friend lists in the quantum communication network instant messaging system was proposed by modifying SIP source code and configuration files.

KEYWORDS: Quantum communication; SIP; instant messaging system; load friend lists

1 INTRODUCTION

Quantum communication originated in confidentiality requirements for communications. Communications security has long been of great importance, especially in military fields. In today's society, with continuous improvements in information technology, the use of the internet, instant messaging, and e-commerce all involve information security, which in turn is closely related to the interests of all users. Encrypting information is an important way to ensure information security. In 1917, G.Vernam proposed the one-time pad (OTP) method (Vernam1926), to encrypt text by using a string of random numbers of the same length, while the recipient decrypts using the same random number. The random number is called a key, which is truly random and used only once. OTP protocols have been shown to be safe (Shannon 1949). However, it is critical that the key has to be long enough and distributed securely and unconditionally in an insecure channel, which is difficult to implement in traditional communications. Later, public-key cryptography, such as the famous RSA protocol, emerged (Rivest et al. 1978). In such protocols, the recipient has a public key and a private key. The recipient sends the public key to the sender, who in turn encrypts the data with the public key before sending it to the recipient. Only with the private key can the data be decrypted. Public key cryptography is widely used. Its security is guaranteed by a mathematical hypothesis that the prime factor decomposition of a large number is very difficult problem to solve. However, the idear ofa quantum computer has changed this view. It has been proved that once a quantum computer is invented, large numbers can be easily decomposed and the widely used cryptographic system can be cracked (Shor 1994). Fortunately, before people realize the power of a quantum computer, quantum key distribution (QKD) technology based on quantum mechanics as been proposed(Bennett & Brassard1984). Quantum key distribution applies the principles of quantum mechanics, and distributes secure keys unconditionally, thereby ensuring the absolute confidentiality of communications when combined with the OTP method.

2 RELATED KNOWLEDGE

2.1 *Quantum secure communication network*

With the development of quantum communication technologies (Johnson et al. 2011), point-to-point communication has become increasingly unable to meet people's requirements. In order to offer secure communications services to customers with limited resources, quantum communication industries have gradually moved in the direction of network development. Quantum communication networks are still at the realization research stage (Patel et al. 2012), and as far as the current tests and research are concerned, quantum communication networks are generally carried out in laboratory or metropolitan settings, and are considered quantum LANs. The quantum WAN is still in the research stage. A quantum LAN includes star topology, ring topology, and bus topology. The experimental platform of quantum communication networks in this research adopted the star topology.

2.2 *SIP protocol*

SIP (session initiation protocol), proposed by the Internet Engineering Task Force (IETF) in 1991, is an IP network -based signalling protocol to realise complex communication applications. Session refers to the exchange of data between users. In SIP-based applications, each session can be a transmission of a variety of data, such as plain text, ordigitized audio and video data. SIP applications have great flexibility.

SIP messages are text-encoded, including the request message and response message. TheSIP request message includes INVITE, ACK, OPTIONS, BYE, CANCEL, and REGISTER. The SIP response message is used to respond to the request message, indicating the success or failure status of a call or registration. Different types of response messages are distinguished by the status code, which contains three integers. The first digit in the status code is used to define the type of response, while the other two are used to make more detailed explanations for the response.

2.3 *XCAP*

XML configuration access protocol (XCAP), is an application protocol that allows the XML format of the application configuration data stored on the server to be read, written, modified, and deleted at the client end. XCAP maps XML document sub-trees and element attributes to HTTP URIs, so that these components can directly use the HTTP protocol clients for access.

3 REALISATION PROCESS

3.1 *Adding the friend list*

To add a friend list, install the mysql database in the kamailio server. Data Sheet xcap is used to manage and maintain the friend list information of the instant messaging end-user. The friend list is generated with the file in the XML format, and uploaded and saved to the list using the curl tool. It is generated and uploaded in the following way:

File name format: <user account> .xml, file content format:

```
<?xml version="1.0" encoding="UTF-8"?>
    <rls-services>
      <service uri="sip:<user account >-list@<sip server>">
        <entry uri="sip:<friend 1 >@< sip server>">
          <display_name>nickname</display_name>
        </entry>
        <entry uri="sip:<friend 2 >@< sip server>">
          <display_name>nickname</display_name>
        </entry>
        <entry uri="sip:<friend 3 >@< sip server>">
<display_name>nickname</display_name>
        </entry>
        <packages>
          <package>presence</package>
        </packages>
        </service>
    </rls-services>
```

Upload the set xml to the xcap server within the signaling system using the curl tool. It is necessary to make sure that the signaling system has been activated before uploading.

The following example illustrates this operation using a user account 1001: the friend list file is 1001.xml. Run the following command in the xml file directory:

"curl -u <user account>:<user password> -T 1001.xml -X PUT http://<sip-serverIP > :5060/xcap-root/rls-services/users/sip: 1001@<sip-server IP>/index"

3.2 *Modifying the kamailio source code*

In order for the server to recognize the tag in the above xml, it is necessary to modify the source code in the rls modules of the signaling server. The modified process is shown in Figure 1. After the signaling server is started, the relative path of the resources, user name and address based on the URI uploaded using the curl tool can be accessed. The friend list can be identified following this path.

The pseudo-code is as follows:

```
int process_list_and_exec()
{
        Traversal node
        if(node->name=="resource-list")
          ……
        else if(node->name=="entry")
          process_uri_name();
        else if(node->name=="list")
        ……
}
int process_uri_name()
```

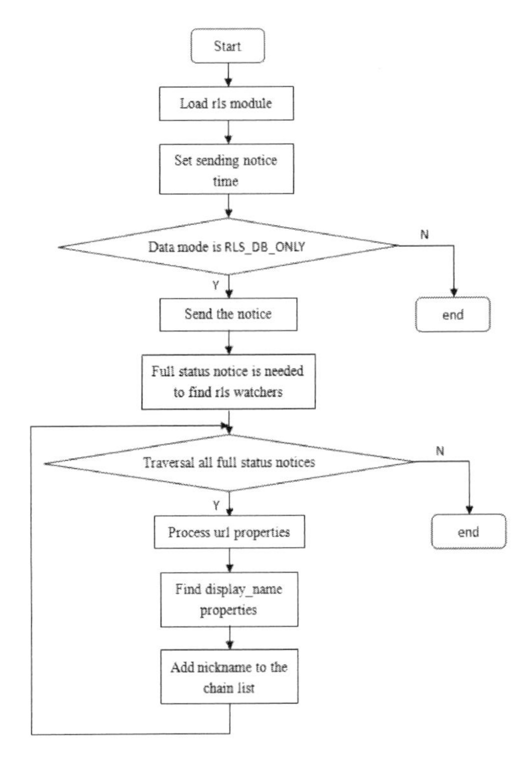

Figure 1. XML tag identification process.

```
{
        Traversal tag
        if(node->name=="display_name")

XMLNodeGetNodeContentByName("display_
name");
        add_resource_tolist(uri,name);
}
```

3.3 *Editing and configuration of the kamailio.cfg file*

At the top of the file, add the definition that supports the XCAP; and define the TCP's lifecycle in a global variable position. In the meantime, load the xcap module and modify the corresponding parameters.

```
#!define WITH_XCAPSRV
tcp_connection_lifetim=3064
#!ifdef WITH_XCAPSRV
Tcp_accept_no_cl=yes
#!endif
#!ifdef WITH_XCAPSRV
loadmodule "xhttp.so"
loadmodule "xcap_server.so"
#!endif
#!ifdef WITH_XCAPSRV
modparam("xcap_server", "db_url", DBURL)
#!endif
#!ifdef WITH_PRESENCE
modparam("presence", "db_url", DBURL)
modparam("presence", "fallback2db", 1)
modparam("presence", "db_update_period", 20)
modparam("presence_xml", "db_url", DBURL)
modparam("presence_xml", "force_active", 0)
modparam("presence_xml",        "integrated_xcap_
server", 1)
#!endif
```

4 CONCLUSION

In this study, a method to load friend lists in a three-node quantum communication network SIP server terminal was realised, by modifying the source code of the SIP server and making use of the curl command. This has laid the experimental foundation for the further development of the client terminal of the instant messaging system. However, this method was used only for loading friends onto the SIP server. Further study is required for adding additional friend lists.

REFERENCES

Bennett, C. H. & Brassard, G. 1984. Quantum cryptography: public key distribution and coin tossing, *in Proceedings of the IEEE International* Conference on Computers, Systems and Signal Processing. Bangalore, India. 1984:175–179. New York: IEEE.

Johnson, M. W., Amin, M. H. S., Gildert, S, et al. 2011. Quantum Annealing with Manufactured Spins. *Nature*. 473: 194–198.

Patel, K. A., Dynes, J. F., Choi, I., Sharpe, A. W., Dixon, A. R., Yuan, Z. L., Penty, R. V. & Shields, A. J. 2012. Coexistence of High-Bit-Rate Quantum Key Distribution and Data on Optical Fiber. *PHYSICAL REVIEW X* 2(4): 041010.

Rivest, R. L., Shamir, A. & Adleman, L. M. 1978. A method for obtaining digital signatures and public-key cryptosystems. *Commun. ACM*, 21: 120–126.

Shannon, C. E. 1949. Communication theory of secrecy systems. *Bell Syst. Tech*. 28:656–715.

Shor, P. W. 1994. Algorithms for quantum computation: discrete logarithms and factoring II. *Proceedings of the 35th Symposium on Foundations of Computer Science. (Edited by Goldwasser, S.),Santa Fe, NM, 20-22 November 1994*:124–134. Los Alamitos, California: IEEE Computer Society.

Vernam, G. S. 1926. Cipher printing telegraph systems for secret wire and radio telegraphic communications. *J Amer. Inst. Elec. Eng*. 45:109–115.

Computer technology

Electronics, Communications and Networks IV – Hussain & Ivanovic (eds)
© 2015 Taylor & Francis Group, London, ISBN: 978-1-138-02830-2

Analysis of a linear, time-dependent process using evolutionary parameters

Abdullah I. Al-Shoshan*
Computer Engineering Department, Qassim University, Saudi Arabia

ABSTRACT: In this paper, for modeling a linear, time-dependent process, we propose a method based on the evolutionary parameters of the process. This approach exploits the fact that the evolutionary bispectrum contains information regarding the time-variation, the phase and the magnitude of the process. In addition, if the process is corrupted by stationary/non-stationary noise with symmetric distribution, the time-dependent coefficients of the process can be identified using the evolutionary parameters. Simulations are given to show the effectiveness of the proposed method.

1 INTRODUCTION

The problem of the analysis of a time-dependent process (TDP) has attracted considerable attention during the past decades due to its large number of applications in diverse fields such as geophysics Landers & Lacoss (1977), biostatistical processing Diggle (1990), radar Kenny & Boashash (1993), medicine (Wood et al. 1990), and many physical applications Percival & Walden (1993). For each application, the analysis of a TDP needs to deal with a set of mathematical equations which can be used to understand the behavior of the process. In recent years, the estimation of the TDP has been considered by various authors (Laurain et al. 2011, Rao 1993). For TDP, a process which has at least one time-dependent parameter, the conventional power spectrum does not reflect the time variation of the process characteristics, hence cannot deal with its variations of its characteristics. With the recent introduction of evolutionary bispectrum (EB) in digital signal processing (Priestley 1988, Al-Shoshan 2011), a new approach to the modeling of a TDP is introduced. In this paper, an algorithm based on the EB to solve these problems is proposed. Finally, we have shown that if the TDP is corrupted by stationary/non-stationary noise with symmetric distribution; the evolutionary parameters of the process can be estimated.

2 TIME-DEPENDENT PROCESS

A discrete TDP, $x(n)$, can be represented as the output of a causal, linear, and time-dependent (LTD) system, with impulse response $h(n,m)$ to a discrete-time, zero-mean, unit-variance white noise process, $e(n)$. According to the Wold-Cramer decomposition, we can represent as follows Priestley (1988)

$$x(n)=\int_{-\pi}^{\pi} H_x(n,w)e^{jwn}\mathrm{d}Z(w) \quad (1)$$

Then the evolutionary spectrum of $x(n)$ can be estimated as

$$\hat{S}_{EP}(n,w)=\frac{N}{M}|\hat{H}(n,w)|^2 \quad 0\le n\le N-1$$

where $\hat{H}(n,w)$ is an estimate of the time-frequency kernel and can be estimated as follows.

$$\hat{H}(n,w)=\sum_{k=0}^{N-1} w_n(k)x(k)e^{-jwk} \quad (2)$$

which uses a window that varies with time and constructed from a combination of orthogonal polynomials (OP) such that

$$w_n(k)=\sum_{i=1}^{M}\beta_i^*(n)\beta_i(k$$

The variation of $w_n(k)$ depends on the value of M, which is the order of the OP set used. If the process $x(n)$ is corrupted with a non-stationary noise, i.e., $z(n) = x(n) + \eta(n)$ the traditional technique (Choi 1992, Nikias & Mendel 1993, Luo & Chaparro 1991) will not be able to remove the noise. However it will simply reduce the effect of non-stationary noise with symmetrically distributed like Gaussian,

* Corresponding author: *drshoshan@gmail.com*

Laplace, uniform, and Bernouli-Gaussian. To remove these types of noise, we use the properties of the EB (Al-Shoshan 2011 & Priestley 1988) such that

$$H_z(n,w) = H_x(n,w) + H_\eta(n,w)$$

which produces the EB such as

$$S_z(n,w_1,w_2) = S_x(n,w_1,w_2) + S_\eta(n,w_1,w_2)$$

where $S_x(n,w_1,w_2)$, the EB of $x(n)$, is defined as

$$H_x(n,w_1)H_x(n,w_2)H_x(n,-w_1-w_2)$$

and an estimate of $H(n,w)$ can be calculated using equation (2). However, since $\eta(n)$ is a non-stationary noise with symmetrically distributed, $S\eta\ (n,w_1,w_2)=0$, which means that the noise has no effect on the EB of the process, therefore

$$S_z(n,w_1,w_2) = S_x(n,w_1,w_2) \tag{3}$$

From equation (3) we reconstruct the time-dependent amplitude $H_x(n,w)$ of the noiseless output $x(n)$ by using the proposed method.

A subclass of discrete TDP is that for which the input and output sequences $e(n)$ and $x(n)$ satisfy a time-dependent difference equation of the form (Laurain et al. 2011)

$$\sum_{k=0}^{p} a_k(n)x(n-k) = \sum_{i=0}^{q} b_i(n)e(n-i) \tag{4}$$

where $\{a_k(n)\}$ and $\{b_i(n)\}$ are functions of time n and $a_0(n) \neq 0$, for all n. In previous work (Laurain et al. 2011, Rao 1993), for modeling a TDP $\{x(n)\}$, it has been assumed that the process is changing "slowly" over time, or the time-dependent coefficients of the process can be represented using orthonormal polynomials such that

$$a_k(n) = \sum_{i=0}^{d} a_{ki}\beta_i(n) \tag{5}$$

where d is the order of the orthonormal polynomials $\{\beta_i(n)\}$. This assumption simplifies the time-dependent modeling problem in such a way that the equations resemble those in the time-independent case. Note that the dimension of the problem increases as we increase the decomposition order, d.

3 TIME DEPENDENT FOURIER COEFFICIENTS

The time dependent Fourier coefficients (TDFC) $\{f_n(\mu,l)\}$ of the EB can be estimated as follow:

$$\hat{f}_n(\mu,\ell) = \left(\frac{N}{2\pi M}\right)^2 \int_{-\pi}^{\pi}\int_{-\pi}^{\pi} \hat{S}(n,w_1,w_2)e^{jw_1\mu}e^{jw_2\ell}dw_1dw_2 \tag{6}$$

where N is the length of the data sequence and M is the order of the EB (Kayhan et al. 1994, Neuman & Schonbach 1974). Using the EB defined as

$$\hat{S}_x(n,w_1,w_2) = \hat{H}(n,w_1)\hat{H}(n,w_2)\hat{H}(n,-w_1-w_2) \tag{7}$$

then we get

$$\hat{f}_n(\mu,l) = \left(\frac{N}{M}\right)^2 \sum_{m_1=0}^{N-1} w_n(m_1)x(m_1)w_n(m_1+\ell-\mu)x(m_1+\ell-\mu) \; w_n(m_1-\mu)x(m_1-\mu) \tag{8}$$

In the stationary case, equation (8) reduces to

$$\hat{f}(\mu,l) = \frac{1}{N}\sum_{m=0}^{N-1} x(m)x(m+l-\mu)x(m-\mu) \tag{9}$$

which is an estimate of the third-order cumulant of $x(n)$. The TVFC $f_n^x(\mu,l)$ obtained from the EB has the following relationship with the TD cumulants (TDC)

$$\begin{aligned} c_n^x(\mu,l) &= \int_{-\pi}^{\pi}\int_{-\pi}^{\pi} H(n,w_1)H(n+\mu,w_2)H(n+l,-w_1-w_2)e^{-j(w_1\mu+w_2l)}dw_1dw_2 \\ &= f_n^x(\mu,l) + \int_{-\pi}^{\pi}\int_{-\pi}^{\pi}[H(n,w_1)H(n+\mu,w_2)H(n+l,-w_1-w_2) \\ &\quad - H(n,w_1)H(n,w_2)H(n,-w_1-w_2)]e^{-j(w_1\mu+w_2l)}dw_1dv \end{aligned} \tag{10}$$

which implies that when the time variation of $H(n,w)$ is small, for small μ and l, $f_n^x(\mu,l)$ is a good estimate of the TDC, i.e., $c_n^x(\mu,l) \approx f_n^x(\mu,l)$.

4 SIMULATION

In this section, we will consider two examples to show the performance of the proposed method. The first example considers continuous TD parameter, and the second example considers TD parameters with some discontinuity. Considering the TD-AR model case with a noisy output, let us have the following examples.

Example 1: Let $x(n)$ be a second order TD-AR process such that

$$x(n) = e(n) - a_1(n-1)x(n-1) - a_2(n-2)x(n-2) + \eta(n)$$

where η(n) is a TD Gaussian noise such that SNR=10 db, and

$$a_1(n) = .99 - \frac{.99n}{N-1} \text{ and } a_2(n) = \frac{.99n}{N-1},$$

In this example, we have two TD parameters, one is decreasing and the other is increasing. Using the TVFC, with 60 Monte-Carlo runs on the input, we observe from Figure 1 that we are able to track the variation of the parameters, and hence to model the process.

We observe in Figure 2 that the variances of the estimates of the parameters are high at the end points. This is also reflected in the modeled process depicted in Figure 3(a), when looking at the mean-square error (MSE) plot shown in Figure 3(b).

Example 2: Let $x(n)$ be a second order TD-AR process such that

$$x(n) = e(n) - a_1(n-1)x(n-1) - a_2(n-2)x(n-2) + \eta(n)$$

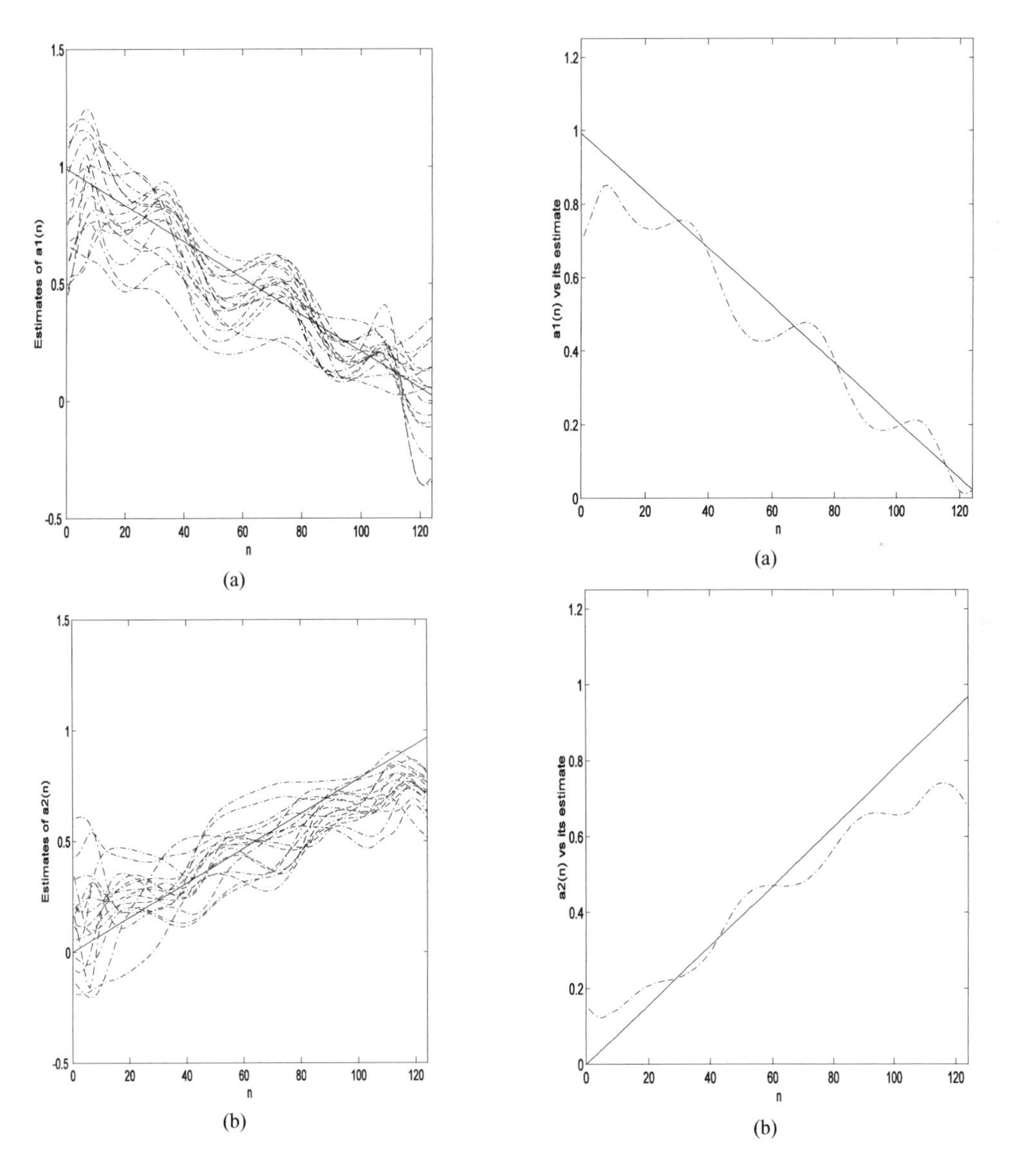

Figure 1. (a) Estimates of $a_1(n)$, (b) Estimates of $a_2(n)$.

Figure 2. (a) $a_1(n)$ and $\hat{a}_1(n)$, (b) $a_2(n)$ and $\hat{a}_2(n)$.

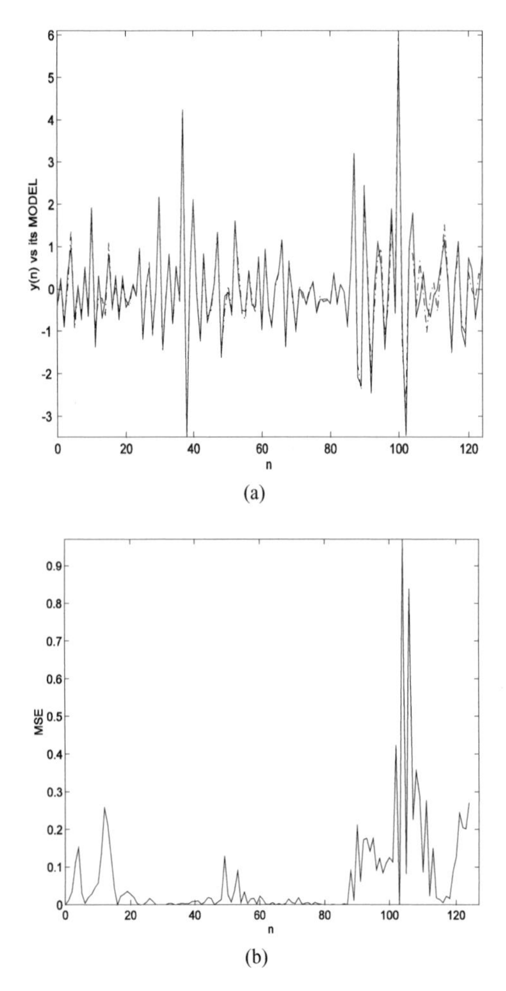

Figure 3. (a) $x(n)$ vs. $\hat{x}(n)$, (b) MSE.

where $\eta(n)$ is a TD Gaussian noise such that SNR=10 db,

$$a_1(n) = 0.5 \begin{cases} 0 < n < \dfrac{N}{4} \\ \dfrac{3N}{4} < n < N-1 \\ 0 \quad otherwise \end{cases}$$

and $a_2(n)=0.5-a_1(n)$. In this example, the parameters have rapid changes at $n=N/4$ and at $n=3N/4$.

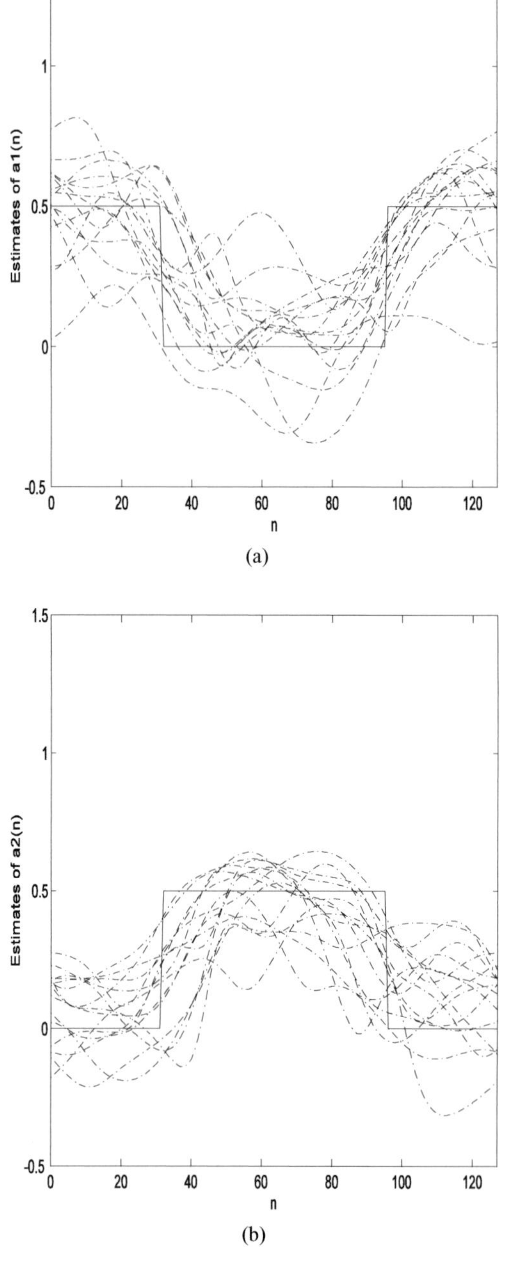

Figure 4. (a) Estimates of $a_1(n)$, (b) Estimates of $a_2(n)$.

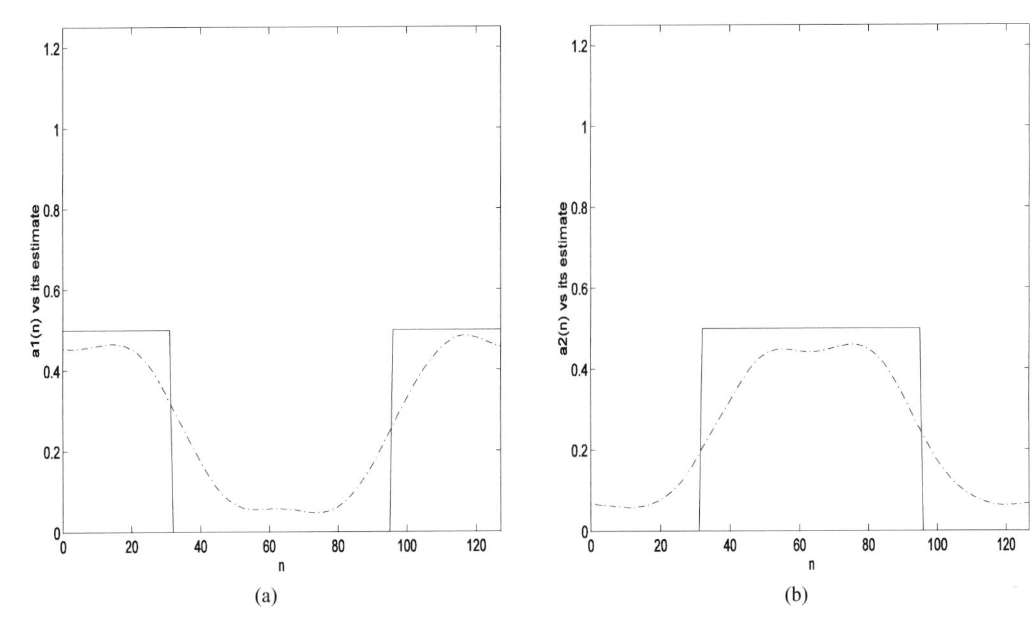

Figure 5. (a) $a_1(n)$ and $\hat{a}_1(n)$, (b) $a_2(n)$ and $\hat{a}_2(n)$.

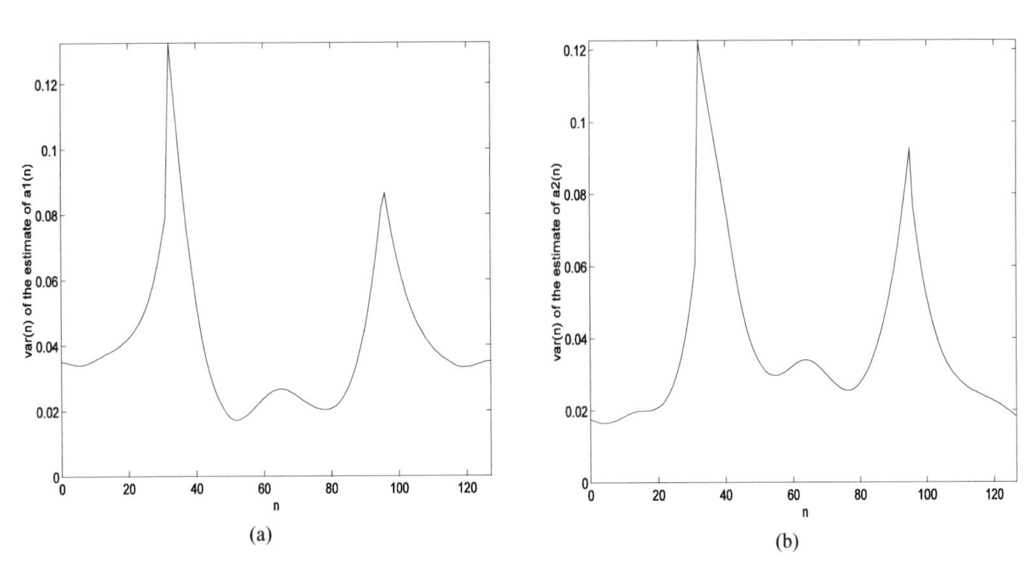

Figure 6. Variances of the estimates of (a) $a_1(n)$, (b) $a_2(n)$.

Because of these rapid changes of the parameters, the variances of the estimated TD parameters are high at these two points of time. Also, we observe the effect of this sudden change on the estimated process reflected by the MSE. We observe from Figure 4 and Figure 5 that the TVFC algorithm does not have problems at the end points. Figure 6 depicts the variances of $a_1(n)$ and $a_2(n)$, and Figure 7 shows the estimate of the TDF $x(n)$ vs. $\hat{x}(n)$ and its MSE.

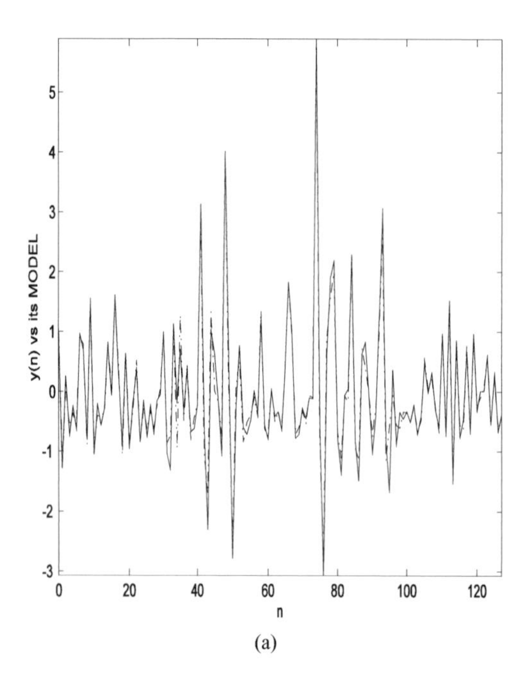

(a)

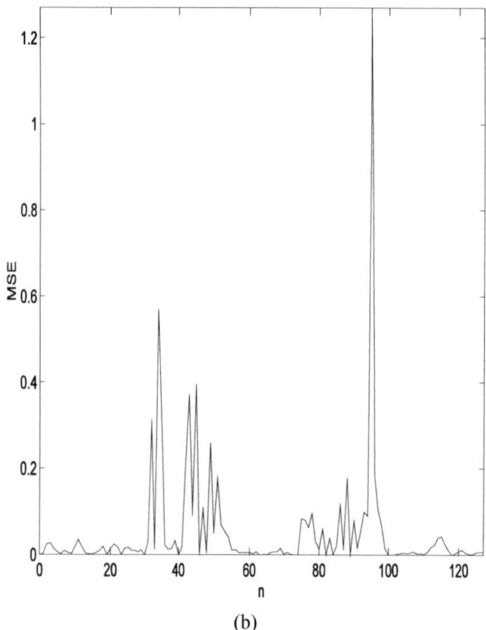

(b)

Figure 7. (a) $x(n)$ vs. $\hat{x}(n)$, (b) MSE.

5 CONCLUSIONS

Although the bispectrum has been applied in modeling of time-independent processes, it requires the assumption of stationarity and restricts the process to have non-symmetric probability density. However, when the characteristics of the process are time-dependent, neither the power spectrum nor the bispectrum can handle this problem because they do not reflect the time variation of the process characteristics. With the recent introduction of EB in digital signal processing, a new approach to the solution of the analysis/synthesis of a time-dependent process has been devised. In this paper, an algorithm based on the EB to solve these problems was proposed, whether the process is clear or corrupted with a symmetric distribution noise, using the TVFC. Some simulation has been depicted to show the powerful of the proposed method. For future work, one may try to apply the EB for finding the direction of arrival of a TDP.

REFERENCES

Al-Shoshan, A.I. 2011. Modeling of a non-stationary process corrupted by a symmetrically distributed noise using the evolutionary bispectrum, *CAINE2011*: 272–277.

Choi, B. S. 1992. *ARMA model identification*. New York: Springer-Verlag.

Diggle, P. J. 1990. *Time series: a biostatistical introduction*. Oxford: Oxford University Press.

Kayhan, A.S., El-Jaroudi, A. & Chaparro, L.F. 1994. Evolutionary periodogram for non-stationary signals. *IEEE Trans. on Signal Processing* 42: 1527–1536.

Kenny, O. P. & Boashash, B. 1993. Time-frequency analysis of backscattered signals from diffuse radar targets. *IEEE Proc. F. Radar Signal Processing* 140(3): 198–208.

Landers, T. E. & Lacoss, R. T. 1977. Some geophysical applications of autoregressive spectral estimates. *IEEE Trans. on Geoscience Electronics* (15): 26–32.

Laurain,, V., T´oth, R., M. Gilson, et al. 2011. Direct identification of continuous-time LPV input/output models. *IET Control Theory and Applications* 5(7): 878–888.

Luo, L. & Chaparro, L. F. 1991. Parametric identification of systems using a frequency slice of the bispectrum. *ICASSP* (E7.14): 3481–3484.

Neuman, C.P. & Schonbach, D.I. 1974. Discrete (Legendre) orthogonal polynomials - a survey. *International J. for Numerical Methods in Engineering* (8): 743–770.

Nikias, C. L. & Mendel, J. M. 1993. Signal processing with higher-order spectra. *IEEE SP Mag*azine: 10–37.

Percival, D. B. & Walden, A. T. 1993. *Spectral analysis for physical applications*. Cambridge: Cambridge University Press.

Priestley, M. B. 1988. Non-linear and non-stationary time series analysis. New York: Academic Press.

Rao, C. R. 1993. Multivariate analysis: future directions, bispectral analysis of non-stationary processes. In *M. B. Priestly & M. M. Gabr (eds), Amsterdam: North-Holand.*

Wood, J. C., Buda, A. J. & Barry, D. T. 1990. Time-frequency transform: a new approach to first heart sound frequency dynamics. *IEEE Trans. Biomedical Engineering* 39(7): 730–740.

Electronics, Communications and Networks IV – Hussain & Ivanovic (eds)
© 2015 Taylor & Francis Group, London, ISBN: 978-1-138-02830-2

Visual feature clustering using temporal, color and spatial information

A.M.R.R. Bandara & L. Ranathunga
University of Moratuwa, Moratuwa, Western Province, Sri Lanka.

N.A. Abdullah
University of Malaya, Kuala Lumpur, Wilayah Persekutuan Kuala Lumpur, Malaysia.

ABSTRACT: Salient Dither Pattern Feature (SDPF) is one of the visual features, which is highly efficient in feature extraction. The incorporated descriptor of SDPF is a lower dimensional spatial-chromatic histogram which is proven to be robust to both rotation and scale variations. The clustering of feature points to semantically meaningful visual segments is an essential step in using SDPF feature points for visual data understanding. Unsupervised clustering methods by using only non-temporal information suffer from lack of reliability. In this study, three clustering methods, namely K-means, Expectation Maximization (EM) and DBSCAN clustering were evaluated on clustering SDPF features to semantically meaningful clusters using motion, SDPF color patterns and spatial information of SDPF points. The results revealed that using k-means clustering with the temporal, color and spatial properties able to cluster SDPF points reliably with 86% of average completeness and 31% of average spatial accuracy error.

1 INTRODUCTION

Compacted Dither Pattern Code (CDPC) feature is one of the pioneer studies of using dither patterns in visual data understanding (Ranathunga et al. 2011). Salient Dither Pattern Feature (SDPF) is a newly derived feature from the concept held in CDPC, which incorporates both color and spatial details of visual data. SDPF feature has showed its stability over various geometrical transformations of visual data (Bandara et al. 2013).

A frame of a video may contains more than one salient object with a background filled with many different colors, textures and shapes. Therefore, applying SDPF and constructing a single descriptor for the whole frame will lead to misclassification. Constructing separate SDPF descriptors for each of the semantically meaningful objects can be used to overcome this problem. The term "semantically meaningful" expresses what human understand by means of compositions of different shapes, color and other visual properties of a video frame. This study focused on segmenting the video frames to semantically meaningful spatial segments.

Practically it is unable to supervise any segmentation method, to effectively segment all the objects in the universe, hence the solution is approximated with different unsupervised segmentation methods in applications which are not built to be domain specific(Arora et al. 2007,Gould et al. 2009).

Besides the low reliability of unsupervised spatial segmentation methods, they are computationally expensive(Vögele et al. 2014) Once the data is being digested by extracting features, the segmentation problem becomes a clustering problem of visual features which is computationally efficient than conventional segmentation methods due to low data density.

Due to the high complexity of visual objects there is no consistent relationship between visual properties of semantically meaningful segments. This fact motivates to use multiple properties for the clustering. A several recent studies (Aallaoui & Gourch 2013,Li et al. 2007)"have introduced a several spatio-temporal segmentation techniques which can segment to homogeneous areas but all they are not capable of segmenting to semantically meaningful objects without an aid of a supervision method. This paper presents a study of clustering SDPF feature points to semantically meaningful visual objects, based on their temporal details extracted from Lucas-Kanade optical flow estimation(Tamgade & Bora 2009), SDPF color patterns (Bandara et al. 2013) and spatial properties. The clustering performance of the selected properties were evaluated with three different clustering methods namely K-Means clustering (Jain 2010), expectation maximization (EM) clustering (Jin & Han 2010) and density based scanning (DBSCAN)(Sander et al. 1998) using CamVid dataset(Brostow et al. 2008).

2 RELATED STUDIES

This section explains SDPF visual feature, the related data clustering techniques and optical flow estimation technique.

2.1 *SDPF*

The study focused on clustering SDPF feature points. SDPF is an adaption of the concept in CDPC which the dithering reduces both color and spatial data density while preserving the visual impression(Ranathunga et al. 2011). The studies of CDPC has shown the effectiveness and the efficiency of reducing the dimensionality of visual descriptors by using dithering techniques(Ranathunga et al. 2010). SDPF adapt the concept in CDPC to include both color and spatial information in a low dimensional descriptor. This adaption ultimately creates a point like feature that extracted based on its saliency over a set of local dither patterns. The detailed steps of extracting SDPF features and its properties are described in(Bandara et al. 2013). In order to self-contain the paper the steps of extracting SDPF will be explained briefly here onward in this section.

- **Step 1**: Apply a filter by averaging neighborhood colors within a square block of pixels.
- **Step 2**: Considering any four adjacent blocks as a pattern as showed in the Figure 1, find the feature strength by calculating the summation of absolute color differences from its all neighbor patterns.
- **Step 3**: Apply the threshold to the feature strengths and finally apply non-maximal suppression filter.
- **Step 4**: the hue color component of four colors in the feature are quantized to 12 levels.

The above four steps produces a point like feature where each feature contains four colors.

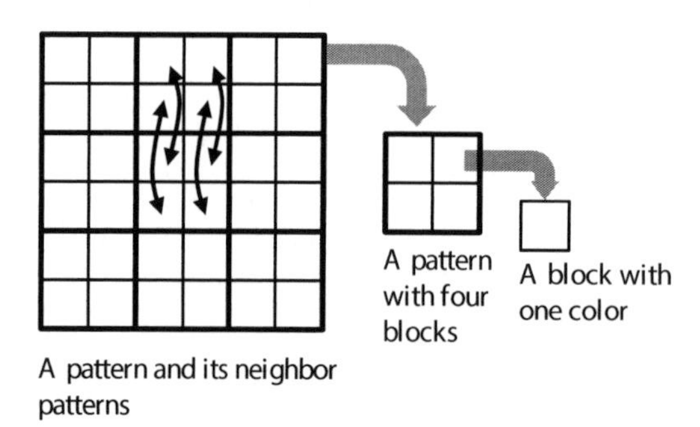

Figure 1. Structure of the dither pattern.

2.2 *Data clustering*

Data clustering is almost a matured field equipped with many supervised and unsupervised clustering techniques including many center based clustering, model based clustering and density based clustering techniques. K-Means clustering, Expectation Maximization (EM) and density based scanning (DBSCAN) are the mostly cited methods from the three types of clustering techniques respectively. K-Means clusters data points into K number of clusters by minimizing the squared error. It always clusters the data points to a one of the clusters whereas DBSCAN avoids noisy points based on their local density. But in DBSCAN it is hard to decide its two parameters namely minimum neighborhood distance and minimum number of points required to form a dense region. EM clustering in other hand is well known for having the capability of handling high dimensionality and robust to noisy data, among many other preferable qualities (Abbas 2008). EM finds the distribution parameters that maximize the log likelihood assuming the data set can be modeled as a mixture of multivariate normal distributions (Jin & Han 2010).

2.3 *Optical flow estimation*

The apparent motion of objects between two consecutive frames which is known as optical flow, has been used as an important detail for image segmentation in many studies (Aallaoui & Gourch 2013,Li et al. 2007) There are two mostly cited methods available for optical flow estimation namely Lucas-Kanade(Tamgade & Bora 2009) and Gunner Farneback's algorithm (Farnebäck 2003). The pyramidal implementation of Lucas-Kanade method can be used for effectively compute the optical flow for a sparse feature set (Tamgade & Bora 2009).

3 PROPOSED METHOD

This section explains how the properties of the feature are selected for the clustering, preparation of feature data, estimation method of number of clusters and the methodology.

3.1 *Selection of feature properties*

The spatial density property of SDPF feature points is necessary for identifying the different clusters, but not sufficient due to two or many semantically different objects can be seen overlapped and share a similar density measurement of feature points. In order to discriminate overlapping objects, color or texture property can be used. Each of SDPF feature points includes a local feature property, namely the dither color pattern. Each pattern contains four elements which are corresponding to the value of the hue, color component of four blocks of pixels around the SDPF point. In the proposed method, these four colors in the pattern are used in addition to the spatial properties to cluster the feature points.

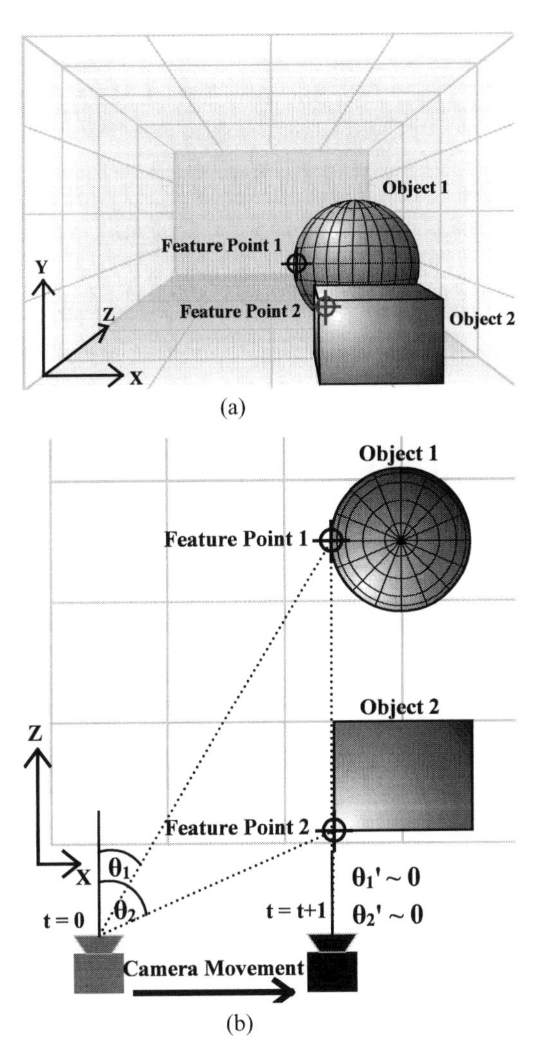

Figure 2. Differentiating two overlapping objects by motion details. (a) 3D Scene of two overlapping objects in XY plane. (b) Top view of the scene with details of camera motion.

The integration of the color property partially solves the problem in clustering features of overlapped objects while introducing a negative effect which over-clusters the feature points due to the occurrences of significantly different colors within a single object. In the proposed method, this negative effect is reduced by using the apparent motion details of the feature points. When the objects are moving, they have significantly different motion details, which is also a strong discrimination factor for accurate clustering. When there are only camera movements the different objects will still have different relative motion details due to the different depths of the locations of the objects, as shown in Figure 2.

In the figure the two objects appear overlapping in 3D perspective and the feature points incident on the camera with the incident angles of θ_1 and θ_2 when the time $t=0$. The geometrical information clearly shows that $\theta_2 > \theta_1$. When the camera moves, at the time $t=t+1$, both the in angles θ_1 and θ_2 become zero, hence the angle variation over the time while the camera movements is not similar. This fact yields that feature points in different objects can be discriminated by using the apparent motion details when objects or the camera is moved.

The apparent motion is captured by Lucas-Kanade optical flow algorithm and then the direction and magnitude of the flow of feature points are calculated.

3.2 *Preparation of feature data*

The preparation of feature points for clustering is done under four steps.

1 SDPF feature points have been extracted from the current frame of the given test video.
2 Calculate the optical flow estimation using the next frame of the video.
3 If the number of feature points found in the next frame is less than the 50% of the number of original SDPF feature points, restart from the step 1 as it clues a shot-boundary. Otherwise, proceed to the step 4.
4 Calculate the flow angle and hypotenuse of each of the feature points found in the next frame. The angle measurement is noisy when the flow magnitude is low; therefore the flow angle is scaled by multiplying with the flow magnitude.

Each of the feature points corresponds to a data vector has 8 elements, namely x coordinate, y coordinate, angle of flow multiplied by the flow magnitude, flow magnitude and four colors in SDPF patterns. This vector with the dimension of 8 is used for the clustering. Both the K-means and EM clustering algorithms require the number of clusters (K). The estimation of K is described in the next section.

3.3 *Estimation of number of clusters (K)*

The K is estimated based on the number of feature points available in the current frame. As the first step the ground truth number of clusters was obtained for 50 random frames by manual inspection and recorded against the number of feature points. Then a curve was fitted to the ground truth number of clusters Vs number of feature points (m) and obtained the equation (1) of the curve. It is then used in the experiment to estimate the K.

$$K = 3.117 \ln(m) - 13.7 \qquad (1)$$

3.4 *Methodology*

In this study, the CamVid dataset (Brostow et al. 2008) is used. It contains video data over ten minutes of high quality 30Hz footage captured from the perspective of a driving automobile. Both the moving and fixed Objects from a set of 7 concepts were selected to test the clustering. As the first step, the feature points were labeled by the selected clustering method. Then spatial centroid was calculated for each of the clusters. Each of these clusters ware labeled with the object on which the spatial centroid is located. This process was carried for 100 visually different frames. To have the objective comparison, a spatial accuracy measure S_{er}(Wollborn & Mech 1998) in each frame is defined as(2).

$$S_{er} = \frac{\sum_{n=1}^{N} |M^{cl}(n) \oplus M^{ref}(n)|}{\sum_{n=1}^{N} |M^{ref}(n)|} \quad (2)$$

Where N is the number of objects in the frame, $M^{cl}(n)$ is the set of points clustered for the nth object, $M^{ref}(n)$ is the set of points that have been clustered by hand to the nth object and $\oplus$ denotes the exclusive disjunction of the two sets.

To objectively compare the completeness of the clustering, a completeness measure S_{compl} is defined as (3).

$$S_{compl} = \frac{\sum_{n=1}^{N} |M^{cl}(n) \cap M^{ref}(n)|}{\sum_{n=1}^{N} |M^{ref}(n)|} \quad (3)$$

The symbol $\cap$ denotes conjunction of the two sets.

4 EXPERIMENTAL RESULTS AND DISCUSSION

Figure 3 shows the results obtained from the experiment for first 18 frames. Figure 3a shows the completeness of K-Means algorithm is the highest for the majority of the frames. The completeness of both K-Means and EM clustering fluctuates similarly over the frames but the completeness of EM lies below the K-Means at most of the cases. Although DBSCAN shows the highest completeness occasionally it is not consistent as the other two algorithms over the frames. DBSCAN shows the worst performance by means of the spatial accuracy measure which is shown in Figure 3b. K-Means keeps the error at the lowest in majority of the cases. Besides the high completeness and high spatial accuracy, using K-Means has a competitive advantage over efficiency since EM clustering requires a costly re-training process as the number of clusters (K) changes. The overall result shows that the average completeness of K-Means clustering is 86% and the average spatial accuracy measurement is as low as 31%. The next best performed algorithm is EM clustering which shows the average completeness of 81% and average spatial accuracy measurement of 43%. DBSCAN is the worst with average completeness of 77% and average spatial accuracy measurement of 54%. The results yield that the proposed scheme with K-Means clustering works well for clustering the SDPF feature points for semantically meaningful visual segments using the temporal, color and spatial details.

(a)

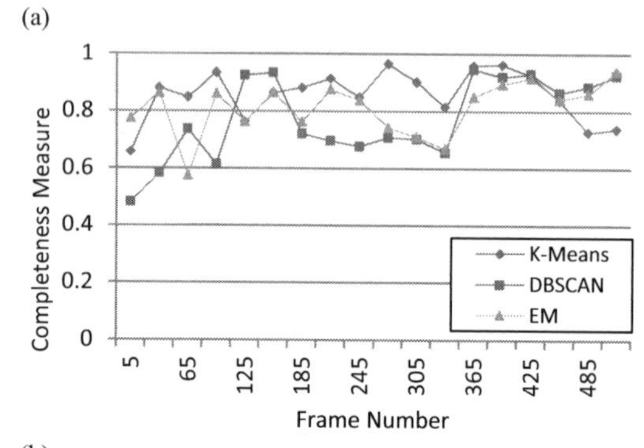

(b)

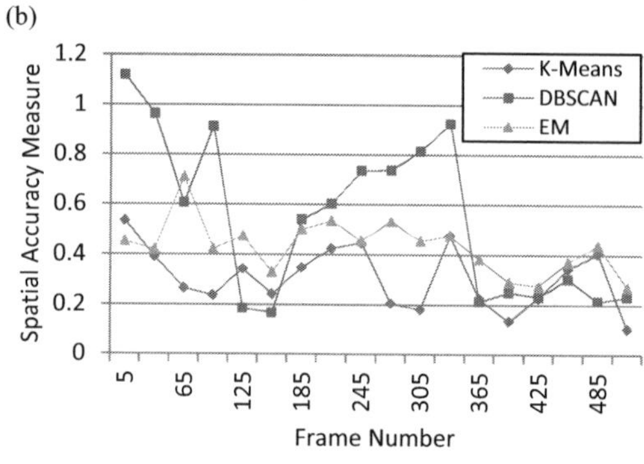

Figure 3. Comparison of the three algorithms with the sequence 16E5. (a) Completeness measure. (b) Spatial accuracy measure.

Figure 4 shows a sample of the clustering process of SDPF points. The visualization of the flow feature points evidences how the proposed method takes the advantage of temporal information of SDPF points. The proposed method shows failures only where the color information, flow information and spatial information have less discrimination such as, clustering points to the fence and road marks. The SDPF points on buildings, cars and trees which are located at different depths, have been successfully clustered, which evidences that the depth of the object also significantly affect the clustering performance.

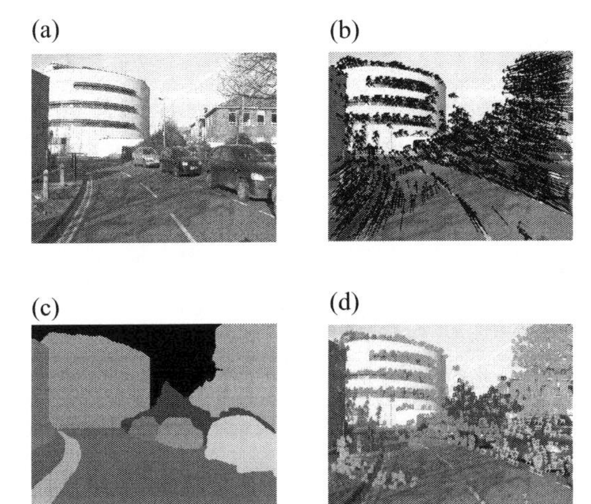

Figure 4. A sample of clustering SDPF points using K-Means -(Frame No. 34). (a) Original. (b) Flow magnitude and angle representation. (c) Resultant frame after clustering. (d) Ground-truth segments.

5 CONCLUSION

This paper presents a study on clustering SDPF feature points to semantically meaningful visual segments using color, spatial and temporal information based scheme. The study has proved that the proposed scheme works best with K-Means clustering. The performance evident that the proposed scheme can be used to cluster the feature points to semantically meaningful visual segments without supervising the clustering method. Since the scheme used local motion details the issue how to utilize global motion estimation for clustering remains. The findings of this study will be used for extracting multiple semantics out of every single frame in video sequences using SDPF feature.

ACKNOWLEDGEMENT

This work was carried out with the support of the National Research Council, Sri Lanka (Grant No. 12-017). The authors also acknowledge the support received from the LK Domain Registry in publishing this paper.

REFERENCES

Aallaoui, M.E. & Gourch, A. 2013.Spatio-temporal segmentation with Mumford-Shah functional. 2013 ACS International Conference on Computer Systems and Applications (AICCSA),Ifrane, 27-30 May 2013: 1–4. IEEE.

Abbas, O.A. 2008. Comparisons Between Data Clustering Algorithms. Int. Arab. J. Inf. Technol. 5: 320–325.

Arora, H., Loeff, N., Forsyth, D. &Ahuja, N. 2007. Unsupervised Segmentation of Objects using Efficient Learning. IEEE Conference on Computer Vision and Pattern Recognition, 2007. CVPR2007. IEEE.

Bandara, R., Ranathunga, L. & Abdullah, N.A. 2013. Invariant Properties of a Locally Salient Dither Pattern with a Spatial-Chromatic Histogram. Eighth International Conference on Industrial & Information Systems 2013 (ICIIS'13). Peradeniya, Sri Lanka.

Brostow, G.J., Fauqueur, J. & Cipolla, R. 2008. Semantic Object Classes in Video: A High-Definition Ground Truth Database. Pattern Recognit. Lett. 30: 88–97.

Farnebäck, G. 2003. Two-frame motion estimation based on polynomial expansion. In Image Analysis. Springer Berlin Heidelberg.

Gould, S., Gao, T. & Koller, D. 2009. Region-based segmentation and object detection. In Advances in Neural Information Processing Systems: 655–663.

Jain, A.K. 2010. Data clustering: 50 years beyond K-means. Pattern Recognit. Lett. 31: 651–666.

Jin, X. & Han, J. 2010. Expectation Maximization Clustering, in: Encyclopedia of Machine Learning: 382–383. Springer ..

Li, R., Yu, S. & Yang, X. 2007. Efficient spatio-temporal segmentation for extracting moving objects in video sequences. IEEE Trans. On Consum. Electron. 53: 1161–1167.

Ranathunga, L., Zainuddin, R. & Abdullah, N.A. 2010. Compacted Dither Pattern Codes Over Mpeg-7 Dominant Colour Descriptor In Video Visual Depiction. Malays. J. Comput. Sci. 23: 68–84.

Ranathunga, L., Zainuddin, R. & Abdullah, N.A. 2011. Performance evaluation of the combination of Compacted Dither Pattern Codes with Bhattacharyya classifier in video visual concept depiction.Multimed. Tools Appl. 54: 263–289.

Sander, J., Ester, M., Kriegel, H.P. & Xu, X. 1998. Density-based clustering in spatial databases: The algorithm gdbscan and its applications. Data Min. Knowl. Discov. 2: 169–194.

Tamgade, S.N. & Bora, V.R. 2009. Motion vector estimation of video image by pyramidal implementation of Lucas Kanade optical flow. 2nd International Conference on Emerging Trends in Engineering and Technology (ICETET), Nagpur, 16–18 December 2009: 914–917. IEEE.

Vögele, A., Krüger, B. & Klein, R. 2014. Efficient Unsupervised Temporal Segmentation of Human Motion.In2014 ACM SIGGRAPH/Eurographics Symposium on Computer Animation. Copenhagen, Denmark, 2014. ACM.

Wollborn, M. & Mech, R. 1998. Refined procedure for objective evaluation of video object generation algorithms. Doc ISOIEC JTC1SC29WG11 M 3448: 1998.

Electronics, Communications and Networks IV – Hussain & Ivanovic (eds)
© 2015 Taylor & Francis Group, London, ISBN: 978-1-138-02830-2

Construction and investigation aircraft control system in a class of one-parameteric structurally stable mappings using Lyapunov functions

Mamyrbek Beisenbi, Gulzhan Uskenbayeva* & Sandygul Kaliyeva
Department of System Analysis and Control, L.N. Gumilyov Eurasian National University, Astana, Kazakhstan

ABSTRACT: The article presents a new approach to building control systems for objects with uncertain parameters in the form of one-parametric structurally stable mappings from catastrophe theory.

KEYWORDS: control systems, robust stability, one-parametric structurally-steady maps, a steady state, Lyapunov function.

1 INTRODUCTION

Currently, control problems are characterised by increasingly complex, high-order systems, requirements for high efficiency and stability, numerous uncertainties and incomplete information. In these systems, uncertainty can occur because of the presence of uncontrolled disturbances acting on the system or because the true values of the parameters of the system are unknown, either initially or as the system changes over time. The main goal in control system design is, in some sense, to provide the best protection against uncertainty in the knowledge of the system.

The ability of a control system to maintain stability in the presence of parametric or nonparametric uncertainties is known as system robustness (Polyak & Scherbakov 2002). In the general, research of system robust stability consists in the indication of restrictions on control system parameters change (Polyak & Scherbakov 2002, Dorato & Yedavalli 1990).

The many papers are devoted to the problem of research of robust stability (Polyak & Scherbakov 2002, Dorato & Yedavalli 1990, Kuntsevich 2006, Liao & Yu 2008). In these works investigated the robust stability of polynomials, matrixes, within the linear principle of stability of continuous and discrete control systems. In works (Kuntsevich 2006, Liao & Yu 2008) absolute stability as robust stability of the control system concerning a choice of nonlinearities from the given (sector) class is considered, i.e. are solving the problems of absolute robust stability. In the practical tasks, connected with development and creation of control systems in technology, economy, biology and other spheres, in the conditions of essential parametrical uncertainty, the increase in potential of robust stability is one of the key factors, which guaranteeing to a control system protection from entry in regime of determined chaos and strange attractors. And guarantees applicability of models and reliability of the designed control systems work.

At present it is conventional that, real control objects are nonlinear, and one of the main properties of nonlinear dynamic systems is functioning in the mode of the determined chaotic traffic (Andrievsky & Fradkov 1999, Nicols & Prigogine 1989, Loskutov & Mikhaylov 2007). In linear dynamic systems it appear in the form of control system's zero steady state stability loss (Beisenbi 2002, Beisenbi 2011, Beisenbi & Erzhanov 2002). In this regard, in the conditions of uncertainty, there was a need for development of models and methods of design control system with rather wide area of robust stability. Which called control systems with increased potential of robust stability (Beisenbi 2002, Beisenbi 2011, Beisenbi & Erzhanov 2002). The concept of creation of a control system with increased potential of robust stability based on results of the catastrophes theory (Gilmore 1984, Poston & Stewart 2001), where the main structural-stable maps are received.

This article is devoted to design of control systems with increased potential of robust stability by dynamical objects with uncertain parameters in a class of the one-parametric structural-stable maps (Beisenbi 1998, 2002, 2011, Beisenbi & Erzhanov 2002, Ashimov & Beisenbi 2000).

Research in recent years has shown that the method of Lyapunov functions (Barbashin 1967, Krasovsky 1959, Malkin 1966) can be successfully used to analyse the robust stability of linear and nonlinear control systems. Usage of Lyapunov's function method for the solution of a set of practical linear or nonlinear tasks is constrained by the lack of a general method for selecting or constructing Lyapunov functions

* Corresponding author: *gulzhum_01@mail.ru*

and difficulties with their algorithmic representation (Barbashin 1967, Malkin 1966). An inappropriate choice of a Lyapunov function or the inability to construct one does not indicate instability of the system, only that a proper Lyapunov function has not been found.

The method of design of Lyapunov vector function (Voronov & Matrosov 1987), on the basis of geometrical interpretation of asymptotic stability theorem and concepts of stability, is offered. Therefore, the origin corresponds to a predetermined condition of the system, the unperturbed state, and the equations of the state are formed concerning perturbations, i.e. in deviations of the perturbed motion from unperturbed (Malkin 1966). Consequently, the state equations express the speed of change of a perturbations vector (deviations) and for steady system is directed toward the origin. And the gradient vector from required Lyapunov function, for stable system, will be always directed to the opposite side. It allows to present Lyapunov function in the form of a potential surface (Gilmore 1984). Research of robust stability of the control system with uncertain parameters are based on ideas of a Lyapunov direct method.

2 MATHEMATICAL MODEL FORMULATION

2.1 *Area of system stability*

System can be written in expanded form:

$$\begin{cases} \dot{x}_1 = -b_{11}x_1^3 + (a_{11} + b_{11}k_1)x_1 + a_{12}x_2 + \ldots + a_{1n}x_n \\ \dot{x}_2 = -a_{21}x_1 - b_{22}x_2^3 + (a_{22} + b_{22}k_2)x_2 + \ldots + a_{2n}x_n \\ \ldots \\ \dot{x}_n = -a_{n1}x_1 + a_{n2}x_2 + \ldots + b_{nn}x_n^3 + (a_{nn} + b_{nn}k_n)x_n \end{cases} \quad (1)$$

where $x(t) \in R^n$ - control object state vector, $\boldsymbol{a_{ij}}$, $\boldsymbol{b_{ij}}$, $i=1,\ldots,n$, $j=1,\ldots,n$ - the elements of the control object.

The control law is given by a vector function in the form of one-parametric structural-stable maps (Gilmore 1984, Poston & Stewart 2001):

$$u_i = -x_i^3 + k_i x_i,\ i = 1,\ldots,n \quad (2)$$

The steady state of the system (1) $\boldsymbol{x_{is}^1 = 0}$, $i=1,\ldots,n$ is defined by the solution of the equations

$$\begin{cases} -b_{11}x_{1s}^3 + (a_{11} + b_{11}k_1)x_{1s} + a_{12}x_{2s} + \ldots + a_{1n}x_{ns} = 0 \\ -a_{21}x_{1s} - b_{22}x_{2s}^3 + (a_{22} + b_{22}k_2)x_{2s} + \ldots + a_{2n}x_{ns} = 0 \\ \ldots \\ -a_{n1}x_{1s} + a_{n2}x_{2s} + \ldots + b_{nn}x_n^3 + (a_{nn} + b_{nn}k_n)x_{ns} = 0 \end{cases} \quad (3)$$

From (3), we receive a stable state of the system

$$x_{is}^1 = 0,\ i = 1,\ldots,n \quad (4)$$

Other stationary states will be defined by solutions of the equation

$$-b_{ii}x_{is}^2 + (a_{ii} + b_{ii}k_i) = 0 \quad x_{js} = 0,$$

$$i \neq 0,\ i = 1,\ldots,n;\ j = 1,\ldots,n \quad (5)$$

Great number of solutions of the equations (5) can be written as

$$x_{is}^{2,3} = \pm\sqrt{\frac{a_{ii}}{b_{ii}} + k_i}, \quad x_{js} = 0,$$

$$i \neq j,\ i = 1,\ldots,n,\ j = 1,\ldots,n \quad (6)$$

Here the system of the nonlinear algebraic equations (3) has the trivial decision (4) and uncommon decisions (6) when $a_{ii}/b_{ii}+k_i>0$, $i=1,\ldots,n$. At negative value $a_{ii}/b_{ii}+k_i<0$, $i=1,\ldots,n$ the equation (5) has imaginary decisions that, can't correspond to any physically possible situation (Nicols & Prigogine. 1989). These decisions are joined with (4) when $a_{ii}/b_{ii}+k_i=0$, $i=1,\ldots,n$. and branch off from it when $a_{ii}/b_{ii}+k_i>0$, $i=1,\ldots,n$, i.e. in a point $a_{ii}/b_{ii}+k_i=0$, $i=1,\ldots,n$, bifurcation happens. It is provided, that the state (4) is globally asymptotically stable for all $a_{ii}/b_{ii}+k_i<0$, $i=1,\ldots,n$, and unstable for $a_{ii}/b_{ii}+k_i>0$, $i=1,\ldots,n$. States (6) also will be asymptotically stable, in other words, branches appears as a result of bifurcation, while the state (4) loses stability, and these branches are stable.

Verification of these statements made on the basis of vector Lyapunov functions ideas (Voronov & Matrosov 1987).

If Lyapunov function $\boldsymbol{V(x)}$ sets in the form of the vector function $\boldsymbol{V(V_1(x),\ldots,V_n(x))}$, then components of the rate vector will be equal (Beisenbi & Uskenbayeva 2014a, 2014b, Beisenbi & Yermekbayeva 2013a, 2013b, Beisenbi & Abdrakhmanova 2013):

$$-\frac{dx_i}{dt} = \frac{\partial V_i(x)}{\partial x_1} + \frac{\partial V_i(x)}{\partial x_2} + \ldots + \frac{\partial V_i(x)}{\partial x_n},\ i = 1,\ldots,n \quad (7)$$

In the equation (7), substituting values of components of the rate vector, we will get

$$\frac{\partial V_i(x)}{\partial x_1} = -a_{i1}x_1,\ \frac{\partial V_i(x)}{\partial x_2} = -a_{i2}x_2,\ldots,$$

$$\frac{\partial V_i(x)}{\partial x_i} = b_{ii}x_i^3 - (a_{ii} + b_{ii}k_i)x_i,\ldots,\ \frac{\partial V_i(x)}{\partial x_{n-1}} = -a_{i,n-1}x_{n-1},$$

$$\frac{\partial V_i(x)}{\partial x_n} = -a_{in}x_n\ i = 1,\ldots,n \quad (8)$$

The time derivatives of the components of the vector Lyapunov function $V(x)$ can be obtained from the state equation (1) using the scalar product of the components of the gradient of the vector Lyapunov function and the components of the state rate vector (Beisenbi & Uskenbayeva 2014a, 2014b, Beisenbi & Yermekbayeva 2013a, 2013b, Beisenbi & Abdrakhmanova 2013), i.e.

$$\frac{dV(x)}{dt} = -\sum_{i=1}^{n}\left(\sum_{j=1}^{n}\frac{\partial V_i(x)}{\partial x_j}\right)\frac{dx_i}{dt} = -\sum_{i=1}^{n}[a_{i1}x_1 + a_{i2}x_2 + \ldots - b_{ii}x_i^3 + \\ +(a_{ii} + b_{ii}k_i)x_i + \ldots + a_{i,n-1}x_{n-1} + a_{in}x_n]^2 i = 1,\ldots,n \quad (9)$$

From (9) follows, that full time derivate from the Lyapunov function will be negative.

From (8), for components of the vector Lyapunov function we will get

$$V_i(x) = -\frac{1}{2}a_{i1}x_1^2 - \frac{1}{2}a_{i2}x_2^2 -,\ldots,-\frac{1}{4}b_{ii}x_i^4 - \frac{1}{2}(a_{ii} + b_{ii}k_i)x_i^2 - \\ -\ldots - \frac{1}{2}a_{i,n-1}x_{n-1}^2 - \frac{1}{2}a_{in}x_n^2, i = 1,\ldots,n$$

We can present Lyapunov function in a scalar form in the view:

$$V(x) = \sum_{i=1}^{n}V_i(x) = \frac{1}{4}x_1^4 - \frac{1}{2}(a_{11} + b_{11}k_1 + a_{21} + a_{31} + \ldots + a_{n1})x_1^2 + \\ \frac{1}{4}x_2^4 - \frac{1}{2}(a_{12} + a_{22} + b_{22}k_2 + a_{32} + \ldots + a_{n2})x_2^2 + \quad (10) \\ \frac{1}{4}x_3^4 - \frac{1}{2}(a_{13} + a_{23} + a_{33} + b_{33}k_3 + \ldots + a_{n3})x_3^2 + \ldots \\ \ldots + \frac{1}{4}x_n^4 - \frac{1}{2}(a_{1n} + a_{2n} + a_{3n} + \ldots + a_{nn} + b_{nn}k_n)x_n^2$$

Function (10) is Lyapunov function and conditions of positive definiteness are defined by inequalities

$$\begin{cases} a_{11} + b_{11}k_1 + a_{21} + a_{31} + \ldots + a_{n1} < 0 \\ a_{12} + a_{22} + b_{22}k_2 + a_{32} + \ldots + a_{n2} < 0 \\ a_{13} + a_{23} + a_{33} + b_{33}k_3 + \ldots + a_{n3} < 0 \\ \ldots \\ a_{1n} + a_{2n} + a_{3n} + \ldots + a_{nn} + b_{nn}k_n < 0 \end{cases} \quad (11)$$

Thus, the area of system stability (1) for the established state (4) is defined by system of inequalities (11).

2.2 *Research of stationary states x_s^2stability*

The equations of the state (1) in deviations in relative stable state x_s^2 (6) can be written as (Beisenbi & Abdrakhmanova 2013, Beisenbi 2002, 2011, Beisenbi & Erzhanov 2002):

$$\begin{cases} \dot{x}_1 = -b_{11}x_1^3 - 3b_{11}\sqrt{\frac{a_{11}}{b_{11}} + k_1x_1^2} + (a_{11} + b_{11}k_1)x_1 + a_{12}x_2 + \ldots + a_{1n}x_n \\ \dot{x}_2 = -a_{21}x_1 - b_{22}x_2^3 - 3b_{22}\sqrt{\frac{a_{22}}{b_{22}} + k_2x_2^2} - 2(a_{22} + b_{22}k_2)x_2 + \ldots + a_{2n}x_n \\ \ldots \\ \dot{x}_n = -a_{n1}x_1 + a_{n2}x_2 + \ldots - b_{nn}x_n^3 - 3b_{nn}\sqrt{\frac{a_{nn}}{b_{nn}} + k_nx_n^2} - 2(a_{nn} + b_{nn}k_n)x_n \end{cases} \quad (12)$$

The full derivative from Lyapunov function $V(x)$, considering the state equations in deviations (12), relative to the stationary state x_s^2 (6), is defined as:

$$\frac{dV(x)}{dt} = \frac{\partial V}{\partial x}\frac{dx}{dt} = -\sum_{i=1}^{n}(a_{i1}x_1 + a_{i2}x_2 + .a_{i3}x_3 + .. \\ -b_{ii}x_i^3 - 3b_{ii}\sqrt{\frac{a_{ii}}{b_{ii}} + k_ix_i^2} - 2(a_{ii} + b_{ii}k_i)x_i + \ldots + a_{in}x_n)^2 \quad (13)$$

Function (13) is the negative function. We can find components of the gradient vector of Lyapunov function

$$\frac{\partial V_i(x)}{\partial x_1} = -a_{i1}x_1, \quad \frac{\partial V_i(x)}{\partial x_2} = -a_{i2}x_2,\ldots,$$

$$\frac{\partial V_i(x)}{\partial x_i} = b_{ii}x_i^3 + 3b_{ii}\sqrt{\frac{a_{ii}}{b_{ii}} + k_ix_i^2} - 2(a_{ii} + b_{ii}k_i)x_i,$$

$$\frac{\partial V_i(x)}{\partial x_{i+1}} = -a_{i,i+1}x_{i+1} \quad \frac{\partial V_i(x)}{\partial x_n} = -a_{in}x_n, i = 1,\ldots,n;$$

From here, we receive Lyapunov function in a scalar form

$$V(x) = \sum_{i=1}^{n}\left(\frac{1}{4}b_{ii}x_i^4 + b_{ii}\sqrt{\frac{a_{ii}}{b_{ii}} + k_ix_i^3}\right) \\ +\frac{1}{2}\sum_{i=1}^{n}(-a_{1i} - a_{2i} - \ldots + (a_{ii} + b_{ii}k_i) - \ldots - a_{ni})x_i^2 \quad (14)$$

Function (14) on the beginning of coordinates addresses in zero, is continuous differentiable function and has the form of variables with odd degrees. Therefore, on the basis of the Morse lemma (Gilmore 1984, Poston & Stewart 2001), function (14) around the steady state x_s^2(6) can be represented as a quadratic form

$$V(x) = \frac{1}{2}\sum_{i=1}^{n}(-a_{1i} - a_{2i} - \ldots + (a_{ii} + b_{ii}k_i) - \ldots - a_{ni})x_i^2$$

From here, positive definiteness of Lyapunov function will be defined by an inequality

$$(a_{ii} + b_{ii}k_i) > a_{1i} + a_{2i} + \ldots - a_{ni}, i = 1,\ldots,n; \quad (15)$$

2.3 *Let investigate stability of a steady states $x_s 3$.*

The equation of the state (1) in deviations in relative steady state x_s^3 (6) can be written as (Beisenbi & Abdrakhmanova 2013, Beisenbi 2002, 2011, Beisenbi & Erzhanov 2002):

$$\dot{x}_i = a_{i1}x_1 + a_{i2}x_2 + .a_{i3}x_3 + .. - b_{ii}x_i^3 +$$
$$+3b_{ii}\sqrt{\frac{a_{ii}+b_{ii}k_i}{b_{ii}}}x_i^2 - 2(a_{ii}+b_{ii}k_i)x_i + ... + a_{in}x_n$$
$$i = 1,...,n; \tag{16}$$

Omitting formal actions for research of stability of stationary states x_s^3(6), similar for a steady state x_s^3(6), we will receive Lyapunov function in a scalar form

$$V(x) = \sum_{i=1}^{n}(\frac{1}{4}b_{ii}x_i^4 - b_{ii}\sqrt{\frac{a_{ii}}{b_{ii}} + k_i x_i^3})$$
$$+\frac{1}{2}\sum_{i=1}^{n}(-a_{1i} - a_{2i} - ... + (a_{ii} + b_{ii}k_i) - ... - a_{ni})x_i^2$$

On the Morse lemma, we will lead (Gilmore 1984, Poston & Stewart 2001) Lyapunov function, by means of the stability matrix, to a quadratic form (Beisenbi & Abdrakhmanova 2013)

$$V(x) = \frac{1}{2}\sum_{i=1}^{n}(-a_{1i} - a_{2i} - a_{3i} - ... + (a_{ii} + b_{ii}k_i) - ... - a_{ni})x_i^2$$

Stability conditions of a steady state x_s^3(6) will be expressed by system of inequalities:

$$(a_{ii} + b_{ii}k_i) > a_{1i} - a_{2i} + ... + a_{ni},\ i = 1,...,n; \tag{17}$$

Thus, the control system constructed in a class of one-parametric structural stable maps will be stable in indefinitely wide limits of change of uncertain parameters of the control object. The steady state x_s^1 (4) exists and is stable at change of uncertain parameters of object in area (11). And stationary states $x_s^2$and x_s^3(6) appear at loss of stability of a state x_s^1 (4), and they are not simultaneously exist. Stationary states $x_s^2$and x_s^3(6) are stable when performing system of inequalities (15) and (17).

3 CASE STUDY

3.1 *Description of dynamics of the aircraft angular motion*

We investigate a task of traffic control of the aircraft by the pitch. Let consider, that aircraft have constants, aprioristic-uncertain parameters, which values are located in a set area. We will notice, that a similar situation can take place, when aircraft were flying on various modes: height, speed and load of the aircraft varies slowly, compared with the rate of angular motion. For the description of dynamics of the aircraft angular motion we use the following linearized equations (Andrievsky & Fradkov 1999, Bukov 1987):

$$\begin{cases} \dot{\alpha}(t) = \omega_z(t) + \alpha_y^{\alpha}(t)\alpha(t) + \alpha_y^{\delta_\beta}\delta_\beta(t) \\ \dot{\omega}_z(t) = -\alpha_{m_z}^{\alpha}\alpha(t) - \alpha_{m_z}^{\omega_z}\omega_z(t) - \alpha_{m_z}^{\delta_\beta}\delta_\beta(t) \\ \dot{\vartheta}(t) = \omega_z(t) \end{cases}$$

Where $\vartheta(t),\omega(t)$ -the angle and the pitch rate, $\alpha(t)$ - the angle of attack, $\delta_\beta(t)$ - angle of a deviation of the rudder height; $\alpha_y^{\alpha}(t)$, $\alpha_y^{\delta}\beta(t)$, $\alpha_{mz}^{\alpha}(t)$, $\alpha_{mz}{}^{\omega}{}_z(t)$, $\alpha_{mz}\delta\beta(t)$- aircraft parameters. Their values depend on the factors stated above and can change over a wide range, depending on the height and the speed of flight. Exact values of parameters a priori not defined. Also, we assume, that dynamics of the executive body it is possible to neglect and consider, that control object is the deviation of rudder $\delta_\beta(t)$.

$$x_1 = \vartheta(t), x_2 = \omega_z(t), x_3 = \alpha(t), a_1 = \alpha_{m_z}^{\alpha}, a_2 = \alpha_{m_z}^{\omega_z},$$
$$a_3 = \alpha_{m_z}^{\delta_\beta}, a_4 = \alpha_y^{\alpha}(t), a_5 = \alpha_y^{\delta_\beta}, u = \delta_\beta(t)$$

Then, the equation of the aircraft angular motion will assume the form:

$$\begin{cases} \frac{dx_1}{dt} = x_2(t) \\ \frac{dx_2}{dt} = -a_1x_3(t) - a_2x_2(t) - a_3u \\ \frac{dx_3}{dt} = x_2(t) + a_4x_3(t) + a_5u \end{cases} \tag{18}$$

As the control law, we will choose

$$u = -x_3^3 + k_3x_3 \tag{19}$$

Thus, the system (18) with the control law (19) will assume the form:

$$\begin{cases} \frac{dx_1}{dt} = x_2(t) \\ \frac{dx_2}{dt} = -a_1x_3(t) - a_2x_2(t) - a_3(-x_3^3 + k_3x_3) \\ \frac{dx_3}{dt} = x_2(t) + a_4x_3(t) + a_5(-x_3^3 + k_3x_3) \end{cases} \tag{20}$$

From the equations (20) we define the established conditions

$$\begin{cases} x_{2S} = 0 \\ -a_1 x_{3S}(t) - a_2 x_{2S}(t) - a_3(-x_{3S}^3 + k_3 x_{3S}) = 0 \\ x_{2S}(t) + a_4 x_{3S}(t) + a_5(-x_{3S}^3 + k_3 x_{3S}) = 0 \end{cases} \quad (21)$$

The system (21) has the following stationary states

$$x_{1S} = 0, x_{2S} = 0, x_{3S} = 0 \quad (22)$$

And other stationary conditions of the system (21) are defined by the solution of the equations

$$(a_3 - a_5)x_{3S}^2 - a_1 + a_4 - (a_3 - a_5)k_3 = 0$$

This equation has nonzero solutions in the form

$$x_{1S} = 0, x_{2S} = 0, x_{3S} = \pm\sqrt{k_3 + \frac{a_1 - a_4}{a_3 - a_5}} \quad (23)$$

3.2 *Research of stationary states (22) stability*

We investigate a stability of the system (20) in stationary points by the method of Lyapunov function. Lyapunov function ***V (x)*** sets in the form of a vector function ***V(V₁(x),..., Vₙ(x))*** then from geometric interpretation of the theorem of asymptotic stability, we will get (Barbashin 1967, Malkin 1966):

$$\frac{\partial V_1(x)}{\partial x_1} = 0; \quad \frac{\partial V_1(x)}{\partial x_2} = -x_2; \quad \frac{\partial V_1(x)}{\partial x_3} = 0;$$

$$\frac{\partial V_2(x)}{\partial x_1} = 0; \quad \frac{\partial V_2(x)}{\partial x_2} = a_2 x_2;$$

$$\frac{\partial V_2(x)}{\partial x_3} = -a_3 x_3^3 + (a_3 k_3 + a_1)x_3;$$

$$\frac{\partial V_3(x)}{\partial x_1} = 0; \quad \frac{\partial V_3(x)}{\partial x_2} = -x_2;$$

$$\frac{\partial V_3(x)}{\partial x_3} = a_5 x_3^3 - (a_5 k_3 + a_4)x_3$$

The full time derivative of the components of the vector Lyapunov function ***V (x)*** can be obtained from the state equation (20) using the scalar product of the components of the gradient of the vector Lyapunov function and the components of the state rate vector, i.e.

$$\frac{dV(x)}{dt} = -\sum_{i=1}^{n}\left(\sum_{j=1}^{n} \frac{\partial V_i(x)}{\partial x_j}\right)\frac{dx_i}{dt} = -(2 + a_2)x_2^2 - \\ -[a_3 x_3^3 - (a_3 k_3 + a_1)]x_3^2 - [-a_5 x_3^3 + (a_5 k_3 + a_4)]x_3^2 \quad (24)$$

From the expressions (24) follows, that the full time derivate from the Lyapunov function will be always negative.

On the basis of the Morse lemma, we will present Lyapunov function in a scalar form in the following view

$$V(x_1, x_2, x_3) = \frac{1}{2}(a_2 - 2)x_2^2 + \frac{1}{4}(a_5 - a_3)x_3^4 - \\ -\frac{1}{2}(a_3 k_3 + a_1 - a_5 k_3 - a_4)x_3^2$$

The conditions of (20) system stability in a steady state (22), we obtain, taking into account the negative definiteness of the functions (24) in the form of a system of inequalities

$$a_2 > 2, \; a_5 - a_3 > 0, \; k_3 + \frac{a_1 - a_4}{a_3 - a_5} < 0 \quad (25)$$

3.3 *Research of stationary states (23) stability*

The equations of system state (20) with respect to deviations of the stationary state (23) is written

$$\begin{cases} \dot{x}_1 = x_2 \\ \dot{x}_2 = -a_2 x_2 + a_3 x_3^3 + 3a_3\sqrt{k_3 + \frac{a_1 - a_4}{a_3 - a_5}}x_3^2 + 2(a_1 + a_3 k_3)x_3 \\ \dot{x}_3 = x_2 - a_5 x_3^3 - 3a_5\sqrt{k_3 + \frac{a_1 - a_4}{a_3 - a_5}}x_3^2 - 2(a_4 + a_5 k_3)x_3 \end{cases} \quad (26)$$

Full time derivative of the Lyapunov function ***V (x)*** with the equation of state (26), with respect to the stationary state (23), is defined as

$$\frac{dV(x)}{dt} = -(2 + a_2)x_2^2 - [a_3 x_3^3 + 3a_3\sqrt{k_3 + \frac{a_1 - a_4}{a_3 - a_5}}x_3^2 \\ +2(a_1 + a_3 k_3)x_3]^2 + \\ [a_5 x_3^3 + 3a_5\sqrt{k_3 + \frac{a_1 - a_4}{a_3 - a_5}}x_3^2 + 2(a_4 + a_5 k_3)x_3]^2 \quad (27)$$

From the expressions (27) follows, that the time derivative of the Lyapunov function will be a negative function. We find the gradient vector components from vector Lyapunov function.

$$\frac{\partial V_1(x)}{\partial x_1} = 0; \quad \frac{\partial V_1(x)}{\partial x_2} = -x_2; \quad \frac{\partial V_1(x)}{\partial x_3} = 0;$$

$$\frac{\partial V_2(x)}{\partial x_1} = 0; \quad \frac{\partial V_2(x)}{\partial x_2} = a_2 x_2;$$

$$\frac{\partial V_2(x)}{\partial x_3} = -a_3x_3^3 - 3a_3\sqrt{k_3 + \frac{a_1 - a_4}{a_3 - a_5}}x_3^2 - 2(a_1 + a_3k_3)x_3;$$

$$\frac{\partial V_3(x)}{\partial x_1} = 0; \frac{\partial V_3(x)}{\partial x_2} = -x_2;$$

$$\frac{\partial V_3(x)}{\partial x_3} = a_5x_3^3 + 3a_5\sqrt{k_3 + \frac{a_1 - a_4}{a_3 - a_5}}x_3^2 + 2(a_4 + a_5k_3)x_3;$$

On a gradient, we will construct Lyapunov's function

$$V(x) = \frac{1}{2}(a_2 - 2)x_2^2 + \frac{1}{4}(a_5 - a_3)x_3^4 + \\ +(a_5 - a_3)\sqrt{k_3 + \frac{a_1 - a_4}{a_3 - a_5}}x_3^3 + \\ +(a_5k_3 + a_4 - a_1 - a_3k_3)x_3^2 \tag{28}$$

By the Morse lemma, from the catastrophe theory, we can replace Lyapunov function (28) with a quadratic form

$$V(x) = \frac{1}{2}(a_2 - 2)x_2^2 + [(a_5 - a_3)k_3 + a_4 - a_1)x_3^2 \tag{29}$$

The condition of positive definiteness of Lyapunov function (28) or (29), we will get in a view

$$a_2 > 2 \quad k_3 + \frac{a_1 - a_4}{a_3 - a_5} > 0 \tag{30}$$

Hence, a necessary and sufficient condition for the stability of the stationary state (23) of (20) system is performance of an inequality (30).

4 CONCLUSION

Thus, the control system of aircraft motion with the increased potential of robust stability constructed in a class of one-parametric structural stable maps provides stability for changes of uncertain parameters of the system.

It appears, the steady state (22) is globally asymptotically stable when performing conditions (25). And unstable at violation of conditions (25). The stability of a steady state (23) requires performance of conditions (30). When $k_3+(a_1-a_4)/(a_3-a_5)=0$ there is a branching, and there are new steady branches.

In other words, branches (23) appear as a result of bifurcation while the steady state (22) loses stability, and these branches are steady. Stationary states (22) and (23) at the same time don't exist. It allows to increase the potential of robust stability of the system in the conditions of uncertainty of parameters.

REFERENCES

Andrievsky, B. R. & Fradkov, A.L. 1999. *The elected heads of the theory of automatic control with application in the Mathlab.* St. Petersburg: Nauka.

Ashimov, A.A. & Beisenbi, M.A. 2000. Robustness of the control systems and the structural-steady maps. *Reports of Kazakhstan Science Academy* 42(6): 28–32.

Barbashin, E.A. 1967. *Introduction in the theory of stability.* Moscow: Nauka.

Beisenbi, M. & Uskenbayeva, G. 2014. The New Approach of Design Robust Stability for Linear Control System. *Proceeding of the International Conference on Advances in Electronics and Electrical Technology, Bangkok, Thailand, 4–5 January 2014:* 11–18. Institute of Research engineers and Doctors.

Beisenbi, M. 1998. A construction of extremely robust stable control system. *Reports of Kazakhstan Science Academy* 42(1): 41–44.

Beisenbi, M.. & Yermekbayeva, J. 2013. The Research of the Robust Stability in Dynamical System. *Proceeding of the International Conference on Control, Engineering & Information Technology (CEIT`13), Sousse, Tunisia, 4–7 June 2013:* 142–147.

Beisenbi, M.A. & Abdrakhmanova, L. 2013. Research of dynamic properties of the control systems with increased potential of robust stability in a class of two-parameter structurally stable maps by Lyapunov function. *Proceeding of the International Conference on Computer, Network and Communication Engineering, 23–24 July 2013.* Atlantis Press.

Beisenbi, M. 2002. *Methods of increasing the potential of robust stability of control systems.* Kazakhstan, Astana: L.N. Gumilyov Eurasian University editorial office.

Beisenbi, M.A. & Erzhanov, B.A. 2002. *Control systems with the increased potential of robust stability.* Kazakhstan, Astana: L.N. Gumilyov Eurasian University editorial office.

Beisenbi, M.A. & Uskenbayeva, G.A. 2014. Research of Robust Stability of linear control systems with m inputs and n outputs by the method of A.M. Lyapunov functions. *Proceedings of the International conference on New Trends in Information and telecommunication technologists. Kazakhstan, Almaty, 20–21 February 2014*: 274–277.

Beisenbi, M.A. & Yermekbayeva, J.J. 2013. Creation of Lyapunov's function in research of robust stability of linear systems. *Scientific journal of Kazakh national technical university* 42(1): 315–320.

Beisenbi, M.A. 2011. *Models and methods of the system analysis and control of the determined chaos in economy.* Kazakhstan, Astana: L.N. Gumilyov Eurasian University editorial office.

Bukov, V.N. 1987. *The adaptive predicting control systems of the flights.* Moscow: Nauka.

Dorato P. & Rama K. Yedavalli. 1990. *Recent Advances in the Robust Control.* New York: IEEE press 3.

Gilmore, R. 1984. *Applied theory of the catastrophe.* Moscow: Mir.

Gregoire Nicols & Ilya Prigogine. 1989. *Exploring Complexity an Introduction.* New York: W.H. Freeman & Co.

Krasovsky, N. N. 1959. *Some tasks of the motion stability.* Moscow: Fizmatgiz.

Kuntsevich, V. M. 2006. Stability analysis and synthesis of stable control systems for a class of nonlinear time-varying systems. *Scientific journal of the Steklov Institute of Mathematics. 255(2): 93–102.*

Liao, X. & Yu, P. 2008. *Absolute stability of nonlinear control systems.* New York: Springer Science. Business Media B.V.

Loskutov, A. & Yu. Mikhaylov, A.S. 2007. *Foundation of the theory of difficult systems.* Izhevsk: Institute of computer researches.

Malkin, I.G. 1966. *The theory of stability of motion. 2-d edition.* Moscow: Nauka.

Polyak, B. & Scherbakov, P. 2002. *Robust Stability and Control.* Moscow: Nauka.

Poston, T. & Stewart, E. 2001. *Theory of catastrophe and its applications (New edition).* Dover Publications Inc.

Voronov, A.A. & Matrosov, V.M. 1987. *Method of vector Lyapunov functions in the stability theory.* Moscow: Nauka.

Electronics, Communications and Networks IV – Hussain & Ivanovic (eds)
© 2015 Taylor & Francis Group, London, ISBN: 978-1-138-02830-2

Virtual sailing master

L. Bortoni-Anzures*, C. Calles-Arriaga, M. Hernández-Ordoñez & Y. Padilla-Moreno
Politechnic University of Victoria

ABSTRACT: This paper does an inventory of the navigation systems under development which promise to be key systems in the following decades. But the main aim of the article is to add some complementary tools of the spot light to the navigation systems to assure they arrive at the port, based on information technologies and social structures. Weather information, conditions of the infrastructure and social movements may be sources of efforts to improve a mature technology. The Global Positioning System shows you the way to get there, but the virtual sailing master shows you how to get there.

1 INTRODUCTION

Undoubtedly, the technology related to navigation assistance has changed much since the days of compasses and sextants in the old medieval vessels compared with the currently employed equipment. Thanks to global positioning system, the development of complex electronic equipment and information management, today's technology is within the reach of most people, from carriers in huge land, sea or air to traveling families by their vehicles and even tourists walking with their cell phones in almost any city on this planet.

In 1957, to study the Doppler Effect, the Soviet Union launched the first artificial satellite, the Sputnik 1, in orbit. Scientists noticed that the Doppler Effect could be used to track an object on the earth's surface, and with that in mind, the U.S. Army began the development of a system called TRANSIT, operated in 1964, and was commercially available in 1967. The system was capable of updating an object position every 40 minutes with certain accuracy. By those decades, atomic incorporated to the satellites watches the increasing precision. (Painter 2013).

Later, in 1973, the U.S. Air Force joined efforts with the Army to deploy the Navigation Technology Program (or NAVSTAR GPS), consisting of a new satellite network with codified signals and a pseudo-random noise, fully functional in April 1995.

As to today, twenty four satellites have been distributed in six orbits at 20200 km over the earth surface. The system has evolved to provide an accuracy of 2.5 to 3 meters, whose rate reaches 95%. By 2017 it is expected to reach GPS3 standards allowing 1 meter precision. (Gurko 2013).

Currently, several world teams (Table 1) are developing their own systems, which will collaborate with the GPS to increase accuracy in certain areas.

Table 1. Current known navigation systems under development.

System	Origin
Global Navigation Satellite System (GLONASS)	Russia
BeiDou or Compass	China
GALILEO	European Community
Indian Regional Navigation Satellite System (IRNSS)	India
Quasi-Zenith Satellite System (QZSS)	Japan

All of these systems focus on improving accuracy and reducing the effect of the error sources, like ionosphere, stratosphere interference, physical obstacles (like mountains or buildings) and numerical errors, which are mentioned little.

2 FOLLOWING THE MAP

Going to a specific place in an unknown city is no longer a problem. Usually, every few miles, the visiting driver has to ask for directions in a gas station or a convenience store. But now it is possible to have a recommendable route from origin place to the final destination. Even if the driver deviates from the suggested route, the system is able to calculate an alternative route (Figure 1).

*Corresponding author: *lbortoni@upv.edu.mx*

Most of the time, there is enough information to get to the destination, but there are some factors that could make the task difficult to accomplish.

Figure 1. Typical GPS screen format.

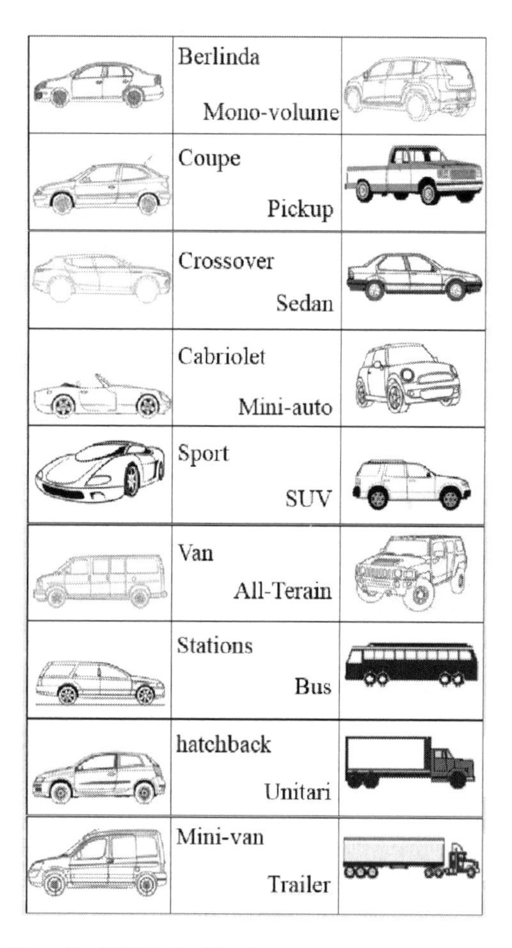

Figure 2. Vehicle classification.

3 FACTORS TO CONSIDER

3.1 *Vehicular configuration*

It is not the same to drive a commercial truck and a trailer or a mini-compact vehicle. The dimensions of the vehicle, weight, weight distribution, and location of the center of gravity, number of passengers, pay load type of suspension, suspended mass, and other factors could make an important difference in order to take a turn at a safe speed, to deal with a road bump, or to cross a bridge. Figure 2 shows the common classification of terrestrial vehicles. Figure 3 represents a simple vehicular model. (Bortoni-Anzures 2009).

And it is not only about safe considerations, local police regulations may apply for some vehicles as well.

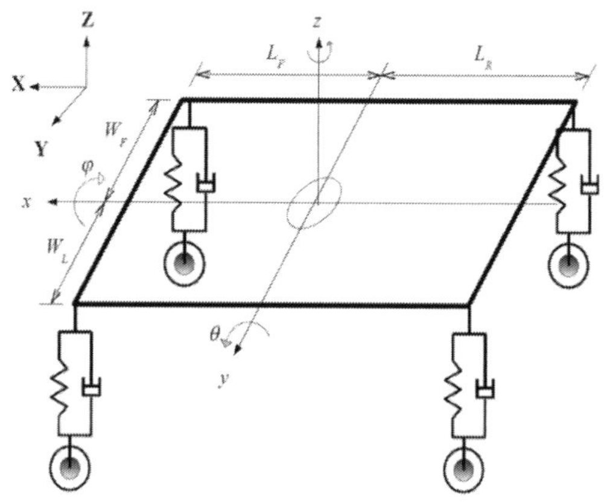

Figure 3. Idealized car suspension.

3.2 *Infrastructure*

All countries have some dangerous spots in their cities or road networks, which are well known by locals, but not for new travelers. Closed turns, damaged pavements and visual obstacles are sources of emergency maneuvers causing a simply unpleasant surprise in the best case scenario, but important damages to vehicles and passengers may also occur. This information could be included in the database to warn the driver on time with minimum computational effort, and the participation of local authorities. Figures 4 to 5 show different infrastructure factors. (Ibtishamiah et al, 2003).

Other simple examples are wedge-plateau bumps in tortuous areas, or road designs without any consideration of the vehicular needs. Most of the factors are included as infrastructure remains in time, so it is important to let the drivers know about them.

Figure 4. Stelvio valtellina, at adigio's valley, italy. Elfinanciero. 2013.

Figure 5. Blind turn in a damaged road. Marrakech, about 20 minutes from col du tichka. Pictured by kevin kl. 2007.

3.3 *Temporary detours*

Other different sources of inconveniences while driving are not permanent, and it is not possible to know their impact or duration.

They depend on two main causes, people's behavior and weather conditions. Figure 6 shows a scene of protestors blocking an important street while they march to the main square plaza. Vehicles have to wait for hours before being able to come out of there. Similar effects are encountered in some labor days, festivities, parades, cultural events, temporal flea market's or sport final-score celebrations.

Sometimes things go out of control, but normally all of these events follow an authorized schedule. So transit authorities can update the data base daily to warn the drivers to avoid those areas, and simply the returning to normal conditions after the event.

Blocking streets is a popular way for protestors. Organizations like unions, peasants, students, political parties, organize an average of 3000 events every year freezing the traffic in Mexico City by blocking about 25,000 streets. (Anonymous 2012).

Eventually, infrastructure needs maintenance. But not only pavements of road surface, but also many city services that run along the streets, like electricity, telephone, tap water and drain, contribute to this endless list, not to mention vehicle accidents. Some are incidentals, but many of them are scheduled and could be advised to drivers in order to take alternative routes.

Figure 6. Down town at mexico city, one of the blocked streets by 15,000 protestors. Elinformador. 2013.

3.4 *Weather*

Detours also occur because of weather events like snowfall and tsunami, but more often for the rain visibility difficulty. Regularly when the rain remains for a considerable time, water will block a few streets or road segments. As a result, geographical configurations may make inner-city lagoons and rivers not suitable for all vehicles to deal with.

Recursively, after about an hour of continuous rain, those areas are flooded, which will remain 45 minutes until the last rain drop. Every local citizen is aware of these areas, but not all car drivers, and not to mention the tourists.

Other examples are tornados and hurricanes, which could cause significant damage to infrastructure. Figure 7 shows a low crossover at the San Marcos River, totally dry all year long, but commonly flooded on rainy days. During "Irene Hurricane" the river overflowed its banks for several days, destroying the pavement.

It took several months for this crossover to be re-opened to traffic, and this history is not exclusive for this town. After two years, New Orleans, USA, has similar consequences of the Katrina hurricane.

As the bad-designed infrastructure remains, situations like Figure 8 have been recurrent for a hundred times. Every year the government limits itself just to fix the damages, rather than build a bridge as it should have done a long time ago.

Cities continue to grow, not always in accordance with a plan. Temporary solutions look like the construction philosophy; it is very frequent to apply a just-for-now fix that at the end remains for decades.

Due to the lack of infrastructure maintenance, climatic change, economic restrictions and many other reasons, road networks are not expanded to keep pace with the demand.

According to the World Health Organization and the National Advice for the Prevention of Accidents, Mexico ranks the 7th in the number of vehicular accidents among countries. 17,000 people die and 50 million people get injured in Mexico's cities, and the number of death reaches 1.3 million per year in road accidents.

Figure 7 & 8. Victoria City's," Av. Rodriguez Inurrigarro" Low crossover on San Marcos River on a normal day (7), and just after the "Irene" hurricane (8).

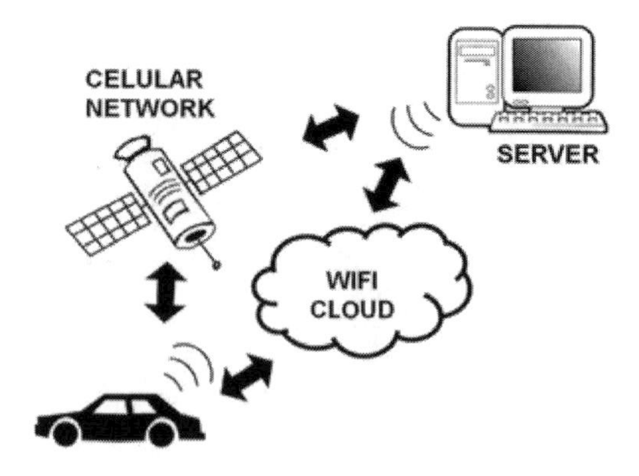

Figure 9. The communication between vehicle and server may use WIFI or cellular network to transfer voice or data information.

Figure 10. Warning signals examples.

4 ABOUT THE SAILING MASTER

4.1 *Proposed smart phone application*

Today's mobile phones have reached high levels of computer power, mass storage capacity. Thanks to the cellular service carriers, an extensive area of the planet is covered with enough signal power to call or transfer information worldwide. (Figure 9).

Due to the long history on the evolution of these devices, there are countless numbers of cell phone platforms. But today, Android and Apple dominate the market, so it is relatively simple to offer this new application to a wide number of users. Currently, the application is under development, so a dummy is presented in its place.

4.2 *Getting started*

Once the application is activated, the phone's GPS will locate current vehicle location. Then, it will download all the possible incidents in a hundred kilometer radius, ensuring enough information to navigate for at least one hour.

The database records with the minimum possible weight reduce data transfer, change the GPS position, incident type, severity (where apply), initial date-time, final date-time plus a comment on the incident. Figure 10 shows an example of a few incident identifications.

4.3 *Standby mode*

During the time, it seems that there are no obstacles ahead. The application will run in the background, so the phone is able to run any other applications, or go to the rest mode to save battery.

Until a flagged GPS location approaches our current path, the phone will wake up and resume the application.

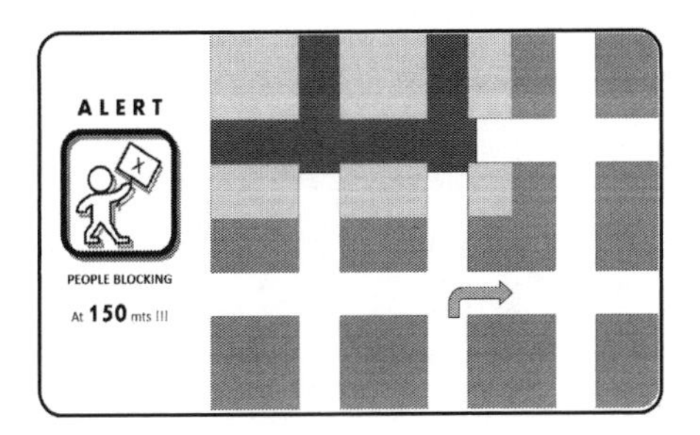

Figure 11. Application's print-screen. On a "People blocking ahead" alert.

4.4 *Situation approaching*

When an incident gets in our 500-meter horizon, the application will enter the active mode, to warn us of the type of incident and the distance to it.

Figure 11 shows the incident alert where blue blocks indicate non-incident area (normal), yellow blocks indicate preventive area (warning), and the red ones refer to the directly affected area (closed). The driver is given enough time to take evasive actions.

Depending on the type of incident, more information could be useful like an event description or a transit authority recommendation, until the obstacle is overcome allowing the device to return to the standby mode. Until getting close to the next incident on the database, an update of the current weather condition on the surrounding area could be useful.

4.5 *Information update*

As can be observed, the system is not too complex, but it has its Aquiles heel. In order to develop the system, it is necessary to have veracity and accuracy on the information database.

Who will be responsible for keeping the information up-to-date properly in order to have only real world-wide records on the system? It is necessary to involve the Bureau of Tourism, Transit Authorities, and Civil Protection Organizations in every country and make them compromise, because getting funds to hire people all over the world seems like a non-feasible solution.

5 CONCLUSION

The automobile is an invention that will never terminate. For more than a century, cars have been everywhere. But each time technology evolves, and new devices and gizmos are invented, they try to get incorporated with the vehicles.

Radio, Television, Computers, improved engines, brakes, tires, materials, GPS, etc. and TI technologies are not different, and the smartphones are mature enough to open the door to a new set of commodities and advantages.

Better communication networks, more affordable service and equipment with tremendous advance in capabilities now make it possible to incorporate new concepts, like infotainment (hardware/software products and systems which provide information-based media content or programming also including entertainment content in an effort to enhance popularity with audiences and consumers).

The Virtual Sailing Master is an effort to complement GPS navigation with local information that could reduce the risk of an accident, or at least reduce the disadvantage of getting caught in traffic.

REFERENCES

Anonymous, 2012. "Wisdom of stairs" How many riots are in a day in the city of Mexico? Mexico *http://www.sabiduriadeescalera.com/?p=3475.*

Bortoni-Anzures & Liborio. 2009. *Multiplatform Instrumentation for Ride Analysis*. Vehicle Dynamics Expo. Messe Stuttgart, Germany.

Gurko, Alexander, 2013. The GLONASS system also can have accuracy. RIA NOVOSTI. Russia. *http://in.rbth.com/.*

Kristen Leigh Painter 2013. Colorado at center of new global positioning technology GPS III. The Denver Post. *http://www.denverpost.com/ci_23320394/colorado-at-center-new-global-positioning-technology-gps.*

Nik Ibtishamiah Ibrahim, Wan Reezal ARIF & Mohamed Rehan Karim 2003. "Road humps as traffic calming devices" *Proceedings of the Eastern Asia Society for Transportation Studies* 4: 1435–1441.

Electronics, Communications and Networks IV – Hussain & Ivanovic (eds)
© 2015 Taylor & Francis Group, London, ISBN: 978-1-138-02830-2

Preliminary design and trial of data reception and transformation system for the East China Sea Seafloor Observation System

Hui Chen, Huiping Xu*, Yang Yu, Rufu Qin, Changwei Xu & Huizi Dong
State Key Laboratory of Marine Geology, Tongji University, Shanghai, China

ABSTRACT: One of the issues concerning undersea observation is data reception and transformation characterized by multi-disciplinary, multi-parameter and weather-independent continuous observations. This paper introduces the overall architecture of the East China Sea Seafloor Observation System (ECSSOS), and preliminary designs a data reception and transformation system for ECSSOS and realizes a prototype. Based on C/S architecture, the system establishes a communication between infrastructure and the observation center using Socket technology. The system for ECSSOS can be divided into four parts called data reception and transmission, data transformation, data storage and data display. And it enables the real-time interaction among the junction box, observation center and users. Through the specific data transformation model, raw data can be converted to standardized scientific data which is the foundation for research. Given the successful trial in Xiaoqushan from August 2013, the solution proposed in this paper could be a reference to the following construction phases of ECSSOS.

1 INTRODUCTION

Seafloor observatory characterized by long-term, continuous and weather-independent observations have been the important technical means to monitor marine environment, submarine earthquake and global climate change. The basic unit is a seafloor observation station composed of various facilities. For example, GEOSTAR mainly concludes an Active Docker, a bottom station, communication system and scientific payloads in Europe. The Active Docker is mainly used to install the bottom station and the communication system, and the communication system is used to transmit data. These represent a new mode for deep sea observation technology (Beranzoli et al. 1998). At present, there are several seafloor observatories in operation or under construction. Victoria Experimental Network Under the Sea (VENUS) is a submarine cable observation network whose observation depth in Canada. It consists of two parts: a 3 km cable node has been set up at a depth of 96 m in the bay of Saanich in 2006; in 2008, two bottom stations in at the depth of 17 m and 300 m respectively in the Georgia Strait has been constructed (Dewey & Tunnicliffe 2003). The Network Operation Center of VENUS remotely controls the whole system by cable, and the data processed little at the shore station will be transmitted to the Data Management and Archive System in Victoria University by Internet, then which can be accessed by users (Dewey et al. 2007). Monterey Accelerated Research System (MARS), built in August 2008, is the first submarine cable connected observation network in the North America, which is located in the Monterey Bay of the central California. Neptune Canada is currently the world's largest submarine observation network put into use from December 8, 2009, whose cable is about 800km. It has 5 nodes from the depth of 17m to 2600m and monitors the north of Juan de Fuca plate in North Pacific (Barnes 2009, Martin 2009). In general, seafloor observatory is characterized by observation platform deep in the seabed and networked energy supply and information transformation.

The Xiaoqushan Coastal seafloor observatory, located between 30°31′44″N, 122°15′12″E and 30°31′34″N, 122°14′40″E, was constructed near the Xiaoqushan Island outside the Hangzhou Bay on the inner continental shelf of the East China Sea in April 2009. It was the first seafloor observatory in China and has been running successfully for more than 2000 days (Xu et al. 2011). The East China Sea shelf is the focus of sea-land interaction. It means the interaction on the East China Sea shelf between the Yangtze River diluted water and the Kuroshio / Taiwan warm current. This region is also by control of seasonal and annual variability of monsoon climate. The overall pattern of the region changes from northwest to southeast gradually and water structure

*Corresponding author: *xuhuiping@tongji.edu.cn*

will stratify by season, which thereby affects various ocean processes of sedimentation, chemistry and biology. ECSSOS will be constructed from the inner shelf of the northeast corner of the East China Sea and outside the Yangtze River Estuary. It will carry out observation which centers on not only earthquake and tsunami, but also marine hydrology, sediment transport and biogeochemistry (Li et al. 2002). Therefore, a data reception and transformation system needs to be developed for ECSSOS to receive the data collected by the instruments. The system will realize data interpretation, storage and display in a real time manner to ensure the whole network's successful operation.

In this paper, we first propose an initial concept for the overall architecture of ECSSOS based on the Xiaoqushan Coastal seafloor observatory. Secondly, we expound the software framework of the data reception and transformation system for ECSSOS and discuss the main functions, then test the prototype in the Xiaoqushan Coastal seafloor observatory. Finally, we summarize and evaluate the overall architecture and the practical performance of this system.

2 OVERALL ARCHITECTURE OF ECSSOS

The overall architecture of ECSSOS mainly concludes science nodes and infrastructure, shore station and observation center. The science nodes and infrastructure present a main junction box and several secondary junction boxes and various instruments connected to them like ADCP, CTD, HD camera. An optical cable enables energy transmission and data communication between the main junction box and secondary junction boxes. The shore station is composed of a power supply device, a data storage device and a command repeater. The electric energy is continuously transported to the main junction box by the optical cable, and the data are transmitted to the storage device and the observation center via the optical cable. The remote control system, data reception and transformation system and monitoring system compose the observation center. Commands from the remote control system are first sent to the main junction box via the shore station and then assigned to the corresponding secondary junction boxes and observation sensors. The raw data and infrastructure information gathered from undersea observatory facilities are transmitted to the data reception and transformation system via the shore station and processed and stored in the database. The monitoring system mainly monitors the operation status of the whole system and sends messages to the remote control system and manager to adjust the commands. The overall architecture of ECSSOS is shown in the figure below:

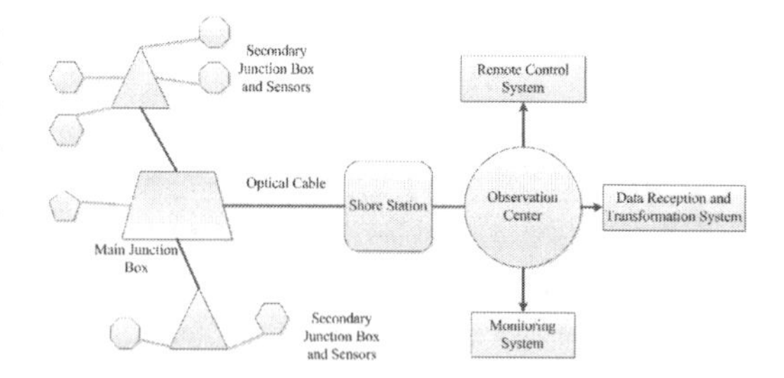

Figure 1. Overall architecture of ECSSOS.

As shown in Figure 1, three function nodes can greatly guarantee the normal operation of ECSSOS. The junction box has 3 functions: 1) the main junction box converts the power from high voltage to medium or low voltage; then the converted power is allocated to the secondary junction boxes or sensors to ensure their normal operation according to the power requirements of the corresponding ports they are connected to in the junction box; 2) send the commands from the remote control system to the corresponding junction boxes or observation sensors via IP and serial ports; 3) and by the same time transmit the data collected from various sensors and the infrastructure information gathered from undersea observatory facilities to the data reception and transformation system. The shore station acts as the intermediate link between the junction boxes and the observation center. The raw data is first stored here to avoid data loss caused by faults on the specific cable or observation center. The existence of the shore station further guarantees the power supply and data security. The observation center is the control terminal and the data center of the observation network.

In the observation center, the data reception and transformation system, as the core module, makes the major parts of the observation networks form a whole from the data aspect (shown in the Fig. 2). And the raw data collected by undersea observation instruments arrive to the data reception and transformation system via the secondary junction boxes and the main junction box (red stream). The data reception and transformation system stores the scientific data that interpreted from the raw data in the database. The monitoring system obtains the interpreted data from the database to make visualizations. Moreover, it will notify the managers if there is data beyond the threshold or other abnormal conditions, and meantime the message will be sent to the remote control system and the data reception and transformation system (green stream). Commands according to the appropriate messages from the remote control system are first sent to the main junction box, then received and processed in the secondary junction box, and lastly assigned to the corresponding observation sensors via serial ports (blue stream). On the other hand, the data reception and transformation system mainly checks whether the interpreting processes execute correctly or not when receiving the backing information from the monitoring system. In terms of the scientific objectives of the observation networks, the primary purpose is to do some researches on the seismic observation, tsunami warning, oceanographic, sedimentary transportation and biogeochemistry, so it is necessary to analyze the scientific data from the observation network. Similarly, the remote control system and the monitoring system also need to access the scientific data to make control and judgments on operational status of observation network. The data reception and transformation system can receive the raw data and convert to the standardized scientific data for post use, so its role is important in ECSSOS.

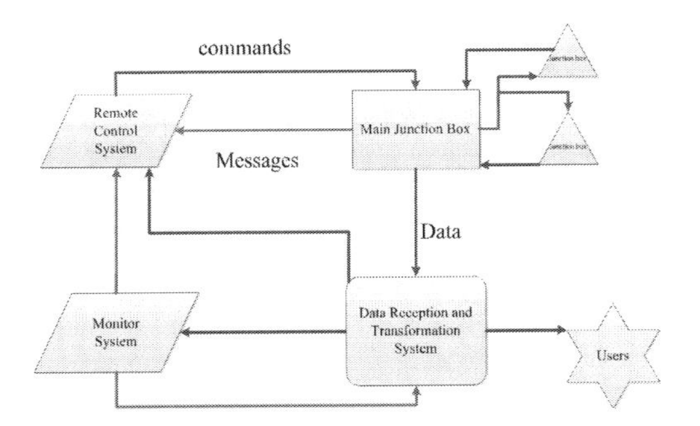

Figure 2. Role of the data reception and transformation system.

3 DESIGN AND TRIAL OF THE DATA RECEPTION AND TRANSFORMATION SYSTEM

From the previous discussion, we can see the significance of the data reception and transformation system in ECSSOS. This section focuses on the basic components of the system and the realization of the basic functions. Based on the design of C/S architecture, the overall system establishes a communication link between the observation infrastructure and the observation center using Socket technology. The data reception and transformation system enables multi-source and heterogeneous data to be received, interpreted, stored and displayed in a real time manner. The interpreted data can be free to users in accordance with their corresponding rights.

3.1 *Architecture design*

The data reception and transformation system concludes the data reception and transmission subsystem, data transformation subsystem, data storage subsystem and desktop and web display subsystem (shown in the Fig. 3). The raw data collected by the observation sensors connected to the junction boxes (green stream) are first received by the data reception and transmission subsystem, and then sent to the secondary data receiving system and the data transformation subsystem. The raw data is interpreted in the data transformation system and the standardized scientific data (blue stream) are stored in the database via the storage module. The desktop and web display subsystem visualize the scientific data by accessing the database. All these processes are accomplished in a real time manner which is the characteristic of ECSSOS. Besides, the scientific data can be accessed by users and the external systems like the remote control system and the monitoring system.

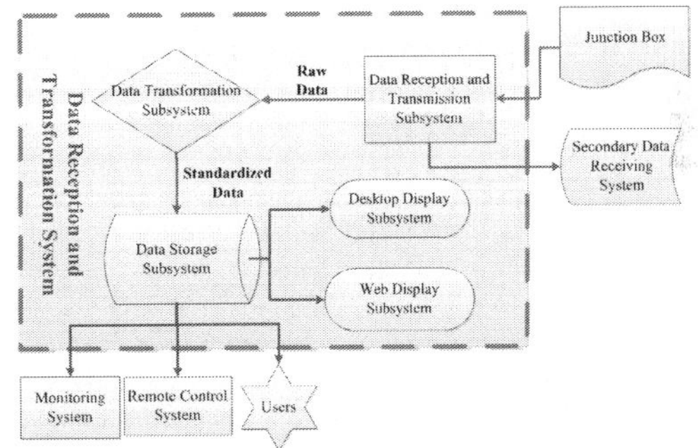

Figure 3. Overall framework of the data reception and transformation system.

We can easily find that the data reception and transformation system are the important bridge between the data occurrence end and the data application end as the raw data are converted to the standardized scientific data. The relationship between this system and the other equipment can be considered as three layers of C/S architecture. Strong interaction capabilities are a major advantage of a C/S model. And a C/S architecture model has a comprehensive client application that provides robust error message prompts. The C/S architecture can effectively assign tasks to the client and server, which will promise to not only reduce the communication overhead between them, but also make full use of the hardware at each end. Therefore, this architecture has strong data manipulation and transaction processing capabilities. The security and integrity of information will be guaranteed by its peer-to-peer structure mode, which is applicable to a more secure protocol network. In the first layer, the client is the data reception and transmission system, and the server is the junction boxes and the instruments connected to them (data occurrence end). Meanwhile, the data occurrence end connects with another client called the remote control system (Yu et al. 2012). For the second-layer architecture, the client is the secondary receiver terminal for the different testing user; the server is the computer in the observation center (the data reception and transmission system). This server is able to establish a communication link with several clients. The different client has different rights so it can acquire the corresponding raw data. The users and the external systems can access the standardized scientific data via the data storage system (the server). This is the third layer of C/S architecture. Therefore, the data reception and transformation system is not only the client but also the server. This architecture makes the system become a bridge in ECSSOS.

3.2 *Functions realization*

As shown in Figure. 3, the functions of the data reception and transformation system consist of real-time reception and transmission, real-time interpretation, data storage and management and real-time visualization. The real-time reception and transmission module and real-time interpretation module are the cores.

The data reception and transmission subsystem establish a communication link with the junction boxes and the secondary receiving terminal based on TCP/IP. The observation data can be received and transmitted between these systems via unique IP and port, according to the established socket communication. Through multithreading mode, this module can transmit the raw data to secondary receiving systems at the same time in a real time way. It can automatically record the observation data, the running state of the system operation and the status of the data transmission. These data will be stored in the computer according to the default format and path. This will provide the basis for the following data backtracking and the program running state examination.

The core of the data transformation subsystem is the data interpretation module which is the most complex one. The data are obtained from the data reception subsystem instead of the data file in the data interpretation procedure. The advantages are as follows: 1) avoid the fault caused by written errors of the data reception and transmission subsystem; 2) reduce the intermediate segment of data storage and read and memory overhead to improve the real-time data interpretation. Thus, this effectively improves the execution efficiency of the program and it also ensures the real time of the entire observation network. Characterized by multi-source, heterogeneous observation data, each type of data will need to have a specific procedure for interpretation, so the repeatability is very low. ECSSOS needs to embrace different types of observation sensors, which are not known in advance and so does the data. Thus, after the data reception and transformation system has been put into use, the data interpretation and storage module should make appropriate changes if the observation equipment is replaced. We propose the concept of "Hot Swapping" interpretation during the data transformation subsystem design, which means that the data reception and transmission and data display module should not be influenced in the process of changing the data interpretation procedure. Moreover, "Hot Swapping" process does not interfere with the other instruments' interpretation and storage. This is like a U disk that can be hot-swappable, but has no other effects on the computer and itself. This module invokes the appropriate interpretation program, according to the specific identifier.

The main function of the data storage subsystem is to store the standardized scientific data to the database which requires to be checked first. The test contents mainly include data repeatability, data length and anomaly field. Meanwhile, the subsystem will manage the scientific data in the database. According to the different users and external systems, it will give them appropriate rights to access the data. The subsystem needs to check the integrity of the data as a result of accidental factors which may lead omission or error data entry. This function will access and interpret the data file written by the data reception and transmission subsystem and compare the data in the database. If the data have not been stored, this module will enter it. The data display subsystem establishes a communication link with the database and gets the scientific data to visualize it in desktop end and web end. This module not only adds in the data reception and transformation system but also operates alone.

3.3 *C. Practical performance*

The data receiving system for Xiaoqushan Submarine Comprehensive Observation and Marine Equipment Test Platform is the prototype of the data reception and transformation system. Since August 11, 2013, the data receiving system has been put into use and working continuously for more than one year. The system has achieved the functions of data receiving, interpretation, storage and display in real time, and historical data download, display and so on. As shown in Figure 4, the lasted, received data packets will show as the scroll of the interface, the contents of the packets can be viewed when double-clicking on them. In the window of the data interpretation, we can see the dynamic libraries added to the system and load, unload and exchange them. Xiaoqushan Submarine Comprehensive Observation and Marine Equipment Test Platform was under maintenance in April 2014 and replaced the "Hydrolab" instrument with the PH instrument. Accordingly, the system has been upgraded to interpret the new data through "Hot

Swapping". The replacement process has no effect on data reception, which achieves the desired objective. The prototype system has realized two core modules called the data reception and transmission and the data interpretation. It lays the foundation of the data reception and transformation system proposed in this paper.

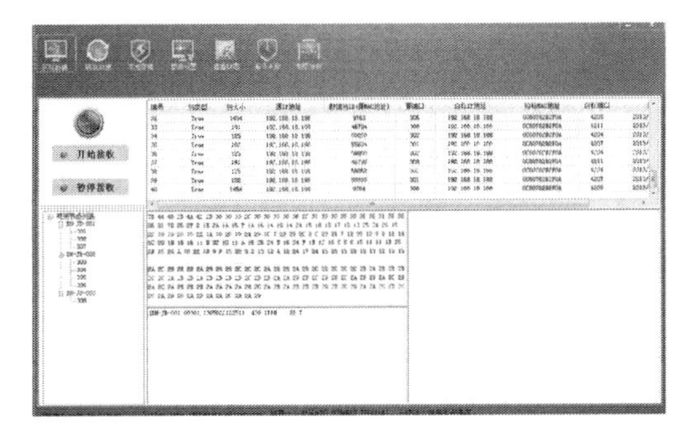

Figure 4. The main interface of the prototype system.

4 CONCLUSION

The data reception and transformation system for ECSSOS establishes a communication link between the junction boxes and users which just likes a bridge. The standardized scientific data provide the foundation for the researches on marine hydrology, sediment transport and biogeochemistry. The overall system adopts C/S architecture and makes the junction box, secondary receiving system and users become an entire network via socket technology. Based on this network, the observation data can be received and transmitted in real time.

This paper mainly introduces the architecture of ECSSOS which concludes the junction boxes and the observation sensors, the shore station and the observation center. There are remote control system, data reception and transformation system and monitoring system in the observation center. The remote control system can change the state of the equipment by sending commands to the junction box and the sensors connected to it. The data reception and transformation system receives and transmits the raw data collected by the observation sensors and the interpreted data will be stored in a database. At the same time, the desktop and the web display subsystem will access the database and visualize the scientific data in real time. The monitoring system can visualize and monitor the status of the junction boxes and sensors by accessing the database. If there is abnormal situation, it will return the messages to the remote control system, the data reception and transformation system and managers by e-mail and short message. The data reception and transformation system is the core system because of the data. This system consists of the data reception and transmission subsystem, data transformation subsystem, data storage subsystem and data display subsystem, which is not only the client but also the server. The main function is to receive the raw data and transform it via specific identifier. The prototype, the data receiving system for Xiaoqushan Submarine Comprehensive Observation and Marine Equipment Test Platform, has completed the most of the prospected functions.

ACKNOWLEDGMENT

The authors feel grateful to other members of our research group and friends for their help in designing and developing the Data Reception and Transformation System for ECSSOS. The study for this paper was funded by the National High Technology Research and Development Program of China (863 Program) (2012AA09A407), 2011 Research Program of Science and Technology Commission of Shanghai (11DZ1205500), and Charity Project of State Oceanic Administration (201105030-2).

REFERENCES

Barnes, C. 2009. Transforming the ocean sciences through cabled observatories. *Aerospace conference, Big Sky Resort, 7–14 March 2009. Piscataway: IEEE.*

Beranzoli, L., De Santis, A., Etiope, G., et al. 1998. GEOSTAR: A geophysical and oceanographic station for abyssal research. *Physics of the earth and planetary interiors* 108(2): 175–183.

Dewey, R, et al. 2007. The VENUS cabled observatory: Engineering meets science on the seafloor. *OCEANS 2007, Aberdeen, 2007. Piscataway: IEEE.* Li, D., et al. 2002. Oxygen depletion off the Changjiang Estuary. Science in China. *Earth Science 45*: 1137–1146.

Dewey, R. & Tunnicliffe, V. 2003. VENUS: Future science on a coastal mid-depth observatory. *Scientific Use of Submarine Cables and Related Technologies, New York, 2003. Piscataway: IEEE.*

Martin T.S. 2009. Transformative ocean science through the VENUS and NEPTUNE Canada ocean observing systems. *Nuclear Instruments and Methods in Physics Research Section A: Accelerators, Spectrometers, Detectors and Associated Equipment* 602(1): 63–67.

Xu, H., et al. 2011. Coastal seafloor observatory at Xiaoqushan in the East China Sea. *Chinese Science Bulletin* 56(26): 2839–2845.

Yu, Y., et al. 2012. A study of the remote control for the East China Sea Seafloor Observation System. *Journal of Atmospheric and Oceanic Technology* 29(8): 1149–1158.

Electronics, Communications and Networks IV – Hussain & Ivanovic (eds)
© 2015 Taylor & Francis Group, London, ISBN: 978-1-138-02830-2

Data calibration and dimensionality reduction for health care system

Kemeng Chen* & Janet M. Roveda
Department of Electrical and Computer Engineering, The University of Arizona, Tucson, Arizona, USA

Richard D. Lane
Department of Psychiatry, The University of Arizona, Tucson, Arizona, USA

ABSTRACT: Wearable technology and mobile platform are becoming more and more popular in clinical context, such as health care or monitoring and a lot of research has been conducted on data processing algorithms. However, very little work has been done on real time feedback data calibration which is critical for training purpose work. In addition, as raw data have been usually noisy, real time data calibration become more and more difficult. Therefore, this paper discusses a Binary Encoded Symbolic Representation (BESR) based calibration algorithm to provide feedback in real time. This new algorithm employs a new pruning method using a Distance Filter with Starting Symbol Requirement (DFSSR) to provide real time feedback. We apply the proposed work on breathing exercise training. Experimental results show that the BESR reduces storage to a factor of 160*X* compared with raw data storage. The new calibration algorithm achieves time complexity *O(1)* for each sample which makes it suitable for real time application.

1 INTRODUCTION

Wearable technology has been widely used in clinical context, such as disorder detection, treatment efficiency assessment, home rehabilitation and other health care research (Patel et al. 2012). Such systems usually collect data from sensors and transmit data to mobile devices via Bluetooth to process (i.e. monitoring, feature extraction). Those data collected from sensor are defined as time-series (Esling et al. 2012). Time series processing algorithms have been studied by many researchers. In (Esling et al. 2012) many time-series processing algorithms have been discussed, including clustering (Lin & Keogh 2005), classification (Xi et al. 2006), segmentation (Keogh et al. 2003), etc. However, very little effort has been put on data calibration especially on real time calibration feedback. Therefore, this paper focuses on fast time-series data calibration algorithm which could be applied to training and provides real time feedbacks. Since raw data is often huge in amount and contains a lot of noise, it may not be the optimal representation for both data process and storage. Thus, we also include a binary encoded symbolic representation based on which we build the proposed real time feedback calibration algorithm.

The contribution of the paper can be summarized as follows: 1) propose a fast time-series calibration algorithm which provides real time feedback; 2) introduce a binary encoded symbolic data representation which is highly storage efficient and noise tolerant to some extent.

The rest of the paper is organized as follows: section 2 introduces the goal of calibration and also defines the mathematical model for time-series data. Section 3 discusses the introduced BESR to save storage and the proposed DFSSR algorithm using examples. Section 4 provides experimental results to demonstrate the effectiveness of the proposed algorithms. Finally, the paper concludes in section 5.

2 CALIBRATION GOAL AND DATA MODEL

This section discusses the goal of calibration and introduces the mathematical model for time series data from sensors. The calibration intends to solve two problems: 1) given a pre-defined pattern how much does the actual data deviate from the pre-defined pattern; 2) how to guide the actual data to follow the pre-defined pattern as close as possible in real time. For example, 4-4-6-2 breathing exercise (Brown & Gerbarg 2012) requires trainees to breathe strictly following a pre-defined pattern. Then, the calibration is responsible to tell if the actual breathing waveform matches the pre-defined pattern, faster or slower than pre-defined pattern and provide instantaneous feedback to help trainees follow the breathing exercise. The breathing exercise can activate the vague nerve, which promotes a relaxation response to help reduce mental stress (Brown & Gerbarg 2012). It is also critical to know how well the trainee is adhering to the prescribed breathing pattern. Figure 1 describes the 4-4-6-2 breathing pattern,

*Corresponding author: *kemengchen@email.arizona.edu*

which consists of four stages: “Inhalation”, “Hold1”, “Exhalation” and “Hold2”. The time duration of each stage is 4:4:6:2 (counts). The horizontal axis is the time coordinate and vertical axis is the breathing depth. We will continue using this example in the rest of the paper to discuss the proposed work.

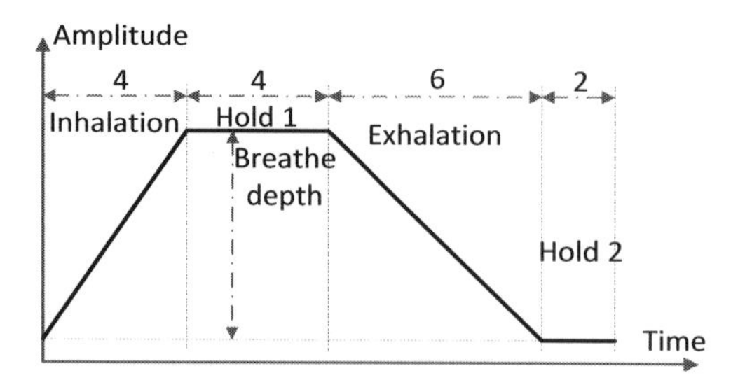

Figure 1. Waveform of pre-defined breathing exercise pattern.

The sensory data is modeled as time series. Let us start with this definition of time-series.

Definition 1: Time series is a set of samples collected at different time point. Each entry contains a real variable s_i to denote sample data and a time stamp t_i to denote time stamp. We can write a time series as:

$$A = \{(t_1, s_1), \cdots, (t_N, s_N)\}, s_i \in \Re, 1 \le i \le N \tag{1}$$

where A is a time series. It includes N entries which is an ordered sequence of time instants whose interval is determined by sampling rate fs. Thus, relationship of sampling rate and the time stamp is

$$\frac{1}{fs} = \Delta t = (t_{i+1} - t_i), 1 \le i \le N-1 \tag{2}$$

Real time attribute refers as each sample s_i in A is only visible during $t_i+\Delta t$. Figure 2 is an example of respiratory waveform data from sensors. The horizontal axis represents time coordinate and vertical axis the amplitude. Each point in this respiratory waveform is an entry of a time series as (ti,s_i). Follow **Definition 1,** for example, we pick up two successive samples (102, 9) and (103, 9) in Figure 2.

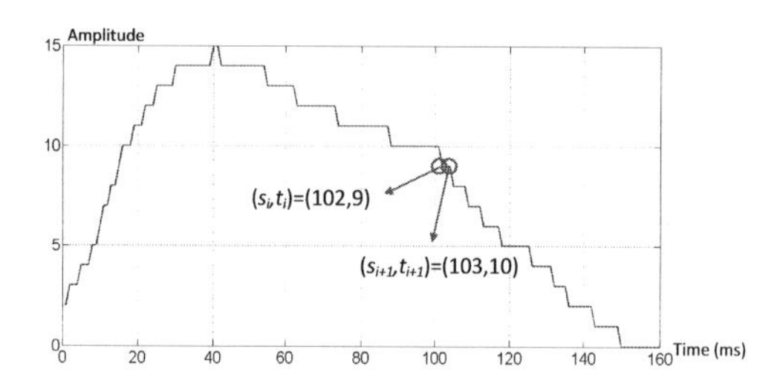

Figure 2. Respiratory waveform data collected in real time from sensor.

102 and 103 are the time stamp and 9 is the sample data in the form of amplitude. The interval of time stamp in Figure 2 is the Δt in equation (2). Respiratory waveform is also periodic data.

3 DATA REPRESENTATION AND REAL TIME CALIBRATION FEEDBACK

This section introduces the BESR based on our data model defined in section 2. We also discuss the proposed real time feedback time-series calibration algorithm and its time complexity.

3.1 *Data representation*

In real time measurement, raw data contains a relatively large amount of samples which lead to higher storage complexity and higher computational complexity so that providing real time feature extraction is difficult. Also, raw data from sensors is noisy, which usually causes difficulties in identifying meaningful features in the data. The goal of data representation is to find a simple form to represent the data in order to reduce data dimensionality and preserve useful features which also excludes noise to some extent. In this work, we propose a Binary Encoded Symbolic Representation (BESR). This new approach extends the classic Symbolic Aggregate approXimation (SAX) representation (Lin & Keogh et al. 2007) with additional pruning and digitization. As a first step, SAX transforms raw data into the (Pairwise Aggregate Approximation) PAA representation (Keogh et al. 2001) which partitions raw data into several segments of equal length and uses the average value of all samples within each segment as the PAA representation. Next, data on PAA representation will be transformed into symbols (i.e. alphabet) by multiple thresholds. The SAX representation computes distance between symbols by using a lookup table which defines the distance between each alphabet. The proposed BESR transforms the SAX into binary number which uses less storage space and also could be used to compute the distance. In this paper, we partition breathing waveform of each cycle (assuming n samples) into k segments. So each cycle C which is a sub sequence of sample set S and can be further represented as

$$C = (s_{x+1}, \cdots, s_{x+n}), 0 \le x \le N-n \tag{3}$$

where x is an index of a starting sample of the breathing cycle. Then, PAA representation of C is

$$C = (\bar{C}_1, \cdots, \bar{C}_k), 1 \le k \le n \tag{4}$$

where each C_i is the mean value of samples in segment i. Each C_i could be computed as

$$\bar{C}_i = \frac{k}{n} \sum_{j=0}^{\frac{n}{k}-1} s_{x+j} \tag{5}$$

where x is the index of starting sample of the ith segment. Next, each value will be transformed into symbols. Figure 3 shows an example of PAA data representation of one cycle respiratory waveform.

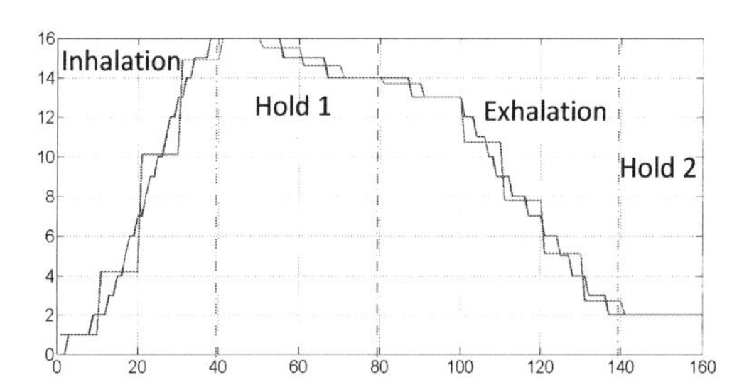

Figure 3. PAA data representation (Red plot: PAA; Blue plot: raw data).

In the example, k is 16 and n is 160. The Horizontal axis is the time axis for the time stamp and vertical axis is the amplitude. To map numerical values into symbols, we use multiple thresholds to identify which symbol should each value be mapped to. The rule to design thresholds is to cover the expected range of data in order to capture more information from raw data. Figure 4 shows an example of numerical value to the symbol mapping for the 4-4-6-2 breathing exercise pattern. Four symbols have been used for this example named as "a", "b", "c" and "d". PAA representation will be mapped into corresponding symbols which it fits into. We also use different thresholds for mapping in different stage in order to capture more information from raw data since the actual waveform is expected to closely follow the pre-defined pattern. As shown in Figure 4, symbolic representation for the pre-defined pattern is $a_1b_1c_1d_1c_2c_2c_2c_2d_3c_3c_3b_3b_3a_3b_4b_4$. Since only four symbols are used in this example, the proposed BESR encodes them using two bits binary numbers defined as a: 00, b: 01, c: 10, d: 11. The binary number takes less space for storage and can be used for distance measurement. Thus, we explained how raw data could be transformed into BESR.

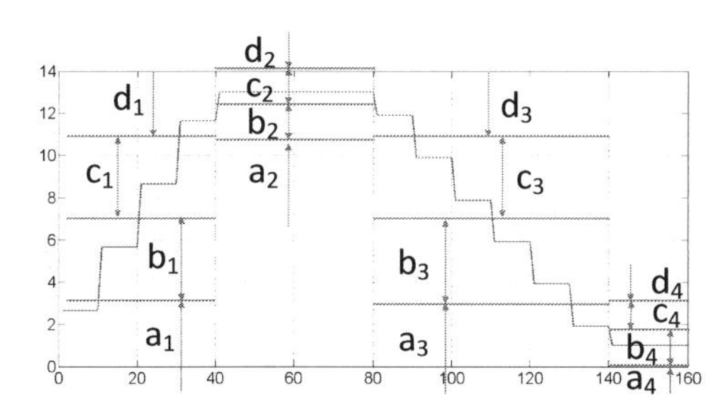

Figure 4. Symbolic transform for pre-defined pattern.

3.2 *Data calibration*

As we stated before, the goal of calibration is to find how much the actual data deviates from a given pattern and how to guide the real data to follow the pre-defined pattern as close as possible. More specifically, matching calibration measures the similarity between actual data (i.e., respiratory waveform from sensor) and pre-defined pattern (i.e. breathing exercise pattern). In some cases, even if the actual data does not follow the pre-defined pattern very well, it still reflects useful information to guide actual data to follow the pre-defined pattern as close as possible. Let's use the breathing exercise as an example. We could partition the actual respiratory waveform into four clusters: "Faster than expected", "Slower than expected", "Match" and "Missed". So, that meaningful information could be used to guide trainees to better follow the exercise. However, as a number of possible scenarios for respiratory waveform in SAX is very large (4^{16} in this work) which may cause a data explosion, creating one to one mapping look up table is not feasible. To reduce calibration complexity, we develop local matching to calibrate each stage independently and use local calibration results to represent the global calibration. Let's use the breathing pattern as an example, for stage 1 and 2, we could have $4^4(=256)$ kinds of scenarios. Similarly, stage 3 and 4 have $4^6(=4096)$ and $4^2(=16)$ each. To further reduce lookup table size, we also assume that a large portion of them will be mapped into the "Missed" cluster. So, we apply a distance filter to filter out scenarios belonging to "Missed" cluster before sending them to look up table. The distance filter works by first exam the starting symbol and then measuring the distance difference between current symbol and next symbol at each step. If any violation step detected, current scenario will be mapped to "Missed" cluster immediately without reaching the end of the current stage. Next, scenarios which are not filtered by the distance filter at the end of each stage will be sent to a hash table to find corresponding calibration results. The DFSSR significantly reduces the number of scenarios the hash table deals with and also save time for calibration. For instance, in Figure 5, the red arrow from c_1 to a_1 is a violation step which will be immediately detected before the end of the entire breathing cycle.

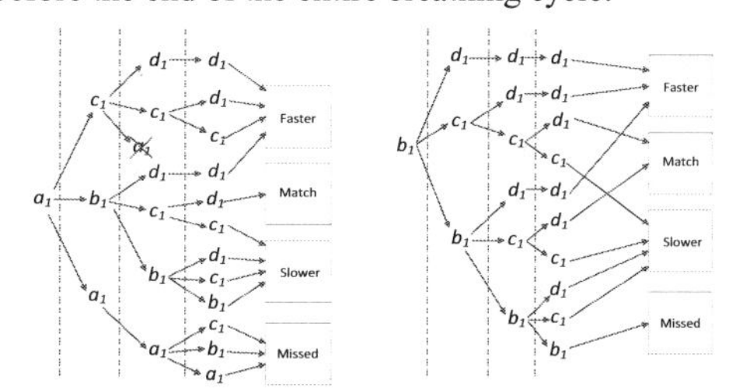

Figure 5. Example of DFSSR for stage 1.

Then, stage 1 of current waveform will be mapped to "Missed" cluster.

3.3 *Data dimensionality reduction and calibration complexity*

Since time-series data could be defined as an N-dimensional vector, data representation should transform the raw data into n-dimensional where $n < N$. This can greatly decrease data storage. Let's assume p number of samples in each segment and 16 segments in each breathing cycle. In this work, the proposed BESR uses only 2 bits for each symbol which end up using 32 bits for each breathing cycle. Comparing with a raw data storage format which is $512 \times p$ bits, assuming the sample is in 32 bit integer format. A comparison table is in section 4.

Calibration complexity determines how fast data could be transformed into meaningful information which is the ultimate goal for the real time feedback or guidance. For real time data, since each data sample is only available during a very short time Δt and the buffer size is limited, each sample must be processed within time Δt. In this work, complexity is reduced by reducing the hash table size, which maps SAX representation data to corresponding meaning. In our application, there can be 4^{16} possible scenarios. Using local calibration reduces the number of possible scenarios to 4624. (For example, stage 1 has four segments and each segment has four possible scenarios summing up to 4^4 scenarios). Then, with the DFSSR, a large number of scenarios which belongs to the "Missed" cluster will be filtered out. As demonstrated in Figure 4, after the DFSSR, only 22 scenarios are left for stage 1. Compared with 256 before DFSSR, reduction rate is very high. Detail comparison of scenario reduction is in section 4.

The algorithm speed complexity is computed using the *Big O* notation (Thomas et al. 2009). Assuming one breathing cycle contains n sample which are partitioned into k segments. A computing mean value for each segment and comparing those mean values with thresholds takes time $(n+k)$ which is $O(n)$ complexity since $k<n$. From Figure 4, time complexity for DFSSR is the number of segments k which is less than n. Thus, complexity for DFSSR is also $O(n)$. Since the time complexity of hash table is $O(1)$, Overall time complexity for the calibration algorithm is $O(n)$. Amortized time complexity for each sample is $O(1)$ and absolute time complexity for each stage is $O(n)$. Therefore, the proposed algorithm is very suitable for a real time application.

4 EXPERIMENTAL RESULTS

Table 1 is the comparison of storage for one respiratory waveform cycle. We assume raw data is in the format of 32 bits integer and each cycle contains 160 samples and each alphabet symbol is a char type which takes 8 bits. The proposed BESR saves storage by a factor of $160X$ which demonstrate its advantage in dimensionality reduction.

Table 1. Data storage comparison.

Raw data	SAX	BESR
5120 bits	128 bits	32 bits

Table 2 is the calibration complexity reduction in terms of number of scenarios for each respiratory waveform stage. The DFSSR achieves a reduction rate of 91.41%, 87.11%, 98.27% and 43.75% for each stage and 97.06% totally. Therefore, size of hash tables used to mapping symbol to the cluster will also be reduced by a ratio of 97.06% which makes it feasible for real time application.

Table 2. Waveform scenario comparison.

Stage	1	2	3	4	Total
SAX	256	256	4096	16	4624
DFSSR	22	33	71	9	136
Rate	91.41%	87.11%	98.27%	43.75%	97.06%

Figure 6 shows the comparison of expected waveform and real respiratory waveform in SAX representation. The first and second stages will be mapped into the "Faster than expected" cluster and the third and fourth stages be mapped into "Match" cluster. Thus, the global calibration would be faster than expected breathing exercise pattern which agrees with our visual calibration.

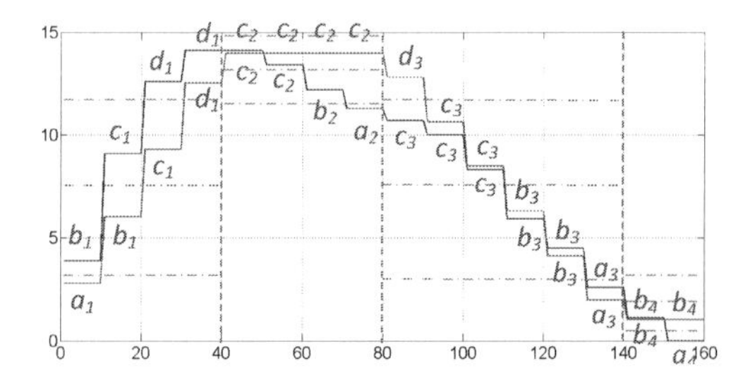

Figure 6. Calibration of breathing exercise pattern.

5 CONCLUSION

This paper proposed a fast time-series calibration algorithm with real time feedback based on the BESR. We theoretically proved that BESR reduced data storage by a factor of $160X$ and DFSSR achieved total hash table reduction rate of 97.06% which makes it suitable for real time application. Finally, experimental results on breathing exercise show that calibration results generated by the algorithm agree with human visual calibration.

REFERENCES

Brown, R. & Gerbarg, P. 2012. *The Healing Power of the Breath: Simple Techniques to Reduce Stress and Anxiety*: 144–148. Shambhala Publications.

Esling, P. & Agon, C. 2012. Time-Series Data Mining. *ACM Computing Surveys* 45(1): 12.

Keogh, E., Chakrabarti, K., Pazzani, M., et al. 2001. Dimensionality Reduction for Fast Similarity Search in Large Time-Series Databases. *Knowl. Info. Syst.* 3(3): 263–286.

Keogh, E., Chu, S., Hart, D., et al.2003a. Segmenting time series: A survey and novel approach. *Data Min. Time Series Databases* 57: 1–21.

Lin, J. & Keogh, E. 2005. Clustering of time-series subsequences is meaningless: Implications for previous and future research. *Knowl. Info.* Syst. 8(2): 154–177.

Lin, J., Keogh, E., Wei, L. & Lonardi, S. 2007. Experiencing SAX: a Novel Symbolic Representation of Time Series. *DMKD Journal* 2007(15): 107–144.

Patel, S., Park, H., Bonato, P., et al. 2012. A review of wearable sensors and systems with application in rehabilitation. *Journal of NeuroEngineering and Rehabilitation* 9(1): 21.

Thomas H. Cormen, Charles E., et al. 2009. *Introduction to Algorithms. Third Edition.* Massachusetts London, England: The MIT Press Cambridge.

Xi, X., Keogh, E., Shelton, C., et al. 2006. Fast time series classification using numerosity reduction. *ICML '06 Proceedings of the 23rd international conference on Machine learning, June* 2006: 1033–1040. ACM.

Electronics, Communications and Networks IV – Hussain & Ivanovic (eds)
© 2015 Taylor & Francis Group, London, ISBN: 978-1-138-02830-2

SIR epidemic model with attribute set rough on exploitation testing

Y.Y. Chen* & X. L. Zhang
Computer Technology Application Key Lab of Yunnan Province, Kunming University of Science and Technology, Kunming, China

ABSTRACT: This paper proposes a Susceptible-Infective-Removal (SIR) infection model with Two-direction Singular Attribute Rough Sets (2-SARS). In this model, every infected node with different possibility could propagate the virus to adjacent nodes in a single infecting period. Considering that information network belongs to scale-free networks, it focuses on the bidirectional rough set to evaluate nodes' states. Theoretical analysis and simulated results show that: for information network, unfixed node state plays a significant role in altering infected density and exploitation spreading velocity.

KEYWORDS: Rough set; SIR; Exploitation testing.

1 INTRODUCTION

The exploitation testing is the most important step of network security. The dynamic evolution of complex network should be studied, so that we understand the changes in the dissemination and safety performance properly. The epidemiological methods in transmission dynamics can describe changes and trend of a node's state. The security's state and evolution can be understood in (Miller et al. 2013, Cao et al. 2013, Masuda et al. 2013). Hattaf et al. (2013) use the technique of Lyapunov functions to establish the global stability of both the disease-free and endemic equilibrium. However, there is an assumption that all cells are exactly the same. But, in fact, the level of a node's infection and the possibility of infection will vary. Liu et al. proposed an improved spread of SIR model, adding D-dead state. But prevention programs and researches of the disease spreading are not involved in. Tiberiu Harkoa et al. proposed the exact analytical solution of the Susceptible-Infected-Recovered (SIR) epidemic model obtained in a parametric form. Ge et al. proposed a network model on transition of strangers and acquaintances. But the cumulative effect in strangers has not been dealt with well. According to the above analysis, there are still two problems about the degree of the node not been solved in the process of exploitation testing:

a. Obviously, the possibility of infection is not fixed because of different node degrees. Information network, usually has a wide distribution, which belongs to scale-free networks. If susceptible nodes contact with the infected individuals, the possibility of infection may be different. Platform, software, service, and configuration may result in different infections. It brings a different possibility of infection.
b. The ratio of the node's degree is a little unstable, because the ration of the infected node is not a fixed value.

Two-direction Singular Attribute Rough Sets (2-SARS) is a good choice (Guo et al. 2010, Wang et al. 2013) to represent discrete state if there are limited states. In this paper, the SIR model of infection, based on 2-SARS, will be proposed. The method that has a heterogeneous distribution of node degree can solve the problem that the possibility of infection is not fixed. Finally, we use the flat-field theory to simulate network exploitation testing. The experimental evidences prove the veracity and validity of the conclusion.

2 SIR MODEL OF INFECTION WITH UNFIXED POSSIBILITY

It is supposed that the network node presents a finite number of discrete states: susceptible state (S), infected state (I), and removed state (R), that is $S \xrightarrow{\beta} I \xrightarrow{\gamma} R$. Every node in information network has an obvious difference in their degrees. The distribution of degree is denoted by$P(k)$. And information network is a typical scale-free network.

*Corresponding author: *dawanmitang@163.com*

Because nodes are different in their system platform, application software, service and configuration, when a node infects its adjacent nodes, the possibility of its success may not be a fixed value. These lead to the different results caused by different susceptible node's infections. We cannot uniform the definitions of the infection rate in the traditional non-homogeneous model.

The infection process in model can be described as follows: the susceptible node would be infected as a possibility β_i when contact i.with infected node i. After the node becomes an infected node, it may recover to a healthy state as a probability γ, and is no longer be subjected to infection. Variable $i_k(t)$ represents the possibility of the infected nodes to all nodes with degree of K. $s_k(t)$ represents the relative density of infected nodes as K degree. $r_k(t)$ represents the relative density of recovered nodes as K degree. They satisfy the normalization condition $i_k(t)+s_k(t)+r_k(t)=1$. An infection model under this condition can be described as equation 1:

$$\begin{cases} \frac{ds_k(t)}{dt} = -\sum_{i=1}^{k} \beta_i k s_k(t)\Theta(t) \\ \frac{di(t)}{dt} = \sum_{i=1}^{k} \beta_i k s_k(t)\Theta(t) - \gamma i_k(t) \\ \frac{dr_k(t)}{dt} = \gamma i_k(t) \end{cases} \tag{1}$$

The proportion of infected nodes in the entire network is $i(t)=\sum i_k(t)p(k)$. $\Theta(t)$ denotes the probability of any given edge connected to an infected node. Its value relates to degree k, $\Theta(t)=\frac{\sum_k kp(k)i_k(t)}{\sum_k kp(k)}=\frac{\sum_k kp(k)i_k(t)}{\langle k\rangle}$. If an effective infection rate is $\lambda=\langle\beta\rangle/\gamma$, then integrate the first equation of the equations, we get equation 2:

$$s_k(t)=\exp(-\int_0^t \sum_{i=1}^{k} \beta_i k s_k(t)\Theta(t)dt$$

$$=\exp(-\frac{\sum_{i=1}^{k}\beta_i k}{\gamma}\int_0^t \Theta(t)dt)$$

$$=\exp(-\frac{\sum_{i=1}^{k}\beta_i k}{\gamma}\frac{\sum_k kp(k)r_k(t)}{\langle k\rangle}) \tag{2}$$

When

$$\varphi(t)=\frac{\sum_k kp(k)r_k(t)}{\langle k\rangle}$$

thus:

$$s_k(t)=\exp(-\frac{\sum_{i=1}^{k}\beta_i k}{\gamma}\varphi(t)).$$

When $t\to\infty, i_k(t)=0$, thus:

$$\varphi(t)=\frac{\sum_k kp(k)r_k(t)}{\langle k\rangle}=\frac{1}{\langle k\rangle}\sum_k kp(k)[1-\exp(-\frac{\sum_{i=1}^{k}\beta_i k}{\gamma}\varphi(t)]\cong\omega(\varphi(t)).$$

In order to obtain the existence of a non-trivial solution, the condition must meet $\frac{d\omega(\varphi(t))}{d\varphi(t)}\Big|_{\varphi(t)=0}>1$. The threshold of the average infection rate in a heterogeneous network environment is $\overline{\lambda_c}=\frac{\langle k\rangle}{\langle k^2\rangle}$. We take the relationship of network nodes into account. So the threshold of the average infection rate is $\overline{\lambda_c}=\frac{1}{\sum_k k' P(k'|k)}$.

3 EVALUATE CELL STATE BASED ON 2-SARS

In the process of exploitation, due to differences in exploitation path and node's ability, there are three states in the transitions. The process of evaluating node is as follows: Firstly, the current state of the cellular is given. Secondly, node starts repair strategy to fix itself when it is infected. At last, resilience and safety value is re-calculated. With the time moving on, quantitative and qualitative changes between three states happen. In particular, around the threshold of state transition, there is a high probability. We can make an evolution simulation on the propagation of weakness by defining the rules of cellular automata changes.

Each node in the network is defined as a cell, and all cells constitute the cellular space (Tiberiu et al. 2014). In SIR model, a cell should be in one of three states {S, I, R}. At any time, a cell's state may be susceptible state (S), or infected state (I), or immune state (R). The state of a cell is not fixed in one state. It may be somewhat unstable randomly when cell X in SIR infection model swings around threshold under the impact of transitions of other nodes F. So it is difficult to judge if a node is in a particular state or is changed to another state. Because there is a certain probability that a cell may not be fully moved into or moved out of a state, 2-SARS can be used to represent this kind of dynamic characteristics. Based on the above discussion, the 2-SARS model is established to evaluate the cell state.

Definition 1: Supposing U describes a set of cell states. $X^{**} \subset U$, is bidirectional S Rough Set about U. Supposing $f \in \underset{\sim}{F}$ and $\overline{f} \in F$, and $X^{**} = X' Y \{x \mid u \in U, u \overline{\in} X, f(u) = x \in X\}$, $X' = X \setminus \{x \mid x \in X, \overline{f}(x) = ux \underset{\sim}{\in} X\}$.

Because $X^{**} \subset U$ may be moved into or moved out of state U according to the probability of node's state.

Definition 2: X^{**} is 2-SARS of U. $(R,F)_{\circ}(X^{**})$ is under approximation of X^{**}. $(R,F)^{\circ}(X^{**})$ is upper approximation of X^{**}. $(R,F)_{\circ}(X^{**})$ and $(R,F)^{\circ}(X^{**})$ is 2-SARS of $X^{**} \subset U$.

4 SIR EXPLOITATION ALGORITHM BASED ON 2-SARS

The importance of a node in information network is closely related to the degree of the node. Algorithm of cell state evaluating is introduced into the SIR infection model. It is assumed that relevant parameters are: transmission rate β, recovery rate γ, period of infection τ, program execute time T, the adjacency matrix $N_{n\times n}$, the domain of discourse U and knowledge equivalent class of cell X. The process of infection is as follows:

Step1: The adjacency matrix consists of all adjacent nodes. Normally, the adjacency matrix $N_{n\times n}$ represents the number of the nodes. All initial value is -1. The cell precursory detection sets are evaluated from the adjacency matrix. The domain of discourse $U$is made of all cells' adjacency matrix.

Step 2: The knowledge equivalent class of cell X that is related to the domain Uneeds to be deduced. The lower approximation and the upper approximation of X can be evaluated. According to the result, the 2-SARS X^{**} is established, and then the set of cell precursory detection is established.

Step 3: Set the every cell state in cell precursory detection X, and that means knowledge equivalent class of a cell. The state of node is related to the state of cell precursory detection. If cells' states could not be evaluated, we should repeatedly process the following steps until the whole cells' states could be evaluated.

Step 4: It is assumed that all nodes' initial states are susceptible. *node*(i).*state* is a row vector and has elements that present every node's state. If a node is not a member of X, then the corresponding value is set to -1. Then we evaluate every node in X: if the value is 0, the corresponding node has not been infected yet. If the value is (0,1], the corresponding node has been contaminated and infected. The value means infection rate β_i: if the value is 2, then the corresponding node is in the state of immune and will be removed from X. The state of immune is a special state. A node's state should be 2, because it could be infected by targeted attack.

Step 5: Re-calculate the lower approximation and the upper approximation of X^{**} according to the result of the transition. $(R,F)_{\circ}(X^{**}_{(\alpha)})$ And $(R,F)^{\circ}(X^{**}_{(\beta)})$ represent minimal set and maximal set after state transitions respectively.

Step 6: Randomly select a node from the N nodes as an infected node. Change the state of the node from 0 to β_i. The process of infection is simulated: if $node(i).state \notin (0,1]$, then skip node i and deal with the next node j; if $node(i).state \in (0,1]$, then node i is recognized as infected node. Search in the adjacency matrix excluding all cells whose state is 2, and the current cell precursory detection set of node i is obtained.

According to the cells added or removed, the state of the set is obtained.

Step 7: If there exists a node that $\lambda_j \geq \lambda_c$, then the node has been infected, otherwise the node remains its state. In order to prevent other susceptible nodes from being infected at the same time, the state of neighbor nodes is marked as 3 rather than β_i. All neighbor nodes are dealt with the same process.

Step 8: Recovery process of node: if $node(i).state \neq 3$, then skip node i and analyze node j; if $node(i).state = 3$, then change the state of the node to 2 after making node i fixed and immune with probability γ.

Step 9: Repeat step from 3 to 8 until τ time.

Step 10: recovery process of node in time $\tau + 1$.

Step 11: Repeat step 5 to step 8 until all nodes' states have been evaluated, and then the algorithm comes to an end.

5 SIMULATION EXPERIMENT

We have analyzed and verified the critical characteristics of the new model through Matlab.

Parameters are set as follows: Max degree $k <= 8$, and repair rate $\gamma = 0.1$. The initial network topology has 16 nodes called $e_1, \ldots, e_{16}$, respectively. The adjacency matrix is a 16*16 matrix. These nodes are divided into three classes, including susceptible state, infected state as well as immune state. The following work takes node e_{13} as an example to verify the process. And $k_{13} = 5$. Initial infection density $\rho = 0.1$;

when $\overline{\beta} = 0.051$, we get the average effective transmission rate $\lambda_c = 0.15$. In the standard model, it is supposed that average neighbor nodes K = 4, critical value $\lambda_c = 0.25$.

With the increase in scale, for example, the value of N is increased from 16 to 5000. When the initial infection density $\rho = 0.1$, critical value is

$$\overline{\lambda_c} = \frac{1}{\sum_k k' P(k'|k)} \approx 0.0604.$$

As shown in figure 1 and figure2, when the effective transmission rate is greater than the critical value, attack will burst in a large scope and spread out. As the infection density increases, more infections happen, and transmission speed is higher.

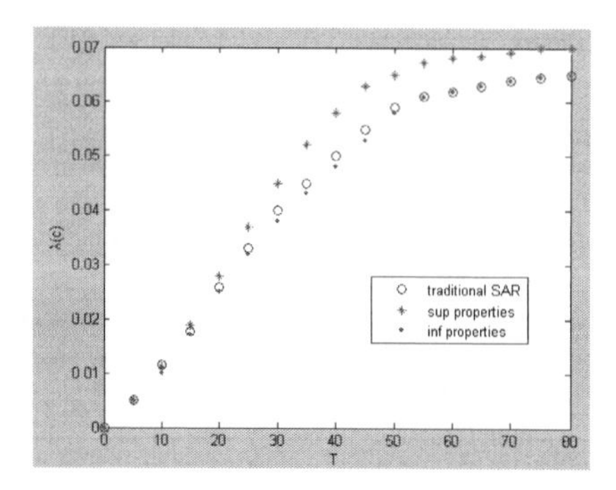

Figure 1. Changes in SIR infected.

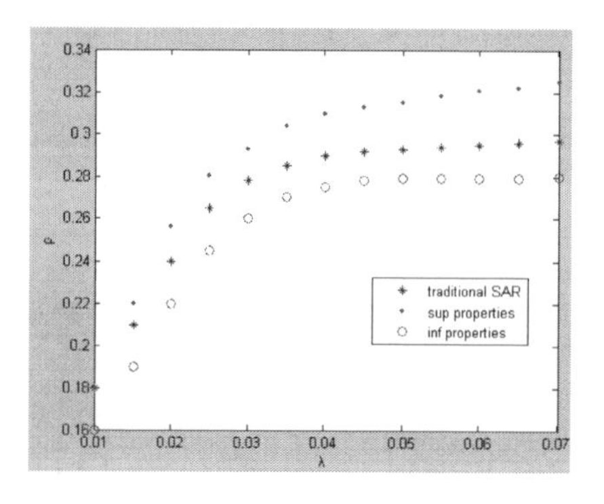

Figure 2. The relationship of density and infection.

6 CONCLUSIONS

In this paper, we propose a novel SIR model that is based on 2-SARS to examine the dynamic process of infected node. The gradual change process of vulnerability's state can be reflected through examining the attack propagation in networks. The simulation results verify that the evaluation node's state has a significant impact on infection density and propagation velocity. The state of nodes can't be easily evaluated when nodes have unfixed costs. For future work, we will further study how different degrees of discrete state influence the infection ratio and the peak value.

REFERENCES

Miller, J. C. & Volz, E. M. 2013. Incorporating Disease and Population Structure into Models of SIR Disease in Contact Networks. *PloS one*, 8(8): e69162.

Cao, J., Wang, Y., Alofi, A., et al. 2014. Global stability of an epidemic model with carrier state in heterogeneous networks. *IMA Journal of Applied Mathematics*, hxu040.

Masuda, N. & Holme, P. 2013. Predicting and controlling infectious disease epidemics using temporal networks. *F1000prime reports*, 5.

Hattaf, K., Lashari, A. A., Louartassi, Y., et al. 2013. A delayed SIR epidemic model with general incidence rate. *Electronic Journal of Qualitative Theory of Differential Equations* 3: 1–9.

Liu, T. Y. 2011. Study of influenza A(H1N1)spreading model on complex networks. *Computer Engineering and Application* 47(15): 222–224.

Harko, T., Lobo, F. S., & Mak, M. K. 2014. Exact analytical solutions of the Susceptible-Infected-Recovered (SIR) epidemic model and of the SIR model with equal death and birth rates. *Applied Mathematics and Computation* 236: 184–194.

Ge, X., Zhao, H. & Zhang, J. 2011. Complex Networks Immune Strategy Based on Acquaintance Immunization. *Computer Science* 38(11): 83–86.

Guo, Z. L. 2010. Two-direction Singular Attribute Rough Sets and Its Properties. *Journal of Shanxi University(Nat. Sci. Ed.)* 33(4):13–518.

Wang, S., Zhu, Q., Zhu, W., et al. 2013. Quantitative analysis for covering-based rough sets through the upper approximation number. *Information Sciences* 220: 483–491.

Electronics, Communications and Networks IV – Hussain & Ivanovic (eds)
 ISBN: 978-1-138-02830-2

Image fusion algorithm based on NSST and PCNN

Guang-qiu Chen*, Jin Duan, Hua Cai & Guang-wen Liu
School of Electronic and Information Engineering, Changchun University of Science and Technology, Changchun, China

ABSTRACT: For enhancing the fusion accuracy of multi-modality images, an adaptive image fusion algorithm based on Non-Subsampled Shearlet Transform (NSST) and Pulse Coupled Neural Network (PCNN) was proposed. First, source images were decomposed to multi-scale and multi-direction sub-bands using NSST. Secondly, local area singular value decomposition in each sub-band was performed to construct a Local Structure Information Factor (LSIF) that served as the linking strength of each neuron in the PCNN. After the fire processing of the PCNN, new fire mapping images of all the sub-bands were obtained, the clear objects of the sub-band images were selected by the compare-selection operator with the fire mapping images pixel by pixel, and then all were merged into a group of clear new sub-bands. Finally, the fused sub-bands were reconstructed to images by Non-Subsampled Shearlet Inverse Transform (NSSIT). Some fusion experiments on two sets of different modality images were performed and objective performance assessments were implemented on the fusion results. The experimental results indicated that the proposed method performed better in subjective and objective assessments than a number of existing typical fusion techniques and obtained better fusion performance.

KEYWORDS: Image processing; Nonsubsampled shearlet transform; shift-invariant; Pulse coupled neural networks; Linking strength.

1 INTRODUCTION

Over the years, image fusion technologies based on multi-scale decomposition (MSD) have attracted the attention of relevant domestic and foreign scholars, who have achieved many outstanding research results. (Yang et al. 2010) Recently, fusion technologies based on MSD, such as DWT (Tang et al. 2012), Curvelet (Candès& Donoho 2004), Contourlet (Do& Vetterli 2005) and pulse coupled neural networks (PCNNs) for combining multi-modality images have become a hot research topic in the fusion field. Easley, et al. (2008) proposed a novel image representation method: the non-subsampled shearlet transform (NSST), which not only has the advantages of Curvelet and Contourlet, but also holds shift-invariant. Therefore, in this paper, the NSST was used as the MSD image tool.

A pulse coupled neural network (PCNN) is a new model of artificial neural network, which is applied to the field of image fusion because of having synchronous excitation, variable thresholds, etc. The singular value (SV) of an image is a good algebraic nature and the SV matrix can represent structural information of the original image. The square of the SV matrix's F-norm represents the energy of the image and reflects the local features of the image. In this paper, a novel image fusion algorithm based on PCNN is proposed.

2 NON-SUBSAMPLED SHEARLET TRANSFORM (NSST)

In dimension $n = 2$, the shearlet system with discrete parameters is defined as follows:

$$S_{AB}(\phi) = \{\phi_{j,l,k}(k) = |\det A|^{j/2} \phi(B^l A^j x - k), (j,l) \in Z, k \in Z^2\} \quad (1)$$

where $\phi \in L^2(R^2)$, A, B are 2×2 invertible matrix and $|\det B| = 1$, j is the scale parameter, l is the direction parameter, and k represents the spatial position. *If* $j \geq 0, -2^j \leq l \leq 2^j - 1, k \in Z^2, d = 0,1$, the Fourier transform of the shearlet in the compact form can be expressed as:

$$\hat{\phi}^{(d)}_{j,l,k} = 2^{3j/2} V(2^{-2j}\xi) W^{(d)}_{j,l}(\xi) \exp(-2\pi i \xi A_d^{-j} B_d^{-l} k) \quad (2)$$

where $V(2^{-2j}\xi)$ is the scaling function, $W^{(d)}_{j,l}$ is the window function localized on a pair of trapezoids, A_d is the anisotropic dilation matrix, and B_d is the shear matrix. The shearlet transform of $f \in L^2(R^2)$ can be computed by:

$$\left\langle f, \phi^{(d)}_{j,l,k} \right\rangle = 2^{3j/2} \int_{R^2} \hat{f}(\xi) \overline{V(2^{-2j}\xi) W^{(d)}_{j,l}(\xi) \exp(2\pi i \xi \mathbf{A}_d^{-j} \mathbf{B}_d^{-l} k)} d\xi \quad (3)$$

*Corresponding author: *guangqiu_chen@126.com*

As can be seen from Equation (3), the shearlet transform of $f \in L^2(R^2)$ is mainly composed of the following two parts:

MSD (multi-scale decomposition): two-channel non-subsampled filter banks (NSFB) are applied to an image for non-subsampled pyramid (NSP), so the multi-scale property of the shearlet is obtained.

Directional localization: implemented by a small-sized shear filter. Given an image is decomposed to low-pass sub-band f^{j+1} and band-pass sub-band g^{j+1} by a NSP, where j denotes the number of scales, j=1,2,...,M. The process of the directional localization can be explained in the following steps.

1 Meyer wavelet function is used to produce Meyer window function $g(\theta)$.
2 In a pseudo-polar grid, $g(\theta)$ is re-sampled to produce the shear filter window $\boldsymbol{W}$.
3 $\boldsymbol{W}$ is mapped into the Cartesian coordinate system from the pseudo-polar grid and a new shear filter $\boldsymbol{W}_{new}$ is obtained.
4 The discrete Fourier of the band-pass sub-band is computed to obtain matrix Fg^{j+1}.
5 $\boldsymbol{W}_{new}$ is operated with Fg^{j+1} to obtain directional sub-bands.
6 The inverse Fourier transform is conducted on each directional sub-band to obtain coefficient $c^{j+1,l}$.

In reference (Easley et al. 2008), the above shearlet is called the non-subsampled shearlet transform (NSST). For removing down-sampled operations in NSSTs, the shift-invariant property is obtained. In the directional localization process, a small shear filter can avoid blocking artifacts conducted by a big filter and reduce the Gibbs-type ringing phenomenon. An image decomposed by NSST with J scales can produce $\sum_{j=1}^{J} 2^{l_j}$ band-pass sub-bands and one low-pass sub-band, where l_j denotes the number of levels at the jth scale.

3 FUSION CRITERION

3.1 *The basic principle of PCNN and the simplified model*

PCNN is a nonlinear dynamic feedback network composed of neuron mutually-linking plurality, which include a receptive field, modulation domain, and pulse generator. Initially, the PCNN model based on mammalian visual system was extremely complex. Because of its complexity, it was modified and simplified in many literatures. In this paper, the discrete mathematic iterative model was used, which is the most commonly used model, as shown in Figure 1.

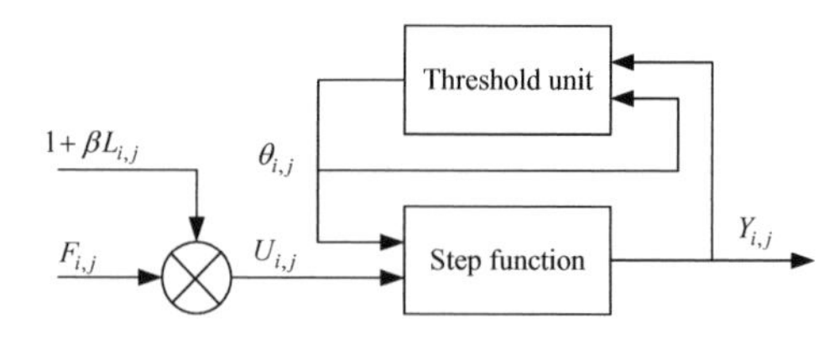

Figure 1. Simplified model of PCNN neuron.

Its mathematical equation is:

$$F_{ij}(n) = I_{ij}(n) ; \tag{4}$$

$$L_{ij}(n) = \exp(-\alpha_L)L_{ij}(n-1) + V_L \sum_{pq} W_{ij,pq} Y_{ij}(n-1) \tag{5}$$

$$U_{ij}(n) = F_{ij}(n) \times (1 + \beta_{ij} L_{ij}(n)) \tag{6}$$

$$\theta_{ij}(n) = \exp(-\alpha_\theta)\theta_j(n-1) + V_\theta Y_{ij}(n-1) \tag{7}$$

$$Y_{ij}(n) = \begin{cases} 1 & U_{ij}(n) > \theta_{ij}(n) \\ 0 & U_{ij}(n) \le \theta_{ij}(n) \end{cases} \tag{8}$$

where (i,j) is the neuron location label, $F_{ij}(n)$ is the feedback input at nth iterated calculation, I_{ij} is the external stimulus input, i.e., pixel gray scale value at location (i,j), $L_{ij}(n)$ is the neuron linking input, β_{ij} is the linking strength, U_{ij} is the internal activity item of the neuron, Y_{ij} is the output at the nth iteration calculation, W is the weighted coefficient matrix, V_L is the amplified factor, θ_{ij} is the output of the variable threshold function, $V\theta$ is the amplified factor of the threshold, α_L and $\alpha\theta$ are time constants, and n is the iterated times. Neurons output one pulse and produce one-time fire when $U_{ij}(n)$ is greater than θ_{ij}. After n times iteration, the firing times of the neuron at location (i,j) represent the feature information of that position. So, after the firing processing of the PCNN, the fire mapping images are formed by the firing times of each neuron in the source images, and the output of the PCNN is obtained.

3.2 *Singular value decomposition of image*

Singular value decomposition (SVD) theorem is described as follows: (Wang 2012)

In matrix $\boldsymbol{A} \in \boldsymbol{R}^{m \times n}$, there are two orthogonal matrices $\boldsymbol{U}$ and $\boldsymbol{V}$, and a diagonal matrix $\boldsymbol{S}$ that make the following Equation hold:

$$\boldsymbol{A} = \boldsymbol{U}\boldsymbol{S}\boldsymbol{V}^T = \sum_{i=1}^{p} \sigma_i u_i v_i^T \tag{9}$$

where $U = [u_1, u_2, \cdots, u_m] \in R^{m\times n}$, $S = diag[\sigma_1, \sigma_2 \cdots, \sigma_p]$,

$U^T U = I$, $V^T V = I$, $p = \min(m,n)$, $\sigma_1 \geq \sigma_2 \geq \cdots \geq \sigma_p > 0$

Equation (9) is called the SVD of matrix A, σ_i is called singular value (SV) of matrix A, and S is called singular value matrix. SV is the nature of the image rather than visual characteristics. The energy of matrix A can be expressed as $E = \| A \|_F^2$, and combining Equation (9), there is:

$$E = \| A \|_F^2 = \| USV^T \|_F^2 = tr[(USV^T)\cdot(USV^T)^T] \\ = tr(S\cdot S^T) = \| S \|_F^2 = \sum_{i=1}^{p} \sigma_i^2 \tag{10}$$

Thus, the energy of image A is focused on matrix S.

3.3 *PCNN fusion criterion based on area SVD*

In the traditional PCNN algorithm, the linking strength of the neurons is a constant that is achieved through experience or experiment. In the human visual system (HVS), the reaction extent of vision on different characteristic image regions is different, i.e., the coupling of different neurons in the visual cortex is different. This suggests that the linking strength β of neurons is different in the PCNN. As can be seen from the internal activity item of the PCNN mathematical equation, the proportion of the clear objects in the source image in the fused image is related to the linking strength β. Therefore, it is believed that β should change on the basis of the image feature, which can represent various region feature information and should not be a constant. The SV of an image contains structure information and concentrates on energy information. In this paper, the mean of local area SVs was used to define a local structure information factor (LSIF), which can represent the content and characteristic change of an image. That is to say:

$$e(i) = (\sigma_1(i) + \sigma_2(i) + \cdots + \sigma_r(i))/r \tag{11}$$

where σ_j is the local area SV and e embodies the basic structure and richness degree of detail information contained in the local area. In this paper, the LSIF was used as the linking strength value in the PCNN to embody feature image information in different areas. In Figure 2, the fusion of two images illustrates the fusion process of the adaptive PCNN fusion algorithm based on SVD.

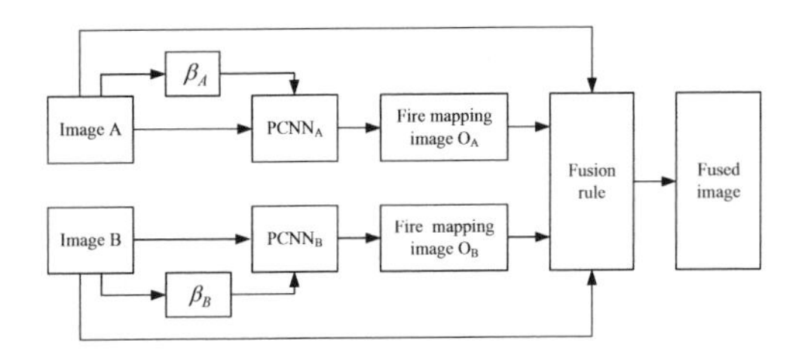

Figure 2. Schematic of the PCNN-based fusion algorithm.

In this paper, two normalized source images were decomposed to multi-scale and multi-direction sub-bands by NSST, respectively denoted $\{IL1, IH1^{l,k}\}$, $\{IL2, IH2^{l,k}\}$, where ILX denotes the low-pass sub-band, $IHX^{l,k}$ denotes the kth band-pass sub-band at the lth scale, and X=1,2. $\{IL1, IH1^{l,k}\}$ and $\{IL2, IH2^{l,k}\}$ were successively set as A and B in Figure 2 and fused sub-bands$\{IL, IH^{l,k}\}$ were obtained. The fused sub-bands were reconstructed to an image by NSSIT.

4 SIMULATION AND RESULTS ANALYSIS

For demonstrating the effectiveness and stability of the proposed method, different fusion methods based on PCNN combining different MSDs were compared with the proposed method, which was denoted as NSST+SVDPCNN.

Method 1: in lifting stationary wavelet (LSWT) domain, sum-modified-laplacian and pixel gray scale values are regarded as external stimulus inputs of PCNN neurons in low-frequency and high-frequency sub-bands, respectively, linking strength β = 0.2; denoted as LSWT+SMLPCNN. (Guo et al. 2010)

Method 2: in Curvelet domain, average rule is used in low-frequency sub-bands and in high-frequency sub-bands, pixel gray scale values are regarded as external stimulus inputs of PCNN neurons, region energy is used as linking strength; denoted as Curvelet+APCNN.(Zhao & Qu 2011)

Method 3: in Contourlet domain, pixel gray scale values are regarded as external stimulus inputs of PCNN neurons, linking strength β = 0.2; denoted as Contourlet+PCNN. (Jin et al. 2012)

Method 4: in NSCT domain, pixel gray scale values are regarded as external stimulus inputs of PCNN neurons, linking strength β = 0.2; denoted as NSCT+PCNN. (Jin et al. 2012)

Method 5: in NSCT domain, spatial frequencies are regarded as external stimulus inputs of PCNN neurons, linking strength β = 0.2; denoted as NSCT+SFPCNN. (Jin et al. 2012)

Method 6: in NSCT domain, pixel gray scale values and LOG energy values are regarded as external stimulus inputs of PCNN neurons in low-frequency and high-frequency sub-bands, respectively, linking strength $\beta = 0.2$; denoted as NSCT+LOGPCNN. (Jin et al. 2012)

The source images consisted of remote sensing and medical images, as shown in Figures 3(a) and (b). The results are shown in Figures 3(c)–(i), and the objective evaluation data is listed in Table 1. The fused image was subjectivity assessed by visual observation and three criterions were used for objective evaluation including the MI (Qu 2002), SSIM(Wang et al.2004) and $Q^{AB/F}$ (Xydeas& Petrovi 2000) criterions. In the proposed method, β was obtained by computing the LSIF of a 3×3 block matrix, with the other parameters of the PCNN set as: each neuron was linked with the surrounding 3×3 neighborhood, i.e., p×q = 3×3, linking nuclear matrix

$$W_{ij} = \begin{bmatrix} 0.707 & 1.000 & 0.707 \\ 1.000 & 0 & 0.707 \\ 0.707 & 1.000 & 0.707 \end{bmatrix}, \ \alpha_L = 0.06931, \quad \alpha_\theta = 0.2,$$

$$V_L = 1.0, \quad V_\theta = 20, \quad N_{\max} = 200.$$

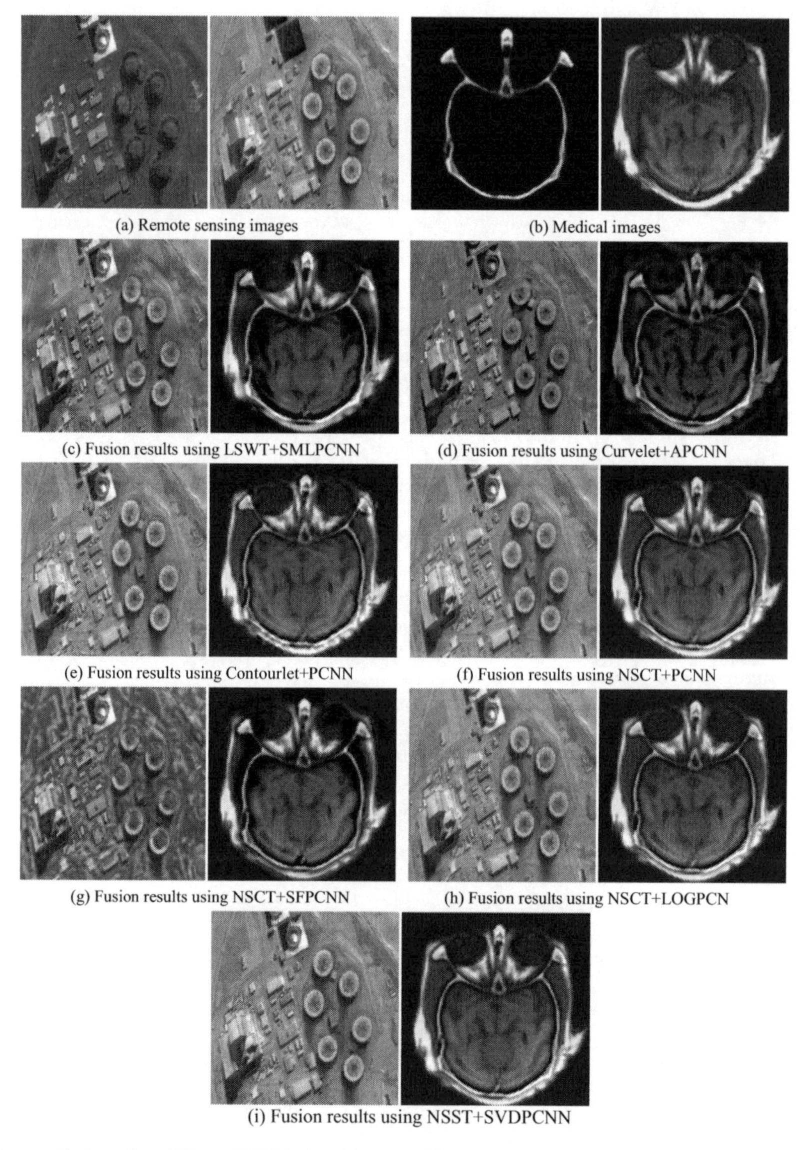

(a) Remote sensing images
(b) Medical images
(c) Fusion results using LSWT+SMLPCNN
(d) Fusion results using Curvelet+APCNN
(e) Fusion results using Contourlet+PCNN
(f) Fusion results using NSCT+PCNN
(g) Fusion results using NSCT+SFPCNN
(h) Fusion results using NSCT+LOGPCN
(i) Fusion results using NSST+SVDPCNN

Figure 3. Fusion results based on different PCNN algorithms in different MSD domains.

Table 1. Fusion results' comparison based on different PCNN algorithms in different MSD domains.

Fusion method	Remote sensing images			Medical images		
	MI	SSIM	$Q^{AB/F}$	MI	SSIM	$Q^{AB/F}$
LSWT+SMLPCNN	1.8828	0.6483	0.5373	1.8867	0.4437	0.7024
Curvelet+APCNN	2.1027	0.6611	0.5168	1.6681	0.3948	0.6426
Contourlet+PCNN	2.9251	0.6447	0.4963	3.5671	0.4779	0.6945
NSCT+PCNN	3.1054	0.6590	0.5481	4.0376	0.4914	0.7785
NSCT+SFPCNN	1.5805	0.4403	0.4464	3.0377	0.4631	0.7296
NSCT+LOGPCNN	3.2252	0.6540	0.5555	4.1194	0.4916	0.7820
NSST+SVDPCNN	3.6186	0.6877	0.5903	4.8420	0.5289	0.8120

As can be seen in Figure 3 and Table 1, the results obtained by the proposed method were optimal in visual observation and objective evaluation data, which indicates that the proposed method is outstanding for multi-modality image fusion.

5 CONCLUSION

Aimed at the defects of existing multi-scale image fusion methods, NSST was introduced to the field of image fusion and a PCNN fusion criterion-based area SVD was proposed. NSST with the flexible multi-resolution and directional expansion can better represent the edges and texture structures of images. The LSIF constructed by a local area's singular value as linking strength were illustrated to represent the characteristics of an image adaptively. Experimental results demonstrated that the proposed method was successful in remote sensing and medical image fusion.

REFERENCES

Candès, E. J. & Donoho, D. L. 2004. New tight frames of curvelets and optimal representations of objects with piecewise C^2 singularities. *Comm. on Pure and Appl. Math* 57(2): 219–266.

Do, M. N. & Vetterli, M. 2005. The contourlet transform: an efficient directional multiresolution image representation. *IEEE Trans. Image Proc.* 14(12): 2091–2106.

Easley, G., Labate, D. & Lim, W.Q. 2008. Sparsedirectional image representations using the discrete shearlet transform. *Applied and Computational Harmonic Analysis* 25(1): 25–46.

Guo, M. Y., Li, H. F. & Chai, Y. 2010. Image fusion using Lifting Stationary Wavelet Transform and Adaptive PCNN. *Opto-Electronic Engineering* 37 (12): 67–74.

Jin, X., Li, H. H. & Shi, P. L. 2012. SAR and multispectral image fusion algorithm based on pulse coupled neural networks and non-subsampled contourlet transform. *Journal of Image and Graphics* 17(9): 1188–1195.

Qu, G. H., Zhang, D.L. & Yan, P.F. 2002. Information Measure for Performance of Image Fusion. *Electronic Letters* 38(7): 313–315.

Tang, Y. Q., Zhang, X. X., Li, X. E., et al. 2012. Image Processing Method of Dynamic Range With Wavelet Transform Based on Human Visual Gray Recognition Characteristics. *Chinese Journal of Liquid Crystals and Displays* 27(3): 385–390.

Wang, Y. Q. 2012. Image quality assessment based on complex number representation of image structure and singular value decomposition. *Journal of Optoelectronics·Laser* 23(9): 1827–1834.

Wang, Z., Bovik, A. C., Sheik, H. R., et al. 2004. Image Quality Assessment: From error visibility to structural similarity. *IEEE Transactions on Image Processing* 13(4): 600–612.

Xydeas, C. S. & Petrovi, V. 2000. Objective Image Fusion Performance Measure. *Electronics Letters*36(4): 308–309.

Yang, B., Jing, Z. L. & Zhao, H. T. 2010. Review of pixel-Level Image Fusion. *Journal of Shanghai Jiao tong University (Science)* 15(1): 6–12.

Zhao, J. C. & Qu, S. R. 2011. A Better Algorithm for Fusion of Infrared and Visible Image Based on Curvelet Transform and Adaptive Pulse Coupled Neural Networks (PCNN). *Journal of Northwestern Polytechnical University* 29(6): 849–853.

Electronics, Communications and Networks IV – Hussain & Ivanovic (eds)
© 2015 Taylor & Francis Group, London, ISBN: 978-1-138-02830-2

Wavelet and pattern trends based co-occurrence features for age group classification of a facial image

Ye-gang Chen
School of Computer Engineering, Yangtze Normal University Fuling, Chongqing, China

ABSTRACT: The paper presents an integrating approach for defining the different age groups of a human. This approach uses wavelets with a new structural approach derived from four distinct binary patterns on a 3×3 wavelet facial image and features derived from Grey Level Co-occurrence Matrix (GLCM). The present paper derived two 3-pixel local patterns Left-Right Diagonal Pattern (LRDP) and Central Horizontal-Vertical Pattern (CHVP).which are derived from four distinct patterns called Left Diagonal Pattern (LDP), Right Diagonal Pattern (RDP), Vertical Centre Pattern (VCP) and Horizontal Centre Pattern (HCP) patterns. For all four patterns the central pixel value of the 3x3 neighborhood is significant. Based on these two CHVP and LRDP, GLCM is computed and four features a evaluated on LRDP-CHVP-GLCM to classify the human age into four age groups, i.e.: Child (0-15), Young adult (16-30), Middle aged adult (31-50) and senior adult (>50). The co-occurrence features extracted from the 4-DLBP provides complete texture information about an image which is useful for classification. The proposed method is experimented with a wide variety of facial images, and exhibited with a higher classification rate.

1 INTRODUCTION

The Age estimation is an important task in facial image classification. Human face embodies rich amount of information usable in many interesting applications, like age classification, face recognition, facial expression etc., These issues has inspired many researchers and leading to a diverse et of solutions. In this respect, more is needed to prove the classifications based on human faces to be worthwhile in more general and realistic scenarios, i.e., in settings acquired in unconstrained conditions.

Recently Human age classification has become an active research topic in computer vision because of its widespread potential real world applications such as electronic customer relationship management (ECRM), security control and surveillance monitoring (Guo et al. 2008, Ramanathan et al. 2006, Patterson et al. 2007), biometrics(Lanitis et al.2004), and entertainment. There are two fundamental problems in designing the techniques, i.e., face image analysis and face image synthesis. In both cases Age of face has also been considered as an important semantic or contextual cue in social networks (Gallagher et al. 2008). Theoretically , Human Aging can be categorized into many phases as age between 1 to 10, 11 to 20, 21 to 50, 51 to 60 and above 60.

Kwon and Vitoria Lobo (Kwon et al. 1999) who proposed a method to classify input face images into one of the following three age groups: babies, young adults and senior adults. Their study was based on geometric ratios and skin wrinkle analysis. Their method was tested on a database of only 47 high resolution face images containing babies, young and middle aged adults.

The changes of face shape and texture patterns related to growth are measured to categorize a face into several age groups. These methods are suitable for coarse age estimation or modeling ages just for young people (Ramanathan et al. 2006). Gray level co-occurrence matrices (GLCM) introduced by Haralick for texture classification (Haralick et al. 1973, Haralick et al. 1979) for extracting the features of textures. The disadvantage of GLCM is high computational cost. The wavelet methods (Laine et al. 1993, Antonini et al. 1992, Van et al. 1999, Montiel et al. 2005) offer computational advantages over other methods for texture classification and segmentation. In our proposed method integrate the GLCM and wavelets for age group classification.

The rest of the paper is organized as follows. Section 2 describes wavelet and pattern trends based co-occurrence matrices and feature extraction method. Experimental results and comparison of the results with other methods are discussed in section 3 and conclusions are given in section 4.

2 GENERATION OF FEATURES OF WAVELET AND PATTERN TRENDS BASED CO-OCCURRENCE MATRIX

The proposed method of the feature extraction method is represented in the following figure 1. The block diagram consists of 6 steps

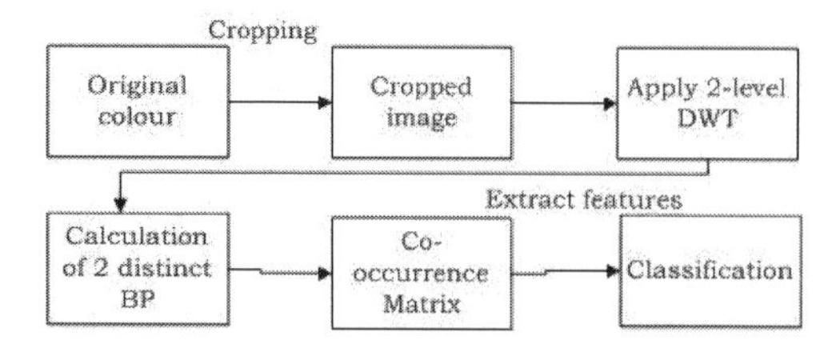

Figure 1. Block diagram of the proposed method.

The original facial image is cropped based on the location two eyes of in the first step as shown in figure 2. In the step 2, apply the wavelet to the image. In the third step, identify the 4 distinct patterns from that calculate 2-distinct pattern binary patterns. In the fourth step, co-occurrence matrix (CM) is formed on 2-distinct pattern LBP's. In the fifth step, statistical features are evaluated on the new CM for age group classification. In the last step using the k-nearest neighbor classifier is used for classification of a facial image (Yazdi et al. 2005).

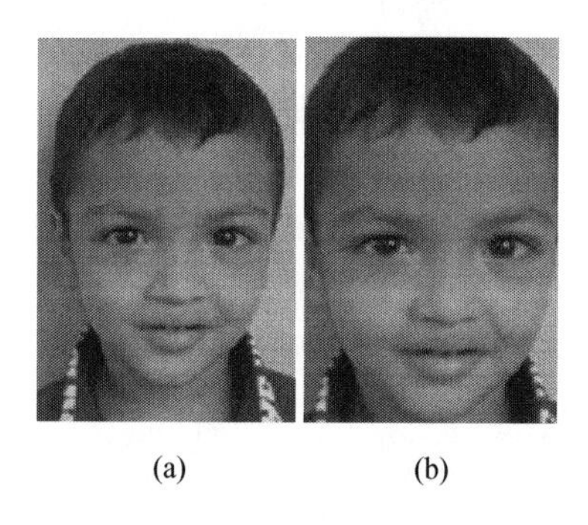

Figure 2. (a) original image; (b) cropped image.

2.1 *Generation of wavelet image*

The word wavelet is due to Morlet and Grossmann in the early 1980s. Today wavelets play a significant role in Astronomy, Acoustics, Nuclear Engineering, Sub-band Coding, Signal and Image Processing (Zhao et al. 2003), Neurophysiology, Music, Magnetic Resonance Imaging, Speech Discrimination, Optics, Turbulence, Earthquake Prediction, Radar, Computer and Human Vision, Data Mining and Pure Mathematics Applications such as Solving Partial Differential Equations etc.

The most commonly used transforms are the DCT, DFT, DWT, DLT and DHT. The present paper adopted DWT techniques to achieve better performance. DWT is a powerful tool of signal and image processing that have been successfully used in many scientific fields such as signal processing, image compression, image segmentation, computer graphics, and pattern recognition. The DWT based algorithms, has been emerged as another efficient tool for image processing, mainly due to its ability to display image at different resolutions and to achieve higher compression ratio. Haar wavelet is one of the oldest and simplest wavelet. Therefore, any discussion of wavelets starts with the Haar wavelet. The Haar, Daubechies, Symlets and Coiflets are compactly supported orthogonal wavelets.

The image is actually decomposed i.e., divided into four sub-bands and subsampled by applying DWT as shown in Figure 3(a). These subbands are labeled LH_1, HL_1 and HH_1 represent the finest scale wavelet coefficients i.e., detail images while the sub-band LL_1corresponds to coarse level coefficients i.e., approximation image. To obtain the next coarse level of wavelet coefficients, the sub-band LL_1 alone is further decomposed and critically sampled. This results in two-level wavelet decomposition as shown in Figure 3(b). Similarly, to obtain further decomposition, LL_2 will be used. This process continues until some final scale is reached. The values in approximation and detail images (sub-band images) are the essential features, which are shown here as useful for image analysis and discrimination. In this paper Haar wavelet, Daubechies wavelets, and Symlet wavelet are used for decomposition.

LL_1	HL_1
LH_1	HH_1

3(a) First level of DWT

LL_2 HL_2 / LH_2 HH_2	HL_1
LH_1	HH_1

3(b) second level of DWT

Figure 3. DWT Decomposition.

2.2 *Generation of LRDP-CHVP-GLCM matrix*

The present research LBP values are evaluated by comparing all 9 pixels of the 3x3 neighborhood with the average value of the neighborhood. The four distinct BP's are grouped into two distinct binary patterns called LRDP and LRDP. The present method derived age classification based on the features derived from a co-occurrence matrix generated from LRDP and CHDP. Most of the statistical methods suffer from the generalization problem due to the unpredictable distribution of the face images in real environment, which might be far different from that of the training face image. The structural method like LBP suffers from illumination effect. To avoid these problems, the present research combined structural and statistical methods based LRDP-CHVP-GLCM features. The calculation of LRDP-CHVP consists of seven steps described below.

Step 1: Convert each 3×3 window of facial image into binary values based on the following equation 1.

$$BM = \begin{cases} 0 \quad if \quad {}_{i,j} < V_0 \\ 1 \quad if \quad X_{i,j} > V_0 \end{cases} \tag{1}$$

Where V_0 is the mean of the 3×3 sub matrix.

Step 2: Convert each 3×3 binary window of facial image into four distinct binary patterns, named as LD-BP, RD-BP, VC-BP and HC-BP, which are shown in figure 4. All the four binary patterns contain three pixels only. For all these patterns involve the central pixel value of the 3x3 neighborhood window. It is significant. That is the reason in the present research pattern values are evaluated by comparing all 9 pixels of the 3x3 neighborhood window with the average value of the neighborhood window.

Step 3: Convert the binary patterns into decimal values by using binary to decimal conversion method. Each of them will have a decimal value ranging from 0 to 7.

Step 4: Based on the step 3 the present method divided the binary pattern units into two categories, i.e. LRDP and CHDP as given in equation 2 and 3 respectively.

$$\text{LRDP} = \text{LDP} + \text{RDP} \tag{2}$$

$$\text{CHDP} = \text{VCP} + \text{HCP} \tag{3}$$

Step 5: Generate the co-occurrence matrix based on values generated in step 5. The proposed GLCM is constructed by representing the LRDP values on the X-axis and CHVP values on the Y-axis as shown in figure 5 (c). This GLCM is named as LRDP-CHVP-GLCM. The LRDP-CHVP-GLCM has the elements of relative frequencies in both LRDP and CHVP as in figure 5(a) & 5(b). The values of the LRDP and CHVP ranges from 0 to 14. Thus the LRDP-CHVP-GLCM will have a fixed size of 15×15.

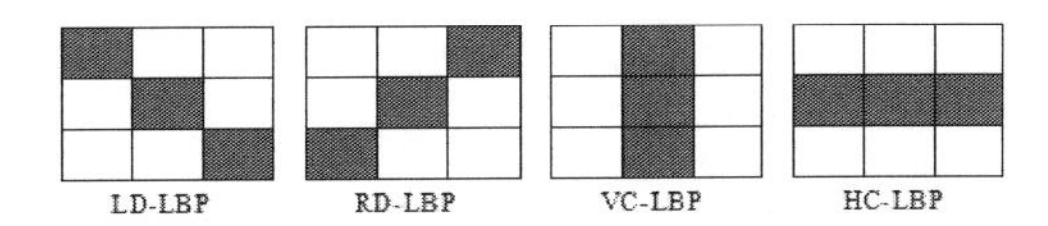

Figure 4. Four different patterns of BP.

LRDP	Fq1
0	
1	
2	
--	
14	

CHVP	Fq2
0	
1	
2	
--	
14	

	0	1	2	--	14
0					
1					
2					
--					
14					

Figure 5. (a) & (b) Frequency occurrences of LRDP and CHVP units; (c) LRDP-CHVP-GLCM matrix.

Step 6: From the generated LRDP-CHVP-GLCM, a set of 4 Haralick features i.e. energy, homogeneity, contrast and correlation are evaluated as given in equations 4, 5, 6 and 7.

$$Engery: \sum_{i,j} P(i,j)^2 \tag{4}$$

$$Homogenity = \sum_{i=0}^{C-1}\sum_{j=0}^{C-1} \frac{1}{1+(i-j)^2} P(i,j) \tag{5}$$

$$Contrast = \sum_{i=0}^{C-1} n^2 \sum_{i=1}^{C} \sum_{j=0}^{C-1} P(i,j) \tag{6}$$

$$Correlattion = \sum_{i=1}^{C-1}\sum_{j=1}^{C-1} \frac{iX_j P(i,j) - \{\mu_x \mu_y\}}{\sigma_x \sigma_y} \tag{7}$$

Step 7: By using K-nn Classifier on the proposed LRDP-CHVP-GLCM features, the facial image is classified as one of the category (Child age (0-15), Young Age(16-30), Middle Age(31-50) and Senior Age (>50).

3 RESULTS AND DISCUSSIONS

The proposed scheme established a database from the 1002 face images collected from FG-NET database and 600 images from Google database and other 700 images collected from scanned photographs. Some of these images are shown in figure 6. This leads to a total of 2302 sample facial images. In the proposed LRDP-CHVP-GLCM method the sample images are grouped into four age groups of Child age(0-15), Young Age(16-30), Middle Age(31-50) and Senior Age(>50). The contrast, correlation, energy and homogeneity features are extracted from LRDP-CHVP -GLCM of different facial images and the results are stored in the feature database. Feature set leads to a representation of the training images. The contrast, correlation, energy and homogeneity features derived from LRDP-CHVP-GLCM of four age groups of facial images are shown in tables 1, 2, 3, and 4 respectively. To evaluate the efficiency of the proposed method, 30 different facial images are considered from the FG - NET database, Google database and scanned photographs as a test database. The present chapter estimated distance between feature database and test database by using Euclidian distances. To classify the test images into appropriate age group the present method utilize K-nearest neighbor classifier with minimum distance. The successful classification results of proposed LRDP-CHVP-GLCM method on the test database are shown in table 5.

Table 1. Feature set values of LRDP-CHVP-GLCM for child age images.

NO	IName	Contrast	Corre lation	Energy	Homog enity
1	001A08	8.1819	0.8082	0.0321	0.5132
2	008A12	8.0168	0.7980	0.0345	0.5062
3	001A14	8.2747	0.8002	00.324	0.5043
4	002A12	8.2534	0.9282	0.0317	0.4909
5	002A04	8.0096	0.7938	0.0322	0.5003
6	002A05	8.1681	0.9123	0.0311	0.4998
7	001A10	8.1823	0.9210	0.0302	0.4989
8	002A07	8.1234	0.9089	0.0340	0.5001
9	002A17	8.2345	0.9093	0.0352	0.4899
10	002A12	8.2534	0.9282	0.0317	0.4909

Table 2. Feature set values of LRDP-CHVP- GLCM for young.

NO	IName	Contrast	Corre lation	Energy	Homog enity
1	001A16	7.3658	0.8331	0.0356	0.5032
2	001A19	7.5204	0.8218	0.0358	0.5072
3	001A29	7.2747	0.8202	00.359	0.5063
4	002A16	7.7747	0.8230	0.0354	0.5032
5	002A18	7.3096	0.8238	0.0364	0.5003
6	002A22	7.2681	0.8223	0.0351	0.5098
7	001A18	7.3823	0.8210	0.0352	0.5089
8	002A07	7.3234	0.8189	0.0357	0.5091
9	002A20	7.2345	0.8293	0.0355	0.5099
10	002A21	7.2534	0.8282	0.0360	0.5009

Table 3. Feature set values of LRDP-CHVP-GLCM for middle age images.

NO	IName	Contrast	Corre lation	Energy	Homog enity
1	001A43	8.6158	0.7330	0.0386	0.5970
2	002A31	8.5204	0.7211	0.0388	0.5972
3	003A29	8.5747	0.7205	00.389	0.5933
4	003A16	8.5647	0.7232	0.0384	0.5942
5	003A18	8.5306	0.7233	0.0381	0.5903
6	003A22	8.5268	0.7222	0.0382	0.5908
7	003A18	8.5323	0.7212	0.0390	0.5909
8	004A07	8.5234	0.7183	0.0383	0.5901
9	001A20	8.5345	0.7291	0.0385	0.5899
10	003A21	8.5234	0.7281	0.0390	0.5930

Table 4. Feature set values of LRDP-CHVP-GLCM for senior age images.

NO	IName	Contrast	Corre lation	Energy	Homog enity
1	001A43	8.6158	0.7470	0.0381	0.5970
2	002A31	8.5336	0.7441	0.0376	0.5951
3	002A38	8.5897	0.7405	00.388	0.5927
4	003A35	8.5947	0.7412	0.0384	0.5923
5	003A47	8.5906	0.7431	0.0380	0.5981
6	003A49	8.5998	0.7421	0.0381	0.6003
7	001A18	8.5993	0.7412	0.0386	0.5929
8	001A37	8.6004	0.7433	0.0382	0.5921
9	001A40	8.6005	0.7401	0.0389	0.5918
10	003A47	8.6004	0.7408	0.0380	0.5913

4 COMPARISON OF THE PROPOSED LRDP-CHVP-GLCM METHOD WITOTHER EXISTING METHODS

The proposed method of age classification is compared with the existing methods (Pullela et al. 2014). The method proposed by M. (Yazdi et al. 2013) classifies the age group using RBF Neural Network Classifier. The age group classification method proposed by Wen-Bing Horng et.al is based on two geometric features and three wrinkle features of facial image. PSVVS Ravi Kumar proposed a method for age group classification by integrating the features derived from Grey Level Co-occurrence Matrix(GLCM) with a new structural approach derived from four distinct LBP's (4-DLBP) on a 3 x 3 image. The percentage of classification of the proposed method and other existing methods are listed in table 6. The graphical representation of the percentage means the classification rate for the proposed method and other existing methods are shown in figure 7.

Table 5. Test vector feature set values of LRDP-CHVP -GLCM for different dataset images.

NO	IName	Contrast	Correlation	Energy	Homogenity	Group	Result
1	Gogle-in01	8.6058	0.8330	0.0383	0.6109	Middle	Success
2	Gogle-in02	8.5904	0.8311	0.0378	0.5979	Child	Success
3	Gogle-in03	8.5947	0.8295	00.381	0.5999	Young	Success
4	Gogle-in04	8.5944	0.8241	0.0379	0.5982	Child	Success
5	Gogle-in05	8.5906	0.8255	0.0377	0.5973	Middle	Success
6	Gogle-in06	8.5968	0.8372	0.0380	0.5998	Serior	Success
7	Gogle-in07	8.6003	0.8382	0.0378	0.6009	Middle	Success
8	Gogle-in08	8.6004	0.8293	0.0389	0.6191	Young	Success
9	Gogle-in09	8.5955	0.8294	0.0382	0.6089	Young	Success
10	Gogle-in10	8.5994	0.8289	0.0384	0.5998	Middle	Success

Table 6. Classification rate of the proposed LRDP-CHVP-GLCM method with other existing methods.

Image dataset	Method[14]	Method[15]	GLCM Method[16]	Proposed LRDP CHVP-GLCM
Google	89.67	90.52	92.23	94.35
FGNET	85.30	81.58	29.50	96.54
Scanned	88.72	85.42	91.50	92.64
Average	87.90	85.84	92.41	94.50

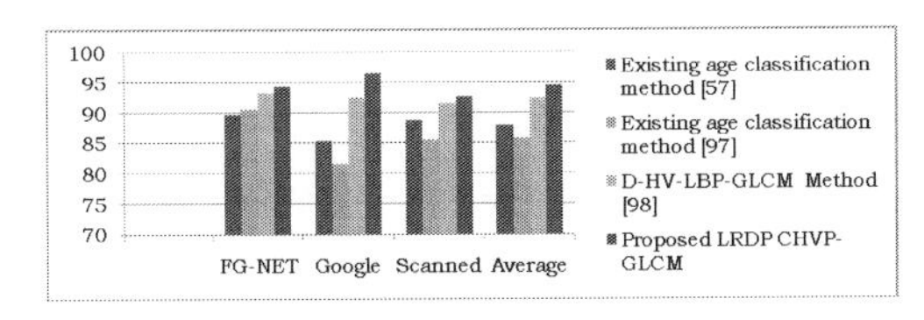

Figure 7. Classification chart of proposed.

5 CONCLUSIONS

The proposed LRDP-CHVP-GLCM reduced the computational time complexity because of the reduced size of the co-occurrence matrix size from 255 to 14. This new method combines the merits of both GLCM and LRDP-CHVP for the effective age classification purpose. The proposed method got high classification rate. The existing method D-HV-LBP-GLCM Method [98] also got good classification results, but the method is extended with wavelets give higher classification rate compare to existing methods.

ACKNOWLEDGEMENTS

Chunhui project by the ministry of education plans to “Wujiang Fuling based on Zigbee to Wulong period of the sewage outlet of real-time monitoring system of research”.

REFERENCES

Antonini, M., Barlaud, M., Mathieu, P. & Daubechies, I. 1992. Image coding using wavelet transform. *IEEE Trans. Image Processing* 1(2): 205–220.

Gallagher, A. & Chen, T. 2008. Estimating Age, Gender, and Identity Using First Name Priors. *Proc. IEEE Conf. Computer Vision and Pattern Recognition(CVPR), Anchorage, AK, 1–8 November 2008*. Piscataway: IEEE.

Guo Guodong, Fu Yun & Charles R. Dyer. 2008. Image-based Human Age Estimation by Manifold Learning and Locally Adjusted Robust Regression. *IEEE Transactions on Image Processing* 17(7): 1178–1188.

Haralick, R.M., Shanmugan, K. & Dinstein, I. 1973. Textural features for image classification. *IEEE Trans. Sysr., Man., Cybern.SMC* (3): 610–621.

Haralick, R.M. 1979. Statistical and structural approaches to texture. *Proceedings of the IEEE*. 64(7): 786–804.

Kumar, P. S., Kumar, V. V., & Venkatarao, R. 2014. Age Classification Based On Integrated Approach. *I.J. Image, Graphics and Signal Processing* 6(6): 50–57

Kwon, Y. H. & da Vitoria Lobo, N. 1999. Age Classification from Facial Images. *Computer Vision and Image Understanding Journal* 74(1): 1–21.

Laine, A. & Fan, J. 1993. Texture classification by wavelet packet signatures. *IEEE Trans. On PAMI*. 15(11): 1186–1190.

Lanitis, A. Draganova, C. & Christodoulou, C. 2004. Comparing different classifiers for automatic age estimation. *IEEE Trans. Syst., Man, Cybern. B, Cybern.* 34(1): 621–628.

Montiel, E., Aguado, A. S. & Nixon, M. S. 2005. Texture classification via conditional histograms. *Pattern Recognition Letters* 26(3): 1740–1751.

Patterson, E., Sethuram,A., Albert, M. Ricanek, K. & King, M. 2007. Aspects of age variation in facial morphology affecting biometrics. *IEEE Conf. on Biometrics: Theory, Applications, and Systems(BTAS), 1–6 July 2007*. Piscataway: IEEE.

Ramanathan, N. & Chellappa, R. 2006. Face verification across age progression. *IEEE Trans. on image Processing* 15(11): 3349–3361.

Ramanathan, N. & Chellappa, R. 2006. Modeling age progression in young faces. *in Proc. IEEE Conf. CVPR*. 1(3): 387–394.

Van de Wouwer, G., Scheunders, P., Livens, S. & Van Dyck, D. 1999. Wavelet Correlation Signatures for Color Texture Characterization. *Pattern Recognition* 32(3): 443–451.
Yazdi, M., Mardani-Samani, S., Bordbar, M. & Mobaraki, R. 2012. Age Classification based on RBF Neural Network. Canadian *Journal on Image Processing and Computer Vision* 3(2): 38–42.
Zhao, W., Chellappa, R., Phillips, P. J. & Rosenfeld, A. 2003. Face recognition: A literature survey. *ACM Computing Surveys* 35:399–458.

Electronics, Communications and Networks IV – Hussain & Ivanovic (eds)
© 2015 Taylor & Francis Group, London, ISBN: 978-1-138-02830-2

An implement of the digital certificate on electrical IC card

Huajun Chen
Southern Electric Power Research Institute Co. Ltd, Guangzhou, Guangdong, China

Yanrong Zhang & Qingqin Fu
National Electric Power Research Institute, Nanjing, Jiangsu, China

ABSTRACT: This paper analyzes disadvantages of electrical IC card traditional authentication method with the symmetric SM1 encryption algorithm, proposes a new IC card authentication method based on the digital certificate. Meanwhile, the paper introduces the digital certificate and chip operating system, and analyzes the mutual identification of electrical IC card using the digital certificate, also describes the trust flow of digital certificate chain. The cipher algorithm of the mutual authentication based on the digital certificate is asymmetric SM2 encryption algorithm. Compared with the traditional symmetric identification, the security level is much higher. Besides, the digital certificate ensures the confidentiality and integrity of both entities in communication. In a word, the implement of the digital certificate on electrical IC Card guarantees the communication of IC card in the electrical field greatly.

1 INTRODUCTION

With the rapid development of the Smart Grid, the security issue of electrical IC card arouses more and more attention. Chip Operating System, referred to as COS, its main function is to control the communication with outside and process the various commands inside. Among them, the exchange of information with outside is the basic requirement of COS. The focus of exchanging information with outside is the safety problem. Through successful authentication the read device can get access to IC card and achieve a certain usage right which can do some security-related operations (Nash et al. 2002).

The IC card security authentication mechanism includes two ways which are symmetric encryption algorithm and asymmetric encryption algorithm. The main algorithm of symmetric encryption is SM1 and DES, while the main asymmetric encryption algorithm is SM2 and RSA (Gao & Fang 2011). In communication symmetric encryption algorithm uses the same key, while asymmetric encryption algorithm uses two keys that are the public key and the private key in pairs. The security level of latter is much higher than the former.

The application of asymmetric public encryption algorithm is mainly digital envelope, digital signature and digital certificate. The digital certificate is a series of data which mark network user identity information (Guan 2002). It is the equivalent of a real-life personal identity, and it can be used as proof of the user identity in the network to provide trust foundation for the two mistrustful communication entities. Through the implement of the digital certificate on IC card, this paper has achieved mutual authentication between electrical IC card and reader device. As a result, the level of security authentication mechanism in COS has been improved a lot.

2 TRADITIONAL AUTHENTICATION METHOD

The electrical IC card traditional authentication uses symmetric encryption algorithm. The IC card and the reader device operate the same random with the same encryption algorithm and then determine the consistency of the two operation results. Through the above process, the legitimacy of the IC card or the reader device can be verified correctly. According to the authentication of the object, the authentication is divided into internal authentication and external authentication. Internal authentication is that the reader device verifies the legitimacy of IC card. External authentication is that IC card verifies the legitimacy of the reader device (Liu et al. 2006).

Such authentication method has its disadvantages. The traditional authentication method uses the fixed symmetric encryption algorithm. The symmetric encryption algorithm is easy to be attacked by the malicious attacks such as SPA and DPA (Yang & Ma 2011). The new method with the digital certificate has higher security mechanism than the traditional authentication method. The security level of the communication will be improved a lot by integrating the digital certificate with IC card.

3 THE DIGITAL CERTIFICATE AUTHENTICATION METHOD

The digital certificate uses asymmetric encryption algorithm. The algorithm used in this paper is SM2. The digital certificate is issued by the third party authority organization which is called CA Centre. The most important information includes the user name, user public key, the signature by CA and algorithm. The most common certificate format is X.509 v3 (Wang 2009). In a large-scale CA authority, the certificate issued work cannot be done by one CA, so the CA hierarchy needs to be established. If the CA among the system is trustful between each other, users can trust the certificate issued by any CA among the system through a certificate chain.

3.1 *CA and certificate chain*

The standard CA hierarchy can be described as an inverted tree. Root CA represents an entity that has special significance of the whole CA hierarchy. The root CA has zero or more layers of intermediate CA which can be called sub-CA. The leaf node corresponding to the non-CA entity is referred to as certificate unity (Wang 2009). The typical CA hierarchy is as follows:

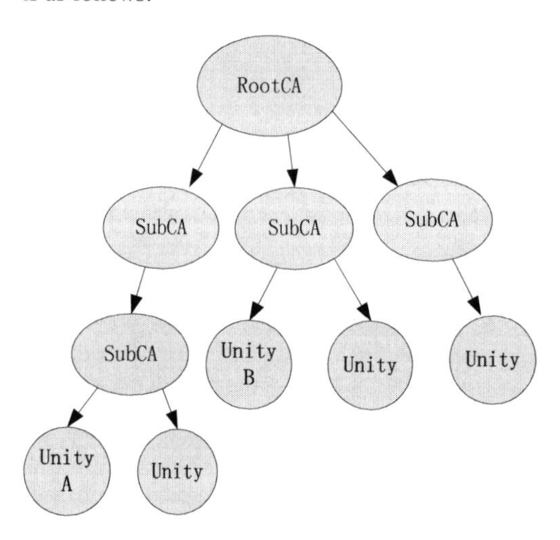

Figure 1. The typical CA hierarchy

- The establishment rule of CA hierarchy:

1. Root CA owns a self-signed certificate.
2. Root CA signs the CA below in order.
3. Bottom CA signs the leaf unities.
4. Each unity needs to trust root CA, not to care about the sub-CA, but its certificate is issued by the bottom CA.
5. Each sub-CA needs to keep two kinds certificates. One is the certificate issued to itself by other CA, the other is that it issues to other CA or unity.

- The trust process of certificate chain:

A certificate chain is a trusted path which consists of a series of interrelated certificates. In the CA hierarchy when a unity sees the certificate which presented by another unity, it can determine whether the certificate holder can be trusted through verifying the certificate path.

The following example describes that the unity A how to trust the unity B in the Figure 1.

1. Unity B contains information of its CA. Along the hierarchical tree until the root certificate, a certificate chain is formed.
2. The verifying process is that from the root unity A turns down to verify each signature of the chain unity B in turn.
3. The root certificate is self-signed with its public key for verification.
4. If all of the signatures are verified successfully, then unity A can make sure that all certificates are correct. If it trusts root CA then it can trust the certificate and public key of unity B.

3.2 *Two-level certificate's design*

IC Card achieves mutual identity authentication with the digital certificate. In this paper, the public encryption algorithm adopted is SM2. The communication unity is in the same level which is secondary. The root CA generates two-level certificate chain for each unity respectively. Figure 2 shows the two-level certificate topology system.

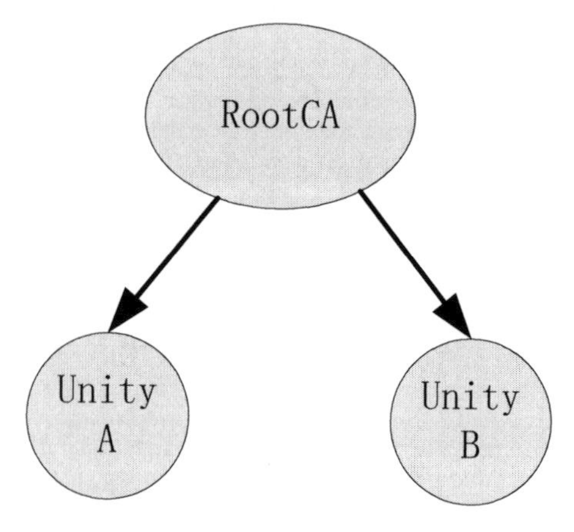

Figure 2. Two-level certificate topology.

- Generating certificates:

1 CA generates a root public key and a root private key in pair, and then issues the root certificate with the signature by its root private key.
2 CA generates a public key and a private key of unity A in pair, and then issues the certificate of unity A with the signature by the root private key.
3 CA generates a public key and a private key of unity B in pair, and then issues the certificate of unity B with the signature by the root private key.

- Verifying and trusting flow:

1 Unity A verifies the root signature with the root public key.
2 Unity A verifies the signature of unity B with the root public key.
3 If the above process is passed then unity A can trust unity B.
4 The process of unity B trusting unity A is the same as above in the converse direction.

3.3 *Two-level certificate's example*

X.509 certificate's data structure is described with ASN.1 which full name is Abstract Syntax Notation One, and coded with ASN.1 syntax. ASN.1 uses some blocks of data to describe the entire data structure. Each data block has four parts: the data type flag, the length, the value and the end flag of data block (Sun & Liu 2010).

In this paper, a certificate with two-level structure is shown in Figure 3. It comprises two certificates which are the root certificate and its certificate.

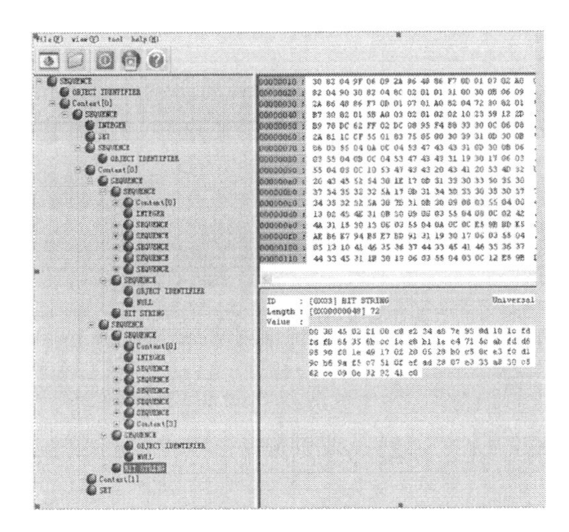

Figure 3. Two-level certificate's example.

3.4 *Comparison of the two methods*

In this paper the implemented method of a digital certificate on electrical IC card is proposed, the new method has much more security advantages compared with the traditional method. The asymmetric encryption algorithm SM2 is much more complex than the symmetric encryption algorithm SM1, and it is not easy to be attacked. The digital certificate uses a pair asymmetric key matched each other for encryption and decryption. Even if a third party tries to intercept communication data, but there is no corresponding private key, the data is unable to be decrypted likewise. This new method protects the communication process greatly. At the same time, the digital certificate ensures the authenticity and uniqueness of the two communication unities.

4 CONCLUSIONS

The electrical development technology based on the chip becomes more and more mature. This paper has applied the digital certificate on electrical IC card to realize the mutual authentication between IC card and outside. This new method not only ensures the authenticity of the electrical communication unities but also protect the confidentiality and integrity of the communication process. In the light of the security advantages of the digital certificate, it will be gradually applied in more electrical IC card application in the future.

REFERENCES

Andrew Nash, William Duane & Celia Joseph. 2002. *Public key infrastructure implementation and management of electronic security* Beijing: Tsinghua University Press.

Gao Liang & Fang Yong. 2011. A smart card-based mutual authentication scheme. *Institute of Information, Sichuan University* 2(44)82–89.

Guan Zhensheng. 2002. *Public key infrastructure and CA*. Beijing: Electronics Industry Press.

Liu Chun, Zhang Qishan & Fan Xiaohong. 2006. The application of smart card in PKI system. *Electronic Information Engineering Institute, Beijing University of Aeronautics and Astronautics* 9(5):122–130.

Sun Xianyou & Liu Ying. 2010. *X.509 identity authentication based on smart card.* Institute of Electronic Technology, PLA Information Engineering University.

Wang Aiying. 2009. *Smart card technology*. Beijing: Tsinghua University Press.Wang Yafei. 2011. A study of smart card password authentication scheme. *Computer Application and Software* 28(9): 295–297.

Yang Li & Ma Jianfeng. 2011. Trusted smart card password mutual authentication scheme. *Journal of University of Electronic Science and Technology of China* 40(1):128–133.

Electronics, Communications and Networks IV – Hussain & Ivanovic (eds)
ISBN: 978-1-138-02830-2

Practical realization of the stationary noise cancellation algorithm in the speech signals on the basis of SMV-canceller

O.V. Chernoyarov* & A.A. Makarov
National Research University "MPEI", Moscow, Russia

D.N. Shepelev
Moscow Financial and Law University "MFUA", Moscow, Russia

B. Dobrucky
University of Zilina, Zilina, Slovak Republic

ABSTRACT: We considered the speech filtration algorithm, which is effective in the presence of stationary disturbances and is based on the distinction in the rate of change of the useful signal and noise spectral densities. Its software and hardware real time realization was implemented. Experimental results confirming reliability and efficiency of the suggested noise canceller were presented.

1 INTRODUCTION

In voice data transmission systems the input signal contains the accompanying noises caused by an environment background (where speech signal is formed), internal apparatus noises and hindrances, which pass to the electro-acoustic converter together with useful signal (Vaseghi 2008). In some cases a noise level can be conspicuous enough, and that results in a speech signal degradation and deterioration of sound and phrase speech intelligibility as a consequence. One of the basic ways of the noisy signal refinement is the filtration (Vaseghi 2008). Unfortunately, many adaptive linear and nonlinear filtration algorithms (that are offered in the known literature and belong to the various signal models, for example – Markovian) depend on a choice of initial model parameters essentially, presuppose a multichannel realization frequently (Gur'ev 1983, Nazarov & Prokhorov 1985) and do not always provide a required signal-to-noise ratio (SNR) (Vaseghi 2008.). These conditions determine the necessity of the designing of most simple, effective and practical algorithms, when the specific signal and noise characteristics are especially considered.

In the present work, we specify the general problem of synthesis, with the focus on a practical important case when the additive mix of a speech signal and stationary noise is made available for the observation. We show that the considered algorithm is universal enough, and that it demands the minimum amount of the prior information and allows for the essential improvement of sounding (perception) quality. We introduce both the software and the hardware realization of this algorithm in real time, effected by means of application package MatLab 8.0 and the digital signal processor (DSP) of the TMS320 family, respectively.

2 SPEECH SIGNALS STATIONARY NOISE CANCELLATION TECHNIQUE

Let the additive mix of the useful signal $s(t)$ and the hindrance $\nu(t)$ passes to the receiver input:

$$x(t) = s(t) + \nu(t). \quad (1)$$

The useful signal $s(t)$ represents the speech transaction that occupies a frequency band from 0 to 4000 Hz and the source of which is not constrained. As hindrance, we will assume stationary random process with any distribution law. It is necessary to separate the useful signal $s(t)$ from observable realization $x(t)$ (1) with the specified level of intelligibility and quality.

The formulated filtration problem becomes complicated due to the fact that it is not obviously possible to construct adequate mathematical model of a speech signal in general. Besides, the received (under the specified model of a useful signal) theoretical results cannot be always generalized on a case of non-Gaussian hindrances. Thereupon, it is necessary to put into practice the approaches to filtration algorithm synthesis based on comparison of the signal and hindrance characteristics, which differ from each other considerably.

*Corresponding author: o_v_ch@mail.ru

In our study, we apply the feature of speech signal nonstationarity for noise cancellation algorithm synthesis. Speech signal nonstationarity appears through its characteristics varying with time, sometimes changing abruptly, while remaining approximately constant on short intervals of 10÷30 ms (a so-called local stationarity feature of a speech signal) (Oppenheim & Schafer 1975, Picone 1993). Hindrance characteristics, due to its stationarity, will remain constants during the whole observation interval.

One of basic parameters, the dynamics of which we can trace easily enough, is a signal energy. Through energy we can characterize the intensity of received realization (1) both in the whole frequency range of a speech signal (from 0 to 4 kHz) and in any frequency band. The last case is especially important, because, as a rule, the hindrances mixing with speech transaction have bandwidth considerably shorter than a useful signal bandwidth. Thus, it is possible to identify the approximate spectral structure of a hindrance with its subsequent removal (filtration) tracing energy variations in various bandwidths. For optimal partitioning of the whole spectral range of a speech signal into subbands, we will consider how speech spectral analysis is carried out by a human ear.

In a number of researches (Nazarov & Prokhorov 1985, etc.) it was established that sound "analog-to-digital conversion" occurs in an internal ear along a membrane plane. Various areas in a cochlea, each of which contains neutral receptors are tuned in various frequency bands. Empirical researches allowed to create a modern representation about critical ranges corresponding to its own area in a cochlea. From the experimental point of view the critical frequency band can be defined as a frequency band, in which sound signal abrupt changes can be subjectively separated. Sensation level of a narrow-band sound signal source in case of the constant level of sound pressure remains constant even if the frequency band will be expanded to critical band and thereafter loudness will begin to amplify. The detection threshold of a narrow-band sound signal source between two masking tones remains constant until the area of frequency division between two tones will lie within a critical frequency band. For the average listener the critical band can be approximated following expression (Picone 1993, O' Shaughnessy 1987):

$$BW_c(f) = 25 + 75\left[1 + 1.4(f/1000)^2\right]^{0.69}. \qquad (2)$$

Function (2) is continuous, however, it is possible to present an ear as a discrete set of the bandpass filters named as a bank of critical band filters on its basis (Picone 1993, O' Shaughnessy 1987).

In accordance with the above, the cancellation procedure of stationary hindrances in a speech signal can be presented in the form of following consecutive operations: time processing, spectral analysis, filtration as shown in Figure 1.

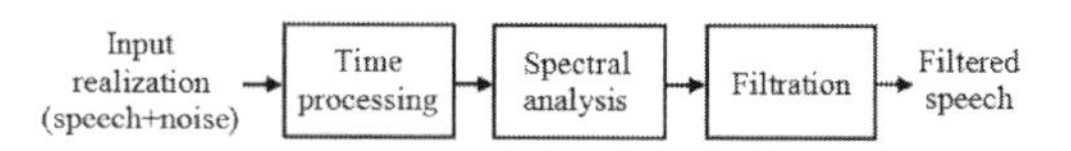

Figure 1. The basic stages of stationary hindrance cancellation in a speech signal.

Procedure of time processing is not trivial on the structure and contains a number of consecutive operations displayed in Figure 2.

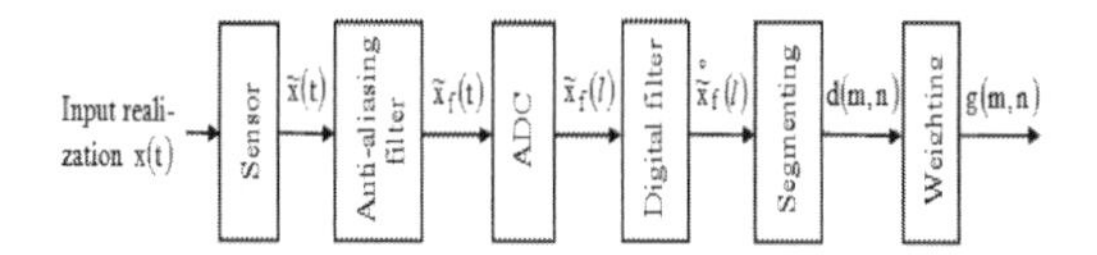

Figure 2. Time processing of the observable data realization.

The first four blocks intend for providing the fullest signal representation in the digital form with the minimum loss of the useful information. For this purpose high-sensitivity microphones with a wide frequency band and low own resistance are usually used as sensors. The anti-aliasing filter requires prevention of the information loss because of aliasing during the signal analog-to-digital conversion (ADC) (Marple 1987). The selection procedure of the digital filter is more important in the scheme presented. Its transfer characteristic should correspond to following two assumptions concerning a speech signal. Firstly, it is supposed that sonant consonants have attenuation in spectral domain (20 dB per decade, approximately) in view of physiological features of speech organs (Markel & Gray 1976). The second assumption consists that the ear is more sensitive to spectral components above 1 kHz (Markel & Gray 1976). In a number of works (Markel & Gray 1976, etc.) it is offered to use an adaptive filtration for the spectral conditioning of a speech signal and amplification of spectral components in the high-frequency range. However, in real systems preference is given to digital filters of the first order with one coefficient ζ_p which values belong to the range from –1.0 to –0.4 (Picone 1993. According to (Boll 1979), the ζ_p value named also as emphasis coefficient was equal –0.8, so

$$\overset{\circ}{\tilde{x}}_f(l) = \tilde{x}_f(l) - 0.8\,\tilde{x}_f(l-1). \qquad (3)$$

Here $\tilde{x}_f(l)$ is the ADC output signal (Figure 2), $\overset{\circ}{\tilde{x}}_f(l)$ is the output signal of the digital filter and l is the discrete time.

Segmenting procedure for calculation of current parameters of the input realization is, as a rule, carried out on the basis of a technique "frame-for-frame"

(Picone 1993). This kind of the analysis is often named as analysis with overlapping, since the part of the data change with each new step only. Mathematically the signal $d(m,n)$ on the segmenting block output (Figure 2) can be presented as follows

$$d(m,n)=\begin{cases} d(m-1,L+n), & 0\le n<D, \\ \overset{\circ}{\tilde{x}}_f(n-D), & D\le n<L+D. \end{cases} \quad (4)$$

Here m is the number of a current segment, n is the sample number in a current segment, L is the length of a modified segment, and D is the overlapping size. The specified segmenting way guarantees tracking of all acoustic phenomena in a speech signal providing smooth enough variation of parameters from frame to frame at the same time.

For capture of fast dynamics of an instantaneous spectrum of input realization, the frame length in the noise canceller (NC) was chosen to be 10 ms, due to the fact that speech has sharp spectral transitions which can end by spectral peaks with shifts on 80 Hz/ms, so it is not recommended to use frames by duration less than 8 ms (Markel & Gray 1976). In that case we get 80 discrete samples by frame (L=80) at an input signal to the standard PCM-format with a sampling frequency 8000 samples/s (the sampling frequency used in telephone communication channels). As the segment size should be small also, so that spectrum details are not to be excessively smoothed, and the relative size of overlapping $D/(L+D)$ should be chosen to be no less than 0.3, so the segment duration was assumed to be equal to 13 ms, that corresponds D=24 and $L+D$=104 samples.

For minimization of the unwanted “tail” effects conditioned by segmenting, each segment is multiplied by weight function (window) $w(n)$ with length of $L+D$ samples:

$$\tilde{g}(m,n)=d(m,n)w(n). \quad (5)$$

Following (Boll 1979), for segment weighing we will use a trapezoidal window of a kind

$$w(n)=\begin{cases} \sin^2\left[\pi(n+0,5)/2D\right], & 0\le n<D, \\ 1, & D\le n<L, \\ \sin^2\left[\pi(n-L+D+0,5)/2D\right], & L\le n<L+D. \end{cases} \quad (6)$$

In the spectral analysis the time array $\tilde{g}(m,n)$ is associated with array in the frequency domain determined by discrete Fourier transformation (DFT). It is well known (Marple 1987), that DFT algorithm is most effective (takes the minimum calculating expenses), if the number of elements of the transformable array can be presented as $2p$ where p is some natural number. Thereupon, the initial array $\tilde{g}(m,n)$ before frequency transformation is preliminary modified by means of zero padding procedure (Marple 1987) as follows:

$$g(m,n)=\begin{cases} \tilde{g}(m,n), & 0\le n<L+D, \\ 0, & L+D\le n<N. \end{cases} \quad (7)$$

Number N in Eq. (7) is chosen as the nearest integer to $L+D$ assuming representation $N=2^p$. Thereby, if $L+D$=104 than we have $N=2^7=128$. Operation (7) allows to realize DFT algorithm using less than $N\log_2 N$ additions and $N\log_2(N/2)$ multiplications of complex numbers (so-called fast Fourier transformation with an exception of the calculating ways containing only zero values) (Marple 1987). The modified segment $g(m,n)$ resulting spectrum

$$G(m,k)=\mathrm{DFT}\left[g(m,n)\right] \quad (8)$$

will contain 128 samples ($k=\overline{0,127}$) and will be symmetric relative to the normalized frequency $\tilde{f}=63.5$. Here as DFT [·] the DFT operator is designated. Hereupon, the further processing can be conducted for first 64 $G(m,k)$ samples only. We will divide the full frequency range (from 0 to 63) into 16 channels corresponding to critical bands of a human ear (Picone 1993). Then low $\tilde{f}_L(i)$ and high $\tilde{f}_H(i)$ normalized frequencies of i-th channel will be defined as follows:

$$f_L(i)=\{2,4,6,8,10,12,14,17,20,23,27,31,36,42,49,56\},$$
$$(9)$$
$$f_H(i)=\{3,5,7,9,11,13,16,19,22,26,30,35,41,48,55,63\},$$

$i=\overline{0,15}$. Here the numbers divided by a comma correspond to different values of i. Further, we will make an energy estimate in frequency bands corresponding to each channel:

$$\tilde{E}_{ch}(m,i)=\frac{1}{\tilde{f}_H(i)-\tilde{f}_L(i)+1}\sum_{k=\tilde{f}_L(i)}^{f_H(i)}|G(m,k)|^2, \quad (10)$$

$i=\overline{0,15}$. In practice, current estimates (10) are passed through the digital filter of the first order in order to receive more robust estimates of channel energies. As a result, we have estimates

$$E_{ch}(m,i)=\alpha_{ch}(m)E_{ch}(m-1,i)+\left[1-\alpha_{ch}(m)\right]\tilde{E}_{ch}(m,i), \quad (11)$$

$i=\overline{0,15}$. Here $\alpha_{ch}(m)$ is the smoothing coefficient of channel energy equal to zero for the first segment and 0.45 for all subsequent ones.

Instead of Eq. (11), for emulation of logarithmic reaction of the human auditory system we will use the logarithm of channel energy times 10, $E_{dB}(m,i)$, which adds up the channel energies in decibels:

$$E_{dB}(m,i)=10\lg\left[E_{ch}(m,i)\right],\ i=\overline{0,15}. \tag{12}$$

Variations of energy values with time can be estimated by means of quantity

$$\Delta_E(m)=\sum_{i=0}^{15}\left|\ E_{dB}(m,i)-\overline{E}_{dB}(i)\ \right|, \tag{13}$$

Where $\overline{E}_{dB}(i)$is the average value of energy in i-th frequency channel expressed in decibels. As the value $\overline{E}_{dB}(i)$ is unknown, we use an estimate $\tilde{E}_{dB}(i)$ instead of it (Boll 1979):

$$\begin{aligned}\tilde{E}_{dB}(i)&=\alpha(m-1)\tilde{E}_{dB}(m-1,i)+\\&+\left[1-\alpha(m-1)\right]E_{dB}(m-1,i).\end{aligned} \tag{14}$$

Here $\alpha(1)=1$ and $\alpha(m)$, $m>1$ is the weighing coefficient characterizing the m-th segment contribution to revaluation of average energy in i-th channel. The value $\alpha(m)$ can be calculated as follows. A total energy of m-th segment in decibels

$$E_{\Sigma}(m)=10\lg\left[\sum_{i=0}^{15}E_{ch}(m,i)\right] \tag{15}$$

is linearly interpolated between their admissible minimum E_L=30 dB and maximum E_H=50 dB values:

$$E_{\Sigma}(m)=\left\{\left[\alpha(m)-\alpha_L\right]E_H+\left[\alpha_H-\alpha(m)\right]E_L\right\}/\left(\alpha_H-\alpha_L\right), \tag{16}$$

where α_L=0.5, α_H=0.99 are weighing coefficients corresponding to linear interpolation borders. Then from Eq. (16) for coefficient $\alpha(m)$ is received

$$\alpha(m)=\alpha_H-\left(\alpha_H-\alpha_L\right)\left[E_H-E_{\Sigma}(m)\right]/\left(E_H-E_L\right), \tag{17}$$

Energy variations $\Delta_E(m)$, calculated according to Eqs. (13)-(17), help us to establish the character of observable realization, namely: if throughout 50 and more segments (with possible breaks no more than in 6 segments) value $\Delta_E(m)$ does not exceed 28 dB, then the realization of the observable data is accounted as a stationary process (i.e., does not contain a useful signal).

If hindrance intensity is rather big, then in the absence of a useful signal the stationarity condition of observable data realization formulated above cannot be always fulfilled. Thereupon, for each processed segment (4), the SNR was also traced. Under small SNR, it is decided that a useful signal is absent in a current segment, and vise versa. Control of SNR for each segment was carried out on the basis of a voice metrics method (Boll 1979). According to this method, for all allocated frequency subbands channel SNRs were calculated

$$\sigma_q(i)=10\lg\left[E_{ch}(m,i)/E_N(m,i)\right],\ i=\overline{0,15}. \tag{18}$$

Here $E_N(m,i)$ is the hindrance energy in i-th subband. Calculated SNRs were limited from below and above by levels 0 dB and 89 dB accordingly, and they were also quantized with step of 1 dB:

$$\sigma_q'(i)=\max\left(0,\min\left(89,\operatorname{round}\left(\sigma_q(i)/0.375\right)\right)\right), \tag{19}$$

$i=\overline{0,15}$. Here round(·) is a rounding of number to the nearest integer. Based on the received quantized SNR values (19), the sum of voice metrics of a current segment $v(m)$ was determined:

$$v(m)=\sum_{i=0}^{15}V\left[\sigma_q'(i)\ \right], \tag{20}$$

where $V(k)$ is k-th element of voice metrics vector V:

$$\begin{aligned}V=\{&2,2,2,2,2,2,2,2,2,2,2,3,3,3,3,3,4,4,4,5,5,5,6,6,7,7,7,8,8,9\\&9,10,10,11,12,12,13,13,14,15,15,16,17,17,18,19,20,20,21,22,23,\\&24,24,25,26,27,28,28,29,30,31,32,33,34,35,36,37,37,38,39,40,\\&41,42,43,44,45,46,47,48,49,50,50,50,50,50,50,50,50,50,50\ \}.\end{aligned} \tag{21}$$

If $v(m)<35$, then SNR for the analyzed segment is considered as small, and the segment oneself – as purely noise.

With the conditions of stationarity or SNR smallness fulfilled, we are allowed to make a revaluation of a hindrance channel energy for the current segment of realization of the observable data by means of the smoothing filter of the first order with coefficient α_n=0.9 (Boll 1979):

$$E_n(m+1,i)=\alpha_n E_n(m,i)+\left(1-\alpha_n\right)E_{ch}(m,i), \tag{22}$$

$i=\overline{0,15}$. There are some techniques for definition of initial value $E_n(0,i)$. For practical realization of the noise cancellation algorithm we calculated the value $E_n(0,i)$ by standard way (Marple 1987) on the previous 40 segments which do not contain a useful signal, as it was supposed.

Values $E_n(m,i)$ allow calculating of the total cancellation level for a current segment:

$$\gamma_n=-10\lg\left[\sum_{i=0}^{15}E_n(m,i)\right]. \tag{23}$$

It is assumed that value γ_n cannot be less –13 dB, i.e. $\gamma_n=\max(-13,\gamma_n)$. Cancellation coefficients for everyone frequency subband were corrected with the help of the found channel SNR as

$$\gamma_{dB}(i) = \gamma_n + 0.39\left[\sigma'_q(i) - \sigma_{q\min}\right]. \tag{24}$$

Here $\sigma_{q\min}$=6 dB is minimally possible channel SNR. For the purpose of optimization of cancellation coefficient value (24) quantized channel SNRs σ'_q(i) were preliminary modified to $\sigma''_q(i)$ by a following rule. If since the fifth channel it is not coming up to five channels with SNR not less than 12 dB in everyone, then all SNRs smaller 12 dB are taken equal to value $\sigma_{q\min}$. In addition, if the condition $v(m) < 46$ is satisfied, then other channel SNRs are equated to minimally possible value $\sigma_{q\min}$ also. Otherwise (if there is not less than five $\sigma'_q(i)$: $\sigma'_q(i) \geq 12$, $i \geq 5$) updating is not made.

Transformation of factors (24) from logarithmic to linear scale is made as

$$\gamma_{ch}(i) = \min\left(1,\ 10^{\gamma_{dB}(i)/20}\right). \tag{25}$$

In Eq. (25) it is taken into account that the cancellation coefficient cannot possess the value big 1.

Using the found cancellation coefficients (25) and according to (Boll 1979), the filtration of a current segment in frequency domain was carried out for each channel:

$$H(m,k) = \begin{cases} \gamma_{ch}(i)G(m,k), & \tilde{f}_L(i) \leq k \leq \tilde{f}_H(i),\ 0 \leq i \leq 15, \\ G(m,k), & 0 \leq k < \tilde{f}_L(0). \end{cases} \tag{26}$$

Further, for transition in the time domain the inverse DFT for array $H(m, k)$ was implemented and sequence of the signal samples cleared from hindrance already was restored. In Figure 3 the block diagram of NC blocks of the spectral analysis and filtration (Figure 1) is shown.

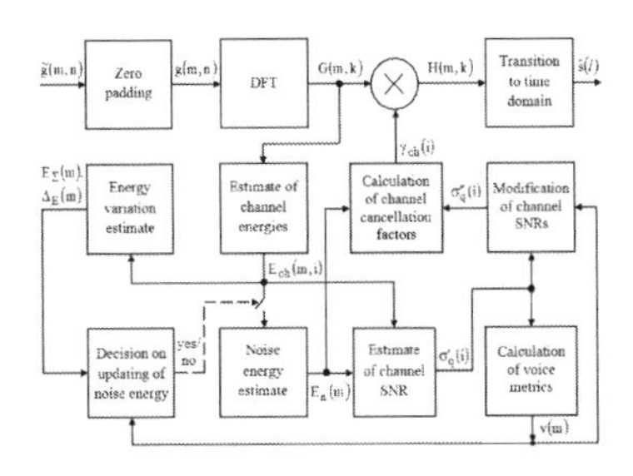

Figure 3. The block diagram of noise cancellator blocks of the spectral analysis and filtration.

3 MODELING RESULTS OF NOISE CANCELLATION ALGORITHM

The considered cancellation algorithm of stationary hindrances in a speech signal was represented as softwired (in MathLab 8.0 system) and hardwired (based on the DSPs of the TMS320 family). As observable data realization the additive mix of a speech signal and engine noise of various transport vehicles was used. Working efficiency of the noise cancellation algorithm was estimated by the SNR value of an NC output:

$$\text{SNR} = 10\lg\left[\sum_l s^2(l) \Big/ \sum_l \left(s(l) - \hat{s}(l)\right)^2\right]. \tag{27}$$

Here $s(l)$ are samples of an initial speech signal, $\hat{s}(l)$ are signal samples on an NC output. In particular, if the input SNR was −3 dB then the output SNR has made ~ 6 dB; if 0 dB, then ~ 9 dB; if 6 dB then ~ 21 dB. The conducted researches have shown that other choice of frequency subbands (9), a voice metrics vector (21) and thresholds used in algorithm do not lead to quality enhancement of noise cancellation, anyway.

In Figures 4–7 the qualitative illustration of NC work is shown. In Figure 4 time diagrams of the speech signal distorted by engine noise of car "Volga" on an NC input and output are presented. In Figures 5–7 similar time diagrams of the speech signal distorted by engine noise of car "Zhiguli" (Figure 5), helicopter MI-8 (Figure 6) and armored personnel carrier BTR-90M (Figure 7) are displayed.

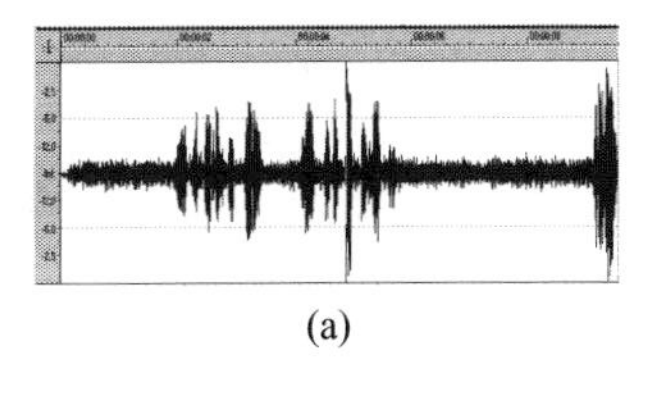

(a)

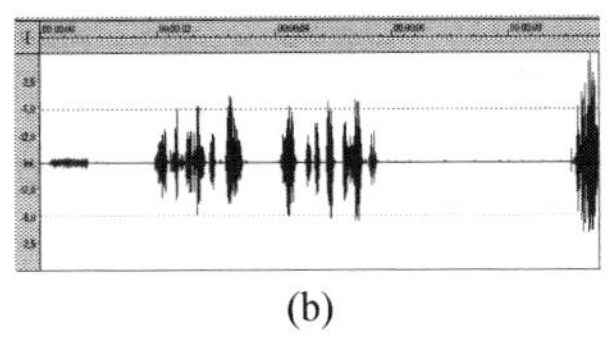

(b)

Figure 4. The test speech signal distorted by car "Volga" engine noise on the noise canceller input (a) and output (b).

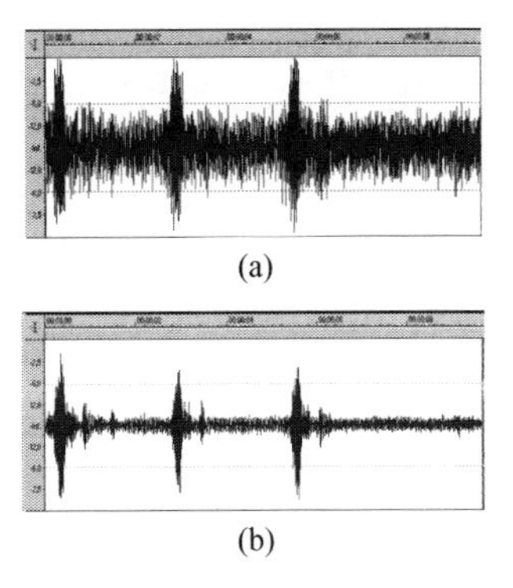

(a)

(b)

Figure 5. The test speech signal distorted by car "Ziguli" engine noise on the noise canceller input (a) and output (b).

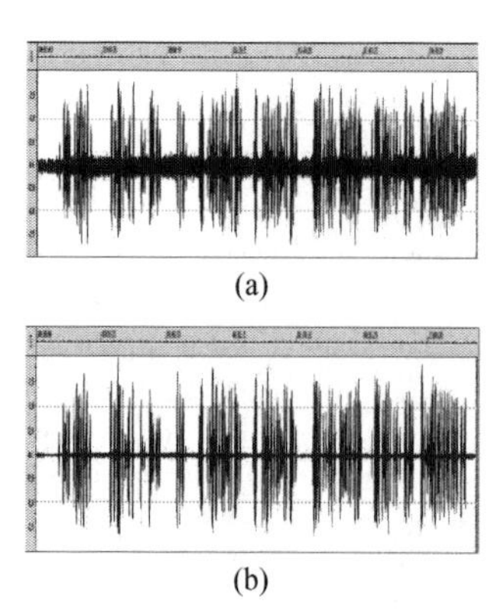

Figure 6. The test speech signal distorted by helicopter MI-8 engine noise on the noise canceller input (a) and output (b).

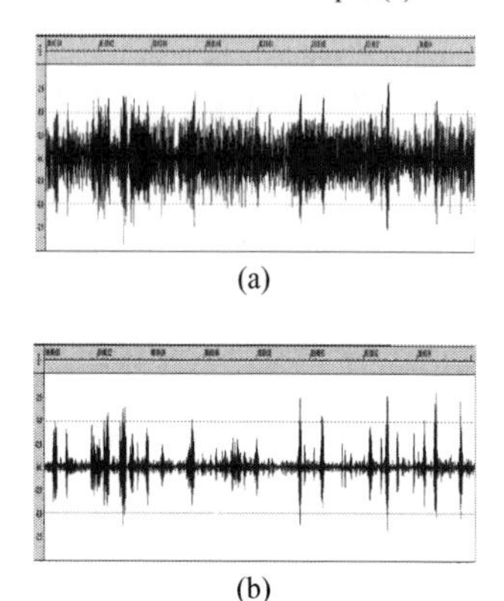

Figure 7. The test speech signal distorted by armored personal carrier BTR-90M engine noise on the noise canceller input (a) and output (b).

As appears from Figures 4–7 the use of the noise cancellation algorithm allows considerable reduction of the noise level in a speech signal and, as consequence, it helps to raise speech intelligibility and sounding comfort.

Thus, it can be drawn a conclusion on availability of considered NC for suppression of additive (quasi) stationary hindrances superimposed on speech messages. According to the received characteristics, the quality of its work can be estimated as commercial (Nazarov & Prokhorov 1985).

4 CONCLUSION

For increase of sound and phrase intelligibility of the speech distorted by stationary noise, random process and with any distribution law and other statistical characteristics, it is possible to use the filtration algorithm, based on distinctions in the spectrum changes of a useful signal and a hindrance, estimated through energy variations in frequency bands, corresponding to critical frequency bands of a human ear. Working capacity, efficiency and practical reliability of a given algorithm in real time is corroborated by the results of computations (MathLab 8.0 application) and hardware (based on digital signal processors of the TMS320 family) modeling.

Results of work have the rather general character and can be used in designing of various adaptive systems for the elimination of an influence of interfering factors, which characteristics are constant or change slowly with time.

ACKNOWLEDGEMENTS

The reported study was supported by the Russian Science Foundation (research project No. 14-49-00079).

REFERENCES

Boll, S.F. 1979. A Spectral Subtraction Algorithm for Suppression of Acoustic Noise in Speech. *Acoustics, Speech, Signal Processing, IEEE International Conference on ICASSP'79, Washington, 2–4 April 1979*. Washington: IEEE.

Gur'ev & Yu.Yu. 1983. Markov Nonlinear Filtration of a Speech Signal from a Mix with Stationary Noise [in Russian]. *Radiotekhnika* (12): 48–51.

Hartmann, W.M. 1998. *Signals, Sound, and Sensation*. New York: Springer Verlag.

Markel, J.D. & Gray, A.H. 1976. *Linear Prediction of Speech*. New York: Springer-Verlag.

Marple, S.L. & Jr. 1987. *Digital Spectral Analysis with Applications*. Englewood Cliffs NJ: Prentice-Hall.

Nazarov, M.V. & Prokhorov, Yu.N. 1985. *Methods of Digital Processing and Passing of Speech Signals* [in Russian]. Moscow: Radio i Svyaz'.

Oppenheim, A.V. & Schafer, R.W. 1975. *Digital Signal Processing*. Englewood Cliffs NJ: Prentice-Hall.

O'Shaughnessy, D. 1987. *Speech Communication: Human and Machine*. New York: Addison Wesley.

Picone, J. 1993. Signal Modeling Techniques in Speech Recognition. *IEEE Proceedings* 81(9): 1215–1247.

Vaseghi, S.V. 2008. *Advanced Digital Signal Processing and Noise Reduction*. Chichester: John Wiley & Sons Ltd.

Electronics, Communications and Networks IV – Hussain & Ivanovic (eds)
© 2015 Taylor & Francis Group, London, ISBN: 978-1-138-02830-2

A stage manager system based on SIP technology

Dejing Cui & Peng Sun
Hengde Digital Choreography Technology Co., Ltd, Qingdao Economic and Technological Development Zone, Shandong, China

Binguo Wang, Yanlei Zhang & Yue Hou
College of Information Science and Engineering, Shandong University of Science and Technology, Qingdao, Shandong, China

ABSTRACT: This paper proposes a stage manager system based on SIP technology. The system uses SIP as the signaling protocol. By utilizing the RTP protocol transmission media and combining with the VoIP soft switching technology, it realizes real-time speech and video communication between departments on stage by one interface. The whole system is a multimedia communication platform based on IP network. Experimental results show that: The stage manager system not only realizes visualization dispatch and centralized control for the stage, waiting area and other areas, but also makes multi-system integration simple and clear. It improves mobile integrated command and dispatching ability of stage managers, directors and other personnel, and the communication is not limited by the distance. That there is no signal interference or signal source blind area can ensure the instructions of system seamless docking.

1 INTRODUCTION

The digital stage manager system has been an important research direction in digital communication technology for more than ten years since it is proposed. With the wider application of digital technology in the theatre and the improvement of people's cultural life, higher quality of the stage show is in demand. At present, the digitization stage manager system becomes a hot research topic, and it has very broad development prospects and practical value.

Stage supervision system is an auxiliary system to serve the theater cast and crew. It mainly aims to meet the stage manager's need of supervision, management, dispatching. Stage supervision system is also known as a stage manager system, which includes internal communications subsystems, lighting tips subsystems, reminders field broadcast subsystems and video surveillance subsystems. (National Theatre Stage Audio Technology Group 2008, Zhou & Xue 2009). There have been some more successful stage manager systems abroad, for example, Telex, Altair and Clear-Com. However, its intercom systems and video supervision systems often exist as a separate. In addition, wired intercom and wireless intercom often have their respective host. They will make the system complicated and cumbersome to operate, prone to error, and inconvenient to maintain. A stage manager system has not appeared at home and abroad, which is the whole IP of wireless intercom signals, wired intercom signals and the video signals.(Li et al. 2014) The existing stage supervision systems can realize reminders field, lighting tips, intercom and other basic functions. However, usually they have many issues such as multi-system and multi-host being independent, standard being not united, devices being not compatible, the pertinence of call being too low, (Li et al. 2014) call being limited by distance, and communication blind area. Therefore, they have not satisfied the high demand of the stage manager.

In order to solve the above problem, this paper proposes a stage manager system based on SIP (Si et al. 2005, Wang et al. 2014)technology. By using SIP as the signaling protocol and the RTP (Qiu & Wang 2006) protocol transmission media, and combining with VOIP(Li et al. 2002, Song et al. 2008) soft switching technology, the proposed system can realize real-time speech and video communication between departments on stage by an interface. The whole system is a multimedia communication platform based on IP network. Experimental results show that: The stage manager system not only realizes visualization dispatch and centralized control for the stage, waiting area and other areas, but also makes the multi-system integration simple and clear. It improves mobile integrated command and dispatching ability of stage managers, directors and other personnel, and the communication is not limited by the distance. That there is no signal interference or no signal source blind area ensures the seamless docking of instructions.

2 VOIP

VoIP (Voice over Internet Protocol), in a nutshell, is an analog signal (Voice) digitization, and the signal is transmitted on an IP network in the form of the data packet. Compared with the traditional business, VoIP is providing more and better service which is able to be widely used on the Internet and global IP interconnection environment.

As displayed in Figure 1, the basic principle of VoIP is to compress the encoded speech data through voice compression algorithm, and then voice data is packaged according to the TCP / IP standard. The data packets are delivered to receiving area through the IP network; afterwards, the voice data packet is strung; then, after decompression it will be restored to the original voice signal, so as to achieve the purpose of transmitting voice from the Internet.

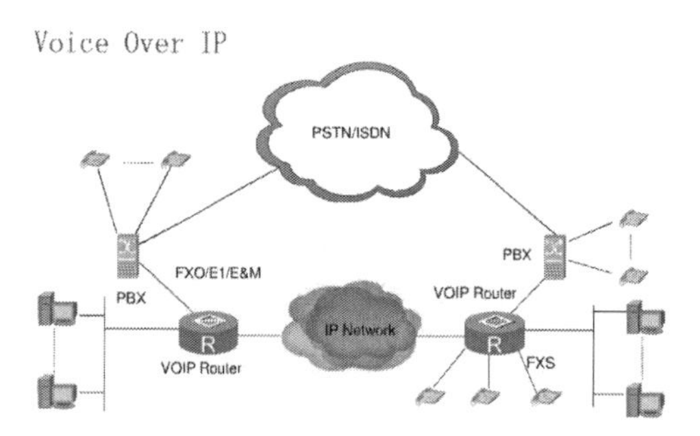

Figure 1. The basic principle of VoIP.

The core and key device of IP phone is IP gateway. It maps the regional telephone area code to the corresponding regional gateway IP address. This information is stored in a database, and the data connection processing software will complete call processing, digital voice package, routing and management and other functions.

3 SIP

SIP (Session Initiation Protocol, RFC 3261) is the Internet conferencing and telephony signaling protocol from the IETF (the development of the Internet Standards Organization). It is an application layer protocol that can establish, modify and suspend multimedia sessions or calls. It is an ASCII-based protocol, and it provides "dating" service on the Internet.

As an application layer signaling protocol, SIP is not a complete communications system solution. It requires other solutions or protocols combined to achieve the entire system. For example, real-time transport protocol (RTP) (RFC1889) is used to transmit audio and video and other real-time streaming data.

Compared with the traditional H.323 protocol, SIP has obvious advantages:

Excellent scalability: greatly improves the processing capability of the system;

Close integration with the Internet: makes communication easier and more convenient;

Remarkable openness: provides not only good support for mobile phones, PDA and other mobile devices, but also benefits online instant communication, voice and video data transmission and multimedia applications.

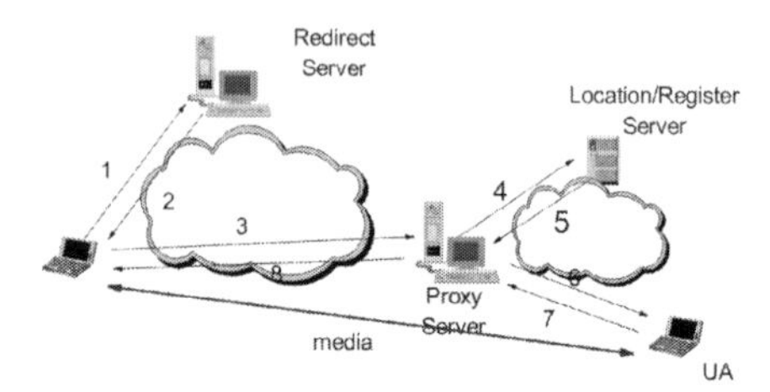

Figure 2. The principle of SIP.

SIP and HTTP (client-server protocol) have a similar structure, as shown in Figure 2. The client sends a request to the server. Then the server processes the request and sends a response to the client. The request and response form a transaction.

4 STAGE MANAGER SYSTEM BASED ON SIP TECHNOLOGY

4.1 *Stage manager system architecture based on SIP*

The stage manager system, based on SIP, comprises a core switch, stage manager server, stage manager sets, video access gateways, video server, wrist-mounted terminal, terminal control room and reminders field terminals. Figure 3 shows the topology map:

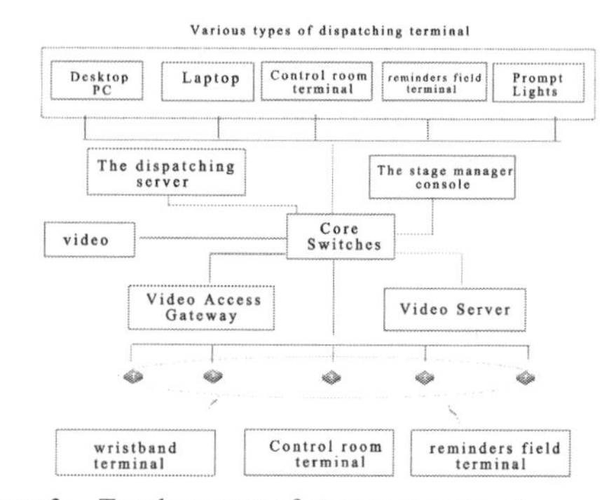

Figure 3. Topology map of stage manager system.

4.2 *Functional description of the main entities*

The stage manager server is the core of the entire system. By using the international standards SIP protocol as the signaling protocol to design and combining with VOIP soft switch technology, it achieves registration authentication for each subsystem of the stage manager system. Logically, it is divided into registration server, proxy server, media processing server and video server.

The Video Access Server is a powerful video proxy gateway. It realizes the seamless switch of implements the traditional monitoring devices and IP soft. The independent monitoring system will be extended to the soft switching subsystems to achieve the convergence of communications monitoring system and soft switch systems. Video Access Server supports standard SIP protocol, which can be used as a standalone IP soft switch server that supports user registration. Locally, registered users can view their video surveillance agents.

Control room terminals are deployed on the stage of the machine room, lighting and sound control room and so on. The control room terminal is the fixed equipment in the control room, and it is used to communicate with the working groups or staff. The control room terminal can display multiple Road surveillance videos and easily control the room scheduling.

Reminders field terminals are deployed in the dressing room, VIP room and so on. The Reminders field terminal is the fixed equipment. It is one-way communication. By playing the stage manager's voice, you can display one or more channel surveillance video. Meanwhile, reminders field terminal can play cable TV signal to enable you to switch freely between the TV signal and reminders field terminal software.

Smart three anti-mobile phones are deployed as mobile wristband terminal, which may be carried out at anytime or anywhere by stage manager dispatching. Mobile workers equipped with Wristband terminal that can carry the job without space constraints. Wristbands terminal can display a video surveillance, and it can enable the communication in work groups.

4.3 *SIP basic call setup process*

SIP basic call setup process as shown in Figure 4:

(A) When a user off-hook to initiate a call, user agents A initiated Invite request to the proxy server of the region;

(B) Proxy server confirms that the user authentication has passed through the authentication /accounting, and then checks whether the request message Via header field already contains its address. If it does, indicates that the occurrence of loopback response and an error is returned. If there is no problem, the proxy server inserts its own address in the Via header field of the request message. And it transmits Invite request to user agents B which is Invite message of Invite To field indicated.

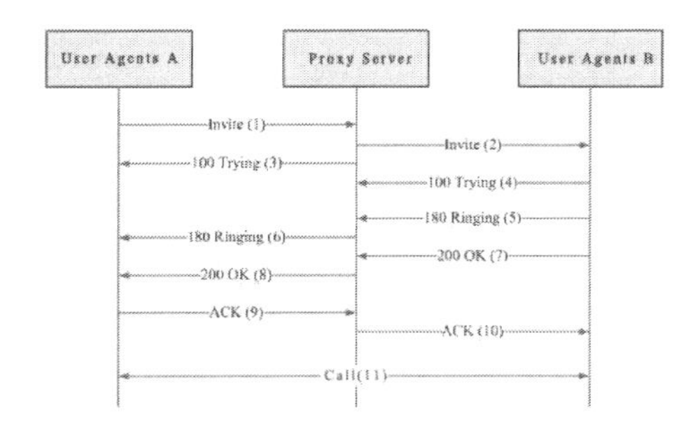

Figure 4. SIP basic call setup process.

(C) Proxy server sends the answering call processing information: 100 Trying to user agents A.

(D) User agents B send the call processing in the response information: 100 Trying to the proxy server.

(E) User agents B instruct the called user ringing. User sends 180Ringing information to the proxy server after ringing.

(F) Proxy server forwards the called ringing formation to user agents A.

(G) The called user goes off-hook, and the user agents B returns a connection with a successful response (200 OK) to the proxy server.

(H) Proxy server forwards the success indicator (200 OK) to user agents B.

(I) User agents A sends an ACK message for confirming after receiving information from the proxy server.

(J) The proxy server forwards the ACK confirmation message to the user agents B.

(K) To establish a communication connection between the calling and called user, start talking.

4.4 *Principle of the system*

The stage manager system is designed based on IP soft switch architecture. It uses SIP as the signaling protocol. The media are transmitted by RTP protocol, and are combined with VOIP soft switch technology. Wrist-mounted audio and video mobile terminal sends the real-time audio streaming to dispatch host. After the dispatch host looking up the routing, the call is routed to the system's built-in conference room. In order to achieve interconnection and interworking among the various terminals and real-time communication, the real-time data acquired by cameras will be transmitted to the video access gateway, and other types of terminal establish a call through video access gateway to real-time view the surveillance video. In the system built conference room. You can adjust the encoding format or transmission on-demand to achieve dynamic grouping. The dispatching host management, the member of conference room, and each terminal can flexibly leave from or join the meeting. During the meeting, the terminal can send orders by the dispatching host to achieve listening or speaking. The terminals upload audio and video data which are packed by RTP protocol, forwarding from the server. In this way, the system can effectively solve the NAT traversal problem.

5 THE RESULTS OF THE EXPERIMENT

In order to verify the correctness of the method, we use C# as the experimental platform to realize the stage manager system. Through several experiments, it can quickly and efficiently realize speech and video call between departments on stage.

Figure 5. Interface of the system.

As displayed in Figure 5, the top right corner of the interface is the main video area. It displays "main video" which is set in the configuration stage manager station. When the stage manager system logs in, the only one main video opens automatically.

Top left corner of the interface is the auxiliary video area. It shows other videos addition to the main video. Auxiliary videos are opened and closed by video surveillance subsystem in the stage manager station operating area. The auxiliary video area can display multiple videos. Every time the videos are turned on or off, the program will automatically adjust to the most appropriate display.

Below part of the interface is the operating area. It includes the title bar, timer, internal communication subsystem, lighting subsystem, reminders field subsystem and video surveillance subsystem.

Delay Test Results: Videos delay within 20ms; voice delay of the terminals wired connection within 20ms; voice delay of terminal wireless access within 300ms; and with the increasing use of time, delay time gradually increases.

6 CONCLUSIONS

At Present, a digitization stage manager system has not appeared at home and abroad, which based on SIP technology. By using SIP as the signaling protocol and the RTP protocol transmission media, and combining VOIP soft switching technology, the proposed system realizes real-time speech and video communication between departments on stage by an interface. The whole system is a multimedia communication platform based on IP network. Although experimental results show that: The stage manager system possesses the advantages of integrated design, low cost deployment, perfect compatibility, high stable performance, low maintenance cost, and being simple and easy to operate. It also can realize the HD and SD hybrid decoding on the wall, remote mobile terminal monitoring, image control, video downloads, video transformation, audio and video dispatching, action with alarm, log query, authority management and other functions.. However, many questions remain insoluble. For example, the system also has a time delay. So, stage manager systems need to be further optimized.

ACKNOWLEDGEMENT

The authors are grateful for the support from National Key Technology R&D Program (2013BAH60F01).

REFERENCES

Li Lin, Wang Xingang & Chai Qiaolin. 2002. Implementation of SIP Protocol in Open VOIP Model. *Computer Engineer* 28(z1): 222–226.

Li Yihua, Zhang Xiaodong & Luan Zhenhui 2014. Study on Stage Video Supervision and Scheduling System Based on IP Network. *Journal of Anhui University of Science and Technology (Natural Science)* 2: 19–22.

Li Zhen, Zhao Xuejun & Yang Qianyi. 2014. Research of Networked Intercom System for Stage Dispatch. *Journal of Communication University of China Science and Technology* 1: 16–18.

National Theatre Stage Audio Technology Group. 2008. Theatre stage manager system. *Art Science and Technology* 3: 45–49.

Qiu Ying & Wang Ku. 2006. The Implementation of Net Video Watch System Base on RTP. *Microcomputer Applications* 27(4): 436–439.

Si DuanFeng, Huan XinHui & Long Qin. 2005. A Survey on the Core Technique and Research Development in SIP Standard. Beijing: *Journal of Software* 16(2): 239–250.

Song Xiuhong, Xiao Zongshui & Wei Benjian. 2008. Analysis and study of DoS attacks on SIP-based VoIP network. *Computer Engineering and Design* 27(10): 2479–2482.

Wang Shuang, Lian DongBen & Kang HongNan. 2014. Central Signaling Control Server Based on SIP. *Computer Systems & Applications* 23(3): 93–97.

Zhou Kesheng & Xue Bin. 2009. JBC City TV studio theater sound and stage monitoring systems design and implementation. Beijing: *International Pro-Audio & Lighting*4: 22–24.

Electronics, Communications and Networks IV – Hussain & Ivanovic (eds)
© 2015 Taylor & Francis Group, London, ISBN: 978-1-138-02830-2

A new Kalman filter model for deconvolution of single channel of seismic record

Xiaoying Deng* & Zhengjun Zhang
School of Information and Electronics, Beijing Institute of Technology, Beijing, China
Department of Statistics, University of Wisconsin, Madison, WI, USA

ABSTRACT: To remove the effects of seismic wavelet's complexity on the resulting seismic record and improve the vertical resolution of the earth, a new model for the Kalman filter on the seismic deconvolution is proposed in this paper, considering that the reflectivity function is not changeable in all recursive computations for single channel of seismic record. The new model and detailed filtering steps are given. Results of experiments show that for the synthetic seismic record with the Ricker wavelet or the damped sine wave, the Kalman filter based on the new model can compress the wavelet to improve the seismic resolution, but the Kalman filter based on the old model cannot do it. So the Kalman filter based on the new model works better than the one based on the old model.

1 INTRODUCTION

In seismic exploration, the seismic record is generally considered as the convolution result of a seismic wavelet with a reflectivity function which corresponds to the layered earth. To remove the effects of seismic wavelet's complexity on the resulting seismic record and compress the seismic wavelet so as to improve the vertical resolution of the earth, many deconvolution techniques are available in time or frequency domain such as Wiener filtering (Robinson & Treitel 1967), Kalman filtering (Crump 1974, Mahalanabis et al. 1981, 1983, Song & Lv 2009, Kumiadi & Nurhandoko 2012), homomorphic filtering (Ulrych 1971), spectral whitening (Coppens & Mari 1984), inverse Q filtering (Wang 2008), Gabor deconvolution (Margrave et al. 2011, Wang et al. 2013), recursive maximum posteriori probability algorithm (Machalanabis et al. 1982), principle phase decomposition (Baziw & Ulrych 2006), and distributed algorithms (Plataniotis et al. 1998). Kalman filter (Kalman 1960) was proposed in 1960, and then was widely applied in tracking, navigation, control, communication and etc. It is an efficient recursive filter which uses the state equation and measurement equation to describe a linear, time-varying system. In seismic exploration, Crump first used the discrete Kalman filter successfully for the deconvolution of seismic signals to generate an estimate of the reflectivity function. Then Mahalanabis proposed fast and adaptive Kalman filter algorithms. Song discussed detailedly how to choose the initial state vector and estimate covariance. Kurniadi combined the Kalman filter with primitive deconvolution.

In the above approaches based on Kalman filter for seismic deconvolution, the same Kalman model was nearly built. In this paper, a new Kalman filtering model for the deconvolution of seismic record is proposed. Not like the old Kalman state equation, we present a new state equation considering that the reflectivity function is not changeable in all recursive computations for single trace of seismic record. Using the new proposed Kalman system model, the simulations on the synthetic seismic records show the proposed model method works better than the conventional Kalman filter.

2 KALMAN FILTER MODEL FOR SEISMIC DECONVOLUTION

For single seismic channel, it is well known that the seismic signal can be modeled as

$$y(n) = \sum_{m=1}^{M} b(m) r(n-m) + e(n), n = 1, 2, \cdots, N, \quad (1)$$

in which $y(n)$ is the discrete seismic signal with a length of N, $b(m)$ is the seismic wavelet with a length of M, $r(n)$ is reflection coefficient, and $e(n)$ is additive noise. According to this convolution model of seismic signal, most of seismic deconvolution methods based on Kalman filter used the following Kalman system model

*Corresponding author: *xydeng@bit.edu.cn*

$$\mathbf{x}(k) = \mathbf{\Phi}(k,k-1)\mathbf{x}(k-1) + \mathbf{g}u(k-1), \tag{2}$$

$$y(k) = \mathbf{H}(k)\mathbf{x}(k) + v(k). \tag{3}$$

In the state equation (1), the state vector $\mathbf{x}(k)=[x(k)\ x(k\text{-}1)\ \ldots\ x(k\text{-}M\text{+}1)]^T$. And the state transition matrix $\mathbf{\Phi}(k, k\text{-}1)$ describes the relationship between the state vectors at two iterations

$$\mathbf{\Phi}(k,k-1) = \begin{bmatrix} \varphi_1(k) & \cdots & \varphi_{M-1}(k) & \varphi_M(k) \\ 1 & \cdots & 0 & 0 \\ \vdots & \vdots & \vdots & \vdots \\ 0 & \cdots & 1 & 0 \end{bmatrix} \tag{4}$$

which is expected to change with depth. Assigning the first line all zeros is equivalent to assuming that the reflection coefficients form a white random sequence. Also someone simply chose the first line all $1/M$, which means $x(k)$ equals the average of the M previous values. Furthermore, $\mathbf{g}=[1\ 0\ \ldots\ 0]^T$, and the random noise $u(k\text{-}1)$ is often supposed as the white noise sequence. In the measurement equation (2), on the assumption that all traces are generated by the same wavelet, the observation matrix $\mathbf{H}(k)=[b(1)\ b(2)\ \ldots\ b(M)]$, that is the seismic wavelet, and $v(k)$ is the measurement noise which is also often assumed as the white noise sequence.

After the above model is completely defined, the seismic data can be sequentially processed by using the standard discrete Kalman filtering equations, which will be list detailedly in the next section. Finally after N iterations, the reflected coefficients can be estimated by extracting the first component of every estimated state vector.

3 NEW KALMAN FILTER MODEL

For single channel of seismic record, the reflectivity function implies the attributes of the vertical layered earth along the center point between the shot and the geophone. So the reflectivity sequence is deterministic and not changeable in the whole recursive processes sample by sample. Let the state vector $\mathbf{x}(k)=[x(1)\ x(2)\ \ldots\ x(N)]^T$ with the same length as the seismic record, instead of the same length as the seismic wavelet used in the past Kalman filter model. Then the following new model is proposed

$$\mathbf{x}(k) = \mathbf{x}(k-1), \tag{5}$$

$$y(k) = \mathbf{W}(k)\mathbf{x}(k) + v(k). \tag{6}$$

In the new measurement equation (6), the observation matrix, $\mathbf{W}(k)=[w(1)\ w(2)\ \ldots\ w(M)]$ and $v(k)$ means the same as the one in (3). Note that $\mathbf{W}(k)$ is cut out from the seismic wavelet sequence extended with $2N$-2 zeros shown as follows

$$B = \begin{bmatrix} \underbrace{0 \ \cdots \ 0}_{N-1} & b(M) & \cdots & b(1) & \underbrace{0 \ \cdots \ 0}_{N-1} \end{bmatrix}$$

and $w(i)=B(N\text{-}k\text{+}1\text{-}(M\text{-}i))$ for $\mathbf{W}(k)$.

3.1 *Initialization of Kalman filter*

Before recursive filtering, the initialization for the variables must be done. We can let

$\mathbf{P}(0) = \sigma^2\mathbf{I}_N, \hat{\mathbf{x}}(0) = \begin{bmatrix} 0 & \cdots & 0 \end{bmatrix}^T, \mathbf{W}(1) = \begin{bmatrix} b(1) & 0 & \cdots & 0 \end{bmatrix}$ in which σ^2 is the variance of the measurement noise which is supposed known or can be estimated by using some statistical methods.

3.2 *Recursive processes*

After the initialization, according to minimum mean square error rule, one can deduce the following filtering equations for iteration k from 1 to N just like other references.

1 Predictions for state vector, measurement vector and state covariance matrix

$$\hat{\mathbf{x}}(k) = \hat{\mathbf{x}}(k-1), \tag{7}$$

$$\hat{y}(k,k-1) = \mathbf{W}(k)\hat{\mathbf{x}}(k,k-1), \tag{8}$$

$$\mathbf{P}(k,k-1) = \mathbf{P}(k-1) \tag{9}$$

2 Inovation and gain

$$\mathbf{S}(k) = \mathbf{W}(k)\mathbf{P}(k,k-1)\mathbf{W}^T(k) + R(k), \tag{10}$$

$$\mathbf{K}(k) = \mathbf{P}(k,k-1)\mathbf{W}^T(k)\mathbf{S}^{-1}(k), \tag{11}$$

where $R(k)$ is the variance of measurement noise $v(k)$.

3 Correction for state vector and state covariance matrix

$$\hat{\mathbf{x}}(k) = \hat{\mathbf{x}}(k,k-1) + \mathbf{K}(k)\left[y(k) - \hat{y}(k,k-1)\right], \tag{12}$$

$$\mathbf{P}(k) = \mathbf{P}(k,k-1) - \mathbf{K}(k)\mathbf{S}(k)\mathbf{K}^T(k) \tag{13}$$

Up to now, an iteration is finished, and then k is added one and go back to 1) for next iteration until $k=N$. Finally $\hat{\mathbf{x}}(N)$will be the desired output—estimated reflectivity sequence.

From the above new model, the state equation has a great change from (2) to (5). The state transition matrix $\mathbf{\Phi}(k, k\text{-}1)$ and state noise $u(k\text{-}1)$ are removed, which is reasonable because we change the state vector into the whole reflected coefficients instead of part of them. In the past, the problem of optimally estimating

$\mathbf{\Phi}(k,\ k\text{-}1)$ is not solved well. Nevertheless, this problem disappears in the new model. So the new model not only removes the boring estimation problem, but also simplifies the followed filtering computations.

4 SIMULATIONS

For investigating the performance of the new Kalman model for single-channel seismic deconvolution, many simulations have been done and compared with the conventional Kalman model.

According to the convolution model (1) of seismic signal, the noisy seismic records with different noise levels are synthesized in our experiments. Two different seismic wavelets are used as shown in Figure 1, in which one is the common Ricker wavelet with the dominate frequency f 30Hz

$$b(t) = (1 - 2\pi^2 f^2 t^2) e^{-\pi^2 f^2 t^2}, \tag{14}$$

the other is a damped sine wave

$$b(t) = e^{-100t} \sin(100t) \tag{15}$$

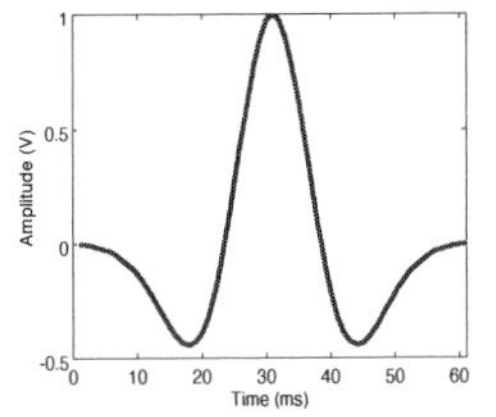

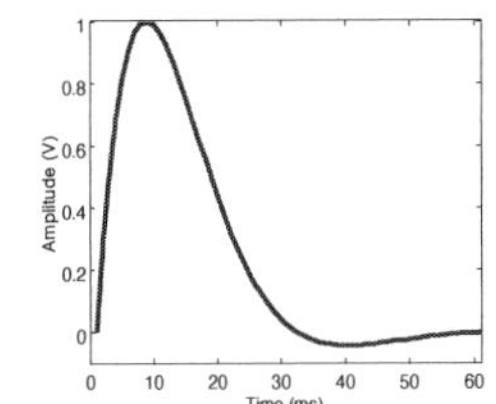

Figure 1. Seismic wavelets: (a) Ricker wavelet; (b) damped sine wave(normalized).

As an example for seismic deconvolution, Figure 2 shows a set of deconvolution result. Note that all plots including the following figures are normalized for easy comparison. Figure 2a gives the reflected coefficients that means there are two reflected interfaces, and Figure 2b shows the synthetic noisy signal polluted by Gaussian white noise with the signal-to-noise ratio (SNR) 30dB. The SNR is defined as follows

$$SNR = 10 \log_{10}\left(\frac{A^2}{\sigma_n^2}\right) \text{dB}, \tag{16}$$

where A is the maximal amplitude of useful signal, and σ_n^2 is the variance of the additive white noise. Figure 2c and 2d show the deconvolution results by using the proposed new model and the conventional Kalman model, respectively. From Figure 2c, we can see that the seismic wavelets are compressed, which means the improvement of seismic resolution. More importantly, the peak-value positions of wavelets are completely accurate. However, from Figure 2d the original peak-value positions almost change into the beginning time of the deconvolution results. Apparently, the new model method is better than the old even if the noise left in Figure 2c is stronger than the one in Figure 2d.

When the noise becomes stronger, the Kalman filter based on the new model still works better than the old model. Figure 3 gives another example for deconvolution with SNR=15dB. From Figure 3b, the noise is so strong that the seismic wavelets are distorted and the background is very bad. From Figure 3c and 3d, the noise becomes weak after Kalman filtering, and the filtered result based on the new model is better than the old model.

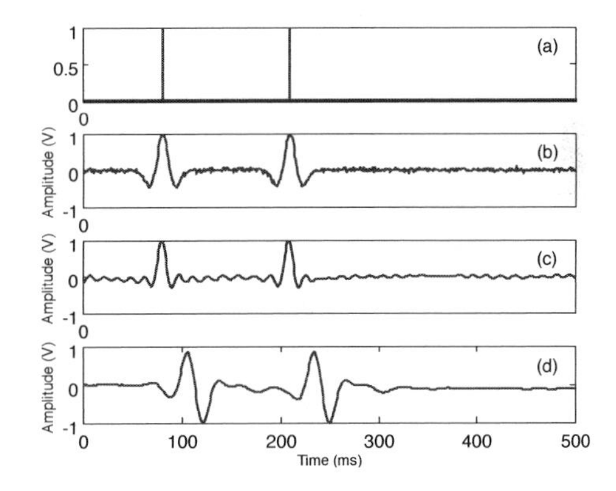

Figure 2. Example for deconvolution with SNR=30dB: (a) reflected coefficients; (b) noisy seismic record; (c) deconvolution result by using the new model; (d) deconvolution result by using the old model.

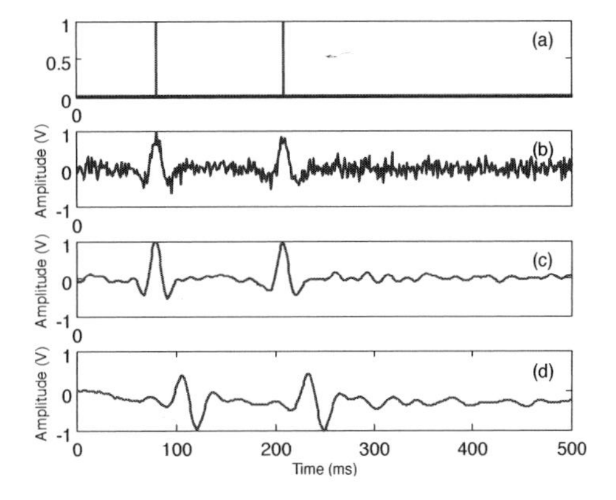

Figure 3. Example for deconvolution with SNR=15dB: (a) reflected coefficients; (b) noisy seismic record; (c) deconvolution result by using the new model; (d) deconvolution result by using the old model.

For testing the ability of Kalman filter to improve the seismic resolution, the simulations for thin layer are done. Figure 4 shows a set of deconvolution results for thin layer with SNR=30dB. Figure 4a shows that there is a thin layer between 208ms and 219ms. From Figure 4b, the thin layer is so thin that the two seismic wavelets are overlapped and cannot be identified. But from Figure 4c, the thin layer can nearly be recognized after using Kalman filter based on the new model because two wavelets can be seen. On the contrary, from Figure 4d, the thin layer still cannot be identified after using Kalman filter based on the old model. So the new model is better than the old.

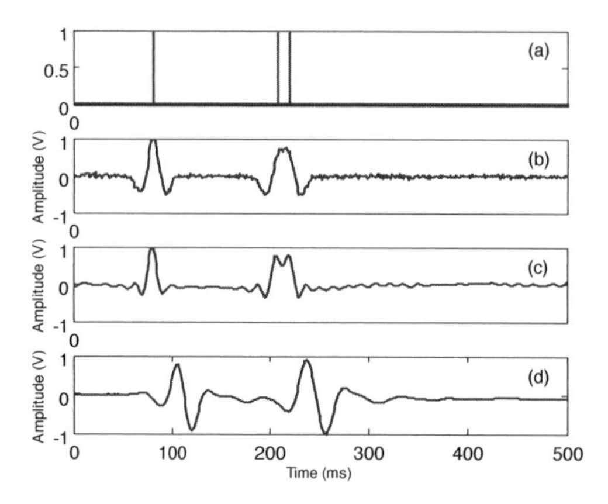

Figure 4. Deconvolution for thin layer with SNR=30dB: (a) reflected coefficients; (b) noisy seismic record; (c) deconvolution result by using the new model; (d) deconvolution result by using the old model.

In the above simulation, the used seismic wavelet is Ricker wavelet which is zero-phase and symmetrical. In order to test on different wavelet, the damped sine wave (15) is also used, which is minimum-phase and asymmetrical. The normalized damped sine wave is shown in Figure 1b. Figure 5 shows a set of deconvolution simulation for the signal with the damped sine waves and a thin layer. Note that in Figure 5b the peak-value positions are not aligned with the reflected points in Figure 5a because the reflected points should be aligned with the centers of the damped sine wavelet, and the third wavelet is not differentiated from the second wavelet. They are overlapped and seem one wavelet. But after Kalman filter based on the new model, the two overlapped wavelets are separated with appearing two spikes as shown in Figure 5c. However, in Figure 5d, the two overlapped wavelets still are almost kept the original state only smoothed, i.e. the noise is removed.

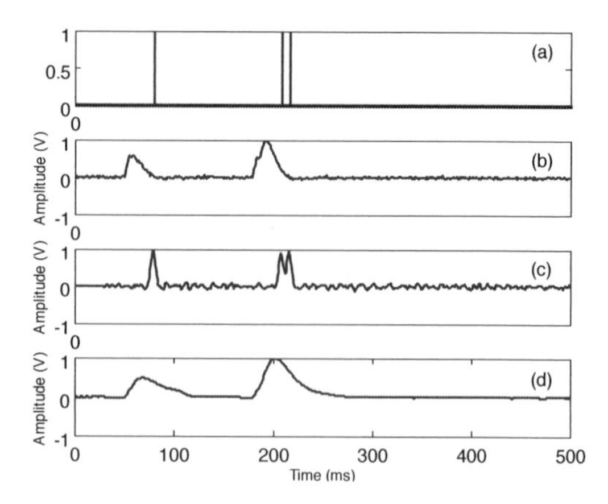

Figure 5. Deconvolution with the damped sine wave and thin layer: (a) reflected coefficients; (b) noisy seismic record; (c) deconvolution result by using the new model; (d) deconvolution result by using the old model.

From the above, for the synthetic seismic record with the Ricker wavelet or the damped sine wave, the Kalman filter based on the new model can obtain the better performance than the old model on positioning the locations of the reflected lays and improving the seismic resolution.

5 CONCLUSIONS

A new model for the Kalman filter on the seismic deconvolution is proposed in this paper. The new model and detailed filtering steps are given. Results of experiments show that for the synthetic seismic record with the Ricker wavelet or the damped sine wave, the Kalman filter based on the new model can compress the wavelet to improve the seismic resolution, but the Kalman filter based on the old model cannot. So the Kalman filter based on the new model works better than the old model. But it is a pity that the denoising performance of the Kalman filter based on the new model is not as good as the one based on the old model. The improvement of the denoising performance will be done further in the future.

ACKNOWLEDGMENT

This work is supported by the National Natural Science Foundation of China (41374114).

REFERENCES

Baziw, E. & Ulrych, T.J. 2006. Principle phase decomposition: a new concept in blind seismic deconvolution. *IEEE Trans. on Geoscience and Remote Sensing* 44(8): 2271–2281.

Coppens, F. & Mari, J.L. 1984. Spectral equalization: a means of improving the quality of seismic data. *Geophysical Prospecting* 32(2): 258–281.

Crump, N.D. 1974. A kalman filter approach to the deconvolution of seismic signals. *Geophysics* 39(1): 1–13.

Kalman, R.E. 1960. A new approach to linear filtering and prediction problems. *Trans. ASME (Series D)-J. Basic Engr.* 82: 35–45.

Kurniadi, R. & Nurhandoko, B.E.B. 2012. The discrete Kalman filtering approach for seismic signals deconvolution. *International conference on physics and its applications, Bandung, Indonesia, 10–11 November 2011*: 91–94. New York: AIP Publishing.

Mahalanabis, A.K., Prasad, S. & Mohandas, K.P. 1981. A fast optimal deconvolution algorithm for real seismic data using Kalman predictor model. *IEEE Trans. on Geoscience and Remote Sensing* GE-19(4): 216–221.

Mahalanabis, A.K., Prasad, S. & Mohandas, K.P. 1982. Recursive decision directed estimation of reflection coefficients for seismic data deconvolution. *Automatica* 18(6): 721–726.

Mahalanabis, A.K., Prasad, S. & Mohandas, K.P. 1983. On the application of the fast Kalman algorithm to adaptive deconvolution of seismic data. *IEEE Trans. on Geoscience and Remote Sensing* GE-21(4): 426–433.

Margrave, G.F., Lamoureux, M. P. & Henley, D. C. 2011. Gabor deconvolution: Estimating reflectivity by nonstationary deconvolution of seismic data. *Geophysics* 76(3): W15–W30.

Plataniotis, K.N., Katsikas, S.K., Lainiotis, D.G., et al. 1998. Optimal seismic deconvolution: distributed algorithms. *IEEE Trans. on Geoscience and Remote Sensing* 36(3): 779–792.

Robinson, E.A. & Treitel, S. 1967. Principles of digital Wiener filtering. *Geophysical Prospecting* 15(3): 311–333.

Song, W.Q. & Lv, S.C. 2009. Seismic deconvolution method based on Kalman optimal estimation theory. *Geophysical Prospecting for Petroleum* 48(4): 342–346.

Ulrych, T.J. 1971. Application of homomorphic deconvolution to seismology. *Geophysics* 36(4): 650–660.

Wang, L.L. Gao, J.H., Zhao, W. ,et al. 2013. Enhancing resolution of nonstationary seismic data by molecular-Gabor transform. *Geophysics* 78(1): V31–V41.

Wang, Y.H. 2008. *Seismic Inverse Q filtering.* Singapore: Wiley Blackwell Publishing.

Electronics, Communications and Networks IV – Hussain & Ivanovic (eds)
© 2015 Taylor & Francis Group, London, ISBN: 978-1-138-02830-2

Tuning quick search string matching algorithms with multi-windows and integer comparison

Hongbo Fan, Shupeng Shi, Li Dong & Jing Zhang*
Department of Computer Science, Kunming University of Science and Technology, Kunming, China
Computer Technology Application Key Laboratory of Yunnan Province, Kunming, China

ABSTRACT: String matching is a fundamental problem in computer science, and has been widely used in many important fields. This paper improves the classic BM type algorithm, Quick Search Using multiple sliding window method, the comparison unit from character to integer is increased in the first comparison of the current window. These two improved point decreases the average cost of the comparison obviously. Thus, a new series of algorithms are presented named QSMI_*k*_*x* (Quick Search with multi-sliding windows and integer comparison for *k* sliding windows and *x* type integer comparison). It is shown in the experiment that the QSMI algorithm is faster than other known algorithms in many cases on our platform, such as the pattern length of 4 and 8 on middle-level or large alphabet.

1 INTRODUCTION

String matching is a fundamental problem in computer science which has been widely used in most fields of text or symbol processing, such as network security, computational biology and information retrieval. Exact single pattern string matching is the basis of string matching which means seeking all the occurrences of a pattern $P = P[0,...,m-1]$ in a text $T = T[0,...,n-1]$ over the same alphabet Σ of length σ. To date, this problem has been researched widely, and hundreds of algorithms have been presented to solve it. This paper concentrates on designing more practical exact single pattern string matching. All the algorithms in this paper are for searching an exact pattern.

The Quick Search (QS) algorithm (Daniel 1990) is a very important suffix type string matching algorithm. The QS algorithm is a simplification of the Boyer-Moore algorithm (Robert & Strother 1977) and it is very easy for implementation. The jump distance of QS can reach $m+1$ which is even better than the length of the pattern. For very large alphabet, QS has outstanding jump capability. However, too much branch operations in the core loop of QS will lead to the low matching speed of QS.

The paper (Lv et al. 2012) presents an algorithm called *k*SW*x*C. This algorithm improves TSW (Amjad et al. 2008) by extending its 2 sliding windows method to multiple sliding windows and using integer as a comparison unit to replace character comparison in the first check in the sliding window. The two improved methods obviously increase the branch performance. *k*SW*x*C is very fast for short patterns and small alphabet.

This paper applies the multiple sliding windows and the integer comparison method on the QS and presents a series of algorithms called QSMI_*k*_*x* (Quick Search with multi-sliding windows and integer comparison with *k* sliding windows and *x* type integer comparison). It is shown in the experimental results that QSMI_*k*_*x* is faster than other known algorithms in many cases for short patterns on middle-level or large alphabet.

2 BASIC DEFINITION AND TERMINOLOGY

For a string $S, X, Y, Z \in \Sigma^*$, if $S = XYZ$, X is called a prefix and Z is called a suffix of S, and X, Y, Z are the factors of S. And $S[i...j]$ is the substring of S from $S[i]$ to $S[j]$. If $S = s_0 s_1 ... s_{k-1}$, we say that $S^{rv} = s_{k-1} s_{k-2} ... s_0$ is a reversal of S.

The QS algorithm is a basic algorithm in string matching field. It uses only the bad-character shift method of Boyer-Moore. Firstly, the algorithm attempts to check whether string matching occurs in the current sliding window ($T[i...i+m-1]$). If string matching occurs, the location of the sliding window will be reported. And then, the window should shift at least one because the current window has been checked. Therefore, the character after the current window ($T[i+m]$) is necessarily involved in the next attempt, and thus can be used for the bad-character shift after the current attempt. The code of QS is listed in Code 1.

*Corresponding author: *270677673@qq.com*

$preqsBc$(char* p, int m, int $qsBc[\sigma]$)
{ **for** $c \in \Sigma$ **do** $qsBC[c] \leftarrow m+1$;
  for $i \in [0,...m-1]$ **do** $qsBC[P[i]] \leftarrow m-i$;}
Quick Search (char* p, unsigned char* t, int m, int n)
{ $preqsBc(p, m, qsBc[\sigma])$; $j \leftarrow 0$;
  while $j < n-m$ **do** *//Searching*
{ $i \leftarrow 0$; **while** $i < m$ **and** $P[i] == T[i+j]$ **do** $i \leftarrow i+1$;
  if $i >= m$ **then** *Output(j)*;
  $j \leftarrow j + qsBC[T[j+m]]$;}}
Code.1 the QS algorithm

3 MULTIPLE SLIDING WINDOWS

In TSW, it presents a 2 sliding windows method. It alternatively uses two sliding windows, which scan the text from both sides to the middle position of the text. The window w_0 utilizes the shift values table built according to P for jumping forward and the window w_1 utilizes the shift values table built according to P^{rv} for jumping backward. The method is expanded to the multiple windows method that is shown as Fig.1 (C) by the paper(Berry & Ravindran 1999).

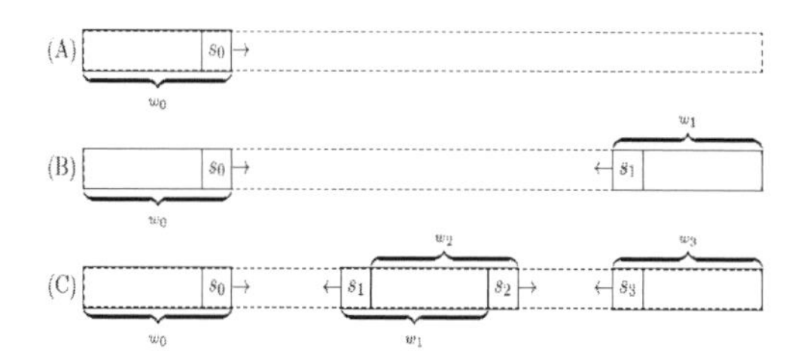

Figure 1. The multiple sliding windows method.

The two or multiple sliding windows method can enhance the algorithm performance greatly, and the reasons are listed as follows:

1 The two or multiple window algorithms can check the cross-border situation of two windows through checking whether two windows are overlapped by one branch operation, that is, the branch operation of the cross-border check is reduced by 50% than the original algorithms.
2 There is no date dependence for each window in the multiple windows method, so when treating, the operations of each window can be parallelly executed on different pipelines on super-scalar processor.
3 There are multiple plates in the memory system. The memory can open more than one plate at the same time to achieve the parallel transmission of data. After applying the multiple windows method, since the data to be accessed may be allocated on the different plates of memory, they can be accessed parallelly to some degree. Meanwhile, there are a lot of delays when opening a closed plate of memory (more than 10000 processor ticks on some platforms (Berry & Ravindran 1999). After the multi-windows method is applied, when a window is beyond the plate boundary, it may still come into another opening plate. It can avoid some of memory plate opening delay, so as to improve memory performance obviously.
4 The table look-up operation needs much time to wait memory to return the results. At this time, the pipeline does nothing but wait. After the multi-windows method is introduced, the operations without date dependence for processing other windows can be executed in the same pipeline when the pipeline is waiting memory response of one memory access. Thus, the average cost of memory access is reduced furthermore.

In fact, the multiple sliding windows method is a general code optimization method of flow handle, and it has attracted the attention of researchers in the string matching field recently. The detail of the method, called the list splitting method, is introduced in Kris (2003), in this book, this method is. The literature (Kris 2003) indicates that because there are four plates of memory on the current platform generally, it is reasonable that a stream is split into 4 or 6 blocks, which has been proved by the experimental data in Reference(Berry & Ravindran 1999), 4 or 6 windows solution is always the fastest one.

The multiple sliding windows method can be easily applied in the QS algorithm and it can also improve the algorithm. For the backward searching, the bad-character jump table is built by the reverse of the pattern.

4 INTEGER COMPARISON METHOD

Branch is a high cost operation of the current pipeline processor. It consumes most of computing time in matching. The cost is dependent on the branch predicted failure rate. If a branch is successfully predicted, this branch only needs one processor tick; but a branch predicted failure need a punishment of dozen processor ticks to empty the pipeline. Reducing the branch prediction failure rate is the primary task to optimize algorithm performance.

Let the punishment of the branch prediction failure be about 40 ticks (it is the results of Intel Pentium IV). In QS, the comparison unit to check whether string matching occurs in the current window is one character, and the average cost of a comparison is $1*(1-\sigma^{-1})+punish*\sigma^{-1}$ ticks. For example, on small alphabet, such as DNA sequence, the average branch cost for character comparison is 10.75 ticks, which is more than the average cost of the sum of all other operations in processing this window.

If the comparison unit is extended from character to some bigger unit, and let the comparison unit include w characters, the branch prediction failure rate will reduce to $1*(1-\sigma^{-w})+punish*\sigma^{-w}$. For example, on DNA sequence, the average branch cost of for comparison with uint16, uint32 and uint64 as the comparison unit would reduce to 3.44, 1.15, 1.00 processor ticks, respectively.

To date, most of current processors have supported unaligned read. In unaligned read, some adjacent characters that may not be in the boundary range of an integer of memory can be read in one instruction with one or two ticks cost. In C/C++ code, let the text be t and the offset of the current window be j, *(*type of integer**) (*t+j*) can unalignedly read the first integer of the current window. Using unaligned read, although it adds a bit of read cost, the comparison unit is extended from character to integer and the average branch cost of the branch reduces obviously. Furthermore, the probability of matching an integer is exponentially lower than the probability of matching a character, using unaligned read can reduce the total number of branches in checking whether string matching occurs in the current window. Thus, using unaligned read can increase the performance of QS.

The probability of the smaller read-in integer being in the range of a real memory access is higher. Thus, the average cost of unaligned read for smaller integer is lower. And since smaller integer can present low enough branch predicted failure rate on very large alphabet, using short integer (uint16) as the comparison unit may be optimal. And for very small alphabet, the branch predicted failure rate is still high for comparing uint16, and using bigger integer such as uint32/uint64 as the comparison unit may be optimal.

The q-grams method has become one of the most important improved points in string matching research field. Most of known researches are focused on how to increase the jump capability by using q-grams. In fact, integer comparison is a kind of q-grams method to reduce the average cost of basic operation in string matching.

The algorithm that applies the multiple windows method and the integer comparison method on QS is called QSMI_*k*_*x* (Quick Search with multi-sliding windows and integer comparison with k windows and x type integer comparison). The code of QSMI_2_uint16 is listed as Code. 2.

```
preqsBc(char* p, int m, int qsBc[σ])
{ for c ∈ Σ do qsBC[c] ← m + 1;
  for i ∈ [0,...m − 1] do qsBC[P[i]] ← m − i;}
QSMI_2_uint16 (char* p, char* t, int m, int n)
{ preqsBc(p, m, qsBc[σ]); preqsBc(p^rv, m, qsBcR[σ]);
  j ← 0 ; j1 ← n − m ; sc ← *(uint16*)p ;
  while j ≤ j1 do //Searching
{ if sc == *(uint16*) ( t + j ) then
  {i ← 2; while i < m and P[i] == T[i + j]do i ← i + 1;
  if i >= m then Output( j );}
  j ← j + qsBc[t[ j + m]];
if sc == *(uint16*) ( t + j1 ) then
{i1 ← 2;
while i1 < m and P[i1] == T[i1 + j1] do i1 ← i1 + 1;
if i1 >= m then Output( j1 );}
  j1 ← j1 − qsBcR[t[ j1 − 1]];}}
```

Code.2 the code of QSMI_ *k*_ *x* algorithm for 2 sliding windows and uint16 comparison

5 EXPERIMENTAL DATA

The following experiment is done based on SMART 13.02(Simon & Thierry 2013), which gives the implements of most known algorithms as of Feb. 2013 and is the algorithm implement of the most comprehensive survey (Faro & Lecroq 2013) on exact single pattern string matching. The hardware platform of this experiment is Intel Core2 E3400 @3.0GHz / Intel P43 / 4G DDR3 RAM. The software environment is Ubuntu 10.10 64-bit edition/ g++4.4.5 with -O3 optimizing parameter. The DNA sequence (*E.coli*), the English text (Bible.txt), the natural language samples text (world192.txt) are tested, which are from (Jürgen 1997), and four 20MB length random texts with alphabet size 2, 16, 64 and 254. We test patterns of length 2,4,8,16,32 and 64.For each matching condition, the 100 patterns in each pattern set are picked from the 100 random selected and non-overlapping positions of the text, and the average matching speed is recorded as the final result among which the highest and the lowest 20% results are ignored. The text is read in memory and a big table is read sequentially before each time of matching to avoid the impact of disk and cache. Only the matching phase is timed by RDTSC* (±30 CPU ticks). SpeedStep** and TurboBoost*** are disabled and the CPU frequency is locked by cpufrequtils****. The network and the unrelated background service are closed to ensure the processor utilization before matching is below to 5%.

This experiment compares all algorithms of SMART 13.02 and adds some algorithms not being included by SMART, such as SBNDM*q*b(Branislav et al. 2010), FSBNDM*q*b(Hannu & Jorma 2011), GSB2b(Hannu & Jorma 2011) , FQHash(Fan et al. 2010), HBF(Zhou et al. 2012), kSWxC(Lv et al. 2012), HGQSkip(Wu et al. 2013), SBNDM_sb(Zhang et al.

* http://en.wikipedia.org/wiki/Time_Stamp_Counter

** http://en.wikipedia.org/wiki/SpeedStep

*** http://en.wikipedia.org/wiki/TurboBoost

**** https://wiki.archlinux.org/index.php/CPU_Frequency_Scaling

2013), Greedy_QF(Chen et al. 2012), BOM*q* (Fan & Yao 2012), SufOM (Fan & Yao 2012), etc. In this experiment, all bit parallel solutions are implemented with the 32-bit edition (i32) and the 64-bit edition (a64), both of which are tested. This experiment complements the unlisted parameters in SMART. If algorithms with different parameters are called different algorithms, more than 800 algorithms are compared, these algorithms cover most of known algorithms. Though dozens of thousand records of the complete record of the experiment cannot be listed all, the fastest three algorithms with their optional parameters and their matching speed (the unit is MB/s) are listed.

It is shown in the experimental results that our algorithms cannot be faster than other algorithms on very small alphabet, and the results on binary text and DNA sequence are not listed. The experimental results are listed as Tables 1~5.

Table 1. Results on English text (Bible.txt), unit is MB/s.

m=2	*m*=4	*m*=8
HBF_f2u16a64 2715.9	**QSMI_4_uint32** **3394.2**	**QSMI_4_uint32** **4747.5**
QSMI_4_uint16 **1971.9**	SSWIC 2945.6	SBNDM2_2sbi32 4162.7
SSWSC 1815.0	SBNDM2_2sbi32 2883.0	SSWIC 3798.8
m=16	*m*=32	*m*=64
SBNDM4_4sbi32 5600.6	SBNDM4_sbi32 6195.1	FSBNDMq_q5f1a64 6400.2
QSMI_4_uint32 5516.5	UFNDM6_a64 6044.7	SBNDM6_sba64 6328.0
Greedy_QF_4_3i32 4752.9 (Chen et al. 2012)	FSBNDMq_q5f0i32 5946.3	BNDM4_a64 6248.5 (Hannu & Jorma 2011)

Table 2. Results on random text of alphabet size 16, unit is MB/s.

m=2	*m*=4	*m*=8
HBF_f2u16a64 3102.2	**QSMI_4_uint32** **3470.3**	SBNDM2_2sbi32 5340.6
QSMI_4_uint16 2197.1	SBNDM2_2sbi32 3347.7	**QSMI_4_uint32** 4986.4
SSWSC 2160.8	SSWIC 3217.3	GSB2b_i32 4894.0 (Hannu & Jorma 2011)
m=16	*m*=32	*m*=64
SBNDM4_sbi32 6229.3	SBNDM4_sbi32 6737.5	SBNDM3_a64 7033.0 (Branislav et al. 2010)
HGQSkip_q3g2c1 5875.9 (Wu et al. 2013)	FSBNDMq_q3f0i32 6425.8	FSBNDMq_q4f1a64 6915.9
BOM3 5794.8 (Fan & Yao 2012)	UFNDM4_a64 6261.4 (Branislav et al. 2010)	HGQSkip_q3g1c1 6730.9

Table 3. Results on natural language samples text (world192.txt), unit is MB/s.

m=2	*m*=4	*m*=8
HBF_f2u16a64 3165.9	**QSMI_4_uint32** **3503.1**	**QSMI_4_uint32** 5034.3
QSMI_4_uint16 **2247.4**	HBF_f2u16a64 3250.4	SBNDM2_2sbi32 4610.4
TVSBS_w4 1286.3 (Berry & Ravindran 1999)	SBNDM2_2sbi32 3007.8	GSB2b_i32 4360.1
m=16	*m*=32	*m*=64
QSMI_4_uint32 **5736.1**	SBNDM4_sbi32 6171.5	SBNDM3_4a64 6510.4
SBNDM4_sbi32 5556.2	FSBNDMq_q3f0i32 5945.6	FSBNDMq_ q4f1a64 6444.8
Greedy_QF_3_4i32 5433.5(Chen et al. 2012)	UFNDM4_a64 5893.9	BNDM4_a64 6270.4

It is shown in the experiment that the QSMI algorithm is faster than other known algorithms in many cases such as pattern length of 4 and 8 on middle-level or large alphabet. The optimal number of sliding windows of QSMI is always 4 because there are only 4 plates on our platform. The optimal number of sliding windows may increase for platform with more plates, but we have not tested that. And the comparison unit longer than uint32 can not speed up further, because using uint64 as the comparison unit can reduce the average branch cost little even on very small alphabet, but lead high probability that the read-in integer crossing the memory access border increases the average read cost of unaligned read.

Table 4. Results on the random text of alphabet size 64, unit is MB/s.

m=2	*m*=4	*m*=8
HBF_f2u16a64 3544.3	**QSMI_4_uint32** **3886.3**	**QSMI_4_uint32** 5601.9
QSMI_4_uint16 **2505.2**	HBF_f2u16a64 3541.9	SBNDM2_2sbi32 4984.2
TVSBS_w4 1893.9 (Berry & Ravindran 1999)	TVSBS_w4 2726.9	GSB2b_i32 4915.9
m=16	*m*=32	*m*=64
SBNDM2_2sbi32 6223.5	SBNDM2_2a64 6893.2	QF_2_6i32 7063.5 (Branislav et al. 2010)
FSBNDM_w2a64 6220.8 (Berry & Ravindran 1999)	FSBNDMq_q2f0i32 6883.7	SBNDM3_a64 7013.8
SBNDM2_3i32 6165.5 (Branislav et al. 2010)	BNDM2_i32 6839.2	BXS3_a64 6994.6 (Branislav et al. 2010)

Table 5. Results on the random text of alphabet size 254, unit is MB/s.

m=2	m=4	m=8
HBF_flu16a64 4959.0	HBF_flu16a64 4883.8	**QSMI_4_uint32** **5868.7**
HGQSkip_q1g2c2b2 2572.0	HGQSkip_q1g2c2b2 4148.0	SBNDM_w2i32 5582.2
QSMI_4_uint16 **2540.9**	**QSMI_4_uint32** **3982.6**	HGQSkip_q1g2c2b2 5431.8
m=16	m=32	m=64
FSBNDM_w2i32 6313.4	FSBNDMq_q2f0a64 7011.4	QF_2_6i32 7100.2
FSBNDM_a64 6241.9	SBNDM2_3_a64 7009.9	BXS3_a64 7022.1
BNDM2_a64 6217.8	BXS2_i32 6965.1	FSBNDMq_q2f0i32 7012.9

6 CONCLUSION

This paper adds the multiple sliding windows method and the integer comparison method into a classic BM type algorithm, Quick Search, and a serial of algorithms called QSMI_*k*_*x* are presented. It is shown in the experiment that the QSMI algorithm is faster than other known algorithms in many cases such as the pattern length of 4 and 8 on middle-level or large alphabet.

To date, many algorithms have reached the lower bound of the average time complexity of string matching. Increasing the jump capability for algorithms is harder and harder to increase the performance of the algorithm. This paper tries to reduce the average cost of basic operations in string matching and gains a very high performance algorithm even based on very classic string matching algorithm such as Quick Search. We will follow this direction in the future work.

ACKNOWLEDGMENT

This paper is supported by the Applied Basic Research project Foundations of Yunnan Province of China (No. 2012FB131).

REFERENCES

Amjad Hudaib, Al-Khalid Rola, Dima Suleiman, Mariam Itriq, et al. 2008. A fast pattern matching algorithm with two sliding windows. *Journal of Computer Science* 4(5): 393–401.

Berry T. & Ravindran S. 1999. A fast string matching algorithm and experimental results. In *Proceedings of the Prague Stringology Club Workshop '99,J. Holub & M.Simanek ed., Collaborative Report: 16–28, Czech Technical University, Prague, Czech Republic.*

Branislav Ďurian,Jan H., Hannu P., et al. 2010. Improving practical exact string matching. *Information Processing Letters* 110(4): 148–152.

Chen Zhiteng, Liu Lijun, Fan Hongbo, et al. 2012. Fast exact string matching algorithms based on greedy jump and QF. In *IET Conference Publications, IET International Conference on Information Science and Control Engineering.*

Daniel M.S. 1990. A very fast substring search algorithm. *Communications of the ACM* 33(8):132–142.

Ďurian Branislav, Hannu P., Leena S.,et al. 2010. Bit-Parallel Search Algorithms for Long Patterns. SEA2010. *Lecture Notes in Computer Science* 6049/2010: 129–140.

Fan Hongbo & Ya Nianmin o. 2012. Q-gram variation for EBOM. In *Lecture Notes in Electrical Engineering,v211 LNEE*:453–460.

Fan Hongbo & Yao Nianmin. 2012. Tuning the EBOM algorithm with suffix jump.In *Lecture Notes in Electrical Engineering,v211 LNEE*: 965–973.

Fan Hongbo, Yao Nianmin & Ma Haifeng. 2010. A practical and average optimal string matching algorithm based on Lecroq. ICICSE'10. *Washington DC: IEEE Conference Publishing Services* 2011: 57–63.

Faro S. & Lecroq T. 2013. The Exact Online String Matching Problem: a Review of the Most Recent Results. *ACM Computing Surveys* 45(2).

Hannu P. & Jorma T. 2011. Variations of Forward-SBNDM. In *Proceedings of the Prague Stringology Conference*: 3–14.

Jürgen A.1997/2014.12. *The Large Canterbury Corpus.* http://www.data-compression.info/Corpora/CanterburyCorpus/.

Kimmo F. & Szymon G. 2005. Practical and Optimal String Matching. SPIRE 2005 LNCS 3772: 376–387. Berlin: Springer-Verlag.

Kris Kaspersky. 2003. *Code Optimization: Effective Memory Usage*. A-List Publishing.

Lv Zhenhong, Fan Hongbo, Liu Lijun, et al. 2012. Fast single pattern string matching algorithms based on multi-windows and integer comparison. *Information Science and Control Engineering 2012 (ICISCE 2012), IET International Conference on:1-5, 7-9 Dec. 2012.*

Ricardo A.B. & Gonnet H.G. 1989. A new approach to text searching. *Communications of the ACM* 35(10): 74–82.

Robert S.B. & Strother J M. 1977. A fast string searching algorithm. Communications of the ACM 20:762–772.

Simon F. & Thierry L. 2012. A Fast Suffix Automata Based Algorithm for Exact Online String Matching. *CIAA 2012, LNCS* 7276:149–158. Berlin: Springer-Verlag.

Simon F. &Thierry L.2013/2014.12.*Smart:String matching research tool*. http://www.dmi.unict.it/~faro/smart/.

Thierry L. 2007. Fast exact string matching algorithms. *Information Processing Letter* 102(6): 229–235.

Wu Wenqing, Fan Hongbo, Liu Lijun, et al. 2013. Fast string matching algorithm based on the skip algorithm. In *Lecture Notes in Electrical Engineering* 236:247–257.

Zhang Jian, Fan Hongbo, Huang Qingsong, et al. 2013. Single pattern string matching algorithm based on unaligned 2-byte reading mechanism.

Zhou Yao, Fan Hongbo, Liu Lijun, et al. 2012.Fast string matching algorithms for very short pattern. In *IET Conference Publications, International Conference on Information Science and Control Engineering.*

Electronics, Communications and Networks IV – Hussain & Ivanovic (eds)
© 2015 Taylor & Francis Group, London, ISBN: 978-1-138-02830-2

Scenario-based security risk assessment

Rong Fu & Xiaofang Ban
China Information Technology Security Evaluation Center, Beijing, China

Xin Huang & Dawei Liu
Xi'an Jiao Tong-Liverpool University, Suzhou, China

ABSTRACT: Traditional assessment method for security risks cannot provide a clear analysis to information systems. In this paper, a scenario-based method is proposed. Scenarios are developed from the standpoint of attackers. Also, impacts on business operations and technique operations are clearly separated. In the past few years, this scenario-based risk assessment method has been used by China Information Technology Security Evaluation Center in many real-world cases. Advantages of this method are: (1) this method can provide a clear description for security risks, and (2) it is easier to achieve an agreement between analysts and corporations under assessment. This scenario-based method is a good addition to the traditional method.

1 INTRODUCTION

Security risk analysis is the foundation of information system security, especially the security of national information infrastructure and corporation key information system. Thus, Chinese government published GB/T 20984 standard (GB/T 20984, 2007) a few years ago. Also, the security standards and studies of financial and many other sectors have also been published in the past few years. (GB/T 22239 2008, Kim & Cha 2012, Innerhofer-Oberperfler & Breu 2006, Breu et al. 2008, Piètre-Cambacédès & Bouissou 2010).

However, most current security risk analysis methods are based on quantitative risk analysis. These methods cannot provide a clear description and a fine-grained analysis in many important aspects, for example, the probability of certain attack, the attack path, and the influence to the business processes. These drawbacks may lead to the following problems. (1) It is difficult to understand the importance of risk evaluation results. (2) The impact on technique processes is more overestimated than the impact on business processes. (3) Attack paths of combination attacks cannot be clearly described.

In this paper, a scenario-based method is proposed. Risk scenarios are firstly developed from the standpoint of attackers. Also, we describe the procedure of security risk assessment using these scenarios. Meanwhile, factors of developing scenarios and template of describing scenarios are specified in order to help security analysts to understand and use this risk assessment method.

The rest of this paper is organized as follows. In section 2, the procedure of the security risk assessment method is introduced. In section 3, factors of the risk scenario are discussed. In section 4, template of describing scenarios is specified. In section 5, survey results are discussed. Finally, some conclusions are made in section 6.

2 METHOD OVERVIEW

2.1 *Procedure*

Using scenario-based risk assessment method, analysts identify assets in one information system, analyse vulnerabilities, describe risk scenarios for serious vulnerabilities, find out probability of each scenario, and evaluate risks. The procedure is as follows.

- Step 1. Identify key assets (Based on GB/T 20984 and ISO 27005)
- Step 2. Identify serious vulnerabilities
- Step 4. Describe risk scenarios
- Step 5. Evaluate the probability of risk scenarios
- Step 6. Identify technique layer impacts and business layer impacts
- Step 7. Evaluate risks

Remark 1. Only risk scenarios corresponding to high and middle level vulnerabilities are important. Because low-level vulnerabilities are always corresponding to low-level risks, they are not used to develop risk scenarios in most cases.

Remark 2. Scenarios and vulnerabilities are not one- to- one mapping. Several vulnerabilities can be related. They should be described in one scenario.

2.2 *Risk synthetic*

Many security risks are caused by multiple vulnerabilities that are exploited by attackers. Thus, it is useful to discuss the risk synthetic.

Risk synthetic requires analysts to analyse the system from the standpoint of attackers. Analysts should find out how attackers can combine multiple vulnerabilities to construct more risky scenarios.

Take a big corporation with many subsidiary corporations as an example. There are data flows between headquarters and subsidiary corporations. There are also data flows among subsidiary corporations. Many different IT systems are working together in order to support these data flows. Whenever one IT system is compromised, other IT systems can be also compromised; the whole corporation is then paralysed.

Therefore, the risk synthetic is useful in complex IT systems. We will discuss the risk synthetic from three aspects below.

2.2.1 *Inter-components synthetic*

Inter-components synthetic focuses on different components in one system.

Use case. There is no physical access control policy to computer rooms. Outside attacker can access computers in these rooms. If some computers do not have passwords, this attacker can log in to them and get sensitive data. There are two components here. (1) Outside attacker can physically access computer rooms. (2) Computers do not have passwords.

2.2.2 *Inter-system synthetic*

Inter-system synthetic focuses on different systems in one corporation.

Use case. Suppose some data stored in one system is maliciously modified. The data provided by other systems that rely on the data stored in this system is also compromised.

2.2.3 *Inter-corporation risk synthetic*

Inter-corporation risk synthetic focuses on different corporations.

Use case. The IT system of headquarters is compromised. Subsidiary corporations that rely on real-time data provided by headquarters cannot function correctly.

2.3 *Characteristics*

Characteristics of this method are two folds.

1. This method highlights key vulnerabilities. Risks are evaluated based on vulnerabilities instead of assets in traditional methods.
2. This method highlights business layer risks. The separation of technique layer impact and business layer impact helps non-IT staff in corporations to manage risks.

3 SCENARIO DEVELOPMENT

Scenario development is the core of our method. A risk scenario is a collection of the following factors.

3.1 *Time*

IT system security status can be divided based on time. (1) Past: the system has already been attacked. (2) Now: what are current potential attacks? (3) Future: what risks are remained in the system? Time is one useful categorization factor that helps security analysts to analyse the system in a comprehensive way.

3.2 *Location*

Location refers to physical location or logical location. Examples of physical locations are XX city, XX building, and XX room. Examples of logical locations are XX corporation network and XX subnet. Location information can help analysts to identify the attack source and potential impact areas.

3.3 *Vulnerability*

Vulnerability refers hardware, software or management flaws. It is the core in one risk scenario. One risk scenario may be developed based on just a single vulnerability or multiple vulnerabilities in one attack path.

3.4 *Threat*

Threat is the means of attacks; it can be categorized as malicious threats and accidental threats based on the motivation. Malicious threats include denial of service and stealing of password. Accidental threats include operation errors and inappropriate authorizations. The motivation helps analysts to understand the cause of threats and the attack procedure.

3.5 *Threat Source*

This factor is used to identify the threat source. Threat source can be virus, people, and IT systems. People can be further categorized as insider and outsider.

3.6 *Probability*

Probability of one risk scenario is closely related to the threat source and the motivation. For example, if the threat is a malicious one, the analyst should focus on technique requirements, equipment requirements, the costs of the attack, etc. However, if the threat is an accidental one, the analyst should focus on the privileges of the employees, operation skills, internal conflicts, etc.

3.7 *Impact*

Impact can be categorized as technique layer impact and business layer impact. Technique layer impact refers to the impact on confidentiality, availability, integrity, etc. The business layer impact refers to the impact on business goals, enterprise honours, etc.

4 RISK SCENARIO TEMPLATE

A clear representation of assessment results is critical. The clear representation can help analysts and the corporations under evaluation to achieve an agreement on system risks. Thus, a template for representing assessment result is provided in Table 1.

Table 1. Risk scenario template.

Factor	**Description**
Risk	Technique layer risk: high/middle/low Business layer risk: high/middle/low
Vulnerability	[Vulnerability 1] [Vulnerability 2] ...
Threat source	[Insider] [Outsider]
Threat	[Threat 1] [Threat 2] ......
Scenario description	[Overview of attack procedure] Example: Outside attackers can visit internal website via a VPN (virtual private network). They get weak password and get some privileges in certain server. They find that there are a large number of sensitive files in the server. Using passwords recorded by web browsers, they log into users' account and get sensitive information.
Probability	[Probability of the scenario]
Impact	[Technique layer impact: impact on confidentiality, integrity, availability, etc.] [Business layer impact: impact on business tasks, enterprise honours, etc.]
Evidence	[List evidences, for example, past attack videos.]
Suggestion	...

5 RESULT DISCUSSION

A questionnaire is designed in order to study the proposed risk assessment method. 11 people answered the questionnaire. Their backgrounds are listed as follows.

- Sex: 9 candidates are male. 2 candidates are female.
- Age: 5 candidates are between 18 and 30. 6 candidates are between 30 and 44.
- Education: 9 candidates get Master degree or Ph.D. degree. 2 candidates are bachelors.
- Job: 2 candidates are related to information security. 6 candidates are related to other IT area. 3 candidates are not related to IT.

Two types of risk description results are given: one is our scenario-based method; the other is the traditional asset-based method. 7 candidates think that the combination of these two methods is a better choice. 4 candidates think that the traditional asset-based method is better.

Therefore, we can claim that this scenario-based method is a good addition to the traditional method. However, it should not be used independently.

6 CONCLUSION

Traditional security risk assessment methods lack clear description and fine-grained analysis to risk scenarios. Scenario-based method can help business staff and technique staff to correctly understand assessment results in technique and business layers. This method can also help them to find out key problems easily. Meanwhile, this method involves a risk synthetic corresponding to combination attacks, and this is an extension to traditional methods.

In the past few years, this scenario-based risk assessment method has been used by China Information Technology Security Evaluation Center in many real-world cases. It has two advantages. (1) This method can provide a clear description for security risks. (2) It is easier to achieve an agreement between analysts and corporations under assessment. In summary, this scenario-based method is a good addition to the traditional method.

ACKNOWLEDGMENT

This work was supported in part by the Natural Science Foundation of Jiangsu Province under Grant BK20140404, by the Jiangsu University Natural Science Research Programme under Grant 13KJB510035, and by the Suzhou Science and Technology Development Plan under Grant SYG201405.

REFERENCES

GB/T 20984-2007, Information security technology-Risk assessment specification for information security.

GB/T 22239-2008, Information security technology-Baseline for classified protection of information system security.

Kim Young-Gab & Cha Sungdeok. 2012. Threat scenario-based security risk analysis using use case modeling in information systems. *Security and Communication Network* 5(3): 1 : 293–300.
Innerhofer-Oberperfler, F. & Breu, R. 2006. Using an Enterprise Architecture for IT Risk Management. In *ISSA* : 1–12.
Breu, R., Innerhofer-Oberperfler, F., & Yautsiukhin, A. 2008. Quantitative assessment of enterprise security system. *Third International Conference on Availability, Reliability and Security ARES 08, Barcelona, 4–7 March 2008* : 921–928. IEEE.
Piètre-Cambacédès, L., & Bouissou, M. 2010. Beyond attack trees: dynamic security modeling with Boolean logic Driven Markov Processes (BDMP). In *Dependable Computing Conference (EDCC), 2010 European Valencia, 28–30 April 2010*: 199–208. IEEE.

Electronics, Communications and Networks IV – Hussain & Ivanovic (eds)
© 2015 Taylor & Francis Group, London, ISBN: 978-1-138-02830-2

Design and implementation of algorithm recommended by new users based on Mahout

Xianwei Gao* & Zhibin Shi
North University of China, College of Computer and Control Engineering, Taiyuan, Shanxi

ABSTRACT: In order to solve the problems of difficult recommendations and low efficiency of recommendations, etc. due to new users having no historical data under the background of big data, the techniques of combining Mahout-based collaborative filtering algorithm with MapReduce-based Top N algorithm is put forward to achieve the recommendation of new users. The framework of the new user recommendation system is constructed, and the collaborative filtering algorithms in the Hadoop Top N algorithm and Mahout are designed and realized. The theoretical analysis and experimental verification show that the algorithm recommended by new user is significantly better than the collaborative filtering algorithm used and recommended by new user alone on recommendation efficiency, scalability for large-scale data processing and recommendation quality.

1 INTRODUCTION

With the rapid development of information technology and network technology, as well as the rise of cloud computing, mobile Internet and Internet of things and other technologies, the data continues to grow and accumulate at an unprecedented speed. It is difficult for people to obtain accurate information meeting their own needs, thus giving rise to the problem of "information overload". Therefore, people put forward the concept of "personalized services"(Zeng et al. 2002, Adomavicius & Tuzhilin 2005) to provide different services or information contents for different users. As an important branch of the research field of personalized service, recommender systems (Adomavicius & Tuzhilin 2005, Ricci et al. 2011, Xu et al. 2009, Liu et al. 2009) help users find the projects which they may be interested in from a lot of data through digging the binary relations between the users and projects, and generate personalized recommendation to meet individual needs.

However, the algorithm of traditional recommender systems recommends based on historical data. New users do not have historical data, therefore, there are three methods commonly used: First, collaborative filtering recommendation methods discover new exclusive interests and do not rely on domain knowledge, but on historical data. Second, new user label recommendation method (Liao et al. 2013), personalized labels indirectly react users' interests; however, there are typical problems, such as scalability, sparsity and cold start, etc. Third, collaborative filtering and small-world model new user recommendation methods make new users find the group more similar to themselves (Hu et al. 2012); however, recommendation quality and effect are not very obvious in the fusion of the algorithm.

Meanwhile, due to the diversity of data sources (Internet logs, sensor data and mobile Internet, etc.) and many kinds of data (structured, semi-structured and unstructured, etc.), the storage and parallel computing of recommendation system data withstand the severest test. Hadoop distributed parallel computing framework solves the problems of large data storage and parallel computing, etc. effectively.

However, achieving the recommender system algorithm is a very laborious thing, and the appearance of Mahout solves this problem. When the algorithm of Apache Mahout runs under the platform of Hadoop, Mahout projects encapsulate commonly used algorithms, such as clustering, classification, recommendation engine and excavation of frequent item sets, etc. and are designed to help developers develop intelligent applications more quickly and easily.

This paper innovatively combines the Top N algorithm and collaborative filtering algorithm, and introduces a new user recommendation algorithm based on a large amount of data, thus solving the problems of the new user having difficulty in the recommendation, being not high efficiency and no matching recommendation, etc. And it also verifies its effectiveness and scalability through recommender system to date.

*Corresponding author: *gaoxianwei@126.com*

2 THE RELEVANT TECHNICAL PRESENTATIONS

2.1 *The main recommendation algorithms*

Recommendation algorithms mainly include (Wang et al. 2012): collaborative filtering recommendation, content-based recommendation, knowledge-based recommendation and hybrid recommendation. Comprehensive comparison of their advantages and disadvantages (Li & Xu 2014) is shown in Table 1.

Table 1. Comparison of main recommendation algorithms.

	Advantages	Disadvantages
Collaborative filtering recommendation algorithm	Discover the novel interest and do not rely on the domain knowledge; The effect becomes better and better with the passage of time and accumulation of data amount; The personalized and automation degree of recommended procedure is high; Be able to handle complex and unstructured objects;	It has scalability, sparsely, cold start and other typical problems; The recommendation quality depends on the historical data set;
Content-based recommendation algorithm	Discover the novel interest and do not rely on the domain knowledge;	Rule extraction is difficult and time-consuming with low personalized degree;
Knowledge-based recommendation algorithm	Consider the non-product attributes; reflect the users' needs; Make up for the lack of knowledge and experience of the users;	It is difficult to obtain the knowledge and experience; The recommendation is static;
Hybrid recommendation algorithm	Solve the lack of single recommendation with better recommendation;	The algorithm is complex and cannot take into account all the circumstances;

2.2 *Distributed and storage technology - Hadoop*

The paper about distributed infrastructure released by Google has a huge impact on the industry, in which the thoughts of MapReduce and GFS, etc. provide a key reference for distributed computing and storage, and Hadoop is its open source implementation.

Hadoop (Yang & Liu 2013) platform is with such characteristics as high scalability, low cost, high reliability and easy-to-use, etc. Its core is HDFS distributed file system and MapReduce framework. The former makes processing massive amounts of data possible under the circumstance of controllable cost; the latter has simplified programming model using divide and conquer strategy and being designed for large-scale distributed parallel data processing, which draws on the idea of functional programming and is uniformly abstracted to two operations, map and reduce, for large-scale data processing task. Because Hadoop cluster can extend dynamically and laterally according to demand, using the Hadoop platform can break through the bottleneck of big data analysis brought by data scale to recommender system to meet the needs of high-performance and highly scalable computing.

2.3 *Machine learning algorithms - Mahout*

Apache Mahout (Zhu & Qian 2013) is a new open source project developed by Apache Software Foundation (ASF), whose main objective is to create a number of scalable machine learning algorithms for developers under the permission of Apache. Mahout contains many implementations, including clustering, classification, Collaborative Filtering (CF) and evolutionary program. In addition, Mahout can be effectively extended to the cloud through the use of the Apache Hadoop library.

A building recommender system with Mahout is a simple yet difficult thing. It is simple because Mahout complete packages "collaborative filtering" algorithm and realizes parallelization, and provides a very simple API interface; It is difficult because we do not know the details of an algorithm, and it is difficult to configure and optimize the algorithm based on the business sense. The common algorithms implemented by Mahout are: user-based collaborative filtering algorithm UserCF, item-based collaborative filtering algorithm ItemCF, SlopeOne algorithm, KNN Linear interpolation item-based recommendation algorithm, SVD recommendation algorithm and Tree Cluster-based recommendation algorithm, etc.

3 NEW USER RECOMMENDATION SYSTEM COMBINING COLLABORATIVE FILTERING ALGORITHMS AND TOP N ALGORITHML

The new user recommendation must have three elements: new user A, recommended object B and evaluation model C.

3.1 *Interpretation of name*

Name 1 (Top N algorithm) uses the direct sorting method or Hash Table method to find the highest (lowest) collection.

Name 2 (new user A) is recommended object as the beneficiaries of recommendation.

Name 3 (recommended object B) is the recommended object, which is recommended to the new user A.

Name 4 (evaluation model C) is the score of new user A on recommended object B.

3.2 *The thought of algorithm recommended by new user*

Top N and collaborative filtering algorithms are comprehensively used to make recommendations for new users. Its algorithm idea is as follows:

Analyzing the raw data, Top N algorithm is used to obtain the ID of N recommended objects B to obtain basic data "Top basic data D"; analyzing the raw data, collaborative filtering algorithm is used to calculate the information of recommended object B and obtain the basic data "recommended basic data E"; new user A conducts the related evaluation on "Top basic data D", and records the evaluation model "evaluation model C"; the "evaluation model C", screen "recommended basic data E" are utilized to identify the evaluation model in "recommended basic data E" and the data "filtering data F" similar to "evaluation model C"; for the "filtering data F", the results calculated by using Top N algorithm are the results of algorithm recommended by new users.

The algorithm process chart is as Figure 1:

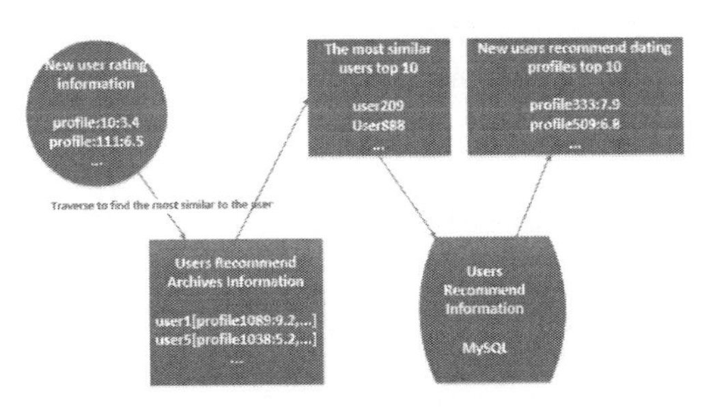

Figure 1. The process chart of algorithm recommended by new users.

3.3 *System design and implementation*

The system is developed by using Spring MVC + MyBatis + Spring framework, and HDFS file system and MySQL database of Hadoop cloud platform are used to store data. Raw data are directly stored into HDFS, and the user profiles calculated recommend basic data into MySQL database.

3.3.1 *System architecture*

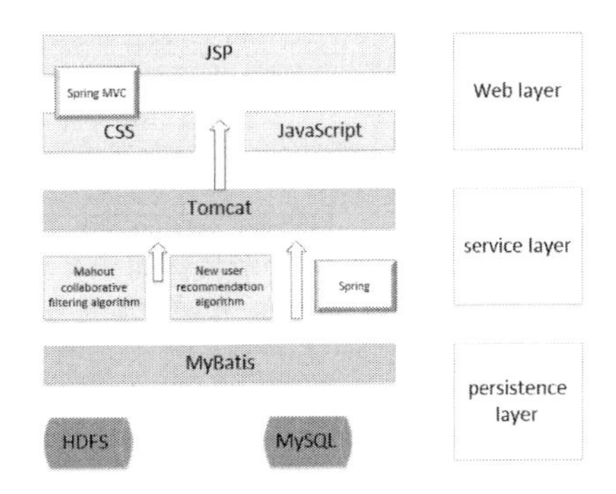

Figure 2. System framework chart.

3.3.2 *System flowchart*

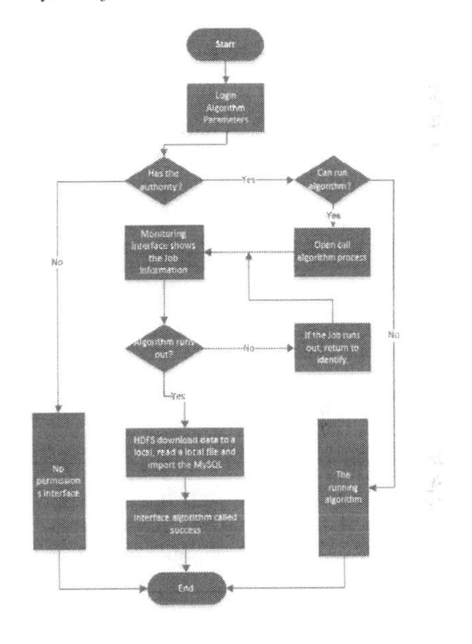

Figure 3. System flowchart.

3.3.3 *Algorithm design and implementation*

1 Top N Algorithm Design

The Hadoop top N algorithm is front N items with the average largest user rating in all statistical files, that is, it needs to calculate the average user rating of each item for the raw user data, and then orders these items in accordance with the order of user rating from large to small, and takes the front N items.

First, the Map Reduce process of the Top N algorithm is analyzed. The format of data input is: [userId, profileId, prefValue], in which userId is not needed, and profileId and the corresponding prefValue are just needed, and recording how many times each item has been evaluated by the users is also needed, which is convenient for average score of the item. The format of Mapper input and output data is shown in Table 2.

Table 2. Input and output format of Mapper.

Input		Output	
key	value	key	value
LongWritable Data location	Text Data	Text File ID	TransData Score and times

a. Mapper Code List:

```
public class TopMapper extends
Mappper<LongWritable,Text,Text,TransData>{
    private String topSplit=",";
    public void setup(Context cxt){
   String tempSplit = cxt.getConfiguration().
get("TOP_SPLIT");
   topSplit = (null ==tempSplit)?topSplit:tempSplit;
   }
public void map(LongWritable key,Text value,Context
cxt)
          throws IOException,InterruptedException{
   String[] line = value.toSting().split(topSplit);
   if(null==line || line.length!=3){return;}
   cxt.write(new Text(line[1]),new TransData(Double.
parseDouble(line[2]),1));
   }
}
```

Mapper converts such data [userld, profileld, prefValue] and then outputs <key, value> being <profileld, [prefValue, 1]>, that is, the output is each file and the score and number of file is once.

Combiner is followed. Combiner mainly works at Map terminal, and it can integrate part of the data, so that data transmitted to reduce terminal will be reduced, which improves efficiency. Combiner respectively adds up all the same keys, that is, profileld is the same score and time, the <key, value> and Mapper output are the same, but the data has changed. Score and time is the sum of the same files in the same Mapper.

b. Reduce Code List:

```
public void reduce(Text key,Iterable<TransData>
value,Context cxt) throws
IOException,InterruptedException{
    double prefs = 0.0,int counts = 0;
    for(TransData v value){
   prefs += v.getPref().get();
  counts += v.getCount().get();
}
if(counts < 10){return;}
if(map.size() <10){
   map.put(key.toString(),new DoubleWritable(prefs/
counts));
    }else{
    insertToMap(key.toString().prefs/counts);
    }
}
public void cleanup(Context cxt) throws
IOException,InterruptedException{
    List<Entry<String,DoubleWritable>> list =
sortByValue(map);
    for(Entry<String,DoubleWritable> l : list){
   cxt.write(new Text(l.getKey()),l.getValue());
   }
}
```

The main work of Reduce is to calculate the average rating of each file, and then order all items in accordance with the rating descending order, and finally take the ID of the front 10 (N Takes 10) files. The core code is shown below.

2 Collaborative Filtering Algorithm Design in Mahout

The component diagram is implemented by collaborative filtering recommendation in Mahout,as Figure 4:

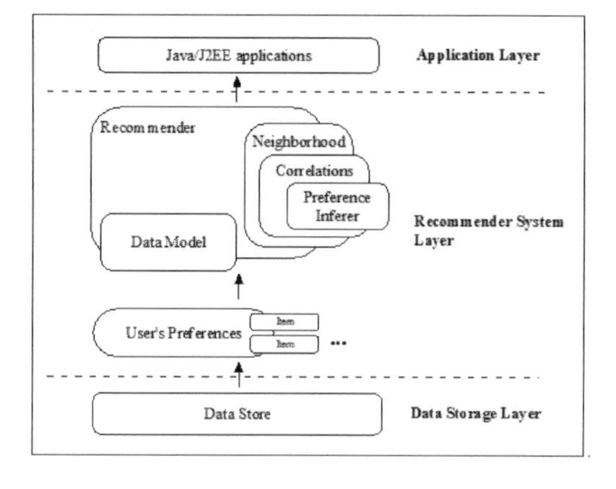

Figure 4. The component diagram implemented by collaborative filtering recommendation in Mahout.

The most classic three recommended strategies of collaborative filtering of Mahout are: User CF, Item CF and Slope One. Date recommender system in the paper uses the Item CF algorithm of Mahout to obtain the basic data recommended.

Realize Item CF based on Mahout:

```
DataModel model = new FileDataModel(new
File("preferences.dat"));
ItemSimilarity similarity = new
PearsonCorrelationSimilarity(model);
Recommender recommender = new
GenericItemBasedRecommender(model, similarity);
```

3 The New User Recommendation Algorithm Design

The algorithm runs Mapper at the beginning. In Mapper's initialization, that is, setup function, needs to read the scores of users on top 10 files, and then read these data into a vector userPrefs. And then calculate the similarity between itself and vector userPrefs in the map function for each data recorded, and then output the user ID and the similarity. The formula of similarity can refer to Equation 1.

$$Similarity = (1 / (1 + sqrt (v1, v2))) \quad (1)$$

where sqrt (v1, v2) represents the mean square error of vector v1 and vector v2.

```
//Calculate the similarity between the two vectors, it is
more similar with bigger similarity
private double calc(VectorWritable userVec){
   double distance =0.0;
   Vector userPrefsVec = userPrefs.get();
   Vector userVecV = userVec.get();
   distance
   userPrefsVec.getDistanceSquared(userVecV);
   if(distance - 0.0 < 0.0000001){
   return Double.MAX_VALUE;
   }
return 1/(1 + distance);
}
```

Considering uniformly processing all data is needed in the following Reduce, key output by Mapper here must be consistent, and the key set and output here is new IntWritable (1). The core Mapper code list is as follows:

After Mapper outputs data, Reduce integrates the same value of key, and uniformly processes (because key output by Mapper are the same, so the data here are all the data). The value format output in Mapper is the defined Writable type.

In fact, the process in Reduce terminal is similar to a 10 element stack maintained by the front top 10 algorithms, the key of Map element in stack is the user ID. At the end of the algorithm, the ID information of date file recommended by new users can be obtained on HDFS file system.

4 EXPERIMENT AND ANALYSIS

Verify the validity of the new user recommendation system and scalability facing big data through implementing the system of recommending date. The data used in the system is downloaded from http://www.occamslab.com/petricek/data/. This data is the rating data of 135,359 anonymous users on the file information of other 168,791 users. Data format is: [1,133,8], which indicates that the score of user 1 on 133 files is 8 points (10 points is full mark).

4.1 *Environment configuration*

Table 3. Hardware configuration.

Environment name	Environment version	Environment description
4 or more cheap PC machine (named as: hp0 to hp3)	Normal PC, only the host	hp0:NameNode,JobTracker hp1-hp3:DataNode,TaskTracker
Router	A TPLink router	Normal router

4.2 *Recommended performance evaluation*

This paper conducts performance analysis and experimental test on the algorithm recommended by new users, which is analyzed from performance evaluation and recommendation quality evaluation (Feng & Huang).

Firstly, performance evaluation is mainly considered from the execution efficiency. The parameter f = T1 / Tn is utilized for comparison. T1 is the execution time of the algorithm when TaskTracker node is 1, and Tn is the execution time of the algorithm when TaskTracker node is n. The author can draw the influence condition of the number of Hadoop nodes on the implementation efficiency of algorithm through the comparison of parameter f (f = T1 / Tn).

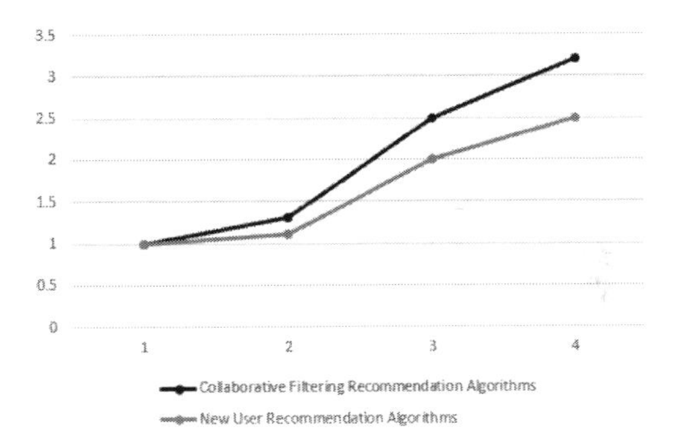

Figure 5. 10M data.

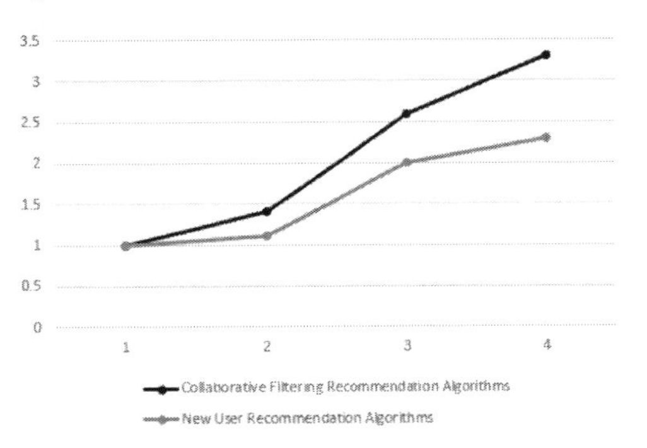

Figure 6. 100M Data.

The x-axis is number of nodes, the y-axis is parameter f (f = T1 / Tn).

From the experimental results shown in Figures 5 and 6, the number of TaskTracker node for the same data set increases, and the efficiency of algorithm recommended by new users is significantly higher than that of collaborative filtering recommendation algorithm. And compared with the calculated data sets, the impact of the size on Hadoop clusters is more obvious.

In general, the recommendation quality assessment method uses mean absolute error MAE for evaluation. When MAE is smaller, recommendation quality is higher, the MAE formula is:

$$MAE=\sum na \in U|Raj - Paj|/N \quad (2)$$

where N is the number of users of recommended items obtained by users, j is 1 , ... , N recommended items, a is user to be recommended.

RecommenderEvaluator in Mahout Algorithm library uses MAE to test whether the recommended items are consistent with the actual situation. But, this is recommended through historical data of users, the new user recommendation lacks of historical data, and there are certain difficulties in all quantitative analysis of the recommended quality evaluation. This paper analyzes the problem of recommended quality evaluation from the perspective of qualitative.

Only collaborative filtering recommendation algorithm is utilized to recommend the new users. It can be seen from the results that it can find their own potential interests and preferences which have not been found yet, as well as their similar interests and preferences can be recommended. But after all, they are similar interests and preferences. Sometimes the recommended is too many, thus making the recommendation quality fail to keep up. While the algorithm recommended by new users makes them orient ownself, together with integrating the concept of similarity of collaborative filtering, thus the interests and preferences recommended are their favorite interests and preferences, and their amount has been featured. This is the charm of algorithm recommended by new users.

5 CONCLUSION

The authors in this paper use the "hybrid" of collaborative filtering algorithm provided by Mahout with Top N algorithm based on MapReduce to achieve new user recommendations. A model recommended by new users is built and the validity and scalability facing large-scale data of the new user recommender system are verified through making recommendation appointment systems. Because the goal of Mahout is a scalable machine learning algorithm library, the distributed parallel computing combined with Hadoop and the processing of large data sets by MapReduce are realized so that Hadoop and Mahout can quickly design and develop enterprise-class recommender systems with powerful data processing capability. Recommendation date system can provide better recommendations for new users, and it is with certain accuracy and feasibility. In the future, the algorithm will be optimized so that the accuracy moves on up.

REFERENCES

Adomavicius, G & Tuzhilin, A. 2005. Personalization technologies: A process-oriented perspective. *Communications of the ACM* 48(10):83–90.

Adomavicius, G & Tuzhilin, A. 2005. Toward the next generation of recommender systems: A survey of the state-of-the-art and possible extensions. *IEEE Trans. on Knowledge and Data Engineering* 17(6):734–749.

Feng Guohe & Huang Jiaxing. Research on Collaborative Filtering Book Recommendation based on Hadoop and Mahout. *Library and Information Service* 20.

Hu, Zhuqing, Liu, Sisi, Liu, Chenguang, et al. 2012. Exploration on the Combination of New User Issues with Small World Network of Collaborative Filtering System. *Silicon Valley* (8): 191–191.

Li, Wenhai & Xu, Shuren. 2014. Design and Implementation of E-commerce Recommender System based on Hadoop. *Computer Engineering and Design PKU* 35 (1).

Liao, Zhifang, Wang, Chaoqun, Li, Xiaoqing, et al. 2013. Recommendation Algorithm of Label Recommendation of Tensor Decomposition and New User Label. *Small Computer and Microcomputer Systems* 34 (011): 2472–2476.

Liu, JG, Zhou, T & Wang, BH. 2009. Personalized recommender systems: A survey of the state-of-the-art. *Chinese Journal of Progress in Natural Science* 19(1):1–15.

Ricci, F, Rokach, L, Shapira, B, et al. 2011. *Recommender Systems Handbook*: 1-842. Berlin: Springer-Verlag.

Wang, LiCai, Meng, Xiangwu, Zhang, Yujie. 2012. Context-aware Recommender System. *Journal of Software* (1): 1–20.

Xu, HL, Wu, X, Li, XD, et al. 2009. Comparison study of Internet recommendation system. *Journal of Software* 20(2):350–362.

Yang, Zhiwen & Liu, Bo. 2013. Collaborative Filtering Recommendation Algorithm based on Hadoop Platform. *Computer System Application* (7): 108–112.

Zeng, C, Xing, CX & Zhou, LZ. 2002. A survey of personalization technology. *Journal of Software* 13(10):1952–1961.

Zhu Qian & Qian Li. 2013. Analysis and Design of Mahout-based Recommender System. *Bulletin of Science and Technology ISTIC PKU* (6).

Electronics, Communications and Networks IV – Hussain & Ivanovic (eds)
© 2015 Taylor & Francis Group, London, ISBN: 978-1-138-02830-2

The automatic planning method for mainstream velocity line

Jianhua Gao, Xueshi Dong & Wenyong Dong
Computer School, Wuhan University, Wuhan, Hubei, China

ABSTRACT: The main velocity line of a channel is a difficult and hot research area in channel route planning. Nowadays, how to plan a main velocity line of a channel with the scientific and reasonable method is the scientific problem in the research of channel route planning. In this paper, geographic information system and machine learning apply to the planning of mainstream line, the cross section data are used for the spatial interpolation, generating the corresponding irregular triangular mesh and DEM data, the ration of velocity data is calculated according to the section data, the mainstream velocity line with the threshold is generated and the prediction and analysis of main velocity line based on these are carried out. The paper firstly provides the automatic programming method of mainstream velocity and implements this successfully, which has a certain referential significance of the study and work in the relevant areas.

KEYWORDS: cross section data; mainstream velocity line; automatic planning method; geographic information system

1 INTRODUCTION

The automatic planning method for the mainstream velocity line has the following two significances: on the one hand, it can save the cost of waterway departments in waterway planning and greatly reduce the cost of human resources. Compared with traditional artificial waterway planning that costs many human resources and materials, these aspects have been greatly improved through this method; on the other hand, it saves a lot planning time and enhances the waterway planning efficiency.

At present, problems related to planning method for the mainstream velocity line include the followings: In the past, it took many financial and material resources and a lot of time to plan the method for mainstream velocity line, which was far from meeting the needs of various waterway departments and relevant departments. In this paper, geographic information systems, machine learning and other technologies are used in the planning method for mainstream velocity line, and carrying out mainstream velocity line planning based on this approach is firstly proposed in the country, which is a big innovation in waterway planning, for this method not only saves financial and material resources but also greatly reduces the operating time, thus having important theoretical and practical significance for relevant work and research.

This paper proposes for the first time the automatic planning method for mainstream velocity line and has some innovations in the sub-model or algorithm, which are shown as follows: first, based on actual experience, it first proposes the empirical formula of the mainstream velocity line model and innovatively adopts linear regression and life cycle prediction algorithm to predict the mainstream velocity line. What's more, it predicts the evolution of the line based on history data by using the geometric principles to convert it into data of prediction points and then connect them. Second, in terms of mainstream velocity line generation, it applies the cubic Bezier curve to the smoothing of mainstream velocity line, getting quite good practical results; and it uses geometry calculation to classify cross-section data which are divided into two categories according to certain principles. Third, it adopts the IDW spatial interpolation method for the interpolation of cross-section data so as to generate TIN and DEM data (Zhang et al. 2012). When it is impossible to interpolate or when the interpolated data is missing during the interpolating process due to the overlarge interval between cross-section data, the author inserts some short-interval cross-section data before interpolating, therefore it obtains better results. There are some papers (Chen et al. 2014, Formetta et al. 2014, Li et al. 2007, Xu & Chen 2003) which could be used for reference.

The Yangtze River (Li 2014) is the unique waterway transportation channel through the east and west, and the main channel flowing through seven provinces and two cities. The main tributaries of Yangtze River run through south and north. So it is the main part of the basin comprehensive transportation system, and plays an important role which can't be replaced because of the huge transport capacity and the important areas it flows. It promotes the formation of the economic belt along the river, and also provides important support for export-oriented economic development.

The article firstly gives the automatic programming method and implements it successfully. In this paper, the history data and materials are utilized to analyze the middle low channel segment and draw relevant rules; and the modern advanced scientific theory and methods are used to simulate the mainstream line; and the main velocity line in the future time is predicted, which can provide exact decision guideline for channel maintenance and management.

This paper consists of five parts: The first part is the introduction; the second part is about the overall structure of the automatic planning method for mainstream velocity line; the third part is the detailed discussion of the various methods adopted in this paper; the fourth part is the effective analysis of the planning method; and the fifth part is concluded.

2 OVERALL STRUCTURE OF AUTOMATIC PLANNING METHOD

2.1 *Basic steps of the study*

The basic steps of the study mainstream line planning are shown in Figure 1.

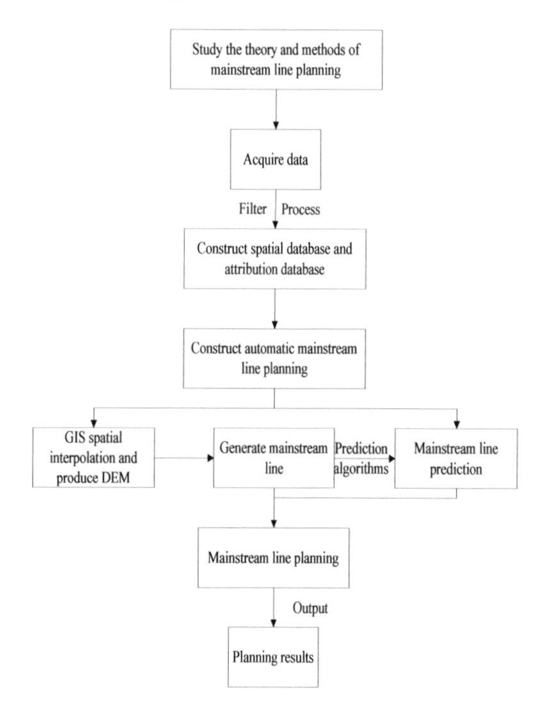

Figure 1. Basic steps of study mainstream line planning.

1 Study the theory and methods of mainstream line planning. To study the field, one needs to acquire knowledge of different areas, and studies the theory and methods of the method at home and abroad.
2 Acquire data, in this step, one needs to obtain many kinds of data, such as data in the form of DXF and TXT and some relevant spatial data and attribution data.
3 Filter and process the data, during the process, one should deal with the initial data, for example, process the missing data and irregular data in order to meet the requirement.
4 Construct spatial database and attribution database, the automatic planning method has two kinds of databases named spatial database and attribution database. The attribution database stores the attribution information and the spatial database saves the data related to the spatial location.
5 Construct automatic mainstream planning, based on the step mentioned above, then the automatic mainstream planning can be constructed, and the planning method includes three steps in the following parts.
6 GIS spatial interpolation and generate DEM data, spatial interpolation is an important operation of GIS, after the interpolation it can generate the DEM data.

   The interpolation of the DEM is operated by using the cross section data during the process. When the distance between the two ones is very long such as more than one thousand meters, it is difficult to interpolate the data or the interpolated data is missing or not perfect. The algorithm utilizes the method that it interpolates some cross section data between the two cross data based on a short distance, then interpolation is done, so that it can obtain the satisfying results.
7 Generate mainstream line, the method can utilize DEM data to generate velocity line, and according to the chosen threshold it can produce the mainstream line.

   In the step, it calculates the ration of velocity data according to section data, and generates the mainstream line with the threshold. The method contains the ways of producing mainstream lines including the threshold of a single cross section and threshold of multi cross section, and the single cross means that the threshold is calculated with the single section data, and the multi cross shows the threshold is computed by using multi section data or all cross section data.
8 Mainstream line prediction, the method contains three prediction models, including models of empirical formula, linear regression model and life cycle prediction model.

   The basic way for mainstream line prediction: choose the representative point of the mainstream velocity line generated by the used history data, and predict the velocity and coordinate of the each point; then connect the points to generate the lines. The lines are the prediction mainstream lines that we want to obtain by using the method.

During the process of prediction, the prediction model is based on the method of prediction line by using the data in the form of a line. The basic theory is choosing the representative points of history mainstream velocity line. This method generates the intersection point set by the lines produced by using cross section data to intersect with the history mainstream velocity line, which can translate the predicting line by using line data to predict point with the data in the form of point.

Mainstream velocity prediction is based on the theory of predicting line by using line data; firstly, according to history data and the mainstream line prediction models, the method generates the mainstream velocity line.

9 Mainstream line planning, based on the works and steps mentioned above, people can do mainstream line planning and make some decisions by using the processed data.
10 Output the planning results, and it can save the planning results in different forms for decision service.

2.2 *Basic steps of data processing*

After the data of the automatic planning method for mainstream line being loaded, the mainstream velocity line is generated, during which the following items are required, the basic steps of data processing are shown in Figure 2:

1 Erasing some invalid data of the measurement point (decision rule undetermined due to the requirement of the information about the water level);

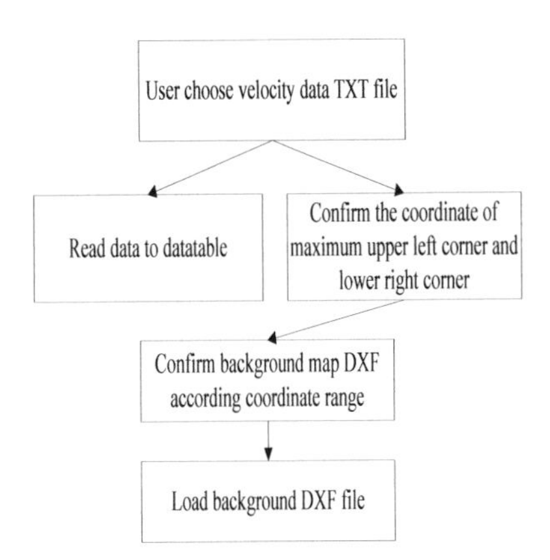

Figure 2. Basic steps of data processing.

2 Inputting a percentage by user. On the basis of this and together with the consideration of the velocity information, two mainstream velocity points can be determined in each cross section (interpolation required and interpolating required to be conducted within the banks);
3 Connecting and smoothing the upper and lower mainstream velocity points of each cross section, respectively, thus forming two mainstream lines.
4 Selecting a style for filling the two mainstream velocity lines;
5 Generating the results into a DXF file which supplements the information of layers (generation time, water level, percentage and other information included).

3 DETAILS OF METHODS

3.1 *Data processing*

During the execution of the DEM interpolation model, the interpolation is conducted on the basis of cross section data.

Digital Elevation Model (DEM) is the solid ground model with a group of ordered numeral arrays to represent ground elevation, which is the division of Digital Terrain Model and other terrain values are generated from this. Generally speaking, DTM describes many kinds of geomorphologic factors, including digital elevation such as slope, aspect and slope change rate and the spatial distribution with the combination of linear and nonlinear, and DEM is the simple zero order and digital surface model, other features such as slope, aspect and slope change rate can be derived based on DEM (Liu et al. 2009) (Lu et al. 2012).

Because DEM describes the ground digital elevation information, it is widely used in survey, hydrology, meteorology, geology, geomorphology, soil, engineering construction, national economy, national defense construction, the humanities and natural sciences. In the engineering construction, it can be used for the calculation of earthwork, visibility analysis and so on; in flood control and disaster reduction, DEM is the foundation of hydrological analysis, such as water network analysis, analysis of rainfall, flood calculation, flood analysis; in wireless communication, it can be used in the base station analysis of cellular telephone etc..

Among various types of existing spatial interpolation methods, such as IDW, Kriging interpolation, Natural Neighbor method, Spline Interpolation, Topo to Raster Interpolation, Trend Interpolation and some other interpolation methods, IDW interpolation method is adopted in this paper.

IDW is a simple and convenient spatial interpolation method which is commonly used. The distance between the interpolation point and sample point is regarded as the weight, on the basis of which the weighted average method is conducted. And the closer the sample point is to the interpolation point, the larger its endowed weight is.

$$Z = \left[\sum_{t=1}^{n} \frac{Z_t}{d_t^2} \right] / \left[\sum_{t=1}^{n} \frac{1}{d_t^2} \right] \quad (1)$$

$$d_i^2 = (X - X_i)^2 + (Y - Y_i)^2 \quad (2)$$

The basic thought is that the closer the target is to the observation point, the larger the weight is and the greater the influence over the observation point is. The disadvantages are that the observation point itself is absolutely exact and can limit the number of the interpolation points. Power can help to determine the degree that the proximate principle influences the results, and search radius can help to control the number of the interpolation points.

$$Z = \left[\sum_{i=1}^{n} \frac{Z_i}{d_i^2} \right] / \left[\sum_{i=1}^{n} \frac{1}{d_i^2} \right] \quad (3)$$

$$d_i^2 = (X - X_i)^2 + (Y - Y_i)^2 \quad (4)$$

Through the average calculation of values of each sample point in the neighboring regions, IDW can acquire the interpolation unit. This method requires that the discrete points should be distributed homogeneously and the density should meet the requirement of reflecting the local surface changes during the analysis.

When one cross section is rather far from the other cross section, above 1,000 meters, it is difficult to interpolate data or the interpolating data is incomplete between the cross sections. In this algorithm, some short-interval cross-section data will be interpolated between the two sections before the performance of the cross section interpolation, so as to obtain a satisfactory result.

3.2 *Forming of main velocity line*

Bézier curve is a kind of mathematical curve applied to stereogram application. Common vectorgraph applications use it to draw precise curves. Bézier curve consists of the segment and node, which accounts for a drawable fulcrum, and segment appears as an elastic rubber band; they are adopted in making vector curves of the pen tools in drawing application. Bézier curve is of great importance in computer graphics, and it also appears in certain developed bitmap applications like PhotoShop. There are not yet complete curve tools in Flash 4, but Bézier curve has been provided in Flash 5.

Bézier curve had been widely published in 1962, by a French engineer named Pierre Bézier, who designed the main body of a car by using Bézier curve. The curve was first developed by Paul de Casteljau in 1959 from de Casteljau algorithm, and obtained by the method of stability number.

Cubic formula of Bézier curve:

P_0, P_1, P_2 and P_3 define the cubic Bézier curve in a plane or three-dimensional space (Wang & Shang 2011). It starts from P_0 to P_1, then from P_2 to P_3. Generally it will not pass by P_1 or P_2, which only provides information of direction. The distance between P_0 and P_1 determines the "length" to P_2 before approaching P_3.

$$B(t) = P_0(1-t)^3 + 3P_1 t(1-t)^2 + 3P_2 t^2(1-t) + P_3 t^3, t \in [0,1] \quad (5)$$

Parameter form of the curve is:

$$B(t) = P_0(1-t)^3 + 3P_1 t(1-t)^2 + 3P_2 t^2(1-t) + P_3 t^3, t \in [0,1] \quad (6)$$

Modern imaging systems such as PostScript, Asymptote and Metafont, use the cubic Bézier curve formed by Bézier spline for describing profile curve.

There are two automatic programming methods of main velocity line in this article: one is to use section data to conduct spatial interpolation for DEM data, based on which to form an equal velocity line and get needed threshold value, according to which a new flow velocity line get formed, and this is the main velocity line; the other way is to sort section data, set relevant rate, which will be the threshold value of main velocity line, and then the main velocity line is got according to the value.

3.3 *Prediction of main velocity line*

Basic way of the prediction of main velocity line: select a certain amount of representative points among main velocity lines formed by history data, predict the flow rate and the coordinate of each point, match the points. The predicted main velocity line is made.

Prediction of main velocity line is conducted based on the principle of line prediction. First, form a model for main velocity line by history lines and save it in formation of TXT, and make predictions based on the TXT by main velocity line prediction model, then save the predicted lines as TXT file; the saved history and predicted main velocity lines can all be visually presented by automatic generation model. During prediction, the model is based on line-to-line method, whose theory lies in choosing typical point data on history main velocity lines, obtaining a set of crossing points through certain distant straight lines and history main velocity lines, and then turning the form of line-to-line into point-to-point.

Multiple main velocity lines formed in different time can be read; main velocity lines of another time are predicted via ways of interpolation and extension. A prediction model for the automatic programming method is shown as follows:

1 Model of Empirical Formula

Model of empirical formula:

$$p_3 = p_2 + w\frac{Q_3 - Q_2}{Q_1 - Q_2}(P_1 - P_2) \tag{7}$$

P_3 is predictive data, P_1 and P_2 are needed history data for the prediction before P_3, Q_3 is the corresponding data for water depth, Q_1 and Q_2 serve as relevant data on water levels of P_1 and P_2, W stands for a coefficient and is set as 1 in the formula. The model can be used for predicting the changes of flow velocity and coordinate. Basic way of the prediction of main velocity line: select a certain amount of representative points among main velocity lines formed by history data, predict the flow rate and coordinate of each point, match the points, and predicted main velocity line is made.

2 Linear Regression Model

The so-called linear regression model (Song & Wang 2012) (Si et al. 2010) refers to that the relation of the dependent and independent variable is a linear one. Linear regression analysis is an analysis of the quantitative relation of things; it is an important statistical method and has been widely applied to research on the influencing factors and relations of social economic phenomena, which is too complicated to be expressed by a single variable. Assume X as an independent variable and Y as a dependent variable, linear regression method is to establish a linear regression equation of X and Y.

$$Y_t = a + bX_t \tag{8}$$

Prediction model of linear regression method:

$$Y_t = a + bX_t \tag{9}$$

In the formula, X_t stands for the value of the independent variable at t; Y_t for that of the dependent value; a and b stand for parameters in the formula. a and b can be obtained through the following formula, of which $\sum$ stands for $\sum_{i-1}^{n}$.

$$a = \frac{\sum Y_i}{n} - b\frac{\sum X_i}{n} \tag{10}$$

$$b = \frac{n\sum X_i Y_i - \sum X_i \sum Y_i}{n\sum X_i^2 - (\sum X_i)^2} \tag{11}$$

Put a and b in the linear regression equation $Y_t = a + bX_t$ and a predictive model is set up.

3 Life Cycle Prediction Model

Life cycle prediction model (Zhang et al. 2008) is suitable for monotone increasing or monotone decreasing non-linear systems, as well as non-linear equation with them both. Assume during the changing of annual runoff as Q(t) according to t, n-th power function proportional to t emerges, and attenuates with the negative index of t. Such a process can be expressed as the formula as follows:

$$\begin{cases} Q(\theta) = A\theta^n e^{-\theta} \\ \theta = (y - y_0)/c \end{cases} \tag{12}$$

In the formula, $Q(\theta)$ refers to the fitting or predictive annual runoff; θ means discrete time; y is the predicted time; y_0 is the predictive starting time, and annual A, n and c are undetermined parameters.

The key lies in the solving of model parameters, $B = y - y_0$, and then the above formula can be shown as $Q(t) = b \times t^n \times e^{at}$.

In the formula, $b = A/c^n$, $t = B$, $a = -1/c$. By applying the least square to get the above parameters and removing the logarithm, a formula comes out:

$$\ln Q(t) = \ln b + n \ln t + at \tag{13}$$

The above formula contains three unknown numbers a, n and b, so a three-element simultaneous equation is needed. Generally, history data of predicted object are required to be divided into 3 groups, through statistical analysis, a simultaneous equation containing the three unknown numbers can be obtained; then the equation is solved, and the value of the unknown numbers is got. Assume total history data are $(m_1 + m_2 + m_3)$, of which first group occupies m_1, second m_2, and the third group m_3, and an equation group is such acquired.

$$\begin{cases} \sum_{i=1}^{m_1} \ln Q(t) = m_1 \ln b + n\sum_{i=1}^{m_1} \ln t_i + a\sum_{i=1}^{m_1} t_i \\ \sum_{i=m}^{m_1+m_2} \ln Q(t) = m_2 \ln b + n\sum_{i=m_1+1}^{m_1+m_2} \ln t_i + a\sum_{i=m_1+1}^{m_1+m_2} t_i \\ \sum_{i=m_1+m_2+1}^{m_1+m_2+m_3} \ln Q(t) = m_3 \ln b + n\sum_{i=m_1+m_2+1}^{m_1+m_2+m_3} \ln t_i + a\sum_{i=m_1+m_2+1}^{m_1+m_2+m_3} t_i \end{cases} \tag{14}$$

Solve the equation, and a, n and b can be known.

4 ANALYSIS ON AUTOMATIC PROGRAMMING EFFECT

The DEM map made by overall structure of automatic programming method for main velocity line is shown in the following Figure 3:

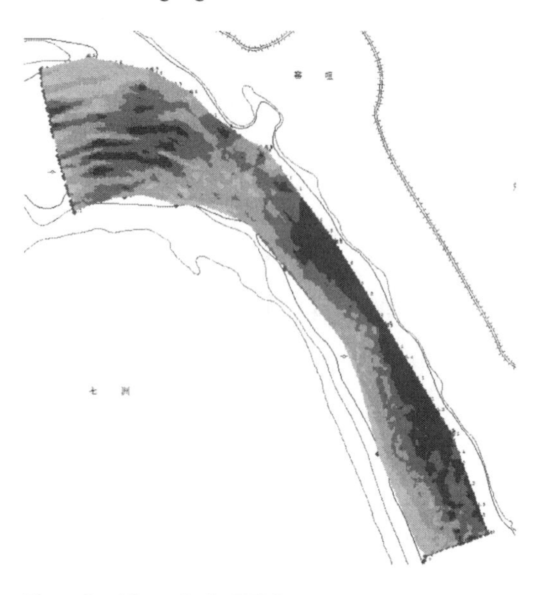

Figure 3. Flow velocity DEM.

Figure 3 is a map data made by DEM, different colors in the figure stand for different velocity flows.

The yellow line in Figure 4 is the formed main velocity line. We export and save the formed result in a widely used file format "DXF".

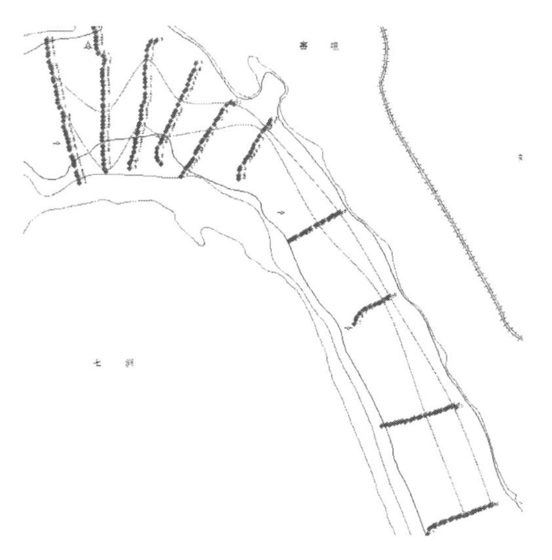

Figure 4. single-section main velocity line.

Before prediction, first a basic process is needed for the predicted data: open main velocity lines formed by history data in turns, and then merge data of each period into one file, which is the needed data for prediction. The main velocity line in Figure 4 is obtained by cubic Bézier curve.

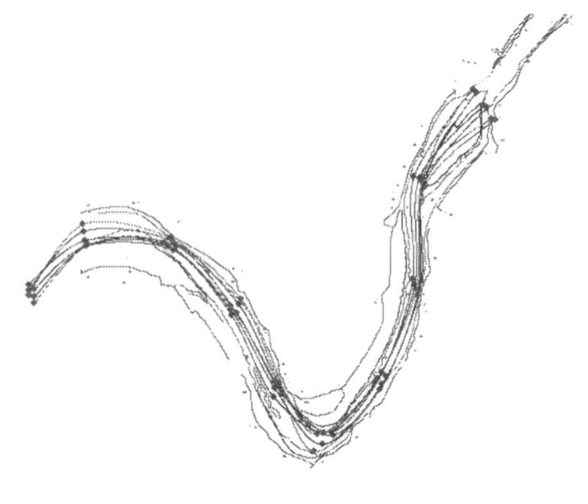

Figure 5. Predictive and analytical with empirical formula.

By loading and opening the history section data in three different days, which are represented by three different colors, user can set some parameters by their experience. Finally, we can generate the prediction data which is shown in Figure 5 by the empirical formula we defined.

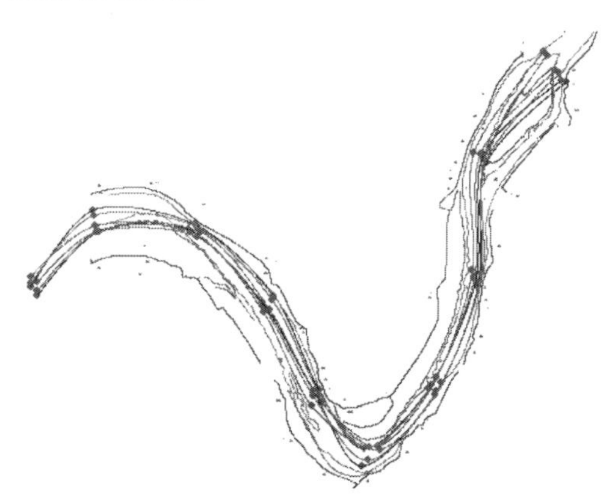

Figure 6. Predictive and analytical data with linear regression.

The predictive and analytical data with linear regression prediction is displayed in Figure 6.

The article automatically programs the main velocity line applications such as geographical information system, machine learning and computational geometry, it greatly increases the efficiency of waterway planning, saves manpower and material resources, and it is of great significance both theoretically and practically on the working and research in related fields. The article has made an analysis on the effect of automatic programming in terms of GIS spatial interpolation, forming and prediction of main velocity line. First, automatic programming method adopts spatial interpolation to supplement missing data between sections, and interpolates remote data based on certain principles; such method has been proven effective. Second, the method has smoothed main velocity line with bigger winding by cubic Bézier curve, and has proven satisfied to practical requirement, with a certain innovation on the algorithm for threshold value. Third, the article has put forward the prediction model of main velocity line by empirical formula; and applied linear regression and life cycle methods, turning line-prediction to point-prediction; then main velocity line is formed by the predictive points, realizing a kind of innovation as to prediction method.

1 Comparison between the effect of automatic programming and artificial programming

   Automatic programming can save manpower and material resources for waterway department. On one hand, traditional programming method requires manual operation, which is in need of a great human resources; at the same time, during operation, man-made operation error may be very likely to appear; on the other hand, traditional programming method makes use of man power, wasting much time; whereas automatic programming method of main velocity line is able to greatly cut down the programming time and increase work efficiency. So it has significance to utilize the method for programming.

2 Section flow data included in the automatic programming curve is larger than the given rate of threshold value

   While forming main velocity line, a rate for section threshold value must be given; section threshold value also corresponds to the flow rate of section data. The flow rate between lines formed by an automatic programming method is smaller than the corresponding one of the threshold value. In spite of that, some may be bigger, which is due to the error by abnormal data and man-made measurement, the mainstream programming method is proven practical.

3 Comparison and analysis of empirical formula, linear regression and life cycle. The parameters of empirical formula are determined by experience, and they can be adjusted according to practical situation till they are relatively accurate; parameters of linear regression model are acquired by applying the least square method, and to solve the parameters, some historial data needs to be trained; life cycle model is rather complicated compared with the former two methods, with more parameters to be trained, and they are also obtained by least square method.

5 CONCLUSIONS

Programming of main velocity line has always been a heated topic in the waterway planning field. The article firstly puts forward the automatic programming method based on section data. The method can not only save manpower and material resources for waterway department, but also greatly save time to increase work efficiency. As for engineering practice, the article has proposed some new empirical ways that can best solve practical problems and hold referential significance of the work and research of related fields.

In order to improve planning method, there are some works we can do for the next work: First, study the more effective and better mainstream line prediction algorithms and the improvement of the algorithms and consider the effective combination with other intelligent algorithms; second, study the more advanced methods of data processing to deal with kinds of data including DEM, TIN and the data from TXT and DXF, which can improve the effectiveness and reduce time of processing; finally, explore other better mainstream line planning methods which can perform more accurately and effectively than the method in this paper.

ACKNOWLEDGMENT

We are gratefully acknowledged the financial support from the National Natural Science Foundation of China under Grant No. 60873114, 61170305.

REFERENCES

Chen, J., Wu, X.D., Finlayson, B. L., et al. 2014. Variability and Trend in The Hydrology of The Yangtze River, China: Annual Precipitation and Runoff. *Journal of Hydrology* 513: 403–412.

Formetta, G., Antonello, A., Franceschi, S., David O. & Rigon Ret al. 2014. Hydrological Modelling with Components: A GIS-based Open-source Framework. *Environmental Modelling & Software* 55: 190–200.

Li, B.D. 2014. Data Resource Planning of The Yangtze River Waterway. *Port & Waterway Engineering* 11: 15–18.

Li, Q.Q., Yang, B.S. &, Zheng, N.B. 2007. An Integrated Spatio-Temporal Data Model for GIS-Transportation and Related Applications. *Geomatics and Information Science of Wuhan University* 32: 1034–1041.

Liu, X.J., Zhang, P. &, Zhu, Y. 2009. Suitable Window Size of Terrain Parameters Derived from Grid-based

DEM. *Acta Geodaetica et Cartographica Sinica* 38(3): 264–271.
Lu, H.X., Liu, X.J., Wang, Y.J., et al. 2012. Noise Error Analysis of Slope Algorithms Based on Grid DEM Derived from Interpolation. *Acta Geodaetica et Cartographica Sinica* 41(6): 927–932.
Si, S.B., Ni, M.N. &, Jia, D.P. 2010. A Different Optimization Method of Two-Echelon Inventory Control System Based on Linear Regression Theory. *Journal of Northwestern Polytechnical University* 28(6): 844–849.
Song, X. &, Wang, C.R. 2012. Linear Regression Based Distributed Data Gathering Optimization Strategy for Wireless Sensor Networks. *Chinese Journal of ComputersHINESE JOURNAL OF COMPUTERS* 35(3): 568–578.
Wang, X.C. &, Shang, J.H. 2011. Concrete Application of Triple Bezier Curve Jointing Model in Method of Curve-rization to Broken-lined Contours. *Science of Surveying and Mapping* 36(2): 192–194.
Xu, Z.L. &, Chen, Y.M. 2003. Design of Simulating and Predicting System on Waterway of Minjiang River Based on GIS. *Computer Engineering* 29: 146–148.
Zhang, J.X., Ma, X.Y., Zhao, W.J., et al. 2008. A Prediction Model for River Annual Runoff Based on Life Cycle-weighted Markov. *Engineering Journal of Wuhan University* 41: 1–15.
Zhang, X.Z., Shi, Y. &, Chen, L.G. 2012. Method of Data Extraction of River Section Based on ArcGIS. *Water Resources and Power* 30: 139–141.

Electronics, Communications and Networks IV – Hussain & Ivanovic (eds)
© 2015 Taylor & Francis Group, London, ISBN: 978-1-138-02830-2

Large pattern online handwritten Chinese character recognition based on multi-convolution neural networks

Mingtao Ge & Yuan Sang
SIAS International University, Zhengzhou University, Zhengzhou, Henan, China

Liwu Pan* & Junhui Liu
Dept. of Information Engineering, Henan University of Animal Husbandry & Economy, Zhengzhou, Henan, China

ABSTRACT: Online handwriting recognition is widely applied in production and life and is focused on pattern recognition. So far, the traditional Convolutional Neural Network (CNN) technique has been applied, with a quite high rate of small scale on-line handwritten Chinese character recognition and a high overall performance. But in the case of large-pattern character sets, it has a low recognition rate. This paper proposes a character recognition method based on multi-convolution neural networks so as to address the above problem and improve the recognition rate of large-model character sets. Stochastic diagonal Levenberg-Marquardt (L-M) is adopted to improve training. Unipen training set test shows online handwriting recognition has an accuracy of 89%, with a promising prospect.

KEYWORDS: Pattern recognition; neural network; convolution; character recognition.

1 INTRODUCTION

With the rapid development of global information technology and automation requirements, handwriting recognition technology has been widely applied. In recent years, mobile phones, tablet PCs and other intelligent electronic products characterized by handwriting functions have been increasingly popular, and masking researches on online handwriting recognition have become the focus. Online handwriting recognition requires timeliness, which is a hot issue in large-scale model recognition. It is necessary to have a quite high feature space dimension in recognition, extensive aspects in feature sample training and multiple support feature values or feature objects (Wu & Zhang 2001, Zhang 2012).

CNN has the advantage that the visual model is obtained directly from the original image in recognition. Handwriting image preprocessing in the design has the following advantages: The matching rate between the image to be detected and pre-established network topology is high; Feature extraction and model classification can be conducted simultaneously; Training parameter is important to the system calculation; CNN can apply weight sharing technology, which can greatly reduce the parameter and make the system structure simpler and the whole system more adaptable(Xu et al. 2013, Lu 2011, Pham 2012, Sermanet et al. 2012).

So far, CNN has been successfully applied to human-computer interaction handwriting recognition, human-computer interaction, vehicle license plate number recognition and face recognition, commonly used in information security. LeCun Y et al. adopted a 4-layer CNN LeNet-5 to conduct Mnist databank recognition test, with a recognition rate of 98.4%, and adopted a 2-layer BP network, with a recognition rate of 87%(Lu 2011, Pham 2012, Sermanet 2012, LeCun et al. 1998). Multiple scholars have made multi-faceted researches on CNN application to online handwritten Chinese character recognition with a quite high recognition rate of small character sets, but a low recognition rate of large-model character sets. This paper introduces CNN basic concepts and a typical CNN structure, and proposes a character recognition model based on multi-convolution neural networks. This author conducts trainings and tests with large-modeled character set Unipen database, and proposes large-model online handwriting recognition, achieving a higher recognition speed and a satisfactory recognition rate.

2 CNN

LeCun Y. et al. (1998) and Patrice et al. (2003) described in detail how CNN ensures image displacement, scaling, and robust distortion performance. A

*Corresponding author: *panliwu@163.com*

typical handwritten Chinese character CNN LeNET structure is shown in Figure 1.

Figure 1 shows, the input layer receives 32*32 handwritten Chinese character image to be verified and simple processing of 2 aspects, namely, size normalization and image grayness, is conducted. The results are taken as an image of the sampling layer.

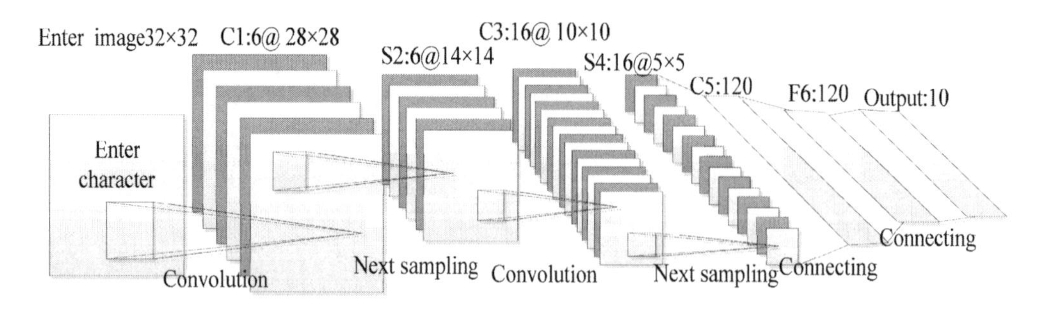

Figure 1. Typical CNN structure.

After the sampling layer is obtained, a kernel which can learn is used for convolution. The result of such convolution can form the neurons of this layer through the output of an activating function Each neuron connects a 5×5 neighbor of the input image, and the first hidden layer is obtained (Layer C1), which consists of 6 feature graphs. Each feature graph has 25 weights (such as directed line segment, end and corner). If the boundary effect is taken into account, the resulting feature map is 28×28, which is less than the input layer (Lu 2011, Pham 2012, Sermanet 2012, LeCun et al. 1998, Patrice et al. 2003). Convolution layer mathematical calculation can be expressed as Formula (1)

$$x_j^l = f\left(\sum_{i \in M_j} x_j^{l-1} * kernel_{ij}^l + b_j^l\right) \tag{1}$$

where, l represents the number of layers, *kernel* represents the convolution kernel, M_j represents a choice of the input feature graphs. Each output graph has a bias b.

The results of each convolution serve are taken as the input of the next sampling layer, which samples the input information. In case of n input feature graphs, the number of feature graphs after the next sampling layer is still n, but the output feature graphs will become smaller (for example, each dimension will become a half of the original). The hidden layer S2 will become a next sampling layer, composed of six 14×14 feature graphs. Next sampling layer calculation can be represented by Formula (2).

$$x_j^l = f(\beta_j^{l-1} down(x_j^{l-1}) + b_j^l) \tag{2}$$

where, $down\ (x_j^{l-1})$ represents the next sampling function, which is a range summation of the size of $n \times n$ of the input image of the layer typically. The output image size is $1/n$ of that of the input image. Each feature graph has its own β and b.

Similarly, C3 layer has 16 convolution layers, consisting of 10×10 feature graphs. Each neuron connects 5×5 neighbor of certain feature graphs of S2 network layer. Network layer S4 is the next sampling layer, consisting of 16 5×5 feature graphs. Each neuron of the feature graphs connects a 2×2 neighbor of C3 layer. Network layer C5 is a convolution layer composed of 120 feature graphs. Each neuron connects 5×5 neighbors of all feature graphs of S4 network layer. Network layer F6 including 84 neurons fully connects network layer C5. The output layer has 10 neurons, consisting of radial basis function (RBF). Each neuron of the output layer corresponds to a character class. RBF output yi calculation is shown Formula (3).

$$y_i = \sum_j (x_j - w_{ij})^2 \tag{3}$$

Multiple researchers have made elastic training of character sets. Test results show the recognition rate on Mnist character set is as high as 99% (LeCun Y et al. 1998, Patrice et al. 2003). CNN main advantage lies in small model sets, with a high recognition rate of collections, consisting of numbers or 26 alphabets. However, recognition of large model sets remains a challenge. It is quite difficult to design an optimal and large enough single network, which needs longer training time. Therefore, this author intends to combine multiple CNNs with a high recognition rate of a certain character set, constitute multi-convolution neural networks and increase the recognition rate of the CNN large model handwritten Chinese character.

3 MULTI-CONVOLUTION NEURAL NETWORKS

3.1 *Multi-convolution neural networks characters recognition*

According to the conventional convolution neural networks and their failure to process large model handwritten Chinese characters, this author proposes a multi-convolution neural network to improve the conventional CNN model and combine a multi-convolution neural network with a high recognition rate of multiple small CNN. Each small CNN has a high recognition rate of a particular character set. In addition, single CNN has a formal output set, and produces an unknown output (namely characters are difficult to identify). In case an input character is not correctly identified, it will be outputted as an unknown character, and then the input model will enter the next CNN for recognition. Finally, judgment will be conducted through a spell checker module to select the best result output. Flow of the system is shown in Figure 2.

Where CNN1 identifies handwritten digits and CNN2 identifies handwritten lowercase alphabets. With a very strong scalability, this model can be supplemented with multiple arbitrary CNN (such as Chinese and Japanese).

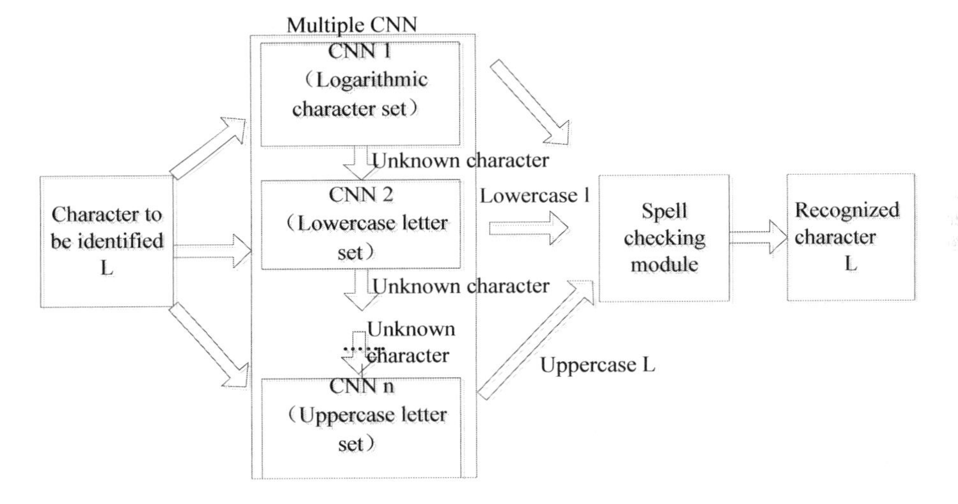

Figure 2. Multiple CNN character recognition diagram.

3.2 *Stochastic diagonal L-M training method*

The traditional structure is quite simple. Single CNN adopts the basic Back Propagation (BP) training network, which often requires several hundred iterations with a low convergence speed. This author adopts stochastic diagonal L-M algorithm proposed by Dr. Le Cun for training, with iterations required by the algorithm significantly less than that of the basic BP algorithm (Lu 2011, LeCun et al. 1998). Stochastic diagonal L-M algorithm formula is:

$$\eta_{ki} = \varepsilon \Big/ (\langle \partial^2 E / \partial w_{ij}^2 \rangle + \mu) \tag{4}$$

where, ε is the global learning rate, with an initial value of 0.01. If the initial value is too large, the network will not converge. If the initial value is too small, the convergence rate will decrease, which will make the network vulnerable to local minima. In the training, it is proper to adopt the heuristic rule to change ε value. This author takes the lowest value of 5e-005.$\langle \partial^2 E / \partial w_{ij}^2 \rangle$ is an estimate. Relying on the size of the training set, it is proper to adjust the number of samples. This author makes stochastic selection of 200 samples to estimate its value. μ is used to avoid too large change of μ_{ki} when $\langle \partial^2 E / \partial w_{ij}^2 \rangle$is too small.

3.3 *Multiple CNN expressions recognition*

Multiple CNN handwritten expressions recognition proposed in this paper can be described by a simple process: Firstly, handwritten image input witnesses preprocessing and segmentation, and then respective recognition is conducted through Multiple CNN module and finally judgment of the recognition results with the word recognition module so as to select the best result output. The process is shown in Figure 3.

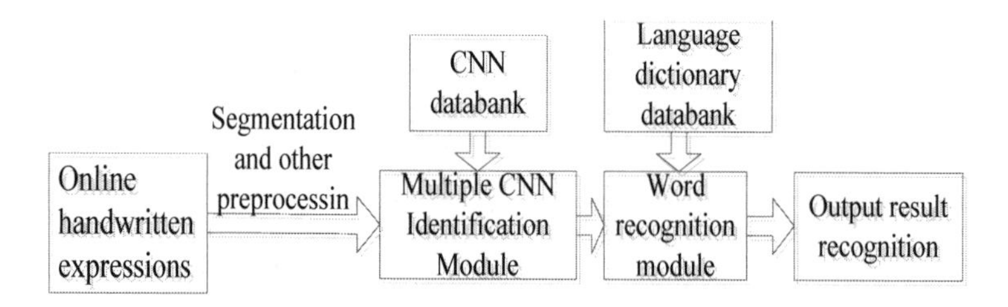

Figure 3. Multiple CNN online handwritten expression recognition process.

Multiple CNN on-line handwritten Chinese character recognition proposed in this paper has eliminated the limitations of traditional CNN recognition to characters. Each CNN serves as a small model, which is easy to train and optimize. More importantly, this program has high flexibility and scalability, which will facilitate parameter adjustment. Each CNN can be reused and supplemented with one or more new networks as needed, without changing or rebuilding the original network.

4 TRAINING AND TESTS

In order to assess the performance of multiple CNN on-line handwritten Chinese character recognition based on large model character sets, the system adopts the 2 different handwritten Chinese character training sets of Mnipen and Unipen for test. Unipen database was proposed and established in IEEE IAPR Conference held in 1992, aiming at creating a large handwritten Chinese character database and providing the foundation of research and development to online handwritten Chinese character recognition, which has received the support of several well-known companies or institutes. And Unipen design specifications have been completed. In data comparison tests, this author adopts Mnist handwritten digital database, applied in multiple studies. Designed by NEC Research Center, the database is a subset of NIST (The National Institute of Standards and Technology) Database. The training integrates a large number of training samples and test cases. This paper tacitly approves the following definition:

$$Recognition\ rate = N/Ts^{*} \times 100\% \quad (1)$$

where N is the Number of correct recognition×100% ; Ts is the total number of samples.

$$Error\ rate = Tw/Ts \times 100\% \quad (2)$$

where, Error rate is the wrong recognition rate; Tw is the total number of wrong recognitions; Ts is the total number of samples×100%.

Experimental tests are conducted on a common desktop computer. The entire recognition prototype system adopts MVC framework as system architecture and operates on Net Framework 4.5 platforms. Tests show Mnist training set correct recognition rate is 99% (LeCun et al. 1998, Zhou et al. 2012), Unipen digital correct recognition rate is 97%, Unipen digital and uppercase letter correct recognition rate is 89% (1a, 1b), and Unipen lowercase letter correct recognition rate is 89% (1c). Figure 4 shows MSE comparison of 3 Unipen lowercase trainings.

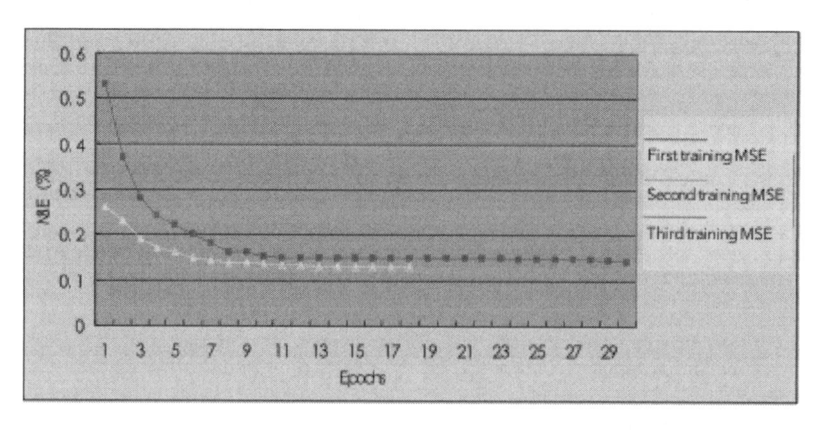

Figure 4. Training error data.

Figure 4 shows, in the first training periods, MSE drops very quickly, and after the first 13 cycles the neural network reaches a stable value of about 0.1485, that is, after the first 13 cycles, the network witnesses a low degree of improvement. After the value of the training error rate is corrected as 0.00045, the second training of the 18th cycle is conducted, with MSE decreasing. After the third training, it tends to be stable, with a Unipen lowercase letter correct recognition rate of 89%. Tests show the stochastic diagonal L-M makes convergence much faster than the basic BP algorithm. After 68 cycles of training, the correct recognition rate reaches 89%.

5 CONCLUSIONS

This paper proposes the online handwritten Chinese character recognition method based on multiple CNN, applying multiple CNN with a high recognition rate and stochastic diagonal L-M method that supports large model online handwritten Chinese character recognition. Experimental data comparison shows this method achieves a high recognition rate of the large model online handwritten Chinese character recognition, with high recognition speed, timeliness and overall effect. Against today's touch-screen applications in production and life, it has a broad application prospect. At the same time, this method provides a sound reference to researches on handwritten Chinese character recognition.

ACKNOWLEDGMENT

This work is supported by the key science and technology attack project of Henan Province of China (Project No. 142102210104) and the key science and technology attack project of Science and Technology Department of Henan Province of China (Project No. 132102210493).

REFERENCES

LeCun, Y., Bottou, L., Bengio, Y. & Haffner, P. 1998. Gradient-based learning applied to document recognition, *Proc. IEEE, 1998. USA:*2278–2324.

Lu Gang. 2011. Multi-font character recognition based on CNN. *Journal of Zhejiang Normal University (Natural Science Edition)* 34 (4): 425–428.

Patrice Y. Simard, Dave Steinkraus & John Platt. 2003. Best Practices for CNN Applied to Visual Document Analysis. *Proc. intern. symp.* On docucument Analysis and Recognition,Edinburgh, 6 August 2003. IEEE Computer Society: Los Alamitos.

Pham D V. 2012. Online handwriting recognition using multi CNN. *Simulated Evolution and Learning Springer Berlin Heidelberg*: 310–319.

Sermanet, P., Chintala, S. & LeCun, Y. 2012. CNN Applied to House Numbers Digit Classification. *Proc. Inter. Symp. of Pattern Recognition, Tsukuba, 11–15 November 2012.* IEEE Computer Society: Los Alamitos.

Wu Mingrui & Zhang Bo. 2001. A neural network algorithm for large-scale pattern recognition. *Journal of Software* 12 (6): 851–855.

Xu Shanshan, Liu An & Xu Sheng. 2013. Timber Defect Recognition Based on CNN. *Journal of Shandong University (Engineering Science)* 43 (2): 23–28.

Zhang Hui. 2012. Massive online handwritten Chinese character recognition database arrangement, statistics and experimental analysis. Guangzhou: South China University of Technology.

Zhou Hui, Ren Haijun, Ma Liang, et al. 2012. Application of MVC design pattern in the development of information system. *Software Guide* (10): 120–122.

Electronics, Communications and Networks IV – Hussain & Ivanovic (eds)
 ISBN: 978-1-138-02830-2

Efficient dynamic detection of data races for multi-core software

Ok-Kyoon Ha*
Engineering Research Institute, Gyeongsang National University, Jinju, Republic of Korea

Yong-Kee Jun
Department of Informatics, Gyeongsang National University, Jinju, Republic of Korea

ABSTRACT: Data races in multi-core software represent the most notorious class of concurrency bugs. For detecting data races, happens-before analysis is precise, but inefficient due to the additional runtime overhead. To detect data races with both precision and efficiency during an execution of multi-core software using multithreads, previous work still requires time and space complexities that depend on the maximum parallelism of the program to partially maintain expensive data structures, such as vector clocks. This paper presents an efficient dynamic data race detector which uses only two epoch clocks instead of full vector clocks in access histories. We implemented our detector on top of the Pin binary instrumentation framework and compared it with three state-of-the-art detection techniques. Empirical results using C/C++ benchmark show that our detector efficiently locates data races with reducing the runtime and the memory overhead to 60% and 5% of the others, respectively.

1 INTRODUCTION

Data races (Netzer & Miller 1992, Pozniansky & Schuster 2003) in multi-core software represent the most notorious class of concurrency bugs that cause non-atomic execution of critical sections. They occur when two concurrent threads access a shared memory location without explicit synchronization, and at least one of them is a write. A multithreaded program for multi-core systems may not exhibit the same execution instance on different runs with the same input. It is difficult to figure out whether a program runs into data races, because there are many possible executions of the program and a lot of data races hard to reproduce. Detecting data races is therefore important, since they may lead to unpredictable results from an execution of the program.

This paper presents an efficient dynamic data race detector which uses only lightweight identifier instead of heavyweight data structures, such vector clocks, in each access history. Our detector requires $O(1)$ runtime and memory overhead to maintain an access history and to locate data races with no loss of precision. We implemented our detector on top of the Pin instrumentation framework (Bach et al. 2010) which uses a just-in-time (JIT) compiler to recompile target program binaries for dynamic instrumentation.

To compare the efficiency of the new detector for locating on-the-fly data races, we also implemented other three state-of-the-art detection techniques, Djit$^+$ (Pozniansky & Schuster 2003), FastTrack (Flanagan & Freund 2009), and AccuLock (Xie & Xue 2011), on top of the same framework. The experimental results on C/C++ benchmark using Pthreads show that our detector reduces the runtime and the memory overheads to 60% and 5% of the others, respectively.

2 RELATED WORK

Dynamic data race detection techniques (Dinning and Schonberg 1991, Savage et al. 1997, Harrow 2000, O'Callahan and Choi 2003, Pozniansky and Schuster 2003, Yu et al. 2005, Flanagan and Freund 2009, Jannesari et al. 2009, Serebryany and Iskhodzhanov 2009, Jannesari and Tichy 2010, Xie and Xue 2011, Ha et al. 2012) reports data races which are merely detected during an execution of multithreaded programs. These techniques dynamically detect data races with still less overhead in storage space than other techniques due to the fact that unnecessary information are removed as the detection advances.

The happens-before analysis reports data races involving the current access and earlier accesses maintained in a special kind of data structure, called *access history*, for a shared memory location during an execution of a multithreaded program. To determine the logical concurrency among concurrent accesses to shared memory locations, this technique widely uses *vector clocks* (VCs) (Baldoni & Raynal 2002) which

*Corresponding author: *jassmin@gnu.ac.kr*

represent the Lamport's happens-before relation (Lamport 1978). Generally, VC is well-known as an expensive data structure because it requires $O(n)$ size of space for maintaining a thread information and $O(n)$ time for VC operations (e.g. creation and comparison), where n designates the maximum number of simultaneously active threads during an execution.

The happens-before analysis using VCs is precise, since it does not produce unnecessary warnings (false positives) and can be applied to all synchronization primitives. However, the technique is quite difficult to be efficiently implemented due to the performance overhead of VCs. Thus, previous work using the happens-before analysis still requires a large amount of time and space costs by VCs.

Recently, a new technique, called FastTrack technique (Flanagan & Freund 2009) replaces heavyweight VCs with a lightweight identifier, called *epoch clocks*, that only uses a pair of the clock value and the thread id. The epoch-based happens-before analysis reduces runtime and memory overhead of almost VC operations from $O(n)$ to $O(1)$ for detecting data races. However, the FastTrack technique still needs VC operations in order to guarantee that there is no loss of precision. Therefore, the overhead problem still potentially exists to dynamically analyze programs with a large number of concurrent threads.

3 DESIGN OF SOFTWARE ARCHITECTURE

3.1 *VC-based happens-before analysis*

The happens-before relation determines the logical concurrency between two thread segments. By the relation, if a thread segment t must happen at an earlier time than a thread segment u, t *happens before* u, denoted by $t \xrightarrow{hb} u$. If neither $t \xrightarrow{hb} u$ nor $u \xrightarrow{hb} t$ is satisfied, we say that t is *concurrent with* u, denoted by $t \parallel u$.

VCs are widely used to precisely analyze the happens-before relation $hb\rightarrow$, because VCs can inform the execution order of thread segments and synchronization order including thread operations and events. A vector clock *VC*: *Tid* → *Nat* records a clock value c for each thread in an execution of the program. Thus, a thread segment t maintains a vector clock $C_t = \langle c_1, \cdots, c_n \rangle$ which has n entries, if the maximum number of active threads in an execution of a multithreaded program is n. The VC of each thread segment is partially ordered ($\sqsubseteq$) point-wise with a minimum element $\langle 0, \cdots, 0 \rangle$ and associated synchronization primitives that define point-wise maximums. For instance, the entry $C_t[u]$ for any thread segment u stores the latest clock value of u that happens before the current synchronization primitive of t.

Using the VCs of each thread segment, we simply analyze the happens-before relation between any two thread segments. If the clock value of a thread segment t is less than or equal to the corresponding clock of another thread segment u, we can conclude that t happens before u. Otherwise, t is concurrent with u. Formally,

$$\begin{cases} t \xrightarrow{hb} u \equiv C_t[t] \leq C_u[t] \\ t \parallel u \equiv (C_t[t] > C_u[t] \vee C_t[u] < C_u[u]) \end{cases}$$

Finally, the happens-before analysis locates a data race during an execution of a multithreaded program, whenever any two events on two concurrent thread segments access a shared memory location with at least one write.

Definition 1 Given two access events e_t and e_u to a shared memory location from two distinct thread segments t and u respectively, if the two events are not synchronized (i.e. neither $C_t \sqsubseteq C_u$ nor $C_u \sqsubseteq C_t$) and at least one of the events is a write, there exists a data race between e_t and e_u

3.2 *Efficient dynamic detection of data races*

We have improved FastTrack technique (Flanagan & Freund 2009) to reduce the time and space overheads even in the worst case through it uses only epochs instead of full VCs to detect data races. For a shared memory location x, our detector defines an access history using two entries: R_x that records two epochs for the two concurrent read accesses of x, and W_x that records only an epoch for the last write access to x.

The dynamic detector reports data races by analyzing $\xrightarrow{hb}$, and simply maintains epochs by updating an access history. For the detector, some notions are used to analyze $\xrightarrow{hb}$ using the epoch. The function $E(t)$ is a shorthand for $c@t$, and $E(t) \preceq VC$ denotes that an epoch $E(t)$ happens before a vector clock VC, where $E(t) \preceq VC$ iff $c \leq VC[t]$.

When an event e_t newly occurs in a thread segment t, the technique for reporting data races and maintaining each entry is the following:

- In case: e_t is a *Read* event

If $R_x \neq E(t)$ then check $W_x \preceq C_t$ to report a data race between earlier write event and current event. Then $E(t)$ is kept in R_x.

- In case: e_t is a *Write* event

If $W_x \neq E(t)$ then check $W_x \preceq C_t$ to report a data race between an earlier write event and current event. If there exists only an epoch in R_x than check $R_x \preceq C_t$ to report a data race between an earlier read event and current event, otherwise check $R_x \sqsubseteq C_t$ for two concurrent epochs maintained in R_x. The previous epoch(s) is removed from R_x, and $E(t)$ is inserted into W_x.

In the case of read, if the epoch of the current e_t is same with R_x, $R_x=E(t)$, then the detector returns with no action, and the write case is similar to this read case.

We provide a detailed view of how our dynamic detector locates data races for concurrent events. The detector using the R_x, which maintains only two epochs instead of a full VC, guarantees that it detects data races without the loss of precision, because it locates at least one of read-write data races.

Lemma 1 If there exist data races between R_x and a current write event w, the detector soundly locates *one or two data races.*

Proof. Two distinct shared read events toward a shared memory location x are kept in R_x. Since two read events are concurrent with each other, we guarantee that

1. If only one of the events kept in $R_x \parallel w$, then the detector reports a data race, because neither the concurrent event in $R_x \sqsubseteq w$ nor $w \sqsubseteq R_x$ for the event is satisfied by Definition 1.
2. If all of the events kept in $R_x \parallel w$, then the detector reports two data races between w and both shared read events, because they satisfies Definition 1.
3. If all of the events kept in $R_x \xrightarrow{hb} w$, then the detector reports no data races.

All write events to x are totally ordered with the assumption that no data races have been detected on x. Thus, the detector records the epoch of the write event in W_x, and it locates a data race between W_x and a later event to x by analyzing the epoch of W_x and the current vector clock of the read/write event, $W_x \preceq C_t$.

Lemma 2 *If there exist data races between W_x and a current read/write event , the detector always locates the data races.*

Proof. When a current event e_t occurs in a thread segment t , the detector analyzes $W_x \preceq C_t$.

1. If $W_x \parallel e_t$, then the detector always reports a data race, because they always satisfies Definition 1.
2. If $W_x \xrightarrow{hb} e_t$, then the detector reports no data races.

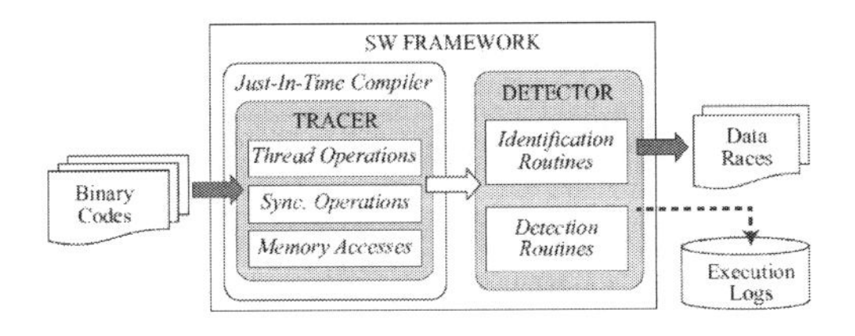

Figure 1. The overall architecture of a data race detector.

Theorem 1 Our dynamic detector efficiently locates data races with no loss of precision, if it maintains only epochs instead of full VCs in R_x and W_x.

Proof. Our new detector offers $O(1)$ time and space overheads to detect data races, because it removes the switching between epochs and VCs for R_x in FastTrack technique by maintaining only two concurrent epochs in R_x. From Lemma 1 and Lemma 2, the detector soundly locates data races because it reports at least one of data races, if there exists any.

4 EVALUATION

4.1 *Implementation & experimentation*

We implemented our dynamic detector and three other dynamic detection techniques on top of the Pin instrumentation framework (Bach et al. 2010) which uses a just-in-time (JIT) compiler to recompile target program binaries for dynamic instrumentation. The three other techniques are: Djit$^+$ (a high performance VC-based happens-before analysis algorithm), FastTrack (a state-of-the-art happens-before analysis algorithm), and AccuLock (a hybrid algorithm that combines FastTrack with the lockset analysis to consider the coverage of thread interleaving by locks).

Figure 1 depicts the architecture the detectors. Each detector consists of a TRACER and a DETECTOR module to report apparent data races during a program execution. The TRACER tracks thread operations and event instances to every shared memory location considering synchronization primitives. The DETECTOR involves the thread identification routines to generate and manage VCs for each active thread segment, and the detection routines to report data races considering the four algorithms.

The thread identification routines employ the VC primitives, such as thread fork/join, acquire lock, and release lock, and the VC primitives are commonly used to analyze happens-before relation in the detection routines for all techniques. Whenever the TRACER catches one of the thread operations or events, it calls either the thread identifier routines or the detection routines to add instrumentation at each interesting point of the running target binaries.

Our experimentation focused on comparing the efficiency of dynamic data race detection in programs with a large number of concurrent threads. To evaluate the fulfillment of the objectives intended by the new detector, we measured the execution time and the memory consumed by the execution instances of a set of C/C++ benchmark using Pthread, and compared the data races reported by each detector. For this purpose, we used X.264 application (Bienia et al. 2008) which targets computer vision (i.e., an open library for encoding video streams). The application in our experimentation were executed with simulation inputs to produce proper runtime overhead and memory consumption.

Table 1. The runtime slowdown and memory consumption of x264 application with six simulation inputs.

				Runtime Slowdown				Memory Overhead			
Inputs	# of Threads	Base Time (Sec)	Base Memory (MB)	Djit+	FastTrack	AccuLock	Our Detector	Djit+	FastTrack	AccuLock	Our Detector
test	1	1.1	14	0.5	0.5	0.5	0.5	0.4	0.4	0.4	0.4
dev	6	1.5	77	0.7	0.7	0.7	0.7	0.2	0.2	0.2	0.2
small	16	1.8	204	1.7	1.6	1.7	1.5	1.0	0.5	1.4	0.2
medium	64	2.2	233	3.9	3.2	3.9	2.8	7.8	3.4	7.8	0.3
large	256	3.1	250	10.8	9.1	10.6	4.7	51.6	25.8	51.6	0.3
native	1024	38.9	534	–	–	–	11.5	–	–	–	1.6
Average				3.5	3.0	3.5	2.1	12.2	6.1	12.3	0.3

The implementation and experimentation were carried on a system with two 2.4GHz Intel Xeon quad-core processors and 32GB of memory under the Kernel 2.6 of Linux operating system. We installed the most recent version of the Pin framework (Ver. 2.12), and the applications were compiled with gcc 4.4.4 for all detectors. We employed a programmed logging method to measure the execution time and the memory consumption of each application. This method uses system files in the proc directory which provides several real-time information of the system. The average runtime and memory overheads of all applications were measured for ten executions under each detector.

4.2 *Results and analysis*

We measured the runtime and the memory consumption of an application under four detectors to compare the efficiency of our detector with others. The comparison used all of six simulation inputs because they lead to increasing thread size of each input frame. Table 1 shows the results of the comparison. The table lists the number of threads, running times, and memory requirements for each simulation input. The "Base Time" and "Base Memory" columns contain the measured execution times and memory consumptions for the original running of the application with each input. The "Runtime Slowdown" and "Memory Overhead" columns show the execution times and memory consumptions with the ratios of the original run under each of the detectors.

In the results, our detector incurred a runtime overhead of 2.1x on average, meanwhile other detectors incurred more than 3x slowdown. Moreover, the detector shows a distinguished performance, reducing memory overhead, because it ran only with the memory overhead of 0.3x in the average case, while the memory overhead of others increased with a factor of more than 95% relatively to our detector's overhead. The application under our detector ran with native input using 1,024 concurrent threads, but all other detectors ran out of memory on the input due to the limitation of 32GB on our system. In that case, our detector required the runtime overhead of 11.5x and the memory overhead of 1.6x for locating 3 data races.

From Table 1, our new detector reduced 9.9x memory overhead and 1.3x speedup over other dynamic detectors, Djit+, FastTrack, and AccuLock. These empirical results show that our new detector is practical for dynamic data race detection due to the fact that it reduces the runtime and memory overhead to 60% and 5% of the others.

5 CONCLUSION

FastTrack is the fastest happens-before analysis technique. However, there is still room for improvement, since the technique partially needs VC operations which require $O(n)$ size of space and time. In this paper, we presented a dynamic detector that uses only epochs in each access history, unlike other happens-before based detectors. This detector is practical to locate data races due to fact that it requires only $O(1)$ runtime and memory overheads. We implemented new detector on top of the Pin instrumentation framework, and compared it empirically with three detection algorithms, Djit+, FastTrack, and AccuLock.

The empirical results using a set of C/C++ benchmark show that our detector is the practical for dynamic data race detection due to the fact that it reduces the runtime and memory overhead to 60% and 5% of the others, respectively. Future work includes additional improvement of new detector to design the hybrid detection technique to exclude the false positive problem, and additional enhancement of the precision to handle more variant synchronization primitives.

ACKNOWLEDGMENTS

This research was supported by the Basic Science Research Program through the National Research Foundation of Korea (NRF) funded by the Ministry of Education (NRF-2013R1A1A2011389), Republic of Korea.

REFERENCES

Bach et al. 2010. Analyzing parallel programs with pin. *Computer* 43(3): 34–41.

Baldoni, R. & Raynal, M. 2002. Fundamentals of distributed computing: A practical tour of vector clock systems. *IEEE Distributed Systems Online* 3(2).

Bienia et al. 2008. The parsec benchmark suite: Characterization and architectural implications. In *Proceedings of the 17th International Conference on Parallel Architectures and Compilation Techniques*: 72-81. Torronto: ACM.

Dinning, A. & Schonberg, E. 1991. Detecting access anomalies in programs with critical sections. *SIGPLAN Not.* 26(12): 85–96.

Flanagan, C. & Freund, S. N. 2010. FastTrack: efficient and precise dynamic race detection. *Commun. ACM*, 53(11): 121–133.

Ha et al. 2012. On-the-fly detection of data races in OpenMP programs. In *Proceedings of the 2012 Workshop on Parallel and Distributed Systems: Testing, Analysis, and Debugging*: 1-10. Minneapolis: ACM.

Harrow, J. J. 2000. Runtime checking of multithreaded applications with visual threads. In *Proceedings of the 7th International SPIN Workshop on SPIN Model Checking and Software Verification*: 331–342. London: Springer-Verlag.

Jannesari et al. 2009. Helgrind+: An efficient dynamic race detector. In *Proceedings of the 2009 IEEE International Symposium on Parallel & Distributed Processing*: 1–13, Rome: IEEE.

Jannesari, A. & Tichy, W. F. 2010. Identifying ad-hoc synchronization for enhanced race detection. In *Proceedings of the 2010 IEEE International Symposium on Parallel & Distributed Processing*: 1-10. Atlanta: IEEE.

Lamport, L. 1978. Time, clocks, and the ordering of events in a distributed system. *Commun. ACM* 21: 558–565.

Netzer, R. H. B. & Miller, B. P. 1992. What are race conditions?: Some issues and formalizations. *ACM Lett. Program. Lang. Syst.* 1: 74–88.

O'Callahan, R. & Choi, J.-D. 2003. Hybrid dynamic data race detection. *SIGPLAN Not.* 38(10): 167–178.

Pozniansky, E. & Schuster, A. 2003. Efficient on-the-fly data race detection in multithreaded c++ programs. *SIGPLAN Not.* 38(10): 179–190.

Savage et al. 1997. Eraser: a dynamic data race detector for multithreaded programs. *ACM Trans. Comput. Syst.* 15: 391–411.

Serebryany, K. & Iskhodzhanov, T. 2009. ThreadSanitizer: data race detection in practice. In *Proceedings of the Workshop on Binary Instrumentation and Applications*: 62–71. New York: ACM.

Xie, X. & Xue, J. 2011. AccuLock: Accurate and efficient detection of data races. In *Proceedings of the 9th Annual IEEE/ACM International Symposium on Code Generation and Optimization*: 201–212. Chamonix: IEEE.

Yu et al. 2005. RaceTrack: efficient detection of data race conditions via adaptive tracking. *SIGOPS Oper. Syst. Rev.* 39(5): 221–234.

Electronics, Communications and Networks IV – Hussain & Ivanovic (eds)
© 2015 Taylor & Francis Group, London, ISBN: 978-1-138-02830-2

Friction compensation and identification for tank line of sight stabilization system

Bin Han*, Tianqing Chang, Kuifeng Su & Rui Wang
Department of Control Engineering, Academy of Armored Force Engineering, Beijing, P.R. China

ABSTRACT: For nonlinear friction disturbances existing in tank stabilized sight systems, effects on the control performance were analyzed and verified. A friction compensation method was proposed based on the simplified Stribeck friction model to improve system stability accuracy. An equivalent current was used to achieve the characteristic curve of frictional disturbance instead of friction torque to overcome the problem of inaccuracy in the motor parameter. A simplified model of friction was obtained by parameters identification of the friction model based on GA. The experimental results showed that the identification method was effective. The tank stabilized sight system performance was improved through the feed-forward compensation.

1 INTRODUCTION

Tank stabilized sight systems use the fixed axis of inertial devices to isolate the interference of the vehicle so it can guarantee steady sight in inertial space. As an important part of a tank's Fire Control System, its performance directly affects the function of the weapons system. Therefore, research on precision tank line of sight stabilization systems plays an important role in improving the accuracy of tank Fire Control Systems. As a major disturbance in tank stabilized sight systems, nonlinear friction has a great impact on the dynamic and static system performance, primarily as low speed jitter and zero crossing error pole, steady-state error, and limit cycle oscillations. Since the tank line of sight stabilization system swings back and forth in the equilibrium position and the operating speed is generally low, the friction disturbance is particularly evident. Therefore, friction compensation in the system is especially important.

Building a friction model that can truly reflect the friction phenomenon is a prerequisite for friction compensation. Practice shows that the classical friction models, such as Coulomb friction and a viscous friction model cannot reflect the real dynamics of friction. At present, the Stribeck friction model and LuGre friction model are used to conduct friction-applied research. Document 1 used particle swarm to do research on friction identification and compensation (Zhang et al. 2007). Document 2 used a disturbance observer to conduct friction compensation based on identification of the Stribeck friction model and achieved good simulation results (Yu et al. 2008). Document 3 analyzed the influence on gun control at low speed made by nonlinear friction and compared the effect of several non-model friction compensations to provide a basis for future studies (Yuan et al. 2007). Regardless of the model, may it be Stribeck or LuGre, for friction compensation, samples of Stribeck friction need to be drawn from torque curves at different speeds. The motor parameters of the stabilized sight system are not known; therefore, the friction torque value cannot be sampled. In view of this situation, an equivalent current was used to achieve the characteristic curve of frictional disturbance instead of friction torque so that the friction model can be obtained for friction compensation.

2 SIMPLIFIED MODEL OF THE FRICTION TORQUE

Selecting the appropriate friction torque model is particularly important for friction compensation. The Stribeck friction model and LuGre friction model reflect the real dynamics well and are widely used. The former is a static model, which is the dynamic model. The applications of the LuGre model are limited because of the large number of parameters and complex models. If the parameter identification is not accurate, it not only is unable to reflect the dynamic friction characteristics correctly, the accuracy of the model is also affected. The Stribeck friction model is widely applied in practical applications, and the static friction properties of the entire model can be approximately 90% (Song). As such, the model can be used as a friction torque model.

*Corresponding author: *han-b08@163.com*

The Stribeck friction model defines the friction torque between two objects as a function of the relative velocity. This relationship is often called the Stribeck curve. The mathematical expressions are as follow (Armstrong-Hélouvry et al 1991):

$$T_f = \begin{cases} T_f(\dot{\theta}), & \dot{\theta} \neq 0, \\ T_m, & \dot{\theta} = 0, \ddot{\theta} = 0, |T_m| < T_s; \\ T_s \operatorname{sgn}(T_m), & \dot{\theta} = 0, \ddot{\theta} \neq 0, |T_m| > T_s; \end{cases} \quad (1)$$

$$T_f(\dot{\theta}) = \left[T_c + (T_s - T_c)\exp\left(-\left|\dot{\theta}/\dot{\theta}_s\right|^2\right)\right] \bullet \operatorname{sgn}(\dot{\theta}) + b\dot{\theta} \quad (2)$$

in which T_s and T_c represent the maximum static friction torque and sliding friction torque, respectively; T_m represents the motor output torque; T_f represents the system friction distance; $\dot{\theta}$, $\ddot{\theta}$, and $\dot{\theta}_s$ represent speed, acceleration, and the Stribeck speed, respectively; b represents the viscous friction coefficient; and sgn is the sign function. According to the principle of inversion symmetry, there are four parameters to be identified: Ts, Tc, $\dot{\theta}_s$, and b.

The Stribeck curve is difficult for calculating and adjusting because of its nonlinear characteristic. Different parameters have different influences on the value of a function, which causes some parameters to be difficult to identify correctly. Therefore, it is necessary to simplify the Stribeck model. There are two simplified methods, the Taylor expansion and the piecewise linear function. Since the changes in the system's operation speed is small, we can use the piecewise linear function to simplify the Stribeck model.(Zhai et al. 2013)

Obtain the first derivative and second derivative of Formula (2) to get Formula (3) and Formula (4), so the direction of the Stribeck curve can be determined.

$$\dot{T}_f(\dot{\theta}) = b\text{-}2*(T_s - T_c)/\dot{\theta}_s^{\,2} * \dot{\theta} * \exp\left(-\left|\dot{\theta}/\dot{\theta}_s\right|^2\right) \quad (3)$$

$$\ddot{T}_f(\dot{\theta}) = 2*(T_s - T_c)/\dot{\theta}_s^{\,2} * \exp\left(-\left|\dot{\theta}/\dot{\theta}_s\right|^2\right) * \left(2*\dot{\theta}^2/\dot{\theta}_s^{\,2} - 1\right) \quad (4)$$

By solving the equation: $\ddot{T}_f(\dot{\theta}) = 0$, we can get $\dot{\theta} = \frac{\sqrt{2}}{2}\dot{\theta}_s$. Since function $\ddot{T}_f(\dot{\theta})$ is monotonically incremental, when $\dot{\theta} < \frac{\sqrt{2}}{2}\dot{\theta}_s$, we have $\ddot{T}_f(\dot{\theta}) < 0$ and $\dot{T}_f(\dot{\theta})$ decreases; when $\dot{\theta} > \frac{\sqrt{2}}{2}\dot{\theta}_s$, we have $\ddot{T}_f(\dot{\theta}) > 0$ and $\dot{T}_f(\dot{\theta})$ increases. For $\dot{T}_f(\dot{\theta})$, the initial value is b, which is above zero. Then the value reduces to $\dot{\theta} = \frac{\sqrt{2}}{2}\dot{\theta}_s$, falling below zero, and then increases. Thus, the Stribeck curve increases first, then decreases, and finally increases again. Consider that when $\dot{\theta} > \sqrt{2}\dot{\theta}_s$ and $\dot{T}_f(\dot{\theta}) \approx b$, the value of the Stribeck curve is near the minimum and the curve simplification results are good. Thus, select Points (0, y_1) and (x_2, y_2) as the first segment, for the second straight line, take (x_2, y_2) as a starting point and k_2 as the slope. Thus, the simplifying piecewise linear function expression (velocity positive) is as follows:

$$T_f(\dot{\theta}) = \begin{cases} (y_2 - y_1)/x_2 * \dot{\theta} + x_1, & \dot{\theta} < \sqrt{2}\dot{\theta}_s \\ k_2 * \dot{\theta} + y_2 - x_2 * k_2, & \dot{\theta} > \sqrt{2}\dot{\theta}_s \end{cases} \quad (5)$$

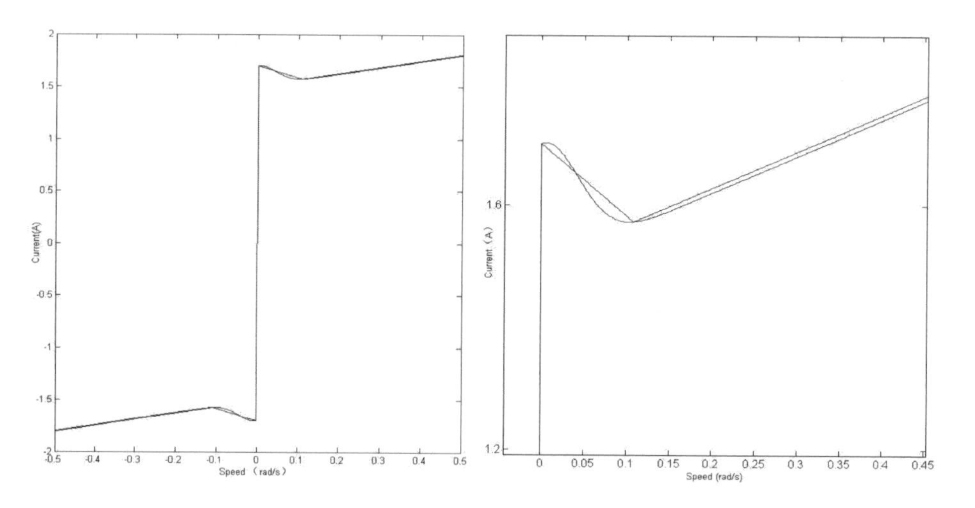

Figure 1. The effect of simplified function.

After conversion, it is shown that the parameters to be identified were (y_1, y_2, x_2, k_2). Compared to the original parameters to be identified, their effect on the function graph is relatively large, thus relatively easier to identify. The effect of the simplified function is shown in Figure 1.

3 FRICTION MODEL PARAMETER IDENTIFICATION

3.1 *The system model to be identified*

Tank line of sight stabilization systems use DC torque servomotors, which is controlled and driven by current directly. The servo system is composed of three control loops: a position loop, speed loop, and current loop. In order to reduce the error of multiple controllers during friction identification, position loop are ignored in mathematical model(Xie 2013). The simplified system model schematic is shown in Figure 2.

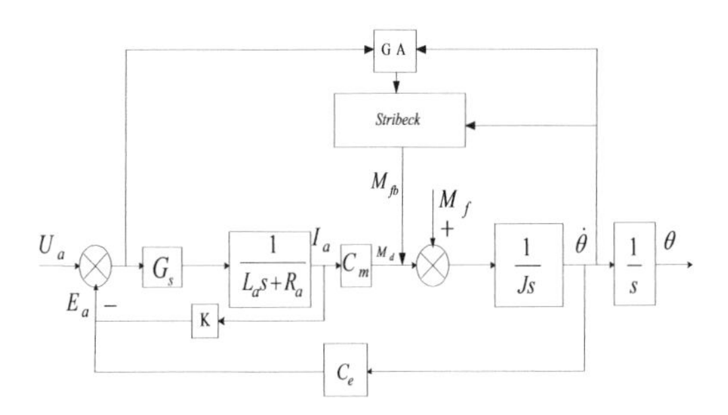

Figure 2. The structure of the simplified system model.

In figure 2, K is the current loop feedback factor with its value set to 1,; M_f is friction interference and M_{fb} is friction compensation; Gs is the controller for current loop PI. The input and output data in the current loop can be sampled through the DSP system and the step response curves can be obtained through cubic spline interpolation processing. The close-loop mathematical model for the current loop is:

$$G(s)=\frac{1}{8.154e-7s^2+6.484e-4s+1} \tag{6}$$

The remaining parameters in Figure 2 including motor torque coefficient Cm = 4.35Nm / A; system inertia J = 0.1kg·m^2.

3.2 *Friction model experimental data acquisition*

A control system and experimental platform was employed to get speed - driving current data on the state of the uniform system. To maximize the accuracy of the collected data, after full consideration of the response bandwidth of the current loop and control system response speed, the data sampling period was 1 ms. Meanwhile, considering the mechanical methods of working stabilized sighting systems, a square wave signal was used to acquire the speeds of the forward and reverse directions corresponding to the current value simultaneously. A DSP system was used to sample input and output data. And in order to achieve filtering effects, data was sampled at eight points simultaneously, using the averages as the output data. Finally, the driving current value corresponding to the system speed was obtained. Figure 3 shows the speed - driving current relationship.

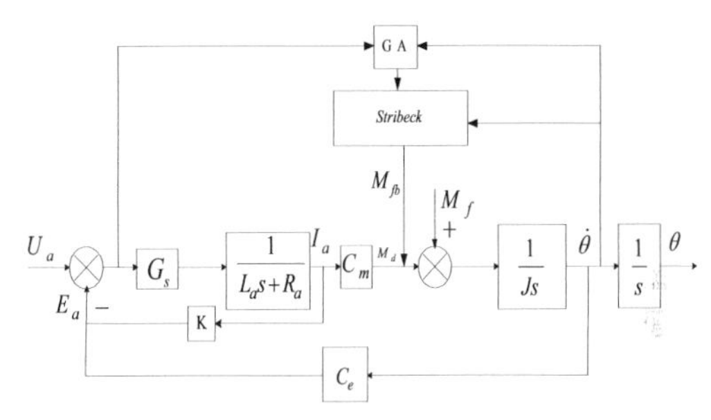

Figure 3. Speed - driving current data diagram.

As can be seen in the figure, the experimental data, except for a few outliers, were in line with the Stribeck friction model, demonstrating the effectiveness of the experimental data.

3.3 *Parameter identification based on GA*

Using the Formula (5) simplified friction model as the pending identification model and the experimental data shown in Figure 3 as the input and output data, the parameter identification of the friction model was conducted based on a genetic algorithm (GA).

Encoded in decimals, five genes represent a parameter and a chromosome contains twenty genes. The population size was set to 100. The absolute value of the errors at all points in the fitness function can be represented as:

$$G=\sum|e(t)| \tag{7}$$

The mutation operator in GA is random. The mutation probability and mutation place ranges adaptively as the evolution algebra ranges(Xie et al. 2012):

$$\begin{cases} p_m=0.1-(0.1-0.001)\dfrac{t}{T} \\ n=N-t \end{cases} \tag{8}$$

To save time, each parameter range was confirmed using the data fitting method first, and then the program was written in accordance with the genetic algorithm. After 200 generation iterations, the results of the identification were (0.0412814, 0.0349667, 0.751121, and 0.0084195) and the model achieved (take the positive speed as an example) as shown:

$$T_f(\dot{\theta})=\begin{cases}-0.008407034*\dot{\theta}+0.0412814, & 0<\dot{\theta}<0.751121\\ 0.0084195*\dot{\theta}+0.02864264 & \dot{\theta}>0.751121\end{cases} \quad (9)$$

The data fitting extent and recognition effect is shown in Figure 4. It can be seen that the effect of the friction model identification was good.

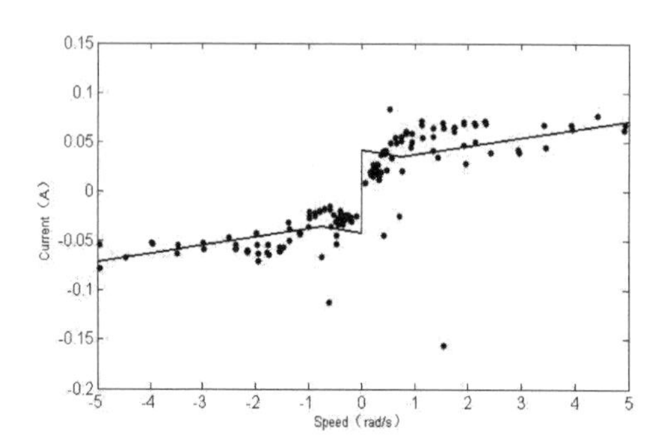

Figure 4. Simplified friction model.

4 FRICTION COMPENSATION BASED ON THE FRICTION MODEL

Using a sinusoidal signal with the amplitude of 1 and the frequency of 1 Hz as an input signal, the speed-tracking signal in an actual DSP system was gathered. The tracking curve is shown in Figure 5.

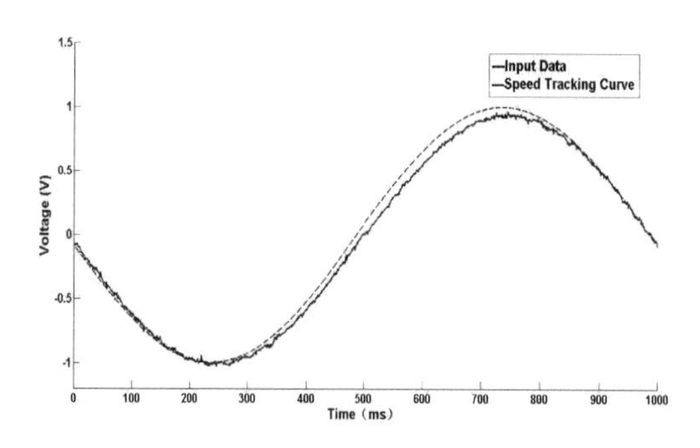

Figure 5. Speed tracking curve.

It can be seen that the speed-tracking curve appeared "zero crossing error pole" and was unable to reach peak. When the speed was zero and as the direction of the speed changed, the friction torque changed suddenly when the speed direction changed; thus, the error of the tracking curve abruptly became large. Subsequently, the error was corrected and the tracking curve went back to normal. The tracking curve could not reach maximum speed because of friction.

Friction compensation can obey friction model or not obey friction model. The kind obeys friction model is feed-forward control acctually, calculate friction disturbance torque according to the friction model anduse the compensation to balance friction torque disturbance. In this study, the friction compensation is calculated based on the fixed friction model.

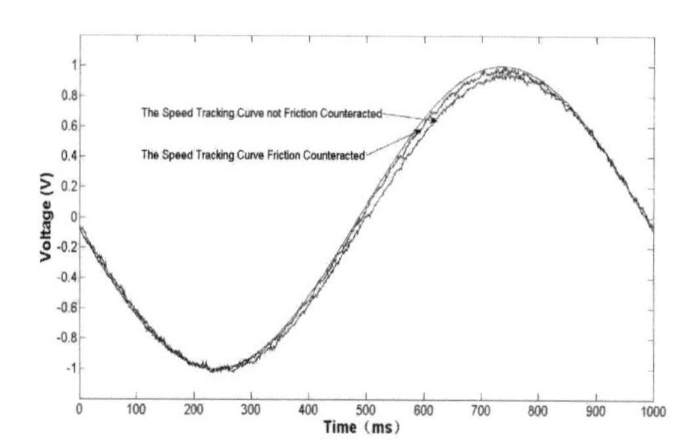

Figure 6. The effect of friction compensation.

Write the friction model identified in stabilized sighting system control algorithms, compensate velocity loop control amount that is current loop input, the result of the friction compensation experiment is shown in Figure 6. By comparing the speed-tracking curve that was friction-compensated with the curve that was not friction-compensated, it can be seen that the effect of speed-tracking improved after friction compensation and errors were reduced.

5 CONCLUSIONS

This paper proposed a friction compensation method based on the simplified Stribeck friction model to improve system stability accuracy. An equivalent current was used to achieve the characteristic curve of frictional disturbance instead of friction torque to overcome the problem of the inaccuracy of the motor parameters. On the basis of determining the friction model, research on parameter identification of a friction model was performed in an actual system.

Considering the real-time response speed and system reliability of the model, friction compensation based on a fixed friction model was studied. The result of the experiment showed that the effect of speed-tracking improved after friction compensation and errors were reduced.

REFERENCES

Armstrong-Hélouvry, B., Dupont, P. & De Wit, C. C. 1991. A survey of models, analysis tools and compensation methods for the control of machines with friction. *Automatica* 38(5): 363–368.

Song Yan. 2010. *Study and realization on key technology for improve velocity stability.* Chinese Academy of Sciences.

Xie Jie. 2013. Design of line-of-sight control system for armored vehicle. *Academy of Armored Force Engineering.*

Xie Min, Zhao Wenlong, Lu Daowang, et al. 2012. Research on mathematical model identification of steering gear based on genetic algorithms. *Computer Measurement & Control* (2): 428–430.

Yu Shuang, Fu Zhuang, Yan Weixin, et al. 2008. Method of friction compensation based on disturbance observer in inertial platform. *Journal of Harbin Institute of Technology* 40(11):1830–1833.

Yuan Dong, Ma Xiaojun, Wei Wei, et al. 2007. Research on friction nonlinearity and low velocity performance of gun control system of tank. *Journal of Academy of Armored Force Engineering* 21(4):57–61.

Zhai Yuanlin, Wang Jianli, Wu Qinglin, et al. 2013. Friction compensation control system design based on stribeck model. *Computer Measurement & Control* 21(03): 85–87, 126.

Zhang Wenjing, Zhao Xianzhang & Tai Xianqing. 2007. Parameter identification of gun servo friction model based on the particle swarm algorithm. *Journal of Tsinghua University(Science and Technology)* 47 (S2): 1717–1720.

Electronics, Communications and Networks IV – Hussain & Ivanovic (eds)
 ISBN: 978-1-138-02830-2

Study of greenhouse intelligent control system based on STM32

Kun Hao*
School of Electronics and Information Engineering, Tianjin University, Tianjin, China
School of Computer and Information Engineering, Tianjin ChengJian University, Tianjin, China

Feilong Zhai
School of Computer and Information Engineering, Tianjin ChengJian University, Tianjin, China

ABSTRACT: Aiming at solving the complexity and low accuracy of manual control methods of greenhouse conditions, this paper proposed and realized an intelligent greenhouse monitoring system based on the STM32. GPRS and sensor technologies were used for monitoring and controlling major environmental factors such as temperature, humidity, and illumination. The test results showed that this system could be easily operated and provided a suitable environment for agricultural crops. Hence, the efficiency of raising crop could be improved.

KEYWORDS: Intelligent control; STM32; Sensor Technology; GPRS.

1 INTRODUCTION

Greenhouses provide suitable growth environments for commercial crops, and because of the outstanding economic and social benefits, greenhouse-cultivating technology has become an important part of modern agriculture. Temperature, humidity, and illumination are major factors that affect the growth of agricultural crops, and how to realize intelligent control of these aspects has become a key issue and research focus in this field. Hence, the search for an intelligent control system for greenhouses is important for the modernization of agriculture.

The intelligent control system of reference (Kang & Yan 2009) adopts a host-terminal mode including one host computer and terminal handling equipment. Although this mode is flexible in the control method, it has disadvantages such as low reliability and high cost. MCU is used as the control unit for the system proposed in reference (Yao & Zhao 2007). It has a simple structure but complicated human-machine interface, and is difficult to operate. Recently, the appearance of embedded systems and wireless sensors provide excellent solutions for monitoring greenhouses. Traditional MCUs such as 32-bit ARM7 and ARM9 are designed for data communication and data processing, and have excellent data processing capabilities, but they have complex structures and poor real-time performance. In 2007, the ST company launched the high-performance and low-cost STM32 (ST Microelectrionics Corporation 2007), which was used as the core unit for an intelligent greenhouse control system proposed in this paper. Accurate measurement and control of greenhouse environment factors could be carried out in real-time using this system, which synthesizes sensors and GPRS (Cao & Dong 2012) technologies to adopt cost-effective temperature, humidity, and illumination sensors. A CAN bus was used for reception of data and transmission of instructions and a serial port was used for data communication with the GSM module, which could immediately send abnormal signals to users.

This paper realized and proposed an intelligent greenhouse monitoring system based on the STM32. GPRS and sensor technologies were used for monitoring and controlling the major environmental factors such as temperature, humidity, and illumination. (Cao & Dong2012) This paper is organized as follows: Section II describes the system overall design. Section III presents the key technologies of this system. Section IV presents the system establishment and testing. Section V concludes this paper.

2 OVERALL DESIGN

The proposed system includes the main control module, the data collection and processing module, and the GSM module. The data collection module includes the temperature, humidity, and illumination sensors. On the monitoring terminal, there is a camera module. Data processing includes the LCD display module and data fusion. When the monitored parameter exceeds the set range, the GSM module will send a message to users. The overall structural diagram is shown in Figure 1.

*Corresponding author: *littlehao@126.com*

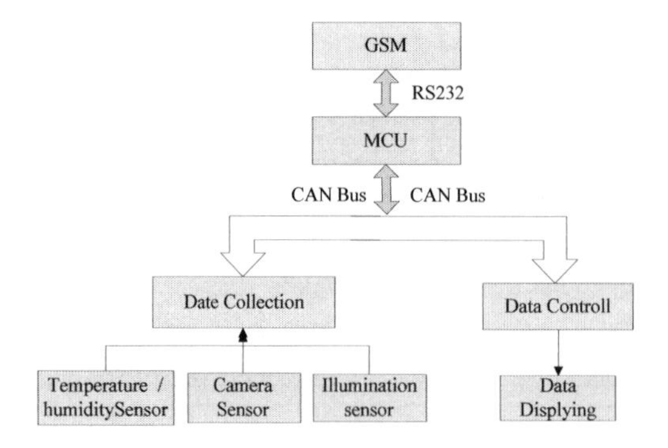

Figure 1. Overall structural diagram of intelligent control system.

As shown in Figure 1, the main control module is the core of the system and is responsible for receiving data from the sensors. The temperature, humidity, and illumination values are converted into electrical signals and the filtered electrical signals are sent to the MCU to finalize the signal collection. The collected signals are compared to the preset threshold value and a warning mechanism will be activated if environmental factors exceed the threshold value. A relevant message will be sent to the user to realize intelligent control of temperature, humidity, and illumination.

3 SYSTEM DESIGN

3.1 *Microcontrol Unit (MCU)*

The STM32 development board uses the high-configuration STM32F103ZETT6 (ST Microelectrionics Corporation 2010) as the MCU. In addition, there is a FSMC (flexible static memory controller) for external expansion of SRAM and connection to an LCD, which can improve the refresh speed of the LCD and reduce its cost.

3.2 *GSM module*

The main function of the GSM module is sending messages to the user; that is to say, if a parameter exceeds the set range, a relevant message will be sent to the user for the purpose of resolving the issue immediately. An EM 310 (Hua wei Corporation 2009) module is used in this system.

The frequency range of the EM310GSM included two frequencies, EGSM900 and GSM1800, and its power supply was 3.5 V. It had functions such as message sending, telephone communication, and GPRS network connection, and its message-sending function was used in the greenhouse control system.

The code for sending instructions to the GSM module in the main() through the serial port is shown in Figure 2. AT is the initial communication instruction, and the return value of OK shows that the connection was successful. The instruction "AT+CMGF=1" set the message format as txt format, and "AT+CMGS=13752582980" set the telephone number of the receiving terminal. After writing the data message, the send instructions were written.

```
void fasong(void)
{
        printf("AT\r\n");
        delay_ms(1000);
        printf("AT+CMGF=1\r\n");
        delay_ms(1000);
        printf("AT+CMGS=13752582980\r\n");
        delay_ms(1000);
        printf("HELLO0x1A\r\n");
        delay_ms(1000);
        USART2->DR=0x1A;
        while(1);

}
```

Figure 2. Sending AT instruction code.

3.3 *Data collection*

3.3.1 *Temperature and humidity sensor module*

A DHT11 (Aosong 2011) which measures the temperature and humidity of a greenhouse environment, was used as the temperature and humidity sensor. It included a NTC component, which was used for the measurement of temperature, and a resistance-type component, which measured the environmental humidity. The DHT11 was connected to the 8-bit MCU, and the collected data was transmitted through the IO port. Its normal working voltage was 5 V and average working current was 0.5m A. It was programmed to measure temperatures ranging from 0-50°and humidity ranging from 20% to 90% RH. If the greenhouse temperature exceeded the preset threshold value (for example, 25° C), the GSM module sent a message to inform the user to adjust the temperature immediately. If the temperature was lower than a preset value (for example, 20° C), a light was turned on to warm the greenhouse.

Before the collection of data, the system needed to detect whether the DHT11 module existed. If the DHT11 was located, the system would operate; otherwise, "DHT11 ERROR" was displayed on the LCD, showing that the sensor did not exist. In the main(), temperature and humidity values were displayed on the LCD, and the temperature was judged for further action. If the temperature exceeded 25° C, a message was sent to the user, and if it was lower than 20° C, a light was turned on to simulate the warming action. In order to realize real-time monitoring, after repeated debugging, each 100 ms the data will be refreshed. Relevant code in the main function is shown in Figure 3.

```
if(t%10==0)
{
    DHT11_Read_Data(&temperature,&humidity);
    LCD_ShowNum(60+40,150,temperature,2,16);
    LCD_ShowNum(60+40,170,humidity,2,16);
if(temperature>25)
{fasong();}
if(temperature<20)
{LED0=0;}

}
delay_ms(10);
t++;
if(t==20)
{
    t=0;
}
```

Figure 3. Code for acquirement of DHT11 data in main function.

3.3.2 *Illumination sensor module*

A BH1750FVI, which is a digital light intensity sensor IC suitable for two-wire serial bus, was adopted, and it adjusted the back-light of the LCD or keyboard according to the received light-intensity data. It was high resolution and could detect light-intensity variations in a wide range.

An IIC (inter-integrated circuit), a two-wire serial bus that includes one data signal wire SDA and one clock signal wire SCL, was used for communication with the MCU for both sending and receiving data. The illumination sensor design was focused on debugging the IIC, the connections and communication between the sensor and the MCU, and the display of certain data on the LCD. The initialization function of the IIC is shown in Figure 4.

```
void IIC_Init(void)
{
    GPIO_InitTypeDef GPIO_InitStructure;
    RCC_APB2PeriphClockCmd( RCC_APB2Periph_GPIOB, ENABLE );

    GPIO_InitStructure.GPIO_Pin = GPIO_Pin_10|GPIO_Pin_11;
    GPIO_InitStructure.GPIO_Mode = GPIO_Mode_Out_PP ;
    GPIO_InitStructure.GPIO_Speed = GPIO_Speed_50MHz;
    GPIO_Init(GPIOB, &GPIO_InitStructure);
    GPIO_SetBits(GPIOB,GPIO_Pin_10|GPIO_Pin_11);
}
```

Figure 4. Initialization function of IIC.

The program for the illumination sensor BH1750FVI was written in BH1750.c, and there were two programmed functions: Function BH1750_Write( ), which was used for writing relevant instructions or signals to the sensor after finding the address of the sensor and communicating with it; BH1750_Read_Data(void), which was used for reading signals from the sensor. In main.c, illumination data from the BH1750 was read and displayed on the LCD, as shown in Figure 5.

```
void  BH(void)
{
    Data=BH1750_Read_Data();
    BH_temp=(Data)/1.2;
    LCD_ShowxNum(190,180,BH_temp,5,16,0)   ;
    LCD_ShowString(60,180,200,16,16,"Light intensity");

}
```

Figure 5. Illumination data of BH1750 is displayed on LCD.

3.3.3 *Camera module*

There is a built-in interface on the STM32 board for connecting with the camera module. In this system, the pictorial sensor OV7670 from OmniVision was used for real-time monitoring of the greenhouse environment. The monitored pictures were displayed on the LCD.

3.3.4 *Display module*

The collected data were displayed on the LCD module, and a TFT-LCD (Shen 2012) was used in this system. It displayed both black-white and 16-bit true color pictures. Compared to the traditional OLED panel that could only display one or two colors, the TFT-LCD penal is more flexible and has higher precision. In this system, a 2.8-inch panel was used.

```
void LCD_WR_DATA(u16 data)
{
    LCD->LCD_RAM=data;
}
u16 LCD_RD_DATA(void)
{
    return LCD->LCD_RAM;
}
```

Figure 6. Functions for writing or reading LCD data.

```
void LCD_DisplayOn(void)
{
    if(lcddev.id==0X9341||lcddev.id==0X6804)LCD_WR_REG(0X29);
    else LCD_WriteReg(R7,0x0173);
}

void LCD_DisplayOff(void)
{
    if(lcddev.id==0X9341||lcddev.id==0X6804)LCD_WR_REG(0X28);
    else LCD_WriteReg(R7,0x0);
}
```

Figure 7. Function for turning the LCD on or off.

```
void LCD_SetCursor(u16 Xpos, u16 Ypos)
{
    if(lcddev.id==0X9341||lcddev.id==0X6804)
    {
        LCD_WR_REG(lcddev.setxcmd);
        LCD_WR_DATA(Xpos>>8);
        LCD_WR_DATA(Xpos&0XFF);
        LCD_WR_REG(lcddev.setycmd);
        LCD_WR_DATA(Ypos>>8);
        LCD_WR_DATA(Ypos&0XFF);
    }else
    {
        if(lcddev.dir==1)Xpos=lcddev.width-1-Xpos;
        LCD_WriteReg(lcddev.setxcmd, Xpos);
        LCD_WriteReg(lcddev.setycmd, Ypos);
    }
}
```

Figure 8. Function for setting the location of the display.

Programs for the TFT-LCD module were defined in lcd.c. LCD_WR_DATA(u16 data) was the function for writing data onto the LCD screen, and LCD_RD_DATA(void) was the function for reading data from the LCD, and they are both shown in Figure 6. LCD_DisplayOn(void) was the function for turning on the LCD display, LCD_DisplayOff(void) was the function for turning the LCD display off, and they are shown in Figure 7. LCD_SetCursor(u16 Xpos, u16 Ypos) was the function for selecting the location for displaying, and x,y were the coordinates of the display location, as shown in Figure 8. LCD_Clear (YELLOW) was the function for changing the display colors, and the color in brackets illustrated the display color.

3.3.5 *Data communication*

The CAN bus was a bus-based serial communication network that operated under several modes. It was easy to constitute a multi-machine control system, and any node on the network could send messages to other nodes. Node information was divided into different priorities according to real-time performance of the system. Short frame structure was adopted by CAN bus, and point-to-point or other data transmission modes were realized with message filtering. The CAN protocol adopted in this system was based on the CAN frame structure and the master-slave network structure. It adopted standard frame format, and supported CAN2. 0A(ST Microelectrionics Corporation 2010) technical regulations.

4 SYSTEM ESTABLISHMENT AND TESTING

4.1 *System establishment*

System connection diagram is shown as Figure 9. The first part is main control module MCU. The second part is GSM module. The third part is LCD display module. The fourth part is BH1750 illumination sensor module. The fifth part is DHT11 temperature and humidity sensor module. The sixth part is OV7670 camera module.

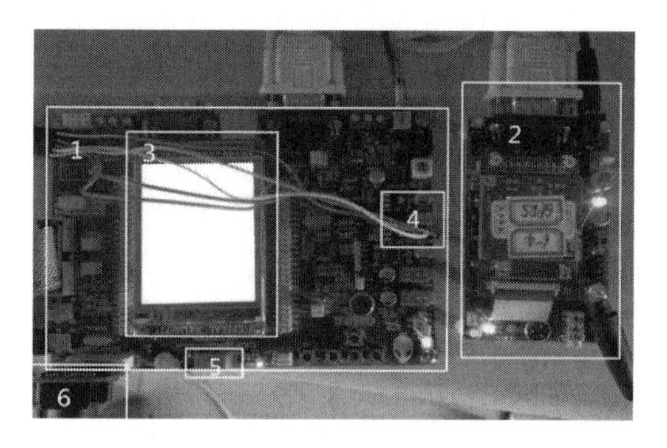

Figure 9. System connection diagram.

4.2 *System testing*

For this system, MDK4.2 (Li 2012) was used for programming software and C language was used for programming. An LCD module was used for displaying the received temperature, humidity, and illumination data, as shown in Figure 10. The pictorial information acquired by the OV7670 module is shown in Figure 11, and the message received on a mobile phone is shown in Figure 12.

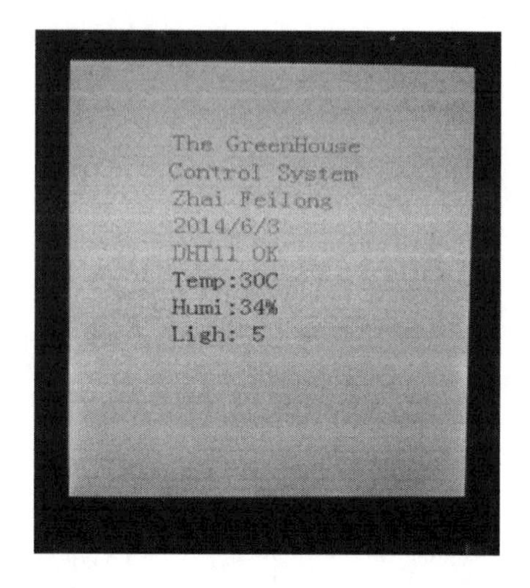

Figure 10. Temperature, humidity, and illumination parameters detected by the system.

Figure 11. Pictorial information acquired by camera module.

The temperature assumed to be suitable for crops is 20-30° C. If the real-time temperature was lower than 20° C, the light was turned on by the system to warm the greenhouse, and if the temperature was higher than 25° C, the red light was turned on and a

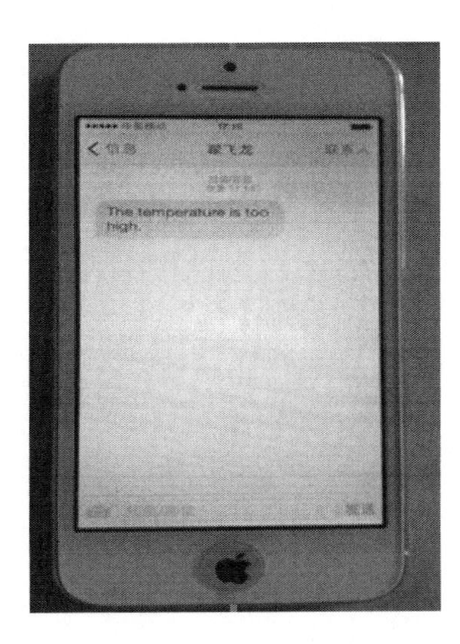

Figure 12. Message received on mobile phone.

message was sent to the user. If the temperature was too low and urgently needed to be raised, more warming lights were turned on manually to quickly warm the greenhouse. If pictorial information was needed, the KEY_UP button was pressed, and the LCD display was switched on. If the reset button was pressed, the LCD display was reversed back. Monitoring of the humidity or illumination was also controlled according to temperature.

5 CONCLUSION

The paper researched a greenhouse control system that could monitor environment parameters such as temperature, humidity, and illumination in real-time. When the temperature fell lower than the preset value, a heat lamp was turned on to warm the greenhouse. When the temperature was higher than the preset value, a buzzer was sounded to raise the alarm. At the same time, the GSM module sent a message to the user. The camera module was used for monitoring real-time local environmental situations in the greenhouse.

ACKNOWLEDGMENT

This research was supported by The Higher Education Science and Technology Planning Project of TianJin (20130419).

REFERENCES

Aosong. 2011. The knowledge of humidity and moisture measuring. http:// www.aosong.com.

Cao Xin& Dong wei. 2012. The intelligent greenhouse monitoring system based on wireless sensor network. *Electronic Design Application* 38(2): 84–87.

Hua wei Corporation. 2009. *Huawei EM310 wireless module AT command manual*.1(1)

Kang dong & Yan Hailei. 2009. The Design of the remote greenhouse control system. *Control Engineer* 16(S4): 8–11.

Li, Ning. 2012. *ARM MCU Development and MDK Technology*. Beijing: Beihang University Press.

Shen Zhiyuan. 2012. *TFT-LCD Techonolgy: Structure, Principle and Manufacturing Technology*. Beijing: Publishing House of Electronic Indursty.

ST Microelectrionics Corporation. 2007. *STM32F103XX Datasheet*.(11).

ST Microelectrionics Corporation.2010. *STM32F103XX Datasheet*.(5).

Yao Youfeng & Zhao Jiangdong. 2007. The design of environmental condition monitoring system based on single chip microcomputer technology. *Measurement and control technology* 31(1): 105–108.

Electronics, Communications and Networks IV – Hussain & Ivanovic (eds)
© 2015 Taylor & Francis Group, London, ISBN: 978-1-138-02830-2

Three-dimensional positioning based on weighted centroid algorithm

Zhenzhen Hao
Institute of Information Science and Technology, Jinan University, Guangzhou, Guangdong, China

Ribin Wang & Yuanliang Huang*
Institute of Electrical Automation, Jinan University, Zhuhai, Guangdong, China

ABSTRACT: In order to suppress effectively the effects of environmental factors and improve the precision of distance measurement, Gaussian model is used to get the final measured Received Signal Strength Indication (RSSI). Then the weighted centroid algorithm is extended from two-dimensional space in three-dimensional space, which forms a weighted centroid algorithm for three-dimensional positioning based on RSSI. The running result shows that this localization algorithm is quite practical for short distance positioning with high precision, and it needn't additional hardware spending.

KEYWORDS: RSSI; Gaussian model; weighted centroid algorithm; three-dimensional positioning.

1 INTRODUCTION

In recent years, the application of wireless sensor network is increasing rapidly. ZigBee (IEEE 802.15.4 standard) is a rising wireless network technology which is of short space, low power consumption, low data rate and low cost (Li & Duan 2008). Wireless location technology based on ZigBee can realize relatively precise positioning of the indoor environment, such as underground parking lot and mine. Therefore, Zigbee will be applied widely in the field of short distance positioning.

So far, positioning algorithms are mainly based on two broad categories of ranging algorithms and algorithms without ranging. The ranging algorithms include Received Signal Strength Indication, Link Quality Indicator, Time of Arrival, and Angle of Arrival and so on while the positioning without ranging algorithms include Centroid Method, and Distance Vector Hop algorithm (Bulusu et al. 2001). Comparatively speaking, the ranging algorithms have higher positioning precision than the without ranging positioning algorithms distance. Among the positioning algorithms, the positioning technology based on RSSI is still widely concerned because of its simple structure without additional hardware device, although it can only provide the rough precision of locating (Yao & Fu 2010).

However, the research on wireless positioning based on ZigBee is limited to two-dimensional space. In order to increase the practicality of ZigBee positioning technology, this paper extends the weighted centroid algorithm from two-dimensional space to three-dimensional space. Finally, the simulation experiment shows that the algorithm can realize high precision and Real-time positioning, which is quite valuable for short distance positioning.

2 THE PRINCIPLE OF CALCULATING DISTANCE AND THE DATA PROCESSING

2.1 *The principle of calculating based on RSSI*

The different loss of wireless signal strength has a great effect on the location accuracy of the localization algorithm based on *RSSI*. There are many models to measure distance based on *RSSI* (Rappaport 1996), such as a Free Space Propagation Model, Logarithm Distance Path Loss Model, Hata Model, Logarithm Normal Distribution Model, etc. We use the Simplified Logarithm Normal Distribution Model which is shown as equation (1).

$$[RSSI]_{\mathrm{dBm}} = [PL(d)]_{\mathrm{dBm}} = -[A + B\lg(d)] \qquad (1)$$

Where A is the value of RSSI when the distance between blind node and beacon node is one meter. B is an experience value which should be measured in the specific environment. In this paper, we get the relationship between *RSSI* and the distance d by filtering and fitting the measured data.

*Corresponding author: *tyoll@jnu.edu.cn*

2.2 *The data processing of Gaussian Model based on RSSI*

In reference (Zhang et al. 2009), Statistical Average Model, Random Model and Gaussian Model are used to correct the measured *RSSI* from the experiment. Finally, they find that the method of data processed by Gaussian Model has the highest accuracy, which can satisfy the needs of the localization of most WSN nodes.

The principle of data processed by Gaussian Model gives up the *RSSI* with small probability and selects the *RSSI* with high probability, then calculates the average of this *RSSI*. This method can eliminate the influence of the *RSSI* with small probability and enhance the accuracy of localization.

We credit the *RSSI* of blind node receiving from the same beacon node to *Beacon_i*[], then adopt Gaussian Model to deal with these *RSSI*. The Gaussian Model is introduced as follows (Sheng & Xie 2009).

$$F(RSSI)=\frac{1}{\sigma\sqrt{2\pi}}\int_{-\infty}^{RSSI}e^{-\frac{(x-\mu)^2}{2\sigma^2}}dx \tag{2}$$

$$P(RSSI_i\le RSSI_{01})=F(RSSI_{01})=\alpha_1,i=1,2,\cdots,n \tag{3}$$

$$P(RSSI_j\ge RSSI_{02})=1-F(RSSI_{02})=1-\alpha_2,j=1,2,\cdots,n \tag{4}$$

Where μ is the average of these *RSSI*, σ is the Standard deviation of these *RSSI*. $RSSI_{01}$ and $RSSI_{02}$ are the upper limit and lower limit of *RSSI* processed by Gaussian Model. α_1 is the probability of the *RSSI* which are less than $RSSI_{01}$, and α_2 is the probability of the *RSSI* which are less than $RSSI_{02}$. Based on past experience, in this paper, α_1=0.3, α_2=0.7. Therefore, if the measured *RSSI* is between $RSSI_{01}$ and $RSSI_{02}$, it will be considered a high probability event. So we should select the *RSSI* from $RSSI_{01}$ to $RSSI_{02}$, then credit it into the *Beacon_i_gauss*[]. At last, we calculate the average of the *RSSI* in *Beacon_i_gauss*[] which is considered as the final *RSSI* from the certain beacon node.

The model can effectively solve the problem that RSSI is easily interfered by environmental factors. So the algorithm we put forward below will adopt the Gaussian Model to process the measured *RSSI*.

2.3 *Error correction of the distance measurement*

Distance measurement cannot be eliminated though this paper uses the Gaussian model to get the final RSSI. So it is necessary to correct the error of distance measurement, which can improve the precision further. Suppose that the real distance between beacon node and blind node is *d* and the measured distance between beacon node and blind node is *d'*. The distance measurement absolute error is

$$e_a=d'-d \tag{5}$$

And the relative error can be calculated through equation (6).

$$e_r=\frac{e_a}{d}\times 100\% \tag{6}$$

In a particular environment, it is not difficult to get average relative error through a lot of experiments. The more experiments, the average relative error is close to the facts more. Therefore, the corrected distance between beacon node and blind node is d″.

$$d''=d\times(1-e_r) \tag{7}$$

3 TWO-DIMENSIONS POSITIONING ALGORITHM BASED ON WEIGHTED CENTROID

In two-dimensional positioning, three beacon nodes which are in the same line can position the blind node. In an optimal situation, take the distances between three beacon nodes and blind node as radiuses to draw 3 circles centered on the 3 beacon nodes respectively, and the three circles will intersect at a point, which is concerned as the blind node. However, the calculated distance between beacon nodes and blind node is always longer than the real one because of the interference of the environment so that the 3 circles can't intersect at one point (Lin & Chen 2009).

As shown in Figure 1, $D_1(x_1, y_1)$, $D_2(x_2, y_2)$, $D_3(x_3, y_3)$ are beacon nodes and $O(x, y)$ is blind node. Take $|D_1M|$, $|D_2M|$, $|D_3M|$, as radiuses to draw 3 circles centered on D_1, D_2, D_3, respectively to get the overlap region. The basic thought of triangle weighted centroid positioning algorithm is described as follows. In figure 1, *E*, *F*, *G* are the intersections of circle D_1, circle D_2, circle D_3. Then take the 3 points as vertexes of the triangle and calculate the triangle centroid through a weighted algorithm. So take triangle centroid as the blind node coordinate. The coordinates of $E(x_e, y_e)$ can be worked out through equations (8).

$$\begin{aligned}(x_e-x_1)^2+(y_e-y_1)^2&\le r_1^2\\(x_e-x_2)^2+(y_e-y_2)^2&=r_2^2\\(x_e-x_3)^2+(y_e-y_3)^2&=r_3^2\end{aligned} \tag{8}$$

Similarly, it is not difficult to calculate the coordinates of $F(x_f, y_f)$,$G(x_g, y_g)$. The coordinates of triangle centroid N, which is taken as the blind node $O(x, y)$, can be worked out through equations (9).

$$x_N = \frac{x_e / r_1 + x_f / r_2 + x_g / r_3}{1/r_1 + 1/r_2 + 1/r_3}$$
$$y_N = \frac{y_e / r_1 + y_f / r_2 + y_g / r_3}{1/r_1 + 1/r_2 + 1/r_3} \tag{9}$$

The selecting rule of the weight is according to the distance, which means that the weight is inversely proportional to distance.

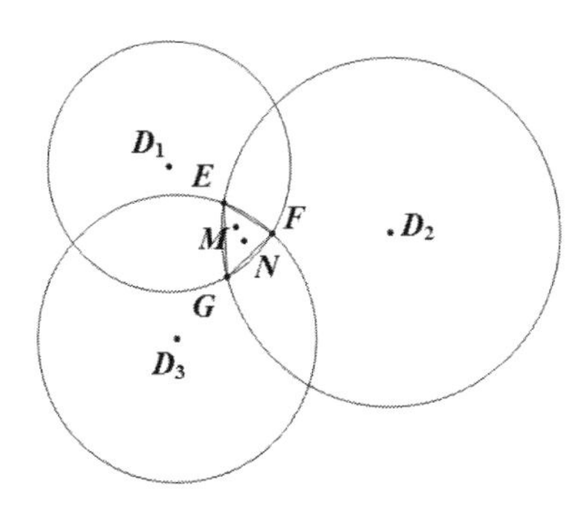

Figure 1. Three positioning circles.

4 THREE-DIMENSIONS POSITIONING ALGORITHM

4.1 *Three-dimensions positioning space model*

As shown in Figure 2, place i beacon nodes, which are recorded as $N_k(k=1, 2, ..., i)$ in three-dimensional Positioning space and ensure that there is enough beacon nodes in *XOY* plane. The coordinates of the K-th is recorded as $N_k(x_k, y_k, z_k)$. Suppose that there is a blind node $M(x, y, z)$, and choose 3 beacon nodes $A_1(x_1, y_1, z_1)$, $A_2(x_2, y_2, z_2)$, $A_3(x_3, y_3, z_3)$ in the *XOY* plane, which satisfies the condition that the projection of blind node $M(x, y, z)$ in the *XOY* plane locating inside triangle $\Delta A_1, A_2, A_3$. Besides, in order to improve positioning accuracy, it is necessary to make the distances between M' and A_1, A_2, A_3 as short as possible when selecting A_1, A_2, A_3. Then choose another beacon node $A_4(x_4, y_4, z_4)$ outside the *XOY* plane. The distances between blind node M and beacon nodes $A_j(j=1, 2, 3, 4)$ can be calculated through equation (10).

$$d_j = \sqrt{(x_j - x)^2 + (y_j - y)^2 + (z_j - z)^2} (j = 1,2,3,4) \tag{10}$$

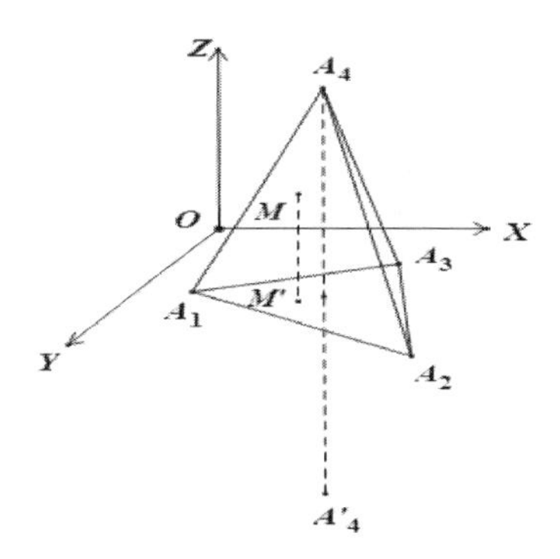

Figure 2. Three-dimensional positioning space model.

4.2 *Description of the three dimensional localization algorithm*

In ZigBee 3D wireless location networks, the distance between blind node and beacon node can be calculated according to RSSI. This paper proposed a 3D wireless localization algorithm based on RSSI values, which extends the weighted centroid algorithm from two-dimension to three-dimension. In figure 2, M' is the projection of blind node M in the *XOY* plane. Take $|A_1M'|$, $|A_2M'|$, $|A_3M'|$ as radiuses to draw 3 circles O_1, O_2, O_3 centered on A_1, A_2, A_3 respectively. It is easy to get the *XY* coordinate of M', concerned as the *XY* coordinate of M, through a weighted centroid algorithm. Then calculate the z coordinate of M(the mean value of $z_j(j=1, 2, 3, 4)$) through equation (10), the beacon nodes coordinates and the *XY* coordinate of M. Finally, the z coordinate of M can be worked out by following.

$$z = \frac{\sum_{j=1}^{4} z_j}{4} (j = 1,2,3,4) \tag{11}$$

In figure 2, suppose that the existence of another point $M''(x'', y'', z'')$ (The distance between it and beacon node A_4 is recorded as d''_4) and node M on the plane *XOY* of symmetry. If $d''_4 > d_4$, the z coordinate of M is positive since this paper selects beacon node A_4 is close to blind node M as much as possible. Otherwise, it is negative. Therefore, it can get the coordinate of $M(x_0, y_0, z_0)$.

5 THE PROCESS OF THE ALGORITHM

After researching on the weighted centroid algorithm for two-dimensional positioning, this paper puts forward a three-dimension location algorithm based on a weighted centroid algorithm. The main steps of the algorithm are as follows.

Step 1: Before measuring *RSSI*, put *i* beacon nodes with certain coordinate in the positioning place in *XOY* plane and out of the *XOY* plane, which is according to the three-dimension location model. The coordinates of these *i* beacon nodes are $A_1(x_1, y_1, z_1), A_2(x_2, y_2, z_2)...A_i(x_i, y_i, z_i)$. Then put a blind node $M(x, y, z)$ according to Section 3.1.

Step 2: After finishing measuring *RSSI*, use the Gaussian model to get the final n *RSSI* from the n beacon nodes. So it can work out the distances between blind node and the *n* beacon nodes, and output them from small to large order. Then select 4 beacon nodes, including 3 beacon nodes in *XOY* plane and another beacon node outside the *XOY*, which meet the requirements of Section 3.1.

Step 3: The blind node $M(x, y, z)$ was projected in *XOY* plane to get the node $M'(x, y, 0)$.Then take the 3 beacon nodes which is chosen in step 1 as the center of the circle and $|A_1M'|$, $|A_2M'|$, $|A_3M'|$ as radius to draw 3 circles. Calculate node M' coordinate according to aweighted centroid algorithm, which is recorded as $M_0'(x_0, y_0, 0)$.

Step 4: Work out the z coordinate through node $M_0'(x_0, y_0, 0)$ and equation (10), (11)so that the coordinate of blind node is $M_0(x_0, y_0, z_0)$. The positioning error e_s can be calculated through equation (12).

$$\varepsilon_0 = \sqrt{(x_0 - x)^2 + (y_0 - y)^2 + (z_0 - z)^2} \qquad (12)$$

6 VALIDATION OF ALGORITHM

This paper adopts the programming of Matlab to validate the correctness of the three-dimensional positioning based on a weighted centroid algorithm, setting the signal is sent from CC2430 and adding noise model to simulate the signal transmission interfered by practical environment. Take (10, 3, 0), (21, 9, 0), (32, 21, 0), (15, 30, 0), (8, 23, 0), (7, 15, 0), (19, 17, 9), (25, 20, 19) as the beacon nodes and (20, 18, 6), (11, 8, 5), (23, 19, 15), (15, 20, 5)as blind nodes. The practical running results according to the algorithm are shown as Table 1.

Table 1. The practical running results.

Blind nodes	Positioning Results	Positioning Error
(20, 18, 6)	(20.39, 18.54, 6.10)	0.68
(11, 8, 5)	(11.79,8.35,5.12)	0.87
(23, 19, 15)	(22.44,19.93,14.76)	1.12
(15, 20, 5)	(15.41,20.14,4.84)	0.47

The practical running results show that this algorithm not only can orientate the blind node quickly but also have high location precision, which is valuable for three-dimensional short distance positioning.

7 CONCLUSIONS

This paper adopts a Gaussian model to sift RSSI, which increases the precision of distance measurement based on RSSI. The weighted centroid algorithm is extended from two-dimension to three-dimension. This means adopting dimension reduction to project the blind node in *XOY* plane. Using weighted centroid algorithm to work out the *X,Y* coordinate and then calculate the *z* coordinate.

Experiments indicate that the positioning precision is high. Besides, the algorithm can also orientate the blind node quickly, reduce hardware cost and save power. Consequently, the algorithm for three-dimensional positioning based on a weighted centroid algorithm is valuable for realizing short distance positioning.

ACKNOWLEDGEMENTS

This research is supported by the fund of Guangdong Province Science and Technology Project (2013B010401019) and Zhuhai Public Platform Project (2013D0501990002).

REFERENCES

Bulusu, N., Heidemann, J. & Estrin, D. 2001. Density adaptive algorithms for beacon placement in wireless sensor networks. *IEEE ICDCS01, AZ, April 2001*. Mesa: IEEE.

Li Wenzhong & Duan Chaoyu. 2008. *Wireless and positioning base on ZigBee*. Beijing: Beijing University of Aeronautics & Astronautics publishing house.

Lin Wei & Chen Chuanfeng. 2009. Wireless sensor networks triangle centroid localization algorithm based on RSSI. *Sensor Technology* 2009(2): 180–182.

Rappaport & Theodore S. 1996. *Wireless communications: principles and practice*. New Jersey: Prentice Hall PTR.

Sheng Zhou & Xie Shiqian. 2009. *Theory of probability and mathematical statistics and its applications*. Beijing: Higher education publishing house.

Yao Junyi & Fu Xuan. 2010. Location estimation based on CC2431 RF transceiver in WSN. *Information and Electronic Engineering* 8(3): 257–260.

Zhang Jianwu, Lu Zhang, Ying Ying, et al. 2009. Research on ZigBee-Based RSSI ranging. *Sensing Technology Journal* 22(2): 285–288.

Electronics, Communications and Networks IV – Hussain & Ivanovic (eds)
© 2015 Taylor & Francis Group, London, ISBN: 978-1-138-02830-2

System architecture model for runtime deployment adaptation

Zhiyong He & Yue Liang
Naval University of Engineering, Wuhan, Hubei, China

ABSTRACT: One of the crucial aspects that influence reliability and availability of control systems is the deployment of software components to hardware nodes. If some hardware nodes or communication lines are failed or destroyed, which is often the case in military environment, the software architecture needs to be reconfigured to a new deployment to keep key function's availability. We introduce an approach to automate the task to find an alternative deployment to reserve some key function of the control system. As distinct to related approaches, which typically need offline reconfiguration, we want to preserve service continuity and assure run-time adaptation process safety.

1 INTRODUCTION

System deployment, i.e. mapping of software components to hardware hosts, is a typical example of a design task (Meedeniya 2011). Control Systems usually distribute software applications on heterogeneous hardware nodes and communication infrastructures. They are made up of a lot of components, each component having its own requirements (both hardware and software). The problem of automatic exploration of deployment alternatives, and reporting a set of near-optimal solutions has already been addressed with respect to various system quality attributes, including latency, performance, security, and resource usage (Belguidoum & Dagnat 2007). However, in some situations, control system can't be suspended as a whole. Runtime adaptation may avoid the lost of system shutdown, and keep availability of key services.

Among the numerous problems that arise in the runtime adaptation to an alternative deployment of components (in a system), we concentrate on two main attributes of dependability problem (Meedeniya 2011). The first one is guaranteeing deployment success, i.e. finding all dependencies of a software component and explicated specify them. The second problem concerns state consistency of the target system during and after adaptation (Figure 1). To reach the goals we mentioned before, we propose a specific model for deployment dependencies attributes. The model extends Meedeniya's work (Meedeniya 2011), to add some important distinguished resource types. We also propose a runtime adaptation approach based on reflection mechanism. This approach modifies meta information of software architecture through meta operations such as "add", "delete", "replace", "migrate", "combine", and get the new runtime architecture of base-level using the reflection protocol (Hughes et al. 2010).

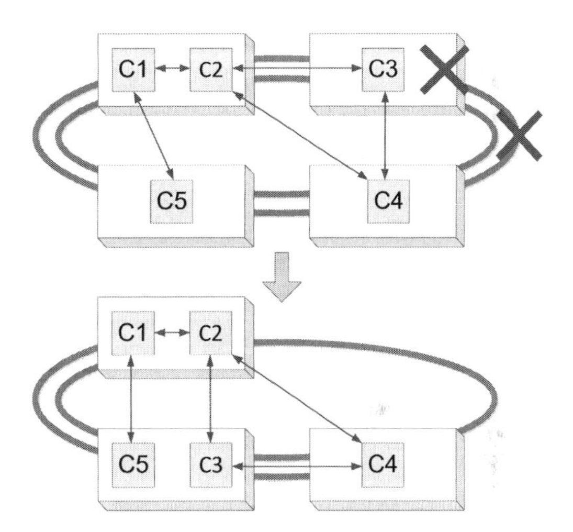

Figure 1. Reconfigure to alternative deployment.

2 SYSTEM ARCHITECTURE MODEL

Control systems interact with the physical environment closely, typically via sensors and actuators. They are deployed on compact self-contained computational units (containing both processing and memory part), known as Electronic Control Units (ECUs). All the hardware modules (ECUs, sensors, actuators) are connected through communication lines, and form the hardware architecture of the system. The software layer of a control system consists of a high number of lightweight components, representing the logical blocks of system functionality. This forms the software architecture of the system. The rest of this section discusses our modeling abstraction of such systems, presenting the software and hardware architecture models, along with the attributes respect to the deployment problem (Su et al. 2011).

2.1 *Software architecture model*

We classify the resources into three types: exclusive, shared, and divisible. An exclusive resource can be used by only one software component. Competition between components to an exclusive resource will produce a failure, for example, a communication port. In other hand, shared resources can be used by many components, like display devices and some sensors. Some resources are divisible. A component just needs a part of them, the rest part can be used by another component. Memory, bandwidth, and computing power are divisible resources, and always has a quantitative attribute. Then we can annotate each component with its resources dependencies (deployment constraints). (1)Divisible resources: (a) Size: memory size of a component used in deployment constraints; expressed typically in KB (kilobytes). (b) Workload: computational requirement of a component for each service; expressed in MI (million instructions). (c) Data size: the amount of data transmitted from software component C_i to C_j during execution of a single service; expressed in KB (kilobytes).(2) Exclusive resources: (a) Port: Listening port used by service in a component, identified by port number. (b) Actuator: required to perform the control orders, express by actuator ID.

2.2 *Hardware architecture model*

The hardware model is described by the same resource type in software model. The hardware nodes and the communication lines connecting them are modeled separately.

1 Nodes: (a) Capacity: memory capacity of an ECU used in deployment constraints; in KB (kilobytes). (b) Processing speed: the instruction-processing capacity of the nodes; expressed in MIPS (million instructions per second). This is used to calculate the execution time, which is a function of processing speed of the node and the computation workload of the service. (c) Actuator and Sensor: the actuator it can control, and the sensor it has access to. (d) Display Capacity: model the display capacity by resolution (1024*768), color mode (24bit), and size (23cm*17cm). A special monitor type maybe required, eradiate screen.
2 Lines: (a) Data rate: the data transmission rate of the bus; expressed in KBPS (kilobytes per second). This is used to calculate the time taken for data transmission, as it is a function of the data rate of the line, and amount of data transmitted during the communication. (b) Security level: what security policy or mechanism can be used to process information with different sensitivities.

2.3 *Self-loading function module*

Self-loading, as the extended loading bootstrap program, is a piece of program code written by the user to realize the function of data transmission from FLASH to the chip compared with the Loader of chip level inside the chip ROM. It solves the problems such as the failure of repeated invocation of the loader program in the chip, the limitation of the length of the loading program, etc. Furthermore, it makes it possible to selectively load the application program when starting the chip, and to realize the remote update and debugging of the system on chip. The basic function of self-loading is to read the program code and data information stored in the designated address of Flash chip and run the program code and data information in RAM of chip through loading according to the mapping relation between the established memory address and run address.

2.4 *Synchronization mechanism of network loading protocol*

In order to ensure the smooth progress of network loading, each node in the network must follow the synchronization mechanism in the process of executing network loading protocol. The foresaid three delay operations in the literature are a kind of simple synchronization mechanism. There are two synchronization mechanisms in the execution process of network loading protocol: network node extension synchronization mechanism and user code starting synchronization mechanism. The network node extension refers to the process of loading the unloaded node through the loaded network. It needs to ensure the new loaded node LINK port enters working state before the coming of the follow-up loading package, so as to avoid the loss of loaded data. The user task starting synchronization mechanism is used to ensure the user code of any node only starts after completing the network loading of the whole processor, so as to avoid the influence of interoperation like visiting, interruption, etc., of user code of multi-node on the loading process. The design of these two synchronization mechanisms is shown as follows: (1) The network node extension synchronization mechanism is shown in Figure 2. The new node can obtain the LINK port used by it when loading through query of the DMA channel status of each LINK port. It can also transmit the "extending node response" information to the ancestor nodes through this LINK port. After receiving the response information, the ancestor node will transmit the tag of new node, and the new node will analyze the tag and execute the remaining loading task. Then, the synchronizing process of network node extension ends. (2) The user code starting synchronization mechanism. After a certain node's completion of transmitting the loaded data, it should firstly wait for the information of "sub-tree loading finish response" returned by all the sub-trees. Then, it can transmit the information of "sub-tree loading finish response" to its ancestor nodes. Later, it should wait for the information of "allowing to exit the loading process" transmitted by the ancestor nodes. After receiving the information, it can transmit the information of "allowing to exit the loading process" to each sub-tree, and start executing its user code. This process should expand to the last "leaf" node. Meanwhile, the application of global signal line can also realize the synchronizing start and execution of network node. Then, the user code starting synchronizing process ends.

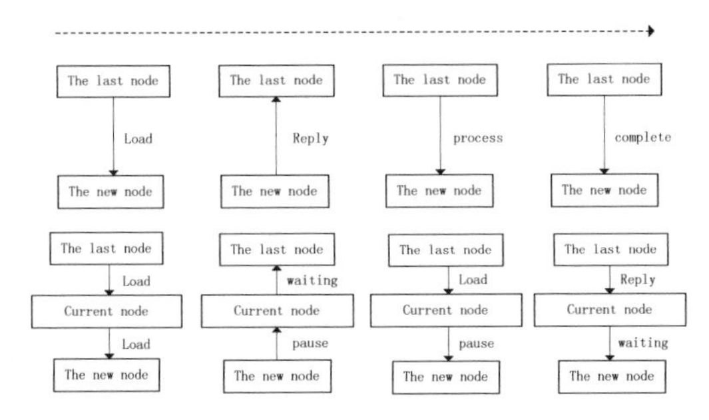

Figure 2. Network node extension and loading sychronization mechanism.

2.5 *Data communication module design*

Data communication module is to read the data collection stored in SRAM which is in reading status of SRAM_A and SRAM_B, and transmit the data collection to industrial personal computer through parallel port. The industrial personal computer will store, process, analyze, show and print the received data. According to the requirement of this system, the transmission rate of parallel port should be higher than 500KBps at least. However, the transmission rate of SPP (standard parallel port) is lower than 350KBps, which cannot meet the system's requirement of transmission. The ideal transmission rate of EPP (enhanced parallel port) can reach 650KBps to 3.2MBps, which can completely meet the system's requirement. Thus, this system chooses EPP to conduct the data communication. It applies the EPP data write operation to complete. The interface principle of chip controller and industrial personal computer is shown in Figure 3. In this system, the C1 and C2 of I/O port of the chip control core are respectively connected with Pin 14 and Pin 11 of the parallel port. The data link Data_A and Data_G are respectively connected with Pin 2-9 of the parallel port. If the chip control core needs to be communicated with the industrial personal computer, it should firstly read the data of SRAM which is in reading status. Then, it can transmit the data to industrial personal computer according to the sequential write operation of EPP mode.

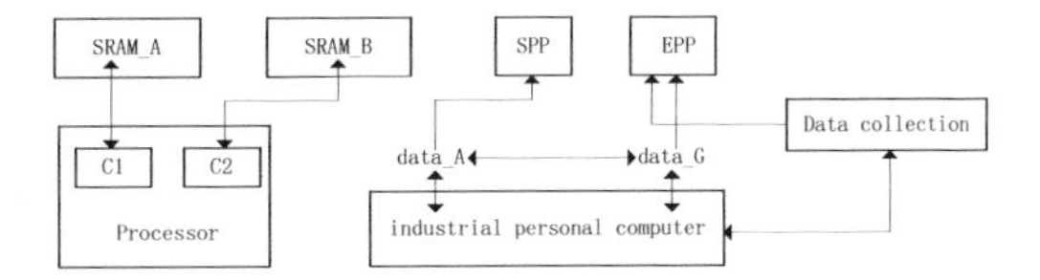

Figure 3. Coomunication mechanism of chip controller and industrial personal computer.

3 TEST RESULTS

This paper realizes the foresaid improved DIT-FFT based on 2 on TMSVC3402 platform. This processor adopts the advanced modified Harvard architecture. The chip totally has 8 buses (1 program memeory bus, 3 data storage unit buses and 4 address buses). The provided function of bit-reversed addressing improves the efficiency and execution rate of applying program memory in FFT algorithm routine. In this kind of addressing mode, the integer N stored in AR0 is one half of FFT points. Another auxiliary register points to the data storage unit. When using the bit-reversed addressing to add AR0 to the auxiliary register, the address is produced in bit-reversed mode from left to right, rather than from right to left. In addition, the processor also provides two accumulatos A and B, and seven auxiliary registers AR0-AR7. The complete utilization of these registers can minimize the number of variables, and in turn save the clock cycle and reduce the unnecessary waste. Table 1 shows the difference between the general DITFFT and improved DITFFT in terms of execution cycle and memory references under some different input points. Obviously, the application of improved DITFFT greatly reduces the memory reference times and execution cycle.

Table 1. Platform test data.

FFT number		Before improved time(s)	After improved time(s)	Reduced time	Reduced memory
First node	64	26.4	24.2	2.2	85.2%
	512	312.3	298.5	13.8	73.1%
	1024	2031.4	1983.5	48.9	71.8%
Second node	64	27.8	24.8	3.0	84.7%
	512	323.7	301.1.5	22.2	75.3%
	1024	2101.6	2053.6	4.8	70.7%
Third node	64	28.4	26.1	2.3	85.7%
	512	335.1	316.2	18.9	72.6%
	1024	2204.8	2112.7	92.1	69.4%

In addition, we also have studied the time spent and the amount of data transmission of DITFTT nodes after improvement as well as the relationship between the error rate and the data amount. Experimental conditions inside are to test three nodes respectively for several times and compute the time spent and error rate of each node under different amounts of data. The concrete computing method is that the error rate equals to the data packet failed in treatment divided by the total data packet. The Figure 4 shows the time spent and error rate before and after improvement. From the figure, it can be seen that the time spent and error rate of nodes after improvement obviously decrease, proving the designed rationality and practicability of the model.

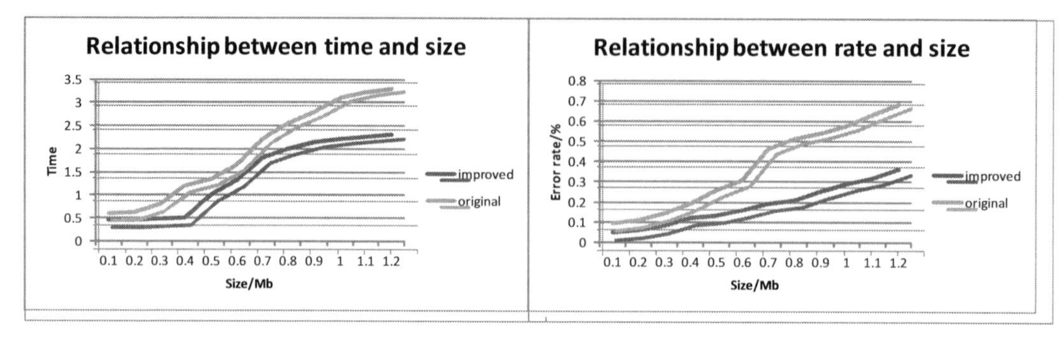

Figure 4. The performance comparison between improved and original.

4 CONCLUSION

In distributed environments, the connections between the hosts on which a software system is running are often unstable. he overall availability of the system decreases. This paper has presented a detailed model and an efficient algorithm for improving a distributed, component-based system's availability via redeployment. In this work we have argued that a global architecture model, describing resources needed by components explicitly, can offer a good support to overcome this issue. Using an interpreted procedural configuration language, it is also possible to address an important requirement of current applications: the need for dynamic reconfiguration and redeployment.The critical difficulty in achieving this task lies in the fact that during the reconfiguration process, application status continuity must be preserved to achieve availability. In our approach, the distributed check-point approach is used to keep the globe status, but there are a number of pertinent questions remain unexplored.

ACKNOWLEDGMENT

This work is supported by State Key Laboratory of Software Engineering (SKLSE)

REFERENCES

Belguidoum, M. & Dagnat, F. 2007. Dependability in software component deployment. Dependability of Computer Systems, 2007 DepCoS - RELCOMEX '07. 2nd International Conference on, Szklarska, 14–16 June 2007: 223–230 . IEEE.

Hughes, D., et al. 2010. A graph based approach to supporting software reconfiguration in distributed sensor network applications. *Journal of Internet Technology* 11(4):561–571.

Meedeniya, I. 2011. Reliability-driven deployment optimization for embedded systems. *Journal of Systems and Software*: 84(5): 835–846.

Pinello, C., Carloni, L. P. & Sangiovanni-Vincentelli, A. L. 2008. Fault-tolerant distributed deployment of embedded control software, *IEEE Transactions on Computer-Aided Design of Integrated Circuits and Systems* 27(5): 906–919.

Su X., Liu, H., Wu, Z., et al. 2011. Sa based software deployment reliability estimation considering component reliability of exponential distribution. *Journal of Software* 6(6): 1140–1145.

Su X.H., Wu, Z., Liu, H., et al. 2011. Reliability analysis of SA based software deployment with consideration of system deployment. *Intelligent Automation and Soft Computing* 17(6): 749–758.

Electronics, Communications and Networks IV – Hussain & Ivanovic (eds)
© 2015 Taylor & Francis Group, London, ISBN: 978-1-138-02830-2

Development and application of software for expanding map symbol database in Oracle Spatial

Yuanrong He
Institute of Computer and Information Engineering, Xiamen University of Technology, Xiamen, Fujian, China
Institute of Urban Environment, Chinese Academy of Sciences, Xiamen, Fujian, China

Yuantong Jiang
Institute of Architecture and Urban Planning, Hunan University of Science and Technology, Xiangtan, Hunan, China

Jiahao Li
Institute of Computer and Information Engineering, Xiamen University of Technology, Xiamen, Fujian, China
Institute of Urban Environment, Chinese Academy of Sciences, Xiamen, Fujian, China

Guoliang Yun
Institute of Computer and Information Engineering, Xiamen University of Technology, Xiamen, Fujian, China

ABSTRACT: In the application of GIS data and the secondary development of GIS, vector data, raster data, attribute data and cartographic symbols not only can't be managed easily, but also reduce the efficiency. Oracle Spatial is the application extension of Oracle database on the storage of spatial data. As it has developed for about twenty years, at present, it provides powerful support for the storage and operation functions of vector data, raster data and attribute data, and also provides import tool and developer components for these three kinds of data above. However, it can't satisfy the needs of the clients for it doesn't provide cartographic symbols management tool and it can't extend cartographic symbol library. In order to realize the integrated storage of vector data, raster data, attribute data and cartographic symbols, this paper elaborates the design and construction of a visual symbol library management system for Oracle Spatial data, which can realize the import and management from SVG format to Oracle. Practical applications show that the tool provides the convenient way for the display of spatial data and web visualization. It also can cooperate with Oracle Builder import tools and Oracle Map server, and becomes an effective complement of a series of Oracle Spatial tools.

KEYWORDS: Oracle Spatial, Map Symbol Database, Share, Java, SVG

1 DEMAND BACKGROUND

In traditional GIS softwares and its application systems, the storage of vector data, raster data and attribute data is separated. For example, ArcGIS uses *.shp file to store graphic data, *.dbf file to store attribute data, *.shx file to connect these two data and *.style file to store map symbols (Chen et al. 2001). The storage method is not only inconvenient to be managed but also reduces the running efficiency. The appearance of object-relational databases makes the integration of storage and network sharing of vector data, raster data and attribute data possible. Taking Oracle Spatial which has developed for more than 20 years as an example, Oracle Spatial possesses a complete data reception tool of vector data, raster data and attribute data. At the same time, it provides the data storage and operation function and front-end development tool (Table 1). However, the lack of map symbol management tool and associated symbols makes even rich data contents appear as black and white, silent movies. As a result, storing map symbols in Oracle, integrating the management of four databases, can not only make the spatial data stored in Oracle be accurately conveyed, but also make the symbol library be carried between systems and shared on the network just like the other three kinds of databases. Therefore, designing and developing a set of generalized management tools which stores and uses symbol library in Oracle can collaboratively apply Oracle Builder import tool and Oracle Map server. Then, it can provide convenient approach for expressing spatial data standardization and visualized Web service, making it an effective complement of Oracle Spatial series tools (Zhang et al. 2007).

Table 1. Oracle spatial space support analysis tools.

Vector Data	Raster Data	Spatial Data Tools	Front-end development tools
Geometry support vector data storage, spatial query, indexing and analysis, data management support and unified storage attributes associated with the object.	Georaster supports raster data storage, spatial query, indexing and analysis.	Provided EasyLoader, GeoRasterLoader, Shp2SDO, Map Builder software tools and Oracle Spatial's Java API, OCCI, ADO application interface receives data into GIS.	Provide built on the underlying XML API, classic Java and JSP API web map service clients, Ajax-based JavaScript API variety of interfaces.

2 DEVELOPMENT AND VALIDATION OF RESEARCH IDEAS

Oracle Spatial starts from 11g version, providing a symbol library for storing data table, and Map Builder provides picker, import, export, and backup tools like transplants, providing support for front-end applications and back-end storage conditions for the symbol library on the target system storage and sharing application specifically as follows (He 2011):

1 Storage Support. Oracle Database provides two data tables, USER_SDO_STYLES and MDSYS.SDO_STYLES_TABLE, shown in Table 2 and Table 3, which can make each point, line, surface and annotation style stored into a table record. Both data sheets contain five fields which are used to store data carrier symbol (DEFINITION field) and graphic description (NAME, TYPE, DESCRIPTION, IMAGE). MDSYS.SDO_STYLES TABLE adds SDO_OWNER field on this basis, and each symbol is used to describe attribution recorded by users. Through these two views, an Oracle database symbol can be uploaded. The main difference is that MDSYS.SDO_STYLES_TABLE contains all the symbols and the system created by users, and USER_SDO_STYLES contains only specifically created symbols by users. For example, suppose that A user (SDO_OWNER is A) creates 10 symbols, B user creates 15 symbols, C users create 20 symbols. Then the symbol resources of MDSYS.SDO_STYLES_TABLE not only contains the system's own symbols, but also owns all 45 symbols of the three symbols A, B and C, which makes it convenient to back up all symbols at one-time into a single *. dat file, which is put into Oracle server of other hosts to realize the transplant share. Therefore, the user's best choice for importing symbol target database table should be MDSYS.SDO_STYLES_TABLE. It can be seen from Table 2,that USER_SDO_STYLES contains five fields. NAME field is used to uniquely identify the name of the symbol, DESCRIPTION field is the further description for symbol style, TYPE field values with six kinds are labeled corresponding to symbol style types, which Marker, Line, Area, Text corresponds to the four kinds of symbol types in GIS software, namely, point, line, surface, annotation by two field data contents DEFINITION (CLOB type, SVG symbols) and IMAGE (BLOB type, image symbol) to achieve.

Table 2. The storage structure of table for Oracle Spatial symbol library USER_SDO_STYLES.

Column Name	Data types	Description
NAME	VARCHAR2	Style name, the name must be unique. Can not be empty.
TYPE	VARCHAR2	Symbol style types, including Area, Color, Line, Marker, Text, Advanced six kinds. Can not be empty.
DESCRIPTION	VARCHAR2	The text description of Symbol style. Can be empty
DEFINITION	CLOB	The XML code of Symbol style. Can not be empty.
IMAGE	BLOB	It is used to store image file. Can be empty.

2 Supporting of style pickup tool. Oracle provides the style pick-up function developed by Map Builder, which provides the main store pickup USER_SDO_STYLES styles and preview display functions as a layer or SQL spatial data sets. However, Map Builder does not provide the additional symbol library for symbol, the function for modifying, which are the extensions that the system needs.

Table 3. The storage structure of table for Oracle Spatial symbol library MDSYS.SDO_STYLES_TABLE.

Column Name	Data types	Description
SDO_OWNER	VARCHAR2	Username, the name must be unique. Can not be empty.
NAME	VARCHAR2	Style name, the name must be unique. Can not be empty.
TYPE	VARCHAR2	Symbol style types, including Area, Color, Line, Marker, Text, Advanced six kinds. Can not be empty.
DESCRIPTION	VARCHAR2	The text description of Symbol style. Can be empty
DEFINITION	CLOB	The XML code of Symbol style. Can not be empty.
IMAGE	BLOB	It is used to store image file. Can be empty.

3 Supporting of backup and migration. Map Builder provides output metadata, which can connect all symbol libraries, themes, and maps, exported together into a dat file on another computer by importing metadata, supporting backup and migration.

Based on the above conditions, developing the symbol library system which supports the expansion in line with MDSYS.SDO_STYLES_TABLE data format and content requirements is feasible. And it is also currently the important content absent in the expansion of space provided by Oracle. At the same time, allowing for that the Map Builder is a pure Java implementation of graphical user interface JAR main pack, target system as an auxiliary tool for Map Builder, uses the same Java development language, similar style interface JAR package program oracle map style extensor. Development and validation of research ideas are as follows:

1 SVG symbols use Java to read, upload, preview, edit, and delete the full functionality (Yu 2008).
2 In order to ensure continuity of software, the data connectivity is developed so that users can connect with the server user connection parameters of Oracle established according to the target system. Meanwhile, use the eclipse plugin jar package to fight Fat jar, pack it into a Jar file to realize cross-platform operation.
3 Share application test results, make the style, size, and color specifications of SVG symbols in line with the "map symbol library to establish the basic requirements" and put them into Oracle's symbol table through the developed symbol library software system, imports the experimental data of topographic maps using Oracle Map Builder, picks up the corresponding symbols, and refers to the single layer of ArcGIS and the show effect of all maps, test the overall show effect of the built SVG symbol library (Wang et al. 2007, Dang et al. 2011, Zuo & Nie 2011, Nass et al. 2011). To test the portability and sharing symbol library, symbol libraries, together with the bulk export and import the database on the server, Fusion Middleware Oracle Map Viewer for map publishing, network analysis maps in front of the display is used (Zhai et al. 2004, Xi & Wu 2008, Robinson et al. 2011).

3 KEY TECHNOLOGY

3.1 *Read and format validation of SVG files*

As the information of symbols is stored in SVG files, so it is needed to load the path and validate the format of the SVG files before saving SVG files (Chang & Park 2006, Peng & Zhang 2004, Huang et al. 2009). The implementation process is as follows:

1 Import the batik jars. Batik is a Java-based toolkit for applications or applets that use SVG format graphics to realize various functions. JSVGCanvas module is a swing component provided by Batik toolkit, and application developers can use the module in the main program to achieve the function of SVG graphic symbol display. Batik1.7 also has the function of checking whether the files loaded are SVG files and whether the version of SVG files is 1.0, and it also coincides with the SVG version supported by Oracle Map Builder, making the upload of SVG symbols more standardized.
2 Loading and getting path of SVG files. Define a class called myJSCGCanvas in the development environment in order to display SVG documents and obtain the path of SVG documents. Create a svgCanvas type which contains internal JSVGCanvas type in the constructor of the class, and then define a button to read the SVG file path, and display them. The core program is as follows:

```
// create object svgCanvas of JSVGCanvas type
JSVGCanvas svgCanvas = new JSVGCanvas();
```

```
button.addActionListener(new ActionListener() {
   public void actionPerformed(ActionEvent ae) {
   JFileChooser fc = new JFileChooser(".");
   int choice = fc.showOpenDialog(c);
   if (choice == JFileChooser.APPROVE_OPTION)
   {
   File f = fc.getSelectedFile();
   try {svgCanvas.setURI(f.toURI().toString());
   // obtain the path of SVG documents

    d = f.getPath();} catch (Exception ex) {
   ex.printStackTrace();
   }}
```

3.2 *Parsing and upload of SVG files*

Oracle Map Builder based on the XML document stores the SVG graphic symbols, so the SVG file needs to be parsed before imported into the database (Santosh et al. 2012). Parsing process roughly consists of the following 3 steps:

1. Create the input stream. According to the already accessed SVG file path, create a FileInputStream object fis to read the original byte stream of SVG files. The code is as follows:
   `FileInputStream fis = new FileInputStream(d);`
2. Create a temporary byte array. Put the raw bytes of data of SVG files into a temporary byte array. The code is as follows:
   `byte[] buffer = new byte[1024];`
3. Read the data. Apply the reading method in FileInputStream to read the data in a temporary array of bytes. The core code is as follows: fis.read (buffer,0,buffer.length)

The essence of the process of uploading symbols is combining the content of SVG documents and other attributes data, and then inserting it into the MDSYS. SDO_STYLES_TABLE table. The SQL inserting string is defined as follows (Guan 2006, Lin & Huang 2001):

```
String sql="insert into
MDSYS.SDO_STYLES_TABLE (NAME. TYPE,
DESCRIPTION, DEFINITION)
values('"+text1.getText()+"','"+a+"','"+text4.
getText()+"','"+s.trim()+ "')";
```

3.3 *Preview display of symbol graphics and code*

Before uploading symbols, in order to enable users to catch a sight, the loaded SVG graphics files are checked. This system provides a preview function of SVG symbols and code. Users can modify the XML code to edit, modify, load SVG files, making the uploading of SVG files more humanized (Liu & Bi 2001). The realization of the symbol graphics and the code previews mainly consists of the following two steps:

1. Preview of SVG symbol graphics. In the process of loading SVG graphics, SVG symbol graphics can be displayed in the interface and read the SVG file path by capturing the SVG file address.
2. Preview of the XML code. In the process of uploading SVG files, the XML documents in SVG files have been gained. Therefore, in order to preview the XML codes, XML codes only need to be displayed in the form of character string. Implementing the preview of XML documents by FileInputStream method, then assigning the XML codes to a TextArea object text2 and displaying it in the interface. The core program is as follows:

```
// generate objects representing input stream
FileInputStream fis = new FileInputStream(d);
// generate a byte array
byte[] buffer = new byte[1024];
// invoke the read method of input stream object
int temp = fis.read(buffer, 0, buffer.length);
String k=new String(buffer);
//assign the XML documents
TextArea text2 = new TextArea(k.trim());
```

As the function interface shown in Figure 1, through the reading, verification, parsing and database insert program mentioned above, it implements the storage of notation style in the symbol library data table MDSYS. SDO_STYLES_TABLE. After connecting Map Builder to Oracle, its display on the list of style resources is shown in Figure 2.

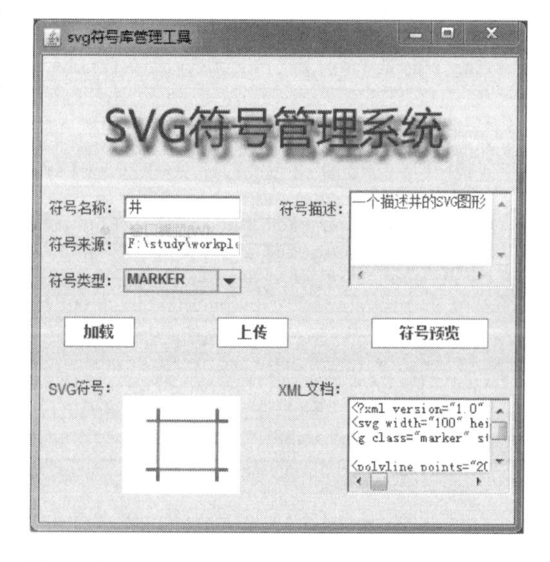

Figure 1. Symbol upload function interface of symbol library system.

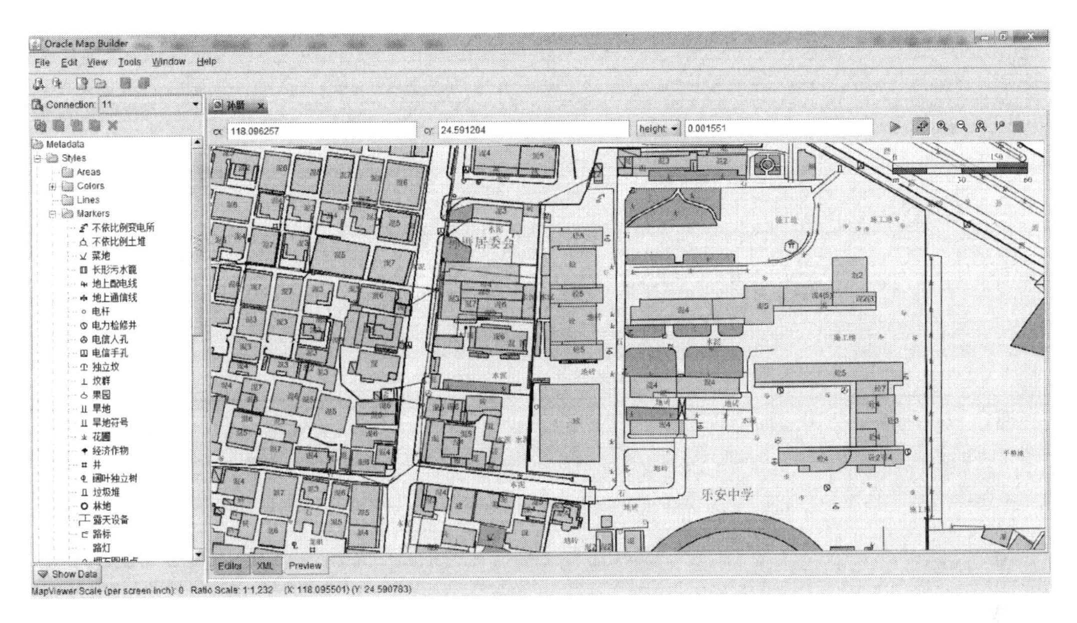

Figure 2. The display effect of symbols within symbol library system in Oracle Map Builder.

4 EXPERIMENTS AND RESULTS

Using Xiamen Qiao Ying Sun Cuo communities New Rural Construction of large scale mapping in ArcGIS as the experimental data, importing the dataset which contains 55 layers into Oracle by Map Builder, the SVG topographic map symbols of Oracle Map Builder picked upload eventually get the result shown in Figure 2. The output is consistent with the ArcGIS maps to verify the availability of the symbol library.

To further test the symbol library portability sharing, following two pairs of dual operation of spatial data and symbol library are imported or exported on the development machine and server:

1 Type "exp sa/sa123 @ orcl file = d: daochu. dmp full = y" in command, export the spatial data tables as *. Dmp file. Notably, where full = y behalf exports the entire database, export data for all users except the users of the system. If the parameter is not specified, then the exported data are incomplete. The basic table information found in the Oracle Map Builder can not be correct in display.
2 Open the Oracle Map Builder. When establishing a connection, selecting "Export Metadata" under the "Tools menu" to export the symbol library to *. Dat file.
3 Type "imp sa/sa123 @ orcl file = d: daochu.dmp full = y" in command, the imported and exported spatial database is collected by the server of the development machine.
4 Select the "Import Metadata" under the selection tool of Oracle Map Builder, and export *. Dat symbol library into the database server.

Based on the spatial database and imported symbol library, the technology of fusion middleware to publish WebGIS in Oracle Map (shown in Figure 3) is used, consistent with the output of ArcMap and display of Map Builder, which verifies the portability of sharing symbol library (Dunfey et al. 2006). At the same time, it also describes the developed symbol library and Oracle provides the ability of collaborative applications for Map Builder and fusion middleware, filling the blank in symbol library management tools for Oracle space expansion tools (Zou et al. 2012.).

5 ACKNOWLEDGMENTS

Funded projects: National Natural Science Foundation of China (No. 41204032), The fourth sub-topics of National Key Technology Research and Development Program of the Ministry of Science and Technology of China (number: 2012BAC16B00), One hundred billion yuan in Guangxi major scientific and technological projects (number: Guikegong 1218017-9D), Guangxi University Scientific Research Projects (number:201103YB151).

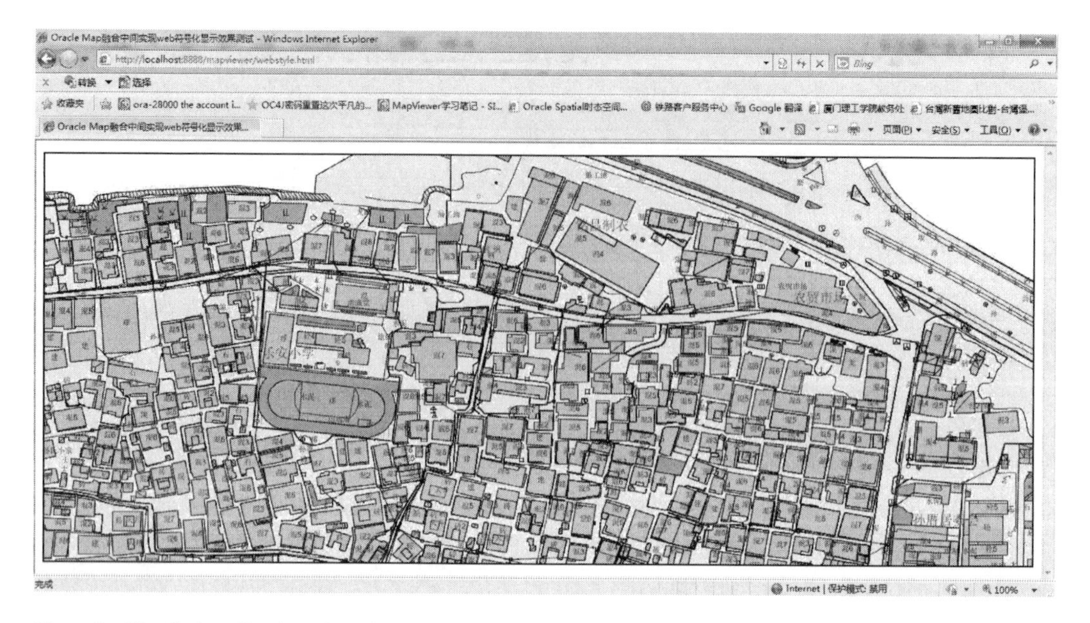

Figure 3. The display effect in WebGIS based on Oracle Map.

REFERENCES

Chang, Y. S. & Park, H. D. 2006. XML Web Service-based Development Model For Internet GIS Applications. *International Journal of Geographical Information Science* 20 (4): 371–399.

Chen, S., Lu, X. & Zhou, C. 2001. *Introduction of GIS*: 28-30. Beijing: Science Publish.

Dang Lina, Dang Gaofeng & Wu Fan. 2011. The Research on Represention and Realization of Map Symbol Based on Text. *Procedia Environmental Sciences* 10: 2342–2347.

Dunfey, R. I., Gittings, B. M. & Batcheller, J. K. 2006. Towards an Open Architecture for Vector GIS. *Computers & Geosciences* 32(10): 1720–1732.

Guan Jihong. 2006. GQL: Extending Xquery to Query GML Documents. *Geo-Spatial Information Science* 9(2): 118–126.

He Yuanrong. 2011. *High Resolution Remote Sensing Monitoring of Mining Area Environment and Its Information Resources Utilization: Methods & Applications*. Changsha: Central South University.

Huang, C. H., Chuang, T. R., Deng, D. P., et al. 2009. Building GML-native web-based geographic information systems. *Computers& Geosciences* 35 (9): 1802–1816.

Lin Hui & Huang Bo. 2001. SQL/SDA: a query language for supporting spatial data analysis and its web-based implementation. *IEEE Transactions on Knowledge and Data Engineering* 13 (4): 671–682.

Liu Xiao & Bi Yongnian. 2001. *The SVG Application Guide Based on XML*. Beijing: China Water & Power Press.

Nass, A., van Gasselt, S., Jaumann, R., et al. 2011. Implementation of Cartographic Symbols for Planetary Mapping in Geographic Information Systems. *Planetary and Space Science* 59(11): 1255-1264.

Peng, Z. R. & Zhang, C. 2004. The Roles of Geography Markup Language (GML), scalable vector graphics (SVG), and Web feature service (WFS) specifications in the development of Internet geographic information systems (GIS). *Journal of Geographical Systems* 6(2): 95–116.

Robinson, A. C., Roth, R. E., Blanford, J., et al. 2011. A Collaborative Process for Developing Map Symbol Standards. *Procedia-Social and Behavioral Sciences* 21: 93–102.

Santosh, K.C., Lamiroy Bart & Wendling Laurent. 2012. Symbol Recognition Using Spatial Relations. *Pattern Recognition Letters* 33(3): 331–341.

Wang Cheng, Li Lin &Yin Zhangcai. 2007. The Area Symbols Design and Implementation Based on SVG. *Science of Surveying and Mapping* (3): 168–145.

Xi Yantao & Wu Jiangguo. 2008. Application of GML and SVG in the development of WebGIS. *Journal of China University of Mining and Technology* 18(1): 140-143.

Yu Zhengwei. 2008. Study on Control SVG File Based on Java. *Computer Programming Skills & Maintenance* (14): 16–18.

Zhai Liang, Li Lin & Tong Xuejuan. 2004. Use of Scalable Vector Graphics in Web Map Issue. *Bulletin of Surveying and Mapping* (5): 38–41.

Zhang Junling, Xiong Weidong & Xia Bin. 2007. Establishment of a Map Symbol Making Tool Based on ArcGIS Engine. *Science of Surveying and Mapping* 32(5): 86-88.

Zou Qiang, Wang Qing & Wang Chengzhong. 2012. Integrated Cartography Technique Based on GIS. *Energy Procedia* 17: 663–670.

Zuo Xiaoqing & Nie Juntang. 2011. Algorithm of Symbol Generation and Configuration of Land Polygons in Present Land-Use Map. *Transactions of Nonferrous Metals Society of China* 21: s743-s747.

Electronics, Communications and Networks IV – Hussain & Ivanovic (eds)
© 2015 Taylor & Francis Group, London, ISBN: 978-1-138-02830-2

A hybrid task scheduling algorithm in cloud computing

Meng Hu* & Yingchun Yuan
College of Information Science and Technology, Agricultural University of Hebei, Baoding, Hebei, China

Boshen Chen
College of Computer and Information, Hefei University of Technology, Hefei, Anhui, China

ABSTRACT: Task scheduling is an essential problem in achieving high efficiency in cloud computing. However, it's a significant challenge to design and implement the efficient scheduling algorithm, since general scheduling problem is NP-complete. In cloud computing, most existing task-scheduling methods, based on clustering of resources features, only consider how to implement resources cluster, rarely further study on how to schedule in each cluster. In order to obtain better performance, we propose a hybrid scheduling algorithm in cloud computing. In this algorithm, firstly, the resource characteristics in the data centers are quantified and normalized based on the fuzzy mathematical theory. These resources will be divided into three clusters. Secondly, in each cluster Min-Min heuristic algorithm is applied for task allocation. Finally, according to the given threshold, the allocation is adjusted for improving the scheduling results. The performance of the algorithm is evaluated using CloudSim toolkit. Experimental result shows that the proposed algorithm has better performance and load balancing.

1 INTRODUCTION

Cloud computing (Foster et a1. 2008) is emerging as a new commercial computing model (Daniel et al. 2009). It is a fusion and further development of distributed computing, parallel computing (Michael et al. 2010), grid computing and other traditional computer technology and network technology. Cloud computing is an on-demand payment model, which can provide useful, convenient and on-demand network access. So users can quickly obtain the shared pool configurable resources (including computing resources, storage resources and network sources) with minimal management effort or very little interaction with the service providers (Rochwerger et a1. 2009).

Task scheduling is a key technology in a cloud computing environment (Chen & Deng.2009), where resource type is various, user requirements are different and tasks numbers are large scale. Therefore, task scheduling algorithm has become an important factor influencing the efficiency of cloud computing. Recently, different models and algorithms (Zuo & Cao.2012, Zhang 2009) are proposed to deal with task scheduling in a cloud computing environment. Most of these methods focus on all the resources in the datacenter. However, with large amount of tasks, overhead time of tasks, choosing resources is increasing, leading to low scheduling efficiency. So it is very much necessary to find a proper way to reasonably divide resources into groups so as to narrow the scope of the optional resource. Clustering is an effective means of classifying targets. Similar work has been done in heterogeneous computing systems. For example, Du et al. (2006) use the fuzzy clustering theory to analyze heterogeneous resource characteristics in grid environment. They present a heuristic scheduling algorithm according to their analysis result. Aiming at independent task scheduling in grid issues, Li et al. (2008) put forward a hybrid clustering method based on both tasks and resources. Yang et al. (2009) implement the task mapping and scheduling of the grid environment with the aid of hypergraph clustering. Above algorithms are mostly for grid environment, but for a cloud computing platform is relatively few. Guo et al. (2013) propose a workflow task scheduling algorithm based on resource clustering in a cloud computing environment. The algorithm, firstly, according to the priority of tasks, constructs the scheduling order list. And then apply the fuzzy clustering to dividing resources into two collections: selection collection and spare collection. According to the task completion time, assign task to resource. The experimental results show that the algorithm improves the performance of workflow task scheduling. Li et al. (2012) put forward a two-level task scheduling algorithm based on the principle of resources equitable distribution. This algorithm,

*Corresponding Author: *humengok@sina.com*

firstly, implements the overall scheduling with the unit of the user, choosing a cloud service provider according to the user overall demand. Then, according to the task preference coefficient of the user submits, choose the resource in the corresponding clustering. The algorithm ensures the overall system throughput while maintaining a high customer satisfaction.

A few of them focus on the scheduling in each cluster after resources clustered. Furthermore, few optimize the utilization of each cluster. Aiming at this shortcoming, a hybrid scheduling algorithm is proposed. The innovation of this paper lies in: divide resources into three groups on the basis of the fuzzy clustering, apply min-min algorithm to each cluster and adjust allocation according to the given threshold, in order to reflect the fairness of task scheduling to the greatest extent and improve the system resource utilization ratio.

The remaining parts of this paper are organized as follows: Section II presents the related works about task model and resource model in the cloud computing. In Section III, a hybrid cloud task scheduling algorithm we proposed is introduced. The implementation and experiments are given in Section IV. Finally, Section V concludes the paper and presents future works.

2 PROBLEM DESCRIPTION

In a cloud computing environment, there have lots of computing resources, network resources, storage resources and other resources. Different tasks have different preferences for these resources. Some computational tasks prefer the high-performance computing resources, and some network interaction tasks prefer adequate network resources to high-performance resources. In order to utilize the power of cloud computing completely, we need an effective and efficient task scheduling algorithm. Task scheduling algorithm is responsible for dispatching tasks submitted by users to cloud provider onto heterogeneous available resources. Therefore, this paper focuses on the efficient tasks scheduling considering the total completion time of tasks, average response time and resource utilization in the cloud environment.

2.1 *Task model*

In a cloud computing environment, there have two categories about the task: independent tasks and related tasks. This paper only considers independent tasks, laying the foundation for later continue to study related task. Assuming we have a set of m tasks ($Cloudlet_1$, $Cloudlet_2$, …, $Cloudlet_m$),they need to be scheduled onto n available virtual machine resources(Vm_1, Vm_2,…,Vm_n). Cloudlet features can be described with one-dimensional vector as follows: Cloudlet = $\{cloudlet_{id,} cloudlet_{userid,} cloudlelength, cloudletPEs, cloudlet_{bw,} cloudlet_{stor,} cloudlet_{ifile,} cloudlet_{ofile}\}$.

Where $cloudlet_{id}$ is task indentifier; $cloudlet_{userid}$ is user indentifier; $cloudlet_{length}$ is a task length (unit:MI,one million instructions);$cloudlet_{PEs,}$ $cloudlet_{bw}$and $cloudlet_{stor}$ respectively denote user's desired number of PE, network bandwidth (unit: MB/s) and the storage size (unit:MB); $cloudlet_{ifile}$ and *cloudletofile* respectively denote the input and output file size.

The Cloudlet computing requirement is expressed as: $cloudlet_{comp} = cloudlet_{length} / cloudlet_{PEs}$

2.2 *Resource model*

In a cloud computing environment, resources are almost virtual resources through virtualization technology presented to the users. Suppose that a resource collection contains n virtual machine resources and the n virtual machine resources have different performance. Resources characteristics of the virtual machine can be described to one-dimensional vector $Vm = \{Vm_{id,} Vm_{userId,} Vm_{CPU,} Vm_{MIPS,} Vm_{bw,} Vm_{stor}\}$.

Where the Vm_{id} is resource indentifier; *VmuserId* is the cloud provider identifier; Vm_{CPU} is the number of CPU; Vm_{MIPS} is the number of executing one million instructions per second; Vm_{bw} and Vm_{stor} respectively denote resources of communication ability and storage capacity.

The Vm computing capacity is expressed as:

$$Vm_{comp} = Vm_{cpu} \times Vm_{MIPS}$$

The greater the value is, the better the performance of the computing resources is.

The comprehensive performance of resource j is calculated by formula (1) .

$$\mathrm{V}m_{pj} = \sqrt{\frac{a(Vm_{comp})^2 + b(Vm_{bw})^2 + c(Vm_{stor})^2}{a+b+c}} \tag{1}$$

Where a, b and c respectively represent as empirical coefficient of three resource performance parameters.

The comprehensive performance of every cluster is obtained from average comprehensive performance of all resources in one cluster:

$$c_{r_{pj}} = \frac{1}{n}\sum_{j=1}^{n} Vm_{pj} \tag{2}$$

3 HYBRID SCHEDULING ALGORITHM

Massive tasks and resources lead to blindly select resources. The blindness influences algorithm execution time. Therefore, we put forward the hybrid task scheduling algorithm. This algorithm is implemented by three phases: resources clustering, task allocation and algorithm improvement.

3.1 *Resource clustering phase*

In cloud computing environment, on the one hand, it is difficult to describe tasks demand for resources; on the other hand, due to the dynamic characteristics of a cloud computing environment, it is difficult to accurately describe the attributes of resources themselves. Therefore, the properly matching process is also fuzzy. This article uses the fuzzy c-means clustering method (FCM) (Qu et al 2009), according to the resource characteristics of multidimensional fuzzy classification. The main steps are as follows:

Step 1: According to n resource attribute values $Vmcomp$, Vm_{bw} and Vm_{stor}, establish an initialization sample matrix $R_{n\times 3}$.
Step 2: According to the formula (3), standardize the matrix $R_{n\times 3}$ and get r'_{ij}. Then according to the formula (4), compress the values to [0, 1] and get r'_{ij}.

$$r'_{ij}=\frac{r_{ij}-\overline{r_j}}{S_j} \tag{3}$$

$$\overline{r_j}=\frac{1}{n}\sum_{j=1}^{n} r_{ij}, S_j=\sqrt{\frac{1}{n}\sum_{i=1}^{n}(r_{ij}-\overline{r_j})^2}$$

Where $\overline{r_j}$ refers to an average values of the resource attribute in the *jth* dimension performance; S_j stands for the standard deviation of the resource attribute in the *jth* dimension performance.

$$r''_{ij}=\frac{r'_{ij}-r'_{j\min}}{r'_{j\max}-r'_{j\min}} \tag{4}$$

Where $r'_{j\min}$ and $r'_{j\max}$ denote the min value and the max value among $r'_{1j}, r'_{2j}, \ldots, r'_{nj}$.
Step 3: According to the performance, resources are divided into three categories, namely three clustering centers. Initialize the membership degree matrix U.
Step 4: According to the formula (5), calculate clustering center c_k.

$$c_k=\frac{\sum_{i=1}^{n} u_{ki}^{m} x_i}{\sum_{i=1}^{n} u_{ki}^{m}}, k=\{1,2,3\} \tag{5}$$

Step 5: According to the formula (6), update matrix U.

$$u_{ki}=\frac{1}{\sum_{p=1}^{c}(\frac{d_{ki}}{d_{pi}})^{2/(m-1)}} \tag{6}$$

Where u_{ki} is the degree of membership of x_i in the cluster k; c_k is clustering center of the fuzzy set of k; $d_{ki}=||c_k - x_i||$, x_i is the *ith* measured data, c_k is the *kth* center of the cluster, and ||*|| is any norm expressing the similarity between any measured data and the center; m is an arbitrary real number greater than 1, the best value range for [1.5, 2.5], generally take $m = 2$.
Step 6: Calculate the value function

$$J(U,c_1,c_2,\ldots,c_c)=\sum_{k=1}^{c} J_k=\sum_{k=1}^{c}\sum_{i=1}^{n} u_{ki}^{m} d_{ki}^{2}$$

If J <θ, or up to the max iteration, or $||U^{(k+1)} - U^{(k)}|| < \varepsilon$, then STOP; otherwise return to step 4 and start the next iteration.

After the fuzzy clustering algorithm, finally n resources are divided into three clusters: computational resource set (CompCluster), the bandwidth resource set (BwCluster) and storage resource set (StroCluster).

3.2 *Task allocation based on Min-Min*

According to the task preference, tasks are assigned to the three clusters generated in the A section. Specific algorithm steps are as follows:

a. For every cloudlet in the cloudletList, according to formula (7), get task preference value tqi and then add it to the corresponding resources queue: computational queue CompCloudletList, bandwidth queue BwCloudletList or storage queue StorCloudletList.

$$t_{qi}=\sqrt{\frac{\alpha(cloudlet_{comp_i})^2+\beta(cloudlet_{bw_i})^2+\gamma(cloudlet_{stor_i})^2}{\alpha+\beta+\gamma}} \tag{7}$$

Where $a,\beta,\gamma\in[0,1]$ respectively represent as empirical coefficient of task preference.

b. According to the Min-Min algorithm, CompCloudletList, BwCloudletList and StorCloudletList are respectively allocated to CompCluster, BwCluster and StorCluster. Specific algorithm steps are as follows:

Step 1: If a task set is not empty go to the step 2, otherwise jump to step 8.
Step 2: For all tasks, calculate the time is expected to finish mapped to the resource set.
Step 3: For all tasks, they are mapped to all available resources on the earliest completion time of $ECT_{ij} = ETC_{ij} + rt_j$. rt_j represents for ready time of the j machine.
Step 4: According to the result of step 3, find out the task t_i and corresponding resource r_j which the earliest time is the smallest.
Step 5: Map task t_i to r_j resources; And remove the task from the task set.

Step 6: Update expectation ready time *(rt_j)* of resources r_j.
Step 7: Update the earliest completion time of other tasks in the resource r_j; Return to step 1.
Step 8: Stop.

3.3 *Algorithm performance improvements*

Min-Min algorithm fails to utilize the resources efficiently, which lead to a load imbalance. In order to avoid the drawback of load imbalance, an adjustment is proposed in this paper to optimize the load balance of Min-Min. It aims to increase the utilization of resources with light load or idle resources in the cluster, which has the minimal completion time, thereby freeing the resources with heavy load in the cluster which has the maximal completion time. Specific steps are as follows:

Step 1: Calculate the makespan of each cluster separately, and the difference D between the max value in the cluster *ii* and the min value in the cluster *jj* for the completion time. Record the maximum and minimum value D_{max} *and* D_{min} respectively.
Step 2: If $D_{max} < \theta$ (a given threshold) , algorithm go to Step 4, or go to Step 3.
Step 3: Find $Cloudlet_i$ that cost minimum execution time on the heavy load resource Vm_j in the cluster *ii*; Find the minimum completion time of $Cloudlet_i$ produced by resource Vm_k in the cluster *jj*.
If such minimum completion time < makespan, reassign task *Cloudleti* to resource *Vmk* and then update the ready time of both *Vmj* and *Vmk*.
Step 4: End the task scheduling and exit the program.

4 EXPERIMENT ANALYSIS

4.1 *Experimental environment*

In accordance with the features of the cloud computing, this algorithm implements on simulation platform—CloudSim. CloudSim(Buyya et al 2009) is a toolkit of simulation announced by the grid laboratory of University of Melbourne. Simulation experiment is respectively combined the Round Robin, Min-Min and improved algorithm based on fuzzy clustering, and using time span(makespan), average response time(RT) and average resource utilization ratio(ARUR) three indicators to evaluate the algorithm.

In the simulation experiments, the tasks and virtual machine resources are respectively randomly generated by CloudletGenerator and VmGenerator. Each experiment, you can specify the scope of the task number and each task attribute values range, as well as the number and the performance value range of resources. The specific experimental environment settings are as follows:

a. The parameter settings of cloud datacenter: cloud system with 10 datacenters, 4 hosts in each datacenter. Each host configuration is as follows: randomly assigned to PE among 1, 2 and 4; processor rate 2000 MIPS; 4 GB memory, disk 1 TB and bandwidth 10 GB/s; datacenter features: the system architecture for x86, an operating system for Linux, the virtual machine for Xen.
b. The parameter settings of tasks: according to customer's requirement to create a cloud task, task length distribution in the interval [500, 4000], communication capacity requirements for [1000, 4000], storage capacity requirements for [500, 3000].
c. The parameter settings of virtual machines: virtual machine number of PE range for {1, 2, 4}, the processor speed for [500, 1000], communication ability for [500, 3000], storage capacity for [512, 2048].

4.2 *Experiment and result analysis*

In this experiment, assume that tasks number m = {100,110,120,130,140,150}, the scope of resources number n is [15, 25]. *CloudletList* and *vmList* are respectively generated by CloudletGenerator and VmGenerator. The comparison is a Round Robin (FCMRR), Min-Min (FCMMM) and improved algorithm based on clustering. From makespan, the average response time and system resource utilization three aspects we compare three algorithms. Comparison results is as shown in Figure 1, Figure 2, Figure 3, where the horizontal axis shows the cloud task number, the vertical axis respectively shows the makespan (unit: s), the average response time (unit: s) and the system resource utilization.

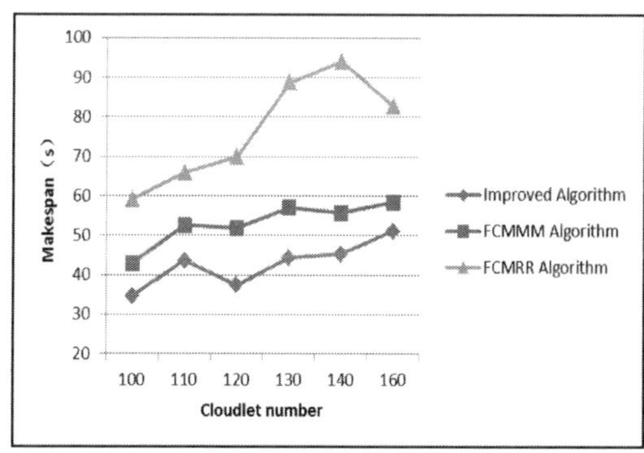

Figure 1. Comparison of makespan.

From Figure 1, we can see the improved algorithm at the completion time, significantly lower than the FCMRR and FCMMM algorithm. Because the improved algorithm is combined with Min-Min, and Min-Min scheduling goal is based on the earliest completion time of task. So task completion time is smaller than the overall FCMRR algorithm. Furthermore, improved algorithms for clustering is adjusted according to the threshold of completion time. Compared with FCMMM algorithm, the completion time is further reduced.

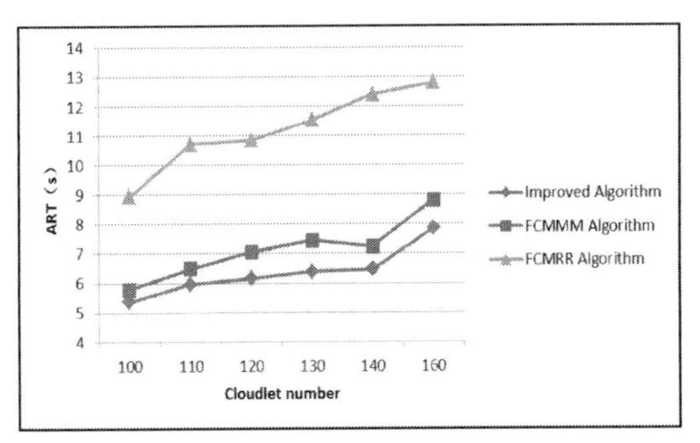

Figure 2. Comparison of average response time.

As can be seen from Figure 2, the improved algorithm makes the average response time of all tasks shortened significantly. For example, when the task number m changed from 110 to 120, average response time of improved algorithm is increased by 3.17%, FCMRR and FCMMM are increased by 6.48% and 6.48% respectively. It can be seen that the average response time of improved algorithm is growing slowly, relatively stable.

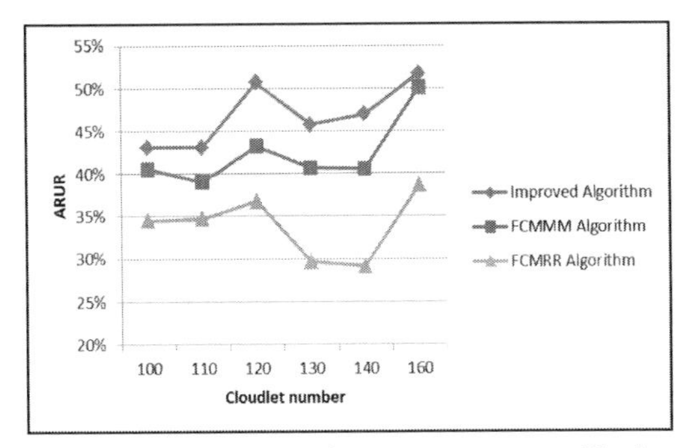

Figure 3. Comparison of average resources utilization ratio.

As can be seen from Figure 3, FCMMM algorithm system resource utilization is below 40%, due to the unbalanced load of traditional Min-Min. When the max completion time difference Dmax between two clusters is greater than threshold value of adjustment, apply the adjustment method. Thus the system resource utilization is over 40%, slightly higher than the FCMRR algorithm, significantly higher than the FCMMM algorithm.

The above analysis shows that in the cloud computing model, the improved algorithm is not only significantly higher execution efficiency, but also can shorten the makespan and the ART of the task collection, to a certain extent. Furthermore, it can improve the ARUR and keep the system load balance.

5 CONCLUSIONS

On the basis of the independent task scheduling algorithm in cloud computing, summarize the characteristics of the traditional scheduling algorithm. Through deeply analyzing the characteristics of resources in cloud data center, a hybrid cloud task algorithm is proposed. Firstly the resource characteristics are quantified and normalized, and then carry on the fuzzy clustering based on these characteristics, significantly reducing the scope of the mission selection resources and thus effectively reducing the execution time of the algorithm. Finally, the experimental analysis shows that, on the one hand, this algorithm effectively shorten completion time and the average response time. On the other hand, from the angle of the system, it improved the average resource utilization ratio of the system, and kept the load balance. The further research will consider dynamic factors of resources and tasks, in order to adapt to the dynamic cloud computing environment.

ACKNOWLEDGMENTS

This work was supported by Natural Science Foundation of Hebei Province of China under Grant No. F201204089. In addition, our research is also supported by the Science and Technology Research Foundation of Agricultural University of Hebei in China under Grants No.LG201407 and No.LG20140703.

REFERENCES

Buyya R,Ranjan R & Calheiros R N. 2009. Modeling and simulation of scalable cloud computing environments and the CloudSim Toolkit: challenges and opportunities.//Proceedings of the 7th High Performance Computing and Simulation Conference.New York,USA: IEEE Press.

Chen, Q & Deng, Q.N. 2009. Cloud computing and its key technology. Journal of computer applications. 29(9): 2572–2567.

Chen, Z.G & Yang, B. 2009. Task Scheduling Based on Multidimensional Performance Clustering of Grid Service Resources. Journal of Software. 20(10):2766–2774.

Daniel Nurmi, Rich Wolski, Chris Grzegorczyk. 2009.The Eucalyptus Open Source Cloud Computing System. Proceeding of the Cluster Computing and the Grid. California:University of California.

Du, X.L. Jiang, C.J. Xu, G.R. et al. 2006. A Grid DAG Scheduling Algorithm Based on Fuzzy Clustering. Journal of Software.17(11): 2277~2288.

Foster I, et al. 2008. Cloud Computing and Grid Computing 360degree Compared. Proceedings of the 2008 Grid Computing Environments Workshop. Washington DC: IEEE Computer Society.

Guo, F.Y. Long Y. Tian, S.W. et al. 2013. Workflow task scheduling algorithm based on resource clustering in cloud computing environment. Journal of computer applications, 33(8):2154–2157.

Li, F.F. Qi, D.Y. Zhang, X.G. et al. 2008. Grid independent task scheduling algorithm based on fuzzy clustering, Computer engineering. 34(22):19–21.

Li, W.J. Zhang, Q.F. Ping, L.D. et al. 2012. Cloud scheduling algorithm based on fuzzy clustering. Journal on communications.33(3):146–154.

Michael A, Armando F, Rean G, et al. 2010. A view of cloud computing *J*. Communications of the ACM,53 (4), 50–58.

Qu, F.H. Cui, G.C. Li, Y.F. et al. 2011 .Fuzzy clustering algorithm and its application . Defense Industry Press.50–58.

Rochwerger, et al. 2009. The Reservoir Model and Architecture for Open Federated Cloud Computing. IBM Journal of Research and Development, 53(4):1–17.

Zhang, C. Y. 2009. Research and implementation of job scheduling algorithm in cloud computing. Beijing: Beijing Jiaotong University.

Zuo, L.Y & Cao, Z. B. 2012. Review of scheduling research in cloud computing. Application Research of Computers. 29(11): 4023–4027.

Electronics, Communications and Networks IV – Hussain & Ivanovic (eds)
 ISBN: 978-1-138-02830-2

Credibility test for frequency estimation of sinusoid using F-test

Guobing Hu, Shanshan Wu*, Yan Gao & Ning Ding
School of Electronic information Engineering, Nanjing college of information technology, Nanjing, Jiangsu, China
College of Computer and Information Science, Ho Hai University, Nanjing, Jiangsu, China

ABSTRACT: Estimation of sinusoid frequency is a key research problem related to radar, sonar, and communication systems. The results of numerous investigations on frequency estimation have been reported in the literature. Nevertheless, to the best of our knowledge, none of them have dealt with credibility evaluation, which is used to decide whether an individual frequency estimation of the sinusoid is accurate. In order to assess the credibility of a frequency estimation of the sinusoid, this paper proposes a method based on a lack-of-fit F-test for linear regression of the correlations between the received signal and the reference signal generated according to the frequency estimation. Simulations show that the proposed method performs well even at low signal-to-noise ratios.Keywords: frequency estimation, credibility evaluation, F-test.

1 INTRODUCTION

Frequency estimation of a sinusoid signal is a critical problem concerning applications related to commercial and military signal processing systems. For some specific methods, estimation of frequency is the precondition for the estimation of the other parameters of sinusoid signals (Rife & Boorstyn 1974) as well as frequency estimation of modulated signals (Peleg 1991, Ghogho et al. 2000). Some algorithms proposed for estimation of frequency from received signals are provided in the literature (Rife & Boorstyn 1974, Jain 1979, Dash & Hasan 2011). Generally, performance evaluation of frequency estimation algorithms is considered from two points of view. Algorithm designers focus on the overall statistical performance that can be evaluated by comparing the mean square error with the Cramér–Rao lower bound (CRLB). On the other hand, users consider the credibility of individual frequency estimation important. In the noncooperative context, especially at low signal-to-noise ratios (SNR), it is important to decide whether an individual frequency estimation is accurate when the frequency is unknown at the receiver side.

Recently, investigations conducted have focused on the confidence evaluation of the blind modulation recognition result aimed at enhancing the reliability of the overall processing units and decreasing the wastage of both software and hardware resources. It has been reported in (Su et al. 2006) that the confidence measurement of modulation recognition becomes the key output information of the signal processing system in military applications that can be used to identify unknown radar signals. According to literature (Pucker 2009), the modulation recognition confidence rating is regarded as additional output information in some civilian signal processing devices such as Agilent's option MR1 for E3238S signals detection and monitoring systems. Fehske (Fehske et al. 2005) defined the half value of the maximum minus the second maximum output of the back propagation (BP)-based classifier as the confidence metric for modulation classifying results in CR. Lin (Lin & Liu 2008) proposed a confidence measurement method based on the information entropy to measure the confidence of modulation recognition results in single-input and single-output (SISO) and multiple-input multiple-output (MIMO) channels for CR. Still, credibility evaluation of blind frequency estimation remains a research problem that has not been adequately addressed in the literature.

The method proposed in this paper is aimed at automatically deciding the credibility of an individual frequency estimation of a sinusoid without any priori knowledge about the parameters of the received signal. Section 2 of this article presents the signal model and the hypothesis test for credibility evaluation. In Section 3, the statistic is defined by calculating the magnitude of the correlation function between the received signal and the reference signal generated according to the frequency estimation, and then a decision rule for credibility evaluation of the frequency estimation of a sinusoid based on a lack-of-fit test for regression is presented. Section 4 reports the simulation results. Finally, Section 5 gives the conclution.

*Corresponding author: *wuss@njcit.cn*

2 SIGNAL MODELS AND BASIC ASSUMPTIONS

2.1 *Signal model*

A complex sinusoid contaminated by noises can be described by the following signal model:

$$\begin{aligned} x(n) &= s(n) + w(n) \\ &= A\exp[j(2\pi f_0 n\Delta t + \theta)] + w(n),\ 0 \le n \le N-1 \end{aligned} \tag{1}$$

Where A, f_0, and θ denote the amplitude, carrier frequency, and initial phase of the sinusoid signal $s(n)$, respectively. Δt is the discrete sampling interval, and N corresponds to the length of the samples. The additive noise $w(n)$ is supposed to be a white complex Gaussian process with a zero mean and variance σ^2 whose real part and imaginary part are independent of each other.

2.2 *Hypothesis model for credibility evaluation*

In the noncooperative environment, the modulation format and the parameters of the signal are both unknown at the receiver side. For a certain processing cycle, credibility evaluation of the frequency estimator is aimed at detecting whether the individual frequency estimation is accurate. In practice, for the widely used fast Fourier transform (FFT)-based estimators, if the signal-to-noise ratio (SNR) is greater than the moderate threshold, the maximum absolute bias of the frequency estimation ($|\Delta f|$) is less than a quarter of the discrete sampling frequency interval (ΔF) (Hu et al. 2013). Therefore, the credibility assessment is described as the following hypothesis test:

$$\begin{cases} H_0 : |\Delta f| \le 0.25\Delta F \\ H_1 : |\Delta f| > 0.25\Delta F \end{cases} \tag{2}$$

3 STATISTIC SELECTION AND ANALYSIS

3.1 *Feature analysis*

Assuming that the observed signal is a single-tone sinusoid, the reference signal can be constructed by the sinusoid model as follows:

$$y(n) = \exp(-j2\pi f_1 n\Delta t), 0 \le n \le N-1 \tag{3}$$

where f_1 is estimated by the maximum likelihood (ML) method or by other suboptimal estimators. The correlation function between the observed signal $x(n)$ and the reference signal $y(n)$ can be expressed as

$$z(n) = \sum_{m=0}^{n} [s(m) + w(m)]y(m) = z_s(n) + z_w(n) \tag{4}$$

where $z_s(n)$ and $z_w(n)$ are the signal part and noise part of $z(n)$, respectively. Consequently, when $\Delta f \Delta t$ is small, the signal part $z_s(n)$ can be further derived as

$$z_s(n) \approx a_s(n) e^{j\beta_s(n)} \tag{5}$$

Where $a_s(n) \approx A(n+1)|sinc[\Delta f \Delta t(n+1)]|$ denotes the magnitude of $z_s(n)$, $sinc(x)=sin\pi x/\pi x$ is the sinc function, and $\beta_s(n)= \pi n\Delta f\Delta t +\theta$ is the phase of $z_s(n)$.

Letting $\delta=\Delta f/\Delta F$, the magnitude $a_s(n)$ can be rewritten as

$$a_s(n) = A(n+1)\,|\,\text{sinc}[\delta(n+1)/N]\,| \tag{6}$$

where δ is the factor of the frequency estimation error.

According to (Rife & Boorstyn 1974), if $0 \le \pi\Delta f\Delta t(n+1) \le \pi/2$, Equation (6) can be further written as

$$\begin{aligned} a_s(n) &\approx A(n+1) - 1.6388A\delta^2(n+1)^3/N^2 \\ &= a_{sl}(n) + a_{sc}(n), \end{aligned} \tag{7}$$

where $a_{sl}(n)$ and $a_{sc}(n)$ denote the linear part and the nonlinear part of $a_s(n)$, respectively.

With moderate SNR conditions, for the appropriate frequency estimator, the absolute estimation error is always less than the discrete sampling frequency interval ΔF, i.e., $0 \le \delta \le 1$. Under hypothesis H_0, assuming that the absolute error is relatively small, i.e., $|\Delta f| \le 0.25\Delta F$ and $\delta \le 0.25$, it follows that

$$a_s(n)\big|_{\delta=0.25} \le a_s(n) \le a_s(n)\big|_{\delta=0} \tag{8}$$

where

$$a_s(n)\big|_{\delta=0} \approx A(n+1) \tag{9}$$

and

$$a_s(n)\big|_{\delta=0.25} \approx A(n+1)\left[1 - 0.1024\left(\frac{n+1}{N}\right)^2\right] \tag{10}$$

From Equation (9), we see that if the estimation error factor equals to zero, then $a_s(n)$ is purely linear. In a similar manner, if the absolute frequency estimation equals to $0.25\Delta F$, i.e., $\delta=0.25$, then the maximum value of the second term of Equation (10) is less than 0.1024, $a_s(n)$ is determined mainly by the linear part, as can be seen from Equation (10). Therefore, with $\delta \le 0.25$, the main part of $a_s(n)$ is the linear part $a_{sl}(n)$ and the nonlinear part $a_{sc}(n)$ is relatively small, implying that $a_s(n)$ can be approximated as a line in the entire range of $0 \le n \le N-1$.

Under hypothesis H_1, because of the low SNR or other possible reasons, the maximum magnitude of the observed signals' spectrums will not be located in the correct point or the discrete frequency interval and the estimation error Δf is larger than $\Delta F/4$, i.e., $\delta \geq 0.25$. For instance, if $\delta=0.5$, it follows from Equation (7) that

$$a_s(n)|_{\delta=0.5} \approx A(n+1) - 0.4097A\frac{(n+1)^3}{N^2} \tag{11}$$

The derivative of Equation (11) can be given by

$$\frac{\partial a_s(n)}{\partial n} \approx A - \frac{1.2291A(n+1)^2}{N^2} \tag{12}$$

When $0 \leq n \leq N/4$, it follows that

$$\frac{\partial a_s(n)}{\partial n} \geq \frac{\partial a_s(n)}{\partial n}|_{n=N/4} = 0.925A \tag{13}$$

From Equation (12) and (13), it can be seen that the slope of $a_s(n)$ decreases slightly when $0 \leq n \leq N/4$. Therefore, it can be approximately regarded as a straight line. When n decreases, the nonlinear part of Equation (11) increases.

For $n=N-1$, it follows that

$$\frac{\partial a_s(n)}{\partial n}|_{n=N-1} \approx 0.29A \tag{14}$$

Hence, the slope of $a_s(n)$ decreases with the increase of n and it is predominantly determined by the nonlinear part when n is large.

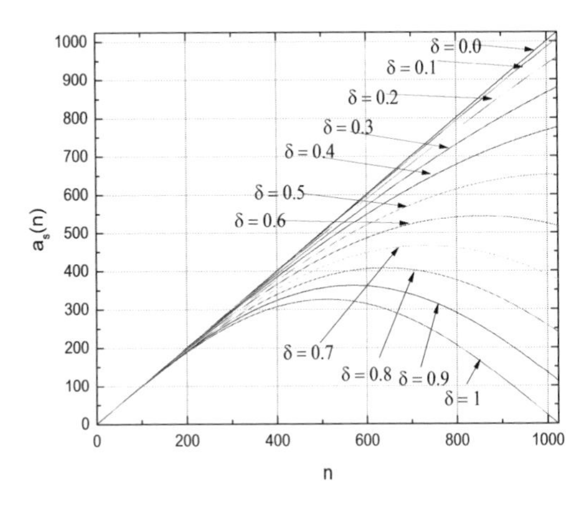

Figure 1. Relationship between $a_s(n)$ and δ.

The relationship between δ and $a_s(n)$ is shown in Figure 1. For $\delta \leq 0.25$, $a_s(n)$ can be regarded as a linear function associated with n and its slope decreases slightly with increasing δ. For $\delta > 0.25$, $a_s(n)$ decreases dramatically with increasing δ. Hence, we can infer that the absolute error of the certain frequency estimation depends on the shape of the wave of $a_s(n)$ and obtains the credibility metric of the frequency estimation.

Next, we consider the effect of the noise on the property of $z_s(n)$ mentioned above.

Recalling Equation (4), the noise part $z_w(n)$ can be rewritten in polar form as $z_w(n)=a_w(n)e^{j\beta w(n)}$, where $a_w(n)$ and $\beta_w(n)$ are the modulus and the angle of $z_w(n)$, respectively. As both the real and imaginary parts of $z_w(n)$ are normally distributed, those of $a_w(n)$ are Raleigh distributed, and those of $\beta_w(n)$ are uniformly distributed in the range of $[0,2\pi)$. From Equation (4) and (5), we arrive at

$$\begin{aligned} z(n) &\approx a_s(n)e^{j\beta_s(n)} + a_w(n)e^{j\beta_w(n)} \\ &= a_s(n)e^{j[\beta_s(n)+\gamma(n)]}\sqrt{\left(1+\frac{a_w(n)}{a_s(n)}\cos\varphi_{sw}(n)\right)^2 + \left(\frac{a_w(n)}{a_s(n)}\sin\varphi_{sw}(n)\right)^2} \end{aligned} \tag{15}$$

where $\varphi_{sw}(n)=\beta_w(n)-\beta_s(n)$ is uniformly distributed in the range of $[0,2\pi)$ and

$$\gamma(n) = \arctan\frac{a_w(n)/a_s(n)\sin[\varphi_{sw}(n)]}{1+a_w(n)/a_s(n)\cos[\varphi_{sw}(n)]}$$

Furthermore, the magnitude of $z(n)$ can be represented by

$$|z(n)| \approx a_s(n)\sqrt{\left(1+\frac{a_w(n)}{a_s(n)}\cos\varphi_{sw}(n)\right)^2 + \left(\frac{a_w(n)}{a_s(n)}\sin\varphi_{sw}(n)\right)^2} \tag{16}$$

Letting

$$\frac{E[a_w^2(n)]}{a_s^2(n)} = \frac{1}{(n+1)SNR[1-0.16605\pi^2\delta^2(n+1)/N^2]^2} \tag{17}$$

and considering that the SNR is moderate, n is relatively large (usually supposing $n \geq 0.3N$) and $\delta \leq 1$, it's obtained that $E[a_w^2(n)]/(a_s^2(n)) << 1$ from the above equations. Therefore, Equation (16) can be approximated as

$$|z(n)| \approx a_s(n)\sqrt{1+2\frac{a_w(n)}{a_s(n)}\cos\varphi_{sw}(n)} \tag{18}$$

Using the Taylor expansion of Equation (18), it can be approximately expressed as

$$|z(n)| \approx a_s(n)\left[1+\frac{a_w(n)}{a_s(n)}\cos\varphi_{sw}(n)\right] = a_s(n) + a_w(n)\cos\varphi_{sw}(n) \tag{19}$$

Therefore, the magnitude of the correlation function $|z(n)|$ can be regarded as a deterministic component $a_s(n)$ contaminated by a noise part $a_w(n)cos\varphi_{sw}(n)$. Under H_0 assumption, the term $a_s(n) \approx A(n+1)$, $|z(n)|$ can be regarded as a straight line contaminated by noise. Under H_1 hypothesis, for the increasing frequency estimation factor δ, the term $a_s(n)$ is a cubic function associated with n and $|z(n)|$ can be regarded as a nonlinear function contaminated by noise when n is relatively large.

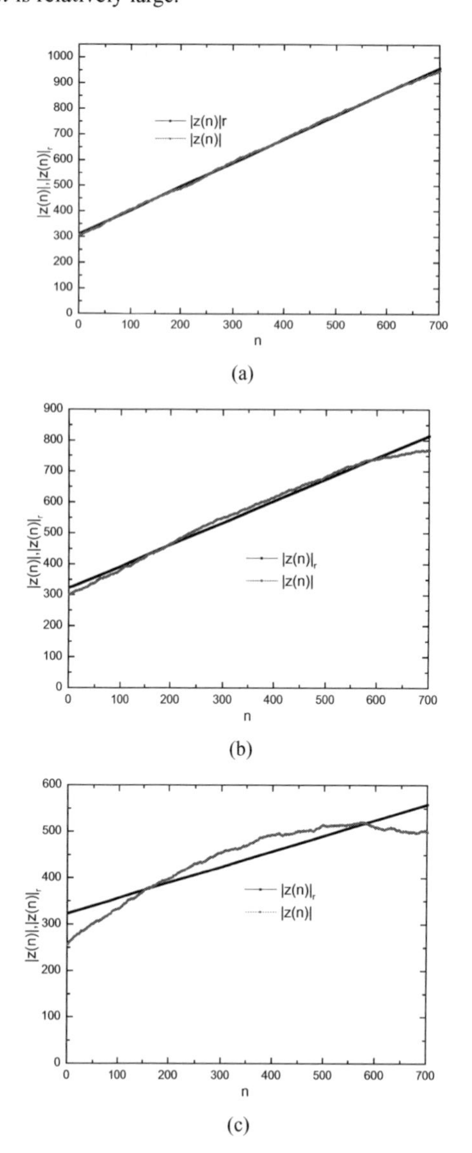

Figure 2. Magnitude of the correlation function $|z(n)|$ and its estimation by simple linear regression under different values of δ when an SNR = 0 dB.
(a) $\delta=0.2$ (b) $\delta=0.4$ and (c) $\delta=0.5$

Figure 2 (a) shows that $|z(n)|$ can be an approximation of a straight line with an identical slope under H_0 assumption. Thus, $|z(n)|$ can be fitted well by a simple linear regression and the regressive error is small. In contrast, under H_1 assumption, as shown in Figure 2 (b) and (c) that each $|z(n)|$ with the frequency estimation error factor of 0.4 and 0.5 cannot be adequately fitted by the linear regression because it is an approximation of a cubic function. Moreover, the degree of the lack of fit is increased by the increase of the frequency estimation error factor.

3.2 *Decision rule and the threshold*

In brief, the credibility test for an individual estimation of a sinusoid signal can be transformed to a lack-of-fit test for the regression of the magnitude of the correlation function $|z(n)|$.

Given the paired observations $(x_i,y_i), i=1,\ldots,M$, the simple linear regression model is represented by (Rife & Boorstyn 1974)

$$y_i = b_1 x_i + b_0 + e_i, i = 1,\ldots,M \tag{20}$$

where e_i is referred to a residual error, b_1 and b_0 are regression coefficients. Using the least-square method to estimate the above two parameters b_1 and b_0, the following regression equation is obtained:

$$y_i^{'} = \hat{b}_1 x_i + \hat{b}_0, i = 1,\ldots,M \tag{21}$$

By defining $p_i=y_i-y_i^{'}$, $i=1,\ldots,M$ as the residual error at xi, the sum of squares for error (SSE) can be expressed by

$$SSE = \sum_{i=1}^{N} p_i^2 \tag{22}$$

A test for lack of fit can be used to check whether the observed data can adequately be regressed by the simple linear regression model defined by Equation (20), and can be given as follows (Rife & Boorstyn 1974):

1 By letting $x_i=i$, $y_i=|z(i)|, i=1,\ldots,M$, and grouping (x_i,y_i) into c clusters of near replicates with the $i\text{-}th$ cluster containing N_i cases, three models can be defined as follows:

$$\begin{aligned} A &: y_{ij} = x_{ij} b_1 + b_0 + e_{ij} \\ B &: y_{ij} = x_{mi} b_1 + b_0 + e_{ij} \\ C &: y_{ij} = \mu_0 + e_{ij} \end{aligned} \tag{23}$$

where $i=1,\ldots,c, j=1,\ldots,N_i$, and x_{mi} is the mean of the horizontal ordinate in the $i-th$ cluster.

2 Fitting the data (x_i, y_i) by using the three regression models defined by the respective equations and calculating the SSE of each model and their degrees of freedom, respectively, the statistic can be obtained:

$$F_1 = \frac{[SSE(B) - SSE(C)]/[df(B) - df(C)]}{[SSE(A) - SSE(B) - SSE(C)]/[df(A) - (df(B) - df(C))]} \quad (24)$$

where $df(\bullet)$ denotes the degree of freedom. From (Christensen 2002), the degree of freedom of each model defined by Equation (23) is then obtained as $df(A)=M-2$, $df(B)=M-2$, and $df(C)=M-c$.

3 For a certain significance level α, the test threshold th is obtained from

$$th = F(1-\alpha; c-2, M-c) \quad (25)$$

where $F(1\text{-}\alpha;c\text{-}2,M\text{-}c)$ represents the critical value of the F-distribution, and the numerator and denominator degrees of freedom are c-2 and M-c, respectively.

4 If $F_1 \leq th$, it's decided that the observed data can be adequately fitted by a simple linear regression model; this also means that the hypothesis H_0 defined by Equation (2) can be accepted. If the statistic is greater than the threshold th, the data cannot be fitted by the simple linear regression, that is to say, the hypothesis H_1 is chosen.

4 SIMULATIONS RESULTS

Monte Carlo simulations are carried out to study the behavior of the proposed algorithm in different environments. The simulations are aimed at detecting whether the absolute error of a certain estimated frequency is greater than a quarter of the discrete sampling frequency. One thousand Monte Carlo trials are performed for each condition. The sampling frequency is set to 100MHz.

Figure 3 shows the detection behavior for the credibility test by using the F-statistic with respect to the false alarm. The carrier frequency of the sinusoid is 19.081MHz, while the sample size is 1024. The initial phase is $\pi/6$, and the frequency estimation error factor is 0.5. The SNR varies from -12 dB to 3 dB in steps of 3 dB. The false alarms are set to 0.05, 0.01, and 0.001, respectively. In the three cases, the probability of detection (P_d) is close to 1 at SNR values greater than 0 dB. The detection probability is enhanced by a greater value of false alarm. With $\alpha=0.05$, the probability of detection is close to 1 for an SNR of -3dB.

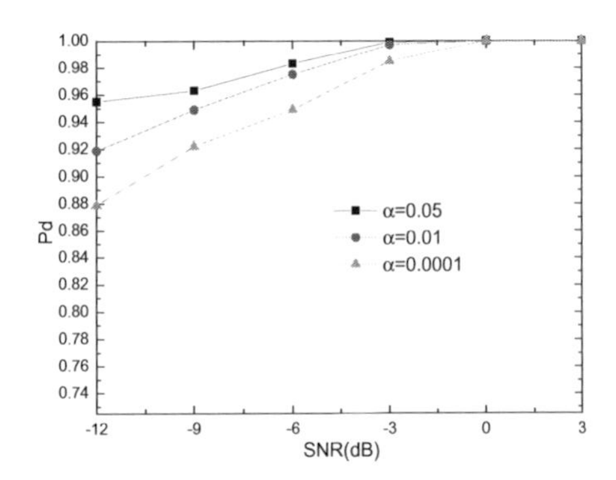

Figure 3. Effect of the false alarm probability on the detection probability obtained by the credibility test of the frequency estimation.

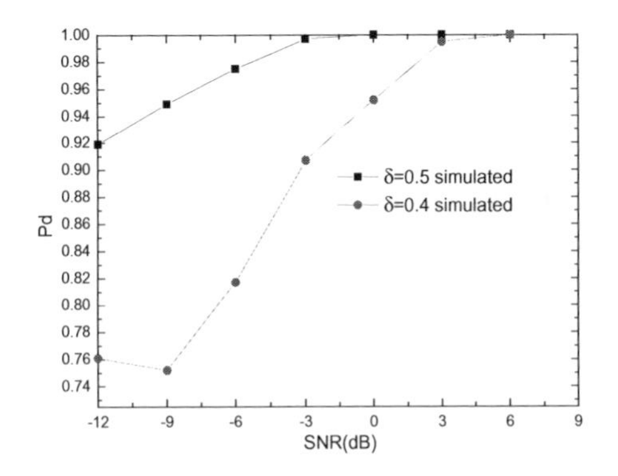

Figure 4. Effect of the frequency estimation error factor on the detection probability obtained by the credibility test of the frequency estimation.

Figure 4 shows the probability of detection with respect to the carrier frequency estimation error. The carrier frequency of the sinusoid is 19.081MHz. The initial phase is $\pi/6$. The carrier frequency estimation error equals to 0.4, 0.5 for a sinusoid with 1024 samples and a probability of false alarm equals to 0.01 respectively. The detection probability is increased by elevating the carrier frequency estimation error factor. For $\delta=0.5$ and an SNR is close to 3 dB, the probability of detection approximately equals to 1.

In fact, a larger frequency estimation error factor enhances the mean of the proposed statistic, and the distinction between non-null and null is easier.

5 CONCLUSION

The paper presents a credibility test algorithm based on the F-statistic for blind frequency estimation of a sinusoid. A credibility assessment testing model is defined to analyze the simple linear regression characteristics of the correlation function between the observed signal and the reference signal under different hypotheses. The test is performed with the proposed threshold based on the F-statistic. Experimental results show a good performance even at a low SNR. The research findings will be useful in electronic warfare signal processing and CR security applications.

ACKNOWLEDGMENTS

This study is financially supported by the 333 Research Project of Jiangsu Province (Project No. BRA2013171) and supported by the Natural Science Foundation of Jiangsu Province (Project No. BK2011837).

REFERENCES

Abatzoglou, T. 1985. A fast maximum likelihood algorithm for frequency estimation of a sinusoid based on Newton's method. IEEE Transactions on Acoustics, Speech and Signal Processing 33(1): 77–89.

Christensen, R. 2002. Plane Answers to Complex Questions:The Theory of Linear Models. New York: Springer-Verlag.

Dash, P. K. & Hasan, S. 2011. A Fast Recursive Algorithm for the Estimation of Frequency, Amplitude, and Phase of Noisy Sinusoid. IEEE Transactions on Industrial Electronics 58(10): 4847–4856.

Fehske, A., Gaeddert, J. & Reed, J. H. 2005. A New Approach to Signal Classification using Spectral Correlation and Neural Networks. First IEEE International Symposium on New Frontiers in Dynamic Spectrum Access Networks: 144–150.

Ghogho, M., Swami, A. & Durrani, T. 2000. Blind estimation of frequency offset in the presence of unknown multipath. IEEE International Conference on Personal Wireless Communications: 104–108.

Hu, G.-B., Xu, L.-Z. & Jin, M. 2013. Reliability testing for blind processing results of LFM signals based on NP criterion. Tien Tzu Hsueh Pao/Acta Electronica Sinica 41(4): 739–743.

Jain, V. K. 1979. High-accuracy analog measurements via interpolated FFT. IEEE Transactions on Instrument and Measurement 28(2): 113–123.

Kay, S. 1989. A fast and accurate single frequency estimator. IEEE Transactions on Acoustics, Speech and Signal Processing 37(12): 1987–1990.

Lin, W. S. & Liu, K. J. R. 2008. Modulation Forensics for Wireless Digital Communications. IEEE International Conference on Acoustics, Speech and Signal Processing: 1789–1792.

Pantazis, Y., Rosec, O. & Stylianou, Y. 2010. Iterative Estimation of Sinusoidal Signal Parameters. IEEE Signal Processing Letters 17(5): 461–464.

Peleg, S. & Porat, B. 1991. Linear FM Signal Parameter Estimation from Discrete-time Observations. IEEE Transactions on Aerospace and Electronic Systems 27(7): 607–615.

Pucker, L. 2009. Review of Contemporary Spectrum Sensing Technologies (For. IEEE-SA P1900.6 Standards Group). http://grouper.ieee.org/groups/scc41/6/documents/white_papers/P1900.6_Sensor_Survey.pdf.

Rife, D. C. B. P. & Boorstyn, R. R. 1974. Single-tone Parameter Estimation from Discrete-time Observation. IEEE Trans on Information Theory 20(5): 591–598.

Stegun, M. A. a. I. A. 1965. Handbook of Mathematic Function with Formulas, Graphs, and Mathematical Tables. New York,Washington I.A. U.S. Department of Commerce.

Su, W. & Yu, M. 2006. Dual-use of modulation recognition techniques for digital communication signals. Systems, Applications and Technology Conference: 1–6.

Wood, D. C. 1980. Fitting Equation to Data. New York: Wiley.

Electronics, Communications and Networks IV – Hussain & Ivanovic (eds)
 ISBN: 978-1-138-02830-2

A simple LLR correction for LDPC coded BICM-ID systems

Ping Huang*, YueHeng Li & MeiYan Ju
Department of Computer Information, HoHai University, Nanjing, Jiangsu, P.R.China

ABSTRACT: In this paper, a linear Log-Likelihood Ratios (LLRs) correction method is proposed for Low-Density Parity-Check (LDPC) coded Bit-Interleaved Coded Modulation with Iterative Decoding (BICM-ID) systems. As the demodulator is sensitive to mismatched decoder feedback LLRs, it is necessary to correct the LLRs to provide the reliable prior information for demodulator. The proposed method searches the scalar factors based on the *consistency condition* for the decoder feedback bit LLRs. Different from the previous reference, the proposed method applies the scalar factors to the decoder output LLRs instead of the demodulator output LLRs. Furthermore, as the probability distribution function (pdf) of Belief Propagation (BP) decoder output LLRs is Gaussian-like, simpler scalar factors for each bit level LLRs with minor complexity can be found compared with the correction based on GMI. Numerical and simulation results verify that the proposed method achieves noticeable performance improvement for BICM-ID systems with LDPC codes.

1 INTRODUCTION

BICM (Zehavi 1992)has been widely adopted in wireless standards thanks to its excellent performance and flexibility. The BICM-ID between demodulator and decoder presented in Reference (Li et al. 2002)can improve error performance both in the Additive White Gaussian Noise (AWGN) channel and Rayleigh fading channel. The soft information exchange between the demodulator and the decoder is usually represented byLLRs.

Under the assumption of an ideal interleaver, the channel decoder treats bits within the symbol as independent. This allows BICM to be modeled as the independent parallel bit channels (Caire et al. 1998). The BICM capacity was defined as the sum of mutual information for the parallel channels. But in the real systems, with limited depth interleaver, sub-optimal demodulator, etc., independent assumption is not satisfied, and the LLRs are described as mismatched LLR in recent work by Reference (Martinez et al. 2009). To improve the achievable rate, LLR correction should be considered. The LLR correction in previous reference (Nguyen et al. 2011)is performed by the rule that corrected LLR should satisfy the so-called *consistency condition*(Tuchler 2004). The other method searches the factors based on generalized mutual information(GMI) maximization for each bit channel(Jalden et al. 2010). These methods firstly find the pdf of the demodulator output LLRs and then numerically scale the LLRs. Both methods require the numerical quadrature because the pdf of demodulator output LLRs is unknown. Moreover, they preclude the on-line correction.

LDPC codes approach the Shannon limit on various channels with relatively low complexity using the BP decoding algorithm. In LDPC coded BICM system, the demodulator is sensitive to imperfect soft information from the BP decoder. It is necessary to correct the LLRs to provide the reliable information for demodulator. Chung (Chung et al. 2001) observed that the output LLRs of BP decoder are Gaussian-like. Based on this assumption, in this paper, a linear LLR correction is considered for LDPC coded BICM-ID systems. The decoder output LLR is modeled as Gaussian variable. According to the *consistency condition* which matched LLR should obey, the scalar factors for the decoder output LLRs are derived. Simulation results verify that the proposed correction method can achieve noticeable performance gains with a minor complexity.

2 SYSTEM MODEL

The BICM system is shown in Figure 1. A block of data bits is encoded by a channel encoder and bit-interleaved by the random interleaver Π. The interleaver may not be required for the LDPC codes, as the interleaver is inherent in the structure of these codes when the parity check parity matrix is constructed randomly. The interleaved sequence is denoted by $\boldsymbol{b}$, m consecutive bits of the sequence $\boldsymbol{b}$ are grouped to form the sub-sequence $\boldsymbol{b}_k = (b_k^0,\ldots,b_k^{m-1})$. Each $\boldsymbol{b}_k$ is mapped to a complex $x_k = \mu\ (b_k)$ symbol chosen from the 2^m-ary signal constellation χ according to the labeling map μ.

*Corresponding author: *huangpinghope@hhu.edu.cn*

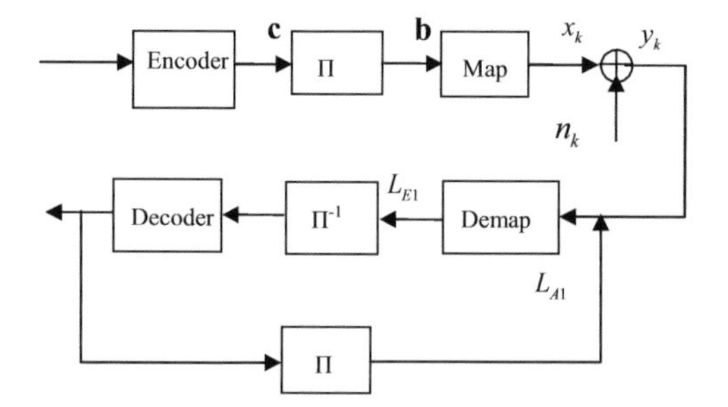

Figure 1. Conventional BICM-ID system model.

At the receiver, the signal is iteratively decoded by mutually exchanging soft information between demapper and decoder. The extrinsic output L_{E1} from the demapper is de-interleaved to become the *a priori* input to the SISO decoder which calculates extrinsic LLRs on the coded bits. The extrinsic information delivered by the decoder is re-interleaved and fed back as *a priori* knowledge L_{A1} to the demapper where it is exploited to reduce the error rate in further iterative procession.

3 LLR CORRECTION BASED ON SCALAR FACTORS

3.1 *Consistency condition*

The channel seen by the channel decoder can be described by the channel transition probability $p(y \mid b)$. Under the assumption of an ideal interleaver and optimal demodulator, the LLR from the decoder is matched to $p(y \mid b)$. In this case, the LLR should satisfy the so-called *consistency condition*(Tuchler 2004)

$$\ln \frac{p^{(i)}_{L_n|b_n}(l|1)}{p^{(i)}_{L_n|b_n}(l|0)} = l, \tag{1}$$

where $p^{(i)}_{L_n|b_n}(l|b)$ is the sufficient statistics of *i*-th level bit channel transition probability.

3.2 *Scalar factors for LLR optimization*

In the real systems, with limited depth interleaver, sub-optimal channel estimation and demodulator, etc., the LLR from the channel decoder is mismatched to the ture channel transition probability, that is

$$\ln \frac{p^{(i)}_{\tilde{L}_n|b_n}(l|1)}{p^{(i)}_{\tilde{L}_n|b_n}(l|0)} = f(l). \tag{2}$$

By adopting a channel adaptor to BICM proposed by Hou in Reference (Hou et al. 2003), the conditional pdf of per bit channel output LLR satisfies *symmetry condition* $p^{(i)}_{\tilde{L}_n|b}(l|1) = p^{(i)}_{\tilde{L}_n|b}(-l|0)$. Then the LLR is exchanged and updated between the demodulator and BP decoder iteratively.

Reference (Chung et al. 2001) and simulation have proved that the output LLRs of BP decoder are Gaussian-like. Then the LLR pdf of decoder output can be described as

$$p^{(i)}_{\tilde{L}_n|b}(l|1) = p^{(i)}_{\tilde{L}_n|b}(-l|0) = \psi(l - \tilde{\mu}_i, \sigma_i^2) \tag{3}$$

where $\psi(l,\sigma_i^2) = \frac{1}{\sqrt{2\pi}\sigma}\exp\left(-\frac{l^2}{2\sigma^2}\right)$. Furthermore, Equation (3) is applied to Equation (2) to derive the scalar factors such as

$$\begin{aligned} \ln \frac{p^{(i)}_{\tilde{L}_n|b_n}(l|1)}{p^{(i)}_{\tilde{L}_n|b_n}(l|0)} &= \ln \frac{\exp\left(-\frac{(l-\mu_i)^2}{2\sigma_i^2}\right)}{\exp\left(-\frac{(l+\mu_i)^2}{2\sigma_i^2}\right)} \\ &= \frac{2\mu_i}{\sigma_i^2} l \\ &= \alpha_i l \end{aligned} \tag{4}$$

Compared with Equation (1), the decoder output extrinsic LLRs should be scaled by the correction factor $2\mu_i / \sigma_i^2$. The factors are called the Gaussian factors. The sample expectation and variance of the absolute value of the LLRs from the decoder are obtained as

$$\mu_i = \frac{1}{N_m}\sum_{j=0}^{N_m-1} |l_{i,j}| \tag{5}$$

$$\sigma_i^2 = \frac{1}{N_m}\sum_{j=0}^{N_m-1} \left(|l_{i,j}| - \mu_i\right)^2, \tag{6}$$

where $l_{i,j}$ is the *i*-th bit level LLRs from the decoder.

3.3 *Complexity analysis*

It is observed that in the signal noise ratio (SNR) region of interest, the estimation accuracy can be successively improved by using multiple samples. But in fact, a couple of code blocks should be enough to get sufficiently accurate estimations of μ_i and σ_i^2 in equations (5) and (6). When N_m approaches a certain value, the Gaussian factors will tend to a constant α, thus the factor does not need to be calculated anymore. The constant can be applied to replace α_i in Equation (4) directly to begin the next iterative processing.

Table 1 shows the scalar factors in the different SNR with 8PSK, SP labeling for (504, 252)LDPC coded BICM-ID system. As shown in Table 1, in the SNR region of interest, the variation of each bit channel correction factors is ignorable.

Table 1. The scalar factors at the different SNR.

Factors	α_1	α_2	α_3
5.2dB	0.53	0.44	0.41
5.4dB	0.56	0.48	0.44
5.6dB	0.60	0.53	0.49

As presented in Reference (Jalden et al. 2010), the GMI $I_{qB_i,Y}^{GMI}$ for the i-th bit channel can be expressed as

$$\begin{aligned} I_{qB_i,Y}^{GMI} &\triangleq \max_{s>0} I_{qB_i,Y}(s) \\ &= \max_{s>0}\left\{1 - E_{X,Y}\left(\log_2(1+\exp(-\mathrm{sgn}(b_i(X))L_{B_i,Y}^{sub}(y)s))\right)\right\} \end{aligned} \quad (7)$$

To get the expectation value of Eq. (7), the GMI optimization method requires the numerical quadrature because the pdf of demodulator output LLR is unknown. Then the factor s is chosen to maximize the per bit channel GMI in Eq. (7).

Compared with the GMI method, in the initial a couple of iterative procession, the proposed method computes the expectation and variance according to Eqs.(5) and (6), and then Eq.(4) is utilized to get the scaling factor for LLRs optimization. It is clear that the computation complexity of the proposed method is much less than the GMI method.

4 NUMERICAL AND SIMULATION RESULTS

In this section, the proposed LLR correction method is applied to LDPC coded BICM-ID system. Regular 1/2- rate LDPC code of block length 504 and 1056(MacKay 2007), 8PSK/16QAM mapping with SP labeling is considered in this paper. At the receiver, a max-log demodulator is applied. The decoder adopts BP decoding and the maximum number of BP iteration is 10. The maximum number of iteration between demodulator and decoder is also set to 10.

Figure 2 shows EXIT curves comparison with/without LLR correction in LDPC coded BICM-ID system. One can see that when applying the proposed method, the demodulator achieves lager output mutual information I_{E1} with the same decoder feedback mutual information I_{A1} than the demodulator in the system without LLR correction. This result shows that the proposed LLR correction can achieve better performance and the result is also consistent with the latter bit error ration(BER) simulation.

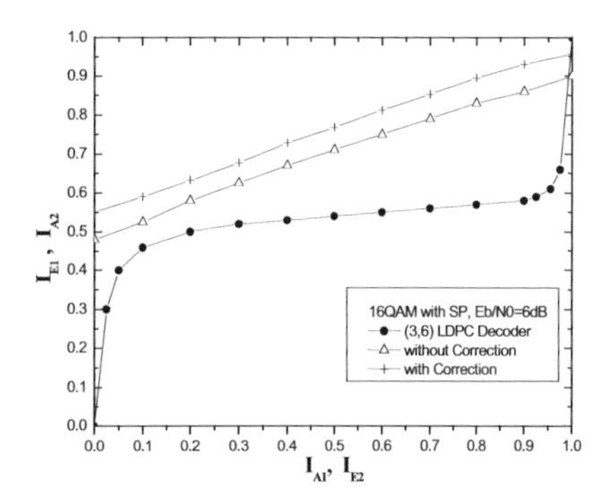

Figure 2. EXIT Curves comparison with/without LLR Correction in LDPC coded BICM-ID system.

Figure 3 and 4 depict the BER performance comparison with/without LLR correction in LDPC coded BICM-ID system. Note that the code 1 represents (504, 252) LDPC code, while code2 denotes (1056, 528) LDPC code. As can be seen from Figs.3 and 4, when

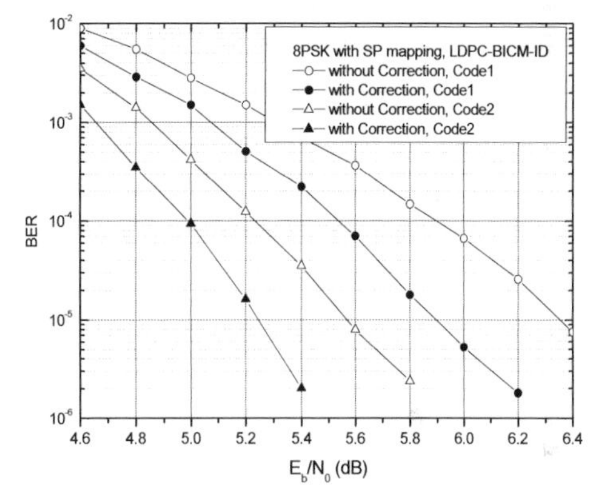

Figure 3. BER comparison with/without LLR Correction in LDPC coded BICM-ID system with 8PSK.

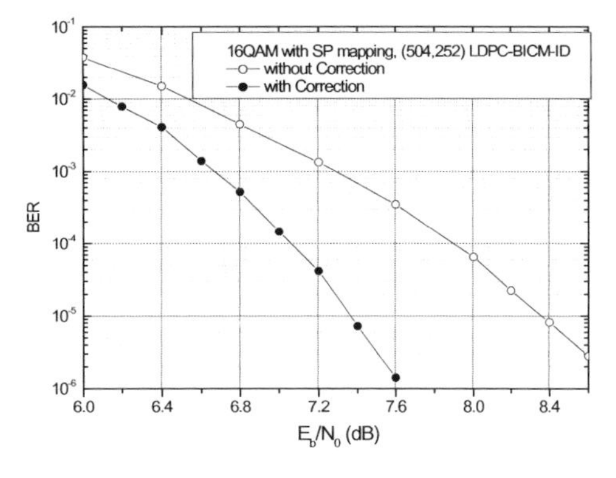

Figure 4. BER comparison with/without LLR Correction in LDPC coded BICM-ID system with16QAM.

the BER is about 10^{-5}, the proposed method achieves 0.6-0.8 dB gains with code 1 and 0.4dB gains with code 2. It is also found that as the codeword length increases, the gains achieved by LLR correction will be reduced, i.e. for (5040, 2520) LDPC coded BICM, LLR correction outperforms the method without correction only 0.2dB. This may be explained as follows, as the codeword length increases, the decoder feedback each bit information will be independent, thus the error propagation between the bits within the same symbol will be reduced.

5 CONCLUSIONS

This paper introduces a simple and efficient LLR optimization method for BICM-ID systems. The proposed method searches the scalar factors to each bit channel LLRs based on the *consistency condition.* The proposed optimization method reduces the computation complexity greatly compared with the other linear correction such as GMI optimization. Numerical and simulation results verify that the proposed method improves error performance especially for BICM-ID systems with LDPC codes.

ACKNOWLEDGMENTS

This work was supported by "Central university base science research project fund" (HoHai university, Contract number 2010B06514).

REFERENCES

Caire G , Taricco G & Biglieri E. 1998. Bit-interleaved coded modulation. *IEEE Trans. on inform. Theory* 44(3): 927–946.

Chung Sae-Young, Richardson T.J. & Urbanke R.L. 2001. Analysis of sum-product decoding of low-density parity-check codes using a Gaussi- an approximation. *IEEE Trans. on Information Theory* 47(2): 657–670.

Hou, J., Siegel, P.H., Milstein, L.B. et al. 2003. Capacity-approaching bandwidth-efficient coded modulation schemes based on low-density parity-check codes. *IEEE Trans. on Information Theory* 49(9): 2141–2155.

Jalden, J, Fertl, P & Matz, G. 2010. On the generalized mutual information of BICM systems with approximate demodulation. In *Proc. IEEE Information Theory Workshop:*1–5. Cairo, Egypt.

Li Xiaodong, Chindapol A. & Ritcey J.A. 2002. Bit-interleaved coded modulation with iterative decoding and 8 PSK signaling. *IEEE Trans. on Commu.* (50) 8: 1250–1257.

MacKay D J C. Encyclopedia of Sparse Graph Codes [Online]. Available: http://www.inference.phy.cam.ac.uk/mackay/codes/data.html. 2007,1.

Martinez A., Fabregas A. G. i, Caire G., et al. 2009. Bit interleaved coded modulation revisited: A mismatched decoding perspective. *IEEE Trans. Inform. Theory* 55: 2756–2765.

Nguyen T. & Lampe L. 2011. Bit-interleaved coded modulation with mismatched decoding metrics. *IEEE Trans. Commun.* 59: 437–447.

Tuchler, M. 2004. Design of serially concatenated systems depending on the block length. *IEEE Trans,Commun.* 2(2): 209–218.

Zehavi E. 1992. 8-PSK trellis codes for a Rayleigh channel. *IEEE Trans. Commun.* 40(5): 873–884.

Electronics, Communications and Networks IV – Hussain & Ivanovic (eds)
 ISBN: 978-1-138-02830-2

An adaptive PI active queue management algorithm based on queue length

Hongcheng Huang*, Fan Yang, Shiwei Wang & Gaofei Xue
School of Communication and Information Engineering, Chongqing University of Posts and Telecommunications, Chongqing, China

ABSTRACT: A Queue Length-based Adaptive Proportional Integral (QL-API) algorithm is proposed to cope with a decline in the performance of the network caused by the mismatching of the fixed PI controller parameters and time-varying network parameters. QL-API is designed according to the theory of TCP/AQM feedback system. The proportion coefficient and integral coefficient that matches with the current network status are calculated by the router queue length and the loss probability. Verified by using NS-2 simulations, QL-API can reduce queue oscillation and achieve faster convergence speed than PI. It also shows better stability and robustness than PI.

1 INTRODUCTION

With the explosive growth of the Internet, the Internet traffic has increased quickly, so that network congestion occurs frequently. Network congestion will directly lead to the degradation of the entire network performance, such as the decrease of network throughput, and the increase of packet loss rate and end-to-end delay. The crash of the network will be caused by serious network congestion.

The AQM (Active Queue Management) techniques (Briscoe & Manner 2014) which adopt the queuing algorithm and pack discard strategy to manage the router buffer, so that it can adjust the data transmission rate to avoid a more serious congestion and improve network performance. Misra, V. established a TCP/AQM cybernetic model (Misra et al. 2000), which allows people to study the AQM by using control theory. By using the instantaneous queue length, which is based on the TCP/AQM model as a measure of the congestion status, Hollot, C. designed the Hollot-PI algorithm (Hollot et al. 2001), which can eliminate the steady-state error effectively and keep the queue length in a remain stable. But as a result of the PI controller with fixed configuration parameters, the system response is slower, queues jitter is larger, and network throughput and bandwidth utilization are low.

As pointed out in the literature (Wang et al. 2012), the PI controller can eliminate the steady-state error of the queue length effectively by integral operation, but the parameters of PI algorithm are sensitive to network scenarios. According to TCP / AQM model, configuration parameters of PI algorithm are affected by the time-varying parameters which include connection number, link capacity and round trip time, so that they will affect the overall performance of the network eventually (Liu et al. 2009). An adaptive PI algorithm is proposed in the literature (Hong & Yang 2006). It sets the parameters of the PI controller by the phase angle magnitude margin. However, the algorithm detects network status by a heuristic based strategy, and poor performance is exhibited in the network whose parameters fluctuate in a small range. Literature (Huang et al. 2013) proposed a new API algorithm, but it needs to configure the PI controller parameters based on the network status.

In this paper, a QL-API (Queue Length-Based Adaptive Proportional Integral) algorithm is proposed, which can analyze the transfer function of the controlled object in the TCP/AQM dynamic feedback system and calculate the appropriate proportional integral factor of the PI controller in real time. Simulation results show that QL-API algorithm has a faster response speed and a smaller queue jitter than PI algorithm. Even in the time-varying parameter network, the QL-API algorithm can increase the bandwidth utilization and reduce the packet loss rate, and it has better stability and robustness.

2 PI CONTROLLER FOR TCP/AQM MODEL

The PI controller for TCP/AQM feedback control model (Zhang & Sun. 2010) is shown in Figure 1, where q_0 means the target queue length, and P is the packet loss rate.

*Corresponding author: *huanghc@cqupt.edu.cn*

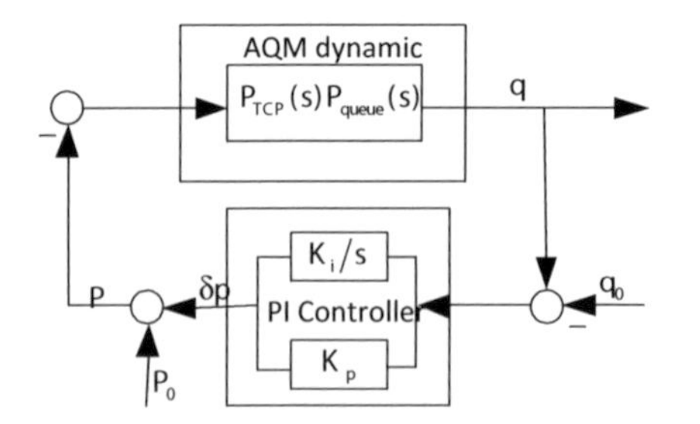

Figure 1. AQM (PI) feedback system.

Let N be the number of system connections. Let C denote the link bandwidth and R denote the round trip time. In this system, the transfer function of PI Controller $G_1(s)$ and the transfer function of AQM Dynamic $G_2(s)$ are as follows:

$$G_1(s)=\frac{1+\tau s}{Ts} \tag{1}$$

$$G_2(s)=\frac{K_m e^{-sR}}{(T_1 s+1)(T_2 s+1)}(K_m=\frac{(RC)^3}{4N^2}, T_1=\frac{R^2C}{2N}) \tag{2}$$

τ and T are set to be determined parameters, $T_2=R$. So the open-loop transfer function of the system is expressed as follows:

$$G(s)=G_1(s)G_2(s)=\frac{1+\tau s}{Ts}\frac{K_m e^{-sR}}{(T_1 s+1)(T_2 s+1)} \tag{3}$$

The formula of PI active queue management algorithm (Arpaci & Copeland 2000) is as follows:

$$p(t)=k_p e(t)+k_i\int_0^t e(t) \tag{4}$$

where $p(t)$ represents the packet loss probability. k_p and k_i are proportional coefficient and integral coefficient respectively. Let $q(t)$ be the length of the current queue, and q_{ref} denote the length of the target queue. $e(t)=q(t)-q_{ref}$ is the length difference between the current queue and the target queue.

The packet loss rate is calculated according to certain sampling time in the implementation of PI algorithm. Let k denote the current sampling time, so the discrete equation of PI active queue management algorithm is as follows:

$$p(k)=p(k-1)+k_p(e(k)-e(k-1))+k_i e(k) \tag{5}$$

According to the definition of stability margin. k_i and k_p can be obtained from the following formulas:

$$k_i=w*z*\left|\frac{jw+\frac{1}{R}}{\frac{C^2}{2N}}\right|, \; w=\frac{\beta}{R}, \; z=\frac{2N}{R^2C} \tag{6}$$

$$k_p=\frac{R^2C}{2N}*K_i \tag{7}$$

In Formula(6), $w=\beta/R$ and $z=2N/R^2C$, where β is the cutoff frequency when the system is stable.

3 QL-API ALGORITHM DESIGN

QL-API algorithm which is based on the PI algorithm analyzes the transfer function of the controlled object module in the TCP/AQM congestion control dynamic feedback system. In the controlled object system, input variable is the packet loss probability $p(t)$, and output variable is the queue length $q(t)$. The equation of $p(t)$ and $q(t)$ can be derived by the method of z-transform. Then, real-time values of $q(t)$ and $p(t)$ are used to calculate the intermediate variables (K_m, T_1 and T_2), which are associated with the current network status. The k_p and k_i can be calculated by using the intermediate variables and can be updated to the PI controller. Figure 2 is the flow chart of QL-API algorithm, and the details will be described as follows.

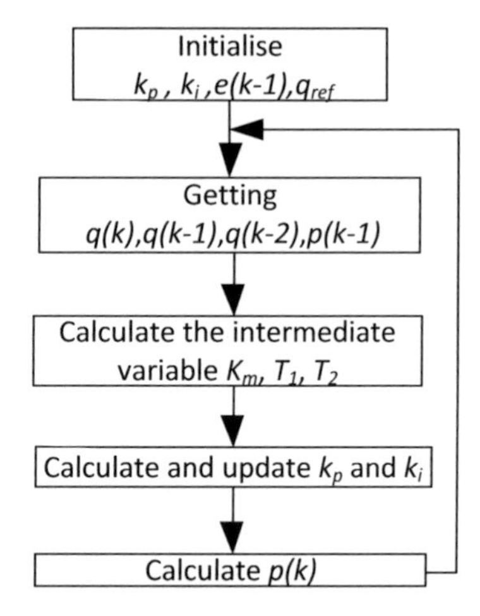

Figure 2. The flow chart of QL-API algorithm.

3.1 *Analysis of TCP/AQM system*

Since the sampling period T is equal to R, and according to the formula of $\lim(1+x)^{1/x}=e(x\to 0)$ and the relationship between the Laplace transform and z-transform, we know $s=(z-1)/zT$, $e^{-sR}=z^{-1}$. Substitute them into Formula(2), the expression of $G_2(z)$ can be achieved:

$$G_2(z)=\frac{Q(z)}{P(z)}=\frac{\sigma_0 z^{-1}}{1-\sigma_1 z^{-1}-\sigma_2 z^{-2}} \tag{8}$$

In Formula (8), $Q(z)$ and $P(z)$ are output function and input function, respectively. Where $\sigma_0=bT^2/(1+a_1T+a_2T^2)$, $\sigma_1=(2+a_1T^2)/(1+a_1T+a_2T^2)$, $\sigma_2=1/(1+a_1T+a_2T^2)$, $a_1=T_1T_2/(T_1+T_2)$, $b=K_m/T_1T_2$. The discrete time domain expression of the controlled object can be obtained by inverse Z-transform from Formula(8):

$$q(k)=\sigma_1 q(k-1)+\sigma_2 q(k-2)+\sigma_0 p(k-1) \tag{9}$$

Formula (9) describes the relationship between input variables p (t) and output variables q (t) in the TCP/AQM system.

3.2 *Calculate the intermediate variance of network status*

According to Formula(9), set up three equations about $q(k)$, $q(k\text{-}1)$, $q(k\text{-}2)$. σ_0, σ_1, and σ_2 can be obtained by solving the three equations. Then, by substituting σ_0, σ_1, and σ_2 into Formula(8), K_m, T_1, *and* T_2 can be obtained:

$$\begin{cases} K_m=\dfrac{\sigma_0}{1+\sigma_2-\sigma_1} \\ T_1=\dfrac{T(\sigma_1-\sigma_2-\sqrt{\sigma_1^2-4\sigma_2})}{2(1+\sigma_2-\sigma_1)} \\ T_2=\dfrac{T(\sqrt{\sigma_1^2-4\sigma_2}+\sigma_1-2\sigma_2)}{2(1+\sigma_2-\sigma_1)} \end{cases} \tag{10}$$

In Formula(10), K_m, T_1, *and* T_2 are the intermediate variables that reflect the system's real-time state, and K_m, T_1, *and* T_2 are also the functions of N, C and R.

3.3 *Calculate the parameters of PI controller*

Based on Formula(6)(7) and Formula(10), k_p and k_i can be expressed as:

$$k_p=w\frac{K_m}{T_1}\sqrt{w^2T_2^2+1} \tag{11}$$

$$k_i=\frac{w\sqrt{w^2T_2^2+1}}{K_m} \tag{12}$$

Substitute Formula(11)(12) into Formula(5), so the packet loss probability of PI controller can be expressed as follows:

$$p(k)=w\frac{K_m}{T_1}\sqrt{w^2T_2^2+1}(e(k)-e(k-1)) + \frac{w\sqrt{w^2T_2^2+1}}{K_m}e(k)+p(k-1) \tag{13}$$

where $e(k)=q(k)\text{-}q_0$. $p(k)$ is the probability of packet loss and $q(k)$ is the queue length at the time k.

When a router receives a packet, it will decide whether to discard the packet according to the results of the Formula (13). Since $p(k)$ is updated in real time based on the network conditions, it avoids the increasing of router queue jitter and the reducing of the bandwidth utilization caused by unreasonable controller parameters. QL-API algorithm can update the parameters of the PI controller in real-time, which are calculated by intermediate variable to satisfy the current network status periodically. So that router can maintain the stability of the router buffer and increase the link bandwidth utilization.

4 SIMULATION AND ANALYSIS OF QL-API ALGORITHM

4.1 *Simulation environment*

Using the NS-2 simulation software, the classic dumbbell network topology and configuration parameters are shown in Figure 3. AQM algorithm is deployed on the router B. All senders establish a TCP connection to send infinite data, and the receivers have enough buffers to receive data. The buffer of bottleneck router is set to 500 packets, and each experiment simulates 150 s of a run.

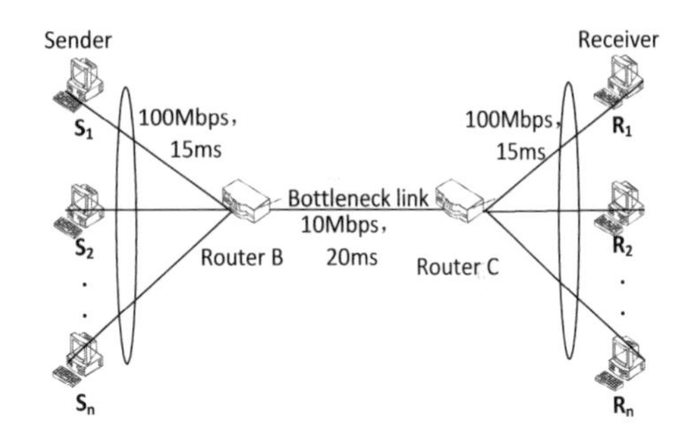

Figure 3. Network topology.

When the TCP/AQM system is in a stable state, the simulation configuration parameters of the two algorithms are as follows:

① PI Algorithm: In the closed-loop system, the gain margin is 3 and the phase margin is $\pi/3$. Set cutoff frequency β as 0.53. Set the value of a as 0.00001822, and the value of b is 0.00001816. Then sampling frequency should be set to 170 Hz. The length of the target-queue is 200 packets.

② QL-API Algorithm: Set cutoff frequency β as 0.53. The value of k_p is 0.00014723, and the value of k_i is 0.0000004277. The sampling frequency is 170 Hz. The length of the target-queue is 200 packets.

4.2 *The number of connections is variable*

Since burstiness is an important feature of network traffic, the simulation is designed to test AQM algorithm performance in the scenario where a sudden change happens in the number of connections. To simulate a sudden change in the number of TCP connections, the number of TCP connection changes every 30 s. After changing, the numbers of connections are 200, 600, 300, 800 and 500 respectively.

As can be seen in Figure 4, though the queue lengths of the router show some fluctuations in both PI and QL-API, QL-API shows a faster response and better stability. After the number of connection changes, it is difficult for a PI algorithm to converge the queue length to 200 pkt in 30 s. QL-API algorithm takes about 10 s to make the queue length into a stable value which conforms to the target queue length.

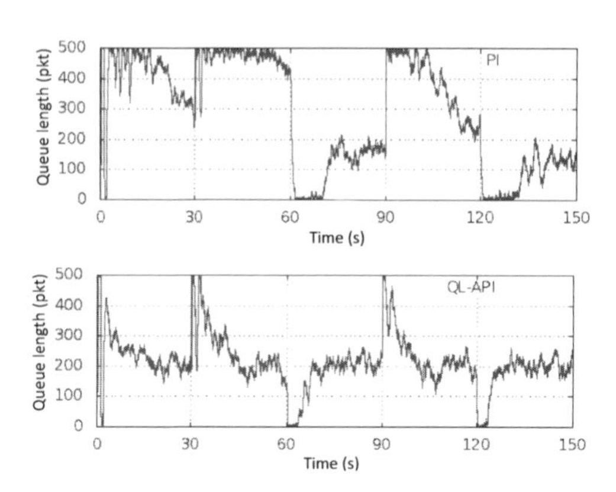

Figure 4. Queue length with variable connection number.

Table 1. Performance comparison.

	Queue length	Queue oscillation	Packet loss rate	Throughput
Algorithm	Packets	packets	%	Mbps
PI	285.38	178.21	12.29	9.755
QL-API	220.42	88.37	8.02	9.921

As shown in Table 1, compared with PI algorithm, the average queue length of QL-API is closer to the target queue length, the queue oscillation is about 90 pkt and the packet loss probability is reduced by approximately 4%. QL-API algorithm shows better stability and robustness than PI algorithm.

4.3 *Various RTT*

This scenario will test the stability and robustness of the QL-API algorithm in the networks of different RTT. The number of TCP connections through the bottleneck link is 400. The initial experimental setup has a 60 ms RTT. At each step, RTT will increase. Thus, for each new configuration, the RTT is increased by 60 ms.

The result shown in Figure 5 indicates that the performance of QL-API algorithm is better than that of a PI algorithm in average captain, queue oscillation, bandwidth utilization and packet loss rate. In the queue oscillation, QL-API algorithm exhibits better stability. With the increase of RTT, link utilization and packet loss rate have shown a downward trend in both PI and QL-API. But the packet loss rate of QL-API is typically 3% lower than that of PI.

Figure 6 shows the queue length from the representative experiments with RTT setting of 180 ms between PI and QL-API. QL-API will converge the queue length to the target value in 12 s, and PI takes more than 50 s.

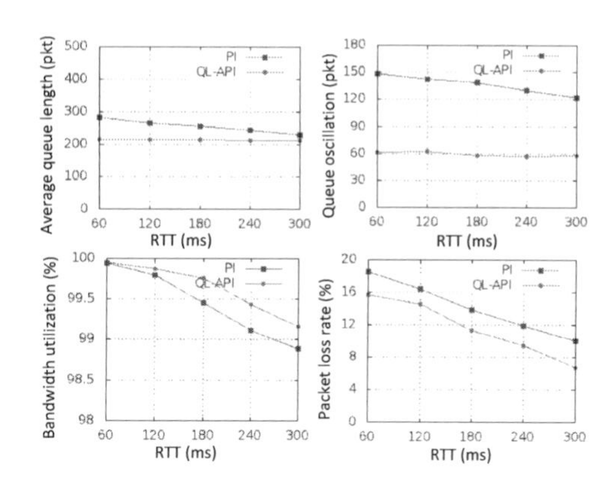

Figure 5. Algorithm performance analysis of various RTT.

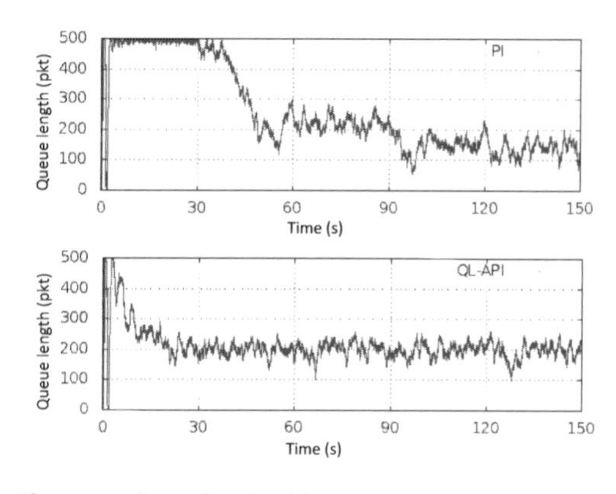

Figure 6. Queue length with RTT = 180 ms.

4.4 *Various bottleneck capacity*

The performance of QL-API algorithm will be investigated in the scenario that the number of TCP connections through the bottleneck link is 300 and RTT is 100 ms. The initial experimental setup has a 5 Mbps bottleneck. For each new configuration, the bottleneck capacity is increased by 10 Mbps.

Figure 7 shows the comparison between QL-API and PI. With the increase of the bottleneck capacity, PI can converge the queue length to the target value (200pkt), but queue oscillation is significantly greater than QL-API. The packet loss rate of QL-API is 2% lower than that of the PI.

When the bottleneck capacity is 25 Mbps, the queue length comparison between QL-API and PI is shown in Figure 8. QL-API will converge the queue length to the target value in 14 s, while PI takes about 40 s. As shown in Figure 7 and Figure 8, QL- the API algorithm performs better stability and robustness than the PI algorithm in the network of different bottleneck capacities.

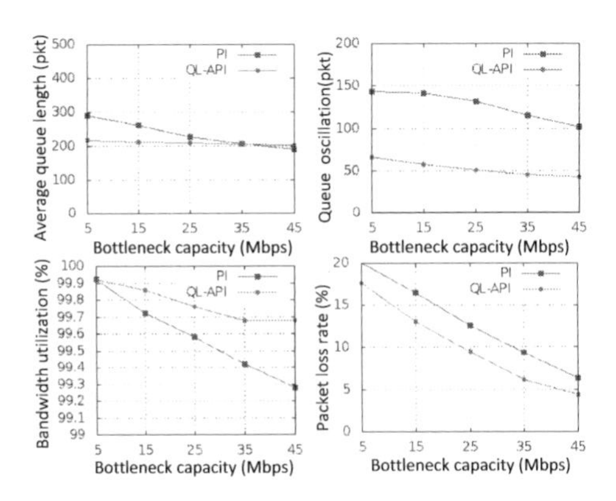

Figure 7. Algorithm performance analysis of various bottleneck capacity.

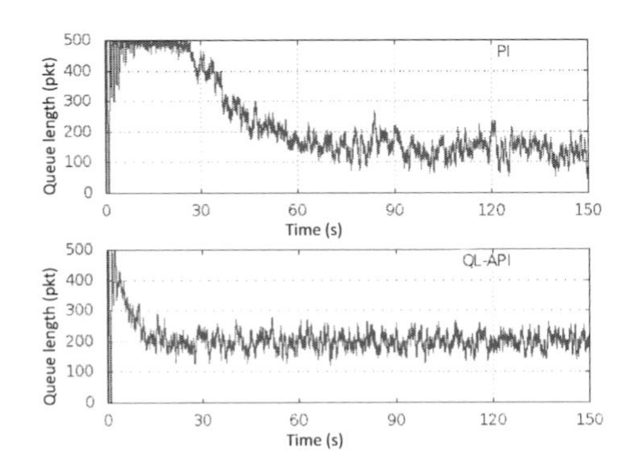

Figure 8. Queue length with C = 25Mbps.

5 CONCLUSION

This paper proposes a new active queue management called QL-API algorithm which is based on the router queue length to solve the problem that the fixed configuration parameters are inapplicable to the network whose parameters are time-varying in PI algorithm. In our experiment, we compare the performance of QL-API algorithm with the PI algorithm in the scenarios which include the different numbers of connections, the different RTT and the different bottleneck capacities. The result shows that QL-API algorithm has a faster response speed and a smaller queue jitter than PI algorithm. It can weaken the congestion of the network effectively. QL-API also has higher bandwidth utilization and lower packet loss rate than PI, and shows a better stability and robustness.

ACKNOWLEDGEMENT

This work was supported by Scientific and Technological Research Program of Chongqing Municipal Education Commission (KJ1400402), and the Foundation and Frontier Research Project of Chongqing Municipal Science and Technology Commission (cstc2014jcyjA40039).

REFERENCES

Arpaci, M. & Copeland, J. A. 2000. An adaptive queue management method for congestion avoidance in TCP/IP networks. *IEEE Global Telecommunication Conference, 27 November-1 December*: 309–315. Piscataway: IEEE.

Briscoe, B. & Manner, J. 2014. Byte and Packet Congestion Notification. *http://datatracker.ietf.org/doc/rfc7141/.*

Hollot, C. V., Misra, V., Towsley, D., et al. 2001. On designing improved controllers for AQM routers supporting TCP flows. *20th Annual Joint Conference of the IEEE Computer and Communications Societies, Anchorage, 22–26 April 2001*: 1726–1734. Piscataway: IEEE.

Hong, Y., & Yang, O. W. 2006. Adaptive AQM controllers for IP routers with a heuristic monitor on TCP flows. *International Journal of Communication Systems* 19(1): 17–38.

Huang, H., Xue, G., Wang, Y., & Zhang, H. 2013. An adaptive active queue management algorithm. *2013 3rd International Conference on Consumer Electronics, Communications and Networks, CECNet 2013, Xianning, 20–22 November 2013*: 72–75. Piscataway: IEEE.

Liu, F., Dang, X. L. & Xu, Z. 2009. An Improved PI Active Queue Management Algorithm Based on Network State Parameters Estimation. *Computer Research and Development* 46(7): 1086–1093.

Misra, V., Gong, W. B., & Towsley, D. 2000. Fluid-based analysis of a network of AQM routers supporting TCP flows with an application to RED. *ACM SIGCOMM*

2000 Conference, Stockholm, 28 August-1 September 2000: 151–160. New York: ACM.
Wang, H., Li, X. M., Yan, W. et al. M. H. 2012. A Novel AQM Algorithm Based on the PI Controller with Minimum ISTE. *Jisuanji Xuebao/Chinese Journal of Computers* 35(5): 951–963.
Zhang, S. & Sun, J. 2010. Adaptive PID active queue management algorithm design. *2010 Chinese Control and Decision Conference, CCDC 2010, Xuzhou, 26–28 May 2010*: 3194–3198. Piscataway: IEEE.

Electronics, Communications and Networks IV – Hussain & Ivanovic (eds)
 ISBN: 978-1-138-02830-2

Comparative analyses of a fast DOA estimation algorithm using a subarray approach for single-snapshot data

Ching Jer Hung, Xiarong Cui & Xiaofei Li
Department of Mathematics and Computer, Wuyi University, Fujian, China

ABSTRACT: In this study, a new Direction Of Arrival (DOA) estimation technique, using subarray processing with application to the uniform linear sensor array in the single snapshot case, is presented. The proposed technique is very suitable for real-time processing, even under non-stationary and coherent interference environments because there is no need for complicated computations and prior knowledge of the number of signals, unlike most DOA estimation algorithms. Results demonstrated that successful DOA estimations could be obtained when the Signal-to-Noise Ratio (SNR) and the Degrees Of Freedom (DOF) of the sensor array were large enough.

1 INTRODUCTION

A vast number of high-resolution direction of arrival (DOA) estimation approaches have been proposed, but real-time tracking capabilities for signals with quick variations, such as suddenly appearing and disappearing, is still challenging due to the following possible issues:

1. A large number of snapshots are required in order to obtain good performance (Macinnes 2004, Xin & Sano 2004, Yang 1995, Yang & Kaveh 1988, Schmidt 1986, Capon 1969, Roy & Kailath 1989, Shan et al. 1985, Williams et al. 1988, Stoica & Sharman1990).
2. A complicated computational burden is required because of the eigendecomposition process or nonlinear optimization algorithm (Schmidt 1986, Capon 1969, Roy & Kailath 1989, Shan et al. 1985, Williams et al. 1988, Stoica & Sharman 1990).
3. A prior knowledge of the number of signals is required in the single snapshot case (Fuhl & Molisch 1995, O'Brien et al. 2005, Radich & Buckley 1997, Zhang et al. 1991).

In practical scenarios, there may be limited snapshots available and no prior knowledge of the number of signals in the fast DOA estimation process. Hence, most of the DOA estimation algorithms are severely limited when the above issues cannot be solved. For example, a particular algorithm found during the literature research (Zhang et al. 1991) contained fewer computations in the single snapshot case, but the estimation performance would be severely degraded due to the number of incident signals not being correctly estimated.

In order to acquire a viable real-time application for fast DOA estimation, a new method, termed Single Snapshot- Direction-of-Arrival (SS-DOA), is proposed in this paper. The SS-DOA algorithm is a non-statistical technique, which uses subarray processing (Shan et al. 1985, Williams et al. 1988) of the uniform linear sensor array in the single snapshot case. Even though the subarray processing application must take into account losing the number of degrees of freedom (DOF), the SS-DOA has the advantage of tracking the DOA of fast varying signals without complicated computations and prior knowledge of the number of signals. The proposed algorithm only uses a single snapshot to estimate DOA of fast varying signals, thereofore, arbitrary probability density functions can be allowed.

2 METHODS AND ANALYSES

2.1 *Proposed SS-DOA algorithm*

Assume the observations vector $\underline{b}(l)$ received from a uniform linear array of m-omnidirectional sensors at l-th snapshot is a linear combination of q narrow-band signals with additive noise. In the general form, it can be modeled as

$$\underline{b}(l) = A(\theta)\underline{s}(l) + \underline{n}(l) \quad (1)$$

$$= \begin{bmatrix} b_1 & b_2 & \cdots & b_m \end{bmatrix}^T$$

where $\underline{s}(l) = \begin{bmatrix} s_1 & s_2 & \cdots & s_q \end{bmatrix}^T$ is a $q \times 1$ vector of signal amplitudes, $\underline{n}(l)$ is a $m \times 1$ vector of additive noise and $A(\theta)=\left[\underline{a}(\theta_1),\underline{a}(\theta_2),\cdots,\underline{a}(\theta_q)\right)$ is a $m\times q$ matrix of the steering vectors whose i-th column $\underline{a}(\theta_i)=\left[1,e^{j2\pi d/\lambda \sin\theta_i},\cdots,e^{j2\pi d/\lambda (m-1)\sin\theta_i}\right]^T$ denote the steering vector of source s_i coming from θ_i, d is the spacing between the sensors and λ is the wavelength of the carrier, $i=1,2,\cdots q$.

In this section, we divide the uniform linear array into high overlapping subarrays of size M, then the sensors $\{1,2,\cdots,M\}$ form the first subarray and the sensors $\{2,3,\cdots,M+1\}$ form the second subarray and so on, i.e., we rearrange the received observations $\underline{b}(l)=\left[\begin{matrix} b_1 & b_2 & \cdots & b_m\end{matrix}\right]^T$ to form the following data matrix:

$$B(l)=\left[\begin{matrix}\underline{b}_1(l) & \underline{b}_2(l) & \cdots & \underline{b}_M(l)\end{matrix}\right]$$

$$=\left[\begin{matrix} b_1 & b_2 & \cdots & b_M \\ b_2 & b_3 & \cdots & b_{M+1} \\ \vdots & \vdots & \ddots & \vdots \\ b_M & b_{M+1} & \cdots & b_m \end{matrix}\right] \quad (2)$$

where M $(M>q)$ is equal to $\lfloor (m+1)/2 \rfloor$ and $\lfloor y \rfloor$ $(y\geq \lfloor y \rfloor)$ is the closest integer to y. Let $\underline{b}_k(l)$ denotes the k-th column vector of the data matrix (2), then we can write

$$\underline{b}_k(l)=\tilde{A}(\theta)F^{k-1}\underline{s}(l)+\underline{n}_k(l) \quad (3)$$

where $\tilde{A}(\theta)$ and F can be expressed as

$$\tilde{A}(\theta)=\left[\underline{\tilde{a}}(\theta_1),\underline{\tilde{a}}(\theta_2),\cdots,\underline{\tilde{a}}(\theta_q)\right]$$

$$\underline{\tilde{a}}(\theta_i)=\left[1,e^{j2\pi d/\lambda \sin\theta_i},\cdots,e^{j2\pi d/\lambda (M-1)\sin\theta_i}\right]^T$$

$$F=diag\left[e^{j2\pi d/\lambda \sin\theta_i}\right]$$

We can rewrite the first column vector of (2) as

$$\underline{b}_1(l)=\tilde{A}(\theta)F^0\underline{s}(l)+\underline{n}_1(l)$$

$$=\left[\underline{\tilde{a}}(\theta_1),\underline{\tilde{a}}(\theta_2),\cdots,\underline{\tilde{a}}(\theta_q)\right]\left[\begin{matrix} s_1 \\ s_2 \\ \vdots \\ s_q\end{matrix}\right]+\underline{n}_1(l)$$

$$=s_1\underline{\tilde{a}}(\theta_1)+\cdots+s_q\underline{\tilde{a}}(\theta_q)+\underline{n}_1(l) \quad (4)$$

then

$$\left[\underline{b}_1-s_1\underline{\tilde{a}}(\theta_1)\right]=\left[s_2\underline{\tilde{a}}(\theta_2)+\cdots+s_q\underline{\tilde{a}}(\theta_q)\right]+\underline{n}_1 \quad (5)$$

Please notice that we skip the snapshot index l for shortness' sake of equation representation. Based on the equation (5), $\left[\underline{b}_1-s_1\underline{\tilde{a}}(\theta_1)\right]$ exclude the desired signal $s_1\underline{\tilde{a}}(\theta_1)$, i.e., only include other $q-1$ desired signal $s_2\underline{\tilde{a}}(\theta_2),\cdots,s_q\underline{\tilde{a}}(\theta_q)$ and noise vector $\underline{n}_1$. If we can find a vector $\underline{w}(\theta_1)=\left[w_1(\theta_1),w_2(\theta_1),\cdots,w_M(\theta_1)\right]^T$ to satisfy $\left[s_2\underline{\tilde{a}}(\theta_2)+\cdots+s_q\underline{\tilde{a}}(\theta_q)\right]^T\underline{w}(\theta_1)=[0]$, then equation (5) will be rewritten as

$$\left[\underline{b}_1-s_1\underline{\tilde{a}}(\theta_1)\right]^T\underline{w}(\theta_1)=[0]+\underline{n}_1^T\underline{w}(\theta_1) \quad (6)$$

Therefore

$$\underline{b}_1^T\underline{w}(\theta_1)=s_1\underline{\tilde{a}}(\theta_1)^T\underline{w}(\theta_1)+\underline{n}_1^T\underline{w}(\theta_1) \quad (6\text{-}1)$$

By the same process, we can then get

$$\underline{b}_2^T\underline{w}(\theta_1)=s_1e^{\phi_i}\underline{\tilde{a}}(\theta_1)^T\underline{w}(\theta_1)+\underline{n}_2^T\underline{w}(\theta_1) \quad (6\text{-}2)$$

$$\underline{b}_3^T\underline{w}(\theta_1)=s_1e^{2\phi_1}\underline{\tilde{a}}(\theta_1)^T\underline{w}(\theta_1)+\underline{n}_3^T\underline{w}(\theta_1) \quad (6\text{-}3)$$

$$\vdots$$

$$\underline{b}_M^T\underline{w}(\theta_1)=s_1e^{(M-1)\phi_1}\underline{\tilde{a}}(\theta_1)^T\underline{w}(\theta_1)+\underline{n}_M^T\underline{w}(\theta_1) \quad (6\text{-}M)$$

Where $\phi_i=j2\pi d/\lambda \sin\theta_i$. Combining (6-1)–(6-M) into a matrix notation gives

$$\left[\begin{matrix}\underline{b}_1^T \\ \underline{b}_2^T \\ \vdots \\ \underline{b}_M^T\end{matrix}\right]\underline{w}(\theta_1)=s_1\left[\begin{matrix}\underline{\tilde{a}}(\theta_1)^T \\ e^{\phi_1}\underline{\tilde{a}}(\theta_1)^T \\ \vdots \\ e^{(M-1)\phi_1}\underline{\tilde{a}}(\theta_1)^T\end{matrix}\right]\underline{w}(\theta_1)+\left[\begin{matrix}\underline{n}_1^T \\ \underline{n}_2^T \\ \vdots \\ \underline{n}_M^T\end{matrix}\right]\underline{w}(\theta_1)$$

$$=\left[\begin{matrix}s_1\underline{\tilde{a}}(\theta_1)^T+\underline{n}_1^T \\ s_1e^{\phi_1}\underline{\tilde{a}}(\theta_1)^T+\underline{n}_2^T \\ \vdots \\ s_1e^{(M-1)\phi_1}\underline{\tilde{a}}(\theta_1)^T+\underline{n}_M^T\end{matrix}\right]\underline{w}(\theta_1) \quad (7)$$

If we use $\begin{bmatrix} s_1\underline{\tilde{a}}(\theta_1)^T \\ s_1e^{\phi_1}\underline{\tilde{a}}(\theta_1)^T \\ \vdots \\ s_1e^{(M-1)\phi_1}\underline{\tilde{a}}(\theta_1)^T \end{bmatrix}$ as an approximation to $\begin{bmatrix} s_1\underline{\tilde{a}}(\theta_1)^T + \underline{n}_1^T \\ s_1e^{\phi_1}\underline{\tilde{a}}(\theta_1)^T + \underline{n}_2^T \\ \vdots \\ s_1e^{(M-1)\phi_1}\underline{\tilde{a}}(\theta_1)^T + \underline{n}_M^T \end{bmatrix}$ under the condition of a sufficiently high signal-to-noise (SNR), then equation (7) can be rewritten as

$$\begin{bmatrix} \underline{b}_1^T \\ \underline{b}_2^T \\ \vdots \\ \underline{b}_M^T \end{bmatrix} \underline{w}(\theta_1) \approx \begin{bmatrix} s_1\underline{\tilde{a}}(\theta_1)^T \\ s_1e^{\phi_1}\underline{\tilde{a}}(\theta_1)^T \\ \vdots \\ s_1e^{(M-1)\phi_1}\underline{\tilde{a}}(\theta_1)^T \end{bmatrix} \underline{w}(\theta_1)$$

$$= \begin{bmatrix} s_1\underline{\tilde{a}}(\theta_1)^T \\ s_1e^{\phi_1}\underline{\tilde{a}}(\theta_1)^T \\ \vdots \\ s_1e^{(M-1)\phi_1}\underline{\tilde{a}}(\theta_1)^T \end{bmatrix} \begin{bmatrix} w_1(\theta_1) \\ w_2(\theta_1) \\ \vdots \\ w_M(\theta_1) \end{bmatrix}$$

$$= s_1 w_1(\theta_1)\underline{\hat{a}}(\theta_1) + \cdots + s_1e^{(M-1)\phi_1} w_M(\theta_1)\underline{\hat{a}}(\theta_1)$$

$$= \alpha\underline{\hat{a}}(\theta_1) \tag{8}$$

where $\underline{\hat{a}}(\theta_i) = \left[1, e^{j2\pi d/\lambda \sin\theta_i}, \cdots, e^{j2\pi d/\lambda (M-1)\sin\theta_i}\right]^T$ and $\alpha = \left(s_1 w_1(\theta_1) + \cdots + s_1 e^{(M-1)\phi_1} w_M(\theta_1)\right)$, then

$$\underline{w}(\theta_1) \approx \alpha \begin{bmatrix} \underline{b}_1^T \\ \underline{b}_2^T \\ \vdots \\ \underline{b}_M^T \end{bmatrix}^+ \underline{\hat{a}}(\theta_1) \equiv \begin{bmatrix} \underline{b}_1^T \\ \underline{b}_2^T \\ \vdots \\ \underline{b}_M^T \end{bmatrix}^+ \underline{\hat{a}}(\theta_1) \tag{9}$$

where $[\ ]^+$ denotes the inverse matrix. Even though the α is skipped, the $\underline{w}(\theta_1)$ retains the same vector direction due to the α being a scalar. Equation (9) is an approximate solution for $\underline{w}(\theta_1)$, so $\underline{w}(\theta_1)$ must be normalized to reduce noise effect. By repeating this process, a weight vector $\underline{w}(\theta_i)$ is obtained, which is orthogonal to steering vectors except for steering vector, $\underline{\tilde{a}}(\theta_i)$. Now, the DOA of the desired signals from the vector $\underline{w}(\theta_i)$ and $\underline{b}_1$ can be estimated as

$$d(\theta_i) = \frac{w(\theta_i)^T \underline{b}_1}{\|\underline{w}(\theta_i)\|} = \underline{\hat{w}}(\theta_i)^T \underline{b}_1 \tag{10}$$

Where $\|\underline{w}(\theta_i)\|$ denote the norm of $\underline{w}(\theta_i)$ and $\underline{\hat{w}}(\theta_i)$ denote the normalized $\underline{w}(\theta_i)$. There are two possible conditions for $d(\theta_i)$ as listed below:

Condition 1) If there is an incident signal in the look direction θ_i, the vector $\underline{\hat{w}}(\theta_i)$will make nulls in the looking direction of another incident signals as

$$d(\theta_i) = \underline{\hat{w}}(\theta_i)^T \underline{b}_1$$

$$= \underline{\hat{w}}(\theta_i)^T [s_1\underline{\tilde{a}}(\theta_1) + \cdots s_i\underline{\tilde{a}}(\theta_i) \cdots + s_q\underline{\tilde{a}}(\theta_q) + \underline{n}_1]$$

$$= s_i \underline{\hat{w}}(\theta_i)^T \underline{\tilde{a}}(\theta_i) + \underline{\hat{w}}(\theta_i)^T \underline{n}_1 \tag{11}$$

Condition 2) if there is no incident signal in the looking direction θ_i, then equation (11) will be rewritten as

$$d(\theta_i) = \underline{\hat{w}}(\theta_i)^T \underline{b}_1 = \underline{\hat{w}}(\theta_i)^T \underline{n}_1 \tag{12}$$

Equation (11), which has an incident signal s_i, with added noise $\underline{n}_1$, is more powerful than equation (12), which contains only noise $\underline{n}_1$. Based on the above description, the DOA of the incident signals can be estimated by searching for local peak values of $P(\theta_i) = d(\theta_i)d(\theta_i)^*$ from all possible incident angles. Where * denotes the complex conjugate.

Table 1. Single snapshot DOA algorithm.

1. Divide the uniform linear array into high over-lapping subarrays of size M as in equation (2).
2. Calculate $\underline{w}(\theta_i)$ for all possible incident angles as in equation (9)
3. Search for local peak values of $P(\theta_i) = d(\theta_i)d(\theta_i)^*$ as the DOA of the inci-dent signal

2.2 *Computer simulation*

To test the proposed method, narrowband farfield DOA estimation examples, using a uniform linear array with 41 sensors and half-wavelength element spacing in the single snapshot case, were carried out. Two correlated equipower signals were impinging on the array from directions of $5^0$and 10^0. The noise at each antenna element was assumed to be additive.

Figure 1 shows that two peaks can correspond to the DOA of the simulated sources using the SS-DOA algorithm under the condition of SNR=20dB.

Figure 2(a) shows the comparison result of mean-square-error (MSE) versus different SNRs. Figure 2(b) shows the probability of success versus different SNRs using a reference (Zhang M. et al., 1991); including 30 independent test runs of the SS-DOA algorithm. When the number of incident signals is not correctly estimated (q=3) at the reference, we can find that the estimation performance of the reference will be severely degraded.

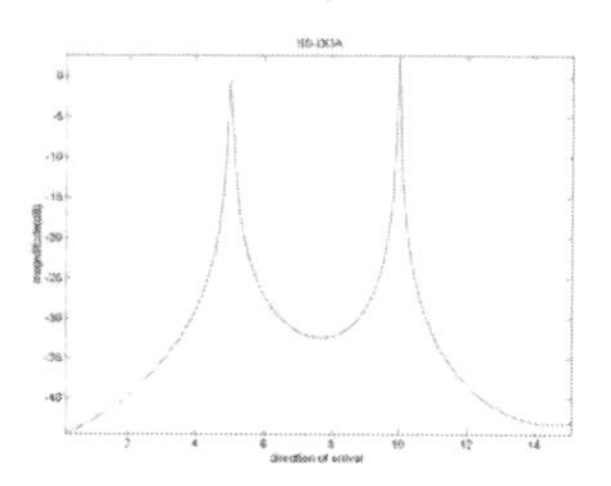

Figure 1. DOA estimation by the SS-DOA algorithm for two correlated signals having SNR of 20dB from 5^0 and 10^0.

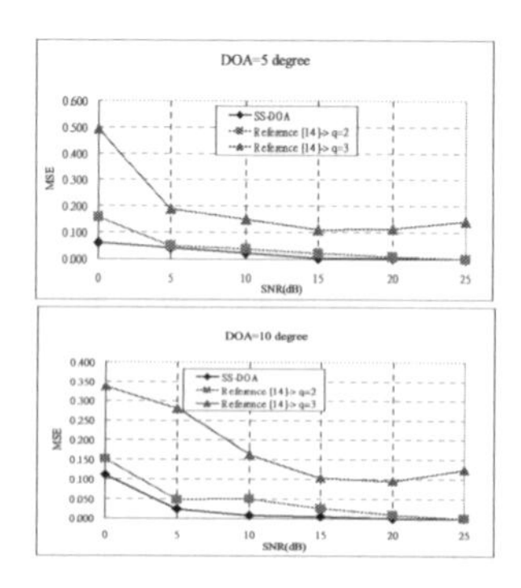

Figure 2(a). Comparison result of MSE by the reference (Zhang et al., 1991) and SS-DOA algorithm for two correlated signals from 5^0& 10^0 under different SNR.

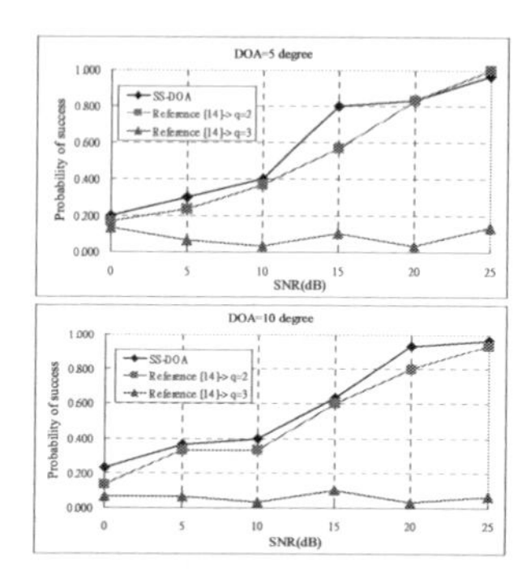

Figure 2b. Comparison result of probability of success by the reference (Zhang et al. 1991) and SS-DOA algorithm for two correlated signals from 5^0& 10^0 under different SNR.

3 CONCLUSION

A new direction of arrival estimation technique, using a subarray processing with application to the uniform linear sensor array in the single snapshot case, is presented. The major contribution in the new DOA estimation algorithm is that it does not require complicated computations. Also, it is very suitable for real-time processes, even in non-stationary and coherent interference environments. Furthermore, it allows for arbitrary probability density functions of incident signals since SS-DOA is a non-statistical technique. Computer simulations show that successful DOA estimation results can be obtained without prior knowledge of the number of signals, whereas the reference (Zhang et al. 1991) does not. Since this new approach improves performance by reducing computational complexity while maintaining sufficient resolution in non-stationary environments, it will have a wider range of prospective applications in real time DOA estimation compared to other methods.

REFERENCES

Capon, J. 1969. High-resolution frequency-wavenumber spectrum analysis. *Proc IEEE* 57:1408–1418.

Fuhl, J. & Molisch, A.F. 1995. Virtual-image-array single-snapshot algorithm for direction-of-arrival estimation of coherent signals. in *Proc. PIMRC'95* 2: 658–663.

Macinnes, C.S. 2004. Source localization using subspace estimation and Spatial Filtering. *IEEE Journal of Oceanic Engineering* 29: 488–497.

O'Brien et al. 2005. Single-Snapshot Robust Direction Finding. *IEEE Trans Signal Process* 53(6): 1964–1978.

Radich, B.M. & Buckley, K.M. 1997. Single-Snapshot DOA Estimation and Source Number Detection. *IEEE Signal Processing Letters.* 4(4): 109–111.

Roy, R. & Kailath, T. 1989. ESPRIT-Estimation of signal-parameters via rotational invariance techniques. *IEEE Tran ASSP.* 37: 984–995.

Schmidt, R.O. 1986. Multiple emitter location and signal parameter estimation. *IEEE Tr. Ant. Prop.* 34: 276–280,

Shan, T. et al. 1985. On spatial smoothing for direction-of-arrival estimation of coherent signals, *IEEE Trans Acoust Speech Signal Process* 33(4): 806–811.

Stoica, P. & Sharman, K. 1990. Maximum likelihood methods for direction of arrival estimation. *IEEE Trans Acoust Speech Signal Process.* 38: 1132–1143.

Williams, R.T., et al. 1988. An improved spatial smoothing technique for bearing estimation in a multipath environment. *IEEE Trans on Acoust., Speech Signal Processing* 36: 4425–4432.

Xin, J. & Sano, A. 2004. Computationally efficient subspace-based method for direction-of-arrival estimation without eigendecomposition, IEEE Trans. *Signal Process* 52(4): 876–893.

Yang J. & Kaveh, M. 1988. Adaptive eigensubspace algorithms for direction or frequency estimation and tracking. *IEEE Trans on Acoustics Speech and Signal Processing* 36:241–251.

Yang, B. 1995. Projection approximation subspace tracking. *IEEE Trans On Signal Proc* 44(1): 95–107.

Zhang M. et al. 1991. New method of constructing the projection matrix for array processing in single snapshot case. *IEE Proceedings F: Radar and Signal Processing.* 138(5): 407–410.

Electronics, Communications and Networks IV – Hussain & Ivanovic (eds)
© 2015 Taylor & Francis Group, London, ISBN: 978-1-138-02830-2

Modern machine learning techniques and their applications

Mirjana Ivanović & Miloš Radovanović
Department of Mathematics and Informatics, Faculty of Sciences, University of Novi Sad, Novi Sad, Serbia

ABSTRACT: During the last several decades Machine Learning (ML) became a mainstay of information technology. One view of ML is that it represents the science of finding patterns and making predictions from data, with the goal of creating systems that learn from experience. ML borders with multivariate statistics, data mining, pattern recognition, and advanced/predictive analytics. Increasing amounts of available data suggest that ML will become an even more pervasive component of technological progress. ML techniques usually belong to one of the main learning styles: supervised, unsupervised, semi-supervised, and reinforcement learning. Relevant methods include decision trees, nearest neighbor techniques, partitional/hierarchical/density-based clustering, Bayesian approaches, kernel methods, neural networks, and deep learning. The first part of our paper will give an overview of most important ML techniques. We will outline common problems: the bias-variance tradeoff, under/over-fitting, high dimensionality, and big data. The second part will highlight ML applications in business and telecommunications.

1 INTRODUCTION

Machine learning (ML) has grown into one of the most important fields of research spanning computer science, statistics, pattern recognition, data mining and predictive analytics. It has also become one of the mainstays from an application point of view, making strong headway into contemporary information technology and practice.

One significant view on machine learning is that it represents the modern science of finding patterns and making predictions from large amounts of data, with the goal of creating systems that learn from experience. Increasing amounts of data made available each day is good reason to believe that machine learning, data mining, and other related fields will become even more pervasive as a necessary component for technological progress.

Our goal is to describe and discuss a representative selection of machine learning techniques and applications of these techniques in the areas of business and telecommunications. Section 2 presents the techniques, by first discussing different learning styles (supervised, unsupervised, semi-supervised, and reinforcement learning), presenting a selection of relevant algorithms, and discussing some significant challenges faced by ML: the bias-variance tradeoff, over/underfitting, high dimensionality, and big data. In Section 3 we will highlight some of the recent applications of ML in areas of business and telecommunications, including predicting customer churn, personalization, recommendation, privacy-preserving data mining, as well as speech, gesture and handwriting recognition.

2 MACHINE LEARNING TECHNIQUES

The field of machine learning (ML) is concerned with the question of how to construct computer programs that automatically improve with experience (Mitchell 1997), where experience is gained in the form of examples coming from a set of collected data. Data mining (DM), also referred to as knowledge discovery from data (KDD), deals with concepts and techniques for uncovering interesting data patterns hidden in large data sets (Han & Kamber 2006). Despite the apparent difference in the central motivation, the two fields share many tasks, techniques, and data representations discussed in this paper.

General-purpose machine-learning techniques, like those for classification and clustering, are usually designed for examples which have a fixed set of nominal (symbolic), discrete (ordinal), or numeric (integer or continuous) features. A data set is then represented by a table where columns correspond to features, and rows are individual examples. For various types of data which do not conform to this format, or contain additional assumptions, there exist numerous ways to transform the data to this simple form, e.g., the "bag-of-words" representation for text data (Sebastiani 2002), feature extraction for images (Nixon & Aguado 2013), etc.

After discussing typical machine learning styles and tasks in Section 2.1, we will present a representative selection of algorithms in Section 2.2. Section 2.3 will examine some of the common challenges faced by machine learning methods.

2.1 *Learning styles*

Machine-learning methods typically fall into one of the following four learning styles: supervised, unsupervised, semi-supervised, and reinforcement learning. In supervised learning, computer programs capture structural information and derive conclusions from examples (also called instances), previously annotated by labels denoting *classes*. This enables supervised learning algorithms to process new examples and apply the conclusions to them. Unsupervised learning deals more with the *analysis* of data, in the sense of capturing relationships between examples and grouping them without relying on outside information. Semi-supervised learning is concerned with ways to combine the two paradigms, predominantly by using unlabeled data to assist supervised learning. Reinforcement learning is relevant to the scenario with an entity (robot, software agent, etc.) immersed in an environment, where the goal of the learning algorithm is to derive correct responses and behavior, data consists of sensory input from the environment, and supervision comes in the form of rewards/punishments for correct/incorrect actions.

ML tasks can roughly be divided into four distinct areas: classification, clustering, association learning, and numeric prediction (Witten et al. 2011). In the machine-learning approach, classification algorithms (classifiers) are trained beforehand on previously sorted (labeled, classified) data, before being applied to classifying unseen examples.

Clustering is a basic unsupervised learning task concerned with finding groups of examples based on some notion of similarity between them. While classification is concerned with finding models by *generalization* of evidence produced by a data set, clustering deals with the *discovery* of models which describe patterns in data, with little or no external guidance.

Association learning can be viewed as a generalization of classification which aims to capture relationships between arbitrary features (also called attributes) of examples in a data set. In this sense, classification captures only the relationships of all features to the one feature specifying the class.

Numeric prediction (also called *regression*, in a wider sense of the word), may be viewed as another generalization of classification, where the class feature is not discrete, but continuous. This small shift in definition results in large differences in the internal workings of classification and regression algorithms. However, by dividing the predicted numeric feature into a finite number of intervals, regression algorithms can generally also be used for classification, while the opposite is not usually possible.

Another important task associated with machine learning and data mining is outlier detection, which is concerned with locating examples which are in some sense different from the majority of others. Besides being valuable in its own right for many applications, outlier detection is also considered an important preprocessing step for application of learning methods discussed above, particularly clustering.

It is widely acknowledged that a large number of features, that is, high dimensionality of a data set, can cause severe problems in many machine-learning applications. These problems are commonly referred to as the curse of dimensionality. A solution that is often employed is to reduce the dimensionality of the data space by selecting only a subset of the feature set, or applying some transformation on data points, projecting them to a feature space of lower dimensionality. Techniques for achieving this task are collectively known as *dimensionality-reduction* methods.

2.2 *Algorithms*

In this section we will present some of the most common and useful machine-learning algorithms belonging to the tasks/styles of classification (Section 2.2.1) and clustering (Section 2.2.2). Due to its specific nature, we will discuss deep learning in a dedicated Section 2.2.3.

2.2.1 *Classification*

Perceptrons and neural networks. The perceptron, originally introduced by Rosenblatt (1959), is a binary classifier which uses the value of the inner product of vectors $\mathbf{w} \cdot \mathbf{x}$ to classify instance $\mathbf{x}$ according to the previously learned weight vector $\mathbf{w}$. If the inner product, summed with some value b (called the bias, or threshold), is greater than or equal to 0, the instance is assigned to the positive class, and vice versa. More precisely, for binary class $c \in \{-1, 1\}$, $c = \text{sg}(\mathbf{w} \cdot \mathbf{x} + b)$, where $\text{sg}(\mathbf{x}) = 1$ if $\mathbf{x} \geq 0$, and $\text{sg}(\mathbf{x}) = 0$ otherwise. This means that $\mathbf{w}$ and b define a hyperplane which linearly separates the vector space, as exemplified in Figure 6(a) for the two-dimensional case.

Figure 1 (adapted from Matthews 2010) shows a schematic representation of the perceptron. Every input x_i represents a component of vector $\mathbf{x}$, and has an associated weight w_i ($i \in \{1, 2, \ldots, d\}$). The bias can be viewed as constant input +1, with the associated weight b. To compute the output of the perceptron, every input is multiplied by its corresponding weight, the sum is taken, and the classification decision made based on its sign (the sigmoid function $1/(1 + e^{-x})$ is often used instead of the sign).

Learning the vector w starts by assigning it a 0 vector (or a vector of small positive weights) and continues by examining each training instance $\mathbf{x}$ one at a time, classifying it using the currently learned $\mathbf{w}$. If the classification is incorrect, the vector is updated: $\mathbf{w} \leftarrow \mathbf{w} \pm \eta\mathbf{x}$, where addition (subtraction) is used

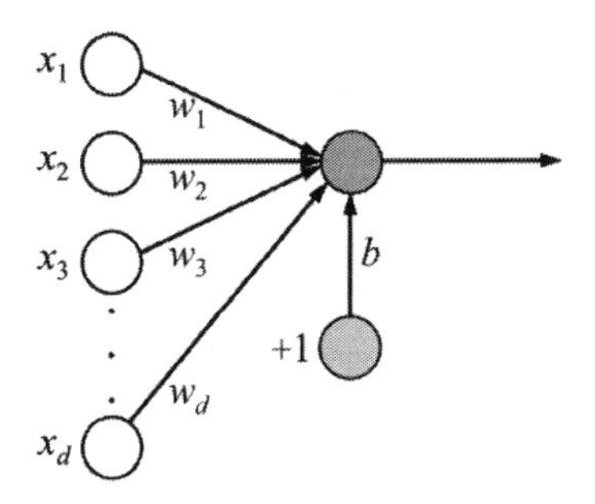

Figure 1. The perceptron.

when **x** belongs to the positive (negative) class, and η is a small positive number – the *learning rate*. The effect of the update is to shift the weights of **w** towards the correct classification of **x**, in proportion to their "importance" signified by the values of weights in **x**. The algorithm iterates multiple times over instances in the training set, until all examples are classified correctly, or some other stopping criterion is met.

Perceptrons are the building blocks of a common type of neural network called the multilayer feed-forward network or multilayer perceptron. Here, perceptrons are joined into a network organized into layers: an input layer (to which data is passed), an output layer (outputting predictions), and zero or more hidden layers in between. Connections between nodes in the network are formed exclusively between layers, with each connection having an associated weight. Training the network is naturally more complex than training a single perceptron, but can be viewed as an extension of this process. One common approach is to start with randomly assigned weights and biases, for an input data instance propagate the outputs of the perceptrons layer by layer from the input layer towards the output, compute the error of the output, propagate this error back towards the input layer, and finally adjust the weights and biases based on the estimated contribution of each node to the error. This approach is referred to as backpropagation (Hastie et al. 2009). Various challenges in training neural networks for classification are surveyed by Zhang (2000).

Support vector machines. One of the most sophisticated classifiers is the support vector machine (SVM) classifier. SVM is a binary classifier, and its main idea lies in using a predetermined *kernel function*, whose principal effect is the transformation of the feature vector space into another space, usually with a higher number of dimensions, where the data is linearly separable. Quadratic programming methods are then applied to find a *maximum margin hyperplane*, that is, the optimal linear separation in the new space, whose inverse transformation should yield a good classifier in the original vector space. Figure 2 (from Sebastiani 2002) shows a graphical representation of the separating hyperplane for a two-dimensional space (after the transformation), where the class is depicted with labels + and °. The hyperplane, in this case a line, lies in the middle of the widest strip separating the two classes, and is constructed using only instances adjacent to the strip: the *support vectors* (outlined by squares in the figure).

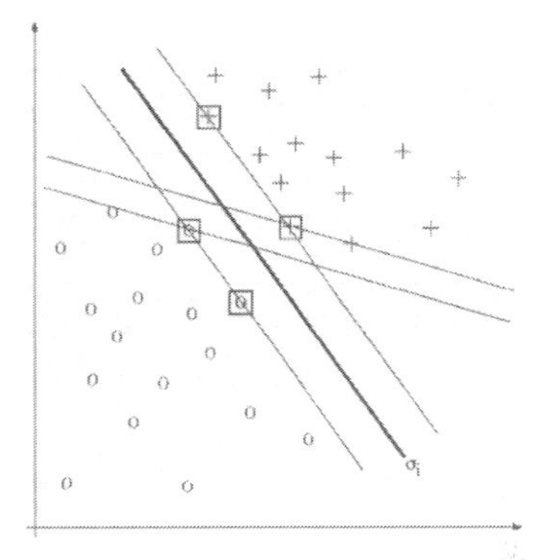

Figure 2. The maximum margin hyperplane separating the two classes, with highlighted support vectors.

Although the theoretical foundations for support vector machines were laid out in the 1970s (Vapnik & Chervonenkis 1974, Vapnik 1979), the computational complexity of various solutions to the quadratic programming problem restricted the use of SVMs in practice. Only relatively recently were approximate solutions derived which enabled feasible and, compared with some other classifiers, superior training times. One solution was proposed by Osuna et al. (1997), improved by Joachims (1999) and implemented in his *SVM^{light}* package. An alternative is Platt's *sequential minimal optimization* (SMO) algorithm (Platt 1999, Keerthi et al. 2001), available, for instance, as part of the Weka machine-learning workbench (Witten et al. 2011).

Support vector machines can handle very high dimensionality, and are not particularly sensitive to overfitting. SVMs belong to a larger family of kernel methods, which has drawn in, through introduction of a kernel function, numerous classic machine-learning algorithms, producing their "kernelized" versions, such as kernel perceptron, kernel K-means, etc. (Shawe-Taylor & Cristianini 2004).

Bayesian learners. The probabilistic approach to modeling data has resulted in several useful machine-learning techniques which can be used on high-dimensional data. One of them is the simple, but effective *naïve Bayes classifier*, and another, more expressive but also more complex and still actively researched – *Bayesian networks*.

Naïve Bayes. The naïve Bayes classifier has been "around" for a long time, but was initially more in the focus of information retrieval, rather than the machine-learning community (Lewis 1998). The main principles of its functioning are as follows. Let random variable C denote the class feature, and A_1, A_2, ..., A_d the d components of the attribute vector. Then, the classification of a specific vector $(a_1, a_2, ..., a_d)$ is

$$c = \underset{c_j \in C}{\operatorname{argmax}} \; \mathrm{P}(c_j | a_1, a_2, \dots, a_d),$$

providing that one class maximizes the expression. Application of the Bayes theorem transforms the expression to

$$c = \underset{c_j \in C}{\operatorname{argmax}} \; \frac{\mathrm{P}(a_1, a_2, \dots, a_d | c_j)\, \mathrm{P}(c_j)}{\mathrm{P}(a_1, a_2, \dots, a_d)}$$

$$= \underset{c_j \in C}{\operatorname{argmax}} \; \mathrm{P}(a_1, a_2, \dots, a_d | c_j)\, \mathrm{P}(c_j)$$

$$= \underset{c_j \in C}{\operatorname{argmax}} \; \mathrm{P}(c_j) \prod_{i=1}^{d} P(a_i | c_j)$$

The last derivation uses the assumption that attributes are mutually independent, which generally does not hold in reality, hence the prefix "naïve." Nevertheless, the assumption has been shown to work in practice. Training involves approximating the values $\mathrm{P}(c_j)$ and $\mathrm{P}(a_i \mid c_j)$ from data. Several approaches exist, depending on the assumed data distribution. In the classification phase, if multiple classes maximize the last expression in the derivation, different strategies may be employed to resolve the ambiguity, e.g., by selecting the class with the highest prior probability $\mathrm{P}(c_j)$, or simply by choosing one of the classes randomly.

Bayesian networks. Note that without the independence assumption in naïve Bayes, estimating the values of $\mathrm{P}(a_1, a_2, ..., a_d | c_j)$ would have been infeasible. However, data attributes usually *are* interrelated, and one way to capture such dependencies is by using Bayesian networks.

Generally speaking, Bayesian networks consist of nodes which are random variables, and vertices representing conditional probabilities between them. Their aim is to offer a computationally feasible and graphically representable way to express and calculate dependencies between events.

Again, it would be computationally infeasible (and not even allowed in a Bayesian network) to calculate dependencies between *all* attributes, especially in high-dimensional data. The objective of Bayesian networks is to express only the dependencies that are necessary (or strong enough to have an impact on the solution to a particular problem), under constraints which ensure the correctness and feasibility of computation. This can be done manually, by supplying the structure of the network – then training a Bayesian network resembles the training phase of the naïve Bayes classifier, with conditionals being estimated from the data set. If estimation of dependencies from data is not possible, training becomes more difficult, with several solutions being available (Mitchell 1997). Learning the *structure* of the network presents a bigger challenge, and is still an area of active research. More comprehensive information can be found in several books devoted to the subject of Bayesian networks (Koller & Friedman 2009, Jensen 2007, Neapolitan 2003).

Nearest-neighbor classifiers. The training phase of *nearest-neighbor* (also known as *instance-based*, or *memory-based*) classifiers is practically trivial, and consists of storing all examples in a data structure suitable for their later retrieval. Unlike other classifiers, all computation concerning the classification of an unseen example is deferred until the classification phase. Then, k instances most similar to the example in question – its *k nearest neighbors* – are retrieved, and the class is computed from the classes of the neighbors. The computation of the class can consist of selecting the majority class of all the neighbors, or distance weighing may be used to reduce the influence of faraway neighbors on the classification decision. The choice of k depends on the concrete data and application – there is no universally best value. The most frequently used distance/similarity measure for determining neighbors is Euclidean distance, with a plethora of other choices available. With richer representations of instances and more complex similarity functions the issue moves into the field of case based reasoning (Mitchell 1997, Kurbalija et al. 2009).

To illustrate the workings of the k-nearest neighbor (k-NN) classifier, for the simple case of $k = 1$ and data dimensionality 2, Figure 3 shows the Voronoi tessellation of the data space by training points (Taskar 2010), where the region surrounding every point represents the area in which the point is the nearest one of all points in the training set. At classification phase, the label of an unseen point is determined as the label of the training point in the corresponding region. Therefore, depending on the labeling of training points, the boundaries between classes for the 1-NN classifier model follow the boundaries between regions.

Decision trees. A decision tree (DT) is a tree whose internal nodes represent features, where arcs are labeled with outcomes of tests on the value of the feature from which they originate, and leaves denote categories. The decision tree constructed from a simple weather data set is shown in Figure 4. Classifying

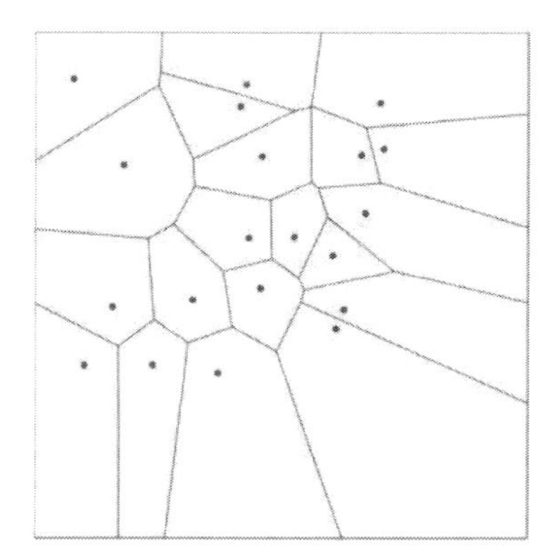

Figure 3. Voronoi tessellation of the data space, for the 1-NN classifier.

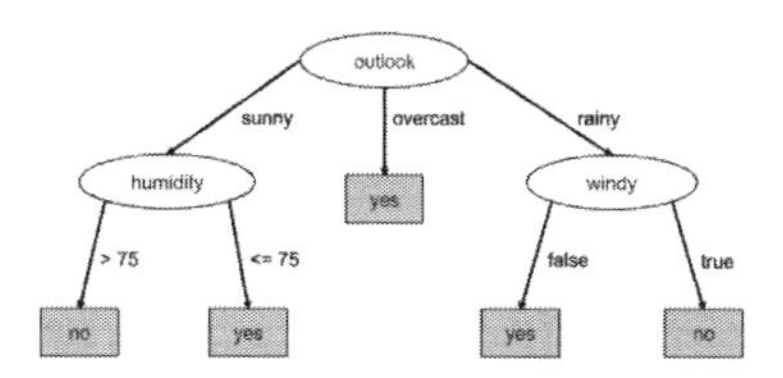

Figure 4. The decision tree generated from weather data.

a new instance using a decision tree involves starting from the root node and following the branches labeled with the test outcomes which are true for the appropriate feature values of the instance, until a leaf with a class value is reached.

Two of the most widely used decision-tree learning algorithms are classification and regression trees (CART) by Breiman et al. (1984), and C4.5 by Quinlan (1993). Learning a decision tree with C4.5 involves selecting the most informative feature using a combination of information-gain and gain-ratio methods for feature scoring, determining how best to split its values using tests, and repeating the process recursively for each branch/test, without considering features which were already assigned to nodes. Recursion stops when the tree perfectly fits the data, or when all features have been used up. The tree in Figure 4 was generated using the C4.5 algorithm.

To avoid overfitting (see Section 2.3.2), pruning can be performed on the learned tree, which reduces its fit to the training data, at the same time attempting to improve its accuracy in the general case. C4.5 performs this task by converting the tree to an equivalent rule form (one rule for each path in the tree from root to leaf), estimating the general accuracy of each, and improving it by removing some tests. Then, rules are sorted in decreasing order of estimated accuracy and used in this form for classification.

Decision trees are especially useful when the workings of the classifier need to be interpreted by humans, offering insight into the structure of data. As for text data, DTs may be unsuitable for many applications since they are known not to be able to efficiently handle great numbers of features. Nevertheless, sometimes they do prove superior, for instance with data sets in which a few highly discriminative features stand out from the many (Gabrilovich & Markovitch 2004).

Hidden Markov models. When a data instance represents a *sequence*, e.g., features in the data table are measurements of a numeric value produced from a single source over time (such as speech), or appearances of different types of molecules in biological sequences (such as DNA), a whole class of sequence learning techniques can be used. Classic examples of these techniques are Markov chains and hidden Markov models (HMMs).

Both types of models are probabilistic, and are composed of states and possible transitions between them. The probability of a state depends only on the previous state, with states emitting symbols once a transition to them is made. A Markov chain can be defined by a set of states with associated symbols, and a matrix of transition probabilities between the states. Given a data set of training sequences, this model can be learned from the data.

When sequences are long, a common task is to associate appropriate class labels to parts of a sequence. This can be accomplished with a hidden Markov model, which could be viewed as containing multiple "hidden" Markov chains (each modeling one class), themselves interconnected with associated transition probabilities. Algorithms for training HMMs include the forward algorithm, the Viterbi algorithm, and the Baum-Welch algorithm (Rabiner 1989, Han & Kamber 2006).

2.2.2 *Clustering*

***K*-means clustering.** The basic K-means clustering algorithm is one of the oldest and simplest clustering algorithms suitable for application to high-dimensional data, which may still produce good results. It involves randomly choosing K points to be the centroids of clusters, and grouping instances around centroids based on proximity. Then, centroids are iteratively recomputed for each cluster, and instances regrouped until there is sufficiently little change in centroid positions. This algorithm depends heavily on the choice of K (which may not be obvious at all for a particular application), and the initial positioning of centroids. Having K-means generate empty clusters is not a rare occurrence.

Instead of explicitly assigning examples to clusters (*hard* assignment), each cluster can be represented by a vector of features and updated on witnessing an example (*soft* or *fuzzy* assignment), based on proximity. That way, representations of clusters are not limited to centroids and may fit some data distributions more naturally.

Hierarchical clustering. These techniques derive a nested hierarchy of clusters, with the extreme of a single cluster containing all instances on one end, and a collection of one-element clusters on the other. One such partition is shown by the *dendrogram* in Figure 5 (from Chakrabarti 2003). The hierarchy can be constructed from the bottom up (*agglomerative* approach), by starting with single-element clusters and merging two of the most similar in each step, and from the top down (*divisive* approach), by repeatedly dividing the cluster with least internal similarity. In both approaches, a concrete set of clusters can be obtained simply by choosing a suitable degree of inter-cluster similarity and "reading off" the clusters from the dendrogram using a horizontal cut-off.

Compared to the divisive approach, the agglomerative approach is relatively straightforward, with the main issues concerning the choice of the inter-cluster distance metric and optimizing the search for most similar clusters. Methods for determining the distance between two clusters include: *single link*, where the distance between two clusters is taken to be the minimum distance (maximum similarity) between two points from the two different clusters, *complete link*, where cluster distance is defined as the maximum distance (minimum similarity) between two points from the two different clusters, and *group average*, where average distance between all pairs of points from the different clusters is used (Tan et al. 2014). The divisive approach, on the other hand, offers a wide variety of ways to split the chosen cluster. One way is to use the basic K-means algorithm multiple times with different choices of starting points, and select the split with the highest overall similarity, which is referred to as *bisecting K-means* (Steinbach et al. 2000).

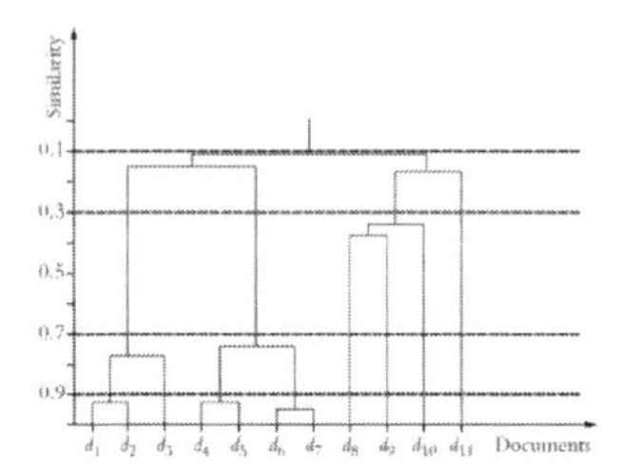

Figure 5. A dendrogram representing a hierarchical clustering of a set of examples (in this case, documents).

Probabilistic clustering. In the probabilistic clustering approach, instances are considered to be generated from a *mixture model* of k probability distributions, by first choosing model j with probability p_j, and then drawing an example adhering to the distribution (Berkhin 2006). Each cluster corresponds to a distribution, with instances gathering around its mean at distances determined by variance. The *likelihood* that a particular data set is drawn from a particular mixture model of k distributions is given by

$$L(X|R) = \prod_i \sum_j p_j P(\mathbf{x}_i | r_j),$$

for instances $\mathbf{x}_i$ and clusters r_j. One probabilistic method, the EM algorithm (Mitchell 1997), is based on alternatively estimating (the "E" step) and maximizing (the "M" step) the expected value of the *log-likelihood* function $\log L(X|R)$.

Benefits of probabilistic clustering include the ability to build clusters using different data sets (because clusters are represented independently from examples), iterative examination of instances, and output of results which are easy to interpret (Berkhin 2006). More details on the probabilistic approach to clustering can be found in (Chakrabarti 2003).

Spectral clustering. Spectral clustering refers to a family of algorithms designed for situations where clusters in the data can have predominantly nonconvex shapes (Bishop 2006). Starting from a matrix of pairwise similarities of points in a data set, an adjacency matrix is formed using, for example, mutual k-nearest neighbors or ε-neighborhoods, resulting in an undirected graph whose edges are weighted by similarities between nodes. Some notion of a graph Laplacian matrix is then computed, and the eigenvectors corresponding to the smallest eigenvalues are found (the set of eigenvalues of a matrix is referred to as "the spectrum," giving rise to the name of the clustering method family). The final step usually involves using some standard clustering algorithm like K-means to find the groups of points.

The most well-known spectral clustering algorithms are described in the works by Shi & Malik (2000), Ng et al. (2002), and Meilă & Shi (2001), with the principal differences being in the type of graph Laplacian adopted (von Luxburg 2007).

The objective behind spectral clustering methods is to identify different groups of data points by finding local neighborhoods within a graph. There exist several viewpoints on how this objective should be achieved, resulting in different approaches to spectral clustering, and explanations as to why spectral clustering methods work: the graph cut point of view (partitioning the graph so that the edges between different groups have a low weights, and the edges within a group have high weights), the random walk point of view (through a stochastic process which jumps from node to node), and the perturbation theory point of view (examining the behavior of eigenvalues and eigenvectors with the introduction of small changes to the matrix, that is, perturbations) (von Luxburg 2007).

2.2.3 *Deep learning*

Deep learning consists of a family of methods belonging to a broader area of *representation learning* (Bengio et al. 2013). Instead of working with a fixed set of features (which was assumed by all methods discussed above), these methods learn the representations themselves in conjunction with building the model for the main concept being learned. Moreover, deep learning approaches derive featural representations in a layered manner, starting from a low level of abstraction and working the way to higher levels of abstraction which rely on the features from the lower levels, before finally reaching the layer responsible for learning the primary concept.

For example, in the domain of text the input to the learner can consist solely of characters, while different layers are able to derive appropriate features denoting words, word groups, clauses and sentences, all at increasing levels of abstraction, until finally reaching the level of the main problem (e.g., document summarization).

Deep learning is inspired by the perceived way in which the human brain works (Arel et al. 2010), and is regarded as breathing new life to the field of neural networks, moving closer to the goal of achieving "true artificial intelligence."

Classic techniques for deep learning include convolutional neural networks (CNN) and deep belief networks (DBN). CNNs represent a family of multi-layered neural networks designed for use on two-dimensional data, such as images and videos, while DBNs are probabilistic models that combine multiple layers of restricted Boltzmann machines, which is a type of neural network (Arel et al. 2010). The current state-of-the-art methods in terms of success are recurrent long short-term memory (LSTM) neural networks and variations of CNNs (Schmidhuber 2014).

For comprehensive overviews of this complex and vibrant field the reader is referred to articles by Deng & Yu (2014), Schmidhuber (2014), and Arel et al. (2010).

2.3 *Challenges*

In order for machine learning to be successful, it is often faced with different challenges. We will discuss some of the most common, namely the bias-variance tradeoff (Section 2.3.1), the related notions of overfitting and underfitting (Section 2.3.2), high dimensionality (Section 2.3.3), and big data (Section 2.3.4).

2.3.1 *The bias-variance tradeoff*

In the supervised learning setting, the error which an algorithm makes while performing prediction can be split into three components: bias, variance and irreducible error (Hastie et al. 2009). While the last component cannot be controlled, the first two can be affected by tuning algorithm parameters and by other means. Bias refers to the notion of how consistently the learned model is "right" or "wrong," compared to the "ground truth." On the other hand, variance expresses how "smooth" the model is, with larger variance indicating that the model is more complex and that small changes in data points can lead to radical changes to outcomes of prediction. Generally, while attempting to increase the accuracy of supervised learning, reducing bias will tend to increase variance and vice versa, with the overall goal becoming that of finding and optimal balance between the two controllable sources of error.

2.3.2 *Overfitting and underfitting*

Overfitting is a related notion to the bias-variance tradeoff within supervised learning, and refers to the notion that a classifier can be trained "too much," in the sense of maximizing its performance on the training set, which may in fact lead to suboptimal performance on a separate test set and real life data. Overfitting may come as a consequence of a small or large number of training instances, noisy data, and/or high dimensionality. Some classifiers are more prone to overfitting than others, and many of them employ complex strategies to avoid it. The philosophical equivalent of the problem lies in the *Occam's razor* principle, which in ML terms translates to preferring a simple model which reasonably fits the data, to a complex one which does so more accurately.

To illustrate, consider one binary classification problem, in a two-dimensional feature space, of separating salmon and sea bass on evidence of their width and lightness of scales, shown in Figure 6 (from Duda, Hart, & Stork 2001). The data is clearly not linearly separable, therefore the linear classifier in Figure 6(a) leaves several misclassified examples on both sides of the boundary. On the other hand, the complex model in Figure 6(b) perfectly fits all examples, making great efforts to "pick up" every fish which strayed deep into the waters occupied by instances of the other species. This leads to whole regions being marked for one class on the evidence of a single example, when, considering all surrounding examples, they have a greater probability of belonging to the other class. In one such region a new example is marked by a question mark in the figure – it will be classified as sea bass although it is more likely to be a salmon, considering the surroundings. In all probability, the linear model will perform the same or better on real-world data than the complex one, at the same time being much simpler to derive, apply, and maintain. In this example, the complex model has lower bias, but higher variance than the simple model.

Underfitting is the opposite extreme of overfitting, where the derived model is too simple, and thus not

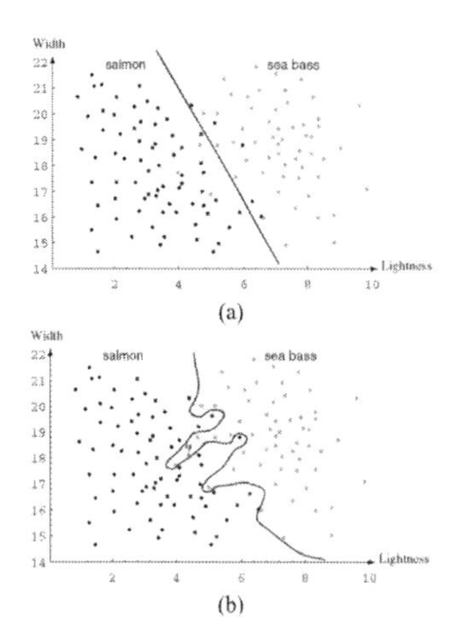

Figure 6. A simple and complex model for binary classification in a two-feature space.

able to accurately describe the learned concept. In this setting the variance is low since the model is simple, but bias is high.

2.3.3 *High dimensionality*

The basic representation in which information is gathered and stored, the data table (data set), can have a large number of rows (instances) and/or a large number of columns (features). The second case is typically referred to as high dimensionality, and is able to cause problems in tasks related to many fields, including machine learning. These problems are usually denoted by a common term the curse of dimensionality originally introduced by Bellman (1961).

The curse of dimensionality can be manifested in many different forms and contexts. Within the field of machine learning, the curse can affect Bayesian modeling (Bishop 2006) by making the estimation of mutual dependencies between features (as random variables) infeasible. Nearest-neighbor prediction is also affected (Hastie et al. 2009), for example, by the exponential raise of the number of required samples of data points to achieve a required sampling density. The search for nearest neighbors may suffer from high dimensionality (Korn et al. 2001), since indexing methods tend to lose their effectiveness in high dimensions (Kibriya & Frank 2007). High dimensionality is also known to reduce the performance of learning methods such as neural networks (Bishop 1996).

Recently it was established that k-NN graphs obtained from high-dimensional data using a distance measure can possess hub nodes similarly to different types of real-world networks with a different origin (Radovanović et al. 2010). This phenomenon, referred to *hubness*, produces effects relevant to the cluster structure of data, as well as algorithms for classification, clustering, semi-supervised learning, outlier detection, and others. Hubness is manifested as (intrinsic) data dimensionality increases, with the distribution of data point in-degrees, i.e., the number of times points appear among the k nearest neighbors of other points in the data, becoming very skewed. This results in hub points that can have in-degrees multiple orders of magnitude higher than expected. Different approaches to reducing hubness in the data, or employ them to improve learning, were proposed (Hara et al. 2015, Schnitzer et al. 2012, Tomašev and Mladenić 2012, Tomašev et al. 2014, Tomašev and Mladenić 2014).

2.3.4 *Big data*

"Big data" refers to data sets whose size is beyond the ability of typical database software tools to capture, store, manage, and analyze (Manyika et al. 2011). Besides sheer volume, other factors are often encompassed by this term, such as data variability (inconsistency), velocity (speed by which new data is being generated) and complexity (of structure, relationships, etc.). The big data setting presents a clear challenge to ML techniques from the computational standpoint, bringing back into focus the simple techniques (such as K-means clustering) which are computationally more efficient, and easier to apply in a parallel and distributed fashion. High-dimensional data can be viewed as one aspect of big data, where "big" does not reside in the number of instances, but rather in the number of features used to describethem.

3 MACHINE LEARNING APPLICATIONS

Data management based on rapid growth of on-line data and use of statistical methods challenged extracting useful insights from different data sources. Key advances in robust and scalable data mining techniques, detection of significant patterns from very large databases, employment of machine learning techniques for business applications are recognized in the last two decades as an emergent research topic.

Different research institutions (Apte 2010) are oriented towards business analytics and optimization in order to help businesses optimize their important decisions. Advanced methods caused a shift in businesses and data management, from the traditional descriptive analytics, to use of predictive and prescriptive analytics. Descriptive analytics gives answers to questions such as: "what happened," "how many times," and "where." To obtain a prospective view on the business it is unavoidable to use predictive analytics and get answers of the kind: "what could happen," "what might happen in the future if ..." Prescriptive analytics could improve business under a given set of predictions and constraints and give an answer to the key question: "what is the set of required actions."

Starting from the early 1990s and first successful machine learning applications (in the manufacturing quality control and computer performance management areas, inside IBM) different research institutions continued with processing immense amounts of business data like: marketing campaigns, financial credit, airline reservation systems data, telecommunications calls, and so on.

In the rest of this section we will focus on the following application areas: business (Section 3.1), telecommunications (Section 3.2), and various types of recognition (Section 3.3).

3.1 *Examples of using ML in the business domain*

For processing big data (Vossen 2014) as large and complex collections it is necessary to apply wide range of methods and techniques, and machine learning algorithms play a very important role. To solve complex business problems, different research institutions have oriented their research efforts towards leveraging more advanced ideas from machine learning. They tried to overcome challenges from special characteristics of the data like: high dimensionality, link information, insufficient number of labeled examples, timely analysis requirements. Numerous successful applications have been developed including credit card fraud detection, marketing optimization, life sciences and healthcare management, and sustainability management.

The interplay between machine learning and optimization represents a new dimension as well. New kinds of decision support systems that incorporate prediction and prescription require coupling predictive modeling and optimization for obtaining adequate solution plans. Reinforcement learning is becoming an unexpectedly useful paradigm for many business applications.

3.2 *ML techniques in the telecommunications industry*

The telecommunications industry is an extremely dynamic area with a large base of customers. An important factor for competition and profitability of a telecommunications company is the development of effective strategies to attract more customers, but also to retain the existing ones. Therefore, it is crucial to reduce the level of customer churn.

Several interesting research results on the application of machine learning and analytical tools in telecommunications and prediction of churn are presented by Xevelonakis (2005), Oseman et al. (2010), Jahromi (2009) and Lu (2005). Xevelonakis (2005) focused on several aspects in proposing strategies for customer retention. More recent work by Oseman et al. (2010) is based on the application of a decision tree classifier for churn prediction using data customer geographical region. A dual-step model building approach is presented by Jahromi (2009). The idea is first to cluster different types of customers and then to learn the classification models separately for each cluster. A somewhat different approach is presented by Lu (2005). Using survival analysis apart from prediction of churners, the author also tried to estimate how soon customers would churn.

To illustrate the use of ML techniques in telecommunications we will outline several particular cases.

CASE 1: ML technologies in telecommunications

Different kinds of human interaction with telecommunication services/networks contain unconnected peculiarities of information. This amount of data could be analyzed and correctly interpreted to give insights into behavior and interaction of people.

Research by Svensson & Söderberg (2008) is concentrated on three general scenarios where ML technologies can be successfully used: recommendations, personalization, and media recognition.

Business scenario 1: recommendations. Recommendation services, based on the analysis of available data, have to provide hints about a user's preference for certain information. In fact it is important to identify the services interesting for a user. A recommender for a given service identifies users with high probability for uptake (based on similarities among service-usage patterns). Apart from that service recommenders may also be used to boost existing services or to identify why users do not adopt some services. In addition recommenders can be used to predict churn – using information derived from the usage patterns of past churners, one can detect changes in other usage profiles.

Business scenario 2: personalization. Personalization is one of the buzz words in contemporary ICT research. In the telecommunications area it is based on adapting operator services and offerings to the needs and preferences of the customer base. ML techniques support automation of process of dividing customers into groups; for example assigning a profile to customers according to calling and messaging behavior. Apart from that, operators can by personalizing their offerings reduce spam, by directing adequate marketing to users with a certain profile.

One of the more frequent business scenarios is personalized (targeted) advertising, where advertisements are tailored to an individual and situation. Operators and advertisers can focus on customers with ads that fit their needs and interests. Also, it is possible to separate the "typical users" associated with a given service.

Business scenario 3: media recognition. Recognition in different media as pictures, sound, video is very important nowadays. Based on a given classification of known patterns, the goal is to identify the category to which previously unseen patterns belong. Svensson & Söderberg (2008) recognized the activities of photo tagging and mobile tagging as particularly important.

Photo tagging. Operators can facilitate a community of users to share different image metadata using collaborative indexing. Using annotations of other subscribers and available image classification system learned to recognize a large set of different images, users can have great benefits.

Mobile tagging. Modern mobile devices can facilitate photo tagging by detecting images and objects. Automatic tagging functions can suggest to the user the class of the photo (s)he has taken and suggest an annotation keyword as well. Such annotations and visual descriptors for the photos are stored in a common repository.

Software systems and architectures. During the long history of use of ML techniques in real domains, different software systems (and architectures) have been developed. Majority of the architectures follow the same basic scheme (e.g., the cross-industry standard process for data mining, www.crisp-dm.org) and comprises of the following components:

A data source collection tier stores raw data and cleans them. This component is essentially based on traditional extraction, transformation and load tools developed by the IT industry.

A data preparation tier supports preparation / aggregation of data (also removes redundant information and applies domain specific knowledge) and make them suitable for analysis by the ML component.

An analysis tier is responsible for applying adequate ML algorithms to data. The idea is to train models that will be used for solving concrete business questions.

A presentation tier, usually using visualization tools, presents derived knowledge in a suitable form.

Nevertheless, to describe tiered conceptual architecture the data-mining process is in nature iterative. To obtain a satisfactory answer to a use case many of mentioned steps must be iterated several times. Like in other domains, in telecommunications to get a lot of advantages from available data resources, suppliers must design systems specifically for the domain. With constantly evolving networks and new protocols/nodes/services, the sources of raw data are heterogeneous and generate huge amounts of data. In order to trust output from the ML system operators must be certain that the correct domain specific adaptation has been made.

CASE 2: Churn prediction in telecommunications
Preventing customer churn, as one of challenges of global companies' competition, is significant concerns in different industries. The telecommunication sector with a churn rate of 30%, is in the first position. Movement from one provider to another is a consequence of better rates, services, or other different benefits that a competitor company offers.

Predictive models are in charge of trying to solve this problem. They can be employed to identify customers who are at risk of churning. Brandusoiu & Toderean (2013) proposed an advanced methodology for predicting customer churn in mobile telecommunications.

Usually companies employ a defensive marketing strategy to keep their customers. So they need a method to identify possible churning customers and use proactive retention campaigns. Thanks an increased performance generated by ML algorithms, churn prediction modeling is highly dependent on the data mining methods. Brandusoiu & Toderean (2013) used support vector machines to build the predictive model, and report on experimental evaluation with accuracy of success of around 80%.

As in the testing set there were 15% of churners and 85% of non-churners present, in order to train the SVM algorithm successfully Brandusoiu & Toderean (2013) cloned the "yes" labels they became nearly equal in number to the "no-s." They proposed four SVM models and after evaluation their three SVM models that use the RBF, linear, and polynomial kernel functions correctly classify 8 out of 10 subscribers as churners. The last proposed model based on the sigmoid kernel function performs a bit worse (7 out of 10). Finally, they concluded that the models based on RBF and polynomial kernel functions perform better than the other two. Therefore, if a mobile telecommunications company wants to send incentive offers, they can easily select, for example, the top 20% subscribers, classify them, contact the predicted churners, and expect that 80% of the contacted subscribers would have in reality become churners.

CASE 3: Privacy preserving data mining
When data mining techniques are used to connect personal identifiers (like, names, addresses) with other person related information then privacy is threatened. On the other hand, society may have benefits from the distilled knowledge from sensitive information.

Therefore, one of recent research directions in data mining is privacy preserving. Huge amounts of available data are maintained in the telecommunications industry: call data, describing the calls over telecommunication networks, network data in relation to the state of hardware/software components, customer related data. Such existing data may involve a threat to the privacy of users (Granmo & Oleshchuk 2007). The authors recognize two essential approaches to achieving privacy preserving in application of different data mining methods:

Data transformation/randomization that are connected to modifying sensitive data. Such modified data loses its sensitive meaning, but still retains statistical properties of interest.

Secure multi-party computation – several different parties possess parts of data and that union of these data represents the input needed to perform the computation. These methods used for privacy-preserving data mining are based on elements as secure sum, secure set union, secure size of set union, etc.

It is necessary to apply different algorithms (Pinkas 2002) in order to combine different sources of data without revealing the content of the data elements themselves but obtain useful and highly appreciated results.

Privacy preserving is obviously a research area that will play an important role in future applications not only in telecommunications. But in this area there are several less explored research directions (Najaflou et al. 2013) that could be attractive in the near future:

Privacy preserving adaptive control assumes learning to control a dynamic system without revealing significant system's properties.

Privacy preserving adaptive resource allocation takes care of allocation of some kind of resources to appropriate entities with main intention to optimize some performances but again without revealing information of these resources.

Privacy preserving adaptive routing takes care of finding routers near to optimal routing strategies, without revealing information about important elements like customers, preferences, traffic patterns.

Privacy preserving multi-dimensional scaling assumes application of privacy preserving, first of all pattern matching techniques in huge amounts of data received from monitoring cameras or from distributed sensor networks.

3.3 *Other applications*

Different ML methods are recognized as extremely productive and applicable in a wide range of domains. For example, hidden Markov models are especially useful and frequently used in temporal pattern recognition like speech, gesture, and handwriting recognition (Rabiner 1989, Liu and Lovell 2003, El-Yacoubi et al. 1999).

3.3.1 *Speech recognition*

Rabiner (1989) highlights several interesting implementation issues related to HMMs like scaling, multiple observation sequences choice of model size and type. Apart from that the author presents several implementations of speech recognizers based on HMMs.

The first presented system is concentrated on use of HMMs to build an isolated word recognizer. They assume to have a vocabulary V of words to be recognized. Each word is modeled by a distinct HMM and for each word from V there is a training set K of occurrences spoken by one or several speakers.

Experiments were performed in the area of recognition of isolated digits in a speaker-independent manner. For this purpose they used several conventional recognizers and recognizers based on HMMs, showing that HMM-based recognition achieves comparable performance (Rabiner 1989).

Next, a more complex application presented by Rabiner (1989), where HMMs have been successfully applied, is devoted to connected word recognition. In this case recognition is based on individual word models. The crucial recognition problem here is to find an optimum concatenation of word models that best matches an unknown sequence of words. Two methods are proposed to perform this task: level building approach and frame (time) synchronous Viterbi search. Again, experiments were performed in order to recognize sequences of connected digits. The proposed model has been tested in 3 modes:

1 Speaker trained using a training set that contained about 500 connected digit strings for each of 25 male and 25 female speakers. They also used appropriate independent testing sets.
2 Multispeaker mode where training sets from item 1 were merged in one big training set. The same has been done for testing sets. A set of 6 HMMs per digit was used.
3 Speaker independent mode where testing and training sets had 113 speakers divided in 22 dialect groups. A set of 4 HMMs per digit was used. Lengths of digit strings in experiments vary from 1 to 7.

The most challenging task in application of HMMs is continuous speech recognition, devoted to recognition of basic speech units smaller than words. In the most advanced systems, theory of HMMs could be applied to the representation of phoneme-like subwords as HMMs. In such systems it is necessary to use a triple embedded network of HMMs. Unfortunately, it is difficult to search and process such large networks. Therefore, it is necessary to invent new and efficient algorithms (in addition to existing stack algorithms, various forms of Viterbi beam searches, etc., Lowerre & Reddy 1980).

3.3.2 *Gesture recognition*

Together with the growing importance of interactive systems in everyday human activities, a new interesting research area appeared: hand gesture detection and recognition. Initial attempts in developing hand gesture recognition systems employed geometric feature and template-based methods, and more recently active contour and active statistical models. Motivated by the success of HMMs in speech and character recognition, some researchers realized that

HMMs could successfully be applied in a framework for hand gesture detection and recognition (Liu and Lovell 2003). In this framework, the gesture is modeled as a HMM. The observation sequence used to characterize the states of the HMM is obtained from the features extracted from the segmented hand image by vector quantization. For experimental purposes, the authors used several different HMMs and training algorithms in order to obtain high recognition rate and low computational complexity.

In the system presented by Liu and Lovell (2003), the hand gesture video is first captured by a digital camera (each video is composed of 25 frames) and skin color segmentation is based on HSV color space. To smooth the image and remove noise authors applied preprocessing (morphological) operations. An improved camshift algorithm has been successfully applied in tracking of the hand. Moment algorithm was applied on binary images of the hand in order to calculate the features of the hand. Then the features were normalized and sent as input to the vector quantizer. The pre-trained vector quantizer outputs the codeword sequence (each frame corresponding to one codeword). After that the input to a HMM is the codeword sequence representing a discrete observation sequence. Authors used and trained 8 different HMMs based on model type, structure and number of states. For experimental purposes the authors designed three poses and six gestures and compared two training algorithms. Both applied training algorithms (traditional Baum-Welch, and the Viterbi path counting method) achieved optimistic and reasonable performance.

3.3.3 *Handwriting recognition*

Handwriting recognition is nowadays one of the more challenging research areas. In spite of the fact that research teams all over the world work on these problems still there is no satisfactory solutions (El-Yacoubi et al. 1999). Several of the most difficult problems in the area are connected to a huge variability of handwriting and include: "inter and intra-writer variabilities, writing environment (pen, sheet, support, etc.), the overlap between characters, and the ambiguity that makes many characters unidentifiable without referring to context" (El-Yacoubi et al. 1999).

A popular basic technique is segmentation of words into basic elements (letters or pieces of letters) but this operation is rather difficult. Segmentation-recognition methods are the most successful. In these approaches a loose segmentation of words into (pieces of) letters is first performed. After that, using dynamic programming techniques, the optimal combination of these units to retrieve the entire letters (definitive segmentation) is applied. Success of HMMs in speech recognition motivated researchers to use these models in handwriting recognition as well. In the majority of cases, authors represent a word image as a sequence of observations and usually use two methods: implicit segmentation similar to speech representation of the handwritten word image, and explicit segmentation where segmentation algorithms are used to split words into basic units. El-Yacoubi et al. (1999) propose an explicit segmentation-based HMM method for recognition of unconstrained handwritten words (uppercase, cursive and mixed).

El-Yacoubi et al. (1999) generate three sets of features. The first set includes features like loops, ascenders and descenders. The second set includes features obtained by an analysis of the bidimensional contour transition histogram for segments. The third set depicts segmentation points between formed units like spaces between letters or words; vertical position of the segmentation points splitting connected letters and so on. According to that, each word was represented by two feature sequences (of shape-symbols and segmentation-symbols) of equal length. To achieve this structure the preprocessing phase consisted of four steps: baseline slant normalization, lower case letter area (upper-baseline) normalization for cursive words, character skew correction, and smoothing. The result of the first two steps is the first feature set. The result of the third step is the second feature set. El-Yacoubi et al. (1999) present results of extensive experiments on unconstrained handwritten French city name images. The learning, validation and test sets were include 11.106, 3.610 and 4.280 elements, respectively. The logarithmic version of the Viterbi procedure has been used for recognition and the Baum-Welch algorithm was used for training. Obtained results showed that HMMs can be successfully employed in development a high-performance handwritten word recognition system.

3.3.4 *Neural network classification*

Another technique which is frequently used in contemporary research, especially for classification purposes, are neural networks. As a consequence of research efforts in different domains it appeared that neural networks are very promising alternative to various other classification methods. There are several advantages of neural networks that qualify them for wider use: they are data driven self-adaptive methods; they are able to approximate a wide range of functions with arbitrary accuracy; as a nonlinear models, they offer significant flexibility in modeling complex real world relationships; they are able to estimate the posterior probabilities and make easy process of defining classification rules and fulfilling statistical analysis. Therefore, neural networks are becoming more attractive for applications in a wide range of real-world tasks (Zhang 2000): speech and handwriting recognition, bankruptcy prediction, medical diagnosis, product inspection, fault detection and so on. In spite of the success of different real-world applications of neural networks still there are some unsolved or incompletely solved issues like: development of more effective and efficient methods in neural model identification, feature variable selection, classifier combination, and uneven misclassification treatment. In addition, recent developments in deep learning (Deng and Yu 2014, Schmidhuber 2014, Arel et al. 2010) have reinvigorated neural network research and applications.

4 CONCLUSION

We hope that through this paper we provided a representative short overview of machine learning techniques and applications focusing on the business and telecommunications sectors. We foresee an increasing impact and application of ML in these areas (particularly novel applications of deep learning), which would open avenues for new practices, as well as research goals.

ACKNOWLEDGMENT

This is an invited paper corresponding to the keynote talk given by M. Ivanović. The authors thank the organizers of CECNet 2014 for this exquisite opportunity. We also thank the Ministry of Education, Science and Technological Development (Republic of Serbia) for support through project no. OI174023, "Intelligent techniques and their integration into wide-spectrum decision support."

REFERENCES

Apte, C. 2010. The role of machine learning in business optimization. *Proceedings of the 27th International Conference on Machine Learning (ICML)*: 1–2.

Arel, I., Rose, D. C., & Karnowski, T. P. 2010. Deep machine learning - a new frontier in artificial intelligence research. *IEEE Computational Intelligence Magazine* 5(4): 13–18.

Bellman, R. E. 1961. *Adaptive Control Processes: A Guided Tour*. Princeton University Press.

Bengio, Y., Courville, A. & Vincent, P. 2013. Representation learning: A review and new perspectives. *IEEE Transactions on Pattern Analysis and Machine Intelligence* 35(8): 1798–1828.

Berkhin, P. 2006. Survey of clustering data mining techniques. J. Kogan and C. N. M. Teboulle (Eds.), *Grouping Multidimensional Data - Recent Advances in Clustering*: 25–71. Springer.

Bishop, C. M. 1996. *Neural Networks for Pattern Recognition*. Oxford University Press.

Bishop, C. M. 2006. *Pattern Recognition and Machine Learning*. Springer.

Brandusoiu, I. & Toderean, G. 2013. Churn prediction in the telecommunications sector using support vector machines. *Annals of the Oradea University, Fascicle of Management and Technological Engineering* 1: 19–22.

Breiman, L., Friedman, J. H., Olshen, R. A. & Stone, C. J. 1984. *Classification and Regression Trees*. Chapman & Hall.

Chakrabarti, S. 2003. *Mining the Web: Discovering Knowledge from Hypertext Data*. Morgan Kaufmann Publishers.

Deng, L. & Yu, D. 2014. Deep learning: Methods and applications. *Foundations and Trends in Signal Processing* 7(3–4): 197–387.

Duda, R. O., Hart, P. E. & Stork, D. G. 2001. *Pattern Classification* (2nd ed.). Wiley.

El-Yacoubi, A., Sabourin, R., Gilloux, M. & Suen, C. 1999. Off-line handwritten word recognition using hidden Markov models. L. C. Jain and B. Lazzerini (Eds.), *Knowledge-Based Intelligent Techniques in Character Recognition*: 191–229. CRC Press.

Gabrilovich, E. & Markovitch, S. 2004. Text categorization with many redundant features: Using aggressive feature selection to make SVMs competitive with C4.5. *Proceedings of the 21st International Conference on Machine Learning (ICML)*: 321–328.

Granmo, O. C. & Oleshchuk, V. 2007. Privacy preserving data mining in telecommunication services. *Telektronikk* 3: 84–89.

Han, J. & Kamber, M. 2006. *Data Mining: Concepts and Techniques* (2nd ed.). Morgan Kaufmann Publishers.

Hara, K., Suzuki, I., Shimbo, M., Kobayashi, K., Radovanović, M. & Fukumizu, K. 2015. Localized centering: Reducing hubness in large-sample data. *Proceedings of the 29th AAAI Conference on Artificial Intelligence*.

Hastie, T., Tibshirani, R. & Friedman, J. 2009. *The Elements of Statistical Learning: Data Mining, Inference, and Prediction* (2nd ed.). Springer.

Jahromi, A. T. 2009. *Predicting customer churn in telecommunications service providers*. Master's thesis, Tarbiat Modares University Faculty of Engineering, Tehran, Iran & Lulea University of Technology, Lulea, Sweden.

Jensen, F. V. 2007. *Bayesian Networks and Decision Graphs* (2nd ed.). Springer.

Joachims, T. 1999. Making large-scale SVM learning practical. B. Schölkopf, C. J. C. Burges, and A. J. Smola (Eds.), *Advances in Kernel Methods – Support Vector Learning*: 169–184. MIT Press.

Keerthi, S. S., Shevade, S. K., Bhattacharyya, C. & Murthy, K. R. K. 2001. Improvements to Platt's SMO algorithm for SVM classifier design. *Neural Computation* 13(3): 637–649.

Kibriya, A. M. & Frank, E. 2007. An empirical comparison of exact nearest neighbour algorithms. *Proceedings of the 11th European Conference on Principles and Practice of Knowledge Discovery in Databases (PKDD), Lecture Notes in Artificial Intelligence* 4702: 140–151. Springer.

Koller, D. & Friedman, N. 2009. *Probabilistic Graphical Models: Principles and Techniques*. MIT Press.

Korn, F., Pagel, B.-U. & Faloutsos, C. 2001. On the "dimensionality curse" and the "self-similarity blessing". *IEEE Transactions on Knowledge and Data Engineering* 13(1): 96–111.

Kurbalija, V., Ivanović, M., & Budimac, Z. 2009. Case-based curve behaviour prediction. *Software: Practice and Experience* 39(1): 81–103.

Lewis, D. D. 1998. Naive (Bayes) at forty: The independence assumption in information retrieval. *Proceedings of the 10th European Conference on Machine Learning (ECML), Lecture Notes in Artificial Intelligence* 1398: 4–15. Springer.

Liu, N. & Lovell, B. C. 2003. Gesture classification using hidden Markov models and Viterbi path counting. *Proceedings of the 7th Biennial Australian Pattern Recognition Society Conference*: 273–282.

Lowerre, B. & Reddy, R. 1980. The HARPY speech understanding system. W. Lea (Ed.), *Trends in Speech Recognition*: 340–346. Prentice-Hall.

Lu, J. 2005. Predicting customer churn in the telecommunications industry - an application of survival analysis modeling using SAS. *SAS User Group International (SUGI27) Online Proceedings*: 114–27.

Manyika, J., Chui, M., Brown, B., Bughin, J., Dobbs, R., Roxburgh, C., & Byers, A. H. 2011. *Big data: The next frontier for innovation, competition, and productivity*. McKinsey Global Institute.

Matthews, J. 2010. The perceptron. *http://www.generation5.org/content/1999/perceptron.asp*.

Meilă, M. & Shi, J. 2001. Learning segmentation by random walks. *Advances in Neural Information Processing Systems* 13: 873–879.

Mitchell, T. M. 1997. *Machine Learning*. McGraw-Hill.

Najaflou, Y., Jedari, B., Xia, F., Yang, L. T., & Obaidat, M. S. 2013. Safety challenges and solutions in mobile social networks. *IEEE Systems Journal* PP(99): 1–21.

Neapolitan, R. E. 2003. *Learning Bayesian Networks*. Prentice Hall.

Ng, A. Y., Jordan, M. I. & Weiss, Y. 2002. On spectral clustering: Analysis and an algorithm. *Advances in Neural Information Processing Systems* 14: 849–856.

Nixon, M. S. & Aguado, A. S. 2013. *Feature Extraction & Image Processing for Computer Vision* (3rd ed.). Academic Press.

Oseman, K., Shukor, S. M., Haris, N. A., & Bakar, F. A. 2010. Data mining in churn analysis model for telecommunication industry. *Proceedings of the Regional Conference on Statistical Sciences (RCSS), Malaysia*: 215–223.

Osuna, E., Freund, R., & Girosi, F. 1997. An improved training algorithm for support vector machines. *Proceedings of the 7th IEEE Neural Networks for Signal Processing Workshop*: 276–285.

Pinkas, B. 2002. Cryptographic techniques for privacy-preserving data mining. *ACM SIGKDD Explorations Newsletter* 4(2): 12–19.

Platt, J. C. 1999. Fast training of support vector machines using sequential minimal optimization. B. Schölkopf, C. J. C. Burges, and A. J. Smola (Eds.), *Advances in Kernel Methods - Support Vector Learning*: 185–208. MIT Press.

Quinlan, J. R. 1993. *C4.5: Programs for Machine Learning*. Morgan Kaufmann Publishers.

Rabiner, L. R. 1989. A tutorial on hidden Markov models and selected applications in speech recognition. *Proceedings of the IEEE* 77(2): 257–285.

Radovanović, M., Nanopoulos, A., & Ivanović, M. (2010). Hubs in space: Popular nearest neighbors in high-dimensional data. *Journal of Machine Learning Research* 11: 2487–2531.

Rosenblatt, F. 1959. The perceptron: A probabilistic model for information storage and organization in the brain. *Psychological Review* 65: 386–408.

Schmidhuber, J. 2014. Deep learning in neural networks: An overview. *Neural Networks*. Forthcoming.

Schnitzer, D., Flexer, A., Schedl, M., & Widmer, G. 2012. Local and global scaling reduce hubs in space. *Journal of Machine Learning Research* 13: 2871–2902.

Sebastiani, F. 2002. Machine learning in automated text categorization. *ACM Computing Surveys* 34(1): 1–47.

Shawe-Taylor, J. & Cristianini, N. 2004. *Kernel Methods for Pattern Analysis*. Cambridge University Press.

Shi, J. & Malik, J. 2000. Normalized cuts and image segmentation. *IEEE Transactions on Pattern Analysis and Machine Intelligence* 22(8): 888–905.

Steinbach, M., Karypis, G., & Kumar, V. 2000. A comparison of document clustering techniques. *Proceedings of the ACM SIGKDD Workshop on Text Mining* 400(1): 525–526.

Svensson, M. & Söderberg, J. 2008. Machine-learning technologies in telecommunications. *Ericsson Review* 3: 29–33.

Tan, P.-N., Steinbach, M., & Kumar, V. 2014. *Introduction to Data Mining* (2nd ed.). Addison Wesley.

Taskar, B. 2010. CIS520: Machine learning. *http://alliance.seas.upenn.edu/~cis520/wiki/*.

Tomašev, N. & Mladenić, D. 2012. Nearest neighbor voting in high dimensional data: Learning from past occurrences. *Computer Science and Information Systems* 9(2): 691–712.

Tomašev, N. & Mladenić, D. 2014. Hubness-aware shared neighbor distances for high-dimensional *k*-nearest neighbor classification. *Knowledge and Information Systems* 39(1): 89–122.

Tomašev, N., Radovanović, M., Mladenić, D., & Ivanović, M. 2014. The role of hubness in clustering high-dimensional data. *IEEE Transactions on Knowledge and Data Engineering* 26(3): 739–751.

Vapnik, V. 1979. Estimation of dependencies based on empirical data (in Russian). *Nauka*. English translation Springer-Verlag, Berlin, 1982.

Vapnik, V. & Chervonenkis, A. 1974. Theory of pattern recognition (in Russian). *Nauka*. German translation Akademie-Verlag, Berlin, 1979.

von Luxburg, U. 2007. A tutorial on spectral clustering. *Statistics and Computing* 17(4): 395–416.

Vossen, G. 2014. Big data as the new enabler in business and other intelligence. *Vietnam Journal of Computer Science* 1(1): 3–14.

Witten, I. H., Frank, E., & Hall, M. A. 2011. *Data Mining: Practical Machine Learning Tools and Techniques* (3rd ed.). Morgan Kaufmann Publishers.

Xevelonakis, E. 2005. Developing retention strategies based on customer profitability in telecommunications: An empirical study. *Journal of Database Marketing & Customer Strategy Management* 12: 226–242.

Zhang, G. P. 2000. Neural networks for classification: A survey. *IEEE Transactions on Systems, Man, and Cybernetics, Part C: Applications and Reviews* 30(4): 451–462.

Electronics, Communications and Networks IV – Hussain & Ivanovic (eds)
© 2015 Taylor & Francis Group, London, ISBN: 978-1-138-02830-2

A hybrid image registering algorithm of improved Powell algorithm and simulated annealing algorithm

Yingtian Ji*, Dengyin Zhang & Lipin Tan
Key Lab of Broadband Wireless Communication and Sensor Network Technology, Nanjing University of Posts and Telecommunications, Ministry of Education, Nanjing, Jiangsu, China.

ABSTRACT: With the development of computer technology and Biomedical Engineering, the image registering technology is widely used in healthcare and the disease diagnosis and clinical care demand higher registering precision and speed. In this paper, against the defect that the Powell method is prone to obtain a local extremum, we propose an improved Powellalgorithm which can improve the registering precision and robustness with the simulated annealing algorithm in a multi-resolution strategy. The experiment demonstrates our method is an effective way to reduce the registering error.

1 INTRODUCTION

The real human body situation can be reflected only by integrating various medical images because the imaging principles of medical imaging equipments are different and the one single image can hardly provide enough information (Kagadis et al. 2002). The premise of medical image integration is image registering. The main purpose of image registering is to remove the geometrical differences between the image to be registered and the reference image including translation, rotation and other kinds of geometrical deformation. The result of image registering is a space transformation through which the image to be registered is able to achieve the most geometrical similarity to the reference page.The registering accuracy is an immediate factor that affects the integration quality and the registering technique has become asignificant aspect ofmodern medical image processing technology.At present, the most widely used algorithms in medical image registering includesimplex algorithm (Zheng 2005), Powell algorithm (Nelder& Mead 1965), simulated annealing (SA) algorithm (Kecun 1998) and genetic algorithm (Xin et al. 2006). Among these methods, the convergence rates of simplex algorithm and SA algorithm are relatively slow. The genetic algorithm tends to be premature due to its randomness. Powell algorithm is a local-search method with powerful search capability and high convergence speed which bases on mutual information. However, in this method, the initial spot has great impact on search result which makes Powell algorithm prone to obtain a local extremum. Consequently, In order to obtain a global result, the Powell method must be used combined with a global search method.

2 TRADITIONAL POWELL ALGORITHM

The Powell algorithm divides the whole search progress into several rounds and conducts 1-dimension search for $n+1$ times in each iteration round where n represents the number of the search function parameters. The error is calculated after each round. If the error is larger than the required convergence accuracy, the next iteration is supposed to be carried out until the error is smaller than the required accuracy. In each iteration round, the search starts from the corresponding initial spot and goes along the directions on the direction set of the current round. That means in the k_{th} round, the 1-dimension search starts from $x^{(k,0)}$ and goes along the direction $d^{(k,1)}$, $d^{(k,2)}$, $d^{(k,3)}$...$d^{(k,n)}$, in turn to get point $x^{(k,n)}$from which another 1-dimension search starts and goes along the connection from $x^{(k,n)}$ to$x^{(k,0)}$. The best-matched point is obtained through this way. If the point meets the convergence requirement, the search is considered to be a success. Otherwise, the initial spot need to be updated and a new corresponding direction set should be constructed in the next round.

The direction set needs to be updated in each round in Powell method and it has great impact on the result and speed of the optimal research. If one of the direction sets is inappropriate or even incorrect, the whole search is likely to fail. The rule of direction set updating is therefore quite important. In the general case, the connection of the start point and stop point in each round is added into direction set as a new direction replacing the first direction in the set without any filtering or restriction. This rule might cause the directions in set become linear correlated which leads to the degradation of search direction and a wrong result.

*Corresponding author: *jiyingtian@foxmail.com*

3 HYBRID ALGORITHM

3.1 *An improved Powell algorithm*

To solve the problem introduce above, we propose an improved Powell method in which the new direction for the next round is judged. If the new direction is a good one, it shall replace an old direction in the direction set, otherwise the set shall remain unchanged. The definition of a good new direction is that the new direction does not make the direction set linear correlated. Furthermore, the discarded old direction is the direction along which the change of the function value is the largest instead of the first direction in the direction set. Through a certain number of iterations, the search can be considered as a success once the error meets the convergence requirement.

The specific steps of the improved Powell method are described as following which is also shown in Figure 1:

3.1.1 *Step 1 - Set the initial direction*

Preset the convergence accuracy $\varepsilon>0$ and the start point$x^{(1,0)}$. Determine$n$, the number of the objective function parameters and construct n linear independent directions. that is a maximum linear independent group in which each direction is a unit vector$d^{(1,1)}$, $d^{(1,2)}$, $d^{(1,3)}$...$d^{(1,n)}$. Set k=1.

3.1.2 *Step 2 - Conduct k_{th} round search*

Conduct 1-dimension search from$x^{(k,0)}$, along the direction $d^{(k,1)}$, $d^{(k,2)}$, $d^{(k,3)}$...$d^{(k,n)}$ in turn. Find out the points $x^{(k,1)}$, $x^{(k,2)}$, $x^{(k,3)}$... $x^{(k,n)}$ respectively corresponding to the function maximums for each direction and make these points fulfill the following requirements:

$$f\left(x^{(k,i-1)}+\lambda_i d_i\right)=max\left(f\left(x^{(k,i-1)}+\lambda_i d_i\right)\right) \quad (1)$$

$$x^{(k,i)}=x^{(k,i-1)}+\lambda_i d_i \quad (2)$$

If $x^{(k,n)}$- $x^{(k,0)}<\varepsilon$,stop the iteration and $x^{(k,n)}$ is the best point. If not, move to Step 3.

3.1.3 *Step 3 - update the start point.*

Along the direction $d^{(k,n+1)}=x^{(k,n)}-x^{(k,0)}$, find the λ corresponding to the objective function maximum.

$$f\left(x^{(k,0)}+\lambda_{n+1}d^{(k,n+1)}\right)=\max_\lambda\left(f\left(x^{(k,0)}+\lambda d^{(k,n+1)}\right)\right) \quad (3)$$

Let $x^{(k+1,0)}=x^{(k,n+1)}=x^{(k,0)}=\lambda_n d^{(k,n+1)}$, find out the parameter values corresponding to maximum in the direction of $d^{(k,n+1)}$.

If $x^{(k+1,0)}-x^{(k,0)}<\varepsilon$, stop the iteration and $x^{(k+1,0)}$ is the best point. Otherwise, move to step 4.

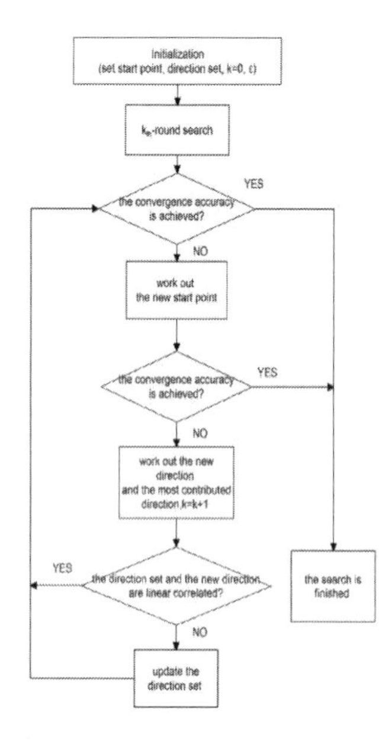

Figure 1. Flow process of the improved Powell method.

3.1.4 *Step 4 - Find the new direction*

Find out the fastest decline direction which fulfills the following equations.

$$f\left(x^{(k,m-1)}\right)-f\left(x^{(k,m)}\right)=\max_{j=1,2,\cdots n}\left\{f\left(x^{(k,j-1)}\right)-f\left(x^{(k,j)}\right)\right\} \quad (4)$$

$$\Delta m=f\left(x^{(k,m-1)}\right)-f\left(x^{(k,m)}\right) \quad (5)$$

3.1.5 *Step 5 - Update the direction set*

Check whether$d^{(k,n+1)}$ is linear correlated with the original direction set (without$d^{(k,m)}$). That is $d^{(k,1)}$, $d^{(k,2)}$,...$d^{(k,m-1)}$,$d^{(k,m+1)}$...$d^{(k,n)}$. Let $f_1=f(x_k(0))$, $f_2=f(x_k(n))$ and$f_3=f(x_k(n+2))=f(2x_k(n)-x_k(0))$. It can be proved that if the two conditions$f_3<f_1$and$(f_1-2f_2+f_3)(f_1-f_2-\Delta m)^2<0.5\Delta m(f_1-f_3)^2$ are satisfied at the same time, the direction $d^{(k,n+1)}$ is linear independent of the original direction set which means th$d^{(k,m)}$ can be replaced with $d^{(k,n+1)}$for the next round as following.

$$d^{(k+1,j)}=d^{(k,j)}j=1,2\cdots\cdots n,\ j\neq m \quad (6)$$

$$d^{(k+1,m)}=d^{(k,n+1)} \quad (7)$$

Otherwise, the original direction set is still suitable for the $k+1_{th}$ round.

3.2 *A hybrid algorithm in the multi-resolution strategy*

In this paper, we combine our improved Powell algorithm with the SA algorithm in the multi-resolution strategy. In this hybrid algorithm, we firstly apply wavelet decomposition respectively on the floating image and the reference image. Then we use the SA method to search the lowest-resolution layer and make the gradient weighted normalized mutual information (GWNMI) function (Rirong 2005) achieve the maximum through constant space transformation. The corresponding parameters are regarded as the initial values for the Powell search in the high-resolution layer. The specific steps are as following, which is also shown in Figure 2.

Step1 - Decompose the image to be registered with *N*-level wavelet and get the decomposed images.

Step2 - Register the N_{th}-layer image with SA method and set the space transformation parameters obtained as T_N.

Step3 - Register the m_{th}-layer ($m<N$) image with our improved Powell method in which the start point is initialized byT_m+_1 and the new space transformation parameters is set asT_m.

Step4–$m=m-1$, if $m<0$, the image registering is finished. Otherwise, move to step3.

4 SIMULATION RESULTS

In order to test and verify the performance of the optimized search method in this paper, we adopt five search methods for registering shown as Table1.

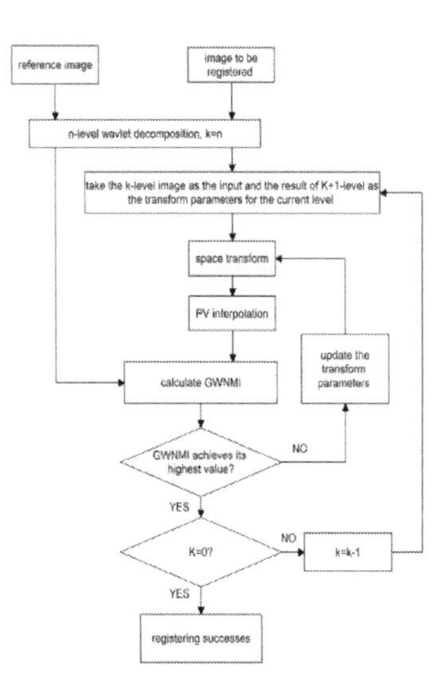

Figure 2. Flow process of the hybrid algorithm.

The images we adopt are shown in Figure3 in which Figure 3(a) and Figure 3(b) are both human brain CT (Zhijie 2006) images and they are already aligned, that is they are already registered very well. We firstlytransform Figure3(b) in space which is a translation 10 pixel to the right and 5 pixel down and arotation 10 degree anti-clockwise. The transformed image is Figure 3(c). Then we register Figure 3(a) and Figure 3(c)with the five methods respectively. The GWNMI function, linear transformation and PV interpolation are applied for the similarity measurement.

Table 1. Adopted search methods and schemes.

Schemes No.	Image Registering search method
A	Powell algorithm
B	the improved Powell algorithm
C	thehybrid algorithm of the improved Powell and SA algorithm
D	the hybrid algorithm of Powell and SA algorithm in multi-resolution strategy
E	the hybrid algorithm of the improvedPowell and SA algorithm in multi-resolution strategy

Table 2. Experimental data of five registering methods.

–	*X* offset	*Y* offset	Rotation
Actual value	10	–5	–10
Scheme A	7.3844	–4.1341	–11.0396
Scheme B	8.8242	–4.3353	–9.2374
Scheme C	9.3643	–5.3432	–10.4043
Scheme D	9.3532	–4.6654	–10.2348
Scheme E	9.6741	–5.1809	–9.9045

Table 3. Performance comparison of five registering methods.

–	ΔX	ΔY	Δθ
Scheme A	2.6166	0.8759	1.0394
Scheme B	1.1758	0.6647	0.7626
Scheme C	0.6357	0.3432	0.4043
Scheme D	0.6468	0.3346	0.2348
Scheme E	0.3259	0.1809	0.0955

The 3-level wavelet decomposition is adopted in this paper and the convergence accuracy $\varepsilon=10^{-4}$. The registering results are shown in Figure 3 (d)-(h).

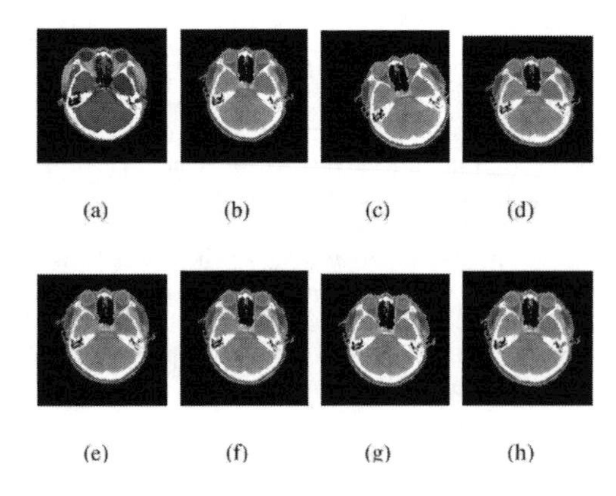

Figure 3. Original images and registering results.

The experimental data of the five methods are shown in Table 2 and the performance comparison in Table 3. The result of Scheme B is obviously more accurate than that of Scheme A which means the improved Powell method in this paper outperforms the original method. Moreover, the performance of Scheme C is better than Scheme A and B which demonstrates the hybrid algorithm is superior to the Powell method.

5 CONCLUSION

In this paper, against the defect of Powell method, we propose an improved Powell algorithm which introduces the judgement rule into direction set updating. At the same time, we propose a hybrid algorithm of our improved Powell method and SA method in multi-resolution strategy. In this hybrid algorithm, the search result of the lower-resolution layer is utilized to the initial start point of the high-resolution layer. Moreover, the SA method applied to the low-resolution layer can prevent the Powell method for obtaining a local extremum. The experiment demonstrates that the hybrid algorithm is an effective way to reduce the registering error.

ACKNOWLEDGEMENT

This research work is supported by the National Natural Science Foundations of P. R. China (NSFC) under Grant No. 61071093, National 863 Program No.2010AA701202, Returned Overseas Project, Jiangsu Province Major Technology Support Program No.BE2012849, Graduate Research and Innovation project No. KYLX_0812.

REFERENCES

Kagadis, C., Kdelibasis. & P. Asvetas. 2002. A comparative study of surface and volume-base techniques for the automatic registering between ct and spect brain images. *Medical Physics* 29(2): 201–213.

Kecun, Z. 1998. *The algorithm and analysis in engineering optimization.*Xi'an: Xi'an Jiao Tong University press.

Nelder, J. &Mead,A. 1965. A simplex method for function minimization. *Computer Journal* 7: 308–313.

Rirong, Y. 2005. Research on medical image registering based on genetic algorithm and maximum mutual information. *Kunming University of Science and Technology Journal* 3: 24–38.

Xin, W., Z. Chunhui& Z. Chaozhu. 2006. Research on improved simulated annealing algorithm in image registering. *Application Technology* 33(1): 10–12.

Zheng, C. 2005. Overview of research on the technology of medical image fusion. *Medical Equipment* 26(3): 48–49.

Zhijie, Y. 2006. Research on medical image registering based on mutual information. *Hebei University of technology journal* 3: 22–30.

Electronics, Communications and Networks IV – Hussain & Ivanovic (eds)
© 2015 Taylor & Francis Group, London, ISBN: 978-1-138-02830-2

Standards and protocols for RFID tag identification

Xiaolin Jia* & Yajun Gu
School of Computer Science and Technology, Southwest University of Science and Technology, Mianyang, Sichuan, China

Quanyuan Feng
School of Information Science and Technology, Southwest Jiaotong University, Chengdu, Sichuan, China

Limin Zhu & Li Yue
Sichuan Institute of Standardization, Chengdu, Sichuan, China

ABSTRACT: RFID technology provides a method to connect nearly any items or objects in the world, and trace, control, and manage them effectively. In this process, RFID tag identification and anti-collision is one of the most important technologies. This paper analyzes the classical anti-collision protocols and the international standards for RFID tag identification. At the same time, an efficient and stable anti-collision protocol, i.e., Collision Tree protocol (CT), is introduced in this paper, whose performance is much better than that of the classical anti-collision protocols used in the international standards for RFID tag identification. Therefore, CT could be used as a replacement protocol of the classical anti-collision protocols.

KEYWORDS: international standard; RFID tag identification; anti-collision protocol; collision tree protocol.

1 INTRODUCTION

Radio Frequency Identification (RFID) is one of automatic identification methods, which is used to wirelessly identifying data stored in a tag's microchip by using RF waves (Finkenzeller 2010, Klair et al. 2010). A typical RFID system consists of a reader, RFID tags, and application servers. The reader is typically a powerful device with ample memory and computational resources. The RFID tags include many types of dumb passive tags in smart, active tags, in which the passive ones become the most popular choice for large scale RFID tag applications due to their low cost and power consumption.

The reader and tags communicated with each other by the RF signal through the air interface. In RFID system, when multiple tags and/or readers transmit their IDs or data simultaneously in the same channel, the signal will interfere in the wireless channel, i.e., the collision occurs, which make the tag identification and date collection failure, even the system uselessness. Therefore, the anti-collision protocol is very important in the RFID tag identification system. At the same time, anti-collision protocol is one of the three supporting technologies in UHF RFID air interface, which includes wireless power transmission, backscattering modulation, and anti-collision mechanism. Anti-collision protocol is also one of the bottleneck technologies of the standard system in the UHF RFID air interface.

RFID system is the most important component of the terminal and the boundary system of Internet of Things (IOT) (Jia et al. 2012a). RFID tag identification technology is the key and core technology in RFID system and IOT system. This paper introduces and analyzes the standards and anti-collision protocols for RFID tag identification. A novel anti-collision protocol is recommended also in this paper, which is called a collision tree protocol (CT). For simple and straightforward to implement with high performance, CT could be used as the replacement protocol of the classical anti-collision protocols used in RFID tag identification.

2 STANDARDS FOR RFID TAG IDENTIFICATION

2.1 *Standard organization*

Standards have a very important influence on the development and application of RFID technology. There are many organizations and institutions for RFID standards, such as ISO/IEC, EPCGlobal, ITU-T, UIDC (Ubiquitous ID Center), AIM (Automatic Identification Manufacturers), IP-X, etc.

The most important international organization for RFID standards is ISO/IEC. There are many technology sub-committees under the ISO/IEC to research and establish related standards. For example, ISO/

*Corresponding author: *my_jiaxl@163.com*

IEC JTC1 SC31, the automatic identification and data capture technology sub-committee, is in charge of the communication interface and parameter standards of the automatic identification and data collection when working under different frequency, i.e., ISO/IEC18000 series standards. ISO/IEC JTC1 SC17, the identification card and identification technology sub-committee, is in charge of the ISO/IEC14443 series standards.

EPCGlobal is another famous standardization organization for RFID technology. EPCGlobal devotes to construct and implement the internet of things (IOT) to share the information from and between every entity by giving a unique identifier to every entity and object and using the Internet, RFID system, and global identification and recognition system, and so on. EPCGlobal has been issued a series of standards and specification, including electronic product code (EPC), RFID tag specification and interoperability, communication protocols between the reader and tags, middleware systems and interface, etc.

2.2 *Standards for RFID tag identification*

The air interface standard is most important in RFID standards, which affect the RFID reader and tags significantly. The main standards of the RFID are established by ISO/IEC and EPCGlobal, such as ISO /IEC 18000 series, ISO/IEC 14443 series, EPCGlobal standard series, and so forth. This part introduces some standards used in RFID tag identification. The anti-collision protocols and standards for RFID tag identification are listed in Table I.

Table 1. Standards and Anti-collision Protocols.

Anti-collision protocols	Standards for RFID tag identification
Query tree protocol (QT)	ISO/IEC 18000-3 Mode 1
Binary tree protocol (BT)	ISO/IEC 18000-6 Type B EPCGlobal Class 0 EPCGlobal Class 1
Dynamic binary search algorithm (DBS)	ISO/IEC 14443-3 Type-A
Q Protocol	ISO/IEC 18000-6 Type C EPCGlobal Class 1 Generation 2
Pure ALOHA protocol (ALOHA)	ISO/IEC 18000-3 Mode 1 Extension
Slotted ALOHA (S-ALOHA)	ISO/IEC 18000-3 Mode 2
Framed slotted ALOHA (FSA)	ISO/IEC 18000-3 Mode 1 Extension ISO/IEC 18000-6 Type A EPCGlobal Class 1 EPCGlobal Class 1 Generation 2
Dynamic framed slotted ALOHA (DFSA)	ISO/IEC 18000-3 Mode 1 ISO/IEC 14443-3 Type-B

ISO/IEC 18000, titled information technology – radio frequency identification for item management, is the most important international standard for RFID tag identification, which describes a series of different RFID technologies. ISO/IEC 18000 includes seven parts from 18000-1 to 18000-7, each of them used in a specific frequency range.

- ISO/IEC 18000-1 describes the architecture and definition of parameters to be standardized.
- ISO/IEC 18000-2 defines the air interface and parameter at 135 kHz.
- ISO/IEC 18000-3 defines the air interface and parameter for RFID communications at 13.56 MHz.
- ISO/IEC 18000-4 defines the air interface and parameter for RFID communications at 2.45 GHz
- ISO/IEC 18000-6 defines the air interface and parameter from 860 MHz to 960 MHz. ISO/IEC 18000-6 is the largest document, which is split into multiple types form type A to type D including a general part.
- ISO/IEC 18000-7 defines the air interface and parameter for RFID communications at 433 MHz.

ISO/IEC 14443, titled identification cards – contactless integrated circuit cards and proximity cards, is another important international standard, which is developed by SC 17 in ISO/IEC JTC 1. ISO/ IEC 14443 defines the proximity cards (also RFID tags) for identification, transmutation, and communication with it. ISO/IEC 14443 includes four parts form ISO/IEC 14443-1 to ISO/IEC 14443-4.

- ISO/IEC 14443-1 defines the characteristics of the physical layer of the proximity cards.
- ISO/IEC 14443-2 defines the radio frequency signal interface and power supply.
- ISO/IEC 14443-3 defines the initialization process and anti-collision protocol.
- ISO/IEC 14443-4 defines the transmission protocol and control specification.

ISO/IEC 14443-3 refers two type cards: Type A and Type B, both of them communicate at 13.56 MHz and use the same transmitting protocol. The main distinctions between them are the modulation methods, the coding schemes, and the initialization procedures.

EPCGlobal focus to achieve the standardization and adoption of EPC technology, and to establish a worldwide standard for RFID to sharing the data and information via the Internet and EPCGlobal network. Many standards are developed or in developing by EPCGlobal, as shown in Figure 1.

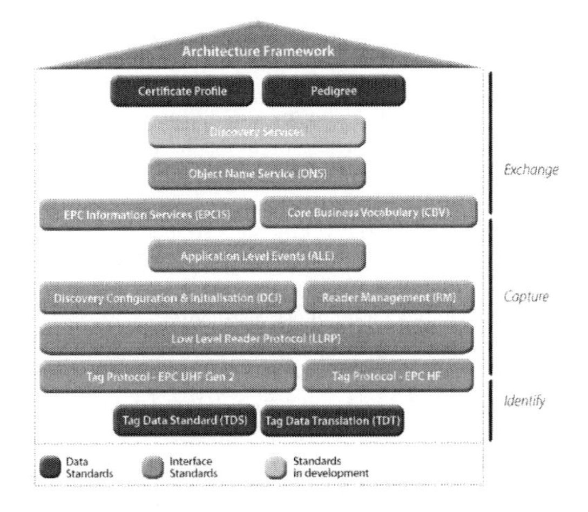

Figure 1. Standards of EPCGlobal (EPCglobal, 2014).

The main standards developed by EPCGlobal are EPC™ Radio-Frequency Identity Protocols EPC Class-1 HF RFID Air Interface Protocol, and UHF Air Interface Protocol Standard Generation 2, i.e., EPC Gen2. The former one defines the physical and logical requirements for a passive-backscatter RFID system of readers and tags working at 13.56 MHz. The latter one defines the physical and logical requirements for a RFID system of readers and passive tags working in the 860 MHz – 960 MHz. EPC Gen2 has been used in many fields as the standard for UHF RFID tag identification, and becomes the core part of the various RFID application systems.

3 CLASSICAL RFID TAG IDENTIFICATION PROTOCOLS

In RFID system, tag identification protocol is also called an anti-collision protocol for the purpose of it is to solve the collision problem in multiple tags identification. The proposed anti-collision protocol for RFID tag identification could be classified into tree-based protocol and ALOHA-based protocol. This section introduces the classical anti-collision protocols of them, for they are used in the international standards at present, as listed in table I, and many other improved or hybrid anti-collision protocols are also based on them.

3.1 *Query tree protocol*

Query tree protocol (QT) consists of many query-response cycles (Law et al. 2000). In every cycle, the reader sends a query command with a parameter called prefix composed of a series of binary "0" and "1", and waits the tag's response. The tags received the query command compare their IDs with the prefix. If the matching is successful, the tag transmits its ID to respond the reader's query. Otherwise, the tags not do anything.

If there has no response, the reader gets another prefix and queries again, or ends the identification if no prefix exists. If no collision happens in the response, the reader identifies one tag successfully, and starts another identifying cycle with a new prefix. If a collision happens in the response, the reader generates two new prefixes by extending the previous prefix with one bit "0" and "1" for the following query. The reader repeats the identifying process constantly, until all tags are identified.

For every time when the collision happens, the reader extends the prefix with only one bit, many idle cycles is occurred, in which there is no tag answer the reader's inquiry. But QT belongs to the class of memoryless protocol, for the tags need not to memorize the state or information of the before query and response process. Therefore, QT could be used in passive RFID tag identification condition.

3.2 *Binary tree protocol*

Binary tree protocol (BT) splits the collided RFID tags into different groups according to the random number generated by the collided tags (Klair et al. 2010). In BT, every tag has a counter to memory the group and identification state. Initially, the value of each counter is set to 0. When received the reader's request command, the tags which the counter value equals 0 transmit their IDs to the reader.

If no collision happens in the response, the reader identifies one tag, and all unidentified tags decrease their counter by 1. If a collision happens, the responded tags generate a binary number 0 or 1, and add it to their counter. The other unidentified tags increase their counter by 1. If there is no response, all unidentified tags increase their counter by 1 also. The reader requests repeatedly until all tags are identified.

Since the number is generated randomly, BT can not avoid the idle cycle problem. Furthermore, the identification performance of BT is very different, even in different identifying process for the same tags set.

3.3 *Binary search protocol*

Binary search protocol (BS) splits and identifies the tags by scanning the tag's ID (Finkenzeller 2010). The reader uses a binary serial number as the parameter of request command. Initially, the serial number is the biggest possible tag ID, in which every bit is 1. The unidentified tag whose ID equals or less than the serial number responses the reader's request with its ID.

When a collision occurs, the reader decreases the serial number according to the collision place in the received ID, and requests the tags again with the new serial number. If no collision occurs, the reader identifies one tag, and requests with the initial serial number continually until all tags are identified.

In BS, both the reader and tags use the whole length binary string as the parameter of request and response command. In fact, both the latter part of the reader's serial number and the former of the tag's response ID are not useful. So in dynamic binary search protocol (DBS), an improved protocol for BS, the reader only transmits the former part of the serial number, and the tags only transmit the latter part of their IDs. Therefore, the transmitted bits could be reduced up to 50% by DBS.

3.4 *ALOHA-based protocol*

ALOHA-based protocol mainly includes the pure ALOHA protocol (PA), slotted ALOHA protocol (S-ALOHA), framed slotted ALOHA protocol (FSA), dynamic framed slotted ALOHA protocol (DFSA), Q-protocol, etc. (Finkenzeller 2010, Klair et al. 2010).

In ALOHA-based protocol, the time is always divided into many equal slots, in which RFID tags could finish a response to the reader's request. The frame is the time interval between two requests of the reader, which is composed of a number of slots. Every tag generates a random number which is not more than the frame size and waits. When the random number is equal to the slot number of the frame, the tag transmits its ID to answer the reader's request. If there is only one tag answered in a slot, the tag could be identified by the reader. If not, the tags need to generate random number again and wait to be identified in the next frame.

For all tags select the slots randomly, the optimal identification performance could be achieved when the number of tags to be identified is nearly equal to the frame size. So, it is important to estimate the number of tags to be identified and adjust the frame size according to the number of tags. In an FSA, the size of the frame is fixed. In DFSA, the size of the frame could be adjusted dynamically. Different estimate algorithms form various DFSA, in which Q-protocol is the simplest. In the Q-protocol, the frame size is adjusted only according to a number between 0 and 1 called Q factor, when collision occurred.

4 HIGH PERFORMANCE ANTI-COLLISION PROTOCOL

As discussed previously, the performance of the classical anti-collision protocols is very limited. For instance, the optimal identification efficiency of them is below 36.8%, even lower in actually. This section introduces a novel anti-collision protocol for RFID tag identification, i.e., collision tree protocol.

Collision tree protocol (CT) adopts Manchester Coding to encode the tag's ID for transmission (Jia et al. 2010, Jia et al. 2012b). So that the reader could obtain the correct bits in the tags' response and trace the collided bits to each bit. The reader sends a query command with a prefix as the parameter. The tags whose ID match the prefix transmit their ID except the same part of the prefix to respond the reader's inquiry.

If no collision happens in the response, the reader identifies one tag, and query again with a new prefix. When a collision happens, the reader generates two new prefixes according to the first collided bit of the received ID, and puts them into the prefix pool. The reader repeats this process until all tags are identified or the prefix pool is empty.

Figures 2 to 5 illustrate the performance and the stability of CT. From Figure 2, the identification efficiency (the ratio between the tags to be identified and the total cycles required to identify the all tags) is much better than the classical protocols. This is because CT eliminates the idle cycle or idle slot in RFID tag identification absolutely.

According to Figure 3, the performance of CT is stable in various distributions of tag IDs (Jia et al. 2012b), such as uniform distribution (S1), continuous distribution in front part (S2), continuous distribution in middle part (S3), continuous distribution in end part (S4). Even in different continuous degrees and different specific experiments, the performance of CT is stable also, as shown in Figure 4 and 5.

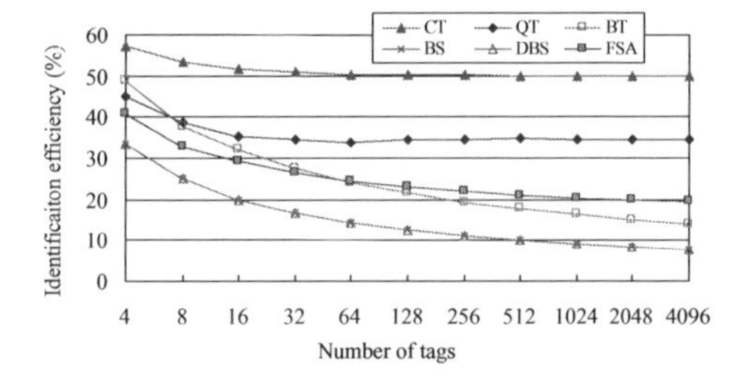

Figure 2. The identification efficiency of CT.

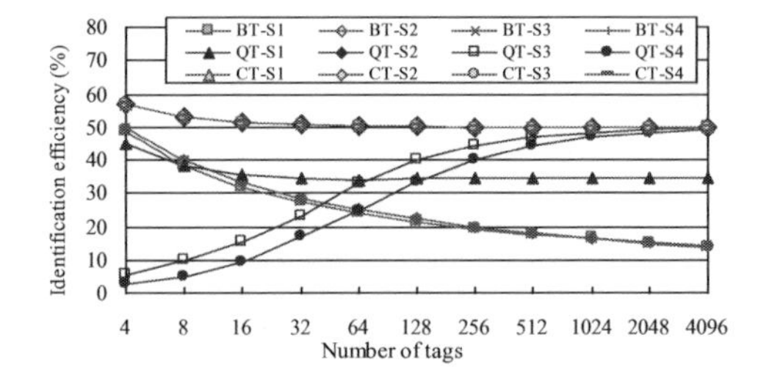

Figure 3. Stability of the identification efficiency of CT in different distributions.

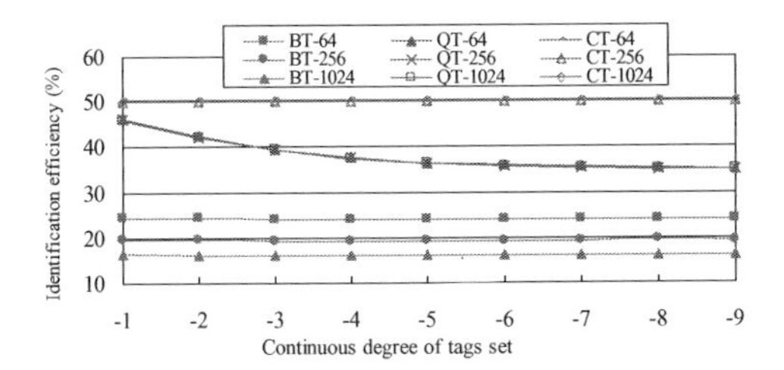

Figure 4. Stability of the identification efficiency of CT in different continuous degrees.

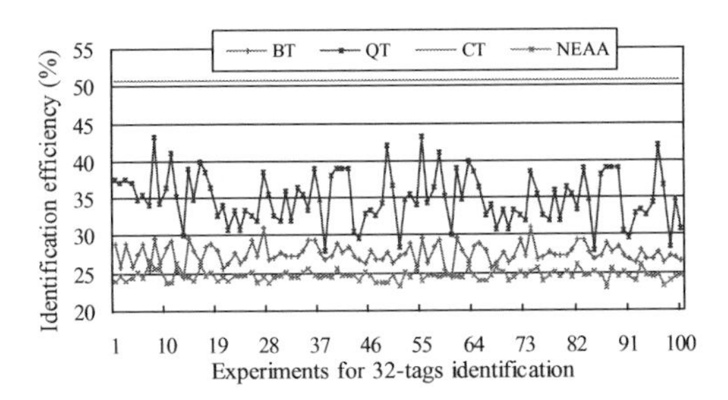

Figure 5. Stability of the identification efficiency of CT in 100 specific experiments.

5 CONCLUSIONS

This paper analyzes the international standards and anti-collision protocols for RFID tag identification, especially introduces the collision tree protocol (CT), which improves RFID tag identification efficiency significantly. For simple, straightforward, and easy implementation with high performance, CT could be used in various RFID systems and as a replacement protocol of the classical anti-collision protocols used in the international standards.

ACKNOWLEDGMENT

This work is supported by the National Natural Science Foundation of China (NNSF) under grant 61471306, 61271090, and the National 863 Project of China under grant 2012AA012305, and Sichuan Provincial Science and Technology Support Project under grant 2014JY0230, 2012GZ0112, and the Scientific Research Fund of the Education Department of Sichuan Province under grant 13CZ00025, and the Key Laboratory of Artificial Intelligence (Sichuan) fund under grant 2014RYY03, and Mianyang Network Integration Laboratory under grant 12ZXWK10.

REFERENCES

EPCglobal, 2014. *EPCGlobal standards overview*, http://www.gs1.org/gsmp/kc/epcglobal.

Finkenzeller, K. 2010. *RFID handbook: fundamentals and applications in contactless smart cards, radio frequency identification and near-field communication (3rd Ed.)*, New York: Wiley.

Jia, X.L., Feng, Q.Y. & Ma, C.Z. 2010. An efficient anti-collision protocol for RFID tag identification. *IEEE Communications Letters* 14(11): 1014–1016.

Jia, X.L., Feng, Q.Y. & Yu, L.S. 2012b. Stability analysis of an efficient anti-collision protocol for RFID tag identification. *IEEE Transactions on Communications*, 60(8): 2285–2294.

Jia, X.L., Feng, Q.Y., Fan, T.H., et al. 2012a. RFID technology and its applications in internet of things (IOT). *the 2nd International Conference on Consumer Electronics, Communications and Networks (CECNet'12)*: 1282–1285.

Klair, D.K., Chin, K.W. & Raad, R. 2010. A survey and tutorial of RFID anti-collision protocols. *IEEE Communications Surveys & Tutorials* 12(3): 400–421.

Law, C., Lee, K. & Siu, K.Y. 2000. Efficient memoryless protocol for tag identification. *Proceedings of the 4th International Workshop on Discrete Algorithms and Methods for Mobile Computing and Communications*. ACM: 75–84.

Electronics, Communications and Networks IV – Hussain & Ivanovic (eds)
© 2015 Taylor & Francis Group, London, ISBN: 978-1-138-02830-2

Development of attention and motion-sensitive behavior training platform for children with ADHD using BCI and motion sensing technology

Taesuk Kihl
Sangmyung University, Seoul, Republic of Korea

Kyungeun Park
Towson University, Towson, Maryland, USA

Min-Jae Kim
Sangmyung University, Seoul, Republic of Korea

Hyoungmok Baek
Cheil Worldwide, Seoul,Republic of Korea

Juno Chang
Sangmyung University, Seoul, Republic of Korea

ABSTRACT: This research introduces the development of the attention and motion-sensitive behavior training platform for the children with Attention-Deficit/Hyperactivity Disorder (ADHD) and weak social skills using NeuroSky's MindWave Brain-Computer Interface (BCI) and Microsoft's Kinect motion sensing technology. Also, the narrative contents combined with fairy tales and corresponding characters, and quests are embedded into the platform. The children with ADHD are to train themselves by completing the quests of the given narrative game of the platform. While playing the game, MindWave biosensor and Kinect motion sensor continuously track a player's brainwave and motions which may restrict the game character's behavior in case the player's attention level decreases or motion level increases. As the training effect, it is expected that the player shows the improved performance, such as decreased number of motions, increased average attention level, and reduced clearing time throughout the entire game session. The first experiment with the proposed platform involves tracking and identifying a variety of skeleton and head motion, and brainwaves of five participating testers. The main contribution of this research is the implementation of the behavior training platform which combines BCI and motion sensing technology with narrative contents and treats the children with ADHD in terms of the behavior management by helping the children increase attention to the game and feel confidence on themselves with the improved test results. This results in developing their social skills as they learn how to control their behavior under specific circumstances.

1 INTRODUCTION

ADHD is a specific disorder which shows deficits in behavioral inhibition, sustained attention, and resistance to distraction, and the regulation of one's activity level to the demands of a situation (hyperactivity or restlessness) (Barkley 1999, 2014). The American Psychiatric Association (APA 2013) states in the Diagnostic and Statistical Manual of Mental Disorders (DSM-5) that 5% of children have ADHD. However, studies in the U.S. have estimated higher rates in community samples. Approximately, 11% of children 4–17 years of age (6.4 million) have been diagnosed with ADHD as of 2011 (CDC 2014). Boys (13.2%) are more likely than girls (5.6%) to have ever been diagnosed with ADHD (CDC 2014). Three to five percent of all children have ADHD, with some estimates as high as 15%. That means that more than a million children in the U.S. have ADHD (NYU CSC 2014).

Barkley (Barkley 1998) analyzed the characteristic behaviors of disorder observed in children with ADHD. The obvious distinctions are impaired behavioral inhibition and self-control which are a critical foundation for the performance of any task. In addition, his research introduces four major mental activities, known as executive brain functions, which are needed to help a person achieve a goal in work or play. As children mature, the brain functions including nonverbal working memory, internalization of self-directed speech, self-regulation of motivation and level of arousal, and reconstitution capability should be internalized and made private in order to inhibit the public performance of these executive functions. In addition, he concludes that the inattention, hyperactivity, and impulsivity of children with ADHD are caused by their failure to be guided by internal instructions and by their inability to curb their own inappropriate behaviors (Barkley 1998).

ADHD can be treated and managed effectively once a child is diagnosed with ADHD. Fortunately, Rief (Rief 2005) introduces that children with ADHD show some common positive characteristics and traits: creative and inventive, artistic, innovative, imaginative, warmhearted, intelligent/bright, willing to take a risk and try new things, observant, intuitive, etc. As a result, children with ADHD are expected to perform much better in behavioral intervention, which is designed to help the children actively participate and train themselves to complete tasks presented on more creative and motivating environment. Those attempts might cause them to have preoccupation or hyperactivity with the given treatment programs. Accordingly, the intervention strategy for children with ADHD adopts attractive and efficient medium and recovers their self-esteem by increasing attention, controlling hyperactivity including emotion, and improving communication skills under specific context.

It is said that the psychological/cognitive functions of fairy tales include the positive influence on the growth of children. Bettelheim (Bettelheim 1989) mentioned indirect experiences from the traditional fairy tales with violent fantasies and separation anxiety allow children to be absorbed in the story and face to analyze and solve the predicaments in their imagination structures. In other words, children learn the characteristics or knowledge of human life through cruel stories, resulting in building problem solving capability, establishing their identity, and improving social skills.

As a result, the motion-sensitive behavior training platform has been proposed to treat the children with ADHD, keeping track of children's brainwaves and the associated behaviors responding to the given narrative contents based on fairy tales. While playing the game, the users are given quests to be able to increase their concentration level, control their hyperactive behavior, and improve their social skills as they continue playing the game.

The platform monitors players' brainwaves, motion, and time information as they interact with narrative contents of the training platform. To achieve this, Unity3D is used as the game development engine to which NeuroSky's MindWave biosensor and Microsoft's Kinect motion sensor are connected as a BCI and a motion sensing input device for the game platform. The MindWave captures user's brainwaves with the real-time attention and meditation eSense meters. Kinect recognizes user's motion at the same time. The captured data are maintained within a MySQL database for analyzing user's behavior and the effectiveness of the system.

The platform includes three major functions: 1) the brainwaves and motion recognition; 2) the interactive intervention with narrative contents; and 3) the sensor and intervention data management. While a player is playing a given narrative game, the player's brainwaves and motion sensing data are transmitted to the platform. This system simultaneously controls the intervention procedure, helping the player concentrate on the story and accomplish quests appropriately. Accordingly, the progress of the system is dynamically influenced by the player's behavior. For example, if the eSence value from MindWave drops below a threshold or the player's excessive motion is recognized, this system has the main character of the prototype stop without further progress.

The narrative contents and scene managers provide a visual interface combined with text and graphic features. This system is deeply related to the behavior management through which players are expected to learn how to behave in a certain context and slowly but surely improve their social skills.

The sensor and intervention data stores dynamically changing attention and meditation scale data and motion records with twenty joints' coordinates. In addition, the real-time interaction data and the platform's intervention data are maintained within the data center. The data will be semantically analyzed to review the effectiveness of the intervention procedures of the platform.

The final goal through this research is to treat children with ADHD in terms of the behavior management. The behavior training game is expected to help the children improve their attention level and control hyperactive behaviors through the narrative game. Moreover, the game helps the children develop their social skills as they learn how to control their behavior under specific circumstances.

Section one and two describe the introduction and research background of this paper. The behavior training platform is introduced in more detail in section three. The next section explains the experiments of the platform and the associated analysis on the sensor and intervention data conducted with five test participants. The paper concludes with future research directions.

2 RESEARCH BACKGROUND

Recently commercialized non-invasive Brain-Computer Interfaces (BCIs) in the form of lightweight headset devices are designed to read users' electroencephalogram (EEG) and analyze their attention and meditation level. This technology enables many developers to make the best use of users' brain activities for them to develop the various applications including medical treatment, education, and entertainment purpose interactive programs. EEG reads scalp electrical activity generated by brain structure as well-suited BCIs to measure the brain's electrical activity (Huang et al. 2014) as one of medical imaging techniques (Teplan 2002). Also, Scherer showed that the Kinect interface is useful to relate hand and feet movements to the specific patterns of the EEG results (Scherer 2013) while playing a virtual ball game in order to establish Kinect and EEG-based functional brain mapping

2.1 *Brain-Computer Interfaces (BCIs)*

As a BCI interface, this research considered using NeuroSky's MindWave EEG biosensor technology by virtue of its brain activity digitization capability which recognizes the wearers' analog electrical brainwaves and analyzes them into different emotional states (See Figure 1).

Figure 1. MindWave headset (NeuroSky).

2.2 *Motion tracking interface*

Generally, a typical motion sensing interface is divided into contact and non-contact interfaces. Contact interfaces require sensor devices to be attached to a user and works depending on the captured motion information. Whereas, non-contact interfaces are based on users' gesture tracked by multiple cameras (Bolt 1980, Shin et al. 2014). The proposed platform adopts Kinect controller of Microsoft as a non-contact motion sensing interface (See Figure 2).

Figure 2. Kinect controller (Microsoft).

Figure 3 shows twenty joints' positions of a tracked person, of which Kinect recognizes the values of the individual coordinates.

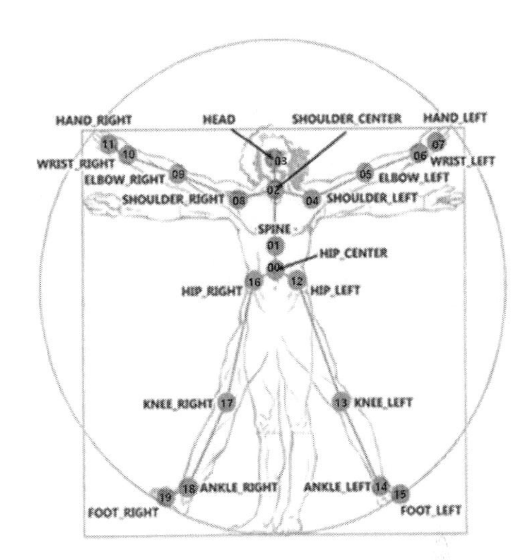

Figure 3. Kinect skeleton structure.

3 BEHAVIOR TRAINING PLATFORM

The proposed behavior training platform is composed of the brainwaves and motion recognition interfaces, the interactive intervention controller with the scene and narrative content managers, and the data center for sensor and intervention data (Figure 4 shows the block diagram of the proposed platform).

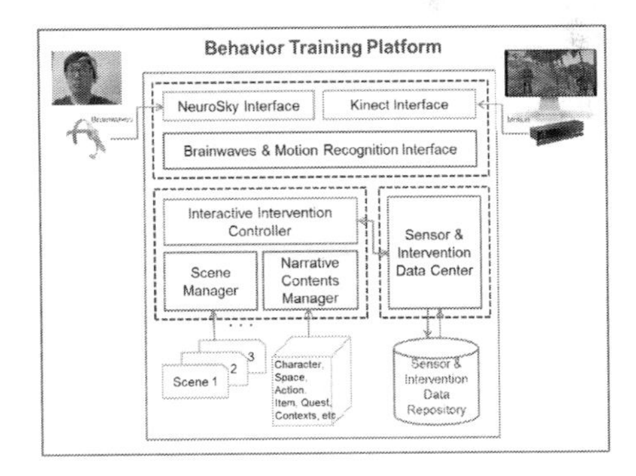

Figure 4. Block Diagram of the behavior training platform.

3.1 *Brainwaves and motion recognition interface*

The brainwaves and motion sensing data are collected through this front-end sensor interface. The motion tracking is accomplished by tracking users with the Kinect skeletal tracking mechanism. By analyzing the tracked skeleton data, a sequence of joints' positions of the twenty recognized human joins in space over time, a specific motion is recognized. Motion tracking has been accomplished in term of skeleton tracking and head tracking (See Figure 5 and 6).

Figure 5. Raising right hand from skeleton tracking..

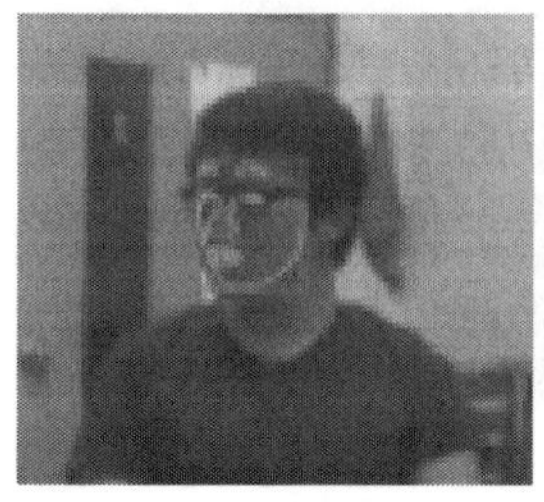

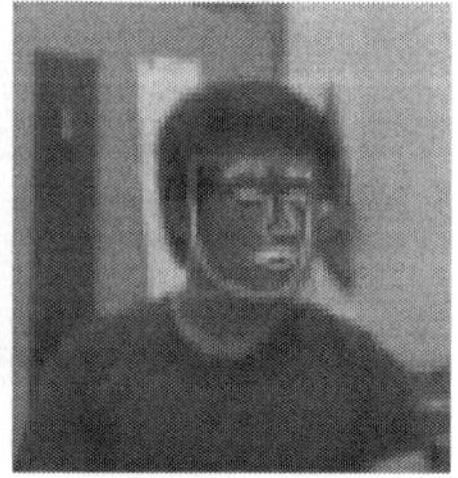

Figure 6. Moving head from right to left from head tracking.

3.2 *Interactive intervention controller*

The interactive intervention controller dynamically controls the game according to the recognized brainwaves and skeleton and head tracking results. It performs the intervention/training functions of the game depending on the player's interactive action, reaction and status. Also the controller processes front-end user interaction information and delivers them to the scene manager, narrative content manager, and sensor and intervention data center. The intervention/training function has been implemented by prohibiting the character of the game from proceeding any further when the player is not concentrating on the narratives or moving in hyperactive motion (See Figure 7).

While a user plays a given narrative game, he/she is given a sequence of quests within individual scenes and guided to appropriately respond to accomplish given tasks in each scene. The system continuously attracts the participants and gives rewards back to them to maximize the effect of the game and increase the level of concentration and calm state.

3.3 *Sensor and intervention data center*

Once completing each game, sensor and intervention data is accumulated in the system data center including the player's brainwaves, skeleton tracking, and head tracking data, and intervention records issued by the platform. The stored is used to reproduce the player's past game and assess the training effect.

1) Normal playing with the player sitting in place

2) Paused upon recognizing the player's motion

Figure 7. Interactive intervention mechanism.

The data repository is built on MySQL database which is interfaced with PHP. The data includes ID, data, time, eSense meters including attention and meditation eSense meters of the brainwaves, motioning tracking skeleton data, and clearing time of the quests per each scene.

4 EXPERIMENTS AND ANALYSIS

First, the prototype training platform has been experimented to see if brainwaves and motion tracking data are successfully collected while playing the narrative game. The experiments were conducted by five college students with MindWave headset on. The results show that twenty two skeletons and six head tracking motions were successfully recognized through the brainwaves and the motion recognition interface of the platform. After then, the data has been analyzed in terms of the number of motions and attention and meditation eSense meters.

4.1 *Motion tracking test*

Table 1 shows the motion tracking capabilities of the prototype platform with the average time to recognize twenty two individual motions made by five different participants. On average, a motion was recognized within 1.6 sec. as a reasonable response time to make the interactive intervention controller either proceed or pause the game according to the players' motion frequencies.

Table 1. Motion tracking test.

Motion	P-1	P-2	P-3	P-4	P-5	Mean (sec)
0	1.11	1.00	1.04	1.23	1.01	1.1
1	1.20	1.29	1.30	1.44	1.21	1.3
2	1.43	1.41	1.41	1.42	1.43	1.4
3	2.80	1.41	1.21	1.33	1.22	1.6
4	1.40	1.20	1.55	1.31	1.24	1.3
5	1.42	1.41	1.31	1.24	1.51	1.4
6	1.60	1.41	1.42	1.41	1.50	1.5
7	1.60	1.52	1.42	1.31	1.20	1.4
8	1.50	1.21	1.33	1.42	1.12	1.3
9	2.09	2.20	2.02	1.95	1.98	2.0
10	1.60	1.55	1.60	2.00	1.98	1.7
11	2.60	2.05	2.10	1.98	2.41	2.2
12	1.85	1.81	1.91	1.94	2.01	1.9
13	1.83	1.51	1.72	1.95	1.80	1.8
14	1.23	1.53	1.11	1.44	1.22	1.3
15	1.29	1.11	1.30	1.29	1.55	1.3
16	2.14	2.22	2.20	1.99	2.09	2.1
17	1.25	1.24	1.11	1.09	1.19	1.2
18	1.44	1.52	1.69	1.20	1.43	1.5
19	1.68	1.77	1.52	1.55	1.44	1.6
20	1.55	1.48	1.60	1.52	1.55	1.5
21	1.78	1.60	1.66	1.77	1.72	1.7
22	1.10	1.05	1.22	1.24	1.15	1.2
Total Time (Mean)	37.49 (1.7)	34.50 (1.6)	34.75 (1.6)	35.02 (1.6)	34.96 (1.6)	35.3 (1.6)

4.2 *Attention variation*

While playing a scene of the game, the five participants' brainwaves were collected (See Figure 8).

4.3 *Game tracking test*

As a next step to test the functions of the platform and its effectiveness, skeleton and head tracking have been tested. At the same time, the five participants' brainwaves were measured and collected through MindWave. Table 2 briefly summarizes the identified motions, including some of the skeleton and head tracking motions, two eSense meters of attention and meditation values, and scene clearing time of individual participants.

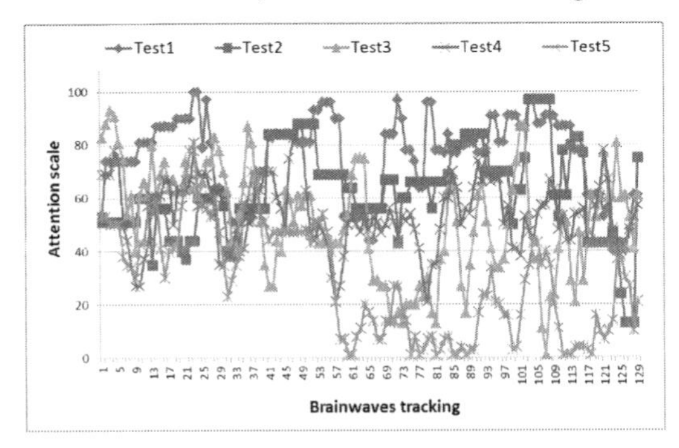

Figure 8. Attention variation of five participants.

Table 2. Skeleton and head tracking, brainwaves, and clearing time.

Sensing & Measurement		P-1	P-2	P-3	P-4	P-5
Skeleton Tracking	Standing	0	2	0	1	1
	Left Hand Left	3	2	1	3	0
	Upper Body Left	2	4	1	3	0
	Left Foot Left	1	1	1	0	0
	...			...		
	Head Right Tilt	0	1	0	1	1
Brain waves	Attention	76	63	51	54	30
	Meditation	67	63	61	55	34
Scene clearing time (sec)		65	112	67	101	61

5 CONCLUSION AND FUTURE RESEARCH

The suggested behavior training platform continuously keeps track of players' brainwaves and motions while they interact with the narrative game. Accordingly, the platform controls the flow of the game so as to help the children with ADHD successfully finish the game and train themselves. For this, a narrative storytelling structure is embedded to the platform in order to attract the children and help them obtain good effects through the game. As a BCI system, MindWave biosensor captures users' brainwaves. The kinect motion sensor is integrated to the platform.

The platform has been tested to see how the brainwaves and the motion recognition interface correctly responds to changing status and the interactive intervention functions. The interactive intervention between the players and the platform can be used to reproduce the past game of a player by referring to the brainwaves and motion sensing data.

The main contribution of this research is the implementation of the behavior training platform which combines BCI and motion sensing technology with narrative contents and treat the children with ADHD in terms of the behavior management. Moreover, the game helps the children develop their social skills as they learn how to control their behavior. The narrative contents will be diversified with wide-range of fairy tales for the children to actively participate in the training program. Furthermore, the platform would be integrated with Big Data framework to effectively handle numerous sensing events and applied to assistance or surveillance solutions by adapting motion sensing technology.

REFERENCES

American Psychiatric Association. 2013. *Diagnostic and Statistical Manual of Mental Disorders, Fifth edition: DSM-5*. Washington: American Psychiatric Association.

Barkley, R.A. 1998. Attention-Deficit Hyperactivity Disorder. *Scientific American, September 1998*: 6–71.

Barkley, R.A. 1999. *Attention Deficit Hyperactivity Disorder: a Handbook for Diagnosis and Treatment, 2nd ed*. New York, NY: Guilford Press.

Barkley, R.A. 2014. Fact Sheet: Attention Deficit Hyperactivity Disorder (ADHD) Topics. *http://www.russellbarkley.org/factsheets/adhd-facts.pdf*.

Bettelheim, B. 1989. *The Uses of Enchantment: The Meaning and Importance of Fairy Tales*. New York: Vintage Books.

Bolt, R. A. 1980. "Put-That-There": Voice and Gesture at the Graphics Interface. *Proceedings of the 7th Annual Conference on Computer Graphics and Interactive Technique, Seattle, US, July, 1980*14(3): 262–270.

Centers for Disease Control and Prevention. 2014. Key Findings: Trends in the Parent-Report of Health Care Provider-Diagnosis and Medication Treatment for ADHD: United States, 2003–2011. *http://www.cdc.gov/ncbddd/adhd/features/key-findings-ADHD72013.html*.

Huang, J., Yu, C., Wang, Y., Zhao Y., Liu, S., Mo, C., Liu, J., Zhang, L. & Shi, Y. 2014. FOCUS: Enhancing Children's Engagement in Reading by Using Contextual BCI Training Sessions. *Proceedings of the CHI Conference on Human Factors in Computing Systems, Toronto, Canada, April 2014*. ACM.

Microsoft. 2014. microsoft.com Tracking Users with Kinect Skeletal Tracking. *http://msdn.microsoft.com/en-us/library/jj131025.aspx*.

Neurosky. 2014. neurosky.com eSense(tm) Meters. *http://developer.neurosky.com/docs/doku.php?id=esenses_tm&s[]=attention&s[]=esense*.

NYU Child Study Center. 2014. Attention Deficit Hyperactivity Disorder (ADHD): Children at Risk. *http://www.aboutourkids.org/families/disorders_treatments/az_disorder_guide/attentiondeficit/hyperactivity_disorder/children_ris*.

Rief, S.F. 2005. *How to reach and teach children with ADD/ADHD: practical techniques, strategies and interventions, 2nd ed*. San Francisco: Jossey-bass.

Scherer, R. 2013. On the Use of Games for Noninvasive EEG-Based Functional Brain Mapping. *IEEE Transactions on Computational Intelligence and AI in Games* 5(2): 155–163.

Shin, S.H., Kim, S.K., Chang, J. & Park, K. 2014. An Implementation of the HMD-Enabled Interface and System Usability Test. *Proceedings of the 1st International Symposium on Simulation & Serious Games, Seoul, Korea, May 2014*.

Teplan, M. 2002. Fundamentals of EEG Measurement. *Measurement Science Review* 2(2): 1–11.

Electronics, Communications and Networks IV – Hussain & Ivanovic (eds)
© 2015 Taylor & Francis Group, London, ISBN: 978-1-138-02830-2

Low-complexity implementation of moving object detection

B.S. Kim, J. Kwon & D.S. Kim
Korea Electronics Technology Institute, Seongnam, Republic of Korea

ABSTRACT: So far, moving object detection is assumed as the scene taken by stationary cameras. In the case of tracking the vehicle, however, the camera mounted on the vehicle moves according to the vehicle's movement, resulting in ego-motions on its background. The scene contains mixed motions and it is difficult to distinguish between the target objects and background motions. Without further improvements on the mixed motion, traditional fixed-viewpoint object detection methods lead to many false-positive detection results. In this paper, a procedure of using with traditional moving object detection methods which relax the stationary cameras restriction is suggested, by introducing additional steps before and after of the detection. It also states about methods for implementation that can alleviate overall complexity. The target application of this suggestion is to use with a road vehicle's rear-view camera system.

1 INTRODUCTION

Recently, road vehicles have been equipped with various sensors and monitoring systems to provide safety related functions, such as a lane departure warning system, parking assistance, and backward collision alarms (Fleming 2012, Okuda et al. 2014). In the case of backward collision prevention, so far, the implemented feature is that it uses active sensors, including laser, ultrasonic, or microwave radar (Cheam & Saman 2014) in conjunction with vision sensors to provide live images to its user. Although those active sensors detect stationary obstacles well based on the distances between objects and the vehicle, we focus on the detection solely from a vision sensor, which is mainly a camera mounted on the back of a vehicle.

In this proposal, as a practical approach, we do not assume the obstacles to be a specific type, therefore, the target objects can be from pedestrians to other vehicles in our suggestion. The detection results may also be further processed with the results from other active sensors to provide combined information in an integrated form such as distance, speed, and direction on the screen if possible.

This paper first examines the previous methods for object detection, followed by details about the suggested procedure. The implementation of the suggestion as a hardware platform is also stated in this paper.

2 PREVIOUS WORKS

From the early years in the computer vision, finding target objects without additional information from other sensing sources has been a challenging job due to its ill-posed characteristics (Poggio et al 1985). Despite this major drawback, various object detection algorithms have been introduced for use with applications such as surveillance cameras, robotics, and intelligent systems. Among the various methods, two major approaches are widely used which are background subtraction and optical flow method.

2.1 *Background subtraction*

Background subtraction is the method of separating moving foreground objects from a stationary background image. The key of the subtraction process is to keep the background image to separate the foreground and background objects. The actual form differs from methods, though, keeping the background information as accurate as possible is always critical. Another common part of background extractions is to hold the pixels relatively consistent across two or more consequent images, while rejecting other pixels that have rapid changes in its value.

Frame difference is one of the earliest suggestions for background subtraction process (Jain & Nagel 1979). It generates a background image by simple frame differences between two consequent images, making the method depend solely on the previous frame. Due to its simplicity, this method can generate the background image with only two input frames, making the method highly adaptive. The major drawback is, however, moving foreground objects may easily become part of the background if the object stop for more than one frame. The interior pixels of an object also may not be distinguished correctly if the inside has uniformly distributed intensity value, leaving no difference between the two frames.

Instead of simple subtraction, approximated median (McFarlane & Schofield 1995) keeps the accumulation of approximated pixel values. As a result, the background image is updated continuously. During the approximation, generated background converges to the median value of all frames. Therefore, background objects stationed across frames may be imprinted better than foreground moving objects which have less opportunity to converge. The foreground objects temporarily stopped or had the plane interior can now better filtered than simple frame differences, unless the object becomes background by appearing in the same position longer than the originally occluded background.

Gaussian mixture models (Friedman & Russell 1997), on the other hand, processes the input frames in different ways. Unlike the mentioned methods above, the background components are accumulated as terms for Gaussian distribution functions. By using its terms, the functions can decide whether a pixel of the input image is part of foreground or background image. This statistical determination is effective for minor changes in the background, such as waving leaves, moving clouds or rain drops.

2.2 *Optical flow*

Another major strategy of finding moving objects is the optical flow of the frames. According to its definition, optical flow is the motion of the apparent movement of brightness patterns in the image, making it sensitive to light sources (Horn & Schunck 1981). Finding the movement can be described as finding the displacement δ which minimizes ε in the following equation

$$\varepsilon(\delta_x,\delta_y)=\sum_{x=u_x-w_x}^{u_x+w_x}\sum_{y=u_y-w_y}^{u_y+w_y}\left(I_1(x,y)-I_2(x+\delta_x,y+\delta_y)\right) \tag{1}$$

with a given point (ux, uy) in the image I. Searching for such value is performed within a small window w. The window should be introduced because it becomes an aperture problem without the window. Adequate size of the window provides hints on the direction of the vectors from searching through neighboring points. The resulting displacements are going through a further classification process so that they can be separated as vector groups, which in turn represent moving objects.

3 COMBINATIONAL METHOD IMPLEMENTATION

In this section, the proposed procedure for moving object detection is presented. To achieve better detection and less false positives than previous standalone approaches, we use both background subtraction and optical flow methods simultaneously. Both background subtraction and optical flow methods have been widely used for applications including intrusion or motion detection from video streams with stationary cameras. Recent studies show the use of grouping algorithms with optical flow also alleviates the limitation of the stationary camera by some level (Kim et al. 2013, Sheikh et al. 2009). It is mainly done by further classification of feature points into two different motion groups.

Instead of using those classification methods, in our suggestion, we combine a background subtraction and an optical flow process to derive moving object regions. The key benefit of using this combinational approach is that the detection results from optical flow and background subtraction can be cross-checked later. This combinational scheme is composed of the following two-step approaches: background motion compensation and object detection.

The proposed algorithm detects the moving object from the images and Figure 1 shows a block diagram of the proposed moving object detection method. The first stage is to scale down the image resolution. The 1280x720 resolution is scaled down to the 320x240 resolution. Original image's bit depth is 8bit, which can have any one of 256 possible values, from 0 (black) to 255 (white). After quantization, the output bit is 4bit values, from 0 to 15. The main filter is used to reduce excessive noise from the image. The proposed architecture adopts a modified 3x3 mean filter for the profit of the hardware implementation. The filter outputs are written to the memory buffer. Second stage is the shifting range determination that calculates the shifting range of a specific area between the current and the previous frame. The

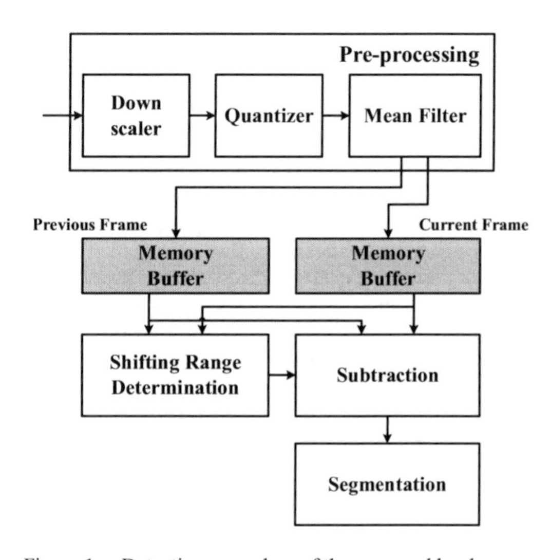

Figure 1. Detection procedure of the proposed hardware.

comparison operation is performed for searching the matching blocks and calculating the shifting range by using the 11x11 blocks in the current frame and the search windows in the previous frame.

The shifting range is applied to the previous frame, and the subtraction operation is started. The subtraction operation is performed by using the 4x4 blocks in the current and the previous frame. The SAD calculation is applied to the whole frame and the comparisons between the SAD results and the threshold is performed. If the SAD result between the present frame and the previous frame is larger than the threshold, proposed architecture marks that 4x4 blocks. The last stage for moving object detection is a segment that is performed to draw the moving object by using marked 4x4 blocks.

3.1 *Down-scaler*

A Down-scaler is used for scaling down the image. In this architecture, 1280 x 720 images are scaled down to 320 x 240 images. Proposed architecture reduces the memory usage by this scaling down step.

3.2 *Quantizer*

A quantizer passes its input pixels through a stair-step function so that many neighboring points on the input axis are mapped to one point on the output axis. Input pixels in the quantizer are 8 bits that can have 256 possible values, from 0 to 255 and after the quantization output bit is 4 bits, from 0 to 15 as shown in Figure 2. This architecture uses a half bit of the original image bit by using the quantization. This can reduce the memory usage and be suitable for low memory hardware implementation.

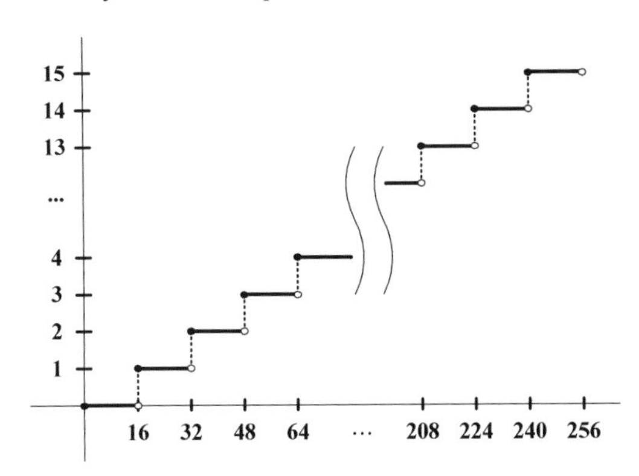

Figure 2. Quantization.

3.3 *Low-pass filter*

Both the Median filter and the Mean filter can be used to remove noise from an image. As the Figure 3 shows a Mean filter is a filter that takes the average of the current pixel and its neighbors. In this architecture, 8-tap mean filter is used to remove the noise of the image. Modified mean filter is only used as adder and shifter for low-complexity design.

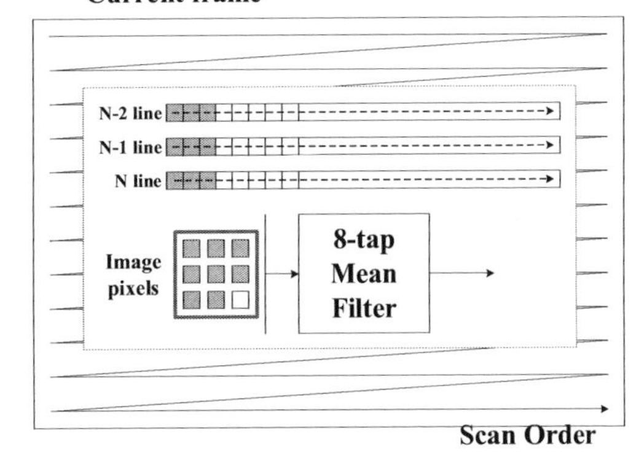

Figure 3. Low-pass filter.

3.4 *Shifting range determination*

The shifting range determination block calculates a shifting range for an accurate sum of absolute difference (SAD) calculation. To compare with one of the blocks on the current frame, C_l is one of the blocks on previous frame and P_l are used for calculating the SAD. After comparison operation, the SAD operation is applied to the search range as shown in Figure 4. The search range is limited from the selected point to 10 pixels toward right and left and from the selected point to 5 pixels toward up and down. The block on the previous frame, which has the least SAD, will be recognized as a similar block with a block on the current frame. Then the average of the difference in the coordinate between the four blocks on current frame and four blocks which have the least SAD on the previous frame is the shifting range. This shifting range is applied to the subtraction block.

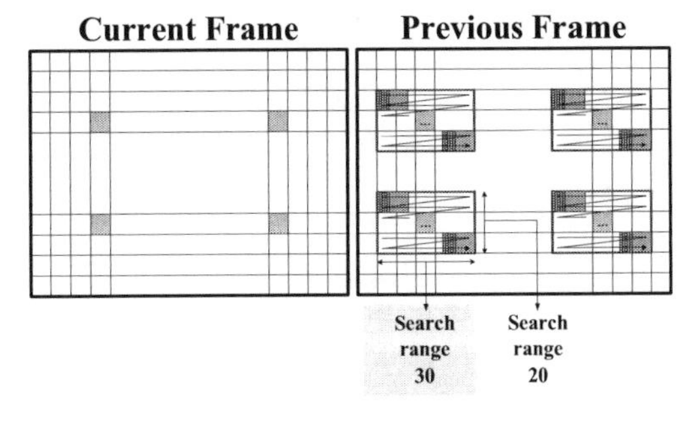

Figure 4. Scan-order of the shifting range determination.

3.5 *Subtraction*

The subtraction block calculates the SAD by using 4x4 blocks on the current and previous frame. The 4x4 blocks are moved to four pixels in the current and previous frame horizontally and vertically, and the SAD is calculated respectively as shown in Figure 5. If calculated SAD is bigger than the threshold, the x and y coordinate of the block is saved in the local register.

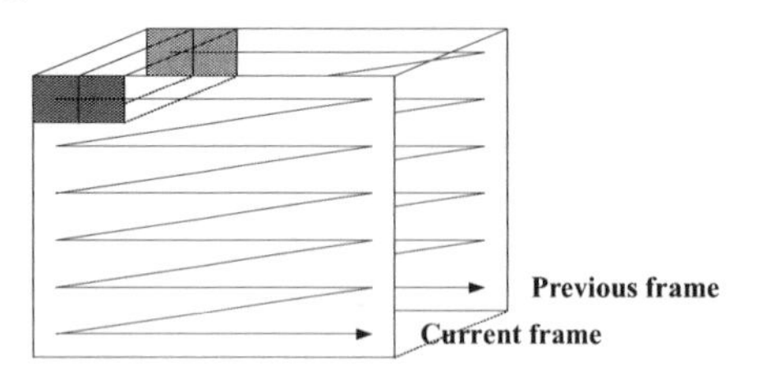

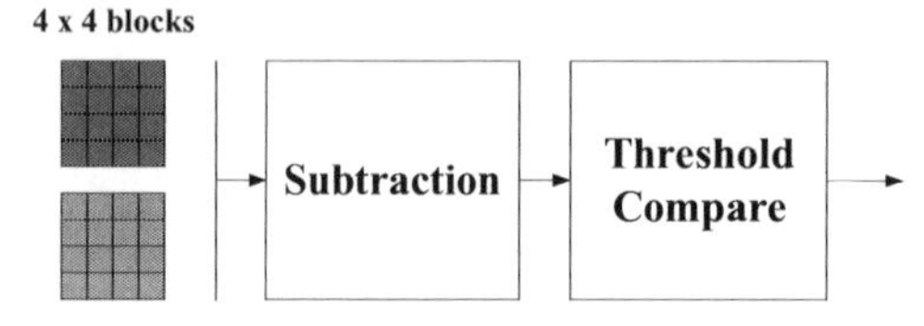

Figure 5. Subtraction.

3.6 *Segmentation*

The segmentation block calculates the coordinate of the moving object and draws the area of a moving object. Adjacent marked blocks are merged into one area. In Figure 6, the segmentation block is divided into 16 areas. First, the blocks on the second line and a block on the third line are united into one block respectively. Then adjacent two blocks on the top and the bottom line are united into one respectively. Last the blocks on the top and the bottom line are united. In the process of combining, detection of the moving object is performed as shown in Figure 7.

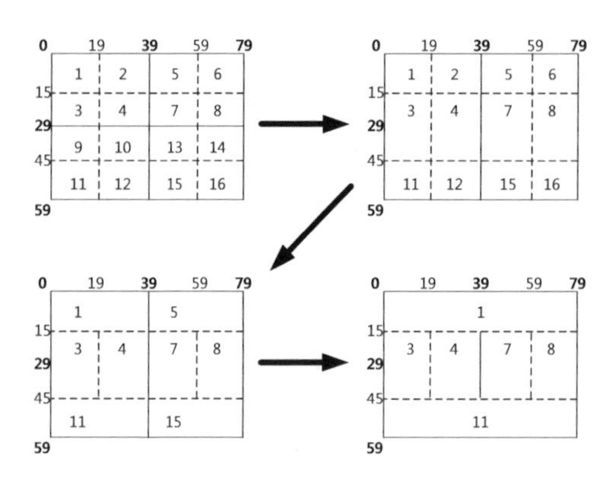

Figure 6. Segmentation.

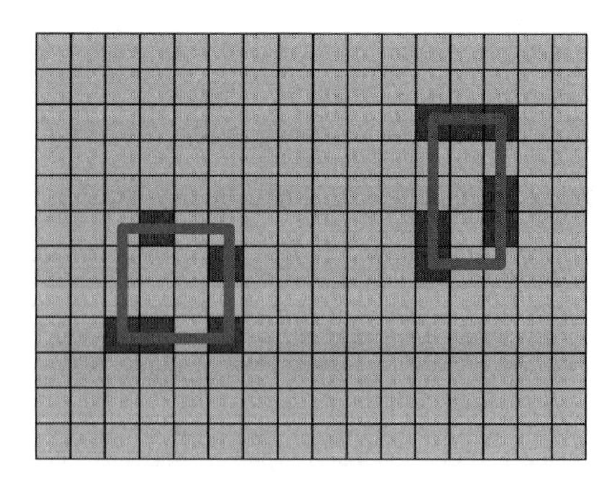

Figure 7. Object detection.

4 EXPERIMENTS

To evaluate our suggestion and compare the result with other methods, two experimental results are presented in this section. In most cases, evaluations of optical flow based methods usually focus on the terms of precision and recall, relative distance, angular error, and computation time. Since it is the accuracy that matters in our target application, we state mainly about the precision and recall terms.

The first sample sequence used in the test consists of 200 consecutive frame sets with three vehicles moving from right to left. The camera also moves from right to left, leaving ego-motion on its background pixels. As shown in the Table 1, rank-constraint model (Sheikh et al. 2009) results in a precision of 0.83 and recall of 0.99 with the sample frames. The suggested procedure gives precision of 0.98 and recall of 0.78.

Table 1. Precision and recall comparison.

Algorithm	Precision	Recall
Rank-constraint	0.83	0.99
Suggested method	0.98	0.78

Another sample video for evaluation has 360 frames taken at an indoor parking lot. The vehicle at this time moves from left to right as the driver intends to make 90-degree turn. The background in the result contains pan motion, and a person who comes into the frame from the left side simultaneously. The video contains the total of 259 ground true movements.

As a comparison, a video of the optical flow method under the assumption of the stationary camera is run (Brox et al. 2004). Because of the ego-motion and changing lights, reported movements are far more than the truth, which include 599 movements with precision of 0.39. Proposed scheme gives the results of precision of 0.80 and recall of 0.74 as described in the following Table 2.

Table 2. Precision and recall comparison with optical flow.

Algorithm	True-positives	False-Positives	Precision	Recall
Optical flow	231	368	0.39	0.89
Suggested method	190	48	0.80	0.74

5 CONCLUSION

In this paper, the two major types of methods for moving object detection and their constraints are discussed. A combinational method of alleviating the overall complexity is suggested. Experimental results show improvements in terms of precision, which is an important feature for using with the applications.

REFERENCES

Brox, T., Bruhn, A., Papenberg, N., et al. 2004. High accuracy optical flow estimation based on a theory for warping. *in Proc. Computer Vision – ECCV 2004, Prague, Czech Republic, 11–14 May 2004* 3024:25–36. Berlin Heidelberg: Springer

Cheam, S.-R. & Saman, A. B. S. 2014. Developing algorithm for object tracking using passive sensors. *5th International Conference on Intelligent and Advanced Systems, June 2014*: 1–5.

Fleming, B. 2012. New automotive electronics technologies. *Vehicular Technology Magazine, IEEE* 7 (4): 4–12

Friedman, N. & Russell, S. 1997. Image segmentation in video sequences: a probabilistic approach. *in Proc. 13th Conf. on Uncertainty in Artificial Intelligence: 175–181, August 1997.* Morgan Kaufmann Publishers Inc.

Horn, B. & Schunck, B. 1981. Determining optical flow. *in 1981 Technical Symposium East, International Society for Optics and Photonics: 319–331, Washington, D.C., 21 April 1981.* International Society for Optics and Photonics.

Jain, R. & Nagel, H.-H. 1979. On the analysis of accumulative difference pictures from image sequences of real world scenes. *IEEE Transactions on Pattern Analysis and Machine Intelligence* (2): 206–214.

Kim, J., Wang, X., Wang, H., et al. 2013. Fast moving object detection with non-stationary background. *Multimedia Tools and Applications* 67(1): 311–335.

McFarlane, N. & Schofield, C. 1995. Segmentation and tracking of piglets in images. *British Machine Vision and Applications* : 187–193.

Okuda, R., Kajiwara, R. & Terashima, K. 2014. A survey of technical trend of ADAS and autonomous driving. *2014 International Symposium on VLSI Technology, Systems and Application, Proceedings of Technical Program: 1–4.* IEEE.

Poggio, T., Torre, V. & Koch, C. 1985. Computational vision and regularization theory. *Nature* 317(26): 314–319.

Sheikh, Y., Javed, O. & Kanade, T. 2009. Background subtraction for freely moving cameras. *IEEE 12th International Conference on Computer Vision*: 1219–1225.

Stauffer, C. & Grimson, W. 1999. Adaptive background mixture models for real-time tracking. *Proc. IEEE Computer Society Conf. on Computer Vision and Pattern Recognition 2 Fort Collins: 246–252, CO, 23 June 1999.* IEEE.

Electronics, Communications and Networks IV – Hussain & Ivanovic (eds)
© 2015 Taylor & Francis Group, London, ISBN: 978-1-138-02830-2

Design of the online temperature monitoring system for distributed high-voltage switchgear

Zi-hua Kong* & Jin Luo
Department of Missile Engineering, Ordnance Engineering College, Shijiazhuang, Hebei, China

ABSTRACT: In allusion to the power equipment failure caused by the electrical connection superheating inside the high-voltage switchgear, this paper proposed a scheme based on Zigbee technology and GPRS mobile communication network on the foundation of current study about the temperature monitoring system of power plant and transformer substation voltage switchgear. The scheme set up the cluster-tree Wireless Sensor Network (WSN) and transmitted the temperature data to the remote monitoring terminal through the embedded gateway and GPRS module. Finally, busbar and connection temperature online monitoring and warning were realized for number of high-voltage switchgear. The whole system has such advantages such as stability and high reliability in operation, it's effectively guarantee the safe operation of power equipment.

1 INTRODUCTION

Switchgear is one of the most important electrical equipment in power system, it's running a state has a significant impact on the reliability of the power system. Traditional measurement methods which measure multi-switchgear by artificial manner causes labor waste and heavy workload, and connections superheat cannot be found timely which is caused by oxidation and corrosion of the electrical connection surface and poor contact or aging of the busbar connection during the long-term operation. If we don't troubleshoot timely, it even cause equipment to be burned, a large area power outages and other accidents (Wang et al. 2009). To avoid such accidents, the real-time temperature monitoring of the cable connection of high-voltage switchgear became an important problem related to the safe operation of the power system. Because the hot spots are in the sealed switchgear with its door is prohibited opening in running, the duty officer could not find the heat defect through normal surveillance method that once the connections superheated seriously, it will lead to serious accidents and affect the safe operation of the system.

At present, there are three main methods of online temperature measurement: IR mode, fiber mode and wireless mode(Qian 2007, Gong et al. 2006), among which, the fiber optic sensors have difficulties in arrangement and easily bend. IR model takes the temperature of measured points by using infrared thermal imager, this non-contact method is less susceptible to electromagnetic interference, but expensive, used by certain people and have inconvenience in measuring temperature inside a small space of the switchgear. Wireless mode is to install a wireless transceiver module on each measured spot, with this module, there is one or more sensors to measure the temperature of the measured spots and the collected data is sent to the data nodes by way of wireless communication. This paper proposed a wireless sensor network (WSN) based on Zigbee technology for the online monitoring of the high-voltage switchgear temperature. For there is difficult in using electricity from high-voltage terminal, the data processing chip CC2430 and MSP430F149 are chose due to they have low power consumption and a variety of power-saving modes coexist. The digital temperature sensor DS1631 with ESD (Electro-Static Discharge) protection is utilized to measure temperature and the GPRS module transmit the local data remotely to realize remote online monitoring for multi- switchgear temperature.

2 ZIGBEE TECHNOLOGY

Zigbee, which emerged in recent years is a wireless networking technology for the low-cost device. It is a kind of wireless communication technology with short distance, low complexity, low power, low data rate and low-cost two-way(Wang & Sun 2011). The Zigbee protocol developed by an IEEE802.15.4 working group and Zigbee Alliance jointly. A complete protocol stack is only 32KB in size. The working frequency of Zigbee is divided into 3 sections: 868MHz, 915MHz and 2.4GHz. Among witch, 2.4GHz is free of charge, application and of the universal frequency range. The Carrier Sense Multiple Access

*Corresponding author: *sky-speed@163.com*

with Collision Avoidance (CSMA/CA) is adopted in communication to effectively avoid collision of the wireless carrier. CCM* mode, the improved model of CCM is adopted by Zigbee network, and security and confidentiality of the data can be ensured via implement 128 bit AES encryption algorithm to encrypt the data frame. In addition, wireless sensor networks using the anti-interference of high-frequency radio signal (especially 2.4GHz) make the data communication security guaranteed effectively; it gives a better solution to the high-voltage isolation of signal transmission. Therefore, it's very suitable for wireless temperature monitoring of high-voltage switchgear.

3 SYSTEM DESIGN

Zigbee supports three kinds of static and dynamic self-organizing wireless network topology structure: Star, Mesh and Cluster-tree. Generally there are dozens of switchgear in small and medium substation, which distribute dispersedly. So using the star network can not meet the requirements. Therefore, Cluster-tree network is adopted in this paper; it has the strongest network robustness and system reliability, more of flexibility and freedom in the self-organizing network path. Although Zigbee is short-distance transmission technology, in the outdoor open area (such as the substation) the transmission distance between adjacent nodes can reach 400 meters. But Zigbee network can accommodate up to 65535 devices, if by communication between the routers and nodes, the coverage of the whole ZigBee network will be expanded indefinitely, which meet the requirement of network arrangement of small and medium-sized substation.

Zigbee wireless sensor network mainly constitutes by network coordinator, router and sensor nodes, of which, network coordinator and router is mainly responsible for organizing WSN, forming a transmission path and sensor node acquiring temperature data. Network coordinator and router are served as Full Function Device (FFD) and sensor nodes are served as Reduced Function Device (RFD). In the wireless networks organized in this paper, there are multiple sensor nodes and router nodes but only one coordinator node. In addition, because of the dispersed layout of switchgear in the transformer substation, it need to set a certain number of routing nodes to ensure the reliability and integrity of the data communication.

A primary equipment of the sealed switchgear is distributed in 3 separate rooms: switch room, busbar room and outlet room. In the switch room of the primary switch, there are 6 main plug and 18 connections flowing load current through. In the outlet room there are 3 main plug which connect with many other electrical connections. Select key parts of these electrical connections to install the temperature sensors. The sensor nodes installed inside the switchgear, the router nodes and coordinator nodes installed outside the switchgear are organized to the wireless monitoring network by Cluster-tree topology structure. Take one switchgear as a cluster and set the router as cluster head. Each cluster number with RF transceiver is responsible for data acquisition, processing and transmitting to the cluster head. The cluster head can transmit the data acquired by cluster number to the nearest network coordinator after data fusion. The cluster head can also broadcast the data packet, which is transmitted by the network coordinator, to the cluster managed by cluster head itself.

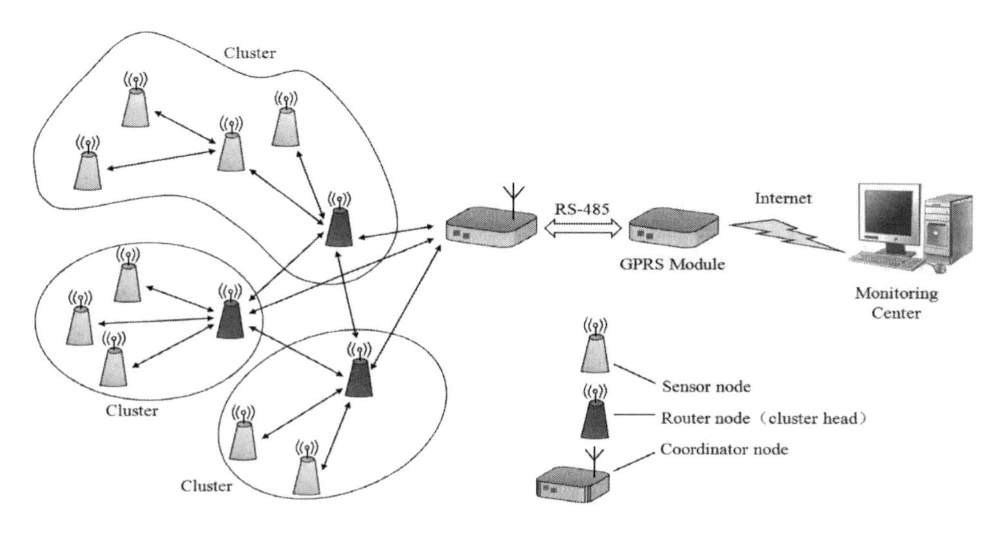

Figure 1. Overall structure of the monitoring system.

The network coordinator and GPRS module is connected by the serial form, so data can be transmitted directly to the monitoring center by GPRS. At the same time, the remote communication between the monitoring center and Zigbee network can be also realized through GPRS public channel and the effective control and management of the locale network is realized by the relevant data acquired. The monitoring system structure is shown in Figure 1.

4 HARDWARE DESIGN

4.1 *Sensor node design*

Sensor node, which consists of data processing module, data acquisition module and power modules etc, is mainly responsible for temperature information collection, data transmission, reception and implementation of control instruction. The core part of the sensor node uses CC2430 chip developed by Texas Instruments, Inc. The chip is based on Smart RF technology, whose interior contains a 8051 MCU of enhanced industry standard working in 32MHz, a RF front-end of 2.4GHz, programmable flash memory of 32/64/128KB, Static RAM of 8KB embedding Z-Stack in. The maximum transmission rate of CC2430 is 250Kbps, and its current is less than 27mA and 25mA respectively under receiving and sending modes, only 0.9mA under sleep mode, less than 0.6μA under standby mode. Therefore, it is particularly suitable for battery powered. The temperature sensors use DS1631 chip developed by Maxim company, which has digital output and I^2C an interface. This chip provides 9～12 bit sampling precision for the user, and its measuring range is between -55°C and 125°C, the measuring error is within ±0.5°C between 0°C and 70°C. Furthermore, it offers ESD protection of 8000V(Wendel 2003) to meet the requirement of the work environment of the system. The temperature information collected by DS1631 is translated into a digital signal that to be processed and transmitted after outputting to CC2430. CC2430 connects DS1631 by I^2C bus which is simulated by I/O port. Generally speaking, the sensor node can survive for about 2 years with two AA batteries, so that the maintenance personnel can replace the batteries during routine maintenance of the electrical equipment. Moreover, Z-Stack don't support a sleeping pattern of the router node, so hardware of the router node should be designed to low power consumption as far as possible. The hardware structure of the sensor node is shown in Figure 2.

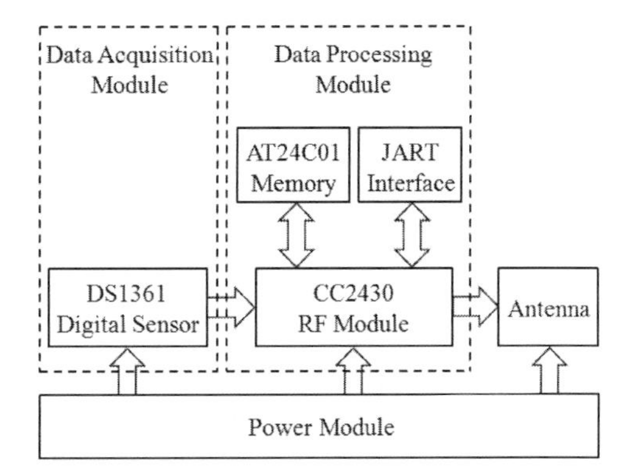

Figure 2. Hardware structure of the sensor node.

4.2 *Coordinator node design*

Coordinator node includes processor module, RS-485 serial communication module, wireless communication module and power module etc. The serial communication module uses a enhance RS-485 transceiver ADM2483 with isolation. The chip adopts i Coupler magnetic isolation technology in stead of optical isolation device whose transfer efficiency is influenced by external and it has the function of TSD and fail safe(Wang et al. 2014). The hardware structure of coordinator node is shown in Figure 3. The function of the coordinator node is: to establish and maintain the wireless network, to receive, process, gather the sensor nodes information and transpond to the monitoring center. All the data transmitted by sensor nodes are gathered to transmit to the GPRS transceiver, and then forwarded to monitoring terminal through the coordinator node.

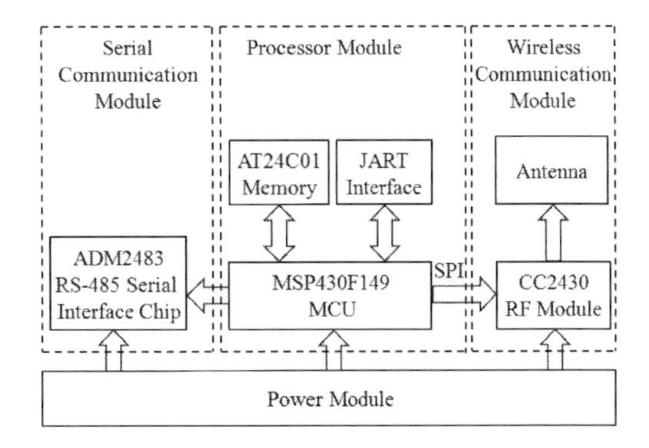

Figure 3. Hardware design of the coordinator node.

4.3 *GPRS module design*

GPRS module, which consists of a processor module, GPRS transceiver module, power modules and other peripheral circuit, is mostly used for transmitting the data from WSN and receive control command from monitoring center. If the distance between the monitoring center and coordinator node is greater than 1000m, it isn't suitable for establishing cable communication channel. So data communication should proceed by dint of the mobile communication network provided by operators. GPRS is now very mature mobile wireless network technology of 2.5 generation, it is convenient to use and install, low cost, wide coverage, reliable operation, less leakage code and bit error rate. GPRS can provide a data transfer rate of 115kbps, and it is suitable for transmitting the low amount of data frequently. The Transmission time interval is larger than the network delay. MC52i, the industrial dual-frequency (GSM/GPRS) module of Cinterion is selected in this paper. Because TCP/IP protocol stack is embedded in the module, the MCU can directly control the module, by AT instruction set, to translate serial data into a data packet of TCP/IP. The hardware structure of GPRS module is shown in Figure 4.

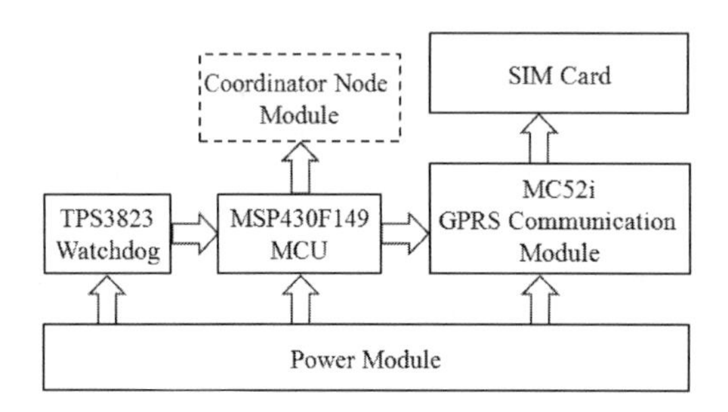

Figure 4. Hardware design of the GPRS module.

5 SOFTWARE DESIGN

5.1 *Software design of sensor node and coordinator node*

After power on, the coordinator node complete the task of Z-Stack initialization and channel assessment, select a free channel automatically and set PAN identifier after scanning channel, and then start a Zigbee network and release broadcast frame to wait for the link request of terminal node. Meanwhile, sensor node scans effective network channel initiatively to seek a suitable coordinator nearby. Once find the broadcast frame, the sensor node will send the network access requirement to the coordinator node and then join the network authentically after confirmed. Next, the coordinator node assigns a unique ID for each sensor node and register MAC address binding with ID to start data transceiving and various operating instructions performing.

After the network is established, the router node and the coordinator node are of the same function. Because replacement of each node battery is difficult, in order to reduce the consumption, prolong the network life cycle. CC2430 of the sensor node using PM2 mode (low power mode in which CC2430 can be awakened in time), that is, sensor nodes enter into PM2 mode just after joining the network successfully. The timing of 30 min achieve, the sensor nodes are awakened to read the data acquired by the temperature sensor DS1361 and to transmit. After finishing the work, CC2430 enters into PM2 mode once again. Nodes program flow is shown in Figure 5(a) and Figure 5(b).

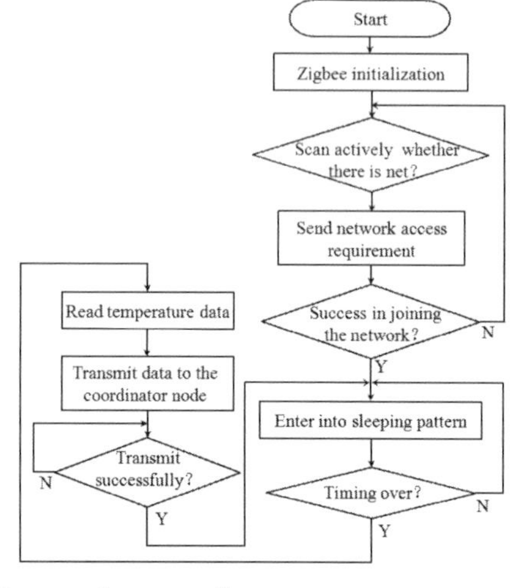

(a) Sensor node program flow

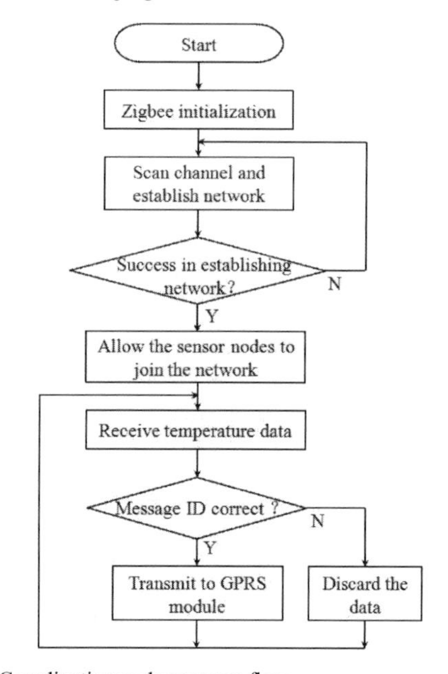

(b) Coordination node program flow

Figure 5. Nodes program flow.

5.2 *Software design of GPRS module*

GPRS module is responsible for transmitting the temperature data gathered by the coordinator node to the monitoring center. Meanwhile, it receives the control command (such as temperature acquisition command) sent from the monitoring center and then parse and forward to relevant sensor nodes through the coordinator node. The connectivity of GPRS module and the Internet can be achieved by installing a SIM card and open web service. Th program flow of GPRS module is shown in Figure 6.

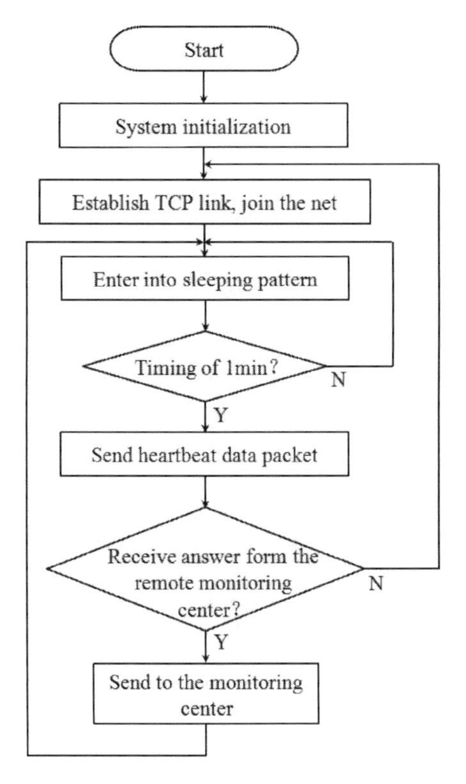

Figure 6. Program flow of GPRS module.

After initializing and joining the network, the system turn into a sleeping pattern (current only 3mA). In order to avoid network interruption, a length of heartbeat packet is sent to the monitoring center server every 5 min of system setting to ensure the GPRS module long-term online. If there is no answer received during a certain time, it is judged to network disconnection and require to re-establish the connection. When the coordinator node transmits the data, GPRS module begins to parse a data frame of Zigbee, from which, extract temperature data and package to the data frame of TCP/IP according to the GPRS standard. Then the data are uploaded to the server via the Internet, according to the IP and port number of the terminal that set by MC52i to accomplish the protocol conversion from Zigbee to GPRS. In the temperature monitoring system, the size of each data frame is 20 bytes. After testing, when data transmission conducts of TCP mode at the speed of 500 frames per second in the spacious open air, there is almost no packet loss, which is stable and reliable communication.

5.3 *Software design of monitoring platform*

System software of the remote monitoring center is based on C/S structure oriented Socket communication mechanism of TCP/IP, Visual C++, object-oriented visual integrated programming system, is adopted as well. The temperature data on the electrical connections inside the switchgear are saved in Access database by means of MFC ODBC, which is convenient for recalling of the historical data. The Monitoring platform provides users with a graphical operation interface, and its main function is as follows:

a. Real-time display of temperature: To take a real-time display of the measured temperature in each high-voltage switch gear and store the data in the database.
b. Nodes management: For the registration information of new nodes added in, monitoring center should add them into the node list of databases to update and it warns managers to enter the annotation of the node installation location to mark the physical location of the node.
c. Temperature prediction and alarm: To alarm for the condition of exceeding the temperature threshold, and to make early warning for not exceeding but violent rise in temperature. To determine the position of the temperature anomaly point by the node's location and then get the diagnosis of the operating state of the important parts such as busbar connection and electrical switches.
d. Historical data query and report output: To query the historical data of each measurement point of the switchgear, plot the temperature curve and finally generate the data report.
e. Parameter setting: To accomplish the setting of remote temperature limit, sample frequency and transmission rate.

6 SYSTEM TESTING

Due to the particularity of contact inside the high-voltage switchgear during actual operation, the accuracy analysis of wireless temperature measuring system based on ZigBee is conducted in the simulation laboratory. Copper row is used to simulate the heating parts such as busbar and contact of the high-voltage switchgear (10-150°C can be simulated). In order to realize temperature measurement of multiple measuring points by WSN at the same time, connect the coordinator to the computer through the serial port line and start the software of a temperature monitoring system. When temperature data output from each measuring point of the monitoring terminal, it indicates that the test platform of temperature monitoring network based on ZigBee can realize the function of networking and online temperature monitoring. By comparing with the actual temperature, system measuring error is not more than ±0.5°C, which can meet the requirement of temperature measurement system of the high-voltage switchgear. The comparative data are shown in Table 1.

Table 1. Comparison of the node measurement data and actual temperature.

time	No.1 node	No.2 node	No.3 node	actual temperature
16:30	27.0°C	27.3°C	26.8°C	27.1°C
16:35	28.9°C	29.1°C	28.7°C	29.2°C
16:40	34.1°C	34.3°C	33.8°C	34.0°C
16:45	38.5°C	38.5°C	38.2°C	38.4°C
16:50	43.0°C	43.3°C	42.8°C	42.7°C
16:55	47.3°C	47.7°C	47.4°C	47.8°C

7 CONCLUSION

The design of the online temperature monitoring system for switchgear in this paper, by virtue of the introduction of wireless communication technology, has solved the problem of the temperature measurement of the switchgear in the environment of sealed space, high voltage, high magnetic field and large current. The measurement arrangement taking advantage of WSN based on Zigbee technology is realized to the online temperature monitoring of multi-switchgear, multi-channel. It has characteristics of low power consumption, thorough isolation, flexible networking and large network capacity. Meanwhile, it incorporates the advantages of wide coverage area, remote transmission of GPRS, field data can be collected to achieve the teletransmission and then by means of the monitoring platform to monitor the real-time temperature and predict an accident or fire caused by local overheating of the equipment. Thus, the system provides a powerful safeguard for the normal operation of the electrical equipment.

REFERENCES

Gong Xiao-feng, Yi Hong-gang & Wang Chang-song. 2006. Reseach On Temperature Monitoring of Isolators in HV Switchgear. *Proceeding of the CSEE* 26(1): 155–158.

Ken Wendel. 2003. Reliability Report for DS1361(Rev A4). Dallas Semiconductor.

Qian Xiang-zhong. 2007. Design of On-line Monitoring System of Temperature for Contact Inside of The High Voltage Switchgear. *Instrument Technique and Sensor* 2: 73–75.

Wang Qi-wu, Zhou Feng-xing & Yan Bao-kang. 2014. Design of Mornitoring System for The Stability of The Base of Voltage Towers. *Application of Electronic Technique* 40(3): 126–29.

Wang Ru-chuan & Sun Li-juan. 2011. Sensor Network and Its Applications. Beijing: Posts & Telecom Press.

Wang Xiao-hui, Su Biao & Rong Ming-zhe. 2009. Development of On-line Monitoring System of Medium Voltage Switchgear. *High Voltage Apparatus* 45(3): 52–55.

Electronics, Communications and Networks IV – Hussain & Ivanovic (eds)
© 2015 Taylor & Francis Group, London, ISBN: 978-1-138-02830-2

A close-form bias reduced solution of source localization based on Chan's algorithm

Qian Li, Junhui Liu, Shuang Liu & Lei Xi*
Henan University of Animal Husbandry and Economy, Zhengzhou, China

ABSTRACT: The paper provides an algorithm for reducing the bias of the famous Weighted Minimum Least Square (WMLS) algorithm proposed by Chan and Ho. The new method only requires the structure of the noise covariance matrix. It creates an augmented vector and an imposing constraint. The analysis shows that the bias is reduced considerably and the MSE performance can achieve CRLB. Furthermore, it can be applied to moving source by using FDOA localization.

1 INTRODUCTION

Source Localization is a fatal problem in next-generation communication that has been applied in military, medical treatment, environment and commercial areas (Kay & Sengupta 1995). Traditionally, accuracy dominates the performance of localization, especially in some applications such as navigation and emergency service. Thus, designing an efficient algorithm to decrease error and bias is encouraged widely. The close-form non-recursive hyperbolic solution, so called WMLS by Chan & Ho (1994) is attractive for its superior performance. It requires neither initial guess nor iteration, and it uses a linear constraint to estimate the TDOA source position and attain Cramer-Rao lower bound (CRLB) in most cases under high SNR and fine geometry condition, especially when the source is distant. However, although its MSE performance can achieve the CRLB, the bias dominates the mean square error when the SNR is low and virtual geometry is poor. Compared to the previous methods, the Bias-Reduction method in this thesis is different:

- The algorithm that proposed in this thesis can be applied in both two dimensional (2-D) and 3-D.
- It can deal with correlated noise. However, most TDOA noise is often assumed to be uncorrelated.
- It does not require numerical search in (Huang et al. 2001).
- The computational complexity is relatively simple and does not need singular value decomposition (SVD) in total-least-square methods.
- The solution is available to reach CRLB under the assumption that SNR is considerably high, and the source is far from sensors.

The simulations corroborate that the new methods can reduce the bias to a large extent in most circumstances.

2 CLOSE-FORM BIAS REDUCTION SOLUTION

2.1 *Discussion of previous work*

In this paper, it adopts Chan's algorithm as the benchmark for new methods that verify the performance. The bias of Chan's algorithm is caused by the noise correlation between regressor and regressand in the computation process of WLS algorithm (Ho 2012). The modified algorithm Weighted Total least square (WTLS) approach (Barton & Rao 2008) uses weighted TLS technique to decrease the bias. It is computationally expensive and the estimation standard deviation is greater than Chan's solution; Newton-Raphson (N-R) iterative (Barton & Rao 2008) brings in the estimate solution of Chan's algorithm as an input. It is an approximate ML method to solve the nonlinear TDOA equations iteratively; Linear-correction least-squares (LCLS) estimator (Huang et al. 2001) imposes a constraint to the unknowns. It makes no assumption about the covariance matrix of measurement errors and the linear approximation. The method is attractive for that it is difficult to verify the assumed error distribution. However, without the second stage iterations of WMLS, it would lead to a big bias. LCLS has relatively small bias and low standard deviation, but it requires numerical search; a previous research by Kovavisaruch & Ho (2005) presents a solution for arbitrary sensors with random sensor position measurement errors. It utilizes several

*Corresponding Author: *chenhong@gcidesign.com*

WLS and does not suffer from ambiguity solution. The accuracy performance of this algorithm is close to CRLB with Gaussian noise and small position error. Although the solutions reduce the bias, compared to WMLS, they increase the computational complexity in different extent.

2.2 *Close-form bias-reduced algorithm*

It assumes that there are M separated sensors in the scenario and u^o denotes the true Cartesian coordinate of source which $(*)^o$ means the noise-free true value. The i_{th} sensors locate at si that are known accurately. N presents the dimension, which means that u^o and s_i are $N\times 1$column vectors. The distance between i_{th} sensor and source is formed below:

$$R_i^{o2} = \left\|\mathbf{u}^o - s_i\right\|^2 = (\mathbf{u}^o - s_i)^T(\mathbf{u}^o - s_i) \tag{1}$$

where ||*|| is Euclidean norm. $ni,1$ is the additive noise. The TDOA measurement is the distance difference between i_{th}($i\neq$hetween e. The TDOA measurement irrived sensor:

$$R_{i,1}^o = \left\|\mathbf{u}^o - s_i\right\| - \left\|\mathbf{u}^o - s_1\right\| \tag{2}$$

$$R_{i,1} = R_{i,1}^o + n_{i,1} \tag{3}$$

$$R_i^o = R_{i,1}^o + R_1^o \tag{4}$$

Assume the measurement vector as:

$$\mathbf{R} = \mathbf{R}^o + \mathbf{n} \tag{5}$$

$$\mathbf{R} = [R_{2,1} R_{3,1} \cdots R_{M,1}] \mathbf{R}^o = [R_{2,1}^o R_{3,1}^o \cdots R_{M,1}^o]$$

where n is zero-mean Gaussian with covariance matrix Q. We are estimating u^o and R.

The Chan's algorithm requires at least N+1 sensor. In Chan's method, the original non-linear equation set will be firstly linearized into linear equations. An initial estimation is preliminarily provided by applying the WLS estimator under the assumption that u^o and R_1^o are independent. In the second stage, it improves the estimation solution of the first stage using the relationship with them. Finally, the improved solution is obtained as a result of the known constraints and extra variable using the second weighted LS. The bias analysis demonstrates that the bias of Chan's algorithm mainly is caused by the bias in the first stage. Bias-Reduced method modifies the first stage of the Chan's algorithm, while stage 2 and 3 maintain the same. Suppose φ_1 is the estimate source position, and $\varphi_1 = [\mathbf{u}^T, \mathbf{R}_1]^T$. Adding both sides of (3) with $R_1^o\left\|u^o - s_1\right\|$, it results in $R_i^o + n_{i,1} = R_{i,1} + R_1^o$. Squaring both sides and ignoring$n_{i,1}^2$, it can be acquired that $2R_i^o n_{i,1} \approx R_{i,1}^2 - s_i^T s_i + 2(s_i - s_1)^T \mathbf{u}^o + 2R_{i,1}R_{i,1}^o$. After integrating all conditions when $i=1,2,3...M$:

$$\mathbf{B}_1\mathbf{n} = \mathbf{h}_1 - \mathbf{G}_1\varphi_1^o \tag{6}$$

$$\mathbf{B}_1 = 2diag\left\{R_2^o, R_3^o, \cdots, R_M^o\right\} \tag{7}$$

$$\varphi_1 = (\mathbf{G}_1^T\mathbf{W}_1\mathbf{G}_1)^{-1}\mathbf{G}_1^T\mathbf{W}_1\mathbf{h}_1 \tag{8}$$

$$\mathbf{W}_1 = (\mathbf{B}_1\mathbf{Q}\mathbf{B}_1^T)^{-1} \tag{9}$$

$$\mathbf{G}_1 = -2\begin{bmatrix} (s_2 - s_1)^T & R_{2,1} \\ (s_3 - s_1)^T & R_{3,1} \\ \vdots & \vdots \\ (s_M - s_1)^T & R_{M,1} \end{bmatrix} \mathbf{h}_1 = \begin{bmatrix} R_{2,1}^2 - s_2^T s_2 + s_1^T s_1 \\ R_{3,1}^2 - s_3^T s_3 + s_1^T s_1 \\ \vdots \\ R_{M,1}^2 - s_M^T s_M + s_1^T s_1 \end{bmatrix} \tag{10}$$

φ_1^o is the true value of φ_1. The close-form solution can be estimated by minimizing the ε:

$$\varepsilon = \left(\mathbf{h}_1 - \mathbf{G}_1\varphi_1\right)^T \mathbf{W}_1\left(\mathbf{h}_1 - \mathbf{G}_1\varphi_1\right) \tag{11}$$

Define an augmented matrix and vector

$$\mathbf{A} = [-\mathbf{G}_1, \mathbf{h}_1], \mathbf{v}^o = [\varphi_1^{oT}, 1]^T \tag{12}$$

(6) and (11)can be described as:

$$\mathbf{B}_1\mathbf{n} = \mathbf{A}\mathbf{v}^o, \varepsilon = \mathbf{v}^T\mathbf{A}^T\mathbf{W}_1\mathbf{A}\mathbf{v} \tag{13}$$

In presence of noise, $A = A^o + \Delta A$. Moving the true value of h1 and G1 from A, ignoring the second-order of h1:

$$\Delta\mathbf{A} = 2[\mathbf{O}_{(M\times 1)\times N}, \mathbf{n}, \widehat{\mathbf{B}}_1\mathbf{n}] \tag{14}$$

$$\mathbf{B}_1 = diag\{R_{2,1}^o R_{3,1}^o ... R_{M,1}^o\} \tag{15}$$

Substituting (14) into (13), acquiring the error function that needs to be minimized:

$$\varepsilon = v^T\mathbf{A}^{oT}\mathbf{W}_1\mathbf{A}^o v + v^T\Delta\mathbf{A}^T\mathbf{W}_1\Delta\mathbf{A}v + 2v^T\Delta\mathbf{A}^T\mathbf{W}_1\mathbf{A}^o v \tag{16}$$

The expectation of ε is obtained because that $E[A] = 0$ and the third term of (16) equals to zero:

$$E[\varepsilon] = v^T\mathbf{A}^{oT}\mathbf{W}_1\mathbf{A}^0 v + v^T E[\Delta\mathbf{A}^T\mathbf{W}_1\Delta\mathbf{A}]v = v^T\mathbf{A}^{oT}\mathbf{W}_1\mathbf{A}^o v \tag{17}$$

In addition, $E[\varepsilon]$ depending on v and when $v=v^o$, $A^o v^o=0$, $E[\varepsilon]=0$ will achieve the minimum. In that case, the first term of (17) will be minimized. Bias-Reduced method minimizes the ε in terms of constraint and the second term of (17), so that $E[\varepsilon]$ can attain the minimum value. Thus, v is found to be:

$$\text{Minimize } v^T \mathbf{A}^T \mathbf{W}_1 \mathbf{A} v \text{ subject to } v^T \Omega v = k \quad (18)$$

$$\Omega = E[\Delta\mathbf{A}^T \mathbf{W}_1 \Delta\mathbf{A}]$$

k can be any value. Varying the value of k does not affect φ_1 but the scaling of v. The constraint minimization problem is proposed to be solved by Lagrange multiplier λ, and the auxiliary cost function is defined as $v^T \mathbf{A}^T \mathbf{W}_1 \mathbf{A} v + \lambda(k - v^T \Omega v)$. Take the derivative in terms of v and set it to zero:

$$\mathbf{A}^T \mathbf{W}_1 \mathbf{A} v = \lambda \Omega v \quad (19)$$

Multiply equation above with v^T and combine it with the constraint, and the cost function that required to be minimized is given. $\lambda = v^T\left(\mathbf{A}^T \mathbf{W}_1 \mathbf{A}\right)v / k$ is the generalized eigenvector of $\mathbf{A}^T \mathbf{W}_1 \mathbf{A}$ and Ω. Bringing (14) into the constraint:

$$\Omega = E\left[\Delta\mathbf{A}^T \mathbf{W}_1 \Delta\mathbf{A}\right] = \begin{bmatrix} \mathbf{O}_{N\times N} & \mathbf{O}_{N\times 2} \\ \mathbf{O}_{2\times N} & \hat{\Omega}_{2\times 2} \end{bmatrix} \quad (20)$$

$$\hat{\Omega} = 4\begin{bmatrix} tr(\mathbf{W}_1 \mathbf{Q}) & tr(\mathbf{W}_1 \hat{\mathbf{B}}_1 \mathbf{Q}) \\ tr(\hat{\mathbf{B}}_1 \mathbf{W}_1 \mathbf{Q}) & tr(\hat{\mathbf{B}}_1 \mathbf{W}_1 \hat{\mathbf{B}}_1 \mathbf{Q}) \end{bmatrix} \quad (21)$$

Partitioning v and $\mathbf{A}^T \mathbf{W}_1 \mathbf{A}$ into:

$$v = [(v_1^T)_{N\times 1} (v_2^T)_{2\times 1}]^T \quad (22)$$

$$\mathbf{A}^T \mathbf{W}_1 \mathbf{A} = \begin{bmatrix} (\mathbf{A}_{11})_{N\times N} & (\mathbf{A}_{12})_{N\times 2} \\ (\mathbf{A}_{12}^T)_{2\times N} & (\mathbf{A}_{22})_{2\times 2} \end{bmatrix} \quad (23)$$

The modified version of (19) is:

$$\begin{bmatrix} \mathbf{A}_{11} v_1 + \mathbf{A}_{12} v_2 \\ \mathbf{A}_{12}^T v_1 + \mathbf{A}_{22} v_2 \end{bmatrix} - \begin{bmatrix} \mathbf{0}_{N\times 1} \\ (\lambda \hat{\Omega} v_2)_{2\times 1} \end{bmatrix} = 0 \quad (24)$$

The first N rows give v_1 subjects to v_2:

$$v_1 = -\mathbf{A}_{11}^{-1} \mathbf{A}_{12} v_2 \quad (25)$$

The rest two rows:

$$(\mathbf{A}_{22} - \mathbf{A}_{12}^T \mathbf{A}_{11}^{-1} \mathbf{A}_{12}) v_2 = \lambda \hat{\Omega} v_2 \quad (26)$$

Ω is positive definite and there is only one solution. There is no need for the Eigen-decomposition calculation. The only solution of v_2 can be obtained by setting:

$$J v_2 = \lambda v_2 \quad (27)$$

where

$$\mathbf{J} = \hat{\Omega}^{-1}(\mathbf{A}_{22} - \mathbf{A}_{12}^T \mathbf{A}_{11}^{-1} \mathbf{A}_{12}) = \begin{bmatrix} J_{11} & J_{12} \\ J_{21} & J_{22} \end{bmatrix} \quad (28)$$

The eigenvalue λ satisfied $det(\mathbf{J}-\lambda\mathbf{I})=0$. Equation can be described as $\lambda^2-(J_{11}+J_{22})\lambda+J_{11}J_{22}-J_{12}J_{21}=0$. The smallest root of the equation is:

$$\lambda_{\min} = \frac{(J_{11}+J_{22}) - \sqrt{(J_{11}-J_{22})^2 + 4J_{12}J_{21}}}{2} \quad (29)$$

Take the value back to (27) and set the second element of v_2 to unity yield, and the value is acquired:

$$v_2 = \left[\left(\frac{\lambda_{\min} - J_{22}}{J_{21}}\right), 1\right]^T \quad (30)$$

Thus, from (25), the first stage φ_1:

$$\varphi_1 = \left[v_1^T, v_2(1:L-1)^T\right]^T / v_2(L) \quad (31)$$

$v_2(1{:}L\text{-}1)$ is the first L-1 element of v_2 in (30) and L is the length of v_2.

The method continues in stage 2 and stage 3 of the Chan's method. Usage of the approximate estimation in solutions and measurements with noise will result in a bias in BiasRed. BiasRed method can be applied with positive definite TDOA noise covariance matrix, and it has no restriction with correlation of noise. Scaling the Q affects λ but does not change the final solution.

3 STIMULATIONS

There are two groups of stimulations provided to corroborate the accuracy performance of BiasRed. The first set stimulation is applied in 2-D scenario, from which the idea of geometry position setting comes. The results evaluate the performance of presented method for TDOA localization in terms of SNR, sensor number and source range/array radius ratio. Then the second set stimulation is assessed in 3D plane. It computes the accuracy using stationary geometries for TDOA. The mean squared error (MSE) and Bias are general methods to assess the accuracy of the estimator. Suppose that u is the estimated source position solution.

$$MSE = \sum_{i=1}^{L} \left\|\mathbf{u} - \mathbf{u}^o\right\|^2 / L$$

$$bias = \left\|\sum_{i=1}^{L} \left(\mathbf{u} - \mathbf{u}^o\right)\right\| / L$$

In order to bound the accuracy performance limit, the definition of CRLB is proposed under the assumption that TDOA measurement errors were i.i.d. zero-mean Gaussian using a far-field source (Although the BiasRed method is not strict with the noise to be uncorrelated). The noise covariance $\mathbf{Q} = E[\mathbf{n}\mathbf{n}^{\mathrm{T}}] = \sigma_d^2[\mathbf{I} + \mathbf{1}\mathbf{1}^{\mathrm{T}}]/2$. σ_d^2 is the TDOA and noise power. TDOA variance equals to $c^2\sigma_r^2$, $c=3\times10^8$ is signal propagation speed. $\sigma_r^2 = 1/8\pi SNR(16\times10^{18})\sec^2$, is noise covariance matrix which is related to the SNR.

$$CRLB = c^2(\mathbf{G}_t^{oT}\mathbf{Q}^{-1}\mathbf{G}_t^o)^{-1}, \mathbf{G}_t = \begin{bmatrix} (\mathbf{u}-s_2)^T/R_2 - (\mathbf{u}-s_1)^T/R_1 \\ \vdots \\ (\mathbf{u}-s_M)^T/R_M - (\mathbf{u}-s_1)^T/R_1 \end{bmatrix}$$

3.1 *Stimulations in 2-D*

Throughout this part, the first sensor is considered as a reference sensor located at the original point and the remaining sensors located around it in a circle with different radius. The number of ensemble run is 10000. The source and sensor position are the same as in. The performance of the variety algorithms is presented in Fig.1 and Fig.2 as the SNR varying across a range of 40 to 10dB. The radius of the sensor array is set to be 20 meters and the source range is 250 meters. Compared with BiasSub method (Ho, 2012), it has two advantages: Only require the structure of noise covariance Q; Avoid the evaluation of the expected bias value. In Fig.1, BiasRed method provides the best performance in low SNR region (-40dB) and BiasSub is almost identical to BiasRed but there is still slight increase. Fig.2 shows the excellent bias performance of BiasRed compared with WMLS. When SNR is quite low, BiasSub shows a rapid decline. BiasRed has a 20dB gain compared with Chan's method across the entire SNR.

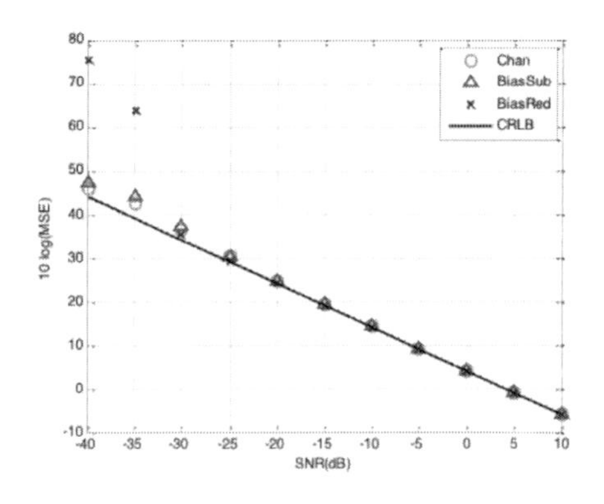

Figure 1. MSE Comparison of three methods.

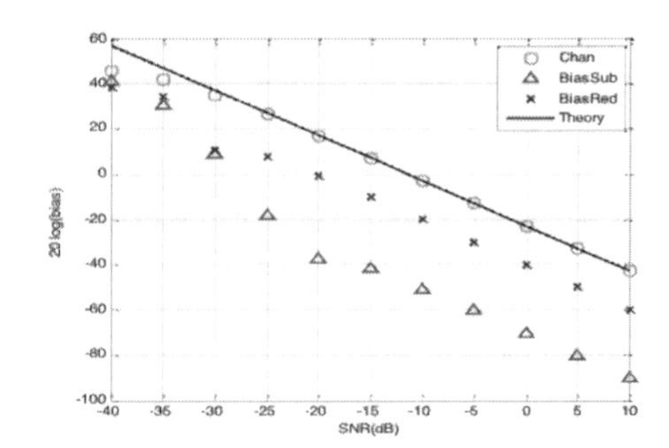

Figure 2. Bias comparisons of three methods.

The performance as the range to array radius ratio increase is shown in Fig.3 and Fig.4. The circle radius of the sensor array is maintained at 20 meters, and the default value of low SNR is -25dB. The simulation performances are measured up with the theoretical analysis. Accordingly, when the source is far and under moderate noise level, BiasRed has significant development in bias.

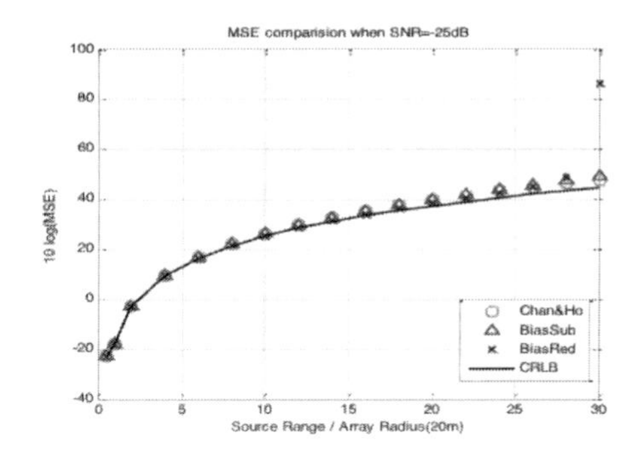

Figure 3. MSE performances under Low SNR.

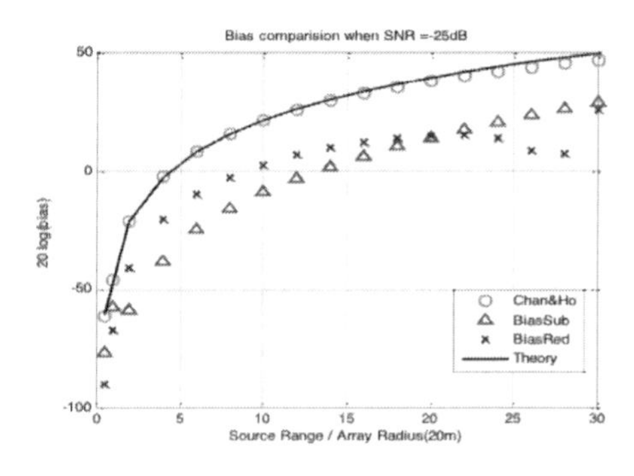

Figure 4. Bias performance under low SNR.

3.2 *Stimulations in 3-D*

In this section, comparisons of performance of the new algorithms in 3-D are provided. The target is located at $u^o=[500,550,550]^T$. Table 1 indicates sensor coordinates. Performance of accuracy is shown as the increase of TDOA covariance.

Table 1. Sensor locations for stationary source.

Receiver *i*	x	y	z
1	-15	15	-13
2	-20	25	-35
3	20	-15	30
4	10	10	60
5	-90	-10	-10
6	35	20	100
7	30	50	20
8	40	15	50

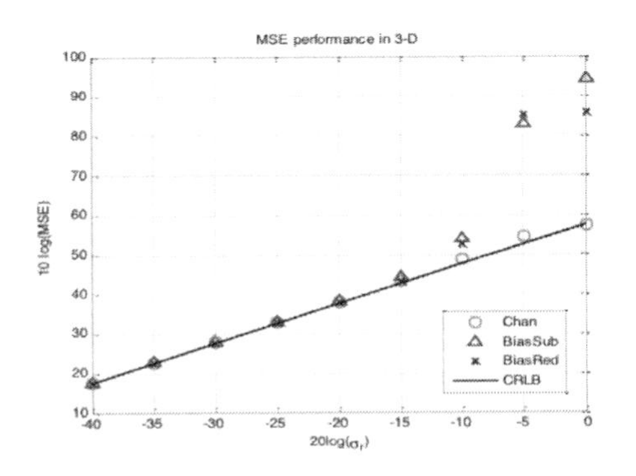

Figure 5. MSE performances for 3-D TDOA localization.

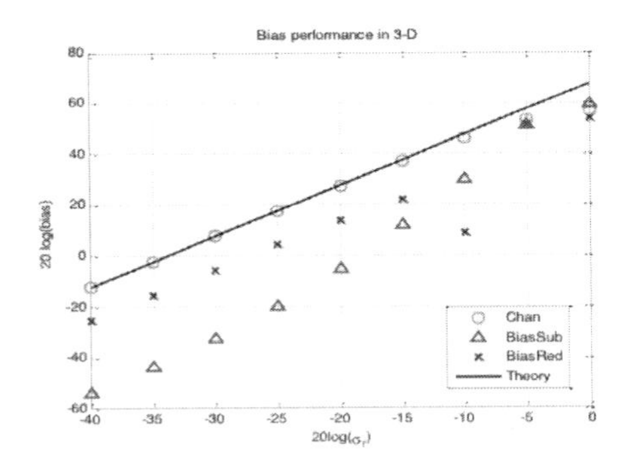

Figure 6. Bias performance for 3-D TDOA localization.

4 CONCLUSION

Evaluating the influence of low SNR, range /array ratio on algorithms is important. This paper shows that BiasRed is quite efficient to reduce the bias in different scenarios. It exploits a quadratic constraint to reduce the bias. In terms of the source range, the MSE of Chan's method is the most robust in three methods. But under the low SNR situation and source range is large, BiasRed method shows up an excellent performance. In this case, BiasRed method is extremely suitable for far field localization with considerable noise level. For practical consideration, it is hopeful to be utilized in long-range UWB location and tracking systems, where the bias dominates the performance and it requires that the algorithm must be stable and should be adaptive to poor environment.

REFERENCES

Barton, R. J. & Rao, D. 2008. Performance capabilities of long-range UWB-IR TDOA localization systems. Eurasip Journal on Advances in Signal Processing 2008: 81.

Chan, Y. T. & Ho, K. C. 1994. Simple and efficient estimator for hyperbolic location. IEEE Transactions on Signal Processing 42(8): 1905–1915.

Ho, K. C. 2012. Bias reduction for an explicit solution of source localization using TDOA. IEEE Transactions on Signal Processing 60(5): 2101–2114.

Huang, Y., Benesty, J., Elko, G. W. et al. 2001. Real-time passive source localization: A practical linear-correction least-squares approach. IEEE Transactions on Speech and Audio Processing 9(8): 943–956.

Kay, S. M. & Sengupta, S. K. 1995. Fundamentals of statistical signal processing: estimation theory. Technometrics, 37(4): 465–465.

Kovavisaruch, L. & Ho, K. C. 2005. Alternate source and receiver location estimation using TDOA with receiver position uncertainties. 2005 Philadelphia, PA, United states. Institute of Electrical and Electronics Engineers Inc.: IV1065–IV1068.

Electronics, Communications and Networks IV – Hussain & Ivanovic (eds)
© 2015 Taylor & Francis Group, London, ISBN: 978-1-138-02830-2

Multi-task object detection based on sharing linear weak classifiers

Yali Li, Shengjin Wang* & Xiaoqing Ding
State Key Laboratory of Intelligent Technology and Systems, Department of Electronic Engineering, Tsinghua University, Beijing, P R China

ABSTRACT: We focus on the problem of applying sharing information into multi-task object detection (*i.e.*, multi-view multi-object detection). Common features among multiple classes are often ignored in traditional object detection methods. To cope with this, a method which learns to share linear weak classifiers among multiple classes is proposed in this paper. Considering the characteristic of exhaustive search for sliding-window based methods, we introduce a new mathematical definition of shared features among multiple classes. A learning framework based on boosting algorithm and genetic programming based feature selection method is proposed. Experimental results show that the application of sharing information can significantly reduce the number of used features for multi-task object detection.

1 INTRODUCTION

The goal of object detection is to determine whether generic objects exist in images and obtain the locations of objects if so. It is an important topic in computer vision and has a wide range of applications, such as video surveillance and content-based image retrieval. Great efforts have been devoted to object detection and much progress has been achieved. However, the problem still remains challenging. It is mainly because that the appearances of object of categories can vary a lot when illuminations and viewpoints change. A straightforward way to deal with appearance variety is to learn multiple classifiers for all possible views. Also in applications we need to recognize multi-class objects simultaneously. Besides, fusing the output of multiple part classifiers is a typical approach to obtain high-performance detectors. These can all be viewed as multi-task object detection since more than one object/non-object classifiers are required. In this work we focus on learning shared parameters among multiple classifiers.

Briefly the object detection methods can be divided into two categories as part model based approaches (Felzenszwalb et al. 2010) and sliding window based approaches (Dalal & Trigges 2005). Part-model based approaches implicitly or explicitly divide the object into parts and build up models of these parts. Object positions can be predicted through part locations with the spatial relations. Whereas for sliding window based approaches, features (*e.g.*, HOG (histograms of oriented gradients)) are combined with classifiers for object/non-object discrimination. On detection, all candidate sub-windows in test images are scanned and tested whether they are object regions or not.

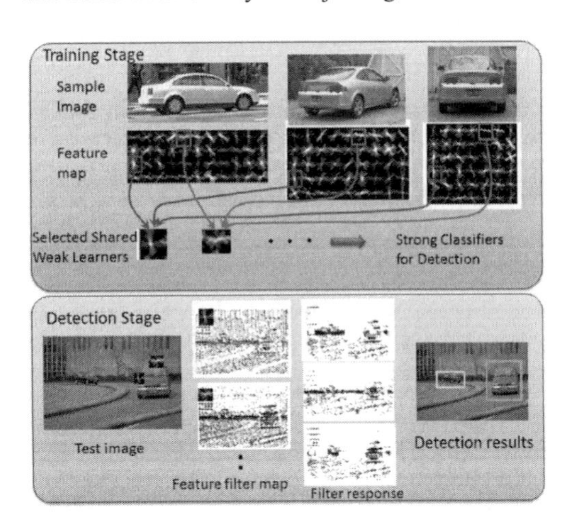

Figure 1. The pipeline of the system.

The intuitional way for multi-task object detection is to learn separate classifiers for each task. Information sharing for multi-task object detection has two advantages. First, for objects with few samples the classifiers are hard to be learned. The samples for learning models can be increased by sharing. Second, sharing methods can also inhibit the number of model/classifier parameters, increasing linearly with the task number. Existing information sharing methods can also be categorized into two kinds:

*Corresponding author: *wgsgj@tsinghua.edu.cn*

shared models based methods and shared features based methods. The former concentrates on learning generic object component models among multiple classes and construct hierarchical models to represent the sharing the information. Fergus et al. (2010) propose a WordNet hierarchical tree method to share semantic labels among many categories. Zhu et al. (2010) propose recursive compositional models to share sub-parts for multi-view multi-object detection. Salakhutdinov et al. (2011) propose a hierarchical classification model to split object samples into subclasses step by step and share visual appearance across categories. The descriptions for shared models are often complex and abundant information. However, learned models becomes more class-specific as it becomes complex. Besides, some research focuses on select shared weak features and combines weak linear classifiers into strong classifiers. Torralba et al. (2007) propose joint-boost algorithm to learn shared weak features of multi-class multi-view object detection. It can significantly reduce the number of used features as well as the runtime for multi-task object detection. However, their work mainly concentrates on obtaining the optimum shared classes subset and they make a strict restrict that the samples belong to all shared classes need to be normalized into the same size. It is difficult to satisfy many applications. On other aspects, sharing information between root filters and part filters for part-model based object detection methods are always ignored, which in fact can be an important application of feature sharing.

In this paper, we focus on sharing weak linear classifiers for multi-task object detection. The pipeline system is shown in Figure 1. Shared weak linear classifiers are learned based on HOG and Boosting is applied to aggregating the multiple sharing classifiers. At each round, genetic programming is applied to finding the locations of features and weighted linear regression is used to get the parameters of weak classifiers. On detection, the outputs of shared weak classifiers are computed. The output response maps of a strong classifier for each task are obtained by combining the outputs of shared classifiers. Section 2 introduces the linear weak classifiers for Boosting. The shared linear weak classifiers learning framework is presented in Section 3 and Section 4 gives the experimental results. The conclusion is given in Section 5.

2 LINEAR WEAK CLASSIFIERS FOR BOOSTING

Most features in object detection such as HoG are represented by a high-dimensional vector. To obtain weak classifiers for Boosting, weighted linear regression is applied to randomly selected low-dimensional data (Parag et al. 2008). Suppose $\boldsymbol{x} = \left[x_1, x_2, \cdots, x_d, 1\right]^T$ is an augmented feature vector and y is the label, we want to find the weight vector $\alpha \in \mathcal{R}^{d+1}$ to minimize $E\left\{\| \boldsymbol{x}^T\alpha - y \|^2\right\}$. The solution is $\tilde{\alpha} = E\left\{\boldsymbol{x}\boldsymbol{x}^T\right\}^{-1} E\left\{\boldsymbol{x}y\right\}$. In Boosting, suppose $X = \left[\boldsymbol{x}_1, \boldsymbol{x}_2, \cdots, \boldsymbol{x}_N\right]$ is a $(d+1)\times N$ matrix whose rows are the feature vectors for samples and $\boldsymbol{y} = \left[y_1, y_2, \cdots, y_N\right]^T$ is a N-vector whose elements are the labels of samples. W is a $N \times N$ diagonal matrix consists of the weights. Then $\tilde{\alpha}$ is denoted as:

$$\tilde{\alpha} = \left(XWX^T\right)^{-1}\left(XW\boldsymbol{y}\right) \quad (1)$$

For multi-task object detection, positive and negative samples are collected to learn multiple classifiers simultaneously. Assume the sample labels for each task are $\pm k, k = 1, \cdots, K$, $\tilde{\alpha}$ can be further denoted as:

$$\tilde{\alpha} = \left(\sum_{k=1}^{K} X_k W X_k^T\right)^{-1}\left(\sum_{k=1}^{K} X_k W \boldsymbol{y}_k\right) \quad (2)$$

where $X_k, k = 1, \cdots, K$ is the feature matrix for each task and $\boldsymbol{y}_k$ is the label vector. Its elements indicate whether the samples are positive/ negative.

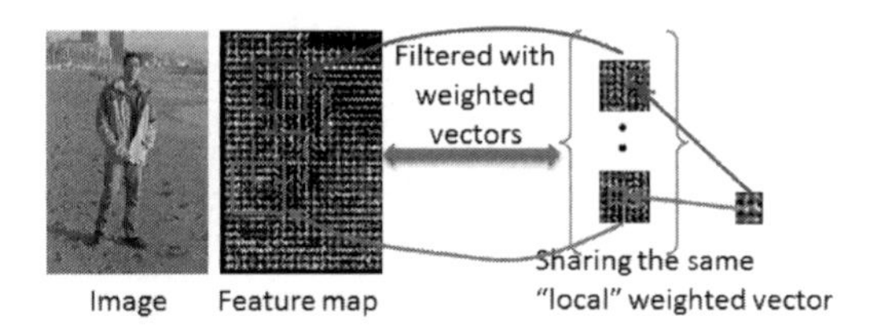

Figure 2. The definition of shared weak classifiers for sliding-window based object detection methods.

3 LEARNING SHARED WEAK CLASSIFIERS

In this Section, the shared weak classifiers learning framework is given. First the definition of shared features for the sliding window based object detection methods is presented. Furthermore a general mathematical model for shared features learning is provided. The learning framework based on genetic programming and Real-Adaboost algorithm is given.

3.1 *The definition of shared weak classifiers*

We first present the definition of shared weak classifiers. As illustrated in Figure 2, the similar local regions of different objects lead to similar local linear classifiers for discrimination. Suppose the local classifiers as $(\boldsymbol{p}, \psi)$, where $\boldsymbol{p} = (ix, iy)$ indicates the location of local filters and ψ denotes the operation of extracting features and computing the filter response. For HOG, $\boldsymbol{p}$ indicates where the histograms of gradients are statistically computed and ψ denotes the extractions of HOGs and obtaining the output of the linear classifiers. We construct our shared classifiers learning framework based on Real-Adaboost(RAB) (Freund and Schapire 1995). Several positive and negative samples of K tasks are collected to learn multiple classifiers. For each task, a positive label k is assigned to positive samples and $-k$ is assigned to the corresponding negative samples. The loss function for multi-task classification is

$$J(F)=\sum_{k} E\{I(|y|=k)\exp[-\operatorname{sgn}(y)F(x)]\} \quad (3)$$

where y is the label of samples and $I(|y|=k)$ is the indication function. It is equal to 1 only if the sample label is equal to $k,-k$. Greedy strategy is applied to minimize $J(F)$. That is, at each round of boosting we try to find the weak hypothesis h_t to minimize Z_t as:

$$Z_t=\sum_{k}\left[\sum_{i} I(|y|=k)D_t(i)\exp(-\operatorname{sgn}(y_i)h_t(x))\right] \quad (4)$$

where $D_t(i)$ denotes the weight of each sample. The samples are split into two parts as shared classes S and non-shared classes. For shared classes, the domain-partitioning hypothesis is applied on the output of weak linear classifiers to obtain $h_t(x)$. While for non-shared classes, a constant is set as weak classifiers at this round. Suppose that

$$W_k^j=\sum_{i:x_i\in X_j\cap y_i=k} D(i)$$
$$W_k^+=\sum_{i:y_i=k} D(i), W_k^-=\sum_{i:y_i=-k} D(i) \quad (5)$$

We obtain the weak classifiers h_t as:

$$h_t(i)=0.5\log\left(\sum_{k\in S} W_k^j / \sum_{k\in S} W_{-k}^j\right), \text{if } x_i\in X_j$$
$$h_t(i)=0.5\log\left(W_K^+/W_K^-\right), \text{if } y_i=\pm k, k\notin S \quad (6)$$

Then the minimum Z_t can be represented as:

$$Z_t=2\sum_{j}\sqrt{\left(\sum_{k\in S}W_k^j\right)\left(\sum_{k\in S}W_{-k}^j\right)}+2\sum_{k\notin S}\sqrt{W_k^+W_k^-} \quad (7)$$

By now we have obtained the minimum optimization function for each iteration of boosting algorithm as Z_t. To simplify the discussion, the subscript t is omitted and Z can be viewed as a function of with variables as shared classes S and different local weak classifiers. With the definition of shared features above, local weak classifiers can be denoted as $\left(\boldsymbol{p}_{S(1)}, \boldsymbol{p}_{S(2)}, \cdots, \boldsymbol{p}_{S(K_S)}, \psi\right)$. Intuitively, this optimization problem is a *NP*-hard combinational problem. To tackle with this, genetic programming is introduced to select the most discriminative shared linear classifiers.

3.2 *Find shared classes and shared linear classifiers*

There are 2^K-1 possible combinations of shared classes. Instead of testing all possible solutions, we choose the shared class subset in a backward elimination manner. For each possible shared linear classifier, we assume that all classes are shared and obtain the parameters of linear classifier. After that, compute the classification error for each class and find the one which has highest error. Remove it from the shared classes and obtain classifier parameters for the updated shared subset. Iteration ends when only one class is left. Only K combinations of shared classes are tested. Although it cannot guarantee that the obtained solutions are best, experiments show that it can achieve better results.

There are two kinds of classifier parameters as locations $\boldsymbol{p}_{S(1)}, \boldsymbol{p}_{S(2)}, \cdots, \boldsymbol{p}_{S(K_S)}$ and linear projection vector $\tilde{\alpha}$. Once the location parameters are determined, the projection vector can be calculated with eq.(2). Possible values of each $\boldsymbol{p}_{S(k)}$ can be obtained by the size of features and samples. To obtain optimum combinations of $\boldsymbol{p}_{S(1)}, \boldsymbol{p}_{S(2)}, \cdots, \boldsymbol{p}_{S(K_S)}$, the genetic programming is applied. Inspired from biological evolution, the genetic programming methods proceed as follows.

3.2.1 *Initialization*

Based on the consideration that the discriminating features for single class are also prone to show higher performance when used for sharing, we compute Z_k for each single class by eq.(7) after obtaining the linear projection vector and attached weak hypotheses. We let $\boldsymbol{p}_{S(k)}$ with smaller $Z\left(\boldsymbol{p}_{S(k)}, \psi\right)$ be with higher possibilities to be chosen to form the individual solutions in the initial generation.

3.2.2 *Producing children individuals for the next generation*

In theory, the optimal criterion as eq.(7) can be reversed to serve as the fitness function. But the computation cost of computing Z is high. To avoid this, we use the weighted fitting error of linear regression function to rank the individuals for each generation. All the individuals are ranked by their fitting error in ascending order. The individuals with higher fitting errors are wiped out while those with lower ones are kept to generate the next generation.

The children individuals of the next generation are produced by operations as crossover and mutation. The crossover operation has three steps: (1) randomly select two different parent individuals; (2) randomly select several element bits; (3) exchange the elements in the selected bits to produce two children individuals. In mutation operation, several element bits are chosen to be changed. Besides, the parent individuals with higher fitness (*i.e.*, lower Z) have a higher opportunity to be chosen to generate individuals of the next generation. This helps accelerate the optimization.

4 EXPERIMENTS

Our proposed framework of sharing linear weak classifiers can be applied to various multi-task object detection problems, including multi-view object detection, multiple object detection and multiple parts based object detection. In the train stage, several object samples and background subregions are collected to train classifiers. In the detection stage, image regions are tested with the classifiers. To test the performance our shared weak classifiers based on Boosting, two groups of experiments are done. The used features are HOG. First, a large number of linear weak classifiers are constructed with randomly selected feature bits in the same blocks for Boosting. By comparison, shared linear weak classifiers are trained with the proposed framework for Boosting.

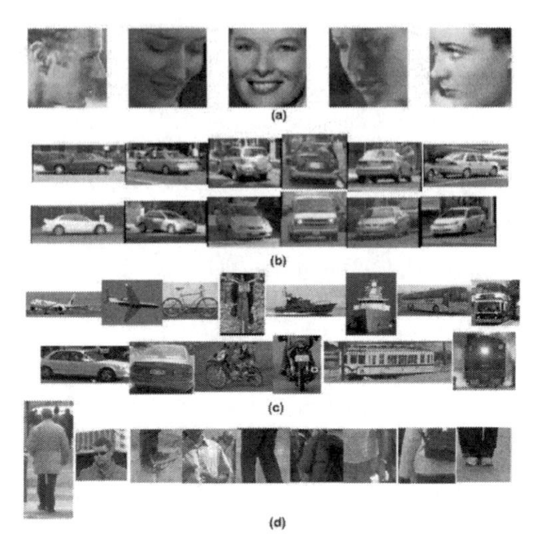

Figure 3. Samples for all classes to train the shared classifier. (a)multi-view face detection. (b)multi-view car detection. (c)Pascal 2007 outdoor objects. (d) Parts of pedestrians.

In the first group of experiments, positive/negative samples for multiple tasks are collected. They are randomly split into the training and test sets. Detection rates and false alarm rates change along with the number of applied weak classifiers are recorded. The ROC curves with $1, 2, \cdots, K$ (*i.e.*, the number of classes) weak classifiers are plotted. Besides, the ROC curves of single class training classifiers are recorded for comparison. Four experiments as follows are done.

Multi-view face: We choose five views of faces as left profile, left half-profile, frontal, right half-profile and right profile (Figure 3(a)). 300 positive samples are collected for each view. Negative samples of the same number are randomly cropped from backgrounds. Among them, 100 positive/negative samples are for training and the left samples are for testing. The ROC curve is shown in Figure 4(a).

Multi-view car: The experiments are done on multi-view car dataset provided by (Fritz et al. 2005). It contains totally 1279 complete car images under 7 views. By flipping, car examples of 12 views from 0~330 degrees at intervals of 30 degrees can be obtained. The first 50 positive samples for each view are used as positive train samples and the rest samples serve as true object samples. Besides, 150 sub-images are randomly cropped from background images without cars and the first 50 ones are for training. The example positives and ROC curve are shown in Figure 3 and 4(b).

Pascal 2007 outdoor objects: The Pascal 2007 VOC dataset (Everingham et al. 2007) contains four groups of objects as person, animals, indoor objects and outdoor objects. Several annotated examples of seven categories as airplane, bicycle, boat, bus, car, motorbike, train consist into the outdoor subset. Considering about the view changes, for each class samples are split into two parts according to the aspect ratio (Figure 3c). In total 14 classes of objects are obtained. The samples in the training set are used for training the classifiers and the ones in validation set are for testing. The same number of train/test negative samples is cropped from other images which do not contain the objects. The ROC curves are in Figure 4(c).

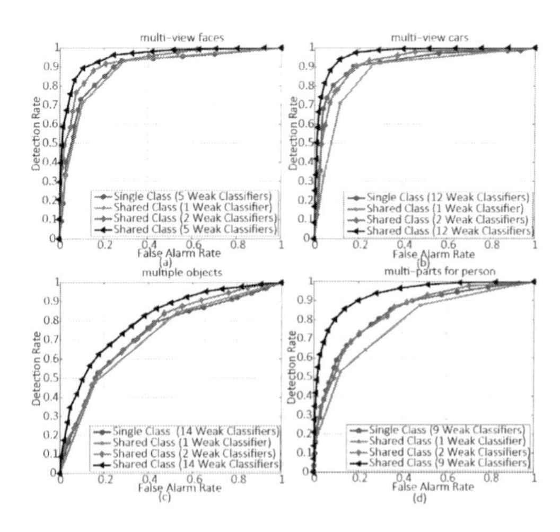

Figure 4. ROC curves on test sets of the share-trained classifiers and single-trained classifiers. (a) multi-view face detection. (b) multi-view car detection. (c) Pascal 2007 outdoor objects. (d) Parts of pedestrians.

Parts of pedestrians: For part based object detection, classifiers for different parts are trained with samples. We test the performance of sharing framework with pedestrian part images obtained from Inria dataset(Dalal & Trigges 2005). The pedestrian images are obtained by annotations and the remaining part images are cropped from detection results with deformable part models(DPM) (Felzenszwalb et al. 2010). 400 sub-images for each part are used for training and the left ones are for testing. The example positives and ROC curve are shown in Figure 3 and 4(d).

Traditionally, if no feature sharing techniques applied, the number of used weak classifiers increases linearly with the number of classes. That is, for K-task (K is the number of classes) object detection, at least K weak classifiers are needed. In Figure 4, the ROC curves with non-sharing techniques are plotted with red lines. Besides, the ROC curves with $1, 2, \cdots, K$ weak classifiers are plotted with magenta, blue and black lines. From ROC curves in Figure 4 we can find that with the same number of weak classifiers, better performance can be obtained. Also the method of sharing weak classifiers breaks the limitation that the classifiers number increases linearly with object classes. With $0.15K \sim 0.2K$ of weak classifiers per class, the performance reaches the same or even better than that of single class training with K classifiers.

Figure 5. Examples of outputs of boost shared classifiers.

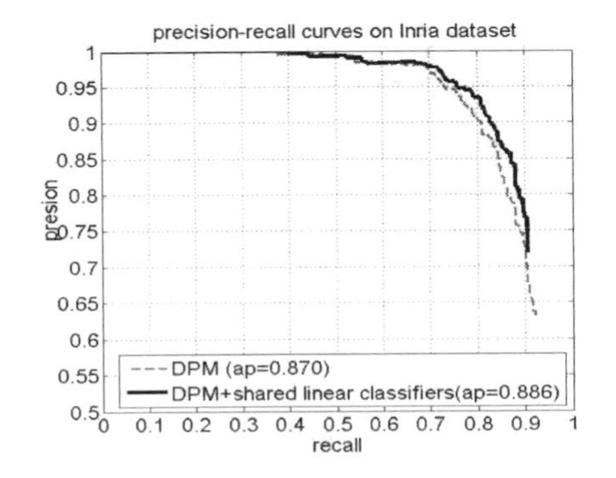

Figure 6. Comparison of PR curves on Inria dataset.

We also integrate the sharing framework with DPM by a cascade structure and test the performance on Inria dataset. Stages of cascaded boost classifiers are trained with sub-images of pedestrians and 8 body parts. For each strong classifier, the weak linear classifiers are shared among the whole pedestrians and body parts. The Boosted shared classifiers serve as a pre-stage filter here. The examples of filtered outputs are shown in Figure 5. Only those regions which can pass the boost shared classifiers need to further calculation of filter response with DPM. The precision/recall (PR) curves of DPM and that with shared-classifiers are plotted in Figure 6 for comparison. With boost shared classifiers, the AP is increased by 1.5%. Besides, the detection time is also recorded. Without considering about the extraction time of HoGs, 3~6 speedup factors can be achieved. That is because the parameters of classifiers can be reduced with the sharing framework.

5 CONCLUSIONS

We investigate on sharing linear classifiers for multi-task object detection. A framework to learn shared linear weak classifiers for multiple classes is proposed in this paper. Boosting is applied to aggregating linear classifiers into strong classifiers for object detection. Genetic programming is applied to finding the optimum locations of weak classifiers. Experiments on multi-view object detection and multi-object detection prove that the framework can significantly reduce the number of used classifiers.

ACKNOWLEDGMENT

This work is supported by the National High Technology Research and Development Program of China (863 program) under Grant No. 2012AA011004 and the National Science and Technology Support Program under Grant No. 2013BAK02B04.

REFERENCES

Dalal N. & Trigges B. 2005. Histograms of oriented gradients for human detection. *ICCV*.

Everingham M., Van Gool L., Williams C. K., et al. 2007. The PASCAL visual object classes challenge 2007 results, *http://www.pascalnetwork.org/challenges/VOC/voc2007/*

Felzenszwalb P. F., Girshick R. B., McAllester D. A., et al. 2010. Object detection with discriminatively trained part-based models. *IEEE Trans. Pattern Anal. Mach. Intell.*32(9): 1627–1645.

Fergus R.,Bernal H.,Weiss Y., et al. 2007. Semantic label sharing for learning with many categories. *ECCV*.

Freund Y. & Schapire R. E. 1995. A decision-theoretic generalization of on-line learning and an application to boosting. *ECCV.*

Fritz M., Leibe B., Caputo B., et al. 2005. Integrating representative and discriminant models for object category detection. *ICCV.*

Parag T., Porikli F. & Elgammal A. 2008. Boosting adaptive linear weak classifiers for online learning and tracking. *CVPR.*

Salakhutdinov R., Torralba A. & Tenenbaum J. 2011. Learning to share visual appearance for multiclass object detection. *CVPR.*

Torralba A., Murphy K. P. & Freeman W. T. 2007. Sharing visual features for muticlass and multiview object detection. *IEEE Trans. Pattern Anal. Mach. Intell.* 29(5):854–869.

Zhu L., Chen Y., Torralba A., et al. 2010. Part and appearance sharing: recursive compositional models for multiview multi-object detection. *CVPR.*

Electronics, Communications and Networks IV – Hussain & Ivanovic (eds)
© 2015 Taylor & Francis Group, London, ISBN: 978-1-138-02830-2

An English sentence pronunciation evaluation system using speech recognition and multi-parametric method

Xinguang Li, Minfeng Yao, Dongxiong Shen, Jiyou Xu* & Junyu Chen
Cisco School of Informatics, Guangdong University of Foreign Studies, Guangzhou, China

ABSTRACT: This paper introduces the design of an English pronunciation evaluation system with voice recognition capabilities based on the multi-parameters method. In this system, the Mel Frequency Cepstrum Coefficient (MFCC) feature and the Hidden Markov model (HMM) algorithm are used to establish a model for speech recognition; at the same time, this paper builds a pronunciation evaluation model with multi-parameters of pronunciation, speed, stress, rhythm and intonation; it can give feedback to the users with pronunciation problems based on experts' knowledge. Experiments have shown that the model has certain credibility and accuracy, and the pronunciation evaluation software based on this model is of great help to improving the efficiency of the learners' oral English learning.

1 INTRODUCTION

A technique named Computer-Assisted Language Learning (CALL) in computer-assisted pronunciation learning for non-native language learners has become a popular research topic. Meanwhile, automatic evaluation on pronunciation quality is the core technology of CALL system(2006). This technology will change the existing language learning environment and teaching method, which will greatly improve the efficiency of language learning. The timely, accurate and objective evaluation as well as feedback can help learners find out the gap between test pronunciation and the standard pronunciation.

Evaluating the learners' English pronunciation objectively is an effective way to guide them to learn. By using voice recognition technology, the intonation of sentence can be accurately evaluated, and the sentence pronunciation quality can be embodied in many ways. This paper describes a design of a multi-parameter objective evaluation system which is equipped with the function of voice-recognition and aims at pitch, speaking speed, stress, rhythm and intonation of the speech. Meanwhile, it can give the users some instruction on pronunciation and help them correct the errors in time through a correction from the experts' knowledge. Experiments have shown the effectiveness of this system.

2 METHOD FOR ENGLISH SPEECH RECOGNITION AND EVALUATION

2.1 *The process of evaluating pronunciation quality*

Studies have shown that in the objective evaluation process of voice quality, different indicators will be expressed with different speech features. The general process of pronunciation evaluation is described in Figure 1.

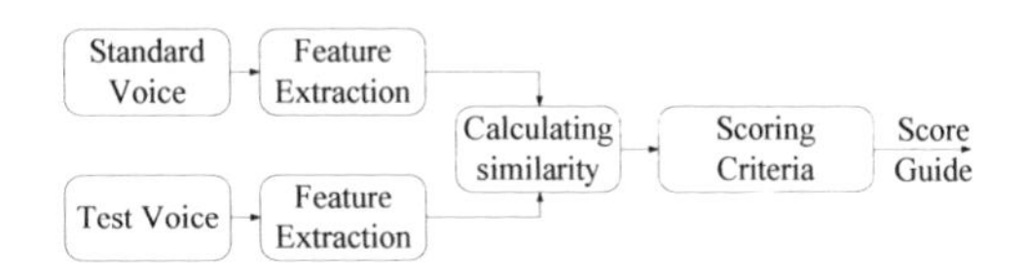

Figure 1. Process of pronunciation evaluation.

2.2 *Selection of indicators and evaluation methods*

This paper adopts the Mel Frequency Cepstrum Coefficient (MFCC) as the pitch evaluation indicators, which is based on human auditory model. However, the pitch indicators consist mainly of the contents of the pronounced sentence; while the pronunciation indicators include reading speed, sound

*Corresponding author: *190142094@qq.com*

duration, rhythm and intonation, which cannot be reflected. The pronunciation quality of an English sentence largely depends on the accuracy. In addition, some prosodic features of the sentence also depend on the sentence itself.

Prosodic, also known as rhythm, from the linguistic point of view, refers to the phonological organization above one segment level. It is a system organization which is able to organize various linguistic units into words or associated discourse chunk. Li & Weiqian(2009) proposed that prosodic realization can not only transmit linguistic information, but also deliver paralinguistic and non-linguistic information (which includes focus conveying, emphasis, phrase grouping, etc.). Besides, it is also equipped with functions like distinguishing words as well as expressing the modal attitude. Prosodic features, also called suprasegmental features, mainly refer to stress, rhythm, tone and intonation, which are separately composed of related dynamic parameters including pitch, intensity and duration etc.(Jin & Gui-zhen, 2013) Thus, the quality of English sentences pronunciation is closely related to its rhythm.

2.2.1 *Pronunciation indicator*

This paper builds a standard speech corpus with the MFCC parameters and HMM algorithm. In the evaluation process, test speech recognition is realized by collecting MFCC parameters of the test voice and using the Viterbi algorithm in HMM. Viterbi algorithm will export the similarity output probability between the test speech and the standard speech. The greater the output probability value, the higher accuracy the phonetic pronunciation will have. The output probability maps to the pitch percentile score, and the conversion method is shown in Eq.1.

$$S_{\text{pronunciation}} = \frac{100}{1 + a(dist)^b} \tag{1}$$

The independent variable *dist* represents the output probability difference which is processed by each HMM model identification. a and b should be obtained by multiplying pairs pronunciation-dist value as samples and calculated with the least square method. Through the experiments, this paper sets a as 0.005 and b as 1.5.

2.2.2 *Speed indicator*

Speed refers to the perception impression on the auditory speed of words. Speed measure is the number of syllables said per second. It is generally measured by a total length of discourse containing standstills. Calculating the average speaking rate can be judged by the length of the effective voice after the endpoint detection. Relative speaking rate coefficients are used as indicators to determine speech as shown in Eq. 2.

$$V_{\text{Relative Speed}} = \frac{S_{pl} - T_{pl}}{S_{pl} + T_{pl}} \tag{2}$$

Among this, S_{pl} refers to the pronunciation duration of the standard sentences, T_{pl} refers to the pronunciation duration of the test sentences.

By setting double threshold of short-term energy and zero-crossing rate, this paper has applied endpoint detection to the input speech to get $V_{\text{Relative Speed}}$ into Eq.3. Through experimental test, take α as 100, and take β as 80.

$$S_{\text{Speed}} = \alpha - \beta \times V_{\text{Relative Speed}} \tag{3}$$

Parameter S_{Speed} represents the percentile score of Speed indicators. Thus, the final score of the speech rate will be mapped to the form of percentile scores.

2.2.3 *Stress indicator*

Stress refers to reading a word or syllable which needs to use larger or heavier strength than the surrounding word or syllables. The listener can often feel that the stressed words or syllables are much louder than their surrounding words or syllables.

Stress distribution is one aspect of sentences' prosodic features and plays a very important role in semantic organization and semantic expression. Therefore, to observe the English sentences' prosodic stress distribution, first of all, it should get the English sentences' stresses divided.

This paper divides a sentence's stress mainly based on three characteristics of the stressed syllable in the English sentences, they are: (1) Read loudly; (2) Long pronunciation; (3) Clear and easy to distinguish.

The stress scoring mechanism includes the following steps:

1 Extract energy values of a sentence. Loud stressed syllables in the sentence will be reflected in the energy intensity of time domain features.
2 Sentences warping. In the process of scoring tested sentences, adjusting the length of the tested sentences in proportion to the standard sentence in a similar level can be beneficial to data processing.
3 This paper uses dual threshold comparison method for stress endpoint detection. After a lot of experiments, this system sets the two thresholds as Eqs.4 and 5:

Stress syllable threshold:

$$T_u = \left(\max\left(sig_{in}\right) + \min\left(sig_{in}\right)\right)/2.5 \tag{4}$$

Non-stressed syllable threshold:

$$T_l = \left(\max\left(sig_{in}\right) + \min\left(sig_{in}\right)\right)/10 \quad (5)$$

Where sig_{in} is the input speech signal energy.

For the stress rating, the evaluation score for the stress is mainly from two aspects: (1)the number of stress pronunciations; (2)stress pronunciation and relative short-length energy, among which the stressed pronunciation length and the sentence rhythm have large correlation, and this will be discussed in detail in Section 2.2.4. Thus, in a stress rating module, mainly the number of stresses and their average short-term energy are considered as indexes. The stress score S_{stress} is calculated as shown in Eqs. 6, 7, and 8:

$$S_{stress} = \omega_1 \times S_{stress1} + \omega_2 \times S_{stress2} \quad (6)$$

$$S_{stress1} = 100 - 100 \times \left|S_{Snum} - T_{Snum}\right| / (S_{Snum} + T_{Snum}) \quad (7)$$

$$S_{stress2} = 100 - b \times \left|S_{Se} - T_{Se}\right| / S_{Se} \quad (8)$$

Where, S_{Snum} and T_{Snum} refer to the number of stresses in a standard sentence and test sentence, respectively; S_{Se} and T_{Se} refer to the average short-term energy values of the stresses in a standard sentence and test sentence, respectively. As in Eq. 9, coefficient b is calculated as 68 according to the analysis of experiments. As for the establishment of coefficients ω_1 and ω_2 in Eq. 7, this paper sets them as 0.3 and 0.7 respectively based on phonetics experts' recommendation.

2.2.4 *Rhythm indicator*

In this hypothesis, the rhythm of the language is defined as "the isochronous repetition time of a language unit fragment". Language is divided into timing and syllable timing.

After that, Low & Grabe (2005) proposed that using successive syllable unit fragment length to calculate Pairwise Variability Index(PVI) parameters can obtain the result of syllable duration differences and can be regarded as a measure of the differences among stress-timed language, syllable-timed language and mora-timed language.

PVI is used to calculate the length of duration variability between adjacent units, and if the variability is smaller, it represents that the unit has isochrony. Low (1998) of the Singapore's Nanyang Technological University first presented the PVI equation in Singapore English-paced study which is used to draw speech rhythm correlation by calculating gaps between the continuous front and rear syllables of stressed and non-stressed vowels. Grabe & Low (2002) from Cambridge believed that the impact of voice speed on the consonant intervals is not so significant, thus they use the original The raw Pairwise Variability Index (RPVI) parameters to calculate the differences of intervals between consonants, as Eq.9:

$$RPVI = \left(\sum_{k=1}^{m-1} | d_k - d_{k-1} | / (m-1) \right) \quad (9)$$

Chiu(2010) promoted this equation, m represents the sum of the speech unit intervals, while d_k represents the duration of the k-th speech unit, which includes the time between vowels, and ignores the standardization of the speech rate.

This paper adopted an improved Distinct Pairwise Variability Index (DPVI) parameter calculating equation. This equation takes into account the characteristics of the English pronunciation unit Durational Variability. It compares as well as calculates the syllable unit fragment length of a standard sentence and the test statements, and makes the conversed parameters available for the system rated basis. It is shown in Eq.10.

$$DPVI = 100 \times \frac{\left(\sum_{k=1}^{m-1} | d1_k - d2_k | + | d1_t - d2_t | \right)}{Len} \quad (10)$$

Where $d1_k(d2_t)$ is the fragment duration of the k-th(t-th) interval in standard(test) speech, m is minutes, Len is the duration of the standard speech.

2.2.5 *Intonation Indicator*

Intonation refers to the changes of the speakers' tone in the level of severity and circumflex. In addition to lexical meaning, English sentences also have intonation meaning expressed by the speaker's attitude or tone. Lexical meaning of sentences plus intonation meaning finally constitute the complete significance.

The purpose of intonation assessment is to show how much difference there is between a speaker's English intonation and the standard by the scoring algorithms.

As for the extraction of English sentence intonation curve, the key lies in extracting pitch to each frame of speech signal in sentences; and the speech material form of the pitch is expressed as the fundamental frequency of the vocal cords. The traditional method of pitch tracking can be divided into two categories: the time domain and the frequency domain.

In this paper, the complexity of operation has been taken into consideration, and the evaluation of the tone index will adopt the autocorrelation function method (ACF) in time domain to select the pitch of each frame corresponding to the English sentence; then smooth the pitch by setting the Median Filter to exclude the unstable speech frame with abnormal pitch values; and use the DTW algorithm to compare the difference between these two, and then work out a score.

Auto-correlation function uses the autocorrelation function to calculate a speech frame s(i) and their similarity, as shown in Eq.11:

$$acf(\tau)=\sum_{i=0}^{n-1-\tau} s(i)s(i+\tau) \quad (11)$$

Where n is the length of a frame of voice data, τ is the amount of time delay. The value which can make acf(τ) within a reasonable specific section is found, then pitch can be calculated.

After extracting the sentence intonation curve, it will use the DTW algorithm to compare differences in tone data so as to calculate the test parameters *dist* between the standard sentence and test sentence. However, *dist* obtained by DTW algorithm cannot directly be served as a pronunciation score; there must be a reasonable mapping from the distance to the score as shown in Eq.12:

$$S_{\text{int}onation}=\frac{100}{1+a(dist)^{b}} \quad (12)$$

Obviously, this equation can map the distance to the [0,100] score range. According to the data experiment, a and b in Eq. 12 are set value as 0.0005 and 2, respectively.

3 ENGLISH SENTENCE PRONUNCIATION EVALUATION SYSTEM

3.1 *System framework*

In this paper, speech recognition and speech evaluation process are shown in Figure 2.

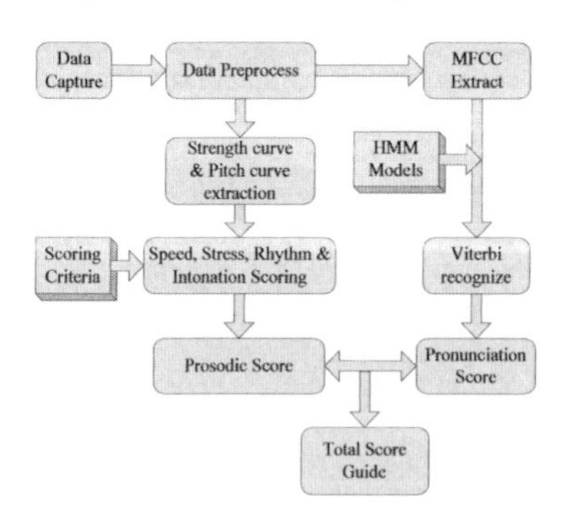

Figure 2. Speech recognition and evaluation process.

3.2 *Scoring method and experiment analysis*

3.2.1 *Scoring method*

Scoring method used in this system is a supervised evaluation. For each test sentence, there must be a corresponding standard pronunciation. In order to reduce the pronunciation differences between men and women which can lead to a deviation of evaluation results, the system has recorded two categories of male and female standard pronunciation for users to select.

3.2.2 *Score basis*

This paper adopts the suggestions of four English phonetic pronunciation experts. Five indicators (pitch, speed, stress, rhythm and intonation) are selected as a basis for evaluation. Among these factors, pitch is relevant to the content of the user's pronunciation. This paper attempts to seek the pronunciation fitness degree between the testing speech and standard speech through the HMM Viterbi decoding algorithm as a pitch reference index. On prosody, it calculates an overall score through the speech rate, stress, rhythm and evaluates four indices weighted. The Prosodic Index weighting methods is shown in Eq.13:

$$Prosodic = A\times 0.1 + B\times 0.3 + C\times 0.3 + D\times 0.3 \quad (13)$$

In Eq. 13, *Prosodic* means the weighted score of the four indicators of the Prosodic, A refers to the speed, B the stress, C the rhythm, D the intonation.

In assessment and evaluation, the above four metrics' weights should be different, and *Prosodic* is obtained by weighting the four indicators. The final score of the sentence is obtained through the weighting of two indicators: *Pronunciation* and *Prosodic* in Eq.14.

$$Total = Pronunciation\times 0.4 + Prosodic\times 0.6 \quad (14)$$

Total is the final score of the sentence, *Pronunciation* is the final score in Eq. 1, and *Prosodic* is the final score of weighting four indicators of the *Prosodic* calculated in Eq. 13.

3.2.3 *Experiment analysis*

This paper carefully selects 50 English sentences in the field of tourism which are suitable for practicing spoken English to serve as a software system corpus. Besides, two English pronunciation linguistics experts (a male and a female) are invited to record 50 standard sentences as the system's standard voice.

Within the recognition module, the HMM training requires a lot of voice samples so that 50 sentences are chosen to train the HMM model. Ten English majors (5 males and 5 females) are invited to record these 50 sentences, totally there are 50 (sentence) × 10 (people) = 500 sentences taken part in the HMM model training. Meanwhile, 10 non-English majored students are invited to record 50 (sentence) × 10 (people) = 500 sentences as the test speech samples of identification and scoring.

In the pronunciation evaluation module, three English teachers are invited to score the 500 voice samples and the score results are divided into four grades: A, B, C and D.

The experiment data in Table 1 indicates the difference between *St*(teachers' score) and *Sm*(machine's

Table 1. Part of experiment data of intonation.

	|St- Sm/			
Indicators	=0	=1	=2	=3
Pronunciation	86.25%	13.33%	0.42%	0%
Speed	77.08%	22.92%	0%	0%
Stress	80.00%	18.75%	1.25%	0%
Rhythm	79.45%	19.50%	1.05%	0%
Intonation	70.42%	26.25%	3.33%	0%
Total	87.50%	12.50%	0%	0%

score). |*St-Sm*| being 0 means that the teachers' score and machine's score are the same. Among the 500 sentences, 86.25% of pronunciation score, 77.08% of speed score, 80% of stress score, 79.45% of the rhythm and 70.42% of the intonation score are the same between teachers and machine.

It can be seen that the consistence between machine's total score and teachers' total score is higher than other indicators, because of the linear weighed equation Eqs.(13) and (14) being trained by the teachers' score data. The difference larger than 1 point only counts for 12.5% with full mark 4 point as total score.

4 CONCLUSION

This paper firstly describes the development status of the speech signal processing technology and the importance of CALL in English education. Secondly, it discusses the speech intonation theory, the pretreatment process, the five parametric indicators principle and specific implementation methods of pronunciation, speed, stress, rhythm and intonation.

Verified by experiment data and pronunciation linguistic experts, the pronunciation evaluation model built in this paper has certain credibility.

There is still space for improvement on this evaluation model. Here is the outlook for the future work:

1. In the experiment data analysis stage, due to limited corpus, the score results are influenced by subjective opinions of the experts to a certain extent. In order to reduce the personal and subjective opinion, the accumulation of corpus will be increased and more will be invited to evaluate the pronunciation. In the future, more experiment data will be made use of to calculate a more objective and reasonable target weight.
2. Currently, in the speech recognition module, due to limited objective conditions, the standard corpus is relatively rare, only 50 sentences are selected to train the model; in the pronunciation evaluation module, there are just 500 sentences for system testing. At the moment, the standard voice has a certain unity, in the future the standard reference and the scoring basis can be more scientific by the accumulation of more standard voice and test voice samples and a more rigorous test can be made to improve the credibility of the model.
3. This system is currently implemented on the PC platform. In order to meet the needs of practicing English anytime anywhere, this system can be ported to the other mainstream platforms such as Android and IOS.

ACKNOWLEDGEMENTS

This work is supported by the Guangzhou science and technology project (2014J4100025), the Ministry of Education humanities and social science project (13YJAZH046) and Guangdong University of Foreign Studies postgraduate innovation research project (14GWCXXM-35).

REFERENCES

Chen Juanwen. 2004. *The comparison of Shanghai mandarin Chinese and mandarin prosodic features*. Hangzhou: Zhejiang University.

Chiu, Yu-Hsiang, Bosco. 2010. *Learning-based auditory encoding for robust speech recognition*. Carnegie Mellon University.

Doh-Suk Kim & Tarraf A.Enhanced. 2006. Perceptual Model for Non-Intrusive Speech Quality Assessment. *IEEE International Conference on Acoustics, Speech and Signal Processing* l: 829–832.

Falk T.H & Wai-Yip Chan. 2006. Non-intrusive speech quality estimation using Gaussian mixture models. *Signal Processing Letters*. 13(2): 108–111.

Grabe, E. Low, E. L. 2002. Durational Variability in Speech and the Rhythm Class Hypothesis. *In Laboratory Phonology VII*. Berlin: Mouton de Gruyter.

Jin C & Gui-zhen W. 2013. The Validity of Rhythm Measurements in Evaluating the Rhythm Proficiencies of Chinese EFL Learners. *Foreign Language in China* 1(60): 64.

Li G, Wei-qian L & Yu-guo D. 2009. Feasibility Study and Practice of Machine Scoring of Repetition Questions in Large-scaled English Oral Test. *Computer-Assisted Foreign Language Education in China* 2:10–15.

Low E. L. 1998. Prosodic Prominence in Singapore English. *D. Unpublished doctoral dissertation*. London: University of Cambridge.

Shao Pengfei. 2009. *The research for comparison of rhythm of mandarin dialects and mandarin and the study of evaluation system*. Jinan: Shandong Normal University.

Wang Guizhen. 2005. *English phonetics tutorial*. Beijing: Higher Education Press.

Electronics, Communications and Networks IV – Hussain & Ivanovic (eds)
© 2015 Taylor & Francis Group, London, ISBN: 978-1-138-02830-2

A transition from microstrip to slotline to the PCB version of SINRD waveguide

Qian Li & Feng Xu*
College of Electronic Science and Engineering, Nanjing University of Posts and Telecommunications, Nanjing, Jiangsu, China

ABSTRACT: As an improvement of the conventional NRD, the Printed Circuit Boards (PCB) version of the Substrate Integrated Nonradiative Dielectric (SINRD) waveguide or metalized dielectric layer shows its excellent performances at the microwave and millimeter-wave frequencies. This paper presented a transition for the SINRD wave guides used the structure of microstrip to slotline. By using two different heights of the Arlon TC600, the structure of microstrip to slotline can be directly fabricated in the thin board. To do this, we can excite the required mode of SINRD near the middle of the dielectric strip of SINRD waveguide which means that we can obtain wider bandwidth and highly integration. Simulation results also confirmed the above views.

1 INTRODUCTION

In 1981, T. Yoneyama and S. Nishida proposed the concept of NRD waveguide (Yoneyama & Nishida 1981). With its excellent performance in high frequency, the structure of NRD waveguide attracted much attention in the field of millimeter-wave circuits and systems. Different from the traditional transmission line (e.g. microstrip), NRD waveguide appears low transmission losses and almost nonexistent radiation at bends and discontinuities. But, the structural defects greatly limit the development of NRD waveguide.

In order to solve the problem of mechanical and assembly errors, the substrate integrated nonradiative dielectric (SINRD) waveguide proposed in (Cassivi & Wu 2004). By extending the width of the dielectric, SINRD become more easy to integrate with other planar circuits. Nevertheless, as the first improvement of the NRD waveguide, SINRD waveguide also need two extra metal plates to cover. It means that the gap still existed between the dielectric and the metal plate. It will affect the performance of the device. Therefore, the second improvement of the NRD waveguide, the PCB version, has been presented in (Xu & Wu 2011). Compare to the SINRD waveguide, the PCB version directly drill the air holes or slots on the PCBs, which can be eliminated alignment problems and mechanical errors of conventional NRD waveguide without extra metal cover. So, this kind of improvement structure, which can be fabricated directly on PCBs or metalized layer will be easy to integrate with other planar circuits and, at the same time, it becomes a potential structure for millimeter-wave hybrid integrated systems.

The problem about how to excite the NRD waveguide is difficult. In consideration of the particularity of the field type of the NRD waveguide, the excitation cannot be directly applied to the port. Thus, a variety of transition structures are applied to the NRD waveguide. Ref. (Han et al. 1996, Wu & Han 1997) introduced two transition structures of microstrip line to NRD waveguide. However, the transition about the PCB version of SINRD has not attracted enough attention yet. According to traditional structure in NRD waveguide, the similar structure applied in the PCB-SINRD waveguide in (Li & Xu 2013). It provided a transition from microstrip line to PCB-SINRD waveguide. But it still can not meet the requirement of the PCB-SINRD waveguide. In addition, the transition described in (Li & Xu 2013) is fabricated in the top or bottom of the SINRD waveguide which cannot be well excite the desired mode and have not enough bandwidth. Otherwise, the transition from microstrip line to PCB-SINRD waveguide needs an extra metal layer as its ground and it will obviously affect the performance of the whole circuit. In this paper, according to the structure described in (Esquius-Morote et al. 2013) which used for exciting the extended substrate integrated waveguide (SIW), we introduced a novel transition from microstrip line to slot to SINRD waveguide. In this scheme, two different heights of Arlon TC600 are used. Thus, the required mode can be excited near the middle of the dielectric strip of the SINRD waveguide. Simulation results verified that this kind of proposed transition has a good performance and wider bandwidth.

*Corresponding author: *feng.xu@njupt.edu.cn*

2 DESCRIPTION OF THE PROPOSED TRANSITION

Considering that the operating mode of NRD waveguide is LSM_{11} (shown in Figure 1), the structures described in (Li & Xu 2013) used the quasi-TEM mode generated by microstrip line through a coupling aperture to excite the PCB version of SINRD waveguide. As shown in Figure 2, the feeding method of the structure mentioned in (Li & Xu 2013) is applied, the transition at the top or bottom of the PCB version of SINRD waveguide which can not excite the LSM_{11} mode effectively.

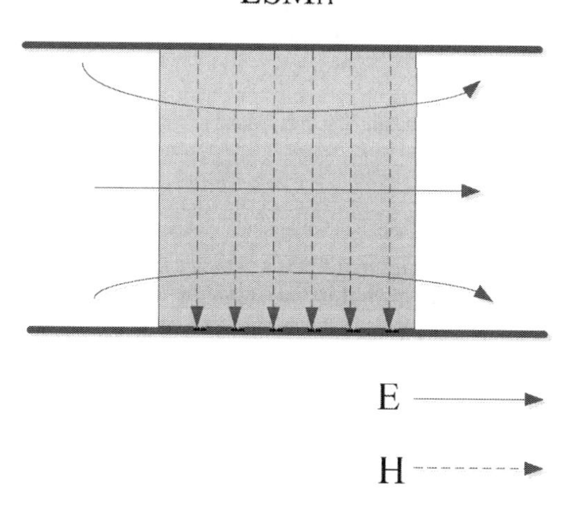

Figure 1. The operating mode of NRD waveguide.

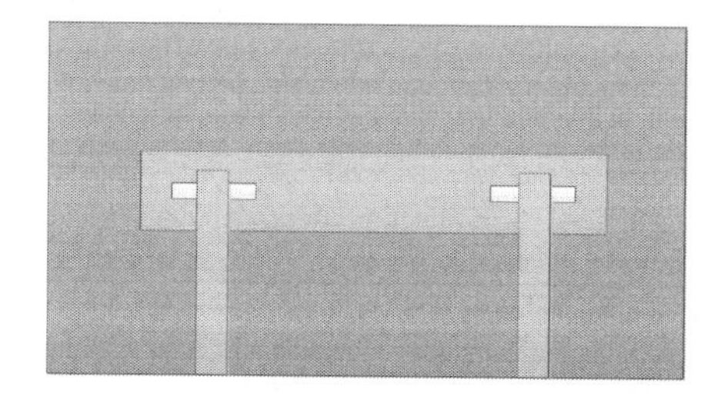

Figure 2. Geometry of the transition from microstrip line to PCB-SINRD waveguide.

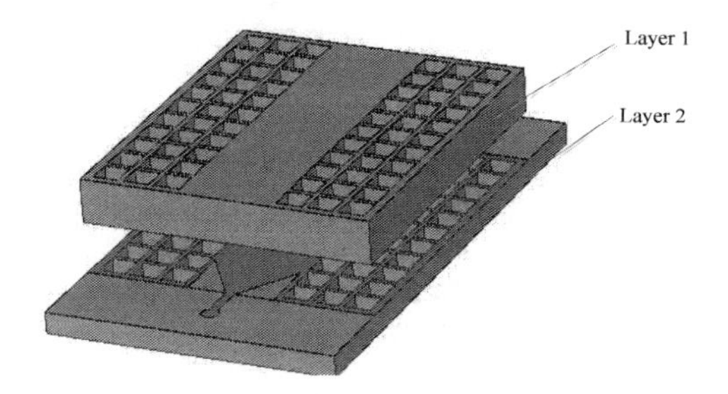

Figure 3. Topology of the proposed transition.

In this segment, a novel transition will be introduced. As is shown in Figure 3, a slotline is designed in a thinner substrate and then sandwiched by another substrate. In addition, we can find that at the bottom of the substrate (shown in Figure 4), two microstrips is positioned to feed the slotline. To compensate the slotline is not placed at the middle of the PCB-SINRD waveguide height, we used linear tapers between the slotline and SINRD waveguide. And we adopted circle at the end of microstrip and slotline to adjust match. Moreover, we utilized two different heights of Arlon TC600 combined together by mechanism at the height of the PCB-SINRD waveguide. By doing so, we can make the slotline excited the desired mode very well. Furthermore, since the two different heights of substrate integrated with mechanism, very well, it is no needs to assemble any unit. It turns out that this scheme of the structure has significant structural advantages.

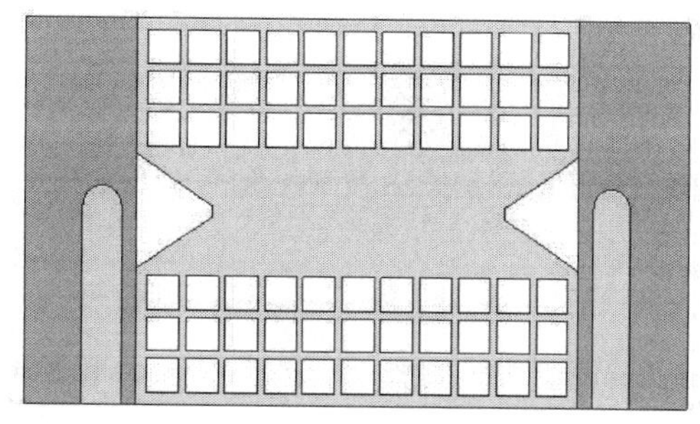

Figure 4. Bottom of the proposed transition.

3 GEOMETRY OF THE PROPOSED TRANSITION AND SIMULATION RESULTS

The topology of the proposed transition is illustrated in Figure 5. In layer 1, the substrate of Arlon TC600 ($\varepsilon_r = 6.25$) is 20×16 mm^2 and has a height of $h_2 =$ 2.54mm. In layer 2, the size of the Arlon TC600 is 30×16 mm^2 with a height of $h_1 =$ 1.27mm and two identical microstrip lines are positioned on the back of it. The strip width of the microstrip line is $w_m =$ 1.75 mm. Besides, the slotline which fabricated in layer 2 has a wide of $w_s =$ 0.5mm. The central channel width $W_{nrd} =$ 5.5 mm, diameter d of the air holes is 1.45 mm and the distance s between them is 0.3 mm. In addition, the circulars used for matching are $r_m =$ 0.89mm and $r_s =$ 0.45mm, respectively. These via-air hole dimensions should carefully designed to minimize the leakage losses from the uncovered metal plate of the PCB-SINRD waveguide.

By using the 3D electromagnetic simulation software, we required the simulation results of S

parameters of the transition from the microstrip line to slotline to the SINRD waveguide. Moreover, with the help of another one 3D electromagnetic simulation software which is based on FEM while the former is based on FDTD method, we verified the accuracy of the simulation result. The comparison of the two simulation results is shown in Figure 6. It can be seen from the Figure 6 that the transition has a central frequency of f_0=23.5GHz with a 21% fractional bandwidth. In the passband, we achieved that the return loss is better than 15dB and the insertion loss is generally smaller than 6dB. Obviously, the bandwidth of the proposed transition is wider than the structure mentioned in (Li & Xu 2013). And the performance of the transition is well.

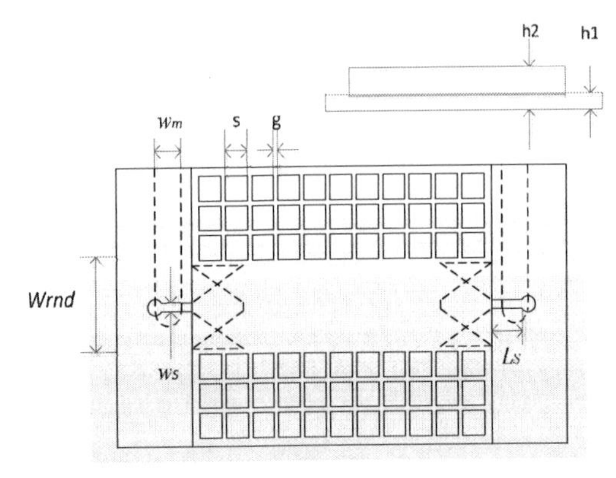

Figure 5. Geometry of the proposed transition.

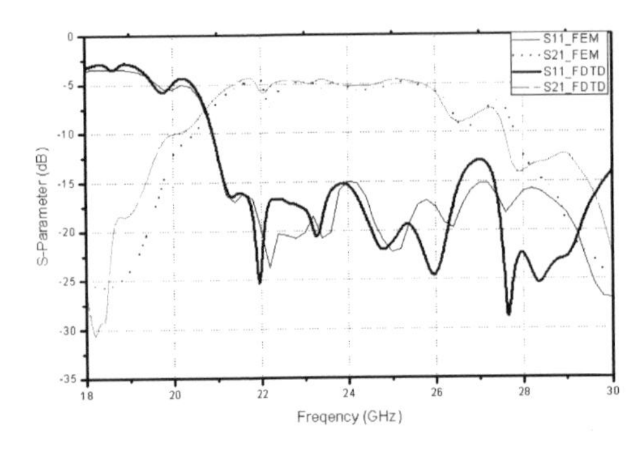

Figure 6. Simulation results of the proposed transition.

4 CONCLUSION

A novel transition of microstrip to slotline to PCB-SINRD waveguide was proposed in this paper. By using two different heights of substrates and designing the slotline sandwiched between them, we realized to excited SINRD waveguide near its middle height. The proposed structure has a very high level integration and can be easily processed. Moreover, this geometry of transition can acquire wider bandwidth.

The simulation results of the proposed transition exhibit that the PCB version of SINRD waveguide was an excellent NRD waveguide replacement and this scheme of transition would be potentially useful for future millimeter hybrid integrated systems.

ACKNOWLEDGEMENT

This work was supported by the specially appointed professor program foundation of Jiangsu province of China, the national natural science foundation of China under grant No. 61171052, the doctoral fund of ministry of education of China under grant No. 20123223110004 and the fund of state key laboratory of millimeter waves of China under grant No. K201414.

REFERENCES

Cassivi, Y. & Wu, K. 2004. Substrate integrated nonradiative dielectric waveguide. *IEEE Microwave and Wireless Components Letters* 14(3): 1639–1642.

Esquius-Morote, M., Fuchs, B., Zurcher, J.- F., et al. 2013. Extended SIW for TEm0 and TE0n Modes and Slotline Excitation of the TE01 Mode. *IEEE Microwave and Wireless Components Letters 23(8): 412–414.*

Han, L., Wu, K. & Bosisio, R. G. 1996. An integrated transition of microstrip to nonradiative dielectric waveguide for microwave and millimeter-wave circuits, *IEEE Tansaction on Microwave Theory and Techniques* 44(7): 1091–1096.

Li, F. & Xu, F. 2013. An integrated transition of microstrip to substrate integrated nonradiative dielectric waveguide based on Printed Circuit Boards. *2013 Proceedings of International Symposium on Antennas and Propagation. Nanjing, 23–25 October 2013.* IEEE.

Wu, K. & Han, L. 1997. Hybrid integrated technology of planar circuits and NRD guide for cost effective microwave and millimeter-wave applications. *IEEE Tansaction on Microwave Theory and Techniques* 45(6): 946–954.

Xu, F. & Wu, K. 2011. Substrate integrated nonradiative dielectric waveguide structures directly fabricated on printed circuits and metallized dielectric layers. *IEEE Tansaction on Microwave Theory and Techniques* 59(12): 3076–3086.

Yoneyama, T. & Nishida, S. 1981. Nonradiative dielectric waveguide for millimeter-wave integrated circuit. *IEEE Tansaction on Microwave Theory and Techniques* 29(11): 1188–1192.

Electronics, Communications and Networks IV – Hussain & Ivanovic (eds)
© 2015 Taylor & Francis Group, London, ISBN: 978-1-138-02830-2

Author index